The New York Botanical Garden Illustrated Encyclopedia of Horticulture

ERRATA

Volume 7
Page 2214, column 3, line 37:
"1772" should read "1768"

Page 2227, column 1, line 22:
Insert "live on decaying organic material in the soil or" after "They"

Page 2233, column 1, line 24:
Add entry "**MORETON BAY FIG** is *Ficus macrophylla*."

The New York Botanical Garden Illustrated Encyclopedia of Horticulture

Thomas H. Everett

Volume 7
Ma-Par

Garland Publishing, Inc.
New York & London

15 14 13 12 11 10 9 8 7 6 5 4 3 2 1

Library of Congress Cataloging in Publication Data

Everett, Thomas H
The New York Botanical Garden illustrated encyclopedia of horticulture.

1. Horticulture—Dictionaries. 2. Gardening—Dictionaries. 3. Plants, Ornamental—Dictionaries. 4. Plants, Cultivated—Dictionaries. I. New York (City). Botanical Garden. II. Title.
SB317.58.E94 635.9'03'21 80-65941
ISBN 0-8240-7237-5

PHOTO CREDITS

Black and White

Bodger Seeds Ltd.: *Narcissus* 'Paper White' forced into early bloom in a mixture of peat moss and perlite, p. 2284. Bulb growers of Holland: Division 7 (Jonquilla): 'Trevithian', p. 2274. Desert Botanical Garden: *Neolloydia conoidea,* p. 2301. Estelle Gerard: *Mitriostigma axillare,* p. 2217. Keline-Wilcox Company: *Monstera deliciosa,* a tub specimen, p. 2228. Kirstenbosch Botanic Garden: *Pachypodium namaquanum* in South Africa, p. 2462. Malak, Ottawa, Canada: Division 1 (Trumpet) (part b), p. 2272; *Narcissus bulbocodium,* p. 2277; Planted in small groups narcissuses can be very effective (part b), p. 2280. Netherlands Flower Institute: Montbretia (*Crocosmia crocosmaeflora*), p. 2229; Division 4 (Double-flowered) (part a), p. 2273; *Narcissus triandrus albus,* p. 2277; Narcissuses naturalized under trees, p. 2280; Planted in small groups narcissuses can be very effective (part a), p. 2280. The New York Botanical Garden: A field of *Manihot esculenta* in the American tropics, p. 2134; *Marshallia obovata,* p. 2149; *Matricaria recutita,* p. 2155; Sow seeds of many sorts of annuals and vegetables outdoors, p. 2162; *Meehania cordata,* p. 2174; *Mikania scandens,* p. 2201; *Mitella diphylla,* p. 2215; *Monarda punctata,* p. 2223; *Myristica fragrans,* nutmegs (seeds) enclosed in a net of mace (outer covering), p. 2264; Narcissuses forced as pot plants (part d), p. 2285; *Nelumbo nucifera* at The New York Botanical Garden, p. 2292; *Nelumbo nucifera* (seed pods), p. 2292; A summer bed of *Nierembergia scoparia,* p. 2327; Night-blooming-cereus (*Hylocereus*), p. 2328; *Nuphar luteum,* p. 2349; Hardy nympheas at The New York Botanical Garden, p. 2352; *Nymphaea marliacea* variety, p. 2353; Day-blooming tropical nympheas (parts a, b, c & d), p. 2354; *Nyssa sylvatica* (leaves), p. 2358; *Nyssa sylvatica* (fruits), p. 2358; *Onoclea sensibilis* (fertile leaves), p. 2394; *Oplopanax horridus,* p. 2403; *Orchis spectabilis,* p. 2427; *Ornithogalum arabicum* p. 2434; *Osmunda regalis,* p. 2443; *Osmunda cinnamomea* (fertile fronds), p. 2444; *Oxalis bowiei,* p. 2453; *Pachysandra procumbens* in bloom, p. 2464; Palms in containers (part b), p. 2475; *Pamianthe peruviana,* p. 2478; *Panax trifolius,* p. 2478. Other photographs by Thomas H. Everett.

Color

Malak, Ottawa, Canada: *Muscari armeniacum,* Trumpet narcissus 'van Wereld's Favourite', Double-flowered narcissus 'Texas', *Narcissus bulbocodium* with grape-hyacinths. The New York Botanical Garden: *Medeola virginica, Mertensia longiflora,* A hybrid *Miltonia, Mitella breweri, Narthecium americanum, Nymphoides peltata, Oenothera xylocarpa.* Other photographs by Thomas H. Everett.

Published by Garland Publishing, Inc.
136 Madison Avenue, New York, New York 10016

Printed in the United States of America

This work is dedicated to the honored memory of the distinguished horticulturists and botanists who most profoundly influenced my professional career: Allan Falconer of Cheadle Royal Gardens, Cheshire, England; William Jackson Bean, William Dallimore, and John Coutts of the Royal Botanic Gardens, Kew, England; and Dr. Elmer D. Merrill and Dr. Henry A. Gleason of The New York Botanical Garden.

Foreword

According to Webster, an encyclopedia is a book or set of books giving information on all or many branches of knowledge generally in articles alphabetically arranged. To the horticulturist or grower of plants, such a work is indispensable and one to be kept close at hand for frequent reference.

The appearance of *The New York Botanical Garden Illustrated Encyclopedia of Horticulture* by Thomas H. Everett is therefore welcomed as an important addition to the library of horticultural literature. Since horticulture is a living, growing subject, these volumes contain an immense amount of information not heretofore readily available. In addition to detailed descriptions of many thousands of plants given under their generic names and brief description of the characteristics of the more important plant families, together with lists of their genera known to be in cultivation, this Encyclopedia is replete with well-founded advice on how to use plants effectively in gardens and, where appropriate, indoors. Thoroughly practical directions and suggestions for growing plants are given in considerable detail and in easily understood language. Recommendations about what to do in the garden for all months of the year and in different geographical regions will be helpful to beginners and will serve as reminders to others.

The useful category of special subject entries (as distinct from the taxonomic presentations) consists of a wide variety of topics. It is safe to predict that one of the most popular will be Rock and Alpine Gardens. In this entry the author deals helpfully and adequately with a phase of horticulture that appeals to a growing group of devotees, and in doing so presents a distinctly fresh point of view. Many other examples could be cited.

The author's many years as a horticulturist and teacher well qualify him for the task of preparing this Encyclopedia. Because he has, over a period of more than a dozen years, written the entire text (submitting certain critical sections to specialists for review and suggestions) instead of farming out sections to a score or more specialists to write, the result is remarkably homogeneous and cohesive. The Encyclopedia is fully cross referenced so that one may locate a plant by either its scientific or common name.

If, as has been said, an encyclopedia should be all things to all people, then the present volumes richly deserve that accolade. Among the many who call it "friend" will be not only horticulturists ("gardeners," as our author likes to refer to them), but growers, breeders, writers, lecturers, arborists, ecologists, and professional botanists who are frequently called upon to answer questions to which only such a work can provide answers. It seems safe to predict that it will be many years before lovers and growers of plants will have at their command another reference work as authoritative and comprehensive as T. H. Everett's Encyclopedia.

John M. Fogg, Jr.
Director Emeritus, Arboretum of the Barnes Foundation
Emeritus Professor of Botany, University of Pennsylvania

Preface

The primary objective of *The New York Botanical Garden Illustrated Encyclopedia of Horticulture* is a comprehensive description and evaluation of horticulture as it is known and practiced in the United States and Canada by amateurs and by professionals, including those responsible for botanical gardens, public parks, and industrial landscapes. Although large-scale commercial methods of cultivating plants are not stressed, much of the content of the Encyclopedia is as basic to such operations as it is to other horticultural enterprises. Similarly, although landscape design is not treated on a professional level, landscape architects will find in the Encyclopedia a great deal of importance and interest to them. Emphasis throughout is placed on the appropriate employment of plants both outdoors and indoors, and particular attention is given to explaining in considerable detail the how- and when-to-do-it aspects of plant growing.

It may be useful to assess the meanings of two words I have used. Horticulture is simply gardening. It derives from the Latin *hortus*, garden, and *cultura*, culture, and alludes to the intensive cultivation in gardens and nurseries of flowers, fruits, vegetables, shrubs, trees, and other plants. The term is not applicable to the extensive field practices that characterize agriculture and forestry. Amateur, as employed by me, retains its classic meaning of a lover from the Latin *amator*; it refers to those who garden for pleasure rather than for financial gain or professional status. It carries no implication of lack of knowledge or skills and is not to be equated with novice, tyro, or dabbler. In truth, amateurs provide the solid basis upon which American horticulture rests; without them the importance of professionals would diminish. Numbered in millions, amateur gardeners are devotees of the most widespread avocation in the United States. This avocation is serviced by a great complex of nurseries, garden centers, and other suppliers; by landscape architects and landscape contractors; and by garden writers, garden lecturers, Cooperative Extension Agents, librarians, and others who dispense horticultural information. Numerous horticultural societies, garden clubs, and botanical gardens inspire and promote interest in America's greatest hobby and stand ready to help its enthusiasts.

Horticulture as a vocation presents a wide range of opportunities which appeal equally to women and men. It is a field in which excellent prospects still exist for capable entrepreneurs. Opportunities at professional levels occur too in nurseries and greenhouses, in the management of landscaped grounds of many types, and in teaching horticulture.

Some people confuse horticulture with botany. They are not the same. The distinction becomes more apparent if the word gardening is substituted for horticulture. Botany is the science that encompasses all systematized factual knowledge about plants, both wild and cultivated. It is only one of the several disciplines upon which horticulture is based. To become a capable gardener or a knowledgeable plantsman or plantswoman (I like these designations for gardeners who have a wide, intimate, and discerning knowledge of plants in addition to skill in growing them) it is not necessary to study botany formally, although such study is likely to add greatly to one's pleasure. In the practice of gardening many botanical truths are learned from experience. I have known highly competent gardeners without formal training in botany and able and indeed distinguished botanists possessed of minimal horticultural knowledge and skills.

Horticulture is primarily an art and a craft, based upon science, and at some levels perhaps justly regarded as a science in its own right. As an art it calls for an appreciation of beauty and form as expressed in three-dimensional spatial relationships and an ability to translate aesthetic concepts into reality. The chief materials used to create gardens are living plants, most of which change in size and form with the passing of time and often show differences in color and texture and in other ways from season to season. Thus it is important that designers of gardens have a wide familiarity with the sorts of plants that lend themselves to their purposes and with plants' adaptability to the regions and to the sites where it is proposed to plant them.

As a craft, horticulture involves special skills often derived from ancient practices passed from generation to generation by word of mouth and apprenticeship-like contacts. As a technology it relies on this backlog of empirical knowledge supplemented by that acquired by scientific experiment and investigation, the results of which often serve to explain rather than supplant old beliefs and practices but sometimes point the way to more expeditious methods of attaining similar results. And from time to time new techniques are developed that add dimensions to horticultural practice; among such of fairly recent years that come to mind are the manipulation of blooming season by artificial day-length, the propagation of orchids and some other plants by meristem tissue culture, and the development of soilless growing mixes as substitutes for soil.

One of the most significant developments in American horticulture in recent decades is the tremendous increase in the number of different kinds of plants that are cultivated by many more people than formerly. This is particularly true of indoor plants or houseplants, the sorts grown in homes, offices, and other interiors, but is by no means confined to that group. The relative affluence of our society and the freedom and frequency of travel both at home and abroad has contributed to this expansion, a phenomenon that will surely continue as avid collectors of the unusual bring into cultivation new plants from the wild and promote wider interest in sorts presently rare. Our garden flora is also constantly and beneficially expanded as a result of the work of both amateur and professional plant breeders.

It is impracticable in even the most comprehensive encyclopedia to describe or even list all plants that somewhere within a territory as large as the United States and Canada are grown in gardens. In this Encyclopedia the majority of genera known to be in cultivation are described, and descriptions and often other pertinent information about a complete or substantial number of their species and lesser categories are given. Sorts likely to be found only in collections of botanical gardens or in those of specialists may be omitted.

The vexing matter of plant nomenclature inevitably presents itself when an encyclopedia of horticulture is contemplated. Conflicts arise chiefly between the very understandable desire of gardeners and others who deal with cultivated plants to retain long-familiar names and the need to reflect up-to-date botanical interpretations. These points of view are basically irreconcilable and so accommodations must be reached.

As has been well demonstrated in the past, it is unrealistic to attempt to standardize the horticultural usage of plant names by decree or edict. To do so would negate scientific progress. But it is just as impracticable to expect gardeners, nurserymen, arborists, seedsmen, dealers in bulbs, and other amateur and professional horticulturists to keep current with the interpretations and recommendations of plant taxonomists; particularly as these sometimes fail to gain the acceptance even of other botanists and it is not unusual for scientists of equal stature and competence to prefer different names for the same plant.

In practice time is the great leveler. Newly proposed plant names accepted in botanical literature are likely to filter gradually into horticultural usage and eventually gain currency value, but this sometimes takes several years. The complete up-to-dateness and niceties of botanical naming are less likely to bedevil horticulturists than uncertainties concerned with correct plant identification. This is of prime importance. Whether a tree is labeled *Pseudotsuga douglasii, P. taxifolia,* or *P. menziesii* is of less concern than that the specimen so identified is indeed a Douglas-fir and not some other conifer.

After reflection I decided that the most sensible course to follow in *The New York Botanical Garden Illustrated Encyclopedia of Horticulture* was to accept almost in its entirety the nomenclature adopted in *Hortus Third* published in 1976. By doing so, much of the confusion that would result from two major comprehensive horticultural works of the late twentieth century using different names for the same plant is avoided, and it is hoped that for a period of years a degree of stability will be attained. Always those deeply concerned with critical groups of plants can adopt the recommendations of the latest monographers. Exceptions to the parallelism in nomenclature in this Encyclopedia and *Hortus Third* are to be found in the CACTACEAE for which, with certain reservations but for practical purposes, as explained in the Encyclopedia entry Cactuses, the nomenclature of Curt Backeburg's *Die Cactaceae,* published in 1958–62, is followed; and the ferns, where I mostly accepted the guidance of Dr. John T. Mickel of The New York Botanical Garden. The common or colloquial names employed are those deemed to have general acceptance. Cross references and synonomy are freely provided.

The convention of indicating typographically whether or not plants of status lesser than species represent entities that propagate and persist in the wild or are sorts that persist only in cultivation is not followed. Instead, as explained in the Encyclopedia entry Plant Names, the word variety is employed for all entities below specific rank and if in Latin form the name is written in italic, if in English or other modern language, in Roman type, with initial capital letter, and enclosed in single quotation marks.

Thomas H. Everett
Senior Horticulture Specialist
The New York Botanical Garden

Acknowledgments

I am indebted to many people for help and support generously given over the period of more than twelve years it has taken to bring this Encyclopedia to fruition. Chief credit belongs to four ladies. They are Lillian M. Weber and Nancy Callaghan, who besides accepting responsibility for the formidable task of filing and retrieving information, typing manuscript, proofreading, and the management of a vast collection of photographs, provided much wise council; Elizabeth C. Hall, librarian extraordinary, whose superb knowledge of horticultural and botanical literature was freely at my disposal; and Ellen, my wife, who displayed a deep understanding of the demands on time called for by an undertaking of this magnitude, and with rare patience accepted inevitable inconvenience. I am also obliged to my sister, Hette Everett, for the valuable help she freely gave on many occasions.

Of the botanists I repeatedly called upon for opinions and advice and from whom I sought elucidation of many details of their science abstruse to me, the most heavily burdened have been my friends and colleagues at The New York Botanical Garden, Dr. Rupert C. Barneby, Dr. Arthur Cronquist, and Dr. John T. Mickel. Other botanists and horticulturists with whom I held discussions or corresponded about matters pertinent to my text include Dr. Theodore M. Barkley, Dr. Lyman Benson, Dr. Ben Blackburn, Professor Harold Davidson, Dr. Otto Degener, Harold Epstein, Dr. John M. Fogg, Jr., Dr. Alwyn H. Gentry, Dr. Alfred B. Graf, Brian Halliwell, Dr. David R. Hunt, Dr. John P. Jessop, Dr. Tetsuo Koyama, Dr. Bassett Maguire, Dr. Roy A. Mecklenberg, Everitt L. Miller, Dr. Harold N. Moldenke, Dr. Dan H. Nicolson, Dr. Pascal P. Pirone, Dr. Ghillean Prance, Don Richardson, Stanley J. Smith, Ralph L. Snodsmith, Marco Polo Stufano, Dr. Bernard Verdcourt, Dr. Edgar T. Wherry, Dr. Trevor Whiffin, Dr. Richard P. Wunderlin, Dr. John J. Wurdack, Yuji Yoshimura, and Rudolf Ziesenhenne.

Without either exception or stint these conferees and correspondents shared with me their knowledge, thoughts, and judgments. Much of the bounty so gleaned is reflected in the text of the Encyclopedia but none other than I am responsible for interpretations and opinions that appear there. To all who have helped, my special thanks are due and are gratefully proferred.

I acknowledge with much pleasure the excellent cooperation I have received from the Garland Publishing Company and most particularly from its President, Gavin Borden. To Ruth Adams, Nancy Isaac, Carol Miller, Melinda Wirkus, and Geoffrey Braine, I say thank you for working so understandingly and effectively with me and for shepherding my raw typescript through the necessary stages.

How to Use This Encyclopedia

A vast amount of information about how to use, propagate, and care for plants both indoors and outdoors is contained in the thousands of entries that compose *The New York Botanical Garden Illustrated Encyclopedia of Horticulture.* Some understanding of the Encyclopedia's organization is necessary in order to find what you want to know.

Arrangement of the Entries

Genera

The entries are arranged in alphabetical order. Most numerous are those that deal with taxonomic groups of plants. Here belong approximately 3,500 items entered under the genus name, such as ABIES, DIEFFENBACHIA, and JUGLANS. If instead of referring to these names you consult their common name equivalents of FIR, DUMB CANE, and WALNUT, you will find cross references to the genus names.

Bigeneric Hybrids & Chimeras

Hybrids between genera that have names equivalent to genus names—most of these belonging in the orchid family—are accorded separate entries. The same is true for the few chimeras or graft hybrids with names of similar status. Because bigeneric hybrids frequently have characteristics similar to those of their parents and require similar care, the entries for them are often briefer than the regular genus entries.

Families

Plant families are described under their botanical names, with their common name equivalents also given. Each description is followed by a list of the genera accorded separate entries in this Encyclopedia.

Vegetables, Fruits, Herbs, & Ornamentals

Vegetables and fruits that are commonly cultivated, such as broccoli, cabbage, potato, tomato, apple, peach, and raspberry; most culinary herbs, including basil, chives, parsley, sage, and tarragon; and a few popular ornamentals, such as azaleas, carnations, pansies, and poinsettias, are treated under their familiar names, with cross references to their genera. Discussions of a few herbs and some lesser known vegetables and fruits are given under their Latin scientific names with cross references to the common names.

Other Entries

The remaining entries in the Encyclopedia are cross references, definitions, and more substantial discussions of many subjects of interest to gardeners and others concerned with plants. For example, a calendar of gardening activity, by geographical area, is given under the names of the months and a glossary of frequently applied species names (technically, specific epithets) is provided in the entry Plant Names. A list of these general topics, which may provide additional information about a particular plant, is provided at the beginning of each volume of the Encyclopedia (see pp. xvii–xx).

Cross References & Definitions

The cross references are of two chief types: those that give specific information, which may be all you wish to know at the moment:
Boojam Tree is *Idria columnaris.*
Cobra plant is *Darlingtonia californica.*
and those that refer to entries where fuller explanations are to be found:
Adhatoda. See Justicia.
Clubmoss. See Lycopodium and Selaginella.

Additional information about entries of the former type can, of course, be found by looking up the genus to which the plant belongs—*Idria* in the case of the boojam tree and *Darlingtonia* for the cobra plant.

ORGANIZATION OF THE GENUS ENTRIES

Pronunciation

Each genus name is followed by its pronunciation in parentheses. The stressed syllable is indicated by the diacritical mark ´ if the vowel sound is short as in man, pet, pink, hot, and up; or by ` if the vowel sound is long as in mane, pete, pine, home, and fluke.

Genus Common Names
Family Common Names
General Characteristics

Following the pronunciation, there may be one or more common names applicable to the genus as a whole or to certain of its kinds. Other names may be introduced later with the descriptions of the species or kinds. Early in the entry you will find the common and botanical names of the plant family to which the genus belongs, the number of species the genus contains, its natural geographical distribution, and the derivation of its name. A description that stresses the general characteristics of the genus follows, and this may be supplemented by historical data, uses of some or all of its members, and other pertinent information.

Identification of Plants

Descriptions of species, hybrids, and varieties appear next. The identification of unrecognized plants is a fairly common objective of gardeners; accordingly, in this Encyclopedia various species have been grouped within entries in ways that make their identification easier. The groupings may bring into proximity sorts that can be adapted for similar landscape uses or that require the same cultural care, or they may emphasize geographical origins of species or such categories as evergreen and deciduous or tall and low members of the same genus. Where the description of a species occurs, its name is designated in ***bold italic.*** Under this plan, the description of a particular species can be found by referring to the group to which it belongs, scanning the entry for the species name in bold italic, or referring to the opening sentences of paragraphs which have been designed to serve as lead-ins to descriptive groupings.

Gardening & Landscape Uses
Cultivation
Pests & Diseases

At the end of genus entries, subentries giving information on garden and landscape uses, cultivation, and pests or diseases or both are included, or else reference is made to other genera or groupings for which these are similar.

General Subject Listings

The lists below organize some of the encyclopedia entries into topics which may be of particular interest to the reader. They are also an aid in finding information other than Latin or common names of plants.

PLANT ANATOMY AND TERMS USED IN PLANT DESCRIPTIONS

All-America Selections
Alternate
Annual Rings
Anther
Apex
Ascending
Awl-shaped
Axil, Axillary
Berry
Bloom
Bracts
Bud
Bulb
Bulbils
Bulblet
Bur
Burl
Calyx
Cambium Layer
Capsule
Carpel
Catkin
Centrals
Ciliate
Climber
Corm
Cormel
Cotyledon
Crown
Deciduous
Disk or Disc
Double Flowers
Drupe
Florets
Flower
Follicle
Frond
Fruit
Glaucous
Gymnosperms
Head
Hips
Hose-in-Hose
Inflorescence
Lanceolate
Leader
Leaf
Leggy
Linear
Lobe
Midrib
Mycelium
Node
Nut and Nutlet
Oblanceolate
Oblong
Obovate
Offset
Ovate
Palmate
Panicle
Pedate
Peltate
Perianth
Petal
Pinnate
Pip
Pistil
Pit
Pod
Pollen
Pompon
Pseudobulb
Radials
Ray Floret
Rhizome
Runners
Samara
Scion or Cion
Seeds
Sepal
Set
Shoot
Spore
Sprigs
Spur
Stamen
Stigma
Stipule
Stolon
Stool
Style
Subshrub
Taproot
Tepal
Terminal
Whorl

GARDENING TERMS AND INFORMATION

Acid and Alkaline Soils
Adobe
Aeration of the Soil
Air and Air Pollution
Air Drainage
Air Layering
Alpine Greenhouse or Alpine House
Amateur Gardener
April, Gardening Reminders For
Aquarium
Arbor
Arboretum
Arch
Asexual or Vegetative Propagation
Atmosphere
August, Gardening Reminders For
Balled and Burlapped
Banks and Steep Slopes
Bare-Root
Bark Ringing
Baskets, Hanging
Bed
Bedding and Bedding Plants
Bell Jar
Bench, Greenhouse
Blanching
Bleeding
Bog
Bolting
Border
Bottom Heat
Break, Breaking
Broadcast
Budding
Bulbs or Bulb Plants

Gardening Terms and Information (Continued)

Types of Gardens and Gardening (Continued)

Miniature Gardens
Native Plant Gardens
Naturalistic Gardens
Nutriculture
Organic Gardening
Rock and Alpine Gardens
Roof and Terrace Gardening
Salads or Salad Plants
Seaside Gardens
Shady Gardens
Sink Gardening
Terrariums
Vegetable Gardens
Water and Waterside Gardens
Wild Gardens

PESTS, DISEASES, AND OTHER TROUBLES

Ants
Aphids
Armyworms
Bagworms
Bees
Beetles
Billbugs
Biological Control of Pests
Birds
Blight
Blindness
Blotch
Borers
Budworms and Bud Moths
Bugs
Butterflies
Canker
Cankerworms or Inchworms
Casebearers
Caterpillars
Cats
Centipede, Garden
Chinch Bugs
Chipmunks
Club Root
Corn Earworm
Crickets
Cutworms
Damping Off
Deer
Die Back
Diseases of Plants
Downy Mildew
Earthworms
Earwigs
Edema
Fairy Rings
Fire Blight
Flies
Fungi or Funguses
Galls
Gas Injury
Gophers
Grasshoppers
Grubs
Gummosis
Hornworms
Inchworms
Insects
Iron Chelates
Iron Deficiency
Lace Bugs
Lantana Bug
Lantern-Flies
Larva
Leaf Blight
Leaf Blister
Leaf Blotch
Leaf Curl
Leaf Cutters
Leaf Hoppers
Leaf Miners
Leaf Mold
Leaf Rollers
Leaf Scorch
Leaf Skeletonizer
Leaf Spot Disease
Leaf Tiers
Lightening Injury
Maggots
Mantis or Mantid
Mealybugs
Mice
Midges
Milky Disease
Millipedes
Mites
Mold
Moles
Mosaic Diseases
Moths
Muskrats
Needle Cast
Nematodes or Nemas
Parasite
Pests of Plants
Plant Hoppers
Plant Lice
Praying Mantis
Psyllids
Rabbits
Red Spider Mite
Rootworms
Rots
Rust
Sawflies
Scab Diseases
Scale Insects
Scorch or Sunscorch
Scurf
Slugs and Snails
Smut and White Smut Diseases
Sowbugs or Pillbugs
Spanworms
Spittlebugs
Springtails
Squirrels
Stunt
Suckers
Sun Scald
Thrips
Tree Hoppers
Virus
Walking-Stick Insects
Wasps
Webworms
Weevils
Wilts
Witches' Brooms
Woodchucks

GROUPINGS OF PLANTS

Accent Plants
Aquatics
Aromatic Plants
Bedding and Bedding Plants
Berried Trees and Shrubs
Bible Plants
Broad-Leaved and Narrow-Leaved Trees and Shrubs
Bulbs or Bulb Plants
Bush Fruits
Carnivorous or Insectivorous Plants
Dried Flowers, Foliage, and Fruits
Edging Plants
Epiphyte or Air Plant
Evergreens
Everlastings
Fern Allies
Filmy Ferns
Florists' Flowers
Foliage Plants
Fragrant Plants and Flowers
Gift Plants
Graft Hybrids
Grasses, Ornamental
Hard-Wooded Plants
Houseplants or Indoor Plants
Japanese Dwarfed Trees
Medicinal or Drug Plants
Night-Blooming Plants
Ornamental-Fruited Plants
Pitcher Plants
Poisonous Plants
Shrubs
State Flowers
State Trees
Stone Fruits
Stone or Pebble Plants
Stove Plants
Succulents
Tender Plants
Trees
Windowed Plants

The New York Botanical Garden Illustrated Encyclopedia of Horticulture

MANDRAGORA (Mandrá-gora) — Mandrake or Love-Apple. One species of the six that belong in this Old World genus of the nightshade family SOLANACEAE is of especial interest, chiefly because of its former importance as a medicinal herb and the folklore that accumulated around it. It is grown in herb and medicinal gardens. The name *Mandragora* was used by Hippocrates and is believed to have alluded to its being harmful to cattle. In North America another plant, the may-apple (*Podophyllum*), is sometimes called mandrake, but it is better to restrict the use of the name to *Mandragora*. Love-apple, another common name of *Mandragora*, was at one time also used for the tomato.

Mandragora officinarum, in fruit

The mandrake (***M. officinarum***) is a stemless, deciduous, herbaceous perennial. It has large, thick, deep, usually forked, fleshy roots and a clump of coarse, deep green, about 1-foot-long, wavy-edged, toothed leaves, the inner ones erect, the outer spreading. The bell-shaped greenish-yellow flowers about 1 inch long are in clusters low down among the leaves. They are succeeded by glossy egg-shaped fruits about as big as large plums and at first green, but maturing yellowish. These are technically berries. Another species, ***M. autumnalis***, a native of southern Europe and North Africa, is similar, but is smaller and has violet flowers. Other species inhabit the Himalayan region. Like so many of the nightshade family, *Mandragora* has slightly poisonous properties and is emetic, purgative, and narcotic. Its fruits have a sweetish, insipid flavor.

Ancient beliefs and superstitions about the mandrake were based on the alleged and sometimes fancifully close similarity of the form of its roots to that of the lower portions of the human body. Enlisting the doctrine of signatures (the belief that each kind of healing herb carries in its shape or markings indications of the conditions it alleviates) the mandrake was esteemed as an aphrodisiac. Old herbals contain intriguing illustrations of the plant, usually depicting the roots as being realistically manlike. Among the many superstitions regarding this plant one of the most common is acknowledged by Shakespeare in "And shrieks like mandrakes torn out of the earth, that living mortals hear them run mad" and "could curses kill as doth the mandrakes groan." The belief referred to is that when removed from the ground mandrakes cry out and that whoever hears the sound dies. Josephus recommends that "he who would take up a plant thereof, must tie a dog thereunto, to pull it up, otherwise if a man should do it, he would surely die in a short space after." The device of first digging around the root and then attaching a dog to it to complete its extraction was generally held to be the best way of obtaining mandrake. The dog, it was said, perished soon afterward. John Donne's "Song" (in 1633) advises "Goe and catche a falling starre, Get with child a mandrake roote, . . ." Pliny and others attest to the high regard in which this plant was held as a restorative and medicine by the Romans. He describes the procedure for gathering the roots whereby the collector drew three circles around the plant with the point of a sword while standing with his back to the wind, poured a libation on the ground, faced the west, and dug the plant out with his sword. Other ancient peoples, including the Greeks and Jews, were superstitious about the mandrake and regarded it with awe. To the Arabs, who regarded the fruits as aphrodisiacal, they were known as devil's-apples.

Garden Uses and Cultivation. The mandrake has no more ornamental value than a clump of horseradish, which it rather closely resembles, but it is of interest as a conversation piece and to include in herb and medicinal gardens because of its former importance and the legends associated with it. It is hardy and easy to grow in any sunny place in well-drained ground, succeeding best in soil that is deep and fertile. Propagation is by seed, which germinate readily, or by root cuttings.

MANDRAKE. See Mandragora and Podophyllum.

MANETTIA (Manét-tia)—Firecracker Vine. Native to the West Indies, Central America, and tropical South America, *Manettia*, of the madder family RUBIACEAE, consists of 130 species, few of which are cultivated. The name commemorates Saverio Manetti, prefect of the Botanic Garden, Florence, Italy, who died in 1785.

These are evergreen, twining vines with opposite leaves. The flowers have a four- or eight-lobed, often leafy calyx, a tubular corolla with four or five lobes (petals), four or five stamens, and one style. The fruits are many-seeded capsules.

Manettia inflata

Manettia inflata (flowers)

The firecracker vine (***M. inflata***, in gardens often misidentified as *M. bicolor*), is best known. Native to Paraguay and Uruguay, this attains a height of several feet, has finely-hairy stems and distinctly stalked, lanceolate leaves 1 inch to 2 inches long, bright green and on their undersides, or at least along the veins there, clothed with short hairs. The solitary, long-stalked blooms come from the leaf axils. They have tubular corollas ¾ inch long, conspicuously swollen at their bases and scarlet with short, bright yellow petals. The style does not protrude. Brazilian ***M. bicolor*** differs from the last in having hairless stems and leaves, the latter almost stalkless, sepals that are erect or spreading, but not reflexed, and protruding styles.

Very attractive ***M. cordifolia*** (syn. *M. glabra*), also a native of Brazil, is distinguishable by its smooth, shining, ovate to nearly round leaves sometimes slightly heart-shaped at their bases and 1½ to 4 inches long. Its flowers, with four-lobed calyxes, are red, funnel-shaped, 1 inch to 2 inches long, and in few-flowered clusters. In its homeland this is used in treating dropsy and dysentery. It is a powerful emetic.

Tropical American ***M. reclinata*** has pointed-ovate to pointed-lanceolate leaves up to 4 inches long and red flowers ¾ to 1 inch long, several together in leafy-bracted clusters. Each bloom has eight narrow calyx lobes. This has been confused with ***M. coccinea,*** a species common through much of tropical America from Mexico southward, that has glossy, hairless, ovate or pointed-elliptic leaves up to about 2 inches long and red flowers solitary or few together about 1 inch long. The calyx lobes of this vary from four to eight.

Garden and Landscape Uses and Cultivation. Manettias are attractive for gardens outdoors in countries where there is little or no frost and may be used as temporary summer plantings elsewhere, although where winters are too harsh for their permanent survival outdoors they are more commonly accommodated in greenhouses and sometimes grown as houseplants. Propagation is very easy by cuttings taken any time. If made in spring or summer from rather soft, somewhat immature shoots they root more quickly than do firmer stems taken in fall or winter. After planting keep the cuttings in a humid atmosphere and shade them from sun until roots are formed. Then plant them individually in small pots in porous, sandy soil. Supports around which the stems can twine must be provided. Brushwood, trellis, or wires are satisfactory.

Manettias can be trained up pillars or to wires stretched under the roof glass of a greenhouse. They are also decorative as pot specimens supported by trellis or brushwood. To encourage bushiness, pinch the tips out of the shoots occasionally. From spring-rooted cuttings excellent examples in 7- or 8-inch pots are obtainable by late summer. They bloom almost continuously. Indoors in winter a night temperature of 55°F is adequate with a five to fifteen degree rise during the day. Good light with some shade from strong direct sun is needed. Their most common pests are red spider mites and aphids.

MANFREDA (Man-frèda) — False-Aloe, Huaco or Texas-Tuberose, Rattlesnake Master. Some botanists include *Manfreda,* of the amaryllis family AMARYLLIDACEAE, in *Polianthes,* but most recognize it as a separate genus. When this is done it comprises about twenty species of perennial herbaceous plants, natives of the United States and Mexico. Its name honors Manfredus de Monte Imperiali, "an ancient writer on simples whose work is in the Parisian library."

Manfredas have fleshy, spindle-shaped roots and underground rhizomes. In rosettes, sometimes more than one from each rhizome, the soft, fleshy, and spineless leaves have smooth or minutely-toothed, frequently wavy or crisped margins. The flowers, in erect racemes or spikes, are usually solitary from each node of the flowering stalk. They have six petals, or more correctly, tepals, joined at their bases into a narrow tube and in fully open flowers recurved. There are six usually long-protruding stamens and one style. The fruits are capsules. From closely related *Agave* this genus differs in its flowers being asymmetrical and in generally branchless spikes or racemes.

The huaco or Texas-tuberose (***M. variegata*** syns. *Agave variegata, Polianthes variegata*) occurs wild in Texas and adjacent Mexico. It has few, mostly lanceolate, deeply-channeled, finely-toothed basal leaves 1 foot to 1¼ feet in length by 1 inch to 2 inches broad, and loosely-flowered stalks 3 to 4 feet tall. The greenish-brown, broadly-funnel-shaped blooms have perianth tubes under ½ inch long and slightly longer petals. The long-protruding stamens are three to six times as long as the petals.

From the last, ***M. maculosa*** (syns. *Agave maculosa, Polianthes maculosa*) differs in its protruding stamens being less than twice the length of the petals. It has brown- or green-blotched, minutely-toothed, basal leaves, under 1 inch wide, and flowering stalks 1 foot to 2 feet tall or taller, with a few purplish bracts, and up to twenty-five individual, nearly stalkless, fragrant, purplish to greenish-white flowers with narrowly-funnel-shaped perianth tubes and slightly spreading petals a little longer than the tubes. The blooms are nearly 2 inches long. This sort is a native of Texas.

False-aloe or rattlesnake master (***M. virginica*** syns. *Agave virginica, Polianthes virginica*) is native from Virginia to Illinois, Missouri, Florida, and Texas. All in a basal rosette, its lanceolate to oblanceolate, nearly flat, fleshy leaves are 6 inches to 1¼ feet long and 1 inch to 2 inches wide. Their margins are smooth or minutely-toothed. From 3 to 6 feet tall, the glaucous flowering stalks have about seven bracts and some thirty loosely arranged, solitary, greenish blooms a little under to a little over 1 inch in length, with erect petals approximately one-half as long as the perianth tubes. The dark-purple-stalked stamens, with spotted-whitish anthers, protrude only slightly. Variety *M. v. tigrina* has foliage blotched with purple.

Garden and Landscape Uses and Cultivation. Manfredas are chiefly adapted for informal landscapes and gardens where native plants are emphasized. They are content with dry soils, although preferring somewhat more moisture than most agaves. They succeed in sun or a little part-day shade. Propagation is easy by seed and by division.

MANGIFERA (Mang-ífera) — Mango. The mango is the only well-known member of *Mangifera,* a genus of forty species of tropical Asian trees of the cashew family ANACARDIACEAE. Planted abundantly wherever it will ripen its delicious fruits, it succeeds in southern Florida, southern California, Hawaii, and in most tropical and warm subtropical regions that are almost or quite frost-free and have sufficient water available. Mango fruits are shipped to northern markets in the United States and Canada and are available at high class fruit stores. The name *Mangifera* is from the common name mango and the Latin *fero,* to bear.

These trees are often very large. They have alternate, stalked, smooth-edged, lobeless, firm, leathery leaves and small flowers with four or five calyx lobes and the same number of petals, in branching terminal panicles. Although the stamens may be as many as five, only one or two are fertile. Male, female, and bisexual flowers are borne on the same tree. The large fleshy fruits, structured like plums, contain a single large stone with a solitary seed.

The mango (*M. indica*) is a handsome evergreen tree up to 90 feet in height. Native to Malaya, Burma, and northern India, it attains a huge spread, sometimes as much as 125 feet, and has semilustrous, rich green, usually slightly curved, lanceolate leaves 6 to 16 inches in length with stalks, swollen at their bases, up to 4

Mangifera indica, young tree

Mangifera indica (foliage and flowers)

Mangifera indica (fruits)

inches in length. The leaves hang vertically and have slightly rippled margins that do not lie in one plane. They are more or less crowded at the ends of the twigs. The flowers, slightly over 1/4 inch in diameter, have only one fertile stamen. They are pinkish-white, yellowish, or greenish and in erect, hairy clusters 1 foot long or longer. At first pleasantly fragrant, their odor becomes unpleasant as they age. The smooth fruits are clustered, the individuals 2 to 6 inches long and weighing 6 ounces to 4 pounds, are variously shaped, according to variety. Oval, ellipsoid, almost spherical, obliquely-heart-shaped, or kidney-shaped and yellowish, reddish, greenish, or yellow suffused with red, at maturity they are suspended on long stalks. They have rather thin skins, juicy pulp, which may be sweet or acid and often has a resinous flavor, and a large, oval, flat stone covered with a fibrous coat. This tree belongs to the same botanical family as the poison-ivy and some people develop a rash if their skin comes into contact with the sap of the leaves or shoots or with the peels of the fruits; the flesh of the fruits is harmless.

Mangoes (the plural may also be spelled mangos) have been cultivated for at least 4,000 years as one of the premier fruits of the tropics. In warm climates they occupy the position of importance that apples do in temperate regions and, like apples, are double-duty trees, ornamental as well as food producing. True, they do not present the floral display of their temperate region counterpart, but for this they compensate in being evergreen and having colorful young leafage. Their noble proportions, handsome foliage, and often highly colored ripe fruits combine to make mangoes first-rate ornamentals.

The mango is sacred to Hindus and Buddhists. It is recorded that the great Mogul emperor Akbar planted 100,000 of them in one place. In the sixteenth or early seventeenth century the Portuguese brought the first mangoes to the New World and established them in Brazil. Since, other importations have been made; in the present century those of Dr. David Fairchild and others have been important. Because the mango has been cultivated so long numerous varieties have been developed. These differ not only in size, shape, and quality of their fruits, but in their adaptability to specific regions. In the continental United States, southern Florida is the prime area for mango cultivation. The first kinds grown there were importations, but programs of breeding and selection of the most desirable kinds and their propagation by vegetative methods rather than by seed are bringing good results. Outstanding among Florida-raised varieties is 'Hayden'. This is a seedling of the first Indian mango to be introduced to Florida, a variety called 'Mulgoba', brought there in 1889. Prior to that the only mangoes in Florida were seedlings with fibrous, poorly flavored fruits. Unfortunately, the 'Hayden' mango does not fruit heavily consistently and is not a commercial success even though greatly esteemed for home gardens. Other kinds of Indian mangoes, the results of systematic breeding, are planted commercially, but their fruits are less appreciated than those of the 'Hayden'. For home gardens, varieties of Philippine and Saigon mangoes have received public acceptance. They are easy to grow, come tolerably true from seeds, and their fruits are almost without fibers. However, they lack the rich spicy flavor and aromatic bouquet of 'Hayden'. Importation of mango varieties into Hawaii has also received attention and there, too, the 'Hayden' is considered to be among the choicest. In the Hawaiian Islands these fruits grow best at low altitudes.

Garden and Landscape Uses. Although fruit production is usually the chief objective when mango trees are planted, one can not overlook the very real virtues of the trees as ornamentals. In bloom and in fruit they are decidedly decorative, and even when not so ornamented the trees are very handsome by reason of their shapely tops and splendid foliage. There is a tremendous difference between the fruits of fine varieties and those of nondescript seedlings. The latter are likely to have a minimum of good flesh, much stringy fiber, and to taste strongly of turpentine. In the best varieties these objectionable features are eliminated, and there is ample luscious flesh of the most exquisite flavor. Mangoes are chiefly eaten as dessert fruits. They are also made into chutney and used in other ways.

As houseplants young mangoes raised from seeds are successful and attractive. Admired for their foliage, they neither flower nor fruit. Seeds from ripe fruits planted individually in 4-inch pots in ordinary well-drained potting soil soon germinate in a temperature of about 70°F if the soil is kept reasonably moist. Their

Young mangoes are attractive houseplants: (a) Removing mango seed from fruit

(b) Planting the seed

(c) After about one year's growth

(d) A plant from seed

subsequent care calls for no more skill than is necessary with most easy-to-grow houseplants. It consists of watering, occasionally fertilizing, and transferring to a larger pot when the roots become crowded. When grown in this way the plants need good light but not necessarily exposure to full sun. Winter night temperatures of 65 to 70°F and moderate humidity are appropriate.

Cultivation. In parts of the tropics where there is abundant moisture throughout the year, mangoes often grow to immense size but fruit sparingly. The check to growth provided by a distinctly dry season seems to be necessary to stimulate fruit production. Another cause of failure to bear well is soil excessively rich in nitrogen; not even alternate wet and dry seasons overcome the effects of such soil. About the cultivation of mangoes in the United States much is still to be learned, and would-be planters are advised to seek the advice and follow the recommendations of the State Agricultural Experiment Station, Gainesville, Florida, and the State Agricultural Experiment Station, Honolulu, Hawaii. Special attention should be given to securing the varieties suggested and to the fertilizing and irrigating practices recommended. In general, fertilizers should be applied in spring and early summer, but not in fall. Because mangoes become large trees, although never such immense specimens in the United States as in parts of the tropics, they need ample space, certainly 35 to 45 feet between individuals. Young specimens are very sensitive to cold and in Florida should have their trunks wrapped in November with several inches of straw or similar material. This is necessary only for the first three or four years. Dry weather in late winter and spring may cause the fruits to drop; this may be prevented by thorough irrigations at about ten-day intervals. Propagation in the United States is chiefly by budding, but in the Old World tropics the much slower approach grafting is often employed. Understocks raised from seeds are used in both cases. The seeds must be sown soon after removal from the fruits because they do not retain their ability to germinate for many weeks.

Pests and Diseases. Scale insects and thrips infest mangoes. The chief disease is anthracnose, which causes the flowers to drop and affects foliage and fruits. It is especially prevalent in humid weather. From two to five applications of a neutral copper spray are recommended to control this, the first made just before the flowers open and the second when the fruits begin to set, with additional applications later. The first two applications are the most important.

MANGO is *Mangifera indica,* the mango melon, *Cucumis melo chito.*

MANGOSTEEN is *Garcinia mangostana.*

MANGROVE. See Rhizophora. The button mangrove is *Conocarpus erectus,* the mangrove trumpet tree, *Dolichandrone spathacea.*

MANIHOT (Máni-hot)—Cassava or Manioc or Yuca or Tapioca Plant. One of the most important food plants of the lowland tropics, especially in the Americas, cassava or manioc belongs to the genus *Manihot,* of the spurge family EUPHORBIACEAE. Native only to warm parts of the New World, *Manihot* consists of 170 species of milky juiced herbaceous plants, shrubs, or rarely trees, often with tuberous roots. Its name is an adaptation of the native Brazilian one, manioc.

Manihots have alternate, usually long-stalked leaves, undivided, but often deeply-palmately- (in hand-fashion) lobed. Unisexual flowers of both sexes are on the same plant. They are without petals and in terminal or axillary panicles. They have often colored, petal-like, bell-shaped calyxes with five short or deep lobes. The males have ten stamens. Female blooms have three styles united on their lower parts and sometimes nonfunctional stamens (staminodes). The fruits are capsules.

The cassava, manioc, yuca, or tapioca plant (***M. esculenta*** syn. *M. utilissima*) is a shrubby species with soft, brittle stems, and clusters, somewhat like those of dahlias, of large, tuberous roots. It is 3 to 15 feet tall, and has somewhat the aspect of its relative, the castor bean (*Ricinus communis*). Its leaves are hairless except minutely on the veins of their glaucous undersides. They are deeply-lobed, the lobes being narrowly-lanceolate to obovate and 3 to 6 inches long. The flowers are under ½ inch in length. The capsules are spherical and about ½ inch in diameter. Numerous varieties of cassava are cultivated, but the one or more wild species from which they have been developed are not known. They fall into two main groups, bitter cassavas and sweet cassavas, the latter sometimes distinguished as *M. dulcis.* Bitter cassavas have leaves with three to seven lobes, seed capsules with six winged angles, and tubers that contain such dangerous amounts of hydrocyanic acid throughout that, eaten without proper preparation, they are poisonous. It is remarkable that primitive American Indians devised methods, consisting of successively peeling, shredding, soaking in water, squeezing until nearly dry, and boiling in fresh water, that effectively rid the tubers of the poison and converts them into useful food. The starch of manioc dried as little lumps is tapioca, ground into flour, is Brazilian arrowroot. Unfortunately cassava is deficient in protein and where it forms a major part of the diet malnutrition often occurs. Sweet cassava varieties have leaves with three to thirteen lobes, seed capsules not winged, and tubers like those of bitter cassavas, but with the hydrocyanic acid confined to their outer layers. These can be eaten without the elaborate preparation needed for bitter cassava, but they are less heavy croppers. A very beautiful ornamental cassava, *M. e. variegata* has leaves liberally variegated with pale yellow.

A field of *Manihot esculenta* in the American tropics

Manihot esculenta variegata

The source of Ceara rubber, and in times past extensively cultivated for that, ***M. glaziovii*** is a tree about 40 feet tall. It has leaves with three to five 4-inch-long, toothless lobes and seed capsules without wings. It is native to Brazil.

Garden and Landscape Uses and Cultivation. The last species described is sometimes planted for ornament in the tropics and warm subtropics. It stands dryish conditions and exposure to sun. Except for its variegated foliaged variety, *M. esculenta* is of practically no horticultural interest except for inclusion in educational exhibits of plants useful to man. It, and other kinds, are best satisfied with fertile, sandy soil. They are propagated by cuttings consisting of short pieces of stem and, less frequently, by seed. The variegated-leaved variety is pretty as an ornament in tropical greenhouses and outdoors in warm countries. It is much subject to infestations of red spider mites.

MANILA. This word occurs as part of the common names of these plants: Manila grass (*Zoysia matrella*), Manila-hemp (*Musa textilis*), manila palm (*Veitchia merrillii*), and Manila-tamarind (*Pithecellobium dulce*).

MANILKARA (Manil-kàra)—Sapodilla. The sapodilla or chewing gum tree, previously known as *Achras sapota* and *Sapota achras*, is *Manilkara zapota*. It is quite distinct from the sapote (*Pouteria mammosa*), which has been known as *Achras zapota*. Consisting of seventy species of trees of the West Indies and tropical America, *Manilkara* belongs to the sapodilla family SAPOTACEAE. Its name is an adaptation of a Malayan native one.

Manilkaras are alternate-leaved, milky-juiced trees with foliage crowded at the ends of the branchlets and flowers in clusters from the leaf axils. The flowers have six, more rarely four or eight sepals, corollas with as many lobes (petals) as there are sepals, and stamens or stamens and staminodes (nonfunctional stamens) equaling in number the corolla lobes. There is one style. The fruits are technically berries. The only important member of the group, the sapodilla is of interest not only because of its delicious fruits, but because until synthetic substitutes were developed, chicle, derived from its sap, was the chief basis of chewing gum.

Manilkara zapota, with fruit

The sapodilla (***M. zapota***), native to Yucatan and Central America and widely cultivated in the tropics, exceptionally attains a height of 125 feet, but more commonly does not exceed 75 feet and often is much lower. It has leathery, hairless, lustrous, evergreen, ovate-elliptic to elliptic-lanceolate, toothless leaves 2 to 5 inches long, sometimes notched at their apexes, and with strongly defined midribs. In crowded clusters among the leaves near the branch ends, the stalked, white blooms develop. They have six-lobed calyxes, hairy on their outsides, and urn-shaped, six-lobed corollas ¼ to ½ inch wide. The three staminodes are more or less petal-like. The thin-skinned, rough-surfaced, rusty-brown fruits, scurfy when fully ripe, contain few to several black seeds surrounded by translucent, brown flesh of good flavor.

Other species sometimes cultivated are these: ***M. bidentata,*** of northern South America, Panama, and the West Indies, is up to 100 feet tall or sometimes taller. It has stalked, pointed-oblanceolate to pointed-oblong-obovate leaves with blades 3 inches to 1 foot long. The flowers, in clusters of ten or more, have lobed corollas (petals). The long-stalked, globose to ellipsoid fruits, each with a short, sharp terminal point, are ½ to 1 inch long. This species is the chief source of the nonelastic rubber called balata. Its lumber, known as bully wood and bullet wood, has many uses. ***M. kauki*** (syn. *Mimusops kauki*), of tropical Asia, has long-stalked, broad-elliptic to obovate-elliptic leaves with blades up to 4 inches long and 2 inches wide with a prominent mid-vein and with white, silky undersides. The ¾-inch-long white flowers are in clusters of usually ten or more on stalks about 1 inch long. The long-stalked, red fruits are about ¾ inch long. ***M. roxburghiana*** (syn. *Mimusops roxburghiana*), of tropical Asia, is a large tree with elliptic to oblanceolate hairless leaves up to 3 inches long and 1¾ inches wide,

Manilkara roxburghiana, with fruits

with a prominent midrib. The flowers are in clusters of four or fewer on stalks about 1 inch long. The fruits are stalked and spherical.

Garden and Landscape Uses and Cultivation. In the American tropics, this frost-tender species is cultivated for its edible fruits. It thrives in ordinary good soils and is especially partial to those of limestone derivation. Established trees require no special care. Propagation is by seed.

In greenhouses devoted to displays of plants useful to man, the "chewing gum tree" or sapodilla is popular as an educational exhibit. It grows readily in a humid atmosphere with the minimum winter night temperature 60°F and by day five to fifteen degrees higher. Some shade from strong sun is needed. Repotting and any pruning needed to shape or restrain the plants is done in spring. At all seasons the soil is kept moderately moist.

MANIOC is *Manihot esculenta*.

MANO DE MICO is *Chiranthodendron pentadactylon*.

MANTIS or MANTID. See Praying Mantis.

MANUKA is *Leptospermum scoparium*.

MANULEA (Man-ulèa). One South African species of this genus of fifty, of the figwort family SCROPHULARIACEAE, is occasionally cultivated. Its name derives from the Latin *manus*, the hand, and refers to a fancied faint resemblance of the divisions of the flower. The genus includes annuals, tender herbaceous perennials, and subshrubs. They have mostly opposite, lobeless, toothed leaves that may or may not be all basal and spikes, racemes, or panicles of asymmetrical, tubular flowers with five-lobed corollas.

A charming annual up to 1 foot tall, ***Manulea violacea*** (syn. *Lyperia violacea*) has pointed-ovate leaves about 1 inch long and slender spikes of ½-inch-long violet flowers.

Garden Uses. As an edging, for the fronts of borders, and in rock gardens, this is attractive. Like many South African annuals, it does not take kindly to extremely hot, humid weather and so its use is limited in many parts of the United States. Where cool summers prevail it is well worth growing. It prefers a sunny location and moderately fertile, well-drained soil.

Cultivation. Seeds may be sown in spring where the plants are to remain and the seedlings thinned to 4 to 6 inches apart, but usually it is preferable to start them indoors in a temperature of about 70°F ten weeks before it is warm enough to transplant the resulting plants to the open ground. This may be done when it is safe to set out geraniums. Seedlings raised indoors are transplanted 2 inches apart in flats or individually into small pots and grown in a sunny greenhouse where the night temperature is 50 to 55°F. One week or ten days before they are to be planted in the garden they are stood in a cold frame or sheltered place outdoors to harden. A planting distance of 6 inches is satisfactory.

MANURES. Since before the beginning of recorded history manures—animal excreta usually mixed with such vegetable matter as straw, leaves, hay, or wood shavings used as bedding for animals—have been employed by gardeners and farmers to fertilize the soil. Indeed until well into the twentieth century they were by far the most commonly used sources of plant nutrients.

More concentrated forms of nitrogen, phosphorus, and potassium (the three chief nutritive elements supplied by manures) came into use toward the end of the nineteenth century and later. These, at first known as artificial manures, later were called fertilizers, by which name they are known today. Although manures are fertilizers in the sense that they increase soil fertility, they are usually not referred to as such. In horticultural and agricultural practice and in common understanding a distinction is made between manures and fertilizers. The latter are discussed in this Encyclopedia under Fertilizers.

Reliance upon fertilizers instead of manures increased as they became more generally available and as crop farming not combined with stock raising replaced mixed farming. A more decisive blow came to dependence on manure with the development and vastly increased use of the gasoline engine and the consequent decline in the horse population. Now, except in those few places where local supplies are available, manures play a minor part in horticultural practice. Most gardeners must substitute for them fertilizers and compost and other sources of bulk organic material.

Manures are extremely valuable soil amendments that vary considerably in quality depending upon a number of factors including the kind and age of the animals, their food, the sort and proportion of bedding material it contains, its degree of freshness, and where and how it has been stored. One of the surest ways of depleting manure of its nutrients is to stock pile it outdoors where it is subject to leaching by rain. Manure in storage should always be covered.

Besides supplying nutrients, although scarcely in well-balanced amounts of each, manures according to kind contribute greater or lesser amounts of organic matter, which as it decays adds valuable humus to the soil.

Cow manure averages about 0.4 percent nitrogen, 0.3 percent phosphoric acid, and 0.45 percent potash. Unlike horse manure it remains cold and wet when piled or stacked. Partially rotted, it is excellent for all soils, especially those of a light, sandy nature. Apply a cubic yard to 200 to 400 square feet.

Horse manure contains on an average approximately 0.75 percent nitrogen, 0.55 percent phosphoric acid, and 0.65 percent potash. Piled when fresh, it generates much heat and unless turned (forked over) occasionally it may as a result lose some of its value. Looser and more fibrous than cow manure and unless half-rotted more likely to harm plants by "burning" their roots, horse manure is excellent on all soils, particularly clayey ones. Apply at the rate of a cubic yard of a fairly compact, half-rotted sample to 250 to 500 square feet.

Pig manure averages about 0.3 percent nitrogen, 0.35 percent phosphoric acid, and 0.45 percent potash. Its qualities are similar to those of cow manure, but because of its consistency and odor it is less pleasant to work with. Best for light, sandy soils, apply a cubic yard to each 150 to 300 square feet.

Poultry manure averages 1.8 percent nitrogen, 1.0 percent phosphoric acid, and 0.5 percent potash. Because it becomes greasy in texture and difficult to handle when wet, store this under cover, preferably mixed with a little dry soil to make it more crumbly and easier to spread, or add it to the compost pile. This is a quickly available source of nitrogen, can easily result in "burned" roots if applied in too great amounts. Use fresh poultry manure at 100 pounds to 200 to 300 square feet on land vacated of plants, at about one-quarter that rate around living plants.

Manures of rabbits and other small animals including pets contain varying amounts of plant nutrients and are usually best managed by adding them to compost piles.

Commercial pulverized and shredded, dried or dehydrated manures are the only ones available to many gardeners. Because of their comparatively high cost per unit of nutrients and because at rates commonly applied the amounts of organic matter they add is small, their use is largely limited to adding in small amounts to potting soil mixes. Commercial dehydrated cow, sheep, and goat manures average 1.2 percent nitrogen, 1.2 percent phosphoric acid, and 2.3 percent potash, the poultry manures average 5 to 6 percent nitrogen, 2 to 3 percent phosphoric acid, and 1.2 percent potash.

Green manures are not manures in the sense the word is employed here. They are crops of living plants, such as buckwheat, winter rye, and soy beans, grown for the express purpose of turning them under when partly grown to add organic matter and improve the texture of the soil. See Cover Crops.

MANZANITA. See Arctostaphylos.

MAPLE. See Acer. Flowering-maple is the colloquial name for *Abutilon.*

MARAH (Màr-ah) — Wild-Cucumber, Big Root, Chilicothe, Man Root. The Pacific Coast region is home to the seven species of *Marah,* of the gourd family CUCURBITACEAE. The name is aboriginal. The genus is closely related to *Echinocystis* and is composed of herbaceous plants with large perennial tubers, and leafy stems that last for only one season. The leaves are alternate, rounded in outline, and with more or less heart-shaped bases. They are three- to nine-lobed or cleft. The tendrils, opposite the leaves, are branchless or three-branched. The flowers are unisexual. They have five-lobed calyxes, or in the male flowers these may be wanting, and wheel-shaped to shallowly-bell-shaped corollas, with commonly five lobes. The males have three or four stamens joined into a spherical head; in the females, the stigma sits on top of a very short style or directly on the ovary. Male blooms are in racemes or panicles and come from the same leaf axils as the solitary females. The fruits are spherical to egg-shaped, usually conspicuously spiny, and contain several seeds.

The chilicothe (***M. macrocarpus*** syn. *Echinocystis macrocarpa*) inhabits dry soils in southern California and Baja California. Its stems are up to 20 feet in length, and have leaves with blades 2 to 4 inches wide and slightly- to deeply-lobed. The flowers are white and somewhat cup-shaped, the males up to ½ inch in diameter, the females somewhat bigger. The densely-spiny, cylindrical fruits are 3½ to 4½ inches long. Similar to the last, but with flatter, cream-colored flowers and 2-inch-wide, spherical fruits with pointed tips is ***M. fabaceus*** (syn. *Echinocystis fabacea*), a native of California.

The man root (***M. oreganus*** syn. *Echinocystis oregana*) ranges from California to British Columbia. Sparsely-hairy or nearly hairless, it has stems up to 20 feet long and roundish leaves with shallowly-lobed blades 3½ inches to 1 foot across. The flowers are white and distinctly bell-shaped, the males almost or quite ½ inch in diameter, the females a little bigger. Ovoid, and tapering at both ends, the fruits, approximately 2 to 3 inches long, have a few weak spines.

Garden and Landscape Uses and Cultivation. In their native regions and places of similar climate, marahs are occasionally cultivated as bank covers, vines for summer screens, and similar purposes. They need well-drained soil and sun or part-shade. They are propagated by seed.

MARANG is *Artocarpus odoratissimus.*

MARANTA (Marán-ta)—Arrowroot, Prayer Plant. In gardens it was for long common practice to apply this name to species of *Calathea* and sometimes to species of *Ctenanthe* and *Stromanthe,* which like *Maranta,* are members of the maranta family MARANTACEAE. Fortunately, this is now less often done. In appearance *Maranta* and *Calathea* are quite similar. The difference lies in recondite technical details of the ovaries of the flowers and in the blooms of *Maranta* being in branched arrangements (inflorescences), whereas those of *Calathea* are generally crowded in often conelike heads. In the wild, *Maranta* is restricted to the American tropics. It comprises twenty-three species, chiefly plants of the forest floor. The name dedicates the group to the sixteenth-century physician and botanist Bartolommeo Maranti. From the roots of *M. arundinacea,* tapioca and West Indian arrowroot are obtained.

Marantas are erect or prostrate, stemmed or stemless, evergreen, sometimes tuberous-rooted, herbaceous perennials with all basal or basal and stem leaves that have stalks that sheathe or partially sheathe the stems or each other. The leaf blades are undivided and without lobes or teeth. The flowers, in panicles that are to a greater or lesser extent branched and have spathe-like bracts, make but little show. They are small and markedly asymmetrical. They have three sepals, three petals (one larger than the others), one fertile stamen, and usually five more or less petal-like staminodes (nonfunctional stamens) of which two are conspicuously bigger than the others. The fruits are capsules.

Arrowroot (***M. arundinacea***) has tuberous roots, and slender, jointed, zigzag-forked stems 1½ to 6 feet tall. Its longish-stalked, pointed leaves are ovate-oblong to ovate-lanceolate or narrowly-heart-shaped and 6 inches to 1 foot long by 1½ to 4 inches broad. Loosely arranged, the little white blooms have an upper broad staminode longer than the petals and notched at its apex. Its foliage variegated with yellow or white, *M. a. variegata* is sometimes known in gardens as *Phrynium variegatum* and *P. micholitzii.*

Very short-stemmed, up to 1 foot tall ***M. leuconeura*** (syn. *M. l. massangeana*), of Brazil, has no tubers. Its broad-elliptic or oblong, blunt or short-pointed leaves up to 6 inches long, are whitish along the veins and to a greater or less extent dark-blotched along their edges. Their undersides are powdery-glaucous or purplish. The little flowers are white and purple. Attractive *M. l.* 'Manda's Emerald' has emerald-green to chartreuse leaves with blotches, dark chocolate-brown on young foliage, that fade to moss-green as the leaf ages. The bright to pale grayish-green leaves of the common prayer plant (*M. l. kerchoveana*) lack the white veins of the typical species. They are clearly marked with two rows of rich chocolate blotches that change with age to deep green. Their undersides are blotched with purplish-red. Similar, and undoubtedly a dwarf form of the common prayer plant, is a smaller-leaved variant that has been called ***M. repens.*** Very beautiful, the elliptic-obovate leaves of *M. l. erythroneura* (syn. *M. fascinator*) are up to 4 inches long or slightly longer. Their downy midribs are narrowly bordered with light green, outside of which is a zone or row of irregular dark blotches on a ground of light green. The side veins and sometimes the basal parts of secondary veins are pink to rose-red, most intensely so on young foliage. The under surfaces of the leaves are pale green with red veins and reddish-purple blotches. The purple-based staminodes of the flowers are rosy-lavender.

Maranta leuconeura kerchoveana

Maranta leuconeura erythroneura

Maranta leuconeura 'Manda's Emerald', in bloom

Maranta repens of gardens

From the common prayer plant and its varieties, ***M. bicolor*** differs in having tuberous roots. From 6 inches to a little over 1 foot tall, it is stemless or almost so, and has elliptic to oblong or ovate, somewhat wavy-margined leaves 3 to 6 inches long. Their upper surfaces are glaucous-green, with a jagged-edged, paler, broad central band bordered with dark green, and with grayish-green along the leaf margins. The flowers are white, spotted and striped violet.

Garden and Landscape Uses and Cultivation. Marantas need warmth, high humidity, and shade. Their soil must be moist, but not saturated and contain generous amounts of organic matter. These are splendid ornamentals for gardens in the tropics and for greenhouses in which minimum winter temperatures are not below 60°F. The common prayer plant finds some favor as a houseplant, but ordinary room conditions are usually too dry for it. It does better in a terrarium or elsewhere where the air is humid. Too intense light causes the leaves to curl. Repotting or replanting is best done at the beginning of a new season of growth, which in North America means spring. Well-rooted specimens respond to fairly frequent applications of dilute liquid fertilizer. Increase is usually by division.

MARANTACEAE—Maranta Family. Perennial stemmed or stemless herbaceous plants totaling thirty genera and 400 species compose this family of monocotyledons. They are natives of the tropics, mostly of the Americas. Included are many ornamentals esteemed for their handsome, often beautifully patterned and colored leaves. Arrowroot is a useful product of *Maranta.* From the ginger family *Zingiberaceae,* the *Marantaceae* is readily distinguished by the presence of a conspicuously swollen joint at the junction of the stalk and blade of the leaves.

Members of this family have rhizomes.

Their leaves are in two ranks with the lower parts of their stalks sheathing each other. Often, but not always they are all basal or apparently so. Undivided, they have linear, elliptic, oblong, or paddle-shaped blades with, angling from the midribs in pinnate-fashion, many parallel lateral veins. The asymmetrical flowers are in spikes or panicles usually with spathelike bracts and arising from the stems or directly from the rhizomes. Each has a three-parted calyx and a corolla of three petals, joined at their bases. There is one often petal-like fertile stamen, up to five petal-like nonfunctional stamens (staminodes), and one style, flat, twisted, and lobed. The fruits are capsules or berries. Genera cultivated include *Calathea, Ctenanthe, Hybophrynium, Maranta, Monotagma, Phrynium, Pleiostachya, Stromanthe,* and *Thalia.*

MARATTIA (Mar-áttia). Sixty species of stately ferns of the tropics and subtropics of the Old and the New World and of New Zealand compose *Marattia,* of the marattia family MARATTIACEAE. The name commemorates the Italian botanist Giovanni Francesco Maratti, who died in 1777.

Marattias have much the aspect of related *Angiopteris,* from which they differ in their spore capsules being fused in rows inside elongated, purselike containers that appear as neat stitches set along or near the margins of the leaf segments at right angles to them. They have massive, short, globose stems or crowns, and large, twice- or thrice-pinnate and pinnately-lobed fronds (leaves) that have long stalks conspicuously swollen, and flanked by stipule-like appendages at their bases where they join the crowns. In countries where these ferns grow wild their succulent crowns, after boiling or baking, were often eaten by the native peoples.

New Caledonia is home to ***M. attenuata.*** This sort has thrice-pinnate leaves up to about 8 feet long and 4 feet wide. The leaflets, 1 inch wide or wider, have toothed apexes. Twice-pinnate fronds, up to 12 feet in length, are typical of ***M. smithii,*** of

Marattia attenuata

Fiji, New Guinea, and Polynesia. Central America is home to ***M. excavata,*** which has fronds three-times-pinnately-divided. The leaflets, 6 to 10 inches long and sometimes sickle-shaped, have toothed segments. The fronds are 3 to 7 feet long. Endemic to Hawaii and the only species native there, ***M. douglasii*** has glossy, dark green, three-times-pinnate fronds up to about 12 feet long. Their ovate to oblong leaflets are toothed and about 1 inch long.

Indigenous to New Zealand, Australia, South Africa, tropical Asia, Tahiti and other islands of the Pacific, ***M. fraxinea*** (syn. *M. salicina*) has basically triangular-ovate fronds up to 15 feet long and 6 feet wide, with stalks as long as 3 feet. Mostly they are twice-, less commonly thrice-pinnate. The primary leaflets may be 3 feet long, their final divisions up to 1 foot long by ¾ inch wide. They have wavy, incurved margins and on their undersides clusters of spore capsules that appear like stitches along their edges.

Garden and Landscape Uses and Cultivation. Most of these highly ornamental ferns give satisfaction only in the humid tropics and in tropical greenhouses. The New Zealand species succeeds under somewhat cooler, subtropical or warm-temperate conditions. Outdoors they are splendid ornaments for watersides and other moist-soil areas. In greenhouses they may be accommodated in ground beds or large pots or tubs in a coarse soil containing abundant decayed organic matter. To assure constant moisture for their roots it is well to keep container specimens standing in shallow water even though, unusual for ferns, their fronds, soon after water becomes available, recover from wilting as a result of temporary dryness. Shade from strong sun is needed. In greenhouses a minimum winter night temperature of 60°F and higher temperatures by day are desirable for most kinds, but the New Zealander is satisfied with five or even ten degrees less. Propagation is usually by scales taken from the bases of the fronds, which are placed on a bed of moist peat moss and sand or sphagnum moss in a humid atmosphere in a temperature of 70 to 75°F. Spores may also be used to secure increase. For more information see Ferns.

MARATTIACEAE — Marattia Family. This family of ferns is by some authorities divided into four lesser families. Considered as one, it comprises half a dozen genera that include a large number of tropical and subtropical species, nearly all with pinnate fronds with joined or separate clusters of spore capsules on their under surfaces. Very exceptionally they have fronds of one leaflet. Only *Angiopteris* and *Marattia* seem to be cultivated.

MARCGRAVIA (Marc-gràvia). Indigenous to Central America, tropical South America, and the West Indies, *Marcgravia* commemorates by its name George Marggraf, a German writer on the natural history of Brazil. He died in 1644. It comprises fifty-five species and belongs in the marcgravia family MARCGRAVIACEAE.

Marcgravias are climbing, shrubby plants that sometimes when young root into the soil, but later exist on trees without attachment to the ground. They take no nourishment from their hosts and so are not parasites, but instead are epiphytes that live in the same way as many orchids and bromeliads. Each individual has two very different kinds of shoots, sterile ones, with small, stalkless leaves in two ranks, that send out roots and attach themselves to tree trunks and other supports, and flowering and fruiting ones without roots and with bigger leaves arranged spirally and with at least short stalks. The flowers are in terminal clusters with the central blooms aborted and their bracts represented by stalked, pocket-like, brightly colored nectaries containing fluid that attracts hummingbirds. The fertile blooms hang upside down on the pendulous flower clusters. They have six sepals, a caplike corolla that falls in one piece, and twelve to many stamens. The many-seeded, more or less fleshy fruits are spherical to egg-shaped. The plant often mistakenly named *M. paradoxa* is *Raphidophora celatocaulis.*

Most common in cultivation, ***M. rectiflora,*** of Puerto Rico and other West-Indian islands, has rooting stems with ½-inch-long, elliptic to ovate leaves and lanceolate-elliptic to elliptic, flowering shoot

Marcgravia rectiflora

leaves 3 to 5 inches long. The flowers are greenish tinged with red. A native of Jamaica, ***M. umbellata*** varies considerably in aspect at different stages of its growth. At first a rather weak vine with short-stalked, heart-shaped leaves, it later develops many slender, drooping branches with two ranks of alternate, nearly stalkless, elliptic leaves 3 to 5 inches long. The flower clusters consist of several normal flowers

Marcgravia rectiflora (fruits)

Marcgravia umbellata

and, at their centers, three to five aborted flower stalks often with a flower or aborted bud at the top.

Garden Uses and Cultivation. In the humid tropics marcgravias can be used as self-clinging vines. They also make very satisfactory plants for humid, tropical greenhouses and grow well in pots or other containers in well-drained soil that contains an abundance of organic matter and is kept evenly moist, but not stagnant. A minimum winter night temperature of 55 or 60°F is satisfactory, with an increase of up to about fifteen degrees by day and warmer conditions at other seasons. Shade from strong sun is needed. Increase is easily had by cuttings and by seed.

MARCGRAVIACEAE. About 100 species distributed in five genera compose this tropical American family of dicotyledons. They are evergreen climbing shrubs or vines, frequently epiphytic, with alternate, undivided leaves and usually pendulous flower clusters, the bracts of which are commonly represented by brightly colored nectaries. The calyxes of the flowers are of four or five sepals, the corollas of the same number of lobes or petals. There are three to many stamens and a short style. The fruits are capsules. Only *Marcgravia* seems to be cultivated.

MARCH, GARDENING REMINDERS FOR. Except in the coldest parts of the north, March brings the beginning of spring or at least an immediate prelude to it. In milder regions that welcome season has already arrived. In either case this is a busy time for gardeners. All aware of the unseemly haste with which winter gives way to spring and soon to summer know full well the importance of "keeping up with the season." Once you fall behind with spring work, it is difficult indeed to catch up.

As early as conditions permit prepare planting sites: (a) Fertilizing

(b) Forking

Tasks that can profitably be undertaken now depend on local climate and taking advantage of every favorable opportunity the weather affords. On days when the ground is in suitable condition for the preparation of planting sites and making ready for seed sowing, and when pruning and other outdoor work can be done, push on, waste no favorable hour. If you "miss the boat" by not sowing peas and sweet peas or planting onion sets at the earliest possible date, or if you delay long after the weather makes possible the readying of areas where lawn grass is to be sown, results will be less satisfactory than if you were "on the ball."

Where climate permits, begin cautiously to remove winter protection from plants that have been covered but do not be in too great a hurry to expose them to harsh winds and bitter cold that may follow spells of mild weather in this treacherous month. In cold parts of the north, leave this task until next month and in any case with rock garden and other low perennials protected with salt hay, branches of evergreens, or similar loose material take it off in two or three stages instead of all at once. Follow the same procedure in taking down soil hilled up to roses. If possible choose dull humid days for this work of uncovering.

Prune deciduous trees and shrubs, including roses in need of this care, but unless you are undertaking a drastic renovation of old, much overgrown specimens of kinds you know will readily renew themselves from the base, do not prune spring-flowering shrubs now. Leave that until after they are through blooming. Also, it is best to delay pruning birch trees and maples until midsummer or at least until they are in full leaf. Then the pressure of rising sap is much less and the cuts will not "bleed" to the horrifying extent any made in spring will. Do not forget to pare smooth the edges of large cuts and to coat

Prune, before new growth begins:
(a) Roses

(b) Other deciduous shrubs and trees

(c) Pare the edges of cuts above one inch in diameter with a sharp knife

(d) Cover with tree wound paint

all more than about an inch in diameter with tree wound paint.

Complete pruning apples and other orchard fruits. Above all hurry to attend to grapes because if these are left late the upward surge of sap causes the wounds to "bleed." If you failed to cut second-year canes out of red raspberries after they fruited last summer, do so immediately.

Complete the pruning and thinning out of red raspberries and tie retained canes to wires

Also, thin out weak new canes and cut off the upper foot or so of stronger ones before tying them neatly to the wires stretched to support them.

Drastic renovation pruning of overgrown or unshapely shrubs and small trees that after severe cutting are capable of sprouting strongly from the stubs is now in order. Kinds that respond include lilacs, rhododendrons, and yews. It takes confidence and a stout heart to "butcher" large specimens to the extent needed for the best results and with flowering sorts this season's bloom must be sacrificed and next year's probably foregone. But if cut to within 6 inches to 1 foot of the ground, fertilized, mulched, and kept watered during dry spells in summer, such plants will renew themselves and if not crowded by others or denied adequate care soon make shapely specimens.

Hedges, deciduous and evergreen, that are very unshapely or overgrown may be treated similarly if they are of privet, rose-of-Sharon, yew, or other plants that grow freely after being cut back drastically, but

Overgrown hedges of privet and of some other shrubs and trees can be renovated by pruning them back severely at this time

to prune hedges of hemlocks or other unresponsive kinds in this way would invite disaster. Hedges that have been reasonably cared for through the years will need no such drastic action, but routine trimming and shaping may be appropriate.

Dormant spraying to control scale insects and other pests may be done when the temperature is 40°F or above and there is no danger of it dropping at night to the freezing point. Complete any needed pruning in advance of spraying.

Plant deciduous trees and shrubs as soon as the ground is workable and in a suitable crumbly state, but if you can, delay planting evergreens until they are just about to start into new growth. If bare-root trees or shrubs are delivered before you are ready to plant, heel them in temporarily.

Plant bare-root trees and shrubs before their new growth begins

Delay no longer than weather necessitates setting in their flowering quarters such spring-flowering biennials as English daisies, canterbury bells, forget-me-nots, foxgloves, pansies, violas, and wallflowers. If these are to be moved from cold frames, harden them gradually and for a couple of weeks before actual transplanting leave them completely without glass sash or other protection day and night.

Work to be done with perennials this month or next, depending upon the for-

Foxgloves: (a) Wintered in cold frames

(b) Transplanted to their flowering stations now

(c) Blooming in June

(c) Pansies

wardness or otherwise of the season in different localities and how soon it is safe to remove winter covering, may include lifting, dividing, and replanting clumps of such vigorous growers as asters, heleniums, and phloxes that have become too big, or if this is not necessary thinning out crowded shoots. When dividing old clumps of perennials, keep for replanting younger, more vigorous parts and discard inner, older portions.

Whether dividing and replanting is part of your spring program or not, apply a dressing of a complete fertilizer to perennial beds and prick it into the surface with a spading fork, taking great care not to damage the shoots of sorts such as balloon flowers, rose-mallows (*Hibiscus moscheutos*), and lilies that are notorious late-starters.

(d) Sweet-williams

Other biennials, or plants grown as biennials, to plant in their flowering quarters include: (a) English daisies

Tidy ground beneath shrubs except strongly surface-rooting ones, such as azaleas and other rhododendrons, mountain-laurels, and many other kinds of the heath family ERICACEAE, by forking it shallowly and if compost or other suitable organic material is available spreading a layer of it as a mulch. If only a limited amount of mulch can be had reserve it for evergreens, especially those of the heath family.

Lawn work that in many parts of the north must be delayed until April may be carried out in March in parts of that region where spring comes relatively early. Consult April, Gardening Reminders For in this Encyclopedia.

In greenhouses there is call for much activity. Longer, sunnier days favor rapid growth and one must step lively to keep up with all that needs doing. Even a few days delay in transplanting seedlings can cause something akin to disaster and sowing dates missed and cuttings taken too late result in less satisfactory plants than planned. Attend promptly to such matters as shading where this is needed to protect tender foliage from being scorched and see that humidity is maintained at desirable levels by wetting paths, floors under benches, and other surfaces sufficiently frequently. On sunny days, but not at other times, most plants benefit from being misted or sprayed very lightly with water, but not so heavily that the soil is wetted or the moisture will not dry within an hour. Never spray so late in the day that the foliage is wet at nightfall.

Ventilating must be done with especial care during this fickle month. Do this on all favorable occasions to admit air and reduce temperatures appropriately, but be conscious always of the dangers of allowing cold drafts to strike the plants.

Watering needs more frequent attention now. On sunny days soil in pots and benches dries more rapidly and roots are taking up more water. This must be replaced without keeping the soil in a state of constant saturation. Certainly it must not be allowed to dry to the extent that foliage wilts, but this phenomenon is not always caused by dryness of the soil (it can indeed be the result of excessive wetness that has caused the roots to rot), but at this time of the year, it may occur when a bright sunny day follows several dull days. Then, lightly spraying the foliage with water is more helpful than soaking soil already wet.

Fertilizing well-rooted, actively growing pot plants does much to sustain vigor and health. Attend to this routinely, weekly or

(b) Canterbury bells

Fertilize and lightly fork over established beds of perennials

Regularly fertilize well-rooted specimens of: (a) Agapanthuses

(b) Calla-lilies

(b) Marigolds

(c) Clivias

(c) Petunias

more frequently if dilute liquid fertilizer is employed, less often with fertilizers that release their nutrients slowly over longer periods. Among plants benefited by this attention are agapanthuses, azaleas, calla-lilies, cinerarias, clivias, hydrangeas, jasminums, schizanthuses, stocks, sweet peas, and other annuals such as asters, calendulas, and clarkias, that are flowered indoors.

Seed sowing of annuals and some vegetables for planting outdoors later demands much attention in March. True, a few slow-growing kinds will have been sown earlier, but now is the time for most sorts. Too early sowing gives plants too big and if in flats too crowded by planting-out time. It is far better then to have specimens that could have remained without harm for another week or two in flats or pots, than to have to set out plants starved and lanky from having been too long cramped. Among the many annuals to sow indoors for later use outdoors are annual phloxes, asters, globe-amaranths, marigolds, petunias, stocks, strawflowers, and zinnias. Vegetable seeds to sow indoors now include those of broccoli, cabbage, cauliflower, celery, eggplant, lettuce, peppers, and tomato.

Also, in readiness for outdoor planting

(d) Straw flowers

Seeds of annuals to sow indoors in March include: (a) Globe-amaranths

(e) Zinnias

Vegetable seeds to sow indoors in March include: (a) Cabbage

(b) Lettuce

(c) Tomatoes

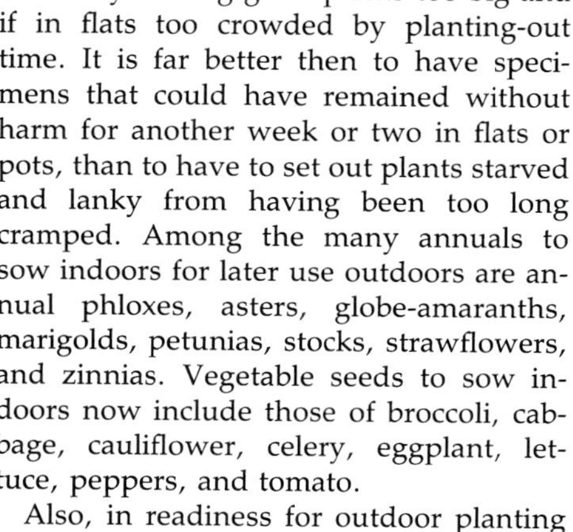

later, start into growth tubers of caladiums, cannas, and tuberous begonias. There is still time to take cuttings of such fast-growers as ageratums, coleuses, and wax begonias to have small plants for setting outdoors in late May or early June, but get these in as early as possible.

Cuttings of a vast majority of greenhouse plants that accommodate to this method of propagation root readily now and the young plants have ample time to achieve respectable size before fall. A brief selection of sorts includes abutilons, acalyphas, aphelandras, begonias, coleuses, chrysanthemums, English ivies, fuchsias, geraniums, *Iboza riparia,* jacobinias, philodendrons, ruellias, *Salvia leucantha,* and stevias. And do not forget leaf cuttings of African-violets, peperomias, pick-a-back plant, and rex-cultorum begonias, as well as such succulents as echeverias, kalanchoes, and sedums.

For summer display start into growth indoors tubers of: (a) Caladiums

(b) Cannas

(c) Tuberous begonias

(d) A begonia tuber

(e) Planting a caladium tuber

Cold frames are extremely useful. Besides affording winter protection for plants on the borderline of hardiness, unheated frames serve as way stations, hardening-off quarters, for plants raised in greenhouses for planting later outdoors. In most parts of the north this holding function will not come into serious use until next month, but a start on the season may be stolen by sowing seeds of transplantable hardy annuals and vegetables, asters, broccoli, cabbages, and lettuce and the like in frames three weeks earlier than outside sowing is practicable and transplanting the young plants outdoors when conditions are favorable.

Propagate many indoor plants now from cuttings: (a) Chrysanthemums

(b) African-violets (from leaf cuttings)

The management of cold frames through this weatherwise unpredictable and often blustery month calls for care. Ventilate those containing plants that have wintered in them freely whenever the air temperature is above the freezing point, but be more gentle with frames in which you have sown early annuals or vegetables or into which you have transferred plants from indoors. In them aim to maintain a cozy 45° to 55°F on sunny days, ventilating carefully to prevent the sun building up temperatures inside the frame high enough to harm the young plants, but at the same time avoiding cold drafts.

Hotbeds are a great boon to gardeners without or hard pressed for greenhouse space. This month they can serve most excellently as places to raise and temporarily accommodate plants to be set outdoors later. Such plants may be in pots or flats or in a bed of soil forming the floor of the hotbeds.

The management of hotbeds is much like that for cold frames detailed above, but with even more regard for the avoidance of cold drafts that might harm delicate foliage. Choose warm days and times of day to attend to watering and other chores that necessitate opening the sash. Covering the sash on cold nights retains heat.

Most houseplants now if not earlier should be showing signs of new growth. If you have not already done so appraise your collection and decide what is to be done with each specimen. Some perhaps you will want to discard after propagations have been secured from them, some undoubtedly will benefit from pruning to shape and seemly size, many will be in need of repotting into bigger containers or after old, worn-out soil is pricked away from their roots into pots of the size they previously occupied. Specimens you do not repot will be helped by top-dressing, pricking off as much surface soil as is easily removable and replacing it with rich new soil with which fertilizer has been mixed. Make certain that the drainage in the pots of all houseplants is functioning properly. If there are suspicions that it is not take the plant out of its pot and if corrections are needed make them.

Repot houseplants in need of this attention

Scrutinize houseplants closely for pests, especially aphids, mealybugs, scale insects, red spider mites, and whiteflies. If confirmatory evidence is found take prompt remedial measures. Often much good can be done without harm to the plants by cutting out shoots and removing leaves that are worst affected and treating the parts that remain with an insecticide. Physical removal of such calamities as mealybugs and scale insects by rubbing with a sponge dipped in diluted insecticide and squeezed nearly dry is effective. A forceful spray of water, directed to the undersides of the leaves, discourages mealybugs and red spider mites.

Cleanse houseplants adaptable to such treatment, of mealybugs and red spider mites by directing a forceful spray of water on them

Plants that appreciate full exposure throughout the winter, but are intolerant of the stronger sun that spring and summer bring need consideration now. Move such kinds as African-violets, sensitive sorts of begonias, ferns, and pick-a-back plants to locations where they will not receive full middle-of-the-day sun, or shade them lightly during that period.

New plants from old ones of many kinds are easy to secure and no season offers better chances of success with numerous sorts than spring. Simple division at potting time suffices with such subjects as aspidistras, clivias, rhizomatous begonias, snake plants, and others of more or less tufted growth. Air layering is a sure and effective way of securing youngsters from crotons (codiaeums), dieffenbachias, dracaenas, rubber plants, and kinds that similarly develop stems that become too tall and leggy. But cuttings are probably most often resorted to as means of propagating houseplants. A vast number of sorts lend themselves to this method of increase, most by leafy stem cuttings, some by leaf cuttings. With all except cactuses and other succulents, shade from direct sun and a more humid than normal atmosphere is necessary for success. By keeping them in a terrarium or under a framework of wire covered with transparent polyethylene plastic these requirements are easily met. Cuttings of many tropical plants root more readily if the perlite, vermiculite, or sand in which they are planted is gently warmed to a little above air temperature. Where practicable, standing the propagating box over a radiator may be helpful.

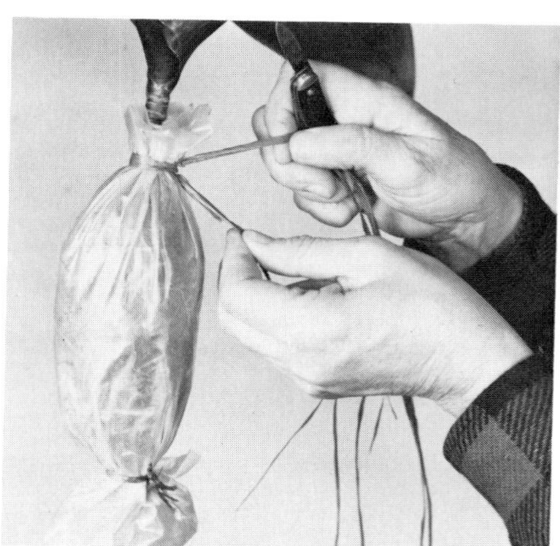
Air layering is a satisfactory way of propagating dracaenas and some other sorts of tall-stemmed houseplants

In the south as soon as azaleas and camellias are through blooming attend to any needed pruning, fertilize them with an acid fertilizer and mulch with bagasse, compost, peat moss, or other suitable material, or omit the fertilizer, and mulch with rotted manure. Should your soil not be sufficiently acid, supplement the manure with an application of iron chelate, sulfur, or aluminum sulfate.

If you neglected the pruning and fertilizing discussed in the Encyclopedia entry February, Gardening Reminders For and you are in the upper south delay no longer. In the deep south, the season is now too far advanced for this type of pruning, but fertilizing may be done and repeat applications made where earlier use was made of kinds that leach fairly quickly from the soil.

Bare-root planting of shrubs and trees that naturally leaf out rather late and of roses is still practicable in early March in the upper south, but not in regions where they are already in foliage. But do not wait a day longer than absolutely necessary. To adopt a legal cliché, time is of the essence. Prune a little more severely specimens set out now than you would similar ones planted earlier.

Trees, shrubs, and herbaceous perennials from containers can be planted with every prospect of success after they are in foliage and in some cases in bloom. Take care not to break their root balls and after they are installed water them copiously and mulch the ground around them.

Evergreens, and some deciduous trees and shrubs, such as azaleas, dogwoods, and magnolias, dug with substantial balls

Transplant, this month or the next, evergreens dug with substantial root balls: (a) A small false-cypress

(b) A larger pine

of earth about their roots kept intact by wrapping and tightly lacing them in burlap or by other effective means, can be transplanted with great success in March. Complete the job by thoroughly soaking the soil with water, and mulching.

In some areas it is still practicable to lift, divide, and replant herbaceous perennials in need of this attention, but unless this is accomplished before new growth is 2 to 3 inches high it is usually advisable to postpone the job until fall. Give first attention to sorts that start early and grow rapidly. Not all perennials need dividing every year or even every second or third and some, such as peonies and oriental poppies, are better left undisturbed as long as they are performing well.

Bulb plants of many kinds, including some with corms or tubers instead of true bulbs, may be planted in most sections in March, but in parts of the upper south planting should be delayed until the following month. Sorts to consider include caladiums, cannas, crinums, dahlias, elephant ears, hedychiums, gladioluses, gloriosas, montbretias, oxalises, spider-lilies, tigridias, tuberoses, and zephyranthes. To assure a long season of bloom successional plantings at two-week intervals may be made of gladioluses, montbretias, and tuberoses. If cannas or dahlias were left in the ground over winter, lift them before new growth begins, wash the clumps of tubers free of earth, and divide them and replant in newly spaded and fertilized soil.

Seed sowing now needs much attention. Annuals in wide variety, excluding only those most sensitive to cool weather, can be sown now in the middle and upper south and so can all the vegetables recommended for sowing in February, Gardening Reminders For. In localities with mild climates, sow the first corn. In the vegetable garden too plant asparagus, horseradish, potatoes, and rhubarb. In the lower south it is time to sow or plant beans, beets, carrots, corn, cucumbers, eggplants, melons, New Zealand spinach, okra, peppers, pumpkins, squash, tomatoes, and watermelons.

Lawns that were oversown with rye grass should be cut frequently to a height of 1 inch to assure the permanent grasses enough light to encourage new growth. New lawns may be made between now and midsummer of Bermuda, centipede, St. Augustine, and zoysia grasses that are grown from sprigs or plugs.

Routine measures for pest and disease control to be initiated include regular spraying of roses and of azaleas to control petal blight.

On the West Coast, March gardening procedures closely parallel those detailed above as appropriate for gardens in the south. They include the completion of planting deciduous trees, shrubs, roses, and herbaceous perennials, in northern parts by the end of the month, earlier southward. Planting avocados, bamboos, citrus fruits, camellias, gardenias, guavas, hibiscuses, and other subtropical sorts may be done this month and into early April. Bulb plants of kinds suggested for planting this month in gardens in the south are also appropriate for setting out in West Coast gardens.

Annuals and vegetables, except in colder sections and except decidedly warm-weather sorts, may be sown now in great variety. Other tasks to give attention to are the pruning of spring-blooming shrubs that need this care and of course measures to keep weeds, pests, and diseases under control.

MARCOTTAGE. See Air Layering.

MARE'S TAIL. See Hippuris.

MARGINATOCEREUS (Margináto-cereus). As *Marginatocereus* those who split genera finely accept one beautiful Mexican species, of the cactus family CACTACEAE. More conservative botanists include it in *Pachycereus* or *Lemaireocereus.* The name alludes to the striking contrast its areole-margined ribs make with the body of the stem. It derives from the Latin *marginatus,* a border, and the name of the related genus *Cereus.*

Known as organ pipe cactus, ***M. marginatus*** (syns. *Pachycereus marginatus, Lemaireocereus marginatus*) commonly branches from its base, becomes treelike, and is up to 25 feet tall. Its erect stems, 5 to 6 inches thick and dark green, have five or six broad, high, triangular ribs with confluent white-woolly areoles forming continuous lines along their crests. Each areole sprouts a cluster of very short, conical spines, at first dark red, later whitish. Each cluster has seven radials not over 1/8 inch long and one or two centrals a little over 1/2 inch long. The flowers, 2 inches long by 1 inch wide, are reddish on their outsides, greenish-white within. The fruits are about 1½ inches across. Variety *M. m. gemmatus* has slender, needlelike spines 1 inch, and pink flowers 1½ inches, long. The name organ pipe cactus is also used for *Marshallocereus thurberi.*

Garden and Landscape Uses and Cultivation. In Mexico this cactus is much planted as living fences. As an ornamental it can be used to good effect outdoors in warm, dry climates and is handsome in pots, tubs, and ground beds in greenhouses given over to collections of cactuses and other succulents. It succeeds in part-shade or full sun in well-drained, porous soil of reasonable fertility and responds to care appropriate for most columnar-type cactuses. For more information see Cactuses.

MARGOSA is *Azadirachta indica.*

MARGUERITE or PARIS-DAISY. This, *Chrysanthemum frutescens,* of the daisy family COMPOSITAE, a native of the Canary Islands, is a popular greenhouse plant and is much grown in suitable climates for spring and summer outdoor display in flower beds, window and porch boxes, tubs, urns, and other containers. Its blooms have some value as cut flowers. The plant called blue-marguerite is *Felicia amelloides.* The name golden-marguerite is applied to *Anthemis tinctoria.*

Marguerites or Paris-daisies are subshrubs, bushy, and 2 to 3 feet tall, with chrysanthemum-like foliage, strong-smelling when bruised. They have numerous, up-facing, yellow-eyed, white, creamy-yellow or light yellow, daisy-like flower heads, 1½ to 2½ inches across, produced in succession over a long period. There are double-, anemone-, and single-flowered varieties.

Marguerites: (a) Double-flowered

(b) Single-flowered

The cultivation of marguerites is simple. They are easily raised from cuttings made from flowerless shoots. This is the usual method of multiplication, but the wild species comes well from seed. Cuttings, which root readily at any time, are usually made in August or September and planted in a lightly shaded greenhouse or cold frame propagating bed or in full sun outdoors under mist. Good small specimens can be had from cuttings taken in January.

When rooted, pot the cuttings individually in small pots and successively into bigger ones as growth makes necessary until the containers in which they are to bloom are attained. For first-year plants, these may be 5- to 7-inch pots. Specimens grown on for a second year may need tubs up to 1 foot or more in diameter.

Any fertile, porous soil suits. To induce sturdy growth pack it firmly about the roots. When the young plants are about 4 inches tall pinch out their tips and repeat this two or three times when the branches attain lengths of 5 or 6 inches.

Greenhouse conditions favorable to marguerites are full sun, a buoyant, airy, not too humid atmosphere, and from fall through spring a night temperature of 45 to 50°F, with an increase of five to ten, or on sunny days in spring, fifteen degrees. Marguerites when well rooted are thirsty subjects, needing generous amounts of water. From the time they fill their final pots with roots until the flowers expand give them weekly applications of dilute liquid fertilizer, biweekly ones thereafter. Keep faded flower heads picked.

MARGYRICARPUS (Margyri-cárpus)—Pearl Fruit. The Andes is home to the ten species of *Margyricarpus*, of the rose family ROSACEAE. Their name is from the Greek *margaron*, a pearl, and *karpos*, fruit, and alludes to the white fruits.

These are low, rigid, branching shrubs, related to *Acaena*, but with blooms solitary or few together in the leaf axils instead of being assembled in heads or spikes. Their leaves are pinnate and have a terminal as well as side leaflets. The flowers have four- or five-lobed calyxes, but are without petals. They have one to three stamens and a short style. The fruits are berries or achenes.

The pearl fruit (***M. setosus***), of from Ecuador to southernmost South America, is evergreen, prostrate, heathlike, and up to 1 foot in height. Its awl-shaped, reflexed leaves, ¾ to 1 inch long and dark green, are divided into linear leaflets up to ⅓ inch long. The solitary, stalkless, greenish, inconspicuous flowers are succeeded by white or pinkish berries, up to ⅕ inch in diameter, freely produced and fairly persistent. They have a pleasing acid flavor. They make a good decorative display.

Garden Uses and Cultivation. This is a choice, ferny-foliaged evergreen of especial interest to rock gardeners, but unfortunately not hardy in the north. It is suitable for parts of California and other places where severe winters are not experienced and may be grown in pans (shallow pots) in greenhouses devoted to alpines. It does not like very hot summers. At home it inhabits dry mountainsides. Its roots range deeply and widely. Under cultivation it must be given thoroughly well-drained soil. One of a sandy peaty character that does not dry completely, but is never for long periods saturated is satisfactory. Propagation is by seed, summer cuttings, and layering.

MARIA is *Calophyllum brasiliense.*

MARIANTHUS (Mari-ánthus)—Marybells. Dedicated to the Virgin Mary, *Marianthus* has a name derived from Maria and the Greek *anthos*, a flower. It belongs to a genus of sixteen Australian and Tasmanian species of the pittosporum family PITTOSPORACEAE.

Nonhardy trailing or climbing subshrubs with twining stems, these have undivided leaves, toothed or toothless, the lower ones sometimes lobed. In terminal or lateral clusters, the white, orange-red, red, or blue flowers have five sepals, five petals united at their bases into a tube, usually five stamens, and one style. The fruits are capsules.

An attractive, woody-based, hairless or hairy climber, ***M. ringens*** has slender stems and rather distantly spaced thin, leathery, evergreen, pointed, lanceolate to elliptic leaves 2 to 4 inches long and ½ to 1 inch wide or a little wider. The flower clusters, 1½ to 3 inches wide, have stalks ½ inch to 2 inches in length. The blooms are ¾ to 1 inch wide. They have yellow corolla tubes ⅝ inch long. The spreading, well-separated petals are light brick-red, and the stamens and long, slender style protrude.

Marianthus erubescens

Garden and Landscape Uses and Cultivation. In warm, dryish climates, such as that of California, the species described here is charming for furnishing trellises and other supports around which its stems can twine. It thrives in full sun in well-drained, reasonably fertile soil, and needs no special care. A little judicious thinning out of old stems to prevent crowding may with advantage be done from time to time. This is also nice to grow in pots in greenhouses and in windows in cool rooms. Night temperatures in winter of 50 to 55°F are satisfactory, with a daytime rise of five to fifteen degrees permitted. Watering to keep the soil always fairly moist is necessary. In summer the plants can be put outdoors with their pots buried to their rims in sand, peat moss, or similar material. Propagation is by cuttings of moderately firm shoots, by layering, and by seed.

MARIGOLD. Without qualification the name marigold as used in America commonly refers to kinds of *Tagetes* and includes African marigolds and French marigolds. The classical marigold (Mary's gold) or pot marigold is *Calendula officinalis*, a plant still called marigold in the British Isles. Other plants the common names of which include the word marigold are bur-marigold (*Bidens*), Cape-marigold (*Dimorphotheca*), corn-marigold (*Chrysanthemum segetum*), desert-marigold (*Baileya*), field-marigold (*Calendula arvensis*), and marsh-marigold (*Caltha palustris*).

MARIJUANA is *Cannabis sativa.*

MARINE-IVY is *Cissus incisa.*

MARIPOSA-LILY. See Calochortus and Erythronium.

MARITIMOCEREUS. See Loxanthocereus.

MARJORAM. Four species of *Origanum* are commonly grown as culinary herbs, for use fresh or dried. They have also been employed medicinally. For descriptions of these and other species see *Origanum.*

Sweet, garden, or knotted marjoram (*O. majorana*), although sometimes perennial, is usually cultivated as an annual; pot marjoram (*O. onites*), which is somewhat more winter hardy, is grown as a perennial as are the even hardier wild marjoram (*O. vulgare*) and winter marjorum (*M. heracleoticum*). The dried leaves of the last two sorts are called oregano. All succeed in well-drained, dryish soil in sunny locations. They need moderately fertile earth, but excessive fertility may reduce their aromatic qualities.

Pot marjoram

Wild marjoram

Seeds are slow to germinate. Sow those of sweet marjoram in early spring in rows 1 foot apart. Make one or two later sowings to provide succession. During their early stages it is well to shade the seedlings lightly. Thin them to 6 inches apart in the rows. Keep down weeds by practicing clean cultivation.

Pot marjoram, wild marjoram, and winter marjorum prosper under the same conditions as sweet marjoram but need a little more space. Rows 1¼ feet apart with 9 or 10 inches between individual plants suit. Propagation may be by seeds sown in a nursery bed and the young plants transplanted, or as for sweet marjoram, sown where the plants are to remain. Division in spring or early fall is an alternative and common way of increasing pot and wild marjorams.

Harvesting to dry for later use is done just before the flowers open. Cut the stems of sweet marjoram close to the ground, those of the others higher so that some stems and foliage are left. Tie the harvested material in bundles and hang them in shade in a cool, dry, airy place. When completely dry rub the leaves free of stems and store in airtight containers.

MARKHAMIA (Mark-hàmia). A near relative of the better known and in warm climates nearly ubiquitous African-tulip-tree (*Spathodea*), *Markhamia* comprises a dozen species of tropical Africa and Asia. They belong in the bignonia family BIGNONIACEAE. From *Spathodea,* markhamias differ technically in the arrangement of their ovaries and in details of the seed capsules. The name commemorates Sir Clements Robert Markham, English explorer and author, who died in 1919.

Markhamias are trees and shrubs. They have opposite, pinnate leaves, with an uneven number of leaflets, and flowers in ample panicles at the branch ends or from the leaf axils. The blooms have a large calyx that splits down one side in spathe-like fashion and a tubular, funnel- to bell-shaped corolla, usually yellow, with red, brown, or purple stripes, or on their insides more rarely lilac. The fruits are long capsules, containing seeds winged at their ends.

West African ***M. lutea*** is a shrub or tree approximately 20 feet tall, with leaves about 1½ feet long, of seven to eleven elliptic to obovate, stalked leaflets up to 7½ inches long. The yellow flowers are a little over 2 inches long, the curved seed pods nearly 2 feet long. With brown-striped, yellow, bell-shaped flowers 1½ inches long or a little longer, and with glandular corolla lobes (petals), African ***M. lanata*** (syn. *M. paucifoliata*) is an attractive shrub or small tree. Its downy leaves approximately 1 foot long, have five to eleven ovate to obovate leaflets about 5 inches long by 2 inches wide. The seed pods are 1 foot to 2 feet in length and downy.

East African ***M. hildebrandtii*** is a tree, in Florida, up to 30 feet tall, or sometimes only a tall shrub. It has ovate leaves, smaller than those of *M. lanata,* and very fragrant, bell-shaped, clear yellow blooms. Its seed pods are 2 feet long. This is highly drought resistant and withstands light freezes.

Garden and Landscape Uses and Cultivation. In the tropics and frost-free or nearly frost-free subtropics, markhamias are excellent for general ornament. They succeed in ordinary well-drained soil in full sun and are increased by seed.

MARL. This naturally occurring mixture of lime and clay or fine silt, where abundant locally, is sometimes used in place of lime as a soil amendment. Because its content of calcium carbonate is comparatively low much larger amounts of marl than lime are needed to achieve equivalent results.

MARLBERRY is *Ardisia paniculata.*

MARMALADE. This word is used as part of the common names of these plants: marmalade-box (*Genipa americana*) and marmalade fruit or marmalade-plum (*Pouteria sapota*).

MARRI is *Eucalyptus calophylla.*

MARROW. This is a type of squash. See Cucurbita.

MARRUBIUM (Mar-rùbium)—Horehound or Hoarhound. About thirty species are recognized in *Marrubium,* of the mint family LABIATAE. They are natives of Europe and temperate Asia. One, the common horehound, is naturalized throughout most of the United States and southern Canada especially in waste places and along roadsides. The name is an ancient Latin one that seems to have been used for several members of the mint family.

These are herbaceous perennials with square stems and opposite leaves. They have tubular, white, markedly-two-lipped flowers. The upper lip is erect and notched at its end, the lower recurved and three-lobed. There are four stamens, the two lower slightly longer than the upper pair, and one style. The fruits, commonly called seeds, are small nutlets.

The common horehound (***M. vulgare***) was as well known to Dioscorides as the candy made from it was to children a few decades ago. Familiar with its habits, he describes it as growing near houses and rubbish of buildings. Dioscorides says it "is a shrub of many branches from one root, somewhat rough, white, four cornered in the rods, but the leaf is equal to the great finger, somewhat round, thick, wrinkled, bitter to the taste. But the seeds on the stalks by distances, and the flowers as the vertebra of the backbone, sharp." It is hard to improve on that homey description. Dioscorides' "rods" are, of course, the stems. They are 1 foot to 2 feet tall and, like the leaves, softly-white-hairy. The latter are ovate, up to 2 inches long, somewhat shorter than Dioscorides' "great finger," perhaps. The lower leaves are stalked, the upper nearly stalkless. The flowers, about ¼ inch long, are densely crowded in the axils of the rather widely spaced upper leaves to produce the "backbone" effect referred to by the Greek.

Marrubium vulgare

Dioscorides attributed to horehound virtues it is still believed to have as well as medicinal ones no longer recognized. Chief of the former is its use to alleviate the distress of coughs and colds of which the Greek physician wrote "The dry leaves of this with the seed sodden in water, or juiced when green is given with honey to the phthisicall, asthmaticall, and to such as cough," a not very different recipe than that for more modern horehound candy. While accepting this use of horehound, we no longer go along with Dioscorides' recommendations that it is an antidote for poisons swallowed and the bites of venomous beasts, or that it cleanses foul ulcers, is a "sight quickener," and is good for ear pains.

Less frequently cultivated, ***M. incanum*** (syn. *M. candidissimum*), of the Balkan Peninsula and Italy, has erect, densely-white-woolly stems 1 foot to 1½ feet tall, and oblong-ovate, round-toothed leaves wedge-shaped at their bases and up to 2 inches long, with gray-green upper surfaces, whitish undersides. The whitish flowers are many together in rather distantly spaced whorls (tiers).

Marrubium incanum

Marrubium incanum (flowers)

Garden Uses. Horehounds are primarily subjects for herb gardens. They are scarcely decorative enough to be commonly planted in flower beds, although in dryish, poorish soils an occasional clump may be used in such places with good effect. Common horehound is used as a home medicine by boiling one cup of fresh leaves or one-half a cup of dried leaves in two cups of water and straining the infusion. This is diluted with twice as much boiling water to make horehound tea. A cough syrup is prepared by thoroughly stirring one part of the infusion into two of honey (shades of Dioscorides). A recommended recipe for horehound candy is to place two cups of sugar and ⅛ teaspoon of cream of tartar in a pan, mix thoroughly, then add a cup of horehound infusion and stir until the sugar dissolves. The mixture is cooked slowly with low heat until when a little is dropped in cold water it hardens satisfactorily. Then it is poured onto a buttered tray and when semihardened is cut into suitably sized pieces.

Cultivation. The cultivation of horehound calls for no special skill. Provided the soil is well drained and porous and the site sunny, it thrives with little care. Its best qualities are brought out in dryish soils not excessively rich in nitrogen. Propagation is by seed sown in spring or by summer cuttings. For drying, the stems are cut in summer, tied in bundles, and hung in a dry, airy, shady place. The dried leaves are rubbed free of the stems and stored in closed jars, bottles, or other suitable containers. The other species described responds to the same culture as the common horehound.

MARSDENIA (Mars-dènia). One or two of the 100 or more species that constitute the European, Asian, and African genus *Marsdenia,* of the milkweed family ASCLEPIADACEAE, are cultivated. The generic name commemorates William Marsden, author of a history of Sumatra, who died in 1836.

Marsdenias are shrubs and twining woody vines with opposite leaves and panicles or umbels of usually small bell- or urn-shaped flowers with five small sepals, a corolla with a slender tube, five spreading petals, and five scalelike lobes forming a crown (corona) in the center. The fruits are winged or wingless, often fleshy follicles.

Himalayan ***M. roylei*** is a robust vine with finely-pubescent young stems and leaves, the latter longish-stalked, pointed-ovate-heart-shaped and 4 to 8 inches long, frequently velvety-hairy on their under surfaces. The orange to brick-red, bell-shaped flowers, a little over ¼ inch long, with pubescent corolla lobes, are in clusters 1½ to 2 inches across. This is not hardy in the north.

Hardy in southern New England, ***M. erecta,*** of southeastern Europe and adjacent Asia, when supported attains heights of 20 to 25 feet. Without support it is low and sprawling. It has stalked, bright green to grayish-green leaves 1½ to 3½ inches long and pointed-ovate with heart-shaped bases. They are hairless except for a minute covering on the veins on the undersides of young ones. The fruits, spindle-shaped and about 3 inches long, contain many seeds each with a 1-inch-long tuft of hairs. The sap is poisonous and may blister the skin of some people.

Marsdenia erecta

Garden and Landscape Uses and Cultivation. In mild climates *M. royleyi* is useful for furnishing arbors, pillars, and other supports appropriate for a vigorous twiner. Of lesser decorative worth, much hardier *M. erecta* may be allowed to tumble over rocks or old tree stumps or to twine up wires or other suitable supports in sunny locations where the soil is well drained. Propagation of marsdenias is by seed or by summer cuttings in a propagating bed in a greenhouse or shaded cold frame or under mist. No special care is needed other than sufficient pruning to keep them shapely and uncrowded. This is best done in spring.

MARSH. As part of their common names the word marsh applies to these plants: marsh fern (*Thelypteris palustris*), marsh-fleabane (*Pluchea*), marsh-mallow (*Althaea officinalis* and *Hibiscus moscheutos palustris*), marsh-marigold (*Caltha palustris*), marsh-pea (*Lathyrus palustris*), marsh pennywort (*Hydrocotyle vulgaris*), marsh-pink (*Sabatia*), and marsh-St.-John's-wort (*Triadenum*).

MARSHALLIA (Marshàll-ia). Ten species of the daisy family COMPOSITAE, natives of eastern North America and the Gulf States, constitute *Marshallia,* named in honor of the American botanist Moses Marshall, who died in 1813.

Marshallias are herbaceous perennials with lobeless basal leaves and sometimes alternate, lobeless stem leaves. Their flower heads, without ray florets, are purple, pink, or white. They have longish, naked stalks and somewhat resemble the flower heads of scabious (*Scabiosa*) of the teasel family DIPSACACEAE. The fruits are seedlike achenes. Marshallias are not closely related to other cultivated members of the daisy family.

Native from Missouri to Texas, ***M. caespitosa,*** about 1½ feet tall, has mostly basal foliage. Its linear leaves are up to 4 inches long, its flower heads are pink or white, about 1 inch in diameter. Foliaged to their middles, the sparsely-branched stems of ***M. trinervia*** have ovate to ovate-lanceolate, three-veined leaves. The usually solitary, purplish flower heads are ¾ inch to 1¼ inches in diameter. This species, up to 2½ feet in height, is native in dryish soils from Virginia to Alabama and Mississippi. Favoring moister soils and woodlands from Pennsylvania to Kentucky and North Carolina, ***M. grandiflora*** is 1 foot to 2 feet high or sometimes higher. Its lower leaves are elliptic to spatula-shaped and up to 7 inches long. The upper ones are smaller and linear-elliptic. The solitary flower heads are ¾ inch to 1¼ inches in diameter. A lower species of the southeastern United States is ***M. obovata.*** This has chiefly basal, obovate or spatula-shaped leaves and white flower heads about 1½ inches in diameter. It attains a height of 1¼ to 1½ feet.

Marshallia obovata

Garden Uses and Cultivation. Although not among the showiest of hardy herbaceous perennials, marshallias possess a grace and quiet charm that please and are appropriate for planting toward the fronts of flower beds and borders, in rock gardens, and in wild gardens. The limits of their hardiness have not perhaps been completely explored but certainly even southern *M. obovata* persists in the vicinity of New York City without special protection and so do other kinds. Marshallias are easy to raise from seeds sown in gritty, porous soil containing an abundance of organic matter and kept evenly moist, but not wet. The seeds may be sown in fall or early spring. The plants grow well in any ordinary, porous garden soil in sun or where they receive a little shade. An alternative way of propagation is by careful division of the plants in early fall or early spring.

MARSHALLOCEREUS (Marshállo-cereus)—Organ Pipe Cactus. By many botanists included in *Lemaireocereus,* and in recent treatments by conservative students of cactuses native to the United States in *Cereus,* the genus *Marshallocereus,* of the cactus family CACTACEAE, inhabits Arizona, Mexico, and the West Indies. It comprises two species, one widely known as organ pipe cactus, a name also applied to *Marginatocereus marginatus.* The name *Marshallocereus* honors William Taylor Marshall, American cactus enthusiast, who died in 1957.

These are big trunkless or very short-trunked cactuses that branch freely from about ground level and send up many erect, fluted, columnar stems suggestive of the pipes of an organ. The many rounded, notched ribs are furnished with clusters of rigid spines. The funnel-shaped flowers have corolla tubes and ovaries, like the fruits clothed with woolly hairs and spines. The spherical to egg-shaped fruits are edible.

Organ pipe cactus (***M. thurberi*** syns. *Cereus thurberi, Lemaireocereus thurberi*) is generally 9 to 20 feet in height, with clusters 6 to 18 feet in diameter of green stems 4 to 8 inches in diameter with twelve to nineteen shallow ribs. The brownish, gray, or black, needle-like spines, up to ½ inch long and in clusters of eleven to nineteen, spread in all directions. The white-edged-petaled flowers are 2½ to 3 inches in diameter. The fruits, approximately spherical and at maturity red and 1¼ to 3 inches in diameter, are densely covered with deciduous spines. Variety *M. t. littoralis,* not exceeding 3 feet in height, has rose-red flowers. Native to Costa Rica, ***M. aragonii*** (syn. *Cereus aragonii*), almost 20 feet high, has stems with six to eight broad, rounded ribs and clusters of eight to ten ½- to 1¼-inch-long, gray spines. The flowers are white, 3 inches long, the fruits about 1½ inches long.

Garden and Landscape Uses and Cultivation. Of striking aspect, these rather slow-growing cactuses are admirable for outdoor landscaping in warm desert regions and for cultivation in greenhouses. They adjust well to domestication, finding agreeable well-drained, slightly acid soils and exposure to full sun. They require little water, are sensitive to frost, and are easily increased by cuttings and seed. For more information see Cactuses.

MARSILEA (Mar-sílea) — Water-Clover or Pepperwort. These four-leaf-clover-like plants look surprisingly unlike the ferns they are. They belong in the marsilea family MARSILEACEAE and grow in mud or shallow water, with their alternate leaves floating or standing above the surface. The number of species of *Marsilea* is estimated as being from sixty to seventy. The genus is widely distributed throughout most of the world and is considered to be an extremely ancient form of plant life. The name commemorates Giovanni Marsigli, an eighteenth-century Italian botanist.

Marsileas have creeping rhizomes from which arise the fronds (leaves). Each has two joined pairs of opposite leaflets terminating a long, slender stalk. The spores are in special purse-like, more or less bean-shaped organs called sporocarps, which are morphologically modified leaflets. Similar to *Marsilea* is *Regnellidium,* but its fronds have only two leaflets. The only other genus of the family, *Pilularia* has bladeless leaves consisting only of grass-like leafstalks.

A hardy species, ***M. quadrifolia,*** native to Europe and Asia, is naturalized in North America. It has thick, sparingly-branched rhizomes and nearly hairless,

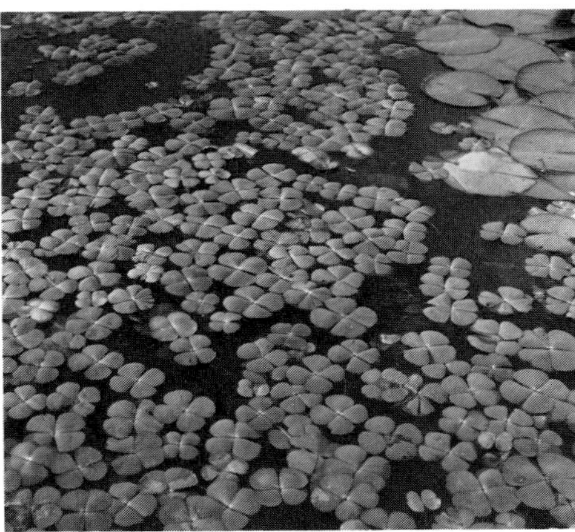
Marsilea quadrifolia

floating leaves with blades ¾ inch to 2 inches wide and stalks up to 1 foot long or a little longer. The stalked, elliptic sporocarps arise from the leafstalks just above their bases and are usually two together, but may be in threes or solitary. From the last, hardy ***M. mucronata*** differs in its sporocarps being shorter-stalked, solitary, and arising from the very bases of the leafstalks. Its leaves are usually hairy.

Australian ***M. drummondii*** has leaves floating or raised above the mud or water

Marsilea drummondii

Marsilea drummondii, showing at its base purse-like sporocarps

in which it grows. They form crowded tufts and have silky-hairy, wavy-edged blades about 3 inches across, which, with age, sometimes lose their hairs. The leaves of ***M. fimbriata,*** of tropical Africa, have blades about 2 inches in diameter. Their upper surfaces are hairless, beneath they have bristly hairs. The sporocarps are also hairy.

Marsilea fimbriata

Garden Uses and Cultivation. Hardy water-clovers, the first two described above are attractive for decorating the surfaces of ponds and pools, but, unless restricted to containers, may become invasive and perhaps difficult to control. They are planted in fertile earth in pots, pans (shallow pots), or boxes set a few inches beneath the surface. The Australian and African species described need subtropical or tropical environments. They do well in pots or pans of rich earth positioned a few inches below the water surface or with the bases of the containers only standing in water. A humid greenhouse, lightly shaded in summer and with a winter night temperature of 55 to 60°F rising five to fifteen degrees by day, suits their needs. Marsileas are easily increased by division and may be raised from spores. For more information see Ferns.

MARSILEACEAE—Marsilea or Water-Clover Family. This family of ferns comprises three genera totaling seventy species and is represented in the native floras of North and South America, Europe, Asia, Australia, and New Zealand.

The sorts of the marsilea family are marsh and aquatic plants with creeping rhizomes and often floating leaves. The leaf blades, not present in *Pilularia,* are of two or four fan-shaped leaflets. Propagation is by spores of two types, one that germinates to give female egg cells, the other male elements. The spores are in bean-shaped structures called sporocarps that develop on the leafstalks or from their bases. The genera are *Marsilea, Pilularia,* and *Regnellidium.*

MARTINEZIA. See Aiphanes.

MARTYNIA (Martýn-ia). The only species of *Martynia* is closely related to the unicorn plant (*Proboscidea*). It differs in having two instead of four fertile stamens and in its sepals being joined in spathelike fashion. Belonging in the martynia family MARTYNIACEAE, this plant takes its name from a professor of botany of Cambridge University, England, Dr. John Martyn, who died in 1768. For other plants previously included in this genus see Ibicella, and Proboscidea.

An annual vine some 10 feet tall, ***M. annua*** is native from Mexico to Central America, the West Indies, and tropical Asia. The entire plant is clammy-hairy, more especially its younger parts. The heart-shaped to triangular, long-stalked, coarsely-toothed leaves have blades up to 1¼ feet long. Their main veins spread palmately (like the fingers of a hand). The bell-shaped blooms are in racemes of ten to twenty. They have a five-cleft calyx, a five-lobed corolla, and four stamens. They are 1¾ inches long, creamy-white, yellowish, rose-pink, or reddish-purple, usually spotted with yellow and marked with purplish blotches. The eight-ribbed, egg-shaped fruits, 1½ inches long, are horned capsules with the horns shorter than the bodies of the fruits.

Garden and Landscape Uses. Only where the growing season is long and the summer warm is this vine likely to attain its full height. Elsewhere it may be grown as a sprawler, without supports up which to climb. It then may not be more than 2 feet tall. Where it makes sufficiently vigorous growth it may be used to clothe posts, walls, and fences. Its flowers do not last when cut.

Cultivation. This is a tropical plant that revels in sun and hot weather and responds to fertile, well-drained soil supplied with adequate moisture. Its needs are those of cucumbers and melons. Propagation is by seed, which may be sown after the ground has warmed in spring where the plants are to remain or, especially in the north is this desirable, be started indoors ten weeks or so before it is safe to set the young plants in the garden (that is, when it is time to plant out geraniums and begonias). The seeds often germinate very irregularly, sometimes taking three weeks or a month to break ground. For indoor germination a temperature of 70°F should be maintained. The seedlings are transplanted individually into coarse, fertile soil in 4-inch pots, and are grown in a sunny, warm greenhouse until a week or ten days before they are to be planted outdoors. They are then stood in a cold frame or sheltered spot outside to harden. Spacing of 1 foot to 1½ feet apart outdoors is generally satisfactory.

MARTYNIACEAE — Martynia Family. As wildlings confined to warm parts of the Americas, the martynia family comprises five genera of dicotyledons totaling thirteen species of often tuberous-rooted, sticky-hairy annuals and herbaceous perennials. Characteristically, they have somewhat fleshy stems and leaves, the latter opposite, becoming alternate toward the branch ends, wavy or lobed.

The flowers, in terminal racemes, are asymmetrical, have a five-cleft or spathelike calyx and a tubular bell- or funnel-shaped, five-lobed, slightly two-lipped corolla, four stamens in pairs of unequal lengths, and one staminode (nonfunctional stamen) or less commonly two stamens and the other pair represented by staminodes. There is one slender style and a two-lobed stigma. The fruits, of peculiar and intriguing forms, are long, prominently beaked capsules. They are used ornamentally and the young ones of some sorts are pickled and eaten. Cultivated genera include *Ibicella, Martynia,* and *Proboscidea.*

MARVEL-OF-PERU is *Mirabilis jalapa.*

MARYBELLS. See Marianthus.

MARYLAND-DITTANY is *Cunila origanoides.*

MASCARENA. See Hyophorbe.

MASDEVALLIA (Mas-devállia). Named in honor of the Spanish botanist and physician Jose Masdeval, who died in 1801, *Masdevallia,* of the orchid family ORCHIDACEAE, native from Mexico and the West Indies to tropical South America, is most abundant in Colombia. Most of its 275 species are found at high altitudes where, for the tropics, comparatively cool conditions along with high humidity prevail throughout the year.

Included are sorts with some of the most extraordinary flowers in a family notable for blooms of unusual form and coloring. They vary in size from decidedly small to almost 1 foot across from tip to tip of the sepals, which are the showy members. Their colors range from pure white through greenish-white, buff, and other sorts of yellow, to brilliant red and almost black-purple, and various combinations of these. Despite these intriguing features, masdevallias have been less popular than they were in the late nineteenth and early twentieth centuries when numerous hybrids of them were raised. A few enthusiasts maintain considerable collections and signs of renewed interest in the group are evident.

Masdevallias are without pseudobulbs, having clusters of not very evident rhizomes from which usually sprout tufts of tiny secondary stems or in a few cases longer and more obvious sprawling ones. The leaves are solitary, usually leathery, and somewhat fleshy, most commonly oblongish to linear or oblanceolate, generally widest above their middles and with sheathing bases. The flowers are most often in ones to threes from near the apexes of the stems, but those of some kinds are in raceme arrangements of several. They are on stalks that come or apparently come from the bases of the plants, and in some sorts these remain green and, if retained, after the blooms fade produce additional flowers later. The calyx, the showy part of the bloom, consists of three sepals united into a tube in their lower parts and two of them often well beyond the tubular part. Often the tips of the sepals are extended as long tails. The two small petals and the short lip are frequently but not always concealed or partly concealed in the calyx tube.

Among cultivated sorts are those now to be described. The measurements given for the flowers are those of the greatest dimension (length) across the face of the bloom from tip to tip of the sepals or, if they have tails, from tip to tip of the tails of the sepals. ***M. attenuata,*** of Panama, Costa Rica, and Puerto Rico, has linear-lanceolate to oblanceolate, stalked leaves usually up to 4 inches in length and solitary flowers on stalks mostly somewhat shorter. Bell-shaped and about 1 inch long, the flowers are white to cream-colored, with the tubes sometimes streaked with red and with ½-inch-long, bright yellow to orange-yellow tails to the sepals. The lip of this easy-to-grow sort is white tipped with brown. ***M. barlaeana,*** of Peru, tufted, has stalked, lanceolate leaves 3 to 4 inches long and solitary blooms on erect stalks up to 8 inches tall. The showy flowers have narrow pink-magenta tubes suffused with coral-red to crimson. The sepals, the upper with a slender tail 1 inch to 1½ inches long, the others united for two-thirds of their lengths and with tails about ½ inch long, are red except for the basal part of the upper one, which is orange-yellow. The petals and lip are minute. ***M. bella,*** of Colombia, has blunt, oblong-lanceolate leaves, 5 to 9 inches long, and large solitary, spidery blooms on drooping stalks mostly shorter than the foliage. Fragrant, pale yellow to brownish-yellow spotted with brownish-crimson and with the bases of their sepals forming a triangle 2 inches across, the flowers have tails to the sepals up to 4 inches long. The petals are yellow spotted with red. The lip is white. ***M. calura,*** of Costa Rica, has 3- to 4-inch-long, stalked, oblanceolate leaves ordinarily shorter than the flower stalks. The solitary blooms, up to 3½ inches long, are chiefly dark red with deeper suffusions. Their sepals have slender, flat, orange-yellow tails, the tail of the upper or dorsal one 1½ to 2 inches long, the others shorter. ***M. carderi*** is a Colombian with lustrous, thinnish, narrow-spatula-shaped leaves 3 to 5 inches long. Its drooping, slender flower stalks, each with one bloom, are up to 7 inches long. The flowers, 2 to 2½ inches long and bell-shaped at their centers, are mainly creamy-white blotched with purplish-brown toward the bases of the sepals, and with their inner surfaces clothed with short hairs. The sepals have tails 1 inch to 2 inches long. The petals are white with a center line of purplish-brown. The lip is white. ***M. caudata,*** of Colombia, has elliptic-oblong to obovate leaves with stalks about as long as the 2- to 3-inch-long blades and equaled or somewhat exceeded by the stalks of the solitary blooms. The latter from tip to tip of the long tails of their sepals may be as much as 8 inches long. The center part of the bloom formed by the blades of the sepals is 1 inch to 1½ inches across. The dorsal sepal is yellow with, except on the tail, red spots and five to seven red veins. Unlike the lateral ones, it is concave and shell-like. The lateral sepals are mauve-purple mottled with white. The tiny petals are white, the lip light mauve. ***M. chestertonii,*** of Colombia, has narrowly-oblanceolate leaves 4 to 7 inches long and solitary blooms on slender, pendulous stalks 6 to 8 inches long. From 2½ to 3½ inches across, they have ovate-oblong sepals ending in 1-inch-long, black-purple tails. The blades of the sepals, keeled on their backs, are greenish-yellow spotted with black-purple. The tiny petals are yellowish-red tipped with black. Comparatively large, the lip is light orange-yellow marked with radiating reddish lines. ***M. chimaera,*** of Colombia, has lustrous, short-stalked, narrowly-oblanceolate leaves 6 to 9 inches long. Its remarkable, but somewhat ill-scented flowers, up to about 9 inches across and with sepals but slightly joined at their bases, are borne singly, but if their stalks are left after the flower fades they usually produce on extensions developed successfully from below the ovaries one to several additional blooms. Varying considerably in size and coloring, the sepals are prevailingly whitish-yellow to light yellow or brownish-yellow, more or less densely spotted with brown to brownish- or blackish-purple. Warty and hairy on their insides and hairy at their margins, they have long, slim tails. The petals and lip are usually white, the latter often marked with yellow, orange, or red spots. Many variants of this species have been given horticultural names. ***M. civilis,*** of Peru, has densely-clustered, linear to oblong-linear, short-stalked, very fleshy leaves 5 inches to nearly 1 foot long by up to ½ inch wide. On stalks much shorter than the leaves, the solitary, ill-scented flowers, about 2 inches wide, are mainly buff-yellow to greenish with purple spots and with the inside of the throat purple. The sepals have rather short tails. ***M. coccinea,*** of Colombia and Peru, forms large tufts and has long-stalked, lustrous, dark green, obovate-lanceolate leaves 6 to 10 inches long in clusters. The flowering stalks, much exceeding the foliage and up to more than 1½ feet in length, terminate in a solitary, thick-textured, waxy bloom 3 to 3½ inches long. In color they come in a

Masdevallia attenuata

Masdevallia chimaera

Masdevallia civilis

wide range of reds from orange-scarlet to crimson-purple and in tones of yellow, orange, and creamy-white to white. Except for its long, slender tail, the dorsal or upper sepal is much smaller than the others. The lateral ones, pointed-ovate and without tails, are united below their middles. The petals and lip are similar and very small. ***M. coriacea,*** of Colombia, has thick, leathery, linear-lanceolate leaves 5 to 7 inches long. On stalks about as long as the leaves its solitary, 3-inch-long flowers have fleshy sepals, the upper or dorsal one triangular with a short, broad tail and whitish-yellow with some purple dots. The yellowish lateral sepals terminate in sharp points. The petals are white, the lip yellowish, all with purple mid-lines. ***M. corniculata,*** of Colombia, has very fleshy, oblong-oblanceolate leaves 6 to 7 inches long. Its solitary flowers, on erect stalks about 3 inches in length, do not expand widely. They are approximately 3 inches long. Mainly they are light yellow mottled with brownish-red. The tail of the upper sepal is about 2 inches in length and like the shorter down-pointed ones of the laterals, yellow. The petals are white tipped with yellow. The lip is yellowish, spotted with purple. ***M. cucullata,*** of Colombia, has very leathery, oblong-lanceolate leaves 8 inches to 1 foot long and solitary flowers on up to 1-foot-long stalks. The hooded blooms, their lower parts enclosed by a big bract, are about 2 inches long. They have maroon-purple sepals with yellowish-green tails, about 1½ inches in length. The petals are white with purple-warted tips. The lip is dark purple. ***M. elephanticeps,*** of Colombia, owes its name to the vaguely elephant-head-like shape of its solitary blooms, rather suggestive of those of certain stapelias, which they resemble also in being malodorous. The flowers do not expand fully. On stalks shorter than the foliage, and 2½ to 4 inches long, they are carried horizontally or hang downward. Their bells have rosy-purple interiors, which are paler and yellowish on their outsides. The petals are yellow, the lip purple.

Masdevallia elephanticeps

The leaves are oblanceolate to narrowly-wedge-shaped and 6 to 10 inches long. ***M. ephippium,*** of Ecuador and Colombia, has elliptic-lanceolate to oblong-ovate, stalked leaves 6 to 9 inches long. Its flowers, solitary or in twos or threes on erect stalks up to somewhat over 1 foot tall, are carried horizontally or sometimes droop. About 8 inches long and thick-textured, they have sepals with yellow tails about 4 inches long. The upper or dorsal one is smaller than the others, concave, and stained with brown on the outside. The laterals are united to form a deep, predominantly dark brown cup. The petals are white, the lip reddish-brown. ***M.erythrochaete,*** of Costa Rica, has somewhat the aspect of *M. chimaera.* Its scarcely fleshy, channeled leaves, 6 inches to 1 foot long, are linear-lanceolate. The short-tubed blooms, usually solitary or rarely in twos, are on slender, erect to pendulous stalks up to 1½ feet long. About 3 inches across, the blooms are white or yellowish-white and yellow spotted with reddish-purple. The sepals have reddish-purple, 2-inch-long tails. Their insides are clothed with white hairs. ***M. estradae,*** of Colombia, is densely tufted and has elliptic to spatula-shaped leaves with blades up to 3 inches long and stalks about as long as the blades and often toothed at their apexes. The solitary flowers on stalks somewhat exceeding the leaves are 2 to 3 inches long. They have a helmet-shaped upper sepal, yellow in its basal half, violet-purple above, and ending in a 1-inch-long, threadlike, yellow tail and a pair of similarly colored, nearly flat lateral sepals with considerably longer slender tails. ***M. floribunda*** is Mexican. It has oblanceolate-oblong leaves 3 to 5 inches long and solitary flowers on erect or spreading stalks generally longer than the leaves. The blooms, which open only partially, are 1 inch long or somewhat longer and bell-shaped. Chiefly buff-yellow dotted with brownish-purple, they have sepals with slender, recurved, reddish-brown tails, that of the dorsal sepal longer than the others. The petals are white, the lip white with purplish- or reddish-brown spots. ***M. gemmata,*** of Colombia, is diminutive, with clustered, 2-inch-long, channeled, linear leaves and solitary flowers nearly 2 inches long on lax, threadlike stalks longer than the leaves. The blooms have a purple-veined, brownish-yellow upper or dorsal sepal, with a very slender, orange-yellow tail, about 1 inch long, and a pair of bigger lateral sepals united into a darker-veined, wine-purple, boat-shaped blade with near its apex two orange-yellow tails. The petals and lip are tiny, purple. ***M. houtteana,*** of Colombia, has clustered, very slender, linear-lanceolate leaves 5 to 7 inches long. Its flowers are solitary at the ends of slender, procumbent stalks 3 to 4 inches long. About 3 inches long, most of which measurement accounted for by the tails of the sepals, the flowers at their centers are creamy-white spotted with purple or red and densely clothed with short white hairs. The tails of the sepals are reddish-purple. The tiny white petals are tipped with a tuft of blackish hairs. The lip is concave, white or pink. ***M. infracta,*** of Brazil and Peru, has lanceolate to oblanceolate leaves 5 to 6 inches long, and at the apexes of erect, three-angled stalks considerably longer than the leaves, one to five flowers. The broadly-bell-shaped, about 3-inch-long blooms have a slender tail ¾ inch to 1¾ inches long to each sepal. They are yellowish- to greenish-white, light wine-purple within. ***M. leontoglossa,*** of Colombia and Venezuela, has lanceolate leaves 6 to 8 inches long. On stalks shorter than the leaves, the solitary flowers, which do not open fully, are greenish on their outsides, rose-red spotted thickly with deeper red within. The tails of the sepals are about 2 inches long. ***M. macrura,*** of Colombia, is robust and has stems about 6 inches long and elliptic-lanceolate to elliptic-oblong, thick leaves sometimes over 1 foot long by up to 3 inches wide. The solitary, fleshy, short-tubed blooms on erect stalks up to 1 foot long, are sometimes 10 inches long. Tawny-yellow suffused on their outsides with brown and studded with blackish-purple warts on their insides, the sepals, yellowish and 4 to 5 inches long, are without warts on their paler, very long tails. The petals and lip are tawny-yellow, the latter spotted with purple and with a warty apex. ***M. melanopus*** is a densely-tufted native of Colombia, Ecuador, and northern Peru. It has oblong-spatula-shaped leaves 4 to 5 inches long and minutely-three-toothed at their apexes. Freely produced, the slender, one-sided racemes of three to eight about 1-inch-long blooms are up to 10 inches tall. The flowers, which do not expand fully, are white lightly sprinkled with purple and have bright yellow, ½-inch-long tails to the sepals. The

Masdevallia infracta

petals are minute, the lip tongue-shaped and tipped with yellow. ***M. militaris*** (syn. *M. ignea*), one of the most splendid and satisfactory of the genus, is a variable native of Colombia. It has stalked, elliptic-lanceolate leaves up to 4 inches long, and topping slender stalks 1 foot to 1¼ feet tall, solitary, brilliant flowers 1½ to 2½ inches across. Their color, typically cinnabar-red suffused with scarlet to crimson, varies to more orange-red tones. The upper or dorsal sepal is narrow and tapers to a down-pointed tail that droops in the separation between the lateral sepals. The latter, joined for more than one-half their lengths, are elliptic-ob-

Masdevallia militaris

Masdevallia militaris (flowers)

long and much broader than the dorsal sepal. Each has three distinct longitudinal veins. The petals and lip are white with purple mid-lines and on the lip an apical stain of orange-yellow. ***M. mooreana,*** of Colombia, and allied to *M. elephanticeps,* has linear-oblong leaves 5 to 9 inches long and solitary, horizontally-held or down-pointing, ill-scented flowers 3 to 3½ inches long on shorter, erect stalks. The narrowly-triangular free parts of the sepals protrude forward, those of the lateral pair united for more than one-half their length and nearly parallel. The upper sepal, yellow with three purple streaks toward its base, has a twisted tail about 2 inches long. The lateral sepals, except for their yellow tails, are wine-red sprinkled with tiny purple-black warts. The petals are white with purple mid-veins. The lip is blackish-purple. ***M. nycterina,*** of Colombia, a smaller-flowered ally of *M. chimaera,* has linear-lanceolate leaves 6 to 8 inches long and solitary flowers on shorter, more or less horizontal or drooping stalks. The triangular blooms are up to 6 inches long. Except for their 3-inch-long, purple-red tails, the sepals are pale yellow spotted with reddish-purple and furnished with tiny white hairs. The petals are yellow with nearly black spots. The concave lip has numerous radiating, raised lines. ***M. pachyantha,*** of Colombia, has oblanceolate leaves 6 to 7 inches long. The solitary blooms, 4 to 6 inches long, are on stalks longer than the leaves. They are pale yellowish-green to yellowish-ochre, the dorsal petal with three brown-purple veins and a 1-inch-long tail, the lateral sepals spotted with rosy-purple and with yellow tails. The whitish petals have brown-purple mid-veins. The lip is nearly black at its tip, brown toward its base. ***M. peristeria,*** of Colombia, has oblanceolate-oblong leaves 4 to 7 inches long. On stalks shorter than the leaves, its 3- to 5-inch-wide solitary blooms spread their sepals widely. They are yellow to yellowish-green spotted with rose-purple and have short, greenish- to tawny-yellow tails to the sepals. The petals are light greenish-yellow, the lip dotted with rosy-purple warts. The petals and white column together suggest, in much the manner of those of *Peristeria elata,* a dove. ***M. platyglossa,*** of Colombia, has stiff, erect, lanceolate-spatula-shaped leaves 4 to 6 inches in length and solitary or rarely paired flowers on shorter, sprawling stalks. Evenly pale green, translucent, and up to about 2½ inches across, the blooms have sharp-pointed, tail-less sepals, each with three conspicuous longitudinal veins. The petals are tongue-shaped, the lip warted at its tip. ***M. polysticta,*** of northern Peru, and densely-tufted, has narrow, spatula-shaped to oblanceolate-linear leaves minutely-three-toothed at their apexes and up to 7 inches long. Much exceeding the leaves, the flowering stalks carry three to nine hooded blooms about 1½ inches long, white to pale lilac, with numerous purple-red spots. The sepals terminate in slender tails about ¾ inch long. Their outer halves are bright orange-yellow or yellow. The petals and lip are minute. ***M. racemosa,*** of Colombia, has elliptic-oblong leaves 2 to 4 inches long spaced ½ to 1 inch apart along creeping rhizomes. The slender, arching flowering stalks, 9 inches to 1¼ feet long, each bear up to fifteen flowers, 2 to 2½ inches long, and bright orange-red suffused with crimson or less often almost yellow. The dorsal sepal is triangular and reflexed. The lateral ones are joined into a blade 1 inch to 1½ inches wide that broadens toward its apex and has two lobes or points, but is without tails. The minute petals and lip are whitish. ***M. radiosa,*** of Colombia, allied to *M. chimaera,* has oblanceolate leaves 5 to 8 inches long and flowering stalks with three or more broadly-bell-shaped blooms up to 7 inches long. Their upper insides are ochre-yellow, the remainder purplish densely covered with blackish warts. The 2- to 3-inch-long tails to the sepals are black-purple paling toward their apexes. The shell-like lip is white with radiating rose-pink ridges. The yellowish column is tipped with black. ***M. reichenbachiana,*** of Costa Rica, has clustered, stalked, pointed-oblanceolate, erect leaves up to about 6 inches long. Its flowers, two to four together or sometimes solitary, are on stalks that somewhat exceed the leaves in length. Almost 3 inches long, they have red-veined, yellowish-white sepals, dark red on their outsides, that contract abruptly into 1½-inch-long slender, recurved tails, those of the down-pointing lateral pair with their tips overlapping. The petals and lip, tiny and hidden within the throat of the bloom, are whitish. ***M. rolfeana,*** of Costa Rica, much resembles *M. reichenbachiana,* but is rather smaller in its parts and has dark purple blooms approximately 2½ inches across. ***M. schlimii,*** of Venezuela, has clustered, lanceolate-spatula-shaped leaves 6 to 10 inches long or longer and considerably longer, erect flowering stalks with three to eight blooms approximately 4½ inches long, the tails of their sepals 1 inch to 2 inches long, prevailingly yellow sometimes lightening to cream on their insides and variously marbled and spotted with brownish-red, and with yellow tails 1 inch to 2 inches long. The strap-shaped lip is rose-pink. ***M. schroederana,*** of Peru, densely-tufted, has erect, oblanceolate leaves some 6 inches long and on stalks slightly exceeding the foliage, flowers with orange-yellow upper sepals with two dark crimson spots and an orange tail 2 to 3 inches long. The other sepals white toward their bases, violet to rose-crimson above, have 2-inch-long, recurved, orange-yellow tails slightly shorter than that of the upper petal. The petals are minute, the lip only slightly bigger. ***M.simula,*** widely distributed in Central America, is diminutive. It has clustered, 2- to 3-inch-long, bright green or purple-tinged, linear to linear-lanceolate leaves. The solitary, hooded flowers, on stalks ½ inch long, are about ½ inch wide. They have yellow sepals spotted and barred with purple and a wine-purple, tongue-shaped lip much bigger than the tiny greenish petals. ***M. perpusilla,*** a Peruvian, is very similar. ***M. tovarensis,*** of Venezuela, willing in cultivation, has clustered, erect, elliptic-spatulate leaves slightly notched at their apexes, and 5 to 6 inches long, and three-angled stalks up to 7 inches long that carry two to five, long-lasting, 3-inch-long, pure white blooms and that if left after blooming will flower the following year. The upper sepal, 1½ to 1¾ inches long, has a triangular base and a slender tail. The other two, united for most of their lengths, have very much shorter tails. The petals are two-lobed, the lip pointed and reflexed. ***M. triangularis,*** a Venezuelan, has slim-stalked, elliptic-oblong to oblanceolate leaves 4 to 6 inches long and single-flowered, erect stalks with broadly-bell-shaped blooms about 5 inches long. Their slightly concave, triangular sepals, the laterals somewhat sickle-shaped, are ochre- to tawny-yellow thickly spotted with brownish-purple or purple. They have very slender, brownish-purple tails up to 3 inches in length. The petals and lip are white, the former three-toothed, the latter spotted purplish-pink to reddish-purple and its turned-back tip with a tuft of blackish hairs. ***M. triaristella*** is a very diminutive Colombian. It has densely crowded, slender, spindle-shaped leaves, grooved down their face sides, up to 2 inches long. Slightly exceeding the leaves in length, the very slender, erect flowering stalks carry one or two nearly 1-inch-long blooms. Each has a short upper sepal with a yellow, ½-inch-long tail. The dark brownish-red lateral sepals are fused for most of their lengths to form a boat-shaped blade with near the tip a pair of short, hornlike, yellow tails. ***M. veitchiana,*** a Peruvian, occurs on the ruins of Machu Picchu and elsewhere. A showy

Masdevallia veitchiana

species, this has clustered linear- to oblong-oblanceolate leaves 4 to 10 inches long. On stalks 1 foot to 1½ feet tall are borne one or less often two showy flowers up to 8 inches long. Chiefly orange to cinnabar-red, they are studded on their faces with crimson-purple hairs. Their reverses are tawny-yellow. The two lateral sepals are joined from their bases to beyond their middles and end in slender tails shorter than the 2-inch-long or slightly longer one of the upper or dorsal sepal. The tiny petals and lip are white. ***M. velifera,*** of Colombia, has erect, linear-elliptic leaves 6 to 8 inches long and solitary, hooded, ill-scented flowers, 2½ to 3 inches long, on stalks about 4 inches long. They are ochre-yellow speckled on their faces with minute reddish-brown spots. The upper sepal has a 2-inch-long tail, the lateral ones are declined and have thickish tails. The petals are greenish-white. The lip is thickly covered with brown-red warts. ***M. ventricularis,*** a native of Colombia and Ecuador, has clustered, lustrous, lanceolate-spatula-shaped leaves, up to 6 inches long, and solitary, brownish flowers, up to 4½ inches long, their sepals with slender, spreading tails. The small, strap-shaped lip is violet-purple. ***M. wagenerana,*** of Venezuela, is a miniature, with spatula-shaped leaves 2 to 2½ inches long and solitary, slender-stalked flowers as long or a little longer. The blooms are buff-yellow dotted and to some extent lined with red. The lateral sepals, which like the concave upper one have yellow tails, are united from their bases to beyond their middles. The hatchet-shaped petals are toothed and yellow, the whitish lip is lozenge-shaped and toothed.

Garden Uses and Cultivation. Masdevallias are fanciers' orchids with a somewhat undeserved reputation of being more or less difficult to grow. True, they tend not to give satisfaction in environments not to their liking, but given a little understanding of their needs and some serious effort to meet them a great many kinds respond favorably. Because they come from widely different natural habitats their cultural requirements vary according to kind, and adjustments must be made to satisfy special needs.

Cool greenhouse treatment suits the great majority of masdevallias, winter temperatures of 50°F at night and a few degrees more by day being adequate. A few sorts appreciate just a little more warmth, intermediate greenhouse temperatures. In many parts of North America hot summers bring trying conditions for these orchids. Then, the greenhouse should be kept as cool as practicable, if possible it should be air conditioned.

Always, high humidity must be maintained without permitting the air to become stagnant. Circulating fans are a great help in summer. Because masdevallias are without pseudobulbs and have no season of rest, they must be watered generously throughout the year. This necessitates sharp drainage in the containers in which they are grown and a very porous rooting medium. Growers vary in their choices. Many successful ones employ mixtures of osmunda fiber, peat moss, and sphagnum moss, with a little chopped charcoal mixed in. For additional information see Orchids.

MASK FLOWER. See Alonsoa.

MASSONIA (Mass-ònia). The genus *Massonia,* of the lily family LILIACEAE, an endemic of South Africa, consists of about forty-five species. Its name commemorates Francis Masson who, as the first plant collector to be sent abroad by the Royal Botanic Gardens, Kew, England, traveled in South Africa, North America, and the West Indies. He died in 1805.

Massonias are deciduous bulb plants of unusual aspect. They have two opposite, flat, broad, ovate, or oblongish leaves that spread on or close to the surface of the ground. The flowers are crowded in a stalkless or nearly stalkless umbel-like head surrounded by an involucre (collar) of membranous bracts. They are erect and have a tubular perianth with six spreading or reflexed lobes (petals). There are six stamens and one style. The fruits are capsules.

Its bulb about 1 inch in diameter, ***M. depressa*** (syn. *M. latifolia*) has thin, broad-ovate, hairless leaves from 4 to 6 inches or sometimes up to 10 inches long. About 2 inches in diameter, the dense head of bloom sits tightly between the leaves. The flowers have perianths about 1 inch long, with linear-lanceolate, reflexed lobes. The stamens have red stalks and yellow anthers.

Massonia depressa

Garden and Landscape Uses and Cultivation. Massonias prosper outdoors in dry, essentially frostfree, Mediterranean-type climates such as prevail in parts of the southwestern United States. There, they are suitable for rock gardens and other intimate plantings. They are also easy to grow in sunny greenhouses in which conditions appropriate for cactuses and other succulents are maintained, and in windows. They respond to reasonably fertile, well-drained soil. During their season of dormancy they should be kept dry; while in leaf watering to keep the soil moderately moist is in order. Propagation is easy by seed and sometimes by offsets.

MASTERWORT. See Astrantia.

MASTIC. This name is used for *Mastichodendron foetidissimum, Pistacia lenticus,* and *Schinus molle.*

MASTICHODENDRON (Masticho-déndron) — Mastic. The genus *Mastichodendron,* of the sapodilla family SAPOTACEAE, was formerly included in *Sideroxylon.* Native to southern Florida, the West Indies, Mexico, Central America, northern South America, Indochina, and southern China, it consists of about nine species of trees and shrubs. The name, alluding to the type of sap, derives from mastic, and the Greek *dendron,* a tree.

Mastichodendrons have milky sap, undivided leaves, and axillary clusters of small, five- or rarely six-parted, more or less bell-shaped flowers, with as many stamens as petals and the same number of scalelike staminodes (abortive stamens). The fruits are egg-shaped or globular berries.

Mastic (***M. foetidissimum*** syn. *Sideroxylon foetidissimum*) is indigenous from central Florida to the West Indies. A variety of it occurs in Mexico and Central America. A stately evergreen in southern Florida and some other parts of its natural range attaining a height of 75 feet but often much lower, it usually develops an irregular head. Its slightly lustrous, alternate, yellow-green leaves, somewhat wrinkled at their margins, have bluntish-elliptic to oblong blades 2½ to 4 inches long by 1 inch to 2¼ inches wide, and stalks ¾ inch to 1¼ inches long. Except when young they are devoid of hairs. The light yellow, short-stalked blooms that develop on young shoots are about ⅜ inch wide and ¼ inch long. Usually in clusters, sometimes solitary, they have a strong, rather unpleasant, cheeselike odor. The petals spread, the stamens are erect. The fruits, ¾ to 1 inch long, and usually with one large seed, are olive-shaped, yellow, and decidedly ornamental. They are edible, but sour and unpleasantly flavored. Animals eat them.

Garden and Landscape Uses and Cultivation. The mastic is planted as a shade and ornamental tree in Florida and other areas with a warm climate. It succeeds in ordinary, even poorish soils, including those derived from limestone, and withstands dry conditions reasonably well. Seed afford means of propagation, but the percentage of germination is often rather low.

MATAI is *Podocarpus spicatus.*

MATCHWEED is *Gutierrezia sarothrae.*

MATE is *Ilex paraguariensis.*

MATELEA (Mat-èlea). The genus *Matelea,* of the milkweed family ASCLEPIADACEAE, comprises approximately 130 species of twining herbaceous and shrubby vines. From related *Gonolobus* they are distinguished by botanical details of the anthers of the flowers, particularly in that they are without dorsal appendages. The genus is endemic to the Americas. Its name is derived from a native one used in Guiana.

Native to moist woodlands and thickets from Maryland to Missouri and southward, finely-hairy ***M. carolinensis*** (syns. *Gonolobus carolinensis, Vincetoxicum carolinense*) has pointed, broadly-ovate to nearly round leaves up to 8 inches long by 5 inches broad. The brownish-purple flowers have widely spreading petals ⅓ to ½ inch long. About 4 inches long, the seed pods are roughened with prominent sharp, warty projections.

Garden Uses and Cultivation. The species described is of use for native plant gardens and for naturalizing in informal areas under conditions approximating those under which it grows in the wild. Seed afford the simplest means of propagation.

MATILIJA-POPPY. See Romneya.

MATRICARIA (Matric-ària) — Matricary, Sweet-False-Camomile. From *Chrysanthemum* this genus of the daisy family COMPOSITAE is distinguished by its achenes (seedlike fruits) being ribbed on their fronts or fronts and margins, but not on their rounded backs. The thirty-five species of *Matricaria* are mostly aromatic annuals, biennials, and herbaceous perennials native to the northern hemisphere and South Africa. The name, derived from the Latin *mater,* mother, was given by early herbalists who employed *Tripleurospermum maritimum inodorum* (formerly *Matricaria inodora*) for treating diseases of the uterus.

Matricarias have alternate, pinnately-, twice-pinnately-, or thrice-pinnately-divided or lobed leaves and from the ends of the branches, clusters of flower heads, which may, like daisies, have centers of tubular, yellow disk florets encircled by petal-like, white ray florets or consist only of yellow disk florets. The disk florets are tubular and four- or five-lobed. For other plants often cultivated under the name *Matricaria,* see *Tripleurospermum* and *Chrysanthemum parthenium.*

Sweet-false-camomile (***M. recutita***) is often misnamed *M. chamomilla.* Native to Europe and western Asia and naturalized in waste places and roadsides in North America, it is a much-branched, sweet-scented annual 1 foot to 2½ feet tall, hairless or nearly hairless and with finely-twice-pinnately-divided leaves up to 2½ inches long. The ¾- to 1-inch-wide flower heads have ten to twenty-five strongly-reflexed, white ray florets about ¼ inch long and a yellow disk or eye.

Three somewhat similar natives of South Africa, occasionally cultivated, are ***M. globifera, M. grandiflora,*** and ***M. suffruticosa.*** All are much-branched annuals about 1½ feet tall with leaves twice-pinnate or those of the first two sometimes three-times-pinnate. Their yellow flower heads are without ray florets. Another South African, ***M. africana*** (syn. *M. capensis*) has flower heads with a yellow eye and white ray florets. It is suspected that plants cultivated under the names of the first three mentioned are often *Chrysanthemum parthenium* and that *Tripleurospermum maritimum inodorum* is often the plant grown as *M. grandiflora.*

Matricaria recutita

Garden and Landscape Uses and Cultivation. Sweet-false-camomile is sometimes grown in herb gardens by sowing seeds in early spring directly where the plants are to remain. Choose a sunny spot and well-drained ordinary garden soil and thin out the seedlings sufficiently to prevent undue crowding. The South African species described respond to similar treatment.

MATRICARY. See Matricaria.

MATRIMONY VINE. See Lycium.

MATTEUCCIA (Mat-teùccia)—Ostrich Fern. One American and one European and two Asian species comprise *Matteuccia.* Only the first is likely to be cultivated in North America. The name commemorates an Italian physicist, Carlo Matteucci, who died in 1868. These are large deciduous ferns of the aspidium family ASPIDIACEAE.

Ostrich ferns are bold in appearance, vigorous, and distinctive. Their leaves (fronds) are clustered. Subterranean runners from which arise new plants are well developed. There are two very different kinds of fronds, sterile ones that begin to grow earlier in spring than the others and are typically fernlike, and spore-producing or fertile fronds that come much later and are surrounded by sterile ones. The leaflets or segments of the fertile fronds are rolled into slender, podlike containers of spore-producing organs.

The American ostrich fern (***M. struthiopteris*** syn. *Pteretis pensylvanica*) is sometimes distinguished from the European population of *M. struthiopteris* as *M. s. pensylvanica*. It inhabits moist woodlands and swamps from Newfoundland to Alaska, Virginia, Ohio, Missouri, and British Columbia. Rather late in starting its spring

Matteuccia struthiopteris

growth, this species first develops sterile leaves that may attain a length of 6 feet, but are often smaller. They have oblong-lanceolate blades from 6 inches to 1 foot broad or a little broader, and of twenty or more alternate, long-pointed primary divisions or leaflets that gradually shorten toward the base of the blade and abruptly toward the apex, and have toothed margins. Summer is at hand before the fertile fronds appear. They are about one-half as long as the sterile ones and have double-comblike, spore-bearing portions about as long as the stalks that hold them aloft. Their segments, not over 2 inches long, become bronzy as they mature, and finally turn brown.

Garden and Landscape Uses and Cultivation. A stately vase-shaped fern useful for bold landscape effects by watersides and in moist and wet soils, the ostrich fern succeeds with little care, but gives of its best only where the soil is deep and fertile. It spreads rapidly and should not be planted where this will discomfort less vigorous neighbor plants. Established specimens need no special attention. They appreciate being mulched occasionally with nourishing compost or similar organic material and, if there is any danger of the soil drying, of copious watering before this occurs. Increase is easily had by division and by spores. For further information see Ferns.

MATTHIOLA (Matthì-ola)—Stock. Fifty-five species of annuals, biennials, and subshrubby perennials of the mustard family CRUCIFERAE constitute *Matthiola*. Two have considerable horticultural importance. Foremost is the common stock once called gilliflower, the other the evening stock. The plant known as Virginian-stock is *Malcolmia maritima*. Other species of *Malcolmia* are called malcolm-stocks. The sorts of *Matthiola* are mostly natives of the Mediterranean region, but a few occur in central Asia and South Africa. The name commemorates Pierandrea Mattioli, an Italian physician and botanist, who died in 1577.

As they grow in the wild, matthiolas are grayish-pubescent, have more or less narrow, oblong leaves, sometimes lobed or with wavy edges, and spikes or racemes of often fragrant, white, lilac, or purple flowers. The blooms have four erect sepals, four spreading petals that narrow at their bases into long claws, four long and two shorter stamens, and one style tipped with a lobed stigma. The fruits are podlike. The hairs responsible for the grayness of the stems and foliage are branched and starlike. In cultivation varieties that are double-flowered or show other improvements are favored. Stocks are closely related to wallflowers (*Cheiranthus*), differing in their seed pods being cylindrical or flattened, but not four-sided, and in their seeds being winged. Like stocks, wallflowers are sometimes called gilliflowers. It is of interest to note that double-flowered stocks do not produce seeds. Those from which they are raised are borne by single-flowered plants especially selected to give seeds that ensure a high percentage of double-flowered progeny.

Matthiola incana, single-flowered

The common stock (***M. incana***), in its wild state, is a branched biennial or perennial with the base of its stem more or less woody and with alternate, long-oblong to oblanceolate, blunt leaves that taper gradually to short stalks and are neither lobed nor toothed. Stems and leaves are felty-gray-pubescent. The fragrant flowers, in loose, erect, terminal racemes that elongate as the blooms open in succession from below upward, are purple, reddish, yellowish, or white. The common stock, a

Matthiola incana, double-flowered

native of southern Europe and The Isle of Wight off the southern coast of England, is naturalized as an escape from gardens in California. Variety *M. i. annua* is less woody and comes into bloom sooner from the time the seed is sown.

Cultivated varieties of the common stock are classified into three groups, ten-week stocks, intermediate or East Lothian stocks, and Brompton stocks. In all there are wide choices regarding flower color, from pure white to deep crimson and including all shades of pink, lilac, and purple as well as pale yellow. Ten-week stocks (*M. i. annua*) are invariably, and intermediate and East Lothian stocks may be, grown as annuals. When accorded this treatment intermediate or East Lothian stocks take longer to bloom than the ten-week varieties. From spring sowings they flower in fall. Brompton stocks are always, and intermediate or East Lothians may be, grown as biennials.

Because stocks do not thrive in extremely hot weather and are killed by temperatures below freezing, the biennial sorts are little grown in most parts of America, being more popular in the British Isles. In America reliance is placed mostly on ten-week stocks, which can be raised and bloomed successfully during favorable periods. Double-flowered plants of all *M. incana* varieties are superior to single ones and reputable seed raisers give much attention to producing strains that give a high percentage of doubles and relatively few weedy, single-flowered ones.

The evening stock (***M. longipetala***) is a loose-growing annual or biennial, often subshrubby, with gray, downy stems and foliage and smaller flowers than those of *M. incana*. Its leaves, at least the lower ones, are narrow, lobed, and up to 3 inches long. The flowers open only at night and are then delightfully and intensely fragrant. They are purplish to lilac. Its long seed pods are cylindrical and have two horns. This is native to southeast Europe. Its variety *M. l. bicornis* (syn. *M. bicornis*) has pods with longer horns.

A border of double-flowered stocks

Garden Uses. These are admirable decoratives, well suited for seaside as well as inland gardens. The common stock is esteemed for beds and borders and for the production of outdoor flowers for cutting. It is extensively grown in greenhouses for cut flowers and to some extent as pot plants. In pots it serves well to embellish conservatories and large greenhouses in which winter and spring flower displays are maintained. The evening stock, without appreciable decorative value, is grown exclusively for the penetrating and delicious fragrance of its nocturnal blooms that perfume the air so pleasantly. A small patch beneath a window open in the evening is a pretty way of using this plant, or it may be located alongside paths apt to be used after dark. Where the Virginian-stock flourishes, that is where summer nights are fairly cool, it is practicable to mix seeds of it with those of evening stock and sow them together. The result is a planting made colorful by the Virginian-stock by day and fragrant by the evening stock after nightfall. Another possibility to relieve the daytime drabness of the evening stock is to interplant it with sweet-alyssum.

Cultivation. For their best development stocks need fertile, well-drained, neutral, slightly alkaline or slightly acid soil in good physical condition, not gravelly, sandy, nor excessively clayey and compact. Sunny locations are preferred, but a little part-day shade is tolerated especially by the evening stock.

To have ten-week stocks bloom outdoors in summer it is usually best to start seeds early indoors. Sow them in a temperature of 60°F about eight weeks before the plants are to be set in the garden, which can be done as soon as there is no longer danger from frost. Transplant the seedlings 2½ to 3 inches apart in flats or individually in 3-inch pots in fertile, porous soil as soon as their second pair of leaves is half-grown. If the soil is acid mix a little lime with it. Experience indicates the strongest seedlings are most likely to be double-flowered, and so at this transplanting it is good practice to discard all weak ones. Grow the young plants in a sunny greenhouse with a night temperature of 50°F and do not allow the day temperature to exceed this by more than five to ten degrees before opening the ventilators. Water so that the soil is allowed to become partially dry between applications. Constant saturation causes roots and stems to rot.

An alternative method is to transplant the young seedlings to a mild hot bed or cold frame, spacing them 3 to 4 inches apart. If this is done take care to grow them as cool as possible short of allowing them to be frosted. Whether in a greenhouse, hotbed, or cold frame, ventilate freely on all favorable occasions.

From the flats, pots, or frames transplant to the flowering locations as early as weather permits so that ample opportunity is given for maximum root growth before really warm weather arrives. It is very important to handle the plants in such a way that their roots are damaged as little as possible and good balls of soil are retained around them. Because stocks tend to have straggly root systems careless handling easily results in most or all of the soil falling away. A final spacing of about 1 foot each way between plants is satisfactory. Although an early start indoors is usually advantageous, where summers are not torrid good results can be had by sowing outdoors as early in spring as possible and transplanting the seedlings or thinning them to 10 inches to 1 foot apart. Bloom may be expected in about twelve weeks from the time seeds are sown.

In greenhouses the finest quality ten-week stocks are produced. Long-stalked, massive spikes of gorgeous, fragrant blooms repay good culture if a good strain of seeds is sown. The plants may be 3 feet or more in height. As cut flowers they are scarcely excelled and, grown shorter and with more flower spikes, are delightful pot plants. Seedsmen offer varieties specially selected for greenhouse cultivation, and it is important to choose from these. They may be grown in ground beds, benches, or pots. For an early spring crop sow seeds in late August, for successional blooming in September or early October. Seeds sown in December or January give plants to bloom in late spring. Plants from these later sowings will not have as long-stemmed or large spikes of bloom as those from earlier seedings.

To ensure fine results in greenhouses, provide coarse, fertile soil sweetened with the addition of a little lime. For cut flowers space plants 6 to 8 inches apart. In pots use as finals, depending upon how early the seeds are sown, containers 4 to 6 inches in diameter. Maintain a night temperature of 50°F, five to ten or perhaps up to fifteen degrees more by day in sunny weather. Water moderately and after the available soil is well filled with roots apply dilute liquid fertilizer at weekly intervals.

A greenhouse crop of double-flowered stocks

Intermediate or East Lothian stocks sown in spring and treated in the manner recommended for ten-week stocks bloom in fall, but succeed only where summers are cool. In regions of mild winters sow seeds in August to give plants that, wintered in cold frames or where little freezing occurs, in outdoor nursery beds, will be ready for transplanting to their blooming quarters in early spring. In cold frames or nursery beds space them 6 inches apart. When moved to their flowering locations allow about 1 foot between the plants.

Brompton stocks are sown in June or July in a cool, lightly shaded cold frame or outdoor bed. Transplant the seedlings 6 inches apart in cold frames or where little frost occurs to outdoor nursery beds and give them their final move to their blooming stations early the following spring or, in regions of very mild winters, in fall. A spacing of about 1 foot each way between plants is satisfactory.

The evening stock is simple in its requirements. Sow seeds outdoors in early spring where the plants are to stay and thin the seedlings to 3 to 4 inches apart. This sort is less demanding as to soil quality than ten-week and intermediate or East Lothian tocks.

Diseases and Pests. Stocks are subject to damping off, bacterial rot, white rust, wilts, and downy mildew as well as the virus disease called curly top. Their chief insect pests are aphids, caterpillars, flea-beetles, and springtails. The latter are responsible for tiny dead spots on the lower foliage and the eventual withering of affected leaves. They are controlled by dusting the soil surface in the evening or early morning with malathion or other suitable insecticide.

MATUCANA (Matu-càna). This genus of the cactus family CACTACEAE is so similar to *Borzicactus* that conservative botanists include it there. Native at high altitudes in Peru, *Matucana* was for long considered to comprise only *M. haynei.* From 1932 on, another dozen species were described, based on such minute and often unstable differences that all may with much justification be considered variants of *M. haynei.* The name *Matucana* is that of a village near which *M. haynei* was discovered.

Matucanas are medium-sized and have solitary or clustered, flattish-globular to cylindrical stems or plant bodies closely furnished with clusters of generally needle-like spines. The day-opening, slender, funnel-shaped blooms come from the tops of the stems. They have very oblique lips or faces. The perianth tubes, ovaries, and fruits are without hairs or scales or sometimes are minutely-hairy. This genus scarcely differs from *Submatucana.*

Attaining up to 1 foot in height and a stem diameter of 3½ to 4 inches, ***M. haynei*** (syn. *Borzicactus haynei*) has a plant body at first spherical, but with age becoming cylindrical. It has twenty-five to thirty slightly-notched, low ribs. The areoles (spine-producing areas) when young are woolly. The formidable needle-like white to grayish spines with darker ends, thirty or more from each areole, intertwine and effectively conceal the plant body. They are up to 1½ inches long, the centrals longer and darker than the radials. The flowers are orange-red to dark red with violet edges to their pointed petals. They are 2½ to 3 inches long, 2 inches wide or slightly wider, and have petals of varying lengths, the outer ones shortest, the others progressively longer toward the center and overlapping like roof shingles. The stamens are pink, the four-lobed stigma yellow.

Garden and Landscape Uses and Cultivation. Matucanas are collectors' plants that respond to the general care suggested for desert kinds under Cactuses. In cultivation they seem to be rather shy about blooming. Propagation is by seed.

MAUGHANIA (Maugh-ània). Herbaceous plants and shrubs, sometimes climbing, belonging to the pea family LEGUMINOSAE, compose *Maughania,* a genus of thirty-five species of the Old World tropics. Formerly they were named *Flemingia.* The name has also been spelled *Moghania.* It is a latinized form of an Indian name.

Maughanias have leaves undivided or of three leaflets, conspicuously veined, and with glandular dots on their undersides. In racemes or large panicles from the leaf axils and terminal, small clusters of red, purplish, yellow, greenish, or whitish flowers are hidden behind large, persistent, folded, leafy bracts. The standard or banner petal has earlike appendages at its base. There are ten stamens, nine united and one free. The fruits are short, oblique, swollen pods.

Naturalized in the southern United States and elsewhere in warm parts of the Americas, ***M. strobilifera*** (syn. *Flemingia strobilifera*), a native of India, Malaya, China, and Taiwan, a slender-stemmed shrub up to 10 feet tall, is often considerably lower. It has undivided, ovate to elliptic leaves 2 to 6 inches long, thinly-silky-hairy on their lower sides. The tiny, greenish-white flowers are in racemes or panicles 3 to 6 inches long. The bracts are heart-shaped and downy, the seed pods under ½ inch long.

Garden and Landscape Uses and Cultivation. Of minor importance horticulturally, the species described succeeds in warm climates, under ordinary conditions of soil and exposure, and is well suited for planting near the sea. It is best propagated by seed. Cuttings may also be used.

MAURANDYA. See Asarina.

MAURITIUS-HEMP is *Furcraea foetida.*

MAXILLACASTE. This is the name of orchid hybrids the parents of which are *Lycaste* and *Maxillaria.*

MAXILLARIA (Maxil-lària). About 400 evergreen species of the orchid family ORCHIDACEAE constitute *Maxillaria,* a genus of chiefly tree-perchers (epiphytes) or on occasion rock-perchers, native from Florida to Argentina. The name is derived from the Latin *maxilla,* a jaw. It alludes to the flowers of some sorts being suggestive of the jaws of an insect.

Maxillarias usually have pseudobulbs. Those of some sorts are flattened and in clusters with each pseudobulb having a single large, oblanceolate leaf. Those of certain other sorts are clustered, scarcely or not flattened, and each has one or two generally strap-shaped leaves. Yet other maxillarias have pseudobulbs spaced at intervals along pendulous, trailing, or erect rhizomes. Some few sorts are without pseudobulbs or at most have inconspicuous ones. Their leaves or bracts are in fans or are two-ranked on canelike stems. The short- to long-stalked flowers, often produced in great profusion, are always solitary. They originate from the bottoms of the pseudobulbs or from the leaf axils and range in size from small to as much as 6 inches across. They exhibit tremendous variety in flower color, from pure white to almost black and with intermediates including greens, yellows, browns, and reds. In aspect the blooms show affinity to those of *Bifrenaria* and *Lycaste,* to which genera *Maxillaria* is obviously kin. The three sepals are separate or the lateral ones may be united at their bases. Mostly a little smaller than the sepals the two petals often extend forward over the lip and parallel the often curved, wingless column. Three-lobed or lobeless, the lip is concave and usually shorter than the petals. More species than are described here are likely from time to time to be included in collections of orchid fanciers.

Largest-flowered and for that reason very worthwhile, ***M. sanderana,*** of Ecuador and Peru, has clustered, compressed, one-leaved pseudobulbs about 2 inches long. The narrowly-oblong leaves are up to 1 foot long or sometimes longer by a little over 2 inches wide. The fragrant flowers, 5 to 6 inches in diameter, have white sepals and petals spotted in their lower parts with purple-red or crimson. The fleshy, creamy-white lip is also marked with purple or crimson near its base.

Others with flattened, clustered pseudobulbs and solitary leaves include these: ***M. grandiflora,*** of Ecuador, has pseudobulbs 1½ to 3 inches tall each with a strap-shaped leaf up to 1 foot long or longer and 2 inches wide. Fragrant and 3 to 4 inches in diam-

Maxillaria grandiflora

eter, the blooms are milky-white with an orange-yellow lip strongly striped with red or wine-purple. ***M. lepidota,*** of Guatemala, has ovoid pseudobulbs up to 1½ inches long and broad-oblongish-elliptic leaves up to 1 foot long and heavily spotted with light gray. The spidery, buff-yellow blooms, up

Maxillaria lepidota

to nearly 5 inches across, have usually brown-tipped sepals. The lip is spotted with black-purple. ***M. luteoalba,*** of Central America and northern South America, has pseudobulbs 1½ to 2 inches tall and broad, strap-shaped to elliptic-lanceolate leaves 9 inches to 1½ feet long. Its flowers, variable in size and color and sometimes fragrant, may be 4 inches across. Typically, they have pale yellow sepals and petals that become tawny toward their apexes. The three-lobed lip is creamy-white or yellow margined with white, its side lobes lined with purple. ***M. venusta*** is a beautiful native of Colombia with pseudobulbs 2 to 3 inches long and oblong-lanceolate leaves 1 foot long or somewhat longer. The fragrant blooms, up to nearly 6 inches wide, have white sepals and petals, the latter markedly the shorter and jutting forward parallel with the column. The even shorter, fleshy, three-lobed lip is buff-yellow on the inside, creamy-white sometimes stained with red along the margins and spotted with red on the back.

Sorts with ovoid pseudobulbs not markedly compressed include ***M. picta*** and ***M. porphyrostele,*** both of Brazil. Those of the first, clustered or somewhat spaced along the rhizomes, are 2 to 3 inches tall and carry one or two thick, narrow, strap-shaped leaves up to 1¼ feet long. The pseudobulbs of the other are about 1¼ inches tall, densely clustered, and usually have two linear leaves not more than 9 inches long. The sometimes fragrant, 2½-inch-wide flowers of *M. picta* have yellow sepals and petals dotted and streaked with purple or chocolate-purple. The lip is white with purple on the side lobes. The blooms of *M. porphyrostele* are 1¼ inches wide and yellow, with a central stripe of

Maxillaria picta

purple toward the bases of the petals and purple streaks on the side lobes of the lip.

Slightly flattened, nearly circular pseudobulbs, about 1 inch in diameter, are characteristic of ***M. brunnea,*** of Colombia. Each has a solitary leaf with a petiole 2 to 3 inches long and an elliptic, leathery blade up to 6 inches long. The flowers are solitary on stalks, clothed with brown scales, 2 to 3 inches long, that sprout from the bases of the pseudobulbs. About 1½

Maxillaria brunnea

inches across, they are yellow and brown with chestnut-brown outsides to the sepals and petals. Miniature ***M. reichenheimiana,*** of Costa Rica, has flattened pseudobulbs. It was described originally, in 1871, as being "not larger than a threepenny piece" (Americans can substitute ten-cent piece or dime). The broad-elliptic leaves, about 4 inches long by 2½ inches wide, are green, conspicuously decorated in the manner of those of *Phalaenopsis schillerana,* with gray spots. The nodding flowers are solitary atop slender stalks 3 to 4 inches long. Approximately 1½ inches across, they are yellow and cream with brown or violet markings on the thick lip. Tightly-clustered pseudobulbs, about ¾

Maxillaria reichenheimiana

inch long, more or less triangular in section, and longitudinally fluted, are characteristic of Brazilian ***M. seidellii.*** From the apex of each sprout two erect, slender-linear leaves 1 inch to 1½ inches long. The solitary, white or whitish flowers, on erect stalks slightly shorter than the leaves, are ½ to ¾ inch wide.

Maxillaria seidellii

Sorts with pseudobulbs spaced along the rhizomes include these: ***M. coccinea,*** of the West Indies and northeastern South America, has flattened pseudobulbs 1½ inches long or longer, each with one strap-shaped leaf about 1 foot long by 1 inch wide. Its flowers, mostly under 1 inch across, are light rose-red or red with a usually darker colored lip. ***M. densa*** (syn. *Ornithidium densum*), of Mexico and Central America, has clustered along long rhizomes flattened, oblongish-egg-shaped pseudobulbs up to 3 inches long. The solitary leaves are linear, lanceolate, or oblanceolate, up to 1¼ feet long by 2 inches wide. The flowers, densely clustered in the axils of the sheaths of the new growths, are greenish-white to yellowish tinged with pink or purple, and about ½ inch in diameter. ***M. elatior,*** of Mexico and

Central America, has rhizomes up to 1½ feet in length with a few scattered one- or two-leaved pseudobulbs. The blooms, sol-

Maxillaria elatior

itary from the leaf axils, are small and fleshy and reddish-yellow to brick-red with darker centers. ***M. juergensii,*** of Brazil, has tufts of erect rhizomes 6 inches or so high, each with several small, ellipsoid

Maxillaria juergensii

pseudobulbs topped with linear leaves some 2 inches long. The ¾-inch-wide blooms are dark reddish-brown. ***M. sanguinea,*** of Central America, has tightly clustered pseudobulbs ½ to ¾ inch long, each with a solitary grasslike leaf 8 inches to 1 foot long by under ⅛ inch wide. The flowers, little over 1 inch across, have reddish-chocolate sepals and yellow petals heavily spotted with the same hue. The purple-brown lip is cream-margined at its apex. ***M. tenuifolia,*** of Mexico and Central America, has erect rhizomes and 1-inch-long, ovoid pseudobulbs, with one leaf up to 1 foot or somewhat more in length. The strongly coconut-scented blooms, 1½ to 2 inches wide, are rather variable in color. Most commonly the sepals and petals are deep red speckled or mottled with yellow and the lip blood-red in its lower part, dark yellow spotted toward its apex with reddish-purple or reddish-brown. ***M. variabilis,*** a native from Mexico to northern South America, has erect to drooping rhizomes up to about 1 foot long with spindle-shaped to subcylindrical pseudobulbs 1 inch to 2½ inches in length, each with a single strap-shaped leaf that may attain a length of 6 inches. Approximately ¾ inch across, the flowers exhibit great variation in color. They may be white or yellow with deep red markings, completely dark red or reddish-brown, or have red-marked white or yellow sepals and petals and a dark red lip.

Maxillaria sanguinea

Maxillaria variabilis

Native from southern Florida to Brazil and the West Indies, ***M. crassifolia*** is less ornamental than most cultivated sorts, but has certain interest because of its inclusion in the native flora of the United States. Its pseudobulbs are inconspicuous or lacking. From 3 inches to 1½ feet long and up to 2 inches wide, its thick leaves are in loose fans. The yellow to orange-yellow flowers, which usually do not open fully, ½ to ¾ inch long, have lips often streaked with purple. Its natural range extending from the West Indies to Central America, northern South America, Brazil, and Peru, variable ***M. rufescens*** has clustered, mostly somewhat four-angled, ovoid pseudobulbs 1 inch to 2¼ inches long, each with a solitary stalkless or very short-stalked, strap-shaped to elliptic-oblong leaf 6 to 8

Maxillaria rufescens

inches long. Its fragrant flowers, on horizontal to erect stalks up to 2 inches long, sometimes do not open fully. From ¾ inch to 1¼ inches wide, they have usually pale to dark pinkish-brown to reddish-brown sepals and somewhat narrower light yellow petals. The erect, three-lobed lip, its center lobe squarish, is yellow spotted with red or brown.

Garden Uses and Cultivation. Maxillarias for the most part are not difficult to grow and are well suited for beginners as well as advanced growers of orchids. However, with a group as vast as this, the members of which come from such varied habitats, accommodation must be made to the particular needs of different kinds, particularly with regard to temperatures. Most succeed in environments appropriate for cattleyas. Some do well under cooler conditions. All need fairly bright light with shade from strong sun. Maxillarias grow satisfactorily in osmunda, tree fern, and other fibers and bark chips that suit most epiphytic orchids. They are sensitive to the rooting medium becoming soggy and sour and should be repotted before this happens. Those with elongated rhizomes need the support of "totem poles" of tree fern trunk or other aids to keep their rhizomes erect, or they can be grown in hanging baskets or otherwise be suspended and allowed to droop. Maxillarias have no definite dormant period. They should be watered throughout the year, but less abundantly when quiescent than when in active growth. During the latter period mild fertilizing is advantageous. For more information see Orchids.

MAY is *Crataegus laevigata* and *C. monogyna.*

MAY-APPLE is *Podophyllum peltatum.*

MAY, GARDENING REMINDERS FOR. One of the most rewarding months in northern gardens, May is also a busy one with a good deal that needs doing to ensure satisfaction now and for the rest of spring and productivity throughout summer and fall. The pressure for immediacy that characterized so many awaiting tasks earlier is now relieved to some extent. Longer days and more favorable and predictable weather make it easier to schedule work and accomplish it at a more leisurely pace. Still there is no time to waste.

Much planting can be done this month. Nearly everywhere there is yet time to move evergreen trees and shrubs, all but the very smallest balled and burlapped or from containers. Be sure after transplanting to water evergreens very thoroughly and mulch around them. Do not transplant bare-rooted deciduous trees and shrubs after they have come into foliage, but specimens of these and of roses and herbaceous perennials from containers can be safely set out in full leaf and even in bloom, thus making instant landscaping feasible. This technique calls for generous watering not only at planting time, but during dry spells that may occur afterward. Late spring, when they are in foliage, is a good time to transplant magnolias and tulip trees. Be sure to take a large intact ball of soil with their roots.

Balled and burlapped evergreens as sold for planting

Planting an evergreen: (a) Filling soil around the root ball

(b) Watering very thoroughly

Removal as soon as their floral display is over and before their foliage has died of bulbs and other early bloomers from beds planted for spring display may be necessitated by need to replace them with summer bedding plants raised indoors in flats or pots. Discard such biennials as English daisies, forget-me-nots, pansies, and wallflowers. Dig out hardy perennials, such as *Arabis albida, Polemonium reptans,* dwarf phloxes, and polyanthus primroses, and plant them permanently elsewhere or carefully divide them and set the divisions in nursery beds to provide furnishings for next year's spring bedding.

Set plants from containers in the garden: (a) Removing the container

(b) Planting

Tulips, hyacinths, and other bulbs to be taken from flower beds before their foliage has died and saved for next year are treated differently. Dig them out very carefully with as much soil as possible held about their roots and heel them in (plant them temporarily) closely together in a place shaded from strong sun. Water them immediately and copiously and allow them to die down naturally. After the tops are brown and dry, lift the bulbs and clean off dead stems and foliage and store the bulbs in a dry, airy, cool cellar or similar place indoors until the fall planting season.

Heeling in tulips lifted from beds after they have finished flowering

Do not remove foliage until it has died completely from crocuses, hyacinths, narcissuses, tulips, and other spring-flowering bulbs that are to remain in the ground. If you do you will seriously reduce the buildup of stored food materials in the bulbs upon which the quality and quantity of next year's bloom depend. Where bulbs of these sorts are naturalized in lawns and meadows the grass must not be cut until the leaves are quite brown and dead, and this will not be for awhile. In the meantime you may relieve the untidy appearance of the sprawling, dying leaves of narcissuses by bundling them and securing the tops of each small bundle with a tie or by plaiting or knotting them.

Annuals and other bedding plants raised indoors or in cold frames for the summer embellishment of flower beds, window and porch boxes, and other containers will mostly be ready for planting in their summer quarters this month, al-

Indoor-raised annuals to be planted in May: (a) Hardening for a week or two in a cold frame or sheltered place outdoors

(b) Planting

though with more tender sorts, such as begonias, caladiums, fuchsias, geraniums, and heliotropes, it may be wise, in cold sections even necessary, to wait until the first week in June. In any case make sure the plants are gradually hardened to inure them to outdoor conditions before they are planted. Tomatoes, peppers, and eggplants raised indoors or in cold frames must also be carefully hardened before planting them outside after the weather is warm and settled. The tomatoes are a little hardier to cold than the others.

Summer-flowering and summer-foliage bulbs, using these terms in the broadest popular sense to include plants with corms, tubers, and similar food-storage organs, may be planted from early May to early June depending upon kind and local climate. With some, such as dahlias and elephant's ears, only one planting is usually made, but with gladioluses, montbretias, and tuberoses, it is common practice to make two, three, or more plantings at ten-day or two-week intervals to ensure as long a succession of bloom as possible. Among sorts you might like to consider are caladiums, lycorises, summer-hyacinths (*Galtonia candicans*), tigridias, and zephyr flowers (*Zephyranthes*).

Make first plantings of bulb plants in warm and settled weather: (a) Gladioluses

(b) Montbretias

(c) Tuberoses

Outdoor seed sowing is a frequent pleasant task in May. Not only are successional sowings of many hardy annuals and vegetables of which sowings were made last month in order, but first sowings of more tender sorts may be made as soon as the weather is reasonably warm and settled. About the time it is safe to sow the first corn it is appropriate also to sow beans, cucumbers, melons, okra, squash, and frost-tender annuals and plants grown as annuals, such as dwarf dahlias,

Bulbs to be planted in May: (a) Summer-hyacinths

(b) Tigridias

(c) Zephyranthes

Sow seeds of many sorts of annuals and vegetables outdoors

portulacas, tithonias, and torenias. Sow nearly all kinds of biennials and short-lived perennials best treated as biennials toward the end of May or at latest in early June. Pansies are an outstanding exception. Their sowing is best deferred until late July. Sorts to sow earlier include aubrietas, columbines, English daisies, forget-me-nots, foxgloves, honesty, verbascums, and wallflowers.

Perennials to be raised from seeds do well from sowings made outdoors or in a cold frame from the middle to the end of May. Among the many kinds satisfactory when raised in this way are coreopsises, delphiniums, gaillardias, geums, liatrises, and pyrethrums.

After-flowering pruning of spring-blooming shrubs and trees is apt to be overlooked by amateurs, yet any pruning they need is best done as soon as blooming is through, in most cases sometime this month or next. Not all spring-bloomers need this attention, but sorts that develop new branches so freely that the bushes become overcrowded and too dense if neglected benefit from annual or biennial pruning. Among such sorts are deutzias, forsythias, and weigelias. But look and appraise all kinds with a critical eye for possible need. Even crab apples, flowering cherries, magnolias, and others that need no regular pruning may call for a little cutting to remove an occasional damaged or ill-placed branch.

Shrubs to prune shortly after they have finished flowering: (a) Deutzias

(b) Forsythias

(c) Weigelias

Herbaceous perennials will be making good growth now. Attend to staking those likely to require support before the need becomes desperate. Stems kept straight and orderly from the beginning assure much more shapely and attractive plants than if you procrastinate and attempt correction after they have been allowed to sprawl or have been beaten down by wind or rains.

Herbaceous perennials staked to prevent their stems flopping: (a) With brushwood

(b) With canes and twine

The scuffle hoe is very efficient for destroying young weeds

Routine matters claiming attention in May include weed elimination, much of which can be accomplished without great physical effort and with much satisfaction by using a scuffle or Dutch hoe. But this is only true if that effective tool is employed while weeds are still small, and so its frequent use is indicated. Even so, a certain amount of hand pulling close to the plants is likely to be necessary. Sheets

A short-handled cultivator can be used to destroy weeds and to loosen surface soil in a rock garden

Sheets of black plastic mulch, here weighted down by stones, suppress weeds

of black plastic mulch are a boon in keeping down weeds between plants in rows as in vegetable gardens and nursery areas. Give timely attention to mowing lawns.

Mow laws regularly (a) With a hand mower

(b) With a power mower, here fitted with a grass catcher, for a heavy cut

If dry spells occur, as is not uncommon in May, water very thoroughly at weekly to ten-day intervals herbaceous perennials, roses, vegetables, and especially recently transplanted trees and shrubs. And keep up with pest and disease control work.

In greenhouses work tends to be less demanding in May than previous months. Crowding is greatly relieved by transfer outdoors or to cold frames of plants for summer display and for the vegetable garden. Furthermore, the display values of such sorts as cinerarias, cyclamens, freesias, primulas, schizanthuses, and a host of others that provided gay blooms in winter and early spring are now over or nearly over and their passing relieves pressure on space.

Warm days necessitate much more generous ventilating of structures housing such cool greenhouse plants as calceolarias, chrysanthemums, cinerarias, cyclamens, and primulas. Whenever outdoor temperatures allow, keep the ventilators open a little by night, much more widely by day. The use of slowly revolving greenhouse fans does much to create the free circulation of air so beneficial to plants of these and similar kinds.

Damp down floors and other surfaces during the day and in sunny weather lightly spray the foliage early enough to ensure that it dries before nightfall. Of the plants mentioned in the last paragraph all except chrysanthemums now need light shade from strong sun.

To have flowering specimens next winter and spring take cuttings of marguerites (Paris-daisies), geraniums, hydrangeas, *Coleus thyrsoideus*, *Erlangea tomentosa*, stevia (*Piqueria trinervia*), and *Plumbago indica*. When rooted, transfer them individually to small pots and successively as growth demands to bigger ones. Pinch the tips out of the stems of the young plants (except hydrangeas) to induce branching. First-year hydrangeas, grown to single stems, need no pruning. Nip out while yet small all flower buds that appear between now and fall on geraniums being grown for winter bloom.

Cuttings for greenhouse pot plants: (a) *Coleus thyrsoideus*

(b) Hortensia hydrangeas

(c) Marguerites (*Chrysanthemum frutescens*)

Poinsettias dormant since shortly after they finished blooming may now be started into growth. Cut them back to heights of from 6 inches to 1 foot, weak stems more severely than strong ones. Remove them from the pots, shake most old soil from their roots and repot in a sandy mix in containers just big enough to comfortably hold the roots. Place in full sun in a warm, humid atmosphere. Do not keep the soil too wet, but spray the tops lightly daily with water to encourage the buds to break and produce shoots suitable for cuttings. An alternative plan is to plant the old plants in a sunny spot in the garden. There they will develop strong shoots that can be used as cuttings.

Poinsettias: (a) Taken from their storage quarters

(b) Pruned, repotted, and put in a warm, sunny place to start into new growth

Seeds of *Primula malacoides*, *P. obconica*, *P. sinensis*, and *P. kewensis* to give plants for winter and spring blooming may be sown in May, but these are very definitely cool-climate plants and in regions of hot summers it calls for much skill to care for the young plants successfully. Where this cannot be given, August sowing is advisable, although resulting plants are smaller than from seeds sown in May. You may now sow pansy seeds to give plants for blooming in winter in a cool greenhouse and seeds of lovely blue-flowered *Coleus thyrsoideus* to have plants in flower at Christmas in an intermediate-temperature greenhouse.

In readiness for planting new carnations and roses next month in greenhouse benches or beds, remove old plants and old soil and before replacing it with new thoroughly cleanse the inside of the structure by washing the glass, scrubbing the woodwork or metalwork, and whitewashing or painting walls. Do not neglect to clean thoroughly beneath the benches.

Rest for a month or so by allowing the soil to dry considerably but not completely roses and gerberas being kept over a second or a third year in beds or benches. About the middle of the month begin to taper off watering calla-lilies and early next month cease completely. Lay the pots on their sides in a cold frame or similar place, there to be kept quite dry until repotting time in late August or September. Treat deciduous South African bulb plants, such as freesias, ixias, lachenalias, ornithogalums, sparaxises, and tritonias, in the same way.

Prune early in May spring-flowering nonhardy shrubs grown in greenhouses. Such sorts include acacias, cytissuses, ericas, and jasminums. Remove thin, weak stems and dead ones and if compact specimens are desired shorten others of last season's growth to within 1 inch of their bases. Cut back less severely if you would like larger plants. After pruning water moderately and spray the tops two or three times a day lightly with water to encourage sprouting. When new shoots are about 1 inch long repot or topdress according to the needs of individual plants.

Cold frames in May serve mainly as hardening-off and holding areas for plants soon to be planted outdoors. Their management is essentially that recommended in the Encyclopedia entry April, Gardening Reminders For, except that freer ventilation often involving the complete removal of the sash, and more frequent watering will be called for, and with shade plants some protection from increasingly strong sun.

Houseplants in general should now show substantial improvement. Trials occasioned by short days, poor light, and most importantly the employment of artificial heat are over, and good growth should in most cases be evident.

Because of the reduction in or cessation of indoor heating, some plants, especially those in moisture-holding soils of high organic content such as are used for begonias and African-violets and other gesneriads, are likely to need less frequent watering, but this will not necessarily be true of well-rooted specimens of other kinds particularly if they are located where breezes or drafts from open windows or doors affect them. But newly potted specimens of all plants that have not yet filled their soil with roots will need less frequent watering than better-rooted ones.

Fertilize plants well-rooted in their containers regularly in accordance with their individual needs, but withhold fertilizer from any recently potted or that, like Martha Washington geraniums and calla-lilies, will soon enter their resting periods.

Cuttings planted earlier and now with roots an inch or two long are ready for potting individually in sandy peaty soil in small pots. Young plants now occupying small pots will in most cases be ready for transfer to somewhat larger ones. Incidentally, this is a good time to increase your stocks of houseplants by purchasing small specimens from specialist dealers. With a full summer and fall ahead they have ample time to grow into sizable specimens and become acclimated before winter.

Propagation of a great variety of houseplants can be achieved by division and cuttings this month and the results by fall will be scarcely smaller than if earlier action had been taken. For suggestions of sorts to consider see the Encyclopedia entry April, Gardening Reminders For. Many other sorts can be easily increased.

In the south, inasmuch as plant growth typifies the season, early summer is at hand or has arrived, bringing with it the need for greater emphasis on such regular maintenance work as weeding, mowing, watering, and spraying or dusting to battle diseases and pests.

Now raise mower blades so that lawns are cut at a height of 1½ to 2 inches. Fertilize established lawns of Bermuda grass and in the middle and upper south sow new ones on soil that has been very thoroughly prepared by turning it over to a minimum depth of 6 inches and incorporating compost, peat moss, or other suitable organic matter and fertilizer. Be alert for the early germination of crabgrass and take measures to control it, using a selective weed killer or hand picking or both.

Weeds now grow with amazing speed. Destroy them while yet small. Between rows in vegetable and flower gardens and nursery areas and between plants in perennial and shrub beds where clean cultivation is practiced this necessitates frequent use of hoe or cultivator and probably a certain amount of hand weeding. Especially admirable for the purpose is the Dutch or scuffle hoe. Mulching is an attractive alternative to keeping the soil around and between plants weed-free by cultivating, hoeing, or hand weeding. It aids too in conserving moisture, keeping the roots relatively cool, and with some sorts of mulch supplying nutrients.

Other tidying and clean-up work includes picking faded flowers promptly, a procedure that prolongs the blooming season of many sorts, and staking and tying plants likely to be damaged by wind and rainstorms.

Make succession sowings of warm weather annuals and vegetables. Among the former are amaranths, castor beans, cosmos, marigolds, moonflowers, portulacas, strawflowers, sunflowers, tithonias, and zinnias. Dwarf dahlias, although not true annuals, do well when treated as

In the south make second sowings of:
(a) Amaranths

(b) Cosmos

(c) Dwarf dahlias

(d) Tithonias

(e) Zinnias

such, and can be sown now. Vegetables to sow include beans, collards, corn, cucumbers, melons, New Zealand spinach, squash, and watermelons. Set out young plants of eggplants, peppers, and tomatoes and rooted slips of sweet potatoes.

Bulb and tuber plants of many sorts may now be set out. Dahlias planted in late May or early June generally give better results than if planted earlier. Other sorts to plant this month include caladiums, cannas, crinums, elephant ears, gladioluses, gloriosas, hedychiums, summer hyacinths (*Galtonia candicans*), montbretias, tigridias, and tuberoses. Plant tropical water-lilies and other tender aquatics as soon as the water is sufficiently warm, not before.

Tropical water lilies are among aquatics to be planted now in southern gardens

Also plant annuals and plants treated as annuals from flats or pots to fill spaces in beds left vacant by the passing of spring blooms.

Evergreens of narrow- or needle-leaved sorts, such as arborvitaes, hemlocks, and junipers that are kept to formal shapes, may be sheared as soon as new growth is completed.

In the lower south cuttings of many sorts of shrubs taken late this month or in June root readily under mist or in a shaded cold frame. Sorts that respond include azaleas, boxwood, and camellias.

Shrubs to root from cuttings in May in the lower south: (a) Azaleas

(b) Boxwood

(c) Camellias

Chlorosis, an unnatural yellowing of foliage with the veins staying green, often may indicate lack of availability to the plants of iron and can usually be rectified by spraying or by applying to the soil iron chelates or sulfate of iron.

West Coast gardens, like those in the south, now call for more attention to routine maintenance. If not already done, prune spring-flowering shrubs that need this attention as soon as they pass out of bloom.

Make successional sowings of annuals and vegetables to assure continued displays and harvests and successional plantings of gladioluses, montbretias, and tuberoses. Other bulb and tuber plants to set out now are those suggested above for planting in gardens in the south. Dig up tuberous anemones and ranunculuses that are through blooming after their foliage has yellowed, dry them in a shady place, and store until fall planting time.

MAYACA (May-áca)—Bog-Moss. The ten species of *Mayaca* that compose the bog-moss family MAYACACEAE are endemic to the warmer parts of the Americas including the West Indies. The name is a vernacular one used in French Guiana.

Mosslike aquatic and bog herbaceous perennials with copiously leafy stems, mayacas have alternate, undivided, slender leaves pinnately-notched at their tips. The flowers, solitary on slender stalks from the leaf axils or several from near the tips of the stems, have three each sepals, spreading petals, and stamens, and one sometimes three-cleft style. The fruits are capsules.

Native from the southern western United States to throughout tropical America, ***M. fluviatilis*** occurs submerged and in bogs. It has clustered or running stems 2 to 9 inches long or longer, closely feathered along their lengths with hairlike, green leaves ¼ to ½ inch long. Solitary and ½ inch or a little more in diameter, its flowers are lilac to light purple or violet, often with white bases to the petals and with yellow anthers. Similar in size and aspect to the last, ***M. sellowiana,*** of Brazil

Mayaca sellowiana

and Paraguay, has ½-inch-wide flowers with lilac, lavender-pink and white, or violet petals and golden-yellow anthers.

Garden Uses and Cultivation. These are interesting plants for aquariums, for wet watersides and bogs in the tropics and subtropics, and for growing in pans (shallow pots) or pots kept standing in saucers of water in warm, humid greenhouses. They succeed when planted in sphagnum moss or in sandy soil containing an abundance of decayed organic matter, such as peat moss. Propagation is by division, cuttings, and seed.

MAYACACEAE Bog-Moss family. Botanically closely related to the spiderwort family COMMELINACEAE, the dicotyledonous bog-moss family MAYACACEAE has the characteristics of its only genus, *Mayaca*.

MAYFLOWER is *Epigaea repens*.

MAYPOP is *Passiflora incarnata*.

MAYTEN is *Maytenus boaria*.

MAYTENUS (May-tènus)—Mayten. About 200 species of trees and shrubs natives of the tropics and subtropics of the Old and the New World compose *Maytenus*, of the staff tree family CELASTRACEAE. The name is adapted from a vernacular one of Chile.

Spiny or spineless, members of this genus have alternate or sometimes clustered, undivided, leathery leaves and in clusters or less often panicles, small white, yellowish, greenish, or red bisexual or unisexual flowers with four to six each sepals, petals, and stamens and one style. The fruits are capsules.

The mayten (***M. boaria***) is a graceful, broad-crowned, pendulous-branched tree, in cultivation 20 to 30 feet in height, with much the aspect of a weeping willow. In some parts of its native Chile it attains a height of 75 feet. Its thin, finely-toothed, lanceolate leaves are about 2 inches long. In early summer a profusion of tiny greenish-white flowers appears. The pea-sized, two-seeded, yellow fruits split when mature to reveal scarlet-cloaked seeds.

Native only to the Canary Islands, ***M. cassinoides*** (syn. *Gymnosporia cassinoides*) is an evergreen shrub, erect and hairless. It has short-stalked, broad-ovate, slightly-toothed or wavy-edged leaves with blades 2 to 3 inches long by up to nearly 2 inches wide. The white flowers are in twos or threes. Ethiopian ***M. serratus*** (syn. *Gymnosporia serrata*) is a sometimes spiny, evergreen shrub 6 feet tall or taller. It has short-stalked, finely-toothed, ovate to elliptic leaves 1½ to 3 inches long, and whitish flowers in branching clusters from the leaf axils.

Garden and Landscape Uses and Cultivation. The mayten, hardy in mild climates only, is esteemed in California as a street and lawn tree. It seems able to adjust to either frequent irrigation or rather sparing supplies of water. It grows rapidly and prospers in a variety of soils. Propagation is usually by seed and suckers. Root cuttings would probably also be successful. Less common in cultivation, the other species described here may be expected to respond to similar environments and care.

MAZUS (Mà-zus). In the wild, ranging from Japan and China to the Himalayas, Tasmania, and New Zealand, *Mazus*, of the figwort family SCROPHULARIACEAE, includes twenty species. Its name, from the Greek, *mazos*, a breast, calls attention to the two protuberances or ridges in the throats of the flowers.

These are low, often carpeting annuals and herbaceous perennials, the cultivated ones perennials. Their leaves are commonly toothed or lobed, the lower ones clustered or opposite, those higher on the stems, alternate. The little flowers, in terminal, more or less one-sided racemes, have bell-shaped, five-toothed calyxes, a tubular, two-lipped corolla with the upper lip erect and two-lobed, the lower markedly larger, three-lobed, and spreading. At the base of the lower lip are two strongly evident longitudinal ridges. There are four stamens, and a style with two stigmas. The fruits are capsules with the persistent calyxes attached.

Most widely grown, and probably native to the Himalayas, ***M. reptans*** presents a tight mat of trailing, rooting stems from which are sent to heights of an inch or two short leafy branches bearing few-toothed, lanceolate to elliptic leaves usually not over 1 inch in length. About ¾ inch long, the broad-lipped flowers, suggestive of tiny lavender-blue snapdragons, in racemes of two to five, freely dot the greensward-like carpet of foliage. Their lower lips are decorated with white, yellow, and purple. In gardens this species is sometimes wrongly identified as *M. rugosus*, which is a synonym of *M. japonicus*. Native from Japan to India, ***M. rugosus*** is without trailing stems and up to 1 foot tall. It has coarsely-toothed, obovate leaves up to 2½ inches long. Its ¾-inch-long blue flowers have the brown-spotted ridges on their lower lips bearded with club-shaped hairs. Japanese ***M. miquelii*** (syn. *M. stoloniferus*) differs in having creeping stems, few-leaved at their bases, up to 6 inches long. The very blunt leaves are obovate, elliptic, or broadly-ovate and 1½ to 3 inches long, including the stalk, by up to a little over ½ inch wide. The loosely-flowered racemes are of few purple-blue blooms with two ridges with yellow-brown hairs on the lip. Variety *M. m. albiflorus* has white flowers.

Mazus reptans between stepping stones

Mazus reptans (flowers)

Two species are natives of New Zealand, *M. pumilio* and *M. radicans*. Also inhabiting Australia and Tasmania, ***M. pumilio*** has subterranean creeping stems from which come short branches closely set with short-stalked, toothed or toothless, obovate-spatula-shaped leaves ¾ inch to 3 inches long. The flowers, up to six on a slender stem, are under ½ inch long, and white with a yellow throat or purple with a white throat. Creeping, rooting stems are developed by ***M. radicans,*** with erect

Mazus pumilio

branches up to 2 inches tall. The closely-set, hairy or hairless, obovate to narrowly-spatula-shaped, toothed or toothless leaves, including their stalks, are ¾ inch to 2 inches long. From ½ to ¾ inch long, the yellow-throated flowers, in racemes of up to three, are white, often with purple markings on the lip.

Garden Uses and Cultivation. The plants of this genus are best suited for rock gardens and similar places. The low ones are well adapted for planting in crevices between paving laid on soil. A persistent spreader, *M. reptans* is too aggressive to neighbor choice small alpine plants, but not for planting in rock gardens where its pleasant, flower-sprinkled carpet of green can be displayed without threat to weaker-growing plants. This kind, hardy in southern New England, sometimes becomes established in lawns, where it suffers not at all from repeated mowing. The degree of hardiness of the other kinds is not precisely known, but the New Zealanders are certainly much more tender to cold than *M. reptans*. Gritty, fairly nourishing soil, not overly dry, and full sun or part-day shadow suit mazuses. They are easily increased by seed, division, and cuttings.

MAZZARD. This is a common name of *Prunus avium*.

MEADOW. The word meadow appears in the common names of these plants: meadow beauty (*Rhexia*), meadow foam (*Limnanthes douglasii*), meadow foxtail grass (*Alopecurus pratensis*), meadow grass (*Poa*), meadow-rue (*Thalictrum*), meadow saffron (*Colchicum autumnale* and *C. speciosum*), Queen-of-the-meadow (*Filipendula ulmaria*), and reed meadow grass (*Glyceria grandis*).

MEADOWSWEET. This vernacular name is used for *Filipendula, Spiraea alba,* and *S. latifolia*.

MEALY FLATA. See Plant Hoppers.

MEALYBUGS. These common pests of plants indoors and out, are not true bugs, but close relatives of scale insects usually found in colonies that look like little puffs of cotton. Most sorts feed on aboveground plant parts, root or ground mealybugs on the roots. All feed by piercing the skins and sucking the juices of the plants they infest. They have soft, segmented, oval bodies covered with particles of white wax that extend from the projections on the threads usually of about equal length, but in long-tailed mealybugs with those directed from the rear longer. Females of most kinds lay up to 600 eggs contained in a waxy sac below the rear of the body. The eggs soon hatch into crawling, pale yellow young. Males change into tiny two-winged flying insects and die after mating. The long-tailed mealybug, which does not lay eggs, bears living young. Mealybugs excrete honeydew, which attracts ants and upon which the disfiguring fungus called sooty mold grows. Control is had by the use of contact insecticides and by washing the insects off with a forceful stream of water.

Mealybugs on the underside of a leaf

MECONELLA (Meconéll-a). The genus *Meconella*, of the poppy family PAPAVERACEAE, consists of three or four species, all natives of Pacific North America. The name comes from the Greek *mekon*, a poppy, and *ella*, a diminutive.

Meconellas are slender-stemmed, poppy-like annuals with opposite or basal lobeless leaves and solitary flowers, with six, or rarely four, soon-deciduous petals, six to twelve or sometimes numerous, stamens, and usually three carpels. The fruits are capsules. The blooms look somewhat like those of cream cups (*Platystemon californicus*) to which *Meconella* is closely related.

Kinds sometimes grown in gardens are *M. linearis*, of California, and *M. oregana*, indigenous from Oregon to British Columbia. Both are pretty and adaptable for rock gardens and edgings. Clothed with stiff, spreading hair, ***M. linearis*** forms tufts of linear, pointed basal leaves up to 2¼ inches long, and leafy flower stems 6 to 8 inches high terminated by pale yellow, poppy-like blooms ½ to 1 inch in diameter. A variety, *M. l. pulchella*, has pale lemon-yellow outer petals and cream-colored or nearly white inner ones. The other species, ***M. oregana,*** is hairless and has very slender stems 2 to 6 inches tall. Its leaves are obovate to spatula-shaped and up to ¾ inch long. Its white flowers are ¼ to ⅓ inch across.

Garden Uses and Cultivation. These plants, useful as edgings and for sowing in drifts in rock gardens, thrive in regions of cool summers. They soon deteriorate if subjected to prolonged hot, humid weather. They thrive in any ordinary garden soil, generally preferring one that is sandy, and in full sun. Under favorable conditions they bloom profusely throughout the summer. Their needs are simple. Seeds are sown in early spring where the plants are to bloom and the seedlings are thinned to 3 or 4 inches between individuals. No further attention is needed other than that necessary to control weeds.

MECONOPSIS (Mecon-ópsis)—Blue-Poppy, Welsh-Poppy, Harebell-Poppy. Belonging in the poppy family PAPAVERACEAE, and numbering forty-five species, *Meconopsis*, with one exception, inhabits the Himalayas and mountains of western China. The exception, the Welsh-poppy, is confined in the wild to western Europe, including Great Britain. Except in a few favored parts of North America where cool, moist summers and relatively mild winters prevail, as in certain regions of the Pacific Northwest, *Meconopsis* is not for American gardeners. The reverse holds true in Great Britain, especially in its northern reaches, where many kinds are cultivated and highly prized. In the 1930s and later, numerous capable and experienced gardeners attempted to grow the blue-poppy (*M. betonicifolia*) as well as other kinds in the east and in other warm-summer regions, with almost total failure. At The New York Botanical Garden many careful trials established that only the comparatively nonalluring Welsh-poppy (*M. cambrica*) could be grown there outdoors. The name is from the Greek *mekon*, the poppy, and *opsis*, like, and alludes to the appearance of the blooms.

Meconopsises are hardy biennials and herbaceous perennials. Their leaves, usually in basal rosettes and on the stems, are without lobes or teeth or are variously lobed or dissected. The yellow, reddish, purple, blue, or rarely white flowers, nodding, facing outward, or upturned, closely resemble those of poppies. They may be solitary, in racemes, or in panicles. They have usually two, sometimes three or four sepals, mostly four, but up to ten petals, many stamens, and an ovary with an usually short-styled stigma. The fruits are erect capsules. In addition to the wild species several hybrids, some of extraordinary beauty, are grown by connoisseurs.

The blue-poppy (***M. betonicifolia*** syn. *M. baileyi*) is probably the best known.

Meconopsis betonicifolia

Good color forms of it are certainly among the most glorious of garden herbaceous plants. A biennial 4 to 5 feet tall, this has reddish-hairy leaves with oblong-elliptic blades 4 to 6 inches long. The sky-blue to pinkish-lavender, crepe-textured blooms, about 2 inches in diameter, are freely produced. The upper ones have five or six, the others four, petals.

Meconopsis betonicifolia (flowers)

Hybrids of *M. betonicifolia* include sulfur-yellow-flowered ***M. sarsonsii,*** the other parent of which is *M. integrifolia*. About 3 feet tall, this has large, outward-facing

Meconopsis sarsonii

flowers on erect individual stalks. Magnificent ***M. sheldonii,*** a 3- to 4-foot-tall hybrid between *M. grandis* and *M. betonicifolia*, is

Meconopsis sheldonii

longer-lived and has larger flowers than *M. betonicifolia*. Its slightly down-facing blooms are in fact larger and, in its finest forms, of a more beautiful clear blue than those of any other meconopsis.

Meconopsis sheldonii (flowers)

The Welsh-poppy (***M. cambrica***) is a perennial 1 foot to 1½ feet tall. It has deeply-lobed, somewhat hairy leaves with the lobes again cut. The basal leaves, including their stalks, are up to 8 inches long.

Meconopsis cambrica

The yellow, or in *M. c. aurantiaca* orange, flowers are on stalks from the axils of the upper leaves. They are nearly 1½ inches across, and have four or more petals. Variety *M. c. flore-pleno* has double yellow or orange flowers.

Other kinds that may be tried from time to time by American enthusiasts include the following: ***M. aculeata*** has irregularly-pinnately-lobed leaves sparsely furnished with pale yellow hairs, and, including the stalks, up to 9 inches long by 2 inches broad. The blue to reddish-purple blooms, about 1½ inches wide, are in racemes. Following blooming the plant dies. ***M. dhwojii,*** a soft yellow-flowered species that dies after flowering, is from 2 to 3 feet tall, with deeply-gashed, bristly-hairy, long-stalked leaves about 1 foot in length. Their hairs have purple-black bases. The flowers, the

Meconopsis dhwojii

Meconopsis dhwojii (flowers)

upper ones solitary, the lower ones few together, are about 1¼ inches in diameter. ***M. grandis*** is a 3-foot-tall, short-lived perennial. Its brownish-purple to dark slaty-blue, long-stalked blooms are solitary from the leaf axils. They have four or more petals, and are up to 5 inches across. The reddish-hairy, toothed leaves, with their stalks, are up to 6 or 7 inches long. They are elliptic-oblong to oblanceolate. ***M. horridula,*** a variable biennial, is 3 to 4 feet tall, with spiny stems and leaves. The latter,

Meconopsis horridula

with their stalks up to 10 inches long, are about 1½ inches wide, narrow-oblong to oblanceolate, and may be lobed. On long spiny stalks the 1½-inch-wide blooms, are light blue to reddish-purple or white. They have four to eight petals and dark-stalked stamens. ***M. integrifolia*** is sometimes called yellow Chinese-poppy. It is up to 3 feet in height, and has linear-lanceolate, lobeless leaves, 2 inches wide, and including their stalks, up to over 1 foot long. When young they are thickly-hairy. The blooms have six to eight yellow, rarely

Meconopsis integrifolia

white petals and are about 2½ inches in diameter. Their stamen stalks are yellow. This kind dies after it has bloomed. ***M. latifolia,*** up to 3½ feet tall and perennial, has pale blue or sometimes white, four-petaled blooms, about 2 inches in diameter, with deep blue stalks to the stamens. Including the stalk about 8 inches long, the

Meconopsis latifolia

ovate to oblong leaves are clothed with pale yellow hairs. ***M. napaulensis*** (syn. *M. wallichii*) dies after flowering. Up to 8 feet tall, its branching flower stems carry many nodding, red, blue, purple, or sometimes white blooms some 3 inches in diameter,

Meconopsis napaulensis

and with four petals. The largest leaves, including their stalks, exceed 1½ feet in length, and are deeply-lobed with the lobes again cut. The foliage is clothed with reddish hairs. ***M. paniculata*** also has nodding flowers, with four, or rarely five, petals. They are about 2 inches wide, numerous, and yellow with stamens with pale yellow stalks. A rare white-flowered variety is known. The upper blooms are solitary, those below two to six on each branch. This species is up to 6 feet tall. The densely-hairy, long-stalked leaves are much cleft. The largest, including the stalks, are up to 2½ feet long. ***M. quintuplinervia,*** the harebell-poppy, is a spreading perennial up to 1 foot high. From ro-

Meconopsis quintuplinervia

settes of densely-hairy, obovate to narrowly-oblanceolate leaves that, with their stalks, are under 1 foot long, it produces usually solitary-flowered stalks with light lavender-blue to purplish blooms 1½ inches in diameter. The flowers have four or rarely more petals. ***M. regia*** forms very beautiful rosettes of narrowly-elliptic, toothed leaves, thickly clothed with silvery or yellow hairs and, with their stalks, sometimes more than 3 feet long. The flower stalks, up to 5 feet in height, have

Meconopsis regia

branches with four or fewer yellow blooms 4 to 5 inches wide, with yellow-stalked stamens and a reddish-purple style. The flowers, which mostly have four petals, do not make as fine a display as those of some other kinds. This dies after blooming.

Garden Uses and Cultivation. As indicated at the beginning of this entry, the cultivation of meconopsises, except *M. cambrica,* is practicable in very limited regions of North America. They are appropriate for cool, open, humid woodlands, where filtered light reaches them, which provides a moist, but not stagnant root run containing an abundance of organic matter. They can also be accommodated in partially shaded beds and borders, and, the smaller ones, in rock gardens. Shelter from wind, which can be provided by nearby evergreens such as rhododendrons, is important. A loose mulch of peat moss or compost is helpful. Propagation is mostly by seeds, which should be sown as soon as they are ripe in sandy peaty soil in pots or pans in a cool greenhouse or cold frame. As soon as the seedlings are large enough to handle easily they are transplanted to flats or small pots in which they are grown until they are large enough to be transplanted to their outdoor flowering sites. A few kinds can be increased by division or by removing side growths.

MEDEOLA (Medèo-la)—Indian Cucumber-Root. The only species of *Medeola,* of the lily family LILIACEAE, is endemic to rich woodlands from Nova Scotia and Quebec to Minnesota, Missouri, Georgia, and Alabama. Its botanical name commemorates Medea, a sorceress of mythology. The common name refers to the flavor of its edible rootstocks.

Indian cucumber-root (***M. virginica***) has a short, tuber-like rhizome and slender, erect, branchless stems 1 foot to 2½ feet tall with two tiers of leaves. The lower consists of five to eleven individuals, oblong-oblanceolate and 2½ to 5 inches long. The three to sometimes five leaves of the upper tier are ovate, 1 inch to 2 inches long. From the center of this tier arises in late spring or early summer clusters of three to nine spreading or nodding greenish-yellow flowers on slender stalks ¾ to 1 inch long. They have six similar petals (more correctly, tepals), quite separate, recurved, and ⅓ inch long; six stamens; and three recurved styles. The fruits, nearly spherical dark purple berries, ripen in fall.

Garden Uses and Cultivation. Although by no means showy, Indian cucumber-root is not without horticultural interest for woodland gardens, informal shaded areas, and native plant gardens. It prospers in moist or wet, moderately acid, woodsy soil and may be propagated by seeds separated from the pulp of the berries as soon as they are ripe and sown immediately in sandy peaty soil in a cold frame.

MEDICAGO (Medic-àgo)—Medick, Calvary-Clover, Alfalfa. Of minor interest horticulturally, this genus of fifty or more species of the pea family LEGUMINOSAE includes one, alfalfa (*Medicago sativa*), of immense agricultural importance. Like it, but to a lesser extent, some others are grown for forage and hay. The group comprises annuals, biennials, herbaceous perennials, and a few shrubs. In the wild it inhabits Europe, North Africa, and western Asia. Several kinds are naturalized in North America. The name is derived from *medice* the ancient name for alfalfa, and indirectly from Media, now part of Iran, and thought to be the original home of alfalfa.

Medicagos have alternate, pinnate leaves of three usually toothed leaflets, and small pea-like, yellow to violet flowers in heads or short racemes from the leaf axils. The blooms have nine stamens united and one separate, and a beardless style. The fruits are small, spiraled or contorted, spiny or smooth pods containing one to few seeds.

Calvary-clover (***M. echinus***) received its vernacular name because its leaflets are usually clearly marked near their centers with a reddish spot, and its spirally coiled, spiny seed pods suggest crowns of thorns. Native of the Mediterranean region, it is a prostrate annual with stems extending up to 3 feet. The leaves have obovate, toothed leaflets up to 1 inch long. The small yellow flowers are crowded in clover-like cylindrical heads up to ¾ inch long. Variety *M. e. variegata* is described as having leaflets with a large blood-red spot.

Medicago echinus

Snail medick (***M. scutellata***) is so called because its nonspiny, coiled seed pods suggest the shells of snails. Native to Europe, this more or less densely-hairy, small annual has pale green leaves with broadly-obovate to elliptic leaflets, toothed in their upper parts, and small yellow blooms, solitary or in racmes of two or three. The spiraled seed pods are about ½ inch across.

Tree-alfalfa or moon-trefoil (***M. arborea***) is an evergreen, sparsely-branched, leafy shrub 3 to 12 feet in height that inhabits rocky areas, especially near the sea, in the Mediterranean region. It has gray-downy stems, and leaves with obovate leaflets, toothed or toothless at their apexes, ¼ to ¾ inch long, and silky-hairy on their undersides. The blooms, in stalked, short, four- to eight-flowered clusters, are yellow to orange-yellow and nearly ½ inch long. They provide a succession, but not very plentiful display, over a long summer period. The coiled pods with a hole in the center of the spiral so that they resemble little doughnuts, contain two or three seeds. This kind is not hardy in the north. A subshrubby species from the Crimea, ***M. cretacea,*** attains a height of up to about 1½ feet, and has leaves with broad-ovate leaflets scarcely over ⅓ inch in length. Its ¼ inch-long yellow flowers are many together in stalked, rather headlike clusters. The compressed, broad-sickle-shaped pods are under ½ inch long. This is perhaps hardy in sheltered places as far north as Philadelphia.

Garden Uses and Cultivation. As a matter of interest calvary-clover and snail medick are occasionally cultivated and are adaptable for sunny slopes and inconsequential areas. They thrive in almost any well-drained soil, preferring those of a neutral to alkaline reaction to acid ones. They are raised by sowing seeds in spring where the plants are to remain, thinning the seedlings to a few to several inches apart.

Tree-alfalfa is a general purpose shrub for sunny locations in California-type climates. It thrives in porous soil, and is easily raised from seeds, and by summer cuttings planted under mist or in a greenhouse or cold frame propagating bed, preferably with a little bottom heat. Lower-growing and hardier *M. cretacea* responds to the same conditions and treatment.

MEDICINAL OR DRUG PLANTS. Throughout the ages plants have been esteemed for their actual or supposed virtues as alleviants or cures of bodily ills. Every primitive people employed some for this purpose. Often their uses were associated with witchcraft or religious practices and those deemed most knowledgable of the kinds of plants to employ and how to use them were accorded status as witch doctors or priests.

Most sorts accepted by the ancients and primitive peoples as having curative properties so far as scientific investigations have proved are without such virtues, but some are sources of extremely efficacious drugs, among them belladonna, digitalis, morphine, and quinine, and others are plants from which are derived bitter aloes, cascara, castor oil, witch hazel, and a goodly number of other medicinal products.

In the middle ages, medicinal gardens developed largely by physicians as a means of assuring supplies of plant products they needed, gave impetus to the establishment of the first botanic gardens. Advancement in medical knowledge and in refining, standardizing, and in some cases synthesizing drugs obtained or previously obtained from plants has made unnecessary, and indeed often dangerous, reliance upon the crude products of field or garden. Only for the mildest of indispositions do moderns ordinarily resort to herb teas and the like, and these are more likely to be comforting than especially efficacious.

Medicinal gardens are occasionally maintained as educational features in botanic gardens and similar places. In them are grown plants that are or have been used as sources of medicines and other healing products. Because in the past many such plants now called herbs were grown as medicines, such medicinal gardens can very appropriately be associated with herb gardens.

Local climate will dictate which of the sorts of plants suitable for inclusion in medicinal gardens can be grown outdoors and which must be kept in a greenhouse or other place indoors. Among medicines and drugs obtained from plants are aconite (roots of *Aconitum napellus*), aloe (leaves of *Aloe barbadensis, A. ferox,* and *A. perryi*), belladonna (*Atropa belladonna*), cascara (bark of *Rhamnus purshiana*), chamomile (flower heads of *Chamaemelum nobile*), cocaine (leaves of *Erythroxylum coca*), codeine (seed capsules of *Papaver somniferum*), colchicum (bulblike corms of *Colchicum autumnale*), colocynth (fruits of *Citrullus colocynthis*), cortisone (parts of species of *Strophanthus*), cubeb (fruits of *Piper cubeba*), curare (various parts of species of *Strychnos* and other tropical genera), digitalis (leaves of *Digitalis purpurea*), ephedrine (all plant parts of *Ephedra*), eucalyptus (leaves of *Eucalyptus globulus*), gentian (roots of *Gentiana lutea*), ginseng (roots of *Panax schinseng* and *P. quinquefolium*), goldenseal (roots and rhizomes of *Hydrastis canadensis*), henbane (leaves of *Hyoscyamus niger*), hoarhound (leaves and flowering parts of *Marrubium vulgare*), ipecac (rhizomes and roots of *Psychotria ipecacuanha*), jalap (roots of *Exogonium purga*), licorice (roots of *Glycyrrhiza glabra*), lobelia (leaves and tops of *Lobelia inflata*), morphine (seed pods of *Papaver somniferum*), nux-vomica (seeds of *Strychnos nux-vomica* and *S. ignatii*), opium (seed pods of *Papaver somniferum*), pennyroyal (leaves and tops of *Hedeoma pulegioides*), podophyllum (roots and rhizomes of *Podophyllum peltatum*), psyllium (seeds of *Plantago psyllium* and other species), quassia (wood of *Quassia amara* and *Picrasma quassioides*), rhubarb (roots and rhizomes of *Rheum officinale* and *R. palmatum*), santonin (flower heads of *Artemisia cina*), senega (roots of *Polygala senega*), senna (leaves of species of *Cassia*), slippery elm (bark of *Ulmus rubra*), squills (bulbs of a white variety of *Urginea maritima*), stramonium (leaves and flowering tops of *Datura stramonium*), valerian (roots and rhizomes of *Valeriana officinalis*), witch hazel (leaves, twigs, and bark of *Hamamelis virginiana*), wormseed (fruits of *Chenopodium ambrosioides*), and wormwood (leaves and tops of *Artemisia absinthium*).

MEDICK. See Medicago.

MEDINILLA (Medín-illa). Botanists are acquainted with many more beautiful flowering plants of the melastoma family MELASTOMATACEAE than are gardeners. This is to be regretted because the group is rich in highly decorative species. That the great majority are tropical places certain limitations on their usefulness in the continental United States, but does not preclude their being tried outdoors in southern Florida and Hawaii or being grown in greenhouses. One old standby fairly commonly grown is *Medinilla magnifica*, and another, of entirely different appearance, *M. sedifolia*, has recently been brought into cultivation. One or two other species of *Medinilla* are seen rarely in choice collections of tropical plants.

All natives of the Old World tropics, the possibly 400 species of *Medinilla* are evergreen shrubs or vines. They have opposite or whorled (in circles of more than two), undivided leaves, or, much more rarely, solitary ones. The pink or white flowers are in panicles or clusters, many or few in each. Sometimes they are accompanied by showy bracts. The persistent calyxes may be without teeth or have four or five. There are four or five, or less commonly six, petals, twice as many stamens as petals, and one style. The fruits are berries. The name honors an early nineteenth century governor of the Marianas Islands, Jose de Medinilla y Pineda.

Medinilla, undetermined species

Native of the Philippines, where more than 100 species of this genus are indigenous, ***M. magnifica*** well deserves its botanical specific designation. It is one of the most gorgeous of tropical shrubs. It has stiff, four-winged branches, attains a height of 3 feet or somewhat more, and when well grown may be as broad as tall. Its broadly-ovate to obovate, opposite leaves, 6 to 10 inches long, with a prominent mid-vein and strong side veins branching from it, are glossy green and decidedly ornamental. From the branches the coral-pink flowers, each about 1 inch across, and with yellow-stalked stamens with purple anthers, hang in grapelike, inverted, pyramidal panicles up to 1 foot long. They are pink and are attended by showy bright pink bracts 4 inches long or longer. Variety *M. m. rubra* has darker pink flowers than the typical species. Those of *M. m. superba* are also slightly deeper colored and are somewhat larger than in the type.

Medinilla magnifica

Quite distinct from the last, ***M. scortechinii*** in the wild often grows as an epiphyte, perched on branches of trees and living there without taking nourishment from its host. Up to 3 feet tall, it has round stems and opposite, pointed, ovate-lanceolate leaves. They have three main longitudinal veins and are 5 to 8 inches long. The repeatedly forked stalks of the axillary, erect clusters of bloom, as well as the flowers, are pink. At each branching point of the clusters, which are 4 to 5 inches in length, are a pair of pink bracts. The flowers have four spreading or reflexed petals. This species is a native of Malaya.

Medinilla scortechinii

Differing markedly from other cultivated kinds, ***M. sedifolia*** is a native of Malagasy (Madagascar). It is a slender, vining plant that in the wild grows as an epiphyte, that is to say that like many orchids and bromeliads, it lives perched on the trunks and branches of trees, but without taking nourishment from them as do parasites. The shoots develop rootlets from the nodes (joints) and, in pairs or singly, broadly-elliptic to nearly round, fleshy leaves, ¼ to slightly over ½ inch long. At maturity the leaves are without hairs. Solitary in the leaf axils, the short-stalked flowers have oblanceolate, round-ended petals about ½ inch long and approximately one-half as broad. Each bloom has two kinds of stamens one of which is decidedly longer than the other. The style is without branches.

Medinilla sedifolia

Garden and Landscape Uses and Cultivation. Because high temperatures and high humidity are essential for the medinillas described above, their usefulness for outdoor cultivation is confined to the moist tropics where they succeed best in light shade. Elsewhere they must be considered for cultivation in tropical greenhouses in which a minimum winter night temperature of 60 to 70°F is maintained. By day and at other seasons the temperatures may with advantage be five to fifteen degrees higher. As much light as the plants will stand without their foliage scorching is desirable; too little inhibits blooming. Coarse, rich soil that contains an abundance of organic matter and permits the free passage of water is conducive to good growth.

Repotting, when needed, is done in late winter. Specimens of *M. magnifica* accommodated in containers 8 inches or more in diameter will not need this attention every year. When done, some of the old soil may be shaken from the roots to permit the addition of a maximum amount of new earth. Any pruning needed to keep the plants shapely is also done in late winter before new growth begins. Well-rooted specimens respond to frequent, liberal applications of dilute liquid fertilizer. For satisfactory results it is important to allow *M. magnifica* plenty of room. Good specimens cannot be had if they are crowded among other plants. Copious watering is the rule at all seasons, except that from October to early February supplies are reduced and only enough is given to prevent the foliage from wilting. At the same time the night temperature is lowered to 60°F. This partial rest encourages the production of flowers. When in bloom the atmosphere may, with advantage, be less humid than at other times. The growing conditions favored by the other species described here are essentially those recommended for *M. magnifica*. Propagation is usually by cuttings, which may be rooted in a greenhouse propagating bench furnished with mild bottom heat. Sand, perlite, and vermiculite are satisfactory materials in which to plant the cuttings. In the case of *M. magnifica* firm shoots of medium vigor make better cuttings than more robust ones. Pinching out the tips of the shoots of young medinillas may be done to promote branching.

MEDIOCACTUS (Medio-cáctus). By some authorities this genus is included in *Selenicereus*. When segregated two, or by those who split the genus more finely, five South American species, of the cactus family CACTACEAE, constitute *Mediocactus*. The name from the Latin *medius*, middle, and cactus, calls attention to these plants being intermediate in aspect between *Hylocereus* and *Selenicereus*.

Mediocactuses are big, bushy, rock-inhabiting or tree-perching (epiphytic) plants with slender, rambling, or drooping, mostly triangular stems that sprout aerial roots. The areoles (specialized locations on the stems of cactuses from which spines and flowers originate) have very tiny, bristle-like spines. The long-tubed flowers open at night. They are white with green outer segments (petals), funnel-shaped, and very big. The fruits have areoles with a felt of hairs. They are spherical, knobby, and spiny, red when ripe.

Most frequently cultivated, ***M. coccineus*** (syn. *M. setaceus*) has clambering stems branching freely from their bases, 2 to 3 feet long by up to 3 inches wide. The one to four spines in each cluster are pinkish to brown or yellowish. The blooms have scaly perianth tubes. They are 10 inches to 1 foot long by approximately as wide. From the last, ***M. megalanthus*** differs in having much slenderer stems, often only up to ¾ inch thick, and blooms up to 1¼ feet long, among the largest borne by any cactus.

Garden and Landscape Uses and Cultivation. These easy-to-grow night-blooming plants thrive outdoors in warm regions and in greenhouses under conditions that suit most epiphytic cactuses. They require full sun and a coarse, porous soil kept moderately moist, but allowed to become dryish between watering. In the fairly humid tropics and subtropics they can be grown perched in the crotches of trees or among the persistent leaf bases attached to the trunks of palms. Propagation is easy by cuttings and by seed. For more information see Cactuses.

MEDIOLOBIVIA (Medio-lobívia). Closely related to *Rebutia*, the genus *Mediolobivia*, of the cactus family CACTACEAE, differs in having usually larger blooms and bigger seeds and in its stems being ribbed. The blooms are in most species smaller than those of *Lobivia*. There are other technical differences. In the wild *Mediolobivia* is indigenous to Argentina and Bolivia. The name, from the Latin *medius*, middle, and that of the genus *Lobivia* alludes to the intermediate aspect mediolobivias present between *Lobivia* and *Rebutia*. By some authorities the kinds of *Mediolobivia* are included in these genera.

Mediolobivias usually form clusters of small plant bodies (stems) with low, spiraled, often notched ribs and clusters of weak spines. The funnel-shaped, day-opening flowers are yellow to orange-yellow. The fruits are bristly. There are sixteen species.

Variable ***M. aureiflora*** (syn. *Rebutia aureiflora*) has clusters of spherical to nearly cylindrical plant bodies (stems) about 1½ inches thick with twelve to seventeen notched ribs and spine clusters of fifteen to twenty ¼-inch-long or shorter radials and three to four ½-inch-long centrals. The 1½-inch-wide, golden-yellow flowers have white throats. Varieties are *M. a. albiseta*, with soft white spines much longer than those of the typical species and light yellow flowers; *M. a. boedekeriana* (syn. *Rebutia boedekeriana*), with spine clusters with two or three rigid, 1-inch-long centrals; *M. a. rubelliflora*, with red flowers; *M. a. rubiflora* (syn. *Rebutia blossfeldii*), with bluish-red blooms; *M. a. sarothroides* (syn. *Rebutia sarothroides*), with blood-red blooms; and *M. a. duursmaiana* (syn. *Rebutia duursmaiana*), which has spine clusters of ten to twelve slender, glassy, white, ¼-inch-long white spines and light yellow flowers.

The tubercles on the ribs of ***M. elegans*** (syn. *Rebutia elegans*) are much smaller than those of *M. aureiflora*. This has spherical, pale green plant bodies about 1½ inches in diameter, with notched ribs. The spine clusters are of about fourteen yellowish-white, bristle-like radials about ⅛ inch in length and three or four darker centrals almost twice as long. The 1½-inch-wide flowers are bright yellow. The fruits are reddish. The stems of *M. e. gracilis* are more slender than those of the typical sort.

Other species cultivated include these: ***M. costata*** (syn. *Rebutia costata*) has stems with eight or nine ribs and spine clusters of eleven or twelve white, ¼-inch-long radials and one or no centrals. The 1-inch-wide or somewhat broader flowers are orange-red. ***M. eucaliptana*** has ¾-inch-wide stems with eight or nine ribs and spine clusters of nine to eleven yellowish, ½-inch-long radials. About 1¼ inches wide, the flowers are light red. ***M. ritteri*** has stems with eight or nine ribs and clusters of ten or more white spines. The flowers, vermilion with a violet-red throat, are 1¾ inches wide. ***M. spiralisepala*** (syn. *Rebutia spiralisepala*) has cylindrical stems about 3½ inches tall by 2¼ inches thick. They have spiraled ribs and clusters of fifteen or sixteen radial spines and four centrals about ¾ inch long. About 1¼ inches wide, the flowers are orange-red.

Garden Uses and Cultivation. Mediolobivias are among the choicest of small cactuses, generally rewarding to grow. In warm, desert climates they are admirable for rock gardens. They are easy to grow in greenhouses and as window plants. Their needs are like most other small desert cactuses. For more information see Cactuses.

MEDLAR is *Mespilus germanica*, the haw-medlar is *Crataegomespilus*.

MEEHANIA (Mee-hània)—Meehan's-Mint. Consisting of one species native to the eastern United States and one, or perhaps more, to eastern Asia, the genus *Meehania* belongs to the mint family LABIATAE. Its name commemorates the distinguished American horticulturist and horticultural editor Thomas Meehan, of Philadelphia, Pennsylvania, who died in 1901.

Meehanias are low, usually trailing, hardy herbaceous perennials with four-angled stems and opposite, stalked, undivided, toothed leaves. Their flowers are in whorls (circles of more than two) in short, erect or ascending terminal spikes. They have a bell-shaped, five-lobed calyx and a corolla with a slender-based tube, an upper lip with two lobes, and a lower lip slightly larger and more deeply cleft into three lobes. There are four stamens pressed against the upper lip, and one two-lobed style. The fruits are of four seed-like nutlets.

Attractively-flowered ***M. cordata*** inhabits rich mountain woodlands from Pennsylvania to Ohio, North Carolina, and Tennessee. It has trailing stems up to 2 feet long and more or less erect branches that bear the flowers. The long-stalked leaves are hairy. They have broadly-blunt,

Meehania cordata

heart-shaped shallowly-round-toothed blades 1¼ to 2¼ inches long. From 1 inch to 1¼ inches long, the erect, lavender-blue to lilac flowers are few together in one-sided spikes 1¾ to 3 inches long. They are distinctly hooded, after the fashion of *Scutellaria,* so that the mouth of the corolla faces to one side. Not known to be in cultivation in North America but promising as being worthy of introduction, ***M. urticifolia*** of China, Manchuria, Korea, and Japan is a quite variable native of habitats similar to those favored by the American species. Generally larger and more vigorous than that sort, it has long trailing stems and flowering stems up to 1 foot tall. Its lanceolate to triangular-ovate or broadly-heart-shaped, toothed leaves have blades from 1 inch to 7 inches long. In one-sided spikes up to 5 inches in length the blue to blue-purple flowers are 1½ to 2 inches long.

Garden Uses and Cultivation. The American species is a useful groundcover for shady places. It has success with minimum attention in well-drained but not dry soil that contains an abundance of leaf mold, peat moss, or other semidecayed humus-forming material. Propagation is easy by seed, division, and cuttings. It may be assumed that the Asian species would prosper under similar conditions.

MEGACARYON (Mega-càryon). The only species of *Megarcaryon,* of the borage family BORAGINACEAE, is native to western Asia. Its name, from the Greek *mega,* big, and *karyon,* a nut, presumably is a not very apt reference to the fruits.

Megacaryon orientale

Closely related to *Echium,* from which it differs in having smooth instead of wrinkled nutlets (commonly called "seeds") ***M. orientale*** is a hairy herbaceous perennial.

Megacaryon orientale (flowers)

Up to 4 feet tall or perhaps sometimes taller, it has stalkless or nearly stalkless, oblong-lanceolate leaves up to 1 foot long, the basal ones in a loose rosette, those on the stem alternate and diminishing in size upward. The numerous flowers, from 1¼ to 1¾ inches long, are in large, loosely-branched, terminal panicles. They have a five-lobed calyx and a slightly asymmetrical tubular corolla with five whitish to rosy-lilac, spreading lobes (petals), the upper two with longitudinal purplish streaks. There are five stamens and one style. The fruits consist of four seed-like nutlets.

Garden and Landscape Uses and Cultivation. Hardy probably only where winters are mild, this moderately ornamental species is appropriate in beds of herbaceous perennials. It thrives in ordinary well-drained soil and is easily raised from seed.

MEGACLINIUM. See Bulbophyllum.

MEGASEA. See Bergenia.

MEGASKEPASMA (Mega-skepásma). The only species of this genus of the acanthus family ACANTHACEAE in gardens has sometimes been misidentified as *Adhatoda cydoniaefolia,* a designation that correctly belongs to a different plant. A handsome ornamental, it is a native of Venezuela. The name is from the Greek *mega,* large, and *skepasma,* a covering. It presumably refers to the floral bracts.

A shrub up to 10 feet tall, ***Megaskepasma erythrochlamys*** is more or less clothed with deciduous, tawny hairs. It

Megaskepasma erythrochlamys in Florida

has stems lined with small glands and opposite, short-stalked, oblong to oblong-elliptic leaves 6 inches to 1 foot long by approximately one-half as wide. From each side of the midrib arise nine to twelve, conspicuous, upcurving veins laced by cross ties of smaller ones. The

Megaskepasma erythrochlamys (foliage and flowers)

stalkless flowers are in panicles of spikes, the terminal spikes up to 8 inches long, the lateral ones smaller. At the base of each bloom is a showy, rose-pink to purple-pink, broad-ovate bract up to 1¾ inches long. The flowers have five-lobed calyxes and slender, tubular, white or delicate pink corollas 2 to 3½ inches long, deeply divided into two lips, the lower lip markedly recurved toward its apex. There are two stamens and a slender style. The fruits are four-seeded capsules.

Garden and Landscape Uses and Cultivation. This is a beautiful shrub for outdoor cultivation in southern Florida, Hawaii, and other places with humid, tropical, or subtropical climates, and for growing in ground beds, pots, or tubs in greenhouses. Indoors it responds to the conditions and care that suit *Adhatoda.* Outdoors it thrives in sun or part-shade in ordinary soil of moderate fertility. It is easily increased by seed and by cuttings. The chief blooming season is winter. In late winter or spring, after flowering is through, it is advisable to prune the bush severely to induce a desirable compact habit and encourage the production of strong new growth.

MEIRACYLLIUM (Meira-cýllium). Two species constitute Mexican and Central American *Meiracyllium,* of the orchid family ORCHIDACEAE. The name, in allusion to the small size of the plants, comes from the Greek *meirakyllion,* a little boy.

Creeping, freely-rooting rhizomes with extremely short side branches each with a broad-ovate to nearly circular, fleshy leaf ¾ inch to 2 inches long are characteristic of ***M. trinasutum.*** Its flowers, ¾ to 1 inch wide, are in short racemes of up to six.

Meiracyllium trinasutum

They are red-purple, often with the deeply-cupped lip darker. Usually somewhat more robust, ***M. wendlandii*** (syn. *M. gemma*), has longer racemes of slightly larger flowers with narrower petals, yellowish-white at their bases.

Garden Uses and Cultivation. These plants appeal to orchid fanciers and succeed best when grown in osmunda fiber on rafts or attached to slabs of tree fern trunk in a humid greenhouse with a minimum winter night temperature of 60°F and shade from strong sun. For more information see Orchids.

MELALEUCA (Mela-leùca) — Bottle Brush, Punk Tree or Cajeput Tree. The myrtle family MYRTACEAE includes as one of its more attractive genera *Melaleuca.* This comprises more than 100 species, all except one endemic to Australia. The exception, *M. leucadendron,* occurs also in the East Indian Archipelago and Malaysia. The name, from the Greek *melas,* black, and *leukos,* white, alludes to the black and white trunks of an Asian variant of this kind. The vernacular designation bottle brush is also applied to related *Callistemon.*

Melaleucas are trees and shrubs with evergreen, leathery, undivided, lobeless, toothless, usually short-stalked, sometimes stalkless leaves. The stalkless flowers, in heads or cylindrical spikes, are succeeded by fruits that form crowded, persistent, woody clusters encircling the bases of the new branches. The latter develop as extensions of those of the previous year, growing out at an early stage from the apexes of the flower clusters. Each small bloom has a cup-

Melaleuca, undetermined species (foliage and flowers)

to bell-shaped torus, five sepals, usually deciduous in the fruiting stage, five longer, concave, orbicular petals, numerous stamens much exceeding the petals in length, arranged in five bundles opposite the petals, unlike those of *Callistemon* with their bases united. There is one style. The fruits are woody capsules. Cajeput oil is processed from the leaves of *M. leucadendron.* Several species in their native Australia supply good lumber.

The punk tree or cajeput tree (***M. leucadendron***) is a variable, large tree, conspicuous because of its very thick, whitish, spongy bark, which, when dry, can be ig-

Melaleuca leucadendron

Melaleuca leucadendron (trunk and branches)

nited and used as punk. Its branches often droop. Elliptic to oblong and tapered at both ends, the leaves are 2 to 4 inches long by ½ to ¾ inch wide. They have three to seven longitudinal parallel veins tied with numerous cross veinlets. The creamy-white blooms, with nearly ½-inch-long stamens, are in cylindrical, terminal spikes 2 to 6 inches long. The globular seed capsules are ⅛ inch in diameter. A kind sometimes grown under this name is *M. quinquenervia.*

A hairless shrub with slender, wandlike branches, ***M. acuminata*** has mostly opposite, pointed-elliptic, sometimes spine-tipped leaves ¼ inch long or a little longer. Its small whitish flowers are in tuftlike clusters of few along shoots of the previous year's growth. Graceful ***M. armillaris,*** 15 to 30 feet tall, has furrowed, gray bark that peels from the lower part of the trunk, pendulous branches, and needle-like leaves ½ to 1 inch long. White, the flowers are in 1- to 2-inch-long, cylindrical spikes. The stamens are about ¼ inch long. This sort becomes picturesque with age. Lilac- to purple-flowered ***M. decussata,*** 9 to 20 feet in height and usually as broad as high, has shredding brown bark. Its branches droop. Its bright green to bluish, opposite, rather heathlike, narrow, elliptic-oblong to lanceolate leaves are crowded and ¼ to ½ inch long. The blooms are in up to 1-inch-long cylindrical to globular terminal heads. Attaining heights of 8 to 15 feet, ***M. elliptica*** has shedding brown bark and broad-elliptic to nearly round leaves approximately ½ inch in length. Its showy red to crimson flowers are in bottle brushes up to 3½ inches long by 2 inches wide or wider. Fast-growing ***M. ericifolia*** is a shrub or tree up to 20 feet tall and eventually 20 to 30 feet wide. It usually has several trunks and wide-spreading, generally pendulous branches. Its up to ½-inch-long, crowded, slender leaves, narrow-linear to nearly cylindrical, often curve backward from their middles. Yellowish-white, the flowers are in more or less globular spikes ½ to 1 inch long. Their stamens are ¼ inch long. This stands alkaline and poorly drained, even swampy soils. Up to 8 feet tall, ***M. fulgens*** is a shrub with very narrow lanceolate leaves, ½ to 1 inch long or a little longer, conspicuously speckled with dark glands and grooved along the upper side. The

Melaleuca fulgens (flowers and fruits)

showy, cylindrical, about 3-inch-long flower spikes are of blooms with closely arranged bundles of rich crimson stamens. The urn-shaped fruits are ⅓ to ½ inch in diameter. Erect and up to several feet tall, ***M. filifolia*** is a shrub with almost thread-

Melaleuca filifolia

like leaves, ¾ inch to 1½ inches or sometimes 2 inches long, and short, roundish heads of flowers each with a bundle of seven to nine red, or perhaps sometimes yellow or white stamens.

A shrub up to 10 feet tall, ***M. hypericifolia,*** when not in bloom, looks much like a *Hypericum*. It has spreading, recurved branches and mostly opposite leaves. Lanceolate to elliptic or oblong, ¾ inch to 1¾ inches long by up to over ½ inch wide, they

Melaleuca hypericifolia

have the mid-veins prominent on their undersides. The brilliant orange-red blooms, crowded in cylindrical spikes about 2 inches long, have stamens at least ¾ inch long. Although tolerant of dry soil and drought, this species is less well adapted than some for growing right at beaches. A graceful, hairless, alternate-leaved shrub 6 to 10 feet tall, with slender, wandlike branches, ***M. lateritia*** has closely set, pointed, flat to rather concave, narrowly-linear leaves ¼ to 1 inch long. The quite large, bright scarlet flowers, with stamens fully ¾ inch long, are very attractive. They are in cylindrical spikes up to 3 inches long. White-trunked ***M. linariifolia*** is a slender-branched large shrub or tree up to 30 feet tall, with shredding bark and branches more erect and rigid than those of the last. Willowy when young, this species later develops an umbrella-shaped head. Its rigid, needle-shaped, light green to bluish-green, opposite leaves, about 1¼ inches long, have midribs prominent on their undersides. The cylindrical spikes, up to 1½ inches long, of small white flowers, their stamens under ½ inch long, are produced in abundance. Attractive ***M. megacephala*** is a large shrub with slender, erect branches furnished with gray-green, obovate, pubescent leaves up to 1 inch long. The spherical flower spikes, about 2 inches in diameter, are of blooms with clusters of tightly-packed white to pale yellow stamens tipped with golden-orange. With yellow-tipped, pink to mauve flowers that fade to nearly white as they age, in crowded, roundish

Melaleuca megacephala

terminal heads about 1 inch in diameter, ***M. nesophila*** is a large, quick-growing shrub or picturesque tree up to 35 feet tall. Its gnarled trunk and stout branches are clothed with thick, spongy bark. The leaves are ovate-oblong, up to 1 inch long by ¼ inch wide. Called pink melaleuca, this is highly tolerant of adverse environments. It survives in dry or moist soils including rocky ones. It withstands high temperatures under desert conditions and locations subject to high winds and ocean spray. A shrub or tree 12 to 18 feet tall, ***M. preissiana*** has hairy young shoots. Its alternate, rather crowded, spreading or recurved, rigid, ½-inch-long, lanceolate to broadly-linear leaves rarely exceeding ½ inch in length, are spreading or recurved. The smallish, whitish or yellowish blooms are in loose spikes 1½ to 4 inches long. They have very short stamens. A spreading shrub up to 3½ feet tall, ***M. pulchella*** has slender, hairless shoots and closely spaced, alternate leaves, under ¼ inch long, blunt and recurved. The rose-pink flowers are solitary or in twos or threes near the shoot ends. Their bundles of stamens are up to ½ inch long.

A tree 20 to 40 feet tall, ***M. quinquenervia*** is often grown as *M. leucadendron*. Of spreading, open habit, this has drooping young branches. Its thick, spongy, tan to

Melaleuca quinquenervia

whitish bark peels in large sheets. The leaves, lustrous green, narrow- to broad-elliptic, about 3 inches long, turn purplish after a touch of frost. Yellowish-white, less commonly pink or purplish, the flowers are in terminal, more or less globular heads 2 to 3 inches long.

A shrub, tall and with wandlike branches, ***M. radula*** has opposite, pointed-linear leaves, concave or with turned-up margins, ¾ inch to 2 inches long. The fairly large white or pink blooms are in pairs arranged distantly or occasionally forming interrupted spikes. A beautiful upright shrub, ***M. squarrosa*** is erect and commonly 6 to 10 feet tall, rarely twice that maximum. Its young shoots are hairy. Mostly opposite or almost so, the rigid, sharp-pointed, ovate- to ovate-lanceolate leaves are usually under ½ inch long. Yellowish-white, the flowers are in tight, slender, cylindrical, terminal spikes 1 inch to 2 inches long. Exceptionally fine ***M. steedmanii,*** from 3 to 6 feet tall, has short, narrow-elliptic leaves and cylindrical flower spikes composed of blooms with bundles of stamens of intense crimson tipped with yellow. Sometimes 40 feet tall, more often considerably lower, ***M. styphelioides*** is a tree of graceful, open habit. Its trunk is clothed with thick, spongy, light tan bark that becomes black with age and peels in papery strips. The rigid, sharp-pointed, light green leaves are up to ¾ inch long by ¼ inch wide. Creamy-white, the flowers are in brushlike clusters 1 inch to 2 inches long. They have stamens ¼ inch long. A shrub, spreading or straggling, ***M. wilsonii*** is 5 to 8 feet tall. It has rough, scaly bark and, closely set on slender branches, opposite, linear-lanceolate to oblanceolate leaves ⅓ to a little over ½ inch long. The flowers, in axillary or terminal 1-inch-wide clusters of two to five are pinkish-magenta.

Melaleuca radula

Melaleuca steedmanii at the Huntington Botanic Garden, California

Garden and Landscape Uses. Melaleucas are excellent landscape furnishings for dryish, Mediterranean-type climates such as that of California. Most succeed even in poor soils, and stand high temperatures, exposure to wind, and salt air. Especially adapted for seaside planting and alkaline soils are *M. armillaris, M. ericifolia, M. hypericifolia, M. leucadendron,* and *M. nesophylla.* The tree types, such as *M. leucadendra* and *M. quinquenervia,* are good street trees. Some, including *M. armillaris, M. hypericifolia,* and *M. nesophylla,* lend themselves for use as sheared hedges.

Cultivation. This is as for Callistemon.

MELAMPODIUM (Melam-pòdium). Melampodiums are tap-rooted annuals, biennials, and subshrubs, restricted in the wild to the Americas, mostly to Mexico and Central America, and belonging to the daisy family COMPOSITAE. There are thirty-six species. The name comes from the Greek *melam,* black, and *pous,* a foot, and alludes to the black stalks.

The leaves of *Melampodium* are opposite, with each pair at right angles to the pairs below and above it. The stems fork freely and from the upper forks develop stalks with solitary flower heads. Like daisies, the flower heads have a central eye of disk florets surrounded by a circle of petal-like ray florets. The bisexual disks are yellow and have five-lobed corollas, the rays are yellow or white, and female. The involucre (collar of bracts behind the flower head) is of two rows. The species described and occasionally cultivated are low, bushy, herbaceous perennials or sometimes subshrubs with creamy-white-rayed flower heads. They are natives of Texas.

Variable, and at its showiest with flower heads 1 inch in diameter, ***M. cinereum*** has rough, gray-hairy stems and foliage and is from 6 inches to 1½ feet tall. Its leaves, up to about 2 inches long by ½ inch wide, are linear-oblong, and mostly have toothed or wavy-lobed margins. The flower heads have small, raised, bright yellow disks, and about eight broad, white ray florets, three-toothed at their apexes.

From the last, ***M. leucanthum*** differs in the bracts of its involucres (collars of bracts at the backs of the flower heads) being united for more than one-half, rather than for less than one-third, of their lengths. Also, its linear-oblong leaves are sometimes pinnately-lobed, and its flower heads are 1 inch to 1½ inches wide. This species favors limestone soils.

Garden and Landscape Uses and Cultivation. Where they are hardy, which probably limits their cultivation pretty much to the west and southwest, melampodiums are suitable for rock gardens and native plant gardens. They are raised from seeds and need well-drained soil and full sun. The plants should be set in their permanent locations early. Because of their tap roots they do not transplant well when large.

MELANDRIUM. See Lychnis and Silene.

MELANTHIUM (Melánth-ium) — Bunch Flower. In its natural distribution confined to North America, *Melanthium,* of the lily family LILIACEAE, comprises five species. Not of sufficient ornamental value to be important garden plants, they are occasionally grown in native plant gardens, bog gardens, and informal moist soil areas. The name, from the Greek *melas,* black, and *anthos,* a flower, has no obvious application.

Melanthiums are perennial herbaceous plants with thick rhizomes and long, sheathing, mostly basal leaves. Their green or greenish, starry flowers, usually bisexual and unisexual on the same plant, are in large terminal panicles. They are small and have six spreading perianth segments (petals or more properly tepals), six stamens, and three short styles. From the blooms of related *Veratrum* they differ in that the petals narrow at their bases into distinct claws, and at the bottoms of their blades have a pair of glands. The fruits are capsules.

Native of wet soils from New York to Indiana, Minnesota, Florida, and Texas, ***M. virginicum*** has a stout stem up to 5 feet in height, and linear leaves, the lower ones up to 1 foot long, and mostly ½ to ¾ inch wide. From ½ to ¾ inch wide, the greenish-yellow flowers are in loose, ovoid, pubescent panicles up to 1 foot long. The lower branches of the panicles bear bisexual blooms, the upper ones flowers that are male or have aborted female parts.

Garden Uses and Cultivation. Possible uses for the bunch flower are suggested above. It grows without trouble in sun or part-shade in moist or wet ordinary soil and is propagated by division and seed.

MELASPHAERULA (Mela-sphaèrula). The solitary spring-blooming species of the iris family IRIDACEAE that constitutes *Melasphaerula* is a native of South Africa related to *Ixia.* Not hardy in the north, it makes an interesting addition to gardens in milder climes and is well adapted for cool greenhouses. Its name, alluding to its small blackish corms (bulblike structures) is from the Greek *melas,* black, and *sphaerula,* a globe.

From its corms, ***M. ramosa*** (syn. *M. graminea*) develops fans of about six narrow, parallel-veined leaves, 6 inches to 1 foot long, and with prominent midribs. The slender, loosely-branching, often zigzagged flower stalks are erect or lax. The branches, sometimes again branched, are nearly at right angles to the main stem and are down-turned at their ends. The rather ill-scented flowers open white and change to dull yellow; they may be flushed purplish or have purple mid-veins to the perianth segments (petals), of which there are six. They are ½ to ¾ inch wide, and nodding, and arise from two calyx-like green bracts. There are three stamens one-half as long as the petals and a three-branched style. The fruits are capsules.

Garden Uses and Cultivation. Although much less showy than its more popular relatives, freesias, ixias, tritonias, and babianas, this graceful plant is worth the attention of gardeners interested in the unusual. In bloom its pattern of flowering brings to mind *Thalictrum* and *Chlorophytum,* to neither of which *Melasphaerula* is related. In climates where freesias and their kin prosper outdoors our plant can be conveniently accommodated at the fronts of flower borders and in rock gardens. It can also be grown in pots in cool greenhouses. Its cultural needs are those of *Ixia.*

MELASTOMA (Melá-stoma)—Indian-Rhododendron. This, the type genus of the melastoma family MELASTOMATACEAE, is little known in American horticulture, although several of its tropical relatives such as *Tibouchina,* as well as native North American *Rhexia,* are grown. In the wild, *Melastoma* is confined to tropical Asia, Indonesia, and Polynesia. It consists of about seventy species of shrubs or rarely small trees. Its name, from the Greek *melas,* black, and *stoma,* a mouth, refers to the staining of the mouth if the fruits are eaten.

Melastomas are evergreen, hairy shrubs, small trees, or rarely herbaceous plants, with more or less leathery or firm, undivided, lanceolate to oblong or elliptic, stalked leaves that have three to seven conspicuous veins running from base to apex and are opposite or rarely in threes. The pink, red, purple, or seldom white flowers, in clusters commonly of three to seven at the branch ends, or rarely solitary, have five- to seven-toothed calyxes and five to seven petals. There are ten to fourteen stamens and a slender, incurved style tipped with an inconspicuous stigma. The fruits are leathery or fleshy berries. The plant once named *M. corymbosum* is *Amphiblemma cymosum.*

Indian-rhododendron (***M. malabathricum***) is a broad shrub 6 to 10 feet tall, a native of Malaya, India, Ceylon, and Sumatra. It is sometimes called Malabar melastome. Its four-angled shoots, leafstalks, flower stalks, and calyxes are clothed with narrow, pointed, bristly hairs. From 2 to 6 inches in length, the pointed, generally toothed leaves are broadly- to narrowly-elliptic, have three to five chief veins running longitudinally, and are clothed on both surfaces with bristly hairs parallel with the surface. Bright pink to violet-mauve, and 1½ to 3 inches in diameter, the short-stalked flowers are in clusters of about five or occasionally are solitary at the branch ends. The berries are red. This is naturalized in Hawaii where in some places it has become a pesky weed. It has been used medicinally and as a source of pink and black dyes.

Native of China, where in some places its wood is used for fuel, ***M. candidum*** is a shrub 3 to 6 feet tall. This has densely-bristly-hairy shoots and foliage. The leaves, broad-lanceolate to elliptic or ovate, are 3 to 5 inches long. They have five to seven chief veins. The red, pink, or white flowers, about 2 inches in diameter, are fragrant. The fruits are purplish-red. Another handsome Chinese, ***M. sanguineum*** (syn. *M. decemfidum*) also ranges into tropical Asia. It is a shrub 5 to 10 feet tall, with leaves up to 6 inches in length and proportionately

Melastoma sanguineum

longer than those of the last. Also, the hairs on its young shoots, leafstalks, and calyxes are much longer and shaggier than those of *M. candidum* so that the calyxes have a thistle-like appearance. The upper leaf surfaces have hairs parallel with their surfaces, but the undersides are hairless except along the veins. The purplish-pink to clear pink blooms are about 3 inches across. Native of Java, ***M. setigerum molkenboeri*** is a shrub or tree up to 20 feet tall. It has hairy leaves about 4 inches long by 1½ inches broad. Its mauve-purple flowers are 2 inches in diameter. The blooms of *M. s. album* are white.

Garden and Landscape Uses and Cultivation. These are as for *Tibouchina.*

MELASTOMATACEAE—Melastoma Family. Trees, shrubs, and herbaceous plants, numbering 4,500 species accommodated in 200 genera constitute this family of dicotyledons. The vast majority are tropical or subtropical, but *Rhexia* extends into temperate North America. Some are vines, some epiphytes that perch on trees without taking nourishment from them.

Melastomes, as members of this family are called, have frequently four-angled stems and opposite, undivided, sometimes toothed leaves with each pair generally at right angles to those above and below. Commonly one leaf of each pair is markedly smaller than the other and may wither early. A very characteristic feature of melastomes is that their leaves have three to nine veins of equal importance that spread outward from the base, converge at the apex, and are linked by parallel secondary veins that diverge from them.

The flowers, often showy and very beautiful, are clustered or grouped in various ways. A very common feature that greatly aids in family recognition is the conspicuous appendages to the anthers. There are generally four or five sepals and usually the same number of petals and as many or twice as many, often bent or down-pointing stamens sometimes of two lengths. There is one style and one stigma. The fruits are berries or capsules. Cultivated genera include: *Amphiblemma, Arthrostema, Bertolonia, Bertonerila, Bredia, Centradenia, Dissotis, Heterocentron, Medinilla, Melastoma, Miconia, Monolena, Phyllagathis, Rhexia, Salpinga, Sonerila, Tibouchina, Tococa,* and *Triolena.*

MELIA (Mè-lia)—Chinaberry or China Tree or Bead Tree or Pride-of-India or Indian-Lilac, Texas Umbrella Tree. Deciduous trees and shrubs of an uncertain number of species, possibly as many as twenty, constitute *Melia,* of the mahogany family MELIACEAE. They are natives of warmer parts of Asia and Australia. The name, an ancient Greek one for the ash (*Fraxinus*), was applied because of some similarity of the foliages.

Melias have large, alternate, once-, twice-, or thrice-pinnate leaves with toothed or toothless leaflets. Their many small, white or purple flowers are in large panicles from the leaf axils. Each has five or six sepals, and the same number of much longer petals. The ten or twelve stamens, united into a tube, are lobed at its top. The long style terminates in a three- or six-lobed, knoblike stigma. The fruits are berry-like drupes. The tree sometimes named *M. azadirachta* is *Azadirachta indica.*

Chinaberry, China tree, bead tree, pride-of-India, or Indian-lilac (***M. azedarach*** syn. *M. sempervirens*) is a wide-topped, densely-branched tree up to 50 feet tall, a native of Asia, and commonly planted for ornament in the south where it has escaped from cultivation and become naturalized as far north as southern Virginia. In Hawaii also, it has established itself as a wildling. It is not hardy in the north. The leaves of the Chinaberry, 7 inches to over 1½ feet long, are divided into twice-pinnately-lobed or -toothed, ovate to ovate-lanceolate leaflets 1½ to 3 inches long. The lilac and purple, fragrant flowers are succeeded by yellow fruits about ½ inch in diameter that remain decorative long after the foliage has fallen. The fruits, appreciated by birds and cattle, are reported to be poisonous to poultry, hogs, and humans. Leaves placed in books are said to repel insects. The Texas umbrella tree (*M. a. umbraculiformis*) is a variety, with crowded, erect branches that form a wide-spreading, much-flattened head and more or less pendulous leaves. Bushy, 6 to 8 feet tall *M. a. floribunda* is especially free-flowering.

Garden and Landscape Uses and Cultivation. In tropical, subtropical, and warm-temperate regions much planted for ornament and shade, the Chinaberry grows rapidly and blooms when quite young. It withstands hot, dry conditions well and provides heavy shade. It has the disadvantages of, for much of the year, dropping either leaves or fruit, and of giving origin to numerous self-sown seedlings. It is rather short-lived. No special care is needed. It grows well in a variety of soils and is easily raised from seed. The coarse-grained, weak wood of this tree, not durable under moist conditions, is used for making musical instruments and some other purposes.

MELIACEAE—Mahogany Family. Mahoganies (*Swietenia mahagoni* and *Khaya senegalensis*) and West-Indian-, Spanish-, or cigar-box-cedar (*Cedrela odorata*) are the most important commercial products of the mahogany family MELIACEAE. The seeds of others are sources of oils, the fruits of some are edible. The Chinaberry tree (*Melia azedarach*) is much planted for ornament in the south. The family consists of 1,400 species of dicotyledons contained in fifty genera of trees, shrubs, and a few subshrubs, often with fragrant wood, natives of warm-temperate to tropical regions throughout the world.

Members of the *Meliaceae* have usually pinnate, rarely undivided leaves without tiny, punctate (translucent) dots usual in the nearly related rue family RUTACEAE. The small, symmetrical flowers, in panicles, have calyxes of three to five lobes or separate sepals and corollas of usually four or five, less frequently as few as three or more than five, generally separate, seldom united, petals. The mostly five or ten, occasionally fewer or more stamens are commonly completely joined into a tube. There is one style and one stigma. The fruits are capsules or berries or are drupe-like. Genera in cultivation are *Aphanamixis, Azadirachta, Cedrela, Dysoxylum, Melia, Nymania, Swietenia, Trichilia,* and *Turraea.*

MELIANTHACEAE — Melianthus Family. Here belong fifteen species of African trees and shrubs allotted to three genera. Dicotyledons, they have alternate, pinnate leaves the midribs of which are often winged, and racemes of bisexual or bisexual and unisexual flowers, with calyxes of five or less often four unequal lobes or sepals, corollas of four or five unequal petals much narrowed at their bases, five, four, or sometimes ten stamens separate or briefly united at their bases. There is one style and one usually four- or five-lobed stigma. The fruits are capsules. An interesting characteristic of the family, which distinguishes it from the related soapberry family *Sapindaceae,* is that before the blooms attain maturity they turn on their stalks through a 180 degree angle. Genera cultivated are *Greyia* and *Melianthus.*

MELIANTHUS (Meli-ánthus)—Honey Bush. This, the type genus of the melianthus family MELIANTHACEAE, is endemic to South Africa. It consists of half a dozen species of nonhardy, evergreen, soft-wooded shrubs or subshrubs and has a name from the Greek *meli,* honey, and *anthos,* a flower, that refers to the abundant nectar secreted by the blooms. From *Greyia,* the other cultivated genus of the family, *Melianthus* is readily recognized by its pinnate leaves.

Honey bushes have attractive foliage that when bruised is ill-scented. The leaves are alternate and have an uneven number of asymmetrical, toothed leaflets and at the bases of the leafstalks, large leafy appendages (stipules). The flowers, asymmetrical and in terminal or axillary racemes, have compressed, five-lobed calyxes with a nectar-producing gland on the inside. There are four normal petals narrowed at their bases into long claws, and sometimes an aborted one, and four stamens. The style is slender and incurved. The fruits are deeply-four-lobed, papery or woody capsules.

Most commonly cultivated, ***M. major*** grows naturally in damp soils. Up to 10 feet or so tall, it spreads vigorously by sucker shoots. Its light bluish-green or glaucous leaves, 1 foot to 1½ feet in length, have seven or nine leaflets approximately 2 to 3 inches long. Between the leaflets the midrib is conspicuously winged. The stipules are united to form one winglike attachment at the base of the leafstalk. From about 1 foot to 1½ feet long, the one-sided racemes of ill-scented, dull, dark red flowers are displayed on long, erect stalks well above the foliage. The green, bladder-like seed pods are as decorative as the blooms.

Another kind sometimes cultivated, ***M. minor*** is a shrub 4 or 5 feet tall, with smaller leaves than those of *M. major* and

Melianthus major at Durban, South Africa

Melianthus major in bloom at the Kirstenbosch Botanic Garden, South Africa

light reddish-brown flowers that nestle under the foliage rather than being carried well above it.

Garden and Landscape Uses. The hardiest species *M. major* survives outdoors in the south of England. It is commonly grown in California and may be expected to persist wherever the ground does not freeze sufficiently to injure its roots. If only the tops are frozen the shrub renews itself from the base. The other kind described here is a little more tender. Both are good-looking foliage plants, and *M. major* is quite striking in bloom and in fruit. They need full sun and are particularly effective near watersides and elsewhere where there is plenty of space for them to spread and display themselves. They are not adapted for crowding among other plants. Honey bushes are also sometimes grown in greenhouses and are occasionally treated as annuals and used as foliage plants for summer beds.

Cultivation. For the best results these plants need a hearty, fertile soil that does not lack for moisture, but is not wet. They can be increased by division, and by spring or summer cuttings in a greenhouse propagating bench or under mist, but as seeds germinate readily and the young plants make rapid growth they are most commonly grown from seed.

MELICA (Mél-ica) — Melic Grass. To the grass family GRAMINEAE belong the forty to fifty species of *Melica*. The name comes from the Greek *meli*, honey. In ancient times it was applied to a kind of sorghum. Melicas are tufted or creeping perennial grasses, natives of North America, Europe, Asia, and North and South Africa. They have flat-linear or rolled leaf blades and usually nodding spikelets of flowers, without awns (bristles), in loose or compact racemes or panicles. The spikelets have two or more flowers.

The most ornamental, ***M. altissima***, of Europe and temperate Asia, is 2 to 4 feet tall or sometimes taller and has creeping rhizomes. Its pointed leaf blades are up to 9 inches long by up to ½ inch broad. The one-sided panicles, 5 to 10 inches long, are crowded above, but toward their bases have the short-stalked spikelets, up to ½ inch long, more loosely organized, with sections of bare stem showing between. Variety *M. a. atropurpurea* has purple spikelets, those of *M. a. alba* are whitish.

Densely tufted, and with slender stems up to 2½ feet tall, ***M. ciliata***, of Europe, adjacent Asia, and North Africa, has leaves with slender, often almost thread-like blades up to 7 inches long by up to ½ inch wide and grayish-green. The dense, cylindrical, spikelike panicles of purplish spikelets are 2 to 6 inches long. The lemmas (bracts) of the spikelets, unlike those of the other species dealt with here, are fringed with long white hairs.

Mountain or nodding melic grass (***M. nutans***), of Europe and adjacent Asia, is 9 inches to 2 feet tall. This kind has thin rhizomes and loosely-clustered, slender stems. From 2 to 8 inches long by up to ¼ inch wide, the bright green leaf blades have short hairs on their upper surfaces. The purplish or reddish-purplish spikelets are in gracefully arching, loose racemes.

Melica nutans

Garden and Landscape Uses and Cultivation. Hardy and easy to grow, melic grasses flourish in ordinary garden soil in sun or part-day shade. They do especially well in limestone soils. They are useful as decoratives in informal landscapes and in beds, and their flowers are suitable for bouquets and for fresh and dried arrangements. For drying they should be harvested a little before they reach their fullest maturity. Propagation is by seed and by division.

MELICOCCUS (Meli-cóccus) — Spanish-Lime or Genip or Mamoncillo. Despite its common name, *Melicoccus* is no relation of the true lime, which is a species of *Citrus*. It belongs, rather, in the soapberry family SAPINDACEAE and is kin to the akee (*Blighia*) and to *Litchii*. There are two species, natives of tropical America. The name, which is also spelled *Melicocca*, is from the Greek *meli*, honey, and *kokkos*, a berry, and refers to the flavor of the fruits.

The Spanish lime, genip, or mamoncillo (***M. bijugatus***) is cultivated in tropical countries, including Hawaii, and in southern Florida, where it succeeds best so far as fruit production goes at Key West. It is a slow-growing tree up to 60 feet tall. Of erect habit, it has shining, pinnate leaves with two pairs of pointed-elliptic to elliptic-lanceolate, hairless leaflets 2 to 4 inches long, and many-flowered panicles about 4 inches long of fragrant, greenish-white blooms. Individual flowers have four or five sepals, the same number of petals, eight stamens, and a two- or three-lobed stigma. They are unisexual or bisexual. By the latter it appears that no viable pollen is produced. Individual trees may bear flowers of one or more than one type. The fruits, almost spherical, green, and about 1 inch in diameter, are in dense, grapelike clusters. Each contains a solitary large seed surrounded by pleasantly flavored, more or less acid, whitish or yellow pulp enclosed in a thick, tough, smooth green rind. Until they are ripe the fruits are astringent. When mature they are juicy with somewhat the flavor of grapes. The genus *Genipa* is also called genip.

Garden and Landscape Uses and Cultivation. In suitable climates the Spanish-lime is cultivated for its fruits, of which both the pulp and roasted seeds are agreeable. They are held in considerable esteem in Key West, the less acid ones especially for eating out of hand. This tree flourishes in ordinary soil. It is propagated by seed and to a lesser extent by air layering. Grafting has not proved successful, which is unfortunate because it would make possible the ready multiplication of heavy-bearing trees with the most desirable fruits. Seedlings vary much in these respects.

MELICOPE (Melicò-pe). This group of possibly seventy species of trees and shrubs is indigenous from tropical Asia to New Zealand. It belongs to the rue family RUTACEAE and has a name, alluding to four nectar-secreting glands at the base of the ovary, from the Greek *meli*, honey, and *kope*, a division.

The leaves of melicopes are alternate or opposite. They are dotted with tiny glands and are of three leaflets or are undivided and have winged stalks. The small, unisexual or bisexual flowers have four minute sepals, four petals, eight stamens, a four-lobed ovary, four styles that often are joined, and a four-lobed stigma. The fruits are dry capsules consisting of four one-seeded lobes.

New Zealand ***Melicope ternata*** is a hairless, freely-branching shrub or tree up to 20 feet tall. Its opposite, stalked leaves have mostly three obovate to elliptic, 2- to 4-inch-long, toothless leaflets, or those of the flowering parts may have two or only one leaflet. The greenish flowers, ⅓ inch wide, are in usually three-branched, often paired, roundish clusters 2 to 2½ inches long. The blooms are unisexual or bisexual. The lobes of the fruits are ¼ inch across.

Melicope ternata

Garden and Landscape Uses and Cultivation. In regions of frost-free or practically frost-free winters the species discussed can serve as a general purpose evergreen. It responds to ordinary soil and sun. Propagation is by summer cuttings and seed.

MELICYTUS (Meli-cỳtus) — Mahoe or Whitey Wood. Native to New Zealand, and the islands of Fiji, Norfolk, and Tonga, *Melicytus* consists of five species of the violet family VIOLACEAE. One kind is planted in California and other mild climate areas. The derivation of the name is not known.

Trees and shrubs, melicytuses have alternate, toothed or wavy-edged leaves and symmetrical, unisexual blooms with the sexes on separate plants. The flowers are in clusters or more rarely are solitary. They have calyxes with five lobes or teeth, five spreading petals, five stamens, and a two- to six-branched style. The fruits are berries containing angled, black seeds.

The mahoe or whitey wood (***M. ramiflorus***) has the range of the genus. Round-headed, white-barked, and up to 30 feet or somewhat more in height, it branches and suckers freely from the base and has many short, brittle branchlets. Its slender-stalked, bluntly-toothed leaves, lanceolate- to elliptic-oblong, are 2 to 6 inches long by up to 2 inches broad and vary considerably in shape. In clusters of two to ten, the slender-stalked, greenish-yellow flowers are borne abundantly below the leafy portions of the stems. They are ⅛ inch or slightly more across. On female trees fertilized with pollen from a male the blooms are succeeded by egg-shaped to nearly spherical dark blue or purplish fruits up to ⅕ inch long and containing three to six seeds. In New Zealand the berries are much esteemed by wild pigeons. Cattle are fond of the foliage. The soft wood of the mahoe was used by the Maoris to make fire by friction.

Garden and Landscape Uses and Cultivation. This attractive species can be used for general landscaping. To assure fruiting it is necessary to have a male plant near females. The mahoe is hardy only where little freezing occurs (it succeeds outdoors in the mildest parts of the British Isles). It grows in ordinary soil and is propagated by seed.

MELILOTUS (Meli-lòtus) — Melilot, Sweet-Clover. Except that they are sometimes planted as cover crops or green manures, the sorts of *Melilotus,* of the pea family LEGUMINOSAE, are without horticultural significance. The genus consists of about twenty-five species, mostly annuals and biennials, but including a very few shrubby kinds. It inhabits Europe, Asia, and Africa. Some kinds are naturalized along roadsides and elsewhere in North America. Some are grown for forage, and are good bee plants. The name is from the Greek *meli,* honey, and *lotos,* some plant of the pea family.

Melilots have leaves of three toothed leaflets and small yellow, white, or violet, pea-shaped flowers in spirelike racemes that come from the upper leaf axils. The fruits are pods. The chief kinds are white melilot or Bokhara-clover (***M. alba***), with white blooms, yellow melilot (***M. officinalis***), with yellow flowers, and yellow-flowered ***M. indica.*** These are all annuals or biennials 1½ to 3 feet tall or taller, naturalized in North America.

MELIOSMA (Meli-ósma). This genus of the sabia family SABIACEAE is named in allusion to its honey-scented blooms from the Greek *meli,* honey, and *osme,* an odor. It consists of about 100 species of deciduous and evergreen trees and shrubs, mostly Asian, but a few tropical American.

With one exception, *Meliosma alba,* of Mexico, all American meliosmas have undivided leaves. The Asian kinds have, according to species, undivided or pinnate leaves with an odd number of leaflets. The flowers are commonly bisexual, sometimes bisexual and unisexual on the same tree. They are in terminal and axillary panicles. Each bloom has five or rarely four sepals, three rounded petals, and two much smaller, often scalelike ones, the latter frequently united to the only two fertile stamens of the five. The fruits are small, usually one-seeded, drupes (fruits structured like plums).

Melilotus officinalis

A deciduous shrub or tree up to 20 feet in height, ***M. cuneifolia,*** native of western China, has undivided, toothed, obovate leaves, pointed at their apexes and hairless except for tufts in the axils of the veins on the undersides. They are 3½ to 6½ inches long and have twenty to twenty-five pairs of parallel veins extending from their midribs. The delightfully scented, yellowish-white blooms become almost white as they age. They are ¼ inch across and are in loose panicles 6 to 10 inches long and almost as broad. They come in summer. The black fruits are spherical and about ¼ inch in diameter.

Another with undivided leaves, ***M. myriantha*** is native to Japan and Korea. This sort is up to 30 feet tall and deciduous. Short, dull yellow hairs clothe the young shoots and undersides of the leaves. The latter are pointed, obovate to oblong, toothed, and 4 to 10 inches long. Greenish-yellow, the ⅛-inch-wide flowers, which come in early summer, are in erect, pyramidal panicles 6 to 10 inches long and almost as broad. About ⅙ inch across, the fruits are dark red.

Pinnate, sumac-like leaves, 1½ to 2½ feet long, with nine or eleven ovate or oblong, rarely distantly toothed leaflets 3½ to 7 inches in length, are features of ***M. veitchiorum,*** a native of central China. This is a deciduous, erect-branched tree 30 to 45 feet tall. Except for the midribs beneath, the leaves are without hairs. The loose, drooping, pyramidal panicles of fragrant, white blooms are up to 1½ feet long by 1 foot broad. They are displayed in late spring. Individual blooms do not exceed ¼ inch in width. The fruits, the size of large peas, are purplish-black. Also Chinese, ***M. pendens*** is a shrub or tree up to about 15 feet in height. It has undivided, pointed-elliptic-obovate, coarsely-toothed leaves 3 to 8 inches long, hairy on the midrib and on their undersides along the veins. The fragrant, white flowers are in pendulous, loose panicles.

Meliosma veitchiorum

Meliosma veitchiorum (leaves)

Meliosma veitchiorum (flowers)

Garden and Landscape Uses and Cultivation. Meliosmas grow satisfactorily in sunny locations in well-drained, reasonably moist soils. With the exception of *M. veitchiorum,* which is hardy about as far north as Philadelphia, they are not hardy in the north. They need no particular care and are propagated by seed, by summer cuttings under mist or in a greenhouse propagating bench, and by layering. They are attractive as single specimens or for inclusion in groupings of shrubs and trees.

MELISSA (Mel-íssa)—Balm or Bee Balm or Lemon Balm. The only cultivated member of *Melissa* shares with an entirely different plant, *Monarda didyma,* the colloquial name bee balm, which when applied to the latter is preferably spelled with a hyphen, thus, bee-balm. This clearly illustrates the difficulty sometimes encountered in being definitive when reliance is had on common names. The bee balm, lemon balm, or just plain balm discussed here is hardy *Melissa officinalis,* one of that great assemblage, so abundant in fragrant herbs of the mint family LABIATAE. There are three or four species of *Melissa,* natives of Asia and the Mediterranean region. The generic name derives from the Greek *melissa,* a bee, and refers to the attraction the flowers have for bees.

These are herbaceous perennials with erect, branched, square stems and broad, toothed leaves. Their asymmetrical, two-lipped, white or yellowish flowers are in clusters in the leaf axils. Their curved, tubular corollas project from long bell-shaped calyxes. Their upper lip is erect and notched, the lower is divided into three lobes and spreads. There are four stamens. The fruits, commonly called seeds, are smooth nutlets.

Classed as a sweet herb, balm has foliage that, fresh or dried, if bruised emits a pervading perfume of lemon, fortified with a soupçon of that of mint. Its fresh leaves may be used to add piquancy to fruit cocktails and salads, more commonly are employed to make a herb tea and as a flavoring for claret cup. The dried foliage is chiefly esteemed for its fragrance.

Balm (***M. officinalis***) is 1 foot to 2 feet in height. Loosely-branched, it has stalked, ovate, crinkly leaves up to about 3 inches in length. Its flowers, about ½ inch long, are borne in summer. It is naturalized in many parts of North America. Variety *M. o. aurea* has leaves variegated with yellow.

Garden Uses. Except in herb gardens and perhaps for tucking away in sunny spots in vegetable gardens, ordinary green-leaved balm makes little claim to garden space. Its appeal, and then only when it is crushed or bruised, is to the nose and palate rather than the eye. The variety with variegated foliage is more decorative and suitable for the fronts of flower beds.

Melissa officinalis

Melissa officinalis aurea

Cultivation. Balm needs a sunny location and a well-drained, porous soil of moderate fertility. When raised from seeds, which is easy to do, it takes two or more years to produce sizable plants. First-year specimens have little foliage, certainly not enough to harvest. Sow seeds outdoors in spring or indoors earlier. Transplant the seedlings in the garden at a spacing of about 1½ feet apart. Propagation can also be by division of old plants in spring and by cuttings in early summer. To harvest balm, cut off the top few inches of the leafy stems several times during the summer, dry them quickly in a shaded, airy place, and store in tightly closed bottles or canisters.

MELITTIS (Melít-tis) — Bastard-balm. The name of this genus of one species is derived from the Greek *melissa,* a bee, and refers to the attraction that its flowers have for that insect. Of the mint family LABIATAE, bastard-balm is native to Europe.

A herbaceous perennial, ***Melittis melissophyllum*** is related to *Stachys* and *Physostegia.* It has an erect, branchless stem up to 1½ feet tall and wrinkled, ovate, toothed, hairy leaves up to 2 inches long. Its attractive blooms are borne in clusters of two to six in the leaf axils. They are up to 2 inches long, and have a two-lip-

Melittis melissophyllum

ped, bell-shapped calyx with the lower lip three-lobed, and a corolla that is creamy-white spotted with purple or pink. The fruits consist of four seed-like nuts. The entire plant is aromatic. In variety *M. m. album* the blooms are white, those of *M. m. grandiflorum* are creamy-white with a purple-red lip.

Garden Uses and Cultivation. Bastard-balm is easily raised from seed and by division. It thrives in any reasonably fertile garden soil, but is most prosperous in deep, rich loam, which is never excessively dry. Light shade from strong summer sun is advantageous. When dried, the stems and foliage retain their pleasant fragrance for a long time. This species is useful for the fronts of perennial beds and for inclusion in herb gardens. It is hardy.

MELOCACTUS (Melo-cáctus)—Melon Cactus, Turk's Cap. The species of *Melocactus,* possibly thirty, are not clearly defined. Belonging to the cactus family CACTACEAE, they are inhabitants, often of saline, sandy soils usually near the sea, of Mexico to tropical South America and the West Indies. The name, from the Greek *melon,* a melon, and *Cactus,* the name of a related genus, alludes to the form of the plant body.

Melocactus, undetermined species, with a well-developed cephalium

Melocactuses are spherical, somewhat flattened-spherical, or shortly-columnar, stout-spined cactuses with several to many usually straight, not spiraled, ribs. Their plant bodies or stems, when the plants reach maturity, are crowned with a topknot of long woolly hairs and bristles that form a cushion-like mass called a cephalium. This may be flat, hemispherical, or cylindrical. From the cephalium sprout the pinkish flowers. These, open by day, are tubular with rather few, spreading perianth segments (petals). The club-shaped, white to scarlet, edible fruits, protruding from the cephalium, retain their withered flower parts. Among the first cactuses taken from America to Europe, melocactuses were illustrated there as early as the beginning of the sixteenth century.

The easiest species to grow, ***M. neryi*** (syn. *Cactus neryi*) seems to be the only one that in the wild occurs away from coastal regions. It is a native of interior Brazil. Up to 6 inches in diameter and somewhat less in height, this has ten broad, low ribs and rather distantly spaced clusters of seven to nine stout, spreading, 1-inch-long spines, and sometimes one or two longer central spines. The flowers, scarcely 1 inch in length, have green-lobed stigmas. The fruits are red.

Melocactus neryi

Other natives of Brazil include these: ***M. azureus,*** up to 7 inches tall and 6 inches wide and of a beautiful glaucous-blue hue of almost frosted aspect, has nine or ten sharp ribs. Its spine clusters are of seven radials, one of which points downward and is up to 1½ inches long, and one, or less commonly up to three central spines, the largest up to 1 inch long. The cephalium, up to 3 inches wide by 1½ inches tall, is of pure white, woolly hair with red bristles intermingled. The flowers are carmine with yellow stigmas. ***M. macrodiscus,*** flattened-spherical, is usually somewhat wider than high and up to 6 inches tall. It has about eleven sharp ribs and spine clusters of up to nine gray or reddish radials, up to ¾ inch long, and sometimes one central of similar color and about as long. The tiny, nonprotruding, pinkish-red flowers are succeeded by violet-pink fruits. ***M. salvadorensis,*** hemispherical and up to 4½ inches in diameter, has ten to twelve ribs. Its spine clusters are of eight or nine awl-shaped, reddish radials that spread horizontally, and sometimes one central spine up to 1 inch long. The flattened-rounded cephalium is 2 to 2½ inches wide and ¾ inch tall. It is furnished with needle-like, red spines. The flowers and fruits are unknown.

Melocactus azureus

Melocactus macrodiscus

Melocactus salvadorensis

Cuban ***M. matanzanus*** has solitary or clustered plant bodies up to 4 inches in diameter and scarcely as high. They have eight or nine ribs. The spines, reddish when young, yellowish when mature, are in clusters of eight or nine radials from slightly more to rather less than ½ inch long, and one longer central spine. The rose-pink flowers are about ¾ inch long.

Endemic to the West Indies, including Puerto Rico, ***M. intortus*** (syn. *M. communis*) is a globose or cylindrical species that may attain a height of 3 feet. It has fourteen to twenty ribs armed with clusters of ten to fifteen yellow to brown spines ¾ inch to 2½ inches long. The pinkish blooms, approximately ¾ inch long, develop from an imposing, cylindrical cephalium of white, woolly hairs and brown bristles sometimes nearly as long as the plant body. Peruvian ***M. trujilloensis,*** with clusters of formidable, recurved spines, attains a height and width of about 4½ inches and is ten-ribbed. Its flowers, under ½ inch wide, are deep carmine-red.

Garden and Landscape Uses and Cultivation. Except for *M. neryi,* melocactuses do not adapt as well as many other cactuses to pot cultivation. Confining their naturally very wide-spreading roots seems detrimental to their well-being. They are more likely to succeed in sharply-drained ground beds in full sun. They require more heat than many members of the cactus family. A minimum winter temperature of about 60°F suits best. Although it is essential to keep the soil much drier in winter than at other seasons, it never should be allowed to become completely dry. For additional information see Cactuses.

MELON. See Muskmelon or Melon. The bitter-melon is *Momordica charantia.* The name preserving-melon is applied to *Benincasa* and *Citrullus lanatus citroides.* The watermelon is *Citrullus lanatus.*

MELON-PEAR or MELON-SHRUB is *Solanum muricatum.*

MELOTHRIA (Melóth-ria) — Creeping-Cucumber. In the broad sense accepted here, *Melothria* comprises about eighty species of mostly tropical vines of the Old and the New World. A more restricted view limits its application to natives of the Americas. The genus belongs to the gourd family CUCURBITACEAE. Its name is derived from *melothron,* the Greek one for *Bryonia.*

Melothrias are slender-stemmed trailers or climbers with branchless, often much-coiled, tendrils. Their usually thin leaves are alternate. The flowers are small and unisexual, the males in racemes or clusters, the females clustered or solitary. They have a five-toothed calyx and a deeply-five-lobed, bell-shaped corolla. The male flowers have three short-stalked stamens, the females a short style usually with three stigmas. The many-seeded fruits, structurally berry-like, are smooth or warty.

Inhabiting dry and moist woodlands and thickets from Virginia to Indiana, Missouri, Florida, Texas, and Mexico, the creeping-cucumber (***M. pendula***) is a variable, slender-stemmed annual or perhaps perennial, hairless or nearly hairless vine that has roundish, five-angled or five-lobed, often ivy-shaped leaves from 1 inch to 3½ inches long and heart-shaped at their bases. Its small flowers, the males in short racemes, the females long-stalked and solitary, are yellow or greenish. The ovoid, juicy fruits, up to ⅓ inch long, when ripe are purple to blackish.

Native to Mexico and similar to the last, ***M. scabra*** has roundish to triangular, more or less five-angled or five-lobed, toothed leaves. Its small flowers are yellow or greenish. The egg-shaped to roundish fruits, greenish to yellowish then becoming blackish, are up to ¾ inch long. Their seeds are reported to be purgative.

An attractive South African species, ***M. punctata*** (syn. *Zehneria scabra*) has a carrot-shaped, tuberous root, and slender, hairy stems and foliage. Its thin, five-lobed and toothed leaves, paler beneath than on their upper sides, are under 2 inches in length and have stalks about as long. The flowers are in groups of seven or eight. The smooth, red, spherical fruits are up to ¾ inch in diameter.

Garden and Landscape Uses and Cultivation. Although not of first horticultural importance, the species described are sometimes cultivated as quick-growing vines for screening and interest. They are generally treated as annuals by sowing the seeds, after all danger of frost has passed and the weather has become warm and settled, where the plants are to remain, and thinning out the seedlings to allow ample room for their growth. An alternative method, advantageous in the north, is to sow the seeds indoors in a temperature of 60 to 65°F some six weeks before it is expected that the weather will be warm enough to set the young plants outdoors, and then plant in the garden.

MEMPAT is *Cratoxylum polyanthum.*

MENISPERMACEAE — Moonseed Family. Dicotyledons numbering 350 species allotted among sixty-five genera compose this family. Mostly twining, woody vines, also included are a few erect shrubs and trees. Natives of warm regions, they have alternate, usually undivided, lobeless, palmately-veined leaves and mostly symmetrical, unisexual flowers generally in racemes or panicles. Less often the leaves are lobed or have three leaflets. The tiny blooms generally have six sepals and usually six petals. In males there are three, six, or an indefinite number of separate or variously united petals. Females have one style or none, a lobed or lobeless stigma, and sometimes staminodes (aborted stamens). The fruits are drupes or achenes. Genera cultivated include *Cocculus, Menispermum, Sinomenium,* and *Stephania.*

MENISPERMUM (Meni-spérmum)—Moonseed. Because their seeds, contained singly in berry-like fruits, are crescent-shaped these plants are called moonseeds and are named *Menispermum* from the Greek *mene,* moon, and *sperma,* a seed. They belong in the moonseed family MENISPERMACEAE and are hairless or slightly pubescent twining vines. There are two species.

Moonseeds have alternate, long-stalked leaves with the stalks attached to the blades a little distance in from the margins. In outline they are broad-ovate to broad-heart-shaped and are usually angled or shallowly-lobed. The small greenish or white flowers, in loose clusters from the leaf axils, are unisexual, have four to eight petals shorter than the four to eight sepals, nine to twenty-four stamens and two to four pistils. The bluish-black fruits resemble currants or small grapes and are strongly suspected of being poisonous.

The common moonseed (***M. canadense***) inhabits rich moist soils from Quebec to Manitoba, Georgia, and Arkansas. Attaining a height of 10 to 15 feet, it has slender stems, woody in their lower parts, and pubescent when young. Its flowers most commonly have six sepals, six petals and,

Menispermum canadense

the males about twenty stamens; the females have six aborted stamens. Not very different is ***M. dauricum,*** of Japan, Korea, China, and Siberia. A more vigorous, hairless plant with smaller leaves, with their stalks attached more in from the margins than in the American kind, it has yellowish flowers, the terminal ones with six sepals, nine or ten petals and approximately twenty stamens, the lateral ones with four sepals, six petals, and about twelve stamens.

Menispermum dauricum

Garden and Landscape Uses. These are neat vines for ornamenting fences, walls, posts, and other vertical surfaces, but because they sucker freely they should not be located near other choice plants. They succeed in any fairly good garden soil that is not excessively dry, in part-shade. Their chief attraction is their foliage.

Cultivation. Moonseeds grow readily from seeds sown in cold frames, protected outdoor beds, or cool greenhouses in fall or spring in any porous soil. The native kind also transplants easily from the wild. Cuttings of ripened shoots afford another means of propagation. Pruning consists of cutting the stems back almost to ground level in fall or early spring.

MENODORA (Meno-dòra). Seventeen species of more or less woody herbaceous plants and subshrubs constitute *Menodora,* of the olive family OLEACEAE. The group is indigenous to the southwestern United States and Mexico, from Bolivia to Chile and Argentina, and in South Africa. The name comes from the Greek *menos,* force, and *doron,* a gift.

Menodoras have alternate leaves or the lower ones may be opposite. Stalkless or nearly so, they are smooth-edged, toothed, or pinnately-lobed. The flowers are solitary at the ends of the stems, in racemes from the leaf axils, or in branched clusters (panicles). They have tubular calyxes with five to fifteen deep lobes, a five- or six-lobed, wheel- to salver-shaped corolla with a funnel-shaped tube, two stamens, and a slender style. The fruits are berry-like capsules containing eight or fewer seeds.

Native from Bolivia to Paraguay, Uruguay, and Argentina, ***M. integrifolia*** is a bushy, somewhat woody, herbaceous perennial 1 foot to 2 feet tall, with much the aspect of a *Helianthemum.* It has hairless, linear to linear-lanceolate leaves up to 1 inch long with smooth, recurved edges. The slender-stalked, about ½-inch-long, yellow, bell-shaped blooms are freely produced in terminal clusters. Variety *M. i. trifida* is a spreading, freely-branching plant up to 1 foot tall. Its leaves are often lobed.

Garden and Landscape Uses and Cultivation. The kinds described are cultivated as ornamentals in mild climates, such as that of California. They are suitable for the fronts of borders and rock gardens. They need well-drained soil and a sunny location. Propagation is by seed, and by cuttings in a greenhouse or under a mist.

MENTHA (Mén-tha) — Mint, Pennyroyal. Mints or their products used to flavor a wide variety of chewables, eatables, and drinkables, from candies and chewing gum to mint juleps and crème de menthe, in one form or another are familiar to almost everyone. They constitute the genus *Mentha,* the type of the mint family LABIATAE. There are twenty-five species, natives chiefly of Europe and Asia, with one of these also indigenous in North America. Others are endemic in South Africa and Australia. In addition, there are numerous varieties and hybrids. The name, from the Greek *Minthe,* the name of a nymph, who was changed into a mint plant by Persephone to protect her from Hades.

Mints are herbaceous perennials, the cultivated kinds, except the Corsican mint, hardy. They are prostrate or erect, often spreading by underground runners (stolons). Their leaves are opposite, undivided, and more or less toothed. The small, asymmetrical flowers are in terminal spikes or heads or axillary clusters. Their calyxes are somewhat two-lipped or equally five-toothed. The corollas are tubular, and have a face of four nearly equal lobes (petals) or have the upper lobe broader than the others and notched at its apex. There are four stamens and one style with two stigmas. The fruits are of four seedlike nutlets. As a result of hybridization and other causes, the European and Asian mints are represented by numerous slightly different varieties. Of most there are kinds with crisped leaves, any of which are likely to be grown as *M. crispa.* Without flowers it is usually impossible to identify these as to species, and not easy even if blooms are available. In like manner many of this group of mints are represented in gardens by variegated-leaved varieties.

Spearmint (***M. spicata***) is the kind most familiar in home gardens. There are many varieties. Selected ones are cultivated commercially for the distillation of oil used for flavoring and in medicine. Spreading aggressively by underground runners, this hairless or nearly hairless kind has erect stems with upright branches. It is 1 foot to 2 feet tall. Its sharply-toothed, pointed-ovate-lanceolate leaves, up to 2½ inches in length, are without stalks or have very short ones. The whitish to lilac or purplish flowers are in whorls spaced to form slender, interrupted spikes. Of uncertain, possibly hybrid origin, spearmint is widely naturalized in Europe and to some extent in North America. From spearmint ***M. longifolia,*** a native of Europe and Asia and naturalized in North America, differs in having pointed-lanceolate to lanceolate-ovate, sharp-toothed leaves, two to three times as long as broad. It has runners that spread vigorously, and branched or branchless stems 1½ to 3 feet tall. Stalkless, its leaves are 2 to 5 inches in length, hairy on their upper surfaces, and with white-hairy undersides. The tiny, purplish flowers are in stoutish spikes with sometimes the lower portion interrupted so that parts of stem show between the tiers of blooms.

Mentha spicata

Mentha suaveolens

Mentha longifolia

Apple or round-leaved mint (***M. suaveolens***) in gardens is often misidentified as *M. rotundifolia.* It has decidedly softly-hairy stems and foliage. From 1½ to 2½ feet tall, it spreads vigorously by runners. Its stems are slender, erect, and branched or branchless. The stalkless, round-toothed leaves, 1 inch to 4 inches long, have more or less heart-shaped bases and blunt or rounded apexes. They are not more than twice as long as wide. The tiny white to reddish-purple blooms, in solid or interrupted spikes, are up to 4 inches in length. This southern European species is naturalized in North America as far north as Maine and Ohio. Variety *M. s. variegata* has leaves variegated with creamy-white.

Mentha suaveolens variegata

Water mint (***M. aquatica***), native to Europe and naturalized in North America, spreads by runners. It has stalked, pointed, lanceolate to heart-shaped, toothed leaves, usually more or less hairy, and up

Mentha aquatica crispa

to 2 inches long. Its rather weak stems are up to 3 feet long. The little, lilac flowers are in compact spikes about 1½ inches in length. The leaves of *M. a. crispa* are crisped.

Peppermint (***M. piperita***), a presumed hybrid between spearmint and *M. aquatica*, closely resembles the first, from which it is distinguishable by its longer-stalked leaves and wider flower spikes. There are

Mentha piperita

many varieties some of which are cultivated commercially for oil production. This is naturalized in wet soils in North America. Bergamot mint or lemon mint (*M. p. citrata* syns. *M. aquatica citrata, M. citrata*) is nearly hairless; spreading by runners, it has procumbent stems up to 2 feet long. Its thin-textured, decidedly stalked, toothed leaves, up to 2 inches in length, are bluntly-elliptic to ovate, or the upper ones may be pointed-lanceolate. The ¼-inch-long, lavender flowers are in short, dense, headlike spikes.

Field mint (***M. arvensis***), a widely dispersed native of wet soils in North America and other temperate parts of the northern hemisphere, is a very variable sort that spreads rapidly by runners. Its erect or ascending stems 1 foot to 2 feet tall or sometimes taller have usually short-stalked, ovate to broad-lanceolate, toothed leaves mostly 2 to 2½ inches long. The widely separated tiers of flowers are in dense clusters in the leaf axils of the upper parts of the stems, not in spikes. Japanese mint (*M. a. piperascens*), a large-leaved variety up to 3 feet tall, is cultivated commercially as a source of oil.

Scotch mint is one of many varieties of ***M. gentilis*** (syn. *M. cardiaca*), a reputed hybrid between *M. spicata* and *M. arvensis*, naturalized in North America. Much like the last, the distinguishing characteristic of *M. gentilis* is that the bracts of the flower spikes are leafy and twice or thrice as long as the corollas. Some varieties are grown instead of spearmint as sources of oil of spearmint. A variety with yellow-veined

Mentha gentilis

Mentha gentilis aureo-variegata

leaves is grown as an ornamental under the name *M. g. aureo-variegata*.

Pennyroyal (***M. pulegium***), not to be confused with American-pennyroyal, which is *Hedeoma pulegioides*, has the specific epithet *pulegium*, meaning fleabane, alluding to a supposed virtue of driving away fleas. Native to Europe and adjacent Asia, it is a freely-branched plant 6 inches to 1½ feet tall. Its stalked, hairy leaves, not over about ½ inch long, smooth-edged or slightly toothed, are round-elliptic and

Mentha pulegium

have only two or three side veins on either side of the midrib. The little bluish-lilac blooms are crowded in tiers in the leaf axils. Related to pennyroyal, but with linear to linear-lanceolate leaves, ***M. gattefossii***, of Morocco, is up to 1 foot tall.

Corsican mint (***M. requienii***) is very distinct and easily recognizable. A miniature creeper, it forms absolutely flat mats scarcely ½ inch high of threadlike stems and round, pubescent, stalked leaves up to ⅛ inch across. Its scarcely visible, almost microscopic bluish-lilac flowers are few together in the leaf axils. When stroked or stepped upon this has a strong odor of pennyroyal.

Garden and Landscape Uses and Cultivation. With one exception, the Corsican mint, menthas are rarely grown other than as flavoring herbs. Spearmint is especially esteemed for flavoring, apple mint for garnishing drinks. All are very easy to grow in sunny or lightly shaded places in reasonably moist, fertile soil. They are invasive and must be restrained from occupying territory reserved for other purposes. For further details regarding their cultivation see Mint, and Pennyroyal. Unlike other kinds dealt with here, Corsican mint is a delightful miniature ornamental for rock gardens and for setting between paving stones where the earth is moist and sufficient shade to moderate strong sun is available. This is less hardy than the taller mints described above. It is unlikely to survive outdoors north of Philadelphia, but even in colder climates it frequently renews itself by self-sown seedlings. It is easily propagated by seed and by division.

MENTZELIA (Ment-zèlia) — Blazing Star. Possibly sixty species of annuals, biennials, and herbaceous perennials, sometimes somewhat woody toward their bases, constitute *Mentzelia*, of the loasa family LOASACEAE. The majority are natives of southwestern United States, Mexico, and the West Indies. They are free-branching and clothed with barbed, but not stinging hairs. Their name commemorates the German botanist Christian Mentzel, who died in 1701.

Mentzelias have usually alternate, rarely opposite, undivided, but often pinnately-lobed, stalked or stalkless leaves. The blooms are solitary, clustered, or in racemes. Those of some kinds open in late afternoon or evening and close the following morning, or later in dull weather. They have five sepals, five to ten petals, numerous stamens, the outer ones of which are sometimes petal-like and sterile, and a style branchless or three-cleft at the apex. The fruits are cylindrical or top-shaped capsules.

Most commonly grown is ***M. lindleyi*** (syn. *Bartonia aurea*), an annual 6 inches to 2 feet tall, and a native of California. This is distinct from *Eucnide bartonioides*, of the

Mentzelia lindleyi

same family, which in gardens is often known as *Bartonia aurea*. Usually freely branched, *M. lindleyi* is rough-hairy, and has stalkless, lanceolate to ovate leaves with slightly stem-clasping bases. They are up to 6½ inches long and generally lobed, with the lobes sometimes toothed. Solitary from the leaf axils or in terminal twos or threes, the fragrant blooms, which open at dusk, are 2 to 3½ inches across. They have five golden-yellow petals with orange-red bases, and at their apexes a little projection.

Blazing star (***M. laevicaulis*** syn. *Bartonia laevicaulis*) is a stout biennial with shining-white stems. From 1½ to 3½ feet tall, it has oblanceolate, deeply-wavy-lobed basal leaves up to 1 foot long and smaller, ovate-lanceolate stem leaves. Terminal, and solitary or in twos or threes, the five-petaled flowers are light yellow and 2½ to 4 inches or somewhat more in diameter. They open only at night. This occurs wild from Montana to Utah, Washington, and California.

An annual or biennial of California and Baja California, ***M. involucrata*** is usually branched from its base and up to 1½ feet tall. Its stems are white and shortly-pubescent. Its leaves are coarsely-wavy-toothed, linear to oblong-lanceolate, and up to 4½ inches long or sometimes longer. The upper ones are stalkless and stem-clasping. The solitary, satiny blooms are terminal, 2½ inches or rarely more across, and have five white petals with reddish bases and veins. Below each bloom is a pair of broad, white, deeply-toothed, papery bracts with green tips. Variety *M. i. megalantha* has flowers 3 to 5 inches across.

Perennial mentzelias bloom when quite young. Some are comparatively short-lived and on occasion behave as biennials or annuals. Notable among them is ***M. albescens*** (syn. *Bartonia albescens*), a native from Colorado to Kansas, Missouri, Oklahoma, Texas, and Mexico, as well as in Argentina and Chile. From 1 foot to 2 feet tall or sometimes taller, it has few-branched stems and coarsely-lobed or toothed leaves, those below linear-lanceolate to obovate, the upper ones narrower. The flowers, which open in late afternoon, are pale yellow, about ¾ inch wide, and in clusters. Handsome ***M. decapetala*** (syn. *M. ornata*) attains a height of 3 feet or sometimes more. It has fragrant, ten-petaled blooms 4 to 5½ inches in diameter. White to yellowish, they open shortly after sunset. The oblanceolate leaves are sinuously-pinnately-lobed and up to 7 inches long. This is a native from Alberta and Saskatchewan to Texas. Wild from Wyoming to California, Texas, and Mexico, ***M. multiflora*** is 1¼ to 2½ feet tall. It has white stems, and lanceolate, sinuously-lobed leaves up to 4 inches long. The flowers, in clusters of three or four at the branch ends, have ten yellow petals. They are 1 inch or a little more in diameter. Its stems but little branched, ***M. nuda*** (syn. *M. stricta*) occurs wild from South Dakota and Montana to New Mexico and Texas. Up to about 3 feet tall, it has linear-lanceolate, toothed leaves, and fragrant, white flowers about 2½ inches wide that open in late afternoon.

Garden and Landscape Uses and Cultivation. Except in dry or dryish climates that approximate those where they grow as natives, perennial mentzelias are not satisfactory as permanent garden plants, but some can be treated as annuals, as can the biennial kinds. They, and the true annuals, are attractive for summer flower beds and for informal use in semiwild areas. In dry climates the perennial kinds can be used similarly in more permanent displays. Mentzelias make good shows of, in some cases fragrant, blooms. They need sunny locations, sheltered from wind, and ordinary, well-drained soil. Because of their taproots mentzelias do not transplant satisfactorily. Seeds are sown, in spring where the plants are to remain, and the seedlings thinned out just sufficiently to prevent crowding.

MENYANTHES (Meny-ánthes)—Buckbean or Bogbean. One hardy deciduous herbaceous perennial plant that occurs naturally in swamps and wet lands throughout the cooler parts of the northern hemisphere is the only representative of *Menyanthes*, of the gentian family GENTIANACEAE. Its name is a variation of an ancient Greek one, *menyanthos*, probably applied to the plant now named *Limnanthemum*. The buckbean contains a bitter ingredient that has been employed medicinally as a tonic; in Sweden the plant is used to flavor beer.

The buckbean or bogbean (***M. trifoliata***) has running rhizomes and alternate, hairless leaves, somewhat reminiscent of those of the broad bean (*Vicia faba*), crowded near the base of the flower stalk. The leaves have stem-sheathing stalks up to 10 inches long and three elliptic to obovate stalkless leaflets 1½ to 3 inches long. The flowers are in upright terminal

Menyanthes trifoliata

Menyanthes trifoliata (foliage)

racemes. They have a five-parted calyx and a corolla with a funnel-shaped tube and five spreading lobes conspicuously hairy on their insides. The bases of the five stamens join the inside of the corolla tube. The stout style has a two-lobed stigma. In some flowers the style is much longer than the stamens, in others the reverse is true. The flowers are white, pinkish, or purplish and about ½ inch long. The fruits are egg-shaped to shortly cylindrical capsules ⅓ inch long. They contain flattened-egg-shaped, shining seeds.

Garden Uses and Cultivation. The buckbean is a useful and interesting plant for the sunny margins of streams and ponds and for garden pools. It prospers in wet ground and where the soil is covered to a depth of two or three inches with water. It will even grow in constantly damp places in ordinary flower borders, but prefers wetter conditions than usually obtained in such places. An acid to neutral soil is most to its liking. The buckbean has attractive foliage, of a distinctive appearance, and in late spring or early summer, pleasing, fragrant blooms. These can be arranged, together with some of the plants' foliage, as charming and unusual table decorations. The easiest way of increasing and establishing this species is to plant 1-foot-long pieces of the rhizomes, each with a terminal bud, in suitable places in spring. New plants can also be had from seeds sown in wet soil. These may be sown as soon as they mature or the following spring; if they are stored they must be kept constantly moist and in a temperature of about 40°F from the time of ripening.

MENZIESIA (Men-zièsia). As now interpreted, *Menziesia*, of the heath family ERICACEAE, consists of seven species of deciduous shrubs of North America and eastern Asia. Other plants previously included are treated under *Daboecia* and *Phyllodoce*. The name commemorates the distinguished English botanist and naval surgeon Archibald Menzies, who accompanied Vancouver on his voyage of discovery. Menzies died in 1842.

Menziesias have alternate, undivided, thin leaves and clusters or short racemes of small, nodding, bell- or urn-shaped or cylindrical flowers at the ends of shoots of the previous season's growth. They have small, four- or five-parted calyxes, and four- or five-lobed, white, greenish, or pink corollas. The stamens, not protruding, equal in number or are twice as many as the lobes of the corollas. The style is slender. The fruits are small capsules.

The native American species in general are less meritorious as ornamentals than the Japanese. Occurring in woodlands in mountains from Pennsylvania to Georgia, ***M. pilosa*** is 3 to 6 feet tall, and has densely-hairy shoots and elliptic to obovate, toothless leaves up to about 1½ inches long, and hairy on both surfaces. The bell- to urn-shaped flowers, about ⅓ inch long, are in clusters of two to several. Usually they are greenish tinged with red, but they vary to white and yellow. They have four sepals and four corolla lobes. This species is hardy in southern New England.

Native from California to Montana and Alaska, erect or somewhat straggling, ***M. ferruginea*** is 3 to 12 feet in height. It has glandular-hairy shoots and toothed, elliptic-oblong to obovate leaves up to 2½ inches in length. Reddish hairs sparsely cover their upper sides and the veins beneath. The four-lobed flowers, their corollas yellow tinged with red, are about ¼ inch long. From the last, ***M. glabella***, native from Minnesota, to Wyoming, Idaho, Alberta, and British Columbia, differs in its leaves being hairless or nearly so. They are toothed and on their undersides are glaucous. The four-parted, bell-shaped flowers are creamy-white and up to ⅓ inch long.

The most attractive species is the Japanese ***M. purpurea,*** which is 3 to 6 feet tall and has hairless shoots. Its short-stalked leaves, rather under 2 inches long, are ovate to elliptic-ovate, and have minute glandular hairs on their upper sides and sometimes on their glaucous undersides. The midrib beneath is hairy. In clusters of four to eight, the tubular-bell-shaped, deep purplish-pink flowers are approximately ½ inch long. Their four corolla lobes are fringed with glandular hairs. This mountain species is hardy in southern New England. So very variable that botanists recognize several varieties, ***M. ciliicalyx*** (syn. *M. tubiflora*) is another Japanese species. About 3 feet tall, it has hairless or sparingly-short-hairy shoots, and in its typical form, ovate-oblong to obovate, minutely-toothed, more or less hairy leaves. Its pale, yellow or greenish blooms have five-lobed corollas not fringed with hairs. This species is about as hardy as the last.

Menziesia ciliicalyx

Garden and Landscape Uses and Cultivation. Menziesias are rather choice shrubs for the collector of the somewhat rare and unusual. They provide displays of bloom in spring and are neat and attractive in foliage. Requiring approximately the same conditions as deciduous azaleas, they prosper in acid soil that never dries excessively and are grateful for light or part-day shade. A mulch of leaf mold, peat moss, or similar organic material is to their liking. No pruning is needed. Propagation is easy by seeds sown in sandy peaty soil kept uniformly moist and by summer cuttings under mist or in a greenhouse propagating bench.

MERCURIALIS (Mercur-iàlis)—Herb Mercury, Dog's Mercury. Eight species constitute the European, North African, and temperate Asian genus *Mercurialis,* of the spurge family EUPHORBIACEAE. Its name derives from its ancient Latin one, *herba mercurialis,* given by the Romans in accordance with their belief that medicinal virtues had been discovered in these plants by the god Mercury. Belief in such qualities persisted until well after the advent of modern medicine, but the genus is not, at the present time, a source of official drugs.

The sorts of this genus are annuals, herbaceous perennials, and subshrubs with opposite, undivided leaves, and unisexual, small, green flowers with three sepals, no petals, eight to twenty stamens, and two styles. The flowers arise from the leaf axils, the males in stalked spikes, the females in short clusters. The fruits are two-seeded capsules.

Herb mercury (***M. annua***), native to Europe, western Asia, and North Africa, is naturalized in North America. A hairless hardy annual of weedy aspect, it has spreading stems 4 inches to 1½ feet long.

Mercurialis annua

Its toothed leaves are lanceolate to ovate-lanceolate and 1½ to 2½ inches long. The spikes of male flowers are longer than the leaves. The fruits are hairy. Formerly this plant was employed as an emollient and for other medicinal purposes. It has been recommended as a pot herb, but there is evidence that it can be poisonous to humans when eaten, although certainly much less dangerous than dog's mercury.

Dog's mercury (***M. perennis***), of Europe, is an ill-smelling, hairy, deciduous, hardy herbaceous perennial about 1 foot tall. It has slender, creeping underground rhizomes and numerous erect, branchless stems with thin, lanceolate leaves. The insignificant flowers appear before the foliage is fully developed, the male blooms in long-stalked racemes. The female flowers, solitary or in twos or threes on long stalks, are hidden among the foliage. The hairy fruits contain a solitary seed. Dog's mercury is dangerously poisonous if eaten. It was formerly used medicinally and as the source of a blue dye.

Garden Uses. The sorts described above are appropriate for inclusion in collections of old-time medicinal plants. Dog's mercury can also be used effectively as an attractive groundcover in shaded places where the soil is not unreasonably dry. It succeeds in even heavy shade. Herb mercury prospers in sun or part-shade and in

Meconopsis sheldonii

Manettia inflata

Massonia depressa

Masdevallia veitchiana

Masdevallia coccinea

Masdevallia militaris

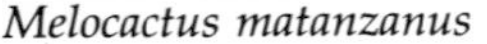

Melocactus matanzanus

Medeola virginica

Mertensia longiflora

Melaleuca steedmanii

ordinary soil. Both sorts are easily propagated by seeds and the dog's mercury, by division as well.

MERCURY is *Chenopodium bonus-henricus*. Dog's mercury and herb mercury are *Mercurialis*.

MERENDERA (Merén-dera). About ten species of *Merendera* range from Europe to Afghanistan. They belong in the lily family LILIACEAE and have corms (bulblike underground organs). The name is from the Spanish *quita meriendas*, for *Colchicum*.

From closely related *Colchicum*, this genus is distinguished by the lower parts of its six petals (more correctly, tepals) coming together to form an apparent tube, but not joined into one, by its six stamens having stalks attached near the bases of the anthers rather than at their middles, and by the three styles being separate to their bases. The fruits are capsules.

Native of mountain pastures in western Europe, ***M. montana*** (syns. *M. bulbocodium, Colchicum montanum*) has corms (bulbs) ½ to ¾ inch long and produces its flowers in fall, one or two stalkless blooms from each bulb. They have lilac-pink, strap-shaped petals, much longer than the stamens, that, when the flower fully opens, spread to form a star. The blunt, linear leaves, about three in number, are 3 to 4 inches long.

Garden Uses and Cultivation. Merenderas are attractive hardy plants for rock gardens. They prosper where they receive slight shade as well as in full sun, and respond to gritty, well-drained, reasonably nutritious earth. They are usually planted and, when necessary, transplanted, 2 to 3 inches deep, in late summer or early fall. They are propagated by seed and by natural multiplication of the corms.

MERINTHOSORUS (Merintho-sòrus). Two species of ferns of the polypodium family POLYPODIACEAE comprise the genus *Merinthosorus*. The name is derived from the Greek *merinthos*, a line or string, and *sorus*, a group of spore-clusters. It alludes to a characteristic of the sori.

Native to Malaysia and Indonesia, these evergreen ferns are close relatives of *Aglaomorpha*. Chiefly tree-perchers (epiphytes), they have horizontal rhizomes and fronds that, except sometimes on young specimens, have the upper, sterile portions of the fronds pinnately-lobed.

Attractive ***M. drynarioides*** has rhizomes densely clothed with long brown hair. Up to a few feet long by 1 foot wide or sometimes wider, the spreading to erect, elliptic to obovate fronds (leaves) are in their upper parts quite deeply-pinnately-lobed after the fashion of a conventional oak leaf.

Garden and Landscape Uses and Cultivation. The species described here is suitable for shady places outdoors and lath

Merinthosorus drynarioides

houses in the humid tropics, and for tropical greenhouses. It needs a coarse, loose rooting mix composed chiefly of such organic material as shredded bark of a kind suitable for orchids, sphagnum moss, peat moss, and partially decayed leaves, with the addition of a little fibrous loam and a generous dash of charcoal broken from peanut to walnut size. The mix may be contained in pans (shallow pots) or hanging baskets or be piled on rafts or slabs of cork bark or rot-resistant wood such as red-cedar or teak. When planting or potting do not bury the rhizomes, instead peg them to the surface of the mix. In greenhouses a winter night temperature of 60 to 70°F is desirable with an increase of five to ten degrees by day. Light shade from strong sun is needed as is a decidedly humid atmosphere. Water to keep the rooting mix always moderately moist without it remaining saturated for long periods. Propagation is easy by division and by spores. For additional information see Ferns.

MERISTEM CULTURE. This highly sophisticated method of plant propagation scarcely comes within the scope of activities of the majority of amateur gardeners, but is of considerable importance commercially, and its use is being expanded. It is therefore important for all gardeners to know about it. Unlike most animals, the majority of plants can be raised from pieces cut from individuals. Propagation by cuttings, leaf cuttings, and root cuttings are familiar examples of advantages gardeners take of this. Success with all these is based on the ability of the meristematic tissues they contain to continue cell division and reconstitute new individual plants. Meristematic or meristem tissue consists of cells that have this peculiar ability. It is located at the growing point or apex center of every shoot and within every growth bud. Only minute amounts occur.

Scientists conjectured that if small portions of meristem could be cut from a plant and kept alive separately they would continue to grow and develop into new individuals. This has proved to be the case, and such culture has been practiced to some extent since the middle of the twentieth century. For success several conditions must be met. First, the removal of the meristem, its division, and transfer to where it is to grow must be performed under conditions at least as aseptic as those of a hospital operating room. If this is not done, contamination with molds and other funguses and bacteria that find the environment prepared for the meristems highly to their liking is inevitable. The next need is to provide the meristem with the nutrients, in their proper proportions, required for growth. These are essentially all nutrients needed by plants. They include nitrogen, phosphorus, potassium, sulfur, calcium, iron, and a number of trace elements. In addition sugar, usually supplied as sucrose, is needed. Experimentally, and sometimes practically, many other additives have been tried. The various nutrients are incorporated in a solution or a jelly, the pH of which has an important bearing on results, as also does temperature and other environmental factors.

Not all cultured meristems develop into new plants. For this to happen it is necessary for differentiation to take place in the multiplying cells so that roots, stems, leaves and other plant parts can be formed. But some meristems under some conditions continue to grow almost in cancerous fashion without differentiation and produce only larger masses of meristem. Much research is under way to determine the causes of differentiation. If this is successful it would theoretically be possible to grow masses of meristem indefinitely, to remove small portions as needed, culture them under conditions that would stimulate differentiation and so become new plants, and keep the stock meristem growing in an undifferentiated condition for future use. To date the greatest practical applications of meristeming, as meristem culture is sometimes called, has been with orchids, many of which are multiplied commercially by this method. Plants raised from meristems are called mericlones.

Gardeners may well ask about the advantages of such a demanding technique over more traditional methods. With many orchids, increase by division, usually the only alternative, of rare species and superior hybrids is distressingly slow, with the result that the products are extraordinarily expensive. Meristem culture can correct this. One of the most important practical uses to which the new technique has been put is the raising of virus-free stock from plants infected with virus diseases. The meristematic cells of such plants are very often free of viral contamination and can be used as starts for new healthy populations. This remarkably useful phenomenon has been put to good use with carnations, chrysanthemums, dahlias, oranges, orchids, strawberries, and other plants.

MERMAID WEED. See Proserpinaca.

MERREMIA (Mer-rèmia)—Wood-Rose. Visitors to Hawaii and other tropical places are often intrigued by the decorative sprays of wood-roses they see in dried flower arrangements. They recognize, of course, that they are not roses, but few, not previously informed, would guess correctly their botanical relationship. They are the fruits of a morning glory relative, a plant so closely akin to those familiar flowers that it was once included with them in the genus *Ipomoea*. Now it belongs in *Merremia*, of the same morning glory family, the CONVOLVULACEAE. Another plant, called the transparent wood-rose, is *Operculina turpethum*. The small-wood-rose is *Argyreia nervosa*. The name *Merremia* commemorates a German naturalist, Blasius Merrem, who died in 1824.

Merremias differ from morning glories chiefly in having smooth instead of spiny pollen grains. This can easily be observed with a good hand lens. They number about eighty species and occur as natives of the tropics and parts of the subtropics throughout the world.

The wood-rose (***M. tuberosa*** syn. *Ipomoea tuberosa*) is a tuberous-rooted, rampant vine capable of climbing into the tops of tall trees. It has hairless stems and foliage. The leaves, 3 to 8 inches across, are fingered in hand-fashion into seven, or sometimes only five, long, narrow, elliptic lobes. The funnel-shaped flowers are bright yellow, 1 inch to 2 inches long and about as wide. Their stalks bear few to several blooms. The flowers have spreading, blunt-ovate sepals that enlarge to about 1 inch in length and become woody and rigid as the fruits develop. The sepals are the "petals" of the "roses." They surround the globular capsule containing angular seeds that forms the center of the "rose." At maturity, capsules and sepals are brown.

Merremia tuberosa (flowers)

Merremia tuberosa (fruits)

Another species, sometimes known as Alamo vine, is ***M. dissecta*** (syn. *Ipomoea dissecta*), a native of South America naturalized from Georgia to Florida and Texas. This is a tuberous-rooted vine with yellow-hairy stems and leaves 2 to 4 inches long that have seven, narrow, toothed leaflets. The flowers, singly or in pairs on the stalks, are broadly-funnel-shaped, 1 inch to 2 inches across, and nearly white with purple throats. Their sepals become much enlarged as the fruits develop. Native to Baja California, ***M. aurea*** is a slender, tuberous-rooted vine with five-lobed leaves and golden-yellow, funnel-shaped blooms with faces 2 to 4 inches across.

Garden and Landscape Uses. As a screening vine where there is ample space for it to develop, the wood-rose is useful in tropical gardens. The other species may be used similarly in warm, essentially frost-free climates. Additionally, the wood-rose is grown for its decorative fruits, which are esteemed by creators of dried flower arrangements. Usually they are varnished to give them gloss or are painted before use.

Cultivation. In hot or warm, humid climates no difficulty attends the cultivation of these plants. They are sun-lovers that enjoy deep, fertile, fairly moist soil. Seed affords the most satisfactory way of increase. Cuttings of the wood-rose can be rooted, but not with ease or certainty.

MERRYBELLS. See Uvularia.

MERTENSIA (Mer-ténsia)—Bluebells, Virginia Bluebells or Virginia-Cowslip. One-half or more of the perhaps fifty species of the northern hemisphere genus *Mertensia*, of the borage family BORAGINACEAE, are natives of North America. The name commemorates the German botanist Franz Carl Mertens, who died in 1831.

Mertensias are hardy, hairy or hairless herbaceous perennials. They have alternate, undivided leaves and terminal racemes or loosely-branched clusters of generally blue to purple, more rarely pink or white flowers disposed along one side of the stalks that bear them. The blooms have a calyx of five lobes or five separate sepals, a corolla cylindrical toward its base, trumpet- or bell-shaped above, and with five short lobes (petals). There are five protruding or nonprotruding stamens and a stigma that may be slightly two-lobed. The fruits consist of four seedlike nutlets.

Virginia bluebells or Virginia-cowslip (***M. virginica***) is by far the best known of cultivated sorts. Native from New York to Michigan, Wisconsin, Kansas, Alabama, Tennessee, and Missouri, this hairless inhabitant of moist and wet woodlands has elliptic to oblanceolate or obovate leaves up to 6 inches long, the lower ones much tapered toward their bases. The nodding, porcelain-blue flowers, ¾ to 1 inch in length, distinctly bell-shaped and in the bud stage pink, are in showy clusters. In early summer, after blooming, the plants become dormant and remain so without above-ground parts until early spring. A white-flowered variety occurs.

Mertensia virginica

With much the same garden and landscape uses as Virginia bluebells, ***M. paniculata*** occurs as a native from Quebec to Oregon and Alaska. From 1 foot to 3 feet in height and hairy or hairless, this inhabitant of wet meadows and other damp places has many lanceolate to ovate leaves and purple-blue, ½-inch-long flowers.

A beautiful dwarf 2 to 10 inches tall, ***M. longiflora*** is hairy to nearly hairless. Its few bluish-green leaves, elliptic to ovate or obovate, are up to 2½ inches in length. The blooms are crinkled bells of pale blue, ¾ to 1 inch long. This sort inhabits lightly shaded or open, dryish places from Montana to British Columbia and northern California. Similar, but with less tuberous roots and commonly better-developed basal foliage and longer, narrower leaves, ***M. oblongifolia*** sometimes exceeds 1 foot in height. It inhabits dryish soils from Nevada to British Columbia. Practically all low alpine species of western North America, many from a garden point of view rather similar, would make splendid additions to rock gardens were it possible to domesticate them.

Garden and Landscape Uses. Virginia bluebells is a spring-blooming ornamental of outstanding merit for naturalizing in shade to almost full sun and for grouping in more formal flower borders. It succeeds and usually increases freely by self-sown seeds in slightly acid to neutral soil that is moist from early spring to at least midsummer and that contains a generous proportion of leaf mold, compost, or peat moss. Because its above-ground parts disappear early, associate it with ferns or other plants that furnish foliage to hide the bare spots it leaves.

Western North American mertensias include several that may from time to time be brought into cultivation, but most, especially those from high altitudes, are distinctly challenging to grow under conditions that differ much from those they know in the wild. They are sometimes grown in rock gardens and alpine greenhouses. Cool summers seem essential for success.

Cultivation. Plant the curiously formed, brittle, tuberous roots of Virginia bluebells approximately 1½ feet apart any time between when the foliage dies and early fall or transplant young seedlings in early spring. Mulch established plants each fall with leaf mold, compost, or peat moss. If the location tends to be dry water freely through the spring season of growth. New plants are easily raised from seeds sown in a cold frame or similar place in woodland-type soil that is constantly damp. Transplant the seedlings to partially shaded nursery beds. They should flower in their third year. Give dwarf alpine mertensias gritty, well-drained soil containing an agreeable amount of humus, kept moist during their period of growth, at other times drier. In rock gardens place them in as cool locations as practicable and provide light shade from intense sun. Propagation is by seed.

MERYTA (Méry-ta)—Puka. Native chiefly of New Caledonia, but with representatives in other Pacific islands in addition to one kind each in Australia and New Zealand, *Meryta* belongs in the aralia family ARALIACEAE. Its name is based on the Greek *merytos*, glomerate, and alludes to the male blooms being in crowded heads. There are twenty-five species.

Merytas are small resinous trees, with leaves alternate or clustered toward the ends of the branches. The small flowers are in panicles or heads. The sexes may be on separate trees, or unisexual and bisexual blooms may occur on the same individual. Male flowers have three to five tiny to minute calyx teeth, and four, five, or sometimes more petals. The styles of the female blooms are usually five-branched. The fruits are fleshy.

The only kind known to be cultivated is the one endemic to certain offshore islands of New Zealand, and there called the puka. It is ***M. sinclairi***, an evergreen, up to 25 feet in height, that has brittle branches and forms a dense crown. Its leaves are crowded at the ends of branchlets. They are undivided, have stalks about 1 foot long, and glossy, oblong blades that may attain 1½ feet or somewhat more in length and 8 inches in width. Their margins have suggestions of shallow lobes. Terminal, erect, and up to 1½ feet long or rather longer, the panicles of greenish blooms are oblongish in outline. The flowers are essentially without calyxes. The males have four petals and the same number of stamens. The females have four or five petals, the same number of staminodes (infertile stamens), and four- or five-branched styles. The succulent, glossy black fruits are ⅓ to ½ inch long.

Garden and Landscape Uses and Cultivation. Not hardy in the north, this species may be in cultivation in gardens on the Pacific Coast. Little is recorded about its cultural requirements, but these probably are not difficult to satisfy. Plants can be raised from seed.

MESCAL-BEAN is *Sophora secundiflora.*

MESCAL BUTTON is *Lophophora williamsii.*

MESEMBRYANTHEMUM (Mesembryánth-emum)—Ice Plant. This genus of annual and nonhardy biennial, succulent, herbaceous plants, of the carpetweed family AIZOACEAE, has had a confused and involved history. At various times more than a thousand species have been ascribed to it. Botanists mostly agree that this vast array comprises a number of separate genera, but authorities have disagreed as to which group of segregates the name *Mesembryanthemum* should apply. Here the view of the great South African student of the genus H. M. Louisa Bolus is accepted, and the name is used to include the group that was previously named *Cryophytum.* There are seventy-four species. Mesembryanthemums grow and reproduce spontaneously in South Africa, the Mediterranean region, the Sahara desert, southwest Asia, and in coastal areas of southern California and Baja California. This has resulted in their being described and accepted as natives of all these places, but such an unusual natural geographical distribution seems unlikely. It is far more probable that the genus was originally confined to South Africa and that its distribution elsewhere is the result of man's activities. If this be so, its introduction to California must have occurred very early. The name, referring to the position of the ovary, comes from the Greek *mesos*, middle, *embryon*, embryo, and *anthemon*, a flower.

Mesembryanthemums are fleshy annuals and biennials usually prostrate or at least with prostrate branches. All or parts of them are furnished with glistening, watery pustules. Their leaves, opposite or the upper ones alternate, are stalkless or short-stalked, flat, and sometimes quite large, or cylindrical. Their solitary flowers are daisylike in appearance, but structurally are very different. In daisies the flower heads consist of numerous tiny flowers called florets, each usually with its own reproductive organs. The blooms of *Mesembryanthemum* have each four or five sepals, two of which are often leafy, many slender, linear petals, numerous stamens, and five or rarely four stigmas. The fruits are capsules containing many small seeds. Of the genera described in this Encyclopedia the following belong in the great complex once included in *Mesembryanthemum.* Some still often known as such are *Aptenia, Carpobrotus, Cephalophyllum, Cheiridopsis, Conophyllum, Delosperma, Dorotheanthus, Drosanthemum, Erepsia, Lampranthus, Malephora, Oscularia,* and *Ruschia.*

The ice plant (***M. crystallinum*** syn. *Cryophytum crystallinum*) is most commonly cultivated. A prostrate, much-branched, very succulent annual with spatula-shaped to broad-elliptic, wavy-edged leaves, it has stems up to 2 feet long. The lower leaves may be 6 inches long by 3 inches broad,

Mesembryanthemum crystallinum

but these usually soon die. The leaves of the flowering shoots are much smaller. All the foliage is thickly covered with glistening, crystalline pustules, which give reason for the common name. The whitish, pink, or reddish blooms open only in bright weather. They are about ¾ inch across. In California and Baja California this species is abundant along the coasts, often in more or less saline soils.

Another kind occurring under similar conditions in the same parts of North America as the ice plant is annual ***M. nodiflorum*** (syn. *Cryophytum nodiflorum*). This differs in having slender, nearly cylindrical leaves ¾ inch to 1½ inches long. Its stems, prostrate or suberect, are up to 8 inches long, its white flowers under ½ inch wide. As a garden plant this is decidedly inferior to the ice plant. Also sometimes cultivated, ***M. aitonis*** (syn. *Cryophytum aitonis*), a native of South Africa, has prostrate stems and rather blunt, flat, lanceolate leaves. Its flowers are small and white.

Garden and Landscape Uses. These are good plants for dry, sunny locations. They thrive in poor soils and bloom freely. In sun the foliage, especially that of the ice plant, glistens as if covered with hoar frost or sprinkled with minute crystals of ice. The floral displays are pleasing and continue for a long period. The chief garden uses of mesembryanthemums are as edgings and for growing in patches or drifts at the fronts of flower borders and in rock gardens. Another good way to employ them is for carpeting areas among spring-flowering bulbs, to supply a groundcover of foliage and flowers after the bulbs have died down. These plants are also attractive when grown in pots, hanging baskets, and window and porch boxes. The ice plant is best suited for these purposes.

Cultivation. No annuals are easier to grow than these. All they need is perfectly drained soil and full sun. Seeds may be sown outdoors in spring where the plants are to bloom or earlier indoors in a temperature of 60 to 65°F, about eight weeks before the young plants are to be transplanted to the garden. The seeds are covered shallowly with soil. Plants from those sown outdoors are thinned to stand 6 to 8 inches apart, those from seeds sown indoors and transplanted 2 inches apart in flats and are later planted outdoors at a spacing of 8 to 10 inches. Planting is done after the weather is warm and settled. For growing in pots individual plants may be started in 3-inch pots and later be transferred to 4- or 5-inch pots. A very coarse, not excessively fertile soil should be used and watering done only when the soil is obviously dry. Excessively wet soil inhibits the development of the glistening, watery pustules responsible for the decorative appearance of the foliage. For further information see Succulents.

MESPILUS (Més-pilus)—Medlar. A hardy, edible fruit-bearing tree little known in America, but common in parts of its native Europe, the medlar is the only species of *Mespilus*, of the rose family ROSACEAE. Kin of apple, pear, and quince, but most closely related to hawthorn (*Crataegus*), it differs from that genus in having solitary flowers as well as in other details. Its botanical name is an ancient Greek one. The plant called snowy-mespilus is *Amelanchier ovalis*.

The medlar (***M. germanica***) is a frequently crooked and picturesque tree up to 20 feet tall. In the wild generally thorny, under cultivation it is usually less so. It has densely-hairy young shoots and finely-toothed, nearly stalkless, lanceolate to ovate, alternate leaves 2 to 5 inches long. Borne at the ends of short, leafy branchlets, the white or pink-tinged, five-petaled blooms are 1 inch to 2 inches in diameter. Their five sepals are longer than the petals and are covered with gray wool. Apple-shaped, with at the apex of the fruit a large, wide-open eye that exposes the ends of the bony ovaries and is surrounded by the persistent calyx, the brown fruits are 1 inch to 2 inches in diameter.

Mespilus germanica (flowers)

Mespilus germanica (fruit)

The medlar is native to continental Europe and occurs spontaneously in southern England, although it is believed not to be truly indigenous there. It has been cultivated for centuries and at one time was much more important as a fruit than it is today. Several horticultural varieties exist including thornless or nearly thornless ones. These commonly have fruits 1½ to 2½ inches in diameter and larger leaves than the wild species.

The taste for medlars is an acquired one. They are not eaten fresh, but are allowed to hang on the trees until after the first frost and then are picked and stored in a cool root cellar until they are "bletted," that is until they have reached a state of initial decay. Alternatively, they can be made into jelly.

Cultivation. Medlars thrive in any reasonably fertile, well-drained soil, such as suits apples and hawthorns. They need a sunny location. Little or no pruning or other special care is required. The typical species is propagated by seed, the horticultural varieties by grafting onto seedlings of medlar or hawthorn. Seeds often do not germinate until the second year from sowing. For information about the curious graft-hybrids or chimeras between medlar and hawthorn see Crataegomespilus.

MESQUITE is *Prosopis juliflora*.

MESSERSCHMIDIA (Messer-schmídia) — Tree-Heliotrope. Three species of herbaceous plants and shrubs constitute *Messerschmidia*, of the borage family BORAGINACEAE. The group inhabits coastal regions in many warm parts of the Old World and the New World. Its name commemorates Daniel Gottlieb Messerschmid, Polish botanist, who died in 1735.

Messerschmidias have alternate, toothless leaves, and usually branched, terminal clusters of small flowers with five-parted calyxes, tubular corollas with five lobes (petals), five stamens, and a branchless style tipped with a usually two-lobed stigma. The fruits, technically drupes, contain up to four stones, and have corky coverings that aid in their distribution by ocean currents.

Tree-heliotrope (***M. argentea*** syn. *Tournefortia argentea*) is naturalized in Hawaii. Its native range extends from southeast Asia to Taiwan, Japan, northern Australia, Polynesia, and Madagascar. Characteristically it inhabits sea coasts, shores and coral strands, often growing so close to the water that at high tide its lower branches are washed. A rather attractive shrub or broad, short-trunked tree up to about 20 feet in height, the tree-heliotrope has its leaves in clusters at the branch ends. They are obovate to obovate-elliptic, 4 to 10 inches long, up to 3½ inches wide, thick, fleshy, and brittle. Like the shoots, they are covered with whitish, silky hairs. In free-branching clusters of coiled spikes at the ends of the branches, the numerous small, white flowers, globose in bud, are borne. They are about ¼ inch across. The glossy, pointed, pea-like, fleshy fruits split into two corky parts, each containing one or two fertile seeds.

Garden and Landscape Uses and Cultivation. In Hawaii and other tropical places the species discussed has considerable merit for planting near the sea. It withstands maritime environments even more unfriendly than those the coconut survives. It grows in pure sand and persists even in close competition with casuarina trees. Young plants are best raised from seed and should be grown in cans or other containers until they are large enough to set in their permanent location.

MESUA (Més-ua)—Ceylon-Ironwood. One of the three species of the Asian genus *Mesua*, of the garcinia family GUTTIFERAE, is grown to some extent in southern Florida and other warm regions. It is the Ceylon-ironwood, a beautiful evergreen tree much planted in Sri Lanka (Ceylon) near Buddhist temples and as a garden ornamental. Its very durable lumber is esteemed for bridge building, shingles, and cabinetwork. The name commemorates Johannes Mesuë, a physician of Damascus, interested in plants, who died in 1857.

The Ceylon-ironwood (***M. ferrea***) is up to 80 feet tall. It has narrow-elliptic, pointed leaves 3 to 5 inches long, with many parallel veins extending outward from the midrib. They are dark green above and waxy-white on their undersides. When young they are deep red. The delightfully fragrant blooms, 2 to 3 inches or more in diameter, are suggestive of single roses. They have four white petals and a central cluster of yellow stamens with orange anthers. Each bloom lasts for only a day, but a succession is maintained over a long period in late winter and spring. The fruits are woody capsules.

Garden and Landscape Uses and Cultivation. In warm, moist climates, the Ceylon-ironwood is an attractive tree for general purpose planting. It thrives in ordinary soils in sunny locations. Propagation is by seed and by cuttings of firm but not woody shoots taken in late spring or early summer and inserted in a propagating bench provided with a little bottom heat, or under mist.

METAKE is *Pseudosasa japonica.*

METASEQUOIA (Metasequò-ia) — Dawn-Redwood. The name *Metasequoia*, from the Greek *meta*, beyond, and *Sequoia*, the name of the redwood tree, alludes to a presumed ancestral relationship of the dawn-redwood to the redwood. Both belong in the taxodium family TAXODIACEAE.

Since its discovery in 1941 *Metasequoia* has often been referred to as a living fossil, but the term is no more appropriate for the dawn-redwood than it is for tree ferns, cycads, ginkgoes, and other elements of our present flora that are well represented in fossil records. Indeed some of these have longer fossil ancestries than the dawn-redwood.

The unusual feature of the discovery and naming of *Metasequoia* is that it was found as a fossil and described and named before living specimens were known to exist, thus reversing the usual sequence of events. A parallel case in zoology was the catching of living specimens of the deep-sea fish the coelacanth in 1938 in waters off the African coast. This fish, previously thought to be extinct, had been known only from fossils.

The scientific history of the dawn-redwood began when a new type of fossil plant, believed to be an extinct species of redwood, was described as *Sequoia langsdorfii.* As time passed other fossil samples from Asia and North America accumulated and were studied by scientists who recognized their relationship to present-day redwoods (*Sequoia*), to the big tree (*Sequoiadendron*), and to the swamp-cypress (*Taxodium*). In the main they were believed to represent extinct ancestral forms of living genera rather than to being identical with them because the fossils differed in important respects from the living trees to which they were so obviously kin. Notably their branches are opposite rather than alternate. Because of this and other differences, in 1941 the Japanese botanist Miki described and named the genus represented by the fossils *Metasequoia.* At the time he had no reason to suppose that living plants of his new genus existed. Then an exciting coincidence occurred. The Chinese botanist Kan discovered that same year in Szechwan a deciduous conifer previously unknown to science that was called by the local inhabitants shui-sha (water-fir). Three years later specimens were collected for scientific study and the facts about the discovery of the new tree were published in 1946. In 1948 it was formally described and named *M. glyptostroboides*, the last part of the name means similar to *Glyptostrobus*, which is another genus of Chinese deciduous conifers.

Shortly after the announcement of the discovery of the dawn-redwood Dr. E. D. Merrill, Director of the Arnold Arboretum, arranged for seeds to be collected and sent to the United States. Dr. Merrill generously shared the seeds he received with botanical gardens and similar institutions in the United States and abroad and young dawn-redwoods soon became familiar to botanists and horticulturists throughout temperate regions of the world.

Since the discovery of the first few specimens, exploration has revealed that the dawn-redwood grows in a number of places in Szechwan and Hupeh scattered over an area several hundred miles long by perhaps twenty to twenty-five miles wide, reaching its maximum development in the Shui-sha valley. It is found chiefly by stream sides, growing in company with *Cephalotaxus*, *Cunninghamia*, and *Taxus*. In its home territory this tree is planted for ornament. Its lumber is not considered of good quality.

The dawn-redwood (***M. glyptostroboides***) is the only known living species of the genus. Growing up to a height of about 150 feet, it has a tapering trunk. When young it is pyramidal in outline, but when mature it develops a broad, rounded top. The dawn-redwood has soft, green, feathery foliage suggestive of that of a swamp-cypress (*Taxodium*). The leaves, flat and linear and about ½ inch long on older trees, but up to two or three times as long on young specimens, are arranged on both sides of short shoots like the teeth of a comb. Together with the short shoots to which they are attached, they shed in fall. Before they drop they turn a warm russet-red. The cones, ¾ to 1 inch long, solitary and pendulous, are at the ends of lateral branchlets. The dawn-redwood has proved hardy in New England.

Metasequoia glyptostroboides, a young specimen

Metasequoia glyptostroboides, an older specimen

Metasequoia glyptostroboides (foliage)

Garden and Landscape Uses. Dawn-redwood is one of the fastest growing trees. In its early years, under favorable conditions, it may increase almost five feet in height in a single season. It is yet too early to evaluate its final worth for garden and landscape use, but every indication is that it will prove a welcome addition to the kinds worth planting. It is excellent as a single specimen and as an avenue tree, and for grouping. It may be sheared regularly to form an interesting hedge. Specimens trained as standards by removing their lower branches while they are young make good clean-stemmed street trees. Because the dawn-redwood is deciduous it is more likely to withstand city conditions than evergreen conifers. Its leaves when they drop shrivel and present no problem of collection and removal.

Cultivation. Dawn-redwood thrives in any reasonably good soil. It grows best where drainage is good, but the earth always fairly moist. It transplants well and is responsive to pruning and shearing. Propagation is easy by cuttings of short growths, preferably with a heel of older wood attached, taken in late summer and planted in a propagating bench in a greenhouse or cold frame or under mist. It can also be increased from cuttings of mature shoots taken in fall and from seed.

METROSIDEROS (Metro-síderos) — Rata, Lehua, New Zealand Christmas Tree, Ironwood, Bottle Brush. Alluding to the heart wood, the name *Metrosideros* is from the Greek *metra*, heart of a tree, and *sideros*, iron. It belongs to a genus of the myrtle family MYRTACEAE represented in the natural floras of New Zealand, Australia, Polynesia, Malaya, South America, and South Africa. The number of species is variously interpreted as from twenty to sixty. Some are highly attractive ornamentals.

This group of aromatic evergreen trees, shrubs, and woody vines has opposite, undivided, gland-dotted, usually leathery leaves. Generally in terminal clusters or racemes, the numerous small blooms of the cultivated kinds make gay showings, their chief decorative feature being their brightly colored, long-protruding stamens. Each flower has five each sepals and petals, many stamens, and a slender style. The fruits are leathery, top-shaped capsules. In gardens, *Metrosideros* has been confused with *Callistemon*, and some of its kinds have been mistakenly grown under that name.

The northern rata (***M. robustus*** syn. *M. floridus*), of New Zealand, attains heights of 80 feet or more. In its homeland it commonly but not always begins life as an epiphyte, starting from a wind-blown seed that lodges on a tree, and developing into a specimen that extends its roots to the ground, eventually envelops and kills its host, and establishes itself as a free-standing specimen. This sort has elliptic-ovate to elliptic to ovate-oblong hairless leaves 1 inch to 2 inches long, with a tiny notch at the apex. In many-flowered terminal clusters, its bright red blooms have stamens 1 inch to a little more in length. Northern ratas that from the beginning of their lives grow in the ground attain little more than shrub size. Native of Hawaii, the lehua (***M. collinus***) also frequently starts life as an epiphyte and develops in the same manner as the northern rata of New Zealand. This very variable kind ranges from shrub size to a massive tree 100 feet tall and has bright red flowers with stamens ½ to 1 inch long. Hawaiians use its flowers in leis. The leaves of this species are broad-ovate to nearly round, downy on their undersides, and 1½ to 2½ inches long.

Metrosideros robustus (leaves and flowers)

The New Zealand Christmas tree (***M. excelsus*** syn. *M. tomentosus*) attains heights of 35 to 65 feet. It has spreading branches and downy young shoots. Its elliptic to oblongish leaves, mostly 2 to 4 inches in length, are clad on their undersides with white hairs. In many-flowered, rounded terminal clusters, the crimson flowers have stamens 1¼ to 1¾ inches long. This kind inhabits coastal forests and sea cliffs in New Zealand. In cultivation in California, it tends to develop as a multi-trunked specimen 20 to 30 feet in height. If by pruning restrained to one trunk, it makes a fine umbrella-shaped head.

Other large tree species native to New Zealand are *M. kermadecensis* (syn. *M. villosus*) and the southern rata or ironwood (*M. umbellatus* syn. *M. lucidus*). With densely-hairy shoots and broadly-ovate to elliptic-oblong leaves ¾ inch to 2½ inches long, clothed beneath with white hairs, ***M. kermadecensis*** has crimson flowers, many together in terminal, compound clusters, with stamens ½ to ¾ inch long. Attaining a maximum height of 60 feet, ***M. umbellatus*** often is only a small, round-topped shrub. Its shoots, like its 1- to 2-inch-long, pointed-lanceolate to elliptic leaves are hairless. Its very showy, light red blooms, in terminal clusters, have stamens about ¾ inch long.

A straggling, sometimes prostrate shrub or slender tree not over 20 feet tall, ***M. parkinsonii***, of New Zealand, has leathery, ovate-lanceolate leaves mostly 1¼ to 2 inches long. Its compound clusters of bright crimson flowers usually come from older parts of the shoots beneath the leaves rather than being terminal. A climber, ***M. fulgens*** (syn. *M. floridus*) has stems with flaking bark and elliptic-oblong leaves 1½ to 2¼ inches long. Its terminal racemes of orange-red to scarlet flowers have stamens ¾ to 1 inch long. Variety *M. f. aurata* has yellow flowers. This is probably the kind listed in California as *M. aurea*.

Pink- to white-flowered ***M. colensoi***, of New Zealand, is a slender-stemmed climber up to 20 feet tall or taller, with hairy shoots and pointed, ovate-lanceolate leaves ½ to ¾ inch long, when young densely-hairy. The flowers, in terminal racemes and small lateral clusters, have stamens not quite ½ inch in length. Another species with pink or white flowers, ***M. perforatus*** (syn. *M. scandens*), of New Zealand, has slender, trailing stems and closely spaced, broad-elliptic to roundish, boxwood-like leaves ¼ to ½ inch long. The flowers, from the leaf axils, are crowded at the branch ends. Their stamens are under ½ inch long. From the last, ***M. carmineus***, a vine with rooting stems, differs in its shoots being hairless or nearly so and in having more leathery, almost or quite hairless leaves. Its white to pink blooms, with stamens up to ½ inch long, are usually in lateral clusters from below the leaves.

Metrosideros perforatus

Garden and Landscape Uses. No metrosideros is hardy, but in mild, essentially frostless climates they prosper outdoors and are accounted among the most beautiful of flowering evergreen trees, shrubs, and climbers. They are grown to some extent in greenhouses and conservatories in pots, tubs, and ground beds. Their flowers are attractive for use in arrangements. The New Zealand Christmas tree is resistant to salt water spray and so is especially well adapted for planting near the sea. The cultivated representatives of this genus are beautiful in foliage and bloom. Depending upon their growth habits, the different kinds are adaptable as specimen trees or shrubs or as climbers, and for draping over walls, rocks, tree stumps, and banks. They succeed best in sunny locations in well-drained, reasonably fertile, moderately moist soils, preferably of a peaty nature. Some, including the New Zealand Christmas tree, adapt well to soils low in fertility. In greenhouses they may be accommodated in pots, tubs or ground beds. Most kinds bloom even when quite small.

Cultivation. Metrosideroses respond to the conditions and care that suit callistemons. In greenhouses it is essential that a cool, airy environment be provided. Winter night temperatures of 45 to 50°F are adequate. By day these may be increased, depending upon light intensity, by five to fifteen degrees. It is beneficial, but not essential, to stand container specimens outdoors in summer. The soil must never become dry, but constant saturation is to be avoided. Well-rooted specimens benefit from regular applications of dilute liquid fertilizer from spring to fall. Propagation is by seed, cuttings, and layering. Pruning, involving the removal of crowded and misplaced branches to keep the plants shapely, is done as soon as flowering is through.

METROXYLON (Metróx-ylon)—Sago Palm. Containing fifteen species and ranging from Thailand to New Guinea, *Metroxylon* belongs in the palm family PALMAE. Its name derives from the Greek *metra,* heart of a tree, and *xylon,* wood, and refers to the large pith of the trunk. Sago palms are so-called because their mature trunks, which attain a height of 30 feet, contain a nutritious starchy food called sago. The chief commercial source of this is *M. sagu.* The sago is extracted just before the blooms appear by cutting the trunks into pieces, crushing the pieces, and washing the crushed material. The name sago palm is also applied to quite different *Cycas revoluta.*

Sago palms have pinnate leaves and large, loose, erect clusters of flowers that stand well above the foliage. They flower and fruit only once and then die. This usually occurs when the tree is eight to fifteen years old.

The principal sago palm (***M. sagu***), native from New Guinea to Malaya, is cultivated in the Asian tropics for its starch. In America it is scarcely known, but it has been cultivated in greenhouses in European botanical gardens. For successful growth, it needs high temperatures and high humidity. It is propagated by seed. As natives these palms grow in swampy lowlands. For additional information see Palms.

MEUM (Mè-um)—Spignel or Baldmoney. One aromatic hardy herbaceous perennial native of mountain meadows of Europe is the sole representative of this genus of the carrot family UMBELLIFERAE. Its name is a modification of its Greek name *meon,* used by Dioscorides.

Spignel or baldmoney (***Meum athamanticum***), 6 inches to 2 feet tall, is a tufted, early summer-flowering plant with rootstocks crowned with a fibrous mass of remains of old leafstalks. The foliage is mostly basal. The oblong leaves, with stalks equaling the blades in length, are several times dissected into numerous threadlike segments, the ultimate ones 1/10 inch long and disposed in all directions. The scarcely-branched stems end in umbels 1 inch to 2½ inches wide composed of six to fifteen smaller umbels, some of which may contain male blooms only. The tiny white to purplish flowers, without sepals, have five pointed petals. The fruits, about ⅓ inch long, are oblong-ovoid and ribbed.

Garden Uses and Cultivation. This feathery foliaged plant is not often cultivated. It is best suited for informal areas and naturalistic plantings. It succeeds with little care in almost any well-drained garden soil in sun and is propagated by seeds sown as soon as they are ripe or in spring and by division.

MEXICAN. The word Mexican occurs as part of the vernacular names of these plants: Mexican-bamboo (*Polygonum cuspidatum*), Mexican black-torch (*Tillandsia punctulata*), Mexican blue palm (*Erythea armata*), Mexican-buckeye (*Ungnadia speciosa*), Mexican-clover (*Richardia scabra*), Mexican fire plant (*Euphorbia heterophylla*), Mexican flame vine (*Senecio confusus*), Mexican foxglove or Mexican-violet (*Tetranema*), Mexican handflower tree (*Chiranthodendron pentadactylon*), Mexican love grass (*Eragrostis mexicana*), Mexican-orange (*Choisya ternata*), Mexican-primrose (*Oenothera berlandieri*), Mexican rubber (*Castilla elastica*), Mexican shell flower (*Tigridia pavonia*), Mexican star (*Milla biflora*), Mexican-tea (*Chenopodium ambrosioides* and *Ephedra*), Mexican tulip-poppy (*Hunnemannia fumariaefolia*), and Mexican twin flower (*Bravoa geminiflora*).

MEXICAN JUMPING-BEAN. Beanlike seeds that when put in a warm place hop around are sold as curiosities as Mexican jumping-beans. They are seeds of species of *Sesbastiana,* of the spurge family EUPHORBIACEAE, that contain grubs of an insect that under the stimulus of warmth becomes active and supplies the motive force that causes the beans to jump. The plants are not cultivated. The jumping-beans sold are collected from wild stands in Mexico.

MEYEROPHYTUM (Meyero-phỳtum). Only one species of *Meyerophytum* is accepted by most authorities, although some recognize more. The genus, endemic to South Africa, is of the *Mesembryanthemum* relationship of the carpetweed family AIZOACEAE. Its name is seemingly a commemorative one.

A cushiony shrublet, ***M. meyeri*** (syn. *Mitrophyllum meyeri*) has slender stems, their not more than about ½ inch long lower parts often clothed with the remains of old leaves, and with, when dormant, pea-sized plant bodies enclosed in gray skins. In fall each plant body erupts into a new growth consisting of two or sometimes three pairs of leaves, the alternate pairs quite different in appearance. The first is nearly spherical and the two fleshy leaves composing it united practically to their tips. The second pair are cylindrical, nearly ½ inch long by less than one-half as wide, and conspicuously divided at their apexes so that the upper portions of the leaves are slender, cylindrical lobes. The flowers, daisy-like in aspect, but not in structure, are red to rose-purple, and approximately 1½ inches across. Each is a single bloom, not as are daisies, heads of numerous florets. The fruits are capsules.

Garden Uses and Cultivation. This is a real gem for collectors of rare and choice succulents. It needs the same conditions and care as *Conophyllum,* but is less amenable to cultivation. It is propagated by seed and by cuttings.

MEZEREON is *Daphne mezereum.*

MICE. Meadow mice and pine mice damage roots, rhizomes, bulbs, and other underground plant parts, frequently reached by following burrows made by moles. They themselves also burrow beneath mulches and feed on bulbs and roots and gnaw the bark of young trees and shrubs. Potted bulbs buried outdoors or in a cold frame for preforcing rooting are also likely to be injured by the depredations of mice. The simplest method of control is to bait the burrows with a ready-to-use commercial mouse bait.

MICHAELMAS-DAISY. See Aster.

MICHAUXIA. See Mindium.

MICHELIA (Mich-èlia)—Banana-Shrub. In technical details only does this Asian genus of deciduous and evergreen trees and shrubs vary from nearly related *Magnolia.* The chief easily observable differences are the much smaller flowers and the fact that they are mostly lateral on the branches and branchlets rather than at their ends. The sepals and petals are similar to each other. The group belongs in the magnolia family MAGNOLIACEAE and comprises about a dozen species. Its name honors the Florentine botanist Peter A. Michel, who died in 1737.

Several species of *Michelia*, including *M. champaca* of the Himalayas and evergreen *M. formosana* of Taiwan, are in their native countries important sources of lumber. The Hindus carve the wood of *M. champaca* into statues of Buddha and into beads and are fond of planting the tree near temples. In Malaya and other parts of Asia, the flowers of *M. champaca* are esteemed for stringing into necklaces and wearing in the hair and for making perfume. The Chinese use the flower buds of *M. fuscata* to perfume hair oil, and its wood is used in Java for handles of the knives called krisses.

In the continental United States the banana-shrub ***M. figo*** (syns. *M. fuscata, Magnolia fuscata*) is the most familiar kind. A large, compact, freely-branched evergreen shrub, its colloquial name refers to the very marked banana-like fragrance of its erect flowers, which are 1 inch to 1½ inches across and brownish-yellow rimmed with maroon or carmine-red. Borne in spring or early summer, they have narrow-oblong sepals and petals. The banana shrub attains a height of 10 to 15 feet and has branchlets clothed with velvety brown hairs and elliptic to broad-ovate, bluntly-pointed, glossy leaves 2 to 4 inches long. This species, a native of China, is not hardy in the north, but is a popular garden shrub in the middle and lower south.

A more tropical kind is ***M. champaca,*** of the Himalayas. This quite tall evergreen

Michelia champaca in the Hope Botanic Garden, Kingston, Jamaica

tree has a smooth gray trunk and narrowly-ovate, wavy, glossy leaves up to 10 inches long. Its yellow to orange, more or less drooping blooms have fifteen to twenty sepals and petals and are 2 to 2½ inches across. They are deliciously fragrant, especially at night. The fruits have reddish seeds and are quite decorative. This species is adaptable to cultivation in tropical and near tropical climates only.

Michelia champaca (flowers)

The hardiest *Michelia* is ***M. compressa,*** an erect evergreen endemic tree of the warmer parts of Japan and the Ryukyu Islands, which lives outdoors in sheltered places in England, but is not hardy in the northeastern United States. It attains a height of 40 feet or more and develops a dense, rounded head. Its leathery, glossy green leaves, 2 to 4 inches long and oblong, taper to slender stalks ½ to 1 inch long. The fragrant, pale yellow blooms are 1½ to 2 inches across and are succeeded by 2-inch-long fruits usually containing three seeds.

Garden and Landscape Uses and Cultivation. The cultivated michelias are attractive in gardens and other landscapes. They are useful as accents and as lawn trees and shrubs. They grow well in any ordinary garden soil with minimum attention. No regular systematic pruning is required. Propagation is by seed and by cuttings of firm shoots, in a greenhouse propagating bed or under mist.

MICONIA (Mi-cònia). Belonging to the melastoma family MELASTOMATACEAE, the genus *Miconia* contains about 1,000 species endemic to tropical America and the West Indies. Few are cultivated. The name commemorates a Spanish physician and botanist, Francisco Mico, who died in 1528.

Miconias are trees and shrubs, often with strikingly handsome foliage. Their undivided leaves, usually opposite, more rarely in whorls (circles of more than two), have a few very distinct longitudinal veins and, connecting them, many lateral ones at right angles to them. The rather small flowers are generally in terminal panicles or clusters. They have calyxes with four to eight, usually five, teeth or lobes, as many spreading or reflexed petals, and usually twice as many stamens as petals. The fruits are several- to many-seeded berries.

Plants known in gardens as *M. magnifica* (syn. *Cyanophyllum magnificum*) and *M. velutina* are *M. calvescens*. They do not differ sufficiently to warrant botanical separation, but the one called *M. velutina* has foliage more bronzy than the other and its leaves are flat rather than arched. It may be worth distinguishing horticulturally as *M. c. velutina*. Native to Mexico, Central America, and South America, ***M. calvescens*** is an evergreen tree up to 50 feet tall. It has reddish-brown shoots and pointed, broad-ovate leaves, up to 2½ feet long. They are lustrous green above, with three narrow, ivory-white veins running lengthwise, and reddish-purple to green on their hairless or minutely-sparsely-hairy undersides. The panicles of the small, stalkless, white flowers are up to 1½ feet long. The berries are purple. Generally resembling the last, ***M. astroplocama,*** of Costa Rica, differs in its leaves, dark green above, without white veins, and paler beneath, being densely clothed with stellate (star-shaped) hairs on their undersides. It differs also in having stalked flowers in panicles with densely-velvety-hairy, stalked branches.

An evergreen shrub 3 to 5 feet tall, ***M. hookeriana*** is an inhabitant of South American rain forests. It has leaves up to 1 foot long, with the center one of the three longitudinal veins marked on each side with a silvery-white band that flares briefly along each side vein. The leaves are up to 1 foot in length, broad-elliptic, and with wrinkled, velvety, dark green upper surfaces. Their undersides are paler and velvety, with short brown hairs. The flowers are white, the fruits, at first red or violet, become black when ripe. Another West Indian and South American evergreen shrub, ***M. racemosa*** attains a height of up to about 8 feet. It has broad-elliptic, short-stalked leaves 4½ to 9 inches long, fringed with short hairs. They have three main longitudinal veins. The tiny flowers,

Miconia racemosa (flowers)

in loose panicles, are white with pink stamens. The fruits, at first violet, change to black as they mature. This species in gardens has been misidentified as *M. pulverulenta*. An attractive sort cultivated in tropical greenhouses and perhaps outdoors in

Miconia racemosa (fruits)

parts of the tropics as ***M. saldanhaei*** is almost certainly misidentified. As grown in greenhouses it attains a height of 3 to 4 feet, or perhaps considerably more, and branches freely from the base. It has decidedly ornamental, pointed-ovate leaves with five conspicuous longitudinal veins.

Cultivated as *Miconia saldanhaei*, this plant is probably misidentified

Garden and Landscape Uses and Cultivation. Miconias are occasionally grown in tropical greenhouses for the beauty of their foliage. When so accommodated they are maintained as low shrubs in large pots or tubs or, in spacious conservatories, planted in ground beds. Even tree kinds are satisfactory when treated in these ways. In the humid tropics they may be planted outdoors and allowed to grow as large as the climate allows. Miconias respond to well-drained, rich soil that does not lack for moisture, but is not wet. In greenhouses they do well under conditions that suit anthuriums, alocasias, philodendrons and many other tropical foliage plants. A highly humid atmosphere is to their liking, and a minimum winter night temperature of 60 to 65°F and five to fifteen degrees more by day. At other seasons higher temperatures are in order. Shade from strong sun is needed. Propagation is by cuttings, air layering, and seed.

MICRANTHOCEREUS (Micrantho-cèreus). The only species of *Micranthocereus*, of the cactus family CACTACEAE, is native to Brazil. By botanists who are conservative in their interpretation of genera of the family, it is included in *Cephalocereus*. The name, alluding to the size of the plant, is from the Greek *mikros*, small, and the name of the related genus *Cereus*.

Forming colonies of stout, cylindrical, erect stems up to 5 feet in height, attractive ***M. polyanthus*** has stems with fifteen to twenty low ribs and clusters of twenty-five to thirty-five white to yellow, slender, needle-like, ½-inch-long radial spines and one to three stouter yellow to brown centrals 1 inch to 1¼ inches long. Its many pale pink to creamy-pink blooms, ½ to ¾ inch long, have perianth tubes naked of scales or hairs on their outsides. The smooth, pink fruits, the size of small peas, contain black seeds.

Garden and Landscape Uses and Cultivation. These are as for *Cephalocereus*. For more information see Cactuses.

MICROBIOTA (Micro-biòta)—This genus of one species, like its near ally *Juniperus*, belongs in the cypress family CUPRESSACEAE. Its name derives from the Greek *mikros*, small, and *Biota*, an obsolete name for *Platycladus*.

A native of eastern Siberia, ***Microbiota decussata*** is a hardy, dwarf, evergreen, flat-topped shrub with crowded, spreading branches and small opposite leaves, most of which are almost scale-like, but some on some branches may be awl-shaped. Individual plants are unisexual. The berry-like fruits, borne of course only by female specimens, have almost woody scales that break apart when the fruits mature. These characteristics of the fruits distinguish *Microbiota* from *Juniperus*.

Microbiota decussata

Microbiota decussata (foliage)

Garden Uses and Cultivation. This rare and choice species makes special appeal to collectors of dwarf conifers. It is well adapted for rock gardens and other intimate plantings and thrives in ordinary well-drained soil in sunny locations or where it receives light shade for part of each day. Propagation is by cuttings and by seed.

MICROCITRUS (Micro-cítrus)—Australian Finger-Lime. This genus of half a dozen or fewer species of generally spiny shrubs and small trees endemic to Australia and New Guinea belongs to the rue family RUTACEAE. The name is derived from the Greek *mikros*, small, and the name of the related genus *Citrus*.

The leaves of *Microcitrus* vary with the maturity of the plant. Those of young seedlings are very small. These are followed by slender juvenile foliage and, as growth continues, by adult foliage typical of the species. Adult-type leaves have short, finely-pubescent stalks, strongly veined, leathery blades, and are without hairs except near the bases of the midribs. The spines, usually solitary, or in some species paired, are sharp and slender. The small flowers, from the leaf axils, have four-, five-, or less commonly three-parted calyxes and corollas, and four or more times as many stamens as petals. The fruits are technically berries, globular, cylindrical, or spindle-shaped.

The Australian finger-lime (***M. australasica***) is a tall shrub or tree up to 30 or 40 feet tall that occurs near the coasts of Queensland and New South Wales. It is the only member of the *Citrus* relationship that has the long, slender, yellowish-green fruits responsible for its colloquial name. These, 2½ to 4 inches long by up to 1 inch in diameter, contain an acid juice. Juvenile foliage is slender and linear and, because of this and their strongly horizontal branching, young plants present much the appearance of small fir trees or araucarias and are decidedly ornamental. Mature leaves are ovate to lozenge-shaped, about

1 inch long by approximately one-half as broad, and inconspicuously toothed. The spines of this species are solitary. The flowers, with petals 1/3 inch long, are usually solitary, rarely in pairs, from the leaf axils. Many are functionally male, some bisexual. The fruits are acid-lemon-flavored. Variety *M. a. sanguinea* has fruits with red pulp.

The remarkable 'Sydney hybrid' has as parents *M. australasica* and globular-fruited *M. australis*. It has vast numbers of twigs, and leaves scarcely larger than those of the finger-lime, but more conspicuously toothed and more pubescent. The ellipsoid to obovoid, yellowish-green fruits, up to 2 inches long by a little over one-half as wide, are small, pungent-aromatic, and acid.

Garden and Landscape Uses and Cultivation. The Australian finger-lime is an interesting ornamental, hardier than the lemon. It and the hybrid thrive under conditions suitable for oranges.

MICROCLIMATES. See Climates and Microclimates.

MICROCOELUM (Microcoè-lum). Two Brazilian species of palm of the palm family PALMAE, previously included in *Syagrus*, constitute *Microcoelum*, the name of which is derived from the Greek *mikros*, small, and *koilos*, a hollow. It refers to the hollow interiors of the seeds.

Popular in cultivation, as an indoor pot plant and for outdoor landscaping in the tropics and frost-free subtropics, ***M. weddellianum*** (syns. *Cocos weddelliana, Syagrus weddelliana*) has a slender trunk, up to 7 feet in height, thickly covered with the remains of old leafstalks that form a net of black fibers. Its gracefully arching pinnate leaves are 1 foot to 4 feet long and have about fifty slender arching leaflets on each side of the midrib. They are dark green above, somewhat glaucous beneath.

Garden and Landscape Uses and Cultivation. These palms are raised from seeds sown in sandy peaty soil in a temperature of 75 to 90°F. When grown in greenhouses they need a humid atmosphere, shade from strong sun, a coarse, fertile, porous soil, and well-drained containers. The pots or tubs should be smallish in comparison to the size of the plants. Watering should be done to keep the soil moderately moist, but not excessively so. Well-rooted specimens benefit from being given dilute liquid fertilizer at biweekly intervals. The minimum night temperature in winter should be 65 to 70°F, higher at other seasons. At all times day temperatures should exceed night ones by five to ten degrees. For additional information see Palms.

MICROCYCAS (Micro-cỳcas)—Corcho. The name of this genus of the cycad family CYCADACEAE, derived from that of *Cycas* and the Greek *mikros*, small, is particularly unsuitable because its only species, the corcho (*Microcycas calocoma*), is one of the three or four tallest of its family. The inappropriateness resulted from it being named by a botanist who had available only a few small leaves which seemed to resemble those of *Cycas*. From closely related *Zamia* this genus differs in having the scales of the male cones flat instead of shield-shaped, and those of the females thick and blunt-ended.

The corcho is limited in the wild to a small region in western Cuba. An evergreen, palmlike plant, ***M. calocoma*** has a trunk, rarely branched, that lifts its head of abundant foliage to heights of 10 to 30 feet. The pinnate, pubescent leaves, up to 3 feet long, have about eighty pairs of linear, slender-pointed leaflets. The narrow cones are up to 2 feet long.

Garden and Landscape Uses and Cultivation. These are as for other cycads. For more information see Cycads.

MICROGLOSSA (Micro-glóssa). Natives of Africa and warm parts of Asia, the ten species of *Microglossa* are shrubs and woody climbers of the daisy family COMPOSITAE. They are closely related to *Aster* and *Erigeron*. The name, from the Greek *mikros*, small, and *glossa*, a tongue, refers to the short ray florets.

Microglossas have alternate, undivided, usually toothless leaves and panicles or clusters of small flower heads with bisexual disk florets (the kind that form the centers of daisies) and short, white or bluish female ray florets (the petal-like ones of daisies). The seedlike fruits are achenes.

Probably the only kind cultivated, ***M. albescens***, as a wildling occurs from China to the Himalayas. It is a pretty, summer-blooming shrub or subshrub up to 3 feet in height with smooth-edged or finely-toothed, lanceolate leaves 2 to 5 inches long, felted with grayish hairs on their undersides. The aster-like flower heads, about 1/3 inch wide, are in terminal clusters 3 to 7 inches across. Each has a yellow disk, and about fourteen narrow, pale blue to almost white ray florets. This species is probably not hardy north of Virginia.

Microglossa albescens

Garden and Landscape Uses and Cultivation. Easily satisfied, *M. albescens* does well in sunny places in moderately fertile, well-drained soil. It is useful for fronting shrub borders, for flower borders, and for rock gardens. It is readily multiplied by seed, summer cuttings under mist, or in a propagating bed in a cold frame or greenhouse, and by division in spring or early fall. Pruning consists of cutting the plants back fairly hard in late winter or early spring.

MICROGRAMMA (Micro-grámma). Twenty species of small tropical ferns constitute *Microgramma*, of the polypody family POLYPODIACEAE. Most are natives of the New World, two or possibly more of Africa. The name derives from the Greek *mikros*, small, and *gramma*, a mark or letter.

Microgrammas are closely related to *Pleopeltis* and like it by some authorities are included in *Polypodium*. They have creeping rhizomes and lobeless, toothless fronds (leaves) of firm to leathery texture and usually of two sorts, the fertile (spore-bearing) ones generally narrower than the sterile ones.

A common native of tropical America and Africa, ***M. lycopodioides*** (syn. *Polypodium lycopodioides*) has long, slender, rooting rhizomes and distantly-spaced, short-stalked, linear to lanceolate or elliptic fronds 1 inch to 4 inches long by 1/3 to 1/2 inch wide, the fertile ones longer and narrower than the sterile.

Native to tropical America, including the West Indies, ***M. vaccinifolia*** (syn. *Polypodium vaccinifolium*) has firm, undivided sterile and fertile fronds of markedly different aspect, under 2 inches long, and widely-spaced along creeping, slender rhizomes. The sterile fronds are roundish to elliptic, the spore-producing ones linear to strap-shaped with big clusters of spore capsules in a single row. Also tropical American, ***M. piloselloides*** (syn. *Polypodium piloselloides*) has wide-ranging creeping rhi-

Microgramma piloselloides

zomes closely furnished with glossy, lobeless, ovate to obovate, leaves up to 2 inches long with the veins not visible. Widely distributed in warm parts of the Americas, ***M. ciliata*** (syn. *Polypodium ciliatum*) has elliptic to ovate-elliptic, ½- to 2½-inch-long sterile fronds, fertile ones about as long, but very much narrower. This is a close relative of *M. piloselloides.*

Microgramma ciliata

Garden and Landscape Uses and Cultivation. These are attractive ferns for cultivating in greenhouses in pans, hanging baskets, or attached to slabs of the trunks of tree ferns. Outdoors in the humid tropics they are suitable for shaded rock gardens and similar sites. There they may be grown in the crotches of trees or on rough-barked palms or the trunks of tree ferns. They succeed in well-drained soil largely composed of coarse organic debris that is always moist, but never water-logged, and may be propagated by division and by spores. In greenhouses a winter night temperature of 55 to 60°F suits, with an increase of five to ten degrees by day allowable. Shade from bright sun and a constantly humid atmosphere are desirable. For additional information see Ferns.

MICROLEPIA (Micro-lèpia). Medium- to large-sized ferns compose this genus of forty-five species of the pteris family PTERIDACEAE. The name, derived from the Greek *mikros,* small, and *lepis,* a scale, refers to the little half-cup-shaped coverings of the clusters of spore capsules. Except for one kind indigenous to the tropics of the western and eastern hemispheres, they are confined in the wild to the tropics and subtropics of the Old World, with extensions to Japan and New Zealand. The group is closely similar to *Dennstaedtia,* the distinction lying in the locations of the spore clusters and the shapes of their coverings. In *Microlepia* they are set in from the leaf margins, and are half-cup-shaped.

Microlepias have creeping, hairy rhizomes, and fronds (leaves) one- or more-times-pinnately-divided, usually set fairly closely together along the rhizomes, and with the ultimate segments toothed. The clusters of spore capsules terminate the veins.

The most widely distributed species in the wild, ***M. speluncae*** occurs in many slightly different forms in the tropics of the Americas and the Old World and as far south as New Zealand. A vigorous grower, it has soft, thin-textured fronds 3 to 6 feet long, three- or four-times-pinnately-divided, and with white hairs. In gardens ***M. hirta*** (syn. *M. speluncae hirta*) is often grown as *M. speluncae.* This strong-growing, somewhat variable native of tropical Asia and Polynesia has rich green fronds as big as those of the species. Their stalks are brown-purple, their blades twice- or thrice-pinnately-divided into deeply-lobed and toothed segments.

Commonest in cultivation, ***M. platyphylla,*** a native of warm parts of Asia, but not of, as is sometimes reported, Japan, has fronds, white-hairy when young, hairless and lustrous when old, 3 to 4 feet long, and three-times-pinnately-divided into broad, bluish-green, deeply-lobed segments. They are firmer in texture than those of most kinds. Endemic to Hawaii, ***M. setosa*** (syn. *M. strigosa hirta*) is there called palai. A lacy-leaved fern that inhabits forest margins, it has underground rhizomes and slender-stalked, slightly hairy fronds 3 to 4 feet tall, and thrice-pinnate into ultimate, toothed, four-sided segments. This fern was used in the worship of Laka, goddess of the hula.

Also cultivated are *M. cristata* and *M. pyramidata* (syn. *M. strigosa cristata*). Native of Assam, ***M. cristata*** has 2½-foot-long soft-hairy, twice-pinnate, light green fronds with white-hairy stalks. The fronds of ***M. pyramidata*** are some 2 feet tall, and lustrous deep olive-green, their extremities usually forked and crested. Somewhat similar ***M. strigosa,*** of Japan, Ryukyu Islands, Taiwan, Korea, China, Malaysia, India, and Polynesia, is variable, but less so than some species. It has short, creeping rhizomes, and evergreen fronds with stalks 1 foot to 4½ feet long and hairy at their bases, and spreading, twice-pinnate, broadly-lanceolate to ovate, firm blades 1 foot to 2 feet long and up to over 1 foot wide. They are olive-green above and paler beneath. The ultimate linear-lanceolate divisions are toothed and ¾ inch to 1¼ inches long.

Microlepia strigosa

Garden Uses and Cultivation. These are as for *Davallia* except that because of their size they are less adaptable for growing in hanging baskets. For additional information see Ferns.

MICROMERIA (Micro-mèria). Approximately seventy species of more or less thymelike plants comprise *Micromeria,* of the mint family LABIATAE. The name, from the Greek *mikros,* small, and *meris,* part, alludes to the small blooms. The genus, widely distributed in temperate and warm-temperate areas and especially abundant in dry, rocky, exposed habitats in the Mediterranean region, is closely related to *Satureja,* and some species previously in *Micromeria* now belong there.

Micromerias are often aromatic, herbaceous perennials, subshrubs, or low shrubs with small opposite leaves, and in spikes, whorls (circles of more than two) from the leaf axils, or rarely in little panicles, small to minute blooms with straight or curved, unequally five-toothed calyxes, asymmetrical, purplish to white, straight-tubed, two-lipped corollas, with the upper lip erect and the lower spreading. There are four stamens in two pairs, and one style. The fruits consist of four seedlike nutlets.

Native to the Canary Islands, ***M. ericifolia*** is a perennial with procumbent stems somewhat woody at their bases and ovate to lanceolate leaves, with rolled-back margins. The small, purple, nearly stalkless flowers have calyxes almost as long as the corollas.

An aromatic shrublet 4 inches to 1¼ feet high with many erect mostly branchless stems, ***M. juliana*** (syn. *Satureja juliana*) has stalkless, ovate to linear-lanceolate, pubescent leaves up to ⅓ inch in length and with inrolled margins. The essentially stalkless, reddish-purple flowers, about as long as the leaves, are in tiers that compose very slender, somewhat interrupted, long spikes. This inhabits dry, stony sites in Europe. From the last, very variable ***M. graeca*** (syn. *Satureja graeca*) differs chiefly in having stalked clusters of flowers that spread outward to form much looser, broader spikes. Native of Portugal and southern Europe this is up to 1½ feet tall.

A southeastern European, freely-branched and about 1½ feet tall, ***M. dalmatica*** has erect or ascending, branchless or branched stems, ovate, pubescent leaves ⅓ to ¾ inch long, and white blooms in loose, many-flowered, stalked clusters. Prostrate stems that become erect at their ends are characteristic of ***M. thymifolia*** (syns. *M. rupestris, Satureja thymifolia*), a southern European with hairless or nearly hairless, ovate to oblong leaves, and lavender-spotted, white blooms two to ten together in short-stalked clusters.

Forming tufts of erect stems 3 to 6 inches tall, ***M. croatica*** (syn. *Satureja croatica*), of Yugoslavia, is a densely-pubescent shrublet 4 to 8 inches high, with ovate to round leaves and stalked, solitary, paired, or clusters of three, rose-violet flowers in compact panicles. Also compact, 3- to 6-inch-tall ***M. piperella,*** of southwestern Europe, has ovate leaves sometimes heart-shaped at their bases and hairless or nearly so. Its purplish-red blooms are in clusters of few.

Garden and Landscape Uses and Cultivation. Micromerias, although not especially showy or important horticulturally, are pleasant enough plants for herb gardens, rock gardens, and similar sites. The majority are not reliably hardy in the north. All need full sun. Poorish, well-drained soil suits best. Too rich a diet results in too lush growth. Propagation is easy by seed, by cuttings, and often by division.

MICROSERIS (Micro-sèris). Plants previously known as *Uropappus* belong here. The genus *Microseris,* of the daisy family COMPOSITAE, has an unusual natural distribution. Of its thirty or somewhat fewer species most are natives of western North America, but one occurs in Chile, and one in Australia and New Zealand. The name, from the Greek *mikros,* small, and the Latin *seris,* chicory, alludes to similarities of the flower heads.

Microserises are milky-juiced annuals, biennials, or herbaceous perennials, with taproots or rhizomes. Generally their foliage is in basal rosettes, but sometimes on the stems. The undivided leaves may be pinnately-lobed or toothed. Borne singly at the ends of long, leafless stalks or branches, the flower heads are of all strap-shaped florets, yellow or white, often striped with red. The fruits or achenes are seedlike.

An annual favoring dry soils and open places from Idaho to Washington, Arizona, and Baja California, ***M. linearifolia*** (syn. *Uropappus linearifolius*) is hairless or sometimes long-hairy and 6 inches to 2 feet tall. Usually it branches low down. Its linear, sometimes lobed leaves are 6 to 8 inches long. The flower heads are solitary and up to 1¾ inches wide.

Garden Uses and Cultivation. This little known annual is useful for adding variety to flower gardens and, within its natural range, for including in collections of native plants. Its culture is of the simplest. It is best adapted to sunny locations and dryish, moderately fertile earth, but it grows satisfactorily in any fairly good garden soil. Seeds are sown in early spring outdoors where the plants are to bloom and are raked into the soil surface. The young plants are thinned so that they are spaced about 6 inches apart. No special care is needed beyond keeping down weeds and removing spent flower heads. This plant produces blooms in succession from early summer until frost.

MICROSORIUM (Micro-sòrium). This genus of the polypody family POLYPODIACEAE is one of a considerable number of segregates from *Polypodium* recognized by many botanists, by others treated as subgenera of *Polypodium.* It consists of about sixty species of the tropics of the Old World and includes kinds that some botanists further segregate as *Phymatodes.* The name comes from the Greek *mikros,* small, and *soros,* a heap, in allusion to the clusters of spore capsules.

Microsoriums are typically tree-perching (epiphytic) ferns of moderate to large size with creeping rhizomes furnished with usually broad scales. The fronds (leaves) are lobeless, pinnately-lobed, or rarely pinnate. Mostly they are hairless, sometimes more or less pubescent. The clusters of spore capsules are usually round.

Its rhizomes clothed with dark brown scales, ***Microsorium punctatum*** (syn. *Polypodium polycarpon*) has hairless, leathery fronds, mostly smaller than their maximums of 3 feet long by 3½ inches wide.

Microsorium punctatum cristatum

Approximately strap-shaped and broadest above their middles, they have pointed apexes, and winged bases. Only the main veins are prominent. This sort inhabits Hawaii and other Pacific islands. The fronds of *M. punctatum cristatum* fan out and are much frilled at their apexes. Native to Taiwan and Indochina, ***M. steerei*** is closely allied to *M. punctatum.* It has practically stalkless, thick, fleshy, strap-shaped to broadly-oblanceolate fronds, with spore cases in numerous, minute, circular clusters.

Microsorium steerei

Native to the same regions as the last, ***M. scolopendria*** (syns. *Polypodium scolopendrium, Phymatodes scolopendria*) has long slender rhizomes and thick, leathery, broad, flat sea-green to dark green fronds up to 3 feet long that are lobeless or deeply-pinnately-cleft in much the manner of the fronds of the hare's foot fern (*Phlebodium aureum*) into several to many bluntish to pointed lobes up to 8 inches long by ¾ inch to 2 inches wide, and with sunken, but not clearly evident veins. The clusters of spore capsules are in one or two rows paralleling the mid-veins of the leaf lobes.

Microsorium scolopendria

Except for the considerable diversity of its leaves, which vary from smooth-margined and about 9 inches long by 2 inches wide to irregularly- or deeply-pinnately-lobed and up to 1½ feet long by 9 inches broad, ***M. diversifolium*** (syn. *Polypodium diversifolium*), of Australia, Tasmania, and New Zealand, much resembles *M. nigrescens.* Its clusters of spore capsules are in single rows. From it, ***M. nigrescens*** (syn. *Polypodium nigrescens*), of the East Indies, Ceylon, and southern India, differs in having darker green, wavy-edged fronds with veins clearly evident. Its clusters of spore capsules are in single rows at the sides of the veins.

Native to southern China and Taiwan, ***M. fortunei*** has creeping rhizomes and short-stalked, spreading to erect, linear-lanceolate fronds 1 foot to 1½ feet long by

Microsorium fortunei

¾ inch to 2 inches wide. Except for a prominent midrib the veins are obscure. The orange-yellow clusters of spore capsules are in single rows, one row or sometimes two on each side of the midrib.

Other sorts cultivated include these: ***M. musifolium*** (syn. *Polypodium musifolium*), of the East Indies, has stout rhizomes and pointed, stalkless, lobeless fronds mostly 2 to 3 feet long by 3 to 4 inches wide. Its very tiny clusters of spore capsules are distributed over nearly the entire under surfaces of the fronds. ***M. normale,*** of southern China, the Malay Peninsula, and India, has rhizomes with black scales and narrowly-ovate, lobeless fronds up to 3 feet long by about 2 inches wide without prominent veins. Large clusters of spore capsules are near the midrib. ***M. pteropus*** (syn. *Polypodium pteropus*) is a semiaquatic that inhabits stream banks and similar places where in its native southern China, Malay Archipelago, and India during the wet season it is beneath water for considerable periods. This has fleshy rhizomes

Microsorium pteropus

and lobeless to deeply-three-lobed fronds up to 1 foot or more in length. ***M. pustulatum*** (syn. *Polypodium pustulatum*), of Australia and New Zealand, has thin, pointed fronds, smooth-margined to deeply-pinnately-lobed, the former up to 9 inches long and under ¾ inch wide, others up to 1½ feet long and 6 inches wide.

Garden and Landscape Uses and Cultivation. These are in general as for the hare's foot fern (*Phlebodium aureum*) and *Drynaria,* but *M. pteropus* is sometimes grown as a submerged aquatic in aquariums. For more information see Ferns.

MICROSTYLIS. See Malaxis.

MIDGES. Tiny two-winged flies called midges include sorts familiarly known as no-see-ums and punkies. The maggots of some sorts damage plants, some causing galls or blisters and some by tunneling as leaf miners through the tissues. Among the most troublesome are the apple leaf-curling midge, catalpa midge, cattleya midge, chrysanthemum gall midge, dogwood club gall midge, honey-locust pod gall midge, rose midge, and spruce gall midge. For controls, which differ with different kinds of midges, consult a Cooperative Extension Agent, State Agricultural Experiment Stations, or other authorities.

MIDRIB. This is the rib or vein that forms the central axis of a leaf or sometimes of a bract, sepal, or petal.

MIGNONETTE. See Reseda.

MIGNONETTE TREE is *Lawsonia inermis.*

MIGNONETTE VINE is *Anredera cordifolia.*

MIKANIA (Mikàn-ia)—Climbing Hempweed. To the daisy family COMPOSITAE belongs *Mikania,* a genus of about 250 species of vines, shrubs, and perennial herbaceous plants, all natives of the Americas. The name commemorates J. G. Mikan, professor at Prague, who died in 1814, or his son J. C. Mikan, a botanical collector of plants in Brazil, who died in 1844.

Mikanias are mostly tropical or subtropical. They have opposite, usually stalked leaves and many small flower heads, each composed of all disk florets (tiny flowers of the kind that form the centers of daisy flower heads). There are no ray florets (those that in daisies look like petals). The fruits are five-angled, seedlike achenes.

Climbing hempweed (***M. scandens***) is a twining, deciduous vine with stems up to 15 feet long. Native from Maine to Michigan, Missouri, Florida, and Texas, and widely distributed in many tropical and subtropical places, it has stalked, pointed-heart-shaped, toothed or toothless leaves from 1 inch to 5 inches long and up to 3½

Mikania scandens

inches wide. Its many tiny heads of white or pinkish florets are in roundish clusters at the ends of long stalks that come from the leaf axils.

Much more attractive, ***M. ternata*** (syn. *M. apiifolia*) is a variable native of Brazil.

Mikania ternata

An evergreen climber, it has purple-hairy stems and softly-hairy leaves, with five to seven leaflets arranged palmately (with their bases spreading from a common point at the top of the leafstalk). The leaflets are blackish-green above, dull purple on their undersides. The central one may be three-lobed, deeply-pinnately-lobed, or coarsely-toothed. The small yellowish flower heads are in broad, loose clusters. Less well known ***M. hemispherica,*** a native of Brazil, is a vine with toothed, satiny-green, arrow-shaped leaves that have silvery veins and are purplish on their under surfaces. The flower heads are flesh-pink.

Garden and Landscape Uses. Climbing hempweed is occasionally cultivated in gardens of native plants and in more or less wild landscapes for covering walls, trellis and other supports, but has little to recommend it for wider horticultural uses. The other kinds discussed are hardy only in warm, frost-free climates. As a greenhouse and room plant *M. ternata* is highly satisfactory.

Cultivation. Climbing hempweed grows without difficulty in ordinary garden soil and is easily raised from seed. Readily increased by cuttings, *M. ternata* grows vigorously in fertile, well-drained soil kept evenly moist, in sunny or lightly shaded locations. It needs a minimum temperature of 55 to 60°F, a moderately humid atmosphere, and a little shade from strong summer sun. Regular applications of dilute liquid fertilizer to well-rooted specimens promote good growth. Pruning to shape and repotting is done in spring. The other Brazilian species described above needs the same growing conditions as *M. ternata*.

MILA (Mì-la). The name *Mila,* an anagram of that of the Peruvian city of Lima, applies to a genus of small, clump-forming, Peruvian spiny cactuses of the cactus family CACTACEAE. About a dozen species are recognized, but few are cultivated.

These plants branch freely from their bases and have cylindrical, often more or less prostrate stems with shallow depressions between the ribs and needle-like spines. Their growth habit is similar to that of *Echinocereus.* The bell-shaped blooms, small and at the tops of the stems, are yellow and have short, scaly perianth tubes and scaly ovaries. There are a few long hairs in the axils of the scales. The shiny, spherical, small green fruits contain black seeds.

The sorts here discussed are about 6 inches tall. The stems of ***M. caespitosa,*** from slightly under to a little over 1 inch in diameter, have about ten ribs and very closely set areoles (regions of the stem from which spines develop). Tipped with brown, the whitish to yellowish spines are up to twenty from each areole, the central one exceeding 1 inch in length, the others about one-third as long. The blooms are almost ¾ inch long. Semiprostrate and with about eleven slightly lumpy ribs on the stems, ***M. kubeana*** has two to four central spines and nine to twelve radial ones from each areole. The former, up to ¾ inch long, are comparatively stout and yellowish with dark tips, the latter are white, bristle-like, and under ½ inch long. With stouter stems, up to 2 inches in diameter, and having eleven to thirteen ribs, ***M. nealeana*** develops three or four yellow central spines, about twelve glassy white radials and yellowish wool from each areole. Its blooms are 1 inch in diameter.

Garden and Landscape Uses and Cultivation. These are the same as for the majority of other small desert cactuses and are discussed under Cactuses.

MILDEW. See Downy Mildew, and Powdery Mildew.

MILFOIL is *Achillea millefolium.* For water-milfoil see Myriophyllum.

MILIUM (Mì-lium)—Wood Millet. Half a dozen species of the grass family GRAMINEAE constitute *Milium,* the name of which derives from the ancient Latin one for millet.

Native to temperate parts of North America, Europe, and Asia, miliums include annuals and perennials with slender to thick stems, leaves with flat blades, and flowers in loose or compact panicles.

A bright yellow-leaved variety of wood millet (***M. effusum***), identified as *M. e. aureum,* is decidedly ornamental. Perennial and 1 foot to 2 feet tall, this forms thick clumps of arching leaves. The flower panicles are yellow.

Milium effusum aureum

Garden Uses and Cultivation. The sort described is attractive for beds and the fronts of borders. It thrives in ordinary soil in sun or part-day shade. It is easily increased by division and seed.

MILK. This word forms part of the vernacular names of these plants: African milk bush (*Synadenium grantii*), milk bush (*Euphorbia tirucalli*), milk maids (*Dentaria*), milk-thistle (*Silybum marianum*), milk tree (*Brosimum galactodendron*), and milk-vetch (*Astragalus coccineus*).

MILK-AND-WINE-LILY is *Crinum latifolium zeylanicum.*

MILKWEED. See Asclepias. Giant-milkweed is *Calotropis gigantea.*

MILKWORT. See Polygala.

MILKY DISEASE. This is a bacterial infection that kills Japanese beetles. Powders containing its spores, prepared under authority of the United States Government, are sold for Japanese beetle control. Apply them as directed by the manufacturers to lawns infested with grubs of the beetles. The bacteria live in the ground indefinitely, but it usually takes a few years for their numbers to increase sufficiently to markedly reduce the population of beetle grubs.

MILLA (Míl-la)—Mexican Star. Belonging to the lily family LILIACEAE, the genus *Milla* consists of six species of bulb plants. It is represented in native floras from the southern United States to Central America. The name commemorates Juliani Milla, gardener in the eighteenth century to the Spanish Court in Madrid. The plant sometimes called *M. uniflora* is *Ipheion uniflora.*

The bulblike organs of millas are technically corms. From them come two to seven linear, parallel-veined leaves and one or rarely two leafless flower stalks each ending in an umbel of up-facing blooms with at the base of the umbel four spreading, papery bracts. The flowers are white with longitudinal green stripes, pink, white flushed with pink, or blue. They have perianths with tubes one-and-a-half to six times as long as their lobes (petals, or more properly, tepals), six stamens, and one style, reaching to or beyond the throat of the perianth tube, tipped with a three-lobed stigma. The fruits are three-compartmented many-seeded capsules.

Mexican star (***M. biflora***), native chiefly in volcanic soils in mountains from Arizona and New Mexico to Guatemala, has narrowly-linear to tubular, glaucous leaves. Its fragrant, white, starry flowers are in umbels of one to six or rarely as many as eight, terminating slender, erect stems 6 inches to 1½ feet tall. The individual blooms are stalkless, but have fleshy perianths with stalklike tubes 2 to 6 inches long or slightly longer, expanded for a short distance below the petals, and up to 2 inches or a little more across the face of the bloom. The oblong-lanceolate, spreading petals are white with a green keel down their back sides. The fruits are elongated.

Garden and Landscape Uses. Mexican star is not hardy in the north. It is useful for late winter and spring display in greenhouses, and in climates where its bulbs are not subjected to freezing, it may be planted permanently outdoors in rock gardens and flower beds. Where winters are too cold for the bulbs to persist outdoors, they may be lifted each fall and stored over winter in a temperature of 35 to 40°F.

Cultivation. Fertile, well-drained soil and a sunny location are needed. For greenhouse display the bulbs are planted several together with a distance equal to from one-half to the full width of a bulb between individuals. They are watered sparingly at first, more freely as growth develops. Throughout the winter a night temperature of about 50°F is satisfactory. By day this is increased by five to ten degrees. Well-rooted specimens are given dilute liquid fertilizer at about weekly intervals. After flowering, watering is continued as long as the foliage remains green. When it begins to die back, watering is reduced and eventually discontinued, and the soil kept dry throughout the dormant period. Repotting is done in early fall. Propagation is by seed, offsets, and bulb cuttings.

MILLET. Several grasses known as millets are identified adjectivally as African-millet (*Eleusine coracana* and *Pennisetum americanum*), broom-corn-millet (*Panicum miliaceum*), foxtail-millet (*Setaria*), Indian- or pearl-millet (*Pennisetum americanum*), and wood millet (*Milium effusum*).

MILLETTIA (Millét-tia). From *Wisteria,* to which it is closely related, *Millettia* differs in its tough pods not splitting readily to release their seeds and in its racemes of flowers being usually branched. Native chiefly of warm parts of the Old World, it consists of 180 species of mostly evergreen trees, shrubs, and twining vines that belong to the pea family LEGUMINOSAE. The name commemorates the nineteenth-century French botanist Charles Millett.

Millettias have pinnate leaves with an unequal number of leaflets and pea-like flowers in racemes from the leaf axils or in terminal panicles. The flowers, greenish-white, pink, purple, or rarely blue, have a four-lobed calyx, nine united and one free stamen, and one style. The fruits are thick, woody or very leathery, flat pods.

A woody vine, native to China and Taiwan, ***M. reticulata*** (syn. *Wisteria reticulata*) is hardy perhaps as far north as Philadelphia. Its hairless, somewhat leathery, evergreen leaves have usually three pairs of lateral leaflets and a terminal one. The leaflets, elliptic-lanceolate to lanceolate, are 2 to 4 inches long and ¾ inch to 2 inches broad. The flowers, pinkish-lavender to purplish-red, fragrant, and about ½ inch long, are in large panicles composed of racemes up to 8 inches long. The brown seed pods are 4 to 6 inches long. This is a good vine for the South and California. Another vine, ***M. nitida,*** of China, has rusty-hairy branchlets and young leafstalks. Its leaves are of usually five ovate to elliptic-oblong to broad-elliptic leaflets 2 to 4½ inches long. In crowded panicles with rusty-hairy stalks the purple flowers have a 1-inch-long, silky-hairy standard or banner petal. Also rusty-hairy, the seed pods are about 4 inches long.

A beautiful, medium-sized, semideciduous or deciduous tree of tidy habit, ***M. ovalifolia,*** of Burma, has proved satisfactory in Florida, where it blooms in spring about the time the new leaves appear. This has slightly drooping branches and pretty foliage and flowers. Up to 1½ feet long, its leaves are hairless and usually of seven broad leaflets 2 to 3 inches long. The ¼-inch-long, lilac and bright mauve blooms, in drooping racemes from the leaf axils, are succeeded by hairless, knobby-surfaced, two- or three-seeded pods 2 to 3 inches long. Also Burmese, ***M. pendula*** is a tree with silky-hairy branchlets and leaves about 6 inches long with seven ovate-oblong leaflets thickly clothed with gray hairs on their undersides. The flowers, their calyxes longer than the petals and the standard or banner petal not hairy, are crowded in short racemes. The knobby-surfaced fruits are 3 to 5 inches long.

Several tree or shrub species native to Africa are cultivated. They are likely to be evergreen or deciduous depending upon climate. South African ***M. caffra,*** up to 30 feet high, has leaves with eleven or thirteen lanceolate-oblong leaflets up to 2½ inches long, slightly silky-hairy on their undersides. Up to 8 inches long, the panicles are of flowers silky-hairy on the backs of their standard petals. The velvety, brown seed pods contain two seeds. Fast-growing, free-blooming, fragrant-flowered ***M. dura,*** of East Africa, is up to 25 or sometimes even 40 feet tall. Its leaves, rusty-hairy when young, have thirteen to nineteen ovate to oblong leaflets, the laterals up to 1 inch, the terminal up to 3 inches, long. In drooping, branched racemes 4 to 8 inches long, the predominantly pink flowers are bluish toward the petal margins and yellow at the bottom of the ¾-to 1-inch-long standard or banner petal. The hairless seed pods are up to 8 inches long. Also East African, ***M. stuhlmannii*** sometimes exceeds 15 feet in height, but often is lower. It has leaves of seven or nine elliptic to somewhat reverse-heart-shaped leaflets 2 to 5 inches long by 1½ to 3½ inches broad, slightly hairy on their undersides. Its light violet or lavender flowers are 1 inch long. One- to four-seeded, the seed pods are 7 to 10 inches long.

Garden and Landscape Uses and Cultivation. In appropriate climates, millettias are, according to species, beautiful vines or ornamental flowering trees. The vines need supports around which their stems can twine and the same care as wisterias. The trees are admirable as lawn specimens and for other locations where they are seen to advantage. All are satisfied with well-drained, not overly fertile soil and need full sun. Propagation is by seed, perhaps also by cuttings.

MILLIPEDES. Mostly scavengers living on decaying or decayed organic material, but seldom initiating decay, millipedes sometimes damage young living roots and seedlings. Called "thousand leggers" (they have two pairs of legs on each of their many body segments) these elongated, cylindrical, hard-shelled insects are 1 inch to 2 inches long or sometimes longer; when at rest they are coiled like springs. They are sometimes confused with the garden centipede (not a true insect and with only one pair of feet to each body segment) and with wireworms (which are flatter than cylindrical, do not coil, and have only six legs). Control consists of sanitary measures to eliminate dark, damp hiding places, and spraying with malathion or Sevin.

MILPASIA. This is the name of bigeneric orchids the parents of which are *Miltonia* and *Aspasia.*

Milpasia 'Gold Star'

MILTASSIA. This is the name of bigeneric orchids the parents of which are *Miltonia* and *Brassia.*

Miltassia 'Flower Drum'

MILTONIA (Mil-tònia) — Pansy Orchid. Some twenty-five species of the orchid family ORCHIDACEAE are accounted for in *Miltonia,* a genus native from southern Mexico to South America where it occurs from near sea level to high in the Andes. In addition to species, there are some natural and many man-made hybrids. The name honors Viscount Milton, a patron of horticulture, who died in 1851.

Botanically most closely related to brassias, odontoglossums, and oncidiums, miltonias are tree-perching (epiphytic) or occasionally rock-perching orchids usually with conspicuous pseudobulbs each with one, two, or very rarely three evergreen leaves from their apexes and others from their bases. The flowers, solitary or in racemes from the bases of the pseudobulbs, have nearly similar spreading sepals and petals, the latter somewhat broader, and a large, showy, lobeless lip sometimes notched at its apex, but usually not narrowed at its base into a shaft or claw. The blooms of many sorts have a striking superficial resemblance to pansies, hence the vernacular designation, pansy orchids.

The flowers of some miltonias superficially resemble pansies

Kinds previously named *M. laevis* and *M. reichenheimii* belong in *Odontoglossum*.

Hybrid miltonias: (a) 'Gaiety'

(b) 'Princess Elizabeth'

Pansy orchids that need or tolerate fairly high temperatures are generally more adaptable to cultivation than cool-temperature miltonias that are natives of high altitudes and resent torrid summers. Warm-temperature kinds, all unless otherwise mentioned believed to be restricted in the wild to Brazil include these: ***M. anceps*** has crowded, compressed pseudobulbs 2 to 3 inches long and with a pair of oblongish-linear, apical leaves up to about 6 inches in length. Its flattened, bracted flowering stalks bear solitary blooms 2 to 2½ inches in diameter that have yellow-green sepals and petals and a rather fiddle-shaped white lip striped and spotted with purple. ***M. bluntii,*** with pseudobulbs and leaves resembling those of *M. spectabilis,* is a natural hybrid between that species and *M. clowesii.* Its fragrant blooms, solitary or few on the flowering stalks, are about 3 inches in diameter. Their sepals and petals are pale yellow somewhat blotched with reddish-brown. The lip is white with a purplish-crimson base. Variety *M. b. lubbersiana* is distinguished by its flowers being bigger, up to 4 inches across, and by having pale yellow sepals and petals extensively decorated with bands and blotches of purplish-brown. The purple-based lip, paler toward its apex, is marked with red-brown lines. ***M. candida*** differs from the last in its apical-two-leaved, 3- to 4-inch-long pseudobulbs being appreciably spaced along rhizomes rather than in close clusters. Its pointed, linear-oblanceolate leaves are 9 inches to 1½ feet in length. The generally two- to eight-bloomed, stout flowering stalks attain lengths up to 1½ feet or more. The waxy blooms are fragrant, a little less or more than 3 inches in diameter. Their chestnut-brown sepals and petals, the latter especially, spotted and tipped with bright yellow, are approximately of equal size. The petals have usually more or less wavy edges as does the obovate to nearly circular lip. The latter is white with pale or less commonly dark violet-purple to purple-brown spots at its base. Its lower part forms a tube around the column. Variety *M. c. grandiflora* has blooms twice as big as those of the typical species, their sepals and petals except for their yellow tips, brown, and the lip white. ***M. clowesii*** has tightly clustered, ovate-oblong pseudobulbs each with a pair of linear-strap-shaped leaves 1 foot to 1½ feet long at their summits. The flowering stalks, 1 foot to 2 feet in length, carry three to ten blooms 2 to 3 inches wide across; the nearly equal sepals and petals are yellow with conspicuous cross-bands of reddish-brown. The somewhat fiddle-shaped, long-pointed lip is rich purple below, white in its outer half. ***M. cuneata*** is a handsome species. It has strong creeping rhizomes and ovoid-oblong, flattened pseudobulbs 3 to 4 inches long and with at their apexes a pair of pointed, narrowly-lanceolate leaves up to 1¼ feet long. The 1½- to 2-foot-tall flowering stalks carry up to eight blooms 2½ to 3 inches in diameter. They have chestnut-brown sepals and petals tipped and sometimes near their bases lightly streaked with pale yellow. The lip, notable for narrowing markedly below to form a distinct claw or shaft, is squarish, has wavy margins, and is white with occasionally a few pink spots near its base. ***M. flavescens,*** of Paraguay and Brazil, has pseudobulbs spaced along rhizomes rather than tightly clustered.

Miltonia flavescens

Conspicuously flattened, they are 3 to 5 inches long and each has at its summit two strap-shaped leaves up to 1 foot long and ¾ to 1 inch wide. The seven- to ten-flowered racemes are up to 1½ feet in length. The fragrant, starry blooms, 2½ to 3 inches or more across, have pointed, light yellow sepals and petals and a shorter, ovate-oblong, wavy-margined, white lip with its lower one-half decorated with a few radiating reddish-purple lines. The sepals, petals, and lip of the blooms of *M. f. stellata* are white. ***M. regnellii*** has compressed, ovoid pseudobulbs 3 to 4 inches tall crowned with two narrowly-strap-shaped leaves about 1 foot in length. From 2 to 3 inches wide, the flowers are seven or fewer on slender stalks up to 2 feet long. They have white sepals and petals sometimes with a suggestion of pink at their bases and a slightly three-lobed, reverse-heart-shaped, white-margined, pink lip streaked with purple-pink and with yellow lines radiating from its crest. The blooms of *M. r. purpurea* have white-edged, light rose-purple sepals and petals and a dark-veined rich magenta lip with a white- and yellow-marked crest. ***M. roezlii,*** of Colombia, has flattened pseudobulbs with one apical leaf and flowering stalks carrying up to half a dozen fragrant, 4-inch-wide flowers with white sepals, white petals with a wine-purple eye, and a yellow-eyed, white lip. Variety *M. r. alba* has flowers entirely white except for some yellow at the bottom of the lip. ***M. russelliana*** has flattened-ovoid-oblong pseudobulbs 2 to 3 inches tall, each with two narrowly-lanceolate apical leaves 6 to 9 inches long. The flowering stalks, up to 2 feet long, may carry as many as nine 2½-inch-wide blooms that have reddish-brown sepals and petals tipped with light yellow and a mainly rosy-lavender lip with its outer one-third white or pale yellow. ***M. spectabilis*** has strong, creeping rhizomes with oblongish-ovoid, flattened pseudobulbs 3 to 4 inches in length at intervals along them. Each pseudobulb has at its top

Miltonia spectabilis

a pair of narrowly-strap-shaped leaves up to 7 inches long. The very flat blooms, some 3 inches across, are solitary on about 8-inch stalks. Somewhat variable, they most usually have sepals and petals that are creamy-white or white, possibly stained with pink toward their bases, and a big, spreading, roundish-ovate, pink- or white-margined, wine-purple lip decorated with six or more darker veins running lengthwise and a usually yellow crest. Its margins are wavy or crisped. Variety *M. s. moreliana* has blooms up to 4 inches in diameter that have plum-purple sepals and petals and a bright rose-purple lip with darker veining. Native to Equador, Colombia, and Peru, ***M. warscewiczii*** has clusters of much-flattened pseudobulbs 3 to 5 inches in length. Each

Miltonia warscewiczii

has at its apex one linear-oblong leaf 5 to 7 inches long. The numerous fragrant, waxy blooms in crowded, nodding, branched racemes that attain lengths up to 2 feet are 1½ to 2 inches across. They have nearly equal, narrow, much-waved sepals and petals somewhat variable in color, but generally are reddish-brown paling to yellow at their tips. The lip, notched at its apex and with its edges recurved, has usually broad, white margins and a rose-purple disk with a light-yellow-brown blotch above its middle. This species is sometimes cultivated as *Oneidium warscewiczii* and *O. weltonii*.

Cool-temperature sorts include *M. endresii, M. phalaenopsis,* and *M. vexillaria.* A native of Costa Rica and Panama, ***M. endresii*** has pseudobulbs up to 2 inches long, with a single linear-lanceolate apical leaf up to 1 foot long. Erect or arching, the up to 1-foot-long flowering stalks carry five or fewer 2½- to 3-inch wide, flat blooms with creamy-white sepals and petals, rose-purple at their bases, and a fiddle-shape lip similarly colored and with a yellow crest. A native of Colombia, ***M. phalaenopsis*** has ovoid pseudobulbs some 1½ inches long, topped by a pair of linear leaves up to 9 inches long or rather longer. Shorter than the leaves, the flowering stalks have five or fewer 2- to 2½-inch-wide blooms with pure white sepals and petals and a white four-lobed lip with a purple-crimson center and a yellow base. The flowers of *M. p. alba,* except for yellow markings at the base of the lip, are white. Very variable ***M. vexillaria,*** of Colombia, has compressed, not very evident, elliptic-oblong psuedobulbs 2 to 2½ inches long and with up to eight two-ranked, bluish-green, linear-lanceolate leaves. Mostly two

Miltonia vexillaria

from each pseudobulb, the arching flowering stalks 1 foot to 2 feet long carry four to ten pansy-like flowers that vary as to size and color on different plants. The best blooms may be 4 inches across and are usually white flushed with pink, rose-pink, carmine-pink, or lavender. They have a nearly circular, flat lip, deeply notched at its apex, and with a yellow crest. The flowers of *M. v. alba* are pure white, sometimes lightly blushed pink at the bottoms of the sepals and petals. Smaller flowers with white-margined, rose-pink sepals and petals are typical of *M. v. rubella*. The blooms of *M. v. rubra* are deep rose-pink with darker veins and three crimson lines radiating from the yellow disk at the base of the lip.

Garden Uses and Cultivation. Among the showiest of orchids, miltonias are prized for inclusion in collections. Those from low and relatively low altitudes, chiefly of Brazilian origin, ordinarily present little difficulty, but the few sorts that are native high in the Andes, unless an air-conditioned greenhouse is available, are less easy to maintain where summers are hot.

All are evergreens without a decided season of rest and so must be watered throughout the year, but less abundantly in winter than at other times. Excessive wetness must be avoided. These plants need a humid but not stagnant atmosphere and temperatures as cool as practicable in summer and at other times, for warm-temperature sorts, about 55°F at night increased five to fifteen degrees by day; for cool-temperature kinds night and day temperatures should be five to ten degrees lower. To flower satisfactorily, as intense light as they will endure without the foliage scorching or bleaching is needed, which means that shade is necessary from spring to fall.

Miltonias are impatient of stagnant or soured conditions about their roots and respond to repotting every year or two. As a potting mix use osmunda or tree fern fiber and chopped sphagnum moss with some crushed charcoal. Many sorts, because of their creeping rhizomes, are better accommodated in pans (shallow pots) or in baskets than standard flower pots. For more information see Orchids.

MILTONIDIUM. This is the name of bigeneric orchids the parents of which are *Miltonia* and *Oncidium*.

Miltonidium 'Aristocrat'

MIMETES (Mi-mètes). Belonging to the protea family PROTEACEAE, the South African genus *Mimetes* consists of sixteen species. Its name derives from the Greek *mimetes*, a mimic.

Evergreen shrubs or small trees, the sorts of this genus have undivided, stalkless, leathery leaves sometimes with three or five short teeth at the apex. The leaves that accompany the flower heads are in some species brightly colored. The flowers are bisexual and are in stalkless, crowded heads of four to twelve in the axils of the upper leaves. Each flower has four perianth segments each with a stalkless anther, and one protruding style. The fruits are thick-walled achenes.

Beautiful ***M. cucullata*** (syn. *M. lyrigera*), in its homeland considered the easiest sort to grow, as a wildling inhabits moist, peaty soil. From 1½ to 4 feet tall, and branching from the base, it has erect branches and stalkless, oblong-oblanceolate leaves, slightly toothed at their tips and 1¼ to 2 inches long. The four- to six-flowered heads of bloom, including the bright red styles, are about 2¼ inches long. The leaves from the axils of which they sprout are bright red toward their apexes, have yellow bases. They are responsible for the showy, flame-like tassels of bloom that are the inflorescences. A rare native in peaty marshes, ***M. hirta*** is an erect-branched shrub 3 to 4½ feet tall or sometimes taller with narrow-ovate-elliptic to oblanceolate, all-green leaves up to 2 inches long. The usually about twelve-flowered, yellow and red heads of bloom are, including the bright red styles, about 2¼ inches long.

Mimetes cucullata

Mimetes hirta

Garden Uses and Cultivation. Perhaps not presently cultivated in North America, these choice plants afford real challenges to the collector and cultivator of the rare and unusual. They may respond to environments and care that satisfy *Leucospermum* and *Protea.*

MIMOSA (Mimòs-a)—Sensitive Plant. The name mimosa has various currency values among gardeners. It is the common name of certain kinds of *Acacia* (yellow-flowered trees and shrubs) and of related *Albizia julibrissin,* a pink-flowered tree commonly planted in the south. The botanists' genus *Mimosa,* with which we are concerned here, is entirely different, although it belongs in the same pea family, the LEGUMINOSAE. Its name comes from the Greek *mimos,* a mimic. It alludes to the sensitivity to stimulation of the leaves of some kinds.

This genus consists of 400 to 500 species of trees, shrubs, and herbaceous plants, mostly natives of tropical and subtropical America, but represented also in warm parts of Africa and Asia. They have alternate, twice-divided leaves, with the leaflets of the ultimate divisions arranged pinnately. The small, more or less tubular flowers, in fuzzy, globular clusters, have protruding stamens. The fruits are flat pods each containing several seeds. The genus is remarkable because of the irritability of the leaves of some species, notably of those of the common sensitive plant. This curious and extremely interesting sort is easy to cultivate and never ceases to astonish those unfamiliar with its responses. The only other species at all common is *M. polycarpa.* A few others are rarely grown.

The common sensitive plant (***M. pudica***), a native of Brazil, has become naturalized in the warmer parts of the south. It is a prickly, subshrubby perennial usually cultivated as an annual. Only a few inches high, it has slender stems and branches and pink or lavender-pink flowers. Its fernlike leaves are very sensitive, especially when the temperature is optimal. Below 60 or above 104°F no movement takes place. The response is most rapid between 75 and 85°F. It results from any mechanical stimulus, such as the touching of its leaflets or of their stalks, a drop of water falling on the leaves, or the foliage being disturbed by wind or by air currents. They also respond to the application of heat locally and if the flame of a lighted match is held near them, the leaflets and leaves collapse. The sensitivity manifests itself first at the point of irritation, at the spot touched by the water drop, finger, or other object, or closest to the flame. The

Mimosa pudica (flowers)

Mimosa pudica: (a) Touching the leaves

(b) The leaves collapsing

(c) The leaves completely collapsed

leaflets fold together quickly and in rapid succession, then the main leafstalk droops, and if the stimulus is sufficiently strong, all or most of the leaves collapse. Recovery is quite rapid if the plants are not further disturbed, but nevertheless takes much longer than does the collapse. After repeated stimuli recovery becomes slower. Young plants are more sensitive than old ones. The movement is made possible by groups of specialized cells located at the bases of the leafstalks, the turgidity of which, as a result of stimuli of the types described, rapidly changes.

South American ***M. polycarpa*** is a shrub 4 to 10 feet tall with leaves less sensitive than those of *M. pudica,* yet it shows obvious movement in response to appropriate stimulation. From *M. pudica* this differs in its leaves having only one pair of main divisions of leaflets up to slightly over ½ inch long. The flowers are rose-purple. Variety *M. p. spegazzinii* (syn. *M. spegazzinii*) has leaves with smaller leaflets, hairy on their upper surfaces.

Garden Uses. In frost-free and nearly frost-free climates and in greenhouses where botanical collections are maintained *M. polycarpa* and its variety are cultivated for ornament. The common sensitive plant is frequently grown in greenhouses as a curiosity. It may even be accommodated indoors on a window sill, although it succeeds better in a terrarium or if kept covered by a glass or transparent plastic bell jar or bowl. It is of great interest to children (and, of course, to adults also).

Cultivation. In favorable climates the outdoor cultivation of these plants presents no difficulties. They grow well in any well-drained ordinary garden soil in sun or light shade. They also respond satisfactorily to greenhouse cultivation in pots or other containers or in beds. Indoors, minimum winter temperatures of 50 to 60°F are satisfactory. The common sensitive plant, like others of its genus, comes readily from seeds sown in a temperature of 65 to 70°F. The young seedlings are transplanted individually to 2½-inch pots, later to pots 4 or 5 inches in diameter. A humid environment where the minimum temperature is 60°F provides suitable growing conditions. The soil must be moist at all times and weekly applications of dilute liquid fertilizer are helpful after the containers are filled with roots. Good light with a little shade from strong sun is required. This plant is easily damaged by drafts of cold air and by a too-dry atmosphere. Its chief pest is mealybugs.

MIMULUS (Mímu-lus) — Musk, Monkey Flower. Chiefly natives of temperate regions and especially numerous in western North America, the 100 or more species of *Mimulus* belong in the figwort family SCROPHULARIACEAE. The majority are herbaceous, a few shrubby. Their name, in allusion to the fanciful grinning faces of the flowers of certain kinds, is a diminutive of the Latin *mimus,* a little buffoon.

Mimuluses are erect or spreading, hairless or hairy, with an often glandular-sticky pubescence. Their undivided, toothed, toothless, or rarely lobed leaves are opposite. Solitary in the leaf axils or in terminal racemes, the mostly showy blooms have persistent, conspicuously five-angled, five-toothed, tubular calyxes and more or less distinctly two-lipped corollas with cylindrical tubes. The upper lip of the corolla is two-lobed and erect or reflexed, the lower spreading and three-lobed. There are four stamens and a two-lobed, slender style. The fruits are capsules.

Herbaceous perennial monkey flowers include several good garden plants, among them the musk (***M. moschatus***). Formerly much cultivated in greenhouses and window gardens, as well as outdoors, musk is less popular than it was, chiefly because it has "lost its scent." This has occasioned much speculation and some mystery. There is no doubt that until the early years of the twentieth century cultivated musk plants had a strong and pleasant musklike odor. It is equally certain that no such examples are now known to exist. It seems probable that the fragrant-foliaged plants popular up to about the time of World War I represented one aberrant clone propagated vegetatively by cuttings and that even then the vast majority of the wild population of *M. moschatus* had no such strong odor. Just why or how the scented form was lost is not known, perhaps the upheaval of the war, or possibly the introduction of new seedling stock, was responsible. There is always the possibility, of course, that a new scented seedling will appear and be perpetuated by vegetative propagation.

Mimulus moschatus

Musk, native in moist soils from the Rocky Mountains to the Pacific Coast and widely distributed either as a native or naturalized inhabitant from Newfoundland and Quebec to West Virginia has creeping, slimy-hairy, rooting stems up to 1 foot long, and oblong-obovate, toothed or toothless, short-stalked leaves up to 1½ inches long. Dotted with light brown, the pale yellow, funnel-shaped, long-stalked blooms are about ¾ inch long. They have petals of nearly equal size and nonprotruding stamens tipped with pubescent anthers. Variety *M. m. longiflorus* has more erect stems and blooms up to 1¼ inches long. Also more upright than the typical species, *M. m. sessilifolius* has mostly stalkless leaves.

Yellow-flowered and a native of Chile, ***M. luteus*** is a hairless, herbaceous perennial of moist soils. It is one parent of the hybrids gardeners call *M. tigrinus.* It has rooting, creeping stems up to 1 foot long, and sharply-toothed, broad-ovate leaves 1 inch to 2 inches long, from the bases of which spread five or seven veins. The upper leaves are stalkless, the lower short-stalked. In loose racemes of few, the long-stalked, usually red- or purplish-spotted, yellow blooms with gaping throats, are 1 inch to 2 inches long. Varieties are *M. l. alpinus,* with stems up to 6 inches tall and blooms with a reddish spot on their lower lips, *M. l. rivularis,* with a large red spot on the center lobe of the lower lip; *M. l. variegatus,* with petals edged with pinkish-purple and the throat of the corolla pale yellow; and *M. l. youngeanus,* with all its petals spotted.

Mimulus luteus

Similar to *M. luteus,* but distinguished by its usually red-brown- or purplish-spotted, yellow blooms having their throats almost closed by two hairy ridges and by its more erect growth and usually more abundant blooms, is ***M. guttatus*** (syn. *M. langsdorfii*), a wet-soil, hardy herbaceous perennial or annual indigenous from Montana to Alaska, California, and Mexico. This very variable kind, up to 1 foot or a little more in height, has mostly branchless stems and ovate to oblong-lanceolate, many-veined, coarsely-toothed leaves ½

Mimulus guttatus

inch to 6 inches long. The lower leaves have long stalks, the upper ones none. The flowers, ½ inch to 1½ inches long, have stalks usually shorter than the corollas. Hybridized with *M. luteus* this species is the source of the showy, popular, large-flowered, varicolored garden hybrids known as ***M. hybridus*** (syn. *M. tigrinus*).

Mimulus hybridus

A low, creeping herbaceous perennial, ***M. tilingii*** (syn. *M. implexus*) has weak stems up to 8 inches long and few broad-ovate to broad-elliptic leaves, with three or five chief veins from their bases. The lower leaves are short-stalked, the upper ones stalkless. The long-stalked flowers, yellow, spotted with red, are up to 1½ inches in length. This species is closely related to *M. guttatus.* It inhabits wet soils at high altitudes in western North America.

A charming small yellow-flowered herbaceous perennial, ***M. primuloides,*** as its name suggests, has something of the aspect of a primrose. Its leaves may be all basal or some basal and some on the up-to-4-inch-tall stems. They are broadly-obovate to wedge-shaped and usually not over 1 inch long by ⅓ inch wide. They may or may not be hairy. The few flowers are mostly solitary and sometimes dotted

Mimulus primuloides

with reddish-brown. They are ½ to ¾ inch long. This kind is native in moist meadows at fairly high altitudes from the Rocky Mountains to Washington and California. Variety *M. p. linearifolius* has linear to spatula-shaped leaves up to 1¾ inches long and flowers ¾ to 1 inch long.

Blue-flowered ***M. ringens*** and ***M. alatus*** are hardy herbaceous perennials of similar aspect. They inhabit swamps and wet woodlands in eastern North America, with the range of the first extending to Oklahoma. From 1 foot to 4 feet tall, their four-angled stems may be branched above; those of *M. alatus* are winged. They have pointed, lanceolate to oblanceolate leaves. Those of *M. ringens* are stalkless and obscurely round-toothed, those of *M. alatus* stalked and more coarsely-sharp-toothed. From 1 inch to 4 inches long, they become progressively smaller upward. On slender stalks, the flowers are 1 inch to 1½ inches long. Plants with pink or white flowers are occasionally seen.

Mimulus ringens

Scarlet or rarely yellow blooms are possessed by ***M. cardinalis*** and rose-red, pink, or rarely white ones by ***M. lewisii.*** Both are herbaceous perennials of western North America, growing in moist places in the mountains. The first has sticky-hairy, weak stems up to 1 foot long and stalkless, sharp-toothed, obovate leaves 1 inch to 4½ inches long. The long-stalked, strongly two-lipped flowers, 1½ to 2 inches long have slightly protruding stamens with bearded anthers. The flowers of *M. lewisii* are less markedly two-lipped than those of *M. cardinalis.* Their more or less hairy-anthered stamens do not protrude. This sticky-hairy species attains heights up to 2½ feet, and has irregularly-toothed, oblong leaves 1 inch to 3 inches long.

Annual monkey flowers cultivated are mostly natives of western North America. This is true of all described below except *M. cupreus,* which hails from southern Chile. In addition, varieties of *M. hybridus* (syn. *M. tigrinus*) are often grown as annuals. The flowers of ***M. cupreus,*** spotted with red in their throats and 1 inch to 1½ inches long, open golden-yellow and change to a brilliant copper hue. About 8 inches tall and freely-branched, this kind is hairless or slightly hairy and has ovate or elliptic, coarsely-toothed leaves 1 inch to 1¼ inches long, with three or five veins spreading from their bases.

Yellow or predominantly yellow flowers are characteristic of *M. brevipes, M. glaucescens,* and *M. nasutus.* Up to 2½ feet tall, and branched or single-stemmed, ***M. brevipes*** has lanceolate to linear leaves up to 3 inches long and beautiful two-lipped, yellow blooms up to 2 inches long. It grows in dry soils. About twice as tall, ***M. glaucescens*** is hairless. Its glaucous, coarsely-toothed leaves are from broad-ovate to nearly round, the lower ones long-stalked, those above stalkless and sometimes joined at their bases. Spotted with red, its yellowish flowers are 1½ inches long. This inhabits moist soil. Up to 2½ feet tall, ***M. nasutus*** is more or less short-hairy, and usually branched. Its round-ovate to oblong leaves are coarsely-toothed. Those below are long-stalked, the upper ones stalkless or nearly so. The yellow flowers are up to nearly 1 inch long. Their lower lips are densely-hairy, and usually marked with a prominent reddish-brown blotch. This is a moist-ground plant.

Red, pink, and reddish-purple are the predominant flower colors of *M. breweri, M. fremontii, M. nanus,* and *M. tricolor.* About 6 inches tall and pubescent, ***M. breweri*** has linear or oblong, scarcely-toothed or toothless leaves ½ inch to 1¼ inches long and funnel-shaped, pink to purplish-red blooms ⅓ inch long. It grows in moist soil. About 8 inches in height, and sticky-pubescent, ***M. fremontii*** is a desert, dry-soil plant, reddish throughout and generally freely-branched. Toothless or nearly so, its stalkless leaves are up to 1¼ inches long. The rose-purple, broadly-funnel-shaped blooms, pubescent on their outsides, are ¾ to 1 inch long. Generally much-branched, and sticky-hairy, 2- to 6-inch-tall ***M. nanus*** is decidedly attractive.

It has elliptic-lanceolate to oblong, toothless leaves up to 1¼ inches long, and magenta-purple or rarely yellow blooms up to ¾ inch long, with patches of white or yellow dotted with purple inside the throats. It favors dryish soils. Attractive ***M. tricolor*** has erect or spreading, glandular-pubescent branches up to 6 inches long and oblanceolate to oblong leaves up to 1 inch long and one-third or less as wide. The flowers, up to 1¾ inches long, have yellow corollas tubes with reddish-purple throats and rose-purple petals, each ususally with a large darker spot near its base. This species prefers moist soils.

Shrubby monkey flowers, bushy kinds with permanent, usually sticky, woody stems, and foliage often with a somewhat varnished appearance, by some botanists are segregated as the genus *Diplacus*. Even when retained in *Mimulus* they form a distinct subgroup. Although they hybridize with some freedom among themselves, they have never been hybridized with the herbaceous perennial or annual kinds. All are natives of California, Baja California, or islands off the coasts of those places, where they are components of the chaparral vegetation that occupies dryish soils.

A well-known shrub monkey flower is ***M. aurantiacus*** (syns. *M. glutinosus, Diplacus glutinosus*). Somewhat straggly and freely-branched, it has sticky stems and foliage and attains heights up to 4 feet. Its oblong leaves are stalkless, more or less toothed, and densely-yellowish- or brownish-hairy beneath. They are up to 2 inches long and ½ inch broad, with rolled-under margins. The rich yellow, funnel-shaped blooms have spreading petals, and are about 1½ inches long. From the last ***M. puniceus*** (syn. *Diplacus puniceus*) differs chiefly in its leaves being less hairy beneath and its flowers being brick-red to dark red. Up to 6 feet tall, it has narrowly-lanceolate, linear, or oblanceolate leaves up to 2½ inches long and blooms up to 1½ inches long. A white-woolly or densely-long-hairy covering of the younger shoots and calyxes is a distinguishing feature of ***M. longiflorus*** (syns. *Diplacus longiflorus, D. speciosus*). Mostly not over 3 feet tall, this has lanceolate leaves up to 3 inches long and ½ inch wide, usually hairy underneath, sometimes toothed, but generally not revolute. The slender-tubed, yellowish-salmon-pink to cream-colored flowers are up to 3 inches long, with the petals wavy- or jagged-edged. The shoots and foliage are sticky.

Shrubby kinds with leaves hairless on both surfaces are *M. aridus, M. flemingii* (syn. *M. parviflorus*), and *M. leptanthus*. These have sticky stems and foliage. About 1 foot tall, ***M. aridus*** has oblanceolate to oblong leaves up to 1¾ inches long, toothed or not, the smaller ones with revolute margins. Its numerous pale buff to deeper yellow flowers, 2 inches or less in length, have short, thick stalks. Brick-red flowers tinged with yellow and up to 1¾ inches long are borne by 2-foot-tall ***M. flemingii.*** This is heavily foliaged with obovate leaves paler beneath than on their upper sides and up to 1¾ inches long. The margins are sometimes rolled under. Thin-stemmed, 3-foot-tall ***M. leptanthus*** has slender-tubed yellow flowers up to 2½ inches long, with deeply-notched petals. Its numerous glossy, oblong to spatula-shaped leaves may be slightly toothed and frequently have rolled-under margins. They are up to 2 inches long and ½ inch wide.

Mimulus aurantiacus

Garden and Landscape Uses. Mimuluses serve various purposes. The annuals and perennials grown as annuals, such as *M. cardinalis, M. guttatus, M. luteus*, and *M. hybridus*, are attractive where winters are cold for summer display in outdoor flower beds, and for winter and spring blooming in garden beds in regions of essentially frostless winters. They do best in fertile, moderately moist soil, with a little shade from the hottest sun. They are also pretty for flowering in late winter and spring in greenhouses. The hardiness of herbaceous perennial monkey flowers varies and generally may be inferred from the regions where they are native. Some, such as *M. moschatus*, and *M. ringens*, are hardy far north, but this is not true of some of the western North American kinds or of those from South America. The taller herbaceous perennials, including *M. cardinalis, M. guttatus*, and *M. luteus*, are appropriate for informal groupings near watersides, in open woodlands and similar places, and in flower borders. Dwarfer kinds, such as *M. fremontii, M. moschatus*, and *M. primuloides*, are for rock gardens and other intimate areas. In the main the herbaceous perennials prefer rich, moist soil and light shade.

None of the shrub monkey flowers, the kinds sometimes named *Diplacus*, is hardy in the north. Where winters are at all severe these must be kept indoors at that season. They are pleasing as pot specimens in greenhouses and sun rooms, and can be planted in outdoor beds, much like geraniums or fuchsias, for summer flowering. In California and other places with climates approximating those under which these plants grow in the wild, shrub monkey flowers are useful permanent garden plants and bloom freely for long periods. They accommodate to sun or part-day shade and to dryish soils of ordinary fertility.

Cultivation. All mimuluses are easily raised from seeds, which, being small, should be covered with soil very shallowly. Multiplication of the herbaceous perennials can also be achieved by division and cuttings, of shrub monkey flowers by cuttings. To have plants of annual mimuluses and of kinds grown as annuals that will bloom in winter and spring outdoors in mild climates, and in greenhouses, seeds are sown in fall or in January in a temperature of 60°F. The seedlings are potted individually in small pots and, for greenhouse cultivation, successively into larger ones. The final containers in which the plants bloom may be 5, 6, or 7 inches in diameter. The greenhouse is maintained with a night temperature of

50°F and, according to the brightness of the weather, five to fifteen degrees warmer by day. On all suitable occasions it is ventilated freely. During their young stages the tips of the shoots are pinched out two or three times to encourage branching. At all times the soil is kept moist, and when the final containers are well filled with roots, regular applications of dilute liquid fertilizer are given.

To have summer flowering plants outdoors, seeds of annuals and sorts grown as annuals are sown in spring in a greenhouse. The young plants are grown in flats or small pots until there is no longer danger of frost and then are planted in the garden. During their time indoors they are treated as advised above for plants grown throughout in greenhouses. Hardy perennial mimuluses give good results from seeds sown in a shaded cold frame in May in soil kept evenly moist.

Shrub mimuluses are raised from seeds sown in a temperature of about 60°F, or by spring or summer cuttings of firm, but not hard shoots planted in sand, a mixture of sand or perlite and peat moss, or perlite, in a greenhouse propagating bench or under similar conditions. The rooted cuttings are potted individually in small, well-drained pots of sandy peaty soil and are transferred successively to bigger pots as growth makes necessary. Occasional pinching out of the shoot tips promotes desirable bushy growth. Old specimens, pruned to size and shape in late winter, are then repotted and restarted into growth in temperatures of 55°F by night and five to fifteen degrees warmer by day. The soil is kept moderately, but not excessively moist. In favorable weather the greenhouse is ventilated freely. In summer the plants may be put outdoors with their pots buried nearly to their rims in a bed of sand, ashes, peat moss, or soil. They are brought inside before fall frost.

MIMUSOPS (Mímu-sops)—Spanish-Cherry. As now understood, *Mimusops,* of the sapodilla family SAPOTACEAE, except for one species indigenous to tropical Asia and islands of the Pacific, is restricted in the wild to Africa, Madagascar, and the Mascarene Islands. New World species once included, such as the balata rubber (formerly *M. balata*), have been transferred to *Manilkara* and other genera. There are sixty species of *Mimusops.* The name, from the Greek *mimo,* an ape, is without obvious application.

Evergreen, milky-juiced trees and shrubs, the members of this genus have alternate, undivided, usually thick and lustrous leaves with only one prominent vein, the center one. Neither large nor showy, the white or whitish flowers are solitary or clustered in the leaf axils. They have a calyx with eight lobes (sepals) in two rows, a corolla with eight lobes (petals) each with a pair of appendages on its back, eight stamens, eight staminodes (nonfunctional stamens), and a slender style. The fruits are few-seeded berries, those of some species edible.

Spanish-cherry (***M. elengi***) a round-topped, evergreen tree up to 50 feet in height, is a native of tropical Asia and islands in the western Pacific. Its young shoots and flower stalks are clothed with rusty hairs. Dark green, the leaves, elliptic to broad-elliptic and 2 to 6 inches long by 1 inch to 3 inches wide, have wavy, upturned margins. Short-stalked and about ½ inch long, the flowers are white and very fragrant. The green, yellow, or orange, egg-shaped fruits are approximately 1 inch long. They are edible but too astringent to be widely appreciated. Variety *M. e. parvifolia* has smaller leaves. A species with larger, broadly-elliptic to obovate, short-stalked leaves and larger, milder-flavored fruits, ***M. commersonii,*** native to Madagascar, is cultivated in Hawaii.

An evergreen shrub or tree up to 45 feet in height, ***M. zeyheri,*** of South Africa, has rusty-hairy young shoots. Its leaves, ovate-lanceolate to obovate-oblanceolate, are 1¾ to 4 inches long and nearly one-half as wide. The fragrant, small, whitish or greenish-white flowers, without decorative appeal, are succeeded by ovoid, ellipsoid, or spherical, deep yellow, edible fruits ¾ inch to 1¼ inches long and up to 1 inch thick. They contain one to four seeds and sweet pulp that clings to the tongue and palate rather unpleasantly. The fruits are rich in vitamin C and in Africa are part of the native diet. Closely related and probably only to be regarded as a warmer-region variety of *M. zeyheri* is ***M. kirkii,*** also of South Africa. South African ***M. caffra*** is a large shrub or tree up to 45 feet tall. It has rusty-hairy young shoots and obovate to nearly reverse-heart-shaped leaves 1½ to 3 inches long. The egg-shaped, usually one-seeded fruits, about ¾ inch long, and red when ripe, are edible. In the wild this inhabits coastal dunes exposed to high winds and salt spray.

Garden and Landscape Uses and Cultivation. In the tropics and warm subtropics the kinds described here are planted for ornament and interest. They thrive without special care in ordinary soils and locations and are propagated by seed and by cuttings.

MINA (Mì-na). The only species of this genus is botanically so closely related to morning glories that many botanists include it with them in the genus *Ipomoea.* Yet it is so different in general appearance from all other kinds that it seems best to accept the opinion of those experts who tend to split genera on minor technicalities and for horticultural purposes to regard this as a separate entity. It belongs in the morning glory family CONVOLVULACEAE. Its name commemorates a Mexican minister, F. X. Mina. It is native to Mexico.

A vigorous perennial, twining, hairless vine 15 to 20 feet tall, ***Mina lobata*** (syns. *Ipomoea versicolor, Quamoclit lobata*) has heart-shaped, three-lobed leaves 2 to 3 inches across. The lobes are pointed, and the center one narrows toward its base. The flowers are in more or less one-sided, curving spikes on branched stalks from the leaf axils. They are ¾ inch long and tubular. Their five sepals are bristle-pointed. The corolla, at first rich scarlet-crimson, soon fades to pale yellow or cream. Its lower part is narrowly cylindrical; above it expands into a pouched or bag-shaped portion with five tiny lobes (petals). The five stamens and style protrude. The fruits are capsules.

Mina lobata

Garden and Landscape Uses and Cultivation. This very distinctive and showy vine is a welcome perennial in tropical and subtropical gardens and in greenhouses and is easily cultivated as a summer-flowering annual outdoors where winters are too severe for it to live over. It can be used to furnish pergolas and pillars and to climb wherever there are supports that its stems can entwine. It responds to fertile, reasonably moist soil and rejoices in sun. In greenhouses this vine succeeds in pots, tubs, and ground beds. A night temperature of 50 to 55°F in winter is adequate with an increase of a few degrees by day permitted. At other seasons both day and night temperatures may be higher. Refurbishing, consisting of pruning, and of top dressing with fresh fertile earth or repotting, as seems wisest, is done in late winter or spring. Generous supplies of water and, after the containers are filled with roots, regular applications of dilute liquid

Mina lobata clothing a fence

fertilizer, are needed from spring through fall. No fertilizer is given in winter, and at that season watering is reduced without permitting the soil to become extremely dry. Propagation is by seeds sown in late winter or spring in a temperature of 60 to 70°F. When the seedlings have their first true leaves fairly well developed pot them individually in 3 or 4 inch pots and later transplant them to larger containers as growth necessitates or, if they are to be planted outdoors, keep them growing indoors until the weather is warm enough to put out such tender plants as dahlias and tomatoes, and then set them out.

MINDIUM (Mín-dium). It is to be regretted that rules of botanical nomenclature make it necessary to substitute *Mindium* for *Michauxia* as the name of this genus of seven species, of the bellflower family CAMPANULACEAE, because *Michauxia* honored Andre Michaux, a distinguished French botanist, who lived in North America for a decade and contributed much to the knowledge of American plants. He died in 1803. The name *Mindium* is said to be "a barbarous corruption by an Arabian writer" of *Medium*, an ancient name for various species of *Campanula*.

Mindiums are biennial or perhaps sometimes perennial herbaceous plants restricted in the wild to southwest Asia. Little known in American gardens, these stately ornamentals have irregularly-lobed and toothed leaves and spikes, racemes, or panicles of attractive large blooms that differ from those of related *Campanula* in having seven to ten each sepals, petals, and stamens. The long, forward-pointing style ends in a seven- to ten-lobed stigma. The fruits are capsules that open at their bases in seven to ten places to release the seeds. Unlike those of many campanulas, the conspicuously stalked to nearly stalkless flowers are not bell-shaped but have long, spreading or eventually reflexed, slender petals and sepals. The upper blooms open first.

The best, ***M. campanuloides*** (syn. *Michauxia campanuloides*) is 3 to 6 feet tall or sometimes taller and has lanceolate, irregularly lobed, bristly-hairy leaves, those of the stems smaller than those below. Its wheel-shaped, nodding blooms, bristly-stalked and in loose, candelabra-like panicles, sometimes attain a diameter of 5 inches, but more usually are about 3 inches across. They are white with purplish tinting, and the corolla is cut almost to its base to form the petals. Native in the mountains of Asia Minor, ***M. tchihatcheffii*** (syn. *Michauxia tchihatcheffii*), 6 to 7 feet tall, has stout, branchless stems and narrowly-oblong to somewhat fiddle-shaped leaves 6 to 7 inches long. The blooms are crowded in cylindrical, terminal spikes up to 2 feet long and 5 or 6 inches wide. They have eight flaring, white petals.

Garden Uses and Cultivation. These imposing, handsome bellflower relatives add dignity to flower borders and are useful for tucking into openings at the fronts of shrub planting. To give of their best they must be afforded generous treatment in the matters of soil preparation and watering. They respond poorly to earths lacking in nourishment or moisture. Adequate drainage is needed. Stagnant water in the soil is not tolerated. Mindiums usually bloom in their second year from seed, but sometimes take three or four years. After blooming they die. They are raised from seeds sown in flats, cold frames, or outdoor seedbeds in spring. As soon as the seedlings are big enough to handle they are transplanted to nursery beds or cold frames where they remain until the fall or spring; then they are transferred to their flowering quarters. In cold climates they may be carried through the winter in cold frames. They need a warm, sunny location, and should be watered freely in dry weather. The hardiness of mindiums is not fully determined. Certainly in regions of cold winters they should be given sheltered locations, and a light winter covering of salt hay, branches of evergreens, or other suitable protective material does not come amiss.

MINER'S-LETTUCE is *Montia perfoliata*.

MING TREES. These are florists' confections constructed of dried plant parts to somewhat resemble living trees dwarfed as bonsai in the Japanese manner. The trunks and branches of ming trees are commonly of a Californian manzanita (*Arctostaphylos*) or other attractive smooth wood. To the branches are wired flat pads of North American *Eriogonum ovalifolium* or Brazilian *Syngonanthus elegans* or *S. niveus*.

MINIATURE GARDENS. See Dish Gardens.

MINIATURE JOSHUA TREE is *Sedum multiceps*.

MINT. Besides being the common name for members of the genus *Mentha*, dealt with in the next entry, the word mint forms part of the names of these plants: Australian mint-bush (*Prostanthera*), coyote-mint (*Monardella villosa*), horse-mint (*Monarda*), lemon-mint (*Monarda citriodora*), mint-bush (*Elsholtzia stauntonii*,) mint-geranium (*Chrysanthemum balsamita*), mountain-mint (*Pycnanthemum* and *Monardella odoratissima*), and stone-mint (*Cunila origanoides*). Meehan's mint is *Meehania cordata*.

MINT. This is the name of species and varieties of *Mentha*. Under that entry descriptions of the various kinds, as well as of the uses and cultivation, of the decorative Corsican mint (*Mentha requienii*) are given. Here we are concerned with the ones grown in gardens for flavoring and garnishing. Chief among these are spearmint and apple mint, but other kinds, including peppermint are grown. All need the same care and cultivation. All are hardy and easy to grow.

Spearmint, ready for picking

Mints succeed in a variety of soils, preferring those that are deep and fertile and that do not dry unduly. They succeed in locations shaded lightly for part of each day as well as in open, sunny ones. If the soil tends to dryness a partially shaded location is advantageous. A difficulty with mints is that they spread by underground runners to the extent that they are given to invading adjacent lawns and paths and

Mints forced into early growth in a sunny window; left, apple mint, right, spearmint

areas intended for other plants. This can be prevented or checked by sinking in the ground old pails or other containers from which the bottoms have been removed and planting inside these.

Although seeds can be used as sources of new plants, mints are so easily increased by division that this method is used almost exclusively. Cuttings root without trouble and if taken early in the season make substantial plants by fall. Although mints persist indefinitely, the finest produce is had from young plantings; therefore it is advisable to lift the roots every second or third year, divide them, and reset strong, small pieces in newly spaded and fertilized soil. The older interior parts of the clumps that have become hard and woody should be discarded. Established mints are improved by a spring application of a complete garden fertilizer and by watering generously in dry weather. A serious rust disease, apparent as orange pustules on the leaves, affects some mints. There is no adequate control. Infected plants should be destroyed.

Mint may be dried for winter use by cutting it before flowers develop, tying it in bunches, and hanging these upside down in a shady, cool, dry, airy place. When thoroughly dry the leaves are stripped from the stems and stored in tightly corked or stoppered bottles or similar receptacles.

To have fresh mint in winter and spring before outdoor supplies are available is a simple matter. In fall dig strong clumps of roots and plant them in porous soil closely together in boxes or other containers about 8 inches deep. Water thoroughly and stand the containers in a cold frame or bury them to their rims outdoors in peat moss, sand, sawdust, or similar material until late December or January. Then bring them into a sunny greenhouse or window where the temperature at night is 50 to 55°F and by day is ten degrees or so warmer. Keep the soil well watered. Soon excellent picking will be ready.

MINUARTIA. See Arenaria.

MIRABILIS (Miráb-ilis)—Four O'Clock or Marvel-of-Peru, Umbrella Wort. The name of this genus of the four o'clock family NYCTAGINACEAE, derived from the Latin *mirabilis*, wonderful, probably alludes to the blooms. Comprising sixty species, *Mirabilis* is a native of chiefly warm parts of North, Central, and South America.

Mirabilises are annuals and nonhardy, often tuberous-rooted herbaceous perennials often grown as annuals. They have stems generally swollen at the nodes (joints) and opposite leaves, the lower stalked, the upper sometimes stalkless. The long-tubed flowers, one or more together from calyx-like involucres (collars of bracts), are without corollas, but have as the showy parts of the blooms corolla-like calyxes with five spreading, petal-like lobes. There are three to six stamens, the bases of their stalks united, and a slender style. The fruits, technically achenes, are leathery and longitudinally ribbed.

The common four o'clock or marvel-of-Peru (***M. jalapa***) received its specific name because its tuberous roots were incorrectly believed to be the source of jalap, a product of *Ipomoea*. The name four o'clock refers to the blooms not opening until late afternoon. Except in dull weather, they close the following morning. Erect, freely-branched, and 1 foot to 3 feet tall, this is by far the most popular kind. A hairless or slightly hairy, perennial, it has pointed-ovate leaves 2 to 6 inches long or sometimes longer with heart-shaped or squared bases, and stalks about 1 inch long. The abundant, showy, white, yellow, red, or striped blooms, about 1 inch across and 1 inch to 2 inches long, are solitary from the involucres. Dwarf varieties and a variety with variegated foliage are available. The four o'clock, native of South America, is naturalized in the southern United States and also in Hawaii. An old garden plant, 2- to 3-foot tall ***M. longiflora,*** of Mexico, is a perennial with pointed-heart-shaped leaves, and from each involucre a cluster of three or more white, pink, or violet flowers that have clammy calyx tubes 5 to 6 inches in length. They are fragrant in the evening. Hybrids between this and *M. jalapa* are named *M. hybrida*.

Mirabilis jalapa

Annual kinds are *M. viscosa* and *M. dichotoma*, the first native of northern South America and Mexico, the other of Mexico. Sticky-hairy and up to 3 feet tall, ***M. viscosa*** has stout stems and broadly-heart-shaped, fleshy leaves up to 4 inches long. The purple, red, pink, or white flowers, usually one from each involucre, are up to ¾ inch long. From 1½ to 2½ feet in height, ***M. dichotoma*** has round-based, ovate leaves and yellow flowers.

Kinds native to the United States are sometimes cultivated. Here belongs the wishbone bush (*M. laevis* syn. *M. californica*), native of California, *M. froebelii*, of the southwestern United States and Mexico, and *M. multiflora*, which is wild from Colorado to Utah and Mexico. A thick-rooted perennial, ***M. laevis*** has many usually repeatedly forked, slender, sprawling, more or less sticky-hairy stems. Its ovate leaves have blades up to 1 inch long or sometimes longer. The short-stalked, solitary flowers, clustered near the branch ends, are bell-shaped, rose-pink to purplish red, and ½ to ¾ inch wide. Three or more deep pink to rose-purple flowers come from each involucre of ***M. froebelii.*** They are funnel-shaped and 1½ inches long or longer, and ¾ inch to 1¼ inches wide. This much-branched, very sticky species has thick, tuberous roots and leaves with broadly-ovate blades 1½ to 3½ inches long. Handsome ***M. multiflora,*** is more or less sticky, and 1 foot to 3 feet tall. It forms clumps of erect or spreading leafy stems and has broadly-ovate to heart-shaped leaves, with blades up to 3 inches long. There are usually more than three rose-pink to magenta-purple flowers from each involucre. They are funnel-shaped, 1½ to 2 inches long, and nearly as wide.

Ranging in the wild from South Dakota to Montana and Mexico, ***M. linearis*** (syn. *Allionia linearis*), up to 2 feet tall or sometimes taller, has sticky-hairy upper parts and nearly to quite stalkless linear leaves up to 4 inches long. Usually three together, the flowers are light pink to purplish-red. Native from Illinois to Colorado, Louisiana, and Texas, and naturalized in eastern North America, ***M. nyctaginea*** (syns. *Oxybaphus nyctagineus, Allionia nyctaginea*), up to 3 feet tall, has stems branched above and pointed-ovate-oblong to triangular-ovate, stalked leaves up to 4 inches long. The pinkish-purplish flowers are about ⅓ inch long, the saucer-shaped involucres about as wide. Up to 4 feet tall or taller, ***M. wrightiana*** (syn. *M. longiflora wrightiana*) is minutely-hairy, scarcely sticky. It has distinctly stalked, heart-shaped to triangular-ovate leaves and white flowers, 4 to 5½ inches long, tinted with pink or purple. This is a native of Texas, New Mexico, and Arizona.

Garden and Landscape Uses. Four o'clocks are easy to grow. The more common kinds are suitable for grouping in flower beds and for planting as ribbon borders, edgings, and low hedges, as well as in window and porch boxes. They bloom for a long period in summer and fall, until killed by frost. The kinds described above that are natives of the United States are suitable for dry regions, especially for inclusion in collections of native plants.

Cultivation. In warm climates where the growing season is long, four o'clocks can be raised from seeds sown directly outdoors, but in the north it is more usual to start them early indoors in a temperature of 65 to 70°F and to transplant the seedlings to flats or small pots. Six to eight weeks later, when it is safe to plant dahlias, tomatoes, and other frost-tender plants outdoors, the young plants are set in the garden. Another plan is to dig the tubers after their tops have been damaged by frost, store them in the same way as dahlias, and replant in spring. Suitable spacing for four o'clocks is about 1 foot apart. They thrive in ordinary soil in sunny locations or where they get a little shade for part of each day. If the ground where they have grown is left undisturbed, it is not uncommon for volunteer seedlings to come up the following year, and these can be readily transplanted.

MIRO is *Podocarpus ferrugineus*.

MIRROR SHRUB is *Coprosma repens*.

MISCANTHUS (Miscán-thus)—Eulalia, Zebra Grass. The genus *Miscanthus*, of the grass family GRAMINEAE, a native of the Old World, includes twenty species represented in native floras from Africa to Japan and the Philippine Islands. Its name, derived from the Greek *mischos*, a stem, and *anthos*, a flower, alludes to the stalked flower spikelets.

Miscanthuses are tall perennials with ample, flat, fanlike or feather-duster-like terminal panicles of slender, flexuous spikes of tiny flowers, with clusters of silky hairs that come from the bases of the flower spikelets and produce a soft, feathery effect. The genus is related to sugarcane (*Saccharum*).

The kinds commonly cultivated are the eulalia (*M. sinensis*) and its varieties. At one time this was included in the botanical genus *Eulalia*. Although now reclassified, the cultivated kind has retained eulalia as its common name and is generally known as such among gardeners and others who use colloquial names.

One of the most popular ornamental grasses, ***M. sinensis*** forms dense clumps of numerous, slender, erect stems 4 to 9 feet in height, with arching or spreading, rough-margined leaves 2 to 3 feet long, about 1 inch wide, with whitish midribs.

Miscanthus sinensis

Miscanthus sinensis (flowers)

Miscanthus sinensis gracillimus

In winter stems and foliage die to the ground. The flowers, in coppery-reddish panicles 6 inches to 2 feet long carried well above the foliage, appear in fall. They add considerably to the beauty of the plants and are appropriate for use in fresh and dried flower arrangements. This species, a native of China and Japan, has escaped from cultivation and naturalized itself to some extent in North America. More commonly cultivated than the plain green-leaved typical kind are two varieties with variegated foliage and *M. s. gracillimus*, which has very narrow (about ¼ inch wide), channeled leaves. In *M. s. variegatus*, the leaves are striped lengthwise with white or pale yellow. The leaves of the zebra grass (*M. s. zebrinus*) are marked with horizontal bands of pale variegation.

Miscanthus sinensis variegatus

Miscanthus sinensis zebrinus

Garden Uses and Cultivation. These plants excel for producing bold foliage effects. They can be used to good purpose in isolated lawn clumps, in waterside plantings (if their roots are not in water-soaked soil), and in masses in front of shrub borders. Their culture is of the simplest. They grow most vigorously in deep, fertile earth, but respond quite satisfactorily to any ordinary garden soil and to locations in full sun. They are hardy and are readily increased by division of the clumps in early fall or early spring. Little or no routine care is required, nor is staking needed. For more information see Grasses, Ornamental.

MISSION BELLS is *Fritillaria biflora*.

MIST and MISTING. The application of water in a very fine mistlike spray to the above-ground parts of plants is misting. A common greenhouse routine, its usual purpose is to prevent the dehydration and wilting that may come from too-rapid transpiration stimulated by dry or moving air, high temperatures, or bright sun. Misting is also practiced to induce the buds of dormant plants being forced to plump up and grow, in gardeners' language to "break." See also Mist Propagation.

MIST FLOWER is *Eupatorium coelestinum.*

MIST PROPAGATION. Unless special precautions are taken to prevent newly planted cuttings losing more moisture by transpiration than their cut ends can replace, they inevitably become dehydrated and perish. Traditional methods of preventing this have included partial defoliation, maintaining a highly humid atmosphere, providing warmth, shading from sun, and in favorable weather of occasionally misting the cuttings with water. Good results are had by these means, but to assure satisfactory ones considerable attention and good judgment as to the degree of shading and the frequency of misting are needed. Furthermore, shading lowers the rate of photosynthesis and consequently to some extent delays growth and root formation, and a too stagnant humid atmosphere favors the development of funguses that cause damping-off and rotting.

Mist propagation: Cuttings in a greenhouse bathed in a mist of water from a row of overhead nozzles

Mist propagation, or propagating under mist, affords superior rooting conditions. No shading is required, and by employing simple devices, the cuttings are prevented from wilting by bathing them in a mist of water. This reduces danger of rotting by preventing atmospheric stagnation and by washing away fungus spores that otherwise might germinate and invade plant tissues.

Mist propagation outdoors with one overhead nozzle

Essential to success are nozzles that under ordinary water pressures deliver a very fine mist and extremely well-drained beds, benches, pots, or other containers filled with sharply porous sand, perlite, or vermiculite that assures the rapid draining away of water and the consequent presence of needed air in the rooting medium. Equipment of the type called fog nozzles is employed. To prevent drift in wind it is well to surround outdoor installations with screens of polyethylene plastic film or other suitable material.

When first popularized this method of propagation involved maintaining uninterrupted mist throughout the hours of daylight, shutting it off at night. This procedure, called continuous mist, is still sometimes employed, but it is now much more usual and certainly more economical of water to arrange for the mist to operate for a few seconds at intervals and to have longer periods between without mist. This may be done by an automatic timer that activates the nozzles at predetermined intervals or by using sensors of various types that turn on the mist when the film of moisture on the cuttings dries.

Although mist propagation as now practiced was not developed until after World War II, its principles were well known and to some extent used much earlier. In Great Britain and probably elsewhere, by the beginning of the twentieth century, what were known as sun frames were used by some professional propagators. These were cold frames fully exposed and without shading. Cuttings, usually planted in pots of sand, were put in the frames, which were kept closed except when it became necessary to open them slightly to prevent the inside temperature rising above about 90°F. On sunny days the cuttings were sprayed with a hand-operated garden syringe sufficiently often to prevent their surfaces from ever becoming dry. The disadvantages of this system were the need for frequent, possibly hourly or more often, attention on sunny days and the possibility of human failure resulting in neglect and likely disaster.

MISTLETOE. Besides its use as the name of the plants discussed in the next entry, the word mistletoe forms parts of the common names of mistletoe cactus (*Rhipsalis*) and mistletoe honeysuckle (*Lonicera quinquelocularis*).

MISTLETOE. Not garden plants, the mistletoe of North America is *Phoradendron serotinum,* that of Europe the related *Viscum album,* both of the mistletoe family LORANTHACEAE. Like other members of the family, they are semiparasites that absorb water and part of the nutrients they need from the trees upon which they grow, but because they contain chlorophyll they are able to manufacture food by photosynthesis as do other green plants.

Popular as Christmas greens, these mistletoes are gathered in great quantities from wild sources. In Europe *Viscum album* is occasionally propagated artificially as something of a curiosity. This is done by squashing berries into crevices on the undersides of branches of apple trees, hawthorns, or other sorts agreeable to the mistletoe, and leaving them to germinate and eventually parasitize the tree. In a suitable climate, this probably could also be successfully done with American mistletoe. Unless a tree is heavily loaded with mistletoe no appreciable harm results from its presence.

Mistletoe (*Viscum album*) growing on hawthorn (*Crataegus*), Chelsea Physic Garden, London, England

MISTOL is *Ziziphus mistol.*

MITCHELLA (Mitchél-la)—Partridge-Berry, Twin-Berry, Squaw-Berry. Named in honor of one of America's early botanists, Dr. John Mitchell, who died in 1772, this interesting genus consists of two species, one American and one Japanese. It belongs in the madder family RUBIACEAE.

Mitchella repens

Mitella diphylla

Only the American kind is common in cultivation.

The partridge-berry, twin-berry, or squaw-berry (***Mitchella repens***) is indigenous from Nova Scotia to Minnesota, Florida, and Mexico. It forms broad, flat mats of trailing stems that hug the ground and root at the joints and has opposite, ovate to nearly round, stalked leaves up to ¾ inch long. Lustrous green and often with white veins, they form a neat pattern. The small, fragrant flowers, pink in bud and white with pink throats when open, in spring erupt in pairs from leaf axils near the shoot ends, and are succeeded by bright scarlet berries that remain attractive throughout the winter. The flowers have four-toothed calyxes, long, funnel-shaped corollas, hairy inside and with four spreading lobes (petals), usually four stamens, and a style with a four-lobed stigma. The berries, about ⅜ inch in diameter, are edible, but flavorless. A white-berried variety, *M. r. leucocarpa*, exists.

Garden and Landscape Uses. The partridge-berry is charming for woodland gardens, rock gardens, and for locating beneath shrubs and trees, especially evergreens, where it receives part-shade and is in acid, fairly moist soil. It is not sufficiently vigorous or aggressive to cover large areas, nor would it be in scale for such purpose, but as a plant to form patches of low carpeting in smaller, more intimate spots it is ideal. Partridge-berry is frequently planted in glass bowls and terrariums for winter interest and decoration indoors. When so used it succeeds best in a cool room where it receives good window light with some shade from strong, direct sun.

Cultivation. Given reasonably favorable conditions, no undue difficulty attends the cultivation of partridge-berry. It transplants from the wild easier than many eastern American woodlanders and is simple to propagate by division, cuttings, and seed. A soil well enriched with organic material, such as leaf mold or peat moss, that is never excessively dry suits it well. The chief routine care consists of preventing other plants from overpowering it or seeding among it. Occasional careful weeding is enough to take care of this.

MITELLA (Mi-téllа)—Bishop's Cap, Mitrewort. The dozen species of *Mitella,* of the saxifrage family SAXIFRAGACEAE, bear a name alluding to the shape of their young fruits. It is a diminutive of the Latin *mitra,* a cap. The genus is native in North America and eastern Asia.

Mitellas are hardy, more or less glandular and hairy, herbaceous perennials with short horizontal rootstocks and mostly basal foliage. In some kinds the stems bear one to three leaves, in others they are leafless. The leaves are more or less shallowly-lobed. The flowers are in branchless racemes. They have saucer- to bell-shaped, deeply-five-lobed calyxes. There are five white, greenish, pinkish, or purplish, pinnately or three-lobed petals, five or ten stamens, and two styles. The fruits are many-seeded capsules or little pods that soon split to reveal their usually glossy black seeds.

North American natives of the eastern and central reaches of the continent, *M. diphylla* and *M. nuda* are most easily recognized as to kind by the flowering stems, which in the first have a single pair of, or very rarely three, leaves, and in the second a solitary leaf or none. From 6 inches to 1½ feet tall, ***M. diphylla*** is pubescent below and glandular in its upper parts. Its long-stalked basal leaves are round-ovate, shallowly three- or five-lobed, round-toothed, and from 1½ to 2½ inches wide. The stem leaves are smaller and three-lobed, with the center lobe bigger than the others. The raceme of short-stalked flowers, 2 to 6 inches long, has white, lacy blooms ¼ inch wide. The tiny petals are deeply cleft. There are ten stamens. This is an endemic of rich woodlands from Quebec to Minnesota, Alabama, and Missouri. The flowering stems of ***M. nuda*** are up to 8 inches tall. The yellowish-green blooms have ten stamens. The leaves are less deeply lobed than those of *M. diphylla.* This species is indigenous entirely across northern North America and southward to Pennsylvania, Michigan, Minnesota, and Montana and in eastern Asia.

Western American natives, except *M. nuda* described above, have flowers with five stamens. They include ***M. caulescens***, a kind unique in that its racemes of flowers open from the top downward. Native to woodlands, meadows, and swamps from Montana to British Columbia, Idaho, and California, this has glandular stems 8 inches to 1 foot tall or a little taller, with one to three short-stalked leaves considerably smaller than the basal ones. The latter are long-stalked and sparsely-hairy and have three- to seven-, but mostly five-lobed blades, roughly heart-shaped in outline. The flower stalks terminate in slender, loose racemes, up to 8 inches long, of yellowish-green blooms. Individual flowers are about ¼ inch across.

Another Westerner, ***M. stauropetala*** has strongly one-sided racemes of flowers with usually three-lobed, spreading, white or purplish petals up to about ⅙ inch long. The racemes of bloom are carried to heights of up to about 1½ feet on leafless stalks. This species, a native of moist, dense woodlands from Montana to Oregon, Colorado, and Utah, has often purplish-tinged, heart- to kidney-shaped, sparsely-hairy leaves, with obscurely or conspicuously five- or seven-lobed, round-toothed margins.

Easily distinguished from other western North American mitellas with flowers that have five stamens, ***M. pentandra*** is distinct in having those organs alternating with, instead of opposite to, the sepals. Generally not over 1 foot tall, its flower stems are naked or carry one or two bracts or very much reduced leaves. The basal leaves, coarsely-glandular-hairy to nearly hairless, are ovate-heart-shaped, obscurely five- to nine-lobed, and doubly-round-toothed. The flowers are less than ¼ inch wide and have greenish petals. This inhabits moist woodlands, often favoring stream sides, from Alberta to Alaska, Colorado, and California.

A high-mountain species, native up to timber line from Alberta to British Columbia, Montana, Idaho, and California, ***M. breweri*** has all basal foliage. Its long-stalked, heart- to kidney-shaped leaves, sometimes 4 inches but usually from 1½ to 3 inches in diameter, are very indistinctly seven- to eleven-lobed, slightly round-toothed, and hairy or hairless. The greenish-yellow flowers are about ¼ inch wide.

Garden and Landscape Uses. Although their displays of blooms are decidedly inferior to those of the showiest members of the related genus *Heuchera*, mitreworts are worth planting in shaded places. They provide much the effect of their close relative the foam-flower (*Tiarella*), which differs from them in its petals not being lobed and its fruits being of two markedly unequal parts. The foliage of mitellas is plentiful and of pleasing texture and forms agreeable patterns. Even without flowers, this makes them acceptable as groundcovers in woodland gardens, along shaded paths, and in similar environments. They prefer moist soil of a woodland type that contains abundant organic matter.

Cultivation. Little or no difficulties attend the cultivation of these plants. Once established, they appreciate an annual top dressing with leaf mold, well-rotted compost, or peat moss, and if the earth is poor in nutrients a light application in early spring of a complete garden fertilizer. In dry weather periodic watering may be needed. Propagation is extremely easy by division in early fall or spring and by seed.

MITES. Mites are not insects. They are more closely related to spiders. They differ from insects in having at maturity eight, or in some species four, instead of six legs, from spiders in having their bodies seemingly in one piece instead of consisting of obvious segments. Small, sometimes microscopic in size, mites of many species infest and seriously damage a wide variety of plants. The harm results from their feeding, which is done by piercing or rasping the plant tissues and sucking the juices.

Red spider mites and allied species commonly cause foliage to become pale, sickly, stippled, or russeted. These mites, which feed on many kinds of indoor and outdoor plants and often spin fine webs on the undersides of the leaves, multiply especially rapidly in hot, dry environments. The cyclamen or strawberry crown mite, which seriously infests not only the plants those common names suggest, but also a wide variety of others, including aconites, African-violets, azaleas, begonias, delphiniums, gerberas, marguerites, marigolds, peppers, tomatoes, verbenas, and zinnias, cripples leaves by curling and distorting them and causing them to become noticeably rigid and brittle. Also, the development of new foliage practically ceases.

Foliage severely damaged by red spider mite

Typical damage to foliage caused by cyclamen mite: (a) African-violet

(b) Gloxinia (*Sinningia*)

(c) *Eupatorium atrorubens*

Bulb mites feed on many kinds of bulbs, corms, tubers, and other underground parts. They are especially fond of rotting tissues, but will burrow into healthy ones. In doing this they may transmit fungal, bacterial, and virus diseases. They are active on bulbs in storage as well as in the ground.

Other kinds of mites attack evergreens, fruit trees, and lawn grasses. Some produce galls (swellings). Notable among these is the maple bladder gall mite, the cause of small, often clustered, red, green, and eventually black galls on the upper surfaces of maple leaves. This mite may also cause puckering of the leaf surfaces. Yet another mite is associated with the development of witches' brooms, common on hackberries, and bud mites cause swelling and destruction of buds of currants and flowering currants.

Successful control of such a varied group of creatures that infest so many different kinds of plants requires measures tailored to particular plants and the habits and lifestyles of the various mites. Most insecticides are not effective against mites and may indeed, as a result of killing certain insect enemies of the mites, bring about substantial increases in mite populations. In some cases, however, systemic insecticides are effective. Special miticides are available from dealers in garden supplies

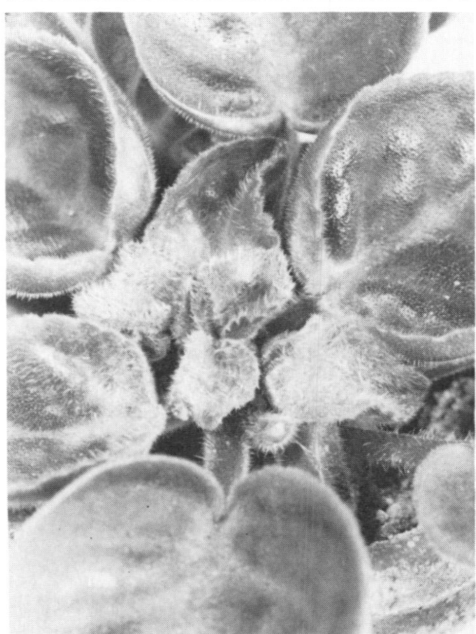

The center leaves of this African-violet clearly show the distortion, hardening, and failure to grow characteristic of cyclamen mite infestation

and, if used according to directions, are effective against many kinds of mites. Immersing infested plants in water at 112°F for twenty minutes gives control in some cases. For more detailed recommendations consult an up-to-date book on garden pests and their control or seek advice from your Cooperative Extension Agent or your State Agricultural Experiment Station.

MITRARIA (Mi-trària). The gesneria family GESNERIACEAE is replete with choice plants of interest to gardeners. Here belong African-violets, episcias, gloxinias, and streptocarpuses, to name but a few, and here also belongs the only species of *Mitraria*. Its name, alluding to the miter-shaped seed capsules, comes from the Greek *mitra*, a miter or cap.

A semitrailing, thin-stemmed, twiggy, evergreen, woody-stemmed shrub, ***M. coccinea*** inhabits dim, humid, rain forests in Chile, where it sends its roots into loose, always damp leaf mold and other organic debris. Its short-stalked, coarsely-toothed, ovate leaves are in twos or occasionally threes. From ½ to 1 inch long or rarely longer, most commonly about midway between, and about one-third as broad, they have lustrous, green upper surfaces, and are paler underneath. The bright orange-scarlet to scarlet, pendent blooms are solitary at the ends of slender, drooping stalks longer than the leaves. The calyx is small and usually has five unequal, lanceolate lobes. The corollas are about 1½ inches long, slightly curved, and tubular, with one side of the tube markedly pouched or bellied, and at the apex five small, spreading lobes (petals). There are four slightly protruded stamens, one awl-shaped staminode (sterile stamen), and one style, also slightly protruded. The fruits are fleshy capsules ½ inch in diameter.

Garden Uses and Cultivation. This is an attractive, relatively easy-to-grow greenhouse plant, suitable for outdoors in regions of mild winters and fairly cool, humid summers. It succeeds in the open in the mildest parts of Great Britain. Without support it is a trailer or dense mound 1 foot to 2 feet tall, but against a wall or among shrubs it is likely to reach a height of 6 feet or more. For its well-being it needs shade from strong sun and a constantly humid atmosphere. A somewhat acid soil composed largely of coarse leaf mold, or containing large proportions of peat moss and sand or perlite, suits. The addition to the soil of chopped charcoal and some shredded dry cow manure is beneficial.

In greenhouses *Mitraria* does well in ground beds, pots, and hanging baskets. It is very easily increased by cuttings and seed and is happy with a winter night temperature of 45 to 50°F, with an increase of five to fifteen degrees by day. At other seasons higher temperatures are appreciated. Unless the plants are in baskets or are to be allowed to sprawl, staking is needed. The soil should be kept moderately moist at all times. Biweekly applications of dilute liquid fertilizer from spring to fall help to maintain health and vigorous growth. Repotting is attended to in late winter or spring and again later if needed.

MITREWORT. See Mitella. False-mitrewort is *Tiarella*.

MITRIOSTIGMA (Mitrio-stígma). The genus *Mitriostigma* of the madder family RUBIACEAE consists of two or three species, natives of Africa. The name, alluding to the caplike stigma, is derived from the Greek *mitra*, a miter, and *stigma*.

Mitriostigmas are thornless, hairless, evergreen shrubs allied to *Gardenia*. They have opposite, rather leathery, pointed, elliptic-lanceolate leaves, and in branched, few-flowered clusters from the leaf axils, funnel-shaped flowers each with a five-lobed calyx, a corolla with five rounded lobes (petals), five stamens, and one style. The fruits are berries.

Mitriostigma axillare

Of compact habit and up to about 5 feet tall, ***M. axillare*** (syn. *Gardenia citriodora*), called in South Africa wild-coffee, has dark green, elliptic-lanceolate leaves 3 to 4 inches long. Its flowers, approximately 1 inch in diameter, are white and very fragrant.

Garden Uses and Cultivation. The species here described has uses similar to *Gardenia* and thrives in environments agreeable to the sorts of that genus. Propagation is easy by seed and cuttings.

MITROPHYLLUM (Mitro-phýllum). This intriguing group of South African succulent shrublets consists of a few more than twenty species. From *Conophyllum* it is separated only by technical differences in the fruits (capsules). Belonging in the *Mesembryanthemum* section of the carpetweed family AIZOACEAE, like other members of that group *Mitrophyllum* has flower heads daisy-like in aspect, but very different in structure. They are single blooms, not heads of numerous florets. Alluding to the appearance of the pairs of joined leaves, the name is derived from the Greek *mitra*, a miter or bishop's cap, and *phyllon*, a leaf.

Mitrophyllums produce alternate pairs of leaves of quite different character. Pairs joined almost to their tops to form subcylindrical, more or less conelike bodies enclose and protect from dehydration other pairs of decidedly longer leaves joined only at their bases. After a time the short, conelike pairs become flabby, lose their green color, dry, shrivel, and become an enveloping skin that splits to reveal a new pair of much longer leaves that divide almost to their bases and spread widely. From the center of these, another conelike pair in turn arises. The uppermost pairs of leaves on flowering shoots are of a third type; they are bractlike. The flowers, white, pink, or in some kinds perhaps yellow, are rare in cultivation. Plants previously named *M. dissitum*, *M. grande*, and *M. proximum* belong in *Conophyllum*.

Most likely to be cultivated is clump-forming, soft-leaved ***M. mitratum***, which attains 1 foot in height and has joined leaf bodies up to 3½ inches in length and notched at their tops. The other leaves, up to 4 inches long and spreading widely, have slightly rounded upper surfaces and rounded undersides. An inch or slightly more across, the white flowers have reddish tips to their petals.

Garden Uses and Cultivation. Rare and choice, these remarkable plants appeal to specialist collectors of succulents. Not hardy, they require the general treatment appropriate for *Pleiospilos*, *Lithops*, and other very juicy, succulent members of the *Mesembryanthemum* group. Their growing period begins in fall. Then they should be watered moderately. During their long season of rest the soil is kept dry. Full sun and well-drained, porous, sandy soil are needed. Propagation is by seed, or with considerably more difficulty, by cuttings. For further information see Succulents.

MIZUTARA. This is the name of orchid hybrids the parents of which include *Cattleya, Diacrium,* and *Schomburgkia.*

MLANJI-CEDAR is *Widdringtonia whytei.*

MOCCASIN FLOWER. See Cypripedium.

MOCK. This word appears as part of the common names of these plants: flame-mock-orange (*Chorizema*), mock-cucumber (*Echinocystis lobata*), mock-heather (*Haplopappus ericoides*), mock-orange (*Philadelphus, Pittosporum tobira, Prunus caroliniana*), and mock-strawberry (*Duchesnea indica*).

MOCKERNUT is *Carya tomentosa.*

MODECCA. See Adenia.

MODESTY is *Whipplea modesta.*

MOGHANIA. See Maughania.

MOHAVEA (Mohàv-ea) — Ghost Flower. Two western American species of the figwort family SCROPHULARIACEAE constitute the genus *Mohavea,* named after the stream along which specimens were first collected. Both are desert annuals, closely related to snapdragons (*Antirrhinum*), from which they differ in having two instead of four fertile stamens. They are erect plants with alternate, lanceolate to ovate-lanceolate leaves and dense leafy spikes of solitary, yellow, two-lipped flowers. The blooms have short, pouched tubes and fan-shaped, lobed lips, the lower one with a conspicuous, hairy palate that closes the throat of the bloom. There are three abortive stamens in addition to the two functional ones. The fruits are capsules.

The ghost flower (***M. confertiflora***) is a sticky-pubescent showy plant 4 inches to 1¼ feet in height. Its leaves are ½ inch to 2½ inches long. The creamy-yellow flowers, 1 inch to 1¼ inches in length, have palates covered with tiny brown or purple-black dots. Similar ***M. breviflora*** has flowers up to ¾ inch long; lemon-yellow, their palates are not conspicuously spotted and their lower lips are more deeply lobed than those of *M. confertiflora.*

Garden Uses and Cultivation. These are delightful annuals for rock gardens, edgings, for the fronts of flower borders, and in their native regions, for wild gardens. They need a sunny, well-drained location and porous soil and succeed in comparatively poor earth provided it is sandy or gravelly. They bloom in early summer and do not withstand well extremely hot, humid weather. Seeds are sown in early spring or in mild climates in fall where the plants are to bloom and are lightly raked into the soil surface. The seedlings are thinned to about 4 inches apart. No special care is required.

MOKARA. This is the name of orchid hybrids the parents of which include *Arachnis, Ascocentrum,* and *Vanda.*

MOLD. This is an imprecise term rarely employed for woodland soils having much the characteristic of leaf mold and resulting largely from the decay of organic matter. It is also more frequently and somewhat loosely used as a name for profuse fungus colonies, with a more or less woolly or felty appearance, that grow on the damp surfaces of plants, soil, and other organic materials.

Many molds are saprophytic, that is, they live on dead material, but some invade living tissues and are mildly to seriously parasitic. Medicinal penicillin is a product of a mold (*Penicillium*) related to the common blue molds that grow on citrus and other fruits and bread.

Most familiar of plant diseases called mold is the gray mold or botrytis blight that affects a wide variety of plants. Sooty molds live on insect exudates called honeydew. They annoy chiefly by making foliage, fruits, and other parts unsightly. Snow mold affects lawn grasses in spring as the snow disappears, causing conspicuous approximately circular patches of light brown dead foliage and sometimes rotting the crowns and roots of the grasses.

The chief precautions against molds lie in avoiding the stagnant, humid conditions that favor them. Control measures include spraying or dusting with sulfur or with copper-based or other fungicides. Sooty molds are checked or eliminated by destroying aphids, mealybugs, scales, and other honeydew-producing insects. Snow mold usually disappears as the season advances and light and temperatures improve and humidity decreases, so that treatment is usually unnecessary. Under exceptional circumstances it may be worthwhile using a mercury fungicide in late fall before the first snowfall.

MOLE PLANT is *Euphorbia lathyrus.*

MOLES. Although credited with destroying grubs and pupae of many harmful insects, moles are generally highly undesirable in gardens, particularly in lawns and rock gardens. In the course of their extensive tunneling they disturb and mechanically damage the roots and other underground plant parts and admit air in amounts sufficient to dry fine feeding roots. It is thought too that they feed to a limited extent on some bulbs, but such damage attributed to moles may be the work of mice that enter mole runs. Besides the damage moles do to plants they cause unsightly upheavals of the surface soil, those of the common eastern North American species forming long, loose ridges, those of the eastern star-nosed mole and the Oregon mole, large mounds. Moles can be combated by using, strictly according to the manufacturer's direction, a commercial poison bait. If properly employed, harpoon-type, scissors-jaw, and choker traps are effective. When handling these it is advisable to use gloves and to pass the traps through a flame to ensure that they carry no human scent. Position the trap by opening and blocking a short portion of a runway with soil. Set the trap above the blocked portion so that when the mole attempts to repair the damage it will spring the trap. Small bulbs are sometimes protected by planting them inside wire mesh baskets.

MOLINERIA RECURVATA is *Curculigo capitulata.*

MOLINIA (Molín-ia). The two hardy perennial grasses that constitute this genus are native to Northern Europe and Asia. One is sometimes grown as an ornamental. The name commemorates Juan Ignacio Molina, who studied and wrote about the natural history of Chile, and who died in 1829. The genus *Molinia* belongs in the grass family GRAMINEAE.

Molinias form dense tufts of erect, slender stems and leaves with flat blades. Their loose or compact flower spikes have one- to four-flowered spikelets.

The purple moor grass (***M. caerulea***), a native of Europe, is naturalized in fields and at roadsides from Maine to Ontario and Pennsylvania. From 1 foot to 4 feet tall, it has stems with their lower leaves much closer together than the upper ones. The stiffish leaves are up to 1 foot long by ¼ inch wide. Its flower panicles are slender and purplish, the branches erect. Variety *M. c. variegata* is about 1½ feet tall

Molinia caerulea variegata

and has leaf blades striped with cream or white.

Garden Uses and Cultivation. The purple moor grass and its variegated variety are well suited for moist, acid soils. The latter is sufficiently ornamental to warrant its use in beds and borders; the typical species is more suited for watersides and informal areas. Both are easy to grow in sunny locations. They are readily increased by division in spring or early fall, and, the typical species, by seed.

MOLOPOSPERMUM (Molopo-spérmum). Fortunately perhaps for gardeners with an antipathy for long botanical names, the only species *Molospermum peloponnesiacum* (syn. *M. cicutarium*) of this genus is rarely cultivated. It belongs in the great carrot family UMBELLIFERAE. Occasionally it is planted as a hardy ornamental. The name, from the Greek *molopos*, a weal, and *sperma*, a seed, alludes to the striped seeds.

Native to the Pyrenees and southern Alps of Europe, ***M. peloponnesiacum*** is kin to lovage (*Levisticum*). It is a stout, hairless, herbaceous perennial, with hollow stems, up to 5 or 6 feet in height. Its leaves, those from the base and lower parts of the stem up to 3 feet in length, are two- to four-times-divided into deeply-toothed, lanceolate leaflets. The tiny white to yellowish flowers, each with a five-lobed calyx and five each petals and stamens, are in umbels with bracts that are sometimes leaflike. The umbels are held well above the foliage. The lateral ones, smaller than the terminal ones, are mostly opposite or in whorls (circles of more than two).

Garden Uses and Cultivation. Informal landscapes provide the best setting for this plant, which is grown chiefly for its bold foliage effects. It is grown without difficulty in ordinary, moderately moist soil, in sun, and is increased by division and by seed.

MOLTKIA (Mólt-kia). This segregate from *Lithospermum* consists of six species, one-half of which are natives from Italy to Greece and the others to Asia Minor. With the exception of *M. aurea*, a kind not known to be cultivated, the sorts of *Moltkia* differ from *Lithospermum* in not having yellow flowers; theirs are blue or purple and are in one-sided, elongated clusters. The genus belongs in the borage family BORAGINACEAE. Its name honors Count Joachim Gadake Moltke of Denmark, who died in 1818.

Moltkias are rough-hairy herbaceous perennials and subshrubs. Their flowers are funnel-shaped and have five lobes that, unlike those of *Lithospermum*, do not spread widely. The nutlets of the fruits may be shiny and smooth or opaque and rough. From nearly related *Lithodora* moltkias differ in their flower clusters being coiled and consisting of more than ten blooms and in the nutlets of the fruits being obviously bent at their centers. From *Buglossoides* they may be told by the absence of vertical lines of glands or hairs within the corolla tube.

Two subshrubby species of this genus, *M. petraea* and *M. suffruticosa*, are cultivated. Attaining a height of 6 inches to a foot, ***M. petraea*** has pale green, leathery, narrow-linear or linear-oblong leaves up to 2 inches long and dense, terminal, sometimes branched clusters of flowers that at first are pinkish-blue, but deepen to rich violet-blue. They are about ½ inch across. Their anthers, which protrude well out of the corolla tube, are blue. A wide-spreading tufted plant, ***M. suffruticosa*** differs from the last in having terminal clusters of pale blue flowers about ½ inch in diameter, with yellow anthers that do not protrude beyond the corolla tube. Its leaves, slender and grasslike, are 2 to 6 inches long. Hybrids intermediate in appearance between these two species are named *M. intermedia* and *M. intermedia froebelii*.

Garden Uses and Cultivation. These, primarily plants for rock gardens, are not well adapted to very cold winters and prosper best where summer temperatures are moderate. They thrive in well-drained, limy soil in sun and are increased by seed, cuttings, and layering.

MOLUCCA-BALM is *Moluccella laevis*.

MOLUCCELLA (Moluccél-la) — Molucca-Balm or Shell Flower or Bells of Ireland. The genus *Moluccella*, of the mint family LABIATAE, is native from the Mediterranean region to northern India. Its name derives from the Molucca Islands, at one time wrongly thought to be its native home. There are four species. The name has been spelled *Molucella*.

Moluccellas are annuals with toothed or lobed, stalked leaves and symmetrical flowers in axillary whorls arranged in tiers. Their prominent calyxes are saucer- or cup-shaped. The corollas, which scarcely if at all exceed the calyxes in length, have an erect, hooded upper lip and a three-lobed lower one, with the center lobe broader than the others and notched. The blooms have four stamens under the upper lip, a two-branched stigma, and an oblique hairy ring within the corolla. The fruits consist of four seedlike nutlets. This genus is easily distinguished from such related ones as *Lamium*, *Phlomis*, and *Leonotis* by the large, undivided, saucer-like calyxes of its flowers.

The most popular kind, the Molucca-balm, shell flower, or bells of Ireland (***M. laevis***) is a native of western Asia admired chiefly for its unusual spires of bloom. Under favorable conditions it attains a height of 3 feet. Long-stalked and nearly circular, its leaves are about 1½ inches across. The

Moluccella laevis

Moluccella laevis (flowers)

small, white or pinkish, fragrant flowers are not very conspicuous except for their large seashell-like calyxes, which are pale green netted with white veins and slightly prickly at their angles. Native to southern Europe and Syria, ***M. spinosa*** is an annual or a biennial, sometimes 3 feet tall. It has brownish stems, deeply-cut, ovate leaves, and flowers in terminal clusters. They have obliquely-two-lipped calyxes, with one long spine above and seven shorter ones below, and white corollas.

Garden Uses. As a garden plant Molucca-balm is more interesting than beautiful. Unless supported it tends to sprawl and, in hot weather, after its flowers are past, it dries rather suddenly and presents an unattractive mass of yellow and brown foliage. Its flower spikes are useful for cutting for indoor arrangements and this species is grown for this purpose. Best adapted for planting in the rear of mixed flower borders is *M. spinosa*. Both kinds need full sun.

Cultivation. Moluccellas thrive in a well-drained, sandy fertile soil. Where the season is long enough, the seeds may be sown in early spring directly outdoors where the plants are to bloom. Alternatively, they may be sown indoors in a temperature of about 60°F about eight weeks before it is safe to set the young plants outdoors, which is about the time that the first sowing of corn can be made; the young plants are grown in pots until setting-out time. Spacing between plants of *M. laevis* should be about 1 foot, one-half as much again should be allowed between individuals of *M. spinosa*. These plants sometimes reproduce themselves from self-sown seeds. For the production of cut flowers in greenhouses, seeds are sown in sandy soil in September, October, or January, and the young plants are planted in small pots or flats. From these, before their roots become crowded, they are planted either in soil beds or benches, spaced 10 to 12 inches apart, or are successively potted into bigger pots until the final containers, 6 or 7 inches in diameter, are achieved. When they have filled their available soil with healthy roots they benefit from weekly or semiweekly applications of dilute liquid fertilizer. Throughout, they are grown in full sun in a night temperature of 50 to 55°F and day temperatures five to ten degrees higher.

MOMBIN. See Spondias.

MOMORDICA (Momórd-ica)—Balsam-Pear or Bitter-Melon, Balsam-Apple. Fast-growing annual and herbaceous perennial vines with branchless or branched tendrils constitute *Momordica*, of the gourd family CUCURBITACEAE. They are natives of the tropics and subtropics of the Old World, with some kinds naturalized elsewhere in warm regions, including the southern United States and Hawaii. The name, from the Latin *mordeo*, to bite, makes reference to the seeds of some of the forty-five species having eroded margins that look as if they have been bitten. In the Orient the fruits of the balsam-pear, cooked before they are fully ripe, are eaten, and the seeds are used in curries. The young shoots and leaves are also used as food in some places, and are employed medicinally. If taken in excess they may be poisonous.

Momordicas have alternate leaves, smooth-edged or lobed, or divided palmately (in hand-fashion) into three to seven leaflets. The yellow or white flowers are unisexual, the males and females on the same or different plants. The males are solitary, the females solitary or in panicles. They have five-lobed calyxes, and broadly-bell- to wheel-shaped, usually deeply-five-lobed corollas. There are three stamens and a long style ending in three, sometimes two-lobed, stigmas. The small to fairly large fruits, often warted or spiny, are spherical to cylindrical. In some species they split from the apex downward into three parts.

Balsam-pear or bitter-melon (***M. charantia***) is a vigorous, heavily foliaged annual with stems up to 10 feet long or longer. Its moderately long-stalked leaves, round to heart-shaped in outline, deeply three- to five-lobed, and coarsely-sharply-toothed, are 1 inch to 4 inches in diameter. The shallowly-bell-shaped, 1-inch-wide, yellow blooms are solitary from the leaf axils. Those of the males have a bract near the middle or at the base of the flower stalk. The short-pointed, cylindrical, longitudinally-furrowed, warty-surfaced fruits, 4 to 8 inches long or sometimes longer, are orange-yellow when ripe. They split into three folded-back segments to disclose the brilliant scarlet coverings of their seeds. A small variety, *M. c. abbreviata*, with narrower leaves, and fruits 1 inch to 3 inches long, is a weed of waste places and gardens in Florida, Hawaii, and other warm regions.

Balsam-apple (***M. balsamina***), a less rambunctious grower than the balsam-pear, has similar but mostly smaller, leaves. An annual, a native of Africa, its solitary flowers, 1 inch wide or a little wider, and shallowly-bell-shaped, are yellow, often with blackish centers. The bract on the stalk of the male bloom is located well above its middle. Orange, and bursting when ripe into three recurved segments to exhibit their bright red-coated seeds, the egg-shaped to nearly round, more or less warty fruits are up to 3 inches long. This species is naturalized in the American tropics.

Less commonly cultivated, slender-stemmed ***M. involucrata*** has leaves with stalks up to 1 inch long and roundish, lobed, bluntly-toothed blades about 2 inches wide. Its white to cream-colored often black-dotted blooms have the bract of the stalk of the male flower near its top. The 2-inch-long fruits, at first sulfur-yellow, change to scarlet as they ripen and then burst and display their seeds.

Garden and Landscape Uses and Cultivation. The chief employments of momordicas in gardens is for screening and, as vines, interesting because of their unusual fruits. They grow without difficulty in moistish, fertile soils in sunny locations and are responsive generally to conditions satisfactory for cucumbers, melons, and gourds. They abhor cold weather and are killed by the first fall frost. Seeds may be sown outdoors in spring where the plants are to remain, but in the north it is better to start them early indoors and to set out sturdy young plants from pots about the time it is safe to plant tomatoes. Supports to which the vines can attach their tendrils are needed.

MONADENIUM (Mona-dènium). None of the about fifty species of the African genus *Monadenium* is well known horticulturally, a few kinds are grown by fanciers of succulents. The genus belongs in the spurge

Momordica balsamina

Metasequoia glyptostroboides

A hybrid *Miltonia*

Mimulus aurantiacus

Monarda didyma variety

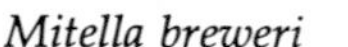

Mitella breweri

Mimulus guttatus

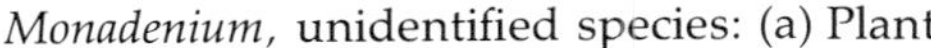

Monadenium, unidentified species: (a) Plant (b) Flowers

family EUPHORBIACEAE and is related to *Euphorbia,* from which it differs in technicalities of its flowering parts. The name, derived from the Greek *monos,* one, and *aden,* a gland, refers to a characteristic of the flowers.

Monadeniums fall into two groups, those with perennial, above-ground, often sparsely-branched and sometimes spiny stems, and those with tuberous rootstocks from which arise annual, usually branchless, stems. The quickly deciduous leaves are alternate and fleshy. They are not divided, but may have toothed or wavy edges. The flowers are mostly in many-branched clusters and are sometimes attractively colored.

A pleasing kind is the erect or somewhat sprawling ***M. coccineum,*** which has five-angled stems 1 foot to 4 feet long and almost stalkless, hairless, oblanceolate to ovate, toothed leaves 1½ to 3½ inches long, with soft prickles on the midribs beneath. The flowers, from near the terminations of the stems, are bright red or red and green. From 1 foot to 2 feet tall, ***M. echinulatum*** has branchless, cylindrical stems. Like the flower stalks and undersides of the leaves, they are thickly beset with fine hairs. The almost stalkless, toothed leaves are 1 inch to 4 inches long, about one-half as wide, and elliptic to elliptic-obovate. The flowers, from the leaf axils, are green, sometimes flushed with pink, and softly spiny on their outsides. Having a tuberous root and stout, erect, cylindrical, spineless stems that sometimes are very shortly-hairy, ***M. lugardiae*** attains a height of 1 foot to 5 feet. Its obovate or spatula-shaped leaves, toothed toward their apexes, are ¾ inch to 3½ inches long and about one-third as wide. They are crowded at the ends of the branches. The short-stalked flowers, in once-forked clusters solitary from the upper

Monadenium echinulatum

Monadenium coccineum

Monadenium lugardiae

Monadenium lugardiae (flowers)

Monadenium ritchiei

leaf axils, are pale green. Its lumpy stems cylindrical, up to 1¼ feet long and ¾ to 1 inch thick, ***M. ritchiei*** has a few elliptic to obovate leaves up to 1 inch long but often smaller. The flowers are pink and stalkless or nearly so. The stems of ***M. schubei,*** at first erect, trail as they lengthen. They are sparingly-branched and thickly covered with prominent tubercles about ⅓ to ½ inch wide and about as high. The stalkless leaves are crowded near the ends of the stems. Oblanceolate and 1½ to 3 inches long, they have crisped edges. The flower clusters, once- or sometimes twice-forked, are short-stalked. The green blooms are sometimes tinged with pink.

Garden Uses and Cultivation. These are collectors' plants. They may be grown outdoors in warm, dry climates and in sunny greenhouses. They require the same conditions as succulent euphorbias. Those with annual stems are kept quite dry during the time of the year when they are without leaves. Propagation is by seed and by cuttings. For information on their general culture see Succulents.

MONANTHES (Monánth-es). A dozen species of the Canary Islands, Salvage Islands, and Morocco constitute this group of dwarf, sometimes minute, herbaceous plants and subshrubs of the orpine family CRASSULACEAE. All but one *Monanthes,* seemingly not cultivated, are perennials. None is hardy in the north. The name, given under a false assumption that the flowers of the first kind named were solitary, is from the Greek *monos,* one, and *anthos,* a flower. Because of the considerable variability of some species and the freedom with which they hybridize in the wild and in cultivation, precise identification often presents problems, even to skilled botanists.

Members of this genus have erect or prostrate stems, branched or not, and alternate or rarely opposite, succulent, cylindrical, club- to egg-shaped leaves often crowded toward the ends of the branches and frequently thickly covered with tiny projections. Usually on comparatively long, slender stalks, the yellowish, greenish, or purplish, small, starry flowers are in racemes or clusters. The six- to eight-lobed calyxes are saucer-shaped. There are as many linear to lanceolate petals as calyx segments and twice as many stamens. The carpels are as many as the petals.

Frequently cultivated, and very variable, ***M. laxiflora*** is easily distinguished by its fat, narrowly-egg-shaped to nearly globular, opposite leaves, up to 1 inch long and up to slightly over ½ inch wide, but often smaller. They are loosely arranged on forking, erect or lax stems that bear at their ends six- to ten-flowered, usually slightly hairy racemes of five- to eight-petaled, greenish-yellow flowers often marked with reddish dots or lines that give a purplish appearance. Varying considerably in size, the flowers are mostly about ⅓ inch wide. They have hairless or slightly hairy calyxes. The leaves are green, purplish, or, under dry conditions, silvery; the sprays of flowers are from very short to up to 3 inches or so long.

A usually solitary, loose rosette of about twenty very fleshy leaves at the top of a short, thick, branchless cylindrical or bulb-like stem or rootstock is characteristic of ***M. brachycaulon,*** which inhabits the Canary Islands and the Salvage Islands. Its blunt-pointed, oblanceolate-spatula-shaped, leaves, flat on top and with rounded un-

Monanthes brachycaulon

dersides, are up to ¾ inch long by ⅕ inch wide. Hairless and mottled with purple, they have a stain at the base and midribs of the same color and are covered with tiny raised projections. The purplish-green flowers, ⅓ inch wide or a little wider, have hairy calyxes and are in racemes of five to seven.

A diminutive, creeping species of considerable charm, ***M. polyphylla*** has slender, prostrate stems. It forms broad cushions of rounded to conical, very dense rosettes of numerous overlapping wedge- to spatula-shaped, fleshy leaves, up to ⅓ inch long, and rounded on both surfaces. The rosettes in the wild are pinkish-gray-green, but in cultivation are usually green or purplish. From their centers arise slender, few-branched stalks, with a flower at the end of each conspicuously hairy branch. Pale greenish to brownish-purple, they have six to eight petals and are somewhat under ½ inch wide. Their calyxes are hairy.

Garden Uses and Cultivation. Collectors of succulents find these somewhat sedum-like plants attractive for growing in pans (shallow pots) and for planting in ground beds among rocks in greenhouses devoted to desert and other plants needing arid conditions. In frost-free, dry climates they can be accommodated in outdoor succulent gardens. They prefer sandy soil, and it must be well drained. Somewhat unusual for succulents, they appreciate a little shade from the full intensity of the summer sun, but this must not be overdone, and at other seasons full exposure is needed. When grown indoors, cool conditions in winter are requisite. Night temperatures of 40 to 50°F are sufficient, and by day they should not be more than five or ten degrees higher. Little water is required in winter and only very moderate amounts from spring to fall. Propagation is very easy by cuttings, leaf cuttings, and seed. For further information see Succulents.

MONARCH-OF-THE-EAST is *Sauromatum guttatum.*

MONARDA (Mon-àrda)—Horse-Mint, Oswego-Tea or Bee-Balm or Fragrant-Balm, Wild-Bergamot. Sixteen species compose the wholly American genus *Monarda.* They belong in the mint family LABIATAE and have a name that commemorates the Spanish physician and botanist Nicholas de Monardes, who died in 1588.

Mostly fairly tall, erect, and aromatic, these annuals, biennials, herbaceous perennials, or rarely shrubs have four-angled stems, and opposite, undivided, ovate to linear, usually toothed leaves. Their asymmetrical, narrowly-tubular flowers are crowded in heads that are solitary and terminal or come from the leaf axils and form interrupted spikes of two or more tiers. The heads usually rest on a collar of often brightly colored bracts. The flowers have five-toothed calyxes with fifteen veins, and two-lipped corollas, with tubes slightly broadened toward their mouths, and hairy on their outsides. The upper lip of the corolla is erect or arching, the lower is three-lobed, with the center lobe longer than the others. There are two generally protruding stamens and sometimes a pair of rudimentary ones. The style has two stigmas. The fruits consist of four seedlike nutlets. Several species were used by American Indians for flavoring foods and sometimes for making teas.

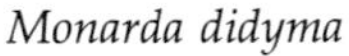
Monarda didyma

Monarda fistulosa

Oswego-tea, bee-balm, or fragrant-balm (***M. didyma***) is a handsome, variable perennial, the progenitor of many improved garden varieties. As a wildling it ranges from Maine to Michigan and North Carolina and is most abundant in moist thickets and woodlands. This has sometimes branched stems, 2½ to 4 feet tall, and ovate to nearly lanceolate leaves 3 to 6 inches long. Its 2-inch-long, showy, bright crimson flowers are in heads with bracts usually tinged red.

Wild-bergamot (***M. fistulosa***), another attractive perennial, is indigenous from Quebec to Manitoba, Georgia, Louisiana, Texas, and Arizona, and in British Columbia. From 2 to 4 feet in height, it has sometimes branched stems, and triangular-ovate to lanceolate leaves 2 to 4½ inches long. The muted lilac to pale purple or sometimes white flowers, about 1½ inches long, are in terminal heads with whitish or purplish-tinged bracts. The upper lip of the corolla is softly-hairy, densely so at its apex. A rather inferior white-flowered variant is *M. f. alba*. Western American *M. f. menthaefolia* (syn. *M. menthaefolia*) differs in its leafstalks being under ¼ inch instead of over ⅓ inch long. In *M. f. rubra* the flowers are rose-red. Closely related, variable, perennial ***M. media*** is thought to be a hybrid between *M. fistulosa* or *M. clinopodia* and *M. didyma*. Its rose-purple flowers are in heads with bracts strongly purple-tinged. The upper lip of the corolla is hairless or very shortly hairy, but has no dense hairiness at the apex. This supposed hybrid is found within the ranges of its conjectural parents.

Inhabiting dry soils from Illinois to Alabama, Louisiana, Arkansas, and Texas, ***M. russeliana*** (syn. *M. bradburiana*) is 1 foot to 2 feet tall and hairless or sparingly-hairy. It has ovate to triangular-ovate or lanceolate, short-stalked, remotely toothed to nearly toothless leaves up to 2¼ inches long. Its purple-spotted, whitish to pinkish flowers, about 1½ inches long, are in solitary, terminal heads.

Spotted bee-balm (***M. punctata***) is a perennial, a biennial, or an annual, commonly up to 3 feet tall. Native from Vermont to Minnesota, Florida, and Texas, it varies so considerably that botanists recognize many varieties. The annual, with purple-spotted, white or yellowish flowers previously called *M. lasiodonta*, is *M. p. lasiodonta*. Spotted bee-balm has stalked, lanceolate to oblong-lanceolate leaves from ¾ inch to 4 inches long. The yellowish to white or rarely pinkish flowers, usually spotted with purple or maroon and about 1 inch long, are in heads tiered to form spikes. The bracts are white, yellowish, or purplish.

Monarda punctata

Lemon bee-balm or lemon-mint (***M. citriodora***) is a variable annual or biennial 1 foot to 2 feet tall, native from Missouri to Nebraska, Utah, Texas, and Arizona. It has lanceolate to oblanceolate leaves up to a little over 2 inches long. The heads of flowers are in tiers. They have equal-sized, purple to green bracts. The yellowish-white to pink corollas, stippled with purple, are about ¾ inch long, and slightly longer than the bracts.

Many splendid horticultural varieties, chiefly variants and hybrids of *M. didyma* and *M. fistulosa* have been raised. These are sturdy plants of compact habit, usually with larger flowers, and flower heads of superior decorative merit. They come in a wide range of colors from white to pink and deepest red and include shades of mahogany and maroon. Here belong such outstanding varieties as 'Adam', compact and with cerise-red flowers; 'Croftway Pink', with lovely large heads of rose-pink flowers; and 'Mahogany', with rich, deep red blooms.

Garden and Landscape Uses. These are handsome summer- and fall-blooming, hardy plants for flower beds, wild gardens, and naturalistic landscapes. Clumps show to advantage near watersides, and the flowers provide useful cutting for arrangements. It is proper, too, to include oswego-tea, lemon bee-balm, and some others in herb gardens.

Cultivation. Few groups of plants so worthy of the attention of gardeners are less demanding of their skills and time than these. Monardas thrive in a variety of

Monarda 'Croftway Pink'

Monarda 'Croftway Pink' (flowers)

soils, preferring those reasonably fertile and moderately moist. They prosper in sun or part-day shade. The perennials are very easily increased by division in early fall or spring, and the species by seed. Seeds of horticultural varieties do not give progeny identical with the parents. Annual or biennial kinds are raised from seeds sown in fall or early spring where the plants are to remain, and the seedlings thinned out to prevent overcrowding.

MONARDELLA (Monard-élla)—Mountain-Mint, Coyote-Mint. The western North American genus *Monardella*, of the mint family LABIATAE, consists of about twenty species of aromatic annuals and herbaceous perennials. The name is a diminutive of *Monarda*, which *Monardella* resembles.

Monardellas have opposite, toothed or toothless, small leaves and terminal heads of flowers, with an involucre (collar) of often purplish bracts. The blooms have narrow, tubular, five-toothed calyxes and tubular slightly two-lipped corollas of five nearly equal lobes. The lower lip extends outward or downward and is three-lobed; the erect upper lip has two lobes. The corolla is pink, purplish, red, yellowish, or white. There are four fertile stamens and a style with two unequal branches. The fruits are of four seedlike nutlets included in a persistent calyx.

An annual, ***M. lanceolata*** is 6 inches to 2 feet tall and has toothless, lanceolate to oblong leaf blades ¾ inch to 2 inches long on stalks up to ½ inch long. Rose-purple or paler, the flowers, in heads up to a little over 1 inch across, are ½ inch long. The bracts are green, often tipped with purple. This kind inhabits dry soils in California.

Mountain-mint (***M. odoratissima*** syn. *M. purpurea*) is a very variable perennial from 6 inches to 2 feet tall. Native from Washington to New Mexico, it branches from a woody base, has short-stalked to nearly stalkless, toothless, lanceolate leaves more or less hairy, and is ½ inch to 1¼ inches long. The light purple to rose-purple flowers are in compact heads up to 2 inches across with broad, sometimes roundish, hairy bracts.

The coyote-mint (***M. villosa***) is a perennial with more or less prostrate stems, up to 2 feet long, woody at their bases, branched or not, and hairy in their upper parts. Its ovate, usually toothless leaves are dark green, sparsely-hairy, and have slender stalks about one-half as long as the blades, which are up to 1 inch long. The compact flower heads, of purple, pink, or whitish blooms, are up to 1½ inches wide.

Scarlet to yellowish flowers are characteristic of ***M. macrantha***, a native of dry soils in California. They are 1 inch to nearly 2 inches long, in loose heads of ten to twenty that somewhat suggest those of ixoras. The heads are up to 1¾ inches wide and have collars of purplish, hairy-margined bracts. This species is rarely over 1 foot tall and often much lower. Its leaf blades, on stalks one-half as long as themselves, are ovate to elliptic, ½ inch to 1¼ inches long, pubescent beneath, less hairy on their upper surfaces. This species is perennial.

Garden and Landscape Uses and Cultivation. Monardellas are most satisfactory in climates of fairly mild winters and moderately dry summers. They prosper in full sun and are readily propagated by seed and division. Annual kinds must, of course, be renewed from seeds sown each year. They are useful for the fronts of flower borders, rock gardens, and in their home territories for native plant gardens.

MONDIA. See Chlorocodon.

MONDO-GRASS. See Ophiopogon.

MONESES (Mon-èses) — One-Flowered-Shinleaf. The charming little one-flowered-shinleaf, sole species of its genus of the heath family ERICACEAE, haunts cool northern parts of all continents of the northern hemisphere. It is interesting botanically because, although not of hybrid origin, it is intermediate between *Pyrola* and *Chimaphila*. Its name is from the Greek *monos*, single, and *hesis*, a delight, and alludes to the blooms being solitary.

The one-flowered-shinleaf (***M. uniflora***) is a delicate, hairless, evergreen perennial herbaceous plant that extends itself by underground stolons, the ends of which emerge to bear small basal clusters of nearly round, toothed to almost toothless, flat, thickish leaves up to ¾ inch long, in twos or threes. The fragrant, nodding, solitary blooms, ½ to ¾ inch wide, top slender, naked stalks ¾ inch to 4 inches tall. There are five sepals and five wide-spreading, roundish, white or sometimes pinkish petals, ten stamens with anthers with two conspicuous horns, and a long-protruding, shield-shaped, five-lobed style. The fruits are small, nearly round capsules. They open from the apex downward. The western American form of the species is sometimes identified as *M. u. reticulata*.

Garden Uses and Cultivation. Appropriate for woodland gardens and rock gardens, this frail denizen of the cool north is not well adapted for growing in warmer places. It needs acid soil containing abundant leaf mold or other satisfying organic debris that is always moist, but not sodden, a humid atmosphere, and shade. A north-facing slope or the shaded side of a rock is likely to afford the most acceptable environment provided the soil is agreeable. A light mulching with decayed leaf mold is desirable, and in regions of sparse snow covering, winter protection is afforded by a few branches of pine or other evergreens positioned over it. Propagation is by careful division in spring and by seeds sown as soon as they are ripe in pots in a shaded cold frame or similar environment.

MONEYWORT is *Lysimachia nummularia*. Cornish-moneywort is *Sibthorpia europaea*.

MONILARIA (Monil-ària). The Latin *monile* means a string of pearls and from it the name *Monilaria* is derived. It alludes to the stems, which in most species suggest rows of beads joined at their points of contact. Monilarias are among the most curious of a curious race, the *Mesembryanthemum* group of small succulents that inhabit South Africa and include the unusual stone and windowed plants, such as argyrodermas, gibbaeums, pleiospiloses, oscularias, and ophthalmophyllums. They belong in the carpetweed family AIZOACEAE.

There are a dozen species of monilarias. They form low clumps, with short, fat stems, that for much of the year show scarcely any evidence of life. In fall, from the end of each stem develops a pair of leaves joined almost to their tips into a tiny hemispherical body. After a while these gape and from between them come two comparatively long, nearly cylindrical leaves joined only at their bases. These remain until the beginning of the next resting season in spring when they wither and remain as dried remnants. The handsome flowers terminate stalks that arise from between the pairs of cylindrical leaves. In apparent form they are daisy-like, but in structure they differ very markedly from daisies for each bloom is a single flower, not a head composed of numerous florets. Very like those of *Lampranthus*, the flowers have numerous slender petals and five to seven stigmas. The fruits are capsules. Thus, throughout their lives, these plants develop alternately pairs of leaves of two distinct types, normally one pair of each kind each year. In this they resemble their close relatives *Mitrophyllum* and *Conophyllum*.

Known as long ago as 1795, ***M. moniliformis*** attains a height of about 4 inches and has branching, beaded stems. Its subspherical leaf bodies are dark green and under ½ inch in diameter. The cylindrical leaves are up to 4 inches or more long and ⅕ inch wide. White with yellow stamens and 1½ inches across, the blooms are on stalks up to 2 inches long. The pairs of short rounded leaves of ***M. chrysoleuca*** are cleft more deeply at their tops than those of the last. Its cylindrical leaves are up to 2 inches long by ⅕ inch wide. This kind is also about 4 inches tall. A little over 1 inch in diameter, its flowers, on stalks about 1½ inches long, are pure white with yellow stamens. Very succulent ***M. globosa*** is only about 1¼ inches tall and has stems densely clothed with old leaves. It has small globose or conical branches. From between the resting pairs of leaves, which are very small, there arise other pairs joined for about one-third of their lengths, and 1½ to 1¾ inches long. The white flowers, 1½ inches across, are on stalks up to slightly over 1 inch long. From 3 to 5 inches tall, ***M. peersii*** has short, papery-surfaced branches and glaucous leaves scarcely or not visible when the plant is dormant, 1½ to 2½ inches long when the plant is in active growth, and then thick and with a flattened upper surface and rounder under one. The flowers, on ¼-inch-long stalks, are white and about ¾ inch across.

Garden Uses and Cultivation. These very rare plants are treasured by collectors of choice, nonhardy succulents. Not a great deal is known about their cultivation, but this appears to be less difficult than was at one time supposed. They should be given conditions appropriate for *Lithops* and other dwarf relatives of the *Mesembryanthemum* clan. It is especially important to respect their dormant season by keeping them dry at that time. Propagation is by seed and careful division. For additional information see Succulents.

MONIMIACEAE—Monimia Family. Of minor horticultural impact, this tropical and subtropical family of dicotyledons, which is most abundant in the southern hemisphere, comprises twenty genera totaling 150 species. Its members are evergreen trees and shrubs, or rarely vines, with most often opposite, less frequently whorled (in circles of more than two) or alternate, leathery leaves, usually aromatic when bruised.

Solitary or clustered, the prevailingly bisexual, but sometimes unisexual, symmetrical flowers have hollowed, cuplike centers and inconspicuous perianths or none. When present they may be of all calyx-like parts or of distinctive sepals and petals. The short-stalked stamens are few to many. There is one short style and one stigma. The fruits are achenes or drupes. Genera sometimes cultivated are *Hedycarya*, *Laurelia*, and *Peumus*.

MONKEY. This word monkey occurs as part of the common names of these plants: monkey bread (*Adansonia digitata*), monkey flower (*Mimulus*), monkey flower tree (*Phyllocarpus septentrionalis*), monkey nut (*Hicksbeachia*), monkey pod (*Samanea saman*), monkey pot (*Lecythis*), monkey-puzzle tree (*Araucaria araucana*), monkey's comb (*Pithecoctenium cynanchoides*), and monkey's hand (*Cheirostemon platanoides*).

MONK'S PEPPER TREE is *Vitex agnus-castus*.

MONKSHOOD. See Aconitum.

MONOCOTYLEDON. Monocotyledons, often called monocots, form one of two groups into which botanists classify angiosperms, a major subdivision of the flowering or seed plants. The other group is the dicotyledons. There is no one constant difference between the two. Separation is based on the sum total of a few varying characteristics. The most constant is the one suggested by their names, that of the first derived from the Greek *monos*, one, and cotyledon, that of the other from *dis*, two, and cotyledon. Monocotyledons nearly always have seeds with a single cotyledon. Other common characteristics of monocotyledons are that their leaves are usually parallel-veined and the parts of the flower, when definite in number, are generally in threes or multiples of three. Also, in those that are trees and shrubs there is no cambium layer beneath the bark, and the trunks and branches do not thicken by the addition of annual rings. Because of the lack of a cambium layer it is rarely possible and certainly not practicable to propagate monocots by grafting. More recondite differences between monocotyledons and dicotyledons concern their roots and pollen. Among the most familiar monocots are those belonging to the arum, iris, lily, and orchid families.

MONOECIOUS. Kinds of plants in which the flowers are unisexual, but both sexes are on the same plant, such as begonias, corn, and pines, are termed monoecious. Compare Dioecious.

MONOLENA (Mono-lèna). About eight to ten species of rather fleshy, stemless herbaceous perennials related to *Bertolonia* that inhabit the humid forests of Central America and western South America constitute *Monolena*, of the melostoma family MELASTOMATACEAE. The name, derived from the Greek *monos*, one, and *olene*, an arm, refers to the solitary projection or spur from the base of each anther.

Monolenas have short fleshy rhizomes and toothed or smooth-edged, elliptic-oblong to nearly round leaves. Their pink, five-petaled flowers are in showy, curved racemes. The fruits are capsules.

The only kind known to be cultivated, beautiful ***M. primulaeflora*** (syn. *Bertolonia primulaeflora*) native from Colombia to Peru, has short, thick, erect, clustered rhizomes.

Monolena primulaeflora

An epiphyte (plant that perches on trees without taking nourishment from its host) 6 inches to a foot tall, it blooms freely, usually several times each year, and is attractive both in foliage and bloom. Its flowers, pink with white centers and yellow anthers, are about 1 inch in diameter. The pointed-elliptic leaves, up to 6 inches long, are bright lustrous green with wine-purple undersides and with five conspicuous veins from base to apex. Their margins are slightly wavy-toothed and fringed with hairs. In one variant the leaves are delicately variegated with dark red.

Garden Uses and Cultivation. This species is for enthusiastic collectors of rare tropical plants. It needs the same growing conditions and attentions as *Bertolonia.*

MONOLOPIA (Monolòp-ia). The genus *Monolopia* comprises four species occurring natively only in California. It belongs in the daisy family COMPOSITAE, and its name, from the Greek *monos,* one, and *lopos,* a covering, refers to the bracts of the involucre (collar of leafy organs behind the flower head) being in one row.

Monolopias are white-woolly-hairy annuals with stalkless leaves, the lower ones opposite, the upper alternate. Their rather large, stalked daisy-type flowers terminate the branches. The tips of the involucral bracts are usually furnished with black hairs. The fruits are seedlike achenes.

Differing from some monolopias in having its involucral bracts joined in their lower halves to form a cup instead of being separated to their bases, ***M. major,*** from 6 inches to 2 feet tall, is usually sparsely-branched. It has oblong or lanceolate leaves and golden-yellow flower heads about 2 inches across with about eight ray florets. Very similar ***M. lanceolata*** tends to be somewhat smaller in all its parts.

Garden Uses and Cultivation. These are attractive annuals for sunny flower beds and borders and for inclusion in wild gardens in their native California. They bloom freely in early summer and flourish on sun-drenched slopes where the ground is porous and well drained. The flowers are attractive for cutting and last well in water. Seeds may be sown outdoors in early spring where the plants are to bloom, or they may be started indoors earlier in a temperature of about 60°F. If the latter procedure is followed the young plants are transplanted 2 inches apart in flats and grown in a sunny greenhouse in a night temperature of 50°F and a day temperature five to ten degrees warmer until all danger of frost is passed and it is safe to transplant them to the garden. Indoor sowing should be done seven or eight weeks before the expected planting out date. In the garden the plants should be spaced about 6 inches apart. Monolopias need no particular care, but weeds must be controlled, faded flower heads removed, and in very dry weather water provided. Staking is not needed. They do not withstand very hot and humid weather well.

MONOPSIS (Mon-ópsis). Relatives of *Lobelia,* and sometimes included in that genus, the eighteen species of *Monopsis,* of the bellflower family CAMPANULACEAE, are annuals and herbaceous perennials of South Africa and northward in Africa. The name, derived from the Greek *monos,* one, and *opsis,* a face, refers to the flowers not being markedly two-lipped. From *Lobelia* the sorts of *Monopsis* differ in the lobes of the stigma being thread-like instead of short and roundish.

Low or prostrate, with spreading branches, the species of this genus have alternate, toothless, toothed, or divided leaves, and solitary, long-stalked blue or yellowish-brown flowers from the leaf axils. The calyx is deeply-lobed. The corolla has a tube split down one side to the base and five spreading lobes (petals), the front ones overlapping. The fruits are capsules.

An attractive, bushy annual of South Africa, ***M. campanulata*** (syn. *Lobelia campanulata*) rarely exceeds 1 foot in height and may be lower. It is hairless. The lower leaves are few-toothed or toothless, ovate-lanceolate, and about ¾ inch long by one-third as wide. Rich blue, and with contrasting bright yellow anthers, the bell-shaped flowers are about ¾ inch across.

Garden and Landscape Uses and Cultivation. The species described is best suited for regions of fairly cool summers. It is appropriate for rock gardens and the fronts of flower borders and succeeds in ordinary, well-drained, moderately fertile soil, preferably in part-day shade, although it will stand full sun. It is raised from seeds sown in spring where the plants are to remain, and the seedlings thinned to prevent undue crowding, or from seeds sown indoors earlier, and the young plants treated like lobelias are when grown as annuals, and planted outdoors after the weather is warm and settled.

MONOPTILON (Monóp-tilon)—Desert Star. Two species of small desert annuals of the southwestern United States and adjacent Mexico are the only representatives of *Monoptilon,* of the daisy family COMPOSITAE. The name is derived from the Greek *monos,* one, and *ptilon,* a feather, and refers to the single feathery pappus hair on the ovary of *M. bellidiforme.*

The commonest kind, ***M. bellioides*** forms a specimen up to 10 inches in diameter and 4 inches in height and has roughly-hairy stems and foliage. The leaves, linear to spatula-shaped, are up to ¾ inch long. The abundant, daisy-like, white-rayed flowers, with yellow or purplish disks, are ½ inch or more in diameter. This differs from much rarer ***M. bellidiforme*** in usually being larger and in not having plumose (feathery) pappus hairs. The fruits of these plants are achenes.

Garden Uses and Cultivation. These are attractive for early summer display and for relatively poor, dryish soils. They do not succeed in very hot, humid weather, nor do they prosper in shade. They are well adapted for edgings, rock gardens, and in regions where they are native, for wild gardens. Although small, the blooms last well when cut and can be used effectively in small arrangements. Their cultural requirements are not demanding. Seeds are sown in open, sunny areas in early spring and are lightly raked into the soil surface. The seedlings are thinned to about 4 inches apart. No special care is needed, though weeds must be kept down. The prompt removal of faded flower heads prolongs the season of bloom.

MONOTAGMA (Mono-tágma). The twenty species of *Monotagma,* relatives of *Calathea,* belonging to the maranta family MARANTACEAE, are natives of South America. The name, from the Greek *monos,* one, and *tagma,* order, has no obvious significance.

Monotagmas are herbaceous perennials with creeping rhizomes and more or less two-ranked, long-stalked basal leaves much like those of marantas and calatheas, and like them, with a swollen joint at the junction of leafstalk and leaf blade. Their solitary flowers, in bracted spikes, differ from those of *Maranta* in having one or rarely no staminodes (nonfunctional stamens) instead of two. There are three sepals, the same number of petals, and one fertile stamen.

About 1½ feet in height, ***M. smaragdinum*** (syns. *Calathea smaragdina, Maranta smaragdina*) has lanceolate to oblong-lanceolate, emerald-green leaves with a longitudinal, dark green center stripe, and paler, downy undersides. The swelling where the long leafstalk joins the blade is also downy. This is a native of Ecuador.

Garden and Landscape Uses and Cultivation. These are as for *Calathea.*

MONOTROPA (Monó-tropa)—Indian Pipe, Pinesap. Monotropas are scarcely cultivated plants in the ordinary sense of the word, but are elements in our American native flora that excite curiosity when found in the wild and are sometimes transferred to gardens. They should be afforded conditions as similar to those under which they grow naturally as possible. By various authorities *Monotropa* is put in the shinleaf family PYROLACEAE, the heath family ERICACEAE, and in a family of its own, the MONOTROPACEAE. Their name comes from the Greek *monos*, one, and *tropos*, turn, and refers to the nodding flowers. There are two species.

Although to nonplantsmen the rather ghostly appearance of monotropas may suggest some strange mushroom-type fungus, these are not funguses (which are without flowers), but true flowering plants. They are parasites on soil funguses that live upon decaying organic matter. From these they obtain their nourishment. Because they have no need to manufacture food from simple elements, as do most plants, they have no need of chlorophyll, and they have none.

Monotropas are erect plants with leaves represented by fleshy scales along the stems. Their flowers, broadly-tubular or urn-shaped, have four or five separate petals and two to five or no sepals. There are eight to ten slender, pubescent stamens and a stout, short style with a shield-shaped stigma. The fruits are capsules.

The Indian pipe (***M. uniflora***) is native throughout much of North and Central America as well as eastern Asia. It is a waxy-white plant, 4 to 8 inches tall, with a shepherd's-crook-shaped stem and a solitary, stiffly nodding, terminal, often sepal-less flower a little under to a little over ½ inch long. It grows in the leaf mold of rich woodlands in deep shade.

Monotropa uniflora

The pinesap (***M. hypopithys***), by some botanists treated as a separate genus and named *Hypopitys monotropa*, is indigenous in acid-soil woodlands throughout much of North America, including Mexico, and is widely distributed in the Old World. Up to 1 foot tall and yellow, pink, or red, its pubescent stems curve at their tops and end in a raceme of few to many fragrant, orange flowers under ½ inch long. The upper blooms are often larger than those at the base of the raceme and have five sepals and petals instead of the four of the blooms below.

MONOTROPSIS (Monó-tropsis)—Sweet-Pinesap. Three species parasitic on soil funguses in the southeastern United States constitute this genus, separated from *Monotropa* by its petals forming a bell-shaped corolla instead of being separate to their bases. The name is derived from that of the genus *Monotropa* and the Greek *opsis*, resembling. This genus belongs in the heath family ERICACEAE or, according to some authorities, to the shinleaf family PYROLACEAE. Yet others assign it to the monotropa family, the MONOTROPACEAE.

The crowded, scalelike leaves overlap. Remarks made under *Monotropa* regarding the family, growth habits, and garden possibilities apply here also.

Native to dry woods from Maryland and Kentucky southward, ***M. odorata*** is a pinkish plant 2 to 5 inches tall.

MONSTERA (Mons-tèra)—Ceriman or Swiss Cheese Plant, Taro Vine. Few tropical plants are more widely used for decoration than *Monstera deliciosa*. Even northerners, far from where its outdoor cultivation is possible, recognize this species and with ease bred of familiarity call it monstera. A favorite for growing in pots and tubs to ornament homes, hotels, places of business, and other indoor areas, it is sometimes called Swiss cheese plant, a designation inspired by the perforations in its large and handsome leaves. The Swiss cheese plant is one of about twenty-five species of its genus, natives of the American tropics, and belonging to the arum family ARACEAE. The name *Monstera* is of unknown derivation and application.

Monsteras are vines that climb by aerial roots growing from their semiwoody stems. They are related to Asian *Epipremnum* and *Raphidophora*, also called taro vines. The differences between these genera lie in details, apparent only by microscopic examination of the ovaries of their tiny flowers. Monsteras have alternate leaves, usually in two rows, with long stalks that with their bases sheathe the stems. Their blades are undivided or deeply-cleft and may be perforated with sizable holes. What are commonly thought of as flowers are inflorescences, assemblages of blooms, and associated parts. The true flowers of *Monstera*, tiny and bisexual, are thickly studded over the entire surface of a spikelike column called the spadix (corresponding to the central yellow organ of a calla-lily inflorescence). From the base of the spadix comes a partially enveloping petal-like bract called a spathe, the equivalent of the white, trumpet-shaped part of the calla-lily inflorescence. In *Monstera* the spathe exceeds the spadix in length and is soon deciduous. The fruits are berries that cohere to form a columnar or conelike body.

The ceriman or Swiss cheese plant (***M. deliciosa*** syn. *Philodendron pertusum*), native from Mexico to Central America, is a tall vine, or it sometimes scrambles over the ground. From its thick stems it sends long cordlike roots that may or may not reach the ground. Its leaf blades, in the adult phase of the plant 1 foot to 3 feet

Monstera deliciosa

Monstera deliciosa (inflorescence)

Monstera deliciosa, a tub specimen

Monstera siltepacana

Monstera friedrichsthalii

long by some two-thirds as wide, are regularly cleft into oblong lobes, and are randomly perforated with holes of various sizes. Juvenile leaves are much smaller and without lobes or perforations. It is plants with such foliage that are sold by florists as *Philodendron pertusum.* When ripe, the soft-fleshed, cone-shaped, pervasively fragrant, compound fruits of the ceriman are delicious. Before maturity they contain crystals of calcium oxalate severely irritating to the mouth and throat. A variant with its foliage unpredictably and irregularly variegated with creamy-white to greenish-yellow is *M. d. variegata* (syn. *Philodendron pertusum variegatum*). This not infrequently reverts and develops plain green leaves. Variety *M. d. borsigiana* is distinguished by its smaller, glossy leaves being more deeply cleft, sometimes to their midribs, and without or with few perforations. At their junctions with their blades, the leafstalks are distinctly corrugated. Much resembling the ceriman, a plant cultivated as ***M. perforoides*** has smaller, grayish-green, lobed and perforated leaves that have stalks with black margins. This name is not recognized botanically. Another sort with perforated leaves, Mexican ***M. siltepacana*** is a vigorous climber with large ovate leaves mostly heart-shaped at their bases and perforated with elliptic to ovate holes of irregular sizes. The inflorescences have a large, club-shaped spadix.

Differing markedly from the ceriman, ***M. standleyana*** has been misidentified in cultivation as *M. guttifera.* The leaves of this fine species are without lobes or holes and have broadly-winged stalks. A native of Costa Rica, as usually seen in cultivation this has thin stems, and dark green juvenile leaves with long stalks and lustrous oblong blades about 6 inches long by under 2 inches wide. Stems with adult foliage are much stouter. The adult leaves are crowded and have stalks about two-thirds as long as the blades, which are oblong-ovate, up to 2 feet in length by slightly over 1 foot wide. The spadix is creamy-white.

Distinctive ***M. friedrichsthalii,*** of Costa Rica, is a vigorous, high-climbing vine of handsome style. Its asymmetrically-oblong-elliptic to ovate leaves, up to about 1 foot long by approximately one-half as broad, are proportionately narrower than those of the ceriman, and their wavy margins are not slashed into lobes. Those of climbing specimens are abundantly perforated with two or three rows of various-sized elliptic to oval holes. Somewhat similar, but with mostly narrower, longer-pointed leaves not always perforated, ***M. obliqua*** is an inhabitant of the Amazon region. It has spadixes up to 2 inches long and spathes not greatly longer. This sort is perhaps not cultivated, but a close relative, ***M. uleana,*** of Peru, is. It has leaves up to 1 foot long and somewhat less wide, irregularly pinnately-cleft into six or seven lobes. Variable ***M. adansonii*** (syn. *M. pertusa*), of tropical America, has leaves with ovate to oblong-ovate or oblong-elliptic blades 2 to 3 feet long by somewhat under 1 foot wide. Markedly asymmetrical at their bases, they have a single row of more or less elliptic holes, that occasionally cut through the leaf margin, on both or only one side of the midrib. This has white spathes up to 8 inches in length and slender spadixes about one-half as long. The long-stalked adult leaves of ***M. dubia,*** of Central America, have oblong-ovate blades 2 feet long by three-quarters as wide, dissected almost or quite to the midribs into twelve to twenty long, narrow, curved lobes or leaflets. Its juvenile leaves are very much smaller and uncleft. The spathes, white on their insides, are up to 1½ feet long.

Other cultivated monsteras include these: ***M. acuminata,*** of Guatemala, like certain kinds of *Raphidophora,* with which it shares the colloquial name shingle plant, has juvenile foliage that presses closely to the support up which the vine grows. These leaves, which overlap like shingles on a roof, are practically stalkless, broad-ovate, and 2 to 4 inches long. The asymmetrically pointed-ovate adult leaves stand free of the supports and are much larger than juvenile leaves. The spadixes are about 2 inches long. ***M. epipremnoides*** (syn. *M. leichtlinii*), of Costa Rica, has adult leaves with ovate to oblong-elliptic blades up to 3 feet long by two-thirds as broad, pinnately-cleft into lobes ½ inch to 1½ inches wide and with two or three rows of holes on each side of the midrib, the outer ones much bigger than the inner and extending to or sometimes cutting through the margin. Juvenile-type leaves are similar but much smaller. About twice as long as the spadixes the white spathes are about 1¼ feet in length. ***M. karwinskyi,*** of Mexico, in gardens often misidentified as *M. egregia,* has twisted, asymmetrically ovate to lanceolate-ovate, lobeless leaves with blades up to 2½ feet long by two-thirds as wide. Those of the adult type are perforated with a few holes. The about 6-inch-long, cream-colored spathes are only slightly longer than the ovoid spadixes. ***M. latiloba,*** of Peru, has leaves with ovate-oblong blades up to 1½ feet long by two-thirds as wide, cleft-pinnately nearly to their midribs into three or four broad lobes. The leafstalks approximate the blades in length. The spadix is orange. ***M. maxima,*** of Ecuador, differ from *M. karwinskyi* in the blades of its leaves being up to almost 2½ feet long, irregularly lobed, and with small perforations along the midrib. The spadix is white. ***M. pittieri*** is Costa Rican. It has thin, slender-pointed, asymmetrically-ovate leaves up to 6 inches long and one-half as wide that display a few large holes. Native to Mexico, ***M. punctulata*** has leaves with white-dotted stalks and asymmetrical, irregularly-perforated, ovate to oblong-elliptic blades up to 4 feet long by one-half as broad. The spathes, up to about 5 inches long, are pinkish outside, white within. ***M. subpinnata,*** of Peru, in cultivation often misnamed *Raphidophora laciniosa,* has leaves with stalks somewhat shorter than the broad-ovate blades, which are up to about 1 foot in length and have three or four pairs of widely-spaced, narrow-oblanceolate, 1-inch-wide lobes. The spathes are muddy-yellow.

Garden and Landscape Uses and Cultivation. Among the noblest and most satisfactory vines of the arum family, monsteras can be used with magnificent effects in gardens in the humid tropics and warm subtropics, as high climbers to clothe tree trunks and similar supports; the thick-stemmed ones, such as the ceriman, can be used as mounded, hummocky specimens without any considerable support. The ceriman is also a popular and very satisfactory plant for growing in pots and tubs for decorating large rooms. For their best comfort monsteras need a reasonably high humidity and high temperatures. Indoors a minimum of 60°F is about right. So long as the air is not excessively dry, considerably higher temperatures are in order. These plants succeed in a wide variety of soils that are reasonably moist and well drained. They do best in soil well supplied with organic matter. Shade from strong sun is needed. Propagation can be accomplished by sowing seeds that are ripe and have not been allowed to dry, by air layering, and by cuttings.

Montbretia (*Crocosmia crocosmaeflora*)

MONTANOA (Montan-òa). This is one of comparatively few genera of the vast daisy family COMPOSITAE that consists of trees and shrubs. It comprises fifty species and is indigenous from Mexico to northern South America. Its name commemorates a Mexican politician, Don Luis Montana. The sorts of *Montanoa* have opposite, lobeless to deeply-lobed leaves. Their daisy-like flower heads consist of both disk and ray florets, the latter white or pink. They are in branching clusters.

About 8 feet in height, ***M. bipinnatifida,*** of Mexico, has deeply-twice-cut, toothed leaves, ovate in outline, and up to 3 feet long. Its flower heads, 2 to 3 inches wide, are in terminal clusters of up to twenty. They have disks up to ¾ inch across and white, obovate ray florets, notched at their ends. This is handsome both in foliage and in bloom. Another splendid Mexican, ***M. grandiflora,*** exceeds 10 feet in height and forms a bush at least as wide as it is tall. It has deeply-lobed and irregularly-toothed leaves. Its flower heads, yellow-centered and with white ray florets, resemble marguerites. They are 3 inches in diameter and are succeeded by decorative, chartreuse seed heads. This beautiful kind blooms profusely in fall. It has a resinous fragrance. Native from Guatemala to Costa Rica, ***M. hibiscifolia*** is up to 25 feet tall. It has leaves 1 foot in diameter, lobed-palmately, in the way of maple leaves, to their middles. The yellow-centered, white-rayed flower heads are 1½ inches across.

Garden and Landscape Uses. These beautiful, tall shrubs are not hardy in the north, but may be grown there as annuals for summer foliage effects outdoors and for fall and winter bloom in greenhouses. In southern Florida, southern California, and other practically frost-free regions, they are hardy and bloom splendidly in late fall or winter. They are striking in the landscape and are deserving of greater popularity. Montanoas are useful for beds, borders, and backgrounds, and succeed in full sun in a variety of soils including limestone ones. They grow rapidly.

Cultivation. Propagation is by seed, root cuttings, or cuttings started in spring. Cuttings root most readily in a greenhouse propagating bench with a little bottom heat. The young plants, potted in sandy soil, make rapid progress. In the north they are kept in a sunny greenhouse with a night temperature of 50 to 55°F and day-time temperatures somewhat higher. By late May or early June those intended for temporary foliage effects outdoors are ready for planting. Those to be bloomed in greenhouses in pots or tubs may be grown under conditions that suit chrysanthemums.

In warm climates some staking or other support may be needed, especially where, as in southern Florida, summer humidity is such that a somewhat lax and drooping habit of growth is encouraged. After blooming or in early spring, established specimens should be cut back severely to encourage the development of vigorous growth and large leaves.

MONTBRETIA (Mont-brètia). As the name of a genus, *Montbretia* is no longer valid. Plants previously so identified, botanically *Crocosmia crocosmaeflora,* represent a swarm of hybrids between *C. aurea* and *C. pottsii.* For these, the plants dealt with here, montbretia is the accepted common name. Montbretias belong in the iris family *Iridiacaea.*

Montbretias are splendid for flower beds and borders and for supplying cut flowers. They look well planted at watersides in soil a foot or so higher than the water level. They come in numerous named varieties, with flowers in many tones in the yellow to copper to nearly vermilion-red range. Growing from small bulblike organs, called corms, they send to heights of 2 to 3 feet erect, wiry stems furnished below with a few narrowly-sword-shaped leaves, and above branching into elegant sprays of slender spikes of blooms each with six spreading petals.

The cultivation of montbretias is simple. Essentially they are treated like gladioluses. They are hardy about as far north as Philadelphia, and even in the region of New York City they persist outdoors if heavily mulched over winter. In cold climates, however, it is usual to dig them up after the first frosts of fall, allowing the tops to dry before removing them, and to store the corms (bulbs) packed tightly together in soil, peat moss, or sand, in a frost-proof cellar, garage, attic, or similar place where the temperature does not go above 50°F. An alternative method is, after lifting, to heel the plants in (plant them closely together) in soil in a well-protected cold frame.

Montbretias stand light shade better than gladioluses, but are equally as happy in full sun. They need well-drained, but not dry, fertile soil, preferably somewhat sandy and enriched with compost, well-rotted manure, or other decayed organic matter. Plant the corms in early spring about 3 inches deep and 2 to 4 inches apart, in groups in flower beds or borders, or for cut flowers in rows 1 foot to 2 feet apart. Summer care consists chiefly of keeping weeds down by frequent shallow cultivating or by mulching. The last is excellent because it also keeps the roots desirably cool. When the plants are half grown apply a dressing of a complete garden fertilizer. During spells of dry weather supply water generously. When cutting the flowers leave as many leaves as possible to build corms large enough to make strong growth the following year. Propagation is by the abundant natural increase of the corms. New varieties are raised from seed.

MONTEZUMA (Monte-zùma)—Maga. In commemoration of the Aztec ruler of Mexico, his name was applied to a plant in the mistaken belief that it was a native of Mexico. Now, in accordance with the rule of priority that governs the application of botanical names, it is the rightful designation of two species, neither indigenous to Mexico. One is endemic to Puerto Rico, the other Cuba. The former is planted as an ornamental in southern Florida and Hawaii. The genus *Montezuma* belongs in the mallow family MALVACEAE. Its members are heavily foliaged evergreen trees with large, handsome blooms. From the closely related *Thespesia* it differs in its flowers having calyxes that break apart and that are not persistent.

The maga (***M. speciosissima*** syn. *Thespesia grandiflora*), native to moist limestone forest lands in Puerto Rico, attains a height of 30 to 50 feet and has stout, warty branchlets. Its long-stalked, alternate leaves have heart-shaped, toothless blades up to 9 inches long by 4 to 6½ inches

Montezuma speciosissima, Fairchild Tropical Garden, Miami, Florida

wide, with lower surfaces paler than the upper. The long-stalked flowers originate from the leaf axils at the ends of the branches. Of the several near each branch termination only one opens at a time. The calyx, green and cup-shaped, sheds when the flower fades. There are five broad, glossy, conspicuously veined, red petals 3 to 3½ inches long and numerous short stamens that form, with bright yellow anthers, a whitish column about 2¾ inches long. The slender style, 2¼ inches long, has three or four yellow stigmas that are joined. Fleshy or leathery, the nearly globular fruits are 1½ to 2 inches in diameter. They do not split.

This decidedly handsome tree is less planted than formerly in Puerto Rico because it serves as alternate host to the pink bollworm that is a pest of cotton and because it has not proved as successful a forestry tree as was hoped. Another strike against it is that in Puerto Rico it is susceptible to damage by a scale insect. None of these handicaps is necessarily significant in other parts of the tropics. The heavy, handsome, durable wood is splendid for furniture, turnery, and musical instruments and is also used for poles and posts.

Garden and Landscape Uses and Cultivation. In the humid tropics and near-tropics this is a handsome ornamental for use as a specimen tree. It grows well in limestone soils and is easily propagated by seeds, provided they are fresh; they do not retain their vitality for more than about a month.

MONTEZUMA-CYPRESS is *Taxodium mucronatum.*

MONTIA (Mónt-ia)—Winter-Purslane or Miner's-Lettuce. Few are likely to connect montias with brilliantly flowered portulacas, yet both belong in the purslane family PORTULACACEAE. The genus *Montia,* named to honor the Italian botanist Giuseppe Monti, who died in 1760, contains about fifty species of hairless, often glaucous, annuals and herbaceous perennials of temperate parts of the northern hemisphere, mountains of tropical Africa, and Australia. It is closely related to *Claytonia,* from which it differs in the number of ovules in the ovary.

Montias have somewhat fleshy leaves, alternate or opposite or all basal. Their small blooms, as buds usually nodding, are in racemes, panicles, or umbel-like clusters. They have two persistent and often unequal-sized sepals, two to five or rarely six petals that in some kinds are joined at their bases, as many stamens as petals, and a three-branched style. The fruits are small spherical or egg-shaped capsules containing one to three often shining seeds.

Winter-purslane or miner's-lettuce (***M. perfoliata*** syn. *Claytonia perfoliata*) is a complex assemblage of variable forms, several of which have been given varietal names. The species ranges as a native from British Columbia to Baja California and is widely naturalized elsewhere. An annual up to 1 foot tall, it has green and sometimes glaucous, long-stalked, rhombic-ovate to oblanceolate or linear basal leaves and at the tops of each of several more or less erect stems, a single pair of semicircular leaves joined at their bases to form a shallow, fleshy, circular cup or saucer ½ inch to 3 inches across. From the centers of the saucers the clusters or racemes of flowers emerge. The blooms are white and ¼ to ⅓ inch wide.

An aquatic or semiaquatic perennial, ***M. chamissoi*** occurs at watersides and in wet meadows, often at high altitudes, from California to New Mexico and Alaska and from Iowa to Minnesota. It has creeping or floating stems with opposite leaves and slender runners that at their ends bear little bulblets. Its branches are more or less erect. The leaves are oblong-spatula-shaped, and ½ inch to 2 inches in length. In three- to eight-flowered racemes at or near the ends of the shoots, the slender-stalked white or pink blooms, ⅓ to ½ inch in diameter, come in summer.

Garden Uses and Cultivation. Winter-purslane is rarely cultivated as an early salad plant, and is mildly interesting as a rather unusual item for semiwild gardens. It is raised from seeds sown in spring in damp or wet soil where the plants are to remain. The aquatic species described is of interest for planting in bog gardens and at pool sides where its roots are in saturated ground and its stems can creep over wet soil or float on the water. It is propagated by seed and bulblets and by division.

MONTRICHARDIA is Pleurospa.

MONVILLEA (Mon-víllea). The South American night-blooming, desert cactuses that comprise this genus of the cactus family CACTACEAE bear a name commemorating M. Monville, a nineteenth-century student of that family. There are fifteen species.

In the wild, monvilleas form thickets, with semierect or sprawling, slender stems, near the ends of which medium-sized flowers develop. The blooms have slender perianth tubes furnished with tiny scales. The outer perianth lobes (petals) are greenish to pinkish, the inner white or yellow. The stamens and style are white. After fading, the withered perianth remains without falling. In this, *Monvillea* differs from related *Cereus*.

One of the most satisfactory night-blooming cactuses, ***M. cavendishii*** branches from the base and is 3 to 10 feet tall. It has stems up to 1 inch or a little more thick, with nine or ten low blunt ridges. The needle-like spines, in clusters of eight to twelve, sprout from specialized areas called areoles spaced less than ½ inch apart. The white flowers have straight, slender perianth tubes 4 to nearly 5 inches long and petals that spread as widely. The stamens and style more or less protrude. The nearly globular fruits are 2 inches or slightly less in diameter. This is a native of Brazil, Paraguay, and northern Argentina.

Erect, with bluish-green, more or less white-spotted stems with three toothed or wavy-edged conspicuous angles, ***M. spegazzinii*** is native to Argentina and Paraguay. Its young branches carry short dark spines with broadly-conical bases, three from each areole. From the areoles on older stems sprout a central spine with five others encircling it. Long, slender, and funnel-shaped, the flowers are 4½ to 5 inches long, purplish outside, their inside petals almost white.

Low and sprawling ***M. phatnosperma,*** of Paraguay, has 1-inch-thick, four- or five-ribbed, bright green stems 3 to 6 feet long. From each areole there are five or six radial spines approximately ½ inch long and sometimes a central one up to 1 inch long. About 5 inches long, the flowers are white.

Garden and Landscape Uses and Cultivation. Interesting for outdoor succulent collections in warm, dry regions, and for inclusion in greenhouse collections, monvilleas respond to treatment appropriate for most cactuses. For additional information see Cactuses.

MOON. This word is included in the common names of these plants: moon-cereus (*Selenicereus*) and moon-trefoil (*Medicago arborea*).

MOONFLOWER is *Ipomoea alba*.

MOONSEED is *Menispermum*. Carolina moonseed is *Cocculus carolinus*.

MOONWORT. This name is applied to *Lunaria* and *Botrychium lunaria*.

MOOSEWOOD, MOOSE BUSH, MOOSEBERRY. Moosewood is a name of *Acer pensylvanicum* and *Viburnum alnifolium*. The last is also called moose bush and mooseberry.

MORACEAE—Mulberry Family. Important as a source of ornamentals and other useful plants, the *Moraceae* numbers among its fifty-three genera 1,400 species of dicotyledons. Natives chiefly of the tropics and subtropics and with a few sorts inhabiting more temperate regions, included are breadfruit, figs, hops, and mulberries, as well as such sorts as the osage-orange that supply lumber and others that are sources of caoutchouc and certain kinds of rubber.

Mulberry family members are mostly evergreen and deciduous trees, shrubs, or less commonly vines. Some few are herbaceous. Except for *Cannabis* and *Humulus*, they have milky sap. Their mostly alternate leaves are usually undivided, often lobed, and sometimes toothed.

The minute, inconspicuous, generally unisexual flowers with the sexes on the same or different plants are in spikes or heads or the males in racemes, or in *Dorstenia* and *Ficus*, on fleshy bodies called receptacles; those of the latter are hollow and mature as fruits of the fig type. The blooms have a perianth of usually four, but from one to six parts not well differentiated as sepals and petals, in the males usually as many stamens, in the females two or less commonly one each styles and stigmas. The fruits are very various. They may be separate achenes, collections of achenes embedded in fleshy perianth parts, as in the mulberry, osage-orange, and breadfruit, or, as in figs, contained in a fleshy receptacle. Cultivated genera include *Antiaris, Artocarpus, Brosimum, Broussonetia, Cannabis, Castilla, Cecropia, Chlorophora, Cudrania, Dorstenia, Ficus, Humulus, Maclura,* and *Morus*.

MORAEA (Mor-aèa)—Peacock-Iris or Peacock Flower. Graceful, iris-like plants numbering about 100 species constitute the genus *Moraea*. They are natives of Africa and Madagascar and belong to the iris family IRIDACEAE. Their name commemorates Robert More, English botanist, who died in 1780, not, as is sometimes reported, the Swede Dr. Johan Moraeus. None is hardy in the north. All described below are South African. A few species previously included in *Moraea* are segregated as *Dietes*. These have creeping rhizomes and fibrous roots. True moraeas have small bulblike, underground parts called corms.

Moraeas have from their bases narrow, sometimes nearly threadlike, parallel-veined leaves and wiry, stiffly erect stems with clusters of lovely iris-like flowers, each cluster with a spathe (bract) at its base. From those of irises, the blooms of moraeas differ in their perianth parts being quite separate instead of being joined into a tube at their bases. There are six perianth parts, the three inner true petals, the outer three technically sepals, but resembling petals. For convenience, we shall refer to the former as the inner petals, and the latter as the outer petals. The outer petals are reflexed (down-turned). The inner petals may be similar, but smaller, or very much reduced in size and three-toothed. The lower parts of the stalks of the three stamens join to form a tube. The style has three petal-like branches ending in bilobed crests, with the anthers pressed against their undersides. The fruits are pods. In most species the individual flowers are open for only a day or two, but those of each cluster expand in succession over a fairly long period.

The three or four leaves of ***M. polystachya*** are up to 2 feet long. From 2 to 3½ feet tall, its stout stems bear loose assemblages of five to twenty clusters, each of three to six short-lived lilac blooms about 1½ inches across. The three oblong-spoon-shaped outer petals have a bright yellow spot at the bottom of their blades. This species is poisonous to livestock.

Moraea spathulata

Perhaps the finest horticultural kind, variable ***M. spathulata*** (syn. *M. spathacea*) in bloom has much the aspect of the yellow flag (*Iris pseudacorus*). Its leaves, up to 4 feet in length and ¾ inch broad, bend over and may sweep the ground. Lifted above the foliage on stems about 1½ feet long, the bright yellow blooms, up to 3 inches across, have darker markings at the bases of their spreading outer petals. The inner petals are somewhat narrower than the outer ones and are suberect. Similar ***M. balenii*** differs in having much narrower, whiplike leaves.

The peacock-iris or peacock flower (***M. villosa*** syn. *M. pavonia*) was given its vernacular names because the brilliantly colored markings of its blooms bring to mind those of a peacock's tail. This kind has a solitary, long, very slender, white-downy basal leaf and stems 1 foot to 2 feet long, with, at their ends, gorgeously colored,

1½-inch-wide flowers. Individual blooms last about four days and do not close at night. The blooms range in color from beige to yellow or orange, or they may be light blue. Their rounded outer petals have at their bases blotches of strongly contrasting hues, usually iridescent peacock-blue or green, margined with yellow, orange, deep purple, or almost black. The small and narrow inner petals form a small central tuft to the bloom.

Individual flowers of ***M. tripetala*** remain open for three or four days. This excellent species has usually one narrowly-linear basal leaf and a slender flowering stem up to 2 feet tall, generally without branches. The blooms are in clusters of up to three. About 2 inches wide, they are dark lilac to turquoise-blue, with a triangular patch of yellow at the bases of the blades of the three outer petals, or they may be plain purple. The outer petals have erect, narrow, stalklike shafts and drooping blades. The very small, narrow inner petals angle outward.

Beautiful ***M. tricuspidata*** (syn. *M. glaucopis*) inhabits Table Mountain and other places near Cape Town, South Africa. It has a small corm and a single slender-cylindrical, sinuous basal leaf. On the lower parts of the usually branchless, 1½- to 2-foot-tall stems are two or three sheathing, smaller leaves. The propeller-shaped blooms, in groups of two or three, are about 2 inches across. They are white with obscure blue veins and with, at the base of each of the rounded blades of the short-shafted, spreading outer petals, a richly contrasting, metallic-blue-black blotch. The inner petals are small, with white or creamy-white crests.

Quite charming ***M. tricuspis*** has a long, linear basal leaf. Its very slender stems, usually branchless, are up to 2 feet tall. The lilac-tinged, white flowers about 1 inch across, have spreading outer petals with roundish blades that reflex sharply from their primrose-yellow, bearded shafts and have, at their bases, a conspicuous blotch of peacock-blue. The three-toothed inner petals, much smaller, narrower, and suberect, have large crests.

Very different from other kinds, ***M. ramosissima*** (syn. *M. ramosa*) has pale green leaves in basal tufts, and candelabra-branched stems up to 3 feet tall that carry pale yellow, 1½-inch-wide blooms at the numerous branch ends. The outer petals of the flowers have at their bases richer yellow blotches margined with gray. Unfortunately the individual flowers last for a few hours only, opening about midday and closing in the evening. A wet-ground plant, this needs plenty of moisture during its growing period, and preferably light shade.

Garden and Landscape Uses and Cultivation. In mild regions where freezing is not a hazard, and especially where, as in California, a Mediterranean-type climate prevails, moraeas are splendid flower garden plants. The flowers of those kinds that remain open for more than a day or so are elegant for cutting. They are also of interest for growing in cool greenhouses. They need open, sunny locations, or some part-day shade, well-drained, fertile soil, and plenty of moisture during their seasons of active growth. Propagation is by natural increase of the corms and by seed. The latter are planted in pots of sandy soil and kept shaded. They germinate in about a month. Two to four years elapse from seed sowing to first blooming.

In greenhouses the bulbs are planted in well-drained pots in fall or early winter, being set fairly close together. Five or six bulbs can usually be accommodated in a 5-inch container. Scarcely any water is given until growth begins. Then, at first sparingly, and more frequently as foliage develops, water regularly until the foliage begins to turn yellow and die naturally. When this happens, the length of the periods between applications is gradually increased. Finally, the water is withheld completely, and the bulbs are kept in the soil, in an airy, shaded dry place. Repotting is done annually, just before new growth begins, in decidedly nutritious, porous soil. During winter a night temperature of 45 to 50°F is adequate, with up to ten degrees increase on sunny days permitted. On all favorable occasions the greenhouse must be ventilated freely. Dank, humid conditions, and shade are highly detrimental.

Moraea tricuspidata

MORAINE GARDEN. See Rock and Alpine Gardens.

MORANGAYA (Moran-gàya). The only species of *Morangaya*, of the cactus family CACTACEAE, is akin to *Aporocactus*, but is larger and differs in botanical details. Its name honors the twentieth century American botanists Dr. Reid Moran and Ed and Betty Gay who collected this cactus in Baja California.

Native to Baja California, ***M. pensilis*** (syn. *Echinocereus pensilis*) has prostrate or drooping stems branched from their bases and up to several feet long by 1 inch to 2 inches thick. They have eight to eleven lumpy ribs and clusters of at first six to twelve, later many more, needle-like, yellowish to reddish, spines from under ½ to 1 inch long. The flowers, 2 to 2½ inches long, are bright red.

Garden and Landscape Uses and Cultivation. This species is attractive for growing in hanging baskets or, in warm, dry climates, in an area where its stems can hang over a wall or a cliff. Its cultural needs are those of *Echinocereus*. For additional information see Cactuses.

MORAWETZIA (Mora-wétzia). One or two species constitute *Morawetzia*, of the cactus family CACTACEAE. The genus doubtfully merits recognition as being distinct from *Oreocereus* and like it is included by conservative botanists in *Borzicactus*. Endemic to the Peruvian Andes, *Morawetzia* bears a name commemorating Victor Morawetz of New York, who provided financial support for collecting expeditions made to South America in the 1930s by the German botanist Curt Backeberg.

The first described and possibly only species, ***M. doelziana*** (syns. *Oreocereus doelzianus, Borzicactus doelzianus*) forms colonies of beautiful, erect, club-shaped, olive-green to grayish-green stems 2 to 3 feet tall, 1¾ to 3 inches in diameter, and branched from near their bases. The branches are sometimes prostrate. The stems have nine to eleven shallow, rounded, notched ribs with white-hairy areoles up to ¾ inch apart, each with a cluster of ten to about sixteen needle-like, grayish to reddish radials approximately ½ inch long and four similar but stouter centrals, three of which are about ¾ inch long, the other up to twice that length. The day-opening, nearly cylindrical bluish-red flowers come from a patch of grayish wool at the apex of the stem. They are up to 3 inches long and have a slightly oblique lip or face. The numerous stamens are shorter than the style, about as long as the petals. The globular, greenish stigma has a dozen or more lobes. Like the perianth tube the fruits are hairy. In *M. d. calva* the areoles are hairless.

Garden Uses and Cultivation. These are as for *Oreocereus*. For more information see Cactuses.

MORETON-BAY-CHESTNUT is *Castanospermum australe.*

MORETON-BAY-PINE is *Araucaria cunninghamii.*

MORICANDIA (Moric-ándia). To the mustard family CRUCIFERAE belongs this genus of eight species. Native from the Mediterranean region to central Asia, *Moricandia* bears a name honoring the Swiss botanist Moise Etienne Moricand, who died in 1854.

Annuals, biennials, and herbaceous perennials, moricandias usually branch freely from their bases. They have somewhat fleshy, undivided, alternate leaves, the upper ones commonly more or less stem-clasping. The blue, purple-violet, or white flowers, in loose racemes, are typical of those of the majority of plants of the mustard family. They have four sepals, four petals that spread to form a cross, six stamens, two of which are shorter than the others, and one style. The fruits are slender, four-angled capsules.

Native to the western Mediterranean region, ***M. arvensis*** is an annual or a biennial 1 foot to 1½ feet tall. Its foliage is glaucous and fleshy. The lower leaves are obovate, toothless, and short-stalked. The upper ones clasp the stems with heart-shaped bases. In racemes of up to twenty, the pale violet-purple flowers are about 1½ inches across. Perennial ***M. ramburii*** has a somewhat woody base, and a basal rosette of toothed or nearly toothless, obovate leaves. The other leaves clasp the stems with deeply-heart-shaped bases. The purple flowers are almost 2 inches wide. The plant sometimes named *M. sochifolius* is *Orychophragmus violaceus.*

Moricandia arvensis

Garden and Landscape Uses and Cultivation. The species described above are less often grown than their merits deserve. The perennial is hardy only in regions of mild winters, where it can be treated as a biennial by sowing seeds in early summer to give plants to bloom the following year. The other does best where summers are moderately cool. In mild regions it can be bloomed in spring from fall-sown seeds. Both kinds respond best to deep, moderately fertile, neutral or somewhat alkaline soil and open, sunny locations. For the best results sow seeds in fall or early spring where the plants are to bloom, and thin the seedlings to about 6 inches apart. No staking is needed. Moricandias stand up well to wind and rainstorms.

MORINA (Mor-ína)—Whorl Flower. The genus *Morina,* sometimes segregated as the morina family MORINACEAE, is retained by more conservative authorities in the teasel family DIPSACEAE. It consists of seventeen species chiefly from high altitudes in the Himalayas. Its species much resemble each other and differ from teasels (*Dipsacus*) in their flowers being in spikes composed of whorls (circles of more than two) of flowers borne in the axils of large, usually spiny bracts; the clear stem shows between the lower whorls. In appearance the flower spikes resemble those often found in the unrelated mint family (LABIATAE). The name commemorates the French botanist Louis Morin, who died in 1715.

Thistle-like herbaceous perennials, morinas have usually spiny leaves in pairs or whorls. Their somewhat two-lipped, gaping-mouthed, tubular blooms, white, yellow, pink, or red, have two-lipped calyxes, five corolla lobes (petals), and two or four fertile stamens, with sometimes one nonfunctional stamen (a staminode). The seedlike fruits are technically achenes.

The most familiar species, the whorl flower, Himalayan ***M. longifolia***, is 2 to 4 feet in height and has evergreen, linear, very prickly-toothed leaves up to 1 foot long by about 1½ inches wide. Its flowers, which open white and change to pink and then crimson, protrude from the axils of spiny bracts and are about 1 inch long and ¾ inch wide. There are two fertile stamens. Native from Greece to the Himalayas, 3 to 4 feet-tall ***M. persica*** has stems

Morina longifolia

Morina longifolia (flowers)

hairy in their upper parts. Its leaves, 6 to 9 inches long are doubly-spiny-toothed. In rather loose, erect spikes, the pink flowers, 1 inch to 1½ inches long, have each two fertile stamens.

Garden Uses and Cultivation. Not adaptable to regions of very cold winters or torrid humid summers, morinas are easily satisfied in regions with more equable climates. They grow well in deep, sandy loamy soils that, without being wet, do not lack for moisture, and are best satisfied with light shade. Shelter from strong wind should be given. Propagation is by seed and by division as soon as flowering is through. These are impressive plants for grouping in flower borders.

MORINDA (Mor-índa)—Indian-Mulberry or Painkiller, Royoc. Representatives of this genus of trees, shrubs, and a few climbers are natives of tropical and subtropical regions of the Old World and New World. One species is indigenous to Florida. The genus *Morinda,* of the madder family RUBIACEAE, consists of eighty species. Its name is derived from the ancient Latin name of the mulberry, *morus,* and *indica,* Indian. The fruits are vaguely remindful of mulberries, although the plants are not related.

Morindas have their leaves opposite or sometimes in threes and small white or red flowers often in tight clusters. Their fruits are formed by the joining of many berry-like parts. Some species yield dyes and some edible fruits.

The Indian-mulberry (***M. citrifolia***), native from tropical Asia to Australia, is an evergreen tree or large shrub up to about 20 feet in height and with squarish twigs, the older ones marked with prominent leaf scars, and pairs of wavy-surfaced, deeply-veined, dark green, broadly-elliptic leaves that may be nearly 1 foot long by 6 to 7 inches wide. They have short stalks, at the bases of which are a pair of leafy appendages about ¼ inch long. These are called stipules. The leaves are without hairs except for small tufts in the axils of the mid-vein on the underside. Their margins are without teeth. The four- to six-lobed tubular white flowers are in globular or ovoid, axillary, stalked clusters. They are about ½ inch long. The clusters of bloom are succeeded by curious, soft, juicy, egg-shaped fruits 3 to 4 inches long and some 2½ inches wide. They are whitish or pale green and have a peculiar, not pleasant, cheesy odor. Because their surfaces are warty and are scored into four- to six-sided, 1-inch-wide segments representing the individual parts of the compound fruit, they have somewhat the appearance of a pineapple without the terminal tuft of leaves. Each segment contains two seeds. Variety *M. c. potteri* has its leaves handsomely variegated with white.

Morinda citrifolia (flowers)

Morinda citrifolia (fruit)

The insipid fruits are eaten by hogs and, reportedly, by some peoples, but probably only when food is scarce. In the West Indies and elsewhere the leaves are used as home applications for relieving painful swellings, headaches, neuralgia, and head colds. Because of this, the colloquial name painkiller is sometimes used for this species. A red dye is obtained from the roots of the Indian-mulberry, and from its leaves a bright yellow one.

The royoc (***M. royoc***) is a shrub or vine with elliptic to wedge-shaped leaves 2 to 4 inches in length. The small white or reddish flowers are several together in heads. They are succeeded by yellow fruits ¾ inch to 1½ inches long. This species is native to Florida, including the Keys.

Garden and Landscape Uses and Cultivation. In the tropics and extreme southern Florida the Indian-mulberry is planted as an ornamental. Because of its large leaves it is conspicuous in the landscape and generally should be restricted to places where bold foliage can be displayed to advantage. It grows rapidly and is sufficiently dense to provide a good screen. Its fruits are objects of interest and curiosity. Any ordinary garden soil will support the growth of the Indian-mulberry, but those of reasonable fertility and fair moisture content give the best results. Propagation is simple by seed and by cuttings.

MORINGA (Moríng-a)—Horse-Radish-Tree. The horse-radish-tree (*Moringa*), is the only commonly cultivated species of the dozen in this genus. From casual inspection, the botanically uncritical might well suppose it to be a member of the pea family LEGUMINOSAE. It is not, but it does belong to a group that some believe to be a connecting link between the pea family and the caper family CAPPARIDACEAE. The name *Moringa* is a modified form of the native Malabar name, *moringo.* It identifies the only genus of the moringa family MORINGACEAE, an assemblage consisting of deciduous trees with alternate, twice- or thrice-pinnate leaves and asymmetrical, bisexual flowers, with five recurved sepals, five petals, one of which is erect and bigger than the others, and three to five fertile stamens alternating with infertile ones. The fruits are slender, angled pods, containing seeds that often are winged. The genus is native from Africa to India.

The horse-radish-tree (***M. pterygosperma*** syn. *M. oleifera*), of India, is naturalized in southern Florida, the West Indies, and other parts of the American subtropics and tropics. It is 20 to 30 feet tall and has corky bark, soft wood, and brittle branches. The lacy, deciduous leaves, of numerous ovate leaflets ½ to ¾ inch long by one-half as wide, are 1 foot to 2 feet long. The terminal leaflets are slightly larger than the others. About 1 inch across, the white or creamy-white, fragrant flowers are many together in panicles from the leaf axils. They are slightly asymmetrical and starry in aspect. The three- or nine-ridged seed pods contain single rows of triangular, three-winged seeds. They are up to 1½ feet long.

The uses of the horse-radish-tree are many. Its common name suggests one. Its pungent roots smell and taste like horse-radish, and are used similarly. The shoots, leaves, flowers, and immature seed pods are cooked and eaten as vegetables. The leaves provide cattle fodder. From the bark a tragacanth-like gum, sometimes used in printing cottons, is obtained. The most important product is a nearly colorless, non-drying oil called oil of ben. This is used in the arts and in perfumery and as a lubricant and salad oil.

Garden and Landscape Uses and Cultivation. In the tropics and warm subtropics, including southern Florida and southern California, the horse-radish-tree is used ornamentally for planting as solitary specimens, groups, and in hedges. It has the disadvantages of not being especially attractive when old and of having weak branches subject to storm damage. In Puerto Rico and some other places it is not recommended for planting because of its marked susceptibility to attack by termites. It thrives in a wide variety of soils and is easily multiplied by seed and by cuttings.

MORINGACEAE — Moringa Family. The characteristics of this family of one genus of dicotyledons are given under Moringa.

MORISIA (Morís-ia). A little-known herbaceous perennial, of the mustard family CRUCIFERAE, is the only representative of *Morisia,* the name of which honors Professor Giuseppe Giacinto Moris, an Italian botanist, who died in 1869.

A delightful little native of sandy places at low altitudes in Sardinia and Corsica, ***M. monanthos*** (syn. *M. hypogaea*) has no aboveground stem, but develops rosettes of dark, glossy, pinnate, ferny leaves 2 to 3 inches long that lie flat or nearly so on

Morisia monanthos

the ground. The golden-yellow, four-petaled flowers, solitary on short slender stems and resembling miniature blooms of wallflowers, are ½ to ¾ inch in diameter. They are borne in spring and early summer. After the flowers fade their stalks bend downward and bury the seed capsules in the soil to complete their ripening.

Garden Uses. Unfortunately, morisia does not withstand severe winter cold nor hot, humid summers. The regions in North America where it is adaptable for outdoor cultivation are therefore limited, the most favorable being the Pacific Northwest. This is primarily a plant for rock gardens and is also appropriate for cultivating in pots or pans in cold frames and cool greenhouses along with alpines and other choice rock plants.

Cultivation. A deep, well-drained, sandy, limy soil that is lean rather than very fertile and a sunny location please this plant. It is well suited for scree conditions. It may be raised from seeds sown directly where the plants are to remain in a deep, soil-filled crevice in a rock garden, or seeds may be sown in a pot or pan in a cool greenhouse or cold frame and the seedlings later transplanted. Root cuttings provide an alternative and easy means of increase. Cut the thonglike roots into pieces 1 inch to 1½ inches long in early spring and plant them in sand, covering them to a depth of about ¼ inch. Set the containers in which the cuttings are planted in a shaded cold frame and water them rather sparingly (but avoid complete dryness) until new leaves are well developed. Then pot them individually in small pots and grow them on until they are of an appropriate size for planting outdoors or in larger pots or pans.

MORMODES (Mormó-des) — Goblin Orchid. Thirty species constitute *Mormodes,* a genus of the orchid family ORCHIDACEAE. Endemic from Mexico to tropical South America, its sorts are most abundant in Brazil. From related *Catasetum* and *Cycnoches* this genus differs in its flowers being always bisexual. Also, unlike those of *Catasetum,* their columns are not winged. The name, in allusion to the unusual appearance of the blooms, is derived from the Greek *mormo,* a phantom.

Deciduous and generally epiphytic (tree-perching), less often growing on rocks or in the ground, these orchids have elongated, tapered pseudobulbs sheathed with the bases of the leafstalks, which after the leaves die and their blades fall away, persist. Each first-year pseudobulb has lengthwise-pleated leaves. The flowers are in racemes of several to many from the sides of the pseudobulbs, usually but not always from their middles or above. Generally quite large and showy, they have nearly equal sepals and petals, spreading widely or some pointing backward or forward. The lip, often lobed, commonly has a rolled-under margin. The column is curiously twisted and so positioned that its apex approaches or touches or nearly touches the lip.

Sorts cultivated include these: ***M. buccinator,*** of Mexico and Venezuela, has pseudobulbs 4 to 8 inches long, narrowish-lanceolate leaves up to 1 foot long, somewhat longer erect or arching racemes of up to a dozen spicily-fragrant blooms

Mormodes buccinator (flowers)

1¾ to 2½ inches long, which are very various as to color, often pale green or brownish-buff and dotted with red-brown, with an ivory-white lip. The lateral sepals point backward. The sides of the trumpet-like lip are strongly recurved. ***M. colossus,*** of Costa Rica, Panama, and Colombia, has pseudobulbs 6 inches to 1 foot long and elliptic-ovate leaves 1 foot to 1¼ feet in length. In arching racemes up to 2 feet long, which mostly originate from below the middles of the pseudobulbs, the fragrant flowers are from 3 to 5 inches or more across. Quite variable in color, they mostly have olive-green, yellowish-brown, or cream sepals and petals and a brown, light brown, or yellow lip, with its side

Mormodes igneum

Mormodes igneum (flowers)

margins strongly recurved. ***M. igneum,*** of Costa Rica, Panama, and Colombia, has narrow pseudobulbs up to 1¼ feet tall, with pointed-lanceolate leaves that may be as long. Sometimes exceeding 2 feet in length, the erect or arching racemes carry few to many fragrant blooms up to 2 inches or more across, often varying in size and color even in the same raceme. Their sepals and petals are yellow, light brown, olive-green, or red, often spotted reddish-brown, or brown. The petals and lateral sepals usually bend backward. The lip, which if spread is nearly circular, has strongly-reflexed margins and is often chocolate-brown with darker spots. ***M. luxatum,*** of Mexico, has pseudobulbs about 9 inches long and leaves 1½ to 2 feet long. The fragrant flowers, in racemes often directed horizontally and up to 2½ feet in length, are of approximately a dozen blooms 3 to 3½ inches wide. They are yellowish-green, with the sepals and considerably broader petals sometimes spotted with purple, and with streaks of the same color on the inside of the lip. Variety *M. l. eburneum* has ivory-white flowers larger than those typical of the species.

Garden Uses and Cultivation. These are as for *Catasetum*. For additional information see Orchids.

MORMON-TEA-BUSH. See Ephedra.

MORNING GLORY. This is the common name of *Ipomoea*. The woolly-morning-glory is *Argyreia nervosa*, the dwarf-morning-glory *Convolvulus tricolor*.

MORRENIA (Mor-rènia). The genus *Morrenia*, of the milkweed family ASCLEPIADACEAE, consists of ten species of twining vines, of South America. Its name commemorates the Belgian botanist, Charles Jacques Édouard Morren, who died in 1886.

Morrenias are milky-juiced, and have opposite leaves. Their flowers, in clusters in the leaf axils, are typical of those of the family. They have five sepals, five petals, a tubular corona or crown, five stamens with anthers that adhere to the stigma, and pollen in waxy masses called pollinia. The fruits are paired, podlike follicles containing many seeds, each with a floss of silky hairs.

Native to Argentina and Paraguay, ***M. odorata*** is a perennial that has long-stalked, halberd- to heart-shaped leaves with two spreading basal lobes. Their blades are 2 to 4 inches long, and at their bases may be as wide as long. The short-stalked, white flowers have a pleasing vanilla fragrance. They are about 1 inch in diameter and have slender, widely spreading corolla lobes and an erect, tubular, five-lobed corona. The leathery fruits are ovate and up to 4 inches long by 3 inches wide. The floss that accompanies the seeds is used in South America for stuffing pillows.

Garden and Landscape Uses and Cultivation. Morrenias are not hardy in the north. In mild climates they may be used where wires, strings, or other supports around which their stems may twine are available. They succeed in ordinary well-drained soil in sun or part-day shade. Propagation is by cuttings, layering, and seed.

MORUS (Mòr-us)—Mulberry. The genus *Morus*, of the mulberry family MORACEAE, comprises up to ten species of mostly variable deciduous trees and large shrubs. It is represented in the native floras of the Old and New Worlds and has as its botanical name its ancient classical one. Mulberries have for long been esteemed for their edible fruits. References to them, under the name sycamine, are made in the Bible. In the Orient their leaves are used to feed silkworms. A nineteenth-century attempt to develop silkworm culture in the United States was accompanied by extensive planting of *M. alba multicaulis*, but the venture was not a commercial success and was abandoned.

Mulberries contain milky juice and have alternate, undivided, variously lobed and toothed leaves with three or five chief veins from their bases. Often there is considerable variation in the size, shape, lobing, and toothing between leaves on the same plant. The flowers, in small, pendulous, catkin-like spikes, have four-parted perianths. They are unisexual, with the sexes on the same or separate trees. Male flowers have four stamens, and females a two-armed style. The latter are succeeded by juicy fruits that look like blackberries.

White mulberry (***M. alba***) is most important horticulturally. It is planted for its fruits and for ornament. Native to China, naturalized in North America, and hardy through much of New England, it is a dense, round-topped tree sometimes 80 feet tall, but generally not much more than one-half that. Its irregularly lobed or toothed, bright green leaves, with lustrous upper sides, are mostly broadly-ovate in outline and 2 to 4 or rarely 6 inches long. Except sometimes on the veins beneath, the leaves are without hairs. Individual trees may be bisexual or unisexual. The sweet, whitish, pinkish, or purple fruits, much esteemed by birds, mature in early summer. Those of selected horticultural varieties are 1 inch to 2 inches long, those of trees raised from seeds often smaller. Weeping mulberry (*M. a. pendula*) has long, slender, drooping branches. Grafted several feet up on the trunk of an erect mulberry understock, it assumes a very distinct cascade form, its branches reaching to the ground. Russian mulberry (*M. a. tatarica*), believed to be the hardiest kind, is smaller. It has lobed or lobeless leaves 1½ to 3½ inches long and dark red fruits not over ½ inch long. The fruitless mulberry, especially recommended for planting where falling fruits of other kinds are likely to be a nuisance, is *M. a.* 'Kingan'. Variety *M. a. pyramidalis* is narrowly-pyramidal. In *M. a. Paciniata* (syn. *M. a. skeletoniana*), the leaves are deeply divided into narrow lobes.

Morus alba (trunk)

The chief source of silkworm forage *M. alba multicaulis* is a tall shrub or sometimes tree that differs from *M. alba* chiefly in having larger, rarely lobed, toothed leaves, sometimes 1 foot long or longer. They are dull on their upper sides and more hairy beneath than those of *M. alba*. The fruits are black and sweet. This is hardy in southern New England.

Black mulberry (***M. nigra***) has dark red, purple, or black fruits up to about 1 inch

Morus alba (leaves and young fruits)

Morus nigra

Morus nigra (foliage and fruits)

long. Most tender of the kinds discussed here, it survives in southern New England only in sheltered locations. This is a dense, dark-foliaged tree up to 30 feet in height. Its dull, thickish leaves have rough upper surfaces and are more or less hairy beneath. They are 2 to 8 inches long and ovate with deeply-heart-shaped bases and toothed margins. They are lobed or lobeless. The black mulberry is a native of western Asia.

Red mulberry or American mulberry (***M. rubra***) is endemic from Massachusetts to Michigan, Florida, and Texas. Broad-headed and up to 60 feet tall, it has pointed, ovate to oblong-ovate, sharp-toothed, lobed or lobeless leaves 3 to 5 inches long or sometimes longer, with more or less rough upper surfaces and softly hairy undersides. The dark purple fruits are up to 1¼ inches long.

Garden and Landscape Uses. Because their fruits do not last long or pack or ship well mulberries are not commercial fruits. Varieties of white and black mulberries are grown in a small way for home production. They make good jellies, jams, and wine. As ornamentals they have not a great deal to offer, except that the white mulberry stands dry conditions and exposed locations well and grows rapidly even in poor soil, soon providing shade. The falling fruits seriously stain pavements, automobiles, clothing, and other surfaces on which they drop. Because of this, it is often most practical to plant *M. alba* 'Kingan', which is fruitless. Because of its distinctive habit of growth the weeping mulberry can be used effectively in some locations.

Cultivation. No special care is needed. Propagation is easy by seed and by leafy summer cuttings and hardwood cuttings. The weeping mulberry is most usually increased by grafting. The only pruning needed is any to shape the trees or limit their size. White flies and scale insects are sometimes troublesome. A bacterial blight disease kills foliage and branches, and prompt cutting out of affected branches is recommended.

Morus rubra, a partly decayed old trunk with aerial roots extending down the rotted portion to the ground

Morus rubra, young shoot with leaves

MOSAIC DISEASES. A group of virus infections, the symptoms of which normally involve mottled foliage and not infrequently stunting or dwarfing, are called mosaics. Sometimes, as with certain mosaic viruses that infect lilies, the symptoms may be masked, not being apparent even though the virus is present and capable of being transmitted to other plants that may then evidence mottling.

Mosaic infections are commonly transmitted by aphids and by such pruning and cutting tools as budding and grafting knives. The virus of tomato or tobacco mosaic is most often transmitted by workers whose hands become contaminated as a result of handling cigarettes or smoking tobacco. In greenhouses it is not unusual for flats, faucets, and doorknobs to be contaminated by gardeners, who have been in contact with infected plants, and for them to serve as sources of future infections. If diseased plants or tobacco have been handled, thoroughly wash the hands with soap before touching uninfected plants.

There are no effective cures for mosaic diseases. Control of aphids and strict sanitary measures to avoid inadvertent transmission of the viruses from infected to healthy plants are employed to limit their spread as is prompt roguing out and destruction of infected plants.

Not all mosaic infections are considered undesirable. A classic example of a benign one is that which causes the leaf variegation of *Abutilon striatum thompsonii.* Now widely cultivated as an ornamental, plants of this variety have all been derived by vegetative propagation from a single plant with variegated foliage that appeared among a batch of green-leaved seedlings in the West Indies about 1868. It is probable that some other variegations considered ornamental are caused by viruses.

MOSCHARIA (Moschàr-ia). One annual, a native of Chile, is the only member of *Moscharia,* of the daisy family Compositae. Its name, which refers to its musky odor, is derived from the Greek *moschos,* musk.

From 1½ to 2 feet tall and diffuse, ***M. pinnatifida*** has stem-clasping leaves, the lower ones pinnate or deeply-divided in a pinnate manner, the upper ones more or less lobed. The daisy-like flower heads, consisting of both disk and ray florets, are white to light rose-pink. About ¾ inch in diameter, they are in loose terminal clusters.

Garden Uses and Cultivation. An attractive plant for flower beds and borders, *Moscharia* has blooms useful for cutting and produces freely from midsummer until frost. Seeds may be sown in early spring directly where the plants are to bloom or, and this is the more usual practice, they may be sown early indoors and the resulting plants set in the garden after all danger of frost has passed. If sowings are made outdoors the seedlings are thinned to 7 to 9 inches apart. Indoors, the seeds germinate well in a temperature of 60 to 65°F, the seedlings are transplanted 2 inches apart in flats and grown in a sunny greenhouse with a night temperature of 50 to 55°F, with day temperatures five or ten degrees higher. Sowing should be done about eight weeks before the plants are to be transplanted to the garden. Appropriate spacing in the garden is about 9 inches between individuals.

MOSCOSOARA. This is the name of orchid hybrids the parents of which include *Broughtonia, Epidendrum,* and *Laeliopsis.*

MOSES-IN-A-BOAT or MOSES-IN-A-CRADLE is *Rhoeo spathacea.*

MOSQUITO. This word appears as parts of the common names of these plants: mosquito fern (*Azolla*), mosquito plant (*Lopezia hirsuta*), and mosquito trap (*Cynanchum ascyrifolium*).

MOSS. As parts of vernacular names of plants that are not true mosses, the word moss is used for these: ball-moss or bunch-moss (*Tillandsia fasciculata*), floating-moss (*Salvinia rotundifolia*), flowering-moss (*Pyxidanthera*), Irish-moss (*Arenaria verna caespitosa* and *Sagina subulata*), moss campion (*Silene acaulis*), moss-pink (*Phlox subulata*), rose-moss (*Portulaca grandiflora*), Scotch-moss (*Arenaria verna aurea* and *Sagina subulata aurea*), and Spanish-moss (*Tillandsia usneoides*).

MOSS, PEAT. See Peat Moss.

MOSS, SHEET. The bright green or yellow-green material known as sheet moss used by gardeners to line hanging baskets before filling them with soil and for other purposes consists of sheets of dried and dead carpet moss (chiefly *Hypnum curvifolium*). This is gathered from the wild, bagged, and sold through garden and florists' supply outlets.

MOSS, SPHAGNUM. This is gathered from the wild and used alive or dried and dead by gardeners and florists for a number of purposes. It is wrapped around sticks to make "totem poles" on which are grown vining philodendrons and other stem-rooting plants, to line hanging baskets, and in the propagation technique called air layering.

As a rooting medium for a wide variety of plants, most especially epiphytes, it is excellent to use by itself or mix with other materials. Ground, dried sphagnum moss, sold as milled sphagnum, is an excellent medium in which to sow seeds. Because of its antiseptic quality it inhibits growth of damping-off disease.

Sphagnum mosses (members of the genus *Sphagnum*) decay slowly under water to form peat moss and, under compression, the fuel lignite. The chief components of peat bogs, they are widely distributed in the wild, often covering extensive areas of acid wetlands.

MOSSES. Plants correctly identified as mosses constitute the subdivision of nonflowering plants called bryophytes of which there are nearly 15,000 species. Certain other plants are known by colloquial names that include the word moss, among them club-moss (*Lycopodium*), Florida- or Spanish-moss (*Tillandsia usneoides*), flowering-moss (*Pyxidanthera barbulata*), Irish-moss (*Arenaria verna caespitosa, Sagina subulata* and a type of edible seaweed), Scotch-moss (*Arenaria verna aurea* and *Sagina subulata aurea*), reindeer-moss (a lichen), and rose-moss (*Portulaca grandiflora*).

True mosses are low and have slender, soft stems bearing small green to brownish leaves and, except in sphagnum mosses, rootlike organs called rhizoids. Sexual reproductive cells develop from the centers of rosettes of leaves at the tips of the shoots, males and females on the same or different plants. The male cells, or sperm, have the ability to swim in the films or drops of water that collect in the cupped rosettes of leaves and in this manner to reach and fertilize the female reproductive cells, or eggs, after which the latter develop into spore-containing capsules. When mature these burst and release their dust-fine contents that, dispersed by the wind and lighting on moist ground, rocks, rough-barked tree trunks, or other suitable moist places, germinate and produce new moss plants.

A moss flourishing in a Japanese garden

Only very rarely in American and European gardens are attempts made to cultivate mosses, but in Japan they are successfully grown as ornamentals and one moss garden in Kyoto is reported to contain ninety species and varieties. Characteristically, they are loose and spongy, of feathery appearance, are capable of absorbing up to thirty-five times their dry weight of water.

MOTH MULLEIN is *Verbascum blattaria.*

MOTH ORCHID. See Phalaenopsis.

MOTHER. The word mother occurs in the common names of these plants: mother-of-thousands (*Saxifraga umbrosa* and *S. urbana*), mother of thyme (*Thymus drucei*), and mother spleenwort (*Asplenium bulbiferum*).

MOTHER-IN-LAW PLANT. See Dieffenbachia.

MOTHERWORT is *Leonurus cardiaca.*

MOTHS. These are the adults of a group of insects related to butterflies that, like butterflies, pass in succession through four well-defined stages. From eggs are hatched caterpillars. These in turn become pupae from which emerge after a period of dormancy the mature insects called moths. The most obvious way in which moths differ from butterflies is that their antennae do not end in knobs. They taper to sharp points or are feathery. Also, when at rest their wings spread horizontally or are held against the body. Those of most butterflies stand vertically above the body. Most, but not all moths fly at night, butterflies are active during the day.

Moths in the adult stage do no harm. Some kinds are helpful in transferring pollen from flower to flower. But the larvae (caterpillars) of many kinds are extremely destructive. Among them are some of the most serious of plant pests. Examples are cankerworms, earworms, tent caterpillars, and the caterpillars of gypsy moths and tussock moths. Practical control measures are chiefly aimed at destroying the caterpillars. See also Caterpillars.

MOUNTAIN. This word forms parts of the common names of these plants: Australian-mountain-ash (*Eucalyptus regnans*), fire-on-the-mountain (*Euphorbia heterophylla*), mountain-andromeda (*Pieris floribunda*), mountain-ash (*Sorbus*), mountain avens (*Geum montanum*, the name mountain-avens is also applied to *Dryas*), mountain-bluet (*Centaurea montana*), mountain bugbane (*Cimicifuga americana*), mountain cowberry or mountain cranberry (*Vaccinium vitis-idaea minus*), mountain-dandelion (*Agoseris*), mountain-ebony (*Bauhinia variegata*), mountain fringe (*Adlumia*), mountain-holly (*Nemopanthus mucronata* and *Prunus ilicifolia*), mountain-laurel (*Kalmia latifolia*), mountain-lily (*Ranunculus lyallii*), mountain mahoe (*Hibiscus elatus*), mountain-mahogany (*Cercocarpus*), mountain-mint (*Pycnanthemum* and *Monardella odoratissima*), mountain-parsley (*Pseudocymopterus montanus*), mountain-pink (*Centaurium beyrichii*), mountain pride (*Penstemon menziesii*), mountain rimu (*Dacrydium laxifolium*), mountain-rose (*Orothamnus zeyheri*), mountain-sorrel (*Oxyria digyna*), mountain supple jack (*Serjania exarta*), mountain tawhiwhi (*Pittosporum colensoi*), snow-on-the-mountain (*Euphorbia marginata*), and Texas-mountain-laurel (*Sophora secundiflora*).

MOURNING BRIDE is *Scabiosa atropurpurea*.

MOUSE PLANT is *Arisarum proboscideum*.

MOUTH. As parts of common names the word mouth appears in these: adder's mouth (*Malaxis*), dragon's mouth (*Arethusa*), and snake mouth (*Pogonia ophioglossoides*).

MOWING. See Lawn Maintenance.

MOXIE is *Gaultheria hispidula*.

MU TREE is *Aleurites montana*.

MUCUNA (Muc-ùna)—Cowitch or Cowage, Florida Velvet-Bean. As here interpreted, *Mucuna* includes the plants previously segregated as *Stizolobium*. Most modern botanists do not accept the split into two genera, based chiefly on characteristics of the seeds, as justified. In the broad sense *Mucuna* consists of 120 species of woody and herbaceous vines, including annuals, of the pea family LEGUMINOSAE. Some have hairs that sting or cause severe itching. Some are cultivated for food and fodder. Some are beautiful in bloom. The name is a Brazilian one, the genus tropical.

Mucunas have leaves of three leaflets, the lateral ones unequal-sided. Their pea-shaped blooms, purple, red, yellow, or orange, are in racemes or umbel-like clusters. They have a short, bell-shaped calyx with four or five teeth or segments, a standard or banner petal shorter than the wing petals with earlike appendages at its base, and ten stamens of which nine are united and one is free. The pods, often large, are leathery and hairy, the hairs of some kinds stinging.

Cowitch and cowage are names applied to several species that have irritating hairs. One, ***M. urens,*** a native of tropical America, is a perennial, woody vine that climbs to heights of 30 feet or so. It has hairless leaves with short-pointed leaflets 3½ to 7 inches long. The purplish-blue flowers with yellowing keels are in clusters of ten to fifteen on slender, pendulous stalks about ten times as long as the leafstalks, and up to 4½ feet long. Their broad standard petals are 1¼ to 1½ inches long and from three-quarters to nearly as long as the wings, which nearly equal the keel in length. The seed pods, 4½ inches to 1 foot long by about 2 inches broad, are thickly clothed with bristly, stinging hairs.

Differing from *M. urens* in its leaves being hairy on their undersides, and its umbel-like clusters of yellow flowers having stalks not more than twice as long as the leaves, ***M. sloanei*** is a native of tropical America, the West Indies, and tropical Africa. The standard petal is about one-half as long as the wings, the bristly pods 4 to 6 inches long by 1½ inches broad. Also yellow-flowered, ***M. fawcettii,*** of Jamaica, differs from *M. sloanei* in the stalks of its flower clusters broadening to almost ½ inch at the apex where the 3-inch-long blooms are attached. The standard petals are about two-thirds as long as the wings. The bristly seed pods are about 4½ inches long by 2 inches broad.

Ornamental-flowered, woody perennials sometimes cultivated in Hawaii and other parts of the tropics include ***M. bennettii,*** considered to be one of the most beautiful of flowering vines. A rampant climber, it has drooping trusses of brilliant orange-red blooms each 3 to 5 inches long. Much like it, ***M. kraetkei*** is distinguished by its shorter calyx lobes. Somewhat smaller, greenish-yellow, less showy blooms are borne by ***M. albertisii.*** Smaller, but darker and brighter red flowers than those of *M. bennettii* are borne by ***M. novaguineensis.***

Annual mucunas without woody stems, previously segregated in the genus *Stizolobium,* include the kinds dealt with now. Called cowitch or cowage, ***M. pruriens*** (syn. *Stizolobium pruriens*) has leaves, hairy on their undersides, with leaflets 3 to 6 inches long and stalks up to 1 foot long. The flowers, dark brown or purplish, less often paler or white, and in racemes, have standard petals one-half as long as the wings which are about 1½ inches in length and are slightly exceeded by the keel. The spotted seeds are in pods 1½ to 3½ inches long and densely clothed with brownish-yellow hairs. This species is native to the tropics.

Florida velvet-bean (***M. deeringiana*** syn. *Stizolobium deeringianum*), probably a native of the Asian tropics, is an annual vine cultivated for ornament as well as for food, fodder, and green manure. Its stems are sometimes 50 feet in length. Its flowers dark purple and about 1½ inches long, hang in pendent clusters of up to thirty. Its black-hairy pods, containing three to five nearly spherical seeds, are narrow and 2 to 3 inches long. Other annuals that have value as forage are the Yokohama-bean (*M. hassjoo* syn. *Stizolobium hassjoo*) and the Lyon-bean (*M. cochinchinensis* syns. *M. nivea, Stizolobium niveum*). Probably native to Japan, ***M. hassjoo*** has dark purple blooms about 1½ inches long, in racemes 4 to 6 inches in length. The white-hairy pods are 3 to 4½ inches long. The flowers of ***M. cochinchinensis,*** a species believed to be a native of southeast Asia and the Philippine Islands, are white or rarely purple, and in clusters with stalks 1 foot to 2 feet long. Its pods are 4 to 5 inches long.

Garden and Landscape Uses and Cultivation. In general, mucunas are rampant and need much room as well as supports around which their stems can twine. Lovers of warmth and needing a long season of growth, they are not suited for the north. Appreciating fertile earth and reasonable supplies of water, they are ordinarily propagated by seed.

MUD-PLANTAIN. See Heteranthera.

MUEHLENBECKIA (Muehlen-béckia)—Wire Plant or Maidenhair Vine. Two of the fifteen species of *Muehlenbeckia* are commonly cultivated in North America. A third plant, often called *Muehlenbeckia platyclados,* is correctly *Homalocladium platycladum.* The group includes plants of various habits and is confined to the temperate and warm-temperate parts of the southern hemisphere including southern South America, New Zealand, Australia, and New Guinea. It belongs in the knotweed family POLYGONACEAE. The name commemorates H. G. Muehlenbeck, a Swiss physician, who died in 1845.

Muehlenbeckias are more or less woody, erect, vining, or creeping plants usually with alternate leaves, but sometimes without any. The leaves have sheathing stipules (basal appendages), a condition common in the knotweed family. The minute flowers have little decorative merit. They are unisexual or bisexual; sometimes the sexes are on different plants, sometimes unisexual and bisexual flowers are borne by the same individual. The flowers are deeply-five-parted and commonly have eight stamens and a three-lobed stigma. The berry-like fruits are nutlets surrounded by parts of the flower that become fleshy.

The wire plant or maidenhair vine (***M. complexa***) is a robust native of New Zealand, creeping, or climbing to a height of several feet, and deciduous or evergreen according to climate. Its thin, wiry, reddish, much interlaced stems are grooved and more or less hairy on their younger parts. They form dense tangles when allowed to develop without support. The stalked leaves, ¼ to ¾ inch long, are rounded, heart-shaped, or contracted at their middles and fiddle-shaped; sometimes they are obscurely lobed. The flowers, the males and females on separate plants, are few together in axillary and terminal spikes. They are greenish-white. The black nutlets are contained in waxy-white cups.

The other commonly cultivated species ***M. axillaris*** (syn. *M. nana*) is also a native of New Zealand. It is one of exceedingly few plants native south of the Equator reliably hardy in the vicinity of New York City. This is much smaller and lower than *M. complexa* and is deciduous. It comes into foliage late in spring. Usually not over 1 foot tall, it has creeping and sprawling, threadlike, much-branched stems and round or blunt-oblong leaves up to ¼ inch long. The yellowish-green flowers are ordinarily solitary or in pairs in the leaf axils, rarely the males are in short terminal spikes. The black nutlets are enclosed in the whitish succulent remains of the flowers.

Muehlenbeckia axillaris as a groundcover

Garden Uses. The wire plant is popular in California and other mild climate regions for training up chimneys, against walls, and on other surfaces to which supports can be affixed. It is also useful for clambering over tree stumps and for forming mounded unsupported masses. In greenhouses it makes a very good hanging basket plant and may also be accommodated in pots or tubs; under warm conditions it retains its foliage throughout the year. The wire plant is said to be root hardy at Philadelphia, Pennsylvania. The maidenhair vine is a charming groundcover for rock gardens and makes a good cover for spring-flowering bulbs of the smaller kinds favored by rock gardeners. It may also be used effectively to clothe banks and other places where a neat carpet of foliage is needed.

Cultivation. These plants thrive without difficulty in any reasonably well-drained soil in full sun. They are easily increased by cuttings and, *M. axillaris*, by careful division.

MUGWORT. This is the colloquial name of the pestiferous weed *Artemisia vulgaris*. The ornamental hardy herbaceous perennial called white mugwort is *A. lactiflora*.

MUILLA (Mu-ílla). Five species of the southwestern United States and Mexico belong in *Muilla*, of the lily family LILIACEAE. Closely related to *Allium*, this genus has a name contrived by spelling *Allium* backward. From alliums, muillas differ in being free of any characteristic onion-like odor and in having minute bractlets on the stalks of the individual flowers.

Muillas have corms covered with fibrous coats somewhat similar to those of crocuses. Their few narrow leaves are nearly cylindrical. The starry flowers are in umbels atop naked stalks. Each has six whitish or greenish perianth parts (commonly called petals, more correctly, tepals), six stamens, and one style. The fruits are capsules.

Native to California and Baja California, ***M. maritima*** occurs in coastal areas and in saline soils inland. It is from 4 inches to 1½ feet tall, with leaves as long or longer than its flower stalks. Its flowers, in heads of up to seventeen, but more usually of five to fifteen, are greenish-white with purplish anthers and about ½ inch across.

Garden Uses and Cultivation. In their native regions these muillas are suitable for inclusion in native plant gardens and occasionally are grown in rock gardens elsewhere. They need full sun and very well-drained soil. Propagation is by seed and by separating natural increases of the corms.

MUIRIA (Mùir-ia). One remarkable nonhardy succulent of the carpetweed family AIZOACEAE is the only representative of this genus. It is nearly related to *Gibbaeum* and belongs with the many genera once included in *Mesembryanthemum*. The name commemorates Dr. John Muir, a twentieth-century collector of South African plants.

Native of the Little Karroo in South Africa, where it blooms in December, ***Muiria hortenseae*** forms clusters of globose to egg-shaped, soft plant bodies about ¾ inch in diameter. Green or tinted red, they are thickly clothed with short velvety down. Each plant body consists of two leaves so completely joined that the only evidence that they are not one is a tiny slit representing an apical aperture between the leaves. Curiously, this is not on top of the plant but at one side. Even more unusual, reportedly the flower does not emerge from this fissure, but ruptures the top of the plant body and pushes upward through the wound. In the entire *Mesembryanthemum* group this procedure is unique to *Muiria*. The white or pinkish blooms, rarely if ever developed in cultivation, are ⅓ to ¾ inch in diameter and have a six-lobed calyx, many entirely separate narrow petals, and six or seven short, thick stigmas. The fruits are small capsules. A natural hybrid between this and *Gibbaeum album roseum* is named *Muirio-gibbaeum muirioides*.

Garden Uses and Cultivation. A plant for collectors of choice succulents, *M. hortenseae* persists under cultivation if afforded the care appropriate for *Lithops* and other very fleshy South African *Mesembryanthemum* relatives. Its growing season, during which water is given in moderation, is short, from July to September. At other times water should be withheld even to the extent that the plant bodies shrivel slightly. Porous, sandy soil containing crushed limestone suits this plant. Propagation is by careful division and by seed. For further information see Succulents.

MUIRIO-GIBBAEUM (Mùirio-gibbaeum). This name belongs to a natural hybrid between *Muria hortenseae* and *Gibbaeum album roseum*. It is composed of the names of its parents, which belong in the carpetweed family AIZOACEAE. A low, nonhardy, clump-forming plant with short, thick branches covered with the remains of dead leaves, *Muirio-gibbaeum muirioides* has plant bodies ¾ to 1 inch long and as much in diameter. With much the appearance of those of *Gibbaeum album*, they are greenish-brown and densely clothed with short, fine hairs. The white to pink flowers are

Mukdenia rossii

produced in South Africa, the homeland of this interesting hybrid, in December.

Garden Uses and Cultivation. These are the same as for *Gibbaeum album*. For further information see Succulents.

MUKDENIA (Muk-dènia). Previously known as *Aceriphyllum*, eastern Asian *Mukdenia*, of the saxifrage family SAXIFRAGACEAE, consists of two species of deciduous herbaceous perennials segregated from *Saxifraga*, in which by some authorities they are included, by technical differences only.

Native to Korea and Manchuria, ***M. rossii*** (syns. *Aceriphyllum rossii*, *Saxifraga rossii*) is an attractive hardy ornamental 1 foot to 2 feet tall. It has short rhizomes and deeply five- to nine-lobed, bronzy-green, toothed leaves rounded in outline. Its numerous small, white, bell-shaped flowers in long-stalked, erect, loose panicles tower above the foliage. It blooms in early May.

Garden and Landscape Uses and Cultivation. Easily raised from seed and by division, this is suitable for fringes of woodlands and other lightly shaded locations where the soil has a fairly high organic content and is not excessively dry.

MULBERRY. See Morus. The French-mulberry is *Callicarpa americana*, the Indian-mulberry *Morinda citrifolia*, the paper-mulberry (*Broussonetia papyrifera*), the sea-mulberry (*Conocarpus erectus*).

MULCHING and MULCHES. The covering of the ground around plants with a layer of material to conserve moisture, prevent erosion, check weed growth, moderate temperatures, keep the surface from crusting and packing, and in some cases serve as an incidental source of nutrients is called mulching, the materials used, mulches. An ancient and admirable garden practice, wisely employed this can greatly reduce labor needed for weed control and watering, and by assuring a favorable climate for root growth promote more vigorous and healthier ornamentals and bigger yields of fruits and vegetables.

Mulching is the adoption and adaptation of a natural process. In forest and meadow, leaves and other plant parts fall to the ground to become a friendly layer of debris that affords protection to roots and as it decays nourishes them. The methodical gardeners' passion for leaf raking, debris disposal, and suchlike tidy horticultural housekeeping is artificial albeit often necessary as for instance in maintaining lawns and in cultivating crops for which the ground must be spaded annually, and often also as a disease and pest control measure. As practiced in gardens, mulching is the selective placement of the equivalent of natural accumulations.

Selective is the key word in the last sentence. The thoughtless or indiscriminate use of mulches can be disadvantageous. For example, applied too early in the season to such warm-weather crops as tomatoes, they can slow growth and delay ripening. Winter mulches installed before the top inch or so of ground freezes may encourage mice to seek homes among them and tunnel with resulting damage or destruction to roots and bulbs. Another serious damage mice attracted to mulches are likely to commit is the gnawing of bark from the bases of the stems and trunks of young trees and shrubs. Some mulches may contain pests or disease organisms or become repositories for these, some may bring with them weed seeds.

Other possible disadvantages that need consideration include high cost. Most commercial mulches are not inexpensive, a factor that should be weighed in deciding whether or not to employ them. Often their use is fully justified for small, intensively maintained areas when expense precludes their more general application. For larger expanses locally available, generally noncommercial materials may satisfy. Another matter to take into account, especially in places accessible to the public, is that some mulches constitute a fire hazard. A carelessly dropped lighted cigarette can initiate a slow burn or conflagration in dry leaves, hay, salt hay, pine needles, and even in dry peat moss or peanut shells. And finally, plants mulched with fresh hay, straw, sawdust, and other materials that contain high proportions of carbon relative to nitrogen are likely to suffer, often severely, from temporary nitrogen deficiency during the period the mulch takes to decompose, since the microorganisms responsible for decay themselves need nitrogen, and if it is in short supply they preempt what is available. The obvious way of staving this off is to apply a nitrogen fertilizer immediately before mulches of this type are spread. With fresh, dry sawdust, for example, 1 pound of nitrogen (an amount supplied by 20 pounds of a 5-10-5 or 10 pounds of a 10-6-4 fertilizer, by 6¼ pounds of nitrate of soda or by 2⅕ pounds of urea) should be applied with each 100 pounds of the sawdust. Lesser amounts are satisfactory with wood chips because being larger these decay more slowly, and still smaller amounts are needed with hay, straw, and similar materials.

The benefits of mulching properly carried out far outweigh possible disadvantages. Especially important is the more uniform environment assured the roots. Extremes of temperature are mitigated. Availability of moisture in the soil remains more constant. Penetration of the soil by rain and irrigation water is greatly improved, loss by run-off being eliminated or minimized. Soils protected by a mulch are not battered by rains. The force of the impact of the drops is absorbed by the mulch, the water trickles gently to the ground. The desirable crumb structure of the soil is encouraged and preserved, and compaction is prevented.

Because they shade the ground, and because most mulches are poor conductors of heat, the temperature of soil beneath them does not rise as high by day in summer nor drop as low at night in winter as with uncovered ground. Best insulation is given by loose, poor conductors of heat, such as bagasse, buckwheat hulls, salt hay, peanut shells, peat moss, and sawdust. Light-colored mulches reflect more heat and so keep the soil cooler than darker ones, a consideration of practical importance when plastic mulches are being considered.

Water loss by evaporation from the soil surface is greatly reduced by mulching. Evaporation is related to vapor pressure gradients, the vapor moving from regions of high humidity to those of lower humidity, at a rate that increases with temperature. Because of this, water loss from soil is reduced when its temperature is lowered and when the relative humidity of the air immediately above it is high, conditions mulches favor. As a result, root development close to and at the soil surface is frequently stimulated by mulching without any reduction in root growth at lower levels, although the actual pattern of root distribution largely depends upon the kind of plant. Such dividends in extra roots result in more vigorous growth.

Choice of mulch depends to a large extent upon its availability and cost as well as its effectiveness, and for use in many landscapes, its appearance. In the past coarse, half-rotted animal manure containing fairly large amounts of straw or other bedding litter was popular, and if there is considerable need for added nutrients as may be true around old-established trees and shrubs, this still offers possibilities as a mulch. A chief disadvantage is the likelihood that it contains seeds of troublesome weeds or even if it does not, of providing a fertile germinating medium for blown in weed seeds. Compost made by piling leaves and other plant wastes, grass mowings, and the like, used in a coarse, partly decayed rather than more decomposed state, also is a good mulch. Although less nutritious than manure, it has the same disadvantages.

Partially rotted manure is a nourishing mulch for shrubs and trees

Coarse compost from plant wastes is an excellent mulch, particularly for shady areas

Mulches are of two types, organic and inorganic. The first, the most numerous, are of plant or less commonly animal origin. They gradually decay and in doing so add some nutrients to the soil. Eventually they become humus. Inorganic mulches, the chief of which are calcined clay, gravel, stones, and polyethylene plastic, are not plant or animal products, do not decay, and do not supply nutrients. The selection of mulches now to be presented is by no means exhaustive, but includes those most commonly used.

Bagasse (dried refuse of sugar cane from which the sap has been expressed) is excellent and particularly suited for plants that prefer acid soil. The roots of azaleas, rhododendrons, and others of this preference spread greedily into this material as it decomposes, and it supplies considerable nutrients. It should be spread 2 to 3 inches deep.

Barks of various kinds, including those of pines and redwoods, broken to egg size or smaller are commercially available. Of earthy-brown colors and rot-resistant, they look well and are long-lasting. A layer 2 to 3 inches deep is adequate.

Buckwheat hulls, the husks of the seeds, rich brown to nearly black and of neat appearance, are much favored for rose beds. Very light, they are sometimes disturbed by high winds and easily by a stream of water directed at them. They are not suitable for slopes. A 1- to 2-inch layer is adequate.

Calcined (heat-treated) clays are inorganic mulches sold under various trade names. They are yellowish-brown to brown and somewhat resemble gravel or crushed crocks. Pleasing in appearance, calcined clays do not decay and do not add nutrients to the soil. They are satisfactory in layers 1 inch to 2 inches deep.

Cocoa shells (pulverized hulls of cocoa beans from which chocolate is made) are attractively colored and light in weight, but have the disadvantage of caking on the surface upon drying after wetting. They pervade the air with a faint, cloying odor of chocolate, which is pleasant or otherwise, depending on personal likes. Nutrients re-

Commercially available mulches: (a) Tree barks processed to suitable sizes

(b) Being spread on rose beds

Mulches of calcined clay are effective and pleasing for some uses

leased into the soil include potash in amounts that if the mulch is used to excess may injure such surface-rooting plants as tomatoes, young azaleas, rhododendrons, lilacs, and maples. A layer from 1 inch to 2 inches thick is sufficient.

Coffee grounds, available locally from coffee processing plants, less commonly in these days of instant coffees from home kitchens, may be used, but the surfaces of mulches of them cake and unless the crust is broken occasionally deny free access of air to the roots. As they decay they add some nutrients to the soil. Spread them not over 1 inch thick.

Grass mowings may sometimes be used with benefit, but generally are less satisfactory than other materials. This is especially true if they are spread when green and succulent. The chief trouble is that as they begin to decay they ferment, heat up, give off a disagreeable odor, and settle to become a more or less slimy layer that prevents an adequate amount of air reaching

Grass mowings, here being raked from a lawn, have some limited use for mulching after being mixed with leaves or other coarse debris

the soil. The best results are had by mixing with them sufficient coarse stuff, such as partly decayed leaves or crushed corncobs, to keep the whole fairly loose and to prevent the mulch from matting. All things considered, unless there is imperative need it is better to add grass mowings to compost piles instead of employing them for mulching.

Ground corncobs form a very good mulch, too light in color when first applied to be exactly pleasing, but darkening as they weather. As they decay they provide nutrients and later when incorporated in the soil improve its structure. A 2- to 3-inch layer serves well.

Ground tobacco stems, in some places available, form a satisfactory mulch that readily admits air and water. This mulch has the advantage of possessing mild insecticidal properties, the disadvantage of being so loose that small leaves that fall from plants are likely to lodge in it; if these harbor diseases as, for instance, black spot of roses, sources of further infection may be established.

Hay, salt hay, and straw are bulky mulches not good looking enough to encourage their general use in ornamental plantings, but useful and popular for or-

Salt hay is excellent and weed-free as a mulch: (a) Shaking it out to form a loose surface layer around plants

(b) Around a rhododendron

(c) On a bank, held in place by strings stretched between stakes

chards and vegetable gardens. To be effective a depth of 3 to 4 inches is needed. As these decay they improve the fertility of the soil. Besides hays of grasses, and these include salt or marsh hay cut from saline meadows and wetlands, there are legume hays, which are the dry stems and foliage of such legumes (plants of the pea family) as alfalfa, clovers, peas, and soy beans. Legume hays are especially high in nutrients that are released slowly. Grass hays other than salt hay have the disadvantage of containing seeds that can give rise to seedlings unwanted in cultivated ground. Seeds of salt hay germinate only in salt marshland.

Leaves of deciduous trees and shrubs are one of the most obvious mulches. For some purposes all that is necessary is to rake them from lawns and other open areas and spread them under trees and bushes, but it usually is better to collect and stack the leaves (compost them) and use them as a 4- to 6-inch-deep mulch when they are half rotted. This prevents them from blowing and eliminates any potential fire hazard. Another way of reduc-

Freshly fallen leaves may be used as mulch where the wind is unlikely to disturb them, but are often better after having been composted until partially decayed

ing these dangers with freshly fallen leaves is to chop them finely in a hammer mill type of grinder before they are spread. All deciduous leaves may be used and all as they decay nourish the soil, but sorts differ considerably in their merits as mulches. Those such as maples, poplars, and horse-chestnuts, which lie flat, pack down, and when wet become layered wads that seal the surface soil from air, are not nearly as good as leaves of oak and other trees that remain reasonably loose and permit free passage of air.

Peanut shells, though light and easy to handle, are a little too pale in color to be as attractive as some mulches. They provide effective insulation and permit ready passage of air and water. Disadvantages are that they sometimes, but rarely, attract mice and, if dry, may burn if a carelessly discarded cigarette lands on them. A depth of 2 to 4 inches is appropriate.

Peat moss (sphagnum peat) available at garden supply outlets in convenient bales is rich brown and good looking. Easy to apply, this is one of the most popular mulches. Other native American peats, mostly fine-textured and darker-colored than sphagnum peat, can also be used but most are not as long lasting as sphagnum. Practically all peats have an acid reaction, so they are particularly suitable for acid-soil plants. If it becomes very dry it takes a great deal of rain or watering to soak peat moss. Because of this, it should not be applied as a layer more than 2 inches deep.

Pine needles or pine straw, as the fallen dead leaves of pine trees are called, are splendid mulch material suitable for applying 1 inch to 2 inches deep. Of agreeable appearance, pine needles decay slowly and have a reaction well suited to the needs of such acid-soil plants as azaleas and rhododendrons. Should they, as the result of long use, cause the soil to become too acid, correction is easily achieved by applying a dressing of ground limestone or lime. Sawdust of any kind may be used in layers not more than 2 inches thick. Fresh, it has a rather raw appearance, but as it ages it darkens and mellows, or it may be piled for a couple of years or more so that it partly decays before being spread. In any case a fairly heavy application of nitrogenous fertilizer (one pound of sulfate of ammonia or nitrate of soda or equivalent for each bushel of sawdust) should be spread immediately before the mulch is put in place. Sawdust decays slowly, eventually becoming humus and enriching the soil.

Seaweed, in coastal areas sometimes available for gathering, is a very good mulch, well suited for vegetable gardens and orchards. It decays fairly rapidly and in doing so appreciably improves the fertility of the soil. Apply it at a depth of 3 to 6 inches.

Spent hops, a residue of breweries, are available in some areas. Because of their high water content they are heavy in relation to their about 12 percent dry weight, which makes long-distant haulage impracticable. When first spread this mulch gives off a sour, pervasive odor, but this disappears after two or three weeks. In hot weather, they are likely to ferment and heat to such a degree that shoots or young trunks in contact with them may be injured. Spread this mulch 3 to 4 inches deep, keeping it a little away from plants it may harm.

Wood chips, obtained by putting branches and other prunings through a chipper machine, are commonly obtainable in regions where trees are plentiful; they are good looking and long lasting. Spread them to a depth of 3 to 4 inches, taking care before doing so to dress the ground with a nitrogen-rich fertilizer.

Machines "chew up" prunings from trees and shrubs to make chips: (a) Chips being transported

(b) Chips being spread

Wood shavings, occasionally employed as mulches, are less satisfactory than wood chips, this because they are lighter and more likely to be blown by wind and, on slopes, disturbed by heavy rains. They also are less pleasing in appearance than wood chips and may be a fire hazard. Their installation should be preceded by an application of nitrogen fertilizer.

Stones, stone chips, pebbles, and gravel are widely available nonorganic mulches especially useful beneath trees, in desert gardens, rock gardens, and other special areas where a permanent, nonnutritious mulch is satisfactory. The patterns and colors of these mulches are often very attractive. They may be used in depths of 1 inch to 3 inches.

Sheets of black polyethylene plastic employed as mulch conserve moisture and by denying light to weeds suppress them. Black plastic also aids by absorbing sun heat, which warms the soil in spring a little ahead of the normal season. This material is most serviceable in vegetable gardens, cut flower gardens, nursery areas, and elsewhere where plants in rows facilitate running strips of plastic between them. Squares of the material with a hole in the center of each and a slit extending from it to one side are practicable for covering the ground around young trees, especially recently planted ones.

Weigh down the edges of the plastic with stones, soil, or boards, and punch some holes through it to give access to rain water.

MULGEDIUM. See Lactuca.

MULLEIN. See Verbascum. The Cretan-mullein is *Celsia cretica,* the mullein-pink *Lychnis coronaria.*

MUNDI ROOT is *Chlorocodon whitei.*

MUNG BEAN is *Vigna radiata.*

MUNTINGIA (Mun-tíngia) — Capulin or Jam Fruit. This is a genus of one species, a tree up to 30 feet tall, native from Mexico to northern South America and in the West Indies. It bears edible fruits and belongs in the elaeocarpus family ELAEOCARPACEAE. Its name honors Professor Abraham Munting of Groeningen, Holland, who died in 1683.

The capulin (***Muntingia calabura***) has alternate, long-pointed, oblong-lanceolate, short-stalked, leaves with markedly unequal bases and coarsely- and irregularly-toothed margins. They are up to 5 inches long and are arranged as in flat sprays in one plane. Their undersides are densely covered with white or grayish hairs of the kind called stellate, which means that the individuals are branched with the branches radiating like the points of a star. From the leaf axils the long-stalked flowers develop either singly or in clusters. Produced over a long season, they are ¾ to 1 inch across and have five sepals and five white petals

Muntingia calabura (foliage and flowers)

Muntingia calabura (fruits)

and many prominent stamens. The fruits are spherical berries, about ½ inch in diameter and ripening red or yellow; they have juicy, sweet pulp and many small seeds. They may be eaten out of hand or used in pies or for jam.

Garden and Landscape Uses and Cultivation. This tree, easily raised from seed, grows with great rapidity and often fruits in its second or third season. Although not of major importance, it is of interest in home gardens and is best adapted to nonalkaline soils; on limestones it tends to deteriorate and die back. It is suitable only for frost-free or nearly frost-free climates. Because its branches grow rapidly, they should be pruned back severely whenever this seems desirable, usually at intervals of two or three years. The trees recover rapidly from this treatment.

MUNTRIES. The fruits of *Kunzca pomifera* in Australia are called muntries.

MURIATE OF POTASH. This, potassium chloride, is an important fertilizer. See Fertilizers.

MURRAEA. See Murraya.

MURRAY-RIVER-PINE is *Callitris columellaris.*

MURRAYA (Mur-ràya) — Orange-Jessamine, Curry Leaf. The spelling *Murraya,* sometimes rendered *Murraea,* is correct for the name of a genus of about a dozen species of eastern Asia, Indomalaya, Australia, and islands of the Pacific that belongs in the rue family RUTACEAE. The name commemorates Johan Andreas Murray, a Swedish student of Linnaeus and a professor of botany, who died in 1791.

Murrayas are nonhardy, spineless trees and shrubs related to the orange (*Citrus*). They have alternate, pinnate leaves with an uneven number of alternate leaflets. In some species the leaves have only one leaflet, and then they look like undivided leaves. The flowers, in rather large, axillary or terminal panicles, have deeply five-toothed calyxes, five petals, ten stamens, and a slender style. The fruits are ovoid or nearly globular berries containing mucilaginous pulp and few to several seeds. The leaves of *M. koenigii* are used in curries.

Orange-jessamine (***M. paniculata*** syns. *M. exotica, Chalcas paniculata*) is a popular ornamental in the tropics and subtropics and is sometimes cultivated in greenhouses. It is greatly esteemed for its handsome evergreen foliage, pleasing displays of deliciously scented blooms, and small, red, ornamental fruits. Native to India, this is a somewhat variable shrub or tree up to about 12 feet in height. Except sometimes when very young, its shoots and shiny foliage are hairless. The leaves have three to nine broad, more or less ovate, round-toothed or toothless leaflets 1 inch to 2 inches long, blunt or sometimes notched at their apexes. The rather bell-shaped blooms, about ¾ inch long, have pointed petals. Ovoid and ⅓ to ½ inch in length, the fruits are bright red. Orange-jessamine often blooms several times a year and frequently is in flower and in fruit simultaneously.

Murraya paniculata

Much less common than the orange-jessamine and of quite distinct appearance, ***M. stenocarpa*** (syn. *Atalantia stenocarpa*) is planted to some extent in southern Florida. A shrub 4 to 6 feet tall, when young this has pinnate foliage, but its later leaves have only one elliptic to ovate-oblong leaflet. Shining deep green, they are 4 to 6 inches long by approximately 2 inches wide. The small, white, fragrant flowers are in clusters of few and are succeeded by red fruits about ½ inch long. This species is native to China.

The curry leaf (***M. koenigii***), widely distributed as a native in warm parts of Asia, is an evergreen tree up to 20 feet tall. Its leaves have five to ten pairs of oblong-lanceolate to ovate leaflets and a terminal one 1 inch to 1½ inches long. The small white flowers are in terminal panicles. They are succeeded by pea-sized, bluish-black fruits. This quite attractive, round-headed, lacy-leaved tree thrives in southern Florida.

Garden and Landscape Uses and Cultivation. Orange-jessamine is an excellent, easy-to-satisfy, general-purpose shrub for tropical and subtropical gardens in areas in which frost presents little or no hazard. It can be used effectively in the vicinity of buildings, as foregrounds for taller plantings, and as backgrounds for lower ones. A good screen or hedge, it stands pruning and shearing well. Ordinary garden soil and sun or part-day shade are to its liking. In pots or tubs, orange-jessamine is suitable for terrace, patio, and greenhouse decoration. In cold climates specimens may be stood outdoors in summer and wintered in a greenhouse. A night temperature of about 50°F is satisfactory, with a daytime rise of five to fifteen degrees. Container specimens respond to rich, well-drained soil kept moderately moist. Monthly or semimonthly applications of dilute liquid fertilizer help keep well-rooted samples healthy and vigorous. Pruning to shape and size is done in late winter or early spring. New plants are easily had from seeds or from cuttings. The other species described above are suitable for planting as ornamentals in warm, essentially frost-free climates. They need no special care and may be increased by seed and by cuttings.

MUSA (Mù-sa) — Banana, Plantain. Despite a popular misconception, bananas do not grow on trees. The apparent trunks of the plants that bear them are formed of the leaf stalks. Banana plants are evergreen herbaceous perennials, though giants among that group. Their genus, *Musa*, belongs in the banana family MUSACEAE and numbers about twenty-five species, natives of the Old World tropics and warm subtropics. Some plants previously included in *Musa* belong in the genus *Ensete*. The name *Musa* commemorates Antonio Musa, physician to the first Roman emperor, Octavius Augustus. He died in 14 B.C.

Musa, undetermined as to sorts: (a) Fruiting in tropical South America

(b) Young plants in Panama

(c) In New Orleans

Musa, undetermined sort (flowers)

Musas have thick bulbous or rhizomatous rootstocks from which arise the leaves and flower stalks. The latter ascend inside the false trunks formed by the leafstalks and emerge from their tops. Unlike those of closely related *Strelitzia*, *Ravenala*, and *Heliconia*, the leaves of *Musa* are arranged spirally, rather than in two ranks, and spread in all directions. They have large, often huge, paddle-shaped blades with stout midribs and numerous lateral veins that spread more or less at right angles to the mid-vein. The leaves tear easily along the lateral veins, and on plants grown in the open, older foliage is usually shredded by the wind. The flower spikes are erect or pendulous and comprise numerous flat, handlike clusters of blooms borne in the axils of broad, colored bracts. The flowers are effectively unisexual, the elements of one sex in each bloom being nonfunctional. Those flowers toward the ends of the spikes are male, those nearer the bases female. Individual flowers have a tubular calyx that soon splits down one side. The corolla consists of a single petal. There are five fertile stamens and one sterile one. Surprisingly, the elongated, often pulpy fruits are technically berries. In wild types they contain many seeds. Cultivated varieties of the common banana and plantain are seedless.

Among the world's most important food crops, bananas and plantains yield much more food per acre than any other plant. They are cultivated throughout the tropics for local consumption, and in selected areas, notably South America, the West Indies, and Canary Islands, bananas are grown in great numbers for export to temperate regions. Plantains and bananas, among the oldest of crops, have been cultivated for at least 4,000 years. Another important product of *Musa* is Manila-hemp or abaca fiber, a product of the leafstalks of *M. textilis*. Other species also yield usable fibers.

Common bananas, their yellow or red fruits suitable for eating raw, and plantains, which have more starchy fruits chiefly employed for cooking, are varieties of *M. acuminata* (syns. *M. cavendishii*, *M. chinensis*) or of *M. paradisiaca* (syn. *M. sapientum*), a hybrid of *M. acuminata* and *M. balbisiana*, the latter seldom cultivated. A variable native of the Asian tropics, Indonesia, the Philippine Islands, Australia, and East Africa, ***M. acuminata*** is up to 20 feet tall or taller, with false trunks usually blotched with brown or black. Ascending or erect, and with blades up to 19 feet long by 2 feet wide, the leaves have green upper surfaces sometimes with darker flecks

Musa acuminata variety

and green or purple undersides. The flowering stalks, at first horizontal become pendulous under the weight of the developing fruits. Numerous, mostly seedless, varieties are cultivated. A flowerless, nonfruiting variety, *M. p. oleracea*, is cultivated in some parts of the tropics for its large tubers, which are cooked and eaten.

The dwarf banana (*M. acuminata* 'Dwarf Cavendish'), also known as the Chinese banana, governor banana, and ladyfinger banana, and one of the most widely cultivated varieties, is the chief sort grown commercially in the Canary Islands, Israel, South Africa, and other subtropical regions. This differs from other varieties of *M. acuminata* in being lower and more compact (its "trunk" does not exceed 7 feet in height) and in its seedless fruits being only 4 or 5 inches long. In bunches of up to two or three hundred, they are yellow and eaten raw. The leaves of the dwarf banana are short-stalked, broad-bladed, and arching and compose a dense crown. Their blades are 2 to 3 feet long and about 1 foot wide. When young they may be blotched with dull red. The short flower clusters are pendulous, their bracts reddish-brown to purplish-red, those of the female blooms deciduous, those of the males persistent. The calyx tubes are yellowish-white, 1 inch long, and five-toothed. The petal is about one-half as long. The dwarf banana is hardier to cold than most other kinds.

Musa acuminata 'Dwarf Cavendish' and palm trees in Florida

Musa acuminata (fruit)

Excellent fruits for cooking are borne by a group of varieties grouped as ***M. fehi*** (syn. *M. troglodytarum*), which is naturalized in Hawaii and cultivated throughout the islands of the Pacific. From 15 to 20 feet in height, it has a blackish "trunk" and violet sap. Its leaves are very large, bigger than those of the plantain, with stalks 1 foot to 1½ feet long. The long flower spikes are erect or nearly so. The blooms are stalkless, in clusters of six to eight, and have calyx tubes with five unequal lobes. There is a short petal. The fruits are yellow and 5 to 6 inches long. When grown at high altitudes they contain seeds, but are seedless when produced at lower altitudes.

Variegated-leaved musas, kinds grown chiefly for the decorative qualities of their foliage, include varieties of the common banana and plantain. One, *M. paradisiaca vittata*, has leaves with white midribs and red margins; the light green body of the leaf is irregularly variegated with bands of milky-green and white. Its chubby, yellow fruits are eaten after cooking. Handsome *M. p.* 'Aeae' (syn. *M. koae*) has "trunks" and leaves, including their stalks and midribs, as well as the young fruits, strongly variegated. The leaf variegation consists of irregular bands of cream, pale green, and dark green. The short, roundish fruits, at maturity uniformly yellow, have yellow flesh. They are eaten after being cooked.

Foliage blotched with dark maroon or purple-red is characteristic of some species. Here belongs ***M. zebrina,*** of Assam, a dwarf species with comparatively long-stalked, bluish-green leaves of satiny texture, strongly variegated above with deep blood-red, and purplish-red on their undersides. The midribs are brownish-red. This kind, often misnamed *M. sumatrana,* retains its foliage color throughout the life of the leaves. It is very tender to cold. Imperfectly known botanically, the plant originally named ***M. sumatrana*** and to which that name properly belongs, rarely exceeds 7 feet in height and has short-stalked, shorter leaves than *M. zebrina.* They are grayish-green blotched with deep wine-red above, and wine-red on their undersides. With age they tend to lose their variegation. This species is more cold-resistant than *M. zebrina.* Less freely variegated and usually not over 6 feet tall, ***M. ornata,*** of India, is a slender, suckering kind with blue-green leaves usually somewhat marked on their uppersides with wine-red, and mostly purplish-red beneath. This species has yellow flowers and pink floral bracts. Its yellowish-green fruits, about 2½ inches long, contain many small black seeds. They are not hairy and do not split open. This is often misnamed *M. rosacea.*

Musa zebrina (leaf)

Musa coccinea (flowers)

Musa velutina

Musa textilis in a conservatory

Brilliantly colored flower clusters are a showy feature of ***M. coccinea,*** a native of Vietnam. The color is provided by large, yellow-tipped, bright scarlet bracts. A suckering species up to 6 feet tall, its long-stalked leaves are bright green. The flower spikes are erect, the blooms yellow. The seedy, hairless, orange-yellow fruits, 2 to 3 inches long, do not split open. A somewhat similar species with erect flower clusters with bright pink bracts is ***M. velutina,*** of Assam. A slender plant with pinkish leafstalks and midribs, it is up to 6 feet tall, and has leaf blades 3 feet long by one-third as wide. Its pink or red, seedy fruits 2 to 3 inches long, are finely hairy and, when ripe, split open.

Manila-hemp or abaca (***M. textilis***), 12 to 20 feet tall or slightly taller, a native of the Philippine Islands, is much cultivated there for its fiber used for ropes, paper, cloth, and other purposes. Its seedy fruits, curved and 2 to 3 inches long, are not edible. A suckering plant, this has bright green, often brownish-spotted leaves with stalks about 1 foot long. The flowers, in drooping spikes, have five-lobed calyx tubes about 1½ inches long with the outer lobes bearing a slender horn near their ends. The male blooms fall early. Probably the hardiest species, ***M. basjoo,*** of the Ryukyu Islands, is also cultivated as a source of fiber. It has a false trunk about 8 feet tall, and green, not glaucous leaves approximately 4½ feet long. The flower spikes are pendulous. The greenish-yellow fruits, barely 3 inches long, have white pulp. There is a variegated variety.

Garden and Landscape Uses. In addition to their importance as food plants, bananas, plantains, and other members of the genus *Musa* are highly ornamental. Few plants other than palms give such a characteristically tropical aspect to landscapes. Wherever they can be grown outdoors they are worth cultivating as decoratives. They are at their best in small groups, with the individuals of each cluster of different sizes. This duplicates the way in which they mostly occur in the wild. They should not be crowded among other vegetation, but should stand alone, perhaps well in front of sheltering trees or towering above lower plants. They are impressive silhouetted against the sky. For best results, they need plenty of light and fertile, moist soil. They are admirable for ground beds and containers in conservatories and large greenhouses. The dwarf banana can be accommodated and fruited even in small greenhouses.

Musa textilis (flowers)

Musa basjoo (flowers)

Musa acuminata 'Dwarf Cavendish' as a container plant

Cultivation. Provided with sufficient warmth, reasonable humidity, fertile soil, and generous supplies of moisture, no difficulty ordinarily attends growing musas. The hardier kinds thrive outdoors in the deep south, the more tender ones appreciate the lowland tropics. The most common means of increase is by suckers. Only certain species bear fertile seeds and can be propagated from seeds. Musas grow well in full sun or part-shade and best where they are sheltered from strong winds. In greenhouses they must be planted in very rich, well-drained soil and repotted promptly when their roots become crowded, at least until they occupy containers 2 feet or more in diameter. Then, regular applications of dilute liquid fertilizer will be needed at weekly or semiweekly intervals. For best results, a minimum winter night temperature of 60 to 65°F is maintained, and by day this is increased by five to fifteen degrees. At other seasons both night and day temperatures may be higher. Provided they are kept growing without check musas flower and fruit in one year to one year and one-half from the time young suckers are removed from the parent plants.

Muscari botryoides

MUSACEAE—Banana Family. Included in the genus *Musa* are such important food plants as bananas and the similar fruits called plantains and the plant from which is obtained abaca or Manila-hemp. Considered broadly, that is, including genera by some authorities segregated as the *Heliconiaceae, Strelitziaceae,* and *Lowiaceae,* the *Musaceae* consists of half a dozen genera of monocotyledons totaling about 150 species, all natives of the tropics and subtropics. Strictly ornamental members of the family include the bird of paradise (*Strelitzia*) and the travelers' tree (*Ravenala*).

Mostly large herbaceous plants, stemless or often of treelike aspect and with branchless, more or less trunklike stems generally sheathed by the bases of the leafstalks, these have alternate, sometimes two-ranked leaves with prevailingly paddle-shaped blades with prominent midveins and pinnately-arranged lateral ones.

In spikes, racemes, panicles, or heads with conspicuous bracts or spathes, the bisexual or unisexual flowers have perianths of six segments of different shapes and sizes, separate or variously united. There are rarely six fertile stamens, more often five and one staminode (nonfunctional stamen), which may be petal-like. There is one style and usually three sometimes branched stigmas. The fruits are berries or capsules. Genera are *Ensete, Heliconia, Musa, Orchidantha, Ravenala,* and *Strelitzia.*

MUSCARI (Mus-càri) — Grape-Hyacinth, Starch-Hyacinth. Among hardy small-flowered bulbs of spring, several kinds of grape-hyacinths rank highly and are popular. There are others worth growing for their beauty or interest, a few for their exquisite fragrance. They belong to the lily family LILIACEAE. The genus *Muscari,* the name comes from the Latin *moschus,* musk, and alludes to the scent of the blooms of some kinds, includes possibly sixty species, that with their varieties have been for long abominably confused as to names in botanical and horticultural literature. The publication of a recent monograph has brought much order to this nomenclatural mish-mash. The genus is native in Europe, North Africa, and western Asia.

Grape-hyacinths are related to hyacinths (*Hyacinthus*), from which they differ in being much smaller and in their flowers usually not having wide-open throats. They have small bulbs and few, all-basal, more or less fleshy, linear leaves. The little urn-shaped to nearly spherical or somewhat cylindrical, ascending, spreading, or nodding blooms, those of most kinds with strongly pursed (constricted) mouths, are in usually crowded racemes or spikes, or rarely are panicled, atop leafless, erect stalks. Except for six usually tiny, teethlike lobes (petals), the perianth parts are united. There are six nonprotruding stamens and one style tipped with a three-lobed stigma. The fruits are three-angled capsules. Generally the upper flowers of the spikes, racemes, or panicles are sterile, of a different color to those below, and in some kinds distinctly showy. Flower colors range through blues and purples to yellowish, brownish, and white.

The common grape-hyacinth (***M. botryoides***) is 6 inches to sometimes 1 foot tall. It has two or three, rarely four, oblanceolate or occasionally linear leaves, erect and usually shorter than the flower stalks and racemes. The sky-blue, spherical flowers have tightly pursed mouths and tiny, recurved, white petals. They are up to 1/6 inch long, in racemes at first tight, but becoming looser as they age. The few pale sterile blooms are cylindrical. This is a native of central and southern Europe. Variety *M. b. heldreichii,* dwarfer and with racemes that remain compact, is an endemic of high mountains in the Balkan Peninsula. Variety *M. b. kerneri* has linear leaves, and flowers in loose racemes. The leaves of *M. b. longifolius* are linear and longer than the flower stalks and racemes. Horticultural varieties include the superb white-flowered *M. b. album,* and a pink-

Muscari botryoides heldreichii

Muscari armeniacum 'Heavenly Blue'

Muscari comosum monstrosum

flowered one called *M. b. carneum,* with flowers of a rather uncertain, washy hue.

Sometimes called starch-hyacinth, early-blooming ***M. neglectum*** enjoys the widest natural distribution of any grape-hyacinth. It is indigenous from England to the Altai Mountains. From the common grape-hyacinth it differs in having three to seven narrowly-linear, limp rather than erect leaves 9 inches to 1 foot in length, and very dark blue to blackish-blue flowers. Variety *M. n. pulchellum,* of southern Europe and Turkey, differs in not producing offset bulbs and in its flower spikes not exceeding 6 inches in height. The old name starch-hyacinth alludes to the blooms having a fragrance thought to be like that of hot starch. The late Louise Beebe Wilder described it as the odor of ripe plums.

Variable ***M. armeniacum,*** of the Caucasus, Turkey, and southeastern Europe, has four to eight linear leaves 4 inches to 1 foot long by up to ⅓ inch wide, and channeled along their upper surfaces. Its flower stalks and dense to somewhat loose racemes are 4 inches to 1 foot long. The spicily fragrant, sky-blue flowers, sometimes slightly more, sometimes less than ¼ inch long, are obovate-urn-shaped and have white, reflexed little petals. There may be as many as ten sterile blooms, paler than the others. An old favorite, variety 'Heavenly Blue', has bright blue, fragrant flowers. It reproduces generously by offsets. Other varieties are *M. a.* 'Cantab', the fragrant blooms of which are clear light blue, and *M. a.* 'Blue Spike', with light blue double blooms.

The feather-hyacinth (*M. comosum monstrosum* syn. *M. plumosum*) is a horticultural variety more frequent in gardens than the tassel-hyacinth (***M. comosum***). The latter, native to Europe, North Africa, and western Asia, has three to six linear to narrowly-strap-shaped leaves 2 to 8 inches long by ¼ to ¾ inch wide. The brownish-greenish, oblong-urn-shaped flowers are ¼ to over ⅓ inch long and have greenish-white or cream-colored petals. They are in loose racemes that, with their stalks, are 8 inches to sometimes much more than 1 foot long. The generally ten to twenty sterile flowers, like their long, erect individual stalks, violet or purplish, are in a usually conspicuous terminal head. In *M. c. monstrosum,* all the flowers are sterile. Similar to *M. comosum* and with tassel-like racemes is ***M. pinardii***. The flowers are grayish with pale blue tips. Their perianths are shredded into numerous dove-purple, hairlike segments curled and twisted at their ends and responsible for the plume-like aspect of the racemes.

Muscari comosum

Dark black-violet fertile flowers and broad leaves characterize ***M. latifolium,*** a native of pine forests in Turkey. This has usually one, sometimes two, leaves from a little under ½ inch to 2 inches wide. In length 3 inches to 1 foot, they have pointed or hooded apexes and bases that sheath the flower stalks. The fertile flowers, oblong-urn-shaped and up to ¼ inch long, have spreading or slightly recurved petals of the same dark hue as their bodies. These contrast strikingly and pleasingly with the up to fifteen soft, pale blue sterile blooms at the apex of the raceme. The racemes, tight at first, but later looser, together with their stalks, are 6 inches to well over 1 foot long. Also with black-violet blooms is early-blooming ***M. commutatum.*** This native of central and southeastern Europe has its flowers in ovoid racemes that with their stalks are 3 inches to 1 foot long. They have strongly pursed mouths, and petals of the same hue as the body of the blooms. The flowers are from somewhat under to a little over ¼ inch long and egg-shaped.

The grape-hyacinth called ***M. tubergenianum*** is perhaps a variety of *M. aucheri.* When in full bloom the upper flowers of its raceme are bright, clear blue, the lower ones dark blue. The buds are turquoise-blue. The species ***M. aucheri,*** native to Anatolia, has two or occasionally up to four narrow-elliptic leaves, the lower parts often sheathing the stalk of the raceme. They are 2 to 8 inches long and from very slender to slightly over ½ inch wide. Their upper surfaces are lighter colored than the lower and glaucous. Spherical and up to ⅕ inch long, the sky-blue flowers, with reflexed, white petals, are in dense ovoid ra-

Muscari tubergenianum

Muscari azureum

cemes that with their stalks are 3 to 10 inches long. Bright blue flowers in dense, ovoid racemes are freely displayed by ***M. azureum.*** They are carried on stems to heights of 4 to 10 inches. Unusual for grape-hyacinths, the blooms are nearly bell-shaped. They are, including the darker-striped, recurved petals, about 1/6 inch long. The smaller, paler, sterile flowers are few. There are two or three narrowly-oblanceolate leaves, often hooded at their apexes, 2½ to 8 inches long. This distinctive species is a native of Turkey. Pure-white-flowered *M. a. album* is especially charming. Variety *M. a. amphibilis* is larger than the typical species and has light blue, more conspicuously toothed flowers. Probably a variety of *M. tenuiflorum*, a species ranging in the wild from Asia to the Near East, the grape-hyacinth called ***M. masseyanum*** has a tasseled raceme of pink-tinged flowers. About 6 inches tall, this is reputed to be a native of Asia Minor.

Musk-scented ***M. racemosum*** (syn. *M. moschatum*) is not showy and can scarcely be called pretty, but its flowers sweeten the air for long distances with their delightful fragrance. Native to Turkey, this kind has four to six linear leaves 4 to 8 inches in length and from ¼ to a little over ½ inch wide. In racemes that with their stalks are 4 to 9 inches long, the narrowly-urn-shaped blooms, at first purplish, later paling to greenish, ivory-white, or yellowish, but never as yellow as some catalog writers infer, have little pale brown petals. There are no sterile flowers. Variants offered as *M. r. majus* and *M. r. minus* are perhaps, respectively, larger and smaller than the typical species. Similar to *M. racemosum*, Turkish ***M. macrocarpum*** (syn. *M. moschatum flavum*) has yellow flowers of a

Muscari racemosum

Muscari racemosum (flowers)

somewhat different shape. Unlike most grape-hyacinths, which produce a new set of roots each year, *M. racemosum* and *M. macrocarpum* have perennial roots.

Often misidentified as *M. racemosum*, a designation that correctly belongs to another species, ***M. atlanticum*** is native to Europe and Asia. It has three to five linear, semicylindrical leaves up to 1 foot long and crowded racemes carried to heights of 6 to 10 inches of ten to thirty, fragrant, egg-shaped, ¼-inch-long dark blue flowers with tiny lighter colored or white petals.

Its name indicating the taxonomic puzzle it has presented botanists, ***M. paradoxum*** has at various times been named *Bellevalia paradoxa*, *Hyacinthella paradoxa*, and *Hyacinthus paradoxus*. Native to Armenia, Iran, and the Caucasus, this attractive species has comparatively large bulbs from which sprout two or three, narrowly-linear, channeled leaves ½ to ¾ inch wide

Muscari paradoxum

and longer than the combined lengths of the racemes of flowers and their stalks, which together are up to 1 foot long. The dense racemes, 1 inch to 1½ inches long, are of twenty to thirty bell-shaped, dark navy-blue to bluish-black blooms a little over ¼ inch long. The flowers of *M. p. album* are white.

Garden and Landscape Uses. Among the most useful spring-flowering bulbs, grape-hyacinths can be effectively naturalized beneath deciduous shrubs to supply in season vast carpets of blue that from some little distance away seem to reflect the sky itself. They are excellent in rock gardens, but there some thought must be given to locating the more prolific kinds, such as *M. armeniacum, M. botryoides,* and *M. racemosum,* lest their proclivity to give rise to self-sown seedlings may result in a takeover of spaces allotted less hearty neighbors. Similar effects can be had in lawns where the first grass cutting can be postponed until after the grape-hyacinth foliage has died. Grape-hyacinths are equally as meritorious for grouping at the fronts of flower beds, for planting as ribbon borders or edges to beds or paths, and for underplanting other spring-flowering bulbs, such as narcissuses, in more formal bedding arrangements. They force easily and in pots or pans (shallow pots) are attractive in greenhouses and windows.

Muscari armeniacum 'Heavenly Blue'

Cultivation. Grape-hyacinths adapt to ordinary soil of fair to good quality, well-drained, but not during their growing season excessively dry. They prefer full sun, but stand light or part-day shade. Planting is normally done in fall, setting the bulbs 2 or 3 inches apart and about as deep, but the best time to transplant perennial-rooted *M. racemosum* and *M. macrocarpum* is as soon as their foliage dies down. Some grape-hyacinths increase by offsets, others do not. All can be propagated by bulb cuttings and, except certain horticultural varieties, by seed. Many kinds multiply by self-sown seeds. Seeds are best sown in cold frames as soon as they are ripe.

MUSHROOMS. This is not a precise term. In the restricted sense employed here, it refers to the commercially cultivated fungus *Agaricus bisporus* and closely similar *A. campestris,* an edible wildling of fields and meadows that is not cultivated. A broader view accepts mushroom as an embracing name for fleshy, spore-bearing bodies (sporophores) of a much wider variety of funguses, mostly of the order *Agaricales* of the *Basidiomycetes,* but including some others. Under this interpretation mushrooms include inedible and even poisonous species often called toadstools. But as no scientific distinction can be made between mushrooms and toadstools, the terms in their broad usages are interchangeable.

The cultivation of mushrooms on a commercial or other large scale is highly sophisticated and exacting, appropriate only for those with considerable experience in this phase of horticulture. Anyone contemplating engaging in it should first consult a State Agricultural Experiment Station or other reliable source of information. But it is quite practicable with fair prospect of success to raise these delicacies for home consumption without the elaborate, carefully controlled set-ups commercial growers find necessary.

Mushrooms, being without chlorophyll, are unable to photosynthesize foods in the manner of plants that contain that remarkable substance, and so have no need for light. The food materials they need are obtained from organic matter, that of the sorts with which we are concerned, from dead and decaying organic matter. Because of this our mushrooms are classified as saprophytes in contrast to some other funguses that grow on living organisms as parasites.

Mushrooms are the reproductive parts of non-flowering plants that reproduce by spores. Their vegetative portions consist of webs of mycelium that thread their way through the material upon which the plant feeds.

To cultivate mushrooms, a cellar, building, cave, or other place that can be kept completely dark and where the temperature can be maintained between 55 and 60°F and the humidity at 70 to 75 percent is employed. Temperature requirements suggest fall as the best season for amateurs to start mushroom beds.

Besides a suitable place, a satisfactory organic compost for the mushrooms to grow in must be provided. Traditionally this has been fresh horse manure, with only the longest straw shaken out from it and that is still highly satisfactory and probably the best medium for amateurs to employ. Its scarcity has forced commercial growers in many regions to adopt other composts, several of which have been devised. In the 1940s composts made from corncobs, straw, and hay, to which were added carefully measured amounts of organic and inorganic fertilizers, were introduced and proved successful. By the 1970s corncobs became less available and research was undertaken to develop new mushroom composts. One such is a mixture of horse manure from racing stables and corncob compost. In Holland, success has been obtained with a compost of wheat straw and chicken manure, with some brewers' grains and gypsum added.

Whichever compost is used, the mix is piled so that it ferments and generates much heat, up to 130°F. During this period it is turned over several times at intervals of a few days. After about two weeks it will have completely changed in appearance and will be without offensive odor; it then is ready for planting.

It is turned into beds about 8 inches deep or filled into boxes as deep called trays. In commercial mushroom houses these are set on racks one above the other, spaced about 1½ feet apart. After the beds or trays are positioned commercial growers pasteurize the compost by maintaining the temperature of the mushroom house at 140°F for five to twelve days, and the compost itself at this temperature for at least four to six hours. This procedure, which usually necessitates blowing live steam into the mushroom house, is obviously impracticable for amateurs. The reason for pasteurization is to free the compost of disease organisms or mites that may damage the crop. It is not absolutely essential.

The compost in beds or trays is packed firmly by tamping with a flat board, and when its temperature has dropped to 75°F it is planted with spawn (mycelium prepared in special laboratories under sterile conditions and grown on a suitable medium). Spawn grown on grain is most popular. This is sown over the surface of the compost and dug in to a depth of an inch or two. Brick spawn is broken into pieces about 2 inches square, which are planted 1 inch to 2 inches deep about 10 inches apart each way. After about three weeks mycelium growing from the spawn will have invaded the whole mass of compost the temperature of which should then be between 60 and 65°F. Then the beds or trays are cased, that is covered with a 1-inch layer of good garden soil of approximately neutral pH reaction and free of stones, weeds, fertilizers, and residues of insecticides, fungicides, or other such materials, and made firm.

In from seven to eight weeks from planting the spawn, the first pin-head-sized mushrooms should come into view. During the waiting time, and subsequently, the beds must never dry nor must they be allowed to be saturated. Use tepid water to keep the surface damp, but not muddy.

The crop comes into harvest in a series of spurts or flushes. Harvest the mushrooms as soon as they reach desired sizes by twisting them while pulling them gently from their sockets. Be careful not to disturb younger ones left in the bed.

Other methods amateurs sometimes employ with some degree of success include the purchase of trays prefilled with a suitable compost and impregnated with spawn. These, if watered and kept in environments suitable for the growth of the fungus as recommended by the purveyors of the trays, provide considerable interest and modest harvests. Less likely to succeed is the plan sometimes advocated of planting pieces of brick spawn about 2 inches in diameter, about 2 inches deep, in lawns with soils containing fair amounts of organic matter. Even if this proves successful the appearance of mushrooms is unlikely the first year, but if moisture and other conditions are favorable crops may be had in the late summer or early fall of several succeeding years.

MUSINEON (Musìn-eon). Restricted as natives to western North America, the four species of *Musineon* belong in the carrot family UMBELLIFERAE. Their name is derived from an ancient Greek one for some member of the family, perhaps *Foeniculum*.

Of minor horticultural interest, musineons are tap-rooted, stemmed or stemless hardy herbaceous perennials with alternate to subopposite, once-, twice-, or thrice-pinnate leaves borne mostly at or near the bases of the plants. The tiny white, cream or yellow blooms are in dense umbels of smaller umbels. They have five-toothed calyxes, five petals, five stamens, and two styles. The fruits are seedlike, ovoid to linear-oblong, and compressed.

Stemless or nearly so, ***M. tenuifolium*** (syn. *Daucophyllum tenuifolium*) has ferny, twice- or thrice-pinnate, glaucous, hairless leaves, their ultimate divisions narrow-linear to threadlike. Its flowering stalks are 4 to 10 inches tall. The blooms are creamy-white to yellow. This sort inhabits dry plains and hills in South Dakota, Nebraska, and Wyoming.

Garden and Landscape Uses and Cultivation. Musineons are best adapted to rock gardens and similar places. They respond to dryish climates and soils and open, sunny locations. They are not likely to succeed where summers are hot and humid. Established specimens do not transplant well. The best procedure is to raise new plants from seeds, pot the seedlings, and later plant them from the pots to their permanent location.

MUSK is *Mimulus moschatus*. The name musk mallow is used for *Malva moschata*, musk-mallow for *Abelmoschus moschatus*. The musk root is *Adoxa moschatellina*.

MUSKMELON or MELON. Here belong the delicious fruits familiarly known as muskmelons, honeydew melons, cassaba melons, cantaloupes, and related kinds, varieties of *Cucumis melo*, of the gourd family CUCURBITACEAE which is native from Iran to central Asia. The kinds commonly called cantaloupes are not the true cantaloupe (*Cucumis melo cantalupensis*) which has a hard, rough, warty, scaly rind and is not grown in North America. Watermelons belong to a different genus. For them see Watermelon.

Muskmelons are frost-tender, vining annuals that delight in a long season of hot weather. Selected varieties can be grown where there are 140 days or more between the last frost of spring and the first one of fall, with at least eighty-five days of decidedly warm weather. Many kinds need longer periods of warmth, so varieties must be chosen to suit local conditions. Cooperative Extension Agents can advise on this.

Soil for muskmelons should be well-drained, fairly loose, and slightly acid to neutral. Cold, clayey, compact earths are unsuitable. They should have a fairly high organic content, be fertile, and not more than very slightly acid. Prepare the ground by spading, rotary tilling, or plowing, if possible at least a month in advance of seed sowing. This gives the soil time to settle. If coarse, undecayed or little decayed vegetation, compost, or manure is to be turned under, do this the previous fall. Only such materials that are well decayed should be incorporated in spring. If they are in short supply, work them into the earth at the individual stations or hills where the plants will stand rather than distribute them over the entire area. The hills may be about 2 feet in diameter and raised above the surrounding ground level or not. Space them 5 by 6 feet apart each way. In addition to organic matter, mix in each hill a generous sprinkling of a complete garden fertilizer.

Seeds may be sown directly in the hills or started earlier in a greenhouse or hotbed to give plants to set out later. Do not sow or set out young plants until warm, settled weather arrives. If Hotkap-type, paper, Glassine, or plastic tentlike covers are used to cover the seeded hills and seedlings, you may sow about three weeks earlier than without such protection. Two or three strong plants are sufficient at each hill, but twice as many seeds should be sown in each to allow for germination failure, accident, and insect injury. Pull out surplus seedlings when they begin to crowd. To have plants to set in the hills, sow seeds in a temperature of 65 to 70°F about six weeks before it is safe to plant outdoors. Grow the young plants in pots or other individual containers in a warm, fairly humid, sunny place and harden them by standing them for about a week in a cold frame or sunny place outdors before planting them in the garden.

Care after sowing or planting outdoors is simple. If Hotkap-type covers are used, be sure to ventilate them so that temperatures inside do not rise high enough to scorch the plants. Keep the soil very shallowly cultivated to admit air and to destroy weeds. Do not water unless quite

A hill of muskmelons after thinning

Muskmelons in vigorous growth

Muskmelons ready for harvesting

A fine greenhouse crop of muskmelons; nets support the fruits

necessary (need for water is indicated by a slight wilting of the foliage), then do so without wetting the foliage. A soil-soaker hose is useful for this purpose.

Harvest promptly as individual fruits ripen, but fruits taken too soon lack sweetness, those left too long lose sugar and become soft. Signs of ripeness are a change from brighter green to graying and slight yellowing of the rinds, the blossom ends of the fruits yield slightly when pressed, and the readiness with which the stalks of the fruits of some varieties break from the vine. This last is not true of all. Fruits of some varieties, including 'Casaba', 'Crenshaw', and 'Honeydew', must be cut from the vine.

Cold frame cultivation of muskmelons is highly practicable and gives good results where the outdoor season is too short. Through the early part of the season and again in fall the covering sash are left in place, with sufficient ventilation applied on sunny days to prevent excessive heat building up inside the structure. During the summer the sash are removed, or if the frame is of the portable or knockdown type, it may be taken away. A shingle, slate, or flat stone placed under each developing fruit helps to keep the fruit clean and perhaps to allow it to ripen more uniformly. Muskmelons were at one time grown by amateurs in greenhouses, but this is scarcely ever done now.

Diseases and pests of muskmelons are those that attack such other plants of their family as cucumbers, gourds, pumpkins, squashes, and watermelons. Consequently, if possible avoid planting them on ground that has carried a crop of any of these or of muskmelons in the previous three years. Buy only the best seed of varieties believed to be disease resistant. Keep the planting free of weeds. Destroy at once any plant that wilts for reasons other than lack of water.

MUSKRATS. Native in many parts of North America, these large rodents feed on such aquatic and waterside vegetation as water-lilies and cat-tails. The disturbance caused by the construction of their extensive burrows can seriously weaken the banks of streams, ponds, lakes, and irrigation ditches. Indications of the presence of muskrats include tracks in mud made by their broad hindfeet and tail, large droppings on land and water, and floating pieces of the vegetation they have fed upon. Also, by day they may be seen swimming.

Entrances to muskrat burrows are just below water level, usually under a clump of sod, cat-tails, or other densely-rooted vegetation. Control measures include trapping, and the use as a bait of rolled wheat or rolled barley treated with Warfarin. If bait is used, be sure to enclose it in special bait boxes that deny access to birds or other desirable creatures.

MUSQUASH ROOT is *Cicuta maculata.*

MUSSAENDA (Mus-saénda). Nearly 200 species of the Old World tropics constitute *Mussaenda,* of the madder family RUBIACEAE. They are mostly erect or somewhat climbing shrubs. A few are small trees, a few herbaceous perennials. Many are remarkable because of their large, brightly colored, or cream or white, petal-like sepals, much like those of *Schizophragma.* The name *Mussaenda* is derived from a native one of Ceylon.

Mussaendas have stalked leaves in twos or threes. The flowers, with white, yellow, or red, tubular corollas, are in terminal clusters. The corollas broaden above and have five wide-spreading lobes (petals). Although pretty, the corollas are not the showiest parts of the blooms. That distinction rests with one, or in some kinds more, of the five calyx lobes of some or all of the flowers. These are much enlarged and petal-like. The blooms have five stamens and a style with two stigmas. The fruits are many-seeded berries.

Native to Africa, ***M. erythrophylla*** is very showy. It is said to climb to heights of 30 feet or more in its home territory, but in cultivation it is an erect shrub a few feet tall. Its leaves are ovate and about 6 inches long. The flowers are 1½ inches long and about 1 inch wide. Their corollas have red tubes, pale yellow petals, and deeper yellow throats. Some flowers of each cluster have one sepal much enlarged, from 2 to 4 inches long, broad-ovate and bright red. The hairy fruits are red and egg-shaped. Another African, ***M. luteola,*** is an erect shrub with ovate-lanceolate leaves up to 2 inches long. Its flowers are yellow. Some of each cluster have one enlarged, cream-colored, oblong to ovate sepal up to ¾ inch long. Tropical Asian ***M. roxburghii*** (syn. *M. frondosa*), a shrub 3 to 5 feet tall, has pointed-elliptic to pointed-ovate leaves

Mussaenda luteola

Mussaenda roxburghii

with blades 2 to 4 inches long, and yellow flowers, often one or more of each cluster with one elliptic to ovate, white, petal-like sepal with a blade 2 to 3 inches long.

A native of the Philippine Islands, *M. philippica* is not as much admired as its magnificent variety *M. p. aurorae,* which was first brought to the United States in 1959. The species ***M. philippica*** is widespread in its homeland. It is a shrub 3 to 10 feet tall, with ovate to elliptic leaves up to ¾ inch long, minutely-hairy beneath and very slightly so on their upper sides. Few together in terminal clusters, the flowers are deep yellow, and some of each cluster have one much enlarged creamy-white sepal. Resembling the species except that one to five of the sepals of all of the flowers are large, white, and petal-like, *M. p. aurorae* is an outstanding ornamental. It has not been known to produce seeds, but it has fertile pollen and has been used as a male parent, with *M. erythrophylla* as female, to originate several horticultural varieties including *M.* 'Dona Luz', in which all five calyx segments are rose-purple and petal-like, and *M.* 'Dona Alicia', in which only one is so enlarged and colored. Variety 'Dona Leonila' has creamy-white calyx segments.

Garden and Landscape Uses and Cultivation. In the humid tropics and subtropics mussaendas are lovely shrubs for garden display. For best results they need full sun and fertile, porous soil. Regular applications of fertilizer are usually required. Because the plants tend to become lanky fairly severe pruning is necessary from time to time. This is best done just before new growth begins. Propagation can be by seeds, but these are not produced by some of the best kinds, such as *M. philippica aurorae.* Plants from seeds from hybrids are

Mussaenda philippica aurorae

Mussaenda 'Dona Leonila'

not identical with their parents. Alternative methods of multiplication are by cuttings and air layering. The latter is highly successful.

As greenhouse plants mussaendas can be successfully grown where the atmosphere is humid and the winter night temperature is from 60 to 65°F. By day this may rise five to fifteen degrees, and in summer considerably higher temperatures are in order. A little shade from strong sun is needed. Ordinary, porous, fertile soil of coarse, rather than fine, texture is suitable. Watering is done to keep it always moderately moist. Well-rooted specimens benefit from regular applications of dilute liquid fertilizer.

MUSTARD and MUSTARD GREENS. Several similar species of cabbage relatives of the genus *Brassica,* of the mustard family CRUCIFERAE, are grown for human consumption under the names mustard and mustard greens. Especially in the south where high summer temperatures render the production of lettuce and spinach difficult or impossible, varieties of *B. juncea, B. rapa,* and *B. hirta* are esteemed as greens and potherbs. Much more commonly cultivated in northern Europe than in North America, black mustard (*B. nigra*) and white mustard (*B. hirta*) are grown to seedling size, often along with cress, for harvesting as salads.

Mustards or mustard greens are easy to grow. They respond to any reasonably fertile, not-too-dry garden soil and sunny locations. Sow seeds very early in spring or even the fall previous, to harvest in May or June. Make successive sowings at two-week intervals so long as there is time for a crop to develop before the onset of hot summer weather. With the arrival of such conditions the plants run to flower and seed. A second sowing in August will produce a fall harvest. The seeds may be broadcast in a bed up to 6 feet wide and the plants thinned to 4 to 6 inches apart, or sown in rows 1 foot to 1½ feet apart and the young plants thinned to 3 to 5 inches apart in the rows. In dry weather watering helps to delay premature bolting to seed. Harvest by picking the young leaves and discarding the stems. Good varieties are 'Florida Broad Leaf', 'Fordhook Fancy', 'Giant Southern Curled', and 'Tendergreen'.

Mustard to be used in its very young seedling stage with seedling cress as a salad is as simple to grow in North America as in Europe. By making successive seedings, indoors during the cold season of the year, outdoors at other times, continuous supplies may be had. The procedure is as follows. Sow the seeds thickly in a flat or other shallow receptacle containing soil, sand, or vermiculite that has just been drenched with water. Press the seeds into it, but do not cover them. Place a piece of glass or polyethylene plastic film over the container. Shade from sun and strong light. As soon as the seeds begin to sprout (in about two days) remove the shade, and a little later the glass or plastic, and expose to full light. Keep the soil moist. During the part of the year when temperatures can be controlled night ones of 50 to 55°F, with a few degrees increase by day are satisfactory. Moderately humid, fairly airy atmospheric conditions are desirable. Harvest before the first true leaves, the second pair of leaves to appear, develop. Do this by snipping the stem with scissors or shearing them with a sharp knife as near to the ground as possible. To have the crops at the right stages for harvesting together sow the mustard three days after the cress. As an interesting gimmick, likely to amuse children, mustard and cress can be grown without soil, sand, or similar material on pieces of flannel or suchlike absorbent fabric laid in the bottom of a saucer or shallow tray and kept uniformly moist.

MUTANT or SPORT. It occasionally happens that among a population of seedlings of a species or variety one departs from the norm. It may, for example, be dwarf or extraordinarily tall or its flowers may differ in form, color, or size from those of the typical kind or its fruits ripen earlier or later. Individuals that exhibit such changes not resulting from such environmental factors commonly recognized as affecting plant growth as soil fertility, water supply, temperature, and exposure and not the outcome of hybridization or of infection with a virus or other living organism are called in lay terms sports or breaks, more technically mutants. By definition a mutant is the result of any inheritable change (mutation) that occurs in the genetic structure of the cells of a plant or an animal. Artificial methods employed by experimental scientists and plant breeders to induce mutation include the use of X-rays and rays given off by radioactive materials, the employment of certain chemicals, such as colchicine, formaldehyde, or mustard gas, or exposure to ultraviolet light or to extremes of heat or cold.

Not all mutants exhibit desirable change. Only those that do interest gardeners and horticulturists sufficiently to warrant their perpetuation by propagation. Most commonly mutation occurs among seedlings, and plant breeders raise and scan for desirable change many thousands of those of kinds with which they are working. Seedling mutants also occur natively among wild populations of species and are occasionally detected by keen-eyed observers and brought into cultivation and multiplied. Sargent's weeping hemlock (*Tsuga canadensis sargentii*) is a good example. Besides seedling mutants there are bud mutants (bud sports), variations evidenced by a single branch or part of a plant displaying change. Many varieties of florists' flowers, such as carnations, dahlias, chrysanthemums, and roses, and some varieties of fruits have originated in this way. In such cases the mutant is commonly multiplied by vegetative propagation, by cuttings or grafting, for example.

MUTISIA (Mu-tísia). Unfamiliar to most American gardeners, *Mutisia,* of the daisy family COMPOSITAE, contains several attractive species. They are not hardy in the north, but are suitable for California, parts of the northwest, and for greenhouses. There are about sixty species, all natives of South America. The name commemorates the Spanish naturalist José Celestino Mutis, who died in 1808 or 1809.

Mutisias are erect or climbing subshrubs and shrubs, with plain-edged, toothed, or lobed leaves, undivided or with more than one leaflet. Commonly they end with a branched or branchless tendril. The long-lasting flower heads, ordinarily thought of as flowers, are solitary, terminal, and composed of a center of disk florets (the equivalent of those that form the eyes of daisies) with a few petal-like ray florets encircling them. The rays are orange, yellow, pink, red, lilac, violet, or white and usually female. The disk florets are bisexual and generally yellow. Each has five protruding stamens and usually a two-lobed style. The involucre (collar of bracts behind the flower head) is cylindrical or bell-shaped and composed of several rows of overlapping bracts.

Five species are reported to be in cultivation in the United States, with ***M. clematis*** the most abundant. This native of the Colombian Andes is a woody climber 6 to 10 feet tall. Its pinnate, evergreen leaves, 2 to 4 inches long, end in a three-forked tendril. Usually they have six or eight leaflets each 1 inch to 2½ inches in length and yellowish-hairy on their under surfaces. The drooping flower heads, 2 inches wide, are on stalks as long or shorter. The disks are yellow, the nine or ten rays bright orange-scarlet to purple-red. The collar of bracts is cylindrical and approximately 2 inches long.

Brilliant orange rays and a central disk of yellow give the flower heads of ***M. decurrens*** great display value. Upturned and 4 to 5 inches in diameter they have about fifteen rays, and something of the aspect of single marigolds, single dahlias, or tithonias. The cylindrical collar of bracts behind each head is 1 inch to 1½ inches long. This is a suckering, hairless, slender-stemmed climber of Chile and Argentina. It has stalkless, undivided, toothed or toothless, evergreen, lanceolate leaves up to 4 inches long by up to ¾ inch wide, each with a two-forked tendril at its end. The bases of the leaves are continued as narrow wings down the stems.

Mutisia decurrens

Mutisia oligodon

Shrubs or subshrubs with erect stems 15 to 18 feet long, *M. ilicifolia* and *M. spinosa,* of Chile and Argentina, have undivided leaves. Those of ***M. ilicifolia*** are evergreen, somewhat leathery, stalkless, and 1 inch long or a little longer by ¾ inch wide. They have margins with ten to twelve teeth along each side and end in a branchless tendril. The undersides are more or less hairy. Upturned, the 2- to 3-inch-wide, short-stalked flower heads have yellow centers and rose-pink, pale mauve to almost white rays. The collar of bracts is cylindrical and up to ¾ inches long. From the last, ***M. spinosa*** differs in its leaves rarely having more than three teeth on each side. If there are more, then the fact that the secondary veins are inconspicuous serves to distinguish this from *M. ilicifolia,* the leaves of which have strongly evident secondary veins. Yellow disks and pale pink rays are characteristics of its up-looking, 2½-inch-wide flower heads. The collar of bracts behind each is bell-shaped and about ¾ inch long.

A low shrub with straggling or ascending branches up to 1 foot high, ***M. oligodon*** hails from Chile and Argentina. Hairless or sparingly-hairy, it has evergreen, stalkless, oblong to elliptic leaves up to 1½ inches long and up to about ½ inch broad; each leaf ends in a branchless tendril and generally has on each side toward the apex three to six teeth. The lower surfaces are densely-hairy, the upper ones hairless. About 2½ inches across, the upturned flower heads, on stalks up to 2 inches long, have yellow disks and six to twelve satiny, bright rose-pink rays. The collars of bracts at their rears are bell-shaped and about ¾ inch long.

Garden and Landscape Uses and Cultivation. These are choice plants for gardeners who appreciate the unusual and are willing to give their treasures a little special attention. It is well worthwhile to go to some little trouble to locate mutisias happily. They prefer loamy, porous, well-drained soil, deep and moderately fertile. Their stems require support. A good way of supplying this is to plant them among shrubs over which their stems can clamber. Alternatively, they can be trained to trellis, wires, or other supports. It is helpful to keep the ground about them mulched with peat moss, compost, or other organic material. In dry weather water regularly. Pruning is limited to cutting out weak and straggly shoots as soon as flowering is through. Increase is by summer cuttings under mist or in a greenhouse propagating bench, or in the case of kinds that develop suckers by taking these off in spring. Seeds also are satisfactory. In greenhouses mutisias do best under cool, airy conditions. Winter night temperatures of 40 to 50°F are satisfactory. By day they are increased by up to about ten degrees. On all favorable occasions the greenhouse must be ventilated freely.

MYCELIUM. The mass of branched filaments (hyphae) produced by many funguses is called a mycelium (plural, mycelia). These are vegetative as distinct from reproductive or spore-bearing parts. The mycelia of mushrooms and many toadstools live in the ground while their fructifications or spore-producing organs rise out of it.

MYCORRHIZA or MYCORHIZA. Some or all of the roots of many plants live in intimate association with funguses of benefit to them and which they benefit. Such a symbiotic combination of root and fungus is called a mycorrhiza (plural, mycorrhizae). For growth under natural conditions of certain kinds of plants, notably orchids, many members of the heath family ERICACEAE, and pines and other conifers, mycorrhizal funguses are essential; with others they are helpful, but not necessary; and yet others do not enter into such symbiotic associations.

Mycorrhizal funguses, which live partly in the soil and partly in the root, may invade the plant cells or grow between them. They serve higher plants by supplying them with water and nutrients they absorb from the soil, especially from the organic matter it contains, and benefit by obtaining food for their own needs by digesting and absorbing the products of grains of starch stored in the cells of the roots of these plants.

The mycorrhizae are often of great importance to plants that grow natively in soils with a high humus content, and the successful cultivation of such sorts generally depends upon maintaining conditions favorable to the well-being and growth of the symbiotic funguses.

MYOPORACEAE—Myoporum Family. Few of the ninety species of the four genera that compose this family of dicotyledons are known to gardners. Natives chiefly of Australia and islands of the South Pacific, a few occur in the native floras of Hawaii, eastern Asia, Mauritius, South Africa, and the West Indies.

Mostly trees, infrequently shrubs, these plants have alternate or rarely opposite, undivided, toothed or toothless leaves often dotted with minute glands or clothed with woolly or scaly hairs, sometimes of much reduced size. Solitary or in clusters from the leaf axils, the usually asymmetrical flowers have a five-cleft calyx, a five-lobed corolla, four stamens in two pairs, and sometimes one staminode (nonfunctional stamen) or less commonly five stamens, one style, and one stigma. The fruits are drupes. Genera sometimes cultivated are *Bontia, Eremophila, Myoporum,* and *Oftia.*

MYOPORUM (Myóp-orum) — Boobyalla, Bastard Sandalwood. The genus *Myoporum,* of the myoporum family MYOPORACEAE, consists of thirty-two species of shrubs and trees and has a natural range extending from Hawaii to eastern Asia, Mauritius, Australia, and New Zealand. Its name, derived from the Greek *myo,* I close, and *poros,* a pore, has reference to the translucent resinous dots on the leaves.

Myoporums have alternate or rarely opposite, toothed or toothless leaves, and usually in clusters, small or medium-sized, most commonly white blooms. The flowers have calyxes with five lobes or sepals, symmetrical or nearly symmetrical bell- to funnel-shaped corollas with usually four, rarely five or six, spreading lobes (petals), and four or rarely five stamens. The fruits are small drupes (fruits of plum-like structure).

A dome-shaped or billowy-headed tree up to 30 feet tall with stout, spreading branches, but often considerably lower and shrubby, New Zealand ***M. laetum*** is densely furnished with stalked, lustrous, bright green, hairless, more or less fleshy, lanceolate to obovate-lanceolate leaves, toothed above their middles and 2 to 4 inches long. In clusters of two to six are borne bell-shaped, four-stamened, purple-spotted, white flowers about ½ inch across. The reddish-purple fruits are ¼ to ⅓ inch long. Variety *M. l. decumbens* has sprawling to prostrate branches and white fruits. Variety *M. l. carsonii* is exceptionally fast growing; it has broader, darker green leaves with less abundant translucent dots.

The boobyalla (***M. insulare***), of Australia, under favorable conditions attains heights of 20 to 30 feet, but usually is lower and often is only a shrub. Its leaves and blooms, except for being smaller, much resemble those of *M. laetum*. The fruits are globular, purplish-blue, and ¼ inch in diameter. Other Australians include *M. debile, M. montanum,* and *M. serratum*. Also native in New Zealand, ***M. debile*** is a trailer 1 foot to 3 feet tall with a spread of 2 to 6 feet or sometimes more. It has reddish-green, lanceolate leaves 2 to 3 inches long and slightly or sometimes toward their bases conspicuously, toothed. Its white, pink, or lavender flowers are ½ inch in diameter. The fruits are reddish. A shrub 6 to 9 feet tall, ***M. montanum*** has toothless, narrow-elliptic to lanceolate leaves up to 2 inches long and white flowers bearded on their insides and purple-spotted in the throat. About ¼ inch in diameter, the fruits are spherical, purple when ripe. Sometimes a tree 20 to 30 feet tall, ***M. serratum*** is often lower and shrubby. It has more or less toothed, oblong to lanceolate leaves and white, ¼-inch-long, ½-inch-wide flowers.

Bastard sandalwood (***M. sandwicense***), of Hawaii, is variable and up to 60 feet high. It has elliptic- to ovate-lanceolate leaves 2 to 6 inches long. About ⅓ inch across, the fragrant, white to pink flowers have five, six, or less commonly seven corolla lobes (petals) and stamens. The fruits are pale yellow to white. They are about ¼ inch in diameter.

Garden and Landscape Uses. Among the most satisfactory shrubs and trees for the seaside and other exposed places, *M. laetum*, and its varieties, and *M. insulare* are fast and satisfactory growers in mild-climate regions, such as California. They are extremely useful for windbreaks and may be sheared to form hedges. Disadvantages are that their roots are invasive and their foliage drops more or less throughout the year. Because of this they

Myoporum sandwicense

are not suitable for locating near pools or in manicured garden areas. They stand full sun and strong winds, and resist the ill effects of blowing sand. They make no effective display of blooms, but their fruits are more colorful. For satisfactory growth *M. laetum* and its varieties need fairly moist soil, *M. insulare* stands drier conditions, and dwarf *M. debile*, which is suitable for planting on banks and in rock gardens in mild climates, is remarkably drought resistant.

Cultivation. Once established in environments reasonably to their liking, myoporums are unlikely to need much attention. In exposed locations severe pruning the third or fourth year after planting is highly desirable. This forstalls the development of heads so large that they become subject to storm damage, and encourages satisfactory branching. Also as a precaution against breakage, it is wise to stake young, newly set-out specimens. When used as hedges periodic shearing is routine. Propagation is easy by cuttings and layering and usually by seed, although the latter are sometimes difficult to germinate.

MYOSOTIDIUM (Myosot-ídium)—Chatham-Island-Forget-Me-Not. Endemic to the Chatham Islands, near New Zealand, the only species of *Myosotidium*, of the borage family BORAGINACEAE, is notoriously difficult to cultivate. Its name reflects its relationship to forget-me-nots (*Myosotis*) and the similarity of its blooms to those of that genus.

A herbaceous perennial, often rather short-lived, ***M. hortensia*** (syn. *M. nobile*) has a stout rootstock and numerous heart- to kidney-shaped, deeply-corrugated, fleshy, basal leaves somewhat reminiscent of those of certain plantain-lilies (*Hosta*). They are hairless and lustrous above, but have minute scattered hairs on their under surfaces. The stout stems, 1 foot to 2 feet tall, have stalkless, oblong leaves smaller than those nearer the ground. The pale to dark blue flowers are in handsome, large, dense terminal clusters the branches of which are longer than the leaves. Individual blooms are about ½ inch across. Wheel-shaped, they have five-parted calyxes, a short corolla tube with five conspicuous scales in the throat, and five rounded corolla lobes (petals). There are five short stamens and a short, stout style. The fruits are four-winged, flattened nutlets (usually called seeds). In the wild, this plant grows on sandy beaches and rocky places near the sea. Because of browsing animals, it is now much rarer than it once was and is largely confined to places unreachable by such animals.

Garden Uses and Cultivation. This is a plant for the few. It certainly cannot be expected to succeed over vast areas of North America. Only where winters are practically frost-free and summers are cool and humid, as in parts of the Pacific Northwest, are there prospects of it prospering. It needs moist, well-drained, sandy peaty soil quite free of any stagnant water. Propagation is by seed.

MYOSOTIS (Myo-sòtis)—Forget-Me-Not. Forget-me-nots are well known to most people who have a reasonable acquaintance with garden flowers, and for long have been popular. The best known kinds are easy to grow, but there are much less compliant ones that test the skills of the most experienced alpine gardeners. The genus *Myosotis* belongs in the borage family BORAGINACEAE and inhabits temperate, alpine, and arctic regions, especially in Europe and New Zealand. It is represented also in Asia, North America, and elsewhere. Of the about fifty species at least thirty-two are endemic to New Zealand. The name is an ancient one used by Dioscorides. It is derived from the Greek *myos*, a mouse, and *ous*, an ear, and refers to the appearance of the leaves of some kinds.

Forget-me-nots are erect or sprawling, more or less hairy annuals, biennials, and perennials. They have alternate, undivided, toothless leaves. The salver- to broadly-funnel-shaped flowers, small but numerous, are usually in terminal, one-sided, generally bractless racemes, coiled or recurved at first, but straightening and lengthening as the blooms open in succession. Rarely the flowers are solitary. The calyxes are small and five-cleft. The corollas have five lobes (petals) and in their throats five small appendages. The five stamens are enclosed in the corolla tube, and the slender style ends in a minute stigma. The fruits consist of four seedlike nutlets.

Biennial in cultivation, but perhaps sometimes perennial in the wild, ***M. syl-***

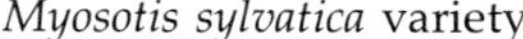

Myosotis sylvatica variety

Myosotis sylvatica (flowers)

vatica is often misnamed *M. alpestris* in gardens. The two are closely related, but the latter, the state flower of Alaska, is not commonly cultivated and the forget-me-not varieties grown for temporary spring displays probably all belong to *M. sylvatica.* That native of Europe and temperate Asia is naturalized in North America. In the wild it is 1 foot to 2 feet tall and softly-hairy, with bluntish, oblong-lanceolate leaves, but garden varieties vary greatly in height, compactness, and other characteristics. Usually they are not more than about 9 inches tall. About ⅓ inch in diameter and usually clear blue, but in some varieties pink or white, the flowers have yellow eyes. Characteristic features of this species are that the stalks of the individual flowers are much longer than the calyxes, and its seeds (really nutlets) are without stalks. From it, ***M. alpestris*** (syn. *M. rupicola*), of the mountains of Europe, arctic Europe and North America, may be distinguished by its smaller blooms having individual stalks not longer than the deeply-cleft calyxes. Also, *M. alpestris* is more likely to behave as a perennial and not to exceed about 8 inches in height. However, dwarf varieties of *M. sylvatica* are not taller. A third similar species, ***M. dissitiflora,*** of Switzerland, can be distinguished from *M. sylvatica* by its seeds (nutlets) having little stalks. This is probably perennial, but the plant usually cultivated under its name is a variety of *M. sylvatica* and is biennial.

Perennial ***M. scorpioides*** (syn. *M. palustris*) is hairy to nearly hairless, and blooms in spring and early summer. Its dwarfer variety *M. s. semperflorens* flowers practically continuously through the summer.

Myosotis scorpioides beside a pond

Myosotis scorpioides (foliage and flowers)

This native of Europe and Asia has stems 1 foot to 1½ feet long, much branched and prostrate at their bases, and then erect. Its nearly stalkless leaves are oblong-lanceolate to spatula-shaped and ½ inch to 2 inches long. The flowers, typically bright blue, but sometimes white or pink, are about ⅓ inch in diameter and have yellow eyes. They are in long, loose, often-paired racemes. A similar perennial, but with much smaller blooms, the calyxes of which are lobed at least to their middles instead of having, like those of *M. scorpioides,* lobes much shorter than the tubes, ***M. laxa*** is a native of North America, Europe, and Asia.

Alpine species from New Zealand include ***M. colensoi*** (syn. *M. decora*), a white-hoary, 2-inch-high gem with white flowers; ***M. traversii,*** a white-hairy kind about 6 inches tall with lemon-yellow to white blooms; and unusual ***M. uniflora,*** which forms a low, compact cushion and has solitary yellow blooms. There are several others that have appeal to alpine gardeners.

Garden and Landscape Uses. The many charming varieties of *M. sylvatica* are grown as biennials for temporary spring displays. They do best in sunny locations where the soil is moderately but not excessively moist, and are popular for interplanting among spring bulbs, especially tulips, in flower beds and in porch and window boxes. After blooming they are discarded and replaced with summer-flowering annuals or plants treated as annuals. These varieties are also sometimes grown to flower in winter and spring in greenhouses as decorative pot plants and as charming flowers for cutting. Perennial *M. scorpioides* and less showy *M. laxa* are useful for decorating the banks of streams and ponds, and other moist-soil areas. They stand part-day shade and flourish with little or no care.

Cultivation. Varieties of *M. sylvatica* are raised from seeds sown in a cold frame or outdoor bed in late spring or early summer. The seedlings are transplanted as soon as they have their second pair of leaves to nursery beds or, in very cold climates, to cold frames. They are spaced 6 to 8 inches apart in soil moderately fertile and well drained, but not wet. The location must be sunny. In cold climates a light protective covering of branches of evergreens, salt hay, or other material that admits air freely is advisable over winter. Transplanting to blooming sites is done in fall or spring in mild climates and, in cold ones, in early spring.

A good way of obtaining new seedlings is to pull up the plants after they are through blooming and strew them over a soil bed in a cold frame covered with open lath shades or similar protection that will keep the plants from blowing away. There, the seeds complete their ripening, are discharged and, if the ground is kept moist, produce a fine crop of youngsters soon ready for transplanting to nursery beds.

In greenhouses, early blooms of the *M. sylvatica* varieties can be had by lifting plants from nursery beds in early fall, potting them in 5-inch pots, and putting them into a cold frame. Later they are brought into a sunny greenhouse where the night temperature is 40 to 50°F and that by day not more than five to ten degrees higher. The plants must be watered moderately and as spring approaches stimulated by applications of dilute liquid fertilizer. If the soil is kept too wet, the plants are likely to rot at their bases.

Even earlier blooms are had by sowing varieties especially selected for greenhouse cultivation (such are offered in seed catalogs) in June or July, potting the seedlings individually in small pots, and into larger ones as growth makes necessary. Until frost threatens the plants are kept in a cold frame or outdoors. Before winter they are moved to a cool, sunny greenhouse and are treated as advised above for plants lifted and potted from nursery beds. It is important to maintain a free circulation of air and a fairly dryish atmosphere. Excessive dampness of soil or air soon brings trouble.

Myosotis sylvatica variety as a pot plant

Perennials *M. scorpioides* and *M. laxa* are easily increased by seed and by division and need practically no attention to establish or maintain them in moist or wet soils. The reverse is likely to be true of the rare and choice New Zealand species. These need careful placing in gritty soil in a rock garden or alpine greenhouse, and attention appropriate for choice primulas and suchlike alpines. They are increased by seed and by division.

MYRCEUGENIA. See Luma.

MYRCIARIA (Myrci-ària)—Jaboticaba. From closely similar *Eugenia* with which it was once united, *Myrciaria*, of the myrtle family MYRTACEAE, differs in having few, usually only two, ovules in each division of the ovary. It consists of sixty-five species of South American and West Indian trees and shrubs with opposite, undivided leaves and usually white or cream-colored blooms. The name may allude to a similarity of the plants to *Myrtus*.

Jaboticaba is the name given in Brazil to ***M. cauliflora*** (syn. *Eugenia cauliflora*) and two or three related species much esteemed for their edible fruits. Uncommon in Florida, *M. cauliflora* is occasionally grown there. An evergreen tree that branches from near the ground and is up to 40 feet tall, in Florida it generally is less than half that height and grows slowly. Its rather leathery, lanceolate or somewhat broader leaves are 1 inch to 2 inches long and not quite half as wide. Similar in form to the flowers of pitanga (*Eugenia uniflora*) and some other eugenias, the small white blooms of *M. cauliflora* are clustered along the trunk and branches. They are succeeded by very dark maroon-purple, grapelike, thick-skinned fruits ½ inch to 1¼ inches in diameter that contain juicy, wine-flavored pulp and up to four seeds. Under favorable conditions, five or six crops of fruit are produced annually.

Other species are grown in warm regions, among them the rumberry (***M. floribunda*** syn. *Eugenia floribunda*), a tree up to 30 feet tall with lanceolate to ovate-lanceolate leaves up to about 3 inches long, white flowers, and red or yellow fruits about ⅓ inch in diameter, and ***M. myriophylla*** (syn. *Eugenia myriophylla*), which is a much-branched shrub with very slender, often crowded leaves up to ½ inch long by about 1/16 inch wide. The flowers are solitary and white. The plant that has been called *M. edulis* is *Eugenia aggregata*.

Garden and Landscape Uses and Cultivation. The cultivated myrciarias require tropical or warm, frostless, subtropical climates. They accommodate to ordinary fertile soils and are increased by seed, grafting, and inarching. For more detailed cultivation of *M. cauliflora* see Jaboticaba.

MYRICA (My-rìca)—Bayberry, Wax-Myrtle, Sweet Gale. Bayberries and some other species of *Myrica*, of the bayberry family MYRICACEAE, are especially useful because of their adaptability to soils and locations ill-suited to most other plants. Some kinds grow well in sandy almost sterile soil and some in saline ones, and some in seaside and other exposed locations. Common bayberry and probably some other species are strongly resistant to the harmful effects of deicing salt that, used to melt snow and ice, often injures or kills roadside plants. The name *Myrica* is an adaptation of *myrike*, an ancient Greek one for some fragrant-foliaged shrub, probably *Tamarisk*. Sometimes named *Myrica asplenifolia*, the sweet-fern is treated in this Encyclopedia as *Comptonia peregrina*.

Except for sweet-fern, all approximately fifty species of the bayberry family are commonly included in *Myrica*, but some botanists segregate sweet gale as the genus *Gale*. Sweet gale differs from other myricas in that its catkin branches drop as soon as flowering or fruiting is through and in its fruits being winged. In these respects it resembles the sweet-fern, but from this it differs in its leaves not being pinnately-lobed.

Alternate-leaved, deciduous and evergreen small trees and shrubs, sometimes prostrate, of cold, temperate, subtropical, and tropical parts of the Old and New Worlds compose *Myrica*. Their short-stalked, toothless or toothed leaves are usually dotted with resin glands. Insignificant in appearance, without sepals or petals, and in dullish catkins, the tiny unisexual blooms, the sexes often on separate plants, are succeeded by small, spherical to egg-shaped, berry-like drupes (fruits structured like plums) or nutlets, often coated with wax or resin.

The foliage of almost all myricas is pleasantly fragrant when crushed. Bayberry candles are made from the waxy secretions of the fruits of common bayberry and of others that develop such deposits. In ancient times, before hops were grown in northern Europe, leaves of sweet gale were used to flavor beer and were employed medicinally. Fruits of some kinds, including those of two not hardy in the north, *M. rubra*, of China and Japan, and *M. faya*, of the Canary Islands, are edible. Following its introduction to Hawaii for reforestation the last named became a troublesome weed.

Common bayberry (***M. pensylvanica,*** in gardens often miscalled *M. carolinensis*) is

Myrica pensylvanica, a native stand

a bushy, billowy, very hardy, deciduous to semievergreen shrub, a native of shores and dry, mostly coastal, uplands from Quebec and Labrador to New Jersey. A good ornamental, it has fragrant, blunt, oblanceolate, obovate, or elliptic, dull green leaves 1½ to 3½ inches long, sometimes finely-hairy on their undersides. Male and female catkins are on separate plants. If a pollinating male is near, the flowers of females are succeeded by considerable clusters of beautiful small fruits thickly coated with bluish, whitish-gray wax. They remain attractive for a long time. Where the southern part of the natural range of common bayberry overlaps the northern limits of the wax-myrtle, intermediate forms with the blunter leaves of common bayberry and the smaller fruits of wax-myrtle occur. These are sometimes distinguished as ***M. heterophylla.***

Wax-myrtle (***M. cerifera*** syn. *M. carolinensis*) may be regarded as a southern representative of the common bayberry. Characteristically it differs from that bayberry, although intermediates occur, in being a tall shrub or tree up to 30 feet in height, with more pointed, evergreen, dark green, mostly oblanceolate leaves at least four times as long as wide, and 1½ to 4 inches in length. They are often toothed above their middles. The fruits, thickly coated with whitish-gray wax, are smaller than those of common bayberry. Much less hardy, the wax-myrtle is native from New Jersey to Arkansas, Florida, Texas, and the West Indies, chiefly in damp or wet, sandy soils.

California bayberry (***M. californica***) is quite different. Esteemed for its lustrous, evergreen, bronzy foliage, and thinly-wax-coated, purple berries, this is an easy-to-grow ornamental shrub or erect, slender tree up to almost 40 feet tall. Native from Washington to California, it succeeds in nearly sterile, sandy soil, but is not hardy in the north. Individual plants bear both male and female flowers. Their lanceolate leaves are up to 4 inches long, the fruits about ¼ inch long.

Sweet gale (***M. gale*** syn. *Gale palustris*) is more botanically interesting than horticulturally exciting. The plants are usually unisexual, but individuals sometimes change sex from year to year. Upright, and up to 5 feet tall, this extremely hardy, deciduous shrub inhabits moist peaty soils and bogs in cold and temperate parts of North America, Europe, and Asia. It has dark green, oblanceolate leaves with usually hairy undersides, toothed toward their apexes. They are 1 inch to 2¼ inches long and glandular. The flowers come from leafless shoots of the previous year's growth. The fruits, in conelike catkins, are resin-dotted.

Garden and Landscape Uses. The myricas discussed above are all worthy ornamentals especially partial to acid soils. Common bayberry and California bayberry are especially well adapted for dry, sandy, nearly sterile places. Wax-myrtle and sweet gale need moister, peaty ones.

Myrica pensylvanica (foliage and fruits)

Myrica gale

All do well near the sea and there and elsewhere are tolerant of wind and sun. Most are salt tolerant. As landscape elements bayberries and wax-myrtle can be employed very effectively in naturalistic groups, as informal hedges, and as more formal backgrounds for flowering and other plants. None is difficult to grow.

Cultivation. Plant myricas in spring or early fall. California bayberry is reported to transplant less readily than others. All should be pruned back severely at planting time, but once established they need minimal attention. If they become too large or straggly, they can be reduced in size and shaped by pruning in late winter or spring. Cutting berried branches for indoor decoration is a rewarding way of keeping the plants lower and more compact than they would otherwise be. Propagation is by seeds taken from the surrounding pulp as soon as ripe and sown promptly in a cold frame or sheltered place outdoors, and by layers. Suckers provide a means of increasing sweet gale.

MYRICACEAE—Bayberry Family. The fifty species of aromatic, deciduous and evergreen shrubs and trees of this family of dicotyledons of wide natural distribution are allotted to two genera. Its members have alternate or subopposite, toothed, toothless, or pinnately-cleft leaves peppered with minute glandular dots.

The flowers are small and mostly unisexual, with the sexes on the same or different plants, or less frequently are bisexual. With regard to sex those borne by an individual plant or particular branch may vary from year to year. In catkin-like spikes from the leaf axils, the flowers, without perianths, usually have associated with them tiny bractlets. Males have from two to twenty, but usually four to eight stamens, their stalks free or united. Females have a pistil with a very short or scarcely evident two-branched style and two stigmas. In bisexual flowers there are three or four stamens and one pistil. The fruits are drupes often coated with wax. The genera are *Comptonia* and *Myrica.*

MYRICARIA (Myric-ària)—False-Tamarisk. Characteristics of flowers and seeds distinguish this genus from tamarisk (*Tamarix*). In tamarisk the stamens are not, or are only very slightly, joined at their bases, there is a definite stalk (style) between the ovary and the stigma, and the tufts of hairs on the seeds are stalked. The reverses of all these are true of *Myricaria,* of which there are ten species. They are natives of southern Europe and temperate Asia and belong in the tamarisk family TAMARICACEAE. The name is an adaptation of *myrica,* which is probably an old one for tamarisk.

False-tamarisks are deciduous shrubs and subshrubs with alternate, scalelike leaves and short-stalked, pink or white small flowers in dense, terminal or lateral racemes. The blooms have five each sepals and petals and ten stamens, united in their lower parts. There are three stalkless stigmas. The fruits are capsules.

Hardy in southern New England, ***M. germanica*** (syn. *Tamarix germanica*), a native of southern Europe and adjacent Asia, is a rather gaunt shrub 6 to 8 feet tall, with erect, wandlike, hairless branches plumed with smaller branchlets furnished with round-pointed, grayish-green, linear leaves up to 3/16 inch long. The pink to whitish flowers, 1/4 inch long, are in slender racemes 3 to 8 inches long. These terminate the upper branchlets. Each flower is in the axil of a bract longer than the bloom.

Garden and Landscape Uses. The species described is less attractive as an ornamental than the best tamarisks. Its chief appeal is to those who maintain botanical collections and to gardeners interested in the rare and unusual. Well-drained soil and full sun are needed.

Cultivation. No special demands are made by this shrub. To keep it shapely, and perhaps to reasonable size, judicious pruning may be done in spring. Propagation is easily effected by hardwood cuttings, taken in fall, and by seed.

MYRIOCARPA (Myrio-cárpa). This genus of the nettle family URTICACEAE consists of fifteen or perhaps more species of alternate-leaved small trees and shrubs of tropical America. Its name, from the Greek *myrios,* many, and *karpos,* a fruit, is self-explanatory.

Myriocarpas have often large, undivided, toothed or rarely nearly toothless leaves, with three chief veins from their bases or with one center vein and pinnately-arranged laterals. The tiny, usually unisexual flowers from the leaf axils or leafless stems are in slender racemes, spikes, or rarely panicles. The males generally have four- or five-parted calyxes and the same number of stamens. Female blooms are without calyxes, but may have two or four small bractlets beneath them. The conspicuous style ends in an oblique, hairy stigma. The fruits are achenes. From *Boehmeria,* with which the species described has been confused, *Myriocarpa* differs in calyxes being absent from the female flowers.

In cultivation ***M. stipitata,*** of South America, has been misnamed *Boehmeria argentea*. It is a shrub or tree up to 30 or 40 feet in height, with usually downy shoots, and pointed-elliptic to broadly-ovate toothed leaves up to 1 foot long by one-half as wide. Their upper surfaces are roughened in finely quilted fashion and are bluish-green liberally brushed with silver toward their margins. The undersides are paler, their veins reddish-brown. The

Myriocarpa stipitata

greenish or whitish flowers hang in long slender streamers, plentiful enough to make a fair display.

Garden and Landscape Uses and Cultivation. In the humid tropics and in tropical greenhouses, the species described is an admirable ornamental, general purpose, evergreen foliage plant. Good specimens can be had in large pots and tubs and in ground beds. It succeeds best in fertile, loamy soil kept evenly moist, but not waterlogged, and appreciates some shade from strong sun. A humid atmosphere is to its liking. Specimens that have filled their containers with roots are much benefited by regular applications of dilute liquid fertilizer. Repotting should receive attention in late winter or spring, and any pruning needed to shape or control the size of the plants may be done then. Specimens in small pots are likely to need repotting again about mid-summer. Propagation is by cuttings of firm young shoots planted in a propagating bench in a humid, tropical greenhouse, by air layering, and by seed.

MYRIOCEPHALUS (Myrio-céphalus) — Poached-Eggs-Daisy. Australia is home to this group of ten species of the daisy family COMPOSITAE. The name *Myriocephalus*, from the Greek *myrios*, many, and *kephale*, a head, alludes to the arrangement of the flowers.

This genus consists of annuals, one of which is sometimes cultivated. Usually woolly plants, they have alternate, undivided leaves. The tiny heads of flowers are crowded into flattish, disklike clusters that look like the central eyes of daisies and nestle among involucres (collars) of several rows of spreading bracts, often with conspicuous appendages that simulate the ray florets of daisies.

Called poached-eggs-daisy in Australia, ***M. stuartii*** (syn. *Polycalymma stuartii*) is so common in places that sand hills are covered with it to the exclusion of almost all other vegetation. It is 1 foot to 2 feet tall and sticky, with linear to lanceolate leaves up to 2 inches long. The flower heads are bright yellow hemispherical buttons ¾ inch to 1¾ inches wide, each surrounded by a collar of white-tipped bracts responsible for their daisy-like aspect.

Garden Uses and Cultivation. Although by no means showy, this species can add interest to flower gardens. To prepare its blooms as everlastings, they should be cut as soon as they expand, be tied in loose bundles, and hung out of the sun, in a cool, airy place until they are completely dry. To raise this plant the seed may be sown outdoors in early spring where the plants are to bloom or indoors some eight weeks before it is expected that the resulting plants will be set in the garden. Plants from outdoor sowings are thinned so that they are spaced about 6 inches apart, those from indoor sowings are transplanted about 1½ inches apart in flats of porous soil and are grown in a greenhouse in a night temperature of 55°F (by day the temperature may rise five or ten degrees above this), in full sun. Care must be taken not to water excessively. Before the young plants are transplanted to the garden, at about the time it is safe to set out tomatoes, they are gradually hardened by standing the flats outdoors for about a week. Spacing between the plants in garden beds or borders may be about 6 inches. For best results, *M. stuartii* should be given a sunny location and a dryish or at least very well-drained and porous, not too fertile soil.

Myriophyllum aquaticum

MYRIOPHYLLUM (Myri-ophýllum)—Water-Milfoil, Parrot's Feather. The genus *Myriophyllum* consists of forty-five species distributed in the wild almost throughout the world. Several kinds are cultivated in water gardens and aquariums. The group belongs in the water-milfoil family HALORAGIDACEAE. Its name comes from the Greek *myrios*, innumerable, and *phyllon*, a leaf, and alludes to the appearance of the much-divided foliage.

Water-milfoils are perennial herbaceous aquatics, with generally slender stems and often two kinds of leaves. Those that develop under water are finely-pinnately-divided into more or less hairlike segments; the above-water ones may be undivided or merely toothed. The species may be unisexual or bisexual or have unisexual and bisexual flowers on the same plant. In such cases the uppermost flowers are male, the next lowest bisexual, and the bottom ones female. They are minute and may or not have a four-lobed calyx. There are four or no petals; in some species the petals are much reduced in size. The stamens number four or eight, and there are four very short styles. Often the flowers are in terminal spikes and raised out of the water. The small nutlike fruits are more or less four-lobed. In the related genus, *Proserpinaca*, they are three-lobed.

The most popular species is the parrot's feather Brazilian ***M. aquaticum*** (syns. *M. brasiliense*, *M. proserpinacoides*). The leaves of this kind, which thickly clothe long trailing floating stems that at their ends rise about 6 inches out of the water, are all similar. They are fresh green, feathery, and are in whorls (circles) of four to six. Each is about 1 inch long and divided into ten to twenty-five hairlike segments. Male and female flowers are borne on separate plants in the axils of submersed leaves.

Several other sorts are grown by fanciers of aquatic plants. All are variable and are practically impossible to identify as to species without careful examination of flowers or fruits. Indigenous from Newfoundland to Alaska, Maine, New York, Michigan, and Minnesota, as well as Greenland, and Europe, ***M. alterniflorum*** has leaves rarely exceeding ⅓ inch long and often shorter than the portions of bare stem that separate successive whorls (circles) of foliage. The flowers, the males with eight stamens, are above water, in spikes up to 2 inches long. The bracts (small leaves associated with the flowers) and flowers, except sometimes the lowest ones, are alternate.

The most common native in the northern United States and Canada, ***M. spicatum exalbescens*** extends across the continent and also inhabits Greenland. This has finely-dissected leaves, all in whorls. Its flowers are in above-water spikes; the males have eight stamens. The bracts, concave and shorter than the flowers, are in circles of three or four. In contrast to those of the last, the long-persistent and eventually reflexed, sharply toothed bracts of the immersed flower spikes of ***M. heterophyllum*** vary in length on different plants, but are much longer than the blooms. The spikes are 2 to 6 inches and occasionally up to 1¼ feet long. The male flowers of this kind have four stamens. This stout-stemmed species has whorled leaves up to 1½ inches long. It is indigenous from New York to Ontario, South Dakota, Florida, and Texas. A western American species, native from California to Washington, ***M. hippuroides,*** has whorls of very finely-dissected leaves up to 1¼ inches long. The bracts of its above-water flower spikes are toothed, but not deeply-pinnately-divided as in *M. verticillatum*. From that species the male flowers of *M. hippuroides* differ in having four stamens. In strong light the foliage of this species becomes reddish. Some or all of the leaves of ***M. pinnatum*** (syn. *M. scabratum*) are alternate. They are ⅓ inch to 1¼ inches long. The flowers are in above-water spikes with toothed bracts, mostly in whorls, and much longer than the flowers, the males of which have four stamens. Above-water leaves, when present, resemble the bracts. This kind is native from Massachusetts to Ohio, Iowa, Kansas, Georgia, and Texas. Deeply-pinnately-divided bracts, often reflexed, and equaling, or the lower ones exceeding, in length the flowers, characterize ***M. verticillatum.*** The leaves are ¾ inch to 1¼ inches long, finely-dissected and in whorls. The flower spikes, above water, are 1½ to 3 inches in length. The male flowers have eight stamens.

Garden Uses. The parrot's feather is a great favorite for outdoor and indoor pools and is hardy in the north provided it is planted below freezing level. It is often displayed to advantage planted in mud by watersides where its elegantly foliaged stems can trail out over the water. It is also satisfactory for use in submerged containers and for planting in pond bottoms. The underwater species are good aquarium plants and can also be accommodated in pools and ponds. They oxygenate the water and provide cover and spawning places for fish. They are best adapted for large aquariums, but rarely or never bloom except in outdoor waters.

Cultivation. Water-milfoils grow freely in mud or, in aquariums, in unwashed river sand to which it is advantageous to add a little clay or loam. Most thrive best in cool water, but the parrot's feather makes good growth only when the water is warm; it luxuriates in humid tropical conditions. The parrot's feather needs full sun, the others discussed thrive in more subdued light. Propagation is easily effected by division and by cuttings.

MYRISTICA (Myrís-tica) — Nutmeg. The most extensive genus of the nutmeg family MYRISTICACEAE, this comprises 120 species. Its members are small to very large trees, natives of the Old World tropics. The name *Myristica* comes from the Greek *myron*, a sweet liquid distilled from plants.

Myristicas have more or less tiered branches and alternate, undivided, leathery, evergreen leaves in flat sprays. Often their undersides are glaucous. The usually little flowers, the sexes commonly on different trees, are mostly in panicles. They have three-, more rarely two- or five-lobed calyxes, but are without petals. The stamens are united in a column, and there is one style. The fruits, usually fairly large, split to reveal their one seed encased in a network-like fleshy or waxy, spicy envelope called the aril.

The nutmeg (***M. fragrans***) is a good-looking tree up to about 70 feet tall, native to the Molucca Islands, formerly called the Spice Islands. It has oblong-lanceolate, yellowish-brown leaves mostly up to 5 inches long. Male and female flowers are on separate trees. The reddish to yellowish fruits, up to about 2 inches long, split to reveal the brown seed and its bright red aril. The latter is the spice mace of commerce. The seeds are nutmegs.

Garden and Landscape Uses and Cultivation. Except in commercial plantations the nutmeg is only occasionally planted in the tropics for interest and ornament. It is sometimes grown in tropical greenhouses where plants useful to man are displayed. It needs fertile soil, plenty of heat, and high humidity. Propagation is by seed and by grafting.

MYRISTICACEAE — Nutmeg Family. Of minimal horticultural importance, this family of tropical dicotyledons comprises eighteen genera totaling 300 species of evergreen trees. *Myristica fragrans* is best known for its products nutmeg and mace.

The sorts of this family have alternate, undivided, toothless, leathery leaves, like the wood usually aromatic when bruised, and often with pellucid (translucent) dots that can be observed by viewing a leaf against the light.

The small, unisexual, symmetrical flowers are variously clustered or in racemes or heads. They have a usually three-lobed, saucer- to funnel-shaped calyx and are without corollas. In males there are two to twenty stamens united as a column. Female flowers have a short style or sometimes none and one stigma. The fruits are drupes. Only *Myristica* is sometimes cultivated.

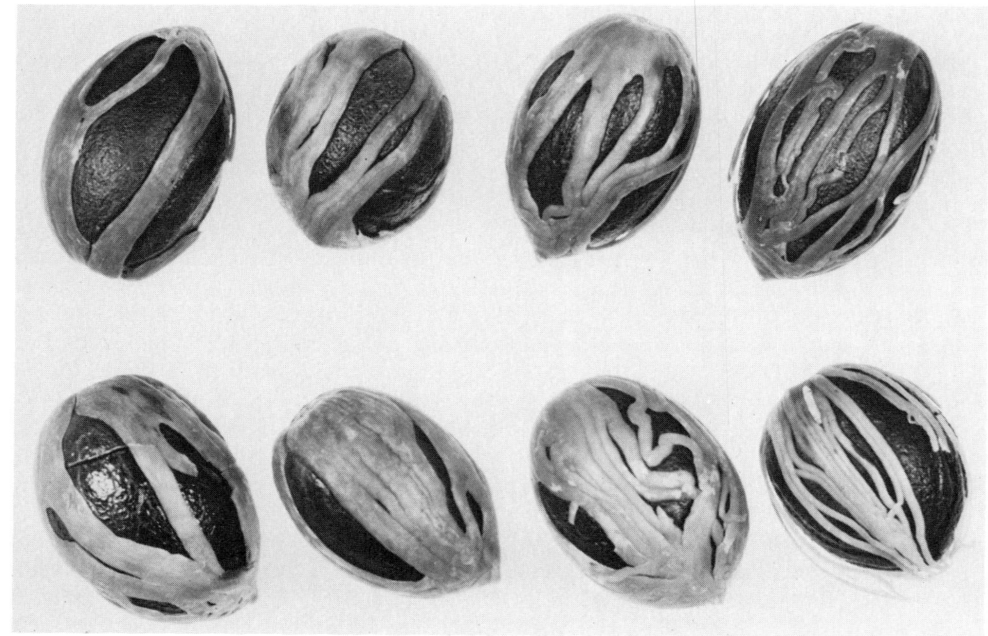
Myristica fragrans, nutmegs (seeds) enclosed in a net of mace (outer covering)

MYRMECODIA (Myrmec-òdia). This curious and interesting group of plants, called myrmecophilous, literally ant-loving, because in the wild fiercely aggressive ants make their homes in communicating hollows and passageways that penetrate the mostly corky tissue of the large tuberous stems. These can make it a highly unpleasant adventure to gather specimens. It has been suggested that the ants, in return for shelter, defend their hosts from voracious animals. It is doubtful, however, that they play any significant part in the economy of the plants or that they influence to any appreciable extent the preservation of the species. There are some forty-five species of *Myrmecodia*, natives of Malaysia, the Solomon Islands, Fiji, New Guinea, and northern Australia; they belong in the madder family RUBIACEAE. The name is from the Greek *myrmekos*, an ant, in allusion to the association of the plants and insects.

Myrmecodia, undetermined species

Small subshrubs, myrmecodias are epiphytes that perch on the trunks and branches of trees in the manner of many orchids and bromeliads and, like them, not extracting nourishment from their hosts. They have much-swollen, tuberous stems and short, thick branches with, clustered near their apexes, opposite, stalked, leathery leaves. Solitary or in clusters of few, the small, stalkless, white flowers have a brief calyx and an urn-shaped, four-lobed corolla. There are four stamens. The fruits are berries.

Native to Malaya, ***M. armata*** (syn. *M. tuberosa*) is an epiphyte with a big irregular

Myrmecodia armata

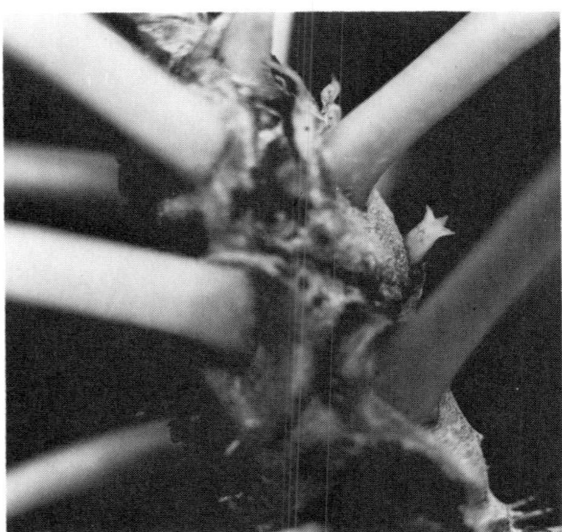
Myrmecodia armata with tiny flowers

tuber furnished with bristle-like prickles and with a few irregular, fleshy branches. Its opposite, stalked, obovate-oblong, lobeless, rather fleshy, toothless leaves have blades 3 to 4 inches long. The stalkless white flowers are succeeded by white fruits. Northern Australia and perhaps New Guinea is the home of ***M. echinata*** (syn. *M. antoinii*). This has a turnip-shaped, very spiny stem 4 to 7 inches in diameter. From its apex develop one to four short, fat, four-angled, spiny branches with at their tops, a few thickish, hairless, obovate leaves 4 to 5 inches long that arise from spiny cushions. The flowers, borne among the foliage, are about ½ inch long.

Myrmecodia echinata

Somewhat similar to the last, ***M. platyrea*** is a native of New Guinea.

Garden Uses and Cultivation. Myrmecodias are plants for lovers of the unusual and collectors of botanical curiosities. In the humid tropics they will grow outdoors, elsewhere they are accommodated in greenhouses devoted to moisture-loving tropical plants. Because they are epiphytes they thrive in environments and with the care that suit many orchids and bromeliads. Of first importance is provision of a rooting medium that admits freely the passage of water and air. A coarse soil, consisting of osmunda fiber or fir bark of the kind used for potting orchids, rough leaf mold or peat moss, and broken charcoal, with a little fibrous topsoil mixed in, is satisfactory. Pot the plants rather loosely, with the bulbous stems above ground. Water when the plants are in leaf to maintain the earth moderately moist, when no foliage is present the soil is kept dry or nearly dry. Myrmecodias grow satisfactorily in greenhouses in which a minimum night temperature of 65 to 70°F is maintained, with higher temperatures by day. Much higher ones in summer are allowed. High humidity is to their liking. Propagation is by seed.

MYROBALAN. This name is used for *Phyllanthus emblica* and *Terminalia catappa*.

MYROSPERMUM (Myro-spérmum). This genus of the pea family LEGUMINOSAE has only one species, *Myrospermum frutescens*, a native of Central America, northeastern South America, Trinidad, and the West Indies. Its name, from the Greek *myrios*, many, and *sperma*, a seed, is self-explanatory.

A deciduous shrub, small tree, or rarely a tree up to 60 feet tall, ***M. frutescens*** has much the appearance of a locust (*Robinia*). Its leaves, marked with translucent dots and lines, are approximately 6 inches long. They have, including a terminal one, about eleven oblong to elliptic leaflets 1¼ to 1¾ inches long by approximately one-third as broad. The white flowers, tinged with pink, and about ½ inch long, are borne in profusion in clustered racemes 2 to 4 inches long from the ends of short shoots. Each has an obovate standard or banner petal, and four very similar smaller, narrower, separate petals. The stamens are not united. The style is awl-shaped. The fruits, which look like one-half of a maple key (fruit) have a broad wing 2 to 3 inches long that margins both sides of the single seed that is located at one end.

Garden and Landscape Uses and Cultivation. Adaptable for outdoor cultivation in the tropics and warm subtropics, this is sometimes planted for ornament. It succeeds under ordinary conditions in well-drained soil, and is propagated by seed.

MYROXYLON (Myr-óxylon) — Balsam-of-Peru Tree, Tolu Balsam Tree. Two species constitute the genus *Myroxylon*, of the pea family LEGUMINOSAE. They are large, evergreen trees of Mexico and Central and South America. The name, alluding to their pleasant fragrance, comes from the Greek *myron*, perfume, and *xylon*, wood.

Myroxylons have alternate, pinnate leaves with a terminal and lateral leaflets. Their more or less pea-shaped, white or whitish blooms are in racemes from the leaf axils or panicles at the branch ends.

They have irregularly-toothed calyxes, a large standard or banner petal, four smaller petals about equal in size, ten stamens that fall with the petals, and one style. The fruits are flattened, stalked pods that have a long, broad wing with a single seed at one end. The balsams for which these trees are exploited are thick, fragrant liquids that flow from incisions made in the trunks. Once believed to have medicinal virtues, they are chiefly employed to perfume ointments, salves, and similar preparations and to flavor cough syrups and other medicines. Balsam-of-Peru chiefly comes from Salvador. It received its misleading name because it was shipped to Spain through the Peruvian port of Callao. Tolu balsam is obtained from *M. balsamum* (syn. *M. toluiferum*), the superior balsam-of-Peru from *M. b. pereirae* (syn. *M. pereirae*). The plant sometimes named *M. senticosum* is *Xylosma conjestum*.

Native to northern South America, ***M. balsamum*** is 75 to 100 feet tall. Its variety, indigenous from Mexico to northern South America, is 50 to 65 feet and sometimes more, high. The leaves of *M. balsamum* are hairless and have five to eleven pointed, oblong leaflets 2 to 3½ inches long, besprinkled with translucent dots. Its fruits, curiously resembling one-half of a maple key (fruit) are 4 or 5 inches long. From the last, *M. b. pereirae* differs in having leaves with nine to thirteen leaflets finely-hairy on their stalks and along the midribs on their undersides.

Garden and Landscape Uses and Cultivation. For interest and ornament, myroxylons are occasionally planted in the tropics and sometimes are grown among collections of plants useful to man in tropical greenhouses. They succeed under ordinary conditions and are usually propagated by seed. They can also be increased by cuttings of firm but not hard shoots planted in a greenhouse propagating bench, preferably with slight bottom heat.

MYRRH is *Myrrhis odorata*.

MYRRHINIUM (Myr-rhínium). Opinions differ as to whether *Myrrhinium*, of the myrtle family MYRTACEAE, should be treated as one or more species. Some authorities recognize five. If more than one are accepted, it is difficult to separate them satisfactorily on the basis of reasonably constant botanical characteristics. The extremes grade into each other. As treated here the genus is composed of one variable species. The name, in allusion to the balsamic gum-resins of these trees being similar to the biblical myrrh, is derived from the Greek *myrrha*.

Indigenous from Ecuador to Argentina, ***M. atropurpureum*** is an evergreen, aromatic shrub or small tree up to about 15 feet high. Its shoots and leafstalks are often hairy. The leaves, opposite, undivided, and without lobes or teeth, are nearly stalkless and 1½ to 4 inches long by up to ¾ inch wide. They are pointed-elliptic, broadly or narrowly so, and have somewhat rolled-under margins. The ¼-inch-wide flowers in crowded 2- to 3-inch-wide clusters arise from older, leafless shoots from the branches and trunk. They have four-lobed calyxes, four fleshy, recurved, pinkish, red, or dark red petals, and usually six, but up to eight long-protruding stamens. The style, slender and red, ends in a slightly two-lobed stigma. The fruits are small, whitish, few-seeded berries, reputed to be edible.

Garden and Landscape Uses and Cultivation. This pleasing shrub is suitable for general landscape planting in mild, essentially frost-free, dryish climates. It succeeds in ordinary soil in sun or part-shade, and is propagated by seed and by cuttings.

MYRRHIS (Mŷr-rhis) — Sweet Cicely or Myrrh. Depending upon the authority chosen, *Myrrhis*, of the carrot family UMBELLIFERAE, consists of one or two species. It is endemic to Europe and western Asia and bears a generic name used by Dioscorides.

Sweet cicely or myrrh (***M. odorata***) is a graceful, erect, finely-hairy, anise-scented, hardy herbaceous perennial, 2 to 3 feet tall, with hollow, grooved stems that branch above. Its two- or three-times-pinnately-divided leaves, whitish on their undersides, have leaflets pinnately-lobed and toothed. The leafstalks sheathe the stems. Tiny and white, the numerous flowers are in compound umbels, of five to ten smaller umbels, up to 2 inches across. The inner blooms of the umbels are male, the outer bisexual. The shining, dark brown to black, sharply-ridged fruits (seeds) are ¾ to 1 inch long. This is not the myrrh of the Bible, that is a product of a species of *Commiphora*.

Garden Uses and Cultivation. In the past sweet cicely was more popular than at present as a salad and flavoring plant. Its roots were boiled and its leaves and seeds used without cooking. It was also employed medicinally. It is entirely appropriate therefore to include sweet cicely in herb and vegetable gardens. In addition to its interest as an edible, it is of ornamental merit and is attractive in informal, naturalistic areas. It is a good bee plant. Sweet cicely thrives in a wide variety of soils in sun or part-day shade and is easily raised from seed. They are sown outdoors as soon as they are ripe or in spring.

MYRSINACEAE — Myrsine Family. Comprising about 1,000 species contained in thirty-five genera, the MYRSINACEAE family of dicotyledons occurs natively in tropical and subtropical regions, including South Africa, New Zealand, Mexico, and Florida. Its members have mostly alternate, usually leathery, evergreen, undivided leaves, often clustered in more or less rosette fashion, and with tiny oil glands or ducts generally visible to the naked eye if a leaf is viewed against the light.

The flowers, usually rather small and in raceme-like groups or panicles, are bisexual or unisexual. They most commonly have a calyx of four or five lobes or separate sepals and a four- or five-lobed or four- or five-petaled corolla. There are as many stamens as corolla lobes or petals, one style, and one sometimes lobed stigma. The fruits are drupes or berries. Cultivated genera include *Ardisia, Hymenandra, Maesa, Myrsine,* and *Suttonia*.

MYRSINE (Myrsì-ne)—African-Boxwood. Five species of African and Asian much-branched shrubs and small trees constitute *Myrsine*, of the myrsine family MYRSINACEAE. By some authorities the genus treated by other botanists and in this Encyclopedia as *Suttonia* is included in *Myrsine*. The name is derived from that of the myrtle (*Myrtus*), an unrelated plant.

Myrsines have mostly alternate, undivided, broad- to narrow-elliptic leaves,

Myrrhis odorata

Myrrhis odorata (flowers)

toothless or toothed toward their apexes. The tiny, scarcely-stalked, unisexual flowers are in clusters of few from the leaf axils. They have four- or five-parted calyxes, short-tubed corollas with four or five lobes or petals, as many stamens as corolla lobes or petals, and a short style ending in a headlike or fringed stigma. The fruits are slightly fleshy, one-seeded, berry-like drupes.

African-boxwood (***M. africana***), a shrub 3 to 6 feet high or rarely higher, is native from the Azores to Africa, China, and Taiwan. Not related to boxwood (*Buxus*), it has short-stalked, glossy, leathery, hairless leaves of somewhat boxwood-like appearance, $\frac{1}{10}$ to $\frac{3}{4}$ inch long, elliptic to lanceolate or nearly round, with short hairs on their undersides, and usually toothed above their middles. The flowers, three to eight together in stalkless clusters from the leaf axils, have four each sepals and petals. The stamens, united at their bases, terminate in large, red, partly-protruding anthers. The tiny fruits are purple-red to blue.

Myrsine africana

Garden and Landscape Uses and Cultivation. African-boxwood is much esteemed in California, and places with similar climates as a general purpose evergreen foliage shrub and hedge. It stands trimming well and adapts to ordinary soils and locations. It is easily propagated by cuttings and seed. In small sizes it is interesting as an indoor window plant.

MYRTACEAE—Myrtle Family. Some 3,000 species of dicotyledons belonging in 100 genera constitute this family of trees and shrubs, some of the latter low creepers. They inhabit warm-temperate to tropical regions and are especially numerous as to kinds in Australia, where they attain their maximum development in the genus *Eucalyptus,* and in South America.

Characteristically, this group has evergreen foliage, which is aromatic or fragrant when bruised. The short-stalked leaves are usually opposite, undivided, and generally lobeless, and with tiny pellucid (translucent) dots visible when viewed against the light. The flowers, solitary in the leaf axils or in racemes or otherwise clustered, have calyxes with tubes more or less united with the ovary and four or five persistent sepals. There are four, five, or rarely no petals, generally numerous stamens often in groups or bundles and with their stalks separate or partly united, and one style. The fruits are berries, drupes, capsules, or nuts.

The family is of great importance as a source of lumber, oils, and gums, with commercial uses, and of allspice, cloves, nutmegs, and some edible fruits. Its members include many fine ornamentals. Cultivated genera are *Acmena, Agonis, Amomyrtus, Angophora, Backhousia, Baeckea, Beaufortia, Callistemon, Calothamnus, Calythrix, Chamelaucium, Darwinia, Eucalyptus, Eugenia, Feijoa, Hypocalymma, Kunzea, Leptospermum, Lhotskya, Lophomyrtus, Luma, Melaleuca, Metrosideros, Myrciaria, Myrrhinium, Myrtus, Neomyrtus, Pimenta, Psidium, Rhodomyrtus, Syncarpia, Syzygium, Tristania, Ugni,* and *Verticordia.*

MYRTILLOCACTUS (Myrtillo-cáctus). Four species constitute *Myrtillocactus,* a Mexican and Guatemalan genus, of the cactus family CACTACEAE. The name is from the Latin *myrtillus,* a little myrtle, and *Cactus,* a genus of the same family. It refers to the small, berry-like fruits having much the aspect of those of myrtle.

Treelike, myrtillocactuses have short trunks, and thick, erect, spiny branches with few strongly defined ribs. Their small flowers, often several from the same areole, open by day.

Myrtillocactus geometrizans

One of the most popular and widely cultivated sorts is ***M. geometrizans*** (syn. *Cereus geometrizans*). Native to Mexico, this is up to 15 feet tall and has numerous upright branches. Its stems, six-ribbed and bluish-green, are about 4 inches thick on mature plants, but thinner on smaller specimens. The younger parts of the stems are especially colorful and handsome. The areoles (places from which spines arise) are approximately 1 inch apart. The spines, about six from each areole, are not always well developed on cultivated plants. The central one is dagger-shaped and often down-pointed, the others, shorter and slightly flattened, bend backward. The fragrant flowers 1 inch to $1\frac{1}{4}$ inches wide, with erect, white perianth segments (petals), are succeeded by pur-

Myrtillocactus geometrizans, cristate variety

plish-black, olive-sized, edible fruits, in Mexico called garrambulla. There is a handsome cristate variety.

With green rather than blue-green stems, *M. cochal* and *M. schenckii* are also cultivated. They attain heights of perhaps 10 feet and have six- to eight-ribbed stems. Those of the latter are much more spiny than those of the former. In ***M. cochal,*** a native of Baja California, the central spine of each group is ¾ inch long and there are five shorter ones. The 1-inch-long blooms,

Myrtillocactus cochal in Baja California

pale green tinged with purple, are succeeded by edible fruits ½ inch in diameter. The spines of ***M. schenckii*** number six to eight from each areole with the central one 1 inch to 2 inches, the others up to 1 inch long. The small flowers are pale yellow to cream with brownish outsides. This is a Mexican species.

Garden and Landscape Uses and Cultivation. The sorts of this genus are attractive for including in desert gardens in California and parts of the southwest where such developments are practicable outdoors, and for inclusion in indoor collections of cactuses. In small sizes they may be grown as window plants. They are easily propagated by cuttings and seed. For more information see Cactuses.

MYRTLE. See Myrtus. Allegheny sand-myrtle is *Leiophyllum buxifolium prostratum;* crape-myrtle, *Lagerstroemia indica;* creeping- or running-myrtle, *Vinca minor;* downy-myrtle, *Rhodomyrtus tomentosum;* fringe-myrtle, *Calythrix;* gum-myrtle, *Angophora lanceolata;* heath-myrtle, *Baeckea;* prickly-myrtle, *Clerodendrum aculeatum;* sand-myrtle or box sand-myrtle, *Leiophyllum buxifolium;* Swan-River-myrtle, *Hypocalymma robustum;* sweet-verbena-myrtle, *Backhousia citriodora;* and wax-myrtle, *Myrica cerifera.*

MYRTUS (Mýr-tus)—Myrtle. The myrtle of antiquity belongs in this genus of possibly 100 species of shrubs or rarely trees, of the myrtle family MYRTACEAE, natives of warmer parts of the Old and New Worlds. The name is an ancient Greek one. This group must not be confused with other genera that have common names that include the word myrtle, for instance crape-myrtle (*Lagerstroemia*), downy-myrtle, (*Rhodomyrtus tomentosa*), gum-myrtle (*Angophora*), running-myrtle (*Vinca minor*), sand-myrtle (*Leiophyllum*), and wax-myrtle (*Myrica cerifera*). The Chilean-guava, previously named *Mrytus ugni*, is *Ugni molinae.* The plant called *M.* luma is *Amomyrtus luma.* New Zealand plants previously included in *Myrtus* are now named *Lophomyrtus* and *Neomyrtus.*

True myrtles (*Myrtus*) have predominantly opposite, undivided, toothless, evergreen leaves that are spicily aromatic when crushed. They have a prominent vein encircling the leaf just inside its margin. The flowers are solitary in the leaf axils or are in clusters of few. They have a tubular, top-shaped, persistent, four- or five-lobed calyx, four spreading petals, and many stamens longer than the petals and separate to their bases. The fruits are ovoid to spherical one- to several-seeded berries with the persistent calyx lobes at their apexes.

Most familiar, the common myrtle has been esteemed since classical times as a symbol of love and peace. The ancients dedicated it to the goddess Venus, and decorated magistrates, poets, playwrights, and victorious athletes with it. It is the myrtle of the Bible, still used by Jews in connection with the feast of the Tabernacle. A traditional bridal flower, in Europe it was, and in some parts still is, the custom to root as cuttings sprigs carried in bridal bouquets and keep them through life as carefully tended pot plants. An oil used in perfumery is obtained from the common myrtle, and high quality charcoal is made from its heavy, fine-grained wood.

Common myrtle (***M. communis*** syn. *M. latifolia*) is a dense, leafy shrub 3 to 12 feet tall. It has glossy, boxwood-like, very aromatic, pointed-egg-shaped to lanceolate, minutely-stalked, leathery leaves ¾ inch to nearly 2 inches long and generally about one-half as wide. They are dark glossy green above, paler on their undersides. The white or pink-tinged, slender-stalked blooms are solitary and ½ to ¾ inch in diameter. When ripe, the nearly spherical berries are bluish-black, or in one variety white. This species is a native of lands surrounding the Mediterranean Sea. Variety *M. c. buxifolia* has elliptic leaves. Those of *M. c. microphylla* are much smaller than those of the species. They are long-pointed, angle toward the tips of the shoots, and more or less overlap. The leaves of dwarf *M. c. minima* and *M. c. nana* are also much smaller than those of typical *M. communis.* The flowers of *M. c. flore-pleno* are double. In *M. c. variegata,* the foliage is variegated with creamy-white. The leaves of *M. c. tarentina* are small, often alternate, and in four ranks. Dried twigs of this variety have been recovered from Roman tombs more than 2,000 years old. Yellowish-white fruits distinguish *M. c. albocarpa.*

Myrtus communis

Myrtus communis (flowers)

Garden and Landscape Uses. As is typical of Mediterranean region plants, myrtles thrive in warm, dryish climates in sunny locations, but not in excessively dry soil. They stand some frost, but not hard freezing. They are admired for their dense foliage effects, for the fragrance of their leaves when lightly crushed or brushed against, for their quite pretty blooms, and for their not very conspicuous fruits. Myr-

Myrtus communis variegata

Mystacidium distichum

Myrtus communis variegata, trained as a standard

tles are good components of shrub beds. They are useful for foundation plantings and sheared hedges. They also are excellent for tubs and other containers used to decorate terraces, patios, steps, and other architectural features, and for espaliering against walls. As smallish specimens they make, especially little-leaved varieties, good window plants for cool rooms, and they are easily grown in greenhouses. They lend themselves well to training as standard (tree-form) specimens.

Cultivation. Little trouble attends the cultivation of myrtles. Outdoors or in, if the environment is at all to their liking, they grow graciously and well. They respond to fertile, well-drained, loamy soils containing fair amounts of such organic matter as peat moss, humus, leaf mold, or compost, and in spells of dry weather, to watering. Pruning to any extent desired to keep them shapely or limit their size is done in late winter or early spring. Hedges and formally shaped specimens may be sheared then and later. Myrtles can be raised from seed, but cuttings 1½ to 3 inches long of moderately firm shoots root so readily that they afford the most common means of multiplication. Cuttings are usually taken in summer or early fall and planted in sand, vermiculite, mixtures of one of these materials, and peat moss, or in sandy soil, in a greenhouse, cold frame, or under a glass jar or similar protection in a shady spot outdoors. Another method of increase is by layering.

MYSTACIDIUM (Mysta-cídium). African orchids of the orchid family ORCHIDACEAE, totaling a dozen species, compose *Mystacidium*. They resemble angraecums and are epiphytes (tree-perchers). The name, from the Greek *mystax,* mustache, alludes to the hairs at the top of the column.

Mystacidiums have short to long leafy stems and leathery or fleshy leaves in two ranks. Solitary or in racemes, the small to medium-sized flowers have sepals and petals of nearly equal size, a spurred lip with or without a pair of small lateral lobes, and a short column.

Native to southeast Africa, charming ***M. capense*** (syn. *Angraecum capense*) has a short, thick stem and 3-inch-long, leathery, strap-shaped leaves notched at their apexes. Loosely-arranged in arching racemes up to 6 inches long, the fragrant, waxy-white blooms, 1 inch to 1½ inches wide, are freely produced. Their narrow-lanceolate sepals, petals, and lip are similar except for the last being somewhat broader, sometimes three-lobed, and with a slender basal spur up to 2 inches long. Tropical African ***M. distichum*** (syn. *Angraecum distichum*) has tufts of stems up to 3 to 9 inches in length with, evenly spaced along their lengths, broadly-elliptic-oblong, sickle-curved, overlapping leaves about ½ inch long. The white flowers are solitary and short-stalked. About ⅜ inch long, they have a hooded, three-lobed lip.

Garden Uses and Cultivation. Collectors' items, these orchids may be grown in much the same way as angraecums. They succeed in intermediate-temperature greenhouses. For additional information see Orchids.

NAEGELIA. See Smithiantha.

NAGELIELLA (Nageli-élla). In gardens, species of *Nageliella,* of the orchid family ORCHIDACEAE, are often grown under the superseded name of *Hartwegia.* There are two or three species, natives of Mexico and Central America. The name honors Otto Nagel, a twentieth-century collector of plants in Mexico.

Nageliellas are good-looking small tree-perchers (epiphytes), or more commonly they grow on rocks or partially in the ground. They are tufted and have stemlike pseudobulbs, often club-shaped above. From the top of each pseudobulb comes a single fleshy, mottled leaf. The racemes, which bear blooms successively for two or more years, arise from junction of pseudobulb and leaf. Their flowers are small and brightly colored.

Its pseudobulbs 2 to 4 inches tall, ***N. purpurea*** (syn. *Hartwegia purpurea*), of Mexico and Central America, has racemes of amethyst-purple flowers reaching upward from them. Its olive-green leaves often purplish on their undersides, are spotted with brown, reddish-brown, or purple. Somewhat triangular, they are 3 to 4 inches long by 1 inch wide or wider. The bell-shaped blooms, ½ inch or slightly more across, are in racemes that have rigidly erect, wiry stems. Similar ***N. angustifolia*** (syn. *Hartwegia purpurea angustifolia*) differs in having narrower, fleshier, linear-lanceolate to oblong-lanceolate leaves and slightly smaller blooms.

Garden Uses and Cultivation. These are pretty little orchids for fanciers of such plants, and being easily grown, are suitable for inclusion in the collections of even beginners in orchid cultivation. They need warm, humid conditions and as much light as they will take without the foliage showing signs of scorching. At no time must their roots be dry, but excessive wetness for extended periods is damaging. They may be kept in pots in firmly packed osmunda fiber or other favored rooting medium or may be attached to slabs of tree fern trunk. Applications of mild liquid fertilizer are of benefit. For more information see Orchids.

NAIBEL is *Hesperethusa crenulata.*

NAILWORT. See Paronychia.

NAMA PARRYI is *Turricula parryi.*

NAMES OF PLANTS. See Plant Names.

NANANTHUS (Nan-ánthus). The highly specialized group identified as *Nananthus* is of special interest to collectors of choice succulents. It consists of thirty South African perennial species of the *Mesembryanthemum* relationship, of the carpetweed family AIZOACEAE. The naming of its species has been variously and confusingly shuffled between *Nananthus, Aloinopsis,* and *Rabiea.* Its name from the Greek *nanos,* dwarf, and *anthos,* a flower, is of obvious application.

These are tufted perennials up to 2 inches tall, with long tuberous roots. They much resemble *Aloinopsis.* They have short stems with three to six pairs of fleshy, opposite leaves, without portions of stem showing between, and with their lower parts sheathing the stems. Erect or spreading, the conspicuously warted leaves have linear-lanceolate to ovate, often more or less spatula-shaped upper surfaces with, above their middles, one side generally wider than the other. The leaf usually ends in a point or bristle. The greatest depth of the leaves, as seen from their sides, is from the middle outward. The solitary flowers, daisy-like in aspect, but not in structure, ¾ inch to 1¼ inches wide, open near midday. They are stalkless or almost so, yellow, sometimes with a middle stripe of red on their petals. They have prominent stamens in an erect cone and are without staminodes (nonfunctional stamens). The fruits are capsules. Because cultivated specimens often differ in appearance from wild ones, and because the species are variable, the identification of plants in this genus is often difficult.

Rosettes of six to eight prominently white-warted, lanceolate to narrowly-lozenge-shaped leaves up to 2 inches long, the older ones spreading, the younger erect, are typical of ***N. aloides*** (syns. *Aloinopsis aloides, Mesembryanthemum aloides*). Their upper surfaces are flat or shallowly-concave, the keel beneath is prolonged to give the leaves a three-angled apex. The 1-inch-wide, short-stalked flowers are yellow. Variety *N. a. latus* has broader leaves. The petals of *N. a. striatus* are striped with dark yellow. The six to ten waxy-gray-green leaves with pale green warts of ***N. pole-evansii*** are about 1¼ inches long by ⅓ inch wide. They narrow slightly to their bases, more markedly to their sharp-tipped apexes. They are keeled, and in side elevation boat-shaped. The angles of the leaves are often reddish. A red line marks the center of each of the yellow petals of the ¾- to 1-inch-wide, bright yellow flowers. Blooms of the same size as those of the last and with a line of red down the center of each light yellow petal are borne by ***N. vittatus*** (syns. *Aloinopsis vittata, Mesembryanthemum vittatum*). This species has six to eight dull green, conspicuously warted, spreading leaves ¾ inch to 1¼ inches long by up to ⅓ inch wide. Asymmetrically lanceolate, thick, and tipped with a brief point, they have broad keels toward their apexes.

Garden and Landscape Uses and Cultivation. Only where the climate approximates that of the South African deserts or semideserts, where these plants grow as natives, are they suitable for outdoor cultivation. There, they may be accommodated in rock gardens. More often grown in greenhouses by keen collectors of rare and choice succulents, because of their tuberous roots, which may be 1 foot in length, they are most conveniently accommodated in pots considerably deeper than those of standard dimensions or in sections of porous clay drain pipes. The soil must be extremely porous. One containing much gravel and sand is recommended. Very little water is needed in winter. In summer the soil is kept moderately moist, but never for long periods wet. It should be nearly dry before applications are made. Propagation is best by seed. Shoots, carefully removed from the parent, may be used as cuttings. Nananthuses need full sun.

NANDINA (Nandì-na)—Heavenly-Bamboo or Chinese Sacred-Bamboo. The generic name of the only species of *Nandina* is derived from the Japanese one *nanten*. The plant is a beautiful evergreen shrub, native from India to China and introduced into Japan, that by some authorities is considered to constitute a family of its own, the nandina family NANDINACEAE, and by others to belong in the barberry family BERBERIDACEAE. Certainly its affinities are with the barberries and podophyllums (the latter are by some authors placed in a distinct family, the PODOPHYLLACEAE, and by others are included in the BERBERIDACEAE).

The heavenly-bamboo or Chinese sacred-bamboo (***N. domestica***) is from 3 to 8

Nandina domestica with smaller plants in front

feet in height and has many erect, branchless stems. Its alternate, somewhat fernlike leaves, 1 foot to 1½ feet long, are twice- or thrice-pinnately-divided into many long-pointed, toothless leaflets 1 inch to 3 inches in length; one to five leaflets constitute each ultimate leaf division. The whitish flowers about ¼ inch in diameter are produced in great profusion in large terminal, pyramidal clusters, but are not very showy. Each bloom has numerous sepals and petals and six stamens with large yellow anthers. From the outside to the center of the flower the parts change gradually from sepals to petals. Much more decorative than the flowers are the long-lasting clusters of bright red, globose berries which are borne in abundance. The berries, each of which contains two seeds, are about ⅓ inch in diameter. Variety *N. d. alba* has white berries. Others with pale

Nandina domestica (fruits)

purple berries are reported from Japan. In the latter country many horticultural varieties are grown, including miniature kinds that exhibit a wide range of foliage characteristics.

Garden and Landscape Uses. The heavenly-bamboo is hardy in sheltered locations about as far north as New York City, but as a landscape shrub it is most useful in the south and on the Pacific Coast. Its very considerable decorative merits commend its use in shrub borders, foundation plantings, and as a specimen in small gardens. It is admirably adapted for inclusion in Japanese gardens. The heavenly-bamboo is also an excellent ornamental for growing in tubs. Much of the beauty of this plant lies in the delicate tracery of its foliage, which is enhanced by its attractive bronzy coloring when young and the red hues its mature leaves assume in fall. The fruit display is also highly ornamental. Nandinas prosper in sun or light shade in any fairly good garden soil that does not become excessively dry.

Cultivation. If soil and site are suitable, no special care is needed to grow this attractive evergreen. It may be transplanted without danger of loss even when old and big. In dry weather regular applications of water are helpful, and it is certainly beneficial to keep the soil around it mulched with compost or other appropriate organic material. Toward the northern limits of regions where it may be grown, winter protection from strong sun and wind enables it to survive where otherwise it might succumb.

Containers for nandinas must be well drained. The soil should be fertile, porous, and contain an abundance of peat moss, leaf mold, rotted manure, or other organic matter. Water must be given often enough to keep the soil always fairly moist and from spring through fall weekly applications of dilute liquid fertilizer are conducive to good growth. In cold climates, tub plants can be stored over winter in a cool, light, frostproof or nearly frostproof place such as a cellar or garage. Propagation is usually by seed sown in sandy peaty soil kept evenly moist. Cuttings, preferably under mist, afford another means of increase, but they root rather slowly. They are chiefly employed in the propagation of horticultural varieties.

NANNORRHOPS (Nán-norrhops). This genus, native from Arabia to northwestern India, includes four species one of which is sometimes cultivated. It belongs in the palm family PALMAE. The name from the Greek *nannos*, dwarf, and *rhops*, a bush, refers to the low stature of the plants.

Palms of the genus *Nannorrhops* are low and usually have branched stems that are more or less subterranean or prostrate. The fan-shaped leaves are not spiny, but have toothed leafstalks. The flower clusters are of bisexual and unisexual, three-petaled blooms, the former with usually six stamens, the males with nine. The spherical to oblong, berry-like fruits contain one seed.

The Mazari palm (***N. ritchiana***, sometimes misspelled *ritchieana*) exceptionally has erect stems and attains a height of 20 feet, but more usually has prostrate, creeping stems. Its rigid, leathery, short-stalked leaves, grayish-green or powdery-white, may be up to 4 feet across. They are deeply divided into eight to fifteen narrow, bifid lobes. The flower clusters, erect and loosely branched, rise high above the foliage. The blooms are creamy-white. The brown or orange-brown, wrinkled fruits are about ½ inch in diameter. In its native regions, this species inhabits dry mountain areas at elevations of up to 5,000 feet, where it is not unusual for snow to lie for many months at a time. The flesh of its fruits is eaten when food is scarce, and its young shoots are used as a vegetable. The leaves are employed for making baskets and sandals and for thatching.

Garden and Landscape Uses. The species described here has much the aspect of the European fan palm (*Chamaerops humilis*), but differs in its spineless leafstalks. It is well adapted for places where space is limited and may be used as a lawn specimen, or grouped with other plants. It is especially appropriate near such architectural features as steps, terraces, and buildings. It may also be grown in large pots or tubs in greenhouses.

Cultivation. The Mazari palm thrives in sun or part-shade in any ordinary garden soil and withstands dry conditions; however, it grows faster and is more lush where the moisture supply is not too restricted and the soil drainage is good. It is possible to propagate it by careful division of the clumps, but it is far better to grow it from fresh seeds sown in sandy peaty soil in a temperature of 70 to 80°F. This palm thrives in greenhouses where the minimum night temperature in winter is 40 to 50°F and the daytime temperature is five to ten degrees higher. The atmosphere should be moderately humid and light shade is necessary in summer. Well-drained pots or tubs are suitable containers or the plants may be set in ground beds. The soil should be fertile and porous. From spring through fall generous watering is needed, in winter the soil should be allowed to become dryish before water is applied. The chief pests are scale insects, mealybugs, and red spider mites. For additional information see Palms.

NANNYBERRY or NANNY-PLUM is *Viburnum lentago.* The name nannyberry is also used for *V. prunifolium.* The rusty nannyberry is *V. rufidulum.*

NARANJILLA is *Solanum quitoense.*

NARCISSUS (Nar-císsus)—Daffodil, Jonquil. Horticulturally one of the most important genera of bulb plants, *Narcissus,* of the amaryllis family AMARYLLIDACEAE, consists of sixty species and a vast number of natural and horticultural varieties and hybrids. Native to Europe, North Africa, and western Asia, and with one variety of Mediterranean-region *N. tazetta* indigenous to China and Japan, the genus *Narcissus* has its chief center of natural distribution in central Europe and the Mediterranean region. The name is a classical Greek one of a mythical youth of such beauty that he became entranced with his own reflection in a pool and was by the gods transformed into a flower.

Narcissuses (some prefer the Latin plural narcissi) have small to comparatively large bulbs and all-basal foliage. The leaves, flat and linear or cylindrical and rushlike, develop in late winter or early spring and die in late spring or early summer. Solitary or in clusters and arising from a membranous spathe (bract) atop a branchless stalk, the prevailingly yellow or white blooms generally are more or less nodding. Their perianths have six spreading petals (more correctly tepals) inside of which is a crown or corona of petal-like texture developed as a trumpet of large size and bold appearance, as a smaller trumpet, or as a shallow ring or cup. There are six stamens and one style ending in a three-lobed stigma. The fruits are capsules.

Narcissus bulbs vary greatly in size according to variety

The history of narcissuses in cultivation is long. Theopharastus, three hundred years before the birth of Christ, made reference to them. In the first century A.D., Dioscorides wrote of *N. poeticus* "Is like in ye leaves to ye Leek, but they are thinner and smaller by much, & narrower; it hath a stalk, empty without leaves, longer than a span. On which is a white flower, but within of a saffron color, & in some of a purple color, but ye root white within, round Bulbous-like." As benefits the fountain-head of classical medicine, Dioscorides, in his Herbal, detailed the healing virtues, as he understood them, of the plant. Among the more intriguing, "the root of this being eaten and drank doth move vomitings," "Being beaten small and laid on with honey, it helps both ye luxations of ye Malleoli, and the long continued griefs about ye joints," "with nettle seed and acetum it cleanseth both sunburnings, & ye vetiligo," "being laid on with Loliacean meal, & honey it draws out splinters."

Although modern findings fail to support Dioscorides' beliefs as to the healing virtues of narcissuses, they do demonstrate their powerfully emetic and even narcotic and somewhat poisonous properties. There is a twentieth-century report of a nonfatal poisoning resulting from narcissus bulbs having been cooked and eaten in mistake for leeks.

The chief interest of gardeners in *Narcissus* relates to the numerous magnificent garden varieties and hybrids of which such splendid selections are offered and described, often with excellent photographic color illustrations, in the catalogs of dealers in hardy bulbs. But enthusiasts, especially rock gardeners, also concern themselves with the wild species and a few of their varieties and simple hybrids that possess daintiness and charm distinct from those of their larger relatives.

Narcissusses are classified in these divisions and subdivisions:

Division 1, Trumpet Narcissuses of Garden Origin. Stalks with only one flower, the trumpet (corona) of which equals or exceeds the length of the perianth segments (petals).

a. Perianth and trumpet colored, the latter at least as deeply as the former
b. Perianth white, trumpet colored
c. Perianth white, trumpet white, not lighter than the perianth
d. Flowers of any other combination of colors

Division 1 (Trumpet): (a) Unidentified variety

(b) 'Mount Hood'

(c) 'W. P. Milner'

Division 2, Long-cupped Narcissuses of Garden Origin. Stalks with only one flower, the trumpet (corona) of which is more than one-third as long, but does not equal the length of the perianth segments (petals).

a. Perianth and trumpet colored, the latter at least as deeply as the former
b. Perianth white, trumpet colored
c. Perianth white, trumpet white, not lighter than the perianth
d. Flowers of any other combination of colors

Division 3, Short-cupped Narcissuses of Garden Origin. Stalks with only one flower, the cup (corona) of which is one-

Division 2 (Long-cupped): (a) 'Fortune'

(b) 'Jaguar'

(c) 'Niagara'

(d) 'Polindra'

Division 3 (Small-cupped): Unidentified variety

Division 4 (Double-flowered): (a) 'Cheerfulness'

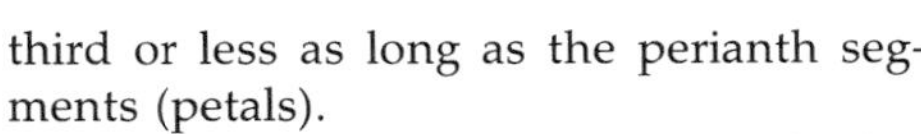

third or less as long as the perianth segments (petals).

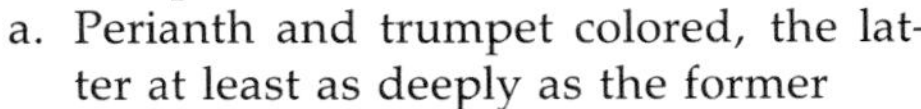

a. Perianth and trumpet colored, the latter at least as deeply as the former
b. Perianth white, trumpet colored
c. Perianth white, trumpet white, not lighter than the perianth
d. Flowers of any other combination of colors

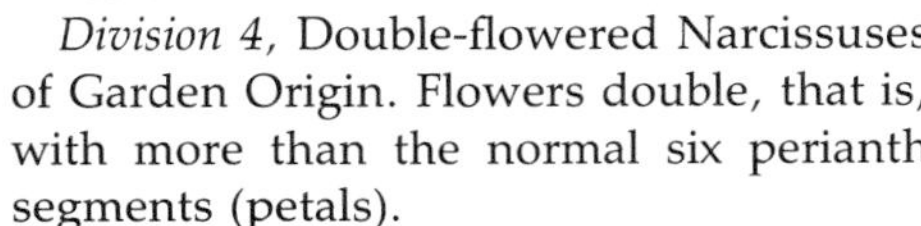

Division 4, Double-flowered Narcissuses of Garden Origin. Flowers double, that is, with more than the normal six perianth segments (petals).

Division 5, Triandrus Narcissuses of Garden Origin. Specific characteristics of *N. triandrus* obviously dominant, stalks with one or more generally white flowers.

(b) 'Golden Ducat'

(c) 'White Lion'

Division 5 (Triandrus): 'Sidhe'

a. Cup (corona) at least two-thirds as long as the perianth segments (petals)
b. Cup (corona) less than two-thirds as long as the perianth segments (petals)

Division 6, Cyclamineus Narcissuses of Garden Origin. Specific characteristics of *N. cyclamineus* obviously dominant, stalks with one flower, the narrow trumpet (corona) of which is as long as the usually

Division 6 (Cyclamineus): 'February Gold'

Division 7 (Jonquilla): 'Trevithian'

backward-pointing, yellow perianth segments (petals).

a. Cup (corona) at least two-thirds as long as the perianth segments (petals)

b. Cup (corona) less than two-thirds as long as the perianth segments (petals)

Division 7, Jonquilla Narcissuses of Garden Origin. Specific characteristics of *N. jonquilla* dominant, stalks with usually two or more scented, yellow blooms with cups (coronas) less than one-half as long as the perianth segments (petals).

a. Cup (corona) at least two-thirds as long as the perianth segments (petals)

b. Cup (corona) less than two-thirds as long as the perianth segments (petals)

Division 8, Tazetta Narcissuses of Garden Origin. Specific characteristics of *N. tazetta* clearly evident, stalks with two to six fragrant blooms, their perianths (petals) usually white and much longer than the pale yellow or white cup (corona).

Division 9, Poeticus Narcissuses of Garden Origin. Specific characteristics of *N. poeticus* without evident admixture of

Division 8 (Tazetta): 'Geranium'

Division 9 (Poeticus): 'Mega'

other types. Flowers fragrant, mostly one on a stalk, with white perianth segments (petals) and a very short, red-rimmed, yellow cup (corona).

Division 10, Species and Naturally Occurring (Wild) Forms and Hybrids. All narcissuses that occur wild (i.e., natively) somewhere.

Division 11, Split-corona Narcissuses. Here belong all narcissuses that have the corona of the flower split for at least one-third of its length.

Division 12, Miscellaneous Narcissuses. All sorts not classifiable in any of the previous eleven divisions.

Species narcissuses, sorts that exist in the wild, often exhibit considerable variation and many deviations from what may be considered norms are recognized as subspecies or varieties. We shall now consider the most important of these species and their variants.

The common daffodil (***N. pseudonarcissus***), largest-bloomed of *Narcissus* species, is ancestor of the large-trumpet horticultural varieties of daffodils. Originally a native of Spain and southern France, it now is naturalized in one or more of its many forms through most of Europe. It has flat, bluntish or short-pointed, bluish-green leaves 6 inches to 1 foot long or longer, ½ to ¾ inch wide. On distinctly two-edged stalks abut equaling the leaves in length, the solitary, drooping, horizontal, or upangled, large-trumpeted blooms, 1½ to 2 inches long, have petals and trumpets of the same or different shades of yellow. The trumpets are usually not spreading or lobed at their margins. Several natural variants have been given varietal names.

Other species, some tall, some dwarf, with large-trumpeted flowers, closely related to the common daffodil and like it with one flower on each stalk, are by some authorities treated as subspecies of it, by others, as is done here, treated as separate species. Among such sorts are these: ***N. alpestris,*** of the Pyrenees, 4 to 6 inches in height, has on scarcely flattened stalks nodding, pure white blooms with narrow, twisted petals that instead of spreading widely angle forward around the slender trumpet. ***N. asturiensis,*** of the mountains of northern Portugal and Spain, is the smallest cultivated Narcissus species. An almost perfect replica in miniature of *N. pseudonarcissus,* it has two or three glaucous leaves 2 to 4 inches long, up to ¼ inch wide, and on stalks as long as the leaves, hanging golden yellow flowers ¾ to 1 inch long, their trumpets contracted at their middles, spreading at their margins. ***N. hispanicus,*** of Spain, has flat, erect, more or less twisted, glaucous

Narcissus asturiensis

leaves 1 foot to 1½ feet tall, by up to ½ inch wide. Its rich golden-yellow flowers, 2 to 2¾ inches long, have a large trumpet with a spreading, deeply-round-toothed margin. ***N. longispathus,*** also of Spain, has erect, glaucous leaves approximately 1 foot long and up to ½ inch wide. The yellow, somewhat up-pointed flowers, usually one, but occasionally two on each stalk, are 1½ to 1¾ inches long. ***N. minor,*** of Europe, is similar to but larger than *N. asturiensis* with which it is often confused.

Narcissus minor

Approximately 6 inches tall, *N. minor* has glaucous, ¼-inch-wide, slightly-channeled leaves and horizontal or nodding flowers 1¼ to 1½ inches long with strongly pleated trumpets not contracted at their middles and deeper yellow than the wavy petals. *N. m. minimus* (syn. *N. minimus*), dwarfer than *N. minor*, has spreading leaves, and flowers scarcely 1¼ inches in

Narcissus minor minimus

length. ***N. moschatus,*** in gardens often misnamed *N. cernuus,* which name properly belongs to a variety of *N. triandrus,* is questionably a native of the Pyrenees. From 8 inches to a little over 1 foot tall, it has erect, somewhat twisted, glaucous leaves ¼ to ½ inch wide, and uniformly pale sulfur-yellow to creamy-white, nodding flowers 1½ to 2 inches long with twisted petals shorter than the trumpets and drooping over them. ***N. nobilis,*** an inhabitant of damp meadows in Portugal and Spain, of moderate height, has more or less horizontal flowers 2 to 2½ inches long with twisted, pale yellow or cream petals and golden-yellow trumpets. *N. n. leonensis* is taller and has similarly colored flowers almost 3 inches long with trumpets sometimes much expanded toward their mouths. ***N. obvallaris,*** the Tenby daffodil, is of uncertain origin. This has erect, flat, bluish-green leaves approximately 1 foot long by up to ½ inch wide. On stalks as long as the leaves, the blooms 1½ to nearly 2 inches in length, have flat, spreading, golden-yellow petals and trumpets of the same color.

The poets' or pheasant's eye narcissus (***N. poeticus***) is a lovely, variable native of southern Europe, the ancestral stock of many splendid horticultural varieties. It has flat, glaucous-green, bluntish leaves up to ⅜ inch wide and flower stalks about as long as the foliage with one sweetly-scented bloom 1½ to 2 inches wide held horizontally or facing upward. The flowers have wide-spreading, broad-obovate, white petals with overlapping bases and a short, flat, yellow cup or corona with a more or less frilled, red margin. The stamens are clearly of unequal lengths and only three protrude. Popular *N. p. ornatus,* which blooms earlier than the typical species, has a red-edged, yellow corona. Very like typical *N. poeticus,* but differing in its flowers having petals not overlapping at their bases and stamens of almost equal length that all protrude, *N. p. radiiflorus* (syn. *N. radiiflorus*) has sweetly fragrant flowers 2¼ to 2¾ inches wide with slightly greenish-white petals and a cup-shaped, red-margined bright yellow corona. The blooms of *N. p. recurvus* (syn. *N. recurvus*), 2 to 2¾ inches wide, have a cup-shaped corona with red margins and greenish centers. The petals, with incurved margins, are reflexed. The upper parts of the ½-inch-wide, glaucous leaves bend downward.

Narcissus poeticus variety at Brooklyn Botanic Garden

Jonquil, a name most properly reserved for graceful ***N. jonquilla,*** in some parts of America is loosely applied to all daffodils. Here it is more appropriately used in its restricted sense. Highly fragrant, the jonquil, a native of Spain and Portugal, is naturalized in France, Italy, and North Africa. It has slender, cylindrical or semicylindrical leaves channeled down their faces (its old name jonquilla derives from the same Latin source as that of the botanical name *Juncus* for rushes) and slender flower

Narcissus jonquilla

stalks up to 1 foot tall topped by clusters of six or fewer small, starry, horizontal or slightly nodding, golden-yellow flowers 1 inch to 1¼ inches in diameter and with round-toothed, cup-shaped coronas less than one-half as deep as the petals are long. Much dwarfer *J. j. minor* has ½-inch-wide flowers.

The campernelle or campernelle jonquil (**N. odorus** syns. *N. campernellii, N. calathinus*), a hybrid of *N. jonquilla* and *N. pseudonarcissus,* has never been found truly wild. Known in cultivation since the sixteenth century, it differs most obviously from the true jonquil in having broader subcylindrical rather than cylindrical leaves channeled down their faces and in its sweetly scented, approximately 2-inch-wide flowers having less slender, ¾-inch-long perianth tubes and much deeper cups.

Species allied to or generally resembling *N. jonquilla* are *N. calcicola, N. juncifolius, N. rupicola, N. scaberulus,* and *N. watieri.* Spanish ***N. calcicola,*** 4 to 6 inches tall, has atop each stalk an umbel of several deep yellow blooms with the aspect of those of the poets' narcissus, but smaller. A native of Spain, Portugal, and southern France, ***N. juncifolius*** (syn. *N. requienii*) has cylindrical, rushlike leaves. Very slender and usually not over 6 inches long, they are somewhat exceeded by the also slender flower stalks each with four or fewer about ½-inch-wide, bright yellow, fragrant blooms carried horizontally or facing upward. They have ½-inch-long perianth tubes and crinkle-edged cups approximately one-half as deep as the lengths of the ovate petals. Much like *N. juncifolius* but with leaves more or less triangular in section and each flowering stalk carrying only one highly-scented, ¾-inch-wide, yellow bloom, ***N. rupicola*** is native to Spain and Portugal. Rich orange-yellow blooms about ¾ inch wide are borne singly or in pairs atop short stalks by ***N. scaberulus,*** a rare, not very hardy native of Portugal. The leaves of this are thin, twisted, and more or less prostrate. Native to North Africa, ***N. watieri*** has blunt, narrow-linear leaves and stalks 4 to 6 inches tall with solitary, scentless, pure white or rarely pink-flushed blooms, 1 inch to 2 inches wide with perianth tubes about ½ inch long.

Winter-blooming ***N. tazetta*** (syn. *N. canaliculatus*), of the Mediterranean region and with one variety in the Far East, is not hardy in the north. It has four to six slightly glaucous leaves ⅜ to ¾ inch wide and approximating in length the 1- to 1½-foot-long stalks that carry clusters of flowers with short lemon-yellow to orange coronas of uniform texture, with neither membranous nor red margins. Its best known variety, *N. t.* 'Paper White', is a favorite for forcing into early bloom indoors. This has clusters of 1- to 1¾-inch-wide flowers with pure white petals and coronas, the latter not edged with red. Variety *N. t. polyanthos,* of southern France, has clusters of ten to twenty flowers, their corollas with lobes (petals) are shorter than the tube. Native to China and Japan, *N. t. chinensis* has white flowers with a yellow corona. Possibly a hybrid between *N. tazetta* and *N. incomparabilis,* the Chinese sacred-lily (*N. t. orientalis*) has clusters of few to many flowers with white or creamy-white petals and yellow, cup-shaped coronas.

Narcissus tazetta

Slightly later-flowering than *N. tazetta* and its varieties, ***N. aureus,*** of southern France, Italy, and the Balkan Peninsula, has green leaves and clusters of ten to twelve blooms 1 inch to 1¼ inches wide with yellow petals and slightly deeper yellow coronas. From it ***N. cupularis,*** of southern France, Italy, and Sardinia, differs in having glaucous foliage and yellow-petaled flowers with deep orange-yellow coronas. Winter-blooming ***N. bertolonii,*** of Italy, much like the last, has somewhat glaucous leaves and flowers almost 1½ inches across with bright yellow petals and coronas. Variety *N. b. algericus,* of Algeria, differs in its petals being bright yel-

Narcissus calcicola

Narcissus tazetta 'Paper White'

Narcissus tazetta orientalis

Narcissus 'Grand Soleil d'Or'

Narcissus triandrus albus

Narcissus bulbocodium

low, the coronas orange-yellow. In all probability the golden-yellow-flowered horticultural variety 'Grand Soleil d'Or' is a derivative of *N. bertolonii* or its variety.

The angel's tears narcissus (***N. triandrus***) is a very lovely variable native of the Pyrenees. From 6 inches to 1 foot high or sometimes taller, it has two to four rush-like, cylindrical leaves channeled down their faces, and slender stalks carrying one to six nodding, 1-inch-long blooms. The flowers have pure or nearly pure white petals and coronas, the former lanceolate, reflexed, and about twice as long as the latter, which are cup- or vase-shaped. Three of the stamens are longer than the others. Variety *N. t. cernuus*, of Spain, Portugal, and Glenans Islands in the Bay of Biscay, have sulfur-yellow or yellowish flowers. Horticultural selections with identifying names include *N. t. albus*, with white flowers; *N. t. loiseleurii*, with larger, creamy-white blooms and leaves that curl toward the ground; *N. t. concolor*, with rather small, light golden-yellow blooms; and *N. t. aurantiacus*, with deeper golden yellow blooms.

The hoop petticoat narcissus (***N. bulbocodium***), so delightfully and appropriately named, is a distinct, charming, highly variable sort. Its native range includes nearly all of Spain and Portugal, a small adjacent part of France, and northwest Africa. Blooming in early spring, this has three or four slender leaves nearly round in cross section, channeled along their upper surfaces, and 4 inches to 1 foot long or longer. On stalks shorter than the leaves, the solitary, asymmetrical flowers, 1 inch to 1½ inches long, have very slender, spreading petals, a large bell-shaped trumpet as long or longer than the petals and toothed or not, and conspicuously curved stamens. Among its many varieties one of the most frequently cultivated is *N. b. conspicuus*, the golden-yellow flowers of which have especially large trumpets and protruding styles. From the last, *N. b. citrinus* is distinguishable by its lemon-yellow blooms 1½ to 2 inches long. Scarcely exceeding 4 inches in height, *N. b. nivalis* has yellow flowers that open very early.

Narcissus bulbocodium conspicuus

Closely related to *N. bulbocodium* and sometimes treated as subspecies or varieties of it are these: ***N. cantabricus*** (syn. *N. clusii*), of Spain, has prostrate leaves and pure white or greenish-white flowers. Variety *N. c. foliosus* has three to eight leaves and white flowers 1½ to 2¼ inches long. It is a native of Morocco. Variety *N. c. petunioides*, probably a native of Algeria, has flowers with wide-spreading coronas giving somewhat the effect of petunia flowers. *N. c. monophyllus*, of Spain and Algeria, as its name suggests, has usually only one leaf; it is not more than 1/25 inch wide. The blooms, resembling those of *N. bulbocodium*, are pure white and have widely open coronas. ***N. obesus*** (syn. *N. bulbocodium obesus*), of Portugal and Morocco, has narrow, prostrate leaves and golden-yellow blooms with very large coronas usually constricted at their throats and of almost globular aspect. ***N. romieuxii***, of North Africa, has large, pale yellow blooms, which in mild climates appear in December. This is much more tender to cold than most sorts of the *N. bulbocodium* alliance.

The cyclamen-flowered narcissus (***N. cyclamineus***) is a very distinct endemic of damp meadows in the northwestern

Narcissus cyclamineus

coastal parts of Portugal and Spain. It has flat, keeled, slender-linear leaves 4 to 8 inches long and, appearing even earlier than those of the hoop petticoat narcissus, solitary, rich, golden-yellow flowers of such unusual appearance that they prompted the imaginative English authority on rock gardens Reginald Farrer to liken them to the head of a hare with the ears laid back as in flight. In truth they do bear some such resemblance or perhaps they would be more appropriately compared to a horse with its ears set back in anger. Whether or not these similes are accepted, the blooms are choice. The toothed trumpet, tubular and about 1 inch long by under ½ inch wide, is nearly equaled in length by the backward-streaming petals, which represent the ears of the hare or horse.

Three autumn-blooming species are cultivated to some extent by conoisseurs of the rare and unusual. None is hardy in the north. One, *N. viridiflorus*, a native of Gibraltar and Morocco, is allied to *N. jonquilla*. Another, *N. elegans* (syn. *N. autumnalis*), which is related to *N. tazetta*, occurs in Italy, Sicily, Sardinia, Corsica, and along the coast of North Africa. The third, *N. serotinus*, which forms a botanically distinct section of the genus, is distributed as a wildling along almost the entire lengths of both shores of the Mediterranean Sea and along the African coast westward of Gibraltar. Blooming in late fall, ***N. viridiflorus*** has one or two subcylindrical leaves. The two to four flowers on stalks 1 foot to 1½ feet long have narrow, olive-green, lanceolate petals and a very short, six-lobed, green corona. Blooming earlier than the last, in September or October, ***N. elegans*** has up to four nearly cylindrical leaves flattened in their upper parts and 4 to 6 inches long that come before the flowers. The fragrant blooms, two to seven on each flowering stalk, of starry aspect, and 1½ to 2 inches across, have spreading pointed-lanceolate petals about ½ inch long and a short, dull-orange to yellow, saucer-shaped corona. Leafless when it blooms in September, ***N. serotinus*** waits until spring to produce its foliage, which consists of a few very slender leaves. The

Narcissus viridiflorus

fragrant blooms, solitary or in twos, are up ¾ inch in diameter and have greenish-white petals and extremely short, wide-open coronas, three- or six-lobed to their bases or sometimes more shallowly.

Hybrids, natural and artificial, between species of *Narcissus* are numerous and some have been given names. Mostly they exhibit characteristics intermediate between those of the parents. One such hybrid, the campernelle (*N. odorus*) has been discussed above. Here are some others, with the names of the parents in parentheses: ***N. bakeri*** (*N. bulbocodium* × *N. pseudonarcissus*), is a natural hybrid with slender, linear, somewhat glaucous leaves and solitary, horizontal, 1½-inch-long, yellow flowers with narrow petals somewhat shorter than the trumpet. ***N. dubius*** (*N. juncifolius* × *N. papyraceus*) is a natural hybrid with narrow, glaucous leaves and clusters of a few white flowers ½ to ¾ inch wide. ***N. gracilis*** (*N. poeticus* × *N. jonquilla*), presumably a garden hybrid and one of the latest narcissuses to flower, has highly scented, saffron-yellow blooms resembling those of *N. jonquilla* in clusters of three to five. ***N. incomparabilis*** (*N. hispanicus* × *N. poeticus*) is a garden hybrid, with glaucous, ½-inch-wide leaves and solitary flowers 2 to 3 inches wide with sulfur-yellow or paler petals and a bright yellow corona. ***N. intermedius*** (*N. jonquilla* × *N. tazetta*) is a natural hybrid with bright green, subcylindrical leaves up to ⅓ inch wide and clusters of up to ten 1½-inch-wide flowers with yellow petals and orange-yellow coronas. ***N. johnstonii*** (*N. pseudonarcissus* × *N. triandrus cernuus*) is a natural hybrid with nearly flat leaves ⅓ inch wide and solitary, uniformly pale yellow flowers about 1½ inches long with more or less reflexed petals about as long as the trumpet. ***N. medioluteus*** (syn. *N. poetaz*) (*N. poeticus* × *N. tazetta*) is represented by sev-

Narcissus johnstonii

eral named horticultural varieties. Their leaves are 1 foot to 2 feet long by about ½ inch wide. In clusters of two to six, the fragrant flowers have white or cream perianth tubes about 1 inch long, white, cream, or yellow petals, and a waxy-margined, yellow or orange corona decidedly shorter than the petals. ***N. tenuior*** (*N. jonquilla* × *N. poeticus*) is a natural hybrid similar to *N. gracilis*, but smaller in all its parts.

Garden and Landscape Uses. The numerous, large-flowered horticultural varieties of daffodils and other narcissuses are among the most satisfactory spring-flowering bulb plants, esteemed for garden decoration and for use as cut flowers and for forcing into early bloom indoors. The species and miniature sorts especially interest collectors of the unusual and rock gardeners. The great majority of kinds are satisfactorily hardy at least in climates not harsher than that of New England, a few, notably *N. tazetta* and related species and their varieties, such as 'Paper White', 'Grand Soleil d'Or', and the Chinese sacred-lily, succeed outdoors only in mild climates.

Narcissuses, except very tiny sorts, are undoubtedly at their best when naturalized, planted informally in cloudlike drifts in meadowland or very thin woodland where the soil is deep and nourishing, and neither too wet nor too dry, and the lay of the land or presence of trees, shrubs, hedges, large rocks, cliffs, or other impedimenta afford shelter from strong, sweeping winds. In such places for all practical purposes if given reasonable care they will persist indefinitely.

But narcissuses can be employed in rather more formal surroundings to great advantage. Groups of twelve to three times as many in perennial and mixed borders or tucked in spaces at the fronts of beds and foundation plantings of deciduous or evergreen shrubs, perhaps among English ivy or other groundcover, can be delightful. They are less well adapted to strictly formal uses and look uncomfortable in serried rows in beds of geometric shapes after the fashion in which hyacinths and tulips are so often effectively used. Nor do they appear at their best strung in ribbons in narrow beds bordering paths, fences, or building foundations. Dwarf and miniature narcissuses are elegant adornments for rock gardens and alpine greenhouses.

Besides their uses outdoors, narcissuses are among the most satisfactory spring-flowering bulbs for forcing into early bloom in greenhouses and window gardens. The earliest varieties 'Paper White',

Narcissuses naturalized in a meadow: (a) In full sun

(b) Along a woodland path beside a quiet stream

(b) In light shade

(c) At a lakeside

Narcissuses naturalized in light shade: (a) Along a woodland path featuring a rock outcrop

(d) On a steep bank

Narcissuses naturalized under trees

Planted in small groups narcissuses can be very effective: (a) A long-cupped variety

(b) A poeticus variety

(c) A double-flowered variety

'Grand Soleil d'Or', and the sort called Chinese sacred-lily, of the *N. tazetta* relationship, can be had in flower by late November and from then until late winter, early varieties of some others can be flowered by Christmas or shortly thereafter.

Cultivation. Because narcissuses survive and even bloom regularly and fairly well without much attention does not mean they are unresponsive to care. Far from that, they repay intelligent management handsomely with improved quality as well as greater quantities of flowers and in more rapid multiplication of the bulbs, which makes possible the extension of plantings without purchases.

Clumps of dwarf narcissuses are attractive in rock gardens: (a) *Narcissus cyclamineus*

(b) *Narcissus bulbocodium* variety

The ideal soil is deep, fertile, crumbly loam that contains enough clay to give it body and assure reasonable water retention, but not so much that it is sticky and nearly impervious to air and water. But soils widely different from the ideal will support, or without too much trouble can be brought into condition to support, satisfactory growth. Very good results are possible on sandy soils, loamy soils, and all but the very stiffest, almost unworkable clays. Some of the tiny rock garden sorts appreciate earth of a sandy peaty character.

Subsurface drainage should be such that the soil in summer is not abnormally wet, certainly not more so than suits the general run of vegetable and flower garden plantings, and drier conditions are advantageous. But from early fall until the foliage dies in late spring, generous moisture without actual waterlogging promotes the finest results. Lack of enough soil moisture is undoubtedly a factor responsible for the flower buds of some late blooming narcissuses, notably those of double-flowered poeticus sorts, blasting (drying and failing to open). All poeticus varieties thrive in soil moist or even quite wet in winter and spring and even stand considerable dampness through the summer. They are among the best sorts for waterside and low-lying areas. Where underground drainage is inadequate, good results are usually possible by planting in beds raised 6 to 8 inches above the normal ground level.

In preparation for planting, unless of course you are setting bulbs in a meadow, lightly wooded area, or other site where the earth is already agreeable to them, turn the ground over to a depth of 8 inches or more with a spade, spading fork, rotary tiller, or plow and incorporate generous amounts of compost or other soil-improving organic matter, except manure, unless this last is so thoroughly decayed that it is little more than humus. A dressing of bonemeal, of superphosphate, or of a fertilizer especially compounded for bulbs well mixed with the soil before

Planting narcissus bulbs in a meadow

(b) Forking the ground deeply

Planting narcissuses: (a) Spreading fertilizer

(c) Planting the bulbs with a trowel

planting is generally advisable. If you bring your ground into a condition that would suit most vegetables and flower garden plants it will suit narcissuses.

When is best to plant purchased bulbs is easily settled. It is as soon as they are obtainable from the supplier, ordinarily in September or October. The earlier they are put into the ground the sooner they begin to root and the more satisfactory will be their spring performance. However, if absolutely necessary planting can be done as late as November or under exceptional circumstances into December with assurance of flowers in spring. Bulbs dug from the garden in late spring and stored will be available earlier and may with excellent prospects of success be planted in August.

Spacing calls for some judgment. Unlike hyacinths and tulips, narcissuses are not seen at their best in even rows or other patterned plantings, or in rectangular, circular, or similar geometrically-shaped groups. Rather, have them sprouting from the ground randomly. Set the bulbs in clusters or drifts with the individuals spaced irregularly and with a few outliers representing chance seedlings slightly further away from the main bodies of the colonies. Within the groups average distances between large bulbs of vigorous growing sorts may be 6 to 8 inches, 4 to 5 inches for medium kinds, and 1½ to 3 inches for small ones.

Depth to plant should relate to size of bulbs and this varies greatly with type and variety. Set big ones such as those of most tall, large-trumpeted daffodils with their shoulders (it is better to measure from there than from the tips of the bulbs because the long necks of some sorts may deceive one into planting too shallowly) 4 to 5 inches beneath the surface, medium-sized bulbs 3 to 4 inches, and small ones 2 to 3 inches. These measurements are those made after the earth has settled and compacted, not when it is newly spaded and loose. All may with advantage be planted deeper in sandy soils and even in heavier ones in good tilth, always provided there is a reasonable depth of fertile soil beneath the bulbs. A common mistake of inexperienced gardeners is planting the bulbs too shallowly.

Immediately following planting, if the ground has been much compacted by treading (which should be avoided, if possible), loosen its upper couple of inches with a fork to prevent hard spots remaining that may obstruct the emergence of the shoots in spring. If the earth is dry or dryish, soak it thoroughly as soon as planting is done. This encourages prompt root development. Although not necessary if the bulbs are set out early, some gardeners mulch newly planted narcissuses, unless of course they are in grass, after the ground has frozen to a depth of an inch or two. This practice is advantageous with late-planted bulbs because it maintains soil temperatures favorable to rooting a little longer. Remove mulches, with care taken not to damage any sprouts that may be showing, in late winter or very early spring.

Routine care of established plantings involves the application in spring as soon as the shoots are safely above ground of a complete fertilizer with a fairly high content of available nitrogen or of bonemeal supplemented with a dressing of nitrate of soda, sulfate of ammonia, or urea. At the same time, in plantings other than those in turf or groundcovers, carefully stir the upper inch of the soil with a cultivator or hoe to incorporate the fertilizer and facilitate the absorption of rain. A second and similar fertilizer application made as soon as the flowers fade is generally advantageous, although not necessary if the ground is quite fertile.

At flowering time and after, two important considerations present themselves, both related to the fact that the prosperity of the bulbs and next year's bloom depend upon the retention of as much foliage as possible for as long a time as practicable. Therefore, when cutting flowers take no more leaves than actually needed and be careful not to take more than a small proportion of those of any individual bulb, and be sure to allow the foliage that remains to die naturally and completely before removing it.

To avoid untidiness: (a) Tie fading narcissus foliage into small bundles

Temporary untidiness resulting from maturing, yellowing foliage can be much relieved by gathering together the leaves of one or of a few neighbor bulbs and loosely plaiting them and tying their tops in a loose knot. If for any reason bulbs must be dug before their foliage has completely withered, lift them with as much soil adhering to their roots as possible, heel them in (plant them shallowly closely together) in a shaded place, and water them well.

Fall attention is rarely given established plantings of narcissuses yet a little care at that time pays rich dividends. This is especially true if rainfall has been deficient. Then one or two deep soakings in late Au-

(b) Alternatively, knot it loosely together

To transplant narcissuses: (a) Dig the bulbs before the foliage has completely died

(b) Clean them of soil and debris

gust or early September does much to encourage the active growth of roots and if the soil is tending to become exhausted of nutrients, as is often true of old plantings, an application of fertilizer rich in potash and phosphate, but comparatively low in nitrogen, can work small wonders. A combination of superphosphate or fine bonemeal and unleached wood ashes works well.

Lifting and replanting the bulbs needs attention from time to time, the frequency depending upon a number of variables. Except when it becomes needful to rearrange plantings or to multiply as expeditiously as possible a choice or otherwise desirable sort, the task need not be done until because of overcrowding and possible soil deterioration the flowers are noticeably fewer and of inferior quality. Commercial growers of bulbs for market usually lift and replant every year.

The best time to lift is in late spring when the foliage has almost completely ripened, but not to the extent that when pulled it separates readily from the bulbs. Use a spading fork in such a way that as few as possible of the bulbs are pierced. This is best assured by carefully digging away the surface soil from one side of the clump and up to within an inch or two of it before driving the fork a few inches further away down and under the bulbs to pry them out.

Lifted bulbs may be cleaned off, sorted to size, dried, and stored in a well ventilated, cool, dry place until fall, but unless a matter of convenience there is no advantage in this. Immediate replanting in ground newly spaded and adequately prepared by incorporating organic matter and fertilizer, brings excellent results.

Forcing narcissuses into early bloom indoors is a very simple horticultural exercise that can be successfully achieved in homes as well as in greenhouses. By spacing plantings and times of introducing the bulbs to congenial warmth a succession of flowers may be had from November until March or April.

(c) Sort them to size

Earliest to bloom are the varieties 'Paper White', 'Grand Soleil d'Or', and the

Paper white narcissus forced in pebbles and water: (a) Plant the bulbs closely together with their tops well above the pebbles

(b) Fill the container with water

(c) In a temperature of 50 to 55°F, growth soon begins

Chinese sacred-lily, all belonging to the *N. tazetta* assemblage of species and none hardy in the north. Because of their susceptibility to damage by low temperatures these are managed for forcing differently from the bulbs of hardy kinds. They may be grown in bowls containing pebbles and water or moist peat moss mixed with fine charcoal or with charcoal and perlite or in pots of soil. If they are to be planted outdoors after flowering (practicable only in mild climates) use soil, otherwise the medium matters not because the bulbs will be of no further use once blooming is through.

To force hardy narcissuses, secure good quality bulbs as soon as they are obtainable in fall and plant immediately in adequately drained pots, pans, or 4-inch deep flats in fertile, porous soil. What are known as double-nosed, as opposed to single-nosed bulbs are often the best buy because each gives two or more blooms and on the basis of number of flowers obtained are usually cheaper than single noses. Except with very small bulbs, such as those of dwarfs and miniatures occasionally cultivated in alpine greenhouses or sometimes forced for rock garden displays at spring flower shows, it will usu-

(d) In four to six weeks, the plants are in full bloom

Narcissus 'Paper White' forced into early bloom in a mixture of peat moss and perlite

Muscari armeniacum 'Heavenly Blue' and *Aurinia saxatilis*

Musa textilis

Mussaenda, undetermined species

Muscari armeniacum

Chrysanthemums mulched with black plastic

Peppers mulched with straw

Nandina domestica, a miniature variety, in fall

Myosotis sylvatica variety

Myosotis sylvatica variety as a pot plant

Myosotis sylvatica variety with pansies

Narcissuses forced for cut flowers: (a) Plant double-nosed bulbs in a flat

(b) A flat in full bloom

ally be desirable to have a portion of the bulb above the soil surface, but it is better that this be not more than one-third. Set the bulbs so that they almost touch each other. The number accommodated in any particular container will depend not only on its dimensions, but also on the size of the bulbs. Three or four of the largest in a 6-inch pot or five to seven in 7- or 8-inch containers will usually be enough.

Firm the soil around the bulbs and level it, leaving room between its surface and the rim for watering. Stand the planted containers in a well-drained site outdoors or in a cold frame, soak them very thoroughly with water, and then cover with 6 inches or so of sand, peat moss, sawdust, or similar material.

After six to eight weeks examine for progress. Good rooting by then should be

Narcissuses forced as pot plants: (a) After planting, stand the pots closely together outdoors or in a cold frame

(b) Cover them with 6 to 8 inches of coarse sand or similar material

(c) After the bulbs are well rooted and have developed short new shoots, move them to where they are to be forced, shading them from direct sun at first

(d) After the young shoots become green, expose to full sun

(e) Full sun and fairly cool temperatures assure sturdy foliage

(f) A well-grown pot of bulbs in full bloom

achieved. If plump shoots an inch or two high are showing, and most certainly before their tips begin to open into separate leaves, remove the pots or flats to a cold frame or greenhouse where they will be cool (a temperature of 40°F suits), but not subjected to frost or to freezing temperatures. For the first week or so shade from strong sun, then expose fully. Keep the soil well watered and to promote the finest blooms give an occasional application of dilute liquid fertilizer. To ensure a succession of flowers over a long period, transfer batches of pots or flats at ten-day or two-week intervals to temperatures of 55 to 60°F.

Bulbs that have been forced are not suitable for forcing the next year, but if little or none of their foliage is cut and proper care is given after flowering they may be planted outdoors, there to recover and gain strength and bloom satisfactorily in future years.

Appropriate care consists of keeping them watered and maintaining them in a light, fairly cool, frost-free place until the foliage dies naturally, then either planting them immediately outdoors or storing the bulbs for fall planting outdoors. Applications of dilute liquid fertilizer given during the period when the foliage is green improves the quality of the bulbs.

Propagation of narcissuses except by breeders of new varieties and in the reproduction of some species is almost exclusively by naturally produced offsets removed when the bulbs are lifted for replanting. A more rapid means of achieving increase of a choice kind is by the little known technique of bulb cuttings. From seeds the plants mostly take five or six years to reach flowering size. The seeds may be sown as soon as they are ripe in sandy peaty soil in a cold frame or cool greenhouse.

Pests and Diseases. In the main narcissuses are extraordinarily free of pests and diseases, but not immune. The most frequently encountered pests are the large and lesser bulb flies the larvae of which bore into and destroy the bulbs. The large bulb fly, which resembles a small bumble bee, lays eggs near the bases of the leaves and necks of the bulbs about the time the foliage is dying. The grubs, usually only one in each bulb, grow to ½ to ¾ inch long. The lesser bulb fly more resembles a housefly. It lays its eggs in the manner of the larger kind, but several to many of its ½-inch- or less-long grubs are usually found in each infested bulb. Other sometimes troublesome pests are bulb mites, millipedes, and nematodes.

Diseases that may occur include crown rot, basal rot, root rot, and fire disease, the last causing spotting of the flowers and premature dying of the foliage. Leaf spot and leaf scorch diseases sometimes affect narcissuses, and they are also susceptible to virus infections called yellow stripe, flower streak, white streak, and mosaic.

If troubles are suspected send samples to a Cooperative Extension Agent, State Agricultural Experiment Station, or other authority for diagnosis and recommendations, and follow the advice given. When lifting bulbs discard and destroy any that are suspiciously soft or mushy.

NARDOSTACHYS (Nardo-stáchys). Native to the Himalayas and China, *Nardostachys* of the valerian family VALERIANACEAE, consists of three species. Its name comes from *nardos,* the ancient Greek name of a fragrant oil derived from *N. jatamansi,* and *stachys,* a spike, in reference to the arrangement of the flowers.

These are hardy herbaceous perennials with fragrant, woody roots and undivided, mostly basal leaves supplemented by one to a few pairs of opposite stem leaves. The tiny red to purplish flowers in headlike clusters have a five-lobed, persistent calyx, five-lobed corolla, four stamens, and one slender style. The fruits are slender achenes.

The spikenard of the ancients and the source of an essential oil, Himalayan ***N. jatamansi*** is 4 inches to 1 foot tall or sometimes taller. It has oblong-lanceolate to spatula-shaped leaves 2 to 4 inches long and crowded heads of ¼-inch-wide, rose-purple flowers.

Garden Uses and Cultivation. The species described is appropriate in herb gardens and rock gardens. It prospers in sunny locations in ordinary, well-drained soil and is easily propagated by seed.

NARTHECIUM (Nar-thècium) — Bog-Asphodel. The species of *Narthecium* are so similar that botanists differ in their evaluations of how many should be recognized. From four to eight are usually accepted. One occurs in western North America, one to three in eastern North America, one in Japan, and the others in Europe. The name is an anagram of that of the genus, *Anthericum,* which, like *Narthecium,* belongs in the lily family LILIACEAE.

Nartheciums are hardy herbaceous perennials with stout rhizomes, grasslike basal leaves, and a few reduced, bractlike leaves on the flower stems. Their greenish-yellow to yellow flowers are in racemes terminating erect stalks. Each bloom has six petals (properly tepals) that wither and enclose the seed pods. There are six stamens with conspicuously woolly-hairy stalks and a short style that more or less merges into the ovary and is topped with a slightly three-lobed stigma. The fruits are capsules.

A rare native of wet pine barrens from New Jersey to South Carolina, ***N. americanum*** has upright, slender-linear, basal leaves 4 to 8 inches long. The flower stalks rise to a height of 1 foot to 1½ feet. The dense racemes of yellowish blooms are 1½ to 2 inches long, the petals up to ¼ inch long. The kind differentiated as ***N. montanum,*** a native of swampy places in the mountains of North Carolina, differs in having looser racemes of slightly larger blooms with the flower stalks shorter than the bracts in whose axils they originate. The western American ***N. californicum*** has basal leaves up to 1 foot long and flower stalks up to 1½ feet tall. Its racemes of yellow blooms are 3 to 6 inches long. This inhabits wet meadows and banks above a 2,500-foot altitude in California.

Japanese ***N. asiaticum*** which favors wet soils, attains heights of 1 foot to 2 feet and has linear leaves up to 10 inches long by ⅓ inch broad. Its flowers are in racemes 3 to 5 inches long. They have yellow perianth segments, and stamens with white-hairy stalks and pale yellow anthers.

Garden Uses and Cultivation. Occasionally cultivated in bog gardens and native plant gardens, bog-asphodels must be provided with wet, acid, peaty soil. They are propagated by seed and by division.

NASTURTIUM (Nastúr-tium)—Watercress. The gaily colored popular garden flowers commonly called nasturtiums do not belong here, but in the genus *Tropaeolum,* under which name they are treated in this Encyclopedia. The *Nasturtium* of botanists is a genus of the mustard family CRUCIFERAE that includes watercress. The name, from the Latin *nasus,* nose, and *torus,* distortion, was originally applied to some pungent species of cress.

Native to Europe, Asia, and North Africa, *Nasturtium* is naturalized in North America. It consists of hairless, aquatic or wet-soil herbaceous perennials with pinnately-divided leaves and small white flowers each with four sepals, four petals that spread to form a cross, four long stamens and two shorter ones, and a single style. The fruits are slender capsules con-

Nasturtium officinale

taining two rows of seeds. Six species are recognized of which ***N. officinale*** (syn. *N. nasturtium-aquaticum*) and a hybrid between it and *N. microphyllum* are widely cultivated as the salad crop watercress. For the cultivation of these see Watercress.

NATAL. This word appears in the common names of these plants: Natal-bottle-brush (*Greyia*), Natal-orange (*Strychnos spinosa*), and Natal-plum (*Carissa grandiflora*).

NATIVE PLANT GARDENS. Areas maintained as gardens devoted exclusively to plants native to particular regions can be delightful, interesting, and educational. It is usual to accept in such developments naturalized as well as strictly native plants, the selections usually being based on kinds admitted in published floras.

Most often native plant gardens are associated with wild life reservations, botanical gardens, public parks, and similar places, but they can be appropriately fitted into home properties. They may be extensive or of quite modest size.

Some gardens of this sort emphasize particular habitats, thus they may be alpine, desert, seashore, or woodland gardens, but in most regions to present effectively a representative cross section of the spontaneous flora calls for a variety of habitats. Where choice is practicable select an area where variations in topography and other details exist. It may for example include woodland, meadow, lake, pond, or stream, high and dryish ground as well as lower and moister places.

Where space is limited it is often necessary to contrive habitats by modifying the soil, planting or removing trees or shrubs to increase or decrease shade or to provide shelter, establish water features, and so on. If such changes are made sympathetically they can be charming.

Naturalistic rather than formal effects should be the aim of developers of native plant gardens. Ecologically the results should be satisfying and convincing. The result should look as though it belonged and should not too obviously betray human interference. But do not let this mislead you into thinking a native plant garden needs little or no care. It is a garden, not a piece of self-sustaining countryside, and as such control must be exercised over the plants allowed to occupy it.

A native plant garden, Santa Monica Botanical Garden, California; *Coreopsis maritima* and *Eschscholzia californica* in bloom

A carpet of *Podophyllum peltatum* in the wild garden of The New York Botanical Garden

As with all gardens, the initial task is to carefully survey the site with the objective of retaining what is good and modifying or removing that which is not. This is likely to involve reducing the number of too-abundant species, removing or transplanting ill-placed ones, and eliminating kinds inadmissable as natives. The introduction of natives not present in the garden area will be needed, a procedure that necessitates careful and sympathetic choice of planting sites.

Maintenance of a native plant garden calls for constant vigilance to make sure that vigorous growers do not overgrow weaker neighbors. There may be need to clear away dead tops or plants that die back each fall and give attention to removal of other trash before it accumulates it too great amounts. Mulching, watering, and other garden procedures but usually not fertilizing, is likely to be desirable for some of the inhabitants of native plant gardens.

Echium in bloom in the wild garden at Rancho Santa Ana Botanic Garden, California

NATURALISTIC GARDENS. The naturalistic style of landscape design is in sharp contrast to the strictly formal, which emphasizes symmetry and uses geometrical patterns. Naturalistic planning seeks to minimize revelation of the designers' dominance over nature, to give the impression that the garden has been achieved with minimal interference with a natural landscape.

To assure this, the competent designer retains all existing features that can be incorporated to advantage. These may include the contours of the land, trees, shrubs, outcropping rocks, ponds, streams, and the like, as well perhaps as such evidences of human occupancy as old fences, stone walls, rural paths, and fields or meadows.

The presence of features such as these simplifies the task of creating a naturalistic garden, but even where none exist, where

Naturalistic gardens: (a) A pond, with deciduous trees and in the foreground Japanese irises

(b) Trees and shrubs, a lawn with a curving path, and a bed of snapdragons

(c) A tree-shaded path with appropriate underplantings

(d) A path, bordered with rhododendrons and groundcover plants, leads to a rustic pergola

the terrain is flat or otherwise uninteresting, the objective can be attained by careful planning and layout.

The first essential is that the developer have a sympathetic understanding of natural landscapes. Without that, the "naturalistic" garden is likely to look contrived and unconvincing. This does not mean that it is necessary for a garden of this type to be as informal as a wild garden, but that as a result of asymmetrical rather than symmetrical balance it fits with easy grace into the surrounding landscape.

Generally, lawns, beds of shrubs and flowers, ponds, and pools will be of easy free-form shapes that flow with the contours, and trees and other prominent features will be strategically located, but not in regular patterns. Formally trained or sheared shrubbery and severely clipped hedges generally are abjured. Rarely will paths be straight, but meaningless curves and wiggles should be avoided. A shrub, group of shrubs, or rising ground may give reason for a pleasing diversionary curve.

Man-made objects need not be excluded but should be subsidiary to the living landscape rather than dominant. Where appropriate, steps of casual design may be introduced, a suitable seat may invite in a shaded spot, a bird bath, sundial, or simple piece of statuary may be featured.

By planning so that the entire garden is not in view at once, the element of surprise is introduced and this can be a very valuable asset in a naturalistic garden.

NATURALIZED. In botanical usage naturalized refers to plants other than original natives that have become so thoroughly established that they maintain and reproduce themselves as elements of regional floras. Examples are the common dandelion, a native of Europe, now naturalized in nearly all temperate parts of the world and the tree-of-heaven (*Ailanthus altissima*), a native of China, now established as a naturalized species in New York City and some other urban regions as well as in some rural areas where it competes successfully with native vegetation.

Gardeners employ the word naturalized more broadly to describe plants that have been sown or planted in places where they may be expected to succeed with minimal care and so displayed that they suggest the informality of a natural landscape rather than more geometrical or orderly patterns. In this sense of the term, crocuses may be naturalized in lawns, daffodils in meadows and open woodlands, Japanese irises at watersides, and so on.

NAUPAKA. See Scaevola.

NAUTILOCALYX (Nautílo-calyx). Relationships between this genus, *Episcia,* and *Alloplectus* are somewhat confused. Further study by botanists may result in some rearrangement. As presently understood, *Nautilocalyx* comprises fourteen species of evergreen, herbaceous perennials of the gesneria family GESNERIACEAE, natives of tropical America. The name comes from that of the marine nautilus, and calyx, and alludes to a rather indistinct similarity in form. Technical differences between the two other genera mentioned above and *Nautilocalyx* chiefly relate to the anther cells and their placement.

Nautilocalyxes are erect to spreading, 6 inches to 3 feet tall. They have opposite leaves, those of each pair of approximately equal size; usually short-stalked, they are broadly-elliptic to ovate. The clusters of flowers, from the leaf axils, are commonly accompanied by leafy bracts. The blooms

Nautilocalyx bullatus

Nautilocalyx bullatus (flower)

Nautilocalyx forgetii

Nautilocalyx lynchii

Nautilocalyx melittifolius

have five-parted, large leafy calyxes and tubular corollas narrowed below and with a distinct spur hidden by one of the calyx lobes, at the base. They are whitish or yellow with sometimes purple-red markings in their throats. There are four stamens. The stigma is shortly two-lobed. The fruits are capsules.

Easily recognized by its pebbly-surfaced, wrinkled foliage, ***N. bullatus,*** of Peru, about 2 feet tall, has olive-green or brownish leaves, vinous red on their undersides, exceptionally 1 foot long by one-half as broad, with short, winged stalks. Up to ten pale yellow blooms are in tight clusters in the leaf axils. Another Peruvian of similar height, ***N. forgetii*** has wavy-edged, elliptic leaves with about 2-inch-long wingless stalks. Above they are lustrous green with brown veining, and on their reverse sides wine-purple. Almost white except for their yellow throats, the blooms have reddish calyxes. Colombian ***N. lynchii*** is another tall kind. It is distinguished by its slender-pointed, oblong-lanceolate, dark maroon leaves, with purplish undersides, and short, winged stalks. They are about 6 inches long by 3 inches wide. The not very showy flowers, two or three together, have reddish calyxes and pale yellow corollas.

Often more sprawling than those described above, ***N. villosus,*** of Venezuela, has purplish stems up to 1 foot long, which, like the foliage and flowers, are thickly clothed with white hairs. Beneath the hairy coating the leaves are pale green. They are pointed-ovate and 6 to 7 inches long by about 3 inches wide. About 2 inches long, the white flowers, flecked in their throats with purple, are few together. Native of the West Indies **and**

South America and sometimes producing small tubers, ***N. melittifolius*** (syn. *Episcia melittifolia*) has squarish, spreading or trailing stems that root from the nodes and lustrous green leaves up to 10 inches long with ovate to heart-shaped, round-toothed blades. Its long-stalked, short-tubed, red flowers, in clusters of up to a dozen, are about 1½ inches in diameter and last for only a day or two.

Much newer in cultivation than any of the foregoing, ***N. picturatus*** was first grown in America in the late 1960s. Out of bloom it has much the aspect of a rather narrow-leaved, beautifully colored florists' gloxinia. Native to Peru and Venezuela, it has stems not over 6 inches in height and very closely spaced, short-hairy, pebble-surfaced, olive-green or brownish leaves with a herringbone of yellowish-green veining on their upper sides. Beneath they are vinous purple with green veins. Slightly recurved, they are up to 6 inches long by 3 inches wide and have 1-inch-long stalks. White with purple streaks, the flowers, in small clusters well down in the leaf axils, are 1¼ to nearly 2 inches long and 1¼ inches across their faces.

Garden Uses and Cultivation. These are primarily plants for tropical greenhouses, but are also successfully grown by fanciers of gesneriads who accommodate them in terrariums, with other special facilities, often including fluorescent lighting, that provide approximately as agreeable conditions. They are lovers of warmth and humidity and appreciate as much light as they can take without their foliage being damaged or scorched, but they will get along with less. Shade from strong sun is necessary. Although the flowers of some species are not without interest, the chief decorative feature is the beautiful foliage. Porous, coarse, fertile soil that contains liberal amounts of peat moss or leaf mold, and is kept evenly moist but not saturated, favors good growth, as do minimum temperatures on winter nights of 60 to 65°F, with daytime temperatures somewhat higher. At other seasons the lowest temperature should be 70°F. Fertilizing should be moderate rather than generous. Propagation is easy by cuttings, leaf cuttings, and seed. For further information see Gesneriads.

NAVELWORT. This is a vernacular name of *Umbilicus rupestris* and *Omphalodes.*

NEANTHE BELLA is *Chamaedorea elegans.*

NECKLACE TREE is *Ormosia monosperma.*

NECTARINES. Nectarines are smooth-fruited peaches for convenience botanically distinguished from peaches (*Prunus persica*) as *P. p. nucipersica.* Besides not being covered with a fuzz of hairs, the fruits of nectarines are usually smaller and more richly flavored than peaches. Nectarines sometimes develop from peach seeds and peaches from nectarine seeds without cross-pollination being involved. Each may also arise from the other as mutations or bud sports. For available varieties consult catalogs of nurseries specializing in fruit trees. The cultivation of nectarines is as for peaches. See the Encyclopedia entry Peach.

Nectarine, trained as an espalier

NEEDLE. This word occurs in the common names of these plants: needle flower (*Posoqueria latifolia*), needle palm (*Rhapidophyllum hystrix*), and shepherd's needle (*Scandix pectenveneris*). The slender leaves of many conifers, for example those of pines, are called needles.

NEEDLE CAST. Diseases of conifers, the symptoms of which are conspicuous shedding of the leaves or needles, are called needle casts or needle blights. They are caused by various funguses, some but not all controllable by timely spraying with bordeaux mixture or other fungicides. For advice in particular cases consult Cooperative Extension Agents, State Agricultural Experiment Stations, or other authorities.

NEEM TREE is *Azadirachta indica.*

NEILLIA (Néil-lia). Approximately a dozen species are included in *Neillia,* of the rose family ROSACEAE. They are deciduous shrubs or rarely subshrubs with much the aspect of *Spiraea.* The genus inhabits Asia from the Himalayas to China, and Java and Sumatra. Its name commemorates Patrick Neill, one-time secretary of the Caledonia Horticultural Society of Scotland. He died in 1851.

The closest relatives of *Neillia* are *Physocarpus* and *Stephanandra.* From the first *Neillia* differs in its follicles (seed pods) opening along one side instead of two and in usually having fewer carpels and from *Stephanandra* in the styles being always lateral instead of terminal and the seeds being rough instead of smooth and glossy.

Neillias are good looking, with arching, usually zigzagged branches and alternate, mostly lobed, double-toothed leaves, with stipules (appendages at the bases of the leafstalks) that generally soon fall. The development of these is one difference between *Neillia* and *Spiraea.* The flowers are in racemes or panicles. They have bell-shaped to tubular calyx tubes, five short, erect sepals pubescent on their upper sides, five roundish, white or pink petals sometimes hairy above, and from ten to thirty, but usually not fewer than fifteen, stamens. There are one or two carpels with slender styles. The fruits are follicles containing several lustrous seeds.

The showiest kinds are closely related *N. sinensis,* of central China, and *N. thibetica* and *N. longiracemosa,* of western China. The first is so variable that half a dozen varieties are recognized, the others are more uniform. The blooms of ***N. thibetica*** are smaller but more numerous than those of ***N. sinensis.*** The calyx tubes of both are longer than broad. Those of *N. sinensis* are hairless or have only a few hairs near their bases, those of *N. thibetica* are pubescent all over. Both sorts are about 6 feet tall, with short-stalked, long-pointed ovate or oblong-ovate, lobed leaves up to 3½ inches long and to a greater or lesser extent hairy. The flowers of *N. thibetica,* about ⅓ inch long, are crowded, from nine to seventeen or sometimes up to twenty-one together in racemes 1¾ to 3½ inches in length, those of *N. sinensis* are longer-stalked, ⅓ to ½ inch long and are mostly twenty-three to sixty, but sometimes as few as nineteen together in more or less nodding, slender racemes 1½ to 6 inches long. From typical *N. sinensis,* which has calyx tubes up to ¼ inch long, *N. s. ribesioides* differs in its calyx tubes being longer, up to ⅓ inch. Up to about 10 feet tall, ***N. longiracemosa*** has downy stems, pointed-ovate, often three-lobed leaves, and peach-pink flowers with hairy calyxes in slender racemes 4 to 6 inches long.

Neillia sinensis

Distinct from all other neillias in having on the branches near the flowers stellate (star-shaped) hairs rising from little protuberances which remain and are obvious even after the hairs fall, ***N. uekii*** is the only species native of Korea. In the wild it is restricted to that country. It has ovate leaves up to 3½ inches long and hairy on the veins beneath. The white flowers, with glandular-hairy calyx tubes longer than broad, are in racemes up to 2 inches long.

Differing from those already discussed in its calyx tubes being as broad or broader than their lengths, ***N. affinis*** is another inhabitant of western China. It has angled young branches and long-pointed, ovate to ovate-oblong, lobed leaves up to 3½ inches long. They are nearly hairless or slightly hairy on the veins beneath. In crowded racemes up to 3½ inches long of ten or more blooms, the pink flowers have calyx tubes clothed with fine hairs. Variety *N. a. pauciflora* has shorter racemes of ten or fewer blooms.

Easily recognized among cultivated neillias by having more than one bud in each leaf axil, ***N. thyrsiflora*** also differs in its flowers usually being in much-branched panicles instead of racemes. From 3 to rarely 6 feet tall, this native of the Himalayas is less showy in bloom than the other kinds discussed here. It has long-pointed, ovate, lobed leaves, sometimes hairy on the veins beneath, and up to 4 inches long and terminal panicles, some 3 inches long, of white flowers.

Garden and Landscape Uses. Neillias are suitable for shrub borders and informal areas. They are graceful and associate well with other shrubbery. With the exception of late summer- and fall-blooming *N. thyrsiflora,* the cultivated kinds flower in late spring or early summer. All except *N. longiracemosa* are hardy at New York City and most are hardy in southern New England. They grow satisfactorily in ordinary well-drained soil in sun or part-day shade.

Cultivation. No special demands are made of gardeners by these shrubs. Pruning to thin out the branches to forestall overcrowding is done with most kinds immediately after the blooms fade, but *N. thyrsiflora,* which blooms on current season's shoots, is pruned in late winter or early spring. Even if its tops are killed back by winter cold, provided its roots and lower parts survive, it flowers the following fall. Propagation may be accomplished by summer cuttings under mist or in a greenhouse propagating case and by seed.

NELIA (Nél-ia). The four species of *Nelia* are endemic to a small region close to the west coast of South Africa. They belong in the *Mesembryanthemum* complex of the carpetweed family AIZOACEAE. The name honors the distinguished South African botanist Dr. Gert Cornelis Nel, who died in 1950.

Nelias are low, tufted, stemless, herbaceous perennials with very succulent leaves and, as is typical of the group of which they are part, flowers that look something like daisies, although completely different in botanical structure. They are not heads of many florets, but individual blooms. The blunt or pointed leaves, in pairs with the bases of each pair joined for a short distance, are nearly cylindrical to three-angled. They have slightly rounded or flat upper surfaces. Beneath, their upper halves are markedly keeled. Mostly under 1½ inches long, they are approximately one-quarter to one-third as broad and thick as their lengths. Bluish- or whitish-green with sometimes a reddish tinge, they are without darker dots. Solitary or in twos or threes, the white, long-lasting blooms, up to ¾ inch in diameter, are open day and night. They have five sepals longer than the numerous linear-spatula-shaped petals. The latter, notched at their apexes and with short hairs at their bases, are in several rows. There are many stamens and staminodes (nonfunctional stamens) and five very small styles. The fruits are capsules.

From 1 inch to 2 inches in height, ***N. pillansii*** has shoots with two or three pairs of bluish- to whitish-green, spreading to erect leaves ¾ inch to 1¼ inches long by up to ⅓ inch wide, and nearly as thick. The short-stalked flowers are approximately ½ inch in diameter.

Garden Uses and Cultivation. Among the easiest to grow of the many small succulents of the *Mesembryanthemum* relationship, nelias are good items for the beginning collector. For more information see Succulents.

NELUMBIUM. See *Nelumbo.*

NELUMBO (Ne-lúmbo)—Lotus or Water Chinquapin or Pond Nut. This aquatic genus, once named *Nelumbium,* consists of one species native of the New World, and one, or according to some botanists two, endemic to the Old World. Its name is a Sri Lankan one for *Nelumbo nucifera.* Nelumbos are not related to the genus *Lotus.* They belong in the water-lily family NYMPHAEACEAE.

Nelumbos have thick rhizomes, and large, circular, bowl-like or plate-like leaves with the stalks joined to the underside centers of the blades. Both leaves and flowers are normally raised high above the water on stout stalks, the flowers mostly overtopping the foliage. The handsome, sometimes almost peony-like, solitary blooms have usually twenty or more sepals and petals, scarcely differentiated one from the other except that the outermost are green. The many stamens closely sur-

Nelumbo (leaf flower, and seed pod)

Nelumbo lutea

Nelumbo nucifera at The New York Botanical Garden, with a reflection of the dome of the conservatory

round the pistils. The funnel-shaped "seed pods" are flat-topped enlargements of the receptacles of the flowers besprinkled with holes; each hole contains a large seedlike nut. The "pods" look like large pepper or salt shakers. Frequently they are dried, bronzed or silvered, and used as decorations. In the Orient the rhizomes, leaves, and seeds of *N. nucifera* are much used as food.

American lotus, water chinquapin, or pond nut, ***N. lutea*** (syn. *Nelumbium pentapetalum*) inhabits quiet ponds and sluggish streams from Ontario to New York, Minnesota, Florida, and Texas. Mostly rising about 3 feet above the water but sometimes floating, its bluish-green leaves, 1 foot to 3 feet wide, have nearly hairless stalks. The blooms, pale yellow and 6 to 10 inches in diameter, are succeeded by "pods" 3 or 4 inches wide, with acorn-like fruits (seeds) nearly ½ inch in diameter.

East Indian lotus (***N. nucifera*** syns. *N. speciosa, Nelumbium nelumbo*) is native from tropical and subtropical Asia to Australia and is naturalized in Hawaii. A sacred flower of India, Tibet, and China, it was taken to Egypt about five hundred years before the birth of Christ, but is not spontaneous there. Although the name is sometimes applied to this, Egyptian lotus is more correctly reserved for *Nymphaea lotus*. The East Indian lotus has rhizomes many feet in length and leaves with rough, somewhat prickly stalks. The leaf blades are 1 foot to 3 feet wide. The fragrant blooms are 4 to 10 inches across. Typically they are pink, but there are varieties with white, red, and striped blooms, varieties with double-flowered ones, and dwarf kinds. Several have horticultural varietal names. Individual blooms remain open for two days. Less frequently cultivated ***N. caspica*** differs from the East Indian lotus in its white flowers having blunter petals and in technical botanical details. It is a native of the Caspian Sea region.

Nelumbo nucifera (seed pods)

Garden and Landscape Uses. Among the most beautiful of aquatics, nelumbos are magnificent for planting in mud bottoms or containers of rich soil where their rhizomes are covered with 1 foot to 3 feet of quiet water and are not subjected to freezing. They need a sunny location and, for their best development, ample space to spread.

Nelumbo caspica

Cultivation. Nelumbos are usually propagated by division. This should not be done until growth begins in spring unless the divisions are to be potted and started in an indoor tank or pool. It is important that the water be warm enough to encourage new growth immediately. When planted in outdoor pools they are laid horizontally in a shallow trench, covered with soil, and kept in place by putting a stone or brick over them. Alternatively, plants started indoors in pots are transplanted to outdoor pools after the weather is warm and settled. Another method of increase is by seeds sown in pots of sandy soil immersed so that the soil is covered with 2 or 3 inches of water, which should be kept at 60°F. The seedlings are potted individually and grown under similar conditions, but in deeper water, until large enough to be planted outdoors.

Where the rhizomes are covered with sufficient water so that there is no danger of them freezing, nelumbos can remain outdoors over winter. Or the pool may be drained in fall and the roots protected with a 3-foot covering of salt hay, straw, leaves, or other loose insulating material. Alternatively, tubs or other containers may be moved to a cellar or other place indoors where the temperature holds at 35 to 45°F. Seeds collected for sowing later should be kept in water. Under favorable conditions they retain their powers of germination for extremely long periods.

NEMASTYLIS (Nemá-stylis). Belonging in the iris family IRIDACEAE and comprising twenty-five species of bulb plants, *Nemastylis* occurs natively from the southern United States to tropical America. Its name, in allusion to the slender styles of its flowers, comes from the Greek *nema*, a thread, and *stylos*, a column.

These plants have long, linear to narrowly-sword-shaped, sometimes half-cylindrical leaves. Their blue or less commonly yellow blooms are few to many from the axils of bractlike spathes (bracts). The flowers have six perianth segments (petals or, more correctly, tepals), three stamens, and a three-branched style with the branches cleft to their bases into two slender arms. The fruits are capsules. This genus is closely related to *Eustylis* and *Salpingostylis*.

Spring-blooming ***N. acuta,*** a native of clay soils and prairies in the lower Mississippi basin, has bulbs about 1 inch in diameter, sparingly-branched or branchless stems 6 inches to 2 feet tall, and linear leaves from 6 inches to over 1 foot long. The blue or purple flowers, nearly or quite 2 inches in diameter, last only a few hours. They open in the morning and close the same afternoon. Fall-flowering ***N. floridana*** inhabits swamps, marshes, and moist woodlands in Florida. It has bulbs about ¾ inch in diameter. From 1½ to 5 feet tall, the stems are slender, branched, or branchless. The narrow, sword-shaped leaves are up to 2½ feet in length. The blooms, pretty, starlike, and 1¼ to 2 inches wide, except for a small white eye are bright violet. They remain closed in the morning, open for a few hours in the afternoon, and then fade.

Garden and Landscape Uses and Cultivation. Rarely cultivated, the sorts described here should be worth attempting by those interested in unusual plants. They would be suitable for informal plantings and, in their natural ranges, for inclusion in native plant gardens. Very little is reported about their cultivation. It is unlikely that they would prove hardy in climates much colder than those in which they grow in the wild. Propagation is by seed and probably by offsets.

NEMATANTHUS (Nemat-ánthus). Endemic to South America where its members most commonly grow on trees as epiphytes, that is, without taking nourishment from their hosts, *Nematanthus* belongs in the gesneria family GESNERIACEAE. Including kinds some botanists segregate as *Hypocyrta*, there are more than thirty species. The name, derived from the Greek *nema*, a thread, and *anthos*, a flower, refers to the long slender flower stalks of some sorts at the ends of which the blooms dangle.

Nematanthus strigillosus

Nematanthuses are close allies of *Columnea*, but most kinds have longer flower stalks and funnel- to trumpet-shaped blooms. Many are comparatively large shrubby plants or vines of often rather ungainly habit and with the older parts of the stems woody. Others are attractive trailers. Their undivided, toothless or toothed, hairless or hairy leaves are opposite, those of each pair of equal or of different sizes. Borne on long or short stalks from the leaf axils, the flowers have deeply-five-parted calyxes, the spreading segments of which are slender and often toothed, and pouched corollas with five short, broad, rolled-back lobes. The anthers of the four stamens are united, the stigma, broad or mouth-shaped, terminates a long style. The fruits are capsules, often berry-like.

Horizontally branched and trailing sorts include *N. gregarius*, *N. strigillosus*, *N. nervosus*, and *N. wettsteinii*, all natives of Brazil. Hairless ***N. gregarius*** (syns. *N. radicans*, *Hypocyrta radicans*) has fleshy, shining, dark green leaves, those of each pair or whorl (circle) of three of equal size, up to 1½ inches long by ½ inch wide. On stalks shorter than the leaves the flowers, solitary from the leaf axils, have orange perianth tubes, about ¾ inch long, and small yellow petals. Its minutely-hairy, very short-stalked leaves, elliptic, up to 1¾ inches long by ¾ inch wide, and those of each pair of similar size, ***N. strigillosus*** (syn. *Hypocyrta strigillosa*) has solitary flowers on stalks shorter than the leaves. They are red-orange to orange, about 1¼ inches long, conspicuously pouched, and much contracted at their mouths. The pairs of finely-hairy, fleshy, elliptic, lustrous leaves of ***N. nervosus*** (syn. *Hypocyrta nervosa*) are of individuals of nearly equal size and up to 2 inches long by ¾ inch wide. Solitary and short-stalked, the flowers are minutely-hairy, red to orange-red, and 1 inch to 1¼ inches long. Broadly-pouched, and narrowing to a tiny mouth, they have small, erect corolla lobes (petals). More strikingly beautiful than the last, ***N. wettsteinii*** (syn. *Hypocyrta wettsteinii*) has slender, trailing stems and ovate, toothless, short-stalked leaves ¾ to 1 inch long by approximately one-half as wide. They have a prominent mid-vein and are so glossy they resemble polished plastic.

Nematanthus wettsteinii

Nematanthus wettsteinii (flowers)

The short-stalked, solitary, very conspicuously pouched, 1-inch-long blooms are tangerine-orange, yellow toward their bases, and with tiny yellow petals. They come from the leaf axils and are produced with great freedom over a long period. This species is reported as having survived brief periods of freezing, but such low temperatures are of course disadvantageous.

Quite distinct and larger than the kinds discussed above, Brazilian ***N. perianthomegus*** (syn. *Hypocyrta perianthomega*) has upright or spreading, rigid stems and thick, glossy, broad-elliptic to obovate, sparingly-hairy, toothless leaves, with one of each pair markedly bigger than the other and up to 4½ inches long. They are recurved at their apexes. The blooms, on stalks shorter than the leaves, are in twos or threes from the leaf axils. About 1½ inches long, they have dull orange calyxes and yellow corollas pouched on their upper sides and with longitudinal dark maroon stripes and a band of the same color beneath the petals.

Upright and with rigid stems, shrubby ***N. fissus*** (syns. *N. selloanus*, *Hypocyrta selloana*), of Brazil, has velvety-hairy, toothed, oblanceolate to elliptic, dull green leaves 3 to 4 inches long by about 1 inch wide. Its short-stalked flowers, solitary or several together from the leaf axils, have densely-white-hairy, orange-red to scarlet corollas, cylindrical at their bases, pouched above, and constricted to a tiny mouth. In habit

Nematanthus perianthomegus

Nematanthus perianthomegus (flowers)

Nematanthus fissus

similar to the last, Brazilian ***N. fritschii*** has leaves with the largest of each pair broad-elliptic to ovate and 2 to 3 inches long, and the other markedly smaller. Their upper sides are hairless; beneath they are clothed with fine hairs and are marked with a conspicuous central red blotch. The 1½- to 2-inch-long, pouched flowers are lavender-pink spotted on their insides with rose-pink. Their stalks are 3 to 4 inches long.

Remarkable for its brilliant red blooms, Brazilian ***N. longipes*** (syn. *Columnea splendens*) is a lax-branched shrub with short-stalked, nearly elliptic leaves slightly broadened above their middles and with pointed apexes. Those of each pair are of equal size. Up to 4 inches long by 1½ inches wide and sometimes indistinctly toothed, the leaves have undersides paler than their upper surfaces or are purplish and furnished with short hairs. The flowers, on stalks 3 to 6 inches long, have ¾-inch-long, toothed calyx lobes, sometimes red at their apexes. From between two of these the 2¼-inch-long, finely-hairy, wide-mouthed, trumped-shaped corollas angle obliquely outward.

Hybrid nematanthuses are fairly numerous and many bloom more freely than most species. The earliest hybrids were raised in the late 1960s in Massachusetts by amateur gardener William Saylor and in New York at Cornell University by Dr. Robert Lee. Saylor's early successes included 'Rio', which more than a decade later retained its reputation for having the truest red flowers of all nematanthuses, and 'Tropicana', with maroon-striped, yellow flowers. In 1967 Dr. Lee introduced 'Stoplight', which inherits its red flowers from its parent, *N. longipes*, and its red-marked foliage from its other parent, *N. fritschii*.

The about thirty hybrids available in the late 1970s include 'Butterscotch', a trailer with deep yellow blooms; 'Encore', remarkable for shining rich green foliage and bright orange flowers; and 'Moonglow', with lemon-yellow flowers striped pale pink. The foliage of 'Golden West' is variegated creamy-yellow.

Garden Uses. The cultivation of these plants and their popularity are much more recent developments than are those of many other members of the gesneria family. Besides providing specimens for specialist collectors of gesneriads, they include attractive kinds for the general embellishment of warm greenhouses. The smaller ones are appealing as houseplants where fairly high humidities and other environmental needs can be met. The trailing kinds are attractive in hanging baskets.

Cultivation. Because they are epiphytes, nematanthuses need a coarse, loose, porous soil or substitute mix that encourages the free passage of water and ready access of air. It should have a high organic content and be maintained in a fairly moist, but not saturated condition. High humidity is important as is good light with shade from strong sun. Minimum winter night temperatures of 60 to 65°F are satisfactory, and by day in winter temperatures five to fifteen degrees higher. At other times of the year even more warmth is to the liking of these plants. Pruning to shape and refurbishing including top dressing with new soil or repotting is done in late winter or spring. Very well-rooted specimes benefit from summer applications of dilute liquid fertilizer. Propagation is simple by cuttings and by seed. For more information see Gesneriads.

NEMATODES or NEMAS. These are minute, predominantly thread- or worm-like creatures, usually whitish, yellowish, and more or less transparent and often of sizes invisible to the naked eye. Formerly they were called eelworms. Many kinds cause serious injury to plants. Some live internally in their tissues, others feed from the outside. The majority of sorts that parasitize plants attack roots, bulbs, and other underground organs, others infest stems, leaves, flower buds, and flowers. The identification of the different kinds is a highly specialized scientific procedure.

As a general rule nematodes are not able to travel far under their own power. Their dispersal among cultivated plants is almost entirely by being carried on shoes and tools or in irrigation water and by shipping infected plants and soil from place to place.

The effects of nematode invasions may be systemic, causing a general unhealthy appearance and stunting including perhaps yellowing and dieback or it may produce more distinctive symptoms, such as galls or in the cases of foliar species blackish patches between the leaf veins. Among those that cause galls are the well-known root-knot nematodes, which afflict many different kinds of plants. Foliar nematodes that attack begonias, chrysanthemums, dahlias, ferns, and some other ornamentals are prevalent.

Control of nematodes consists of sanitary measures to avoid transference of infested plants or soil; crop rotation; avoiding so far as is practicable overhead wetting of plants subject to foliar nematodes; treating infected bulbs and other plant parts by soaking them in hot water for a predetermined period, the water temperature and length of subjection being dependent upon the type of plant and kind of nematode; and steam or chemical sterilization of infected soil. For more specific recommendations consult a Cooperative Extension Agent, State Agricultural Experiment Station, or other authority.

NEMESIA (Nemès-ia). Excellent showy ornamentals are included in *Nemesia*, an African genus of fifty species of the figwort family SCROPHULARIACEAE. The generic name is one used by Dioscorides for a snapdragon (*Antirrhinum*).

Nemesias, related to toadflaxes (*Linaria*), include annuals, nonhardy herbaceous perennials, and subshrubs. Those cultivated are mostly annuals. They have four-angled, grooved stems and stalkless or nearly stalkless, lanceolate leaves in pairs, those of each pair spreading at right angles to the leaves of the pairs above and below. Toward the tops of the stems the leaves gradually are reduced in size. The flowers, in terminal, often compact racemes, have a five-parted calyx and a tubular, two-lipped corolla with a short pouch or spur. The upper lip is four-cleft, the lower one is notched and sometimes has a palate-like development at its base. There are four stamens and one style. The wide color range includes white, cream, yellow, orange, brown, pink, red, and blue, the hues often brilliant. The fruits are capsules.

The most commonly cultivated annual ***N. strumosa,*** of South Africa, has given rise to many horticultural varieties to which the group name *N. s. suttonii* is applied. Mostly from 1 foot to 2 feet tall, they have lanceolate to linear, toothed leaves

Nemesia strumosa variety

up to 3 inches long. Their flowers are pouched, but not spurred, and come in a very wide range of colors. A dwarf sort named 'Blue Gem' has forget-me-not-blue flowers, and is about 9 inches high.

Nemesia versicolor variety

The only other species cultivated to any considerable extent is ***N. versicolor,*** also South African and an annual, but differing from *N. strumosa* and its varieties in having flowers with spurs and in being lower, normally not exceeding 1 foot in height. There are two callosities and some pubescence in the throat of each bloom, the spur about equaling the lower lip in length. The flower colors are various and include good blues, as well as lilacs, yellows, and whites. A selected, compact, free-flowering strain is called *N. v. compacta.*

A few other species are occasionally cultivated. The following are all South African: ***N. azurea,*** similar to *N. strumosa,* has blue flowers; ***N. barbata,*** up to 20 inches tall, has coarsely-toothed, lustrous leaves and spurred, blue and white flowers; ***N. chamaedrifolia,*** up to 2 feet tall, has pale pink, short-spurred, solitary blooms in the leaf axils; ***N. floribunda,*** less attractive than many, has loose clusters of creamy-white or white-and-yellow, fragrant flowers; ***N. lilacina,*** glandular-pubescent, has lilac-and-white flowers with purple stripes on the upper lip and yellow on the lower.

Garden Uses. Nemesias are among the most showy of garden annuals, but unfortunately they do not succeed where hot humid summers prevail. Only where temperatures are moderate and nights are cool can they be relied upon to behave well insofar as summer display is concerned. In the British Isles they are favorites for outdoor planting but in most of North America their chief merits are as pot plants for spring blooming in greenhouses, and for this they are superb. They have limited uses as cut flowers.

Cultivation. If nemesias are to be grown in the outdoor garden, seeds should be sown in a greenhouse in a temperature of about 60°F in well-drained containers of porous soil about eight weeks before it is safe to transfer the plants outdoors, which can be done as soon as all danger of frost has passed. The seedlings are transplanted 2 inches apart in flats and are grown on in a sunny greenhouse in a night temperature of 50 to 55°F and a day temperature five to ten degrees higher. They may be pinched once to induce branching. In the garden they need a fertile, reasonably moist, slightly acid soil and a sunny location. They should be spaced 4 to 7 inches apart according to the vigor of the variety. Because they are rather brittle special care should always be used in handling nemesias when transplanting or working among them. Through the summer they require no special attention beyond the routine care given most outdoor annuals.

In greenhouses nemesias are grown to a limited extent for cut flowers and as decorative pot plants. Seeds are sown in well-drained containers of sandy soil from September to January, the seedlings are transplanted individually to 2½-inch pots, and later planted out in ground beds or benches or potted into larger containers. The kinds used for these purposes are almost exclusively varieties of *N. strumosa suttonii,* but other annual nemesias respond to the same treatment. When planted in beds or benches, 6 inches between plants allows ample space for growth. The final pots may be from 4 to 6 inches in diameter depending upon the vigor of the variety and the time of sowing. Plants from early fall seedings need larger accommodations than those from seeds sown later. The soil should be fertile, porous and rather coarse, and kept always fairly moist, but not constantly saturated. As soon as the final pots are filled with roots, regular feeding with dilute liquid fertilizer should be instituted. Ideal growing conditions are those of a sunny greenhouse where the winter night temperature is held at 50°F and the daytime temperature does not exceed this by more than five to ten degrees. The greenhouse should be ventilated as freely as weather permits to maintain an airy, moderately dry, buoyant atmosphere rather than a stagnant, oppressively humid one. Stake before the stems become twisted or broken. Nemesias are not much troubled by diseases and pests.

NEMOPANTHUS (Nemo-pánthus)—Mountain-Holly. There is only one species of this holly relative of the holly family AQUIFOLIACEAE. An inhabitant of bogs, swamps, and wet ground from Nova Scotia to Ontario, Wisconsin, and Virginia, it differs from holly (*Ilex*) in its flowers having strap-shaped petals not joined at their bases and tiny sepals that soon fall. Its name stems from the Greek *nema,* a thread, *pous,* a foot, and *anthos,* a flower, and alludes to the flower stalks.

The mountain-holly (***Nemopanthus mucronatus***) is a hardy, deciduous, hairless shrub up to 10 feet tall. It has many slender branches and alternate, slender-stalked, slightly toothed or toothless, elliptic to oblong leaves 1 inch to 1½ inches long. Its whitish blooms, on very slender stalks from the leaf axils, are about ⅙ inch in diameter. They have four or five sepals and the same number of petals and stamens as sepals. The stigma is lobed. Individual plants may bear flowers of one sex or a mixture of unisexual and bisexual blooms. The berry-like fruits, which ripen in summer, are ¼ to ⅓ inch in diameter and dull red.

Garden and Landscape Uses and Cultivation. Mountain-holly is attractive in fruit and when its foliage turns yellow in fall. It is well adapted for partly shaded places where the soil is fairly moist or wet. It lends itself best for use in casual landscapes. It calls for little or no regular care, and the only pruning needed is any deemed desirable to keep the plants shapely. Propagation is by seeds, sown as soon as they are ripe in a cold frame or outdoors, and by summer cuttings under mist or in a greenhouse propagating bench.

NEMOPHILA (Nem-óphila) — Baby Blue Eyes, Five Spot. About thirteen species of North American annuals compose *Nemophila,* of the water leaf family HYDROPHYLLACEAE. Most are Western. The name, alluding to the natural habitats of some kinds, is derived from the Greek *nemos,* a grove, and *phileo,* to love.

Nemophilas are generally freely-branching plants of lax or prostrate habit, hairy or not. Their leaves are stalked, and opposite or sometimes alternate. They may be pinnately-divided, lobed, or merely toothed. The slender-stalked, solitary, wheel- to broadly-bell-shaped flowers originate in the leaf axils or opposite the leaves. Blue, purple, or white, they are sometimes conspicuously spotted. The calyxes are deeply-five-parted and usually have a projection that looks like an additional smaller sepal at the base of each cleft. These enlarge as the fruits develop. The corollas are deeply-five-lobed and usually have ten scaly appendages inside their throats. The stamens do not protrude. The style is two-lobed. The fruits are globular or ovoid capsules.

Baby blue eyes (***N. menziesii***) is a variable kind. Of lax growth, it has more or less succulent, branching stems that are usually pubescent, indistinctly angled or winged, and 4 inches to 1 foot long. The opposite leaves are deeply divided into usually nine to eleven, but sometimes less or more, toothed lobes, except in variety *N. m. integrifolia,* in which the upper leaves have only shallow, toothless lobes. In outline the leaves are ovate to oblong and, including their stalks, are 2 to 4 inches long. The flower stalks are longer than the leaves and end in a bowl-shaped bloom with narrow calyx lobes with projections between them. The blooms are ¾ inch to 1½ inches or more in diameter. Typically they are bright blue with paler centers, but vary to pale blue or white and may be variously spotted, blotched, or striped with darker veins. In *N. m. atomaria* the flowers are white with conspicuous black spots radiating from the center almost to their edges.

The five spot (***N. maculata***) differs from baby blue eyes in technical characteristics of the seeds. It is more or less prostrate, rarely exceeding 6 inches in height, and has hairy stems up to 1 foot in length. Its opposite leaves, oblong to broad-elliptic in outline, and hairy, are, including their stalks, up to 3 inches long and pinnately-cleft usually into five to nine lobes, but the upper leaves may be three-lobed. The thick flower stalks, 1 inch to 2 inches long, each end in a white bowl-shaped bloom ¾ inch to 1¾ inches wide, purple-dotted, and with a conspicuous dark purple spot at the end of each corolla lobe.

Garden and Landscape Uses. Nemophilas are pretty, easily grown annuals for flower beds, rock gardens, and within their natural range, native plant gardens. They are also useful for providing spring bloom as pot plants in greenhouses, but are of little use as cut flowers. Because they are intolerant of hot weather seed should be sown so that the plants come into bloom before high summer. They are happiest and bloom longest where they receive a little dappled shade or part-day shade. Moist soil of average garden fertility is required. Because nemophilias are rather fragile and subject to storm damage, locations least subject to high winds and other disturbances should be chosen.

Cultivation. For outdoor display, seeds should be sown in spring or, in regions of moderately mild winters, in fall where the plants are to remain, and covered with soil to a depth of not more than ¼ inch. The seedlings are thinned to 6 to 8 inches apart, the lesser distance being appropriate on poor soils and for spring-sown plants. Frequent shallow surface cultivation discourages weeds and promotes growth. It may be advisable to support the plants lightly by inserting among them short pieces of brushwood. This helps to reduce the danger of damage by heavy rains.

For greenhouse decoration, seeds are sown in September or October in a temperature of 55 to 60°F. The seedlings are transplanted individually to 2½-inch pots and later transferred singly to 4- or 5-inch pots or set three or more together in pans (shallow pots) of 6-inch size or larger. The containers must be well drained, and porous, fertile soil used. Water with care. It is fatal, especially in the early stages before the plants are well rooted, to permit the soil to remain saturated for long periods. A sunny location in a well-ventilated greenhouse where the night temperature is 45 to 50°F, and the day temperature is not more than five to ten degrees higher, encourages sturdy growth. In late winter, as the days lengthen and the sun becomes stronger, more generous watering may be initiated and occasional applications of dilute liquid fertilizer given. Neat staking keeps the plants tidy and controllable.

Nemophila maculata

NEOBENTHAMIA (Neo-benthàmia). The only species of this genus of the orchid family ORCHIDACEAE has a name commemorating the English botanist George Bentham, who died in 1884.

Native to tropical East Africa, ***Neobenthamia gracilis*** is a beautiful, easily cultivated evergreen orchid that grows in the ground rather than perched on trees. It is without pseudobulbs and has slender, branching stems 4 to 6 feet tall fitted along their lengths with narrow, lustrous, stem-clasping, soft leaves 6 to 8 inches long. Many together in crowded, half-rounded clusters at the stem ends, the fragrant, long-lasting flowers are about 1 inch across. Mostly they do not open widely. They are white, with, on the lip, a bright yellow stripe spotted on each side at its base with purplish-red.

Garden Uses and Cultivation. A suitable soil for this worthy orchid consists of a mixture of two parts turfy loam and one each of osmunda fiber and coarse sand or broken crocks or brick, together with a generous spiking of dried cow manure. To accommodate the vigorous roots, the pots should be fairly large for the sizes of the plants, and well drained. The soil must be kept moist at all seasons, but less moist in winter. Summer applications of dilute liquid fertilizer are beneficial. Repotting is done in spring following the chief season of bloom. A humid atmosphere and tropical temperatures, the minimum at night in winter 60°F, are needed as is shade sufficient to prevent the foliage from being scorched. For more information see Orchids.

NEOBESSEYA (Neo-bésseya). Modern botanists with conservative leanings include *Neobesseya*, of the cactus family CACTACEAE, in *Mammillaria*. As a segregated genus it comprises half a dozen or fewer species native to the western United States and Mexico. Its name honors the American botanist Dr. Charles Edwin Bessey, who died in 1915.

Neobesseyas are small cactuses with solitary or clustered, spherical or somewhat flattened stems or plant bodies having spiraled or irregularly-disposed, nipple-like tubercles, usually grooved along their upper sides. The latter do not persist as woody knots after the spines drop. The spines are in clusters at the tips of the tubercles. Borne close to the tops of the plant bodies, the comparatively large and attractive flowers open by day. The bright red, spherical fruits contain very dark brown to black seeds.

Here are sorts likely to be cultivated: ***N. asperispina,*** of Mexico, has turnip-shaped plant bodies 2 to 2½ inches wide and clusters of nine to ten radial spines with one or no centrals. Its 1-inch-wide flowers are greenish-yellow. ***N. missouriensis*** (syn. *Mammillaria missouriensis*) has usually clustered, nearly spherical plant bodies 2 to 4 inches in diameter. The spine clusters are of ten to seventeen radials up to ½ inch long and sometimes one central. From 1 inch to 2 inches wide and with fringed petals, the flowers are greenish, yellow, bronze, or bronze streaked with pink. This is native from North Dakota to Montana, Oklahoma, and Texas. ***N. notesteinii*** (syn. *Mammillaria notesteinii*), of Montana, has solitary or clustered plant bodies scarcely exceeding 1 inch in diameter and with short tubercles with clusters of twelve to eighteen pubescent spines, one of which is usually a central, the other radials. The flowers are ashy-gray lined with pink. ***N. rosiflora*** (syn. *Mammillaria rosiflora*) forms small clumps of nearly spherical plant bodies generally broader than tall and up to 2¾ inches across. They have about ¼-inch-long, bristle-like spines, one of which is a central, the other twelve to fifteen radials. About 1½ inches in diameter, the flowers of this endemic of Oklahoma are pink without stripes or bands of other colors. ***N. similis*** (syn. *Mammillaria similis*), of Texas, has plant bodies identical with those of the last, but 2 to 4 inches wide. The spine clusters are of ten to seventeen slender radials and sometimes one stouter central. The flowers, 1 inch to 2 inches in diameter, have slender, sharply-pointed petals that may be greenish, yellow, yellow streaked with pink, or bronze, generally with many colors in the same bloom.

Garden Uses and Cultivation. These are essentially as for mammillarias. Neobesseyas are generally easy to manage in well-drained, slightly acid soil. Some seem to appreciate slight shade from the full intensity of the summer sun. For additional information see Cactuses.

NEOBRACEA (Neo-bràcea). The dogbane family APOCYNACEAE contains *Neobracea,* a genus of four species endemic to the Bahamas and Cuba. The name commemorates Lewis Jones Knight Brace, a collector of the flora of the Bahamas. He died in 1938.

The only species known to be cultivated is *N. bahamensis,* a shrub 3 to 4 feet tall planted in southern Florida. This has short-stalked, firm-textured, oblong-lanceolate leaves, blunt or slightly notched at their apexes, and concentrated near the branch ends. They are 2 to 4 inches long by up to 1 inch wide and have rolled-under margins. Their upper surfaces are glossy-green. Beneath, they are densely-white-hairy and clearly netted with green veins. White with reddish throats, and hairy on their outsides, the ¾-inch-wide blooms are solitary or in pairs from the upper leaf axils. They have a deeply-five-cleft calyx, a corolla with a short tube broadening above and five obliquely-ovate, spreading lobes (petals), five stamens, and one style. The fruits are paired, hairy, pendent pods up to 6 inches long.

Garden and Landscape Uses and Cultivation. Not a great deal of information is recorded about this species in cultivation. In its home islands, it grows in exposed coastal areas. Propagation is by seed and presumably by cuttings.

NEOBUXBAUMIA (Neo-buxbáumia). The Mexican genus *Neobuxbaumia,* of the cactus family CACTACEAE, consists of four species closely related to *Cephalocereus.* Its name honors a distinguished twentieth-century German student of cactuses, Dr. Franz Buxbaum. The plant grown as *N. euphorbioides* is *Rooksbya euphorbioides.*

Neobuxbaumias are tall, columnar, or treelike. They have branchless or branched trunks up to 50 feet tall with many narrow ribs and needle-like spines. The branches are stout and erect. Coming from the sides of the stems near their tops, the funnel- to bell-shaped flowers open at night. Their ovaries and the fruits are scaly.

Kinds of neobuxbaumias cultivated include these: ***N. mezcalaensis*** (syn. *Cephalocereus mezcalaensis*) has light green solitary stems up to about 20 feet tall that have clusters of six or seven ½-inch-long radial spines and one slightly longer central. The greenish-white flowers are 2 inches long. ***N. polylopha*** (syn. *Cephalocereus polylophus*) has solitary, columnar, light green stems up to 45 feet in height and 1 foot or more in diameter with clusters of seven to ten slender, brown-tipped, yellow radial spines ½ to ¾ inch long and one central about as long. The 2½-inch-long blooms are pink. ***N. scoparia*** (syn. *Cephalocereus scoparius*) develops a large head of numerous erect branches. It grows about 25 feet tall and has dark gray-green stems with clusters of five to seven bent, ½-inch-long radial spines and one stiff, 1-inch-long or a little longer central, when young black, later gray. The flowers are bell-shaped, reddish, and about 1½ inches in length. ***N. tetetzo*** (syn. *Pilocereus tetetzo*) attains heights of 25 to 50 feet and usually branches sparingly from the ground, more freely above. Its erect, grayish-green, many-ribbed stems have clusters of eight to ten stout, stiff, dark-tipped, white radial spines ½ to 1 inch long and one to three centrals, sturdier and up to 2 inches or sometimes more in length. Old specimens bloom freely from the tops of their stems. The blooms are funnel-shaped, 2½ inches long, white inside, and reddish-green on their outsides. Variety *N. t. nuda* has clusters of not more than three spines.

Neobuxbaumia scoparia

Garden and Landscape Uses and Cultivation. These are as for *Cephalocereus.* For more information see Cactuses.

NEOCHILENIA (Neo-chilènia). As indicated by its name a native of Chile, *Neochilenia* comprises fewer than twenty species, by some botanists included in *Neoporteria,* of the cactus family CACTACEAE.

Of small to medium size and often with large tuberous roots, neochilenias have plant bodies with notched ribs and clusters of weak to very stout spines. Their flowers, open by day, have perianth tubes and ovaries like the fruits clothed with wool, hair, and bristles. The blooms are broad-funnel-shaped in contrast to the more cylindrical ones of *Neoporteria.*

Kinds likely to be found in cultivation include these: ***N. aerocarpa*** (syn. *Chileorebutia aerocarpa*) has cylindrical, brownish plant bodies with closely set clusters of about ten ¼-inch-long, gray to black spines sprouting from white-woolly areoles. The crimson flowers are 1½ inches in diameter. ***N. chilensis*** (syn. *Neoporteria chilensis*) has cylindrical plant bodies with about twenty notched ribs and clusters of up to as many as thirty off-white, strong, reedlike spines ½ inch to 1¼ inches long. The rose-red flowers are 2 inches in diameter. ***N. esmeraldana*** (syn. *Chileorebutia esmeraldana*) has dark brown, squattish to spherical plant bodies with brown spines in clusters of about fourteen. Its pinkish to greenish-yellow flowers are 1½ inches across. ***N. fusca*** (syn. *Neoporteria fusca*) has brown-tinged plant bodies at first spherical, later cylindrical. From 3½ to 5½ inches across, they have about twelve strongly-notched ribs and clusters of about seven awl-shaped, ½-inch-long radial spines at first black, becoming gray, and generally one longer central ¾ to 1 inch long. Greenish on their outsides, yellowish within, the flowers are about 1½ inches in diameter. ***N. hankeana*** is much like *N. fusca,* but has light green plant bodies and longer spines. Its 1¼-inch-wide blooms are yellowish-white. ***N. jussieui*** (syn. *Neoporteria jussieui*) is much like *N. fusca*. It has dark green, reddish-brown, or nearly black plant bodies with thirteen to sixteen strongly-notched ribs and white spines eventually becoming yellowish in clusters of seven to fourteen radials and one or two thicker centrals up to 1 inch in length. The 1½-inch-wide flowers are pink with a dark stripe down the center of each petal. They have green stamens, and stigmas with six reddish lobes. ***N. mitis*** (syn. *Neoporteria mitis*) has brownish plant bodies scarcely 1½ inches in diameter and about ¾ inch tall. They have closely set clusters of six to eight extremely short spines. The pink blooms are 1½ inches across. ***N. napina*** (syn. *Neoporteria napina*) has a very thick taproot and a grayish-green plant body up to 3½ inches high by rarely over 2 inches wide, with fourteen spiraled ribs and clusters of three to nine very short, black spines. About 1½ inches across, the light yellow flowers have reddish stigmas. ***N. nigriscoparia*** has solitary plant bodies about 1½ inches tall, broader than high, with conspicuous tubercles and clusters of blackish-brown spines about 1½ inches long. About 1 inch wide, the blooms are pink. ***N. occulta*** has solitary, spherical to ovoid, brownish-green plant bodies 2 to 3½ inches tall. They have fourteen much-notched ribs. In clusters of up to ten or sometimes solitary or lacking, the spines, except that they vary in length from under ½ inch to 1½ inches long, are all similar. The 1-inch-wide flowers are yellow, their outer petals with a reddish streak down their centers. ***N. odieri*** (syn. *Neoporteria odieri*) has reddish-brown, spherical plant bodies with thirteen ribs with pronounced tubercles. Its thin, ¼-inch-long spines, all radials, are in clusters of six to ten. The flowers, white to pink, are 2 inches in diameter. From the typical species *N. o. mebbesii* differs in being larger in all its parts and green. Also, the spine clusters frequently have one central. ***N. odoriflora*** (syn. *Pyrrhocactus odoriflorus*) has dark gray-green plant bodies with ten to sixteen ribs. The greenish-brown spines in clusters of seven to ten are up to ¾ inch long. About 2 inches wide, the flowers are brownish-green. ***N. paucicostata*** (syn. *Pyrrhocactus paucicostatus*) has bluish-gray plant bodies with eight to twelve conspicuously notched ribs and clusters of five to eight ½-inch-long radial spines and one to four stouter centrals at first black later becoming whitish-gray. The 1½-inch-wide blooms are white flushed with red. Variety *N. p. viridis* has green plant bodies. ***N. reichei*** (syn. *Chileorebutia reichei*) is very attractive. It has brownish-red spherical to somewhat cylindrical and sometimes branched plant bodies 1½ to 2½ inches in diameter with thirty or more ribs and clusters of seven to nine tiny, pale grayish spines pressed against them that spread in two directions like the teeth of combs. The yellow flowers, rather sparingly produced and 1¼ inches in diameter, have red styles and stigmas. ***N. taltalensis*** (syn. *Neoporteria taltalensis*) has dark green plant bodies with thirteen ribs divided into chinlike tubercles. Its twisted spines ½ to 1 inch long, at first dark brown, later whitish, are in clusters of seven to sixteen. The purple flowers are about 1 inch wide. ***N. wagenknechtii*** (syn. *Pyrrhocactus wagenknechtii*), somewhat variable, has gray-green plant bodies with eighteen to twenty-two ribs and clusters of six to eight radial spines, up to ¾ inch long, and one to four longer centrals. Approximately 1½ inches wide, the flowers are greenish- to brownish-yellow or brownish-red.

Garden and Landscape Uses and Cultivation. Neochilenias are attractive for outdoor plantings in warm, dry climates and for including in greenhouse collections of cactuses. They increase in size slowly, but for the most part are easy to manage, responding to care and environments that suit the majority of desert cactuses. Most benefit from a little shade from the fiercest sun of summer, among them *N. napina*. This sort often thrives best when grafted onto other strong-growing cactuses. For more information see Cactuses.

NEODAWSONIA (Neo-dawsònia). Three Mexican species by conservative botanists included in *Cephalocereus* constitute *Neodawsonia*, of the cactus family CACTACEAE. The name honors the American twentieth-century botanist and cactus authority E. Yale Dawson.

Neodawsonias are columnar cactuses with branched or branchless stems 9 to 25 feet tall that have numerous ribs and many slender spines. Arising from a woolly-hairy cephalium or band that surrounds the top of the stem like a collar, the flowers are small, funnel-shaped, and pink. Their perianth tubes and ovaries and the fruits are scaly and hairy.

Dark bluish-green and 3 to 10 feet tall, ***N. apicicephalium*** has stems branched from their bases, 3 to 4 inches in diameter, and with twenty-two to twenty-seven ribs. The spines are in clusters of nine to twelve radials ½ to ¾ inch long and two to six centrals ¾ inch to 1½ inches long. The cephalium is white. About 1¼ inches across, the blooms are pink suffused with yellow. From 6 to 10 feet tall, ***N. nizandensis*** has stems 4 to 5 inches thick that branch in their upper parts and have twenty-five to twenty-eight ribs. The spines are in clusters of about sixteen radials ½ to ¾ inch long and six stouter centrals. The cephalium is golden-brown. About 1½ inches long, the flowers are pink. Erect, branchless, gray-green stems 10 to 25 feet high and 4½ to 6 inches in diameter are characteristic of ***N. totolapensis***. They have about twenty-eight ribs and spine clusters of eight to thirteen ¼ to ½-inch-long radials and three to six stouter centrals. The cephalium is yellowish toned with brown. The flowers, 1½ inches long and about 1 inch wide, are pink tinged with yellow.

Garden and Landscape Uses and Cultivation. These are as for *Cephalocereus*. For more information see Cactuses.

NEODYPSIS (Neo-dýpsis). The fifteen species of *Neodypsis*, of the palm family PALMAE, are endemics of Malagasy (Madagascar). From *Chrysolidocarpus* they differ in technical details only. The name derives from from the Greek *neo*, new, and *Dypsis*, the name of another genus of Madagascan palms.

About 30 feet tall, ***N. decaryi*** has a single trunk up to 1½ feet in diameter, brown and roughened with fissures and leaf scars. Its graceful, pinnate leaves, arched at their extremeties and up to 15 feet long, are one above each other in three distinct vertical rows, so that viewed from above they are seen to spread in only three directions with 120 degree angles between. When seen from normal eye level, only two of the three rows of leaves are visible and these spread like feathers of a gigantic fan. The overlapping leaf bases covered with fuzzy brown hairs form a triangular pattern. Similar brown fuzz occurs on the leafstalks and branches of the flower clusters. The latter come among the foliage. Each has two large spathes, the outer of which is persistent and twice as large as the other. The clusters are thrice-branched and include both male and female flowers with the usual arrangement being one female flanked by two males, however, in the majority of groups, the female flower does not develop. Male blooms have six stamens. The fruits, egg-shaped and olive-green, are about 1 inch long.

Its trunk not over 9 inches in diameter, but up to 40 feet tall or taller, ***N. lastelliana*** has a crown of ascending leaves with stalks the red-scurfy bases of which sheathe each other to form a distinct crownshaft (apparent extension of the trunk). The leaf blades, up to 10 feet or more in length, have nearly 100 leaflets on each side of the midrib. The branched flower clusters up to 3 feet or so long are among or just below the foliage. The fruits are about ¾ inch long.

Garden Uses and Cultivation. Little experience has been had with these plants in cultivation, but judging from specimens of *N. decaryi* at the Fairchild Tropical Garden, Miami, Florida, that species is likely to prove well adapted for warm subtropical and tropical regions and greenhouses. It is interesting in habit of growth and decidedly ornamental. At the Fairchild Tropical Garden this succeeds on porous, oolitic limestone in full sun and part-shade without irrigation or supplemental watering.

Neofinetia falcata

few, erect, rigid, spiny-edged leaves. The flowers, much like those of *Billbergia*, but symmetrical, are in spikes and have six perianth parts (petals), six stamens, and three stigmas. The fruits are berries. The leaves are the source of a strong fiber exploited commercially in Brazil.

Slender and almost whiplike, the few nearly cylindrical, overlapping, gray-green leaves of ***Neoglaziovia variegata***, from 2 to 15 feet long, are marked with distinct horizontal bands of light gray. The spikes of bloom, sometimes as tall as the leaves, have coral-red stalks and bracts. The about 1-inch-long flowers are bright purple and darken with age. There are thirty to forty in each spike.

Garden Uses and Cultivation. This rare bromeliad can be grown outdoors in well-drained, porous soil in sunny locations in Florida, California, and other warm regions. It does well under quite dryish conditions, such as suit dyckias and hechtias and is propagated by division and seed. Indoors it is treated as is *Billbergia*. For more information see Bromeliads or Bromels.

At Daytona Beach, Florida, it has withstood temperatures of 25°F without injury. Plants in full sun mature earlier than those in shade. The other species described here will probably respond to conditions and care that suit *N. decaryi*. For further information see Palms.

NEOFINETIA (Neo-finétia). Consisting of one variable, charming species, ***Neofinetia***, of the orchid family ORCHIDACEAE, was formerly included in *Angraecum*. Native to Japan and Korea, it has a name compounded of the Greek *neo*, new, and a now discarded name, *Finetia*, of this orchid.

Epiphytic, that is tree-perching instead of growing in the ground, ***N. falcata*** (syn. *Angraecum falcatum*) is without pseudobulbs. It has densely leafy stems 2 to 3 inches long, which often branched from near their bases. The leaves, their bases overlapping, are pointed-linear, fleshy, usually curved, and up to 3 inches long. Very fragrant, especially at night, the pure white flowers are in two- to seven-flowered racemes shorter than the leaves from the axils of which they arise. They are about 1¼ inches across, and have a curved, slender spur 1½ to 2 inches long. Variety *N. f. variegata* has leaves with cream-colored longitudinal stripes. Other named varities are *N. f. albomarginata*, 'Seito Fukurin', and *N. f. a.* 'Toko Fukurin'.

Garden Uses and Cultivation. These delightful orchids are adapted for pots and hanging baskets. They thrive, planted in osmunda, tree fern fiber, or other well-drained rooting medium, under the cool, humid conditions that suit odontoglossums. For more information see Orchids.

NEOGARDNERIA (Neogard-nèria). Three tree-perching (epiphytic) species of the orchid family ORCHIDACEAE are the only members of *Neogardneria*. Endemic to Brazil and Guiana and once included in *Zygopetalum*, this genus, on the basis of technical details of its flowers, is now treated separately. Its name commemorates George Gardner, Superintendent of the botanic garden, Peradeniya, Ceylon. He died in 1849.

Sometimes cultivated ***N. murrayana*** (syn. *Zygopetalum murrayanum*), of Brazil, has clusters of ovoid pseudobulbs about 2½ inches long. Each has two or more pointed, strap-shaped-lanceolate leaves up to 1½ feet long by 1½ inches broad. The erect or nodding flowering stalks up to 1 foot long carry up to half a dozen fragrant blooms about 2 inches wide. They have green sepals and petals and a white lip with a yellow patch with five purple-brown stripes, and a suffusion of the same hue extending from it.

Garden Uses and Cultivation. These are attractive orchids for inclusion in collections. They respond to conditions favorable to warm-temperature lycastes. For more information see Orchids.

NEOGLAZIOVIA (Neoglaz-iòvia). This Brazilian genus, of the pineapple family BROMELIACEAE, contains two species, one occasionally cultivated. The name commemorates the French landscape architect Auguste F. M. Glaziou, who superintended the public gardens at Rio de Janeiro, Brazil late in the nineteenth century. He died in 1906.

Unlike many members of the family, Neoglaziovias do not perch on trees, but grow in the ground. They have rosettes of

NEOGOMESIA (Neo-gomèsia). Those who prefer not to split genera extravagantly include the only species of *Neogomesia*, of the cactus family CACTACEAE, in closely similar *Ariocarpus*, from which it differs in its flowers having long perianth tubes and its fruits being visible from their beginnings. Known from one limited locality in Mexico, this species has a generic name that honors Marte Gomez, a Governor of the State of Tamaulipas, Mexico.

Superficially this cactus resembles *Ariocarpus trigonus*, but has much longer tubercles with conspicuous areoles well back from their apexes. Low and with a thick root and a short, greenish-brown, flat-topped stem or plant body 1½ to 3½ inches in diameter, ***N. agavioides*** (syn. *Ariocarpus agavioides*) has leaflike tubercles mostly four times as long as broad, not erect, and often flaccid. A woolly, spineless areole is present a short distance below the tip of each tubercle. The magenta flowers, 1½ inches in diameter, come from the centers of the plants. The 1-inch-long, club-shaped, red fruits like the perianth tubes and ovaries are without scales or hairs.

Garden Uses and Cultivation. A choice item for collectors, this needs conditions and care appropriate for *Ariocarpus*. For more information see Cactuses.

NEOLEPTOPYRUM. See Leptopyrum.

NEOLITSEA (Neolít-sea). Native to eastern Asia, Malaysia, and India, this genus of some eighty species belongs to the laurel family LAURACEAE. The name *Neolitsea* is that of related *Litsea*, in which it was previously included, combined with the Greek *neo*, new.

Neolitseas are evergreen small trees and shrubs with alternate leaves, usually three-veined from their bases, rarely pinnate-veined. The small flowers, in stalkless umbels with bracts at their bases, are unisexual, the sexes on different plants. They have four deciduous perianth segments, commonly called petals, and in male flowers six or rarely eight fertile stamens. The stigma of the female flowers is shield-shaped. The fruits are red or black berries with their bases enclosed in the inflated perianth tubes. This genus differs from *Litsea* in its flowers having fewer perianth segments and stamens and in its leaves generally not being pinnate-veined.

Native to Japan, the Ryukyu Islands, and Korea, ***N. sericea*** (syn. *Litsea glauca*) is an evergreen tree with greenish branchlets and leathery, oblong to ovate-oblong leaves, 3½ to 6½ inches long, with stalks from ¾ to a little over 1 inch long. They have three chief veins from their bases and when young are clothed with yellow, silky hairs. Their under surfaces are white or silvery and sometimes glaucous. The flowers are in yellow-hairy, dense umbels. The subspherical fruits are up to ½ inch long, and red.

Garden and Landscape Uses and Cultivation. Of minor horticultural importance, the species described is planted to some extent for variety in regions where little or no frost occurs. It prospers in ordinary garden soil in sun or part-shade. Propagation is by cuttings under mist or in a greenhouse propagating bed and by seed.

Neolloydia conoidea

NEOLLOYDIA (Neo-llóydia). About eight species of the cactus family CACTACEAE constitute *Neolloydia*, a genus related to *Thelocactus*. It is native in the southern United States, Mexico, and Cuba. The name honors Francis Ernest Lloyd, an American for many years professor of botany at Montreal, Canada. He died in 1947. By some authorities *Glandulicactus* and *Gymnocactus*, treated separately in this Encyclopedia, are included in *Neolloydia*.

Neolloydias are small cactuses with usually clustered, spherical to cylindrical stems or plant bodies that have distinctly tubercled (lumpy), spiraled ribs. The tubercles are grooved on their upper sides, and each one ends in a cluster of many spreading radial spines and one or more sturdier, longer centrals. The comparatively large pink to purple blooms come from axils of the tubercles near the tops of the plants. They open by day. The scaly or few-scaled fruits, flattened-spherical, later become dry and papery.

Mexican ***N. ceratites*** has more or less egg-shaped, grayish-green stems about 4 inches tall by one-half as wide. Its spine clusters are of about fifteen grayish radials somewhat over ½ inch long and five or six stout centrals ¾ to 1 inch long. Straight, they are gray with black tips. The purple blooms are about 1¼ inches across. Native to Mexico and Texas, ***N. conoidea*** (syn. *Echinocactus conoideus*) usually forms clusters of gray-green short-cylindrical stems 3 to 4 inches long by 2 to 3 inches thick. Their blunt tubercles are woolly in their axils and densely covered with numerous spines in clusters of ten to sixteen white radials ¼ to ½ inch long and one to five black centrals up to 1¼ inches long. From 2 to 2¼ inches in diameter, the flowers are purple. The Texan form of this species, rather smaller and with longer spines than more typical specimens from Mexico, has been distinguished as ***N. texensis,*** but modern botanists tend not to accept the segregation. Similar in habit to *N. conoidea*, but its spine clusters of about twenty-five radials and one or two black centrals or none, ***N. grandiflora,*** of Mexico, has stems about 4 inches tall by about 3 inches wide. The purple blooms are approximately 2½ inches wide. Mexican ***N. matehualensis*** has gray-green, cylindrical stems about 6 inches long by 2 inches wide. The spines are in clusters of ten to twelve grayish radials, some ½ inch long, and two or three light brown centrals ¾ inch long. The blooms of this are not known.

Garden Uses and Cultivation. These are interesting plants for collectors. Except for *N. conoidea*, most can be rather balky under cultivation. They appreciate full sunlight and an acid, very sandy soil mixed with chips of shale or stone. Moderate amounts of water are given from spring to fall, none or scarcely any should be given in winter. For more information see the Encyclopedia entry Cactuses.

NEOMARICA (Neo-márica)—Apostle Plant or Twelve Apostles. Entirely tropical and subtropical American, *Neomarica*, of the iris family IRIDACEAE, comprises fifteen species. Previously named *Marica*, and still often called that in gardens, its present designation is formed of its older one (the name of a nymph), preceded by the Latin *neo*, new. The change became necessary after it was discovered that *Marica* was a synonym of the related genus *Cipura*. The kinds usually grown are often known as apostle plants and twelve apostles, because it is popularly believed that blooms develop only on plants with fans of twelve leaves.

Neomaricas are iris-like, herbaceous perennials. They have short rhizomes and erect fans of all-basal, sword-shaped leaves. The flower stalks are erect and terminate in a large leafy spathe (bract). The flower stalk is short, apparently lateral, and often nodding. It bears few to several beautiful white, blue, blue-and-white, or yellow, iris-like blooms, each of which ordinarily lasts only one day. Their tubeless perianths have six segments (commonly called petals), the spreading, broadly-ovate, outer three very different from the erect, fiddle-shaped, much smaller inner ones. There are three short stamens, one style, three-angled in its upper half and with three crested branches that are opposite the stamens. The fruits are oblong capsules.

Widely cultivated ***N. caerulea,*** a native of Guinea and Brazil, is 3 to 5 feet tall. Its sword-shaped leaves, without definite midribs, are 2 to 3 feet long and up to 1½

Neomarica northiana

Neomarica northiana (flower)

inches broad. The flower stalk is as broad as the leaves and has bractlike leaves 2 to 3½ inches long. The bright blue flowers, 3 to 4 inches wide, are marked at the bases of their petals with brown crossbars. From 2 to 3 feet tall, ***N. gracilis,*** indigenous from Mexico to northern Brazil, has leaves up to 2½ feet long by ¾ inch wide, with a distinct midrib. The flower stalk, conspicuously winged in its upper part, has a terminal spathe about 1 foot long, and with a distinct midrib. The pretty white or blue flowers are about 2 inches in diameter. Vigorous ***N. northiana,*** of Brazil, is up to 3 feet tall. Its leaves may be 2 feet long and 2 inches in width. The fragrant flowers, 4 inches across, have white outer petals mottled with yellow and purple at their bases, and sky-blue inner petals similarly marked low down. Yellow-flowered ***N. longifolia,*** of Brazil, about 2 feet tall, has leaves about 1 inch wide. Its blooms, crossbarred with brown, are 2 inches wide.

Neomarica longifolia

Garden and Landscape Uses. Neomaricas are not hardy, but are splendid garden adornments where little or no frost is experienced. They are suitable for using in beds and borders in the manner of irises and are appropriate for cultivation in sunny greenhouses and windows. Because of their short lives the blooms do not serve well as cut flowers. Left to develop on the plants the flowers succeed each other in flushes and give a fine show over fairly long periods.

Cultivation. Neomaricas grow gratefully in any well-drained, not-too-dry, nourishing soil. They need full sun and are readily increased by division and seed. Indoors they succeed where the night temperature in winter is 55 to 60°F and that by day and at other seasons somewhat higher. Dividing and repotting is done in late winter, following which water is given rather sparingly until new roots are well established, then water is applied freely, except that from November to February only sufficient is given to prevent the rhizomes from shriveling.

NEOMOOREA (Neo-moòrea). The one attractive species that constitutes this genus, of the orchid family ORCHIDACEAE, is occasionally cultivated. The name, a replacement for *Moorea,* honors Sir Frederick Moore, distinguished botanist and Director of National Botanic Gardens, Dublin, Ireland, who died in 1949.

Native to Panama and Colombia, ***Neomoorea wallisii*** (syn. *N. irrorata*) in the wild grows in the ground or perched on trees. It has flattened, furrowed, egg-shaped pseudobulbs, 3 to 5 inches in length, each surmounted by two elliptic-lanceolate, stiffish, rather thick, pleated leaves up to almost 3 feet long by 4 to 6 inches wide. The erect racemes of ten to twenty 2- to 2½-inch-wide, fragrant, fleshy blooms rise to a height of nearly 2 feet. The sepals and petals are reddish-brown with white bases. The deeply three-lobed lip has its middle part pale yellow dotted with red, its side lobes yellow banded with brownish-purple.

Garden Uses and Cultivation. An outstanding orchid, this has much the aspect of a *Stanhopea*. Easy to grow under humid, tropical conditions, it does well in pots in osmunda or tree fern fiber with which a little fibrous loam, coarse sand, and chopped charcoal has been mixed. From spring to fall ample supplies of water are needed, and then well-rooted specimens may be fertilized regularly. In winter drier conditions at the roots are in order, but complete dryness must be avoided. Unless made quite necessary because of deterioration of the rooting medium, repotting should not be done. Root disturbance is very inimical to this orchid, but then so is stale, waterlogged, unaerated potting compost about its roots. Reasonable shade from bright sun is necessary. For more information see Orchids.

NEOMYRTUS (Neo-mýrtus). The only species of this New Zealand genus was formerly included in *Myrtus*. Its name combines that of that genus with the Greek *neo*, new. It is a member of the myrtle family MYRTACEAE.

An evergreen shrub or tree up to 20 feet tall, ***Neomyrtus pedunculata*** (syn. *Myrtus pedunculata*) has four-angled branchlets and short-stalked, obovate-oblong leaves from ½ to nearly 1 inch long by two-thirds as broad. The flowers, solitary from the leaf axils, slender-stalked, and about ¼ inch in diameter, are white. They have a more or less top-shaped calyx with five lobes (sepals), the same number of petals and many stamens, and one style. The fruits are red or yellow, broadly-egg-shaped berries about ¼ inch long.

Garden and Landscape Uses and Cultivation. This has much the same horticultural uses as the common myrtle (*Myrtus communis*). It is probably best suited for use in North America in mild parts of the Pacific Northwest. Propagation is by seed and by cuttings.

NEOPANAX (Neo-pánax). This is a genus of six species of evergreen trees and shrubs previously included in *Nothopanax*. It belongs in the aralia family ARALIACEAE and is native only in New Zealand. Its name is from the Greek *neos*, new, and *Panax*, another genus of the same family. A notable characteristic is that the foliage may show great diversity of form at various stages of development of the plants. This is so true that juvenile and adult trees of some kinds have been described and named by competent but not fully informed botanists as different species.

Neopanaxes are shrubs or trees with leaves of several leaflets that spread from the top of the leafstalk in handlike fashion, or in adult-type foliage, that may consist of only one leaflet. In compound umbels (umbels of smaller umbels) or rarely in simple umbels, the tiny flowers have a minute calyx, five petals, five stamens, and two styles. The fruits are berry-like.

A usually unisexual, stout-branched, hairless shrub or tree up to 20 feet tall, ***N. colensoi*** (syn. *Nothopanax colensoi*) has leaves of three to five leathery, narrowly-obovate to elliptic-oblong, stalkless or nearly stalkless leaflets, 2 to 6 inches long, and coarsely-toothed above their middles. The flowers, in crowded terminal umbels of smaller umbels, are succeeded by more or less spherical, nearly black fruits up to ¼ inch in diameter. From the last, ***N. arboreus*** (syn. *Nothopanax arboreus*) differs chiefly in having leaves with five to seven distinctly-stalked, toothed, ovate leaflets up to 8 inches long and 1 inch to 3 inches wide. It is a somewhat taller, round-headed, unisexual tree that in the wild generally is found at lower elevations than *N. colensoi*. Its fruits are black and up to ⅓ inch in diameter. A shrub or tree up to 20 feet tall, ***N. laetus*** has thick, leathery leaves with five to seven elliptic leaflets 6 inches to 1 foot long that in their upper two-thirds are toothed. The leafstalks and mid-veins of the leaflets are purplish-red. The fruits are dark purple.

Neopanax laetus

Most variable is ***N. simplex*** (syn. *Nothopanax simplex*), a shrub or tree up to 25 feet in height. The usual progression is that as a young seedling it has long-stalked, ovate, toothed leaves. Next, five-leafleted leaves are produced, with the leaflets stalked and linear or pinnately-lobed. The next change is to leaves with three stalkless, lanceolate leaflets, and finally, at maturity, undivided, more or less toothed leaves, 2 to 5 inches long, are produced. In short umbels composed of few smaller umbels, the flowers are borne in the leaf axils and from the shoot ends. Male and female flowers are on the same plant. The fruits are up to ⅙ inch across. Variety *N. s. sinclairi* (syn. *Nothopanax sinclairi*) passes quickly from the seedling stage to the form with leaves with three leaflets and then retains that characteristic throughout life.

A shrub up to 10 feet tall, ***N. anomalus*** (syn. *Nothopanax anomalus*) is very distinct. It has densely-forking branches and juvenile foliage of wing-stalked leaves of three, sometimes pinnately-lobed, leaflets. The leaves of its adult phase, small and undivided, have flattened stalks and obovate to rounded, toothed blades ½ to ¾ inch long. The umbels, each of up to ten minute flowers, are on short stalks. Up to 30 feet in height, ***N. kermadecense*** (syn. *Nothopanax kermadecensis*) is a tree with slender-stalked leaves of usually seven pointed, elliptic to somewhat obovate, stalked leaflets and umbels of flowers more compact than those of *N. arboreus*.

Garden and Landscape Uses and Cultivation. In warm, mild climates where little frost is experienced neopanaxes are interesting and useful for outdoor planting. They provide diversity and are something of conversation pieces. They grow well in part-shade or sun in ordinary well-drained soil and are propagated by seed and air layering.

In greenhouses they succeed where the winter night temperature is 45 to 55°F and daytime temperatures are a few degrees higher. Moderately humid conditions are appropriate and some summer shade is desirable. Water should be given rather sparingly in winter, moderately at other times. Well-rooted specimens benefit from spring-to-fall regular applications of dilute liquid fertilizer.

NEOPORTERIA (Neo-portèria). As understood by those who split genera of the cactus family CACTACEAE finely, *Neoporteria* embraces thirteen species, all natives of Chile. Conservative botanists also include in *Neoporteria* sorts treated in this Encyclopedia as *Neochilenia*. The name honors Carlos Emilia Porter, a nineteenth-twentieth century Chilean entomologist.

Neoporterias have spherical to long-cylindrical stems or plant bodies with notched, usually straight ribs and clusters of generally very strong, stout spines. The funnel-shaped flowers, open by day, have perianth tubes, ovaries, and fruits furnished with a little felt of hairs. The fruits are small, spherical, and dry.

Neoporterias cultivated include these: ***N. castaneoides*** (syn. *Chilenia castaneoides*) has usually solitary, spherical to cylindrical, green or grayish-green stems up to 6 inches tall by 3½ inches thick with about fourteen ribs, and at their tops whitish wool. The spine clusters are of about twenty white radials, up to ¾ inch long, and six or seven honey-yellow to light brown, thicker centrals somewhat over 1 inch in length. The 3-inch-long blooms, with pink to red outer petals and white inner ones, have white stamens and a white style, tipped with an eight-lobed, reddish stigma. ***N. clavata*** (syn. *Arequipa clavata*) has a grayish-green stem, spherical at first, later cylindrical or club-shaped, and at its top, 8 inches wide. Up to 5 feet tall, it has about ten prominently notched ribs with spine clusters of four to six thick straight or curved radials up to 1¼ inches long and one central. The nearly cylindrical red blooms have petals a little over ½ inch long. Variety *N. c. grandiflora* has bigger blooms. ***N. gerocephala*** (syn. *N. senilis, N. nidus senilis*) is much like *N. nidus*, except that its shorter, slenderer, more twisted spines are all creamy-white to white. ***N. heteracantha*** (syn. *Chilenia heteracantha*) has flattish-spherical plant bodies approximately 4½ inches wide, with about nineteen ribs, and spine clusters of about twenty bristly radials almost 1 inch long and about six centrals nearly twice as long. The flowers are carmine-pink with white bases to the petals. ***N. littoralis*** has semispherical, gray-green plant bodies up to 3½ inches across with fourteen to twenty-one ribs. The clusters of yellowish to blackish spines are of about thirty hairlike radials, up to ¾ inch long, and eight to twelve centrals, up to 1¼ inches long. Approximately 1 inch long by ¾ inch across, the blooms are carmine-pink. ***N. mamillarioides*** (syn. *Malacocarpus mamillarioides*) has subspherical to short-cylindrical, bright green plant bodies with fourteen to sixteen strongly-notched ribs, clusters of ten to twelve short, slender radial spines, about ⅓ inch long, and four centrals arranged crosswise. The yellowish-red flowers are 1½ inches wide. ***N. nidus*** has solitary, grayish-green plant bodies at first spherical to ovoid, but later cylindrical, 2 to 4 inches thick. They have sixteen to eighteen ribs and are thickly covered with curved spines in clusters of about thirty. The radials are slender, white, hairlike, and up to ¾ inch long, the six to eight centrals, thicker, more twisted, and up to 2 inches long. The pinkish to reddish flowers are about 2½ inches long. ***N. nigrihorrida*** (syn. *Chilenia nigrihorrida*) has flattish-spherical stems, up to about 2½ inches tall by 4 inches wide, with sixteen to eighteen ribs. The spine clusters are of about sixteen often twisted radials, up to a little over ½ inch long, and six or seven awl-shaped centrals that may exceed 1 inch in length. Gray, they turn black when wet. The flowers, carmine-red with white interiors, are about 1½ inches long. ***N. subgibbosa*** (syn. *Chilenia subgibbosa*) has spherical to short-cylindrical stems up to 5 inches high by 3½ inches wide with fourteen to sixteen ribs and clusters of about twenty-eight amber-yellow spines, which later turn gray. The blooms are 1½ inches long. ***N. villosa*** has stems at first spherical, later cylindrical, up to 5 inches tall and 3½ inches wide, grayish to reddish-green to almost black and with thirteen to fifteen ribs. The spine clusters are of twelve to sixteen dark-tipped, yellowish, bristly radials, up to ¾ inch long, and about four stouter, darker-colored centrals that may exceed 1 inch in length. The many flowers, 1 inch to 1¼ inches long, have pink petals with white bases, white stamens, and a white style.

Garden and Landscape Uses and Cultivation. Neoporterias are good-looking plants for gardens outdoors in warm dry regions and for inclusion in greenhouse collections of cactuses. They respond to environments and care that suit *Cephalocereus* and similar desert cactuses. For more information see Cactuses.

NEORAIMONDIA (Neo-raimóndia). The remarkable genus *Neoraimondia*, of the cactus family CACTACEAE, consists of four species, natives of Peru and northern Chile. Its name honors the nineteenth-century Peruvian naturalist and geographer Antonio Raimondii. These are massive, columnar cactuses with erect, prominently-ribbed, stout stems, and strong spines.

Branching freely from its base, ***N. arequipensis*** (syn. *N. macrostibas*) forms bushes with stems 6 to 12 feet high. The unusual feature of the plant is its greatly enlarged areoles (specialized areas on cactus stems from which spines, hairs, and flowers develop). These sit like globose or elongated brown cushions, up to 4 inches high, studded along its about eight widely spaced ribs. From each areole arises a dozen or more spines of unequal lengths, the biggest probably the largest of any in the cactus family and up to 1¼ feet long, but usually shorter. The solitary or paired, greenish-white, funnel-shaped blooms are 1 inch to 1½ inches long and under 1 inch wide. Egg-shaped to spherical and up to 3 inches across, the purple fruits have brown-woolly areoles with short spines. Robust and about 25 feet tall, ***N. gigantea*** (syn. *N. arequipensis gigantea*) has dense clusters of stems with four or occasionally five ribs. Its blooms are purplish-pink. Not over 6 or 7 feet in height, ***N. roseiflora*** has stems with usually five or six, rarely four, ribs. Its blooms are pink.

Garden and Landscape Uses and Cultivation. These are those of other columnar cactuses of the *Cereus* relationship. For more information see Cactuses.

NEOREGELIA (Neo-regèlia) — Fingernail Plant. This horticulturally important genus, of the pineapple family BROMELIACEAE, is named in honor of Eduard Albert von Regel, one-time Director of the botanical garden at what was then St. Petersburg, Russia, and distinguished editor of the German magazine *Gartenflora*. He died in 1892. The group now named *Neoregelia* has suffered several name changes. At various times it has been dubbed *Karatas, Regelia*, and *Aregelia*. Its present name was first applied by the distinguished American student of the family Dr. Lyman B. Smith. There are forty species. The genus is essentially Brazilian.

Neoregelias are generally epiphytes (plants that, like many orchids, perch on trees without taking sustenance from them). Some sometimes grow in the ground. They are stemless, evergreen perennials of varying dimensions and forms, but compared with bromeliads in general are mostly medium in size. Their commonly prickly-toothed leaves form rosettes typically cupped at their bases to hold water. Depending upon kind, the foliage ranges from very rigid to leathery or softer-textured, from scaly to smooth and shining. It may be clear green or various tones of bronzy-red to maroon, in some kinds tipped with red or variegated with bands, marblings, or spots differing in hue from the ground color. The small flowers, not individually showy and mostly blue, purple, or white, are in compact heads that nestle low in the centers of the rosettes and are surrounded by bractlike, often highly colored, shorter leaves. Each bloom has three sepals generally united by their bases, three spreading petals, their lower parts joined into a tube, and six stamens longer than the style, which has three twisted branches. The fruits are many-seeded berries. In addition to the natural species, many hybrid neoregelias have been raised. The most easily observable difference between neoregelias and nidulariums is that the flower heads of the former are not branched and are so low in the cups of the rosettes that their stalks are not visible; also, the petals of their flowers spread widely during the few hours before they fade, and when they fade the flowers become twisted. In *Nidularium* the flower heads are definitely branched and except in one or two rare species the blooms remain tubular, do not spread their petals, and do not twist when they die.

One of the most beautiful and popular species is ***N. carolinae*** (syns. *N. marechalii, Aregelia marechalii*). Somewhat variable, this has thickish, strap-shaped, glossy, bright green, finely-spine-toothed leaves 1 foot to 1½ feet long by approximately 1½ inches wide. The fiery hearts of the rosettes are of often blue-tinged, red bracts that develop as the flowering stage nears. These form long-lasting decorative foils for the slightly raised heads of blue-purple blooms. Variety *N. c. meyendorffii* is perhaps not distinct from the typical species. The leaves of *N. c. tricolor* are beautifully variegated with longitudinal stripes of ivory-white that in mature specimens are suffused with pink. Similar to *N. carolinae*, but with stubbier, deeper-colored leaves, ***N. farinosa*** is equally as attractive. Under conditions of good light its foliage becomes coppery-olive-green to purplish. The central bracts, as the flowering stage approaches, become brilliant crimson.

The fingernail plant, as it is sometimes called, ***N. spectabilis*** (syn. *Aregelia spectabilis*) earns that designation by reason of its strap-shaped, slightly concave, leath-

Neoregelia carolinae

Neoregelia ampullacea

Neoregelia ampullacea (flowers)

Neoregelia carolinae meyendorffii

Neoregelia carolinae tricolor

ery, erect leaves being tipped with brilliant cerise-red "fingernails." There are other kinds that also display this feature. Approximately 1 foot long by 1½ to 2 inches broad, the leaves of *N. spectabilis* are green above, faintly to definitely horizontally banded with ashy-gray on their undersides. The outer ones may be more or less spiny, the inner ones smooth-edged. The bracts surrounding the head of blue flowers are not highly colored. They are banded with purple. In cultivation this species shows some variation, especially in the foliage assuming tints of rosy-pink. One such varient is named 'Pinkie'. Another species with red "fingernails" tipping its leaves is ***N. cruenta*** (syn. *Aregelia cruenta*). This has rosettes of blunt, strap-shaped, yellowish leaves about 1 foot long by 3 inches wide, red-spine-margined and ending in a brief point. Bluish bracts encircle the head of blue flowers.

Miniature ***N. ampullacea*** (syns. *Aregelia ampullacea, Nidularium ampullaceum*), unlike the great majority of neoregelias, produces numerous suckers. Its tubular rosettes are of few slender-pointed, linear leaves 4 to 5 inches long, red-tipped and banded and flecked with reddish-brown, most plentifully on their undersides. The blue and white flowers, in heads of about a dozen, are surrounded by green bracts. Variety *N. a. tigrina* is more brightly variegated than the species. Of comparable size, ***N. pauciflora*** differs from *N. ampullacea* in its foliage being speckled instead of cross-banded with mahogany-brown.

Other species of *Neoregelia* cultivated include those now to be described. Additional kinds may be expected in the collections of enthusiasts and are likely to be offered in the catalogs of specialist dealers. There, too, will be found listed a variety of attractive hybrids. ***N. babiana*** has tubular rosettes of bright green, glossy leaves, the innermost with red upper surfaces. The flowers have white, blue-tipped petals. This grows slowly. ***N. carcharodon*** has gray-green to green, red-tipped leaves blotched and banded on their undersides and spotted above with purplish-maroon. The white flowers have lavender-apexed petals. ***N. concentrica*** (syns. *Aregelia concentrica, Nidularium acanthocrater*) forms tight, thick-set rosettes of blunt, spreading leaves, fresh green on their upper surfaces, grayish-streaked beneath, margined with black spines and, the young leaves particularly, spotted with dark brown. The bracts surrounding the flower heads are yellowish-white suffused with violet or are purplish-carmine. An intermediate hybrid between this and *N. johannis* is named 'Vulcan'. ***N. fosterana*** has in dense rosettes broad coppery leaves blotched with grass-green. The tips of the leaves and,

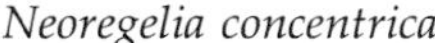

Neoregelia concentrica

Neoregelia concentrica (flowers)

Neoregelia marmorata 'French Hybrid'

Neoregelia marmorata 'French Hybrid' (flowers)

Neoregelia mooreana

when flowering time approaches the bracts, are burgundy-red. The blooms are pale blue. ***N. johannis*** has shining, dark green, thin but firm, leathery leaves, fluted at their bases and more or less clothed with grayish scales. Beneath they are purple with gray lines. The heads of bright blue flowers are in the lavender-violet centers of the rosettes. ***N. marmorata*** (syn. *Aregelia marmorata*) is rarely cultivated. The plant usually grown under its name is a hybrid between it and *N. spectabilis*. The true species is robust and has pale green leaves magnificently marked on both surfaces with reddish-brown spots that run together. They are tipped with bright red. The hybrid has olive-green foliage attractively blotched with maroon. Its leaf apexes are red, its flowers lavender. ***N. mooreana*** has tubular rosettes 8 to 10 inches tall of arching, leathery leaves that taper to slender, more or less recurved, pointed apexes. Often marked with pale longitudinal lines, they have margins furnished with black spines. The flowers are white. ***N. pineliana*** (syn. *Aregelia pineliana*) has gray-scaly, coppery-green, narrowly-strap-shaped leaves and red bracts surrounding blue flowers. ***N. princeps*** (syn. *Aregelia princeps*) has short, broad, glaucous-green leaves with prickly margins. They are up to about 1 foot long by 1½ to 2 inches wide. The heads of violet-blue blooms are displayed against collars of brilliant orange-scarlet bracts. ***N. sarmentosa*** has slender rosettes of channeled, linear leaves up to about 1 foot long by 1¼ inches wide, with a brief point at their rounded apexes. There are few spines along their margins. The heads of few blue flowers are surrounded by red bracts. ***N. tristis*** has red-tipped, prickly-edged leaves 6 to 7 inches long by about ¾ inch wide, spotted especially toward their bases with brownish-red, their undersides with grayish crossbands. The flowers are lilac. ***N. zonata*** is low and compact. It has olive-green leaves conspicuously banded and marbled on both surfaces with purple-red. The flowers are white, with blue apexes to the petals, or pale blue.

Garden and Landscape Uses. Neoregelias are splendid for outdoor cultivation in the humid tropics and in tropical greenhouses. Some are fairly satisfactory houseplants. If the atmosphere is excessively dry, they will not prosper permanently, but they are able to adapt pretty well to not exactly ideal conditions, remaining attractive for many weeks or even months. They are well suited for growing in terrariums. For details of their cultivation see Bromeliads.

NEOTRELEASIA. See Setcreasea.

NEOWERDERMANNIA (Neowerder-mánnia). One or two species by some authorities referred to *Gymnocalycium* belong in South American *Neowerdermannia*, of the cactus family CACTACEAE. The name honors Dr. Erich Werdermann, Director of the botanic garden, Berlin, Germany. He died in 1959.

Endemic high in the Andes from Bolivia to Argentina, ***N. vorwerkii*** has much the aspect of a *Mammillaria,* but is distinct in its blooms coming from the ends of the tubercles rather than from depressions between them. Globular, green, and with about sixteen ribs, this cactus develops a deep, thick taproot. Its tubercles (protrusions from the plant body) are triangular and flat-topped. At their tips are clusters of seven to ten or more ½- to ¾-inch-long, spreading, curved spines and one about twice as long and hooked. When young, the spines are purplish-brown, later they become yellowish or light gray. The largest are about 2 inches in length. The funnel-shaped blooms, 1 inch long by ¾ inch wide, have spreading, pointed-lanceolate, white perianth lobes (petals) with lilac midribs. When mature the fruits protrude from the surface of the areole. The flowers of *N. v. gielsdorfiana* have blunt, pinkish-lilac petals. It is smaller than *N. vorwerkii* and has much shorter spines.

Garden and Landscape Uses and Cultivation. These are as for mammillarias. For more information see Cactuses.

NEPENTHACEAE—Nepenthes Family. This family of dicotyledons has the characteristics of what is usually accepted as its only genus *Nepenthes.*

NEPENTHES (Nep-énthes)—Pitcher Plant. Carnivorous plants, those that obtain part of their nourishment from small creatures they trap, have a fascination of their own. There are many different kinds, belonging to various families and widely distributed in many parts of the world. Not least interesting are the tropical Old World pitcher plants belonging to the genus *Nepenthes.* These are quite distinct from American pitcher plants (*Sarracenia*) and belong in a different, but related family. The chief, and according to many botanists, only representative of the nepenthes family NEPENTHACEAE, the genus *Nepenthes* occurs in the wild in tropical Asia, Malaysia, New Caledonia, and northern Australia. It consists of more or less woody, sprawling or vining plants that grow in the loose, organic debris of the forest floor or sometimes where it has collected in the crotches of tree branches and similar places. Some botanists recognize a species native of the Seychelles Islands as a second genus of the nepenthes family and name it *Anurosperma.*

The greatest horticultural interest in *Nepenthes* occurred in the last half of the nineteenth and the early part of the twentieth century. As early as 1830 problems associated with growing these denizens of the tropics in greenhouses had been solved, although their successful cultivation remained challenging. After the middle of the nineteenth century, hybrids, which proved generally easier to grow than the wild species and had more and as beautiful or more beautiful pitchers, were developed. Nor were all these hybrids simple ones between two species; some were complex kinds involving three or four parent types. Such was the fascination of *Nepenthes* that fine hybrids were raised in Great Britain, France, and the United States, and pampered in the tropical greenhouses of botanical gardens and wealthy amateur horticulturists as well as of nurserymen who specialized in catering to the latter's interests. Now, a century or so after the major excitement that attended the early cultivation and breeding of *Nepenthes,* they are comparatively rare as cultivated plants. Occasionally orchid fanciers grow a few as curiosities (they are not, of course, orchids, but their environmental needs are similar to those that suit many tropical orchids) and quite extensive collections are maintained at Longwood Gardens, Kennett Square, Pennsylvania, and at the Royal Botanic Gardens, Kew, England.

Nepenthes greenhouse, Royal Botanic Gardens, Kew, England

The genus *Nepenthes,* the name of which comes from the Greek *nepenthes,* without care or removing sorrows, consists of nearly seventy species in addition to some splendid garden hybrids. In their native forests the more vigorous kinds have stems that attain lengths of 70 feet, but in cultivation they are ordinarily restrained to from 2 to 3 feet. The leaves are alternate and have flat blades from the ends of which protrude tendril-like portions of the midribs that terminate in the truly remarkable, and often quite beautiful, lidded pitchers. Misconceptions exist about these. They are not, as nonbotanical people sometimes surmise, flowers. They are portions of the leaves highly modified to serve a special function. Nor do their lids close or other parts move to trap their prey, as happens in the Venus's fly trap (*Dionaea*). Pitchers of *Nepenthes* are passive traps, baited to attract insects and so constructed that when these or other small creatures enter they are unable to climb out because of the smooth surfaces and inrolled rim. Perforce they must remain to drown in the abundant digestive fluid the pitchers hold and add their bodily remains to the nutritive brew from which the plant absorbs part of the nitrogen and other elements it needs. The flowers of *Nepenthes* are in long slender racemes or panicles, usually from longish shoots. They are small, unisexual, and the sexes are on separate plants. They have no petals. The males have four yellowish, green or reddish sepals and ten to twenty stamens. The females have the same number of sepals and a short-stalked or stalkless, four-lobed stigma. Neither are showy and both secrete abundant nectar, which together with the rather evil odor of some kinds, attracts insects that carry pollen from male to female. The fruits are greatly elongated, many-seeded capsules.

Kinds from the wild that may be cultivated include slender ***N. gracilis,*** which has pitchers not over 4 inches long by 1 inch wide, often smaller and pale green dotted or flecked with purple. This is native to Malaya, Borneo, and Sumatra. In contrast, ***N. hookerana,*** a native of Borneo and probably a natural hybrid, exhibits

Nepenthes hookerana

greater vigor. One of the easiest to grow, it has handsome, richly colored, pale green, blotched-with-purple pitchers up to 5 inches long by 3 inches wide. Another kind amenable to cultivation is ***N. maxima,*** of the Celebes Islands, Borneo, and New Guinea. This, the parent of several fine hybrids, has green pitchers beautifully splotched with purple. Native from the Malay Peninsula to Sumatra and Borneo, ***N. rafflesiana*** is a straggling climber. Its lower pitchers are urn-shaped, its upper ones trumpet-shaped and 8 to 10 inches long. They are greenish-yellow, conspicuously mottled with purplish-brown. From

Nepenthes rafflesiana

Hybrid nepenthes: (a) *Nepenthes chelsonii*

(b) *Nepenthes dicksoniana* (flowers and pitcher

(c) *Nepenthes mixta sanguinea*

(d) *Nepenthes superba*

(e) *Nepenthes* 'F. W. Moore'

(f) *Nepenthes ratcliffiana*

the Philippine Islands comes ***N. ventricosa,*** which succeeds in somewhat lower temperatures than most. It has pitchers up to 6 inches long by almost one-half as wide.

Hybrid nepenthes in cultivation at Longwood Gardens in 1965 were *N. chelsonii, N. dicksoniana, N. dormanniana, N. edinensis, N. intermedia, N. mastersiana, N. mixta sanguinea, N. superba, N. williamsii, N.* 'Lt. R. B. Pring', *N.* 'Marcoz', and *N.* 'St. Louis'. Other hybrid sorts in cultivation include *N.* 'F. W. Moore', and *N. ratcliffiana.* It would serve no useful purpose to attempt to describe these hybrids here. Those particularly interested will find helpful descriptions of many of these and other hybrids as well as of a long list of species in *The Standard Encyclopedia of Horticulture* by L. H. Bailey.

Garden Uses and Cultivation. These are choice plants for the collector of rare tropicals. For their successful growth, the cultivated kinds of *Nepenthes* need the hot, steamy atmosphere associated with lowland, tropical jungles. It is possible, judging from conditions under which some of the higher altitude species reportedly grow naturally, that there are kinds that prosper under rather cooler conditions, but these are not in cultivation. A shaded greenhouse with a constant relative humidity of 75 percent or higher, and a minimum winter temperature of 65 to 70°F is ideal. On bright winter days the temperature may be increased by five or ten degrees and in the spring to fall season of active growth may be allowed to reach 90°F or higher. Nepenthes are grown in hanging baskets, preferably of cedar, teak, or other rot-resisting wood. The soil should be highly organic and very loose and porous so that water and air pass through readily. A mixture consisting of a little turfy loam (the fibrous roots of partially decayed grass sods shaken partially free of soil), and a much larger proportion of coarse orchid peat (the roots of osmunda fern), fibrous peat, or similar material, together with some sphagnum moss, and, to keep the whole open and sweet, some chopped charcoal, as well as some fir bark of the type in which orchids are potted, and some coarse sand, is satisfactory.

Planting new and refurbishing old baskets is done in early spring. At that time, too, second-year and older plants are pruned by shortening their stems to 9 inches to 1 foot from their bases. New growth is encouraged by frequently spraying the plants with a mist of water. This is most easily done by an automatic misting system. When the new shoots have made four to six leaves, their tips are pinched out. This concentrates the energy of the shoot on the development of large pitchers instead of further growth. Only if flowers are needed to secure seeds should the shoots be allowed to grow without pinching. Following planting, watering should be done with some caution, but once roots have penetrated the rooting medium generous supplies are needed to keep the roots bathed in moisture, but not deprived of air (hence the need for porous soil and perfect drainage). Well-rooted plants benefit greatly from weekly applications of dilute liquid fertilizer. It is not necessary or desirable to "feed" the plants by supplying them with organic matter in the form of insects, bits of meat, or the like, placed in the pitchers. This can be very harmful. In practice it rarely pays to keep plants of *Nepenthes* after their second year; young specimens give superior results. The best means of propagation is by cuttings of firm shoots taken in December or January. The stem of each cutting is pushed through the hole in the bottom of an inverted, 2½-inch pot, which is almost buried in a bed of sphagnum moss or coarse vermiculite in a greenhouse propagating bench where the temperature is 80 to 90°F. The atmosphere should be highly humid and the cuttings shaded from sun. When roots about 1 inch long have developed, the pots are broken and the rooted cuttings potted in small well-drained pots. Later they are transferred to 4-inch pots and subsequently to baskets.

Nepenthes are easy to raise from seeds sown in well-drained pots or pans (shallow pots) filled with a similar, but less coarse soil than that used for larger plants. In a temperature of 80 to 90°F, they germinate in four or five weeks. When the seedlings have produced four or five leaves they are potted individually in small pots and treated in the same manner as rooted cuttings.

NEPETA (Né-peta)—Catmint or Catnip, or Catnep. Annuals and herbaceous perennials, natives of Europe, Asia, North Africa, and the mountains of tropical Africa, to the number of 250 species, constitute *Nepeta*, of the mint family Labiatae. The name, of uncertain origin, is perhaps derived from that of the Etrurian city Nepete. The plant previously called *Nepeta hederacea* is *Glechoma hederacea*.

Nepetas are mostly aromatic. They have opposite, incised or toothed leaves, and whorls (circles of more than two) of blue, lavender, or white, asymmetrical flowers in more or less elongated spikes or panicle-like arrangements. The flowers have tubular, five-toothed, fifteen-ribbed calyxes and two-lipped corollas, with the upper lip straight and the lower one spreading and three-lobed. The center lobe is larger than the others and cupped. There are four stamens in pairs, the upper pair longer than the lower, and one style with two stigmas. The fruits are of four seedlike nutlets. Catmint has been used in infusions to relieve colds, fevers, and other ailments and as a flavoring in cooking.

Catmint, catnip, or catnep (***N. cataria***) is an old-fashioned garden plant native of Europe and western Asia, widely naturalized in North America. Its common names and botanical specific testify to the extraordinary appeal this species has for cats. It also is a good bee plant. An erect, stately, hardy perennial 2 to 3 feet tall, catmint has densely-grayish-hairy stems and foliage. Its ovate, round-toothed leaves, 2 to 3 inches long, whitish on their undersides, have deeply heart-shaped bases. The dense clusters of small, violet-dotted, white flowers are in heads 1 inch or so long at the ends of stems, branches, and branchlets. Their upper lips are erect.

Nepeta cataria

Nepeta faassenii

Most familiar perhaps to American gardeners is ***N. faassenii***. This is commonly misidentified as *N. mussinii,* although actually it is a hybrid between that species and some other, most probably *N. nepetella,* but perhaps *N. grandiflora*. The hybrid shows some variation. Typically 1 foot to 1½ feet tall, and more erect than *N. mussinii,* it has narrow-ovate or lanceolate, round-toothed leaves, more or less wedge-shaped at their bases. It flowers so profusely that its blooms are presented as clouds of lavender-blue above masses of grayish foliage. Each is about ½ inch long. They are attractive to bees. Cut sprays are delightful in flower arrangements. True ***N. mussinii*** is less erect than *N. faassenii*. It has gray-hairy, round-toothed ovate leaves proportionately broader than those of the hybrid, with heart-shaped bases. Up to 1 foot tall, this species is of loose, lax habit. Its lavender-blue flowers set seeds freely and from these new plants are easily raised.

Fairly common in gardens, ***N. sibirica*** (syns. *N. macrantha, Dracocephalum sibiricum*) is 1½ to 3 feet tall. It has oblong-lanceolate, toothed leaves 2 to 3½ inches long, and minutely-hairy on their undersides. The 1-inch-long or slightly bigger lavender-violet to bluish flowers are in distinctly separated whorls (tiers) assembled in spikelike arrangements 5 inches to nearly 1 foot long. This is a native of Siberia and China. Less well known than the last, ***N. nervosa,*** of Kashmir, is a handsome sort 1 foot to 2 feet tall with conspicuously-veined, linear-lanceolate, toothed leaves, the lower ones with short stalks, the upper ones stalkless. In dense, cylindrical, terminal spikes 1 inch to 3 inches long, the flowers are clear blue or yellow. Native to the Caucasus, ***N. grandiflora,*** 1½ to 3 feet tall, has ovate, round-toothed leaves up to 2 inches long, hairless or with short hairs, with heart-shaped bases. In interrupted spikes the lavender blue to blue flowers are ¾ inch long.

Nepeta sibirica

Nepeta nervosa

Nepeta grandiflora

Garden and Landscape Uses. One of the most useful of edging plants, *N. faasenii* is ideal where a low bordering plant 2 feet or more wide can be accommodated. It is equally as appropriate for grouping at the fronts of perennial beds and for planting over the tops of retaining walls. The other kinds considered above can be used in perennial beds. Catmint is also proper in herb gardens.

Cultivation. No difficulties attend the cultivation of nepetas. Seeds remain viable for several years and germinate satisfactorily, but usually rather slowly, when sown in fall or early spring outdoors or in a cold frame, or in spring in a greenhouse. Most kinds can be increased by division at those same seasons, and by cuttings taken in summer and planted in a greenhouse propagating bench, cold frame, or under a glass jar or little polyethylene tent in a shaded spot outdoors. Nepetas are plants for sunny places and well-drained, dryish soil not too rich in nutrients. Like many herbs of the Mediterranean region they give of their best in environments a little trying for plants that come from areas where milder, moister summers prevail.

NEPHELIUM LITCHI is *Litchi chinensis.*

NEPHRODIUM. See Dryopteris.

NEPHROLEPIS (Nephró-lepis)—Sword Fern or Boston Fern. Few nonhardy ferns are as well known to gardeners and houseplant fanciers as certain members of the genus *Nephrolepis*. Here belong the distinguished Boston fern and its vast array of associate and derivative varieties, as well as a few other kinds cultivated in greenhouses and, in warm climates, outdoors. They represent a small minority of perhaps thirty species that constitute this botanically puzzling genus, which is widespread in the wild throughout the tropics and subtropics, extending to Florida, Japan, and New Zealand. It belongs in the davallia family DAVALLIACEAE and has a name that comes from the Greek *nephros,* a kidney, and *lepis,* a scale, that alludes to the shape of the coverings (indusia) of the spore-case clusters.

There are both terrestrial and epiphytic sword ferns; some kinds grow in both ways. Terrestrial sorts root in the ground, epiphytic perch on trees in the fashion of many orchids without abstracting nourishment from their hosts. Nephrolepises usually have short, erect stems and send out stolons or runners. Their fronds (leaves) in the wild kinds are typically once-pinnate, but in horticultural varieties are often more-times-divided and variously lobed, crested, or frilled. In general they are long and comparatively narrow, erect, arching, or pendulous, and are crowded in tufts or clumps. The arrangement of the spore-case clusters differs according to species. They may form lines along the leaf edges, or appear as roundish bodies on the backs of the fronds. Most familiar of the genus is *N. exaltata,* of which the Boston fern and its kin are variants, but before we deal with that great horde it may be well to consider other species cultivated.

Common throughout the tropics, ***N. biserrata*** (syn. *N. acuta*) has fresh green fronds up to 6 feet long or sometimes longer and 6 inches to sometimes as much as 1 foot wide. They arise from a short stock furnished with numerous scales edged with fine hairs. Their many leaflets are leathery, slightly toothed, and so spaced that they do not touch each other, and may be as far apart as their own width. Spore-bearing leaflets are narrower than barren ones, the latter being ½ to ¾ inch wide. The location of the circular spore-case clusters, well in from the edges of the leaflets, and the numerous spreading, narrow scales at the bases of the leafstalks are distinguishing characteristics of this species. It produces runners freely, and in the wild grows in open or lightly shaded habitats on the ground or on palm trunks or rocks. Variety *N. b. furcans* is distinguished by having the ends of its leaflets conspicuously forked. Another native of the tropics, ***N. hirsutula*** is a vigorous grower with erect to drooping fronds 2 to 4 feet long, up to 6 inches wide, and with the stalks, midribs and undersides conspicuously furnished with soft, tan, hair-

Nephrolepis hirsutula

like scales. The clusters of spore cases are along the margins of the finely-toothed leaflets.

Very different ***N. acuminata*** (syn. *M. davallioides*), a native of Sumatra to the Celebes Islands, has a short stock with the scales near its apex dark brown, and springing from it bright green fronds 3 to 6 feet long by up to over 1 foot wide. They have numerous leaflets, up to ¾ inch wide, notched along their margins into little lobes. The upper leaflets are joined to form a lobed terminal one. Fertile leaflets have a single round spore-case cluster on the underside of each lobe. This species has long, slender, rooting runners. It is commonly an epiphyte.

A handsome native of tropical mountains and of lower altitudes in the subtropics, and extending to Japan and New Zealand, ***N. cordifolia*** (syn. *N. tuberosa*) has short, erect or slanting, scaly stems and neat, erect or nearly erect fronds up to 2½ feet long by up to 2½ inches wide. The edges of the leaflets are toothed. The rounded spore-case clusters are midway between the leaflet edges and their midribs. Many runners are produced and these frequently bear small tubers. Variety *N. c. duffii* (syn. *N. duffii*) never bears spores. Coming from New Zealand and islands of the Pacific, it presumably is a natural mutant (sport) of *N. cordifolia*. Compact and of pleasing appearance, this very

Nephrolepis cordifolia in California

Nephrolepis cordifolia duffii

distinctive kind has erect fronds, usually forked or crested at their tips, strung with rounded, toothed leaflets of such size that the fronds are only about ½ inch wide. They are up to 2 feet long. Other varieties are *N. c. compacta*, with fronds leafy to their bases and closely arranged, dark green, toothed leaflets about 1 inch long, and *N. c. plumosa* (syn. *N. c. tessellata*), which has stiff, dark green, narrowish fronds, the leaflets of which are toothed or finely divided at their ends, and runners with tubers. Related to *N. cordifolia* and much resembling it, but lacking the small tubers common to that species, and with the bases of the ultimate divisions of its leaves with small instead of usually more prominent little lobes or ears, ***N. pectinata*** is a native of tropical America.

Extensively creeping or scrambling ***N. radicans*** (syn. *N. volubilis*) is most easily identified by its rather few, long wiry runners, which, unlike those of other kinds that develop new plants at their ends, have them at irregular intervals along their lengths. Also characteristic of this kind are the round-toothed apexes of its leaflets. Native to tropical Asia, *N. radicans* is terrestrial or epiphytic. It has short stems furnished with blackish scales and fronds ordinarily 2 to 2½ feet long or sometimes

Nephrolepis radicans

longer and 2 to 3½ inches broad. The leaflets, although not touching, are close together. The spore-case clusters, close to the leaflet margins, are round.

Native only to the Americas, where it occurs in the tropics and other warm regions including Florida, or according to some authorities including populations of the Old World, ***N. exaltata*** in its numerous varieties is the most commonly cultivated sword fern. The typical species

Nephrolepis exaltata in Florida

somewhat resembles *N. biserrata*, but has the leaflets of its erect, stiff fronds, which spread at their ends and are 2 to 5 feet long by 3 to 6 inches broad, close together. The rounded spore-case clusters are midway between the midribs and margins of the leaflets. This species has been the source of numerous mutants (sports), which are much more commonly cultivated than the wild type. Some are so very different in appearance, having fronds very much divided, crested, and in other ways modified, that it is not easy to relate them to *Nephrolepis*. However, nearly always they produce occasional more typical fronds that make identification easy and certain.

The Boston fern (*N. e. bostoniensis*) was the first mutant of *N. exaltata* to receive

Nephrolepis exaltata bostoniensis

considerable attention. Introduced to horticulture in 1895, it soon became popular. The Boston fern is vigorous, and its once-pinnate fronds droop more than do those of *N. exaltata*. They are up to 3 feet long by 6 to 8 inches broad, and have toothed or toothless, more or less wavy leaflets about ¾ inch broad. Dwarf Boston fern (*N. e. b. compacta*), as its name suggests, is a more compact, dwarfer edition. The Boston fern and its sports have given rise to a long series of mutations many of which have been named; they are grown as horticultural varieties.

These varieties have fronds once-pinnate or essentially so. *N. e. childsii* is a slow-growing dwarf kind with much-curled or crisped, closely set leaflets. *N. e. dreyeri* has slender, strap-shaped, light green, spreading fronds. *N. e. falcata* is similar to *N. e. scottii*, but the ends of its fronds are forked. *N. e. giatrasii* is a bushy sport of the Boston fern. *N. e. gretnai* has strap-shaped fronds 3 feet or more long, with the tips of the leaflets often forked. *N. e. hillsii* is a vigorous grower with wavy or deeply-lobed, crisped, overlapping leaflets. *N. e.* 'New York' is a vigorous, handsome variety with fronds shorter than those of the Boston fern; their leaflets are wavy. *N. e. randolphii* has broad fronds with wavy leaflets. *N. e. rooseveltii* is an improved Boston fern with fronds with wavy leaflets. *N. e. r. plumosa* has feathery fronds up to 3 feet long with leaflets lobed or crested at their tips. *N. e. scottii* is a compact sport of the Boston fern with spreading fronds. *N. e.* 'Teddy Junior' is a compact sport of *N. e. rooseveltii*. Its fronds have wavy leaflets. *N. e. victori* has wavy, frilled leaflets to its short, erect fronds. *N. e. viridissima* has stiffly erect fronds with curled and twisted leaflets. *N. e. wagneri* is a dwarf with stiff fronds closely furnished with wavy leaflets.

Varieties with all or most of their fronds twice-pinnate include the following. *N. e.* 'Anna Foster' has low, spreading, eventually drooping, fronds. *N. e. barrowsii* has fronds up to 2 feet long or longer, with wavy leaflets. *N. e.* 'Colorado' grows strongly, and has erect, rigid fronds with

Nephrolepis exaltata elegantissima

flattish, toothed leaflets. *N. e. elegantissima* has lacy fronds, two- or three-times-pinnate and up to 1½ feet long or a little longer and 6 to 8 inches wide. *N. e.* 'Fluffy Ruffles' has stiffish erect, dark green fronds under 1 foot long. *N. e.* 'M. P. Mills' has mostly twice-pinnate fronds among which quite often develop once-pinnate ones. It grows slowly. *N. e.* 'Ostrich Plume' has lacy fronds, rather coarser than those of *N. e. whitmanii*, and two- or three-times-divided. It not infrequently produces once-pinnate leaves. *N. e. piersonii* is much like the Boston fern, but its

Nephrolepis exaltata whitmanii

fronds are mostly twice-pinnate, with some partly or entirely once-pinnate. *N. e. scholzelii* has fronds a little over 1 foot long by up to 5 inches wide. *N. e. splendida* has fronds exceeding 3 feet in length. The leaflets toward the forked tips of the fronds are much crested. *N. e.* 'Trevillian' has very lacy fronds, almost ball-like in form, and very compact. *N. e. whitmanii* has lacy, light green, arching or drooping, sometimes three-times-pinnate fronds up to 1½ feet long. *N. e. wicheri* has fronds forked at their ends. Their upper leaflets are crested.

The fronds of the following kinds are three- or more-times-pinnate. *N. e. magnifica* is of loose growth and has lacy fronds about 1¼ feet long. *N. e. muscosa* has fronds up to 8 inches long and is much like *N. e. superbissima*, but dwarfer. *N. e. norwoodii* is an elegant variety with broad fronds. *N. e. scholzelii tripinnata* is much like *N. e. scholzelii*, but its fronds are more-times-divided. *N. e. smithii* is a lacy variety with fronds four-times-pinnately-divided, and mostly over 1 foot long. Their leaflets are very crowded. *N. e. superbissima* has three-times-pinnate fronds 1 foot long or a little longer, with irregularly twisted leaflets. Twice- and once-pinnate leaves are often produced. *N. e.* 'Verona' has short, drooping, lacy fronds.

Garden and Landscape Uses. As plants for greenhouses and room decoration, sword ferns attain their most extravagant horticultural popularity, but in warm, not excessively arid regions, where frost seldom if ever occurs, they are esteemed as outdoor underplantings and for growing as epiphytes on the trunks and branches of palms and other trees that afford root-

Nephrolepis ferns are attractive in hanging baskets: (a) Lining the basket with sheet moss

(b) Planting the basket

(c) The basket after a few months growth

Nephrolepis ferns grouped in a porch box

Nephrolepis exaltata bostoniensis thrives in a lighted window

Nephrolepis ferns summering in a shaded place outdoors

Nephrolepis fern propagation: (a) Divide the ferns into individual rooted crowns

(b) Pot sepatately in small containers

holds. Lacy-leaved varieties are not suitable for outdoor plantings; only kinds with once-divided leaves should be used.

Indoors, all kinds are satisfactory, and there is a wide variety of sizes and forms from which to choose. Those with drooping fronds are especially well adapted for hanging baskets. Others are of service in window and porch boxes. In large conservatories the simple-leaved species and varieties are good groundcovers for use beneath trees or shrubs. Because of the need to keep their foliage dry, lacy-leaved varieties are not adaptable for this purpose. With few exceptions, sword ferns are extraordinarily adaptable and very tolerant of less than ideal environments. It is not unusual to find huge specimens of the Boston fern, and sometimes other kinds, that have lived and prospered for decades in a large pot or tub in a window of a house or store. Such examples usually are where temperatures are not excessively high; a sunroom, closed porch, or bay window in an older house is more to their liking than most city apartments. Good light, but not necessarily strong sun, although they stand considerable sunlight, is essential for best success, as is a fairly free circulation of air. Dank, dark, excessively humid conditions are not liked by sword ferns. If subjected to them their fronds become weak and straggly and the plants lose the sturdiness and crispness, and profuseness of foliage that are so much part of their beauty.

Cultivation. Propagation of sword ferns is almost entirely by offsets and division. Many horticultural varieties do not have spores. However, when spores are produced, they may be used as a means of increase, even though because of the hybrid or other irregular origin of many of these plants the offspring may not duplicate the plants from which the spores were taken. Although division of old plants, selecting more vigorous outer portions of the clumps each consisting of several crowns, for replanting, and discarding the old, worn-out inner parts, is satisfactory where only a few specimens are needed, for rapid, plentiful multiplication, the best results are had by starting with single plantlets, each consisting of a young crown with its cluster of small leaves. Plants raised in this way usually develop into more uniform, shapely specimens than those from larger divisions. They are started in late winter or spring and, if grown in a humid greenhouse where the minimum temperature is 55 to 60°F and, once growth is well started, is ventilated fairly freely on all suitable occasions, they make rapid growth, and by fall the more vigorous kinds will be handsome specimens in 6-inch pots. Less robust varieties by then should occupy containers 4 or 5 inches in diameter. It is important to afford these ferns all the light they will stand without their foliage yellowing or scorching. This means that light shade is needed in summer, but not too much. Exposure to weak direct sun is beneficial. The soil for sword ferns should be more loamy and less woodland in character than for many ferns. It should, nevertheless, contain a fair proportion of peat moss, leaf mold, or compost and most certainly must be porous enough to allow the free passage of water; stagnant, waterlogged conditions about the roots, soon cause deterioration. A hearty, fertile soil of coarse structure is best for sword ferns. Their containers must be adequately drained. In bright weather, spraying the foliage once or twice a day with water, provided the moisture dries within about an hour, is beneficial. Watering of most kinds, including all of the *N. exaltata* varieties, is done to keep the soil evenly moist at all times, but a few of the tuber-bearing species and varieties may be deciduous or nearly so and they are kept dryish, but never completely dry, during their season of winter rest. Well-rooted specimens benefit greatly from regular applications of dilute liquid fertilizer. Throughout the winter established plants do well where the night temperature is about 55°F, and by day not over five to ten degrees higher, but they will stand conditions both somewhat cooler and somewhat warmer. The chief pests of these ferns are scale insects, mealybugs, whiteflies, and, under humid conditions, slugs and snails. For additional information see Ferns.

NEPHROPHYLLIDIUM. See Fauria.

NEPHTHYTIS (Nephthȳ-tis). The genus *Nephthytis*, of the arum family ARACEAE, consists of five species of tropical West Africa. It is named after the mythical goddess Nephthys. The name is often misapplied in gardens, the plant often called *N. liberica* is *Syngonium podophyllum* and the one known as *N. picturata* is *Rhektophyllum mirabile*.

Nephthytises have thick, tuberous rootstocks and long-stalked, evergreen leaves with arrow-shaped blades. Their flowers have the form of and are borne in the way of calla-lilies. Individually tiny, they are clustered closely together along an erect central column called a spadix from the base of which a broad-ovate, petal-like bract called a spathe sprouts. Collectively spadix, spathe, and supporting stalk is an inflorescence.

Commonest in cultivation, ***N. afzelii*** has knobby, spreading rhizomes and leaves with stalks 1½ feet long or somewhat longer, and blades about 1 foot long. The inflorescence has a stalk up to 1 foot long. The leaf-green spathe is much longer than the spadix. The lower part of the latter is

Nephthytis afzelii (inflorescences and fruits)

ellipsoid and consists of green female flowers. The upper part is cylindrical, tan, and has only male flowers. The berry-like, egg-shaped fruits are bright orange at maturity.

Other kinds occasionally cultivated are ***N. gravenreuthii,*** with broad, yellow-green leaves with darker veins, a dotted spathe, a white spadix with green dots, and orange fruits, and ***N. poissonii,*** with rich green leaves and elliptic, green spathes finely-spotted with brown.

Nephthytis poissonii

Garden and Landscape Uses. In the humid tropics nephthytises are useful for growing in shady places outdoors. They are also admirable for pots and beds in tropical greenhouses, responding to the same conditions and care that suit dieffenbachias and similar tropical foliage plants.

Cultivation. These plants revel in high temperatures, high humidity, filtered light with shade from strong sun, and fertile soil containing an abundance of organic matter, which is always kept moist. Use coarse soil of a sort that will not break down quickly, become compact, and impede free access of air. To assure this, mix with it broken crocks or brick, coarse sand, and charcoal. When repotting is needed do it in spring. Specimens that are pot bound, having filled their containers with roots, may with advantage be given dilute liquid fertilizer at weekly intervals from spring through fall. Propagation is by division in spring and by seed. The seeds, freed from the pulp that surrounds them and sown while fresh in sandy, peaty soil in a temperature of 65 to 75°F, soon germinate, and the young plants grow quickly. A minimum winter night greenhouse temperature of 60°F is needed for the best growth of these plants, but they will stand temperatures five degrees or so cooler. At other seasons, and by day in winter, a temperature of 70°F or higher is desirable.

NEPTUNE-FERN. See Air-Fern or Neptune Fern.

NEPTUNIA (Nep-tùnia). Chiefly native to the tropics and subtropics of the Americas, Asia, Australia, and Africa, but extending into the southern United States, *Neptunia* consists of fifteen species. It belongs in the pea family LEGUMINOSAE. The name, applied because some species are aquatics, derives from that of Neptune, god of the sea. In parts of the tropics, *N. prostrata* (syn. *N. oleracea*) is used as a pot herb.

Neptunias are herbaceous perennials with erect, sprawling, or prostrate stems those of some floating kinds are surrounded by a thick layer of spongy air-filled cells. The leaves are twice-pinnate and more or less sensitive when touched. Their primary segments, of two to eleven pairs, each have from eight to more than forty pairs of leaflets. The small, yellow or greenish flowers are in densely-crowded, spherical, ovoid, or short-cylindrical, fuzzy heads resembling those of the sensitive plant (*Mimosa pudica*). Each of the tiny flowers that compose the heads has five each sepals and petals, and at least the upper flowers, ten stamens. The fruits are stalked, flat pods up to 2 inches long.

Indigenous to Oklahoma and Texas, prostrate or nearly prostrate ***N. lutea*** has hairy stems and foliage. Its leaves have three to six pairs of primary divisions, each with seven to seventeen pairs of oblong leaflets up to ⅓ inch long. The flowers, in ellipsoid heads ½ to ¾ inch long, are yellow. Terrestrial, or with its roots anchored in soil and its branches floating, ***N. plena*** has erect or prostrate stems up to 3 feet long. Its leaves have two to four or rarely five pairs of primary divisions, each with nine to nearly forty pairs of leaflets. The leaves, like those of *Mimosa pudica,* fold together rapidly when touched. The yellow flowers are in ellipsoid heads 1 inch to 1½ inches long. This sort is native from Mexico to Brazil and the West Indies.

Garden Uses and Cultivation. Because of its sensitive leaves, *N. plena* is occasionally grown as a botanical curiosity, but it does not take kindly to cultivation. It re-

Neptunia plena

quires a humid tropical environment and is best accommodated in wet waterside soil where its stems can float out over the surface of a pond or pool. A sunny location is needed. Propagation is by seed and perhaps by cuttings. The other species described, which requires similar treatment, is perhaps easier to grow.

NERINE (Ne-rìne)—Guernsey-Lily. There are about thirty species of *Nerine,* bulb plants of the amaryllis family AMARYLLIDACEAE. All are natives of South Africa, and in the northern hemisphere all bloom in fall. None is hardy outdoors in the north. The name is that of the water nymph Nerine. Nerines have deciduous, strap-shaped to narrowly-linear leaves that come with or just after the usually short-tubed, funnel-shaped flowers. The latter, on solid, leafless stalks, are in umbels of two to many. The perianth is divided nearly to its base into six segments, commonly called petals, but more correctly, tepals, that are erect or widely spreading and often waved at their margins. In some kinds, none of which is described below, there is a corona of scales or frayed membranes in the throats of the blooms, often joined to the stamens. There are six stamens, three longer than the others, and a style with a three-branched or three-lobed stigma. The more or less spherical, three-lobed fruits are capsules that split along three longitudinal lines to free the fleshy, spherical seeds.

The name Guernsey-lily, applied to *N. sarniensis* is somewhat misleading. It is not native to the English Channel island, the name of which it bears, but was brought there as the result of a Dutch or English ship carrying bulbs foundering off its coast. The bulbs, washed ashore, took root and became firmly established as a naturalized element in the island flora. Because the ship had sailed from Japan it was long believed that the Guernsey-lily was native to that country, but since the genus *Nerine* is confined in the wild to South Africa, it is fair to assume that the ship had called at the Cape of Good Hope, a common practice in the eighteenth century, on its way from Japan.

The Guernsey-lily (***N. sarniensis***) is a highly variable species that has been hybridized with *N. flexuosa, N. undulata, N. pudica,* and possibly other species, to give progeny of great charm and decorative value. Typically, it produces three to five, blunt, strap-shaped, green or glaucous leaves up to 1 foot long by ½ to ¾ inch wide. Umbels of six to eight, occasionally fewer or more, scarlet, vermilion, deep rose-pink, or paler pink flowers top each stalk. They are 1¼ to 1½ inches long or slightly longer and have erect, flaring and recurved petals with slightly wavy margins. The stamens are 1½ to 2 inches long and like the style are erect, protruding, and nearly straight. The stalks of the individual flowers, and the ovaries, are without hairs. Variety *N. s. corusca* has glaucous leaves broader than those of the typical species and crossbarred between the veins. Its comparatively large flowers are salmon-red. The blooms of *N. s. maxima,* (syn. *N. s. major*) are crimson. Those of *N. s rosea,* a name frequently misapplied to *Lycoris radiata,* are rose-red.

Nerine sarniensis

The hardiest species ***N. bowdenii*** differs from the last in having down-pointing stamens and styles, upturned at their ends, and 2 to 2½ inches long. Each mature bulb produces about eight strap-shaped leaves, 1¼ feet long by ½ to ¾ inch wide. The flower stalks, up to 2 feet tall, carry in a horizontal position seven to ten blooms on hairless individual stalks. The petals, bright pink with a darker mid-vein, wavy margins, and 2 to 2½ inches long, at first form a funnel-shaped flower, but later spread and recurve into a flatter bloom. Variety *N. b. alba* has almost white flowers.

Nerine bowdenii

Narrow petals spreading widely from their bases so that the flowers are not funnel-shaped are characteristic of ***N. flexuosa***. They are slightly wavy at the margins, 1¼ to 1½ inches long, and pale pink with a darker mid-vein. Each umbel of nine to eleven blooms tops a flexible stalk up to 2 feet long or longer. The ovaries and stalks of the individual flowers are hairless. There are about four strap-shaped, bright green leaves 1½ feet or so long by about ¾ inch wide. Variety *N. f. alba* has white flowers. The plant previously named ***N. excellens*** seems to be an

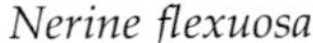
Nerine flexuosa

Nerine filifolia

Nerine curvifolia

Nerine humilis

Nerine undulata

especially large-flowered form of *N. flexuosa.* There are hybrids between the latter and *N. sarniensis* and others.

Quite distinct from those previously described, ***N. filifolia*** has up to ten very slender, linear, grass-green leaves 6 inches to 1 foot in length and very finely-hairy. The umbels of up to a dozen flowers, with individual stalks and ovaries densely-hairy, are held aloft on slender stalks 9 inches to 1½ feet long. The narrow, wavy-margined, rose-pink petals, 1¼ to 1½ inches long, spread widely and are recurved. The down-pointing stamens are 1 inch to 1½ inches long.

Other species cultivated include ***N. curvifolia,*** which has curved, glaucous leaves up to 1½ feet long, and in umbels of up to a dozen, scarlet blooms with nearly straight stamens and a nearly straight style almost equaling the petals in length. Popular *N. c. fothergillii* (syn. *N. fothergillii*) is more robust and very free-flowering. Up to nearly 2 feet tall, ***N. falcata*** has slightly twisted, sickle-shaped leaves. Its flowers, their petals white with recurved pink apexes, are in umbels of eighteen to twenty-five. The stamens point downward. Round-tipped, somewhat channeled, oblong-linear leaves are characteristic of ***N. humilis***. Its light to deep rose-purple flowers, in umbels of six to twenty that terminate stalks longer than the leaves, have slender, somewhat wavy petals. Its flowers with down-pointing stamens, ***N. lucida*** (syn. *N. laticoma*), 6 to 8 inches tall, has linear leaves and narrow-petaled, red flowers with long stalks in umbels of twenty to forty. Very slender leaves are typical of 6- to 9-inch-tall ***N. masonorum.*** Its rose-pink flowers have a darker center line to each recurved, wavy, ½-inch-long petal. The stamens, shorter than the petals, point downward. Nod-

Narcissuses naturalized in a meadow

Trumpet narcissus 'van Wereld's Favourite'

Narcissuses naturalized in open woodland

Double-flowered narcissus 'Texas'

Narcissus bulbocodium with grape-hyacinths

Nemophila menziesii

Neillia sinensis

Nemesia strumosa varieties

Nemophila maculata

Neochilenia fusca

ding flowers about ¾ inch long and with pale pink petals about equaling the down-pointed stamens in length, are characteristic of *N. **undulata.*** The blooms are in umbels of a dozen or fewer. The leaves are linear, 1 foot to 1½ feet long by about ½ inch wide.

Garden Uses. Nerines are very beautiful ornamentals for outdoor beds and borders and rock gardens in areas where they survive winters and for cool, sunny greenhouses. They supply admirable flowers for cutting that last exceptionally well in water.

Cultivation. In regions of nearly frost-free winters and dry summers, or where mild winters occur and the bulbs can be protected from excessive moisture during their dormant summer period, nerines are excellent for sunny locations and porous soils where they can remain undisturbed. In the south of England, *N. bowdenii* succeeds at the foot of south-facing walls and would undoubtedly be hardy in similar locations in climates not harsher than that of Washington, D.C. Other species are somewhat more tender.

In greenhouses nerines are best grown in pots because they bloom most surely when their roots are confined. They are not difficult to manage, but are apt to be a little erratic as to flowering, not always blooming as freely as could be wished. One point is certain, they resent root disturbance. Because of this, repotting is done only at intervals of several years. In intervening years some of the surface soil is removed each September, just as new growth begins, and is replaced with rich earth. Potting or repotting, when necessary, is done in August when the bulbs are dormant. The receptacles should be rather small in comparison to the size of the bulbs. It is a matter of choice whether they are planted singly or three or four bulbs in a large pot. Many gardeners prefer the former plan at first potting and, if they wish eventually to have pots containing more than one bulb, to allow offsets to develop and remain and to transplant the whole group to a bigger pot later. If it is desired to maintain solitary bulbs, offsets that develop are taken off from time to time and potted separately. Nerines need a porous, loamy earth. One consisting of good topsoil with which has been mixed one-fourth part by bulk each of coarse sand and dried cow manure or, if the latter is not available, as a substitute for it two-thirds peat moss and one-third commercial dried sheep manure, is satisfactory.

Following the annual top dressing or less frequent potting, the soil is soaked with water. Little leaf growth develops until the flower spikes are well advanced, and until then care must be taken not to water excessively, but without allowing the soil to dry completely. Light overhead sprayings with water on sunny days are beneficial. At this time a sunny environment, airy and as cool as can be maintained without dropping below 50°F, is needed. To assure this, on all favorable occasions ventilate the greenhouse freely. Throughout the winter a night temperature of 50°F is adequate, with daytime temperatures not more than five to ten degrees higher whenever they can be held that low. Too much warmth is damaging, but temperatures appreciably below 50°F are a likely cause of nerines not flowering freely even though they produce abundant foliage. As the foliage develops supplies of water are increased and overhead spraying is continued, taking care never to spray so late in the day that the foliage does not dry by nightfall. Fertilizing established, well-rooted specimens regularly is important. Without this, potbound plants cannot be expected to build strong bulbs that will flower well. From the time the foliage is well above ground until late spring, when it begins to die naturally, weekly applications of dilute liquid fertilizer are in order. When the foliage does begin to yellow, intervals between soakings are lengthened, overhead spraying is stopped, and finally water is withheld completely. From then until it is again time to start the bulbs into growth, the pots are laid on their sides in a greenhouse or cold frame where the bulbs can be exposed to full sun, but protected from rain.

NERIUM (Nèr-ium)—Oleander. This Old World genus of three species is best known because one, the oleander, is commonly planted in warm countries outdoors, and elsewhere as a tub plant to be wintered indoors. Like many members of the dogbane family APOCYNACEAE, neriums are poisonous. Pliny, who died in A.D. 79, called attention to this. Not only are all parts of the plants deadly if eaten, but some people develop a skin irritation upon contact with them. Animals have been killed by eating the foliage, but cats and dogs, not being leaf eaters, are not likely to be harmed. The name *Nerium* is the ancient Greek one for the oleander.

Neriums are erect, evergreen, milky-juiced, tall shrubs or small trees with leathery leaves usually in whorls (circles) of three, but sometimes in fours or twos. In large, showy, branching clusters, their handsome blooms are borne over a long period in summer. They have deeply-five-cleft calyxes, furnished at the bases on their insides with many glands, and funnel-shaped corollas with normally (in double-flowered varieties there are more) five spreading lobes (petals) that are twisted and overlap to the right in the bud stage. In the throats of the blooms are five teeth. There are five short-stalked stamens that do not protrude. Their anthers, with two tails at their bases, are tipped with long appendages, and are adherent to the stigma. The fruits are paired, elongated follicles (pods) containing seeds with tufts of hairs at their apexes.

The common oleander (***N. oleander***), a native of the Mediterranean region and naturalized in warm parts of the United States, has been cultivated for centuries and was brought to North America by early settlers. It attains a height of about 20 feet and forms a billowy, much-branched shrub that, by pruning, can be persuaded to become treelike. It has nar-

Nerium oleander

Nerium oleander, double-flowered

Nerium odorum of gardens

Nerium oleander variegatum

rowly-oblong, dark green or grayish-green, short-stalked leaves 4 inches to 1 foot long with prominent mid-veins and undersides paler than the upper sides. The satiny, tubular blooms, with wheel-shaped faces 1½ to 3 inches across and toothed appendages in their throats, range in color from white through cream to yellow, pink, and deep red. There are both single- and double-flowered varieties to which some of the names indicative of the colors of their flowers, such as *album* (white), *atropurpureum* (purple), and *roseum* (pink), are applied. Sorts with foliage variegated with yellow or white are identified as *N. o. variegatum.* The seed pods are pendulous and 4 to 7 inches long.

Sweet-scented oleander is a name applied to varieties of *N. oleander* often identified as *N. indicum* and *N. odorum,* ordinarily less robust than the common oleander and with more widely spaced, slender, linear-lanceolate leaves in threes and up to 6 inches long. The flowers occur in about the same color range as those of the common oleander. They are fragrant, about 2 inches wide, and single or double.

Garden and Landscape Uses. Oleanders are lovers of summer heat and brilliant sunshine. Without these they do not bloom profusely. They withstand dry conditions and are good seaside and city plants. Although they are at their best in fertile soil, they succeed surprisingly well in a wide variety of earths, even poorish ones, provided good drainage is assured. They endure several degrees of frost so long as it is not of long duration. Survival after exposure to 18°F is recorded. Oleanders are quite stunning as solitary specimens and are beautiful for avenues and driveways and as windbreaks, backgrounds, screens, and informal hedges. They grow rapidly and need minimum care. As tub plants they are esteemed for decorating patios, terraces, steps, and similar places. Their blooms are not satisfactory for cutting, but their seed pods are used in dried arrangements.

Cultivation. The outdoor cultivation of oleanders, in warm climates, is extremely easy. Routine care calls for little more than such pruning as is needed to keep the bushes shapely, tidy, not overcrowded with thin, weak shoots, and of suitable size. The flowers are borne on new shoots of the current season's growth and pruning is done in late winter before new growth begins.

As tub plants oleanders are partial to fertile, porous soil that is well drained. In the north they must be sheltered during the winter by storing them in a light shed, cellar, cool greenhouse, or similar place that is cool and well ventilated but where they will not be subject to freezing. Temperatures in the 35 to 50°F range are adequate. Throughout the winter they are given little or no water. In late winter or early spring the plants are pruned by removing all thin and crowded shoots and shortening the sturdier old ones that bloomed the previous year. Then the plants are top dressed, or retubbed if they are to occupy larger containers, and are given more warmth, 50 to 55°F at night and five to ten degrees higher by day. Normal watering is then resumed and the tops of the plants are sprayed lightly with water on sunny days. From the time spring growth starts it is essential that the plants receive maximum light; then a cool, sunny greenhouse or its equivalent is ideal. After the weather has warmed sufficiently for tomatoes to be planted outdoors, tubbed oleanders may be stood outside where it is warm and sunny.

Propagation is very easily accomplished by cuttings consisting of sections or terminal pieces of firm shoots taken in late summer and planted in sand or sandy soil, outdoors in nearly frost-free climates, indoors elsewhere. Cuttings are also easily rooted by standing them in containers of water. Young plants may be grown in pots or other containers or planted outdoors in nursery beds. The latter plan can be followed successfully in the north, as well as areas where oleanders are winter hardy, but where winters are cold the plants must be lifted and potted before fall frost and be wintered indoors.

Pests and Diseases. The chief pests of oleanders are scale insects and mealybugs. Aphids are also sometimes troublesome. Oleanders are subject to leaf spot diseases and to a bacterial gall disease, which causes wartlike growths on the aboveground parts. For this, the suggested control is to cut out affected parts.

NERTERA (Nertè-ra)—Bead Plant. Of the dozen species of the madder family RUBIACEAE included in *Nertera,* only one appears to be cultivated and that but rarely. The group is native from southern China and Taiwan to Australia, New Zealand, many islands of the Pacific, and parts of South America.

This genus consists of creeping plants with small, opposite leaves, inconspicuous, solitary flowers, and fleshy, berrylike fruits each containing two seeds. The blooms have toothless or four- or five-toothed calyxes, tubular, four- or five-lobed corollas, four or five stamens, and two slender styles. The name, referring to the lowly habit of the plants, is from the Greek *nerteros,* low down.

The bead plant ***N. granadensis*** (syn. *N. depressa*), of South America, New Zealand, and Tasmania, forms a dense, quite prostrate carpet of trailing stems and minute, broad-ovate leaves. The extremely small greenish flowers are without decorative importance, but the translucent bright orange fruits, which under favorable circumstances besprinkle the mat of foliage and contrast most pleasingly with its bright green, are highly ornamental. They are about 1/4 inch in diameter and remain attractive for a long period.

Garden Uses. In mild climates such as are typical of parts of California, the bead plant is a delightful groundcover for rock gardens and other choice areas. It is also charming in pans in cool greenhouses.

Cultivation. Sandy soil, moderately moist but permitting the easy passage of water, and shade from strong sun are important requirements. The bead plant does not succeed where temperatures are excessively high or the atmosphere is oppressively humid. Even when vegetative growth is satisfactory, this plant does not always fruit well, which is maddening to the hopeful cultivator. New plants of *N. granadensis* are easily had from divisions or seed. Sow the latter in spring in a cool greenhouse or cold frame in pots or pans of sandy peaty soil. Transplant the young seedlings to other shallow containers in similar, but slightly more nourishing soil and grow them under cool, but frostproof, shady conditions. A winter night temperature of about 50°F is adequate, with a daytime rise of five or ten degrees allowed.

NESAEA VERTICILLATA is *Decodon verticillatus.*

NESTEGIS (Nes-tègis). Closely related to *Olea,* and by some authorities included there, *Nestegis,* of the olive family OLEACEAE, consists of four species native to New Zealand and *N. apetala* of New Zealand and Norfolk Island. From *Olea* this genus differs in its flowers having no corollas. The name, of uncertain application, derives from the Latin *ne,* not, and *tego,* covered.

Nestegises are trees and shrubs with opposite, toothless leaves, those of adult plants distinctly different from those of juvenile specimens. Bisexual or unisexual, with the sexes on the same or different plants, the flowers are in racemes. They have a four-cleft calyx, two short-stalked stamens, and a short style. The fruits are red or orange drupes (fruits structured like plums).

The tallest species ***N. cunninghamii*** (syn. *Olea cunninghamii*) reaches 65 feet in height. On juvenile specimens, the narrowly-linear leaves are 6 to 10 inches long by up to 3/4 inch wide. Adult specimens have lanceolate to ovate or elliptic-lanceolate leaves 3 to 6 inches long by 3/4 inch to 2 inches wide. The flowers are in slender racemes with stout, densely-hairy stalks. The red fruits are about 1/2 inch long. Approximately 45 feet tall, ***N. lanceolata*** (syn. *Olea lanceolata*) has leaves 2 to 4 1/2 inches long, on juvenile trees narrowly-linear, on adults ovate-lanceolate to narrowly-elliptic and up to 3/4 inch wide. The slender flower stalks are hairless or nearly so. The red or orange fruits are up to 1/2 inch long. Both juvenile and adult leaves of ***O. montana,*** a tree up to 30 feet in height, are linear, or in adults narrowly-lanceolate. The stalks of the flower clusters are hairless or nearly so. The fruits are red and 1/3 inch or slightly more in length.

Garden Uses and Cultivation. These evergreens are suited for use as ornamentals

Nertera granadensis

Nertera granadensis (fruits)

Neviusia alabamensis

in essentially frost-free climates. They require the same general conditions and care as the olive (*Olea europea*), but thrive also under more humid conditions.

NET BUSH. See Calothamnus.

NETTLE. See Urtica. The dead-nettle is *Lamium*, the hedge-nettle, *Stachys*, the nettle-tree *Laportea*.

NEVIUSIA (Neviùs-ia) — Snow Wreath. There is only one species of *Neviusia*, a native of and indeed as a wild plant confined to Alabama. Despite this, it is quite hardy in much of New England and as far north as central New York. The snow wreath belongs in the rose family ROSACEAE and is related to *Kerria* and *Rhodotypos*, although its general aspect is quite different. Its name honors its discoverer, Ruben Denton Nevius, who found it growing on sandstone cliffs near Tuscaloosa about the middle of the nineteenth century. It was brought into cultivation about 1880. Nevius died in 1913.

A deciduous shrub 3 to 6 feet in height, the snow wreath (***N. alabamensis***) is remarkable for the very attractive display it makes in summer of flowers that have no petals but that depend upon their conspicuous fluffy pompons of white petal-like stamens for effect. The blooms suggest those of meadow-rues (*Thalictrum*), but the plants are not, of course, related. The snow wreath has slender, branching stems and alternate, doubly-toothed, ovate leaves somewhat resembling those of *Stephanandra;* they are up to 3 inches long. The pure white flowers, about 1 inch in diameter, have numerous stamens and are occasionally solitary, but more commonly occur in clusters of three to eight. They are borne on short axillary shoots from shoots of the previous year's growth. The fruits are drupelike achenes.

Garden and Landscape Uses. Here is an exquisite and choice shrub that deserves much wider appreciation than it presently receives. It may be used with good effect in beds, borders, and on the fringes of woodlands. It thrives in warm locations.

Cultivation. The snow wreath needs porous, well-drained soil on the dryish side and of moderate fertility and should not be watered except during periods of extreme drought. Pruning to prevent its stems becoming overcrowded should be done immediately after the flowers fade, and be restricted to removing obviously old, worn-out and weak stems so that new shoots receive sufficient light to develop strongly. Propagation is easy by division, cuttings, and seed.

NEW-CALEDONIAN-PINE is *Araucaria columnaris.*

NEW-JERSEY-TEA is *Ceanothus americanus.*

NEW SOUTH WALES CHRISTMAS BUSH is *Ceratopetalum gummiferum.*

NEW ZEALAND. This appears as parts of the names of these plants: New Zealand black-pine (*Podocarpus spicatus*), New Zealand bur (*Acaena microphylla*), New Zealand Christmas tree (*Metrosideros excelsus*), New-Zealand-edelweiss (*Leucogenes*), New-Zealand-flax (*Phormium tenax*), New-Zealand-honeysuckle (*Knightia excelsa*), New Zealand lacebarks (*Hoheria*), New-Zealand-laurel (*Corynocarpus laevigata*), New-Zealand-passion-flower (*Tetrapathaea tetrandra*), New-Zealand-privet (*Geniostoma ligustrifolium*), New-Zealand-spinach (*Tetragonia tetragonioides*), New Zealand tea tree (*Leptospermum scoparium*), New Zealand white-pine (*Podocarpus dacrydioides*), and New Zealand wineberry (*Aristotelia serrata*).

NEW-ZEALAND-SPINACH. Quite distinct from ordinary spinach, New-Zealand-spinach is a good substitute for that popular vegetable. It delights in hot weather and so is available for harvesting when ordinary spinach is not. New-Zealand-spinach, botanically *Tetragonia tetragonioides,* of the carpetweed family AIZOACEAE, and thus related to the ice plant (*Mesembryanthemum*), is a lush, vigorous annual with prostrate stems and somewhat fleshy leaves. The edible parts are the 3- or 4-inch-long shoot ends with a few leaves attached. These are cooked like ordinary spinach. Following each picking new shoots develop, which can be harvested to provide a succession throughout the warm months.

Among the easiest of crops to grow, New-Zealand-spinach prospers in any ordinary garden soil of reasonable fertility.

New-Zealand-spinach

In poor soils and those not supplied with adequate moisture, the plants are stunted and the crops are of inferior quality. A few plants will ordinarily suffice for home garden needs. Each plant spreads to form a circle 3 feet or more in diameter, so at least that distance must be allowed between individuals. When grown in considerable quantities space the plants 3 feet apart in rows 4 feet asunder.

Delay sowing until the weather is warm, until it is time to sow corn, or sow indoors early, raise the young plants in pots, and plant them in the garden after settled warm weather is assured. Before sowing soak the seeds in water for twenty-four hours. Other than watering generously in dry weather and keeping weeds under control this crop needs no particular attention.

NEYRAUDIA (Ney-raùdia). The name of this genus of the grass family GRAMINEAE is an anagram of *Reynaudia*, of the same family. Consisting of two species of reed-like perennials of tropical East Africa, Malagasy (Madagascar), and Asia, *Neyraudia* is cultivated to some extent as an ornamental in southern Florida and southern California. It has tall, erect, more or less woody, stout stems, long, linear, flat or rolled leaf blades, and densely-branched flower panicles. The flattened spikelets have four to eight flowers.

Native from tropical East Africa to Madagascar and India, ***N. arundinacea*** (syn. *N. madagascariensis*) is 8 to 12 feet in height. Its leaf blades are 2 feet long or longer by 1 inch wide. Its flowers are in great decorative panicles up to 3 feet in length. In this species the lowest bract of the flower spikelet is barren and without hairs. Very similar, ***N. reynaudiana*** (syn. *N. madagascariensis zollingeri*), native from Japan to China, Malaya, and India, differs in the lowest bract of the spikelet of flowers being fertile and hairy.

Garden and Landscape Uses and Cultivation. Suitable for bold decorative effects in the tropics and subtropics, neyraudias thrive in deep, rich soil, in sun. They are propagated by division and by seed.

NICANDRA (Nicán-dra)—Shoo-Fly Plant or Apple-of-Peru. One species from Chile and Peru comprises *Nicandra*, of the nightshade family SOLANACEAE. It is closely akin to *Physalis* from which it differs chiefly in having larger and more ornamental blooms and in its fruits being nearly dry rather than fleshy berries. Its name commemorates Nikander, a poet of ancient Colophon who wrote of plants about 100 B.C.

The shoo-fly plant or apple-of-Peru (***N. physalodes***) is an old garden favorite, especially in the south. In some places in North America it has escaped from cultivation and became established as something of a weed. A sturdy, freely-branched annual up to 4 feet tall, this has alternate, oval leaves, sinuately-toothed at their margins and up to about 6 inches long. The nodding, blue flowers are solitary, shallowly-bell-shaped, and 1 inch or more in diameter. They have five stamens and one style. The many-seeded fruits are enclosed in markedly five-winged, inflated calyxes.

The name shoo-fly plant alludes to its supposed virtue of being repellent to flies. It is called apple-of-Peru because of the appearance of its fruits and its place of origin. Variety *N. p. violacea* has larger flowers and leaves. Its stems are dark violet and its leaves covered with small purple pustules from which arise violet hairs. Its flowers are violet with white centers.

Garden Uses. Nicandras are suitable for flower beds and borders and are especially appropriate in unsophisticated, simple garden settings, especially those of an old-fashioned character. Branches with their decorative fruits can be dried and used as decorations indoors. These plants are attractive for their bold foliage, pretty flowers, and ornamental fruits. They thrive in sunny locations in any ordinary garden soil.

Cultivation. In the south and other regions where the growing season is long, seeds are sown outdoors, but in the north it is better to start them indoors about eight weeks before the young plants are to be transplanted to the garden, which should be about the time it is safe to set out tomatoes. Success can be had in the north from outdoor sowing, but plants so raised come into bloom and fruit later than those given an early start inside. Seeds sown indoors should be germinated in a temperature of 60 to 65°F and the young plants grown under slightly cooler conditions in a sunny greenhouse or approximately similar environment. Until the time comes to transplant them to the garden they may be grown 3 inches apart, in flats, but a better plan is to accommodate them individually in 3- or 4-inch pots. Outdoors they are spaced 2 to 3 feet apart. They stand storms well and ordinarily need no staking; no other special care is required.

NICOBAR-BREADFRUIT is *Pandanus leram*.

NICODEMIA. See Buddleia.

NICOLAIA (Nicol-àia) — Torch-Ginger. Many of the changes in plant names that sometimes inconvenience or annoy gardeners result from the botanist's logical law of priority. Briefly, this asserts that, unless upon the recommendation of a special committee, an international botanical congress has approved the retention of a later name as part of the nomina conservanda (conserved names), the oldest name under which a plant was acceptably described shall be its correct one. In accordance with this rule the torch-ginger popular in Hawaii and other tropical parts becomes *Nicolaia elatior* instead of the long familiar *Phaeomeria magnifica*. The genus, a member of the ginger family ZINGIBERACEAE, has a name honoring Czar Nicholas I of Russia. It consists of about twenty-five species, of which only the one described here seems to be cultivated.

The torch-ginger (***N. elatior***) is one of the most ornamental tropical herbaceous plants. An evergreen, it forms clumps of arching, leafy stems 10 to 20 feet high. Its oblongish leaves are in two rows and alternate. They are 1 foot to 2½ feet long by 4 to 6 inches broad. The distinctive flower heads, shaped like classical torches, are carried at the tops of leafless stalks 2 to 6 feet tall that come directly from the roots. They are highly decorative and brilliantly colored. From 6 to 10 inches wide, the glossy, waxy heads are brilliant red or pink and are composed of numerous bracts, the lower ones large, recurved or spreading, petal-like, narrowly-white-edged, and usually without flowers in their axils. The more numerous and smaller ones are arranged spirally in a tight cone that nestles in the center of the others. In the axils of the bracts of the central cone are the small flowers, the only visible parts of which are erect, yellow-margined red lips. Each flower has a single, fertile, short stamen and no staminodes (abortive stamens). This species is a native of the East Indies.

Garden and Landscape Uses and Cultivation. As screens and for massing in borders, beds, or groups in sun or half-shade, the torch-ginger is highly satisfactory in the humid tropics. Its blooms are excellent for cutting and associate well with other tropical flowers and foliage in arrangements. This plant is grateful for deep fertile soil and fair supplies of moisture. It is readily increased by division.

NICOLLETIA (Nicol-létia). There are three species of *Nicolletia*, of the daisy family COMPOSITAE. In the wild restricted to the southwestern United States and Mexico, they are herbaceous perennials or perhaps sometimes annuals, with alternate, pinnately-lobed leaves with narrow segments. Their daisy-type flower heads have central disks of bisexual florets and around them a few petal-like, flesh-colored to purplish female ray florets. The involucre (collar) behind the flower head consists of a single row of about eight bracts. The seed-like fruits are achenes. The name commemorates Jean Nicholas Nicollet, an explorer of the American Southwest, who died in 1843.

A pretty, herbaceous perennial, ***N. edwardsii*** is slender-stemmed, hairless, and 4 to 8 inches tall. Branching freely from near its base, it has leaves divided into three or five linear to nearly threadlike divisions, each with a gland-tipped apex. The short-stalked, rosy-pink to lavender-pink or purple flower heads are 1 inch wide or a little wider. This inhabits sandy and limestone soils.

Garden Uses and Cultivation. The species described may be planted in native plant gardens in regions where it is indigenous and in rock gardens there and in places with similar climates. Little information about its cultivation is recorded, but it may be expected to need thoroughly

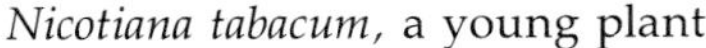
Nicotiana tabacum, a young plant

Nicotiana tabacum, in flower

drained, dryish soil and a sunny location. It can be raised from seed and probably can be increased by division.

NICOTEBA BETONICA is *Justicia betonica.*

NICOTIANA (Nico-tiàna)—Tobacco, Flowering Tobacco, Tree Tobacco. Pronounced nico-sheàna, this genus of the nightshade family SOLANACEAE comprises approximately sixty species, natives of North and South America, Australia, and Polynesia. The name commemorates Jean Nicot, French Consul to Portugal who introduced tobacco to the royal courts of Portugal and France. He died in 1600.

An important commercial crop, common tobacco is cultivated in numerous varieties in many parts of the world for its leaves, which are dried and slightly fermented to produce cigarette, cigar, pipe, and chewing tobacco and snuff. This and *N. rustica* were cultivated and ritually smoked by Indians long before the discovery of America by Europeans. Tobacco is depicted in relief carvings of about A.D. 100 on temples in Mexico.

The genus *Nicotiana* consists of annual and perennial, often clammy or sticky, herbaceous plants or less commonly sorts more woody and shrublike or even treelike. They have alternate, short-stalked or stalkless, undivided, occasionally wavy-edged leaves. The flowers are in one-sided, terminal racemes or in panicles. They have white, greenish, yellow, pink, purple, or red flowers, those of some sorts partly closed during the day and opening fully toward evening and remaining expanded and deliciously fragrant throughout the night. Each has a five-lobed tubular to bell-shaped calyx and a corolla with a long slender tube and five lobes (petals) that spread to form a usually oblique face to the bloom. There are five stamens and one style. The fruits are many-seeded capsules.

All species are believed to contain narcotic poisons. That of common tobacco and of *N. rustica* is nicotine, in its pure form one of the deadliest of all plant products. A single drop placed on the skin of a human being causes death.

Common tobacco (***N. tabacum***), cultivated as an annual, sometimes endures for more than one year in mild climates. A native of tropical America and 6 to 8 feet tall, it has hairy, slightly sticky stems and foliage. The stalkless, oblong-lanceolate leaves are often 1 foot long or longer. From 1½- to 2-inch-long and open during the day, the funnel-shaped blooms are rose-pink to red and have pointed petals. Variety *N. t. variegatum* has foliage beautifully variegated with creamy-white.

White-flowered nicotianas cultivated as ornamentals are *N. alata grandiflora, N. sylvestris,* and *N. suaveolens.* The first is a large-flowered horticultural variety of a species native of Brazil, Uruguay, and Paraguay. The next is native of Argentina, the third of Australia. A nonhardy perennial cultivated as an annual, ***N. alata grandiflora*** is 3 to 5 feet tall and glandular-hairy. It has erect stems and ovate to oblong-lanceolate leaves up to 8 inches or more in length. In loose racemes, its flowers, with yellowish corolla tubes 3 to 3½ inches long and pointed petals, are 1½ to 2½ inches across their faces. Closing on bright days, they are fully expanded and extremely fra-

Nicotiana tabacum (flowers)

Nicotiana alata grandiflora

Nicotiana alata grandiflora (flowers)

Nicotiana alata grandiflora, dwarf variety

Nicotiana sylvestris

Nicotiana sanderae

grant at night. From the last, ***N. sylvestris,*** which is as tall or taller, differs in its pendulous, fragrant flowers having slender, pure white corolla tubes about 3½ inches long and in the blooms being in short, headlike panicles. About 1½ inches across, they have pointed petals and remain open on cloudy days. The fiddle-shaped leaves of this are up to 1 foot long. Annual ***N. suaveolens,*** 1 foot to 2 feet tall, has smaller flowers than the other white-flowered sorts and usually somewhat sticky stems and foliage. The spatula-shaped to lanceolate, hairless leaves have rather glossy upper surfaces and are downy on their undersides. The nodding 1- to 2-inch-wide, fragrant blooms are in very loose racemes. They open at night and partly close during the day.

Red flowering tobacco (***N. sanderae***) is a hybrid raised early in the twentieth century at Sander and Sons nursery in England. Its parents are *N. alata,* or perhaps *N. alata grandiflora,* and red-flowered *N. forgetiana.* Seeds of the latter had been collected in Brazil in 1900 by Louis Forget, botanical collector for Sander and Sons. Subsequently *N. forgetiana* was lost to cultivation, but was discovered again in the wild in 1945. In the meantime the hybrid *N. sanderae* had produced progeny exhibiting great diversity in the intensity of the colors of their blooms, and excellent horticultural varieties that come quite true to type have been developed from it. In the main, these hybrids resemble *N. alata grandiflora,* but have bright red to rich crimson blooms.

The tree tobacco, ***N. glauca*** of Argentina and Bolivia is naturalized in the southwestern United States and the Mediterranean region. A shrub or tree 10 to 30 feet tall, of small ornamental merit, this has thickish leaves with ovate to lanceolate leaves up to 10 inches long. In loose or

Nicotiana glauca

crowded panicles, its narrow-tubular, yellow flowers 1¼ to 1¾ inches long have nonprotruding stamens. This sort is poisonous to livestock.

Garden and Landscape Uses. Normally a field crop, common tobacco is cultivated in educational demonstrations of commercially useful (economic) plants and is by no means without merit as a bold component in mixed flower beds. However, as ornamentals, the others described, with the exception of *N. glauca,* are more widely planted. Practically without value as cut blooms, they are elegant for flower beds and borders and are especially esteemed where their night fragrance can be appreciated. Nicotianas prosper in sun or part-day shade in ordinary garden soils. Modern varieties are available with flowers of pink, red, and lime-green as well as white. They are sometimes grown in pots in greenhouses.

Cultivation. Invariably grown as annuals, nicotianas are among the simplest of such plants to raise. They give of their best in fertile soil that is moist, but not wet. It is usual to start seeds indoors in a temperature of 55 to 60°F some ten weeks before it is safe to set the resulting plants in the garden, which may be done when all danger of frost is over. When the seedlings from indoor sowings are big enough to handle, transplant them individually to small pots or spaced 2 to 3 inches apart in flats. Throughout their period indoors grow them in full sun where the night temperature is 50°F, that by day up to fifteen degrees higher. They appreciate airy conditions.

For flowering in pots in greenhouses sow from September to January for spring bloom. Pot the plants successively into larger containers until they attain those in which they are to bloom, which may be from 6 to 8 inches in diameter. Use coarse, nourishing, porous soil and pack it moderately firmly. After the final pots are filled with roots, apply dilute liquid fertilizer at weekly intervals. Keep the soil always moderately moist.

NICOTINE. The principal alkaloid of tobacco, nicotine (*Nicotiana tabacum*) is an extremely toxic colorless liquid. Its compound nicotine sulfate (black leaf 40) is a contact insecticide especially effective for controlling aphids, mealybugs, thrips, and certain other insect pests. Also a deadly poison, nicotine sulfate is highly soluble in water. Because of this it is quickly removed from plant parts by rains or washing and leaves no residue. As an insecticide nicotine is also employed in dusts and as a fumigant.

NIDULARIUM (Nidu-làrium). The name *Nidularium,* derived from the Latin *nidus,* a nest, and alluding to the form of the plants, is that of a genus of approximately twenty species of bromeliads, plants of the pineapple family BROMELIACEAE. All are endemic to tropical South America, nearly all to Brazil where they grow in the ground or perch on trees as epiphytes (plants that grow without taking nourishment from their hosts). In addition to the natural species there are a number of horticultural hybrids.

Mostly of medium size, nidulariums have rosettes of evergreen, sword- or strap-shaped or linear leaves the bases of which are expanded so that the center of the rosette forms a crater or tank that holds water. The leaf margins are furnished with spines or prickles. The small flowers commonly in tight, short-stalked, usually dense headlike panicles are generally well down in the centers of the nests of leaves, less commonly lifted out on longer stalks. They have three separate sepals, the same number of petals with their bases united to form a tube, six stamens, and a slender style capped with a three-parted stigma. The fruits are many-seeded berries. Unless otherwise stated the sorts described here are endemics of Brazil.

One of the most popular and attractive species, ***N. fulgens*** occurs natively at alti-

Nidularium fulgens

tudes up to 3,300 feet. It has rosettes of fifteen to twenty strongly-toothed, strap-shaped leaves 9 inches to 1 foot in length by nearly 2 inches broad. They are pliable and lustrous green with a mottling of darker green. The green-tipped bracts surrounding the small heads of bloom are predominantly of an intense cerise hue, which later slowly changes to lavender. The deep blue flowers have petals margined with white.

Especially fine, shade-loving, and variable, ***N. innocentii*** in the wild mostly grows in the ground or among moss or debris on the bases of the trunks of trees and large shrubs. This has stolons and typically rosettes of about twenty approximately strap-shaped, finely-toothed leaves, broadest toward their middles and metallic-purple with glossy undersides. They are up to 1 foot long or somewhat longer. Bright rose-pink to rusty-red or brighter red bracts surround the dense heads of white flowers cosetted in the centers of the rosettes. A smaller variant with very dark leaves is known. Named varieties are *N. i. lineatum,* with light green leaves that are clearly striped lengthwise with very numerous white lines; *N. i. paxianum,* its leaves green with a center band of white; *N. i. striatum,* with coarser and fewer ivory-white or yellowish lines or bands than those of *N. i. lineatum* decorating its leaves; *N. i. viridis,* with mottled pea-green leaves and floral bracts with carmine tips; and *N. i. wittmackianum,* which has plain green foliage. The stemless species ***N. paxianum*** forms a handsome rosette of wide-spreading, strap-shaped, green leaves with finely-toothed margins. The bracts associated with the whitish blooms, nestled in the center of the rosette, are red in their upper parts, green below.

Other sorts include these: ***N. billbergioides*** (syn. *N. citrinum*) in the wild perches on trees and limestone rocks. It has stolons and erect rosettes of inconspicuously spined leaves some 1½ feet long by 1½ inches wide and dark green. On a stalk up to about 9 inches long, the heads of white flowers, surrounded by burnt-orange bracts, are lifted well above the centers of the rosettes. ***N. burchellii*** is a tree-percher with long, wiry stolons and dull, minutely-toothed leaves up to 9 inches long by 1½ to 2 inches wide with purple-red under surfaces and green upper ones. On stalks that raise them slightly above the centers of the rosettes, the white-petaled flower heads have orange bracts. ***N. deleonii*** inhabits forests in Colombia. It has funnel-like rosettes of soft foliage and deep within them the heads of pale blue blooms. Up to about 1½ feet long, the leaves are clothed on their under surfaces with white scales. ***N. microps*** is a stolon-producer with rosettes of lustrous, 1½-foot-long, 1-inch-wide, erect, toothed, green leaves, in the centers of which develop the flower heads surrounded by deep purple bracts. In *N. m. bicense* the foliage is deep burgundy-red. ***N. procerum*** occurs natively as a ground-growing species and also grows on rocks. A robust sort, its broad, copper-tinged, green leaves margined with tiny spines are 1½ to 3 feet long. The floral bracts surrounding the heads of bright red blooms are rusty- to brighter-red. In variety *N. p. kermesianum,* the leaves are not over 1½ feet in length. ***N. regelioides*** as a native occurs at altitudes of nearly 6,000 feet. It has shapely, compact rosettes of fine-toothed, glossy-green leaves with darker green mottlings. They approximate 1 foot in length by 2½ inches in width. The heads of deep-orange to red flowers are surrounded by rose-pink to rusty-red bracts. ***N. scheremetiewii*** is a tree-percher with compact rosettes of sharply-toothed, bright green, channeled leaves about 1 inch broad and up to 1½ feet long. The heads of violet flowers are nestled in collars of brilliant scarlet bracts. ***N. seidelii,*** of vigorous habit, forms rosettes of dark green, practically smooth-edged leaves about 2 feet long by 1½ inches wide. The flowers, in long spikes that rise above the foliage, have stalks with boat-shaped bracts that, like the blooms, are bright yellow-green.

Hybrid nidulariums include ***N. chantrieri,*** which has rosettes of red-tipped, lustrous green leaves clearly marked with darker mottlings, with pinkish undersides. Its heads of white flowers are surrounded by brilliant red bracts. The parents of this handsome sort are *N. fulgens* and *N. innocentii.* Of similar origin, except that one parent is *N. i. viridis,* robust *N.* 'François Spae' has, except for the brilliant orange-red bracts that set off the heads of flowers, plain green foliage. A sort cultivated as *N. maureanum,* is perhaps ***N. morrenianum,*** a hybrid between *N. innocentii* and *N. ferdinando-coburgii.* This has rosettes of purplish-olive-green foliage, with the undersides of the leaves lustrous purple, and a head of white flowers set in a collar of brownish-pink bracts.

Garden and Landscape Uses and Cultivation. These are as for *Aechmea.* For more information see Bromeliads or Bromels.

Nidularium innocentii lineatum

Nidularium paxianum

NIEREMBERGIA (Nierembérg-ia) — Cup Flower. The genus *Nierembergia* grows natively from Mexico to Argentina and includes thirty-five species of annuals, herbaceous perennials, and subshrubs. It belongs in the nightshade family SOLANACEAE. Its name honors the Spanish Jesuit, Juan Eusebio Nieremberg, first Professor of Natural History at the University of Madrid, who died in 1658.

Nierembergias creep or have slender, erect or semierect, much-branched stems, and alternate leaves with toothless margins. The white, bluish, or violet flowers have five-parted calyxes; saucer- or bell- or funnel-shaped corollas, tubular below and shallowly-five-lobed; five stamens, four of which are in two pairs and the other small; and one style. The fruits are capsules. None of the cup flowers is reliably hardy at New York, although *N. repens* will persist through some winters in sheltered locations. In southern New Jersey *N. scoparia* is hardy outdoors. There is confusion in the identification of the kinds commonly grown, all of which are perennials.

The white cup (***N. repens*** syn. *N. rivularis*), a creeper with spoon-shaped or oblong leaves up to 1 inch long, is an outstanding species. This native of Argentina forms mats of bright green foliage 2 to 4 inches high and in summer is besprinkled with upturned, delicate creamy-white, 2-inch-wide saucers, each with a golden center. Sometimes the flowers are tinged faintly with pink or blue. Another native of Argentina, ***N. hippomanica*** is not known to be in cultivation, but *N. h. violacea*, often misidentified as *N. hippomanica* and *N. caerulea*, is popular. Up to 1 foot tall, this has many somewhat lax, much-

Nierembergia hippomanica violacea

branched stems and linear leaves up to ¾ inch long. Its flowers, about 1 inch wide, are lavender-blue to blue-violet, deepening in color toward their centers. Yet a third Argentinian is nearly prostrate ***N. gracilis.*** This is 6 to 8 inches tall, with linear leaves up to ½ inch long and 1-inch-wide, white flowers tinged and veined with purple toward their centers, with yel-

Nierembergia scoparia

low throats. Variety *N. g. crozyana* blooms more profusely and has violet-tinted blooms. Argentinian and Uruguayan ***N. scoparia*** (syn. *N. frutescens*) is stiffer, erect, and 2 to 3 feet tall. It has slender, wiry stems. Its linear leaves are 1 inch long or longer, its flowers about the same in diameter. Typically, they are delicate blue-lilac with yellow throats, but *N. s. albiflora* is a white-flowered, yellow-throated variety. *N. s. atroviolacea* has deep purple flowers with yellow throats, and *N. s. grandiflora* has extra large blooms.

Garden and Landscape Uses. Low-growing *N. repens* is useful for rock gardens, for crevices between paving, and as a groundcover. It can be a fairly aggressive

Nierembergia repens in crevices in between paving

spreader and should not be planted near choice, weaker growing plants. The species *N. scoparia, N. gracilis,* their varieties, and *N. hippomanica violacea* are excellent for summer beds, window boxes, and porch boxes. They bloom continuously and have the advantage of not needing to have dead flowers removed as these dry without leaving any mess. They are also pretty as summer blooming pot plants.

Cultivation. Porous, well-drained, reasonably moist soils of moderate fertility suit cup flowers. They need full sun and are easy to increase. Division affords the

Nierembergia repens

A summer bed of *Nierembergia scoparia*

simplest method with *N. repens.* Leafy cuttings of the others root readily at any time, but are usually taken from outdoor plants in late summer or fall, and in January or February from stock plants wintered in a greenhouse. Seeds sown indoors in February in a temperature of 55 to 60°F produce flowering plants the same year, and by following this method of increase, nierembergias can be treated as annuals. The seedlings are transplanted about 2 inches apart in flats, where they remain until they are transplanted outdoors at the time it is safe to plant petunias, verbenas, and other tender annuals. Spacing outdoors should be 5 to 6 inches apart for *N. hippomanica violacea* and *N. gracilis,* 9 inches to 1 foot for *N. scoparia.* Whether grown from cuttings or seed, the shoots of the young plants should be pinched two or three times to encourage branching. This is especially necessary with *N. scoparia.* For summer blooming in pots, pot on the young plants as root growth necessitates until they occupy 5- or 6-inch containers (*N. gracilis* can be finished in 4-inch pots) and grow them in a cool, airy, greenhouse with just a little shade from very strong sun. They need liberal watering, and when their final pots are filled with roots, they benefit from weekly or semiweekly applications of dilute liquid fertilizer. Bushier specimens of *N. scoparia* result if three seedlings or three cuttings are planted in each pot.

NIGELLA (Nigél-la)—Fennel Flower, Love-in-a-Mist or Devil-in-a-Bush, Black-Cumin. Natives of the Mediterranean region and western Asia, the twenty species of *Nigella* are annuals of the buttercup family RANUNCULACEAE. The generic name is derived from the Latin *niger,* black, in reference to the color of the seeds. The seeds of the species called black-cumin are used for seasoning.

Nigellas are erect, branching plants with finely-cut, alternate leaves often divided into such threadlike segments that the foliage has a lacy appearance. The flowers, blue, purple, yellowish, or white, are at the ends of the branches and branchlets and in some kinds are surrounded by threadlike bracts that form the mist alluded to in the common name love-in-a-mist. They have five to eight notched petals, an indefinite number of stamens, and most usually five to ten pistils joined at their bases. The fruits are inflated capsules from which the seeds escape through holes at the top.

Love-in-a-mist or devil-in-a-bush (***N. damascena***) is up to 2 feet tall and much-branched, with fine, lacy, bright green foliage and solitary blooms 1½ inches in diameter that are powder-blue or white and appear to float in green clouds of long, threadlike involucral bracts. This is a delightful native of southern Europe. The variety 'Miss Jeykll', with semidouble, blue or white flowers, is most commonly cultivated. Quite as beautiful is ***N. hispanica,*** of Spain and North Africa. It grows about as tall as love-in-a-mist, but has very narrow-linear rather than threadlike leaves and faintly fragrant, solitary or paired blooms about 2½ inches in diameter. These are without the involucres (collars of bracts) so conspicuous in *N. damascena.* The flowers are blue with contrasting red stamens. Variety *N. h. atropurpurea* has purple blooms, those of *N. h. alba* are white. From southern Europe, North Africa, and nearby Asia comes ***N. arvensis.*** Up to 1½ feet tall, it has bluish-white

Nigella damascena

flowers about 1 inch across without conspicuous involucres. Its foliage is finely divided. Attractive ***N. integrifolia*** (syn. *N. diversifolia*), of Turkestan, up to 1 foot high, has undivided, oblong lower leaves and stem leaves cut in handlike fashion into linear leaflets. The small pale blue flowers are in clusters. Occasionally offered in seed catalogs, but no great beauty, ***N. orientalis,*** of Asia Minor, about 1 foot in height, has dark green, deeply-divided leaves and ½-inch-wide, curiously shaped, yellowish-green flowers. Black-cumin (***N. sativa***), of the Mediterranean region, attains a height of 1 foot or more and has narrowly-linear, but not thread-like leaflets and blue flowers about 1½ inches across, without involucres.

Garden Uses. Nigellas are dainty and charming plants for flower beds and borders, and their flowers are elegant for cutting. The seed pods of kinds with conspicuously inflated fruits, such as *N. damascena, N. hispanica,* and *N. sativa,* are used for dried flower arrangements. To prepare them for this the plants are cut when the pods are fully developed, tied in small bundles, and hung upside down in a dry, airy place until they are completely dry. Love-in-a-mist is an easy plant to grow for late winter and early spring bloom in greenhouses and no doubt other species would respond to the same treatment. It is appropriate to include black-cumin in herb gardens.

Cultivation. No annuals are easier to grow. They thrive in any ordinary fertile garden soil and need an open, sunny location. Seed is sown by scattering it broadcast or along drills (shallow trenches) spaced 1 foot to 1½ feet apart in early spring or, where winters are mild, in fall. The seeds are covered with soil to a depth of about ¼ inch. Seedlings from broadcast seeds are thinned to 8 or 9 inches apart on good soils, somewhat closer on less fertile ones; those from seeds sown in rows may be thinned to about 6 inches apart. Flowering from spring-sown seeds begins in early summer and slightly earlier from fall sowings. In regions where cool summers are the rule, a longer succession of flowers can be had by making a few sowings in spring spaced two or three weeks apart, but elsewhere it is best to rely upon the one sowing made as early as it is possible to work the ground so that the plants will mature and bloom before really torrid weather comes.

Greenhouse cultivation of love-in-a-mist for cut flowers is done by sowing seeds in pans or flats of porous soil from late August to January and carefully transplanting the seedlings to 2½-inch pots. Later they are planted about 6 inches apart in benches or beds or are potted into 4- or 5-inch pots in which they are grown to blooming size. Alternatively, three plants may be accommodated in a pot 6 inches in diameter. Whether for beds, benches, or pots, the soil should be coarse, porous, and fertile. Nigellas need full sun and a well-ventilated greenhouse with a night temperature of 50°F, and daytime temperatures some five or ten degrees higher. Water with care; too wet soil causes rotting. These plants need no pinching, but do require staking or other support.

NIGHT. As parts of the common names of plants the word night appears in these: night-flowering-jessamine (*Cestrum nocturnum*), night-jasmine (*Nyctanthes arbor-tristis*), and night-phlox (*Zaluzianskya capensis*).

NIGHT-BLOOMING-CEREUS. This name is applied to several kinds of cactuses, most commonly perhaps to *Hylocereus undatus,* but also to *Nyctocereus serpentinus, Selenicereus pteranthus,* and some others with large flowers that open at night and close the following morning. Cactus fanciers cultivate these, outdoors in warm climates, and in greenhouses, appropriately identified, as they are in this Encyclopedia, by their correct names. But many amateurs grow a night-blooming-cereus without further identification. Such specimens are often quite large and old and excite considerable attention when they bloom. It is not unusual for local newspapers to publish articles about them, often accompanied by a picture of the plant and the proud owner.

As houseplants all these kinds respond to the same care. A well-drained pot or tub is a suitable container. The soil should be coarse, fertile, and contain enough sand, small pieces of broken brick, or other nondecayable, gritty material so that it remains porous enough to allow the free passage of water. Do not use a too-large pot or tub. One into which the root mass fits, with just a little space around it, is

Night-blooming-cereus (*Hylocereus*)

Night-blooming-cereus (*Selenicereus*)

adequate. Do not set the plant too deeply. If you do it is likely to rot at its base. Rather let the basal portions of the stems, and even parts of the uppermost roots, show at the soil surface. Young specimens may need to be repotted every two or three years. Larger plants, in containers 8 to 10 inches or more in diameter, go for much longer periods without this attention. The best accommodations for house-grown night-blooming-cereuses are a sunny window in a fairly cool room from fall to spring, and a partially shaded place outdoors or on a porch in summer. Watering is simple. Soak the soil thoroughly, then let it become almost dry before soaking again. Once-a-week applications in summer will be about right. In winter much longer periods between waterings are in order. Give plants that have filled their pots or tubs with healthy roots, but not others, a houseplant fertilizer each month from spring to early fall. Because the stems of these plants are weak or trailing, stakes or other supports must be provided. Most likely pests are scale insects (they appear as nonmoving, dark-colored little bumps) and cottony-looking mealybugs. The last can be flushed away with water from a hose (if the plant can be taken outdoors) or by dabbing them with a swab dipped in insecticide or in alcohol, taking care not to get this on the stems. The most practical way of eliminating scale insects is to scrub them off with a soft toothbrush dipped in soapy water or in a mild insecticide. Night-blooming-cereuses are easily propagated by cuttings. For further information see Cactuses.

NIGHT-BLOOMING PLANTS. The flowers of most plants stay open day and night, those of others close at night and expand by day. Yet a third group, the blooms of which open fully only at night, ordinarily remain closed or partially closed throughout the day; in dull weather, however, they may stay open well into the morning or sometimes later.

These last, called nocturnal or night-blooming plants, are most numerous in the tropics and many of their sorts are not in cultivation. Several night-bloomers, however, are useful garden plants in temperate regions. Because most are pollinated by moths or other nocturnal insects, or in the tropics by bats, their flowers are usually easily visible white or pale yellow and often are highly fragrant at night.

Night-blooming plants are especially appropriate for locating where they are most likely to be appreciated after dusk, near dwellings, along paths leading to a front door, and fragrant ones especially, beneath windows open on summer nights.

The most familiar night-bloomers include a number of cactuses with blooms of imposing size and aspect that last for only one night. Here belong species of *Hylocereus, Nyctocereus, Monvillea,* and *Selenicereus*. Equally as short-lived are the individual blooms of the cannon ball tree (*Couroupita guianensis*), cultivated in Hawaii and other tropical parts and in southern Florida as a curiosity. Also nonhardy

Night-blooming plants: (a) *Couroupita guianensis*

(b) *Oenothera missourensis*

(c) *Mirabilis jalapa*

(d) *Gladiolus tristis*

night-bloomers, appropriate for gardens in mild climates and greenhouses, are night-flowering-jessamine (*Cestrum nocturnum*) and night-jasmine (*Nyctanthes arbor-tristis*).

Other night-blooming plants include evening-primroses (*Oenothera*), evening stock (*Matthiola longipetala bicornis*), flowering tobacco (*Nicotiana*), four o'clock (*Mirabilis jalapa*), *Gladiolus tristis*, moonflower (*Ipomoea alba*), night-phlox (*Zaluzianskya capensis*), *Schizopetalon walkeri*, and some sorts of tropical water-lilies (*Nymphaea*).

NIGHTSHADE. Several plants belonging in the nightshade family SOLANACEAE are called nightshades. See Atropa and Solanum. For enchanter's-nightshade see Circaea. The Malabar nightshade is *Basella*.

NINEBARK. See Physocarpus.

NIPA. See Nypa.

NIPHAEA (Níph-aea). Because of its great richness in attractive species, the gesneria family GESNERIACEAE has long interested gardeners, and special emphasis on collecting and growing its many kinds expanded greatly in America at the beginning of the second half of the twentieth century. This led to the introduction into cultivation of many genera and species not previously grown as well as some cultivated in the past that had been lost to gardens.

The genus *Niphaea*, known to science and named as early as 1841 was not grown by American gardeners until 120 years later when plants were brought from Chiapas, Mexico. It comprises from three to five species that, unusual but not unique for the family, have flat or nearly flat rather than pronounced long-tubular blooms. In this it resembles African-violets (*Saintpaulia*) and *Ramonda*. The name comes from the Greek *nipha*, snow, and alludes to the pure white flowers. Niphaeas have scaly rhizomes similar to those of closely related *Phinaea*. From the latter *Niphaea* differs in the stalks of its four stamens being straight and shorter than the

(e) *Nymphaea leucantha*

anthers. This genus inhabits Mexico, Central America, and the West Indies.

The only species cultivated ***N. oblonga***, up to 1 foot tall, has fairly stout, erect stems with usually four or five pairs of heart-shaped, rough-hairy, coarsely-toothed leaves about 4 inches long by 3 inches broad. The upper surfaces are impressed with dark purple veins. The veins on the undersides are prominently raised and purplish. In clusters from the upper leaf axils, the white flowers on slender hairy stalks have small calyxes and five, rounded, spreading corolla lobes (petals)

Niphaea oblonga

that form a face to the flower about 1 inch to 1½ inches in diameter, with a small yellow center to each bloom. There are four fertile stamens. The fruits are capsules.

Garden Uses and Cultivation. The importance of this species is as a greenhouse ornamental and to a lesser extent perhaps as a houseplant for collectors and reasonably skilled cultivators of gesneriads (members of the gesneria family). It requires the same treatment as *Achimenes*. For additional information see Gesneriads.

NIPHIDIUM (Niph-ídium). The genus *Niphidium*, of the polypody family POLYPODIACEAE, comprises ten rather similar species, natives of warm parts of the Americas and except for the one discussed below, all of quite restricted natural distribution. The name derives from the Greek *niphas*, snowflake, and *idium*, like.

Niphidiums are tree-perching (epiphytic) or terrestrial ferns with creeping, scaly rhizomes and undivided, lobeless fronds. They are distinct from polypodiums in having the clusters of spore capsules in single rows between clearly evident lateral veins.

A large, usually tree-perching species, ***N. crassifolium*** (syns. *Pessopteris crassifolia*, *Polypodium crassifolium*) has short creeping rhizomes clothed with scales and hairy roots. Its leathery, narrow-oblanceolate fronds (leaves) suggest those of a giant hart's tongue fern (*Phyllitis scolopendrium*). Up to 3 feet in length, they have blades 1 foot to 2 feet long by up to 6 inches wide. The clusters of spore capsules are large and circular.

Garden and Landscape Uses and Cultivation. These are as for tropical species of *Polypodium*. For additional information see Ferns.

NIPPLE FRUIT is *Solanum mammosum*.

NIPPON BELLS is *Shortia uniflora*.

NITELLA (Ni-télla)—Stonewort. The genus *Nitella*, of the stonewort family CHARACEAE, consists of submerged aquatic algae. It is sometimes cultivated in aquariums especially in those used as breeding tanks for fish.

Nitellas are branched green plants having a superficial resemblance to equisetums. They occur natively in many parts of the world in still or slowly moving, usually mildly acidic water. The species most commonly grown are *N. flexilis* and *N. gracilis*.

NITRATE OF AMMONIA. This, ammonium nitrate, is an important fertilizer. See Ammonium Nitrate, and Fertilizers.

NITRATE OF LIME. This, calcium nitrate, is an important fertilizer. See Calcium Nitrate, and Fertilizers.

NITRATE OF POTASH. This, potassium nitrate, is an important fertilizer. See Potassium Nitrate, and Fertilizers.

NITRATE OF SODA. This, sodium nitrate, is an important fertilizer. See Sodium Nitrate, and Fertilizers.

NITRIFICATION. In the process of decay, whereby complex organic substances that form the bodies of animals and plants are after death reduced to simpler compounds and elements, nitrification plays an important part. It consists of the conversion of ammonia into nitrates and is effected by specialized groups of microorganisms that perform only when temperatures are favorable and moisture, air, and a mineral base to neutralize the nitric acid produced are available. The process involves two groups of microorganisms, one of which converts ammonia into nitrites, the other nitrites into nitrates. For practical purposes this last is the only form in which plants are able to absorb the nitrogen essential for their well-being and growth.

Many soil management practices, including the use of nitrogenous fertilizers, manuring, green manuring, and compost making, are dependent for their success on satisfactory nitrification. Liming, by supplying the necessary mineral base, favors the process. Where large amounts of organic matter accumulate and nitrification is retarded or prevented, as in bogs by insufficient air and lack of a mineral base, acid or "sour" conditions unfavorable to the growth of most plants result.

NITROGEN. This is a colorless, odorless gas that forms four-fifths of ordinary air. Although an essential component of all living organisms, except by a few very lowly sorts it cannot be utilized directly but most plants of the pea family LEGUMINOSAE are able through the agency of bacteria that live in small tubercles on their roots, to fix nitrogen obtained from the air into a form they use as a nutrient. Most plants secure their supplies from the soil in the form of nitrates. Nitrogen is an important plant nutrient. The percentage contained in commercial fertilizers is indicated by the first of the three numbers of the formula given on the containers. See Fertilizers.

NIVENIA (Niv-énia). Sometimes included in closely related *Aristea*, South African *Nivenia*, of the iris family IRIDACEAE, consists of eight species. The name commemorates James Niven, a gardener who collected plants in South Africa. He died in 1826.

Low subshrubs, nivenias have iris-like foliage, and clustered wheel- or star-shaped flowers with six petals (more correctly, tepals), three stamens, and one style. The fruits are capsules.

The fans of slightly glaucous-green, sword-shaped leaves of ***N. corymbosa*** sprout from erect, branched stems about

Nivenia corymbosa

1½ feet tall. About ¾ inch in diameter, the bright blooms, with a spot at the base of each petal, are in roundish-topped branched clusters.

Garden and Landscape Uses and Cultivation. These are as for *Aristea.*

NOBLEARA. This is the name of orchid hybrids the parents of which include *Aerides, Renanthera,* and *Vanda.*

NODE. Often called a joint, a node is the place on a stem from which a leaf, in some cases rudimentary, has arisen. Often, as with bamboos, begonias, and the common geranium, the nodes are swollen and thicker than the spaces (called internodes) between them, but this is not always true. Nodes have especial importance to gardeners because cuttings of most plants root most readily from nodes to regions near them, and it is common practice to make their basal cuts just below a node. Pruning cuts are usually made just above a node because this allows for new shoots to sprout from the node without leaving stubs above it that will die back.

NOLANA (Nolàn-a)—Chilean-Bellflower. These natives of Chile and Peru are tender herbaceous perennials, but in gardens are almost invariably grown as annuals raised from seeds each year and discarded at the end of the blooming season. They belong in the nolana family NOLANACEAE. There are eighty species of *Nolana.* The name is from the Latin *nola,* a little bell, and refers to the shape of the flower. The fruits are nutlets.

Nolanas are prostrate or reclining. They have angular, much-branched stems, upturned for several inches at their ends and in some kinds clothed with sticky hairs. Their spoon-shaped leaves are often somewhat fleshy and are without lobes or teeth. In the vegetative parts of the plants they are alternate, but in the blooming regions they are in pairs. The solitary, angled, or lobed, bell-shaped flowers come from the leaf axils. Mostly blue or purple, rarely pink or white, they expand only in sunshine.

The Chilean-bellflower (***N. paradoxa*** syns. *N. atriplicifolia, N. grandiflora*), 6 to 10 inches tall, has stems up to over 1 foot long spotted and streaked with purple. The lowermost of its fleshy leaves are long-stalked, those above stalkless or nearly so. The flowers up to 2 inches in diameter are blue with white throats and yellow on their insides. White flowers are borne by *N. p. alba.* The blooms of *N. p. violacea* are violet. Very similar to *N. paradoxa,* but with flowers about ¾ inch across, ***N. humifusa*** (syn. *N. prostrata*) is a native of Peru. Its flowers are blue with white throats veined with purple. The plant cultivated as *N. lanceolata* is likely to be *N. paradoxa.* True ***N. acuminata*** (syn. *N. lanceolata*), of Chile, is white-hairy and has lanceolate leaves 4 to 6 inches long. Its blooms, up to 2 inches in diameter, are azure or dark blue with yellowish-white throats spotted with yellowish-green. This is a native of Chile. Chilean ***N. tenella*** is sticky-hairy and has slender stems, pointed-ovate leaves. Its white-throated, pale violet flowers have each of their five lobes tipped with a broad point.

Garden Uses. Nolanas are excellent plants for summer display in sunny locations and are especially suited for rather poor, dryish soils. They grow splendidly near the sea. Throughout the summer their convolvulus-like blooms are pro-

Nolana humifusa (flower)

Nolana acuminata

Nolana humifusa

Nolana acuminata (flowers)

duced in profusion, but open only in sun and are useless as cut flowers. These plants are good for edging paths, the fronts of flower border, slopes, and for hanging baskets, ornamental urns, and porch and window boxes.

Cultivation. Sow seeds in early spring where the plants are to bloom and cover them with soil to a depth of up to ¼ inch. Thin the seedlings while yet small to 9 inches to 1 foot apart. Alternatively, seeds may be sown earlier indoors in a temperature of about 60°F, and the young plants grown in flats until the weather is suitable for transplanting them outdoors, which may be done at the time it is appropriate to set out such half-hardy annuals as petunias and verbenas. No special summer care is needed. Weeds must, of course, be kept down. Only under exceptionally dry conditions is watering necessary.

NOLANACEAE—Nolana Family. Endemic to the west coast of South America, this family of dicotyledons consists of two genera totaling eighty-five species of herbaceous plants, subshrubs, and low shrubs. Its members have alternate, undivided leaves in their lower parts, opposite ones in their flowering parts. Solitary from the leaf axils, the blooms have a five-parted calyx, a broad, bell- to funnel-shaped, five-lobed corolla, five stamens of two different lengths, and five united or separate carpels. The fruits are of five separate or united nutlets. The only genus cultivated is *Nolana*.

NOLINA (No-lìna)—Bear-Grass. This genus, of the lily family LILIACEAE, or if an alternative classification is used, of the agave family AGAVACEAE, is sometimes interpreted as including *Beaucarnea*. That is not done here. So limited, *Nolina* consists of twenty-five species, natives of desert and semidesert regions of the southwestern United States and Mexico. The name commemorates P. C. Nolin, an eighteenth-century Frenchman who wrote about agriculture.

Nolinas have much the aspect of yuccas, but their flowers, some of which are unisexual, some bisexual, are much smaller. They have woody stems, subterranean or emerging as trunks, sometimes of considerable height and sometimes much expanded at their bases. The numerous leaves, narrow-linear, evergreen and clustered, widen markedly at their bases. The whitish flowers are in tall, branched panicles. They have perianths of six persistent segments ordinarily called petals. Male flowers have six stamens, females a short style capped with a three-lobed stigma. The fruits are capsules.

Native in Arizona, southern California, and Baja California *N. **bigelovii*** has a branched trunk 2 to 3½ feet tall. The branches are crowned with great mop

Nolina, undetermined species, in bloom

heads of numerous, glaucous leaves up to 4 feet long by ½ inch to 1¼ inches wide, not or scarcely toothed, but with shreds of brown fibers along their margins. The dense panicles, with their stalks, are 3 to 10 feet long. Similar to the last, but with the leaf margins finely-sharp-toothed instead of being fringed with separate fibers, ***N. parryi*** is endemic to southern California. Its leaves are 1½ to 3 feet long, grayish-green, and concave. By some authorities accepted as a separate species, *N. p. wolfii* (syn. *N. wolfii*), of southern California, differs from typical *N. parryi* in its green leaves being flat and 1 inch to 1½ inches wide.

Other kinds with sizable trunks include ***N. beldingii***, of Baja California. Up to 25

Nolina parryi, with flower panicles in bud

Nolina parryi

feet tall, this branches near its top and has slightly-glaucous leaves about 3 feet long by ¾ inch wide. Mexican ***N. longifolia***, about 10 feet in height, has a trunk swollen at its base, branched at its top. Its rough-margined leaves are 3 feet long or longer by 1¼ inches wide.

Without an evident trunk, ***N. microcarpa***, of Arizona, Texas, and northern Mexico, has rough-edged, finely-toothed leaves 2 to 4 feet long by up to ½ inch wide, ending in a tuft of fibers several inches long. The minute flowers are in erect, plumelike panicles 3 feet long or longer and carried well above the foliage. Also without an above-ground stem, southern Californian ***N. interrata*** has glaucous leaves very little if at all more than ½ inch wide by up to 3 feet long, margined with minute teeth of two sizes. The panicles are up to 6 feet long.

Garden and Landscape Uses and Cultivation. Nolinas are adapted for outdoor gardens in essentially frost-free desert and semidesert regions and for including in greenhouse collections of succulents. Kinds with trunks are also grown as container plants for indoor decoration.

Propagation by seed is easy. The plants grow slowly without special care in porous, well-drained soil watered rather

Nolina parryi wolfii

sparingly but not, if in containers, with such deprivation that the leaf ends become brown. The general cultivation is that of desert yuccas.

NOLTEA (Nól-te-a) — Soapbush. Perhaps the only species of its genus, the soapbush (*Noltea africana*) belongs in the buckthorn family RHAMNACEAE. Its botanical name honors a German professor, Ernst Ferdinand Nolte, who died in 1875. Its common one refers to its macerated twigs and leaves having been used by natives of Africa for washing.

The soapbush (***N. africana***) is a South African, evergreen, hairless shrub up to 12 feet tall, with erect, wandlike branches. It has alternate, stalked, undivided, elliptic to oblong-lanceolate, toothed leaves 1 inch to 2½ inches long. Glossy-green above, they have paler undersides. The whitish, unisexual and bisexual blooms, ⅙ inch across, are in small axillary and terminal panicles that at the ends of the branches are crowded to form larger panicles several inches long. They have bell-shaped, five-lobed calyxes, five obovate, hooded petals shorter than the sepals, the same number of stamens, and an undivided, three-angled style. The fruits are three-lobed, obovoid, three-seeded capsules about ⅓ inch wide.

Garden and Landscape Uses and Cultivation. Although of no outstanding ornamental value, this shrub is sometimes planted for variety in California and regions of similar mild climates. If kept sheared it makes a fairly good hedge. It succeeds without special care under ordinary garden conditions and is multiplied by seed and by summer cuttings under mist or set in a greenhouse or cold frame propagating bed.

NOMENCLATURE. See Plant Names.

NOMOCHARIS (Nomó-charis). So closely related to lilies (*Lilium*) and *Fritillaria* is this genus that at one time or another several of its sixteen species have been classified in those genera. Little known to American gardeners, *Nomocharis* is more familiar in the British Isles and other parts of Europe. It belongs to the lily family LILIACEAE and is native, mostly at high altitudes to western China and the Himalayas. The name, alluding to the natural habitats and aspects of the plants, is from the Greek *nomos*, a pasture, and *charis*, grace or charm.

Nomocharises have bulbs composed of overlapping scales and upright, deciduous, leafy stems with one to several more or less nodding, open-faced blooms. The parallel-veined leaves, alternate or in whorls (circles of more than two), are linear to lanceolate. The flowers have six separate, spreading perianth segments (petals, or more correctly, tepals) more or less fringed at their margins, the inner three usually with a basal gland divided by the mid-vein. There are six stamens with stalks sometimes much inflated at their bases, and a club-shaped style tipped with an obscurely three-lobed stigma, or sometimes the styles are missing. The fruits are capsules.

Nomocharis farreri

Kinds cultivated by connoisseurs include ***N. farreri*** (syn. *N. pardanthina farreri*), of Burma. From 2 to 3 feet tall, this has whorls of lanceolate leaves and light pink flowers spotted with red toward the bases of the three inner petals and sometimes sparsely so on the outer ones. From 1½ to 4 feet in height, beautiful ***N. mairei,*** of Western China, has pointed, lanceolate to ovate-lanceolate leaves, often in whorls, and up to 4 inches long. Its pendulous, white flowers, 3 to 4 inches wide, are sometimes suffused with purple on their outsides and spotted with rose-purple within. Their three inner petals are considerably broader than the outer ones and are fringed. The stalks of the stamens are much swollen at their bases. Variety *N. m. leucantha* has spotted blooms; those of *N. m. candida* are without such markings. Another from western china, ***N. aperta,*** 1½ to 3 feet tall and with sharp-pointed, lanceolate leaves, has stems with up to six pink to rose-purple blooms spotted and blotched with crimson, and styles longer than the ovary. From the last, ***N. pardanthina,*** of western China, differs in the maroon-red spots and small blotches on the inner petals of its 2- to 4-inch wide, down-facing flowers being confined to their bottom halves. The stalks of the stamens are dilated at their bases. This, 1 foot to 2½ feet tall, has often whorled leaves. Native to western China and Burma, ***N. saluenensis*** has broader leaves than the other kinds discussed here and up- or outfacing, bowl-shaped flowers about 3½ inches wide, sprinkled with fine dots rather than blotched with crimson, and stamens with stalks not conspicuously di-

Nomocharis farreri (flowers)

Nomocharis mairei

Nopalea cochenillifera

lated at their bases. The style does not exceed the ovary in length.

Garden and Landscape Uses and Cultivation. The hardiness of these rare and choice plants is not precisely known, but it is certain that over large areas of North America hot, dry summers rather than winter cold preclude their successful cultivation. The comparatively cool, humid summers and mild winters characteristic of much of the Pacific Northwest are more to their liking. Where such prevail nomocharises are more likely to prove tractable. They can be used with lovely results in seminaturalistic surroundings, such as suit many lilies. They require light shade from strong sun, but cannot compete with the roots of trees. Soil deep and somewhat acid, well-drained but not dry, and fat with decayed organic matter such as leaf mold or rich compost is needed. Seeds afford the best means of propagation. Plants so raised take three to six years to attain flowering size. During this time they may be grown in deep beds of agreeable soil in cold frames or outdoors. Measures to protect them from slugs and snails, which are extraordinarily fond of these plants, are likely to be needed. Once established and doing well, it is inadvisable, unless absolutely necessary, to transplant bulbs of flowering size.

NOPALEA (No-pàlea)—Cochineal Plant. The nine species of the cactus family CACTACEAE that constitute *Nopalea* are natives of Mexico and Guatemala. They closely resemble prickly-pears (*Opuntia*) and by some authorities are included in that genus. From *Opuntia* they differ in having flowers that do not open fully, but have erect perianth segments (petals) that form almost tubelike blooms with long-protruding, nonsensitive stamens and styles. The name is derived from the Mexican *nopal*, applied to species of *Opuntia*.

Nopaleas are much-branched and often tall. They have cylindrical trunks and flattened branches composed of fleshy, often narrow pads or joints that have minute, nearly cylindrical leaves that soon fall. The stems may or may not by spiny. Spines and a few glochids (minute, barbed spines) arise from specialized small areas on the stems called areoles. The areoles are white-woolly. Juicy and edible, the usually spineless fruits contain many flat seeds.

Until analine products superseded it, the dye cochineal was of great commercial importance. Nopaleas were among the cactuses basic to its production. These were cultivated in large orchards and upon them the cochineal insects were placed, and there they fed and reproduced. After about four months they were collected and dried to form commercial cochineal. Although unknown in the Old World until the discovery of America by Europeans, cochineal had been produced by the Indians since prehistoric times. It became one of the first exports of the Spaniards from Mexico. Cortez, in 1523, was instructed to obtain and ship to Spain as much as he could. Later its production spread to other parts of the world, notably the Canary Islands, and plantations of nopaleas, introduced from America, were established to support the industry.

The cochineal plant (***N. cochenillifera***) is native to tropical America and Jamaica. Erect, with a trunk up to 8 inches in diameter, and nearly or quite spineless stems, this species is up to 15 feet in height and has oblongish stem joints that may be 1½ feet or even more long. Their areoles bear early-deciduous glochids. The flowers, including the stamens, are 2 inches long or longer. They are abundant from near the tops of the stem segments and are pink to red, with pink stamens. The fruits are red and about 2 inches long.

Other species include ***N. auberi***, a treelike kind of Mexico, which attains a height of 30 feet. It has bluish-green, spiny or not spiny branches with thick segments up to 1 foot long. The spines are solitary or in pairs. The pink flowers, including the stamens, are 3½ inches long. The Mexican and Central American ***N. dejecta*** has straggling, lax branches and is up to 6 feet tall. Its trunk is very spiny with spines in clusters of up to eight. Its branches are bright green and have usually paired, yellowish to pinkish spines that gray with age. Including the stamens and styles the dark red blooms are about 2 inches long.

The Mexican ***N. inaperta*** may attain a height of 20 feet, but is frequently lower and shrublike. It is very spiny. On young stems the spines are three to six from each areole, more on older parts of the plant. With the stamens and style the blooms are up to 1¾ inches long. They are red, as are the small fruits.

Garden and Landscape Uses and Cultivation. These are the same as for warmth-loving kinds of opuntias. In greenhouses, minimum winter night temperatures of 50 to 55°F are satisfactory. For general information and suggestions see Cactuses.

NOPALXOCHIA (Nopalx-òchia). Native from Mexico to Peru, *Nopalxochia* belongs to the cactus family CACTACEAE. Its name is a latinized form of its Aztec one, *nopalxochitl,* which identifies a picture published in 1651 in a Spanish work describing the plants of "New Spain." So these plants have been known for a long time. The genus consists of three species.

From closely related *Epiphyllum* this genus differs chiefly in its flowers having perianth tubes shorter than the free parts of the petals. The older, main portions of its stems are cylindrical, but the many and more obvious round-toothed branches are flat and, in appearance and function, leaflike. They have evident mid-veins and lateral veins and along their margins areoles (specialized areas on cactuses from which spines, bristles or hairs, and flowers sprout). The areoles are bristly rather than spiny and are most prominent on young shoots. The flowers, which remain open during the day, have many spreading, pointed petals. The fruits are fleshy.

Frequently cultivated ***N. phyllanthoides*** has arching, eventually pendulous stems the branches of which are frequently tinged red. Its rose-pink flowers, about 2 inches wide, have perianth tubes ½ inch long. They are produced in great profusion. This is believed to be a native of southern Mexico. Brought from Europe to Mexico in 1824, ***N. ackermannii,*** which shares with *Epiphyllum* the common name orchid cactus, has somewhat arching stems up to about 3 feet in length and thin, dark green branches. Its red flowers are 4 inches or more in diameter.

A hybrid between *N. ackermannii* and *Heliocereus speciosus* has been and often still is confused with *N. ackermannii.* It differs from that species in having often three-angled, fleshier branches and in the areoles on the perianth tubes of its blooms being felted and spiny. In these respects it resembles its *Heliocereus* parent. Because the hybrid, correctly identified as ***Heliochia violacea*** and which is a surviving representative of a group of the same parentage and name raised in Europe early in the nineteenth century, is more vigorous and freer-flowering than the species, it practically replaced the latter in cultivation; it masqueraded under the name *Epiphyllum ackermannii* for long and sometimes still does.

The approved name for hybrids between *N. phyllanthoides* and *Heliocereus speciosus* is *Heliochia vandesii* and so the plant long grown as *Phyllocactus jenkinsonii* becomes *Heliochia vandesii jenkinsonii.* The sort widely grown as *N.* 'Deutsche Kaiserin' is believed to be of hybrid origin.

Garden Uses and Cultivation. These are as for *Epiphyllum.* For more information see Cactuses.

NORFOLK-ISLAND-PINE is *Araucaria heterophylla.*

NORMANBYA (Normán-bya)—Black Palm. The Australian genus *Normanbya,* consisting of one species of the palm family PALMAE, is closely related to and by some authorities is included in *Ptychosperma* from which it may be distinguished by its large fruits and peculiar grouping of the leaflets. Its name honors a Marquis of Normandy.

Native in humid forests in a limited area in northern Queensland, the black palm (***N. normanbyi***) is a unisexual, feather-leaved tree with a single trunk somewhat enlarged at its base and up to 60 feet tall or taller. The nearly stalkless leaves, green above and ashy-gray on their undersides, are 6 to 12 feet long. They have numerous narrow leaflets with blunt, often oblique, ragged apexes. The bases of the leafstalks are sheathing and form a distinct crownshaft that looks like a continuation of the trunk. The leaflets, in groups of two to several with the bases of those in each group joined, are scattered along the midribs. The much-branched flower clusters arise from below the crownshaft. The fruits, ovoid to broad-pear-shaped and 1½ to 2 inches long, are peach-colored and have a pleasing, fruity odor. The black palm takes its name from the hard outer wood of its trunk, which is ebony-black with small red flecks and is used for making walking canes.

Garden and Landscape Uses and Cultivation. Comparatively little is recorded about the horticultural use and cultivation of this rare palm. It is graceful and undoubtedly could be used as effectively in landscapes as *Ptychosperma.* It may be expected to prosper in warm, humid climates, such as those of southern Florida and Hawaii, and in tropical greenhouses and to respond to the cultural care recommended for *Ptychosperma.* For more information see Palms.

NORONHIA (Noròn-hia) — Madagascar-Olive. Closely related to the olive (*Olea*) and belonging to the olive family OLEACEAE, this genus comprises forty species, natives of Madagascar and other islands of the Indian Ocean. One is occasionally planted in Hawaii. The name commemorates the Spanish naturalist and traveler Fernando de Noronha, who died in 1787.

Cultivated ***Noronhia emarginata*** (syn. *Olea emarginata*) is an evergreen tree or a large shrub with opposite, leathery, short-stalked, toothless, obovate leaves up to about 6 inches long by one-half as wide. They have midribs conspicuously raised on the lower surfaces, and rolled-under margins. Their apexes are rounded or indented. The small, yellowish, fragrant, four-petaled flowers are in panicles in the leaf axils. The edible, purplish, fleshy, olive-like, sweet fruits are globular to egg-shaped, and about 1 inch long. This is called Madagascar-olive.

Garden and Landscape Uses and Cultivation. Little experience has been had with the cultivation of this species. It may be expected to survive in the tropics and warm subtropics in open, sunny locations, and to thrive near the sea. Propagation is by seed and by cuttings.

NOTHOCALAIS (Notho-càlais). This horticulturally unimportant genus, closely related to *Agoseris,* consists of four species of the daisy family COMPOSITAE. It is confined in the wild to western North America and has a name derived from the Greek *nothos,* false, and *Calais,* of Greek mythology, who had scales on his back.

Sorts of *Nothocalais* are hardy perennials with milky juice and thick taproots. Their leaves, all basal, are in rosettes. Often irregularly curled at their edges, they may be hairless or have opaque white hairs on the midribs and margins. Similar hairs may be present on the flower stalks. The flower heads are solitary, and like those of dandelions, are of all strap-shaped florets. The fruits are seedlike, beakless achenes.

Native from Wisconsin to Montana, Missouri, Texas, and New Mexico, ***N. cuspidata*** (syn. *Agoseris cuspidata*) has somewhat fleshy and glaucous, linear-lanceolate leaves 4 to 6 inches long by ¼ to ¾ inch broad, with margins that become very wavy when the leaves dry and are furnished with minute curled hairs. The flower stalks, 4 inches to 1 foot long, terminate in yellow flower heads about 1 inch in diameter, with each petal-like floret five-toothed at its apex.

Garden and Landscape Uses and Cultivation. The remarks made on these matters under *Agoseris* apply here also.

NOTHOFAGUS (Notho-fàgus)—Southern-Beech. This genus is the southern hemisphere representative of the northern hemisphere beeches (*Fagus*). Its name, indicating the relationship, comes from the Greek *nothos,* false, and *Fagus,* but perhaps *notos,* south, was intended for *nothos.* There are thirty-five species, natives of New Zealand, Australia, Tasmania, New Guinea, New Caledonia, and temperate South America. They belong in the beech family FAGACEAE.

Nothofaguses are deciduous and evergreen trees or sometimes shrubs, some very large. From true beeches they differ in the male flowers being solitary or in spikes of up to three, or less commonly five, instead of in heads of many, and in having considerably smaller, closely spaced leaves. The female blooms are solitary or in twos or threes. The two-ranked leaves are alternate, undivided, and toothed or toothless. The fruits are usually three-angled nuts similar to those of *Fagus*, but smaller. Among southern hemisphere genera, *Nothofagus* ranks second only to *Eucalyptus* as a source of lumber.

Evergreen species numbering five are endemic to New Zealand. Two, *N. cliffortioides* and *N. solandri,* have toothless leaves densely-hairy on their under surfaces. Typically a tree 20 to 50 feet tall, at high altitudes ***N. cliffortioides*** is reduced to shrub size. Its very short-stalked, broad-elliptic to roundish leaves, ¼ to ½ inch long, are spaced up to ¼ inch apart. They have four or five pairs of rather obscure veins. Much like the last, but sometimes 75 feet tall, ***N. solandri*** differs in its leaves being blunter, narrow- to elliptic-oblong, and usually conspicuously veined.

The other New Zealand natives have toothed leaves, hairless or sparsely-hairy on their undersides. Here belongs ***N. menziesii,*** distinguished by the double toothing of its leaves and its male flowers having thirty to thirty-six stamens. A tree sometimes 100 feet in height, this often has a buttressed trunk. Its broadly-triangular-ovate to roundish leaves are rather obscurely veined. From the last, *N. fusca* and *N. truncata* differ in having single-toothed leaves, that is to say, the teeth are not again toothed. Both attain extreme heights of 100 feet and have leaves ¾ inch to about 1¼ inches long. Those of ***N. fusca*** are thinnish, broadly-ovate to ovate-oblong, and sharply and rather deeply-toothed. The leaves of ***N. truncata*** are thicker, broadly-ovate to roundish, and more shallowly-bluntly-toothed. The male flowers of *N. fusca* have eight to eleven stamens, those of *N. truncata* ten to thirteen.

Tasmanian ***N. cunninghamii*** is an evergreen that in its homeland attains heights of up to 200 feet. It has leaves ¼ to a little over ½ inch long, broadly-ovate to triangular or lozenge-shaped, bluntly-toothed in their upper parts. Except for their very short stalks they are hairless. Native to Argentina and Chile, evergreen or sometimes nearly deciduous (at least in southern England) ***N. dombeyi*** in its homelands attains a very large size, sometimes exceeding 130 feet in height. It has hairless, toothed, ovate to lanceolate leaves ¾ to 1 inch long, with undersides lighter than the upper, and sprinkled with fine black dots.

Deciduous species native of Chile include *N. antarctica, N. obliqua,* and *N. procera.* The first two in their homeland attain heights of 100 feet, the last occasionally sometimes exceeds 80 feet. Its young

Nothofagus dombeyi at Bariloche, Argentina

shoots very hairy, ***N. antarctica*** has round-ended, broadly-ovate to triangular leaves ½ inch to 1¼ inches long and nearly as broad. They are minutely-toothed, downy only on the undersides of the midribs, and ¼ to ¾ inch apart on the twigs. From the last, ***N. procera*** differs in having blunt, ovate-oblong, minutely-toothed leaves 1¼ to 4 inches long, with fourteen to eighteen pairs of veins. They are downy on the undersides of the midribs. The young shoots are hairy. Probably the hardiest species, ***N. obliqua*** has hairless young shoots and irregularly-toothed or at the base shallowly-lobed, broadly-ovate to oblong leaves 1½ to 3 inches long.

Nothofagus obliqua at the Royal Botanic Gardens, Kew, England

Garden and Landscape Uses and Cultivation. It is to be regretted that these very handsome trees are hardy only in mild climates. The South American kinds described above are much more cold-resistant than the others, but none survives in the north. Climates such as those of California and the Pacific Northwest are to the liking of nothofaguses. They are suitable for planting as single specimens and in small groups. Nothofaguses prosper in ordinary, well-drained soil, but seem not to like those of a limestone nature. Once established, they need no special attention. Propagation is by seed and layering. Some kinds, including *N. procera,* can be increased from cuttings.

NOTHOLAENA (Notho-laèna)—Cloak Fern. The genus *Notholaena,* of the polypody family POLYPODIACEAE, consists of about sixty species of mostly rock-inhabiting ferns, natives chiefly of arid regions in the warmer parts of the Americas and one species in the Old World. The name is derived from the Greek *nothos,* false, and *chlaina,* a cloak, in allusion to the incomplete indusiums (coverings of the clusters of spore capsules).

Notholaenas are small ferns closely related to *Cheilanthes* and by some authorities included in that genus. From it they differ chiefly in that the margins of the leaves are usually flat and not turned under to cover the clusters of spore capsules, or at most are only slightly reflexed. Characteristically notholaenas have pinnately-divided fronds (leaves) and are mealy, hairy, or clothed with chaffy scales. The clusters of spore capsules, are near the leaf margins and at or near the ends of the veins. They are sometimes partly covered by indusiums.

Native from Texas to California and central Mexico, ***N. sinuata*** is a variable sort that commonly grows in alkaline soils.

Notholaena sinuata growing as a "volunteer" in a pot with a cactus

Scaly but not mealy on their undersides, the leaves have twelve or more pairs of broad-oblong to triangular-ovate, lobed or lobeless leaflets up to nearly 1 inch long and ½ inch wide. Their upper surfaces are sage-green besprinkled with white, stellate (star-shaped) hairs; their undersides are brown. Occurring natively from southern Europe and North Africa to China, ***N. marantae*** is of tidy habit. It has many spreading, arching, and erect, twice-pinnate fronds up to about 1 foot long by

Notholaena marantae

1½ to 3 inches wide and with densely-scaly scaly stalks.

Garden Uses and Cultivation. These are attractive plants appropriate for rock gardens and other outdoor locations where they are hardy, and for greenhouses. The sorts described here prosper indoors in a winter night temperature of 45 to 50°F with a ten to fifteen degree increase by day. They need a buoyant rather than excessively humid atmosphere and well-drained soil. For *N. marantae* the soil should be always fairly moist, but *N. sinuata* stands conditions considerably drier and will even grow in environments that suit desert cactuses and other succulents. While light shade from strong summer sun may be beneficial it is important not to shade too heavily. Notholaenas are impatient of this as they are of too-high temperatures and excessive wetness. They are propagated by division and by spores.

NOTHOLIRION (Notho-lírion). Native from Iran to western China, *Notholirion,* of the lily family LILIACEAE, consists of six species, closely resembling lilies (*Lilium*), except that their bulbs are covered with smooth skins instead of overlapping scales and the stigmas of their flowers are deeply-three-cleft rather than three-lobed. Also, after flowers are borne the bulbs die, although usually not without having produced bulblets. The flowers have six petals (more correctly, tepals), six stamens, and one style. The fruits are capsules. The name, indicating the relationship, is from the Greek *nothos,* spurious, and *leirion,* a lily.

Native to the Himalayan region, ***N. thomsonianum*** (syn. *Fritillaria thomsoniana*), up to about 3 feet tall, has lower leaves approximately 1 foot along by ¾ inch wide. Its fragrant, pale to medium pink, funnel-shaped flowers, sometimes twenty or more on a stem, are carried horizontally. They are about 2 inches long and have straight stamens. From the last, ***N. macrophyllum,*** also from the Himalayas, differs in having 1-foot-tall, pliable

Notholirion thomsonianum

stems that carry not over half a dozen purplish-pink blooms 1½ to 2 inches long; these have curved stamens. Western Chinese ***N. bulbiferum*** (syn. *N. hyacinthinum*), 3 to exceptionally 6 feet tall, has lower leaves up to 1½ feet long and 1 inch wide. Those above are smaller. The many lavender to rosy-purple, 1-inch-long blooms, funnel-shaped and carried horizontally, have decidedly curved stamens.

Garden and Landscape Uses and Cultivation. Notholirions are for connoisseurs of the choice and unusual. Only moderately hardy, they are most likely to give satisfaction outdoors in regions of mild winters and cool summers, such as parts of the Pacific Northwest. They need well-drained but not dry soil, light shade, and shelter from strong winds. Offset bulblets and seed afford ready means of propagation. Seedlings of *N. thomsonianum* usually bloom when about four years old.

Notholirion bulbiferum

NOTHOPANAX. Plants previously known by this name are included in *Neopanax* and *Polyscias*.

NOTHOSCORDUM (Notho-scórdum) — False-Garlic, Crow Poison. This group of onion relatives belongs in the lily family LILIACEAE. It comprises about thirty-five species, natives of the Americas. The name comes from the Greek *nothos,* false, and *skordion,* garlic, and alludes to a resemblance and relationship. From onions, garlic, and other members of the *Allium* genus *Nothoscordum* differs in being without the characteristic odor of that clan, as well as in technical details. It consists of perennial, herbaceous bulb plants with all basal, linear leaves and erect, leafless stalks, each ending in an umbel of blooms with at its base two papery bracts. The starry flowers have six similar perianth segments (petals, or more correctly tepals), six stamens, and a slender style capped with three minute stigmas. The fruits are capsules.

Often favoring moist soils, ***N. bivalve,*** sometimes called crow poison, inhabits open woodlands, prairies, and barrens from Virginia to Ohio, Illinois, Kansas, Nebraska, Florida, Texas, Mexico, and Central and South America. It has a globose bulb up to 1 inch in diameter and a few leaves ⅙ to ¼ inch wide and 4 inches to 1 foot long. The umbels top stalks 4 inches to 1 foot long and are of five to twelve pale straw-colored to greenish blooms, each with a slender stalk up to 2 inches long. The flowers are ¾ to 1 inch in diameter and keeled down their backs with red, purple, or green. This kind blooms in spring and often again later.

Nothoscordum bivalve

Heliotrope-scented, white flowers, slightly blushed pink and lined with darker pink, are borne by ***N. inodorum*** (syn. *N. fragrans*). The nativity of this kind is uncertain, but is probably the West Indies and northern South America. It is naturalized in the southern United States, Hawaii, and Bermuda. This has a yellowish bulb a little bigger than that of *N. bivalve* and blunt, linear leaves 8 inches to 1 foot long. The flower stalks, up to 1½ feet tall, carry umbels of eight to twenty blooms with purplish stamens in spring and early summer.

Nothoscordum inodorum

Garden and Landscape Uses and Cultivation. Although not generally accounted hardy in the north, the species described here live outdoors satisfactorily in the vicinity of New York City. They are best in well-drained, not too fertile earth, and in sun. Because they make no great show they are better adapted to the needs of the avid collector of plants than to the gardener whose prime interest is colorful displays. In mild climates they can be slightly pestiferous because of the freedom with which they spring from self-sown seeds, but this can be prevented by the prompt removal of all faded flower heads. Nothoscordums are better adapted to semiwild areas and to rock gardens than to more formal plantings. They are very easily grown in pots in greenhouses in a winter night temperature 45 to 50°F, with a few degrees increase by day permitted. Propagation is by seed and offsets.

NOTOCACTUS (Noto-cáctus)—Ball Cactus. The genera *Brasilicactus, Eriocactus,* and *Wigginsia* (formerly *Malacocarpus*), by conservative botanists included in *Notocactus,* are dealt with separately in this Encyclopedia. Notocactuses belong to the cactus family CACTACEAE. They are natives of Brazil, Uruguay, and Argentina. There are fifteen species, often variable and not always easy to identify with certainty. The name, referring to the geographical distribution of the genus, comes from the Greek *notos,* southern, and cactus.

The stems or plant bodies of this genus are solitary or clustered, and are spherical, flattened-spherical, or short-cylindrical. They have more or less lumpy (tuberculate) ribs with often densely-felted areoles (spine-producing cushions). The blooms come from near the tops of the stems. The yellow to red flowers usually have red to purplish, sometimes yellow, stigmas. Shortly funnel- to bell-shaped, they have spreading petals, and open by day. Their calyx tubes have small scales with much wool in their axils and dark bristles. The small, dry or fleshy fruits, to which parts of the withered flowers cling, are felted with hairs and bristly.

Its stems usually solitary and thickly covered with yellow to brownish or reddish, soft spines, ***N. apricus*** of Uruguay is spherical and about 2 inches in diameter. It has fifteen to twenty low ribs, with to their tops, spines in clusters of eighteen to twenty radials approximately ½ inch long and four centrals exceeding 1 inch in length. The yellow flowers, up to 1¼ inches long, have toothed inner petals. The stigma is thirteen-lobed. Of much the same aspect, ***N. concinnus*** (syn. *Malacocarpus concinnus*) is about 2½ inches tall and 4 inches wide. Its low, glossy green ribs are lumpy in their upper parts and have except toward their tops clusters of ten or twelve bristly radial spines about ¼ inch long and four centrals a little over ½ inch long. Up to 3 inches long and canary-yellow becoming carmine-red at their outsides, the flowers have red, ten-lobed stigmas. This sort is a native of southern Brazil and Uruguay. Other kinds of much the same habit are *N. muricatus* and *N. tabularis.* Up to 6 inches high by 4 inches in diameter, ***N. muricatus*** ordinarily has sixteen to twenty low, lumpy ribs. The ¼-inch-long radial spines are fifteen to twenty in each cluster, the slightly longer centrals, three to four. Pale yellow and up to 1½ inches long, the blooms of this native of southern Brazil have nine- to eleven-lobed stigmas. Spherical to shortly columnar and more or less glaucous, ***N. tabularis,*** a native of Brazil, has sixteen to eighteen ribs. The spine clusters are of sixteen to eighteen radials and four centrals. The yellow flowers are 2 inches or more in length. Spherical or shortly-club-shaped, ***N. werdermannianus,*** of Uruguay, is up to 2¾ inches tall and 2¼ inches wide. It has about forty ribs with chin-like projections (tubercles). The spines are in clusters of sixteen yellowish-white radials about ⅕ inch long and four centrals, the longest ½ inch or slightly more long and pointing downward. The 3 inch-wide glossy flowers are sulphur yellow.

Notocactus werdermannianus

Stouter and longer spines than those of the species considered above are characteristic of *N. mammulosus* (syn. *Malacocarpus mammulosus*), *N. submammulosus* (syn. *Malacocarpus submammulosus*), *N. s. pampeanus* (syn. *Malacocarpus pampeanus*), and *N. floricomus* (syn. *Malacocarpus floricomus*). With nearly spherical, glossy green plant bodies about 3 inches in diameter with eighteen to twenty-five ribs, ***N. mammulosus,*** of Brazil, Uruguay, and Argentina, has spine clusters of twenty to thirty radials and two to four centrals slightly more

Notocactus mammulosus

than ½ inch long. The blooms are yellow and about 1½ inches long. From it, Argentinian ***N. submammulosus*** differs in having fewer, low ribs, and its spines in clusters of six radials and one central, the latter up to ¾ inch long. Up to 4 inches tall and 2½ inches in diameter, the spherical to subcylindrical plant bodies of *N. s. pampeanus* have about twenty-one ribs, with spine clusters of seven to ten ½-inch-long radials and one or two strong central spines. The yellow flowers have toothed inner petals. At maturity 7 inches in height and 5 inches in diameter, the solitary, flattened-spherical to short-cylindrical plant bodies of ***N. floricomus,*** of Uruguay and Argentina, have twenty ribs with red-based white to gray spines, the radials numbering fifteen to twenty, the centrals four to five and up to 1 inch long. The 2½-inch-long flowers have toothed inner petals and an eight-lobed, red stigma. Variety *N. f. velenovskii* (syn. *N. velenovskii*) has spine clusters of twenty-five glassy white, bristly radials and five to nine slender, brown centrals about ¾ inch long.

Popular, variable ***N. scopa,*** of southern Brazil and Uruguay, has stems, at first spherical but becoming elongated with age, completely covered with slender, rather soft spines. They have thirty to thirty-five notched ribs. Each spine cluster is of up to forty white radials about ¼ inch long and three or four stronger, brown centrals. The canary-yellow blooms, 1 inch

Notocactus scopa

to 1½ inches wide, have toothed petals and a stigma with ten to twelve red lobes. Variety *N. s. candidus* has all white spines. The centrals of *N. s. ruberrimus* are crimson.

One of the most distinct, attractive, and variable species, ***N. ottonis*** has usually freely-clustering, spherical or flattened-spherical, bright green plant bodies 2 to 4½ inches in diameter, with ten to thirteen ribs. The short, needle-like spines are in clusters each of three or four 1-inch-long central spines and several shorter yellow radials. The blooms are golden-yellow with the outer petals reddish, on their outsides brownish-woolly, and 1½ to 2½ inches across. This species is native to southern Brazil, Uruguay, and Argentina.

Notocactus ottonis

Garden and Landscape Uses and Cultivation. Deservedly popular among cactus fanciers, notocactuses are generally easy to manage and are delightful for rock gardens and similar locations in warm, dry climates and as greenhouse and window plants. They flower freely, the blooms lasting for four to six days. They succeed under conditions favorable to most small cactuses, such as echinocactuses and mammillarias. Well-drained, slightly acid soil is to their liking and they are readily propagated by offsets, seed, and sometimes by grafting. For more information see Cactuses.

NOTOSPARTIUM (Noto-spártium)—Pink-Broom. The pea family LEGUMINOSAE is not well represented in the native flora of New Zealand, but it does provide a few genera, some of which occur only there, and most of which are unusual. Consisting of three species, *Notospartium* is an endemic that in its native land is becoming rare. Its members are shrubs, leafless except when young, and having arching branches and flattened, pendulous branchlets that serve as leaves. The blooms, pea-shaped and in racemes, are very like those of *Carmichaelia.* They are succeeded by flattened, straight or curved pods, which, unlike those of *Carmichaelia,* do not open spontaneously. The name comes from the Greek *notos,* south, and *Spartium,* the name of a related genus.

The pink-broom (***N. carmichaeliae***) is a graceful and lovely plant from 4 to 12 feet tall, freely-branched, and in summer almost smothered with lilac-pink blooms each ⅓ inch long, in 1- to 2-inch-long, dense racemes. The leaves, which are few except on young plants, are small and rounded. The slender pods, up to 1 inch long, are constricted between the three to eight seeds. A round-headed tree 15 to 30 feet high, ***N. glabrescens*** has, except for the uppermost that are erect, slender, pendulous branches and rather loose racemes up to about 2 inches long of purple-veined or purple-flushed, rose-pink flowers almost or quite ½ inch long. Up to 1 inch in length, the slender pods contain eight or fewer black-dotted, red seeds.

Notospartium carmichaeliae

Notospartium glabrescens

Garden and Landscape Uses and Cultivation. Not hardy in the north, these quite charming shrubs can be grown only where mild winters prevail. They stand but a few degrees of frost. Well-drained, sandy, peaty soil and sunny locations are most to their liking. The most satisfactory means of propagation is by seed.

NOTOTHLASPI (Noto-thláspi). In the wild found only in New Zealand, the two species of this genus of the mustard family CRUCIFERAE are small, deep-rooted herbaceous perennials that occur in rocky and stony places in the mountains. Fleshy, and branched or not, they have more or less spatula-shaped leaves and white flowers in racemes or clusters. Each bloom has four sepals and four spreading, white petals that narrow markedly at their bases. Of the six stamens four are longer than the others. There is one style. The name is from the Greek *notos,* south, and *Thlaspi,* the name of a related genus. The fruits are flat, broadly-winged capsules containing many seeds.

With crowded rosettes of more or less toothed leaves with stalks and blades up to ¾ inch long, but often shorter, ***Notothlaspi rosulatum*** has branchless stems that, as the plant passes into seed, elongate to about 9 inches. Earlier, they carry a crowded, pyramidal raceme of stalked blooms each a little under ½ inch in diameter. The heart-shaped seed capsules are ½ to 1 inch long.

Garden Uses and Cultivation. This rare plant is suitable only for gardeners well experienced in the cultivation of capricious alpines, and success even then is not likely without the convenience of an alpine greenhouse. Although reported to be a perennial in the wild, this plant is likely to behave as a biennial in cultivation. Little experience is reported about growing it. It may be raised from seed.

NOTOTRICHIUM (Noto-tríchium) — Kului. Endemic to Hawaii, *Nototrichium* consists of three or four closely similar species of shrubs and trees of the amaranth family AMARANTHACEAE. One is sometimes cultivated. The name perhaps comes from the Greek *noton*, back, and *trichion*, a small hair. Its application is not apparent.

Nototrichiums have forking branches and opposite leaves. The tiny, conical, pubescent flowers, in hairy-stalked, terminal and axillary spikes, have perianths deeply divided into four equal lobes, stamens slightly united at their bases, and a slender style. The fruits, technically utricles, are enclosed in the perianths.

The kului (***N. sandwicense***) is a shrub, or tree, up to 20 feet tall, with slender branches covered with yellowish hair. Its short-stalked, pointed-elliptic to ovate leaves, up to 3 inches long, have silvery-hairy undersides. The flower spikes, usually drooping in threes or fives from the branch ends, are ovoid to cylindrical and ¾ inch to 1½ inches long.

Garden and Landscape Uses and Cultivation. In its home islands, California, and other mild climate regions, the kului is planted to some extent as an ornamental. It succeeds under ordinary garden conditions and is propagated by seed.

NOTYLIA (Notýl-ia). Consisting of some sixty species, *Notylia*, of the orchid family ORCHIDACEAE, inhabits Mexico, Central America, tropical South America, and the West Indies. Its name, from the Greek *noton*, back, and *tylos*, a hump, alludes to a peculiar protuberance on the column of the flower.

Plants of this genus roost on trees (are epiphytes) or on cliffs and rocks. Some have small pseudobulbs each with a single flat leaf, the leaves of others overlap and have folded bases that sometimes envelop small pseudobulbs. Their little, usually white, off-white, greenish, or yellow, often fragrant blooms are generally in many-flowered racemes. They have erect or spreading, similar narrow sepals and petals and a lip with a triangular or arrowhead-shaped blade.

Among sorts cultivated are these: ***N. barkeri***, native from Mexico to Panama, up to 1 foot tall, has ellipsoid pseudobulbs each with one strap-shaped leaf up to 8 inches long and nearly 1 inch wide or somewhat wider. Its slightly fragrant flowers in arching or drooping racemes up to 1 foot long have greenish sepals and white petals mottled with yellow. The lip is white. ***N. bungerothii***, of Central America, has comparatively large pseudobulbs and leaves up to 9 inches long by 2 to 3 inches wide. The blooms, closely arranged in many-flowered racemes, have narrow, sickle-shaped, yellowish to white petals with an orange spot at the base. The lip is small and white. ***N. sagittifera*** (syn. *Pleurothallis sagittifera*) has 1-inch-long pseudobulbs each with one elliptic-oblong leaf up to 6 inches long and 1½ inches broad. The

Notylia bungerothii

Notylia sagittifera

½-inch-wide flowers loosely strung along pendent racemes up to 6 inches long have apple-green sepals, pale yellowish-green petals with two yellow spots on the lower half, and a white triangular lip. ***N. xyphophorus***, of Ecuador, has fans of small, pointed-ovate to sword-shaped leaves and on slender, pendulous stalks terminal loose, umbel-like clusters of individually long-stalked, purplish-lilac flowers with narrow sepals and petals.

Garden Uses and Cultivation. Of interest to collectors of orchids, notylias prosper affixed to pieces of trunks of tree fern. They need humid tropical conditions, such as suit cattleyas, and require shade from bright sun and abundant moisture. For more information see the Encyclopedia entry Orchids.

Notylia xyphophorus (flowers)

NOVELTIES. This somewhat ambiguous and not infrequently misapplied term in its best usage is reserved for newly introduced horticultural varieties, sorts developed by plant breeders or those obtained by selecting and multiplying desirable chance seedlings or vegetative mutants (sports). Species newly introduced from the wild should not be called novelties, nor should horticultural varieties after their first year of introduction into commerce without the year of their introduction being given.

Many seed and nursery catalogs list novelties annually, most often priced substantially higher than older, standard varieties of the same kinds of plants. A small proportion only of those promoted achieve permanence in gardens. Most disappear from the trade within a very few years. But some, of course, persist as definite improvements over older varieties, or at least they are sufficiently different to make them worthwhile.

As a rule it is best to rely upon well tried and proved varieties for general plantings and to try on a somewhat experimental basis the most appealing novelties, discounting to some extent the loud praises heaped upon them by their introducers and the blandishments of the catalogs.

NOVEMBER, GARDENING REMINDERS FOR. In the north this month heralds the beginning of winter and brings need for preparations for its arrival. Inexperienced gardeners are apt to close the season's work too soon, after a little cleaning up following the first killing frost to "call it a day," but knowledgeable ones know that much else can and should be done to ensure the well-being of many sorts of plants and to lay the groundwork for a successful garden next year.

Fall clean-up of course needs attention. It includes pulling or cutting the tops off annuals and vegetables that have ended their usefulness and off most herbaceous perennials as soon as they are killed by frost. The foliage of the few perennials that dries rather than rots may be left until spring as natural protection for the roots. Most of the debris can be safely put on the compost pile. Exceptions are plant parts known or strongly suspected of hosting pests or diseases that live over in soil. These, if added to compost become sources of future troubles; however, the great majority of pests and diseases will not survive in compost. Among serious ones that do are nematodes and crown rot disease.

Final cleanup of the garden in November. (a) Pull up annuals that are through blooming

(d) Transport by wheelbarrow

(f) Add to free-standing compost pile

(b) Clear away old tops of perennials

(e) Or by other means

(g) Or to a compost bin

(c) Collect the tops of string beans and other harvested vegetable garden crops

Stakes and other temporary plant supports need cleaning and carefully bundling or stacking in dry storage in readiness for future use. So do empty pots, tubs, flats, hanging baskets, and similar appurtenances. Most important of all, thoroughly clean and store in a dry place all tools and equipment, such as mowers and rotary tillers, as soon as their season's usefulness is over. Wet days give good opportunity to attend to this.

Leaf raking, or rather leaf collection because in large gardens fallen leaves are often dealt with by assembling them into windrows with blower machines and then sucking them into a truck body with a giant vacuum cleaner type apparatus, is a traditional fall chore. Add collected leaves of deciduous trees to the compost pile or stack them separately to make leaf mold. Do not include leaves of evergreens in leaf mold piles, include only in small amounts if any at all in general compost piles.

In woodland and semiwoodland areas, in shrub borders and similar places, it is often preferable to let fallen leaves remain as a natural mulch instead of collecting and stacking them. If you do this check once in a while that they have not gathered or been blown into layers too thick for the well-being of herbaceous perenni-

Rake leaves as they fall; do not allow them to lie on lawns for long periods

In woodland areas, fallen leaves serve as a natural mulch

als that may be underneath. If they have, level them out with a fork or rake.

Never allow masses of leaves to lie for long periods on lawns. They weaken the grasses by denying them light and establish conditions favorable to the growth of funguses that may rot the turf. Do not forget to free gutters and drains that may be blocked.

Complete fall planting without delay. In cold areas it is generally better to postpone moving evergreens until spring, but in most parts there is yet time to set out deciduous trees and shrubs, including roses and most herbaceous perennials. But do not delay, and be prepared before the full intensity of winter comes to afford special protection by wrapping trunks of newly transplanted trees with strips of burlap or paper made especially for the purpose and by mulching or winter covering those needing such protection. Early November is the opportune time to plant tulips. Hyacinths, narcissuses, crocuses, and other spring-flowering bulbs that should preferably have been set out earlier will nevertheless give acceptable results if planted without delay.

Propagation of hardy trees and shrubs from cuttings can be done this month. A phase of horticulture much less practiced by amateurs than it should be, this makes it possible to raise at home young stock for new plantings and for replacing excessively old and worn-out specimens, as well as the multiplication of rare and unusual sorts not available from nurseries. Not all kinds of trees and shrubs are susceptible to this means of increase, but a very considerable number, including some evergreens as well as deciduous sorts, are.

Cuttings of deciduous trees and shrubs taken when leafless are easily rooted without greenhouse facilities. All that is necessary is to cut shoots of the current season's growth into pieces 6 to 9 inches long, tie them in bundles, and bury them beneath 6 inches or so of sand or peat moss outdoors or in a cold frame, then in spring take them out, separate them, and plant in nursery rows, there to grow into new plants. Among the numerous kinds that respond are deutzias, flowering currants, forsythias, honeysuckles, hydrangeas, mock-oranges, plane trees, privets, rose-of-Sharon (*Hibiscus syriacus*), roses, spireas, viburnums, vitexes, and weigelas.

Evergreens that can be increased by cuttings at this time include aucubas, barberries, boxwood, English ivy, hollies, leucothoes, junipers, and yews. In general, firs, hemlocks, pines, spruces, and other tall tree types do not lend themselves to

Plant tulips in November in beds and in small groups in borders

A small group of tulips in bloom

Propagate appropriate shrubs and trees by hardwood cuttings: (a) Cut shoots of current season's growth into pieces 6 to 10 inches long

(b) Tie them into small bundles and label, then bury them in a cold frame or outdoors; in spring dig them up and plant them, with only their tips above ground, in rows outdoors

Shrubs easy to propagate by hardwood cuttings taken in November: (a) Deutzias

(b) Flowering currants (*Ribes*)

(c) Forsythias

(d) Mock-oranges (*Philadelphus*)

(e) Rose-of-Sharon (*Hibiscus*)

(f) Roses

(g) Viburnums

(h) Weigelas

this method of propagation, but dwarf spruces do. Take the cuttings before severe freezing weather comes and plant them in a propagating bench in a cool greenhouse. These cuttings are, of course, leafy.

Cuttings of many hardy evergreens root readily in greenhouse propagating case in November

Cuttings rooted in a propagating case

Winter covering, strewing the ground and covering the tops of low plants with materials that mitigate the effects of alternate freezing and thawing and give pro-

Evergreens that may be rooted from cuttings taken in November: (a) Aucubas

(b) Barberries

(c) Boxwoods (*Buxus*)

(d) Leucothoes

(e) Junipers (*Juniperus*)

(f) Yews (*Taxus*)

tection from sun and wind will need attention this month or next. It substitutes for the continuous deep snow cover that in parts of the north affords protection.

Materials used must be loose enough to admit air freely and not pack unduly under the influence of rain and snow. Among those commonly employed are recently fallen leaves, corn stalks, salt hay, littery manure, and branches of evergreens (those of discarded Christmas trees serve well).

Do not winter cover too early. If you do you may encourage mice and other rodents to make homes beneath coverings and damage plants there. Wait until the ground is frozen to a depth of 2 to 3

For winter protection: (a) Salt hay spread as a loose cover over perennials

(b) Keep salt hay in place by laying woody branches across it

inches. And do not use too thick a layer of covering. In most cases one just heavy enough to stay in place and adequately shade the ground is sufficient. This admonition not to cover too thickly applies especially to such plants as basket-of-gold, evergreen candytuft, dianthuses, and helianthemums, which retain their foliage through the winter. Take care not to cover deeply the centers of canterbury bells, foxgloves, and similar rosette-type plants.

Winter protection of roses in cold sections is best done by hilling soil around their bases and, after it has frozen to a

Hill soil around roses to protect them against the rigors of winter

depth of 3 inches or so, applying a layer of winter covering that covers the mounds and fills the hollows between.

Where ground remains frozen to a considerable depth for protracted periods, evergreens on the borderline of hardiness, such as boxwoods in many places as well as newly planted evergreens and some newly set out deciduous trees, can be seriously injured by exposure to sun and sweeping winds. Spraying two or three times at monthly intervals with an antidesiccant, beginning just after Christmas, is helpful, but even more positive aid is achieved by erecting around exposed

Protect specimen boxwood: (a) Build a frame around them and surround them with burlap

(b) The finished job

specimens screens of burlap or by sticking into the ground around individual plants branches of pine or other suitable evergreens to afford shade and break the force of wintery blasts. Be aware of the danger to evergreens of snow slides from roofs. Where these threaten, install snow guards on the roofs or protect endangered plants by erecting platform-like covers supported on stakes or poles over them. Even in climates as cold as that of New York City, figs may be wintered successfully by tying their branches together, surrounding them with suitable insulation, and wrapping them in waterproof roofing paper or polyethylene plastic.

Greenhouses in November will normally be chock-a-block full, and there will be little opportunity or indeed need to engage in many tasks other than routine watering, damping down, management of ventilators, and the like. Still, there will be enough to keep one busy on days too inclement for outdoor activities, which when weather permits must be given preference.

Weaker sunlight and shorter days call for the removal of permanent shading from greenhouses. Even plants most sensitive to strong sun have no need of protection after the end of November and many that appreciated light shade in summer and early fall may be exposed earlier.

In New York City, a fig wrapped for winter protection

Propagate varieties of: (a) *Begonia hiemalis*

(b) From leaf cuttings

Colder weather brings increased use of artificial heat and that, in turn, dries the atmosphere, occasioning more frequent need, especially where tropical plants are grown, for damping down paths, floors under benches, and benches between pot plants. Little of this is needed in cool greenhouses in which such plants as carnations, chrysanthemums, and cactuses and other succulents are accommodated, but on very bright sunny days a damping down in the morning is of benefit.

If inclined to be heavy-handed with the watering can or hose, and a great many amateurs are, use considerable restraint from now throughout the winter. Give plants actually in need generous soakings, but refrain from wetting the soil of those moist enough to get along without watering until the next day. And do not splash water on foliage. Smooth-leaved tropical plants, such as crotons, dracaenas, and ficuses, may perhaps be lightly misted overhead on bright sunny days, but only if there is every expectation that the moisture on the foliage will dry before nightfall. Most other plants get along better at this season without this.

Tulips, hyacinths, and narcissuses, Dutch and Spanish irises, and other hardy spring-flowering bulbs for forcing into early bloom indoors may still be potted or if needed for cut flowers planted in flats, but as time is running out, the work should be delayed no longer. Now, too, hardy lilies, except Madonna lilies (which should be planted much earlier), can be potted, put in a cold frame, and covered heavily with dry leaves, salt hay, or similar material, there to establish good root systems before being brought indoors later for blooming in late spring. Pot Easter lilies late this month or early next and carry them in a greenhouse with a night temperature of 60°F or a few degrees cooler for the first three or four weeks.

Plant propagation takes little of gardeners' time in November, but to have fine specimens of winter-flowering begonias of the *Begonia hiemalis* group, leaf cuttings should be taken in late November or December and planted in a propagating bench with a temperature of 60 to 70°F to give plants for next year.

Care of cold frames now includes making sure their occupants will be fairly snug for the next few months, but avoid too much coddling. On sunny days when the temperature inside rises above 35 or 40°F, ventilate frames a little or more generously depending upon the outside temperature. Remember the objective is to keep the plants dormant or nearly so, certainly not to encourage top growth.

Strew salt hay loosely over biennials in cold frames for winter protection

It is very helpful to strew over biennials and hardy perennials housed in frames a light layer of salt hay, straw, or dry leaves. This, besides giving some insulation against extremely low temperatures, checks loss of moisture from foliage and keeps the soil at a more uniform temperature.

Plants rather tender to severe cold, such as English wallflowers in many areas, which without extra protection would not survive, can often be wintered successfully if the frames are well banked around with soil or insulating material such as dry leaves, straw, corn stalks, or salt hay, and on cold nights are covered with thick mats or wooden shutters. To contain the insulation a temporary framework of boards, its walls standing 9 inches to 1 foot away from the cold frame proper and the space between packed with insulating stuff, is effective. For the best success with sweet peas sow seeds this month in pots of sandy soil and bury them to their rims in a cold frame in a bed of sand or peat moss, then cover them with 1 inch to 2 inches of peat moss. On the first occasion in spring when the ground is workable, remove the plants from their pots and without disturbing the roots plant them in a sunny place in the garden in ground that has been deeply spaded and conditioned by mixing in compost or manure.

An insulated cold frame; on mild days, raise the sash to assure good ventilation

Houseplant management during November calls for little more than watering sufficiently often to keep the soil evenly moist, but not soggy-wet. Increasing use of artificial heat dries the atmosphere. This can be partially allieviated by misting foliage with water a couple of times a day and standing the plants on broad trays filled with pebbles, moss, or other material kept

Water houseplants adequately, but not excessively, during November

Sheets of paper positioned on cold nights between window plants and the glass protect against excessively low temperatures

(b) Coreopsises

(c) Cornflowers (*Centaurea*)

(d) Gaillardias

(e) Mignonette (*Reseda*)

constantly damp from which moisture rises to humidify the air around the plants.

Only very well rooted leafy specimens of such plants as dracaenas, fatshederas, ferns, rubber plants, and philodendrons should be fertilized in winter, and they not more than once or twice a month with dilute liquid fertilizer. Under no circumstances fertilize plants that have not filled their containers with healthy roots.

Inadequate light from now on is a frequent cause of such sun-loving plants as geraniums becoming thin and spindly. The trouble is compounded if temperatures are much higher than the minimum necessary for decent survival. No doubt about it, a temperature in the 70°F-plus range that most humans find comfortable is too high for any but the most tropical houseplants. But some relief is attainable. Close to windows, the best locations for most plants, temperatures are usually appreciably lower than elsewhere in a room. And it helps tremendously to lower the thermostat a few degrees at night. Finally, illuminating plants with artificial light as a supplement to daylight helps to counteract short days and poor natural light.

Because of possible danger of temperatures close to windows falling too low on very cold nights, it may be desirable to pull shades, place sheets of paper between the plants and glass, or move the plants away from the window.

Other dangers to the winter well-being of houseplants stem from proximity to radiators or other sources of dry heat and exposure to cold drafts.

November in the south is a busy time. Put off no longer placing orders for trees, shrubs, and vines, ornamental and edible-fruited, deciduous and evergreen, and roses, needed for planting, and proceed with reasonable haste in preparing planting sites. The ideal time for planting differs somewhat with latitude and other variables, but in areas where deciduous trees and shrubs are common, the heavy falling of their leaves may be taken as a signal that planting time has arrived.

Now, too, you may set out most kinds of herbaceous perennials, those obtained from nurseries as well as divisions of specimens dug from one's own or friends' gardens.

There is still time to plant many sorts of bulbs as well as tuberous anemones and ranunculuses. Among true bulbs are Dutch and Spanish irises, grape-hyacinths, narcissuses, scillas, snowdrops, and lilies other than Madonna lilies (for which it is now too late). Tulips may be planted in northern parts of the south toward the end of November, but in the middle and lower south it is better to wait until December and then to set only bulbs that have been chilled for a few weeks in a refrigerator. In the deep south plant freesias.

In the deep south and middle south sow now for late winter bloom seeds of many sorts of annuals, such as anchusas, calendulas, California-poppies, coreopsises,

Annuals to sow outdoors in November in regions of mild winters: (a) California-poppies (*Eschscholzia*)

(f) Sweet peas

cornflowers, gaillardias, larkspurs, mignonette, phloxes, poppies, and sweet peas. To have pansies for spring bloom sow in a lightly shaded cold frame seed that has been kept for three or four weeks in a refrigerator.

Vegetables to sow in November, the sorts likely to succeed depending upon the part of the south in which you garden, include beets, broccoli, brussels sprouts, cabbage, celery, collards, endive, kale,

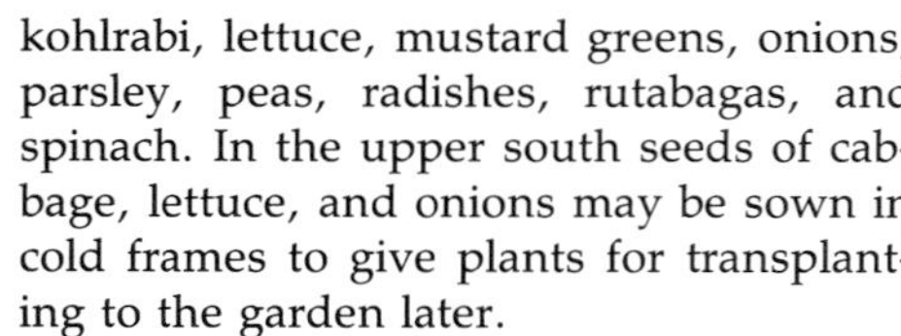
kohlrabi, lettuce, mustard greens, onions, parsley, peas, radishes, rutabagas, and spinach. In the upper south seeds of cabbage, lettuce, and onions may be sown in cold frames to give plants for transplanting to the garden later.

Begin mowing lawns of rye grass or those that have been overseeded with rye grass as soon as the new grass is 2 inches tall. Be sure the mower cuts sharply.

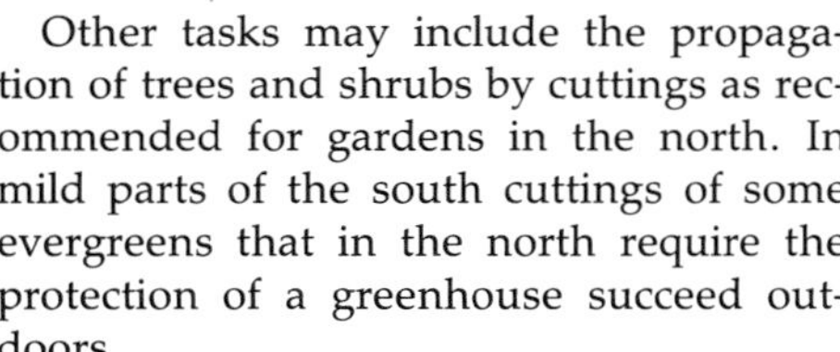
Other tasks may include the propagation of trees and shrubs by cuttings as recommended for gardens in the north. In mild parts of the south cuttings of some evergreens that in the north require the protection of a greenhouse succeed outdoors.

West Coast gardeners in colder sections will busy themselves with fall clean-up work and soon with affording protection

(c) Cabbage

(f) Peas

Vegetable garden crops to sow in November in regions of mild winters:
(a) Beets

(d) Lettuce

(g) Radishes

(b) Brussels sprouts

(e) Parsley

(h) Spinach

to plants likely to be harmed by low temperatures. Cut back snapdragons that are to be left for blooming a second year.

Preparations for planting should now be in full swing. Whenever weather is favorable make ready sites for immediate and for spring installations. In many areas this is the ideal time to set out deciduous and evergreen trees, shrubs, vines, and roses. Continue to plant spring-flowering bulbs and in mild regions such bedding plants as cinerarias, pansies, Paris-daisies, primulas, and stocks, for late winter and spring display.

In favorable areas sow seeds of a wide variety of annuals as well as such cool-weather vegetables as beets, lettuce, onions, peas, radishes, rutabagas, spinach, and turnips.

Other matters to look to include taking cuttings for rooting in a greenhouse or cold frame of coleuses, geraniums, heliotropes, lantanas, and numerous other plants. If winter rains are late in coming, be sure to keep camellias, roses, and all other plants that are evergreen or in active growth well watered.

NUN'S ORCHID is *Phaius tankervilliae.*

NUPHAR (Nùph-ar)—Spatterdock, Cow-Lily, Yellow Pond-Lily. Spatterdocks, less effective ornamentals than water-lilies (*Nymphaea*), are more often established in natural ponds and other quiet waters than grown in garden pools, yet they are not without beauty and interest and afford variety in collections of aquatics. They belong in the water-lily family NYMPHAEACEAE and number twenty-five species, all of north temperate regions and mostly natives of North America. The name *Nuphar* derives from the Arabic name for water-lilies, *naufar.* The rhizomes of these plants were used as food by American Indians.

The spatterdocks differ from water-lilies in that the showy parts of their flowers consist not of petals, but of five to twelve, leathery, yellow or purplish, concave, petal-like sepals; the true petals are minute and, except that they have no anthers, resemble the very numerous stamens. The stigma is disk-like and has five to many rays. The blooms vaguely resemble huge buttercups and are usually held above the water. The egg-shaped to cylindrical fruits generally mature above water. These are vigorous, often rampant plants with stout rootstocks that grow in mud beneath fairly shallow water in lakes, ponds, and along the margins of slow-moving streams. Their leaves are nearly round to lanceolate and are deeply heart-shaped at their bases; often some are submersed, some float, and some stand out of the water. The submerged leaves are thin and fragile.

The common spatterdock (***N. advena***) occurs naturally from Maine to Nebraska, North Carolina, Kentucky, Arkansas, and Texas. All or nearly all of its leaves are commonly above water; they rarely float or are submersed. This distinguishes it from several other eastern American species. The leaves are nearly round and 8 to 16 inches across. The flowers are yellow, sometimes flushed with purple, and the three inner sepals tipped with green. Up to 2 inches in diameter, they usually have six sepals. Formerly included in *N. advena,* sometimes under the name *N. a. variegata,* a similar plant that normally has floating and some submersed leaves, is now distinguished as ***N. variegatum***. Its floating leaves are usually 6 to 10 inches long and two-thirds as wide; their stalks are flattened on their upper sides, whereas those of *N. advena* are rounded. The flowers of *N. variegatum* have sepals that are usually red in their lower halves. This species ranges from Quebec and Newfoundland to British Columbia, New Jersey, Ohio, Indiana, and Iowa. A quite different American species is ***N. sagittifolium,*** indigenous from Virginia to South Carolina. It has both floating and submersed leaves. They are narrowly-oblong, 6 to 16 inches long by 2 to 4 inches wide. The yellow flowers are 1 inch or slightly more in diameter. It has cylindrical fruits. A hybrid between this and *N. advena,* intermediate in character and named *N. interfluitans,* occurs in Virginia. The western American ***N. polysepalum*** occurs as a native from South Dakota and Colorado to California and Alaska. It has floating or erect, broadly-ovate leaves, 4 to 16 inches long, and flowers with seven to nine yellow sepals, often tinged with red. The blooms are 3 to 4 inches in diameter or sometimes larger.

Nuphar luteum

A European spatterdock, the yellow pond-lily (***N. luteum***) has smaller blooms than *N. advena.* They have five sepals and are somewhat fragrant; the scent, according to some people, is that of an alcoholic beverage for which reason the plant is sometimes called brandy bottle. The leaves float or are raised slightly above the water; they have triangular stalks. Variety *N. l. sericeum* has larger blooms. A native of Europe and Asia, ***N. pumilum*** is a similar, but smaller plant with translucent underwater leaves and deeply-lobed floating ones. The Japanese ***N. japonicum*** has narrow leaves like those of *N. sagittifolium,* but its flowers are 2 to 3 inches in diameter and its fruits are flask-shaped. Variety *N. j. rubrotinctum* has orange-red flowers with red-tipped stamens. Variety *N. j. variegatum* has leaves mottled with cream.

Hybrid spatterdocks of interest are ***N. spenneranum,*** which has as its parents *N. luteum* and *N. pumilum* and is a much tidier and smaller plant than *N. luteum,* and an unnamed hybrid between *N. pumilum* and *N. japonicum,* which has the large flowers of the last named and the translucent underwater foliage of *N. pumilum.*

Garden and Landscape Uses. The spatterdocks have some appeal as subjects for planting at the margins of natural ponds and slow streams. Those geographically appropriate may properly be included in collections of native plants. They are less useful in more sophisticated water gar-

Nuphar pumilum

dens, although they may occasionally be used in a small way to provide variety. They need full sun.

Cultivation. Spatterdocks are planted and grown in the same way as hardy water-lilies (*Nymphaea*) and are propagated by division and by seed in the same way as water-lilies.

NURSERY. A nursery is an area of ground devoted to the care of young plants until they are transplanted to their permanent locations. Propagation is usually an important part of the work of nurseries, but may be dispensed with and reliance placed upon bought young stock. Plants of the smallest size suitable for planting in rows outdoors, called liners, are often chosen for purchase from nurseries specializing in propagation.

Besides land, a nursery generally comprises appurtenances and structures necessary for its operation. There may include greenhouses, lath houses, cold frames, storage sheds, and an office. Retail nurseries usually have a display and sales area, often of the garden center type. Some have landscape departments and offer planning services as well as the laying out and planting of gardens and other landscape developments.

In addition to being important as sources of plants and supplies, nurseries are fascinating places for home gardeners and others to visit. From them one can learn a great deal about the appearances, characteristics, hardiness, and adaptability of plants in particular localities, and from friendly and knowledgeable nurserymen, much about the selection and proper management of plants. But remember, specimens in nurseries are mostly young, and some sorts, especially evergreen and deciduous trees, assume quite different forms and become larger as they mature. Before purchasing sorts with which you are not familiar it is a good idea to look around the neighborhood and appraise specimens that have been established in permanent locations for several years.

A small nursery, or at least a nursery bed, is a desirable adjunct to home gardens, especially younger ones still in process of development. It may be associated with a vegetable garden, cut flower garden, or other crop production area or be by itself a little apart and screened from ornamental parts of the garden.

Such a feature is of great interest to manage and of inestimable value in supplying at remarkably small cost young trees, shrubs, and other plants for extending and refurbishing plantings or if a surplus is produced for giving to friends or disposing of in other ways.

NUT. The word nut is employed as part of the common names of a wide variety of plants not all of which have true nuts as fruits. Among such sorts are Australian nut-palm (*Cycas media*), banucalad nut (*Aleurites trisperma*), Barbados nut (*Jatropha curcas*), betel nut palm (*Areca catechu*), bladder nut (*Staphylea*), Brazil nut (*Bertholletia excelsa*), Chilean nut (*Gevuina avellana*), clearing nut (*Styrychnos potatorum*), goat nut (*Simmondsia chinensis*), grass-nut (*Triteleia laxa*), gympie or macadamia nut (*Macadamia ternifolia*), Jesuits' or water nut (*Trapa natans*), lucky nut (*Thevetia peruviana*), monkey nut (*Hicksbeachia*), paradise or sapucaia nut (*Lecythis zabucajo*), physic nut (*Jatropha curcas* and *J. multifida*), pistachio nut (*Pistacia vera*), pond nut (*Nelumbo lutea*), Queensland nut (*Macadamia integrifolia*), singhara nut (*Trapa bispinosa*), soap nut (*Sapindus mukorossii*), tallow nut (*Ximenia americana*), and tiger nut (*Cyperus esculentus sativus*).

NUT and NUTLET. In their strictly botanical sense, a nut is a one-celled, one-seeded fruit with a hard, bony outer cover that does not split spontaneously and a nutlet is a tiny nut. Examples of nuts are acorns and hazelnuts. Typical nutlets are the seedlike fruits, usually called seeds, of coleuses, monardas, salvias, and other plants of the mint family LABIATAE. In common parlance many fruits not botanically nuts are called nuts. These include almonds, coconuts, hickory nuts, peanuts, pecans, and walnuts.

NUTMEG is *Myristica fragrans*. The California-nutmeg is *Torreya californica*.

NUTRICULTURE. See Hydroponics or Nutriculture.

NUTS. Edible fruits popularly called nuts include sorts that under strict botanical interpretation are not nuts. Here the word is employed in its broader, colloquial sense, and includes the coconut and peanut, immensely important as sources of edible oils as well as for eating, and others, of which almonds, butternuts, chestnuts, filberts, hazelnuts, hickory nuts, litchis, macadamia nuts, pecans, pignuts, pistachios, and walnuts are examples, that are esteemed as nutritious foods for consumption raw or cooked.

Although some kinds of nuts are cultivated commercially none is of much importance in home gardens. Anyone contemplating planting them for their fruits should ask for information from their State Agricultural Experiment Station. For further information consult entries in this Encyclopedia under the names of the nuts and under the names of the genera to which they belong.

NYCTAGINACEAE—Four O'Clock Family. This family of dicotyledons includes thirty genera totaling 290 species, the majority tropical or subtropical and mostly American. Belonging are trees, shrubs, woody vines, and herbaceous plants with alternate or opposite leaves, those of pairs often of unequal size. The bisexual or unisexual flowers are in branched, sometimes panicle-like clusters with large bracts that in some sorts simulate a calyx, that in others are brightly colored and petal-like. There is a tubular, often petal-like calyx, no petals, one to many stamens, and one slender style. The fruits, frequently glandular, are achenes enclosed in persistent calyxes. Cultivated genera include *Abronia, Boerhaavia, Bougainvillea, Mirabilis,* and *Pisonia.*

NYCTANTHES (Nyct-ánthes) — Tree-of-Sadness or Night-Jasmine or Harsinghar. In applying two of the common names of the genus *Nyctanthes* it is necessary to avoid confusion with those of species of other genera for which somewhat similar vernacular designations are used. These are the sorrowless tree (*Saraca indica*) and night-flowering-jessamine (*Cestrum nocturnum*). A somewhat straggling shrub or tree of the vervain family VERBENACEAE, the tree-of-sadness, night-jasmine, or harsinghar is one of two species of its genus. Its name, from the Greek *nyx,* night, and *anthos,* a flower, alludes to its nocturnal blooming.

Native to India, the tree-of-sadness (***N. arbor-tristis***) attains a height of up to 30 feet. Rough all over with stiff, white hairs, it has quadrangular shoots, opposite, very short-stalked, pointed-heart-shaped leaves up to 4½ inches in length, sometimes with a few teeth. The exceedingly fragrant flowers, which open at night and drop the following morning, have four-lobed calyxes. Their corollas have cylindrical, orange tubes, and four spreading, white lobes (petals). There are two stamens, and a two-lobed style. The blooms, ¾ to 1 inch wide, are in few-flowered clusters with collars of bracts at their bases, arranged in large terminal panicles. The fruits are flat, nearly round, papery capsules about ¾ inch long. When ripe they separate into two one-seeded parts. In Burma a dye made from the flowers is used to color the robes of Buddhist monks, and there and elsewhere in the Orient the blooms are employed in religious exercises. The wood is esteemed as fuel.

Garden and Landscape Uses and Cultivation. Hardy only in regions of little or no frost, this species is suitable for planting where it is likely to be appreciated after dark, beside a window or near a frequented path perhaps. It is often of ungainly habit and makes no fine display. The fragrance of its blooms is its only recommendation to gardeners. Occasionally this species is cultivated in greenhouses. It thrives in ordinary, well-drained, reasonably fertile soil in sun or part-day shade. A greenhouse winter night temperature of 55 to 60°F, with a daytime increase of five to fifteen degrees, is appropriate. The atmosphere should be fairly humid. Pruning to shape and any repotting needed is done in late winter or spring. At all times the soil is kept moderately moist. Well-rooted specimens benefit from regular applications of dilute liquid fertilizer from spring to fall. Propagation is easy by seed and by cuttings.

NYCTERANTHUS. See Aridaria.

NYCTERINIA. See Zaluzianskya.

NYCTOCEREUS (Nycto-cèreus). The seven species of cactuses that constitute *Nyctocereus* are native to Mexico and Central America. Clambering, trailing, or rarely erect, they belong to the cactus family CACTACEAE. The name, derived from the Greek *nyktos,* night, and *Cereus,* the name of a related genus, alludes to their nocturnal blooming. They are commonly called night-blooming-cereuses.

The thin stems of nyctocereuses are ribbed and very spiny. Their white, funnel-shaped flowers open only at night. They have longish calyx tubes with small scales and tufts of conspicuous bristly spines and woolly hairs. The perianth segments (petals) spread widely. Spherical to ovoid and with deciduous spines, the red fruits are about 1½ inches long. Embedded in their red, crystalline flesh are a few large, smooth, black seeds.

Common in cultivation and variable, ***N. serpentinus*** (syn. *Cereus serpentinus*) has clambering, creeping, or pendulous, ten- to thirteen-ribbed, cylindrical stems up to 10 feet long by ¾ inch to 2 inches in diameter. The ribs are low and rounded. They have felted areoles (spine-producing cushions) close together. White to brownish, usually with darker tips, the needle- or bristle-like spines are in clusters of about a dozen. The largest spines are a little over 2 inches in length. From toward the ends or at the ends of the stems the flowers develop. They are 6 to 8 inches long, about 3½ inches wide, and have long, bristly spines on their calyx tubes. The fruits are similarly spiny.

Its prostrate or pendulous stems not over 1 inch wide, and with ten rather acute ribs with clusters of seven to nine radial spines about ⅓ inch long and three to five centrals, the longest ¾ inch long, ***N. hirschtianus*** is a native of Nicaragua. The white to pinkish flowers, 2 inches long or slightly longer have narrow petals. Usually more or less upright and with arching or sometimes prostrate or creeping stems up to 2½ inches thick, ***N. guatemalensis*** is a native of Guatemala. The spines on the stems are in clusters of about ten radials and three to six longer, up to 1¼-inch-long, central spines. The fragrant, white blooms have yellow to red-brown outer segments (petals). They are up to 7 inches long. The spiny fruits are about ¾ inch long.

Garden and Landscape Uses and Cultivation. Nyctocereuses grow without trouble under conditions appropriate for most thin-stemmed cactuses, outdoors in warm, dryish regions, and in greenhouses. They may be trained to supports or allowed to sprawl. For more information see Cactuses.

NYMANIA (Ny-mània). The only species of *Nymania* belongs in the mahogany family MELIACEAE, or according to botanists who accept the segregation, in the aitonia family AITONIACEAE. It was named to honor the Swedish botanist Carl Fredrik Nyman, who died in 1893. It is endemic to South Africa.

A shrub or small tree 3 to 10 feet or more rarely up to 20 feet in height, known in its native land as klapperbos and Chinese lanterns, ***N. capensis*** (syn. *Aitonia capensis*) has rigid branches, and alternate, often apparently clustered, linear to linear-oblong, hairless leaves up to 2 inches long. The solitary, axillary flowers are rose- to carmine-pink. They have four-lobed calyxes and four erect petals about ½ inch long. The eight protruding stamens are united at their bases. The style is long-exerted. Bladder-like and 1 inch to 1¾ inches long, the inflated, four-lobed, papery seed capsules are mottled and suffused with carmine. Each lobe contains usually two seeds.

Nymania capensis in South Africa

Nymania capensis, in fruit

Garden and Landscape Uses and Cultivation. An unusual looking shrub that can be used to add variety to landscape plantings in warm, dry climates, such as those of parts of California, *Nymania* succeeds in well-drained soil in full sun. It is propagated by seed and by cuttings.

NYMPHAEA (Nymph-aèa)—Nymphea or Water-Lily or Pond-Lily, Egyptian Lotus. This best-known genus of showy-flowered aquatics includes fifty species of the water-lily family NYMPHAEACEAE and numerous varieties and hybrids of great diversity in size, foliage pattern, flower color, and other horticulturally important characteristics. Some open their blooms by day, others only at night. The name *Nymphaea* is derived from that of Nymphe, one of the lesser deities of mythology called water nymphs. The plants popularly called giant water-lilies belong to the separate genus *Victoria.*

As natives, nympheas are widely distributed in temperate, subtropical, and tropical regions, inhabiting ponds, lakes, and quiet backwaters. As ornamentals they are of great antiquity. The Egyptians admired and depicted them four thousand years before the birth of Christ and included them in offerings to the dead. Petals of the Egyptian blue lotus and white lotus have been identified in the funeral wreaths found in the tomb of Rameses II, the Pharaoh at the time of the Israelite captivity.

Nympheas are deciduous or evergreen perennials with tubers or horizontal to erect, thick rootstocks that grow in muddy bottoms beneath water from a few inches to sometimes as much as 6 feet deep. They have nearly circular to oval leaves that float or when crowded may toward the centers of the plants rise slightly above the water. The leafstalks adjust in length to the water's depth. The blades are according to sort from 2 inches to 2 feet across. They may have smooth, crinkled, or toothed edges. Those of some kinds develop plantlets. The flowers, from 1 inch to 1 foot or more in diameter and white, yellow, pink, red, purple, violet, or blue, have each four sepals and numerous petals and stamens. There are many carpels and up to about thirty radiating stigmas. The blooms float, or are lifted some distance above the water, then as the fruits develop sink beneath the surface and mature there. Although the numerous splendid hybrid water-lilies, the ancestries of many not clear, are the most commonly grown, the natural species should not be overlooked as attractive ornamentals.

Hardy nympheas at The New York Botanical Garden

White-flowered natives of North America include the fragrant water-lily (***N. odorata***), indigenous from Newfoundland to Manitoba, Michigan, Florida, and Louisiana. This has nontuberous rootstocks and round to slightly oblongish, smooth-margined, thickish leaves 4 to 9 inches wide, with reddish to purple undersides and stalks not striped. The 2- to 4½-inch-wide, strongly fragrant blooms, open during the mornings and closing about noon, white or rarely pink, have usually more than twenty-four petals. They have from fifty to over 100 stamens and ten or more stigmas.

Nymphaea odorata

Somewhat larger-flowered and with leaves upturned at their edges, *N. o. gigantea* inhabits the coastal plain from Delaware to Florida. From the fragrant water-lily, ***N. tuberosa*** differs in having leaves generally with striped stalks and usually green undersides and nonfragrant white blooms that open in the afternoon. Except along the northern boundary of the United States, this is the only species native west of the Appalachians. It occurs from Lake Ontario to Michigan, Minnesota, Kentucky, and Oklahoma and occasionally from Vermont to New Jersey. Also with scentless white flowers that open after noon, the pygmy water-lily (***N. tegragona***) extends as a native northward from Maine to Michigan and Idaho and occurs also in eastern Asia. From the previous two species it differs in having elliptic leaves not over three-quarters as broad as long and in having flowers with fewer than ten stigmas. Mexican ***N. flavovirens*** has nearly circular leaves approximately 1½ feet in diameter, their margins irregularly round-lobed, their undersides green. Open from morning until late afternoon, the 6- to 8-inch-wide, white flowers have fifteen to twenty petals. This is probably not hardy in the north.

Other white-flowered species are the European white water-lily (*N. alba*) and the Egyptian white lotus (*N. lotus*). Native to Europe and North Africa, ***N. alba*** is reported to be sparingly naturalized in New England. It has nearly circular, smooth-margined leaves up to 1 foot in diameter, with reddish undersides, and scentless white flowers 4 to 8 inches in diameter that have usually fewer than twenty-four petals. They remain open until late afternoon. Variety *N. a. rubra* has rosy-red blooms. Nonhardy, ***N. lotus*** differs from sorts previously discussed in having sharply-toothed leaves and blooms that expand in the evening and remain open all night. The leaves are nearly round, brownish and with a fuzz of short hairs on their undersides. The creamy-white flowers 5 to 10 inches across have about twenty petals and almost or quite 100 stamens. Central African *N. l. dentata* has pure white blooms 8 to 10 inches wide with narrower petals than those of the species. They stay open until about noon of the day following their opening. West African ***N. micrantha*** has roundish-elliptic leaves suffused with purplish-brown and flowers 6 inches across, lifted well above the water and with both bluish-white and white pointed petals. The blue-stalked stamens have bright golden yellow anthers.

Nymphaea micrantha

Blue-flowered water-lilies are all natives of warm regions. Most familiar are the blue lotus of Egypt, the blue lotus of India, the blue water-lily of South Africa, sometimes called the Cape blue water-lily, and the Australian water-lily. Considering these in turn, the Egyptian blue lotus (***N. caerulea***) is a vigorous sort with leaves 1 foot to 1½ feet across blotched with purple on their undersides. Its not especially showy blooms, 3 to 6 inches in diameter and pale blue with whitish bases to the fourteen to twenty petals, open in the mornings. They have fifty or more stamens. The blue-lotus of India (***N. stellata***) also occurs in southeast Asia. A much finer ornamental than its Egyptian counterpart, it differs chiefly in its leaves being not over 6 inches wide and having rich blue-violet under surfaces and toothed edges. Open in mornings and early afternoons, the 3- to 7-inch-wide blooms have eleven to fourteen light blue petals with white bases and thirty-five to a few more than fifty stamens. The blue water-lily, of South Africa (***N. capensis***), has conspicuously-toothed, roundish-ovate leaves from 1 foot to nearly 1½ feet across. Its sky-blue flowers, whitish toward the bases of their petals and 6 to 8 inches in diameter, are open from morning until late afternoon. They have twenty to thirty petals and about 150 stamens. Variety *N. c. zanzibariensis* has usually smaller leaves with more rounded basal lobes and flowers up to 1 foot in diameter with more stamens. The

Nymphaea gigantea

Nymphaea gigantea alba

Australian water-lily (***N. gigantea***) has leaves up to 1½ feet in diameter with their undersides brownish-pink or purplish. The blooms, open by day and approximately 9 inches in diameter, are lifted on stout stalks 1 foot to 1½ feet above the water. They have pale blue petals with darker apexes and a central cluster of incurved yellow stamens. Each bloom lasts three or four days. White-flowered *N. g. alba,* except for color, has similar blooms, which are reported to remain open by night as well as day. A less well-known, blue-flowered water-lily, ***N. colorata,*** of East Africa, has flowers much deeper colored at their centers than elsewhere. Because it produces clusters of tiny tubers it is easy to propagate.

Nymphaea colorata

Yellow-flowered, hardy ***N. mexicana*** has tuber-like rootstocks and spreading rhizomes (runners). Its slightly wavy-edged leaves, 4 to 8 inches across, are blotched with brown or brownish-red on their undersides. From 2 to 5 inches in diameter, the bright yellow blooms, standing a few inches above the surface of the water, open from noon to late afternoon.

Nymphaea mexicana

Purple-red-flowered, nonhardy ***N. rubra,*** of India, has almost circular, sharply-toothed leaves 1 foot to 1½ feet in diameter, green above and with their undersurfaces covered with a reddish-brown scurf. The blooms, 6 to 10 inches across, expand in the evening and close about midday. They have fifteen to twenty petals.

Hybrid water-lilies in general greatly surpass the natural species in size of bloom and range of flower color. Practically all are creations of plant breeders of the past 100 years. The great work began with the efforts of M. Bory Latour-Marliac, of Temple-sur-Lot in southern France, who raised his first hybrid, ***N. marliacea rosea,*** in 1879. The water-lilies depicted in Claude Monet's famous paintings are Marliac hybrids. They include pink-flowered *N. marliacea rubra punctata* and yellow-flowered *N. m. chromatella*. Latour-Marliac continued his highly successful efforts for many years. He produced, among many notable varieties, a group he named the *N. laydekeri* or the Laydeker hybrids in honor of his son-in-law, Maurice Laydeker. These, probably having *N. tetragona* as one parent, are especially well adapted for small pools. A charming miniature with tiny leaves and flowers that is well adapted for small pools, ***N. pygmaea helvola*** (syn. *N. helvola*) is even smaller than *N. tetragona,* and hardy. Its parents are *N. tetragona* and *N. mexicana*. It has brown-blotched leaves and canary-yellow flowers that open in afternoons.

Nymphaea marliacea variety

Nymphaea pygmaea helvola

If to France belongs the chief credit for the production of most of the finest hardy water-lilies that grace our gardens, certainly the United States has led the world in breeding tropical water-lilies. Foremost among American breeders of water-lilies, distinguished horticulturist George H. Pring, formerly Superintendent of the Missouri Botanical Garden at St. Louis, Missouri, conducted his hybridizing at that garden. Other American breeders of note include Tricker of Saddle River, New Jersey, and Independence, Ohio; Richardson of Ohio; Dreer's of Philadelphia, Pennsylvania; and Three Springs Fisheries, Lilypons, Maryland.

Specialist nurseries offer a very wide selection of water-lilies and to their catalogs one can turn for descriptions and often colored illustrations of the very beautiful sorts available. They include a large number of well-tested old varieties as well as newer introductions. The listings are commonly grouped under the chief headings of hardies and tropicals, with the latter broken down into day-bloomers and night-bloomers. Among the hardies the flower colors available run from purest white to sulfur-yellow and through a vast assortment of pinks, apricots and the like, to deepest reds. There are no hardy blue-flowered water-lilies. But that color in various shades and associated tones, as well

Hardy nympheas: (a) Yellow-flowered 'Sunrise'

(b) Pink-flowered 'Pink Sensation'

Day-blooming tropical nympheas: (a) Blue-flowered 'Blue Beauty'

(b) Blue-flowered 'Bob Trickett'

(d) Pink-flowered 'Shell Pink'

(c) Pink-flowered 'Pink Pearl'

as shades of lavender, violet, and purple, are well represented among the tropicals, which also include varieties in the color spectrum afforded by the hardies.

Garden and Landscape Uses. Of all aquatics, water-lilies are undoubtedly the most widely appreciated for their beauty and suitability for decorating pools, ponds, lakes, and such other waters as the margins and backwaters of streams. Aquatic gardens without nympheas are barely imaginable. In most they are the chief floral features. They are adaptable for expansive landscapes and small home gardens where their accommodation is a miniature pool or perhaps a tub.

Water-lilies need exposure to full sun and still or very sluggish water. They do not prosper in streams with any appreciable flow or in pools with fountains splashing into them. Waters to accommodate these plants can have natural mud bottoms or be constructed of various artificial materials (see Water and Waterside Gardens). It is entirely practicable to accommodate small and miniature sorts often listed in dealers' catalogs as pygmy water-lilies in tubs, cut-down barrels, or similar containers sunk into the ground. These can be charming garden features.

Cultivation. Few difficulties attend the cultivation of water-lilies and the sunny summers experienced throughout most of North America strongly favor their growth. For cultural purposes they are divided into two groups, hardy kinds, which even in cold climates can remain outdoors all winter, and tropical water-lilies, which will not survive winters outdoors except in tropical or warm subtropical climates. These last are highly successful outdoors in the north if treated in much the manner of certain other popular nonhardy perennials, such as fuchsias, geraniums, and lantanas, by setting them out each late spring or early summer as young plants raised in greenhouses and discarding them at the end of the first season.

Hardy water-lilies luxuriate in water that covers them to a depth of from 8 inches to 1½ feet. Only the most vigorous species and varieties are likely to give satisfactory results in deeper water. They are gross feeders, needing for their best development rich soil preferably on the heavy (clayey) side. In natural ponds the native soil can be successfully enriched, if the pond can be drained, by mixing cow manure or a slow-release fertilizer into it. For constructed pools a mix consisting of four or five parts by bulk of fertile topsoil and one part rotted or dried cow manure to which bonemeal has been added at the

Night-blooming tropical nympheas:
(a) Creamy-white-flowered 'Missouri'

(b) Red-flowered 'H. C. Haarstick'

rate of one pint to each bushel is ideal. Do not use peat moss, leaf mold, or other organics commonly mixed with potting soils for other plants and do not substitute other manures for cow manure. If the latter is not available rely upon fertile topsoil enriched with bonemeal and a slow-release fertilizer. It is pointless to add sand, perlite, or other materials designed to improve porosity and drainage to soils that in any case will be constantly saturated.

The soil may be spread over the bottom of the pool to a depth of 5 to 8 inches or be confined in tubs, baskets, or other containers of wood, concrete, or wickerwork. The first method has the merits of ensuring a more extensive root run and of making it possible for the plants to remain longer without replanting. It has the decided disadvantage that if it is necessary to enter the pool to groom the plants or for other purposes the soil is churned and the water becomes muddy. All in all, in artificial pools the container method is usually the best. Even for less vigorous sorts the tubs should hold at least a bushel of soil and two to four times that amount is better, especially for robust varieties.

Planting is best done in early spring just as new growth begins. Be sure that plants received from nurseries, and others being transplanted, are kept constantly wet. Either submerge them or if that cannot be done put them in a shady place and cover them with wet leaves, straw, burlap, or other material. In natural ponds and similar waters and in large soil-bottomed pools the best method is to plant in soil-filled wire or wicker baskets and, after lacing a few sticks across the tops of these or covering them with chicken wire to prevent the rhizomes from floating out, to sink them where they are to grow. Containers to be located in constructed pools may be planted and then gently lowered into the water, but since they weigh a great deal, a much easier method is to empty the pool, locate the containers, and then plant. Position the rootstocks correctly according to kind, and only one in each receptacle. Those of some sorts are spreading rhizomes that should be set horizontally with the top of the rhizome and growing point barely exposed. Other sorts, notably the Marliac hybrids, have vertical rootstocks with at one end a distinct crown or collar that should be just visible at the soil surface.

After planting cover the surface of the soil with an inch or two of clean sand or better still grit or small pebbles. This to keep bits of manure and other debris from floating away. Unless quite necessary do not at first cover the newly set plants with the full depth of water. From 3 to 4 inches over them is enough. As they make growth gradually run in more water until the required depth is reached.

Summer care consists chiefly of keeping dead leaves and faded flowers picked. Fertilizing once a month is very helpful. To do this, tie in small pieces of cheesecloth about an eggcupfull of a garden fertilizer, preferably one of the newer slow-release kinds and certainly not a rapidly available sort, then carefully reach down through the water and plant four or five of these packets in the soil of each water-lily.

Fall brings cessation of growth and eventually dying back or killing of the foliage by frost. If the depth of water is adequate to protect the rhizomes from freezing, the plants may be left outdoors over winter. However, in regions of cold winters, small pools are usually drained then and provision must be made for preserving the water-lilies. This may be done by leaving them in the empty pool and covering them with enough leaves, straw, hay, or similar material to prevent freezing or by moving the containers into a cool, frost-free place and keeping them covered with leaves, straw, hay, moss, or burlap kept constantly moist.

Replanting and possibly propagation are matters that need attention from time to time, the first at intervals of two to five years, the other when additional plants are needed. Plants in small containers require replanting more often than those with more root room. The need is clearly signaled by poorer, less vigorous foliage and fewer, smaller blooms. Propagation of hardy water-lilies is most commonly done by dividing the rhizomes at replanting time. It is simple to do and without hazard to the plants. An alternative method applicable to species, but not practicable for multiplying garden varieties true to type, is by the use of seeds. Sow these in pots or pans (shallow pots) of sandy soil and cover them with 1 inch of sand. Immerse the containers beneath 2 to 3 inches of water maintained at about 60°F. As soon as the seedlings have floating leaves, transplant them individually to small pots and as growth demands to bigger ones.

Tropical water-lilies are mostly large plants that for successful growth need plenty of room, sufficient to spread their leaves over a circle of 6 to 8 feet in diameter. The minimum amount of soil needed by each for the best results is a bushel. Three or four times that amount is better for strong-growing sorts. It should be of the same character as that recommended for hardy kinds and be covered with 1 foot to 1½ feet of water. Planting must not be done until the water temperature is 70°F and is not expected to go lower. Colder water sets the plants back to the extent that they become dormant and recovery is slow. Set only one plant in each container or at each station. More destroys the beautiful radial disposition of the leaves, which is a chief charm of these plants. Care must be taken not to break the root ball at planting time. Merely put it into a scooped-out hole and firm into place. Set the plant with its crown at such a level that even after the soil has been covered with an inch or two of coarse sand or gravel it is still just above the surface. Do not be concerned if immediately after planting the foliage is beneath the surface. Within twenty-four hours or so the leafstalks lengthen sufficiently to permit the leaf blades to float. Not more then a foot of water should cover the soil surface at planting time. Later it may, but not necessarily must, be deepened by running more water into the pool, but never should the crowns of the plants be more than 1½ feet beneath the surface. When adding water take care the pool is not chilled to below the requisite 70°F.

Propagation of tropical water-lilies can be by seed, but with natural species or when breeding for new varieties, this is less practicable than obtaining increase by plantlets that develop naturally on the leaves of some kinds, or by tubers. Seedlings of choice garden varieties are likely to be inferior to their parents. If seeds are to be sown do this in summer or fall in pans (shallow pots) in soil covered after the seeds are placed with about an inch of sand, and immersed a couple of inches beneath the surface of water kept 75 to 80°F in a sunny greenhouse. As soon as the seedlings are big enough to handle conveniently transplant them individually to small pots and keep them growing under the same conditions, transferring them to larger pots later, through the winter and spring until it is warm enough to plant them outdoors.

Increase by plantlets is had by removing these from the leaves on which they develop during summer and planting them individually in small pots. Submerge them 3 to 4 inches beneath the water surface of an outdoor pool until the coming of cold weather. Then transfer them to an indoor pool or tank where they will receive full sun and the water temperature will not drop below 70°F. To prevent them developing into undesirable, multiple-crowned plants, when they are very young take a sharp knife and carefully slice the plants into separate single crowns, each with roots attached, and pot each division separately.

Tuber propagation is very satisfactory. To do this, after the first frost dig the plants up and wash away soil that adheres to their roots. Then search around their bases, and with some kinds around the tops of the old tubers, for new, baby tubers. These will usually be about the size of walnuts. Remove them and store in slightly moist sand in closed containers in a temperature of about 55°F. In early spring plant the young tubers about 2 inches deep in sandy soil and submerge them in water kept at 70°F in a sunny place. As soon as new growths have their first floating leaves, run finger and thumb down the stemlike shoots to their base and nip them off where they join the tuber. Pot the baby plants immediately into 3-inch containers and, without letting them dry, return them to the comfort of their 70°F water. Growths that develop in succession from the tubers may be removed similarly. When well rooted in the 3-inch pots, transfer the infant plants to 4- or 5-inch containers. They can stay in these until they are planted outdoors. When potting, finish with the crown of the plant slightly above soil level and do not press the root ball too firmly in case the rather brittle roots are broken. Cover the soil with 1 inch of coarse sand or grit.

NYMPHAEACEAE—Water-Lily Family. This assemblage of dicotyledons consists of more than 100 species of often showy and beautiful aquatic plants distributed among nine genera. Some authorities divide the group into half a dozen families, but that view is not accepted in this Encyclopedia. The sorts of the NYMPHAEACEAE are herbaceous perennials, some hardy, others not, usually with tuberous rootstocks. Most have long-stalked, undivided, heart-shaped to circular floating leaves sometimes of great size, but *Cabomba* has as well as floating leaves submerged ones that are much dissected, and most of the leaves of *Nelumbo* stand well out of the water. The flowers of this family float or are lifted above the water. Solitary, they have four to six sepals and except in *Brasenia* and *Cabomba* numerous petals and stamens with often a gradual transition from one to the other. There are from two to many separate or united carpels that develop into the often large fruits, those of *Nelumbo* of peculiar form and decidedly ornamental. Cultivated genera include *Brasenia, Cabomba, Euryale, Nelumbo, Nuphar, Nymphaea,* and *Victoria.*

NYMPHOIDES (Nymph-òides)—Floating Heart, Water Snowflake, Banana-Plant. The botanically noncritical may be excused for supposing that these plants, at one time named *Limnanthemum,* belong to the water-lily family NYMPHAEACEAE. Botanists recognized this similarity in over-all appearance by bestowing the name *Nymphoides,* which is the Greek suffix *oides,* meaning resembling, attached to a part of the name *Nymphaea,* that of the water-lilies. But over-all appearances, which in this case apply only to the leaves and manner of growth, can be deceiving. An examination of their blooms quickly dispels any belief that *Nymphaea* and *Nymphoides* are kin. The latter belongs in the gentian family GENTIANACEAE. Among wet soil and water plants its "blood" relation is the buckbean (*Menyanthes trifoliata*), although, apart from its floral structure, it looks less like the buckbean than a water-lily.

All aquatics, the species of *Nymphoides* number about twenty, mostly tropical, and natives of both the Old and New Worlds and of Australia. Some occur naturally in temperate climates, in North America as far north as Canada. They have underwater, perennial rhizomes from which are sent long leafstalk-like stems. The sterile stems bear at their tops a solitary, broadly-heart-shaped, floating leaf; the flowering stems produce one or more similar leaves and an umbel of yellow or white blooms, which have a five-lobed calyx and a shallowly-bell- to almost wheel-shaped, five-lobed corolla. There are five stamens and a stout style tipped with a broad, two-lobed stigma. The fruits are capsules.

Among hardy floating hearts, the largest flowered and most beautiful is vigorous ***N. peltata,*** a native of Europe and Asia naturalized in some parts of the United States. It has stout stems with, at the top of each that bears flowers, a pair of usually unequal, wavy-margined, opposite leaves, mottled with reddish-brown, and an umbel of bright yellow blooms. Often the stem continues beyond the flowers and bears additional umbels. The

Nymphoides peltata

nearly round leaves are 2 to 5 inches across. The flowers are 1 inch or slightly more in diameter and have slightly fringed corolla lobes. Variety *N. p. bennettii* has plain green leaves. These are hardy plants.

The water snowflake (***N. indica***) is a beautiful tender aquatic of less vigorous growth than *N. peltata.* A native of the tropics, it is sparingly naturalized in warm parts of North America. This kind has round leaves, 2 to 8 inches in diameter, and pure white blooms with yellowish centers. The corolla lobes are fringed and densely covered with white hairs. Individual blooms remain open for only one day.

Nymphoides indica

Other kinds sometimes cultivated, but less showy than those discussed above, are *N. cordata* (syn. *N. lacunosa*) and *N. aquatica,* the latter offered by dealers in aquarium plants under the name banana-plant. In the wild ***N. cordata*** is found from southeastern Canada to New York and Connecticut, ***N. aquatica*** from New Jersey to Florida and Texas. They are similar except that *N. aquatica* is larger and its calyx lobes are flecked with purple. Both have slender stems, and white or cream flowers about ½ inch in diameter. They often produce tubers among the umbels of flowers.

The flowering stems have a single broadly-ovate, terminal leaf, 1½ to 3 inches long, that is above the cluster of blooms.

Endemic to Tasmania, ***N. exigua*** grows natively in wide swards in wet, sometimes slightly brackish mud bordering lakes and lagoons. It has basal rosettes of leaves with ovate to rounded blades up to ¼ inch long that terminate slender stalks twice that length. Long runners from the rosettes bear masses of tangled leafy stems. Slightly more than ¼ inch in diameter, the starry, pale-yellow flowers contrast pleasingly with the rich green foliage.

Nymphoides exigua

Garden Uses. As plants for pools and other still waters these are delightful. Under favorable conditions they give little trouble and provide long displays of bloom. The patterns formed by their leaves are decidedly decorative. They may be grown in containers or with their roots in soil bottoms. The latter method best suits *N. cordata* and *N. aquatica* but *N. peltata* spreads so rapidly that if unconfined it soon occupies a considerable space and may become difficult to control or eradicate. It is usually best to grow this species in a tub. The water snowflake is adapted to either tub or pool-bottom planting. It survives winters outdoors in mild climates only. Dealers in aquarium plants sometimes recommend that *N. aquatica* be floated on the surface without its roots having access to soil. Although this permits observation of the interesting banana-like clusters of little tubers, it does not make for the well-being of the plants, which when so grown deteriorate and eventually die. All kinds of *Nymphoides* need full sun.

Cultivation. These plants give of their best in rich, fertile, rather heavy soils such as suit water-lilies (*Nymphaea*). Their planting and cultural requirements are those appropriate for water-lilies. Planting is best done in spring, early for hardy kinds, but not until the weather and water are reasonably warm for the water snowflake. Propagation is by division in spring and by separation of plantlets in fall with those kinds that develop young plants on their leaves. Seeds germinate readily if sown and cared for in the manner recommended for water-lilies. From the time of ripening until they are sown the seeds must be kept in water or at least constantly moist (in wet moss, for instance); if they dry they lose their power to germinate.

NYPA (Nỳ-pa)—Nipa Palm. One species, *Nypa fruticans,* native of brackish swamps and tidal estuaries in Malaysia, India, Ceylon, the Philippine Islands, certain Pacific Islands, and northern Australia, constitutes this genus of the palm family PALMAE. The name is a native Moluccan one. Travelers along the klongs of Bangkok, the delta of the Ganges, river mouths in the Philippine Islands, and other places where this palm flourishes can scarcely fail to notice its huge, erect leaves rising directly from soil or water level and towering like gigantic fern fronds to heights of 20 to 30 feet or more.

The most truly aquatic palm, ***N. fruticans*** grows where its base is submerged for at least part of each day in slightly saline water. It is strictly a plant of marsh and swamp lands where fresh and salt water meet and mingle. A vigorous grower, it has thick, prostrate subterranean branches from the ends of which the huge pinnate leaves originate. It forms great clumps that play an important part in building new land and protecting swamp lands from erosion by the water. The flowers of the nipa palm are in crowded heads atop 3- or 4-foot, golden-brown stalks that arise from the leaf axils. The female flowers are in rounded clusters surrounded by leafy bracts from which catkin-like spikes of tiny bright yellow male flowers protrude. The knobby, rough, mahogany-brown fruits about 1 foot in diameter and approximately globular, are at the ends of the flower stalks. The seeds are as big as hens' eggs.

In their home lands nipa palms are put to many uses, chief of which is supplying leaves to thatch buildings. The foliage is also employed for making baskets, mats, and wearing apparel, the leafstalks are used for fuel. From the flowering stems sap is obtained for making palm wine, alcohol, and vinegar and is used as a source of sugar. In the Philippine Islands areas of otherwise useless swamp land have been planted with nipa palms to provide these useful products.

In the 1930s this species was introduced to the United States Plant Introduction Garden and to the garden of Colonel Robert H. Montgomery, both at Miami, Florida, by seeds collected from specimens growing in the botanical garden at Georgetown, British Guiana. It is surmised that the plants were brought to Georgetown by the British toward the end of the nineteenth or early in the twentieth century. The Florida plants have successfully withstood hurricanes and innundations of sea water as well as quite severe freezes without permanent damage.

Landscape Uses and Cultivation. It is probable that nipa palms could be used effectively in landscaping suitable sites in the warmest parts of Florida and in Hawaii. They can be raised from fresh seeds sown in sandy, peaty soil kept constantly wet in a temperature of 80 to 90°F. Their cultural needs can be surmised from a consideration of the conditions under which they grow naturally. For additional information see Palms.

NYSSA (Nỳs-sa)—Sour Gum or Black Gum or Pepperidge, Cotton Gum or Tupelo, Ogeechee-Lime. Ten species of Asian and North American deciduous trees of the nyssa family NYSSACEAE constitute *Nyssa.* The name, that of a water nymph, was applied in recognition of the wet soil habitats of some species.

Nyssas have alternate, undivided leaves, smooth-edged or with a few distantly-spaced teeth, and unisexual or unisexual and bisexual flowers on the same tree. The minute, five-petaled flowers are greenish-white, in clusters or solitary in the leaf axils. Male and bisexual blooms have five to twelve protruding stamens. The one-seeded fruits resemble tiny plums.

This genus includes some of the handsomest American trees. The lumber of *N. sylvatica, N. s. biflora,* and *N. aquatica,* called tupelo wood, is esteemed for boxes, crates, plywood, flooring, paper pulp, and other uses. From the fruits of *N. ogeche* delicious preserves are made.

Sour gum, black gum, or pepperidge (***N. sylvatica***) is native from Maine to Michigan, Florida, and Texas. Pyramidal, with branches slightly pendulous, it exceptionally attains a height of 100 feet and has blunt, lustrous, obovate to elliptic, leathery leaves 2 to 5 inches long that in fall become brilliant scarlet or orange. At that

Nyssa sylvatica in summer

Nyssa sylvatica in winter

Nyssa sylvatica (fruits)

Nyssa sylvatica (leaves)

season sour gums illuminate the forest with fiery displays, rivaling those of sugar maples and scarlet oaks. The sour gum is a wet soil species, but will grow away from swamps if the soil is not excessively dry. Its female flowers, rarely solitary, usually are in twos or threes. The berry-like fruits are ½ inch more or less long and dark blue.

Tupelo or cotton gum (***N. aquatica***), plentiful in swamps from Virginia to Illinois, Florida, and Texas, differs from the sour gum in having solitary flowers, 1-inch-long fruits, and pointed leaves. It attains heights of up to 100 feet.

The Ogeechee-lime (***N. ogeche***), an inhabitant of wet lands from South Carolina to Florida and not hardy in the north, rarely is over 60 feet tall. It has a bushy head and young twigs clothed with red hairs. The leaves are blunt, elliptic to elliptic-oblong, lustrous and slightly hairy above. The female flowers are solitary. Sour and juicy, the red fruits are ½ to ¾ inch long. They make excellent preserves.

Garden and Landscape Uses. Picturesque and colorful, nyssas are esteemed for their wonderful fall coloring and attractive branching habits, the latter seen to best effect in winter when the trees are bare of foliage. The branches of old specimens droop much like those of the pin oak. Because nyssas are naturally gregarious they are appropriate for planting in groups to be viewed from a distance. They are well suited for watersides. The sour gum stands city conditions remarkably well and thrives near the sea. It endures considerable shade.

Cultivation. Established specimens need no special care, but successful transplanting to new locations presents problems. Nyssas do not transplant easily. Losses among those moved from the wild are likely to be discouragingly high. As a rule only nursery-grown trees that have been transplanted every two or three years and are not over 10 feet tall are worth moving. Dig these with a comparatively large ball of soil and thin out or shorten the branches to reduce the amount of foliage the newly transplanted tree will be called upon to support the first summer after moving. It is very important that the soil be not allowed to dry out for the first year or two after transplanting.

Diseases and Pests. Nyssas are not much subject to diseases or pests. Diseases that sometimes affect them are cankers, leaf spots, and rust. Insect enemies are leaf miner, azalea sphinx moth, and San Jose scale.

NYSSACEAE—Nyssa Family. This is a family of three genera of dicotyledons totaling twelve species represented in the native floras of eastern North America, eastern Asia, Malaysia, and the Himalayan region. Its members are deciduous trees with alternate, undivided, toothed or toothless leaves. The tiny unisexual or bisexual flowers are in terminal or axillary heads. They have minute calyxes of five tiny teeth, or the calyx may be wanting, and small corollas with five or more petals, or are without corollas. There are up to a dozen stamens, twice as many as the petals when those are present. The slender style is sometimes divided. Genera in cultivation are *Camptotheca, Davidia,* and *Nyssa.*

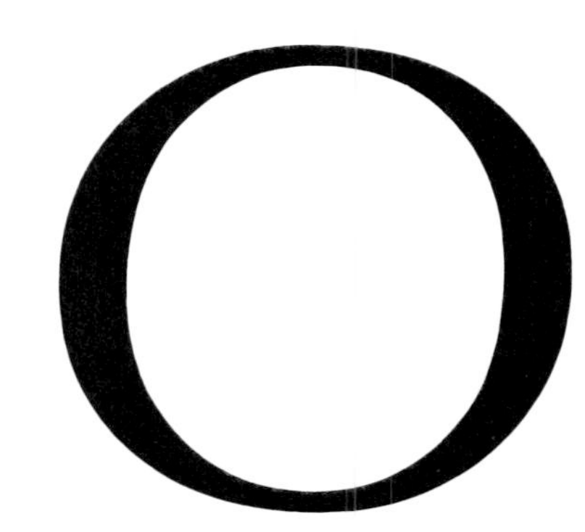

OAK. See Quercus. Other plants have common names including the word oak are these: French- or Spanish-oak (*Catalpa longissima*), Jerusalem-oak (*Chenopodium botrys*), oak fern (*Gymnocarpium dryopteris*), oak leaf fern (*Drynaria quercifolia*), oak-leaved fern (*Polypodium quercifolium*), poison-oak (*Rhus diversiloba*), she-oak (*Casuarina*), silk- or silky-oak (*Grevillea robusta*), tanbark-oak (*Lithocarpus densiflorus*), and white silky-oak (*Grevillea hilliana*).

OAK APPLES. These are more or less spherical galls, usually the size of small marbles, that develop on the leaves of some oak trees. Caused by insects, they do no appreciable harm.

OAT. See Avena. Sea-oats is *Uniola paniculata*.

OBCORDATE. Reversed cordate (heart-shaped) in shape, with the widest part near the apex.

OBEDIENT PLANT is *Physostegia virginiana*.

OBLANCEOLATE. Reversed lanceolate in shape, with the widest part near the apex.

OBLONG. Used in reference to leaves and other plant parts, oblong lacks the precision of its strict geometrical application. It means the sides of the part referred to are approximately parallel for an appreciable portion of their lengths. Parts that are oblong are proportionately wider than those identified as linear.

OBOVATE. Reversed ovate. Shaped like the outline of a hen's egg with the broad end forming the apex.

OBREGONIA (Ob-regònia). The only species of *Obregonia* is a Mexican plant related to *Echinocactus* and belonging in the cactus family CACTACEAE. By some authorities it is included in *Strombocactus*. Its name honors Don Alvaro Obregon, a President of Mexico.

This cactus, ***O. denegrii,*** has much the appearance of *Ariocarpus*. It has a stout taproot and a globular or top-shaped body up to 5 inches in diameter with spirally arranged leaflike extensions called tubercles, which give to the plant something of the apperance of a grayish-green artichoke. The tubercles are triangular, sharp-pointed, keeled on their undersides, and up to slightly over ½ inch long. They end in specialized areas (areoles) at first woolly, furnished with a few soon-deciduous, weak spines about ⅜ inch in length. The blooms, white or pale pink, about 1 inch long and up to ¾ inch across, come from the centers of the plants. The fleshy, white fruits dry on the plants and remain attached for a long time. They contain black seeds.

Garden Uses and Cultivation. Not particularly difficult to grow, this cactus is basically a collectors' plant suitable for greenhouse cultivation and for outdoors in warm, dry climates in moderately fertile, slightly acid soil. It grows slowly and should not be watered overhead. A sunny location is needed. Propagation is by seed. Seedlings grafted onto columnar cactuses grow more quickly than specimens on their own roots. For general cultural information see Cactuses.

OCEAN SPRAY is *Holodiscus discolor*.

OCHAGAVIA (Ocha-gàvia). As here accepted *Ochagavia* consists of four species of the pineapple family BROMELIACEAE, three native of Chile, one of the island of Juan Fernandez. Some sorts are by some botanists treated separately as *Rhodostachys*. The name commemorates Sylvestris Ochagavir, a former Minister of Education of Chile.

Ochagavias are evergreen, short-stemmed or stemless pineapple-like perennials that grow in the ground or on cliffs, not like so many bromeliads, perched on trees. They have clusters of rosettes of spiny-margined narrow leaves, low in the centers of which develop compact, many-bracted, spherical to ovoid heads of little pink or yellow stalkless blooms. The flowers have three each sepals and petals, and six stamens. The fruits are berries.

Chilean ***O. lindleyana*** (syn. *O. carnea, Rhodostachys andina*), of noble aspect, has rosettes of numerous spreading and recurved, strongly-spined leaves about 1½ feet long by 1 inch wide, tapered to sharp points. Their undersurfaces are furnished with rows of grayish, mealy, tiny scales. The quite striking short-stalked flower head, about 4 inches in diameter, has rose-red bracts and many blooms with lavender petals and protruding bright yellow stamens.

Ochagavia lindleyana

Garden Uses and Cultivation. Adapted for outdoor cultivation in dry, essentially frost-free climates only and for inclusion in greenhouse collections of succulents, the species described grows readily in porous, well-drained soil in full sun. It is easily propagated by offsets and seed.

OCHNA (Óch-na). The genus *Ochna* of the ochna family OCHNACEAE consists of eighty-five species, native to tropical and subtropical Africa and Asia. The name comes from the Greek *ochne,* used by Homer for the wild pear. The leaves of *Ochna* were thought to somewhat resemble those of the pear.

Ochnas are shrubs or trees with alternate, finely-toothed, lustrous, leathery leaves. Their yellow, or sometimes greenish flowers, solitary or in racemes, have five persistent sepals, five to twelve spreading petals, many stamens, and one style. In cultivated kinds the fruits are even more ornamental than the flowers. They have fleshy disklike receptacles to which are attached a few stalkless, seedlike drupes that look like tiny plums and contain the seeds.

Most frequent in cultivation and often misnamed *O. multiflora,* South African ***O. serrulata*** is a shrub 3 to 5 feet tall or sometimes taller. It has very short-stalked, elliptic to elliptic-ovate, sharply-toothed leaves, with prominent midribs, and up to about 2 inches long by nearly one-half as wide. The solitary or rarely paired, fragrant flowers, freely produced and about ¾ inch wide, are at the ends of short lateral shoots. The sepals, at first about ¼ inch long, increase considerably in size and become bright red as the fruits develop. The obovate petals are yellow. Attached to the bright red disk and enhanced by the showy, recurved sepals of the same color are usually five egg-shaped fruits nearly ⅓ inch long, at first bright green, but ripening to a glossy black. They make a colorful display over a long period. Heavily fruited bushes are highly ornamental.

Ochna serrulata

Other species cultivated include these: ***O. jabotapita,*** of Ceylon, is a shrub or small tree with lanceolate, toothed leaves 3 to 7 inches long, by up to 2 inches wide. At the ends of short branchlets long-stalked, 1-inch-wide, five-petaled flowers are borne in clusters of six to twelve. The fruits have five to seven drupes. ***O. kirkii,*** of East Africa, up to 15 feet tall, has oblong-elliptic to narrowly-obovate, often wavy-margined leaves up to 4 inches long by 1¾ inches wide. The 1½- to 2-inch-wide flowers, in short panicles on short branchlets, have calyxes that are bright red and highly ornamental in the fruiting stage. The fruits have up to a dozen drupes. ***O. mossambicensis,*** of Mozambique, East Africa, attains a height of about 10 feet. Its obovate to oblanceolate, toothed leaves are up to 9 inches long by one-third as broad. The flowers, about 1¼ inches wide, are in loose panicles on short branchlets. The fruits have persistent red calyxes and eight to ten drupes. ***O. obtusata,*** of Sri Lanka and India, is a small tree, hairless and with spiny-toothed leaves up to 5 inches long and oblanceolate to obovate or elliptic. About 1½ inches in diameter, long-stalked, and seven- to twelve-petaled, the flowers are in umbel-like panicles at the ends of short shoots. The name *O. squarrosa* (correctly a synonym of *O. jabotapita*) is sometime misapplied to this species. ***O. pumila,*** of the Himalayan region, is a subshrub with oblanceolate to obovate, spiny-toothed leaves up to 6 inches long. It has fragrant, long-stalked, 2-inch-wide flowers with petals much shorter than the sepals. The fruits have greenish drupes. ***O. thomasiana*** is an East African much resembling *O. kirkii,* but usually with smaller leaves decidedly eared at their bases and less abundant, smaller flowers.

Garden and Landscape Uses. Very lovely *O. serrulata* is delightful for planting outdoors in warm, nearly frost-free climates and is also very satisfactory in greenhouses. It blooms while quite small, and in flower and fruit makes a colorful display over an extended period. It is suitable for general-purpose planting and responds to well-drained, fertile soil and a sunny location. The other sorts described here have similar uses.

Cultivation. Seeds germinate reliably and quickly. Cuttings can also be used as a means of increase. The only pruning needed is any required to keep the specimens shapely. This may be done in early spring. In greenhouses a sunny location or one slightly shaded in summer is ideal. The minimum night temperature in winter should be 55 to 60°F. By day at that season it may be five to fifteen degrees higher. Higher temperatures in summer are in order. A moderately humid atmosphere and watering to keep the soil, which should be porous and fertile, moderately moist at all times are satisfactory. Spring-to-fall applications of dilute liquid fertilizer benefit specimens in pots.

OCHNACEAE—Ochna Family. The forty genera of this tropical family of dicotyledons total 600 species of mostly trees and shrubs. They have alternate, undivided or extremely rarely, pinnate leaves. In panicles, racemes, branched clusters, or false umbels, the flowers have five sepals, four to ten petals, four to many stamens, sometimes some staminodes (nonfunctional stamens), and one style that may be cleft at its apex. The fruits often are of several separate, berry-like drupes clinging to an enlarged, fleshy receptacle or are solitary drupes or capsules. The only cultivated genera seem to be *Ochna* and *Ouratea.*

OCHROMA (Och-ròma)—Balsa. The only species of *Ochroma,* of the bombax family BOMBACACEAE, is native throughout much of tropical America and the West Indies and has a name that comes from the Greek *ochros,* pale yellow, in allusion to the color of the blooms. Its common name is an aboriginal one for raft and alludes to the wood of this tree being commonly used by natives of regions where it is native for constructing rafts. From logs of it the famous raft Kon-Tiki, on which Thor Heyerdahl made his notable voyage from South America to Easter Island, was constructed. In World War I 80,000 logs of balsa were used as floats for the 250-mile-long mine barrier across the North Sea. In that conflict too, as in World War II, balsa wood (the lightest commercial lumber) was designated a strategic material and was used for life preservers and rafts. In peacetime it is used as fishnet floats, insulation, model airplane construction, boxes, and novelties. Ropes have been made from fibers in the bark of *Ochroma,* and the hairs surrounding the seeds have been used, like kapok, for stuffing pillows and mattresses.

Balsa (***O. pyramidale*** syn. *O. lagopus*) is a fast-growing, evergreen tree 50 to 80 feet in height, with a few-branched, spreading, open crown. It has alternate, broadly-heart-shaped to nearly round leaves with five- to seven-lobed or -angled blades 8 inches to 1 foot or more in diameter, and toothed or not. The long leafstalks are reddish. The tubular-bell- to trumpet-shaped blooms, on long stalks near the ends of the shoots, yellowish or brownish-white, are 5 to 6 inches long by 3 to 4 inches broad. They have five-lobed calyxes, below which are three bracts and five petals. The many stamens are joined in a column that surrounds the pistil. The long style ends in five spiraled stigmas. The fruits are slender, ten-angled capsules up to 1 foot long, with many seeds embedded in dense masses of tawny-woolly hairs.

Garden and Landscape Uses. The balsa tree is of minor service as an ornamental. In the tropics and frost-free parts of the subtropics it is occasionally planted as a matter of interest, and young specimens are sometimes included in greenhouse collections of plants useful to man. It thrives in ordinary, not-too-dry soil, and is easily propagated by seed and cuttings. A minimum greenhouse temperature of 55 to

60°F is satisfactory. The atmosphere should be humid. Repotting in ordinary, fertile, porous, coarse soil is done in late winter, and at that time any pruning needed to shape or limit the plant to size should receive attention.

OCHROSIA (Och-ròsia). Occurring as natives in Australia, New Caledonia, Malagasy (Madagascar), Hawaii, and other Pacific islands, *Ochrosia* consists of thirty species of the dogbane family APOCYNACEAE, although it is generally represented in cultivation by only one kind. The generic name, from the Greek *ochros,* pale yellow, refers to the flowers.

Ochrosias are milky-juiced trees with undivided, lobeless, toothless leaves in pairs or whorls (circles of more than two). Their flowers, in stalked, axillary or terminal clusters, have deeply-five-parted calyxes, five-lobed, tubular corollas, and five stamens alternating with the spreading corolla lobes (petals). There is one style. The hard fruits contain one or two seeds.

The kind best known is ***O. elliptica*** (syn. *O. parviflora*), a native of Australia, New Caledonia, and islands of the Pacific. A spreading tree, in the wild it may be 30 to 40 feet tall, but in cultivation in North America it is usually about 20 feet in height. Its glossy, leathery, hairless, elliptic to oblong-obovate, short-stalked leaves have numerous parallel veins from midrib to margin. From 3 to 7 inches long, they are in twos, threes, or fours. The dense, short-stalked, branched clusters of individually stalkless, ½-inch-long, fragrant, cream flowers develop from the axils of the upper leaves and are succeeded by cerise-scarlet, pointed, plumlike, glossy fruits 1 inch to 2½ inches long and usually in pairs. The fruits are poisonous. When crushed they have a violet-like fragrance. Each contains one or sometimes two narrowly-winged seeds. In Florida this tree is sometimes misidentified as *Kopsia arborea.* True *K. arborea* has purple-violet to black fruits.

A rare Hawaiian native, ***O. sandwicensis*** differs from *O. elliptica* in having longer-pointed, smaller leaves and yellow fruits. Its bark and roots supplied a yellow dye formerly used for decorating tapa cloth.

Garden and Landscape Uses and Cultivation. In climates not colder than that of southern Florida *O. elliptica* is a good ornamental with attractive foliage and fruits. Its blooms make no appreciable display. It succeeds in a variety of soils and, being resistant to salt spray, does well near the sea. For short periods it successfully withstands temperatures as low as 25°F. It grows rather slowly and is well suited for cultivating in large containers. Propagation is by seed and by cuttings set in a greenhouse propagating bench, preferably with a little bottom heat.

OCIMUM (Òcim-um)—Basil. Except for common or sweet basil, which is grown as a fragrant culinary herb, and its purple-leaved variety sometimes as a flower garden ornamental, *Ocimum,* of the mint family LABIATAE, is not much known to gardeners. It consists of about 150 species of aromatic annuals, herbaceous perennials, and shrubs. The name is an ancient Greek one. This genus inhabits warm regions.

Ocimums have opposite leaves, and whorls (circles of more than two) of usually six small flowers, arranged in terminal, often panicled, racemes. The asymmetrical flowers have unequal-toothed calyxes, usually as long as the tubes of the two-lipped corollas. The upper corolla lip has four lobes. There are four down-pointing stamens and a shortly two-cleft style. The fruits are of four small nutlets.

Ocimum basilicum

Common or sweet basil (***O. basilicum***), 1 foot to 2 feet tall or sometimes taller, is a bushy, leafy annual, or perhaps in the wild sometimes a short-lived perennial, with stalked, ovate, often purplish, hairy or hairless leaves 1 inch to 2 inches long, and with toothed or smooth margins. Its little, white or purplish-tinted blooms, up to ½ inch long, have slightly protruding stamens with white anthers, and are in fairly crowded racemes. This is a native of tropical Asia, Africa, and islands of the Pacific. Bush basil (*O. b. minimum* syn. *O. minimum*) does not exceed 10 inches in height, and is more compact and smaller-leaved than common basil. The foliage of *O. b. purpurascens* is beautiful rich purple. That of *O. b. citriodorum* is lemon scented. A variety of *Perilla frutescens* is sometimes misnamed *O. b. crispum.*

Holy basil (***O. sanctum***), of warm parts of Asia, in the Hindu religion is held sacred to the deities Krishna and Vishnu. It is frequently cultivated by Hindus, near their dwellings, in areas where they pray. For common basil this species differs in its toothed leaves being somewhat lozenge-

Ocimum basilicum purpurascens

shaped and having blunt apexes, in the stalks of its individual blooms being as long as or longer than the calyx, and in the anthers being yellow. The stems of holy basil are whitish-hairy. Both surfaces of the leaves are hairy. The flowers are purplish, pink, or white.

Tree basil (***O. suave***), of Africa, is a nonhardy, much-branched shrub up to 8 feet in height. It has densely-woolly-hairy, ovate to ovate-lanceolate, toothed leaves 2 to 4¼ inches long and up to 2 inches broad. Its ¼-inch-long whitish to purplish flowers are set closely in dense, panicled racemes up to 9 inches long. They have slightly protruding stamens. This species closely resembles ***O. gratissimum*** of India.

Garden Uses and Cultivation. Tree basil may in warm, nearly frostless climates occasionally be planted for interest. It needs well-drained soil and a sunny location. Propagation is by seed or by cuttings. For the uses and cultivation of other kinds see Basil.

OCONEE BELLS is *Shortia galacifolia.*

OCOTILLO is *Fouquieria splendens.*

OCTOBER, GARDENING REMINDERS FOR. In northern gardens one of the most satisfying months, October is also a busy one. Preparations for winter and the next season must proceed apace for much needs doing before time runs out. Now weather is mostly pleasant, but cold nights following warm days forewarn of harsher times to come.

Give lifting, planting, and transplanting high priorities. The first involves taking from the ground and putting into storage all nonhardy bulb plants. Unless their foliage has yellowed it is best to delay this until it is touched by the first frost. So long as the leaves are green they work to fatten the bulbs, corms, or tubers with stored food needed for next year's growth.

When lifting, use a spading fork and take care not to spear or break the bulbs or parts to be saved. Wounds afford entry for

After the first light frost: (a) Dig up dahlias

Before the ground freezes: (a) Dig gladioluses

Plant or transplant deciduous trees and shrubs in October: (a) A young, bare-rooted tree

(b) Label and store the clumps of tubers

(b) Clean off their tops

(b) An older, balled and burlapped tree

organisms that cause rotting. Drive the prongs nearly vertically into the ground close to but not immediately over the bulb parts and pry them gently upward and out. With clumps such as those of montbretias, tuberoses, and dahlias, it will be necessary to dig away carefully some of the soil from around them before attempting prying. With ismenes or Peruvian daffodils, it is especially important not to break the roots more than is absolutely necessary.

Inspect carefully all lifted bulbs, corms, and tubers. Discard any that are diseased as well as those too small to be worth keeping. Keep sorts and varieties well separated and clearly labeled. If you do not there will be confusion when planting time comes. After the bulbs, corms, or tubers are dug, some sorts will need spreading out in a dry place where they are safe from frost until their foliage dies and dries and they can easily be cleaned of it. Store on shelves or trays or in net bags or old stockings suspended from a ceiling as may be most convenient.

Planting and transplanting practically all types of deciduous trees and shrubs, including fruit trees and except in the coldest parts of the north roses, can be carried out with maximum prospect of success from the first shedding of leaves until the ground freezes several inches deep, but the sooner this is done after leaf fall begins the more time there is for new roots to push into still relatively warm earth before freezing halts their growth. Fall planting assures an early start next spring. Exceptions among deciduous trees, sorts that move better in spring than fall, include those such as magnolias, which have fleshy roots, and any known to be somewhat tender to cold as *Albizia julibrissin* is likely to be.

(c) Store them in a cool, airy, dry place

(c) Transporting a large tree

(d) A newly transplanted tree securely guyed against storm damage and with a "saucer" around it to facilitate watering

Many kinds of hardy bulbs can be planted in October: (a) Chionodoxas

(c) Fritillarias

(e) Narcissuses

(b) Crocuses

(d) Lilies

(f) Scillas

With evergreens it is different. October is getting late for transplanting these in the north and unless quite necessary it is usually better to wait until spring. If you must move evergreens now do so as early as possible, and take substantial balls of soil with their roots. When planting is completed water them very thoroughly and after the ground has frozen a couple of inches deep spread a layer of mulch around them. Also, sometime before Christmas or immediately after and again a month later, spray them with an antidesiccant to check loss of moisture from the foliage.

Push on with planting hardy bulbs. Only with tulips is it advisable to wait until November. Other sorts, chionodoxas, crocuses, fritillarias, hyacinths, lilies, narcissuses, scillas, and the rest, are better for early planting.

Hardy herbaceous perennials of all kinds can be safely transplanted now, even those such as iris that for certainty of the best flower display next year would have been better moved earlier. Not only can they be transplanted, they can be divided and so multiplied. This all makes it practicable, indeed desirable, to manage perennial beds and borders by completely remaking a portion of them each year, and October is a fine time to do this.

Begin by digging up all plants, except possibly clumps of a very few such as aconites or peonies that you know are better left undisturbed, pack them closely together in an out-of-the-way corner, and cover them with leaves or other material to keep them moist. Then spade the ground deeply and mix in compost or other organic matter, bonemeal or superphosphate, and if the soil is too acid, a dressing of ground limestone. Complete the job by dividing and resetting the plants. Select for replanting pieces from the vigorous outer parts of old clumps. Unless these are in short supply discard tired, old inner portions. Set out also young plants from nursery beds to add to the variety of the planting.

In the vegetable garden carrots, Jerusalem artichokes, leeks, parsnips, and salsify may be left in the ground all winter and

October is an excellent month in which to plant bulbs of most lilies

Prepare planting sites for bulbs by forking in fertilizer

Also plant small bulbs: (a) Chionodoxas, planted relatively shallowly

Prepare for planting herbaceous perennials: (a) By spading

Plant or transplant hardy herbaceous perennials early enough for root growth to take place before the ground freezes

(b) Plant large bulbs, such as hyacinths, deeper

(b) By forking

(c) The hyacinths in bloom in spring

(c) Or by deep rotary tilling

dug for use as needed, but this will become impossible if the ground freezes deeply; therefore cover the rows with a thick layer of straw or leaves held down with boards or chicken wire before this happens.

Alternatively, these crops may be lifted and stored in a cool root cellar or similar place where the atmosphere is humid and the temperature is kept above 32°F, but whenever outdoor conditions permit not over 40°F. Other crops to store similarly are beets, cabbages, celery, potatoes, turnips, and winter radishes. Onions keep best in a dry atmosphere in about the same temperature range. For pumpkins and squashes the air should also be dry, but the temperature between 40 and 60°F.

Yet another way of holding celery for later eating is to lift the plants from the garden with goodly root balls, pack them closely together in a deep cold frame, water them thoroughly, keep them shaded from all light and protected from freezing by banking leaves or other material around the sides of the frame and covering its top with straw or mats. On warm days ventilation is needed.

Chives and parsley may be carefully dug, potted, and watered well and then buried to the rims of their pots in sand or peat moss in a cold frame, later to be brought indoors and gently forced to give winter supplies of fresh leaves. Or they may be planted in a bed of sandy soil in a cold frame for spring picking before supplies from outside are available.

If you plan to force rhubarb or whitloof chicory for winter and spring use, dig the roots now and store them with sand, sandy soil, or moist leaves packed around them in a root cellar, protected cold frame, or equivalent accommodation.

Clean-up work calls for much attention in October. Raking or other ways of collecting the bounty of fallen leaves will be in full swing. And bounty it is, for whether used as a direct mulch or in other ways to afford winter protection, or composted to give a nourishing product to add to the soil later, the annual cast of foliage is one of the most valuable of garden "crops." Fortunately the burning of leaves, so often practiced in the past, is now forbidden in enlightened communities because of the pollutant smoke it causes.

Be sure to keep lawns cleanly raked of fallen leaves. If allowed to collect and mat down they may do considerable damage by obstructing light and maintaining such humid conditions that the grasses rot. If the grass continues growing this month give it a last cutting at a height of 2 to 2½ inches.

Winter protection of various plants now needs serious thought, but except in the extreme north and elsewhere where winter comes early it is yet too soon to do it. Do not therefore be stampeded into in-

Keep free of fallen leaves: (a) Lawns

(b) Other areas

(c) Do not contribute to air pollution by burning leaves; instead, turn leaves into excellent leaf mold or compost

stalling protection early. In most cases it is wisest to wait as long as reasonably practicable, but do not delay locating and assembling supplies that will be needed for the purpose.

Mulches put in place before the ground has frozen to a depth of an inch or so are likely to attract mice and other creatures that may feed through the winter on your choicest plants or gnaw the bark of shrubs and young trees. To discourage this last, which is commonly done by rabbits as well as mice, surround endangered specimens with small-mesh chicken wire or wire netting.

Hilling soil about the bases of roses just before the ground freezes and later covering with leaves, straw, or other loose winter covering is a traditional and effective method of affording bush types that bloom on shoots of the current year adequate winter protection. The soil for hilling may be drawn up from between the roses if that can be done without exposing the roots, but many gardeners prefer to bring in additional soil (to be removed in spring) for the purpose, and this may be the only way if the bushes are planted closely.

Tree roses (also called standards) and, in regions of severe cold, climbing roses are best protected over winter by laying their stems horizontally along the ground and covering them with soil.

Pruning to the extent of somewhat shortening extra long shoots of bush roses, raspberries, and other plants with canelike growths likely to be badly whipped about by winter winds gives some protection from tearing and breakage.

To prevent damage from winter storms, shorten excessively long canes of roses and other plants likely to be harmed

Garden tools and equipment, including furniture, hoses, and of course such items as lawn mowers, fertilizer spreaders, and sprayers, need to be cleaned; oiled, if appropriate; and safely put in dry storage as soon as their seasonal usefulness is over. Be sure to wipe all unpainted steel parts of spades, forks, hedge shears, pruning shears, and other tools with an oily rag before putting them away.

Construction work of many kinds can be done most conveniently in fall. The coming of cooler weather and lessening of routine maintenance make building rock gardens, making terraces and patios, installing paths, and similar projects practicable. But inevitably even colder weather will call a halt to such activities so again the word is to push on.

Only after their season's usefulness is through, clean and, as needed, oil before putting away: (a) Tools

(b) Implements

Greenhouses are now chock-a-block full, not only with permanent occupants and such one-season crop plants as cinerarias, cyclamens, and poinsettias kept indoors throughout their briefer lives, but also with sorts wintered in greenhouses and displayed outdoors in summer. These may include fuchsias, geraniums, heliotropes, and lantanas. The passing out of bloom of greenhouse chrysanthemums will soon relieve pressure on space, but temporarily at least greenhouses are likely to be crowded.

Fortunately it is now easier to maintain congenial growing conditions, especially for such cool-climate plants as calceolarias, carnations, cinerarias, cyclamens, cymbidiums, and primulas, than it was during the hot summer months. To do so it is necessary to ventilate cool greenhouses freely on all favorable days and even perhaps to leave the ventilators open just a crack at night, but that of course depends upon outside temperatures.

Before the end of the month shading from most greenhouse roofs should be completely removed. If heavy, it is better to do this in two operations so that the plants become gradually accustomed to more intense light, rather than being suddenly exposed.

Greater care in watering, overhead spraying, and wetting down paths and other interior surfaces is now necessary. An excessively humid atmosphere encourages mildew on chrysanthemums, primulas, and some other plants and constant wetness resulting from excessive watering can rot these and others. In warm greenhouses where tropical plants, including African violets and other gesneriads, begonias, and poinsettias, are housed more humidity is needed, but it can in the main be achieved by wetting the floors, the benches between pots, and other surfaces instead of by overhead spraying of foliage.

Shortening days and less intense light slows growth and reduces need for water, therefore intervals between waterings will likely be longer than a month or two ago. Still, be sure when using the hose or watering can to thoroughly saturate the soil. So far as practicable get done with watering before noon. Only in cases of quite obvious need give water late in the day. And in dull weather take care not to wet foliage unnecessarily. To do so favors fungus diseases.

Ease up on fertilizing permanent greenhouse occupants and all others that have not yet filled their available soil with roots, but continue applications to chrysanthemums until their flower buds announce their intentions of opening by "showing color." And give dilute liquid fertilizer, on a regular basis until Christmas, to begonias, cyclamens, poinsettias, and other early winter-bloomers that have filled their containers with roots.

Repotting now needed includes transferring cinerarias, cyclamens, and primulas intended for late winter and early spring display to bigger containers. It is fatal to best results to allow such plants to become really pot-bound before attending to this. Do not move cyclamens that are to be bloomed by Christmas or early in the new year but do pick off any flower buds that develop too early so that they will be at their best at the main blooming time.

Seeds of many annuals of sorts recommended in the entry September, Gardening Reminders For can still be sown with every prospect of having nice-sized plants, albeit a little smaller than from sowings made a month earlier, for blooming in late winter and spring. Young snapdragon plants from October sowings will be ready to follow chrysanthemums in soil beds or benches. Calendulas and stocks from seeds sown now provide good flowers for March cutting and make attractive pot plants for displaying then.

Cool greenhouse, nonhardy bulb plants, nearly all natives of South Africa, include a wide selection of sorts. Among the most familiar are freesias, lachenalias, and nerines. Others are babianas, ixias, and sparaxises. Maximum sunlight and cool, airy conditions are musts for these. Do not maintain the night temperature at more than 50°F. Ventilate freely on days when outdoor temperatures much exceed that. Keep humidity on the low side. Until frost becomes a real threat, cold frames will give adequate protection for plants of these sorts, but very soon the greater security of a greenhouse will be needed.

Hardy bulbs for forcing into early bloom in greenhouses will mostly be potted, or if being grown for cut flowers, planted in flats this month. Chief among them are hyacinths, narcissuses, and tulips. Others include Dutch and Spanish irises and such sorts as chionodoxas, crocuses, grape-hyacinths, scillas, and snowdrops, as well as many lilies.

Make certain the containers are adequately drained and the soil is porous, not necessarily rich. After planting hardy bulbs (narcissuses of the 'Paper White', 'Grand Soleil d'Or', and Chinese sacred-lily sorts are not hardy) bury them under sand, peat moss, or similar material outdoors or put them in a dark cellar where the temperature is approximately 40°F for a rooting period of a few weeks, in most cases until January or later, before bringing them into the greenhouse.

An important task in relation to greenhouses is to make sure adequate supplies of topsoil, leaf mold, sand, and suchlike "makings" for potting and planting mixes are stored under cover where they will be accessible when hard freezing makes outdoor supplies impossible or at least difficult to get.

Cold frames require essentially the same care as in September with, in cold sections, a little extra watchfulness when night temperatures drop below freezing to see that no harm comes to their occupants.

Houseplants are about to enter a difficult period. Shorter days, poorer light, and most importantly a drier atmosphere, the result of more steam, hot water, or hot air heat being used bring definite stresses. Combat such adverse environmental conditions as well as you reasonably can by moving plants to your brightest windows and if desirable and practicable supplementing natural light with artificial; by standing the pots on broad shallow trays filled with pebbles, sand, or moss kept always moist; and especially with such sorts as fuchsias and geraniums, which prefer cool conditions, by keeping temperatures, especially at night, as low as you reasonably can consistent with human comfort.

Ease up on fertilizing; the majority of houseplants are better if given no or at most very little such stimulation for the next three months or so. Unlike the same sorts in greenhouses, houseplants may require as frequent or even more frequent watering in winter than summer, this because of the drying effect of the more arid atmospheres of most homes. So be alert to their needs and satisfy them without keeping the soil over-saturated. Even cactuses and other succulents suffer if their soil remains parched for long periods.

Christmas cactuses are likely to bloom more surely if rested by moving them to a fairly cool room and keeping their soil nearly dry from early October to mid-November. Plant hardy bulbs for later forcing as described in the recommendations made above for greenhouses and also nonhardy narcissuses of the 'Paper White', 'Grand Soleil d'Or', and Chinese sacred-lily sorts for earliest bloom.

In parts of the south, October often brings serious dry spells, and gardeners must be alert that azaleas, camellias, and other plants likely to be harmed are not allowed to suffer. Periodic deep watering is likely to be needed. Pay special attention to newly seeded or planted lawns and recently set out trees and shrubs.

Sow rye grass for temporary winter lawns and oversow established lawns of Bermuda grass to ensure greenness through the winter. Use approximately one-half as much seed for oversowing as is used for all new rye grass lawns. New lawns can now be made by planting plugs of St. Augustine, centipede, and zoysia grasses and by sprigging with Bermuda grass. Fertilize established lawns.

Spring-flowering bulb plants of sorts suggested in the entry September, Gardening Reminders For can still be planted as well as those of poppy anemones, ranunculuses, and in the lower south alstroemerias, gladioluses, gloriosas, and 'Paper White', 'Grand Soleil d'Or', and Chinese sacred-lily narcissuses.

Cuttings of camellias and most other evergreens taken now and planted in a propagating bed in a greenhouse or cold frame are likely to root with facility. Those of azaleas and other plants made earlier

Bulb plants to be planted in regions of little or no frost in October: (a) Poppy anemones

(b) Turban ranunculuses

(c) Alstroemerias

(d) Gloriosas

(e) 'Paper White' narcissuses

Gladioluses can be planted in the lower south in October

may now be sufficiently rooted for them to be potted individually into 3-inch pots in sandy peaty soil. Water thoroughly and plunge (bury to the rims of their pots) in sand or peat moss in a cold frame after potting.

Except in the lower south, where a fertilizing followed by mulching is in order, stop fertilizing roses. They should now be making satisfactory displays. To assure the finest long-stemmed blooms on hybrid tea varieties disbud them to one flower to each stem.

In the south: (a) Remove all but one flower bud from each stem of hybrid tea roses

(b) To attain the finest blooms for cutting

Herbaceous perennials in need of dividing and replanting may be given attention now, and young perennials set out. For both, before planting, improve the ground by spading it deeply and incorporating manure, compost, or other organic matter, an application of a complete fertilizer, and if the ground is too acid a dressing of ground limestone or lime. Tidy perennial beds that are not being remade by cutting off and taking away to the compost pile if they are free of pests and diseases that survive composting (nematodes and crown rot disease are chief examples of these), otherwise to some place where they will not contaminate garden soil.

Biennials and plants grown as such may be transplanted to their flowering sites in October

Biennials and plants grown as such, including canterbury bells, pansies, and wallflowers, as well as such annuals as calendulas and snapdragons intended for early spring or in some sections winter flowering, may now be taken from nursery beds and planted where they are to bloom.

Set out young plants of broccoli, cabbage, cauliflower, and kale and toward the end of the month sow seeds of beets, broccoli, cabbage, carrots, kale, lettuce, onions, spinach, Swiss chard, and other cool-weather vegetables, also those of cool-weather annuals for blooming in late winter and spring. Seeds of perennials that reproduce true to type when raised in this way may be sown in cold frames or nursery beds.

Prepare for planting to be done in November and December by placing orders for trees, shrubs, roses, and other plants you may need and by getting the planting sites into agreeable condition. Make ready too soil in places where next month you will sow sweet peas.

For West Coast gardeners, October is a time for much planting. Evergreens, herbaceous perennials, and many bulb plants, including where hardy, such sorts as babianas, freesias, gloriosas, ixias, and watsonias, that will not endure much cold,

Set out young plants in parts of the south in October: (a) Calendulas

(b) Snapdragons

may be set out now. Plant deciduous trees and shrubs as soon as they begin to shed their foliage. It is also time to transfer from nursery beds to their flowering quarters such spring-blooming bedding plants as English daisies, forget-me-nots, pansies, and polyanthus primroses and in mild sections young plants of hardy annuals, such as snapdragons and stocks. Other appropriate work at this time includes cleaning up and other routines discussed earlier in reference to gardens in the north and in the south.

Setting out broccoli plants in southern gardens in October

Sow seeds of cool season vegetables and annuals and give attention to making new lawns. Renovate established ones of Bermuda grass by raking out accumulated thatch, then top dressing and watering generously.

Many deciduous shrubs including roses are easily propagated by hardwood cuttings 6 to 9 inches long taken after leaf-fall and planted vertically in rows in sandy ground with only their tips exposed. The soil must of course be kept moist, but must not be waterlogged.

OCTOMERIA (Octo-mèria). Not frequent in cultivation, this tropical American and West Indian genus of 130 species of the orchid family ORCHIDACEAE is sometimes represented in the collections of orchid enthusiasts. Its name, from the Greek *okto*, eight, and *meris*, a part, refers to the number of pollen masses of each flower, a feature that distinguishes *Octomeria* from related *Pleurothallis*, which has four masses of pollen to each bloom.

Octomerias perch on trees (are epiphytes) or rocks. They vary much in aspect, but generally are low and often tufted. They are without pseudobulbs, but produce from branching rhizomes bracted stems that persist for several years. Their leathery, evergreen leaves are oblong, roundish, or nearly cylindrical. Mostly white or yellowish, the small flowers are usually in dense clusters or heads or sometimes are solitary or in pairs, from the axils of the leaves. They have sepals and petals much alike, although the latter are often the shorter, and a small lip joined to the column.

Native of the West Indies and Trinidad, ***O. graminifolia*** has rooting rhizomes from which arise at intervals stems about 1½ feet tall. Its leaves, often suffused with red, are up to 2½ inches long. Not opening widely, the solitary blooms, ¼ inch or so long, are yellow with a red spot on the lip. With clustered stems 5 to 7 inches tall, ***O. grandiflora,*** of tropical South America and Trinidad, has solitary, narrowly-strap-shaped, often purplish leaves 4 to 6 inches long, and yellowish-white flowers about ¾ inch long, with a yellow lip that like the column is reddish at its base.

Other kinds cultivated include these: ***O. erosilabia,*** of northern South America, has three- to five-jointed stems up to 6 inches high or a little higher and erect, oblong-linear leaves up to 6 inches long and with rolled-back margins. The flowers, solitary and clustered, are creamy-white to yellow. ***O. surinamensis,*** of tropical South America, has tufts of stems 3 to 4 inches long, and of about the same lengths, leathery leaves often reddish or maroon-red on their undersides. The pale yellow flowers are in clusters of up to five. ***O. tunicola,*** of Brazil, has low, grasslike tufts of foliage and, from the leaf axils, clusters of little pale yellow flowers.

Garden Uses and Cultivation. Only collectors are likely to concern themselves with these small orchids, which are most likely to succeed under conditions that suit cool-climate odontoglossums. For further information see Orchids.

OCTOPUS TREE is *Brassaia actinophylla.*

ODONTADENIA (Odontadèn-ia). Thirty species of tropical America and the West Indies are included in *Odontadenia,* of the dogbane family APOCYNACEAE. The name, derived from the Greek *odontos,* a tooth, and *aden,* a gland, alludes to the cup-shaped group of toothed glands below the pistil. Only *O. grandiflora* seems to be cultivated.

Like its relatives the vining allamandas, ***O. grandiflora*** (syn. *O. speciosa*) is a tall-growing, vigorous climber. Its woody stems twine counterclockwise and like its leaves and other parts contain a milky latex. The glossy, lanceolate, opposite leaves, up to 1 foot long or longer by one-half as wide, often are smaller. The handsome, fragrant, 3-inch-wide flowers have apricot-colored,

Odontadenia grandiflora

funnel-shaped corollas streaked with yellow on their tubes. The spreading lobes (petals) overlap each other slightly to the right and are twisted to the left. The blooms are in ample terminal and axillary panicles usually borne over a long period in summer, or sometimes this vine blooms more than once each year. Fruits are rarely produced. They are woody pods, about 2 inches long and normally in pairs, although one member of the pair may abort. The numerous brown seeds have parachute-like clusters of hairs, which aid in their distribution by winds. They do not remain viable for long. Another species, ***O. nitida,*** with smaller, scentless flowers and beautiful shining foliage is worth cultivating. It is a native of Trinidad.

Garden and Landscape Uses and Cultivation. These are essentially as for allamandas. The species described flowers best in full sun. In temperate climates it requires the protection of a greenhouse where high humidity and a minimum temperature of 60 to 70°F is maintained. Container-grown specimens benefit from regular applications of liquid fertilizer from spring through fall. Pruning consists of thinning out weak shoots and shortening others in late winter or spring. Attention must be given to keep the plants free of mealybugs, scale insects, and red spider mites. Cuttings root readily and layering forms a sure means of propagation.

ODONTIODA. This is the name of orchid hybrids between the genera *Odontoglossum* and *Cochlioda*.

Odontioda variety

ODONTOGLOSSUM (Odonto-glóssum). The genus *Odontoglossum*, of the orchid family ORCHIDACEAE, consists of about 250 species natives from Mexico to tropical South America, with the majority Andean. It includes also many handsome hybrids. The name, from the Greek *odontos*, a tooth, and *glossa*, a tongue, alludes to the toothed crests on the lips of the blooms.

Odontoglossums are epiphytes, in their natural environments growing on trees, but not extracting nourishment from them. Usually they have spherical to ovoid, one- to three-leaved pseudobulbs spaced on short rhizomes. The flowers of most sorts are showy and are in slender racemes or panicles of few to many that originate from the bottoms of the pseudobulbs and are generally erect or arching. Those of a few sorts are pendulous. The blooms have three sepals and generally similar to them, but sometimes broader, two petals. Sepals and petals spread widely. The lip is vaguely to decidedly three-lobed with the center lobe variously shaped according to kind and with a fleshy, toothed crest close to its base. The club-shaped column is parallel to the lower part of the lip and sometimes joined to it. This distinguishes *Odontoglossum* from nearly related *Oncidium*, in which the lip diverges from the bottom of the column.

Among sorts cultivated are these: ***O. bictoniense***, of Mexico and Central America, has compressed, ovoid pseudobulbs up to about 7 inches long with three or fewer elliptic-oblong to pointed-linear leaves up to 1¼ feet long by up to 1 inch wide. As much as 4 feet tall and generally branchless, the racemes are of regularly spaced, scented or scentless blooms 1½ to 2 inches in diameter. Their spreading sepals and petals are yellowish-green conspicuously banded with chocolate-brown.

Odontoglossum bictoniense

The large and prominent white to pink or lavender-pink, broadly-heart-shaped lip is crisped at its edges. ***O. cariniferum***, of Central America and northern South America, has flattened, ovoid to elliptic-oblong pseudobulbs 3 to 5 inches long. Its pointed-linear leaves are 1 foot to 1½ feet long by 2 to 3 inches broad. Erect or arching, 2 to 4 feet long, and usually branched, the flowering stalk has many 2-inch-wide blooms with pointed-lanceolate, prevailingly brown sepals, petals generally tipped and edged with yellow or brown markings, and a white to yellow lip, purple-red to violet at its base. ***O. cervantesii***, of Mexico and Guatemala, has one-leaved, clustered, compressed, ovoid pseudobulbs 2 to 2½ inches tall. The thinnish, stalked, ovate-lanceolate to oblong-elliptic leaves may be 6 inches long. Erect to somewhat drooping, the flowering stalks, usually under 1 foot long, bear half a dozen or fewer fragrant blooms 1½ to 2½ inches wide. They have white to pinkish sepals and petals marked with many concentric chocolate-brown lines in their lower thirds. The white to pinkish, more or less heart-shaped, concave lip has a yellowish callus, purplish-striped at its base. ***O. constrictum***, of Venezuela, has compressed-ovoid pseudobulbs about 3 inches long, each with a pair of pointed, linear-strap-shaped leaves up to 1 foot long. The arching, usually branched, flowering stalks, 3 to 5 feet long, carry many 1½-inch-wide blooms, their spreading, pointed-lanceolate sepals and petals, lemon-yellow spotted toward their bases with brown and the disk of the broadly-fiddle-shaped, white lip with a pair of rose-pink spots. *O. c. majus* has larger blooms with a long, arrow-head-shaped lip. *O. c. pallens* has spotless, pale yellow sepals, petals, and lip. *O. c. sanderanum* has flowers 2½ inches wide that have yellow sepals and petals blotched

Odontoglossum cariniferum

Odontoglossum constrictum sanderanum

with light chocolate-brown and a white lip with a rose-purple spot. ***O. convallarioides***, of Mexico and Central America, is similar to *O. pulchellum*, but its pink- or lavender-flushed, white flowers rarely exceed ¾ inch in diameter and have concave instead of deflexed or recurved lips. ***O. cordatum***, of Mexico and Central America, has flattened-ovoid pseudobulbs up to about 3 inches long, each with one elliptic-lanceolate to elliptic or oblong leaf up to 1 foot long. The arching to erect flowering stalks 1 foot to 2 feet long have few to

Odontoglossum cordatum

Odontoglossum cordatum (flowers)

many starry blooms 2½ to 3 inches wide, with spreading, sharp-pointed sepals and petals that are whitish, yellowish, or greenish heavily spotted, blotched, or banded with chocolate-brown or the sepals almost solid brown. The lip has a blade heart-shaped toward its base and brown-spotted there and at its long-pointed apex. ***O. crispum,*** one of the most beautiful and variable odontoglossums, is grown in many varieties. Native to Colombia, this has flattened-ovoid pseudobulbs up to 4 inches long. The two or three narrow-strap-shaped leaves may exceed 1 foot in length. Generally arching, rarely branched, the flowering stalks, up to 1½ feet in length, carry five to twenty rather closely set blooms variable in size, coloring, and other details. Mostly they are 2½ to 3½ inches wide and white to light pink, blotched or spotted with reddish-brown. They have broad-elliptic sepals and petals with wavy, crisped, or toothed margins. The lip, usually white with a few red dots, has a large yellow patch near its center. ***O. edwardii,*** of Ecuador and Peru, has ovate pseudobulbs 3 to 4 inches tall with pairs of strap-shaped leaves up to 2 feet long by up to 1½ inches wide. In semierect panicles up to 3 feet long or sometimes longer the numerous 1-inch-wide, fragrant flowers are borne. They range in color from magenta to reddish-purple with toward the bottom of the lip a patch of bright yellow. ***O. grande,*** a native of Mexico and Guatemala, has clusters of more or less compressed, ovoid to roundish pseudobulbs up to about 4 inches long by rather over one-half as wide. They have usually two, stalked, pointed, elliptic to lanceolate, glaucous leaves that may be as much as 1¼ feet long by nearly 3 inches wide.

Odontoglossum grande

Odontoglossum grande (flowers)

Up to 1 foot long and usually erect, the flowering stalks rarely bear seven or eight, usually fewer waxy blooms that may be more than 6 inches wide. They have yellow sepals cross-branded with cinnamon-brown; the petals are brown in their lower halves, yellow above. The lip is cream-colored spotted with brown around the edges. ***O. krameri,*** of Costa Rica and Panama, has clusters of light green, flattened, spherical to ovoid pseudobulbs with solitary, elliptic-lanceolate leaves up to 9 inches long by 2 inches wide. The erect or horizontal 6- to 7-inch-long flowering stalks have three or fewer blooms about 1¾ inches in diameter. Their sepals and petals are ivory-white suffused with lilac or blush-lilac-pink. The bright pinkish-violet to violet lip, spotted with yellow and purple, has a pair of dark brown lines near its base. ***O. laeve*** (syn. *Miltonia laevis*), of Mexico and Guatemala, has flattened pseudobulbs up to 5 inches long, each with two or occasionally three more or less strap-shaped leaves up to 1 foot or so long by approximately 1½ inches wide. The starry flowers, in panicles 2 to 3 feet long, are up to 2¼ inches across, have yellow to greenish-yellow, similar sepals, petals blotched or barred with reddish-brown, and a fiddle-shaped lip, white in its outer half, purple-violet toward its base. ***O. pendulum*** (syn. *O. citrosmum*), a native of Mexico and Guatemala, has flattened, ovoid to nearly spherical pseudobulbs 3 to 6 inches long, each with a pair of oblong-strap-shaped leaves up to 1 foot long by up to 2½ inches wide. The pendulous flowering stalks, shorter than the leaves, carry fifteen or fewer 2- to 3-inch-wide, white or creamy-white, sometimes pink-tinged flowers with pale pink to deep rose-pink lips. The pointed-ovate sepals and petals are similar. The kidney-shaped lip, notched at its apex, narrows at its base to a shaft or claw with a red-dotted, yellow callus. The flowers of *O. p. album* are

Odontoglossum laeve

Odontoglossum pendulum

Odontoglossum pendulum album

pure white except for the lip which has a yellow claw. Those of *O. p. punctatum* are pale pink dotted with purple. The flowers of *O. p. rosellum* are rose-pink with a yellow-clawed lip. In *O. p. roseum* the lip of the flower is deeper pink than that of the typical species. ***O. pulchellum,*** of Mexico and Central America, has flattened, oblongish pseudobulbs 2 to 3 inches tall with two or occasionally three pointed-linear leaves approximately 1 foot long. The fragrant blooms, on weak stalks that carry ten or fewer, are from 1 inch to 1½ inches across. They are white with, at the bottom of the oblong lip, which is sharply recurved or deflexed at its apex, a red-dotted patch of yellow. ***O. reichenheimii*** (syn. *Miltonia reichenheimii*), of Mexico, has flattened, two-leaved pseudobulbs. The leaves are oblong-lanceolate, up to 1 foot long. In many-flowered panicles up to 2 feet in length, the showy, 2-inch-wide blooms have similar sepals and petals, bright yellow heavily marked with bands of chocolate-brown. The somewhat fiddle-shaped lip is pale lemon-yellow with a half-crescent-shaped pink mark. ***O. rossii,*** of Mexico and Central America, has much-flattened, curved pseudobulbs each with a single leaf up to 6 inches long. Its predominantly white, pinkish, or pale yellow blooms are two to four on stalks up to 8 inches tall. Their spreading sepals, bases of their petals, and deep yellow, vaguely heart-shaped lips are more or less spotted with reddish-brown. ***O. schlieperanum,*** of Costa Rica and Panama, has pseudobulbs and foliage like those of *O. grande.* Its two- to eight-bloomed, erect flowering stalks are up to 9 inches long. The flowers, up to 3½ inches across, are yellow to greenish-yellow with their parts cross-banded and blotched with reddish-brown. The lip is somewhat fiddle-shaped.

Odontoglossum pulchellum

Odontoglossum pulchellum (flowers)

Odontoglossum schlieperanum

Garden Uses and Cultivation. Among the most beautiful of cultivated orchids, odontoglossums are extremely worthy of inclusion in collections and if fairly treated respond with good displays of bloom. They are, however, decidedly more difficult to manage in regions of torrid summers than many orchids. The great majority respond best to cool greenhouse environments where night temperatures are 50°F in winter and are permitted during the day to rise five to ten or on sunny days up to fifteen degrees above the night levels. Some few sorts, those that in the wild inhabit lowland regions, prefer temperatures five to ten degrees higher. In summer the greenhouse should be kept as cool as outdoor temperatures permit. At all seasons as much ventilation as can be given without lowering the greenhouse temperature below acceptable levels is needed. A humid, but not dank atmosphere is necessary as is moderate, but not excessive shade from strong sun.

In general odontoglossums succeed best when their roots are rather confined. It is a mistake to put them in pots too big in relation to the sizes of the plants. Repotting is best done annually, and at that time plants too large should be carefully divided. As a potting medium a mixture of equal parts of osmunda fiber and chopped sphagnum moss is satisfactory, and most growers recommend topping off the surface of the root balls with a layer of living sphagnum. Most sorts have no decided rest periods and watering is necessary throughout the year, but less frequently when the plants are not in active growth than when they are. During their active periods they benefit from weekly or biweekly applications of dilute liquid fertilizer. For further information see Orchids.

ODONTONEMA (Odonto-nèma). Tropical America and the West Indies is home to the forty species of *Odontonema,* of the acanthus family ACANTHACEAE. A few of its showier members are cultivated, sometimes under the now displaced name of *Thyrsacanthus.* The name, derived from the Greek *odous,* a tooth, and *nema,* a thread, alludes to the stamens.

These are erect evergreen shrubs or herbaceous perennials with opposite, undivided leaves and flowers in loose or crowded, erect or pendulous, terminal panicle-like clusters, the bracts of which, unlike those of many of the acanthus family, are small and inconspicuous. The blooms are tubular, symmetrical or somewhat two-lipped and have five tiny corolla lobes (petals). There are two fertile stamens and two vestigial ones (staminodes). The fruits are capsules, normally with four seeds.

Commonly planted in the tropics and warm subtropics, ***O. strictum*** (syn. *Jacobinia coccinea*) is a handsome shrub 5 to 12 feet in height. Its pointed-ovate or oblongish, short-stalked, wavy-margined, leaves are up to 10 inches long and are smooth except for a slight hairiness on the veins beneath. The slender, bright red flowers, in erect, crowded, narrow, spikelike racemes carried well above the foliage from the branch ends, open a few at a time over a long period. They are about 1 inch long. This species is indigenous to Mexico and Central America. Variety *O. s. variegatum* has grayish or bluish-green leaves irregularly variegated along their margins with white.

The slender-stemmed, pendulous panicles of red flowers of ***O. schomburgkianum***

Odontonema strictum variegatum

(syn. *Thyrsacanthus rutilans*) form delicate cascades 1 foot to 3 feet in length. A native of Colombia, this shrub is up to 6 feet tall and has oblong-lanceolate leaves 3 to 8 inches long. Its narrow blooms are about 1½ inches in length. Less common, ***O. callistachyum,*** of Mexico and Central America, is a more or less hairy plant some 2 feet tall, with wrinkled, oblong-ovate leaves up to 1 foot long and erect, branched, crowded, spikelike clusters of red or pink flowers, 1¼ inches long and two-lipped, with the lower lip deflexed.

Garden and Landscape Uses. In the humid tropics and warm subtropics these are attractive plants for garden beds and borders. They succeed in sun in ordinary, moderately moist, well-drained soil. They are also satisfactory for cultivation in tropical greenhouses.

Cultivation. Outdoor-grown specimens need no particular care other than occasional pruning to shape them, which is best done at the beginning of a new growing season. At that time, too, the application of a slow-acting complete fertilizer is of value. Propagation is easy by cuttings.

In greenhouses these plants need a minimum winter night temperature of about 60°F, a humid atmosphere, good light with just a little shade from strong summer sun, and rich, well-drained soil kept moderately moist without becoming stagnant. New plants are commonly raised from cuttings in spring, and at that time old ones are pruned, top dressed or repotted, and started on a new cycle of growth. In spring and early summer the shoots of both new and old specimens should have their tips pinched out once or twice to induce branching.

ODONTONIA. This name is applied to bigeneric orchid hybrids the parents of which are *Miltonia* and *Odontoglossum.*

ODONTOPHORUS (Odontó-phorus). Half a dozen species of succulent plants of South Africa are the only representatives of *Odontophorus,* of the carpetweed family AIZOACEAE. The genus, one of the many once included in *Mesembryanthemum,* takes its name from *odus,* a tooth, and *phorus,* a bearer, presumably referring to the toothed leaves.

Odontophoruses are miniature fleshy-rooted shrublets with prostrate or upright branches, each with one or two pairs of opposite, soft, fleshy, gray-green, velvety-downy, bluntly-keeled leaves roughened with minute wartlike tubercles, and toothed. The white or yellow flowers are solitary. They have eight to ten stigmas. Although superficially resembling daisies, they differ in being single blooms instead of being composed of many florets.

Best known to fanciers of succulents is ***O. primulinus.*** This has short stems with clinging to them the dried remains of old leaves. The live leaves are white-hairy, subcylindrical, and about 1½ inches long by approximately one-half as broad, or at their bases somewhat wider. Those of each pair spread at about a forty-five degree angle. Toward their apexes they are three-angled, the angles furnished with short, thick teeth. The pale yellow blooms are about 1½ inches wide. Similar to the last, but smaller and with white flowers is ***O. nanus.*** The short stems of ***O. marlothii*** (syn. *Mesembryanthemum marlothii*) carry two or three pairs, set at right angles to each other, of white-warted, three-angled leaves about 1¼ inches long by ¼ inch wide and deep, toothed along their margins. Those of each pair are united at their bases. From the leaf axils longer shoots, at first upright, later prostrate, develop. The blooms, on stalks about ½ inch long, are yellow and 2 inches or a little more in diameter.

Garden and Landscape Uses and Cultivation. Choice subjects for inclusion in fanciers' collections of succulents, odontophoruses may be grown in greenhouses and, in desert and semidesert regions where conditions approximate those under which they grow in the wild, in rock gardens. They are very sensitive to excessive moisture and must have porous, coarse soil. One of a sandy, gravelly nature is best. Crushed brick is excellent to mix with the soil for specimens grown in pots. Pans (shallow pots) are the most suitable containers for the indoor cultivation of these plants. They must be very well drained. Any suspicion of soil stagnancy brings trouble. In fall and winter the earth is kept almost completely dry. When new growth starts, in late winter or spring, water is given regularly but always with caution, the soil being allowed to approach dryness before being wetted. This procedure is followed until fall. Indoor night temperatures in winter should be about 50°F, with increases of up to about ten degrees by day permitted. Propagation is easy by cuttings and seed.

ODONTOSORIA. See Sphenomeris.

ODONTOSPERMUM. See Astericus.

OEDEMA. See Edema.

OEMLERIA (Oem-lèria)—Osoberry or Skunk Bush or Bearberry. The hardy deciduous shrub called osoberry and sometimes skunk bush belongs in the rose family ROSACEAE. Its only species, allied to *Prunus* from which it differs in having chiefly unisexual flowers with five pistils, is indigenous from British Columbia to California where it lives in moist, rich woodlands. The common name osoberry represents a mongrel marriage of convenience between the Spanish word for a bear, *oso,* and the English word berry. Often the plant is called bearberry, although that name is more commonly used for quite different *Arctostaphylos uva-ursi.*

The osoberry (***Oemleria cerasiformis*** syn. *Osmaronia cerasiformis*) is a bushy, suckering plant 3 to 15 feet in height with slender, straight stems and alternate, lobeless, toothless, oblong to oblanceolate,

Oemleria cerasiformis

short-stalked leaves up to 4 inches long, and rank-scented (hence the name skunk bush) flowers that expand at the dawn of the new growing season, sometimes as early as January. They have lobed, bell-shaped calyxes, and five spoon-shaped, white or greenish-white petals alternating with the calyx lobes. Those of male flowers have spreading, those of the female erect, petals. The males have fifteen stamens, the females five pistils and abortive stamens. About ⅓ inch across, the flowers are in pendulous panicles at the ends of leafy branches. They are pollinated by flies for whom the stink of the blooms is presumably an attraction. The young foliage when crushed is also odoriferous, with a scent that has been described as similar to that of watermelon rind. The one-seeded, flattish, berry-like fruits are about ½ inch long. Bluish-black with a heavy waxy coating or bloom, they are insipid and bitter but edible. Their stalks are often orange or red.

Garden and Landscape Uses and Cultivation. The osoberry is an attractive landscape shrub with good-looking foliage. It lends itself for grouping and is thought well of because it leafs out and blooms early. It is hardy in sheltered locations in southern New England, although it may be injured there sometimes by late frosts. Given a moistish fertile soil and some shade, it grows without difficulty and spreads. It is easily propagated by seeds sown as soon as they are ripe, or stratified and sown in spring, and by division, cuttings, and layering. Transplanting is best done in spring or early fall. Pruning needed to shape the plants or to contain them to size should be done as soon as they are through blooming. Ordinarily it consists of a moderate thinning out of weak, crowded, and old branches. If necessary, severe pruning may be carried out without harm to the bush.

OENOTHERA (Oen-othèra) — Evening-Primrose, Sundrop, Golden Eggs. Belonging to the evening-primrose family ONAGRACEAE, the genus *Oenothera* consists of approximately eighty species. Although not of outstanding horticultural importance, the group includes some popular ornamentals and others, less well known, worthy of trial. Especially intriguing, but probably not amenable to cultivation except in arid regions, are a few desert species, notably *O. californica, O. deltoides,* and *O. primiveris.* This genus is endemic to the New World, but a few sorts are well established as weeds elsewhere. It is not botanically related to true primroses (*Primula*). The name is derived from one used by the ancient Greeks, probably for an *Epilobium.*

Oenotheras include annuals, biennials, herbaceous perennials, and more rarely, subshrubs. They have alternate, stalked or stalkless, undivided leaves with smooth, lobed, or toothed margins. Mostly they have showy blooms solitary from the axils of leaves or in terminal racemes. Each flower has four usually strongly reflexed sepals, four oblanceolate to obovate petals, eight stamens sometimes of two lengths, and a style with an undivided, lobeless stigma or one that is four-lobed or four-parted. The fruits are capsules. The blooms of some sorts open toward evening and, unless the weather is dull, close the next morning; those of others remain expanded through the day.

Evening-primroses, the flowers of which expand in the evening and close the next morning, include these sorts: ***O. acaulis*** (syn. *O. taraxifolia*), of Chile, a perennial or sometimes a biennial, is stemless when young, when older has prostrate stems up to 6 inches long. Its stalked, deeply-pinnately-divided, dandelion-like, or coarsely-

Oenothera acaulis

toothed, oblanceolate leaves are 4 to 8 inches long. The 2- to 3-inch-wide blooms, white when they first open, become pink as they age. They have stamens of unequal lengths and four-lobed stigmas. ***O. argillicola,*** a biennial or perennial known in the wild only from shale barrens in the southern Appalachian Mountains, accommodates in cultivation to a great variety of well-drained soils. During its first season this forms stemless rosettes of remotely-toothed, linear-lanceolate to oblanceolate leaves 3 to 7 inches long. The following summer it develops more or less sprawling, leafy flowering stems 2 to 3 feet long.

Oenothera argillicola

Freely produced and making a brave show, the bright yellow flowers are 2½ to 4 inches in diameter. ***O. biennis,*** of eastern North America, is a biennial often of rather weedy aspect but some variants are sufficiently attractive to be accorded a place, at least in native plant gardens. From 2 to 6 feet tall, this has basal rosettes of shallowly-toothed, lanceolate leaves up to 6 inches long and many smaller stem leaves. The flowers, 1½ to 2½ inches wide, are clear yellow becoming more golden yellow as they age. The young shoots are sometimes used in salads; the roots may be cooked and eaten as a vegetable. ***O. caespitosa*** is a very variable perennial or biennial native from South Dakota and Nebraska, westward and

Oenothera caespitosa

southward. Stemless or practically so, it forms rosette-like clusters of hairy, narrowly-lanceolate to elliptic, spatula-shaped, or obovate, more or less dandelion-like, toothed leaves up to 1 foot long. The white to pink flowers, which deepen in color as they age, are 1½ to 3 inches in diameter. ***O. californica,*** of desert parts of California and Baja California, has much the aspect of *O. deltoides* but it has running, underground rhizomes and is perennial. Its leaves are oblanceolate to spatula-shaped or lanceolate and from nearly toothless and lobeless to quite coarsely toothed. They are ¾ inch to 2¼ inches long. In clusters of several, the flowers,

Oenothera californica

from 1½ to 2¼ inches across, have obovate-circular, white petals that change to pink as they age. ***O. campylocalyx*** is an annual or a biennial native of Bolivia. It is 1½ to 2½ feet tall, branched or branchless. Its stalkless or nearly stalkless, toothed,

Oenothera campylocalyx

linear-lanceolate leaves, green with a central white stripe, are spaced all along the stems, to the tops of the terminal flowering portion. The 1-inch-wide flowers are orange-red with yellowish edges to the petals. ***O. deltoides,*** of desert regions in Arizona, California, and Baja California, is an erect annual or biennial up to 1 foot tall with a few decumbent branches extending from its base. In a loose rosette, the 1- to 3-inch-long lower leaves are obovate to oblanceolate or lanceolate and toothed or

Oenothera deltoides in its native range

nearly toothless. Solitary from the upper leaf axils, the 1½- to 3-inch-wide flowers have white petals that change to pink as they age. ***O. drummondii*** is a 1- to 2-foot-tall native of Texas. Although probably a biennial in the wild, this is often grown as an annual. Softly-hairy, it has lanceolate-oblong to oblong-ovate, smooth-edged or slightly-toothed leaves. The decorative 2- to 3-inch-wide flowers are bright yellow. ***O. erythrosepala,*** believed to be of garden origin, is naturalized in some parts of North America. A biennial, this forms basal rosettes of large wavy-toothed leaves and in its second year develops erect, branched, red-warted stems 3 to 4 feet tall bearing nearly stalkless, crinkled, oblanceolate to ovate leaves. The yellow flowers are 2½ to 3½ inches across. Those of *O. e. rubricalyx* (syn. *O.* 'Afterglow') have rich

Oenothera erythrosepala

red calyxes. ***O. lavandulaefolia,*** native from Wyoming to Texas and Arizona, is handsome. A tufted perennial 4 to 8 inches high, from a woody rootstock this develops many erect or ascending, slender stems bearing gray-hairy, linear to oblanceolate or narrowly-obovate leaves mostly under 1 inch long. The long-tubed flowers, two to three to a stem and 1½ to 2 inches wide, are yellow becoming reddish with age. ***O. missourensis,*** is native to dry,

Oenothera missourensis

rocky, limestone soils from Illinois to Nebraska, Colorado, Missouri, and Texas. A perennial with decumbent to erect stems occasionally 1½ feet tall, usually lower, it has softly-hairy, thick, pointed-narrowly-elliptic to lanceolate, distantly-toothed or toothless leaves 4 to 5 inches long. Its broad-petaled flowers are 3 to 5 inches across. They have stamens of two different lengths and a four-cleft stigma. ***O. primiveris,*** of deserts in Utah, Texas, and California, is a stemless or nearly stemless annual or biennial. Hairy throughout, it has leaves with blades up to 3 inches long, deeply pinnately-cleft into lobes which are again lobed or toothed. The leafstalks are about as long as the blades. The flowers, 1 inch to 2 inches across, have yellow petals that become pinkish to orange-red as they age.

Sundrop is the common name of a few yellow-flowered day-bloomers including these sorts: ***O. fruticosa,*** native in dryish soils from Massachusetts to Florida and Alabama, is perennial or perhaps sometimes biennial. It has erect or spreading, hairy stems from a few inches to 2 feet tall. Its stalked basal leaves have ovate to spatula-shaped blades. Those on the stems are stalkless or nearly stalkless, linear to lanceolate, and rarely over 2¼ inches long.

Oenothera fruticosa

The clear yellow blooms 1 inch to 2 inches across have usually shallowly-toothed petals and a four-lobed stigma. The seed capsules are hairless or nearly hairless. ***O. perennis,*** of eastern North America, is a slender, erect perennial or biennial 6 inches to 2 feet tall. Its lower leaves are oblanceolate to spatula-shaped and up to 2¼ inches long, those above tend to be linear-lanceolate. The flowers, which nod in the bud stage, are from ½ to ¾ inch across. ***O. pilosella*** (syn. *O. pratensis*) is a perennial native of the central United

States. It has erect, softly-hairy stems 6 inches to 1½ feet tall. Its lower leaves are obovate to oblanceolate, those above lanceolate to elliptic-lanceolate and 1 inch to 4 inches long. Mostly clustered at the tops of the stems, the yellow flowers are 1½ to 2½ inches across. The seed capsules are hairy. ***O. tetragona*** (syns. *O. glauca, O. fruticosa youngii, O. youngii*) is a variable perennial up to 3 feet tall. It occurs usually on dry soils from Nova Scotia to Michigan, South Carolina, Alabama, and Louisiana. Similar to *O. fruticosa,* from that species *O. tetragona* can be distinguished by its seed capsules usually being furnished with glandular hairs. Its flowers are generally not over 1½ inches wide. Those of *O. t. fraseri* are 2 to 2½ inches across. Variety *O. t.* 'Highlight' is exceptionally free-flowering.

Other day bloomers include these: ***O. berlandieri*** (syns. *O. speciosa childsii, O. tetraptera childsii*), of Texas and Mexico, is sometimes called Mexican-primrose. More or less prostrate and much like *O. speciosa,*

Oenothera berlandieri

it has very slender stems up to 6 inches long, leaves 1 inch to 2 inches long, and pink blooms 1 inch to 2 inches wide. ***O. bistorta*** is an annual or sometimes longer-lived native of coastal California. From 1 foot to 2 feet tall, its hairy stems bear spatula-shaped to lanceolate leaves 1½ to 3 inches long, the lower ones stalked, those above stalkless. The 1-inch-wide flowers are yellow, usually with a reddish-brown spot at the bottom of each petal. More slender *O. b. veitchiana* is native inland from the coast. ***O. ovata*** (syn. *Camissonia ovata*), sometimes called golden eggs, is a good-looking native of California. Usually perennial, it has a deep taproot, stemless rosettes of leaves, and ¾- to 1-inch-wide, golden-yellow blooms. Up to about 8 inches long, its toothed or toothless leaves are ovate to oblong-lanceolate. ***O. speciosa*** is an attractive native of an extensive region westward and southward from Missouri into Mexico. Usually perennial, but

Oenothera ovata

sometimes perhaps biennial or even annual, this sort is 1 foot to 2 feet tall. It has more or less erect stems and linear to lanceolate or oblanceolate leaves 1½ to 4 inches long. Their edges are distantly-toothed or wavy, or those of the lower leaves may be pinnately-lobed. The up-facing blooms, white or more rarely pink when they first open, become pinkish as they fade. They are 2 to 3½ inches across and come from the axils of the upper leaves. They have stamens of two different lengths and a four-lobed style.

Garden and Landscape Uses. The cultivated oenotheras are suitable for displaying in sunny flower beds and the low ones in rock gardens and similar environments. Where geographically appropriate they may be used in native plant gardens. All are sun-lovers that need well-drained soil and accommodate well to dryish ones, although a few sorts not generally grown in gardens prefer moister conditions. Most but not all kinds described are hardy. Of those that will not survive winter cold or winter wet in the northeast, *O. acaulis* and *O. caespitosa* are satisfactory when treated as annuals and *O. berlandieri* is suitable for use as a cool greenhouse pot plant.

Cultivation. Few plants are easier to grow than evening-primroses and sundrops. All come readily from seed and the perennials from cuttings and some kinds by division. Biennials are managed by sowing the seeds from May to July, transplanting the seedlings to nursery beds to make their first season's growth and to their flowering sites in fall or early spring. Sow seeds of annuals in early spring where they are to bloom and thin the seedlings enough to prevent undue crowding. The perennials where hardy need no special care beyond curbing the invasive tendencies of any such as *O. speciosa* that spread by underground runners.

OFFSET. As gardeners use the term, an offset is a fairly easily separable young bulb or plant originating from at or near the base of an older one. Detached offsets afford ready and certain means of propagation of many sorts of plants.

OFTIA (Óf-tia). The meaning of the name *Oftia,* a genus of two South African species of the myoporum family MYOPORACEAE, is not known.

Oftias are nonhardy evergreen shrubs with toothed leaves that are alternate, or the lower ones opposite or whorled (in circles of more than two). Produced from the axils of the upper leaves, the small, stalkless or short-stalked, white flowers have a five-parted calyx, a corolla with a slender cylindrical tube, five spreading lobes (petals), and four stamens in two pairs. The fruits are berry-like drupes.

Up to about 3 feet tall, ***O. africana*** has somewhat succulent stems and stiffish, pointed-ovate leaves margined with unequal teeth, the lower ones opposite, the upper alternate. The starry flowers are about ½ inch wide. Their petals are white with a blue streak at the bottom of each. The spherical, purple fruits are ¼ inch in diameter.

Oftia africana

Garden and Landscape Uses and Cultivation. The species described is a useful outdoor shrub in the tropics and subtropics and is interesting for cultivation in greenhouses where it succeeds in environments appropriate for begonias. Propagation is easy by cuttings and by seed.

OGEECHEE-LIME is *Nyssa ogeche.*

'OHE'OHE is *Tetraplasandra kauaiensis.*

OIL. This word forms parts of the common names of these plants: castor oil plant (*Ricinus communis*), China wood oil or tung oil tree (*Aleurites fordii*), Japan wood oil tree (*Aleurites cordata*), oil palm (*Corozo oleifera* and *Elaeis guineensis*), and Poona or Poonga oil tree (*Pongamia pinnata*).

OKRA or GUMBO. Much more commonly cultivated in the south than elsewhere, but adaptable to northern conditions, okra or gumbo is *Abelmoschus esculentus,* a native of the Old World tropics. It belongs in the mallow family MALVACEAE. As a vegetable it is notable as the name ingredient of much esteemed gumbo soup and is also cooked in other ways. Its edible parts are its nutritious young seed pods. These are used fresh, dried, canned, and frozen.

Okra

Okra requires essentially the same conditions as corn. Because of the height of many varieties it is well to locate them where they will not shade lower crops. A warm weather plant, okra will not tolerate cold, or early in the season, wet weather. Deep fertile soil, slightly acid to neutral, is best to its liking. Sow the seeds after the weather is settled and warm where the plants are to stay. Vigorous varieties should be in rows 4 to 5 feet apart with the final spacing of the plants 1 foot to 3 feet apart in the rows. Dwarfer, less robust kinds may stand 9 inches to 1¼ feet apart in rows 3 feet asunder. Cover the seeds to a depth of ½ inch. Because often they germinate rather poorly, it is well to sow up to a dozen seeds to each 1 foot of row, and thin out surplus seedlings. Germination is aided if the seeds are soaked in water for twenty-four hours before sowing. Seedling okra plants do not transplant readily, but this can be done to fill gaps in the rows if the plants are taken with a good ball of soil and well watered and shaded after they are set in their new locations.

Care through the growing season involves weed control, most conveniently done by frequent shallow cultivations, aided perhaps after hot weather arrives by mulching. Okra withstands hot, dry weather well, but is greatly helped through such periods by periodic deep watering. After the first harvesting some gardeners pinch out the tip of the main stem to induce branching and prevent too tall growth, others prefer to let the plants run.

Harvesting is continuous until frost. Even if not needed for use, remove the pods promptly to prevent them draining the plants of strength that would be better employed in supporting new growth and a succession of young pods. Harvestable pods, the best those 2 to 3 inches long, should snap when bent between the fingers. If they only bend they are too old, tough, and stringy. Pick all pods that have attained harvesting size every day or two.

Popular varieties are 'Clemson Spineless', 'Dwarf Green Longpod', 'Perkin's Longpod', and 'White Velvet'. Troubles that may come are fusarium wilt, for which crop rotation affords the best prevention, chlorotic yellowing, the result of excessively acid or alkaline soil, and bud dropping, which occurs when the soil is poorly drained or excessively dry.

OLD MAN is *Artemisia abrotanum.* The old man cactus is *Cephalocereus senilis,* the old man's beard, *Clematis vitalba.*

OLD WOMAN is *Artemisia stellerana.*

OLDENBURGIA (Olden-búrgia). Named in honor of Fredrick Peter Oldenberg, who died in the latter part of the eighteenth century, this genus of three species of African shrubs and subshrubs belongs to the daisy family COMPOSITAE. Its members have leathery, undivided, toothless leaves, alternate or in rosettes, that are densely-hairy beneath and have one prominent mid-vein. The large heads of flowers are solitary in the leaf axils, or are solitary or in twos or threes at the ends of the branchlets. They have disk florets (corresponding to those in the center of a daisy) and petal-like ray florets. The former are five-lobed, the latter two-lipped with the outer lip long and strap-shaped, the inner one minute. The seedlike fruits are achenes.

Suggesting by its habit of growth and overall appearance a large-leaved rhododendron, ***Oldenburgia arbuscula*** (syn. *O. grandis*) attains a height of up to 8 feet and may be almost as broad as tall. Its broad-elliptic to obovate, thick, leathery, stalkless or nearly stalkless leaves are 6 inches to 1½ feet long and up to 8 inches broad. When young they are covered with silvery-white hairs, but these soon disappear from the upper surfaces, which become

Oldenburgia arbuscula in South Africa

Oldenburgia arbuscula showing young foliage

Oldenburgia arbuscula, with young flower buds

Olea europaea as a lawn specimen in California

Olea europaea by a roadside in California

glossy dark green. The large purple and white flower heads have the outer lips of the ray florets three-lobed. The collar or involucre is composed of pointed bracts, which, like the stalks that carry the flower heads, are densely-woolly.

Garden and Landscape Uses and Cultivation. Very little experience has been had in North America with the cultivation of the handsome species described, but it is to be expected that it would grow satisfactorily in California and elsewhere in dryish, sunny climates free or nearly free of frost. It serves well as a single specimen and as a component of shrub groupings. Propagation is by seed and by cuttings.

OLEA (Ò-lea)—Olive. The genus *Olea,* of the olive family OLEACEAE, contains twenty species of evergreen trees and shrubs and is widely distributed as a native of the eastern hemisphere. Its name is the classical one of the common olive. The most important species of the genus, the common olive is widely grown in Mediterranean-type climates everywhere for its fruits, which are greatly esteemed for food and as a source of oil. Since time immemorial this species has been so cultivated.

Oleas have opposite, usually toothless leaves, scurfy with scales on their undersides. Their little white flowers, bisexual or unisexual, are in panicles or clusters from the leaf axils or terminal on the branchlets. They have a short, four-toothed calyx, a corolla with a short tube and four spreading lobes (petals), two stamens, and a short style. The fruits are drupes (fruits of plumlike structure).

The common olive (***O. europaea***), believed to be a native of Asia Minor but now naturalized elsewhere in the Mediterranean region, includes the thorny-branched wild species distinguished as *O. e. oleaster,* and the much better known, thornless cultivated varieties derived from it, which are grouped as *O. e. sativa* (syn. *O. e. communis*). The common olive is generally not over 25 or 30 feet tall, but sometimes reaches twice this. It is extremely long-lived, its possible life span certainly being in excess of 1,000 years. Old specimens are decidedly picturesque. Their trunks, frequently twisted and gnarled, are often hollow. The leaves of the common olive are elliptic, oblong, or lanceolate and 1 inch to 3 inches long. Their upper surfaces are dark gray-green, their undersides densely-silvery-scaly. In panicles shorter than the leaves, the fragrant flowers are borne. The fruits are the familiar olives of commerce, green at first, becoming black and glossy when ripe, and ½ inch to 1½ inches long.

African ***O. africana*** (syn. *O. chrysophylla*) differs from the common olive chiefly in its smaller flowers and fruits. A round-headed, slender-branched tree up to about 30 feet in height and a native of semidesert habitats, it has leaves 2 to 4 inches long, narrow-elliptic to lanceolate, bright green on their upper sides and with brownish scales beneath. The flowers are axillary. The fruits are spherical, black, and ¼ inch in diameter. The black ironwood, of South Africa (***O. laurifolia***), which sometimes attains an 80-foot height, is often lower and has a smallish crown. Its lustrous, ovate to elliptic leaves are up to 4 inches long by 1½ inches wide and have wavy margins. The flowers, in terminal clusters, are creamy-white and fragrant. The edible, dark purple fruits are about ⅓ inch long.

Garden and Landscape Uses and Cultivation. Oleas are not hardy in the north. They survive only where winters are mild and hard freezing does not occur. Sunny climates with dry summers and moist winters best suit the common olive and the South African species. Under such conditions they fruit satisfactorily. In moister climates they may live, but not fruit well. The common olive and other cultivated members of the genus *Olea* are good ornamentals, admired for their often picturesque forms, their frequently somber evergreen foliage, and some for the fragrance of their flowers. They are handsome as solitary specimens and for closer setting as elements in mixed plantings and screens. They need well-drained, porous soil, and sunny locations. They stand exposure to wind well. Propagation is by seed and cuttings. For cultivation of the common olive see Olive.

OLEACEAE — Olive Family. The twenty-nine genera of this widely distributed family of dicotyledons account for 600 species of trees, shrubs, and occasionally woody climbers. Especially abundant in temperate and tropical Asia, its species nearly always have opposite, very rarely alternate or whorled leaves that may be undivided or pinnate, and toothless or toothed. Bisexual or rarely unisexual, the small flowers are in panicles or branched clusters. They may be without calyxes, but more often have small, usually four-lobed ones. The corollas are of four or more separate or partly united petals or may be lacking. There are two or very seldom four stamens and one style capped with a two-lobed stigma. The fruits are berries, drupes, capsules, or samaras. Genera cultivated include *Abeliophyllum, Chionanthus, Fontanesia, Forestiera, Forsythia, Fraxinus, Jasminum, Ligustrum, Menodora, Nestegis, Noronhia, Olea, Osmanthus, Osmarea, Phillyrea, Syringa,* and *Ximenia.*

OLEANDER. See Nerium. The yellow-oleander is *Thevetia peruviana.*

OLEARIA (O-leària) — Daisy Bush, Daisy Tree, Tree-Aster. Despite its 100 or more species being shrubs and trees instead of herbaceous plants, *Olearia* is very closely allied to *Aster.* Unlike that familiar genus, which in the wild is restricted to the northern hemisphere, *Olearia* is native only south of the equator, chiefly in Australia and New Zealand, with a few extensions into New Guinea and Lord Howe Island. It belongs to the daisy family COMPOSITAE and has a name derived from *Olearius,* the Latinized form of the name of Adam Olschlager, of Germany, who died in 1671.

Olearia gunniana

Olearia gunniana (flowers)

It is not derived, as is sometimes stated, from *Olea*, the olive. For the plant previously named *O. insignis* see Pachystegia.

Daisy bushes are evergreens with usually alternate, less commonly opposite leaves, usually clothed on their under surfaces with a felt of white or yellowish hairs. The small to fairly large flower heads are solitary or in clusters or panicles. Each consists of few to many bisexual disk florets similar to those that form the centers of daisies and usually but not invariably encircled by a single row of white, blue, or purple, female petal-like, ray florets. The fruits are seedlike achenes.

One of the most attractive daisy bushes, ***O. gunniana,*** a Tasmanian endemic, is a much-branched shrub 5 to 10 feet in height, with heavily white-felted young shoots. It has alternate, short-stalked, oblong to narrowly-obovate leaves with pointed apexes and shallowly-toothed margins. From ½ inch to 1½ inches long, and about one-quarter as broad, they are clothed with a felt of white hairs on their undersides. The yellow-centered flower heads, very aster-like in appearance, have ten to sixteen white, pink, mauve, blue, or purple ray florets. From 1 inch to 1½ inches across, they are borne with great freedom in slender-stalked clusters 4 to 8 inches in diameter. This species has been confused with *O. stellulata*, an Australian kind with much larger leaves, perhaps not in cultivation.

Its leaves rather holly-like, ***O. macrodonta,*** of New Zealand, is not infrequently cultivated as *O. dentata*, a name that correctly belongs to another New Zealander. A shrub or tree up to about 20 feet high, *O. macrodonta* has short-stalked, more or less pointed, opposite leaves that when bruised give off a musky odor. They are ovate-oblong to oblong and 2 to 4 inches long by rather less than one-half as broad and white-hairy on their undersides. They have coarsely-sharp-toothed, wavy margins. Slightly under ½ inch wide, the flower heads are in clusters 3 to 6 inches across. They have reddish centers and eight to twelve florets. Similar to the last, ***O. ilicifolia*** differs chiefly in its pointed leaves being linear to lanceolate or narrowly-oblong instead of broader and in having more definitely spiny teeth. Their undersides are thinly clad with yellowish-white hairs. It is a native of New Zealand.

Olearia ilicifolia

The large panicles of flower heads of ***O. traversii*** are less attractive than those of most other kinds discussed here, but the form and foliage of the shrub or tree, which is sometimes 35 feet tall, are agreeable. The flower heads, a little over ¼ inch wide and without ray florets, are of five to fifteen white disk florets. The opposite, leathery, short-stalked, toothless leaves, ovate-oblong and mostly 1½ to 2½ inches long, are silky-hairy on their undersides. This species is endemic to the Chatham Islands.

Ranging in height from 8 to 20 feet, ***O. avicenniaefolia*** is a shrub or small tree with white-scurfy young shoots and alternate, pointed or bluntish, elliptic-lanceolate, toothless, lustrous, leaves 2 to 4 inches long and approximately one-half as wide. Their upper sides are grayish-green, their undersurfaces clothed with white or yellowish felt. The 2- to 3-inch-wide rounded clusters of numerous ¼-inch-long flower heads have long slender stalks. Each flower head has two or three florets, one or two of which may be of the ray type. This is native to New Zealand. Another from that land, ***O. moschata*** is a shrub up to 12 feet in height, its flowers with a distinct musklike fragrance. It has rather crowded, grayish-green, obovate-oblong leaves up to ¾ inch long, clothed beneath with white hairs. About ¼ inch wide, the white flower heads, in clusters

Olearia avicenniaefolia

Olearia moschata

Olearia moschata (flowers)

Olearia haastii

Olearia albida

Olearia erubescens

of twenty to thirty, have each ten to twenty florets, of which a dozen or fewer are ray florets.

The hardiest daisy bush and one of the most ornamental, ***O. haastii*** is a natural hybrid between *O. avicenniaefolia* and *O. moschata*. A much-branched shrub 4 to 10 feet tall, this has white-hairy young shoots and crowded, alternate, short-stalked, oblong to ovate, toothless leaves, glossy-green above and white-felted on their undersides. The flower heads, about ⅓ inch across and with yellowish centers and three to five white ray florets, are in flattish, long-stalked clusters 2 to 3½ inches in diameter. Another hybrid, ***O. oleifolia,*** has as parents *O. avicenniaefolia* and *O. odorata*, the latter a shrub of little merit. From 5 to 10 feet tall, *O. oleifolia* has leathery, lanceolate to narrowly-oblong leaves 1½ to 3 inches long and white-felted on their undersides. Its ¼- to ½-inch-long white flower heads are many together in leafy clusters.

Other sorts include these: ***O. albida*** is a shrub up to 10 feet tall or sometimes in its native New Zealand a small tree. It has toothless, narrowly-oblong to somewhat broader leaves, white-felted on their undersides and 1 inch to 4 inches long. The white flowers are in terminal and axillary clusters 2 to 3 inches wide. The leaf margins of *O. a. undulata* are markedly undulate. Native to South Australia and Tasmania, ***O. erubescens*** is 3 to 5 feet tall. Its glossy-green, broad-elliptic to oblong, toothed, leathery leaves, ½ inch to 1½

Olearia furfuracea

inches long, on their undersides are brownish-downy. The 1-inch-wide, starry flower heads, with yellow centers and a few long, narrow, white disk florets, are in long, narrow, leafy panicles. ***O. furfuracea*** is a New Zealand shrub or tree up to 20 feet tall with alternate, leathery, ovate to elliptic-oblong leaves 2 to 4 inches long, toothless or slightly toothed. At maturity their upper surfaces are hairless, their undersides are furnished with silvery or brownish hairs. The yellow-centered flowers with two to three white rays are up to ½ inch wide. They are in branched, panicle-like clusters 3 to 5 inches across. ***O. scilloniensis,*** a hybrid between *O. gunniana* and *O. lyrata,* is a handsome, rounded shrub up to 5 feet tall. It has lanceolate, dark green, wavy-edged leaves and, in massive, crowded clusters a great abundance of pure white, ¾-inch-wide flowers.

Garden and Landscape Uses and Cultivation. Unsuitable for cold climates or those characterized by hot, dry summers, daisy bushes are excellent for more salubrious, equable conditions such as are found in California and the Pacific Northwest. They serve admirably as general purpose shrubs and small trees, and in general do well in coastal regions and cities. They need sunny locations and rather light, well-drained soil of a peaty character and free of lime. Propagation is by cuttings, made from firm but not hard ends of young shoots, planted in a cold frame or greenhouse propagating bench, and by seed. No regular pruning beyond any to keep them shapely is ordinarily needed. This is done immediately after flowering or in spring. If the bushes become too tall and straggly they may be pruned hard back. They soon renew themselves with new growth.

OLEASTER is *Elaeagnus angustifolia.*

OLIVE. Plants other than the one discussed in the next entry that have common names including the word olive are these: Bermuda olive-wood bark (*Elaeodendron laneanum*), black-olive (*Bucida buceras*), false-olive (*Elaeodendron orientale*), Madagascar-olive (*Noronhia emarginata*), Russian-olive (*Elaeagnus angustifolia*), spurge-olive (*Cneorum tricoccon*), sweet-olive (*Osmanthus fragrans*), and wild-olive (*Cordia boissieri*).

OLIVE. One of the world's most important sources of edible oil, the olive (*Olea europaea*), presumably a native of Asia Minor, has been cultivated there since prehistoric times. A very long-lived evergreen tree that with age becomes decidedly picturesque, it was introduced from the area of its provenance to other lands bordering Mediterranean by the Phoenicians, Greeks, and Romans and since has become enormously important to the agricultural and horticultural economies of the region. Later it was transported further afield and is now extensively cultivated wherever Mediterranean-type climates prevail, in California and the southwestern United States, Mexico, parts of South America, South Africa, and Australia. The tree also succeeds in more humid warm climates such as that of the Gulf states; under such conditions it does not fruit, being of value only as a handsome ornamental. Not only are the fruits of the olive valued for the oil pressed from them, but also for eating green or ripe after pickling.

Olive (*Olea europaea*)

At maturity olive trees are likely to be about 25 feet tall and somewhat wider. In orchard plantings they are often set 25 to 30 feet apart, but 35 to 40 feet is better. With such plantings it is usual to start with young trees, dug bare-rooted from nursery rows, to prune them heavily at planting time, and to whitewash their stems to reflect light and heat and so reduce danger of sunscald. Amateur gardeners are more likely to begin with container-grown specimens, which eliminate the need for such precautions. Larger specimens can be safely transplanted if good balls of earth are kept intact about their roots. In any case generous watering should be done after planting.

Olives adapt to a wide variety of soils so long as they are not waterlogged; they even succeed in ground not fertile enough for many other trees. Their need for nitrogen is fairly minimal. Although they persist where water supplies are meager and are accounted highly drought-resistant, for the most satisfactory cropping they must receive about as much water and as often as needed as citrus fruits and other evergreen trees.

Routine care is not onerous. Besides irrigation, a certain amount of pruning and of thinning of young fruits may be done to improve the size of those that remain and to help to discourage a tendency to bear heavily in alternate years and sparsely between. Except on very infertile soils, fertilizing is not usually required, but as needs vary in different localities it is well to consult the local Cooperative Extension Agent about this. If hand thinning is practiced reduce the load to three or four fruits to each foot of shoot. Thinning can also be done by hormone-type sprays. Ask the advice of the Cooperative Extension Agent about this procedure.

Harvesting for pickling for home use may be done as soon as the fruits begin to change from green to pale yellow or pinkish. Handle the fruits with care to avoid bruising. This can be avoided by dropping them into pails containing water instead of into baskets. Pickling, rather a tedious process, consists of first soaking the fruits in a solution of lye and later in brine. Several changes are necessary. Details are obtainable from State Agricultural Experiment Stations in such states as California where olives are grown commercially.

Varieties most commonly grown include 'Ascolano', 'Manzanillo', 'Mission', and 'Sevillano'. They are most commonly propagated by cuttings, which root with facility if made of pieces of shoot 4 to 5 inches long taken from a little way below their apexes and with all except the uppermost pair of leaves removed. Even larger cuttings of older wood can be rooted. Other methods of securing increase are by taking off and planting suckers that often develop from the bases of trees, and sometimes by grafting.

The chief pests are scale insects. Diseases include one called peacock spot caused by a fungus, for which the control is spraying with bordeaux mixture, verticillium wilt, and a bacterial disease called black knot, which necessitates cutting out affected shoots and being careful not to spread the infection by cutting healthy shoots with tools that have cut diseased ones without first sterilizing them by dipping in alcohol.

OLIVERANTHUS. See Echeveria.

Neoregelia carolinae tricolor

Nephrolepis exaltata in a hanging basket

Nerium oleander variety

Nertera granadensis

Nerium oleander, double-flowered variety

Nidularium innocentii striatum

A hardy nymphea

Nierembergia hippomanica violacea

Nymphoides peltata

A tropical nymphea

OLMEDIELLA (Olmedi-élla)—Guatemalan-Holly. There is only one species of this botantically and horticulturally interesting genus. It has an interesting history, having been cultivated for more than seventy-five years before its provenance was known. Then, in 1932, it was discovered growing wild in the mountains of Guatemala, a finding that aroused considerable interest among botanists because it brought to light a completely new addition to the known flora of the Americas. The only near relatives of *Olmediella* are natives of the Old World. Since the 1932 finding, it has been discovered also in Honduras and Mexico. First mistaken for an Ilex of the holly family AQUIFOLIACEAE, and named accordingly, *Olmediella* belongs instead to the quite different flacourtia family FLACOURTIACEAE. Its name is a diminutive of *Olmedia,* a genus of plants named after the eighteenth-century botanist Vincent de Olmedo.

A handsome broad-topped, evergreen tree that in the wild sometimes attain a height of 90 feet, but in cultivation is commonly a shrub or small tree, ***O. betschlerana*** has handsome, alternate, short-stalked, elliptic leaves up to about 6 inches long by 2½ inches wide. Those on young trees and the lower parts of older ones are coarsely- and sharply-spine-toothed, those higher on sizable trees are much more sparsely-toothed or toothless. The flowers are unisexual and without petals. The males have about fifteen sepals and numerous stamens, the females seven to nine sepals, nonfunctional stamens (staminodes), and a short style with six to eight stigmas. The fruits are spherical berries.

Garden and Landscape Uses and Cultivation. In frostless and nearly frostless climates, such as that of California, *Olmediella* is a good general purpose evergreen. It prospers in sun or part-day shade in well-drained, fertile soil. Propagation is by seed and by cuttings under mist or in a greenhouse propagating bed.

OLNEYA (Olnèy-a)—Desert-Ironwood. The only member of its genus, *Olneya tesota* is endemic to California, Arizona, and Mexico. Belonging in the pea family LEGUMINOSAE, it has a name that commemorates the American botanist, Stephen Thayer Olney, who died in 1878. A broad-headed tree 15 to 25 feet tall, ***O. tesota*** has gray-hairy leaves with four to twelve pairs of leaflets and sometimes a terminal one. The leaves are 1½ to 4 inches, the leaflets ¼ to ¾ inch long. Below each leaf is a pair of stout spines. The flowers are few together in axillary clusters and appear before the new crop of leaves. They are pea-like, pale rose-purple, and under ½ inch in length. The upper or standard petal is deeply notched. There are nine stamens joined and a tenth separate from the others. The upper part of the style is bearded. Glandular-hairy, the slender seed pods are 1 inch to 3 inches long. They contain up to eight black seeds and are usually somewhat constricted between them. The remarkably hard wood of this tree was used for tool handles and arrow heads by the Indians, and its seeds are eaten by them.

Garden and Landscape Uses and Cultivation. In its home territory, and where similar warm, dry climates prevail, the desert-ironwood may occasionally be planted for general landscape purposes. It is difficult to transplant. Young specimens should be raised from seed and grown in containers until they are planted permanently. Well-drained soil is essential.

OMBU is *Phytolacca dioica.*

OMPHALODES (Om-phalòdes) — Navelwort or Navel Seed, Creeping-Forget-Me-Not or Blue-Eyed Mary. Forget-me-not-like annuals and hardy or moderately hardy herbaceous perennials of Europe, Asia, and Mexico make up *Omphalodes,* of the borage family BORAGINACEAE. There are nearly thirty species. The name, inspired by the shape of the nutlets of the fruits, stems from the Greek *omphalos,* a navel, and *oides,* resembling. The name blue-eyed Mary used for one species is also applied to *Collinsia verna.*

Navelworts have hairless or minutely hairy stems and foliage. Their leaves are undivided, the basal ones long-stalked, those on the stems few and alternate. The usually long-stalked, small blue or white flowers are in loose racemes sometimes with a few bracts at their bases. The blooms have a five-parted calyx, a corolla with a very short tube with scales in its throat and five broad lobes (petals), and five stamens. The fruits consist of four nutlets.

The only annual kind ordinarily cultivated, ***O. linifolia*** hails from Spain and Portugal. Erect, up to 1 foot in height, grayish-glaucous, and practically hairless, this has spatula-shaped basal leaves 1 inch to 2 inches long, and stalkless, linear-lanceolate stem ones with a few hairs along their margins. In long racemes, the about ½-inch-wide flowers are white. Those of *O. l. caerulescens* are bluish or pinkish.

Creeping-forget-me-not or blue-eyed Mary (***O. verna***) is a sparsely-hairy European perennial 2 to 8 inches tall that spreads by underground stolons and in mild climates is evergreen. Its short-pointed basal leaves, ovate to ovate-heart-shaped, are 1 inch to 3½ inches long. Similar in shape, the few shorter-stalked stem leaves are smaller. Few together in short racemes or solitary, the ½-inch-wide blooms are blue with white throats, or in *O. v. alba,* white. Slightly taller than the creeping-forget-me-not, and without sto-

Omphalodes verna

lons, perennial ***O. cappadocica,*** of Asia Minor, has creeping rhizomes and erect stems. Densely-hairy, its heart-shaped, prominently veined basal leaves are up to 4 inches long. Similarly shaped, those on the stems are smaller and almost stalkless. In long racemes, the long-stalked, bright blue, white-eyed flowers are ⅓ inch across. Japanese ***O. japonica*** is a hairy perennial with stems 4 to 8 inches tall and oblanceolate to broadly-oblanceolate basal leaves 4 to 6 inches in length that narrow to stalklike bases. The few stem leaves are smaller. The blue flowers, nearly ½ inch across, are in loose racemes. Variety *O. j. albiflora* has white flowers. A taller perennial, ***O. nitida*** (syn. *O. lusitanica*) may attain a height of 2 feet. Its lustrous leaves, the lower stalked, oblong-lanceolate, and 6 to 10 inches in length, are smooth and lustrous above, downy on their lower surfaces. About ⅓ inch wide, the white-centered blue blooms are in long, loose racemes. This is a native of Portugal.

Considered difficult to grow, or at best capricious, ***O. luciliae*** is a hairless native of Greece and Asia Minor. Tufted and from 3 to 6 inches high, it has ovate or oblong basal leaves ½ inch to 1¼ inches in length, and smaller, stalkless ones on its stems. In loose racemes, its flowers open pink and change to a beautiful sky-blue. They are about ½ inch wide. In *O. l. alba* the blooms are white.

Garden and Landscape Uses. Rock gardens afford suitable habitations for all navelworts described here, except that *O. luciliae* may accommodate better, albeit grudgingly, to an alpine greenhouse or cold frame. Its whims are best ministered to by dedicated and experienced growers of alpine plants. Contrariwise, beautiful blue-eyed Mary adapts to lightly shaded places where the soil is of a woodland character, with great ease and spreads willingly, but not obnoxiously. So far as ease of cultivation goes, the other perennial kinds are likely to prove somewhat intermediate between these extremes, with *O. cappadocica* often tricky to satisfy. The annual kind comes freely from seed and not infrequently perpetuates itself by self-distributed seeds.

Cultivation. To have *O. linifolia* all that is necessary is to sow seeds in spring where the plants are to stay and thin out the seedlings to prevent undue crowding. The perennials are also easy to raise from seeds, preferably sown in pots or pans (shallow pots) of light soil in a cool greenhouse or cold frame. Division in early fall or spring is practical with most kinds. Except *O. luciliae,* for which a light, well-drained, moist soil containing lime and exposure to full sun are recommended, these plants respond to woodland-type soils, dampish rather than dry, and dappled shade, or light shadow at least during the heat of the day.

OMPHALOGRAMMA (Omphalo-grámma). So closely related to *Primula* is *Omphalogramma,* of the primrose family PRIMULACEAE, some botanists have included it in that much better-known genus. It differs in its seeds being markedly flattened and winged and in its slightly asymmetrical, solitary blooms being without bracts and having six corolla lobes (petals). The flowers of some species appear before the leaves. Alluding to the peculiar seeds, the name is from the Greek *omphalos,* a navel, and *gramma,* writing. Omphalogrammas are herbaceous perennials with basal rosettes of toothed or toothless leaves. Their primrose-like flowers have calyxes with five to eight lobes. The corollas have slender tubes and wide-spreading lobes. There are five stamens. The fruits are capsules. There are fifteen species, natives from the mountains of western China to the Himalayas.

A variable species, ***O. vinciflorum,*** native to western China, has broad-ovate to oblongish leaves up to 3½ inches long and glandular-hairy on both sides. Its 1-inch-long flowers, on stalks 6 to 9 inches long, come later than the foliage. They are 1¾ inches in diameter, and have violet petals notched at their apexes and darker-colored at their bases. Three stamens are erect and three reflexed.

Garden Uses and Cultivation. Omphalogrammas are notoriously difficult to grow. Their cultivation is to be attempted only by skilled growers of choice alpines. They cannot be expected to live where summers are hot, and elsewhere are primarily plants for pots in alpine greenhouses and cold frames. They need a cool, humid atmosphere, shade, and a porous soil containing an abundance of leaf mold or peat moss, that is kept moist, but not constantly saturated. Monthly applications from spring to late summer of dilute liquid fertilizer are of benefit. Propagation is by seed and division.

ONAGRACEAE—Evening-Primrose Family. Some 640 species allotted among twenty-one genera constitute this family of dicotyledons that occurs natively in many parts of the world, but is especially abundant in the Americas. Its members are chiefly herbaceous perennials, a few shrubs, or trees. Among the herbaceous sorts aquatics are well represented. The usually undivided leaves are alternate, opposite, or in whorls (circles of more than two). The flowers are solitary from the axils of the leaves or in spikes, racemes, or panicles. Most commonly they have four each calyx lobes and petals, but sometimes there are fewer or more, or the petals may be lacking. There are as many or twice as many stamens as petals, one style and one to four stigmas. The fruits are usually capsules, less frequently berries or nuts. Genera cultivated include *Boisduvalia, Circaea, Clarkia, Epilobium, Fuchsia, Gaura, Lopezia, Ludwigia, Oenothera,* and *Zauschneria.*

ONCIDASIA. This is the name of orchid hybrids the parents of which are *Aspasia* and *Oncidium.*

ONCIDENIA. This is the name of orchid hybrids the parents of which are *Macradenia* and *Oncidium.*

ONCIDETTIA. This is the name of orchid hybrids the parents of which are *Oncidium* and *Comparettia.*

ONCIDIODA. This is the name of orchid hybrids the parents of which are *Oncidium* and *Cochlioda.*

ONCIDIUM (Oncíd-ium) — Butterfly Orchid, Dove Orchid. The extensive genus *Oncidium,* of the orchid family ORCHIDACEAE, consists of not fewer than 350 species and numerous handsome hybrids. It occurs natively from southern Florida to the West Indies, Mexico, and through Central and South America to Argentina. Alluding to the crests on the lips of the flowers, its name comes from the Greek *onkos,* a swelling.

Oncidiums belong in the *Brassia, Miltonia,* and *Ondontoglossum* relationship. Most are epiphytes, plants that perch on trees without extracting nourishment from them, but a few inhabit rock cliffs and still fewer are at times terrestrial, growing in the ground. Among the sorts and to a lesser extent within them, variation is considerable. This makes classification and identification a matter of some complexity and the presentation of a short, embracing description of the genus difficult.

Most oncidiums have prominent pseudobulbs, but those of some sorts are much reduced and in a few are absent. Frequently they are borne at well-separated intervals along rhizomes. The pseudobulbs are topped by one, two, or sometimes more leaves that vary in shape and size according to kind. The flowers, solitary or in racemes or panicles frequently of great length, are often showy. Their prevailing colors are yellows and browns with white, green, magenta, and red occurring less frequently. The blooms have three sepals, two lateral petals, and representing a third petal, a three-lobed lip bearing a prominent crest, cushion, or warty development. The sepals and petals are usually sharply constricted into shafts or claws in their lower parts. A useful identifying characteristic is the short, generally winged column swollen below its apex. Besides the species described here others are likely to be grown by specialists and from time to time be offered in dealers' lists. There too, are to be found descriptions of hybrids.

The butterfly orchids *O. papilio* and *O.*

Oncidium papilio

Oncidium varicosum

krameranum are remarkable examples of floral mimicry. Their blooms astonishingly resemble giant butterflies with long, forward-pointing antennae and proboscis, brilliantly colored bodies and wings. Native to northern South America and Trinidad, ***O. papilio*** has tight clusters of compressed to roundish pseudobulbs about 2 inches long each with usually one elliptic-oblong leaf 6 to 9 inches long by 2 inches wide or somewhat wider and mottled, especially on the underside, with reddish-purple. The slender, jointed flowering stalks, flattened above and 2 to 5 feet in length, produce blooms in succession and continuously for many months or even years. The flowers, 4 to 5 inches wide or wider, have a dorsal sepal and petals that are slender, and reddish-brown marked sparingly with yellow; they point upward. These represent the proboscis and antennae of the "butterfly." The pair of wavy, oblong side sepals, not united at their bases, curve downward and are cinnamon-brown with narrow, yellow crossbands. The lip has two small red-spotted, yellow sides lobes and a large fiddle-shaped bright yellow center one edged with reddish-brown and with a red-spotted, white crest. Its apex is shallowly-notched. Except in technical details ***O. krameranum*** closely resembles *O. papilio* and like it produces blooms for several years from the same stalks. The blooms tend to be a little smaller than those of *O. papilio* and the "antennae" and "proboscis" are broader. Also, the lateral sepals are blotched or spotted rather than banded with cinnamon-brown. This is a native of Central America and northern South America.

The golden butterfly orchid (***O. varicosum***), of Brazil, differs from those just discussed in its "butterflies," displayed in great swarms, being much smaller and without long antennae and proboscises. But the flowers are remindful of exquisite butterflies. One of the most beautiful oncidiums, this has clustered, compressed pseudobulbs 2 to 4½ inches in length, each with two or occasionally three rigid, lanceolate-strap-shaped leaves 6 to 9 inches long. The arching, branched flower stalks 3 to 5 feet long carry numerous blooms, 1 inch to 1½ inches in diameter with inconspicuous, reddish-brown-barred, dull sepals and petals and a showy, brilliant yellow lip sometimes with a toothed crest and a patch of reddish-brown. The center lobe of the lip is kidney-shaped and three- or four-lobed, the side lobes are much smaller than the center one and roundish.

Impressive and free-blooming ***O. tigrinum,*** of Mexico, and Brazilian ***O. marshallianum*** have more or less spherical pseudobulbs 3 to 4 inches tall with two or rarely three blunt, linear-oblong leaves up to 1 foot long. The usually upright, loosely-branched, flowering stalks of *O. tigrinum,* 2 to 3 feet tall, have many fragrant flowers 3 inches across with richly brown-blotched, bright yellow sepals and petals and a very large brilliant yellow lip sometimes suffused with brown at its waist, notched at its apex. The flowering stalks of *O. marshallianum,* 3 to 5 feet tall, carry many 2½-inch-wide blooms with brown-banded, greenish-white, concave sepals, and larger brown-dotted golden-yellow petals that at their bases narrow, as does the fiddle-shaped, wavy lip, to claws or shafts. The lip is yellow, spotted on its crest with red. Its big center lobe is notched at the apex, the side lobes are small. Variety *O. m. sulphureum* has clear yellow flowers.

Its numerous small flowers in panicles suggesting a bevy of ballerinas, ***O. flexuosum,*** of Brazil, Paraguay, and Uruguay, has 2-inch-long, flattened pseudobulbs, each with one or two linear-oblong leaves 6 to 9 inches long. The slender-stalked

Oncidium flexuosum

panicles are 2 to 3 feet long; the flowers, ¾ inch to 1¼ inches across, have small yellow sepals and petals decorated with crossbars of cinnamon-brown, the lateral sepals united at their bases, and a sparsely-red-spotted, yellow lip, narrowed to a shaft or claw at its base, with small side lobes and a big kidney-shaped middle one notched at its apex.

The dove orchid (***O. ornithorhynchum***) is very different from any of the sorts so far dealt with. Occurring in the wild from Mexico to Salvador, this has two-leaved,

Oncidium ornithorhynchum major

flattened pseudobulbs 1½ to 2 inches long. The leaves, up to about 9 inches in length, are linear. The slender-stalked, pendulous panicles are of numerous ¾-inch-wide, rosy-lavender blooms with wavy sepals, broader petals, a little fiddle-shaped lip with a warty, yellow crest, a middle lobe notched at its apex, and two much smaller side lobes. The flowers of *O. o. major* are a little larger. The blooms of *O. o. albiflorum*, except for a yellow crest, are white.

Other species and natural hybrids cultivated include these: ***O. altissimum,*** of the West Indies and South America, forms huge clusters of 4-inch-long pseudobulbs with one or two strap-shaped leaves up to

Oncidium altissimum

1 foot in length. Its flowers are in panicles, sometimes over 6 feet long, with short branches bearing 1- to 5-inch-wide blooms. They have greenish-yellow-edged sepals and petals barred and blotched with brown and a lemon-yellow lip, notched at its apex. ***O. ampliatum,*** of Central and

Oncidium ampliatum

South America and Trinidad, has brown-mottled pseudobulbs with one or two leaves up to a little over 1 foot long by 3½ inches wide and arching flowering stalks 2 to 4 feet long. The numerous 1½-inch-wide blooms have small yellow sepals blotched with red-brown, and yellow petals, white on their backs. *O. a. majus* has blooms 2 inches across. ***O. ansiferum*** (syn. *O. lankesteri*) of Central America, has much-flattened pseudobulbs 2½ to 5 inches tall, with one or two oblong-elliptic

Oncidium ansiferum

leaves up to over 1 foot long and 2 inches broad. The panicles, up to 5 feet long, are of 1-inch-wide blooms with heavily brown-spotted sepals and petals margined and tipped with yellow and a bright yellow and fiddle-shaped lip with a patch of chocolate-brown and notched at its apex.

O. anthocrene (syn. *O. powellii*), of Colombia and Panama, has somewhat widely spaced, flattened, oblong-ovoid pseudobulbs up to 6 inches long by 2 inches wide. They have solitary, oblong-lanceolate to strap-shaped leaves that may exceed 1 foot in length and be 2 inches broad. Arching and branchless or sometimes branched, the flowering stalks have many waxy blooms 1¼ to 2½ inches wide with cinnamon- to chocolate-brown sepals and petals margined and cross-banded with yellow. The fiddle-shaped lip has a center lobe of bright yellow with a chestnut-brown waist and small, generally brown-spotted, yellow side lobes. ***O. barbatum*** is Brazilian. It has one-leaved pseudobulbs 2 to 4 inches long. The leaves, up to 4 inches in length, are linear to ovate-oblong. Sparingly-branched, and 1 foot to 2 feet in length, the flowering stalks carry a few blooms 1 inch to 1½ inches across. They have yellow sepals and petals with chestnut-brown blotches and a yellow lip, its crest spotted with red. ***O. baueri,*** of the West Indies and Central and South America, has pseudobulbs 3 to 7 inches long with usually paired, strap- to sword-shaped leaves 1 foot to 3½ feet in length, 1 inch to 2½ inches wide. From 5 to 9 feet long, the freely-branched, arching to pendulous panicles are of blooms up to 1 inch wide. Their sepals and petals are yellow to greenish-yellow spotted and barred with brown. The yellow lip has a brown shaft or claw and a whitish crest. ***O. carthagenense*** inhabits Central America, northern South America, and the West Indies. Its pseudobulbs are so small they are barely recognizable as such. Each has a solitary elliptic leaf up to about 1½ feet long and pinkish-purple-spotted white blooms in panicles up to 5 feet long by 1 inch wide. ***O. cavendishianum,*** of Mexico and Guatemala, is without obvious pseudobulbs. It has solitary, oblong-elliptic leaves 1 foot to

Oncidium cavendishianum

2 feet long by up to 8 inches broad and many-flowered panicles 2 to 3 feet tall or taller of fragrant, 1½-inch-wide blooms with greenish-yellow to yellow sepals and petals heavily blotched and spotted with cinnamon-brown. The lips are bright yellow. ***O. cheirophorum,*** of Costa Rica and Panama, has roundish, compressed pseudobulbs about 1 inch long, each with a strap-shaped leaf up to 6 inches long. Its

Oncidium cheirophorum

½- to ¾-inch-wide, fragrant flowers, in rather crowded, narrow panicles longer than the leaves, have similar bright yellow, rounded sepals and petals and a three-lobed lip with a whitish crest. ***O. concolor*** is Brazilian. It has clustered pseudobulbs up to 2 inches long with two or sometimes three about 6-inch-long, lanceolate-strap-shaped leaves. The fifteen or fewer blooms of the 1-foot-long pendulous racemes are 1½ to 2 inches across and bright yellow. ***O. cornigerum,*** of Brazil and Paraguay, has 2- to 4-inch-tall pseudobulbs with solitary, elliptic-oblong leaves up to 6 inches long or longer. In many-flowered, arching or drooping panicles up to about 2 feet in length, the ¾-inch-wide blooms have golden-yellow sepals and petals spotted and barred with dark red. Their lips are bright yellow. ***O. crispum*** is a Brazilian with clustered psuedobulbs 3 to 4 inches long, each with two or three oblong-lanceolate leaves 6 to 8 inches long by 1 inch to 2 inches wide. The blooms, in erect or arching, many-flowered panicles 3 to 4 feet long are 2 to 3 inches in diameter. Their very wavy, much crisped sepals, petals, and lips are rich chestnut-brown, sometimes edged and spotted with yellow. ***O. divaricatum*** is Brazilian. It has much-flattened pseudobulbs 1½ inches tall, with solitary leaves up to 1 foot long by 2 inches wide. The freely-branched panicles, 4 to 6 feet tall, are of flowers usually under 1 inch wide with chestnut-brown sepals and petals with a yellow blotch near the tip of each and a yellow lip spotted with the same brown. ***O. excavatum,*** of Ecuador and Peru, has clusters of flattened pseudobulbs 3 to 5 inches long with usually one linear-strap-shaped leaf up to 2 feet long. The 1½-inch-wide blooms in many-flowered pyramidal panicles 2 to 3 feet long have very wavy sepals and petals, the latter much the bigger. The sepals are yellow with two or three cross-bands of brown, the petals yellow with one or two basal spots of reddish-brown. The lip is bright canary-yellow with small red-brown side lobes. ***O. falcipetalum,*** of Venezuela, has clusters of 3- to 4-inch-long pseudobulbs, each with usually two strap-shaped to somewhat lanceolate leaves up to 1¼ feet long by 2 inches wide. Its irregularly-branched, flexible-stalked panicles, up to 10 feet long, have many 2½- to 3-inch-wide blooms with yellow-edged, reddish-brown sepals, much smaller sickle-shaped, yellow petals spotted with brown toward their bases, and a purplish-brown, reflexed lip. ***O. forbesii,*** of Brazil, has clusters of compressed pseudobulbs 2 to 3 inches tall with one or more often two oblong-strap-shaped leaves up to 1 foot long. Up to 3 feet in length, the erect or arching, many-flowered panicles are of blooms 2½ inches in diameter. Their sepals, crisped-margined petals, and the center lobe of the lip are bright chestnut-brown narrowly edged with yellow and the side lobes of the lip are generally bright yellow. ***O. gardneri,*** a Brazilian with clustered, compressed pseudobulbs 2 to 3 inches tall, has from each a pair of linear to oblong-lanceolate leaves some 6 inches to 1 foot long. The many-flowered panicles, 2 to 3 feet in length, are of 2-inch-wide blooms. Their sepals and much bigger petals are brown, the former with cross-bands of yellow, the latter with yellow margins. The lip has small yellow side lobes marked at their bases with red-brown and a big yellow center lobe with a marginal band of red-brown spots. ***O. haematochilum,*** a natural hybrid between *O. lanceanum* and *O. luridum* that occurs in Colombia and Trinidad, is without evident pseudobulbs. It has leathery leaves 9 inches to 1 foot long. Similar in its vegetative aspect to *O. lanceanum,* it has many-flowered, arching to erect, paniculate inflorescences up to 2 feet long or longer. Its very fragrant, long-lasting blooms, about 2 inches in diameter, have yellowish-green sepals and petals generously blotched with brown, and a spreading, crimson and rose-pink lip edged with yellow. ***O. incurvum*** is a native of Mexico. It has compressed, ovoid pseudobulbs 3 to 4 inches long with two or three narrowly-strap-shaped leaves 1 foot to 1½ feet long. Erect or arching and 3 to 5 feet long, the panicles are of lilac-pink to rose-pink, fragrant flowers spotted and blotched with white and 1 inch to 1½ inches wide. ***O. isthmi,*** of Central America, much resembles *O. baueri,* differing

Oncidium haematochilum

most noticeably in its blooms having bigger lips with more constricted waists and its pseudobulbs, leaves, and flowering stalks being shorter than the maximums attained by that species. ***O. jonesianum,*** of southern Brazil, Paraguay, and Uruguay, has practically obsolete pseudobulbs, each with a solitary, cylindrical, pendulous leaf up to 1¼ feet long, by under ½ inch wide. The racemes, up to 2 feet long, are of from five to fifteen handsome blooms 2 to 3 inches wide. They have whitish- or greenish-yellow sepals and petals spotted with chestnut-brown and a lip with a white center lobe spotted near its base with red and very small yellow and red side lobes. ***O. lanceanum*** inhabits northwestern South America and Trinidad. It has stout rhizomes and practically obsolete one-leaved pseudobulbs. The leaves, erect and elliptic-oblong, are spotted with purple and up to 1½ feet long by 5 inches wide. In racemes or panicles 1 foot to 1½ feet tall, the flowers are few to many. Fragrant and 2 to 2½ inches across, they are somewhat variable in color. The sepals and petals are yellow to yellow-green, sparingly- to densely-spotted with chocolate-brown. The very large lip is rich magenta to rose-purple or whitish. ***O. leucochilum,*** of Mexico, Guatemala, and Honduras, has more or less flattened, oblong-ovoid pseudobulbs 3 to 5 inches tall with usually a pair of strap-shaped leaves 9 inches to 2 feet long. The much-branched flowering stalks, up to 7 feet long, carry many blooms 1¼ inches across or often smaller. The sepals and petals are usually light yellowish-green with cross-bands of greenish- to purple-brown. Generally the lip is white with a magenta-pink crest. Its margin, notched at the apex, is sometimes wavy. ***O. longifolium*** (syn. *O. cebolleta*), indigenous from Mexico to South America and the West Indies, has stout rhizomes and very small, one-leaved pseudobulbs. The often reddish or red-spotted leaves, very slender and nearly cylindrical, taper to a sharp point. They are 6 inches to 2 feet long. Carrying few to many 1- to 1½-inch-wide blooms, the branched or branchless flowering stalks are up to 2½ feet long. The blooms, 1 inch to 1½ inches across, have yellow petals spotted with reddish-brown and a comparatively large yellow lip. ***O. longipes*** is Brazilian. It has clusters of 1-inch-long pseudobulbs, each with usually two linear-oblong leaves up to 6 inches long by up to ½ inch wide. The zigzagged flowering stalk, up to 6 feet in length, bears many flowers 1 inch to 1½ inches in diameter. They have light red-brown wavy sepals and petals tipped and streaked with yellow, and a canary-yellow lip often toward its base suffused with brown. ***O. luridum,*** of southern Florida, the West Indies, and Mexico to South America, like *O. carthagenense,* to which it is very closely related, is almost without pseudobulbs. From robust rhizomes it

Oncidium lanceanum

Oncidium lanceanum (flowers)

Oncidium leucochilum

Oncidium longifolium

Oncidium macranthum

Oncidium maculatum

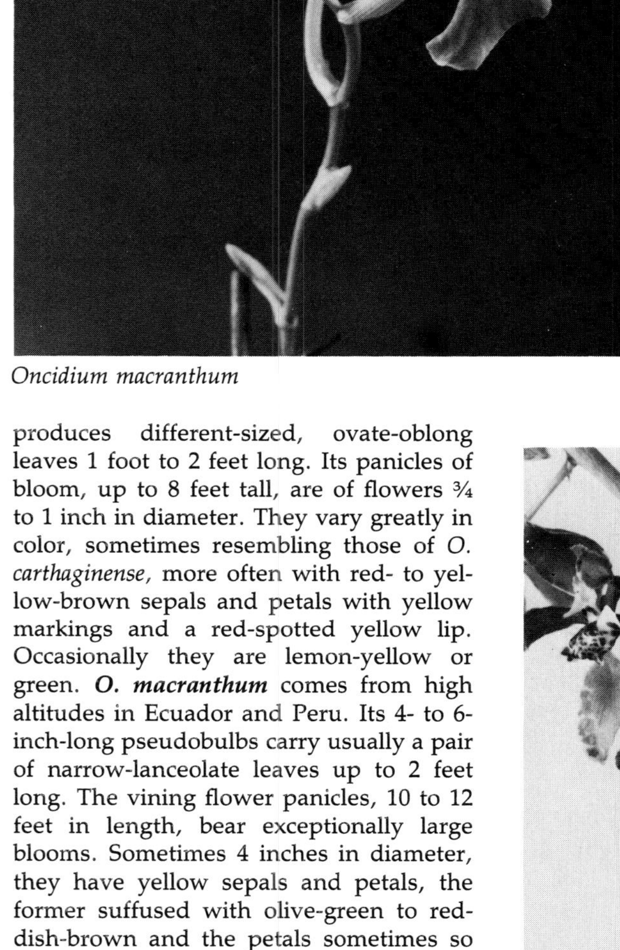

produces different-sized, ovate-oblong leaves 1 foot to 2 feet long. Its panicles of bloom, up to 8 feet tall, are of flowers ¾ to 1 inch in diameter. They vary greatly in color, sometimes resembling those of *O. carthaginense*, more often with red- to yellow-brown sepals and petals with yellow markings and a red-spotted yellow lip. Occasionally they are lemon-yellow or green. ***O. macranthum*** comes from high altitudes in Ecuador and Peru. Its 4- to 6-inch-long pseudobulbs carry usually a pair of narrow-lanceolate leaves up to 2 feet long. The vining flower panicles, 10 to 12 feet in length, bear exceptionally large blooms. Sometimes 4 inches in diameter, they have yellow sepals and petals, the former suffused with olive-green to reddish-brown and the petals sometimes so flushed. The small yellow lip has a white middle lobe and violet-purple side ones. ***O. maculatum*** of Mexico has flattened, ribbed pseudobulbs 3 to 4 inches long with two narrow leaves 7 to 10 inches long. In arching, sometimes branched racemes 1 foot to 1½ feet in length, the 2 inch-wide flowers have light green, yellowish, or yellow sepals and petals conspicuously spotted or blotched with brown. The pointed, oblong-ovate lip is cream or white and yellow with pinkish or red veinings. ***O. microchilum,*** of Mexico and Guatemala, has clusters of more or less flattened, ovoid to nearly round pseudobulbs, 1½ to 2 inches tall, with solitary, oblong to elliptic-oblong leaves up to 1 foot long by 3 inches wide. Its stout flowering stalks, up to 4 feet tall and usually with numerous short branches, are of blooms ¾ to 1 inch in diameter with brown sepals marked with yellow and narrower, chestnut- to purplish-brown

Oncidium microchilum

petals edged and banded with yellow. The lips have rounded, white side lobes with purple spots at their bases and a small white one representing the center lobe. ***O. onustum*** of Panama has small, ovoid pseudobulbs and short, oblong-elliptic leaves. Its flowers, about 1 inch in diameter, are golden yellow with darker lines on the lip. They are displayed in arching, rather one-sided racemes about 1 foot long. ***O. panamense,*** endemic to Panama,

Oncidium onustum

has flattened-ovoid pseudobulbs up to 6 inches long by more than 2 inches wide with leafy bracts below, and each with a pair of linear-lanceolate leaves as much as 2½ feet in length and 1½ inches in width. The flowering stalks, up to 12 feet long and erect, arching, or pendulous, are of ¾-inch-wide blooms with yellow sepals and petals heavily blotched and banded with olive-brown. The yellow, fiddle-shaped lip is marked with a patch of yellowish-brown below its crest. ***O. phymatochilum,*** native from Brazil perhaps north to Mexico, has one-leaved, clustered pseudobulbs. The leaves, narrow-elliptic-oblong and pointed, are up to a little over 1 foot long by 1½ inches broad. The slender panicles, 3 to 6 feet in length, are of many 2-inch-wide blooms with more or less reflexed sepals and petals, that are light yellow with spots and bands of brown or less frequently are ivory-white with orange-red spots. The lip is shorter and is white spotted with red around the yellow crest at its base. ***O. pubes,*** of Brazil and Paraguay, has subcylindrical, tapered

Oncidium pumilum

Oncidium pusillum

pseudobulbs 2 to 2½ inches long with usually two, sometimes one, narrowly-oblanceolate, pointed leaves 3 to 5 inches in length. Many together in somewhat crowded racemes and variable in color, the blooms are generally under 1 inch wide. Most often the sepals and petals are reddish-brown spotted and cross-banded with yellow, and the lip has a yellow-edged, red-brown to purplish center lobe and slender, yellow, reflexed side lobes. ***O. pulchellum,*** of the West Indies and the Guianas, is without pseudobulbs. Generally in twos or fours and forming two ranks, the linear-lanceolate, keeled leaves are up to about 6 inches long. Up to twenty or more blooms, rather variable in size and coloring, are borne on flowering stalks up to 1¼ feet long. White, lightly or deeply suffused with pink, they generally have an ochre-yellow spot in front of the crest on the large, squarish, four-lobed lip that hides the united lateral sepals. ***O. pulvinatum,*** of Brazil, Paraguay, and Argentina, has roundish-oblongish pseudobulbs 1½ to 2 inches tall with usually a solitary, erect leaf up to 1 foot long by up to 3½ inches wide. The slender-stalked, arching to erect, loose panicles, 4 to 9 feet long, are of many nearly 1-inch-wide flowers with similar yellow or yellowish sepals and petals with reddish-brown bases. The lip, pale yellow dotted with red, has a red-spotted, white crest. Its side lobes are fringed. ***O. pumilum,*** Brazilian, has tiny one-leaved pseudobulbs, the leaves elliptic to ovate and 3 to 4 inches long. In many-flowered, erect or arching, spire-shaped panicles, the flowers, about ⅓ inch wide, have similar brownish-red-spotted, yellow, spatula-shaped sepals and petals. The kidney-shaped, brownish-yellow lip is three-lobed. ***O. pusillum,*** native from Mexico to Bolivia, Brazil, and Trinidad, is a miniature without pseudobulbs. Its sword- to sickle-shaped leaves, 1 inch to 2½ inches long, form iris-like fans from which come in season up to half a dozen blooms about ¾ inch across. They have bright yellow sepals, petals of the same color cross-banded with brown, and a fiddle-shaped, brown-blotched, bright yellow lip with a white disk spotted with reddish-orange. ***O. raniferum,*** a Brazilian species, has clustered, oblong-conical, flattened pseudobulbs 1 inch to 2 inches high with a pair of slender-linear leaves 5 to 8 inches long. The few-branched, many-flowered panicles attain lengths of up to 10 feet. The blooms, somewhat more than ½ inch across, are bright yellow with a prominent orange-red crest on the lip. Their sepals and petals are reflexed. The specific name of this species alludes to the form of the crest being remindful of a sitting frog. ***O. sarcodes,*** of Brazil, has rather club-shaped, somewhat flattened pseudobulbs with two or occasionally three strap-shaped leaves up to 1 foot long by 2 inches wide. The panicles, slender-stalked and 3 to 6 feet long, are of many blooms 1½ to 2 inches wide. They have rich chestnut-brown sepals, petals edged and otherwise marked with yellow, and a bright yellow lip with a few red-brown spots bordering the crest. ***O. sphacelatum,*** of Mexico to Central and northern South America, has stout rhizomes with more or less widely spaced, flattened-oblongish pseudobulbs 4 to 7 inches long by up to 3 inches wide. Each bears two or three strap-shaped leaves that may exceed 2 feet in length. On numerous short branches, the flowers about 1 inch across are in many-bloomed panicles up to 5 feet long. They have brown-dotted, yellow sepals and petals and a somewhat fiddle-shaped, golden-yellow lip banded with red-brown in front of the crest. In the bud stage the tips of the sepals and petals recurve like small horns. ***O. sphegiferum,*** of Brazil, has broadly-egg-

Oncidium sarcodes

Oncidium sphacelatum

Oncidium splendidum

shaped to nearly round pseudobulbs up to about 1½ inches long and solitary, erect, pointed-elliptic-oblong leaves up to 8 inches in length by nearly 1¾ inches broad. From 3 to 4½ feet long, the many-flowered panicles are of nearly 1-inch-wide, bright orange blooms stained at the bases of the sepals, petals, and usually paler lip with reddish-brown. ***O. splendidum*** (syn. *O. tigrinum splendidum*), of Guatemala, has flattened, roundish pseudobulbs 1 inch to 2 inches long with one rigid, fleshy, ovate leaf. In loosely-branched panicles 2 to 3 feet long, the gracefully disposed, butterfly-like flowers 1½ to 2 inches wide, have similar yellow sepals and petals heavily barred and blotched with reddish-brown and a creamy-yellow lip with a large, nearly circular center lobe and two much smaller, linear side lobes. ***O. stipitatum*** is a Central American with scarcely evident pseudobulbs mostly not over ¼ inch long and flat-topped. Its solitary, pointed, slender-cylindrical leaves, 9 inches to 2 feet in length, are at first erect to spreading, later drooping. The horizontal to upright, many-bloomed flowering stalks approximate the leaves in length. Generally ¾ to 1 inch wide, the blooms have yellow sepals and petals generously marked with reddish-brown and a fiddle-shaped, bright yellow lip with some reddish-brown markings and more or less wavy margins. Its slender side lobes are often more or less erect. ***O. tetrapetalum,*** of Colombia, Venezuela, and the West Indies, is without pseudobulbs. It has creeping rhizomes and leaves in fanlike tufts of three to five. Erect, three-angled, and up to 6 inches long, they have a channel along one side. Up to three feet in length, the flowering stalks are branchless or sparingly-branched. They have many about 1-inch-wide blooms with chestnut-brown sepals and petals cross-banded with yellow and a white lip with a reddish-brown blotch in front of the crest on the kidney-shaped, notched center lobe and a pair of hornlike side lobes. ***O. triquetrum,*** of Jamaica, is without pseudobulbs. In tufts of three or more, its three-angled leaves are channeled along one side, up to 5 inches long, but often shorter. The slender flowering stalks, mostly under 7 inches in length, have fifteen or fewer ½-inch-wide blooms with purplish-green sepals and wavy, white petals tinted with pale green and spotted with purple. The lip, heart-shaped and with a small orange-yellow crest, is white-streaked and spotted with reddish-purple. ***O. variegatum,*** of the West Indies, has long rhizomes and scarcely evident pseudobulbs. Spaced at intervals along the rhizomes are fans of rigid, recurved, lanceolate, toothed leaves 1 inch to 3 inches long. Its rarely-branched flowering stalks, up to 1 foot tall, have few to many about 1-inch-wide blooms of various coloring. Most commonly they are white to greenish-white suffused with brown or red-purple, and the fiddle-shaped, toothed lip has some yellow spots and a yellow crest. Its side lobes are reflexed. ***O. wentworthianum*** is native to Mexico and Guatemala. It has clusters of ovoid-oblong, slightly-flattened pseudobulbs, up to 5 inches long, each with one or more commonly two strap-shaped leaves up to 1 foot long. The distantly-branched flowering stalks, up to about 10 feet long, have blooms approximating 1 inch wide. Their wavy-edged sepals and petals are yellow, blotched except near their tips with reddish-brown. The lip is yellow with the crest margined with a zone of brown spots.

Garden Uses. Many splendid orchids well adapted for cultivation by amateurs as well as experts are found among oncidiums. They please by the grace of their blooms and their cheerful colors as well as by their profusion. The nearly continuous blooming of many kinds and the often long-lasting quality of their flowers, both on the plants and when cut, are additional merits. For floral decorations many oncidiums are superb.

Cultivation. As is to be expected in such a large and diverse genus, the cultural requirements of oncidiums differ considerably and allowances must be made for this when caring for sorts with different needs. A great many succeed in pots under conditions that suit cattleyas, potted in osmunda or tree fern fiber packed firmly although rather less so than for cattleyas. Some growers mix sphagnum moss with the fiber. Sorts with pendulous stems are best fitted for growing in suspended baskets or rafts, still others for attaching to slabs of tree fern trunks. Let the habit of growth suggest the most suitable accommodation for individual kinds.

Oncidiums need a resting period, usually several weeks, following the completion of growth of the new pseudobulbs or stems. Then the roots must be kept dryish. At other times water freely. Like most orchids these resent a soured, stale condition of the rooting material. To guard against this provide good drainage in the pots and repot if the fiber shows signs of staleness. Temperature requirements must be adjusted to the needs of various kinds. For additional information see Orchids.

ONCIDPILIA. This is the name of orchid hybrids the parents of which are *Oncidium* and *Trichopilia.*

ONCOBA (Ón-coba). As here understood, including sorts sometimes segregated as *Xylotheca,* the genus *Oncoba,* of the flacourtia family FLACOURTIACEAE, comprises thirty-nine species of tropical and subtropical, African and Arabian shrubs and trees. The name is derived from the Arabian vernacular *onkob,* applied to *O. spinosa,* the dried fruits of which are used as ankle rattles by African dancers.

Oncobas have alternate, undivided leaves and white, yellow, or reddish, clustered or solitary, unisexual and bisexual blooms, both on the same plant. The flowers have three to five sepals, five to ten petals, many stamens, one style, and a disklike stigma. The fruits are leathery, berry-like capsules.

A shrub or tree up to about 20 feet tall, ***O. spinosa*** is a native of tropical Africa. It has lustrous, dark green, pointed-elliptic, toothed leaves up to 4 inches long. About 3 inches in diameter and with a central cluster of bright yellow stamens, its blooms somewhat resemble those of *Stewartia* and *Franklinia.* Solitary or in pairs, they are white, fragrant, and are produced even by young plants. Individual blooms remain attractive for several days. They first appear in spring and open in succession over a long period. From the last, closely similar tropical African ***O. routledgei*** differs most obviously in its flowers being only about 2 inches in diameter.

Sometimes called African-dog-rose, South African ***O. kraussiana*** (syn. *Xylotheca kraussiana*) is a rounded, evergreen shrub or sometimes a tree up to 20 feet in height. It has lustrous, short-stalked,

Oncoba kraussiana

broad-elliptic to elliptic-oblong, toothless leaves 2 to 3½ inches long by about one-third as wide that, when young, like the young shoots are hairy. The slightly fragrant flowers, solitary or in twos or threes from the leaf axils or on the branches, resemble miniature flowers of *Magnolia stellata.* They are 2 to 3 inches in diameter and have a central cluster of numerous stamens. The style ends in four stigmas. Variety *O. k. glabrifolia* differs from the typical species in having its shoots and young leaves hairless.

Garden and Landscape Uses and Cultivation. The species described are satisfactory for general landscape purposes. They are content with ordinary soil and sunny locations, generally responding to environments that suit gardenias. In California, *O. spinosa* is reported to have survived temperatures of 20°F. Propagation is by seed and summer cuttings.

ONE-SIDED-BOTTLE-BRUSH. See Calothamnus.

ONION, SEA- is *Urginea maritima.*

ONIONS. Besides the common onion there are others worth cultivating, all esteemed for their alliaceous flavor and healthful qualities as foods. Varieties of the common sort are derivatives of *Allium cepa,* a native of western Asia.

Onions of this most familar type are biennials that in their second full season of growth flower, produce seeds, and die. But unless being grown for seeds, the crop is harvested well before this occurs, as mature bulbs late in their first full growing season or as scallions earlier.

Seeds are the only means of increase. These are sown early in the year the crop is to be harvested, or late in the previous year to give tiny bulbs called sets that are stored in dormant condition indoors over winter and planted outdoors in spring. Sets are sold commercially and provide an easy way for home gardeners to raise scallions and onions, but bulbs raised in this way do not keep as long after harvesting as do onions grown from seeds and kept growing until they mature.

Soil for onions needs to be well-drained and fertile. Exposure to full sun is essential. Spade, plow, or rotary till the ground deeply and mix into it generous quantities of partially rotted manure, compost, or other suitable organic material. If the soil is acid lime it to bring it to near neutral (pH 6 to 6.5). Supplement this with a dressing of complete fertilizer forked or harrowed into the upper 3 or 4 inches a week or two before seeding or planting. If easily available, incorporate at the same time a heavy sprinkling of wood ashes.

The physical condition of the ground is of special importance in growing onions. See that it is thoroughly tilled, large stones removed, lumps broken. Work the earth into a pleasant crumbly condition. If cloddy or sticky it will not suit onions. Should it be clayey, and excellent crops can be grown on such soils as well as sandier ones, do not attempt to work it when wet. Onions like a firm root run, therefore before sowing or planting compact the soil moderately by treading (tramping) across it or by rolling, but do this only when it is so dry that it does not stick to shoes or roller. Immediately before sowing or planting, rake to a fine, level surface.

Sow seeds outdoors in drills 1 foot to 1¼ feet apart and about ½ inch deep as early in spring as the soil can be brought into suitable condition, of if extra large onions are wanted for a flower show or other special purpose sow in a greenhouse in January and grow the young plants in flats or pots until it is safe to plant them outside soon after the last frost. In mild climates sowing outdoors in October for spring or summer harvesting is practicable.

Thin out seedlings from outdoor sowings as soon as they are 2 to 3 inches tall, but do not remove all the surplus ones at one time. At first take out only enough to allow for a little growth and complete the thinning in one or two more operations, using the thinnings for eating as scallions. Final spacing should depend upon the fertility of the ground and the size of the particular variety; from 2 to 5 inches is usually appropriate. It is of great importance when thinning onions not to disturb the plants that remain, or if this is inadvertently done to push back the loosened soil and make it firm about their bases.

Onion sets, purchased or raised the previous year, may be planted at the first opportunity spring affords, as soon as the first peas can be sown. To plant, make drills about ½ inch deep and 1 foot to 1¼ feet apart, and at intervals of 1 inch to 2 inches push the sets into the bottom of the drills. Do not cover with soil. When they attain suitable size and before they crowd unduly thin them out as recommended for seedlings, using the thinnings as scallions. To raise onion sets for planting the following year, sow seeds in spring in wide drills in soil on the lean rather than fertile side and as free as possible of weeds. Sow about ten times as thickly as for normal crop production. Harvest in July and dry and store as with crop onions. Onions

Fertile soil and good cultural care are essential for the production of prize-quality onions

Newly planted onion sets

Pushing the tops of onions over encourages the bulbs to ripen

A crop of onions nearing maturity

started indoors and planted outside will usually be of large exhibition varieties that benefit from spacing 6 to 9 inches apart.

Routine care during the growing season calls for strict attention to controlling weeds. Never allow these to acquire much size. Frequent shallow stirring between the rows with a scuffle hoe or other cultivator, coupled with hand-pulling in the rows, achieves this. Occasional side dressings of fertilizer given until the foliage begins to ripen and die and soaking with water at weekly to ten-day intervals during dry spells promote growth.

Onion bulbs increase most rapidly in size toward the end of their growing season, when the food manufactured in the leaves is being stored in the bulbs at the fastest rate. For this reason and because onions harvested prematurely will not store well, be sure to allow those to be kept for winter use to ripen thoroughly before storing.

In August when growth has nearly finished and the leaves have well begun to die back from their tips and to fall down, push them over with the back of a rake so that they lie along the ground. It is a convenience if the tops of alternate rows are directed toward each so that every other row is clear of foliage. This, by exposing the bulbs to all possible sun, promotes ripening.

Harvest the crop as soon as the foliage has completely died by lifting the bulbs and spreading them in single layers on a dry surface where free air circulation is assured, outdoors if the weather is dry, or under cover.

Trays stood on top of each other and propped to allow air space between facilitate moving the crop when drying is completed and makes it possible to protect the bulbs by covering them with plastic sheeting should rain come during the few days they are drying. When quite dry remove the roots and, unless they are to be braided, the tops also, and store the bulbs in a cool, but frost-proof, dry place on slat-bottomed shelves or trays, in net bags, or in ropes made by braiding the tops, suspended from the ceiling.

After harvesting, onions to be stored benefit from being dried in the open for a few days

Varieties of common onion are fairly numerous and selections should be based on suitability for the purposes for which they are required: flavor (mild or pungent), sizes at maturity, and if they are to be stored, keeping qualities. For pickling 'White Portugal' or 'Silverskin' is a favorite. One of the best for keeping long in storage is 'Minnesota Globe'. Other commonly grown varieties include 'Southport White Globe', 'Southport Yellow Globe', 'Southport Red Globe', 'White Bermuda', and 'Yellow Bermuda'. The variety 'Ebenezer' is mostly grown from sets. In addition to these and other varieties listed and described in seedsmen's catalogs, there are a few called bunching onions, which are grown from seed almost exclusively for use as scallions. Among these are 'Beltsville Bunching', and 'White Bunching'. The rather similar 'Japanese Bunching' is a variety of *A. fistulosum.*

The top or Egyptian tree onion (*Allium cepa viviparum*) is a hardy perennial that instead of blooms bears at the tops of its flowering stalks clusters of small onions chiefly suitable for pickling. Plant full-sized bulbs in spring 5 to 6 inches apart in rows 1 foot to 1½ feet asunder. If small bulbs are planted they probably will not crop the first season.

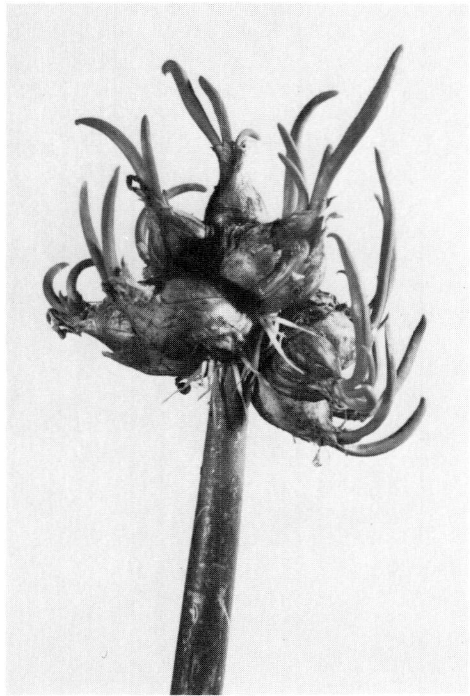

The top or Egyptian tree onion

The multiplier or potato onion, a variety of *A. cepa aggregatum,* has multiple-part bulbs that if planted without splitting into its parts develops several scallion-like clusters of leaves, which may be harvested at any stage and used as a salad and for flavoring. However, it is better to separate the bulbs at planting time into their constituent parts and plant these, merely pressing them into the surface of fertile, well-drained ground without covering them with soil at 1 inch to 2 inches apart in rows about 1 foot apart. By thinning out until the plants that remain stand about 6 inches apart, using excess plants as scallions, adequate space is assured for those left to mature, and some of these may be held over for planting the following season.

Shallots, although the name is sometimes used for any onion grown to be harvested and eaten while young and immature as a "bunching onion" or scallion, is most correctly reserved as the name of a particular variety, probably of Asian origin, of *A. cepa aggregatum*. A hardy perennial, this rarely produces seeds. It is propagated in the same way as the multiplier onion by planting single bulbs. By midsummer or a little later each will have formed a cluster of five or six or more new bulbs, which as soon as the foliage dies may be harvested and the clusters separated into single bulbs and stored. But in the United States it is usual to pull all except sufficient to provide bulbs for planting the next year and use them as scallions. In Europe they are more often left to mature before they are harvested. Shallots are excellent for pickling in vinegar.

Welsh onion in bloom

Shallots

The Welsh onion is of different origin from other sorts discussed here. It is a variety of *A. fistulosum*, a native of temperate Asia. A very leafy hardy perennial without obvious bulbs, the Welsh onion is esteemed for its leaves, used chiefly for seasoning.

Pests and Diseases. The chief pests are the onion maggot, which burrows into the bulbs, and thrips, which cause a grayish to yellowish discoloration of the leaves. The chief diseases are smut, caused by a fungus that lives over in the soil and which attacks young seedlings, and downy mildew, which in cool, damp weather causes the foliage to become yellow and die. Refrain from planting onions in ground where onion smut has occurred during the last two or three years. For controls of the other troubles consult the local Cooperative Extension Agent or your State Agricultural Experiment Station.

ONOARA. This is the name of orchid hybrids the parents of which include *Ascocentrum, Renanthera, Vanda,* and *Vandopsis.*

ONOBRYCHIS (Onó-brychis) — Sainfoin. Of the 120 species accounted for in this genus of the pea family LEGUMINOSAE, very few have horticultural appeal and none is much grown for ornament. Sainfoin is widely cultivated for fodder. Native to Europe, North Africa, and western Asia, *Onobrychis* bears a classical name, presumably derived from the Greek *onos,* an ass, and *brycho,* to eat greedily and of obvious significance.

To *Onobrychis* belong annuals, herbaceous perennials, and spiny subshrubs. They have pinnate leaves with an odd number of leaflets and axillary, stalked racemes or spikes of pea-like blooms. The flowers, white, pink, purple, or rarely yellow, have obovate or reverse-heart-shaped standard or banner petals, short wing petals, and blunt keels as long or longer than the standards. The bell-shaped calyxes have five nearly equal teeth. There are ten stamens, one free, the others joined. The fruits are stalkless, flattish, few-seeded pods.

Sainfoin, saintfoin, or holy-clover (***O. viciaefolia***) is a more or less hairy perennial 1 foot to 2 feet in height with sparingly-branched stems and leaves with eleven to twenty-nine usually ovate to oblong leaflets ½ inch to 1½ inches long. The flowers, pale pink with purplish veins, are up to ½ inch in length. They are in crowded racemes from the upper leaf axils. The seed pods are approximately ¼ inch in length and have toothed margins. Probably originally native of central Europe, sainfoin now occurs beyond its original range. It is naturalized to some extent in North America.

Two annuals, *O. crista-galli* and *O. caput-galli,* are occasionally cultivated. A native of northern Asia, ***O. crista-galli,*** approximately 1 foot tall, has pink flowers about ¼ inch long on stalks not longer than the leaves. Its toothed seed pods are ¼ to ½ inch in length. Native to the Mediterranean region, ***O. caput-galli,*** up to 3 feet in height, has leaves with nine to fifteen obovate to linear leaflets and reddish-purple flowers ⅓ inch long in clusters on stalks as long as the leaves. The seed pods, with long teeth at their edges, are ¼ inch long or a little longer.

Garden Uses and Cultivation. These are plants for semiwild areas. Sainfoin succeeds in dryish soils and especially limy ones. All kinds are raised from seed sown in spring where the plants are to stay. The seedlings are thinned sufficiently to avoid crowding.

ONOCLEA (Onoc-lèa) — Sensitive Fern. One species of fern common in meadows, open woodlands and swamps from Newfoundland to Ontario, Minnesota, South Dakota, Florida, and Texas, and in northern Asia is the only member of this genus of the aspidium family Aspidiaceae. Its generic designation is an ancient Greek one, *onokleion*, for some plant unknown. Its colloquial name refers to the sensitivity of its foliage to frost; it blackens at the first touch.

The sensitive fern (***Onoclea sensibilis***) has vigorous branching, creeping rhizomes from which arise scattered fronds, the sterile ones up to 3 feet tall including the deeply-pinnately-lobed blade that may be 15 inches long. The lobes are opposite and number eight to ten pairs; they are coarsely-lobed or shallowly-toothed. The stiffly-erect fertile leaves, shorter than the sterile ones, have much reduced blades and leaflets, the latter with beadlike segments with rolled margins.

Garden Uses and Cultivation. This coarse fern cannot be rated highly as a horticultural subject. It has limited uses in native plant gardens and areas of lesser importance where tallish groundcover is desirable and it may be used with good effect at watersides. It needs moist or wet neutral or acid soil and stands light shade. Division affords a ready means of increase; it can also be propagated by spores. For further information see Ferns.

ONONIS (On-ònis) — Rest Harrow. The genus *Ononis* consists of herbaceous perennials and low shrubs indigenous to Europe, North Africa, and western Asia. They number seventy-five species and belong to the pea family Leguminosae. The name is one used by Theophrastus.

Ononises have mostly leaves of three usually toothed leaflets, but sometimes there is only one leaflet and rarely there are more than three. At the bases of the leafstalks are appendages, called stipules, joined to the stems. Solitary or two or three in the leaf axils, or in racemes or panicles, the yellow, pink, purple, or white flowers are pea-like and have deeply-divided calyxes. They have ten stamens, joined, and an incurved style. They are succeeded by cylindrical or swollen pods containing one to many seeds.

A shrub up to 3 feet high or somewhat taller, much branched, and deciduous, ***O. fruticosa*** has finely-glandular-hairy young shoots and clustered, nearly stalkless and hairless, leathery, grayish-green leaves, almost all with three, spatula-shaped to oblong-lanceolate leaflets, toothed in their upper parts and usually under 1 inch long. It has long, terminal, leafless branching flower panicles. The blooms, ½ to ¾ inch long, are light pink, veined with deeper pink or purple. The upper petal is hairy. The seed pods, about ¾ inch long, contain about four seeds. This native of dry, rocky,

Ononis species

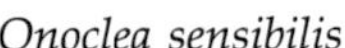

Onoclea sensibilis

Onoclea sensibilis (fertile leaves)

often mountain habitats in Spain and France is hardy in sheltered locations about as far north as New York City.

Another Spaniard, but much more tender, ***O. speciosa*** is perhaps the most beautiful of the genus. A shrub about 3 feet tall, its stems are densely clothed with short, glandular hairs. Its leaves are hairless, but are covered with sticky glands. They have three firm, elliptic to nearly round leaflets under 1 inch long. The golden-yellow, short-stalked blooms, ¾ inch long, are in crowded, cylindrical panicles. Distinguishable from the last by its leaves not being glandular, ***O. aragonensis,*** of the Pyrenees and Spain, is not hardy in the north. A shrub 6 inches to 1 foot tall, with often contorted branches, it is densely-hairy above, with sometimes some of the hairs short and glandular. Its leaves have three leathery leaflets, under ½ inch long, and elliptic to nearly round. The yellow, ½- to ¾-inch-long flowers are in terminal panicles. They are succeeded by seed pods, about ⅓ inch long, that contain one or two seeds.

Subshrubby to nearly shrubby rest harrows include several of merit. Very variable, ***O. natrix*** is chiefly southern European, but extends into northern France. Much-branched and erect, 9 inches to 2 feet tall, and woody at its base, it has densely-glandular-hairy stems. Its leaves have usually three leaflets, but the lowest ones sometimes have more, arranged pinnately. Yellow, often with red or violet veins, the flowers, ¼ to ¾ inch long, are in leafy, lax panicles. Up to 1 inch in length, the seed pods have four to ten seeds. The plant sometimes grown as *O. hispanica* is a botanical variety of this species.

A chiefly mountain species of southern Europe, ***O. rotundifolia,*** up to 1½ feet in height, is a neat, hardy, branched shrub or subshrub with hairy-glandular shoots and leaves of three elliptic to rounded, coarsely-toothed leaflets up to 1 inch long. The middle leaflet is long-stalked. The white to pinkish flowers, ½ to ¾ inch long, are followed by seed pods up to 1 inch or slightly more in length that contain ten to twenty seeds. From southern Europe and North Africa comes ***O. minutissima,*** woody at its base and 4 inches to 1 foot in height. Its stems, usually procumbent and often rooting, are nearly hairless. The leaves have three leaflets not over ¼ inch long and oblanceolate to obovate. The yellow flowers, in dense terminal racemes are under ½ inch long, have calyxes as long as the corollas.

Herbaceous perennial kinds include a few occasionally cultivated. Here belongs ***O. cenisia*** (syn. *O. cristata*), a species with procumbent stems, and from 3 inches to a little over 1 foot in height. It has glandular-hairy stems and leaves with three leaflets. The latter, less than ½ inch long, are more or less oblanceolate. The pink blooms are approximately ½ inch long. Containing about five seeds, the pods are less than ½ inch in length. This species is indigenous to southern and central Europe.

Ononis rotundifolia

Closely related are *O. spinosa*, *O. repens*, and *O. arvensis* (syn. *O. hircina*). A highly variable subshrub, ***O. spinosa*** is 6 inches to 3 feet tall, hairy or not, and with usually spiny stems. Mostly its leaves are of three leaflets and are of various shapes. In lax racemes, the flowers, ¼ to ¾ inch long, are pink or purple and generally have calyxes much shorter than the corollas. The seed pods, with one to few seeds, are less than ½ inch long. From the last ***O. repens*** differs in having often procumbent, rooting stems and in being spineless or with very soft spines. Up to 3 feet tall, it has leaves of one or three leaflets. At its maximum somewhat taller than the two last, ***O. arvensis*** grows erectly. It has mostly leaves with three elliptic to ovate leaflets and dense terminal racemes of paired, pink flowers ½ to ¾ inch long. About ⅓ inch long, the pods have one to three seeds.

Garden and Landscape Uses. Rest harrows are deserving of more attention from American gardeners. The taller ones are appropriate for perennial beds and naturalistic areas, and those of more restrained growth for rock gardens and other intimate plantings. All are children of the sun and glory in dryish, sharply drained soil of considerable depth so that their roots can strike deeply and obtain sufficient moisture for the plants' needs, even when the surface layers are parched. Some, perhaps all, are fond of limestone soils. Exact information about their hardiness is not available, but it is probable that the distinctly shrubby ones described, for the most part, will survive as far north as Washington, D.C. or even New York City, and that the herbaceous kinds are even hardier. Most should prosper in the drier, milder climates of the West.

Cultivation. Rest harrows are easy to grow. They can be raised from seed and cuttings, and some by division. Because they do not transplant well it is important to keep the young plants in pots until they are set in their permanent quarters. No special routine care is needed.

ONOPORDUM (Onopórd-um) — Thistle. The name thistle is commonly used for several genera besides this, which contains some of the largest and boldest sorts known by that name. To *Onopordum*, of the daisy family COMPOSITAE, belongs the Scotch thistle, a kind that despite its col-

Onopordum acanthium (foliage)

Onopordum acanthium (flowers)

loquial name occurs but rarely in Scotland and in all probability only as a long-established immigrant rather than an original native. Be that as it may, there are many who hold that it is the true heraldic emblem of Scotland, whereas others favor the bull thistle (*Cirsium vulgare*). The genus *Onopordum*, sometimes spelled *Onopordon*, consists of forty European, North African, and western Asian species, natives of the Mediterranean region. Its name is a Latinized form of the old Greek name *onoporde*.

Onopordums are coarse, usually freely-branched biennials, often white-woolly, with alternate, pinnately-lobed, spiny-toothed, often large leaves the bases of which are continued down the stems as usually very conspicuous wings. The large flower heads, typically thistle-like and without ray florets, are solitary or few together in tight clusters at the ends of candelabra-like branches. They are bright purple, blue-violet, rose-pink, or sometimes white. From most other genera of thistles and related plants *Onopordum* differs in not having bristles on the receptacle (the part of the flower head to which the florets are attached).

The Scotch thistle (***O. acanthium***) is the most likely species to be cultivated. From 3 to 9 feet tall and much branched, this vigorous kind is covered with white-cottony hairs and has oblong, lobed, prickly-toothed leaves up to 1 foot in length. Its flower heads, usually but not always solitary, are 1½ to 2 inches in diameter and light purple or sometimes white. This sort is naturalized over much of North America. Very different ***O. acaulon*** (syn. *O. acaule*), of Spain and western North Africa, as its name suggests is stemless or nearly so, the stalks rarely exceeding 4 inches in length so that the flower heads nestle at the centers of the basal rosettes of foliage or reach scarcely above them. The narrowly-oblong, short-stalked leaves are 6 inches to 1¼ feet long and deeply dissected into spine-toothed lobes, gray- or white-hairy on both surfaces. The flower heads are very large and are white.

Others that may sometimes be cultivated include these: ***O. bracteatum,*** native to Asia Minor, is white-hairy, up to 10 feet tall, and has large, yellow-spined leaves and huge purple flower heads. ***O. illyricum,*** of southern Europe, North Africa, and the Near East, up to 10 feet in height, is white- or gray-hairy, with the stems branched only in their upper parts. Its spiny leaves are twice-pinnately-parted. Its flower heads are medium-sized and purple. ***O. nervosum*** (syn. *O. arabicum*), of Spain and Portugal, 6 to 9 feet tall, has markedly winged stems, comparatively shallowly-lobed, spiny leaves, and bright rose-purple flower heads.

Garden Uses. There may be prejudice against admitting thistles to the garden, chiefly perhaps because of the well-deserved reputations of some such as the Canada thistle (*Cirsium arvense*) being obnoxious weeds, also because most are too coarse and vigorous for refined, sophisticated landscapes. Onopordums, being strictly biennials, are not invasive, but are coarsish plants better adapted for semi-wild and informal areas than formal flower beds and borders. Planted in naturalistic groups in appropriate settings, they can be quite effective. They thrive in full sun or part-shade in any ordinary garden soil, perferring heavy ones.

Cultivation. Sow seeds outdoors in May or June and transplant the seedlings in rows 1½ feet apart with 9 inches between individuals in the rows, or simply thin them out to 9 inches to 1 foot between individuals. In early fall or early the following spring they are transferred to their blooming sites. The taller ones are spaced 2 to 5 feet apart according to the vigor of the species (3 feet is satisfactory for the Scotch thistle) and the low-growing *O. acaulon* about 1 foot apart.

ONOSMA (On-ósma)—Golden Drop. Annuals, biennials, herbaceous perennials, and subshrubs of the borage family BORAGINACEAE compose *Onosma*. Native from Europe to the Himalayas, it bears a name derived from the Greek *onos*, an ass, and *osme*, smell, alluding to a fancied animal-like odor. There are 150 species.

Onosmas are hairy and have alternate, undivided, lobeless and toothless leaves with conspicuous mid-veins. Borne in one-sided racemes or clusters, the tubular or urn-shaped, yellow, purple or white flowers each has a bract at its base. Their persistent calyxes have five lobes, which increase in size as the fruits develop. Unlike those of some members of the borage family, the corollas are without scales in their throats. They have five small lobes (petals). There are five stamens, usually not protruding, but sometimes briefly exserted. The style is slender. The fruits consist of four seedlike nutlets.

Golden drop (***O. echioides***) has roots that yield a red dye used locally for coloring foods and beverages. Native to Europe, it is a herbaceous perennial with erect stems 1 foot to 1½ feet tall that

Onosma stellulatum

Onosma stellulatum (flowers)

branch above and bear pendulous yellow blooms about ¾ inch long. Up to 3½ inches long, the lower leaves are linear. The upper ones are smaller and lanceolate. All are clothed with bristly, yellow hairs, swollen at their bases but without smaller hairs sprouting around the swollen parts. From the golden drop ***O. arenarium*** differs in having more but decidedly smaller blooms in crowded pyramidal clusters. Its lower leaves are oblong-spatula-shaped, those above oblong. The hairs on the undersides of the leaves sometimes have smaller hairs around their bases. This European is perennial or biennial.

White, bristly hairs, almost all with up to ten smaller hairs spreading in starlike fashion from their swollen bases are a characteristic of ***O. stellulatum.*** This native of southeastern Europe is a herbaceous perennial with a slightly woody base and stiff, erect, branchless stems 6 inches to 1 foot tall. Its leaves are oblong-linear. The racemes of blooms may be branched or not. The short-stalked, pale yellow flowers are finely-hairy on the outsides of their corollas and about ¾ inch long. From the last, ***O. tauricum*** (syn. *O. stellulatum tauricum*) differs in its flowers being stalkless and the lower surfaces of its leaves, except for the mid-veins, being without bristly hairs and having only stars of spreading short hairs. White flowers that become deep pink or red and finally purplish are borne by ***O. alboroseum,*** of Asia Minor. This is about 6 inches tall and has the habit of growth and type of hairiness of *O. stellulatum*. Its leaves are oblong-spatula-shaped. The blooms, velvety-hairy on their outsides, are about 1 inch long.

Onosma alboroseum (foliage)

Other species include these: ***O. albopilosum,*** of Asia Minor, is subshrubby and under 1 foot tall. It has blunt-oblong, gray-downy leaves, the upper ones stalkless and velvety-hairy. Its flowers, white at first, blushing pink as they age, are about 1 inch long. The flowers of *O. a. roseum* are pink. ***O. cassium,*** of Syria, is a handsome herbaceous perennial 1 foot to 1½ feet tall that has blunt, oblong leaves with

Onosma albopilosum

Onosma albopilosum roseum

Onosma helveticum

a few bristly hairs. Hairless and about ¾ inch long, its yellow flowers have slightly protruding stamens. ***O. helveticum,*** of the European Alps, is 1 foot to 1½ feet tall. It has oblong-spatula-shaped, short-hairy leaves, up to 3 inches long, and pubescent, 1-inch-long, pale yellow flowers. ***O. nanum*** (syn. *O. decipiens*) is a brownish-hairy, native of Asia Minor. Up to 5 inches tall, it has rough hairy, linear-lanceolate leaves and hairless, yellow flowers about ¾ inch long that change to blue as they age. ***O. rupestre,*** of the Caucasus, is 6 to 9 inches tall and has stalkless, soft-hairy, linear-lanceolate leaves up to 2 inches long. Its red-stalked, ivory-white flowers, pale yellow in bud, are narrow-urn-shaped.

Garden Uses and Cultivation. Onosmas are primarily rock garden plants. They revel in sun, and in deep, mellow, extremely well-drained, sandy soil. Constant wetness around their roots is disastrous. Their resistance to cold varies and is not reliably known for all species. At New York City, *O. echioides, O. stellulatum,* and *O. tauricum* have proved satisfactory and they probably can withstand colder climates. Propagation is by seed and summer cuttings planted in a shaded cold frame or greenhouse propagating bench.

ONYCHIUM (Oných-ium)—Claw Fern or Cliff-Brake. Old World ferns, the exact number of species uncertain because of diverse interpretations of different authorities, but certainly several, constitute this genus of the pteris family PTERIDACEAE. Its name comes from the Greek *onychion,* a little claw, and alludes to the shape of lobes of the leaves. An American fern sometimes called *Onychium densum* is *Aspidotis densa.*

Onychiums have short, compact, or creeping rhizomes and fronds (leaves) three-times- or more-pinnate, leathery or not, with the final segments narrow and hairless. The clusters of spore capsules are continuous along the margins of the leaf segments.

Widely distributed as a native from Japan, Korea, China, and Taiwan, to Malaysia, and India, evergreen ***O. japonicum*** has sparsely-leafy, creeping rhizomes and hairless fronds with triangular-ovate blades 6 inches to nearly 1½ feet long by 3½ to 8 inches wide. They are dark green on their upper sides, paler beneath, and three- or four-times-pinnately-divided into final narrowly-lanceolate segments, those of sterile fronds, frequently toothed.

Garden and Landscape Uses and Cultivation. Of dainty, lacy appearance, *O. japonicum* is an excellent fern for outdoor cultivation in mild climates, for greenhouses, and as a houseplant. It succeeds in well-drained, humus-rich soil kept moderately moist, but not saturated, where light is good, but shade from strong sun is had. Cool growing conditions are most favorable and a fairly humid atmosphere. Indoor winter night temperatures of 45 to 50°F are adequate, and day temperatures may rise five to fifteen degrees. Plants that have filled their containers with healthy roots benefit from spring-to-fall biweekly applications of dilute liquid fertilizer. Increase is by division and spores. For additional information see Ferns.

OOPHYTUM (Oophýt-um). Closely related to *Conophytum* and once included there, *Oophytum* consists of two or three species of South African, nonhardy, perennial succulents of the *Mesembryanthemum* relationship, belonging in the carpetweed family AIZOACEAE. Its name, derived from the Greek *oon,* an egg, and *phyton,* a plant, is pleasantly descriptive, because the plants suggest clusters of tiny eggs standing on end.

Oophytums form clusters of small plant bodies each consisting of a pair of fleshy leaves completely joined except for a small cleft at their apexes. The flowers, daisy-like in aspect, but not in structure, are solitary. They have a five- to seven-lobed calyx, many separate petals in two or three rows, numerous stamens, and five or six awl-shaped stigmas. The fruits are capsules.

The plant bodies of ***O. oviforme*** are ½ to ¾ inch long, olive-green or reddish. In the fissure the solitary white and purplish-pink flowers, nearly 1 inch wide, are borne. Green plant bodies up to ¾ inch long by ¼ inch wide and with much-rounded apexes are characteristic of ***O. nanum.*** About ⅜ inch across, the flowers are white with reddish margins to the petals.

Garden Uses and Cultivation. Although these plants grow without difficulty under conditions that suit conophytums, they do not always bloom reliably in cultivation. They are easily increased by division and cuttings, as well as by seed. For further information see Succulents.

OPHIOCAULON. See Adenia.

OPHIOGLOSSACEAE — Adder's Tongue Family. Four genera totaling seventy species of small, often fleshy-foliaged ferns comprise this family, represented in the native floras of temperate to tropical regions in the Old World and the New World. Some tropical sorts are epiphytes (tree-perchers). These ferns have short stems and solitary or few fronds (leaves) of two types, sterile and fertile. The latter may bear spores over their entire surfaces or portions of them may not develop spores. Genera cultivated include *Botrychium, Helminthostachys,* and *Ophioglossum.*

OPHIOGLOSSUM (Ophio-glóssum)—Adder's Tongue Fern. To the untutored eye, the species that constitute *Ophioglossum,* of the adder's tongue family OPHIOGLOSSACEAE, scarcely look like ferns. They have more the aspect of small lily relatives, but unlike lilies and other blooming plants, they have neither flowers nor seeds, they reproduce in typical fern fashion from spores. The number of species is variously estimated as from thirty to fifty. The name comes from the Greek *ophis,* a serpent, and *glossa,* a tongue, and alludes to the fertile portions of the fronds. The genus is nearly cosmopolitan.

Ophioglossums have small, subterranean stems and one or more leaves that divide to form a sterile green blade and a fertile, slender, spikelike portion that carries, in two rows, the fleshy clusters of spore cases. Usually the spike is without branches, but in *O. palmatum* it branches.

Common adder's tongue fern (***O. vulgatum***), a native of moist meadows, woodlands, and occasionally beaches from Nova Scotia to Alaska, Florida, Texas, Arizona, and Mexico, and in Europe and Asia, has usually solitary, stalked leaves up to 1¼ feet in length. The sterile blunt blade is oblong-elliptic to ovate, 1½ to 4½ inches long by ¾ inch to 2 inches wide. The fertile spikes, 1 inch to 2 inches long, are on stalks up to 7 inches long. Indigenous from Virginia, to Illinois, Kansas, Florida, Arizona, and Mexico, ***O. engelmannii*** is hardy and has usually two to five leaves about 9 inches in length with pointed sterile blades generally 2 to 3 inches long by ¾ inch to 1¼ inches broad, but sometimes somewhat bigger. From ½ inch to 1¼ inches in length, the fertile spikes terminate stalks up to 4½ inches long. This adder's tongue fern favors limestone soils in light woodlands and pastures.

Occasionally cultivated in greenhouses, ***O. reticulatum,*** of South Africa, East Africa, and parts of Asia, has usually one or two leaves 6 inches to 1 foot tall. The fertile portion, 1 inch to 1½ inches in length, terminates a stalk 2 to 4 inches long. The sterile segment is heart-shaped at its base and 2 to 3½ inches long. Widely dispersed throughout the tropics and subtropics, including Florida, and sometimes occurring spontaneously in tropical greenhouses, ***O. petiolatum*** differs most obviously from other sorts in the leaf blade being lanceolate and decidedly narrower and in the lower stalk-like portion being proportionately longer.

Ophioglossum reticulatum

Ophioglossum petiolatum

Garden Uses and Cultivation. Interesting for fern collections, rock gardens, and native plant gardens, and the nonhardy sorts occasionally cultivated in greenhouses, ophioglossums are not too easy to grow. So far as possible attempts should be to duplicate as closely as practicable conditions under which the plants thrive in the wild. Moist, peaty soil and partial shade are to their liking. For *O. engelmannii* the addition of crushed limestone to the soil may be helpful. Propagation is by spores and very careful division. For further information see Ferns.

OPHIOPOGON (Ophio-pògon)—Lilyturf, Mondo-Grass. The common name lilyturf of this group of twenty species of eastern Asia is applied also to *Liriope.* Both genera belong to the lily family LILIACEAE. From *Liriope* the plants we are concerned with here differ in having apparently inferior (located mostly below the petals) ovaries and stamens with stalks shorter than the pointed anthers. The reason the ovary appears to be inferior is because the lower part of the perianth tube is joined to it and the perianth lobes (petals) spread from near its top. In appearance ophiopogons and liriopes are similar. The name *Ophiopogon* comes from the Greek *ophis,* a serpent, and *pogon,* a beard, and is an interpretation of the Japanese vernacular name. An older, discarded generic name is *Mondo.*

Ophiopogons are clump- or turf-forming, evergreen, low herbaceous plants with narrow leaves, and spikes or racemes of small white to blue-purple flowers that have a perianth of six tepals (usually called petals). There are six stamens. The fruits are berry-like and contain one seed.

The white lilyturf or jaburan (***O. jaburan*** syn. *Mondo jaburan*), of Japan, is a clump-forming, nonspreading kind sometimes 2 feet high, with cordlike, but not tuber-forming, roots and many dark green arching leaves 1 foot long or longer by ¼ to ½ inch wide. The white flowers, up to ½ inch long, are in short, somewhat one-sided spikes carried above the foliage. The fruits are blue-violet. Varieties *O. j. argenteo-variegatus* and *O. j. aureo-variegatus* have their leaves striped longitudinally, respectively, with white and yellow. Those named *O. j. variegatus, O. j. vittatus,* and *O. j. javanensis* are perhaps not different. The white lilyturf is frequently confused with *Liriope muscari.*

Ophiopogon jaburan aureo-variegatus

Dwarf lilyturf (***O. japonicus*** syn. *Mondo japonicum*) is very different. Native to Japan and Korea and only a few inches high, it has tuber-bearing roots, and stolons that spread rapidly and produce a dense, coarse turf of dull, dark green, grasslike leaves up to 1 foot long and not over ⅜ inch wide. The loose racemes of nodding, pale lilac to violet, ⅜-inch-long flowers are mostly hidden among the foliage and make no great show. They are succeeded by pea-sized blue fruits containing white seeds.

Japanese ***O. planiscapus*** has thickened, spindle-shaped roots and often slender stolons. Its leaves, up 1 foot long or longer and ⅛ to ¼ inch wide, are linear, several-veined, and deep green. The racemes of nodding, pale purple or white, ¼-inch-long blooms are in racemes that, with their stalks, are 1 foot to 1½ feet tall. The fruits are dull blue. More common in cultivation than the typical species, dwarfer *O. p. arabicus* (syn. *O. arabicus*) has nearly black leaves and flowering stalks, pinkish flowers, and blue-green fruits. The variety is hardy in New York City.

Garden and Landscape Uses. The white lilyturf and its variegated varieties are handsome, easily grown plants not reliably hardy in the north, but admirable in the south for beds, borders, edgings, rock gardens, and other purposes. They are splendid pot plants for window gardens and greenhouses. In regions where winters are too cold for them to live permanently outdoors, they may be used to good purpose in summer beds and be taken up before the advent of fall frosts and wintered in pots or flats of soil, kept moderately moist in a greenhouse or other light, fairly cool, frost-proof place. Dwarf lilyturf is one of the very finest groundcovers for gardens in the deep south. It is hardier than white lilyturf, but just to what degree is not fully determined. Certainly it is reliable as far north as Washington, D.C., and in reasonably sheltered locations perhaps to New York City or even beyond. It forms a delightful, uniform carpet of grasslike foliage, but, it must be confessed, in regions where winters are too harsh for its liking, it becomes shabby before spring. This, however, is soon repaired by the development of new growth. Dwarf lilyturf may be used for edgings and in rock gardens, but because of its habit of spreading rapidly it is less useful than white lilyturf for these purposes. It is a splendid erosion-checking cover for banks and slopes. Lilyturfs succeed in a wide variety of soils, in shade and provided the soil is not excessively dry, in full sun.

Ophiopogon japonicus

Cultivation. Ophiopogons are easily raised from freshly collected seeds, which should be freed of surrounding pulp and sown about ½ inch deep in sandy peaty soil in flats, cold frames, or the open ground. They germinate in about four weeks and under favorable conditions make plants big enough to set in their permanent locations in about two years. More commonly, division is employed as the means of multiplication. With dwarf lilyturf, this is simply a matter of digging up sods, pulling them apart with the hands, and planting each small rooted piece. With white lilyturf more difficulty may be encountered, especially if the clumps are old and have developed woody centers. Then a heavy knife, cleaver, or axe may be needed to obtain pieces suitable for planting. Divisions consisting of a single tuft of leaves with a few roots attached are to be preferred to larger units. In the lower south the best time to divide is early fall, further north early spring is preferable. For edgings and groups white lilyturf may be spaced 1 foot to 1½ feet apart. Closer spacing, 6 to 8 inches each way between individuals, is appropriate for dwarf lilyturf. Under good conditions in fertile soil this will give complete coverage in two growing seasons. During that time, by mulching or cultivating, weeds must be kept down, but very little attention will be needed from then on. The question is sometimes raised whether ophiopogons will stand mowing and the answer is that *O. japonicus* will, at least to a limited degree, and has been known to persist in lawns for many years and even extend itself. However, in general, they should not be regarded as mowable, but should remain uncut except for any attention of this kind needed at the end of winter.

As pot plants in greenhouses and window gardens ophiopogons grow with great ease. They are best suited with a night temperature in winter of 40 to 50°F, with a daytime rise of a few degrees permitted. At other seasons normal outdoor temperatures are appropriate. Any porous, fertile soil is satisfactory, and it should be kept moderately moist. Well-rooted specimens respond favorably to applications of dilute liquid fertilizer. Dividing and repotting, when needed, are best done in late winter or spring, or in summer as soon as blooming is through, but it may be done in any season.

OPHRYS (Óph-rys) — Bee Orchid. Little known in cultivation in North America, *Ophrys* consists of about thirty often quaint species of terrestrial orchids, of the orchid family ORCHIDACEAE, confined in the wild to Europe, North Africa, and western Asia. Its members usually have small tubers that are replaced annually as the old tubers shrivel. The stems may have leaves only at their bases or throughout their lengths. The blooms, which often strikingly resemble insects in appearance, are in terminal racemes. They have three nearly equal-sized, petal-like sepals larger than the two petals and a large three-lobed or lobeless, inflated, hairy lip. The column is often beaked. There are two stalked pollinia. Unlike closely related *Orchis*, the flowers have lips without spurs. The name is from the Greek *ophrys*, an eyebrow, referring to fringes on the sepals.

The bee orchid (***O. apifera***), of Europe and North Africa, is so called because its blooms are shaped, colored, and hairy so that they look somewhat like bees. It inhabits pastures and other grassy places and thickets and is up to 1½ feet tall. It has a spherical tuber from which develops in fall a basal rosette of foliage. The four to seven leaves are oblong to lanceolate-elliptic. There are two to seven blooms with usually pink sepals greenish on their outsides, greenish, velvety petals, and a velvety-hairy, spherical, red-brown, lip patterned with yellow and with two yellow spots near its tip. Varieties with white or creamy petals are known. This species hybridizes freely with others. Native to the eastern Mediterranean region, ***O. reinholdii*** is mostly from 8 inches to 1½ feet tall. It has oblong-lanceolate, chiefly basal leaves up to 4 inches long by 1 inch wide and on each stem three to eight bee-like flowers. They have rose-violet to pink or greenish-white sepals approximately ½ inch long and shorter and pinkish petals. The about ½ inch-long, velvety, three-lobed lip is dark purple or brownish-purple marked with two white spots. Variety *O. r. straussii* (syn. *O. straussii*) of Turkey and Iran differs in the lip marking being more obvious and horseshoe shaped. Other species include the fly orchid (*O. muscifera*), the wasp or yellow bee orchid (*O. lutea*), and the bumble bee orchid (*O. bombyliflora*), all European kinds described in floras of that continent.

Ophrys reinholdii strausii

Ophrys reinholdii strausii (flower)

Garden Uses and Cultivation. These are collectors' plants best adapted for growing in pans (shallow pots) in cold frames, or in favored spots in rock gardens. They are fairly hardy and respond to excellent drainage. The recommended soil consists of equal parts of good top soil, peat moss, leaf mold, and coarse sand with some crushed limestone added. Ample supplies of water are given during the season of active growth, much less when the plants are dormant. Propagation is by careful division in spring. Repotting is done about every third year also in spring. For further information see Orchids.

OPHTHALMOPHYLLUM (Ophthalmo-phýllum). This intriguing and variable genus, of the carpetweed family AIZOACEAE, much resembles *Conophytum*, from which it differs in its calyxes not being tubular. It is of the *Mesembryanthemum* relationship. Its nineteen species, dwarf, windowed, pebble plants, are natives of South Africa. The name, from the Greek *ophthalmos*, eye, and *phyllon*, a leaf, refers to the fanciful likeness of the windows to eyes.

Ophthalmophyllums are without stems. Their plant bodies, solitary or few, are cylindrical to top-shaped. Each consists of two very fleshy, often finely-hairy, juicy leaves united practically to their tops, and with at their flat or convex apexes a window of translucent tissue that admits light

Ophthalmophyllum, undetermined species

to the chlorophyll-containing cells below. This is an important survival adaptation. It permits ophthalmophyllums to grow in their native desert habitats with their bodies buried in the ground and only their tips exposed. This they do and thus are protected against desiccation by sun and wind, and to a considerable extent from browsing animals. Their close mimicry of the small stones and pebbles among which they so often grow is further insurance against depredations by hungry and thirsty creatures. The flowers of *Ophthalmophyllum*, from somewhat less than ½ inch to a little over 1 inch across, and from white to pink or red-violet, are typical of *Mesembryanthemum* and its relatives. Superficially they resemble daisies, but there is no botanical similarity, for each is a single bloom rather than a head of many florets. The fruits are capsules.

Kinds likely to be found in cultivation include the following. Its red-tinged to coppery-red, cylindrical plant bodies 1 inch to slightly more in length and approximately one-half as wide, ***O. friedrichiae*** is deeply fissured between the rounded, very clearly windowed tips of its leaves. Its white blooms are from a little under to a little over ½ inch in diameter. Solitary or in clusters of several, the plant bodies of ***O. herrei*** are soft, velvety, and olive-green. From ¾ inch to about 1½ inches long, the leaves have gaping extremities separated by a deep cleft. Opening only by day, the somewhat fragrant white to light pink blooms are up to a little over 1 inch in diameter. The white flowers of ***O. maughanii*** are ½ inch wide. Its semi-translucent plant bodies, solitary or in groups of several, are 1¼ to 1¾ inches in length and up to 1 inch in diameter. They are divided by a fairly deep cleft into a pair of gaping rounded lobes with pale yellowish-green windows that have along their outside edges, large transparent dots. Solitary, nearly globose, shallowly-cleft plant bodies sprinkled with small translucent dots, and 1 inch long and one-half or more as thick are characteristic of ***O. schuldtii.*** Its dull, dark purple leaf tips are rounded and have well-marked windows. This has white blooms 1 inch or somewhat more in diameter. Solitary, top-shaped plant bodies fissured into two flattish lobes are features of ***O. triebneri.*** They are ½ to ¾ inch long and generally broader than tall. Ochre-yellow above, they are reddish to purplish or bluish below, with large translucent dots as well as terminal windows. The slightly fragrant, wide-petaled, white flowers are 1 inch in diameter.

Garden Uses and Cultivation. These choice plants are for collectors of rare and unusual succulents. Their needs are essentially those of conophytums, lithops, and other South African stone or pebble plants, but they are less easy to satisfy than many. Perfectly drained soil is a must, and great care must be taken that it does not remain wet for lengthy periods. During dormancy, which lasts for several months, watering is not needed. Then, the plant bodies shrivel. In late summer or fall new plant bodies develop within the old, and the latter shrink to become pieces of dry, skinlike tissue. As soon as new growth begins, watering is resumed, but always with considerable restraint. These plants need a minimum temperature of 55°F and a sunny location. They are easily raised from seeds and are best accommodated in very well-drained pans (shallow pots). Repotting, when needed, is done at the beginning of the growing season.

OPITHANDRA (Opith-ándra). One of the five or six species of this Asian genus of the gesneria family GESNERIACEAE was long known in gardens as *Oreocharis primuloides*. It now appears that no representative of *Oreocharis* is cultivated in North America and that the plant previously grown under that name is *Opithandra primuloides*. This is endemic to Japan.

In general resembling *Ramonda*, ***O. primuloides*** differs in its flowers having two instead of four or five fertile stamens. The foliage consists of rosettes of evergreen, wide-elliptic to broadly-heart-shaped, longish-stalked, coarsely-toothed, densely-hairy, rather fleshy leaves with blades up to 2½ inches long. The leafless flower stalks rise to heights of 4 to 8 inches or more and are topped by clusters or umbels of up to twelve lilac, purple, white, or pink flowers. The blooms have calyxes with five short, linear lobes, and tubular corollas ½ to ¾ inch long, with wide-spreading lobes (petals), and two-straight-stalked, anther-bearing stamens. The fruits are slender capsules.

Garden Uses and Cultivation. A plant for alpine greenhouses, and outdoor cultivation in regions of mild winters and fairly cool summers, *O. primuloides* responds to the same conditions and care and has the same uses as *Ramonda*, but is not quite as hardy. Leaf cuttings planted in sphagnum moss and seed are the means of multiplication. For further information see Gesneriads.

OPIUM POPPY is *Papaver somniferum*.

OPLISMENUS (Oplís-menus) — Basket-Grass. A few species and varieties of *Oplismenus* are grown as foliage plants in greenhouses and in warm climates outdoors, often under the misapplied name of *Panicum*. They are members of the grass family GRAMINEAE. There are fifteen species, natives of shady places in warm parts of the Old and New Worlds. The name, from the Greek *hoplismos*, armed, alludes to the bristle-tipped spikelets. There are annual and perennial species.

These grasses have slender, leafy, trailing or ascending stems, and leaves with spreading, flat, lanceolate to ovate blades. The flowers are in one-sided, spikelike racemes with few to several on a common stalk. Solitary, paired, or clustered, the short-stalked spikelets have two flowers,

Oplismenus hirtellus variegatus

Oplismenus hirtellus variegatus (foliage)

an upper bisexual one and a lower male or sterile one. The spikelets fall without disintegrating.

Native to Africa, Asia, Polynesia, and northern Australia, ***O. compositus*** is a perennial trailer or creeper that roots freely from the nodes (joints). It has pointed, narrowly-elliptic to narrowly-ovate, thin, somewhat hairy leaf blades 1½ to 6 inches in length and up to 1 inch wide. The racemes of up to 4-inch-long spikelets in loose assemblages of up to about ten are 1 foot in length or shorter. The spikelets are lanceolate or lanceolate-elliptic and about ⅛ inch long. From the last, ***O. hirtellus,*** native from Texas to Argentina and in the West Indies, differs in having more compact, slightly spreading racemes of spikelets not over 1 inch long and three to seven on a common stalk. The more or less hairy leaf blades at their maximum are

Oplismenus hirtellus variegatus in bloom

somewhat smaller than those of *O. compositus*. Popular *O. h. variegatus* has white-striped leaves and, like the variegated-leaved variety of *O. imbecillus,* is often misnamed *Panicum variegatum* in gardens.

Often misnamed *Panicum variegatum* in gardens, *O. imbecillis variegatus* has white-striped leaves sometimes tinged with pink. Native of New Caledonia, it is a variant of ***O. imbecillis,*** a creeping or trailing kind with ascending stems and linear-lanceolate to lanceolate, pointed, thin, hairless leaf blades 1½ to 4 inches long and up to a little over ½ inch wide. The up to half dozen loose racemes of spikelets on each common stalk are not over ¾ inch long and are in assemblages up to 6 inches long. This species occurs in Australia, New Zealand, and Polynesia.

Garden and Landscape Uses and Cultivation. These attractive, quick-growing trailers are most used in hanging baskets, urns, vases, porch and window boxes and for decorating fronts of benches in conservatories and greenhouses. They grow with little or no trouble and are very easily multiplied by cuttings and division and, the green-leaved kinds, by seed. Ordinary soil, reasonably well drained, suits them. It should be kept moderately moist. Established plants benefit from regular applications of dilute liquid fertilizer. A humid atmosphere and a minimum temperature of 55°F are to their liking. Shade from strong sun is advantageous. Occasional pinching off the tips of the shoots promotes bushiness. Repotting and general rehabilitation of specimens that have become straggly may be done in spring, but young plants are so easily obtained from cuttings that usually it is better to rely on these and discard those that have outgrown their usefulness. A simple plan is to plant the cuttings directly in pots of sandy soil, three in a 3-inch, or five in a 4-inch size. In a humid atmosphere in a shaded place where the temperature is 60 to 70°F, or more they root in three or four weeks. When new growth is well begun the tip of each plant is pinched out to encourage branching.

OPLOPANAX (Opló-panax)—Devil's Club. The one American species of the two or three that constitute *Oplopanax* (once named *Echinopanax*) is the only kind likely to be cultivated. The others inhabit Japan and Korea. All are deciduous, very prickly shrubs of the aralia family ARALIACEAE. Their name, from the Greek *hoplon,* a weapon, and *Panax* an allied genus, alludes to their formidable spines.

Oplopanaxes have branchless or sparsely-branched stems and alternate, long-stalked, palmately- (in hand-fashion) five- to seven-lobed leaves. The greenish-white blooms are in small umbels aggregated in dense terminal clusters or panicles. Each tiny flower has an indistinctly toothed calyx, five each petals and stamens, and two styles. The fruits are small, flattened, berry-like, red drupes (fruits structured like plums).

Devil's club (***O. horridus***), which ranges in the wild from California to Alaska, in the east is hardy about as far north as southern New England. Up to about 12 feet tall, it has stout, erect stems and somewhat maple-like leaves, hairless above and slightly hairy on their undersides, with rounded ovate, sharply-toothed blades 6 inches to almost 1 foot in diameter. The stems, leafstalks, veins of the leaves, and stalks of the flower clusters are furnished with slender, sharp, yellow prickles. The numerous flowers are in terminal clusters or dense spikes shorter than the leaves. Those of the lower parts of the spikes are mostly males, those above mostly females. The fruits, red when mature, are ¼ inch long. The other species differ in minor ways only from *O. horridus.* Japanese ***O. japonicus,*** an inhabitant of coniferous woods, is about 3 feet tall. Its leaves are more deeply lobed than those of *O. horridus.*

Garden and Landscape Uses and Cultivation. The bold foliage and bright red fruits of the devil's club are attractive, and the stems form formidable barriers. In gardens, unless the plants are suitably located, the spines can be troublesome; wounds caused by them swell and cause much discomfort. In naturalistic landscapes this species can be used effectively in groups. It thrives near the sea, and

Oplopanax horridus

in sun or part-shade, in ordinary soils. Propagation is by seed and by division.

OPOPANAX is a vernacular name for *Acacia farnesiana.* It is also the botanical name of a genus, not cultivated, of the carrot family UMBELLIFERAE. Both are sources of gums used in perfumery.

OPUNTIA (O-púntia)—Prickly-Pear, Cholla, Indian-Fig. Although *Opuntia* is one of the largest and most widely distributed genera of the cactus family CACTACEAE, most of its members, all natives of the Americas, are less appreciated as ornamentals than many other cactuses. This in part because the tiny glochids so abundant in the majority of kinds make them difficult to handle without danger of annoyance and irritation, and also because many are too similar for more than a few to appeal to other than the most avid collectors. Yet in total a considerable number of the approximately 250 species and in addition some varieties are cultivated. A representative selection is described here. The name *Opuntia* is a Greek one for some entirely different plant that grew near the ancient town of Opus.

Edible fruits called tunas, Indian-figs, and prickly-pears of some opuntias, esteemed for eating out of hand, are grown, especially in Mexico, commercially. The young pads of certain prickly pears are cooked as a favorite food in Mexico and parts of Texas. Great quantities of older pads, after their spines have been removed by burning, are used as a cattle feed. The dried stems of suitable chollas are made into various trinkets and travelers' souvenirs. In Australia introduced species (none is native there) have spread widely and become serious weeds of range and pasture land. Some kinds are freely naturalized in dry parts of Africa and some others in desert and semidesert regions as well.

Some authorities include as subgenera of *Opuntia* several groups of species less conservative botanists segregate as distinct genera and the modern tendency, at least in America and England, is toward such consolidation. However, for technical reasons, chiefly because some species have not been sufficiently studied and by adequate publication transferred to *Opuntia* from the genera under which they were described, it is botanically unacceptable to apply *Opuntia* binomials to all kinds at this time. Furthermore, collectors of cactuses and nurseries specializing in them usually grow their plants under the name of the several genera into which the genus is divided by the "splitters," at least they do with reference to species not native in the United States. But species native north of Mexico are in modern floras of the region invariably included in *Opuntia* instead of being segregated according to kind as *Opuntia, Cylindropuntia,* and *Corynopuntia.* Purely then as a matter of convenient reference, cylindropuntias and corynopuntias are treated here as *Opuntia* (with the synonyms in parenthesis), while *Austrocylindropuntia, Consolea, Grusonia, Nopalea,* and *Tephrocactus,* which other segregate genera conservatives retain in *Opuntia,* are listed separately in this Encyclopedia.

Opuntias are succulent perennials, often more or less woody, varying in habit from low mat-forming sorts to large bushes and small trees. Their leaves like those of most cactuses neither strongly evident nor important functionally, are found only on the youngest parts and soon fall. They are small and awl-shaped, or cylindrical. Photosynthesis of food, in most plants the function of the leaves, is mostly done by the cholorphyll-containing stems. These consist of a series of well-defined joints, which may be broad, flat pads or nearly or quite cylindrical. They are without ribs. The areoles (specialized cushion-like portions of the stems of cactuses from which spines and flowers come) may be without spines, but more often have one to fifteen, rarely more. In addition, they develop tiny to minute, barbed bristles or hairs called glochids. These are easily detachable; if brushed against they become embedded in the skin and because of their barbed tips cannot be withdrawn. Shaving is likely to give the best relief.

Produced on the edges of the joints developed during the previous year, the flowers have short perianth tubes, with many spreading perianth parts with a gradual transition from the outer to inner from sepals to petals. There are many stamens more or less sensitive to touch and one style. The fruits when mature are dry or fleshy, spiny or spineless. Technically they are berries.

Opuntias may be sorted rather neatly (there are a very few intermediates) into two distinct groups, those in which segments of the stems are flat pads (the bo-

Opuntias in the Mojave Desert

Opuntia (left) in a botanical garden conservatory

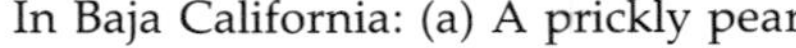

In Baja California: (a) A prickly pear (b) A cholla

tanical group *Platyopuntia*) and those in which they are approximately circular in section (the botanical group *Cylindropuntia*). In the United States the former are generally known as prickly-pears, the tall, upright kinds of the others as chollas (pronounced chòy-as). The *Cylindropuntia* group is often further split into *Cylindropuntia* proper, in which the stem joints are cylindrical, and *Corynopuntia,* in which they are club-shaped.

Prickly pears, sorts that those who favor splitting the genus into several genera retain in *Opuntia,* include the kinds now to be described. They are presented here in alphabetical sequence according to their botanical names.

The Beaver tail cactus (***O. basilaris***), native from Utah to Mexico, develops clumps up to 6 feet across and usually 6 inches to 1 foot tall of stems with broad-ovate to nearly circular, bluish-green to purplish, usually velvety-pubescent pads 2 inches to 1 foot long. Without spines, they have depressed areoles and rusty-brown glochids. The rose- to magenta-pink or less often yellow blooms are 2 to 3 inches long. The fruits, dry when mature, are spineless. Variety *O. b. treleasei* (syn. *O. treleasei*) has narrowly-elliptic to obovate pads with a few 1- to 1¼-inch-long spines.

A native of Mexico and Guatemala, ***O. decumbens*** is a low creeper or sometimes is up to 1¼ feet tall. Its velvety pads are up to 8 inches long and are oval to oblongish. They have yellow areoles with a red spot and either one or more slender, yellow spines up to 1½ inches long or

Opuntia basilaris

Opuntia basilaris in bloom

Opuntia decumbens

Opuntia ficus-indica

Opuntia erinacea

Opuntia ficus-indica (fruits, with one cut open to show interior)

longer. The deep yellow flowers are succeeded by juicy, purple fruits.

The Mojave prickly-pear (***O. erinacea***) includes several varieties chiefly inhabitants of sandy and gravelly soils in the Mojave Desert region. Forming clumps up to 3 feet wide or wider and 1 foot tall, it has thinnish, bluish-green pads, elliptic-oblong to obovate-oblong, mostly 4 to 6 inches long, sometimes shorter or longer, and 2½ to 3½ inches wide. The white to gray, somewhat flexible, irregularly curved spines, most commonly about 2 inches in length but sometimes considerably longer or shorter, are in clusters of four to nine. The rose-pink or yellow blooms are 1¾ to 3½ inches in diameter. Dry when mature, the brownish, densely-spiny obovoid fruits are 1 inch to 1¼ inches long. Grizzly bear cactus (*O. e. ursina*) has slenderer, down-turned, extremely flexible spines 3 to 4 inches long.

Indian-fig (***O. ficus-indica***) is a shrubby or treelike, freely-branched sort. From 10 to 20 feet high, it has thick, smooth, elliptic to oblongish, slightly bluish, gray-green pads 9 inches to 1½ feet long or sometimes a little longer; mostly spineless, they occasionally have single white spines. The many yellow glochids drop early. The yellow flowers, 3 to 4 inches in diameter, are succeeded by edible, fleshy, purple to red, pear-shaped fruits 2 to 3½ inches in length. Now widespread in cultivation and as a spontaneous element in natural floras, this species is of uncertain geographical origin. Variants of it that are cultivated include 'Burbank's Spineless', which is almost free of spines and glochids, and one called mission cactus, often grown as *O. megacantha*. Its specific name, alluding to the ease with which the segments of its stems are detached, ***O. fragilis*** is a native from Michigan to British Columbia, Texas, and Arizona. It forms mats 1 foot wide or wider, 2 to 8 inches

Opuntia humifusa

Opuntia humifusa (flowers)

tall, of bluish-green, ovoid to obovoid or circular pads mostly up to 2 inches long and one-half or more as thick but sometimes up to 4 inches long. The spines are barbed, in clusters of up to nine or sometimes solitary, of varying lengths, up to 1¼ inches. The yellow or greenish blooms, 1½ to 2 inches wide, are succeeded by dry, usually spiny, obovoid fruits up to ½ inch long.

One of the hardiest and the most widely distributed of North American cactuses, variable ***O. humifusa*** (syn. *O. compressa*) occurs on shores, dunes, and rocks from Massachusetts to Montana, Florida, and Texas. Sometimes misnamed *O. vulgaris*, a name which properly belongs to a quite different South American species, the North American sort forms large mats of prostrate stems and obovoid to nearly circular pads 2 to 6 inches long that are spineless or with some of the areoles producing a single spine, rarely more. In winter they often become reddish, wrinkled, and more or less flabby. The bright yellow flowers, 2 to 4 inches wide and sometimes with red centers, are produced in abundance. They are followed by red or red-purple, fleshy, edible fruits 1¼ to 2 inches long. Native from Delaware to Florida and Mississippi, *O. h. austrina* has tuberous roots and erect stems up to 3 feet tall. The upper areoles have usually one or two twisted spines, at first white, becoming gray later.

Treelike Mexican ***O. leucotricha*** attains heights up to 15 feet. From 4 to 10 inches long, its elliptic-oblong to nearly circular pads are pubescent. The spines on young pads are solitary or in two or threes and short, but on older ones they are more numerous and up to 3 inches in length. The 3-inch-wide, yellow flowers are followed by edible, 2-inch-long, white to red fruits.

Sometimes called thimble cactus, ***O. longispina corrugata*** (syn. *O. corrugata*) is a variety of an Argentinian native with mostly white instead of the chiefly purple spines of the typical species. Low and spreading and about 4 inches tall, this has thick, ovate to elliptic segments ¾ inch to 1¼ inches long with white areoles and yellow glochids. Somewhat under ½ inch in length, the spreading, white spines are in clusters of six to eight with sometimes one or two longer and yellowish. The orange-peach flowers, about 1 inch long, are succeeded by red fruits.

Bunny ears (***O. microdasys***) is a favorite. A native of Mexico and often forming wide clumps 1 foot to 2 feet in height, this has stems with oblong to almost circular, spineless, velvety-pubescent pads 3 to 6 inches long decorated with tufts of golden-yellow glochids sprouting from numerous areoles. The 2-inch-long, yellow, usually reddish-tinged blooms with white-stalked sepals and white styles are succeeded by spherical, fleshy, red fruits. Cinnamon-brown glochids are typical of the cinnamon cactus (*O. m. rufida*), those of the angel wing cactus (*O. m. albispina* syn. *O. m. albata*) are white.

A highly variable native of the southwestern United States and Mexico, ***O. phaeacantha*** is by some authorities separated into several species that by conservatives are considered varieties. Characteristically these are prostrate or sprawling, form clumps 2 to 20 feet wide and 1 foot to 3 feet tall. They have bluish-green, obovate pads 4 to 6 inches long or some-

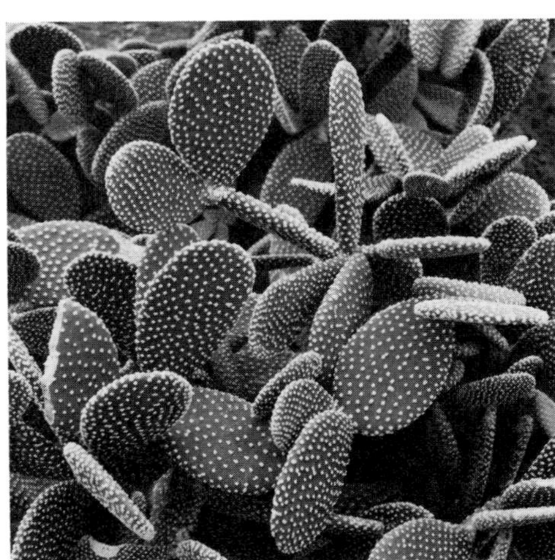
Opuntia microdasys albispina

times longer that in cold weather become reddish. In clusters of up to 9 or solitary, the brown spines are 1 inch to 2½ inches long. The flowers are yellow, sometimes with red bases to the petals. They are 2½ to 3¼ inches in diameter. The fruits are fleshy and wine-colored. The plant known as *O. mojavensis* is *O. p. major*, which has bigger, rounder pads with spines in clusters of not more than three or sometimes solitary. For long mistakenly identified as *O. engelmannii*, variety *O. p. discata* is the largest native prickly-pear of the southwestern United States. Its pads are bigger than those of the typical species and broad-elliptic to nearly circular. The spines are white or ashy-gray.

The plains prickly-pear (***O. polyacantha***) forms clumps up to several feet in diameter and 6 inches to 1 foot high. Its bluish-green, obovate to almost circular

Opuntia polyacantha

Opuntia tuna monstrosa

pads are 2 to 6 inches long and freely decorated with white to brown or gray, mostly down-pointing spines in clusters of five to eleven, of which one is usually conspicuously longer than the others and up to 2¼ inches long. The flowers are yellow, sometimes tinted with pink or red, and 2 to 2½ inches wide. They are succeeded by dry, spiny, obovoid fruits, light to deeper brown when mature and, ¾ to 1 inch long. This quite variable species is widely distributed as a native throughout much of the western United States and adjacent Canada.

Bushy Mexican ***O. setispina,*** up to 3 feet tall, has erect branches. Its roundish, bluish-green pads, about 2 inches across, have yellow glochids and slender, brownish-yellow spines in clusters of four to ten of which a few are up to 1½ inches long, the others much shorter. The blooms are yellow, the fruits purple.

Spreading or prostrate ***O. soehrensii,*** of Peru, Bolivia, and Argentina, forms mounds or cushions up to 3 feet wide or wider, and 8 inches high. Its ovate to almost circular, often purplish, spiny pads are 1½ to a little over 2 inches long. In clusters of eight or fewer, the yellowish to brown, slender spines are up to 2 inches long. From light yellow to orange-red, the flowers are about 2 inches wide. The red fruits, a little over 1 inch long, are without spines.

Forming clumps several feet wide, popular ***O. sulphurea,*** inhabits Argentina and Bolivia. Up to about 1½ feet in height and a rather loose bush, it has ovate to elliptic pads 4 to 10 inches long. They have warted surfaces and little areoles with yellowish-red glochids and clusters of up to eight straight, curved, or twisted, rigid, awl-shaped spines up to ½ inch long. The sulfur-yellow blooms are up to 1¾ inches wide.

One of the first cactuses to be cultivated, ***O. tuna,*** of the West Indies, is a shrub 2 to 3 feet tall with obovate to circular pads approximately 4 inches across, sometimes longer. They have brownish areoles with yellow glochids and usually three to five spreading, yellow spines. The flowers are yellow more or less tinged with red. The fruits are obovoid, red, and about 1¼ inches long. The pads of *O. t. monstrosa* develop numerous small, slender branches.

Purple prickly-pear (***O. violacea***) is chiefly a native of deserts of the southwestern United States and Mexico. Generally a spreading shrub 2 to sometimes 7 feet in height, but at times with a distinct trunk and treelike, this always has purplish to lavender pads that in extremely dry or cold weather become red. They are broadly-obovate to rounder, and 4 to 8 inches in diameter. They may be spineless or have a few spines, solitary, in twos or threes, chiefly on the upper edges of the pads and 2 to 7 inches long. The flowers, with red-based, yellow petals, are 2½ to 3½ inches across. Red to purple, the fleshy fruits are 1 inch to 1½ inches long by about one-half as wide. Sometimes called dollar cactus, *O. v. santa-rita* (syn. *O. santa-rita*) has circular pads 6 to 8 inches in diameter. Called the flapjack cactus, *O. v. gosseliniana* (syn. *O. gosseliniana*) is another variant.

Opuntias with cylindrical or club-shaped stems, those called chollas, are next described. They are presented in alphabetical sequence.

The teddy bear cholla (***O. bigelowii*** syn. *Cylindropuntia bigelowii*), native from Nevada to Arizona, California, and Mexico, has much the aspect of *O. fulgida* and like it is sometimes called jumping cholla. Attaining heights of 3 to 9 feet, it has a single trunk topped by a compact head of tubercled branches that branch only once or twice. The spindle-shaped joints, 3 to 5 inches long by 1¼ to 2½ inches wide and green or bluish, are easily detachable. The clusters of six to ten pinkish or reddish-brown spines, much more strongly barbed than those of *O. fulgida* and spreading in all directions, are ½ to 1 inch long. From 1 inch to 1½ inches wide, the blooms are light green or yellow streaked with lavender. The fleshy, tubercled fruits, up to ¾ inch long and wide, persist for many weeks but unlike those of *O. fulgida,* not for years, nor do new flowers and fruits develop from older ones. Mexican ***O. bulbispina*** (syn. *Corynopuntia bulbispina*) has thick turnip-like roots and forms cushions up to 4 feet across by about 4 inches high of very glaucous, ovoid to club-shaped, warty joints, branched at their ends. The spine clusters are of eight to twelve white, bristle-like radials, about 1/10 inch long, and four, stronger, reflexed, brown to whitish centrals, ½ to ¾ inch long, markedly bulbous at their bases. The flowers and fruits are unknown. Not over 3 inches tall, ***O. clavata*** (syn. *Corynopuntia clavata*) forms mats 3 to 6 feet wide; this is a very distinct native of sandy soils in the Southwest. Its stems are of conspicuously lumpy, club-shaped joints markedly narrowed to their bases, mostly 1½ to 2 inches long by ¾ to 1 inch in diameter. The clusters of ashy-gray, mostly down-

Opuntia echinocarpa

Opuntia echinocarpa (flowers)

pointing, approximately 1-inch-long, flattened spines consist of ten to twenty individuals one of which is stronger than the others and dagger-like. The yellow flowers are 1¼ to 2 inches across.

The silver or golden cholla (***O. echinocarpa*** syn. *Cylindropuntia echinocarpa*), of the Mojave and Colorado deserts, is a freely-branched shrub with a definite trunk. Mostly 2 to 6 inches long, sometimes exceeding 1 foot in length, its conspicuously tubercled joints are ¾ inch to 1½ inches in diameter. The silvery, straw-colored, or golden-yellow spines ¾ inch to 1½ inches long and spreading in all directions, are in clusters of three to twelve. The greenish-yellow blooms, their outer parts tinged or streaked with red, are 1¼ to 2½ inches wide. Changing as they mature from green to pale brown or straw-yellow, the spiny fruits, ¾ inch to 1¼ inches long, are top-shaped to almost hemispherical.

Jumping cholla (***O. fulgida*** syn. *Cylindropuntia fulgida*) is a shrub or small tree 3 to 15 feet tall with one or more distinct trunks up to 3 feet in height ending in much-branched heads of cylindrical to more or less spindle-shaped, easily detached, lumpy-surfaced joints 2 to 6 inches long by 1¼ to 2 inches wide. The needle-like, barbed spines, in groups of two to twelve, are ½ to 1 inch long and spread in all directions. Pink or white streaked with lavender, the ¾-inch-wide blooms have recurved petals. The usually sterile, green, obovoid fruits, 1 inch to 1½ inches long during their first season, but becoming somewhat bigger with age, may remain attached for twenty years or longer. New flowers and fruits develop from older ones and in time from long, branched, pendulous chains. The boxing glove cactus (*O. f. mamillata* syn. *O. mamillata*), 2 to 4 feet tall, has drooping younger parts, more prominent tubercles, and fewer, shorter, less conspicuous spines than the typical species. A crested variant, *O. f. m. cristata*, is cultivated. The name jumping cholla is also sometimes used for *O. bigelowii*.

The chain-link cactus (***O. imbricata***), a native from Colorado to Mexico, is tree-

Opuntia imbricata

like, up to 12 feet tall, and has spreading branches. Its stem joints, up to 8 inches long by about 1 inch thick, have prominent tubercles. In clusters of eight to thirty, the about 1-inch-long spines are brown to whitish. Approximately 3 inches in diameter, the pink to purple flowers are succeeded by yellow fruits. Native to Baja California, ***O. invicta*** (syn. *Corynopuntia*

Opuntia imbricata (flowers)

invicta) is a spreading plant up to 1½ feet tall with egg-shaped joints 3 to 4 inches long by 3 inches wide. The areoles bear a few whitish glochids and clusters of up to eighteen or more spines, of which ten or more are curved, flat or four-angled centrals about 1¼ inches in length, red or red and white when young, later gray. About 2 inches in diameter, the flowers are yellow, as are the fruits. The latter have many glochids, but no spines.

The desert Christmas cactus (***O. leptocaulis*** syn. *Cylindropuntia leptocaulis*) is a shrub 1 foot to 3 feet tall with from its base, stemlike, pencil-thick or thinner, slightly lumpy branches of joints up to 1 foot long or longer and shorter side branches of smooth joints 1 inch to 3 inches long. The gray or sometimes pink-tinged, wide-spreading to down-turned spines are solitary and 1 inch to 2 inches long. From a little under to a little over ½ inch across, the green, yellow, or bronzy blooms are succeeded by obovate to more cylindrical, bright red, fleshy fruits about ½ inch long. This is native to the southwestern United States and Mexico. Mexican ***O. moelleri*** (syn. *Corynopuntia moelleri*) has much the aspect of *O. clavata*, but its stem segments are usually shorter and thinner and more conspicuously warted. Each areole has numerous short, bristle-like radial spines and three or four needle-like centrals up to slightly over ½ inch long. The blooms are yellow to salmon. A natural hybrid betwεen *O. bigelowii* and *O. hoffmannii*, intermediate between its parents, ***O. munzii*** is an upright shrub 6 to 12 feet tall and nearly as wide. Its ultimate joints often are slightly drooping to pendulous. Mostly up to 6 inches long, occasionally longer, they are 1½ to 2 inches thick. The areoles are furnished with tan glochids and ten to twelve needle-like, yellow spines ½ to ¾ inch long. The flowers are yellowish-green, tinted with red and about 1½ inches wide. The fruits are without fertile seeds. Native to California, prostrate or bushy ***O. parryi*** is up to 4 feet tall. Its stems consist of ¾-inch-thick joints 3 inches to 1 foot long. Of unequal size, the spines are in clusters of up to twenty or occasionally are solitary. The fruits are yellow tinged with red.

Opuntia munzii

Opuntia parryi

The diamond cholla ***O. ramosissima*** (*Cylindropuntia ramosissima*), of the southwestern United States and Mexico, forms mats, bushes, or shrubs 6 inches to 5 feet tall and very freely-branched. Its pencil-like stem joints, grayish-green and 2 to 4 inches long by about ¼ inch thick, show a distinctive diamond pattern of flattened, cushion-like tubercles. The mature spines are solitary, needle-like, and 1 inch to 2¼ inches long. On short side branches, the ½-inch-wide blooms are apricot-colored to brown suffused with lavender to red. Native to Argentina, the mule's ears or lion's tongue (***O. schickendantzii***) is 3 to 6 feet tall. Its grayish-green, somewhat flattened to cylindrical stem segments are up to 8 inches long by ¾ inch to 1½ inches wide. The areoles have white glochids and one or two spines up to ¾ inch long. The yellow blooms are 1½ inches wide. The more or less spherical to ovoid fruits, green sometimes with a red blush, are about ½ inch long. The densely-spiny, brownish, burlike, dry fruits are about ¾ inch long. ***O. stanlyi*** (syn. *Corynopuntia stanlyi*) resembles *O. clavata*, but is 6 inches to 1 foot in height and has joints up to 6 inches long and spines at most very faintly grooved or ridged. Variety *O. s. parishii* (syn. *O. parishii*) has smaller joints than the typical species, yellow or red flowers and fruits without spines. Its chief natural range being central Mexico, Ecuador, and Peru, but occurring also locally in Texas, ***O. tunicata*** (syn. *Cylindropuntia tunicata*) is a compact bush 1 foot to 2 feet tall. It has erect stems of prominently tubercled, cylindrical to club-shaped, readily detachable joints 2 to 6 inches long. There are many crowded branches. The white-woolly areoles have clusters of six to ten spines, of which three to six are spreading centrals 1 inch to 2½ inches long and stouter, two to four smaller radials. The pale greenish flowers, about 2 inches wide, are succeeded by spherical to broad-club-shaped, yellowish-green fruits.

Garden and Landscape Uses. Besides the interest collectors of cactuses have in opuntias for outdoor plantings in warm, dry regions and for growing in greenhouses, gardeners with no special affection for the cactus family find some sorts useful and worth exploiting for their decorative qualities. This is especially true in desert and semidesert regions and where long dry summers prevail. But in much colder climates *O. humifusa* is splendid in rock gardens, on banks, for planting near the sea and in other places where it has perfect soil drainage and full sun, and a few kinds succeed in humid, warm areas such as Florida.

In warm, dry regions where, without much attention to watering, other trees are scarce, tall chollas and prickly pears are often prized as ornamentals. They give height and distinctive character to the

A formidable hedge of *Opuntia ficus-indica*

landscape and prosper with little or no attention. As hedges, some opuntias, notably *O. ficus-indica* and its variants, are highly satisfactory in mild, dry regions. These and certain others are also esteemed for their edible fruits. As a safety measure take care not to plant opuntias where passers-by are likely to be injured by their spines or annoyed by their pestiferous glochids.

In greenhouses, windows and other places indoors if they are not watered excessively opuntias are extraordinarily tenacious to life. They put up with, but are not improved by neglect better than almost any other plants. They even persist for very long periods in dim light, and little harm befalls even if left without water for a few weeks while the gardener vacations. In effect they are nearly indestructible.

Cultivation. This is of the simplest. The chief need is for sharp soil drainage and, if they are to flower well, exposure to full sun. Excessive watering is the chief threat to success. Increase is very easy by cuttings and by seed. The type of soil seems not to matter much, but a fertile one with only a small organic content is best. For further information see Cactuses.

ORACH. A rarely grown, spinach-type vegetable or potherb, orach (*Atriplex hortensis*) is an eastern Asian relative of the common pigweed or lamb's quarters. A vigorous, somewhat branching annual in bloom 3 to 5 feet tall, it has long-stalked, triangular-ovate leaves, those of the typical kind green, those of *A. h. rubra* an attractive red hue. The variety is often grown as an ornamental.

For culinary purposes basal leaves harvested before the flowering stalks begin to shoot up are used. Once the flowering stems lengthen, the usefulness of the plants as an edible is over.

Orach, like spinach, is a cool season crop that supplies welcome greens before hot weather arrives. Make the first sowing in rows in a sunny location in fertile soil as early in spring as the ground is workable. One or two later sowings at two-week intervals may be made to provide succession. Have the rows 1 foot to 2 feet apart and thin the seedlings early to stand 8 inches to 1½ feet asunder. In dry weather water freely. Orach may be cooked in the same way as spinach. Because of its bland flavor, sorrel or dandelion are sometimes cooked with it.

ORANGE. Besides being the name of the familiar fruits discussed in the next entry, the word orange is used as parts of the common names of these plants: African-cherry-orange (*Citropsis*), flame-mock-orange (*Chorizema*), Mexican-orange (*Choisya ternata*), mock-orange (*Philadelphus, Pittosporum tobira,* and *Prunus caroliniana*), Natal-orange (*Strychnos spinosa*), orange hawkweed (*Hieracium aurantiacum*), orange-glow vine (*Senecio confusus*), orange-jessamine (*Murraya paniculata*), orange trumpet vine (*Pyrostegia venusta*), osage-orange (*Maclura pomifera*), trifoliate-orange (*Poncirus trifoliata*), and wild-orange (*Prunus caroliniana*).

ORANGE. Most important of citrus fruits, oranges fall into three major categories, sweet oranges, sour oranges, and mandarin, tangerine, and satsuma oranges. Within each major group are subordinate ones and varieties. The fruits are too well known to need description. The trees that bear them are described in this Encyclopedia under Citrus.

Orange: (a) Flowers

Believed to have originated in subtropical and tropical Asia and the Malay Archipelago, oranges have been cultivated for so long that ancestral stocks of most sorts are not surely known. Botanically sweet oranges are grouped as *Citrus sinensis*, sour oranges as *C. aurantium*, mandarins, including tangerines and satsumas, as *C. reticulata*. The plant sometimes called 'Otaheite' orange and more properly identified as the 'Otaheite' lemon, is treated in this Encyclopedia under Lemon. The trifoliate-orange is *Poncirus trifoliata*.

The history of oranges in cultivation and of their distribution from their ancestral homes to virtually every part of the world where climate favors their growth is not documented from its beginning, but sweet oranges were grown in China centuries before Europeans knew of them and there is some evidence that they had been introduced to Italy as early as the first century A.D. Certainly by the fifteenth century, sweet oranges, and considerably earlier sour oranges, were cultivated in southern Europe. Those of the mandarin group came very much later. Although these last had been cultivated for centuries in China and Japan, it was not until 1805 when the first to reach Europe arrived in England. From there, propagations were sent to

(b) Fruits

Malta and eventually other Mediterranean countries and the Americas.

On his second voyage to the New World, Columbus, in 1493, stopped at the Canary Islands and took aboard seeds of useful crops including oranges for transporting to Hispaniola where a Spanish colony was being established. The settlement of St. Augustine, Florida, by Spaniards began in 1565, and fourteen years later an official report stated that great amounts of figs, pomegranates, oranges, and grapes were being grown there.

In the eighteenth and nineteenth centuries orangeries became popular among wealthy people in Europe and some few were constructed in the United States. These were usually massive conservatory-like structures with solid rather than glazed roofs and with glass sides. In them orange trees and sometimes other subtropical kinds, accommodated in tubs and stood outdoors in summer, were wintered.

Sweet oranges fall into four subgroups, the most important of which is the common or blonde orange of which there are numerous horticultural varieties, among them 'Dillar', 'Hamlin', 'Pineapple', and 'Valencia'. Blood oranges, distinguishable because under favorable conditions, but otherwise not dependably, the flesh and juice of the fruits are pink or red, and the skins reddish, have a distinctive flavor. There are several variable varieties of blood oranges, among them 'Maltese Blood' and 'Spanish Sanguinella'. Navel oranges, of which by far the most important is 'Washington' and its variant 'Frost Washington', include also 'Atwood', 'Baianinha Piracaba', and 'Gillette'. They are characterized by the fruits almost without exception having a navel (a tiny rudimentary second fruit embedded at the apex of the major one). Some other oranges occasionally bear a few fruits of this type. The fourth group of sweet oranges, the sugar or acidless varieties, have insipid fruits and are of minor importance.

Sour, or bitter oranges include the common sort used chiefly as understock for grafting and its fruits for marmalade, the bittersweet sort, which has less acid fruits, and kinds classed as variant sour or bitter varieties cultivated as ornamentals and for the production of neroli oil. Noteworthy among the variants is 'Bouquet', a small-foliaged variety excellent as a hedge and for other decorative landscape uses. Myrtle-leaf oranges, sorts with very small, usually lance-pointed leaves of which there are a few varieties, are also classed as belonging with the sour orange group.

Mandarin or tangerine oranges, distinguished from all others by their loose, easily removable skins, and among the most highly esteemed citrus fruits, include a goodly number of varieties, some with fruits with typically tangerine-colored rinds, others paler. The mandarin satsuma varieties, of which there are several, are the most resistant to cold of commercially cultivated oranges. 'Clementine', 'Dancy', and 'King' are well-known varieties of mandarins. Closely similar to the mandarins, variety Temple' is the most cold-sensitive of oranges. Its fruits often have a small navel.

Garden and Landscape Uses. In the United States, oranges are restricted as a commercial crop chiefly to California and Florida with less extensive acreages in Arizona and Texas and still smaller ones in Louisiana and Mississippi. Hardier than grapefruits, lemons, and limes, they are grown for their fruits and as ornamentals in home gardens in some other areas, along the Atlantic seaboard as far north as South Carolina.

As pot and tub specimens, oranges do well and are suitable for decorating terraces, patios, and similar places and for cultivation in greenhouses and even as window plants where space permits.

Choose sites for oranges with some care. Full sun is desirable, exposure to high winds and in some areas high summer temperatures, detrimental. Special consideration must be given to air drainage. Frosts occur more frequently in valleys and other low places than on higher ground since during periods of low temperatures cold air flows down into them. In regions subject to frost choose elevated or protected sites. In Florida, locations near lakes or other large bodies of water are likely to be favorable.

Soils satisfactory for oranges are of various types, but some few, including alkali ones, will not do. Avoid shallow soils unless the underlying stratum can be broken to allow deep root penetration, and do not plant on compact, sticky clay ground or other land that does not drain well. Unless irrigated, excessively dry soils are unsuitable.

Cultivation. Oranges make several flushes of growth each year alternated by periods of semidormancy. They may be successfully transplanted whenever they are not in active growth, but in practice, winter is generally the favored season. For establishing new groves, bare-root trees one or two years old from the time of budding on understocks four or five years old are preferred. Take great care not to allow the roots to dry. If the trees cannot be planted immediately upon receipt, unpack them and heel them in (plant them temporarily closely together) in a shaded place and water them. Exposing the roots to sun and wind is extremely damaging. As an alternative to bare-root trees, container specimens may be planted.

Home gardeners and others concerned with immediate landscape effects often set out trees considerably larger than those used to start commercial groves. Such big specimens must be dug and transplanted with an intact ball of soil around their roots and be either burlapped or boxed. Reduce the size of head of large specimens by pruning before the trees are dug.

Oranges resent being planted too deeply. They must not be set lower than they were previously; in fact, they may well be planted slightly higher to allow for settling. When planting bare-root trees spread the roots, positioning them as they have been growing, work soil between them, taking care not to leave voids, and pack it firmly. Complete the operation by soaking the ground with water to more than the full depth of the roots, then mulch with hay, straw, grass, leaves, or other suitable material.

Prune to whatever degree is desirable immediately planting is completed. This is important to reestablish an acceptable balance between the root system, which has inevitably been reduced in the digging operation, and the amount of top growth the roots will be called upon to support. A common fault of inexperienced gardeners is a reluctance to cut as severely as is best for the trees. Do not hesitate to be bold. By thinning out and shortening branches, remove from one-third to one-half the total top growth of large specimens. Young, single-stemmed trees may be cut back to within 1½ feet of the ground. Those branched at or a little above this height may have the main stem shortened and their branches trimmed so that each is left with only two or three buds.

If the trees will be exposed to strong sun protect their trunks from sunscald by winding around them tree-wrap paper, burlap, or similar material or by wrapping them in straw or moss. Mulching after planting is beneficial.

Adequate supplies of moisture in the soil are essential throughout the year for orange trees to prosper. To achieve this, greater amounts of irrigation water are normally needed in summer than winter. Without sufficient water the growth lacks vigor, the leaves curl and shed, and the young fruits drop. Progressively the trees become debilitated, more subject to pests and diseases, and eventually die. Frequency of watering or irrigation must be adjusted to the specific needs of different climates and soils.

Fertilizing regularly to supply generous amounts of nitrogen, usually by applying some organic material such as animal manure or compost or by turning under cover crops, or by chemical fertilizers, is necessary for best success with oranges.

Pruning established trees demands very little attention. It consists chiefly of the cutting back or the removal of any branches that die as a result of frost injury

Opuntia basilaris

Onopordum acanthium

Oplismenus hirtellus variegatus

Oenothera xylocarpa

Onions harvested and drying in mesh-bottomed trays before storage

Odontonema strictum variegatum

Origanum vulgare

Opuntia humifusa

Ornithogalum thyrsoides

Opuntia humifusa

Ornithogalum umbellatum

or other cause and occasionally a little thinning to prevent overcrowding and to admit more light to the interior of the trees.

Protection against damage by frosts and freezes is necessary or desirable in many areas where oranges are cultivated. Forewarnings are issued over the radio and by other media in orange-growing regions. The most effective way of assuring survival of the trunk (if other parts are killed they can be pruned back and new shoots that grow from the live trunk will replace them) is to mound soil around them in early November before they have been even lightly frosted. Use clean soil free of debris (subsoil is good for the purpose) and clear all weeds and trash away from the vicinity of the trunks before installing it. Because loose earth settles, add more three or four weeks after the first banking and if necessary repair the mounds whenever they show need. In spring when there is no more danger of frost remove the soil mounds.

As an alternative to banking, the trunks of young trees can be effectively protected by wrapping several thicknesses of newspaper around them, tying the wrap securely, and heaping a little soil around the base to make sure that not even a small part of trunk between the ground and bottom of the wrap remains exposed.

Heating by burning oil or other fuel in strategically placed burners is a common method of protecting against damage by cold. Contrary to a once popular belief, blanketing the orchard with a smudge of smoke and water vapor is of slight help, if any. To be effective, sufficient fuel must be burned to maintain the temperature of the air surrounding the trees at above the danger level. This is likely to require for each hour from 10 to 20 gallons of oil or equivalent heat-producing amounts of other fuels for each acre to be protected. Enough fuel should be on hand to allow for at least twenty-four hours continuous operation, with replacement supplies conveniently available.

As container specimens oranges give little trouble in fertile, well-drained, loamy soil kept always moderately moist, never for long periods wet. In cold climates they may be kept the year around in a greenhouse or an equivalent accommodation or kept there from fall to spring and summered outdoors. Winter night temperatures indoors of 40 to 50°F, and daytime ones five to fifteen degrees higher, suit. From spring to fall night temperatures of 60 to 70°F or more, rising five to fifteen degrees by day, are satisfactory. Generous watering and fertilizing from spring through fall is in order. In winter less watering and no fertilizing is proper.

Specimens in small pots may need repotting every two to four years, those in containers 2 to 4 feet or more in diameter less frequently. So long as fertilizing is given regular attention, oranges with roots confined thrive for long periods, those in containers 5 or 6 feet wide and 3 to 4 feet deep often indefinitely.

When repotting is needed, do the work in late winter or early spring before new growth begins and at the same time prune the plants to shape and size. In years when repotting is not needed, top-dress by pricking away some of the surface soil and replacing it with a rich mix, and attend to pruning.

Propagation of oranges is most commonly done by budding or grafting onto two-year-old seedling understocks, usually those of sour oranges, sometimes of sweet orange or grapefruit, or where other understocks do not thrive on rough lemon. Shield budding in fall or winter is the procedure most commonly employed. Oranges are also easily raised from cuttings inserted in a greenhouse propagating bench or, in warm climates, under mist outdoors but this is rarely done except to produce ornamentals for growing in containers. The root systems of specimens propagated from cuttings are likely to be shallower, and so less capable of anchoring trees in the open ground than are those of seedlings. The trifoliate-orange (*Poncirus trifoliata*), rarely employed as an understock, usually dwarfs varieties worked onto it. In general it lowers the quality of the fruits, but is useful for specimens grown in containers.

Pests and Diseases. The most serious pests are scale insects, mealybugs, red spider mites, thrips, and whiteflies. A wide range of diseases may develop, including melanose fungus, which causes rough spots on twigs, foliage, and fruit as well as distortion of the leaves, gummosis, stem-end rot, scab, canker, sooty blotch, and several fruit rots. State Agricultural Experiment Stations and Cooperative Extension Agents in orange-growing regions can supply up-to-date information about pests and diseases and the most practical ways of controlling them.

ORANGEQUAT. See Calamondin, and Fortunella.

ORBIGNYA (Orbígny-a) — Babassu Palm, Cohune Palm. Twenty-five species of tropical American pinnate-leaved palms are included in *Orbignya*. Some are of commercial importance. They belong in the palm family PALMAE. Their name honors the French naturalist Alcide Dessalines d'Orbigny, who died in 1876. Very closely related to *Attalea* and *Scheelea*, orbignyas differ in botanical details of their flowers and seeds.

The babassu palm (***O. barbosiana***) is a Brazilian native, the source of babassu oil used in soaps and margarines. It attains a height of 50 to 60 feet and has leaves 15 to 25 feet long. The oil contained in the kernels of its extremely hard seeds is extracted after their shells are removed. This difficult and laborious task is done by hand by natives of the forests where the palms grow. Despite repeated efforts, no machine satisfactory for shelling the nuts has been devised. The cohune palm (***O. cohune*** syn. *Attalea cohune*), a native of Honduras, is 40 to 60 feet tall. It has a stout trunk and erect, arching leaves up to 20 or 25 feet long with numerous dark green leaflets up to 1½ feet long. The broad-ovoid fruits, about 3 inches long, contain kernels that are the source of cohune oil, a solid yellow fat similar to coconut oil and used for manufacturing margarine and soap.

Garden and Landscape Uses and Cultivation. These palms are suitable for cultivation under humid tropical or essentially tropical conditions only. They may be grown outdoors in southern Florida and Hawaii, but chiefly they are for special collections. They are appropriate for inclusion in displays of plants useful to man, such as are sometimes maintained by botanical gardens outdoors or in greenhouses. They require the same cultural treatment as *Areca*. For further information see Palms.

ORCHARD. A fruit garden or orchard can be an extraordinarily rewarding adjunct to a home and if properly planned and cared for will supply a great deal of healthful food for eating out-of-hand and cooking. The key phrase is "if properly planned and cared for." An ill-conceived or neglected orchard is more likely to be a liability than asset.

Among advantages that come from a well-managed orchard is the enjoyment of higher quality fruit in greater variety than comes to market. Commercial growers concentrate on varieties that mature evenly, and so can be harvested with the fewest possible pickings, that ship well, and that have fruits of even size and attractive eye appeal. Such sorts do not necessarily include the finest flavored nor the best for cooking. Moreover, much commercial fruit of necessity is gathered before it is completely ripe, and variety for variety it does not compare to that left on tree, vine, or bush until perfection is attained.

Besides the varieties of popular fruits grown in vast quantities in commercial orchards there are others available, although fewer alas than half a century and more ago, of superior quality or with other virtues that commend them to the home gardener. Certain ones are offered in catalogs of nurseries specializing in fruits, others can be located through State Agricultural Experiment Stations. There are also some fruits, such as improved varieties of the native persimmon, not grown commercially that may be considered for planting.

The home fruit garden need not, indeed generally should not, be large. Very few or sometimes only one tree of a variety is likely to produce enough fruit of its particular kind for an average family. The same is true of a few bushes or vines of other sorts.

Choice of site is obviously limited by the property, from generally a narrow one on ordinary lots to wider opportunity on farms and homes with acreage. For all fruits full sun is needed and the soil must be well drained, of fair depth, and reasonably fertile or capable of being made so. Hilltops and the bottoms of hollows are much less suitable for fruits than level land or part way down slopes. The former are too exposed to wind, the latter to cold air draining and collecting with the result that spring blossoms are killed.

What to grow must be determined largely by local climate and personal preference. Quite certainly it is impracticable to grow oranges or mangoes in New York or currants or apples in Florida. Nevertheless the home cultivation of many sorts of fruit can succeed where attempts at commercial production would result in total failure. Within the limits of New York City many residents, especially those of Italian ancestry, enjoy each year crops of delicious home-grown figs, and the successful cultivation of grapes, peaches, cherries, or plums is by no means limited to areas in which they are grown for market.

Important matters to consider before deciding whether and what to plant include these: Tree fruits, such as apples, pears, peaches, plums, and cherries, unless grafted on dwarfing understocks require the availability of a sprayer of sufficient power to reach their tops with disease- and insect-controlling sprays that in some cases must be applied several times each year, usually at precisely scheduled times. This is likely to make reliance upon hired spraying services impracticable and to necessitate investment in equipment. Birds, especially in built-up areas, are much more likely to damage the fruits of solitary or few specimens of cherries, plums, and other soft fruits than the same sorts in more extensive plantings. Some few fruits require very specialized conditions, as for instance blueberries, for which a highly acid soil is a must. Certain varieties of fruit trees are self-sterile; they will not crop unless another variety, and that a compatable one, is planted nearby.

Before planting an orchard give serious consideration to the fact that such an undertaking is a fairly long-time project, that except for certain bush and vine fruits, it will not come into full production for several years, and that maintenance necessitates annual timely, intelligent attention to such tasks as spraying, pruning, and harvesting. Should you contemplate dwarf trees keep in mind that these call for more precise attention to pruning than standard ones of the same varieties. Above all, if inexperienced in fruit growing, from the beginning seek the advice of your Cooperative Extension Agent or other local authority. For discussion of individual sorts of fruits consult entries in this Encyclopedia under their names such as Apples, Cherries, Grapes, Peaches, Pears, and Plums.

ORCHID. For common names of orchids see the end of the entry titled Orchids. Some plants that are not orchids have the word orchid as parts of their common names. These include Hong Kong orchid-tree (*Bauhinia blakeana*), orchid cactus (*Epiphyllum*), orchid- or shell-ginger (*Alpinia mutica*), orchid tree (*Bauhinia variegata*), orchid vine (*Stigmaphyllon ciliatum*), poorman's-orchid (*Schizanthus*), and water-orchid (*Eichhornia*).

ORCHID BASKET. Hanging containers made of strips of wood in which epiphytic orchids, especially those with pendulous flowers, and sometimes other plants are grown are called orchid baskets.

The strips, preferably of teak, cypress, redwood, or other rot-resistant lumber, are square in section with ½- to ¾-inch sides. To make a basket, cut the strips into even lengths and bore a small hole through each end of each piece. Push pieces of stout copper wire through the holes of two strips and fasten their ends securely underneath. Then thread onto the wires two more strips connecting the first two to form a square and with their ends overlapping at right angles. Repeat this procedure to form openwork sides for the basket, and when their height is sufficient gather together the free ends of the four wires and twist them together into a hook. Then lay other strips to form the bottom of the basket.

An orchid basket

Orchid baskets of this sort are especially suited to the needs of such epiphytes as nepenthes and many bromeliads as well as orchids because they permit ready access of air to the roots and sharp drainage. For such orchids as catasetums, which send their flower spikes downward through the rooting medium, they are especially appropriate.

ORCHID PEAT. See Osmunda Fiber or Osmundine.

ORCHIDACEAE—Orchid Family. Variously accepted by different authorities as comprising from 400 to 800 genera and 15,000 to 35,000 species, the orchid family consists of monocotyledons widely distributed throughout the world with the greatest concentration of its species in the tropics. Its members, called orchids, are essentially herbaceous, evergreen or deciduous, occasionally vining perennials. Most are epiphytes (tree-perchers), a goodly number terrestrials (grow in the ground), a few are saprophytes, and still fewer parasites on funguses.

Some orchids have underground tubers, a great many have pseudobulbs (thickened, aboveground, bulblike stems). In a few cases the vegetative parts consist only of a mass of roots. The flowers, solitary or in spikes, racemes, or panicles, vary greatly in size according to kind, and often are of extraordinary, sometimes grotesque shapes, in general related to the methods by which pollination is effected. Usually bisexual, rarely unisexual, the blooms have perianths of typically six segments, the outer three sepals, the inner three comprising two petals and a lip, the latter a modified petal. The sepals are often petal-like in texture and color. A striking characteristic of this family is that the style, stigma, and stamens of the flowers are variously united to form a complex organ called the column. If there is only one stamen it is at the apex of the column. If two (or rarely three) they are lateral. The pollen is generally in waxy masses called pollinia, of which there are commonly from two to eight. The fruits are capsules containing numerous minute seeds without endosperm (a starchy nutrient).

Horticulturally extremely important as a source of cut flowers and of plants treasured by collectors and fanciers the orchid family is almost without other values. Natural vanilla, now largely superseded by its synthetic substitute, is a product of *Vanilla planifolia* and to a lesser degree of *V. pompona* and *V. tahitenseis*. Various orchids have been used as native medicines, very few as foods, some as sources of gums or glues. It is estimated that by the beginning of the third quarter of the twentieth century, 60,000 sorts of interspecific hybrids had been produced by orchid breeders.

Besides natural genera (sorts that exist in the wild) numerous artificial genera of orchids (intergeneric hybrids raised by plant breeders) are cultivated. Most of these have identifying names, frequently formed, as with *Laeliocattleya* from those of the parent genera, which in the example given are *Laelia* and *Cattleya*, but sometimes without such association. Here is a list of the natural genera and many of the artificial genera treated in this Encyclopedia: *Acampe, Acanthephippium, Acineta, Ada, Adaglossum, Adioda, Aeridachnis, Aerides, Aeridocentrum, Aeridofinetia, Aeridopsis, Aeridostylis, Aeridovanda, Aganisia, Angraecopsis, Angraecum, Anguloa, Angulocaste, Anoectochilus, Anoectonaria, Anota, Ansellia, Ansidium, Aplectrum, Arachnis, Arachnopsis, Arachnostylis, Aranthera, Arethusa, Armodachnis, Arpophyllum, Arundina, Ascocenda, Ascocentrum, Ascofinetia, Asconopsis, Ascorachnis, Ascotainia, Ascovandoritis, Aspasia, Aspasium, Batemannia, Bateostylis, Beaumontara, Bifrenaria, Bifrenlaria, Bletia, Bletilla, Bloomara, Bollea, Bothriochilus, Brapasia, Brassavola, Brassia, Brassidium, Brassocattleya, Brassodiacrium, Brassoepidendrum, Brassolaelia, Brassolaeliocattleya, Brassophronitis, Brassotonia, Bromheadia, Broughtonia, Bulbophyllum, Burkillara, Burrageara, Calanthe, Calopogon, Calypso, Campylocentrum, Catamodes, Catanoches, Catasetum, Cattleya, Cattleyopsisgoa, Cattleyopsistonia, Cattleytonia, Caularthron, Caulocattleya, Cephalanthera, Ceratostylis, Charlesworthara, Chrondrobollea, Chondropetalum, Christieara, Chysis, Cleistes, Cochleanthes, Cochlenia, Cochlioda, Coelia, Coelogyne, Coeloplatanthera, Colax, Comparettia, Corallorhiza, Coryanthes, Cycnoches, Cycnodes, Cymbidium, Cyperocymbidium, Cypripedium, Cyrtopodium, Dactyleucorchis, Dactyloceras, Dactylorhiza, Dekensara, Dendrobium, Dendrochilum, Dendrophylax, Dewolfara, Diabroughtonia, Dialaelia, Dialaeliocattleya, Dialaeliopsis, Dillonara, Diplocaulobium, Disa, Doricentrum, Doriella, Doriellaopsis, Doritaenopsis, Doritis, Dossinia, Dossinimaria, Ellanthera, Elleanthus, Encyclipedium, Epicattleya, Epindendrum, Epidiacrium, Epigenium, Epigoa, Epilaelia, Epipactis, Epiphronitella, Epiphronitis, Epitonia, Eria, Eriopsis, Ernestara, Erythrodes, Eulophia, Eulophidium, Eulophiela, Eurychone, Fujiwarara, Galeandra, Garayara, Gastrochilus, Gauntlettaria, Giddingsara, Gilmourara, Gomesa, Gongora, Goodyera, Grammatocymbidium, Grammatophyllum, Greatwoodara, Gymleucorchis, Gymnadeniorchis, Gymnanacamptis, Gymnaplatanthera, Gymnoglossum, Gymnotraunsteinera, Habenaria, Haemaria,*

Orchidantha maxillarioides

Harrisiella, Hawkesara, Herbertara, Herminorchis, Hexisea, Holffumara, Houlletia, Huntara, Huntleanthes, Huntleya, Ione, Ionettia, Ionocidium, Ionopsis, Isotria, Iwanagara, Jacquinparis, Jumellea, Kagawara, Kamemotoara, Kirchara, Laelia, Laeliocatoria, Laeliocattkeria, Laeliocattleya, Laeliopleya, Laeliopsis, Laelonia, Laeopsis, Laycockara, Leptotes, Lewisara, Liparis, Lissochilus, Listera, Lockhartia, Lycaste, Lymanara, Macludrania, Macodes, Macradenia, Macradesa, Macragraecum, Macroplectrum, Malaxis, Masdevallia, Maxillacaste, Maxillaria, Meiracyllium, Miltonia, Mizutara, Mokara, Mormodes, Moscosoara, Mystacidium, Nageliella, Neobenthamia, Neofinetia, Neogardneria, Neomoorea, Nobleara, Notylia, Octomeria, Odontioda, Odontoglossum, Oncidasia, Oncidenia, Oncidettia, Oncidioda, Oncidium, Oncidpilia, Onoara, Ophrys, Orchis, Ornithocephalus, Ornithochilus, Osmentara, Paphiopedilum, Paraphalaenopsis, Pattoniheadia, Plexia, Peristeria, Perreiraara, Pescatoria, Phaius, Phalaenopsis, Pholidota, Phragmipedium, Phymatidium, Physosiphon, Pleione, Pleurothallis, Podangis, Pogonia, Polystachya, Ponthieva, Promenaea, Pterygodium, Reinikkaara, Renaglottis, Renancentrum, Renanthera, Renanthoceras, Restesia, Restrepia, Restrepiella, Rhynchostylis, Rhyndoropsis, Rodriquezia, Rodriopsis, Rumrillara, Saccolabium, Sarcanthus, Sarcochilus, Satyrium, Schomburgkia, Scuticaria, Sigmatostalix, Sobralia, Sophronitis, Spathoglottis, Spiranthes, Stanhopea, Stelis, Stenoglottis, Thrixspermum, Thunia, Tipularia, Trichocentron, Trichoceros, Trichoglottis, Trichopilia, Trigonidium, Vanda, Vandopsis, Vanilla, Xylobium, Zygocolax, and *Zygopetalum.*

ORCHIDANTHA (Orchid-ántha). The name of this genus of perhaps eight species means orchid flower (from the English orchid, and the Greek *anthos*, a flower). Yet *Orchidantha* belongs not to the orchid family, but to the banana family MUSACEAE, or according to some botanists to a family of its own they call the LOWIACEAE. But the flowers of *Orchidantha* indeed bear a superficial resemblance to some orchids. The genus is endemic to southern China and Malaysia.

The only species believed to be in cultivation, ***O. maxillarioides*** (syn. *Lowia maxillarioides*) is a stemless, tufted plant with somewhat the appearance of a compact aspidistra. Its pointed-elliptic, leathery leaves have blades 7 to 9 inches long and stalks of about the same length. Few together in loose panicles shorter than the leaves, the blooms are in successive pairs. They have tubular calyxes with three spreading, dark purplish-brown lobes 1¼ to 1½ inches long, and three petals the upper two of which are very small, and the third, ¾ to 1 inch long, elliptic to obovate, and green with purple variegations. There are five very short stamens,

Orchidantha maxillarioides (flower)

and a short style with three brief branches. The fruits are capsules containing pea-sized seeds.

Garden and Landscape Uses. By no means common in cultivation, *O. maxillarioides* makes a shapely, ornamental pot specimen and in tropical and near-tropical regions is useful for rock gardens and planting beneath trees and elsewhere in shade. Its decorative virtues are those of its manner of growth and its foliage. Its flowers are interesting, but are mainly hidden by the leaves.

Cultivation. The species described thrives with minimum care. A rather slow grower, it prefers well-drained, moderately moist, fertile soil. In greenhouses a minimum winter night temperature of 60°F, rising appreciably by day, and at other seasons higher, gives good results. A humid atmosphere and shade from strong sun are needed. Repotting is necessary only at fairly long intervals. Propagation is easy by division and by seed.

ORCHIDS. All plants of the botanical family ORCHIDACEAE are correctly identified as orchids. A vast assemblage, this includes many thousands of garden hybrids as well as the large number of species and few hybrids that occur in the wild. Plants not of the orchid family that have common names including the word orchid are the orchid tree (*Bauhinia variegata*) and the poor-man's-orchid (*Schizanthus*).

Horticultural concern for the group is high, as evidenced by the very great number of books, periodicals, and other publications devoted to it and by the existence and vigor of numerous societies dedicated to promoting interest in orchids and their cultivation. Such societies are to be found in nearly all countries in which plants are grown for ornament. Among the chief ones in English-speaking countries are the American Orchid Society and the Orchid Society of Great Britain.

The history of orchids in cultivation extends much further back than is possible to pinpoint by detailed documentation and as is true of so many cultural activities leads to China. A Japanese authority refers to a treatise on cymbidiums and their cultivation written in China in the tenth century, and several other Chinese writings on orchids appeared before the end of the seventeenth century. In Japan too, interest in orchids began early. The first treatise dealing with their cultivation there of which we have certain knowledge was not produced until 1728, although there are references to earlier writings on the subject. None of these early Oriental productions dealt of course with more than a tiny proportion of the kinds of orchids now in cultivation.

In Europe it is probable that native orchids were grown by herbalists and in botanic gardens before the middle of the sixteenth century, but it was 1698 before the first tropical species bloomed there, in Holland. It was *Brassavola nodosa.* A colored plate published in England in 1732 depicts the first tropical orchid to flower in that country. Obtained from the Bahamas the previous year, this was *Bletia purpurea.* By 1737, and for how long before is not certainly known, hardy *Cypripedium calceolus pubescens* and *C. reginae,* both natives of North America, were in English gardens.

In 1768 the first Asian orchids reached Europe. These, *Cymbidium ensifolium* and *Phaius tankervilliae,* had been brought to England from China.

The cultivation of exotic orchids, other than as occasional curiosities, did not progress much until the early 1800s, but then interest expanded rapidly, first and most especially in England, closely followed by similar enthusiasm in continental Europe and the United States.

English interest was greatly stimulated by a number of factors not the least of which was the founding of the Horticultural Society of London in 1809. But even earlier, British seafarers including such notables as Captain Cook, discoverer of Hawaii, and Captain Bligh of H. M. S. *Bounty* fame had transported from abroad living plants of orchids to their homeland. By 1813 nearly fifty species of exotic orchids were in cultivation at the botanic garden at Kew as compared with the fifteen grown there in 1789.

The first flowering in cultivation of *Cattleya labiata,* an introduction from the Organ Mountains of Brazil, gave tremendous stimulus to interest in orchids. This event, which took place in 1818 at the establishment at Barnet near London of William Cattley, an importer of tropical plants, began what became a highly fashionable enthusiasm among the wealthy for collecting and growing orchids. Further impetus and prestige was given the vogue when in 1833 the Duke of Devonshire after seeing astonishing and beautiful *Oncidium papilio* in bloom at a horticultural exhibition began a collection at Chatsworth, his stately home in Derbyshire. The Duke's participation included the purchase of desirable sorts at auctions of imported plants then becoming popular (at one of these the Duke paid 100 guineas, the equivalent of 5,000 or more 1975 dollars, for one plant of a white-flowered *Phalaenopsis*), and of sending collectors to the Asian tropics to secure plants. Within a decade the Chatsworth collection was the most extensive in existence.

The hobby of collecting and growing orchids eventually almost attained the proportions of the "tulipmania" that swept Holland in the seventeenth century, and as orchid collectors vied with each other for the honor of first flowering new kinds and often less commendably for having such sorts named after themselves, prices of orchid plants rose precipitously, and continued to do so through the nineteenth century and beyond.

Commercial orchid growing was initiated in 1812 by Messrs. Conrad Loddiges of Hackney near London, England. Soon others engaged in the business and the names of some of those early entrepreneurs have become almost legendary in the history of the development of the art of domesticating and cultivating orchids.

To satisfy the rapidly increasing demand from connoisseurs and aspiring beginners for previously unknown species and natural varieties, as well as flowering-size specimens of older ones, it was found highly profitable to send collectors to the tropics to secure native plants for shipment to Europe. And this professional collectors did with such tremendous enthusiasm and effectiveness and on such a vast scale that before the end of the nineteenth century entire regions in South America and some other tropical places were stripped of desirable sorts and certain species practically eliminated. This plunder of the orchid-rich tropics was carried out ruthlessly and with ruinous results to the native vegetation. To obtain the orchid plants, most of which grew high in the crowns of tall trees, forest giants over vast areas were felled and the countryside devastated. In one attack alone, 4,000 trees in Colombia were sacrificed to obtain 10,000 plants of *Odontoglossum crispum,* and other regions were described after the orchid collectors had finished with them as appearing to have been cleaned out by forest fires. The scope of these activities is to some extent indicated by the fact that in 1894 Messrs. Sanders of England and Belgium had twenty collectors actively engaged in Brazil, Colombia, Equador, Peru, Mexico, Burma, India, Malaya, New Guinea, and Malagasy (Madagascar).

In the United States cultivating exotic orchids began sometime before 1829. By then John Wright Boott, a businessman, was growing a few plants in his small greenhouse in Boston, Massachusetts. His modest collection was augmented in 1829 by plants shipped from England by his brother James and later by other importations. Following Boott's death, about 1845, his orchids became the property of John Amory Lowell of Roxbury, Massachusetts. Lowell built a greenhouse to accommodate them and over the next decade continued to expand the collection by importations.

The first man-made hybrid orchid bloomed at Veitch's Royal Exotic Nursery in England in 1856. Attempts made nearly a decade earlier at artificial matings had resulted in the production of viable seeds, but the seedlings soon died. The Veitch achievement, its parents *Calanthe furcata* and *C. masuca,* was appropriately named *C. dominii,* in honor of John Dominy, who had made the actual cross three years ear-

Tropical and subtropical orchids: (a) *Bifrenaria harrisoniae*

(b) *Brassavola cordata*

lier and had sown the resulting seeds in 1854. Dominy followed this success by raising other orchid hybrids, in all twenty-five before he died in 1891. Others attempted hybridizing, but it was not until 1871, fifteen years after *C. dominii* first bloomed, that anyone other than Dominy achieved a flower. During that period Dominy had raised nineteen hybrids involving the genera *Aerides, Anoectochilus, Calanthe, Cattleya, Dossinia, Goodyera, Haemaria, Laelia, Macodes, Paphiopedilum,* and *Phaius,* and the Royal Exotic Nursery was the only place in the world where artificial hybrid orchids were being created and flowered.

Horticultural interest in tropical and subtropical orchids remains high and widespread, and the successful cultivation of numerous species and vast numbers of their hybrids is engaged in by amateurs practically throughout the world. Some sorts are even adaptable as houseplants. In addition, the commercial production of cut flowers of such kinds as cattleyas, cymbidiums, paphiopedilums, and vandas, most popular for that purpose, has become big business.

The simplification of cultural procedures, and as a result of modern methods of propagation, the much greater availability of superior sorts at moderate cost are responsible for the great upsurge of interest in collecting and growing orchids, but this would not have happened if these tropical orchids had not possessed beauty and other intriguing characteristics.

Hardy orchids, terrestrial sorts native to temperate regions have gained no such widespread favor with gardeners. This not

(c) *Brassia longissima*

because they lack appeal or beauty, but because the great majority have not proved easy to grow; indeed they are frequently excessively finicky, and rapid and sure means of propagating them have not been worked out. Still, enthusiasts cultivate some kinds where conditions are reasonably favorable and a few, especially na-

(d) *Calanthe vestita*

(e) *Catasetum pileatum*

(f) *Cattleya percivaliana*

(g) *Coelogyne cristata*

(h) *Cymbidium erythrostylum*

(i) *Dendrobium densifolium*

(j) *Laelia finckeniana*

(k) *Paphiopedilum* 'Onyx'

(l) *Vanda amoena*

Hardy orchids: (a) *Cypripedium calceolus*

(b) *Dactylorhiza elata*

tive species, are practicable for woodland gardens and other shaded areas, bog gardens, wild gardens, and rock gardens.

Associated with the roots of all orchids in the wild are funguses that maintain mutual beneficial relationships with the orchid plants. The combinations of the roots and these symbiotic funguses are called mycorrhizae. Early failures with germinating orchid seeds and keeping alive any seedlings that did appear were often the direct result of the absence of a specific mycorrhizal fungus from the sowing medium. Following the discovery in 1909 of this requirement of orchids as they grew in the wild, a method of germinating their seeds in an artificial culture medium into which the fungus was introduced was worked out. This brought some success, but was far from reliable.

A remarkable breakthrough resulted from investigations by the American plant physiologist Lewis Knudson, who, in 1917, adapted to orchids a technique he had developed for growing other plants from seeds. This pure culture or asymbiotic method as it is called was based on including in a culture medium free of mycorrhizal fungus, sugar as a substitute for the foods that under natural conditions would be supplied by the fungus. The introduction of this method revolutionized the production of orchids from seeds and made possible the extensive hybridizing that has become one of the mainstays of the vast commercial orchid growing industry that now exists.

Another giant step forward came with the discovery, in 1956, of the practicability of raising new orchid plants by culturing minute pieces of meristem tissue taken from the apexes of vegetative growths. Such meristems, as they are called, have the advantage of usually being free of any virus disease with which the mother plant may be infected. By dividing and redividing the protocorm-like bodies that develop from the excised tissue it is theoretically possible to obtain within one year in excess of 4,000,000 orchid plants genetically identical with the original. In practice slower rates of multiplication are accepted, but even so, desirable plants are multiplied with such astonishing speed compared with distressingly slow older methods and in such profusion that prices of all except the newest and scarcest are moderate, and many fine inexpensive kinds are available. This method is only practicable in laboratories.

Tropical and subtropical orchids fall into two broad groups, the epiphytes or tree-perchers and the terrestrials or ground orchids. Representatives of both are cultivated, their needs, especially with regard to rooting mediums, differing somewhat. Gardeners also adopt a convenient classification based on optimum temperature requirements. They group the sorts they grow as cool-, intermediate-, and warm-greenhouse orchids.

Except in the tropics and subtropics where they can be cultivated outdoors or in lath houses, nonhardy orchids are grown chiefly in greenhouses. Keen amateurs have also had considerable success with a limited number of kinds as houseplants, sometimes in terrariums, sometimes under artificial lights. Such achievements confirm what experienced growers have long known, that many orchids are adaptable to a much wider range of environments than most people suspect, that they are "tough," not delicate. Prevailing temperatures determine to a very large degree the kinds of orchids that succeed outdoors and in lath houses in frost-free regions. This is illustrated by the great success had with cymbidiums in southern California and with vandas and other warm-greenhouse sorts in Hawaii.

To grow nonhardy orchids satisfactorily calls for an understanding of their basic needs and the regulation of environmental factors to meet these within acceptable limits and in many cases to satisfy changing seasonal demands. This last is especially important with kinds that have definite resting (dormant) periods. Clues to environmental factors most congenial to different orchids can be gleaned from their habits of growth and types of foliage. Unfortunately temperature requirements are not signaled. Knowledge of these must be obtained elsewhere, by reading, inquiry, observation of the plants' responses, and to some extent, by trial and error.

The possession of pseudobulbs strongly suggests need for one or possibly two periods of dormancy or partial dormancy each year. If the pseudobulbs are well developed and the leaves are thin-textured and deciduous, as, for example, are those of *Catasetum warczewitzii* and some calanthes, assume that plenty of water during the growing season, complete withholding during the resting period, and moderately heavy shade are necessary for satisfactory results.

Sorts with strongly developed pseudobulbs and thicker more leathery leaves than those of the last group respond best to brighter light, but with some shade from all but very weak direct sun. If the leaves are deciduous, water generously during the season of growth, not at all when the plants are leafless. If the foliage is retained throughout the year give only sufficient water through the period of partial dormancy to prevent the pseudobulbs from shrinking, at other times water freely.

Strongly developed pseudobulbs of *Coelogyne cristata*

Orchids without pseudobulbs and with comparatively soft, fleshy, evergreen foliage require shade, quite heavy from spring to fall, from all direct sun; the rooting medium must be kept moist throughout the year.

Slender pseudobulbs of *Dendrobium nobile*

Paphiopedilums are without pseudobulbs

Those without pseudobulbs and with flat, leathery, evergreen leaves need good light with some shade from strong sun, and plenty of water during their period of active growth, lesser amounts at other times. Kinds without pseudobulbs that have evergreen, cylindrical to subcylindrical leaves require little or no shade, generous supplies of water when in growth, moderate amounts when partially resting.

Temperature is obviously important to the successful cultivation of orchids. Cool greenhouse kinds, such as cymbidiums and paphiopedilums, other than those with tessellated (mottled) foliage and white-flowered species and hybrids, prosper where night temperatures in winter are in the 50 to 55°F range; intermediate greenhouse sorts, for example, cattleyas, and laelias, prosper where temperatures on winter nights are between 55 and 60°F; and warm or tropical greenhouse sorts require night temperatures in winter above 60 and up to 70°F. In all cases daytime temperatures may rise five to fifteen degrees above night levels in direct relation to increased brightness of the day, and both day and night temperatures may properly be higher from spring to fall than in winter. In many parts of North America normal outdoor summer temperatures are too high for best success with certain cool greenhouse orchids, such as coelogynes, disas, odontoglossums, and most masdevallias, and best results are had only in air-cooled greenhouses or with some sorts by summering the plants in a lath house or other cool, shaded place outdoors.

Light requirements of different kinds of orchids vary markedly. Some, such as phalaenopsises and ludisias, prefer comparatively low intensities, whereas at the other end of the preference scale are such sorts as vandas that for their best success need bright light; indeed, when grown outdoors in the tropics, they thrive in practically full sun. Cattleyas and the majority of other sorts fall somewhere between these extremes.

Inexperienced growers are much more given to shading too heavily than not enough. By reducing light intensity rich green foliage that suggests health and vigor is developed, but is probably deceiving. Knowledgeable orchid growers are aware that plants that exhibit such coloring are less likely to bloom well than less verdant examples exposed to brighter light. A good rule is to provide the maximum exposure the plants will tolerate without the leaves being scorched. A slight yellowish appearance is more likely to be advantageous than harmful. Seasonal differences in natural light are responsible for the varying degrees of shade needed at different times of the year. For a period in winter, its length depending upon latitude, no shade at all is required.

Humidity maintained at favorable levels is of importance to successful orchid management. Practically all kinds require a moderately high level of relative humidity, forty to fifty percent, and some respond to considerably moister atmospheres. But dank, stagnant air is to be avoided. Provide for reasonable air circulation. Slowly revolving fans located inside and near the ridge of the greenhouse are of great help, especially in hot, humid weather. A too moist atmosphere can result in fungus diseases spotting the flowers of cymbidiums and some other orchids, as well as other troubles. Keep the air sufficiently humid by wetting paths, floors beneath the benches, and other surfaces at whatever intervals are needed and on bright days when foliage will dry within an hour or less by spraying the plants themselves, when not in bloom, lightly with water.

Watering plays a very significant role in determining success with growing orchids. Inexperienced gardeners generally err on the side of keeping the roots too wet. This soon results in too rapid deterioration of the rooting medium, rotted roots, and poor growth.

Although some understanding of water requirements can be obtained from reading and from the admonitions of successful growers, in the final analysis personal experience must guide the gardener. In making use of this, correct interpretations of responses plants have made to previous watering procedures must form the basis for future practices.

Plants in plastic pots require less frequent watering than those in clay pots, this because plastic is not porous. It takes some judgment to accurately determine just when the rooting medium has reached the stage of dryness when watering should be done. Experienced growers evaluate this by visual inspection and their knowledge of when the plant was last watered. If in doubt, it helps to thrust a finger into the compost at the edge of the pot. A cool, damp feeling indicates adequate moisture, a dry warm one, a low moisture content. Lifting the pot to test its weight also gives some idea of the moisture content of the rooting medium.

The quality of the water may be of significance. If it has been heavily chlorinated, it should be allowed to stand for a few hours in an open vessel before use. The pH is also important. For most orchids an approximately neutral or slightly acid reaction is best; too high alkalinity results in distress. Where water is excessively hard, as it is in some limestone regions, it must be softened either by adding a drop or two of phosphoric acid to each gallon or by passing it through a water softener of a sort that does not affect the water adversely for plants.

Fertilizing has been a matter of considerable controversy. For long the consensus was that it was unnecessary and probably harmful to most epiphytic orchids. Many growers, depending chiefly upon osmunda with perhaps the addition of sphagnum moss as a rooting medium, achieved splendid results without fertilizing. But experience with newer, inert potting materials emphasized that orchids, like other plants, absorb nutrients through their roots and a modest uptake in this way is necessary for their well-being. Cautiously at first and then with more confidence, it became accepted that fertilizers carefully used can benefit most if not all orchids, but they must be used with caution and sparingly.

Service only well-rooted specimens during their periods of active growth, and use very dilute solutions frequently rather than stronger ones infrequently. Organic or inorganic fertilizers may be used, diluted to about one-half the strength appropriate for most other pot plants. Many growers perfer slow-release pelleted fertilizers that sprinkled on the surface of the potting material supply minute amounts of nutrients each time the plant is watered.

Orchids are grown in pots, pans, and hanging baskets, on rafts, and attached to slabs of trunks of tree ferns, cork bark, or even sections of tree branches. A few sorts are in warm climates planted in ground beds. Of prime importance is that air circulates freely about the roots of epiphytic sorts. To further this, it is advisable that pots or pans containing them be stood on shelving or staging made of wire mesh or narrow strips of rot-resistant wood spaced 1 inch to 1½ inches apart. The staging should be raised 6 inches to 1 foot or more above the greenhouse bench.

Formerly it was the practice to cover the benches with a layer of coke, coarse cinders, charcoal, or similar material that when wet down several times a day absorbed moisture which it slowly yielded to humidify the atmosphere. Now it is usual to depend upon automatic humidifiers for this.

In the past it was fairly common to facilitate access of air to the roots by using pots having sides pierced with many holes. That is rarely done now, and indeed experience indicates that if watering practices are properly adjusted the great majority of orchids grow as well in nonporous plastic pots as in those of porous clay.

An orchid in a pot

Potting mediums of many sorts give satisfaction, but except for such ground orchids as calanthes and cymbidiums for which mixes containing a proportion of preferably turfy loam are used, soil is not employed.

Osmunda fiber, also known as orchid peat and osmundine, was for many years practically the only material used for potting epiphytic orchids, and it is still fa-

An orchid in a hanging basket

An orchid on a slab of cork bark

vored by many growers for some, and by a few for all kinds. In America osmunda is commonly used alone, in Europe frequently after mixing a small amount of sphagnum moss with it.

Because of increased demand and the greater difficulty of obtaining supplies, the cost of osmunda fiber rose to an extent that stimulated active search for acceptable substitutes. Many, including gravel and other inorganics, were tried and under carefully managed procedures some gave good results. The most universally acceptable proved to be chips of bark of Douglas-fir, fir, pine, or redwood graded to size and sterilized before marketing. More recently, expanded light-weight materials, called orchid gravels, have become popular for potting.

Bark has the great advantage of being much easier to pot with than osmunda, and it is easier to determine when it has reached the stage of dryness that plants growing in it need watering. It comes in several sizes, the smaller grades being used for seedlings and for bigger plants of thin-rooted orchids, the coarser grades for sorts with thicker roots.

For terrestrials such as paphiopedilums (often inaccurately called cypripediums), the rooting medium should be somewhat finer textured than for epiphytes. Various mixes are employed by different growers and for different sorts of orchids. For cypripediums, osmunda fiber mixed with gravel is used successfully and so are small bark chips mixed with about one-sixth part by bulk of clean, washed sand. Phalaenopsises are grown in coarse bark used alone and in a great variety of mixes employing some humus-rich soil or small pieces of partially decayed grass sod with some of the soil shaken out, and varying proportions, depending upon the preferences of individual growers, of osmunda fiber, tree fern fiber, redwood fiber, bark, coarse peat moss, partly decayed leaves, gravel, clean coarse sand, perlite, coarse charcoal, dried cattle manure, and a little hoof and horn meal or bonemeal. An excellent mix for calanthes consists of walnut-sized pieces of old grass sod, coarse sand or perlite, and about ⅛ part by bulk dried cattle manure.

The precise choice of ingredients and the proportions in which they are used matters less than the physical composition of the mix. It must provide for very free passage of water and admission of air and should not decay or become soured too quickly. For orchids with thick roots the ingredients of the potting mixes should be coarser than for fine-rooted ones.

When to repot depends upon several factors. With a few kinds, for example, calanthes that have deciduous foliage and roots that renew themselves each year, annual attention to this gives the best results; this is done at the conclusion of the dormant period just as new growth is about to start. The majority of orchids have perennial roots and are better not disturbed oftener than necessary. For them repotting, rarely advisable more often than every two, three, or even four years, is done just as new growth starts.

Reasons for repotting (or rebasketing) orchids with perennial roots are (1) that the old potting medium is approaching a stage of decay that impedes rapid drainage of water and easy admission of air, (2) that the plant has become obviously too big for its container and its new growths overlap

This cattleya should have been repotted before its new roots grew over the side of the pot

to a distressing degree the rim of the container, or (3) that you wish to divide the specimen to secure additional plants or to forestall it becoming bigger. The beginning of a new cycle of growth is the most favorable time to repot.

The technique of potting depends upon the medium. With terrestrial orchids it does not differ materially from that employed with most other plants grown in soil, but the mix is not pressed very firmly and the finished surface is mounded slightly. With epiphytic sorts somewhat different procedures are used. To assure free passage of water it is helpful to enlarge the drainage holes of standard pots and pans to about twice their normal sizes. There is no need to do this with special orchid pots that have holes through their sides at their bases as well as in their bottoms. Be sure the receptacles are clean and dry. Put into them considerably more drainage material than is usual when potting other sorts of plants, filling them to one-fourth to one-third their depths with crocks, pebbles, coarse gravel, pieces of coke or charcoal, or large chips of bark.

To facilitate the removal of plants to be potted from their containers, it is often helpful to slide a sterilized knife blade around the inner surface of the pot to the full depth of the roots. Tease out or brush off from the roots all spent and decayed portions, cut off any dead roots, and strip the pseudobulbs of any dead foliage and loose scales, taking care not to damage growth buds. Inspect the plants carefully for scale insects and if these or other pests are detected clean the plants of these before potting.

Correct positioning of the plants in the pots is important; it is related to the two modes of growth, monopodial and sympodial, that orchids exhibit. Monopodial sorts, without rhizomes or pseudobulbs, have very short to very long, erect or trailing stems that elongate throughout their lives from terminal buds, a growth pattern botanists call indeterminate. Sympodial orchids, their sorts much more numerous than monopodials, have rhizomes, those of a few kinds including dendrobiums very short, those of many sorts extensive, from which sprout branches the bases of

A monopodial orchid (*Vanda tricolor suavis*)

A sympodial orchid (*Laelia albida*)

which frequently are developed as pseudobulbs. These branches attain certain lengths and then cease to grow. In botanists' language, they are determinate. When potting sympodial sorts that make appreciable lateral growth make allowance for this. Set the plants to one side of center with the rear of the rhizome almost touching the side of the pot and its growing point at approximately its middle, thus allowing for the extension of the rhizomes that will take place before the next repotting. Be sure that the upper part of the rhizome is above the surface of the potting medium.

Potting with bark mixes and lightweight orchid gravel is done by tucking the material among and around the roots, tapping the pot a few times on the workbench or table to settle it, and firming the surface with the fingers or by tamping. Because specimens newly potted in bark or gravel are less securely held than if osmunda fiber is used, tall ones may need staking. This is most conveniently done with wires made to clip to the rims of the pots.

An orchid potted in bark

Potting with osmunda fiber is a quite different technique. First soak the material in water and allow it to dry until pliable but not wet. Next cut it into egg-sized, or for pots of less diameter than 5 inches smaller pieces and push some of these under and among the roots. Position the plant in the pot and holding it in one hand stick additional pieces of fiber between the root mass and the sides of the container.

Potting in osmunda fiber: (a) A cattleya to be repotted

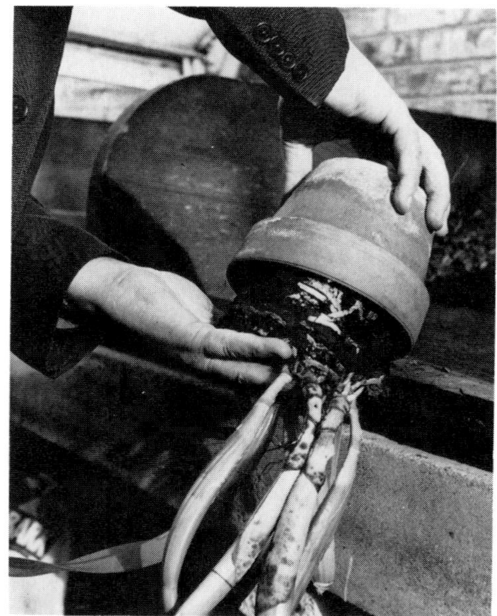

(b) Removing the plant from its container by tapping the rim of the pot on the edge of the potting bench

(c) Cutting away unwanted back bulbs

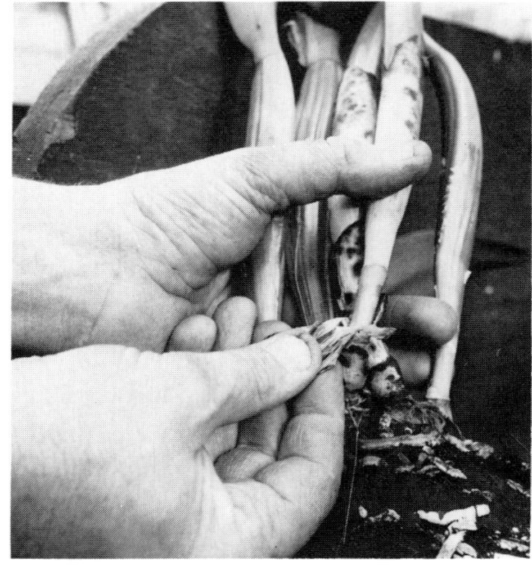

(d) Stripping loose scales from the bases of the pseudobulbs

(e) Cutting osmunda fiber into usable-size pieces

(f) A generous amount of crocks promotes good drainage

As these are placed, push a metal or wooden potting stick down the inner face of the pot and using its lower end as a fulcrum, lever the fiber toward the center. Add more lumps of fiber and repeat the

(g) A layer of osmunda fiber is placed over the crocks

(h) Positioning the plant

(i) Packing the osmunda fiber with a potting stick

levering action until the mass is so tight that it is impossible to make room for more. Have the finished surface slightly mounded from its circumference, which

(j) Staking and tying to afford the newly potted plant support

(k) Cropping the surface of the fiber to give a neat finish

should be approximately ½ to 1 inch below the top rim of the pot, to its center. Give it a tidy finish by cropping with shears. It is unwise to water newly potted orchids for two or three days following the operation, this so that cut ends of roots will have time to heal. During this period mist them lightly with water fairly frequently. A little additional shade for about a month is also advisable.

Installing orchids in baskets and on rafts calls for minor variations only of the techniques used in potting. To attach them to blocks of tree fern trunk, cork bark, and similar materials on which many kinds, especially the smaller ones, can be grown and displayed very satisfactorily, tie them with nylon thread or soft copper wire. It is sometimes helpful to place a small wad of osmunda fiber or osmunda fiber mixed with sphagnum moss between the plant and the support.

The simplest way of propagating most orchids is by division. This is most conveniently done with a thin-bladed, keen knife, which, to avoid possible transmission of virus diseases, should be dipped in a 1 to 10 solution of Clorox and water or in alcohol between each cut. The most propitious season varies with the kind of orchid and to some extent the preferences of the gardener, but most often late winter and spring, which assure prospects of good growing weather ahead, are favored. Some growers divide only just as new growth begins, others wait until after flowering. It is a mistake to make divisions too small. Best results are had if each consists of at least three or four growths. Pot the divisions in receptacles only just big enough to comfortably contain their roots and afford them aftercare appropriate for recently repotted specimens.

A method of division that involves the use of "back bulbs" as sources of new plants is practiced, but because growths from these are often slow to start and weak, it is usually applied only when it is especially important to obtain increase from a particular plant. Back bulbs are old pseudobulbs, those left when the forward portions of rhizomes including three, four, or more pseudobulbs are cut off and repotted. To obtain new plants, lay them on a layer of sphagnum moss or vermiculite kept moist in a congenially warm temperature. Eventually most will develop new shoots from latent buds and then they may be potted.

Back bulbs can often be improved as propagation material if about a year before their removal the rhizome between four, five, or more of them and the forward portion is cut through about one-third of its diameter. This stimulates new growth from the back bulbs and when the portion of the plant consisting of this is separated and repotted it becomes reestablished with minimum shock and delay.

Young plantlets develop from the stems of some orchids, notably dendrobiums and epidendrums with reedlike stems. They are most profuse on plants kept in humid atmospheres with rather less light than is conducive to their best flowering and can easily be removed, potted, and established as new plants. Orchids of these kinds can also be propagated by cutting their stems into lengths of about 4 inches and laying them on a bed of vermiculite, perlite, or peat moss where the atmosphere is humid, the temperature is a little warmer than best suits the growing plants, and there is shade from direct sun. This can also be done with old flower stalks of phalaenopsises. Indeed, left on the plants after the blooms have faded, these will often develop plantlets spontaneously.

Orchids that produce roots freely from tall stems, as do vandas and some other monopodials, are easily increased by cutting the top off a stem at a point below some of the roots and treating the severed part as a cutting. A modification and ofter better way is to air layer the plant. To do this, cut about one-third through the stem at a point at least four leaves below the apex. Paint the incision with bordeaux mixture or dust it with fermate to minimize damage that may result from rot-causing funguses and then wrap the part with sphagnum moss or osmunda fiber tied into place. Keep this moist and after enough roots have formed cut off the top and pot it separately.

Raising orchids from seeds to flowering takes several years and in its early stages is a highly specialized procedure usually embarked upon only by highly skilled hybridists and breeders. Amateurs without the time, facilities, or patience to master the techniques are advised to purchase flasks, or better, community pots of seedlings from specialists. Some, however, may want to start from the beginning, which may prove fascinating as well as adventurous.

Briefly, seeds are sown in sterilized glass flasks or test tubes on a sterile medium containing nutrients necessary for the growth of the young plants. Formulas used by different growers and for different orchids vary somewhat. The original one of Knudson, discoverer of the asymbiotic method, consists of distilled water, one liter; agar, 17.50 grams; sucrose, 20 grams; calcium nitrate, 1.00 gram; ammonium sulfate, 0.50 gram; magnesium sulfate, 0.25 gram; and ferric phosphate, 0.05 gram.

The ingredients are heated together long enough to dissolve the agar, but not longer. The solution is then tested and if necessary brought to an acidity reading, depending upon the formula adopted, of from pH 4.5 to 5.5 by adding a few drops of a very dilute solution of sodium hydroxide to make it more alkaline or a small amount of dilute nitric or sulfuric acid to increase its acidity. Before it cools it is poured into flasks or test tubes. The latter are tilted to assure the largest practicable surface for seed sowing. The flasks or tubes are then plugged with cotton, cellulose, or specially designed stoppers to exclude bacteria and fungus spores, but permit passage of air.

The next step is to sterilize the prepared receptacles and the seeds. The first is accomplished, taking care not to use temperatures high enough to caramelize sugar in the medium, in a pressure cooker or an autoclave. There are several ways of sterilizing the seeds. A convenient one is to dissolve in 140 cubic centimeters of water, 10 grams of sodium hypochlorite (caustic soda), filter the solution, and place a few drops of it along with the seeds in a test tube, which is then shaken for fifteen to twenty minutes.

To reduce danger of contamination do the actual sowing where there are no drafts, over a wide pan containing water boiling gently. Take in one hand the vessel to be sown and the tube containing the sterilized seeds. Remove the plug from the former and with a sterilized spatula transfer the seeds and scatter them over the surface of the sowing medium. Before replacing the plug pass it through a flame to destroy bacteria or fungus spores that may have alighted. Then cover the replaced plug with metal foil or plastic and put the containers in an environment favorable for germination, where a temperature of about 70°F, a relative humidity of sixty to seventy percent, and shade from direct sun and very strong light is assured. Germination may be expected in ten to sixty days, depending upon the kind of orchid. The first developments, called protocorms, are tiny roundish to top-shaped bodies with slender, rootlike organs called rhizoides. In time each produces a first or cotyledon leaf and later regular leaves typical of its kind.

First transplanting is usually delayed until the seedlings have well-developed normal leaves, but if they come up too thickly it is quite practicable to transplant while they have only cotyledon leaves. Whenever done, the transfer is made to flasks containing a medium similar to that used for seed sowing, but with a lower sucrose content, with strict measures taken to prevent contamination. Professionals achieve this by making the actual transfer inside a sterile glass dome having two armholes fitted with cloth sleeves through which the operator works, but it is possible to contrive other ways to assure the necessary sterility. The tiny plants are spaced to allow for a reasonable period of growth.

The next move or pricking off is to community pots or pans or trays that accommodate many seedlings. Its timing depends upon a number of variables, including the kind of orchid and speed of growth, but generally it is done before another year elapses. The medium in which to prick off may be very fine bark chips or other favored sort. It should be not more than 1 inch deep and overlying a substantial layer of drainage material. All traces of the agar medium in which the seedlings have been growing must be carefully washed away and the plants are dipped in an anti-damping-off fungicide before planting them in the new mix. When the young plants begin to crowd they are transplanted again, to community pots that hold four to six. Care from now on is not markedly different than for older plants of the same kinds, except that greater emphasis must be placed on keeping the atmosphere sufficiently humid and to protecting from drafts. Both objectives are furthered by keeping the community pots, and young plants potted from them, for a period, in a propagating case inside a greenhouse. Marked improvement in the rate of growth, with consequent reduction in the time taken to come to first flowering, occurs if the day length for seedlings is extended to sixteen hours by using artificial light.

Hardy orchids have not been domesticated in the sense that they are commonly grown and propagated garden plants, but some native sorts are fairly frequently planted in wild gardens, rock gardens, bog gardens, and other special areas where conditions they favor in the wild can be approximated. For such uses and for naturalizing in congenial environments many are superb. Among those worth trying are species of *Arethusa*, *Calopogon*, *Cypripedium*, *Habenaria*, *Liparis*, *Pogonia*, and *Spiranthes*.

Almost everyone who attempts to grow hardy orchids seeks to duplicate as closely as possible the environmental conditions under which they grow in the wild. In the main this makes good sense. Certainly one would be foolish to install a plant of acid-soil woodlands in a limestone rock garden, or an inhabitant of bogs on a dry bank. But there are indications that with some sorts at least there is room for maneuverability, that at times greater success may be had by diverging somewhat from what one believes to be slavish imitation of "natural conditions," provided of course that basic needs are not neglected.

Supporting evidence for this is supplied by the success had by gardeners in the British Isles in cultivating under climatic conditions decidedly different from those of their homelands sorts not native to the British Isles, including species of *Orchis* and other genera. At The New York Botanical Garden native species of *Spiranthes* grown in pots in greenhouses from which they were never removed and watered regularly with city water, flourished and flowered abundantly for ten years or more. And cypripediums, including the notoriously difficult pink lady slipper (*Cypripedium acaule*), have been known to give far better results in ground abundantly enriched with partially rotted cow manure than in similar soil not so enriched. It thus seems there is some opportunity for testing and experimenting to determine the best means of growing hardy orchids.

Concern for supplying water and in sufficient and timely amounts to keep the ground agreeably moist is undoubtedly of major importance in the successful cultivation of hardy orchids, and so in many cases is provision of generous amounts of organic matter in the soil. Mulching to keep the roots evenly moist and relatively cool is often very helpful.

The majority of native orchids planted in gardens are dug from the wild, a practice to be undertaken with sympathetic understanding of the need to preserve wild populations and not to be indiscriminatingly encouraged. Collecting is especially reprehensible in the case of rare species. Sometimes plants can be obtained from sites about to be destroyed as homes for orchids by road building or other construction. The best time to transplant most sorts is as soon as flowering is through, and it is always advisable to take with the plants a goodly amount of native soil, which possibly contains microorganisms necessary for the well-being of the orchids. Cultivated specimens of some kinds can be increased by carefully dividing them, but unless additional plants are greatly needed, it is usually best to leave flourishing specimens untouched. Propagation from seed has not so far proved practicable.

Orchids are far from immune to pests and diseases, and only vigilance and adequate precautionary measures prevail against them. Among the most serious diseases are virus infections that manifest themselves by often not easily identifiable symptoms; these include streaking, striping, or spotting of foliage, distortion of young growths and flowers, and discoloration of the latter. Destroy affected plants uncompromisingly. If in doubt send a portion of the suspected plant to your State Agricultural Experiment Station for diagnosis. Other diseases include several rots and leaf spots, as well as a petal blight and a rust. Strict sanitary measures and avoidance of excessive wetness and poor air circulation, together with provision for adequate light are the best preventatives and controls for these.

The most frequent pests include, perhaps the most serious, scale insects, and in addition aphids, mealybugs, red spider mites, slugs, snails, thrips, weevils, and cockroaches, as well as pests called cattleya or orchid fly and orchid midge. The larvae of the first feeds on the pseudobulbs and other parts, those of the midge on the roots.

The names of genera of orchids treated separately in this Encyclopedia are given under the entry Orchidaceae. Among the most familiar vernacular names used for them are these: bamboo orchid (*Arundina graminifolia*), bee orchid (*Ophrys apifera*), bucket orchid (*Coryanthes*), bumble bee orchid (*Ophrys bombyliflora*), butterfly orchid (*Oncidium krameranum* and *O. papilio*), chain orchid (*Dendrochilum*), Christmas orchid (*Cattleya trianaei*), clam shell orchid (*Epidendrum cochleatum*), cradle or tulip orchid (*Anguloa*), crane-fly orchid (*Tipularia discolor*), dove orchid (*Oncidium ornithorhynchum* and *Peristeria elata*), Easter orchid (*Cattleya mossiae*), fly orchid (*Ophrys muscifera*), foxtail orchid (*Rhynchostylis*), fringed orchid (*Habenaria*), giant or queen orchid (*Grammatophyllum speciosum*), goblin orchid (*Mormodes*), golden butterfly orchid (*Oncidium varicosum*), golden chain orchid (*Dendrochilum filiforme*), grass-pink orchid (*Calopogon*), hyacinth orchid (*Arpophyllum*), jewel orchid (*Anoectochilus, Dossinia marmorata*, and *Haemaria*), leopard orchid (*Ansellia*), moth orchid (*Phalaenopsis*), nun's orchid (*Phaius tankervilliae*), pansy orchid (*Miltonia*), Punch and Judy orchid (*Gongora*), ram's head orchid (*Cypripedium arietinum*), rattlesnake orchid (*Pholidota*), rein orchid (*Habenaria*), spider orchid (*Brassia*), swan orchid (*Cychnoches*), tulip or cradle orchid (*Anguloa*), violet orchid (*Ionopsis*), virgin orchid (*Caularthron bicornutum*), and wasp orchid (*Ophrys lutea*).

ORCHIS (Órc-his). The name *Orchis*, alluding to the shape of the tubers of some species, comes from the Greek *orchis*, a testicle. It identifies a group of about thirty-five species of north-temperate and warm-temperate regions of the northern hemisphere, chiefly of the Old World, but with two native in North America. Belonging in the orchid family ORCHIDACEAE, it differs from related *Habenaria* in having the glands of the flowers enclosed in small pouches. A few species fairly commonly cultivated in Europe that formerly belonged in *Orchis* are now included in *Dactylorhiza*.

Orchises have fleshy roots or tubers not cleft or divided into finger-like lobes and one to few usually basal leaves. The flowers, in terminal spikes or racemes, and commonly pink, purple-pink, or red, are borne in the axils of bracts. They have three sepals, separate or joined, and approximately similar to the two petals, and a three-lobed or lobeless lip.

Orchis spectabilis

The showy orchis (***O. spectabilis***) ranges in the wild from New Brunswick to Nebraska, Georgia, and Arkansas. Inhabiting rich woodlands, it has a pair of thick, lustrous, narrowly-obovate to broadly-elliptic leaves 3 to 6 inches long and an angled stalk 4 to 8 inches tall that carries up to eight 1-inch-long blooms that have magenta-pink or paler sepals and petals joined into a hood and a lobeless, white lip, or less commonly the flowers are entirely pink or entirely white. From the last, ***O. rotundifolia***, a native of swamps and wet woodlands from Greenland to the Yukon, New York, Michigan, Minnesota, and British Columbia, differs in having only one elliptic to obovate leaf up to about 4 inches long. Slender and 4 to 10 inches long, the stalk carries a few flowers with pink-spotted, white, three-lobed lips, and magenta-pink sepals and petals, the lateral petals wide-spreading.

European species include the soldier orchid (***O. militaris***), 1 foot to 2 feet tall, which has dense cylindrical spikes of more or less spotted, pink to rose-purple blooms, and the early purple orchid (***O. mascula***), 6 inches to 2 feet tall. This latter has an ovoid cylindrical spike of dark crimson-purple flowers with heavily spotted, three-lobed lips. A drink called salep is made from this species in southeastern Europe.

Garden Uses and Cultivation. Orchises are choice plants, challenging to grow, and best suited for naturalistic gardens and rock gardens. They need deep, rich soil and environments that approximate those under which they live in the wild. Ample moisture and light shade is to the liking of most kinds. They are impatient of transplanting. When this is necessary, early fall is the best time to do it. Increase is by careful division in early fall. For further information see Orchids.

OREGON-BOXWOOD is *Paxistima myrsinites*.

OREGON-GRAPE is *Mahonia aquifolium*.

OREGANO. This name is applied to *Origanum vulgare* and *O. heracleoticum* and to *Coleus amboinicus*.

OREOCALLIS (Oreo-cállis). Natives of Australia, the Malay Archipeligo, and South America, the five species that compose *Oreocallis* belong in the protea family PROTEACEAE. The name, alluding to the mountain habitats of some sorts and their beauty, derives from the Greek *oreos*, a mountain, and *kallos*, beautiful.

Closely related to *Embothrium*, oreocallises are evergreen shrubs or trees with alternate, undivided or pinnate leaves and solitary, paired, or clustered racemes of bloom. The flowers have a tubular, four-lobed perianth, four stamens, and one style. The fruits are follicles containing many winged seeds.

Australian ***O. pinnata*** (syns. *Embothrium wickhamii, E. w. pinnata*), because its flowers resemble those of the waratah (*Telopea speciosissima*), is sometimes called the waratah-tree. It is a small tree usually with both undivided and pinnate leaves, the latter with seven to nine leaflets, on the same plant. Its long-stalked, red flowers crowded at the branch ends make a brave show.

Garden and Landscape Uses and Cultivation. These are as for *Embothrium.*

OREOCARYA. See Cryptantha.

OREOCEREUS (Oreo-céreus) — Mountain Cactus. Conservative botanists find no sufficient reason for maintaining *Oreocereus,* of the cactus family CACTACEAE, as a separate genus and include it in *Borzicactus.* Those who favor recognizing it as a separate entity do so chiefly because its members nearly always have abundant long hairs on their stems, at least in their upper parts. These are absent from plants they accept in *Borzicactus.* The name, from the Greek *oros,* mountain, and the name of the related genus *Cereus,* alludes to the native habitats of these plants.

The five species of *Oreocereus* are endemics of the South American Andes. From 2 to 10 feet tall, they have ribbed, spiny stems and usually branch more or less freely from or nearly from ground level. The spines are needle-like to awl-shaped. The red flowers have narrow perianth tubes; they open during the day. The fruits, small, dry, and more or less spherical, have scales, but not spines.

Beautiful ***O. neocelsianus*** (syn. *Borzicactus celsianus*), of Bolivia, Peru, and Chile, has usually erect stems 3 to 4 feet long by up to 5 inches thick. They have nine to seventeen slightly lumpy ribs bearing conspicuously white-woolly areoles approximately ½ inch apart, with one to four strong central spines and a number of shorter, yellowish-brown radials. From the areoles on the upper parts of the stem sprout numerous white, silky hairs, 1¼ to 1½ inches long. These almost hide the spines. Brownish-red on their outsides and paler inside, the flowers come from near the tops of the stems. They are 2¾ to 3¾ inches long and have their ovaries and outsides of the perianth tubes densely clothed with bristles and hairs.

Up to 10 feet in height and with the nine to fourteen ribs of its 3-inch-thick stems more deeply notched than those of the last, ***O. fossulatus*** (syn. *Borzicactus fossulatus*) is a native of Bolivia and Peru. This has loose, white hairs and sturdy, needle-like spines. The not particularly attractive greenish-pink blooms are 2 inches long. From 2 to 3 feet in height, ***O. trollii*** (syn.

Oreocereus trollii

Oreocereus trollii grafted onto *Hylocereus undatus*

Borzicactus trollii) hails from Bolivia. It forms clusters of branchless, light green stems 2½ to 4 inches thick, on their young upper parts about nine-ribbed, below ribless. The abundant whitish or light gray, woolly hairs arising from the areoles are matted around the stems. From each areole springs one to three central spines and about seven shorter radials. The latter are approximately ¾ inch long. At first reddish-brown or reddish-yellow, the spines later become white. The blooms are like those of *O. neocelsianus*.

Its stems up to 4 feet tall and 4 inches wide and with long lustrous hairs at first golden-yellow, changing to whitish, ***O. hendriksensianus*** (syn. *Borzicactus hendriksensianus*), of Bolivia and Peru, differs from *O. fossulatus* in not being as high, in having thicker stems that branch lower, and in its spines being stouter. Also the brownish-pink flowers are more hairy.

Garden and Landscape Uses and Cultivation. The uses of oreocereuses are those common for columnar cactuses discussed under Cactuses. This genus is not difficult to grow. Its members prefer soils that contain some calcareous material, such as limestone, crushed oyster shells, or tufa rock. For additional information see Cactuses.

OREOCHARIS. See Opithandra.

OREODOXA. See Roystonea.

OREOPANAX (Oreó-panax). As its name indicates, this genus of 120 tropical American species is related to *Panax.* The prefix oreo comes from the Greek *oreos,* a mountain, and alludes to the habitats of some species. Belonging in the aralia family ARALIACEAE, the group consists of trees and shrubs with leaves lobed palmately (in handlike fashion) or divided into separate leaflets that spread from a common point. The small flowers are in heads, racemes, or panicles. Typically they are greenish.

A shrub or small tree native of Mexico and Central America, ***Oreopanax xalapensis*** (syn. *O. thibautii*) has good-looking schefflera-like foliage. Its long-stalked leaves have eight to ten, or on young shoots five to seven, obovate, slightly toothed, somewhat drooping leaflets up to 1 foot long, of leathery texture, and with mid-veins slightly paler than the body of the leaflet. The flower heads, about ½ inch in diameter, are in racemes 1 foot long. The fruits are black. A Mexican and

Oreopanax xalapensis

Oreopanax peltatus

Oreopanax peltatus (with panicle of flower buds)

Guatemalan species, ***O. peltatus*** (syn. *O. salvinii*) is a tree with leaves entirely different from those of the last. They are long-stalked, broadly-heart-shaped in outline, but instead of being divided into separate leaflets are deeply-five- to seven-lobed with the lobes again lobed or toothed, and the undersides of the leaves woolly. The veins are paler than the body of the leaf.

Garden and Landscape Uses and Cultivation. The handsome, bold foliage of these plants is useful for creating strong patterns in outdoor landscaping in the tropics and subtropics and for providing lush effects in greenhouses and conservatories and that of pot and tub plants for decorating large rooms. The cultivation of oreopanaxes is simple. Well-drained, fertile soil, never excessively dry, suits. They stand full sun and endure considerable shade. Indoors they are at their best in humid atmospheres and in minimum temperatures of about 60°F, but they will grow where temperatures are sometimes somewhat lower and the air is dryish. Repotting is done in spring, at intervals of a few years. Established plants respond favorably to occasional applications of dilute liquid fertilizer. In the north, in summer, container-grown specimens may be stood outdoors, preferably in partial shade. Propagation is by seed, cuttings, and air layering.

ORGANIC GARDENING. The system of plant cultivation known as organic gardening is based on soil management practices that reject the use of fertilizers other than those approved as being of "organic" origin and of all pesticides or at least all notably toxic to warm-blooded creatures.

Accepted as fertilizers are materials derived more or less directly from animals and plants, such as manures, compost, leaf mold, seaweed, fish meal, cottonseed meal, dried blood, bonemeal, and wood ashes, as well as such chemically untreated products as ground phosphate rock, granite dust, and greensand, the first a source of phosphorus, the last two of potassium.

Fertilizers organic gardeners classify as chemical are not employed and so superphosphate, obtained by treating phosphate rock with sulfuric acid to make the nutrient phosphorus more readily available is rejected, as are such fertilizers as sulfate of ammonia, nitrate of soda, urea, muriate of potash, and sulfate of potash.

For pest and disease control, organic gardeners prefer to rely on methods that include the encouragement of such predators as ladybugs, and praying mantids that feed on harmful insects and, in a few instances, the promotion of disease among the enemy, such as using milky disease preparations to control Japanese beetles. The employment of crop rotations to minimize buildup of populations of damaging pests and disease organisms is favored, and so is setting out plants that discourage or are believed to discourage certain pests (soil nematode populations are sharply reduced near marigolds). In addition, trapping, hand-picking, and hosing-off with water are practiced and some organic gardeners approve spraying, dusting, or baiting when necessary with such nonpoisonous (to warm-blooded creatures) materials as pyrethrum, rotenone, methaldehyde, and sabadilla.

The positive practices organic gardeners advocate are mostly not new. Many are centuries old, some have been followed almost since mankind first cultivated plants. Nor are organic gardening's negative aspects, the what-not-to-dos of the routine, really new. Until late in the nineteenth century "chemical" fertilizers, sprays, and dusts were not available in significant amounts, and it was well into the twentieth century before these proliferated and were used in vast quantities. Until then, gardeners and farmers, of necessity, could not use most fertilizers and pesticides now employed.

Organic gardening then largely represents a return to pre-twentieth-century methods of soil management and pest and disease control. In its modern understanding it was practiced and advocated in Europe and India for a number of years before being introduced as a viable concept to the United States by Jerome I. Rodale, of Pennsylvania, in the 1940s. Rodale, rightly regarded as the founder of the organic gardening movement in North America, met great success in promoting its principles and practices and soon acquired a considerable band of enthusiastic followers. He wrote and lectured extensively and founded the very successful magazine *Organic Gardening*. Certainly the publicity that attended the popularization of the movement called attention to the excesses then being committed in the employment of pesticides and to an extent foreshadowed the concern for protection of the environment that developed two and three decades later.

No competent horticulturist questions the validity of most practices advocated by the organic school of gardening. Every sensible gardener does all he can to maintain or improve the fertility of the soil by methods that include returning to it generous amounts of organic matter that will decay to form humus. Concomitant with that is the need to collect and conserve all suitable organic wastes.

Those who do not accept in totality the precepts of organic gardening, and this includes the vast majority of professional horticulturists, disagree not so much about what organic gardeners do as about what they will not do. There are no scientific reasons for believing nitrogen, phosphorus, potassium, and other elements contained in decayed organic matter are in any way different from the same elements taken from chemical additives to the soil, nor does science uphold the contention that organically grown foods are more nutritious or healthful than others. Despite the protestations of organic gardening enthusiasts, adequate pest and disease control is not practicable by the methods they approve alone.

Although in its extreme manifestations organic gardening has assumed some of the characteristics of a cult, which has included such faddish procedures as sprinkling decoctions of herbs on compost piles and breeding earthworms to introduce to humus-poor soils, yet many of its procedures could with advantage be adopted by those gardeners all too ready to rely wholly or nearly so on "chemical" fertilizers and pesticides. Let the uncommitted gardener follow a middle-of-the-road course, abjuring the excesses of the extreme sprinkle-, squirt-, and spray-chemicals school as well as the complete banishment of such controls advocated by organic enthusiasts. In that way perhaps lies the best of both worlds.

ORGANIC MATTER. Throughout this Encyclopedia frequent mention is made, with especial reference to soils, of organic matter. In gardening and as used here, this term is less encompassing than its fullest meaning implies. It is limited to dead organic matter, or in the case of green manure crops, organic matter that soon will be dead. Actually all living things, including gardeners, are composed of organic matter. And all organic matter decays to form humus and eventually simple chemical compounds and elements. This is of importance to gardeners. It is essential to understand that, unlike the rock or mineral particles of soils, the stones, gravel, sand, silt, and clay particles, organic matter is not stable, it is not inherently unchangeable. Under conditions favorable to the growth of most plants it undergoes constant change, in the process releasing plant nutrients, and itself gradually decaying and disappearing. Because of this, except where the soil is largely composed of organic matter, as are muck and peat lands, the efforts of gardeners must be directed to maintaining a reasonable content in the soil. This they do by adding and incorporating such materials as manure, compost, peat moss, commercial humus, leaf mold, decayed sawdust, and seaweed, in considerable bulk and by growing and turning under such green manure crops as winter rye, cow peas, and soy beans. Other materials of organic origin applied to soils add humus. These include such fertilizers as dried blood and tankage, but these are used in such small amounts that the total organic matter supplied is negligible. They serve chiefly as more concentrated sources of nutrients than the bulky organic materials just mentioned.

Now it is quite possible to grow plants without organic matter. This is done in hydroponic culture where the rooting medium is water, sand, gravel, or other inorganic material. But for most soil-supported plants a reasonable organic content of the soil is desirable or essential for good results. There are extremists, dedicated organic gardeners, who follow the age-old practice of depending entirely upon organic matter as additives to their soils to nourish their crops. Success can be had in this way, usually by expending considerably more money or labor or both than if a program employing intelligent use of nonorganic fertilizers as well as organic matter were followed. There is no scientific evidence that "organically" grown plants are superior to those grown with the help of nonorganic, often man-made nutrients.

Besides supplying nutrients, organic matter serves other important purposes. It improves soil structure by keeping clayey earths more open, porous, and crumbly, thus encouraging the free passage of water and air, and in a well-decomposed condition by cementing together the particles of sandy soils to make them more cohesive and retentive of moisture. Organic matter also tends to darken light-colored soils, and dark soils, being more absorbent of sun heat, warm in spring earlier than light-colored ones.

Other advantages of organic matter in the soil are that it reduces baking and forming of surface crusts that interfere with the emergence of seedlings and the absorption of water. It checks erosion by wind and water and increases water-holding capacity of the soil, which is especially important with sandy ones. It is a food or a source of nourishment of many organisms from earthworms to microorganisms.

Organic matter in the form of humus serves as a buffering agent. It checks rapid changes in soil pH and aids in the retention of plant nutrients that otherwise might be lost by leaching. Used as mulches, organic materials conserve moisture, eliminate the need for surface cultivation, reduce the need for weeding, and aid in stabilizing soil temperatures. Many plants grow better when mulched than under clean cultivation.

ORIGANUM (Orígan-um) — Marjoram, Oregano, Dittany-of-Crete. As here accepted *Origanum*, of the mint family LABIATAE, includes plants sometimes segregated on the bases of minor botanical differences as *Amaracus* and *Majorana*. There are thirty to thirty-five species, natives, chiefly, of dry, rocky environments from Europe, especially the Mediterranean region, to central Asia. The name is modified from *origanon*, applied by the ancient Greeks to several species of the mint family, and said to mean joy of the mountains. Most sorts are good bee plants. The culinary herbs dried and sold as Greek and Italian oregano are *O. heracleoticum* or varieties of *O. vulgare*.

Origanums are aromatic annuals, biennials, herbaceous perennials, and subshrubs. They have opposite, toothed or toothless leaves and small, asymmetrical, tubular flowers in erect, cylindrical to rounded, bracted spikes. The blooms have

Origanum vulgare

two-lipped, top- to bell-shaped, toothless or five-toothed calyxes or calyxes with only one lip and deeply slit down one side. The corollas are two-lipped, the upper lip erect and often indented, and the lower three-lobed. There are four stamens in pairs of unequal lengths and a usually unequally-two-lobed style. The fruits are of four seedlike nutlets.

Wild marjoram (***O. vulgare***), native throughout most of Europe and naturalized in North America, is an erect, woody, more or less hairy-stemmed, hardy herbaceous perennial 1 foot to 2½ feet tall, with nearly horizontal rootstocks and pointed, broadly-ovate, stalked, shallowly-toothed or toothless leaves up to 2 inches long. Its flowers, about ¼ inch long, are in rounded to cylindrical, terminal and axillary clusters that form loose panicle-like aggregations. The blooms have five-toothed calyxes and rose-purple to white corollas and are accompanied by tiny purplish, pink, or nearly white, hairless bracts. All or two of the stamens protrude. Many variants of this highly variable kind have at times been described as separate species. Variety *O. v. aureum* has yellow foliage.

Origanum vulgare aureum

Winter marjoram (***O. heracleoticum***) is very like wild marjoram, differing mostly in the bracts of its flower spikes being thickly covered with glands, instead of the glands there being few or absent. Also, the floral bracts of *O. heracleoticum* are usually green, whereas those of *O. vulgare* are generally purple.

Sweet, garden, or knotted marjoram (***O. majorana*** syn. *Majorana hortensis*), in the wild annual, biennial, or perennial, is bushy and 1 foot to 2 feet tall. It is usually cultivated as an annual. This has stalked, toothless, blunt-elliptic leaves, grayish-hairy on both sides and ¼ to ½ inch long. In stalked clusters of three to five displayed in loose, erect, panicle-like assemblages, the globular to oblong spikes of purplish to white flowers decorate the branchlets in summer. The individual blooms, about ⅙ inch long, are accompanied by hairy, broad-ovate, rounded, overlapping bracts almost as long as the flowers. The blooms have one-lipped, shell-like calyxes deeply slit down one side; funnel-shaped, two-lipped, five-lobed corollas; four protruding stamens; and a slightly two-cleft style.

Closely similar to sweet marjoram and not infrequently mistaken for it, ***O. majoricum,*** a rare native of southwest Europe, is probably a natural hybrid between sweet marjoram and *O. virens,* the latter apparently not in cultivation. The hybrid, if such it be, differs from sweet marjoram in its pink flowers having two-lipped, bell-shaped calyxes with the upper lip deeply-three-toothed and the lower with two teeth. This sort is shrubby, up to 2 feet tall or taller, and has ovate-elliptic, short-hairy leaves up to 1 inch long. It rarely if ever produces seeds.

Pot marjoram (***O. onites*** syn. *Majorana onites*), of the Mediterranean region, is a subshrub about 1 foot in height with scarcely-branched, hairy stems, and ovate, stalkless, slightly-toothed, hairy leaves. Its whitish flowers, a little bigger than those of sweet marjoram, are in dense, roundish or flattish clusters 2 inches or so across composed of small ovoid spikelets. The calyxes of the flowers are one-lipped and deeply slit down one side. This species, native to southeast Europe, Asia Minor, and Syria, is not hardy in the north.

Origanum onites

Flowers with toothless calyxes not slit down one side distinguish ***O. microphyllum*** from all other species considered in this treatment. Native to Crete, this is a dwarf shrub with many usually branched, four-angled stems. Its ovate to oblong leaves are up to ⅓ inch long and more or less hairy. The purple flowers, about ⅕ inch long, are in loose terminal panicles. An intermediate natural hybrid of this species and *O. heracleoticum* named *O. minoanum* occurs locally in Crete.

Frequently grown as *O. maru,* which name has also been used for *O. microphyllum,* handsome ***O. syriacum*** is a common element in the native floras of Israel and other lands to the east of the Mediterranean. A gray-hairy subshrub 1½ to 3 feet tall, it has broad-elliptic to ovate or ovate-oblong leaves ½ inch to 1¼ inches long and tiny purplish flowers in woolly-bracted, oblong spikes disposed in loose panicles. Rare ***O. amanum,*** of Asia minor, when happily accommodated, blooms freely throughout the summer. A compact, thin-stemmed shrublet 3 to 4 inches tall, it has opposite, heart-shaped leaves eyelashed with hairs and ¼ to ½ inch long. The pink flowers, from the axils of purple bracts, have slender corolla tubes about 1½ inches long. Successful cultivators of this generally difficult-to-grow and usually short-lived species recommend cutting it back severely as soon as flowering is through.

Origanum amanum

Dittany-of-Crete (***O. dictamnus*** syn. *Amaracus dictamnus*), of Crete, may survive outdoors in sheltered locations not colder in winter than that of Washington, D.C. Subshrubby, it has procumbent or ascending, conspicuously white-woolly stems up

Origanum dictamnus

to 1 foot long and short-stalked, densely-white-woolly, broadly-ovate to nearly round, toothless, sometimes purplish-mottled leaves up to 1 inch long. The ½-inch-long, pink to purplish flowers are in spikes that look like little hops and are ½ inch long or slightly longer. Their overlapping, broadly-ovate bracts change to rich rose-purple as seeds are formed.

Less commonly cultivated than dittany-of-Crete, subshrubby ***O. sipyleum*** (syn. *Amaracus sipyleus*) is essentially hairless. It has more or less sprawling stems and erect flowering branches up to about 2 feet tall, with ovate to ovate-lanceolate leaves about ¾ inch long. Its pinkish flowers, in hop-like heads with bracts smaller than those of dittany-of-Crete, become bright plum-red. They have two-lipped calyxes with the upper lobe nearly toothless, the lower one shallowly-toothed. This is a native of Asia Minor.

A popular handsome hybrid between dittany-of-Crete and *O. sipyleum* often misnamed *O. dictamnus*, is ***O. hybridum*** (syn. *Amaracus hybridus*). In general it resembles *O. sipyleum*, but has hairy, not white-woolly, stems and foliage. Its leaves are ovate and ordinarily 1 inch long or a little longer. The pinkish flowers are in hoplike spikes that are solitary or in clusters of three, with bracts that turn purplish-red as they mature.

Reportedly recently introduced to American horticulture, ***O. tyttanthum***, of the Soviet Union, 1 foot to 2½ feet tall, has many slender, erect stems and elliptic-ovate, hair-fringed leaves. Its tiny white to rosy-violet flowers with toothed, bell-shaped calyxes are in loosely clustered spikes.

Garden Uses. The chief uses of the cultivated sorts of this genus are as components of herb gardens. A few known as marjorams are fairly widely grown as culinary herbs. In addition, dittany-of-Crete, *O. hybridum*, and *O. sipyleum* are sometimes grown in rock gardens and as pot plants in greenhouses and windows. With the exception of *O. vulgare*, origanums are either not hardy or not reliably so in the north.

Cultivation. For the cultivation of sorts ordinarily regarded as culinary herbs, see the entry Marjoram in this Encyclopedia. Where winters are not excessively cold other sorts succeed outdoors in gritty, very well-drained soil, preferably of a limestone character and not overrich, and dryish rather than dampish. Full sun is needed. Indoors they do well where winter night temperatures are 40 to 50°F and day temperatures only a few degrees higher. They can also be wintered in well-protected cold frames to which, whenever outdoor temperatures rise above 50°F, as much ventilation as practicable should be given. Excessive moisture, especially in winter, is likely to be harmful. Propagation of these charming and interesting plants is usually by cuttings and by division in spring. These presumably are the only methods applicable to *O. hybridum* and *O. majoricum*. The species can also be raised from seed.

ORIXA (Or-íxa). Consisting of one species endemic to Japan, South Korea, and China, *Orixa* belongs in the rue family RUTACEAE. Its name is Japanese.

A deciduous, broad shrub, 10 feet tall, with foliage somewhat ill-scented when bruised, ***O. japonica*** is hardy in southern New England. It has shoots slightly short-hairy when young and lustrous, bright green, alternate, short-stalked, obovate leaves 2½ to 5 inches long that are finely-short-hairy, especially on the veins when young. They have tiny translucent dots, visible when viewed against the light. Their margins are smooth or may have a few irregular indentations. The pale green flowers are unisexual, the sexes on separate plants. Rather inconspicuous, they appear in spring, below the current season's leafy shoots, from wood of the previous year. The females are solitary, the males in racemes ¾ inch to 1½ inches long. The latter have four each sepals, petals, and stamens, the latter shorter than the petals. The females have four sepals, four petals, four nonfunctional stamens, and a short style tipped with a four-lobed stigma. The fruits are of four flattened carpels joined at their bases, each containing a solitary black seed.

Garden and Landscape Uses and Cultivation. In its homelands an inhabitant of woods and thickets at low mountain altitudes, *O. japonica* has not much to recommend it in the way of flowers or fruits likely to endear it to gardeners, but it is of graceful habit and is accounted a very satisfactory foliage plant for backgrounds, screens, and similar purposes. It grows well in ordinary well-drained soil in part-day shade or full sun and is easily increased by seeds sown as soon as ripe, or stratified and sown in spring. It may also be increased by summer cuttings planted under mist or in a greenhouse or cold frame propagating bed, and by root cuttings. The only pruning needed is occasional thinning of old and crowded branches, sufficient to keep the bush vigorous and shapely. This may be done in late winter.

ORLEANS-PLUM is *Chrysophyllum pruniferum*.

ORMOSIA (Ormòs-ia)—Necklace Tree or Bead Tree. The common names given to one of more species of the fifty of *Ormosia*, of the pea family LEGUMINOSAE, refer to the use of their seeds for ornament. The generic name, from the Greek *hormos*, a necklace, has a similar implication.

Ormosias are alternate-leaved, evergreen trees and shrubs of the tropics of the Old World and the New. They have often large, pinnate leaves with an uneven number of leaflets and white, lilac, blue, or dark purple, pea-like blooms generally in terminal panicles. The fruits are flattened pods, usually thick and leathery with, in some species, beautifully and brilliantly colored seeds. The flowers have ten separate, unequal-sized stamens.

The necklace or bead tree (***O. monosperma*** syn. *O. dasycarpa*), a native of the West Indies, is a timber tree of large dimensions in the wild, but often smaller in cultivation. Its leaves, 6 to 8 inches long, have seven to eleven pointed-oblong leaflets 3 to 4 inches long. The blue flowers are ¾ inch long and in large, rusty hairy panicles. The hairy pods mostly have only one bright scarlet seed, with a conspicuous black blotch.

Native to Puerto Rico, Hispaniola, Guadeloupe, and Dominica, ***O. krugii*** is an evergreen, broad-crowned tree 30 to 60 feet tall. Its dull green leaves have five to nine broad-oblong leaflets 3 inches to 1 foot long and 1½ to 8 inches wide. They have strong, slightly sunken, parallel veins running outward from their midveins. The panicles of bloom are 6 inches to 1 foot long. The short-stalked flowers have hairy, brown calyxes and deep violet corollas, a little over ½ inch in length, with the standard or banner petal spotted with white or yellow. The brown pods, 2 to 4 inches long and up to ¾ inch broad, contain one to five seeds.

Native to Panama, ***O. panamensis*** is a vigorous, tall tree. Its leaves have seven to eleven, widely spaced, broad-elliptic leaflets 3 to 8 inches long. The small white flowers, in terminal spikes, are succeeded by pods containing bright red seeds.

Garden and Landscape Uses and Cultivation. The species discussed are adapted for the tropics and near tropics. They succeed in ordinary soil, and are propagated by seed. Under favorable conditions they are fast growers.

ORNAMENTAL-FRUITED PLANTS. Trees, shrubs, and some other plants that make considerable displays of decorative fruits are commonly referred to as ornamental-berried, but as the fruits of cotoneasters, hawthorns, crab apples, and many others that belong are not technically berries the designation ornamental-fruited seems better.

The grouping is convenient because it identifies a wealth of plants of great use to landscapers, gardeners, and flower arrangers. Many of its sorts provide food for birds.

In northern gardens, especially, but by no means exclusively, ornamental-fruited plants have much to offer. The majority make their greatest displays in fall and associate splendidly with the brightly colored foliage of the season. Some, flowering dogwood and crab apples for example, are dual-season plants, delighting with their flowers in spring, their fruits later.

Fruits vary in size from tiny berries, such as those of bush honeysuckles (*Lonicera*), to quinces and oranges. Usually, especially if small, they are numerous. Their colors include the nearly pure white of snowberry (*Symphoricarpos albus*) and the pure white of lovely *Callicarpa japonica alba*; the pink of coralberry (*Symphoricarpos orbiculatus*); the bright reds of ardisias, cotoneasters, hollies (*Ilex*), honeysuckles (*Lonicera*), Brazilian pepper tree (*Schinus terebinthifolius*), California pepper tree (*Schinus molle*), viburnums, and many more; the yellow and orange hues of some kinds of hollies, viburnums, roses and calamondins, hardy-orange (*Poncirus trifoliata*), and oranges; the rich blue of sapphire berry (*Symplocos paniculata*); and the bright purple-violet of several sorts of beauty-berry (*Callicarpa*). There are a goodly number of black-fruited plants, but most are scarcely showy enough to be grown for the decorative merits of their fruits. A few are, including some privets (*Ligustrum*) and buckthorns (*Rhamnus*).

The lasting quality of the fruits varies greatly. Those of some varieties of holly color in fall and remain attractive until May. Aucubas and certain cotoneasters hold their fruits in fine condition throughout the winter. Pearly-white snowberries remain only until severe freezing causes them to blacken, but those of *Callicarpa japonica alba* last much longer. The majority of ornamental fruits, such as crab apples and firethorns (*Pyracantha*) are displayed for shorter periods and a few, either because of the fondness birds have for them or for other reasons, are extremely short-lived. Such a one, unfortunately, is the sapphire berry (*Symplocos paniculata*).

Ornamental-fruited plants in addition to those already mentioned, some suitable only for mild climates, include *Ampelopsis brevipedunculata*, barberries (*Berberis*), bittersweets (*Celastrus*), buffalo-berry (*Shepherdia argentea*), camphor tree (*Cinnamomum camphora*), castor-bean (*Ricinus communis*), Chinaberry (*Melia azedarach*), choke-cherries (*Aronia*), Christmas-berry (*Heteromeles arbutifolia*), *Clematis*, *Daphne* (some sorts), dogwoods (*Cornus*), *Elaeagnus*, elderberries (*Sambucus*), firethorns (*Pyracantha*), golden dewdrop (*Duranta repens*), heavenly-bamboo (*Nandina domestica*), loquat (*Eriobotrya japonica*), *Mahonia*, matrimony vine (*Lycium halimifolium*), mountain-ash (*Sorbus*), Natal-plum (*Carissa grandiflora*), *Pernettya mucronata*, *Phillyrea*, *Photinia*, *Prinsepia sinensis*, *Prunus* (some sorts), roses (*Rosa*), *Sarcococca*, sea-buckthorn (*Hippophae rhamnoides*), service berries (*Amelanchier*), and *Skimmia*.

ORNITHIDIUM. See Maxillaria.

ORNITHOCEPHALUS (Ornitho-céphalus). Fifty species of the orchid family ORCHIDACEAE constitute *Ornithocephalus*, which is wild from Mexico to tropical South America and in the West Indies. Its name, alluding to a fanciful resemblance of the flowers to the head of a bird, derives from the Greek *ornis*, a bird, and *kephale*, a head.

Epiphytes (tree-perchers), in the wild mostly roosting on small branches and twigs, these are small plants without pseudobulbs, with iris-like fans of somewhat fleshy leaves. The usually little flowers, in racemes, have almost equal, spreading sepals and petals and, continuous with the column, a three-lobed lip.

Native from Mexico to Panama, ***O. inflexus*** has lengthwise-folded leaves, their bases sheathing each other, 3 to 4 inches long. Greenish to whitish-green, the 3/16-inch-wide flowers have practically lobeless lips, are in slender, often slightly zigzagged racemes up to 4 inches long. As a wildling ranging from Mexico to Guatemala, ***O. tripterus*** has pointed-sword-shaped leaves 2 to 4½ inches long and slender racemes up to about 6 inches long of two-ranked, evenly spaced, greenish-white flowers ¼ inch or a little more in diameter, with a tongue-shaped lip.

Garden Uses and Cultivation. Of interest to orchid collectors, the sorts of this genus described here are most likely to succeed when attached to suspended slabs of tree fern trunk and grown in an environment favorable to cattleyas. Their roots must be kept always moderately moist. A humid atmosphere and shade from strong sun are necessary. For more information see Orchids.

Ornithocephalus tripterus

ORNITHOCHILUS (Ornitho-chìlus). The Asian genus *Ornithochilus*, of the orchid family ORCHIDACEAE, is occasionally cultivated by orchid fanciers. It consists of one or perhaps two species of evergreen epiphytes (plants that perch on trees without taking nourishment from them) and has a name, alluding to the shape of the lip of the flower, derived from the Greek *ornis*, a bird, and *cheilos*, a lip.

Native from southern China to the Himalayas, ***O. fuscus*** has rather fleshy, oblong-elliptic leaves 5 to 8 inches long and about one-third as broad. Its ½-inch-wide, fragrant blooms are many in arching or pendulous racemes or panicles up to 1½ feet in length. They have similar reddish-striped, greenish-yellow sepals and petals and a spurred, three-lobed lip with the center lobe red and the side ones yellowish.

Garden Uses and Cultivation. The collectors' item described here responds to the same conditions and treatment as *Phalaenopsis*. It is best accommodated in hanging baskets rather than in pots. For additional information see Orchids.

ORNITHOGALUM (Ornithó-galum)—Star-of-Bethlehem, Chincherinchee. Hardy and nonhardy bulb plants, several of great charm, constitute *Ornithogalum*, of the lily family LILIACEAE. The majority of its 150 species are endemics of South Africa, the remainder natives of other parts of Africa and of Europe and Asia. The name, derived from the Greek *ornis*, a bird, and *gala*, milk, is of uncertain application. It probably alludes to the milky-white flowers of some sorts.

In North America, the star-of-Bethlehem and *O. nutans* are naturalized, the former sometimes to the extent of being regarded a weed. But if weed it be at least it is a pretty one. As long ago as 1760 an advertisement in a Boston, Massachusetts, newspaper offered seeds of the star-of-Bethlehem for sale and undoubtedly it was well known long before then. Surely early settlers brought the "starre flower," as it was called in England at the time, to America, for it was popular in England and its bulbs were easy to transport. But pretty though the star-of-Bethlehem is, it cannot compare with the exquisite beauty of Mediterranean *O. arabicum* or that of South African *O. lacteum*, *O. saundersiae*, or *O. thyrsoides* and the latter's golden-flowered variety.

Ornithogalums have small to large bulbs and all-basal foliage. The leaves are linear to strap-shaped, few to several from each bulb. On short to tall, erect, branchless flowering stalks, the few to many blooms are displayed in umbel-like clusters or racemes. The white, greenish, yellow, orange, or reddish, more or less starry blooms have six petals (more correctly, tepals) separate to their bases, six stamens with usually dilated stalks, and one style terminating in a more or less three-lobed stigma. The fruits are three-lobed capsules containing a few black or nearly black seeds.

The star-of-Bethlehem (***O. umbellatum***) has a vernacular name that in New England and some other parts of North America is applied to quite unrelated,

Ornithogalum umbellatum

Ornithogalum umbellatum (flowers)

nonbulbous *Campanula isophylla*. Native to the Mediterranean region, the ornithogalum so called often escapes from cultivation and makes itself thoroughly at home in yards, meadows, and similar places where it can find foothold. From about 1-inch-wide, ovoid, white bulbs come tufts or sheafs of six to nine linear leaves up to 1 foot long by up to ⅓ inch wide and erect flowering stalks of approximately the same length topped by open, flat-topped, umbel-like clusters of few to thirty glistening white flowers 1 inch to 1½ inches wide with the reverses of their petals green with white margins. The stamens are considerably shorter than the petals. The blooms open fully in sun, close at night and in dull weather. This species is believed by some to be the dove's dung of the Bible.

Also naturalized in North America, Europe, and western Asia ***O. nutans*** has racemes of nodding flowers. Its ovoid bulbs exceed 1 inch in diameter. The leaves, limp, linear, and about 1½ feet long by ½ inch wide, soon die after the blooms fade. The flowering stalks, 1 foot to 2 feet tall, have loose racemes of usually ten or fewer

Ornithogalum nutans

bell-shaped blooms nearly 1 inch long, white on their insides, on their outsides silvery-gray-green edged with white. Variety *O. n. boucheanum* has denser racemes of larger blooms.

Beautiful ***O. arabicum,*** of the Mediterranean region, has rather pear-shaped bulbs 1 inch to 1½ inches in diameter and several long-tapering, glaucous green leaves 1 foot to 1½ feet long by ¾ to 1 inch

Ornithogalum arabicum

wide. Its pearly-white, up-facing flowers, 1½ to 2 inches wide and with stamens tipped with yellow anthers, are in short, compact, roundish to pyramidal racemes of usually eight to twelve atop stalks 1½ to 2 feet long. They have the rich, mouth-watering fragrance of ripe apples. About 1½ inches across, each has at its center a conspicuous, large, shining black ovary. The stamens, less than one-half as long as the petals, have stalks dilated in their lower parts. The sort called *O. corymbosum*, treated by some botanists as a separate species, is by others considered to be an especially free-flowering form of *O. arabicum*.

Restricted as a native to the eastern Mediterranean region, ***O. pyramidale*** (syn. *O. narbonense pyramidale*) has bulbs approximately 1 inch in diameter and four to six strap-shaped, channeled, glaucous green leaves 1 foot to 1½ feet long by up to ½ inch wide. From 1 foot to 2 feet tall or taller, the erect flowering stalk carries a 4- to 9-inch-long, slender, cylindrical raceme of twenty to fifty starry blooms ¾ to 1 inch wide with petals pure white on their insides, green-banded on their outsides. Less robust and smaller-flowered, ***O. narbonense,*** of the Mediterranean region, is otherwise similar. Much like *O. pyramidale* and wild through much of southern and central Europe, ***O. pyrenai-***

Ornithogalum thyrsoides

Ornithogalum saundersiae

cum differs in its flowers being greenish-white and turning yellow when dried.

The chincherinchee (***O. thyrsoides***), its common name an onomatopoeic reminder of the rustling sound its dry stems and flowers make when stirred by the wind, is a well-known South African. The blooms

Ornithogalum thyrsoides (flowers)

of this when cut last in good condition for several weeks and even before air transport became available were shipped from South Africa to England for sale by florists. The chincherinchee has spherical bulbs with six or fewer lanceolate leaves up to 1 foot long and ¾ inch wide. The dense, many-flowered, pyramidal racemes are lifted on erect stalks to heights of 1 foot to 2 feet. The blooms are 1 inch to 1½ inches across and white to creamy-white. The style is shorter than the ovary. Formerly considered a variety of *O. thyrsoides*, less robust South African ***O. miniatum*** (syn. *O. aureum*), up to 1 foot tall, has leaves 3 to 5 inches long by up to 1 inch wide. In umbels of few to many, its 1¼-inch-wide flowers are yellow to brilliant orange, less often white.

Variable ***O. lacteum,*** of South Africa, has spherical bulbs 1 inch to 1½ inches in diameter and up to ten somewhat fleshy, strap-shaped to lanceolate leaves, usually reddish toward their bases, fringed with fine short hairs. They are up to 1 foot long. Exceeding the leaves in length, the erect flowering stalk supports a dense, cylindrical raceme up to about 6 inches long of twenty to fifty persistent, pure white to milky-white blooms 1 inch to 1½ inches wide, with styles as long or longer than the ovaries.

Another native of South Africa, ***O. saundersiae*** resembles Mediterranean *O. arabicum* in its blooms having prominent and decorative, beadlike ovaries, in this species greenish-black. Its bulbs are large and white, its several to many strap-shaped leaves limp and 1 foot to 1½ feet long. The stout, 2- to 3-foot-tall flowering stalks terminate in dense, many-flowered clusters of 1-inch-wide, white flowers lightly tinged with green on their outsides, with grayish anthers. Reports from South Africa indicate this species poisons grazing animals.

Slender spires of blooms that open from below upward in long succession and terminate long stalks are characteristic of quite similar ***O. caudatum*** and ***O. longebracteatum,*** of South Africa. These have ovoid bulbs 3 to 4 inches in diameter and four to eight rather fleshy leaves 1½ to 2 feet long by ¾ inch to 1½ inches wide. The flowers, in cylindrical racemes of fifty to 100 terminating 1- to 2-foot-long stalks, are white with green keels down the backs of the petals. The stalks of the stamens of *O. caudatum* are alternately lanceolate and four-sided at their bases, those of *O. lon-*

Ornithogalum caudatum

Ornithogalum caudatum (flowers)

gebracteatum are alternately linear and lanceolate.

Garden and Landscape Uses. The star-of-Bethlehem is suitable for naturalizing in drifts along sunny fringes of woodlands and shrub borders and in other informal areas where on bright days in spring its glistening stars will carpet the ground. Do not admit it to rock gardens or other places where its proclivity to spread rapidly will interfere with choicer plants. The nodding star-of-Bethlehem, as *O. nutans* has been called, is altogether more seemly. It too is adapted for naturalizing and does well in places with light part-day shade as well as in sun. It is very much less aggressive than the star-of-Bethlehem and is much more likely to "stay put" and not threaten delicate neighbors.

The other sorts described here are not reliably hardy in the north, although in choice, sheltered locations it is always possible that some from the Mediterranean region will survive at least for a few seasons if protected by covering them well in winter with leaves, salt hay, branches of evergreens, or similar material. Still, in general, they, like the South African kinds, are adapted for outdoor planting only in mild climates and especially those characterized by dry summers. In such regions nonhardy ornithogalums are splendid for flower beds and borders and kinds such as the chincherinchee, *O. arabicum*, and *O. lacteum* as sources of cut flowers.

Indoors these same splendid sorts are admirable for greenhouses and may even be bloomed in sunny windows in cool rooms and sun porches. For this latter purpose, although less splendid in bloom, *O. caudatum* and *O. longebracteatum* are even better adapted.

Cultivation. Ornithogalums are easy to manage, responsive to ordinary soils, and not finicky about other environmental conditions nor seriously subject to pests or diseases. They need well-drained soil, preferably fertile, and are best left undisturbed as long as they are growing and blooming satisfactorily. Outdoors, set the bulbs, depending upon their size, with their tips 2 to 4 inches beneath the surface.

Indoor cultivation is also simple. Plant the bulbs in early fall in well-drained pots of fertile soil, setting those of *O. caudatum* and *O. longebracteatum* singly in the containers. Other kinds are displayed to better advantage if three or more are put in each pot with space between them equal to about one-half the diameter of the bulb. Let the tips of the bulbs show at the soil surface. Pack the soil firmly.

Grow in a sunny greenhouse or in a window in a cool room. A night temperature of 45 to 50°F is suitable for all except *O. miniatum*, which does better at 55°F. Day temperatures may exceed those maintained at night by five to fifteen degrees depending upon the brightness of the weather. Water with some caution at first, increasing supplies as foliage develops and when in full leafage keeping the soil constantly decidedly moist. After flowering is through and the foliage begins to die naturally, gradually increase the intervals between soakings and finally stop watering and keep the soil dry until the beginning of the next growing season. Fertilizing during the period of active growth from when the pots become well filled with roots until the flower buds begin to open is rewarding. Use dilute liquid fertilizer at about weekly intervals. Propagation is easy by means of offset bulbs, which most sorts produce freely. Bulb cuttings are also satisfactory and seeds afford a ready means of increasing stock although two, three, or more years will elapse before the first blooming of seedlings.

ORNITHOSTAPHYLOS (Ornitho-stáphylos). Closely related to *Arctostaphylos* and by some authorities included there, the genus *Ornithostaphylos* belongs in the heath family ERICACEAE. Its name, presumably referring to the inflorescences, derives from the Greek *ornis*, a bird, and *staphyle*, a bunch of grapes.

The only species is an evergreen shrub with erect, rigid, much-branched stems and undivided, narrow, very short-stalked leaves with toothless, rolled-back margins. The flowers have a persistent, usually five-lobed calyx, spherical to urn-shaped, a usually five-lobed corolla, and usually ten nonprotruding stamens. The fruits are small, dry drupes (fruits of plumlike structure).

Endemic to southernmost California and Baja California ***O. oppositifolia*** is a tall shrub with narrow-linear leaves 1 inch to 3¼ inches long and more or less whitish-pubescent on their undersides. Freely produced in loose, somewhat drooping panicles, the little flowers are succeeded by fruits up to ¼ inch long.

Ornithostaphylos oppositifolia

Ornithostaphylos oppositifolia (foliage)

Garden and Landscape Uses and Cultivation. These are as for western North American dry region species of *Arctostaphylos*.

ORONTIUM (Orónt-ium)—Golden Club. There is scarcely a possibility of mistaking this beautiful aquatic plant for any other. Native to swamps and shallow waters in coastal plain regions from southern New York and Massachusetts to Florida, Mississippi, and Kentucky, the only species of *Orontium* belongs in the arum family ARACEAE and so is kin to the jack-in-the-pulpit (*Arisaema*) and calla-lily (*Zantedeschia*). Its name is an ancient one for some water plant, probably a kind that grew in the Orontes River in Syria.

From stout deep-penetrating rhizomes the golden club (***O. aquaticum***) develops floating or ascending leaves with stalks 4 inches to 1½ feet long, swollen at their bases. Their waxy, velvety, bluish-green blades, up to 1 foot long, are elliptic, without a distinct midrib, and have longitudi-

Orontium aquaticum

Orontium aquaticum (flower spike)

nal parallel veins joined by connecting horizontal veinlets. Produced abundantly, the flower spikes or spadixes are the "clubs" that inspire the common name. Terminating white stalks 6 inches to 3 feet long and flattened toward their tops, they are slender, cylindrical, erect, golden-yellow, and 1 inch to 2 inches long. At the base of each "club" (analogous to the central spikelike column of the jack-in-the-pulpit and calla lily) is a spathe (the equivalent of the large, enveloping, petal-like parts of the jack-in-the-pulpit and calla-lily), but in the golden club these are not showy and soon drop off. The tiny flowers, clustered along the central column, are bisexual. The fruits are utricles.

Garden and Landscape Uses. The golden club is admirable for sunny bogs, shallow pools, and still, shallow waters near the shores of ponds and lakes, planted either in soil bottoms or in containers. Once established where its rhizomes can run freely, it may be difficult to eradicate. In water not over a very few inches deep the foliage stands well above the surface, in decidedly deeper water it floats and only the flower spikes are lifted above the water.

Cultivation. The golden club may be planted in submerged beds or in large pots or tubs of rich soil where the tops or crowns of the rootstocks will be covered with 1 foot to 1½ feet of water. The best planting time is early spring. If they are to root directly into the bed of a pond, lake, or slow-moving stream a simple way to plant is to set the rootstocks in soil contained in wicker baskets or bags of coarse burlap, weight them with stones, and sink them in position. The surface soil in pots or tubs should be covered after planting with 1 inch of clean sand before submerging so that the soil will not muddy the water nor debris float to the surface. Once established, the golden club needs little attention, but it is advisable to divide and replant specimens in pots or tubs every two or three years. The containers should be at least 1 foot wide and deep, larger sizes are advantageous. Propagation is by division of the rootstocks in early spring.

OROSTACHYS (Oro-stáchys). Genera of the orpine family CRASSULACEAE are notoriously difficult to delimit and so *Orostachys,* by some botanists included in *Sedum,* has in the past been accommodated in *Umbilicus* and *Cotyledon.* It consists of twenty species, natives from Japan to South Korea and the Ural Mountains. The name, relating to the habitats of some kinds and the manner of flowering, comes from the Greek *oros,* a mountain, and *stachys,* a spike.

These are biennial or perennial succulent herbaceous plants with dense sempervivum-like, basal rosettes of foliage. Except in *O. malacophylla,* the rosettes that have not yet bloomed wither back in fall to become tight winter buds. In spring their leaves elongate from their bases, carrying with them as callused tips the remains of the previous year's growth. In the summer or fall of their second or third year, the rosettes bloom, then die. The *Sedum*-like flowers are in dense terminal spikes, racemes, or panicles. Each has five sepals, five often red-spotted, white, pinkish, yellowish, or yellow, more or less spreading petals, ten stamens, and five pistils. The fruits are small follicles.

Solitary or clustered, the rosettes of ***O. japonica*** (syn. *Sedum japonicum*), of Japan and Korea, are ½ inch to 4 inches wide. They are of up to twenty thick, oblong-lanceolate leaves generally not over 1/10 inch wide, the outer ones harshly-spine-tipped, the inner ending in softer spines. The many white flowers, in erect, slender spires 2 inches to 1 foot long or somewhat longer, are ¼ to ½ inch in diameter. The flowers of ***O. spinosa*** (syn. *Sedum spinosum*) are yellow and in spikelike panicles. Its clustered rosettes are of wedge-shaped leaves tipped with soft, white spines. The outer rows of leaves are notably bigger and longer than the inner ones. This is native through much of central Asia.

Garden Uses and Cultivation. Not reliably hardy at New York City, these are attractive plants for rock gardens in milder, drier climates, and for inclusion in indoor collections of succulents. They grow well in well-drained, sandy soils in sunny locations. Increase is easy by seed. For further information see Succulents.

OROTHAMNUS (Oro-thámnus) — Mountain-Rose. The only species of *Orothamnus,* of the protea family PROTEACEAE, is endemic to a very limited region in South Africa. Its name, referring to its habitat and form, derives from the Greek *oros,* mountain, and *thamnos,* a shrub.

Probably the rarest species of the protea family, ***O. zeyheri*** is accepted as endangered, a species threatened with extinction. Known in its native land as mountain-rose, it is a slender, evergreen shrub with one to few branches up to 9 feet tall, in their upper parts clothed with overlapping, pointed-broad-ovate leaves fringed with silvery hairs. The flowers, in groups of several at the tips of the stems, are accompanied by broad, promegranate-red bracts folded around each other in the fashion of the petals of a rose, to produce a nodding inflorescence about 3 inches long and 2 to 3 inches wide.

Orothamnus zeyheri

Garden Uses and Cultivation. Perhaps never successfully cultivated away from its native land, and even there considered extremely difficult, the South African mountain-rose is likely to prove a very real challenge to even the most experienced cultivator. It is truly only for the most enthusiastic of collectors. It is described as growing in the wild on or among granite bolders and rooting into a 1 foot-deep, spongy mass of decayed reeds with water seeping through. The suggestion has been made that from ripening until they are sown the seeds should be kept in water.

OROXYLUM (Oró-xylum). This genus of the bignonia family BIGNONIACEAE consists of one or two species. It is a native of India, Malaya, and other warm parts of

Asia. Its name sometimes spelled *Oroxylon,* derives from the Greek *oros,* a mountain, and *xylon,* wood, and alludes to its habitat.

Cultivated in southern Florida and other warm regions, ***Oroxylum indicum*** (syn. *Bignonia indica*) is a deciduous tree of curious and ungainly aspect. Its habit of growth is peculiar. Young specimens extend rapidly and without branching to a height of 10 to 15 feet and then produce their first blooms. Then they almost cease to grow, but later from buds lower on the stem stiffly erect branches develop. The tree may reach an eventual height of 60 feet, but is often lower. In Florida old specimens are often without branches. The enormous leaves of *O. indicum* are clustered at the top of the trunk in the manner of those of a papaya or at the ends of the branches. They are 3 to 5 feet long and two-, three-, or sometimes four-times-pinnately-divided into ovate, toothless, very short-stalked leaflets 3 to 8 inches long. On stout stalks the evil-scented, bisexual blooms are in erect, terminal racemes. They are bell-shaped, whitish to lurid purple, and 2 to 3 inches long by 2½ to 3½ inches wide. The corollas have five crinkly-edged lobes. There are five protruding stamens and a slender style. The fruits are flat, woody capsules 1½ to 3 feet long by about 2½ inches wide and containing winged seeds 2 to 3 inches long. Opening at night, but not closing the following day, the blooms in its home lands attract bats, which are the pollinating agents. The fruits and bark are used for tanning and as a mordant in dyeing. The seeds are purgative and, like other parts, have been used medicinally.

Garden and Landscape Uses and Cultivation. The strange and rather grotesque tree described above is an item for collectors of the unusual, something of a conversation piece. It succeeds in ordinary soil, but preferably one decidedly fertile, in sun or part-shade, and is increased by seed and by cuttings. The latter may be rooted in a propagating bench furnished with mild bottom heat. The plant sometimes called *O. flavum* is *Radermachera pentandra.*

OROYA (Or-óya). This genus consists of one variable and poorly defined species of the cactus family CACTACEAE that, by some botanists, is divided into six separate species. It takes its name from a village in Peru near which it was first discovered. The genus *Oroya* is endemic at high elevations in the Peruvian Andes where its members grow embedded in the earth with only their flat tops visible above the surface.

Flattened-spherical and usually solitary, the plant bodies of ***O. peruviana,*** 4 to 5 inches in diameter, have generally twenty-one blunt ribs formed of low bumps or tubercles. The spine-producing areas (areoles) have eighteen to twenty yellowish radial spines with darker bases and red apexes, of varying lengths up to ¾ inch, and sometimes up to four stouter, longer, red, central spines. Approximately ¾ inch in length, the short, broadly-bell-shaped, variously colored flowers typically have reddish outer segments (petals) and pink, yellow-based inner ones. Neither stamens nor style exceed the petals in length. The latter is capped by a lobed stigma. The fruits are obovoid and slightly fleshy.

Garden Uses and Cultivation. Cactus hobbyists find this species of interest. In cultivation it is best to have the plant body standing above the ground surface. If more or less buried, as in the wild, it is likely to rot. For more information see Cactuses.

ORPINE is *Sedum telephium.*

ORTHEZIA. See Greenhouse Orthezia.

ORTHOCARPUS (Ortho-cárpus)—Owl's-Clover, Cream Sacs. Except for one indigenous in the Andes, all twenty-five species of *Orthocarpus,* of the figwort family SCROPHULARIACEAE, are confined as wildlings to western North America. They are annuals and differ from closely related *Castilleja* in that the lower lip of the flower is as long and bigger than the helmet-like upper one. The name, derived from the Greek *ortho,* straight, and *karpos,* fruit, refers to the symmetrical seed capsules.

Erect or lax, members of this group have alternate, stalkless, narrow leaves, undivided or pinnate. The flowers, crowded among prominent bracts, are in dense, terminal spikes. The more or less bell-shaped calyx is divided into four parts or into two parts with the divisions lobed. The narrowly-tubular corolla is conspicuously two-lipped, the upper lip without lobing or teeth and the lower one three-toothed. There are four stamens. The fruits are capsules.

The owl's-clover (***O. purpurascens***) inhabits fields and grassy slopes in California, Arizona, and Baja California. Its reddish or purplish stems, usually branched from the base, are erect and 6 inches to 1 foot tall or slightly taller. The deeply-pinnately-lobed leaves are up to 2 inches long. The stems terminate in dense spikes, up to about 4 inches long, of light red to purple blooms with white-tipped lower lips and purple or yellow markings. They are set among palmately-lobed, rose-purple-tipped, greenish-purple bracts. The flowers are ½ inch to 1¼ inches long. In variety *O. p. ornatus* the ends of the lower lips of the flowers are bright orange. Variety *O. p. pallidus* has white or yellow lower lips.

Cream sacs (***O. lithospermoides***), native from Oregon to California, is erect and has stoutish stems with upright branches 9 inches to 2 feet tall. Its lanceolate leaves are ¾ inch to 2½ inches long, the upper ones pinnately-lobed and glandular-pubescent. The dense flower spikes have palmately-lobed bracts and clear yellow, shortly-pubescent flowers with two purple spots at the base of the bottom lip. The blooms are up to 1 inch long. The flowers of *O. l. bicolor,* at first white, become pink as they age.

Native from Idaho to Washington and British Columbia, ***O. tenuifolius*** is about 8 inches tall and has hairy, purplish stems and leaves cut into slender, threadlike segments. Its flowers, ¾ inch long, are yellow, sometimes tipped with purple.

Garden and Landscape Uses and Cultivation. These are attractive annuals for flower beds and borders and as cut flowers. They are easily raised from seeds sown in spring where the plants are to remain or, in mild climates, may be sown in fall. They require well-drained, moderately fertile soil and an open, sunny place. Young plants of owl's-clover and cream sacs are thinned to about 6 inches apart, those of *O. tenuifolius* to about 4 inches.

ORTHOPHYTUM (Ortho-phỳtum). Endemic to Brazil, *Orthophytum,* of the pineapple family BROMELIACEAE, comprises ten or more species. Its name, derived from the Greek *ortho,* straight, and *phytum,* a plant, alludes to the erect flower stalks.

Orthophytums are small to medium-sized, semisucculent, stemmed or nearly stemless evergreen perennials that develop clusters of rosettes and when not in bloom have much the aspect of dyckias or aloes. In their native homes they grow on sunny cliffs and rock ledges with their roots in whatever decayed rock and soil is available. They do not, as do so many members of the pineapple family, perch on trees. Their green to coppery leaves are margined with soft spines. The flowers, accompanied by bracts, are white and variously disposed, those of some sorts in crowded heads low in the centers of the rosettes, those of others on tall stalks.

Orthophytum saxicola

Orthophytum vagans

Orthophytum vagans (close-up of a rosette)

Each little bloom has three each sepals and petals, six stamens, and one style. The fruits are berries.

One of the smaller sorts, ***O. saxicola*** has rosettes, usually under 6 inches across, of rich coppery leaves and clusters of flowers on stalks that lift them only about 2 inches above the centers of the rosettes. Longer, trailing stems and narrower leaves distinguish otherwise similar ***O. vagans*** from *O. saxicola*.

Garden and Landscape Uses and Cultivation. Interesting to bromeliad fanciers and collectors of unusual plants, the sorts of this genus are easy to grow outdoors in warm, dry climates and in greenhouses where conditions satisfactory to cactuses and other succulents are maintained. They need full sun and porous, very well-drained soil. Propagation is easy by offsets and by seed.

ORTHOPTERUM (Orthóp-terum). Two species native near the eastern tip of South Africa compose *Orthopterum*, of the carpetweed family AIZOACEAE. They belong to the *Mesembryanthemum* complex of that group. Their flowers suggest to the botanically untrained those of daisies, but are of very different structure. Each is a single bloom, not a head of numerous florets. The name, from the Greek *orthos*, erect, and *pteron*, a wing, alludes to the prominent wings on the cells of the fruiting capsules.

Orthopterums are very succulent, short-stemmed, clump-forming, evergreen perennials, closely related to and much resembling *Faucaria*. Each shoot or short branch carries three or four pairs of opposite, more or less erect, dark-dotted, light green leaves, those of each pair rather unequal in size. Linear to lanceolate, they have deep, ovate-lanceolate sides. Their upper surfaces are flat or slightly concave, their lower ones rounded, keeled toward the apex. The margins and keel have each one or two indistinct, recurved teeth. Stalkless, the flowers, which expand in the afternoon, are golden-yellow often with reddish undersides to the many petals. They have five sepals, three or four rows of petals, a central bunch of erect stamens, and five or six stigmas. The fruits are capsules.

Sometimes cultivated in collections of choice succulents, ***O. coegana*** has leaves 1 inch to 1½ inches long, those of each pair slightly united at their bases. They are ⅓ inch thick and slightly wider. The blooms are about 1¾ inches in diameter.

Garden Uses and Cultivation. These are as for *Faucaria*. For further information see Succulents.

ORTHOTHYLAX (Ortho-thŷlax). One species native to Australia constitutes *Orthothylax* of the philydrum family PHILYDRACEAE. Its name derives from the Greek *orthos*, right, and *thylax*, a bag or pouch.

A tufted, hairless, evergreen herbaceous perennial about 3 feet tall, ***O. glaberrimus*** has sword-shaped, dull green, leathery leaves 2 to 4 feet long and 1 inch to 1½ inches wide with a prominent mid-vein and their lower margins folded inward. The little creamy-white flowers are many together in erect, plume-like panicles with upward-pointing branches. They have three to five quite separate perianth parts, one stamen with a flattened stalk, and one style. The fruits are three-angled capsules.

Garden and Landscape Uses and Cultivation. This species of bold aspect has merit for outdoor beds and borders in warm, dryish climates, and for ground beds and tubs in large conservatories where the night temperature in winter is 45 to 50°F and that by day ten degrees or so higher. It responds to deep, fertile, well-drained soil in a sunny location or where it receives part-day shade. Propagation is by division, best done when new growth is just starting, and by seed.

Orthothylax glaberrimus

Orthothylax glaberrimus (panicle of flowers)

ORTHROSANTHUS (Orthros-ánthus). Relatives of the blue-eyed-grass (*Sisyrinchium*), the eight species of *Orthrosanthus* belong in the iris family IRIDACEAE. Three are natives of Mexico or the Andes, the others of Australia. The name, from the Greek *orthros*, morning, and *anthos*, a flower, alludes to the blooms being open only in the early part of the day.

These are tufted plants with short, woody rootstocks and slender, grasslike or iris-like, mostly basal foliage. The short-stalked flowers, in clusters of two to many, form loose, spikelike, branched or branchless panicles at the tops of erect stalks that have a few, small, distantly spaced leaves. They have short perianth tubes, six spreading perianth lobes (tepals), and three stamens. The fruits are capsules.

A western Australian species, ***O. multiflorus*** has flower stalks and leaves 1 foot to 1¼ feet long. The sky-blue blooms nearly 1½ inches across, are in many-flowered clusters; individual blooms open in succession. Each lasts for only a few hours and only one of each cluster is open at a time. In sunny weather they fade by noon, on cloudy days later, but well before nightfall.

Native from Mexico to Bolivia and Peru, ***O. chimboracensis,*** 1½ to 3 feet tall, has two-ranked, grasslike leaves 1 foot long or a little longer by up to ½ inch or less wide. Taller than the foliage, the flowering stalks carry clusters of two or three light violet-blue flowers approximately 1½ inches wide and with yellow anthers and purple style branches.

Garden and Landscape Uses and Cultivation. These plants succeed outdoors in California and places with similar climates. They make pretty displays in late spring and early summer and succeed in light shade in moist soil that contains an abundance of compost, leaf mold, or humus. In California *O. multiflorus* has survived, with some damage to its foliage, temperatures of 20°F. They may also be grown in ground beds or pots in freely ventilated greenhouses where a winter night temperature of 40 to 50°F is maintained. Indoors they do well in well-drained, sandy peaty soil, watered moderately, but never excessively. Seed and division afford easy means of propagation.

ORYCHOPHRAGMUS (Orycho-phrágmus). Sometimes confused with closely related *Moricandia*, the genus *Orychophragmus*, of the mustard family CRUCIFERAE, consists of two annual or biennial species of eastern Asia. The name, referring to a recondite feature of the seed pods, comes from the Greek *oryche*, a pit, and *phragmos*, a septum.

From *Moricandia* the plants of this genus differ in never developing subshrubby characteristics (woody bases), in their leaves being thin and sometimes with hairs, and in the seed pods not being winged. The flowers have four sepals, four petals that spread to form a cross, four long and two short stamens, and one style. The fruits are slender capsules.

Orychophragmus violaceus

Chinese ***O. violaceus*** (syn. *O. sonchifolius*) is 1 foot to 2 feet tall, has a basal rosette of reverse-fiddle-shaped, deeply-pinnately-lobed leaves and much smaller stem leaves. In racemes terminal on the stems and branches, the blue-lavender flowers, from 1 inch to 1¾ inches in diameter, have four spreading petals. In China this species is eaten as a vegetable.

Garden and Landscape Uses and Cultivation. About as hardy as English wallflowers, the species described is best treated as a biennial by sowing seeds in late spring or early summer. In the vicinity of New York City the plants will not survive winters outdoors, as they do in somewhat milder climates, but carry over satisfactorily in a well-protected cold frame.

Transplant the seedlings as soon as their second pair of leaves has developed, about 6 inches apart, in rows 1 foot to 1½ feet apart in a sunny outdoor nursery bed or, where they are not winter hardy, 6 inches apart each way in cold frames. Transfer them to their flowering quarters in early spring.

Attractive specimens can be had in flower in a cool greenhouse in late winter and early spring. To achieve this, sow seeds in late summer and transplant the seedlings individually to 3-inch pots. As growth makes necessary, repot successively until they occupy 5- or 6-inch containers. Use porous, fertile soil. A winter night temperature of 45 to 50°F, with a daytime rise of from five to fifteen degrees, depending upon the brightness of the weather, is adequate from fall through spring. At all times maintain a good circulation of air by ventilating the greenhouse on all favorable occasions. In February, after the days begin to lengthen, apply dilute liquid fertilizer, at first at two- or three-week intervals, later more frequently.

ORYZA (Orỳ-za)—Rice. Chiefly an agricultural crop, rice is occasionally cultivated in small quantities in educational exhibits of plants useful to man. A species of *Oryza*, of the grass family GRAMINEAE, it is one of fifteen to twenty species of its genus. Wetland and aquatic, annual and perennial grasses, oryzas are widely distributed as natives of the tropics. None is of ornamental importance. The name *Oryza* is the ancient Greek one for rice. The wild-rice of North America belongs to the related genus, *Zizania*. Rice paper is a product of a very different plant, *Tetrapanax papyriferum*.

Rice (***O. sativa***) has been cultivated since before 2600 B.C. An erect annual, 2 to 6 feet in height, it has long, flat leaf blades. Its flower panicles, 6 inches to 1½ feet long, and curved to one side or nodding, have branches with many compressed spikelets about ⅓ inch long, each containing one primitive flower with usually six stamens, and each developing a single seed. The seed or grain, free inside the husk, is yellow or red. The whiteness of rice familiar as food results from polishing the grains. Unfortunately this removes most of the vitamins and proteins so that polished rice is a far less adequate food than unpolished. Rice flour cannot be baked. The grain is commonly prepared for eating by boiling. Rice wine or saki is produced by fermenting the grain.

Oryza sativa

Cultivation. For its satisfactory growth rice requires tropical or warm subtropical conditions. Most varieties succeed only if the soil is flooded to a depth of a few inches during much of the growing season, and in the Orient paddies (the fields where rice is grown) are constructed so that they can be irrigated to meet this need. There are upland varieties that do not have such exacting water requirements, but these are comparatively poor yielders. Most of the rice cultivated in North America is of the upland type. For demonstration in tropical greenhouses, where a minimum temperature of 70°F is maintained, devoted to economic plants (sorts exploited for their uses to man) rice is easy to grow in large pots or tubs. In late winter or spring fill these nearly to their rims with fertile soil and sow the seeds thinly over the surface. Cover with 1 inch of coarse sand and submerge the containers so that the sand is covered with 1 inch to 2 inches of water. When the seedlings appear thin them out as needed for them to stand 3 to 4 inches apart. When the plants are about 1 foot tall the containers may be raised so that the sand surface is 3 to 4 inches above the water level but this is not absolutely necessary.

OSAGE-ORANGE is *Maclura pomifera.*

OSBECKIA (Os-béckia). About sixty species constitute *Osbeckia,* a genus of the melastoma family MELASTOMATACEAE, native from Africa and southern Asia to Australia. Its name honors the Swedish naturalist Peter Osbeck, a clergyman, who died in 1757.

Herbaceous perennials, subshrubs, and shrubs, many of which are attractive ornamentals, constitute this horticulturally neglected genus. Its members have opposite, stalked or stalkless, undivided, toothed or toothless leaves with three to seven prominent longitudinal veins. The generally pink, violet, or reddish showy flowers, in heads or panicles, are less often solitary. They have a bristly-hairy, urn-shaped, to nearly spherical, usually five-lobed calyx, five or less frequently four petals often hair-fringed at their margins, usually ten stamens, and one style. The fruits are capsules.

Native to Ceylon and naturalized in Mauritius, ***O. octandra*** is a low shrub, the stems with appressed hairs, the three-veined leaves sparsely with fine hairs. The calyx tube is usually without hairs and the five petals, ⅔ to ¾ inch long, are pink to mauve or purple. The epithet *octandra,* which means with eight stamens, is misleading, since there are ten. Native to southern Asia from India eastward, ***O. stellata crinita*** is a shrub 1 foot to 4 feet tall with finely-hairy stems and sparsely-hairy five-veined leaves. The calyx tube is bristly with branched hairs and the four petals, ⅔ to 1 inch long, are pale pink to white or red, blue, or purple. The eight stamens have long tubular beaks.

Landscape Uses and Cultivation. Osbeckias require humid, tropical or subtropical environments. Where such conditions prevail outdoors they are excellent for flower beds and the fronts of shrub borders. They are also delightful greenhouse plants. Any reasonably fertile, well-drained, coarse soil suits. Specimens in pots that have filled their containers with healthy roots benefit from regular applications of dilute liquid fertilizer. Propagation is by seed or by cuttings. The latter root readily in sand, perlite, or in a mixture of either of these and peat moss, in a humid atmosphere and a temperature of about 70°F. Old specimens may be pruned back and, if needed, repotted in spring.

OSCULARIA (Oscu-lária). Once included in *Mesembryanthemum,* the five species of small succulent subshrubs that constitute *Oscularia* are natives of South Africa. They belong in the carpetweed family AIZOACEAE and, characteristic of that assemblage, have blooms that, although very different structurally, have much the aspect of daisies, but differ from daisies in each being a single flower instead of a head of many florets. The name *Oscularia,* derived from the Latin *osculum,* a little mouth, alludes to the fanciful mouthlike appearance of the pairs of toothed leaves.

Oscularias have spreading or erect, branching stems and opposite, short, gray-green leaves somewhat joined at their bases, alternate pairs at right angles to each other. The leaves are three-angled. They narrow toward their bases and have keels that are often toothed. In threes, and about ½ inch in diameter, the short-stalked, fragrant blooms, borne in spring and summer, are pink to red. The fruits are capsules.

Very charming, blue-gray or sometimes reddish ***O. deltoides*** (syn. *O. muricatum*) forms small, compact mounds. Its glaucous, triangular leaves, toothed along all three edges are under ½ inch long by at least one-half as wide. The flowers are pink. Variety *O. d. majus* has slightly bigger blooms. Similar, but usually larger, up to 1½ feet tall, and with red stems, ***O. pedunculata*** has pink blooms slightly under ½ inch across. Its light gray, waxy-coated leaves are about ¾ inch long and triangular. They curve inward and are toothed along their margins and keels. With spreading reddish branches, ***O. caulescens*** has incurved, triangular leaves up to ¾ inch long by about one-half as wide. They may be toothless or have a few teeth on the margins toward the apex and sometimes on the keels. They are covered with a waxy, powdery, whitish coating. The pink flowers are ½ inch in diameter.

Garden Uses and Cultivation. Oscularias appeal to collectors of choice succulents. Their resting season is winter. Then they should be kept dry. At other times very moderate watering is appropriate. Easy-to-grow and free flowering, they respond to conditions that suit *Lampranthus* and are readily propagated by cuttings and by seed. For further information see Succulents.

OSIER. See Salix.

OSMANTHUS (Osmán-thus) — Sweet-Olive, Devil Wood. These olive tree relatives number about thirty species of evergreen trees and shrubs of eastern and southeastern Asia, Hawaii, New Caledonia, and North America. They belong in the olive family OLEACEAE. The name *Osmanthus,* of obvious application, comes from the Greek *osme,* fragrance, and *anthos,* a flower.

Osmanthuses have opposite, essentially hairless, undivided, lanceolate, elliptic, or obovate, toothed or toothless leaves, usually with thickened margins. Their tiny, usually fragrant flowers in axillary clusters or panicles, are white, cream-colored, or rarely yellow. They have four-lobed calyxes, four-lobed, bell-shaped or less often cylindrical corollas, two, or in a few kinds

Osbeckia stellata crinita

three or four stamens, and a short style topped with a usually slightly two-lobed stigma. Most cultivated kinds have male and bisexual flowers on separate plants. The fruits, structured like plums, technically are drupes. Those unobservant of botanical detail are likely to confuse some osmanthuses with hollies (*Ilex*). They are easily told apart by the arrangement of their leaves, those of osmanthuses being opposite, those of hollies alternate.

The sweet-olive (***O. fragrans***) is an old favorite for outdoor gardening in mild climates, in greenhouses, and as a window plant. In eastern Asia it has been cultivated for many centuries as an ornamental, especially near temples, and Japanese botanists have recognized several variants as distinct species. Its dried flowers are used in China for scenting tea, and it is said, for keeping insects away from clothes. This is a shrub or tree 6 to 40 feet in height. It has dark green, usually broadly-lanceolate leaves, rarely toothed at their margins, and hairless, except sometimes for very short hairs on the stalks. The leaves, normally 3½ to 4½ inches long, are occasionally shorter or longer, and ½ inch to 2 inches broad. Sometimes toothed, their margins are slightly thickened. Nearly always very fragrant, the white or yellow blooms are in clusters. Variety *O. f. aurantiacus* (syn. *O. aurantiacus*) is distinguished by bright orange flowers and typically, lanceolate leaves with few or no teeth.

Osmanthus fragrans

Much esteemed as an ornamental, ***O. heterophyllus*** (syns. *O. ilicifolius, Olea ilicifolia, Olea aquifolium*), 6 to 25 feet tall, is dense and bushy. It has thick, ovate to elliptic or rarely obovate, holly-like leaves 1½ to 2½ inches long by 1 inch to 1½ inches broad, commonly with three to five prominent, strong, spiny teeth along each side and one at the apex. On tall specimens the upper leaves are likely to be without spines. The midribs on the top sides of the leaves are finely-hairy. The blooms, which come in fall, are fragrant, pure white, and in clusters. The dark purple fruits are a little over ½ inch long by two-thirds as wide. This native of Japan and Taiwan includes a variant, by some authorities distinguished as *O. integrifolius*, in which the leaves are essentially spineless. A number of interesting horticultural varieties are cultivated. They include *O. h. aureus* (syn. *O. ilicifolius aureomarginatus*), with leaves margined with yellow; *O. h. myrtifolius* (syn. *O. ilicifolius myrtifolius*), with leaves 1 inch to 1¾ inches long and without spines except at their apexes; *O. h. purpureus* (syn. *O. ilicifolius purpurascens*), with purplish-black young foliage and older leaves tinged purple; *O. h. rotundifolius* (syn. *O. ilicifolius rotundifolius*), a slow-growing dwarf with usually spineless, rounded to obovate leaves 1 inch to 1½ inches long; and *O. h. variegatus* (syns. *O. ilicifolius variegatus, O. i. argenteo-marginatus*), which has leaves edged with creamy-white.

Osmanthus heterophyllus at Princeton, New Jersey

Osmanthus heterophyllus variegatus

A popular hybrid, ***O. fortunei*** is essentially intermediate between its parents, *O. fragrans* and *O. heterophyllus*. A shrub, it has thick leathery, broad-elliptic, spine-tipped leaves 2 to 3 inches long or somewhat longer by 1 inch to nearly 2 inches wide. They have thickened edges with up to a dozen, short regularly spaced teeth along each side. The small, white fragrant flowers are all males.

Three allied Chinese species are *O. armatus, O. serrulatus*, and *O. yunnanensis* (syn. *O. forrestii*). A shrub or tree up to 30 feet tall, ***O. armatus*** has very thick, leathery, short-stalked, narrowly-lanceolate to lanceolate-elliptic leaves 2½ to 4½ inches long by ¾ inch to 1½ inches wide. They have thickened margins with nearly always six to fourteen, large, strong teeth along each side. The flowers are white, fragrant, and in clusters. Violet-black, the fruits are little over ½ inch long and slightly under ½ inch wide. Some 40 feet in height, ***O. serrulatus*** has thickish, oblanceolate to narrowly-obovate leaves, toothless or with thirty-five or fewer small, forward-pointing teeth along each side, sometimes confined to the upper halves of the leaves. The clustered flowers are white and fragrant. Differing from *O. serrulatus* in having lanceolate to ovate or rarely elliptic leaves usually covered with tiny raised dots and with, when present, larger teeth, ***O. yunnanensis*** is a shrub or tree up to 45 feet tall. It has white, cream-colored, or yellow flowers. Its deep purple fruits, up to a little over ½ inch long, are covered with a waxy bloom.

Another Chinese, ***O. delavayi*** has by some botanists been segregated as *Siphonosmanthus delavayi*. About 6 feet tall, this is a twiggy shrub with opposite, pointed, ovate, toothed, lustrous, hairless leaves ½ to 1 inch long. Their undersides are dotted with tiny glands. In groups of few from the leaf axils, the fragrant, waxy-white flowers, which appear in spring, are about ½ inch in length. They are succeeded in summer by bluish-black, berry-like fruits about ½ inch long.

Devil wood (***O. americanus*** syns. *O. floridanus, Amarolea americana*), native from Virginia to Florida, Mississippi, and southern Mexico, is a shrub or tree up to 45 feet tall. It has short-stalked, elliptic to lanceolate-oblong, toothless, lustrous leaves 2½ to 7 inches long. The fragrant flowers, in panicles, come in spring. The egg-shaped, dark blue fruits are almost 1 inch long. Variety *O. a. megacarpus* has larger fruits. In *O. a. microphyllus* the leaves are smaller than those of the typical species.

Endemic to the Hawaiian Islands, where it occurs at low altitudes in dryish places in forests, ***O. sandwicensis*** occasionally attains heights of up to 60 feet, but usually is considerably smaller. It has pointed-elliptic, leathery leaves 3 to 8 inches long by 1 inch to 3 inches wide, lustrous on their upper sides and very pale beneath; their margins are often wavy. The light yellow bisexual flowers have four stamens. They are succeeded by bluish-black, egg-shaped, one-seeded fruits ½ to ¾ inch long. From its wood, Hawaiians fashioned digging sticks and spears.

Garden and Landscape Uses. Osmanthuses are admirable ornamentals, esteemed for their shapeliness and handsome foliage as well as for the fragrance of their decoratively unimportant flowers. With the exception of *O. heterophyllus* and *O. americanus*, which may survive outdoors in sheltered locations about as far north as New York City, they are not hardy in the north. In the south and west they are employed as single specimens, and in beds and foundation plantings, as well as for informal and formal screens and hedges. They are at their best where they receive a little shade from the fiercest sun, but will grow without this if the earth is not too dry. They need adequately drained, reasonably fertile soil. The sweet-olive, and sometimes other kinds, are grown as pot plants.

Cultivation. Outdoors osmanthuses require no special care. Unless trained as hedges or in formal shapes no regular pruning is required, but it may be desirable to shorten any long, straggly branches and to thin out crowded ones in spring.

As greenhouse and window plants the sweet-olive and other kinds succeed in fertile, porous, loamy soil in well-drained containers. They are watered to keep the soil always pleasantly moist without being constantly wet. Specimens in containers filled with roots are greatly benefited by dilute liquid fertilizer given every week or two from spring through fall. Cool, airy growing conditions are essential. In summer it is a good plan to stand the plants outdoors on a terrace, beneath a tree, or elsewhere where they can be cared for conveniently, or better still, where their pots can be buried to their rims in a bed of sand, peat moss, or other material that will aid in keeping the roots cool and evenly moist. Before frost the plants are brought indoors. Throughout the winter, night temperatures of 40 to 50°F are adequate, with daytime reading not exceeding night limits by more than five to ten degrees. Repotting older specimens is likely to need attention about every third year, in spring. Younger plants in their formative stages respond to annual potting. Cuttings provide the most usual means of increase. Made from firm shoots, and 3 to 4 inches long, these root fairly easily in late summer either under mist or in a greenhouse or cold frame propagating bed, preferably with a little bottom heat. Seeds and layering are also satisfactory modes of multiplication. The pests of these plants are scale insects and aphids.

OSMAREA (Os-màrea). The name *Osmarea* belongs to any hybrids between *Osmanthus* and *Phillyrea*, both of the olive family OLEACEAE. The one known, *O. burkwoodii* is intermediate between its parents *Osmanthus delavayi* and *Phillyrea decora.*

Not very common, ***O. burkwoodii*** is an evergreen, spring-blooming shrub, 3 to 6 feet tall or taller, that originated before 1928. It has lustrous, short-stalked, elliptic to oblong-ovate, finely-toothed but not spiny, leathery leaves, ¾ inch to 2 inches long, and terminal and lateral clusters of five to seven unisexual and bisexual, fragrant, small, ivory-white flowers. They differ from those of *Osmanthus delavayi* in having slender stalks ¼ to ⅓ inch long, and in the corolla lobes (petals) being as long as the corolla tube. From its other parent they differ in the corolla tube being as long as the corolla lobes and in the anthers protruding only very slightly from the throats of the flowers.

Garden Uses and Cultivation. This attractive hybrid is suitable for general landscape planting in sun or part-day shade in deep, fertile, well-drained, but not dry soil. It will not withstand extreme cold, but may be expected to survive in sheltered locations about as far north as New York City. No pruning is needed other than the occasional removal or shortening of an ill-placed branch to promote shapeliness. This is best done immediately after flowering. Propagation is by summer cuttings made from the ends of side shoots and planted under mist or in a propagating bed in a greenhouse or cold frame.

OSMARONIA. See Oemleria.

OSMENTARA. This is the name of orchid hybrids the parents of which include *Broughtonia, Cattleya,* and *Laeliopsis.*

OSMORHIZA (Osmo-rhìza)—Sweet Cicely. To *Osmorhiza*, of the carrot family UMBELLIFERAE, belong about fifteen species of thick-rooted, perennial herbaceous plants with umbels of small, white, greenish-yellow, or purplish flowers. They inhabit North and South America and eastern Asia. The name, from the Greek, *osme,* an odor, and *rhiza,* a root, was applied in recognition of their agreeable odors. The vernacular name sweet cicely is also applied to *Myrrhis odorata.*

Osmorhizas are erect. Their lower leaves are stalked and stem-clasping, the upper ones stalkless. Each leaf is divided into three parts, each part consisting of three toothed or pinnately-lobed, ovate to lanceolate leaflets. The five-petaled flowers are without sepals. They have five stamens and two petals and are in few-branched umbels. The little seedlike fruits are slender and nearly cylindrical to narrowly-club-shaped.

Species native to North America, the only ones likely to be cultivated, are generally similar, differing mainly in technical details not easily observable. They are 1 foot to 3 or 4 feet tall, have slender to fairly stout stems, and are more or less hairy. Inhabitants of open woodlands, they favor moist soils. Four species occur in eastern North America and five in the West. They are occasionally grown in gardens. Eastern species include *O. claytonii* and *O. longistylis*, Western species *O. occidentalis* and *O. brachypoda.* One, *O. chilensis,* occurs both in the east and the west and in South America.

Garden and Landscape Uses and Cultivation. Osmorhizas are suitable for naturalistic and native plant gardens. Hardy, they succeed in deep, moist, fertile soil. They are raised from seed.

OSMUNDA (Osmún-da)—Royal Fern or Flowering Fern, Cinnamon Fern, Interrupted Fern. The osmunda family OSMUNDACEAE, to which this group belongs, consists of three genera of geologically very ancient ferns (the others are *Todea* and *Leptopteris*). Only *Osmunda* has its spores on specialized portions of the fronds; in the others they are on the backs of ordinary leaves. There are half a dozen species of *Osmunda,* a genus widespread over the world. The origin of the name is not clear; it may come from Osmunder, the Saxon name of the god Thor.

Osmundas are rather coarse, vigorous ferns of considerable beauty. Those described here are hardy. They have stout, scaleless rhizomes, covered with the persistent bases of old leaves, and fibrous

Osmunda regalis

roots. The more or less hairy, scaleless, pinnately-divided or lobed fronds (leaves) are long-stalked and erect. The spores are on separate fertile fronds or fertile portions of partly sterile fronds. The spore cases are comparatively large and in more or less globose clusters.

Royal fern or flowering fern (*O. regalis*) has, under ideal conditions, fronds fully 6 feet long, but more often 3 to 4 feet long. They have pinkish stalks shorter than the twice-pinnate, broadly-ovate blades. There are four or five pairs of primary leaf divisions, the largest up to 1 foot long by one-half as wide. The rather widely-spaced leaflets are oblong, up to 3 inches long by less than 1 inch wide, and finely-toothed. Sterile and fertile fronds are borne. The lower parts of the latter are similar to the sterile fronds; their upper parts are erect panicles of numerous clusters of spore cases, at first green, later brown. In appearance they suggest groups of tiny blooms of a flowering plant and are responsible for the vernacular name flowering-fern sometimes applied to this species. The royal fern has coarser-patterned leaves than the next two. It increases in size and spreads slowly.

Cinnamon fern (*O. cinnamomea*) has separate fertile and sterile fronds. The former, in groups of up to four, show earlier in spring than the latter. Erect, and 2 to 5 feet tall, they have upper parts that are erect-branched, slender panicles of clusters of cinnamon-brown spore cases. The arching, erect, sterile fronds surround the fertile ones in vase-shaped fashion. They are 2 to 5 feet long by 5 to 10 inches broad, and lanceolate to oblong-lanceolate, with pinnately-divided blades with numerous narrow-lanceolate, pinnately-lobed, primary divisions. This very beautiful fern, with conspicuous spires of warm-brown, fertile fronds, that at a glance look much like the inflorescences of a flowering plant, is native from Labrador to Minnesota, Florida, Texas, and New Mexico, and in tropical America.

Osmunda cinnamomea

Osmunda cinnamomea (young shoots or fiddle heads)

Osmunda cinnamomea (fertile fronds)

The interrupted fern (*O. claytoniana*) is indigenous from Newfoundland to Minnesota, Florida, New Mexico, and Mexico, and in Asia. From the other osmundas discussed it differs in its fertile parts being at the middles of some of the fronds, with sterile portions above and below. The fronds are 2 to 6 feet long by 9 inches to 1 foot wide. They are pinnate, with fifteen to twenty pairs of primary divisions. The sterile leaflets are up to 6 inches long by 1¼ inches wide, oblong-lanceolate, and deeply-pinnately-lobed. There are one to five pairs of fertile leaflets, smaller than

Osmunda claytoniana

the sterile ones, and they soon wither. Their spore-bearing organs are dark brown. Variants with more pinnately-lobed leaflets occur. The interrupted fern inhabits subacid and neutral soils in open woods and at the edges of swamps.

Garden and Landscape Uses. Osmundas, strong-growing, vigorous, hardy, deciduous ferns of noble aspect, form handsome background for lower plants and are adapted for establishing in groups, and sometimes as single specimens, at watersides, in beds, and in informal areas, where the soil is acid, deep, and fertile. For its best development the royal fern needs swampy earth; that for the cinnamon fern should be constantly damp. The interrupted fern succeeds in somewhat drier, but never dry, earth. All grow in full sun or partial shade.

Cultivation. Osmundas are easy to cultivate. Like most ferns they are propagated by spores or sometimes by careful division. Once established, their care is minimal. An annual mulch with compost, leaf mold, peat moss, or other suitable organic debris is helpful. The dead fronds are left over winter to provide natural protection and are removed in early spring. For more information see Ferns.

OSMUNDA FIBER or OSMUNDINE. This, also called orchid peat, consists of the fibrous, wiry roots of species of the fern genus *Osmunda,* chiefly those of *O. cinnamomea* and *O. claytoniana.* An excellent material in which to pot epiphytic orchids and bromeliads, it is collected from the wild and sold by dealers in supplies for orchid growers.

OSMUNDACEAE—Osmunda Family. This family of three genera of terrestrial ferns, consisting of nineteen species, is native to

Osteospermum fruticosum 'Burgundy Mound'

Osmunda cinnamomea

Orontium aquaticum

Oxydendrum arboreum

Oxalis rubra

Pachycereus pringlei in bloom

Pachysandra terminalis

Pachystachys lutea

Oxalis oregana

temperate to tropical parts of the Old and the New World. Its mostly large members have creeping or upright rootstocks and one- to three-times-pinnately-divided or pinnately-lobed fronds (leaves). The naked spore capsules are borne on fronds that differ markedly in appearance from the sterile ones or are on the undersides of more normal-looking fronds. The genera cultivated are *Leptopteris, Osmunda,* and *Todea.*

OSOBERRY is *Oemleria cerasiformis.*

OSTEOMELES (Osteo-mèles). Two attractive species of possibly fifteen recognized by some botanists as comprising *Osteomeles,* of the rose family ROSACEAE, are cultivated. The group is indigenous to Central America, the Andes, Pacific islands, and eastern Asia. Its name comes from the Greek *osteon,* bone, and *meles,* apple. It refers to the fruits.

These are attractive evergreen or partially evergreen shrubs with alternate, pinnate, ferny leaves, each with many small leaflets. The white, hawthorn-like flowers are in clusters at the ends of short leafy shoots. They have five each sepals and petals, fifteen to twenty stamens, and five separate styles. The fruits are small pomes (fruits with the structure of those of apples and mountain-ashes). They contain five nutlets (seeds) and retain the calyx at their tips. A hybrid between *Pyracantha* and *Osteomeles* is *Pyracomeles.*

Native to western China, ***O. schweriniae*** is a graceful, partially evergreen shrub, reliably hardy about as far north as New Jersey. Up to 10 feet tall, its branchlets, undersides of its leaves, and leafstalks are clothed with grayish hairs. The leaves have narrowly-winged midribs and are 1 inch to 2 or rarely 3 inches long. They have seven to fifteen pairs of elliptic to obovate-oblong leaflets and a terminal one. They are ¼ to ½ inch long and not toothed. The ½-inch-wide flowers are in clusters 2 to 2½ inches across. They are succeeded by subspherical fruits up to ⅓ inch long and not pubescent, which change from red to blue-black as they ripen. Variety *O. s. microphylla* has leaves with fewer leaflets. It has smaller, more compact flower clusters. It is said to be hardier than the species. Also, it is dwarfer.

From Hawaii and other islands of the Pacific comes ***O. anthyllidifolia,*** an evergreen shrub that has been reported to be hardy in New York. Almost surely this is not true. There has been confusion between this species and the last and the kind that survived outdoors at New York was most probably *O. schweriniae* or its variety *O. s. microphylla.* The Hawaiian native varies from a low spreading shrub not more than a foot or two high to a small tree. It has silky-hairy young shoots and foliage. The leaves, of thirteen to nineteen toothless, oblong leaflets, up to ½ inch long, lose the hairs from their upper sides as they age. The ½-inch-wide flowers are followed by nearly spherical, pubescent, white fruits that are about ⅓ inch across and have sweet, white to purplish pulp.

A less well-known kind, ***O. subrotunda,*** a native of China, is a low, compact, slow-growing evergreen with tortuous branches that when young are downy. The leaves, up to 1½ inches long, have nine to seventeen obovate to oblong leaflets and are hairy, especially on their lower sides. The flowers are about ⅓ inch wide. It is probably as hardy as *O. schweriniae.*

Garden and Landscape Uses and Cultivation. These are attractive general purpose shrubs for mild and moderately mild climates and are also satisfactory in cool greenhouses. They can be espaliered with facility and look especially well when so trained. Of tidy appearance, they bloom when quite small. They are easily restrained to size by pruning and they succeed in any ordinary, well-drained soil in sunny locations. Propagation is by seed, summer cuttings, and layering. Seeds are sown as soon as they are ripe or are mixed with slightly damp peat moss and stored in a temperature of about 38°F for three or four months, and then sown. In greenhouses a winter night temperature of 40 to 50°F is adequate, with a rise of a few degrees by day permitted. Container-grown specimens benefit from being stood outdoors in summer.

OSTEOSPERMUM (Osteo-spérmum). Of the three genera *Dimorphotheca, Castalis,* and *Osteospermum* that once composed *Dimorphotheca,* this is by far the largest. Consisting of about seventy species, it is indigenous from South Africa to Arabia. Belonging to the daisy family COMPOSITAE, it has a name, alluding to the hardness of its fruits, derived from the Greek *osteo,* bone, and *sperma,* a seed. Osteospermums differ from *Dimorphotheca* and from *Castalis* in the disk florets of their flower heads being sterile and only the ray florets producing fruits.

Herbaceous plants, subshrubs, and shrubs compose *Osteospermum.* They have usually alternate, rarely opposite leaves that may be lobeless and toothless or pinnately-lobed or toothed. The solitary flower heads are like those of daisies. They have a central eye of tubular disk florets encircled by petal-like ray florets. In *Osteospermum* the ray florets are in a single row and are yellow, mauve, violet, or white with violet-stained undersides. The fruits are seedlike achenes.

The commonest species cultivated, ***O. ecklonis*** (syn. *Dimorphotheca ecklonis*) is very beautiful. A branching subshrub or shrub 3 to 4 feet tall, with distantly-

Osteospermum ecklonis in California

Osteospermum ecklonis (flowers)

toothed, glandular-pubescent, obovate to oblanceolate leaves that narrow to their stalks, it is 2 to 4 inches long by ¼ to ¾ inch wide. The flower heads, solitary or loosely clustered, terminate long stalks from the branch ends. They have centers of azure blue and shining white ray florets with blue or violet undersurfaces often bordered with white.

Other kinds sometimes cultivated include these: ***O. barberae*** (syn. *Dimorphotheca barberae*) is a somewhat shrubby, minutely-glandular perennial with oblong to spatula-shaped, toothless or few-toothed leaves, the lower ones short-stalked and up to 3 inches long, those higher on the stems shorter and stalkless. About 2½ inches in diameter, the long-stalked flower heads have deep purple centers and ray florets deep purple above and with duller undersides. ***O. fruticosum*** is a herbaceous to somewhat woody perennial with long, more or less prostrate stems and alternate, obovate, stalkless leaves, at first glandular-hairy, later nearly hairless. On stalks not over 4 inches long, the about 1¾-inch-wide flower heads have dull violet-purple centers, ray florets that are white above, violet, lilac, or rosy-purple on their undersides. ***O. hyoseroides*** (syn. *Tripteris hyoseroides*), about 2 feet tall and aromatic, is a more or less glandular-hairy annual with alternate, toothless or nearly toothless leaves, the lower ones oblong to oblanceolate, wavy-toothed, and up to 4 inches long, those above oblong-linear to oblanceolate. The flowers have yellow disks brushed with violet, and orange-yellow ray florets. ***O. jucundum*** (syn. *Dimorphotheca jucunda*) is a glandular-hairy herbaceous perennial, up to 1½ feet tall, with erect or sprawling stems. Its leaves are alternate, nearly or quite stalkless, linear-elliptic to elliptic or oblanceolate, and few-toothed. About 2 inches across, the solitary flower heads have a dark purple disk and cerise ray florets with paler undersides.

Garden and Landscape Uses. Beautiful *O. ecklonis* blooms over a very long period and is splendid for flower gardens in California and other places with Mediterranean climates. It needs porous, moderately fertile, well-drained soil and full sun. It is also admirable for pots and tubs in greenhouses and can be trained as a standard (in tree form). The other kinds described are suitable for outdoor flower beds in climates and under conditions that suit *O. ecklonis*.

Cultivation. Where planted outdoors, routine attention needed by osteospermums consists of the removal of faded blooms, periodic watering, and once or twice a year a light dressing of a complete fertilizer. Any pruning to shape needed is done after flowering is through or at the beginning of a new season of growth. Increase is easy by seed and by cuttings of firm but not woody shoots.

In greenhouses *O. ecklonis* responds to a winter night temperature of 50°F, increased five to fifteen degrees during the day. On all favorable occasions the structure must be ventilated freely. In summer, specimens in pots and tubs benefit from being put outdoors with their containers buried nearly to their rims in sand, ashes, peat moss, soil or some similar material. They are brought inside before fall frost. Watering is done to keep the soil always comfortably moist, but not wet. Well-rooted specimens benefit from regular applications of dilute liquid fertilizer. Tree-form or standard specimens are easily trained. A strong young plant is selected and transferred successively from smaller to larger pots as its root growth allows. The tip of its stem is carefully preserved, but all side shoots are nipped off as soon as they are big enough to be seized between thumb and forefinger. The single stem is kept tied to a stake and when it has reached a height of about 2 feet, its tip is removed and from near its top a few side shoots are allowed to develop. When these are 4 or 5 inches long their ends are pinched out and this procedure is repeated until a sizable head is formed. Then, pinching ceases and flowers are allowed to develop.

OSTRICH FERN. See Matteuccia.

OSTROWSKIA (Ostròw-skia)—Giant-Bellflower. The specific epithet *magnifica* of the only species of *Ostrowskia* is well deserved, for in truth the giant-bellflower, well grown and in good condition, is magnificent. Unfortunately in most parts of North America it is difficult or impossible to grow. Only where fairly mild winters and cool summers prevail is it likely to do well. This pretty much restricts it as a garden possibility to the Pacific Northwest and a few other favored places. The genus belongs in the bellflower family CAMPANULACEAE. From *Campanula* it differs in having leaves in whorls (circles) of four or five, and a greater number of sepals and corolla lobes (petals) than the five commonly found in *Campanula*. The name commemorates Michael Nicolajewitsch von Ostrowsky, a Russian patron of bot-

Osteospermum hyoseroides

Osteospermum jucundum

any, who died in 1880. The plant is a native of Turkestan.

The giant-bellflower (*O. **magnifica***) has a tuberous root and stout, erect stems 4 to 6 feet tall, furnished with rather distant whorls of ovate-lanceolate, toothed leaves 4 to 6 inches long. Its pale lilac blooms, suffused with white, form few-flowered terminal racemes. They are bell- or trumpet-shaped and have about eight broad, spreading, corolla lobes and long, slender, spreading sepals. The blooms measure 4 or 5 inches across the mouth of the bell.

Garden Uses and Cultivation. This is a challenging plant for the dedicated gardener. It resents root disturbance and does not divide or transplant well. Seeds afford the best means of propagation. They should be sown as soon as they are ripe. Even under favorable conditions, seedlings do not bloom until they are three or four years old. A deep, well-drained, sandy soil well admixed with plenty of leaf mold or other nourishing organic material is favorable to growth and a sunny location should be chosen. The plant blooms early and dies down before midsummer. From then on it should be protected from excessive moisture. Winter cover in the form of cut branches of evergreens or some similar material laid over the ground surface is advisable in cold climates.

OSTRYA (Ós-trya)—Hop-Hornbeam or Ironwood. Hop-hornbeams are closely related to hornbeams (*Carpinus*), from which they differ primarily in the involucral bracts at the bases of the female flowers. In *Ostrya* these form a baglike husk, open at the top until the flower is fertilized by pollen from a male, when it closes and becomes considerably bigger. They develop into pale green, thin-walled bladders that completely enclose the nutlets (fruits) and overlap to form the hoplike clusters that are responsible for the vernacular name. When not in fruit hop-hornbeams can be distinguished from hornbeams by their lateral buds being erect instead of lying flat against the twigs, by the veins of their leaves sometimes branching at their ends, and by their male catkins developing in fall. Hop-hornbeams belong in the birch family BETULACEAE, or if that group's division into three families be accepted, to the hornbeam family CARPINACEAE. The name comes from *ostrys*, its ancient Greek name. There are ten species.

Hop-hornbeams have a curious natural distribution. Two species occur in North America, one in southern Europe and Asia Minor, one in eastern Asia, the others in Central America. They are deciduous trees with rough, scaly bark and alternate, usually ovate to oblong leaves with double-toothed margins and parallel veins angling upward from the midrib to the leaf margins. They bear a general resemblance to elm leaves, but are thinner and more symmetrical. Male and female flowers are borne on the same tree, the former in slender, drooping catkins, the latter in slender, upright ones. The individual flowers are minute and without sepals or petals. They have no decorative value. The fruits are ribbed nutlets surrounded by a bladder development of the involucral bracts. The cultivated species are quite handsome ornamentals, attractive in foliage and fruit, but without any special displays of color.

The American hop-hornbeam or ironwood (***O. virginiana***), which ranges from Cape Breton to Minnesota, Florida, and Texas, attains heights of up to 60 feet and a trunk diameter of up to 2 feet; often it is lower. Sometimes called ironwood (a name also applied to *Carpinus caroliniana*) because of the extreme hardness of its durable lumber, it is a round-topped or pyramidal, slow-growing species with leaves 2½ to 4½ inches long and slightly heart-shaped at their bases, with eleven or more pairs of veins and stalks under ¼ inch long. The nutlets are spindle-shaped. Another American hop-hornbeam (***O. knowltonii***) is limited in its natural distribution to northern Arizona. Up to 35 feet tall, it differs from *O. virginiana* in having leaves up to 2 inches long, with up to eight pairs of veins.

Ostrya virginiana

The European hop-hornbeam (***O. carpinifolia***) is very similar to *O. virginiana*, but has leaves with rounded instead of somewhat heart-shaped bases, and ovoid nutlets. Much like the American and European species, but more obviously distinct, is the Japanese hop-hornbeam (***O. japonica***), a native of Japan and northeast Asia. It differs in having leaves less markedly double-toothed than the others, and conspicuously hairy on their undersides; those of the American and European kinds are but sparsely-hairy or are without hairs. The leaves of the Japanese hop-hornbeam have nine to twelve pairs of veins. In its homeland this tree attains heights of 80 feet.

Garden and Landscape Uses. The hardiest species is the American hop-hornbeam. The others are satisfactorily hardy in southern New England, and the Japanese kind is reported to succeed at Hamilton, Canada. Although not spectacular, these trees deserve more consideration from landscapers than they have received. Of pleasing form, they have clean-looking foliage of good texture and color and interesting fruit. Because they grow slowly and are of moderate size they are well suited as ornamentals and shade trees for small properties. They are little affected by pests or diseases. Perhaps a little more difficult to transplant than some more popular trees, if the operation is carefully done they can be moved bare-rooted or with a ball of earth with little danger of loss. They prosper in any ordinary garden soil and are well suited for dry conditions and exposed locations.

Cultivation. Hop-hornbeams need no special care. They are propagated by seeds sown as soon as they are ripe or stratified and sown in spring. If kept dry over winter and sown in spring, the seeds usually do not germinate until the following year. Increase of rare kinds can be had by grafting onto the American hop-hornbeam or the hornbeam (*Carpinus*).

OSWEGO-TEA is *Monarda didyma*.

OTAHEITE. This word is included in the common names of these plants: Otaheite-apple (*Spondias cytherea*) and Otaheite-gooseberry (*Phyllanthus acidus*). The variety of lemon named 'Otaheite' is also known as the otaheite-orange.

OTANTHUS (Ot-ánthus)—Cottonweed. A coastal and seashore plant of western Europe and the Mediterranean region, the only species of this genus was formerly named *Diotis candidissima*. Now *Otanthus maritimus*, and commonly named cottonweed, it is a member of the daisy family COMPOSITAE. Its name derives from the Greek *otos*, an ear, and *anthos*, a flower, and takes cognizance of the ear-shaped corolla lobes of the florets.

Cottonweed (***O. maritimus***) is a somewhat tufted herbaceous perennial of the *Achillea* relationship with a creeping woody rootstock and densely-snowy-white-felted stems and foliage. From 6 inches to 1½ feet tall, it has erect stems and alternate, stalkless, oblong to spatula-shaped leaves rather under ½ inch long. The flat-topped clusters of few, subspherical, short-stalked heads of yellow flowers are borne in summer. The flower heads are without ray florets (the kind that in daisies resemble petals). They are of all disk florets (the type that form the daisy's eye). The involucre (collar of bracts behind the flower head) is densely felted with white hairs. The absence of ray florets distinguishes this genus from *Achillea*.

Garden Uses and Cultivation. Cottonweed is admirable for rock gardens, the fronts of borders, and edgings, and is especially well adapted to seaside gardens and sandy, dry soils. It revels in brilliant sun. Provided it is not subjected to long wet periods its cultivation presents no problems. Propagation is easy by seed, division, and cuttings. This species is better suited to the dryish climates of western North America than that of the east. It stands considerable frost, but its exact hardiness is not determined.

OTHONNA (Oth-ónna)—Little Pickles. A few species of this African genus of about 150 are perhaps grown by specialist collectors of succulents, but only one is common in North America. Known as little pickles, because of the form of its leaves, it is a charming and easily grown trailer. The genus *Othonna* belongs in the daisy family COMPOSITAE. Many of its members, but not little pickles, have tuberous roots. Some are subshrubs, others herbaceous. Not all are succulent. Many are too weedy to be acceptable as garden plants. The name is an ancient Greek one for a plant of some other genus in which some species are succulents.

Othonnas generally have fleshy, basal or alternate stem leaves, the latter in clusters at the branch ends. Usually yellow, the daisy-type flower heads are solitary or clustered. The tubular disk florets (those that form the eyes of the flower heads of daisies) are sterile. The petal-like ray florets are fertile. The seedlike fruits are achenes.

Little pickles (***O. capensis*** syn. *O. crassifolia*) is a nearly hairless, evergreen perennial with slender, trailing or pendulous, branching stems up to about 3 feet long that become woody toward their bases and have scattered or clustered, fleshy, sharp-pointed, cylindrical to club-shaped, slightly glaucous-green, soft leaves, those on the stems up to 1 inch long, the basal ones sometimes twice as long, and often purple-tipped. Solitary or few together on slender, erect stalks 2 to 6 inches long, the bright yellow flower heads, with reflexed ray florets, are ½ to 1 inch in diameter. The rays spread only in sun.

Othonna capensis

Very different ***O. quercifolia,*** of South Africa, is a small hairless shrub with, mostly toward the ends of its stout, fleshy branches, succulent elliptic leaves 2 to 4 inches long by up to 1 inch wide and with usually two to four lobes on each margin.

Othonna quercifolia in South Africa

In terminal branched clusters 3 to 6 inches long the yellow flowers have about five ray florets. Endemic to southwestern Africa, ***O. euphorbioides*** is a deciduous subshrub up to 4 inches tall or perhaps a little taller. It has stout, succulent stems and, crowded toward the branch ends, white-waxy, spatula-shaped leaves ¾ to 1 inch long. The two- to several-times-forked spines are nearly as long as the leaves. The solitary, yellowish flower heads terminate stalks about 1 inch long.

Garden and Landscape Uses and Cultivation. Little pickles is delightful for pots and hanging baskets in greenhouses and window gardens and, in dry, frostless and nearly frostless regions, as a permanent outdoor groundcover for rock gardens, banks, and similar areas. Where winters are more severe it may be planted outdoors in summer, and rooted portions taken up in fall and wintered indoors. It needs full sun and good soil drainage. For preference the earth should be sandy and moderately nutritious. Indoors a dryish atmosphere and cool, airy conditions are most to the liking of little pickles. Winter night temperatures of 40 to 50°F are adequate. They may be increased five to fifteen degrees by day. Watering should be done moderately, in winter with considerable care not to have the soil wet for long periods. Propagation is done with extreme ease by division and cuttings. Seeds may also be used. The other species described above is appropriate for desert gardens in arid and semiarid, frost-free regions and for including in collections of succulents in sunny greenhouses. It needs very well-drained soil and watering rather sparingly, with the soil allowed to become nearly dry between applications. Cuttings and seed provide ready means of propagation. For further information see Succulents.

Othonna euphorbioides

OTO. See Xanthosoma.

OURATEA (Our-àtea). The genus *Ouratea,* of the ochna family OCHNACEAE, consists of about 200 species, natives of the tropics, found most abundantly in the Americas. The name is derived from a native one of Guiana.

Ourateas are shrubs and trees with alternate, undivided leaves and prevailingly yellow flowers generally in panicles or racemes. Each flower has five each sepals and petals, and ten stamens. The fruits consist of up to five small, fleshy drupes attached to a fleshy receptacle.

Native to Puerto Rico and to St. Thomas in the Virgin Islands, ***O. littoralis*** is an evergreen shrub or tree up to 30 feet tall with short-stalked, obscurely-toothed or toothless, elliptic to ovate leaves 2 to 4½ inches long. Fragrant and in panicles up to 4 inches long, the about ½-inch-wide, golden-yellow flowers have fan-shaped petals. The fruits are red and blue. Brazilian ***O. olivaeformis*** (syn. *Gomphia olivaeformis*) is an evergreen shrub, sometimes tree-like, up to 15 feet tall, with short-stalked, broad-elliptic to oblong-elliptic, finely-toothed, glossy-green leaves 3 to 7 inches long. In much-branched terminal panicles, its strongly-scented, golden-yellow flowers are approximately ½ inch in diameter.

Garden and Landscape Uses and Cultivation. The sorts described are useful decoratives for gardens in the tropics and warm subtropics and are sometimes

Ouratea olivaeformis

grown in greenhouses where a minimum winter night temperature of 60°F is maintained. They do well in ordinary well-drained soils of medium fertility, in sun or part-shade. Water specimens in greenhouses sufficiently to keep the soil moderately moist at all times and from spring through fall give those that have filled their containers with roots biweekly applications of dilute liquid fertilizer. Do any pruning needed to shape the plants or restrain them to a desired size in late winter or spring. Propagation is easy by seed, and by cuttings planted in sand, perlite, or in a mixture of either of those with peat moss. Mild bottom heat to keep the rooting medium five to ten degrees warmer than the atmosphere is advantageous.

OUR-LADY'S-THISTLE is *Cnicus benedictus.*

OUR-LORD'S CANDLE is *Yucca whipplei.*

OURISIA (Our-ísia). Rare in gardens in North America, *Ourisia,* of the figwort family SCROPHULARIACEAE, comprises twenty species of chiefly low herbaceous perennials. A few, woody at their bases, qualify as subshrubs. The group inhabits New Zealand, Tasmania, southernmost South America, and the Andes. Its name commemorates Governor Ouris, of the Falkland Islands, who died in 1773.

Ourisias have creeping rhizomes and opposite, chiefly basal, stalked, toothed or toothless leaves. Their white, pink, purplish, or scarlet flowers are in terminal clusters or racemes or are solitary from the leaf axils. Each has a five-lobed calyx, a short-tubed corolla with five spreading lobes (petals), two pairs of stamens, a minute staminode (rudimentary stamen), and a long style tipped with a headlike stigma. The fruits are capsules.

Native to Chile, ***O. coccinea,*** 6 inches to 1 foot tall, has unevenly toothed, broad-elliptic to oblong leaves above which rise stalks with panicled clusters of drooping, scarlet flowers about 1½ inches long, and with protruding stamens tipped with cream-colored anthers. New Zealand ***O. macrocarpa*** has hairless, ovate-oblong leaves, up to 5 inches long, and stout flower stalks occasionally 2 feet tall, usually lower. Its flowers, white with yellow centers, and about 1 inch wide, are in whorls (tiers). Differing in its less leathery, downy, ovate to elliptic-ovate leaves up to 3½ inches long, and its more slender flower stalks 3 inches to 1 foot tall, ***O. macrophylla*** also comes from New Zealand. It has white blooms ½ to 1 inch wide in whorls or in a terminal umbel on stalks 6 inches to 2 feet tall. The ovate to orbicular-oblong, mostly long-stalked leaves, toothed and rather thin have blades 1 inch to 5 inches long. Perhaps only a variety of the last, ***O. crosbyi*** is a smaller native of New Zealand. Not over 10 inches tall and often lower, it has very thin membranous leaves and white flowers ⅓ to ½ inch wide.

Ourisia crosbyi

Ourisia microphylla

Native to Argentina, ***O. microphylla*** is a low alpine with slender, branching stems and tiny overlapping leaves in four rows. The broad-funnel-shaped flowers, borne near the tips of the stems, are rosy-lilac.

Garden and Landscape Uses and Cultivation. Ourisias are suitable only for mild, humid climates where summer heat is not oppressive. The Pacific Northwest suggests itself as a region in North America best adapted to their cultivation. They are suitable for shaded locations in rock gardens and woodland gardens, where the soil is mellow, moist, but not wet, and contains a generous organic content.

OUTENIQUA YELLOW-WOOD is *Podocarpus falcatus.*

OVATE. Shaped like the outline of a hen's egg with the narrow end forming the apex.

OVER-POTTED. A plant in a pot too large for its reasonable needs is said to be over-potted. This is likely to result in a sickly appearance, poor growth, and sometimes death. The reasons are the near impossibility of maintaining the roots at the center of the soil mass desirably moist without the surrounding soil being watered so much that it remains for long periods too wet, with the result that it does not contain sufficient oxygen to encourage roots to invade it. Under such conditions changes in the soil unfavorable to the growth of roots take place and nutrients are leached. As the condition progressively worsens the over-potted plant sulks, its roots may rot, and it may die. In the early stages, relief is sometimes had by being extremely careful not to water too often, and by maintaining the soil on the dry side so that the roots are encouraged to search for the moisture they need by traveling into the new earth. But once the soil has soured, as gardeners term it, this will not do. Then the only hope is to remove the plant from its container, take away all soil not permeated by healthy roots, and carefully repot into a well-drained receptacle, no more than just big enough to hold the roots comfortably, in fresh, porous soil, preferably sandier than would normally be used for the particular kind. Following this, water with more caution than usual and keep the plant in a fairly humid atmosphere, a little warmer than before. In growing pot plants it is the practice to make transfers from small to successively bigger containers. The increase at each shift depends upon the rooting vigor of the particular type of plant and upon its known preferences for root room. Some plants do better when pot-bound than when afforded more generous accommodation. Over-potting is a common fault of amateurs. See Potting and Repotting.

OWL'S-CLOVER is *Orthocarpus purpurascens.*

OWN-ROOT. When referring to plants such as roses that are commonly grafted or budded onto understocks of distinctly different species or varieties, but sometimes are grown from cuttings or by other methods than grafting or budding, specimens raised in these other ways are known as own-root plants. The term is also used for plants multiplied, as is sometimes convenient to do, by grafting or budding scions of a plant onto roots of its own kind. An advantage of own-root plants is that they do not produce from below the graft or bud union shoots different from those of the desired variety.

OX-EYE. See Buphthalmum.

OXALIDACEAE — Wood-Sorrel Family. About ten genera totaling perhaps more than 900 species are admitted in the OXALIDACEAE, a family of dicotyledons that occurs natively throughout much of the world, especially in the tropics and subtropics. It includes annuals, herbaceous perennials, subshrubs, shrubs, and trees. Characteristically, but not exclusively, its members have acid juice and all basal or alternate pinnate or palmate leaves. The solitary flowers are in umbel-like or branched clusters or, rarely, in raceme-like arrangements. They have calyxes of five sepals or lobes, five petals, ten stamens, and five styles with headlike or slightly two-lobed stigmas. The fruits are capsules. Cultivated genera are *Averrhoa, Biophytum,* and *Oxalis.*

OXALIS (Óxal-is)—Wood-Sorrel. Although the vast majority of the 800 species of *Oxalis,* of the wood-sorrel family OXALIDACEAE, are easily recognizable as such at a glance, their identification as to species is often more difficult. The name is derived from the Greek *oxys,* sour, and refers to the acid taste of the foliage. The genus is cosmopolitan in its distribution, but is most abundant in South Africa and South America. Several species are pesky weeds, one of the most familiar of which is *O. corniculata* (syn. *O. repens*), common in greenhouses and in lawns and other places. Others are beautiful ornamental plants.

The genus *Oxalis* consists of annual and perennial, mostly low herbaceous plants and a few subshrubby and shrubby ones. The herbaceous kinds may have well-developed, aboveground stems or be essentially stemless. Many species, chiefly natives of regions where long, dry summers occur, have underground bulbs or tubers, those of some edible. The leaves of most *Oxalis* have three leaflets and are clover- or shamrock-like, but others have from four to twenty or more leaflets, arranged in palmate (handlike) fashion with their bases arising from a common point; in some kinds phyllodes (flattened stems resembling leaves) take the place of true leaves. The leaves of many kinds are light sensitive, at night "going to sleep" by folding their leaflets downward. The flowers of *Oxalis* are solitary, or in clusters atop leafless stalks, and are white, pink, red, purplish, yellow, or orange-yellow. Mostly they open wide only in sun or strong light. Each bloom has five sepals, five petals, ten stamens, of which five are longer than the others, and five styles. The fruits are dry capsules containing numerous seeds that are distributed by being shot to some distance from the mother plant when the capsules rupture.

For convenience of recognition, cultivated oxalises may be divided into two groups, those with evident stems and those without. The latter may have underground stems and in some cases the ends of these emerge slightly above ground, nevertheless to all appearances they are stemless. Another way of considering them is as kinds hardy outdoors in the north and kinds that survive winters outdoors only in milder climates or indoors.

Hardy oxalises grown in gardens are comparatively few, and those native to the southern hemisphere are not very hardy; they cannot be expected to live through winters as harsh as those of New York City, for instance. Quite hardy kinds include ***O. corniculata atropurpurea*** (syn. *O. tropaeoloides*), a variety of a common, northern hemisphere, green-leaved weedy species distinguished by its rich reddish-purple stems and foliage that contrast pleasingly with the small yellow flowers. It is a creeping, mat-forming perennial with slender, prostrate, rooting stems.

Native to prairies and dry upland woods from Massachusetts to Florida, Michigan, and Texas, ***O. violacea*** is a pretty spring-flowering perennial that from a scaly bulbous base produces hairless leaves of three leaflets and distinctly longer flower stems, the latter up to 8 inches tall and each with two to several rosy-violet or rarely white blooms about ¾ inch long. It is distinguished by two small callosities at the end of the mid-vein of each leaflet. From the above the wood-sorrel (***O. acetosella***) differs in having solitary flowers on stalks up to 6 inches tall that only slightly surpass the foliage. Its flowers are white, veined

Oxalis violacea

with pinkish-purple. This perennial, which grows from a slender, scaly rhizome and has long-stalked leaves with three sparsely-hairy leaflets, inhabits Europe and Asia as well as North America, where it is found in rich, moist woods from Quebec to Saskatchewan, New York, Michigan, Wisconsin and, in the mountains, to North Carolina and Tennessee. The American plants are sometimes segregated as a separate species under the name *O. montana.* Western North American ***O. oregana*** is more robust. Its flower stems are up to 10 inches tall and its leaflets mostly 1 inch long, about twice as big as those of *O. acetosella;* its flowers are rosy-pink or occasionally white.

Three choice southern hemisphere species, hardy in fairly cold but not extreme climates, are *O. adenophylla, O. enneaphylla,* and *O. magellanica.* The first is a native of Chile, the second of the Falkland Islands, and the third of southern South America, southern Australia, and New Zealand. The Chilean endemic (***O. adenophylla***) is a stemless perennial with a non-scaly, bulbous base. It attains a height of 4 to 6

Oxalis adenophylla

inches and has long-stalked, hairless, silvery, gray-green, or bluish leaves, each with twelve to twenty-two obcordate leaflets that form a single whorl (circle); each leaflet is ¼ inch long and wide. The flowers, solitary on stalks equaling the leaves, are about 1 inch long, bell-shaped, and lilac-pink. Rather similar, but with slender, creeping, horizontal, scaly rhizomes, and having the nine to twenty leaflets of its gray-green, short-hairy leaves usually in two rows or whorls, is ***O. enneaphylla.*** Its solitary blooms are white with bluish centers or sometimes pink throughout. From both of these, ***O. magellanica*** differs in having leaves of three obcordate leaflets. They arise to a height of 1 inch to 1½ inches from slender scaly rhizomes and, except for a few on their stalks, are without hairs. The leaflets are up to ¼ inch

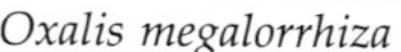
Oxalis megalorrhiza

Oxalis peduncularis

long and the short-stalked, solitary, white or whitish flowers about ½ inch long. In effect this is a miniature version of *O. acetosella.*

Tender oxalises with evident stems that are at all commonly cultivated are less numerous than kinds without stems. Attractive ***O. megalorrhiza*** (syn. *O. carnosa*) inhabits western South America and the Galapagos Islands. This produces bright yellow flowers over a long period. Its thick, fleshy, erect stems tend to sprawl, but rise to a height of 4 inches or so from spreading, horizontal rhizomes. Usually branchless, they bear toward their ends long-stalked leaves of three fleshy, grayish-green, obcordate leaflets that are glossy above, but on their undersides appear frosted. The leaflets are about ½ inch long and wide. The flowers, ¾ inch across, have petals joined at their bases and are borne two to five together on stalks about as long as the leaf stalks, and up to 4 inches long. A distinguishing characteristic is that two of the sepals differ markedly in shape and size from the others. Also yellow-flowered, but otherwise quite distinct, is ***O. peduncularis,*** native from Ecuador to Peru. This has thick stems 1 foot tall or taller their older parts rough with the remains of old leaf stalks. The flower stalks are about twice as long as the leaves and at their bases have small hairy appendages called stipules. The flowers are many together in crowded clusters. They have bright yellow petals and contrasting reddish-purple sepals. Chilean ***O. valdiviensis*** has much the same appearance, but its leaves are without stipules and its petals and leaflets are more broadly-obovate. Although usually grown as an annual, *O. valdiviensis* is perhaps perennial. Very distinct because of it fish-tail-shaped leaflets, of which each leaf has three, is ***O. ortgiesii,*** of Peru. This has erect, purplish, pubescent, leafy stems up to 1½ feet tall and long-stalked leaves purplish on their undersides. The leaflets, 1½ to 2 inches long, are deeply cleft to form two pointed lobes. The dark-veined, yellow flowers are few together in slender-stalked clusters from the leaf axils, and about ½ inch wide.

Pink-flowered oxalises with aboveground stems are *O. rosea*, of Chile, and the South African *O. hirta*. The Chilean species (***O. rosea***) is the only stemmed oxalis in cultivation with pink flowers in clusters. In gardens a quite different, stemless species, *O. rubra*, often masquerades under its name. The true *O. rosea*, 1 foot tall or slightly taller, has stalked leaves of three obcordate leaflets and loose, long-stalked clusters of dark-veined, white-throated, pink flowers, each about ½ inch long. With bulblike tubers, stems that die to the ground each year, stalkless leaves, and solitary flowers from its leaf axils, ***O. hirta*** is distinct from other cultivated species. It is a variable kind with slender stems that branch above and unless supported tend to flop to the ground. The leaves have three leaflets. The flowers, about 1 inch in diameter, appear in fall and early winter, and are pink; they are on stalks 1 inch to 1½ inches long. As its botanical designation *hirta* indicates, this is a hairy plant. A native of southwest Africa, tufted, deciduous ***O. incarnata*** has thin, procumbent to erect stems and long-stalked, light green leaves of three reverse-heart-shaped leaflets. Its lavender-pink-tinged flowers have yellow corolla tubes and are about ½ inch wide. Small tubers form freely at the crowns of the plants.

Oxalis hirta

Oxalis incarnata

Oxalis alstonii

A few woody-stemmed kinds, miniature shrubs, are grown in greenhouses. The best known, the fire-fern (***O. alstonii***), first described under that name in 1978, had previously been misidentified as a variety of *O. hedysaroides*, which name properly identifies a related plant that differs in the pubescence of its stems, leafstalks, and flowering parts. A beautiful shrublet native of shady river banks in Brazil, the fire-fern has slender, erect, branched, woody stems, and at least in cultivation, it is not usually over 6 or 8 inches tall, although under favorable conditions it possibly exceeds this. The leaves, of three stalked leaflets that are heart-shaped with their broad ends attached to the stalks, are intense maroon-red. They are especially beautiful when viewed against the light. The flowers are small and bright yellow. A native of tropical west Africa, ***O. dispar*** is a beautiful hairy shrublet with a stout, branching stem, leaves with three lanceolate, conspicuously-veined leaflets, and nearly stalkless, small yellow flowers in

Oxalis dispar

Oxalis fruticosa

compact clusters atop quite long stems arising from the upper leaf axils. Of considerable botanical interest is ***O. fruticosa.*** What appear to be its leaves are really flattened leaf stalks that serve the purposes of leaves (this phenomenon occurs in some other plants, as in many acacias). These phyllodes or cladophylls, as they are called, sometimes reveal their morphological status by producing from their tips three tiny leaflets. They are narrow-elliptic and parallel-veined in the manner of a bamboo or grass leaf. The small yellow flowers are close to the stems in the leaf axils.

Stemless oxalises not hardy in the north and generally adapted to outdoor cultivation only in regions where little frost is experienced are numerous. They are not always easy to identify. By stemless is meant without obvious aboveground stems; often these kinds have underground stems and in some cases their tips peep out of the ground. In form the subterranean stems are usually tubers or bulbs, but sometimes they are scaly or woody rhizomes. A convenient division of these tender, stemless oxalises is into kinds with yellow flowers and kinds with blooms of other colors.

The Bermuda-buttercup (***O. pes-caprea*** syn. *O. cernua*) is a popular winter-blooming kind. It is the only tender, yellow-flowered, stemless oxalis in cultivation that has clusters of flowers; the others have solitary blooms. Despite its common name, this species is a native of South Africa, but is naturalized in Bermuda, Florida, and California. It has scaly bulbs, a deep, thick taproot, and very long-stalked leaves with three obcordate leaflets, fringed with hairs and ¾ to 1 inch across. Its somewhat nodding flowers, 1 inch to 1½ inches wide, are in clusters atop long stalks that carry them well above the foliage. They are bright yellow. In variety *O. p. flore-pleno* the flowers are double. Chilean ***O. lobata,*** a tuberous kind that grows to a height of about 4 inches, has leaves of three obovate leaflets not fringed with hairs and sometimes cleft so deeply that they are almost two-lobed. Its flowers are yellow dotted and lined with red; they are solitary.

Oxalis pes-caprea

Nonhardy stemless oxalises with pink, purplish, red, or white flowers are more numerous. Commonest is the 'Grand Dutchess' series of color forms of ***O. purpurea*** (syn. *O. variabilis*). Their flowers are white, pink, or rose-red, with yellow bases to the petals. They are solitary, about 2 inches across, and very showy. These plants have bulbs. Their leaves have three obcordate leaflets not notched at their apexes but fringed with hairs. Very handsome ***O. bowiei*** (syn. *O. bowieana*) is the hardiest South African species. In very sheltered places it survives winters outdoors at New York City, but does not

Oxalis purpurea in South Africa

Oxalis purpurea 'Grand Dutchess'

Oxalis rubra

Oxalis bowiei

flourish there and tends to deteriorate from year to year. This kind has glandular hairs as well as sometimes non-glandular hairs on its leafstalks, flower stalks, and calyxes. It has thick roots and large, scaly bulbs. The rather short-stalked leaves of three broadly-obovate, slightly notched leaflets, 1 inch to 1½ inches in diameter, are much overtopped by the flower stalks, each of which carries aloft three to twelve rose-pink to purplish-pink blooms 1½ to 2 inches across. It blooms in summer and fall. Commonly grown as a window plant and in greenhouses, sometimes under the name *O. floribunda*, very free-flowering ***O. rubra,*** a native of Brazil, has tuberous, but not bulblike roots and woody crowns that in old plants rise somewhat from the soil and from which spring leaves with long, slender stalks and three obcordate leaflets ¾ inch long and wide. The foliage is covered with fine-silky hairs that lie flat. Several to many attractive satiny flowers, ½ to ¾ inch long and with pink to rose-pink or sometimes white petals, constitute each flower cluster. They are carried well above the foliage on slender stalks and are borne in great profusion over a long summer period. This species is somewhat naturalized in California. Growing from bulbs and having long-stalked leaves each with eight to ten narrowly-wedge-shaped leaflets, ***O. lasiandra*** is an attractive light crimson-bloomed species. A native of Mexico, its flowers are in crowded clusters. Another Mexican, ***O. deppei,*** has scaly bulbs and long-stalked leaves of four obovate or nearly round leaflets about 1½ inches in diameter and not heart-shaped at their apexes. The red flowers, ½ to ¾ inch long, are in clusters of five to twelve at the tops of long stalks. From the above, ***O. tetraphylla*** differs in having the usually four leaflets of each leaf definitely heart-shaped at their apexes. The flowers, few together in clusters, are lilac or rosy-pink.

Other kinds include these: ***O. brasiliensis,*** native to Brazil, about 6 inches tall, bulbous, and stemless, has leaves with three obcordate leaflets, each about ½ inch across. Its bright rosy-red flowers with darker veins and yellowish throats are 1 inch across and in few-flowered clusters or occasionally solitary. ***O. corymbosa aureo-reticulata*** has clustered scaly bulbs from which arise many prostrate and ascending short stems. The leaves, of three obcordate leaflets, are about 3 inches across and beautifully veined with yellow. Its long-

Oxalis corymbosa aureo-reticulata

stalked flowers are carmine with darker veins and white throats. The species ***O. corymbosa*** (syn. *O. martiana*), native to tropical America, is represented in cultivation only by the variety. ***O. flava,*** native to South Africa, is a bulbous, stemless kind of fleshy character. Its long-stalked leaves have five to twelve narrow, finger-like leaflets. Its large, solitary flowers are bright yellow, they are carried scarcely above the foliage. ***O. gigantea*** is a shrub 3 to 6 feet tall or taller with slender, wand-like stems, fleshy leaves of three tiny obcordate leaflets, and solitary, golden-yellow flowers ¾ inch wide. In its native Chile this sort inhabits sea cliffs and dry slopes. ***O. herrerae,*** of Peru, is similar to *O. peduncularis*, but smaller. It has thick, branched stems and leaves with long, fleshy stalks and three small, notched, fleshy leaflets. Its yellow flowers, in compact clusters, are long stalked. ***O. melanostica,*** of South Africa, is stemless, low, and with firm leaves of three leaflets covered with glistening white hairs. It has solitary yellow flowers carried just above the foliage. ***O. regnellii*** is South African. It has woody rhizomes and long-stalked leaves with three leaflets that appear to have been sheared to produce very wide apexes indented at their centers with a shallow notch. Its white blooms are in long-stalked clusters. ***O. vulcanicola*** (syn. *O. siliquosa*), of Costa Rica, El Salvador, and Panama, is a loose, lax kind with a profusion of thin red stems and coppery-reddish leaves each with three stalkless leaflets. Its bright golden-yellow flowers, ¾ inch across, are in clusters.

Oxalis gigantea

Garden Uses. Of the hardy species, the wood-sorrel *O. acetosella* is suitable for partially shaded locations and moistish soil in native plant gardens, woodland gardens, and rock gardens. The other kinds considered here generally need full sun. Suitable only for rock gardens and alpine greenhouses and needing a connoisseur's care are *O. adenophylla, O. enneaphylla,* and *O. magellanica.*

Oxalis vulcanicola

The tender kinds native of South Africa, Mexico, and other warm-temperate and subtropical regions are hardy outdoors in California and many parts of the south and are good greenhouse and window plants, but the truly tropical kinds, such as *O. ortgiesii* and the fire-fern (*O. alstonii*) need really humid, tropical greenhouse conditions or at least the accommodation afforded by a terrarium.

Three kinds, as follows, can be successfully treated as annuals and used in the north and elsewhere for summer display in rock gardens and as edgings to flower beds and paths and, the first-named, in the crevices of flagstone or other open-joint paved surfaces. Unfortunately quite pretty *O. corniculata atropurpurea* (syn. *O. tropaeoloides*) is apt to spread by self-sown seeds and can become a decided pest. The others that flower well the first year from seeds are *O. rosea* and *O. valdiviensis.* Where they are hardy these tender kinds are excellent for the fronts of flower beds and borders and for rock gardens and bordering paths; they lend themselves for use in window and porch boxes and other outdoor containers. Some kinds, such as *O. rubra, O. pes-caprea,* and *O. variabilis,* are quite stunning when accommodated in hanging baskets.

Cultivation. With very few exceptions oxalises are easy to grow. Those most likely to present the gardener in many parts of the United States with problems are delightful little rock garden species that inhabit southern South America and the Falkland Islands, *O. enneaphylla, O. adenophylla,* and *O. magellanica.* These do best where winters and summers are temperate, they are impatient of torrid heat and severe cold. The Pacific Northwest is the most promising region for their successful cultivation. Soil that is gritty, well-drained, and contains a high proportion of leaf mold or other organic matter is agreeable. Where high temperatures and brilliant sun characterize summers, a location on a north-facing slope, possibly where a little shade from a nearby tree is received during the hottest hours of the day, is to be sought for these gems.

The woodland species *O. acetosella, O. oregana,* and *O. violacea* give no trouble if accommodated in soil that contains an abundance of organic matter, such as leaf mold, humus, or peat moss, and is moderately moist. Dappled shade, or overhead shade with good light from the sides, provides an encouraging environment.

Very different is *O. corniculata atropurpurea.* Although pretty, do not introduce it where the self-sown seedlings it produces freely may endanger choicer plants or engender much toil removing them. This kind flourishes in almost any soil, but for best results a lean earth low in available nitrogen is recommended because it encourages the plants to remain low and compact. For the same reason excessive moisture is to be avoided. Limestone soil is especially favorable. Although perennial, *O. corniculata atropurpurea* can be grown as an annual from seeds sown outdoors in early spring, and the young plants transplanted or thinned out to 5 or 6 inches apart. The other two kinds commonly grown as annuals, *O. rosea* and *O. valdiviensis,* are treated essentially in the same way. The former withstands dry conditions and some shade remarkably well.

Tender oxalises with perennial, aboveground stems fall into two groups: the tropical kinds, such as *O. dispar, O. fruticosa, O. ortigiesii,* and the fire-fern (*O. alstonii*), which need comparatively high temperatures and humidity, and the kinds that succeed under cooler, less moist conditions, such as *O. carnosa, O. peduncularis,* and, in semidesert environments, *O. gigantea.* The first-named should be accommodated in a humid greenhouse with a minimum temperature of 60°F or in an approximately similar environment. Shade from strong summer sun is needed. The pots in which these plants are grown must be well drained and on the small side rather than over-large in comparison to the size of the plant; this is especially important in the case of *O. fruticosa.* Most of these oxalises revel in fertile soil that contains an abundance of organic matter and is kept evenly moist but not constantly saturated, but for *O. gigantea* a less organic, drier soil is necessary.

Tender bulbous and tuberous species, nearly all of which are deciduous and need a definite season of rest, although a few, such as *O. rubra,* lose their leaves or are evergreen depending upon the availability of moisture, require quite different treatment. In regions where they thrive outdoors permanently and where there is little frost and where wet and dry seasons alternate, most grow without difficulty in full sun in any well-drained, reasonably fertile earth. They are not plants for the constantly humid tropics, but rather for Mediterranean-type climates, such as those of California and the southwest.

When grown indoors, these kinds need well-drained, nourishing soil and full sun. They are best accommodated in pots or pans. Six to nine bulbs is about the right number for a container 6 inches in diameter, the exact number depending upon their size and the vigor of the species. They are potted at the beginning of their growing season when they begin to break dormancy naturally. The season varies according to kind, but generally fall-blooming species start into growth in July or August, spring-blooming ones in September or October, and summer-blooming kinds in spring. They should be removed from the old soil and replanted in new every year, with the bulbs set about 1 inch beneath the surface. Immediately after planting the soil is drenched, but from then until new roots are strongly developed water is given sparingly so that the earth dries considerably, but not completely, between applications. As roots and leaves increase the need for moisture becomes greater; enough is then provided to prevent the soil from ever drying out. After roots fill the containers weekly applications of dilute liquid fertilizer are beneficial. When the blooming season ends and natural yellowing of the leaves signals the approach of the dormant season, stop fertilizing and lengthen the intervals between waterings until finally moisture is withheld altogether. The bulbs are then kept quite dry until the beginning of the next growing season. During this period they may remain in the soil or be cleaned, sorted to size, and stored in paper bags or other suitable containers. In either case the storage must be dry and cool, temperatures between 40 and 50°F are suitable. When in growth these oxalises need comparatively cool conditions. From fall to spring the greenhouse in which they are growing should be held to a night temperature of 45 to 50°F, and the day temperature should not then exceed 55 or perhaps 60°F before ventilation is given. These oxalises do not flourish in hot, humid environments. When they are grown as houseplants, a cool sunroom or similar location offers the best chance of success. The attempt should be made to approximate as nearly as possible desirable greenhouse conditions.

Propagation of most oxalises is easily achieved. The bulbous or tuberous kinds normally proliferate freely by offset bulbs or tubers, which are readily separated from the mother plant and potted or planted separately. Cuttings of stemmed kinds are generally easy to root in sand, vermiculite, or perlite in a greenhouse propagating bench or even in a window under an inverted glass jar, ventilated a little, and shaded from direct sun. Cuttings of woody-stemmed kinds, such as *O. alstonii, O. dispar*, and *O. fruticosa*, do not root as easily as those of other kinds. For their surest success they should be treated with a hormone-rooting preparation and be placed over a little bottom heat so that the rooting medium is five to ten degrees warmer than the atmosphere.

Diseases and Pests. Two fungus rust diseases, one with corn as its alternate host and one with *Mahonia*, attack some species of *Oxalis*, and they are also sometimes infected with fungus leaf spot diseases and with a disease called red-rust as well as, occasionally, by root rot and by curly top virus disease. Aphids and mealybugs sometimes infest them.

OXERA (Oxér-a). The genus *Oxera*, of the vervain family VERBENACEAE, is endemic to New Caledonia. It consists of twenty-five species of which one is sometimes cultivated. The name, of uncertain derivation, is probably from the Greek *oxys*, acid, in allusion to the sour taste.

Oxeras are mostly climbing shrubs with opposite, leathery, undivided leaves and branched clusters of white or yellowish flowers of various forms, but always with a four- or five-parted calyx, a more or less bell- or trumpet-shaped, wide-throated, four-lobed corolla, four stamens in pairs of different lengths, and one style. The often four-parted fruits are drupes.

Sometimes called royal climber, ***O. pulchella*** has corky stems and leathery, dark green, oblong-lanceolate leaves up to 5 inches long, often with scalloped margins. The fragrant, ivory-white, curved, bell-shaped blooms, 2 inches long or longer, are in forked clusters of as many as forty at the branch ends. They have prominent pale green calyxes and protruding stamens and styles. This beautiful vine blooms over a long period or sometimes several times a year in California.

Garden and Landscape Uses and Cultivation. A quite rare climber, *O. pulchella* is worthy of a choice location. In California it thrives in sun or part-shade in well-drained, somewhat acid soil that is never excessively dry. It is also interesting for greenhouses and prospers where the minimum winter night temperature is 55 or 60°F and the day temperature a few degrees higher. At other seasons higher temperatures are in order if the greenhouse is ventilated freely. This species needs a rich, loamy, well-drained soil. From spring through fall apply water more freely than in winter, but at no time let the soil become dry. Slight shade from strong sun is desirable. Supports to which the branches can be tied must be provided. Propagation is easy by seed and by cuttings.

OXLIP is *Primula elatior*.

OXYANTHUS (Oxy-ánthus). The name of the African genus *Oxyanthus*, of the madder family RUBIACEAE, derived from the Greek *oxys*, sharp, and *anthos*, a flower, alludes to the sharp-pointed calyx and corolla lobes of its blooms. Akin to *Gardenia*, although to the nonbotanist this is scarcely apparent, *Oxyanthus* more closely resembles tropical American *Posoqueria*, and by some authorities has been included there. The genus consists of fifty species.

Evergreen trees and shrubs, the sorts of this genus have opposite, undivided leaves and, in racemes or panicles from the leaf axils, flowers usually white or yellowish, remarkable for their very long, slender corolla tubes tipped with a star of five spreading, pointed lobes (petals). The calyx has five short teeth. There are five stamens and one spindle- or club-shaped style capped with a two-lobed stigma. The fruits are technically drupes.

Native to moist soils in woodlands in South Africa, ***O. natalensis*** is an attractive shrub or a small tree with short-pointed, elliptic-oblong to ovate-lanceolate, leathery leaves 6 to 8 inches long by about one-half as wide. In loose racemes of up to about twenty, its very fragrant, creamy-white flowers, about 4 inches long, have reflexed petals and a slender, protruding style.

Oxyanthus natalensis

Garden and Landscape Uses and Cultivation. The species described is well adapted for outdoor landscapes in southern Florida, Hawaii, and other places with warm, humid climates, and for greenhouses that afford similar environments. Its cultural requirements are those of gardenias.

OXYBAPHUS. See Mirabilis.

OXYCOCCUS. See Vaccinium.

OXYDENDRUM (Oxy-déndrum)—Sorrel Tree, Sourwood. There is only one species of *Oxydendrum*, of the heath family ERICACEAE, a familiar native of moist and dry open woodlands from Pennsylvania to Indiana, Florida, and Louisiana. Its name, derived from the Greek *oxys*, acid, and *dendron*, a tree, alludes to the sour taste of the leaves and character of the plant.

A lovely deciduous tree, sourwood (***O. arboreum***), sometimes 75 feet tall, is usually considerably lower. It has deeply-fissured bark and a pyramidal crown of slender, spreading branches. Its leathery, pointed-oblong-lanceolate to elliptic leaves have slender stalks, toothed or smooth margins. They are 4 to 7 inches long, hairless except for a few hairs on the midribs and veins beneath. Their upper surfaces are lustrous dark green, their undersides paler. In fall they turn brilliant crimson-scarlet above, but the undersides remain pale green. The flowers are arranged along one side of slender branches of 6- to 10-inch-long, pendulous panicles that remind one somewhat of those of *Leucothoe*. White, they are about ⅓ inch long. They have a deeply-five-parted calyx, a narrow-ovoid, tubular corolla with five short, spreading or recurved lobes (petals) ten nonprotruding stamens and a slender style. The sourwood blooms in July or August. Its flowers are succeeded by small, brown, five-angled dry capsules containing numerous very small seeds.

Garden and Landscape Uses. This is a choice garden and landscape subject, outstanding for the quality of its foliage from spring through fall and charming when in bloom. It associates well with rhododendrons, mountain-laurels, leucothoes, and similar acid-soil plants and provides a rich backdrop for lower plants. It is a slow grower, but starts to bloom when quite small and flowers regularly thereafter. For home landscapes and for groups along parkways and in parks it is ideal. The sourwood is hardier than the limits of its natural distribution suggest. It is satisfactory in most of New England, although, as is to be expected, it grows more slowly in these northern parts and does not become as tall as it does further south.

Cultivation. The sourwood will not tolerate limestone or other alkaline earths. It thrives in slightly acid soils, but will grow in those with a neutral reaction. It does best in full sun, but will stand a little shade. Good drainage with a reasonable amount of moisture is needed. This tree transplants fairly well, but is likely to be rather slow in reestablishing itself after being moved. A mulch of compost, leaf-mold, or other agreeable organic material maintained over the area occupied by its roots is very helpful. No regular pruning is needed. Propagation is by seeds sown in sandy peaty soil in spring and kept evenly moist. The sourwood is sometimes infected with leaf spot diseases and by twig blight.

OXYLOBIUM (Oxy-lòbium). Endemic to Australia, *Oxylobium*, of the pea family LEGUMINOSAE, contains about forty species. The name, from the Greek *oxys*, sharp, and *lobos*, a pod, refers to the sharp-pointed fruits.

Oxylobiums are evergreen shrubs or rarely subshrubs with very short-stalked, undivided, usually opposite or whorled (in circles of more than two), more rarely alternate leaves. Generally their margins are smooth, but those of some species are spiny-lobed or toothed. The red, yellow, or red and yellow, pea-shaped flowers are in terminal and axillary racemes or dense clusters. Their ten stamens are not joined. The fruits are ovoid or oblong pods. From related *Chorizema* this genus differs in the keels of the flowers being as long, rather than much shorter than, the wing petals.

A shrub 6 to 15 feet tall, ***O. callistachys*** (syn. *Callistachys ovata*) has angular shoots, hairy when young, and toothless, sharp-pointed, ovate-oblong to lanceolate or linear, leathery leaves 1½ to 5 inches long. They are mostly in circles of three, and on their undersides are silky-hairy. The flowers, yellow with red bases, and ½ to ¾ inch wide, are crowded in terminal racemes 2 to 6 inches long.

Garden and Landscape Uses and Cultivation. In California and other regions of warm, dry climates, the species described is a useful general purpose shrub for sunny, well-drained locations. It is propagated by seed and by cuttings.

OXYPETALUM (Oxy-pétalum). Sometimes named *Tweedia*, this South American genus, of the milkweed family ASCLEPIADACEAE, consists of 150 species of herbaceous plants and subshrubs. Its name, from the Greek *oxys*, sharp, and *petalon*, a petal, alludes to the petals.

Oxypetalums have opposite leaves and blue, purplish, yellowish, or white blooms

Oxydendrum arboreum

Oxydendrum arboreum (flowers)

Oxypetalum caeruleum

Oxyria digyna (flowers)

in clusters or umbels. The flowers have five-parted calyxes and a short-tubular corolla with five, often slender, lobes (petals). The fruits are pods.

Subshrubby toward its base and with twining stems up to 3 feet tall or taller, ***Oxypetalum caeruleum*** has short-stalked, oblongish leaves with heart-shaped bases, hairy on both surfaces. The flowers, pinkish in bud, in stalked, loose, up-facing clusters of three to four from the upper leaf axils, are about 1 inch across and very attractive. Their petals spread widely and are sky blue; they darken as the blooms age. The protruding corona is of erect, dark blue, fleshy scales, notched at their tips.

Garden and Landscape Uses and Cultivation. Only in fairly dry climates that are almost or quite frostless can the species described here be cultivated permanently outdoors. It can, however, be raised as a flower garden annual by sowing seeds early indoors and having young plants in pots to set in sunny places after all danger from frost has passed, or by sowing directly outdoors in early spring and thinning the seedlings out so that they do not crowd. In greenhouses where the minimum winter night temperature is 55 to 60°F, and that by day somewhat higher, and the atmosphere is humid, this plant can be grown permanently. It succeeds in well-drained, fertile soil in a sunny location. Old specimens are pruned to shape and repotted in spring. It is advisable to pinch the tips out of the shoots of young specimens once or twice to induce branching, and staking is usually needed. Propagation is easy by seed, and by cuttings in a greenhouse propagating bench in spring.

OXYRIA (Oxý-ria)—Mountain-Sorrel. The buckwheat family POLYGONACEAE contains the only species of *Oxyria*. It is the mountain-sorrel, an arctic, subarctic, and mountain plant indigenous in North America, Europe, and Asia. Its name, referring to the acid flavor of the stems and leaves, comes from the Greek *oxys,* sour. From nearly related docks or sorrels (*Rumex*) this plant differs in having flowers with four instead of six perianth segments and two styles. There are six stamens. The fruits, ⅙ inch broad, notched at their apexes, and with two broad wings, are mostly not enclosed by the calyxes.

Mountain-sorrel (***O. digyna***) is a stout-rooted, hairless, pale green, slightly succulent, herbaceous perennial, generally not more than 1 foot tall. Its leaves, nearly all basal, are stalked, kidney- to heart-shaped, and usually up to 1½ inches across. The few stem ones are alternate. The erect stalks, often branched above, carry narrow panicles of minute, bisexual, slender-stalked, often red-edged, greenish flowers.

Oxyria digyna

Garden Uses and Cultivation. The mountain-sorrel may properly be included in collections of herbs. It has scurvy-preventing properties and may be used as a potherb. It thrives in moistish soils, in sun, and is raised from seed.

OXYTROPIS (Oxý-tropis)—Locoweed. Although a number of handsome species are among the approximately 300 that compose *Oxytropis*, of the pea family LEGUMINOSAE, they are not well known to gardeners. A main reason is that many are difficult or almost impossible to tame. The group consists of low, sometimes somewhat spiny, perennial herbaceous plants and subshrubs having much the aspect of *Astragalus,* to which *Oxytropis* is closely related, and from which it differs in its flowers having corollas with keels that end in a sharp beak. Oxytropises are natives chiefly of northern and mountain regions in North America, Europe, and Asia. The name, alluding to the keel of the flower, is from the Greek *oxys,* sharp, and *tropis,* a keel. Some species are poisonous to livestock and share with certain species of *Astragalus* the common name locoweed.

Oxytropises generally have deep taproots and branching stems. Their leaves have a number of toothless leaflets including a terminal one. The pea-like flowers are in spikes or racemes from the leaf axils or bases of the plants. White, yellowish, violet, or purple, they have five-lobed tubular or bell-shaped calyxes, an erect standard or banner petal, wing petals usually notched or two-lobed, and a keel terminating in a narrow appendage. There are ten stamens, nine joined and one free. The fruits are ovoid to cylindrical pods.

Native to North America, Europe, and Asia, ***O. campestris*** is a variable, woolly-hairy herbaceous perennial 4 to 8 inches tall. Its leaves have twenty-one to thirty-one elliptic to lanceolate leaflets up to ¾ inch long. Yellow to white, often tinged with purple, the flowers are in crowded egg-shaped clusters of up to fifteen. From the last, European ***O. halleri*** differs mainly in its flowers, in longer-stalked racemes that lengthen after the blooms fade, being usually blue or purple, with darker keels. Similar, but having racemes that do not elongate after the blooms fade, ***O. uralensis*** is a native of Europe and Asia.

A locoweed, ***O. lambertii*** is found in dry soils from Minnesota to British Columbia, Missouri, Texas, and Arizona. From densely-hairy to nearly hairless, it has very short stems, and leaves 4 to 6 inches long, with eleven to seventeen usually erect, nearly opposite, linear to nearly oblong leaflets up to 1 inch long. The spikes of blue-purple flowers, 1½ to 4 inches in length, are carried on stalks to heights of 6 inches to 1 foot. The cylindrical pods are up to 1 inch long.

Oxytropis lambertii

Densely-silky-hairy, *O. splendens* and *O. sericea* are occasionally cultivated. A locoweed, ***O. splendens*** occurs from Minnesota to Manitoba and British Columbia and in the mountains to New Mexico. Up to 1½ feet tall, it has leaves with pointed, narrowly-lanceolate leaflets, often in groups of three or four along the main axis or midrib, and up to ¾ inch long. The ½-inch-long, deep blue flowers are crowded in spikes up to 3½ inches long. Native from Nebraska to Wyoming and Colorado, ***O. sericea*** has leaves with seven to nineteen elliptic to oblong leaflets. Up to 1 foot tall, its flower stalks terminate in racemes of pale purple to nearly white blooms.

Garden and Landscape Uses and Cultivation. The plants considered here are best adapted to cultivation in sunny rock gardens. They need deep, very well-drained soil. Because they do not transplant well young plants should be raised in pots and then planted permanently with a spacing of 9 inches to 1 foot between individuals, or seeds should be sown directly where the plants are to grow. The last procedure is practicable only in regions especially favorable to these plants.

OYSTER PLANT. For the vegetable called oyster plant see Salsify, Oyster Plant, or Vegetable Oyster. The Spanish oyster plant is *Scolymus hispanicus.* Another plant sometimes called oyster plant is *Rhoeo spathacea.*

OYSTER SHELLS. See Lime and Liming.

P

PACHGEROCEREUS (Pachgero-cèreus). A presumed bigeneric hybrid between *Pachycereus pringlei* and *Bergerocactus emoryi* that occurs sparingly as a wilding in Baja California is named *Pachgerocereus orcuttii* (syn. *Pachycereus orcuttii*). Its name is coined from parts of the names of the supposed parents, which belong to the cactus family CACTACEAE.

Shrubby and up to 12 feet tall by one-half as wide, ***P. orcuttii*** has a short trunk and several to many upright branches 3 to 4 inches thick, narrowed at places that mark the end of each season of growth, and with fourteen to eighteen well-defined ribs. The areoles (cushion-like areas in cactuses from which arise stems and blooms), which touch each other toward the tops of the stems, are more widely spaced below. White to gray, each has twenty-five to over 100 swollen-based, needle-sharp spines from quite small to up to 2½ inches long. The flowers, each open for one day only, are funnel-shaped, pale yellow, and 2 to 2½ inches long by nearly as wide. They have many stamens. The densely-spiny and felted, globular fruits, about 2 inches in diameter, develop fertile seeds that give rise to seedlings that are likely to differ somewhat from the typical first-generation hybrid.

Garden and Landscape Uses and Cultivation. This extremely interesting plant is handsome for inclusion in collections of succulents, outdoors in warm, dry climates, and in greenhouses. It requires treatment appropriate for most columnar cactuses. For more information see Cactuses.

PACHIRA (Pac-hìra) — Guinea-Chestnut. The genus *Pachira,* of the bombax family BOMBACACEAE, consists of two species of evergreen trees endemic to the American tropics. The name is an adaptation of a native South American one. In Florida deciduous *Pseudobombax ellipticum* is often misidentified as *Pachira fastuosa.*

The trunks of pachiras are sometimes buttressed. Their leaves have five to nine stalked or stalkless leaflets 3 to 8 inches long that spread palmately (in hand-fashion) from the tip of the leafstalk. Their margins are smooth or wavy and sometimes recurved. Grouped toward the ends of the branchlets, the mostly solitary flowers are remarkable for the 200 to 700 stamens that are their chief attraction. The stamens are joined at their bases into a tube that separates some distance up into five branches, which divide again at different levels into numerous unequal-sized, long, separate stalks, the ultimate separations each tipped with an anther joined near its end to the stalk. There is a lobeless, tubular calyx and five linear, recurving petals. The fruits are woody capsules containing seeds ½ inch or more in diameter that may be associated with some hairs, but are not surrounded by wool.

In the wild commonly found along river banks and in moist places, the Guinea-chestnut (***P. aquatica*** syn. *P. macrocarpa*), 15 to 70 feet tall, has a spreading crown. Native from southern Mexico to Brazil and Peru, it has hairless leaves with usually oblanceolate leaflets 4 inches to 1 foot in length. Commonly solitary, rarely in twos or threes, the flowers are 7 inches long to twice that length. The narrowly-strap-shaped petals are purplish or pinkish, the stamens red. The fruits are ovoid and 4 inches to 1 foot long by approximately one-third as wide. Unlike the last, ***P. insignis,*** of the West Indies and northern South America, occurs in dry soils. Up to 60 feet in height, it has leathery, broad-obovate leaflets with rounded or slightly notched apexes. The fragrant flowers, up to 1 foot in diameter when expanded, have light red to crimson-purple petals longer than the numerous white stamens. The stalks of the stamens fork and refork almost to their ends. The style is red.

Pachira aquatica

Garden and Landscape Uses and Cultivation. These are highly ornamental flowering trees for sunny locations in the tropics. As indicated by the sites they favor in the wild one needs moist or wet soil, the other drier conditions. Propagation is by seed and by cuttings.

PACHISTIMA or PACHYSTIMA. See Paxistima.

PACHYCARPUS (Pachy-cárpus). This African genus of thirty species closely related to *Asclepias* is rare in cultivation. It belongs in the milkweed family ASCLEPIADACEAE and consists of perennial herbaceous plants with tuberous or fleshy roots. The name, from the Greek *pachys,* thick, and *karpos,* a seed, is of obvious application.

Widely distributed through South Africa, variable ***Pachycarpus grandiflorus*** is 1 foot to 1½ feet in height and pubescent; it has oblong to lanceolate, usually wavy-edged leaves and flowers mostly one to four together from the axils of the upper leaves. Its cup-shaped blooms, greenish-yellow spotted with dark purple-brown, are usually under 2 inches in diameter. They have five triangular corolla lobes and a corona of five lobes with curving, tonguelike tips.

Garden Uses and Cultivation. The sorts of this genus are not hardy. The species described here is suitable for flower beds and borders in dryish, frost-free and nearly frost-free climates, such as are typical of parts of western North America.

PACHYCEREUS (Pachy-cèreus)—Cardon, Hairbrush Cactus. Treelike cactuses, often massive, with erect, thick, ribbed stems constitute *Pachycereus.* The genus, belonging in the cactus family CACTACEAE, is native to Mexico. The name, from the Greek *pachys,* thick, and the name of the related genus *Cereus,* is of obvious application.

Pachycereuses, of which there are five species, have day-open, short-tubular, funnel- to bell-shaped blooms of small to moderate size. They rise from thickly-felted areoles (areas on the stems of cactuses from which spines and flowers develop) and have numerous, not protruding stamens. The bristly fruits are large, dry, and subspherical. The plant formerly named *P. orcuttii* is believed to be a natural bigeneric hybrid correctly identified as *Pachgerocereus orcuttii.*

The cardon (***P. pringlei***), of Baja California, is a huge species rivaling and perhaps sometimes exceeding in size the sahuaro (*Carnegiea gigantea*), which, in aspect, it resembles. This has a stout trunk that usually divides into few to several thick, upright, columnar, eleven- to seventeen-ribbed branches that not uncommonly attain heights of 30, 40, 50 feet or sometimes more. The spine clusters, from brown-felted areoles, are of twenty or more spines on young parts of the stems, but often may be absent from older parts. The glistening white blooms, about 3 inches in diameter, are borne for a considerable distance from the tops of the stems downward, usually on their sunny sides. The fruits look something like big chestnut burs.

Pachycerus pringlei in bloom

Pachycereus pringlei and other desert vegetation in Baja California

Pachycereus pringlei in Baja California

The hairbrush cactus (***P. pecten-aboriginum***) owes its vernacular and botanical names to a reported use made by Indians of parts of the large spiny fruits as combs or hairbrushes. A stout-trunked species 15 to 30 feet tall, this has a trunk 3 to 6 feet long and a head of erect branches with about ten ribs. Its spine clusters are of about eight awl-shaped, ½-inch-long radials, at first brown, later gray with darker tips and one similarly colored central 1½ inches long. The white flowers are 3½ inches long by about 2½ inches across. Native to the extreme south of Baja California, this is more sensitive to cold than *P. pringlei.* Up to 35 feet high, ***P. weberi*** (syn. *Lemaireocereus weberi*) has a short, stout trunk and a broad candelabrum head of many strictly erect, blue-green, ten-ribbed branches with clusters of six to twelve needle-like, ¾-inch-long radial spines thickened at their bases and one flattened central about 4 inches long. All are reddish-brown to black. The yellowish-white flowers are 3½ to 4 inches long.

Garden and Landscape Uses and Cultivation. These imposing cactuses are not difficult to grow. As large specimens they give importance to collections of succulents because of their great size and upright growth. When small they are very attractive. They thrive under conditions that suit thick-stemmed, columnar desert cactuses, reveling in sun and warmth. They are most easily propagated by seed. For additional information see Cactuses.

PACHYCORMUS (Pachy-còrmus) — Elephant Tree. Its vernacular name alluding to its obese trunk, the elephant tree (***Pachycormus discolor***), an outstanding endemic of Baja California and the only representative of its genus, belongs to the cashew family ANACARDIACEAE. Its name, of obvious application, is derived from the Greek *pachys,* thick, and *kormos,* a tree trunk.

Pachycormus discolor

The elephant tree inhabits hot deserts, usually favoring rocky, well-drained soil. It has one or more grotesque, elephantine trunks, with peeling bark, that a few feet from the ground may be 3 feet in diameter. Rarely over 15 feet tall, but sometimes twice that, the rounded head, leafless for most of the year, is of usually crooked branches that taper rapidly from base to tip. The leaves are pinnate and up to 3½ inches long by ¾ inch to 1¼ inches wide. They have an uneven number of broad-elliptic to obovate leaflets. The flowers, small, pink or white, fragrant, and in panicles, are displayed when the tree is leafless. They have five sepals and petals, ten stamens, and three styles. The fruits, technically utricles, are small and bladdery.

Garden Uses and Cultivation. Occasionally grown by fanciers of desert plants, this unusual and rare species may be accorded the conditions favored by warm-region cactuses and other succulents. It is propagated by seed.

PACHYPHYTUM (Pachý-phytum). Closely related to *Echeveria* and by some authorities included there, *Pachyphytum,* of the orpine family CRASSULACEAE, consists of a dozen Mexican species. They are nonhardy, succulent, subshrubby perennials much like echeverias. The name, from the Greek *pachys*, thick, and *phyton*, a plant, alludes to the thick stems and leaves. Hybrids between *Pachyphytum* and *Echeveria* are named *Pachyveria*. The plant once named *P. amethystinum* is *Graptopetalum amethystinum*.

Pachyphytums differ from echeverias in having a pair of small but obvious appendages at the bases of each of the five stamens that are joined to the insides of the petals. Only as abnormalities do similar scales occasionally occur in *Echeveria*. There are other technical differences. The leaves of pachyphytums are alternate, often in elongated rosettes from the more or less branched, frequently short stems. They are from flattish to nearly cylindrical, gray-green, or covered with a whitish, waxy coating. The branchless flowering stalks, from the leaf axils, are at first nodding, later more or less erect. They have beneath the lowermost bloom few to several fleshy bracts. The blooms have five sepals united for a short distance from their bases, five petals, ten stamens of which five are attached to the petals, and five separate carpels, each with a short style.

Semicylindrical to much flatter and broader leaves, and flowers with very brief undivided stalks and white petals longer than the markedly unequal sepals, with a red band below their tips, characterize the section of the genus identified as *Pachyphytum*. These sorts belong here. The 1-foot-tall ***P. bracteosum*** has few-branched stems and round-ended, oblanceolate to spatula-shaped leaves 3 to 4½ inches long by 1 inch to 2 inches wide. Covered with considerable waxy bloom, they are often reddish toward their margins. Up to 2 feet in length, the branchless flowering stalks have a few soon-deciduous bracts and carry ten to about twenty blooms. Not crowded, thickish, glucous, often purplish, oblanceolate leaves, 2½ to 4 inches long or slightly longer by ¾ to 1 inch wide and channeled on their undersides, characterize ***P. longifolium.*** Its 9-inch to 1¼-foot-tall flowering stems carry ten to fifty blooms. Short-stalked ***P. oviferum*** has crowded rosettes of slightly obovate to oblong-obovate leaves, nearly cylindrical in section except for their slightly flattened upper sides. Up to 2 inches long by ¾ to 1 inch broad, they are greenish to reddish and hoary with a waxy coating. The reddish flowering stalks, approximately 4 inches long, carry in one-sided racemes a few predominantly purple-carmine blooms with petals whitish on their backs and white-edged. Distinctly separated rather than crowded, the thickish, green to purplish-red leaves on stems ¾ inch to 1¼ inches thick of ***P. viride*** are elliptic-oblong

Pachyphytum oviferum

and 1 inch to 2 inches long by ¾ inch to 1¼ inches wide. From ten to twenty-two flowers are displayed on each 9-inch- to 1¼-foot-long stalk. Its stems about ½ inch thick, ***P. werdermannii*** has distinctly spaced rather than crowded, glaucous, elliptic leaves 1½ to 4 inches long by ¾ inch to 1¼ inches broad. Ten to twenty-two flowers are displayed on each 6- to 10-inch-long stalk.

Spindle-shaped to nearly cylindrical leaves, and flowers with petals of uniform color or with darker tips distinctly longer than the sepals pressed against them, distinguish the section of the genus named *Diostemon*. To this section belong descriptively named *P. compactum*, and *P. hookeri*. With stems up to 4 inches long or longer and about ¼ inch in diameter, ***P. compactum*** has densely crowded, cylindrical leaves slightly flattened on their upper sides, ¾ inch to 1¼ inches long by ½ inch wide, dark green to reddish, and glaucous. The few-bracted flowering stalks, up to 1¼ feet long, terminate in one-sided racemes of three to ten ⅓-inch-long blooms with nearly equal sepals and green-tipped, reddish petals. Its little-branched stems up to 4 feet long by ⅜ to ¾ inch thick, and with few small bracts, ***P. hookeri*** (syns. *P. aduncum, P. uniflorum*) has blunt, spindle-shaped to club-shaped leaves flattened on their upper sides and rounded on their backs, green or glaucous, and up to 2 inches long by ¾ inch wide. Up to 1 foot in length, the reddish flowering stalks terminate in one-sided racemes of usually three to eighteen, rarely one or two, bell-shaped blooms approximately ½ inch long, the lower ones on individual stalks as long or longer than the petals. They have unequal-sized sepals and yellowish petals flushed with red. A crested variety is *P. h. cristatum*.

Flat leaves, and flowers with unequal sepals, that at least in the flowers that open later than the earliest have petals uniform in color, characterize the section of the genus called *Ixiocaulon*. Here belongs *P. fittkaui* and *P. glutinicaule*. With ½- to 1½-inch-thick stems, ***P. fittkaui*** has blunt, slightly-glaucous elliptic-oblanceolate leaves 1 inch to 4 inches long by ¾ inch to 1½ inches broad. On stalks up to 1¼ inches long are displayed twelve to twenty-five flowers with very short, undivided stalks and pink petals. Sticky stems ½ inch or slightly more in diameter are typical of ***P. glutinicaule.*** The leaves of this are obovate, 1 inch to 2½ inches long by ¾ inch to 1¼ inches wide, and glaucous. The flowers, borne six to twenty-two on stalks up to 10 inches tall, are red. A presumed hybrid of this species, 'Cornelius Hybrid' has bluish-gray, often pink-tinged, red-tipped leaves and red flowers.

Garden Uses and Cultivation. These are as for *Echeveria*.

PACHYPODIUM (Pachy-pòdium)—Ghost Men. Desert plants are notoriously often of curious aspect, but none more so, surely, than certain species of *Pachypodium*, of the milkweed family ASCLEPIADACEAE. This group of twenty species of shrubs and miniature trees inhabits dry parts of South Africa and Malagasy (Madagascar). Its most outstanding members have somewhat the appearance of the equally improbable-looking boojum tree (*Idria*), of Baja California. The two genera are not related; *Pachypodium* is kin to *Adenium*. Its name is derived from the Greek *pachys*, thick, and *podos*, a foot, and alludes to the thickened roots and stems.

Pachypodiums are fleshy, spiny-stemmed plants well designed to conserve such moisture as they can abstract from soil that is arid for long periods. Some have tall, swollen trunks with or without slender branches. In other kinds the fat main stems are low, but have knoblike elevations for branches or more slender branches. The stipules (appendages at the bases of the leaf stalks of many plants) are represented by a pair of spines. The leaves are alternate (unusual in the milkweed family) and arranged in spiral fashion. They drop at the beginning of the dry season. Often colorful and attractive, the white, yellow, or red blooms, which develop from the ends of the trunks or stems, are short-stalked and quite showy. They have small, deeply-five-parted calyxes and tubular corollas with five lobes (petals). There are five stamens, and two styles. The fruits are follicles.

The ghost men (***P. namaquanum***), of South Africa, is an extraordinary species. Its usually branchless erect and slightly curved trunk, somewhat like an inverted, spiny carrot, attains a height of 3 to 5 feet and in season is topped by a tuft of leaves 4 to 5 inches long and 2 inches wide or slightly wider. The numerous reddish-brown, velvety blooms are striped with yellow on their insides. In its native land

Pachypodium namaquanum in South Africa

the upper parts of the trunks (or should it be bodies?) of ghost men are always inclined toward the north, which, south of the Equator is toward the sun. Having a similar trunk, but with ascending, ridged, slender branches, ***P. giganteum,*** another African, attains a maximum height of about 20 feet. It is armed throughout with very stiff spines arranged in threes. Its stemless or nearly stemless leaves are long-ovate, 4 to 4½ inches long by 1½ to 1¾ inches wide. Their margins are fringed with hairs. The pleasingly-scented, wavy-edged, white flowers, about 1¾ inches in diameter, are several together from the ends of the stems. This species has very much the aspect of the American *Idria columnaris.* South African ***P. saundersii*** has a stout, branched stem 3 to 5 feet high, with long, needle-like and elliptic leaves. Its flowers are white striped with red.

Pachypodium saundersii

Treelike ***P. geayi,*** of Malagasy (Madagascar), has a comparatively slender, very spiny trunk up to 30 feet in height and

Pachypodium geayi

Pachypodium rutenbergianum

branchless or sparsely-branched at its summit. Crowded at the ends of the trunk and branches are short-stalked, linear to narrow-lanceolate leaves. The flowers of this are unrecorded. Another Malagasian, ***P. rutenbergianum*** has a tapered trunk up to 15 feet tall, at its base up to about 1½ feet wide. Its short branches are furnished with spines less than ½ inch in length. The leaves have short stalks and narrow-oblong-lanceolate blades 3 to 5 inches long. The flowers are white. Also native to Malagasy, ***P. lamerei*** has a fleshy, sparingly-branched, conical stem 1½ to 15 feet tall. Its short-stalked leaves have blades up to about 10 inches long by 1½ inches wide. The flowers are white.

Pachypodium lamerei, a young specimen

Pachypodium bispinosum

Quite different from the sorts described above, African ***P. bispinosum*** (syn. *P. lealii*) has a massive turnip-shaped root, which is the water storage organ, and slender branches, hairy on their young parts, as are both surfaces of the lanceolate leaves. The pink, purplish-pink, or white, bell-shaped flowers are hairy in their throats. A native of Malagasy, ***P. densiflorum*** is squat. It has a brief stem and short,

Pachypodium densiflorum

Pachypodium succulentum

thick, cylindrical branches. Its spines are solitary, its leaves narrowly-elliptic and white-felted on their undersides. The flowers are five or six together in dense clusters. Tuberous rooted ***P. succulentum,*** of South Africa, has a short, thick stem from which sprout erect or straggling slender branches armed with long, needle-like spines. Its leaves are lanceolate, its flowers pink.

Garden Uses and Cultivation. These are rare, choice plants for keen collectors of succulents. They are suitable for outdoor planting in warm desert and semidesert regions only and in greenhouses with a minimum winter night temperature of 55°F. They require conditions similar to those for *Fouquieria.* For more information see Succulents.

PACHYRHIZUS (Pachy-rhìzus)—Yam-Bean, Jicama. This remarkable genus of non-woody, perennial vines of the tropics of the Old World and the New World belongs in the pea family LEGUMINOSAE. There are half a dozen species. They have enormous, starchy, tuberous roots. Some kinds are grown in the tropics and warm subtropics as food, both their roots and the young seed pods being eaten. Reputedly the seeds are poisonous, but any harmful properties possessed by those of young pods are apparently destroyed by cooking. The name, of obvious application, is from the Greek *pachys,* thick, and *rhiza,* a root.

Pachyrhizuses have twining stems and leaves of three often angled or sinuously-lobed leaflets. The pea-shaped flowers, clustered in long racemes from the leaf axils, are followed by long, narrow, flattened seed pods. Not uncommonly their tuberous roots are 5 to 8 feet long and may weigh 50 to 70 lb or even more. From those of some kinds an arrowroot-like starch is prepared.

The yam-bean (***Pachyrhizus erosus***), of both the New World and Old World tropics, is a high-climbing vine with hairy younger parts. Its commonly long-stalked leaves have leaflets with wedged-shaped bases, and in their upper halves toothed lobes. The racemes of reddish-purple, 1-inch-long blooms, are 6 inches to over 1 foot long. The seed pods are 4 to 9 inches long. Vigorous and high climbing, *P. e. palmatilobus* (syn. *P. palmatilobus*), of tropical America, has hairy or hairless stems and mostly long-stalked, pubescent leaves with conspicuously deeply-lobed leaflets. The flowers are purplish. This has smaller roots than the typical species.

The jicama (***P. tuberosus***) differs from the yam-bean in having much bigger roots, leaflets obscurely or not at all lobed, and seed pods 8 inches to 1 foot long. This is a native of tropical America.

Garden and Landscape Uses and Cultivation. Except in the tropics, these plants are grown only as novelties, sometimes in educational collections of food plants. They thrive in deep, fertile, well-drained, soil and need generous supplies of water. When grown for their roots, all flower buds are kept picked. Propagation is by seed. Plants can also be raised from cuttings and root divisions.

PACHYSANDRA (Pachysán-dra). Three Asian and one American are the only species of *Pachysandra,* of the box family BUXACEAE. One of the former and the American appear to be the only sorts commonly cultivated. The name, from the Greek *pachys,* thick, and *aner,* man, alludes to the stout filaments of the stamens.

Pachysandras are evergreen or deciduous perennial herbaceous plants or subshrubs that spread, quickly or slowly according to kind, by slender underground rhizomes. They have erect branches leafy toward their tops, and leaves alternate or clustered. The flowers are in spikes with male blooms occupying the upper, females the lower portions of each spike. They have four sepals, no petals, four stamens, and two or three spreading styles. The fruits are capsules or drupes.

Most familiar is the Japanese-spurge (***P. terminalis***), in America one of the most

Pachysandra terminalis: (a) Foliage

(b) In bloom

(c) Flowers

extensively used low evergreen groundcovers. From 6 to 10 inches tall, this forms dense, spreading mats of neat evergreen foliage that completely hides the ground with an attractive pattern of greenery of uniform height and appearance. It spreads vigorously, but not so aggressively that it is a threat to most neighbor plants. It is, however, too strong a grower for close association with delicate alpines or other choice rock garden plants, and its roots are too matted and too possessive to permit most bulbs to be planted among it with much chance of permanent success. Japanese-spurge is at its best beneath and between trees and shrubs, including evergreens. There it hides bare ground, keeps down weeds, and reduces need for mowing and other maintenance. In addition to being a work-saver, when thrifty it has considerable beauty. The leaves of the Japanese-spurge, glabrous, thick, leathery, 2 to 4 inches long, obovate, and coarsely-toothed above their middles, are wedge-shaped at their bases. The terminal spikes of white, fragrant flowers, 2 inches long, are borne in spring, usually not very abundantly. They are interesting, but do not make any very great show. A variety with foliage variegated with creamy-white is sometimes cultivated. This, *P. t. variegata*, is somewhat less vigorous than the green-leaved plant.

Pachysandra procumbens

Pachysandra procumbens in bloom

The Allegheny-spurge (***P. procumbens***) is a rather local native of rich woods from Kentucky to Florida and Louisiana. It differs from the familiar Japanese species in having duller foliage and in its flower spikes not being terminal on the shoots, but arising from their bases. It is a much less vigorous spreader than its Oriental relative. Its rootstocks extend slowly, and it poses no threat even to delicate neighbors not actually crowded against it. Under favorable conditions it may increase by self-sown seeds, but this is an event to be grateful for rather than to cause concern, for the progeny can be transplanted with ease and a few extra plants are usually welcome. Allegheny-spurge forms a clump up to 1 foot tall with soft gray-green leaves indistinctly mottled with brownish-green. The leaves are ovate to roundish, 2 to 4 inches long, and coarsely-toothed above their middles and narrowed below to a long leafstalk. The flowers appear in earliest spring. They are white with dull pink anthers, which give an over-all pale smokey-pink appearance to the spikes. This species is hardy far north of its natural range, and certainly into New England. In the south it is evergreen, but in the north it normally loses its leaves in winter, although there appears to be some variation among individuals regarding this. Certainly in the vicinity of New York City occasional strains retain their foliage, whereas others are deciduous.

Pachysandra axillaris

Chinese ***P. axillaris*** grows from 6 inches to 1 foot in height. At the tops of its stems it has three to six ovate, coarsely-toothed leaves that are wedge-shaped or rounded at their bases. Its white flowers are borne in spring in axillary spikes up to 1 inch long. This is probably as hardy as the American species. Much resembling the last, Chinese ***P. stylosa*** differs from *P. axillaris* chiefly in its more robust habit and more elongated flower spikes with the female flowers stalked. In the original description of *P. stylosa* its leaves are said to be without teeth, but that is not true of plants grown in gardens under that name. A closely similar sort, also with few-toothed leaves, is cultivated as *P. s. glaberrima*.

Pachysandra stylosa

Pachysandra stylosa glaberrima

Garden and Landscape Uses. The overwhelmingly most important use for Japanese-spurge is as an evergreen groundcover. The Allegheny-spurge and Chinese-spurges are interesting as single specimens or in small groups in wild gardens and rock gardens and can be employed effectively for more extensive groundcover

A shaded bank planted with *Pachysandra terminalis*

operations. Pachysandras thrive in any fairly good soil that is moderately moist, but free of stagnant water. They grow as well in slightly alkaline as in neutral or somewhat acid soils. For their most satisfactory growth they need some shade from strong sun, dappled shade from high-branched trees being ideal, but they also grow well where they receive ample north light without direct sun. They will not prosper in dry, sandy soils or in places where they are exposed to strong sun or cold sweeping winds.

Cultivation. Plantings of pachysandra that are well established benefit from fertilizing with a complete fertilizer each spring and watering heavily in times of drought. Take care not to step on the plants because injuries caused to the stems and leaves invite entry of a fungus that causes leaf blight. These plants are easily increased by division and by leafy shoot cuttings taken in late summer. Japanese-spurge is also susceptible to increase by cuttings made from its long, underground, rootlike rhizomes. These, chopped into pieces ½ to 1 inch long in spring, scattered over the surface of sandy peaty soil in a cold frame, and covered with ½ inch of a mixture of peat moss and coarse sand kept evenly moist soon root, develop shoots, and become established as new plants. Allegheny-spurge is easily raised from seed sown in well-drained, sandy peaty soil kept evenly moist.

Diseases and Pests. Japanese-spurge is subject to a quite serious leaf blight, the best control for which is had by cutting the stems of affected patches to the ground and spraying several times at ten-day intervals with a fungicide. It is also liable to infestation with a leaf-tier, scale, mites, and the peanut root knot nematode. It is by no means unlikely that the Allegheny-spurge and perhaps the others may be susceptible to the same diseases and pests.

PACHYSTACHYS (Pachý-stachys)—Cardinal's Guard. To the chiefly tropical acanthus family ACANTHACEAE belongs the genus *Pachystachys*. It has five species, all natives of warm parts of the Americas. One, the cardinal's guard, is a popular garden plant in Hawaii, Florida, and other tropical and subtropical regions. It is native of northern South America and Trinidad. The genus *Pachystachys* is akin to *Jacobinia*, differentiated by technical details of the pollen and by the stamens being joined to the base of the corolla tube instead of near its middle. The name is derived from the Greek *pachys*, thick, and *stachys*, a spike, and alludes to the flower clusters.

The genus consists of shrubs with opposite, lobeless leaves and spikelike clusters or heads of white, yellow, red, or purple, two-lipped, tubular flowers with the lips or lobes of the corollas markedly unequal. The individual blooms are solitary, or in twos or threes, in the axils of large bracts. There are two stamens.

Pachystachys coccinea

The cardinal's guard (***P. coccinea*** syns. *P. cardinalis, Jacobinia coccinea, Justicia coccinea*) is a husky shrub up to 7 feet tall. It is hairless and has stout stems noticeably narrowed immediately above each node (joint). Its ovate-elliptic or, in the case of smaller ones, ovate-lanceolate, short-pointed leaves are 4 to 8 inches long and strongly veined. Dense heads, 4 to 5 inches long, of bright scarlet blooms terminate the stems. They have large, pointed-ovate, green bracts and flowers about 2 inches long, with corollas cleft almost to their middles and the lower lips three-lobed. The upper lip of the bloom is slightly arched, the lower curves downward.

Beautiful ***P. lutea,*** of Peru, was introduced to cultivation in North America in 1964 by T. H. Everett, of The New York Botanical Garden. A subshrub, 2 to 6 feet tall, this has elliptic leaves up to 7 inches long by 1¾ inches wide, narrowing to brief stalks and with the upper surface

Pachystachys lutea

sparsely-hairy and the veins beneath pubescent. Its flower spikes have conspicuous butter-yellow, ovate bracts and flowers with curved white corollas about 2 inches long.

Garden and Landscape Uses. Although technically a shrub, the cardinal's guard is not very woody and is often thought of as and used more as a herbaceous perennial. It is tender to frost, but will live through temperatures down to 20°F for brief periods, and although its tops are killed, it recovers and sends up new shoots that bloom the first year. In summer its handsome heads of bloom create a blaze of color. This is a grand plant for providing a gorgeous display at that season. For its best development this species needs fertile soil that does not lack for moisture, yet is not poorly drained, and light shade. It is also satisfactory as a warm greenhouse plant. The lovely *P. lutea* thrives in greenhouses with the same treatment as cardinal's guard. It has not been tested outdoors. Since it grows in the open at Lima, Peru, it might be expected to prosper wherever the cardinal's guard succeeds.

Cultivation. Cardinal's guard is easily propagated by cuttings or seed. Established specimens may, and indeed should, be cut back severely in late winter or early spring. If this is not done the plants tend to become lanky and unkempt. They should be fertilized in spring and again later if the soil be a hungry one. When grown in greenhouses this species and *P. lutea* prosper under conditions that suit justicias.

PACHYSTEGIA (Pachy-stègia). This endemic genus of New Zealand consists of one species. Belonging in the daisy family COMPOSITAE, it is distinguished from closely related *Olearia* by technical characteristics of the parts of its flower heads. The name is from the Greek *pachys*, thick, and *stege*, a roof.

A spreading, thick-branched shrub up to 6 feet tall, ***Pachystegia insignis*** (syn. *Olearia insignis*) has densely-hairy shoots and very thick, leathery, lobeless and toothless leaves clustered at the branch

Pachystegia insignis

ends. The leaves have thick stalks up to 2 inches long or longer and ovate-oblong to obovate-oblong blades 3 to 7 inches long and about one-half as wide. At maturity their shining upper surfaces are hairless, beneath they are densely clothed with soft, white, buff, or rust hairs. The solitary, daisy-like flower heads, about 3 inches across, have a central disk of tubular yellow female florets and a surrounding circle of many petal-like, white, bisexual ray florets. The seedlike fruits are achenes. Variety *P. i. minor* is smaller and more slender.

Garden and Landscape Uses and Cultivation. Not hardy in the north, this attractive shrub is suitable for planting in California and places with like climates. It has the uses of, and needs the conditions and care that suit, Olearia.

PACHYSTIGMA. See Paxistima.

PACHYVERIA (Pachy-vèria). This is the name of hybrids between *Pachyphytum* and *Echeveria* of the orpine family CRASSULACEAE. Of garden origin, they are intermediates between their parents, have the same uses, and respond to the same conditions and care. Among the best known is *Pachyveria clavata*, its parents *Pachyphytum bracteosum* and *Echeveria rosea*. A crested variety of this is *P. c. cristata*. The one named *P. pachyphytoides* has as parents *Pachyphytum bracteosum* and *Echeveria gibbiflora metallica*.

PACO is *Diplazium esculentum*.

PAEDERIA (Paed-èria). The genus *Paederia*, of the madder family RUBIACEAE, consists of fifty kinds of woody or semiwoody vines, frequently with foliage that is malodorous when bruised. The name, in allusion to ill-scented *P. foetida*, derives from the Latin *paedor*, a bad odor. The genus is native to the tropics and subtropics.

These plants have twining stems and opposite, long-stalked, undivided, thinnish leaves. Their small flowers are in axillary and terminal panicles. They have a five-toothed or five-parted persistent calyx, a tubular corolla usually hairy in its throat and with four or five recurved lobes (petals), five stamens, one style, and two twisted stigmas. The fruits are small berries.

Native from China to the Malay Archipelago and the Himalayas, and naturalized in Florida, ***P. foetida*** has ovate to heart-shaped-ovate leaves 1½ to 6 inches long by ½ inch to 2½ inches wide, and downy or hairless on the veins of their undersides. In slender, twisted panicles 9 inches to 1½ feet long, the lilac-purple flowers are about ⅓ inch long. The glossy, flattened-spherical fruits are green. Japan, Korea, and China is the native territory of deciduous ***P. scandens,*** a species hardy in southern England. From 10 to 20 feet tall,

Paederia scandens with flower buds

this has pointed-ovate to ovate-lanceolate or oblong-elliptic leaves 2 to 6 inches long by up to 3¼ inches wide, often downy on their undersides. In panicles 6 inches to 1½ feet long, the flowers are white with a purplish eye. The ¼-inch-wide, spherical fruits are orange.

Garden and Landscape Uses and Cultivation. These are interesting vines for clothing walls, pillars and other surfaces strung with vertical wires around which the stems can twine. They thrive in ordinary soil in sun or part-shade and may be propagated by seed and by cuttings.

PAEONIA (Pae-ònia)—Peony. Peonies are old and respected inhabitants of gardens, too well known it would almost seem to need description, yet in addition to the popular horticultural varieties of herbaceous kinds and the somewhat less familiar ones of tree peonies, there is a goodly group of species, unimproved by the skills of horticulturists or plant breeders, that merit attention. All are easy to grow and are well adapted to American gardens. The genus *Paeonia*, chiefly Asian, is represented less abundantly in the natural floras of Europe and North America. It comprises about thirty species, and belongs to the buttercup family RANUNCULACEAE, or according to those who favor splitting that group, to the peony family PAEONIACEAE. The name, employed by Theophrastus, is believed to have been given in honor of Paeon, a physician reputed to have first used the plants medicinally.

Peonies fall into two distinct groups, herbaceous perennials, which are without woody stems and die to the ground each year, and tree peonies, which are not trees at all, but shrubs with permanent woody stems. Peony roots are thick or tuberous. The quite large leaves are alternate, and of three divisions or pinnately-divided or dissected. Mostly solitary, but sometimes clustered, the usually large and showy blooms are red, purple-red, pink, white, or yellow. They have five persistent sepals, five to ten, or in horticultural varieties more, petals, numerous stamens, and two to five carpels that in fruit become large follicles (podlike structures that contain the seeds).

Seed pods of *Paeonia*

Garden varieties of herbaceous peonies, of which the number of handsome ones is almost legion, are mostly derivatives of ***P. lactiflora*** (syn. *P. albiflora*). This native of

China and Siberia has thick, fleshy roots and clusters of erect stems 1½ to 3 feet tall that usually divide above into two to five branches, each ending in a white or pinkish bloom 3 to 4 inches wide, with leafy outer sepals, inner sepals with prominent midribs extending their entire lengths, eight or more large, wide petals, yellow stamens, and four or five hairless, erect or somewhat spreading carpels. The leaves are of three divisions each deeply divided into three nearly separate elliptic to lanceolate, sometimes lobed leaflets with minutely-rough edges.

From the last, less hardy ***P. emodi,*** of Kashmir, differs in its leaves having minutely-hairy, decidedly impressed veins and smooth margins and in its white flowers, 3 to 5 inches wide, having one or rarely two densely-yellow-hairy, spreading carpels. Related species that differ from *P. emodi* in the terminal divisions of their largest leaves being dissected into a dozen or more lobes are ***P. anomala*** and ***P. veitchii,*** the latter by some authorities considered to be a variety of the former from which it differs most obviously in its stems carrying two to four blooms instead of one. Also, the three to five pistils of the flowers of *P. anomala* are hairless, the two to four of those of typical *P. veitchii* densely-hairy. Native from the Ural Mountains to Siberia and central Asia, *P. anomala,* 1½ to 2 feet tall, has bright red blooms 3 to 4 inches wide. More commonly cultivated *P. a. intermedia* differs from the typical species in the carpels of its flowers being hairy. The hybrid *P. smouthii* not infrequently masquerades as *P. anomala* and *P. a. intermedia.* Native of Japan, Sakhalin, China, and Manchuria, ***P. obovata*** is 1½ to 2 feet tall and hairless or sparingly-hairy. Twice divided into threes,

Paeonia obovata

its leaves generally have an obovate terminal leaflet, the other leaflets ovate to oblong. About 3 inches in diameter, the flowers of the cultivated kind are white, but in the wild pink-flowered forms occur. The stigmas are coiled.

Fern-leaved ***P. tenuifolia*** is very distinct and highly decorative in foliage as well as in bloom. Native from southeastern Europe to the Caucasus, it has thick roots and spreading underground stolons. Its 1- to 2½-foot-tall, densely-leafy stems are usually branchless and one-flowered. The twice-divided leaves, hairless above and pubescent on their undersides, are dissected into numerous segments up to ⅕ inch wide. Dark crimson to purple-red, the cupped blooms, 2¼ to 3¼ inches wide, have eight to ten petals and two or three hairy carpels. Variety *P. t. plena* has double flowers. A hybrid between *P. tenuifolia* and *P. anomala,* named ***P. hybrida,*** is distinct from the plant of unknown origin that has been grown as *P. tenuifolia hybrida.* The leaves of the former are hairless, those of the latter strongly pubescent on their lower surfaces. Another hybrid, ***P. smouthii,*** the parents of which are *P. tenuifolia* and *P. lactiflora,* has been misnamed *P. anomala intermedia* and *P. anomala.* This has hairless foliage. Hybrids between *P. tenuifolia* and *P. mlokosewitschii* are ***P. saundersii.*** These have foliage with minutely-pubescent under surfaces. Their flowers are yellowish-salmon-pink.

Native to southern Europe and adjacent Asia, ***P. officinalis*** is 2½ to 3 feet tall. It

Paeonia officinalis

has usually branchless stems. Its leaves are hairless above, paler and somewhat hairy on their under surfaces. The lower ones are three times divided into seventeen to thirty narrowly-oblong segments. The upper ones are less divided into fewer leaflets. The solitary crimson flowers, up to 5 inches in diameter and with red-stalked stamens, have two or three usually densely-white hairy carpels. Varieties in which the blooms are pink or white, and in some cases double, are cultivated. This species, which is reported to be poisonous to livestock, has medicinal properties. Native to eastern Europe and Asia Minor, variable ***P. mascula*** is much like *P. officinalis* but its lower leaves are mostly twice divided into nine to sixteen narrow-elliptic to broad-elliptic leaflets 1 inch to 1½ inches long and hairless on their undersides. The upper leaves are once-divided into fewer leaflets. The rose-red flowers, 3 to 5 inches across, have three to five usually hairy carpels. From the typical species *P. m. arietina* (syn. *P. arietina*) differs chiefly in its leaves being hairy on their undersides, and the lower ones with twelve to fifteen leaflets. Its flowers are rose-red to dark red. Its leaves are bright red.

Paeonia mascula arietina in fruit

Beautiful rich yellow flowers 4 to 5 inches in diameter, with about eight petals, numerous stamens, a purple stigma, and three hairy carpels, are borne by ***P. mlokosewitschii,*** of the Caucasus. From 1½ to 2 feet tall, this has hairless stems and foliage. Its dark bluish-green leaves have paler, glaucous, short-hairy undersides and are of three divisions each of three leaflets. Some authorities are of the opinion that this is a variant of *P. daurica* (syn. *P. triternata*), which comes from the same general region and has rose-red blooms. Paler yellow flowers are borne by ***P. wittmanniana,*** of the Caucasus. This has leaves with leaflets longer and more pointed than those of the last, and with long hairs on their under surfaces. Its blooms are 4 to 5 inches wide. More common in gardens than the typical species, *P. w. nudicarpa* differs from the species in having hairless carpels.

Native American peonies are *P. brownii* and *P. californica.* The first inhabits dry slopes from California to Wyoming, Nevada, and British Columbia, the other is endemic to California. From 9 inches to 1½ feet in height, ***P. brownii*** has mostly branchless stems and somewhat glaucous,

Paeonia wittmanniana nudicarpa

rather fleshy leaves of three short-stalked divisions of three leaflets each. The petals, up to ½ inch long and shorter than the inner sepals, are brownish-red with yellowish or greenish margins. There are usually five carpels. Taller ***P. californica*** differs from *P. brownii* in having nonglaucous leaves, with the three primary divisions practically stalkless, and flowers with pink-margined, blackish red petals about ¾ inch long that are longer than the inner sepals.

Tree peonies include a wide range of horticultural varieties derived from *P. suffruticosa* (syn. *P. moutan*), a native of China. In the Orient this species and its varieties have long been cherished as ornamentals. They are also grown in American and European gardens. A shrub 3 to 6 feet tall, ***P. suffruticosa*** has twice-pinnate lower leaves, hairy on their undersides, and with the leaflets sometimes two- to five-lobed. The flowers, usually solitary, and up to 6 inches wide or sometimes wider, have eight or more broad, concave, rose-pink to white petals, with at their bases a red-margined, deep magenta blotch. Their apexes are toothed or notched. The spreading carpels are densely-hairy. Horticultural varieties include kinds with flowers of various shades of red and pink as well as white. Some have double flowers. Chinese ***P. delavayi*** is 3 to 5 feet tall. Its foliage is hairless. The lower leaves have three primary divisions, each of three-lobed or toothed leaflets. Dark red, the cup-shaped flowers are 2½ to 3½ inches in diameter. Below the calyx is a collar of up to a dozen green bracts. There are five hairless carpels. Related to the last, but without the collar of bracts beneath the calyx, ***P. lutea,*** of China, has yellow blooms, sometimes stained with purple at the bottoms of the petals, and almost 3 inches in diameter. This hairless shrub is about 4½ feet tall. By some authorities *P. lutea* is considered to be a variety of *P. delavayi* as is Chinese ***P. potaninii*** (syn. *P. delavayi angustiloba*). This last is up to 5 feet tall and has red blooms 2 inches or a little more in diameter, without a collar of bracts below the calyx. The foliage is hairless and more finely divided than that of *P. delavayi*. In *P. p. alba* the flowers are white, those of *P. p. trollioides* have yellow petals that do not spread widely. To hybrids between *P. lutea* and *P. suffruticosa* belongs the group named *P. lemoinei*.

Paeonia suffruticosa breaking into new leaf

Paeonia lutea

Garden and Landscape Uses and Cultivation. Information about the garden and landscape uses of peonies and their cultivation is provided under Peony.

Paeonia potaninii

PAGODA TREE, JAPANESE is *Sophora japonica*.

PAINKILLER is *Morinda citrifolia*.

PAINT and PAINTED. These words are employed as parts of the common ones of these plants: Devil's paintbrush (*Hieracium aurantiacum*), Flora's paintbrush (*Emilia*), paintbrush or painted cup (*Castilleja*), Japanese painted fern (*Athyrium goeringianum pictum*), painted leaf (*Euphorbia heterophylla*), and painted tongue (*Salpiglossis*). For painted-daisy see Pyrethrums.

PAK-CHOI. See Chinese Cabbage.

PALAFOXIA (Palafóx-ia). One of the two species of *Palafoxia*, a North American genus of the daisy family COMPOSITAE, is sometimes cultivated. The generic name honors the Spanish General José de Palafox y Melzi, who died in 1847.

A handsome garden annual, ***P. hookerana*** (syn. *Polypteris hookerana*) is native in sandy soils from Nebraska to Texas. From 1 foot to 4 feet in height, bushy, and sticky-hairy, it has rough, lanceolate leaves, mostly three-veined below. The daisy-form heads of flowers, about 1 inch in diameter, have eight to ten deeply-three lobed, rose-pink or rose-red, petal-like florets. The fruits are seedlike achenes.

Garden and Landscape Uses and Cultivation. This little known, but showy annual is attractive in beds and borders and has flowers useful for cutting that can be dried as everlastings. It needs a sunny location and a well-drained soil. One of a sandy nature is ideal. Sow seeds in early spring directly outdoors where the plants are to bloom, or indoors six to eight weeks earlier to give young plants to grow inside until it is nearing time to plant petunias and tomatoes outdoors. Then harden the young plants by standing them for a week or ten days in a cold frame or in a sheltered spot outside before planting them in the garden. Space them about 1½ feet apart. Summer care is minimal. Keep down weeds by shallowly cultivating the surface soil or by mulching. Water in dry weather. To prepare the flowers as everlastings, cut them as soon as they are fully open, tie their stems together, and hang them upside down in a cool, airy place, out of direct sun, until quite dry.

PALAI is *Microlepia setosa.*

PALAQUIUM (Pal-àquium)—Gutta Percha Tree. Several species of *Palaquium,* a genus of more than 100 species of milky-juiced trees of the sapodilla family SAPOTACEAE, yield gutta percha and some are used in warm countries as ornamentals. The genus in the wild is confined to Malaya, Malaysia, and Polynesia. Its name is a modification of the Philippine Islands plant named *palak-palak.*

Palaquiums have alternate, spirally arranged leaves and stalkless clusters of short-stalked blooms in the leaf axils or on the shoots behind the leaves. The small, tubular flowers have six each sepals and petals, and usually twelve, but varying from eight to thirty-six, stamens. The fruits are fleshy, pointed-oblong berries with the persistent sepals at their bases. They contain one to three seeds.

The original source of gutta percha, ***P. gutta,*** has been largely exterminated in the wild as a result of commercial exploitation, but is cultivated for interest and ornament. It is an evergreen tree recorded as occasionally being 100 feet high, but more commonly is under 50 feet tall. It has an open, conical, or rather flat-topped head of branches and foliage. Its leaves, more or less obovate, dark green above, and covered with coppery-brown silky hairs on their undersides, are 3 to 9 inches long and 1 inch to 3 inches wide. The pale green flowers, about ½ inch across, have a sickly odor suggesting that of burnt sugar. The fruits are ¾ inch to 1½ inches long. A similar native of the Philippine Islands, ***P. philippense*** is cultivated in Florida.

Garden and Landscape Uses and Cultivation. The chief decorative virtue of these trees is their handsome foliage. They are also of interest because of their usefulness as sources of gutta percha. Fresh or dried branches and leaves are very pleasing for flower arrangements. For their best growth palaquiums need humid tropical or near tropical conditions and fairly fertile soil. Often they are of slow growth. Propagation is by seed and by cuttings.

PALI-MARA. See Alstonia.

PALICOUREA (Palicoùr-ea). So closely related to *Psychotria* that it is nearly impossible to completely separate all species of the two genera by easily observable characteristics, *Palicourea,* of the madder family RUBIACEAE, comprises about 200 species of shrubs and trees native from Mexico to South America, and in the West Indies. Its name honors Le Palicour, of Guiana.

Palicoureas have opposite or whorled (in circles of more than two), undivided leaves and most commonly terminal panicles of small tubular flowers that nearly always are enlarged at the bases of their usually yellow, orange, red, purple, or blue, rarely white, corolla tubes. Mostly the panicles have brightly colored stalks. The blooms have five-lobed or lobeless, top-shaped or hemispherical calyxes, tubular, five-lobed corollas, five stamens, and a two-branched style. The fruits are little berry-like drupes (fruits similar in structure to plums).

Widely distributed in the West Indies and South America, ***P. crocea*** is a shrub or tree 4 to about 20 feet tall. It has short-stalked, hairless or nearly hairless, slender-pointed, satiny, ovate-lanceolate to ovate leaves 4 to 8 inches in length, fairly narrow to distinctly broad in proportion to their lengths, and with conspicuous parallel veins extending outward from the midribs. The reddish-orange or sometimes bright yellow flowers, many in loose, pyramidal, orange- or red-stalked panicles, are ⅓ to ½ inch long. The fruits, about ⅕ inch in length, are purple-black. A variable shrub or tree 5 to 35 feet tall, ***P. guianensis*** is a native of South America. Nearly hairless, it has short-stalked, elliptic-oblong to elliptic, pointed leaves 5 inches to over 1 foot long by about one-half as wide, with prominent, parallel side veins extending outward from the midrib. The numerous, yellow to orange-red blooms, ½ to ¾ inch long, are in orange-stalked, dense panicles, the floral portions of which are about 6 inches long. The ribbed, egg-shaped fruits, ⅕ inch long, are blue-purple. Related ***P. barbinervia*** (syn. *P. guianensis barbinervia*), of the West Indies, Trinidad, Venezuela, and Brazil, differs from *P. guianensis* in the veins on the undersurfaces of the leaves being hairy.

Garden and Landscape Uses and Cultivation. Suitable for planting for ornament in the humid tropics and warm subtropics, the palicoureas described here succeed in moist soils of ordinary quality. They need no special care, and do well in partial shade. They are increased by seed.

PALISOTA (Palis-òta). A few handsome foliage plants of the twenty-five species of this tropical West African genus are cultivated. Named after the French botanist and traveler J. Palisot de Beauvois, who died in 1820, *Palisota* is a member of the spiderwort family COMMELINACEAE.

This genus consists of nearly stemless to quite long-stemmed herbaceous perennials with foliage mostly basal or crowded near the apexes of the stems. The longitudinally-parallel-veined leaves have red or gray hairs at their edges, and when young all over. The many small flowers, in usually crowded, terminal or apparently terminal clusters, are whitish, pinkish, or purple. They have three each sepals and petals of much the same size and color, three stamens, and two or three staminodes (rudimentary stamens). The fruits are red or blue or other colored berries.

Nearly stemless, ***P. barteri*** has lustrous, dark green, pointed, obovate-oblong to broad-elliptic or elliptic-lanceolate leaves from 9 inches to 2 feet long by 4 to 4½ inches wide, when young clothed with shaggy hairs. Its light purplish flowers, in short-stalked clusters 2 to 3 inches long from the centers of the rosettes of foliage are succeeded by showy crowded spikes of bright orange-red, pointed berries. This species is native to Fernando Po. West African ***P. mannii,*** usually stemless but sometimes with a stem 1 foot to 6 feet

Palisota mannii

high, has shaggy-hairy younger parts. Its lanceolate to obovate leaves 1 foot to 4½ feet long by up to 4½ inches wide, except for their hairy margins, are when mature, nearly hairless. The very numerous white or pinkish flowers, crowded in dense, oblongish to rounded heads, are succeeded by fruits white or green when young changing to yellow, orange, red, or purple-brown at maturity. Differing from *P. barteri* in having an irregular longitudinal pale yellow band down the center of each leaf, ***P. pynaertii elizabethae*** (syn. *P. elizabethae*) is a short-stemmed native of equatorial West Africa. Its leaves are 1½ to 2 feet or more long by up to 6 inches wide. Obovate-lanceolate, they taper very gradually to thick, longish stalks and have reddish hairs along their margins. The whitish flowers are crowded in erect, cylindrical or long-ellipsoid, spikes.

Palisota pynaertii elizabethae

Paliurus spina-christi

Garden and Landscape Uses and Cultivation. Palisotas, strictly tropical, are at their best where the humidity is high. They favor shade from strong sun and are excellent for planting among palms and other trees, as well as for locating in front of shrubs. Single clumps at pool and stream sides and in rock gardens are effective, but must not be planted where the soil is constantly wet. In addition to outdoor uses in the tropics, palisotas are splendid greenhouse plants, succeeding where the minimum night temperature in winter is 60°F and at other times and seasons higher. A humid atmosphere and shade from strong sun are necessary. These plants succeed in ground beds, pots, and tubs. Coarse, fertile, well-drained soil is to their liking and should be always moderately moist. Well-rooted specimens are much benefited by regular applications of dilute liquid fertilizer. Propagation is easy by division at the beginning of a new season of growth and by fresh seeds sown in sandy peaty soil in a temperature of 70 to 80°F.

PALIURUS (Pali-ùrus) — Christ-Thorn or Jerusalem-Thorn. Extending in the wild from southern Europe to China and Japan, this genus, of the buckthorn family RHAMNACEAE, consists of eight species of shrubs and small trees, one of which is commonly planted in mild climates. The name is an ancient Greek one.

The leaves of *Paliurus* are alternate, stalked, and undivided, but sometimes toothed, and are in two ranks. The appendages at the leafstalk bases, called stipules, often gradually harden and become spines. The small, bisexual blooms, in axillary or more rarely terminal clusters, have five each sepals, petals, and stamens and an ovary with three stigmas. The small fruits are globose, with a rim or wing surrounding the central part so that they resemble Chinese cymbals or, as one authority suggests, "a head with a broad-brimmed hat." Unlike the fruits of related *Rhamnus* and *Zizyphus,* they are dry rather than fleshy.

Christ-thorn or Jerusalem-thorn (***P. spina-christi*** syn. *P. aculeatus*) is believed by some to be the biblical crown of thorns. Others consider *Zizyphus spina-christi* to be that plant. Native from southern Europe to Japan, *P. spina-christi* is deciduous. It varies from being a procumbent shrub to a tree up to 20 feet in height. It is spiny with one of each pair of spines straight and pointing upward, and the other curved and down-pointed. The hairless, ovate, blunt, three-veined leaves are finely-toothed and ¾ inch to 1½ inches long. Their upper sides dark green, beneath they are paler. Greenish-yellow, in umbel-like clusters, the flowers are followed by brownish-yellow fruits ¾ to 1 inch across.

Garden and Landscape Uses. Christ-thorn is a moderately ornamental tree or shrub for well-drained soils in sunny locations. It is not reliably top-hardy north of Washington, D.C., but even in southern New England its roots survive in sheltered locations and new shoots come up each year. It is, perhaps, appropriate for including in collections of biblical plants.

Cultivation. Seeds sown as soon as ripe or in spring after being stratified through the winter give satisfactory results. Other means of propagation are root cuttings and layers. No special cultural care is needed. Any pruning to shape or restrict the tree may be done in spring.

PALM-LILY is *Yucca gloriosa.*

PALMA-CHRISTI is *Ricinus communis.*

PALMA DE COYER is *Aiphanes acanthophylla.*

PALMA PITA is *Yucca treculeana.*

PALMAE. This family of monocotyledons, also known as the ARECACEAE and its members called palms, consists of approximately 2,500 species distributed in somewhat more than 200 genera. Among its sorts are many of great usefulness to man. With very few exceptions, its members in the wild are confined to the tropics and subtropics; very few inhabit warm-temperate regions. For further information about this important group of plants see Palms.

Genera of the *Palmae* dealt with as separate entries in this Encyclopedia are *Acanthophoenix, Acoelorrhaphe, Acrocomia, Aiphanes, Allagoptera, Archontophoenix, Areca, Arecastrum, Arenga, Arikuryroba, Astrocaryum, Attalea, Bactris, Bismarckia, Borassus, Brahea, Brassiophoenix, Butia, Calamus, Caryota, Ceroxylon, Chamaedorea, Chamaerops, Chambeyronia, Chrysalidocarpus, Coccothrinax, Cocos, Colpothrinax, Copernicia, Corypha, Cryosophila, Cyrtostachys, Daemonorops, Deckenia, Desmoncus, Dictyosperma, Elaeis, Euterpe, Gastrococos, Gaussia, Geonoma, Hedyscepe, Heterospathe, Howea, Hydriastele,*

Hyophorbe, Hyphaene, Iriartea, Juania, Jubaea, Jubaeopsis, Kentiopsis, Korthalsia, Latania, Licuala, Linospadix, Livistona, Lodoicea, Metroxylon, Microcoelum, Nannorrhops, Neodypsis, Normanbya, Nypa, Orbignya, Phoenicophorium, Phoenix, Phytelephas, Pinanga, Polyandrococos, Prestoea, Pritchardia, Pseudophoenix, Ptychosperma, Raphia, Reinhardtia, Rhapidophyllum, Rhapis, Rhopaloblaste, Rhopalostylis, Rhyticocos, Roscheria, Roystonea, Sabal, Scheelea, Schippia, Serenoa, Syagrus, Synechanthus, Trachycarpus, Thrinax, Trithrinax, Veitchia, Verschaffeltia, Wallichia, Washingtonia, and *Zombia.*

PALMATE. This term means with three or more parts radiating in open-hand-like fashion from a common basal point and is used to describe leaves with leaflets so disposed and systems of veins arranged in this way. A leaf or other part lobed in this manner is palmately-lobed.

PALMELLA is *Yucca elata.*

PALMERELLA. See *Lobelia dunnii.*

PALMETTO. See Sabal, and Serenoa.

PALMS. This noble group of plants is one of the most interesting and important in the world. It includes about 2,500 species, several of immense commercial importance, many others of significant economic use and numerous fine ornamentals attractive to horticulturists. Inevitably palms are associated with the tropics. The popular conception of a palm is that of a tree with a tall, branchless, columnar trunk topped with a large crown of feathery or fan-shaped leaves and probably coconut-like fruits. To complete the picture, sunny, tropical beaches with breakers, or thatched native huts and possibly monkeys are envisaged. This is not too unreal a visualization, but it is by no means adequate. Many, indeed most palms approximate the conventional concept, others do not. Whereas most are tropical or at least warm-subtropical, palmettos (*Sabal*) are native as far north as North Carolina, the European fan palm (*Chamaerops*) is wild in southern Europe, the Nikau palm (*Rhopalostylis*) occurs in cool parts of New Zealand, the Chilean wine palm (*Jubaea*), grows in southern Chile, and *Trachycarpus* is native in temperate China. In the Andes, wax palms (*Ceroxylon*) ascend to altitudes well over 13,000 feet.

Tall, single-trunked palms: (a) *Trachycarpus fortunei*

(b) *Cocos nucifera*

Not all palms have tall, solitary trunks. Many are multiple-trunked; the trunks of the doum palm (*Hyphaene*) branch freely. Many sorts are low and shrubby; some are without trunks. Others are tall vines with thin stems hundreds of feet long. Palms exhibit tremendous variability in outward appearances and in their uses and adaptability to cultivation. This endears them to landscape gardeners, greenhouse gardeners, florists, houseplant fanciers, and others interested in growing plants. Among palms are to be found the plant with the largest known leaves (*Raphia farinifera*), the plant with the largest clusters of flowers (*Corypha umbraculifera*), and the plant with the biggest seeds, the double-coconut (*Lodoicea maldavica*).

Multiple-trunked palms: (a) *Chamaedorea erumpens*

(b) *Rhapis excelsa*

Palms constitute the botanical family Palmae, an ancient group not closely related to most other plants now living. Cretaceous fossils, 120 millions of years old, testify that they were among the earliest of flowering plants to appear in an evolving world and that they have remained remarkably constant in form throughout the eons. They belong in the subdivision of the plant kingdom known as monocotyledons and are thus nearer kin to grasses, cat-tails, lilies, orchids, tulips, and irises than to magnolias, oaks, pines, and most familiar trees. Except for palms, there are few monocotyledonous trees, among them are species of *Strelitzia, Pandanus,* and *Dracaena.*

Being monocotyledons, palms have trunks that are structually very different from those of trees not of that group. Unlike dicotyledons, they do not add growth rings of wood and bark to the outsides of the trunks each year. There is no cambium layer of multiplying, growing cells beneath a layer of bark. With most palms the only area of active new stem growth is the terminal bud, and if this is seriously injured or destroyed, the plant dies. Similarly, the roots of palms do not thicken by adding layers of new tissue as do those of dicotyledonous plants, but grow only at their extremeties. Individual roots do not remain active indefinitely, they are comparatively short-lived and are constantly replaced by new ones that develop from the basal part of the trunk. Such increase in diameter of a palm trunk as occurs, including the enlargement of such kinds as the bottle palm (*Hyophorbe*) and the Cuban belly palm (*Colpothrinax*), is chiefly the result of increase in the size of original cells formed early in the life of the tree. Palm trunks are not covered with bark like those of dicotyledonous trees, but with a hard, tough rind similar to that of bamboos.

The foliage of palms is evergreen. The leaves are of two main types, pinnate in the case of feather-leaved palms and palmate in fan-leaved palms. Pinnate leaves have a central axis or midrib with leaflets

A feather- or pinnate-leaved palm (*Chrysalidocarpus lutescens*)

A fan- or palmate-leaved palm (*Bismarckia nobilis*)

along both sides in the manner of a feather. Palmate leaves have the form of a fan or of a hand with separate leaflets representing spreading fingers. A modification of the pinnate leaf is the pinnatifid. In this there is a central axis, but the divisions of the blade are not deep enough to form completely separate leaflets. Palms with leaves of this kind are included in the feather-leaved group. A modification of the palmate leaf is the palmatifid. In this the blade is deeply lobed in palmate fashion but is not divided into separate leaflets. Palms of this type are included in the fan-leaved group. Some few palms have simple or entire leaves with the blades not deeply-lobed or divided. In these the patterns of the veins are either distinctly pinnate or palmate, and the plants are classed as pinnately-veined or palmately-veined, accordingly.

In some palms the bases of the leafstalks sheath each other to form a cylinder that appears to be a continuation of the trunk. This false trunk is called a crownshaft. According to species, the leaves that die may fall from the trunks cleanly, leaving a smooth or ring-marked surface, they may drop, but leave the bases of the stalks or "boots" attached to the trunks where they persist for long periods or indefinitely, or they may droop and remain to form a skirt or petticoat around the trunk that lasts for

In southern California, these specimens of *Washingtonia filifera* have well-preserved skirts of old leaves

very long periods until, eventually, they decay and the older parts are lost.

The flowers of palms are rarely individually showy, but not infrequently they are borne in such numbers that they attract attention. They are in clusters, usually branched, and often of great size. It has been estimated that a large flower cluster of the talipot palm (*Corypha umbraculifera*) may contain sixty thousand individual blooms. The small flowers are usually greenish or whitish, rarely reddish or purplish, and typically have three sepals, three petals, and six stamens, but there is considerable variation in the number of stamens. The flowers of some kinds may have a thousand. The clusters may be terminal, in the axils of the leaves, or arise from the trunk below the foliage. The flowers may be male, female, or bisexual. In different species, individual trees may bear blooms of one or more of these types of flowers. The flower clusters are commonly enveloped in and protected when young by two or more large bracts or spathes that are typically boat-shaped and may be persistent or early deciduous. Some palms act like the century plant (*Agave americana*). They grow for many years without flowering, then bloom once and die. The talipot palm and the sugar palm (*Arenga pinnata*) do this.

Some palms die after their first flowering and fruiting; here in fruit: (a) *Arenga pinnata*

(b) *Corypha umbraculifera*

Fruits of palms vary tremendously from small and berry-like to the familiar dates and coconuts and the astonishing double-coconut (no close relative of the ordinary coconut), which contains the largest seed of any known plant. Palms may be spiny, some of them viciously so, or spineless. In some species the spines are confined to the trunks, leafstalks, or both, in others they are prominent on the leaf blades, sheaths, and other parts.

Although many studies have been made in recent years, the botanical classification of palms is by no means complete. It is to be expected that further work will result in the discovery of new species and in the reevaluation of and sometimes the renaming of some kinds already known.

Commercially the most important palm is the coconut (*Cocos nucifera*). This is the chief source of vegetable oil used mainly in margarines and soaps. The second most important oil source is the African oil palm (*Elaeis guineensis*). An extremely important producer of food for millions in desert and semidesert parts of the Middle East and North Africa is the date palm (*Phoenix dactylifera*). From the betel palm (*Areca catechu*) are harvested the betel nuts that serve as a masticatory or chewing stuff throughout much of tropical Asia. The starchy pith of the trunk of the sago palm (*Metroxylon sagu*) is a nutritious food, and from the carnauba palm (*Copernicia prunifera*), of South America, and other species valuable waxes are derived. Several palms are esteemed as sources of sugar, and their fermented and unfermented saps are used as beverages. Included here are the sugar palm (*Arenga pinnata*), the palmyra palm (*Borassus flabellifer*), the toddy palm (*Caryota urens*), and the coconut and date palms. Raffia fiber is a product of *Raphia farinifera*, of Malagasy (Madagascar). The South American piassava palm (*Attalea funifera*) produces stout bristle-like fibers used in street-sweeping brooms. These are but a few of the many kinds of palms useful to man and but a few of their uses. In tropical lands palms provide materials for thatching, basket- and mat-making, cordage, and other purposes too numerous to list. The palmyra palm alone is reputed to have more than eight hundred uses.

Sources of important commercial products: (a) Coconut palm (*Cocos nucifera*)

(b) African oil palm (*Elaeis guineensis*)

(c) Date palm (*Phoenix dactylifera*)

The palmyra palm (*Borassus flabellifer*) is a source of sugar

Palms alone create this attractive Florida landscape

In agreeable climates palms can and should be an important feature in planting design. They are adaptable to small home grounds as well as to more extensive areas, such as estates, parks, and parkways, and as street trees. Under favorable conditions they may be used with good effect to dominate the landscape picture. Mostly, with their vertical trunks and conspicuous crowns of bold leaves, they are strong elements in the landscape, but some shrubby kinds, such as *Chamaedorea* and *Rhapis*, provide softer effects and meld more completely with other plants. Because of their tropical appearance palms associate effectively with other tropical and subtropical plants, especially those with large bold leaves, such as anthuriums, strelitzias, bananas, and pandanuses, and those with brightly colored foliage, such as crotons (*Codiaeum*), acalyphas, and dracaenas. They also "go well" with such warm-climate flowering plants as poinsettia, hibiscus, royal poinciana, frangipani, and oleanders.

Basically palms can be employed in two ways: as solitary specimens or grouped

A single-trunked *Acrocomia aculeata* dominates this scene in the Fairchild Tropical Garden, Miami, Florida

A twin-trunked *Chrysalidocarpus madagascariensis* adds distinction to these beds of flowering plants in Hope Botanic Gardens, Jamaica, West Indies

A group of Washingtonia palms, in association with other trees, dominates this southern California landscape

with others of the same species or other species or with plants other than palms or with a mixture of palms and other plants. Single plants of many palms are admirable as lawn specimens; this is especially true of multitrunked kinds, many of which are comparatively low. Such plants can also be used to good effect widely spaced at regular intervals and with the space between them carpeted with grass or other groundcover to line driveways and avenues. Sometimes pairs of handsome specimens are used as accents or to flank gateways and other features. One of the most spectacular uses of single-trunked palms, such as the royal palm (*Roystonea*) and many others, is to line avenues. Such sorts are also attractive as single specimens. The use of palms in groups calls for special consideration of their silhouettes, of their growth, and type, size, color, and texture of foliage so that combinations that will remain pleasing through the years are

An avenue of royal palms (*Roystonea*) in Panama

achieved. One special type of group is the informal hedge or screen, a narrow band of planting with the plants set so closely that they eventually touch and form a masking mass of foliage. Kinds especially adapted for this use are the multistemmed and suckering palms, such as *Caryota mitis, Chamaedorea, Chrysalidocarpus lutescens, Ptychosperma macarthuri,* and *Raphis.*

For satisfying results it is necessary to know something of the rates of growth and of the likely ultimate sizes and appearances of plants set out. Specimens that appear small, charming, and entirely appropriate for the location at planting time can develop into quite lovely, but overpowering monsters within ten or twenty years. Gardeners unacquainted with kinds of palms should try to see mature specimens of types in which they may be interested before purchasing or planting. This may be possible to do in private gardens, public plantings, and nurseries locally, or a visit to a botanical garden or other special palm collection may be necessary. When designing a planting that is to include palms, the relative scale of the various elements in the picture, including buildings, trunks, foliage, and distances, must be given careful consideration. Only when this is done and the planting reflects the conclusions reached, can palms make their most satisfying contribution to the landscape. In the United States palm collections are established at the Fairchild Tropical Garden, Miami, Florida; the United States Plant Introduction Station, Miami, Florida; the Huntington Botanical Gardens, San Marino, California; Los Angeles State and County Arboretum, Arcadia, California; and at the Foster Botanical Garden in Honolulu. There is also a palm collection at the Federal Experiment Station, Mayaguez, Puerto Rico.

For outdoor planting the kinds of palms chosen should obviously depend upon climate. Roughly, very roughly, they fall into two groups: those that must have tropical conditions and stand little or no frost and those that tolerate several degrees of frost, at least for brief periods. Within these broad groupings are subgroups that for their well-being require very humid conditions or tolerate relatively dry atmospheres. A knowledge of the conditions that prevail where the plants grow as natives is a good basis for projecting their requirements under cultivation, but many palms are much more adaptable than a study of them at home indicates. Thus *Chamaerops humilis,* native to sunny North Africa and southern Europe, thrives in the open in Edinburgh, Scotland. Among the hardiest palms are *Acrocomia totai, Allagoptera campestris, Arecastrum romanzoffianum, Butia capitata, Caryota urens, Chamaedorea cataractarum, C. costaricana, C. erumpens, Chamaerops humilis, Erythea armata, E. brandegii, E. edulis, E. elegans, Jubaea chilensis, Jubaeopsis caffra, Livistonia australis, L. chinensis, L. mariae, Phoenix canariensis, P. dactylifera, P. reclinata, P. roebelinii, P. sylvestris, Rhapis excelsa, R. humilis, Rhapophyllum hystrix, Rhopalostylis sapida, Sabal causiarum, S. etonia, S. minor, S. palmetto, S. texana, Serenoa repens, Thrinax microcarpa, Trachycarpus caespitosus, T. fortunei, Trithrinax acanthocoma, T. brasiliensis, Washingtonia filifera,* and *W. robusta.*

The natural habitats of palms vary considerably not only as to climate, but also in the types of soil they favor, degree of sun or shade they receive, and other details. Under cultivation most show considerable adaptability. As a group they are sun-lovers, although a few, such as *Chamaedorea, Rhapis,* and *Geonoma,* are best with shade. They respond to deep, fertile, porous soils that drain well and are always reasonably moist. Even kinds that grow naturally under less benign conditions make more thrifty growth when treated generously. For most sorts any fairly good garden soil, well prepared at planting time, is satisfactory. Preparation consists of excavating a hole much larger than the root ball of the plant to be set in it, improving the undersoil if it is poor in organic matter by mixing with it generous amounts of compost, rotted manure, peat moss, decayed leaves, grass, or other suitable humus-forming material, and having on hand sufficient good topsoil with which has been incorporated similar organics and a slow-acting, complete fertilizer, to pack around the root ball and fill the hole. Whenever possible the planting pits should be dug and the backfill soil prepared several weeks before planting. This provides the opportunity for bacterial action to ameliorate the soil and to bring it into more perfect condition to encourage root growth. Most palms can be successfully planted at any time, but just prior to the beginning of a new growing season is considered ideal.

Palms transplant readily, even when large, if reasonable care is taken in moving them and in providing aftercare. Unlike the roots of dicotyledons, to which the vast majority of other trees belong, the roots of palms when severed do not sprout side roots, but gradually die back to their bases. For the establishment of the plant in its new location, a palm is entirely dependent upon the roots it retains without their tips being cut off and the new ones it develops later. The new roots push out from the base of the trunk. It is important when digging palms to take with them intact balls of soil. The size of the ball should bear a decent relationship to the height and massiveness of the tree. If it is conspicuously too small it will be difficult or impossible to secure the newly transplanted specimen against disturbance by winds and storms, and it is likely to be blown down or at least to have its new roots torn as a result of movement. Another reason for taking good root balls and being careful that they are not broken is that the newer short roots they contain do not reach to the outside of the ball and hence do not have their ends cut off in the process of digging. Upon these roots the plant depends for water until new roots develop.

Because cutting off many roots inevitably temporarily reduces the palm's ability to absorb water, even though plenty is available in the ground, it is wise to take steps to reduce loss of moisture from above ground parts. This is done by cutting off some of the older green leaves, preferably before digging the plant, alternatively immediately afterwards. The root balls are set a little deeper in the planting holes than they were originally and the soil used to fill around them is packed firmly without leaving voids. If there is any likelihood that the newly set specimen will be disturbed by wind, the trunk should be guyed or braced with planks or poles as soon as planting is finished. Guying is done by stretching wires taut from some distance up the trunk to hold-fast pegs in the ground or to other firm anchors. The parts of the wires that encircle the trunks should be threaded through pieces of garden hose or in some other way be prevented from cutting into the tissues of the palm, equivalent precaution should be taken with plank braces where they contact the trunk. Newly planted palms must be watered very thoroughly as soon as planting is completed and as often afterwards as necessary to keep the earth reasonably moist. Mulching with compost, peat moss, bagasse, or other organic material is very helpful.

Routine care of established palms may involve periodic watering in droughts and fertilizing each spring with a slow-acting fertilizer. In some regions where palms are grown, pale foliage or other abnormalities may result from deficiencies of various elements in the soil, such as potash, iron, boron, and manganese. If deficiencies are suspected, consult the local Cooperative Extension Agent or other knowledgeable person with local gardening experience and be guided accordingly.

Pruning is minimal. Chiefly it consists of taking off dead and unsightly leaves and old flower and fruit clusters. Because they may become a fire hazard, the skirts of old dead leaves of washingtonias are often removed by pruning or controlled burning. Although in some places this may be necessary, it is always regrettable because it seriously detracts from the full beauty of the trees. With palms that produce heavy fruits, such as coconuts, it is advisable to cut off the clusters before the fruits become large enough to menace people or property by dropping on them if the plants are located where this is likely to happen. In public areas it is often advisable to remove the vicious spines from the trunks of such palms as *Acrocomia* and *Veitchia,* as well as those on the leafstalks of *Phoenix.* This is a precaution against accidental injury to the public. Cluster-palms that have trunks that die after they have bloomed and fruited once, such as species of *Arenga* and *Caryota,* should have the trunks removed as soon as they have finished bearing and ceased to be attractive. With other cluster-palms, such as *Chamaedorea* and *Rhapis,* it is often desirable to limit their size by cutting away new stems that develop at their perimeters and occasionally to thin out older stems within the clumps to allow for the development of new sucker shoots without undue crowding.

Many palms are highly satisfactory for growing in greenhouses and conservatories and as decorative plants in homes, offices, stores, and other places indoors. Some lend themselves for cultivating in containers, which are placed outside in summer, thus enabling inhabitants of regions of cold winters to enjoy a flavor of the tropics outdoors during the warm season. Such plants must be stored indoors over the winter. Container-grown palms can be used effectively to ornament patios, terraces, steps, and similar areas.

Palms in containers: (a) *Chamaedorea elegans*

(b) *Howea forsterana*

(c) *Chamaerops humilis*

Except in conservatories, indoor palms are commonly accommodated in pots or tubs. In large greenhouses they may be planted in ground beds. All kinds can be grown in containers and in their early stages most are, but some are much more beautiful than others when so grown and some prosper only in greenhouses. The number handsome and durable enough to withstand room conditions is limited. Among the best are the many kinds of *Chamaedorea, Chamaerops humilis, Chrysalidocarpus lutescens, Coccothrinax argentata, Dictyospermum album, Hedyscepe canterburyana, Heterospathe elata,* species of *Howea, Licuala grandis, L. spinosa, Livistonia chinensis, L. rotundifolia, Paurotis wrightii,* species of *Phoenix, Pinanga kuhlii, Ptychosperma elegans, P. macarthuri, Reinhardtia gracilis, Rhapis excelsea, R. humilis, Thrinax parviflora, Trachycarpus fortunei,* and *Veitchia merrillii.*

Palms grown indoors are not choosey as to soil so long as it is porous enough for water to pass through freely. A fairly heavy medium containing a goodly amount of organic matter is desirable, and it should be comparatively coarse rather than sifted finely. If the topsoil in the potting or planting mixture is so fine and clayey that with the passage of time and under the influence of repeated watering it is likely to become compact to the extent that air does not readily reach the roots, add to the potting mixture a generous amount, say one-eighth part by bulk, or even more, of a porous non-organic material, such as broken crocks, crushed brick, or broken sandstone that will aerate the soil, but not later decay, as do leaf mold, peat moss, humus, and other organic additives. A soil consisting by bulk of four parts good topsoil, two parts leaf mold, peat moss, or humus, two parts coarse sand, one part dried cow manure (or one-third as much dried sheep manure), with coarse bonemeal added at the rate of a pint to each bushel, and a complete garden fertilizer, such as a 5-10-5, at one-half a pint to the bushel, is generally satisfactory. The amount of sand may be decreased if the topsoil is sandy or if a substantial amount of broken crocks, bricks, or sandstone is added to the mixture. Palms do not need repotting at frequent intervals. When past the seedling stage they usually go several years without this attention. They thrive better when their roots are crowded than when too much soil surrounds them. If the plant is thrifty, it is often better to sustain growth and health by fertilizing regularly than by shifting to larger containers. This is especially true once palms occupy pots or tubs of substantial size. It is beneficial to top dress each spring those that are not repotted. This is done by picking away as much of the old surface soil as possible without undue injury to the roots and replacing it with fresh, fertile soil.

Whether planted in ground beds or in pots or tubs, free drainage is essential. Waterlogged soil that does not admit air freely spells death to the roots of most palms, certainly to all commonly cultivated. Pots and tubs must have their drainage holes covered with a generous layer of broken crocks, clinkers, or other material to promote the free passage of surplus water.

When potting palms, the earth should be packed very firmly and care taken not to leave voids. The finished surface should be sufficiently below the rim of the container to leave ample room for watering. Because palms are usually repotted at long intervals only, it commonly happens that the soil of the old ball is exhausted of nutrients and has a poor physical structure at potting time. As much of it as possible, without doing excessive damage to the roots, should be picked out with a pointed stick before the plant is set in its new container. Specimens that have occupied their pots or tubs for many years sometimes become raised in their containers as a result of roots twisting around the bottom of the pot or tub, which makes watering difficult. When this has happened, the bottom of the root ball must be cut off with a sharp heavy knife, axe, or machete at repotting time.

Newly potted palms recover from the shock of transplanting most quickly when provided with a highly humid atmosphere, shade from sun, protection from air currents, and a temperature a few degrees higher than normal for several weeks following potting. This is especially true of kinds, such as howeas, that are natives of the humid tropics, it is less true of semidesert palms, such as phoenixes. If the roots have been much disturbed or substantial portions of them cut off it is especially necessary to provide the environmental conditions suggested.

Watering palms in containers calls for keeping the soil in a satisfactory moist condition at all times without it remaining muddy and waterlogged. The rule when watering is to apply enough to thoroughly saturate the entire root ball and then to give no more until the soil definitely approaches dryness. The lengths of the intervals between waterings depends upon several variables, such as how well rooted the plant is, the prevailing temperature, the relative humidity of the atmosphere, and growth activity. The soil may be allowed to become somewhat drier before water is given in the winter than during the active spring-to-fall growing period. It is sometimes difficult to completely saturate a palm ball that consists of almost a solid mass of roots by watering from above. In such cases the best plan, if the plant is small enough, is to immerse its container to its rim in water for a period of five or ten minutes. Container-grown palms that have filled their available soil with healthy roots benefit from being given dilute liquid fertilizer (any complete soluble fertilizer compounded for houseplants and greenhouse plants is satisfactory) at about two-week intervals from spring through fall. This "feeding" program may be omitted in winter or followed at monthly intervals only.

All palms in pots or tubs succeed best in a humid atmosphere. Excessive dryness of the air results in the leaves browning at their ends, a condition frequently observed in palms, in hotels, banks, stores, homes, and elsewhere where the atmosphere is too dry, especially in winter when the artificial heat used to keep the air temperature acceptable for humans lowers the relative humidity. Any measures that increase humidity benefit palms. It is helpful to wet their foliage with a fine spray of water two or more times a day. They should be kept out of drafts.

Indoor palms need good light, but not necessarily direct sun; indeed they should be protected from bright summer sun, but may be exposed in winter. As a group they thrive where the minimum temperature is 60 to 70°F, or even a little higher; some do well where the minimum is lower than 60°F. Recommended temperatures are given under the entries for each genus.

Propagation by vegetative means is limited by the fact that the stems have no cambium layer. Because of this palms cannot be increased by cuttings, grafting, or budding. Division, including the separating of sucker shoots, is practicable with some cluster-stemmed kinds, but these are comparatively few. In rare instances air layering is possible with thin-stemmed palms that naturally produce aerial rootlets, such as some kinds of *Calamus.* The vast majority of palms have a single stem with one growing point at the apex and can be propagated only from seeds.

Seedling palms

The seeds do not retain their vitality for long; therefore it is important to secure freshly harvested ones. Seeds of palms native to regions of even temperature and rainfall or of swampy tropical areas often fail to grow if they are not planted within two or three weeks of ripening. Others, native where changes of seasonal temperature or availability of moisture are more marked, may grow after being stored for two or three months or in some cases longer. Usually the percentage of success decreases with the age of the seeds. As soon as palm fruits are gathered, the fleshy outer layer that surrounds the seeds of most kinds is removed, the seeds are washed and air-dried and then are sown or stored. If they are to be stored they should be dusted lightly with a powdered fungicide and mixed with very slightly dampened peat moss, vermiculite, or sphagnum moss to assure moisture enough to prevent drying, but not to induce germination. Storage temperatures

Paeonia mollis

A single-flowered herbaceous peony

Paeonia obovata

Pansies are popular spring blooms

A double-flowered herbaceous peony

Papaver nudicaule variety

Paphiopedilum insigne

Papaver orientale variety

should be considerably lower than those recommended for sown seeds, which are given in this Encyclopedia under the generic names of the various palms.

Well-drained pots, pans, or flats are suitable receptacles in which to sow. A mixture of one part topsoil, two parts peat moss, and two parts sand with a little powdered charcoal mixed in is a satisfactory seed soil. It is advantageous to heat-sterilize (pasteurize) it before use. Press the soil moderately firm and level its surface, then scatter the seeds, press them in lightly, and cover to about their own depth with additional soil. Next, water thoroughly with a fine spray and place the containers in a greenhouse or other place where a favorable temperature (optimums for germination are indicated in the discussion of each genus of palms in the Encyclopedia) is maintained. Do not permit the soil to become dry at any time, but avoid maintaining it in a soggy condition.

Most palm seeds germinate fairly uniformly, but some, especially if a little old, start erratically and several months may elapse between the appearance of the first and last seedling of a batch. As soon as the first leaf is fully grown and the second shows signs of appearing, transplant the young plants individually to small pots, using a sandy, peaty soil. To delay beyond this stage increases the danger of damage or even death of the seedlings. At this stage of growth the move can be made with minimum damage to fine roots and while the plant is still drawing from foodstuffs stored in the seed and is not entirely dependent upon what its roots absorb for nourishment. In their young stages all palms, including those that at later ages thrive in full sun, grow best in shade. They abhor drafts and dry atmospheres and respond to temperatures a little higher than those recommended for longer-established specimens of the same kind.

There are few exceptions to the general suggestions made above. Coconuts and double-coconuts (*Lodoicea*) are not covered to the depth of the seed with soil, but are planted horizontally with their upper surface just showing above the soil. A few palms, such as *Hyphaene* and *Jubaeopsis*, have the curious habit of sending down from the seed a primary rootlike organ or hypocotyl that may extend to a depth of three feet or more and upon which develops at a considerable distance below the surface the new shoot growth. Sowing seeds of these palms in shallow containers is disastrous, very deep ones, such as a drainpipe stood on end, or ground beds are the only practical way to provide for these. The need for deep containers is noted in directions for cultivation given in the discussions of each palm genus to which it applies.

The dividing of palms, such as certain kinds of *Chamaedorea* and *Rhapis* that spread by underground runners and have numerous stems, is comparatively simple and not basically different from division as applied to many other clump-forming plants. The beginning of a new growing season, normally early spring, is the best time for the operation. The plant should be carefully split into pieces with as little damage to the roots as possible and with care taken to retain as many roots with each division as practicable. The roots must not dry. The divisions are potted into well-drained containers just large enough to hold the roots without great crowding, in very sandy, peaty soil. Thorough watering follows and the propagations are placed in a greenhouse or other place where they are shaded from direct sun and are in a highly humid atmosphere and a temperature a little higher than is considered optimum for their kind. There they remain until they are again well rooted.

The division of other multistemmed palms that make offsets or suckers close to the mother plant, but do not spread by underground runners is more tricky. Yet date palms, which belong in this group, are commonly propagated in this way, and others can be. The percentage of success varies with the kind of palm as well as with the skill of the operator. Briefly, the procedure is this. Dig around the sucker it is contemplated to separate to discover whether it has roots of its own. If it has, it may be removed without further preparation. If it has not, cut partly through the junction of the sucker and its parent and then heap around and over the wounded part sandy, peaty soil. Keep the mound constantly moist and occasionally fertilize it with dilute liquid fertilizer. The combination of being partly cut off from the mother plant as a source of food and of moisture and nutrients being available around the wounded stem encourages the production of roots from the sucker.

Rooted suckers are carefully severed from the parent, dug with as little damage to the roots as possible and, if practicable with an intact ball of earth surrounding them, and potted and treated as advised above for clump-forming palms that have underground runners. To check loss of moisture from the foliage, it is advisable to remove some of the older leaves from suckers at the time they are taken from the parent. This reduces the area from which loss of moisture occurs.

In regions where palms flourish outdoors, divisions may be planted directly in the ground as soon as they are severed from the parent, but if it is possible to provide greenhouse care chances of success are increased. Outdoors, newly set out divisions should be kept well watered, be afforded temporary shade, and be sheltered from wind. Spraying the foliage with water several times a day is helpful. A tentlike covering of polyethylene plastic film erected over newly planted divisions, in effect a miniature greenhouse, conserves moisture, assures a humid atmosphere, and protects against wind. Care must be taken, if there is danger of the temperature rising too high under the tent, to ventilate to prevent this.

Diseases and pests of palms include several rots, wilts, and leaf spots. These, commonest on outdoor grown plants, are much less likely to affect indoor specimens. If disease is suspected, adequate samples of affected parts should be sent to a Cooperative Extension Agent or to the State Agricultural Experiment Station for examination, diagnosis, and recommendations. The chief pests of palms are scale insects, which are troublesome both indoors and outdoors. A keen watch must be maintained for the first signs of infestation and prompt control measures, such as spraying with a scalecide, instituted. Mealybugs infest many palms and can be controlled by insecticidal sprays. Yet a third group of insect enemies are certain aphids, also easy to eliminate by spraying with insecticide. Other insects that may be troublesome to outdoor palms are leaf skeletonizers and other caterpillars, bagworms, ambrosia beetles, palmetto weevils, royal palm bugs, thrips, mites, and termites. In all cases it is advisable to consult local authorities or the State Agricultural Experiment Station regarding the latest controls.

The names of palm genera treated separately in this Encyclopedia are given under the entry Palmae. Here are the common names of many kinds with their species or genus names in parentheses: African oil palm (*Elaeis guineensis*), Alexandra palm (*Archontophoenix alexandrae*), American oil palm (*Elaeis oleifera*), Arikury palm (*Arikuryroba*), Assai palm (*Euterpe oleracea*), Australian cabbage palm (*Livistona australis*), babassu palm (*Orbignya barbosiana*), bangalow palm (*Archontophoenix*), barbel palm (*Acanthophoenix*), barrel or belly palm (*Colpothrinax wrightii*), Beatrice or step palm (*Archontophoenix alexandrae beatricae*), betel or betel nut palm (*Areca catechu*), black palm (*Astrocaryum standleyanum* and *Normanbya normanbyi*), bottle palm (*Colpothrinax wrightii* and *Hyophorbe lagenicaulis*), cabbage palm (*Euterpe oleracea* and *Roystonea oleracea*), cane palm (*Calamus*), carnauba palm (*Copernicia prunifera*), cherry palm (*Pseudophoenix*), Chilean syrup palm (*Jubaea*), Chinese fan or fountain palm (*Livistona chinensis*), Christmas, Manila, or Merrill palm (*Veitchia merrillii*), coconut palm (*Cocos nucifera*), cohune palm (*Orbignya cohune*), coquillo palm (*Astrocaryum alatum*), coquito palm (*Jubaea chilensis*), corozo palm (*Acrocomia*), coyure palm (*Aiphanes*), date palm (*Phoenix*), doum palm (*Hyphaene*), European fan palm (*Chamaerops humilis*), everglade palm (*Acoelorrhaphe wrightii*), fan-leaved cluster or zombi palm (*Zombia*), fish-tail palm (*Caryota*), Franceschi palm (*Brahea elegans*), gingerbread palm (*Hyphaene thebaica*), grigri palm (*Aiphanes corallina*), gru-gru palm (*Acrocomia*), Guadalupe palm (*Brahia edulis*), hesper palm (*Brahia*), ivory-nut, vegetable ivory, or tagua palm (*Phytelephas*), king palm (*Archontophoenix*), lady palm (*Rhapis*), latan palm (*Latania*), loulu palm (*Pritchardia*), manac palm (*Euterpe*), mazari palm (*Nannorrhops ritchiana*), Mexican blue palm (*Brahia armata*), murumuru palm (*Astrocaryum murumuru*), needle palm (*Rhapidophyllum hystrix*), nikau palm (*Rhopalostylis sapida*), nipa palm (*Nypa fruticans*), overtop palm (*Rhyticocos amara*), palmetto palm (*Sabal* and *Serenoa*), Palmyra palm (*Borassus flabellifer*), parlor palm (*Chamaedorea elegans*), peaberry palm (*Thrinax*), peach palm (*Bactris gasipaes*), petticoat palm (*Copernicia macroglossa*), piassava palm (*Attalea funifera*), piccabeen palm (*Archontophoenix cunninghamiana*), pignut palm (*Hyophorbe*), princess palm (*Dictyosperma*), Puerto Rican hat palm (*Sabal causiarum*), queen palm (*Arecastrum romanzoffianum*), raffia palm (*Raphia*), rattan palm (*Calamus* and *Daemonorops*), rock or soyal palm (*Brahea*), royal palm (*Roystonea*), sago palm (*Metroxylon sagu*), sealing wax palm (*Cyrtostachys*), seamberry palm (*Coccothrinax*), silver palm (*Coccothrinax argentata*), silver-saw palm (*Acoelorrhaphe wrightii*), silver-thatch palm (*Coccothrinax jamaicensis*), spindle palm (*Hyophorbe verschaffeltii*), sugar palm (*Arenga pinnata*), talipot palm (*Corypha umbraculifera*), thief palm (*Phoenicophorium borsigianum*), toddy palm (*Caryota urens*), trash palm (*Scheelea curvifrons*), tucum palm (*Astrocaryum vulgare*), umbrella palm (*Hedyscepe canterburyana*), walking stick palm (*Linospadix*), wax palm (*Ceroxylon* and *Copernicia*), windmill palm (*Trachycarpus fortunei*), wine palm (*Caryota, Jubaea,* and *Pseudophoenix*), and yatay palm (*Butia*).

Plants that are not palms but that have vernacular names of which the word palm forms a part include the Australian nut-palm (*Cycas media*), sago-palm (*Cycas revoluta*), snake-palm (*Amorphophallus rivieri*), and travelers'-palm (*Ravenala madagascariensis*).

PALO VERDE. See Cercidium and *Parkinsonia aculeata*.

PAMIANTHE (Pami-ánthe). Named to honor the English horticulturist Major Albert Pam, who died in 1955, *Pamianthe,* of the amaryllis family AMARYLLIDACEAE, is closely related to spider-lilies (*Hymenocallis*) and *Leptochiton.* From both it differs in having numerous winged seeds in each compartment of the ovary. Like the basket flower or ismene group of *Hymenocallis,* but unlike other sorts of *Hymenocallis* and unlike *Leptochiton,* the basal parts of the leaves of *Pamianthe* sheath each other to form a thick false stem. There are three species of *Pamianthe,* endemic to western South America.

These plants have roundish bulbs and linear leaves with rounded keels. The flattened, leafless flower stalk emerges from the top of the false stem. It has up to four white flowers with six perianth lobes (petals) and a bell-shaped corona or staminal cup that joins the basal parts of the six stamens together and gives a somewhat daffodil-like appearance to the blooms. The free parts of the stamens extend from the rim of the staminal cup and are incurved.

Native to Peru, ***P. peruviana*** is deciduous. It has a bulb about 2 inches in diameter. At the top of the false stem, which is up to 1 foot long, the five to seven leaf blades form a more-or-less two-ranked tuft. They are pointed-linear and up to 1 foot long by ¾ inch wide. The fragrant flowers have slightly curved green perianth tubes, about 4½ inches long. The petals, 4 to 4½ inches long by 1 inch broad, have a yellowish-green band down their centers. The staminal tube is 3 inches long and has six short, petal-like lobes. It measures about 2 inches across.

Pamianthe peruviana

Garden and Landscape Uses and Cultivation. These are the same as for basket flowers, the ismene group of *Hymenocallis.*

PAN. See Air and Air Pollution. See also Pots and Pans.

PANAMA. As part of the common names of plants, Panama is included in these: Panama hat plant (*Carludovica palmata*), Panama rubber (*Castilla elastica*), and Panama tree (*Sterculia apetala*).

PANAMIGO is *Pilea involucrata.*

PANAX (Pàn-ax) — Ginseng. The tropical and subtropical shrubby plants previously known by this name are *Polyscias.* The genus *Panax* consists of eight species of herbaceous perennials of North America and Asia. It belongs to the aralia family ARALIACEAE and has a name alluding to the supposed virtues of some species, derived from the Greek *pantos,* all, and *akos,* a cure, a panacea.

Panaxes have deep-seated, aromatic, tuberous roots from which are sent up branchless stems bearing at their tops a single circle of usually three leaves with their leaflets spreading from the ends of their stalks in handlike manner (palmately). Above the foliage rises a longish-stalked, terminal umbel of tiny white or greenish flowers, each with a five-toothed calyx, five petals, five stamens, and two or three styles. The fruits are flattened or angled berries containing two or three seeds.

American ginseng (***P. quinquefolius***) is a scarce inhabitant of rich woodlands from Quebec to Minnesota, Georgia, and Oklahoma. It has spindle-shaped tubers. From 8 inches to 2 feet tall, its stems bear leaves of usually five-pointed, oblong-obovate to obovate, coarsely-toothed, prominently stalked, leaflets 1½ to 6 inches long. The center leaflet is the largest, the two basal ones smallest. The greenish-white flowers, with usually two styles, are succeeded by bright red berries ⅓ inch or slightly more in diameter. Dwarf ginseng (***P. trifolius***) has globose tubers, stems up to 9 inches tall, leaves of three or five stalkless or nearly stalkless, lanceolate, elliptic, or oblanceolate, finely-toothed leaflets up to 3½ inches long, and white to pinkish flowers with usually three styles. The yellow berries are ⅕ inch in diameter. This occurs in rich woodlands from Nova Scotia to Minnesota, Georgia, and Indiana. Asian gin-

Panax trifolius

seng, ***P. pseudoginseng*** (syn. *P. schinseng*), a native of Korea and Manchuria, is similar to American ginseng, differing chiefly in its leaves being more finely-toothed.

Garden Uses and Cultivation. As ornamentals the species described are appropriate for woodland gardens and other shady places where the soil contains considerable decayed organic matter and is moist, but not wet. They make no great display, but are pretty in an unassuming way, and interesting. They benefit from being mulched annually with leaf mold or well-decayed compost. Propagation is by seed and by division. For the cultivation of ginseng commercially see Ginseng.

PANCRATIUM (Pan-cràtium). Pancratiums are little known in North America. Plants that pass under that name usually belong in the related genus *Hymenocallis*. The chief difference is that each of the three compartments of the ovary in *Pancratium* contains many ovules one above the other, and in *Hymenocallis* each has usually two, occasionally up to eight, side by side. The genus *Hymenocallis* is entirely a New World one. As natives, pancratiums are confined to warm parts of the Old World. They belong in the amaryllis family AMARYLLIDACEAE. The name of the genus is from the Greek *pan*, all, and *kratos*, powerful, and alludes to supposed medicinal qualities.

Pancratiums are stemless plants with large bulbs, linear or strap-shaped leaves, and few to many white flowers in umbels atop leafless stalks. The blooms have cylindrical perianth tubes that broaden at their tops and six narrow, spreading petals (properly tepals). A membranous cup, uniting the stamens, forms a corona in the center of the flower, and the free parts of the stamens project from its margin. The style ends in a slightly three-lobed stigma. The fruits are three-compartmented capsules. There are fifteen species.

Native to the Mediterranean region, ***P. maritimum*** has evergreen, linear, glaucous leaves 2 to 2½ feet in length, and flattened flower stalks 1 foot to 2 feet tall. There are five to ten deliciously fragrant blooms in each umbel, each with a short stalk. Their perianth tubes are 2 to 3 inches long, the petals 1½ to 1¾ inches long. The staminal cup is prominent and has triangular teeth between the ¼-inch-long free parts of the stamens, which gives a frilled effect to the cup. This species blooms in summer. From southern Italy, Sicily, and Sardinia comes ***P. illyricum,*** which has deciduous, broadly-strap-shaped, glaucous leaves, a little over 1½ feet long, and flattened flower stalks about the same length. There are six to twelve short-stalked, very fragrant blooms in each umbel. They develop in late spring and have 1-inch-long, green perianth tubes and narrowly-lanceolate, spreading petals almost 2 inches long. Their staminal cups are shorter than those of *P. maritimum*. The Canary Island ***P. canariense*** has glaucous, sword-shaped leaves, 1½ to 2 feet long, and flattened flower stalks of approximately the same length. The six to ten blooms in each umbel have long individual stalks. Their perianth tubes are about 1½ inches in length, the petals rather shorter, and the staminal cup much shorter than the petals. Between the free parts of the stamens are two teeth. This is a late summer bloomer. Native to tropical Asia, ***P. zeylanicum*** has lanceolate leaves not exceeding 1 foot in length and shorter flowering stalks each with a single bloom with a 2-inch-long perianth tube, petals about as long, and a staminal cup approximately ¼ inch deep.

Garden and Landscape Uses and Cultivation. Pancratiums have the same garden and landscape uses as spider-lilies (*Hymenocallis*) and respond to essentially the same care. Propagation is by offsets, and by seed in soil not kept too wet.

PANDA PLANT is *Kalanchoe tomentosa*.

PANDANACEAE—Screw-Pine Family. This Old World family of evergreen monocotyledons comprises three genera, and 700 species, most inhabitants of the tropics, a few of subtropical and warm-temperate regions. Palmlike in relationship and aspect, these are generally shrubs or trees, less often climbers. They have stems or trunks, frequently freely-branching and producing aerial roots that upon reaching the ground often establish themselves as props or stilts that support the trunk. The leaves, mostly long and swordlike, sometimes proportionately wider, are rigid and leathery, those of most sorts with sharp spines along their margins and sometimes on the undersides of the midribs. They are spirally disposed in tufts of many at the branch ends. The reproductive organs, the stamens and pistils, are separate, not in definite flowers. The fruits are large cone- or ball-like assemblages of carpels. Cultivated genera are *Freycinetia* and *Pandanus*.

PANDANUS (Pan-dànus)—Screw-Pine. The botanical name of this tropical and subtropical genus is derived from its Malayan one. Its popular name alludes to the screwlike, strongly spiraled arrangement of its leaves, which somewhat resemble those of pineapples. Estimates of the number of species range from 200 or fewer to 600. All are natives of Africa, Asia, Australia, or islands of the Pacific and Indian oceans. Together with two other genera, *Pandanus* constitutes the screw-pine family PANDANACEAE.

Screw-pines are evergreen trees and shrubs with erect or more rarely prostrate trunks, branched or not. Each trunk and branch ends in a crown of long, stalkless, usually swordlike, toothed or more rarely toothless, parallel-veined leaves. Very commonly the trunks are braced against wind and other disturbing influences by stilt or prop roots that originate several feet above the ground, angle downwards and outwards, enter the earth, and become firmly anchored. Male and female flowers are in dense spikes on separate plants. Those of males are branched in panicle-like fashion, and are terminal or axillary, female spikes are always terminal and may be branched or not. In neither sex are sepals or petals present. Male flowers have numerous stamens, and female blooms sometimes aborted stamens (staminodes) in addition to a solitary ovary. The ovaries of neighbor flowers may be separate or united. Female flowers are succeeded by angular, woody or fleshy fruits called drupes, tightly clustered and sometimes united in large, pendulous, ball- or pineapple-like aggregations that are ordinarily thought of as the fruits.

Pandanus, undetermined species

Extensive stilt roots of an undetermined species of *Pandanus*, Castleton Gardens, Jamaica, West Indies

Fruits of an undetermined species of *Pandanus*

Stilt roots of *Pandanus leram*

In tropical countries, screw-pines have many economic uses and play important parts in the lives of the people. The fruits of some, notably those of the Nicobar-breadfruit (***P. leram***), are eaten, and both roots and leaves are sources of valuable fibers used for ropes, cordage, baskets, mats, nets, thatching, and other purposes. A native of the Nicobar Islands, the Nicobar-breadfruit is a large, much-branched tree with many stilt roots and long, bluish-green, spiny-margined, pendulous leaves. This species is notable for its big fruits composed of wedge-shaped drupes about 5 inches long.

Native of Hawaii, other islands of the Pacific, southeast Asia, and Australia, variable ***P. odoratissimus*** (syn. *P. tectorius*) is a picturesque, well-branched tree up to 20 feet tall or taller, with few to many prop roots. Its prickly, spine-tipped leaves are 3 to 6 feet long. The male flowers are fragrant and in panicles about 1 foot long surrounded by edible bracts. The spherical to ovoid yellow to red fruits are about 8 inches long. The leaves of a smooth-leaved variety, *P. o. laevis,* are esteemed in Hawaii for making mats, baskets, and other articles.

Popular as an ornamental, ***P. utilis,*** of Madagascar, is a reddish-trunked, stoutly-branched tree 50 to 60 feet tall or taller, and with many prop roots. Its erect, glaucous-bluish-green, red-spined leaves are about 3 feet long by 3 inches wide. The solitary, long-stalked, pendulous fruits are 6 to 8 inches in diameter.

Pandanus utilis

Stilt roots of *Pandanus utilis*

Pandanuses with variegated foliage, much cultivated as ornamentals but because they do not flower or fruit are of uncertain botanical assignations, are *P. veitchii* and *P. sanderi* and their variants. Native to Polynesia, ***P. veitchii*** has thinnish, recurved, long-pointed, small-spined, glossy, green leaves up to 3 inches broad, with wide marginal bands of creamy-white. A compact and especially desirable variety, *P. v. compactus* has denser heads of stiffer, darker green leaves with clearer white variegation. The plant grown in Hawaii and elsewhere in the tropics as ***P. variegatus*** may belong here. Vigorous,

Pandanus veitchii

and with its youngest foliage deep yellow, and the more mature bright green leaves marked longitudinally with bands of yellow, ***P. sanderi,*** a native of the island of Timor, has finely-toothed leaves up to 4 feet long by 2 inches wide or wider, which usually bend and hang downward at or above their middles. Similar *P. s. roehrsianus* is perhaps more strongly variegated.

Without spines on its gracefully arching leaves, ***P. baptistii*** is a short-trunked native of New Britain Islands. Its bluish-green, 1-inch-wide leaves are rather indistinctly striped with one or more longitudinal lines of creamy-white or yellow. Very different are the broad, stubby, bright lustrous green, deeply-channeled leaves, blunt at their ends except for a tail-like apex, and with marginal spines, of ***P. dubius*** (syn. *P. pacificus*). This is a short-trunked, compact native of islands of the

Pandanus dubius

Pacific. Shrubby ***P. polycephalus,*** often misnamed *P. pygmaeus,* is about 15 feet in height and has prickly leaves about 3 feet long by 1 inch wide. Usually not appreciably over 2 feet in height, ***P. pygmaeus*** (syn. *P. graminifolius*), of Madagascar and

Pandanus dubius, a young specimen

Mauritius, spreads widely by means of horizontal, rooting stems. It has long, slender, white-spined leaves, glaucous on their undersides.

Garden and Landscape Uses. Variegated-leaved screw-pines and young specimens of a few others are commonly grown as greenhouse plants and in pots and tubs for indoor decoration. For this last use, the thorny leaves of most kinds are disadvantageous. It is wise to locate such sorts where they are unlikely to be brushed by passersby. In the tropics and subtropics both green and variegated foliaged screw-pines are stunning for creating bold, palmlike effects and may be grouped or used singly to good purpose. They thrive near the sea as well as inland and are effective for holding shores and banks against erosion.

Cultivation. Plenty of heat and high humidity favor the best growth of screw-pines, yet they get along with moderately humid atmospheres if temperatures are congenial and if their roots have access to fair supplies of moisture. Full sun or light shade suits. Indoors a minimum night temperature in winter of 60°F is acceptable. Day temperatures then, and night and day temperatures at other seasons, should be higher. Ordinary, fertile, well-drained soil satisfies. In containers, well-rooted plants, fertilized regularly with dilute liquid fertilizer, remain healthy for several years without repotting. Old screw-pines tend to raise themselves out of their containers so that when they are repotted it is necessary to chop off the basal portion of the root ball. Plants subjected to this treatment, should, when potting is completed, be placed in a highly humid atmosphere, kept warm, and watered only moderately until new roots are well developed. Watering of container specimens should be done by drenching the soil each time it approaches dryness without maintaining a constantly saturated condition. Propagation is commonly and easily done by offsets or by suckers and cuttings. Seeds, when procurable, are also satisfactory. Before sowing they should be soaked in water for about twenty-four hours. These are planted in pots just large enough to conveniently hold them and buried almost to the rims of the pots in a tropical propagating greenhouse or in a similar environment. If the bed is warmed to a few degrees higher than air temperature it is an advantage.

PANDOREA (Pandòr-ea)—Wonga-Wonga Vine, Bower Plant. Eight nonhardy, evergreen, perennial woody vines, natives from Malaysia to Australia, belong in *Pandorea,* of the bignonia family BIGNONIACEAE. The name commemorates Pandora of Greek mythology.

Having neither tendrils nor aerial roots, pandoreas climb by twining their stems around suitable supports. Their opposite leaves have a number of leaflets including a terminal one. Pink or white, their beautiful flowers, in terminal panicles, have small, five-lobed calyxes, funnel- to bell-shaped corollas with five spreading lobes (petals), four nonprotruding stamens, and a slender style with a two-lobed stigma. The fruits are oblong pods containing broad-elliptic, winged seeds.

The wonga-wonga vine (***P. pandorana***), of Australia and New Guinea, climbs high and is a fast grower. When mature it has leaves of three to nine or occasionally more, ovate to lanceolate, toothed or smooth-edged, glossy leaflets 1 inch to 3 inches long and, for a short period in spring, many flowered panicles of ¾- to 1-inch-long, ½-inch-wide, yellowish-white blooms spotted in their throats with violet, red, or brown. Variety *P. p. rosea,* has pale pink blooms. Young plants of wonga-wonga vine have leaves completely different from those of older specimens. They are finely-divided and almost fernlike. Plants in this stage have been called *P. filicifolia,* a name not accepted botanically.

The bower plant of Australia (***P. jasminoides***) grows to heights of 15 to 40 feet. It has leaves of five to nine ovate to lanceolate, almost stalkless, toothless leaflets 1 inch to 2 inches long. The blooms, 1¼ to 2 inches long and about 2 inches across, are produced in few-flowered panicles over a long winter, spring, and summer season. They are white, usually with pink throats, in variety *P. j. rosea,* pink with deeper-colored throats, in variety *P. j. alba,* pure white. Plants previously known as *P. brycei* and *P. ricasoliana* belong in *Podranea.*

Garden and Landscape Uses. Pandoreas are first-class vines for tropical and subtropical climates and for large greenhouses and conservatories. The wonga-wonga vine is highly successful near the sea. In California it withstands for short periods temperatures down to 26°F without appreciable harm. The bower plant will endure 24°F. Both are excellent for covering walls, pergolas, archways, pillars, and other supports. Wonga-wonga vine may also be used as a groundcover for level land and banks. The flowers of the bower vine are showy and handsome, those of the wonga-wonga vine make little display. The chief attraction of this last is its foliage.

Cultivation. Pandoreas need rich earth and full sun. In soils of low fertility they grow poorly, are likely to have sickly, light-colored foliage, and tend to be bare of leaves on their lower parts. The wonga-wonga vine withstands considerable drought, but grows more vigorously where the moisture supply is ample and constant. Pruning to keep the plants within bounds and their stems from becoming too densely crowded should be done as soon as flowering is through. Propagation is easy by seed, cuttings, and by layering. The chief pests are mealybugs and scale insects.

PANIALA is *Flacourtia cataphracta.*

PANICLE. A panicle is a branched, frequently more or less conical, natural disposition or arrangement (inflorescence) of stalked flowers with the blooms of the branches generally in racemes or less elongated clusters.

PANICUM (Pánic-um) — Broom-Corn-Millet, Switch Grass, Witch Grass. Some plants popularly called *Panicum* in gardens belong to other genera. The one known as *P. variegatum* is *Oplismenus hirtellus variegatus.* The one called *P. palmifolium* is *Setaria palmifolia.* The genus *Panicum* includes more than 400 species of the grass family GRAMINEAE, natives of tropical, subtropical, and temperate regions. Kinds cultivated as agricultural crops include broom-corn-millet (*P. miliaceum*) and several species of value for forage. The name is an ancient Latin one for millet.

Panicums are hardy and nonhardy annual and perennial grasses of diverse appearance. They have linear to elliptic, flat or rolled leaf blades and loose or dense flower panicles. The spikelets that compose the panicles are stalked and without awns (bristles). There is one fertile and one sterile flower in each spikelet.

Switch grass (***P. virgatum***), native from southern Canada to central America, a perennial 3 to 7 feet tall, spreads by underground rhizomes. Its pointed-linear, flat leaves, 6 inches to 2 feet long, and ⅛ to a little over ½ inch wide, are nearly hairless. Up to 2 feet in length by nearly one-half as wide, the diffuse panicles of flowers have stiff, spreading, divided branches. They are attractive in late summer.

Witch Grass (***P. capillare***), an annual 6 inches to 2 feet in height, is native from southern Canada to Florida and Texas.

Panicum virgatum

Tufted, it has spreading or erect, often branched stems, and flat, linear to narrowly-lanceolate, stiffly-hairy leaf blades up to 1 foot long by ¼ to nearly ¾ inch wide. The green or purple flower panicles are large and loose. They come in late summer.

Garden and Landscape Uses and Cultivation. The species described above are grown as garden ornamentals and for their panicles of bloom, which, fresh or dried, are useful for bouquets. For such use they must be cut before they mature, otherwise the spikelets fall. Panicums are easily cultivated in ordinary soil in sunny locations and are raised from seed sown outdoors in spring, or the switch grass, by division in spring or early fall. Broom-corn-millet is occasionally included in educational displays of plants useful to man. An annual about 4 feet tall, its seeds are sown outdoors in spring where the plants are to remain.

PANS. See Pots and Pans.

PANSIES and GARDEN VIOLAS. Few plants are more commonly grown or generally appreciated than pansies, but bedding violas, which are similar in general aspect, because of their positive need for cool, humid summers are little known in North America. Visitors to northern Europe are invariably impressed with the splendid displays of magnificent violas presented throughout the summer in parks and gardens there. In addition there is a group of small-flowered garden violas or tufted pansies that succeed better in somewhat warmer climates.

Pansies as we know them are not especially old-timers. It was not until the second decade of the nineteenth century that hybridizing simple species and making selections from their offspring and then recrossing and selecting again began. The project was started by an Englishman named Thompson, gardener to Lord Gambier. Within ten or fifteen years a few hundred varieties of pansies were being cultivated. Later, French horticulturists engaged in the work and modern pansies are largely derived from the sorts they raised. And the word pansy is a modification of the French pensee, thoughts. It derives from the same Latin root as pensive. In the once popular "language of flowers," pansies mean thoughts, and so poor mad Ophelia of Shakespeare's Hamlet refers to an ancestral species as she plucks the flowers from her nosegay and says "There's rosemary, that's for remembrance: pray you love remember: and there is pansies, that's for thoughts."

The European heartsease or johnny-jump-up (*Viola tricolor*) provided the starting point and most important element in this breeding program, but other species, *V. lutea* and probably *V. altaica,* were involved. To the hybrid swarm we now grow the name *V. wittrockiana* (syn. *V. tricolor hortensis*) is applied. Except for the pretty facelike markings of most pansy flowers, it is difficult to relate these often splendid blooms 2 to 4 inches wide to the miniatures of their forebears. Modern pansies come in an extraordinary range of pale and rich colors and combinations including white, cream, yellow, orange-yellow, pinkish-purple, purple, blue, brownish-red, and maroon.

An ancestral parent of the pansy (*Viola tricolor*)

Pansies grow well in a wide variety of garden soils, but are at their best only in moderately fertile, well-drained, cool, and reasonably moist ones. They are intolerant of heat and dryness. Just a little shade from the strong sun of late spring is appreciated. Because of these needs, in all except regions of cool, humid summers, pansies are for practical purposes spring-bloomers raised to give a fine display from the end of winter, or in warm climates during winter, until debilitated by the stresses of late spring or early summer weather. Then they are discarded.

Modern pansies: (a) In the garden

(b) As cut flowers

A narrow border of pansies

Quite marvelous companions for hyacinths, tulips, and other spring-flowering bulb plants, pansies are splendid in patches at the fronts of mixed flower beds, in beds and borders interplanted with bulb plants or by themselves, and for window and porch boxes, urns, and similar containers. Their blooms lend themselves for picking for use in simple arrangements.

A flower bed in spring featuring pansies edged with lamb's ears (*Stachys byzantina*)

Pansies are pretty pot plants

In greenhouses selected strains of pansies with long-stalked blooms are well worth growing for winter cut blooms and regular strains as winter-blooming pot plants.

For best spring blooming grow pansies as biennials. There are annual strains offered and even regular sorts can be flowered as annuals if seeds are started in January or February in a greenhouse to give plants for setting in the garden in early spring. But it is far better to start seeds in a cold frame or nicely prepared bed outdoors in late summer or early fall to produce plants for blooming in spring, or in mild climates through the winter.

The time of sowing is in cold regions very important. The objective is to have sizable plants before growth ceases in fall, but not ones that are too big. Lush, overgrown examples are likely to die in winter or at best be difficult to transplant and floppy and unsatisfactory after they are moved to their flowering quarters. In the vicinity of New York City the last week in July is the ideal time to sow. Where growing seasons are longer later sowings are in order.

The seed bed, in a cold frame or outdoors, should be in a cool, shaded place on the north side of a wall or building perhaps or where shade from tall trees moderates the most intense light and heat. Failing that, other shade must be provided.

Have the soil mellow and well drained. Mix with it sensible amounts of compost, leaf mold, or peat moss and coarse sand or grit. Sow broadcast or in drills and cover with soil to a depth of two or three times the diameter of the seeds. Keep the bed evenly moist.

Transplant the seedlings as soon as their second pair of leaves are well developed to nursery beds or cold frames, spacing them 6 to 7 inches apart. In mild climates the pansies are transplanted to the flowering sites in fall, in cold climates the move is delayed until early spring. From then on routine care consists of shallow surface cultivation to promote growth and control weeds and in dry weather soaking as needed to keep the ground reasonably moist. Picking faded blooms before seeds are formed greatly helps to prolong the flowering season.

Winter protection for outdoor beds is advisable or necessary in regions with winters as cold or colder than those of New York City. This can be given by covering the beds after the ground has frozen to a depth of an inch or so with salt hay, cut branches of evergreens, or some similar loose material. Cold frames containing pansies must be ventilated on all favorable occasions through the winter. It is important to keep the plants sturdy by not stimulating too early growth.

In greenhouses, whether for winter blooms or as annuals to be set in the garden in spring, pansies must, to use the gardeners' expression, be grown cool. Winter night temperatures of 45 to 50°F, with a daytime increase of five to ten degrees, are adequate. For blooming indoors choose early-flowering strains. Sow the seeds outdoors in June or July. Transplant the seedlings to a shaded outdoor nursery bed, and in September, pot the plants individually and bring them into the greenhouse or plant them in greenhouse beds or benches of soil, spaced 8 to 9 inches apart. To raise pansies in greenhouses for setting outdoors in early spring sow in January or February and transplant the seedlings individually to small pots. Managed in this way, pansies are likely to bloom longer into the summer than when grown as biennials, but their early spring display is less impressive.

Bedding violas, or tufted pansies as they are sometimes called, are for convenience identified as *Viola williamsii*. They include such splendid varieties as 'Archie Grant', 'Blue Dutchess', and 'Maggie Mott'. They resulted from crossing pansies with *V. cornuta*. The first such hybrids were raised in Scotland about 1863. And it is to Scotland and places with similar cool summers that one must go to see bedding violas at their best. They differ most noticeably from pansies in their flowers not being marked with facelike patterns. From a batch of seedlings raised in 1887 in Scotland was selected a compact plant, the flowers of which lacked the conspicuous dark lines that typically radiate from the centers of the blooms of bedding violas. This and seedlings with similar blooms have been called violettas.

A bedding viola variety

Under favorable conditions bedding violas and violettas bloom continuously from spring to fall. In climates where they succeed they are much used for summer bedding. They may be raised from seed, but more commonly are propagated by cuttings made in August and planted in cold frames where the plants remain until they are transferred to their flowering quarters after displays of hyacinths, tulips, polyanthus primroses, English daisies, wallflowers, and other spring bedding plants are finished.

A small-flowered viola (*Viola visserana*)

Small-flowered garden violas resulting from hybridizing *Viola cornuta* and *V. gracilis* have the group name of *V. visserana.* Much more amenable to most American gardens than large-flowered bedding violas, these, if afforded a fairly cool location, moist but not wet soil containing ample decayed organic matter, and just a little shade from hot sun, bloom with great freedom over a long spring and summer season. They are easily increased by division and cuttings. Belonging are such sorts as 'Catherine Sharp', 'Floraire', 'George Wermig', 'Jersey Gem', and 'John Wallmark'. For information about other violas see entries in this Encyclopedia under Viola and Violets, Sweet or Florists'.

PAPAVER (Pap-àver)—Poppy. This is the name of the genus of the true poppies. Other plants having popular names with the word poppy part of them belong elsewhere. For example, California-poppy in *Eschscholzia,* prickly-poppy in *Argemone,* blue-poppy in *Meconopsis.* The genus *Papaver* includes about 100 species, chiefly natives of the Old World, but with a few indigenous to western North America, and one to South Africa and Australia. This genus is the name genus of the poppy family PAPAVERACEAE. Its name is an ancient Latin one for the poppy.

Poppies are milky-juiced annuals, biennials, and herbaceous perennials, often with more or less bristly hairs, and sometimes glaucous. They have rather fragile-looking, brightly colored, showy blooms. Their foliage may be all basal or consist of basal and stem leaves, the latter alternate. The leaves are pinnately lobed or divided, or toothed. Sometimes their bases embrace the stems. In panicles, racemes, or solitary, the flowers, demurely nodding in the bud stage, up-facing later, have two or three sepals that soon fall, four or rarely five or six petals (more in double-flowered horticultural varieties), and numerous slender or club-shaped stamens. There is no style. The stalkless stigmas are dotted

A field of opium poppies (*Papaver somniferum*) in Asia

Papaver somniferum

around the edges of the top of the ovary, which later develops into the seed capsule. One species, the opium poppy (***P. somniferum***), native from Greece to the Orient, is an extreme example of a plant being both a boon and a curse to mankind. The first because of the tremendous part its products morphine and related derivatives have played and play in relieving human suffering, a curse because of the devastation that opium and its derivative heroin have wrought among addicts. This species is also the source of poppy seeds used in breads, confectionery, and for other culinary purposes. Because it is illegal in the United States to cultivate the opium poppy except under special license, no further discussion of it will be given here. Sometimes called tulip poppy, *P. glaucum* is closely related to *P. somniferum.* For convenience, cultivated poppies can be considered in two groups, annuals and biennials (the life spans of these are largely determined by when the seeds are sown) and perennials.

Best known of annual poppies is the horticulturally developed race called Shirley poppies. These were bred from the common field poppy or corn poppy (***P. rhoeas***), the kind immortalized in World War I poetry as the poppy of Flanders fields. Not only in Flanders, but over much of temperate Europe and Asia does this brilliant-flowered species grow in abundance. It is naturalized in North America. Erect, slender, and branching, *P. rhoeas* is 1 foot to 3 feet tall. Occasionally nearly hairless, more commonly it is clothed with shaggy hairs. More or less grouped at the bases of its flower stalks, the leaves are nearly always irregularly lobed and toothed. Typically bright cinnabar-red, but varying to scarlet, purple,

Papaver rhoeas

white, and white with reddish-edged petals, each with a black blotch at its base, the blooms are 2 inches wide or wider. Shirley poppies differ from the wild forms of *P. rhoeas* in having larger blooms in a much greater diversity of colors, and in being completely devoid of the black anthocyanin pigments present in wild populations. The field poppy is highly variable and enthusiastic botanists have named several of its variants as species. Closely related kinds perhaps cultivated are ***P. commutatum,*** which has the black blotches of its petals chiefly near their centers instead of at their bases, and ***P. lacerum,*** which has a variant with its petals so dark that they are almost black. A native of western North America, orange-flowered ***P. californicum*** much resembles *P. atlanticum,* but differs in being an annual rather than a perennial. Hairless to sparingly-hairy and 1 foot to 2 feet tall, it has pinnately-divided leaves with lobed or toothed segments. Its 1½- to 2-inch-wide flowers have red petals with a green spot at the base of each.

Papaver orientale

Oriental poppies exist in many magnificent varieties. They are derived from ***P. orientale*** (syn. *P. bracteatum*), a native of Turkey, the Caucasus, and Iran. From 2 to 4 feet tall and clothed with stiff hairs, *P. orientale* has evenly-pinnately-lobed leaves up to 1 foot in length with the lobes again lobed and toothed. The flowers, 4 to 6 inches or sometimes more in diameter, have four to six petals, or in double-flowered horticultural varieties more. In the wild form the petals are usually blood-red, commonly with a dark violet to almost black base. The blooms of *P. o. album* are white. Garden varieties include kinds with blooms in a range of reds and pinks.

Iceland poppies belie their popular name. They are derived from a group of related species, for want of clarification commonly lumped together as ***P. nudicaule,*** that chiefly inhabit Asian mountains, with extensions into subarctic and

Papaver nudicaule

arctic regions. These kinds are difficult to differentiate. In their horticultural manifestations they show considerable variation, especially in their range of flower colors, which include white and many shades of yellow, orange, and pink, often with the petal bases colored differently from their upper parts. Double-flowered varieties are grown. Treated in its broad sense, *P. nudicaule* is a more or less hairy, stemless or almost stemless perennial with all basal foliage. Its leaves, often somewhat glaucous, are stalked, once-pinnately-lobed or cleft, and 1 inch to 3 inches long. The erect, wiry flower stalks, 1 foot to 2 feet tall, carry solitary, fragrant blooms 1 inch to 3 inches wide, or in cultivated varieties larger. Their two inner petals are smaller than the two outer ones.

The alpine poppy, in a broad sense ***P. alpinum,*** is divided by critical students into half a dozen species not always easy to relate to cultivated alpine poppies. It looks much like a miniature edition of the Iceland poppy, but differs in its usually up to 1-inch-long leaves being commonly two- or three-times-pinnately-lobed. On stalks of 6 inches to 1 foot tall, the yellow, orange, pink, or white flowers are 1 inch to 1½ inches in diameter. Names applied to segregates of this species are *P. burseri, P. kerneri, P. pyrenaicum, P. rhaeticum, P. sendtneri,* and *P. suaveolens.*

Densely-hairy foliage and orange to brick-red flowers are typical of a group of perennial species a few of which are met in gardens. In some the blooms are solitary, in others the flowers are in racemes. Typical of the first is ***P. atlanticum,*** of Morocco. In gardens this is often misidentified. Much of the material cultivated as *P.*

Papaver atlanticum

californicum belongs here. From 6 inches to 2 feet tall, *P. atlanticum* has stems branched near their bases, and coarsely-round-toothed, oblong-oblanceolate leaves, the largest lower ones stalked and up to 10 inches long, those above progressively smaller and shorter-stalked or stalkless. The flower stalks, up to 1½ feet long, carry blooms 1¼ to 1½ inches across. The upper surfaces of the leaves and the flower buds are densely-hairy. Very similar ***P. rupifragum,*** a native of Spain, differs in the only hairs of its leaves being on the veins beneath and in the flower buds being hairless. Also to this group belongs ***P. lateritium,*** which inhabits Turkey and the Caucasus. Differing scarcely at all from *P. atlanticum,* it can be most readily identified by its bigger seed capsules, which are bluish-black. Those of *P. atlanticum* are green or brownish.

Flowers in racemes, and seed capsules with rounded instead of tapered bases distinguish ***P. pilosum*** from the kinds discussed in the previous paragraph. The blooms are orange with white stamens. Erect and branched, this perennial, so far as is known, is limited in the wild to one mountain, Ulu Dag, in Turkey. Its foliage and 1½- to 2-foot-tall stems are hairy. The leaves, the basal ones long-stalked, are oblong to ovate-oblong and round-toothed or more or less lobed. The orange to red blooms have six or seven stigmas and are about 3 inches across. Also with flowers in racemes, ***P. spicatum*** (syn. *P. heldreichii*), of Asia Minor, is a densely-white-hairy perennial 1½ to 2½ feet tall, with long-stalked, round-toothed, oblong basal leaves and orange-red flowers in spikelike racemes. Differing from the last in having hairless to sparsely-hairy foliage, ***P. apokrinomenon*** (syn. *P. heldreichii sparsipilosum*), of Turkey, is often cultivated, usually under the name of *P. heldreichii.*

Garden and Landscape Uses and Cultivation. For these see Poppy.

PAPAVERACEAE — Poppy Family. Some 200 species accommodated in twenty-six genera constitute this family of dicotyledons, the majority inhabitants of the north temperate zone. By far the greater number are herbaceous plants, but *Dendromecon* and some species of *Bocconia* are shrubs or trees. The family is notable for the possession of milky or yellowish sap or latex. Its leaves are alternate, or the upper ones sometimes opposite or in whorls (circles of more than two). Solitary, clustered, or in racemes, the flowers have two or less often three, soon deciduous sepals, four to six or sometimes more petals, or the petals may be wanting, numerous stamens, and one or six to twenty pistils with short styles or none. The fruits are capsules or sometimes technically nuts. Genera in cultivation include *Argemone, Bocconia, Chelidonium, Dendromecon, Dicranostigma, Elsholtzia, Eomecon, Eschscholzia, Glaucium, Hunnemannia, Hylomecon, Macleaya, Meconella, Meconopsis, Papaver, Platystemon, Pteridophyllum, Roemeria, Romneya, Sanguinaria, Stylomecon,* and *Stylophorum.*

PAPAYA. This is the preferred common name for *Carica papaya* and its fruits. The latter are sometimes called pawpaws, a confusing practice because that name is also used for completely different *Asimina*

Papaya (*Carica papaya*) in fruit

triloba, a hardy deciduous tree native of North America. Here we are dealing with the tropical species, described in this Encyclopedia under Carica. Papayas are rather strange-looking, but stately and handsome, mostly branchless trees with large, terminal crowns of long-stalked, deeply-lobed leaves and small, fragrant, yellow flowers. Individual plants are usually unisexual, so that it is necessary to have a male growing in proximity to females for the latter to develop their edible, melon-like fruits. Occasional trees carry both male and female blooms and may fruit without another tree being nearby. The edible fruits of papayas are highly esteemed. Their bland flavor is improved, to the taste of most Americans, by drizzling lemon or lime juice over the flesh. Besides being eaten like muskmelons, papayas are used in salads, sliced and served with cream and sugar, candied, and made into sherbets, jellies, pies, and pickles. A confection is made by boiling their flowers in syrup, and the fruits are sometimes cooked and eaten. In parts of Asia the tender young shoots and leaves are cooked as a vegetable.

Garden and Landscape Uses and Cultivation. Papayas are essentially tropical. They stand light frosts for brief periods, but not more intense exposures. In southern Florida, Hawaii, and other warm-climate regions, they are grown outdoors, often in small groups in home gardens, where they are decidedly decorative as well as useful for supplying the table. They are also raised commercially and are occasionally cultivated for interest in greenhouses, especially in those devoted to educational displays of plants useful to man.

For their best growth papayas need plenty of deep, fertile soil, well-drained, but not excessively dry, and sunny locations sheltered from high winds. They are usually raised from seed, but seedlings show much variability. Grafting and cuttings of young shoots planted in a greenhouse propagating bench with bottom heat are also sometimes used as methods of increase. In favorable environments, papayas grow very rapidly and reach fruiting size in their second year. For the best results ample water and nutrients must be supplied, otherwise the fruits develop poorly and are not well flavored. Sometimes papayas carry so many fruits that, unless thinned in their early stages, they injure each other by pressing one against the other; overcrowding also affords ideal hiding places between the fruits for harmful insects. It is important that the fruits have sufficient room to develop fully and hang free. Early thinning results in more nourishment being available to the fruits that remain so that they develop as superior samples. In greenhouses papayas are easily managed where high humidity and high temperatures are maintained. They may be accommodated in large tubs or ground beds in rich, well-drained soil kept evenly moist. Generous fertilizing is needed.

Paphiopedilums at a flower show

PAPER. As parts of the common names of plants the word paper appears in these: paper bush (*Edgeworthia papyrifera*), paper-gardenia (*Tabernaemontana divaricata*), paper-mulberry (*Broussonetia papyrifera*), and rice paper plant (*Tetrapanax papyriferum*).

PAPHIOPEDILUM (Paphio-pèdilum)—Lady Slipper. For long gardeners have known and grown the orchids of this genus under its older name *Cypripedium*. Many still employ that for them. Its contraction "cyps" is freely, almost affectionately used as the vernacular for paphiopedilums. But it is better to identify paphiopedilums as *Paphiopedilum* and so avoid confusion with true *Cypripedium,* which name is properly reserved for a related group of hardy orchids including some very fine native Americans. Both genera and *Phragmipedium* are called lady slipper orchids.

Belonging to the orchid family ORCHIDACEAE, and comprising some fifty species, *Paphiopedilum* is restricted in the wild to warm regions of eastern Asia and Indonesia. Its name comes from the Greek *paphia,* which relates to Paphos, on the island of Cyprus, which was the chief seat of worship of Aphrodite (Venus), and *pedilon* a slipper, in reference to the form of the lip.

These are mostly terrestrial orchids. A few sorts are tree-perchers (epiphytes) or cliff-perchers. In addition to the natural species, there is a vast number of garden hybrids and varieties. From *Phragmipedium,* a related genus of the American tropics, *Paphiopedilum* differs in its flowers having one-celled ovaries.

Paphiopedilums have no pseudobulbs. Their thick, evergreen, strap-shaped, oblong, or elliptic leaves are all basal. They have long flowering stalks with one or more blooms each with three sepals, the uppermost or dorsal one large and showy, the two laterals united into what appears to be one lower sepal. There are two spreading or drooping petals and a pouched slipper-like lip, the margins of which, unlike those of the lips of *Cypripedium,* are rarely inrolled. The column has a flattened staminode in front of and between the two anthers.

Probably best known of the species, variable ***P. insigne*** (syn. *Cypripedium insigne*) is a native of India. This has approximately 1-foot-long leaves, ¾ inch wide and plain green. The flowers, on densely-hairy stalks mostly considerably shorter than the leaves, are 4 to 5 inches across. Generally there is only one, rarely two on each stalk. Typically the broad-oval upper sepal, its edges somewhat rolled back, is green, spotted and striped with brown in its lower part, white above. The yellowish-green, brown-veined, wavy-edged petals spread horizontally. The smooth, glossy lip is yellow-green suffused with brown. But many color variations occur in this species and its numerous hybrids. Popular *P. i. sanderae* has flowers with light green stalks, the lower part of the uppermost sepal primrose-yellow with tiny brown dots, the upper part white. Its petals and lip are yellow.

Other species with plain green leaves not mottled or checkered in two shades include these: ***P. chamberlainianum*** (syn. *Cypripedium chamberlainianum*), of Sumatra, has recurved, strap-shaped leaves up to 1

Paphiopedilum insigne

Paphiopedilum insigne variety

Paphiopedilum insigne sanderae

foot long by 1½ inches wide. Considerably longer, the grayish-brown flowering stalks carry three to five 3½- to 5-inch-wide blooms. The nearly circular uppermost sepal has the lower parts of its margins reflexed and fringed with white hairs. It is greenish, shaded in its middle and lower parts with blackish-purple. The 1¼-inch-long, narrow, twisted petals spread widely. They are greenish-pink with lines of tiny purple spots. About as long as the sepals, the prominent, broad lip is green freely sprinkled with violet or reddish dots. ***P. charlesworthii,*** a native of Burma, has broadly-linear leaves 6 to 9 inches long by about 1 inch wide. Its solitary, waxy flowers are on stalks 4 to 8 inches long. From 2½ to 3 inches in diameter and with purple hairs on their ovaries, they have a very wide, nearly orbicular, flat uppermost sepal flushed with rose-red and veined with purple, and horizontally spreading, greenish to reddish-brown, flat or nearly flat, oblong petals. Helmet-shaped, the polished, yellowish-brown to greenish lip is comparatively small. The staminode is white and hairless. ***P. fairieanum*** (syn. *Cypripedium fairieanum*), of the Himalayan region, is a low sort with light green, linear-oblong leaves 5 to 6 inches long by ½ to 1 inch wide. The solitary 2½-inch-wide flowers, on stalks 4 to 7 inches long, have a large, wavy-margined, broad-elliptic uppermost sepal, white except for its yellowish-green base, and netted and lined with violet. The markedly sickle-shaped, wavy-edged petals, usually upcurved at their ends, are whitish striped with violet. The greenish-purplish lip is purplish-veined. ***P. lowii*** (syn. *Cypripedium lowii*) in the wild usually perches on trees or limestone cliffs. It has strap-shaped, grass-green leaves up to 1¼ feet long. The flower stalks, up to 3 feet long, are usually arching or nodding. They have two to six hair-fringed blooms 3 to 4 inches wide. The uppermost sepal is pointed-ovate, yellowish-green striped toward its base with purple. The twisted, slender-spatula-shaped petals, which angle slightly downward, have black-dotted, yellow lower halves, violet-purple upper halves. The coppery-brown lip, lighter colored on its underside than above, is marked with purple lines. ***P. rothschildianum,*** of Borneo and Sumatra, has blooms of striking appearance, from two to five on each strong stalk. Of robust growth, this has more or less upright leaves up to 2 feet long and 3 inches wide.

Paphiopedilum rothschildianum

The minutely-pubescent, violet-purple flower stalks are a little longer than the leaves. Some 5 inches from top to bottom and twice that distance from petal tip to petal tip, the blooms have a slightly hair-fringed, pointed-elliptic, purple-lined, yellow uppermost sepal and long, narrowly-triangular, horizontally-spreading yellowish-green petals spotted with reddish-purple. The reddish-brown lip is yellow in its throat. ***P. spiceranum*** hails from India and Assam. It has linear-lanceolate to linear-oblong leaves up to 1 foot long and 2 inches wide. The 2½- to 3-inch-wide, solitary blooms have stalks about as long as the leaves. The chiefly white, uppermost sepal has strongly reflexed margins; a crimson center band is sprinkled with green at its base. Somewhat sickle-shaped and angled downward, the markedly wavy-edged, greenish-brown petals have hairy bases. The lip is violet with a greenish base. The staminode is almost spherical. ***P. villosum*** is Burmese. It has strap-shaped leaves 9 inches to 1½ feet long by up to nearly 2 inches wide. The stalks, carrying solitary flowers, are 6 inches to 1 foot long or somewhat longer. Almost 6 inches in diameter, the blooms have a rounded, broad-elliptic to obovate uppermost sepal, olive-green with a narrow white band around its edges. The wavy-margined petals are yellowish-brown with brown-purple mid-veins. They pale toward their bases. The lip is brownish-yellow. The staminode is clothed with white hairs.

Species with leaves more or less distinctly mottled or checkered with two shades of green include these: ***P. argus*** (syn. *Cypripedium argus*), of the Philippine Islands, has pointed, mottled leaves up to 8 inches long by 1½ inches wide. The flowering stalk, longer than the leaves and with brown hairs, has one or rarely two blooms 3 to 4 inches wide. The uppermost sepal, broad-ovate and pointed, is white with green and brown veins and is sometimes spotted with dark purple near its base. The sharply-down-angled, strap-shaped, hair-fringed petals have white lower parts veined with purple. Toward their apexes they are light reddish-purple. They have many dark purple warts. The lip is dark brownish-purple, becoming paler and veined with green on its under surface and with some purple spots. ***P. barbatum*** is a native of Malaya and Thailand. It has leaves up to 6 inches long, from 1¼ to 2 inches wide, checkered in two shades of green. Longer than the leaves, the pubescent flower stalks each with one, or less commonly two 2½- to 4-inch-wide blooms with purple-haired ovaries. The uppermost, nearly circular sepal is whitish with conspicuous purple veins and a greenish base. The narrowly-strap-shaped, down-pointing, hair-fringed petals, purplish at their bases, have little warts along their margins. The brownish-purple lip has warts on its side lobes. ***P. bellatulum*** (syn. *Cypripedium bellatulum*), of Burma and Thailand, has oblong-elliptic leaves 6 to 9 inches long by up to 1½ to 3 inches wide, deep green mottled with lighter green, and purplish beneath. The solitary, saucer-like or slightly cupped flowers have hairy, purplish stalks very much shorter than the leaves. From 2 to 2½ inches wide, they have a nearly circular, concave uppermost sepal and very broad-ovate petals, all white or cream-colored and spotted and blotched with purple-maroon. The lip, which is colored similarly, but is less heavily spotted, is shorter than the petals. ***P. callosum,*** of Indochina and Thailand, has comparatively few leaves, usually up to 9 inches long and not over 2 inches wide and green, blotched or marbled with darker green. The flowers, solitary or sometimes two together, have pubescent stalks considerably longer than the leaves. From 3 to 4 inches across, they have a nearly circular, white uppermost sepal, clouded or striped lengthways with crimson and with a green base. The greenish, sometimes purplish-tipped, down-pointing petals are fringed with hair and have several prominent warts along their margins. The lip is brownish-purple. ***P. concolor,*** of Burma, Thailand, and southern China, of dwarf habit, has leaves up to 6 inches long by

Paphiopedilum concolor

1½ inches wide, dark green, marbled and mottled with paler green, purple-spotted on their undersides. On pubescent stalks shorter than the leaves, the 2- to 3-inch-wide, flat or nearly flat flowers, solitary or rarely paired, are pale yellow with many tiny purple dots. ***P. curtisii*** (syn. *Cypripedium curtisii*) is a native of Sumatra. It has comparatively few mottled leaves up to about 8 inches long and 3 inches wide, and on pubescent stalks longer than the leaves, solitary blooms 3 to 4½ inches in diameter. The broadly-ovate, greenish uppermost sepal is veined with bright green and edged with white. The down-pointed, strap-shaped petals are greenish-pinkish at their bases, and become paler toward their recurved apexes. Their margins are hairy and have black warts. The lip is purplish-maroon. ***P. exul*** (syn. *Cypripedium exul*), of Thailand, has usually slightly mottled leaves, strap-shaped and up to 10 inches long by 1 inch wide. Its 3-inch-wide, solitary flowers have stalks with purple hairs. The uppermost sepal, broadly-ovate and yellowish-green, is narrowly edged with white and spotted with brown. The petals, yellowish-green and longer than the sepals, broaden toward their apexes; they are dark-spotted near their bases. The lip is yellow. ***P. godefroyae*** (syn. *Cypripedium godefroyae*) is much like *P. bellatulum,* but has shorter, narrower leaves and the lips of its flowers longer than the petals. It is a native of Cochin-China. ***P. hirsutissimum*** (syn. *Cypripedium hirsutissimum*), of Assam, has obscurely marbled leaves up to 6 inches long by ¾ inch wide. The solitary flowers on blackish-hairy stalks shorter than the leaves are 4 to 5 inches across. The uppermost sepal is almost circular, green heavily bespattered in its central and lower parts with blackish-purple. The broadly-spatula-shaped petals, black-hairy toward their bases, spread nearly horizontally and have wavy edges. They are green stained with purple in their lower parts, bright rose-purple toward their apexes. The minutely-warted lip is dull green suffused with purple. ***P. lawrenceanum*** (syn. *Cypripedium lawrenceanum*), of Borneo, has mottled leaves 8 to 10 inches long by up to 2½

Paphiopedilum lawrenceanum

inches broad. On downy stalks longer than the leaves, the solitary or less often paired flowers are 4 to 5 inches wide. They have a very broad, rounded uppermost sepal that is white with prominent, glossy, violet-purple lines. The hair-fringed, strap-shaped, purple-tipped, purple-lined, green petals spread horizontally or angle slightly downward and have along each margin five to ten prominent warts. The lip is dull brown-purple. ***P. niveum*** in size and aspect much resembles *P. concolor,* but its flowers have stalks as long as or longer than the leaves. The leaves have purple undersides. From 2½ to 3½ inches in diameter, the blooms, flat or nearly flat, are solitary or occasionally two on a stalk. This is native to Malaya. ***P. venustum*** (syn. *Cypripedium venustum*), of the Himalayan region, has leaves about 6 inches long mottled with two shades of green and beneath heavily blotched with violet. About as long as the leaves, the purple-hairy flowering stalk carries a solitary bloom about 3 inches across. The uppermost sepal is broadly-ovate and white, clearly lined with green. The oblanceolate petals are greenish, with darker lines, purple tips, and small, dark warts. The green-veined, yellow-green lip is suffused to some extent with pink.

Paphiopedilum leeanum

Paphiopedilum maudiae

Cultivated paphiopedilums include not only species, but a vast number of selected horticultural varieties as well as natural and artificial hybrids. Those available are described and often illustrated in the catalogs of specialist dealers. Among hybrids with Latin-form names, ***P. leeanum,*** its parents *P. insigne* and *P. spiceranum,* has 4-inch-wide flowers with a green-based, white dorsal sepal spotted and lined with purple, a green lower sepal, greenish petals heavily suffused and dotted with brown, and a green and purple-brown lip. Another such hybrid, ***P. maudiae,*** the parents of which are *P. callosum sanderae* and *P. lawrenceanum hyeanum,* has mottled leaves. Its flowers have a longitudinally-green-veined, white or greenish-white dorsal sepal, green-based, white petals, and a yellow-green lip. The parents of ***P. nitens*** are *P. insigne maulei* and *P. villosum.* The hybrid has flowers 4½ inches across with a green-based, white dorsal sepal thickly spotted with purple-brown, a light green lower sepal, and green petals generously mottled and suffused with purplish-brown and with violet-purple apexes. The lip is greenish-yellow suffused with brown or with purple-brown spots.

Garden Uses. Among the most distinctive and useful orchids, paphiopedilums are excellent as cut flowers and as greenhouse decoratives. Their blooms are much used for corsages and in other florists' confections. They last well in water.

Cultivation. Paphiopedilums are not difficult to grow. Under reasonable conditions they bloom regularly and reliably. With respect to their cultural needs they fall neatly into two groups. The easiest to manage are sorts with plain green foliage.

Paphiopedilum nitens

These gardeners call cool orchids. They respond to temperatures from fall to spring of 50 to 55°F at night, about ten degrees or in sunny weather perhaps fifteen, degrees higher by day. Kinds with mottled or tessellated foliage and those with white flowers need a minimum winter night temperature of 60°F, with corresponding increases by day.

All prosper in very well-drained pots in a variety of rooting mixes packed reasonably firmly. A satisfactory one consists of one part turfy loam, two parts osmunda fiber, and one part live sphagnum moss.

Garden varieties of *Paphiopedilum*: (a) 'Bit of Sunshine'

(b) 'Cimarose'

(c) 'Golden Fleece'

Many growers like to surface the medium in which these plants grow with living sphagnum. Do not use pots too big for the sizes of the plants. Paphiopedilums are

(d) 'Golden Queen'

(e) 'Onyx'

best when their roots are somewhat cozily accommodated.

Because they are without pseudobulbs or other water-storage organs these orchids have no season of definite rest. It is necessary that the medium in which they grow be damp at all times. This does not, of course, mean constantly soaked. Too-wet conditions turn the rooting medium sour and cause the roots to rot. Gentle fertilizing from spring to fall of well-rooted specimens promotes health and good growth, but must not be overdone. Insufficiently diluted fertilizers can do serious harm. For further information see Orchids.

PAPOOSE-ROOT is *Caulophyllum thalictroides.*

PAPRIKA. See Capsicum.

PAPYRUS is *Cyperus papyrus.*

PARA-CRESS is *Spilanthes oleracea.*

PARA-PARA is *Pisonia umbellifera.*

PARABENZOIN (Para-bénzoin). From *Lindera,* in which its two species are by some botanists included, the eastern Asian *Parabenzoin,* of the laurel family LAURACEAE, differs in having dry fruits that split irregularly into five or six parts and so expose the single seed, rather than one-seeded berries that do not split. The name is derived from the Greek *para,* near, and *Benzoin,* an old generic name of *Lindera benzoin.*

Parabenzoins are deciduous shrubs or small trees with alternate leaves and umbels of unisexual flowers similar to those of *Lindera.*

Japanese ***P. praecox*** (syn. *Lindera praecox*) is similar to *Lindera benzoin,* and like it has pinnate-veined, deciduous leaves, but the leafstalks are longer and more slender. Its winter buds are covered with shining brown scales and its fruits are yellowish or reddish-brown. It attains a maximum height of about 25 feet and has bright yellow foliage in the fall. This is hardy about as far north as Long Island, New York. A native of mountainous parts of southern Japan and China, ***P. trilobum*** (syn. *Lindera triloba*) is less hardy than the last. A shrub, it has usually three-cleft, triangular-ovate leaves up to 5 inches long, with three veins diverging from the base.

Garden and Landscape Uses and Cultivation. These are as for *Lindera.*

PARADISE NUT is *Lecythis zabucajo.*

PARADISEA (Para-dìsea)—St.-Bruno's-Lily. The only species of *Paradisea,* of the lily family LILIACEAE, was at one time included in *Anthericum,* to which the St.-Bernard's-lily (*A. liliago*) belongs. From the St.-Bernard's-lily, *Paradisea* differs in having leaves as long or longer than its flower stalks, and blooms with corolla segments (usually called petals) that do not spread widely to form a starlike flower. The name commemorates Count Giovanni Paradisi of Modena, Italy, who died in 1826.

The St.-Bruno's-lily (***P. liliastrum***) is a deciduous hardy herbaceous perennial native to the European Alps and the Pyrenees. It has a very short underground rhizome and fleshy, clustered roots. Its linear, all-basal, channeled leaves are green. The leafless, green, branchless, erect flower stalks 1 foot to 2 feet long terminate in few- to several-flowered, moderately dense racemes. The blooms somewhat resemble those of plantain-lilies (*Hosta*), but are smaller. Sweet-scented, funnel-shaped and 1½ to 2 inches long, they have a corolla of six separate white petals, or more correctly, tepals. There are six stamens and one style. The fruits are capsules. Taller than the typical species, *P. l. major* also has larger flowers.

Garden and Landscape Uses. St.-Bruno's-lily is a charming addition to gardens. It is

Paradisea liliastrum

suitable for flower borders and for colonizing at the fronts of shrub beds where root competition does not make this impracticable. It is also appropriate for open places beneath trees where some sun reaches. Its needs are simple. They are a rather light, not excessively compact soil, with a fair organic content, and a not-too-dry location in partial shade or, provided moisture is adequate, in sun. Seeds provide an easy means of achieving increase or the plants may be carefully divided in early spring.

PARADRYMONIA (Para-drymònia). Central and northern South America is the native territory of the few species of *Paradrymonia,* of the gesneria family GESNERIACEAE. Its members are by some authorities included in *Episcia.* The name, from the Greek *para,* near, and *Drymonia,* the name of a related genus, alludes to the relationship.

Paradrymonias are tree-perchers (epiphytes) or ground-inhabiting herbaceous perennials with creeping or more or less upright, rooting stems and opposite leaves, those of each pair equal or of different sizes. Many together from the leaf axils, the flowers have a five-lobed calyx, a tubular, funnel- or trumpet-shaped corolla with five spreading lobes, four nonprotruding stamens, and one style. The fruits are fleshy capsules.

Tree-perching ***P. decurrens*** (syn. *Episcia decurrens*) has upright to somewhat procumbent red-hairy stems, and thickish, hairy, pointed-elliptic leaves up to 10 inches long by 4 inches wide, with their blades tapered to the stalks. Cream-colored and 1½ inches long, the flowers have curved corolla tubes, narrow below, widening above. They are red-hairy outside, yellow in their throats, and are held close in the leaf axils among the tangles of slender bracts and sepals. This shrubby species produces shoots freely from the base. Recently introduced from Ecuador to the United States by the Marie Selby Botanical Gardens, Sarasota, Florida, ***P. hypocyrta*** has proved less willing to bloom than other sorts in cultivation. Of erect habit, this spectacular species has narrow, glossy leaves, reddish when young. The flowers have a persistent, brilliant red calyx and a white corolla that differs from those of other cultivated paradrymonias in being strongly pouched and having a very constricted mouth.

Other sorts in cultivation include ***P. ciliosa*** of Brazil, Venezuela, and Peru. From vining, branching, rooting stems furnished with dull magenta hairs this sprouts erect leaves with long, dull wine-purple stalks and broad-elliptic to obovate blades 4 to 9 inches long. From all other species in cultivation this differs in its yellow-throated, white flowers up to 2½ inches long, occurring in clusters, having collars conspicuously fringed after the fashion of those of *Episcia dianthiflora.* The flowers are clustered.

Garden Uses and Cultivation. These are as for *Episcia.*

PARAHEBE (Para-hè-be). Believed to be endemic to New Zealand, *Parahebe,* of the figwort family SCROPHULARIACEAE, contains about a dozen species. Its name, in reference to its close botanical relationship with *Hebe,* derives from the name of that genus and the Greek *para,* near.

The sorts of *Parahebe* are low, more or less prostrate subshrubs or less frequently herbaceous perennials. They have opposite, undivided, generally toothed, stalkless or short-stalked leaves and in squat or long racemes, asymmetrical, white, pink, or blue flowers. These have a calyx with four lobes and occasionally a fifth much smaller one, and a short- to long-tubed corolla with four or five unequal lobes (petals). There are two stamens and a style tipped with a headlike stigma. The fruits are capsules.

Species sometimes cultivated include these: ***P. canescens*** (syns. *Hebe canescens, Veronica canescens*) is a tiny herbaceous perennial that forms mats of creeping stems and broadly-ovate leaves approximately ⅛ inch long and loosely clothed with grayish-white hairs. The flowers are solitary, pale blue, about ⅓ inch across. ***P. catarractae*** (syns. *Hebe catarractae, Veronica catarractae*) is a subshrub up to 2 feet tall. It has prostrate and ascending, purplish stems and toothed, linear-lanceolate to elliptic or ovate leaves ½ inch to 2 inches long or sometimes longer. In racemes of many, the flowers have pink- to purple-veined, white flowers about ⅓ inch in diameter. ***P. decora,*** low and mat-forming, has ovate to nearly circular leaves under ¼ inch long, sometimes lobed at the base.

Parahebe lyallii

The ⅓-inch-wide, white to pink flowers are in racemes of ten or fewer. ***P. linifolia*** (syns. *Hebe linifolia, Veronica linifolia*) is a hairless subshrub up to 10 inches tall with sprawling to erect stems and ¾- to 1-inch-long, blunt, toothless, linear to linear-lanceolate leaves. Mostly in racemes of two to four, the ½-inch-wide flowers are white to light pink. ***P. lyallii*** (syns. *Hebe lyallii, Veronica lyallii*), low and freely-branched, has prostrate, rooting stems up to 1½ feet in length and ovate to linear-obovate to subcircular leaves up to ½ inch long. The white to pink flowers, about ⅓ inch in diameter, are in loosely arranged many-flowered, erect racemes about 6 inches tall.

Hybrid ***P. bidwillii*** (syns. *Hebe bidwillii, Veronica bidwillii*), its parents *P. decora* and *P. lyallii,* has procumbent stems and usually three- or five-lobed, roundish leaves. In racemes of about a dozen, the flowers are white to pink.

Garden and Landscape Uses and Cultivation. Where winters are not excessively severe and summers not extremely hot and humid, parahebes are attractive for rock gardens and similar special areas. They prosper in sunny locations in well drained, not-too-rich soils and are readily propagated by cuttings and seed, and some sorts by division.

PARAMIGNYA (Para-mígnya). Possibly twenty species of the rue family RUTACEAE constitute *Paramignya,* a genus closely similar to *Atalantia* and *Luvunga,* and native to Indomalaya. The name, of no obvious application, is derived from the Greek *paramignunai,* a mixture.

Paramignyas are difficult to distinguish as to species. They are commonly high-climbing, evergreen lianas with hooked spines and generally zigzagged twigs. Their short-stalked leaves are of one leaflet or blade. The flowers, up to 1 inch in diameter in some kinds, are white and fragrant. They have four- or five-parted calyxes and corollas, and twice as many stamens as petals. The thick-skinned, spherical to obovoid fruits, unlike those of their rather distant relative, the orange, are gummy rather than juicy inside.

Introduced to the United States for experimental work with *Citrus,* Indian ***P. monophylla*** is a climber, with alternate, elliptic, oblong, or rarely obovate leaves up to 4½ inches long by 1½ inches broad. The slender-stalked flowers, solitary or few together from the leaf axils, have calyxes with four or five lobes and corollas with four or five petals. The stamens are twice as many as the petals. The lower part of the style is pubescent. Ovoid or obovoid and somewhat angled, the pubescent to nearly hairless fruits are 1 inch to nearly 1½ inches long by almost as wide.

Garden and Landscape Uses and Cultivation. The sort discussed here is occasionally grown for its interest and to add variety to collections. It succeeds under conditions appropriate for oranges, and may be increased by seed, and by cuttings of firm but not hard shoots.

PARAMONGAIA (Para-móngaia) — Cojomaria. First brought into cultivation in the United States from its native Peru in 1949, and again introduced from there in 1965, *Paramongaia* is a genus of one species. It belongs to the amaryllis family AMARYLLIDACEAE. The name derives from the Greek *para,* near, and Paramonga, a locality in Peru near where the plant grows. The vernacular name is apparently the Peruvian Indian one of this plant.

Cojomaria (***P. weberbaueri***) is clearly related to the *Ismene* section of *Hymenocallis* and to *Pamianthe.* It is a deciduous bulb plant with foliage that develops with or later than the blooms, which appear in late fall. The six or seven narrowly-strap-shaped leaves are exceeded in length by the leafless flower stalks, which rise to a height of 2 feet and are topped with a solitary bloom 7 inches in diameter. In aspect suggesting huge daffodils, the flowers are butter-yellow. They have six spreading petals (more properly, tepals), a large trumpet-like crown or corona, six stamens, and one style. The fruits are capsules.

Garden Uses and Cultivation. A very few specialist fanciers of choice and rare plants have had opportunity of testing this beautiful species. Its horticultural possibilities and cultural needs have not yet been adequately explored. It appears to have possibilities as a permanent outdoor plant in regions of mild winters and for growing in greenhouses and conservatories. Success with it has been had at Longwood Gardens, Kennett Square, Pennsylvania, where it was grown in pots in a well-drained, slightly acid mixture of soil, peat moss, leaf mold, and sand or perlite with the addition of a little superphosphate and lime, in a temperature during the growing season 65°F at night, five to ten degrees higher by day. Fairly high humidity and a constant circulation of air must be maintained. The plants are grown in full sun and watered generously. Like many of its relatives, *Paramongaia* blooms most profusely when somewhat pot-bound. Well-rooted specimens should benefit from applications of dilute liquid fertilizer given while they are in active growth. Some six months after growth begins the plants enter their dormant period. Water is then gradually withheld and the bulbs are stored dry in the soil they grew in or in vermiculite until the following late summer or fall when they are repotted or top-dressed and started into growth again. This species propagates readily by offsets and is easily raised from seed. Offsets bloom in two or three years. Seedlings take longer.

PARANA-PINE is *Araucaria angustifolia.*

PARANOMUS (Parán-omus). All fifteen species of this genus are South African. They belong in the protea family PROTEACEAE. Their name, from the Greek *para,* not in accordance with, and *nomos,* law, alludes to some species having two kinds of leaves, the lowermost pinnately-dissected, the upper undivided. Evergreen shrubs with flower heads of four small blooms attended by four bracts, arranged in terminal spikes, these plants are not well known in America, but at least one is cultivated in California.

Paranomus spicatus in South Africa

From 3 to 7 feet tall, ***Paranomus spicatus*** has many erect, downy stems clothed with finely-twice-pinnately-divided leaves about 2 inches long. The flowers, pinkish-lavender to yellowish-white, in spikes 2½ to 3 inches long, are borne in late winter and spring.

Garden Uses and Cultivation. These are similar to those detailed under *Protea.*

ENCYCLOPEDIA OF WORLD ART

Vol. XI

PAKISTAN – REMBRANDT

ENCICLOPEDIA UNIVERSALE DELL'ARTE

Sotto gli auspici della Fondazione Giorgio Cini

ISTITUTO PER LA COLLABORAZIONE CULTURALE
VENEZIA-ROMA

ENCYCLOPEDIA
OF
WORLD ART

McGRAW-HILL BOOK COMPANY
NEW YORK, TORONTO, LONDON

ENCYCLOPEDIA OF WORLD ART: VOLUME XI

Paper for plates and text supplied by Cartiere Burgo, Turin — Engraving by Zincotipia Altimani, Milan — Black-and-white and color plates printed by Tipocolor, Florence — Text printed by "L'Impronta," Florence — Binding by Stabilimento Stianti, San Casciano Val di Pesa, Florence — Book cloth supplied by G. Pasini e C., Milan

Printed in Italy

Library of Congress Catalog Card Number 59-13433
19468

INTERNATIONAL COUNCIL OF SCHOLARS

Production and Technical Direction, English and Italian Editions

CASA EDITRICE G. C. SANSONI

Florence

ABBREVIATIONS

Museums, Galleries, Libraries, and Other Institutions

Antikensamml. — Antikensammlungen
Antiq. — Antiquarium
Bib. Nat. — Bibliothèque Nationale
Bib. Naz. — Biblioteca Nazionale
Brera — Pinacoteca di Brera
Br. Mus. — British Museum
Cab. Méd. — Cabinet des Médailles (Paris, Bibliothèque Nationale)
Cleve. Mus. — Cleveland Museum
Conserv. — Palazzo dei Conservatori
Gall. Arte Mod. — Galleria d'Arte Moderna
IsMEO — Istituto Italiano per il Medio ed Estremo Oriente
Kunstgewerbemus. — Kunstgewerbemuseum
Kunsthist. Mus. — Kunsthistorisches Museum
Louvre — Musée du Louvre
Medagl. — Medagliere
Met. Mus. — Metropolitan Museum
Mus. Ant. — Museo di Antichità
Mus. Arch. — Museo Archeologico
Mus. B. A. — Musée des Beaux-Arts
Mus. Cap. — Musei Capitolini, Capitoline Museums
Mus. Civ. — Museo Civico
Mus. Com. — Museo Comunale
Mus. Etn. — Museo Etnologico
Mus. Naz. — Museo Nazionale
Mus. Vat. — Musei Vaticani, Vatican Museums
Nat. Gall. — National Gallery
Öst. Gal. — Österreichische Galerie
Pin. Naz. — Pinacoteca Nazionale
Prado — Museo del Prado
Rijksmus. — Rijksmuseum
Samml. — Sammlung
Staat. Mus. — Staatliche Museen
Staatsbib. — Staatsbibliothek
Städt. Mus. — Städtisches Museum
Tate Gall. — Tate Gallery
Uffizi — Uffizi Gallery
Vict. and Alb. — Victoria and Albert Museum
Villa Giulia — Museo di Villa Giulia

Reviews and Miscellanies

AAE — Archivio per la Antropologia e la Etnologia, Florence
AAnz — Archäologischer Anzeiger, Berlin
AAV — Archivo de Arte Valenciano, Valencia
AB — Art Bulletin, New York
AbhAkMünchen — Abhandlungen der Bayerischen Akademie der Wissenschaften, Munich
AbhPreussAk — Abhandlungen der preussischen Akademie der Wissenschaften, Berlin; after 1945, Abhandlungen der Deutschen Akademie der Wissenschaften zu Berlin, Berlin
ABIA — Annual Bibliography of Indian Archaeology, Leiden
ABMAB — Anales y Boletín de los Museos de Arte de Barcelona
ABME — Ἀρχεῖον τῶν Βυζαντινῶν Μνημείων τῆς Ἑλλάδος, Athens
AC — Archeologia Classica, Rome
ACCV — Anales del Centro de Cultura Valenciana, Valencia
ActaA — Acta Archaeologica, Copenhagen
ActaO — Acta Orientalia, Leiden, The Hague
AD — Antike Denkmäler, Deutsches Archäologisches Institut, Berlin, Leipzig
AE — Arte Español, Madrid
AEA — Archivo Español de Arqueología, Madrid
AEAA — Archivo Español de Arte y Arqueología, Madrid
AEArte — Archivio Español de Arte, Madrid
AErt — Archaeologiai Értesitö, Budapest
AfA — Archiv für Anthropologie, Brunswick
AfO — Archiv für Orientforschung, Berlin
AfrIt — Africa Italiana, Bergamo
AGS — American Guide Series, U.S. Federal Writers' Project, Works Progress Administration, Washington, D.C., 1935–41
AIEC — Anuari de l'Institut d'Estudies Catalans, Barcelona
AIEG — Anales del Instituto de Estudios Gerundenses, Gerona
AJA — American Journal of Archaeology, Baltimore
AM — Mitteilungen des deutschen archäologischen Instituts, Athenische Abteilung, Athens, Stuttgart
AmA — American Anthropologist, Menasha, Wis.
AmAnt — American Antiquity, Menasha, Wis.
AN — Art News, New York
AnnInst — Annali dell'Instituto di Corrispondenza Archeologica, Rome
AnnSAntEg — Annales du Service des Antiquités de l'Egypte, Cairo
AntC — L'Antiquité Classique, Louvain
AntJ — The Antiquaries Journal, London
AnzAlt — Anzeiger für die Altertumswissenschaft, Innsbruck, Vienna
AnzÖAk — Anzeiger der Österreichischen Akademie der Wissenschaften, Vienna
APAmM — Anthropological Papers of the American Museum of Natural History, New York
AQ — Art Quarterly, Detroit
ArndtBr — P. Arndt, F. Bruckmann, Griechische und römische Porträts, Munich, 1891 ff.
ARSI — Annual Report of the Smithsonian Institution, Bureau of Ethnology, Washington, D.C.
ArtiFig — Arti Figurative, Rome
ASAtene — Annuario della Scuola Archeologica Italiana di Atene, Bergamo
ASI — Archivio Storico Italiano, Florence
ASWI — Archaeological Survey of Western India, Hyderabad
AttiCongrStAMed — Atti dei Congressi di Studi dell'Arte dell'Alto Medioevo
AttiDeSPa — Atti e Memorie della Deputazione di Storia Patria
AttiPontAcc — Atti della Pontificia Accademia Romana di Archeologia, Rome
Atti3StArch — Atti del III Convegno Nazionale di Storia dell'Architettura, Rome, 1938
Atti5StArch — Atti del V Convegno Nazionale di Storia dell'Architettura, Florence, 1957

AZ — Archäologische Zeitung, Berlin

BA — Baessler Archiv, Leipzig, Berlin

BABsch — Bulletin van de Vereeniging tot bevordering der kennis van de antieke Beschaving, The Hague

BAC — Bulletin du Comité des Travaux Historiques et Scientifiques, Section d'Archéologie, Paris

BAcBelg — Bulletin de l'Académie Royale de Belgique, Cl. des Lettres, Brussels

BACr — Bollettino di Archeologia Cristiana, Rome

BAEB — Bureau of American Ethnology, Bulletins, Washington, D.C.

BAER — Bureau of American Ethnology, Reports, Washington, D.C.

BAFr — Bulletin de la Société Nationale des Antiquaires de France, Paris

BAmSOR — Bulletin of the American Schools of Oriental Research, South Hadley, Mass.

BArte — Bollettino d'Arte del Ministero della Pubblica Istruzione, Rome

BAT — Boletín Arqueológico Tarragona

BBMP — Boletín de la Biblioteca Menéndez Pelayo, Santander

BByzI — The Bulletin of the Byzantine Institute, Paris

BCH — Bulletin de Correspondance Hellénique, Paris

BCom — Bullettino della Commissione Archeologica Comunale, Rome

BCPMB — Boletín de la Comision Provincial de Monumentos de Burgos

BCPML — Boletín de la Comision Provincial de Monumentos de Lugo

BCPMO — Boletín de la Comision Provincial de Monumentos de Orense

Beazley, ABV — J. D. Beazley, Attic Black-figure Vase-painters, Oxford, 1956

Beazley, ARV — J. D. Beazley, Attic Red-figure Vase-painters, Oxford, 1942

Beazley, EVP — J. D. Beazley, Etruscan Vase-painting, Oxford, 1947

Beazley, VA — J. D. Beazley, Attic Red-figured Vases in American Museums, Cambridge, 1918

Beazley, VRS — J. D. Beazley, Attische Vasenmaler des rotfigurigen Stils, Tübingen, 1925

BEFEO — Bulletin de l'Ecole Française d'Extrême-Orient, Hanoi, Saigon, Paris

BerlNZ — Berliner Numismatische Zeitschrift, Berlin

Bernoulli, GI — J. J. Bernoulli, Griechische Ikonographie, Munich, 1901

Bernoulli, RI — J. J. Bernoulli, Römische Ikonographie, I, Stuttgart, 1882; II, 1, Berlin, Stuttgart, 1886; II, 2, Stuttgart, Berlin, Leipzig, 1891; II, 3, Stuttgart, Berlin, Leipzig, 1894

BHAcRoum — Bulletin Historique, Académie Roumaine, Bucharest

BICR — Bollettino dell'Istituto Centrale del Restauro, Rome

BIE — Bulletin de l'Institut de l'Egypte, Cairo

BIFAN — Bulletin de l'Institut Français d'Afrique Noire, Dakar

BIFAO — Bulletin de l'Institut Français d'Archéologie Orientale, Cairo

BIFG — Boletín de la Institución Fernán González, Burgos

BInst — Bullettino dell'Instituto di Corrispondenza Archeologica, Rome

BJ — Bonner Jahrbücher, Bonn, Darmstadt

BM — Burlington Magazine, London

BMBeyrouth — Bulletin du Musée de Beyrouth, Beirut

BMC — British Museum, Catalogue of Greek Coins, London

BMCEmp — H. Mattingly, Coins of the Roman Empire in the British Museum, London

BMFA — Museum of Fine Arts, Bulletin, Boston

BMFEA — Museum of Far-Eastern Antiquities, Bulletin, Stockholm

BMImp — Bullettino del Museo dell'Impero, Rome

BMMA — Bulletin of the Metropolitan Museum of Art, New York

BMN — Boletín Monumentos Navarra

BMQ — The British Museum Quarterly, London

BNedOud — Bulletin van de Koninklijke Nederlandse Oudheidkundige Bond, Leiden

BPI — Bullettino di Paletnologia Italiana, Rome

BRA — Boletín de la Real Academia de Ciencias, Bellas Letras y Nobles Artes de Córdoba, Córdoba

BRABASF — Boletín de la Real Academia de Bellas Artes de San Fernando, Madrid

BRABLB — Boletín de la Real Academia de Buenas Letras de Barcelona

BRAH — Boletín de la Real Academia de la Historia, Madrid

BrBr — H. Brunn, F. Bruckmann, Denkmäler griechischer und römischer Skulptur, Munich

Brunn, GGK — H. Brunn, Geschichte der griechischen Künstler, 2d ed., Stuttgart, 1889

Brunn, GK — H. Brunn, Griechische Kunstgeschichte, Munich, I, 1893; II, 1897

BSA — Annual of the British School at Athens, London

BSCC — Boletín de la Sociedad Castellonense de Cultura, Castellón de la Plana

BSCE — Boletín de la Sociedad Castellana de Excursiones, Valladolid

BSEAAV — Boletín del Seminario de Estudios de Arte y Arqueología, Universidad de Valladolid

BSEE — Boletín de la Sociedad Española de Excursiones, Madrid

BSEI — Bulletin de la Société des Etudes Indochinoises, Saigon

BSOAS — Bulletin of the School of Oriental and African Studies, London

BSPF — Bulletin de la Société Préhistorique Française, Paris

BSR — Papers of the British School at Rome, London

Cabrol-Leclercq — F. Cabrol, H. Leclercq, Dictionnaire d'archéologie chrétienne et de liturgie, Paris, 1907

CAF — Congrès Archéologique de France, Paris, 1841–1935

CahA — Cahiers Archéologiques, Fin de l'Antiquité et Moyen-Age, Paris

CahArt — Cahiers d'art, Paris

CAJ — Central Asiatic Journal, Wiesbaden

CAUG — Cuadernos de Arte de la Universidad de Granada

CEFEO — Cahiers de l'Ecole Française d'Extrême-Orient, Paris

CEG — Cuadernos de Estudios Gallegos, Santiago de Compostela

CIE — Corpus Inscriptionum Etruscarum, Lipsiae

CIG — Corpus Inscriptionum Graecarum, Berolini

CIL — Corpus Inscriptionum Latinarum, Berolini

CIS — Corpus Inscriptionum Semiticarum, Parisiis

Coh — H. Cohen, Description historique des Monnaies frappées sous l'Empire Romain, Paris

Collignon, SG — M. Collignon, Histoire de la sculpture grecque, Paris, I, 1892; II, 1897

Comm — Commentari, Florence, Rome

Cr — La Critica, Bari

CRAI — Comptes Rendus de l'Académie des Inscriptions et Belles-Lettres, Paris

CrArte — La Critica d'Arte, Florence

CVA — Corpus Vasorum Antiquorum

DA — N. Daremberg, N. Saglio, Dictionnaire des antiquités grecques et romaines, Paris, 1877–1912

Dehio, I–V — G. Dehio, Handbuch der deutschen Kunstdenkmäler, Berlin, I, Mitteldeutschland, 1927; II, Nordostdeutschland, 1926; III, Süddeutschland, 1933; IV, Südwestdeutschland, 1933; V, Nordwestdeutschland, 1928

Dehio, DtK — G. Dehio, Geschichte der deutschen Kunst, 8 vols., Berlin, 1930–34

Dehio-VonBezold — G. Dehio, G. von Bezold, Die kirchliche Baukunst des Abendlandes, Stuttgart, 1892–1901

DissPontAcc — Dissertazioni della Pontificia Accademia Romana di Archeologia, Rome

EA — Photographische Einzelaufnahmen, Munich, 1893 ff.

EAA — Enciclopedia dell'Arte Antica, Rome, I, 1958; II, 1959; III, 1960; IV, 1961

EArt — Eastern Art, London

EB — Encyclopaedia Britannica

'ΕΕΒΣ — 'Επετερίς 'Εταιρεία Βυζαντινῶν Σπουδῶν Athens

'Εφημ — 'Αρχαιολογικὴ 'Εφημερίς, Athens

EI — Enciclopedia Italiana, Rome, 1929 ff.
EphDR — Ephemeris Dacoromana, Rome
'Εργον — Τὸ ἔργον τῆς ἀρχαιολογικῆς ἑταιρείας, ed. A. K. Orlandos, Athens
ES — Estudios Segovianos, Segovia
ESA — Eurasia Septentrionalis Antiqua, Helsinki
Espér — E. Espérandieu, R. Lantier, Recueil général des Bas-Reliefs de la Gaule Romaine, Paris
EUC — Estudis Universitaris Catalans, Barcelona
FA — Fasti Archaeologici, Florence
FD — Fouilles de Delphes, Paris
Friedländer — Max Friedländer, Altniederländische Malerei, Berlin, 1924–37
Furtwängler, AG — A. Furtwängler, Antiken Gemmen, Leipzig, Berlin, 1900
Furtwängler, BG — A. Furtwängler, Beschreibung der Glyptothek König Ludwig I zu München, Munich, 1900
Furtwängler,KlSchr— A. Furtwängler, Kleine Schriften, Munich, 1912
Furtwängler, MP — A. Furtwängler, Masterpieces of Greek Sculpture, London, 1895
Furtwängler, MW — A. Furtwängler, Meisterwerke der griechischen Plastik, Leipzig, Berlin, 1893
Furtwängler Reichhold — A. Furtwängler, K. Reichhold, Griechische Vasenmalerei, Munich
FWP — U.S. Federal Writers' Project, Works Progress Administration, Washington, D.C., 1935–1941
GBA — Gazette des Beaux-Arts, Paris
GJ — The Geographical Journal, London
HA — Handbuch der Archäologie im Rahmen des Handbuchs der Altertumswissenschaft . . . , herausgegeben von Walter Otto, Munich, 1939–53
HABS — Historic American Buildings Survey, U.S. Library of Congress, Washington, D.C.
HBr — P. Herrmann, F. Bruckmann, Denkmäler der Malerei des Altertums, Munich, 1907
Helbig-Amelung — W. Helbig, W. Amelung, E. Reisch, F. Weege, Führer durch die öffentlichen Sammlungen klassischer Altertümer in Rom, Leipzig, 1912–13
HIPBC — A Handbook for Travellers in India, Pakistan, Burma and Ceylon, London, 1955
HJAS — Harvard Journal of Asiatic Studies, Cambridge, Mass.
Hoppin, Bf — J. C. Hoppin, A Handbook of Greek Black-figured Vases with a Chapter on the Red-figured Southern Italian Vases, Paris, 1924
Hoppin, Rf — J. C. Hoppin, A Handbook of Attic Red-figured Vases Signed by or Attributed to the Various Masters of the Sixth and Fifth Centuries B.C., Cambridge, 1919
HSAI — J. H. Steward, ed., Handbook of South American Indians, 6 vols., Bureau of American Ethnology, Bull. 143, Washington, D.C., 1946–50
IAE — Internationales Archiv für Ethnographie, Leiden
IBAI — Bulletin de l'Institut Archéologique Bulgare, Sofia
IFAN — Institut Français Afrique Noire, Dakar
IG — Inscriptiones Graecae, Berolini
ILN — Illustrated London News, London
IPEK — Ipek, Jahrbuch für prähistorische und ethnographische Kunst, Berlin
ITTM — Instituto Tello Tellez de Meneses, Palencia
JA — Journal Asiatique, Paris
JAF — Journal of American Folklore, Lancaster, Pa.
JAOS — Journal of the American Oriental Society, Baltimore
JAS — Journal of the African Society, London
JBORS — Journal of the Bihar and Orissa Research Society, Patna, India
JdI — Jahrbuch des deutschen archäologischen Instituts, Berlin
JEA — Journal of Egyptian Archaeology, London
JhbKhSammlWien — Jahrbuch der kunsthistorischen Sammlungen in Wien, Vienna
JhbPreussKSamml — Jahrbuch der preussischen Kunstsammlungen, Berlin
JHS — Journal of Hellenic Studies, London
JIAI — Journal of Indian Art and Industry, London
JIAN — Journal International d'Archéologie Numismatique, Athens
JISOA — Journal of the India Society of Oriental Art, Calcutta
JNES — Journal of Near Eastern Studies, Chicago
JPS — Journal of the Polynesian Society, Wellington, New Zealand
JRAI — Journal of the Royal Anthropological Institute of Great Britain and Ireland, London
JRAS — Journal of the Royal Asiatic Society, London
JRS — Journal of Roman Studies, London
JS — Journal des Savants, Paris
JSA — Journal de la Société des Africanistes, Paris
JSAH — Journal of the Society of Architectural Historians, Charlottesville, Va.
JSAm — Journal de la Société des Américanistes, Paris
JSO — Journal de la Société des Océanistes, Paris
KbNed — Kunstreisboek voor Nederland, Amsterdam, 1960
Klein, GrK — W. Klein, Geschichte der griechischen Kunst, Leipzig, 1904–07
KS — Communications on the Reports and Field Research of the Institute of Material Culture, Moscow, Leningrad
Lippold, GP — G. Lippold, Die griechische Plastik (W. Otto, Handbuch der Archäologie, II, 1), Munich, 1950
Löwy, IGB — E. Löwy, Inschriften griechischer Bildhauer, Leipzig, 1885
MAAccIt — Monumenti Antichi dell'Accademia d'Italia, Milan
MAARome — Memoirs of the American Academy in Rome, Rome, New York
MAF — Mémoires de la Société Nationale des Antiquaires de France, Paris
MAGWien — Mitteilungen der anthropologischen Gesellschaft in Wien, Vienna
Mâle, I — E. Mâle, L'art religieux du XIIe siècle en France, Paris, 1928
Mâle, II — E. Mâle, L'art religieux du XIIIe siècle en France, Paris, 1925
Mâle, III — E. Mâle, L'art religieux de la fin du moyen-âge en France, Paris, 1925
Mâle, IV — E. Mâle, L'art religieux après le Concile de Trente, Paris, 1932
MALinc — Monumenti Antichi dell'Accademia dei Lincei, Milan, Rome
Mattingly-Sydenham — H. Mattingly, E. Sydenham, C. H. V. Sutherland, The Roman Imperial Coinage, London
MdI — Mitteilungen des deutschen archäologischen Instituts, Munich
MdIK — Mitteilungen des deutschen Instituts für ägyptische Altertumskunde in Kairo, Wiesbaden
Mél — Mélanges d'Archéologie et d'Histoire (Ecole Française de Rome), Paris
Mem. Junta Sup. Exc. — Memoria de la Junta Superior de Excavaciones y Antigüedades, Madrid
MemLinc — Memorie dell'Accademia dei Lincei, Rome
MGH — Monumenta Germaniae Historica, Berlin
MIA — Material and Research in Archaeology of the U.S.S.R., Moscow, Leningrad
Michel — A. Michel, Histoire de l'art depuis les premiers temps chrétiens jusqu'à nos jours, Paris, 1905–29
MInst — Monumenti dell'Instituto di Corrispondenza Archeologica, Rome
MLJ — Modern Language Journal, St. Louis, Mo.
MnbKw — Monatsberichte über Kunstwissenschaft
MPA — Monumenti della pittura antica scoperti in Italia, Rome
MPiot — Fondation Eugène Piot, Monuments et Mémoires, Paris
MPontAcc — Memorie della Pontificia Accademia Romana di Archeologia, Rome
NBACr — Nuovo Bullettino di Archeologia Cristiana, Rome
NChr — Numismatic Chronicle and Journal of the Royal Numismatic Society, London
NedKhJb — Nederlandsch Kunsthistorisch Jaarboek, 1945ff.
NedMon — De Nederlandse Monumenten van Geschiedenis en Kunst, 1911 ff.
NIFAN — Notes de l'Institut Français d'Afrique Noire, Dakar

NR — Numismatic Review, New York
NSc — Notizie degli Scavi di Antichità, Rome
NZ — Numismatische Zeitschrift, Vienna
OAZ — Ostasiatische Zeitschrift, Vienna
OIP — Oriental Institute Publications, Chicago
ÖJh — Jahreshefte des Österreichischen archäologischen Instituts, Vienna
ÖKT — Österreichische Kunsttopographie, Vienna
OMLeiden — Oudheidkundige Mededeelingen van het Rijksmuseum van Oudheten te Leiden, Leiden
OpA — Opuscola Archaeologica, Lund
OTNE — Old Time New England; the Bulletin of the Society for the Preservation of New England Antiquities, Boston, Massachusetts, I, 1910
OudJb — Oudheidkundig Jaarboek, Leiden
Overbeck, SQ — J. Overbeck, Die antiken Schriftquellen zur Geschichte der bildenden Künste bei den Griechen, Leipzig, 1868; reprint, Hildesheim, 1958
Oxy. Pap. — The Oxyrhynchus Papyri, by B. P. Grenfell, A. S. Hunt, H. I. Bell, et al., eds., London, 1898 ff.
ΠΑΕ — Πρακτικὰ τῆς ἐν Ἀθήναις Ἀρχαιολογικῆς Ἑταιρίας, Athens
PEQ — Palestine Exploration Quarterly, London
Perrot-Chipiez — G. Perrot, C. Chipiez, Histoire de l'art dans l'Antiquité, Paris, I, 1882; II, 1884; III, 1885; IV, 1887; V, 1890; VI, 1894; VII, 1898; VIII, 1903; IX, 1911
Pfuhl — E. Pfuhl, Malerei und Zeichnung der Griechen, Munich, 1923
Picard — C. Picard, Manuel d'Archéologie, La Sculpture, Paris, I, 1935; II, 1939; III, 1948; IV, 1, 1954
PL — J. P. Migne, Patrologiae cursus completus, Series Latina, 221 vols., Paris, 1844–64
PM — B. Porter and R. L. B. Moss, Topographical Bibliography of Ancient Egyptian Hieroglyphic Texts, Reliefs and Paintings, 7 vols., Oxford, 1927–51, 2d ed., 1960 ff.
Porter — A. Kingsley Porter, Romanesque Sculpture of the Pilgrimage Roads, Boston, 1923
Post — Charles Post, A History of Spanish Painting, 10 vols., Cambridge, Mass., 1930 ff.
ProcPrSoc — Proceedings of the Prehistoric Society, Cambridge
PSI — Pubblicazioni della Società Italiana per la ricerca dei papiri greci e latini in Egitto, Florence, 1912 ff.
QCr — Quaderni della Critica, Bari
RA — Revue Archéologique, Paris
RAA — Revue des Arts Asiatiques, Paris
RAAB — Revista de la Asociacion Artístico-Arqueológica Barcelonesa, Barcelona
RABM — Revista de Archivos, Bibliotecas y Museos, Madrid
RACr — Rivista di Archeologia Cristiana, Rome
RArte — Rivista d'Arte, Florence
RArts — Revue des arts, Paris
RBib — Revue Biblique, Paris
RCHS — Records of the Columbia Historical Society, Washington, D.C., I, 1897
RE — A. Pauly, G. Wissowa, Real-Encyclopädie der classischen Altertumswissenschaft, Stuttgart, 1894 ff.
REA — Revue des Etudes Anciennes, Bordeaux
REByz — Revue des Etudes Byzantines, Paris
REE — Revista des Estudios Extremesios, Badajoz
REG — Revue des Etudes Grecques, Paris
Reinach, RP — S. Reinach, Répertoire des Peintures Grecques et Romaines, Paris, 1922
Reinach, RR — S. Reinach, Répertoire des Reliefs Grecs et Romains, Paris, I, 1909; II and III, 1912
Reinach, RS — S. Reinach, Répertoire de la Statuaire Grecque et Romaine, Paris, I, 1897; II, 1, 1897; II, 2, 1898; III, 1904; IV, 1910
Reinach, RV — S. Reinach, Répertoire des Vases peints, grecs et étrusques, Paris, I, 1899; II, 1900
REL — Revue des Etudes Latines, Paris
RendAccIt — Rendiconti della R. Accademia d'Italia, Rome
RendLinc — Rendiconti dell'Accademia dei Lincei, Rome
RendNapoli — Rendiconti dell'Accademia di Archeologia di Napoli, Naples
RendPontAcc — Rendiconti della Pontificia Accademia Romana di Archeologia, Rome
RepfKw — Repertorium für Kunstwissenschaft, Berlin, Stuttgart
REthn — Revue d'Ethnographie, Paris
RhMus — Rheinisches Museum für Philologie, Frankfort on the Main
RIASA — Rivista dell'Istituto Nazionale d'Archeologia e Storia dell'Arte, Rome
RIE — Revista de Ideas Estéticas, Madrid
RIN — Rivista Italiana di Numismatica, Rome
RlDKg — Reallexicon zur deutschen Kunstgeschichte, Stuttgart, 1937 ff.
RLV — M. Ebert, Real-Lexicon der Vorgeschichte, Berlin, 1924–32
RM — Mitteilungen des deutschen archäologischen Instituts, Römische Abteilung, Berlin
RN — Revue Numismatique, Paris
RNA — Revista Nacional de Arquitectura, Madrid
Robert, SR — C. Robert, Die antiken Sarkophag-Reliefs, Berlin, 1890 ff.
Roscher — W. H. Roscher, Ausführliches Lexikon der griechischen und römischen Mythologie, Leipzig, 1884–86; 1924–37
RQ — Römische Quartalschrift, Freiburg
RScPr — Rivista di Scienze Preistoriche, Florence
RSLig — Rivista di Studi Liguri, Bordighera, Italy
RSO — Rivista degli Studi Orientali, Rome
Rumpf, MZ — A. Rumpf, Malerei und Zeichnung (W. Otto, Handbuch der Archäologie, IV, 1), Munich, 1953
RUO — Revista de la Universidad de Oviedo
SA — Soviet Archaeology, Moscow, Leningrad
SAA — Seminario de Arte Aragonés, Zaragoza
SbBerlin — Sitzungsberichte der preussischen Akademie der Wissenschaften, Berlin
SbHeidelberg — Sitzungsberichte der Akademie der Wissenschaften zu Heidelberg, Heidelberg
SbMünchen — Sitzungsberichte der bayerischen Akademie der Wissenschaften zu München, Munich
SbWien — Sitzungsberichte der Akademie der Wissenschaften in Wien, Vienna
Schlosser — J. Schlosser, La letteratura artistica, Florence, 1956
Scranton, Greek Walls — R. L. Scranton, Greek Walls, Cambridge, Mass., 1941
SEtr — Studi Etruschi, Florence
SNR — Sudan Notes and Records, Khartoum
SPA — A Survey of Persian Art, ed. A. U. Pope and P. Ackerman, Oxford, 1938
SymbOsl — Symbolae Osloenses, Oslo
Tebtunis — The Tebtunis Papyri, B. P. Grenfell, A. S. Hunt, et al., eds., London, 1902 ff.
ThB — U. Thieme, F. Becker, Künstler Lexikon, Leipzig, 1907–50
TitAM — Tituli Asiae Minoris, Vindobonae, 1901–44
TNR — Tanganyika Notes and Records, Dar-es-Salaam
Toesca, Md — P. Toesca, Il Medioevo, 2 vols., Turin, 1927
Toesca, Tr — P. Toesca, Il Trecento, Turin, 1951
TP — T'oung Pao, Leiden
UCalPAAE — University of California, Publications in American Archaeology and Ethnology, Berkeley
USMB — United States National Museum, Bulletin, Washington, D.C.
Van Marle — R. van Marle, The Development of the Italian Schools of Painting, The Hague, 1923–38
Vasari — G. Vasari, Vite, ed. Milanesi, Florence, 1878 ff. (Am. ed., trans. E. H. and E. W. Blashfield and A. A. Hopkins, 4 vols., New York, 1913)
Venturi — A. Venturi, Storia dell'Arte Italiana, Milan, 1901 ff.
VFPA — Viking Fund Publications in Anthropology, New York
Vollmer — H. Vollmer, Allgemeines Lexikon der bildenden Künstler des XX. Jahrhunderts, Leipzig, 1953
Warburg — Journal of the Warburg and Courtauld Institutes, London
Weickert, Archaische Architektur — C. Weickert, Typen der archaischen Architektur in Griechenland und Kleinasien, Augsburg, 1929
Wpr — Winckelmannsprogramm, Berlin
WürzbJ — Würzburger Jahrbücher für die Altertumswissenschaft, Würzburg
WVDOG — Wissenschaftliche Veröffentlichungen der Deutschen Orient-Gesellschaft, Leipzig, Berlin

ZäS — Zeitschrift für ägyptische Sprache und Altertumskunde, Berlin, Leipzig
ZfAssyr — Zeitschrift für Assyriologie, Strasbourg
ZfbK — Zeitschrift für bildende Kunst, Leipzig
ZfE — Zeitschrift für Ethnologie, Berlin
ZfKg — Zeitschrift für Kunstgeschichte, Munich
ZfKw — Zeitschrift für Kunstwissenschaft, Munich
ZfN — Zeitschrift für Numismatik, Berlin
ZMG — Zeitschrift der deutschen morgenländischen Gesellschaft, Leipzig
ZSAKg — Zeitschrift für schweizerische Archäologie und Kunstgeschichte, Basel

Languages and Ethnological Descriptions

Alb. — Albanian
Am. — American
Ang. — Anglice, Anglicized
Ar. — Arabic
Arm. — Armenian
AS. — Anglo-Saxon
Bab. — Babylonian
Br. — British
Bulg. — Bulgarian
Chin. — Chinese
D. — Dutch
Dan. — Danish
Eg. — Egyptian
Eng. — English
Finn. — Finnish
Fr. — French
Ger. — German
Gr. — Greek
Heb. — Hebrew
Hung. — Hungarian
It. — Italian
Jap. — Japanese
Jav. — Javanese
Lat. — Latin
Mod. Gr. — Modern Greek
Nor. — Norwegian
Per. — Persian
Pol. — Polish
Port. — Portuguese
Rum. — Rumanian
Rus. — Russian
Skr. — Sanskrit
Sp. — Spanish
Swed. — Swedish
Yugo. — Yugoslav

Other Abbreviations (*Standard abbreviations in common usage are omitted.*)

Abh. — Abhandlungen
Acad. — Academy, Académie
Acc. — Accademia
Adm. — Administration
Ak. — Akademie
Allg. — Allgemein
Alm. — Almanacco
Amm. — Amministrazione
Ann. — Annals, Annali, Annuario, Annual, etc.
Ant. — Antiquity, Antico, Antiquaire, etc.
Anthr. — Anthropology, etc.
Antr. — Antropologia, etc.
Anz. — Anzeiger
Arch. — Architecture, Architettura, Architettonico, etc.; Archives
Archaeol. — Archaeology, etc.
Arqueol. — Arqueología, etc.
attrib. — attributed
Aufl. — Auflage
Aufn. — Aufnahme
B. — Bulletin, Bollettino, etc.
b. — born
Belg. — Belgian, Belga, etc.
Berl. — Berlin, Berliner
Bern. — Berner
Bib. — Bible, Biblical, Bibliothèque, etc.
Bibliog. — Bibliography, etc.
Bur. — Bureau
Byz. — Byzantine
C. — Corpus
ca. — circa
Cah. — Cahiers
Cal. — Calendar
Cap. — Capital, Capitolium
Cat. — Catalogue, Catalogo, etc.
Chr. — Chronicle, Chronik
Civ. — Civiltà, Civilization, etc.
cod. — codex
col., cols. — column, columns
Coll. — Collection, Collana, Collationes, Collectanea, Collezione, etc.
Comm. — Commentaries, Commentari, Communications, etc.
Com. — Comunale
Cong. — Congress, Congresso, etc.
Cr. — Critica
Cron. — Cronaca
Cuad. — Cuadernos
Cult. — Culture, Cultura, etc.
D. — Deutsch
d. — died
Diss. — Dissertation, Dissertazione
Doc. — Documents, etc.
E. — Encyclopedia, etc.
Eccl. — Ecclesiastic, Ecclesia, etc.
Ep. — Epigraphy
Esp. — España, Español
Est. — Estudios
Et. — Etudes
Ethn. — Ethnology, Ethnography, Ethnographie, etc.
Etn. — Etnico, Etnografia, etc.
Etnol. — Etnologia
Eur. — Europe, Europa, etc.
ext. — extract
f. — für
fasc. — fascicle
Fil. — Filologia
Filos. — Filosofia, Filosofico
fol. — folio
Forsch. — Forschung, Forschungen
Gal. — Galerie
Gall. — Gallery, Galleria
Geog. — Geography, Geografia, Geographical, etc.
Giorn. — Giornale
H. — History, Histoire, etc.
hl. — heilig, heilige
Holl. — Hollandisch, etc.
Hum. — Humanity, Humana, etc.
Ill. — Illustration, Illustrato, Illustrazione, etc.
Ind. — Index, Indice, Indicatore, etc.
Inf. — Information, Informazione, etc.
Inst. — Institute, Institut, Instituto, etc.
Int. — International, etc.
Ist. — Istituto
J. — Journal
Jb. — Jaarboek
Jhb. — Jahrbuch
Jhrh. — Jahreshefte
K. — Kunst
Kat. — Katalog
Kchr. — Kunstchronik
Kg. — Kunstgeschichte
K.K. — Kaiserlich und Königlich
Kunsthist. — Kunsthistorische
Kw. — Kunstwissenschaft
Lett. — Letteratura, Lettere
Lib. — Library
ling. — linguistica, lingua, etc.
Lit. — Literary, Literarische, Littéraire, etc.
Mag. — Magazine
Med. — Medieval, Medievale, etc.
Meded. — Mededeelingen
Mél. — Mélanges
Mém. — Mémoire
Mem. — Memorie, Memoirs
Min. — Minerva
Misc. — Miscellany, Miscellanea, etc.
Mit. — Mitteilungen
Mnb. — Monatsberichte
Mnbl. — Monatsblätter
Mnh. — Monatshefte
Mod. — Modern, Moderno, etc.

Mon. — Monuments, Monumento
Münch. — München, Münchner
Mus. — Museum, Museo, Musée, Museen, etc.
Muz. — Muzeum
N. — New, Notizia, etc.
Nac. — Nacional
Nachr. — Nachrichten
Nat. — National, etc.
Naz. — Nazionale
Notit. dign. — Notitia Dignitatum
N.S. — new series
O. — Oriental, Orient, etc.
Ö. — Österreichische
obv. — obverse
öffentl. — öffentlich
Op. — Opuscolo
Pap. — Papers, Papyrus
Per. — Periodical, Periodico
Pin. — Pinacoteca
Pr. — Prehistory, Preistoria, Preystori, Préhistoire
Proc. — Proceedings
Pub. — Publication, Publicación
Pubbl. — Pubblicazione
Q. — Quarterly, Quaderno
Quel. — Quellen
R. — Rivista
r — recto
r. — reigned
Racc. — Raccolta
Rass. — Rassegna
Rec. — Recueil
Recens. — Recensione
Rech. — Recherches
Rel. — Relazione
Rend. — Rendiconti
Rép. — Répertoire
Rep. — Report, Repertorio, Repertorium
Rev. — Review, Revue, etc.
Rl. — Reallexikon
Rom. — Roman, Romano, Romanico, etc.
rv. — reverse
S. — San, Santo, Santa (saint)
S. — Studi, Studies, etc.
Samml. — Sammlung, Sammlungen
Sc. — Science, Scienza, Scientific, etc.
Schr. — Schriften
Schw. — Schweitzer
Script. — Scriptorium
Sitzb. — Sitzungsberichte
s.l. — in its place
Soc. — Social, Society, Società, Sociale, etc.
Spec. — Speculum
SS. — Saints, Sante, Santi, Santissima
St. — Saint
Sta — Santa (holy)
Ste — Sainte
Sto — Santo (holy)
Sup. — Supplement, Supplemento
s.v. — under the word
Tech. — Technical, Technology, etc.
Tecn. — Tecnica, Tecnico
Tr. — Transactions
trans. — translator, translated, etc.
Trav. — Travaux
u. — und
Um. — Umanesimo
Univ. — University, Università, Université, etc.
Urb. — Urban, Urbanistica
v — verso
VAT — Vorderasiatische Tafeln
Verh. — Verhandlungen, Verhandelingen
Verz. — Verzeichnis
Vf. — Verfasser
Wien. — Wiener
Yb. — Yearbook
Z. — Zeitschrift, Zeitung, etc.

NOTES ON THE ENGLISH EDITION

Standards of Translation. Contributors to the Encyclopedia, drawn from the outstanding authorities of over 35 different countries, have written in many languages — Italian, Spanish, French, German, Russian, etc. To ensure faithful translation of the author's thought, all articles have been translated into English from the original language, checked for the accuracy of technical terms and accepted English forms of nomenclature by English and American art historians, and correlated with the final editorial work of the Italian edition for uniformity and coherence of the over-all presentation. Naturally the McGraw-Hill Book Company assumes full responsibility for the accuracy and completeness of all translations. Those articles written in English appear in the words and style of the authors, within the bounds of editorial attention to consistency and stylistic and organizational unity of the work as a whole. Article titles are in most cases parallel to those in the Italian edition, though occasionally they have been simplified, as *Dravidian Art* for *Dravidiche Correnti e Tradizioni.*

New Features. Although generally the English-language edition corresponds to the Italian version, a small number of purely editorial changes have been made in the interest of clear English-language alphabetization and occasional deletions or amplifications solely in the interest of clarity. Three major differences between the two editions do exist, however:

A considerable number of cross-references have been added in many places where it was felt that relating the subject under consideration to other pertinent articles would be of value to the reader.

A more extensive article on the Art of the Americas was projected for Volume One of the English edition with an entirely new text and many plates in black and white and color. This article was designed to give the completest possible coverage within the existing space of some 100,000 words to a subject which, because of its interest to the English-speaking public, was entrusted to a group of well-known American scholars, each expert in his respective area.

Some 300 separate short biographies have been added to the English edition to provide ready access to data on the lives, works, and critical acceptance of certain artists identified with schools, movements, and broad categories of historical development that are treated in the longer monographic articles. These articles are unillustrated, but works of the artists are represented in the plates accompanying the longer articles.

Bibliographies. The bibliographies of the original Italian edition have been amplified at times to include titles of special interest to the English-speaking world and English-language editions of works originally published in other languages.

In undertaking these adaptations of the Italian text and preparing original material for the English edition, the publisher has been aided by the generous advice and, in many cases, collaboration of the members of the Editorial Advisory Committee.

CONTRIBUTORS TO VOLUME XI

Bianca Maria ALFIERI, University of Rome
Martín ALMAGRO BASCH, Director, Instituto Español de Prehistoria, Madrid; University of Madrid
Sir Gilbert ARCHEY, Auckland, New Zealand
Rudolph ARNHEIM, Sarah Lawrence College, Bronxville, N.Y.
Edoardo ARSLAN, University of Pavia
Bernard ASHMOLE, Emeritus Professor, Oxford University
Richard D. BARNETT, Keeper, Department of Western Asiatic Antiquities, British Museum, London
Eugenio BATTISTI, University of Genoa
Kurt BAUCH, Professor Emeritus, University of Freiburg im Breisgau
Hermann BAUMANN, Institut für Völkerkunde der Universität, Munich
Giovanni BECATTI, University of Florence
Ernst BERGER, Basel
T. S. R. BOASE, President, Magdalen College, Oxford
Stefano BOTTARI, University of Bologna
Cesare BRANDI, University of Palermo
Giuliano BRIGANTI, Rome
Alessandro BRISSONI, Accademia di Arte Drammatica, Rome
Anna Maria BRIZIO, University of Milan
Gaetano BRUNDU, Cagliari
Geoffrey H. S. BUSHNELL, Curator, University Museum of Archaeology and Ethnology, Cambridge, England
Mario BUSSAGLI, University of Rome
Michelangelo CAGIANO DE AZEVEDO, Università Cattolica del Sacro Cuore, Milan
Marco CHIARINI, Florence
Pio CIPROTTI, Pontificia Università Lateranense; University of Camerino, Italy
Peter COLLINS, McGill University, Montreal
René CROZET, University of Poitiers, France
Bernhard DEGENHART, Staatliche Graphische Sammlung, Munich
Marguerite Marie DENECK, Musée Guimet, Paris
Sergio DONADONI, University of Rome
Gillo DORFLES, University of Trieste
Dario DURBÉ, Rome
Gustav ECKE, Curator of Chinese Art, Honolulu Academy of Arts; University of Hawaii, Honolulu
Lorenz EITNER, University of Minnesota, Minneapolis
Leopold D. ETTLINGER, University College, London
H. J. EYSENCK, University of London
S. Lane FAISON, Jr., Williams College, Williamstown, Mass.
Oreste FERRARI, Rome
Sydney J. FREEDBERG, Harvard University, Cambridge, Mass.
George Van Derveer GALLENKAMP, New York City
Kenneth GARLICK, Assistant Curator, The Barber Institute of Fine Arts, Birmingham, England
Decio GIOSEFFI, University of Trieste
Antonio GIULIANO, University of Rome
Hermann GOETZ, University of Heidelberg
Luigi GRASSI, University of Rome
Paolo GRAZIOSI, University of Florence
Anna GRELLE, Rome
Francis HASKELL, King's College, Cambridge, England
Egbert HAVERKAMP-BEGEMANN, Yale University, New Haven, Conn.
Wilhelm HOLMQVIST, Superintendent, Historiska Museet, Stockholm
William I. HOMER, Cornell University, Ithaca, N.Y.
Henry R. HOPE, Indiana University, Bloomington
Christopher HUSSEY, London
Carol Stevens KNER, New York City
Stefan KOZAKIEWICZ, Warsaw
Vittorio LANTERNARI, University of Bari
Oliver W. LARKIN, Emeritus Professor of Art, Smith College, Northampton, Mass.
Michael LEVEY, National Gallery, London
Stanislaw LORENTZ, Director, National Museum, Warsaw
Bates LOWRY, Director, Museum of Modern Art, New York City
Arthur MCCOMB, Boston, Mass.
Fernanda DE' MAFFEI, Rome
Roland MARTIN, University of Dijon
Paolo MATTHIAE, Rome
Tullio DE MAURO, University of Rome
A. Hyatt MAYOR, Metropolitan Museum of Art, New York City
Alejandro MIRÓ QUESEDA GARLAND, University of San Marcos, Lima
Thomas MUNRO, Cleveland Museum of Art; Western Reserve University, Cleveland
Peter MURRAY, Courtauld Institute of Art, London
Francesco NEGRI ARNOLDI, Rome
Karl NOEHLES, Biblioteca Hertziana, Rome
Louis-René NOUGIER, Director, Institute of Prehistoric Art, University of Toulouse
Pedro DE PALOL, University of Valladolid
Enrico PARIBENI, Soprintendenza, Foro Romano e Palatino, Rome
Alessandro PARRONCHI, Urbino, Italy
Colette PICARD, Paris
Gilbert-Charles PICARD, Correspondant de l'Institut; Sorbonne, Paris
Antonio PRIORI, Rome
Sir Herbert READ, President, Institute of Contemporary Arts, London
Luís REIS-SANTOS, Director, Museu de Machado de Castro, Coimbra; University of Coimbra
Hans RHOTERT, Director, Linden-Museum, Stuttgart
Gisela M. A. RICHTER, Honorary Curator, Metropolitan Museum of Art, New York City
Jean M. Roger RIVIÈRE, Instituto de Antropología y Etnología, Madrid
Robert ROSENBLUM, Princeton University
Luigi SALERNO, Soprintendenza ai Monumenti, Rome
Bianca SALETTI ASOR ROSA, Rome
Margaretta M. SALINGER, Metropolitan Museum of Art, New York City
Francesco SANTI, Soprintendenza ai Monumenti e alle Gallerie dell'Umbria, Perugia
Gian Roberto SCARCIA, Accademia dei Lincei, Rome
Curtis SHELL, Wellesley College, Wellesley, Mass.
Phoebe B. STANTON, The Johns Hopkins University, Baltimore, Md.
Doris STONE, President of the Board of Directors, National Museum of Costa Rica
Peter C. SWANN, Keeper, Department of Eastern Art, Ashmolean Museum, Oxford
Jacques THUILLIER, University of Dijon
W. F. VOLBACH, Zentralmuseum, Mainz, Germany
Allen S. WELLER, University of Illinois, Urbana
Alex R. WILLCOX, Johannesburg
David M. WILSON, London
Heinrich Wilhelm WURM, Göttingen, Germany
Bruno ZEVI, University of Rome

ACKNOWLEDGMENTS

The Institute for Cultural Collaboration and the publishers express their thanks to the collectors and to the directors of the museums and galleries listed below for permission to reproduce works in their collections and for the photographs supplied.

The Institute also acknowledges the kind permission of H. M. Queen Elizabeth II to reproduce works belonging to the Crown.

Aachen, Cathedral Treasury
Abbeville, France, Musée Boucher de Perthes
Aegina, Greece, Museum
Aleppo, Syria, Musée National
Amsterdam, Rijksmuseum
Argos, Greece, Museum
Athens, National Museum
Auckland, New Zealand, Museum

Baghdad, Iraq Museum
Barcelona, Museo Arqueológico
Barcelona, Museo de Arte Moderna
Baroda, India, Museum and Picture Gallery
Basel, Schweizerisches Museum für Volkskunde
Bassano, Italy, Museo Civico
Beirut, National Museum
Belvoir Castle, Leicestershire, England, Duke of Rutland Coll.
Benares, India, A. Boner Coll.
Bergamo, Italy, Accademia Carrara
Berlin, Kupferstichkabinett
Berlin, Staatliche Museen
Bern, Kunstmuseum, Paul-Klee-Stiftung
Bikaner, India, Coll. Maharaja of Bikaner
Bologna, Museo Civico
Bologna, Pinacoteca Nazionale
Boston, Museum of Fine Arts
Bremen, Kunsthalle
Brno, Czechoslovakia, Moravian Museum
Brookline, Mass., Historical Society
Brunswick, Germany, Herzog-Anton-Ulrich-Museum
Bucharest, National Museum
Buffalo, N.Y., Albright-Knox Art Gallery

Calcutta, Indian Museum
Cambridge, England, Pepysian Library
Cambridge, England, University Library
Cambridge, Mass., Peabody Museum of Archaeology and Ethnology
Carthage, Tunisia, Musée National
Chicago, Natural History Museum
China, Li Family archives
Cleveland, Museum of Art
Cologne, Diözesan-Museum
Copenhagen, Nationalmuseet
Copenhagen, Ny Carlsberg Glyptotek
Cyrene, Libya, Archaeological Museum

Delphi, Greece, Archaeological Museum
Detroit, Institute of Arts
Dijon, France, Musée des Beaux-Arts
Dresden, Albertinum, Skulpturensammlung
Dresden, Gemäldegalerie

Ferrara, Italy, Museo Archeologico
Florence, Convento di S. Marco
Florence, Museo Archeologico
Florence, Museo Nazionale
Florence, Museo dell'Opera del Duomo
Florence, Pitti
Florence, Uffizi
Florence, Uffizi, Gabinetto dei Disegni e Stampe
Frankfort on the Main, Städelsches Kunstinstitut

Ghent, Belgium, Musée des Beaux-Arts
Graz, Austria, F. Böck Coll.

Haarlem, Netherlands, Frans-Hals-Museum
The Hague, Mauritshuis
Halmyros, Greece, Museum
Heidelberg, University, Psychiatric Clinic
Honolulu, Academy of Arts

Istanbul, Archaeological Museums

Kassel, Germany, Gemäldegalerie
Kyoto, Jingoji

La Spezia, Italy, Museo Archeologico Lunense
Leiden, Rijksmuseum van Oudheden
Leningrad, The Hermitage
León, Spain, Cathedral Archives
London, British Museum
London, Gernsheim Coll.
London, National Gallery
London, Tate Gallery
London, Victoria and Albert Museum
Lyons, Musée Historique
Lyons, University, Laboratoire de Géologie

Madrid, Museo Arqueológico Nacional
Madrid, Museo Nacional de Ciencias Naturales
Madrid, Prado
Mainz, Germany, Altertumsmuseum
Melbourne, National Gallery of Victoria
Mexico City, M. Aleman Valdes Coll.
Milan, Coll. A. Borletti

MILAN, Brera
MILAN, Coll. Brissoni
MILAN, Coll. M. Strudthoff Brissoni
MILAN, Museo Poldi Pezzoli
MILAN, Museo Teatrale alla Scala
MONTPELLIER, France, Musée Fabre
MOSCOW, Tretyakov Gallery
MOSUL, Iraq, Museum
MUNICH, Alte Pinakothek
MUNICH, Bayerisches Nationalmuseum
MUNICH, Bayerische Staatsbibliothek
MUNICH, Staatliche Antikensammlungen
MUNICH, Theater-Museum

NANTES, France, Musée des Beaux-Arts
NAPLES, Museo di Capodimonte
NAPLES, Museo Nazionale
NARA, Japan, Hōryūji
NAUPLIA, Greece, Museum
NEW DELHI, National Museum
NEW HAVEN, Conn., Yale University Art Gallery
NEW YORK, Frick Coll.
NEW YORK, Metropolitan Museum
NEW YORK, Museum of Modern Art
NEW YORK, Pierpont Morgan Library
NORTHAMPTON, Mass., Smith College Museum of Art
NÜRNBERG, Germanisches National-Museum

OAKLY PARK, Ludlow, England, Earl of Plymouth Coll.
OISO, Japan, K. Sumitomo Coll.
OLYMPIA, Greece, Archaeological Museum
OSTIA ANTICA, Italy, Museo Ostiense
OXFORD, England, Ashmolean Museum

PALERMO, Galleria Nazionale della Sicilia
PALERMO, Museo Archeologico
PALERMO, Museo Pitrè
PARIS, Bibliothèque Nationale
PARIS, Louvre
PARIS, Musée d'Art Moderne
PARIS, Musée Carnavalet
PARIS, Musée Guimet
PARIS, Musée de l'Homme
PATNA, India, Museum
PERUGIA, Italy, Galleria Nazionale dell'Umbria
PERUGIA, Italy, Coll. Ranieri
PHILADELPHIA, Museum of Art
PHILADELPHIA, Pennsylvania Academy of the Fine Arts
PRETORIA, Old Museum

REGGIO EMILIA, Italy, Musei Civici
ROME, Coll. L. Albertini
ROME, Capitoline Museum
ROME, Galleria Doria Pamphili
ROME, Galleria Nazionale
ROME, Galleria Nazionale d'Arte Moderna
ROME, Galleria Spada
ROME, Museo Barracco
ROME, Museo Nazionale Romano
ROME, Museo Pigorini
ROME, Museo di Villa Albani
ROME, Museo di Villa Giulia
ROME, Palazzo dei Conservatori
ROME, Palazzo dei Conservatori, Pinacoteca Capitolina
ROME, Torlonia Coll.
ROME, Vatican Library
ROME, Vatican Museums
ROTTERDAM, Museum Boymans-Van Beuningen

SAINT-ÉTIENNE, France, Musée d'Art et d'Industrie
SAINT-GERMAIN-EN-LAYE, France, Musée des Antiquités Nationales
SALEM, Mass., Peabody Museum
SALERNO, Italy, Museo Provinciale
SANSEPOLCRO, Italy, Pinacoteca Comunale
SÃO PAULO, Brazil, Museu de Arte
SÃO PAULO, Brazil, Museu de Arte Moderna
SEATTLE, Art Museum
SPARTA, Greece, Museum
STOCKHOLM, Mr. and Mrs. H. Hammarskiöld Coll.
STOCKHOLM, Statens Historiska Museum
STRASBOURG, Musée des Beaux-Arts
SYRACUSE, Sicily, Museo Archeologico

TARANTO, Italy, Museo Nazionale
TARASCON-SUR-ARIÈGE, France, Coll. R. Robert
TEHERAN, Archaeological Museum
TEHERAN, Coll. Gen. M. H. Firouz
TRIER, Germany, Rheinisches Landesmuseum
TRIPOLI, Libya, Archaeological Museum
TRIVANDRUM, India, Sri Chitra Art Gallery
TÜBINGEN, Germany, Vorgeschichtliche Sammlung der Universität
TULSA, Okla., Thomas Gilcrease Institute of American History and Art
TUNIS, Musée du Bardo
TURIN, Galleria Sabauda
TURIN, Museo Egizio

URBINO, Italy, Galleria Nazionale delle Marche
UTRECHT, Netherlands, Aartsbisschoppelijk Museum
UTRECHT, Netherlands, Centraal Museum

VENICE, Accademia
VIENNA, Kunsthistorisches Museum
VIENNA, A. Razumowski Coll.

WASHINGTON, D.C., Freer Gallery of Art
WASHINGTON, D.C., National Gallery
WELLINGTON, New Zealand, Dominion Museum
WICHITA, Kans., Art Museum, R. P. Murdock Coll.
WILLIAMSBURG, Va., Abby Aldrich Rockfeller Folk Art Coll.
WUPPERTAL, Germany, Städtisches Museum

ZAGREB, Yugoslavia, Gallery of Primitive Art
ZURICH, Kunsthaus

PHOTOGRAPHIC CREDITS

The numbers refer to the plates. Those within parentheses indicate the sequence of subjects in composite plate pages. Italic numbers refer to photographs owned by the Institute for Cultural Collaboration.

Abbott, Berenice, New York: 139 (2)
Academy of Arts, Honolulu: 230 (4)
A.C.L., Brussels: 443 (1)
A.G.I.P.A.N., Tunis: *127*
Alinari, Florence: 29 (1); 30 (1–3); 31 (1); 34 (2); 35 (1, 2, 4); *37*; *38*; *39*; 41; *49*; 64 (3); 65 (2); 81 (1); 82 (1); 86 (1); 89 (1); 91 (1); 100 (1); 102 (1, 2); 104; 110; 113 (3); 114 (2); 115 (2); 119 (1, 2); 120 (1); 121 (1–3); 124 (4); 149 (1); 151 (1, 2); 152 (2); 160; *161*; 167; 169; 170; 179; 181 (1); 182 (1); 183 (1, 2); 184 (1, 2); 193 (1); 211 (1–3); 212; *217* (4); 218 (1, 4); 221 (2); 222 (1); *225* (1); 237 (1, 2); 241 (1, 3); 303 (1); 323 (4); 324 (2); 325 (1, 2); 326 (1, 2); 327 (1); 356 (1); 357 (1); 360 (1); 365; 376 (1); 423; 424 (1, 2); *428*; 430 (1); 450 (1, 2); 453 (3)
Anderson, Rome: 34 (1); 47 (2); 67 (2); 84 (1); 99 (1); 105; 116 (1); 119 (3); 147 (2); 172 (2); 177; 186; 189; 216 (4); 221 (1); 303 (3); 327 (2); 345; 376 (2); 429; 431; 432 (1); 436; 450 (3); 452 (4)
Antikvarisk Topografiska Arkivet, Stockholm: 328 (1–4)
Arte e Colore, Milan: *175*; *217* (2); *330*; *401* (2); *421*; 455; *466*
Auckland Museum, Auckland: 203 (2)

Baker, Oliver, New York: 446 (3)
Balić, Branko, Zagreb, Yugoslavia: 337 (1)
Bayerische Staatsgemäldesammlung, Munich: 220 (1); 425 (2)
Benevolo, Rome: 354 (2)
Bernice P. Bishop Museum, Honolulu: 195 (1, 2)
Bildarchiv Foto Marburg, Marburg, Germany: 65 (1); 108 (2); 194 (1, 2); 312 (1)
Brady, M. B.: 137 (2)
Breckon, A. N., Auckland: 205 (2, 3)
Bredol-Lepper, Ann, Aachen: 314 (1)
Brieke, Frankfort on the Main: *18*
Bulloz, J.-E., Paris: 377 (4); 440 (1)
Busch-Hauck, Frankfort on the Main: 460 (1)

Caisse Nationale des Monuments Historiques, Paris: 75 (3); 96; 125 (1); 129 (2); 247 (1); 249 (2, 3); 251 (2); 255; 263; 264; 269; 271 (1); 307 (1–3); 308 (1, 3); 309; 310 (2); 311 (1, 2); 358 (1); 439 (2); 442 (1)
Callaghan, Providence, R. I.: 140 (2)
Cameron, J. M.: 138 (1)
Camponogara, Lyons: 404 (1–3)
Capogiro, P., Winthrop, Mass.: 140 (3)
Carrano, Taranto, Italy: 69
Centre d'Études Médiévales, Poitiers: 310 (3); 311 (3)
Chomon-Perino, Turin: *180*
Crea, Rome: 23 (1–3); 24 (1, 2); 378 (1–3); 379 (1–4); 380 (2); 381 (1, 2); 385; 386 (1, 2); 387 (1, 2); 391 (1); 392 (1, 2); 393; 394 (1); 395 (3); 399; 447 (1–3); 448 (1, 2)
Crozet, R., Poitiers: 310 (1)
Czako, E. M., Athens: 80 (4); 190

De Antonis, Rome: *40*; *50*; 68 (1, 2); *98*; *142*; 145; *156*; *159*; *162*; *208* (2); *226*; *231*; *235* (1); *301*; *302*; *308* (2); *332* (4); 337 (2); *355* (1); *361* (3); *374*; *388*; *427*; *456* (1, 2)
De Harport, David, Cambridge, Mass.: 206
De Liberali, Tripoli: 358 (2); 359 (2)
Department of Archaeology, Government of India, New Delhi: 3 (1); 84 (2); 298 (2); 413 (2)
Derocles, Tunis: 85 (1)
De Rota, Trieste, Italy: 368 (2)
Deutsche Fotothek, Dresden: 47 (1); 48; 433
Deutsches Archäologisches Institut, Athens: 61 (2); 62 (3); 66 (3); 67 (1); 80 (4); 190
Deutsches Archäologisches Institut, Rome: 51 (1); 123 (2); 124 (2); 210 (2, 3); 241 (2); 357 (3)
Dingjan, A., The Hague: 220 (2); 461 (1)
Dominguez Ramos, Madrid: 288 (1, 2)

Fiorentini, Venice: 32 (2)
Fleming, R. B. & Co., London: 153 (1); *155*; *176*; *185*; 235 (2)
Foto Alberto Luisa, Brescia, Italy: 323 (1)
Fotofast, Bologna: 122
Foto-Kittel, Quedlinburg, Germany: 312 (3)
Fototeca ASAC Biennale, Venice: 371 (2)
Frequin, A., The Hague: 461 (3)
Frobenius Institut, Frankfort on the Main: 10 (2); 14 (1)

Gabinetto Fotografico Nazionale, Rome: 83 (1); 100 (2); 101 (2); 148; 171; 173 (1, 2); 210 (1); 215 (1); 303 (2); 344; 360 (2); 362 (2); 363 (2); 449; 452 (1, 2); 453 (1)
Giacomelli, Venice: 33; 371 (2)
Girard, Abbeville, France: 242
Giraudon, Paris: 43; 76; 90; 116 (2); 129 (3); 131 (2); 141; 144 (3); 147 (1); 150 (1); 152 (1); 154; 178 (1–3); *198* (1); *208* (1); 216 (3); 221 (3, 4); 227; *232*; 233; 234 (1); 236 (1); 313 (3); 332 (1); 333 (3); 335 (1); 382; 395 (1); 403 (1); *437*; 438; 440 (2); 441 (1, 2)
Graziosi, Florence: 11; 259; 274; 275 (1, 2); 276; 279; 280; 295 (3, 4); 296

Hell, Hellmut, Reutlingen, Germany: 245 (1)
Hermes, Paris: 396 (1, 2)
Hirmer Verlag, Munich: 63 (1, 2); 66 (1); 71; 75 (1, 2); 78; 125 (2); 126; 132 (1, 2); 187; 188 (1, 2); 191 (1); 314 (2); 375 (1)
Hurault, Saint-Germain-en-Laye, France: 243 (2); 248 (1); 249 (1); 250 (1, 2); 253 (1, 2); 256 (1, 2); 311 (4)

Istituto Centrale del Restauro, Rome: 164 (1)

Kaufmann, Munich: 72: 73 (1, 2); 77 (1, 3)
Kempter, Munich: *111*; *422*

Landesbildstelle, Rhineland: 312 (2)

Magnum Photos, New York: 140 (4)
Mannini, Rome: 295 (2)
MAS, Barcelona: 46 (1); 134; 135; 219 (2); 282; 292 (2); 316 (1–3); 317 (1, 2); 318 (1, 2); 321 (1–5); 322; 376 (3)

Mazda: 53 (1)
Monumenti, Musei e Gallerie Pontificie, Vatican City: 426 (1, 2); 430 (2); 434 (1, 2); 435
Mori, Rome: 17

Museo della Civiltà Romana, Rome: 357 (2)

Naldoni, Ettore Augusto, Rome: 412
Niépce, Nicéphore: 137 (1)

Oriental Institute, University of Chicago: 355 (3, 4)

Phillips Studio, Philadelphia, Pa.: 338 (2)
Popper, Paul, Ltd., London: 25 (1, 2); 26
Pozzar, Trieste, Italy: 324 (3); 326 (3)
Pozzi e Bellini, Rome: *124* (3); *215* (3)

Rheinisches Bildarchiv, Cologne: 313 (4)
Rigamonti, R., Rome: 362 (1)
Robert, Tarascon, France: 251 (1); 266 (1, 2); 268; 270; 277; 278

Savio, Oscar, Rome: *28*; *29* (2); *31* (2); *32* (1); *35* (3); *36*; *70* (1, 2); *87*; *97*; *118*; 120 (2); *157*; *163*; *168*; *192*; *213*; 215 (2); *216* (1); *222* (4); *225* (2); *240*; *257*; *299*; *300*; *320*; *329*; 331 3, 4); *339*; *347*; *351* (1–4); *352* (1–4); *353*; *369* (1); *373*; *377* (1); *383*; *384*; *389*; *390* (1, 2); *391* (2); *394* (2–4); 397 (1, 2); *401* (3); 406 (1–3); 446 (4); *452* (3); *454*; *464* (1–3)
Scala, Florence: *158*; *224*
Schmidt-Glassner, Helga, Stuttgart: 108 (1)
Sellerio, Palermo: *400*
Ševčík, Václav, Prague: 398
Skira, Albert, Geneva : 258; 267
Soprintendenza alle Antichità Egittologia, Turin: 355 (2)
Soprintendenza alle Antichità della Tripolitania, Tripoli: 359 (1)
Soprintendenza alla Galleria Nazionale d'Arte Moderna, Rome: 444 (1)
Soprintendenza alle Gallerie, Florence: 44 (1, 2); 92 (1); 93 (1, 2); 99 (3); 106; 114 (1); 115 (1); 164 (2); 174 (1, 2); 181 (2); 349 (1, 2); 350; 370 (1)
Soprintendenza alle Gallerie, Naples: 149 (2)
Soprintendenza alle Gallerie dell'Umbria, Perugia: 115 (3)
Soprintendenza ai Monumenti, Milan: 324 (1)
Sparrow, Auckland: 197 (2); 198 (4); 201 (3); 202; 207 (2)
Sri Chitra Art Gallery, Trivandrum, India: 366 (3)
Staatliche Kunstsammlungen, Kassel, Germany: 462 (1)
Stefani, Bruno, Milan: 323 (3)
Steichen, E.: 138 (2)
Steinkopf, Walter, Berlin (Dahlem): 461 (2)
Stickelmann, Bremen: 371 (3)
Stieglitz: 139 (1)
Strand, P., Seine et Oise, France: 140 (1)
Sunami, Soichi, New York: 143 (3); 144 (1, 2); 335 (2); 336 (1, 2); 338 (1); 372 (2)

Tombazi, V. & N., Athens: 61 (3); 74

University of Melbourne, Australia: 305 (2)

Vaiani, Reggio Emilia, Italy: 244 (2)
Verlag Gundermann, Würzburg: 109
Villani, A. & Figli, Bologna: 46 (2); 103; 107; *112*; 452 (5)

Willcox, Johannesburg: 27 (1, 2)

Zancolli, La Spezia, Italy: 332 (5)

CONTENTS - VOLUME XI

PAKISTAN. The Republic of Pakistan consists of two provinces more than 1,000 miles apart; West Pakistan (bordering on the Indian Union, the disputed territory of Kashmir, Afghanistan, Iran, and the Arabian Sea) and East Pakistan (bordering on the Indian Union, Burma, and the Bay of Bengal). Although Pakistan, predominantly Moslem, has been politically separate from the Indian Union since 1947, its art history is inextricably bound up with that of the rest of the Indian subcontinent (see INDIAN ART) and must therefore be considered in relation to that of the Indian Union (see INDIA). Some areas of Pakistan, in fact — the Indus Valley and Gandhara, for instance — are among the most important sites of Indian civilization (see GANDHARA; INDUS VALLEY ART).

SUMMARY. Historical survey (col. 1): *West Pakistan; East Pakistan; Colonial and modern periods.* Monumental centers and archaeological sites (col. 5): *West Pakistan; East Pakistan.*

HISTORICAL SURVEY. *West Pakistan.* The northwestern part of India that became West Pakistan in 1947 has always been a zone of transit for invaders from western and central Asia and hence a crossroads of different civilizations. The geographical reasons are obvious: the mountain passes — of which the two most important are the Khyber Pass between Kabul and Peshawar and the Bolan Pass near Quetta — provided the only feasible access to the subcontinent from the north; and the delta of the Indus offered access from the Arabian Sea to the Indus Valley.

It is in the Sohan Valley, in the Punjab, that the oldest prehistoric deposits in the entire Indian subcontinent have come to light, dating from the second interglacial period. The earliest period of the Sohan Valley culture, called Pre-Sohan, has yielded some Clactonian flakes of chipped quartzite pebbles. The second period (Early Sohan), perhaps 400,000–200,000 B.C., is characterized by single-edged choppers of quartzite pebbles. The third, or Late Sohan, period, extending into the third and fourth interglacial periods, is marked by an improved technique of producing flakes, involving careful preliminary shaping of the core. The second and third periods have also yielded a relatively small quantity of double-edged tools of the Chellean-Acheulean type.

Protohistoric civilization is particularly well represented in Pakistan, starting with the beginning of the third millennium, in the then fertile valleys of Baluchistan. Numerous centers have revealed cultural developments that are classified as the Quetta, Amri-Nal, and Kulli civilizations of southern Baluchistan and the Zhob Valley civilization in the north. These civilizations, differentiated by their pottery, are related to those of Iran (see ASIATIC PROTOHISTORY; IRANIAN PRE-SASSANIAN ART CULTURES) and link the latter with India.

The Indus Valley, or Mohenjo-daro, or Harappa civilization (see INDUS VALLEY ART) lasted from approximately 2500 to 1500 B.C. Its two main centers were Harappa on the Ravi and Mohenjo-daro on the Indus; their layout (similar although they were 375 miles apart) indicates an advanced stage of urban civilization. This civilization apparently had contacts with Mesopotamia, where seals from the Indus Valley have come to light, particularly at Kish and Tell Asmar. Numerous inscribed, but as yet undeciphered, seals indicate that writing was known. The rather coarse pottery and uninscribed seals found in the upper strata of the Harappa civilization sites at Jhukar and Chanhu-daro indicate an eventual decline of this culture.

At some time between about 1500 and 1200 B.C. the Aryans descended through the northwest passes and invaded the Indus Valley, first settling in the Punjab. No archaeological remains have survived from this period. All that is known about these invaders is derived from their sacred literature, the Vedas, transmitted orally for many centuries. These Indo-European peoples, related to the Iranians, had a pastoral and agricultural culture, probably inferior to that of the Indus Valley populations. After an undetermined time they infiltrated into the Ganges Valley.

While the Indian civilization was evolving in the Ganges Valley, a new invasion from the west — that of the Achaemenid Persians — reached the northwestern regions during the 6th century B.C., intensifying the Iranian influence already interwoven in the traditions of the area (see INDO-IRANIAN ART). Cyrus the Great (600?–529 B.C), founder of the Achaemenid dynasty, included Bactria, Gandhara, and Arachosia among his satrapies; but it was Darius I (521–486 B.C.) who reached the Indus Valley. According to Herodotus, he had had the course of the river explored by the Greek Skylax, about 519 B.C.; and two inscriptions, at Persepolis and Naksh-i-Rustam, list Sind (and perhaps a portion of the Punjab) among his possessions.

The Persians apparently ruled these regions until the defeat of Darius III in 330 B.C. His conqueror, Alexander the Great, took possession of his empire; crossing Bactria and Lampaka, he reached Gandhara and entered Taxila, going as far as the Hyphasis (mod. Beas) River before descending the Indus. His expedition, although a rapid one, left its mark on the cultural traditions of the regions he traversed, especially in Gandhara.

A few years later the successors of Alexander the Great were obliged to relinquish these Indian conquests to the founder of the Maurya empire, Candragupta. The emperor Aśoka, who reigned from about 274 to about 232 B.C., introduced Buddhism into these regions; his edicts were inscribed in Kharoshthi characters at Shahbazgarhi and at Mansehra. Taxila especially was to become an important Buddhist center and a focus of the Greco-Buddhist school of sculpture.

The Indo-Greek princes who had settled in Bactria (see BACTRIAN ART) took advantage of the fall of the Maurya empire to invade the Indus Valley at the beginning of the 2d century B.C. Menander, known in India under the name of Milinda, settled in the Punjab. His successors were driven out a hundred years later by the Scythians, or Śakas, who in turn were supplanted, in the first century of our era, by the Kushans (see KUSHAN ART), bearers of a Greco-Iranian tradition that flowered in the art of Gandhara (q.v.). The Kushans established one of their capitals at Peshawar and other urban centers at Udegram and Sirsukh (Taxila). Important remains of Kushan architecture are found in the Jandial Temple at Taxila and at Jhelum, Shāh-jī-kī-Ḍherī (Peshawar), and Takht-i-Bahi. Since the Kushans controlled the routes from Turkistan to India and from China to the Mediterranean, West Pakistan was an important commercial center for about two centuries until the middle of the 3d century, when, according to the inscription at Naksh-i-Rustam, Shāpur I conquered Peshawar and Sind (see SASSANIAN ART). The last of the Kushans were driven out of Gandhara in the middle of the 5th century by the Ephthalites (White Huns), who also took the Punjab from the Guptas, who had held it for a century. There are remains of Gupta monuments (see GUPTA, SCHOOL OF) at Mirpur Khas, Charsadda, and Taxila.

The Arabs invaded Sind in 711; the two independent principalities that resulted from the division of Arab possessions in the Indus Valley chose for their capitals Multan in the Punjab and Brahmanabad (mod. Mansura), the Abbaside capital (see ABBASIDE ART), in Sind. Later Moslem invaders included Maḥmūd of Ghazni (see GHAZNEVID ART), who came through the northwest passes at the beginning of the 11th century and annexed to his Afghan kingdom the Arab principalities of Sind as well as the Punjab and the northwest frontier areas; and the Rajput Muḥammad of Ghor, who took Ghazni and conquered the Punjab and Sind in the late 12th century and whose successor, Quṭb-ud-Dīn Aybak, founded the Delhi Sultanate in 1206. Lower Sind, however, was soon seized by a local Rajput dynasty and was ruled from Tatta by the Sammas, and later the Arghuns and Turkhans, until the Moghul conquest. Indo-Moslem art (see INDO-MOSLEM SCHOOLS) made its appearance in West Pak-

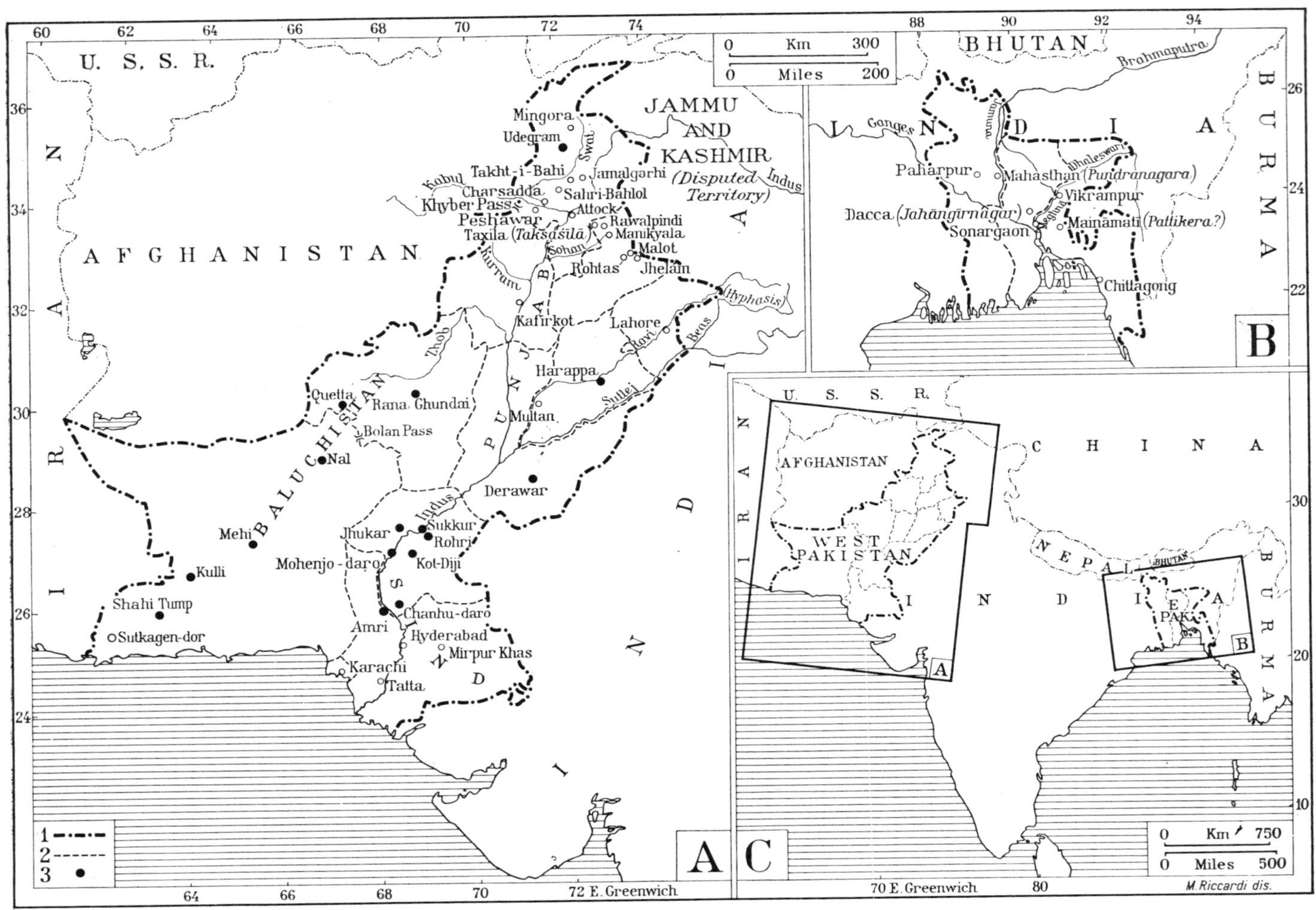

Pakistan, principal archaeological sites and centers of artistic interest. (*A*) West Pakistan; (*B*) East Pakistan; (*C*) geographic positions of West and East Pakistan. *Key:* (1) National boundaries; (2) division boundaries; (3) sites of protohistoric cultures and civilizations.

istan with the Delhi Sultanate; its monuments are found especially in the three capitals, Mansura (ruins of mosques), Multan (mausoleums), and Tatta (the Dabgīr Masjid and the mausoleum of Niẓāmud-Dīn).

The Moghuls (see MOGHUL SCHOOL), after having taken Delhi, united all northern India under their rule; Sind was recovered by Akbar in 1592, but a century later the Kalhōras revolted and established their authority over part of it. The most important concentration of Moghul art in West Pakistan is at Lahore, with its great monuments of Akbar, Jahāngīr, and Shah Jahān. As the Moghul empire disintegrated in the 18th century, Sind and the Punjab were occupied by the Afghans of Aḥmad Shāh Durrānī; eventually small Sikh kingdoms were established in the Punjab.

At the beginning of the 19th century the British established their protectorate over a few Sikh kingdoms between the Jamuna and the Sutlej rivers; in 1843 they annexed Sind, and in 1849 the Punjab and a portion of the northwest frontier states; Baluchistan was peacefully occupied in 1876.

East Pakistan. The great alluvial plain formed by the deltas of the Ganges and Brahmaputra rivers in East Pakistan, or East Bengal, was once inhabited by a people known as the Vaṅgas, mentioned as a non-Aryan people in the *Aitareya Āraṇyaka* and in the later epic texts (the *Rāmāyaṇa* and *Mahābhārata*). They were an enterprising people who traded with distant countries through the delta harbors. The ports were known to the Greeks: Ptolemy refers to them in his *Geography*, and they are also mentioned in the *Periplus maris Erythraei*. The Puṇḍras, inhabitants of the plateau region, are mentioned in the *Aitareya Brāhmaṇa*, but as a savage people.

Little is known about the history of the area for a period of about one millennium. It seems to have formed part of the Maurya empire of Aśoka, as evidenced by an inscription, in Brahmi characters resembling those of the edicts of Aśoka, that was found at Mahasthan, the ancient Puṇḍranagara, capital of the Puṇḍras. Another inscription, on the Allahabad pillar, indicates that the kings of East Bengal maintained relations with the kings of the Gupta dynasty, who eventually annexed East Bengal (see GUPTA, SCHOOL OF). The most important example of Gupta art in East Pakistan is the Viṣṇu temple excavated at Mahasthan.

The fall of the Gupta empire was followed in Bengal by a period of anarchy that lasted until the foundation of the Pala empire by Gopāla, about the middle of the 8th century. The rulers of this empire were Buddhists, and Buddhist monasteries multiplied and flourished under their protection; the ruins of the monastery at Paharpur are an important example of Pala architecture and sculpture (see PALA-SENA SCHOOLS). The Pala dynasty reached the apex of its power under Dharmapāla (ca. 770–810) and experienced a progressive decline during the 9th and 10th centuries. East Bengal then fell under the domination of another Buddhist dynasty, the Chandras, who were in turn replaced by the Hindu Varmans and Senas.

The Sena territories were conquered in the early 12th century by the Delhi Sultanate, which ruled East Bengal for the next two centuries. After having been attached to the Delhi Sultanate, East Bengal reacquired its independence in 1338 under the rule of the Moslem governor of the region and soon after was reunited with West Bengal in an independent Moslem kingdom, which was conquered by the armies of the Moghul Akbar in 1575. The Moghul conquest was followed by more than a century of peace and prosperity, with the capital at Dacca. The middle of the 19th century marked the decline of the Moghul empire, which left the imprint of its art (see MOGHUL SCHOOL) particularly in Dacca; and East Bengal became part of the British Empire until the formation of Pakistan in 1947.

Colonial and modern periods. The architecture of the British period, represented by government buildings in the larger cities (Karachi, Lahore, Peshawar, Chittagong, Dacca) and by numerous public works, is of a composite character, resulting from the grafting of European influences onto local architectural styles derived largely from the Indo-Moslem and Moghul traditions. Of direct European derivation are such buildings as the Neo-Gothic Frere Hall in Karachi and the Government College and Provincial Assembly Building in Lahore. Since World War II a more autonomous architectural expression has come into being, derived from international architectural styles but often adapting them to traditional motifs, particularly in public buildings and industrial complexes. Modern urban planning is an important factor in the creation of new residential and industrial areas in the larger cities and is exemplified especially in the new capital under construction at Islamabad.

In the field of painting (see ORIENTAL MODERN MOVEMENTS) the Moghul tradition strongly influenced the earlier generation of 20th-century artists such as M. A. Rahman Chughtai (X, PL. 411), Fyzee Rahamin, and Haji Sharif. The realistic school that followed had more in common with Western developments in art; and the field has been dominated by abstraction and experimentation in many directions since Pakistan became independent after World War II. Many of the postwar artists studied abroad, and some of the sculptors and painters have worked and exhibited in London and Paris.

The Pakistan Art Councils in the larger centers have been important in promoting interest in art and in sponsoring exhibitions. The chief institutions for art education are the National College of Arts in Lahore, headed by Shakir Ali, the senior abstract painter in Pakistan, and the Art Institute in Dacca, under the direction of Zainul Abedin, one of the early neorealists.

BIBLIOG. *General:* J. W. McCrindle, Ancient India as Described by Megasthenes and Arrian, Calcutta, 1877; S. Beal, Si-yu-ki: Buddhist Records of the Western World, Translated from the Chinese of Hiuen Tsang, 2 vols., London, 1884; J. W. McCrindle, Ancient India as Described in Classical Literature, Westminster, 1901; T. Watters, On Yuan Chwang's Travels in India, 2 vols., London, 1904–05; The Cambridge History of India, I, III, IV, Cambridge, 1922–37; A. Cunningham, Ancient Geography of India (new ed., S. N. Majumdar Sastri), Calcutta, 1924; J. Sion, L'Asie des moussons (Géographie universelle, IX), II, Paris, 1929; F. J. Richards, Geographical Factors in Indian Archaeology, Indian Antiquary, LXII, 1933, pp. 235–43; L. Renou and J. Filliozat, L'Inde classique, 2 vols., Paris, Hanoi, 1947–53 (Eng. trans., P. Spratt, 2 vols., Calcutta, 1954–57); R. E. M. Wheeler, Five Thousand Years of Pakistan, London, 1950; R. E. M. Wheeler, ed., The Indus Civilization, London, 1953; A. L. Basham, The Wonder That Was India, London, 1954; B. C. Law, Historical Geography of Ancient India, Paris, 1954; HIPBC; K. A. Rashid, Military Architecture, Pakistan Q., V, 4, 1955, pp. 25–35, 60–62; Z. H. Syed, Be Your Own Architect and Engineer, Karachi, 1955; Sheikh Mohamed Ikram and T. G. P. Spear, ed., The Cultural Heritage of Pakistan, London, 1956; S. Swarup, The Arts and Crafts of India and Pakistan, Bombay, 1957; R. E. M. Wheeler, Early India and Pakistan to Ashoka, London, 1959; K. Rao, Oxford Pictorial Atlas of Indian History, Oxford, 1960; R. C. Majumdar, Classical Accounts of India, Calcutta, 1961.

West Pakistan: A. Foucher, Notes sur la géographie ancienne du Gandhâra, BEFEO, I, 1901, pp. 322–69; A. Foucher, L'art gréco-bouddhique du Gandhâra, 2 vols. in 4, Paris, 1905–51; A. Foucher, La vieille route de l'Inde de Bactres à Taxila, 2 vols., Paris, 1942–47; V. D. Krishnaswamy, Stone Age India, Ancient India, III, 1947, pp. 11–57; H. Deydier, Contribution à l'étude de l'art du Gandhâra, Paris, 1950; S. Piggott, Prehistoric India, Harmondsworth, 1950 (repr. London, 1962); W. W. Tarn, The Greeks in Bactria and India, 2d ed., Cambridge, 1951; R. E. M. Wheeler, ed., The Indus Civilization, London, 1953; A. L. Basham, The Wonder That Was India, London, 1954; H. Ingholt, Gandhāran Art in Pakistan, New York, 1957; A. K. Narain, The Indo-Greeks, Oxford, 1957; M. Cappieri, L'India preistorica, Florence, 1960; J. Marshall, The Buddhist Art of Gandhāra, Cambridge, 1960; K. M. W. Khan, Sikh Shrines in West Pakistan, Karachi, 1962.

East Pakistan: Government of Bengal, Department of Public Works List of Ancient Monuments in Bengal, Calcutta, 1896; M. Chakravarti, Notes on the Geography of Old Bengal, J. As. Soc. of Bengal, N.S., IV, 1908, pp. 267–91; Abdu'l Wali, Notes on Archaeological Remains in Bengal, J. As. Soc. of Bengal, N.S., XX, 1924, pp. 489–521; N. K. Bhattasali, Iconography of Buddhist and Brahmanical Sculptures in the Dacca Museum, Dacca, 1929; R. D. Banerji, Eastern Indian School of Mediaeval Sculpture (Archaeol. Survey of India, Imperial Ser., XLVII), Calcutta, 1933; N. N. Das Gupta, The Buddhist Vihāras of Bengal, Indian Culture, I, 1934–35, pp. 227–33; R. C. Majumdar and J. Sarkar, History of Bengal, 2 vols., Dacca, 1943–48; B. C. Law, Ancient Historic Sites of Bengal, Ann. Bhandarkar O. Research Inst., XXVI, 1945, pp. 177–91; R. E. M. Wheeler, Five Thousand Years of Pakistan, London, 1950.

MONUMENTAL CENTERS AND ARCHAEOLOGICAL SITES. *West Pakistan.* Amri. At this archaeological site traces of an early civilization, estimated to date from the fourth millennium and named after the site, were discovered under an Indus Valley stratum. The fine wheel-turned pottery found here, characteristically decorated with black or brown geometric designs painted on a buff-colored ground, includes many large, footed jars.

BIBLIOG. N. G. Majumdar, Explorations in Sind (Mem. Archaeol. Survey of India, XLVIII), Dehli, 1934, pp. 24–33; S. Piggott, Prehistoric India, Harmondsworth, 1950, pp. 75–95; D. H. Gordon, The Pottery Industries of the Indo-Iranian Border: A Restatement and Tentative Chronology, Ancient India, X–XI, 1955, pp. 157–89.

Attock. The fortress of Attock, south of the junction of the Kabul and Indus rivers, was built by the Moghul emperor Akbar and was completed in about 1585. Its surrounding stone wall is reinforced by 18 bastions, all circular except for a single rectangular one. A narrow vaulted gallery protected the defenders of the fortress. The gates are restored; only the Kabul Gate dates back to the original construction.

Outside the fortress are the ruins of the Begam-ki Serai, a brick and sandstone shelter for travelers probably mostly built during the early 18th century. It consists of a walled enclosure with small cells or rooms around the inside perimeter. A corner tower reinforcing the wall has a ribbed dome typical of early Moghul architecture. In the southwestern corner of the courtyard is a small triple-arched building popularly believed to have been a mosque, although this has been questioned.

BIBLIOG. Khan Bahadur Maulvi Zafar Hasan, Punjab: Muhammadan and British Monuments, Ann. Rep. Archaeol. Survey of India, 1928–29, pp. 18–26 at 24; R. E. M. Wheeler, Five Thousand Years of Pakistan, London, 1950, pp. 93–95; HIPBC, pp. 505–06; K. A. Rashid, An Old Serai near Attock Fort, Pakistan Q., XIII, 3, 1964, pp. 50–55.

Chanhu-daro. Excavations at Chanhu-daro (see INDUS VALLEY ART) have revealed three separate cultures. First there was a long occupancy by people of the Indus civilization; it is at this level that the workshop of a bead maker was discovered, which has made it possible to study some of the techniques used by this people. After the site had been abandoned (perhaps as a result of an invasion) new occupants settled in the ruins of the huts, which had brick foundations; this level is that of the Jhukar culture, which was less highly developed than its predecessor. The third occupancy was that of a little-known primitive people who used pottery typical of the Jhangar culture.

BIBLIOG. N. G. Majumdar, Explorations in Sind (Mem. Archaeol. Survey of India, XLVIII), Delhi, 1934, pp. 35–38; E. J. H. Mackay, Chanhu-daro Excavations 1935–36, New Haven, 1943; S. Piggott, Prehistoric India, Harmondsworth, 1950, pp. 221–26.

Charsadda. Charsadda, northeast of Peshawar, has been identified as Puṣkalāvatī, capital of Gandhara at the time of the invasion by Alexander the Great. It was situated on the trade route that connected Balkh (Bactria) with India by the Kabul River and gave access also to the Arabian Sea through the Indus Valley. Vestiges of Buddhist art in early Gupta style, with characteristic Greco-Buddhist temple construction of brick and stucco, have been discovered. There are also ruins of monuments dating from the Moslem period.

BIBLIOG. A. Foucher, Notes sur la géographie ancienne du Gandhâra, BEFEO, I, 1901, pp. 322–69; J. Marshall and J. P. Vogel, Excavations at Charsada, Ann. Rep. Archaeol. Survey of India, 1902–03, pp. 141–81; H. Deydier, Contribution à l'étude de l'art du Gandhâra, Paris, 1950, pp. 30–31; R. E. M. Wheeler, Five Thousand Years of Pakistan, London, 1950, pp. 50–51; R. E. M. Wheeler, Chārsada: A Metropolis of the Northwest Frontier, London, 1962.

Derawar. At this archaeological site of the Harappa civilization, situated beside the Ghaggar River (the Sarasvatī of Indian texts), a necropolis has been discovered. A dozen similar sites extend along the river, at one of which pottery with a black design on a buff-colored ground, perhaps comparable to the Quetta pottery, has been found at a level below that of the Harappa civilization. Pottery from a higher level at two other sites is related to that of cemetery H at Harappa.

Harappa. The site of Harappa on the left bank of the Ravi, a tributary of the Indus, had been known for a long time before it was identified; excavations were begun in 1921. This city, the northern center of the Indus civilization (see INDUS VALLEY ART), was founded on the site of a village whose inhabitants had a pottery comparable to that of the Zhob Valley (Rana Ghundai III). The city plan is the same as that of Mohenjo-daro, with the citadel (FIG. 7) built to the west on an artificial platform. To the north of the citadel were discovered two blocks of artisans' houses, near which were some grain mills consisting of rows of circular baked-brick platforms; in the center of some a few kernels of grain were found. A little farther on were the communal granaries. To the west of the city are a few traces of a dike, and to the south of the citadel is cemetery R37, in which the bodies were buried in an extended position, usually surrounded by pottery and sometimes also jewelry and toilet articles. The fortifications appear to have been reinforced during a later period by the erection of corner towers and fortified entrance gates.

Objects similar to those of Mohenjo-daro and typical of the Indus civilization have been recovered; these include inscribed steatite seals, terra-cotta figurines, bronze tools and weapons, flint blades, jewelry of gold and precious stones, and red pottery decorated in black, often with a design of interlocking circles.

Superimposed on a portion of the site are remains of a foreign culture, that of the people of cemetery H, to the northeast of ceme-

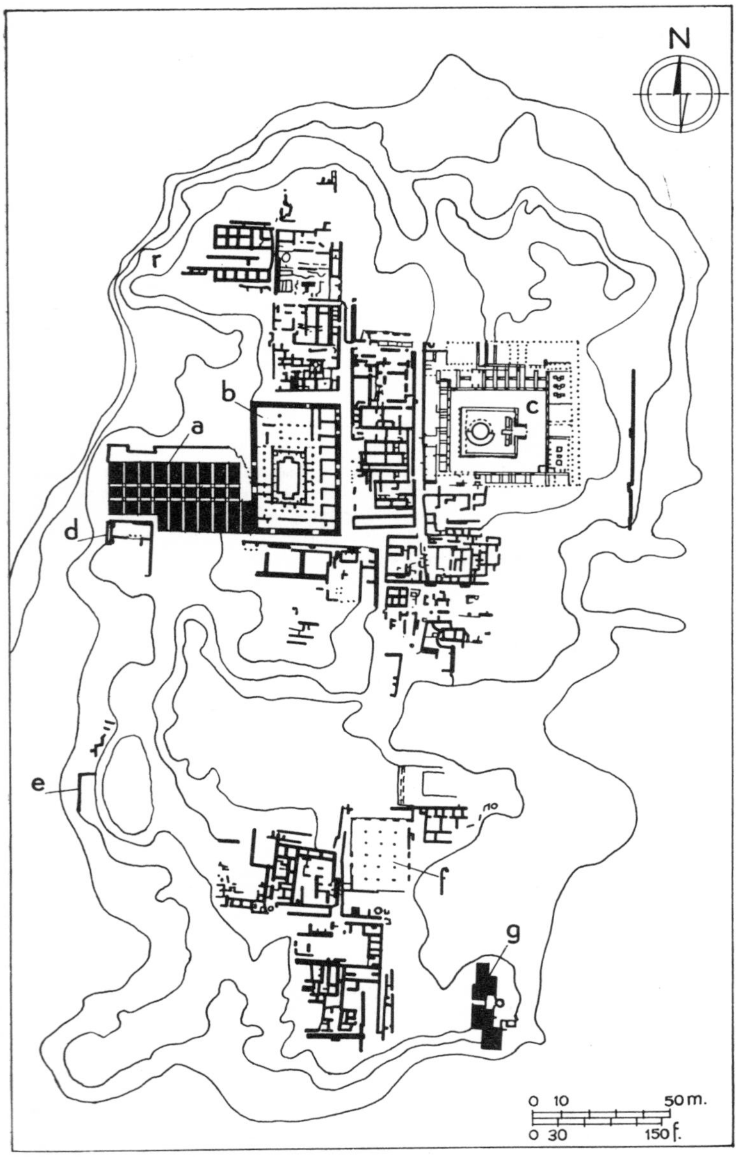

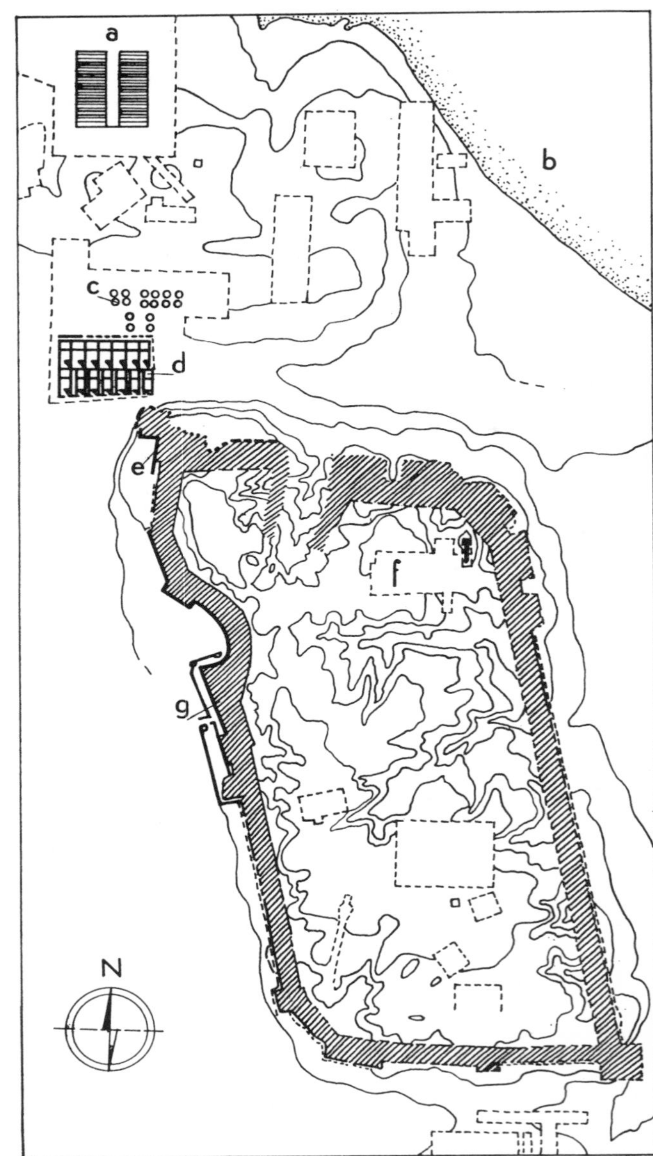

Left: Mohenjo-daro, plan of citadel. (*a*) Granary; (*b*) Great Bath; (*c*) stupa; (*d*) stairway; (*e*) tower; (*f*) hypostyle hall; (*g*) fortifications. *Right*: Harappa, plan of citadel zone. (*a*) Granaries; (*b*) river bed; (*c*) mills; (*d*) craftsmen's houses; (*e*) northwest tower; (*f*) Islamic tomb; (*g*) western gates and terraces of the citadel (*from R. E. M. Wheeler, Early India and Pakistan, London, 1959*).

tery R37. In the oldest tombs the bodies were buried with slightly flexed legs and were surrounded by pottery; the later tombs contain fleshless remains in covered urns. The pottery has a geometric and stylized zoomorphic or plant design painted in black on a shiny red ground, in friezes or panels.

Much of this material is displayed in the site museum.

Bibliog. A. Cunningham, Report of Operations during 1872–73 in Panjab (Rep. Archaeol. Survey of India, V), Calcutta, 1875, pp. 105–08; M. S. Vats, Harappā, Ann. Rep. Archaeol. Survey of India, 1926–27, pp. 97–107; M. S. Vats, Excavations at Harappā, 3 vols., Delhi, 1940; R. E. M. Wheeler, Harappā 1946: The Defences and Cemetery R.37, Ancient India, III, 1947, pp. 58–130; S. Piggott, Prehistoric India, Harmondsworth, 1950, pp. 132–213; R. E. M. Wheeler, ed., The Indus Civilization, Cambridge, 1953. (See also bibliog. of INDUS VALLEY ART.)

Hyderabad (Haidarabad). This city was founded at the end of the 18th century around a fortress built in 1768 by Ghulam Shāh Kalhōra; it came under English rule in 1843. There are important mausoleums of the Kalhōras and the Tālpūrs, the finest being that of Ghulam Shāh Kalhōra, built of marble, with frescoed walls. Hyderabad is an important industrial and commercial center with a number of noteworthy public buildings, among which are the Central Hospital and the University, in traditional Pakistani Islamic style.

Bibliog. H. Cousens, The Antiquities of Sind (Archaeol. Survey of India, Imperial Ser., XLVI), Calcutta, 1929, pp. 131–37; HIPBC, pp. 471–72; M. M. Memon, The City of Wind Catchers, Pakistan Q., VII, 4, 1957, pp. 14–17.

Islamabad. Construction of this city, the eventual national capital of Pakistan, was begun in 1961. The first major building to be completed was Pakistan House, a residence and hotel for members of the National Assembly, with walls of Quetta onyx. By 1965 the Secretariat buildings were nearing completion, as well as about 5,000 private houses and various commercial buildings. The complex built to house the Pakistan Institute of Nuclear Science and Technology is a particularly interesting example of the adaptation of the international architectural style to local requirements and traditions and the integration of landscaping into the architectural scheme.

The housing program, for a projected population of 60,000, has been planned to reflect socioeconomic goals, with services and facilities provided locally in each sector. Landscaping has been planned to complement the architecture, with open spaces in each sector and several large parks with recreational facilities.

Jamalgarhi. At this site, a Buddhist center in Gandhara, a group of monastic buildings around a main stupa has been discovered. Unlike the usual monastery with cells around a courtyard, this consisted of many small buildings, each with its own court and small stupa. The refectory and assembly hall, however, located to the southeast of the main stupa, were communal. Numerous Greco-Buddhist sculptures and low reliefs have been discovered.

BIBLIOG. H. Hargreaves, Excavations at Jamal-Garhi, Ann. Rep. Archaeol. Survey of India, 1921–22, pp. 54–64; G. Deydier, Contribution à l'étude de l'art du Gandhâra, Paris, 1950, pp. 34–35.

Jhangar. This archaeological site in Sind gave its name to a little-known civilization later than that of the Indus Valley, characterized by a black-and-gray pottery with incised decoration.

BIBLIOG. N. G. Majumdar, Explorations in Sind (Mem. Archaeol. Survey of India, XLVIII), Delhi, 1934, pp. 68–70.

Jhukar. Jhukar has given its name to a culture that followed that of the Indus Valley at three Harappan sites: Jhukar, Lohumjo-daro, and Chanhu-daro. This culture is characterized by a buff-colored pottery with stylized plant or geometric motifs painted in black and red, round seals without inscriptions, and bronze shaft-hole axes.

BIBLIOG. N. G. Majumdar, Excavations at Jhukar, Ann. Rep. Archaeol. Survey of India, 1927–28, pp. 76–83; N. G. Majumdar, Explorations in Sind (Mem. Archaeol. Survey of India, XLVIII), Dehli, 1934, pp. 5–18.

Kafirkot. The fortresses of Kafirkot, near the confluence of the Indus and Kurram rivers, dominate the route that leads from the Kabul plateau to the Indus River. The stone wall surrounding the northern fortress is flanked by circular and rectangular bastions. Although its age is undetermined, the construction of the lower portion of the southwestern bastion seems typical of the Kushan period. Inside are five Brahmanic temples of similar plan: a square cell surmounted by a shikara that, although built of stone, has the appearance of a brick structure. From their style it would appear that these temples were erected between the 8th and 10th centuries.

The southern Kafirkot fortress, near the village of Bilot, is on a plateau dominating the Indus. Its massive walls are built of stone blocks and are flanked by 22 circular bastions. Inside are nine Brahmanic temples, similar in plan to those of the northern fortress and probably of the same period.

BIBLIOG. M. A. Stein, Report of Archaeological Survey Work in the North-West Frontier Province and Baluchistan, Peshawar, 1905, pp. 10–17; M. A. Stein, Annual Report of the Archaeological Survey of India Frontier Circle for 1911–12, Peshawar, 1912, pp. 7–9; H. Hargreaves, Inspection and Conservation Notes on the Brahmanical Temples at Bilot-Kafir Kot, Lahore, 1914; R. E. M. Wheeler, Five Thousand Years of Pakistan, London, 1950, pp. 57–60; P. Brown, Indian Architecture: Buddhist and Hindu Periods, 4th ed., Bombay, 1959, p. 161.

Karachi. Karachi was a citadel of the Tālpūr emirs from 1725 until 1838, when it was occupied by the British; the modern city was for twelve years the capital of Pakistan. Its growth dates from the second half of the 19th century; a modern metropolis has grown up around the old town, and it is now Pakistan's most important port and largest city.

In the early phases of urban expansion the English colonial and Pakistani Islamic architectural styles prevailed, as in the buildings of the former Legislative Assembly, the Secretariat, and the Law Courts. European revival styles are represented in the Catholic Cathedral of St. Patrick, Trinity Church and St. Andrew's, and Frere Hall (1865, Neo-Gothic), which houses the National Museum, the Karachi General Library, and the Pakistan Institute of International Affairs. Other noteworthy buildings are the General Post Office, the City Railway Station, the D. J. Sind College, the United States Embassy, the Y.M.C.A., and the President's House (former Governor-General's House; rebuilt, 1947). The persistence of traditional architectural motifs is seen in many other buildings erected between the two world wars and even in later structures, such as the Hotel Karachi International and other hotels and banks and the new Corporation Building. In the international style are the Karachi Airport, the headquarters of the Pakistan Industrial Development Corporation (1955), the State Bank of Pakistan and other commercial structures on McLeod Road, and the University (1959–62; architects, M. Ecochard, P. Riboulet, and C. Thurnauer). The harbor, built in 1854–83 from plans by Bartle Frere and James Thomas Walker, was enlarged in 1957.

Museums include the National Museum (eventually to be transferred to Islamabad, the new national capital), with particularly rich collections of Moghul porcelains and manuscripts, and the Central Museum, opened in 1965.

BIBLIOG. M. B. Pithawalli, An Introduction to Karachi, Karachi, 1950; A. Duarte, Karachi, Pakistan Q., II, 3, 1952, pp. 39–43; HIPBC, pp. 467–69; M. Ecochard et al., La nouvelle Université de Karachi, Arch. d'aujourd'hui, XXX, 86, 1959, pp. 63–65.

Kot Diji. Excavations at Kot Diji have revealed 18 levels of successive occupancy. The six most recent levels are typical of the Indus Valley civilization and are separated by a thick layer of ashes from the lower levels, which extend down to the rocky soil on which once stood a fortress and a town. The surrounding wall, which was reinforced by bastions at regular intervals, was made of undressed stones held together by mud mortar and was topped with sun-dried bricks. These lower layers belong to the Kot Diji culture, which is characterized by a fine, wheel-turned pink pottery, sometimes covered with a buff-colored slip. The simple designs in the more recent levels show patterns that also appear in the Harappa civilization. At these same levels some stone leaf-shaped arrowheads and flint blades were discovered. There was apparently no metal, and writing seems to have been unknown,

The great interest of the Kot Diji culture is that it not only preceded the Indus culture on the same site, but also appears to have had some connection with it. It has not yet been determined, however, whether Kot Diji represents a formative period of the Indus culture or merely came under the influence of that culture as already developed elsewhere.

BIBLIOG. F. A. Khan, Kot Diji, Pakistan Q., VIII, 1, 1958, pp. 13–19; R. E. M. Wheeler, Early India and Pakistan to Ashoka, London, 1959, pp. 106–07; J. M. Casal, L'archéologie pakistanaise: Les fouilles de Kot-Diji, Arts Asiatiques, VII, 1960, pp. 53–60.

Kulli. This site in southern Baluchistan has given its name to a civilization that was partly contemporary with that of the Indus Valley. By virtue of its trade relations with both regions, Kulli furnished a link between Sumer and the Indus Basin.

Remains of stone constructions have been discovered, with traces of a wooden floor and also a stairway. The Kulli civilization is characterized particularly by its pottery and terra-cotta figurines. The pottery has a certain similarity to that of the Iranian Makran, Seistan, and Fars (see CERAMICS; IRAN; IRANIAN PRE-SASSANIAN ART CULTURES); it has a black design on a buff-colored ground. A frieze of animals — mostly humped cattle or felines in a floral setting — encircles the vessels; this frieze is sometimes framed by two red horizontal bands. There is also some red pottery with a black design, similar to that of Harappa. The terra-cotta figurines represent either humped cattle with striped bodies or bird-headed women whose hands rest on their hips.

BIBLIOG. M. A. Stein, An Archaeological Tour in Gedrosia (Mem. Archaeol. Survey of India, XLIII), Calcutta, 1931; S. Piggott, Prehistoric India, Harmondsworth, 1950, pp. 96–118, 142–44; D. H. Gordon, The Pottery Industries of the Indo-Iranian Border: A Restatement and Tentative Chronology, Ancient India, X–XI, 1955, pp. 157–98.

Lahore. This city, the capital of West Pakistan, is often considered the spiritual center of Pakistan because of its Islamic university and its great mosques. In the 8th century it was for a time under the rule of King Lalitāditya of Kashmir (see KASHMIR ART). The city was occupied by Maḥmūd of Ghazni in 1013, but only during the Moghul period did it become an important commercial center. The emperor Akbar built the walls surrounding the city, which were rebuilt by the Sikh Rañjīt Singh in 1812. Now surrounded by the remains of these walls, Lahore has an irregular street plan with buildings dating from various periods and built in a great variety of styles. The modern city is divided into a commerical center, with the Bazaar of Anārkalī, the native quarter of Mian Mir, and the European quarter, known as Donald Town, which was enlarged during the lieutenant-governorship of Sir Charles Rivaz (1902–07).

In the northwestern corner of the walled area Akbar erected a fort, which later became the royal residence. The earliest structure, of brick and Mathura sandstone, dates from the time of Akbar and Jahāngīr (see MOGHUL SCHOOL). Notable within the fort are the eastern gate, flanked by semioctagonal towers; the throne room, with the throne resting on sandstone consoles and covered by a marble canopy; the Jahāngīr quadrangle, surrounded by gardens and with a pool in the center; the Dīwān-i-Am, or public audience hall, with forty pillars; and the Elephant Gate (Hāthī Pol Darwāza), which was begun by Jahāngīr and finished by Shah Jahān. The outside wall of the fort is decorated with glazed-tile mosaics representing court scenes, polo games, animal fights, birds, flowers, and geometric designs. Shah Jahān built the Kwābgāh, an open marble pavilion with polylobed arches inlaid with semiprecious stones; the Mōtī Masjid, surmounted by three onion domes; and the Shīsh Mahal, or Hall of Mirrors, the interior of which is decorated with pieces of colored glass and blue-and-white glazed ceramic tiles inlaid in a network of stucco. This hall is separated from a square court by polylobed arches inlaid with a pattern of flowers and vines and resting on a double row of pillars.

Aurangzēb erected a monumental gate (the Aurangzēb Gate) with fluted towers capped by kiosks, which leads from the fort to the Bādshāhī Masjid (Imperial Mosque). This mosque, said to be the largest in the world, is perhaps the most important monument of Aurangzēb's reign. Built of red sandstone inlaid with white marble, it is surmounted by three onion domes, and the walls are decorated with painted floral panels. The mosque was restored after damage caused in 1840 by an earthquake.

Other monuments of the Moghul era in Lahore and its vicinity are the hexagonal mausoleum of Anārkalī, erected by Jahāngīr in 1615; the mausoleum of Jahāngīr himself, at Shahdara, which stands on the bank of the Ravi in a large garden; the Mosque of Maryam-az-Zamānī, with painted decorations, built by Jahāngīr's mother; the Mosque of Wazīr Khān (1634), with typically Persian exterior

Lahore, plan. Monuments of the Moghul period: (1) Anārkalī Bazaar quarter; (2) cenotaph of Rañjīt Singh; (3) Bādshāhī Masjid; (4) Aurangzēb Gate; (5) fort; (6) city walls; (7) Mosque of Wazīr Khān; (14) to the mausoleum of Jahāngīr; (28) to the Shalimar Gardens. Other buildings: (8) railway stations; (9) Government College; (10) King Edward Medical College; (11) Punjab University; (12) Town Hall; (13) Central Museum; (15) General Post Office; (16) Anglican cathedral; (17) National College of Arts; (18) High Court; (19) Roman Catholic cathedral; (20) Provincial Assembly; (21) Masonic Lodge; (22) Government House; (23) Mayo Gardens; (24) Punjab Club; (25) Aitchison College; (26) Jinnah Gardens; (27) Mian Mir quarter. (*from HIPBC*).

decoration of glazed ceramic mosaic (kashi); and the Mosque of Dāi Ānga, the only one in Lahore in which such ornamental mosaics adorn the interior also.

The Shalimar Gardens (VIII, PL. 446), built by Shah Jahān, are a beautiful example of Moghul landscape architecture. The gardens comprise three descending terraces with numerous pools and waterways, as well as pavilions of red sandstone and white marble. A brick wall with polygonal corner towers surrounds the gardens, entered through gates decorated with a floral pattern of glazed ceramic mosaic.

Buildings of the early English colonial period include the Roman Catholic cathedral, the Anglican cathedral (1884–87, by Gilbert Scott), Lawrence Hall (1862), Montgomery Hall (1866), the Punjab Club (1866), Aitchison College, Nedou's Hotel (now housing government offices), and the Central Museum (opened in 1894, with important archaeological collections). Most of these follow an eclectic style, mingling Neo-Gothic with Moghul motifs, like the later buildings of the High Court (in 15th-century Pathan style), the Provincial Assembly, and the General Post Office. Also of architectural interest are the Neo-Gothic Government College, the Institute of Chemistry (in Pakistani Islamic style), Punjab University (which also reflects the indigenous style), the National College of Arts, and the Government Model School. Among the buildings inspired by more modern international tendencies are Forman Christian College, the Masonic Lodge, the Deputy Commissioner's Court, and the King Edward Medical College and Hospital.

BIBLIOG. S. M. Latif, Lahore: Its History, Architectural Remains and Antiquities, Lahore, 1892; J. P. Vogel, Historical Notes on the Lahore Fort, J. Pañjab H. Soc., I, 1911, pp. 38–55; H. R. Goulding and T. H. Thornton, Old Lahore, Lahore, 1924; Khan Bahadur Maulvi Zafar Hasan, Punjab: Muhammadan and British Monuments, Ann. Rep. Archaeol. Survey of India, 1928–29, pp. 18–26; M. H. Kuraishi, The Punjab and the North-West Frontier Provinces, Ann. Rep. Archaeol. Survey of India, 1935–36, pp. 11–16; R. E. M. Wheeler, Five Thousand Years of Pakistan, London, 1950, pp. 75–93; K. A. Waheed, Lahore, Pakistan Q., I, 6, 1951, pp. 97–102; Muḥammad Bāqir, Lahore: Past and Present, Lahore, 1952, pp. 330–432; Zafar Hasan, Moti Masjid or the Pearl Mosque in the Lahore Fort, J. Pakistan H. Soc., I, 1953, pp. 15–23; HIPBC, pp. 484–94; P. Brown, Indian Architecture: Islamic Period, 3d ed., Bombay, 1960, pp. 33, 100.

Malot. Malot was probably the ancient Singapura mentioned by the Chinese pilgrim Hsüan-tsang. It contains a temple in Kashmir style, built probably in the middle of the 8th century after King Lalitāditya of Kashmir had annexed the northern portion of the Punjab. The temple, of red stone, is square in plan and is adorned on each side with a three-lobed bay framed by fluted columns.

BIBLIOG. M. A. Stein, Archaeological Reconnaissances in North-Western India and South-Eastern Iran, London, 1937, pp. 47, 58; R. E. M. Wheeler, Five Thousand Years of Pakistan, London, 1950, pp. 55–57; HIPBC, p. 497; P. Brown, Indian Architecture: Buddhist and Hindu Period, 4th ed., Bombay, 1959, p. 161.

Manikyala. Remains of a surrounding wall and stupas were discovered at this Buddhist site. In one of the stupas three cylindrical reliquaries in gold, silver, and copper were arranged in a niche covered by a flat stone with an inscription. The gold reliquary contained four gold coins of Kaniṣka; the silver one, seven Roman silver coins; and the copper one, eight copper Kushan coins. This stupa has been identified as the "stupa of the body offering" mentioned by Hsüan-tsang.

BIBLIOG. A. Cunningham, Report of Operations during 1863–64 (Rep. Archaeol. Survey of India, II), Calcutta, 1872, pp. 152–72; A. Cunningham, Report of Operations during 1872–73 in Panjab (Rep. Archaeol. Survey of India, V), Calcutta, 1875, pp. 75–79; A. Cunningham, Report of a Tour in the Punjab in 1878–79 (Rep. Archaeol. Survey of India, XIV), Calcutta, 1882, pp. 1–7; A. Foucher, L'art gréco-bouddhique du Gandhâra, I, Paris, 1905, p. 55; E. Lamotte, Histoire du Bouddhisme indien des origines à l'ère Śaka, Louvain, 1958, pp. 342–43.

Mansura. Built on the site of the ancient Hindu city of Brahmanabad, Mansura became the capital of the Arab kingdom of Sind in 871. Of the earlier city there remain only a few fragments of brick walls and some shards of pottery. From the Arab city there are ruins of mosques, numerous coins, and pottery, including some celadon ware of Chinese origin.

BIBLIOG. H. Cousens, Brāhmanābād-Mansūra in Sind, Ann. Rep. Archaeol. Survey of India, 1903–04, pp. 132–44; R. E. M. Wheeler, Five Thousand Years of Pakistan, London, 1950, p. 63.

Mehi. Mehi is an archaeological site of the Kulli civilization in which traces of settlement, such as stone or sun-dried-brick dwellings, and a small necropolis were discovered. The bodies were cremated and the bones placed either in an urn or directly into the soil. Pieces of buff-colored pottery of the Kulli type have been found, along with red pottery of the Harappa type, terra-cotta figurines, and some bronze objects. Among the latter are a mirror about 5 in. in diameter, with a handle representing a female figure comparable to the terra-cotta ones, and a pin with a lapis lazuli head. Some cylindrical and square containers, divided into four compartments, were probably used as ointment jars; some of these were decorated with an incised chevron pattern. Similar vessels have been found in Seistan (Iran) and Mesopotamia. Weights identical to those of the Indus civilization were also discovered. Like Kulli, Mehi formed a link between the east and the west, between Syria (q.v.) and the Indus Valley.

BIBLIOG. S. Piggott, Prehistoric India, Harmondsworth, 1950, pp. 96–99, 112–13; D. H. Gordon, The Pottery Industries of the Indo-Iranian Border: A Restatement and Tentative Chronology, Ancient India, X–XI, 1955, pp. 157–89.

Mingora. This was the site of probably the most important of the Buddhist monasteries in the Swat region (anc. Udyana), where the Chinese pilgrim Fa-hsien mentions some five hundred; it is almost certainly the Mêng-chieh-li of Hsüan-tsang. The city of Mingora, in the Swat Valley — an important Buddhist center as early as the

Maurya period — was dominated by a citadel. The remains of the monastery surround a stupa decorated with elegant Gandharan low-relief sculptures representing the life of the Buddha. Excavations have uncovered more than sixty lesser stupas, eight of which contained gold reliquaries; in one of the reliquary caskets was a coin dating from the first quarter of the 1st century B.C. In all, nearly 300 pieces of sculpture, throwing light on the development of the Gandharan tradition, have been found at this site; some of this material is in the museum at Saidu Sharif (the capital of Swat, 2 miles away) and some in the Peshawar Museum.

BIBLIOG. H. A. Deane, Notes on Udyāna and Gandhāra, JRAS, XLVIII, 1896, pp. 655–75; E. Chavannes, Voyage de Song Yun dans l'Udyana et le Gandhâra, BEFEO, III, 1903, pp. 379–441; M. A. Stein, Report on Archaeological Survey Work in the North-West Frontier Province and Baluchistan, Peshawar, 1905; M. A. Stein, Serindia, I, Oxford, 1921, pp. 1–24; G. Tucci, Travels of Tibetan Pilgrims in the Swat Valley, Calcutta, 1940; P. R. T. Wright, Graeco-Buddhist Art in the Swat Valley, Indian Art and Letters, XIV, 1940, pp. 1–15; H. Deydier, Contribution à l'étude de l'art du Gandhâra, Paris, 1950, pp. 114–24; D. Faccenna, Mingora, Site of Butkara I, in Reports on the Campaigns 1956–58 in Swat (Pakistan) (Rep. and Mem. ISMEO, Centro Studi e Scavi Archeologici in Asia, I), Rome, 1962; D. Faccenna, Sculptures from the Sacred Area of Butkara I (Swat, Pakistan) (Rep. and Mem. ISMEO, Centro Studi e Scavi Archeologici in Asia, II, 2 and 3), Rome, 1962–64; G. Tucci, La Via dello Swat, Bari, 1963; D. Faccenna, A Guide to the Excavations in Swat (Pakistan), 1956–62, Rome, 1964.

Mirpur Khas. Mirpur Khas, east of Hyderabad, is the site of a brick Buddhist stupa of the early Gupta period. The domical portion of the stupa rested on a high square terrace comprising a basement, or ground story, in which there were some chapels with true arched vaults. Such vaulting, almost unknown in pre-Moslem India, probably indicates Iranian influence. The stupa was adorned with painted terra-cotta statues of the Buddha.

BIBLIOG. H. Cousens, The Buddhist Stūpa at Mirpur Khās, Ann. Rep. Archaeol. Survey of India, 1909–10, pp. 80–92; H. Cousens, The Antiquities of Sind (Archaeol. Survey of India, Imperial Ser., XLVI), Calcutta, 1929, pp. 82–97; R. E. M. Wheeler, Five Thousand Years of Pakistan, London, 1950, p. 58; P. Brown, Indian Architecture: Buddhist and Hindu Period, 4th ed., Bombay, 1959, pp. 44–45.

Mohenjo-daro (see INDUS VALLEY ART). The remains of the citadel of Mohenjo-daro were discovered in 1922 underneath the ruins of a Buddhist stupa. This site, one of the two chief cities of the Indus civilization, was founded during the third millennium on the west bank of the Indus, although the river is now some 3 miles off: the Indus flooded the city several times, and the remains of a dike have been discovered. The ruins of the city reveal a well-organized urban plan, very similar to that of the contemporaneous city of Harappa in the Punjab. The citadel to the west (FIG. 7), which covered a rectangular area of about 1,170 × 585 ft. on a north–south axis, was erected on a platform of sun-dried brick covered by baked brick and was about 30 ft. high. Its most important monuments are the Great Bath (VIII, PL. 64), reached by a stairway; the Granary; and a vast hypostyle hall. In the lower part of the city, outside the citadel, were baked-brick dwellings and shops with stairs leading to the upper stories, laid out along wide streets running either parallel or perpendicular to the axis. There were bathrooms, wells, and drainage systems, and an underground sewage system ran beneath the streets (VIII, PL. 63).

Many of the objects recovered are indicative of a fairly luxurious way of life. Gold and silver jewelry and bronze tools and weapons have been found but no iron, and flint blades continued to be used. Numerous square steatite seals are adorned with naturalistically depicted animals, frequently represented in profile and surmounted by pictographic characters that have not as yet been deciphered (II, PL. 18; VIII, PL. 69). Outstanding are the bust of a bearded man whose garment is decorated with a trefoil pattern (VIII, PL. 67) and a copper female dancer (II, PL. 7) of a type recalling the terra-cotta figurines from Kulli. Some of the numerous terra-cotta figurines (II, PL. 7; VIII, PL. 70) were probably children's toys. The ordinary pottery was plain, but there was also a ware painted in black over a red ground (III, PL. 121; VIII, PL. 70). A few scraps of cotton material prove that both the cotton plant and weaving were known. The use of minerals not found in Sind is evidence of trade relations with other peoples. The several human types found seem to indicate a mixed population. Occupation of the site appears to have lasted at least 700 years, but it has not been possible to reach the level of virgin soil because of the height of the water table. The occupation ended suddenly, probably because the city was conquered by invaders, as can be surmised from the skeletons found in the streets bearing signs of violent death.

Much of the material recovered here is on display in the site museum.

BIBLIOG. R. P. Chanda, Note on Prehistoric Antiquities from Mohenjo-daro, Calcutta, 1924; J. Marshall, The Indus Culture: Mohenjo-daro, Ann. Rep. Archaeol. Survey of India, 1926–27, pp. 51–60; D. R. Sahni, Mohenjo-daro, Ann. Rep. Archaeol. Survey of India, 1926–27, pp. 60–89; J. Marshall, Mohenjo-daro and the Indus Civilization, 3 vols., London, 1931; E. J. H. Mackay, Further Excavations at Mohenjo-daro, 2 vols., Delhi, 1937–38; S. Piggott, Some Ancient Cities of India, Oxford, 1945; R. E. M. Wheeler, ed., The Indus Civilization, Cambridge, 1953; J. van Lohuizen de Leeuw, Note sur un groupe de constructions se trouvant dans la zone HR de Mohenjo-daro, Arts Asiatiques, II, 1955, pp. 145–49. (See also bibliog. of INDUS VALLEY ART.)

Multan. Careful excavations at the site of Tulamba, in the Multan district, have documented its continuous occupation for at least 2,000 years, from before its conquest by Alexander the Great in the 4th century B.C. until its abandonment in the 16th century of our era. The earliest culture levels yielded only unpainted utilitarian pottery, but from about the 2d century B.C. there was stamped and molded pottery related to Gupta designs of the 4th century of our era. Many terra-cotta human and animal figurines have also been recovered, as well as a necklace of 32 carnelian and crystal beads with a silver pendant representing a three-faced Śiva. Glazed pottery from the upper levels, after the Moslem invasion, is similar to that of 11th–13th-century levels elsewhere in Sind and Baluchistan.

The city of Multan, conquered by the Arabs in 712, was the capital of an Arab principality from 871 until the invasion by Maḥmūd of Ghazni in 1005. Five mausoleums are examples of the pre-Moghul Moslem architecture. The oldest, the tomb of Shah Yūsuf Gardīzī, probably dates from 1152; it is built of brick and faced with colored tile. The mausoleum of Bahā-ud-Dīn Zaqariā, of the 13th century, is square in plan, with sloping walls, and is surmounted by a dome. The most important is the mausoleum of Rukn-i 'Ālam (1320–25), which has an octagonal ground plan and is made of brick, with the sloping walls characteristic of the Tughlaq style and a minaret at each corner; it is topped by a low dome. The exterior is decorated with dark-blue, light-blue, and white glazed tile arranged in geometric patterns; only a few traces of the interior decoration remain.

BIBLIOG. A Cunningham, Report of Operations during 1872–73 in Panjab (Rep. Archaeol. Survey of India, V), Calcutta, 1875, pp. 114–36; Khan Bahadur Maulvi Zafar Hasan, Punjab: Muhammadan and British Monuments, Ann. Rep. Archaeol. Survey of India, 1928–29, pp. 18–26 at 25; M. H. Kuraishi, The Punjab and the North-West Frontier Provinces, Ann. Rep. Archaeol. Survey of India, 1935–36, pp. 11–16 at 13; R. E. M. Wheeler, Five Thousand Years of Pakistan, London, 1950, pp. 65–66; HIPBC, pp. 474–75; P. Brown, Indian Architecture: Islamic Period, 3d ed., Bombay, 1960, pp. 33–35.

Nal. An important necropolis has been discovered at Nal, among the ruins of a site related to Rana Ghundai III. Most of the burials seem to have been of fleshless bodies, but in some graves intact bodies had been placed in a rectangular brick tomb. The only well-preserved skull found here is of the same Mediterranean type as the skulls of Mohenjo-daro, a type associated in western Asia with the first known farming communities. Judging by some of the objects found in the tombs — bronze tools, a steatite seal of irregular shape, biconical beads of agate and lapis lazuli — Nal appears to have been in contact with the Indus civilization. The pottery found usually has a polychrome animal or plant design in black or brown on a buff-colored ground with red, yellow, blue, or green highlights. Terra-cotta figurines discovered here seem to belong to the Rana Ghundai III site.

BIBLIOG. H. Hargreaves, Excavations in Baluchistan (Mem. Archaeol. Survey of India, XXXV), Calcutta, 1929; S. Piggott, Prehistoric India, Harmondsworth, 1950, pp. 80–82, 124–25; D. H. Gordon, The Pottery Industries of the Indo-Iranian Border: A Restatement and Tentative Chronology, Ancient India, X–XI, 1955, pp. 157–89.

Nundara. This archaeological site in southern Baluchistan seems to belong to the pre-Harappan civilization known as Amri-Nal. Remains of buildings show stone foundations and walls of stone and sun-dried brick. The pottery is of a fine quality, with geometric and sometimes zoomorphic decorations painted usually in black with red highlights on a buff-colored ground.

BIBLIOG. S. Piggott, Prehistoric India, Harmondsworth, 1950, pp. 76–80, 85–89; D. H. Gordon, The Pottery Industries of the Indo-Iranian Border: A Restatement and Tentative Chronology, Ancient India, X–XI, 1955, pp. 157–89.

Peshawar (Puruṣapura). This city, once the capital of Gandhara, was probably founded by the first Kushans as a fortress controlling the Khyber Pass. It was named Peshawar (meaning "frontier city") by the Moghul emperor Akbar.

To the southeast, on the outskirts of the city, the stupa of Shāh-jī-kī-Dherī, described in the 7th century by the Chinese pilgrim

Hsüan-tsang, was erected to commemorate the conversion of Kaniṣka to Buddhism. Excavations in 1908 revealed the base of the stupa with its masonry wall and stucco decorations representing standing Buddhas between pilasters. The famous casket of Kaniṣka (Peshawar Mus.), containing a crystal reliquary with Buddhist relics, was discovered here (see KUSHAN ART). The lid of the reliquary is surmounted by a group consisting of the Buddha flanked by two bodhisattvas sculptured in the round. The figure of Kaniṣka standing in front of a frieze of cupids carrying garlands decorates the lower part of the reliquary, while a frieze of flying geese adorns the upper part. The inscription, in Kharoshthi characters, mentions the name of Kaniṣka twice.

The original structure of the Bālā Hissār Fort, at the northwestern edge of the city, was erected in the early 16th century, under the Moghul emperor Bābur; it was reconstructed by the Sikhs who ruled Peshawar from 1791 to 1849. Another Moghul monument is the Mosque of Mahabat Khān, built in 1630, in Shah Jahān's reign.

The city of Peshawar came under English rule in 1848. Enclosed by a ring of high walls and defended by two forts, the ancient urban nucleus retains the aspect it assumed in the 18th and 19th centuries, with narrow irregular streets flanked by small baked-brick houses. Noteworthy buildings include the former Governor's Residence, in colonial style, the church and hospital of the Church Mission Society (1853), Edwardes College, the Zenana Hospital, the Victoria Memorial, and the University (in modern Pakistani Islamic style).

The Peshawar Museum (in the Victoria Memorial) has important collections of Gandharan art.

BIBLIOG. A. Foucher, Notes sur la géographie ancienne du Gandhâra, BEFEO, I, 1901, pp. 322–69; D. B. Spooner, Handbook to the Sculptures in the Peshawar Museum, Bombay, 1910; R. E. M. Wheeler, Five Thousand Years of Pakistan, London, 1950, pp. 48–49; HIPBC, pp. 507–08.

Quetta. Situated near the Bolan Pass, which links the Kandahar region (anc. Arachosia) with the Indus Valley, Quetta is the oldest known culture site in Baluchistan. The remains discovered, probably those of a small village of sun-dried-brick structures, have yielded an interesting type of pottery with black geometric designs on a buff-colored ground. It is comparable to the pottery of Tell-i-Bakun, near Persepolis in the Iranian province of Fars, and presumably dates from before the third millennium. Painted pottery dating from the 6th–7th century, with designs showing Sassanian influence, has also been found in the Quetta region.

Once the capital of Baluchistan, Quetta was occupied by the English in 1839 and became an important city toward the end of the 19th century. Almost all the buildings erected by the English administration were destroyed by an earthquake in 1935, and the town has been largely rebuilt in a functional modern style.

BIBLIOG. S. Piggott, A New Prehistoric Ceramic from Baluchistan, Ancient India, III, 1947, pp. 131–42; S. Piggott, Sassanian Motifs on Painted Pottery from North-West India, Ancient India, V, 1949, pp. 31–34; S. Piggott, Prehistoric India, Harmondsworth, 1950, pp. 73–75; W. A. Fairservis, Excavations in the Quetta Valley, West Pakistan, APAmM, XLV, 1956, pp. 169–402.

Rana Ghundai. Rana Ghundai is an important archaeological site in the Zhob Valley where it has been possible to study the stratification of successive cultures. Five main periods can be distinguished. The first (Rana Ghundai I) shows intermittent occupation by a seminomadic people who had an unpainted pottery and flint blades and used domestic animals: remains of the humped ox, the ass, the sheep, and even the horse have been found. After a long interval the site was occupied by a new people (Rana Ghundai II), who lived in houses and made a fine pottery decorated with friezes of stylized animals and geometric designs painted in black on a ground that ranged from pink or buff to dark ocher. This period does not seem to have lasted long. In the next period (Rana Ghundai III), however, three separate levels of dwellings can be distinguished. The pottery is derived from that of the preceding period, but the decorations are in two colors, black and red, on a red ground. No terra-cotta figurines have been discovered here, although some have been found at Moghul Ghundai and Periano Ghundai, two contemporaneous Zhob Valley sites of this culture. Traces of fire mark the end of this period: the site was probably destroyed by invaders about 2000 B.C. The next period (Rana Ghundai IV), which was also terminated by fire, indicates a less developed culture, and the pottery with its painted decoration is much coarser. The final occupants appear with Rana Ghundai V; they had a coarse monochrome pottery with encrusted decoration.

BIBLIOG. E. J. Ross, A Chalcolithic Site in Northern Baluchistan, JNES, V, 1946, pp. 284–316; S. Piggott, Prehistoric India, Harmondsworth, 1950, pp. 119–29; W. A. Fairservis, Archaeological Surveys in the Zhob and Loralai Districts, West Pakistan, APAmM, XLVII, 1959, pp. 277–448.

Rawalpindi. The ancient city of Rawalpindi was occupied by the British in the middle of the 19th century. It has only recently acquired economic importance, which has brought about an urban expansion. Among the noteworthy buildings are the Anglican, Roman Catholic, and Scottish churches, government buildings, Flashman's Hotel, and the very modern American Missionary Hospital. There are also large parks and sports installations. Rawalpindi is the provisional capital of Pakistan; Islamabad, the future capital, is under construction in the vicinity.

BIBLIOG. HIPBC, pp. 500–01.

Rohri. This proto-Neolithic site was discovered in 1866. Its industry corresponds to the third category of the neighboring site of Sukkur and shows similarities to the industry of the earliest level of Mohenjo-daro.

The modern city is said to have been founded at the end of the 13th century. A mosque, the Jāmi' Masjid, decorated with glazed tiles, was built, according to a Persian inscription, by Fateh Khān, an officer of Emperor Akbar, about 1583.

BIBLIOG. H. Cousens, Antiquities of Sind (Archaeol. Survey of India, Imperial Ser., XLVI), Calcutta, 1929, pp. 156–57; H. de Terra and T. Paterson, Studies on the Ice Age in India and Associated Human Cultures (Carnegie Inst., Pub. 493), Washington, 1939; V. D. Krishnaswamy, Stone Age India, Ancient India, III, 1947, pp. 11–57; S. Piggott, Prehistoric India, Harmondsworth, 1950, p. 39; HIPBC, pp. 472–73.

Rohtas. The fortress of Rohtas, located northwest of Jhelum, was built by the Afghan usurper Shēr Shāh in the middle of the 16th century. The surrounding wall is flanked by 68 bastions and pierced by 12 gates, of which the most important are the Sohal, the Khawāss Khān, and the Langar Khān. Within the northern part of the fortress are the remains of a palace built at the end of the 16th century by Mān Singh of Jaipur.

BIBLIOG. M. A. Stein, Archaeological Reconnaissances in North-Western India and South-Eastern Iran, London, 1937, p. 22; R. E. M. Wheeler, Five Thousand Years of Pakistan, London, 1950, pp. 74–75; HIPBC, p. 499.

Sahr-i-Bahlol. Remains of fortifications dating from the Kushan period have been discovered at Sahr-i-Bahlol. There are also ruins of Buddhist stupas and monasteries; schist and stucco Greco-Buddhist sculptures from these are in Peshawar Museum.

BIBLIOG. D. B. Spooner, Excavations at Sahri-Bahlol, Ann. Rep. Archaeol. Survey of India, 1906–07, pp. 102–18; D. B. Spooner, Excavations at Sahri-Bahlol, Ann. Rep. Archaeol. Survey of India, 1909–10, pp. 46–64; M. A. Stein, Excavations at Sahri-Bahlol, Ann. Rep. Archaeol. Survey of India, 1911–12, pp. 95–119; H. Deydier, Contribution à l'étude de l'art du Gandhâra, Paris, 1950, pp. 31–33; R. E. M. Wheeler, Five Thousand Years of Pakistan, London, 1950, pp. 49–50.

Shahi Tump. The lower layers of this site indicate occupancy by people of the Kulli culture; the dwellings had stone foundations and sun-dried-brick walls. The pottery is similar to that of Kulli. In the ruins of the most recent level is a necropolis that seems to belong to the end of the Harappan period, later than 2000 B.C., and to be connected with migrations from the west. The dead were buried intact, and among the grave goods was found pottery with a geometric decoration comparable to that of Khurab in Iran. One of the tombs appears to be that of a warrior; it contains a copper lance of the Harappan type and a copper shaft-hole ax that seems to derive from a Sumerian prototype.

BIBLIOG. S. Piggott, Prehistoric India, Harmondsworth, 1950, pp. 215–20.

Sukkur (Sakhar). Numerous flint blades have been discovered at this proto-Neolithic site on the Indus, some of them long and thin and comparable to those of the Indus culture and others that appear to be older. Three separate categories have been established: the third one corresponds to Rohri and to the oldest level of Mohenjo-daro, while the second shows a Levalloisian technique that gives it the appearance of paleolithic artifacts.

BIBLIOG. H. de Terra and T. Paterson, Studies on the Ice Age in India and Associated Human Cultures (Carnegie Inst., Pub. 493), Washington, D.C., 1939; V. D. Krishnaswamy, Stone Age India, Ancient India, III, 1947, pp. 11–57; S. Piggott, Prehistoric India, Harmondsworth, 1950, p. 39.

Sur Jangal. A necropolis of about 20 mounds, containing bones of cremated bodies, was discovered at this site in the Zhob Valley. The site also yielded pottery of the Rana Ghundai III type, as well as terra-cotta animal figurines and pottery fragments of Rana Ghundai II.

Bibliog. S. Piggott, Prehistoric India, Harmondsworth, 1950, pp. 213–25; W. A. Fairservis, Archaeological Surveys in the Zhob and Loralai Districts, West Pakistan, APAmM, XLVII, 1959, pp. 277–448.

Sutkagen-dor. Sutkagen-dor is the westernmost site at which remains of the Indus civilization have been found. It was fortified, as evidenced by a wall of stone blocks surrounding a rectangular area, with traces of an entrance, probably flanked by towers, at the southwest corner. There are ruins of baked-brick constructions inside as well as outside the enclosing wall. Three funerary urns discovered buried among the ruins of the fortifications seem to belong to the Harappan period. Because of its position on the coast, this site is believed to have played an important role in commercial relations between the Indus and western Asia.

Bibliog. M. A. Stein, An Archaeological Tour in Gedrosia (Mem. Archaeol. Survey of India, XLIII), Delhi, 1931; M. A. Stein, Archaeological Reconnaissance in North-Western India and South-Eastern Iran, London, 1937, pp. 70–71; S. Piggott, Prehistoric India, Harmondsworth, 1950, p. 172.

Takht-i-Bahi. The ruins of the Buddhist monastery of Takht-i-Bahi date from the beginning of the Christian era. This important monastic complex has a main stupa in the center of a court. Around the court are chapels containing votive stupas and statues; the pillars of the chapels are surmounted by Corinthian capitals. A lower terrace is filled with votive stupas and gives access by a stairway to the monastery proper, with the monks' cells and the meeting hall. In secondary courts are various buildings, votive stupas, and the vestiges of a row of colossal Buddhas. Numerous sculptures in the round and in relief (many in Peshawar Mus.), of schist and stucco in Greco-Buddhist style, testify to the wealth of decoration.

Bibliog. D. B. Spooner, Excavations at Takht-i-Bahi, Ann. Rep. Archaeol. Survey of India, 1907–08, pp. 132–48; H. Hargreaves, Excavations at Takht-i-Bahi, Ann. Rep. Archaeol. Survey of India, 1910–11, pp. 33–39; M. A. Shakar, A Short Guide to Takht-i-Bahi, Peshawar, 1946; R. E. M. Wheeler, Five Thousand Years of Pakistan, London, 1950, p. 49; P. Brown, Indian Architecture: Buddhist and Hindu Periods, 4th ed., Bombay, 1959, pp. 33–157.

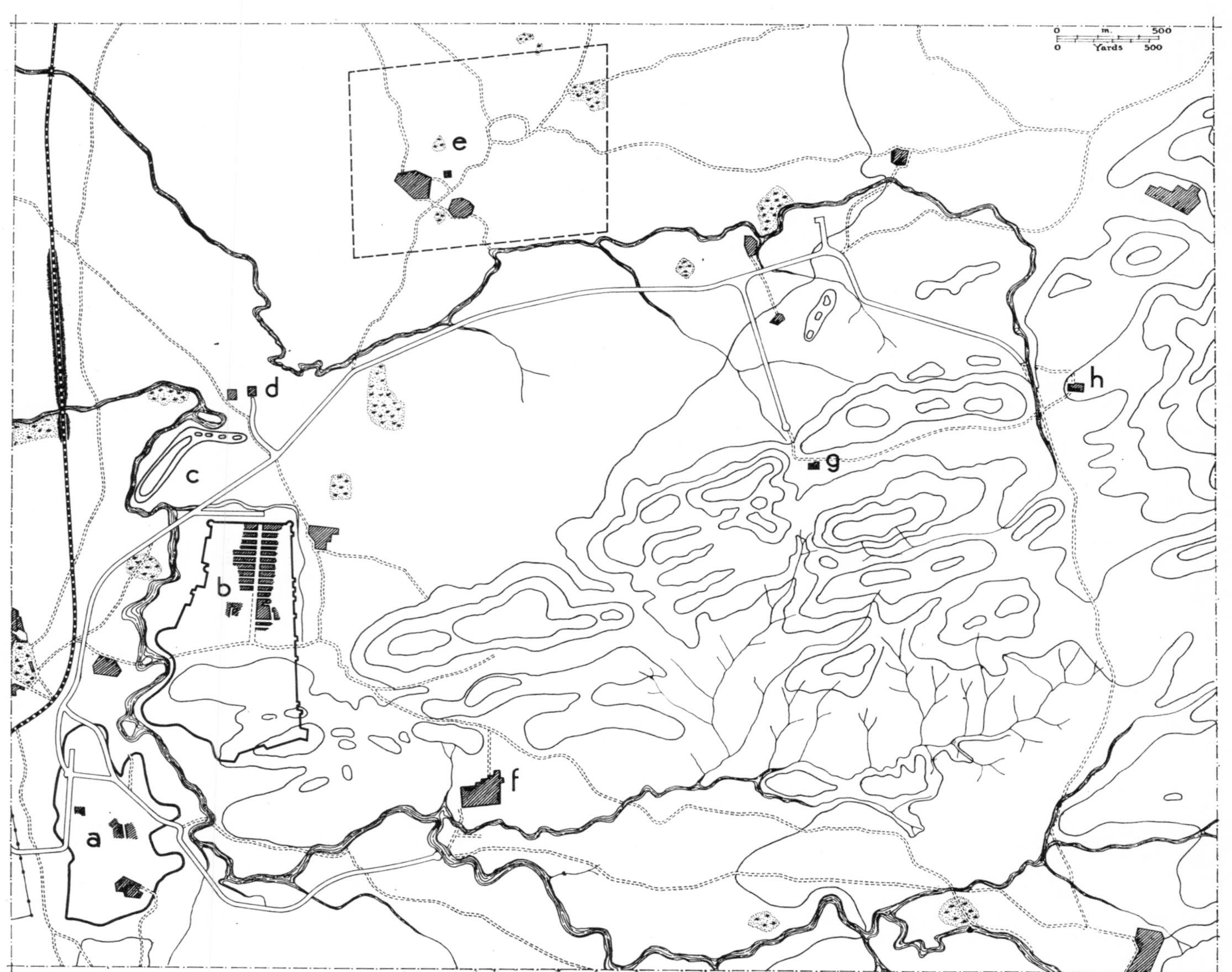

Taxila area, plan. (*a*) Bhir Mound; (*b*) Sirkap; (*c*) Kachchakot; (*d*) Jandial Temple; (*e*) Sirsukh (probable boundaries indicated by broken line); (*f*) Dharmarājika stupa; (*g*) monastery of Mohṛā Morādu; (*h*) monastery of Jaulian (*from J. Marshall, 1951*).

Tatta. Tatta, on the Indus, was from the middle of the 14th century to the beginning of the 16th the capital of the Sammas, and later of the Arghuns and the Turkhans, before the annexation of Sind by the Moghul emperor Jahāngīr in 1625. From the Turkhan period (1588) is the Dabgīr Masjid, now in ruins. This mosque was built of brick; it was decorated inside with painted tiles and had a stone mihrab embellished with low reliefs. Another mosque, the Jāmi' Masjid, was begun in 1644 by Shah Jahān. Also built of brick, but on a high stone basement, it has a rectangular court surrounded by arcades and a hundred-domed roof. Its glazed-tile decorations show the floral and geometric motifs typical of the 17th century.

In the vicinity are stone and brick mausoleums, the oldest of which, in the Makli necropolis, dates from the beginning of the 15th century. The mausoleum of the Samma ruler Niẓām-ud-Dīn was built in the early 16th century and is adorned with sculptured stones that came from Hindu temples.

Bibliog. H. Cousens, The Antiquities of Sind (Archaeol. Survey of India, Imperial Ser., XLVI, Calcutta, 1929, pp. 110–30; M. A. Husain, The Monuments of Tatta, Pakistan Q., I, 2, 1949, pp. 54–58; R. E. M. Wheeler, Five Thousand Years of Pakistan, London, 1950, pp. 68–71; HIPBC, p. 470; B. Ansari, The Forgotten City of Sind (Thatta), Pakistan Q., VII, 4, 1957, pp. 21–28; P. Brown, Indian Architecture: Islamic Period, 3d ed., Bombay, 1960, pp. 124–25.

Taxila. Taxila, the ancient Takṣaśila, was an important Buddhist center. Its university was already renowned by the end of Emperor Aśoka's reign.

Several successive cities were built here (FIG. 17). The oldest, on the Bhir Mound site, was a city of irregular plan, with houses of undressed stone assembled with a mortar made of mud and chips of stone or brick. The city was conquered by the Maurya emperor Candragupta, and the future emperor Aśoka became its governor. The Greco-Bactrians recaptured the city from the Mauryas at the beginning of the 2d century B.C. but abandoned the site for that of Kachchakot, of which only a few ruins remain.

In the middle of the 1st century B.C. the Śakas built a third city, known as Sirkap, which was founded partly on the ruins of the Indo-Greek city but extended farther south. Sirkap was surrounded by ramparts flanked by rectangular bastions along the sides and polygonal ones at the corners. It comprised an upper city to the south and a lower city to the north. This city was of regular plan, with a principal street intersected at right angles by secondary streets; its buildings were of better construction than those of the preceding cities. The two principal buildings were a temple and a palace. The temple was Buddhist and consisted of a porch, a nave that must have been decorated with sculptures, an apse that probably enclosed a stupa, and an ambulatory. A row of monks' cells and two stucco-decorated stupas surround the temple. The palace had two courts and two audience halls.

The Kushans, probably at the end of the 2d century of our era, settled in Sirsukh, in the plain to the northeast of Sirkap and flanked the surrounding wall with semicircular bastions. Within the enclosure excavation has revealed an important building in which the rooms were arranged around two courts. Traces of violent destruction seem to indicate that the last city was invaded by the Ephthalites (White Huns) in the middle of the 5th century.

The environs of these four cities are strewn with remains of the temples and monasteries that formed the famous Buddhist university. The best known are the Jandial Temple (VII, FIG. 942) between Sirsukh and Sirkap, the Dharmarājika stupa southeast of Sirkap, and the Jaulian and Mohṛā Morādu monasteries. They were decorated with typically Greco-Buddhist sculptures in stone and particularly in stucco.

Bibliog. A. Cunningham, Report of Operations during 1872–73 in Panjab (Rep. Archaeol. Survey of India, V), Calcutta, 1875, pp. 66–75; J. Marshall, Excavations at Taxila, Ann. Rep. Archaeol. Survey of India, 1912–13, pp. 1–52, 1914–15, pp. 1–35, 1915–16, pp. 1–38; A. Foucher, Les fouilles de Taxila, JA, XIV, 1919, pp. 311–20; J. Marshall, Excavations at Taxila, the Stūpa and Monastery at Jaulian (Mem. Archaeol. Survey of India, VII), Calcutta, 1921; J. Marshall, Taxila, Ann. Rep. Archeol. Survey of India, 1926–27, pp. 110–19; J. Marshall, A Guide to Taxila, 3d ed., Delhi, 1936; H. C. Beck, The Beads from Taxila (Mem. Archaeol. Survey of India, LXV), Delhi, 1941; S. Piggott, Some Ancient Cities of India, Oxford, 1945; G. M. Young, A New Hoard from Taxila (Bhir Mound), Ancient India, I, 1946, pp. 27–36; A. K. Ghosh Taxila (Sirkap) 1944–45, Ancient India, IV, 1948, pp. 41–84; H. Deydier, Contribution à l'étude de l'art du Gandhâra, Paris, 1950, pp. 125–54; J. Marshall, Taxila, 3 vols., Cambridge, 1951.

Udegram. The site of Udegram (the Ora of Arrian) in the Swat Valley appears to have been inhabited from protohistoric times until Genghis Khan's invasion in the 13th century. Some rock carvings representing wild and domestic animals are from the protohistoric period. The Udegram fort seems to date from the time of Alexander, but it was enlarged and remodeled later; the different methods of construction make it possible to identify the different periods. The fortified city, which stretched along the valley and up the sides of the mountains, was destroyed by Maḥmūd of Ghazni in the 11th century, as was the Raj Gira Castle nearby, of which the massive central building, now in ruins, dates from the Kushan period.

Bibliog. A. Cunningham, Ancient Geography of India (new ed., S. N. Majumdar Sastri), Calcutta, 1924, pp. 93–94; M. A. Stein, On Alexander's Track to the Indus, London, 1929; M. A. Stein, An Archaeological Tour in Upper Swat and Adjacent Hill Tracts (Mem. Archaeol. Survey of India, LXII), Calcutta, 1930; E. Bargen and P. R. T. Wright, Excavations in Swat and Explorations in the Oxus Territories of Afghanistan (Mem. Archaeol. Survey of India, LXIV), Delhi, 1941; G. Gullini, Marginal Report on the Excavations at the Castle of Udegram, East and West, IX, 1958, pp. 329–48; ISMEO, Reports on the Campaigns 1956–58 in Swat (Pakistan), Rome, 1962.

East Pakistan. Chittagong. According to Tibetan sources, the Chittagong region was, during the 7th century, an important center, with Brahmanic temples and Buddhist monasteries. Of the latter, the most important was the Paṇḍita Vihara. Numerous Buddhist images in stone, wood, and bronze have been discovered and confirm the existence of Mahayana Buddhism in the region. Certain bronzes, comparable to those of Nalanda, bear inscriptions that make it possible to date them to the 11th or 12th century.

In modern times the port city of Chittagong, after some centuries of Moslem rule, came under Portuguese domination from 1538 to 1668, when it was recaptured by the Moghuls, who ceded it to the British East India Company in 1760. Among the Moslem monuments are the Qadam-i-Mubārak Mosque (1136) and the Jami' Masjid (1670). The main building of the Islamic Intermediate College, originally an arsenal, dates from the Portuguese period. The older public buildings are in the British colonial style; later buildings of interest include the Court Building, housing Provincial government offices and courts, the Mitford Hospital, the Chamber of Commerce, and the modern port installations.

Bibliog. S. Chandra Das, A Note on the Antiquity of Chittagong, Compiled from the Tibetan Works Pagram Jom Zan of Sunpa Khonpo and Kahbab Dun-dan of Lama Tara Nātha, J. As. Soc. of Bengal, LXVII, 1898, pp. 20–28; D. B. Spooner, Excavations and Explorations: Eastern Circle, Ann. Rep. Archaeol. Survey of India, 1921–22, pp. 81–83.

Dacca. Under the name of Jahāngīrnagar, Dacca was the capital of East Bengal 1608–1704. Of the many monuments of Moghul architecture, the most important are the brick Baṛā Kaṭrā, or Great Market (1644; apparently intended for a royal palace), with a high central portal and octagonal corner towers; and the brick Lāl Bāgh Fort, begun in 1678 by Muḥammad Azam, third son of Aurangzēb, but never finished. Within the enclosure is the mausoleum of Parī Bībī ("fairy lady"), daughter of the governor of Bengal, who died in 1684. The square mausoleum is unusual in that it is built entirely of imported stone: black basalt from Bihar, sandstone from Chunar, and marble from Rajasthan. On the western side of the mausoleum is a small three-domed mosque erected at the same time as the fort. Other important mosques, of predominantly Moghul provincial style, are the Chowk Mosque (1676), the Satgumbad Mosque (1680), the Kar Talab Khān Mosque (1709), and the Star Mosque.

Occupied by the British about the middle of the 19th century, Dacca was the capital of East Bengal 1905–12 and since 1947 has been the provincial capital of East Pakistan as well as the second capital of the Republic of Pakistan. The present city comprises the old city, characterized by domes and minarets, and the European quarter to the east, laid out in the second half of the 19th century, which contains government offices and residences. Outstanding are Government House, the High Court, the Collegiate School, and Curzon Hall (in Pakistani Islamic style); as well as the Anglican church, the Roman Catholic church of Our Lady of the Rosary (16th cent.), the Dacca Museum (opened in 1913), the Medical College and its hospital, the Central Library (which contains valuable manuscripts), and the University (1921–27), all reflecting local traditional architectural motifs. After World War II international architectural trends inspired the design of buildings such as the Circuit House (1954), the Malaria Institute of Pakistan (1954–56), and numerous factories in the industrial area.

Museums include, besides the Dacca Museum (Hindu, Buddhist, and Moslem sculptures and paintings, Arabic and Persian calligraphy), the Balda Museum of Art and Archaeology, rich in old coins and in metal and ivory antiquities; the Arts Institute, with collections of local folk art and of modern East Pakistani painting; and the University Museum (sculpture, painting, and coins).

Bibliog. Sayid Aulad Hasan, Notes on the Antiquities of Dacca, Dacca, 1904; R. E. M. Wheeler, Five Thousand Years of Pakistan, London, 1950, pp. 104, 120–22; HIPBC, pp. 513–14; G. Ruddock, Dacca: Capital of East Pakistan, Pakistan Q., VIII, 1, 1958, pp. 49–58.

Mahasthan. Mahasthan (anc. Puṇḍranagara) was the capital of the Puṇḍras and already existed during the Maurya period, as is evidenced by an inscription found at the site. Excavations have also revealed a city of the Gupta period, to which belongs a Viṣṇu temple bearing traces of restoration, which is adorned with ornamental brick and sculptured terra-cotta panels. Attributed to the same era is a building with a base composed of rectangular earth-filled components, which originally supported a stupa. The stupa was replaced by a temple during the reign of the Senas. Some brick fortifications seem to belong to the Pala period. A mausoleum and a small mosque built in 1718 are in typical Moghul architectural style.

Bibliog. Government of Bengal, Department of Public Works, List of Ancient Monuments in Bengal, Calcutta, 1896, pp. 186–89; N. K. Dikshit, Excavations at Mahasthan, Ann. Rep. Archaeol. Survey of India, 1928–29, pp. 87–97; P. C. Sen, Mahasthan and Its Environs, Rajshahi.

1929; G. C. Chandra and K. N. Dikshit, Explorations in the Eastern Circle, Ann. Rep. Archaeol. Survey of India, 1930–34, pp. 128–29; D. R. Bhandarkar, Mauryan Brahmi Inscriptions of Mahasthan, Ep. Indica, XXI, 1932, pp. 83–91; N. G. Majumdar, Exploration in Bengal, Ann. Rep. Archaeol. Survey of India, 1934–35, pp. 40–42; N. G. Majumdar, Excavations at Gokul, Ann. Rep. Archaeol. Survey of India, 1935–36, pp. 67–69; T. N. Ramachandran, Excavations at Mahasthan, Ann. Rep. Archaeol. Survey of India, 1936–37, pp. 51–54; R. E. M. Wheeler, Five Thousand Years of Pakistan, London, 1950, pp. 102–03; B. C. Law, Historical Geography of Ancient India, Paris, 1954, p. 234.

Mainamati. Near the village of Mainamati have been discovered the ruins of Buddhist monasteries built of brick and decorated with terra-cotta reliefs and ornamental bricks typical of East Bengal sculpture. The layout of these monasteries seems to have been the same as that of Paharpur: a sanctuary in the center of a large court, surrounded by walls and cells. From the style of the sculptures these monasteries are tentatively dated to the 8th–9th century. Later excavations have yielded rich finds, including gold, silver, and copper coins, bronze figures of the Buddha, a bronze votive stupa, copper jars and cooking pots, and an inscription in proto-Bengali script — as yet undeciphered — on a 14-in. copper plaque bearing a circular emblem representing a humped bull. Much of this material is displayed in the site museum.

Bibliog. N. G. Majumdar, Inscriptions of Bengal, III, Rajshahi, 1929, pp. 1–9; N. N. Das Gupta, The Buddhist Vihāras of Bengal, Indian Culture, I, 1934–35, pp. 227–33 at 231; T. N. Ramachandran, Recent Archaeological Discoveries along the Mainamati and Lalmai Ranges, Tipsera District, East Bengal, B. C. Law Commemorative Volume, III, Poona, 1946, pp. 213–31.

Paharpur. About the year 800 Dharmapāla founded a monastery, the Somapura Vihara, now known under the name of Paharpur. This important monastery comprised more than 170 cells, which opened onto a court in the center of which stood a large temple or stupa. The stupa had a cruciform plan and receding tiered levels that gave it a pyramidlike shape. Only three levels remain; at the base of each is a passage for the circumambulation (*pradakṣiṇā*). A stairway on the north side provides access to the upper levels.

The walls of the two lower levels are decorated with friezes of sculptured panels, partly of stone but mostly of terra cotta, representing religious and profane subjects; they give evidence of a local folk art independent of academic tradition and typical of East Bengal art of the Pala period (see PALA-SENA SCHOOLS). The site museum displays many of the finds at Paharpur.

Bibliog. A. Cunningham, Report of a Tour in Bihar and Northern Bengal during 1879–80 (Rep. Archaeol. Survey of India, XV), Calcutta, 1882, pp. 117–20; D. B. Spooner, Paharpur Excavations by Calcutta University, Ann. Rep. Archaeol. Survey of India, 1922–23, pp. 115–19; R. D. Banerji, Exploration at Paharpur, Ann. Rep. Archaeol. Survey of India, 1925–26, pp. 107–13; K. N. Dikshit, Excavations at Paharpur, Ann. Rep. Archaeol. Survey, of India, 1929–30, pp. 138–42; G. C. Chandra and K. N. Dikshit, Excavations at Paharpur, Ann. Rep. Archaeol. Survey of India, 1930–34, pp. 113–28; K. N. Dikshit, Excavations at Paharpur, Bengal (Mem. Archaeol. Survey of India, LV), Delhi, 1938; R. C. Majumdar and J. Sarkar, History of Bengal, I, Dacca, 1943, pp. 489–93, 504–11, 525–32; R. E. M. Wheeler, Five Thousand Years of Pakistan, London, 1950, pp. 98–101; P. Brown, Indian Architecture: Buddhist and Hindu Period, 4th ed., Bombay, 1959, p. 151.

Sonargaon. This ancient town was the seat of the first Pala dynasty and was the capital of East Bengal at the time of the Moghul conquest in 1575. There are ruins of 17th-century monuments as well as a small three-arched mosque. The mosque was rebuilt in 1700, but the mihrab dates from the 15th century.

Bibliog. R. E. M. Wheeler, Five Thousand Years of Pakistan, London, 1950, p. 105.

Marguerite Marie Deneck

(With the assistance of Francesco Negri Arnoldi for the modern sections.)

Illustrations: 4 figs. in text.

PALA-SENA SCHOOLS. In the sphere of the artistic culture of medieval India (see INDIAN ART) the schools of Bihar and Bengal, which take their names from the Pala and Sena dynasties, are distinct by virtue of their attempt to translate into symbolic forms the abstruse concepts of Buddhist esoterism (see BUDDHISM). This attempt was successful owing to the elaboration of a rich repertory of complex iconographic types. To the uninitiated the arcane meanings of these types are undecipherable, but stylistically the types are very fine. Pala-Sena art became known even outside India and nourished the artistic traditions of almost all the Buddhist countries.

Historical and cultural foundations. *Geographical area.* The Pala-Sena schools of Bihar and Bengal occupy an exceptional position in medieval Indian art, because of the peculiar geographic situation, structure, and ethnic character of these regions. The area consists mainly of a vast alluvial plain dotted with marshes and swamps, periodically flooded by the Ganges (and its tributaries, the Son and Gandak), Brahmaputra, Jamuna, and Padma rivers, as well as innumerable minor rivers and channels. The plain is cut off from northern India by the heavily forested mountains between Bihar and Orissa. Nepal and Tibet are accessible through the northern part of the area; Farther India from the east, via Assam, Manipur, Tripura, and Chittagong; and in the south there is a narrow coastal link with Orissa. The delta has many excellent harbors, and intensive traffic with Burma, Malaya, Java, Cambodia, and China has always been maintained.

In the past, Bihar south of the Ganges was called Magadha and north of it Videha or Tirabhukti (Tirhut). The part where the Ganges bends around the easternmost spurs of the Vindhya Mountains was called Anga, farther south Radha and Suhma. The area between the Ganges and Jamuna was Bengal proper (Vanga); in the north (above the junction of the rivers), Pundravardhana; in the south (between Hooghly and Padma), Vyaghratati; in the east, Kamarupa (western Assam), Harikela (Mymensingh and Sylhet), and Samatata (around Dacca).

Settlements, except in the mountainous areas, were built on artificial mounds, and were protected by dikes and drains. The normal building materials were brick and wood, stone (black basalt, quarried at Rajmahal) being used only for statues, mihrabs, applied decorations, and the like.

The indigenous population, which had a strong Mongol strain, was conquered by the Aryans very late. Bihar had been the cradle of Buddhism; Bengal, too, became a bulwark of that anti-Brahmanical religion. It was only under the Sena kings that a Brahmanical aristocracy was installed in Bengal. As in Orissa and Assam, the pre-Aryan indigenous cults, especially of the bloody, orgiastic mother goddess, engendered a special brand of Mahayana Buddhism, the Vajrayana, and later a special brand of Hinduism, Shaktism.

Historical outline. Beginning in the 7th century B.C. Bihar was the center, successively, of the great Indian powers, the Haryaṅkas, Śaiśunāgas, Nandas, and Mauryas. By the beginning of the Christian Era the hub of Indian civilization moved to the upper Ganges plains. In the 4th century, under the Guptas, Bihar again became the center of the empire. From the 5th century on Bihar was increasingly pushed into conflict with the great powers ruling in Ujjain, Kanauj, and southern India, gradually becoming dependent on its hinterland, Bengal, until in the 10th and 11th centuries Bihar was reduced to a frontier march of Bengal. More and more isolated, Bihar was, under the Pala dynasty (ca. 765–1162), the last bulwark of Buddhism, which had disappeared everywhere else in India. Its political and cultural influence extended instead to Nepal, Tibet, China, Burma, upper Thailand, Sumatra, and Java. The Senas (ca. 1050–1230) brought Bengal back into the fold of orthodox Hinduism, but their supremacy lasted no longer than a century, because their suppression of the Buddhists made it easy for the Moslems to "liberate" the country.

The history of the Pala dynasty is divided into three periods, which also represent different art styles. The first two centered on Pataliputra (Patna) and then Srinagar (Bihar-i Sharif) in Bihar, up to the end of the 11th century (especially the late

11th cent.); the third centered on Ramavati (Chandipur near Gaur) and Vikramapura (Rampal) in Bengal, and covered the 12th century. After the middle of the 8th century the last successor states of the glorious Gupta empire disappeared, and the empire of Kashmir, which under Lalitāditya Muktāpīḍa (ca. 725–56) had controlled most of India for two decades, disintegrated. Gopāla (ca. 755/65–70), a devout Buddhist, was elected king of Bengal and firmly established the new Pala dynasty internally as well as externally against the attempts of Jayāpīḍa to restore Kashmir's suzerainty. Gopāla's son Dharmapāla (ca. 770–810/15), in the wake of the collapse of Kashmir, subjugated the whole of northern India, but encountered strong resistance by the Pratīhāras of Rajasthan, Vatsarāja and Nāgabhaṭa II, who tried to capture the old imperial capital of Kanauj. The Rashtrakutas of the Deccan, Dhruva and Govinda III, attacked both the Palas and the Pratīhāras. Devapāla (ca. 810/15–850/54) expanded his empire to Nepal and Assam, controlled Orissa, defeated the Rashtrakuta king Amoghavarṣa, but lost Kanauj and parts of northern India to the Pratīhāras Nāgabhaṭa II and Bhoja I. The next Pala rulers, Vigrahapāla I (ca. 850/54–57), who became a monk, and Nārāyaṇapāla (ca. 857–908/11), were weak, and the vassal kingdoms reasserted their independence; the Pratīhāras expelled Nārāyaṇapāla even from Bihar and western Bengal. The Rashtrakuta victories over the Pratīhāras after the death of Mahendrapāla enabled the Palas to recover Bihar. However, the Candellas of Khajuraho made raids into Pala territory, and the Pala kings Rājyapāla (ca. 908/11–35), Gopāla II (ca. 935–92), and Vigrahapāla II (ca. 992) ruled only a small part of northern India. Even at the zenith of its power the Pala empire — like all medieval Indian states — was loosely knit. Now, almost all Bengal was divided into independent states under local dynasties, of which the Kāmbojas (ca. 911–92), Chandras (ca. 950–1050), and Śūras (ca. 950–1110) were the most important. Moreover, the Tibetans raided the northern Indian plains during the winter wherever and whenever the political situation offered an opportunity.

The invasions of Sultan Maḥmūd of Ghazni and the crisis in the Candella empire that ended in the rise of the Haihaya-Kalacūrīs of Chedi enabled Mahīpāla I (ca. 992–1040) to restore the Pala kingdom, and the raids of the southern Indian Chola emperor Rājarāja I gave him the opportunity to reimpose his suzerainty on the rebellious former vassals, who also were invaded. But under Nayapāla (ca. 1040–55), Vigrahapāla III (ca. 1055–81), Mahīpāla II (ca. 1082), and Śūrapāla II (ca. 1083) the disintegration of the kingdom set in again; first the Kalacūrīs and then later the Gāhaḍavālas of Kanauj occupied most of Bihar. Rāmapāla (ca. 1084–1126) again restored the empire, even controlling Assam and Orissa. However, the vassals had in the meantime become powerful, and after the death of Kumārapāla (ca. 1126–30) and the murder of Gopāla III (ca. 1130), Madanapāla (ca. 1130–61) little by little lost Bengal. The dynasty lingered on in southern Bihar until 1167.

The Sena dynasty, which succeeded the Palas, had immigrated from the Deccan; at first they were officials, then feudatories, in Radha. Sāmantasena (ca. 1050–75), Hemantasena (ca. 1075–97), and Vijayasena (ca. 1097–1159) had been among the most powerful nobles of the late Pala court, the others being the Kaivartas (responsible for the murder of Mahīpāla II), Śūdrakas, Varmas (Varmans) or Yādavas, and Mānas. Under Rāmapāla's weak successors Vijayasena eliminated his competitors and finally also Madanapāla, and he even started expanding into Assam and Orissa. Ballālasena (ca. 1159–85) conquered Bihar. Lakṣmaṇasena (ca. 1185–1206) defeated the Gāhaḍavāla king Jayacandra of Kanauj. But just at the time the Sena court reached the zenith of its splendor, internal disintegration set in. The Senas, all zealous orthodox Hindus, suppressed the Buddhists and imported a new Brahmanical aristocracy. Toward the end of the 12th century the provinces began to reassert their independence, and in 1202 the Moslems easily took the capital, Nadia; Lakṣmaṇasena escaped at the last moment. The dynasty survived for another half-century in southeastern Bengal. In that eastern corner the little kingdom of Pattikera (Mainamati) flourished; very important late Buddhist ruins are found there. Both Pattikera and the last Senas were overthrown by King Dāmodara of the Deva dynasty, whose successor had to submit to Sultan Balban of Delhi in 1283.

Cultural orientations. The Pala-Sena kingdom was a medieval semifeudal state, and the cultural life was dominated by the court, the great nobles, and the priesthood. The Palas supported Mahayana Buddhism against both the Hinayana Buddhists and the Hindu cults. They enlarged the Buddhist monastic university at Nalanda (between Patna and Rajgir) and founded universities at Odantapuri (Uddandapuri = Srinagar, mod. Bihar-i Sharif) and Vikramasila (Patharghata near Bhagalpur). Students flocked to these institutions from Nepal, Tibet, China, Burma, and Indonesia. In the reign of Devapāla the Śailendra king Bālaputradeva erected a monastery at Nalanda for students from Suvarnadvipa (Sumatra). The Palas repaired and embellished the Buddhist places of pilgrimage, such as Sarnath near Benares, Bodhgaya, Kapilavastu (Rummindei), Vaisali (Basarh), and Kusinagara (Kasia). They built numerous monasteries and schools and sent missionaries to Tibet. The most famous monastery was at Somapuri (mod. Paharpur); others were at Traikutaka, Devikota (Bangarh), Pandita, Sannagara, Jagaddala, Phullahari, and Vikramapura. Famous Buddhist scholars, such as Śāntarakṣita, Haribhadra, Buddhajñānapāda, Jinarakṣita, Dharmaśrīmitra, Aśokaśrīmitra, Dīpaṅkara Śrījñāna (also known as Atīśa), Nāropā, and Tilopā, flourished during the Pala dynasty.

Tantric Buddhism was the dominant element in the religious life under the Palas. At first the teachings of the Mādhyamika school, as expounded in the *Aṣṭasāhasrikā Prajñāpāramitā*, prevailed, next the philosophy of the Yogācāra school, then the magico-mystical doctrines of the *Guhyasamāja Tantra*, *Kṛyā Tantra*, and *Yoga Tantra*; finally, under Mahīpāla I the *Kālacakra Tantra* was introduced. Many of these scriptures were translated into Tibetan, Burmese, and Chinese. The Siddhācāra school of Nāropā and Tilopā exercised an immense influence on Tibet. Among the non-Buddhist works written in the period were the *Veṇī Saṃhāra* of Bhaṭṭa Nārāyaṇa (8th cent.), the bucolics of Yogeśvara (ca. 900), and the *Caṇḍakauśikam* (under Mahīpāla). Sandhyākara Nandī composed the *Rāmacarita*, a historical poem, during the reign of Rāmapāla. When the Pala kingdom collapsed, many learned monks emigrated to Nepal, Tibet, and Farther India. Later, when the Moslems were victorious, more monks fled to these territories.

Under the Senas Hinduism became powerful. Vijayasena inaugurated a policy of temple building. Ballālasena called in many Brahman families (whose descendants still form the aristocracy in Bengal), and established the social system of Kulinism. He was a scholar, learned in the Purāṇas and Smṛtis, and the author of several legalistic compendia, such as the *Dānasāgara* and *Adbhūtasāgara*. Lakṣmaṇasena turned from Saivism to Vaishnavism. He was an eminent poet and the patron of some of the greatest Sanskrit poets: Śrīdharadāsa, Halāyudha (author of the *Brāhmaṇasarvasva*), Dhoyī (author of the *Pavanadūta*), and especially Jayadeva, author of the world-famous *Gīta Govinda* (translated into most European languages, the basic inspiration for the great Vaishnava reform of the 15th and 16th centuries), and his colleague Bilvamaṅgala (author of the *Kṛṣṇakarṇāmṛta*).

The suppressed Buddhists turned to several crypto-Buddhist cults, such as that of Dharmaṭhākura. Many were later converted to Vaishnavism by the great mystic Caitanya (1485–ca. 1534), others to Islam (in fact, today East Bengal is a province of Pakistan). But the memory of the Buddhist kings still is alive in the saga of Gopīcandra (Govinacandra) and in the works of the historians of the Tibetan Lamaistic religion, especially Tāranātha and Bu-ston.

Pala Buddhism. The Buddhism of Nepal, Indonesia, and to some extent Burma and the Lamaism of Tibet and Mongolia had as their prototype the Buddhism of the Palas. The path of renunciation (*nivṛttimārga*) of Hinayana Buddhism was intended to lead the individual from this sorrowful samsara via the meditation on its twelve causes (*nidāna*) to nirvana. In place of this negative aim the Mahayana substituted a positive one: the compassion of the Bodhisattva for all suffering beings. In the Yogācāra school the path of renunciation was transformed into a phenomenal cosmogony, proceeding from the void (*śūnyatā* = nirvana) to the formless (*arūpa*) world, where the "body of the law" (*dharmakāya*; i.e., all the buddhas in the nirvana) is active, then to the world of pure forms (*rūpa*), where the "body of bliss" (*saṃbhogakāya*; i.e., the divine bodhisattva saviors) is active, and finally to the samsara world of lust and death (*kāmaloka*), in which the "illusory body" (*nirmāṇakāya*, consisting of the great religious masters incarnated on this earth) is active. The *Guhysamāja Tantra* added to each Dhyani Buddha and his respective bodhisattva a female counterpart (*śakti*, "energy," or *prajñā*, "wisdom"). The Vajrayana, finally, transformed the *śūnyatā* = nirvana into a supreme divinity (Vajrasattva or Ādibuddha, i.e., the archetypal Buddha), attributing to it a positive consciousness (*bodhicitti*), formed of wisdom and compassion (*karuṇā*), and a condition of bliss (*makāsukha* = *ānanda*). In a further development everything in the world was conceived as the dwelling place of a divinity and as evolving from the mystical syllable (*bīja*) corresponding to that deity. Through meditation the devotee would find his way back first to the archetypes and at last to the underlying *śūnyatā*. Though this system was never developed to its ultimate consequences, an immense pantheon of higher and lower deities, each with its various aspects, *prajñās*, and attendants, from Vajrasattva down to the demons, devils, and evil spirits, was evolved. This pantheon was grouped in ritual diagrams (mandalas) into mystical "families" (*kula*) of the five Dhyani Buddhas, embodiments of the Five Wisdoms (*pāramitā*) of the *dharmadhātu* (= *trikāya*). These deities are classified in a number of texts, such as the *Sādhanamālā*, *Niṣpannayogāvalī* of Abhayākaragupta, *Heruka Tantra*, *Hevajra Tantra*, *Vajravārāhī Tantra*, *Kṛyāsamuccaya*, and *Yoginījāla Tantra*.

With the exception of certain terrible deities taken over in Tibet from the indigenous Bon religion, practically the whole system was developed in Bengal and neighboring Assam and Orissa, partly by a subtle distinction of more and more subtypes of the leading deities, partly by the deification of abstractions, and partly by the absorption of Hindu gods and goddesses (mainly Saivite: Śabarī, Ḍombī, Caṇḍālī) and of primitive snake and mother goddesses. The relative influence of each of these successive trends of religious thought in the course of the centuries cannot yet be defined. They seem to have been formulated, in principle, by the 8th century; however, a survey of the extant sculptures reveals a great expansion of the iconography in the 10th and 11th centuries (mainly subtypes, Dhāraṇīs, and popular deities). Nor can the exact relationship between this eastern Indian center of Tantric Buddhism and the western in Udyana-Gandhara (Swat Valley) be determined. It seems that the former was responsible especially for the erotic-orgiastic features, the latter for magic (Mantrayana) and the witch (*ḍākinī*) cult. It is equally difficult to say to what extent the symbolism of blood and sex was accepted at its face value in the Mahāmudrā initiations and the Six Cruel Rites (*ṣaṭkarma*) or how far the shocking terminology of the "twilight language" (*sandhyābhāṣā*) served as a disguise for the sublimest mysteries. In Tantrism both the darkest superstition and the loftiest idealism are present.

Pala Buddhist iconography. The late Mahayana pantheon is immense, and only the barest outline can be presented here. The transcendent reality (*śūnyatā*) for a long time was indicated by a bell (*ghaṇṭa*) as the source of sound from which the *bījas* emerge and only much later as Vajrasattva (iconographically a variant of the Bodhisattva Vajrapāṇi). There are five constituents of human personality (skandhas): form (*rūpa*), sensations (*vedanā*), notions (*saṃjña*), predispositions (*saṃskāra*), and consciousness (*vijñāna*). The five skandhas through which the universe emerges from *śūnyatā* are represented by the Dhyani Buddhas, each with his complement of bodhisattvas in their various aspects, *śaktis* or *prajñās*, human (Mānuṣi) Buddhas, bodhisattvas, apostles, saints, and attendant deities. The aspects, especially of the bodhisattvas and of their *śaktis*, are determined by their respective functions.

Several deities are associated with more than one family, even more when the minor *śaktis* of the attendant gods are counted. Several of the figures are shadowy, theological constructions rather than real objects of veneration. In practice, only a limited number of these deities were commonly venerated, but under an increasing number of aspects, especially Avalokiteśvara (PL. 4), Mañjuśrī (II, PL. 384), Tārā (PLS. 2, 5, 7, 8; II, PL. 384), and secondarily Hevajra, Mārīcī (PL. 3), Prajñāpāramitā, and Cuṇḍā. The veneration of the historical Buddha (PL. 1) continued, but as part of a more general cult revolving around the Mānuṣi Buddhas, with their respective female counterparts or *śaktis* (i.e., wives before their renunciation of the world), the mortal bodhisattvas, and with Maitreya, the Buddha of the future, now living as a god. The historical Buddha Śākyamuni (Śākyasiṃha) was treated mainly as a mythical being, during his Enlightenment wearing the crown of a Bodhisattva, or working miracles, especially that of the self-multiplication at Sravasti. The great power attributed to spells, charms (mantras), and supernatural power (*siddhi*) made itself felt not only in the cult of Avalokiteśvara as protector and of Mahākāla as destroyer but also of the *ḍākinīs* (bringers of enlightenment and *siddhi*), the Five Protectresses (Rakṣas), snake deities, the Hindu goddess of wisdom Sarasvatī, Vajragāndhārī (connected with Kurukullā), Vasudhārā (consort of Jambhala, god of riches, PL. 8), and the twelve Dhāraṇīs.

In order to distinguish between all the deities, a very complicated iconography had to be evolved, following, however, certain definable principles: lotus seat or stand (as a sign of divine beings); one to twelve heads (benign, majestic, angry, demoniac, or animalic); two to a thousand arms (showing the various mudras and carrying symbols or weapons); occasionally also more than two legs. Some standing (*sthāna*) or sitting (*āsana*) postures were: yoga seat for the most abstract and mantric deities and easy seat, with one leg on the seat and the other dangling down (*lalitāsana*, PL. 8), for saviors and deities easily accessible to humans; attacking posture (*ālīḍhāsana*) for terrible deities; wild dancing for demoniac beings. Deities in creative action are shown in sexual intercourse with their *śaktis* or *prajñās*; "victorious" deities, triumphing over the forces of evil, stand on the corpses of heretics or on Hindu gods. Animal vehicles (*vāhanas*) such as are found in Hindu art are rare (e.g., the peacock of Mahāmāyūrī). The high and benign deities are of noble appearance and wear royal costumes; the terrible ones have ugly faces and demoniac bodies with flying hair and fat bellies. The emanations of *śūnyatā* carry a thunderbolt (vajra), those of wisdom the sword of discernment, the saviors a lotus, the bodhisattvas the image of their respective Dhyani Buddha on their crown. The skin color is determined either by the connection with the respective Dhyani Buddha or by the function of the deity: white, yellow, and green for benign deities, red for powerful and active ones, and blue and black for destructive ones. The benign deities are surrounded by a halo of light, the terrible ones by flames. Often the emanations or subaspects are grouped around their basic type. Thus, endless symbolic combinations proved possible. The *Sādhanamālā* describes visions (*sādhanas*) of 312 deities and their different aspects on which meditation turns; the *Niṣpannayogāvalī* describes 26 mandalas, each with innumerable deities.

In contrast to the exuberant speculation of Tantric Buddhism, the other cults developed very little. Jainism (q.v.), though it was the religion of a very small minority but was disseminated throughout India, regained some importance only in its homeland, Bihar, especially at Rajagriha (Rajgir), and in western Bengal (Burdwan-Vardhamana-Bhukti), Bahulara, Bankura, Chota Nagpur), and that only during the crisis of the Pala empire in the 10th and early 11th centuries. Hinduism (q.v.), on the other hand, continued to flourish as a folk religion even during the period when late Buddhism was most popular.

The great Buddhist temple of Somapuri (Paharpur) is covered with terra-cotta reliefs of all the popular Hindu deities: Śiva in several forms, Gaṇeśa, Kṛṣṇa and episodes in his life (including Rādhā), the heroes of the *Rāmāyaṇa*, Brahmā, Indra, Yama, Agni, Kubera, Yamunā, harpies (*kinnaras*), heavenly musicians (gandharvas), and amorous couples (*mithunas*). Hindu cult images of the early Pala period were rather rare but became more numerous during late Pala times and, of course, predominated under the Senas, though not in such great numbers as might be expected. Instead, the Hindu gods infiltrated en masse into the Tantric pantheon, in the disguise of Buddhist gods and goddesses, as minor rulers of the universe and protectors of the Buddhist faith, or even as defeated antagonists attending on or trampled under the feet of Tantric deities. The Sena pantheon reveals little individuality, possibly because of its consciously orthodox character. However, some characteristics may be noted: the most common Saiva sculpture represents Śiva caressing Pārvatī on his lap (*āliṅganamūrti*, PL. 9); Naṭarāja dances not on a dwarf but on his bull; *mukhaliṅgas* are known but are not common. Devī appears for the most part as the majestic divine queen, rarely as Cāmuṇḍā, Mahīṣamardinī, or Vajrayoginī, hardly ever as Kālī (which nowadays is her dominant aspect). Mātṛkā images are rare. Gaṇeśa generally is represented dancing. Viṣṇu rides on Garuḍa, as in Nepal and Indonesia, or is represented as Lakṣmī-Nārāyaṇa (with Lakṣmī on his lap). His best-known avatars are Varāha, Narasiṃha, and Vāmana-Trivikrama (not Kṛṣṇa or Rāma); a mother and child group has also been interpreted as the birth of Kṛṣṇa or as Sadyojātā (Devī). The other deities most commonly depicted are Sarasvatī, Sūrya (PL. 6), and Manasā (a snake goddess). In their composition, minor details, and ornamentation these images are very similar to the late Buddhist ones. Only those of the last phase, in eastern Bengal, reveal a richer iconography and freer composition.

ARCHITECTURE. Of the ancient architecture of Bengal and Bihar disappointingly little has survived; however, this may be partly due to the fact that only a few mounds have as yet been opened. Votive sculptures in the shape of temples, illustrations of famous sanctuaries in Buddhist manuscripts, architectural fragments, and the monuments of Farther India inspired by Bengali prototypes make it possible to reconstruct the main trends of evolution to a fairly satisfactory degree.

Palaces and towns, of course, have disappeared. But their sites, covering areas up to 8 square miles, are still recognizable by the huge rectangular earth platforms (*murā*), dikes (circumvallations), dam-enclosed ponds or lakes (*dīghī*, *pukar*), and temple mounds (*deul*). Among the sites are Ramavati (Chandipur) near Gaur, Vikramapura (Rampal), Vijayapura (near Rajshahi), Srinagar (Bihar-i Sharif, where at least the moats are discernible), Sabhar, Pundranagara (Mahasthan) with the Pala citadel, the Gaṛh, Devikota (Bangarh), Nadia, and Pattikera (Mainamati). The ruins of the ancient royal palaces exist but have not yet been excavated; where the rains have laid bare parts of them, huge brick walls and platforms have come to light. It appears probable that the upper stories of these buildings were constructed of wood, as they were in other parts of India.

In the Pala period stupas (*dhātugarbhas*) had been more or less superseded by temples, but they still played an important role as decorative elements, as small votive examples, and as key parts of more comprehensive structures. The function of the *dhātugarbha* was practically reversed. From a horizontal tumulus it had been transformed into the visible vertical shell of the *dharmadhātu*, in which images or holy scriptures, when there were any, took the place of the relics. The components of the stupa — the high pedestal, the high drum with niches for the images of the Dhyani Buddhas, the "pot" (*kumbha*), and the balustrade (*harmikā*) — were gradually amalgamated into a bell-shaped structure, the symbol of *śūnyatā* (*gaṇḍhakṛti* or *kalaśakṛti* type). Representing a reality even more venerable than the buddhas themselves, this type of stupa could be placed on an open lotus flower (as the buddha images were in Gandharan art).

As an ornamental motif the *dhātugarbha* served as the upper half of the temple spire, or its pot-shaped finial (*kalaśa*); it was used in decorative reduplications of the spire (*upaśṛṅga*) or as the upper end of pillars, doorjambs, and so on; sometimes it was arranged in friezes along the roof edges of the sanctuary. In all these cases the bell-shaped type was the rule.

The votive *dhātugarbhas*, however, found in great numbers at all the Buddhist places of pilgrimage (Sarnath, Bodhgaya, Nalanda, Rajgir, etc.), in most cases retained the older form, well articulated and decorated with four niches for the Dhyani Buddha images or for Tantric deities. In some cases one of the niches was hollowed out into a real cella for a separate cult idol, and an entrance frame, even a porch with its own roof, was added. Sometimes the *dhātugarbha* was placed in the center of a big lotus, or was supported, on a lotus, by four deities (especially in the copper or bronze reliquaries, but also in those of stone). In the latter instances the Dhyani Buddhas were sometimes arranged around the *harmikā*, their bodhisattvas in the ordinary image niches and other deities around the pedestal. In the final stage of development eyes were engraved on the four sides of the *harmikā*, as in the Sambhunatha and Bodhnath (X, PL. 303) stupas in Nepal.

The stupa tower of Gandhara and of the Gupta period (see GUPTA, SCHOOL OF) was still used in early Pala times, at least in the case of time-honored memorials, as, for example, in the final stages of the Dharmarājika and Dhāmekh stupas at Sarnath (9th cent.) or in the stupa of the Goose Relic at Giriyek (at the eastern end of the Rajgir range). The greater part of the tower consisted of the original platforms and circumambulation paths (*pradakṣiṇāpatha*), raised higher and higher and reached by one staircase or four. As time passed this type of tower was more and more influenced by the Hindu temple tower, with its wall projections and recessions, its supporting platform or terraced pyramid, and its surrounding subsidiary temples (*pañcāyatana* type). The stupas at Ushkur and Parihasapura (Paraspor) in Kashmir (which between ca. 734 and 756 governed Gauda, in Bengal) rise on 12-cornered (*dvādaśāṅga*) double platforms with a flight of steps on each side. The great stupa at Nalanda (No. 3; II, PL. 384) in its fifth phase still had one vast platform with four subsidiary corner stupas; in the sixth phase it was a pyramid; and in the last (seventh) phase the central stupa was replaced by a temple. The same development occurred at the great monastery of Somapuri, though to a lesser extent: the central stupa on the pyramid of cross platforms (actually 20-cornered, *viṃśatikoṇa*) was preserved, but the four image niches around it were expanded into monumental chapels surrounded by a *pradakṣiṇāpatha*. The later development is known mainly from monuments outside India inspired by Pala art, such as the Shwesandaw stupa south of Pagan and the Sulamani and Mingalazedi stupas in Pagan itself (see BURMA; BURMESE ART). The Shwesandaw (XIII, PL. 243) is a five-storied pyramid supporting a bell-shaped stupa and was erected during the reign of King Anawratā (1044–77). The Mingalazedi (II, PL. 411), built in 1274, consists of four square terraces crowned with corner stupas and an octagonal terrace bearing the huge central stupa.

The ultimate consummation of these typological tendencies was the mandala stupa. It represented a horizontal expansion of the type prefigured in 8th-century Kashmir on the model of the votive stupa placed in the center of a lotus. After the Yogācāras had introduced the yoga concepts of the wheels (*cakras*) in lotus shape, it was but natural to conceive of the Dhyani Buddhas — already represented in the four chapels around the stupa drum — and of their attendant deities as being present in the petals of an imaginary lotus projected into the stupa. The only such stupa known in the Pala territories is the 60-cornered structure excavated at Lauriya Nandangarh in Bihar, but the top stories unfortunately are missing. The most perfect and best-known example is the Borobudur in Java, of *viṃśatikoṇa* type, rising in 9 (1 + 5 + 3) terraces, of which the three topmost ones bear respectively 32, 24, and 16 subsidiary *dhātugarbhas* surrounding the great central stupa, not to speak of the innumerable miniature stupas crowning every image niche, arch, and so forth (III, PL. 485; VIII, PL. 30; see also VIII, col. 69).

Temples (*prāsāda*) underwent an equally complicated evolution. Whereas the forms of the Hindu temple had become more or less fixed by the 8th century, the shape of the Buddhist sanctuaries remained fluid until at least the 11th century, and the tower (shikhara) temple did not predominate until the

10th century. The reasons for this peculiar situation were the influence of the wooden pagoda and of the stupa, forms which, however, seem closely connected.

The first type, a tower of superimposed pyramidal roofs, still survives in the Himalaya (especially in Nepal and Kulu), in Burma (where it is called *pyatthat* = *prāsāda*), and in Bali (where it is often called a "meru"). Its origins are obscure, but it may possibly have been derived from the set of umbrellas (*chattrāvalī*) erected on top of the stupas and also above the chaityas of various local deities. In the 7th century, at least, it was common in northern India and was one of the starting points for the development of the Hindu temple. The roofs of many early medieval temples are mere copies in stone of such roof pyramids. In the vimana type of temple (in Orissa called *piḍa deul*) these roof pyramids continued in use throughout the medieval period, but when used on top of the sanctuary they were reduced, from the Pratīhāra period on, to low profiles forming a beehive-shaped shikhara. In the alluvial plain of Bengal, especially in Pundravardhana and Samatata, the original wooden type persisted much longer and is depicted in miniatures of the 11th and 12th centuries. A large number of richly carved wooden pillars, brackets, and images have been discovered, well preserved in the waterlogged subsoil.

The influence of the stupa appeared in the elevation of the sanctuary on top of a high pyramid of square or polygonal terraces; in the cross plan (*caturmukha*) of the sanctuary (adopted also by other ancient cults), inspired by the four chapels of the Dhyani Buddhas; in the multitude of small temples surrounding the central temple (like the votive stupas around a much-venerated large stupa); in the prominence given to the circumambulation passage; in the fusion of the shape of the temple proper with that of the stupa; and in the trefoil arches of the cella entrance (apparently inspired by the halo of the buddha image). Finally the stepped-roof pyramid and the terraced pyramid became so similar that the first served as the substructure for a second story.

Only the foundations of most of the temple ruins in Bengal and Bihar have been excavated. The older ones (up to the early Pala period) are massive sanctuaries on a platform with corner projections and a flight of steps, and sometimes also with a pillared porch (e.g., Nalanda, Satya Pīr's Bhīṭā at Paharpur, Bangarh). Concerning the upper structure nothing specific is known; however, it is probable that it may have been the prototype for such Javanese temples as the Pawon (VIII, PL. 34), Kalasan, and Mendut candis with their receding roof stories crowned by small stupas, and also indirectly for numerous other temples in Siam, Champa (Annan), and Cambodia.

The only great temple preserved in Bengal is that of Somapuri (Paharpur; FIG. 30), the nucleus of which is a stupa rising on three *viṁśatikoṇa* terraces, but so well hidden behind four big sanctuaries and a covered circumambulation passage that only its *chattrāvalī* (probably resembling a shikhara) remained visible. Vestiges of similar temples have been traced at Mahasthan (Gokul Math mound) and Mainamati. In Burma the temples of Pagan, erected in the 10th–13th century, form a veritable museum of later Pala types: Bidagat Taik library, with a pyramidal roof ending in a small stupa; Shwegugyi, similar, but with a mandapa and a high shikhara stupa; Ānanda, with four mandapas; Gawdawpalin, two-storied; and Thatbyinnyu, similar, but with an additional mandapa (II, PLS. 412, 414–416). In Cambodia (Koh Ker, Baksei Chamkrong, Phimeanakas, Phnom Bakheng, Takeo), Java [Lara Jongrang, Sewu (VIII, PL. 34, FIG. 62), Singhasāri (VIII, FIG. 78) candis], and Thailand (Lamphun, Sawankhalek, etc.) the temples are raised on terraced pyramids, giving rise to a new development that ended in the famous funeral temple of Angkor Wat (VIII, PL. 369).

The late Gupta shikhara temple (with only a few stories and big chaitya windows) was known, as the Tārā temple at Bodhgaya and several relief representations attest. The famous Mahābodhi temple nearby (VII, PL. 133), a *pañcāyatana* temple on a high platform, in its present form probably represents an early Pala enlargement of a late Gupta design; it was repeatedly copied because of its special sanctity [Pagan, the Mahābodhārāma near Chieng Mai in Thailand (XIII, PL. 20), Lalitapattana in Nepal; a very late echo is the Śivadole at Sibsagar in Assam].

The fully evolved medieval shikhara temple (called Gandhakuṭī by the Buddhists) was introduced from Orissa, Khiching, and the Chedi country during the decline of the Pala empire in the late 10th century. The oldest examples probably are the Śaileśvara (Salleśvara) and Sāreśvara at Dehar (Bankura district), the Ananta Nārāyaṇa temple at Joypur (Bankura), the Siddheśvara at Bahulara (12 miles from Bankura), the Jaṭār Deul in the Sundarbans, and the four "Begūnia" (eggplant-like) temples at Barakar, also in western Bengal; their relationship

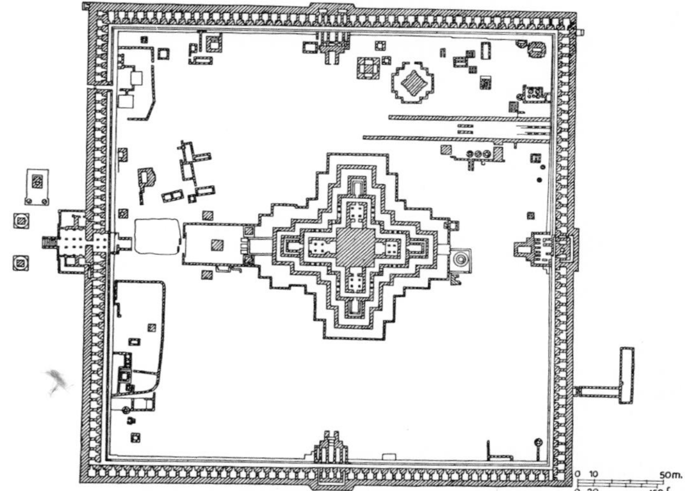

Paharpur, ground plan of the temple (*from B. Rowland, The Art and Architecture of India, Harmondsworth, 1953*).

to the earlier Orissan temples is obvious. A later stage is represented by the Ichaighoṣ temple at Gaurangapur (Burdwan district); in contrast to the preceding sanctuaries, which were made of stone, this is a brick structure. The Kalyāneśvara at Burdwan and the big temple at Konch near Bodhgaya might be adduced as further examples of this style, but they have been so much restored in later times that a correct evaluation will have to wait for a more detailed examination. Characteristic of most of these temples are the exceptionally high and slim shikhara, which rises vertically at first and curves only in the last third; the rather small but high cushion-shaped knob (amalaka); the pronounced *kalaśa*; windows above the entrance; and occasionally even genuine arches. All these temples are Saivite, witnesses of the return to orthodox Hinduism.

Buddhist temples of a similar type must have existed, to judge by architectural fragments and miniature models (Patna Mus.; Calcutta, Indian Mus.; palace of the Maharaja of Dinajpur). They seem to have been dedicated mainly to Tantric deities (the two temples near the Sambhunatha stupa in Nepal still preserve this late Pala type). Many probably were enclosed by a circumambulation passage from the stepped roof of which only the central shikhara emerged. The early doorframes are still of the Gupta type, with the sitting Śākyasiṃha on the coping stone and the Dhyani Buddhas on the lintel frieze above; later the trefoil arch on bulging columns became the rule.

The excavations at Nalanda, Sarnath, Paharpur (Dharmapāla Mahāvihāra), Mainamati, and other places reveal that the earliest monasteries retain the late Gupta plan: they are often several stories high, the rectangular court is surrounded by cloisters and monks' cells, and there is a monumental entrance at one end and a chapel at the other. Soon this chapel was detached by means of a circumambulation passage and finally it was moved into the center of the court, following a model already set in Gandhara and Kashmir.

SCULPTURE. As far as can be deduced from the excavated ruins of the temples in Burma and Nepal that were influenced by Pala art and from developments in later local art, the sculptural decoration of the Pala — and probably also of the Sena — monuments was quite different from that in other parts of India. The walls were of plain brick, interrupted by niches in which terra-cotta plaques and sometimes also sculptured stone slabs were mounted. These terra-cotta plaques have been discovered in great quantities, ranging in date from the Gupta period to the Moslem invasion. They are generally in an uninhibited folk style, crude, summary, without any regard to the canons elsewhere dominating medieval sculpture but very lively and often also most expressive. The subjects are variegated: scenes from life, from the Jātakas, from epics, primitive demons and godlings, Hindu as well as Buddhist deities. Their correct interpretation, however, is most difficult, because the subjects generally are indicated merely by one or two essential traits. Other than this, the decoration was restricted to wood carvings, such as are still seen in Nepal, and to doorframes, perforated window screens, and the like, of basalt or other stone brought by ship from Rajmahal or Orissa. The shikhara temples of western Bengal, wholly constructed with dressed stone, also have almost no exterior sculptural decoration; inside, however, the decoration seems to have been richer. At Nalanda some chapels had life-size clay statues, covered with stucco and painted and gilded in the manner known from the cave temples in the Tarim Basin (see ASIA, CENTRAL), from Tun-huang (q.v.), and from many Tibetan monasteries. With the destruction of the buildings that protected them, such sculptures were, of course, soon washed away. In other places stone and bronze figures were arranged on brick benches or in niches.

Consequently, almost all that is known of high-quality Pala-Sena sculpture consists of cult figures, usually of basalt but sometimes of bronze and occasionally of silver. Their general form is completely stereotyped: an oblong slab, which at the bottom broadens into a pedestal and at the top ends in a semicircle, an elliptic or parabolic curve, or a pointed arch (VII, PL. 462). The divine figure sits or stands on a throne and is surrounded by minor deities and symbols of its legend. On one or both sides of the socle the devout donors are shown. On closer examination, however, an endless variety of elaborations is discovered. The pedestal may be merely a rounded projection or may jut out stepwise; it can be treated as the socle of a temple façade or of a stone altar, or as a (sometimes multiplied) metal platform resting on four legs, or as the box of a chariot, or as the seat of a throne; it can dissolve into rock scenery, or into a lotus rising, on its stem, from the waters, or into a whole group of lotuses. In addition, it may be decorated with the wheel and deer (symbols of the Buddha's First Sermon at Sarnath), lions supporting the *Siṃhāsana*, or nagas holding lotuses. The image proper rises in front of or at the back of the throne of a chakravartin, decorated with all the paraphernalia of sovereignty, or at the entrance of a cave or temple, or in front of a halo of flames. The halo (*prabhāmaṇḍala*) behind the head first was circular, then oval, then an inverted *kumbha*, and finally a scroll that emerged from the mouth of a *kīrtimukha*. A tree behind a lady represents the Buddha's birth in the Lumbini grove; a tree behind the Buddha in yogasana, his Enlightenment at Bodhgaya; the hood of a cobra, his protection by the naga Mucalinda; nude girls and archers on both sides, the temptation by Māra; a reduplication of his figure (on lotuses), the miracle of Sravasti; a stupa behind his recumbent figure, the *parinirvāṇa* (PL. 1). Often the chief events in the Buddha's life — his birth, Enlightenment, First Sermon, miracles, and *parinirvāṇa* — are arranged around his dominating figure. Above Tārā and the bodhisattvas the respective Dhyani Buddhas are depicted, to the sides their attendant deities and *prajñās* (PL. 9). The upper corners generally are filled with flying godlings, gandharvas, and siddhas. In the case of Hindu or Tantric Buddhist deities such an arrangement is even easier, because it is in harmony with a Pan-Indian practice: in the middle the deity flanked by attendants, on the socle the *vāhanas*, on top gandharvas and siddhas. Certain Buddhist conventions, such as the naga holding the stem of a lotus, were taken over by Hindu art.

According to the Tibetan historian Tāranātha, the so-called "Eastern School" developed during the reigns of Dharmapāla and Devapāla under the influence of two master artists, Dhīmān and Bitpālo. As a matter of fact, the stylistic differentiation between an eastern and a western school had already been made under the Gupta emperors and had had a literary counterpart in Daṇḍin's distinction between the Gauda and Vidarbha styles. The western provinces, which had to defend themselves against invasions and had also absorbed many of the foreign barbarians, evolved a simple and rather dry, matter-of-fact style; the opulent eastern provinces, much less affected by barbarian devastations, created a soft, hypersensitive, and exuberant style. But it is surely correct to affirm that under the first two Pala conquerors the art of Bihar and Bengal dissociated itself not only from the late Gupta tradition but also from the other medieval styles derived from the same source. It is equally true that the transformation of Pala-Sena art in the 11th century, though in accord with contemporary trends in other parts of India, was much more incisive.

Early Pala sculpture had all the characteristics of an art after a social revolution: the figures are plump and fleshy, the tender modeling lacks an underlying structure, and the faces often have an insipid smile; yet there radiates from them an earthy, warm, and good-natured humanity. The composition is a simple juxtaposition of parts; the steles are rather short and softly rounded on top. From the late 9th century a change set in: the modeling of the figures became harder, the costumes and jewelry richer, and the background more ornate; the composition of the steles is clearly divided into three sections: socle, central group, halo and top ornament. In the 10th century a sort of new classicism was achieved: elegant vigorous bodies full of energy, powerful movements, beautiful costumes and ornaments, a broadly conceived composition, and an atmosphere of solemn grandeur. From the 11th century on a new style emerged: the technique is like the finest metalwork; the body forms are petrified and become excessively slender and consciously graceful; the movements and postures are exaggerated and affected, as in a ballet (excessive *atibhaṅga* and *tribhaṅga* poses); the involved compositions are of a sumptuous exuberance and an accuracy calculated in every detail; the backgrounds are overladen, often with extraneous details; and the steles are slim and pointed. New court fashions were introduced: the hems of the gorgeously decorated costumes are frizzled; the high crowns are covered with flowers or capped with an amalaka-like top; the hair of the goddesses is drawn backward into a broad, flat chignon or to one side in the form of a mango or a Phrygian cap; the jewelry is elaborated in the most delicate designs. The background too was transformed into costly thrones sculptured with leogryphs (*śārdūlas*), geese or swans (*haṃsas*, *kinnaras*, *makaras*, and *kīrtimukhas*). Nothing of the homely humanity or the spiritual grandeur of the preceding periods was left; it had become a thoroughly frivolous, elegant court art. Under the Senas sculpture acquired more substance, a sweeping rhythm, a physiognomic individuality, and a maximum of sensuousness and sumptuousness and became a composition in depth on three receding planes. At this point the limit of the possible was reached. The elongated bodies degenerate into stiff dolls; the faces become expressionless masks with doubly curved, feelerlike eyebrows, lotus eyes, voluptuously curved lips; the movements are exaggerated; the costume is reduced to scribbles or fluttering scarves; the backgrounds are a jungle of scrollwork or a chaotic multitude of figures. Thereafter Sena sculpture quickly disintegrated into shapeless imitations and disappeared.

Metal images must have been very common in a country where stone is a rarity. The images that have been preserved are more numerous and perhaps more beautiful than the much better known southern Indian bronzes of the same period. Southern Indian procession images exist, but none are extant from the Pala-Sena era. Knowledge of the Buddhist pieces comes mainly from two hoards buried at the time of the Moslem invasion by the monks of Nalanda and Kurkihar (PLS. 2, 3, 8; II, PL. 384). The Kurkihar treasure, discovered in 1930, consists of over 240 images. Minor hoards have been found at Mainamati (from Rupban Mura) and Jhewari (Chittagong district). Hindu pieces have been found all over the country, but especially in eastern Bengal. The most common material was "bronze" (i.e., *aṣṭadhātu*, an 8-metal alloy), which consisted of 83 per cent copper, 13 per cent tin, lead, iron, and other metals. Many of the bronzes are inlaid with gold, silver, and copper or reveal vestiges of gilding. Their stylistic evolution is more or less the same as that of the sculpture in stone. However, with the cire-perdue process much more delicate modeling could be achieved and the shining metal could act as a real *prabhāmaṇḍala*, so that in many cases the background is formed by a blank metal disk or by an arch, framed by richly chiseled borders of tongues of fire. The oldest Buddhist bronze (ca. 780) is a Vāgīśvara from Kurkihar. The most beautiful Buddhist image may be the sitting Tārā (9th cent.) from Kurkihar (New Delhi, Nat. Mus.), exquisitely modeled and the very embodiment of compassion. Iconographic types, found so far only in bronze, are Padmanarṭteśvara (an aspect of Avalokiteśvara) and Vajratārā (an aspect of Tārā), represented in the center of a closed lotus; figures of minor deities are placed on the petals of the lotus if it is open. The oldest Hindu bronze (ca. 700) is the Devī dedicated by Queen Sarvvanī of the Khaḍga dynasty (Dacca Mus.). Noteworthy Hindu images are the 10th-century Viṣṇus from Rangpur (near Dacca) and the delightful silver Viṣṇu from Churain (Sonarang) near Dacca, now in Calcutta.

The most comprehensive collections of Pala-Sena sculpture are in the Indian Museum of Calcutta, the Dacca and Patna museums, and the Varendra Research Society Museum in Rajshahi. There are numerous pieces in all the Buddhist places of pilgrimage; others are found in Nepal, Burma, Thailand, and Java. Many of the leading museums in Europe and America possess some examples.

PAINTING AND DECORATIVE ARTS. The vast brick walls of Bengali architecture naturally invited mural decoration. Though nothing has survived in India except a small fragment at Nalanda, because of the humid climate, extensive cycles of wall paintings inspired by Pala art exist in the temples of Pagan (Kubezatpaya, Kubyaukkyi, Payathonzu, Kyanzittha) and in Nepal and Tibet. In many of these pictorial cycles the wall surface is subdivided into numerous rather small panels, probably a result of the corresponding breaking up of the walls into niches. Almost certainly the crowded mandala compositions, with smaller and smaller groups arranged around one or a few big painted or sculptured figures, go back to this period. Paintings on cloth have found an echo in the tankas of Tibet.

Numerous illustrated manuscripts executed under the Pala and Varma kings have been preserved, mainly in Nepal, as well as early Nepalese manuscripts in the Pala style, but none of those executed under the Hindu Senas have come down to us. Only two of these manuscripts are slightly earlier than the 11th century (Br. Mus., Or. 6902, ca. 965, and Cambridge Univ. Lib., Add. 1464, ca. 985). This may be due to the fact that during the bitter struggles of the late 10th and early 11th centuries the famous monastery of Somapuri and probably other monasteries as well were burned to the ground. (For a list of the Indian and Nepalese illuminated manuscripts that have been preserved, see X, cols. 176–77.)

The manuscripts are on palm leaves. Generally there are two miniatures (ca. 1 ⁷/₈ × 1 ¹/₈ in.) at both ends, near the holes for the strings that hold the pages together. Somewhat larger compositions cover the inner sides of the wooden book covers, and they are generally a little lighter in color than the miniatures proper and with brighter backgrounds. The paintings on the covers usually represent scenes from the legend of the Buddha, the Jātakas, or the Avadānas, whereas the miniatures depict the most venerated deities, especially those of the famous places of pilgrimage (not only in the Pala territory but outside as well), sitting or standing in their respective temples. A good idea of what much of the lost Pala architecture of Bengal and Bihar must have looked like can be gained from these miniatures, but caution has to be exercised in interpreting the architectural features, because they are treated merely as background and are on a much smaller scale than the figures of the deities. The composition of these miniatures is closely related to that of the contemporaneous sculpture, but their repertory is more varied. As in the sculpture, the Gupta heritage of forms and types is still evident, but reduced to a system of fixed formulas and mannerisms, beautifully combined in complicated symphonies. Its evolution cannot be followed, because only works of the 11th and 12th centuries have been preserved. But even in these there are great differences. The work of the early 11th century often has an almost impressionistic freedom, a fluid line, shading, and rich color variations (X, PL. 86); that of the 12th century is hard, flat, and full of absurd mannerisms, which is also true of the Nepalese art. Since some of the work is executed rather carelessly, it is a question of how much is due to different stylistic phases and how much to provincial manifestations of the free folk style known from the terra cottas.

Of the decorative arts very little is known. The sculptures and paintings reproduce beautifully printed or embroidered textiles, exquisite jewelry work, and richly carved furniture. What has been preserved includes finely carved seals, copperplate grants with the Pala coat of arms (the *dharmacakra*, or Wheel of the Law, between two deer), some rather primitive depictions of Viṣṇu's Footprint (*viṣṇupādas*), standard disks with *à jour* reliefs, a small ivory mandala of Vairocana (Seattle Art Mus.), a bronze cornucopia (Nalanda Mus.), lamps and lotus-shaped incense burners, and shards of a polychromous pottery.

THE SPREAD OF PALA-SENA ART. The influence exercised by Pala-Sena art coincided more or less with that of Buddhism, then on the decline in India but on the ascent in central, eastern, and southeastern Asia. The centers of diffusion were especially the three great universities at Nalanda, Odantapuri, and Vikramasila and to a lesser degree Somapuri and Jagaddala. Students flocked to these institutions from all the Buddhist countries; the Mahayana scriptures were translated into Tibetan, Chinese, and other languages; and missionaries were trained here. During the political crises, such as the first eclipse of the Palas, later the collapse of the dynasty, and finally the Moslem conquest, refugee scholars and artists settled in friendly neighboring or overseas countries. Pala influence, therefore, was felt only in two Indian countries.

Orissa, under the Bhaumakara kings (8th–10th cent.), was one of the birthplaces of the Vajrayana, and its Buddhist art had close affinities with the early Pala style. But under the Somavaṃśa and Ganga dynasties Orissa aligned itself with the contemporary Hindu art of northern India. When the Sena kingdom collapsed under the Moslem attacks, Bengali influence showed up in the basalt figures (but not in the other sculpture) of the late Orissan temples at Konarak, Puri, and even Bhuvaneshwar.

Kashmir maintained close relations with Bengal from the 8th century on, but as Buddhism continued to lose ground there it is difficult to define the extent of Pala influence, especially since most of the late Buddhist works have disappeared. The Buddhist and Hindu bronzes from Kulu show that at least in Dharmapāla's time it was under Pala influence (see KASHMIR ART).

Pala influence was most felt in Nepal (adjoining Bihar and in the Buddhist fold since the time of Aśoka). After its liberation from the Tibetan yoke in the early 9th century it became practically a province of Pala culture, and when many monks and artists fled before the Moslems at the end of the 12th century the valley became the cultural heir of Pala art, faithfully preserving it up to our own time (see NEPALESE ART).

King Sroṅ btsan sgam po of Tibet sent his minister T'on mi Saṃbhoṭa to Magadha to study; after the king's marriage in 639 to a Nepalese princess (deified as the Green Tārā) he introduced Mahayana Buddhism to Tibet. The two most influential Buddhist missionaries in Tibet came from the Pala empire: Padmasaṃbhava from Nalanda in 749 and Atīśa about 1040. Tibetan art (q.v.), although it absorbed much of the art of Kashmir, Nepal, and China, and made its own contribution to Lamaistic iconography, remained essentially an offshoot of the Pala school.

In Ceylon the Pala style gained a foothold with Mahayana Buddhism, but neither proved a match for the conservative Hinayana and the already established late Śātavāhana-Gupta art tradition. Only some ruins of monasteries at Anuradhapura and a set of beautiful bronze images in the Colombo Museum bear witness to the short-lived Pala influence (see CEYLONESE ART).

In contrast, Pala influence in Burma proved very strong, although the Tantric Buddhists (Āri) have left only one monument near Pagan, the Payathonzu temple at Minnanthu. But as Burma had only a rather weak (Gupta) art tradition, King Anawratā (1044–77), though introducing a reformed Hinayana from Ceylon, availed himself of Bengali artists who may have left their homes because of the Kalacūrī and Somavaṃśa invasions, which Nayapāla was able to repulse only with the greatest difficulty. The link was the little kingdom of Pattikera in eastern Bengal, the Candra dynasty of which was wiped out by Karṇa Kalacūrī. Thus the new capital of Burma, Pagan, was in the course of three centuries embellished with huge stupas, temples, libraries, and other structures, all of which represent variations of Pala architecture. Some show this descent even in their names, for example, the Mahābodhi temple (a copy of that at Bodhgaya) and the Gawdawpalin (Gauḍapālin). Original Pala sculptures and bronzes have been found which were the models for the later Burmese buddha type; also of Pala origin are the terra-cotta reliefs, stone images, and murals, though in their new milieu they soon assumed Mongolian features.

Farther east, among the various Mon and Thai kingdoms, Pala influence was less conspicuous, because it had to compete with local Śātavāhana, Gupta, and Pallava influences. The bell-shaped stupas on terraced pyramids, so common in Farther India (see INDIA, FARTHER), go back to the same source. The National Museum in Bangkok possesses a bronze of the Buddha invoking the earth goddess, a stone Lokeśvara from Jaya on the Isthmus of Ligor, and a Lokeśvara bronze from southern Siam; in Chieng Mai there is a stone relief of the Buddha with the elephant Nalagīri. These are all original Pala works. Both the crowned and uncrowned type of Siamese buddha images (see SIAMESE ART) are derived from the Pala tradition, though, of course, they soon lost the characteristics of the Pala style.

The strongest Pala influence was in Java (see INDONESIAN CULTURES), where Mahayana Buddhism, already entrenched under the Śailendra dynasty, soon turned Tantric and at last fused with Saivism. The Kelurak inscription of 782 mentions a Gauda monk as royal chaplain. Many learned monks from "Gaudidvipa" visited the island, such as Dharmapāla, Vajrabodhi, Amoghavarṣa, Atīśa, and Kumāraghoṣa. The Pala pre-Nagari script superseded the Pallava Grantha; in the old Javanese *Ādi-Parwan* the influence of Bhaṭṭa Nārāyaṇa's *Veṇī Saṃhārā* (8th cent.) is evident. That the famous Borobudur represents the Pala mandala stupa has already been pointed out. The late Gupta–early Pala temple type is preserved in the Mendut, Pawon, and Kalasan candis. In the Sewu, Parambanam, Singhasāri, and Jago candis the temple on a terrace, often four-sided, has been imitated. In the reliefs of Borobudur Pala influence is also strongly felt, and even more in the bronzes, Hindu as well as Buddhist, of the 9th–11th century (also later in the Plaosan and Singhasāri candis). Similar bronzes have been discovered in a cave at Sambos in eastern Borneo; Pala sculptures and stupas (at Padang Lawas, etc.) have been found in Sumatra. In eastern Java late Pala and Sena influence can be traced. The famous funerary image of King Erlaṅga (Airlangga) at Belahan, for example, is inspired by a Bengali Viṣṇu on Garuḍa. However, in all the areas of Indian "colonization" the Pala impact operated with a considerable time lag and, moreover, was diluted by other influences: the surviving Gupta and Pallava ones and new eastern Chalukya and Chola trends.

Via Sumatra (Srivijaya) Pala influence reached southwestern China (Yunnan), as is proved by an Avalokiteśvara type persisting there up to the 13th century. Pala influence is absent in Cambodia and Champa (Annam), though the pre-Nagari script was introduced there and the terraced temple reached its most glorious evolution in these lands. However, although it is possible that these countries had some acquaintance with the temples of Bengal, this development seems to have been more or less autonomous.

In the Far East Buddhism had produced an artistic revolution, introducing a quite new iconography and typology. During the early T'ang dynasty the late Gupta pictorial style was taken over, but T'ang art soon evolved its own individual style (see CHINESE ART), and although the Buddhist universities of Bihar continued to be held in high esteem, their artistic influence became restricted to the iconographic field. There is strong reason to believe that Pala art furnished the model for many of the mandalas in Chinese Buddhist painting. Later, under the Yüan dynasty, Pala types again infiltrated into China via Nepal, especially through Kublai Khan's court artist A-ni-ko.

Pala-Sena art indirectly influenced classical Moghul art (see MOGHUL SCHOOL). The cusped arches resting on columns in the form of bulbous lotuses, introduced by Shah Jahān, recall those of the minor medieval temples, and they are shown in many votive reliefs and miniatures. The links are to be sought in the architecture of the Bengal sultanate.

BIBLIOG. *Historical and literary characteristics of the period*: R. Pischel, Die Hofdichter des Laksmanasena, Göttingen, 1893; M. Chakravarti, Sanskrit Literature in Bengal during Sena Rule, J. Asiatic Soc. of Bengal, N.S., II, 1906, pp. 157–76; H. Sastri, Literary History of the Pala Period, JBORS, V, 1919, pp. 171–83; H. Sastri, The Nālandā Copperplate of Devapaladeva, Ep. Indica, XVII, 1923–24, pp. 310–27; H. C. Ray, Dynastic History of Northern India (Early Mediaeval Period), I, Calcutta, 1931; P. L. Paul, Rāmapāla: The Last Great Pāla King, Indian H. Q., XIII, 1937, pp. 37–43; P. L. Paul, Relations between Pālas and Senas, Indian H. Q., XIII, 1937, pp. 358–60; N. N. Dasgupta, The Literature of the Age of Dharmapala, Indian Culture, VI, 1939–40, pp. 327–38; S. K. De, Sanskrit Literature under the Pala Kings of Bengal, New Indian Ant., II, 1940, pp. 263–82; B. C. Sen, Administration under the Palas and Senas, Indian Culture, VII, 1940–41, pp. 203–19, 305–26; R. C. Majumdar, The History of Bengal, I, Dacca, 1943.

Religious doctrines: F. A. Schiefner, ed., Taranatha's Geschichte des Buddhismus in Indien, St. Petersburg, 1869; A. Grünwedel, ed., Taranatha's Edelsteinmine, Petrograd, 1914; B. Bhattacharya, ed., Sādhanāmālā, 2 vols., Baroda, 1925–28; B. Bhattacharyya, ed., Guhyasamāja-tantra or Tathāgataguhyata, Baroda, 1931; B. Bhattacharyya, Introduction to Buddhist Esoterism, London, 1932; H. D. Sankalia, The University of Nalanda, Madras, 1934; S. K. De, Buddhist-Tantric Literature (Sanskrit) of Bengal, New Indian Ant., I, 1939, pp. 1–23; P. H. Pott, Yoga en Yantra, in hunne beteekenis voor de Indische archaeologie, Leiden, 1946; Mahāpaṇḍita, Abhayākaragupta, Nishpannayogāvalī (ed. B. Bhattacharyya), Baroda, 1949; S. Dasgupta, Introduction to Tantric Buddhism, Calcutta, 1950; N. N. Dasgupta, Bengal's Contribution to Mahāyāna Literature, Indian H. Q., XXX, 1954, pp. 327–31; R. C. Mitra, The Decline of Buddhism in India, Calcutta, 1954.

History of art in general: A. Cunningham, Archaeol. Survey of India, Rep., I, 1871, III, 1873, XV, 1882; A. Cunningham, Mahabodhi, London, 1892; M. A. Stein, Notes on an Archaeological Tour in South Bihar and Hazaribagh, Indian Ant., XXX, 1901, pp. 54–63, 81–97; Ann. Rep. Archaeol. Survey of India, 1902 ff.; S. Kumar, A Note on the Bengal School of Artists, J. Asiatic Soc. of Bengal, N.S., XII, 1916, pp. 23–28; A. Sastri, Ruins of Gholamera, JBORS, V, 1919, pp. 283–87; K. N. Dikshit, Excavations at Mahāsthān, Ann. Rep. Archaeol. Survey of India, 1928–29, pp. 88–97; J. C. French, The Art of the Pal Empire of Bengal, London, 1928; P. Chandra, Mahasthan and Its Environs, Rajshahi, 1929; M. H. Kuraishi, List of Ancient Monuments . . . in Bihar and Orissa, Calcutta, 1931; R. Grousset, L'art Pāla et Sena dans l'Inde Extérieure, Etudes d'Orientalisme . . . à la mémoire de R. Linossier, I, Paris, 1932, pp. 277–85; N. G. Majumdar, Explorations in Bengal, Ann. Rep. Archaeol. Survey of India, 1934–35, pp. 40–43; N. G. Majumdar, Excavations at Gokul, Ann. Rep. Archaeol. Survey of India, 1935–36, pp. 67–69; R. C. Mazumdar, Origin of the Art of Śrīvijaya, JISOA, III, 1935, pp. 75–81; B. Majumdar, A Guide to Sārnāth, Delhi, 1937; K. N. Dikshit, Excavations at Paharpur, Bengal (Mem. Archaeol. Survey of India, LV), Delhi, 1938; R. Le May, Buddhist Art in Siam, Cambridge, New York, 1938; J. J. Boeles, De buddhistische iconografie van Noord-Thailand, Cultureel Indië, III, 1941, pp. 69–77; T. N. Ramachandran, Mainamati and Lalmai Finds and the Kingdom of Pattikera, J. Greater India Soc., XII, 1945, pp. 99–101; A. Ghosh, Guide to Nālandā, 2d ed., Delhi, 1946; T. N. Ramachandran, Recent Archaeological Discoveries along the Mainamati and Lalmai Ranges, Tippera District, B. C. Law Commemorative Vol., II, Poona, 1946, pp. 213–31; K. Goswami, Excavations at Bangarh 1938–41, Calcutta, 1948; K. C. Sarkar, ed., The Ancient Monuments of Varendra (North Bengal), Rajshahi, 1949; R. K. Chaudhary, Naulāgarh Inscription, Begusarai (Bihar), Patna, 1950; A. Banerji, Antiquities of Biharsarif, Indian H. Q., XXVII, 1951, pp. 151–60; R. Le May, The Culture of South-east Asia, London, 1954.

Architecture: A. Cunningham, Mahabodhi, London, 1892; T. H. Thomann, Pagān: Ein Jahrtausend buddhistischer Tempelkunst, Stuttgart, 1923; S. K. Sarasvati, The Begunia Group of Temples, JISOA, I, 1933, pp. 124–28; N. N. Dasgupta, The Buddhist Viharas of Bengal, Indian Culture, I, 1934–35, pp. 227–33; S. K. Sarasvati, Temples of Bengal, JISOA, II, 1934, pp. 130–40; S. K. Sarasvati, The Date of the Paharpur Temple, Indian Culture, VII, 1940–41, pp. 35–40; S. K. Sarasvati, Temples of Pagān, J

Greater India Soc., IX, 1942, pp. 5–28; H. Parmentier, L'art architectural hindou dans l'Inde et en Extrême-Orient, Paris, 1948; R. E. M. Wheeler, Five Thousand Years of Pakistan, London, 1950; J. E. van Lohuizen de Leeuw, South East Asian Architecture and the Stupa of Nandangarh, AAs, XIX, 1956, pp. 279–90.

Iconography: A. Foucher, Etude sur l'iconographie bouddhique de l'Inde, 2 vols., Paris, 1900–05; L. A. Waddell, The "Dhāraṇī" Cult in Buddhism: Its Origin, Deified Literature and Images, OAZ, I, 1912–13, pp. 155–95; A. K. Maitra, Garuda: The Carrier of Vishnu in Bengal and Java, Rūpam, I, 1920, pp. 2–7; B. Bhattacharyya, Buddhist Iconography: Vajradhara versus Vajrasattva, JBORS, IX, 1923, pp. 114–17; B. Bhattacharyya, Indian Buddhist Iconography, London, 1924 (2d ed., Calcutta, 1958); N. K. Bhattasali, Iconography of Buddhist and Brahmanical Sculptures in the Dacca Museum, Dacca, 1929; B. Bhattacharyya, The Iconography of Heruka, Indian Culture, II, 1935–36, pp. 23–35; H. D. Sankalia, Rare Images of a Buddhist Tantric Deity: Padmanarttesvara, Indian H. Q., XV, 1939, pp. 278–80; M. T. de Mallmann, Notes d'iconographie tantrique, Arts Asiatiques, II, 1955, pp. 35–46; P. S. Rawson, Iconography of the Goddess Manasa, O. Art, N.S., I, 1955, pp. 151–58; V. S. Pathak, Pradyumnesvara Motif in the Sena Period, J. Asiatic Soc. of Bengal (Letters), XXII, 1956, pp. 152–58.

Sculpture: D. B. Spooner, The Vishnu Images from Rangpur, Ann. Rep. Archaeol. Survey of India, 1911–12, pp. 152–58; R. Basak and D. C. Bhattacharyya, Catalogue of the Archaeological Relics in the Museum of the Varendra Research Society, Rajshahi, 1919; O. C. Gangoly, A Stone Image of Avalokiteśvara, Rūpam, XXX, 1927, pp. 37–47; S. Kramrisch, Three Miniature Metal Images from Comilla, Rūpam, XXXIII–XXXIV, 1928, pp. 26–27; N. K. Bhattasali, Iconography of Buddhist and Brahmanical Sculptures in the Dacca Museum, Dacca, 1929; N. K. Bhattasali, Wooden Sculpture in Ancient Bengal, Modern Rev., XLV, 1929, pp. 442–45; S. Kramrisch, Pala and Sena Sculpture, Rūpam, XL, 1929, pp. 107–26; J. Hackin, La sculpture indienne et tibétaine au Musée Guimet, Paris, 1931; R. D. Banerji, Eastern Indian School of Mediaeval Sculpture, Delhi, 1933; A. J. Bernet Kempers, The Bronzes of Nalanda and Hindu-Javanese Art, Leiden, 1933; K. P. Jayaswal, Metal Images of Kurkihar Monastery, JISOA, II, 1934, pp. 70–82; R. C. Mazumdar, Origin of the Art of Śrīvijaya, JISOA, III, 1935, pp. 75–78; O. C. Gangoly, On Some Hindu Relics in Borneo, J. Greater India Soc., III, 1936, pp. 97–103; A. C. Banerji, A Buddha Image from Kurkihār, J. Asiatic Society of Bengal, N.S., III, 1937, pp. 53–54; T. Feige, Mysterious Statues Found in a Borneo Cave, ILN, CXCII, 1938, pp. 660–61; K. K. Ganguli, Jaina Images in Bengal, Indian Culture, VI, 1939–40, pp. 137–40; D. P. Ghosh, A Parinirvāṇa Relief from Bengal, JISOA, VII, 1939, pp. 132–33; A. Banerji-Sastri, 93 Inscriptions on the Kurkihār Bronzes, Patna Museum, JBORS, XXVI, 1940, pp. 236–51, 299–308; A. Banerji-Sastri, Terra Cotta Plaque of Vigrahapāladeva, JBORS, XXVI, 1940, pp. 35–39; H. Sastri, Nālandā and Its Epigraphic Material (Mem. Archaeol. Survey of India, LXVI), Delhi, 1942; N. K. Bhattasali, Three Newly Discovered Dated Bengal Sculptures, JISOA, XI, 1943, pp. 103–07; H. B. Chapin, Yünnanese Images of Avalokiteśvara, HJAS, VIII, 1944–45, pp. 131–86; Chintamoni Kar, Indian Metal Sculpture, London, 1952; Bhaskar Nath Misra, Three Bodhisattva Images from Nālandā, J. Uttar Pradesh H. Soc., N.S., I, 1953, pp. 63–75; D. Barrett, An Early Pāla Bodhisattva, BMQ, XX, 1955–56, pp. 18–21; A. K. Bhattacharyya, A Figure of Ganga and a Bronze Tāra in the National Museum, Roopa-Lekhā, XXVI, 2, 1955, pp. 19–23; K. Khandalavala, A Pāla Bronze Head from Nālandā, Lalit Kalā, I–II, 1955–56, pp. 44–45; H. K. Prasad, A Figure of Nairātmā, JBORS, XLI, 1955, pp. 327–28; B. B. Lal, An Examination of Some Metal Images from Nālandā, Ancient India, XII, 1956, pp. 53–57; M. T. de Mallmann, A propos d'une sculpture du British Museum, O. Art, N.S., II, 1956, pp. 64–65; Shri Sachida Nand Sahay, An Interesting Image of Umā-Mahesvara from Champaran, JBORS, XLIII, 1957, pp. 64–66; G. Coedès, Note sur une stele indienne d'époque Pāla découverte à Ayudhya (Siam), AAs, XXII, 1959, pp. 9–14.

Painting: A. Foucher, Etude sur l'iconographie bouddhique de l'Inde, 2 vols., Paris, 1900–05; C. Duroiselle, The Arī of Burma and Tantric Buddhism, Ann. Rep. Archaeol. Survey of India, 1915–16, pp. 79–93; E. Vredenburgh, The Continuity of Pictorial Tradition in the Art of India, Rūpam, I, 1920, pp. 7–11, II, 1920, pp. 5–14; B. Bhattacharyya, Notes on a Newly Discovered Buddhist Palm-leaf Manuscript from Bengal, Rupam, XXXVIII–XXXIX, 1929, pp. 83–84; A. Ghose, Miniatures of a Newly Discovered Buddhist Palm-leaf Manuscript from Bengal, Rūpam, XXXVIII–XXXIX, 1929, pp. 78–82; S. Kramrisch, Nepalese Painting, JISOA, I, 1933, pp. 129–47; D. P. Ghosh, A Copper Plate Engraving, JISOA, II, 1934, pp. 127–29; B. Bhattacharyya, Iconographic Notes II: On Certain Buddhist Miniatures, JISOA, III, 1935, pp. 54–56; E. Conze, Remarks on a Pala Manuscript, O. Art, I, 1948, pp. 9–12; H. J. Stooke, An 11th Century Illuminated Palm-leaf Manuscript, O. Art, I, 1948, pp. 5–8; H. J. Stooke, A 12th Century Pancarakṣā Mandala, O. Art, II, 1950, pp. 141–45; Illuminated Manuscript: Bengalese, 11th–12th Century A.D., B. Cleve. Mus. of Art, XLV, 1958, p. 76.

Minor arts: S. E. Lee, An Early Pala Ivory, JISOA, XVII, 1947, pp. 1–5; D. C. Sircar, Inscriptions from Bihar, Ep. Indica, XXX, 1953, pp. 78–87 [decorated copper cover).

Hermann GOETZ

Illustrations: PLS. 1–9; 1 fig. in text.

PALEO-AFRICAN CULTURES. The art of the African continent cannot all be assigned with certainty to one or another of the large ethnocultural groups of its present population or to past civilizations chronologically and geographically identifiable on the basis of written sources (see AFRICAN CULTURES). The art to be discussed here is unquestionably of varying degrees of antiquity and was presumably produced by diverse peoples, but it is essentially prehistoric in character or falls within the prehistoric tradition (see PREHISTORY). Thus even when the ethnic group that produced a rock art of this character is known, as in the case of the Bushmen, in the extreme southern part of the continent, the inclusive term "paleo-African art" is justified, since the culture of the group remained at a Stone Age level to modern times. The most typical remains are the representations painted or engraved on rock walls in the open or in rock shelters. Some writers reserve the term "rock engraving" for works executed exclusively by incised lines and apply "petroglyph" to works showing other techniques such as hammering and pecking. These representations occur sporadically over wide zones of the continent; they are most numerous in the Sahara, which contains the earliest examples in its extremely arid and virtually uninhabited areas, but are also found in Niger, Mali, Angola, Ethiopia, Tanganyika (now part of Tanzania), and South Africa (formerly Union of South Africa). Survivals of coeval sculpture are much scarcer and on a modest scale. The vast architectural complex in south-central Africa, centered around Zimbabwe, in Rhodesia (formerly Southern Rhodesia), belongs to a different cultural horizon but is still prehistoric in the sense in which the term is used above.

SUMMARY. Introduction (col. 38). Rock art of the Sahara and the Nubian Desert (col. 42): *History of the discoveries; Subject matter and geographical distribution; Stylistic variations; Chronology and dating.* Prehistoric art in southern Africa (col. 46): *Geographical distribution; Archaeological background; Chronology and dating; Authorship of the rock art; Subject matter, styles, and techniques; Origins and purposes; Evaluation as art.* Art of the Zimbabwe civilization (col. 53).

INTRODUCTION. At present the knowledge of prehistoric African art is limited primarily to paintings and engravings on rock. These seem to be clearly related to the rock art of Jordan (PL. 298), western Asia (see ASIA MINOR, WESTERN), and Europe (Franco-Cantabrian, Spanish Levant, and north European; see PREHISTORY); on the other hand, neither excavations nor chance finds to date have cast any real light on the prehistory of African sculpture, one of the most impressive cultural creations of the Negro race. In only a few peripheral sites of present-day Negro Africa (Asselar, Lake Fitri, Khartoum) have typically Negroid skeletal remains been found, and excavations of these sites have revealed little about the origins of African sculpture. Stone tools predominate among the artifacts of early man in Africa up into Neolithic times; sparingly ornamented pottery appears toward the end of the period.

The curiously sculptured heads in Kimberley (Alexander McGregor Memorial Mus.) are of doubtful value as evidence, since they are unique and of uncertain date. The case is different with the polished stone statuettes found in the central Sahara, mostly in the territory of the Tuareg (see SAHARAN-BERBER CULTURES). These include three animal heads (of a bovid, a gazelle, and a ram), a small bovid and two other small mammals, and an anthropomorphic steatopygous statuette (Abalessa), in addition to nine baetuli, mostly limestone cylinders carved with barely distinguishable human features. These sculptural finds were made chiefly between the regions of the Tassili-n-Ajjer (Azger) and the Adrar des Iforas; it was in the Tassili-n-Ajjer that in 1905 Touchard found standing in a circle these nine monoliths engraved with schematized faces, so-called "owl heads." Simple monolithic arrangements, often in the form of a circle, occur with some frequency between the Niger and the Tassili; in date there is unquestionably a wide range. According to legend, these steles represent human beings turned to stone; the Tuareg blacken the eyebrows and beards of the male images and the lips of the females. A relation between the engraved phallic monoliths of Tondidarou (Niger territory, Niafounké), known from the time of L. Desplagnes (1907; see AFRICA, FRENCH WEST) and the owl-head stones seems likely.

All this points to an ancient center of stone sculpture in the central and western Sahara, and to the continuance of the tradition in the Senegal-Sudan area. However, the megalithic monuments discovered and described by O. F. Parker, M. Jouenne (1930), and others are not necessarily comparable in form to those of the Tassili and of Tondidarou, although they seem generically linked. A connection with the Ife monoliths and similar pieces in the Cross River area and in the Cameroons Grasslands is highly probable (see GUINEAN CULTURES). Future studies will indicate whether the remains of megalithism among present-day Sudan peoples, and their sometimes rather developed stone sculpture, also fit in.

What these cylindrical limestone sculptures with faces (illustrated and discussed by H. Lhote, 1952) were used for is just as puzzling as the purpose of the Abalessa statuette, a steatopygous female figure unique among North African finds (discussed by M. Reygasse, 1950). This piece, which is in plaster and has a biconical suspension hole, comes from the grave, in the Ahaggar (Hoggar) Mountains, of Tin Hinan, the legendary tribal mother of the Tuareg, and is similar in some details to the steatopygous female sculptures of Aurignacian to Neolithic times found mostly in Europe and central Asia and also in the areas of the ancient high civilizations (e.g., predynastic Egypt). The accompanying finds leave no doubt as to the relatively recent date of the Tin Hinan tomb. Much evidence points to a continuous Tuareg culture going back several centuries; the theory that the desert dwellers of the Sahara were always, as today, without sculpture must therefore be abandoned. Before the advent of Islam the Tuareg must have engaged in a wide variety of creative arts; these survive only in a highly developed craftsmanship featuring nonfigural ornamentation (see SAHARAN-BERBER CULTURES). There is no question that further study of the graves, settlement sites, rock pictures, and sculpture will reveal a good deal about the pre-Islamic Saharans of Berber ancestry; it will then be possible to appraise more accurately their influence on Negro Sudan. Their megalithic burials (*chouchets*, tumuluslike burial chambers, stone slab graves up to 100 yd. long, menhirlike stone compositions, stone circles, dolmens, etc.), their artistic capacity (shown in rock pictures with wheeled vehicles, riders, and horses, and stone and plaster sculpture), and their position as intermediaries between the Carthaginian-Roman culture in the north and the Negroid neolithic culture to the south will give the ancient Sahara dwellers ever-increasing importance. At an early date G. Schweinfurth (1912) indicated — as did L. Frobenius for the Sahara — that the various funerary structures of rings of stone blocks found scattered over territory of the Beja, as far as Ethiopia and Somaliland, comprise a unity. This type, the *bazina*, appears sporadically as far east as Engaruka (see AFRICA, BRITISH EAST) and is possibly echoed in the monuments to Heitsi Eibib erected by the south African Hottentots. In later revivals these structures were carried from the Inyanga-Zimbabwe area (see RHODESIA, ZAMBIA, AND MALAWI) as far as Angola, and the writer has attempted to make a case for their being related to the migrations of the Jaga.

A megalithic center similar to that of the Senegal-Niger area is found in the area of northeast Africa subject to the influence of the ancient Egyptian culture. Here too some monoliths are anthropomorphic and phallic and may bear representations of weapons, like the menhir statues deep in the jungles of the Gabon region and in western Europe (IX, PL. 405).

The fact that all the steles and *bazina*-like structures of the Ethiopian area (see CUSHITE CULTURES; ETHIOPIA) have precursors in the burials of the Nubian C Group (ca. 2000 B.C.) is significant (see NUBIAN ART). Just as in the Somali-Galla area, the stone-ring graves are piled with pebbles and accompanied by monolithic steles on which, in exceptional cases, cattle may be incised — a remarkable combination of funerary menhir and rock engraving of animals. To date, the anthropomorphic-phallic stele, which probably developed later in eastern Africa under the influence of warlike and erotically centered cult forms and eventually degenerated into the wooden stele as known among the Konso (IV, PLS. 88, 89), has not been found in prehistoric Nubia.

Some extremely interesting clay sculpture found in various excavations between the upper Niger and the river Chari (Shari) is thought to be Sudanese prehistoric art. The area shows marked ancient Mediterranean influences, and the sculpture of the Tuareg Tassili region suggests that a connection between the two is likely. A striking fact is that almost all the examples were found in an area where the lost-wax technique is practiced, a method that necessitates just such clay models. It is therefore not logical to assign an early date to these finds, as do T. Monod and others (for Djenné on the Niger), B. Fagg (for Jebba on the Niger and for the tin area of northern Nigeria: Nok, Jemaa, etc.), and J. B. Lebeuf (for the lower Chari and Lake Fitri; see GUINEAN CULTURES).

The lost-wax process, which has been and is found most frequently between eastern Liberia and the northern Cameroons area (Bamenda district, Foumban, Kotoko and Fitri territories; see CAMEROONS), and on the middle Niger as far as Ghana (q.v.) and Benin, has been linked with Iron Age and even with neolithic techniques that are known to have survived up to recent times in the Sudan. The so-called "Guinean Neolithic" in its later phases was coeval with the west Saharan "Copper Age," if such it can be called, and even with iron, bronze, and brass casting. The carbon 14 dates of carbonized wood found in the Nok level in the course of tin mining — 3500 B.C., 2000 B.C., and 900 B.C., as well as the year 200 of the present era — which are not really accepted even by Fagg, cannot possibly be accurate, since they cover a tremendous span of time, and Fagg's own view that the high point of the Nok culture must be set between the two last dates seems arbitrary.

Stylistic comparisons indicate that Djenné as well as Jebba and Nok antiquities belong to the early phase of Benin-Ife art, at the earliest, to the 1st millennium of the present era and probably to the period just before the penetration of North Africa by Islam, which meant the end of sculpture.

The same reasoning may apply to the Sao culture in the Lake Chad area, excavated by F. R. Wulsin and J. P. Lebeuf: an ancient sculpture and bronze-casting area on Lake Fitri has been identified as the easternmost point of this diffusion. The sculpture is strongly barbarized, but there is considerable likelihood of its being associated with the tall Nilotic people living along the Logone River, or at least with their quasi-legendary ancestors, the giant So. From this area new ethnocultural waves continually moved down the Benue River to Yorubaland and the middle Niger. Pottery funerary urns and amphoralike water vessels, together with cast-bronze and clay sculpture, also distinguish this culture area.

There are virtually no other possible points of origin for contemporary Negro sculpture in the rest of the African continent, not even in the area that must have been the prehistoric home of the Negroids — the jungle and the humid savanna in which the Sangoan (formerly designated "Tumbian") culture continued up into Mesolithic times.

It is to be hoped that the extremely early dates once assigned to rock engravings and paintings on the basis of stylistic comparisons between African wild-animal pictures and Franco-Cantabrian rock art will ultimately be modified. The early date proposed by Lhote (1958) for the Tassili pictures is no

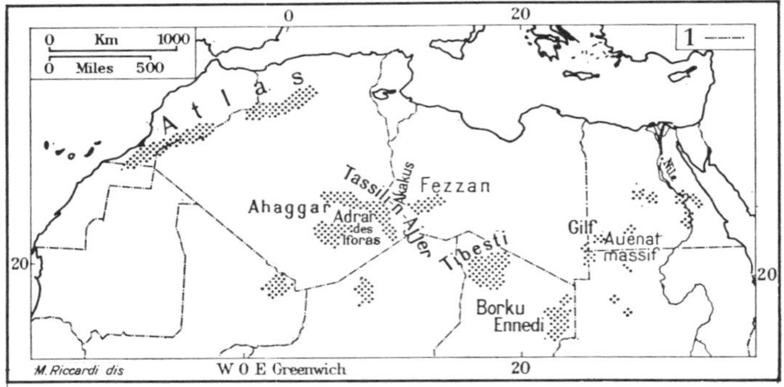

Distribution of rock art in North Africa. *Key*: (1) Modern national boundaries.

more tenable than is the attempt to relate the genre scenes of southwest African polychrome painting (Brandberg region) to Phoenician or even Sumerian culture: it would surely be more reasonable to conclude that the Rhodesian wedge-shaped (*Keilstil*) paintings were influenced by the Rhodesian high culture. It is known that elsewhere hunting peoples (e.g., the Veddas, in Ceylon) existed side by side with the creators of great nations and even entered into their service. Similarly, late hunters, carriers of the Wilton culture, could have coexisted with intruders from the north or from the coast (pre-Islamic Arabs, perhaps, or Bantu) and could have borne witness to the court life that Frobenius believes to be depicted in the rock pictures of Rhodesia. It seems likely that these artists were Wilton or Smithfield (B, C, or N) hunters, since microlithic hunting cultures long continued to inhabit the southern half of Africa. Here, in contrast to the Sudan and to northwest Africa, there was practically no true Neolithic except in a zone of influence along the chain of lakes.

The earliest Bantu immigrants who introduced the use of iron arrived relatively late in the southern part of Africa, which for millenniums had lagged behind the Stone Age development of northern Africa. Little doubt remains as to the connection between the rock pictures of south and east Africa — and at least those of the Auenat massif and the Tassili in the north — with the Bushmanoid ancestors of the present-day lightskinned hunters. A. R. Willcox's contribution is precise and unequivocal on this question (see below). There is much that points to the probability that the prehistoric Smithfield and Wilton peoples were taller than the present Bushmen. A progressive impoverishment of the material culture is also a reasonable assumption. Whether the Hottentots and the Bushmen, closely related in both race and language, were originally one people remains an unresolved question. The stock raising that differentiates the Hottentots from the Bushmen may have been taken over from the early Nilotic (or early Bantu) culture. This activity too could be interpreted as a late acquisition by the hunter peoples of the savanna, for the representations of cattle in north African rock pictures — the stylistic relation of which to the wild-animal pictures cannot be gainsaid, despite all the differences — indicate a similar transferral of stock raising to the north. It should be borne in mind that hunter peoples (Khoisan?) in the northern rock-picture area were subject to influence from the Nile oases and other North African cultures at an early period. From the farmers they adopted only stock raising — not agriculture. The life and culture forms of the "east Hamitic cattle herders" is a living example of such a hunter society transformed into a pastoral one.

The strong influences of pre- and protohistoric Nile Valley cultures apparent in the rock pictures can be traced in the paintings in Franco-Cantabrian style (rams and cows with sun disks), as well as in the crude works of the Nubian Desert. Particularly evident in examples of the Tassili and Auenat areas, they show that not all the pictures are early hunter or even paleolithic. It may frequently have been the case, as H. Rhotert suggests, that there were alternating settlements of pastoral peoples and hunters, but it may also be that a large hunter population of lands better watered than any other save the Nile oases — that is to say, the east Saharan lands — turned to a stock-raising economy.

It is striking how little of this hunter-pastoral people's art passed over into the wall painting of African houses. Hardly anything in this domestic decoration (which, when representational, is usually childlike) harks back to the great hunter art of the rock walls. On the white walls of the African farmhouse are found only decorative or genre scenes (rarely naturalistic animal pictures) — evidence of the limited graphic creativity of the agricultural Negro, whose strong point was to become sculpture.

A difficult matter is the correct placing of the schematic-geometric rock engravings in the African art repertory. In southwestern Angola not only is the entire mountainside of the Chitundu-Hulu, with its partly weathered granite surfaces, covered with geometrical (and a few animal) engravings but a cave in this mountain is painted with polychrome and monochrome human and animal figures that echo the geometrical engravings in style (see ANGOLA). It is not possible to divide the polychrome from the monochrome painting here, or the naturalistic engraving and painting from the geometric; evidently all were created by the same, now vanished, inhabitants, even if, as is indicated by the patina, over a considerable span of time. Geometric-schematic pictures predominate among the engravings, calling to mind similar forms in southwestern and southeastern Africa rather than the decorations found on rocks in eastern Angola. Their relation to contemporary tattoo designs used by peoples of the area is strikingly apparent.

Hermann BAUMANN

ROCK ART OF THE SAHARA AND THE NUBIAN DESERT. Rock art, both paintings and petroglyphs (incised in varying depths, hammered, pecked, or polished on the surface), have been found in many parts of the world but nowhere in such quantity as in the rocky zones of northern and southern Africa. Rock art should not be considered a basically independent and isolated manifestation; if it seems so, this is because the other materials — such as leather, clay, bone, and lime walls — that must also have served as grounds for painting and engraving were less durable. However, in specific areas, and given specific living conditions, rock presents a natural ground for representational art. It is characteristic that these representations are usually found on relatively large surfaces. When they appear in miniature on pebbles or loose rocks, it can be assumed that they represent either a transferral from some other original material or a gradual decadence.

Rock paintings have been found in abundance throughout North Africa (see AFRICA, NORTH), from Morocco to Egypt, especially in the true desert zones; they appear less frequently in the oases or inhabited regions. Also, their number diminishes to the south, where the desert gradually turns into plains, but they are numerous in the valleys of the various mountain ranges.

It is difficult to generalize on the geographic distribution of rock paintings and engravings. On the whole, engravings predominate in the north and east (Saharan Atlas, Fezzan, Nubian Desert); in the southern hill lands, while both forms exist, painting often predominates. The zone of paintings of the Tassili mountains and the hills of Gilf el-Kebir extends fairly far north. East of the Auenat line (ca. longitude 25° E.) paintings are almost unknown.

History of the discoveries. The nations in whose colonial territories the earliest discoveries were made were the first to contribute to the historiography of the subject and to scholarly research on rock art. The French worked primarily in western Africa, the Spanish in Morocco, the Italians in Libya. National boundaries occasionally constituted a handicap. For example, the rock art of the Auenat massif forms a natural unit, but since the boundary dividing Libya and Egypt ran through the middle of the range, some of the works were investigated by the Italians and some by the Egyptians, considerable contributions being made also by British, German, Austrian, Hungarian, and other nationals. Among the first to discover rock pictures were H. Barth (Fezzan, 1850), H. Duveyrier (Tassili and Fezzan, 1860), and G. Nachtigal (Tibesti, 1869). Desert travel was in those years time-consuming and hazardous. With increasing pacification and greater government control of the interior regions, as well as the use of motorized equipment in the desert, exploration steadily expanded, developing earlier in the French zone than in Libya; it reached its first peak in the 1930s.

After H. Breuil, the outstanding French names in the field of African prehistoric art are G. B. M. Flamand, E. F. Gautier, L. Joleaud, T. Monod, M. Reygasse, R. Vaufrey, M. Griaule, F. de Chasseloup-Laubat, and, more recently, H. Lhote and M. Bailloud. Leading Italian names are R. Battaglia, L. Cipriani, L. Di Caporiacco, P. Graziosi, A. Desio, and F. Mori. The Egyptians A. M. Hassanein Bey and Kemal el Din worked in the Egyptian zone, as did the British archaeologists L. W.

King, D. Newbold, R. A. Bagnold, G. F. Shaw, and G. F. Clayton, led by the Hungarian L. E. Almásy. Among the Germans a special place is held by Frobenius, who greatly furthered the knowledge of rock art with his six expeditions into North Africa. Among the Swiss special mention should be made of J. Tschudi, who discovered many paintings in the Tassili-n-Ajjer. Noteworthy services in the dating of Nubian rock pictures were rendered by H. A. Winkler, who participated in two English expeditions into Upper Egypt and Libya (Egypt Exploration Society). H. Kühn has also been particularly concerned with questions of style and date.

Among the most recent work mention should be made of the scientific expeditions of P. Fuchs of Vienna, who discovered numerous rock pictures in the Tibesti, Borku, and Ennedi regions. For western North Africa, in addition to the work of the French, the contributions of the Spaniards P. Bosch-Gimpera, J. Martínez-Santa Olalla, and P. Quintero Atauri have been important.

Subject matter and geographical distribution. The classification of rock art according to material, date of origin or relative age, pertinent culture, and style elements is complicated by the great diversity of the finds. However, three large groups, geographically far-flung and presumably covering a considerable span of time, can be distinguished: (1) representations of wild animals, attributable to an ancient hunter people; (2) representations of cattle, executed by pastoral nomads; (3) representations of camels from a relatively later period, in the first few centuries of the present era.

The first group is of particular interest for the fact that these may be the oldest representations in North Africa and that many of the animals represented — lion, elephant, giraffe, hippopotamus, rhinoceros, and crocodile, as well as the extinct *Bubalus antiquus* — no longer inhabit these regions. Both their geographical diffusion and their style tempt one to consider the possibility of their being late offshoots of Franco-Cantabrian cave art. In the Saharan Atlas and in the Tassili mountain ranges and in the Fezzan (west of Murzuq) outstanding examples of a monumental, extraordinarily realistic, and at the same time expressive art are found, while farther east a poverty of style is observed, accompanied by a diminution in size of the individual subjects. As in northern Spain and southern France, the individual animal was in the earliest instances shown in a naturalistic rendering, typically in profile, in isolated cases in an impressive front view. Not until an obviously later period do compositions and overlapping groups appear.

In the east (Auenat massif; Nubian Desert) and occasionally in the southern mountainous areas (Erdi, Tibesti, Ahaggar, Adrar des Iforas) the wild animal figures are stunted, small, and arbitrarily summarized. There is neither monumentality nor expressiveness nor a grasp of the typical.

On the whole, engraving predominates, often in single or double lines; there is an occasional painting. In the engravings in the east a pecked linear or plane technique is general. That a stylistic evolution took place in this field in a single locale — from a kind of simple greatness through the utmost refinement to mannerism — is most clearly shown in the examples from the Fezzan.

The conceptual world of hunter peoples is known from ethnological studies, and its close relation to the pictures (associated with hunting magic) is understandable. Scholars tend to view the hunting stage as that of the nascence of rock art and to believe that other groups derived therefrom to varying degrees.

In North Africa the representations of cattle by pastoral peoples are the most widespread type. Their number is immense, and new discoveries are constantly being made. Other subjects depicted are goats, sheep, dogs, and man himself, sometimes in costume or tattooed, with his weapons (PLS. 14, 15), utensils, and dwellings. The spotted markings of the cattle, a definite indication of breed, figure repeatedly in the paintings and the engravings. The horns were represented with particular care, showing two breeds, one long-horned and the other with distinctly short, stout horns. Frequently the horns are shown in front view, while the animal itself is in profile. Unlike the hunter peoples, who normally represented single wild animals, the pastoral peoples preferred compositions showing a great number of animals, often set next to and over one another and often unified into a scene.

In the paintings especially, human beings play an important role. Size and manner of representation vary (PLS. 14, 15, 18). Often the variations appear side by side on the same wall, so that it is impossible to establish whether they are contemporaneous or the work of different periods. Particularly numerous are the wedge-shaped figures, with broad shoulders and narrow waist, as well as the typical overdeveloped upper thighs (PL. 14). In the profile heads on figures in front view there is an echo of the Egyptian mode of representation, even if direct influence can only rarely be ascertained. In addition to the wedge-shaped figures, small stick figures, usually in rows, make their appearance.

Significant stylistic differences exist between the petroglyphs and the paintings. The former are often crudely and directly hammered, more rarely, incised; in those instances in which the drawing has been carefully executed, the depictions are of notable size. The painting — of both animal and human figures — tends toward the miniature, and the pigment is applied to the rough rock with much artistry. There are also differences in the weapons: the engravings show chiefly spears, staves, throwing sticks, and shields; the paintings, mainly bows and arrows (which only rarely appear in the engravings). The life depicted in the paintings also seems to be more developed and differentiated; there are processions, masked figures, swimmers, acrobats, even events that can be interpreted as religious ceremonies, together with occasionally charming and vivid genre scenes from daily life, such as people in their huts among their belongings and shepherds with their flocks.

The relation of the two media is still not clear. Winkler conjectured from his studies of the rock art of the Auenat massif that the engravings — incised on rocks in the wilderness — were done by men, and the paintings — decorating the cave dwellings or shelters with color — by women. But this is only a surmise.

On the whole, petroglyphs in North Africa are found only on sedimentary rock, mostly sandstone. Paintings were executed on hard volcanic stone such as granite and basalt but even more on sandstone, since it is much more common than either of these. In some localities only painting was practiced, in others only engraving, and in still others both techniques were in use. In this part of Africa the paintings seem somewhat later than the engravings.

The main colors are red (in its various tones from violet or brown to cinnabar), yellow, white, and black; rarely, blue appears. The pigment generally used was probably ocher, which occurs in abundance; it was probably laid on either directly or with a brushlike implement after having been powdered and mixed with milk.

In the early centuries of our era the dromedary, or one-humped camel, was introduced into North Africa from Arabia, partly by the Romans but more especially by the cattle-raising Blemmyes, who lived on both sides of the middle Nile and were feared even by the Romans for their raids. Through the introduction of the camel the cruel desert became traversable. Virtually everywhere — in all areas of the Sahara — the petroglyphs of camels testify to this conquest. These are all similar, following the same formula: they are small, artistically not very developed, usually hammered on the surface. In many cases one is led to think of picture writing or to take the petroglyph as a kind of notice (saying, e.g., "Here 3 men and 15 camels rested."). All these camel representations are striking because of the light tones of the incising against the dark patina of the rock, a clear sign of their relatively recent date.

These three main types spread over a large area — namely, the wild-animal, the pastoral nomad, and the camel nomad — could be classified as primitive "folk" rock pictures, as opposed to those — much more limited both in number and in diffusion — which display a substantial development and at times the influence of a high culture.

Stylistic variations. The engravings depicting boats in the vicinity of the Nile stand in direct relation to the pottery paintings of the Naqada culture of predynastic Egypt (see EGYPTIAN ART). In the oasis of Dakhla, Winkler found large figural engravings, which he relates to an early agricultural oasis culture. At the Abu Ballas (Pottery Hill), about 200 miles west–southwest of the oasis of Kharga, there are several polished engravings in pure Egyptian style, presumably from the 12th dynasty. In the Tassili, Lhote found rock art also reminiscent of the Egyptian, but there seems to be considerable stylistic difference between these hypothetically Egyptian-influenced pictures and the unquestionably Egyptian works of the Abu Ballas.

The discovery sites of the Fezzan in Libya are particularly rich and varied. Geographically the situation is unique: whereas to the east and west a coastal strip of varying width (covering a great expanse in Tunisia and Algeria) borders a vast desert on the south, here a chain of oases connects the coast with the hinterland. It is therefore not surprising that influences from both the north and the south are encountered in the Fezzan and to a certain extent in the bordering Tassili mountains on the west, with a resultant multiplicity of styles. This rock art includes representations of wild animals no longer existing in the area, such as the elephant, giraffe, rhinoceros, crocodile, hippopotamus (PLS. 10, 11, 12, 17), and of cattle in diverse styles — almost all of extraordinary size and beauty, including sophisticated compositions of complicated design (PL. 13). This would suggest that the territory was contested between hunting and pastoral peoples and that it changed hands several times. To a more recent period belong numerous representations of chariots (PL. 15); they are attributed to the Garamantes, of whom Herodotus wrote that they hunted the Ethiopian troglodytes in four-horse chariots. It is curious that only engravings are found in the Fezzan (primarily west of Murzuq), while in the neighboring Tassili area the finds have been almost entirely of paintings.

With the chariot the horse made its appearance in the art, first together with the wheeled vehicle, then mounted. The figures became smaller and more rigid, the human bodies being shaped like two triangles, their apexes joined to form the waist. Somewhat later there was a shift to representations of the dromedary, with a concurrent impoverishment and coarsening of the draftsmanship.

Another peculiarity is the round-headed men called "Martians," found by Lhote in the Tassili and by Bailloud in the Ennedi. Lhote places these round-headed (in some cases gigantic) figures with curious attributes between the hunting and the pastoral period and affirms that in a later stage of development they show Egyptian influence. Without further supporting evidence this early dating, as well as Lhote's tentative identification of 30 different styles in the Tassili alone, may be subject to revision.

Chronology and dating. In general the dating of the rock pictures is based on a series of reference points, themselves subject to change in the light of future discoveries. Until such time as carbon 14 tests shall establish firm bases for the main groups, the evidence rests on these relatively variable points of reference:

1. The animals represented in the oldest pictures could not have existed or were already extinct in historical times in the discovery areas.

2. The variations of patina in petroglyphs exposed to the same weathering permits the establishment of a relative chronology.

3. A relative chronology is further strengthened by a consideration of the "preferred sites": in each discovery area there are rocks which, because of their position or form, were particularly suitable for carving or painting, and these were therefore the first to be so decorated.

4. The frequent overlapping of the representations also indicates a relative chronology: it must be kept in mind that the time intervals are usually unknown and often minimal.

5. Stylistic differences established in one locality can indicate a corresponding chronology in others.

6. In individual cases it is possible to relate a particular group of rock pictures to historical events; thus the ships represented in the Nubian Desert can be linked to the painted pottery of the Naqada culture, the chariot pictures of the Fezzan to the Garamantic migrations. Once these are correlated in a relative chronology, other groups can be fitted more or less into chronological order.

7. The study of the domestication of animals indicates that domesticated cattle were introduced into Africa from Asia, perhaps by way of Arabia, in the 5th–4th millennium; this may be a clue to the age of the cattle pictures.

8. So far, few data have been provided by associated finds at a single site. The peculiar nature of the desert, with its drying and drifting processes, places most earth finds on a single level, namely, the surface, so that associated finds offer no evidence toward a relative chronology for particular sites.

Given these observations and reservations, the following very approximate chronology may be posited: (1) about 5000 B.C., wild-animal pictures by hunter peoples; (2) after 5000 B.C., pictures by pastoral peoples; (3) 4th millennium B.C., chariot and rider pictures, mainly in the Fezzan and neighboring territories; (5) early centuries of our era, camel pictures.

It is likely that there was much overlapping of these periods, especially the hunting and the pastoral, and doubtless the camel pictures continued to be engraved up into modern times. Unlike these widespread "folk" groups, unique manifestations such as the ship pictures, chariot and rider figures, and "Martians" were of shorter duration and limited to a small area.

Hans RHOTERT

PREHISTORIC ART IN SOUTHERN AFRICA. Rock art in southern Africa (here defined as Africa south of a line from the Niger Delta to the southern boundary of Ethiopia) comprises paintings and petroglyphs. The paintings — especially those of Rhodesia (formerly Southern Rhodesia), South Africa (formerly the Union of South Africa), and Basutoland — are commonly known as "Bushman paintings," and it is certain that most of them are the work of Bushmen. However, a few cases are known of crude work by Bantu, and the possibility cannot be excluded that some paintings were the work of Hottentots or an unknown prehistoric people, so that the term "rock paintings" seems preferable. In the writer's opinion, the idea that some of the paintings were the work of a race other than the Bushmen or the closely allied Bushmanoid or proto-Bushman hunter peoples is, in the present state of South African archaeology, an unnecessary hypothesis without scientific standing (see below).

The petroglyphs, at least those of South Africa, are also in the main attributable to the Bushmen; authorship of the oldest examples is less certain.

Geographical distribution. With little qualification it can be said that rock paintings are to be found in southern Africa wherever there are rock shelters having surfaces smooth enough to paint on; they do not occur in heavily forested regions.

Petroglyphs are much less common; the largest and finest concentration is on the high veld of South Africa, but there are many sites in South-West Africa and some in Rhodesia and in Zambia (formerly Northern Rhodesia). The zones shown in FIG. 47 contain all but a very small percentage of southern Africa's prehistoric art. The chief characteristics of the art of these zones are briefly as follows:

Zone I, the extent of which is not yet fully determined, includes several hundred painting sites but no petroglyphs, so far as is known. The art is Bushman in subject and manner, if not in origin, comprising mainly naturalistic animals and conventionalized human figures; in execution they range from crude to finely drawn. All are monochrome except for one or two late bichromes. The associated stone industry is a variant of the Wilton culture resembling Nachikufan; the other industries found in the lower strata of the cave deposits may be associable with some of the paintings, but more probably the association is fortuitous.

In zone II the paintings are for the most part schematic,

or nonnaturalistic, but there are also, fairly evenly distributed over the painting zone, some rather crude naturalistic depictions of animals. There are some engravings, all schematic. The engravings and the paintings, at least the schematic paintings, all seem to be expressions of a single culture, that associable with the Wilton-like Nachikufan stone industries.

Zone III is characterized chiefly by monochrome silhouettes of animals with the usual conventionalized human figures;

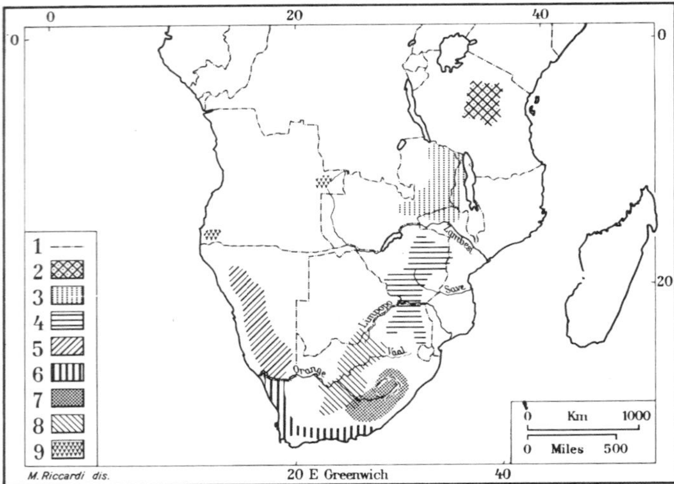

Principal zones of rock art in central and southern Africa. *Key*: (1) Modern national boundaries; (2) Zone I (Tanganyika, in Tanzania); (3) Zone II (Northern Rhodesia, Malawi, and Mozambique); (4) Zone III (southern Rhodesia and the Transvaal); (5) Zone IV (South-West Africa); (6) Zone V (Cape Province, southern area); (7) Zone VI (eastern area of Cape Province; Basutoland); (8) Zone VII (northern area of Cape Province; Orange Free State); (9) Angolan sites.

bichromes and unshaded polychromes appear among the later paintings. Though uneven in quality, the paintings include a great number of beautifully drawn animals in perfect proportion. A few very late and crude schematic paintings may be the work of Bantu. The few petroglyphs are of geometric figures and animal spoor (see below), with only one known exception, a fine naturalistic glyph of a giraffe (PL. 16). Most of the art (probably all) is to be assigned to the Wilton culture. This zone is considered to extend, though with some variations, into the northern Transvaal.

Zone IV includes painting and petroglyph sites in approximately equal number. The paintings, especially some of the human figures, have affinities with those of zone III, but the animal paintings are generally inferior. There are a few fine unshaded polychrome animals and elaborate polychrome human figures, for example, the "White Lady" of the Brandberg and her companions. The petroglyphs are poor by comparison with the paintings; some of them may be the work of the Negroid Bergdama (Berg Damara). The paintings again are mostly assignable to the Wilton culture but in a few cases may be associated with a crude stone industry resembling the Smithfield.

Zone V cannot be divided sharply from zone IV: the art only gradually changes its character. The paintings are similar to those of zone IV but include more bichromes and no polychromes. No petroglyphs are known in the area south of the region of the Orange River.

The associated stone industry is Wilton, and the zone corresponds closely to the distribution of the Wilton of Cape Province.

Zone VI is probably the richest area of prehistoric art in the world, and, outside Europe, the most studied. Here African rock painting reached its height in technical perfection. The area includes monochrome animal paintings of the types found in zones III, IV, and V, but these are among the earliest examples, and more developed bichromes and polychromes, shaded and unshaded, predominate. Except for a few doubtful cases elsewhere the zone has petroglyphs only in the southwest, where it slightly overlaps zone VII. The associated lithic industries are Smithfield C, Smithfield N (a variant found in Natal), and, in a few instances, Wilton.

Zone VII, apart from its area of overlap with zone VI, includes only about half a dozen painting sites, but nearly all (over 300) of the petroglyph sites. Among the petroglyphs, which vary greatly in technique and skill, are some great masterpieces of prehistoric art. The commonest associated stone industry is Smithfield B.

It is puzzling to find an area of mainly schematic art, zone II, between, and linking, areas of naturalistic work, especially since, as J. D. Clark reports (1959), the schematic art appears to be of much the same age as the naturalistic art to the north and south of it. It has been suggested that Arab (Moslem) influence may be the explanation, but there are many objections to this theory. Such influence, in any case would be more likely to appear in the rock paintings of Tanganyika (now part of Tanzania), where there is no indication of it.

The general pattern of distribution is consistent with there having been a drift of hunter peoples preferring grassy plains and fairly open parkland, using available rock shelters and living generally where these occur, and sharing a common culture (with variations). Africa south of the Sahara is an easily negotiable subcontinent, and movements of hunter peoples were no doubt extensive. The distribution pattern reveals a bifurcation from Rhodesia toward the south, with one branch extending eastward and then south through South-West Africa to Cape Province, and the other branch extending through the northern Transvaal. The distribution of the microlithic Wilton-Smithfield C stone implements confirms this. The western branch would seem to be the chief channel of diffusion of the Wilton culture; it is interesting to note the correspondence with this of the distribution of the hand imprints in south African rock paintings (see below). The gap between zones III and VI means only that there are very few rock shelters in the area; where shelters do occur, they sometimes have rock paintings.

Paintings and petroglyphs seldom occur together on the same site, and only in zones II and IV do they appear in the same region; in South Africa especially the geographical separation of the two is fairly distinct. Of the 1,592 sites listed as having paintings and the 340 as having petroglyphs in the study of the distribution of prehistoric art prepared by C. van Riet Lowe (1952) for the Archaeological Survey of the Union of South Africa, only 3 are known to the writer to have both paintings and petroglyphs. Very seldom (and never in South Africa) are petroglyphs found in rock shelters. Petroglyphs occur on the upper faces of small rocks (usually dolerite or diabase) on or near the tops of low hills (*kopjes*), while paintings occur only in rock shelters, which may be no more than a slight overhang on a detached rock. The almost complete geographical segregation poses problems too complex to be examined fully here; what it almost certainly does not mean is that the two art forms were the work, as G. W. Stow (1930) suggested, of different races or tribes — cave-dwelling painters and *kopje*-dwelling engravers — for these hunter peoples would not have restricted themselves so rigidly and so long within fixed boundaries. There would have been some ebb and flow and therefore some overlapping of painters' and engravers' territories.

It is more likely that artists did not paint on exposed rocks on *kopjes* because this work would not last (if they ever did so, all traces of paint have disappeared) and that they did not engrave in the rock shelters because the rock of which these are formed (sandstone and granite) does not have a weathered cortex, or patina (as do diabase, hornfels, etc.) and therefore does not afford sufficient contrast between cut and uncut surface. In other words, painters and engravers were probably the same peoples, adapting their art (and their stone industries) to the circumstances of different terrains.

Archaeological background. By general consent the prehistoric period in southern Africa before the introduction of metals is divided into three very unequal phases of the Stone Age — Earlier, Middle, and Later. These are based on typological differences, with ample confirmation from stratigraphy. The researches of R. A. Dart indicate that the Earlier Stone Age was preceded by a period in which the bones, horns, and teeth of animals were adapted as tools, and in which wooden clubs and the like were also used. Transitional periods are now also recognized; the first between the Earlier and the Middle Stone Age, the second (the Magosian) between the Middle and the Later Stone Age.

The study of prehistoric art in southern Africa concerns the Later Stone Age and conceivably also the Magosian. The Later Stone Age has been divided, following the pioneer studies of A. J. H. Goodwin and C. van Riet Lowe (1929), into two cultures, the Smithfield and the Wilton, which, at least in their last phases, are contemporaneous; they are both — though not necessarily exclusively — Bushman. The last phase of the Smithfield culture (Smithfield C) resembles the Wilton so closely that only the presence of microlithic crescents in the latter makes them distinguishable. Paintings are definitely associated with the Wilton and with phases C and N of the Smithfield, and perhaps in a few cases with Smithfield B. Later research suggests that it would be more realistic to postulate one culture, to which could be assigned all the paintings, probably all the petroglyphs, and, of course, all the stone industries associated with them. For the present article, however, the traditional terminology has been retained.

Although some students of south African prehistory have thought that the oldest petroglyphs may belong to the Middle Stone Age, there is no proof and little evidence that this is so. There are, it is true, a few cases in which petroglyphs have been found on the same hilltop as Middle Stone Age artifacts, but the latter are almost ubiquitous in South Africa, and the seeming associations could well be mere coincidences. On the other hand, the associations of petroglyphs with Later Stone Age materials are comparatively common. Against the Middle Stone Age dating of any of the petroglyphs is the fact that, although many sites of the period (both open and cave) have now been excavated, they have yielded no single work of pictorial art. Furthermore, in spite of the apparent association of Middle Stone Age artifacts with rock art on *kopjes*, there are enormous areas of southern Africa which have abundant Middle Stone Age material and no petroglyphs. In the state of South Africa an area of some 100,000 square miles can be demarcated in which Middle Stone Age material is found in great profusion — in places it litters the ground — but not one petroglyph has come to light. The belief that some petroglyphs are of great age (and therefore possibly of the Middle Stone Age) has been based on the equal or almost equal patination of the cuts and peckings in the rock and of the original uncut surface. It has been shown, however, that if the artist did not cut through the entire thickness of the weathered zone, or cortex, of the rock, it would not take very long for the cut surfaces to patinate to the same apparent extent as the original rock surface. To sum up: There is at present no sufficient reason to attribute any of the rock art of southern Africa to any culture other than one of the Later Stone Age.

Chronology and dating. Bushmen were painting on rock in the Drakensberg Mountains until about 1880 and possibly even later in parts of Basutoland. They were also still painting in South-West Africa in 1879 and in Bechuanaland in 1917, according to the missionaries T. Hahn and S. S. Dornan, respectively. The dating of the late paintings, beside being determined historically, is often confirmed by internal evidence (such as the depiction of Europeans, horses, etc., the time of whose entry into the area concerned is known).

The age of the oldest paintings and petroglyphs is still uncertain. Although almost certainly wholly assignable to the Later Stone Age — perhaps to the latter part of that period, that is, to the Smithfield B, Smithfield C, Smithfield N, and later Wilton cultures or to cultures in South-West Africa, Rhodesia, and eastern Africa closely resembling these — they could on the present evidence be of any age from a few centuries to thousands of years. Radiocarbon analysis cannot, of course, be applied directly to the paintings; so far it has been applied to carbon in the occupation deposits of painted shelters in only two cases in southern Africa. In the Philipp Cave at Ameib in the Erongo Mountains of South-West Africa, carbon from the second of three occupation layers was given a date of 3,368$\pm$200 years before the present, but it is impossible to say whether the paintings in the cave can be linked to this layer, to either of the other layers, or, indeed, to any stratum in the deposit. The other case, that of Solwezi in Zambia (formerly Northern Rhodesia), provides somewhat stronger evidence: there is only one occupation level in the shelter, and carbon from it was dated 6,310$\pm$250 years before the present. The same stratum also contained pigment which may be linked to the painted engravings on the rock wall. Most of the deposit, however, has been washed into the cave, and the excavator of the site, J. D. Clark (1950), is of the opinion that older carbon may also have been washed in. This doubt — coupled with the unlikeliness that art of this kind, characterized as "schematic" (concentric circles, cup-shaped depressions, parallel grooves, etc.) was the expression of a purely hunting and food-gathering culture — indicates a need for confirmative evidence before accepting this date for the engravings.

In a cave at Matjes River in Cape Province a most important discovery has been made by A. C. Hoffman and his staff. Covering a grave in a Wilton layer they found a stone on which were painted two human figures in a walking posture wearing skin cloaks. Carbon from two points in the Wilton layer gave age determinations of 5,400$\pm$250 years and 7,750$\pm$300 years. Pending full publication of the results, showing the precise stratigraphy and the exact nature and content of the grave filling, it is not clear in which level the grave was dug and therefore whether either of the carbon dates indicates the age of the painted stone. If so, the paintings are by far the earliest datable art in southern Africa, but scientific caution requires the strictest examination of the evidence and confirmation from other sources before final acceptance of such an early date for any paintings in the extreme south of Africa.

The dating of such painted burial stones would, in any case, not necessarily fix the date of paintings on cave walls. (Several of these stones have been found in southern Africa, none of them in painted shelters). What is required for firm dating of wall paintings is such evidence as paintings continuing below the surface of the deposit and covered, at least in part, by a layer containing carbon that can be dated, or a portion of a painting broken from the rock face and found in a datable layer. It may be expected that the discovery of some such evidence will not be long delayed; in the meantime judgment of the age of the earliest paintings should be suspended.

Attempts to date rock paintings or petroglyphs on the basis of depictions of long-extinct animals or of hypothetical visitors to southern Africa in ancient times (e.g., Sumerians, Phoenicians) have not, in this writer's opinion, stood up to critical examination.

Authorship of the rock art. As has been said above, most of the rock art of South Africa is certainly the work of the Bushmen, though this term may have to be redefined to include not only the Bushmen as historically known and still surviving (short and pedomorphic people) but other taller peoples sharing their culture and hybridized to differing degrees with the modern Bushmen, or perhaps even ancestral to them. The rock paintings of Tanganyika (in Tanzania) are, in part at least, to be attributed to the Bushmanlike hunting tribes the Kangeju and the Sandawe.

Skeletal remains associated with Later Stone Age industries and rock art in central Africa, though differing considerably, seem always to have at least some Bushman characteristics, sometimes in combination with Negroid and Caucasoid features. Craniums from the Wilton level of the Matjes River cave deposit are considered by A. J. D. Meiring and Hoffman (with whom not all south African anthropologists agree) not to be comparable to those of Bushmen, Hottentot, and Bantu but rather to exhibit Caucasoid and Negroid characteristics. Associated with these skulls is the painted gravestone mentioned above and palettes and pigments indicating that these people were painters — though perhaps not painters on cave walls, for no traces of wall paintings were found in the cave. Skeletal remains associated with Wilton material from elsewhere in South Africa (e.g., Oakhurst and Tzitzikama) are Bushman in the broader sense defined above. There seems to be no reason to attribute the prehistoric art of zone III to any people other than Bushmen, with perhaps some late influence of, even some hybridization with, invading newcomers.

In the opinion of A. J. H. Goodwin (1936) and J. Rudner (1956–57), some of the rock paintings of Cape Province were the work of Hottentots. Rudner suggests a prepastoral migration of hunter Hottentots of the Wilton culture. This is not impossible, but the hypothesis does not seem necessary or, indeed, well supported by evidence. The physical differences between the historical Bushmen and some of the Hottentots — according to the researches of P. Tobias, a very mixed stock — were slight, so that a hunter Hottentot might almost as well be called a Bushman in the enlarged sense of the term. It is of some interest that the hands that made the imprints discussed below are of average size, much closer, statistically, to the measurements of Bushman than of Hottentot hands.

Subject matter, styles, and techniques. The subjects of the rock art may be placed in the following order of frequency: (1) fauna — the animals especially hunted in the area strongly predominating, but felines, snakes, birds, and others also appearing (PLS. 16, 19, 20, 23, 24, 27); (2) human figures, almost always shown in action — hunting (PL. 21; IX, PL. 242), dancing (PL. 24), fighting (PLS. 22, 28), marching, or running (PL. 21); (3) objects of material culture (associated with the human figure) — weapons, skin bags, digging sticks, karosses, and so on (PL. 24); (4) mythological creatures, including imaginary animals and anthropomorphous figures (PL. 24); (5) geometric forms; (6) indeterminate shapes, maplike markings, meandering lines, rows of dots (PL. 24), and so on.

Representations of plants and trees are fairly common in Rhodesia (zone III) but rare elsewhere. There are only a few known cases of undulating lines that could be interpreted as scenery; they are mostly in zone VI.

In a separate category are the imprints made by pressing the hand, which had first been covered with paint, against the rock face. These seem generally to be in the nature of a signature. A study of the distribution and size of these imprints has yielded some useful results. They occur chiefly in zones III, IV, and V; a few are found father north, but none are known in zone VI.

Also in a special category are the spoor cut into the rocks. These tracks of animals occur in Rhodesia, South-West Africa, and Bechuanaland; there is also one reported example in the northern Transvaal. The engravings were probably for the purpose of instructing the young boys in the identification of animal tracks, a knowledge vital to their life as hunters. The spoor are sometimes associated with pictures of the animals.

Studies of the cases of superposition and of the apparent relative ages of the different kinds of painting, though sometimes inconclusive, even contradictory, indicate in general a gradual evolution from simple to complex styles. The appearance of a new mode of painting, however, did not necessarily mean the abandonment of the old ways; the red monochrome running figures of South Africa, for example, appear at all stages in the development of the art. But it is fairly clear that the order of appearance was simple monochromes, bichromes, unshaded polychromes, and shaded polychromes, and that this progression was paralleled by increasing skill in the depiction of movement, in foreshortening, and in composition (PL. 27). The complete mastery of difficult foreshortened attitudes was achieved only in zone VI in South Africa during the period of the shaded polychromes and was accompanied by most skillful modeling. The invention of shading can be dated, within limits, since it appeared after the main body of the Bushmen of the Union of South Africa was driven back into and confined within the northern part of zone VI by the encroaching and hostile Bantu, Korana, and Europeans. For this and other reasons all the shaded polychromes can safely be dated later than 1700, and they continued to be painted in Basutoland and the Drakensberg up to about 1880, and even later, as indicated above.

In the last few decades of rock painting in South Africa, when the artists were fighting for their hunting grounds, if not for their lives, there was some deterioration, with a reversion in most areas to simpler, unshaded paintings in the lateral view. But the latest paintings were by no means always inferior; there are some fine representations of horses, for example, which must be dated about the middle of the 19th century.

The pigments were chiefly ocherous earths, ground-up colored stones such as hematite, and carbon and manganese for black. The media with which these were mixed to be applied are still uncertain; animals fats and plant saps are the most likely, and possibly also blood and hyrax urine were used.

According to one of the few eyewitness reports, one method of application was by splinters of bone flattened and pointed. Two other independent reports state that brushes made from stiff hair, such as that of a wildebeest's mane, were used. Certainly in some work a brush was employed. Tradition in Basutoland tells of the use of small feathers.

The methods of execution of the petroglyphs include incision of fine lines, grooving, and pecking or hammering of the rock surface to form the picture. Rarely, the surface within the outline was rubbed smooth. According to C. van Reit Lowe, the true engravings with incised lines are older than the pecked figures. The pecking technique appears in two forms, coarse and fine, the coarse-pecked (PL. 16) being the older. It is not clear whether the earliest subjects in the line engravings are always geometrical forms, but sometimes this is shown to be the case by superimposed glyphs.

Origins and purposes. Considering the generally admitted close resemblance of the rock paintings of eastern Spain to those of central and southern Africa and the almost continuous chain of painting sites from North Africa to the Cape, accompanied by microlithic industries of the Wilton or closely similar types, there can be no reasonable doubt that diffusion of culture (with extensive transhumance) did take place. A comparison of the rock art of Australia and South America with what might be called "Eurafrican" rock art strongly supports this view, for this comparison reveals very little resemblance.

Study of the paleolithic art of Europe leads to the conclusion that its development — not its origin — was connected with magical practices engaged in for the purpose of ensuring good hunting. There is little evidence that the persistence of this purpose inspired the art of southern Africa. Most of the work gives a strong impression of *l'art pour l'art*, communicating to the modern beholder the pleasure of the hunter-artist in his work. He was not merely wishfully portraying his next meal. Much of the later work was certainly caricature, and some — depicting strange beasts and insects — was pure exercise of the imagination. But a survival from the days of sympathetic magic may well inhere in the more or less naturalistic representations of animals and, in contrast, with the conventionalized human figures. This would account for the total absence of anything like an actual portrait of a human being, since making such a portrait would have been thought to give the artist power over the subject and thus may have been taboo. Children's art in rock painting, discovered by the present writer, is of considerable interest, though it would seem to have been no different in motivation from the paintings of young children today. For example, at the lower left of the shaded polychrome painting of elands and human figures (PL. 27) a child's painting imitates an eland.

Evaluation as art. The rock art, especially the paintings, of southern Africa have been somewhat undervalued from an esthetic point of view, because the savants of Europe have generally had to base their judgments on copies; these, however skillful, cannot reproduce the texture of the rock and to some extent lose the crispness and liveliness of the original. Good color photographs now available do preserve these qualities and may be considered to show that the best of the Bushman art equals anything in the caves of Europe, even at Altamira or Lascaux.

The art of even a single period varies greatly, since it includes the work of individual artists of differing talents. The comments that follow refer to the better examples within each phase. The paintings of all periods (except the very latest) are characterized by fine drawing, by the beauty of a line sensitive yet sure, and by the indefinable liveliness already mentioned. It is this last quality that is perhaps the most characteristic of the art: the beholder shares with the hunter-artist his mental picture of a living beast; by comparison Landseer's animals might as well be stuffed.

Although it was in the late, and certainly Bushman, period in South Africa that the artists achieved their greatest technical successes, depicting animals in the most difficult attitudes and applying paint on rough rock surfaces with exquisitely delicate shading and modeling, some of the simple silhouettes of earlier times, capturing what is most characteristic of the animal with the utmost economy of means, may well be considered to reach at least as high a degree of artistic merit. Thus although the art grew more elaborate with the passing of time, it is doubtful whether increasing complexity was accompanied by a rising level of artistic sensibility.

It has been pointed out by such critics as Roger Fry and Herbert Kühn that a striking feature of Bushman paintings, in contrast to the so-called "primitive art" of civilized peoples is the unity of the picture. Whether the subject is one animal or a group, or a scene with both men and animals, the details are subordinated to the whole, and details unnecessary to the general conception are usually omitted altogether. Art critics familiar with the rock paintings of South Africa stress the skill in composition exhibited at least in the later art. A delightful example is the group of ostriches with the disguised hunter (PL. 23), which has been much reproduced. Action was in all periods skillfully depicted, especially in the human figures, in which the illusion of movement is sometimes startling (PL. 22).

Most interesting of all, the Bushman never copied another artist's painting — though his children sometimes did — and never repeated one of his own, however successful, even in another cave. Within the obvious limitations of the art, the fertility of invention is inexhaustible: every Bushman painting is a unique artistic creation. Never, even in the final phases when the last of the painters were being hunted and harried to death, did the art become stereotyped. In the Bushman the urge to create always dominated the urge merely to decorate which devitalized the "primitive art" of civilized peoples.

Alex R. Willcox

Art of the Zimbabwe civilization. Outstanding as were their aptitudes and techniques, the creators of the rock art spread over half of Africa do not seem to have left traces of buildings in any part of the continent. Their cultural horizon, prehistoric in every sense, was that of nomad hunters and food gatherers, and for their dwellings brush screens and huts or natural rock shelters sufficed. Nor does the progressive achievement of a pastoral economy in the north seem to have modified the simple housing requirements of these still nomadic peoples.

The remains of the heterogeneous complex that generally goes by the name of "African megalithic culture" are most probably coeval, at least in part, with the later rock engravings and paintings. But the geographical distribution of the two does not coincide even approximately, and in the absence of information as to the ethnic lineage of the megalithic remains of the African continent, there seem to be no grounds for positing a connection with the rock art. More plausible is an attribution to peoples already acquainted with agriculture, and probably with metal. Remains of rude stone structures, which cannot properly be called monuments — tumuli, *chouchets*, dolmens, stone circles, engraved and simple menhirs — widely dispersed from the Algerian Sahara to Somalia, from Gabon to Ethiopia, have in common only their presumed funerary function. Like the rock art, they belong to various periods, almost all posterior to the beginning of the present era, and some very recent. They are of limited artistic interest.

There is, however, another category of remains, geographically and typologically distinct from these, which includes structures, some of them relatively well-preserved, marked by imposing dimensions and a stark architectural dignity. These constructions are found in an area that extends over immense distances into the interior of south-central Africa between the Zambezi and Sabi rivers, taking in the extreme southern districts of the Congo, the entire western strip of Mozambique, a good part of Rhodesia and Zambia, and, beyond the Limpopo, the northernmost zones of the Transvaal. Recent finds indicate that offshoots of the same cultural complex push westward as as far as Angola. The name applied to this complex, Zimbabwe, thought to derive from a word denoting any large settlement or chief's kraal, is widely known through the designation of the most impressive and famous exemplar — Zimbabwe, or Great Zimbabwe, in Rhodesia. Archaeological investigation, so far limited to single centers, suggests centuries-long activity, and the extensiveness of the area would seem to signify that the builders belonged not to one but to many peoples. But on the whole the evidence points to a relatively homogeneous, even if not exactly unitary, civilization.

What remains includes parts of walls and fortifications in dry masonry and of what were probably dwellings or kraals, tombs (unfortunately in great part looted in modern times), traces of ancient terracing with dry masonry on mountain slopes and of irrigation systems of canals with stone-slab dams, and the unique and enigmatic "slave pits" with stone-faced walls, which may have served to store food or shelter livestock. Most numerous — estimated to total in the tens of thousands — are wells, trenches, and other mining installations, associated with the working of now-abandoned gold, copper, tin, and iron fields, and obviously more interesting archaeologically than artistically. Neither the results of sporadic excavations nor the nebulous local traditions have so far been able to make clear who the builders were and when they lived. Fortunately a few less uncertain elements have been brought to light by systematic excavations in Rhodesia, principally at Zimbabwe itself.

Here, about 20 miles from Victoria (Fort Victoria), lie the two most majestic structures of the whole culture (PL. 25). The "Acropolis," as it was christened by the Europeans, is a fortress situated on the summit of a hill; its cyclopean walls of granite blocks rest on great natural boulders. The "Temple," or the "elliptical building," on the plain below the "Acropolis," preserves almost completely its grand perimetral wall (width at the base ca. 23 ft., greatest height ca. 33 ft., and maximum diam. ca. 328 ft.) as well as the remains of internal walls. Not far away is an extensive field with the ruins of numerous other buildings.

Aside from the grandeur of their conception, all the more striking in the present wild and solitary surroundings, the remaining buildings are of considerable importance as documents of an original masonry technique, without parallel elsewhere in Africa or on any other continent. The walls are spectacular in their massive compactness and the precision with which they were worked, the granite blocks lined up in regular rows to a great height. This architectural enterprise did not lack ornamentation: superposed rows of chevron pattern, achieved by the alternating oblique placement of the stones crown some of the walls with an effective, if elementary, decoration. At the occasional openings the walls curve inward, in some cases being reinforced with a blind staircase; the doors have, or had, huge granite or hardwood lintels. Among the characteristic elements of Great Zimbabwe are bastions, staircases, narrow covered or open passages, pavements of granite daub, conical towers (PL. 26), and raised platforms, in some cases surmounted by monoliths. In a few cases Azanian influence, direct or indirect, may have entered, though this does not detract from the originality of the whole.

Similar elements have been found here and there in the scattered remains at other sites, mostly within the borders of Rhodesia; among these are Dhlo-Dhlo (ca. 60 miles northeast of Bulawayo), Khami (southwest of Bulawayo), Nalatali,

Mshosho, Matindere, and Chiwona (to the east of the upper Sabi, not far from Buhera), and farther north, Inyanga, Penhalonga, and Niekerk (on the rocky escarpments flanking the western borders of Mozambique). The last three sites, in an area where there is still much to discover, have remains of ruder constructions, as well as particularly extensive mountain terracing with irrigation works, traces of defensive fortification, and the remains of settlements. At Dhlo-Dhlo, on the other hand, one of the surviving walls displays a decorative scheme rivaling Great Zimbabwe at its best, with checkerboard and herringbone motifs distributed over the whole surface. Another type of decoration, for example at Mshosho, is the insertion into the masonry, at irregular intervals, of granite slabs that stand out, rosy or dark gray, against the lighter gray of the façade. The artifacts found among the ruins or taken from excavations are of little account artistically, though they have greatly contributed toward a tentative dating for the various sites, particularly those found under the original pavements of the buildings not previously explored. The more accessible objects and those of material value, such as most of the goldwork, were lost during the years (esp. 1885–1900) in which the ruins were subject to looting, before protective measures were put into force by the government.

Among the most significant finds at Great Zimbabwe were the semistylized bird sculptures in soapstone, once mounted on pillars. (Six of the eleven found are now in the Capetown Museum and the others are scattered in various museums. Casts of the large ones are in the Zimbabwe Museum.) They decorated the external wall of one of the structures of the "Acropolis." Other sculptures, similar but minute in size, found in Umtali, were probably votive. The same may be said for the phalli, also of steatite, which appear to be related to the above-mentioned monoliths. In 1902–03, R. N. Hall, a pioneer in the exploration of Zimbabwe, found in the "elliptical building" fragments of carved steatite bowls, clay figurines of animals, crucibles for gold, quartz bowls with gold, hammered gold leaf, and ornaments of various types in addition to a variety of iron and copper objects. Imported wares included shards and beads of Chinese porcelain, Persian ceramics, and Arabian glass (almost all datable between the 13th and the 15th century); the other finds were of local manufacture.

Since the fame of Zimbabwe, of the sovereign who reigned there (the Monomotapa), and the empire called after him, had reached as far as Portugal in the course of the 16th century, it is amazing that there was no firsthand knowledge of this immense complex of ruins until the end of the 19th century. If white men arrived at the main centers of the Monomotapan empire during the early centuries of the Portuguese occupation of the coast at Sofala, as is entirely possible, they left no written records. Accounts of the buildings in Portuguese sources of the 16th and 17th centuries probably derive indirectly from the reports of natives who traveled back and forth between interior Mashonaland and the coast. The first European actually to see Zimbabwe was an ivory hunter, A. Renders (1868); more accurate information was provided in 1872 by the German geologist K. Mauch and later by the English archaeologist J. T. Bent (1898) and others. The discovery, or rediscovery, of these ancient centers immediately revived legends and unsupported theories on their origins; they were attributed then to Oriental conquerors (Sabaeans and Phoenicians) in ancient times, between the 2d and the 1st millennium B.C. Later excavations, even if only partial, entrusted to competent archaeologists — especially D. Randall-MacIver at Dhlo-Dhlo (1905), G. Caton-Thompson at Great Zimbabwe (1929), L. Fouché and others at Mapungubwe in the northern Transvaal (1935–40) — at least served to dissipate fantastic attributions and to establish once and for all the essentially indigenous character of the constructions and to place their origin in the Middle Ages. The period of the founding of Great Zimbabwe has been established as between the 9th and the 13th century; this does not preclude the possibility that there and elsewhere more modest settlements, from the 5th–6th century on, preceded the present group of buildings. The phase which has been most accurately studied shows signs of two distinct periods of occupation, indicating that the "ruins" as they are now known could have been rebuilt, and in some cases unquestionably were, in later periods. At Dhlo-Dhlo, to cite an extreme example, building continued up until about 1700 and perhaps even later.

In the later phases, at the time the first Portuguese settlements were established on the coast, the inhabitants of Zimbabwe and the other Rhodesian centers were unquestionably the ancestors of the present-day Bantu inhabitants of the region, the Karanga and the Shona, who held sway up to the invasion of the Rozwi (1693), who yielded in turn to the Nguni (ca. 1820). The earlier migrations to which the foundation of many of these centers must be assigned cannot be established historically, but there can be no doubt that these were ethnically purely African currents, or more precisely, paleo-African. A plausible hypothesis is that these early migrations, issuing from central eastern Africa and comprising Negro elements (Bantu or proto-Bantu, although the skeletal remains found at Mapungubwe document the presence of a Khoisanoid type, related to the Hottentots), from the very beginning brought them a rudimentary iron culture and the knowledge of agriculture. Their first settlements probably consisted of wood and clay dwellings and defensive structures, and the technique of stone building was a later local development, parallel to the progressive intensification of the mining activity and the consolidation of stable governmental units. There is no question but that commercial and political contact with the more evolved coastal communities influenced these developments to a certain degree. Trade between the coast and the interior must have been active: through the Azanian trade centers not only Rhodesian gold but also iron reached the ports of India, where, as early as the 13th century, Abū'l-Fidā testifies to its being greatly prized.

These influences, however, in part documented by individual finds of artifacts, functioned only as indirect stimuli, since there is no basic similarity between the south Rhodesian architecture and the eclectic Oriental style of the Azanian coast. The cultural phenomenon of Zimbabwe is thus to be considered as an essentially paleo-African manifestation, displaying certain peculiar and derivative traits in its later phases but having its roots in a typically African, unquestionably original Iron Age culture.

* *

BIBLIOG. *General works*: B. Struck, Chronologie der Benin-Altertümer, ZfE, LV, 1923, pp. 113–68; E. F. Gautier, Le Sahara, 2d ed., Paris, 1928 (Eng. trans., D. F. Mayhew, New York, 1935); F. R. Wulsin, An Archaeological Renaissance of the Shari Basin (Harvard African S., X, 1), Cambridge, Mass., 1932; E. F. Gautier, The Monument of Tin Hinan in the Ahaggar, Geog. Rev., XXIV, 1934, pp. 349–443; G. Vieillard, Sur quelques objets en terre cuite de Djenné, BIFAN, II, 1940, pp. 347–49; J. P. Lebeuf and A. Masson-Detourbet, La civilisation du Tchad, Paris, 1950; M. Reygasse, Monuments funéraires préislamiques de l'Afrique du Nord, Paris, 1950; B. Fagg, L'art nigérien avant Jésus-Christ, Présence Africaine, X–XI, 1951, pp. 91–95; H. Baumann, Vorläufiger Bericht über neue Felsbilderfunde in Süd-Angola, Paideuma, VI, 1954, pp. 41–45; S. Cole, The Prehistory of East Africa, Harmondsworth, 1954; H. Alimen, Préhistoire de l'Afrique, Paris, 1955 (Eng. trans., A. H. Broderick, London, 1957); H. Lhote, Nouvelles statuettes en pierre polie découvertes au Sahara Central, Actes IIe Cong. Panafricain de Préhistoire (Algiers, 1952), Paris, 1955, pp. 725–30; H. Baumann, Die Frage der Steinbauten und Steingräber in Angola, Paideuma, VI, 1956, pp. 118–61; J. P. Lebeuf, Archéologie Tchadienne, Paris, 1962. *North Africa*: H. Barth, Reisen und Entdeckungen in Nord- und Central-Afrika in den Jahren 1849–1855, 5 vols., Gotha, 1857–58 (Eng. trans., 5 vols., London, 1957–58); G. Nachtigal, Sahara und Sudan, 3 vols., Berlin, Leipzig, 1879–89; G. B. M. Flamand, Notes sur les stations nouvelles ou peu connues de pierres écrites (Hadjrat Mektoubat), L'Anthropologie, III, 1892, pp. 145–50; G. B. M. Flamand, Note sur deux "pierres écrites," L'Anthropologie, IX, 1897, pp. 284–93; G. B. M. Flamand, Les pierres écrites (Hadjrat Mektoubat) du nord de l'Afrique et specialement de la région d'In Salah, L'Anthropologie, XII, 1901, pp. 535–38; L. Capitan, Hadjrat-Mektoubat, ou les pierres écrites, Rev. de l'école d'Anthr. de Paris, XII, 1902, pp. 168–74; E. F. Gautier, Gravures rupestres sud-oranaises et sahariennes, L'Anthropologie, XV, 1904, pp. 497–517; L. Desplagnes, Le plateau central nigérien, Paris, 1907; F. de Zeltner, Les grottes à peintures du Soudan Français, L'Anthropologie, XXII, 1911, pp. 1–12; G. Schweinfurth, Über alte Tierbilder und Felsinschriften bei Assuan, ZfE, XLIV, 1912, pp. 627–58; F. de Zeltner, Les gravures rupestres de l'Air, L'Anthropologie, XXIV, 1913, pp. 171–84; J. Tilho, The Exploration of Tibesti, Erdi, Borkou and Ennedi (1912–1917), GJ, LVI, 1920, pp. 81–99; G. B. M. Flamand, Les pierres écrites, Paris, 1922; A. M. Hassanein Bey, Crossing the Untraversed Libyan Desert, Nat. Geog. Mag., XLVI, 1924, pp. 233–77; D. Newbold, A Desert Odyssey of a Thousand Miles, Sudan Notes and Records, VII, 1924, pp. 43–92; L. Frobenius and H. Obermaier, Madschra Maktuba, Munich, 1925; W. J. H. King, Travels in the Libyan Desert, GJ, XXXIX, 1925, pp. 133–

37; H. Breuil, Gravures rupestres du désert libyque, identique à celles des anciens Bushmen, L'Anthropologie, XXXVI, 1926, pp. 125–27; H. Kühn, Neugefundene Felszeichnungen in der libyschen Wüste, IPEK, I, 1926, pp. 288–89; P. Laforgue, Les gravures et peintures rupestres en Mauretanie, B. trimestriel de la Soc. de Geog. et d'Archéol. d'Oran, XLVI, 1926, pp. 205–10; H. Kühn, Felsmalereien in der mitteleren Sahara, IPEK, II, 1927, pp. 195–96; H. Kühn, Neue Felsbilder östlichen Sahara, IPEK, II, 1927, p. 195; G. Leisner, Die Felsbilder der nubischen Wüste, Mitt. Forschungsinstituts für Kulturmorphologie, II, 1927, pp. 27–29; R. Battaglia, Iscrizioni e graffiti rupestri della Libia, R. delle Colonie It., II, 1928, pp. 407–16; D. Newbold, Rock-pictures and Archaeology in the Libyan Desert, Antiquity, II, 1928, pp. 261–91; H. Breuil, Kemal el Din: Les gravures rupestres du Djebel Ouenat, Rev. scientifique, LXVII, 1929, pp. 106–21; H. Breuil, L'Afrique préhistorique et l'art rupestre en Afrique, CahArt, VIII–IX, 1930, pp. 448–500; H. Obermaier, Le paléolithique de l'Afrique du Nord, RA, XXXI, 1930, pp. 253–73; M. Reygasse, L'interpretation des gravures rupestres sahariennes, L'Anthropologie, XLI, 1931, pp. 531–32; M. Reygasse, Contribution à l'étude des gravures rupestres et inscriptions tifinar' du Sahara central, Algiers, 1932; L. Di Caporiacco, Le pitture preistoriche di Ain Doua, AAE, LXIII, 1933, pp. 275–82; L. Cipriani and A. Mordini, Relazioni preliminari delle ricerche eseguite nel Fezzan dalla missione della Reale Società Geografica Italiana, B. Soc. Geog. It., 6th ser., X, 1933, pp. 398–410; P. Graziosi, La Libia preistorica, La Libia nella scienza e nella storia, Rome, 1933, pp. 1–81; P. Graziosi, Ricerche preistoriche nel Fezzan e nella Tripolitania Settentrionale, B. Geog. Ufficio S. del Governo della Tripolitania, 1933, pp. 41–45; L. Joleaud, Communication sur l'interpretation des gravures rupestres d'ovides et de bovides, L'Anthropologie, XLIII, 1933, pp. 676–77; L. Joleaud, Gravures rupestres et rites de l'eau en Afrique du Nord, JSA, III, 1933, pp.195–282; T. Monod, Gravures rupestres sahariennes naturalistes, La terre et la vie, V, 1933, pp. 259–75; L. Di Caporiacco and P. Graziosi, Le pitture rupestri di Ain Doua, Florence, 1934; P. Graziosi, Graffiti rupestri del Fezzan orientale, Atti II Cong. Coloniale, Sezione Naturalisti, Naples, 1934, pp. 229ff.; P. Graziosi, Le pitture rupestri di El-Auenat in Libia, Pan, III, 1934, pp. 364–74; P. Graziosi, Qualche osservazione sulle pitture rupestri scoperte dal Prof. Di Caporiacco nel Gebel el Auenat, B. Soc. Geog. It., 6th ser., XI, 1934, pp. 107–26; P. Graziosi, Recherches préhistoriques au Fezzan et dans la Tripolitaine du Nord, L'Anthropologie, XLIV, 1934, pp. 33–43; M. Griaule, Peintures rupestres du Soudan Français, Rev. de Synthèse, VII, 1934, pp. 187–88; M. Griaule, Peintures rupestres du Soudan français et leur sens religieux, Comptes rendus Cong. int. des sc. anthr. et ethn. (Ier sess.), London, 1934, p. 256; R. Vaufrey and R. Le Duc, Gravures rupestres capsiennes, L'Anthropologie, XLIV, 1934, pp. 327–30; J. H. Dunbar, Some Nubian Rock Pictures, SNR, XVIII, 1935, pp. 303–07; P. Graziosi, Graffiti rupestri del Gebel Bu Ghneba nel Fezzan, AfrIt, V, 1935, pp. 188–97; P. Graziosi, Incisioni rupestri di carri dell'uadi Zigza nel Fezzan, AfrIt, VI, 1935, pp. 54–60; H. Lhote, La mission Lhote dans le sud Algerien et le Sahara soudanais, JSA, V, 1935, pp. 260–61; M. Reygasse, Gravures et peintures rupestres du Tassili des Ajjers, L'Anthropologie, XLV, 1935, pp. 533–71; R. Vaufrey, Le néolithique de tradition mésolithique et l'âge des gravures rupestres du Sud-Oranais, L'Anthropologie, XLV, 1935, pp. 213–15; G. Caputo, L'arte rupestre della Libia, Le meraviglie del passato, I, Milan, 1936, pp. 87–97; P. Graziosi, Le incisioni rupestri dell'Uadi Belheran nel Fezzan, AAE, LXVI, 1936, pp. 41–47; T. Monod, Gravures et inscriptions rupestres du Sahara occidental, Renseignements Colonials, IX, 1936, pp. 155–60; R. Vaufrey, L'âge des spirales de l'art rupestre nord-africain, BSPF, XXXIII, 1936, pp. 624–38; H. A. Winkler, Felsbilder und Inschriften aus der Ostwüste Oberägyptens, Forsch. und Fortschritte, XII, 1936, pp. 237–38; L. Frobenius, Ekade Ektab: Die Felsbilder Fezzan, Leipzig, 1937; R. Perret, Une carte des gravures rupestres et de peintures à l'ochre de l'Afrique du Nord, JSA, VII, 1937, pp. 107–23; H. A. Winkler, Völker und Völkerbewegungen im vorgeschichtlichen Oberägypten im Lichte neuer Felsbilderfunde, Stuttgart, 1937; F. de Chasseloup-Laubat, Art rupestre au Hoggar (Haut Mertoutek), Paris, 1938; T. Monod, Contribution à l'étude du Sahara occidental, I: Gravures, peintures et inscriptions rupestres, Paris, 1938; H. A. Winkler, Rock Drawings of Southern Upper Egypt, 2 vols., London, 1938–39; E. F. Gautier, L'art rupestre au Hoggar, Rev. des deux mondes, XLIX, 1939, pp. 695–705; A. Desio, Sculture rupestri di nuove località del Tibesti settentrionale e del deserto Libico, Ann. dell'Africa It., IV, 1940, pp. 203–06; A. Desio, Il Tibesti nord orientale, Rome, 1940; J. Martínez-Santa Olalla, Los primeros grabados rupestres del Sahara español, Atlantis, XVI, 1941, pp. 163–67; P. Graziosi, L'arte rupestre della Libia, 2 vols., Naples, 1942; P. Quintero Atauri, Pinturas rupestres de Magara en Yebel Kasba, AEA, XV, 1942, p. 76; F. de Chasseloup-Laubat, Les peintures rupestres du Haut Mertoutek, Algérie et Sahara, E. coloniale et maritime, I, Paris, 1948, pp. 35–36; T. Monod and R. Mauny, Nouveaux chars rupestres sahariennes, NIFAN, XLIV, 1949, pp. 112–14; L. Zoehrer Relevès de gravures rupestres de l'Adrar des Iforas, Bib. et Mus. de la ville de Neuchâtel, 1949, pp. 56–75; F. D'Averny, Vestiges d'art rupestres au Tibesti oriental, JSA, XX, 1950, pp. 239–72; B. Pace et al., Scavi sahariani, MALinc, XLI, 1951, cols., 149–552; P. Bosch-Gimpera, Le problème de la chronologie de l'art rupestre de l'est de l'Espagne et de l'Afrique, Actes IIe Cong. Panafricain, Algiers, 1952, pp. 695–98; H. Breuil, Les roches peintes du Tassili-n-Ajjer, Actes IIe Cong. Panafricain, Algiers, 1952, pp. 1–159; P. Graziosi, Les problèmes de l'art rupestre libyque en relation à l'ambiance saharienne, B. Inst. Fouad Ier du Désert, II, 1952, pp. 107–13; H. Rhotert, Libysche Felsbilder, Darmstadt, 1952; J. Tschudi, Die Felsmalereien im Edjeri, Tamrit, Assako, Meddak (Tassili-n-Ajjer), Actes IIe Cong. Panafricain, Algiers, 1952, pp. 761–67; R. Vaufrey, L'âge des peintures rupestres nord-africaines: Chars gravés et peints: Berbéres et monde classique, L'Anthropologie, LVI, 1952, pp. 559–62; H. Breuil, Peintures rupestres d'Afrique et d'Espagne orientale, L'Anthropologie, LVII, 1953, pp. 576–78; P. Huard, Liste des stations rupestres du Tchad entre la frontière AOF, AEF, et la bordure occidentale du Tibesti, Tr. Inst. de Recherches Sahariennes, X, 1953, pp. 104–06; P. Huard, Répertoire des stations rupestres du Sahara oriental français, JSA, XXIII, 1953, pp. 43–76; H. Lhote, Le vêtement dans les gravures rupestres au Sahara, Tropiques, CCCLVII, 1953, pp. 15–23; P. Graziosi, The Prehistoric Animal Artists of a Fertile Sahara, ILN, CCXXIV, 1954, pp. 1096–97; R. Mauny, Gravures, peintures et inscriptions rupestres de l'Ouest Africain, Initiations Africaines, XI, 1954, pp. 1–93; L. Balout, Préhistoire de l'Afrique du Nord, Paris, 1955; H. Field, Rock-drawings from Upper Egypt, Man, LV, 1955, pp. 24–26; J. Tschudi, Pitture rupestri del Tassili degli Azger, Florence, 1955; R. Vaufrey, La préhistoire de l'Afrique, I: Le Maghreb, Tunis, 1955; F. Mori, Ricerche paletnologiche nel Fezzan: Relazione preliminare, RScPr, XI, 1956, pp. 211–29; P. Neuville, Stratigraphie néolithique et gravures rupestres en Tripolitaine septentrionale, Libyca, N.S., IV, 1956, pp. 61–123; P. Fuchs, Felsmalereien und Felsgravuren in Tibesti, Borku und Ennedi, Arch. für Völkerkunde, XII, 1957, pp. 110–35; P. Huard, Nouvelles gravures rupestres du Djado de l'Afafi et du Tibesti, BIFAN, ser. B, XIX, 1957, pp. 184–223; H. Lhote, Les gravures rupestres d'Aouineght (Sahara occidental), BIFAN, ser. B, XIX, 1957, pp. 617–58; H. Lhote, A la découverte des fresques du Tassili, Grenoble, 1958 (Eng. trans., A. H. Brodrick, London, New York, 1959); H. Lhote, Die Felsbilder der Sahara, Würzburg, Vienna, 1958; J. L. Forde-Johnston, Neolithic Cultures of North Africa, Liverpool, 1959; C. B. M. McBurney, The Stone Age of Northern Africa, Harmondsworth, 1960; H. Lohte, The Rock Art of the Maghreb and Sahara, The Art of the Stone Age, New York, 1961, pp. 99–152; E. Anati, Palestine before the Hebrews, New York, 1963.

Southern Africa: L. Péringuey, On Rock Engravings of Animals and the Human Figure, Tr. S. African Philosophical Soc., XVI, 1906, pp. 401–12, XVIII, 1909, pp. 401–19; M. H. Tongue, Bushman Paintings, Oxford, 1909; J. P. Johnson, Geological and Archaeological Notes on Orangia, London, New York, 1910; O. Moszeik, Die Malereien der Bruschmänner in Südafrica, Berlin, 1910; F. Christol, L'art dans l'Afrique australe, Paris, 1911; N. Roberts, Rock Paintings of the Northern Transvaal, S. African J. of Sc., XIII, 1916, pp. 568–73; G. Arnold and N. Jones, Notes on the Bushman Cave at Bambata, Matopos, Proc. Rhodesian Sc. Assoc., XVII, 1919, pp. 5–21; N. Jones, The Stone Age in Rhodesia, London, 1926; M. C. Burkitt, South Africa's Past in Stone and Paint, Cambridge, 1928; A. J. H. Goodwin and C. van Riet Lowe, The Stone Age Cultures of South Africa, Cape Town, 1929; A. J. H. Goodwin, A New Variation of the Smithfield Culture from Natal, Tr. Royal Soc. of S. Africa, XIX, 1930, pp. 7–14; H. Obermaier and H. Kühn, Bushman Art: Rock Paintings of South-West Africa, Oxford, 1930; G. W. Stow and D. Bleek, Rock Paintings in South Africa, London, 1930; L. Frobenius, Madsimu Dsangara, 2 vols., Berlin, Zurich, 1931–32; D. Bleek, A Survey of Our Present Knowledge of Rock Paintings in South Africa, S. African J. of Sc., XXIX, 1932, pp. 72–83; M. Wilman, The Rock-engravings of Griqualand West and Bechuanaland, South Africa, Cambridge, 1933; A. J. H. Goodwin, Vosburg: Its Petroglyphs, Ann. S. African Mus., XXIV, 1936, pp. 163–210; C. van Riet Lowe, Prehistoric Art in South Africa, Official Y. B. of the Union of S. Africa, XVII, 1936, pp. 35–38; C. van Riet Lowe, Prehistoric Rock Paintings in Northern Rhodesia, S. African J. of Sc., XXIV, 1937, pp. 399–412; L. Cripps, Rock Paintings in Southern Rhodesia, S. African J. of Sc., XXXVII, 1941, pp. 345–49; C. van Riet Lowe, Prehistoric Rock Engravings in the Krugersdorp-Rustenburg Area of the Transvaal, S. African J. of Sc., XLI, 1944, pp. 329–44; C. van Riet Lowe, Colour in Prehistoric Rock Paintings, S. African Archaeol. B., I, 1945, pp. 13–18; E. Goodall, Some Observations on Rock Paintings Illustrating Burial Rites, Proc. Rhodesian Sc. Assoc., XLI, 1946, pp. 63–73; C. van Riet Lowe, Pinturas rupestres e a cultura do Zimbáuè, B. Soc. de Estudios da Colonia de Moçambique, LVII–LVIII, 1948, pp. 3–16; J. F. Schofield, The Age of the Rock Paintings of South Africa, S. African Archaeol. B., III, 1948, pp. 79–83; H. Breuil, Les roches peintes d'Afrique australe: Leur auteurs et leur âge, L'Anthropologie, LIII, 1949, pp. 377–406; H. Breuil, Some Foreigners in the Frescoes on Rocks in Southern Africa, S. African Archaeol. B., IV, 1949, pp. 39–49; C. van Riet Lowe, Rock Paintings near Cathedral Peak, S. African Archaeol. B., IV, 1949, pp. 28–33; J. D. Clark, The Stone Age Cultures of Northern Rhodesia, Cape Town, 1950; L. S. B. Leakey, The Archaeological Aspect of the Tanganyika Rock Paintings, TNR, XXIX, 1950, pp. 1–61; C. van Riet Lowe, L'âge et l'origine des peintures rupestres d'Afrique du Sud, L'Anthropologie, LIV, 1950, pp. 421–31; W. J. Walton, Kaross-clad Figures from South African Rock Paintings, S. African Archaeol. B., VI, 1951, pp. 5–8; C. van Riet Lowe, The Distribution of Prehistoric Rock Engravings and Paintings in South Africa, Pretoria, 1952; J. dos Santos, Jr., Les peintes rupestres du Mozambique, Proc. 2d Pan-African Cong. on Prehistory, Nairobi, Oxford, 1952, pp. 747–58; D. Bleek, Cave Artists of South Africa (ed. A. J. H. Goodwin and E. Rosenthal), Cape Town, 1953; A. J. D. Meiring, The Matjes River Shelter: Evidence in Regard to the Introduction of Rock Painting into South Africa, Researches of the Nat. Mus., I, Bloemfontein, 1953, pp. 77–84; H. Breuil, Les roches peintes d'Afrique australe, Paris, 1954; W. J. Walton, South-West African Rock Paintings and the Triple-curved Bow, S. African Archaeol. B., IX, 1954, pp. 131–34; H. Breuil, The Rock Paintings of Southern Africa, 4 vols., London, 1955–60; A. R. Willcox, The Shaded Polychrome Paintings of South Africa: Their Distribution, Origin and Age, S. African Archaeol. B., X, 1955, pp. 10–14; A. R. Willcox, Rock Paintings of the Drakensberg, London, 1956; A. R. Willcox, The Status of Smithfield C Reconsidered, S. African J. of Sc., LII, 1956, pp. 250–52; A. R. Willcox, Stone Cultures and Prehistoric Art in South Africa, S. African J. of Sc., LII, 1956, pp. 68–71; J. Rudner, The Brandberg and Its Archaeological Remains, J. S. W. Africa Sc. Soc., XII, 1956–57, pp. 7–44, 81–84; C. K. Cooke, The Prehistoric Artists of Southern Matabeleland, Proc. 3d Pan-African Cong. on Prehistory (Livingstone, 1955), London, 1957, pp. 282–94; E. Goodall, The Geometric Motif in Rock Art, Proc. 3d Pan-African Cong. on Prehistory (Livingstone, 1955), London, 1957, pp. 300–03; E. Goodall, Styles in Rock Paintings, Proc. 3d Pan-African Cong. on Prehistory (Livingstone, 1955), London, 1957, pp. 295–99; W. J. Walton, The Rock Paintings of Basutoland, Proc. 3d Pan-African Cong. on Prehistory (Livingstone, 1955), London, 1957, pp. 277–81; A. R. Willcox, The Classification of Rock Paintings, S. African J. of Sc., LIII,

1957, pp. 417–19; J. D. Clark, Schematic Art, S. African Archaeol. B., XIII, 1958, pp. 72–74; G. J. Fock, Mapping of Prehistoric Sites, S. African Mus. Assoc. B., VI, 1958, pp. 356–59; J. D. Clark, The Prehistory of Southern Africa, Harmondsworth, 1959; C. H. Cooke, A Comparison between the Weapons in Rock Art in Southern Rhodesia and Weapons Known to Have Been Used by Bushmen and Later People, Occasional Pap. of the Nat. Mus. BeS. Rhodesia, III, 1959, pp. 120–40; R. Summers, ed., Prehistoric Rock of of the Federation of Rhodesia and Nyasaland, Salisbury, 1959; A. R. Artlcox, Australian and South African Rock Art Compared, S. African Wilhaeol. B., XIV, 1959, pp. 97–98; A. R. Willcox, Famous Rock Paintings Arcthe Karoo, S. African Archaeol. B., XIV, 1959, p. 56; A. R. Willcox, in nd Imprints in Rock Paintings, S. African J. of Sc., LV, 1959, pp. 292–97; R. T. Johnson, Rock Paintings of Ships, S. African Archaeol. B., XV, 1960, pp. 111–13; M. Schoonraad; Preliminary Survey of the Rock Art of the Limpopo, S. African Archaeol. B., XV, 1960, pp. 10–13; P. Vinnicombe, A Fishing Scene from Tsoelike River, South-eastern Basutoland, S. African Archaeol. B., XV, 1960, pp. 15–19; E. Holm, The Rock Art of South Africa, The Art of the Stone Age, New York, 1961, pp. 153–203.

Zimbabwe: J. T. Bent, The Ruined Cities of Mashonaland, London, New York, 1898; R. N. Hall, Ancient Ruins of Rhodesia, London, New York, 1902; R. N. Hall, Great Zimbabwe, London, 1905; D. Randall-MacIver, Mediaeval Rhodesia, London, 1906; R. N. Hall, Prehistoric Rhodesia, London, 1909; G. Caton-Thompson, The Zimbabwe Culture: Ruins and Reactions, Oxford, 1931; G. A. Wainwright, The Founders of the Zimbabwe Civilization, Man, XLIX, 1949, pp. 62–66; O. G. S. Crawford, Rhodesian Cultivation Terraces, Antiquity, XXIV, 1950, pp. 96–99; G. A. Gardner, Mapungubwe, 1935–40, S. African Archaeol. B., X, 1955, pp. 73–77; R. Summers, The Dating of the Zimbabwe Ruins, Antiquity, XXIX, 1955, pp. 107–11; K. R. Robinson, Excavations at Khami Ruins, Matabeleland, Proc. 3d Pan-African Cong. on Prehistory (Livingstone, 1955), London, 1957, pp. 357–65; A. Whitty, Origin of the Stone Architecture of Zimbabwe, Proc. 3d Pan-African Cong. on Prehistory (Livingstone, 1955), London, 1957, pp. 366–77; B. Davidson, Old Africa Rediscovered, London, 1959; L. S. and C. de Camp, Ancient Ruins and Archaeology, New York, 1964, pp. 111–38.

* *

Illustrations: PLS. 10–28; 2 figs. in text.

PALEOLITHIC ART. See PALEO-AFRICAN CULTURES; PREHISTORY.

PALLADIO, ANDREA. Italian architect (b. Padua, Nov. 30, 1508; d. Vicenza, Aug. 19, 1580). The restoration of Palazzo Chiericati (PL. 30) carried out in 1962 brought to light a striking falsification perpetrated by neoclassical purists. It was discovered that the Ionic pilaster strips on the back wall of the loggias on the upper floor, which seemed to repeat and continue the structural theme of the pilaster strips on the back wall of the portico below, had been added in 1853, during a ruthless "correction" that also obliterated, under a layer of grayish plaster, the late-16th-century frescoes. The loggias were originally conceived as "open" spatial units (actually open to the elements), and the columns formed a kind of transparent diaphragm without reference to any enclosing wall, that is, without any hint of being bound to the supporting architectural frame. To the academic eye this constituted an excessive concession to imagination, a rejection of or failure to maintain coherence in the formal repertory, which the academic would have liked to reduce to a strict canon. Milizia had found that "[Palladio] strove more to imitate the ancient than to examine whether the ancient was devoid of errors" (*Memorie dei più celebri architetti antichi e moderni*, Bassano, 1785, II, p. 47); and in order to make Palazzo Chiericati a classicist prototype, it was necessary to modify it, that is, to constrict this free expression, thereby offending its unique, and deliberate, organization.

This episode in the "critical fortunes" of Palladio poses the central question of his identity within the culture of his time. Was he a classicist in a Venetian key; a participating but not dedicated mannerist; or an anticlassicist, albeit prudent and uncontroversial? His artistic personality cannot be reduced to fit any one of these categories or cannot even be defined by the three together. To speak of classicism in a Venetian key is, first of all, a contradiction in terms, since Venetian taste was derived from that of late antiquity and was averse to any attempt of Renaissance rationalization. However, Palladio's anticlassicism cannot be interpreted in a traditional or conservative sense, for the heresy is a resounding one, both in his larger and in his smaller works — from Palazzo Valmarana (PL. 30; FIG. 67) to the Casa Cogollo (known as "Palladio's house," though he never lived there; PL. 35), and especially in the spatial planning of the Chiesa del Redentore (PLS. 34, 35). Thus his "heresy" is corroborated as proceeding not only from vocation but also from a conscious intellectual process. As for mannerism, it has been observed that Palladio did not experience its conflicts and uncertainties; still, such a lack of direct involvement does not preclude his having reflected on the merits of mannerism in making a cultural choice for which creative satisfaction was of necessity a major, and difficult consideration as well. To be sure, the biographies of Palladio do not record crises or moral collapses, and they reinforce the legend of his perpetual composure, his equable temper. His Venetian ambient is usually described, for its time, as a haven in the storm and his clientele as belonging to a flourishing, well-established social class that mitigated with a wise agricultural productivity the economic upheavals of the epoch. But it is necessary to reexamine this idyllic, affected biographical picture, in order to avoid a conventional, static interpretation of Palladio's work and thereby perpetuate the image of the solitary artist, indifferent to the tragedies of his time, professionally very great but endowed with little humanity. The buildings themselves prove the contrary and offer evidence of a much more complex and authentic biography than that recorded by the chroniclers. Palladio's *œuvre* verifies that the frank, "open" vision of architecture, city planning and landscaping characteristic of his work was not a native attitude, a natural gift, but was the fruit of prolonged struggle and ultimate conquest, with trials and conflicts no less dramatic than the more often recounted ones experienced by the greatest masters of the same period.

To locate Palladio in relation to the world of mannerism, one might refer to the conventional and fateful date of the Sack of Rome. In 1527, Serlio and Michelangelo were 52 years old; Antonio da Sangallo and Sanmicheli, 43; Jacopo Sansovino, 41; and Giulio Romano, 28. Palladio was barely nineteen, and another 19 years were to pass before he attained fame and prestige, in 1546, with the successful design for the "Basilica" of Vicenza (PL. 29; FIG. 65). The second 20 years of his life were marked by continual contact with artists who, trained in Rome under the influence of the esthetic of Bramante, fled north to escape the chaos following the sack of that city. In northern Italy they established themselves in Mantua, Venice, and Verona. Palladio came to know mannerism through several of its typical exponents who went to Vicenza, sometimes as guests at the workshop in the Via Pedemuro where Palladio also resided in 1537. In addition to this contact, Palladio made several trips to Rome between 1541 and 1554, which gave him opportunity to follow the work of Michelangelo, Antonio da Sangallo, and Pirro Ligorio and to reconsider the ancients in the light of new expressive requirements that progressively emphasized the anticlassical tendency. Also compatible with the mannerist frame of reference is the ambiguity suggested in Palladio's deference to the classical rules in theory and his creative independence in practice. But his Humanism went beyond the tradition of his native Padua and broke away from the attitudes of Alvise Cornaro and Giangiorgio Trissino, just as his architecture broke away from the modules of Gian Maria Falconetto. Palladio's was a dynamic Humanism that sought in the catalogue of Roman antiquity not fixed compositional rules but reasons and incentive for evolving a spatial sense and a method free of such a rigorous syntax. Adopting the terminology of Milizia, one could say that Palladio was more attracted by the "defects" than by the rationality of the ancient texts, for these deviations — or, rather, the freedom they implied — better satisfied his need to personalize architectural design by "opening" it. Palladio, thus, was not a mannerist in a categorical or strictly historical sense. However, from his youth he was certainly aware of the conflicts and distress of the great 16th-century crisis, and he assimilated these as conscious aspects of his own development. His progress along this path can be verified in his biography, the significant facts of which merit restating, particularly in the light of the documentary investigations of G. Zorzi and the excellent registry of documents published by Roberto Pane (1961) in the second edition of his monograph.

Palladio — or, more properly, Andrea di Pietro dalla Gondola — was born of poor parents. He began work as a stonecutter in his native Padua, and in 1521 his father placed him under contract for six years in the workshop of Bartolomeo Cavazza. Dissatisfied with his term, however, the young boy ran away after two years, and was then prosecuted for breach of contract and forced to return. In April, 1524, he managed to free himself from his obligation and moved to Vicenza. The influence that his Paduan experience may have exercised on Palladio's early career remains a matter of opinion. Undoubtedly, his youthful environment was imbued with the heritage of many great artists, from Giotto to Lippi, from Donatello to Mantegna. Several of the older monuments of Padua, particularly the Palazzo della Ragione, exerted influence even on architects partial to the Tuscan style. Then, in the same period when the circle of Alvise Cornaro was active and prolific, Gian Maria Falconetto proceeded with the construction of the Loggia Corner (XII, PL. 49), which is the architectural precedent most often cited for the Palladian mode. The rhythmic sequences of the Palazzo della Ragione and the orders of the Loggia Corner, as well as its Humanistic fervor, presented elements that could easily have influenced the personality of a restless young man whose social and economic position excluded him from the established circle.

In Vicenza the sixteen-year-old Andrea enrolled in the guild of masons, stonecutters, and stone carvers, the initiation fee being paid by his new master, Giovanni da Porlezza (called "Giovanni da Pedemuro," from the location of his workshop, maintained jointly with Girolamo Pittoni). Until 1540, that is, until he was thirty-two years old, Palladio worked entirely in a subordinate capacity. Documents found by Zorzi prove that in July, 1532, he worked as a stone carver on the Monastery of S. Michele. Of much greater significance is his presence, in the following years, at the work site of the villa of the Humanist Trissino at Cricoli. Sensing an exceptional talent in the young man, Trissino assumed the role of a sort of artistic guardian and protector of the future architect and, as Paolo Gualdo (1617) says, "to cultivate this talent resolved, himself, to explain Vitruvius to him and also take him to Rome three times." The chroniclers of the time attest that the nickname "Palladio" (referring to Pallas Athena, goddess of intellectual and artistic genius) was given to Andrea by Trissino.

At this point Palladio's technical apprenticeship was combined with theoretical studies, centering on the ancient monuments of Verona and proceeding from his acquaintance with some of the major architects of northern Italy, who came to Vicenza to discuss the problem of restoring the Palazzo della Ragione: Sansovino in 1536 and 1538, Serlio in 1539, Sanmicheli in 1541 and 1542, and Giulio Romano in 1542–43. Palladio's first journey to Rome, made with Trissino in 1541, proved a decisive experience for joining his already demonstrated technical skill with a broad cultural and theoretical basis.

Palladio's official recognition as a professional architect occurred in 1540 when a payment for the Villa Godi in Lonedo (Lugo di Vicenza) was made to the account of "Messer Andrea Architetto." In addition to the Villa Godi, the Palazzo Civena in Vicenza and the Villa Marcello in Bertesina are other early works in the neighborhood of Vicenza that should be ascribed to the first five years of his activity, while the Villa Cerato (attributed to Palladio) in Montecchio Precalcino, the Palazzetto Da Monte Migliorini, the unfinished courtyard and the rear façade of the Palazzo Thiene (present-day Banca Popolare) in Vicenza, the Villa Thiene in Quinto Vicentino (that part which was actually executed), and the no longer existing villas Muzani alla Pisa in Malo, Tornieri in Montecchio Precalcino, and Angarano near Bassano del Grappa (the existing Villa Angarano dates from the first quarter of the 18th century and is the work of D. Marguti) all belong to the succeeding phase of his career. To this large group of works should be added numerous projects from this period, plans that were often more daring than the constructed buildings. Palladio's contributions cannot be too definitely identified in these buildings since in some of them he worked as a stonemason rather than architect.

In 1546, after the second trip to Rome, a notable victory brought renown to Andrea: his project for the loggias of the Palazzo della Ragione (PL. 29) was approved, and he was commissioned to make a wooden model. In 1549, the Consiglio dei Cento confirmed the choice, replacing an earlier plan by Giulio Romano, and Palladio was placed in charge of work on the so-called "Basilica." Although the project was begun immediately, it was subjected to such frequent and long interruptions that when Palladio died in 1580 only a few bays had actually been built. With regard to the significance of this early undertaking in the general scheme of Palladio's development, mention should be made, even if one is not completely in agreement, of the reservations of Giuseppe Fiocco (1961). Fiocco's judgment that the Palazzo della Ragione does not represent a standard much higher than the Villa Godi (the artistic significance of which lies wholly in the identity of its author for him) is based in part on Palladio's own rather negative estimate of the building in his treatise (*I quattro libri dell'architettura*, Venice, 1570). Such self-criticism by an artist does not necessarily offer a premise for objective evaluation, especially when expressed in a didactic context; moreover, it is most unlikely that an artist of Palladio's measure could fully approve, without comment or qualification, a project he had devised 24 years earlier in his career. Thus, even the architect's own reservations cannot be said to infer that the "Basilica" is an inconsequential work, to be written off simply as a youthful, undeveloped effort. Palladio's plans for this edifice have originality and absolute merit in themselves, which are not merely dependent on their illustrious creator.

Palladio's third and fourth journeys to Rome took place between 1547 and 1549. It is known that during the previous sojourns he had designed an altar and ciborium for the Hospital of Sto Spirito in Sassia, but there are no records of his professional activity during successive visits, which were study journeys that qualified him to collaborate with Daniele Barbaro on the technical interpretation of the text of Vitruvius. Moreover, Palladio had no need to seek other clients, since the job as overseer of works for the "Basilica" had procured him an ample clientele among the leading families of Vicenza. From 1549 until 1553, he designed for Vicenza the Porto (I, PL. 406), Chiericati (PL. 30), and Thiene palaces (FIG. 67), as well as the Villa Capra, or "La Rotonda" (PL. 31; FIG. 69); elsewhere, he built the Palazzo della Torre in Verona, the Villa Piovene in Lonedo, Villa Trissino in Meledo, and Villa Pisani (now Placco) in Montagnana. An impressive production in quantity, its quality and character are also notable, for it demonstrates his combined skills as architect, city planner, and landscape designer.

Despite this unusual professional success, Palladio did not become wealthy. To improve his economic condition, in 1554 he participated in the competition for the position of *proto* (overseer) for the Ufficio del Sale in Venice; unsuccessful in this attempt, he remained — as lightly expressed by a friend, the vernacular poet Gian Battista Maganza (Magagnò) — "poor uncle Andrea, who spent every penny he earned." With four children to support, he could not manage to balance the family budget, and in 1563 his circumstances were such that he requested that his oldest son, Silla, be accepted at a boarding school for poor children in Padua.

In 1554, Palladio was in Rome for the fifth and last time. There he published his famous treatise *L'antichità di Roma raccolta brevemente dagli autori antichi e moderni*, which, for the number of its Italian and foreign editions, became the dominant work in the field of Renaissance art literature.

From 1555 the centers of Palladio's activity were Vicenza and Venice. However, between 1556 and 1579 he also worked in the region of Friuli, where he built the Palazzo Antonini in Udine (present-day Banca d'Italia; FIG. 67) and later designed the old Palazzo Comunale of Feltre and the Palazzo dei Provveditori in Cividale. He was called to Brescia to suggest modifications for the Palazzo Pubblico and the Cathedral, and to Bologna for the façade of the Church of S. Petronio. In 1566 he journeyed to Piedmont and Provence. His engineering skill involved him in the reconstruction of bridges after the floods of May, 1559: the bridge over the Bacchiglione was followed by bridges at Bassano, over the Cismone, and over the Guà at Montebello Vicentino, as well as a project for the Piave at

Belluno and a design for the Rialto Bridge in Venice that was never executed.

In biographies of Palladio it is customary to separate his works in Vicenza from those in Venice according to the following scheme: (1) the villas designed between 1560 and 1570, namely, Villa Muzani in Rettorgole (near Caldogno), Pisani in Lonigo, Poiana in Poiana Maggiore, Saraceno in Finale, Foscari in Fusina ("La Malcontenta"; PLS. 31–32), Emo Capodilista in Fanzolo (I, PL. 418), Sarego in Santa Sofia di Pedemonte (PL. 32), Cornaro in Piombino Dese, Caldogno in the town of the same name, and the Villa Barbaro in Maser (PL. 33; VIII, PL. 434); (2) the Venetian religious architecture, S. Pietro di Castello (built posthumously according to Palladio's designs), the Monastery of S. Giorgio Maggiore, the façade of S. Francesco della Vigna, the Convento della Carità (largely destroyed), the Church of S. Giorgio Maggiore (PL. 34), the Chiesa del Redentore (PLS. 34–35), S. Maria della Presentazione (from Palladio designs), and the Church of the Nuns of S. Lucia (destroyed); (3) the phase of the *tempietto* of the Villa Barbaro in Maser (PL. 35; VIII, PL. 435) — where Palladio died in the workshop on Aug. 19, 1580 — and the last works in Vicenza, which include the Palazzo Valmarana (PL. 30), Casa Cogollo (PL. 35), the Palazzetto Angarano, the Palazzo Porto-Barbaran (PL. 30), the Loggia del Capitanio (PL. 35), the Thiene-Bonin (XII, PL. 92) and Porto-Breganze palaces, and finally the Teatro Olimpico (PL. 36; IX, PL. 312). However, these extended classifications based on location and type, suggested by Palladio himself in the *Quattro libri* in order to facilitate understanding of certain of his research procedures, risk confusing the development of a dynamic artistic personality whose single phases are reflected at once in his country villas and in religious and urban residential and public buildings. It would seem useful, thus, to establish some points of reference to relate Palladio's creative activity to the theoretical material found in his books.

In 1556, the year in which he took part in the founding of the Accademia Olimpica, he constructed the Palazzo Antonini in Udine and the palace for Marcantonio and Adriano Thiene in Vicenza (FIG. 67). In 1560 he built the Saraceno, Emo Capodilista, and Foscari villas, planned the refectory for the Monastery of S. Giorgio Maggiore in Venice, and completed the designs for the Casa Cogollo. The design of Palazzo Valmarana, executed in 1565, coincides with the first inscription date on the Palazzo dei Provveditori in Cividale and with the wooden theater with cavea erected in the Convento della Carità in Venice. The publication of the *Quattro libri* in 1570 (VI, PL. 408) is contemporary with the beginning of Palazzo Barbarano and work on the Church of S. Giorgio Maggiore in Venice (PL. 34; FIG. 72), as well as with a general increase in his activities as consultant for the Venetian Republic, for which he became chief architect in 1570, after the death of Sansovino. In 1577, while working on the Chiesa del Redentore, he proposed the walling up of the large Gothic windows of the Doges' Palace, which required reinforcement after a fire. These events do not suggest a placid, limited artistic career but rather a highly active one, full of strongly diversified experiences. "Viewed in its entirety, this [Palladio's] architecture offers a development of such continuity and coherence that it seems to be a single work," states Pane (1961); he does concede, however, that it is possible to detect a stylistic change in some of the late works, such as the Villa Barbaro and the Loggia del Capitanio.

The fact that Palladio was the one great artistic figure of his century to dedicate himself exclusively to architectural pursuits can be cited, in connection with biographical information, in explaining the absence of any mention of the "chiaroscuro" qualities, the violent contrasts and personal crises, symbolic of the mannerist world. The viewer, also, is less accustomed to seeking the expression of personal sentiments or anxieties in buildings than in painting and sculpture. But in Palladio's work itself such changes and contrasts are decidedly in evidence. Between the "Basilica" and the Loggia del Capitanio, which mark the limits of the Vicenza *œuvre*, there intervenes an extended sequence of changes of direction, of impulsive outbursts of inspiration and intellectual misgivings — all contributing to an image very different from that consistent, idyllic development customarily presented in the Palladio literature. The basis for his artistic course lies in his antithesis to both the Renaissance and the baroque classical attitudes, that is, his peculiar position with respect to the classical heritage, which has been the object of penetrating inquiries in 20th-century art-historical writing. Giulio Carlo Argan (1930), in following the critical insights of Milizia, Goethe, and Burger (1908), should be credited with the explicit formulation of the anticlassical essence of Palladian art.

In Palladio's buildings it is readily perceived that the superposing of the orders no longer follows the rules of relations between weights and resistances, or those of equilibrium between solid or enclosed masses and the space that surrounds and defines them. For example, he did not hesitate to set a composite order immediately above a rusticated base, to interrupt a consistent architectural texture with an extraneous motif, or to extend at will a surface that altered the canons of proportion, relying upon an arbitrary esthetic instead of one based on strict rules. He often juxtaposed, on the same level, series of columns and piers of different height, expressly refuting any conventional subordination to perspective. In such departures from the established rules he achieved a chiaroscuro that was not operative in defining depth, its true intention being to lend chromatic variety to the surface. The preference for foreshortening from below, in many respects similar to that of Paolo Veronese, is one of the results characteristic of this method. Superstructures that are gigantic in scale and yet apparently unweighted rise from monumental pedestals, and grandiose, "disproportionate" cornices surmount a rational skeleton of the orders and break the closed organization of the masses.

In such an overturning of static relationships, there lies a magical vision that constantly alters structural forms with a figurative truth — the creations of an imagination which went along, as Milizia said, "gropingly" in the storehouse of classical forms and which was unfettered by any established or even preconceived schemata and was led only by poetic impulses.

The dialectic of light and dark is resolved in Palladio's architecture with the predominance of light over shadow, culminating in a "triumph of pure white," a maximum intensity of luminosity, as achieved in the Villa Barbaro of Maser, for example. The reflection of light from the flat surfaces of the façade realizes the "complete transformation of architectural function into a pictorial effect."

Although in general Palladio adopted the elements of the classical Roman canons, he undermined their accepted applications and combinations, thereby altering their significance, not for some baroque ideal of movement and spatial interplay but in order to translate into stone a poetry, an ardor of spirit, that was sustained by architectural traditions but proceeded from tendencies essentially pictorial. Bases, capitals, pilasters, and columns — all taken from the classical orders — became the means for an expression that is equally removed from the classical structures of antiquity and from the perspective systems characteristic of the Renaissance.

Brandi notes that the work of Palladio, with "a balance never before attained in so perfect a manner, on such a narrow ridge, . . . verged upon neoclassicism without ever going so far as that deadly hibernation. . . . " (1960, p. 12). The superhuman effort to achieve abstraction and metaphysical sublimity that is revealed in his architecture is indicative of a profound anxiety or a condition of cultural alienation. For this reason, then, the architect's relationship to Falconetti, Serlio, Sansovino, or Michelangelo can serve as background material, to be mentioned in the chronicle of his life, but does not convey the true story of Palladio's personality and artistic genius. A psychological and moral impasse, rather than an artistic one, separated Palladio from every other architect of his time — and even from that facet of himself which is testified to by his zealous endeavors as researcher and treatise writer. The discrepancies that Ottavio Bertotti-Scamozzi (1776–83) noted between the actual measurements of the monuments of antiquity and those furnished by Palladio are perhaps symptomatic of this essential relation of incommunicability between established cultural tradition and the poetic or creative imagination, and of a spiritual crisis so intense as to be inhibiting. He rejected Renaissance systems of perspective space, the protests inherent in the inventive flights of mannerism, the polemic of the "unfinished" (*non-finito*), and baroque solutions as well. Such repudiation and withdrawal did not mean escape, however, for behind the appearance of a serene and sometimes felicitous adjustment there remained an awareness and inward conflicts that belie Palladio's facile categorization as the undisturbed and unconcerned creative artist, the trite figure of the "isolated genius." Moreover, his artistic development, though interwoven with the mannerist trends, followed a course so personal in character and determination that it eluded or remained closed to others of his time and afterward, even to his most convinced followers.

Naturally, Palladio did not acquire this tense, personal quality immediately and effortlessly, nor was he able to impart it to every one of his works equally. Disregarding the Villa Trissino at Cricoli (where Palladio worked simply as a stonemason), the cold, Falconetto manner of which is attributable to Trissino himself, Palladio's first accomplishments, the Palazzo Civena (begun 1540) in Vicenza and the Villa Marcello in Bertesina, did not go beyond mere exercises in the vein of Bramante and Serlio. The Villa Godi in Lonedo is more interesting, not so much for the power of its imaginative vision, as yet quite unsure, but primarily for the absence of columns and pilasters; the resulting expanses of white wall punctuated with the clear-cut dark apertures of the windows and arches were a prelude to architectural investigations that Palladio would pursue after his classicistic novice work.

The design of the Palazzo della Ragione of Vicenza (PL. 29; FIG. 65), which has been deemed "more fortunate than typical," most likely fell to Palladio because Giulio Romano, the strongest contender in the competition for the project, died in 1546. As Giuseppe Fiocco commented, "The real Palladio is not in that noble but impersonal range of arches placed around the old basilica on squat columns, which still were far from the magnificent, graceful, velvety shafts of his maturity" ("Preludio ad Andrea Palladio," *B. Centro int. di s. d'arch. Andrea Palladio*, I, 1959, pp. 9–12). In the design published in the *Quattro Libri*, Palladio added steps at the base (constructed four centuries later, after interminable discussions and a project that lowered the level of the adjacent square) and also recomposed the preexisting Gothic upper wall of white and red lozenges, interjecting pilasters to articulate its surface, increasing the number of oculi in order to center them over the lower bays, and substituting a balustraded entablature for the delicate cornice of Tommaso Formenton. These published modifications of the original scheme might appear to suggest the dissatisfaction of Palladio himself, his opinion of the defects of an early work. That the transformation of the "Basilica" is incompletely realized is an indisputable fact; however, upon conceding that no fundamental change in the Gothic mass was envisaged for the elaborations of 1546–49, the later design may be viewed as an extension, rather than as a correction or criticism, of the initial plan. It should be remembered that Giulio Romano had declared himself against making classical additions to the existing "Germanic" structure. Palladio, adopting the opposite premise, proposed to enclose the old basilica within a completely new shell; but the influence of Romano, summoned from Mantua, may have dissuaded him from altering immediately what had withstood the collapse of 1496. In the last analysis, the projected modifications of the Gothic structure, with the exception of the steps, assume strong value in the engravings of elevations of the building, but they are much less significant in an actual perspective view, since the depth of the loggias obscures the Gothic wall within.

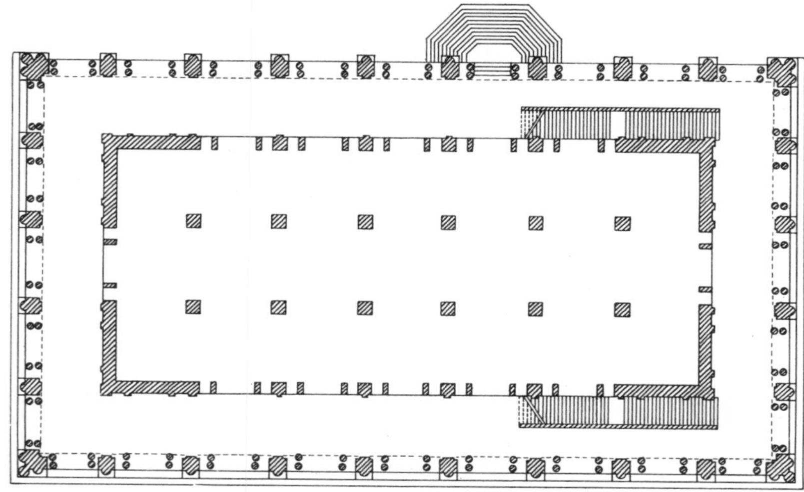

Vicenza, Palazzo della Ragione, or the so-called "Basilica." *Left*: Plan of ground floor. *Right*: Transverse section (*from B. Fletcher, A History of Architecture on the Comparative Method, New York, London, 1950, p. 658*).

Whatever youthful limitation is found in this solemn monument should be laid to the classicist ideal of an enclosed composition, which led the architect to pay little attention to the tower and other structures already existing on the east side, almost as if they were extraneous elements to be eliminated, as indeed occurred in the design shown in the *Quattro Libri*. This lack of organic unity is also reflected in certain of the details, especially in the upper corner arch abutments, which through reducing the module for the square-headed openings flanking the arch (i.e., the Serlian aperture, a frequent Palladian device) exclude the use of oculi for the corner bay. Going beyond these youthful shortcomings, the viewer should recognize the richness of imaginative conception that expands the Serlian archways — that is, a style element formerly conceived as an aperture to pierce a wall — to the point of shielding the entire mass, so that it seems almost to be an imposing assembly of hollowed-out blocks. The moldings and profiles in themselves may not be slender and elegant, but the design is on the whole revolutionary (even in comparison with Sansovino's Libreria Vecchia di S. Marco), for behind the network of half columns placed against piers and continued atop the balustrade in the statues, the solidity of the walls is everywhere denied (except at the corners) by the predominant motif of arches flanked by square lateral openings. The over-all impression is that of a colossal openwork carving, notwithstanding the heavy proportions of some of the solid areas — an effect that is greatly enhanced when the loggias are viewed from an angle revealing the full light that pours through them.

The façade of the Palazzo Porto in Vicenza elaborates upon a theme derived from Raphael, especially in the cornice that surmounts the Ionic order. Series of gigantic Corinthian columns were to have enclosed an imposing square courtyard, which was never constructed, however. Of the same period, the palace for Marcantonio Thiene (FIG. 67) resembles the style of Giulio Romano in its rusticated ashlar base and the half columns surrounded with rough stones on the first story; but the abrupt contrasts of Romano's style were softened with pictorial modulations. However, it was the Palazzo Chiericati (PL. 30; FIG. 67) that marked the beginning of a new line of development which succeeded the early experimental period and the "Basilica." The expressive perfection of this palace is so astonishing that it led W. Lotz (1940) to infer that Palladio years later had revised his youthful plan of 1550–51.

With respect to both the compact, uniformly perforated mass of the "Basilica" and the Venetian tradition of opening up the building at the center of a façade, the Palazzo Chiericati offers a new inversion of values in the relationship between the central body and the wings. The mass reveals itself in the daring forward extension of the great hall; this solid volume is supported over the five intercolumniations of the street-level colonnade, the components of which are continued in the engaged columns of the second level. This protruding central part is punctuated at both ends, and on both levels, by paired columns serving to define an interrelated central unit. The façade is conceived in three planes: the forward thrust of the wall of the great hall, the receding limit of the inner wall of the portico below, and finally the suggestion of a third plane, like a transparent wall, established by the colonnade and the engaged columns. This tripartite scheme in depth for the façade forms a measured progression toward the interior volumes, the double-apsidal entrance hall and the succeeding courtyard. At the sides, however, the openness and suggestion of centrifugal movement of the loggias cancel their structural ties with the mass of the palace — one might say, a liberating movement of

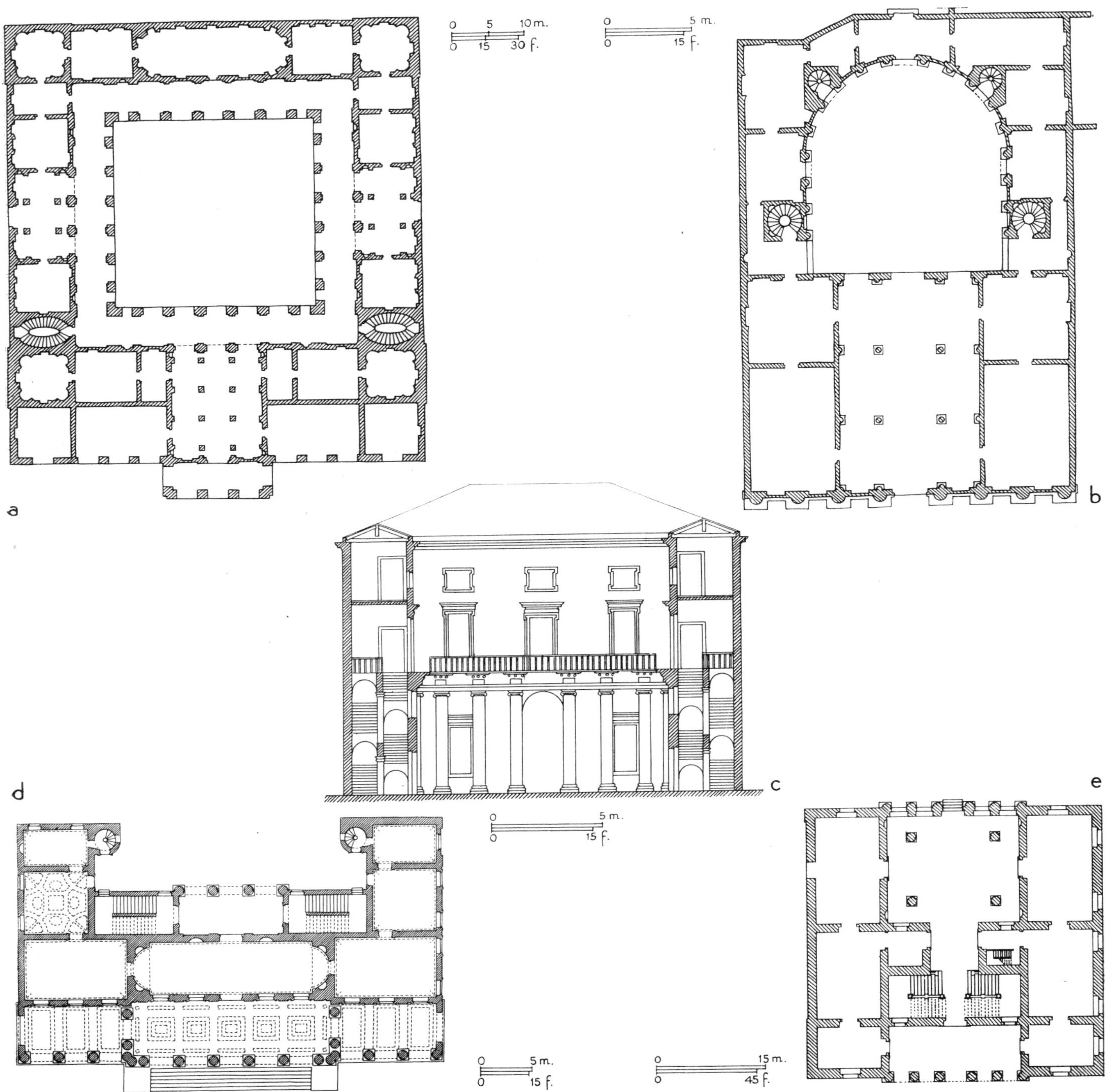

Urban palaces: (*a*) Vicenza, Palazzo Thiene, plan of ground floor; (*b*) Vicenza, Palazzo Porto-Breganze, plan of ground floor; (*c*) Vicenza, Palazzo Valmarana, longitudinal section on the courtyard; (*d*) Vicenza, Palazzo Chiericati, plan of ground floor; (*e*) Udine, Palazzo Antonini, plan of ground floor.

structural form that polarizes the ambient space of the Piazza dell'Isola. Therefore, the design constitutes a radical change from those of the Palazzo Porto and the Palazzo Thiene, since the façade strongly reflects the interior motif of the courtyards, thus coupling interior and exterior elements in a way that graduates the passage from the building to its urban setting, and vice versa. An original device in the organization of the colonnades is that of the arched partitions at the ends of the portico and the loggia, which by inserting an incongruous note in the simple entablature create a faintly visible termination for these essentially open spaces.

Similar arched panels, set perpendicular to the colonnades and clearly separated from them, screen the four symmetrical porticoes of "La Rotonda" (PL. 31; FIG. 69), resolving a problem of composition that involves the entire complex of the villa. The "heart" of the architectural image here is not the mass of an ensemble of walls but the circular hollow of the domed hall. The expression of this fundamental space is carried over to the exterior in four pronaoi with Ionic colonnades. The arched partitions serve specifically to channel and intensify the suggestion of centrifugal force immanent in the central domed core. Within this dominant scheme is introduced the secondary motif of the four rectangular halls, revealed in prismatic form at the corners, then continued in the simple, undecorated angularity of the high pediments of the pronaoi. Clearly, without the arched sections a purely prismatic form would be dominant from the base to the attic level; the pronaoi would become merely superadded elements without any valid structural or formal connection; the precise directional containment of the flights of stairs would be unexplainable; and above all, there would result a closed plan in which mass would simply contain interior space instead of continuing it — or freeing it — into the landscape.

Palazzo Chiericati and "La Rotonda" marked the opening of an area of inquiry that Palladio was to explore for more than ten years, until the "crises" of his late years. At this time there began the development of the unprecedented "open" vision of architecture that found its greatest lyrical success in the Villa Foscari (PLS. 31, 32), the Palazzo Valmarana (PL. 30), and the Chiesa del Redentore (PLS. 34, 35) — three works that, in extraordinary architectural imagery, epitomize many and disparate creative experiments.

The series of villas planned or constructed after 1560 offers many variations on the theme of "La Rotonda." The Mocenigo, Muzani, Pisani, Emo Capodilista, Sarego (at Miega), Valmarana (at Lisiera), and Cornaro villas reveal planning that is in each case personalized but not entirely dissimilar, since all have façades centralized with columned porticoes, which are often surmounted by loggias. Frequently the portico is flattened against the façade in such a way that the principal volume is clearly discernible through its confines and makes a smoother transition into lateral wings or adjoining subsidiary structures, which are arranged to follow the topography of the sites. For the most part, except for those of the Villa Emo Capodilista, these projected annexes were never constructed or have been destroyed. Sometimes, as in the Saraceno and Caldogno villas, a range of arches was substituted for the columned portico.

Occasionally, again taking up the inspiration of the Villa Godi, in the front elevation he depends upon sharply etched windows, rather than the motif of arches and columns, for his artistic effect, as in the Villa Zeno at Cessalto. The Villa Poiana remains within the general scheme of the villas of this period in its plan; its front elevation, however, distinguished by a bare Serlian treatment that incorporates a series of five oculi, makes its design quite original. The Villa Thiene at Quinto Vicentino and the Villa Repeta at Campiglia dei Berici, on the other hand, go beyond the standard plan of the other examples in their spatial organization, since the central core is no longer a solid structure but an open space, a court surrounded by living units. The Villa Sarego at Santa Sofia di Pedemonte is another exception to the rule, because of its semicircular exedra and its majestic columns sheathed in rusticated masonry (PL. 32).

Perhaps each of these villas would have reached truly poetic heights (to look beyond their incomplete or adulterated condition) had it not remained arrested in a stage of architectural allusions, relying too heavily on devices that, nonetheless, are extraordinarily stimulating in their unique invention. "La Malcontenta," on the contrary, apart from any romantic evocation, furnishes a splendid synthesis of the master's many artistic idioms. Its spatial epicenter does not consist in a round, square, or rectangular room, nor in a court or stairwell, but in a cruciform volume that, extending beyond these pure and elementary geometric shapes, governs through its arms the full stereometric expression. On the side overlooking the Brenta River, this focal space bursts forth in a pronaos set on a high base. On the opposite elevation is a large tripartite semicircular window (PL. 31), derived from Roman baths, which serves to focus the variously shaped apertures, the cornices, the broken pediment, and even the attic story and the smoothly embossed pattern of the walls. Thus, the freer and surer handling of apertures creates variety in the elevations: the sides are discreet, almost inhibited, in contrast to the other faces; the pronaos overlooking the river is solemn; the rear façade is striking and strange in the arrangement of the many openings, which are proportioned to the rooms they illuminate. The arched partitions of "La Rotonda," repeated in the Villa Piovene in Lonedo, have no reason to exist here. The spaces of the central core no longer seek to burst forth into porticoes and through

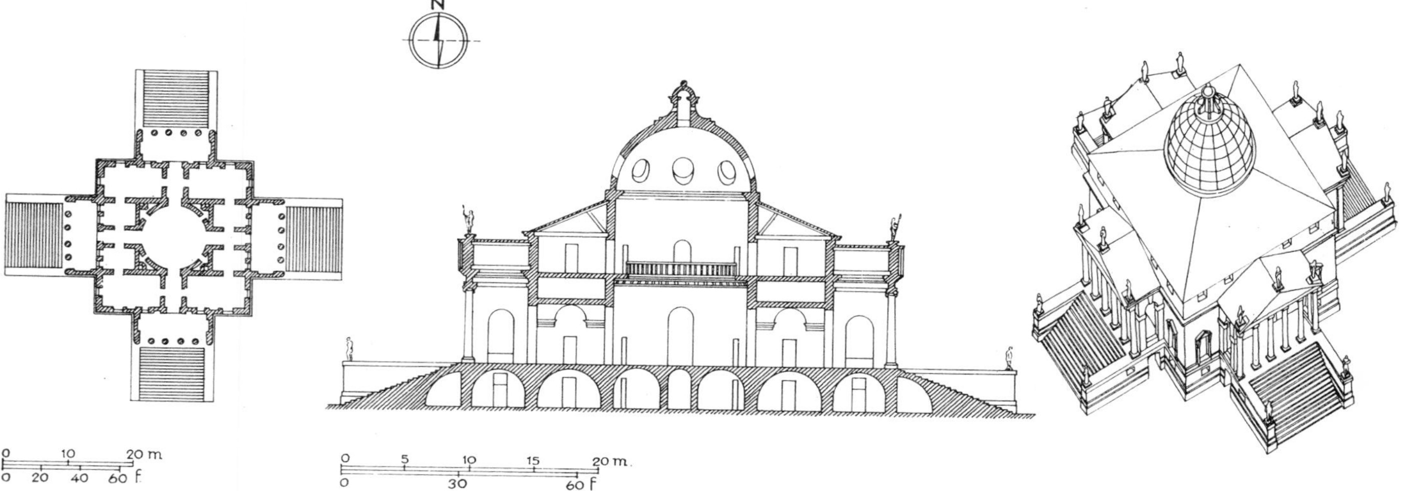

Vicenza, Villa Capra (called "La Rotonda"). *Left*: Plan. *Center*: Section. *Right*: Axonometric view (*from R. Pane, 1948*).

partly enclosed and directional spatial links. They are already in direct contact with the exterior, since the mass of the structure has been so thoroughly perforated with windows — almost as if by the force of light.

The Palazzo Valmarana in Vicenza has been compared with the Villa Barbaro at Maser because of the lively chiaroscuro of its wall surfaces, but its plan is the result of a development begun in the Palazzo Chiericati and interrupted only in the late production: in the Palazzo Porto-Barbaran, the Thiene-Bonin and Porto-Breganze palaces, and the Loggia del Capitanio. Palladio's aim was "open" composition, works that were "unfinished," yet in a manner less outwardly turbulent but no less daring and troubled than those of Michelangelo. His treatment of corners, that is, his manner of resolving the ends of façades, offers one key for tracing the various stages of his progress. In the Porto and Thiene-Bonin palaces the colonnades are extended to the ends of the buildings, giving slight prominence to the corners. The Roman-style design of the courtyard of the Palazzo Thiene seems incongruous, because the corner piers solidify what should be opened up (FIG. 67). Nor is the solution of the Palazzo Antonini in Udine satisfactory, which repeats the scheme of the villa's having a central body strongly emphasized in contrast to the smooth-surfaced wing. Palladio was motivated by a twofold need: to bind together the superposed Renaissance orders, as did Michelangelo, but at the same time not to close off the horizontal components with massive terminal elements that would detract from the unity of pictorial effect. Palladio's various solutions for the problem include such effective devices as the corner column found in the plan of the *palazzetto* for Giacomo Angarano and the double column used in the Palazzo Porto-Barbaran (PL. 30); yet it is only in the Palazzo Valmarana (present Palazzo Braga) that one finds a conception of overpowering boldness.

Raised on a majestic pedestal, which imposes a foreshortened view upon anyone who approaches the palace along the narrow roadway, the façade is dominated by gigantic pilasters, which seem to constrain the superposed Renaissance members, and culminates in a grandiose entablature. At first glance, the composition of the recessed order seems to dissolve into an intricate play of lines and planes toward the center; but then this order abruptly reemerges and becomes more prominent toward the sides, where it is stressed by means of two imposing figures in high relief. Their inclusion is forceful and peremptory; the entablature projects beyond the face of the pilasters but not over the caryatids, while, on the contrary, the broken cornice separating the stories does not protrude at the center but stands out beneath the statues. The desire to overcome every set compositional scheme is obvious elsewhere in the structure (e.g., the attic overhanging well beyond the main body) and is also reflected in the anomalous quality of many secondary elements, such as the truncated ends of the entablature. If one cites as sources for this treatment the wall of the walk above the cavea of the Roman theater in Verona, the caryatids of Giulio Romano for the Palazzo del Te in Mantua, or Michelangelo's piers on the Capitoline buildings (IX, PL. 542), it must be concluded that Palladio surpassed in boldness both the ancient and the contemporary masters, breaking the conventional order in all manner of ways and with departures that even Michelangelo and Giulio Romano had not dared. It has been customary to say that all this was accomplished in purely chromatic or textural terms, that is, without bold plastic or structural innovations; but one should guard against attributing to these so-called "pictorial" effects a merely rhetorical significance, unrelated to structural requirements or means. Perhaps more than in any other mannerist creation, the anticlassical break is resoundingly evident in this design. It is not suprising that neoclassicists criticized the lack of strength in the terminal abutments, since this was a fundamental element in Palladio's unorthodox vocabulary of forms. Nor is it surprising that Burckhardt found the whole to be unsuccessful. If this negative verdict did not regard the artistic worth of the building but the moral and psychological attitudes underlying its creation — in other words, Palladio's *Weltanshauung* — it would be an acceptable judgment, for in no other monument does one sense to such a degree the determining influence of an unhappy and alienated human condition, a terrifying and resigned testimony to an unsatisfied rational conception of the world.

The Chiesa del Redentore (FIG. 72; PLS. 34, 35) is the apotheosis of the Palladian "open" space. Preceded by numerous Venetian accomplishments — the façade designs for S. Pietro di Castello and S. Francesco della Vigna; the enchanting Refectory of S. Giorgio, which should be visualized as flooded with light, with its large Roman thermal windows (now walled up) reinforcing the division of the chamber into three vaulted bays; the Convento della Carità, of which Goethe wrote that it seemed to him that he had never seen anything of greater perfection; and finally, S. Giorgio Maggiore — the Redentore summarizes all of these in a version so taut and resplendent that it approaches the metaphysical. The ready comparison with S. Giorgio Maggiore, though made regularly by scholars, is nonetheless valuable and continues to yield unexpected rewards.

While new and daring, the organization of S. Giorgio Maggiore finds its own limits in the very clarity of the scheme. The sweep of the nave and aisles (which follow the usual 1:2 relationship) is interrupted after only three bays to accommodate the space below the dome, the focal point of the longitudinally oriented structure; this climactic space is extended into transepts terminating in apses. Beyond the piers supporting the dome, there is the raised square area of the sanctuary, at the back of which two flanking pairs of columns provide a screen for the apsidal choir. The contrast between an organization based on multiple spatial divisions arranged in perspective succession and an intention of achieving (by moving the dome forward) a centralized emphasis remains obtrusive, however, although the generous, uniform distribution of light somewhat softens the effect. The front elevation of the church (XII, PL. 49) is even more unresolved, primarily because the four columns are raised on high pedestals and the spacious piazza does not compel a foreshortened view, then because of the half pediments at the sides, which meet the central element at a very indefinite point on the outer columns, and finally because of the crowning pediment, which does not succeed in relating to the mass of the church behind and below it and which is reinforced by weak, incongruous additions of wall section.

In the Chiesa del Redentore the lateral aisles are narrowed to the point of being reduced to a series of almost self-enclosed chapels that, raised on platforms as well, emphasize and delimit the great central space of the nave. With its length not much

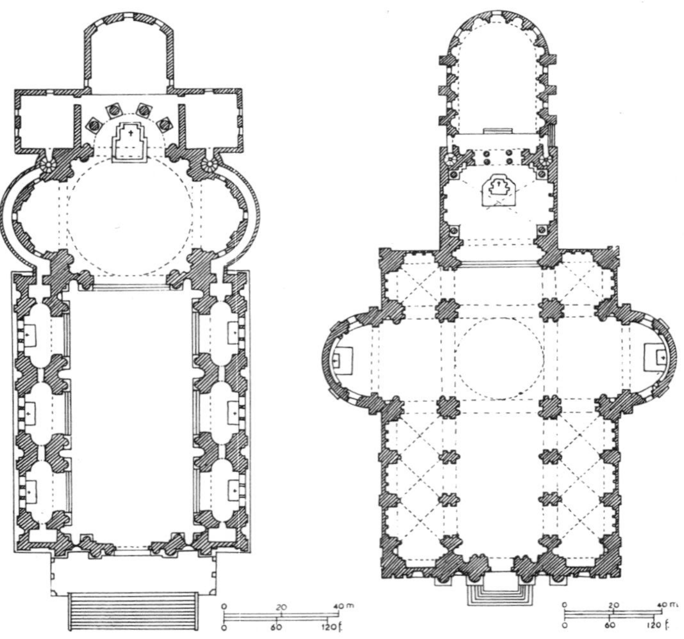

Venetian churches. *Left*: Chiesa del Redentore, plan. *Right*: S. Giorgio Maggiore, plan.

more than twice its width, the nave does not lend itself to the interruption of a dome; the transept and the presbytery, which are distinct in S. Giorgio Maggiore, are united in a single space — instead of the conventional sequence of expansion and contraction of spaces — with flanking apses and a semicircular colonnade closing the far end. The lighting is well differentiated: the nave is cut crosswise by rays coming from the recessed semicircular windows; the space below the dome has a uniform brightness, but beyond the colonnade this burst of light is diffused and seems drawn into infinity. In this composition, the fitting together of the variously dimensioned orders of the Palazzo Valmarana is translated into an interplay of spaces, and the architectural elements mark off separate zones of illuminated space.

Such a coherent and poetic arrangement of interior spaces was meant to be reflected in a perfect mass on the exterior. In the front elevation of the Redentore all the uncertain handling so apparent in S. Giorgio has disappeared; the pairs of engaged columns and pilasters rest directly on the podium for the whole structure, not on individual socles, thereby refuting any intention of foreshortened views; the narrower side panels of the façade, topped by half pediments, form wings that contain and appear to buttress the central section; the motif of repeated pediments serves to define the successive planes and to unify the various elements of the façade. Moreover, there are strong links, or mutual exchanges of function, between the decorative and the structural components: the section of wall above the half pediments of the façade constitutes the first of a series of paired buttresses that rhythmically divide the side walls of the church and, continuing downward into pilasters that flank niches, enliven these lateral surfaces. The immaculate expanses of the walls of the apses are appropriate exterior counterparts of the bare domical vault within. The atmosphere around the dome is vitalized by the thrust of the slender spires, especially toward the rear of the edifice. Thus the Palladian idiom found its supreme affirmation in this absolute, rarely achieved integration of space, natural lighting, structural and decorative elements, and urban setting.

From Palladio's mature period are several minor but exceptionally poetic works. The most notable of these are the Church of S. Maria della Presentazione (or delle Zitelle) in Venice and the Casa Cogollo in Vicenza (PL. 35). The splendid Venetian church restates the theme of the Redentore in a drier, more concise version. In his Casa Cogollo (called "del Palladio"), a most original composition, Palladio set over the columns of a Serlian aperture fluted pilasters intended to frame a fresco, the absence of which cannot be regretted at all since the bare central panel — as was well appreciated by Goethe — seems to polarize so much light in itself that the addition of a painting would merely be superfluous.

Contrary to the customary biographical treatment, it is at this point in the Palladian development that the Teatro Olimpico (PL. 36; IX, PL. 312) should be considered. Although actual construction did not begin until 1580, the germinal form had evolved much earlier: during the architect's studies, in the years 1540 to 1556, of Roman theaters and of the commentary relating to them in Vitruvius, which Palladio interpreted together with Daniele Barbaro; in 1561, when Palladio constructed a temporary theater in the salon of the "Basilica" of Vicenza; and in 1565, when he erected another cavea theater, this one in the Convento della Carità in Venice — "a half theater of wood, for use as a colosseum," as Vasari described it. However, when the members of the Accademia Olimpica acquired the site of the old prisons for a permanent theater on Feb. 25, 1580, they decided to build it "according to the model already made by their colleague Palladio and according to the design of the perspectives also made by him." This statement confirms that the plans had already been completely drawn up. Exactly when the design was worked out is not known, but it cannot be ascribed to the late period — at least with respect to his choice of type — since it is difficult to comprehend in that context, whereas it fits very well with the attitudes and taste of his earlier mature period.

Considering his long apprenticeship in this field, Palladio was certainly aware of accomplishments in theatrical design of the time, particularly with regard to two striking characteristics of the new structures: the rectangular area for the spectators and the proscenium arch, which delimits a space devised to further illusions of perspective. But he rejected these solutions and, attempting to restore the Vitruvian conception of a theater, built an elliptical cavea surmounted by a magnificent Corinthian colonnade, which confronted a fixed stage composed of imaginary foreshortenings of streets in Vicenza. This choice was unusual and anachronistic and had a serious failing: proscenium and cavea meet without being well integrated, as can be seen in the unsatisfactory expedient of statues and niches painted at the sides and in the poor juncture made between the two ceilings. "Academic cultural pretensions contributed to the realization, in the Olimpico, of an archaeologically inspired compromise," wrote Pane (1961), justly describing this theater as "a masterpiece *manqué*." It is interesting to explore, on the basis of studies by L. Magagnato, the inner motives for the architect's choice of structural type. For Palladio, pictorial values were an integral part of architectural expression. The classical proscenium, the broad transparent wall of a unified space, could satisfy him, whereas the proscenium arch of current theatrical practice divided the space into two parts separated by a curtain, and thus lost the possibility of organic pictorial effects. For this reason Palladio even wanted to close the gap between the tiers of seats and the stage, a break that was still apparent in the theater designs of Peruzzi and Serlio, and he conceived of the proscenium as "a screen and not as a frame for perspectives," assigning to it a function like that of the semicircular partitions of columns preceding the choir of the Chiesa del Redentore. Furthermore, rather than noticeably dividing the area of the audience from that of the actors, the unified Vitruvian scheme allows for an unlimited opening since the cavea, originally an open-air plan, retains the expressive potential of an uncovered auditorium. Although all this does not justify the shortcomings of the work, including the abrupt transition between the two ceilings, it does explain why the inorganic, static plan of the theater of his day was unacceptable to Palladio and also why the Olimpico, notwithstanding its defects, remains a unique achievement even with respect to its direct descendants — the theater of Vincenzo Scamozzi at Sabbioneta, that of Giovanni Battista Aleotti in Parma, and the one of Inigo Jones at Oxford. It is impossible to duplicate the vast and sudden upward sweep of the classical cavea in a pillared auditorium, and especially to achieve that "indefinable space" in which acoustics, replacing light, assume a fundamental architectural role.

Palladio's late period is characterized by a virulent anticlassicism that exaggerated the elements of the late-antique tradition and shattered the conventional relationships between them. The Loggia del Capitanio in Vicenza (PL. 35), in front of the "Basilica" in the Piazza dei Signori, is composed in an imposing, extravagant idiom that combines heavy Michelangelesque ornament with reminiscences of the Arch of Septimius Severus in Rome and the triumphal arch of Orange (V, PL. 470). The simple arched panels at the sides of the porticoes of the Palazzo Chiericati and "La Rotonda" are here enlarged and dominate the entire side wall, in which the second story, centered about a Serlian window, loses any logical relation to the ground floor. On the main façade the distended semicolumns of colossal order overpower the pilasters behind; the slight relief of the springing lines of the arches serves only to underline the intentional imbalance of the plan. The fact that two orders of different heights — the gigantic columns of the façade and the smaller ones of the sides — do not converge or correspond but remain juxtaposed without interrelation on orthogonal faces is even more astounding than the fragility of the balcony cornices in comparison with their enormous corbels or the disputable fragmenting of the architraves. In this incomplete, badly restored work there is testimony of a senile anticlassical rage that uses architectural members "as if they were castoff elements, forcibly put together in a new and different rhythm," lacking in any compositional logic (Pane, 1961). Pane rightly adds that the Michelangelesque elements were adopted only for "pictorial" purposes; that is, they were stripped of any function and fragmented. In no other work, not even as seen in the spectral remains of the Palazzo Porto-Breganze, did Palladio attain the same paradoxical degree of formlessness.

Similarly, for psychological effect, the Villa Barbaro at Maser (PL. 33; VIII, PL. 434; FIG. 75) was planned as a low, elongated structure set in an immense park. The vertical section of Bertotti-Scamozzi illustrates the scenographic development on varied levels that culminates in an exedra cut into the hillside. The façade is disengaged from the mass of the

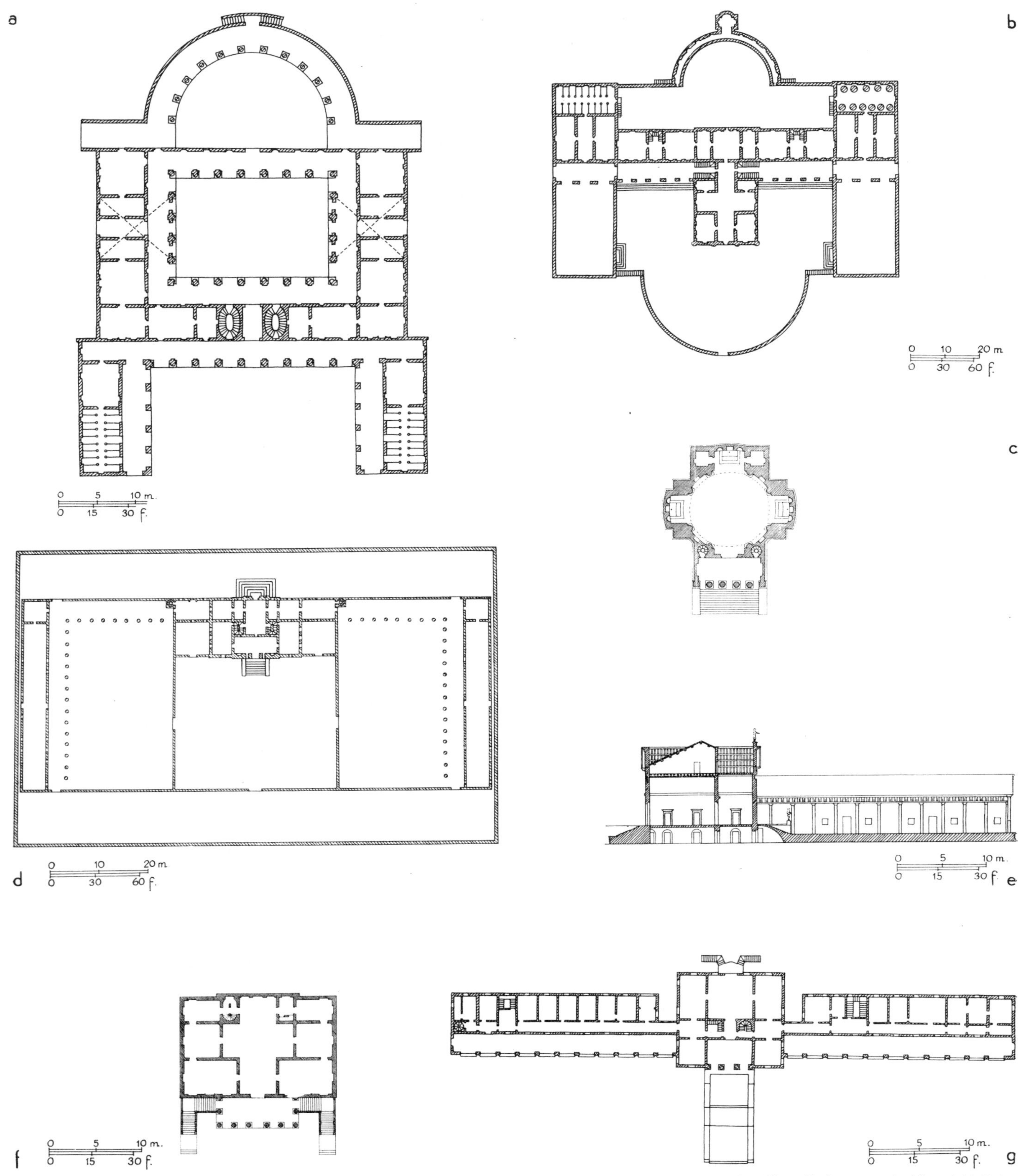

(*a*) Santa Sofia, Villa Sarego, plan; (*b*) Maser, Villa Barbaro, plan; (*c*) Maser, *tempietto* of the Villa Barbaro, plan (*from R. Pane, 1948*); (*d*) and (*e*) Poiana Maggiore, Villa Poiana, plan and section; (*f*) Villa Foscari ("La Malcontenta"), plan (*from Bertotti-Scamozzi, 1776–83*); (*g*) Fanzolo, Villa Emo Capodilista, plan (*from L'Architettura, no. 12*).

building behind, as in the Loggia del Capitanio, and is crowned by a ponderous broken pediment that is further complicated with thick-proportioned illusionistic stuccoes. The circular *tempietto* of the villa (PL. 35; VIII, PL. 435), which can be regarded as the last work of the master, is inspired by the Pantheon only in superficial aspect. In reality, the *tempietto* departs from the Roman model conspicuously in several ways: in the overbearing scale of the pronaos compared with the cella; in the excessive protrusion of the side chapels; in the addition of small belfries, which are not well related to either the intercolumniation below or the dome behind; in the lowered interior arches, which create awkward perspective effects; and in the disorder of the ornament. Pane (1961) maintains that "this is a work that Palladio would not have hesitated to repudiate." Still, the plan was his, and it reflects the aggressive, alienated attitudes of advanced age — the product of a will that is no longer freely imaginative and exalting, but rather of a spirit vitiated and nihilistic toward architecture.

The phases of Palladio's architectural development outlined above are confirmed in his ideas of urban planning and of landscape, which have been thoroughly analyzed by Sergio Bettini (1961). In these fields as well, Palladio differed from the other major artists of the 16th century. At first glance, he would seem less interested in city planning than were Serlio, Scamozzi, or Bartolommeo Ammanati, since he never wrote a treatise on the subject of cities. In substance, however, his work went beyond theirs, for he consistently approached architecture within the context of the city or in relation to a humanized landscape. He did not plan ideal cities: there were no Sforzindas or Palmanovas among his creations. Yet, it is precisely in this refusal to propose such abstract images, strictly geometrical and unrealistic plans, that one can identify a stern, though not explicit, critical position with respect to the culture of his time. Palladio could not accept the closed, symmetrical, "pure" form that characterized the star-shaped, polygonal, and checkered layouts of the ideal Renaissance cities, since this implied an alteration of constructed masses and open spaces in a strictly controlled perspective sequence, that is, an organization so inflexible and axiomatic that it would not be susceptible of easy social living. Granted that the modern concept of urban life and planning had its point of departure in those schemes, since they served to break down the medieval social system of isolated castles and a closed, compartmentalized structure grounded in a hierarchical scale of power; yet they dealt with a Neoplatonic utopia, a product that was best appreciated in the planning stage but quite sterile with respect to the problems of serving an economically dynamic community. To medieval urban development — which at least had the advantage of an organic growth and accommodation to the topography — was opposed a rationality equally if not more dogmatic, one that consisted of empty and solid blocks defining or composing squares, streets, and building complexes. Palladio himself could not subscribe to such an abstract, geometric principle that ignored the more practical considerations. And perhaps, as Bettini believes, Palladio's Venetian experience was a determining factor in his mistrust of the concept of the ideal city, since late-antique Venice had not made use of the defensive military layout characteristic of medieval European cities and therefore was not a likely site for propagation of Florentine idealism. Its watery thoroughfares and open spaces were the connective tissue of an urban conformation that was civilized and unitary, yet not rigidly crystallized.

Also in the sphere of city planning, Palladio's anticlassical sensibility is revealed in his choices as a student of antiquity. The large buildings of the late Roman Empire, especially the baths, became the object of his enthusiastic description (IV, PL. 199). As Zorzi noted, even in the way Palladio sketched the Roman ruins there is a significant transition from the meticulous drawings in the Falconetto style to the suggestive chiaroscuro effects that emphasize, more than the individual architectural elements, the feeling and over-all tone of the spatial interplay. In this cultural orientation there was an insistent manneristic component. In Venice, as in the monuments of the late Empire, Palladio found the passage from a rigorously logical "idea" of space to that encompassing "experience" of space which qualified the crisis of mannerism.

Palladio's attitude toward town planning is clearly evident in his urban buildings, his villas, and his churches, without distinction as to their type. It has been mentioned how expertly the Palazzo Chiericati is fitted into the vista of the piazza in front of it and how it seems to polarize the light from this broad space. The façade of the Palazzo Valmarana, contained within two levels that meet in a barely perceptible obtuse angle which is not orthogonal with respect to the compositional axis, is rich in expedients intended to take advantage of the inclination of the street. But if one thinks of Palladio's uncompleted buildings such as the Convento della Carità and the Palazzo Thiene and of the "Devices according to Different Sites," published in the seventeenth chapter of the second of the *Quattro libri*, and if several very effective isometric reconstructions worked out by Pane for his monograph (1961) are examined closely — in the sequence, juxtaposition, and connection of buildings, courts, and exedrae one can readily discern cadences that reflect and express the evolution of both the scholar and the architect.

The same progress is evident in the villas. The long description that Palladio devoted to the site dominated by "La Rotonda" offers but an early sample of his feeling for landscape, since the villa, despite its dynamic and centrifugal plan, is still clearly an autonomous object of double symmetry. Over the course of years, the classic isolation of the "country houses" was gradually challenged by the architect. The organism already taking form in the Villa Thiene in Quinto Vicentino was further developed in (to mention a completed example) the Villa Emo Capodilista in Fanzolo, later to be culminated in the Villa Barbaro at Maser, where the flattened structure serves as a kind of white "comment" on the hilly landscape. In his method of dealing with the countryside, one also sees a change from a closed and static classical conception to an "open" vision, a contained space that bursts forth through the porticoes and subsidiary constructions into the surrounding nature, at the same time deriving from it subtle topographical incentives and transforming them; ultimately there is achieved the immersion of the architecture into the very fabric of the adjoining landscape.

Palladio's planning, however, was not limited to the relation of the building or the building complex to the urban or rustic setting. The full assimilation of the surrounding landscape into the architectural expression modifies the villas, palaces, and churches. For example, "La Malcontenta" seems more closed and isolated than the Villa Pisani at Bagnolo or the Villa Zeno at Cessalto, but actually its receptivity to light has been sensitized as much as possible. It seems to draw the surroundings into its pronaos, which is imposing in scale but not heavy, into the shadowy areas cut out by the windows and doors, by the very pattern of its rustication, and with wall surfaces that graduate the intensity of reflected light differently. More important is that the setting penetrates the interior spaces in such a way that the exterior panorama is never meant as a "picture" to be admired but as a contributing element to the interior organization.

Site planning — considering and taking full advantage of the relationship between the building and its setting, including absorption of natural light and incorporation of the landscape — became a permanent and prime aspect of Palladio's artistic expression, which was eventually to triumph in the façade (or rather, in the superimposed façades) and in the screened recesses of the Chiesa del Redentore. Bettini wrote: "With a shift comparable to that pointed out by the Viennese art historians — from Riegl to Swoboda to Zaloziecky — in the transition from the architecture of mass of the early Empire to the middle and late Roman view [intended] from a distance, we see in the work of the mature Palladio façades fitting into façades; façades expanding outward and withdrawing inward; spaces inflating alongside spaces, or rising behind spaces. And each one of these spaces, as far as the image is concerned, is not to be defined so much in its plastic properties, immobilized in a checkerboard perspective, but by the timbre of its chiaroscuro.

It is clear, definite, and precise, with the full value of its situation, but it is always related, both on the exterior and in the interior, to the images of the spaces behind and alongside it. These others become the broad backgrounds for the spatial figuration of the foreground; but as we proceed farther, these themselves make a configuration set against other spaces." This variety of functions assumed by the hollows in being presented alternately as subjects or modifiers of the spatial ensemble constitutes final proof of the anticlassical and antiperspective character of Palladio, in brief, of his essential genius. For him, space and mass shape each other at the time when man experiences them; their very existence is justified, or even realized, in their social intent, in the act of perceiving.

These site-planning distinctions of Palladio's mature period cannot be extended indiscriminately to other phases of his development. His sensitivity to architectural setting was less keen during his youth, as is shown by the "Basilica," which was conceived not in firm relationship to the existing constructions about the piazza but almost as if it were a dominant and isolated personage: As for his late years, Palladio's disorder can be noted also in the area of site planning. The later palaces, exaggerated and weakened in their orders, seem to desecrate the surrounding spaces of the city as well. They betoken the architect's overwhelming revolt — his strong reaction against society, against his profession, or perhaps simply against the human condition itself.

Yet these qualities are often forgotten in favor of an idyllic and pallid portrayal of a personality who was, instead, one of the most varied and most dramatic in architectural history. It has already been remarked that the "pictorial" quality of Palladian art should not be interpreted as synonymous with an elegiac detachment, for it is the consequence of a postmannerist and prebaroque sense of alienation. Of this disoriented situation, culturally unique and painful, the last works offer proof also in regard to landscape, because one senses anger and indignation in their relation to it, a desire to offend. His didactic and creative mission evidently completed, Palladio then became incommunicative and turned inward. In this solemn moment, the evaluation he made of the world and of life no longer induced him merely to contest the rules and myths but also to denounce their inconsistency, in the solitary protest embodied in his Loggia del Capitanio.

This unaccustomed interpretation of Palladio finds, too, a natural confirmation on historical and sociological grounds. The Venetian society of the 16th century was troubled and ambiguous; Padua and Vicenza were also involved in its sphere of influence. The exhausting war against the Turks from 1463 to 1479, the prolonged economic stagnation, aggravated by a population increase that created an imbalance in the countryside, the chain of bankruptcies and banking failures in the 16th century, the precipitous decline of the Venetian fleet, and the progressive crisis of trade and markets were grave and complicated phenomena that the apparent prosperity of the second half of the century could not satisfactorily assuage. The memory of lootings, brigandage, and political and religious repression undermined public tranquillity and individual peace of mind; and the plague of 1576 injected a further tragic note into the scene. The opulence of city life, the extravagance of the upper classes, and even the agricultural prosperity of the Vicenza region were but the reverse side of a socioeconomic decline, since the monetary solvency essential for property investments and land reclamation entailed withdrawal from the competition for international markets and eventually brought about bitter losses. Prosperity without security may lead to a sense of alienation, and the megalomaniacal plans of Palladio's unexecuted or incomplete buildings are perhaps one sign of this. It was a society that Fernand Braudel has compared with that of Europe after World War II, when the prosperity was also marred by fear and the realization of having lost a preeminent place in history. Certainly the comparison should not be forced, but the "formlessness" of the late Palladian projects seems fit reflection of an analogous sociological situation.

Nevertheless, it is not necessary to resort to the proof of external events to establish that the Palladian inspiration was complex and troubled in its personal character and, because of this, was an art not to be duplicated. One need only examine the work of both close and distant disciples: Borella, Muttoni, Massari, and Pizzocaro in Italy; Inigo Jones, Colen Campbell, the Earl of Burlington, and John Vardy in England; the Dutchmen Jacob van Campen, Pieter Post, and Justus Vingboons; and Tilman a Gameren, Jean de Bodt, and Johann de Witte, who worked in Poland. These are but a few of the names associated with Palladio subsequently, in an international sphere that, over the centuries, has extended to the most varied cultures and tastes, from America to Russia. Although it would be unfair to undervalue the cultural influence that Palladianism has exercised on so much of the world, it must be recognized that not a single disciple has fully absorbed and expressed Palladio's elusive esthetic and psychological tension or has thoroughly mastered his secret of "open" space. His vital *œuvre* was thereafter reduced to a formula and translated equally into baroque, neoclassic, and functional veins, often in noble and correct forms, but without the authenticity and compelling lyrical outbursts of the originator.

The explanation for this failure can be easily comprehended. These architects were looking backward at Palladio and his work, searching to derive fixed compositional precepts from him, whereas Palladio in his own time was a troubled, even a destructive artist. His attitude in regard to the restoration of the Doges' Palace in Venice, the façade of S. Petronio in Bologna, and the Rialto Bridge documents his faith in an artistic idiom that was new not only with respect to Renaissance art but also with respect to mannerism. As frequently happens, the true heirs of Palladio were not those inspired by his forms — even if Bernini was among them — but those who reexamined his work from unorthodox viewpoints. Approached in this sense, the Palladian source has proved inexhaustible over the centuries, continuing even today, after the rationalistic and functional trends and abstract tendencies, to serve as inspiration for a liberating conception of "open" architecture.

Bibliog. P. Gualdo, Vita di Andrea Palladio, 1617 (ed. G. Zorzi), Saggi e memorie di Storia dell'Arte, II, 1958–1959, pp. 91–104; T. Temanza, Vita di Andrea Palladio, Vicenza, 1762; O. Bertotti-Scamozzi, Le fabbriche e i disegni di Andrea Palladio, 4 vols., Vicenza, 1776–83; T. Temanza, Vita de' più celebri architetti e scultori veneziani, Venice, 1778; O. Bertotti-Scamozzi, Le terme dei Romani disegnate da Andrea Palladio . . . , Vicenza, 1785; F. Burger, Nel 4° centenario della nascita di Andrea Palladio, Mnh. für Kw., I, 1908, pp. 914–15; G. Zorzi, Sull'anno in cui nacque Andrea Palladio, La Provincia di Vicenza, Oct. 15, 1908; W. Arslan, Ritorno al Palladio, Architettura, VI, 1926, pp. 336–47; G. C. Argan, Palladio e la critica neoclassica, L'Arte, N.S., I, 1930, pp. 327–46; H. Willich, ThB, s.v. (bibliog.); G. Fiocco, Andrea Palladio Padovano, Ann. Univ. di Padova, 1932–33, pp. 29–39; F. Franco, La scuola architettonica di Vicenza: Palazzi minori dal secolo XV al XVIII, Rome, 1934; G. Fiocco, Fortune e sfortune del Palladio, Padova, IX, 2, 1935, pp. 7–16; G. Giovanni, L'urbanistica italiana del Rinascimento, in Saggi sull'architettura del rinascimento, Milan, 1935, pp. 265–306 at 290; F. Franco, Classicismo e funzionalità della Villa Palladiana 'città piccola', Atti I Cong. Naz. di Storia dell'Arch. (1936), Florence, 1938, pp. 249–53; J. Lees-Milne, Palladio's Vicenza, Arch. Rev., LXXXV, 1939, pp. 55–60; H. Pée, Die Palastbauten des Andrea Palladio, Würzburg, 1939; W. Lotz, Literaturbericht über die Baukunst des Cinquecento in Italien, ZfKg, IX, 1940, pp. 224–30; R. Pane, Palladio artista e trattatista, Palladio, VI, 1942, pp. 16–24; N. Pevsner, An Outline of European Architecture, Harmondsworth, 1942 (6th ed., 1960); A. Dalla Pozza, Palladio, Vicenza, 1943; R. Wittkower, Principles of Palladio's Architecture, Warburg, VII, 1944, pp. 102–22, VIII, 1945, pp. 68–106; H. Leclerc, Les origines italienne de l'architecture théâtrale moderne, Paris, 1946; R. Zürcher, Stilprobleme der italienischen Baukunst des Cinquecento, Basel, 1947, pp. 79–101; R. Pane, Andrea Palladio, Turin, 1948 (2d ed., 1961; bibliog.); J. Reynolds, Andrea Palladio and the Winged Device, New York, 1948; E. Rigoni, Padova, città natale di Andrea Palladio, Atti Ist. veneto di sc., lett. e arti, CVII, 1948–49, pp. 67–72; R. Wittkower, Architectural Principles in the Age of Humanism, London, 1949 (3d ed., 1962); G. Zorzi, Ancora della vera origine e della giovinezza d'Andrea Palladio secondo documenti, Arte veneta, III, 1949, pp. 140–52; L. Magagnato, La mostra dei disegni del Palladio a Vicenza, Emporium, LVI, 1950, pp. 121–30; R. Wittkower, Giacomo Leoni's Edition of Palladio's Quattro libri dell'architettura, Arte Veneta, VIII, 1954, pp. 310–16; G. C. Argan, L'importanza del Sanmicheli nella formazione del Palladio, Venezia e l'Europa (1955), Venice, 1956, pp. 387–89; F. Franco, Piccola urbanistica della 'casa di villa' palladiana, Venezia e l'Europa (1955), Venice, 1956, pp. 395–98; R. Pane, Andrea Palladio e la interpretazione dell'architettura rinascimentale, Venezia e l'Europa (1955), Venice, 1956, pp. 408–12; N. Pevsner, Palladio and Europe, Venezia e l'Europa (1955), Venice, 1956, pp. 81–94; W. Lotz, Architecture in the Later 16th Century, College Art J., XVII, 1957–58, pp. 129–39; G. Fiocco, Le arti figurative, La civiltà veneziana del Rinascimento, II, Florence, 1958, pp. 177–95; G. Zorzi, I disegni delle antichità di Andrea

Palladio (pref. by G. Fiocco), Venice, 1958; B. Centro int. di s. d'arch. Andrea Palladio, Vicenza, 1959 ff.; B. Zevi, Palladio, 1508–1580, Les architectes célèbres, II, Paris, 1959, pp. 78–79; C. Brandi, Perché Palladio non è neoclassico, B. Centro int. di s. d'arch. Andrea Palladio, II, 1960, pp. 9–13; W. Timofiewitsch, Die Palladio-Forschung in den Jahren von 1940 bis 1960, ZfKg, XXIII, 1960, pp. 174–81 (bibliog.); M. Zocca, Le concezioni urbanistiche di Palladio, Palladio, N.S., X, 1960, pp. 69–83; S. Bettini, Palladio urbanistica, Arte veneta, XV, 1961, pp. 89–98; G. Chierici, Palladio, Milan, 1961; G. Fiocco, Incunaboli di Andrea Palladio, Saggi di storia dell'arch. in onore del Prof. V. Fasolo (Q. Ist. di storia dell'arch., 31–48), Rome, 1961, pp. 169–76; R. Pane, 1961 (see above, Pane, 1948); R. Strandberg, Il Tempio dei Dioscuri a Napoli: Un disegno inedito di Andrea Palladio nel Museo nazionale di Stoccolma, Palladio, N.S., XI, 1961, pp. 31–40; G. Zorzi, Le opere pubbliche e i palazzi privati di Andrea Palladio, Vicenza, 1965.

Bruno Zevi

Illustrations: PLS. 29–36; 5 figs. in text.

PALMA VECCHIO.

PALMA VECCHIO. Venetian painter [b. 1480 (?); d. Venice, July 30, 1528]. Known as Palma Vecchio, Jacopo (or Giacomo) Negreti — Palma is merely a sobriquet — must have been born in 1480 if Vasari was accurate in saying that he lived to the age of 48, since it is known for certain that he died in Venice on July 30, 1528. He was a native of Serinalta, in the region of Bergamo. In 1513 (a date supported by two instances of documentary evidence) he became a member of the Scuola di S. Marco. Tradition says that he was a pupil of Giovanni Bellini, and the testimony inherent in his work reveals that he was influenced by Giorgione, Lotto, and Titian (qq.v.). There exist a number of documents in which Palma is cited as a witness or is mentioned in connection with business transactions and family affairs, but these throw little light on his artistic career.

Many paintings begun by Palma were finished by other hands; at the time of his death, the inventory of his possessions listed no fewer than 62 unfinished works. Moreover, Palma never signed or dated his pictures. Certain recorded dates, however, help to fix various paintings roughly in time and to give some general idea of his development. For example, it is known that his most famous work, the altarpiece of St. Barbara in S. Maria Formosa, was ordered sometime after Christmas, 1509. L'Anonimo Morelliano saw the celebrated *Three Sisters* (Dresden, Gemäldegal.) in the house of Taddeo Contarini in 1525. In the same year Palma and some members of the Scuola di S. Marco petitioned the Council of Ten to allow for the commission of a new altarpiece, representing the death of St. Peter Martyr, for the Church of SS. Giovanni e Paolo, to be executed by one of the leading Venetian artists and to be paid for out of the petitioners' private funds. (This was the first of the proposals which finally resulted in Titian's painting of 1530, destroyed by fire in 1867.)

The general development of Palma's style is marked by a progression from a Quattrocento manner toward an increasingly lyric-idyllic softness and dreaminess. In the second decade of the century his work became more "painterly" and was very blond in tonality. He followed the prevailing taste in his fondness for half-length figures, which were extremely popular about 1505–10. In all these qualities he remained true to his Giorgionesque training. The influence of Giorgione is strikingly apparent in a work such as the *Music Party* (Bowood Park, Calne, Wiltshire, Coll. Marquess of Lansdowne), but is evident to some degree in almost all of Palma's production.

The *sacra conversazione* was a favorite theme of the artist, and it became rather a specialty of his. In general, in treating it, he preferred the oblong (horizontal) to the vertical format.

The following paintings by Palma (with the approximate dates that close study of his style allows one, by more or less general consent, to assign them) should also be mentioned: *Madonna with the Baptist and St. Catherine* (early; Dresden, Gemäldegal.); *Portrait of a Poet* (early; London, Nat. Gall.); the so-called "Violante" (1507–08; Vienna, Kunsthist. Mus.); *Portrait of a Woman* (ca. 1508–10; Lyons, Mus. B. A.); *Madonna with SS. Roch and Mary Magdalene* (1512–15; Munich, Alte Pin.); *The Adoration of the Shepherds* (ca. 1515; Louvre); *Young Woman with a Lute* (ca. 1522; Northumberland, Alnwick Castle, Coll. Duke of Northumberland); *Sacra Conversazione* (ca. 1525; Venice, Accademia); *Portrait of a Lady* (ca. 1525; Milan, Mus. Poldi-Pezzoli); *Portrait of Francesco Querini* (late; Venice, Gall. Querini Stampalia); see also VIII, PL. 213; XII, PL. 68.

BIBLIOG. M. von Boehn, Giorgione und Palma Vecchio, Bielefeld, Leipzig, 1908; J. A. Crowe and G. B. Cavalcaselle, A History of Painting in North Italy . . . from the 14th to the 16th century (ed. T. Borenius), III, pp. 351–90, London, 1912; György Gombosi, Palma Vecchio, Berlin, Stuttgart, 1937.

Arthur McComb

PANAMA.

PANAMA. The Republic of Panama, occupying an area of approximately 30,000 sq. mi., is situated at the eastern end of the Central American land mass that forms the narrowest span between the Atlantic and Pacific oceans. The Panama Canal, in traversing the isthmus, also serves as a distinct east–west dividing line for the country.

The shores of what is modern-day Panama were reached by Columbus on his fourth voyage, in 1502, at which time he established a short-lived colony at Belén. The first settlement of some permanence, Santa María la Antigua del Darién, was founded in 1510, and shortly thereafter, in 1513, Balboa crossed the isthmus and discovered the South Sea — the Pacific Ocean — in the name of the Spanish king. In 1519 the city of Panama was founded on the Pacific coast, later to be linked with Portobelo as the Caribbean terminus. After periods of incorporation in the Spanish viceroyalties of Peru and Nueva Granada, Panama became a province in the Colombian Federation. Following a series of unsuccessful rebellions over a span of more than fifty years, the country declared its final independence from Colombia in 1903.

Panama contains noteworthy remains of pre-Columbian cultures as well as interesting examples of colonial and modern architecture. Archaeologically, the country is divided into four culture areas: Darién, in eastern Panama, and Coclé, Veraguas, and Chiriquí, in western Panama. In spite of extensive trade between these areas, each maintained its own distinct cultural and artistic traits.

PRE-COLUMBIAN PERIOD. *Eastern Panama.* Eastern Panama, called Darién, was dominated at the time of Columbus's arrival by the Cueva Indians, who were found as far westward as Chame Point. There is historical documentation that other peoples, including Caribs, at various times inhabited Darién, but the whole area was under Cueva influence.

Little archaeological work has yet been done in this area, and thus far only one well-defined culture, the Darién, has been identified. The typical pottery of this culture is globular and often has tall, annular bases. The decoration consists of modeled and incised elements, filleting, and applications of red and black paint outlined by incised lines. All but this last characteristic are common in the pottery of lower Central America, which has stylistic connections with northern South America. Historical sources mention elaborate gold objects such as helmets and pectorals, but very little goldware has been found.

Ware recalling that of Santarém, in Brazil, has been discovered in caves in the Canal Zone. Finds at Venado Beach on the Pacific coast show links with areas both to the east and west. Eastern characteristics and types include pedestal bases on the pottery, effigy and globular vessels, and shell animal effigies. Certain pottery characteristics — such as the use of two delineating colors, black line painting on a red ground (as in the earliest Coclé polychrome styles) — and the hammered-gold, cast-gold, and *tumbaga* (gold and copper alloy) figures are more typical of the west. The working of gold is known to have been introduced into the isthmus from South America about 700, after which it continued to spread westward. As with all the cultures of Panama, it is difficult to specify the origins of the Darién culture, but it does seem to be derived primarily from lowland South America.

Western Panama. a. Coclé. The Coclé culture area begins southwest of the Canal Zone and extends through the Azuero Peninsula; it includes the Coclé, Herrera, and Los Santos provinces. Within this zone there are early culture sequences in the Gulf of Parita region: Monagrillo, Sarigua, and Santa Maria. No evidence of transitional phases has been found that would relate these cultures to the rich development associated with the Coclé culture, as known from the finds at Sitio Conte, dating from 1300 to 1500. A fourth sequence, known as El Hatillo, seems in part contemporaneous with the early Coclé. Associated with this sequence are mound and urn burials, shell animal effigies, carved bone objects, greenstone and hollow gold beads, gold-plated and sheet-gold figures, and ceramics. The El Hatillo pottery is characterized by modeled forms, usually of animals and flying birds, and frequently has pedestal bases. Bold outlines enclose designs painted in fine lines, and there is some parallel-line painting, Slips may be rose, buff-tan, or tan-orange with red, black, and purple decoration.

Sitio Conte ceramics, which displayed some of these same elements, were widespread throughout the area and were important trade articles. They have been found to the west in the Diquís region (Costa Rica) and to the east in the Pearl Islands. An early and a late period have been distinguished in this ware.

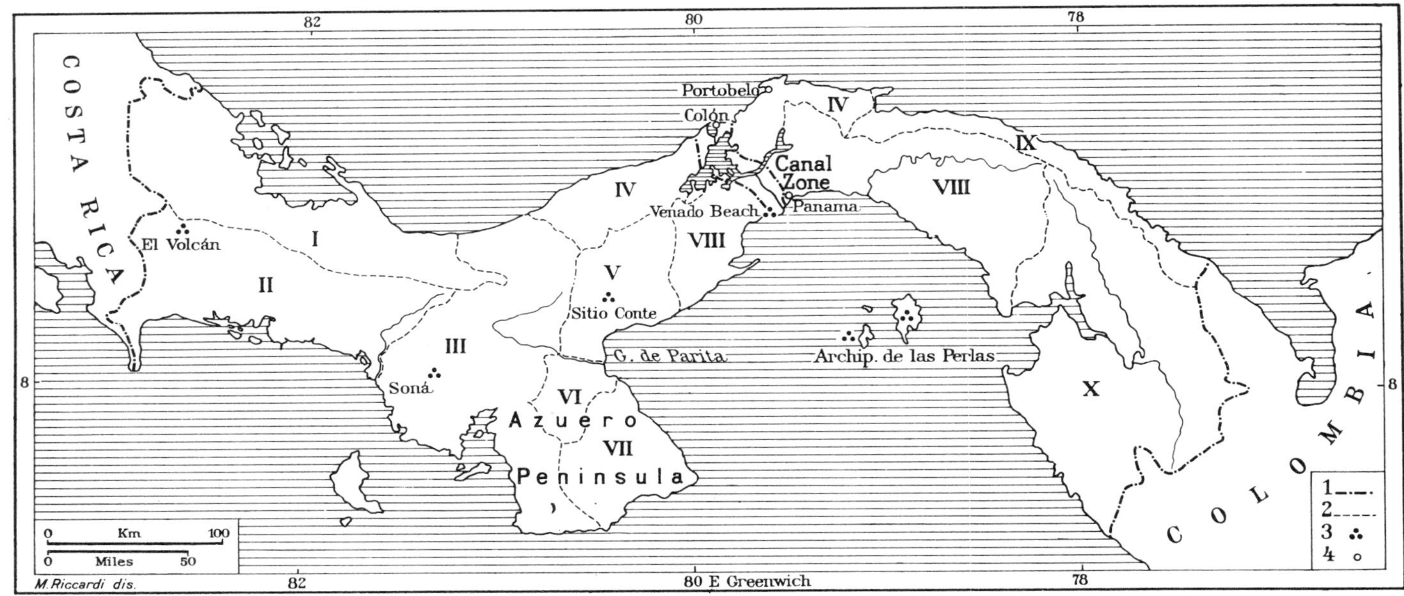

Republic of Panama. *Key:* (1) Modern national boundaries; (2) provincial boundaries; (3) archaeological sites; (4) modern cities. Provinces: (I) Bocas del Toro, (II) Chiriquí, (III) Veraguas, (IV) Colón, (V) Coclé, (VI) Herrera, (VII) Los Santos, (VIII) Panama, (IX) Comarca de San Blas, (X) Darién.

Forms unique to the early Coclé period are slightly curved plates and globular or angular carafes with tall flaring necks. Spouted effigy vessels are common in that phase. In the late period there are pedestal-supported platters; another, more strictly local shape of the late period is a small deep bowl with flaring sides. In both periods, there are almost no naturalistic elements, filleting, incising, or modeling, except on effigy vessels. Typical decorative elements of the early period are two delineating colors, scroll patterns, combinations of animals such as crocodile-headed birds, bird and turtle motifs, and parallel panels. Scroll designs continue through the late period, but fish, claws, and dancing animals are the principal motifs. Distinctive features of Coclé art are eyes depicted with the outer end curved downward to a point, curvilinear patterns, forceful lines, and areas of flat color outlined in black. All but the last characteristic bear resemblance to South American ceramics. Polychrome ware generally has a white slip decorated in black, brown, light and dark red, green, and purple. The completed vessels were covered with a protective resin varnish.

Megalithic columns topped with male or animal figures crudely cut in the round are associated with the early Coclé, but they also continued into the late period. Sharply incised features recall, in their technique and conception, the peg-supported statues of southeastern Costa Rica and southern Honduras.

Jewelry of gold, copper, whale-tooth ivory, agate, quartz, serpentine, bone, and shell is typical of the Coclé culture, with some minor stylistic changes according to period. Metal ornaments include greaves, disks, pendants resembling curly-tailed animals, and gold hats and headbands. Techniques used in metalworking were hollow- and clay-core castings, hammering, soldering, welding, and *mise en couleur* gilding; all of these were South American methods which had been adopted as far north as Costa Rica. Unique Coclé products and techniques were resin-filled gold, the combination of ivory and gold, and animal sculptures with cast-gold heads and emerald, quartz, or agate bodies. It is evident that the Coclé area became a center for the working of precious stones and metals: raw materials such as emeralds were carried from Colombia to Coclé, worked there, and then sent to Oaxaca, Mexico, and gold objects are known to have been exported to Yucatán.

b. Veraguas. The Veraguas culture area lies between the Coclé and Chiriquí regions. Stylistically however, it is linked with northeastern South America and the Costa Rican Atlantic watershed.

Finely executed grinding stones with projecting panels were undoubtedly inspired by religious beliefs and motivations similar to those of Costa Rica. Four-legged grinding stones (made in the likeness of jaguars, with heads and tails protruding beyond the rims) are identical in conception with those from Chiriquí and Costa Rica but are larger and thinner in dimensions. Animals and crocodiles in bas-relief, found on legs of grinding stones resembling those from Barriles in the Chiriquí area, are another link with regions to the west. Trade connections are indicated by the presence of Olmec jade, which has also been found in Central America at Nicoya and Línea Vieja, in Costa Rica.

The Veraguas ceramics are characterized by plain ware with filleting, incising, and modeling; effigy vessels with appliqué features; and globular bowls on tall pedestals. Gutter-rimmed effigy bowls appear in the Veraguas area, as they do elsewhere, in Peru and at Santarém, Brazil. Looped tripod supports recall those of Línea Vieja. Peculiar to the Veraguas culture are enormous flaring strap handles, globular vessels with flat bases, small horizontal barrels with a long protruding tubular neck on one side, and large open-mouth jars without slip but painted black and red. The jars have been tentatively dated 230 B.C. ($\pm$ 60 years).

In Veraguas metalwork, open-back castings in *tumbaga*, usually with *mise en couleur* gilding, are prevalent. Motifs used on pendants include birds, animals, fish, frogs, and human figures. Bells forming protruding eyes were unique to this culture.

c. Chiriquí. The Chiriquí culture area commences at the Tabasará River, the western boundary of Veraguas, and continues into the Diquís region. Certain of its elements are to be found in the Talamanca and Línea Vieja areas of Costa Rica.

Chiriquí stone human figures are of two types. One is a female figure cut in the round and characterized by free limbs and sharply defined features, including eyes with raised rims, a straight nose, and hair portrayed in a blocklike form. These figures are often depicted wearing a patterned loincloth and a headband. Hands are placed on the abdomen above the loincloth. The slightly flexed knees and the completely free legs suggest the figures from the Atlantic watershed of Costa Rica; the squared shoulders, the position of the hands, and the angular spaces freeing the upper arms from the torso recall the slablike peg-supported figures of the Diquís region. The second type is usually peg-supported, and is carved in high relief from a stone block; generally the limbs are not free, and one hand is sometimes placed on the breast. The figures are similar to the squat rounded ones of the Pacific coastal region to the northwest.

Appearing in the Chiriquí area, but characteristic as well of the stone sculpture of Costa Rica, are circular offering tables with carved animal or human heads protruding from the rim and with openwork pedestal bases or Atlantean supports carved so as to create a sense of movement; generalized representations of animals in the outline style of the Diquís region; and jaguar-effigy grinding stones of the type found also in the Veraguas area.

Chiriquí pottery includes many common Central American forms, such as the small, thick-walled globular vessels and the conical, pointed, bulbous, and mammiform tripods. Elongated tripods are similar to certain ones from South America and other parts of Central America. Negative painting in geometric patterns suggestive of textiles and occasional monkey, alligator, and octopus motifs are characteristic of this ware, with a wide distribution extending to both watersheds of Costa Rica. White-slipped alligator ware, with red and black stylized alligator designs, and armadillo ware are equally characteristic of southeastern Costa Rica. Armadillo vessels, often highly polished, with their graceful irregular and globular shapes, minute modeled figures on the shoulders or on the tripod legs, and an economy of detail, are of exceptional quality.

Cast-gold figures are numerous and are framed by decorative elements similar to those of southeastern Costa Rica. There is evidence of extensive trade in Veraguas metal objects, and Nicoyan jadeite pendants and Coclé and northern Colombian agate beads and animal effigies are all found in the Chiriquí region.

Near the volcano Chiriquí (El Volcán) is Barriles, a ceremonial center named for a distinctive art form, the barrel-shaped stones with monkeys in bas-relief on the upper end. These also occur in the Diquís region, where, as at Barriles, they are associated with pictographs on a boulder. Unique to Barriles is a series of realistic basalt statues on pedestal bases of a life-size male nude, who carries on his shoulders a person wearing a conical hat and, often, representations of gold ornaments and holding a human head, vessel, or ceremonial object. These paired figures and large statues of single figures are cut in the round and have few details.

The grinding stones of the Chiriquí region are supported by columns adorned with human figures in relief or by Atlantean supports. The conventionalized human heads decorating the raised rims of the grinding stones differ from those of Costa Rica and Veraguas only in their gigantic size.

Pottery has been found in caches, not in tombs. Unpainted and monochrome ware, tripod vessels with appliqué figures similar to those of Caribbean Costa Rica, and negatively painted inside surfaces of bowls are common. Types peculiar to Barriles are red and yellow vessels, large globular urns with flaring rims and deep necks boldly incised with conventionalized birds and animals painted red and orange. Except for a comparative lack of detail, they recall the Marajó ware of Brazil.

Bibliog. *a. General works*: G. Fernández de Oviedo y Valdés, Historia general y natural de las Indias, islas y tierra firme de la mar océano, 4 vols., Madrid, 1851–55; M. Solà, Historia del arte hispano-americano, Barcelona, 1935; S. K. Lothrop, Cuatro antiguas culturas de Panamá, Actas y trabajos cientificos, XXVII Cong. int. de Americanistas, I, Lima, 1941, pp. 205–09; S. K. Lothrop, The Archaeology of Panama, HSAI, IV, 1948, pp. 143–67; M. W. Stirling, Exploring the Past in Panama, Nat. Geog. Mag., XCV, 1949, pp. 373–99; A. Rubio, Panamá: Monumentos históricos y arqueologicos, Mexico City, 1950; M. W. Stirling, Exploring Ancient Panama by Helicopter, Nat. Geog. Mag., XCVII, 1950, pp. 227–46; M. W. Stirling, Hunting Prehistory in Panama Jungles, Nat. Geog. Mag., CIV, 1953, pp. 271–90; H.-R. Hitchcock, Latin American Architecture, New York, 1955; Pan American Union, Visit Panama, Washington, 1958; S. K. Lothrop, The Archaeological Picture in Southern Central America, Actas XXXIII Cong. int. de Americanistas, I, San José, Costa Rica, 1959, pp. 165–72. *b. Eastern Panama and the Canal Zone*: S. Linné, Darien in the Past: The Archaeology of Eastern Panama and Northwestern Colombia, Göteborg, 1929; T. H. Bull, Excavations at Venado Beach, Canal Zone, Archaeol. Soc. of Panama, I, 1, 1949–58, pp. 6–14; D. Marshall, Archaeology of Far Fan Beach, Panama Canal Zone, AmAnt, XV, 1949, pp. 124–32; S. K. Lothrop, Suicide, Sacrifice and Mutilation in Burials at Venado Beach, Panama, AmAnt, XIX, 1954, pp. 226–34. *c. Western Panama*: W. H. Holmes, Ancient Art of the Province of Chiriqui, Colombia, Ann. Rep. Bureau of Am. Ethn., VI, 1884–85, pp. 3–187; W. H. Holmes, The Use of Gold and Other Metals among Ancient Inhabitants of Chiriqui, Isthmus of Darien (B. Bureau of Am. Ethn., III), Washington, 1887; G. G. MacCurdy, A Study of Chiriquian Antiquities, New Haven, 1911; A. H. Verrill, Excavations in Coclé Province, Panama, Indian Notes, Heye Foundation, IV, 1, 1927, pp. 47–60; C. Osgood, The Archaeological Problem in Chiriqui, AmAnt, N.S., XXXVII, 1935, pp. 234–43; S. K. Lothrop et al., Coclé: An Archaeological Study of Central Panama, 2 vols., Cambridge, Mass., 1937–42; J. A. Mason, New Excavations at the Sitio Conte, Coclé, Panama, Proc. 8th Am. Sc. Cong. (1940), II, Washington, 1942, pp. 103–07; S. H. Wassén, Some Archaeological Observations from Boquete, Chiriqui, Panama, Etnografiska Museet, Etnologiska S., XVI, 1949, pp. 141–201; S. K. Lothrop et al., Archaeology of Southern Veraguas, Panama, Cambridge, Mass., 1950; E. R. V., Reliquias históricas y joyas indigenas en el Museo Nacional, Siete, I, 12, 1953, pp. 10–12; G. R. Willey, C. R. MacGimsey III, and R. E. Greengo, The Monagrillo Culture of Panama, Cambridge, Mass., 1954; G. R. Willey and T. L. Stoddard, Cultural Stratigraphy in Panama: A Preliminary Report on the Girón Site, AmAnt, XIX, 1954, pp. 332–43; C. R. MacGimsey III, Cerro Mangote: A Preceramic Site in Panama, AmAnt, XXXII, 1956, pp. 151–61; R. H. Mitchell, Recent Discoveries in Northern Panama, Archaeology, XIV, 1961, pp. 198–204.

Doris Stone

Colonial and modern periods. Among the numerous monuments surviving from the colonial period, the military constructions along the coast and in the interior of Panama, such as the massive forts near Colón and Portobelo, are of particular interest. There are also a large number of stone churches in the colonial baroque style with ornamental façades flanked by bell towers. In addition to the large churches of Panama City, Colón, and Portobelo, those in Taboga, Natá de los Caballeros, and David are among the oldest in the New World. While many cities in the interior still preserve their original colonial character, the rapid development of the coastal cities in this century has resulted in the construction of modern buildings of international style alongside older structures.

Modern centers. Panama City (Sp., Panamá). Capital of the republic since 1903, Panama City consists of two nuclei, the earlier of which, Old Panama, is situated on the Peninsula of Santa Ana. In the old city, which was sacked and burned by the buccaneer Henry Morgan in 1671 and which remains in ruins, there survive the following noteworthy evidences of the colonial past: the bell tower of the Cathedral (built of wood in 1535 and rebuilt ca. 1650); the ruins of the monasteries of S. Domingo, La Merced, and S. José; the Jesuit chapter house (with Spanish Renaissance architectural elements); the Chapel of S. Ana; the King's Bridge (Puente del Rey); the paved Calle la Empedrada, one of the principal thoroughfares of the city; and remains of numerous private houses with patios in the Spanish style. Rebuilt in 1673 on a small promontory about 5 miles outside the old city, the new settlement of Panama was provided with an efficient system of defenses and with many new public and private edifices built entirely of stone. After a long period of decline, the declaration of independence in 1903 and the opening of the Panama Canal in 1914 effected a swift economic revival, with a resultant increase in building activity.

Monuments of the colonial period include the remains of the fortifications (Las Bóyedas); baroque churches such as the Cathedral, built with material salvaged from the preceding cathedral of Old Panama, with lateral bell towers; La Merced (damaged in a fire of 1963), built from material recovered from the preceding church of the same name; S. Francisco; S. José, within which is a golden altar, a masterpiece of Spanish colonial art; S. Ana; the ruins of S. Domingo, with its famous "flat arch"; and the Jesuit church, also in ruins. Later secular architecture worthy of note includes the presidential palace, with a Moorish patio and small columns decorated with mother-of-pearl, the Ministry of Foreign Affairs, the Municipal Building, the National Theater, and the Palace of Justice. The monuments to Balboa and to the French who attempted to build a canal across the isthmus are also noteworthy. The modern Panama City is laid out on a regular plan and, together with its suburbs, contains interesting examples of international-style architecture, such as the Channel Two (Canaldos Dos) Building by Ernesto de la Guardia III, the Social Security Hospital, University City, La Salle College, and Hotel El Panamá, designed by the American architect Edward Stone. Many new residential quarters, such as San Miguelito, have been laid out in the environs of the city, in accordance with modern architectural styles and techniques and large scale urban planning.

Bibliog. C. P. Kimball, Old Churches of Panama, Arch. Record, XXXIX, 1916, pp. 536–50; J. B. Sosa, Panamá la vieja, Panama City, 1919; J. de la Cruz Herrera, Panamá la vieja, Panama City, 1926; A. Rubio, La ciudad de Panamá, Colon, 1950; M. L. Vidal, La catedral de Panamá, Buenos Aires, 1955; D. Angulo Iñiguez, Historia del arte hispanoamericano, III, Barcelona, 1956, pp. 223–28.

Colón. Situated on the Atlantic coast at the entrance to the Panama Canal, Colón is an important commercial center and a predominantly modern city. The Cathedral is in a colonial baroque style. In the vicinity are Fort S. Lorenzo, built to protect the formerly important town of Chagres and its harbor, and Fort Davis, the largest in the region.

Portobelo. A port city on the Atlantic coast founded in 1597, it enjoyed remarkable prosperity in the colonial era. Impressive reminders of the period are provided by the forts of S. Jeronimo, Santiago de la Gloria, and S. Fernando and by the 17th-century customs house. In the wake of the economic revival of the mid-20th century, like other older cities of Panama, Portobelo has been provided with modern public and private buildings and with planned residential quarters.

BIBLIOG. D. Angulo Iñiguez, Planos de monumentos arquitectónicos de América y Filipinas existentes en el Archivo de Indias, I, Seville, 1933, pp. 18–20; M. J. Buschiazzo, Panamá la vieja, Lasso, V, 2, 1937, pp. 68–73; F. Monasterio, Portobelo: Investigaciones históricos, Panama City, 1940.

Francesco NEGRI ARNOLDI

Illustration: 1 fig. in text.

PAOLO UCCELLO. Italian painter (b. Florence, 1397; d. Florence, Dec. 10, 1475), son of Dono di Paolo and Antonia di Giovanni Castello del Beccuto. In 1407 and 1412, under the name Paolo di Dono, he appeared in a list of young artists helping Lorenzo Ghiberti (q.v.) finish his first doors (on the north side) for the Baptistery in Florence. He joined the Arte dei Medici e Speziali in 1415 and in 1424 became a member of the Compagnia di S. Luca. The next information concerning him is dated Aug. 5, 1425, when he drew his will and left for Venice to work on the mosaics of S. Marco. In 1427 he was not in Florence, but somebody (his mother, perhaps) answered the land registry officials for him, saying, "He went with God over two years ago and is now in Venice." He probably went back to Florence in 1430, for his presence is proved by the registry entries of 1431 and 1433. Possibly on Paolo's own request, and after having made inquiries about his workmanship in Venice, the Opera del Duomo commissioned him to create the frescoed funeral monument of the British captain Sir John Hawkwood, known as Giovanni Acuto. When the work was done, the Opera ordered him to modify parts of it, and he immediately made the alterations (Aug. 31, 1436). There is proof that between 1443 and 1445 he received money for more work in the Cathedral, on the "sphere of the hours" (the clock face) and on cartoons for stained-glass windows. In 1445 he went to Padua, summoned (Vasari says) by Donatello (q.v.), who that year started to work on the *Gattamelata*, his bronze equestrian monument to Erasmo da Narni (IV, PL. 249). It is believed that during his stay in Padua Paolo executed a series of paintings (now lost) of "illustrious men," or "giants," in the Casa Vitaliani. The registry entries of 1442 (or perhaps 1446) and 1457 document his presence in Florence in those years. In 1452 he married Tommasa di Benedetto Malifici, who in the following year bore him a son, Donato, and in 1456 a daughter, Antonia. Documentary evidence shows that he was in Urbino with his fifteen-year-old son between 1467 and 1469. A predella by Paolo, *The Legend of the Profanation of the Host*, is still to be found in that town. His name appears for the last time in the Florentine registry in the entry of Aug. 8, 1469, six years before his death. He was buried in Sto Spirito, in Florence. Critics who have tried to establish the chronological order of Paolo's paintings have had only this meager documentary information to work with.

Paolo belonged to the first generation of the great innovators, including Filippo Brunelleschi, Donatello, and Lorenzo Ghiberti (qq.v.), who were all his seniors; Masolino, Masaccio, Fra Angelico, and Domenico Veneziano (qq.v.), who were more or less his age; and Filippo Lippi (q.v.), his junior; yet even among these artists his individuality stands out. His most marked characteristic has always been thought to consist in his passion for the study of the laws and definitions of the science of perspective.

Despite the high quality of his work (in the assessment of which it must be taken into account that about sixteen paintings or cycles have been lost), there exist in it a variety and an apparent discontinuity that present difficulties for critical interpretation, even in the paintings almost unanimously attributed to him. The list of his works remains highly controversial to the present, and while many scholars more or less accept Vasari's enumeration, others tend to expand it considerably.

For lack of pertinent works and documents, nothing is certain about Paolo's early training, though he must have given proof of great talent at an early age. In any case, he must have received his training during the second decade of the 15th century, when the predominant forms were those of Lorenzo Monaco (q.v.), the visionary Sienese who revived the purest 14th-century style. At that same time, however, many young artists, including Paolo, were having their initial experience under the guidance of Lorenzo Ghiberti, who was then working on his first doors for the Baptistery. In 1425 Paolo executed a frescoed Annunciation for the Carnesecchi Chapel in S. Maria Maggiore. Vasari speaks of it with admiration as being the first noteworthy attempt to render space.

Of the works that have survived, it is possible to attribute to Paolo tentatively, as previous to his stay in Venice in 1425 (Horne, 1905), the first large lunette with scenes from Genesis for the Chiostro Verde of S. Maria Novella in Florence (now removed to the adjoining refectory), that representing the creation of the animals and of Adam (PL. 38), the creation of Eve, and the Fall. This terre-verte fresco, in spite of the basic Gothic flavor, clearly reveals the influence of Lorenzo Ghiberti, especially in the sinopias, while its painting technique is reminiscent of Masolino, an artist who had been Paolo's fellow student in Ghiberti's studio. During his stay in Venice Paolo designed a mosaic of St. Peter (1425) for the façade of S. Marco (Salmi, 1950); it has been suggested that he may also have contributed there to the execution of the architectural elements of *The Visitation* (Longhi, 1926) and *The Birth of the Virgin* and *The Presentation of the Virgin in the Temple* (Pudelko, 1934), in the Mascoli Chapel of S. Marco, both signed by Michele Giambono, as well as to the decoration of the floor of the church (Muraro, 1955). It is likely that he was asked to do this work not as an expert in the technique of making mosaics, for this art was never extinct in Venice, but rather as an inventor of decorative plans, perhaps on the recommendation of Ghiberti during his own stay in Venice in 1424–25.

While Paolo was away from Florence, Masaccio rapidly carried expression of the new ideals of painting to an amazingly full development. In this achievement he was aided by Brunelleschi's clarification of the perspective rendering of space and by Donatello's strong affirmation of the naturalistic ideal in such sculptures as the *St. George* (ca. 1416; IV, PL. 241) and the prophets (ca. 1415–20; IV, PL. 243). To these contemporaneous influences there must have been added the remote but exciting example of ancient art, which Brunelleschi, Donatello, Ghiberti, and (Vasari says) even Masaccio himself had studied in Rome. The study of antique art caused the revival of the orders and of "musical proportions" in architecture (see PROPORTION); while the study of the marbles, cameos, medals, and other remains of antiquity brought about the rediscovery of that sense of nature which scholastic teaching proposed as the artist's ideal. In view of the solemn acceptance of the achievement of Masaccio — confirmed by the veneration that for centuries painters have felt for his unfinished frescoes in the Brancacci Chapel (Florence, S. Maria del Carmine) — the artistic ideal pursued by Paolo Uccello was bound to remain isolated. This isolation also arose because Paolo was already fully formed when his contemporaries were engaged in their first and fiercest experimentations, and because, far more than by the antique, he was fascinated by Gothic subjects, a taste which must have grown stronger during his northern travels. These elements also explain why his works include neo-14th-century motifs (it is better to consider them this than Gothic), which make his work seem self-contradictory in the sense that, while he pursued the formal representation of the concept of perspective even to excess, he maintained a residual taste for archaic patterns of composition. Of color he undoubtedly made a very original, imaginative use; typical are his alternately matching or clashing reds, pinks, and greens. Yet, in contrast to the free atmospheric flow of color in the paintings of Masaccio, Paolo's color is constantly contained within the limits of outlines.

The definition of his neo-14th-century style has always posed a problem, especially since this definition modifies the traditional interpretation given to his work and the opinion that he was a naturalistic painter (Berenson, 1896) and an innovator, the artist who introduced elements and principles of Renaissance art into Venetian painting (Fiocco, 1927). Since he was a deeply inventive spirit, he may well have contributed to the development of the tendency toward greater accuracy in construction in the works of the artists of the Veneto region, where in about 1425–30 the paintings of Jacobello del Fiore, Niccolò di Pietro, and Michele Giambono were still imbued with Gothic style. His work is much more closely related to theirs, however, than to the courtly and sumptuous pictures of Gentile da Fabriano and Pisanello (qq.v.). Also, it is significant that on Paolo's return from the north in the fourth decade of the century he imported once more into Tuscany the elements of International Gothic style, which Gentile da Fabriano had previously introduced in 1423–25 and which Domenico Veneziano was later to reaffirm in Tuscany, though combined with a sense of form influenced by Masaccio's work. Paolo Uccello lived with the new approach, while his spirit was still allied with the past, perhaps partly because of his devotion to Lorenzo Ghiberti, his master.

The fresco of Sir John Hawkwood, in the Cathedral of Florence (XII, PL. 27), is the first instance of a frescoed monument of the equestrian type. It bears in its signature the appellation given Paolo because of his love of animals in general and birds in particular: "Pauli Uccelli Opus" (Vasari). The suggestion for this composition came from the freestanding statues of Alexandrian horses above the portal of S. Marco in Venice. Paolo obtained an effect of abstract solemnity through his use of strictly geometrical drawing and his balance of volumes, through the severity of the color scheme (terre-verte on a porphyry-red background), and by his use of several different perspectives, which create the impression that the viewer is looking up at the monument while the base on which it rests is actually at eye level.

The same solemn and abstract effect may be seen in the four heads of prophets, or rather Evangelists, at the corners of the "sphere of the hours" (1443) and in the cartoons for *The Resurrection* and *The Nativity* (1443), which Bernardo di Francesco and Angelo di Lippo executed summarily in stained glass for two oculi of the drum in the dome of the Cathedral [another, of the Annunciation (1444), has been lost]. These, too, are sharply foreshortened designs in which Paolo took into account the angle from which they would be seen. In the 1440s his interests probably centered on the study of volume and space, subjects which were then major concerns of the Florentine artists but which Paolo sometimes solved with disconcerting preciousness and fastidiousness. The fragmentary *Scenes from the Lives of Monastic Saints*, in the cloister of S. Miniato al Monte, which were rediscovered following a suggestion by Marangoni (1930), were probably created in a mood of experimentation. Unlike his other works, which are crowded with details, these scenes concentrate on very few elements — one or two figures and a few vividly colored touches of landscape, which Vasari considered irrelevant to the subject represented. The intent to create spatial depth is here combined with formal relationships conceived without depth. One distinguishing trait of Paolo's draftsmanship was his purely intuitive and sometimes desultory recourse to the conception of spatial perspective which Filippo Brunelleschi had scientifically established in two experimental paintings (see BRUNELLESCHI), which Masaccio and Donatello had reelaborated, and which Leon Battista Alberti (q.v.) had codified in *Della pittura*. The present writer's interpretation (1957) of Paolo's work, starting from this observation, reveals a series of parallels between Paolo's paintings and a different set of laws — not the laws of perspective for artists' use (*perspectiva artificialis*) but those of visual perspective (*perspectiva naturalis*, or *communis*).

Rejecting Brunelleschi's "mechanical" solution, Paolo applied himself to investigating and drawing figural deductions from the many theorems accumulated in treatises from antiquity through the 14th century [Euclid, Ptolemy, al-Kindi, Alhazen, Roger Bacon, Erasmus Vitello (Erazm Ciołek), Biagio Pelacani] and adopted in his own work that "eye-reason" (*ragione dell'occhio*) which Ghiberti had applied in his door for the Baptistery called the Porta del Paradiso and had expounded in his third *Commentario*, citing those treatises. Thus, for Paolo the problems of rendering visual reality offered an ever-fresh source of suggestion and an invitation to an infinite variety of solutions which, despite the seemingly scientific character of his work, drew their real vitality from the profound imaginative resources of the artist. Perhaps this difference from his contemporaries accounts for the fact that Paolo's name was not included among those artists credited by L. B. Alberti (introduction to *Della pittura*, 1436) or by Domenico Veneziano (letter to Piero di Cosimo de' Medici, Apr. 1, 1438) for notable advances in Florentine art, as well as for the time lag ascribed to Paolo's development by Longhi (1939–40, 1952).

Paolo's imagination was stimulated by the very difficulty of the visual problems, which he strove to resolve in experimental perspective drawings (known from a few extant examples). According to Vasari, Donatello reproved Paolo for spending his time on such "useless" exercises, and Vasari himself considered them a deviation from the true tasks of a painter. Among these drawings are a study of a *mazzocchio*, a many-faceted type of headgear (Uffizi, Gabinetto dei Disegni e Stampe, No. 1756 A), and a chalice (No. 1758 A; IV, PL. 267), which has now been reascribed to Piero della Francesca (q.v.) on the evidence of a passage in Vasari (*Lives*, II, p. 491). The Uffizi also has a study for the monument of Sir John Hawkwood (No. 31 F); a wonderful mounted knight, which is perhaps a study for a St. George (No. 14502 F); a fragmentary angel (No. 1302 F); and a profile of a man (No. 28 E), which has also been attributed to Masaccio. According to Vasari, Paolo proved himself very skillful in the drawing of figures, especially in a cycle of frescoes, now lost, in the cloister of the Camaldolite Convento degli Angeli.

The panel representing St. George freeing the princess from the dragon (Paris, Mus. Jacquemart-André) was perhaps executed at about the time that Paolo was painting his experimental *Scenes from the Lives of Monastic Saints*, which seem so out of step with the general concepts of the early 15th century. In this panel the lines of the upward-sloping plain in the background are drawn at different angles and meet the horizon at two different levels. Furthermore, the construction revealed by the sinopia for *The Nativity and Annunciation to the Shepherds* (detached fresco; Florence, S. Martino alla Scala; PL. 93) appears to be completely anomalous. Here the preparatory perspective plan with distance points produces a composition that strangely diverges in two different directions. One tentative explanation of this oddity is that it was an attempt at solving the problem of binocular vision (Parronchi, 1957), a subject frequently discussed in Paolo's day. Two intersecting vanishing points and other inconsistencies of perspective are discernible in the fourth lunette painted for the Chiostro Verde in S. Maria Novella, with scenes from the life of Noah. This was probably painted in 1447–48, though many critics believe it to be a much later work (Meller, 1960). *The Flood* (PLS. 37, 41) is Paolo's masterpiece, the work that places him among the greatest creative spirits of all times. The difference in style between this fresco and the first lunette (PL. 38) is obvious. In the fourth lunette, having stripped himself of all previous habits, Paolo shows a deep and completely independent consciousness of figural values. He envisions the various themes — the Flood, the receding of the waters, Noah's sacrifice and drunkenness — completely afresh, ignoring all iconographic traditions, and treats them as events of universal life, indelible pages in the history of mankind.

The perspective, however, instead of resulting from a unified scheme of lines intersecting plane and profile, presents a fragmentary demonstration *ad assurdum* of various "laws of vision," constituting a series of optical illusions (*deceptiones visus*, as they are called in optical treatises; cf. Erasmus Vitello, *De perspective*, IV, p. 28). Thus, in the space between the two arks, the Biblical catastrophe offers a spectacle of fantastic disorder in nature with an immense variety of effects. In *The Flood* appears a *mazzocchio*, of the type used later as a common exercise in the practice of perspective. In the lower left-hand corner the sacrifice takes place in a serene atmosphere: here, in the perfect semicircle of the rainbow, is the reversed, foreshortened figure which, on the basis of Biagio Pelacani's explanation (in *Quaestiones perspectivae*) of the phenomenon of mirage, was once considered to be the Lord but now is interpreted as the reflection of Noah sacrificing in the "concave mirror of the clouds" (Parronchi, 1957). Beside the sacrifice, in *The Drunkenness of Noah*, seen as if from a much closer range, the figures of Noah's three sons are grouped around their father (whose foreshortened figure is now lost) — a composition that is Paolo's most classical.

Apart from the paintings mentioned in the sources, there is a distinct group of works attributed to Paolo by only a few critics (Longhi was the first, 1939–40). The group comprises works that have been variously ascribed to the Master of the Quarata Predella (Salmi, 1934–35), or the Karlsruhe Master (Pudelko, 1935), or the Prato Master (Pope-Hennessy, 1950) and includes paintings for which a plausible dating has yet to be found but which are stylistically indistinguishable from those allotted with certainty to Paolo. A saint with two children (Florence, Coll. Contini Bonacossi) is perhaps the earliest of them. Because of its close resemblance to the style of Masaccio, it should be dated no later than the 1430s. The *Madonna and Child* in Dublin (Nat. Gall. of Ireland) bears a similarity to this picture and is not far removed from the early style of Filippo Lippi, though it is also related to the most important cycle in the group — the frescoes comprising *The Birth of the Virgin, The Presentation of the Virgin in the Temple*, and *The Dispute of St. Stephen*, in the Chapel of the Assumption in the Prato Cathedral. This first part of the decoration of the chapel (which was certainly completed by Andrea di Giusto, who died in 1455), the "strangeness" of which was at one time taken for weakness, shows the peculiar characteristics of Paolo's art both in the acuteness of the physiognomic study and in the imagination displayed in the architectural parts, which seem almost to caricature Brunelleschi's works and recall the architectural elements of the above-mentioned mosaics in the Mascoli Chapel in S. Marco, Venice.

Other works of this group are related to the Prato frescoes, which, though not all on the same level, still show a fecundity and an unrelenting dedication characteristic of Paolo. One of these is the predella in S. Bartolomeo a Quarata, with *The Adoration of the Magi* (PL. 44), *St. John on Patmos*, and two kneeling saints (Marangoni, 1932). In the scene of St. John the stormy sky is rendered with a freedom of brushwork that anticipates the technique of Leonardo da Vinci (q.v.). Another of the group is the *St. George and the Dragon* (London, Nat. Gall.; PL. 40), in which the distortions and the weird enchantment of the color appear to be the effect of twilight changes in the atmosphere. Completing this group of attributions are the *Adoration of the Child* (with SS. Jerome, Mary Magdalene, and Eustace) in the Staatliche Kunsthalle, Karlsruhe, with its almost heraldic delicacy of line; the *Crucifixion* in Lugano (Thyssen-Bornemisza Coll.); a Madonna (Berlin, Staatliche Mus.; in storage; Ragghianti, 1938); a canvas with a so-called "Thebaid" (*Scenes from Monastic Legends*; Florence, Accademia), intended to create a visual symbology based on *De oculo morali*, an edifying 13th-century text (Parronchi, 1957); a Madonna with St. Francis in adoration (formerly, New York, Kress. Coll.); *Christ Bearing the Cross*, in the Congregazione di Carità S. Filippo Neri, Parma; a predella from Avane bearing the relatively late date 1452; and, in a private collection in Florence, a Madonna that seems curiously archaistic (Berti, 1961). All these might well be described, in Vasari's terms, as "little things" by Paolo to be found "in many houses in Florence." Most likely belonging to the second half of his career as a painter, they attest to his highly varied style and inventiveness.

Among Paolo's major works are the three panels representing the Battle of San Romano, in which, in 1432, the Sienese suffered a defeat at the hands of the Florentines. Once in the "apartment of Lorenzo" in the Palazzo Medici-Riccardi and now divided among the National Gallery, London, the Uffizi, and the Louvre (PLS. 39, 42, 43), they were first thought to be an early work (Cavalcaselle, 1864), and this error created a misconception as to the historical development and meaning of Paolo's work. After Horne (1901) hypothetically placed them at a more likely date (ca. 1455–60), it was realized that they were not an anticipation but an elaboration or, better still, an almost enigmatic complication of the principles of perspective. The panels were hung high up on the walls, at the bases of three lunettes. Two panels (those in London and Florence) were on the back wall and one (Louvre) on a side wall (Baldini). Because the pictures were to be hung high on the walls, the background hills fill the height of the panels, cutting off the sky. The panels present a battle theme which achieves an effect of fabulous unreality. The preternatural stress of the warriors enclosed in their armor, which was covered with silver leaf and must have shone like mirrors (a device by which the artist obtained, particularly in the Louvre panel, bewildering effects of reflection and multiplication of images); the dark groves with roses and orange trees; the tangle of spears; the crisscrossing lines that create a labyrinth in which the eye of the viewer is enmeshed; the color, warm yet nocturnal, occasionally illuminating an all but hidden face — all these elements combine to create a work which, despite its strict stylistic discipline, enters the realm of sheer fantasy.

Between 1467 and 1469 Paolo painted the Urbino predella, *The Legend of the Profanation of the Host* (PL. 44), above which Justus of Ghent (q.v.) is supposed to have completed his panel *The Communion of the Apostles* in 1473–74. The predella consists of a series of different scenes — interiors designed with varying perspective and outdoor scenes imbued with the mysteriousness of the countryside at night. An apparently later work is the *Night Hunt* (PL. 44), set in the shadows of a pinewoods, where the hunters, the racing horses, the hounds, and the stags seem reduced to the proportions of elegant animated puppets, and the dominant suggestion is that of the distant sounding of horns and the belling of hounds under the moonlit branches.

The problem of the attribution of portraits is one that involves the accurate definition of the style not only of Paolo Uccello but of all the painters of the 15th century. More or less generally accorded to Paolo are the following: a much-faded panel in the Louvre with five portraits identified by Vasari as Giotto (but more probably Masaccio), Brunelleschi, Donatello, Paolo Uccello, and Antonio (or Giovanni) Manetti (according to Lányi, the panel is a late-15th-century elaboration of elements from Masaccio's lost *Sagra*); a portrait of a woman (New York, Met. Mus., Bache Coll.); another, quite similar one, in the Isabella Stewart Gardner Museum, Boston; and the portrait of a young man which bears the motto "elfin fatutto," in the Musée Benoit-Molin, Chambéry, Savoie.

In his biographies of both Paolo Uccello and Raphael Vasari says that Paolo's work declined markedly in his old age, that is, presumably the 1450s, when, in a period alien to his ideals, Paolo was left without followers, except in so far as his poetically decorative quality was echoed by some of the more imaginative *cassoni* painters. However, the spirit of his work was to survive and to find followers at the end of the century.

In order to realize the possibilities for future development in Paolo's style, critics have had to rid themselves of various preconceptions. Vasari criticized Paolo's perspective investigation as too intellectualistic, preferring that of Masaccio, who, in his opinion, realized the tremendous possibilities of such studies for the rendering of atmosphere and light. This fundamental judgment has variously influenced critical opinion to this day, despite the recognition of Paolo's notable ingenuity, though from the 15th to the 18th century the so-called "primitives" were little understood.

Baldinucci emended Vasari's biography and contributed a learned passage on the gait of the horse in the monument to Sir John Hawkwood. Cavalcaselle (1892) wrote an admirable chapter on Paolo. Berenson (1896) focused his study on the interpretation of Paolo's work, but he accepted as naturalism the characteristics which, however dependent on a careful study of objective reality, are better considered as an effort toward pure form. In 1898 Loeser called the attention of scholars to Paolo again, and Horne, in 1901, was the first to stress that the three panels representing the Battle of San Romano must have been late works. Roger Fry, in various studies, contributed some useful observations. Longhi (1914) recognized that Paolo's style anticipated that of Piero della Francesca and, after a temporary attribution of the frescoes of Prato Cathedral to Giovanni di Francesco (1928), finally ascribed them to Paolo (1939–40). Longhi later wrote that Paolo was primarily characterized by the complexity of his genius and pointed out that he lagged behind his contemporaries in their new approach to the representation of space. Lionello Venturi (1930) raised the problem of Paolo's portraits, which he interpreted as fundamentally ingenuous. Marangoni (1930, 1932) attributed to him the frescoes of S. Miniato al Monte and the predella in S. Bartolomeo a Quarata. Schlosser (1933) gave a low rating to Paolo's conceptions while avant-garde critics (Soupault, 1929) were extolling him as the forerunner of modern aspirations. The fascination that Paolo's work has exercised on modern writers — such as Schwob (1896), Giovanni Pascoli (*Poemi Italici: Paolo Uccello*, Bologna, 1911) — is remarkable. Even more notable is his influence on modern artists (e.g., Carrà, 1916) — cubists, surrealists, and abstract painters — who have, because of the irrational and vehement aspect of his style, made him a symbol of intellectual freedom.

Paatz (1934) attributed to him the *Nativity* in S. Martino alla Scala, giving an eminently dramatic interpretation of his work. Pudelko's essays (1934, 1935), because of the associations and comparisons they point out, are of major importance. Ragghianti (1938) studied the importance of the lost frescoes of the Casa Vitaliani indirectly, through other works of art in Padua. Salmi (1938) made the first complete study of Paolo's works. Boeck (1939) divided the works into rigid stylistic periods, reconstructing Paolo's *œuvre* on the basis of an erroneous chronology. Boeck's analysis, however, has some very profound observations. Pope-Hennessy's book (1950), in general agreement with Salmi, is rich in detailed information. Pittaluga (1946), Sandberg-Vavalà (1948), and Carli (1954) also have produced studies of more than popularizing value. A later study by Salmi (1950) contributed well-grounded conjectures concerning Paolo's lost works, and Sindona (1957) considered his production in connection with the philosophy and political history of his time. This approach, supported by much more historical and documentary information, was chosen by Meller (1960) for his interpretation of the Noah frescoes. Gioseffi (1958) rejected the present author's major conclusions (1957) and laid his stress on the inconsistencies of Paolo's draftsmanship, giving an interpretation that is based on technical considerations. However, a careful evaluation of Paolo's work on the basis of the theoretical preparation that he must have received through the rules of perspective would appear to be a logical way of accounting for his marked intellectualism. This viewpoint brings into focus his medieval love of the exception before the rule — although this may also have been partly a consequence of his polemic stand against the unifying rationalism of Brunelleschi. Also, this type of analysis may explain the presence in him of that spirit of investigation (typical of the early 15th century, when the natural philosophy of Aristotle prevailed) which, by way of stylistic elaboration, was to survive throughout the second half of the century and finally break through, humanized and universalized, in the work of Leonardo da Vinci. More recent studies have emphasized Paolo's divergence from the mainstream of 15th-century painting initiated by Masaccio. Here precisely lies the essence of Paolo's art, which consists, not in a final and unchanging technical achievement, as does that of Piero della Francesca (which was itself a development of Paolo's), but in diverse and problematic aspects. The fact that Paolo's production was exceptional shows not only that he was out of touch with his time but also that he applied himself unrelentingly and selflessly to the study of whatever field — animal and human figures, optics, costume — was required in his work. Even in late Gothic painting there was a similar variety of interests, but with Paolo this variety acquired an imaginative unity appropriate to his own times. Far from striking an arrogant attitude, however, he submitted to all the limitations that his profession imposed upon him with that keen sense of responsibility which sets the function of the artisan before the personality of the artist.

SOURCES. A. Averlino (Filarete), Trattato di architettura (ca. 1451–64, ed. W. von Oettingen), Vienna, 1896; G. Rucellai, Il Zibaldone quaresimale (1457–81, ed. A. Perosa), London, 1960, p. 24; C. Landino, La Divina commedia di Dante Alighieri, col commento . . . , Florence, 1481; G. Santi, Cronaca (1490–95, ed. H. Holtzinger), Stuttgart, 1897; A. Manetti, Le vite de XIV uomini singhulary in Firenze dal 1400 innanzi (1494–97, ed. P. Murray), BM, XCIX, 1957, pp. 330–66; F. Albertini, Memoriale di molte statue et picture . . . di Florentia, Florence, 1510 (ed. H. P. Horne, London, 1909); Il libro di Antonio Billi (1516–30, ed. K. Frey), Berlin, 1892; Anonimo Magliabechiano (1520–40, ed. K. Frey), Berlin, 1892; M. Michiel (Anonimo Morelliano), Notizia d'opere di disegno (1521–43, ed. G. Frizzoni), Bologna, 1888; G. Vasari, Le vite, Florence, 1550; R. Borghini, Il Riposo, Florence, 1584, pp. 309–11; F. Bocchi, Le bellezze della città Fiorenza, Florence, 1591 (rev. ed., G. Cinelli, Florence, 1677); Nota di pitture, sculture et fabbriche notabili della città di Firenze (1600–06, ed. P. Galletti), Riv. fiorentina, I, 1, 1908, pp. 27–38, 81–97; F. Baldinucci, Notizie de' professori del disegno, II, Florence, 1686; G. Richa, Notizie istoriche delle chiese fiorentine, 10 vols., Florence, 1754–62; T. Patch (A. Cocchi) and F. Gregori, La porta principale del Battistero di S. Giovanni, Florence, 1773, in E. Müntz, Les archives des arts, I, Paris, 1890, pp. 15–22; J. W. Gaye, Carteggio inedito d'artisti, I, Florence, 1839, pp. 146–47.

BIBLIOG. C. Pini, La scrittura di artisti italiani (ed. G. Milanesi), I, Florence, 1876, no. 47; G. Frizzoni, Die Ausstellung von Handzeichnungen alter Meister im Palazzo Poldi-Pezzoli zu Mailand, Kchr, XVI, 1881, cols. 625–30; H. Ludwig, Leonardo, Das Buch von der Malerei, III, Vienna, 1882, pp. 176–77; A. Schmarsow, Das Abendmahl in St. Onofrio zu Florenz, JhbPreussKSamml, V, 1884, pp. 207–31; E. Portheim, Mantegna als Kupferstecher, JhbPreussKSamml, VII, 1886, pp. 214–26; H. Weizsäcker, Das Pferd in der Kunst des Quattrocento, JhbPreussKSamml, VII, 1886, pp. 45–167; A. de Witt, La pietà di Paolo Uccello, Florence, 1886; J. A. Crowe and G. B. Cavalcaselle, Storia della pittura in Italia, V, Florence, 1892, pp. 19–65; B. Berenson, Florentine Painters of the Renaissance, London, New York, 1896; M. Schwob, Vies imaginaires, Paris, 1896; E. Calzini, Urbino e i suoi monumenti, Rocca S. Casciano, 1897; C. Loeser, Paolo Uccello, RepfKw, XXI, 1898, pp. 83–94; H. P. Horne, The Battle-place by Uccello in the National Gallery, Monthly Rev., V, 1, 1901, pp. 114–38; G. Gronau, Paolo Uccello, RepfKw, 1902, p. 318; B. Berenson, The Drawings of the Florentine Painters, 2 vols., London, New York, 1903 (2d ed., 3 vols., Chicago, 1938); J. de Jongh, Un nouveau portrait de Dante, GBA, XXX, 1903, pp. 313–17; H. P. Horne, Andrea del Castagno, BM, VII, 1905, pp. 222–31; P. Schubring, Matteo de' Pasti, Kunstwissenschaftliche Beiträge August Schmarsow gewidmet, Leipzig, 1907, pp. 103–04; G. Gamba, Di alcuni quadri di Paolo Uccello e della sua scuola, RArte, VI, 1909, pp. 19–30; G. Poggi, Il Duomo di Firenze, Berlin, 1909, pp. 142–47; L. Testi, Storia della pittura veneziana, 2 vols., Bergamo, 1909–15; E. G. Campani, Uccello's Story of Noah in the Chiostro Verde, BM, XVII, 1910, pp. 203–10; R. Fry, On a Profile Portrait by Baldovinetti, BM, XVIII, 1910, pp. 311–12; J. A. Crowe and G. B. Cavalcaselle, A New History of Painting in Italy (ed. R. L. Douglas and G. de Nicola), IV, London, 1911, pp. 106–25; W. Bombe, Die Kunst am Hofe Federigos von Urbino, Mnh. für Kw., V, 1912, pp. 456–74; E. Calzini, Di un'alcova dei Montefeltro nel Museo di Urbino, Rass. bibliog. dell'arte it., XV, 1912, pp. 157–58; W. Kurth, Die Darstellung des Nackten in dem Quattrocento von Florenz, Berlin, 1912; A. Schmarsow, Domenico Veneziano, L'Arte, XV, 1912, pp. 9–20, 81–97; G. J. Kern, Das Dreifaltigkeitsfresco in S. Maria Novella, JhbPreussKSamml, XXXIV, 1913, pp. 36–58; C. Seidel, Paolo Uccello, Arte e storia, XXXII, 1913, pp. 97–102; R. Fry, Three Pictures in the Jacquemart-André Collection, BM XXV, 1914, pp. 79–85; R. Longhi, Piero dei Franceschi e lo sviluppo della pittura veneziana, L'Arte, XVII, 1914, pp. 198–256; L. Venturi, Nella Galleria nazionale delle Marche: L'Alcova di Federico da Montefeltro, BArte, VIII, 1914, pp. 326–27; L. Venturi, Studi nel palazzo ducale di Urbino, L'Arte, XVII, 1914, pp. 415–73 at 456; B. Haendcke, Der französisch-niederländische Einfluss auf die italienische Kunst, RepfKw, XXXVIII, 1915, pp. 28–91 at 75; G. J. Kern, Der Mazocchio des Paolo Uccello, JhbPreussKSamml, XXXVI, 1915, pp. 513–38; P. Schubring, Cassoni (sup.), Leipzig, 1915; C. Carrà, Paolo Uccello costruttore, La voce, VIII, 1916, pp. 375–84; P. Kristeller, Truhen und Truhenbilder der italienischen Frührenaissance, Kunst und Kunsthandwerke, XIX, 1916, pp. 145–64; G. Wolff, Mathematik und Malerei, Leipzig, 1916; C. Carrà, Pittura metafisica, Florence, 1919 (2d ed., Milan, 1946); M. Marangoni, Osservazioni sull'Acuto di Paolo Uccello, L'Arte, XXII, 1919, pp. 37–42; R. van Marle, De ontwikkeling der italiaansche schilderscholen (VI), Oude K., V, 1919–20, pp. 228–34; C. Phillips, Florentine Painting before 1500, BM, XXXIV, 1919, pp. 209–19; R. Fry, The Art of Florence, Vision and Design, London, 1920, pp. 117–22; P. Toesca, Vetrate dipinte fiorentine, BArte, XIV, 1920, pp. 1–6; O. H. Giglioli, I disegni della Galleria degli Uffizi, Rass. d'arte, VIII, 1921, pp. 331–42; G. Fiocco, Un affresco di Paolo Uccello nel Veneto, BArte, N.S., III, 1923–24, pp. 93–96; F. Antal, Gedanken zur Entwicklung der Trecento- und Quattrocento-malerei in Siena und Florenz, Jhb. für Kw., 1924–25, pp. 207–39; E. Moltesen, Paolo Uccello, Copenhagen, 1924; E. Panofsky, Die Perspektive als symbolische Form, Vorträge der Bibl. Warburg, 1924–25, pp. 258–330; F. Antal, Studien zur Gotik im Quattrocento, JhbPreussKSamml, XLVI, 1925, pp. 8–14; G. Fiocco, Il rinnovamento toscano dell'arte del musaico a Venezia, Dedalo, VI, 1925–26, pp. 110–18; H. S. Ede, Florentine Drawings of the Quattrocento, London, 1926; R. Longhi, Lettere pittoriche a G. Fiocco, Vita artistica, I, 1926–27, pp. 127–39; E. Benkard, Das Selbstbildnis, Berlin, 1927; G. Fiocco, L'arte di Andrea Mantegna, Bologna, 1927; R. Longhi, Piero della Francesca, Rome, 1927 (2d ed., Milan, 1946); R. Longhi, Un ritratto di Paolo Uccello, Vita Artistica, II, 1927, pp. 45–49; J. Mesnil, Masaccio et les débuts de la Renaissance, The Hague, 1927; P. Soupault, Paolo Uccello, L'amour de l'art, VIII, 1927, pp. 177–80; R. Longhi, Ricerche su Giovanni di Francesco, Pinacotheca, I, 1928, pp. 34–48; Van Marle, X, 1928, pp. 203–50; R. van Marle, Eine Kreuzigung von Paolo Uccello, Pantheon, I, 1928, p. 242; R. A. Parker, Paolo Uccello: A Precursor of Modern Art, Int. Studio, LXXXIX, 3, 1928, pp. 21–27; M. Salmi, Gli affreschi nella Collegiata di Castiglione Olona, Dedalo, IX, 1928–29, pp. 17–29; H. Beenken, Zum Werke des Masaccio, ZfbK, LXIII, 1929, pp. 112–19; G. Fiocco, Dello Delli scultore, RArte, XI, 1929, pp. 25–42; P. Soupault, Paolo Uccello, L'art et les artistes, N.S., XVIII, 1929, pp. 289–94; M. Marangoni, Gli affreschi di Paolo Uccello a S. Miniato al Monte, RArte, XII, 1930, pp. 403–17; A. Schmarsow, Zur Masolino-Masaccio Forschung, Kchr. und Kunstliteratur, ZfbK, LXIV, 1930–31, pp. 1–3; L. Venturi, Paolo Uccello, L'Arte, N.S., I, 1930, pp. 52–87; W. Boeck, Die "Erfinder" der Perspektive, RepfKw, LII, 1931, pp. 145–47; W. Boeck, Ein Frühwerk von Paolo Uccello, Pantheon, VIII, 1931, pp. 276–81; W. Bombe, Zur Kommunion der Aposteln von Josse van Gent in Urbino, ZfbK, LXV, 1931–32, pp. 68–72; B. Berenson, Italian Pictures of the Renaissance, Oxford, 1932; G. Gronau, Zu Paolo Uccello's Schlachtenbildern, Pantheon, IX, 1932, p. 176; M. Jirmounsky, Une nouvelle predelle de Paolo Uccello, GBA, VIII, 1932, p. 64; M. Marangoni, Una predella di Paolo Uccello, Dedalo, XII, 1932, pp. 329–47; R. van Marle, Eine unbekannte Madonna von Paolo Uccello, Pantheon, IX, 1932, pp. 76–80; A. Stix and L. Fröhlich-Bum, ed., Albertina Katalog, Vienna, 1932; W. Boeck, Drawings by Paolo Uccello, Old Master Drawings, VIII, 1933, pp. 1–3; W. Boeck, Uccello-Studien, ZfKg, II, 1933, pp. 149–75; R. Oertel, Die Frühwerke des Masaccio, Marburger Jhb. für Kw., VII, 1933, pp. 191–289; G. Poggi, Paolo Uccello e l'orologio di S. Maria del Fiore, Misc. in onore di I. B. Supino, Florence, 1933, pp. 323–36; J. Schlosser, Kunstler Problem der Frührenaissance, Vienna, Leipzig, 1933; E. Somarè, Repertorio critico dell'opera di Paolo Uccello, L'esame, V, 1933, pp. 10–34; V. Giovannoni, Note su Giovanni di Francesco, RArte, XVI, 1934, pp. 337–65; W. Paatz, Una Natività di Paolo Uccello e alcune considerazioni sull'arte del Maestro, RArte, XVI, 1934, pp. 111–48; G. Pudelko, The Early Works of Paolo Uccello, AB, XVI, 1934, pp. 231–59; M. Salmi, Paolo Uccello, Domenico Veneziano, Piero della Francesca e gli affreschi del Duomo di Prato, BArte, XXVIII, 1934–35, pp. 1–27; W. Boeck, Malerei, Italien: XV. Jahrhundert, ZfKg, IV, 1935, pp. 253–55; G. Fiocco, I giganti di Paolo Uccello, RArte, XVII, 1935, pp. 385–404; G. Pudelko, Florentiner Porträts der Frührenaissance, Pantheon, XV, 1935, pp. 92–98; G. Pudelko, Der Meister der Anbetung in Karlsruhe: ein Schuler Paolo Uccello's, Festschrift für A. Goldschmidt, Berlin, 1935, pp. 123–30; G. Pudelko, The Minor Master of the Chiostro Verde, AB, XVII, 1935, pp. 71–89; G. Pudelko, Paolo Uccello peintre lunaire, Minotaure, VII, 1935, pp. 33–41; J. Lipman, The Florentine Profile Portrait in the Quattrocento, AB, XVIII, 1936, pp. 54–102; G. Pudelko, An Unknown Holy Virgin Panel by Paolo Uccello, Art in Am., XXIV, 1936, pp. 127–34; C. L. Ragghianti, Casa Vitaliana, CrArte, II, 1936–37, pp. 236–50; M. Salmi, Paolo Uccello, Andrea del Castagno, Domenico Veneziano, Rome, 1936 (2d ed., Milan, 1938); R. Oertel, Wandmalerei und Zeichnung in Italien, Mitt. der kunsthist. Inst. in Florenz, V, 1937–40, pp. 217–314 at 297; R. W. Kennedy, Alesso Baldovinetti, New Haven, London, 1938; G. Pudelko, ThB, s.v.; C. L. Ragghianti, Intorno a Filippo Lippi, CrArte, III, 1938, pp. 22–25; M. Salmi, La Madonna dantesca del Museo di Livorno e il Maestro della Natività di Castello, Liburni Civitas, XI, 1938, pp. 217–56; W. Boeck, Paolo Uccello, Berlin, 1939; R. Longhi, Fatti di Masolino e di Masaccio, CrArte, IV–V, 2, 1939–40, pp. 145–91; M. S. Bunim, Space in Medieval Painting and the Forerunners of Perspective, New York, 1940; M. Wackernagel, Paolo Uccello, Pantheon, XXVII–XXVIII, 1941, pp. 102–10; J. Lányi, The Louvre Portrait of Five Florentines, BM, LXXXIV, 1944, pp. 87–95; V. Moschini, Gli affreschi del Mantegna agli Eremitani di Padova, Bergamo, 1944; C. L. Ragghianti, Argomenti lippeschi e uccelleschi, Miscellanea minore di critica d'arte, Bari, 1945, pp. 69–76; M. Pittaluga, Paolo Uccello, Rome, 1946; E. Somarè, Paolo Uccello, Rome, 1946; R. G. Mather, Documents Mostly New Relating to Florentine Painters and Sculptors of the 15th Century, AB, XXX, 1948, pp. 62–64; E. Sandberg-Vavalà, Uffizi Studies, Florence, 1948; B. Degenhart, Italienische Zeichnungen des frühen 15. Jahrhundert, Basel, 1949; J. Pope-Hennessy, ed., The Complete Work of Paolo Uccello, London, New York, 1950; M. Salmi, Riflessioni su Paolo Uccello, Comm, I, 1950, pp. 22–23; R. Longhi, Il Maestro di Pratovecchio, Paragone, III, 35, 1952, pp. 10–37; I. Toesca, Gli 'Uomini famosi' della Biblioteca Cockerel, Paragone, III, 25, 1952, pp. 16–20; U. Baldini, Restauri di dipinti fiorentini, BArte, XXXIV, 1954, pp. 226–40; F. Canuti, Paolo Uccello in Urbino, Urbania, 1954; E. Carli, Tutta la pittura di Paolo Uccello, Milan, 1954; M. Muraro, L'esperienza veneziana di Paolo Uccello, Atti XVIII Cong. int. di storia dell'arte, Venice, 1955, pp. 197–99; E. Micheletti, Paolo Uccello, Novara, 1956; C. Volpe, In margine a un Filippo Lippi, Paragone, VII, 85, 1956, pp. 38–45; A. M. Fortuna, Andrea del Castagno, Florence, 1957, pp. 40–44; R. Pallucchini, L'arte a Venezia nel Quattrocento, in La civiltà veneziana del Quattrocento, Florence, 1957, pp. 147–77; A. Parronchi, Le fonti di Paolo Uccello, Paragone, VIII, 89, 1957, pp. 3–32, VIII, 95, 1957, pp. 3–33; A. Schmidt, Paolo Uccello: Entwurf für das Reiterbild des Hawkwood, Mitt. der kunsthist. Inst. in Florenz, VIII, 1957–59, pp. 125–30; E. Sindona, Paolo Uccello, Milan, 1957; J. White, The Birth and Rebirth of Pictorial Space, London, 1957; D. Gioseffi, Complementi di prospettiva, CrArte, N.S., V, 1958, pp. 102–49; M. Davies, Uccello's St. George in London, BM, CI, 1959, pp. 309–14; A. Parronchi, Le due tavole prospettiche del Brunelleschi (II), Paragone, X, 109, 1959, pp. 3–31 at 14; E. Sindona, Elementi critici e fantastici nell'arte di Paolo Uccello, L'Arte, N.S., XXVI, 1959, pp. 293–97; P. Meller, Ritratti e politica nelle storie di Noè di Paolo Uccello, communication in Urbino, Dec. 16, 1960; L. Berti, Una nuova Madonna e degli appunti su un grande maestro, Pantheon, XIX, 1961, pp. 298–309; M. Davies, The Earlier Italian Schools, 2d ed., London, 1961, pp. 525–33; G. Francastel, Découverte de Paolo Uccello, XXe siècle, N.S., XXIII, 1961, pp. 21–30; A. Parronchi, Le "misure dell'occhio" secondo il Ghiberti, Paragone, XII, 133, 1961, pp. 18–48; U. Procacci, Sinopie e affreschi, Milan, 1961; C. Shell, The Early Style of Fra Filippo Lippi and the Prato Master, AB, XLIII, 1961, pp. 197–209; A. Parronchi, Una Nunziata di Paolo Uccello: Ricostruzione della cappella Carnesecchi, S. urbinati, XXXVI, 1962, pp. 1–38; R. W. Sciller, Uomini famosi, B. van het Rijksmuseum, X, 1962, pp. 56–67.

Alessandro PARRONCHI

Illustrations: PLS. 37–44.

PARAGUAY. A landlocked country in the central part of South America, Paraguay has an area of approximately 157,000 sq. mi. The Paraguay River divides the country into two quite distinct geographical regions. The western sector, the Chaco region, is on the whole a vast plain that slopes toward the east, where there are found alternating dense forests and lowland plains. Eastern Paraguay is a high, uneven plateau with hills and mountain ridges, the outer slopes of the central Brazilian plateau.

Paraguay has been an independent republic since 1811. The colonization of the country, which was first explored in the 1520s by Spanish adventurers, was chiefly the work of the Society of Jesus, which founded missions there during the first part of the 17th century. (These missions were closed by order of King Charles III of Spain in 1767.)

The mission architecture of Paraguay is probably the most significant achievement of the country's art; the artistic activity of the indigenous peoples is, for the most part, limited to pottery production and weaving.

INDIGENOUS ARTS. When the first Spanish colonists arrived in Paraguay (ca. 1536), they found nomadic and semisedentary forest-dwelling peoples whose only possessions of any artistic interest were personal ornaments, clothing, and ceramics. Only the Guaraní and groups now identified with the Mbayá-Caduveo, as well as certain Arawak tribes such as the Guaná, had any ceramics of artistic merit; these ceramics are the only source for the study of prehistoric Paraguayan art, but their date of origin cannot always be determined. The survival of many of the above-mentioned groups, and the fact that they have continued to retain vestiges of their cultural and artistic heritage, make it difficult to distinguish the prehistoric examples from the historic. The pottery bearing cord impressions that has been recovered from archaeological sites along the Paraguay River is similar to the work of the modern Caduveo tribe, which has been studied by the Italian artist and ethnographer Guido Boggiani. (The Guaraní, however, have not maintained their old artistic traditions in painted ceramics. The artistic development of the Jesuit missions, also known as Guaraní missions, does not derive from indigenous sources, except for the local manpower utilized by the artists and artisans among the Jesuits.)

In their modern-day pottery making, the Caduveo press their designs into fresh clay with string, thereby creating geometric, rectilinear, and curved border designs. They color the surface of the vases

Paraguay, modern centers and distribution of principal ethnic groups (in slanted boldface capitals). *Key*: (1) Modern national boundaries.

red, black, or white. The designs that the Caduveo and related groups apply to their faces and bodies are of singular artistic interest. They consist of spontaneous motifs, usually asymmetrically arranged, and do not follow a set pattern, though to some degree they are based on ancestral esthetic norms. Levi-Strauss (1944) noticed a similarity between the mirror-image pairing or duplication of figural and other design elements in Caduveo art and certain forms used in ancient Chinese art and in the art of the Pacific coast of North America.

The indigenous population of Paraguay, especially those groups considered typical of the culture of the Chaco region, are expert weavers of the caraguata fiber, which is used in the making of sacks and shirtlike garments. These fabrics are neatly decorated over their entire surface with geometric motifs that suggest patterns used in basket weaving.

The Chamacoco and Zamuco produce intricate personal ornaments of featherwork.

Colonial and contemporary architecture. In Paraguayan architecture of the 17th and 18th centuries, the predominant style was Spanish colonial baroque, although there are some examples of Renaissance and Mudejar styles. The architects of this period, of whom the most noteworthy was Giambattista Primoli, made extensive use of indigenous craftsmen, whose influence has been claimed to be evident in details of the ornamentation. Of the numerous communal settlements (reductions) created by the Jesuits before the dissolution of their missions and their expulsion, a few churches have survived largely intact, including S. Roque in Yaguarón, the best preserved, and the churches of the towns of Piribebuy and Paraguarí; there are important remains of other churches such as those of Humaitá, Santa Rosa, and San Cosme. The large church of the town of Jesús, in Spanish baroque style with Mudejar elements, remained unfinished after the expulsion of the Jesuits. There are important ruins, as well, of the Church of SS. Trinidad in Trinidad (ca. 1740; architect, G. Primoli), which was built entirely of stone without mortar and which undoubtedly is the most noteworthy religious construction in Paraguay. Among the significant monuments built in the 19th century, after the proclamation of independence, are various government buildings and the Cathedral in Asunción. Since World War II, contemporary international techniques, materials, and styles have been adopted for new building complexes in the capital and other major cities. Extensive city planning has been carried out, with a view toward decentralization of the older urban centers and creation of new industrial and residential quarters.

Bibliog. E. Gancedo, Un enterratorio payaguá del siglo XVI, La Prensa, Aug. 10, 1930; M. Schmidt, Nuevos hallazgos, prehistóricos del Paraguay, Rev. Soc. científica del Paraguay, III, 3, 1932, pp. 81–102, III, 5, 1936, pp. 133–36; M. Solà, Historia del arte hispanoamericano, Barcelona, 1935; G. Furlong Cardiff, La arquitectura de las misiones guaraníticas, Estudios Acad. literaria de La Plata, 1937, pp. 86–100; M. Schmidt, Los Guisnais, Rev. Soc. científica del Paraguay, IV, 2, 1937, pp. 1–35; M. Schmidt, Los Tapietés, Rev. Soc. científica del Paraguay, IV, 2, 1937, pp. 36–67; C. Levi-Strauss, Le dédoublement de la représentation dans les arts de l'Asie et de l'Amérique, Renaissance, II–III, 1944, pp. 168–86; G. Boggiani, Os caduveos, São Paulo, 1945; J. Giuria, La arquitectura en el Paraguay, Buenos Aires, 1950; P. Hanson, Guides to the Latin American Republics, New York, 1950; H. Busaniche, La arquitectura en las misiones jesuiticas guaranies, Santa Fe, Argentina, 1955; Pan American Union, Visit Paraguay, Washington, 1955; D. Angulo Iñiguez, Historia del arte hispanoamericano, III, Barcelona, 1956, pp. 665–718; G. Pendle, Paraguay, 2d ed., London, New York, 1956; A. Serrano, Manual de la cerámica indígena, Córdoba, 1958; F. A. Plattner and E. Lunte, Kunst im "Jesuitenstaat von Paraguay," Das Münster, XII, 1959, pp. 407–14.

Principal art centers. Asunción. Founded in 1537 on the eastern bank of the Paraguay River by Juan de Salazar, Asunción became the capital of the province of Paraguay as early as 1618. After the expulsion of the Jesuits, there began a period of decline that lasted until the middle of the 19th century, when the dictator Dr. Francia ordered the city demolished and then rebuilt on a regular checkerboard plan. Important architectural monuments are the Palacio del Congresso; Palacio de Gobierno, built in the second half of the 19th century under the dictator Francisco Solano López; the Cathedral (completed 1850), with a magnificent high altar; the Museo Nacional de Bellas Artes; and the Church of La Encarnación. Several buildings are based on famous European models, such as the Oratorio de Nuestra Señora de la Asunción (or Panteón Nacional de los Héroes; 1937), derived from the Hôtel des Invalides in Paris, and the Teatro Municipal (later adapted for government offices), inspired by the Teatro alla Scala in Milan. The Banco del Paraguay is neoclassic in style. There are many parks and gardens, including the Parque Carlos Antonio López, the Parque Caballero, and the Jardín Botánico, where the López villa has been turned into the Museo de Historia Natural y Etnografia. Since World War II important public works, such as the new university center of Yta-Pyta-Punta, have been built in contemporary styles.

Bibliog. F. R. Moreno, La ciudad de la Asunción, Buenos Aires, 1926; R. de Lafuente Machain, La Asunción de antaño, Buenos Aires, 1942; Ecole secondaire modèle à Assunçao, Arch. d'aujourd'hui, XXVI, 62, 1955, pp. 72–73; Teacher Training School: Experimental College Paraguay-Brazil, Asunción, Arch. Rec., CXIX, 1956, pp. 247–50.

Encarnación. The Jesuit mission founded in 1614 under the name of Itapúa developed into the city of Encarnación during the 19th and 20th centuries. Rebuilt after it was destroyed by a tornado in 1926, it has become the most modern city in Paraguay. Construction along contemporary architectural lines is particularly marked in the industrial sectors, as exemplified by the hydroelectric plant (La Usina) and the office building of the Central Eléctrico.

Yaguarón. Outstanding among the monuments of Yaguarón, which was founded in 1539, is the Church of S. Roque (1670), the largest and oldest of the region. Built of stone and brick in the Guaraní mission style and perfectly preserved, the church has sculptured portals and, in the interior, baroque altars, paintings, and a fine retable, the work of indigenous artists.

Bibliog. P. Alborno, A Colonial Church in Paraguay, B. Pan American Union, LXVI, 1932, pp. 700–09.

Francesco Negri Arnoldi

Illustration: 1 fig. in text.

PARMIGIANINO (Francesco Mazzola or Mazzuoli, known as Il Parmigianino). Emilian painter of the 16th century, the most important of a family of painters of Parma (b. Parma, Jan. 11, 1503; d. Casalmaggiore, Aug. 24, 1540). On the death of his painter father, Filippo, in 1505, Francesco became the ward of his uncles Pier Ilario and Michele Mazzola, also painters. However, it was not the example of his family's provincial and *retardataire* workshop but the art of Correggio (q.v.) in Parma that determined the formation of Francesco's early style. A considerable activity in Parma as a painter of both fresco and easel works was followed by a period in Rome, beginning in the latter half of 1524 and ending just after the sack of that city in May, 1527. In this period Parmigianino became prominent as one of the younger generation of the post-Raphaelesque school of Rome and was apparently in close contact there particularly with Perino del Vaga and Il Rosso. From mid-1527 to early 1531 he was settled in Bologna; after that he returned to his native city. In this second Parmesan period his principal commission was the decoration of S. Maria della Steccata; his failure to fulfill his contract resulted in his dismissal in December, 1539. It is fairly certain that in these years Parmigianino was much absorbed in alchemy, to the point of neglecting his commitments as a painter. That occult pseudo science seems to have afforded him the spiritual experience which the organized religion of the time did not offer a person of his highly special temperament. A few months before his dismissal from the Steccata he had taken refuge from the legal troubles arising from his breach of contract by retiring to nearby Casalmaggiore, which was ouside Parmesan territory, and it was there, after about a year's residence, that Parmigianino died in his thirty-seventh year.

The vocabulary of Parmigianino's earliest surviving picture, the *Baptism of Christ* (ca. 1519; Berlin, Staat. Mus.), is clearly bound to the earlier tradition of the pre-Correggesque Emilian school of the 16th century and is on this account an anomaly within his early *œuvre*. Nevertheless, this picture indicates his native disposition to attenuated, cursive forms and to refinement of execution and expression. These qualities are more explicit in the *Marriage of St. Catherine, with St. John the Evangelist and John the Baptist* (1521; Bardi, Church of S. Maria). In this altarpiece, still full of juvenile awkwardness, the fluent and gracile style of Parmigianino is already distinctly recognizable: his mode of painting had assumed the Correggesque cast of surface that, in varying degree, it was to retain for many years to come. From the inescapable example of Correggio in Parma in these years Parmigianino became involved not only with the painterly manner of that master but with the basic formal problems of Correggesque style — which he understood more profoundly than any other painter of the Parma school. His frescoes in S. Giovanni Evangelista (1522; first and second chapels on the left) develop, each successive one in a bolder way, the suggestion of the illusionistic saints done by Correggio in the pendentive frescoes in the same church. The Parmigianino frescoes depend largely on Correggesque precedent, but even as he masters Correggio's repertory of ideas and forms he further affirms his own peculiar personality, arriving at a very different quotient of calculated ornamental grace. The chief production of Parmigianino's early period in Parma is the decoration (1523), with the legend of Diana and Actaeon, of the *saletta* of the Sanvitale castle at Fontanellato, near Parma. Within the framework of an illusionistic arbor based on Correggio's precedent in the Camera di S. Paolo (III, PL. 465), Parmigianino develops devices of illusion as ingenious as Correggio's and combines them with patterns, woven on the figures, of swift-moving, fluid rhythm. An atmosphere of refined sensuality — lighter, more precious, and more intimate than Correggio's — emerges from the decoration of this room.

From the first clear manifestation of his personality in the Bardi altar, Parmigianino's affinity in style and personality with the other early mannerists had been apparent. A demonstration piece he took with him to Rome from Parma to present to Clement VII, the Prado *Holy Family* (1524; PL. 46), was certainly recognizable to his contemporaries — despite the foreign, Correggesque character of its surface — as a work entirely in the new postclassical style. Exposed in Rome not only to the masterworks of High Renaissance accomplishment in Raphael and Michelangelo (qq.v.) but also to the mannerist excursions of his contemporaries, similar in temper to his own, Parmigianino soon developed into a leading representative of the post-Raphaelesque Roman school. His large altarpiece *The Vision of St. Jerome* (1526–first half of 1527; PL. 45) epitomized his experience in Rome and the nature of the contribution he was prepared to make to Roman art. It derived ideas of form and movement, and its impressive Roman scale, from the art of Michelangelo; the character of content and, in part, the composition are closer to the precedents of Raphael. A Correggesque accent remains in the quality of painting, in a more apparent sensibility to effects of texture and of light than was common in the Roman school. This work transforms all its derivative elements into a wholly consistent and now maturely powerful mannerist style, both of detail and of structural principles. Grand in form and movement as they are, the figures are nonetheless revised from nature into appearances of refined rhythmic artifice; the composition (and the very shape of the altar) attenuates the balances of classical design into a more finely calculated equilibrium that is more inherently active and imminently unstable. Equally, the character of content of the actors seems finer, almost to the point of exquisiteness, and considerably less profound than in classical example. A self-conscious estheticism dictates the human meaning of the picture as well as the conception of form.

The *St. Jerome* altar marks not only the personal maturation of Parmigianino but also — together with the contemporaneous S. Felicita altar of Pontormo in Florence (IX, PL. 288) — the coming to maturity of the mannerist style (see MANNERISM). The decade following this artistic landmark is dedicated to further exploration and refinement of its stylistic ideas. Two remarkable works, executed early in Parmigianino's stay in Bologna — the *St. Roch* altar (latter half of 1527; PL. 46) and the *Conversion of St. Paul* (ca. 1528; Vienna, Kunsthist. Mus.; attribution confirmed by the discovery of a preparatory study by Parmigianino) — developed more daringly a concept of *mouvementé* design, in both these works asymmetrical and almost precarious in balance; in the *Madonna with St. Margaret and Other Saints* (1528–29; Bologna, Pin. Naz.) the composition is entirely an exercise of swift-patterned movement. Away from Rome, Parmigianino's technique began to reassert its Corregesque origins, but no longer in a derivative way, through his rendering — as subtle as Correggio's and often more brilliant — of effects of sheer optical sensation. In the *St. Margaret* altar these optically experienced effects are combined, as they are not in Correggio, with substantial revisions of form, of the most arbitrary kind, for purposes of rhythm.

Color assumes, in these pictures of the Bologna phase, a peculiar intensity and a richness of texture, strong and at times somewhat dissonant, tending toward yellows and greens. Refinement of surface, not only in elaboration of rhythmic pattern but in the tissue of colored light woven over it, reaches an unsurpassable climax in two of the latest of Parmigianino's Bolognese productions, the *Madonna of the Rose* (1528–first half of 1530; PL. 47) and the *Madonna and Child, the Infant St. John, the Magdalen, and St. Zacharias* (ca. 1530; PL. 49); in these pictures he fuses draftsmanly and painterly values, both of extraordinary fineness. The character of such paintings is the consequence of his searching after ever-heightened exquisiteness of esthetic sensation; and in the process of this search the physical appearance and the psychological constitution of the persons represented have, equally, become exquisite, of a subtle artificial loveliness. In form and content he creates a climate of the highest poetic hedonism, deliberately excluding the whole wider and very different range of human experience that his Florentine contemporaries Pontormo and Il Rosso (qq.v.) particularly explored. The culmination of Parmigianino's quest for refined and singular sensation is found in his so-called "Madonna dal Collo Lungo" (ca. 1535; XII, PL. 81). No 16th-century work of art goes farther in the arbitrary re-formation of humanity into an image of artificial grace, at once precious and grand. This picture is an ultimate demonstration of one of the major attitudes possible to mannerism: it is a revelation of an improbable and quasi-abstract neoclassic beauty.

The fresco decoration of the east vault of S. Maria della Steccata in Parma (design begun 1531 but executed mostly after 1535, only partly by Parmigianino himself) combines into an original scheme ideas and motifs long before observed in works by Raphael and Giulio Romano (q.v.) in Rome, and it achieves a typical late-classical effect of opulence and dignity, such as is seen in the loggias of the Vatican and the Villa Madama (PL. 431; VIII, PL. 206). The relation to classical style is more explicit than in any of Parmigianino's previous works: the Wise Virgins, the principal figures, have the gravity, and nearly the fullness of proportion, of antique classical images or of figures of the Raphaelesque High Renaissance. The mood of this monumental decoration as a whole, as well as of its large individual forms, is no longer that of the "Madonna dal Collo Lungo": a more profound seriousness of purpose, distinct from the artist's earlier febrile esthetic hedonism, had begun to manifest itself. This increased gravity of form and content is apparent in the tempera *Holy Family with the Infant St. John* in Naples (Mus. di Capodimonte; 1535–39), but it is most evident in the severe, even hieratic, structure and deeply serious atmosphere of the latest surviving painting by Parmigianino, the *Madonna and Child with St. Stephen, John the Baptist, and a Donor* (latter half of 1538–40; PL. 48). The temper of this painting corresponds to the change in the artist's personality that Vasari records (V, 233), and its austere religiosity seems to anticipate the impact of the imminent Counter Reformation on Italian cultural history.

Parmigianino had an extensive practice as a portraitist, and his works in this field, though not of such wide formative influence in Italy as his altar paintings, are of comparable merit. His youthful *Self-portrait in a Convex Mirror* (first half of 1524; X, PL. 388) is an exceptional optical and esthetic experiment, but all his subsequent portraits reveal the same concern with direct visual comprehension of the sitter and his ambience of light and atmosphere; in these he indulges in relatively little of the poetic license he allows himself in the rendering of reality in religious works. His mature portraits, such as the Uffizi *Self-portrait* (latter half of 1527–28), refine the appearance of features, hands, and bodily form and arrange them in a disciplined equilibrium of cursive movement. The effect of this pictorial pattern — one of animated, refined grace held in a precise restraint — corresponds with the interpretation of the sitter's personality: sharply alive, refined to the point of elegance, yet constrained in his communication with the spectator. In psychological tone the portraits of Parmigianino coincide remarkably with those of his mannerist contemporary in Florence, Bronzino (q.v.). Aside from the Uffizi *Self-portrait*, the most notable demonstration of Parmigianino's capacities in this area is the so-called "Antea" (1535–37; PL. 50); in this portrait of a Parmesan noblewoman there is, together with the almost abstract loveliness of the face, the monumental dignity of over-all design of the artist's later style.

Parmigianino was one of the most prolific draftsmen of his generation. It is clear that he drew with prodigious ease, in a mode suggesting the cursive swiftness of calligraphy, and with a brilliant differentiation of rhythmic impulses and accents. In his drawings he was less concerned than were the contemporary mannerists of central Italy with the data of appearance; his draftsmanship is at once more free and more artificial — and more directly a projection of the artist's patternmaking sense. The drawings that remain to us include not only a considerable body of the studies he made in preparation for known paintings but many diverse themes apparently sketched for sheer pleasure in the imaginative and manual exercise of drawing. There are also numerous designs for prints, in particular for chiaroscuro woodcuts, conceived during the Bologna period, and executed by artists more or less regularly attached to Parmigianino's workshop. The most important single group of chiaroscuro woodcuts of the early 16th century is based on his designs. Parmigianino himself — rarely, but with significant effect — also worked as a graphic artist (IV, PL. 428) and appears to have been the first Italian practitioner of etching. Fifteen plates are assigned to him by Bartsch, of which only a half dozen are indisputably his, the finest of them being the *Entombment* (B.5). It is even more to Parmigianino's drawings and prints than to his paintings that the vast influence of his style on 16th-century art, both within Italy and beyond its borders, is due.

MAJOR WORKS. *a*. Religious and allegorical paintings: *Baptism of Christ*, ca. 1519, Berlin, Staat. Mus. – *Marriage of St. Catherine*, 1521, Bardi, Church of S. Maria. – Fresco decoration, *SS. Lucy and Apollonia, St. Agatha, Two Deacon Saints, St. Isidore Martyr* (?), 1522, Parma, S. Giovanni Evangelista, 1st and 2d chapels on the left. – *St. Barbara*, 1522, Madrid, Prado. – *Circumcision* (copy of lost original), 1522, Detroit, Inst. of Arts. – *Marriage of St. Catherine* (copy of lost original), 1522, Parma, Gall. Naz. – Fresco decoration, legend of Diana and Actaeon, 1523, Fontanellato, Sanvitale castle. – *Holy Family*, 1523, formerly Richmond, Surrey, Cook Coll. – *St. Catherine* (?), 1523–24, Frankfurt am Main, Städelsches Kunstinst. – *Holy Family*, 1524, Madrid, Prado (PL. 46). – *St. Cecilia* and *David* (finished and restored by others), 1524, Parma, S. Maria della Steccata. – *Adoration of the Shepherds* and *Madonna and Child*, 1524–25, Rome, Gall. Doria Pamphili. – *Marriage of St. Catherine* (copy?), 1525–26, Somerley, Ringwood, Hampshire, Coll. Earl of Normanton. – *Vision of St. Jerome*, 1526–27, London, Nat. Gall. (PL. 45). – *St. Roch and a Donor*, 1527, Bologna, S. Petronio (PL. 46). – *Conversion of St. Paul*, 1527–28, Vienna, Kunsthist. Mus. – *Madonna with St. Margaret and Other Saints*, 1528–29, Bologna, Pin. Naz. – *Madonna of the Rose*, 1528–30, Dresden, Gemäldegal. (PL. 47). – *Madonna and Child, the Infant St. John, the Magdalen, and St. Zacharias*, ca. 1530, Florence, Uffizi (PL. 49). – *Amor*, 1531–34, Vienna, Kunsthist. Mus. – "Madonna dal Collo Lungo," ca. 1535, Florence, Uffizi (XII, PL. 81). – Fresco decoration, Wise Virgins, Moses, Aaron, Adam and Eve (executed with assistants), 1531–39, but mostly after 1535, Parma, S. Maria della Steccata, east vault. – *Holy Family with the Infant St. John*, 1535–39, Naples, Mus. di Capodimonte. – *Madonna and Child with St. Stephen, John the Baptist, and a Donor*, latter half of 1538–40, Dresden, Gemäldegal. (PL. 48).

b. Portraits: *Portrait of a Priest* (?), 1523, Wrotham Park, Hertfordshire, Coll. Earl of Strafford. – *Self-portrait in a Convex Mirror*, 1524, Vienna, Kunsthist. Mus. (X, PL. 388). – *Gian Galeazzo Sanvitale*, 1524, Naples, Mus. di Capodimonte (PL. 47). – *Lorenzo Cybo*, 1524–26, Copenhagen, Statens Mus. for Kunst. – *Self-portrait*, 1527–28, Florence, Uffizi. – *Allegorical Portrait of Charles V* (unfinished, the head of Charles perhaps by another hand), 1529–30, formerly Richmond, Surrey, Cook Coll. – *Young Man*, 1530–31, Hampton Court, Royal Colls. – *Young Woman in a Turban*, 1530–31, Parma, Gall. Naz. – *Countess Gozzadini* (fragment, unfinished), 1530–31, Vienna, Kunsthist. Mus. – *Pier Maria Rossi, Count of San Secondo*, 1533–35, Madrid, Prado. – *Countess San Secondo* (with assistants), 1533–35, Madrid, Prado. – "Antea," 1535–37, Naples, Mus. di Capodimonte (PL. 50). [Not included in this catalogue are two portraits formerly attributed to Parmigianino by the author (1950): *G. B. Castaldi* (?) and *Girolamo de Vicenti* (?), both Naples, Mus. di Capodimonte.]

Parmigianino is to be considered one of the prime innovators of the mannerist style — along with the Florentines Pontormo and Il Rosso, and with Giulio Romano — and one of the most accomplished of the first generation of mannerists. It is with his style, in fact, that the concept of mannerism has been longest identified. The basic principles of mannerist style (as modern criticism tends on the whole to conceive it) are perhaps more readily recognizable in Parmigianino's works than in those of his fellow innovators; yet it is his variant of mannerism, more than those of his contemporaries, that preserves something of the ideals, as well as the outward forms, of the antecedent classical tradition. No other painter among the early mannerists is so much concerned with the creation of a pictorial realm of suavely harmonious beauty (as in one aspect of Raphael), and none of his contemporaries intends a like fluid purity of form: there is in Parmigianino's art, even more than in that of Raphael's direct pupils, the atmosphere of a neoclassicism. In the observation (made even before Vasari's time) that "the spirit of . . . Raphael had passed into the body of Francesco" (Vasari, V, 223–24) there was more than casual truth. This relation to the style of the High Renaissance was, however, only in part affirmative; in each element in which Parmigianino depends on classical example there is found at the same time a creative deformation of its classical character and meaning. His famed grace of form and feeling is rarified and precious beyond any precedent in the style of the High Renaissance, and the beauty he conceives is far more exquisite and artificial — and often drastically arbitrary in its measure of abstraction from the look of nature. Also, his deliberate rejection, except in his latest works, of all values not having to do with refined and pleasurable sensation is a restriction on the range of art that no personality of comparable stature in the High Renaissance would have entertained.

It was in part this very limitation in Parmigianino's works that gave them their enormous popularity and consequent vast influence. What he had to say was at once irresistibly enchanting and, at least superficially, imitable. The superb quality both of his poetic imagination and of his hand was only rarely approached by imitators, but his style became the model for much of the art of his own generation and still more of the next, not only in the school of Emilia but beyond, in Italy and in France. The art of the *maniera* descends, posthumously, more nearly from Parmigianino than from any other single painter.

SOURCES. M. Biondo, Della nobilissima pittura, Venice, 1549 (repr. Quellenschriften für Kg., V, Vienna, 1873), XIX; G. Vasari, Le vite, Florence, 1550 (2d ed., 3 vols., Florence, 1568); P. Lamo, La graticola di Bologna, ca. 1560 (ed. G. Zanotti, Bologna, 1844); R. Borghini, Il Riposo, Florence, 1584; G. P. Lomazzo, Trattato dell'arte della pittura, Milan, 1584; G. B. Armenini, De' veri precetti della pittura, Ravenna, 1587; C. Ruta, Guida . . . della città di Parma, Parma, 1739; A. M. Zanetti, Raccolta di varie stampe, Venice, 1749; I. Affò, Vita del pittore Francesco Mazzola, Parma, 1784; C. M. Metz, Imitations of Drawings by Parmigianino, London, 1790. See also E. Faelli, Bibliografia Mazzoliana, Parma, 1884.

BIBLIOG. For full bibliog. prior to 1950, see A. O. Quintavalle, Il Parmigianino, Milan, 1948; S. J. Freedberg, Parmigianino: His Works in Painting, Cambridge, Mass., 1950. See also: L. Fröhlich-Bum, Parmigianino und der Manierismus, Vienna, 1921; G. Copertini, Il Parmigianino, 2 vols., Parma, 1932; L. Burroughs, A Drawing by Francesco Mazzola, il Parmigianino, BMMA, N.S., VII, 1948, pp. 101–07; G. Copertini, Nuovo contributo di studi e ricerche sul Parmigianino, Parma, 1949; G. Copertini, Postilla prima all'arte grafica del Parmigianino, Parma, I, 1952, pp. 118–22; Consiglio direttivo del Comitato parmense "per l'Arte," Nota sul Parmigianino acquafortista e xilografo, Parma, III, 1953, pp. 111–16; L. Fröhlich-Bum, Due dipinti attribuiti al Parmigianino, Parma, III, 1953, pp. 109–10; L. Fröhlich-Bum, Latest Literature on Parmigianino, GBA, XLII, 1953, pp. 327–30; A. E. Popham, Un disegno del Parmigianino, Parma, III, 1953, pp. 107–08; A. E. Popham, The Drawings of Parmigianino, London, 1953; F. Bologna, Una Pentecoste del Parmigianino, Paragone, VI, 63, 1955, pp. 31–33; F. Bologna, Il Carlo V del Parmigianino, Paragone, VII, 73, 1956, pp. 3–16; A. O. and A. Quintavalle, Una nuova opera giovanile del Parmigianino, Paragone, VII, 83, 1956, pp. 5–7; I. Toesca, Miscellanea di disegni a Venezia, Paragone, VII, 77, 1956, pp. 50–55; L. Fröhlich-Bum, Additions to a Corpus of Drawings of Parmigianino, GBA, LI, 1958, pp. 9–14; R. Longhi, Un nuovo Parmigianino, Paragone, IX, 99, 1958, pp. 33–38; P. Rotondi, Ipotesi sui rapporti Cambiaso-Tibaldi, BArte, XLIII, 1958, pp. 164–73; A. Quintavalle, Un quadro a tre mani, Paragone, X, 109, 1959, pp. 60–61; R. Wark, A Sheet of Studies by Parmigianino for the Madonna dal collo lungo, AQ, XXII, 1959, pp. 244–48; L. Fröhlich-Bum, Five Unpublished Drawings by Parmigianino, Pantheon, XVIII, 1960, pp. 236–41; L. Fröhlich-Bum, An Unknown Portrait by Parmigianino, Pantheon, XVIII, 1960, pp. 114–16; A. O. Quintavalle, Cinque inediti disegni del Parmigianino, Aurea Parma, XLIV, 1960, pp. 224–26; A. O. Quintavalle, Correggio e Parmigianino a Londra, Aurea Parma, XLIV, 1960, pp. 55–59; A. E. Popham, Dessins du Parmesan au Musée des beaux-arts de Budapest, B. Mus. nat. hongrois de beaux-arts, XIX, 1961, pp. 43–58; L. Fröhlich-Bum, Some Unpublished Drawings by Parmigianino, Apollo, N.S., LXXVI, 1962, pp. 693–96; I. Fenyö, Some Newly Discovered Drawings by Parmigianino, BM, CV, 1963, pp. 144–49.

Sydney J. FREEDBERG

Illustrations: PLS. 45–50.

PARRHASIOS (Παῤῥάσιος). Greek painter. According to most ancient authors (Harpocration, *Parrhasios*, p. 241; Pliny, *Naturalis historia*, XXXV, 67; Athenaeus, *Deipnosophistae*, XII, 543C; Strabo, XIV, 1, 25), Parrhasios was born in Ephesos, although a few say he was from Athens (Seneca Rhetor, *Controversiae*, X, 5; scholia of Acron and Porphyrion on Horace, *Carmina*, IV, 8, 6), either because he became an Athenian citizen or because he lived in Athens for many years. Pliny dates the peak of Parrhasios' career in the 95th Olympiad, about 400 B.C., and that of Euenor, his father and teacher, in the 90th Olympiad, about 420 B.C. (op. cit., XXXV, 60–61). Both these dates seem somewhat early and would more likely correspond to the artists' late phases. Pausanias (I, 28, 2) and Athenaeus (op. cit., XI, 782B) mention as early works of Parrhasios the drawings for the relief executed by Mys on the shield of Phidias's *Athena Promachos* and those for a skyphos executed by Mys (ca. 440 B.C.). At the end of Parrhasios' activity would belong the portrait of Philiskos (Pliny, op. cit., XXXV, 70), which, on the basis of its subject, could not be placed much before 380 B.C.; this date remains uncertain, however, because of doubts concerning the identity of Philiskos himself.

Evidence for his design on the shield of the *Athena Promachos* is scarce; some scholars see echoes of it in a relief in Athens (Nat. Mus., no. 3676), and in a cup in Boston (Mus. of Fine Arts) by the vase painter Aristophanes (Beazley, *ARV*, p. 842, no. 2; E. Langlotz, *Phidiasprobleme*, Frankfurt am Main, 1947, p. 75, no. 11). Of the Ilioupersis designs for the skyphos by Mys there remain no traces whatever.

A painting with the personification of the Athenian Demos (Pliny, op. cit., XXXV, 69) can perhaps be dated between 415 and 410 B.C., if it is indeed the one referred to by Xenophon in his account of the dialogue between Socrates and Parrhasios concerning the representation of character. The most disparate feelings were combined in the same figure; although it has been conjectured that these various emotional qualities were represented by a series of adjoining figures, it is more likely that they were conveyed by an asymmetrical treatment of the facial features, as was done in stage masks.

Ancient sources provide little other chronological data. Euripides's tragedy *Hippolytos* (1. 1004 ff.) mentions a painting of an erotic nature done by Parrhasios, thereby fixing a point of reference of 428 B.C. on the basis of what is known about the stage work. Other pictures by Parrhasios of this type, called *libidines*, are noted by Pliny (op. cit., XXXV, 72) and Seutonious (*Tiberius*, XLIV, 2), particularly one of Atalante and Meleagros that was acquired by the emperor Tiberius for a considerable sum. The mythical huntress Atalante, it should be remarked, was often the heroine chosen for lascivious pictures. This whole group of works may well belong to a distinct phase of the artist's career, to be placed about 430–425 B.C. Along with the Atalante painting, Tiberius bought (for 6,000,000 sesterces) another work, which portrayed an *archigallos*, or castrated high priest of Cybele. Quintilian is perhaps alluding to this painting when he criticizes the interest shown by artists in soft and effeminate bodies instead of sturdy, virile forms (*Institutio oratoria*, V, xii, 21). It would therefore be logical to assign several other paintings to this same phase of the *libidines*: for example, the *Megabyzos* mentioned by Johannes Tzetzes (*Chiliades*, VIII, 398), which is possibly to be identified with the above-mentioned *archigallos* portrait (*Plinio il Vecchio*, ed. S. Ferri, Rome, 1946, p. 155, note), as well as the *Phaidra in Love*, which is perhaps reflected in a Pompeian painting (Naples, Mus. Naz.) from the House of the Fatal Love (house IX.5.18; Picard, 1955).

If later anecdotes about the artist are to be trusted, a painting of Prometheus that existed in Athens (Seneca Rhetor, ibid.) should be dated about 420 B.C., on the evidence that Parrhasios is said to have used as a model a prisoner from Olynthos whom he was to torture. To be placed a few years later — that is, shortly after 415 B.C. — is his self-portrait in the guise of Hermes, which Parrhasios signed with a pseudonym (Themistios, *Orationes*, II, p. 29c). This use of a pseudonym seems to indicate that it occurred near the time when Alkibiades was accused of impiety for having himself portrayed among divinities, a period when the charge against Phidias, that he had put his portrait on the shield of the *Athena Parthenos*, was also still fresh in the people's memory.

Another fixed date for Parrhasios' early period is his activity in Rhodes, for the city of Rhodes proper, where Parrhasios worked, was founded by joint colonization in 408 B.C. The paintings mentioned as being on the island should then be assigned after this date, most likely created within a few years after the founding of the city. As for the *Herakles* he painted for the city of Lindos (Pliny, op. cit., XXXV, 71; Athenaeus, op. cit., XII, 543F), it was destroyed in the fire that devastated the acropolis there about the middle of the 4th century. In this work, an epigram tells us, Herakles was represented as Parrhasios saw him in his dreams; perhaps there is some connection with the rites of induced dreams at the sanctuary of Asklepios in Kos. Other paintings by the artist are known to have been in the temple in Lindos and were destroyed at the same time, though their subjects are not related (ibid., XV, 687B). A painting Parrhasios executed for the city of Rhodes, representing Meleagros, Herakles, and Perseus (three-figure compositions were extremely popular at the end of the 5th century), was hit by lightning three times without suffering damages and was still *in situ* at the time of Pliny (op. cit., XXXV, 69).

Another group of paintings is difficult to date precisely, although connections with other painters show that they must have been late works. This group includes a realistic picture on a curtain painted at Athens, in competition with Zeuxis (ibid., 64–65), a technical tour de force; a painting of Theseus (ibid., 69, 129; Plutarch, *Moralia*, 346A, *Life of Theseus*, IV), once in Athens (perhaps in the Stoa of Zeus Eleutherios, alongside works by Silanion and Euphranor representing a like subject) but later taken by Sulla to Rome, where it was destroyed in the fire on the Capitoline hill in A.D. 70; *Odysseus Pretending to Be Mad* (Plutarch, *Moralia*, 18A), painted for the city of Ephesos in competition with Euphranor, whose work was better recorded (Pliny, op. cit., XXXV, 129); *The Struggle for the Arms of Achilles*, painted for Samos in competition with Timanthes, who won the prize, to Parrhasios' great annoyance (ibid., 72; Aelian, *Variae historiae*, IX, 11; Athenaeus, op. cit., XII, 543E; Eustathios, *Ad Homeri Odysseam*, 1698, 61). The first of these works, the realistic curtain, cannot be dated later than 397 B.C., the year of Zeuxis' death. The works painted in competition with Euphranor and Timanthes cannot be from before 380 and 390 B.C., respectively, since these artists would have been too young prior to that time; these paintings must therefore belong between 400 and 380 B.C. There are no references to Parrhasios' life or work after 380 B.C. By then he was quite old, and it can be assumed that he died soon after.

Many of Parrhasios' works cannot be dated, even roughly: one such is *Philoktetes at Lemnos* (*Anthologia Palatina*, XVI, 111–13), echoes of which have been seen in a south Italian painted vase in Syracuse (Mus. Arch. Naz.; Pace, 1922–23, col. 521, note) and in the Philoktetes cup, one of a pair of silver skyphoi by Cheirisophos found at Hoby, Denmark (PL. 51; K. F. Johansen and G. Rodenwaldt, *AAnz*, 1937, col. 237; Bianchi Bandinelli, 1950, p. 59; Picard, 1957), as well as in a lekythos in New York (Met. Mus.; Rumpf, 1951, p. 2) and in a fragment of Arretine ware from Saint-Bertrand-de-Comminges (K. T. Johansen, *ActaA*, 1960, pp. 185–90). Other undatable works are *Telephos Cured by the Rust from the Sword of Achilles*, a scene in which Agamemnon and Odysseus were also present (Pliny, op. cit., XXXV, 71); *Aeneas between Kastor and Polydeukes* (ibid.); *Dionysos* (Suidas, *Lexicon*, s.v. οὐδὲν πρὸς τὸν Διόνυσον), executed for Corinth in a competition won by Parrhasios; *Thracian Nurse with a Child* (Pliny, op. cit., xxxv, 70), which can be related to a figure on a vase fragment in the British Museum (Pfuhl, II, p. 695), although the vase painting might be earlier than Parrhasios' work (Rumpf, 1951, p. 7); *Priest with Assistant* (Pliny, op. cit., xxxv, 70); *Running Hoplite* (ibid., 71); *Hoplite Taking Off His Armor* (ibid.); *Two Youths* (ibid., 70); *Naval Commander in a Cuirass* (ibid., 69); a painting of unknown subject mentioned in an inscription from Delos (Vallois, 1912, p. 289); and, finally, many sketches on wooden panels and parchment, used as models by other artists (Pliny, op. cit., xxxv, 68).

Most of what is actually known about Parrhasios' work — that is, some appraisal of his distinctive gifts — comes from this famous passage in Pliny: "He was the first to give norms of proportion to painting and the first to lend vivacious expression to the countenance, with elegant arrangement of the hair and beauty of the mouth; indeed, it is admitted by other artists that he won the palm in the drawing of outlines. This in painting is the high-water mark of refinement: to paint the bulk and surface within the outlines, though no doubt a great achievement, is one in which many have won distinction, but to render the contours of the figures and make a convincing boundary where the painting within finishes is rarely attained even in successful artistry. For the contour ought to round itself off and so terminate as to suggest the presence of other parts beyond it also, and disclose even what it hides" (ibid., 67–68).

The central problem in Parrhasios' art therefore seems to have been the representation of a figure in relation to the space around it, by means of articulating the volumes suggested by the contour line. This characteristic of his style is reflected in a group of handsome white-ground lekythoi (PLS. 51, 52) attributed to the Achilles Painter, the Canneto Painter, and the Tumulus Painter.

Parrhasios' basic style seems to have had connections with the artistic circle of Phidias (q.v.), but he developed a distinct artistic personality of his own and in his independent traits was ahead of his time, anticipating the work of Skopas (q.v.). Nonetheless, Parrhasios remained quite close to the life and fashions of his period, both in artistic matters, as is shown by his use of the popular three-figure group, and in his presentation of the gay, at times even exhibitionistic and frivolous, way of life then current, the chief exponent of which was Alkibiades. It was this outlook on life which led to his impatience when others were placed ahead of him and which dictated the following boastful self-appraisal (Athenaeus, op. cit., xii, 543E): "Even if you hear accounts that seem incredible, I affirm that the extreme limit of the art of painting has already been reached by my hand. I have set a boundary that no one will ever be able to surpass. But nothing mortals do is beyond blame."

BIBLIOG. Pliny citations and extracts, Loeb Classical Library edition of the Naturalis historia, London, Cambridge (Mass.), 1952; Overbeck, SQ, nos. 1692–1730; R. Vallois, Les πίνακες déliens, Mél. Holleaux, Paris, 1912, pp. 289–99; Brunn, GGK, II, pp. 97–120; B. Pace, Vasi figurati con riflessi della pittura di Parrasio, MALinc, XXVIII, 1922–23, cols. 521–98; Pfuhl, II, pp. 689–95; A. Reinach, Recueil Millet, Paris, 1927, nos. 257–301; A. De Capitani d'Arzago, La grande pittura greca, Milan, 1945, pp. 50–54; R. Bianchi Bandinelli, Storicità dell'arte classica, 2d ed., I, Florence, 1950, pp. 45–61; G. Lippold, RE, s.v.; A. Rumpf, Parrhasios, AJA, LV, 1951, pp. 1–12; C. Picard, Le motif de Phèdre amoureuse et de la nourrice, RA, XLVI, 1955, pp. 72–74; S. Papaspiridi Karuzos, Scherbe einer attischen Weissgründigen Lekythos, Antike und Abendland, V, 1956, pp. 71–74; C. Picard, D'un tesson arrétin trouvé à St-Bertrand-de-Comminges à l'un des skyphos d'argent du Trésor d'Hoby (Copenhague), Hommages à W. Deonna, Latomus, XXVIII, 1957, pp. 371–84; E. Tomasello, Rappresentazioni figurate del mito di Penteo, Siculorum Gymnasium, N.S., XI, 1958, pp. 219–41; M. Cagiano de Azevedo, EAA, s.v.

Michelangelo CAGIANO DE AZEVEDO

Illustrations: PLS. 51–52.

PARTHIAN ART

PARTHIAN ART. Modern scholarship, especially since the studies of M. Rostovtzeff, tends to grant the status of a definite style to the art that developed and flourished in Iran during the period (ca. 250 B.C.–3d cent. of our era) when the Iranian nomads known as Parthians or Arsacids overran and came to rule the territories formerly held by the Seleucidae. This distinct style, justifiably called Parthian, should not be considered merely a revival of the traditional Iranian art, which had been transformed under Hellenistic domination, but may be included in the trends marking the at least partial decline of Hellenistic-Roman naturalism and illusionism. The prevalence in Parthian art of frontality and of increased conventionality or rigidity in figures, with a linear, almost abstract treatment of drapery (all related to Central Asian traits, which perhaps had appeared earlier in Iranian art), constitutes a series of significant stylistic phenomena. Roughly contemporaneous manifestations of similar tendencies can also be observed in other Eastern schools (see BACTRIAN ART; GANDHARA; INDO-IRANIAN ART; KHWARIZM; KUSHAN ART).

In examining the complex problems concerning its origin and development, the widely diverse influences to which it was subject, and its diffusion and the developments in turn influenced by it, it becomes evident that Parthian art followed a separate course from the concurrent development of the art of the Iranian world proper (see IRANIAN PRE-SASSANIAN ART CULTURES), especially when it is considered in the light of the ethnic group and political system that engendered it.

ORIGINS OF PARTHIAN ART. It should be borne in mind that at the time of its maximum expansion the Parthian kingdom covered an area far greater than that of Iran proper, and included the Indian subcontinent, Mesopotamia, Armenia, and some of the regions where Indian and Iranian influences overlap. The people who created this empire — or, rather, this feudal framework united more by ties of race and custom than by a centralized and solid political structure — showed in their social organization and in various aspects of their folkways and customs traces of the nomadic world of Central Asia, to which it seems to have been related in various ways. According to classical sources (cf. Strabo XI, 1–12) the Parthians were a tribe of the Parni or Aparni, who belonged to the large tribe of the Dahae; their name is clearly related to that of Parthava, the first Iranian region conquered by them. With the exception of Indian documents, which give them the name Pallava (Pahlavāh), nonclassical sources — whether Iranian, Armenian, Syrian, Chinese, or Arabic — prefer to call these people Arsacids, using the dynastic name derived from that of the eponymous sovereign Arsaces (Arshak), who, according to some classical authors, was of Bactrian origin. The north-to-south migration that began circa 250 B.C. and led to the full establishment of Parthian power in Iranian and neighboring territories about 140 B.C. was a historical phenomenon which ultimately resulted in the elimination from this area of every vestige of Hellenistic political structures dating from the invasion of Alexander the Great. The reawakened independence of the Iranian peoples was not, however, guided by the traditional centers of local culture, namely, Persis and Susiana; its leadership came, unexpectedly, from the northern frontier to dominate Iran, in an invasion that seems to have been related in some respects to the movements of the Sarmatians and of the more obscure Yüeh-chih tribes. The Parthians shared many aspects of culture and various folkways with the Kushans, notwithstanding the hostility that sometimes existed between the two groups.

Parthian art reflects some of these historical phenomena in both its evolution and its forms. The changing relationship with Hellenism is evident in the portrayal on coins of the successive kings (see COINS AND MEDALS; PORTRAITURE). There is a progressive departure from Hellenistic forms toward stylization and frontality, though not without an occasional wavering or backward glance clearly related to the attitudes and tastes of individual sovereigns. Similar tendencies are indicated by archaeological finds testifying to the Parthians' interest in Greek art and classicism [e.g., the copies of Greek works found at the first Parthian capital, Nysa (Nisa), and the satyr heads of Hellenistic type from Dinavar]. On the other hand, a substantial series of commemorative and other distinctly Parthian ornamental works testifies to the existence of a marked anticlassical trend, which does not seem to relate to differences in social level but rather to a general coexistence of two antithetical, yet complementary tendencies. Iranian art of the Arsacid period therefore may be considered the particular result of the encounter between local tradition and Hellenistic art; and as D. Schlumberger (1960) states, it was a strain of that Greco-Iranian art which was not limited solely to the Parthian school but extended variously (and with varying degrees of Hellenistic influence) over other regions and civilizations, from Commagene to Mathura and from Auranitis to Khwarizm. Although evincing the wide-

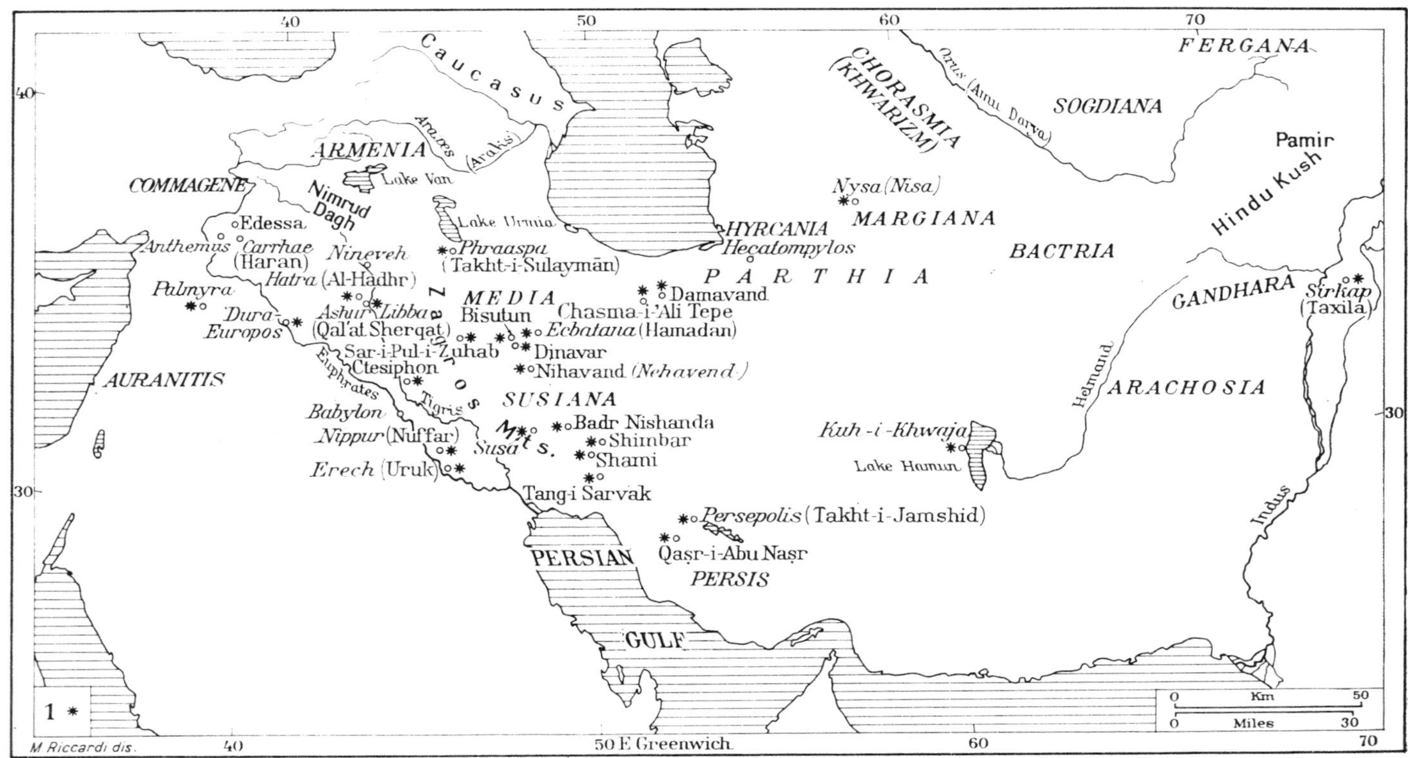

Principal sites of historical and archaeological interest within the Parthian Empire. *Key*: (1) Localities with remains of the Parthian period.

spread and persistent hold of Hellenistic-Roman classicism and Greek culture, which fluctuated with region and epoch, Parthian art never renounced certain figural tendencies of its own and never entirely repudiated others of Iranian origin.

Rostovtzeff has defined among the fundamental characteristics of Parthian art its frontality, spirituality, hieratic order, linearity, and comparative realism. D. Schlumberger considers the first two qualities basic to its character, establishes an antithesis between static scenes and movement, reducing the hieratic order seen by Rostovtzeff to a matter of appearances, and considers the linearity and realism obvious and unquestionable. In some works, however, the realistic intent is almost lost in a linearity in which the images approach an abstract or ornamental rendering of expression, as can be seen in the marble group of veiled women in the frieze over the door of the Temple of Bel at Palmyra (PL. 56). Parthian realism, then, should be assessed as such in the light of its adherence to natural reality only in certain subjects, of its interest as portraiture, and of its lack of concern with fantastic and composite images; but it did not require the artist to make use of illusionistic means in achieving realism. Thus its aims and effects were notably different from the realism of the classical world.

The extensive — one might even say essential — appearance of frontality in Parthian art may be the result of a conscious attempt to express classic forms and attitudes, as peculiarly conceived within a local tradition or otherwise modified for a variety of possible reasons; or it may be the expression of an artistic preference common to ethnic groups linked with Central Asia. In other words, if its basis lies in Greco-Roman classicism, it may be the consequence of a technical decline from the level of illusionistic rendering of the human figure attained in the classical world proper; or it may be a deliberately rigid representation proceeding from a taste which had already found expression, for example, in the Luristan bronzes (see ASIATIC PROTOHISTORY; IRANIAN PRE-SASSANIAN ART CULTURES) and which had been widely dispersed by the Iranian peoples of the East. The frontality occurring so frequently in the Luristan bronzes contrasts with the figural conventions of the ancient Near East and, in a well-founded hypothesis, is connected to Medo-Cimmerian migrations originating more or less to the northeast. The preference for frontality is also affirmed in varied fashion in both the art of Gandhara (q.v.) and in Kushan official art (see KUSHAN ART), both of which closely resemble Parthian art in forms and style. In Sarmatian art as well, frontality has special importance, as is demonstrated by a number of Scytho- Sarmatian works that are in general accord with Parthian figural tendencies.

On the basis of present knowledge, therefore, the origin of Parthian art may be linked to a classical influence entering Iranian territory in the wake of historical developments; but it has not been ascertained whether the principal characteristic of this Iranian school — namely, its frontality — is, in fact, of Western (i.e., Hellenistic-Roman) origin or whether it demonstrates the influence of the Central Asian taste manifested much earlier in the Luristan bronzes. Moreover, because the art of the Steppes (see STEPPE CULTURES), in its phases following the spread of Achaemenid influence, makes but rare use of frontality, it becomes even more important to attempt to trace the true origin of Parthian frontality. Perhaps the origins may lie in the most ancient Iranian tradition, as exemplified in the vase paintings of Susa and Tell-i-Bakun — provided, of course, that these were not at some time influenced by the volumetric experiments of Karasuk, Tagar, and Ordos (q.v.). But the difficulty of exact dating and the combined workings of diverse factors have hindered the search for its precise derivations, and the primal origins of Parthian art remain obscure.

Achaemenid figural art, however, and Greco-Iranian and Greco-Syrian art (Auranitide and Commagene) certainly influenced Parthian art, enriching its expression and imaginative qualities and contributing to its differentiation from the arts of other peoples who had assimilated similar general influences (e.g., Kushan art).

Naturally, the geographical situation of the different centers, together with the prevailing social structures, also influenced their artistic activity. Thus the art of the western caravan cities (e.g., Palmyra) reflects the taste of the rich merchant class and their Hellenic contacts, with marked Greco-Roman influence; and in the eastern centers, contact with the Indian world also produced distinct, but different results, which were nonetheless important from the standpoint of art history. The Hellenistic contribution, uneven and variable though it was, had continuing effects upon various local styles of Parthian art; moreover, this alternately strong, then diminishing classical influence was accompanied by renewals of anticlassical tendencies, which were only in part traditionally Iranian and which perhaps originated in Central Asia or, more probably, belonged to the greater area of Iranian culture.

Stylistic phases. R. Ghirshman distinguishes in Parthian art an initial phase, lasting from about 250 B.C. to the accession of Mithradates II, which was followed by a second, properly Parthian period, extending to the fall of the Arsacids and the rise of the Sassanian Empire.

In the first phase traditional, but by then greatly impoverished, Iranian styles were conserved along with Hellenistic and Greco-Iranian tendencies that had been more lately introduced — a forced coexistence that mirrored the decline of Seleucid power. Throughout the second phase, or the truly Parthian period, an influx of Parthian artistic traits was increasingly in evidence within the area subject to Arsacid political influence, from Dura-Europos and Palmyra to Ashur (Assur), and including the ancient capital Nysa (Nisa) and Sirkap, at or near the site of Taxila in the Indus Valley (part of mod. Pakistan). This division into two phases — though in a sense artificial, being based upon political events — also corresponds to stylistic modifications, for expressive values and traits associated with the Parthians' former nomadic habitat were diminished in the production of the second phase in favor of established Iranian elements, prevalently Achaemenid or Greco-Iranian of Seleucid type.

Architecture and city planning. The liwan (fig. 111), which appears in Parthian architecture, from the very beginning is possibly related to the nomad tent, also open on one side. In the so-called "maison carrée" of Nysa there are four liwans surrounding a central courtyard. A classical influence, possibly Roman, may be seen in New Nysa, where, contiguous with the preexisting city wall, there stands an edifice fronted by a portico, certainly a structural type foreign to Iranian taste. Although foreign or variant elements appeared very early in Parthian architecture, these can be distinguished from recurrent, even constant, features such as the façade characterized by a series of engaged columns and dominated by a large central arch (fig. 112).

In city planning it was necessary to consider the continuous presence of the military, because of the very structure of the feudal, fragmented, and often discordant Parthian society. The Parthian cities, like European counterparts of the Middle Ages, were in varying degree isolated and self-sufficient entities, prepared to assume defensive or offensive positions. They were laid out either on a circular plan (Merv, Phraaspa, Hatra) derived from ancient Central Asian traditions, perhaps supplemented by Mesopotamian and Assyrian practice, or on a rectilinear plan (e.g., the Parthian rebuilding of Sirkap; fig. 110) of vaguely Hellenistic derivation.

Individual structures were sometimes built on the "Babylonian" plan, with a central courtyard, as in the palace of Ashur (fig. 113), or on the "Iranian" plan, with a free adaptation of the liwan, as in the palace at Hatra (fig. 112). The palace of Nippur (fig. 115) shows a somewhat unusual combination of

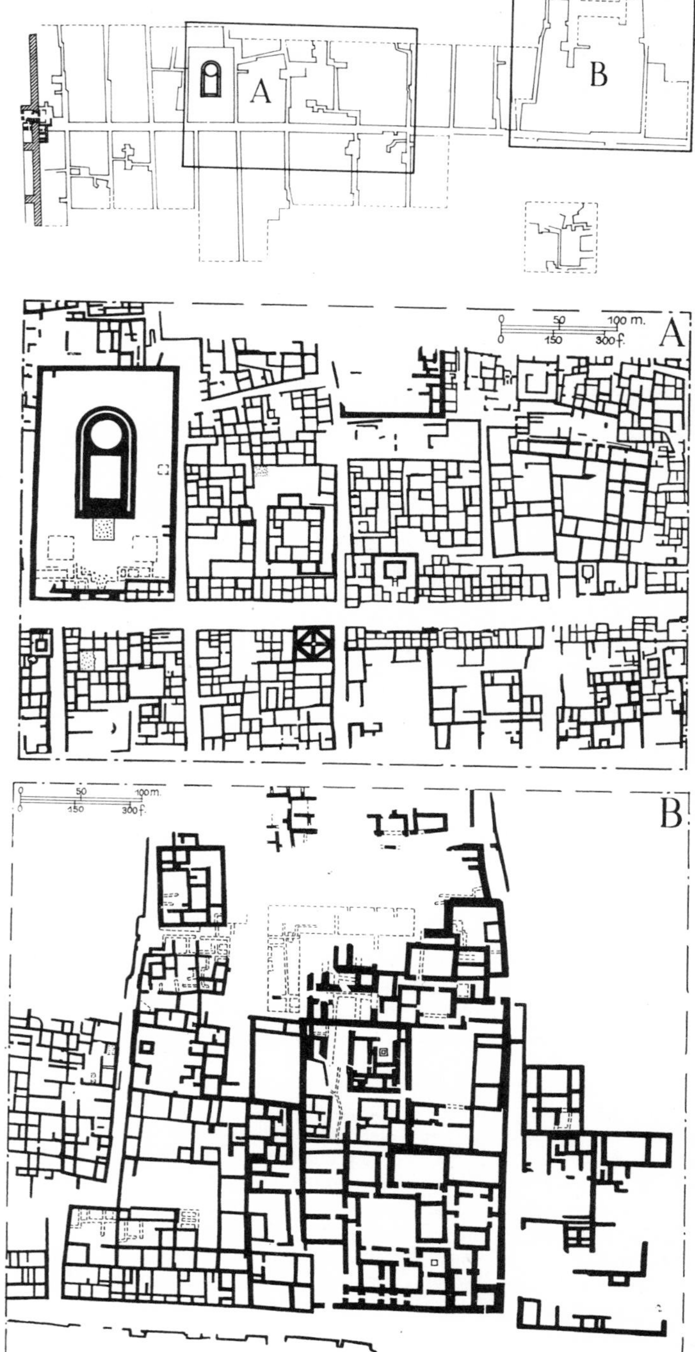

Sirkap (Taxila), plan of the excavated zone. *Inset A*: Temple area. *Inset B*: Palace area (*from Marshall, 1951*).

Greek and Mesopotamian elements. The prevalent use of fired brick did not preclude applied decoration, usually in stucco; nor did it preclude, in a later period, the use of painted decoration on an even greater scale. Various sources — including some Chinese — mention a type of decoration obtained by means of a facing of glazed squares of vitreous paste, with a particularly interesting treatment of color and light.

Sculpture. The Tang-i Sarvak reliefs, attributed to a prince of Elymaïs, seem to introduce into Parthian taste a hitherto unknown art form — the continuous narrative — which undoubtedly emanated from the Hellenistic-Roman world, perhaps reaching the Parthians via Greco-Buddhist art of the East. In these reliefs, carved in rock, it is utilized with schematic and

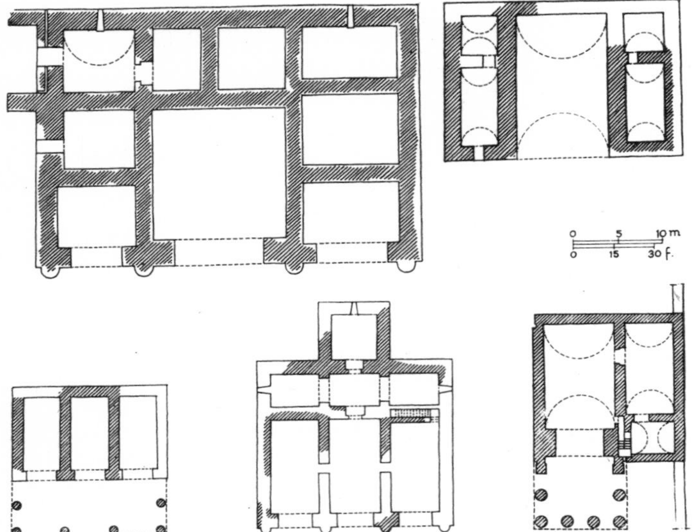
Hatra, various types of liwan structures (*from SPA, I*).

severe forms, in a manner which supports the assumption that by the end of the 2d century of our era the art of the peripheral Parthian world was already pervaded with anticlassical tendencies, even when it employed a Greco-Roman compositional scheme such as continuous narrative. At the farthest remove from the colossal figures of Nimrud Dagh (mid-1st cent. B.C.), which combine both Achaemenid and Hellenistic elements, the Tang-i Sarvak reliefs represent something quite distinct from classical tendencies in their characteristic and unquestionably Parthian rigidity, frontality, and ingenuously employed hieratic proportions. The same characteristics are evident in the art of Palmyra, which marks the encounter of Parthian and classical Roman styles (cf. PL. 54), and in which frontality became a constant factor.

In this zone of overlapping influences there also arose the need — not only among the sovereigns and highest aristocracy, but for all of Palmyrene officialdom — for portraiture, albeit primarily commemorative and funerary portraiture. The narrative genre was not entirely ignored, but it remained secondary to the portrait and symbolic tendencies in religious relief. The high relief widely employed at Palmyra was comparatively rare on the Iranian plateau, where there was a special interest in statuary, both in bronze (e.g., the famous figure of a prince from Shami; VIII, PL. 141) and in marble and other stone: for example, the important portrait statues of the sovereigns and princesses of Hatra, brought to light during the archaeological expeditions of 1951–52, effigies that often reveal an extraordinary expressive power (PLS. 54, 55). In the art of Palmyra, extensive contact with the Roman world was responsible for the remarkable incidence of classical influence in the treatment of drapery. The result of changes in both taste and technique, this treatment certainly goes beyond mere adoption of styles in dress inspired by the fashions of the Roman world. In the statues of Hatra, where the ornamental motifs of sculptured drapery often differ greatly from those most commonly used in Palmyra, the modulation of the folds readily assumes abstract, schematic forms (as in the statue of Ubal; PL. 54). A boldly abstract treatment of drapery is also evident in the above-mentioned group of veiled women in the Temple of Bel in Palmyra, which is rendered with the linearity characteristic of Parthian art. Similar tendencies (i.e., interest in portrait statuary and relief, schematic and linear representation) are found in the official art of the Kushans, as exemplified in the royal statues of Surkh Kotal and Mathura (q.v.), testifying to an affinity of taste deriving from the Iranian origins common to the two civilizations. Extensive use of stucco, a general subordination of decorative sculptural forms to their architectural context, and a broad range of production of the craft level in bronze, terra cotta, and stone are further characteristics of Parthian sculptural activity. In stucco wall decoration with geometric motifs, there is evident the intention to achieve, through manipulation of the projecting and sunken portions of a surface, coloristic effects in a contrast of light and dark that is suggestive of the use of chromatic values in painting. A further connection can be established between sculptured stucco decoration and wall painting, since the former may originally have been a polychrome technique.

PAINTING. Because of the presence of easily recognizable Hellenistic motifs, the wall paintings of the Parthian palace on the island of Kuh-i-Khwaja in Lake Hamun would seem to present significant Occidental inspiration; but taken as a whole, they are decidedly Iranian. The importance of the drawing and of contour lines, the flatness and relative stiffness of the figures (with the exception of the queen, who bends affectionately toward the king at her side) are all Iranian elements, clearly in accord with conventions outside the classical tradition, as is the treatment of eyes and faces. Not all critics, agree, however, on the attribution of these paintings to the Parthian period. D. Schlumberger, for example, excludes any such hypothesis; others think the attribution to be highly probable, if not entirely certain.

Classical inspiration is much more marked in the great mural paintings of Dura-Europos, particularly in the so-called "fresco of Conon," which can be dated to the 1st century of our era. In this fresco, even though the contours are still strongly delineated and there is a symbolic elongation of certain images, the treatment of the faces and the arrangement of drapery folds are evidence that a classical influence was assimilated and modified by the painter, who conveyed profound and obvious feeling in the expressions of his personages. The wall paintings and certain votive paintings in the Mithraeum at Dura are treat-

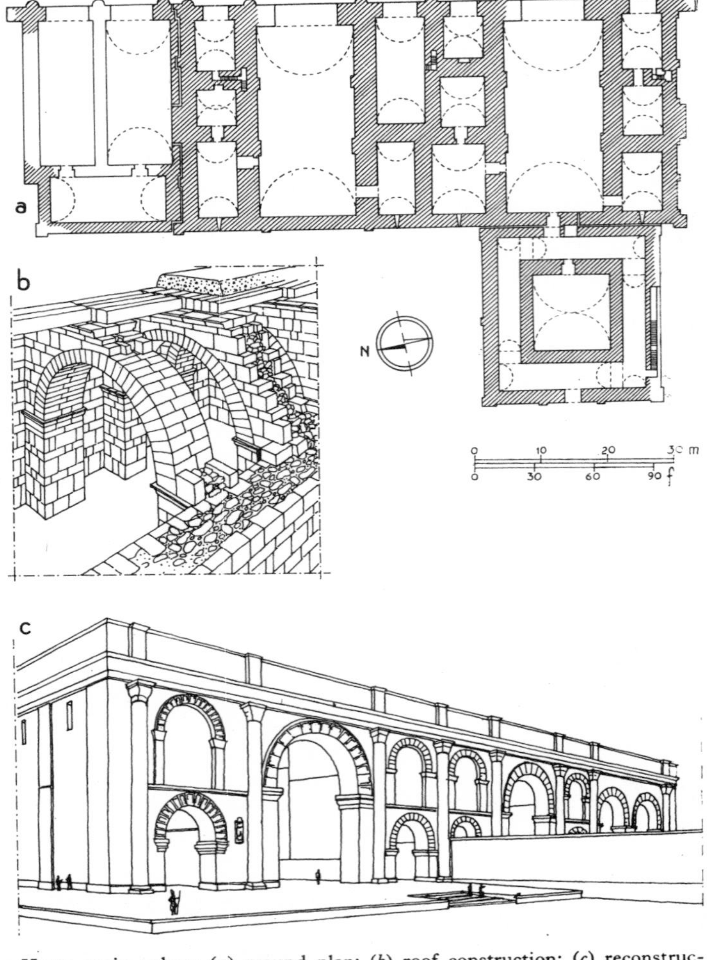

Hatra, main palace: (*a*) ground plan; (*b*) roof construction; (*c*) reconstruction of the façade (*from SPA, I*).

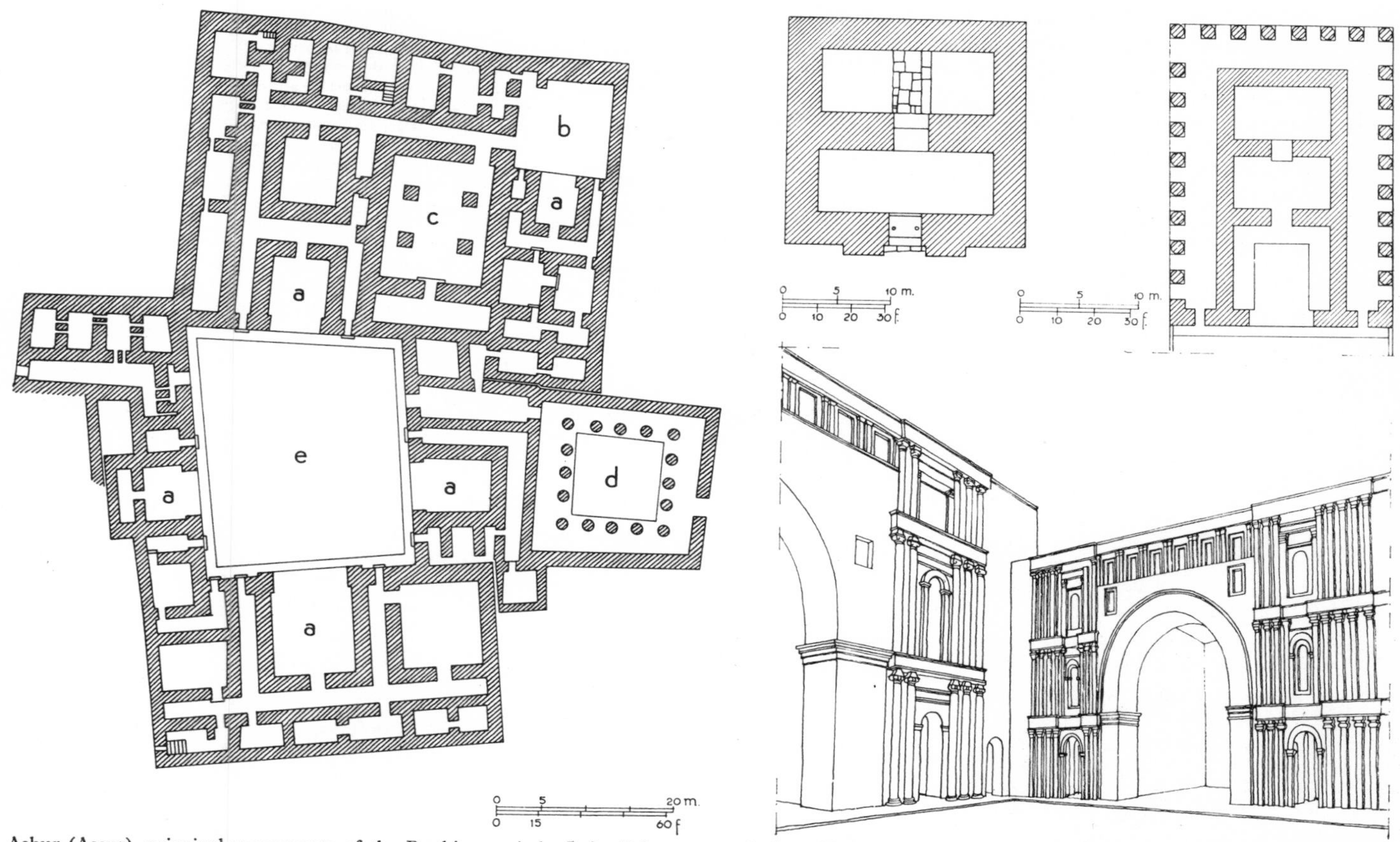

Ashur (Assur), principal monuments of the Parthian period. *Left.* Palace, ground plan: (*a*) liwan; (*b*) courtyard; (*c*) pillared hall; (*d*) peristyle; (*e*) main courtyard. *Above right.* Plan of Temple A and of the "Peripteros." *Below right.* Reconstruction of an interior section of the main courtyard (*from SPA, I*).

ed differently; in these the traditional (i.e., of Achaemenid derivation) Iranian component is much more evident. The paintings in the Synagogue, executed by an Iranian painter (VIII, PLS. 333, 335–337), belong to yet another trend, and the graffiti, in a more popular vein but equally interesting, display the truly Iranian characteristics also present in the large-scale paintings and votive works.

DECORATIVE ARTS. The coexistence of traditional Iranian and Hellenistic schools is exemplified in the terminal protomas, and in the human images sometimes substituted for them, on numerous ivory rhytons found at Nysa. However, the mythological scenes adorning their outer rims are almost wholly Greek, though some typically Iranian ornament of human masks occasionally appears as a complementary motif. In composition and representational techniques, these scenes closely resemble those of the art of Gandhara.

Entirely different in style and technique are some mother-of-pearl figural fragments, possibly once part of a jewel case, found at Shami. In these the images are flat, and the linearity and emphasis on hard contour is extreme. The disproportionately large eyes and the entire arrangement of the figures, as well as the treatment of costume, are a clear expression of Parthian taste in its most markedly anticlassical vein. Both frontal and profile depiction are utilized, with a notable prevalence of the latter. Ghirshman discerns, in the figure of a queen or goddess that constitutes the main fragment, a fairly obvious reminiscence of the art of Palmyra. Small bone and ivory figures of the so-called "nude goddess" (VIII, PL. 239) found in various tombs complete the range of types of Parthian works in these materials.

Examples of Parthian artistic production in precious metals are characterized by a tendency to elongate certain animal images (e.g., the panther on a bronze incense burner; Cleve., Mus. of Art), by the polychrome effects of jewelry settings (some of enamelwork), and by the presence of Central Asian motifs and stylizations not unlike those of Sarmatian art. A winged Eros from the Nihavand hoard rides a fantastic animal, whose body and paws were originally incrusted with precious stones, conceived in the same spirit as Sarmatian figures. The use of gold wire inlaid in silver and the taste for polylobed cups and for earrings shaped like clusters of grapes are also characteristic of Parthian gold- and silverwork, which in many ways appears to be closer to Achaemenid than to Sassanian style.

In Parthian fictile art there can be seen, in addition to the widespread production of small votive images of divinities, a penchant for creating genre figures (horsemen, warriors in repose, female figures) that occasionally recalls the Hellenistic emphasis upon aspects of character, at times even approaching the grotesque. The large sarcophagi and ossuaries found widely in Central Asia are of Egyptian derivation. There are also small plaques and seals with hunting scenes (PL. 57) and symbolic-religious scenes. As for the glazed ceramics, apart from a few very beautiful elegantly stylized rhytons, finds include amphoras, cups, pitchers, and tripods (III, PL. 139). Decoration is usually incised; but in many works of greater value small molded figures are added, and in some amphoras masks of classical derivation are applied at the base of the handles. The technique of green glazing, obtained with copper oxide, is of Western origin.

DIFFUSION OF PARTHIAN STYLE. If Parthian art did not evolve its frontality from a basis of Hellenistic-Roman classicism but from a characteristic preference of the greater Iranian and Central Asian culture areas, the question also arises as to what influence it may have exercised on the Western world during the time of the "great regression" of classical influence. Whatever the answer to this question may be, the influence of Parthian art can definitely be seen in the art of its eastern Greco-Iranian neighbors, although it is difficult to distinguish precisely between those elements in the art of Gandhara and in the school of Mathura which emanate from a taste common to Iranian and other Central Asian peoples and those distinctly Parthian elements which arrived in the wake of commercial activities and historical events. The artistic trends of the classical West and of the Indian cultural sphere in turn influenced the evolution of Iranian art in the Arsacid period and modified its character, especially in the peripheral regions.

If one recognizes the fact that Parthian missionaries played an important role in the spread of Buddhism in China (where they introduced the Amidist movement, or the cult of Amitābha, the Buddha of Boundless Light and Eternal Life), there is further justification for maintaining that the spread of Iranian artistic influence must be closely considered in the light of developments in the surrounding civilizations. For example, from Arsacid Iran there derived characteristic motifs and themes that influenced the art of the Steppes, of India, and of eastern Europe; and the elongated, stylized animal form was diffused throughout every region subject to Parthian influence. It can also be established that classical motifs of Hellenistic and Roman origin found their way toward the Far East by way of Parthian Iran, which, paradoxically, oftentimes constituted both an insurmountable barrier and an important link between the two worlds, East and West.

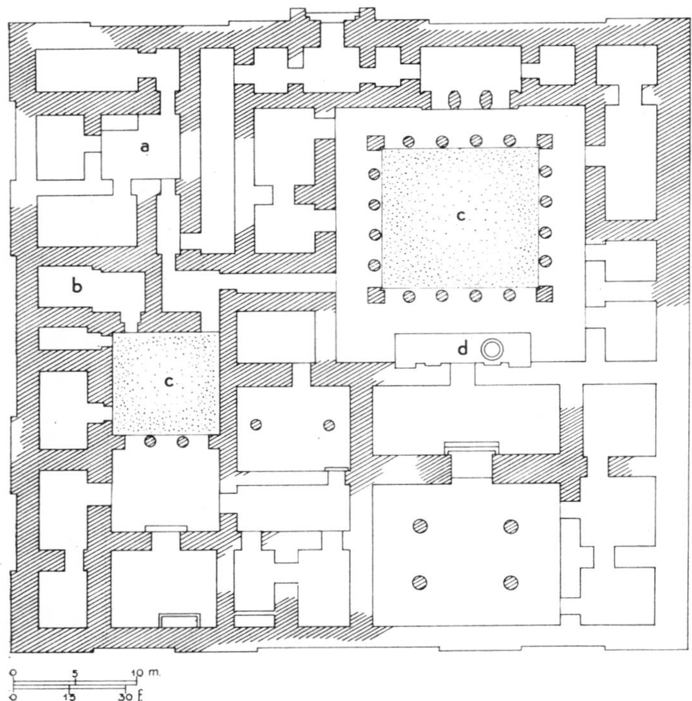

Nippur, plan of Parthian palace: (*a*) kitchen; (*b*) bathroom; (*c*) courts; (*d*) altar (*from SPA, I*).

Bibliog. M. Dieulafoy, L'art antique de la Perse: Achéménides, Parthes, Sassanides, 5 vols., Paris, 1884–89; E. Herzfeld, Am Tor von Asien: Feldsdenkmale aus Irans Heldenzeit, Berlin, 1920; F. Sarre, Die Kunst des Alten Persien, Berlin, 1922; F. Cumont, Fouilles de Doura Europos 1922–1923, 2 vols., Paris, 1926; M. Rostovtzeff, L'art gréco-iranien, RAA, VII, 1931–32, pp. 202–22; W. Andrae and H. Lenzen, Die Partherstadt Assur, Leipzig, 1933; N. C. Debevoise, Parthian Pottery from Seleucia on the Tigris, Ann Arbor, Mich., 1934; H. Seyrig, Antiquités syriennes, 5 vols., Paris, 1934–58; M. Rostovtzeff, Dura and the Problem of Parthian Art, Yale Classical S., V, 1935, pp. 157–304; C. Hopkins, Aspects of Parthian Art in the Light of Discoveries from Dura-Europos, Berytus, III, 1936, pp. 1–30; C. Hopkins, A Note on Frontality in Near Eastern Art, Ars Islamica, III, 1936, pp. 187–98; A. Godard, Les statues parthes de Shami, Āthār-e-Īrān, II, 1937, pp. 285–305; M. Rostovtzeff, Dura Europos and Its Art, Oxford, 1938; D. N. Wilber, The Parthian Structures at Takht-i-Sulayman, Antiquity, XII, 1938, pp. 398–410; SPA, I, pp. 406–92; R. du Mesnil du Buisson, Les peintures de la synagogue de Doura Europos, Rome, 1939; E. Herzfeld, Iran in the Ancient East, London, New York, 1941; G. Contenau, Arts et styles de l'Asie antérieure, Paris, 1948; R. Ghirshman, L'Iran: Des origines à l'Islam, Paris, 1951; J. Marshall, Taxila, 3 vols., Cambridge, 1951; W. B. Henning, The Monuments and Inscriptions of Tang-i-Sarvak, Asia Major, N.S., II, 1952, pp. 151–78; J. Starcky, Palmyre, Paris, 1952; H. Ingholt, Parthian Sculptures from Hatra: Orient and Hellas in Art and Religion (Mem. Conn. Acad. of Art and Sc., XII), New Haven, 1954; E. Will, Le relief cultuel gréco-romain, Paris, 1955; M. E. Masson and G. A. Pugatchenkova, Parfianskie ritony iz Nisy (Parthian Rhytons from Nisa), Moscow, 1956; M. E. Masson and G. A. Pugatchenkova, Mramorniia statui parfianskogo vremeni iz staroi Nisi (Marble Statues of the Parthian Period from Ancient Nysa), Yezhegodnik Inst. istorii iskusstva Akad. Nauk SSSR, VII, 1957, pp. 460–89; L. Vanden Berghe, Archéologie de l'Iran ancien, Leiden, 1959; E. Will, Art parthe et art grec: Etudes d'archéologie classique, Ann. de l'Est, Mém., XXII, 1959, pp. 125–35; D. Schlumberger, Descendants non méditerranéens de l'art grec, Syria, XXXVII, 1960, pp. 131–66, 253–318; R. Ghirshman, Iran: Parthes et Sassanides, Paris, 1962 (Eng. trans., S. Gilbert, Persian Art: The Parthian and Sassanian Dynasties, New York, 1962). See also bibliog. for IRANIAN PRE-SASSANIAN ART CULTURES.

Mario Bussagli

Illustrations: PLS. 53–58; 6 figs. in text.

PA-TA-SHAN-JÊN. Chinese painter (b. Nan-ch'ang, Kiangsi, 1625 or 1626; d. ca. 1705). His real name was Chu Ta, but Pa-ta-shan-jên is the best known of the pseudonyms with which he signed his works. Not much is known about details of his background and life, but he seems to have been distantly related to the Ming imperial house. When the Ming dynasty finally collapsed in 1644, he probably found it politically expedient to efface himself and escape the attention of the incoming Manchu rulers. Thus, like others in a similar situation, he became a Buddhist monk. His father died at about the same time (i.e., when the artist was twenty), and it has been suggested that this strengthened his resolve to live a monastic life. However, other biographers mention that he married and had children and that he only entered a monastery when he was nearly forty.

Dumbness seems to have run in the family (his father, a minor painter, was dumb). Tradition has it that this congenital failing also afflicted the son and that he wrote the character for "dumb" on his door and thereafter shunned society. This is probably an exaggeration, however, for he seems to have enjoyed congenial company and is known to have maintained correspondence with friends such as the painter Shih-t'ao (Tao-chi).

It is often said of Chinese painters whose work shows seemingly hasty execution and great boldness of brushwork that they were heavy drinkers. This is said, for example, of Hsü Wei (1521–93), whose work influenced Pa-ta-shan-jên. It seems to have been true also of Pa-ta-shan-jên, and many stories relate how he would paint only when alcohol had sufficiently inspired him. He did not paint for financial gain and gave his works to friends and drinking companions. What we know of the disorder of his life and his eccentricities would suggest that he was more than just Bohemian. In his later years he may even have been mentally deranged. If these accounts are reliable, an immediate Western parallel to his life springs to mind in that of Vincent van Gogh.

While little is known of Pa-ta-shan-jên's personal mien, his work — most of which was done after he was fifty — is among the most easily recognizable of all Chinese painting and falls into two main categories: the depictions of birds, flowers, insects, trees, and fish, and the landscapes. For the first, he used mainly the album-leaf format or small scrolls (PLS. 59, 60). These are done mostly in ink, sometimes with the addition of faint color; a few summary strokes in very sensitively shaded ink produce a mouse balancing on a huge gourd, a bird at the end of a branch, or a lone fish swimming in the depths of a pond, as seen from above. Almost invariably his animals have quite human expressions, and they have the capacity of making the spectator feel he shares their world. His use of empty space, always a distinctive feature of fine Chinese painting, is masterly and evocative. Brilliant and original though he generally is, there is little doubt that he owes much to Hsü Wei in subject matter, brushwork, and over-all atmosphere; but his approach is more lyrical, more gentle, and altogether more humorous than that of Hsü Wei.

Much has been said of Pa-ta-shan-jên's presumed madness and loneliness, but this is hardly borne out by his work, which gives evidence of a thoughtful, balanced man full of interest in life and quick to select what amused him in the natural world about him. The power of his work often proceeds from a strength of brushwork combined with a simple, almost insignificant subject: for instance, a spider tucked away in the corner of an empty sheet of paper, tense and predatory, seems somehow to fill the whole space as it awaits its prey. Everything Pa-ta-shan-jên touches becomes imbued with a life of its own, and he continually exercises a power to make the spectator share his amusement and savor the almost human character of his animals.

In his small works Pa-ta-shan-jên is perhaps the supreme master of the technique of *hsieh-i* (literally, the "writing of ideas"). The essential principle of this type of ink painting is for the artist to communicate his artistic thought with maximum speed. The mark of his skill lies in the vitality of brushwork and the abbreviation of thought. This, of course, owes much to the Buddhist (particularly Zen) philosophy of sudden enlightenment, of swift realization of the essentials of a composition, and of recognition of the existence of a life essence in every aspect of nature.

Pa-ta-shan-jên's (or Chu Ta's) landscapes, sometimes on album leaves but usually on fairly large hanging scrolls, are perhaps his more ambitious works (III, PL. 290), in which he shows himself to be wholly familiar with the painting styles of the Sung and Yüan dynasties. He often painted landscapes in emulation of his great predecessors such as Ni Tsan (q.v.) and Tung Yüan. Although these works reveal a conscious archaism in composition, the brushwork is highly individualistic, with free use of lines, dots, and washes, applied swiftly and with great assurance. Obviously he was deeply concerned with translating the ideals of the past into his own more radical idiom. The luminous quality of his ink, which unfortunately is not conveyed in reproductions, lends a particularly effective interplay of light and shade to all his work. No better illustration exists of how, in Chinese painting, honored traditions persist and are successfully revitalized.

Pa-ta-shan-jên belonged to a group of painters including Hsü Wei, Shih-tʻao, Kʻun-tsʻan (Shih-chʻi), and Kung Hsien who enlivened painting at a time when its inspiration was in danger of running dry. The famous sketchbooks such as the *Mustard Seed Garden Compendium* had summed up the techniques of many past generations of Chinese painters and had reproduced their brushwork in simplified form. From their elements, anyone who could write might assemble facile paintings. Moreover, the gentlemen scholars exercised tight controls over what was considered the proper method painting. As a result of these restraining influences, an artist of genius understandably might be driven to violent means to affirm that Chinese painting had not reached a dead end. Pa-ta-shan-jên's striking and original qualities gained rapid appreciation and influenced many subsequent painters; because of this continuing esteem, perhaps, forgeries of his work have been numerous.

BIBLIOG. W. Cohn, Chinese Painting, London, 1948; A. Giuganino and J. P. Dubosc, Pitture cinesi delle dinastie Ming e Chʻing, Rome, 1950 (cat.); W. Speiser, Chinesische Gemälde der Mings und Chʻings Zeit, Cologne, 1950; L. Sickman and A. Soper, The Art and Architecture of China, Harmondsworth, 1956, pp. 198–99; O. Sirén, Chinese Painting: Leading Masters and Principles, London, New York, 1958, V, pp. 149–56, VI, pls. 381–87; P. C. Swann, Chinese Painting, London, Paris, 1958; A. Giuganino, La pittura cinese, Rome, 1959, I, pp. 233–34, II, pls. 476–80; Wên Fong, A Letter from Shih-tʻao to Pa-ta-shan-jên and the Problem of Shih-tʻao's Chronology, Arch. Chinese Art Soc. of Am., XIII, 1959, pp. 22–53; J. Cahill, Chinese Painting, Geneva, 1960; Boston Museum of Fine Arts, Portfolio of Chinese Painting (Yüan to Chʻing Periods), Boston, 1961; J. P. Dubosc, Introduction to an Exhibition of the Work of Pa-ta-shan-jên and Tao-tsi, Paris, 1961 (cat.).

Peter C. SWANN

Illustrations: PLS. 59–60.

PATINIR (PATENIER, PATINIER), JOACHIM. Flemish painter (active by 1515; died no later than 1524). Nothing is known with certainty about the early life of Patinir except the fact that he was born in or near Dinant, perhaps at Bouvignes. In 1515 he was registered as a free master in the painters' guild in Antwerp, where he died nine years later, presumably at a youthful age. It is not recorded where he received his training, but it has been suggested that before transferring to Antwerp he may have spent some time in Bruges. There he could have become familiar with the works of his considerably older contemporary Gerard David (q.v.), who had been made a free master in Bruges in 1484 and who seems to have worked in Antwerp as well. In any case, borrowings from David are clearly to be seen in Patinir's paintings. It is even possible that he was in Italy with David, who was working in Genoa about 1512–15.

In his *Book of Painters* (1604), Karel van Mander stated that Patinir collaborated with Joos van Cleve, providing the landscape background for a *Virgin and Child* by that artist. It is probable that he worked in this way with a number of other Flemish painters; an inventory of the Escorial, made in 1574, describes a *Temptation of St. Anthony* (Prado) as a work by Patinir with figures by Quentin Massys (q.v.).

No more than a half dozen of Patinir's paintings are signed. Notable among his authentic works are the *Baptism of Christ* in Vienna (Kunsthist. Mus.), the *Rest on the Flight into Egypt* in Antwerp (Mus. Royal des B.A.), the *St. Jerome* in Karlsruhe (Staat. Kunsthalle), and the *Passage of the Styx* in Madrid (IX, PL. 12). In Spain — where Patinir's work seems to have found its way at an early date — there remain five important pictures, all in the Prado; among these is the *St. Christopher Ferrying the Christ Child* (XII, PL. 378), a work recently transferred from the Escorial, which may have been based on a Dürer drawing. In the center panel of an imposing triptych (Met. Mus.) is depicted the penitence of St. Jerome, with the Baptism of Christ and the temptation of St. Anthony on the wings.

Albrecht Dürer (q.v.), who made the acquaintance of Patinir when visiting Antwerp in 1520–21, described him as a painter of landscape. Although it cannot be denied that Patinir was a specialist in this field, and a great innovator, his pictures, like those of Pieter Bruegel, always include a certain number of figures to illustrate the story or event that forms their real subject, which cannot therefore be construed as pure landscape. Patinir formulated and brought to a high peak a technique for the representation of receding distances that had its origins a century before him in the paintings of Jan van Eyck and the miniatures associated with the names of Hubert and Jan van Eyck. The tonal system that Patinir evolved customarily represented the foreground as brown, the middle distance as green, and the far distance (which usually recalls the craggy valley of the Meuse he probably admired in his youth) as greenish blue (V, PL. 296). Within these general divisions of color there are rich variations, and the total effect is one of poetry and grandeur.

BIBLIOG. G. J. Hoogewerff, Joachim Patinir en Italie, Rev. d'art, 1928, pp. 117–34; M. J. Friedländer, Die Altniederländische Malerei, IX, Berlin, 1931, pp. 101–88; H. B. Wehle and M. Salinger, A Catalogue of Early Flemish, Dutch, and German Paintings (Met. Mus.), New York, 1947, pp. 115–17; M. Davies, ed., Early Netherlandish School (Nat. Gall. Cat.), 2d ed., London, 1955, pp. 119–21; A. Bengtson, A Painting by Joachim Patinir, Figura, N.S., I, 1959, pp. 88–94; Le siècle de Bruegel: La peinture en Belgie au XVIe siècle (cat.), Brussels, 1963, pp. 144–58 (bibliog.).

Margaretta M. SALINGER

PATRONAGE. The love of arts and letters is often expressed through active encouragement and assistance of artists and writers by enlightened governments or by rich and powerful individuals. Such activity, which concerns the history of taste, of criticism (q.v.), and of collecting (see MUSEUMS AND COLLECTIONS), has exerted significant influence on the development of artistic production in many periods of history. In the modern usage of the term "patronage," there is implied a sympathetic comprehension of art as an independent entity with its own inherent values. Thus, ideal patronage implies a personal relationship between the patron and the artist — the development of a respect and an enlightened sense of obligation for the artist's welfare on the part of the patron, regardless of his interest in obtaining a particular work of art.

When the subject of patronage is considered historically, however, it becomes clear that such an ideal is a quite abstract, not to say optimistic, definition of the term. The sensitive concern with "art for art's sake" that is basic to such a concept of patronage is a comparatively modern notion. Throughout history, even during the grandiose patronage periods of the Renaissance and the 17th and 18th centuries, it is difficult to separate the genuine personal taste for art from a generosity conditioned primarily by utilitarian motives and engendered by the desire to assert power, social prestige, religious convictions, and so on.

Within the broader compass of the subject as presented historically, one must examine then all the conditions (historical, cultural, socioeconomic) that have been known to bear upon art patronage. Moreover, the essential interest in tracing the varied manifestations of patronage does not lie in merely listing or describing the activities of munificent individuals but in establishing their incentives and their effect (positive or negative) on art, that is, the resulting interplay between the person who commissions the work of art and the artist who creates it — just as, in a wider frame of reference, the artist has, in most historical epochs, been significantly affected by the attitude of his society in general toward his work.

Any study of patronage should therefore take into consideration these fundamental elements: incidence and degree of appreciation of artists by patrons; sources and motivations of patronage; behavior and reaction of the artist toward the patron. The major object of such a historical examination is an appraisal of the effect of patronage on the nature (type and content) and esthetic quality of the work of art. In making such an appraisal, artistic commissions arising totally or in part from reasons extraneous to a genuine interest in art for its own sake merit closer attention, and yield more general applications, than the isolated personal commissions, the occasional instances of pure patronage. Such individual cases are of concern in a general treatment only in so far as they come to exercise broader influence, those instances when pure patronage made possible the work of great artistic personalities who would otherwise have faced serious obstacles in being accepted by the society of their times: for example, the patrons of Caravaggio and of the French impressionists, who in their patronage went counter to current trends.

There were scattered periods in ancient times during which the production of art involved a love of beauty and luxury on the part of sovereigns and nobles, along with a personal interest in the artists and their work. The development of a stylistic unity influenced by the ideals and by the very physical appearance of the sovereign — such as the figural arts of the Tell el 'Amarna period in Egypt, during the reign of King Akhenaten (IV, PLS. 365, 374, 375) — is one possible outcome of large-scale official patronage. This particular development, which suggests the direct encouragement of Akhenaten, the builder of a new capital and himself a poet, is quite different from the general undertaking of official aggrandizing works by other Egyptian kings such as Thutmosis III, Amenhotep III, and Ramses II. Although the monuments and sculpture erected by these rulers were often far more impressive (at times gigantic in scale) than those of Akhenaten, their architectural and artistic activity was but an indispensable part of a political and religious program conceived along traditional lines or was a means for the display and affirmation of power, rather than an expression of personal taste or philosophy.

The relationship between patron and artist in archaic and classical Greece paralleled the gradual evolution of a sense of human individuality in thought and political institutions, in both literary and artistic creation. In considering the accomplishments of certain enlightened tyrants of the archaic period (Polycrates of Samos) and of the classical era in Greece (Pericles in 5th-century Athens, who has sometimes been compared to Lorenzo the Magnificent), it is possible to speak of patronage. In order to understand the diverse attitudes toward patronage exemplified by Pericles and Lorenzo de' Medici, it should be emphasized that in the Renaissance the name of the patron was closely linked with that of the artist to whom he brought renown, while in the classical world even Phidias was no more than a contractor and fine craftsmen in the eyes of Pericles and the ruling class. In the courts of ancient sovereigns, patronage was largely a question of political expediency and a conscious sign of luxury and wealth. Alexander the Great and the Hellenistic rulers (Ptolemy, for example, founder of the famous Library and Museion in Alexandria) surrounded themselves with poets, learned men, and artists who found ample opportunity for enrichment at court. Their work was regarded as indispensable to the establishment of official iconographies, which might either be treated allegorically or be drawn in terms of everyday court life (see GENRE AND SECULAR SUBJECTS).

In Rome, beginning in the 3d century B.C., art was used to serve political and ethical purposes, both for opulent official display (quantitative as well as qualitative, as evidenced by the attitude toward booty amassed by conquerors) and for private collecting. Possession of objects of artistic worth characterized their owners as men of refinement, with an intelligence and culture above the average. The emphasis on such refinements eventually led to the denunciation of art as a luxury that corrupted morals and customs. It is interesting to note that in the Rome of Maecenas, whose name has become synonymous with munificence and elevated social conscience, patronage was extended only to poets and writers, not to painters and sculptors. The creators of visual art remained, perhaps to an extent even greater than in ancient Greece, the modest executors of commissions. Artistic production became limited chiefly to copies and replicas of Greek originals or to unimaginative works of Greek derivation.

Nevertheless, during the imperial epoch — apart from an ever-increasing demand for architects and sculptors to celebrate the power of Rome and the emperors — there were other signs of public and private munificence that testified to personal interest and passion for art and culture, notably the undertakings of the emperor Hadrian. Further evidence of such personal initiative is furnished by the luxuriously finished palaces and private dwellings of the late imperial period. Moreover, the continuing patronage of men of letters, the growing humanistic interest in scholarship, and a notable increase in libraries certainly affected the figural arts, through the development of manuscript illumination, with a subsequent influence on the other arts (especially mosaics) that was incalculable.

* *

Patronage in the early Middle Ages was largely inspired by religion and, thus, was promoted mainly by the great monastic houses, which could best provide the tranquillity and resources conducive to artistic creation. Early in this period splendid manuscripts were illuminated in the monasteries of England and Ireland; but since the monks themselves were primarily responsible for producing such works for their own use, one cannot properly use the term "patronage" in speaking of the creation of these great books.

In architecture and sculpture, however, there existed a somewhat different situation. In France, one of the great centers of medieval civilization, the Benedictines were the most active builders during the 11th century; the rich decoration they introduced into their abbeys toward the end of that period gave to sculpture its first strong impulse since late antiquity. Research has shown that, though a great deal of the building and decoration was carried out by the monks themselves, they also employed professional "architects" (or master builders) and sculptors to a greater extent than was formerly recognized, and that their patronage brought forth a class of highly gifted artists, some of whom became celebrated personalities. Thus Gislebertus, the sculptor who worked on the Cathedral of St-Lazare in Autun (constructed in the mid-12th century on Benedictine lines), was actually allowed by the bishop to sign his name across the center of his masterpiece, the magnificent tympanum.

Generally, however, it was the patron who took credit for the splendor of his church. This was the case with Abbot Suger (1081–1151) of the Abbey of St-Denis, the one outstanding patron of the 12th century about whom there exists a primary source of information. The abbot himself wrote an account of his patronage, *De rebus in administratione sua gestis* — one of the only such documents extant — wherein he delights in the richness of the decoration he commissioned and justifies it on Neoplatonic grounds, as did the Latin inscriptions he placed on the great doors of gilt bronze: "Whoever thou art, if thou seekest to extol the glory of these doors/ Marvel not at the gold and the expense but at the craftsmanship of the work./ Bright is the noble work; but, being nobly bright, the work/ Should brighten the minds so that they may travel, through the true lights,/ To the True Light where Christ is the true door./ In what manner it be inherent in this world the golden door defines:/ The dull mind rises to truth, through that which is material/ And, in seeing this light, is resurrected from its former submersion."

This reasoning allowed Suger to expend on his abbey all the munificence and skill at his command; from all over France he summoned stonemasons, carpenters, painters, and goldsmiths. But while he held that windows, sculpture, furnishings, and jeweled objects could all help the soul visualize the celestial hierarchies, he was also prompted by another motive, one which has often played a leading role in patronage: the desire to perpetuate his own fame. Everywhere in the church he so splendidly decorated he had his name prominently displayed, and in the embellishment of St-Denis one encounters the first notable instance of that strange combination of personal ambition and reverence which was to recur frequently thereafter. The interplay of these two seemingly disparate factors can be traced also in the convention of the "donor painting," which appeared early in the Middle Ages and through which the patron might choose to be represented in the sacred story only in the person of his name saint. In other examples, the donor himself appears, sometimes with his wife and family, but on a very much smaller scale than the Virgin and Child. Later he assumed a part fully as important as that of the divine personages (cf. V, PL. 218); and before the conventional form finally disappeared in the secularized cultures of the 17th century, it was not unusual to find the place of honor given to the donor, with the sacred figures taking wholly secondary roles.

Not all the medieval religious orders were important patrons of the arts. In Abbot Suger's own lifetime, the Cistercians (founded in 1098) practiced a deliberate austerity that offered few opportunities to sculptors and painters, and St. Bernard of Clairvaux fiercely denounced the decoration of churches with idle ornament. Indeed, each new order began its existence with a return to primitive simplicity and built churches in which architectural style was dictated more by the varying exigencies of function (e.g., special attention to the high altar or the pulpit, display of some special relic) than by any conscious desire for ostentation or beauty. Even the Franciscans, who at Assisi inspired a remarkable humanization of art that heralded the Renaissance, elsewhere showed little interest in such matters. However, such resolutions to maintain simplicity invariably ceded to time and increasing wealth.

While some might attribute these capitulations to a permanent defect in human character, they can be explained more satisfactorily by other, more concrete reasons. Above all, it must be remembered that, though patronage in the Middle Ages was mainly inspired by religion, it was not only churchmen who were responsible for it. Kings, nobles, and craftsmen were all anxious to save their souls, and a lavish gift to some poor religious order or burial near some sainted relic was looked upon as one of the best ways of doing so. Often an initially austere church was turned into a splendid edifice under the impact of royal patronage, sometimes in direct conflict with the intentions of the clergy. One of the most frequent ways whereby color and luxury entered a simple church was through the presence of a royal tomb. The Cistercians at first refused burial to those outside their order, but were later compelled to admit interment of nonreligious personages, and the whole character of their churches was thus changed. Throughout the Middle Ages tombs were among the most permanent sources of patronage for artists of all kinds and were prepared with immense care. A few contracts survive, among them that for erection of the tomb of Louis XI in 1482: "Master Colin d'Amiens, it is necessary that you make a portrait of the King our lord. He must be shown kneeling on a cushion, as here below, with his dog beside him.... Moreover, he must wear buskins, not hose, the most perfect you can make, and be dressed in hunting attire. In particular, you must give him the most handsome face possible, young and smooth. The nose, as you know, is long and a little upturned; and do not make him bald. The nose should be aquiline and the hair longer in back...." Such an explicitly detailed commission serves, incidentally, as a warning for those who look to the sculpture even of this relatively late period for accurate portraiture.

As cathedrals sprang up in growing towns, the orders of chivalry endowed lavish chapels, and craftsmen's guilds took an increasing part in their decoration.

Nonetheless, patronage of a religious nature was no more the sole concern of the laity than it was of the clergy. (For instance, as early as 1153 it was noted that in France certain bishops' palaces contained "walls painted with Trojans clad in purple and gold.") As a measure of security returned to Europe, royal castles and even bourgeois houses (e.g., Hôtel Jacques-Cœur in Bourges; V, PL. 381) began to rival the religious foundations in architectural luxury and lavish furnishings. In the late 14th and the 15th century there were formed the first secular collections that are recognizable as such by modern standards. The Duc de Berry and other princes commissioned Books of Hours from artists of established reputation. Jan van Eyck (q.v.) painted easel pictures for his patrons in the fantastically rich and ostentatious court of Burgundy, so that art works could be taken along on their journeys or displayed in their palaces.

The most significant development of this sort of private patronage took place in Italy. By the third quarter of the 15th century Giovanni Rucellai, a rich Florentine merchant, could claim that "in our house there are many works of sculpture, of painting, of inlay and mosaic, from the hands of the finest masters who have been active from times long past till now, not only in Florence but in all of Italy." His pride was justified, for he refers specifically to paintings and sculpture by Domenico Veneziano, Fra Filippo Lippi, Paolo Uccello, and Andrea del Verrocchio (qq.v.), among others. Since all the artists mentioned were contemporaries, it can be assumed that many, if not all, of these works were specially commissioned; thus Rucellai's proud catalogue of ownership documents the development of an art patron in the modern sense, though the species had certainly been in existence for many generations.

Credit for the most enlightened patronage of the Renaissance is usually given to the Medicis of Florence. Though later research has substantially altered the estimate of their achievements, it remains true that, in the varying attitudes toward art of Cosimo I de' Medici ("Pater Patriae") and the two succeeding generations of his family, there emerged an entirely new outlook and concept of patronage.

Cosimo was brought up in a tradition of patronage that differed greatly from the idea of patronage in France. In Italy (and in Germany and the Low Countries) the early development of free towns had led to an outburst of municipal rivalries that often found expression in great works of art. Guilds of all kinds, from those of bankers to those of the humblest craftsmen, contributed to the erection of great public buildings and churches. In these circumstances artists as a class earned repute very early, and the spirit of communal patronage perhaps can be illustrated by quoting from a document of 1334 (admittedly, not typical of ordinary practice elsewhere) by which the Republic of Florence appointed Giotto its chief architect: "In order that the works which are being undertaken in the city of Florence and are to be carried out for the benefit of the commune may proceed in the most perfect manner, which is not possible unless an experienced and eminent man is chosen as the leader in those works; and as in the whole world there is to be found none better qualified for that, and for much besides, than Master Giotto di Bondone, the painter of Florence, he shall therefore be named in his native city as *Magnus Magister* and publicly regarded as such, so that he may have occasion to abide here; for by his presence many can have the advantage of his wisdom and learning, and the city shall gain no small honor because of him. . . ."

Though the free communes lost power throughout Italy during the 14th century, the republican form of government survived in Florence, and its spirit was proudly adhered to. Cosimo was particularly anxious not to offend these sentiments and, especially in his early years, usually acted well within the communal framework. He imposed his own tastes by offering contributions to a common fund which were somewhat higher than those of his colleagues and by suggesting a particular artist for a project, rather than by the dictatorial methods that afterward became common among rulers more secure of their position. Much of his patronage was for religious foundations — S. Marco in Florence; the Badia of Fiesole; altarpieces by Fra Angelico, Fra Filippo Lippi, and others in various churches — and was probably prompted by any guilt he might have incurred through his business activities, for taking interest was still considered a sin.

Cosimo also commissioned a palace, and for it a number of works of art — among them, the *David* by Donatello (q.v.; XII, PL. 15) and Paolo Uccello's *Battle of San Romano* panels (PLS. 39, 42, 43) — which confirm that he was among the first really important private patrons since the collapse of the Roman Empire. Moreover, it was written about him that "he frequented painters and sculptors; he was very knowledgeable of them and had in his house works from the hands of excellent masters. He was particularly informed about sculpture. He especially encouraged sculptors and all worthy arts. He was a great friend of Donatello and of all the painters and sculptors. . . ." Cosimo even drew up certain architectural plans himself. It seems likely that before he died he was already conscious of his art patronage as an estimable independent activity.

His son Piero was a personal friend of a number of artists and showed keen interest in the techniques of their craft. He showed more sympathy for the ornate, delicate qualities of the lingering International Gothic style than for the more austere tradition associated with his father, and Benozzo Gozzoli's *Procession of the Magi* for his family chapel is the most outstanding testimony of his patronage (VIII, PL. 193).

The designation of Lorenzo de' Medici as the greatest art patron of his century has met with strong reservations, and it has become clear that he was of far greater importance to the writers than to the artists of his era. His most notable contribution to the development of Italian art lay not in the artists he employed personally but in those he sent out of Florence — Leonardo da Vinci to Milan, Verrocchio to Venice, Benedetto da Maiano to Naples, Botticelli and possibly Signorelli (qq.v.) to Rome. Fully aware of the propaganda value of spreading Florentine cultural achievement abroad, he thus inaugurated a type of official patronage that was to become important in later centuries. Lorenzo's personal attitude toward the arts was that of a cultivated dilettante — a type more familiar in the 18th than in the 15th century. He delighted in the small bronzes of Bertoldo di Giovanni, the one artist with whom he was on close terms; he himself indulged in a little amateur architecture; and he enjoyed his role as arbiter of taste, paying homage to the great pioneers of the past and taking a keen but detached interest in certain of the achievements of the present.

If the three generations of the Medici family varied greatly in their approach to the arts, so too did the nature of the painting produced in Florence during the period of their rule. The contrast between the austere, republican, bourgeois world of Cosimo and the aristocratic, almost courtlike atmosphere that surrounded his grandson Lorenzo has inevitably been cited as accounting for the differences between a Masaccio (q.v.) and a Botticelli. Some historians have argued from this premise that the patronage of an "advanced bourgeoisie" ultimately encourages a classic-realist style, while that of an aristocracy tends toward the fanciful, the Neo-Gothic, and the unrealistic. Evidence drawn from the 17th and 18th as well as from the 15th century, though inconclusive, can be marshaled to lend support to this point of view; but an equally important and much more defensible conclusion from such evidence is the implicit recognition that the influence of patrons can be, and has been, very great indeed. The painter who worked independently and sold his pictures as best he could was virtually unknown outside Flanders until the 17th century, and then only rarely until the 19th. Painters worked under contract (i.e., precisely worded commissions), and the few written records that survive show the sort of control which could be exercised. Besides the subject to be painted, the artist was told the size of the work, the number of figures, and often the color and quality of the paints to be used. Moreover, until the 16th century, when Rome became a magnet to artists from all over Italy and Europe, close local contacts between the artists and their patrons offered further opportunities to exert influence.

Some patrons looked upon the artist primarily as a mechanical interpreter of their own ideas or of those of their Humanist friends and advisers. Thus Isabella d'Este, who between 1490 and 1510 decorated her *studiolo* in Mantua with allegorical pictures by the leading artists of Italy, laid down the most stringent programs for Mantegna, Perugino (qq.v.), and Lorenzo Costa. Only her instructions to Perugino (who was living outside Mantua) have survived; in these she gives the most minute details as to the subject to be depicted ("una battaglia di Castità contro di Lascivia"; PL. 116), the size of the figures, and so on, and ends with the recommendation that he is "free to make fewer [figures], but not to add anything else." Clearly, the tyranny of the new secular patrons over the artist far exceeded that imposed on him by traditional religious iconography in preceding centuries.

Yet, despite their strict requirements, it was the very enthusiasm of these patrons and the consequent rise in prestige of the artist that helped change his former anonymous or servile condition. Isabella d'Este herself was prepared to be accommodating when she approached a Leonardo ("if you are satisfied, we shall leave the subject and the time to you"). Also of interest is the story of her relations with Giovanni Bellini. After some negotiations with the aged artist, her agent was forced to write: "The theme that Your Excellency writes I should find for the drawing would better be left to the imagination of him who is to execute it and who feels that many specific instructions do not suit his style, he being accustomed, as he says, always to range freely at his will in his paintings, so that all which is in him may satisfy the beholder. . . ." This expression of the artist's prerogative is a vital landmark in the history of patronage.

Not all artists took this attitude, however. Twelve years later Titian (q.v.) wrote to Alfonso I, brother of Isabella d'Este, about that duke's plans for a picture: "I am firmly convinced that the greatness of the art of the old painters was in large part, or rather altogether, helped by those great princes who very wisely gave them prescriptions, by means of which they achieved fame and praise. Therefore, if it be the will of God that I may in some way satisfy Your Lordship's expectations, who knows but that I may not receive praise? And, nonetheless, in this I will have given only the body, while Your Excellency has given the spirit, which is the worthiest part of a picture." Titian was always a very able courtier, but this letter should not be read as mere flattery. Although the idea that the patron was the true creator was already becoming old-fashioned by then, it had been sincerely upheld for centuries, and as late as 1537 Benvenuto Cellini evoked this comment from one of the Pope's courtiers: "It is our lot to be the inventors, and yours to be the executors."

One man dramatically altered the balance of relation between artist and employer as a result of the immense prestige he acquired. When Michelangelo (q.v.) was only 29, he quarreled with the 53-year-old Pope Julius II, the greatest patron and one of the most powerful figures of the Italian Renaissance, who, in demanding the artist's return from Florence, whither he had fled, wrote: ". . . we are not angry with him, knowing the minds of men of his kind. . . ." Later, the young Federico II Gonzaga wrote of Michelangelo, from whom he desperately wanted some work, to his agent in Florence, Giovanni Borromeo: ". . . and for this there is nothing that we would not do for him, if we knew what would please him . . . we desire and we commission you to try to find him and acquaint him of our good opinion and kindly disposition toward him and ask him in our name, in the most effective and friendly way you deem suitable, if he be willing to do me this honor, to give me some work of his hand, either of painting or sculpture, whichever he prefers, for we do not favor one over the other, so long as it is from his own hand. And if by chance he were to ask you what subject we desire, you will tell him that we seek and desire nothing more than a work of his own creation. . . ."

Michelangelo's fame and the legend he inspired were so great that the figure of the artist as an inspired creator, superior to the petty world in which his patron lived, became through him a deeply rooted concept in the European imagination. "Be it known that men like Benvenuto, unique in their profession, are not to be subjected to the law," said Pope Paul III, according to the account of Benvenuto Cellini. Though the words clearly are fanciful, they express an attitude which persisted and which was revived, sometimes with disastrous results, notably in the 19th century.

The great achievements of Italian art had broad consequences for patronage. Elsewhere in Europe rulers began to recognize the prestige deriving from support of the arts, and they soon took to competing with the Italians. Francis I of France and the Holy Roman emperors Maximilian I and Charles V helped to establish an idea that was to remain important for many centuries, namely, that a prince could gain great glory not only for himself but also for his country by supporting the artist. Venice was one of the states that recognized this possibility early, and the post of "Painter to the Republic" had long been an official, paid, and coveted job. As a result, in the 16th century Venice surpassed all the states of Europe in the rewards she reaped from munificent patronage. The appearance of competing patrons further raised the status of Italian artists — the only ones, apart from Dürer (q.v.) in Germany, thought to be worth considering. The practice of giving titles to artists became more widespread, and biographers celebrated their triumphs with stories such as that of Charles V picking up Titian's paintbrush. Vasari himself took more practical (though, in the long run, probably less effective) steps to consolidate the social gains won by Michelangelo, by helping to found in 1562 the first art academy, the Florentine Accademia del Disegno, headed by Michelangelo himself and Cosimo I de' Medici.

The Reformation and the Counter Reformation radically affected the patronage of religious art. In the Protestant countries commissions for altar paintings and church decoration, which had until then constituted the overwhelming majority of orders for artists, virtually ceased. In the Catholic countries, on the other hand, such commissions vastly increased. Newly constituted or reformed religious orders such as the Jesuits, the Oratorians, and the Theatines rivaled each other in the splendor of their commissions. Churches and cathedrals — foremost among them St. Peter's in Rome — were built and decorated on a scale unknown since the height of the Middle Ages. Fortunes were to be made by painters, sculptors, and architects, at some cost, however, to their independence and social status. The enormous freedom that certain artists had attained during the Renaissance (e.g., there is reason to believe that even in the Sistine Chapel Michelangelo was given a comparatively free hand) was now restricted and attempts were made, even if only sporadically, to impose a uniform iconography. Books were published to expand upon the vague decrees issued by the Council of Trent. There are reports, admittedly rare, of certain works being rejected as unsuitable, and it is quite clear that control over the artist was tightened. An overemphasis of these requirements and proscriptions produced the since-discredited theory that certain orders, notably the Jesuits, were directly responsible for specific artistic styles. While it is true that before the end of the 16th century the Jesuits had claimed responsibility for inaugurating the great fresco cycles of tortures, in general their impact on patronage, apart from such iconographic innovations, was limited to influences that were of a kind too imprecise to be discussed in this context.

Extensive secular patronage continued throughout the 17th century, following on an even greater scale the examples laid down by the popes, Francis I, and Charles V. The aim of establishing strongly centralized monarchies, which met with varying success, led to the artist's becoming increasingly tied to the court. At the same time, the view came to be universally accepted that art patronage was an essential attribute of monarchy. Colbert remarked, "Nothing gives better proof of the greatness and intelligence of princes as do buildings, and all posterity measures them by the yardstick of the superb buildings they raised in their lifetimes." On another occasion he elaborated that he had told His Majesty that, even if he did not take pleasure in beautiful things, "a great prince must pretend that he loves them, and cause all such works of art to be made."

It is no coincidence or mere accident that popes Urban VIII and Alexander VII, kings Charles I of England and Philip IV of Spain, and cardinals Richelieu and Mazarin, as well as King Louis XIV of France and many other rulers, were among the greatest art patrons of their century. In Rome a policy of extravagant patronage of the arts was deliberately adopted to bolster declining prestige, and the popes were fully aware of the part that great artists could play in achieving their aims. Urban VIII is said to have exclaimed to Bernini (q.v.) on becoming pope, "It is your great fortune, oh noble sir, to see as pope Cardinal Maffeo Barberini, but ours is much greater in that the *cavalier* Bernini lives during our pontificate." Between them, these two men transformed Rome.

In France the same policy was pursued to assert the emerging ascendancy; and in England and Spain there were the art-loving kings Charles I and Philip IV, whose courts were the hub of great creative activity. Rulers inevitably acquired a virtual monopoly of the work of favorite artists. Velázquez (q.v.) rarely, if ever, worked outside the court circles once he was established there; Bernini was forbidden by the popes to accept outside commissions without special authorization, which was granted only when it suited their diplomatic purposes. Painters were lodged in the royal palaces and became part of the king's retinue. In 1686, testifying to these conditions, Lebrun wrote: "It has been more than 20 years since, following orders I received from the King, I abandoned the works I was engaged upon and directing for private individuals and undertook exclusively to work on and direct those which were being made for His Majesty at the Gobelins."

It was in France that the political implications of art patronage were most fully realized. Through control of specially created artistic commissions, a thorough dictatorship over the arts was exercised by Louis XIV, his minister Colbert, and Charles Lebrun, "premier peintre du roi." "I entrust to you the most precious thing on earth — my fame," Louis XIV is reported (no doubt apocryphally) to have told his Academy, and for the first 45 years of his reign all the most talented artists in his kingdom concentrated their abilities on celebrating it. With the completion of Versailles, France replaced Italy as the artistic center of Europe; but the cost in talent as well as in money had been considerable. The avowed propagandist motive behind the patronage of Louis XIV and the fact that one artist, Lebrun, was given such great power in administering it meant that — in one of the major instances in the history of patronage — stylistic uniformity was imposed to the detriment of individual genius. Hitherto even those courts which had been most anxious to monopolize the patronage of art had shown greater catholicity of taste.

Nevertheless, the 17th century, which witnessed such a compete royal monopoly of the arts in France, was also the period when collecting and patronage first spread widely outside the range of the ruling circles. In the third decade of the century Giulio Mancini, a Sienese doctor living in Rome, drew up some advice for potential patrons, *Alcune considerazioni appartenenti alla pittura come di diletto di un gentilhuomo. . . .* Quite clearly addressing himself to a whole new class of people who might turn to patronage, Mancini informs them how to set about buying pictures, what prices should be paid (pointing out, incidentally, that the great prefer to reward the artist with a present rather than haggle over money), how to judge quality, and where to hang the pictures once obtained: "Devotional subjects should be placed in the bedchamber; cheerful and secular works in the salon. It should be remembered that among the sacred subjects, the small ones should be placed at the head of the bed and over the kneeler; Christ, the Virgin, and other similar subjects facing the entrance, so that he who enters is reminded that this is a place reserved for devotions. This is for private dwellings."

This new class of patrons was certainly responsible for the great impulse given to the production of small easel pictures (still a comparative novelty, despite their introduction in early 16th-century Venice) and also to a whole new range of subject matter — the still life and the genre scene in particular. Even more important were certain rich and highly placed men whose choice of artists and themes was not dictated by political preoccupations and who therefore were able to indulge in a distinguished and selective patronage of the arts, which in turn gave opportunities to painters who were not fitted by temperament or special gifts to meet the demands of an autocratic court. Such patrons included, in Rome, Vincenzo Giustiniani and Cassiano dal Pozzo (both, incidentally, great collectors of antiquities), to whom so much was owed by artists such as the northern followers of Caravaggio, Nicolas Poussin (qq.v.), and Pietro Testa. It is significant that Poussin, Dal Pozzo's favorite painter, should have been among the first of the artists who worked almost exclusively for private patrons, both in France and in Italy.

During the first half of the 17th century, concurrent with the emergence in France and Italy of the private collector of independent taste, the extremely close, almost feudal links that heretofore characterized the relationship between artist and patron were somewhat loosened. Many painters deliberately refused the opportunity of living in a palace as one among a retinue of officials and servants and chose instead to set up independent studios. There they would sometimes sketch in the outlines of a picture and wait for a prospective purchaser before filling in the details. Some artists — Salvator Rosa is the most conspicuous instance — rejected entirely the established system of patronage. Rosa refused to take an initial down payment for his works or agree upon a price in advance since, as he said, it was impossible to tell how the picture would turn out; furthermore, he claimed that in any case he was unable to paint except when inspired. Such an attitude struck at the very roots of established practice, for it denied the patron that control (and still more, that assurance of results) which he had always expected. The way was in fact open to the complete breakdown of traditional patronage that occurred with the advent of the romantics. For the most part, however, Rosa's contemporaries were more puzzled by his behavior than ready to follow his example. Only a few had an inkling of the real significance of his attitude; a biographer wrote, "The place he had made for himself in the profession was one of esteem, because he knew how to act sagaciously and, moreover, made himself be sought after and importuned." Another factor that helped to change the relationship between patron and artist was the development of the exhibition, originally merely the adjunct of a saint's day, into a regular tribunal where an artist could win a reputation and impress a wide public.

The fundamental breakdown in the aristocratic and official patronage system occurred in the Netherlands. Here, where the Protestant churches showed no interest in painting and the ruling classes consisted largely of newly prosperous businessmen, the arts were treated far more commercially than elsewhere. Dealers, exhibitions, and auctions were essential elements in the artistic scene, rather than the marginal factors that they were in Italy. Pictures were bought and sold for speculation, and John Evelyn wrote in 1641 that "the peasants were so rich that they were looking for investments, and often spent 2,000–3,000 florins for pictures." This extremely wide range of patronage is reported by another English traveler of the time: "As for the art off Painting and the affection of the people to Pictures, I thincke none other goe beyond them, there having bin in this Country Many excellent Men in that Facullty, some att presentt, as Rimbrantt, etts, All in generall striving to adorne their houses, especially the outer or street roome, with costly pieces. Butchers and bakers not much inferiour in their shoppes, which are Fairely sett Forth, yea many tymes blacksmithes, Coblers, etts, will have some picture or other by their Forge and in their stalle. Such is the generall Notion, enclination and delight that these Countrie Natives have to Paintings. . . ."

Despite their popular appeal, Dutch artists did not live very well. Guild rules continued in force long after they had fallen into disuse in Italy, and no artist could sell his work unless he was a member of the guild. In theory he was not allowed to sell privately at all. Moreover, the impact of the Humanists, who in Italy and France had done so much to elevate the status of artists, had been slight in the Low Countries. Certain painters were bound by contract to consign their entire output to specific dealers, and many of these agents, such as Johannes de Renialme and Hermann Becker, became extremely rich and influential. Of true patrons in the Italian sense, there were but few, discriminate collectors such as Jan Six and Constantyn Huygens, the friends of Rembrandt (q.v.).

The insignificance of religious and court patronage and the wider market for painting had important effects on the very nature of Dutch art. The large-scale history painting — the only other kind given serious consideration elsewhere in Europe — was of little interest to businessmen and peasants, who generally preferred smaller pictures that recorded the minutiae of their daily lives.

Apart from this intense commercialization of art — unparalleled elsewhere in Europe until the 19th or even the 20th century — the most conspicuous contribution of the Dutch to art patronage was the group commission. There had, of course, been many precedents for this throughout medieval Europe, in the guilds and later in the Venetian "scuole," which were associations of laymen who banded together for charitable purposes and who commissioned great religious decorations such as the canvases by Tintoretto (q.v.) for the Scuola di S. Rocco. There had also been many official commissions (e.g., the so-called "Opera del Duomo") charged with contracting for and supervising the execution of work on great religious establishments. In the Netherlands, instead, the associations were made up of professional men (army officers, surgeons, governors of hospitals etc.), who preferred to have their own achievements rather than those of the saints recorded by the artist. Their commissions, thus, were usually confined to the group portrait, in which Frans Hals (q.v.) and Rembrandt excelled.

After the decline in power of the popes and the death of Louis XIV, the extremely centralized court patronage of the arts that had been so characteristic of the 17th century greatly diminished in Western Europe. And though private collectors became very active and important, there remained a gap that could never quite be filled. Throughout the 18th century there were laments for the past, and there soon developed a controversy that has lasted until the present: Which came first — the artist or the patron? "Nor would excellent artists ever be lacking in any of the arts if recognition and prizes were forthcoming, for the world always has men who can attaint excellence, but either they remain unknown because held to little account, or else they are not active because they despair of any reward." That was the case for one side, the view that "some mute, inglorious Milton here may lie." The answer came soon enough. "Let there arise once again the Apelleses, the Raphaels, and the Titians — and the Alexanders, the Charleses, and the Leos will not be lacking," said Francesco Algarotti, and the words were repeated a thousand times all over Europe as a stick with which to belabor contemporary painting and justify its neglect.

In reality artists' complaints about the decline in patronage had some justification. In France the royal finances were far too unstable to permit extensive support, and in England the court was altogether marginal in the country's artistic life. But more important than any financial difficulties was a new and pervasive psychological attitude, for by the end of the 17th century everyone felt that the golden age of art (i.e., the High Renaissance, with an afterglow in the Carraccis and their followers) had vanished irrevocably. The battle between the "ancients" and the "moderns" was won by the "moderns" as the very moment when people began to lose faith in them — at least as far as the visual arts were concerned.

For the first time a really serious contradiction between collecting (of antiquities and old masters) and patronage began to bear influence. In Rome, at the beginning of the 17th century, Vincenzo Giustiniani and Cassiano dal Pozzo had both had the finest existing collections of antiquities and of contemporary pictures. By the late 17th century this sort of catholic collection was almost impossible to find. Francesco Algarotti had some difficulty in persuading Augustus III of Poland to patronize his contemporaries; that passionate art lover was far more interested in obtaining works by the great artists of the past. Yet it was only the Eastern European sovereigns — those of Prussia, Saxony, and the other German states, of Poland and Russia — who attempted to continue the large-scale patronage earlier characteristic of the Western European rulers.

In France and England the most influential patrons were not the kings, but the amateurs, the connoisseurs, the dilettanti. In France, especially, a very close relationship developed between a number of artists and enlightened collectors, a rapport that profoundly affected the nature of 18th-century painting, sculpture, and architecture. The heroic and allegorical themes that had always been favored by rulers were replaced by a more intimate kind of art reflecting a whole new range of sensibilities. Pictures came to be admired more for their esthetic qualities than for the subjects they portrayed; "la belle matière," the nervous brushstroke, or the significant detail was picked out of context and praised. The artist's sketches were collected by amateurs, flattered by the illusion of thus sharing in his creative rapture. In some cases the patrons themselves took an active part in executing the works of art they stimulated. In England Richard Boyle, Earl of Burlington, collaborated with William Kent, and often took the initiative, in a number of important buildings; in France Louis XIV's great grandson, the Duc d'Orléans, who was the most important collector and patron of the early years of the 18th century, produced some erotic illustrations for an edition of *Daphnis and Chloe*.

The most notable example of this new type of patron was the agent of the Duc d'Orléans, the fabulously wealthy banker Pierre Crozat (1665–1740), who devoted most of his life to collecting and patronage. His influence has been immortalized in the work of Jean Antoine Watteau (q.v)., who lived in his house. Crozat showed a sympathetic understanding of the special talents of the artists whom he cultivated. He was able to introduce them to the particular masterpieces from his own collection that would encourage their development. With his friends and clients, he could make artistic reputations, notwithstanding the neglect of those traditional sources of patronage, the church and the court. His friend Pierre Jean Mariette wrote, "Monsieur Crozat did not delight in his drawings for himself alone; on the contrary, he took pleasure in showing them to art lovers whenever they requested it." Crozat organized weekly meetings at which scholars, amateurs, and artists looked at his collections and discussed them. Probably for the first time the artist met his clients, his critics, and his protector on equal terms, and it is not surprising that in France, where this new relationship developed, more than anywhere else, every nuance of current fashions was reflected in painting, interior decoration, and architecture.

Such close contacts between artists and public — maintained also through the Salon exhibitions and by means of newspaper criticism — sometimes meant that the painter was subjected to a tyranny surpassing anything he had suffered when still considered a social inferior. The Comte de Caylus was strongly attacked for his ruthlessness with painters, and Jean B. Greuze (q.v.) once caricatured his patroness Mme Geoffrin as a schoolmistress with rod in hand. Mme Geoffrin inaugurated a practice of weekly dinners with artists. Every Monday evening she was at home to all the leading painters, sculptors, and architects of the day — François Boucher, Hubert Robert, Maurice Quentin de Latour, Etienne Maurice Falconet, Jacques Germain Soufflot (qq.v.), Jean Baptiste van Loo, Joseph Marie Vien, Edme Bouchardon, and many others. Though she herself was no great connoisseur, she transmitted to them the ideas of other guests such as Denis Diderot and Baron Melchior von Grimm, who came with the *philosophes* on Wednesdays. Her influence with patrons and collectors was overwhelming; one visitor wrote, "Mme Geoffrin's position was that of a kind of administrator [*police*] for taste." Another said she was "a veritable tyrant, claiming to dominate men of letters and artists." All over Eastern Europe, especially in Russia and Poland, her word was law and no one would think of buying a French picture without asking her advice.

Curiously enough, however, the cosmopolitan 18th century was also the era in which national tastes in art patronage asserted themselves most strongly. For the first time no single country dominated the European taste. The French showed little interest in contemporary Italian art, the English still less in French art, and neither Frenchmen nor Italians were aware that any painting at all was being produced in England. If one is to believe the artists of the time, this neglect extended to the English themselves. William Hogarth (q.v.) was venomous about the typical English patron, who liked anything so long as it was foreign. In reality, English patrons were quite prepared to support English art as long as it kept to the only two branches of painting in which they were interested: portraits and landscapes. In vain did Sir Joshua Reynolds (q.v.) preach the need for a national school of history painting; like every other successful artist he had to earn his living by portraiture. The only serious attempt made by an English patron to put Reynolds's words into practice was that of Alderman John Boydell, who in 1786 proposed a scheme whereby the leading artists of the day were invited to paint a series of subjects from the plays of Shakespeare, paintings which were then to be engraved and issued in volumes. Although a large number were painted, it is characteristic that within a few years the scheme ended in failure and its promoters were reduced to bankruptcy.

Throughout the 18th century, the English were by far the greatest patrons and collectors, and they carried the quest to satisfy their tastes to the Continent, forcing up prices everywhere. Charles de Brosses's complaint that Canaletto was "gaté par les Anglais" was widely echoed — and was perfectly justified. All over Italy artists turned to the previously neglected field of landscapes and *vedute* to satisfy the rich "milordi" whose extravagance became proverbial. History painters such as Pompeo Batoni were compelled to take up portraiture, and one Italian amateur grumbled that "if Bonarotti were alive today to paint his *Last Judgment*, he would find it worth his while to depict Christ Judging as some Milord."

Still, though tastes and patrons changed, about the middle of the 18th century the basic elements of the system itself were essentially what they had been several hundred years before. In Italy especially, there is often no important difference between a contract drawn up in medieval times and one of 1750: the price was still arranged in advance, and the number of figures and sometimes even the colors to appear in the work were dictated to the artist. However much the social position of the artist had improved, he was still looked upon as an infinitely superior craftsman who, once he had learned the rules of his art, could be relied upon to turn out a work of standard quality. Michelangelo's unique genius, as well as sporadic protests by independently minded artists, had not had any lasting effect. At the end of the 18th century, however, an entirely new attitude toward the artist gradually developed, one that was to change the very nature of patronage.

Formulated first in Germany and eventually adopted all over Europe as the romantic movement spread, this theory proclaimed that the artist was a genius whose very nature was, of necessity, different from and superior to that of the society in which he lived. The genius owed it to himself to give expression to his inner convictions and, if necessary, to reject any of the standards, either artistic or moral, that society imposed on him. Clearly, once this view was accepted, the old system of patronage was bound to break down. How could the patron dare to instruct the artist in his art? How could the artist himself decide in advance what his picture was to be like, when its quality was bound to vary with the nature of his inspiration?

The full implications did not become clear until much later, but glimmerings of the great change were already apparent in the attitudes of a number of artists working at the end of the

18th century. In Spain, for instance, Goya (q.v.) said that he was painting a large number of uncommissioned works, for in them he was able "to make observations for which commissioned works give no room and in which fantasy and invention have no limit." Certainly many artists had done this before — especially in their drawings — but it was only now that the issues began to be stated so clearly.

However, the tension between the best artists and the public probably would not have been so great as it was during much of the 19th century had not the patrons themselves changed as drastically as did the claims of the artist. In France the power of the ruling class was broken by the Revolution. The aristocracy and — more important for the arts — the *fermiers-généraux*, those rich, cultivated, easy-going, and leisured patrons who had meant so much to Fragonard (q.v.), for instance, vanished from the scene after 1789. Social changes in England were as profound, if less dramatic. The industrial revolution brought to the fore a class of enterprising manufacturers whose outlook on life was quite different from that of the landed gentry, and isolation from the Continent during the Napoleonic Wars merely strengthened already innate prejudices against cosmopolitanism. Many of these men, such as Robert Vernon and John Sheepshanks, were active patrons, and it can be said that the Victorian period was one of unprecedented prosperity for those artists who conformed to the prevailing taste. It was the standard of taste that succumbed; patronage was as extensive as it had ever been.

In both France and England the state took the initiative in a number of great artistic enterprises. In Paris Delacroix (q.v.) painted extensive decorations in the Palais Bourbon, in the Luxembourg Palace, and elsewhere; in London the rebuilding and decoration of the Houses of Parliament after the fire of 1833 constituted what was probably the greatest official artistic commission ever given in England. But all these schemes were designed to celebrate old institutions. The new industrial society then emerging took little account of the artist, who for the first time found not only himself but also his work considered a nonessential, irrelevant luxury. "Courbet," wrote Sainte-Beuve, "has the idea of making the vast railway stations serve as the new churches to be painted, to cover those immense walls with a thousand perfectly suitable subjects: advance views of the great places to be visited; portraits of great men associated with the cities along the route; picturesque, moral, industrial, and metallurgical themes — in a word, the saints and miracles of modern society. . . ." Unfortunately, such opportunities were not given him, and painting and the arts generally were more and more looked upon as something set apart, with vague ceremonial or spiritual associations but no practical value.

The industrialists who took this line often sought to build private collections themselves, but their choice of artists for patronage appears to have been singularly unfortunate. This failing is difficult to explain — even more so, if one considers that, until the 19th century, there are astonishingly few cases of an artist's contemporary reputation having been reversed by posterity. But once the utilitarian value of art had been rejected, the only basis for satisfactory patronage lay in the amateur approach of the 17th and 18th centuries, in which the patron desired to identify himself with the artist. This the average 19th-century bourgeois refused to do, though he had very definite ideas as to what he wished of the artist. He wanted his picture not only to have but also to emphasize as many signs of solid workmanship as did the product of his factory. He refused to see merit in "unfinished" work, and often he looked upon changes in style as a political or social affront. It is worth mentioning that the impressionists met with far greater appreciation in the more fluid society of the United States than in France or England. As the artist placed more and more value on his own independence and originality, a clash was inevitable. Starting with Gustave Courbet, who defiantly declared that he despised patrons ("je méprise les mécènes"), most of the great artists of the 19th century had difficulty in finding any public at all; however, the subsequent legend that they deliberately flaunted their nonconformity is wholly untrue. With very rare exceptions, the great masters of the 19th century were genuinely puzzled and deeply hurt by the lack of support from their natural patrons, the *bourgeoisie*, whose achievements they so memorably celebrated. The "Bohemian" was an exception, and virtually never a first-rate artist until the very end of the century. Patronage in the old sense of dictation to or collaboration with the artist was by now out of the question, and from the 19th century onward the term "patron" generally meant little more than a collector of contemporary work. Even these were rare enough, though men such as Victor Choquet, Père Tanguy, Gustave Caillebotte, and Georges Charpentier — all of whom bought early works by the impressionists and their successors — should be remembered.

In these critical conditions, however, there emerged a new type of patron — the enlightened dealer. Paul Durand-Ruel, who supported Renoir, Monet (qq.v.), and so many others during their years of hardship, established the type and was followed in succeeding generations by Theo van Gogh, Ambroise Vollard, Daniel-Henry Kahnweiler, and many others. Not only did Durand-Ruel give generous financial backing to the artists who worked for him at a time when he himself had suffered greatly because of his interest in their unpopular works, but he also constantly arranged exhibitions and took steps to ensure that his favorite painters received regular publicity. This type of dealer — one who led rather than followed public opinion — was unprecedented, and Durand-Ruel and his successors can be called patrons more legitimately than many of those to whom the name had been given in previous centuries. In a very real sense, they helped to make possible the triumphs of late-19th-century French painting, and often the artists responded by painting their portraits, much as Van Dyck had painted Charles I or as Velázquez had depicted Philip IV more than two hundred years before. Durand-Ruel, whose extremely reactionary political views contrasted strangely with his daring artistic tastes, understood the role he was assuming and as early as 1869 he wrote: "A true dealer should also be an enlightened art lover, ready, when necessary, to sacrifice his apparent immediate interests in favor of his artistic convictions, and to fight against the speculators rather than share in their machinations." The great period of patronage by the art dealers lasted from about 1870 until about 1910; after that date the general public, perhaps inspired by the acceptance and celebrity of the impressionists, began to take an increasing interest in contemporary work, and the "misunderstood artist" became a rarer phenomenon.

Francis Haskell

In the 20th century genuine patronage, involving the direct commission of art works or regular subsidies to artists as opposed to the collecting of finished works, has come to be limited largely to government, philanthropic, and industrial sources. The purchases of individual collectors have little determinative influence on the character of specific works of art. Government commissions are quite often intended for commemorative purposes; the artistic encouragement and grants of philanthropic foundations are frequently attributable to economic motives such as tax exemptions; and industrial commissions are usually inspired by reasons of publicity, as well as by tax benefits.

Whereas the commissions and subsidies of modern governments have in general tended to support conservative tastes (witness the typical uninspired public memorial, or the social realism favored by totalitarian regimes for propaganda purposes), there have been a number of exceptional official programs that have supported more progressive trends in contemporary art. Notable among these have been the Federal Art Project of the Depression era in the United States and the Arts Council of Great Britain.

In the United States the nationwide program of art subsidy was administered through regional committees and competitions until American entrance into World War II; moreover, as an unusual measure, government support was extended to easel painting and graphics as well as to the more customary forms of large-scale public sculpture and murals. Begun as a relief program for the unemployed, it was a means of encouragement and recognition for many promising and avant-garde artists. Although so extensive a program has never been resumed and official commissions have once more veered toward the academic tradition, there has been continued legislative interest in again establishing some kind of art subsidies. In the meantime, the officially sponsored traveling exhibitions of American art sent abroad, though on a more limited scale, have demonstrated an admirable willingness to display some of the more adventurous currents in American art. The Fulbright program of grants for travel and study abroad, making use of funds accrued from war debts, is another government measure for support of the arts that has found recipients among the more avant-garde American painters and sculptors of the present day. Of the private philanthropies apportioning funds to American artists, the most ambitious in scale are the Ford and Guggenheim foundations.

The Arts Council of Great Britain is a semiofficial body created to support and further the arts. In its activities this group, too, has tended to give substantial recognition to the more modern trends in British art.

Under certain totalitarian regimes in the 20th century, the state not only has insisted on a far greater control over artistic style and subject matter than would ever have occurred even to Louis XIV but also has taken active steps to prevent the creation of art it does not approve of, including works intended for private collectors. The hostility vented upon the German expressionists by the Nazis, as well as their systematic persecution of Jewish artists, exemplifies this type of reprehensible action which arises from political and racist motives, an inverse and destructive exploitation of the propaganda values of art. Similarly, under other authoritarian governments, abstractionist and other avant-garde tendencies have been ruthlessly suppressed as "decadent" or "corrupt," in favor of a prosaic social realism conveying the political and economic aims of the state. Looked upon as insidious by these governments, the bold and free-thinking expression that ordinarily characterizes such movements represents a spirit that is to be suspected and rooted out. This systematic, large-scale "negative patronage" developing in the 20th century has no true comparison in the various historical epochs treated in this article.

* *

The incidence of a genuine, considered patronage that is well grounded in tradition is greatest in cultures rich in — and markedly appreciative of — individual artistic and intellectual endeavor, societies of long refinement such as that of China. Patronage developed in China in the first centuries of our era, following close upon the emergence of a critical attitude toward art. The recognition and protection of artists was limited to painting, the only art regarded as the fruit of genuine inspiration, and sculpture remained the product of anonymous artists. The court became a highly refined center of cultural life in which artists and craftsmen held important positions — such as Han Kan, who gained renown as an imperial painter of horses. Li Yü, the last of the T'ang emperors in Nanking (938–78), was one of the first great patrons and also a collector of earlier art, besides being an accomplished amateur poet. His interest and influence made a direct contribution to the rich artistic activity at his court, where painters were given official status. Many subsequent rulers were themselves painters and poets. (See CHINESE ART.)

Under the Sung dynasty the first imperial academies of men of letters, painters, and calligraphers were established. The Sung emperor Hui-tsung (r. 1101–26), himself an artist, exerted a particular influence on the academy members, who were officially graded in several ranks. Of one notable artistic family, the Ma, five generations participated in the Sung imperial academy. Neglected afterward in the Yüan period, the imperial academy was revived by the first Ming emperor, as a reflection of both the artistic splendor and the political efficacy of the Sung reigns. The painter emperor Hsüan-tsung (r. 1426–35) was one of the principal forces in this Sung revival. Imperial favor again became of prime concern to the artists, and in some instances its loss resulted in a death warrant for the artist. Artistic commissions and collections of precious objects gradually came into existence outside the imperial court, too. Apart from court circles, religious communities in particular, according to their means, supported artists and provided the incentive for many of their works.

In Japan, within a feudal organization that existed until fairly modern times, there early developed a tradition of official patronage and a hereditary artist caste, supported by imperial circles, the shoguns, the daimios, and the warrior class in general. The commission of art and the regular employment of artists was one means of demonstrating political power and acquiring prestige. As early as the 6th century, Chinese and Korean artists had been officially invited to Japan to execute works for Japanese sovereigns; and in 604 it was recorded that painter clans were called to the imperial court, where they were greatly esteemed and were declared exempt from taxes. Later, in 808, a bureau of painters (the E-dokoro) was established to regulate all matters relating to painting for the imperial court, with the power of appointing official painters (*e-shi*). It was during this time, in the Heian period (794–1185), that the members of the court adopted a way of life more distinctly Japanese, and through their patronage the change was clearly reflected in the painting of the period. This Japanese style, called Yamato-e (q.v.), was clearly evident by the 10th century. (See JAPANESE ART.)

In the Muromachi period (1334–1573), the Zen priesthood — men of letters as well as of religious vocation — became an important source of patronage for artists, who were commissioned to draw *sumi-e* scenes to accompany the Zen poetry, giving rise to a form called the "poem-picture roll" (*shigajiku*). Artists who distinguished themselves in religious commissions were given honorary titles (*hokyō, hōin*) by the monasteries they had embellished. Zen priest-artists themselves occupied official posts as government artists. Thereafter the *sumi-e* developed by the priesthood took hold among other groups such as the warrior class, and in the process what was originally a contemplative, literary art gradually changed into a style more suited to the wants of military patrons, with a more secular and decorative tone.

The long tradition and the vicissitudes of Japanese official patronage are best traced in the varying fortunes of the great artistic families of the Kanō and Tosa schools (qq.v.), whose members were widely acclaimed in court circles and among the military caste.

The society of the Edo period (1615–1867) was, in its earlier phases, strictly feudal in organization but Confucian in its philosophy and ethics. In accordance with this Confucian emphasis, the former boldness of the Kanō style was turned toward more gentle harmonies that also better suited the conservative tone of the ruling caste. Once again, distinguished artists were given great recognition and official court status, a practice that continued until the fall of the Tokugawa regime in 1867. On fixed days of the month, senior court artists (*oku-eshi*) assembled to work in a chamber called the "painters' room" (*o-e-beya*) in Edo Castle, where they restored older works and painted their own, in addition to giving expert advice on the official collections.

The patronage of the central government was emulated by provincial administrators, and thus the Kanō style gained wide diffusion throughout Japanese territories. Apart from the local patronage of the daimios, the spread of the style was also due to the fact that the Kanō school had readily adjusted to the needs and taste of the dominant military caste, becoming virtually the artistic expression of the warrior class. Unfortunately, in time the official recognition seems to have affected the quality of the art adversely, and afterward Kanō painters in particular tended toward a hard academicism, with a lack of original development.

While this discussion has centered about painting, it should be noted that lacquer and decorative metalwork were two other artistic forms extensively supported by the ruling classes. Just as there were official court painters, there were officially appointed lacquerers and weapons decorators. In these crafts also, an eventual decline is to be noted in the Edo period, a lack of expressive freedom that perhaps reflects the severe regimentation of all aspects of life during that era in Japan.

BIBLIOG. A. J. Dumesnil, Histoire des plus célèbres amateurs italiens, Paris, 1855; A. J. Dumesnil, Histoire des plus célèbres amateurs français, 3 vols., Paris, 1856–58; A. J. Dumesnil, Histoire des plus célèbres amateurs étrangers, Paris, 1860; Documenti inediti per servire alla storia dei musei d'Italia, 4 vols., Rome, 1878–80; E. Müntz, Les artes à la cour des papes pendant le XVe et le XVIe siècle, 3 vols., Paris, 1878–82; E. Bonnaffé, Dictionnaire des amateurs français au XVIIe siècle, Paris, 1884; E. Müntz, Les collections des Médicis, Paris, 1888; C. Yriarte, Isabella d'Este et les artistes de son temps, GBA, XIII, 1895, pp. 13–32, 189–206, 382–98, XIV, 1895, pp. 123–38, XV, 1896, pp. 215–28, 330–46; G. Clausse, Les Sforza et les arts en Milanais 1450–1530, Paris, 1909; H. Hantsch and A. Scherf, Quellen zur Geschichte des Barocks in Franken unter dem Einfluss des Hauses Schonborn, 2 vols., Augsburg, Munich, 1931–35; L. Venturi, Les archives de l'impressionnisme, 2 vols., Paris, 1939; E. Panofsky, ed., Abbot Suger on the Abbey Church of St.-Denis and Its Art Treasures, Princeton, 1946; M. S. Briggs, Men of Taste from Pharaoh to Ruskin, London, New York, 1947; J. Evans, Art in Medieval France 987–1498: A Study in Patronage, London, 1948; F. H. Taylor, The Taste of Angels: A History of Art Collecting from Rameses to Napoleon, Boston, 1948; G. Ricci, Relazioni artistico-commerciali tra Roma (Italia) e la Grecia negli ultimi secoli della Repubblica e nel I dell'Impero, Antichità, II, 1950, pp. 33–87; N. Lieb, Die Fugger und die Kunst, 2 vols., Munich, 1952–58; D. Sutton, Christie's since the War, 1945–1958: An Essay on Taste, Patronage and Collecting, London, 1959; E. H. Gombrich, The Early Medici as Patrons of Art, Italian Renaissance Studies: A Tribute to the Late Cecilia M. Ady, London, 1960, pp. 279–311; F. Jenkins, Architect and Patron, London, New York, 1961; J. Lees-Milne, Earls of Creation: Five Great Patrons of 18th Century Art, London, 1962; D. M. Fox, Engines of Culture: Philanthropy and Art Museums, Madison, Wis., 1963; F. Haskell, Patrons and Painters: A Study in the Relations between Italian Art and Society in the Age of the Baroque, New York, 1963. See also bibliogs. for DEALING AND DEALERS; MUSEUMS AND COLLECTIONS.

* *

PEALE, CHARLES WILLSON. American painter, naturalist, and inventor (b. Queen Annes County, Md., Apr. 15, 1741; d. Philadelphia, Pa., Feb. 22, 1827). Originally a saddle maker and a decorator of signs and coaches, Charles Willson Peale took lessons in painting from John Hesselius. After a visit to New England in 1765, he began to work as a portrait painter in Virginia. He studied with Benjamin West in London (1767–69) and then worked in Annapolis, Baltimore, Williamsburg, and Philadelphia, where, after having served as a militiaman in the Revolution, he settled down to a long and varied career, which included the opening of a museum of painting and natural history and the promotion of an academy of fine arts.

An early work of Peale, the *Portrait of William Buckland* (1770–74; New Haven, Yale Univ. Art Gall.), imitates the firm clarity of Copley; his full-length portrait of Washington (1778–79; Philadelphia, Pennsylvania Acad. of the Fine Arts) shows the subject as a gangling figure in an awkward pose. The *Staircase Group* (1795; Philadelphia, Mus. of Art; I, PL. 100) presents two of the painter's sons in a composition of great force and originality. In *Exhuming the Mastodon* (1806; Baltimore, Peale Mus.), he produced a faithful documentary record of an important event in which he had played the principal part. At the age of eighty-one he proved his zest for unusual and difficult problems by painting the *Lamplight Portrait of James Peale*, a study of his brother (1822; Detroit, Inst. of Arts). Painting was but one interest among many for "the ingenious Mr. Peale," who made engravings, practiced taxidermy, and invented moving transparencies for his museum. Never as brilliant as Copley nor as suave as Stuart, he relied on disciplined drawing, quiet color, patient modeling, and penetrating, sympathetic observation to lend his sitters a solid presence and vigorous characterization. (Cf. AMERICAS: ART SINCE COLUMBUS, col. 284).

Peale encouraged several members of his large family to become artists. His brother James (1749–1831) was a competent miniaturist, and his self-portrait with his family (ca. 1795; Philadelphia, Pennsylvania Acad. of the Fine Arts) is a charming conversation piece. Among the sons of Charles, Rembrandt (1778–1860) was an uneven painter, fresh and forceful in *Thomas Jefferson* (1805; New York, Mus. of the New-York Historical Soc.) but oversmooth in his several "porthole" portraits of George Washington; Raphael (1774–1825) painted exquisite still lifes (I, PL. 100), and his *After the Bath* (1823; Kansas City, Mo., Nelson Gall. of Art and Atkins Mus.) shows an eye-deceiving realism; Titian (1799–1885) made scientific drawings while on western expeditions; Rubens (1784–1865) turned to still-life painting when he was in his seventies. The Peale dynasty included three daughters of James, all specialists in still life: Anna Claypoole (1791–1878), Margaretta Angelica (1795–1882), and Sarah Miriam (1800–85).

BIBLIOG. C. C. Sellers, The Artist of the Revolution, I: The Early Life of Charles Willson Peale, Hebron, Conn., 1939; W. Born, Still-Life Painting in America, New York, 1947; C. C. Sellers, Charles Willson Peale, II: Later Life, Philadelphia, 1947; C. C. Sellers, Portraits and Miniatures by Charles Willson Peale, Philadelphia, 1952.

Oliver W. LARKIN

PELOPONNESIAN ART. The Peloponnesos, together with Attica and the Aegean islands, was the region having the most intense and continued artistic activity of the territories of ancient Greece. A large area, of varied topography and the home of diverse ethnic groups (some supplanting former groups, others, coexisting at different cultural levels), the Peloponnesos produced artistic styles and forms of great variety and of uneven quality. Busy harbors and seaports, chief among them Corinth, were open to foreign trade and receptive to foreign influences; but there were also quiet agricultural areas, such as Messenia, and mountainous regions, such as Arcadia, that were relatively isolated. In the appellation Peloponnesian there is also an implication of the cultural insularity of the whole region, which was so often in opposition to the rest of the Hellenic world.

It has been possible to discern and to trace the development of characteristic styles in several centers in the peninsula; but in others, general stylistic peculiarities are less evident, and in many cases some single fundamental element must satisfy for conveying their Peloponnesian character.

Systematic investigation of many of the important centers has resulted in a substantially reliable classification of the minor arts, including ceramics, terra-cotta pieces, and to some extent small bronzes. Uncertainties exist, nonetheless, especially for large-scale sculpture in which the artist's individuality prevails over regional characteristics. Considering the traditional mobility of Greek artists, no one may presume to decide, for example, whether the sculptures of the Temple of Apollo Epikourios at Bassai were executed by local or imported artists; the same is true of those in the temple at Skillous, and certainly of the marble sculptures of the Temple of Zeus at Olympia. Uncertainties remain even for series of small monuments commonly attributed (to some extent, on the basis of the area where they are found) to one or another important center of production. Embossed bronze laminae, mostly from shields, must still be described with a double nomenclature (e.g., Argive-Corinthian) despite the wealth of finds and the inscriptions on them. Caldrons with griffin protomas and statuettes of peplos-clad figures used as mirror stands have not been satisfactorily identified with any of the various artistic centers. Even the Doric temple — the supreme invention of Peloponnesian genius — has been assigned to Argolis only by tacit agreement, in the absence of any monumental documentation.

Attempts at establishing specific characteristics for the major artistic centers of the Peloponnesos are complicated by a remarkable ambiguity in the archaeological finds, because of the proximity of the centers and the close contacts that must certainly have existed between them. Except for Sparta, all the great Peloponnesian art centers seem to have been closely situated in the northeastern part of the region. Corinth and Sikyon are only 10 miles apart, and subjective reasoning is often the basis for the attribution of small bronzes found on Corinthian territory to Corinth itself, to Sikyon, or to the more problematic school of Kleonai. According to literary sources, the Sikyonian school of painting was closely connected with the Corinthian school. Moreover, the Argive school of bronzeworkers was in contact for centuries with the Sikyonian school; hence, the ambiguous traditional designation of "Argive-Sikyonian" school. Even more tenuous is the attribution of works known only through Roman copies in marble. (See ATTIC AND BOEOTIAN ART; GREEK ART.)

Peloponnesos, principal ancient centers and archaeological sites.

SUMMARY. Argos (col. 137). Corinth (col. 147). Sikyon (col. 157). Elis (col. 161). Messenia (col. 166). Arcadia (col. 167). Sparta (col. 170). Aegina (col. 176).

ARGOS. The oldest and noblest of Greek cities, Argos is the artistic center of the Peloponnesos about which least is known, in spite of abundant literary references and extensive archaeological exploration. The Sanctuary of the Argive Hera (FIG. 139), which must have been the heart of Argolis, has been substantially investigated, and the settlement of Argos itself has been largely excavated, revealing the arrangement of its magnificent monuments; but finds that are of value in studying the artistic production of the region have been extremely rare. No other city had such a wealth of tradition and of sacred images linked with mythical kings and heroes who were reputed to have lived and fought there. Traditionally, Argos was the location of the xoanon of Hermes by Epeios and of a votive offering dedicated by Hypermnestra, the only Danaïd who refused to kill her husband while he slept; it was also said to have contained the tombs of Tantalos, of Linos, the son of Apollo, and of Sthenelos, among many others. Argos is credited as the site of the throne of Danaos and the xoana of Apollo

Likeyos, of Artemis, and of Zeus, all given by Danaos (Pausanias, II, 19, 3–7). The only tangible remains of its artistic splendor, however, are fragments of Hellenic temples that were built in a simple, refined style upon the ruins of Mycenaean strongholds. These temples bear evidence of the dramatic replacement of one culture by another without however, showing such initial signs of Hellenic resurgence seen in the necropolises.

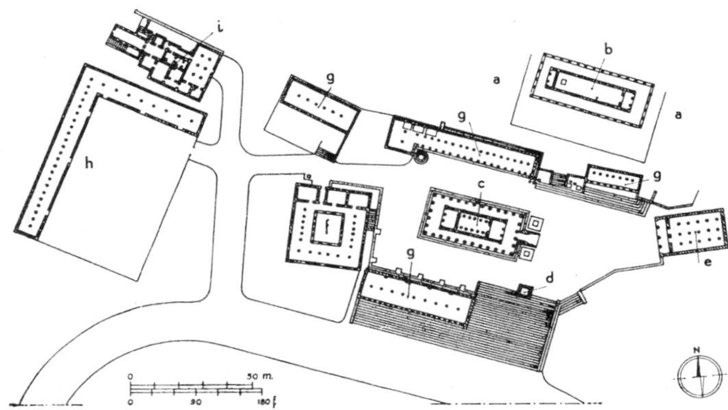

Argos, plan of the Sanctuary of Hera (Heraion). *Key*: (*a*) Terracing in polygonal masonry; (*b*) temple from first half of 7th century B.C.; (*c*) temple from the end of the 5th century B.C.; (*d*) altar; (*e*) hypostyle room; (*f*) banquet hall (*g*) stoa; (*h*) gymnasion(?); (*i*) Roman baths (*from C. Waldstein, 1902*).

The deposits from the tombs of Argos and Asine have proved more valuable than those of the Heraion in helping to estimate the extent and importance of Argive ceramics. The most significant examples are large craters (kraters), with and without the high conical stands or feet, that were developed with remarkable structural compactness and balance. Although many scholars have mentioned their heavy structure and their dependency on Attic forms (observations that are undeniable to some extent), the positive qualities of the Argive Geometric idiom must also be considered. Matz's remark about the "textile" quality of Argive ceramics points up the tight unity and balance of the decorative systems used. Neither the animal nor the human element is an essential of these decorative schemes, though the representations of a hero with a horse that are repeated in the metopelike areas have a monumental quality. These characteristics are even more apparent in some early sub-Geometric fragments of votive shields from Tiryns. In the most complete one (Nauplia, Archaeol. Mus.), which depicts the combat between Achilles and Penthesileia, the figures are rough and inharmonious, and yet they have a quality of grandeur, even in the heavy, summarily drawn filler motifs.

Evidence of the continuation of Argive ceramics in later periods is meagre and uncertain. The hypothesis attributing the Orientalizing craters from the Fusco necropolis, near Syracuse, to Argos has been discredited in view of other links revealed by material from Ithaca. The dating and attribution of the crater of Aristonothos (Rome, Pal. dei Conservatori) are uncertain, and the large conical crater support with mythical figures from the Heraion (Athens, Nat. Mus.) has been assigned to Attica. Apart from the location of the find, the more recently discovered fragmentary crater depicting the blinding of Polyphemos (PL. 66) has been assigned to Argos almost solely by a process of elimination. The few fragments that can be associated with this masterpiece, which is comparable to the great proto-Attic amphora with the same subject from Eleusis (I, PL. 339), are evidence of the importance and the individuality of this ware. On the Argos crater and on a fragment from the Agora in Athens (Mus. of the Agora), the figures are arranged against large expanses of color, as they are on the famous fragmentary plate of Praisos (Heraklion, Archaeol. Mus.). Isolated finds of later date, such as the crater fragment with a black-figured Gorgon from Mycenae, reflect a strictly local style that was apparently dependent upon Attic formulas from the time of Kleitias and Sophilos.

Traditionally, Argos was renowned for the mastery achieved in bronzeworking there. Though finds have yielded no positive data, the more elaborate bronze tripods from the Geometric and sub-Geometric periods, such as those found at Olympia, have been attributed with reasonable certainty to Argos. The distinctive Argive Geometric idiom, with its well-integrated and clear rhythms, was applied with a singular decorative effect to these bronzes. Lofty effects are obtained in the scenes of the metopelike spaces; among these are found some of the earliest representations of myths, such as the one depicting the struggle between two warriors for the tripod (VI, PL. 107). Argos was also traditionally praised for its fine weapons, and recent excavations have uncovered helmets and Geometric breastplates of remarkable beauty. Although these arms were intended to meet practical needs, they may be regarded as works of art in consideration of the effect produced by their rhythmic fundamental lines. The mysterious breastplate from Olympia still awaits sure attribution, but many of the embossed shield laminae that have also come from the Altis can be assigned to Argive workshops.

The large cast-bronze caldrons with griffin protomas (PL. 61) have been generally attributed to Argos, but the wrought metal ones must be regarded as having come from the East. The attribution of these works has been purely on the basis of deductive reasoning and consensus, since the finds are lacking in any definite data. Jantzen's study shows that the greatest number of such finds have been made in Samos, another center of bronze production. Though the Samos protomas form a singularly unequivocal and compact group, the griffin heads, found in abundance at Olympia (PL. 61; I, PL. 340) and considered to be Argive, are some of the finest examples of 7th-century art because of their varied detail and incisive forms. Some of the heads have a monumental character that is quite independent of their actual dimensions, which are quite large; the organic coherence and the remarkable balance attained between fantasy and lucid form show the vigor and individuality of the Hellenic contribution, even though based on an ancient Eastern formula. A similar development can be seen in a type of related bronze "appliqués," which are an enlightened Greek translation of the mechanical, formalized pattern of the *Assurattaschen*. Some authorities attribute this admirable clarification to Argive workshops. In this light, a bust in Athens (Nat. Mus.) and another in Boston (Mus. of Fine Arts) would also appear to be Argive, and their incisive forms and quiet authority rank them close to the head of Apollo from Amyklai (Athens, Nat. Mus.).

It is commonly agreed that monumental sculpture began to be produced in Greece about the middle of the 7th century B.C. As has been observed, Geometric and sub-Geometric plastic art, because its forms were summary and simplified and its characteristics spontaneous and immediate — one might say improvised — was ill-suited to the essential conditions of quiet meditation and slow execution involved in modeling or carving large figures. Technical considerations such as the difficulty of large-scale casting, as well as the customary use of materials (e.g., precious woods, ivory) that by nature dictated a relatively small scale, were other likely reasons for the nonmonumental character of the earliest Greek sculptural art. The idea of creating large statues in the round in carved marble or in cast or hammered bronze seems to have become established toward the middle of the 7th century B.C., at the time of Nikandre (VII, PL. 12), and has come to be associated with the traditions relating to the "inventions" and the personality of Daidalos and his school.

Cretan influence was said to have entered the Peloponnesos through families of artists who were related to or at least were connected with Daidalos, the legendary Cretan sculptor who was absorbed into Attic myth at a later period. On this tradition is based the position that the Cretan artistic developments led the field and that the new forms and idioms emanated from Crete and then spread to the Peloponnesos and the rest of the Greek world. The characteristic features of the Daedalic style seem to have derived from abstract formal principles, in contrast to the spontaneous naturalism of the Geometric style, and to indicate a striving for clear and concise form free of the ardor and improvisational spirit that had dominated in the preceding age. The development was from a multiplicity of elements, which the Geometric mentality set down in apparently uncontrolled fashion, to simplified and unitary compositions, which were no longer transitory and improvised in character but were subject to rules of formal and rhythmical arrangement.

More recently the analysis which seeks to explain the renascence of art in Greece as a definite Cretan importation has been strongly contested (Homann-Wedeking). Primary examples of Daedalic sculptural art have been found in all parts of Greece, particularly in the islands (as can be seen from the surprising discoveries on Ionian Samos), and in many ways the fundamental premises of the Daedalic idiom are only the effects of the new requirements of monumental sculpture. The problems involved in dealing with large blocks of marble or in constructing colossal metal figures inevitably led to a greater concentration on formal structure and on a unitary composition determined through impetus and pause as in music.

Very probably the revolutionary movement toward the attainment of monumentality and the formal reductiveness of the Daedalic idiom did not develop in so simple a manner as tradition would have it. Certainly it involved the collaborated efforts of many Greek artistic centers, particularly the contributions of the marble-producing islands; and Argolis was undoubtedly one of the dominant forces in the development of Greek monumental sculpture during the 7th century B.C.

In connection with the traditional migration of Cretan artists, Dipoinos and Skyllis — who were considered to be closest to Daidalos and who possibly came from Gortyna — are repeatedly mentioned in connection with Argolis. Pausanias mentions ebony and ivory statues of the Dioskouroi on horseback, accompanied by their wives and children, as being in Argos itself (II, 22, 5–6). A statue of Herakles at Tiryns was attributed to these two sculptors, and the same city possessed a xoanon of Apollo and an ancient image of Hera made from pearwood that was either executed or dedicated by Peirasos of Argos. A votive offering in the Heraion was attributed to Daidalos, and the statue of Hera was the work of Smilis of Aegina, the artist who created the famous statue of Hera in Samos. Perhaps this singular image influenced the numerous terra cottas from the sanctuary, some of which are truly remarkable; in these, the small, sharply modeled head dominates the rather formless mass of the traditional idol, over the flat surfaces of which necklaces and diadems could be spread.

The oldest examples of monumental sculpture in the region are metope fragments from Mycenae. Some scholars dispute their description as such; the alternative suggested is that these might be fragments of a pediment. Whatever their purpose, they may be regarded as some of the most authoritative evidence for the Daedalic tradition from approximately the third quarter of the 7th century B.C. The most nearly intact and identifiable fragment (PL. 66), showing the upper half of a veiled goddess, reveals an extraordinary structural sense and subtle harmony. In contrast to the slender shapes and aloof aspect of these examples are two large-scale statues in Delphi from the late 7th or the beginning of the 6th century B.C., which represent the Argive heroes Kleobis and Biton and which are signed by a [Poly?] medes of Argos (I, PL. 342). These figures are heavy and powerful, with thick muscles and short necks. It is quite possible that this accentuated physical force was intended to depict the effort made by the two brothers when, through filial devotion, they yoked themselves to a cart and drew their priestess mother from Argos to the Sanctuary of Hera, 30 stadia away. The virility of these figures and the marked contrast between their vigorous attitudes and the elaborate surface carving places them among the most singular interpretations of the kouros theme. The vitality of the Kleobis and Biton statues seems also to be reflected in the heads of certain fictile statuettes from the Heraion.

Argos and its territory, however, have proved unusually poor in yielding examples of significant monumental works from the 6th century B.C. as a whole. The head of a kouros from Epidauros (Athens, Nat. Mus.), even if regarded as a local work, is a marginal creation that reflects some contact with the sculpture of the islands. The only appreciable evidence for this whole period appears to be the fragmentary bronze statuette of a kouros found during recent excavations of the Heraion. This statuette, which may be dated to about the middle of the 6th century B.C., seems to anticipate the vigorous and sensitive modeling subsequently associated with Argive sculpture.

The Argive school did not achieve a decided supremacy until the time of Hageladas (Ageladas, Hagelaidas) — that is, in the period of the mature archaic style, which was particularly vital and revolutionary in its formal achievements. No work of this artist is extant, and no other traces have been discovered to substantiate what must have been his great spiritual expression. Even the available data about his activity seem to be contradictory; thus, some scholars have been led to suggest the existence of a second Hageladas, while others have shifted his activity to the second half of the 5th century B.C. A pedestal at Olympia bears the signature of his son Argeiadas, who was also a sculptor, inscribed in characters that are from the period 485–480 B.C. This evidence seems to show conclusively that the great Argive master flourished in the mature archaic period and in the early years of the severe style. His most celebrated cult statues, the *Zeus Ithomatas* and the *Herakles Alexikakos*, are beyond description even through indirect documentation, but the possibility remains that Hageladas was the artist who established the motifs of Zeus hurling the thunderbolt and of Poseidon and Herakles in combat. Many small Peloponnesian bronzes executed in these attitudes have been found in the Sanctuary of Zeus at Olympia and at Dodona.

Examples are more numerous from the end of the 6th century on, and to this period may be assigned a fragmentary kore from the Heraion (Athens, Nat. Mus.). Though of the usual island type, it is apparently of local workmanship. The more refined and fully developed groups of the extensive series of nude youths on patera handles of bronze may also be assigned to this period. A splendid bronze hydria (Met. Mus.), with a female bust and an Argive dedication upon it, has led to the assumption that at least some of this type came from Argos. A series of small bronzes, mostly of athletes, characterized by their compact bodies and generous but slightly differentiated musculature, is also attributed to Argos. Of this type are the famous small bronze athlete from the Acropolis (Athens, Nat. Mus.) and the small bronze runner from Olympia (Archaeol. Mus.).

The style of all the above-mentioned works contrasts with the great vitality and angular, tense rendering of hollowed muscles that may be regarded as characteristic of Sikyonian and Corinthian sculpture. The bodies of the Argive examples are full and supple, and the facial expressions are serene and somewhat detached. Confirmation of these distinguishing qualities is provided by the classification and attributions proposed by E. Langlotz for several important series of figurines from mirror supports, which depict Aphrodite wearing the peplos. The group assigned to Argos on the basis of resemblances to the Tiryns terra cottas is marked by heavy but softly rendered draperies, which contrast with the tense lines and the narrow, mechanical fluting of those of the Sikyonian group. One of the main objections to the grouping of Argive works Langlotz proposed lies in the inclusion of a series of small bronzes that, because of their provenance and formal characteristics, should have been assigned to Arcadia. Without excluding the possibility that most of the small bronzes found in Arcadian territory were executed locally, one must admit that influence from neighboring regions was strongly evident. Consequently, within the rude local style of the Arcadian finds can be discerned a line of development that has Argive or near-Argive characteristics, along with others that display Corinthian and Laconian characteristics. The discoverers of the bronze head of a youth from Glanitsa in Arcadia (*BCH*, LXXV, 1951, pp. 224–232) have suggested an Argive origin because of its remarkably noble form, notwithstanding a certain dryness in the modeling. Similarly, there is no alternative but to trace the softer, more tender representations of Arcadian shepherds, with their ample and relaxed bodies, to Argive traditions; the representations with sharp outlines and clear-cut features belong rather to Corinthian traditions. Among other notable works, the small, seated *Demeter* found at Tegea (Athens, Nat. Mus.) must be associated with the great school of Argos as well, because of the calm grandeur of its expression, the softness and unity of its modeling.

There are records of the names and an occasional mention of the works of Argive artists. From the late archaic period are such artists as Eutelidas, Argeiadas (the son of Hageladas), and Chrysothemis; closer to the severe period are Dionysios, Glaukos, Atotos, and Asopodoros. The school of Argos in the generation before Polykleitos the Elder can be studied through representations of athletes, such as the bronze *Diskobolos* in New York (Met. Mus.) and the *Ephebos* from Lygourion (PL. 77). Possibly these sturdy, muscular figures inspired the clear rhythm and simple modeling achieved in the *Ephebos* copied by Stephanos (X, PL. 262). The typical female figure of this period is depicted wearing the heavy but softly draped peplos, as may be seen in many terra cottas and bronze mirror stands. A monumental version of this theme may be seen in Roman statues of the *peplophoros*, for example, in one from Castel Sant'Angelo (Rome, Mus. Naz. Rom.) and in another in Copenhagen (Ny Carlsberg Glyptotek, no. 292). Few scholars continue to maintain that the Olympian sculptures can be traced to the Peloponnesos and to Argos in particular, since these works are too essentially marmoreal to have come from a center of notable bronzeworkers. A respect for traditional opinion, in the absence of more certain evidence, encourages the assignment to the Peloponnesos of the so-called "Blond Boy" (III, PL. 343), which may be looked upon as an anticipation of the style of Polykleitos the Elder.

The great Polykleitos (q.v.), who has come definitely to be regarded as a classical sculptor, is described in most sources as Argive, but his relations with the school of Sikyon are repeatedly mentioned in ancient sources as well. There can be no doubt that from the beginning of his activity — the *Kyniskos* (PL. 191) is believed by some to date from about 460 B.C. — Polykleitos the Elder seems to have attained an equilibrium between the ample, softly modeled volumes of the Argive tradition and the clear-cut forms and precise outlines of the Sikyonian tradition. Though the school of Argos may be regarded as having enjoyed a position of absolute predominance during the generation of Hageladas, in the full flowering of the classical period the art of Polykleitos does not bear comparison with the originality and striking innovations of Phidias (q.v.). Polykleitos the Elder was an incomparable worker in bronze and an admirable student of athletes' bodies, but toward the end of his career, he seems to have acknowledged his artistic limitations by moving somewhat closer to his great rival both in technique and in spirit. In his works (PLS. 189–193; III, PL. 359), Polykleitos reveals the same preoccupation with harmonious and perfectly balanced structure that he set down in his *Canon*. The spiritual detachment or torpor that is so noticeable in the figure of the Argive severe style is transformed in his work into a spiritual serenity. The figures assume attitudes in rhythms arising from a highly calculated counterbalancing of the supporting limbs with those which, as if having discharged their energy, are relaxed. The inclined heads seem to suggest not only a physical or formal necessity of form but some transcendent weight, perhaps fate, and that tragic feeling for life which the Greeks expressed particularly in their poetry. It seems necessary to link the last phase of the activity of Polykleitos with influences from Attica, as suggested by many scholars, especially in regard to the spiritual enrichment shown in his later work, for example, in the *Diadoumenos* (PL. 190). Polykleitos also sculptured divinities, and his last work was a colossal chryselephantine statue of Hera, the great goddess of Argos, for the Heraion. The cult statue of the Argive Hera seems to have been created in response to and considered comparable to the colossal chryselephantine statues erected in many places in Greece (including the Peloponnesos) by Phidias. The later activity of Polykleitos seems to have included, after his prolonged devotion to the athletic and heroic ideal, various female themes. In addition to his necessarily heroic and severely rhythmical *Wounded Amazon* (PL. 193), Polycletan prototypes have been associated with works of a more definite femininity: the *Aphrodite* of Fréjus (Louvre), with drapery of beaten metal, and the Barberini *Suppliant*.

Polykleitos the Elder was also known as the great master of a flourishing school of artists, some of whom depended upon his artistic personality to a very great degree. Only tentative suppositions provide the grounds for attributing statues of a discus thrower (Vat. Mus.), a charioteer (Pal. dei Conservatori), and Hermes (Florence, Pitti) to Polykleitos's brother Naukydes. Through Polykleitos and his wide circle of disciples, the tradition of the school of Argos became increasingly fused with the other great Peloponnesian strain, that of the school of Sikyon. Some considered Polykleitos to have been a Sikyonian, and it is certain that he acquired at least part of his artistic education in Sikyon. Among his pupils are recorded Asopodoros, Perikleitos (or Periklytos), and the latter's pupil Antiphanes, who were Argives; Kanachos and Daidalos, who were Sikyonians; and Athenodoros and Dameas, who were from Kleitor in Arcadia. In general, the Argive school remained linked to the athletic tradition and to the victories in the Panhellenic games at Olympia. Numerous statues by Naukydes and Patrokles, among others, have been recorded. The works of Deinomenes include statues of the heroines Io and Kallisto and of the hero Protesilaos, who was honored with sanctuaries and *manteia* on the shores of the Propontis and in Asia Minor.

At Delphi, only pedestals remain of the great monuments donated by Argos or worked on by Argive sculptors: the *Seven against Thebes*, the *Epigoni*, those of the kings and queens of Argos, and the celebrative monument for the victory of Aigospotamoi. These records afford but a glimpse of a wealth of figural art that, unfortunately, can no longer be evaluated. The hypotheses that authorities have advanced are few, limited to a rather doubtful proposal to identify a statue of an attacking hero (Met. Mus.) with the *Protesilaos* of Deinomenes and to recognize an *Aphrodite* (Athens, Nat. Mus.) from Epidauros as the younger Polykleitos's *Aphrodite* that was dedicated at Amyklai. Statues of divinities and heroes by Antiphanes are recorded in the Arcadian donations at Delphi.

In the generation after Phidias, there was an influx of Attic elements, and an artistic koine largely dominated by Attica emerged even in the Peloponnesos, which hitherto had been a stronghold of autochthonous tradition. This development is substantiated by the remains of the most illustrious sculptural monument in Argolis, the pediments and acroteria of the Temple of Asklepios at Epidauros (III, PLS. 368, 371). It is not certain whether Timotheos, who is indicated as the dominant personality in this ambitious creation, was actually an Argive, but the airy Amazons and winged nymphs with light draperies evidence such a command of marbleworking that they must be linked with the post-Phidian tradition of Agorakritos and Kallimachos. Quite independent works, such as the above-mentioned *Aphrodite* from Epidauros, show unquestionably that these features were not simply personal qualities of Timotheos's art. Other works attributed on reasonable grounds to Timotheos, especially the *Leda* (Rome, Coll. Torlonia; Mus. Cap.) and the Rospigliosi *Athena* (Rome, Coll. Pallavicini), reveal an even more clearly Attic idiom. Soon after this period, the Argive school seems definitely to have lost importance in comparison with the school of Sikyon. The most illustrious continuer of the Polycletan spirit was Lysippos (q.v.) of Sikyon, and almost all the names known from this period are those of Sikyonian artists. Only one statue from the Hellenistic period by an Argive, a sculptor named Kallikrates, is recorded at Epidauros.

ARTISTS. *Argos. a. Architects*: Eupolemos (Εὐπόλεμος) of Argos, last decades of 5th cent. B.C.: the new Temple of Hera, built soon after the destruction of the old temple by fire in 423 B.C. (W. B. Dinsmoor, The Architecture of Ancient Greece, London, 1950, p. 183; G. Caputo, EAA, s.v.). – Sosikles, or Sokles (Σοκλῆς), of unknown origin, 4th cent. B.C., perhaps the custodian of the Heraion: signature found in the Heraion on clay roofing tiles that were perhaps used in a water conduit (C. Wickert, ThB, s.v.). – *b. Sculptors: Legendary times and 7th-6th cent. B.C.*: Peirasos of Argos, considered one of the legendary founders of the Argive school, but perhaps active in 7th cent. B.C.: executed or dedicated most ancient xoanon of Hera at Tiryns, carried to the Heraion at Argos after the destruction of Tiryns in 468 B.C. (Picard, I, p. 494; Lippold, GP, p. 25, note 4). – Daidalos (Δαίδαλος), legendary figure: votive offering in the Heraion. – Dipoinos and Skyllis of Gortyna (?) (see below, *Sikyon*). – Epeios (Ἐπειός) of Phocis: xoanon of Hermes. – [Poly]medes ([Πολυ]μέδης), or [Aga?]-medes, of Argos, ca. 600 B.C.: statues of Kleobis and Biton, signed, discovered at Delphi (Lippold, GP, p. 25; G. von Kaschnitz-Weinberg, Die ungleichen Zwillinge, S. Presented to D. M. Robinson, I, St. Louis, 1951, pp. 525–31; J. Marcadé, Recueil des signatures de sculpteurs grecs, I, Paris, 1953, no. 115). – Smilis (Σμῖλις) of Aegina (see below, *Aegina*). – Eutelidas (Εὐτελίδας) of Argos, active toward end of 6th cent. B.C.: statues of Damaretos and Theopompos, executed in collaboration with Chrysothemis of Argos at Olympia (Lippold, GP, p. 89; M. B. Marzani, EAA, s.v.). – Hybrisstas (Ὑβρίσστας) of Argos (?), 6th or 5th cent. B.C.: possibly a small signed bronze of Zeus hurling a thunderbolt, probably from Epidauros (Picard, II, pp. 162, 916; Lippold, GP, p. 89; G. A. Mansuelli, EAA, s.v.). – *5th cent. B.C.*: Hageladas (Ἀγελάδας), or Ageladas (Ἀγελάδας), or Hagelaidas (Ἀγελαίδας), of Argos, ca. 520–480 B.C., father of Argeiadas of Argos, said to be the master of Myron, Polykleitos, and, according to one source, Phidias: image of Herakles Alexikakos in Athens; offering of the Tarentines at Delphi, executed before 474 B.C., remains of base with inscription; two statues of athletes and a quadriga at Olympia; Zeus Ithomatas, on the acropolis at Messene, reproduced on coins; Muse with barbiton, by conjecture at Sikyon, from a group of three Muses, the others being by Kanachos and Aristokles, both of Sikyon; a young Zeus, reproduced on coins, and a beardless Herakles, both at Aigion (Lippold, GP, p. 88; P. Orlandini, EAA, s.v.). – Chrysothemis (Χρυσόθεμις) of Argos, beginning of 5th cent. B.C., probably an epigonos of the school of Hageladas of Argos: two statues of athletes at Olympia, executed with Eutelidas of Argos (Lippold, GP, p. 89; L. Guerrini, EAA, s.v.). – Aristomedon (Ἀριστομέδων) of Argos: offering of the Phocians at Delphi. – Argeiadas (Ἀργειάδας) of Argos, son of Hageladas of Argos, first half of 5th cent. B.C.: collaborated with Asopodoros and Atotos, both of Argos, on the offering by Praxiteles of Syracuse and Camarina at Olympia, of which the signed base remains (Lippold, GP, p. 103; P. Orlandini, EAA, s.v.). – Asopodoros (Ἀσωπόδωρος) and Atotos (Ἄτωτος), both of Argos: collaborated with Argeiadas on the Praxiteles offering at Olympia (P. Orlandini, EAA, s.vv.). – Glaukos (Γλαῦκος) of Argos, ca. mid-5th cent. B.C.: images of Amphitrite, Poseidon, and Hestia for the offering by Mikythos of Rhegion at Olympia (Lippold, GP, p. 103; G. Cressedi, EAA, s.v.). – Dionysios (Διονύσιος) of Argos, active at Olympia, mid-5th cent. B.C.: offering by Phormis of Mainalos, in collaboration with Simon of Aegina; offering by Mikythos at Olympia, executed with Glaukos (C. Ant; and G. Cressedi, EAA, s.v., no. 1). – Dorotheos (Δωρόθεος) of Argos, first or second half of 5th cent. B.C.: signed base, assigned to an ex-voto (perhaps a quadruped) of Demeter Chthonia offered by Aristomenes, son of Alexias, at Hermione; signed base at Delphi (Lippold, GP, p. 170; J. Marcadé, Recueil des signatures de sculpteurs grecs, I, Paris, 1953, nos. 30–31; L. Guerrini, EAA, s.v., no. 1). – Polykleitos the Elder of Argos (q.v.). – Naukydes (Ναυκύδης) of Argos, possibly the son of Patrokles of Sikyon or of Mothon of Argos, brother of Daidalos of Sikyon, master of Alypos of Sikyon and of the younger Polykleitos or brother of Polykleitos, second half of 5th cent. or ca. 400 B.C. (see POLYKLEITOS): chryselephantine statue of Hebe in the Heraion, reproduced on coins: bronze Hekate near the Temple of Eileithyia; two statues of the Argive Cheimon (victor in wrestling, 448 B.C.), one at Olympia and one at Argos, later in Rome; diskobolos, recognized through copies and coins; figure (perhaps Phrixos) sacrificing a ram, perhaps to be connected with a group on the Acropolis in Athens, probably belonging with a base signed N; statue of Hermes; Eukles of Rhodes, at Olympia, the signed base of which remains; image of Baukis of Troizen at Olympia; portrait of the poetess Erinna of Lesbos or Telos (G. Lippold, RE, s.v., no. 2; Lippold, GP, p. 199). – Phradmon (Φράδμων) of Argos, probably second half of 5th cent. B.C.: Amazon statue in the Artemision of Ephesos, said to have been executed in competition with Polykleitos, Phidias, and Kresilas; portrait of Amertas of Elis (victor in 444/420 B.C.) at Olympia; twelve bronze heifers in the Temple of Athena Itonia in Thessaly; the athlete Pythodelos (or Pithodoros), at Delphi, from which the inscribed base (fragmentary signature) remains. Other attributed works: statues of Sisyphos I, from the Daochos group at Delphi, of Narcissus, and of Dionysos, from Tivoli; bronze head from Benevento, in the Louvre (Lippold, GP, pp. 169–70; P. C. Sestieri, Alla ricerca di Phradmon, AC, III, 1951, pp. 13–32; J. Marcadé, Recueil des signatures de sculpteurs grecs, I, Paris, 1953, nos. 87–88). – Kresilas (Κρησίλας) of Kydonia (Crete): so-called "Diomedes of Cumae," executed for Argos, as indicated by reproductions on coins (P. Orlandini, EAA, s.v.). – Phrynon (Φρύνων), pupil of Polykleitos, perhaps the real name of Dainon (Lippold, GP, p. 216). – Deinomenes (Δεινομένης), perhaps of Argos, of the Polycletan school, ca. end of 5th cent. B.C.: statue of Protesilaos, location unknown, recognized through copies and dated ca. 430 B.C.; Pythodemos, location unknown; Io and Kallisto, on the Athenian Acropolis, attributed also to a namesake of the Hellenistic period (Lippold, GP, p. 203; L. Guerrini, EAA, s.v.). – Deinon (Δείνων), pupil of Polykleitos, possibly an abbreviated form of the name Deinomenes of Argos (?) or perhaps a variant of the name Phrynon (Lippold, GP, p. 217; L. Guerrini, EAA, s.v.). – Alexis (Ἄλεξις), pupil of Polykleitos, ca. end of 5th cent. B.C., identified apparently wrongly as the father of Kantharos of Sikyon (W. Amelung, ThB, s.v.; P. Orlandini, EAA, s.v.). – Argeios (Ἀργεῖος), artist of the Polycletan school whose existence has been doubted but perhaps was the great-grandson or in the family of Hageladas of Argos (W. Amelung, ThB, s.v.; L. Guerrini, EAA, s.v.). – *4th cent. B.C.*: Asopodoros (Ἀσωπόδωρος) of Argos, of the Polycletan school, end of 5th to mid-4th cent. B.C. (Lippold, GP, p. 216; P. Orlandini, EAA, s.v.). – Perikleitos, or Periklytos (Περίκλυτος), of Argos, pupil of Polykleitos and master of Antiphanes of Argos, beginning of 4th cent. B.C. (G. Lippold, RE, s.v.; Lippold, GP, p. 216). – Antiphanes (Ἀντιφάνης) of Argos, pupil of Perikleitos of Argos and master of Kleon of Sikyon: collaborated in the period 405–369 on offerings at Delphi from Argos, Sparta, and Arcadia or Tegea. – Polykleitos the Younger (Πολύκλειτος) of Argos, or Polykleitos II, perhaps son of Patrokles of Sikyon and nephew of Polykleitos the Elder (q.v.), pupil of Naukydes of Argos, first half of 4th cent. B.C., identified with an architect and engraver of the same name: Zeus Meilichios, perhaps reproduced on coins; Zeus Philios in the Temenos of Demeter and Kore at Megalopolis, after 367 B.C.; Aphrodite as a tripod support, an offering at Amyklai for the victory of Aigospotamoi; group of Apollo, Artemis, and Leto in the Temple of Artemis Orthia on Mount Lykone, also attributed to Polykleitos the Elder; six statues of athletes at Olympia, some of which have been attributed to Polykleitos the Elder. Other attributed works: portrait of Hephaistion, a friend of Alexander, also attributed to Lysippos of Sikyon; bronze statues of Hekate for the Hekateion at Argos, possibly executed with Naukydes but more generally attributed to Polykleitos the Elder (Lippold, GP, pp. 216–19; G. Lippold, RE, s.v. no. 11). – Androkydes (Ἀνδροκύδης) of Argos, first half of 4th cent. B.C.: base signed with Polykles at Hermione (J. Marcadé, Recueil des signatures de sculpteurs grecs, II, Paris, 1957, no. 106; L. Guerrini, EAA, s.v., no. 2). – Praxiteles (q.v.) of Athens: group of Leto beside Chloris, in the Temple of Leto, reproduced on coins. – Skopas (q.v.) of Paros: marble Hekate, in the Hekateion at Argos. – ***Epidauros. a. Architects***: Theodotos (Θεόδοτος), first half of 4th cent. B.C., also identified with his namesake of Phocis; doubtful that he was also a sculptor: Temple of Asklepios, ca. 380–370 B.C. (W. Züchner, ThB, s.v.; W. B. Dinsmoor, The Architecture of Ancient Greece, London, 1950, p. 218). – Polykleitos the Architect (q.v.). – *b. Sculptors*: Thrasymedes (Θρασυμήδης) of Paros, 4th cent. B.C.: enthroned chryselephantine Asklepios. – Timotheos (Τιμόθεος), perhaps of Epidauros or from northern Argolis, ca. 370–350 B.C.: typoi (probably reliefs) and acroteria (Nike and two Aurai) on the west side of the Asklepieion (ca. 370 B.C.); reliefs on the south side of the mausoleum at Halikarnassos; statue of Hippolytos at Troizen; Artemia at Rhamnous, later in Rome; Ares at Halikarnassos, attributed also to Leochares of Athens; reliefs of seated Asklepios, near the Asklepieion; figure of a young girl at Hermione; the so-called "Burlington girl"; a nymph (Vat. Mus.); Leda (or Nemesis); the Rospigliosi Athena (Lippold, GP, pp. 219–22; J. F. Crome, Die Skulturendes Asklepiostempels von Epidauros, Berlin, 1951; G. Leoncini, Contributo all'interpretazione dei "typoi" di Timotheos, Aevum, XXX, 1956, pp. 20–29; A. M. Tamassia, Ancora su "typoi" di Timotheos, AC, XIII, 1961, pp. 124–31). – *c. Collaborators on the Asklepieion*: Hektoridas (Ἑκτορίδας): one of the two pediments (perhaps the eastern one displaying the Ilioupersis), perhaps after a design by Timotheos, model of a lion's head for the cyma (M. Bieber, ThB, s.v.; M. B. Marzani, EAA, s.v.). – Theo[n], or Theo[dotos], or Theo[timos] (Θεω[. . .]), whom some identify with the architect Theodotos: acroteria of the eastern façade, from which a draped female torso remains (M. Bieber, ThB, s.v.). – Agathinos (Ἀγαθῖνος): collaborated on the sculptural decoration (W. Amelung, ThB, s.v.; P. Orlandini, EAA, s.v., no. 1). – Spoudias (Σπουδίας) of Athens: votive offering, recorded by its fragmentary base with a dedication to Asklepios by Damar[atos] of Corinth (ThB, s.v. Spudias). – *d. Painter*: Pausias of Sikyon (see below, *Sikyon*). – ***Phlious. a. Sculptor***: Laphaes (Λαφάης) of Phlious mid-6th to beginning of 5th cent. B.C.: cult statue in the Sanctuary of Herakles at Sikyon; xoanon of Apollo at Aigeira, in Achaia (M. Bieber, ThB, s.v.; Lippold, GP, p. 88; L. Guerrini, EAA, s.v.). – *b. Painter*: Sillax (Σίλλαξ) of Rhegion, first half of 5th cent. B.C.: Room of the Polemarchs. – ***Hermione. a. Sculptors***: Dorotheos of Argos (see above, *Argos*). – Kresilas (Κρησίλας) of Kydonia, second half of 5th cent. B.C.: signed base, in Doric dialect and Argive script, assigned to a milch cow offered by Alexias to Demeter Chthonia (P. Orlandini, EAA, s.v.). – Androkydes of Argos (see above, *Argos*). – ***Kalaureia. a. Sculptor***: Pison (Πίσων) of Kalaureia: figure of the soothsayer Agias, an offering of the Spartans at Delphi. – *Tiryns. a. Sculptors*: Dipoinos and Skyllis of Gortyna (?) (see below, *Sikyon*). – Argos ("Αργος) of Argolis, 7th cent. B.C.: possibly executed the earliest xoanon of Hera (Lippold, GP, p. 25, note 4). – Peirasos of Argos (see above, *Argos*). – *Troizen. a. Sculptors*: Hermon ("Ερμων), 6th cent. B.C.: cult statue in the Temple of Apollo Thearios; xoana of the Dioskouroi, probably reproduced on local coins (ThB, s.v.; Lippold, GP, p. 26; G. Cressedi, EAA, s.v., no. 2). – Kalon: xoanon of Athena Sthenias on the acropolis, reproduced on coins. – Timotheos (see above, *Epidauros*).

CORINTH. The most ancient and authoritative traditions concerning Peloponnesian painting are linked with the area lying between Corinth and Sikyon. Pliny (XXXV, 15) mentions first Corinth and then Sikyon as the place of origin of this art. The Corinthian Kleanthes, according to Pliny (XXXV, 16), was the inventor of painting, and his historicity is elsewhere assured. Strabo (VIII, 343) records having seen two of the paintings of Kleanthes in the Temple of Artemis Alpheionia near Olympia, one of which depicted the birth of Athena, a theme that cannot be older than the second half of the 7th century B.C. More recondite names of Corinthian painters are those of Aridikes, Aregon, and Ekphantos. Telephanes and Kraton appear to have been Sikyonians, whereas Kimon, who invented the famous *katagrapha* (or *obliquae imagines*), came from nearby Kleonai and is in many respects the most discussed of the early innovators. The development of the classical and postclassical schools of painting, however, seems to have centered about Sikyon.

The sources mention the considerable activity in Corinth of Cretan artists, who might be either historical or legendary figures, though to a lesser extent than in Sikyon and Argos. A xoanon of Herakles at the Temple of Athena Chalinitas was ascribed to Daidalos himself; it was said to be nude, almost as if in deliberate contrast to the draped Eastern type. Works by Dipoinos and Skyllis are also recorded at Kleonai, in Corinthian territory. Corinth has yielded only a few Geometric bronzes, and those are of little distinction; but the same city has been established as the center of production of a class of tripods which have been studied through the numerous examples found at Olympia and which have been documented in areas of Corinthian influence such as Zakynthos and Ithaca.

Traditionally, the earliest and most famous Corinthian sculpture was a colossal image of Zeus, of beaten gold, offered by the Kypselids at Olympia. This statue may be dated about the third quarter of the 7th century B.C. Its exceptionally large dimensions are characteristic of this period of ancient sculptural art, for when once the concept of monumentality had been realized, every effort was made to explore its possibilities fully in creating numerous gigantic statues to be placed in and around the temples. The use of the peculiar beaten-gold technique suggests that the statue was draped. Moreover, it not only was monumental but was minutely ornamented, like the fragments of chryselephantine statues found under the Sacred Way at Delphi. Perhaps there may be some relation between this technique, which was used both for monumental statuary and for small ornamental metalwork and jewelery, and that found in a number of sheet-gold reliefs from Corinth (VI, PL. 110), which may date from the end of the 8th to the middle of the 7th century B.C. Depicted on the reliefs are figures of a Potnios (Posis), animals, and stories of Theseus, executed in a style that seems to have been a forerunner of the earliest so-called "Argive-Corinthian" bronzes.

An example of sculpture of true Corinthian provenance is a refined perirrhanterion (lustral basin; Corinth, Archaeol. Mus.) from the Isthmian sanctuary of Poseidon. Monuments of this kind, which consist of three female figures (probably representing the great goddess of the animals, Potnia Theron) standing on lions and supporting a basin, are known from other sanctuaries in Greece, for example, the sanctuaries of Rhodes and Samos. A fragmentary perirrhanterion at Olympia can be assigned to Sparta because of the type of marble used and the rugged solidity of the forms; but the one from Isthmia, which has a lighter structure and a wide basin harmoniously balanced by the supporting figures, may well be Corinthian sculpture.

Works of a more definite Corinthian origin are found in ceramics and in terra cotta. As Payne has observed (1931, p. 233 ff.), the most continuous and more precisely datable Corinthian heads are the modeled examples adorning certain proto-Corinthian aryballoi and a series of Corinthian pyxides. A notable aryballos from Thebes (PL. 62), which cannot date later than the middle of the 7th century B.C. because of the painting, has a Daedalic face. Its clear outlines and refined modeling somewhat resemble the style of the famous veiled goddess of the metope of Mycenae. To be placed chronologically between this tiny masterpiece and the female heads on the pyxides — all are from the Corinthian period, about 600 B.C. — are some of the sphinx and siren heads. These heads possess the same superb modeling, and they gradually developed the simple vitality that remained typical of Corinthian work. Not many noteworthy clay statues from the archaic period in Corinth have survived — the Corinthian sanctuary of Hera at Perachora has yielded much richer finds — but mention should at least be made of a small, seated female figure (Corinth, Archaeol. Mus.) from the second half of the 7th century that shows, in a more ample and monumental manner, the qualities which are implied by the modeled ornamentation on the vases. The figure from Corinth has compact and fully realized forms; the head, less intense in its modeling than the above-mentioned Theban aryballos, is enlivened by color and has a quality of candor and quiet lucidity that is characteristic of Corinthian art.

Boutades of Sikyon is another almost legendary "inventor" typical of this period, which was regarded by the ancients as the dawn of their great artistic history. Boutades worked mainly at Corinth, another indication of the similarities and interdependence of Sikyon and Corinth. He has been described as the originator of terra-cotta sculpture and the creator of a relief that is said to have been in the nymphaion at Corinth until the city was destroyed by Lucius Mummius (146 B.C.). Boutades is also said to have been the inventor of figural antefixes. This information helps to confirm the position of absolute dominance that Corinth enjoyed in the production and diffusion of fictile temple decorations, not only in Greece but also in Sicily and the Italian mainland. Some of the Daedalic antefixes in the Temple of Apollo at Thermon (Thermos) (VII, PL. 42) have been dated to the second half of the 7th century B.C.; these, together with the series in Corfu and Kalydon, present evidence of a development extending through almost two centuries, beginning with the Daedalic faces and ending with the masks of satyrs and bearded heroes from the end of the archaic period. Some of the great examples of fictile temple decoration in Sicily and Magna Graecia should also be traced back to Corinth. The great Gorgon faces, for example, which must have decorated the pediments of temples in Selinous (mod. Selinunte), Gela, and Syracuse, can be regarded as Corinthian works or as replicas of a Corinthian innovation. The Gorgon from the Athenaion in Syracuse (VII, PL. 61) may be placed at the beginning of a very long and important series of fictile acroteria (mainly sphinxes), which include a superb head from Kalydon (VII, PL. 41), another from Thebes (Louvre), and heads from Halai and Delphi; this series dates from the beginning of the 6th century to the height of the severe period.

The most important group of monumental stone sculptures of undoubted Corinthian character does not occur in Greece itself but in the Corinthian colony of Corfu. The island possesses the great recumbent lion from the cenotaph of Menekrates (VII, PL. 43), which seems proto-Corinthian in appearance even though it should be dated about 600 B.C. The pediment of the Temple of Artemis, with the Gorgon (Potnia Theron; I, PL. 343), is undoubtedly the oldest and most magnificent of all archaic pediments and serves to justify Pindar's statement crediting Corinth with the invention of the decorated pediment (cf. also VII, PL. 44). The pediment showing Potnia Theron is dated to the early decades of the 6th century on the basis of the development of the type of Gorgon, which in this example is tamed and contrasts with the more terrifying 7th-century images. This does not, however, preclude an even earlier date: indeed, it is possible that the sculptor who carved it and who introduced so many other innovations was also the first to moderate the horrifying appearance of the goddess, who in this work also represents Artemis. Next to the goddess are her son Chrysaor, whose face and angular body are defined by means of hard, simplified planes, and two large panthers arranged in a flat, heraldic fashion at the sides. The massive, smooth lion of Menekrates and the panthers on the pediment are earlier works than the lions of Loutraki (ca. mid-6th cent.) in Copenhagen (Ny Carlsberg Glyptotek) and in Boston (Mus. of Fine Arts). The Loutraki lions are vibrant and highly decorative, like those of the Chimera Painter of Corinth.

Though destroyed by the Romans, Corinth has yielded extremely significant fragments of archaic sculpture. The most important is a fragment of a large head, made of poros (*AJA*, XXXIV, 1930, p. 450; LIV, 1950, p. 266), which has been completed by the addition of clay serpents fitted into the holes in the crown among the curls and which, therefore, was a colossal Gorgon at the center of a pediment. Among the remarkable metope fragments is one with a chariot, or quadriga, depicted frontally, which shows a significant relation to the well-known metope from Selinous (Palermo, Mus. Naz. Arch.) and appears to confirm the correspondences between sculptural art from Corinth and from Sicily.

There seems to be no justification for not attributing to Corinth the one large marble statue found in the area — the *Apollo*, or *Kouros*, from Tenea (I, PL. 350). Not only are the similarities to the fictile heads from Corinthian pyxides strongly apparent (Payne, 1931, pp. 237–38), but the sharpness of the outlines and the clear-cut lines and compact volumes all appear to reflect the artistic traditions of the region between Corinth and Sikyon.

Corinth was famous for its bronzeworking, and a good number of the statuettes and sumptuous cult objects (vases, basins, tripods) found at Olympia and Dodona may be ascribed to Corinthian workshops. Among the bronzes discovered at Olympia are examples that, because of their exquisite workmanship and monumental character, are fully entitled to designation as great works of art. Of the bronzes from the late 7th century, mention should be made of a beaten-metal plaque depicting a mother griffin with her young (Olympia, Archaeol. Mus.), an object that may have served as a metope on a sacred building. Some of the most refined bronze strips (I, PL. 340) have been ascribed to Corinth, including a Bellerophon and Chimera in proto-Corinthian style and an even more monumental depiction of the departure of Amphiaraos (PL. 63). It has not been established as to what degree the long series of so-called "Argive-Corinthian" metal strips is of Corinthian origin. The nucleus of this find comes from Olympia, together with helmets and arms that were included in the great trophies dedicated in the Altis. The presence of fine arm rings for shields has been cited as evidence favoring their attribution to Argos, which was renowned for its beautiful weapons. Several of the relief sheets, however, must have been used in the decoration of storage chests, particularly the series from Eleutherai (Br. Mus.). Certain of the scenes from this series, for example, that depicting the ransoming of the body of Hector, are also found on a mirror handle undoubtedly of Corinthian type (Berlin, Staat. Mus.). H. Payne (1931) has recognized decorative motifs and typically Corinthian forms in at least some of these designs; and independently of what has emerged from the finds at Olympia, the figural idiom of these small works is related more to the Corinthian gold plaques in Berlin (Staat. Mus.) showing stories of Theseus and to the relief sheets from Olympia than to such typically Argive monuments as the strips from Prosymna depicting Kassandra and Klytemnestra. The Corinthian influence — in tracing which, one may go back to the famous chest of Kypselos, which bore a kind of summary of the myths of the Peloponnesos executed in ivory, gold, and precious wood — must have been predominant in the series, for the narrative element emerged in more precise and extended form.

A large portion of the cast ornamental pieces on bronze vases and the male figures forming patera handles should be ascribed to Corinthian bronzeworking shops. Payne (1931) has observed that the peculiar heaviness of Corinthian terra-cotta vases in the foot, the handles, and the moldings on the edge reveals their derivation from bronze vessels, which presented the same difficulty in joining the parts cast in strong relief with the walls of simple beaten metal. Some scholars have asserted, on the evidence of the Gorgons attached to vessels from Perachora and the horseman figure from Dodona, that even peripheral works such as the craters and tripods from Illyria and, in particular, the rich finds at Trebeniste are to be assigned to Corinth. A series of bronze plates with handles, some of which are very decorative, also seem to be of Corinthian manufacture; the most noted examples of these plates, depicting Tritons and Gorgons in repose, are in Naples (Mus. Naz.). A remarkable freestanding Corinthian statuette is a modest kouros from Dodona, with a dedication from Ethymokleidas to Zeus in Corinthian characters.

The Corinthian female figural type from the middle archaic period is exemplified by the statuette of a goddess in the Fitzwilliam Museum (Cambridge, Eng.). The symmetrically flared bell-shaped skirt with two stiff vertical pleats on the sides is found in all other Corinthian statues from this period and in those derived from Corinthian designs, even in wooden statuettes from Palma di Montechiaro (Sicily) and in a series of terra cottas from Locri (Magna Graecia). A statuette of the goddess Artemis from Olympia (Athens, Nat. Mus.) also seems to be Corinthian; the precisely pleated Ionian peplos is resolved into unitary surfaces having no folds. A differently developed version is presented in a series of figurines of Aphrodite designed as mirror stands; in these the garment adheres to the body in minute pleats and the hair falls in compact masses of curls.

A bronze statuette of Herakles from Perachora (Athens, Nat. Mus.), dating from the turn of the 5th century B.C., is a vivid and harmonious work that shows Herakles wielding a club; the violent action is resolved in long and flowing lines. This posture of attack, so much in favor at the end of the archaic period, probably has as an archetype the *Zeus Ithomatas* by Hageladas. Payne's acute comparison (*JHS*, LIV, 1934, pp. 163–74) of this work with the bronze *Herakles* in the Benaki Museum (Athens), which is probably Argive, reveals the individuality of Corinthian artists and their flexible and decorative concepts, which contrast with the more ponderous and compact structure typical of Argive bronzes. The assignment of works from the end of the archaic period is even more uncertain. Probably a great part of the group that Langlotz intuitively attributed to Sikyon and Kleonai should be redistributed between Corinth and Argos, but it is not clear to what extent the bronze mirror stands with peplos-clad figures and the other minor bronzes should be ascribed to Corinth. In regard to the major works, it is not easy to decide whether the *Poseidon* from Livadostra (anc. Kreusis; Athens, Nat. Mus.), with its strongly carved face and decisive gesture, should be assigned to Corinth or to Sikyon.

Corinth has yielded precious fragments of large sculpture from the beginning of the 5th century. A head of a bearded man, rendered in rugged and vigorous planes, may be related to the later heads of Aeginetan warriors; another, less substantial fragment of a head, beardless and with hair that seems to have been cut like a skull cap, has been preserved. The most consistent example, though it has a damaged surface, is a torso of a peplophoros shown in an attitude of advancing; a remarkable vigor is conveyed by its angular structure and canted pose. One of the most famous statuettes from Dodona, which depicts Zeus hurling a thunderbolt (VII, PL. 45), has been assigned to Corinth, but it is executed with a sense of physical form very different from that of the small *Herakles* of Perachora. The *Zeus* from Dodona gains effect not through firmly stressed outlines but because of the broad, less clearly defined musculature.

Works in marble and in bronze are relatively rare, uncertain, and without interrelations, particularly when compared with those in terra cotta. An extensive series of fictile statuettes, heads, and reliefs has been found at Corinth; within limitations, it reveals a remarkable variety and includes some works of genuine artistic interest. The series from Perachora is even more remarkable, for its continuity and the great refinement of many of the pieces. Since a design not unlike that of the Argive Hera type was quite generally adopted for the goddess, it is not surprising that along with the Corinthian finds from Perachora there appear numerous Argive terra cottas, modeled with greater vigor and in sharper relief. A series of terra cottas with quite different qualities and decidedly monumental features has come to light in later years. The fragments of a magnificent Amazonomachy from Corinth (Corinth, Archaeol. Mus.), which is related in conception to the superb sphinxes from Kalydon and Thebes, have suggested the existence of a great body of fictile sculpture about the end of the 6th century B.C.

The sculpture groups from Olympia may be considered to include some of the loftiest and most vivid artistic expressions from the early severe period. A head of Zeus characterized by generous modeling and strong outlines has been connected with a complex group representing the rape of Ganymede (III, PL. 345); moreover, this head may be related to the bronze *Poseidon* from Livadostro. The group of warriors and Athena must have been even more monumental and astonishing, for the clearly outlined and powerfully modeled bodies of the heroes reveal a totally new and different conception. These vigorous, engaged forms already contain a hint of the *Tyrannicides* (III, PL. 347), and simple drapery is arranged over the active, tense relief of the bodies. The head of Athena (III, PL. 342), in its clarity of construction and its vital intensity, is the most lucid and consistent Peloponnesian female figure known in the early severe style.

Corinthian sculpture made less important contributions in the classical and postclassical periods. Kallimachos seems to have been a Corinthian, but as an artist he belonged to the Attic school, close to the style of Phidias. He is traditionally regarded as the inventor of the Corinthian capital; thus, he may be said to have introduced a factor in the Corinthian predominance with regard to the decoration of religious edifices. Euphranor, whose personality is not clearly known to historians of ancient art, was probably a Corinthian, or at least an Isthmian. The fact that he was also a painter provides further reason for linking him with the Corinthian tradition.

The most vivid and personal notes in Corinthian sculptural art are found in the carefully executed series of raised-relief mirrors; the greatest number of these should be assigned to Corinth, and the later, more developed works to places that were artistically dependent upon Corinth, such as southern Italy. Particularly noteworthy are some delicate female heads representing Aphrodite, in which the slender features are enclosed within a cloud of ringlets. These figures seem to be associated with the coins of Syracuse belonging to the Euainetos series (III, PL. 384). Even more representative of the pictorial tendencies of Corinthian art are the engraved mirrors, a few valuable examples of which have been preserved. Depicted in one of the most celebrated examples (Louvre) is Korinthos, the personification of the city, seated in an attitude similar to that of Zeus, while Leukas, a nymph personifying the Corinthian colony, places a crown on his head. Of even higher quality, because of their refined delineation and idyllic freshness, are those in which Aphrodite is accompanied by a little Eros carrying a bow or by a tiny crouching Pan (Br. Mus.).

Various kinds of sculpture that served to adorn the agora and other public places in Corinth during the Roman period have been preserved, but apart from the superb group of imperial portraits, one seeks in vain for a reflection of the former artistic glory of the city. Some interesting "baroque" heads have brought forth suggestions of influence from Pergamon, but their idiom is too commonly found in much of Greece and the surrounding islands to justify ascribing its origin to such a remote province — especially since in the Roman era the dominant influence came from Attica. It is sufficient proof of this to mention the reliefs from the theater (Corinth, Archaeol. Mus.), which follow the motifs of the Gigantomachy and the Amazonomachy on the shield of the Athena Parthenos; the fragments of a relief frieze depicting the slaying of the Niobids, from the Temple of Poseidon at Isthmia (perhaps copied from the throne of Zeus at Olympia); and the fragments of a replica of the reliefs of dancing maenads by Kallimachos (Corinth, Archaeol. Mus.).

It has often been observed that the Corinthian spirit was expressed in its liveliest and most individual forms in painting, a fact that emerges clearly enough from historical records of the traditions of the Corinthian-Sikyonian school. The large architectural terra cottas are substantially pictorial; parts of them, in fact, are no more than flat surfaces enlivened by vigorous color effects. The same is true of the figural antefixes. The great sphinx acroteria and the monumental groups from Corinth and Olympia are infinitely more colorful than any statue, whether of bronze or of marble. It is not, of course, simply by chance that more examples of truly great painting have survived here than in any other Hellenic region. The painted metopes from Thermon (I, PL. 336; VII, PL. 39) and those from Kalydon provide the most monumental examples of ancient painting extant. In these, mythical scenes, such as those of Perseus fleeing, a hunter (perhaps Herakles or Orion), and Chelidon and Philomela, alternate with large Gorgon heads, executed in a manner very much like that of the Argive-Corinthian bronze strips. The simple, angular forms of Perseus and Herakles are similar to those of Chrysaor and Zeus on the pediment from Corfu (VII, PL. 44). The female figures display sumptuous color in their garments, which are notable for their elaborate borders.

A series of painted altars lately discovered at Corinth are mature works, and nothing comparable is to be found in Corinthian ceramic art, which had severely declined as early as 550 B.C. From these altars an attempt has been made to identify an artistic personality, known as the Corinthian Altar Painter, who had a singular gift for depicting mythical scenes and impressive animals. The pinakes from Pitsa, painted wooden tablets that were fortuitously preserved in a cave in Sikyonian territory, are indisputably works of pure painting in all respects. An incomparable and rather mysterious aspect of Corinthian art formerly known only from literary sources, these include, besides the most complete painting, which depicts a solemn sacrifice, fragments of mythical scenes, some of considerable size.

The Pitsa pinakes appear to date from about 550 to 530 B.C., but the abundant series of tablets from Penteskouphia (PL. 64) spans a period from the last quarter of the 7th century to 480 B.C. This series is related both to the Pitsa pinakes and to the painted vases; thus, it forms the most legitimate and convincing bridge in tradition and in development between large-scale painting and vase painting. The Penteskouphia pinakes come from a sanctuary of Poseidon, the supreme deity of the Isthmian region. Poseidon and Amphitrite are depicted on most in static poses; the generously draped divinities are usually shown standing either facing each other or side by side. In contrast with this elevated and solemn world of hieratic figures, other pinakes depict ordinary human beings at work, some apparently in a quarry (VI, PL. 56), others at a potter's or coroplast's kiln. Pinakes with mythical scenes are more rare, but among those found are the stories of Herakles; the birth of Athena; the Homeric episode of the aristeia of Diomedes, with Athena on a chariot and Teukros drawing the bow; and some figural compositions of larger dimensions.

Kunze has observed that in view of certain later finds at Olympia, the Corinthian vases seem to be related — albeit of lesser quality — to the superb bronze reliefs of the utensils. Nevertheless, such was the authority with which painting was applied to almost every type of monument or artifact in Corinth that this medium should not be underestimated, even in comparison with the highly refined bronzes of Olympia. The basic element that facilitated the transition from one medium to another was a feeling for narrative, which was a firmly established feature of Corinthian art. Although the narrative genre and the depiction of myths are found in practically all the art centers of Greece, it was in Corinth that there flourished a continuous vein of narrative of an almost epic character. In fact, the study of Corinthian ceramics enables one to follow the artistic development in Corinth with uninterrupted continuity through almost four centuries (PLS. 62, 63, 70; III, PL. 135; VI, PL. 100).

The last phase of sub-Geometric Corinthian ceramics, which constitutes the early proto-Corinthian period, occurred about the second half of the 8th century B.C. The individual qualities characterizing this period are not so much the novelty or richness of the decorative formulas but the sobriety and sense of balance with which even the humblest and most obvious motifs are employed. The large vases seem less individual and more related to the work of other centers, but the forms of the skyphoi and aryballoi remain distinctive. The development of proto-Corinthian and Corinthian ceramics raises the most difficult problems of dating encountered in the study of Hellenistic artistic history. As is well known, the chronology of the foundation of Greek colonies in Italy is formulated from the study of subtle variations in the shapes of the oldest proto-Corinthian aryballoi. The rather archaic character of the proto-Corinthian aryballoi discovered in Syracuse seems to indicate that this city was founded before Cumae. That such delicate distinctions can be made is further proof of the highly concentrated and systematic development by Corinthian artists of subtle variations in form and in decorative schemes. Even in the Geometric period, Corinthian ceramics did not entirely lack true figural motifs, though these did not reveal the more subtle communion with the world of men and animals found, for example, in Attic works, and to a lesser extent in works from Argos and Sparta. As early as the beginning of the Orientalizing period, the aryballoi, though heavily globular, were enlivened with strange animals, birds and protomas of birds, grazing deer, and fish with fantastic and heraldic shapes. These subjects should be considered individually, as primary rather than accessory elements of a larger pattern. On the larger vases, such as the oinochoai from Cumae and Tolosa (mod. Toulouse), broader and more unusual effects were introduced, such as a large pattern of robust and freely related curves covering the entire surface of the vessel.

Vase production reached a high level in the second proto-Corinthian phase, which, as is generally agreed, lasted from about 700 to 650 B.C. A small, flat-bottomed oinochoe (Corinth, Arch. Mus.) ascribed to the Hare Painter (Hasenmaler) is in itself a highly finished work, having a compact form and superbly balanced decoration. Its most remarkable features are the refined shape and the decorative system emphasizing and enhancing it; the swift, summarily rendered little hares add a rhythmic accent. The almost contemporaneous works of the Hound Painter have superb graphic qualities and are concentrated on an unusual motif, generally a running dog. These vases, by contrast, are more directly related to the realm of large-scale painting. In about the second quarter of the 7th century B.C., large skyphoi with broad surfaces were preferred; their fine, elaborate decorative structure provided a delicate framework upon which large animals were depicted with strong and expansive curves. A little later, still about the middle of the century, the Bellerophon Painter of Aegina made use of these formal achievements in depicting Bellerophon and Pegasos, with the Chimera, the Gorgon, or a Siren, which are the first mythical themes encountered in Corinthian figural art. These images were not introduced as accessory elements but as central motifs executed in decidedly monumental form; the emphasis is less upon violent action than upon the inherent drama of the solemn confrontation of the two adversaries. The smallness and frailty of the hero in comparison with the huge monster are also significant in the bronze strips from Olympia. The portrayal of the heroism of Bellerophon has a unique freshness, but the artist seems to have been most concerned with conveying the power and terrifying aspect of the monster, which has been rendered in firm, flowing lines; the monster advances almost ceremoniously, and its great open maw seems inescapable. The vitality of Corinthian art is due more than anything else to this expression of wonder toward the natural subjects, particularly toward animals. This characteristic freshness survived even in some of the later, mechanically executed continuous friezes.

Apart from the more imposing examples by the Hound Painter and the Bellerophon Painter, most of the painted pottery produced tended to have miniaturistic qualities. The more notable examples are found in the extended series of perfume containers, the globular aryballoi, which were developed in more slender forms before becoming ovoid and then pyriform. Many of these are among the most imaginative masterpieces of Greek art; the spare, refined decoration is organized in carefully balanced patterns, but in no way restricts the freedom and freshness of the figures. The examples from the middle proto-Corinthian period are the most refined of the whole series. The principal motif has been placed within a slightly larger frame or frieze, which is in turn enclosed by an elaborate ornamentation of palms and lotus flowers on the shoulders and a minor frieze at the base. During this period, besides the animal and monster themes, there also appeared authentic mythological scenes, such as Herakles and the centaurs, the abduction of Helen, hoplites in combat, from episodes of the epic cycle, and lion and boar hunts. The vivid figures are usually shown in tense attitudes or in motion, and the lively hare hunts constituting the minor frieze are fanciful additions. The scene of Bellerophon and the Chimera on the kotyle from Aegina (Aegina, Mus.) has traditionally been contrasted with the tiny spear-bearing warrior who confronts the fantastic Chimera with a human head mounted on its back seen on an aryballos in Boston (Mus. of Fine Arts), a work that has been ascribed to the Boston 397 Painter (Benson's Löwenmannmaler).

Numerous individual painters have been distinguished, though satisfactory agreement about them has not always been achieved. The loftiest formal qualities are accompanied by a narrative intensity in the works of the painter of the sacrificial scene on the Macmillan aryballos (Br. Mus.) and especially in the work of the painter of the "Chigi oinochoe" (PL. 70), identified by Benson with one of the "forerunners" known to tradition, the almost mythical Ekphantos. On the Chigi oinochoe, the mythical stories succeed each other in a continuous flow that reveals a strong feeling for narrative, and the formal repertory is of an extreme lightness, even in the large scenes, which have a controlled, orchestrated quality. The Chigi oinochoe is the supreme example of that narrative artistry mentioned in traditional sources as being found in the decorations on the chest of Kypselos, which are known in our era only through modest examples of Corinthian vase painting. A prime difficulty in following the developments in vase painting in general, and in Corinthian art in particular, has arisen from the establishment of a relation between the Chigi oinochoe and the aryballoi of the so-called "style magnifique" (Johansen, Payne); consequently, the Chigi oinochoe must be dated shortly after the middle of the 7th century B.C. Some authorities, who have assumed that this date entails too sharp a break in the substantially unified and harmonious evolution of Greek art, have therefore been led to claim for the Corinthian development an extremely advanced and, to some degree, isolated position. This early date for the Chigi oinochoe is no longer a cause of serious disagreement, since there has been a return to a relatively earlier dating for sculpture belonging to the Daedalic tradition (Homann-Wedeking, 1950; Buschor, *Altsamische Standbilder*, IV, Berlin, 1960; V, 1961). In fact, the most mature and independent Corinthian ceramic figurines fit easily into the line of development of contemporaneous works, such as the ivory figure of a youth, perhaps the support of a throne, from Samos (VII, PL. 32). Beginning with the Chigi oinochoe, the prevailing interest in narrative may be traced in the aryballoi with battle scenes, in skyphoi, in the cups belonging to the Gorgoneion group, and particularly in the large column craters (PL. 63; III, PL. 135) belonging to the mature phase of Corinth.

The logical development represented by Corinthian ceramics does not signify a decadence in comparison with the proto-Corinthian stage. Evidently, the great success and wide distribution of this ware throughout the Mediterranean basin resulted in a certain degree of mechanization in its production. On the one hand, the vitality and freshness of vision so frequently found in the older artifacts diminished as time went on. "Inventions" were of necessity less frequent, and mechanical repetition led to abandonment of efforts to achieve variation in the long rows of animals in the continuous friezes by means of dramatic pairings and implied or actual encounters. On the other hand, the Corinthian taste for narrative continued to develop in the style of the Chigi oinochoe, and further extraordinary effects were produced. On small vessels, the figures of an animal, a sphinx, or other monstrous beasts continued to be depicted with freshness and vigor. One of the most significant achievements of the Corinthian style was the adaptation of a single subject to the curving sides of a vase with complete freedom, without using the system of friezes and interruptions. The gaiety and spontaneity of Corinthian painting and the imaginative clarity of forms and colors should not obscure its underlying adherence to principles of compositional harmony. One might say that concern for proportion and balance prevented Corinthian artists from accepting the grand, rather intemperate style of 7th-century Attic vase painting, in which the large scale of the animals seems to exceed the physical limitations of the walls of the vase. Such an antinomy between the subject and the form of the vase did not exist in Corinth. On the contrary, the Corinthian accommodation to a demanding compositional form led to a stage at which the mechanical fillers began to deprive the central motif of its rightful dominance.

The recognized artistic level of Corinthian vase painting is verified by the appearance of signatures on some of the works. Scholars have made attempts both to distinguish individual personalities and to trace the development of certain workshops. Few artists' names have been preserved, and not all of these have been valuable for modern research. Timonidas signed a superb pinax (PL. 64), and the signature of Milonidas appears on a fragmentary pinax in the Louvre; but Chares was an unskilled painter to whom no one would have paid any attention had he not left his signature. Kalikleas, who painted a conical oinochoe (so-called "candlestick") discovered at Ithaca, was probably a local artist. One of the Pitsa pinakes bears an unclear signature with the additional qualification of "Corinthian." Among the artists whose work scholars have made efforts to individuate are the Amphiaraos Painter, the Tydeus Painter (PL. 63), the Andromeda Painter, and the Chimera Painter.

Narrative scenes prevail on the large vessels, especially on the craters, but a number of the small globular aryballoi show well-developed battle scenes. A late Corinthian aryballos (Paris, Cabinet des Médailles) shows the wooden Trojan horse, from which the Greek heroes are emerging. Many Corinthian cups have epic scenes depicted on them. In Spartan painting the chief scene is placed in the rounded part of the interior, with only decorative motifs on the exterior. In Corinthian painting, the important figures occur on the frieze on the outside (PL. 62), and generally a Gorgoneion or sometimes one or two female heads are found on the inside (PL. 63). Though epic themes are characteristic of this style of painting, there are also works with more modest themes. An oinochoe from Veii (mod. Veio), to be dated slightly later than the Chigi oinochoe, may be considered almost as perfect a work of art, even though it has only two animal figures, a bull and a lion, incised on a black ground. The Corinthian painters developed to a supreme level the ability to centralize a single motif, using as subjects an animal, a monster, or one of the grotesque dancers (komasts) which presage the theater. The most accomplished examples of this technique are the plates by the Chimera Painter, who was one of the most prominent exponents of the so-called "heavy style." This master, who belonged to a relatively mature phase, the middle Corinthian, fulfilled the exacting task of depicting a wholly enclosed figure — capturing, moreover, its very spirit — with lines at once sharp and sensitive.

One of the great achievements of Corinthian artists was that, even though they did not invent them, they codified the narrative methods which were later adopted generally in Attica and which finally became typical of the whole Greek world. According to ancient tradition, almost half of the known iconographic motifs may be traced to the chest of Kypselos, which was kept in the Temple of Hera at Olympia. Within the extant works alone, such as the bronzes from Olympia, the Argive-Corinthian strips, and the vase paintings, there is sufficient proof of the strong Corinthian interest in certain aspects of the figural tradition that in other art centers (in the Hellenic East, for example) were given but occasional importance. This is so essential a characteristic of Corinthian art that it seems impossible not to assign to Corinth at least some of the small ivory friezes from Delphi which are generally assigned to the Ionic sphere (I, PL. 341), particularly the ones depicting the departure of Amphiaraos, the rout of the Harpies, and the birth of Athena, which are executed according to typical Peloponnesian schemes, with compact and outlined forms for which one searches in vain among works from the East Greek area. Vases decorated in relief may be regarded as links between bronze reliefs and painted vases; noteworthy examples of vase reliefs have been discovered at Corinth. One of the oldest and most important is a fragment from Argos with proto-Corinthian animals (*BCH*, LXXX, 1956, pp. 372–73, fig. 19). Some fragments from Corinth seem to prove that an extensive series of Sicilian basins and pithoi, with stamped borders depicting chariot races and winged genii, are of Corinthian derivation. The most notable of these bears a representation of Perseus with the Gorgons in pursuit, which seems to attest that this type of art was included in the common narrative interest.

Preserved in Corinthian ceramic art are stories of Herakles and Perseus, episodes from the Trojan epic, and myths of lesser-known Argive kings and heroes such as Amphiaraos and Adrastos. The narrative is usually subdued and sustained in tone, and is without clashes, dramatic emphases, or interruptions, even when the subject seems to call for such effects; crucial moments are stated simply and with a quiet dignity. The tastes of the Corinthian painters tended chiefly toward the ceremonial and sumptuous aspects of the ancient tales. The scene of the wedding of Paris and Helen (Met. Mus.), which is one of the earliest examples representing the bridal couple's departure in a chariot, came to be frequently repeated. Also depicted were solemn games and competitions, scenes of wooing, banquets, mounted processions, and hunts. Gay and varied colors were used. Garments were ornamented, and horses displayed brown trappings or were presented in alternating shades. Even red and brown human figures are found alternating with white ones.

During the late Corinthian phase, which lasted almost until the first decades of the 6th century B.C., a determining factor was the interchange of influences between Corinth and Athens. Of the two centers, Athens probably profited the more in forms, modes of expression, and figural schemes. In some instances, it is difficult to ascertain which of the two centers should be credited with one form or another, together with its accompanying decorative system; doubt exists, for example, regarding the origin of the paneled amphoras and the lekythoi. Certainly, however, the important development of the Attic cup originated in the Corinthian cup. The rather crude series of Attic komast cups is an obvious importation from Corinth, as the forms and the typology of the komasts both show. As for the Athenian painters, they may well have influenced the unrestrained Corinthian painters to cultivate a more austere style, as seems to be suggested by the skyphoi belonging to the Boston group and by the late black-figured plates. Perhaps these frequent and continual exchanges led to the eventual replacement of Corinthian ceramics by Attic ceramics throughout the ancient world. Various explanations for this development have been advanced, but the decline and substantial disappearance of Corinthian ware was probably due to the irresistible superiority of Attic techniques and products. Some wells explored at Corinth have been found to contain mainly Attic black-figured and red-figured pottery from the second half of the 6th century B.C. on. In Corinth itself, only modest utilitarian pots with rapidly executed decorative motifs continued to be produced locally. The few isolated examples of later outline painting or imitations of Attic red-figured technique only serve to confirm the ultimate exhaustion of the art form that had been the glory of Corinth.

ARTISTS. *Corinth. a. Architect.* Spintharos (Σπινθάρος) of Corinth, ca. mid-4th cent. B.C.: supervised reconstruction of the Temple of Apollo at Delphi. *b. Sculptors.* Daidalos (Δαιδαλος), legendary figure: wood statue of Herakles in the Temple of Athena Chalinitis. – Eucheiros (Εὔχειρος) of Corinth, pupil of Chartas and Syagras, both of Sparta, master of Klearchos of Rhegion, mid-6th cent. B.C., has been identified with the coroplast Eucheir of Corinth (Lippold, GP, p. 28; M. B. Marzani, EAA, s.v., no. 1). – Chionis (Χιονις), or Chion (Χιων), of Corinth, beginning of 5th cent. B.C.: figures of Artemis and Athena, from group offered by the Phocians at Delphi after 485 B.C. – Collaborators on the same Phocian offering: Diyllos (Δίυλλος) and Amyklaios (Ἀμυχλαῖος) of Corinth, who worked in bronze: images of Herakles, Apollo, and Leto. – Kallimachos (Καλλίμαχος) of Athens or Corinth(?): said to have invented the Corinthian capital (L. Guerrini, EAA, s.v.). – Euphranor (Εὐφράνωρ), perhaps a Corinthian (G. Bendinelli and M. Floriani Squarciapino, EAA, s.v.). – Lysippos (q.v.). *c. Painters.* Aridikes (Ἀριδίκης) of Corinth, probably 7th cent. B.C.: employed a monochromatic and linear technique emphasizing anatomical details of the figures, next to which, as in vase painting, their names were written (Rumpf, MZ, p. 32; L. Guerrini, EAA, s.v.). – Kleanthes (Κλεάνθης) of Corinth, 6th cent. B.C., possibly the inventor of painting executed on the basis of outlines: *Capture of Troy* and *Birth of Athena*, both in the Temple of Artemis Alpheionia near Olympia (ThB, s.v.; Rumpf, MZ, p. 16). – Ekphantos (Ἔκφαυτος) of Corinth, period of activity uncertain: reputed to be among the first to use color to fill in the ground of a figure (B. Sauer, ThB, s.v.; M. B. Marzani, EAA, s.v., no. 2). – Aregon (Ἀρήγων) of Corinth, perhaps 6th cent. B.C.: figure of Artemis on a griffin, in the Temple of Artemis Alpheionia near Olympia (B. Sauer, ThB, s.v.; L. Guerrini, EAA, s.v.). – Iphion (Ἰφίων) of Corinth, probably first half of 5th cent. B.C. (ThB, s.v.; L. Guerrini, EAA, s.v.). *d. Coroplasts.* Eucheir (Εὔχειρ), or Eucheiros (Εὔχειρος), of Corinth, 7th cent. B.C.: fled from the tyranny of Kypselos (ca. 650 B.C.) to Etruria in the retinue of Damaratos; identified with a relative of Daidalos having the same name, said to be the inventor of painting, as well as with the sculptor Eucheiros of Corinth and, with less likelihood, with the father of the ceramic artist Ergotimos of Athens (Lippold, GP, p. 28; G. Cressedi, EAA, s.v., no. 2). – Diopos (Δίοπος) and Eugrammos (Εὔγραμμος) of Corinth: accompanied Eucheir of Corinth to Etruria and worked there (G. Cressedi, EAA, s.v. Diopos; M. B. Marzani, EAA, s.v. Eugrammos). – Boutades of Sikyon (see below, *Sikyon*). – *Kleonai*: *a. Sculptors*: Dipoinos and Skyllis of Gortyna (?) (see below, *Sikyon*). *b. Painter*: Kimon (Κίμων) of Kleonai, ca. 500 B.C.: reputed to be inventor of the *katagrapha*, in which a figure is depicted according to several different perspectives at once (S. Ferri, Plinio il Vecchio, Rome, 1946, p. 142; Rumpf, MZ, pp. 71–80; M. Cagiano de Azevedo, EAA, s.v.).

SIKYON. Besides possessing undeniable authority in the traditions of painting, Sikyon enjoyed nearly as much fame from being connected with the legendary beginnings and development of a school of sculpture. According to Pliny (*Naturalis historia*, XXXVI, 9), Sikyon was a center for spreading the influence of Cretan sculpture in the Peloponnesos: "Hi [Dipoinos et Scyllis] Sicyonem se contulere, quae diu fuit officinarum omnium talium patria." ("They [Dipoinos and Skyllis] made their way to Sikyon, which was for a long time the motherland of all such industries.") Among the earliest examples mentioned by tradition are a group by Dipoinos and Skyllis with Apollo, Herakles, Artemis, and Athena, presumably a representation of the contest for the tripod; xoana of Tyche Akraia and of the Dioskouroi (Pausanias, II, 7, 5); a xoanon of Artemis Mounychia; and, in the market place, a seated male figure said to be of markedly archaic character.

In the development of sculpture, as in painting, reciprocal influence beween Corinth and Sikyon was so appreciable that it is extremely difficult to make a clear distinction between the two areas of artistic influence. For example, though famous ancient bronze workshops were located at Corinth, the raw materials had to be obtained from Sikyonian territory. Further evidence, derived from traditional sources, is that Boutades of Sikyon, the reputed "inventor" of terra-cotta sculpture and relief, went to Corinth to practice his art; similarly, Laphaes, an early archaic sculptor from Phlious, was reputed to have worked at Sikyon, where he executed a statue of Herakles. Such reports give the impression of a continuing collaboration and of a substantial unity of development; thus, it is not so surprising that the only examples of true painting from ancient times, namely, the Pitsa pinakes, come from Sikyonian territory but bear the signature of a Corinthian artist and reveal a Corinthian formal idiom.

The oldest example of sculpture from the Sikyonian area is a bronze statuette of a female divinity, wearing a long bell-shaped garment and a polos with a crown of leaves, who, like the Artemis Orthis, is known only from a drawing by Eduard Gerhard. This isolated example is followed, well before the middle of the 6th century B.C., by a series of metopes recovered in the foundations of the so-called "Treasury of the Sikyonians" at Delphi (PL. 65); these are of great importance because of their originality of form. Though a certain relationship to Corinthian forms may be seen in the metopes representing Europa on the bull and Phrixos on the ram, the metopes of the cattle raid (I, PL. 343) and of the Argonauts contain bold solutions of difficult spatial problems. In all of these the statement is firm and clear, and the elongated and clear-cut forms, together with the particular tenderness shown in depicting the animals, witness the extremely high level of development achieved in Peloponnesian sculptural art of this phase. The only extant example of archaic sculpture of definite Sikyonian provenance is a limestone female head from the middle of the 6th century B.C. (Boston, Mus. of Fine Arts).

Sikyon assumed a position of central importance at the time of Kanachos, in the mature archaic period; and from this time on, the close ties that came to exist between the school of Sikyon and the school of Argos can be documented. The special bond between them, formed first at the time of Polykleitos the Elder and again later at the time of Lysippos, seems to be foreshadowed by the collaboration of Kanachos with Hageladas, the most illustrious member of the Argive school, on a group of three Muses. Kanachos is also the earliest representative of the school of Sikyon whose historical and artistic character can be substantiated to some degree. He seems to have been an artist faithful to tradition, expert in the ancient techniques of carving precious materials; Pausanias (II, 10, 5) has recorded a seated statue of Aphrodite by him, made of gold and ivory, which was located in a sanctuary of the goddess at Sikyon. He also made the cedarwood image of Apollo Ismenios at Thebes, a near duplicate of the colossal bronze of Apollo Philesios he made for the Didymaion near Miletos. Pliny (op. cit., XXXVI, 42) states that Kanachos also carved marble statues. There is no doubt, however, that his greatest work was done in bronze, in the traditional Peloponnesian technique, and that, because he was an exponent of quite exceptional artistic concepts and techniques, he was called to as distant a country as Ionia to make the cult statue for one of the most famous of Eastern sanctuaries, the Temple of Apollo Philesios at Didyma. This image of Apollo Philesios must have been his masterpiece, and it is believed that smaller replicas of it exist; at least some similarities can be found by comparing the ancient descriptions with certain bronze statuettes of the same type, such as the Payne Knight *Apollo* (Br. Mus.) and an *Apollo* from Pompeii (Naples, Mus. Naz.). Such data provide fairly certain grounds for placing this artist between the end of the archaic period and the beginning of the severe style. After the head of the *Apollo Philesios* was discovered represented with great purity in a relief on a marble chest at Miletos, attempts were then made to identify characteristics of the Sikyonian school in such works as the *Poseidon* from Livadostro (anc. Kreusis; Athens, Nat. Mus.), a kouros (Apollo?) from Piombino (PL. 76), and a late archaic bronze head from the Villa of the Papyri, near Herculaneum (Naples, Mus. Naz.). Kanachos has also been associated with a superb marble torso from Miletos (Louvre); this powerful work, with tensed, vibrant muscles, is a unique find from Ionia. The brother of Kanachos, Aristokles, is recorded as having been almost as good a sculptor; he was the creator of the statue of a Muse that presumably accompanied the group carved by Kanachos and Hageladas.

As previously mentioned with regard to Argos, during the activity of Polykleitos the Elder, in the mature classical period, the two great schools of the northern Peloponnesos seem to have fused more completely than during the time of Kanachos. Polykleitos himself has sometimes been described as a Sikyonian, and some of his pupils, such as Daidalos and Kanachos the Younger, were also said to be Sikyonians. The schools became more definitely Sikyonian under the dynamic influence of Lysippos (q.v.; III, PL. 382; IX, PLS. 219–226). The ancient sources refer to the extraordinary accuracy and prolificacy of Lysippos — 1500 pieces of sculpture are mentioned by Pliny (op. cit., XXXIV, 36). The genius of Lysippos influenced his epoch more than any other artist and provided the background from which the new achievements of the Hellenistic age were to emerge. As the official portraitist of Alexander the Great, he is also credited with the distinction of having introduced one of the most impressive figures known to posterity (IX, PL. 224) into the formal world of ancient art; for centuries afterward, gods and heroes continued to bear the face of Alexander. Lysippos was surrounded by a host of lesser artists, among whom were his brother Lysistratos; his sons Boidas, Euthykrates, and Daippos; and Teisikrates. But the only pupil of Lysippos who remains more than a mere name is Eutychides of Sikyon, to whom is ascribed the "Tyche of Antioch" (VII, PL. 157), a seated female figure lightly and superbly rendered in a number of contrasting planes. Other Sikyonian artists from the Hellenistic period were Pasias and his brother Aiginetes, Kantharos, Xenokrates, and Thoinias, who can be traced back to Teisikrates and Eutychides.

Some of the anonymous Sikyonian works that are believed to be known through Roman replicas are the Townley *Apollo* (Br. Mus.), the Giustiniani Hestia group (Rome, Coll. Torlonia), a statue of the so-called "Aspasia type," and a peplophoros of the Iraklion type. Langlotz, in his attempt to classify the bronze mirror-stand peplophoroi, has established the existence of two fairly cohesive and profoundly differentiated groups. He has assigned to Sikyon the more refined and vibrant group, characterized by its strongly defined lineaments and attenuated drapery. Some of the figurines were found on Sikyonian territory, and others on Corinthian.

The Sikyonian school of painting, according to traditional data, reached its peak development at a comparatively late date, toward the end of the 5th century B.C. The substantial interruption, or at least reduction, of activity in the wake of the famed preceding artists and until this late period might be connected with the influence of Polygnotos and the consequent shift of artistic interest toward Attica and northern Greece. The literary sources seem to indicate that in this period the school of Sikyon began to be associated to a significant degree with the schools of Attica and Asia Minor but retained its great traditions and stayed in the front rank in this period of great formal achievements. The most illustrious representative of the Sikyonian tradition at the beginning of the 4th century B.C. was, according to Pliny (op. cit., XXXV, 75), Eupompos. A record exists of his disciple Pamphilos, who came to Sikyon from his native city of Amphipolis, as well as notices of Pausias, Nelanthios, and Apelles (q.v.), who seems to have abandoned the Asian traditions at a certain point in order to enter the Peloponnesian school. Timanthes, perhaps a native of Kythnos, also came to Sikyon. Pausias, a specialist in encaustic painting, was followed by his son Aristolaos and by a pupil named Nikophanes. Among the names of other artists of the Hellenistic period is included that of the painter Eutychides; apparently he is to be identified with the sculptor Eutychides, who was a pupil of Lysippos.

The one work that can be suggested as an example justifying the vivid impression left by this school, the traditions of which were continuing and to which artists were attracted from all over Greece and even from rival schools, is a superb mosaic pavement discovered at Sikyon; its intricate circlets and gigantic flowers recall other, more numerous mosaics from northern Greece. A singular votive relief with pictorial and landscape elements (Munich, Antikensamml.), believed to be from Corinth, is the sculptured product of a predominantly pictorial tradition.

ARTISTS. *Sikyon. a. Sculptors*: *7th–6th cent. B.C.*: Dipoinos (Δίποινος) and his brother Skyllis (Σκύλλυς) of Gortyna(?): group with Apollo, Artemis, Herakles, and Athena, possibly depicting the contest for the Delphic tripod, from which the Herakles figure can perhaps be identified with that at Tiryns, and the image of Athena with the one at Kleonai; xoanon of Artemis Mounychia, perhaps same as the Artemis from the group at Sikyon; ebony and ivory statues of the Dioskouroi and their descendants, at Argos; xoana of Tyche Akraia; seated male figure in the agora. – Laphaes of Phlious (see above, *Phlious*). – Mousos (Μοῦσος) of Sikyon, either of the archaic period, if the same as a namesake mentioned in a Sikyonian inscription, or of the Hellenistic age, in which case he is of unknown origin: statue of Zeus on a bronze base, an offering at Olympia by the Corinthians (M. Bieber, ThB, s.v.; G. Lippold, RE, s.v.). – *5th cent. B.C.* Aristokles (Ἀριστοκλῆς) of Sikyon, brother of Kanachos of Sikyon and perhaps father of Kleoitas, 6th–5th cent. B.C.: figure of Muse with lyre from a group of three Muses, the others by Hageladas of Argos and Kanachos (Lippold, GP, pp. 86–87; M. T. Amorelli, EAA, s.v., no. 2). – Kanachos (Κάναχος) of Sikyon, 6th–5th cent. B.C.: Muse with shepherd's pipe, for a group of three Muses, the others by Hageladas of Argos and Aristokles of Sikyon; bronze image of Apollo Philesios, originally in the Didymaion near Miletos, in the right hand of which was a stag with a mechanism that made it rock back and forth, recognized in copies on coins and gems; cedarwood Apollo Ismenios at Thebes, which may have been a duplicate of the Apollo Philesios; a chryselephantine Aphrodite for the sanctuary at Sikyon; boys on running horses, attributed also to Kanachos the Younger; other works in marble, but not identified (ThB, s.v., no. 1; G. Carettoni, L'Apollo della fonte di Giuturna e l'Apollo di Kanachos, BArte, XXXIV, 1949, pp. 193–97; Lippold, GP, pp. 86–87; G. Carettoni, EAA, s.v., no. 1). – Kalamis (q.v.): a bearded chryselephantine statue of Asklepios, at Sikyon. – Alypos (Ἄλυπος) of Sikyon, pupil of Naukydes, end of 5th cent. B.C.: seven statues of nauarchoi for the Spartan offering at Delphi after the victory of Aigospotamoi (405 B.C.), from which the signed base remains; four statues of athletes at Olympia (Lippold, GP, pp. 216–18; J. Marcadé, Recueil des signatures de sculpteurs grecs, I, Paris, 1953, no. 3; P. Orlandini, EAA, s.v.). – Collaborators with Alypos of Sikyon on the same Spartan offering: Patrokles (Πατροκλῆς) of Sikyon, perhaps father of Daidalos and Naukydes: ten statues of nauarchoi executed with Kanachos the Younger; athletes (G. Lippold, RE, s.v., no. 8). Kanachos (Κάναχος) the Younger of Sikyon, pupil of Polykleitos, end of 5th–early 4th cent. B.C.: ten statues executed with Patrokles; the athlete Bykelos of Sikyon at Olympia (ThB, s.v., no. 2; L. Guerrini, EAA, s.v., no. 2). Tisandros (Τίσανδρος), or Teisandros (Τείσανδρος), of the Polycletan school: eleven statues of nauarchoi, from which one signed base remains (J. Marcadé, Recueil des signatures de sculpteurs grecs, I, Paris, 1953, no. 104). – *4th cent. B.C.*: Daidalos (Δαίδαλος) of Sikyon, son of Patrokles, brother of Polykleitos the Younger and Naukydes of Argos, first half of 4th cent. B.C. At Delphi: signed base, perhaps for the statue of Glaukon, son of Taureas, datable ca. 400 B.C.; Nike and Arkas for the anathema of the Tegeans. At Olympia: trophy for the Elians; five statues of athletes; a signed base. At Ephesos: a signed base assigned to the statue of one Eusthenes. At Halikarnassos: a base signed in very late characters (perhaps a copy); statues of two youths with strigils, known from copies and derivations (Lippold, GP, p. 217; J. Marcadé, Recueil des signatures de sculpteurs grecs, I, Paris, 1953, nos. 22–24; G. Cressedi, EAA, s.v., no. 1). – Damokritos (Δαμόκριτος), clearly the same as Demokritos (Δημόκριτος) of Sikyon, first half of 4th cent. B.C., pupil of Pison of Kalaureia and therefore of the school of Kritios: signed base in Rome, clearly a copy from the imperial era, assigned to a statue of Lysis of Miletos; portrait of Hippos of Elis at Olympia; statues of philosophers (Lippold, GP, pp. 247–48; L. Guerrini, EAA, s.v., no. 1). – Kleon (Κλέων), pupil of Antiphanes of Argos, first half of 4th cent. B.C.: signed base, at Delphi, assigned to the statue of Agesipolis (d. 380 B.C.), son of Pausanias, the king of Sparta; statues of philosophers; at Olympia, a signed base belonging to one of two statues of Zeus executed by Kleon, a signed base assigned to a statue of Kritodamos, four statues of athletes, and an image of Aphrodite (Anadyomene?) (Lippold, GP, p. 217; J. Marcadé, Recueil des signatures de sculpteurs grecs, I, Paris, 1953, nos. 60–61; EAA, s.v.). – Skopas (q.v.) of Paros: statue of Herakles in the gymnasion. – Olympos (Ὄλυμπος) of Sikyon, mid-4th cent. B.C.: portrait of the athlete Xenophon, Olympic victor between 400 and 360 B.C. (G. Lippold, RE, s.v., no. 33; Lippold, GP, p. 248). – Lysippos (q.v.) of Sikyon. – Lysistratos of Sikyon, brother and perhaps also pupil of Lysippos, height of his activity ca. 328 B.C.: appears to have been the first to make use of plaster casts to copy statues (see ATTIC AND BOEOTIAN ART). *b. Painters*: *Archaic period*: Kraton (Κράτων), considered by some the inventor of painting: pinax painted in white with profile outlines of a man and woman (Pfuhl, I, p. 497; ThB, s.v.). – Telephanes (Τηλεφάνης), probably 7th cent. B.C.: noted for a linear technique that emphasized anatomical details of the figures, next to which their names were written (ThB, s.v.; Rumpf, MZ, pp. 16, 32). – *4th cent. B.C.*: Eupompos (Εὔπομπος), beginning of 4th cent. B.C., founder of the Sikyonian school, master of Pamphilos of Amphipolis: statue of victorious athlete with the palm (A. de Capitani d'Arzago, La grande pittura greca dei secoli V e IV a.C., Milan, 1945, pp. 60–61; Rumpf, MZ, p. 120; G. Cressedi, EAA, s.v.). – Timanthes, perhaps a native of Kythnos, of the Sikyonian school (see GREEK ART, AEGEAN). – Pamphilos of Amphipolis (see GREEK ART, NORTHERN). – Bryes (βρύης), perhaps of Sikyon, first half of 4th cent. B.C., father and master of Pausias (Rumpf, MZ, p. 132; L. Guerrini, EAA, s.v.). – Pausias (Παυσίας) of Sikyon, son of Bryes, mid-4th cent. B.C., pupil of his father and of Pamphilos, master of his son Aristolaos and of Nikophanes, most proficient in encaustic: scene of the sacrifice of oxen, later found in Rome; portrait of Glykera, called by some "stephanoplocos" and by others "stephanopolis" (girl making, or selling, wreaths), a copy of which was acquired by Lucullus; depictions of Eros, who has cast down his bow and arrows, with a lyre, and Methe drinking wine from a cup, both for the Tholos of Epidauros; the portrait of a youth, Hemeresios; restoration of the frescoes of Polygnotos at Thespiai; decoration of faces on ceiling panels; lascivious subjects (G. Lippold, RE, s.v.; Rumpf, MZ, p. 132). – Melanthios (Μελάνθιος), pupil of Pamphilos, mid-4th cent. B.C.: author of a treatise on painting; portrait of the tyrant Aristratos of Sikyon alongside a chariot conducted by Nike, executed with Apelles and other Sikyonian painters (G. Lippold, RE, s.v.; Rumpf, MZ, p. 132; L. Guerrini, EAA, s.v.). – Apelles (q.v.): completed his artistic apprenticeship in the school of Pamphilos of Amphipolis. – Nikophanes (Νικοφάνης), of unknown origin, second half of 4th cent. B.C., pupil of Pausias: libidinous subjects, one of which (probably showing a courtesan) was perhaps in the stoa poikile at Sikyon (ThB, s.v.; G. Lippold, RE, s.v.). – Aristolaos of Sikyon, son and pupil of Pausias, active in Athens (see ATTIC AND BOEOTIAN ART). – Nikomachos, of Thebes, second half of 4th cent. B.C.: monument of the poet Telestas for the tyrant Aristratos of Sikyon (see ATTIC AND BOEOTIAN ART). – Sokrates (Σωκράτης) of Sikyon, pupil of Pausias, second half of 4th cent. B.C.: group of Asklepios with his offspring; scene with Oknos holding a rope gnawed by an ass (both works should probably be attributed to Nikophanes; G. Lippold, RE, s.v., no. 8; ThB, s.v.). – Athenion of Maroneia (Thrace), second half of 4th cent. B.C.: pupil of Glaukion of Corinth, perhaps of the Sikyonian school (see ATTIC AND BOEOTIAN ART). – Arkesilaos (Ἀρκεσίλαος), perhaps of the Sikyonian school, end of 4th cent. B.C., probably the same as Arkesilas, son of Teisikrates: portrait of the strategos Leosthenes (d. 323 B.C.) with his sons, in the sanctuary of Zeus Soter and Athena Soteira at Piraeus (Pfuhl, II, p. 734; L. Guerrini, EAA, s.v.). – Thales (Θαλῆς), perhaps of the Sikyonian school, presumably active during 4th cent. B.C. (G. Lippold, RE, s.v., no. 6; ThB, s.v.). – Perseus (Περσεύς): disciple to whom Apelles adressed his book on painting (G. Lippold, RE, s.v., no. 8). – Ktesilochos (Κτησίλοχος) of Kolophon, son of Pytheas and brother of Apelles; author of a parody on the birth of Dionysos (ThB, s.v.). *c. Coroplast*: Boutades (βουτάδης), of Sikyon, 7th–6th cent. B.C.: first to have reproduced a human face in clay, which was then offered as an ex-voto to the Nymphaion at Corinth; first to have painted reliefs in red or to mold a red clay directly; also said to have invented the antefix.

ELIS. One of the most common difficulties encountered in attempting to establish some scheme, some line of continuity, in the development of Greek art lies in the fact that the historical data provided by written tradition and by the evidence of finds not only do not coincide but often appear contradictory. A borderline example of this confusing situation is Elis, a region with no real city as its center but containing the enormous Panhellenic sanctuary of Olympia (VI, FIG. 925; I, PL. 331), where the styles of many Greek states were represented. Since it was not an ordinary community, the permanent population of Olympia must have been quite small; and Elis itself was only a "synoikismos," a settlement of Elians that dated from 471 B.C. Nevertheless, it seems reasonable that, beneath the superstructure of the Panhellenic contributions, there must have been a local art with distinct and continuing features of its own.

It is admittedly risky to attempt a classification of Geometric bronzes according to centers of production; yet it is important to remember that in early archaic times the local population must have formed the majority of the devout, and that a wholesale importation of votive art was unlikely. Thus it would seem plausible to attribute a large part of the small Geometric and sub-Geometric votive bronzes found at Olympia to local workmanship. These consist mainly of animal figures (PL. 61), such as might be expected from an agricultural and pastoral population. E. Kunze has quite convincingly proposed to assign a rather peculiar series of bronze warriors (PL. 61) to local craftsmen. This series extends with great consistency from the Geometric period to the beginning of the archaic period. In Kunze's view these are images of Zeus, represented with arms and helmet, which suggest a type not unlike the Apollo of Amyklai. Among the later work, it becomes extemely difficult to distinguish the modest local products in the ever-increasing, more varied deposits of the archaic period found at Olympia. Still, the discovery in the Altis of furnaces for casting statues and the definite knowledge of Phidias's establishment of a workshop for constructing the colossal chryselephantine statue of Zeus encourage the belief that there was continuous artistic activity at Olympia itself. Whatever the evidence, it remains likely that a great religious center such as Olympia had the purpose of an entropôt and a center of diffusion for artistic influences, since it was so hallowed by tradition and because there was, consequently, steady activity of erection, decoration, restoration, and dedication of buildings and votive offerings.

Apart from Olympia, the site in Elis presenting the most significant finds has been Bambes, where a small sanctuary has yielded remarkable examples of sculpture. Among these are a fragment of a poros statue, probably a male subject, from before the middle of the 6th century B.C., and a refined fictile relief, from the end of the archaic period, representing a beardless hero with a lyre in an attitude of repose. This last work shows typological links with the earliest examples from the Taranto series. Another terra-cotta work is a monumental *perirrhanterion* (lustral basin), minutely decorated with elaborate impressed motifs. The exquisite bronze *Artemis* donated by Chimaridas (Boston, Mus. of Fine Arts) is from another Elian sanctuary, near Mazi. Although this statuette, tightly enclosed within a plain long garment, has been assigned to a Spartan workshop, the serenity of the facial expression would suggest a Corinthian origin as more logical.

The central problem concerning Olympia and its relation to Elian art is presented by the figural decoration on the Temple of Zeus (PLS. 74, 75), which poses one of the most vexing questions in the history of Greek art. In later years, the various explanations of its origin have been resolved into two apparently irreconcilable theories: one emphasizes the Ionic elements, the soft modeling and tender effects, and the affinities with the *Nike of Paros*; the other assumes that these sculptures, because of their isolated and unique qualities, represent the Peloponnesian sculpture of the great century that is known only through literary sources. Moreover, as if to emphasize the profoundly autochthonous character of the Olympia marbles, there are scholars who assert that these works were created within the very confines of the sanctuary by a mysterious Elian school.

The only substantial support for a hypothesis of extensive artistic activity in Elis is provided by the superb series of coins that begins about the middle of the 5th century B.C., with examples representing Nike running and carrying a wreath, and others bearing the magnificent head of Zeus, usually traced to the *Zeus* of Phidias. The names of Elian artists from this period are also recorded. For example, Pausanias speaks of a certain Kolotes from Herakleia in Elis but states that others maintain he was from Paros; whatever his origin, Kolotes seems to have been fairly well known and was active at Olympia as a pupil or assistant of Phidias. A specialist in the chryselephantine technique, Kolotes had had ascribed to him the statue of Athena in Elis, that of Asklepios in the Elian city of Kyllene, and the elaborate ivory table, with statues of divinities arranged in triads, upon which the Olympic victors' crowns were laid. There is also mention of a Kalon of Elis at Olympia; he is said to have executed a great donation for Messana, which depicted a *choros* of 35 children accompanied by a fluteplayer and trainer.

ARTISTS. *Elis. a. Architects*: Agamedes and Trophonios: Thesauros of Augeias (see ATTIC AND BOEOTIAN ART). – Libon (Λίβων) of Elis, active a little before mid-5th cent.: Temple of Zeus at Olympia (C. Weickert, ThB, s.v.; W. B. Dinsmoor, The Architecture of Ancient Greece, London, 1950, pp. 149–53; P. Romanelli, EAA, s.v.). – *b. Sculptors*: Smilis of Aegina (see below, *Aegina*). – Phidias (q.v.): image of Aphrodite Ourania. – Kolotes (see below, *Herakleia*). – Kalon of Elis: statues of Hermes and a *choros* of children, at Olympia. – Praxiteles (q.v.): Dionysios image, in the sanctuary of the god. – Skopas (q.v.): statue of Aphrodite Pandemos, seated on a he-goat, in the sanctuary of the goddess. – *c. Painters*: Panainos of Athens: paintings on the shield of Athena by Kolotes (see ATTIC AND BOEOTIAN ART). – Pyrrhon (Πύρρων) of Elis, philosopher, son of Pleistarchos, active in second half of 4th cent.: figures of racers with torches, in the gymnasion of the city (Pfuhl, pp. 700, 781). – *d. Coin engraver*: Euth[ymos?] (Εὐθ[. . .]), active at Elis in mid-5th cent., identified with a namesake active at Syracuse: signed didrachma (L. Forrer, Biographical Dicionary of Medallists, London, 1902 ff., s.v. Εὔθυμος; Weil, ThB, s.v. Euth[ymos]). – *Herakleia. a. Sculptor*: Kolotes (Κολώτης), probably of Herakleia, but according to a less reliable tradition, of Paros and pupil of Paciteles (perhaps two different artists), active in second half of 5th cent., considered to have been a pupil of Phidias and skilled in the chryselephantine technique: at Olympia, collaborated on the Zeus sometimes ascribed to Phidias and executed an offering table; chryselephantine figure of Athena for Elis, generally ascribed to Phidias; statue of Asklepios at Kyllene (Picard, II, pp. 544–45; Lippold, GP, pp. 142–89; M. T. Amorelli, EAA, s.v., nos. 1–2). – *Kyllene. a. Sculptor*: Kolotes (see above, *Herakleia*). – *Olympia. a. Architects*: Pothaios (Πόθαιος), active in second half of 6th cent.: Thesauros of the Syracusans, erroneously ascribed to the Carthaginians (C. Weickert, ThB, s.v.; H. Riemann, RE, s.v.). – Megakles (Μεγακλῆς; E. Fabricius, RE, s.v., no. 10; EAA, s.v.) and Antiphilos (Ἀντίφιλος; H. Thiersch, ThB, s.v., no. 1; G. A. Mansuelli, EAA, s.v., no. 1) collaborated on this thesauros. – Pyrrhos (Πύρρως), father of Lakrates (Λακράτης) and of Hermon (Ἕρμων), with whom he built the Thesauros of Epidamnos in the 6th cent. B.C. (ThB, s.v. Hermon; C. Weickert, ThB, s.v. Lakrates, no. 1, s.v. Pyrrhos; G. A. Mansuelli, EAA, s.vv. Lakrates, Hermon). – Libon of Elis (see above, *Elis*). – Leonides (Λεωνίδης) of Naxos, active in second half of 4th cent. B.C.: perhaps author of a treatise on architectural proportion; dedicated and built the Leonidaion (C. Weickert, ThB, s.v.; W. B. Dinsmoor, The Architecture of Ancient Greece, London, 1950, p. 251; G. Pesce, EAA, s.v.). – *b. Sculptors*: *6th cent. B.C.*: (see above, *Sikyon*). – Smilis of Aegina: enthroned Horai, perhaps chryselephantine, in the Heraion (see below, *Aegina*). – Dorykleidas of Sparta: Themis, perhaps chryselephantine, in the Heraion (see below, *Sparta*). – Dontas of Sparta: offering of the Megarians of wood and gold representing the contest between Herakles and Achelous (see below, *Sparta*). – Medon of Sparta, perhaps to be identified with Dontas: statue of Athena in the Heraion. – Telestas and Ariston, probably Spartans: Zeus, offered by the citizens of Kleitor as a tithe from the conquered cities (see below, *Sparta*). – Hegylos and Theokles of Sparta: Herakles, Atlas and the Hesperides around an apple tree, a grouping (pedimental?) of cedar wood for the Thesauros of Epidamnos. The Hesperides were later placed in the Heraion (see below, *Sparta*). – Dameas (Δαμέας) of Krotona, probably active in the 2d half of the cent.: Milon of Kroton, many times victor (also in 532), of which there remains the base, with a fragmentary inscription (Lippold, GP, p.92; L. Guerrini, EAA, s.v., no. 1). – Kratinos (Κρατῖνος) of Sparta: statue of Philles of Elis, victor prior to 484 (M. Bieber, ThB, s.v., no. 1). – Patrokles (Πατροκλῆς) of Kroton, son of Katillos, worked probably between the mid-6th and the beginning of the 5th cent.: boxwood statue of Apollo with gilded head, offered by the inhabitants of Lokroi Epizephyrioi (G. Lippold, RE, s.v., no. 7; Lippold, GP, p. 92). – Eutelidas (Εὐτελίδας) of Argos, active at the end of the cent.: statue of Damaretos of Heraia, victor in 520 and 516, and of his son Theopompos, victor in 504 and 500, placed on a single base (Lippold, GP, p. 89; M. B. Marzani, EAA, s.v.). – Chrysothemis (Χρυσόθεμις) of Argos, active in the 6th and the beginning of the 5th cent., collaborated on this same work (L. Guerrini, EAA, s.v.). – Askaros (Ἄσκαρος) of Thebes, pupil of a Sikyonian master (Kanachos or Aristokles?), active 6th–5th cent.: Zeus in bronze, offered by the Thessalians after a victory over the Phocians (Picard, I, pp. 440, 491, 511; P. Orlandini, EAA, s.v.). – Aristokles (Ἀριστοκλῆς) of Kydonia in Crete, active 6th–5th cent.: Herakles and the queen of the Amazons, of which there remains a fragment of the inscription, executed for Evagoras of Zankle prior to 494 (Lippold, GP, p. 98; M. T. Amorelli, EAA, s.v., no. 3). – *5th cent. B.C.*: Hageladas of Argos: Anochos of Taras, victor in 520; quadriga of Kleosthenes of Epidamnos, victor in 516; Timasitheos of Delphi (d. 507). – Philesios (Philesias?, Φιλήσιος) of Eretria, beginning of the cent.: bronze bull offered by the inhabitants of Eretria, probably prior to 490, of which a signed base and some fragments remain (G. Lippold, RE, s.v., no 3; Lippold, GP, p. 73, note 4). – Serambos of Aegina; statue of Agiadas of Elis, prior to 480 (see below, *Aegina*). – Theopropos of Aegina: attributed to him a bull dedicated by the inhabitants of Kerkyra (see below, *Aegina*). – Stomios (Στόμιος), beginning of the cent.: Hieronymos of Andros, victor 480 or earlier (G. Lippold, RE, s.v.; M. Bieber, ThB, s.v.). – Glaukias of Aegina: Gelon in his quadriga, dedicated for his victory in 488, of which part of the inscription remains; Glaukos of Karystos, recognized in a small bronze; Philon of Kerkyra, victor in 492 and 488; Theagenes of Thasos, victor in 480 and 476 (see below, *Aegina*). – Athanodoros (Ἀθανόδωρος) of Achaia: offering of one Praxiteles of Syracuse and Camarina, executed in 485–80, of which the signed base remains (Lippold, GP, p. 103; M. T. Amorelli, EAA, s.v. Athenodoros I). – Argeiadas, Asopodoros, and Atotos, all of Argos, collaborated on the same work (see above, *Argos*). – Pythagoras: Astylos of Kroton, victor in 488, 484, 480; Euthymos of Lokroi, victor in 484, 476, 472; Mnaseas of Cyrene, victor in 456; Leontiskos of Messana; victor in 456 and 452; Kratisthenes, son of Mnaseas, victor in 448; Dromeus of Stymphalos; Protolaos of Mantineia; "nude with apples," probably one of the preceding victors (see GREEK ART, WESTERN). – Onatas of Aegina: Hermes, executed with his son or pupil Kalliteles (see below, *Aegina*), an offering of the inhabitants of Pheneos; Herakles, an offering by the inhabitants of Thasos either before 492 or between 478 and 463; group of 9 ancient heroes on a semicircular base and a tenth, Nestor, on a separate base, offered by the Achaians; offering of Deinomenes, perhaps in 465, for a victory of Hieron of Syracuse, a quadriga with a charioteer on which Kalamis collaborated. – Kalamis (q.v.): *choros* of praying boys, offered by the inhabitants of Agrigento for a victory over Motye (480–76 or 451–50); two boys mounted on race horses for the offering of Hieron of Syracuse on which Onatas collaborated (465); Nike Apteros, an offering of Mantinea. – Myron (q.v.): Ladas, victor in 476, probably at Olympia and later at Rome; Chionis of Sparta, athlete; Timanthes of Kleonai, victor in 456; Lykinos of Sparta, twice victor, the first time, in 448, and twice depicted by Myron; Philippos the Azanian, victor perhaps in 444. – Kleoitas (Κλεοίτας), of Doric origin, son of Aristokles (of Sikyon?), father and master of Aristokles, worked in the 1st half of the cent., appears to have invented the starting barrier (*hippaphesis*) of the hippodrome; warrior on the Acropolis at Athens (Lippold, GP, pp. 104–05; G. Carettoni, EAA, s.v.). – Ptolichos of Aegina: statue of Epikradios of Mantinai, victor between 492 and 484; statue of Theognetos of Aegina, victor in 476 (see below, *Aegina*). – Anaxagoras of Aegina: bronze Zeus, offered after the battle of Plataia (479), of which the base remains (see below, *Aegina*). – Akestor of Knossos: statue of the athlete Alexibios of Heraia (see GREEK ART, AEGEAN). – Philotimos of Aegina: with Pantias of Chios, the group of Xenodikos of Kos on horseback next to his father Xenombrotos. There remains an epigram which would seem to be of later date (ca. 550). – Glaukos of Argos: Amphitrite, Poseidon and Hestia, an offering of Mikythos in the temple of Zeus. It is doubtful that the statues of Iphitos and Ekecheiria belong to this group. – Dionysios of Argos: offering in behalf of Mikythos of Rhegion composed of various statues, some of which were carried off to Rome by Nero. These stood on the north side of the temple of Zeus. Fragments of an inscribed base remain; the statue of Orpheus is perhaps recognizable in a copy. An offering, in collaboration with Simon of Aegina, of a horse and charioteer for Phormis of Mainalos was probably done a little earlier. – Polykleitos the Elder (q.v.): Kyniskos of Mantinea; Pythokles of Elis, attributed also to Polykleitos the Younger, as have been the following works: Xenokles of Mainalos; Aristion of Epidauros; Tersilochos of Kerkyra. – Aristokles (Ἀριστοκλῆς), son and pupil of Kleoitas, perhaps of Doric origin, active at mid-cent.: a group of Zeus and Ganymede executed for the commission of Gnathis of Thessaly (W. Amelung, ThB, s.v., no. 4). – Phidias (q.v.): chryselephantine Zeus, for the temple of the god; Anadoumenos. – Kolotes, probably of Herakleia (Elis): worked on the chryselephantine Zeus with Phidias and Panainos; for the Heraion, a table for the offering of the crowns, with twelve divinities and scenes referring to the games. – Phradmon of Argos: statue of Amertas of Elis, victor in 444–20. – Naukydes of Argos: statue of the Argive Cheimon, victor in wrestling in 448. – Paionios (Παιώνιος) of Mende, active mid- or 2d half of 5th cent.: acroteria of the temple of Zeus; marble Nike (preserved together with its signed base) offered by the inhabitants of Messene and Naupaktos, and placed in front of the temple of Zeus. Pausanias erroneously attributed to Paionios the east pediment of this temple (G. Lippold, RE, s.v.; Lippold, GP, p. 205). – Lykios of Eleutherai: Achilles and Memnon, votive offering of the inhabitants of Apollonia in Illyria (see ATTIC AND BOEOTIAN ART; L. Guerrini, EAA, s.v.). – Pantias (Παντίας) of Chios, son and pupil of Sostratos, active in the last decades of 5th cent.: equestrian statue of Xenodikos for the group executed with Philotimos (see above); Nikostratos of Heraia; Aristeus of Argos, victor ca. 420 (G. Lippold, RE, s.v.; Lippold, GP, p. 207). – Kalon (Kallon?, Κάλων) of Elis, active probably in the last quarter of 5th cent.: signed base which upheld a Hermes offered by Glaukias of Rhegion; *choros* of 35 children with trainer and aulos player offered by the people of Messana, a little later in time than the other work (Picard, I, pp. 505–06; Lippold, GP, p. 203; A. Giuliano, EAA, s.v.). – Alypos of Sikyon: statues of Symmachos of Elis, Neolaidas of Pheneos, Archidamos of Elis, and Euthymenes. – Naukydes of Argos: Eukles of Rhodes, of which the signed base remains; Baukis of Troizen. – Nikodamos of Mainalos (Arcadia): Athena, offering of the inhabitants of Elis; Herakles fighting the Nemean lion, an offering of Hippotion of Taras, on the wall of the Altis; Androsthenes of Mainalos, victor for the first time in 420; Damoxenidas of Mainalos, athlete, of which the signed base remains; Antiochos of Lepreon, athlete. – Kallikles (Καλλικλῆς) of Megara, son of Theokosmos and father of Apelleas, worked at the end of 5th cent. according to literary tradition, but at the end of the 4th according to the epigraphic characteristics of his signature: statue of Diagoras of Rhodes, probably erected about 420 B.C., but the base (restored?) on which it rests ought to be dated to 300 B.C.; statue of the athlete Gnaphon of Dipaia. In other localities, uncertain: philosophers; one series of works to which his name is attached should be attributed to another artist (Lippold, GP, p. 204; G. Carettoni, EAA, s.v., no. 3). – *4th cent. B.C.*: Daidalos of Sikyon: trophy offered by the inhabitants of Elis for a victory over the Spartans, about 400 B.C., Eupolemos of Elis, victor in 396; Aristodemos of Elis, victor in 388; Aisypos on horseback and his father, Timon; Narykidas (Tharykidas) of Phigalia, victor in 384(?), of which part of the base remains; signed base of an unknown work (see above, *Elis*). – Apelleas (Ἀπελλέας; or Apellas) of Megara, son of Kallikles, active at the beginning of the cent.: bronze horses in the temple of Zeus, and quadriga with charioteer and Kyniska, both offerings of Kyniska, daughter of Archidamos I of Sparta, for her Olympic victories (probably 396–92), of which there remain parts (one inscribed) of the bases with impressions of statues much smaller than life size. In other localities, uncertain; women adorning themselves (Lippold, GP, p. 222; P. Orlandini, EAA, s.v.). – Kleon of Sikyon: signed base (390–80) assigned to one of two statues of Zeus executed by Kleon and which, with other similar works, was placed between the Metroon and the Stadium; signed base (ca. 350) of a statue of Kritodamos of Kleitor; statues of the athletes Alketos of Kleitor, Deinolochos of Elis, Lykinos of Heraia, Hysmon of Elis; Aphrodite in bronze, in the Heraion. – Polykleitos the Younger of Argos: Antipatros of Miletus (388), attributed by some to Polykleitos the Elder; Agenor of Thebes, dated after 371 or 339–38, or after 316 B.C. Works more often attributed to Polykleitos the Elder: Xenokles of Mainalos, of which the inscribed base remains; Aristion of Epidauros, of which the base remains; Thersilochos of Kerkyra. – Polymnestos of Athens: signed base, reused in 40(?) B.C. (see ATTIC AND BOEOTIAN ART). – Damokritos of Sikyon: the athlete Hippos of Elis. – Kanachos of Sikyon: the athlete Bykelos of Sikyon. – Praxiteles (q.v.): Hermes with the infant Dionysos, in the Heraion, where it was discovered. – Silanion, Athenian or perhaps Megarean: statues of Telestas and Damaretas, both of Messene, erected after 368; statue of Satyros of Elis, after 335–34, identified in a head from Olympia (see ATTIC AND BOEOTIAN ART). – Hippias (Ἱππίας), active at mid-cent.: statue of the athlete Douris who was much later tyrant of Samos, or of his son Skaios (ThB, s.v.; Lippold, GP, p. 288; G. A. Mansuelli, EAA, s.v.). – Sthennis of Olynthos, active at mid-cent.: the athletes Pyttalos and Choirilos (see ATTIC AND BOEOTIAN ART; GREEK ART, NORTHERN). – Olympos of Sikyon: statue of Xenophon, victor between 400 and 360. – Leochares of Athens: chryselephantine statues of Philip, Olympias, Amyntas, Eurydike and Alexander (perhaps the so-called "Rondanini Alexander" is recognizable here), executed after 338 and placed in the Philippeion (see ATTIC AND BOEOTIAN ART; P. E. Arias, EAA, s.v., no. 1). – Lysippos (q.v.): Kairos on a sphere, perhaps at Olympia but sometimes placed at Sikyon, reproduced on gems and reliefs; two athletic statues of Pythes of Abdera; Poulydamas of Skotoussa, of which the base with reliefs survives, probably from Lysippos' workshop; Troilos of Elis, of which a signed bronze tablet remains; Cheilon of Patras, identified also with the *Apoxyomenos*; Kallikrates of Magnesia-on-the-Maeander; pancratiasts, identifiable in a marble head. – Pyrilampes of Messene; statues of the athletes Pyrilampes of Ephesos, Zenon of Lepreos and Asamon of Elis. – Asterion (Ἀστερίων), son of Aischylos, active probably in 4th cent.: the athlete Chaireas of Sikyon (W. Amelung, ThB, s.v.; P. Orlandini, EAA, s.v.). – Theomnestos (Θεόμνηστος) of Sardis, perhaps 4th cent.: the athlete Ageles of Chios; signed base with Dionysios of Chios; athletes, warriors, hunters and sacrificers (M. Bieber, ThB, s.v.; Picard, III, p. 194). – *Of uncertain period*: Onaithos (Ὄναιθος) and Phylakos (or Psylakos; Φύλακος, or Ψύλακος), brothers, executed with their sons a statue of Zeus offered by the Megareans (M. Bieber, ThB, s.v. Onaithos). – Somis (Σῶμις): the athlete Prokles of Andros M. Bieber, ThB, s.v.; J. Marcadé, Recueil des signatures de sculpteurs grecs, I, Paris, 1953, no. 96). – *c. Painters*: Kleanthes of Corinth: birth of Athena, in the Temple of Artemis Alpheionia near Olympia. – Aregon of Corinth: Artemis on a griffin, in the same temple. – Panainos of Athens, collaborator of Phidias: decoration of the wooden barrier placed between the legs of the throne of Zeus (Herakles and Atlas; Theseus and Piritous; Hesperides; Hellas and Salamis; Herakles and the Nemean lion; Ajax and Kassandra; Hippodameia and Sterope; Prometheus and Herakles; Achilles and Penthesileia); decorations of the garments of Zeus (see ATTIC AND BOEOTIAN ART). – Aetion (Ἀετίων), also a sculptor, probably Ionic, 2d half of 4th cent.: exhibited at Olympia a painting, perhaps the nuptials of Alexander the Great and Roxana (327), and as a result was able to marry the daughter of the Hellenodikes Proxenidas. Other works, location not established: Tragedy and Comedy; Dionysos; Semiramis exalted from domestic servant to queen; old woman with a torch in a nuptial cortege (perhaps the same as the nuptials of Alexander and Roxana); the New Spouse, which was perhaps the same as the Semiramis scene (Rumpf, MZ, pp. 147–48; M. Cagiano de Azevedo, EAA, s.v., no. 1). – *d. Coin engraver*: Po[. . .] (Πο[. . .]), perhaps the same as Polykr[. . .] of Elis: signature on local coins between 420 and 416 (J. H. Jongkees, The Kimonian Dekadrachms, Utrecht, 1941, pp. 70–72). *Pisa. a. Sculptor*: Daidalos: Herakles (see GREEK ART).

MESSENIA. Both tradition and archaeological exploration have shown that Messenia was a vast agricultural region without a true artistic character of its own. Sparta, in its desire for expansion was early attracted by Messenia's fertility and richness, and the result was the Messenian Wars, which continued for almost two centuries and did much to impoverish the country. The gradual conquest by Sparta must inevitably have brought the country under Spartan artistic domination, but as late as the end of the archaic period the famous statue of Zeus Ithomatas, erected on Mt. Ithome overlooking Messene, was the work of the great Argive sculptor Hageladas.

An attempt has been made to reevaluate certain important finds in Messenia in order to establish the existence of a local school, at least for smaller bronzes. Later Swedish explorations have concentrated on prehistoric sites, and there have been no outstanding finds of the classical period. But very different finds were made at the sanctuary of Apollo Korynthos (mod. Longa) which have produced a series of remarkable bronzes. The largest and most refined of these is a statuette of a warrior (Athens, Nat. Mus.), from the middle of the 6th century, with a long, bitter, and melancholy face and a chiton decorated with deep embossed borders; it seems to fit easily into the current of Laconian art. Also apparently Laconian is a small Apollo (Athens, Nat. Mus.) with hair falling to his chest; his somewhat vacant facial expression links this work with the Artemis of Chimaridas. A small seated citharist and the small Apollo, in severe style and the most significant Swedish find, are modest by comparison.

The glories of Messenian art are to be found in later periods, at the time when the noted sculptor Damophon was engaged in his remarkable work of restoring and renovating large religious statues all over the Peloponnesos. His technical ability is proven by the fact that he was given the task of restoring Phidias' celebrated statue of Zeus at Olympia. Pausanias expressly stated that Damophon was the only artist worthy of note from Messenia; there is also mention of a Meter and an Artemis Laphria by him at Messene, an Asklepios and his children in the Asklepieion, also at Messene, a Tyche, the City of Thebes, Artemis Phosphoros, and probably an Apollo with the Muses and a statue of Herakles. For Aigion in Achaia he executed an acrolith of Eileithyia and statues of Asklepios and Hygieia; for Megalopolis he made two cult statues of Demeter and the Kore, one of which was acrolithic, with two lesser statues bearing baskets on their heads. For the Temple of Aphrodite in the same city he carved the acrolith of the goddess, a statue of Hermes, and a table decorated with figures of the Horai with Pan, Apollo with the Lyre and the Nymphs tending Zeus. There are also notable remains of his colossal group of cult statues for the Temple of Despoina at Lykosoura, including the heads of Demeter, Artemis, and the Titan Anytos (VII, PL. 187), together with part of the drapery of Despoina decorated with minute reliefs of a mystic character. The images seem to possess that baroque quality which emerged from the dominating classicism characteristic of a slightly later period. This find has made it possible to place Damophon firmly in the last phase of the Hellenistic, about the middle of the 2d century B.C.; he is undoubtedly to be linked with the Neo-Attic movement and that general trend towards a rediscovered classicism which academic writers, echoed by Pliny, regarded as the resurrection of art. Damophon must certainly have enjoyed the stature of a leader of a school, whatever the value of such a position during that period of weakened artistic vitality. His sons Straton and Xenophilos, who collaborated with him on the Lykosoura statues, are perhaps the same as two sculptors of the same name who worked at Argos. Damophon worked for a long time at

Megalopolis, and later artists such as Phileas, Zeuxippos, Nikippos, and Aristeas, whose signatures have been found in the city, may be conveniently said to have belonged to his school. Some scholars have also attributed to Damophon the great stele of Kleitor, probably representing Polybios.

ARTISTS. *a. Sculptors*: Hageladas of Argos: Zeus Ithomatas on the Acropolis (Mount Ithome), executed for Messenian fugitives at Naupaktos, dated 460–55, reproduced on coins. – Pyrilampes (Πυριλάμπης) of Messene, active in the 4th cent. (perhaps during the 2d half, since Messene was founded in 370): three statues of athletes at Olympia (ThB, s.v.; Lippold, GP, p. 248). – *b. Painter*: Omphalion (Ὀμφαλίων), son of Nikomedes and pupil of Nikias (perhaps his slave and favorite), active probably during last decades of 4th cent.: painting of the legendary kings of Messenia in the opisthodomos of the temple of Messene (ThB, s.v.; G. Lippold, RE, s.v., no. 4).

ARCADIA. Arcadia is an isolated and mountainous region the natural features and poor soil of which long defended it from Spartan expansionism. Besides the traditional conservatism of a mountain people, it also possessed a heritage of ancient autochthonous traditions, which in many ways must have separated it from the rest of the Peloponnesos. The glory of Arcadia lay in its ancient sanctuaries, particularly those of Zeus Lykaios and Hermes (who in many respects is an Arcadian god), and it was natural that a certain amount of artistic activity, however sporadic and uneven, should occur in connection with them. The richest and most characteristic art form, the small votive bronzes, seem to be only partly autonomous in character; the rest of the production is dependent upon or linked with the more active and progressive art of the adjoining regions, particularly Argos, Sparta and Corinth.

According to the tradition an ancient statue of Herakles, attributed to Daidalos himself, was erected in very remote times on the borderlands between Arcadia and Messenia. Though it is not easy to establish how much truth lies in this vague, legendary reference, there seems to be more reliability in a report about another Cretan artist, Cheirisophos, active in Tegea, who made a statue of Apollo and a portrait sculpture of marble. Two seated statues, from Hagiorgitika (near Tegea) and from a location near Asea, spectral figures (badly preserved and much worn) of considerable archaicism, have been regarded as examples of this Cretan influence. Even more lost in the mists of fable is an altar from Tegea, donated by a legendary personage, Melampos, son of Amythaon, and decorated with stories of the childhood of Zeus. But by an extremely fortunate circumstance it has been possible to connect an archaic torso of a kouros found in Phigalia with a reference to an athlete of ancient times whose statue stood in the agora of that city. This, the two seated statues from Hagiorgitika and Asea, and the singular acroterium with a Gorgon from the Temple of Artemis Knakeatis at Tegea complete the picture of sculpture in Arcadia in the first half of the 6th century B.C. Their sole common feature seems to be an attenuated and bloodless plasticity.

It is much more difficult to determine in what measure the votive bronzes recovered in the Arcadian sanctuaries, particularly in that of Zeus Lykaios, are of local workmanship. In the deposits which have been explored, works of extreme refinement and of the highest quality have been found side by side with other statuettes, the primitive quality of which are due solely to a lack of culture. E. Langlotz's proposal to abolish the idea of an Arcadian school of art and to distribute objects found in the region between its neighbors, Argos and Sikyon, has, however, been very badly received. Actually not a few of the pieces which may be described as local products possess an attractive directness and virility, for all their undeniable modesty as works of art; their very themes, which are definitely pastoral and alpine, differentiate them from more advanced works of the other art centers in the Peloponnesos. This may be said of the numerous series of Hermes as a shepherd bearing an offering, which runs from the middle of the 6th through a large part of the 5th century B.C. By a process of assimilation that is not at all uncommon in Greek art, the god of the shepherds dresses as a shepherd, in a tunic and heavy mantle, and often carries a lamb on his shoulders. These figures so often have an immediacy closely linked with the realities and circumstances of life in Arcadia that it would be difficult to assign them to any art center outside Arcadia. But it is a different matter when these figures, crude, but with realistic features, are found together with figures of Hermes Kriophoros conceived as a shepherd prince, his garments decorated with borders and studs and sometimes with feathers on his wide-brimmed hat and his boots. For two of these in particular (Boston, Mus. of Fine Arts), which Langlotz attributes to the mysterious art of Sikyon, a place can not easily be found among the picturesque and somewhat crude Arcadian works. The same kind of reasoning is used, it seems, in proposing Argive archetypes, or at least dominant Argive modes of expression for certain fleshy and tender nudes found amongst the shepherd figures. The harmony of the alpine, idyllic world of Arcadia may be said to have ended on an intense and strident note — that of an already classical figure of Pan, with a violently animal goat head and a gross and extended belly upon which hairy legs seem to be grafted.

Arcadian artistic production seems to have become even more reduced and impoverished as it proceeded toward classicism. The rough Nekrodeipnon of Tegea, a somewhat mutilated and worn relief typologically related to the Laconian reliefs, is thus quite isolated. Tradition has handed down the names of some Arcadian artists from the very end of the 5th century B.C., such as Athenodoros and Dameas, who are mentioned as disciples of Polykleitos. Later, the magnificent votive offering dedicated by the Arcadians at Delphi after a victory over Sparta includes the name of an Arcadian sculptor, Samolas, together with those of numerous and well known artists from Argos and Sikyon. It is impossible, however, to deduce from these facts any certain basis for ascribing to local artists the sculptured decorations on the Temple of Apollo at Phigalia. It is known that the architect of the temple (VII, PL. 399) was a famous Attic artist, Iktinos (q.v.), who also designed the Parthenon; therefore it is reasonable to assume that a sculptor, too, was invited from another country. On the other hand, it is undeniable that these sculptures cannot be traced to an Attic origin; they have, moreover, markedly individual features, which are remote from the customary adaptation of the Atticizing formulas in the Argive Heraion, of the pediment from the temple at Phlious and even from those at Epidauros. Peculiar features on the frieze at Phigalia (III, PL. 370) are the heavy, fleshy figures over which the garments are arranged in free ornamental arabesques, as if they were extraneous elements laid over the figures by chance, without achieving any basic accord. Similar characteristics may also be recognized in the singular Mantineia stele, reputed to be of Diotima; it represents a solemn figure wearing a peplos and examining an open liver for auguries.

In the succeding generation the dominant artistic koine favored an even greater inflow of Attic artists throughout the Peloponnesos and even into the remote province of Arcadia. The Temple of Athena Alea at Tegea (FIG. 169) acquired sumptuous sculptural decoration at the hands of Skopas (q.v.; III, PL. 378), and at Mantineia, Praxiteles (q.v.) erected a group of cult statues representing the gods of Delos, the base of which, with figures of the Muses (PL. 241) and the strife of Apollo and Marsyas, has survived intact.

The most noteworthy work of Hellenistic sculpture in Arcadia may be said to be the great warrior stele from Kleitor, which some have identified with one of the many images, statues and reliefs erected at Olympia and other places in the Peloponnesos in honor of the historian Polybios. A large discolored head from an acrolith of Hygeia in Pheneos and signed by an Athenian named Attalos — apparently the same as the sculptor to whom Pausanias attributes the statue of Apollo Lykeios at Argos — is a further example of how the poorest and most featureless kind of Neo-Attic koine prevailed also in Arcadia.

ARTISTS. *Arcadia. a. Legendary artists born or active in Arcadia*: Daidalos: Herakles at the border between Messenia and Arcadia, (see GREEK ART, AEGEAN). – Agamedes and Trophonios: architects either father and stepson from Arcadia or brothers from Boeotia (see ATTIC AND BOEOTIAN ART). – *b. Sculptor*: Samolas: collaborated on an offering by the Arcadians at Delphi, dedicated for a victory

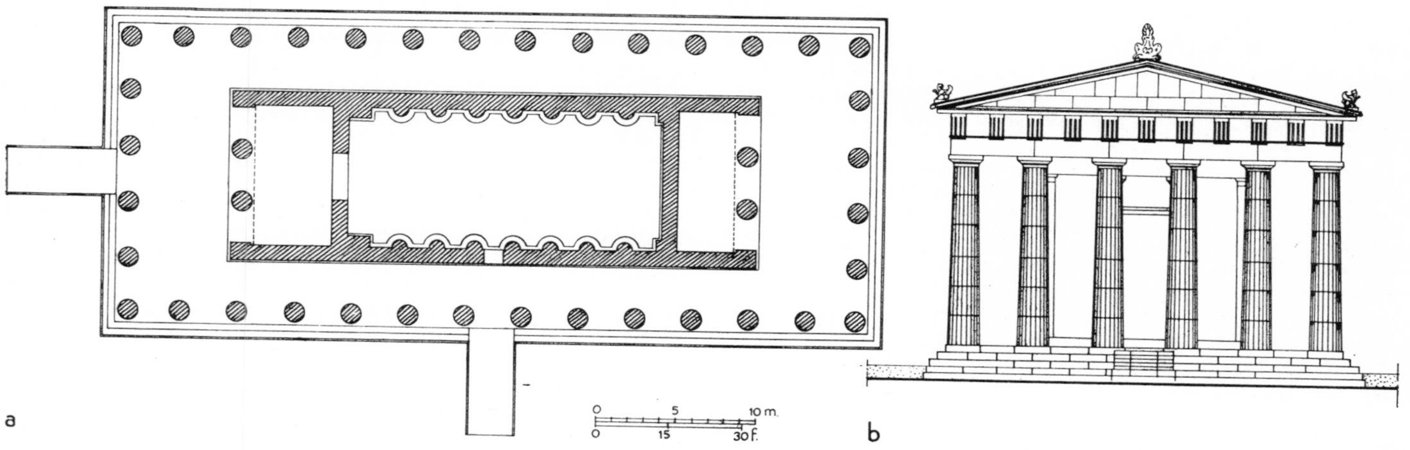

Tegea, Temple of Athena Alea, reconstruction. *Key*: (*a*) Ground plan; (*b*) elevation (*from C. Dugas, J. Berchmans, and M. Clemmensen, Le Sanctuaire d'Aléa Athéna à Tegée au IVe siècle, Paris, 1924*).

over the Spartans (369–68 B.C.), with statues of Triphylos and Azan. Inscribed bases remain (see GREEK ART, NORTHERN). – *c. Engraver of gems, perhaps active in Arcadia*: Olympios (Ὀλύμπιος), 4th cent., perhaps same as the engraver of an Arcadian League coin of 370 (see below): signed carnelian (W. Müller, ThB, s.v.; J. Sieveking, RE, s.v., no. 57). – *Aliphera. a. Sculptors*: Hypatodoros (Hekatodoros?) of Thebes, active probably ca. 450 B.C.: a colossal Athena (see ATTIC AND BOEOTIAN ART; G. A. Mansuelli, EAA, s.v.). – Sostratos (Σίοσ-τρατος) of Chios, collaborator on the same work, father and master of Pantias, active ca. 450–420 B.C. (M. Bieber, ThB, s.v., no. 1). – *Gortys. a. Sculptor*: Skopas (q.v.), of Paros: Asklepios and Hygieia, in the temple of the god. – *Kleitor. a. Sculptors*: Athanadoros, of the Polycletan school, active at end of 5th and in first half of 4th cent., perhaps same as the sculptor of statues of famous women: Zeus and Apollo, an offering of the Spartans at Delphi for Aigospotamoi (see GREEK ART, NORTHERN). – Dameas, of the Polycletan circle, collaborated on the same offering: Artemis, Poseidon and Lysandros. – *Mantineia. a. Architects*: Agamedes and Trophonios: the original wooden temple of Poseidon at Mantineia (see ATTIC AND BOEOTIAN ART). – *b. Sculptor*: Praxiteles (q.v.): Leto and her children, in the Temple of Asklepios, of which three carved slabs of the base remain (according to some, of doubtful attribution); Hera enthroned between Athena and Hebe in the Heraion. – *Mainalos. a. Sculptor*: Nikodamos (Νικόδαμος) of Mainalos, active ca. 400 B.C.: offering of Hetairichos at Delphi, of which the signed base remains; Athena, Herakles wrestling with the Nemean lion, and three statues of athletes, at Olympia (Lippold, GP, p. 218; J. Marcadé, Recueil des signatures de sculpteurs grecs, I, Paris, 1953, no. 84). – *Megalopolis. a. Architect*: Aristandros (Ἀρίστανδρος), period of activity not clear: stoa of the city's agora, of which only traces survive (H. Thiersch, ThB, s.v.; G. A. Mansuelli, EAA, s.v., no. 1). – *b. Sculptors*: Polykleitos the Younger of Argos: Zeus Philios in the Temenos of Demeter and Kore, reproduced on coins after 367 B.C. – Xenophon of Athens: Artemis Soteira and a personification of Megalopolis, in the temple of Zeus Soter, done in collaboration with Kephistodotos (see ATTIC AND BOEOTIAN ART). – *c. Engravers of coins*: Chari[. . .] (Χαρι[. . .]), perhaps Charikles, active during first half of 4th cent., his name appears on a didrachma of the Arcadian League of Megalopolis (370–350 B.C.) and has been completed and interpreted in various ways (C. Seltman, Greek Coins, 2d ed., London, 1955, p, 165; A. Stazio, EAA, s.v. Charikles). – Olym[pios?] (Ὀλυμ[πιος?]), conjectured engraver of coins active during the first half of 4th cent., has been identified with the gem engraver Olympios: didrachma of the Arcadian League on which the signature has been interpreted in various ways (W. Schwabacher, RE, s.v.; C. Seltman, Greek Coins, 2d ed., London, 1955, p. 165). – *Phigalia. a. Architect*: Iktinos (q.v.): Temple of Apollo Epikourios. – *b. Sculptor*: Onatas of Aegina: the Demeter, in the antron, which replaced the ancient xoanon that had been destroyed. – *Tegea. Sculptors*: Cheirisophos (Χειρίσοφος), originally of Crete, probably active in the first half of 6th cent. B.C.: gilded xoanon of Apollo in the temple of that god; stone statue, perhaps a portrait of the sculptor, near the xoanon (Lippold, GP, pp. 23–24; L. Guerrini, EAA, s.v., no. 1). – Endoios: Athena Alea, of ivory, in her temple. Augustus carried it to Rome and placed it in his forum (see ATTIC AND BOEOTIAN ART; P. Orlandini, EAA, s.v.). – Skopas (q.v.): directed the reconstruction of the Temple of Athena Alea after its destruction by fire (395–94 B.C.); its sculptured decoration is also attributed to him by some. Fragments remain. For the same temple he also executed the cult statues of Asklepios and Hygieia. – Pandios (or Pandeios) worked on the sanctuary of Tegea (see ATTIC AND BOEOTIAN ART). – Zeuxippos of Megalopolis: base, signed with Phileas, assigned to the statue of Polias, son of Theon.

SPARTA. It is now beyond dispute that the traditional picture of Sparta as the enemy of the arts and culture, a region rigidly imprisoned in the worship of the state and content merely with military and political success and lost in admiration of its own virtues, is largely the creation of the propaganda of Attic writers, however unconsciously and undeliberately they may have given this impression. Moreover, such an image of Sparta, even if recognized as strained and caricatural, only holds good in comparatively late times — at earliest from the beginning of the 5th century B.C., when the reforms attributed to Lykurgos had penetrated deeply and had produced their damaging effects — for, beyond the traditional, militaristic Sparta of the classical period, there was a Sparta of the archaic periods, which for vitality, interior richness, spirit of enquiry and broadness of vision had little to learn from the more celebrated centers of Peloponnesian art and culture, Argos and Corinth. This "new image" of Sparta has been built up from the evidence gathered during the exploration of the city, which revealed elements of a refined and enlightened cultural life.

From the present state of knowledge, it cannot be said that there was any particularly significant production of Geometric bronzes in Sparta, and the same may be said of Geometric ceramics. The oldest find, which unexpectedly reveals the maturity and high level of Laconian culture, is the well known Sub-Geometric head of Apollo from the Amyklaion (Athens, Nat. Mus.). This is a terra-cotta piece of remarkable size, a warrior god, as were the figures of Zeus at Olympia until the 7th century B.C. and also the Boeotian Apollo with a dedication by Mantiklos (Boston, Mus. of Fine Arts). E. Kunze was the first to point out that this small figure already has monumental characteristics, and he also noted the intense vitality and enlightened intelligence of the expression. Its spiritual qualities flow from a lucid and incisive vision, which sharpens its painted lineaments and reduces the modeling of the face to rather dry and rough planes. The Amyklaion head is on the same level as the Kerameikos ivory statuettes and the Apollo of Mantiklos (the latter is possibly the later work). Thus, with these works it can be seen to belong to the boldest creations of Greek sculptural art in the early archaic age. At the same time, the directness of the carving and the sagged outlines, in which no regard is given to any transitional effects, are sure signs that these are typical Laconian works. The bronze female statuette from the Menelaion (Sparta, Mus.), a modest work when compared with the Apollo of Amyklai, shows that a few decades later the same taste for sharpened forms and hard edges and angles still prevailed; here the hair is gathered into a single curled mass separated from the shoulders, creating a harsh gap.

It was the Daidalic tradition with its clear formal organization which brought a sense of repose to these bold but tormented figures. There is no lack of evidence to link the autochthonous Spartan school with the ancient stories about Cretan influence. Though Dipoinos and Skyllis, the earliest Daidalian artists, do not appear to have worked in Laconia, their immediate disciples, Dorykleidas and Dontas, Hegylos and Theokles were solidly linked with that area. The first two, perhaps brothers, natives of Sparta, are remembered as the creators of numerous works also outside of Sparta, particularly at Olympia, where Pausanias saw a statue of Themis near the Heraion and a group representing Herakles and Acheloos in the Thesauros of the Megarians. Also at Olympia there was a group by Hegylos and his son Theokles representing Herakles, Atlas, and the Hesperides. Both these artists and the more obscure Kallikrates were primarily carvers of ivory and precious woods rather than sculptors; but there is no doubt that the local bluish marble must have been used in Sparta from very early archaic times, as is indicated by the Daidalic statuette from Mistra, which is perhaps from a perirrhanterion, and by the small female torso from the Menelaion. A certain number of xoana of legendary antiquity are reported from Sparta, side by side with works associated with wandering Cretan artists. An example is the statue of Artemis Orthia, identified with the Artemis Taurica carried off by Iphigeneia and dedicated by Orestes; another is a xoanon of Aphrodite Morpho. Amongst the other rare marbles dating from the end of the 7th century B.C. is the so-called "Menelaos and Helen" base (PL. 68), which possesses the enclosed vigor of a very archaic "hieros gamos." The point reached by the long series of terra cottas and bronzes produced by Spartan Daidalic sculptural art may be seen in a highly refined fictile mask (Sparta, Mus.) with immense eyes and a small pointed chin. This is a limpidly conceived and executed image whose quiet authority seems to sum up the formal dogmas of a whole age.

Knowledge of the most archaic sculptors and carvers of precious woods and ivory in Sparta is extensive because of the rich finds of carved ivory, and later of sculptured bone, in the sanctuary of Artemis Orthia, ranging from the middle of the 7th century B.C. to the end of the archaic period (PL. 67). No other Hellenic sanctuary, not even those on the Ionian coast, has so far given so abundant a yield; this alone is sufficient to demonstrate the extent of Spartan artistic contacts during the archaic period. Although normally, artistic dispersion necessarily followed the routes taken by the material used, the Spartan ivories show relatively few signs of Oriental derivation. In defense against severe judgments upon the Artemis Orthia ivories, such as those expressed by E. Homann-Wedeking, one may point to their remarkable formal independence from the immense series of Oriental antecedents. In fact, many of them possess that same brutal immediacy, that aggressive adherence and fidelity to the subject represented, which is not infrequently found and is often admired in many Proto-Corinthian archaic vase paintings from the time of the Ajax Painter and other similar artists. For a long time the ivories retained that violent and somewhat adolescent character which was a constant feature of Spartan art. Their manner of expression was always rapid and immediate, their forms decisive and their movements impetuous. Even in the carving of ivory, which is a soft material and might call for other modes of expression, there was a predilection for rapid and decisive effects, clean-cut color contrasts and sharpened forms, such as are found in vase painting. The great mass of these works are mainly of documentary interest, arising from the typology and modes of figural expression and the archaic world which they reveal, but there are also a number of works of quite high artistic quality. A good many of these carvings come from combs or were quadrangular shields of large fibulas, feminine toilet objects and ornaments which were the most natural and pleasing gifts to offer to the goddess. It was also natural that in most of these plaques, or pinaces, the great nature goddess herself should be depicted in her traditional aspect — a winged standing figure, holding up, with rigid heraldic symmetry, two birds which she has seized by the necks, or two small lions struggling to free themselves. This type ranges from the most primitive depictions, in which a compact and schematic Artemis has no other body than her crossed wings, down to maturer and more knowledgeable representations from the end of the 7th century B.C. There also appears a masculine double, a Potnios, who also holds birds and lions; in one of the latest pinaces, he is shown fighting with a griffin and a winged lion. On one of the most sophisticated and most clearly Daidalic plaques, a youth stands between two female divinities; this most probably depicts the usual Delian trinity. There are, at least on the latest plaques, also mythical scenes, such as the Labors of Herakles (including those of the Hydra and the Centaur), Perseus and the Gorgon, Prometheus, and a mourning scene, which may represent the mourning for Hector. A woman being carried off near a ship, executed according to an old Geometric pattern, and a comb with a Judgment of Paris on it, both reflect definitely Spartan stories. Toward the end of the 7th century, bone took the place of ivory and the production of reliefs, engraved blades and statuettes seems to have declined. The most consistent and artistically better part of this production seems to be that made up of numerous and peculiar images of the goddess, consisting of a small face surmounted by a crown of leaves and a residuary cylindrical body.

It is probably not incorrect to assign the great head of Hera at Olympia to Sparta (PL. 71); its colossal dimensions and high artistic level make it one of the most remarkable of Peloponnesian sculptures. Here also the goddess wears a polos with leaves as Orthia, and the severe expression and strongly incised features are close in feeling to Laconian works. E. Langlotz proposed the name of Medon as the creator of this work, on the basis of Pausanias' statement that the Heraion contained another work by this artist (Medon, however, is sometimes identified with Dontas, who also worked at the Heraion.)

Other notable early archaic works, also in the dark local marble, have been found in Sparta. The singular goddess shown giving birth, naked, kneeling and supported by two tiny attendants, is a rude and violent piece of sculpture and has no iconographic parallels. In the first half of the 6th century there began the series of votive reliefs of the Chrisafa type, representing a divine couple enthroned. These have lately been rescued from their old ambiguous label of "funerary and heroic steles." A perhaps earlier relief from a recently explored temple at Pallantion seems to have some association with these works by reason of its distended planes and its prevalently graphic character. From the second half of the 6th century down to the fully classical period there is a series of reliefs with Dioskouroi, but the most refined and most characteristic example of sculpture is a sharp and vivid Gorgon face, which was perhaps an acroterium on a small temple in Sparta.

Sparta also produced a noteworthy amount of bronze work, of which little has survived. The colossal Apollo of Amyklai was perhaps a xoanon sheathed in beaten metal, and there is a record of a large statue of Zeus Hypatos, made of beaten bronze by that puzzling figure, Klearchos of Rhegion, who is sometimes called a pupil of Daidalos. But the central figure in this art was Gitiadas, an inspired and versatile artist, who according to the sources, made revolutionary experiments in the bronze technique. Bronzeworking techniques were mainly associated with Argos and were perhaps introduced into Sparta by its ally Samos. Gitiadas made the cult statue and the beaten bronze decorations for the temple of Athena Chalkioikos on the acropolis of Sparta. He is remembered for two tripods with statues of Aphrodite and Artemis, both donated to the sanctuary of Apollo at Amyklai.

Other Laconian sculptors whose names have survived were Syagras and Chartas, who are also mentioned as the masters of Eucheiros and who thus suggest interesting relations with the northern Peloponnesos, and Ariston and Telestas, who erected a colossal 18-ft. statue of Zeus at Olympia as a votive offering of the city of Kleitor. No information exists for dating this statue, but it may be presumed that the taste for gigantic statues fits in with that of an early archaic period still in contact with the generations that discovered monumental sculpture and wished to explore all its possibilities. If this is the case, it is possible that the bronze hydria now at Mainz and which bears the name of Telestas is a signed work by this artist, dating from about 600 B.C. During the latter half of the 6th century the names of Laconian artists are replaced by foreign names — Klearchos of Rhegion, and Bathykles of Magnesia (who was responsible for the monumental and highly elaborate altar of Apollo at Amyklai). The constant and friendly relations with Sparta's ally, Samos, are most clearly indicated by the fact that Theodoros, the most illustrious Samian artist, who flourished in the first part of the 6th century, came to Sparta to erect a building known as the Skias, about which no details are extant.

From about the middle of the 6th century B.C., still during Gitiadas's generation, is the date for a bronze head from Sparta (PL. 67). It is about half life-size and its clear and dry structure makes it an extremely significant example of Laconian sculptural art. It is very probable that to Sparta is owed, if not the invention, at least the codification of two of the most characteristic and expressive sculptural types of the late archaic period — namely, the nude Aphrodite as a mirror stand, and the hoplite. The female nudes are light and rhythmic in their neat and elongated lines, and represent an unexpected element in Peloponnesian archaic art. To the hoplite series, besides the local finds, there is every reason to add the splendid warrior statuette from Longa, which is possibly an armed Apollo; another, equally notable, from Dodona, and perhaps even the small warrior found in Southern Arabia (*BSA*, XL, pp. 83–84). Moreover even if they are considered only for variety of motifs, the Laconian bronzes constitute an exceptionally interesting group (PL. 61). The series include pieces with fresh immediacy, such as the boy carrying a hydria, from Phoiniki, and the old man with a bald head making an offering, from Olympia. A confirmation of the exceptional interest aroused by this genre is provided by the considerable export of Laconian bronzes, even to relatively remote places. Judgment on the question whether the Vix crater (I, PL. 354) is Laconian or not may be suspended, but a statuette of a nude Venus belonging to the category of Spartan mirror stands has been discovered at Erice in Sicily: in all probability it was connected with the Sanctuary of Aphrodite on Mount Eryx. In Dodona and in the artistically active provinces of northern Greece the most numerous imports, second only to those from Corinth, consisted of small Laconian bronzes. The highly original and vivid running maiden from Dodona (VII, PL. 45) may serve as an example of all of these, and it may be associated with the advancing maenad figure from Tetove in Illyria (Belgrade, Nat. Mus.), in order to appreciate the extent of the culture and of the influence diffused from Spartan territory. In connection with the Vix crater, it is necessary only to mention the superb relief pithoi of Sparta and the report of the large crater which the Laconians had destined for Croesus but which was donated to the Heraion on Samos.

To return to the question of the warrior figures, it should be recalled that the grandest archaic creation of this kind, the statue of Samos, has been tentatively assigned to Sparta. And, in fact, the type of breastplate, the clean and incisive forms, and the generous and highly elaborate modeling of the hair, may all be regarded as typical of Laconia. Moreover, this type may be traced through a series of monuments in Sparta and at Olympia; the refined fragments of a shield from the acropolis at Sparta, the two heads from Olympia (a fragment which may be referred to one of these, the so-called "Eperastos," had a very fine relief on the shield showing Phryxos on the ram), and finally the disturbing "Leonidas" (PL. 68). There is always the same harsh and severe face, earlier seen in the Charilos, but here enriched with heroic passion. It is impossible to decide whether the face of the warrior expresses a desperate fury or the impassioned concentration of a hoplitodromos. The hoplitodromos of Tubingen may also be reassigned to Spartan territory. A small bronze (lost) of a similar pattern and even more lively came from Cyrene, a city which was considered to be a daughter-city of Sparta.

Figures of divinities seem to be rarer, though a remarkable group of Athena figurines may be assignable to Sparta, as may a series of Apollos with Thyreatic crowns of leaves. The last of these to be discovered is outstanding for its refined workmanship, but especially for its softly modeled and ample structure and well-defined muscles; it seems to give evidence of one of the rare contacts with Argos.

The more refined female types from the late archaic period include an exquisite bronze kore in Berlin, which has finely chiseled features and a clinging garment covered with fine pleats. A trace of East Greek influence in such works has been suggested; this would presumably have been brought to Laconia by Bathykles. The sanctuary of Amyklai has unfortunately yielded only a figured anta capital and some exquisite friezes of leaves and palmettes comparable to those of the Ionian thesauroi at Delphi. From the Amyklaion also comes a fragment of the stele of the discus thrower Ainetos, mentioned by Pausanias. From it, it is possible to reconstruct a pattern of discus throwing which is akin to that on the coins of Kos (III, PL. 393), a design that has a strong suggestion of flight. The superb stele of Ainetos was followed by a noteworthy series of steles and votive reliefs that continued at least through the whole of the 5th century B.C. One of the most interesting works of art, and perhaps the last really valid one to be found on Spartan territory, is the figure of a boy from Geraki (Geronthrai), who, with bowed head, weeps over the brevity of his life.

Laconian ceramic production also contains much that is of interest. The Geometric period did not contribute any distinctive individual features, and in fact it seems to have been dependent on Argive and Corinthian motifs. In the Orientalizing phase the panorama became wider and more varied, and the figural repertory was enriched by an imaginative use of volutes and palmettes, and heavy borders of the pomegranate-row pattern and the check pattern give a fairly precise idea of the effects of vigorous contrast that the Laconian artists sought to achieve. Occasional contacts with Cycladic ceramics may be discerned in certain kinds of free and loose circlets, and particularly in the use of female heads painted on the necks of large vases. There seem to have been more consistent contacts with proto-Corinthian and Corinthian ceramics, seen in the animal processions and brilliant use of graffito. But the shapes, such as the lakaina, remain exclusively Spartan; the decorative motifs and the design are the products of strong individuality. The garlands of myrtle and pomegranates, which are characteristic of the borders, remained in favor for the duration of this ware.

The two fairly primitive cups from Taras, with tondos of tuna fish and dolphins (PL. 69), belong to the end of the 7th century and reveal remarkable compositional formulas. These are undoubtedly some of the most imaginative examples of art from that phase, and at the same time strengthen the possibility that the concept of the round interior of the cup being reserved as a broad space to accommodate a circular, or even a spirally rotating motif, is a Laconian characteristic or, possibly, invention. Thus, there are found circles of tuna fish, dogs running along a belt of parallel lines (which reveal links with the miniaturistic Ionic cups from Samos), and banqueters in a continuous strip surrounding elaborately interwoven palmettes and lotus flowers. The common gorgoneion motif, with its continuous twisting pattern of serpents, seems to be linked with this rotating motif. How deeply ingrained was this feeling for circular composition is also revealed by the absolute incapacity of these artists to resolve satisfactorily a normal figural design on right-angled axes and with an indicated base line. Hence the difficulties in the relation between principal figures and the exergue, which is almost always overdeveloped. In the over-all design it is also the cause of the interruptions, resumptions, and sometimes the continuation of the story in the exergue, as in the small Polyxena in the scene of Achilles in ambush: hence, the absurdly reduced proportions of the figures on the edges and, for example, the arbitrary way in which half of the boar on a cup in the Louvre (no. E 670) has been left out, or, in a well-known cup in Berlin (no. 3404), how the heroes solemnly carrying the body of a fallen warrior are cut in half. Laconian vases lack that feeling for diffused and continuous narration of central importance for the Corinthian and the Attic vase painters. The Laconian scenes are generally summed up in a rapid image that lacks development. It seems necessary to distinguish between those designs made for home consumption, on which are depicted the gods and heroes of the Peloponnesos, Atlas and Prometheus (PL. 67), and those for distant markets, which sometimes deal with themes not found in other examples, such as Arkesilas (I, PL. 338) in his legendary realm, the garden of the Hesperides, and Cyrene struggling with a lion. From this emerges the fact of how widely diffused and how much in contact with distant parts was this Laconian ware. It is worth mentioning that for many years the possibility was discussed that this ware should be assigned to Cyrene, because of the subjects and the large number of Laconian imports at Naukratis.

During their finest period, from the end of the 7th century B.C. to the end of the third quarter of the 6th, Laconian ceramics possessed remarkable artistic qualities, and consequently deserved the favor they enjoyed in distant lands such as Africa, Magna Graecia, and Etruria. In the best examples the drawing has a rude vigor, and curiously vivid color effects are produced by the contrasts between broad areas of black glaze and the white slip, highlighted by abundant retouchings in a somewhat strident purple and with heavy and decisive incising. In these ceramics certain Laconian characteristics are summed up; the harshness of the brush strokes and the abrupt separations seem well-adapted to depictions of that heroic world which is so frankly evoked and which seems to lead back to the angular clarity of the helmeted Apollo from the Amyklaion.

ARTISTS. *Sparta*: *a. Architects*: Theodoros of Samos: *skias*, or assembly hall (see GREEK ART, EASTERN). – Gitiadas (see below). – Kleon (Κλέων) of Sparta, son of Perikleidas, worked in a period not clearly described: one inscription on an architrave, probably discovered west of the theater (C. Weickert, ThB, s.v.; G. A. Mansuelli, EAA, s.v.). – *b. Sculptors*: [. . .]kis ([. . .]κις), perhaps a sculptor of the 7th or first half of the 6th cent.: (signed?) fragmentary inscription on a block of the sanctuary of Artemis Orthia (Lippold, GP, p. 30, no. 8). – Gareas (Γαρέας) of Laconia, probably active before or during 6th cent.: signature on the torso of a horse in the sanctuary of Artemis Orthia (Lippold, GP, p. 30, no. 8; G. Lippold, RE, sup. VIII, s.v.; M. B. Marzani, EAA, s.v.). – Klearchos of Rhegion: Zeus Hypatos, in the temple of Athena Chalkioikos, a work attributed also to one Learchos (see GREEK ART, WESTERN; P. Romanelli, EAA, s.v.). – Chartas (Χάρτας) and Syagras (Συάγρας) of Sparta, active at beginning of 6th cent., masters of Eucheir of Corinth (G. Lippold, RE, sup. VIII, s.v. Syagras; L. Guerrini, EAA, s.v. Chartas). – Dorykleidas (Δορυκλεῖδας) of Sparta, brother of Medon, pupil of Dipoinos and Skyllis, active in first half of 6th cent.: chryselephantine Themis in the Heraion at Olympia (Lippold, GP, p. 30; L. Guerrini, EAA, s.v.). – Medon (Μέδων; perhaps to be identified with Dontas) of Sparta, pupil of Dipoinos and Skyllis, brother of Dorykleidas; active in first half of 6th cent.: statue of Athena in the Heraion at Olympia (G. Lippold, RE, s.v., no. 12; Lippold, GP, p. 30; L. Guerrini, EAA, s.v. Dontas). – Telestas (Τελέστας) and Ariston (Ἀρίστων), brothers from Laconia and probably Spartans, regarded as active during first half of 6th cent. but also dated to first half of 5th: Zeus at Olympia (M. Bieber, ThB, s.v. Telestas; Lippold, GP, p. 32; P. Orlandini, EAA, s.v. Ariston, no. 1). – Hegylos (Ἡγυλος) and his son Theokles (Θεοκλῆς) of Sparta, pupils of Dipoinos and Skyllis, active in mid-6th cent.: wooden group of Herakles, Atlas, and the Hesperides for the Treasury of Epidamnos at Olympia (M. Bieber, ThB, s.v.; Lippold, GP, p. 30; M. B. Marzani, EAA, s.v. Hegylos). – Gitiadas (Γιτιάδας) of Sparta, also architect and poet, active in mid-6th cent.: Temple of Athena Chalkioikos on the acropolis, probably rebuilt by him; cult statue of Athena from the same temple, reproduced on coins; bronze reliefs (Labors of Herakles and the Dioskouroi, the birth of Athena, Perseus and the nymphs, Hephaistos setting free Hera, etc.) which decorated the walls of this temple; two tripods at Amyklai (Lippold, GP, p. 32; G. Pesce, EAA, s.v.). – Gorgias (Γοργίας) of Sparta, if the name Lakon, which follows his in Pliny's list (*Naturalis historia*, XXXIV, 49), is to be interpreted as an indication of his origin and not as the name of some other artist (which is more likely), active at end of 6th cent.: five signed bases from Athens (Lippold, GP, p. 81; M. B. Marzani, EAA, s.v. Gorgias; L. Guerrini, EAA, s.v. Lakon). – Kratinos (Κρατῖνος) of Sparta: Philles of Elis, victor prior to 484, at Olympia (M. Bieber, ThB, s.v.; Lippold, GP, p. 32). – Kallikrates (Καλλικράτης) of Sparta, archaic or Hellenistic era: works in miniature, such as a quadriga with auriga that could be concealed under the wings of a fly or ant (ThB, s.v.; Lippold, GP, p. 30). – Kyranaios (Κυραναῖος), perhaps a Spartan, first half of 5th cent.: fragments of an inscribed base in the Sanctuary of Apollo Hypertеleatas, assigned to an offering by the Queen of Cyrene (Lippold, GP, p. 106, no. 2; L. Guerrini, EAA, s.v.). – *c. Painter*: Adaios or Idaios (Ἀδαίος, writer?), 4th cent.: decorated the harnesses of the horse given by Agesilaos to the son of Pharnabazos (396–94 B.C.; ThB, s.v. Idaios; L. Guerrini, EAA, s.v. Adaios). – *Amyklai*. *a. Architect*: Bathykles (see below). – *b. Sculptors*: Gitiadas: two tripods, with statues of Aphrodite and Artemis. – Bathykles (Βαθυκλῆς) of Magnesia-on-the-Maeander, active at end of 6th cent.: his throne for a colossal image of Apollo, executed with assistants, was more than a chair; it was in the form of a building and was ornamented with friezes, columns, acroteria, and the like. Probably he also executed the statues of Artemis Leukophryene and the Graces which were consecrated at the completion of the work (Lippold, GP, pp. 55–56; M. T. Amorelli, EAA, s.v.). – Kalon of Aegina: bronze tripod, with Demeter and Kore, placed next to two others by Gitiadas, erroneously regarded by Pausanias as an ex-voto for the first Messenian war. – Aristandros (Ἀρίστανδρος) of Paros, probably father of Skopas, active toward end of 5th cent.: female figure with a lyre, the support of a great tripod consecrated by the Spartans after Aigospotamoi (405 B.C.; Lippold, GP, p. 230; P. Orlandini, EAA, s.v., no. 2). – Polykleitos the Younger of Argos: Aphrodite under a tripod, an offering for the victory at Aigospotamoi; it has been identified as a type of the Aphrodite of Epidauros. – *Sellasia*. *a. Sculptor*: Eumythi[s] (Εὔμυθι[ς]), presumably 6th cent.: signed base of a small ex-voto of which he was either sculptor or donor (W. Amelung, ThB, s.v.; G. Fogolari, EAA, s.v.).

AEGINA. The earliest traditions concerning the history of Aegina all speak of close ties with the Peloponnesos, and even of an immigration of colonists from Epidauros. But data from the earliest monuments seem to indicate continual and almost exclusive relations with Attica, after an unexplained interruption of about three centuries, from the end of the Mycenaean period. The earliest evidence of life on the island in Hellenic times is provided by the importation of Attic proto-Geometric vases, beginning in the 10th century B.C. Later there was a considerable importation of Argive ware, which was, however, inferior to the Attic ware. During the Orientalizing period, from the beginning of the 7th century B.C., Corinthian ceramics, which invaded the field with a mass of products of exceptional quality, were supreme. At least half of the fundamental pieces for the history of proto-Corinthian and Corinthian vase painting come from Aegina, and it is significant that A. Rumpf proposed that Aegina itself may have been the center of production of proto-Corinthian ware, which some scholars regarded as being distinct from Corinthian ware. True Aeginetan ware, however, seems to be limited to a few fragments painted in an ambiguous style betraying both Corinthian and Attic influences. Although far outweighed by the mass of Corinthian fragments, outstanding examples of Attic vase painting have come from Aegina. The famous conical support from Berlin, showing Menelaos and the Greek princes, bears an inscription in Aeginetan characters; some authorities have tentatively classified it as a local imitation of Attic ware.

Aegina was poor in agricultural resources and early in its history had to depend upon distant trading and seafaring, eventually attaining great power and prosperity at sea. Just before the Persian invasion, Aegina's fleet was the third largest in Greece, after those of Athens and Corinth. Knowledge of the immensity of the area covered by her trade is derived not so much from the variety of her imports (though there were imports from the Cyclades, such as the so-called "Bird bowls" from Rhodes and some remarkable later fragments from Chios) as from the coins showing a sea turtle (III, PL. 393) that are found throughout the Mediterranean and in the Black Sea region. The Treasure of Aegina (Br. Mus.) can probably be related to the period of expansion, when, among other enterprises, Aegina took part in the joint foundation of Naukratis in the Nile Delta. The provenance of the treasure is not without its uncertainties, and the dates assigned to the material vary between the Mycenaean period and the 7th century B.C. In particular, the gem with Hathor heads and the well-known pendant with an Orientalizing Potnios would fit easily into a composite, indiscriminate tradition as would have existed in a large commercial emporium such as Aegina. Among the earliest sculptural works found on the island is a little Geometric bronze horse from a tripod and a peculiar fictile female statuette with a flattened body.

Aegina is best known for its great school of sculpture; its artists are met with in all the large sanctuaries, and particularly at Olympia, where their athlete statues are found side by side with those from other Peloponnesian schools. According to Pliny (and therefore according to his technical sources), Aeginetan bronze was the product of an excellent formula, second only to that of Delos. It is not possible to say whether Smilis, the earliest Aeginetan artist, was also a bronzeworker, but at any rate it is significant that he was called to work in a large center of bronzeworking (viz., Samos), in order to give Hera, the Great Goddess of the island, a human guise. (Before that, she had been represented by a "sanis," a mere piece of wood.) It is believed that this figure is recognizable on coins of Samos as a broad, static and frontal figure with flowing garments, and it is known that she was dressed as a bride, in royal garments that could be changed. Some scholars have been less certain about the report that Smilis was the sculptor of the statue of Hera at Argos. Nevertheless, as if to confirm that there existed a special link between the artist and the goddess, it was in the Temple of Hera at Olympia that the only other works known to have been made by him were found, namely, some statues of Horai. Kalon, on the other hand, was definitely a bronzeworker and was linked with the school of Tektaios and Angelion, artists whose origin is unknown but who are known to have been disciples of Dipoinos and Skyllis. Kalon is a historical personality, and he worked at the same time as Kanachos of Sikyon and Hageladas; there is a signed base by him on the Acropolis. He appears to have been an artist of exceptional stature, and was active in all the great centers of Greece. He is known to have made a bronze tripod at Amyklai, placed beside those made by Gitiadas, and a cult statue at Troizen representing Athena Sthenias and carved in wood according to the old traditional techniques. Glaukias, who may also be dated about the beginning of the 5th century B.C., is known even more definitely as the sculptor of statues of athletes dedicated at Olympia. The same sanctuary contained a chariot presented in commemoration of a victory of Gelon at Syracuse in the year 488 B.C.; part of its base and part of Glaukias's signature still exist. The central figure of the Aeginetan school appears, however, to have been Onatas, active about the year 480 B.C. His activity is documented by a votive sculpture on the Acropolis, of which part of the inscription in early-5th-century characters has survived. As far as can be judged from what remains, it seems to have been a modest work, a statuette of a horse given as a votive offering to Athena by Timarchos; it suggests the work of an artist who had not yet reached maturity and fame. The sumptuous votive offerings of Delphi and Olympia are on a very different plane. For Olympia Onatas made a colossal statue of Herakles, donated by Tasos, a Hermes Kriophoros, donated by the people of Pheneos, a chariot presented by Deinomenes, and the well-known series of Greek heroes before Troy, arranged in a half circle about the venerable Nestor. At Delphi he collaborated with Kalynthos on a votive sculpture offered by the Tarentines, which included many individual figures in various attitudes, among them Taras and the hero Phalanthos. A colossal Apollo, perhaps brought from Aegina, is documented as being at Pergamon, and at Phigalia there was a famous black statue of Demeter with a horse's head, also by him. There is far less information about other Aeginetan artists, such as Anaxagoras, who erected a colossal statue of Zeus at Olympia as a votive offering for the battle of Plataia, and the tripod supported by a column of serpents (part of which survives in the Hippodrome of Istanbul) which may be tentatively associated with this figure. The same may be said about Theopropos; all that is known about him is that he made the bronze bull of Kerkyra (mod. Corfu) for Delphi; or about Aristonous, who erected the Zeus of Metapontion at Olympia. In the period of the severe style, the fact that Simon of Aegina worked with Dionysios of Argos and that Synnoon and his son Ptolichos were associated with Aristokles of Sikyon, seems to indicate increasingly close relations with artistic centers in the Peloponnesos, certainly in opposition to Athens, which was the enemy of Aegina. The last artist whose name has survived is Philotimos, who made the statue of the athlete Xenombrotos of Kos, dedicated at Olympia about the middle of the 5th century B.C. In 488 B.C. Aegina was conquered by Athens, and in 431 the entire population of the island was deported.

Bronzes of undoubted Aeginetan origin are extremely rare and are generally too modest to give an adequate idea of this renowned art form. A female statuette clad in brief trousers and used as a mirror support, may be dated to about the middle of the 6th century B.C.; however, the attribution by E. Langlotz and others of certain korai figures in Boston (Mus. of Fine Arts) and in Leningrad (The Hermitage), as well as the Corfu pugilist, to Aegina is provisional and uncertain. The provenance of the striding figure, possibly a giant hurling a stone, in New York (Met. Mus.) as from Cyrene is acceptable, and therefore this piece is most probably to be assigned to Sparta.

Among the particularly archaic marbles is a female statuette with a broad *paryphe* (woven border), which apparently must be linked with the Samian tradition; a refined female torso from the middle of the 6th century B.C. may be associated with a torso from Delos (T. Homolle, *De antiquissimis Dianae simulacris Deliacis*, Paris, 1885, pl. 4) and perhaps also with torsos with minutely carved linea from Chios. A weak kouros or sphinx head with the locks of hair over its forehead, a fragmentary Herakles with a bow, also with an Atticizing quality, are both perhaps to be dated from about the middle of the 6th century B.C. The same tendency seems apparent in a fragmentary stele, with seated and a standing female figures, which dates from as late as the last decade of the 6th century. But some fragments from the Temple of Apollo dating from soon afterward, such as the head of a wounded soldier with his hair cut like a cap, and a vigorously rendered horse's head, recall the Parian style and the Thesauros of Siphnos.

The pediments of the Temple of Aphaia (FIG. 178; I, PL. 332) are in complete contrast with this varied and incomplete body of evidence. Even after all the wear and restorations to which it has been subjected, this temple provides one of the finest

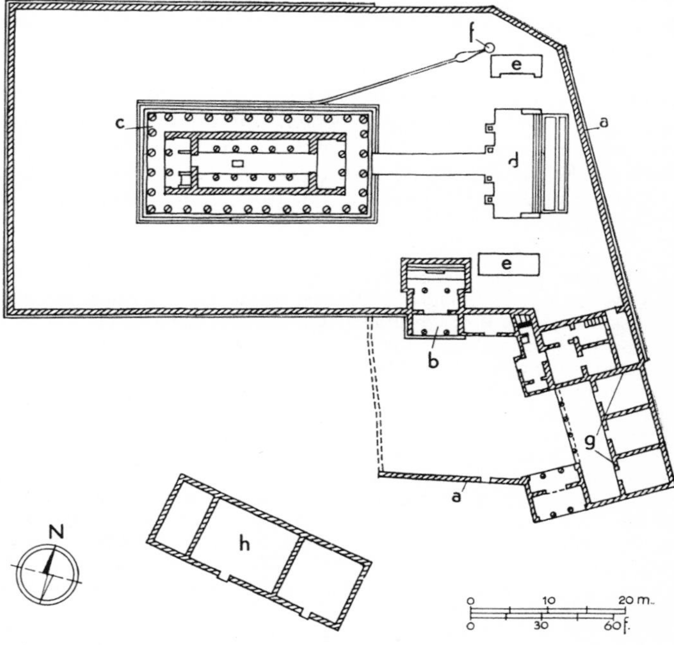

Aegina, ground plan of the Sanctuary of Aphaia at the beginning of the 5th century. *Key*: (*a*) Terrace retaining walls; (*b*) propylon; (*c*) temple; (*d*) altar; (*e*) votive base; (*f*) archaic cistern; (*g*) priests' dwellings; (*h*) southern edifice (*from G. Welter, 1938*).

and most complete examples of Greek sculpture from the end of the archaic period (PLS. 72, 73; I, PL. 371). Enough of the pediments has survived to allow a quite definite reconstruction of the two large, closely knit battle scenes. Each figure has a statuary character and is completely in the round; they recall those on the pediment of Eretria (I, PL. 369), which, because of its theme and the masterly working of the marble, seems to have been the most immediate precedent for the Temple of Aphaia. The two pediments have a surprising unity of tone in theme and in the organization of the groups of combatants alternating with figures of the fallen, into which the more detached figures of the archers are inserted. It may be presumed that the two battles represented glorious episodes in the history of the Aeacides; moreover, since the only recognizable figure is that of Herakles as an archer, in the eastern pediment, it is believed that this battle depicts Herakles and Telamon attacking Laomedon, and that the story on the western side concerns Troy, presumably an aristeia of Ajax, son of Telamon, and Achilles. E. Welter suggests that the inscription referring to the ancient and obscure Cretan goddess Aphaia and her enclosure could refer to a contiguous site or possibly to an earlier phase of the sanctuary. At all events, Athena replaced the older goddess on both pediments, and it is not improbable that the temple is the same as the Athenaion mentioned by Herodotus (III, 59). Undoubtedly an interval must have occurred between the building of the older, west pediment, which perhaps dates from about 500 B.C., and the east pediment, which belongs to the time of Marathon. But between the groups and single figures there is not so much a basic change as an advance, a deepening of the same fundamental theme. The figures of the attackers are disposed more freely, in less stiffly graphic and open ends, and the nudes are more loosely, less deeply modeled; their very faces are more intense and absorbed. One has only to compare the complacent roundness of the face of the Athena on the west pediment with the tenuous, rather ethereal quality of the one on the east pediment.

It is extremely difficult to form a judgment regarding the Aegina sculptures, precisely because they are so famous and because it was through them that modern Europe was first put in contact with archaic Greece. A whole structure of assumptions is built upon them. They do not represent so much a lofty individual message from a great Greek creative center as a rigid expression of that typical hieratic archaism which has been known chiefly through archaicizing sculpture. It is therefore necessary to free these works from such visual distortions and to overcome the obstacles created by the restorations and the smoothing of the surfaces in order to understand their superb formal conception. This is a deeply virile art, primarily expressed by the sharpened structure of tensed muscles and by faces that seem to have become matured through reserve and self-control. On the other hand, the gentle but terse female head of the sphinx acroterium (Munich, Antikensammlungen) and of the so-called "Artemis" reflect isolated moments of peaceful tenderness.

ARTISTS. *a. Sculptors and bronzeworkers*: Smilis (Σμῖλις) of Aegina, regarded as son of Euklides and contemporary of Daidalos, but also assigned to first half of 6th cent.: Hera at Samos, perhaps reproduced on coins; Hera at Argos; the Horai on a throne, perhaps chryselephantine, in the Heraion at Olympia; worked at Elis. He is erroneously named as architect and collaborator with Rhoikos and Theodoros in the construction of the Lemnian labyrinth, probably identifiable with the Heraion of Samos (S. Ferri, Plinio il Vecchio, Rome, 1946, p. 260; Lippold, GP, p. 34; N. Valmin, Skelmis oder Smilis oder beide, AAnz, 1955, cols. 33–40). – Kalon or Kallon (Κάλων; Καλλων) of Aegina, pupil of Angelion and Tektaios, active in last quarter of 6th cent.: signed base from the Athenian Acropolis assigned to ex-voto of a citharist; xoanon of Athena Sthenias at Troizen; bronze tripod at Amyklai (ThB, s.v.; Lippold, GP, p. 97; A. Giuliano, EAA, s.v.). – Theopropos (Θεοπρόπος) of Aegina, active at beginning of 5th cent.: a bronze bull offered by the Kerkyrans at Delphi, of which remains of the base survive; perhaps also a bull offered by the Kerkyrans at Olympia (Lippold, GP, p. 98; J. Marcadé, Recueil des signatures de sculpteurs grecs, I, Paris, 1953, no. 106). – Serambos (Σήραμβος) of Aegina, active at beginning of 5th cent.: at Olympia, statue of Agiadas of Elis, victor before 480 B.C. (M. Bieber, ThB, s.v.; Lippold, GP, p. 98). – Glaukias (Γλαυκίας) of Aegina, perhaps the same as Glaukides, active in first quarter of 5th cent.: Gelon in a quadriga and statues of athletes at Olympia; the *Charioteer* of Delphi is attributed to him (Lippold, GP, p. 97; P. Orlandini, EAA, s.v.). – Synnoon (Συννοῶν) of Aegina, pupil of Aristokles of Sikyon, worked during early years of 5th cent. (M. Bieber, ThB, s.v.; Lippold, GP, p. 98). – Onatas ('Ονάτας) of Aegina, son and probably pupil of Mikon, active from before 480 to about 465 B.C.: colossal statue of Apollo, probably at Aegina and then at Pergamon, of which the base survives; Demeter in the antron at Phigalia; a group of warriors surrounding King Opis at Delphi, executed with Kalynthos on behalf of the Tarentines; signed base, discovered at Athens in a silt deposit, assigned to an ex-voto of Timarchos. At Olympia: Hermes, executed with Kalliteles; Herakles; a group of ten ancient heroes; ex-voto for a victory of Hieron of Syracuse, executed with Kalamis (G. Lippold, RE, s.v., no. 1; Lippold, GP, pp. 98, 102–03). – Kalynthos (Κάλυνθος), also interpreted as Kalliteles, Kalamis, or Kalon, perhaps from Aegina, active ca. 470; collaborated with Onatas on the ex-voto of the Tarentines at Delphi (ThB, s.v.; Picard, II, p. 220; Lippold, GP, p. 98; L. Guerrini, EAA, s.v.). – Kalliteles (Καλλιτέλης), regarded as from Aegina, son or pupil of Onatas with whom he worked: Hermes Kriophoros at Olympia, an offering by the inhabitants of Pheneos (Picard, II, pp. 71, 662; Lippold, GP, p. 98; G. Carettoni, EAA, s.v.). – Philotimos (Φιλότιμος) of Aegina, active probably in first half of 5th cent.: collaborated on the group of Xenodikos and Xenombrotos at Olympia (G. Lippold, RE, s.v., no. 4; Lippold, GP, p. 98). – Ptolichos (Πτόλιχος) of Aegina, son and pupil of Synnoon: two statues of athletes at Olympia (Lippold, GP, p. 98; G. Lippold, RE, s.v., no. 1). – Simon (Σίμον) of Aegina: horse with auriga at Olympia; archer with dog, regarded by some as doubtful in attribution (Picard, II, p. 71; Lippold, GP, p. 98). – Anaxagoras ('Αναξαγόρας) of Aegina, active in first half of 5th cent.: Zeus at Olympia; offering of Praxagoras, son of Lykaios (Lippold, GP, p. 97; P. Orlandini, EAA, s.v., no. 1). – Myron of Eleutherai: xoanon of Hekate, probably before 451 (see MYRON). – Aristonous.: ('Αριστόνους) of Aegina, active probably in first half of 5th cent.: bronze Zeus at Olympia, an offering by the inhabitants of Metapontion (Lippold, GP, p. 98; P. Orlandini, EAA, s.v.). – Ablion ('Αβλίων), perhaps a sculptor, son of Altialos (Altimos?), worked probably in first half of 5th cent.: his name appears on a block, perhaps the base of a statue or part of an altar, with a dedication to a deity (E. Löwy, Inschriften griechischer Bildhauer, Leipzig, 1885, no. 448; L. Guerrini, EAA, s.v.). – *b. Painter*: Elasippos ('Ελάσιππος), active late in 5th cent., perhaps of Aegina: one of his works, according to its inscriptions, either was painted at Aegina or depicted Aegina, daughter of Asopos; he was among the first to adopt the encaustic technique (S. Ferri, Plinio il Vecchio, Rome, 1946, p. 189; M. B. Marzani, EAA, s.v.).

BIBLIOG. *General*: E. Buschor, Die Skulpturen des Zeustempels zu Olympia, Marburg, 1924; E. Langlotz, Frühgriechische Bildhauerschulen, 2 vols., Nürnberg, 1927; E. Kunze, Die Anfänge der griechischen Plastik, AM, LV, 1930, pp. 141–62; Picard, I; Deutsches archäologisches Institut, Bericht über die Ausgrabungen in Olympia, I–II, Berlin, 1936–38; R. J. H. Jenkins, Dedalica, Cambridge, 1936; V. H. Poulsen, Der strenge Stil, ActaA, VIII, 1937, pp. 1–148; E. Homann-Wedeking, Die Anfänge der griechischen Grossplastik, Berlin, 1950; Lippold, GP; F. Matz, Geschichte der griechischen Kunst, Frankfurt am Main, 1950; Rumpf, MZ. *Argos*: A. Furtwängler, Eine argivische Bronze, Wpr., 50, 1890, pp. 125–53; Furtwängler, MW, pp. 42, 78, 370, 380; C. Waldstein et al., The Argive Heraeum, 2 vols., Boston, 1902–05; A. Frickenhaus, Tiryns, I, Athens, 1912, pp. 19–25; S. Papaspiridi-Karouzou, *'Ανασκαφὴ τάφων τοῦ Ἄργους, 'Αρχαιολογικὸν Δελτίον*, XV, 1933–35, pp. 16–53; H. G. Beyen and W. Vollgraff, Argos et Sicyone, The Hague, 1947; G. Kaschnitz-Weinberg, Die ugleischen Zwillinge, S. Presented to D. M. Robinson, I, St. Louis, 1951, pp. 525–31; P. Amandry, Observations sur les monuments de l'Héraion d'Argos, Hesperia, XXI, 1952, pp. 222–74; P. Amandry, Manches de patère et de miroir grecs, MPiot, XLVII, 1953, pp. 47–70; U. Jantzen, Griechische Greifenkessel, Berlin, 1955; J. Marcadé, Sculptures argiennes, BCH, LXXXI, 1957, pp. 405–74. *Corinth*: F. Johansen, Les vases sicyoniens, Paris, 1923; H. Payne, Necrocorinthia, Oxford, 1931; H. Payne, Protokorinthische Vasenmalerei, Berlin, 1933; M. Z. Pease, A Well of the Late 5th Century at Corinth, Hesperia, VI, 1937, pp. 257–316; S. S. Weinberg, Remains from Prehistoric Corinth, Hesperia, VI, 1937, pp. 487–524; M. T. Campbell, A Well of the Black-figured Period at Corinth, Hesperia, VII, 1938, pp. 557–611; H. Payne, Perachora, I, Oxford, 1940; E. Langlotz, Die Bedeutung der neuen Funde in Olympia, Das neue Bild der Antike, I, 1942, pp. 153–71; American School of Classical Studies at Athens, Corinth; Results of Excavations, VII, 1: The Geometric and Orientalizing Pottery, Princeton, 1943; D. A. Amyx, Corinthian Vases in the Hearst Collection at San Simeon, Univ. of Calif. Pub. in Classical Archaeol., I, 9, 1943; O. Broneer, The Corinthian Altar Painter, Hesperia, XVI, 1947, pp. 214–23; E. Lapalus, Le fronton sculpté en Grèce, Paris, 1947; C. Picard, Histoire et archéologie: sur les dates des Cypsélides, RA, XXVII, 1947, pp. 88–90; T. J. Dunbabin, The Early History of Corinth, JHS, LXVIII, 1948, pp. 59–69; F. Villard, La chronologie de la céramique protocorinthienne, Mél, LX, 1948, pp. 7–34; R. J. Hopper, Addenda to Necrocorinthia, BSA, XLIV, 1949, pp. 162–257; E. Capps, Gleanings from Old Corinth, AJA, LIV, 1950, pp. 265–66; P.

Devambez, La fin de la céramique corinthienne et les origines du style attique à figures rouges, CRAI, 1950, pp. 27–31; D. Levi, *'Ηκρήτη καὶ ὁ Κόρινθος*, *Κρητικά χρονικά*, IV, 1, 1950, pp. 129–92; S. Hersom, A Fragment of an Archaic Vessel with Stamped Decoration, Hesperia, XXI, 1952, pp. 275–78; J. L. Benson, Die Geschichte der Korinthischen Vasel, Basel, 1953; S. S. Weinberg, Corinthian Relief Ware: Prehellenistic Period, Hesperia, XXIII, 1954, pp. 109–37; E. Will, Korinthiaka, Paris, 1955; J. L. Benson, Some Notes on Corinthian Vase-painters, AJA, LX, 1956, pp. 219–30; U. Jantzen, Griechische Griff-Phialen, Berlin, 1958. *Sikyon*: A. Orlandos, *'Ανασκαφὴ Σικυῶνος*, *ΠAE*, XC, 1935, pp. 73–83; T. B. L. Webster, Plato and Aristotle as Critics of Greek Art, SymbOsl, XXIX, 1952, pp. 8–23. *Arcadia*: R. J. H. Jenkins, Archaic Argive Terracotta Figurines to 525 B.C., BSA, XXXII, 1931–32, pp. 23–40; E. Kunze, Die Bronzen der Sammlung Helene Statatos (Wpr, 109), Berlin, 1953. *Elis*: BCH, LXIV–LXV, 1940–41, pp. 245–46 (Temple of Scillunte); E. Kunze, Zeusbilder in Olympia, Antike und Abendland, II, 1946, pp. 95–113; *"Εργον*, 1954, p. 41, 1956, p. 85. *Messenia*: J. Mertens, Quatre statuettes messéniennes, AntC, XVIII, 1949, pp. 39–54. *Sparta*: A. J. B. Wace, Laconia: Geraki, BSA, XI, 1904–05, pp. 91–123 at 103; M. N. Tod and A. J. B. Wace, A Catalogue of the Sparta Museum, Oxford, 1906; W. von Massow, Die Stele des Ainetas in Amyklai, AM, LI, 1926, pp. 41–47; R. M. Dawkins, The Sanctuary of Artemis Orthia at Sparta, London, 1929; E. A. Lane, Lakonian Vase-painting, BSA, XXXIV, 1933–34, pp. 99–189; H. Michell, Sparta, Cambridge, 1952; A. Rumpf, Zum Krater von Vix, B. van de Vereeniging tot bevordering der kennis van de antike beschaving, XXIX, 1954, pp. 8–11; B. B. Shefton, Three Laconian Vase-painters, BSA, XLIX, 1954, pp. 299–310; B. Segall, The Arts and King Nabonidus, AJA, LIX, 1955, pp. 315–18; M. Andronikos, *Λακωνικὰ ἀναγλυφα*, *Πελοποννησιακά*, I, 1956, pp. 253–314; G. Hafner, Die Hydria des Telestas, Charites, Bonn, 1957, pp. 119–26; C. Karusos, Ein lakonischer Apollon, Charites, Bonn, 1957, pp. 33–37; E. Homann-Wedeking, Von spartanischer Art und Kunst, Antike und Abendland, VII, 1958, pp. 63–72; P. Pelagatti, Kylix laconica con Eracle e le Amazzoni, BCH, LXXXII, 1958, pp. 482–94; Chronique des fouilles en 1958: Pallantion, BCH, LXXXIII, 1959, pp. 625–27; C. Rolley, Le peintre des Cavaliers, BCH, LXXXIII, 1959, pp. 275–84; H. Riemann, BrBr, 776 ("Leonidas"). *Aegina*: L. Lange, Die Composition der Aegineten, Berichte der sächsischen Gesellschaft der Wissenschaft, XXX, 2, 1878, pp. 1–94; Furtwängler, MW, p. 720; A. Furtwängler, Aegina: das Heiligtum der Aphaia, Munich, 1906; P. Wolters, Aeginetische Beiträge, SbMünchen, 1912, 5; H. Schrader, Die Anordnung der äginetischen Westgiebels, ÖJh, XXI–XXII, 1922–24, pp. 83–95; T. B. L. Webster, The Temple of Aphaia at Aegina, JHS, LI, 1931, pp. 179–83; G. Welter, Aegina, Berlin, 1938; G. Welter, Aeginetica XIII–XXIV, AAnz, 1938, cols. 480–540; E. Homann-Wedeking, Zu Meisterwerken des strengen Stils, RM, LV, 1940, pp. 196–218; L. H. Jeffery, Comments on Some Archaic Greek Inscriptions, JHS, LXIX, 1949, pp. 25–38; A. Raubitschek, Dedications from the Athenian Akropolis, Cambridge, Mass., 1949, p. 85, no. 236, p. 521; W. Kraiker, Aigina: Die Vasen des 10. bis 7. Jahrhunderts vor Christus, Berlin, 1951; G. Welter, Aeginetica XXV–XXXVI, AAnz, 1954, cols. 28–48.

Enrico PARIBENI

Vera Bianco, Mariano Cajano, Giovanni Colonna, Mirella Fantoli, Ambretta Mattei, Giovanna Quattrocchi, Franco Panvini Rosati, Maria Panvini Rosati Cotellessa, Giovanni Scichilone, Sandro Stucchi, and Anna Maria Tamassia assisted in the preparation of the lists of artists.

Illustrations: PLS. 61–78; 4 figs. in text.

PERRAULT, CLAUDE. French scientist and architect (b. Paris, Sept. 25, 1613; d. Paris, Oct. 9, 1688). From 1663 to 1681 this amateur architect played an extremely influential role in the design and construction of several of the principal monuments of the French crown and formulated a body of original architectural theory in his lengthy annotated editions of Vitruvius of 1673 and 1684 and in the *Ordonnances des cinq espèces de colonnes selon la méthode des anciens* (Paris, 1683), the tenets of which were instrumental for the development of architectural thought throughout the 18th century. In addition, he was a respected member of the Académie des Sciences, actively seeking to answer what were then considered to be the critical scientific questions. He published the results of his principal line of research in *Mémoires pour servir à l'histoire naturelle des animaux* (Paris, 1671–76) and *Essais de physique* (Paris, 4 vols., 1680–88).

Although he was equipped for scientific work by his training as a doctor, his architectural activities during the reign of the youthful Louis XIV came about primarily because of the position of his younger brother Charles (1628–1703), who from 1663 was the principal intermediary between Jean Baptiste Colbert and the world of arts, letters, and science. From his position of influence, Charles Perrault was able to put forward Claude's architectural designs, the most famous of which was that for the principal entrance façade of the Louvre, the Colonnade (II, PL. 150). Although Perrault was recognized as its designer by his contemporaries, in 1693, five years after his death and twelve years after Charles had lost favor with Colbert, a long-time antagonist, the critic Nicolas Boileau, charged that the Colonnade actually had been designed by Louis Le Vau (q.v.). Despite a later retraction (1700) of the original charge, this essentially political and personal controversy has continued to obscure Perrault's authorship of this unique façade, a design that in the mid-18th century became a model for monumental civic edifices, such as Ange-Jacques Gabriel's (q.v.) Ministère de la Marine in Paris. The question has been kept alive because the principal documentation of his authorship of the Colonnade — as well as of his general ability as an architect — was lost when the two albums of his original designs (collected by Charles in 1693 for presentation to the King) were destroyed with the burning of the Tuileries in 1871. All historians who had seen the drawings, however, considered them conclusive evidence of Perrault's authorship of the Colonnade, as well as clear proof of his architectural talent.

In 1667 Perrault designed the Observatoire (Paris; later remodeled), the seat of the Académie des Sciences. His design (1669) for a triumphal arch at Porte St-Antoine honoring Louis XIV was chosen in competition with Le Vau and Lebrun (q.v.), but construction was never completed. Recently discovered designs by Perrault for the reconstruction of Ste-Geneviève (1675?) in Paris, the present-day Panthéon, demonstrate his influence on the design by Jules Hardouin Mansart (q.v.) for the royal chapel at Versailles (1698; IX, PL. 319) and, like the Colonnade design, reveal his importance for the new architectural style that developed in France about 1750.

BIBLIOG. R. Blomfield, History of French Architecture, 1661–1774, 2 vols., London, 1921; A. Hallays, Les Perraults, Paris, 1926; M. Petzet, Soufflots Sainte-Geneviève, Berlin, 1961.

Bates LOWRY

PERRET, AUGUSTE. French architect and constructor (b. Feb. 12, 1874; d. Feb. 25, 1954). Auguste Perret was born at Ixelles, near Brussels, where his father, Claude Marie Perret, had been living in exile since the Communard uprising of 1871. In 1881 the family returned to Paris, and Claude Marie, a master mason, reestablished his building firm there. Auguste Perret, like his younger brothers Gustave and Claude, was associated with this firm from the time he left school, and after their father's death, in 1905, the brothers carried on the business under the name of Perret Frères. Auguste and Gustave Perret also studied architecture for several years at the Ecole Nationale des Beaux-Arts in Paris, to which Auguste was admitted in 1891. Neither obtained a diploma (which would have legally prevented them from continuing as building constructors), but Auguste had a distinguished academic record, and he was particularly influenced by the teaching of his master and friend Julien Guadet.

Auguste and Gustave Perret did not undertake construction of the reinforced-concrete buildings on which their fame was to rest until shortly after their father's death, but they had already made one notable contribution to the architecture of reinforced concrete in 1903, when they designed for themselves a reinforced-concrete apartment building in Paris, 25 bis Rue Franklin (V, PL. 102). While not the first reinforced-concrete framed structure, nor even the first reinforced-concrete framed apartment building, this was undoubtedly the first in which a conscientious attempt was made to express externally the character of the structural system used — although the concrete was sheathed in tiles and thus was not visible from outside. The first truly revolutionary building designed and constructed by the Perret brothers was the Church of Notre Dame at Le Raincy, about 9 miles east of Paris (I, PL. 305), in which the reinforced concrete was left uncovered both inside and out and the walls were composed of precast concrete panels filled in with glass. From then on, Auguste became particularly concerned with the search for a valid and urbane way of expressing reinforced concrete; in a number of buildings constructed in Paris from 1928 onward, his almost obsessive zeal for undisguised expression of concrete structure and for refinement of concrete surfaces is very apparent.

The long career of Auguste Perret culminated with his appointment, immediately after World War II, to the post of chief architect for the reconstruction of Le Havre, where he supervised the general design (elaborated by former pupils) and was personally responsible for the Church of St-Joseph and the City Hall. His later works have generally been regarded by art historians as stylistically reactionary, since he had no interest in inventing striking and highly original sculptural shapes and seems to have been more concerned with achieving a harmony between the structural integrity of his own buildings and the masonry buildings surrounding them. Nevertheless, he occupies an important place in architectural history as one of the few idealists who have effectively put into practice the theories of such 19th-century rationalists as Viollet-le-Duc (q.v.). Indeed, it was principally as a disciple of Viollet-le-Duc that he saw himself, despite the fact that the nature of reinforced-concrete frame construction inevitably led him to create forms more in accordance with the French classical tradition.

Principal works. 1902–03: 25 bis Rue Franklin, Paris. – 1905: Garage, 51 Rue de Ponthieu, Paris. – 1911: Théâtre des Champs-Elysées, Paris. – 1922–23: Church of Notre Dame, Le Raincy. – 1925: Observation tower, Grenoble. – 1927: Braque's house, Rue du Douanier, Paris. – 1929: Apartment block, 51–55 Rue Raynouard (plans), Paris; Research laboratories for Marine Nationale, Boulevard Victor, Paris; Concert hall for Ecole Normale de Musique, Rue Cardinet, Paris. – 1930–31: Nubar villa, Garches. – 1935: National furniture storehouse, Rue Croulebarbe, Paris. – 1937: Musée des Travaux-Publics (present Palais du Conseil Economique et Social), Paris. – 1948; Station tower, Amiens; Atomic Energy Research Center, Saclay. – 1952: Church of St-Joseph and City Hall, Le Havre. – 1953: Foundation David-Weill, Boulevard Berthier, Paris.

Bibliog. P. Jamot, A.-G. Perret et l'architecture du Béton Armé, Brussels, 1927; E. Rogers, Auguste Perret, Milan, 1955; B. Champigneulle, Perret, Paris, 1959; P. Collins, Concrete: The Vision of a New Architecture, London, New York, 1959; M. Dormoy, Souvenirs et portraits d'amis, Paris, 1963.

Peter Collins

PERSPECTIVE. The study of the techniques of representing the optical configuration of space is of basic importance in the history of the visual arts. Artists and theorists, especially from the Renaissance onward, have sought theoretical and practical solutions to the problem of conveying spatial relations through lines and volumes and with light, shade, and color, both when representing depth on a plane surface, as in painting, and when arranging volumes in relation to a multiplicity of viewpoints, as in sculpture and architecture.

There are many theories concerning the perspective ideas that were possibly developed in various cultures — or by individual artists — where such spatial concepts were not explicitly formulated and expressed verbally by the artists themselves. For these notations of perspective, definite or implied, many scholars have evolved variant hypotheses.

The first section of this discussion sets forth the generally accepted premises of perspective studies, as well as some fundamental terminology and definitions. The remaining sections summarize the historical development of the various theories of perspective and present the author's interpretation of their influence on works of art, along with comments on theories of artistic perspective propounded by other authorities.

Introduction. "Perspective" originally meant optics. The term was used, instead of the Greek ὀπτική, as early as Boethius, who regarded the subject as a branch of geometry, in keeping with the predominantly geometrical concept of optics held by the ancients. The word kept this meaning in the Middle Ages.

According to Antonio Manetti (1927), the Florentine painters of the 15th century were the first to use the term — Italianized to *prospettiva* — to indicate the newly discovered (or rather, rediscovered) manner of representing objects in accordance with the scientific principles of optics, or *perspectiva.* From then on, to avoid misunderstanding, optics was often called *perspectiva naturalis,* and what is today called "perspective" was variously qualified as *perspectiva artificialis, prospectiva pingendi* (Piero della Francesca), *prospettiva grammica, prospettiva prattica, perspectiva pictorum et architectorum* (Andrea Pozzo), *prospettiva accidentale* (Leonardo da Vinci), and eventually, in English, "linear perspective" (B. Taylor).

In texts written in languages other than Latin the original meaning of the term "perspective" gradually fell into disuse. In Leonardo's writings, natural perspective was still substantially related to optics, that is, to direct vision and the laws governing it; but by the 17th century it had come to mean the view, and perspective began to be equated with prospect and with pictorial composition. G. Troili (1683), for example, designated such prospects as villages, mountains, the sea, islands, valleys, fortresses, cities, squares, and houses as "natural perspectives."

Even apart from this extension, and with reference only to representation, the term "perspective" continues to be used with considerable latitude. In its restrictive sense it means the graphic representation of solid figures in linear diagrams, generally on a plane surface and using the central-projection method. In another sense, not nearly so easy to define, it refers to any system of two-dimensional representation used to indicate the three-dimensional nature of space and the location of objects in it, even the most tentative and reticent allusions to space.

Perspective in the strict sense is now regarded as belonging to what is known as either projective or descriptive geometry, according to whether it is concerned with the methods used (projections) or the purpose intended (to represent, or describe, on a plane surface the location of geometric volumes in space). Considered from the geometrician's point of view, therefore, perspective is a mathematical abstraction, similar to the other abstractions that descriptive geometry has evolved with the object of translating the problems of solid geometry into terms of plane geometry. This being so, its utility is not diminished if, when the construction is complete, it is not viewed from the center of projection. When it is so viewed (i.e., from the center and with one eye), the geometrically constructed perspective image will give a representation of the objects concerned — or rather, of the geometric diagrams inherent in them — that will be identical with the representation which can be perceived and traced on a windowpane when one eye is closed. The picture in perspective, which intersects the cone of projection lines, can be defined as an intersection or cross section of the visual cone.

The central-projection method therefore consists in representing solids by choosing a center of projection from which the significant points and lines are projected onto a picture plane, or plane of projection. This plane may lie behind the objects, as when shadows are projected by means of a point source of light or as in geographic projection; or it may lie in front of the object, between it and the center of projection, or station point (fig. 185, 1 and 2), as in pictorial perspective, where the picture plane is imagined to be transparent. Obviously, the rear plane and a plane through the object can receive the projection equally well.

Each projection point is thus identified by the point where its projection line pierces the picture plane. The projection line is the straight line from the center of projection to that point on the object which is to be projected; in perspective drawing it is preferable to call it the visual ray. The center of projection (O) is also known as the "station point" (the term commonly used), the "eye," the "viewpoint," or occasionally as "point of sight" (although this more often refers to a point on the picture plane). The basic elements of reference in the system are, then, the visual ray perpendicular to the picture plane (often called the central visual ray) and two intersecting planes

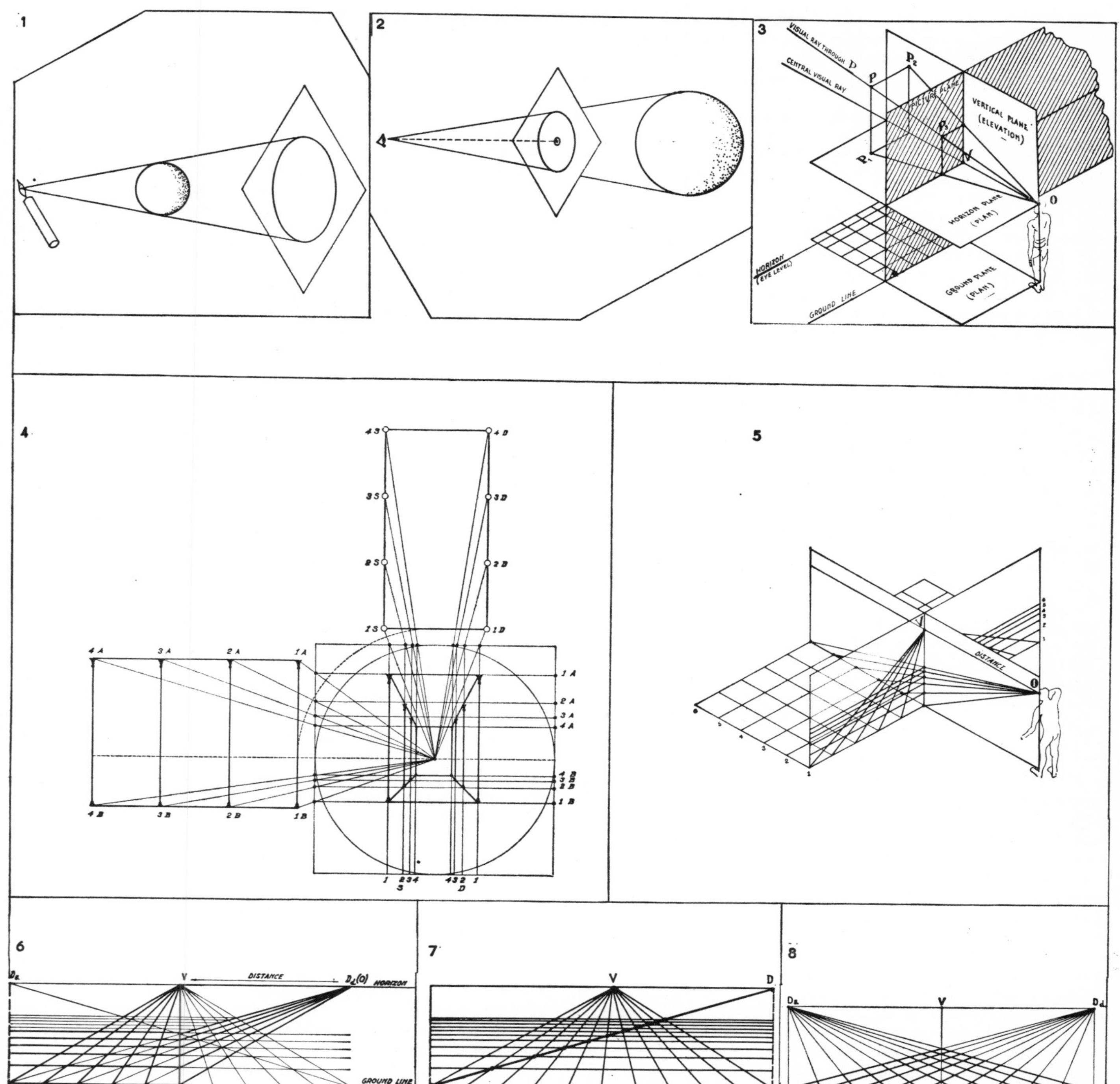

(1) Perspective projection with picture plane behind object; (2) perspective projection with picture plane between station point and object; (3) perspective of a given point — system of reference and analysis; (4) method of direct projection obtained by intersecting visual rays in plan and in elevation, as applied to construction of an "architectural" perspective; (5) "spatial" analysis of the sequence of transversals in perspective, illustrated by two planes interlocked at right angles along the vertical axis through *V*; (6) picture plane with profile plane rotated on it to establish the distance point (*D*); (7) distance-point method; (8) perspective with two distance points for lines at an angle of 45° with the picture plane.

perpendicular to the picture plane as well as to each other, preferably along the central visual ray; these planes are generally taken as horizontal and vertical (FIG. 185, 3).

The horizontal plane (called the "horizon plane") containing the central visual ray meets the picture plane along a straight line called the "horizon" (or "eye level"). This line contains the point of sight, or center of vision (V), which is the intersection point of the central visual ray from which the distance between the eye and the picture plane (OV) is measured.

If the three planes are considered separately — the vertical plane, on which the object will be drawn in elevation; the horizon plane, on which the same object will be plotted in plan; and the picture plane — it is obvious that by drawing lines on them, one can easily find the point in perspective ($P3$) that, for a given station point, corresponds to any points (*P*) whose position in plan ($P1$) and in elevation ($P2$) is already known.

It should be noted that in perspective drawing the plan is usually assumed to be located on a plane at a lower level (viz., the ground plane) that in fact corresponds to the level of the ground or floor on which the observer is standing; the ground line is the intersection of the ground plane and the picture plane. The level at which the plan is located is actually irrelevant, as is the lateral location of the elevation, and does not make any difference to the drawing; but a vertical plane and a horizontal plane that intersect along the central visual ray have the advantage that projection lines drawn on them are in their own form, instead of being orthographic projections of themselves as when they are drawn on the ground plane or on laterally displaced elevations.

If, then, it is decided that the visual rays concerned shall be intersected in plan and in elevation by a straight line representing the picture plane as seen, respectively, from above and from the side, an exact perspective of any three-dimensional object can be constructed, irrespective of its orientation. For "architectural" (or cubical) bodies having one face parallel to the picture plane (FIG. 185, 4), the procedure is elementary. Intersecting the rays viewed in plan gives the sequence of perspective diminutions in depth (foreshortening) of vertical areas (intercolumniations, wall bays, etc.), and the horizontal sequence (diminutions of floor units) is derived from the elevation. No vanishing point is used in this procedure.

However, a vanishing point, defined as the convergence point of parallel lines represented in perspective, is automatically found by extending the orthogonal perspective lines (the lines parallel to the central visual ray that pierces the picture plane at the center of vision opposite the eye). In the procedure illustrated, there are four of these (two stylobates and two architraves), and they meet at a point that has been made to coincide with the center of projection, or the viewing eye, in relation to both the plan and the elevation, which have been suitably juxtaposed. In any event, the vanishing point obviously coincides with the center of vision (V). The center of vision opposite the eye is therefore also called the "principal vanishing point," or simply the "principal point," being the apparent meeting point of all the orthogonals. The intersecting point of a visual ray on the picture plane is the perspective of the whole length of the ray to an infinite distance and of the points at infinity of all lines parallel to it.

If one or more vanishing points are decided on in advance, the procedure of perspective drawing is greatly simplified. By using the center of vision as the vanishing point of the orthogonals, the basic problem of perspective — how to diminish a horizontal plane divided into squares, such as a checkerboard floor — is quickly solved. The ground line need only be divided into equal parts and the perspectives of the orthogonals connected to the vanishing point. Then the rays are intersected, in elevation only, to obtain the receding sequence of transversals, or lines parallel to the horizon (FIG. 185, 5).

If the vertical intersection is used as a hinge and the profile plane rotated to either the right or the left, the fan of rays already drawn in elevation will coincide with the path of the diagonals of the squares drawn in perspective in the picture, and the station point will coincide with the convergence point of these diagonals ($O = D$), so that the distance (OV) from the eye to the principal vanishing point on the picture plane will be accurately registered on the horizon between the vanishing point and the convergence point of the diagonals and can therefore be measured on the picture plane. This convergence point (FIG. 185, 6), which is the vanishing point of all horizontal lines running at an angle of 45° to the picture plane, toward the right or the left, is accordingly called the "distance point" (D).

In a perspective construction using the distance point (or points), the intersecting of the visual rays is therefore unnecessary; it is sufficient to establish on the picture plane the relative positions of two lines and two points: the ground line and the horizon, and on the horizon the principal vanishing point and one of the distance points. After the ground line has been divided into equal parts and the perspectives of the orthogonals converging at the principal point have been completed, a single diagonal drawn to the distance point cannot fail to give the diminishing sequence of transversals and hence a perfect perspective of the checkerboard floor (FIG. 185, 7).

By using both the left and the right distance points and ignoring the principal vanishing point, a special case involving angular orientation can be solved: the case that occurs when the vertical planes to be projected form an angle of 45° with the picture plane (FIG. 185, 8).

The aspect of perspective considered so far, dealing with frontal representations where one face of the reference cube is parallel to the picture plane, is described by various terms, none very precise. The most common term is "one-point perspective"; but only in an abstract scheme using no line that is not parallel to one of the three main axes of the system is it true that there is only one vanishing point. "Parallel perspective," another term in use, risks being confused with parallel projection, in which lines of projection are parallel to one another. There are other, less common terms: "central perspective" — but any accurate construction is central, since it entails central projection; "normal perspective," meaning that the lines converging at the principal vanishing point are normal or perpendicular to the picture plane — but the term might lead to the conclusion that any other representation or orientation is abnormal; and "perpendicular perspective," which though unambiguous is inexpressive. It would be better to speak of "frontal perspective," but the term is not established.

In regard to angular orientation of the objects, perspective is qualified as two-point, angular, or (though rarely) bifocal. The term "accidental perspective" is sometimes used for angular situations other than the special case of a 45° orientation considered above, occasionally called "oblique perspective."

To construct the perspective in an accidental or casual orientation, where lines are neither perpendicular nor at a 45° angle with the picture plane in plan, the same procedures could be used as in one-point perspective, provided that the key points in the figure had first been framed in a plan containing the intersection of an orthogonal and a 45° line. The quickest method (and now the most common), however, is to find the two vanishing points (F_1 and F_2) — sometimes called "accidental points" — on the horizon at the intersections of the visual rays that are parallel to the oblique lines of the figure (FIG. 189, 1 and 2).

The introduction of a third vanishing point (which, again, would be found by tracing the visual ray parallel to the third oblique axis) somewhat complicates the procedure but makes it possible to solve the problems connected with the more general case of freely oriented views, in which all three coordinate axes are oblique to the picture plane. This is known as "three-point," or (less commonly) as "rational," perspective. Two-point perspective, then, is a special case of three-point perspective, and one-point perspective is a very special case.

In modern practice, however, speedier methods are preferred. These are derived from the classic methods of direct projection and the use of one, two, or three vanishing points. To avoid any auxiliary constructions on planes other than the picture plane, and especially on the ground plane, the method of the perspective plan can be used, with the help of other measuring points suitable for measuring the progressive shortening of lines oriented in diverse directions, in the same way as the principal vanishing point is used to measure the shortening of lines parallel to the picture plane, and the distance point to measure the diminishing lengths on the orthogonals. To ensure that the measuring points or other essential parts of the drawing do not fall outside the drawing paper, the fractional system or other devices such as parallel scales can be used.

Various types of scalar guide grids are available, but these restrict the draftsman to a predetermined and therefore limited range of vanishing and distance points. To some extent, these scalar grids may be likened to the many aids that were invented and used through the centuries for drawing in perspective without knowing its rules.

In its broad sense, the term "perspective" is also extended to methods of linear representation other than central projection; thus parallel, or axonometric, projection is often described as perspective with no vanishing point. The methods of projection using parallel projection lines may be considered a special case of central projection in which the station point is assumed to be at infinity. This means that in practice, although the assumption of an observer at an infinite distance is absurd, the smaller or the farther away the object to be represented, the more closely will its image as a parallel projection resemble a true perspective. It follows that the customary parallel projections of buildings, objectively scaled down to a given degree and suitably foreshortened, should be regarded as enormously enlarged when considered as perspective views of the buildings. Hence they should be viewed from a distance of several dozen yards if the observer is to receive an impression of accuracy, and these parallel projections resemble views through a telescope or photographs taken with a telephoto lens. On the other hand, parallel projections of individual objects are unquestioningly accepted as realistic when they represent only a small angle of the visual cone (or a large angle if it is included in a strictly perspective context), because within such limits the eye cannot really apprehend the convergence of the lines toward the vanishing point. One need only see a parallel projection of a cube beside a perspective of the same cube, drawn near the edge of the picture (angular distance from the station point about 25° and perspective distance 2 ft.), to realize how plausible is the illusion created by parallel projections done in this way.

Parallel projection is covered by Pohlke's theorem that, given any three straight lines radiating from a point in any three directions, there is always a plane on which, for a given axis of projection, the lines constitute the projection of a trihedron of three right angles. Parallel projection may be either orthographic or oblique (FIG. 189, 3), depending on whether the projection lines are perpendicular to the plane of projection.

In this discussion of orthographic projection, multiview

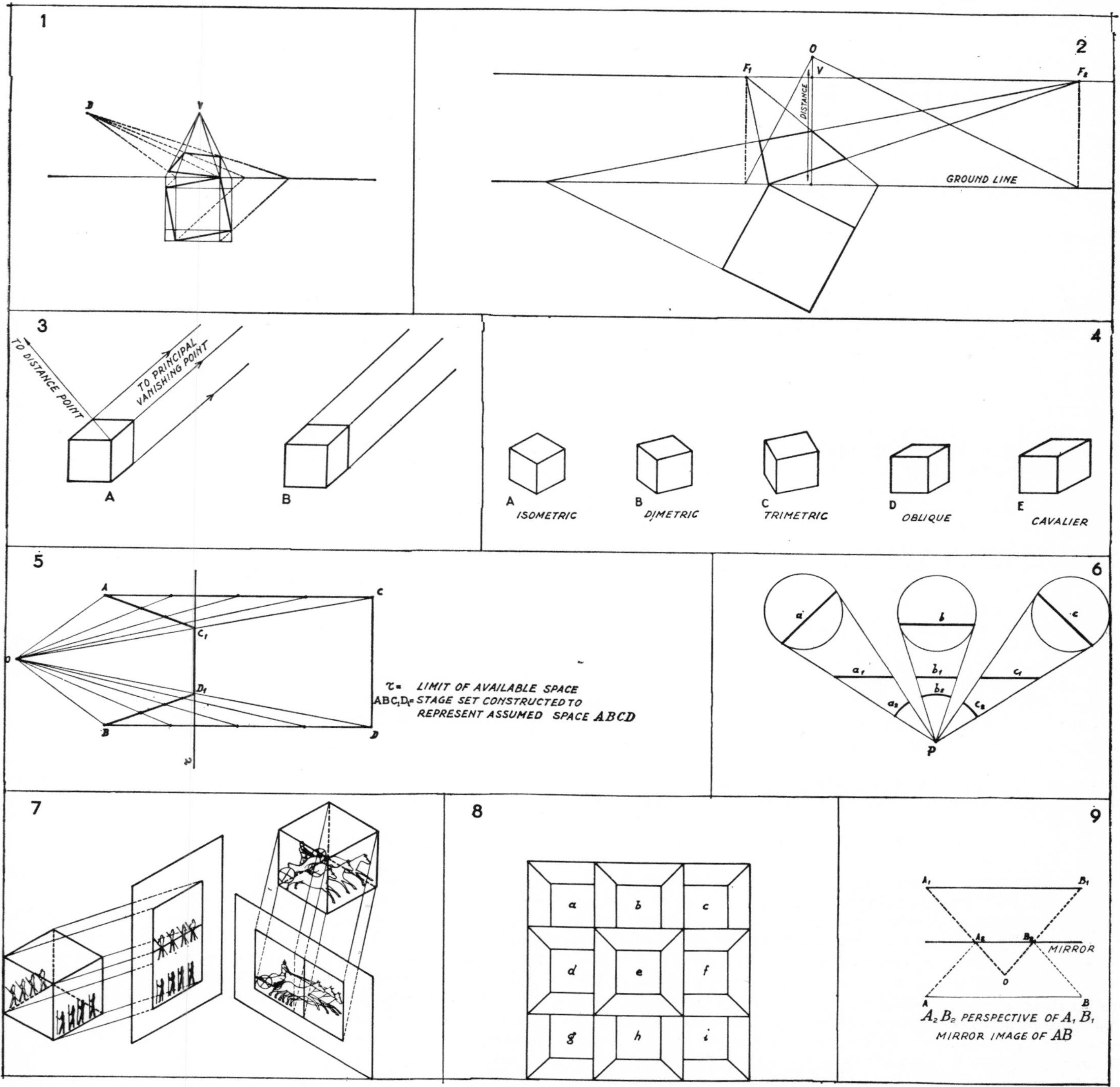

(1) Oblique perspective adapted to one-point system with the principal vanishing point and distance point (V and D); (2) oblique perspective based on the vanishing points (F_1 and F_2) of oblique lines in the figure to be projected; (3) perspective projection (A) and oblique projection (B) of a cube; (4) axonometric (A, B, C) and oblique (D, E) projections; (5) relief perspective for construction of a stage set; (6) discrepancy between ocular-retinal images (a_2, b_2, c_2) and perspective images (a_1, b_1, c_1); correspondence of retinal images both to the object and to the perspective of the object (column a and its perspective a_1 both appear as a_2 to the eye at p); (7) vertical and horizontal superposition as approximations of an oblique perspective; (8) "pigeonhole" arrangement obtained by combining a simplified "herringbone" scheme in two directions; (9) intersection of the virtual pyramid, according to Ptolemy.

projection is not considered but only what is known as "normal axonometric projection" — often called simply "axonometric projection." The reason for this is that multiview projection in which one face of the reference cube is parallel to the plane of projection is not readily assimilable to perspective; from it are derived the plans and elevations used by descriptive geometry in Gaspard Monge's method and the dimensioned-plane method. There are three types of normal axonometric projection; these are known as "isometric," "dimetric," and "trimetric" views, depending on whether the ratio of foreshortening is the same for all three directions or axes, or the same for two axes and different for the third, or different for all three (FIG. 189, 4).

In oblique projection (the second type of parallel projection), the only case considered is that in which two axes are parallel to the plane of projection, which means that one face of the cube will be shown unaltered on the plane; the third axis, inclined at an arbitrary angle, will be foreshortened in a ratio that will also be selected arbitrarily. Oblique drawings are often constructed intuitively rather than mathematically. The particular type of oblique drawing in which the third axis is taken as being inclined at 45° to the plane of projection, and is therefore shown full size or full scale, is known as "cavalier projection" (FIG. 189, 4). When the third axis is shown at half scale, the drawing is called a "cabinet projection."

The term "perspective" can also be legitimately extended to complete representation, including what is known as "aerial perspective." This extension of meaning is implicit in the definition of a perspective picture as a cross section of the visual cone, because this cone is imagined as consisting of rays of various colors and intensities. Color and chiaroscuro ought therefore to be found when taking a cross section, inasmuch as each point on the object should be matched by a point in the perspective — a minute area of paint having the same color intensity. All this can be achieved satisfactorily by color photography, which is a photomechanical and photochemical process; but in practice aerial perspective is not given to that mathematical precision to which propositions dealing with linear perspective can be reduced. This is true even though in theory a system is conceivable — comparable to that used in transmitting photographs by wire — in which the image would be rendered in pointillist fashion by a suitably large number of small squares classified in accordance with a limited scale of colors and gradations, so that a literal and unequivocal description of even *The Night Watch* or the *Mona Lisa* would be possible.

In any case, the so-called "theory of shadows" was developed for chiaroscuro quite early. The phenomena involved were so carefully studied and systematized that chiaroscuro, as a pure science rather than an empirical study, became a fundamental reference point for the practice of painting. Nor was there a dearth of attempts to rationalize aerial perspective and to evolve a theory of chiaroscuro and color. [For example, De Saint-Morien's *La perspective aérienne soumise à des principes puisés dans la nature* (1788) received favorable mention from Leopoldo Cicognara in 1821.]

It is questionable whether it is possible to justify the extension of the term "perspective" to any other empirical or conventional system or method of representing or signifying space or simply of alluding to the three-dimensional nature of objects by means of lines drawn on a plane surface — possibly even reducing such a representation to a strictly plan view. One sometimes speaks of Egyptian, Byzantine, or cubist perspective, even though the arrangement of lines and marks in the picture is far from what would be obtained by central projection and is perhaps even its systematic antithesis. These representations are not true views, but they are always, or nearly always, assemblages of fragments of views or partial views of objects or parts of objects considered separately, and thus to some extent can be traced back to the notion of view and so of perspective. In such instances, however, the use of the term must be recognized as metaphorical.

All the applications connected with central projection onto a nonplane surface are to be included under perspective, even though they may be considered separately. These include manifestations of the arts of painting, sculpture, and architecture, ranging from the simplest case of multiplane perspective to true relief perspective. To obtain a unified perspective construction involving any two or more planes, it is necessary only to make all the station points coincide.

From this practice is derived perspective on irregular surfaces, as applied in particular to the decoration of vaults. This could be constructed geometrically, by the processes taught by Abraham Bosse (1653) and others, which became more complicated as the complexity of the vault increased; but in practice it was done with various kinds of mechanical aids. The commonest method, used by Pozzo (1693–1700), was to stretch a square-mesh net flat under the vault and place a light source at the station point. The net's shadow was outlined by brush and gave the required distorted grid, so that the composition planned on a flat surface could be transferred to the vault from the grid sketch.

The simpler form of theater perspective (see SCENOGRAPHY) is comparable to multiplane perspective, at least in preparing the "funnel-shaped" stage with its floor inclined upward and its side flats inward, which gives an illusion of much greater depth than it really possesses; its floor, ceiling, and two side walls form four trapezoids that the audience sees as receding rectangles (FIG. 189, 5).

The geometrical method by which, given the true depth of the space and the depth to be simulated, the inclination to be assigned to the receding planes was deduced constituted the basis for various experiments in relief perspective. Examples include the false choir of S. Maria presso S. Satiro in Milan (II, PLS. 339, 340), the Teatro Olimpico in Vicenza (IX, PL. 312), and Borromini's perspective gallery in the Palazzo Spada in Rome (II, FIG. 554). According to theorists such as Gérard Desargues and Abraham Bosse (1648), this method, in which the foreshortening factor was increased as much as possible, should have been standard practice for sculptors, who in this way could have made perspective bas-reliefs that receded according to strict rules, rationalizing those flattening methods which had been used empirically from the time of Donatello. This kind of scientific bas-relief remained a dream, though it may have been a schematic reference in the minds of late baroque sculptors and stucco artists. The same trend is evident in the architecture of the period, and the extreme solutions of the Palazzo Spada and the Scala Regia of the Vatican (II, PL. 270), while not in the main stream, had some influence.

PREHISTORY AND THE ANCIENT WORLD. The historical cultures of the ancient Near East were not generally familiar with optical foreshortening such as that in focal perspective, where the diminution is dependent on distance; nor did they have even the most elementary notion of the axonometric foreshortening that is inevitable when a feature is oblique to the picture plane. In their art, therefore, there is no perspective. The images drawn by the artists are not really views but are instead the abstractions of plan and profile and the most varied combinations of the two, and the location of objects in depth is hinted at by ranging the individual objects vertically or horizontally in rows or layers.

The interpretation of vertical superposition as representing a view from above and lateral arrangement as representing a three-quarter view (FIG. 189, 7) is disputed by Erwin Panofsky (1924–25), who prefers to speak of rhythmic linear series. This interpretation, however, needs to be reconsidered and partially revised. First of all, ignorance of foreshortening in those cultures is not constitutional but is connected with a definite *Kunstwollen.* Axonometric foreshortening was practiced in the prehistoric art that reached its zenith in the Magdalenian period, as witness the animals at Altamira (PLS. 268, 277, 278) and Lascaux (PLS. 79, 258, 263–265, 267), and in some cases undeniably focalized solutions were achieved, as in the impressive bull at Font-de-Gaume. In prehistoric painting (see PREHISTORY), however, there is little or no concern with composition. The individual figures (and in this context closely knit groups such as mare and foal can be regarded as single figures) have no relation to one another and seem to share the same wall by chance.

The disintegration of rigidly systematized foreshortening of single views appears only with the development of attempts at composition, inherent in the decorative use of art and in the architectonic organization such a use involves. Narrative art, too, in its need to communicate clearly and unambiguously, is better served by a fixed system of conventional stereotypes than by the spontaneous naturalism of individual episodes. This development can be followed in studying Egyptian art (q.v.). The typical Pharaonic stylization evolved, helped by the concurrent development of ideographic writing, from what was originally a much less constricting figural system. Thus the loose compositions of the predynastic tablets were subsequently reduced to a rhythmic series, and the profiles of objects were rendered geometrically to make them yield their "type" — the final, absolute schematic form. Even the human figure was reduced to an assembly of separate parts that were represented in their individual form, or at least in the distinct form that pictographic tradition had assigned them.

In reality, for any object that is not essentially planimetric, its own particular form is an unattainable abstraction and is therefore replaced by the most significant and typical view. This is the side view for the face, arms, legs, and feet and the front view for the eye and chest. Since, in one way or another, the main "types" all involve the notion of a view, it is impossible to eliminate a residual substratum of perspective. Consider, for example, the way hands are represented. There are so many views that are expressively significant and functionally necessary to the action represented that the image of the closed hand imposes itself alongside the image of the open hand (or the object "fist" alongside the object "hand"). This indicates that not even Egyptian art denied that modicum of perspective spatiality which is essential for representing the fist as a cube and the fingers foreshortened.

A fact perhaps of greater interest concerns the way the parts are put together. The pressure of ideographic custom admittedly encouraged, along with the stereotyping of the figure symbols and the substantially typographic layout of the sequences, the transition from the notion of a series of images to the notion of a series of symbols, with the possibility of interchanging or even confounding the two notions. This did not happen, however, and Egyptian painting remained distinct from writing. The assembling of the various parts of a figure in the general design and the assembling of the individual episodes of a complex scene are done with some thought for naturalistic arrangement in terms of an integrated representation. Thus it cannot be denied that the arrangement of figures in vertical layers or zones also suggests an oblique overhead view, or that the optical model for horizontal series is a three-quarter side view at any level (FIG. 189, 7).

In these models one obviously considers not the focal aspect of the view from close up but the axonometric aspect, which broadly approximates a distant view. The two cases just examined are a special type of oblique projection in which the third axis, objectively perpendicular to the picture plane, is also shown in the picture as an extension of one of the other two. In this way — and above all, in the theoretical case where projection lines are at a 45° angle to the plane of projection — both plan and elevation can be transferred to the picture in their own form and in their own measurements. This makes possible a layout in superimposed or juxtaposed spatial zones, in keeping with the requirements of a planimetrically typographic composition. In practice, it is quite possible to achieve indiscriminate combinations of plan and elevation and the mixed use of vertical and horizontal arrangements, as shown in the famous scene depicting Tutankhamen triumphing over Asian enemies (IV, PL. 368); here, depth is shown by a horizontal overlapping sequence for each pair of chariots and the horses of each chariot, and the relative distances between the several pairs are indicated by vertical superposition. This does not mean that normal visual arrangement was entirely and systematically rejected by the Egyptians: on the contrary, the development of the spatial box, as conceived in modern times in Monge's method and as observed by Panofsky in a Theban representation of a garden, is an extreme case that Egyptian art considered only occasionally and used only for the special requirements of certain topographic descriptions (and then not always).

Mesopotamian art (see ASIA, WEST) is freer than Egyptian inasmuch as, although making use of the same principles of representation, it does not abide by these rules with the iron consistency of Pharaonic art. Nevertheless, even in the most archaic phases, there are daring assemblages of typical views: in the rendering of a phalanx, the Stele of the Vultures of Eannatum, from Telloh (I, PL. 506), represents a truly acrobatic combination of plans, profiles, and front views. Yet a glance at an isometric reconstruction of the presumed layout shows that the assembly is not wholly unrelated to the standard rules for a view and, in effect, is partly connected with the impression of a phalanx as a whole through what is almost exactly a front view, only slightly to the left of and below the center.

In the victory stele of Naram-Sin, King of Akkad (I, PL. 507), the naturalistic freedom of the composition is striking. The lines on which the feet rest are arched in accordance with the curvature of the terrain, and the vertical superposition of the figures in scattered order suggests a unity of space that is found much later in the crater from Orvieto by the Niobid Painter (II, PL. 47).

In Assyrian art the formal decorations of a courtly or ceremonial kind are of less interest for perspective reasons than are the animal studies from life, which also involve an indication of space. Looking at the well-known reliefs showing Ashurbanipal at the chase (17th cent. B.C.), one does not receive an impression that the asses shown higher up are flying but mentally places them correctly in a position behind the other asses depicted at the base of the frieze.

Even in its naturalism, pre-Hellenic art still followed similar schemes; as regards perspective, the striking freshness of the bucolic scenes on the Vaphio cups (IV, PL. 73) marks no advance on the rendering of space by the collateral Mediterranean cultures in the long passage of centuries separating Zoser from Ashurbanipal — to say nothing of Altamira. The next steps toward a more realistic representation of optical space were taken in classical Greece, where true perspective represented by the central-projection method was first discovered.

In the older cultures of central, southern, and eastern Asia, there is no substantial departure from the conventions of the Mediterranean Near East. Various combinations of typical views in plan and elevation, disposition in horizontal and vertical layers, and registration in varied zones all gradually emerge wherever the phase of compositional anarchy and the prehistoric dispersed mode has been passed. This is true of Indian primitive art, although, as far as can be judged from what has survived of the works of the Maurya period (see INDIAN ART), it gradually tended to the Mesopotamian solutions spread by the Achaemenid culture of ancient Persia. It is true also of early Chinese art (engraving on stone or brick; Han funerary paintings), in which there can already be observed a brilliant application to individual figures and compact groups of an axonometric type of foreshortening, as well as convincing indications of the ground plane by means of varying colors.

Pre-Columbian art (see MIDDLE AMERICAN PROTOHISTORY) shows the most drastic disintegration of organic form, due to a combination of symbolic and decorative motivations; this strangely resembles the disintegration of form peculiar to the most ancient Chinese cultures, the Shang and Chou (see CHINESE ART). At the same time, particularly in works that can be attributed to the more refined cultures — Toltec, Zapotec, and, above all, Mayan (III, PL. 1; IV, PL. 396) — there are naturalistic figures, well drawn in accordance with a complicated and well-understood system of foreshortening. The multifigured Mayan murals most often use the system of composition in isolated spatial zones; the vertical superposition and horizontal juxtaposition found in Egypt and Mesopotamia seem to have been unknown. These murals also employ a type of unitary composition that observes a strictly orthographic projection, which in this case is permitted and suggested by the vertical arrangement of human figures on the steps of pyramids (X, PL. 55).

CLASSICAL ANTIQUITY. As the idea of perspective (the genesis of foreshortening) gradually gained acceptance in preclassical Greece, it was inevitably reflected in the vase paintings and reliefs. The basic conventions are identical with those found in the ancient Near East; division into spatial zones, use of typical views, and superposition and juxtaposition. The outstanding advance is that for the first time there is a systematic departure from frontality within a single typical view, that is, foreshortening in the literal sense of cutting short, in the form of simple axonometric foreshortening along a plane oblique to the picture plane.

At first the warriors' shields and the chariot wheels were drawn elliptical instead of round; then the artists learned how to make a convincing three-quarter view of the human figure; and finally entire groups (the quadriga is a typical example) were presented in a way that involved shortening the most representative contour (the profile, in the case of the quadriga). It is this type of foreshortening which Pliny refers to as κατα-γράφα or *imagines obliquae*, attributing its invention to Kimon of Kleonai, a painter of the last decades of the 6th century B.C.

The increasing use of *imagines obliquae* in vase painting throughout the first half of the 5th century B.C. is thus consistent with the literary evidence; but after this time the painted vases and the reliefs are no longer a reliable index — and not only because vases and sculptural friezes are intrinsically unsuitable as "windows through which one looks." In the course of the archaic period and the preclassical period, these art forms had in many ways developed as far as they possibly could within the limits of the conventions evolved, and the introduction of a greater degree of spatial illusion would have disrupted them.

There is still scope for fruitful research along the lines begun by J. White (1956), whose detailed studies of red-figure vases are aimed at tracing the advances in naturalism achieved by the foreshortening of particular objects (chairs and stools) and at establishing within the decade 430–420 B.C. the emergence of what he calls a "foreshortened frontal setting" (though the example he gives of a lekythos in the style of the Meidias Painter seems to be an authentic, if freehand, one-point perspective rather than a simple oblique projection). One must not, however, overlook the fact that such incidental details are by no means the best examples of perspective achieved by this date. These can be only a feeble and distant echo of the successes obtained in other media (and this is even more true of the attempts at perspective aediculae on the Italiote vases of the following century).

Similar observations apply to the perspective seen in certain episodes on the low-relief frieze from the Heroon at Trysa (VII, PL. 254). These serve to bridge the upper and lower registers of the frieze (the division into registers being of ancient Near Eastern inspiration). It is to be taken simply as a trick of perspective introduced into an otherwise purely traditional context — all the more notable in that it demonstrates an unexpected mastery in the handling of two-point perspective late in the 5th century B.C.

Since there remain no large-scale Greek paintings, it seems appropriate to pass over here the derivative arts of vase painting and relief carving and to consider the scant though explicit literary evidence before dealing with the surviving paintings of the Roman period. That the σκηνογραφία (*scaenographia*) of the ancients was nothing other than perspective in the modern sense of central projection (to be rediscovered in Italy in the 15th century) can be inferred from two much-quoted passages in Vitruvius. The first (I, 2, 2) observes that "ichnography" gives the plan and "orthography" the elevation and defines "scenography" as "the shading of the front and the retreating sides, and the correspondence of all lines to the vanishing point, which is the center of a circle" (*Vitruvius on Architecture*, ed. and trans. F. Granger, I, London, Cambridge, Mass., 1931, p. 25). Reference to the graphic demonstration of the direct method based on plan and elevation supports the relevance of this definition to modern perspective.

The second passage (VII, praef., 11) repeats and clarifies the concept already quoted and gives some fundamental chronological details. "In the first place Agatharchos, in Athens, where Aeschylus was presenting a tragedy, painted a scene and left a commentary about it. This led Democritus and Anaxagoras to write on the same subject, showing how, given a center in a definite place, the lines should naturally correspond with due regard to the point of sight and the divergence of the visual rays, so that by this deception a faithful representation of the appearance of buildings might be given in painted scenery, and so that, though all is drawn on a vertical flat façade, some parts may seem to be withdrawing into the background and others to be standing out in front" (Vitruvius, *The Ten Books on Architecture*, trans. M. H. Morgan, II, Cambridge, Mass., 1914, p. 198).

The explicit mention of the center from which the lines are drawn and the reference to the work of Democritus (who, according to Diogenes Laërtius, wrote a treatise on painting, Περὶ ζωγραφίης; an essay on projections, Ἐκπετάσματα; and another on drawing using rays, Ἀκτινογραφίη), are sufficient proof that the method known to Vitruvius could be none other than central projection, which obviously also involved a diminishing scale for receding objects of equal size parallel to the picture plane. (Euclid, in fact, had to give express warning that, in direct views, parallel features of equal size appear unequal in direct proportion to their distance, since their apparent size depends on the angle at the apex.) A clear idea of at least the principal vanishing point (if only as an observation of fact) was also implied. Further indication of the ancients' familiarity with the procedure of taking pyramidal sections are given by the works of Euclid and Ptolemy (FIG. 189, 9) on catoptrics; the knowledge of central-projection methods is abundantly attested by the use of stereographic and gnomonic perspective projections in geography and astronomy by Hipparchus as well as Ptolemy.

It may be assumed that from the mid-5th century B.C., after the work of Agatharchos and the theoretical achievements of Anaxagoras and Democritus, the tendency in Greece was to extend the use of perspective to easel painting. By the end of the century this process culminated in the work of Apollodoros, an Athenian painter, whom Hesychios called the Shadow Painter (σκιαγράφς) and also the Perspective Painter (σκηνογράφος). It was to him that Plutarch attributed the introduction of realistic painting achieved by means of tonality. According to Pliny, his were the first paintings that fixed the eye of the beholder upon the center of vision.

Of the paintings surviving from the Roman period (see HELLENISTIC-ROMAN ART; LATE-ANTIQUE AND EARLY CHRISTIAN ART) very few reveal, or even hint at, the application of strict rules of perspective. The rarity of such in relation to the vast number of paintings discovered might lead one to doubt their value as evidence. Paintings of the imperial period are not all to be treated on the same level, however, and one pertinent fact serves to dispel any doubt: the few pictures that (for reasons at least partly connected with perspective) are rightly considered to be faithful copies of or derivations from 4th-century B.C. or Hellenistic originals, or at least to be in the direct line of Hellenistic tradition, are precisely those which show a formal system of perspective or spatial organization (explicit or easily reconstructed).

Three widely known examples will serve to illustrate the use of perspective in figure painting of classical antiquity. The first is the pinax with Niobe from Pompeii (Naples, Mus. Naz.), probably a copy of an original of the 4th century B.C. With allowances for certain inaccuracies in the freehand copying, this slab shows the correct use of a single vanishing point for the architectural background. The second example is the mosaic depicting the Battle of Alexander (PL. 81), after an original attributed to Philoxenos of Eretria, which he painted for King Kassandros (r. 316–297 B.C.). Clearly, the composition of the painting that the Alexander mosaic derives from was based on the same perspective method. It is interesting to note that the hypothetical perspective scheme of the Alexander mosaic corresponds to Paolo Uccello's sinopia for S. Martino alla Scala in Florence (PL. 93) and probably to his plans for the three paintings of *The Battle of San Romano* (PLS. 42, 43). The third example, an episode from the wanderings of Odysseus (PL. 82; for others of the same cycle, see VII, PL. 180; presumably these are after a late Greek original), would seem to disprove the view that perspective in the modern sense did not exist in the ancient world (Panofsky, 1953; Richter, 1955). Assuming that the Ithacans are not drawn to the same scale as the gigantic Laestrygones, the scale of the figures is so closely proportionate to the distance as to imply the use of a scale of heights similar to that taught later by, for example, Troili (1683).

The examples of correct use of perspective in scenographic wall painting usually cited (since the publication of Beyen's fundamental work in 1939) are the paintings in the House of the Labyrinth and the Villa of the Mysteries at Pompeii and those at Boscoreale — all from the earliest phase of Pompeian Style II. In the Villa of the Mysteries the rooms of particular interest are cubiculum No. 16 (with a double alcove), oecus No. 6 (containing the most splendid perspectives yet discovered), alcove B of cubiculum No. 8 (unique in that the parts below the horizon converge uniformly on the principal vanishing point), and the fragment with coffered ceilings from cubiculum No. 14 (in which the diminution of the orthogonals is correctly measured from the diagonal).

From the House of the Labyrinth at Pompeii and the Villa of P. Fannius Synistor at Boscoreale the most frequently cited examples are the *macellum* scene and a comedy scene (repeated in both places with few variations). The two versions of the *macellum* are noteworthy in that a single vanishing point is used not only for the view itself but also for the architectural framework (both the receding and the projecting parts, as laid down by Vitruvius). Unfortunately, the comedy scene in the House of the Labyrinth is much deteriorated. The one at Boscoreale has been cited by some authorities as an example of dissociated and fragmentary perspective because many of the episodes do not show a principal vanishing point; nevertheless, this scene is worthy of mention, since it is clearly derived from a strict basic scheme of perspective common to all three bays. Not all scholars, however, have accepted this thesis, on the grounds that the errors and aberrations that nearly all the wall paintings display cannot be considered fortuitous. The fact that the lower parts are not correlated to the general scheme, it is argued, shows a basic lack of understanding of perspective on the part of those who devised the scheme, rather than errors in execution or in copying (Sampaolesi, 1960).

The paintings discovered in Rome in 1961 by G. Carettoni in the Room of the Masks (PLS. 83, 87), near the House of Livia on the Palatine, should remove all doubt on this matter. In themselves they are fairly modest achievements, painted by a humble team of decorators like those working in Pompeii, reproducing standard motifs taken at third or fourth hand from the great models of Hellenistic scenography and decoration. However, the exceptional breadth of the perspective and the entirely convincing rendering of the receding colonnade, as well as the way in which all (or most) of the many orthogonals converge strictly on the single vanishing point, both above and below the horizon, could only be achieved by artists who were thoroughly conversant with the laws of perspective.

The few errors in the podia, though serious ones, can be regarded as slips of the hand. In the first place, it is obvious that the painter was confused by the flat angle of perspective, through working with a relatively low vanishing point at a considerable distance. [In similar circumstances, 17th-century Italian scenographers had recourse to a second perspective plan at a lower level (cf. Pozzo, 1693–1700), as did Galli Bibiena in the 18th century.] Secondly, the freehand insertion of details not contemplated in the simple original scheme added a further complication. It is evident, for example, that the addition of bases to the columns flanking the central aedicula made it necessary to alter the orthogonals of the podium. The original lines, still recognizable on the plane on which the columns rest, are identifiable from the orthogonal of the outer side of the shaft without its base and from the bottom of the verticals separating the adjoining walls; these lines converge accurately on the principal vanishing point and the guidelines used to mark out the base of the lateral aediculae may well have served this purpose also. In the third place, it must be remembered that in ancient painting the gridded plane used for measurements and for locating receding objects was not necessarily the floor plane (in fact, it was never so in scenography). More often it was the ceiling plane, where the coffering could supply the required basic scheme of reference provided in Renaissance painting by the checkered floor. This also meant that further difficulties were encountered in constructing the perspective of the lower parts because of the inconvenient plane of reference. The reconstruction of the assumed scheme of perspective by a diagram affords sufficient evidence that these paintings involved the use of a gridded plane to determine diminishing sizes, which implies a thorough grasp of the theory of perspective based on the concept of projection, at least on the part of those responsible for the drawings undoubtedly used by the decorator.

The Room of the Masks is the most important example of ancient perspective yet discovered, if not indeed the oldest surviving example of illusionistic wall painting. Another notable feature of the paintings is the systematic use of shadow, which, though uncertain and incomplete, cannot be taken as an improvised dabbling in impressionism but reflects a deliberate attempt to construct shadow geometrically. This suggests the possibility, never previously entertained, that the ancient theorists had also elaborated some fragmentary and rudimentary theory of shadows.

The later development of wall painting entailed a gradual departure from the strict canons of perspective, as shown in the fanciful inventions of Pompeian Styles III and IV. The transcendental effect achieved by leading the eye to a single vanishing point was replaced by an interest in content in subsequent elaborations, in which unity of the wall was obtained by planimetric composition and by the coloristic texture of the surface. The reckless virtuosity of the decoration and the proliferation of the episodes rendered a strict application of perspective rules extremely difficult, while the search for gaiety, exuberance, and exciting spatial rhythms eventually made them superfluous.

There was thus a return to approximative perspective methods such as those already adopted on the Italiote vases; for example, the so-called "herringbone" perspective, in which, outside a small central area, converging lines do not lead to the vanishing point. Only the central area has a true vanishing point, while those of the sides have two planes converging (floor and ceiling) and two parallel, and the remaining cells (any above or below the central cell) are in oblique projection. Thus, on either side of the central area a series of sections could be laid out to any length in a sort of "pigeonhole" arrangement (FIG. 189, 8). This system, which was to have great success, spelled the end of true perspective construction in ancient art (though the occurrence of striking illusionistic effects in Pompeian Style IV reveals a still undiminished grasp of perspective). At least during the first century of the Empire the question was clearly one of taste, of a deliberate choice dictating the decorator's work. Only later can this lack be identified in the more complex social phenomenon of a less ready awareness of space, which affected the technical capabilities of the painters.

The disappearance of perspective from wall painting does not necessarily imply its final disappearance from all the arts in the ancient world. It is hardly conceivable, for example, that the imperial forums, the Pantheon, or Hadrian's Villa could have been the work of architects unable to make preliminary perspective drawings of their projects. From the 4th century of the Christian era onward, however, it is likely that critical conditions throughout the Empire affected the professional training of architects to such an extent that creative individual art, together with all interest in speculative theory, gave way to the anonymous practice of guild craftsmen (Bianchi Bandinelli, 1956) — though even then there were exceptions such as Anthemios of Tralles and Isidorus of Miletus in the 6th century.

Thus in late antiquity the most notable survivals of the perspective tradition are generally to be found in the more conservative products of the pagan-influenced court style. These, however, are only in the nature of classical revivals, and even when they most brilliantly succeed in evoking the dead past (as in the altar seen at an angle in a leaf of the Symmachorum and Nicomachorum Diptych; I, PL. 301), they are merely painstaking copies of the outward likeness of ancient models, without their spirit or their original technique.

Late-antique and Byzantine art. Perspective illusionism, attacked by Plato as a deceit, was involved in the Neoplatonic condemnation of naturalism and, in the name of a mystical interpretation of the universe, was rejected for its rationalism (see late-antique and early christian art). Thus Plotinus nostalgically supported the absolute quality of Egyptian art, in which things were shown honestly in their true proportions and not deceitfully foreshortened. Christian spirituality was interwoven with similar strands of mysticism, and the spread of Christianity coincided with the revival of Eastern cultures, which during the flowering of Greece and Rome had been relegated to the fringes of the civilized world; but Neoplatonism must be considered a contributory rather than a determining influence for Christian art. A total disregard of space, in the Egyptian or preclassical sense, had become inconceivable. Eight centuries of classical art had left their mark on man's visual imagination; by the end of classical times it was impossible to escape from foreshortened views, the perspective units that were now imprinted on the mind's eye.

The development that began with the dissociated perspective of the later Pompeian styles thus continued, leading in the end to a fragmentation of spatial unity in which single episodes were arranged in rhythmic patterns similar to those of Egyptian or Mesopotamian art. But the scale was changed: where the scheme was once determined along an order of eye-arm-foot-trunk, the episodes were now organized according to a grouping or a unit scale of figures-building-landscape, so that each unit or episode had the indelible mark of the classical tradition, each constituting a niche, so to speak, of individual perspective. The juxtaposition or assembling of these units into a broader composition necessitated those distorted-perspective links which some scholars (Wulff, 1907; Grabar, 1945; Stefanini, 1956), have seen as systematically inverted perspective but the true purpose of which is to fill in the vacant spaces between the individual niches.

Perhaps the most eloquent example is the lunette mosaic of the *Offerings of Abel and Melchizedek* in S. Vitale, Ravenna (pl. 86). Here the altar may be taken as an inverted-perspective link in the form of a table between the spatial "niches" containing the figures; but in reality it is a more complex device. The table itself is a composite view with different vanishing points (with the startling result that the foot of the back leg appears in two different places), the whole being bound together with wedge-shaped segments. One of these has its apex on the right front leg of the table.

Another factor that contributed to the use of dissociated perspective in Early Christian art was the effect of spatial ambiguity inherent in mosaics made with glass tesserae. Such mosaics were already being made in Pliny's time, but they became standard practice in the Christian decoration of later centuries. Since the glass tesserae cannot be polished smooth and their reflecting surfaces are set at different angles, they produce a diffused and ever-changing glitter playing over the surface of the mosaic as the observer moves. This glitter disrupts the solid surface of the walls (especially where the background is gold-colored) and is a constant reminder that between the real space of the building and the transcendental space of the decorated surfaces there is a boundary which can be crossed but which is always there.

This ambiguity, which had long been noted and was to be deliberately exploited in the pre-Byzantine period, led eventually to the development of two different mosaic techniques, wall mosaics of glass and floor mosaics of marble. The difference is dramatically exemplified by two almost contemporary works from the period between the late 4th and the 6th century. The first, the floor mosaic of the Great Palace of the Emperors in Istanbul (II, pl. 437), emphasizes the chiaroscuro effect and sculptural isolation of the figures in the true classical tradition. In the second, the wall mosaics in the Rotunda of St. George in Salonika (II, pl. 438), the volumes emptied of all solidity and related by formal linkages of surface composition — though still conforming faithfully to accepted schemes of traditional perspective — undeniably fall within the abstract tendencies of Byzantine art.

In the mosaics of the Rotunda of St. George the curved aediculae with their columns (Sampaolesi, 1960) are reminiscent of architectural features found in S. Lorenzo Maggiore in Milan, SS. Sergius and Bacchus in Istanbul, S. Vitale in Ravenna, and St. Sophia; a similar motif is found in the mosaics of the Baptistery of the Orthodox in Ravenna, where, however, the technique had somewhat deteriorated. It should be pointed out in this connection that the exceptionally convincing use of perspective in these examples (I, pl. 411; II, pls. 442, 443; IX, pls. 44, 45, 72) is at least partly due to the domical form of the surface on which the mosaic was applied; the central position of the observer made it unnecessary to use those inverted-perspective links which the eye requires in decoration on a flat vertical surface.

This also implies that at this time the fragmentation of perspective space was not yet the rule, as it was to become in medieval Christian art as a result of the renunciation of naturalistic space and solidity. The notions of space and solidity inherited from the imperial age did not all disappear immediately but lingered on spasmodically, marginally, and in clumsily executed form. They were characteristic of the middle period of Byzantine art, where they found vague and sporadic expression in the form of dissociated perspective and in the highlighting technique that was an ill-adapted attempt at chiaroscuro, until they dwindled away altogether. This occurred in the painting of the period in the European West more properly known as Romanesque, in cultural circumstances of great complexity that do not exclude influences from more remote areas of the East, perhaps even from the Far East.

Although a Byzantine work of art is fairly recognizable even to the layman, mainly because of its typical handling of space, this is not to say that the Byzantine concept of space is definable easily, if at all. In essence it was a concept of "non-space," which is to be seen as the converse of the perspective space achieved in the Hellenistic painting of late antiquity. Its initial development before the age of iconoclasm (which includes the pre-Byzantine period) has already been mentioned. At this time the same methods and expedients reappeared and in subsequent stages became mixed indiscriminately. Examples of episodes rendered in their "real" form (in orthographic projection) continued to occur alongside rotated views and multiple viewpoints ("cubism"); attempts at focal perspective with formal links in inverted perspective are associated with vaguely axonometric effects and an occasional example of the better-understood oblique projection. This very mixture of incompatible techniques sometimes has the effect of obliterating all lingering traces of spatial illusion.

The vitality of the Hellenistic tradition up to and even within the period of iconoclasm is shown by the Islamic mosaics in the Great Mosque of Damascus, with landscapes of temples and gardens (II, pl. 441; VII, pl. 407; VIII, pl. 145; X, pls. 379, 380, 386), and by the frescoes of evangelical scenes in S. Maria, at Castelseprio, in Lombardy (II, pl. 445). Compared with the classical models from which they derive, both cycles show a somewhat wider use of oblique projection. Whether this preference sprang from an acquaintance with the civilizations of Central Asia and the Far East, either direct or indirect, it is difficult to say: the growing use of oblique projection was already apparent in book illustration of the 10th century (accepted as Hellenistic), no less than in the 11th-century miniatures (which are agreed to have been inspired from the East).

The steady displacement of what had survived of focal perspective by oblique projection went on everywhere, and there are large-scale examples of inverted perspective in this period that were not simply the result of having to fill in the gap between two passages of focal perspective. Where oblique projection was used in place of perspective, individual elements (e.g., cubic masses such as buildings at the sides of the picture) could still be incorporated in a single unified composition, provided that the oblique projections lay at the same angle (as in Chinese painting) or else converged or diverged symmetrically. When they were convergent the final result might resemble a focal perspective, or at least be assimilated to a scheme tending to be focal, based on late classical models (as is shown, for example, on a page of the Paris Psalter, reproduced by O. Wulff, *Altchristliche und bizantinische Kunst*, Munich, 1914, p. 520, fig. 449). When they were divergent, the central sector necessarily took on the aspect of an inverted perspective.

Architectural elements are all but absent from the larger 11th-century Byzantine mosaic cycles, and landscape elements are rare. In scenes where some kind of furniture or object was necessary, however, oblique projection was still the chosen method for rendering an effect of space. Architectural components reappeared in large numbers in the Venetian and Sicilian mosaics of the 12th century, but their purpose was apparently to serve as a simple rhythmic accompaniment, punctuating the flow of the composition, rather than to achieve the effect of perspective. Thus the flattening of the buildings in the Nativity mosaics on the walls of the north transept of S. Marco in Venice (achieved by orthographic projection or by rotated views and continued lines) constitutes an almost complete negation of space. The mosaics in the Cathedral of Monreale, Sicily (II, PL. 286), which abound in architectural scenes, present different effects. The setting of framing elements and backgrounds is often so convincingly done as to recall in some ways the paintings at Pompeii; but these isolated or partial perspectives are so flatly contradicted in other parts as to place the artist (or artists) of Monreale well within the main stream of the middle Byzantine antiperspective tradition.

The reappearance of these architectural backgrounds, of course, may also have owed something to the contemporaneous neo-Hellenistic style that was gaining preference in the Balkan frescoes of the 12th and 13th centuries (Nerezi, Mileševa, and Sopočani; XIII, PL. 53). In this trend, however, the approach to the problem of space was basically different. It led, in the first half of the 14th century, to the last "renaissance" — the mosaics in the Church of the Chora in Istanbul (II, PL. 464) and the frescoes at Mistra, in Laconia. In these the drawing, chiaroscuro, and color once again combined, as in the most fluent *compendiaria* manner of late Roman painting, to produce a true three-dimensional effect.

At one time it was held that these achievements sprang from the innovations of Giotto (q.v.). When this theory was disproved, opinion moved to the other extreme, and it was argued that these were entirely original and revolutionary achievements in which neither Giotto nor the development he initiated played any part. The paintings later discovered at Castelseprio restore the balance; their close derivation from similar works (even though miniaturistic or peripheral) that can be dated to about the iconoclastic period seriously modifies the previous critical estimate of the innovatory character of this recovery of spatial illusion, a development that seems to have been assimilated rather than actively undertaken.

EUROPEAN MIDDLE AGES. In the long development of Byzantine art, vestiges of the classical tradition were almost continuously in evidence. The most conservative of the various media was miniature painting (see MINIATURES AND ILLUMINATION). There can be no doubt that late classical and proto-Byzantine miniatures (whose influence was protracted for an extraordinarily long time by the almost perfect copies made in the Macedonian era) provided the bases for the Carolingian and Ottonian revivals in the West (see CAROLINGIAN PERIOD; OTTONIAN PERIOD), as well as for the final outburst of neo-Hellenistic art under the Comnenus and Paleologus dynasties. What occurred was the reemergence of a perspective treatment of space, convincing but elusive, and a groping effort toward the landscape effects of the *compendiaria* style of classical art with their loose and imprecise spatial links. The same relation is apparent between painting of the Roman school of the 11th and 12th centuries and the surviving examples of imperial Roman decoration (the spatial and perspective implications of which were no longer understood); in the Romanesque works the figures, backgrounds, and architectural features were equally disposed over a single plane.

During the 13th century this situation underwent considerable change as a sense of material solidity was reawakened, finding diverse expression in the works of a Cimabue or a Cavallini (qq.v.). From this renewed awareness of corporeality grew a need to make the settings more convincing, and in Cavallini's mosaics in the apse of S. Maria in Trastevere in Rome the figures (e.g., *The Presentation in the Temple* and *The Adoration of the Magi*) are set distinctly in front of the buildings, though not yet within them. It was not until Giotto (in the "Isaac Master" period) that the architecture in paintings began to seem habitable. In his work the aedicula, with its parallel vertical faces and converging horizontal planes, seems to have been modeled on the typical layout of Pompeian styles III and IV (and not solely because of the extremely elongated small columns).

An even closer adherence to Greco-Roman perspective methods is shown in the architectural settings of the St. Francis cycle in Assisi, in which the influence of Pompeian Style II can be clearly seen by comparing one of the Assisi bays with, for example, the comedy scene from Boscoreale. It is likely that Giotto's knowledge of models of the Roman imperial age explains why he took such care with the perspective of at least the upper parts of the structures he depicted. This is true of the series with triple arches (the *Pentecost*, the *Approval of the Rule*, and the *Sermon before Honorius*; PL. 89) — an apparently progressive sequence in so far as an entire section is here drawn with a single vanishing point; and also of the exercises in coffered ceilings (*Dream of Gregory IX*, *Wounded Man of Lerida*), whether by Giotto himself or by his assistants, in which the crossbeams are correctly laid off from the diagonal.

In Padua, Giotto showed no inclination to pursue the application of geometrical rules any further, though the diagonal method was still used for the ceilings of the three nuptial scenes in the Scrovegni Chapel; instead he preferred to experiment with greater flexibility in his directional lines in order to obtain a more amenable setting (hence the use of asymmetric one-point perspectives, and two-point perspectives almost parallel to the picture plane). On the chancel arch, however, there is the remarkable achievement of the two painted-perspective chapels that seem actually to pierce the wall. Though the foreshortening was done by eye — copying, one imagines, from reality — the whole is an impressive anticipation of the 15th-century concept of "a window through which one looks."

Further advances in perspective are to be found in Giotto's later work. In the *Dance of Salome* in Florence (Peruzzi Chapel in Sta Croce) the complex architectural setting has a convincing two-point perspective. The *Approval of the Rule* in the Bardi Chapel in Sta Croce has a one-point perspective scheme in a depiction of several rooms, almost perfectly executed in the upper parts and fairly accurately applied to the whole composition. Not to be overlooked are the frescoes of the childhood of Jesus in the right transept of the Lower Church at Assisi. The work of Giotto's pupils, but presumably based on drawings by Giotto himself, these contain the two most consistent and proficient examples of perspective in the entire 14th century (the *Presentation* and *Christ among the Doctors*).

The Siena painters occupy a place apart in 14th-century Italian painting. As long as the *Annunciation* of Ambrogio Lorenzetti (q.v.; IX, PL. 198) was considered the earliest example of uniform one-point perspective, they were believed to be even more advanced than the Florentines; and in any case, they had produced elaborate and picturesque city views as early as Duccio di Buoninsegna (q.v.). In point of fact, the deviations from the mainly focal perspective of Giotto's school toward landscape drawn in terms of parallel projection show a relation to the typical oblique projection of Chinese art that is not merely coincidental. One need only look at the perspective of the town in the *Temptation of Christ* (New York, Frick Coll.) from Duccio's *Maestà* in Siena or at Ambrogio Lorenzetti's allegories of Good Government and Bad Government in the Palazzo Pubblico in Siena (IX, PLS. 194, 195) to discern a Far Eastern influence, and the same is true of contemporaneous work in this genre by Simone Martini (q.v.).

The main line of Western development in the 15th-century revival of perspective runs especially through the Po Valley. The empirical perspective of Giotto's feigned chapels opens up the actual wall surface for a fresh approach to everyday reality as seen in the streets and piazzas of the town. Indeed, the intuitive perspectives of Altichiero and Avanzo (q.v.) in the Oratory of S. Giorgio in Padua (I, PL. 73), and occasionally even the deliberately archaicizing paintings of Giusto de' Menabuoi, achieve a credibility that could hardly be bettered without recourse to scientific theory. Not even the latest courtly Gothic can surpass these works in veracity: neither the Lombards nor Gentile da Fabriano nor Pisanello (qq.v.); neither the analytical realism of the Franco-Flemish painters, with the works of the Limbourgs (q.v.), nor the first fruits of the northern revival (at least in the period of the Master of Flémalle and Hubert van Eyck) — none of these can be considered more modern in their perspective than Altichiero. As for the mature work of Jan van Eyck (q.v.), from the Ghent altarpiece of 1432 to the Rolin *Madonna* (PL. 90), its objective realism (based on exhaustive formal analysis using the convex mirror) remains an unparalleled achievement. Though the dates of these works do not altogether rule out the possibility that Van Eyck had some knowledge of Masaccio (q.v.), even indirect, the fragmentary and still approximate nature of the perspective — and in particular, the fact that the orthogonals do not all meet at the vanishing point — clearly reveals an ignorance of the strict science of projection.

FIFTEENTH CENTURY. *Theory of linear perspective*. The rediscovery of the central projection method is due to the curiosity and scientific interests of Filippo Brunelleschi (q.v.), in connection with the renewed Humanistic interest in ancient texts on optics and catoptrics. Brunelleschi's anonymous biographer (identified as Antonio di Tuccio Manetti) is explicit on this point, saying that "he expounded and practiced what present-day painters call perspective . . . (Manetti, 1927).

Manetti then reports Brunelleschi's first experiment, performed "in his youth" (according to further information given by the author, the date can be placed between 1401 and 1409). Brunelleschi made a panel (lost) with a view of the Baptistery in Florence, which was not to be viewed directly but as reflected in a mirror. In the reflection, seen through a small aperture in the back of the panel (designed to place the eye of the observer at a point exactly corresponding to the vanishing point of the perspective), the picture was seen right way round. In this panel the sky was not painted but was silvered and polished, so that it in turn reflected the actual sky and the passing clouds. Richard Krautheimer, in his study of Ghiberti (1956), was the first to suggest that the whole panel may have been first silvered and polished, so that the perspective as drawn could be checked against the reflection from life. Krautheimer's book draws support from the opinion of A. Averlino (Filarete; 1910), who says that by such mirror experiments Brunelleschi found a form of perspective "that had not been used in other times." But the mirror hypothesis cannot be left halfway, and it can well be believed that the image was really traced on the reflecting surface after the principal vanishing point had been fixed by taking a sight. After all, Ptolemy had long since shown how points corresponding to the objects seen could be marked on a mirror, and elsewhere he had shown how to find intersection points for a given plane.

Since the convergence of parallel lines was already an accepted fact, it is obvious that while Brunelleschi was using a mirror to try to draw the Baptistery he already knew his objective. This was to discover how Ptolemy's schematic construction, which considered only two points on the object and the plane formed by them and the station point, could be applied when dealing with a geometrically complex piece of actual architecture. First of all, he needed to obtain an accurate perspective in one way or another, so that he could analyze it. He chose the octagon of the Baptistery, together with the adjacent buildings, because this would give him the required data in relation to the coordinate axes (orthogonals and diagonals lying in horizontal planes above as well as below) inherent in any architectural situation. Thus he did not fail to note that the "45° lines" converged at two points on the horizon equidistant from the principal vanishing point. But it was still not clear just how a perspective could be drawn from the plan and the elevation, and this problem must have particularly intrigued him if he already had leanings toward architecture. So it can well be believed that he checked his discovery, first in plan and then in elevation, by the catoptric method (Ptolemaic, but also Euclidean) of the intersection of the visual rays. This explains the concurrent and almost immediate spread of both the distance-point method and the legitimate-construction method or projection, and the primacy given in the treatises to the latter as that which contains the proof of the proposition.

No record remains of the many studies and exercises that Brunelleschi must have worked out on paper before and after he drew the Baptistery; but Manetti, echoed by Vasari, also mentions another lost panel, a larger one with a view of the Palazzo Vecchio (Palazzo della Signoria) and the Piazza della Signoria. This was to be viewed directly, not as reflected in a mirror, and was therefore constructed according to the rule, by then firmly grasped; the right-angled grid of the sandstone strips dividing the red brick surface was an invitation to test the rule by comparing the result with the original. Manetti remarks that from the assumed station point in the actual piazza, two faces of the Palazzo "are seen in their entirety." From this remark many students, forgetting that the orientation of the picture plane is what decides whether the view from a given station point will be in one-point or two-point perspective, conclude that the drawing was a two-point perspective. Yet it is evident that this is not so from further information given by the same author, who deduced the assumed station point from the picture; and he could never have inferred that its position must be either outside the piazza "or looking almost along the façade of S. Romolo" unless this façade had been shown in the picture, and shown nearly in section.

A schematic reconstruction of Brunelleschi's panel (PL. 91), taking into account all the data provided by Manetti and no other data, is therefore illuminating — the more so because it turns out that the cubic division of space in Donatello's relief *St. George Slaying the Dragon* (IV, PL. 242), made for the base of his statue of St. George (1416), is modeled on it. This is the earliest surviving monument in which there is a tentative application of Brunelleschi's perspective, constructed from the central vanishing point and the diminished grid of the reference plane. It also serves to date Brunelleschi's view of the Piazza della Signoria to the early years of the second decade of the 15th century and to confirm the view of the Baptistery as being of the first decade.

The next stage in the spread of the new methods in the visual arts appears in the triptych in S. Giovenale at Cascia di Reggello (PL. 92), dated 1422, which L. Berti (1962) correctly ascribes to Masaccio himself, although the artist may have been assisted on the side panels at a later stage. Despite its appearance of steepness, due to the high station point, the altarpiece of Cascia di Reggello is a geometrically perfect perspective, not least in the way that the three panels are subordinated to a single vanishing point and in the proportional diminution in accordance with a distance equal to the spatial depth of the entire triptych. This perfection, which can be easily proved by a summary perspective analysis, shows that the artist had not only fully grasped the theoretical basis of perspective but also had considerable practical skill and that he knew the geometry of the circle and could draw it in perspective — quite a while before Alberti (q.v.), treating of plane surfaces, taught how arcs could be derived from angles.

Alberti's treatise *Della pittura*, written in Italian in 1436 on the basis of his somewhat earlier Latin text, is the oldest text to expound a method of perspective drawing in accordance with the principle of projection, namely, the sectioning of a pyramid. This method, at that time called "legitimate construction," was substantially clarified by Panofsky (1915) and is practically identical with the procedure described above, being based on prior establishment of a vanishing point; then only one auxiliary construction, in elevation, will be needed to give the sequence of transversals. Writing for the instruction of painters, Alberti gives a very precise and full explanation, not of the general principles of this procedure but of a particular and personal practical case, saying "I found this the best method."

This was not intended to deny the validity of other versions of the same method, nor does it imply that he knew no other methods.

The finer points of the procedures used in this "best method" have not always been fully understood, so its individual stages are worth recalling briefly. The first stage (FIG. 205, 1) is performed directly on the wall or panel where the painting is to be done. Alberti first draws a rectangle and divides its base into equal parts whose length is one-third the planned height of the foreground figures. These parts thus correspond to the unit of measurement known as the *braccio*, since the height of an "ordinary man" is 3 *braccia*. The ground line so divided thus becomes the sill of the perspective window. Therefore, irrespective of how and where the station point is located, the scale lateral view of the visual cone. This is a partial view, as is also the case when the direct method is used nowadays, because it does not extend the rays below the picture's ground plane down to the true ground plane (FIG. 205, 5). Alberti takes "a small space" (he does not say whether on a separate sheet, the back of the panel, or the margin of the squared area on the wall), draws a line, and divides it into equal parts proportional to the divisions already marked on the ground line. He then marks a station point on this line, at the same height of 3 *braccia* as the height of the principal vanishing point above the ground line, and from this station point he draws the visual rays to each division marked on the line which has just been drawn.

In the third and final stage (FIG. 205, 3), the distance from the station point to the picture plane is chosen and is set up

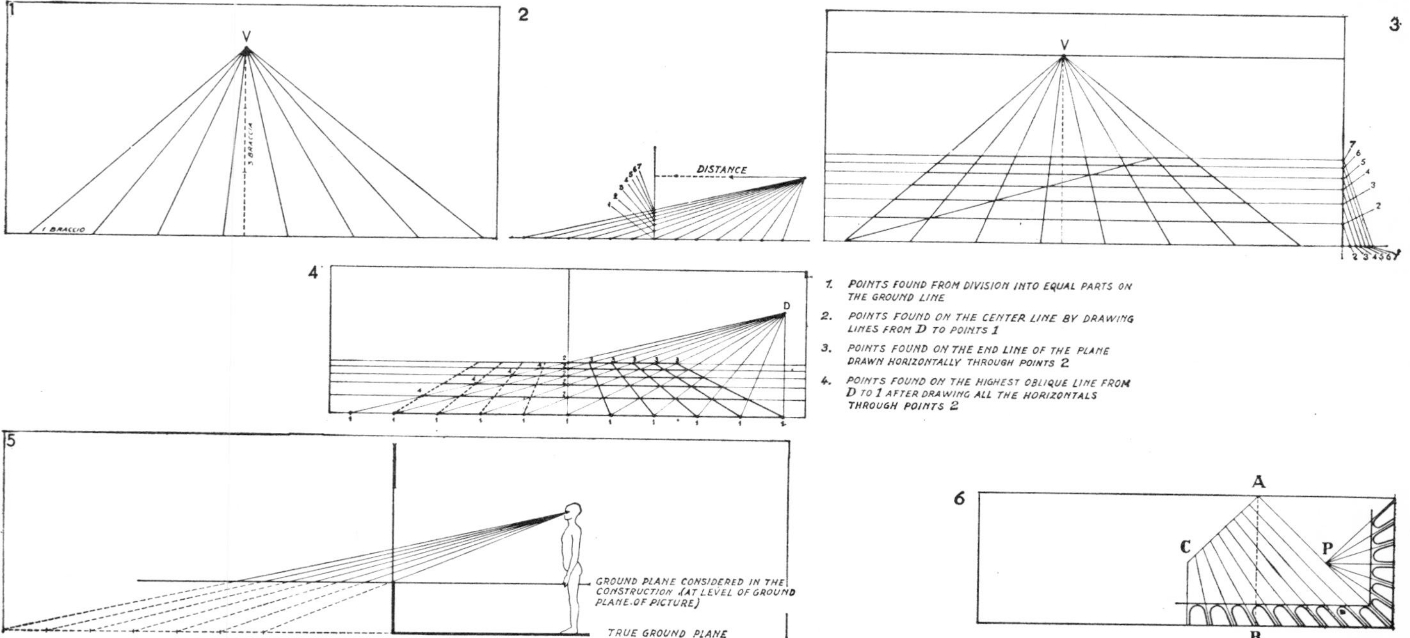

(1, 2, 3) The three stages of Alberti's "best method"; (4) Pomponio Gaurico's method (*after Klein, 1961, p. 218*); (5) Alberti's auxiliary construction stops at the ground plane of the picture, there being no need to consider the true ground plane; (6) perspective system with multiple vanishing points, according to Viola-Zanini (*from J. Schulz, 1961, p. 101*).

ground line will be the perspective image, proportionally reduced, of the first transversal that can be seen through the perspective window on a hypothetical plane at true ground level, "proportionale a quella ultima quantità, quale prima si traversò inanzi" — proportional to that linear quantity which, lying on the true (or assumed true) ground plane, is the last and greatest quantity that can be contained by the fan of rays tracing the ground line on the picture plane, while it is the first and nearest of the visible transversals.

The principal point is then located arbitrarily ("dove amme paia") inside the rectangle. However, there are two provisos. First, the position of the principal point in elevation is restricted by the fact that the central visual ray will have to strike there; it will therefore have to be at the height of a man. This is easily achieved when the picture is on a panel, by hanging it at the required height after it is finished. Second, the principal point, to be properly located, must be at no greater and no less a height above the ground line than the height of the man to be depicted — in other words, 3 *braccia*. The orthogonals are then drawn to the principal point and the first phase is complete. Obviously, if Alberti had to paint on a wall he had to reverse the procedure; first he would mark the piercing point of the central visual ray; then, having selected the scale of the figures, he would draw the ground line one full man's height lower down, divide it into thirds of that dimension, draw the converging lines, and complete the rectangle.

The second stage (FIG. 205, 2) relates to the spacing of the transversals and involves the separate construction of a small-

by dropping a perpendicular (the picture seen in section) at the chosen spot. The intersections of the visual rays with this perpendicular provide the required spacing, which, when plotted at the scale of the picture, will be transferred onto it for drawing the transversals. The plan in perspective will thus be shown as divided into 1-*braccio* squares; and if the construction is correct, the diagonals of the rows of foreshortened squares opposite the vertex will link up into a straight line. Lastly the horizon is drawn to complete the preliminary procedure. Obviously the auxiliary construction could also be done at full scale, but it is more practical to draw it on a sheet of paper to a smaller scale. The spacing as plotted is then transferred in the simplest way: geometrically by the method of Thales of Miletus or else by multiplying the individual quantities by the original reduction factor. Moreover, if the auxiliary construction was scaled to one-tenth or any other submultiple of the picture scale, the multiplication could be done by compass.

Whether the spacing in the picture is reproduced at the side or along any other vertical is a matter of indifference, so Alberti does not mention this detail; but it is highly probable that his demonstrations (which amazed his friends, who called them miracles) were founded on spatial interlocking of the planes considered, and hence the spacing would be marked on the perpendicular dropped from the principal vanishing point of the picture. In any case it seems certain that Alberti's construction derives from some spatial device of this kind.

The legitimate-construction method enjoyed a long period of favor in treatises and was the only one used by Piero della

Francesca to foreshorten the basic field, although the internal divisions were regularly scaled on the diagonal. Later it was carefully described by Pacioli (1494), Serlio (1545), and Dürer (1532). It was through Dürer (q.v.) that Italian methods were definitively introduced into the practice of northern Europe, where, with Dirk Bouts and perhaps even Petrus Christus (qq.v.), there had already been some evidence of a familiarity with the convergence of orthogonals at a single point. Along with the established method of the distance point, the legitimate-construction method was also described by Barbaro (1569) and by Vignola and Danti (1583), and it was not forgotten in the treatises of the following centuries.

Legitimate construction, besides affording a proof of the proposition, has the advantage that the distance is manifest from the beginning; whereas in directing the diagonals to the relevant vanishing point there is the risk, particularly for the expert, of completing a perfectly accurate construction without really knowing just what distance has been used. Moreover, it is much easier to choose relatively long distances, since in the auxiliary construction, done to a smaller scale on paper, they can be adopted without too much risk of practical complications. However, it was known from the start, by Brunelleschi at least, that the construction with a distance point derived from the same spatial model of interlocking planes, rotating on the picture plane the sagittal cross section of the visual cone. This is proved by the earliest graphic and literary references to the method.

The sinopia underneath a Nativity scene by Paolo Uccello in the cloister of S. Martino alla Scala, Florence (PL. 93), shows a preparatory perspective consisting of a diminished grid in which the diagonals are carefully carried to the proper vanishing points. It also gives full prominence to that central vertical line which is the axis of rotation and corresponds to the perpendicular in Alberti's auxiliary construction, the sections of the ground line being measured one by one from its foot. To avoid misunderstanding, it should be made clear that, as is well seen in a clearer outline of the perspective trace, the subsequent spacing of the transversals (beyond a short initial stage equal to five rows of squares) is no longer on the central vertical line of this sinopia, where the diagonals and the visual ray intersect; instead, just as somewhat later with Pélerin (1505), it is determined by the intersection of a single diagonal on the fan of orthogonals converging at the principal vanishing point.

The same vertical is also the basis of the distance-point method expounded by Pomponio Gaurico (1504), who had probably learned it from Mantegna. This method (FIG. 205, 4) does not involve explicit use of a principal vanishing point.

The distance-point method, in the form still taught today, was defined by Pélerin, but without a proof and without an explicit statement that the distance from the principal vanishing point to the distance points, is equal to the distance from the station point to the picture plane. Thus, contrary to the general belief, while it is certain that users of Alberti's method were always aware of the distance, it is less certain that use of the distance-point method always implied this awareness. The fact that the plane was rotated was not clearly stated until Commandino's commentary on Ptolemy's planisphere (1558), which put an end to the earlier ambiguities and misunderstandings. Even Piero della Francesca, in the only case in which he explicitly used the distance point for diminishing a plane figure, correctly measured the distance from that point to the principal vanishing point (*De prospectiva pingendi*, XXIII), but then he went astray in the proof, becoming confused with Alberti's rule.

From that false proof there was derived, through Baldassare Peruzzi, Serlio's false rule (1545), according to which the distance was sought along the ground line from the station point to the edge — instead of to the middle — of the picture. This error, however, did not affect the correctness of the construction and was soon righted by Giacomo da Vignola. But even before the posthumous publication of Vignola's work, with a commentary by Danti (1583), came the publication of Jean Cousin's *Livre de perspective* (1560). For diminishing the plane, he explains only Pélerin's distance-point method, but he anticipates Danti and Vignola in defining the accidental vanishing points as the points of convergence of groups of parallel lines in any horizontal direction.

It fell to Guidubaldo del Monte, Commandino's pupil and one of Italy's greatest scientific thinkers, to formulate a full convergence-point theory, to the effect that a group of parallel lines in perspective meet at the point where the visual ray parallel to the given lines pierces the picture plane. Thus, two centuries after the pioneer experimenters, his *Perspectivae libri sex* (1600) represents the first full mathematical recognition of the more notable corollaries implicit in the revolutionary concept of the picture in perspective as a cross section of the visual cone. These books also taught how the perspective of any significant point on a regular solid could be located by establishing its dimensioned projection on the ground plane, and they laid the foundations of the theory of shadow and of theatrical or relief perspective.

Perspective in art. From the beginning, the application of perspective to the visual arts — including the experiments already mentioned — has taken different directions according to the specific interests of the people concerned. The interest of Masaccio (q.v.) in the subtleties of geometric construction, for instance, was so limited that the suggestion of Kern (1913) that the perspective of the architectural framework in his *Trinity* (IX, PL. 345) may have been drawn by Brunelleschi seems plausible; for Masaccio's central problem was how the volumes, made substantial as never before by a directional light, should be fitted inside a certain space in the easiest relationship between the figures and their surroundings. His instinct for perspective was so ready and live, however, that it need fear no comparison, even where, as in the *Tribute Money* (IX, PL. 348), the architecture is reduced to a bare minimum and the delineation is summary. In contrast to this instinctive gift is the scant spatial consistency in the works of Masolino (q.v.), who completely failed to scale his figures convincingly in depth, even within an academically correct linear architectural setting, as in the *St. Catherine and the Philosophers* (IX, PL. 378), in S. Clemente, Rome.

In the works of Donatello (q.v.), perspective can be studied in its greatest and richest variety. As early as 1427 he was at ease in handling the more usual one-point perspective, in his *Feast of Herod* relief for the Siena Baptistery (IV, PL. 242). Later he was ready to approach the more difficult problems of two-point perspective and even of three-point perspective in an oblique view, such as the tondo with the assumption of St. John in the Old Sacristy of S. Lorenzo in Florence (ca. 1440), as well as of accidental perspective, as in the right-hand building in the scene with the miracle of the healed foot from the altar of the Church of S. Antonio in Padua (ca. 1445–50; PL. 94). It is possible that the treatment of this last structure was suggested to him by Paolo Uccello's Noah's arks (PL. 41) in the Chiostro Verde of S. Maria Novella in Florence, though probably Brunelleschi had been the first to learn how to place the accidental points on the horizon, in the view of the Piazza della Signoria already discussed.

In 1436 Paolo Uccello (q.v.) did the convincing upward perspective on the lower base of the John Hawkwood monument in the Cathedral of Florence (XII, PL. 27), in which, by order of the Operai of the Cathedral, the horse and rider were redone in one-point perspective, which was incongruous for the base in question. Paolo's temperament no doubt led him to take an abstract delight in the mathematical subtleties of perspective for its own sake, achieving a sort of geometric mosaic in which the realistic scope of the perspective method is diverted to solutions with a metaphysical tendency. Thus the distance points on the margin of or even inside a picture sometimes induced him (e.g., on the predella of Urbino; PL. 44), to seek deliberately certain anamorphic distortions toward the edges, which assume lifelike proportions only when viewed from an angle; these seem to have aroused his gift for expression more for the opportunities of caricature that are inherent in the theoretically erroneous front view of the individual details. Surely only Paolo can have supplied Michele Giambono with the cartoons for the Gothic mosaic of the Mascoli Chapel in S. Marco, Venice. Here one of the two buildings is shown in a masterly orientation along the diagonals, in the way taught by Pélerin more than half a century later; the other is crowned by hemispherical domes drawn in perspective by the faceting method used in Paolo's perspective study of a *mazzocchio* (Uffizi, Gabinetto dei Disegni e Stampe).

One should not reject out of hand the theory that Paolo Uccello (to whom Vasari attributes certain experiments in multiplane perspective in a chapel of S. Maria Maggiore, Florence), in the Noah scenes of the Chiostro Verde, may have made some attempt at stereoscopy in complementary colors, to be observed through small circles of red and green glass, based on Ptolemy's by then commonly known theory. The cartoons supplied by him for the Cathedral's stained-glass windows, which were being prepared in just those years, required many such circles.

Two notable early perspective painters who played a fundamental part in spreading the method of perspective are Angelico and Filippo Lippi (qq.v.). G. C. Argan (1946) describes Angelico's work on the predella of Cortona (Mus. Diocesano) as the first perspective of an actual landscape, Lake Trasimeno, although in that case it cannot have been drawn geometrically but must have been taken from life by means of Alberti's *velo* (a reticulated net) or with a mirror or sheet of glass. These two painters often used relatively short distances, which account for their rather steep perspectives.

Domenico Veneziano (q.v.) certainly knew Alberti well, using his *velo* or glass for the unrivaled spatial verity of the landscape in his Berlin tondo. He also introduced the systematic use of relatively longer and more plausible distances. Domenico di Bartolo and the Master of the Barberini Panels both drew upon Domenico Veneziano, as did, later on, Andrea del Castagno (q.v.), and, to some extent, Antonio Pollaiuolo and Andrea Verrocchio (qq.v.); and Piero della Francesca (q.v.) was Domenico's direct disciple. The 50-ft. distance used by Andrea del Castagno for his *Last Supper* in S. Apollonia, Florence (I, PL. 242), and the long distances used and recommended by Piero della Francesca also probably derive from Domenico Veneziano.

As a theorist, Piero della Francesca was not an innovator, unless it be for his systematic practice of rotating the ground plane about the ground line onto the picture plane and for his being the first to assert the need for a true method of establishing geometrically the position of all the significant points on an object. He does, however, represent the culmination of 15th-century perspective, as typified in Florentine rationalism, in which reality was not copied but was reconstructed architectonically, each solid being related to the ideal model of regular bodies. This also lies behind the predilection for frontal views and for frontal illumination, which attenuates the chiaroscuro and emphasizes the prime importance of linear definition. Piero used two-point perspective in the exercises in his writings, without using accidental points in the construction, but he avoided it in his paintings because it would have disturbed the architectonic relationships with the rectangular framework. Nor was Piero greatly worried by the inaccessibility of the station point, as in the upper registers of his frescoes for S. Francesco in Arezzo (PL. 160), since the relationships are so implicit in the internal pattern of the lines that the significant spatial structure is completely discernible without any illusionistic approach.

One of the first to show a concern with the observer's actual position, even when the picture was high up, was Andrea Mantegna (q.v.) in his *Martyrdom of St. James*, in the Church of the Eremitani in Padua (XII, PL. 140). Whether in such a case the station point should be placed as low as the observer's position requires was still being discussed at the height of the mannerist period by, among others, Pellegrino Tibaldi, Palladio, Vignola, Vasari, and G. B. Bertani, as Bassi relates in his *Dispareri* (1572). In Mantegna's Camera degli Sposi (X, PL. 487) it is a question not so much of looking upward as of a reunification of the whole surrounding space through a single station point. Technically at least, this assumption was compulsory, through Giulio Romano and Correggio, for all the large-scale illusionistic decoration of the baroque age.

SIXTEENTH CENTURY. *Leonardo and the crisis of linear perspective.* Perspective painting was able to develop in the 15th century with the support of geometric layouts deduced from the linear hypothesis and its corollary that the visual image can in turn be schematized in accordance with certain fields of equal light and equal color, which roughly correspond with Alberti's doctrine concerning surfaces. This conception is symbolized by the clear, sharply defined areas in inlay work. Wood inlay, in fact, remains one of the hallmarks of perspective virtuosity; it is sufficient but to recall the names of Giuliano and Benedetto da Maiano, Lorenzo and Cristoforo da Lendinara (who worked from cartoons by Piero della Francesca), Giovanni da Verona, and Damiano Zambelli (da Bergamo), who worked well into the 16th century from cartoons by Bramantino and Vignola and whose tints and patinas ended by freeing the genre from its inherent flatness and transforming it into a true picture.

With Leonardo da Vinci (q.v.) the outline, so fundamental to the 15th-century concept of perspective, was roundly rejected. He denied it in theory, both explicitly and by going on from linear perspective to define aerial perspective in its dual aspect of color and diminishing clarity of vision. He likewise denied it in practice by enhancing the atmospheric effects of softness of focus and of haziness depending on distance (see OPTICAL CONCEPTS). This did not mean that doubts were cast on the validity of Brunelleschi's perspective as defined by Alberti. Even Leonardo saw a picture first and foremost as a cross section of the visual cone, in the sense that for each point on the object there must be a point on the image. But thenceforth it was no longer enough to isolate a few special points in order to define what might be called an architectonic framework of space; each point was regarded as having the same worth and as carrying the same minute but inescapable charge of color and light. Italian painting was thus ready to employ the results of the vigorous empiricism of the Flemish painters in the field of chiaroscuro and color, yet without abandoning its own fundamental rationale. Without Leonardo, moreover, it would be impossible to conceive of the tones of Giorgione or of the subsequent development — from Titian to the impressionists — of applying paint in juxtaposed dots or patches of solid color.

Leonardo's experimenting went beyond the mathematical intransigence of his predecessors, especially in investigating the conditions needed for creating the perspective illusion. As a result he became the first to grasp the fact that the ocular image is spherical, and so to realize that it cannot be superimposed on the plane perspective image (FIG. 189, 6). He was thus better able to confirm the true nature of anamorphosis and the complexity involved in multiplane and other presentations connected with projection onto anything but a plane surface. In the light of this knowledge, he posed the alternatives of a strictly orthoscopic view, fixed by means of a sighting device, or the selection of distances that increased as the probability of accurately centered observation decreased. In the end he came to a sort of antiperspective illusionism, as in a painted depiction of a statue; an example is his work in the Sala delle Asse of the Castello Sforzesco in Milan, in which he disguised the vaulting columns as tree trunks and the pendentives as treetops (covering the resulting spherical trihedron with fronds).

Michelangelo (q.v.) may be placed within a similar order of spatial ideas, for he tends to assimilate painting to relief. He paints subjects rather as though following the picture plane and does not try to create the illusion of spaces going far beyond the surface of a ceiling.

Mannerist decoration and architectural perspectivism. Mannerist decoration (see MANNERISM) thus had two aspects, alternating between illusionistic painting and the simpler system of letting the architecture speak for itself, at most embellishing it with separate paintings. This was the course followed by Raphael (q.v.) for the ceilings of the Vatican Stanze; yet it is worth noting that the trend and the practice of illusionistic painting developed in the school of Raphael. The requirement of an integral spatial composition with a broad sweep was taken by Raphael from the tradition of Piero della Francesca, through Perugino, Melozzo da Forlì, and Bramante (qq.v.), and was confirmed in his contacts with his contemporary Baldassare Peruzzi (q.v.). In connection with the revival of the Pompeian illusionistic art of grotesques, and perhaps also with Leonardo's precepts and Paolo Uccello's already familiar attempt in S. Maria Maggiore in Florence, this tradition permitted the extension of illusionistic painting to the vaults in the Vatican Logge merely on the basis of a rather summary calculation of the viewpoints.

Giulio Romano's stay in Mantua, under the influence of the layout of Mantegna's Camera degli Sposi, resulted in the hyperbolic illusionism of the *Fall of the Giants* (IX, PL. 298), which, together with the lost paintings in the Palazzo Ducale, can be regarded as prefiguring Tibaldi's work in the Palazzo Poggi (now Palazzo dell'Università; PL. 100; IX, PL. 298) in Bologna, as well as the Venetian current of the Rossi (or Rosa) brothers in Brescia, who did the lost ceiling of S. Maria dell'Orto in Venice (1556–59). A little later (1559–60) they did the simulated architectural perspectives around Titian's *Sapienza*, on the ceiling above the grand staircase of the Biblioteca Vecchia di S. Marco in Venice.

Still in the vein of Piero, Peruzzi, and Serlio was Giacomo da Vignola (q.v.), the painter of the architectural perspectives that were to frame the scenes painted by Primaticcio at Fontainebleau. About 1562 Vignola painted, in the Palazzo Farnese at Caprarola, the outdoor perspective in the "Camera Tonda," or Guardroom, the columns in the Sala del Concilio, and the greatly admired trompe-l'oeil colonnades and exedras in the Sala di Giove. Thus Vignola's influence met with that of Tibaldi in establishing that scheme in the form of a simulated loggia which, by way of painters of Vignola's circle (Tommaso Laureti's lost ceiling in the Palazzo Sanguinetti, formerly Vizzani, in Bologna, done sometime before 1562 and depicted in an engraving by Danti in his commentary on Vignola's *Due regole*; Ottaviano Mascherino's Sala della Bologna in the Vatican, 1575), was to become the canonic example for architectural perspective painters of later centuries, beginning with Agostino Tassi (PL. 101) and receiving its main impulse from Agostino Mitelli and Angelo Michele Colonna.

SEVENTEENTH AND EIGHTEENTH CENTURIES. *Perspective theory.* Developments in the theory of perspective after Guidubaldo del Monte are of only marginal interest for the visual arts. Numerous painters looked to the theorists, however, who for their part believed they were providing better methods than those already in use. The subject is therefore very briefly discussed, with the warning that in the 17th and 18th centuries the number of books explaining perspective for painters became so great that even a list would be too long for inclusion here.

The mathematical bases of perspective were studied afresh in the 17th century, particularly by Gérard Desargues of Lyons (1593–1662). It was he who finally defined the convergence point as the perspective of a point at infinity. This notion is irrelevant to the practical aims of perspective painting, but it is useful to mathematicians and is one of the foundations of parallel projection. Desargues also laid the foundations for normal axonometric projection, establishing a scalar method involving the use of the three Cartesian axes, which enabled him to determine the perspective of any figure even without using vanishing points. The scientific intransigence of Desargues and his disciple Abraham Bosse, an engraver and teacher at the Royal Academy of Painting and Sculpture under Colbert, caused bitter disputes with the academicians, led by Charles Lebrun, who were tied to the established conventions in painting. Lebrun himself was bound to the tradition of the Carraccis and favored decoration with individual canvases applied to walls and ceilings, as in the Hall of Mirrors at Versailles (II, PL. 212). Desargues also deserves credit for having considered the problem of projection onto an irregular surface in his usual strictly scientific manner and for having founded relief perspective as a separate discipline.

The most notable 18th-century writers on perspective are an Englishman, Brook Taylor (*Linear Perspective*, 1715), and a Swiss, J. H. Lambert (*Freye Perspektive*, 1759). These two men freed perspective from the tyranny of the ground plane, which involved drawing plans or elevations as a first step. Perspective then moved more definitely into the field of pure mathematics and became ripe for inclusion in modern descriptive geometry in the 19th century, following the renaissance inspired mainly by Gaspard Monge and the systematic application of the algebraic concepts peculiar to analytic geometry. Eventually perspective came to be identified primarily with the method of central projection, subsequently playing a part in the modern offshoots of photogrammetry and perspective axonometry, of which Desargues is a remote ancestor. Painters, however, settled for less, and even specialists did not hesitate to take liberties. Inexact methods were even recommended on occasion in the textbooks; thus G. Troili, for example, writing in 1683, says that round balls can be drawn with a compass, ignoring rigorous precepts.

Perspectivist schools. Large-scale decoration in the 17th and 18th centuries took as its starting point Annibale Carracci's gallery in the Palazzo Farnese, Rome, even though in that example the areas of architectural perspective painting and the areas occupied by individual paintings were still largely independent; and its development took the direction of a gradual merging of the architectural-perspective trend with that of free figures painted in empty space (see PERSPECTIVISTS). After Guercino's *Aurora* (II, PL. 197) and Pietro da Cortona's ceiling in the Palazzo Barberini, this fusion was completed in a scheme of acrobatic complexity, which nonetheless fully respects the single station point for the whole space, in Pozzo's masterly *Triumph of St. Ignatius* (II, PL. 172). Himself a treatise writer, Pozzo did not hesitate to speak out against the liberties that had invaded the perspectivist tradition. His was an intransigence that did not shun even the most extreme distortion, provided that the view from the prescribed point ensured its optical correction.

In regard to the liberties mentioned, it may be recalled that Vignola had already used a false distance to make the "Camera Tonda" at Caprarola seem higher than it was, and Danti praised him for it. The Rosa brothers had used multiple vanishing points to correct the effect of rooms that were too long and low and had earned Viola-Zanini's approval (Schulz, 1961; FIG. 205, 6). Pozzo's integrity was thus a beacon, but it did not give rise to a school of disciples. The high point of the amalgamation of architectural-perspective painting with figural depiction is found in the middle decades of the 18th century, in the association of Giovan Battista Tiepolo (q.v.) and Girolamo Mingozzi (Colonna), each an excellent perspectivist in his own special field.

Scenographers and architectural perspectivists were soon competing with the teachers in academies and painters of architecture and ruins in inventing ever more complex and stimulating schemes. These professions often alternated in a family or were practiced by the same person. Theatrical perspective revived the view from an angle, on which Ferdinando Galli Bibiena theorized (PL. 95). This view was developed for the spatial expansion suggested by having two vanishing points and for its practical advantage of disguising the illusionistic inadequacy of the perspective background for spectators at the sides, at least preventing their seeing at once that the visual and perspective axes did not coincide (as they would have with a one-point perspective).

The painting of views could not have reached the photographic accuracy of a Canaletto or a Bellotto (qq.v.) had not the two principal streams joined forces: the documentary (genre and interior painting, *bambocciate*, Dutch landscape painting) and the geometric (architectural *capricci*, scenography). And in truth the unitary credibility of Canaletto's views, mostly montages from partial views taken separately with the camera lucida, could hardly have been achieved without the perspective competence achieved in theatrical scenography. Canaletto's competence was transmitted to his heirs, for the views planned by Bellotto (his nephew) have lessons even for the greatest virtuosos of the genre and the nephew was even more faithful than his uncle to the data provided by the camera lucida.

Relief perspective. The theory of relief perspective was developed particularly to meet the requirements of those elements of stage setting (see SCENOGRAPHY) which had to be practicable (e.g., entries and arches under which performers had to walk or stand). Relief perspective was also fundamental to certain architectural effects that were typical of the baroque period and were prescribed by Desargues and Bosse as normative for low and medium relief. At least in its origins, however, relief perspective was connected with sculpture. It first appeared in Donatello's *stiacciato* (or *schiacciato*: flattened relief) panels, which — unlike the handling of the relief perspective in Ghiberti's second door for the Florence Baptistery (VI, PL. 173) — represent an attempt to keep the index of flattening constant for near and for distant objects, even in the most protuberant parts, as in such later works as his ambos in S. Lorenzo, Florence (IV, PL. 251). These were followed by the notable 15th- and 16th-century attempts inspired by Mantegna (e.g., the triumphal arch of Alfonso of Aragon in the Castel Nuovo in Naples; III, PL. 390) or by Bramante (e.g., tapering-box reliefs by the Mantegazzas and by G. A. Amadeo, showing the influence of the false choir in S. Maria presso S. Satiro, Milan).

Bramante's false choir (II, PL. 340), a slight recess hollowed out with trompe l'oeil to represent an entire choir, remains an isolated instance in the architecture that stemmed from the 15th century in the tradition of Brunelleschi. In this connection it should be stressed that perspective in Brunelleschi's buildings consisted in rejecting any attempt at spatial illusionism. By the deliberate use of equal spacing, it was possible to make the entire building readily measurable to the eye, so that the cubic design was obvious at a single glance and did not need to be assumed to represent any space other than itself. There is a trace of such illusionism in Michelangelo's treatment of the Piazza del Campidoglio (I, PL. 390), which was opened out like a stage but with sides diverging instead of converging, so as to make the piazza appear short, whereas in fact it is too long.

Relief perspective reached its height in the practicable elements that Vincenzo Scamozzi (q.v.) added to the stage of the Teatro Olimpico in Vicenza (IX, PL. 312); and illusionistic tricks based on this technique were to become part of the stock in trade of 17th- and 18th-century architects, hallmarks of a distinctive style and taste. This widespread adoption is seen in various explicit borrowings from the theater for special purposes, such as Borromini's illusionistic gallery in the Palazzo Spada, Rome, and Bernini's Scala Regia in the Vatican (II, PL. 270); and it is even more evident in a subtler manipulation of the techniques of feigned perspective. The antipodes of this tendency are evident in the flattened, almost low-relief façades done by Maderno (q.v.) for S. Susanna (II, PL. 137) and St. Peter's (I, PL. 403) in Rome and in scenographic urban projects such as Bernini's for St. Peter's Square in Rome.

Perspective oddities. The 17th century was an age of delight in perspective tricks and enjoyment of the marvels and bizarre effects of anamorphosis — a taste that went along with the contemporaneous vogue for irregular perspective and relief perspective. After the early experiments of Paolo Uccello and the few known essays in this field by Leonardo, however, interest in such effects of anamorphosis seems to have shifted to the north (if the further attempts attributed to Leonardo or to the painters of his circle, or to later Lombard imitators such as Ambrogio Figino, mentioned by Lomazzo, are disregarded). The oldest known print showing anamorphic representations (dated to 1531–35) is one by Erhard Schön, a pupil of Dürer. From 1546 there is the anonymous distorted portrait of Edward VI of England, with its affinities to the portrait style of Hans Holbein the Younger, whose anamorphic skill in the foreground of *The Ambassadors* (VII, PL. 304) should also be mentioned.

Turning to practical methods of drawing, Danti (in his commentary on Vignola's *Due regole*) and Marolois (1633) taught how to stretch out an image without regard to the degree of divergence required to correct for the obliqueness of the viewpoint. As early as 1569, however, Barbaro had described a mechanical method of anamorphic drawing, using a point source of light to project the pricked outline of a normal drawing onto an oblique screen; and in 1612 Salomon de Caus described a perfectly satisfactory geometrical method. The fashion reached its heyday and interest in it was most widespread during the middle decades of the 17th century, which saw the publication of works on the subject by Niceron (1638, 1646), Kircher (1646), and Dubreuil (1642–49), among others. Interest in anamorphic tricks waned during the 18th century but revived somewhat when the romantic movement started; the popular prints of the Second Empire portraying royalty and erotic subjects relate this last flowering of the genre to its distant origins (Baltrušaitis, 1955). It is at least conceivable that the elongated necks and faces of Amedeo Modigliani (q.v.) in the 20th century may have owed something to these last essays in anamorphosis (whether it was that his eye fell on some belated example, serving as a curiosity for schoolboys, or that he occasionally drew with his cheek resting on the drawing board).

NINETEENTH AND TWENTIETH CENTURIES. The teaching of perspective (though of a much less daring kind than that practiced by the decorators and stage designers of the 18th century) was one of the bases of academic teaching in the 19th century. The romantically inspired revaluation of the primitives led to the recognition (at least in theory, and on the level of historical criticism) of all types of representation as legitimate and of nonperspective art as having its own esthetic validity. At the same time, however, the encroachment of Western-style perspective upon the art of non-European lands was proceeding remorselessly, often with unhappy results, eroding the conventions and systems evolved from centuries of indigenous art.

About the middle of the 19th century the role of perspective in Western painting seemed to be on the decline. Realism and impressionism joined forces, aiming their shafts particularly at the artificial conventions of the school of historical painting — that favorite of the academic diehards — based on the assumption of a spatial box, drawn in perspective and acting as a kind of stage setting for the picture. As the general principles of impressionism (q.v.) gained the day and it became accepted that art was concerned with a more direct copy or impression of everyday reality (*Primamalerei*), artists gradually abandoned geometric construction. This did not imply, however, a deliberate rejection of perspective. For one thing, these copiers from nature were by then too well schooled in the basic principles of perspective, and for another, they now had in photography an even handier and surer yardstick. They no longer went in search of subjects purely as studies in perspective or manipulated them so as to bring out the effects of foreshortening, recession, or diminution; neither did they alter the objective facts as they chanced to find them.

The first notable systematic denials of perspective are found in the still lifes of Cézanne (q.v.; PL. 96), in which the horizontal surfaces are tilted toward the observer and the dishes look as if their contents were about to slip off. These pictures imply more than just the distorting possibilities of memory, as demonstrated in the experiments of H. Thouless (an oblique view of a circle, copied by eye, is less elliptical than the image actually recorded on the retina; Gombrich, 1960, p. 302), a tendency that could still be accepted as an extension of the poetic repertory of impressionism; they represent the deliberate and explicit use of mental images or patterns (*Gestalt*) that serve as a basis for the psychological organization of visual data as perceived by the retina (and hence often coincide with the use of typical views in primitive art). At the same time, they prepare the ground in other respects for the theory of significant distortion which underlies caricature and which corresponds with the expressionist phase of European art, including Fauvism (see FAUVES).

With cubism and its collateral movements (see CUBISM AND FUTURISM), the rejection of perspective took on an openly polemic character. This became increasingly violent as the influence of photography made itself felt and traditional perspective began to disintegrate. As noted by Ragghianti (1962), this disintegration was promoted by use of the conventions of modern technical drawing (as taught by Monge), while a conscious return to the remote historical precedents of Egypt and Byzantium also played its part. In this phase, however, closer study reveals the survival of a number of compositional elements that had been developed and accepted in the traditional school of perspective (Ragghianti). It is also to be noted that an attitude of antagonism toward perspective at least implied a recognition that perspective existed, if only as a premise to be attacked. As the process of dissociation continued, it led eventually to a complete disintegration of organic forms and an impromptu, abstract arrangement of lines on the plane.

The second half of the 19th century also opened the way to a more informed critical history of perspective in art, with the twofold purpose of tracing its vicissitudes and appraising its esthetic significance. The debate was started by Guido Hauck's famous essay (1879; see OPTICAL CONCEPTS). Hauck's aim was to elucidate the reasons for the so-called "optical corrections" described by Vitruvius in connection with Doric architecture. He was also personally convinced that perspective, in so far as it was a central projection onto a plane surface, was inherently against the truth of nature. He therefore attempted to show that the ancients used a different system of perspective based on a sort of pseudo development of the spherical surface. The theory was immediately rejected by other scholars (e.g., Delbrück, 1889) but was later revived with honor. The philological contributions of Kern (1938) and Döhlemann (1904, 1905) in the 20th century were particularly concerned with the intuitive, approximate perspective of ancient times (Kern) and with tracing the antecedents of Renaissance perspective in early Italian and Flemish art.

In Mesnil's studies of Masaccio (1914, 1927), attention was shifted from the philological to the esthetic. Perspective was considered — in the sphere of theories of "pure visibility" — as an element of style within the work of art. Mention should also be made of the further attempt to demonstrate that the usual system of perspective was not true to nature, and also of the claim of Borissavliévitch (1923) to have discovered an optical and physiological perspective. A year later came Panofsky's fundamental study "Die Perspektive als 'symbolische Form'" (1924–25), which remains the fount and mainspring for all later research. In Panofsky's opinion, perspective cannot be isolated and taken as a criterion by which to judge the worth of art but must be viewed simply as a distinguishing element of style, changing with time and place throughout the history of art. In this it resembles one of the "symbolic forms" of Cassirer's theory, according to which a particular spiritual value is associated with a concrete and palpable symbol and becomes its specific attribute. Panofsky reexamined the whole question from this standpoint and formulated in all its details a postulate which had already been guessed at and which is substantially true: that perspective in a work of art is a relative quality. However, his adoption of Hauck's theory, involving the belief that Renaissance perspective was in itself both relative and arbitrary sometimes led to unhistorical conclusions.

The existence of central perspective in antiquity continued to be denied by Little (1936) and Richter (1937, and even later, after Beyen's energetic defense in 1939). Further studies under the influence of Panofsky include those of Nicco Fasola (1942–43), dealing with Piero della Francesca and the Renaissance treatises; Bunim (1940), on the treatment of space in medieval painting and the precursors of perspective; John White (1957), who systematized the curvilinear perspective of Hauck and Panofsky and produced his own theory of synthetic perspective as a constant alternative to artificial perspective throughout the history of Western art; and Pierre Francastel (1951), who extended Emile Durkheim's theory of social space to art criticism and tried to define a new concept of figural space.

In the author's opinion, though the use of perspective in a work of art is a stylistic feature that must be seen against its historical and cultural background, it does not follow that its technical development is irrelevant or immaterial to the art historian. Nor is it obligatory that the scientific concept of a single valid perspective should give way before the relative, stylistic concept according to which there are as many versions of perspective as there are artistic styles — if not indeed, works of art themselves.

With regard to the value of studying the technical development of perspective, the results are a sufficient justification, if only for their use as dating evidence (outlined as long ago as 1905 by Döhlemann). With regard to the two apparently antithetical concepts of stylistic and scientific perspective, much can be done to reconcile them by applying the concept of language evolved by Benedetto Croce's school of criticism — a concept that was only touched on and never fully absorbed in the symbolic theories of Cassirer and Panofsky or in the semantic theory of Francastel (1951).

In effect, when figural styles are considered as historically constituted semantic systems (languages) — with full awareness of the preeminence attributed by this hypothesis to the conventional elements — the perspective of a painting, whether linear or spatial, may be regarded as a structural element inherent in that style (or language), regardless of whether it is based on a geometrically accurate model of perspective or not. In that sense, and even with reference to a correct method of projection, the perspective of the ancients is not the same as that of today. But neither is 15th-century perspective the same as 18th-century perspective; nor can Masaccio and Piero della Francesca be equated in this respect, nor Canaletto and Tiepolo; nor, for that matter, can any two works by Piero, Masaccio, Tiepolo, or Canaletto themselves.

To assume that Apollodoros of Athens, had he wished, might have painted perspectives which followed the geometric scheme of reference to the same degree as Piero della Francesca's *Flagellation* in Urbino and in which the scheme itself was conceived in the same manner as by Piero in that painting would seem tantamount to believing that a 5th-century Greek might have awakened one morning with a burning desire to speak in Italian or German.

Thus the statement, so essential in Panofsky, that Polygnotos could have neither wanted to nor been able to paint a naturalistic landscape, for the simple reason that "such a representation would have belied the immanency of 5th-century Greek art," is beyond criticism up to a point. But it errs in being too categorical. If the same statement were applied to Giotto, for example, it would preclude the possibility of his having painted those feigned chapels in Padua with the spatial naturalism that he indeed achieved.

The fact is that perspective is not only the end product of a process of historical development, the codification of a certain artistic vocabulary, but is also a starting point, a fact of common experience, in that it is an attempt to convey what is actually seen. To some extent, therefore, it can be employed in any system or language of representation, at any stage in its development (though to what extent can be learned only from knowledge of the original "texts," albeit indirect).

For Polygnotos to paint a hypothetical naturalistic landscape would not have been quite the same as to speak German. To have inserted a background of green hills would not have been (to continue the metaphor) the same as saying *Hügel* or *Bäume* but, rather, more like pointing to the actual hills and trees. Such a gesture would have been perfectly clear to anyone speaking another language (be he Dürer or Altdorfer). True, the sense of immanency in 5th-century Greek art might have restrained the gesture, but how far this was so can only be conjectured from current sketchy ideas of that century and its ideals, in the absence of any original paintings.

In this, as in all other cases, the reconciliation of the scientific and the stylistic concepts of perspective can be studied only with reference to the work itself, or at least with the ample aid of philology. Thus philological research in many fields and a more thorough and accurate understanding of the originals are essential preliminaries for adding to the body of objective fact, which alone can serve as a basis for convincing and unambiguous theories.

THE ORIENT AND ISLAM. A conception not far removed from that of Early Christian art (particularly in its proto-Byzantine phase) can be found in the earliest surviving paintings in India (the Ajanta cave paintings, 1st cent. B.C.–7th cent. of the Christian era; see AJANTA). Sources of influence in Syria

(Dura-Europos and Palmyra) and Iran undoubtedly contributed during this period to the spread of Hellenizing fashions that emanated locally from the Greco-Buddhist center of Gandhara (q.v.).

No examples of Gandharan painting have survived, but it is assumed to be the essential antecedent of subsequent developments, in Central Asia as well as in India. The evidence of reliefs, however, can be taken to hold good for painting as well. One significant example, a Buddhist carving of the second half of the 3d century (Karachi, Nat. Mus.), should suffice as proof of how thoroughly the Greco-Roman concept of perspective was assimilated by the Gandharan artists, if only for the Donatello-style foreshortening of the two coffered barrel vaults, which converge fairly accurately from either side on a single vanishing point.

At Ajanta, however, the naturalistic idiom is less consistent and tends to be episodic and fragmentary. Vaguely spatial niches, reminiscent of the later Pompeian styles, are inserted into compositions that seem predominantly two-dimensional, despite the robustly sculptured figures and the crowded scenes teeming with life.

In the cave paintings of Tun-huang (q.v.) in western China, the most noteworthy part of which can reasonably be dated between the 6th and 9th centuries, it can be seen how the spreading of Indian culture in China introduced a naturalistic treatment of space that was Greco-Roman in its distant origins, and how during the T'ang dynasty Chinese sensitivity took over and converted this to its own ends (see CHINESE ART). Thus was born the typical distant prospect, which produces a convincing intuitive rendering of perspective when applied to passages of landscape actually in the distance and foreshadows the naturalistic handling of space by the great Sung landscape painters (III, PLS. 262–267, 269, 270). At the same time a method of parallel projection for buildings developed.

That the Chinese artists also used focal (though not mathematical) perspective for depicting buildings is amply demonstrated by occasional, though none the less eloquent exceptions to the rule. Typical is a well-known representation of a sanctuary (dated by some authorities as early as the 10th cent.) in which the faces of the building are shown in convincing perspective but are nevertheless tied to a completely conventional plan. It might almost be inferred paradoxically that the Chinese used perspective in technical drawing (where parallel projection would be used in the West) and adopted parallel projection in their paintings. The facts, however, are more complex. The Chinese normally preferred parallel perspective not because it seemed more realistic to their eyes but because it was better suited to the unrolling vision and continuous narrative entailed by their vertical and horizontal scrolls (Jap., *kakemono* and *makemono*). In Far Eastern art these take the place of Western easel paintings and permit the observer to immerse himself in the picture, almost to mingle with the characters in the story, to see mountains and villages, palaces and cities through the eyes of the protagonists, as in a film.

It follows that in Oriental painting the commonest and most suitable way of rendering architectural space was oblique projection, but this is not to say that two-point perspective was never used: on the contrary, there are numerous examples of it in Japan, even in the earliest scrolls of the Heian period (784–1185), and it was common in the later Ukiyo-e (q.v.) prints and pictures. This method lent itself to a great variety of treatments and sometimes produced what was to all effects an accurate isometric projection.

The forbidding of figural art by the Koran prevented the development of any continuous pictorial tradition in Islamic art. Ommiad painting (see OMMIAD SCHOOLS) was essentially linked to late classical models and local Syro-Christian precedents, to the extent that the Pompeian-style mosaic perspectives in the Great Mosque of Damascus (II, PL. 441; VII, PL. 407; VIII, PL. 145; X, PLS. 379, 380, 386) are valuable mostly as evidence of the state of secular and aniconic Byzantine painting at the threshold of the iconoclastic age; whereas the Abbasside figure paintings (see ABBASSIDE ART) found in the Samarra palaces are totally devoid of any attempt at perspective and represent a revival of the Sassanian tradition (see SASSANIAN ART) of strictly two-dimensional heraldic stylization.

A certain amount of background detail and settings (though in the form of simple silhouettes) reappears in the earliest Persian Islamic miniatures (see MINIATURES AND ILLUMINATION). These, from the so-called "Baghdad" school, date from the 13th century and appear to owe less to Byzantine and Ommiad traditions than to the influence of the Manichaean Mesopotamian and Turanian miniatures, which were well known to the Seljuk invaders (see SELJUK ART).

In the middle of the 13th century Hūlāgū's Mongols introduced into Iran numerous examples of Chinese painting, and with them the taste for landscape and for more realistic settings. This taste was tempered over the following decades so that the great flowering that took place in the 15th century under the Timurids (see TIMURID ART) coincided with the advent of an original local style. This still retained the narrative style and the settings following the axonometric conventions of Chinese painting; but the last traces of illusionism disappeared under the vivid colors, the gilding, and the damascening applied over the composition with the true Islamic genius for carpet patterns and arabesque decoration. The Herat school with Bihzād (q.v.) marked the zenith of Timurid miniature painting. From it evolved the Safavid style (see SAFAVID ART) in Persia itself and the highly important school of Moghul art (see MOGHUL SCHOOL) in Moslem India. The spatial concepts of the Timurid miniatures can therefore be taken as typical for the entire movement.

Western influence. A first wave of Western influence on Islamic art affected Ottoman miniaturist painting (see OTTOMAN SCHOOLS) toward the end of the 15th century (one of the factors being Gentile Bellini's pioneering visit to the court of Muḥammad II in Istanbul) but soon receded, to leave the traditional Persian style unchallenged. In Persia itself, under the Safavids, Western fashions appeared in the school of Isfahan (see MINIATURES AND ILLUMINATION) as early as the time of Riẓā-i-'Abbāsī (q.v.) and were continued by his 18th-century followers, who showed particular interest in the naturalistic novelties of 17th-century Dutch painting.

During the 18th century a perspective treatment of space in the Western fashion is also frequently encountered in the exquisite Moghul miniatures of Islamic India, although the more conservative Hindu tendencies (Rajput, especially Pahari) remained substantially unaffected by foreign influences (see MINIATURES AND ILLUMINATION; MOGHUL SCHOOL; RAJPUT SCHOOL). Even more striking was the penetration of China and Japan, which were less likely to be harmed by the Western gift.

In China (see CHINESE ART) there were frequent attempts under the Manchu dynasty to initiate Chinese painters in Western methods. Suffice it to mention the Jesuit Giuseppe Benedetto Castiglione, court painter in the Ch'ien-lung period (1736–95), who also painted in the Chinese style (though using perspective) and who taught European painting methods to numerous pupils. A further important contribution to the dissemination of this knowledge was made by the lacquer workers and potters. Producing for export, mainly through the East India Company, they copied or interpreted English, Dutch, and French models supplied by their clients.

In Japan (see JAPANESE ART), in the Tokugawa period (1615–1867), when only the port of Nagasaki was open (and only to the Dutch), the products of Western art were nevertheless known. Through copies made by such minor 18th-century artists as Hiraga Gennai, Shiba Kōkan, and Aōdō Denzen the knowledge of perspective was transmitted to some of the great Ukiyo-e masters such as Toyoharu, Hokusai, Hiroshige (qq.v.), and Kuniyoshi (see UKIYO-E; I, PL. 123; II, PL. 30; IV, PL. 447; VII, PLS. 237–241, 296–300; XIV, PL. 236). In the hands of these painters the new acquisition, when used, was made to harmonize with traditional taste and style. It can therefore be assumed that it was not so much perspective as the arrival of photography and the chromolithographic prints, with their banal perspectives, that dealt the final blow to local traditions at the end of the 19th century.

BIBLIOG. *Technical and scientific works*: G. Loria, Metodi di geometria descrittiva, 3d ed., Milan, 1924; P. Reina, La prospettiva, Milan, 1940; W. Abbott, The Theory and Practice of Perspective, London, 1950; C. L. Martin, Architectural Graphics, New York, 1952; R. Arnheim, Art and Visual Perception, Berkeley, Calif., 1954; E. H. Gombrich, Art and Illusion, New York, 1960; G. Seller, Geometria descrittiva, Milan, 1960; G. Chiesa, Prospettiva, 8th ed., Milan, 1961; S. B. Parker, Linear Perspective without Vanishing Points, Cambridge, Mass., 1961.

Historical works: A. Comolli, Bibliografia storico-critica dell'architettura civile ed arti subalterne, III, Rome, 1788–92, pp. 142–224; L. Cicognara, Catalogo ragionato dei libri d'arte e d'antichità posseduti dal conte Cicognara, Pisa, 1821 (anastatic repr., Cosenza, 1960); N.-G. Poudra, Histoire de la perspective ancienne et moderne, Paris, 1864; M. Chasles, Aperçu historique sur l'origine et le développement des méthodes en géometrie, 2d ed., Paris, 1875; P. Riccardi, Di alcune opere di prospettiva di autori italiani omesse nella "Histoire de la perspective" di M. Poudra, Bib. mathem., N.S., III, 1889, pp. 39–42; F. J. Obenrauch, Geschichte der darstellenden und proiectiven Geometrie, Brünn, 1897; W. De Grüneisen, La perspective: Esquisse de son évolution dès origines jusqu'à la Renaissance, Mél, XXXI, 1911, pp. 393–434; G. Wolff, Neue Perspektive für die Geschichte der Perspektive, Z. math. und nat. Untersuchungen, XLVI, 1914, pp. 263–69; U. Cassina, La prospettiva e lo sviluppo dell'idea dei punti all'infinito, Periodico di matematiche, 4th ser., I, 1921, pp. 326–37; G.Loria, Storia della geometria descrittiva dalle origini sino ai nostri giorni, Milan, 1921; E. Panofsky, Die Perspektive als symbolische Form, Vorträge der Bib. Warburg, 1924–25, pp. 258–330; M. Borissavliévitch, Les théories de l'architecture, Paris, 1926 (2d ed., 1951); G. M. A. Richter, Perspective Ancient, Mediaeval and Renaissance, Scritti in onore di B. Nogara, Rome, 1937, pp. 381–88; G. Nicco-Fasola, Svolgimento del pensiero prospettico nei trattati da Euclide a Piero della Francesca, Le Arti, II, 1942–43, pp. 59–71; T. Viola, Sulle origini della prospettiva, Il Filomate, I, 4, 1948; P. Francastel, Peinture et société, Lyon, 1951; S. Sulzberger, La perspective picturale, Rev. Univ. de Bruxelles, N.S., VIII, 1955–56, pp. 239–43; D. Gioseffi, Complementi di prospettiva, CrArte, N.S., IV, 1957, pp. 468–88, V, 1958, pp. 102–39; D. Gioseffi, Perspectiva artificialis, Trieste, 1957; J. White, The Birth and Rebirth of Pictorial Space, London, 1957; P. Sanpaolesi, Studi di prospettiva, Raccolta Vinciana, XVIII, 1960, pp. 188–202.

Problems of perspective and their individual aspects: J. Kepler, Appendix Hyperaspistis 19 (Opera omnia, ed. C. Frisch, VII, Frankfurt am Main, 1868, p. 279); G. Hauck, Die subiektive Perspektive und die horizontalen Curvaturen des dorischen Styls, Stuttgart, 1879; K. Döhlemann, Raumkunst und Illusionsmalerei, Beilage zur Allg. Zeitung, 168, 1904, part 3, pp. 161–64; K. Döhlemann, Die Verwertung der Linienperspektive zur Datierung von Bildern, Munich, 1905; E. Sauerbeck, Ästhetische Perspektive, Z. für Ästhetik und allg. Kw., VI, 1911, pp. 420–55; R. Müller, Über die Anfänge und über das Wesen der malerischen Perspektive, Darmstadt, 1913; G. Wolff, Mathematik und Perspektive, Leipzig, 1916 (2d ed., 1925); M. Borissavliévitch, Découverte de la perspective optico-physiologique, B. Ecole spéciale d'arch., 3, 1923; G. Schöne, Die Entwicklung der Perspektivbühne von Serlio bis Galli-Bibiena nach der Perspektivbüchen, Leipzig, 1933; W.M. Ivins, Jr., Art and Geometry, Cambridge, Mass., 1946; A. Fornari, Quarant'anni di cubismo, Rome, 1948; P. Francastel, Espace génétique et éspace plastique, Rev. d'esthétique, IV, 1948, pp. 349–80; M. Zanetti, Una proposta di riforma della prospettiva lineare, L'Ingegnere, XXV, 9, 1951, pp.1–9; E. Bock, Binokularperspektive als Grundlage einer neuen Bilderscheinung, Alte und neue K., II, 1953, pp. 101–14; B. Schweitzer, Vom Sinn der Perspektive, Tübingen, 1953; E. Bock, Perspektive Bilder als Wiedergaben binokularer Gesichtvorstellung, Alte und neue K., III, 1954, pp. 81–110; J. Baltrušaitis, Anamorphoses ou perspectives curieuses, Paris, 1955; D. Gioseffi, La falsa preistoria di Piet Mondrian e le origini del Neoplasticismo, Trieste, 1957; D. Gioseffi, La cupola vaticana, Trieste, 1960; C. F. Graumann, Grundlagen einer Phänomenologie und Psychologie der Perspektivität, Berlin, 1960; E. Panofsky, Renaissance and Renascences in Western Art, Stockholm, 1960; L. Brion-Querry, Jean Pélerin (Viator): Sa place dans l'histoire de la perspective, Paris, 1962; T. K. Kitao, Prejudice in Perspective: A Study of Vignola's Perspective Treatise, AB, XLIV, 1962, pp. 173–94; C. L. Ragghianti, Mondrian e l'arte del XX secolo, Milan, 1962.

Antiquity: Plutarch, Moralia 346A, 393C, 395D, 436B, 725C; Pliny the Elder, Naturalis historia, XXXV (see also S. Ferri, ed., Plinio il Vecchio: Storia delle arti antiche, Rome, 1946); Vitruvius, De architectura (ed. F. Krohn, Leipzig, 1912; see also S. Ferri, ed., Vitruvio: Architettura, dai libri I–VII, Rome, 1960); R. Delbrück, Beiträge zur Kenntnis, der Linienperspektive in der griechischen Kunst, Bonn, 1889; H. Schäfer, Von ägyptischer Kunst, 2d ed., Leipzig, 1922; E. Pfuhl, Malerei und Zeichnung der Griechen, 3 vols., Munich, 1923; A. M. G. Little, Scaenographia, AB, XVIII, 1936, pp. 407–18; H. G. Beyen, Die pompejanische Wanddekoration, I, The Hague, 1938; G. J. Kern, Jahreszeitenmosaik der Münchener Bibliothek und die Skenographie bei Vitruv, AAnz, 1938, cols. 245–64; H. G. Beyen, Die antike Zentralperspektive, AAnz, 1939, cols. 47–72; G. B. Levy, The Greek Discovery of Perspective: Its Influence on Renaissance and Modern Art, J. Royal Inst. of Br. Architects, L, 1942–43, pp. 51–57; F. Enriques and M. Mazziotti, Le dottrine di Democrito di Abdera, Bologna, 1948; H. A. Groenewegen-Frankfort, Arrest and Movement, Chicago, London, 1951; R. Bianchi Bandinelli, Il problema della pittura antica, Florence, 1953; P. W. Lehmann, Roman Wall Paintings from Boscoreale in the Metroploitan Museum of Art, Cambridge, Mass., 1953; H. Kenner, Das Theater und der Realismus in der griechischen Kunst, Vienna, 1955; G. M. A. Richter, Ancient Italy, Ann Arbor, Mich., 1955; R. Bianchi Bandinelli, Organicità e astrazione, Milan, 1956; R. Bianchi Bandinelli, Osservazioni storico-artistiche a un passo del "Sofista" Platonico, S. in onore di Ugo Enrico Paoli, Florence, 1956, pp. 81–95; H. A. Stützer, Die Kunst der Etrusker und der römischen Republik, Munich, 1956; J. White, Perspective in Ancient Drawing and Painting, London, 1956; P. H. von Blanckenhagen, Narration in Hellenistic and Roman Art, AJA, LXI, 1957, pp. 78–83; G. Carettoni, Due nuovi ambienti dipinti sul Palatino, BArte, XLVI, 1961, pp. 189–99.

Middle Ages: O. Wulff, Die umgekehrte Perspektive und die Niedersicht, Kunstwissenschaftliche Beiträge August Schmarsow gewidmet, Leipzig, 1907, pp. 1–40; G. J. Kern, Die Anfänge der centralperspektivischen Construktion in der italienischen Malerei des XIV. Jahrhunderts, Mitt. des kunsthist. Inst. in Florenz, II, 1912–17, pp. 39–65; G. J. Kern, Perspektive und Bildarchitectur, RepfKw, XXXVI, 1912, pp. 27–64; H. Berstl, Das Raumproblem in der altchristlichen Malerei, Bonn, 1920; E. Tea, Witelo: Prospettico del secolo XIII, L'Arte, XXX, 1927, pp. 3–30; C. Cennini, Il libro d'arte (ed. and trans. D. V. Thompson, Jr.), 2 vols., New Haven, London, 1932–33; E. Panofsky, The Friedsam Annunciation and the Problem of the Ghent Altarpiece, AB, XVII, 1935, pp. 433–73; G. J. Kern, Die Entdeckung des Fluchtpunktes, Berlin, 1937; G. J. Kern, Die Entwicklung der centralperspektivischen Construktion in der europäischen Malerei, Forsch. und Fortschritte, XIII, 1937, pp. 181–84; E. Panofsky, Once More the Friedsam Annunciation, AB, XX, 1938, pp. 419–42; M. S. Bunim, Space in Mediaeval Painting and the Forerunners of Perspective, New York, 1940; A. Grabar, Plotin et les origines de l'esthétique médiévale, CahA, I, 1945, pp. 15–34; R. Longhi, Giotto spazioso, Paragone, III, 31, 1952, pp. 18–24; E. Panofsky, Early Netherlandish Painting, 2 vols., Cambridge, Mass., 1953; L. Stefanini, La prospettiva tolemaica, Riv. di estetica, I, 1956, pp. 97–106.

Renaissance: *a. General*: W. M. Ivins Jr., On the Rationalization of Sight (Met. Mus. Pap., VIII), New York, 1938; J. White, Developments in Renaissance Perspective, Warburg, XII, 1949, pp. 58–79, XIV, 1951, pp. 42–69; M. Walcher Casotti, Jacopo Barozzi da Vignola nella storia della prospettiva, Periodico di matematiche, 4th ser., XXXI, 1953, pp. 73–103; W. J. Ong, System, Space and Intellect in Renaissance Symbolism, Bib. d'humanisme et Renaissance, XVIII, 1956, pp. 222–39; A. Chastel, Art et civilisation à Florence aux temps de Laurence le Magnifique, Paris, 1959; M. Walcher Casotti, Il Vignola, 2 vols., Trieste, 1960; D. Gioseffi, La prospettiva: Calcolo e scienza, Terzo Programma (Radiotelevisione it.), Q. trimestrali, 3, 1961, pp. 220–29; J. Schulz, A Forgotten Chapter in the Early History of "Quadratura" Painting: The Fratelli Rosa, BM, CIII, 1961, pp. 90–102. *b. Brunelleschi and Masaccio*: A. Averlino (Filarete), Trattato di architettura (ed. W. von Oettingen), Vienna, 1896; G. J. Kern, Das Dreifaltigkeitsfresko von S. Maria Novella, JhbPreussKSamml, XXXIV, 1913, pp. 36–58; J. Mesnil, Masaccio et la théorie de la perspective, Rev. de l'art ancien et moderne, XXXV, 1914, pp. 145–56; A. Manetti, Vita di Filippo di Ser Brunellesco (ed. E. Toesca), Florence, 1927; J. Mesnil, Masaccio et les débuts de la Renaissance, The Hague, 1927; R. Longhi, Fatti di Masolino e di Masaccio, CrArte, IV–V, 2, 1939–40, pp. 145–91; G. C. Argan, The Architecture of Brunelleschi and the Origins of Perspective Theory in the 15th Century, Warburg, IX, 1946, pp. 96–121; P. Sampaolesi, Ipotesi sulle conoscenze matematiche statiche e meccaniche del Brunelleschi, Belle arti, 1951, pp. 25–54; R. Wittkower, Brunelleschi and Proportion in Perspective, Warburg, XVI, 1953, pp. 275–91; G. C. Argan, ed., Brunelleschi, Milan, 1955; R. Krautheimer, Lorenzo Ghiberti, Princeton, 1956; C. G. Lemoine, Brunelleschi et Ptolémée: Les origines géographiques de la "boîte d'optique," GBA, LI, 1958, pp. 281–96; A. Parronchi, Le due tavole prospettiche del Brunelleschi, Paragone, IX, 107, 1958, pp. 3–32, X, 109, 1959, pp. 3–31; L. Berti, Masaccio a S. Giovenale di Cascia, Acropoli, II, 1962, 2, pp. 149–65. *c. Alberti*: A. Bonucci, ed., Opere volgari di Leon Battista Alberti, 5 vols., Florence, 1843–49; H. C. O. Staigmüller, Kannte Leon Battista Alberti den Distanzpunkt?, RepfKw, XIV, 1891, pp. 301–04; E. Panofsky, Das perspektivische Verfahren K. B. Alberti, Kchr, N.S., XXVI, 1915, cols. 505–16; H. Wieleitner, Zur Erfindung der verschiedenen Distanz-Konstruktionen der malerischen Perspektive, RepfKw, XLII, 1920, pp. 249–62; L. B. Alberti, Della pittura (ed. L. Mallé), Florence, 1950; A. Parronchi, Il ' punctum dolens ' della "costruzione legittima," Paragone, XIII, 145, 1962, pp. 58–72. *d. Ghiberti, Paolo Uccello*: L. Ghiberti, I Commentarii (ed. J. von Schlosser), 2 vols., Berlin, 1912; G. J. Kern, Der Mazzocchio des Paolo Uccello, JhbPreussKSamml, XXXVI, 1915, pp. 13–38; A. Parronchi, Le fonti di Paolo Uccello, Paragone, VIII, 89, 1957, pp. 3–32, 95, 1957, pp. 3–33; A. Parronchi, Le "misure dell'occhio" secondo il Ghiberti, Paragone, XII, 133, 1961, pp. 18–48. *e. Piero della Francesca, Luca Pacioli*: L. Pacioli, Summa de Arithmetica, Geometria, Proporzioni e Proporzionalità, Venice, 1494; G. Libri, Histoire des sciences mathématiques en Italie, IV, Paris, 1841, note III; Piero della Francesca, De prospectiva pingendi (ed. C. Winterberg), Strasbourg, 1899; G. Pittarelli, Intorno al libro "De prospectiva pingendi" di Piero dei Franceschi, Atti Cong. int. di sc. storiche, XII, Rome, 1904, pp. 251–66; Piero della Francesca, De prospectiva pingendi (ed. G. Nicco-Fasola), Florence, 1942; R. Wittkower, The perspective of Piero della Francesca's Flagellation, Warburg, XVI, 1953, pp. 292–95. *f. Mantegna, Gaurico*: P. Gaurico, De sculptura, Florence, 1504 (new ed., H. Brockhaus, Leipzig, 1886); P. Kristeller, Andrea Mantegna, London, 1901; G. Fiocco, Andrea Mantegna, Milan, 1937; R. Klein, Pomponius Gauricus on Perspective, AB, XLIII, 1961, pp. 211–30; G. Castelfranco, Note su Andrea Mantegna, BArte, XLVII, 1962, pp. 23–39. *g. Leonardo*: J. P. and I. Richter, The Literary Work of Leonardo da Vinci, 2d ed., 2 vols., New York, London, 1939; E. Panofsky, The Codex Huygens and Leonardo da Vinci's Art Theory, London, 1940; M. H. Pirenne, The Scientific Basis of Leonardo da Vinci's Theory of Perspective, Br. J. for the Philosophy of Sc., III, 1952, pp. 169–85; A. M. Brizio, Il trattato della pittura di Leonardo, Scritti di storia d'arte in onore di Lionello Venturi, I, Rome, 1956, pp. 309–20; Leonardo da Vinci, Treatise on Painting, (ed. and trans. A. P. McMahon), Princeton, 1956; G. Castelfranco, Leonardo, Terzo Programma (Radiotelevisione italiana), Q. trimestrali, 3, 1961, pp. 229–37. *h. Dürer*: A. Dürer, Institutionum geometricarum libri quattuor cum figuris, Paris, 1532; E. Panofsky, Dürers Kunsttheorie, vornehmlich in ihrem Verhältnis zur Kunsttheorie der Italiener, Berlin, 1915; E. Panofsky, Albrecht Dürer, 4th ed., Princeton, 1955.

Treatises 1500–1800: J. Pélerin (Viator), De artificiali perspectiva, Toul, 1505 (2d ed., 1509; 3d ed., 1521); H. Rodler, Perspectiva, Simmern, 1531; S. Serlio, Il libro secondo di prospettiva, Paris, 1545; V. Theriaca, Discorso e ragionamento di ombre, Rome, 1551; F. Commandino, Ptolemaci Planisphaerium, Jordani Plan, F. Commandini in Pl. Commentarius . . . , Venice, 1558; J. Cousin, Livre de perspective, Paris, 1560; C. Bartoli, Del modo di misurare le distanze, le superfici, i corpi, le provincie, le prospettive . . . , Venice, 1564; D. Barbaro, La pratica della prospettiva, Venice, 1569; J. Cousin, Livre de pourtraiture, Paris, 1571; M. Bassi, Dispareri in materia d'architettura e prospettiva con pareri di eccellenti e famosi architetti, che li risolvono, Brescia, 1572; E. Danti, La prospettiva di Euclide . . . aggiuntovi la prospettiva di Eliodoro Larisseo Greco, e Latino, Florence, 1573; J. Androuet Du Cerceau, Leçons de perspective positive, Paris, 1576; J. Barozzi da Vignola, Le due regole della prospettiva pratica, con commentari del P. Egnazio Danti, Rome, 1583; G. P. Lomazzo, Trattato dell'arte della pittura, V, Milan, 1584; L. Sirigatti, La pratica di prospettiva, Venice, 1596; Guidubaldo del Monte, Perspectivae libri sex, Pesaro, 1600; S. de Caus, La perspective avec la raison des ombres et miroirs, London, 1612; F. d'Aguilon, Opticorum libri VI, Antwerp, 1613; P. Accolti, Lo inganno degli occhi, prospettiva pratica, trattato in acconcio della pittura, Florence, 1625; H. Hondius, Instruction en la science des perspectives, The Hague, 1625; G. Viola-Zanini, Dell'architettura, Padua, 1629; S. Marolois, Opticae, sive Perspectivae partes quattuor, Amsterdam, 1633; S. Stevin, Traité d'Optique, Oeuvres mathématiques, Leiden, 1634; S. Marolois, La perspective conténante tant la théorie que la pratique rémise en volume plus commode que auparavant, 2d ed., Amsterdam, 1638; J. F. Niceron, La perspective curieuse, Paris, 1638; J. Dubreuil, La perspective pratique, 3 vols., Paris, 1642–49; A. Kircher, Ars magna lucis et umbrae, Rome, 1646; J. F. Niceron, Thaumaturgus opticus, Paris, 1646; A. Bosse, Manière universelle de M. Desargues pour pratiquer la prospective par pétit-pied, comme le géometral, Paris, 1648; A. Bosse, Moyen universel de pratiquer la perspective sur les tableaux, ou surfaces irregulières, Paris, 1653; A. Bosse, Traité des pratiques géometrales et perspectives enséignées dans l'Académie Royale de la Peinture et Sculpture, Paris, 1665; G. Troili (Paradosso), Paradossi per praticare la prospettiva senza saperla, Bologna, 1683; A. Pozzo, Perspectiva pictorum, et architectorum, 2 vols., Rome, 1693–1700; F. Galli Bibiena, L'architettura civile preparata sulla geometria e ridotta alle prospettive, Parma, 1711; W. J. 's Gravesande, Essai de perspective, The Hague, 1711; B. Taylor, Linear Perspective, London, 1715; F. Galli Bibiena, Direzioni ai giovani studenti del disegno dell'architettura civile e della prospettiva teorica, Bologna, 1725; G. Galli Bibiena, Architetture e prospettive dedicate a S. Maestà Carlo VI Imperatore de' Romani, Augsburg, 1740; G. F. Costa, Elementi di prospettiva per uso degli architetti e pittori, Venice, 1747; F. Jacquier, Elementi di prospettiva secondo i principi di Brook Taylor: con aggiunte spettanti all'ottica e alla geometria, Rome, 1755; J. H. Lambert, Freye Perspective, Zurich, 1759; E. Zannotti, Trattato teorico-pratico di Prospettiva, Bologna, 1766; De Saint-Morien, La perspective aérienne soumise à des principes puisés dans la nature, ou nouveau traité de clair-obscur et de chromatique à l'usage des artistes, Paris, 1788. See also the bibliog. for OPTICAL CONCEPTS.

Decio GIOSEFFI

Illustrations: PLS. 79–96; 3 figs. in text.

PERSPECTIVISTS. Those painters whose work relied mainly on the use of either "artificial" (i.e., linear) or aerial perspective to create illusionistic effects can be grouped under the name "perspectivists." Instances of this kind of painting appeared as early as the 15th century, but the technique was coordinated into a genuine system by the so-called "quadratura painters," or painters of illusionistic architecture, who came into prominence in the 17th century, within a distinct trend that became widespread among the various tendencies in baroque art (q.v.). Illusionistic perspective (*quadratura*) painting was also the basis of scenographic developments (see SCENOGRAPHY). This decorative style flourished in Italy until about the 1770s and spread to other countries, especially to Austria, where it acquired particular characteristics. (The following discussion does not include the *vedute* painters, who painted real or imagined landscape or architectural views, since their work, although it very often depicts architectural structures in perspective, does not emphasize illusionistic effects.) See also OPTICAL CONCEPTS; PERSPECTIVE.

INTRODUCTION. *General criteria of illusionistic perspective.* Illusionism in painting first appeared in the early Renaissance, contemporaneously with the initial theoretical and practical studies of "artificial perspective." At first there were few, often isolated examples (see below), but with the rise of mannerism it became a dominant element of painting in general, and of interior decoration in particular. Among the various and complex reasons for this change in taste were the increasing interest in scenography resulting from the popularity of theatrical performances, the attention stirred by the rediscovery of Vitruvius's *De architectura*, especially concerning the principles of perspective expounded in the chapter on scenography; and the application of the Aristotelian unities of time, place, and action in the theater of the time. All these related elements found visual expression in illusionistic painting. Reversing the objective character of Renaissance perspective, which imposed its laws upon the viewer, illusionistic painting made perspective a subjective quality, dependent on the viewer. It did away with the intersection of the visual cone and substituted for it a unitary spatial continuum that joined observer and painting into a unicum. Not all the theorists agreed on how to obtain this effect, and different principles were developed (cf. M. Bassi, 1572). Nonetheless, the basic measure of this perspective is man, and the objects, especially architectural structures, represented in illusionistic paintings must develop around him in proportion; otherwise the spatial continuity between the viewer and the objects represented becomes absurd and unreal. In wall decoration, perspective conceived in this manner took the form of so-called "sfondati," or perspective views, which create the illusion of space beyond the wall; on vaults there developed the so-called "sottinsù" paintings, which were designed to be seen from below and which concealed the actual shape of the room, simulating nonexistent structures and creating the illusion of reality.

Painted architecture played an important part in this kind of perspective, and initially architects planned and directed the work, even if the actual execution was by a painter. The architect, knowing the five orders to perfection, was able to draw them in perspective; moreover, he was acquainted with the effects of light and shadow, and of the shadows of objects in particular, and if necessary he could simulate architectural structure to correct the proportions of a room.

Illusionistic decoration soon became very fashionable and was also applied to façades. Daniele Barbaro objected to this practice and, with Serlio, maintained that the most suitable surfaces for paintings of architectural structures in perspective were interiors.

Whereas the decoration of a wall might seem a relatively simple matter, vault decoration was far more difficult. In order that the lines appear convincing to the eye when seen from a distance, the artist painting a vault had to draw lines that seemed exaggerated on closer inspection. The difficulty of distributing light and shadow on vaults was such that Egnazio Danti (Vignola's *Le due regole . . .*, 1883, p. 90) suggested following the method proposed by Ottaviano Mascherino, that is, building a model of the architectural structures that the artist proposed to paint on a vault, in order to study natural lighting effects on it.

According to Serlio (1545, IV, p. 191r), artists should choose appropriate subjects for vault paintings (viz., celestial beings and flying creatures rather than earth-bound things). Better still, they should do what Raphael did in the Loggia of the Farnesina in Rome (VIII, PL. 101), where to avoid the "harshness" of *sottinsù* techniques, and despite the fact that he was a master of foreshortening, he painted a blue drapery on the vault and on that painted the banquet of the gods as a *quadro riportato* (i.e., an easel picture usually applied to, sometimes painted directly on, the ceiling). In fact, Serlio affirms that foreshortened figures on vaults can seem very short and monstrous even though, when seen from the proper distance, they appear to be in proportion. Thus the perspectivist was faced with the difficult problem of optical illusion, in regard to which Danti insists that, in decorating vaults, the artist "must keep his eye at the station point and shorten things until they appear to be right to the eye." (op. cit., p. 89). Theory and practice had to be made to agree in order that illusionism might achieve effects as close to reality as possible — that is, as convincing as possible.

This new manner of painting in perspective resulted in a division of labor among painters: The accomplished perspectivist (often an architect) was responsible for planning the architectural perspective elements of the decoration; the figure painter depicted the often life-size figures that animated the architectural background; the ornamentist simulated stuccoes and various other decorative elements; and the landscapist painted views of broad landscapes in simulated apertures. The figure painter often rivaled the perspectivist in his effects, especially in vault painting, since his ability to foreshorten figures allowed him to create an illusion of space by means of them alone. The ability to proportion foreshortenings, by a logical and visual process, so that they would correspond to reality was a basic requisite of the figure painter. In addition, the figure painter had to know aerial perspective, which had been introduced on a scientific basis by Leonardo da Vinci, theorizing on the law of optics according to which colors grow gradually paler and diminish in tone as they recede. Unlike architectural perspective, aerial perspective does not require a knowledge of mathematical laws but a command of tonal gradations and lighting effects.

Architectural perspective (quadratura) *painters.* In the 17th century, as emphasis changed from theoretical to practical criteria, the perspectivist-architect who had prevailed as over-all planner of decorative schemes in the previous century ceded his role to the *quadratura* painter, an artist specialized in *quadratura*, or architectural perspective. The origin of the conventional term *quadratura* may be found in the words *quadro* (square) and *quadrettatura* (division into squares), for, as Serlio said and all the perspectivists maintained, the basis of perspective painting is the perfect foreshortening of a square, from which all other lines will originate (op. cit., II, p. 19).

The *quadratura* painter was essentially a skilled practitioner of a specialized craft, not so learned as the architect in geometric theorems and principles, but master of those quick-to-apply empirical tricks which permitted facile execution of illusionistic decoration of any kind — interior decorations, stage scenery, the large painted canvases that were set up at the city gate or the entrance to a palace courtyard on the occasion of a visit by some great personage, and even the so-called "mortori," or funerary decorations. The ability and inventiveness of these artists was often remarkable, and their methods were sometimes ingenious. Cristoforo Sorte recounts that Giulio Romano made use of a mirror marked off in squares to paint foreshortened figures (*Osservazioni sulla pittura*, Venice, 1580, p. 16). It is known that Lomazzo taught the use of a *telaro*, a full-scale transferring device of cloth (1844 ed., VI, pp. 135, 143); that Danti suggested the use of threads for vault painting (op. cit., pp. 87, 91); and that Andrea Pozzo projected the shadow of a rope net on the wall by means of a bright lamp, placed behind the center of the net in order to produce squares in perspective on the surface to be painted, without resorting to mathematical calculations (A. Pozzo, 1692).

The aim of the *quadratura* painter was to create a general effect of unity in the work, including both wall and ceiling decoration, as seen from some distance and usually from the center of the room. In the *quadratura* technique details were ignored; forms, especially those above the horizon plane and on the vaults, were made disproportionate and often gigantic; and the foreground colors became more and more vivid, as all the elements contributing to the over-all effect were proportioned in order to simulate reality from the point of view of the observer. The *quadratura* painter worked with large brushes attached to long handles, as the scenographer does even now; he painted in sweeping strokes, with strong contrasts of light and shade. In the 17th century it was he, and no longer the architect, who usually directed the work; his assistants always included a figure painter and, often, an ornament painter. Thus there emerged celebrated pairs of artists (see below) who decorated Italian villas and palaces, especially in Emilia, Tuscany, Lombardy, and Piedmont, and who drew inspiration from the contemporary architecture, elements of which they transferred to interiors. *Quadri riportati*, wooden coffers, and grotesques disappeared from the ceilings with the advent of this kind of unified decorative scheme, in which *sfondati* and *sottinsù* offered ideal techniques for historico-celebrative representations.

Illusionistic painting in religious buildings. With its power to create the illusion of infinite space, this mode of painting offered the Church a forceful means to bring before the eyes of the faithful the apotheosis of her triumph after the great upheaval of the Reformation.

The type of illusionistic perspective painting that prevailed in religious buildings differed from nonreligious decoration in that, of necessity, it invariably included figures — divine personages, angels, and saints. The great perspectivist decorators of the churches of the 17th and 18th centuries (obviously, Rome was the center of the taste) were figure painters, who made use of *quadratura* mainly to relate their images to the structure of the building, creating celestial visions and then, by means of architectural perspective, relating them to the atmospheric space of the church, almost as if God and the saints looked down miraculously from the illusionistic sky opening up above the faithful. To create this effect, the figure painters further elaborated the aerial perspective of Correggio and studied, each according to his taste and each using a preferred color range, the relationship between light and shadow and the diminishing of tonal values, as well as the increase in light absorption, as colors recede from the eye. With the organization according to measurable architectural space replaced by an illusionistic atmospheric space, the finite became infinite, the rational irrational, the transcendent returned to the world of man, and eternity was drawn into time. It is understandable, therefore, that such illusionistic painting became popular not only in Italy but also in such other highly religious areas as Austria and Bavaria, where it flourished in the 18th century and developed characteristics very different from those of Italian illusionism.

History of perspectivists. *Italy. a. Forerunners of illusionism.* The earliest existing evidence of a considered attempt to achieve illusionistic perspective effects (two panels by Brunelleschi are known only from written sources) is presented by Masaccio's fresco of the Trinity in the Church of S. Maria Novella, Florence (IX, pl. 345). This fresco, executed with vanishing point and horizon plane lowered to the viewer's level, opens into an architectural *sfondato* simulating a small chapel with a coffered barrel vault. Masaccio also created the first oblique view, but in it he rendered space objectively. In 1436, Paolo Uccello did the opposite with the first painted perspective "statue," in the Cathedral of Florence (XII, pl. 27). The illusionistic perspective makes the figures seem to project from the wall, as if placed on a supporting shelf like a true statue. Vasari relates that this artist also painted, in the Church of S. Maria Maggiore in Florence, an *Annunciation* (lost) in which some columns concealed the arch of the vault, and that in another fresco (lost) he drew the round columns in a corner of the wall in perspective to disguise the corner. Andrea del Castagno painted the equestrian monument to Niccolò da Tolentino in the Cathedral of Florence (1455; I, pl. 246) in imitation of the Uccello work. In decorating the Villa Pandolfini in Florence (frescoes in Cenacolo di S. Apollonia, Florence), Andrea del Castagno painted a series of large, open doors through which the figures of famous men and women appear to be stepping into the room itself (I, pl. 245). In the portrait by Piero della Francesca (q.v.) of Sigismondo Malatesta (1451), in the Tempio Malatestiano in Rimini, the pictorial space is developed in three planes within a landscape background that is seen between two Corinthian columns decked with a flower garland.

It was not until the last half of the 15th century, however, that illusionistic perspective became established as a distinct method. Mantegna's frescoes in the Ovetari Chapel in the Church of the Eremitani in Padua and his altarpiece for the Church of S. Zeno in Verona (1569; IX, pls. 322, 323), with their strictly consistent perspective, are clear antecedents of the Camera degli Sposi, which he later executed in the Palazzo Ducale of Mantua (1574; IX, PLS. 328, 329). The painted doors in the Camera degli Sposi seem to break the bounds of the walls and make the room into an open pavilion covered with a vault — a ceiling that Mantegna's artistry cleverly transformed from flat to domed, opened in the center by a circular oculus (IX, PL. 329). Thus Mantegna intentionally took upon himself the task of the architect, imposing upon the room the shape he wanted. He did so discreetly and unobtrusively, like the "prudent and wise" painter he was (Serlio, op. cit., V, p. 190r); nevertheless, he employed all the laws of perspective, setting the station point in the middle of the room. In order to achieve his architectural effects, he also calculated the light perfectly and, the better to simulate reality, closed some of the archways with painted curtains of cordovan leather.

The development of illusionistic painting on the basis of firm theoretical principles, rather than through the preferences of individual artists, was most successfully fostered in Urbino, where in the last quarter of the 15th century the studies of Piero della Francesca were bearing fruit. Furthermore, scenography must have been an important art in the Montefeltro court, where there existed one of the first specially built theaters. In the *studiolo* of Federigo da Montefeltro (1476), there is a series of inlaid wooden panels in perspective that was intended by the artist who executed them (or provided the designs) to represent "the room as if the prince has just stepped out after taking his favorite books from the shelves, removing his suit of arms and scattering the pieces here and there, after preparing the musical instruments, opening the scores . . . and making his presence felt in the very disorder he had left behind him" (Rotondi, *Il Palazzo Ducale di Urbino*, I, pp. 337–38; VIII, PLS. 80, 100). The artist's strictly mathematical application of the rules of perspective; the preference for geometric shapes, which are an obvious and accurate means of rendering a flat plan in perspective; and the emphasis on contrasts of light and shadow obtained by use of wood of different colors indicate authorship by an architect rather than a painter, at least in the design and general supervision of the work. This architect has been identified by some as Francesco di Giorgio, who was active in Urbino and wrote a treatise on architecture.

From the same milieu came the great perspective works of Melozzo da Forlì (q.v.; 1438–94). In the fresco representing Platina, in the Vatican Library, Melozzo created a superbly constructed *sottinsù* (VII, PL. 274). Later, in the domes of the Sanctuary of the Holy House, Loreto, and the Feo Chapel in the Church of S. Biagio, Forlì (for both of which, perhaps, he provided only the designs), he covered the existing structures with simulated ribs and coffers and with admirable upward foreshortened figures (prophets and angels in Loreto, 1484?; prophets and cherubs in Forlì, probably executed by Marco Palmezzano, PL. 99). The painter Donato Bramante (1444–1514) developed in the same milieu. The few paintings surviving from the time of his stay in Lombardy (1477–79) indicate a relationship between his work and work for the Montefeltro court. The philosophers represented in the few remaining fragments from the façade of the Palazzo dei Rettori in Bergamo (transferred to the Palazzo della Ragione) are clearly reminiscent of the portraits of famous men in the *studiolo* of the Palazzo Ducale in Urbino. The complex structure of the quadrangular niches, recalling the strictly geometric plans of the wooden panels in the Duke's *studiolo*, and the fact that Bramante painted his philosophers above the horizon plane because they were to be placed high on the façade testify to his interest in matters of perspective. The depictions of soldiers of the Panigarola house (fragments in the Brera) that decorated the interior were set within simulated semicircular niches meant to lend depth to the walls and were painted with masterful command of light and shade.

That these principles of perspective aroused the interest of artists and commissioners, started a new taste, and caused it to become established is clearly demonstrated not by monumental work but rather by a series of miniatures that illustrated manuscripts. Most noteworthy is the folio with an Annunciation scene and saints in the Gradual D, begun in 1486, preserved in the Cathedral of Cesena. The figures are placed in a huge, open, three-story building that reveals elements typical of Francesco di Giorgio and is designed with the horizon plane lowered to the level of the man (perhaps the artist himself) who stands in the middle of the vast supporting arch forming the base. Under the arch at the left, a compass and T square testify to the means used in construction of this bold architectural plan.

b. Perspectivists in 16th-century Rome. The important achievements of illusionistic perspective in the first decades of the 16th century were realized in Rome, where the best artists had been summoned to the papal court. In 1508, Michelangelo started work on the vault of the Sistine Chapel. Confronted with an immense surface to decorate, he circumscribed it and gave it a compact unity, sectioning the vault transversely with a series of equidistant, seemingly projecting pillars that form the trellis which supports the gigantic nudes, prophets, and sibyls that seem to burst into the space of the chapel (I, PL. 381; II, PL. 291; III, PL. 312). He placed between the pillars a series of *quadri riportati*, with scenes from Genesis, that recall the eye from a vision of infinity and make it move and survey the whole structure.

Raphael, instead, was attracted to the principles of the Urbino artists, especially to those of Bramante. In the socle in the Stanza della Segnatura of the Vatican the artist had painted a series of illusionistic caryatids (1508). Later, probably unaware of the influence that his perspective motif was to have, he planned two open *sottinsù* perspectives with a square colonnade on two of the small vaults in the Vatican Logge (1516; probably executed by an assistant). The colonnade opens onto the sky, but upward movement is restrained by two vault crossings adorned with five *quadri riportati* (PL. 431; II, PL. 291). Soon afterward, the artist applied illusionistic perspective to cover a whole room, the Loggia of the Farnesina in Rome. Raphael, always a balanced painter, avoided excessive deformations and forced station points in this work.

Baldassarre Peruzzi (q.v.) was bolder in his approach to problems of perspective. In the Farnesina in Rome, in 1519, he intentionally enlarged the space of a room by painting on three of its walls a double-colonnaded loggia with an exquisite landscape background (PL. 118; the basic motif is to be found in a small panel attributed to Francesco di Giorgio, Urbino, Gall. Naz. delle Marche). Peruzzi's "scenographic" qualities derived from his experience as a painter of stage scenery; this can be inferred from certain of his drawings (PL. 99). The fact that in the Farnesina he set the station point between the two doors rather than in the middle of the room (criticized by Danti, op. cit., p. 87) was, perhaps, a consequence of his scenographic experience.

c. Illusionistic aerial perspective in Emilia. In the wake of these achievements of Roman perspectivists, Correggio (q.v.) introduced a new kind of illusionistic perspective, one that was based solely on foreshortened figures and aerial perspective. He first employed it in Parma for the dome of S. Giovanni Evangelista (1623; III, PL. 470) and then for the dome of the Cathedral (1526; III, PL. 471). The originality of these frescoes consists in the fact that Correggio negated the measurable volume of the underlying architectural structure and created the illusion of infinity by depicting *sottinsù* figures and by increasing the areas of light in the upper regions (so that the most distant figures are ethereal and all but absorbed by the light). He may have learned this distribution of light from the works of Leonardo. The same technique was used by Gaudenzio Ferrari on the dome of the Sanctuary of the Madonna dei Miracoli in Saronno, though with denser groups of figures and with some personalized touches indicating that he had been in contact with the artistic ambient of Milan.

In the meantime, the motif of Mantegna's oculus was developed by Benvenuto Tisi, known as "Il Garofalo" (1481?–1559). Scholars (Toesca, *Affreschi decorativi*, Milan, 1917, pl. 130) have attributed to him the decoration of a quadrangular hall in the palace of Ludovico Il Moro, Ferrara (1505?), in which a balustraded balcony, with people looking out and Persian carpets hung as if for a feast, borders the whole ceiling above the cornice (the *tondo* that interrupts the expanse of sky is a later addition). In 1519 the artist repeated the same motif in an octagonal oculus in the Palazzo del Seminario in Ferrara. The influence of the mural decoration of Mantegna's Camera degli Sposi can be seen in a delightfully decorated room of the Palazzo Salvadego in Brescia (frescoes removed during World War II), which Il Moretto (Alessandro Bonvicino; ca. 1498–1554) painted to look as if it opened onto a garden enclosed by low balustrades, in which the richly dressed ladies of the Salvadego family stand. The whole is completed with spacious landscapes in the background.

d. Mannerist perspectivists. The illusionistic perspective works that had the greatest influence on the development of mannerist painting were the Roman works of Michelangelo, Raphael, and Peruzzi. This influence, apparent in the decorations of the Sala di Costantino in the Vatican (executed by Giulio Romano and assistants such as Perino del Vaga and Giovan Francesco Penni), is even more clearly discernible in the decorations of the Sala del Consiglio (Sala Paolina) in Castel Sant'Angelo, which Perino del Vaga frescoed before 1523. In these frescoes elements of diverse nature are crowded together without any unified program, and there is a general effect of confusion. Supporting caryatids seem to project from the walls between simulated columns, recalling elements of various works of Raphael, Michelangelo, and Peruzzi; there are also simulated doors, one of which is open to reveal a figure looking in. Notwithstanding their lack of unity, however, these frescoes give clear evidence of the fact that illusionism prevailed in the most varied forms and that it was especially favored for the decoration of nonreligious buildings. The illusionistic trend was to have two outstanding exponents in Giulio Romano and Girolamo Genga, both active in the 1520s. Giulio Romano decorated the Palazzo del Te in Mantua (1524–25; XII, PLS. 53, 82), and Genga decorated the Villa Imperiale in Pesaro (1526–38). Architects as well as painters, these two artists reaffirmed in their activities that principle which the theorists maintained: that the creator, if not the actual executor, of illusionistic perspectives must be an architect rather than a painter.

In Mantua, Giulio Romano broadened in his development. The influence of Mantegna and of Correggio is evident in his frescoes — Mantegna's in the Sala dei Cavalli, with large, simulated windows opening onto painted landscapes, Correggio's in the central section of the Sala di Psiche. To avoid the harshness of *sottinsù*, the artist found a compromise solution for his foreshortening in the other paintings of the Sala di Psiche, rotating the plane obliquely by the use of a mirror and distributing over it the inclined figures and perspective views. The frescoes in the Sala dei Giganti are notable as illusionistic paintings not so much for their esthetic quality as for their bold execution (IX, PL. 298). Here, wall and vault decorations are fused (perhaps for the first time in the history of painting; Mantegna, for instance, always preserved the distinctions between structural elements, even if only painted). From the blocks and columns crumbling on the giants who have dared to climb Mount Olympus, the artist draws attention upward to the sky, from which the gods look down and where Jupiter's throne, topped by an eagle, stands under a dome with painted panels supported by columns rising from a balustrade and forming a perfect circle. The relationship between vanishing point and station point (set in the middle of the room) is perfectly calculated. Rinaldo Mantovano worked with Giulio Romano on these decorations. Alone, Mantovano decorated one of the chapels of S. Andrea in Mantua, the small dome of which he transformed into a flowery arbor with putti — a work indicative of the influence of Mantegna and of Mantovano's long collaboration with Giulio Romano (PL. 99).

In the decoration of the Villa Imperiale in Pesaro, the situation of Girolamo Genga (1476–1551) was somewhat different. Working with the help of artists such as Raffaello dal Colle, Bronzino, the two Dossi brothers, and Camillo Mantovano (judging from Vasari's account of the life of this artist), Genga had the assistance of more noteworthy personalities than those with whom Giulio Romano worked in Mantua. Unfortunately, since that of Bernhard Patzak (*Die Villa Imperiale in Pesaro*, Leipzig, 1908), no important study has been made of this notable monument; thus the attributions of the various parts are still doubtful. The illusionistic architectural treatments of the various halls, however, seem to show the master planning of no less an artist than Genga. These solutions appear to have been inspired by the Roman works of various artists (especially Raphael and Peruzzi), with the addition of elements probably inspired by the portrait of Sigismondo Malatesta by Piero della Francesca. The predominant feature is that of vast landscapes visible not between the columns of a loggia, as in Peruzzi's decoration of the Farnesina, but between pilaster strips topped by Corinthian capitals, as in the fresco by Piero della Francesca in Rimini. In the so-called "Camera dei Semibusti" there is also found the motif of a garland festooning a simulated window. In the main hall (Sala del Giuramento) a frescoed curtain is held by flying putti, a motif that Genga probably derived from Sodoma. At the edge of the ceiling, smooth columns with Corinthian capitals (reminiscent of Raphael's small vault), which parallel the pilaster strips, support a balcony where flying putti hold up a tapestry (a motif used by Raphael in the Farnesina) narrating the feats of Francesco Maria Sforza, Duke of Urbino. In the Camera degli Amorini, a simulated 16th-century garden is made up of elements derived from Giovanni da Udine and elaborated after the manner of Peruzzi (e.g., the ceiling of Belcaro Castle, Siena). Similar elements are present also in the so-called "Camera delle Forze di Ercole," where the motif of a garden with an arbor of groined arches appears beyond the caryatids. In the Camera delle Cariatidi, which Patzak attributes to the brothers Dossi, the motif of Raphael's loggia in the Farnesina is joined with figures of caryatids emerging from the grass, which the artist must have derived (judging from the drawing that has survived) from the 4th-century decoration of the vault of S. Costanza in Rome. The general effect of the room is pastoral; the ceiling is covered with drapery. In the Sala della Calunnia (where, perhaps, the hand of Raffaello dal Colle can be seen in the figures), the painted structures consist of four large exedrae, one on each side, open at the top and surrounded with a balustrade. In the foreground is a series of smooth columns; two open doors, one on either side of each exedra, show two other columns and the landscape beyond. This painted architecture, although influenced by Peruzzi, is based on designs by Genga. Thus, the cities of Mantua and Pesaro witnessed the first two large-scale achievements in secular illusionistic decoration, and although provincial in comparison with more advanced centers of the time, they had a significant influence on subsequent developments. In the more important centers, illusionism was still not a firmly established trend, and much use continued to be made of grotesque decorations and *quadri riportati*.

In a room in the Palazzo Vecchio in Florence, Bronzino preserved the architectural structure of the vaulted ceiling, into which he inserted moderately foreshortened figures on an oblique plane. Francesco Primaticcio was far bolder in designing an Olympus (not executed) for the Uffizi Gallery, Florence. In flowing lines he drew the half-naked figures of the gods *sottinsù* in a circle around Jupiter and Juno; the perspective is slightly above the horizon line. Generally speaking, however, illusionistic paintings were used only for wall decoration, and the most fashionable manner was that employed by Perino del Vaga in the Sala Paolina in the Castel Sant'Angelo (cf. the decorations of Palazzo Farnese, Rome, and Palazzo Vecchio, Florence, by Francesco Salviati; VII, PL. 268; IX, PL. 294), a style that was occasionally used also in ceiling decoration (e.g., the ceiling of Palazzo Ruspoli, Rome, by Jacopo Zucchi, and that of Palazzo Capponi, Florence, by Bernardino Pocetti).

In Rome, the decoration of the Salone dei Cento Giorni in the Palazzo della Cancelleria, frescoed in 1542 by Vasari and assistants, is outstanding. (In this palace there is also an interesting room attributed to Peruzzi — later unfortunately repainted — with columns that disguise the corners of the room). In the Salone dei Cento Giorni, large painted scenes narrating the historical events of the time alternate with simulated quadrangular niches with statues (a general plan reminiscent of that of the frescoes by Andrea del Castagno in Villa Pandolfini, Soffiano). Each scene (all show the influence of Raphael and Peruzzi) is illusionistically linked to the floor by painted steps, crowded by Vasari with closely overlapping figures so as to create an impression of reality. The device is unquestionably clever, but because each scene and staircase is unrelated to the next one the perspective lacks unity, and the many station points fragment the decoration. Later Poccetti made use of the motif of the staircases in his *Massacre of the Innocents*, which he painted in one of the halls of the Ospedale degli Innocenti, Florence; he unified his work, however, by establishing a single station point for a whole wall surface.

e. The new theorists. While artists continued to produce works in illusionistic perspective, the theorists — first Daniele Barbaro, then Vincenzo Scamozzi — codified the principles upon which the subjective-illusionistic genre was based. Then the architect Giacomo da Vignola applied these principles in his work, providing models conceived according to the criteria expounded by the theorists. After having done perspectivist architectural drawings for Primaticcio at Fontainebleau in 1541–42 (Louvre, Cabinet des Dessins), he worked from about 1556 to 1573 in Caprarola, where he designed the villa and executed the illusionistic columns in the Sala del Concilio and the architectural perspectives on the walls of the Sala di Giove. Perhaps the most important of the artist's decorations in Caprarola is the *sfondato* that he executed in the "Camera Tonda," or Guardroom, where to make the ceiling look higher he painted a small balustrade around the edge, against a background of open sky with birds. It should be noted (cf. M. Walcher Casotti, *Vignola*, Trieste, 1960, p. 45) that in this decoration the station point is below the viewer's feet and the viewer is himself within the visual cone, so that even if one walks away from the center of the room the columns do not seem to fall. True to his profession as architect, Vignola would have no figures in this work. In it, he established the criteria for the centered illusionistic perspective (introduced by Giulio Romano in the Sala dei Giganti of the Palazzo del Te, Mantua) that made it possible to decorate walls and ceiling without a break. This was the kind of perspective that the *quadratura* painters were to favor (see below).

f. Venetian perspectivists. At this time another great architect and treatise writer, Andrea Palladio (q.v.), was active in the Veneto region of Italy. Palladio chose a number of excellent painters to fresco the villas that he built inland from Venice, and these artists developed a particular style of illusionistic painting suited to Palladio's architectural style. Critics have tried to trace this style of decoration back to the lost frescoes by Giuseppe Salviati in Villa Priuli at Treville in Castelfranco Veneto, but from other examples of this painter's work it seems apparent that, except for a few elements, the decorations of Palladio's villas are developed in a completely different manner. The earliest substantial illusionistic decoration which survives is that executed before 1555 in the Palazzo Porto Colleoni at Thiene (Vicenza) by Gian Battista Zelotti and Antonio Fasolo. The decoration simulates a large Corinthian loggia, of the type used by Peruzzi; within it are represented episodes from ancient Roman history, and it opens into a vast landscape. A high point in this style was reached in the decorations of Villa Barbaro (presently Villa Volpi) in Maser, by Paolo Veronese (1561), who had already worked on frescoes (lost) with Zelotti in the Palazzo Porto Colleoni and Villa Soranza. Every room of Villa Barbaro, which Palladio designed to fit into the surrounding countryside, seems to open onto a panorama seen through large simulated windows set between Corinthian columns. Thus, architectural criteria influenced not only the exterior but also the interior plan and the disposition of the rooms, as well as the decorative scheme. On the ceilings the painter preserved Palladio's structural divisions but treated them as if they were trellises, through which the sky could be seen; he scattered figures in inclined perspective on the painted clouds, repeating the oblique perspective that Giulio Romano had used in Mantua. In the Sala dell'Olimpo, the most important room, a painted balustrade with people looking over its edge forms the transition to the frescoed sky. Rather than from Mantegna's work, this arrangement may be derived from the ceiling that Garofalo painted in the Palace of Ludovico il Moro in Ferrara. In the Stanza di Bacco, genuine *sfondati* with pergolas laden with grapes appear on the vault. It is not known how much of the Villa Barbaro decoration was done at the suggestion of Palladio (strangely enough, Palladio's treatise on architecture omits the name of Veronese in connection with Villa Barbaro, though it lists various painters who frescoed his other villas). However, there is no question that Veronese revealed himself an absolute master of architectural perspective in this work, even though some elements are reminiscent of the Villa Imperiale in Pesaro and of the work of Mantegna, Giulio Romano, and Perino del Vaga. After the Villa Barbaro, Veronese did not attempt any similar perspective paintings, but preferred to decorate ceilings with *quadri riportati* representing figures on oblique planes. Nonetheless, with the decorations of the Villa Barbaro, in which color is treated in the bright, fresh tones typical of the Venetian colorists, Veronese established a particular style of illusionistic decoration that became immensely popular with the rich families of Venice and environs.

Gian Battista Zelotti continued this style successfully. He decorated the Villa Godi (Valmarana) in Lonedo (1557), the Villa Emo in Fanzolo di Vedelago (prior to 1565), the Villa Foscari ("La Malcontenta"; 1561), and the Castle of Cataio in Battaglia (ca. 1570). How much he was influenced by Veronese, with whom he worked at first, is readily apparent, but his less original artistic personality permitted him to assimilate Roman mannerist elements, as can be seen, for instance, in occasionally heavy decorative elements over doors (Villa Godi), in enormous caryatids and figures whose legs dangle from where they are seated (Villa Emo), and in his celebrative paintings (Castle of Cataio). Giovanni Antonio Fasolo (1530–72) often collaborated with Zelotti. Fasolo's talent for spatial compositions and for balancing painted sculptural decorations and figures was often uneven, however pleasant the over-all effect (e.g., Villa Caldogno, 1570). Antonio Vicentino, Girolamo del Pisano, and Alessandro Maganza and his son Giambattista were all assistants of Zelotti and Fasolo and followed the style of their masters uninventively. Domenico Riccio, known as Brusasorci (1516–67), Paolo Farinato (1524–1606), Battista Angelo del Moro (1514–75), and Bernardino India (1528–90) formed another group active in the area of Verona under the influence of Veronese. The works of these artists indicate also that they had contacts with nearby Mantua; for example, as is seen in the *Fall of the Giants* in Palazzo Porto in Vicenza, a fresco attributed to Riccio (cf. L. Crosato, *Gli affreschi nelle ville venete del Cinquecento*, Treviso, 1962).

Though from Brescia, Lattanzio Gambara (1530–74) came under the influence of the Venetian painters, especially in the frescoes of villas around Verona (e.g., Villa Di Rovero, Caerano di S. Marco). He was more faithful to mannerism in his frescoes in palaces in Brescia. Neither Veronese's brother Benedetto Cagliari (1538–98), who frescoed Villa Giusti at Magnadola, nor his son Carletto, who frescoed Villa Loredan (Paganizza) in Sant'Urbano, near Padua, created anything comparable to the works of the master. The last notable fresco painter of the Venetian villas was Ludovico Pozzoserrato (Lodewyck Toeput; 1550–1635), a Fleming who, adapting his art to the local style, painted landscapes remarkable for their detail and precision (e.g., Villa Chiericati in Longa).

g. Bolognese perspectivists in Rome. When Gregory XIII of the Bolognese family Bentivoglio became pope (1572–85), a group of artists from Bologna, including Ottaviano Mascherino, Tommaso Laureti (a Sicilian educated in Bologna), Lorenzo Sabatini, Matteo Zaccolini, and the artist-theorist Egnazio Danti, went to Rome. There they executed illusionistic perspective decorations that anticipated *quadratura* painting, basing their work on newly codified principles as well as on their earlier experience. At that time, the faculties of mathematics and optics of the University of Bologna were renowned. Giacomo da Vignola had studied there, and in 1549 Pellegrino Tibaldi painted perspective views on two ceilings in Palazzo Poggi (the present-day Palazzo dell'Università; PL. 100). These views are a large-scale reelaboration of the scheme that Raphael used for the Vatican Logge. Bologna was a meeting ground for artists and theorists, a place where they might formulate and come to agree on the general principles of illusionistic perspective.

It was the papal court in Rome, however, that offered the greatest opportunities to apply these new principles of perspective painting (IX, PL. 294). The architectural structures painted in perspective on the small vaults in the loggias of Gregory XIII, with the intention of making the vaults seem higher, are clear evidence of the presence in Rome of Bolognese artists. One of these works is a faithful reproduction of the dome painted by Giulio Romano in the Sala dei Giganti in the Palazzo del Te, Mantua, except that the central spherical vault in the Vatican version simulates an orchard. Another repeats the colonnade motif of the Sala della Calunnia in the Villa Imperiale of Pesaro. The outstanding work of this group, however, is the Sala della Bologna, where Mascherino transformed the curved-edge ceiling into a gallery, which, with its pairs of columns joined by arches, is reminiscent of that painted by Laureti in the Palazzo Vizani (now Palazzo Sanguinetti) in Bologna (drawing in Danti, 1883, p. 88) and, at the same time, echoes the loggias of Gregory XIII. The figures of astronomers by Lorenzo Sabatini recall the bold, foreshortened figures by Tibaldi in the Accademia delle Scienze of the Palazzo dell'Università in Bologna. In the center of the vault four putti hold up a drapery with an ornamental representation of the zodiac. The station point in the Sala della Bologna is in the center of the room at the level of the viewer, and the precise calculation of the passage from light to shadow confirms Danti's statement that Mascherino made a model of a quarter section of the vault.

The perspective views around the walls of the Sala dei Palafrenieri in the loggias are by Danti himself. In the Sala degli Svizzeri, smooth columns with Ionic capitals rise from a high socle and frame large arched windows and quadrangular niches with painted statues (perhaps by Federico Zuccari). In both halls the corners are concealed by painted trophies. Later followers of this style of perspective painting were Giovanni Alberti (1558–1601), from Borgo San Sepolcro, whose brother Cherubino (1553–1615) assisted in the figure painting, and Agostino Tassi (ca. 1580–1644), from Perugia, both of whom were active during the reign of Gregory XIII. The fame of Giovanni Alberti rests mainly on the decorations of the Sala Clementina in the Vatican (named after Pope Clement VIII, during whose reign it was completed in 1598). In this room, only the fireplace wall is decorated with illusionistic architectural structures, forming a unified whole with those of the vault. The influence of Peruzzi is evident in the slight extension of the painted loggia onto the side walls; Alberti's execution is the more complex however, especially on the vault, where a painted balcony is supported by sturdy painted corbels reminiscent of those in the drawing by Laureti that Danti published. The balustrades and columns arranged in a semicircle at the center of each side (a motif that appears also in the small vaults of the loggias of Gregory XIII) open toward the sky, where a group of angels bearing the insignia of the pontiff create the illusion of additional height. The artist managed to solve the problem posed by the great expanse of the vault only by dividing it into four segments and painting each with a different station point. Prior to this, Alberti had decorated one of the halls in the Galleria delle Antichità in Sabbioneta with simpler and more convincing illusionistic perspectives. More coherent, also, is his decoration of the vault of S. Silvestro al Quirinale (PL. 97), in which the corbel motif in the central oval is a refinement of Laureti's design. Angels painted by Cherubino Alberti, after the manner of Correggio (both brothers had lived in Emilia for a time), fly through the prominent moldings that suggest the structure of the dome. The aerial perspective inspired by Correggio also appears in the ovate painting of Christ in Glory with angel musicians. In the Sagrestia dei Canonici of St. John Lateran (PL. 100) Giovanni Alberti resorted to the much-used scheme of the foreshortened oculus (Domenico Beccafumi had also used it in the Palazzo del Governo, Siena, in 1521); Cherubino adorned the oculus with playing putti, perhaps in imitation of those by Mantegna.

Agostino Tassi also profited from the experience of the Bolognese perspectivists, and in his first Roman work, the Salone dei Corazzieri in the Palazzo del Quirinale (executed with Carlo Saraceni and Giovanni Lanfranco), he used the motif of the loggia with groined arches for the upper band of the walls (1611–18). Summoned to Bagnaia to work at Villa Lante, he used pale, almost transparent colors in the frescoes of the loggia, in which he transformed the oculi into aviaries (PL. 101). Later, in Rome, he worked with Guercino on the decorations of the Palazzo Lancellotti in Via dei Coronari, in a hall for which he reelaborated the motif of Laureti's vaults with spiral columns, while for the decoration of another hall he took inspiration from Tibaldi's double-colonnade motif in the frescoes at the University of Bologna. He repeated similar motifs in the Palazzo Costaguti, Rome, but his boldest perspective is undoubtedly that which frames the chariot of Aurora in the Villa Ludovisi (cf. II, PL. 197). Here, instead of the usual columns, he painted the almost-perpendicular walls of a room, the ceiling of which seems to have collapsed — the right wall, in fact, appears to be crumbling (an anticipation of the 18th-century paintings of romantic ruins) — and beyond the walls he painted the tops of trees in a garden.

h. Illusionistic figure painting. The presence in Rome of the Carracci brothers (q.v.), the fame that Annibale Carracci's decorations of the Galleria in the Palazzo Farnese immediately acquired, the eclecticism with which this painter mixed motifs derived from Michelangelo, Raphael, and Correggio to create illusionistic effects with figures rather than with architectural structures, along with the iconographic program imposed by the Council of Trent — all contributed to the decline of *quadratura* in Rome. The painters of the new Bolognese school then took the place of those of the time of Gregory XIII and, following the example of the Carracci brothers, sought their models in the works of the figure painters of the Renaissance and based illusionistic painting, as much as possible, on figures. This can be seen, for instance, in the frescoes by Domenichino in the Church of S. Andrea della Valle, Rome, which are inspired by Michelangelo, and in the paintings by Guido Reni in the Chapel of the Annunciation (ca. 1610) in the Palazzo del Quirinale. On the ceiling of the Church of S. Maria della Pace (before 1616), Francesco Albani imitated the motif of Raphael's Loggia in the Farnesina. Giovanni Lanfranco imitated his master, Agostino Carracci, in a vault decoration in the Casino Borghese in Rome, but in his religious paintings (dome of the Church of S. Andrea della Valle, Rome, 1621–25; dome of the Chapel of St. Januarius in the Cathedral of Naples, 1641) he returned to the aerial perspective of Correggio. His first innovation occurred in the nave of the Certosa di S. Martino, where in each segment of the vault he resorted to the oblique-plane technique and spread the garments of some figures over the transverse arches. He seems to have aimed at a tight atmospheric unity between the actual architectural space and the painted illusionistic space, a trend that was soon afterward developed in the works of Pietro da Cortona, particularly on the vault of the main hall in the Palazzo Barberini (1635–40; II, PL. 176).

Pietro da Cortona's principles of perspective were to be resumed and developed in a highly personal manner by the Genoese Giovanni Battista Gaulli, known as Bacciccio (1639–1709), on the ceiling of the Church of the Gesù in Rome, where he painted the *Triumph of the Name of Jesus* (II, PL. 171). While in his perspective Pietro da Cortona had preserved an architectural framework, however slight, and had set the vanishing point at its center, Gaulli, using an oval canvas applied to the wooden ceiling of the nave, opened a kind of elliptical oculus in which the figures are distributed along the beams of light radiating from the monogram of Christ.

A contemporary of Gaulli, Andrea Pozzo (q.v.) led a revival of *quadratura* painting in Rome. Pozzo's guides according to the frontispieces of his treatise, were the theoretical works by Palladio and Vignola. The artist, once more basing his work on rules rather than on practical experience, blended *quadratura* and aerial perspective in the Church of S. Ignazio in Rome, introducing a monumental style of decoration that was to be further developed in Austria (see below).

i. Bolognese school of quadratura painting. Illusionistic architectural perspective continued to flourish in Bologna and northern Italy. The treatises of Sebastiano Serlio (1545) and Vignola, especially the latter's *Le due regole . . .*, with commentary by Danti, were readily available to the Bolognese artists, who were interested in architectural perspective also because of the special studies then being made at the university. Thus *quadratura* painting emerged from the Bolognese milieu with a manner of its own, a fully distinct trend among the many schools of baroque painting. It first appeared in the *Coronation of the Virgin* painted by Carlo Bononi (1569–1632), a fresco on the dome of the Church of S. Maria in Vado, Ferrara (PL. 103); the illusionistic loggia painted about the drum recalls Laureti's motif.

The true founder of the *quadratura* school, however, was Girolamo Curti, known as Dentone. According to Malvasia, Dentone showed no ability for figure painting, and having devoted himself to the study of the treatises on perspective of Serlio and Vignola, he became unsurpassed in the art of *quadratura* painting. Moreover, Dentone's inability to paint figures obliged him to join forces with figure painters (he favored the work of Angelo Colonna; see below), and often with ornament painters; thus began the practice of specialized painters working in pairs, each having a clearly defined task, but with the *quadratura* painter the predominant member. The hall painted by Dentone in the Palazzo Chigi (now Palazzo Odescalchi; PL. 98) gained him more praise than had been given the Sala Clementina.

Dentone was influential in the artistic development of all the Bolognese *quadratura* painters of the next generation, among them Angelo Michele Colonna (1600–87) and Agostino Mitelli (1609–60). Among the works that Colonna — also a *quadratura* painter — executed before the beginning of his association with Mitelli (1632), there remain the frescoes of the Casino Malvasia at Trebbo (1625; PL. 101). With Mitelli he completed the decorations in the halls of the Museo degli Argenti, Florence, which had been started by Giovanni da San Giovanni, another able perspectivist and figure painter. The two artists created imaginative *quadratura* paintings that seem to enlarge greatly the size of the rooms by introducing sudden, impressive views of the sky. The renown Colonna and Mitelli enjoyed in Tuscany gained them commissions for the decoration of villas in the countryside around Lucca, including that of the Villa Arnolfini (Marchi) in Gragnano and of the Villa Gardi in Collodi, where flowers, fruits, putti, and allegorical figures appear in rich and vividly colored architectural settings. In their later years Colonna and Mitelli were called to the court in Madrid, where they did not, however, find the favor they had hoped for. Among their many disciples in Italy, Bartolommeo de Santi (ca. 1700–55) continued their manner in Tuscany, introducing in his work a few motifs remotely derived from contemporaneous Roman work (e.g., Sala dei Palafrenieri), as is evidenced by the decoration of the ballroom of the Villa Burlamacchi (Rossi) in Gattaiola. In Bologna, Antonio Rolli (1643–96), with his brother Giuseppe (1645–1727) collaborating as figure painter, painted the ceiling of the Church of S. Paolo (PL. 104), in which the architectural elements have become heavier and there is an excess of ornamentation. In northern Italy (Brescia, Milan, Turin, etc.) there was not a villa or palace of the time that did not have some decoration in *quadratura*, at least in the main hall. In Genoa ornament painters predominated and developed a particular style, with an abundance of painted ornamental motifs such as *rocaille*, that can be seen in the ceilings by Domenico Piola (1627–1703) and Anton Maria Haffner (1654–1732) in the Palazzo Bianco and the Palazzo Rosso. Haffner also worked on the decoration of the Palazzo Bianco in collaboration with Gregorio de' Ferrari (1644–1726) — who was still under the influence of Correggio — and, later, with his brother Enrico Haffner (1640–1702) as figure painters.

Although *quadratura* painting was still much in favor in the first half of the 18th century (e.g., Stefano Orlandi's *quadratura* painting, with figures by Vittorio Bigari, in Casa Tacconi, Bologna; PL. 107), figure painters gradually gained ascendancy in the execution of perspective frescoes and decorations. This was, to a great extent, a result of the influence exercised by the Neapolitan school [cf. C. Lorenzetti, "Interferenze della pittura napoletana con la pittura veneziana: L'origine del settecento pittorico a Napoli," *Atti del XVIII Congresso di Storia dell'arte, Venezia,* 1955 (1956)]. The leading artist of this school was Luca Giordano (1632–1705), who was active also in Spain. His aerial views and his clear, transparent colors heralded a new type of illusionism, no longer bound to determined station points. Followers of Giordano were Giacomo del Pò (1652–1726) and Francesco Solimena (1657–1747). The new Neapolitan style of perspective painting strongly influenced Venetian painters, especially Sebastiano Ricci (1659–1737), who also worked in Rome and Florence (PL. 106).

It was in Venice that this new illusionism was elaborated upon during the 18th century. Giovanni Antonio Fumiani (1643–1710; PL. 102), Giovanni Battista Piazzetta (1683–1754; q.v.), Giovanni Battista Crosato (1697–1756; PL. 110), and, most important of all, Giovanni Battista Tiepolo (q.v.) opened their ceilings onto blue skies crowded with figures. Some painters, such as Crosato and Tiepolo, employed as assistants *quadratura* painters, whose work was limited to the frames of the *sfondati* on the wall — so much like those created by Paolo Veronese and his followers in Venetian villas during the 16th century [e.g., frescoes by Tiepolo in Palazzo Labia, for which the *quadratura* painter was G. Mingozzi (called Colonna)].

Thus optics triumphed over mathematics, bringing to an end the period of *quadratura* painting, which was thereafter employed only by scenographers and by the *vedute* painters, who applied the principles of *quadratura* painting to small panels in which space once again became objective and in which real or imaginary architectural views were the main subjects (cf. PL. 105).

Germanic regions: a. Illusionistic perspective painting beyond the Alps. The earliest attempts at illusionistic perspective in the southern Germanic regions, which appear in palaces and on the façades of houses, date from about the end of the 16th century, when the taste for perspective decoration was introduced by Italian artists active in those areas; this taste was also encouraged by increasing interest in the theater. In 1570, an unknown artist painted a room in the castle in Hannoversch Münden with simulated Doric colonnades and statues in niches recalling those found in Roman frescoes. In 1580, Alessandro Scalzi (known as Paduano) painted one of the staircases — the "Narrentreppe," or madmen's staircase — in the castle at Trausnitz, near Landshut, with illusionistic masks.

The most consistent and monumental early work, however, is that in the Residenz in Munich, where before the end of the 16th century a group of painters — anonymous but obviously trained in the Bolognese school, and revealing a familiarity with the style of Guercino (q.v.) — painted the wooden coffered ceilings with figural compositions in oblique-plane perspective that recall the manner of Giulio Romano in the Sala di Psiche of the Palazzo del Te in Mantua. Architectural *sottinsù* views clearly inspired by the Bolognese masters were painted on the ceiling of the Schwarzen Saales of the Residenz in Munich by Hans Werl, who imitated the motif of Laureti's loggia in his fresco (lost; sketch in Staat. Graphische Samml., Munich). Correggio's aerial perspective also had imitators, as is evident from a sketch of an Ascension scene (private coll.; Tintelnot, 1951, fig. 11, p. 31) attributed to Christoph Schwarz (1545–92), in which the *sottinsù* treatment of the figures seems excessive. In 1615 another Italian painter, the Florentine Servite friar Arsenio Mascagni (1550–1636), introduced in Austria the purest principles of Roman and Emilian *quadratura* painting; the influence of Peruzzi and of Veronese can be seen in Mascagni's wall and ceiling decoration of the Festsaal in the Castle of Hellbrunn, near Salzburg. The Thirty Years' War (1618–48) and the subsequent threat of the Turks (1683) both contributed to delaying the development of illusionistic perspective painting in Germanic regions. Only toward the end of the 17th century was there a noticeable revival of this art, both in Austria and in the southern parts of Germany, where local circumstances offered perspectivists their greatest opportunities. Contacts with nearby Italy were renewed by young Germanic artists who went to Rome and Venice for training and by Italian artists (painters, architects, stucco-workers, etc.) who were commissioned by Germanic nobles to decorate the luxurious palaces and splendid castles being built or restored for them at various places in Austria and Germany. These painters were often honored with the appellation of *Hofmaler* (court painter).

In regions such as Austria and Bavaria, which had remained Catholic in the face of the great changes produced elsewhere by the Reformation, a new fervor for the construction of churches gave impetus to the development of illusionistic perspective, a genre most appropriate for grandiose depictions of the Church Triumphant. Perspectivists became so much in demand that panel painters were almost ignored (contrary to trends in the northern Germanic countries, in England, and especially in France, where the prevailing rationalism rejected the irrational and visionary). Furthermore, illusionistic painting was used to illustrate the absolute power of the king, who was often identified with a mythical hero or god (e.g., Hercules or Apollo). In the Germanic regions that had retained their Catholicism, the figure of the emperor was included in religious representations, in which he occupied a place next to the divinity.

Unlike Italian illusionistic frescoes, the German paintings, generally limited to the ceiling, are executed according to a precise and detailed plan, in which allegory and emblematic symbolism are important and contribute to the iconological character of the painting. These programs, which were provided by men of letters and poets (for secular buildings) and by theologians (for church decorations; those prescribed by the Jesuits, who had earlier provided Pozzo with such a program of decoration for the ceiling of S. Ignazio, Rome, are notable), influenced the artistic treatment of the frescoes also, in so far as the allegorical-doctrinal aim exceeded an interest in perspective based on mathematical principles. Figures were no longer represented in compact groups but, instead, were loosely distributed along oblique axes, arranged at different angles and often intersecting. The distribution of light was used to unify the perspective composition, so that, in place of the architectural spatial continuum, there was an atmospheric continuum. Although this use of light originated with the aerial perspective of Correggio, its treatment became increasingly different from his, since it was rendered not by a gradual paling of the colors as they progressed upward but by strong contrasts and backlighting effects, through which the vivid colors of the figures are made to stand out.

Only in the Germanic regions, and especially in Bavaria (e.g., the pilgrimage church at Wies, near Munich), was there achieved a perfect illusionistic blend of architecture, sculpture, and painting. The result is a kind of objective illusion: that is, the observer cannot fail to perceive the painted perspectives as if they were real space — not as extensions of his own world, however, but as components of another world that imposes its own spatial reality on the room from above and still preserves the objective laws of that space, keeping them distinct from those of the actual architectural structure enclosing it. Though it has not been determined to what extent the prevailing enthusiasm for scenography and the theater contributed to the development of this objective illusionism, it should be remembered, for instance, that every princely residence included a small theater.

Tiepolo expressed the local taste for intricate spatial illusionism in his decoration of the Kaisersaal in the Residenz in Würzburg, one of his finest works (X, PL. 312). The development of German architecture along lines that were soon to change from baroque to rococo (qq.v.) especially favored this type of illusionism. The rococo wall structures, which are clearly delineated and which develop asymmetrically in a play of concave and convex lines interrupted by elegantly modeled but busy stuccoes leading the eye rapidly from one motif to another, prepare the viewer for the fantastic and miraculous scenes that are depicted in the simulated skies of vaults and ceilings, set against backgrounds of gardens, broad landscapes, perspective city views, and rough seas.

b. Italian perspectivists active in Germanic regions. Because cultural exchanges between Italy and the Germanic territories were facilitated by their geographical proximity, and since the fame enjoyed by 18th-century Italian painters had extended into foreign lands, it was natural that Germanic princes should call Italian artists — mainly Lombards and Venetians — to decorate their castles and residences. Thus, there arose a development similar to that of the International Gothic style, in which itinerant painters retained personal stylistic qualities while contributing to the formation of a unifying taste dominant throughout Central Europe. Patrons such as the Schönborns and the Liechtensteins encouraged this development, and often successive generations of artists from families of Lombard origin (e.g., the Carlones) worked in the Germanic regions. (Many of these, however, also worked in Italy at various times.) Italian painters, along with German painters trained in Italy (see below), were undoubtedly responsible for the revival of the taste for illusionistic painting in Austria and Germany. As early as 1660, Carpoforo Tencalla (1623–85) again introduced illusionistic perspective panels in ceiling decorations (e.g., Passau Cathedral).

The arrival in Vienna of Fra Andrea Pozzo marked the beginning of a period of new achievements in illusionistic perspective. The German translation (1702) of his treatise aroused interest in perspective painting among local artists, and the practical schemes he proposed were very popular, especially in the Tirol, through much of the 18th century (cf. H. Hammer, 1912). Among the Italian artists active in Germany and Austria who felt Pozzo's influence were the Lombard Giovanni Francesco Marchini, court painter for the Schönborns [frescoes in the Church of St. Martin (1702) in Bamberg, in the Castle of Weissenstein in Pommersfelden, and in Bruchsal and Mainz], and Luca Antonio Colombo (1661–1737), a *quadratura* painter as well as scenographer born near Lugano, who worked, among other places, in Zwiefalten, Schöntal, Fulda, Mainz, and Ettlingen and was the *Hofmaler* of the Duke of Württemburg. On the ceiling of the Ordenssaal in Ludwigsburg Palace, another *quadratura* painter and scenographer, Giovanni Baronzio, repeated in a more baroque manner Andrea Pozzo's *quadratura* for the Palace of Liechtenstein. Carlo Antonio Carlone (b. Scaria, near Como, 1686) was outstanding among the Italian figure painters active in Germanic regions. After having studied with the Friulian painter Giulio Quaglio and then in Venice and Rome (with Francesco Trevisani), he was commissioned to work in Passau, in Vienna (Belvedere Palace, where he was assisted by the *quadratura* painter Gaetano Fanti), and in Paura, near Lambach. In Paura, he painted domes and ceilings in a manner that was a personalized synthesis of Correggio's art, executed in the colors of the Venetian school. In Ludwigsburg Palace, his assimilation of the influence of Giovanni Antonio Pellegrini (see below) can be seen in the gradual disappearance of strong contrasts of light and shadow and in the distribution of groups of figures along oblique axes. In later works in Ansbach, Passau, Weingarten, and Schönberg, the tone of his palette became still lighter, his figures were more elongated, and the perspective planes intersected more markedly, creating more convincing effects of light and a greater sense of spatial amplitude. This trend in his art explains his susceptibility to the influence of Tiepolo on his return to Lombardy.

A style that was decidedly more original and less eclectic, as well as of higher quality, was introduced into Germany by Giovanni Antonio Pellegrini (1657–1741), who, after leaving Venice at a very early age, worked in various capitals of Europe. Active in Düsseldorf in 1713, he executed painted panels and splendid ceiling decorations of secular subjects (e.g., the *Fall of Phaëthon* in Bensberg Palace) also in Mannheim, Dresden, and Bamberg. The perspective effects of Pellegrini's decorations arise mainly from his beautifully clear colors and delicate effects of backlighting, which made his figures, often placed in loose groups on diagonal intersection planes, seem to blend into the atmosphere. His skies, somewhat reminiscent of those of the Venetians, provided local artists with examples of the fundamental value of light in illusionistic perspective painting.

Jacopo Amigoni (1675–1752), a Neapolitan-born and Venetian-trained artist, worked in Bavaria nearly contemporaneously with Pellegrini. Amigoni's clear palette indicates the influence of Luca Giordano; his sharply outlined figures resemble those of Antonio Bellucci (1654–1726), who worked in Vienna for the Prince of Liechtenstein. Ceiling frescoes by Amigoni at Nymphenburg Palace and in Benediktbeuern and Ottobeuren recall the preciosity of Venetian color. In the central hall of Schlösschen Lustheim at Schleissheim (PL. 108), decorated with the assistance of an anonymous but certainly Bolognese-trained *quadratura* painter, the Italian character of his painting is apparent; however, evidence of his interest in the contemporary Bavarian school also emerges in this hall, where, after the manner of Johann Baptist Zimmermann (see below), he seems to raise the ceiling from the hall and, with figures standing upright on the cornice without any perspective architectural setting, to present the events of the Trojan War against landscape backgrounds, thus conforming to the taste for the " unreal " and the fantastic that was the distinctive characteristic of Germanic illusionism.

Born in Milan in 1701, Giuseppe Appiani left a remarkable number of works in Germany, including those in Seehof Palace near Bamberg, in the pilgrimage church of Vierzehnheiligen (X, PL. 314), in Ober-Marchtal, and in Mainz, where he was court painter and where he died in 1796. A clever interpreter of the themes dictated by his commissions, Appiani painted almost sculptural figures in bold brush strokes. His highly personal style, not unlike that of Rubens, is also based on vivid contrast between the foreground shadows and the sudden, almost diaphanous light of the farther planes, a contrast that accentuates the depth of his paintings.

In the second half of the 18th century the most notable Italian artist active in the Germanic countries was certainly Giovanni Battista Tiepolo, who was called to Würzburg in 1550 to fresco the Kaisersaal and the ceiling over the large staircase in the Residenz. The architect was Balthasar Neumann (q.v.), and the collaborating stuccoworker was the Italian Antonio Bussi. The two perspective scenes from the life of Barbarossa in the Kaisersaal, painted at the base of the ceiling beyond simulated damask curtains, are set within gold-filleted stucco frames, almost as if they were stage scenes, set apart from the actual architectural space of the room but linked by the sky motif to the central oculus of the vault, where the coach of Apollo, drawn by white horses, bears Barbarossa's bride to him.

In the staircase ceiling, Tiepolo used the cornice as a support for lively figures painted against background landscapes and architectural structures that evoke the four corners of the world. Tiepolo's partner, the *quadratura* painter Girolamo Mingozzi (Colonna), collaborated with the master on this work, as he had done elsewhere in Germanic regions. The work of many local painters witnessed the influence of Tiepolo's manner of modeling figures. His unique painted skies, in which mathematical principles of perspective art were boldly applied with a truly poetic inspiration, defied successful imitation, however.

The last important Italian painter active in Germanic regions was Gregorio Guglielmi (1714–73), Roman by birth and training. His masters were Sebastiano Conca and, very likely, Francesco Trevisani. After a short stay in Prague (1734), where his style was not appreciated, he worked in Vienna; there, in 1753, following an iconographic program prepared by Metastasio (the apotheosis of Francis I of Austria and Maria Theresa, surrounded by the Four Faculties), he frescoed the ceiling of the library of the new university (the present-day Academy of Sciences) with groups of figures against a background of luxuriant landscapes and imaginary architectural settings that are reminiscent of Pietro da Cortona. All in rich color, these works exercised some influence on Maulbertsch (see below). In Schönbrunn Castle, in 1762, he represented the theme of the power of the Hapsburgs on the ceiling of the small gallery; the military figures in this decoration show evidence of Tiepolo's influence. After a short stay in Turin, in 1764 he was in Berlin, where, commissioned by Frederick the Great, he painted some ceilings in what is now the university; in these his style appears to be freer and tends toward the rococo. Later he worked in Augsburg and in Russia, where he died in 1773.

c. German perspectivists. The first German artists to go to Italy at the beginning of the 18th century came from the Tirol (Austria); their goal was Rome. They were often painters by family tradition (e.g., the Schors and the Waldmanns), and their Italian experience is clearly evident in the works they executed after their return to Austria. When Egid Schor (trained in the school of Pietro da Cortona) returned to the Tirol in 1666, he frescoed, among others, the church of Schabs (near Brixen), that of the monastery in Wilten (near Innsbruck), and the monastery in Stams. At Schabs his talent as a painter of ornament is more apparent than his ability as a figure painter; this can be seen in the lively, elaborate *rocailles* which frame his slightly oblique perspectives as if they were *quadri riportati*. Kaspar Waldmann (1657–1720), a more gifted artist, enlivened his late baroque ornamentations in the summer house of the Foundation of the Royal Ladies at Hall (Tirol) with figures of angels, more freely and boldly foreshortened and better proportioned than Schor's. Although these repeat the schemes of Pietro da Cortona, they also indicate an acquaintance with the works of the Emilian *quadratura* painters (perhaps Mitelli and Colonna). In the decoration of the Church of the Servites in Rattenberg, by Johann Josef Waldmann (1676–1712), there are instead obvious points of similarity with Lanfranco's domes in the manner of Correggio, although it is done in a heavier style and with stronger color.

The first great innovator in German illusionistic painting and originator of the Bavarian school was Cosmas Damian Asam (see ASAM, THE BROTHERS), who often collaborated with his brother Egid Quirin, an architect and stuccoworker. The singularity of his personality is evident in his decorations of the Benedictine church in Weingarten (begun 1717), where, while accommodating his work to the structure of the bays, he opened a bold *sottinsù* in each of them, displaying his talent for architectural perspective (related especially to the work of Pozzo). In other scenes, however, Cosmas Asam shows a tendency toward the unreal and the imaginary, along with a gradual brightening of color through effects of light. In his later works in Innsbruck (Church of St. Jakob; frescoes destroyed), Forstenfeld, Weltenburg, Einsiedeln, and Wahlstatt, he began to dissociate his *sfondati* from the actual architectural structure and eventually arrived at completely irrational solutions, as in the elliptical dome of the Benedictine church of Weltenburg (1721) and in the so-called "Weinachtskuppel" in the abbey at Einsiedeln, Switzerland (1724–26), in which Asam painted on the dome the earthbound scene of the Birth of Christ and set it against a background of mountains and trees. Stucco angels flying about the frame point to the central scene. It is possible that Asam may have been inspired by a similar scene painted by Michael Willmann in the apse of the abbey church of Grüssau (1692–95). In the nave of the Church of Legnickie Pole in Wahlstatt (1731), visionary perspective is raised to a consummate art, and the space of this visionary world is brought directly into the real space of the church. This scenic control reached its apex in the Church of St. John Nepomuk in Munich (1733–46). The same principles, when applied in secular frescoes (e.g., ceiling of the Rittersaal in Mannheim), marked a decisive step in the breaking away of German-Bavarian late baroque painting from Italian styles.

In Austria illusionistic painting developed along different lines, with the emergence of that trend known — from its origin in the court of Vienna — as the "Hapsburgian." The dogmatic themes prescribed by the court often involved the introduction of the emperor and the aristocracy between the human and divine elements as a kind of link between the people and God. This aggrandizing purpose resulted in a monumental illusionistic style with distinctive characteristics. The first German representative of this trend, which in its beginning was markedly Italianate, was Johann Michael Rottmayr (1654–1730), who had been a disciple of Johann Carl Loth in Venice. He was active in Salzburg (Winter Riding School), at Frain Palace in Mähren, and at Schönbrunn Castle; Rottmayr's art was decidedly academic, however. In Wrocław, Poland, he was commissioned to depict the Adoration of the Holy Name on the nave ceiling of the Church of Świętego Marcina; around the four sides he painted a balcony, on which the emperor, members of the aristocracy, and prelates stand in accordance with a vertical perspective axis. The background opens onto a large ellipse of sky, where the saints and the blessed are grouped asymmetrically around a central coach, drawn by the symbols of the four Evangelists. From the coach shines the light of the name of Christ. The perspective axis of the heavenly figures is markedly oblique. Banks of clouds overlapping the frame (reminiscent of Gaulli) link the heavenly vision with the space of the balcony below and diffuse a warm, golden light. The observer's eye moves from above to below in this space (contrary to the movements in Pozzo's ceiling in S. Ignazio, Rome), whereas the figures on the balcony are gazing upward; this contrast in visual movement is made more conspicuous by their astonished expressions and their hands outstretched in prayer. Rottmayr also used this double movement on the immense dome of the Karlskirche in Vienna, on which pairs of stucco angels between the windows offer laurel wreaths to St. Charles, who is being admitted to the glory of heaven. In the nave of the abbey church in Melk, on the Danube, the *sottinsù* of all the bays (unlike those by Pozzo in the former Universitätkirche in Vienna; 1712–18) are unified by means of the angels and clouds which float over the transverse arches and which seem to become part of the space below. Some of the figures fly upward, while others are emphatically cast downward. (Here, again, Gaulli's influence seems likely to this author.)

Daniel Gran (1694–1757) had been a student of Solimena in Naples, but his academic tendency also reveals the influence of the Venetians and, even more, that of Bellucci. Gran decorated the vault of the National Library in Vienna (1730; architectural elements painted by Fischer von Erlach) on commission from Charles VI. The artist, instructed to represent the Emperor as a patron of the arts and sciences, took Rottmayr's fresco in Wrocław as a model and painted a balustrade around the base of the dome, with figures of philosophers, poets, men of letters, and scientists. (Some of these are reminiscent of the ones by Raphael in the Stanza della Segnatura.) In the sky above, amid transparent effects of backlighting — typical of this painter and probably derived from Solimena — he painted allegories of the sciences and arts in various groups (one group holds a medallion with the portrait of Charles VI), which seem to be descending into the world to enlighten it.

Bartolomäus Altomonte (b. Warsaw, 1702) was the son of Martino Altomonte, who had had his training in Naples. Bartolomäus himself studied in Bologna and Rome (1717–23). On his return to Austria, he became one of the favorite painters of the monastic orders; his earliest work (St. Florian), executed in collaboration with Italian *quadratura* painters, was still somewhat academic, containing elements obviously derived from his Italian education. His personality is best revealed in the frescoes in the choir of the collegiate church of Spital am Pyhrn (1741), for which he took as a model of architectural *quadratura* one of the etchings for a *theatrum sacrum* in Pozzo's treatise; but he painted the figures descending from the top of the small dome almost down to the altar, around which stand the Apostles. In effect, the Virgin of the Assumption seems to rise from the altar to be admitted by the Father and the Son to the glory of heaven. In representing the Triumph of the Church, at Wilhering (near Vienna), Altomonte appears colder and more academic in style; the decoration is based on the same concept as that of Rottmayr in the dome of the Karlskirche, but the theme is not rendered with an equally forceful and suggestive use of light.

Paul Troger (1698–1762) received his education in Venice, was a student of the painter Alberti (1640–1716) at Cavalese, and then went to Venice, Rome, Naples, and Bologna and studied with Piazzetta, Conca, Marco Benefial, and Solimena. He returned to Vienna as a scenographer (1728) and later became the director of the Vienna Academy (1753). His early works (Marble Room of the abbey at Melk, on the Danube; frescoes in the library there, in which Fanti did the *quadratura* paintings; and frescoes in the library of Zwettl, 1732–33) are rather academic, with elements after the manner of Gran, while the figures indicate the strong influence of Giovanni Battista Pittoni. In the dome of the Stiftskirche at Altenburg, however, his hitherto cool and stiff representations were imbued with dramatic movement, emphasized by vivid and contrasting colors, by unreal effects of light, and by twisted trees just inside the frame. The dazzling light that emanates from the asymmetrically placed figures of God and the Virgin seems to create a dramatic tension in this apocalyptic scene. This dramatic power is a fundamental element in Troger's many later frescoes, in which the figures became increasingly isolated, partly perhaps as a consequence of renewed Venetian influence (Marco Ricci, Pittoni, Federico Bencovich, and Pellegrini). The contacts Troger had with Bavarian painters (Johann Zimmermann and Johann Evangelist Holzer) are manifest in his last works (e.g., library in Seitenstetten; Altenburg), though their influence entered into his highly personal vision, with qualities typical of the Viennese baroque and presaging the rococo.

Swabia and Franconia, bordering on Austria and Bavaria, were naturally influenced by the spread of Austrian and Bavarian late baroque illusionistic painting in churches and, especially, in palaces. Bamberg, Würzburg, and Ansbach preserve the best examples of this art. In the Bamberg Residenz the decorations are by Melchior Steidl (d. 1727), who faithfully repeated the plans proposed by Pozzo (as he had already done in the Church of the Foundation of St. Florian). Gabriel Schreyer (1666–1730), the court painter of Bayreuth, and Paul Decker (1677–1713) also relied on Pozzo's projects. Johann Rudolf Byss (1660–1738) worked at Pommersfelden with the Italian *quadratura* painter Marchini. Outstanding among these painters was Johann Georg Bergmüller (1688–1762), director of the Augsburg academy, an expert on architectural perspective, and a decorator of palace façades. In the choir of the church of Diessen (1742), around the stuccoes framing the ceiling, he painted illusionistic visions like those of Asam, which have no relation to the architectural structure of the building. Likewise unrelated to the building structure is his decoration of the old abbey church in Steingaden, also painted with an asymmetry that was clearly rococo. Among the group of painters who worked in Konstanz, Franz Joseph Spiegler (1691–1757), whose works were fully rococo, is notable. His masterpiece is the decoration of the church of Zwiefalten (begun 1747). The theme, the ecstasy of St. Benedict's vision of the Virgin and Child, is rendered boldly, almost expressionistically, especially on the ceiling of the nave, where from the stucco frame there rises an imaginary wall enclosing the pilgrims awed by the vision. From the Virgin's bosom originate rays that pierce the heart of St. Benedict.

The Scheffler brothers (Thomas Christian, 1700–56; Felix Anton, 1701–60) developed an original, though somewhat provincial style, marked by a horror vacui. An example is Thomas Scheffler's fresco in the nave of the Paulinkirche in Trier.

French rococo was introduced to Charlottenburg through the perspective works of Antoine Pesne (1683–1756). The taste for *chinoiserie* was initiated by Thomas Huber (1700–79), who decorated the Teehaus of Sanssouci with frescoes of monkeys and exotic birds perched over the heads of the Chinese and court ladies looking out over the traditional painted balcony.

In Bavaria, rococo illusionism had an outstanding exponent in Johann Baptist Zimmermann, who brought to illusionistic decoration that sense of the idyllic, the lyric, and the bucolic which had already inspired the works of Asam — with the difference that Zimmermann (under French influence) tried to break up, almost to dissolve, the sometimes heavy and weighty constructions of the baroque. On the ceiling of the Church of Steinhausen (1731), the artist transformed the traditional motif of the balcony into an elaborate balustrade, with potted flowers and putti that seem to be molded in ceramic; beyond the balustrade he created a garden with stately trees and flowing fountains. Figures representing the four parts of the world stand in this light-filled garden, contemplating the vision of the Virgin in Glory at the center of the sky. In the St. Michael Hofkirche in Berg am Laim, he abandoned all previous conventions and painted an asymmetrical procession of prelates, princes, and commoners to the grotto of the archangel Michael; the figures recall the 18th-century ceramics of Franz Anton Bustelli. In the Festsaal of Nymphenburg Palace (1757), Zimmermann brought the pastoral motif to poetic heights. The subject of this fresco (executed with assistants) — the kingdom of Flora — was particularly suited to a poetic treatment. Zimmermann created illusionistic paintings of a reality different from the dramatic, visionary, and didactic worlds presented by his contemporaries. His was an Arcadian and dreamlike world more suited to the refined society of his time.

The works of Christian Wink (1738–97) are more dramatic and often have a folk character: for example, the decorations of the parish church of Lohe and those of the Church of St. Leonhard, near Dietramszell, in which horses, oxen, and peasants in contemporary costume appear together with saints. Wink decorated the dining hall of Schleissheim Palace with great imagination and was also adept at architectural perspectives, as is demonstrated by many of his drawings (Munich, Staat. Graphische Samml.). Very similar to Wink in temperament was Joseph Magges (1728–69; Stiftskirche at Altomünster). Johann Holzer (1709–40) returned, instead, to illusionistic architectural structures (e.g., gallery in the Residenz, Würzburg), and his lively, realistic figures (Church of St. Anton, Partenkirchen) seem to anticipate the works of Tiepolo in Würzburg. He was a skilled decorator of palace façades, as was also Franz Zwinck, whose fresco in the form of a *theatrum sacrum* on the façade of the so-called "House of Pilate" in Oberammergau is outstanding. Gottfried Bernhard Göz (1708–74), of the school of Augsburg, framed his allegorical compositions in *quadratura* paintings of elegant *rocailles* in contrasting colors (e.g., Klosterkirche in Birnau) and often made use of the central station point, as found in earlier *quadratura* painting, even though his works (especially the late ones) have elements that are characteristic of rococo style. The frescoes of Franz Martin Kuen (1719–71) give evidence of his contacts with Zimmermann, whose influence is particularly evident in the ceiling of the library at Wiblingen.

Matthäus Günther (1705–88) came from the workshop of the Asams and was active in Swabia, Franconia, Bavaria, and especially in the Tirol. During his stay in Italy, Günther had gained extensive acquaintance with architectural treatises, including the works of Pozzo, and he was not without sympathy for the teachings of Holzer and Tiepolo. Günther distributed his architectural *sottinsù* asymmetrically (parish church at Götzens), according to the rococo taste, and accentuated the contrast between foreground and background planes with strong contrasts of light and shade. He followed the teachings of Asam more closely in his decoration of the parish church of Wilten; in the Stiftskirche of Rott am Inn the influence of Troger is apparent. Johann Zick (1702–62) was influenced by Pozzo and, in the modeling of his figures, by Tiepolo. From Günther he derived his use of contrasts of light and shade in the foreground and background planes of his *sottinsù* paintings. In Grafenrheinfeld, he employed the chiaroscuro technique to create the dramatic effect of a landscape populated with figures recalling those of Rembrandt. Later artists were more or less mannerists and repeated the same themes somewhat perfunctorily.

In Austria, rococo illusionistic painting remained more firmly rooted in *quadratura* painting; therefore, the visual transition from the actual architectural structure of a room to the perspective painting on the ceiling was generally effected by introduction of an "attic" level, simulating either a balcony in the manner of the local late baroque or a *quadratura* painting of stuccoes and *rocailles* (generally executed by Italian artists).

Troger had many disciples in Austria, including Johann Bergl (1718–89), in whose compositions the landscape element is predominant (Klain-Mariazell, 1764–65; University church in Budapest, 1766). In the Sommerpavillon of the abbey at Melk (1763–64), in Schönbrunn Castle (Vienna, 1765), and in the Castle of Ober-Sankt-Veit (1773), he covered the walls with flowers and exotic plants (perhaps deriving his motifs from tapestry), as if the room were a garden opening on all sides to sunny views of palaces and other gardens, thus giving elegant expression to the perspective principles that Paolo Veronese had enunciated in his frescoes at Maser. Johann Lucas Kracker (1717–79), whose style echoes Viennese trends and who belonged to the school of Troger, painted his best-known work in the Church of St. Nicholas in Malá Strana, Prague.

In the last quarter of the 18th century, when neoclassicism began to prevail in Europe, Austria still had a number of painters — such as Martin Knoller, Franz Sigrist, Johann Martin Schmidt, and Januarius Zick — who remained faithful to illusionistic perspective painting, which was already showing signs of decadence. Familiar schemes were being repeated over and over again, with little or no conviction and feeling; there were no new sources of inspiration, perhaps because those absolutist principles which had made possible its development became completely outdated after the French Revolution. Still, the highest note of Austrian rococo was struck during this time by Franz Anton Maulbertsch (1724–96), whose art drew from the most varied sources — from Rembrandt, from Italian painters such as Piazzetta, Bencovich, Ricci, and Bazzani, and perhaps even from Gregorio Guglielmi (1714–73). These diverse influences enabled Maulbertsch, at the very moment of the decline of illusionistic painting, to develop a dramatic technique based on the strong effect of colors against light, brightened by splashes of almost transparent red and touches of white. His architectural perspectives prove that he must have been well acquainted with the works of Pozzo, but at the same time they appear transfigured by color and immersed in a fiery, vibrant atmosphere. Delightful genre details are introduced into his paintings of religious themes. His figures are almost never clearly outlined but are formed of strokes of color either in full light or in shadow (XII, PL. 159).

BIBLIOG. *Treatises*. S. Serlio, Il libro secondo di prospettiva, Paris, 1545; D. Barbaro, La pratica della prospettiva, Venice, 1569; M. Bassi, Dispareri in materia d'architettura e prospettiva con pareri di eccellenti e famosi architetti, che li risolvono, Brescia, 1572; C. Sorte, Osservazioni sulla pittura, Venice, 1580; J. Barozzi da Vignola, Le due regole di prospettiva pratica [1583], con commentari di P. Egnazio Danti, Rome, 1883; G. P. Lomazzo, Trattato dell'arte della pittura, Milan, 1584; G. Viola Zanini, Della architettura libri 2, Padua, 1629; A. Putei (Andrea Pozzo), Perspectiva pictorum et architectorum, Rome, 1692 (Eng. ed., London, 1707); E. Zanetti, Trattato teorico, pratico di prospettiva, Bologna, 1766.

General works. a. Italy: L. Lanzi, Storia pittorica d'Italia, 1789, Bassano (6 vol. ed., 1809); A. Colasanti, Volte e soffitti italiani, Milan, 1915; P. Toesca, Affreschi decorativi in Italia fino al secolo XIX, Milan, 1917; T. Gerevich, Questioni sull'arte barocca e sulla pittura bolognese, Atti X Cong. int. di storia dell'arte, Rome, 1922, pp. 285–90; W. Arslan, L'eredità di Melozzo, Melozzo da Forlì, I, fasc. I, 1937, pp. 19–23; F. Würtenberger, Die manieristiche Deckenmalerei in Mittelitalien, Römisches Jhb. f. Kg., IV, 1940; E. Feinblatt, Jesuit Ceiling Decoration, AQ, X, 1947; R. Bossaglia, Affreschi dei Galliari nelle ville lombarde, Arte lombarda, III, 2, 1958, pp. 105–13, IV, 1, 1959, pp. 131–44; A. F. Blunt, Illusionistic Decoration in Central Italian Painting, J. Royal Soc. of Arts, CVII, 1959; A. M. Romanini, Quadraturisti milanesi della prima metà del XVIII secolo, Storia di Milano, XII, Milan, 1959, pp. 748–51; R. Bossaglia, Riflessioni sui quadraturisti del Settecento lombardo, CrArte, N.S., VII, 1960, pp. 377–98; W. Schöne, Zur Bedeutung der Schrägsicht für die Deckenmalerei des Barock, Festschrift Kurt Badt, Berlin, 1961; L. Crosato, Gli affreschi delle ville venete del '500, Treviso, 1962; P. Charpentrat, Baroque Italie et Europe centrale, Fribourg, 1964.

b. Germanic regions: J. von Sandrart, Iconologia Deorum oder Abbildung der Götter, Nürnberg, 1679; H. Tietze, Programme und Entwürfe den Grossen österreichischen Barockfresken, Jhb. der Kunsthist. Samml. des allerhöchsten Kaiserhauses, XXX, 1911–12; A. Feulner, Christian Wink, Der Ausgang der kirchlichen Rokokomalerei in Südbayern, Munich, 1912; H. Hammer, Die Entwicklung der barocken Deckenmalerei in Tirol, Strasbourg, 1912; A. Feulner, Süddeutsche Freskomalerei, Münchner Jhb. der bildenden K., X, 1916–18; M. Dvorak, Zur Entwicklungsgeschichte der barock Deckenmalerei in Wien, Vienna, 1920; A. Feulner, Die Zick, deutsche Maler des 18. Jahrhunderts, Munich, 1920; K. von Garzarolli-Thurnlackh, Findlinge zu "Programme und Entwürfe" zu den grossen österreichschen Barockfresken, Belvedere, 3, 1923, pp. 118–23; M. Riesenhuber, Die Kirchliche Barockkunst in Österreich, Linz, 1924; B. H. Röttger, Malerei in Untenfranken, Augsburg, 1926; A. Feulner, Skulptur und Malerei des 18. Jahrhunderts in Deutschland, Wilpark-Potsdam, 1929, p. 146 ff.; H. Ginter, Südwestdeutschen Kirchenmalerei des Barock, Augsburg, 1930; A. Feulner,

Bayerisches Rokoko, Munich, 1932; G. Adriani, Die Klosterbibliotheken des Spätbarock in Österreich und Süd-deutschland, Graz, 1935; K. L. Schwarz, Zum ästhetischen Problem des "Programms" und der Symbolik und Allegorik in der barocken Malerei, Wiener Jhb. f. Kg., II, 1937, pp. 77–88; H. Tintelnot, Deckenmalerei, in O. Schmitt, Reallexikon zur Deutschen Kunstgeschichte, 1937 ff., III, col. 1158 ff.; H.Tintelnot, Barocktheater und barocke Kunst, Berlin, 1939; H. Tintelnot, Die barocke Freskomalerei in Deutschland, Munich, 1951; M. B. Heinhold, Süddeutsche Fassaden Malerei, Munich, 1952; N. Lieb, Barockkirchen zwischen Donau und Alpen, Munich, 1953; E. Guldan, Die jochverschleifende Gewölbedekoration von Michelangelo bis Pozzo und in der bayerisch österreichischen Sakralarchitektur, Dissertation, Göttingen, 1954; H. Tintelnot, Barock Freskomalerei in Schlesien, Wiener Jhb. f. Kg., XV, 1954–55; B. Grimschitz, R. Teuchtmüller, and W. Mrazek, Barock in Österreich, Basel, 1960; H. Bauer, Zum Ikonologischen Ail der süddeutschen Rokokokirche, Münchner Jhb. der bildenden K., 12, 1961, pp. 218–40. Additional references for regional and specific works and for individual artists are to be found in the bibliographies for BAROQUE ART; GERMAN ART; ITALIAN ART; PERSPECTIVE; SCENOGRAPHY.

Fernanda DE' MAFFEI

Illustrations : PLS. 97–110.

PERU. The Republic of Peru (Republica del Peru), independent since 1821–24, is situated in a zone of the Central Andes which, together with the Bolivian Altiplano, was in the 3d millennium B.C. the seat of one of the most advanced indigenous American cultures. The period of the highest development of Andean civilization was that between the 3d and 8th centuries of our era, when numerous regional cultures of a high artistic level flourished (see ANDEAN PROTOHISTORY). With the Spanish Conquest and the destruction of the Inca Empire in the 16th century, indigenous arts and cultures declined or disappeared completely. In the succeeding four centuries the development of Peruvian art was decisively conditioned by imported Spanish art, which in turn, however, absorbed traditional local elements. After the declaration of independence in the early 19th century, the country became more receptive to diverse European influences.

SUMMARY. Historical background (col. 243). Cultural and artistic phases (col. 244). *Pre-Columbian era*: *a. Chronology; b. Geographical distribution of sites; c. Stylistic traditions in the minor arts. Colonial and modern periods*: *a. Architecture; b. Painting and sculpture.* Art centers and archaeological sites (col. 257). *Highlands: a. Northern; b. Central; c. Southern. Coastal area*: *a. Northern; b. Central and southern.*

HISTORICAL BACKGROUND. To outline a geographical history of Peru for the period prior to Inca domination is virtually impossible. With the rise of the Cuzco dynasty, not only Peru but also a large part of the neighboring states were united under the scepter of the Inca to constitute the empire known as Tahuantinsuyu (Of the Four Regions). Four roads linked Cuzco, the political and religious capital and the "center" of the empire, with the four parts of the state: namely, the Chinchasuyu, which included almost all of central and northern Peru as well as the territory comprising modern Ecuador; the Cuntisuyu, along the coast between Ica and Moquegua; and Antisuyu, which began on the eastern slopes of the Andes, but the boundaries of which are difficult to establish because of the nature of the terrain and the hostile attitude of the inhabitants of the Amazonian forests; and the Collasuyu, the largest region, which included the whole of the basin of Lake Titicaca, a large part of Bolivia, the north Argentine plateaus, and northern Chile.

Nothing is known of the antecedents of the large and small coastal kingdoms that were formed along the Peruvian coast in about the 13th century. The most important was that of the Chimu, the boundaries of which during the Inca period reached from Tumbez in the north to the vicinity of Lima in the south. The capital was Chan-Chan, near Trujillo. Other smaller kingdoms were the Chincha, with a coastal area corresponding roughly to that of the modern department of Ica, and the Cuismancu kingdom, between the Chancay and Rímac valleys. All the coastal kingdoms, although they continued to retain their own sovereigns, were subjugated by the Topa Inca Yupanqui (1471–93).

The origin of the Inca dynasty is lost in myth; chronicles from the time of the Conquest mention a list of 13 monarchs, but only from the time of the ninth of these rulers, Pachacuti (1438–71), a great conqueror and a political and religious reformer, are the data historically acceptable. The Inca achieved national unity in a very short time — little more than fifty years — by means of military campaigns and peaceful treaties. They dominated their immense territory by means of a complex and efficient bureaucratic system, by the imposition of sun worship as the state religion, and by the adoption of Quechua (the language of the most numerous and most important tribe subjugated by the Inca) as the universal language. There was no independent history for the four regions of the Tahuantinsuyu, which truly constituted a political, religious, and cultural entity.

The Spaniards of Pizarro destroyed this empire when it was less than a century old; only a small Inca kingdom in the Cuzco region survived until 1572.

The modern republic of Peru has been divided into 21 departments, but these do not in any way represent the divisions existing in the Inca or pre-Inca periods. The names of the departments are chiefly of colonial origin (in only a few is it possible to find an ancient Quechua or Aymar root), and the divisions reflect the political and administrative requirements of modern times. However, several cities that grew up on or near ancient sites bear precolonial names.

* *

CULTURAL AND ARTISTIC PHASES. *Pre-Columbian era.* Peru and highland Bolivia together form the Central Andes, a region in which there developed one of the most advanced cultures of ancient America. The two areas have many cultural features in common, and during two periods they were united to some extent politically, as well as artistically by so-called "horizon styles" (Tiahuanaco and Inca). At other times there was a diversity of styles among the different districts, particularly in pottery production; this variation was largely the result of complex and broken topography. Late in the 15th century the Inca Empire spread outward to include northern Chile, northwest Argentina, and a large part of Ecuador; before that time, however, the general artistic development of these areas differed sufficiently from that of the Central Andes for them to be treated as separate entities.

The Central Andean area falls into three zones, one of which — the forested lowlands east of the Andes — can be ignored for present purposes, since no notable civilizations ever flourished there. Of the other two zones, the coastal plain is a narrow strip of rocky and sandy desert, crossed from east to west by a series of irrigated river valleys running from the mountains to the sea. These valleys were the cradles of many Peruvian cultures; the deserts separating them were formidable barriers, but some of the more powerful coastal states were able to group several of these fertile areas under their rule. East of the coastal plain lies the zone of the high Andes, with its snow-covered peaks, bleak plateaus, high passes, and deep valleys, all of which constituted obstacles to communication that frequently led to diversities in art styles.

a. Chronology. It has been found convenient to classify Peruvian cultures in the following periods: preceramic (ca. 3000 B.C.–1000 B.C.); preclassic (ca. 1000 B.C.–A.D. 250); classic (ca. A.D. 250–750); postclassic (after A.D. 750). The preceramic period was marked by agricultural activity on a small scale in the coastal valleys, but maize, later the staple plant, was not yet grown; and as the name demonstrates, there was no pottery. The preclassic period is taken to begin with the introduction of decorated pottery and other practices, including the cultivation of maize, probably from Mexico; plain pottery had appeared somewhat earlier, but its development does not seem to have coincided with any notable difference in the way of life of the people. During the preclassic period the Peruvians used all the known pre-Columbian techniques of agriculture, weaving, pottery, and metallurgy, except the manufacture of bronze. In the classic period these techniques reached their highest level, permitting the production of technically perfect works of art, particularly in weaving and pottery. At the end of this period, the art style associated with Tiahuanaco (see BOLIVIA) was dispersed over certain other highland districts and most of the coast. The postclassic period is distinguished by far-reaching political developments and by something akin to mass production of pottery, textiles, and metalwork, which in the case of pottery resulted in artistic decadence. The period ended with the meteoric rise of the Incas, beginning about 1440, and the organization of their empire on a pan-Peruvian scale.

Peruvian art as a whole does not rise to the heights of inspiration reached by the Maya and the Mexican cultures (see MIDDLE AMERICAN PROTOHISTORY). In pure craftsmanship it is unsurpassed — and this is the more remarkable in that the Peruvians, like other New World peoples, depended on great manual skill, for they used the simplest possible tools, such as the Peruvian belt loom — but it has the limitations of the craftsman's art. As Kroeber pointed out long ago, the craftsman's feet were mired in technology, and he felt with his hands rather than with his emotions. Nevertheless, Peruvian art produced many objects of great charm and interest.

Remains indicate that the keynote of most Peruvian architecture was simple massiveness, lacking even that concern with symmetrical

Peru, principal centers of archaeological and artistic interest. *Key*: (1) National boundaries; (2) archaeological sites.

arrangement of great masses in space which is found in Middle American ceremonial centers such as Teotihuacán (see MEXICO) and in those of the Maya (see MIDDLE AMERICAN PROTOHISTORY). Normal materials used in the highlands were stone, mud, or a mixture of the two; on the coast, mud was customary, either in sun-dried bricks (adobe) or in tapia. The wall surfaces of ceremonial or royal buildings were sometimes relieved by painting or by designs in mud plaster (I, PL. 161).

b. Geographical distribution of sites. Though much broken by natural obstacles, the area of the highlands can, for purposes of description, be divided into three sections — northern, central, and southern.

The known monuments in the northern highlands are in the Cajamarca region and in and around the Callejón de Huaylas. Near Cajabamba, south of Cajamarca, is Marca Huamachuco, where the Cerro del Castillo is crowned by a defensive double wall of coursed rubble masonry about 40 ft. high, with the space between the walls formerly divided by floors. Nearby are smaller circular fortresses of the same sort. Within the perimeter of the Cerro del Castillo are the ruins of houses in the form of long narrow galleries of two or more stories (with masonry that resembles sections of the outer wall), some of which are grouped around courts and the ruins of other buildings. The main buildings appear to date from the classic period.

At a lower level in the same neighborhood are the remains of Viracochapampa, which has been interpreted as a town into which a populace was gathered under Inca domination. Built of masonry of local type similar to that of Marca Huamachuco, it was quite different in layout, being built on a rectangular plan. (For plans of these sites, see T. D. McCown, 1945.)

In the Valley of the Marañón, east of Cajamarca, some great walled towns with mummies encased in the walls have been reported near Kuélape; these appear to belong to the classic or early postclassic period. In the same region are curious little gabled stone burial houses, painted red and white, which are built in virtually inaccessible positions against cliffs.

In the Callejón area, by far the most important site is Chavín de Huántar (east of the Cordillera Blanca), which consists of a rectangular grouping dominated by a massive platform honeycombed with stone-lined galleries and rooms on three stories, connected by stairways and slopes, and supporting the remains of rectangular houses. A stairway cutting the east wall leads down to a terrace, beyond which is a sunken court with platforms on either side. The river lies beyond. Other buildings have been partly buried by landslides. The buildings are faced with fine granite masonry, in which a course of thick slabs alternates with one or two courses of thin ones. A cornice, mostly in ruined condition, once included carved slabs, and a row of massive human and feline heads was tenoned into the wall below it. In one of the rooms of the main building, there is a prism-shaped stone column carved with a feline head doubtless representing the god worshiped there (I, PLS. 166, 167). Stones from the cornice carved in low relief represent the same creature, and other show condors with feline fangs and eyes (I, PL. 166). Two cylindrical columns (a rare feature in Peruvian architecture) bearing similar motifs have been discovered and have been set up flanking an entrance. (For plan of Chavín de Huántar, see Bennett, 1944, fig. 25.)

Though still little explored, Kuntur Wasi, in the province of Cajamarca, promises to be an important and imposing Chavinoid site.

In the Callejón de Huaylas itself, the most notable building is a three-storied structure (perhaps a temple) at Wilkawain, about 8 miles north of Huaraz. About 33 × 49 ft. in area and about 30 ft. high, it is of rough masonry with a low-pitched gabled roof of slabs, on which earth and stones were piled to produce a domed outline. Below the eaves is a recessed zone or cornice, under which carved stone heads were formerly tenoned.

Each floor of this building, which is probably early postclassic in date, is divided into seven rooms. In the same area are poorly preserved remains of lesser stone buildings, including houses of one to three stories and some subterranean stone-lined galleries, at least some of which, since they have yielded pottery of Recuay style, belong to the

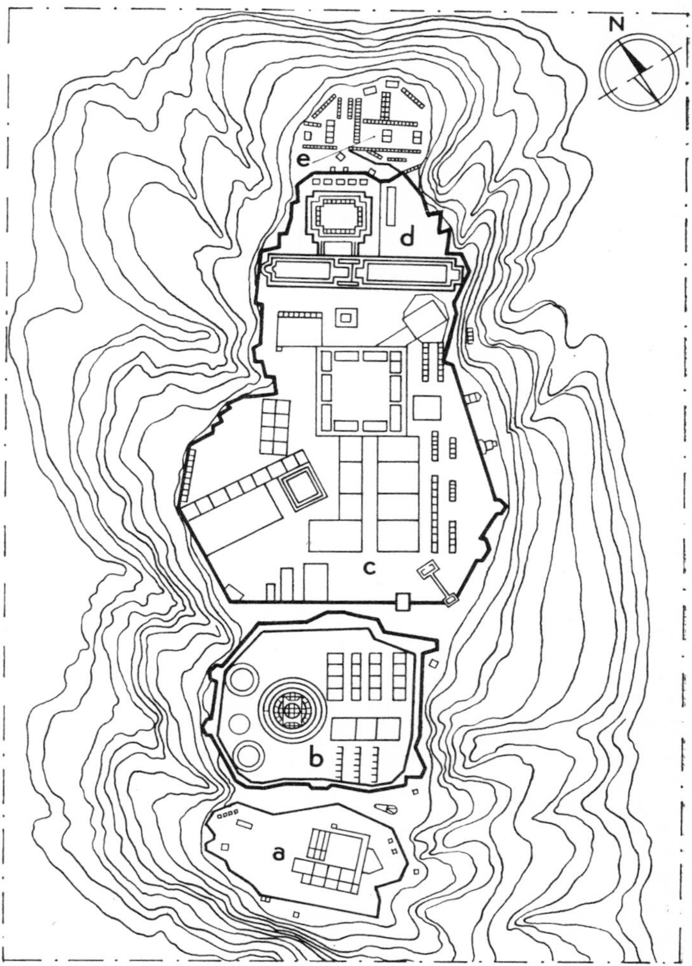

Plan of the ruins of Marca Huamachuco. (*a*) Cerro Viejo; (*b*) Cerro de la Monja; (*c*) Cerro de la Falda; (*d*) Cerro del Castillo; (*e*) Cerro Amaro.

classic period. At Huaraz is a terraced truncated pyramid (poorly preserved) containing interior galleries.

The region of the central highlands is little known, but in its northern part, at Huánuco, is a large Inca city with masonry of the standard types (see below), and another important Inca site had been reported at Bonbón, at the north end of Lake Junín. In the Mantaro Basin, a number of hilltop towns which contain many rough stone house foundations and which are surrounded by multiple defensive walls have been reported in the Jauja district. The large and important town of Huari, about 16 miles north of Ayacucho, lies in the same basin. It covers an area at least a mile square, and most of the constructions are terraces and walls of rough stone which, though much disturbed by cultivation, show no signs of regular planning originally. Amid the general confusion are a few groups of subterranean dressed-stone chambers. The pottery found relates the site both to classic Tiahuanaco and to coastal manifestations of its style. (For plan details, see Bennett, 1953.)

At many sites in the province of Canta, in the Andes east of Lima, there are cylindrical stone burial towers, or chullpas. These are apparently pre-Inca in date but are so well preserved that they are unlikely to be very much older than the Inca Empire. As architectural forms, these structures expand somewhat toward the top; within is a chamber with a central column to support the roof. These chambers are generally found empty, but mummy bundles are buried in smaller chambers beneath the floors. There are good examples of chullpas at Canta Marca, where there are also some rectangular Inca buildings. In the western part of the same province, at Chipprak and other sites, there are rectangular towers characterized by one or more large trapezoidal niches which extend the whole height of the façade and within which are small doorways, also trapezoidal in form. A small structure of the same general type is sometimes annexed to the main one. These buildings have subterranean chambers, some containing mummies; the main chambers, however, often have wall niches containing crude pottery and a fireplace with a chimney, elements which have suggested to some scholars that they may be primarily dwellings. Low rectangular buildings with rows of large trapezoidal niches which have been described as funerary temples, are found at the same sites. (The only source for this area is P. E. Villar Córdova, 1935.)

Monuments abound in the southern highlands, which encompass the homeland of the Incas. The oldest known site with notable architectural remains is Pucára, northwest of Lake Titicaca, which belongs to the late preclassic period and is thus older than the main buildings of Tiahuanaco (see BOLIVIA). The focus of the site is a sunken court about 49 ft. square, under which are two burial vaults lying in a horseshoe-shaped enclosure of finely dressed stone walls that contain small compartments opening toward the court. Each compartment contains one or two altarlike slabs. Associated stone carvings consist of steles and slabs showing geometrical and fish designs, as well as stone statues resembling the rare, more naturalistic Tiahuanaco type. The pottery is clearly related to that of Tiahuanaco, as is particularly evident in the treatment of animal heads, which have eyes divided vertically into black and white halves. (See *Handbook of South American Indians*, Julian H. Steward, ed., Washington, D.C., II, 1946, pls. 37, 38; V, 1949, p. 41.)

The Titicaca Basin is the principal locality for chullpas, the finest being at Sillustani. They may be round or square and either faced with fine, dressed masonry, or entirely built of rough stones. Most of them contain a burial chamber, but some of rough stone are solid and have the burial chamber underneath. The doorway generally faces east. They are believed to have been built by the Aymara both before and after the Inca conquered them. The best examples have a plain projecting cornice and a corbeled vault (I, PL. 158).

Piquillacta lies about 16½ miles southeast of Cuzco. This site is about 2/3 of a mile square and is built on a rectilinear plan, with rough masonry resembling that of Viracochapampa in the north. Evidence of plastered floors and walls has been reported. Some walls over 30 ft. in height remain, and there are signs that some buildings had at least two stories. There is a notable absence of doors or windows. No pottery has been found, but there are Inca buildings nearby, and it has been suggested that the town was built and occupied by a people subject to the Inca, possibly *mitimaes* (forced immigrants) from Cajabamba.

The greatest concentration of Inca ruins is found around and in Cuzco, where Inca walls were often used to form the foundations of Spanish buildings, and where Spanish heraldic doorways were set in Inca walls, making a combination of unique interest. All the Inca work visible is believed to date from the imperial Inca period (15th cent.), and differences in the type of masonry point to differences of function and not of age. Dominating the city is the great fortress of Saccsaihuaman (Sacsahuaman), with its triple zigzag rampart of "megalithic" masonry (I, PLS. 153–155) composed of stones that reach individual heights of over 25 ft. Masonry of this type was used chiefly for retaining walls of platforms and terraces. Within the ramparts are remains of buildings and the foundations of a large round tower that contained a reservoir. Nearby, at Kenko, is an artificially enlarged cave situated beneath limestone outcrops that were carved into steps and platforms. This cave is believed to have been the burial place of the emperor Pachacuti. Behind it a pointed rock, undoubtedly sacred, emerges from a platform built of fine regular masonry of the type reserved for the most important structures; the platform and sacred rock constitute the focus of an amphitheater of similar masonry (I, PL. 156). Pucára, which lies in a small valley running up from this area, is a group of buildings of polygonal masonry of small stones, with rectangular dressed stones framing the doorways. Pucára means "fortress" in Quechua, and this may well have been its purpose. Higher still is situated Tampu Macchai (Tampu Machay; I, PL. 156), a group of terraces with fine polygonal masonry retaining walls and trapezoidal niches. A small stream descends from terrace to terrace in narrow stone-lined channels.

Cuzco itself, like other Inca cities, is built on a rectilinear plan somewhat modified by the topography. Its chief group of temples was on the site of the Dominican convent, where substantial remains of a series of halls of squared masonry are grouped round the main cloister and a curved wall of masonry of the finest quality forms part of the apse of the church (I, PL. 152). The famous "stone with twelve corners" (I, PL. 153) is part of the syenite retaining wall of a platform called Jatunrumiyocc, on which a Spanish house now stands; another Inca wall faces it across the narrow street. The regular, squared type of masonry is well exemplified by the boundary wall of the Accla Huasi (House of the Chosen Women; now the Convent of S. Catalina), opposite which is a wall (I, PL. 152) of similar work belonging to the reputed Palace of Huayna Capac. From the many other Inca remains, one more may be singled out for particular mention; namely, Colqampata, on the outskirts of Cuzco toward Saccsaihuaman, the principal feature of which is a wall with a row of fine trapezoidal niches (I, PL. 152).

Beyond Cuzco it is possible to note only the most prominent Inca sites. To the northwest, in the Urubamba Valley, are the ruins of Machu Picchu (I, PLS. 147–151), with important ceremonial buildings, a city situated amid innumerable agricultural terraces on a saddle falling steeply on either side to the river almost 1,500 ft. below. One remarkable feature of this site is the way in which some of the finer buildings seem almost to grow out of the rocks on which they are founded; another is a series of stone basins or baths, set one above the other on a steep hillside, with stone gullies connecting them. Machu Picchu also contains the finest known example of an *intihuatana*, a stone with a flattened top, from which there emerges a kind of gnomon projecting upward. These forms, found in several places (cf. I, PL. 156), appear to have been sundials, or solar "observatories," traditionally connected with sun worship. Several smaller sites discovered hidden in the scrub around the Urubamba Gorge, near Machu Picchu, repeat features found at the major site, such as terraces (magnificent examples at Inti Pata and Wiñay Wayna), stairways, baths, shrines, and houses. As at Machu Picchu and other sites, the masonry varies from the finest rectangular and polygonal types, with stones closely fitted, to that composed of roughly dressed fieldstones set in mud and chinked with smaller stones. Only the "megalithic" type, the variety of polygonal masonry using very large stones that is so prominent at Cuzco, is absent or rare.

Higher up the valley is Ollantaytambo, which has terraces and buildings of the usual rather standardized Inca character but which is distinguished by a high terrace with a unique retaining wall made of six immense rectangular blocks set on end with a narrow vertical course of smaller blocks set between each two large ones. A notable feature is a stone basin backed by a stone carved with an unusual stepped design (I, PL. 157).

Still higher up the same valley is Pisac, where a group of Inca constructions, including a fine *intihuatana* and some adobe houses, looks down on a great range of agricultural terraces.

At Moray, about 3 miles north of Maras, are three enormous artificial circular depressions, each nearly 500 ft. across, at the head of a tributary valley of the Urubamba. Lined with concentric stone-faced terraces, these are a most impressive sight and must represent a vast amount of labor.

Southeast of Cuzco, at Racche (near Sicuani), is a great wall pierced by trapezoidal openings and rising to a height of over 40 ft. (I, PL. 163). The lower part, about 8 ft. high, is of fine Inca masonry, and the remainder is of adobe. It forms the best-preserved part of a group called the Sanctuary of Viracocha (Wiraqocha).

There are extensive Inca remains of standard type on Titicaca Island.

The coastal area, like the highlands, may also be considered in sections, from north to south. The extreme northern part, from the Ecuadorian frontier, to the Sechura Desert, can be ignored, since it could never have been thickly populated and lacks important monuments. The area from the Rio Leche to the Rio Virú, however, includes important sites. The remains observed in this area fall into five main classes: fortresses, ceremonial centers, cities, towns, and aqueducts. Owing to the ravages of time, many of them appear to the modern-day visitor to be shapeless masses of adobe or scattered remnants of foundations.

The known fortresses are in the Virú Valley and belong to the late preclassic and classic periods. Placed in commanding positions, these seem to have served both religious and military purposes. The best surviving example is El Castillo de Tomaval, a steep terraced adobe pyramid of castlelike outline adjoined by lesser buildings, all within a rough stone wall and crowning a spur in a strategic position overlooking the valley.

Ceremonial centers consist of pyramids or groups of pyramids, to which neighboring dwelling sites, if present, are subordinate. Most belong to the classic period, but some in the more northern valleys may be early postclassic. They usually consist of great steep rectangular adobe blocks broken, if at all, only by narrow terraces and approached either by a direct ramp or by one running round the pyramid. The most famous of these are the Huaca del Sol and the Huaca de la Luna in the Moche Valley. The Huaca del Sol ("Pyramid of the Sun"; I, PL. 163) consists of two rectangular platforms with sides broken by five narrow setbacks, the smaller platform being approached by a ramp and the larger bearing the remains of a stepped pyramid about 335 ft., square and 75 ft. high. The platforms may have been built on a natural hillock. The main structure of the Huaca de la Luna is a great six-stepped platform without a pyramid. It is built against the slope of a hill, along with lesser buildings, in one of which there are remains of wall paintings on soft white plaster in flat colors (red, yellow, light blue, pink, and brown), with the main outlines incised and then painted black. These paintings show combats between men and personified weapons, with the weapons the apparent victors. The style is that of the Mochica pottery, but with a less restricted color scheme than that of the pots. There are several cemeteries around the Huaca.

Similar examples, in various stages of decay, are El Brujo in the Chicama Valley, Dos Cabezas in the Jequetepeque Valley, Pampa Grande in the Lambayeque Valley, and Batán Grande in the Leche Valley. At the last-named site, one of a group of pyramids has a polychrome painted frieze, apparently of early postclassic style. Chotuna, in the Lambayeque Valley, is unusual in that it has a courtyard alongside its pyramid; this contains the remains of a painted frieze in relief, also probably postclassic.

Another type of pyramid rises like a great flight of steps on one side and falls sheer on the others; a notable and well-preserved example located near Etén, in the Lambayeque-Etén Valley, has three terraces of 10, 20, and over 50 ft. high.

The monuments classed as cities and towns are population centers of late postclassic date belonging to the Chimu state, both before and after its conquest by the Inca. The towns are composed largely of undifferentiated house ruins, and are mostly situated where large numbers of people were needed for the protection of irrigation canal intakes and for agricultural developments. The cities contain great compounds enclosed by tapia walls, up to 30 ft. high, within which are structures of several types. The finest of these structures, which have walls covered with designs in molded plaster (I, PL. 161), were probably the dwellings of the nobility (who had the compounds built in order to contain and overawe the common people). The compounds are carefully planned, and may include pyramids, sunken gardens, rows of houses, and small cell-like rooms (probably stores), and stone-lined reservoirs. The greatest of these cities, the Chimu capital Chan-Chan (near Trujillo), contains ten such compounds, with minor buildings, cemeteries, and irrigated areas between them (FIG. 250; I, PL. 160).

Pacatnamú (La Barranca), a smaller city in the Jequetepeque Valley, also contains pyramids that may be of earlier date. El Purgatorio, in the Leche Valley, is remarkable for its compounds surrounded on all sides by an almost 50 ft. drop, which takes the place of the normal enclosure walls.

Some of the larger irrigation canals have aqueducts that are still impressive constructions. A notable example, believed to be of the classic period, carries a canal across the mouth of a side valley of the Chicama near Ascope and is about 2/3 of a mile long and almost 50 ft. high.

The valleys of Santa, Casma, and Nepeña form the next great area of archaeological interest. Exceptionally strong highland influence is apparent in these valleys. It is shown in the use of stone in the structures called *castillos*, which have been reported in all three areas, but these are difficult to date because little pottery has been found. What pottery there is is mainly a utilitarian type with simple incised and reed-stamped decoration, which has not been related to the better-known successions to the north and south. The most striking of the *castillos* is Chancaillo in the Casma Valley, which has three concentric walls containing two circular towers, all of undressed but well-fitted stones. The doorways are designed for defense and

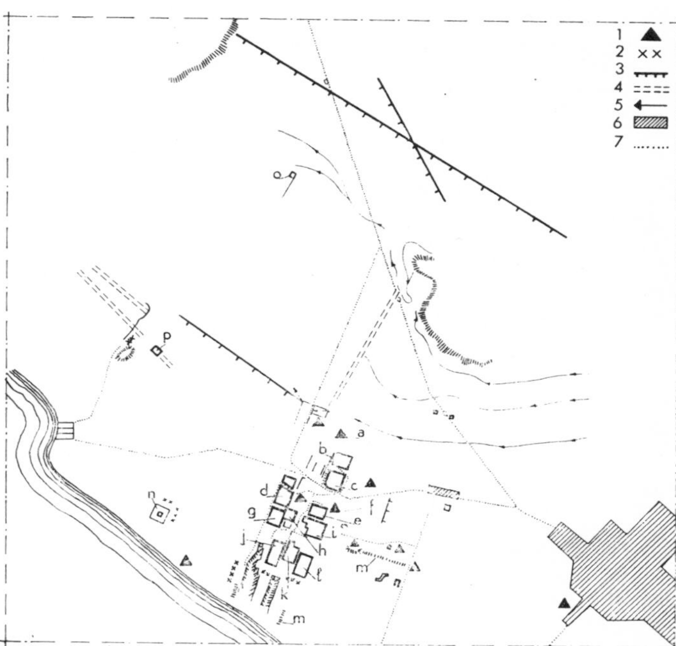

Archaeological plan of the city of Chan-Chan. *Key*: (1) Huacas; (2) irrigation canals; (3) walls; (4) pre-Hispanic streets; (5) necropolises; (6) modern settlements; (7) present-day roads. Principal monuments: (*a*) "Obispo" huaca; (*b*) "La Monjas" complex; (*c*) "Gran Chimú" complex; (d) "Velarde" complex; (*e*) "Bandelier" complex; (*f*) "Toledo" huaca; (*g*) "Labyrinth"; (*h*) "Tello" complex; (*i*) "Uhle" complex; (*j*) "Rivero" complex; (*k*) "Tschudi" complex; (*l*) "Chayhuac" complex; (*m*) dikes; (*n*) "Calvario de los Incas" complex; (*o*) silos; (*p*) storehouses (*from a drawing by H. Horkheimer*).

have wooden lintels. Chancaillo, which seems to be a refuge fortress, has been proved to be of late preclassic date.

Information about buildings of later date is scant. Pañamarca in Nepeña appears to be of the classic period. Consisting of terraces stepped against a natural outcrop, it has remains of very large mural paintings, not in the Mochica style. It is thought to resemble Pachácamac, south of Lima, more than the northern sites. There are no postclassic sites to compare in size with the cities of the north, although some small ones contain buildings of good quality. Punkurí Alto in Nepeña has a group of well-planned rooms and terraces that follow the natural contours of a hill, and there are two adobe friezes in which simple designs are produced by omitting certain bricks.

Nepeña and Casma are best known for their early preclassic Chavinoid temples, which are in themselves evidence of highland contacts. Each has some strongly individual features that were not apparent before partial excavation, owing to their weathered state — itself a measure of their age. Punkurí and Cerro Blanco, in Nepeña, which at first appeared to be merely weathered mounds, proved to contain buried rooms, clay columns, stairs, and clay-plastered walls decorated by carving and polychrome painting. Punkurí has a central stair on which are a modeled cat's head and paws; the stair covers a grave containing the remains of a decapitated woman and a stone vessel with carved Chavinoid designs. Cerro Blanco has rooms on a stepped platform, with clay-plastered walls decorated with Chavín feline eyes and fangs carved and painted in red and greenish-yellow.

Pallka, Moxeke, and Sechín (Cerro de Sechín) are in the Casma Valley. At Pallka the chief ruin is a stepped pyramid of stone and adobe, with four or five broad terraces on one side and narrow steps on the other, resembling to some extent the much later type found at Etén in the north. This pyramid is adjoined by many remains of rude stone rooms. Moxeke has a great stepped platform of stone and conical adobes, from which there rises a pair of low stepped pyramids and some lower platforms. In niches on the façade of the main platform are the remnants of a row of demifigures (probably human) modeled in clay and painted black, white, red, blue, and green; on some are found snakes like those on carvings at Chavín. Sechín seems to have had two terraces, with a central stairway from one to the other. The upper terrace is coated with clay plaster painted with feline designs. The stair is flanked on either side by a row of carved slabs which formed a facing of the lower terrace (I, PL. 166). The upright slabs, which carry a figure, a geometrical design, or, in one case, two columns of faces, alternate with pairs of square slabs bearing heads recalling the trophy-heads of Nazca art. The style of carving does not resemble that of Chavín, or indeed any other, but the general character of the site is

Chavinoid, and small finds from other parts of the coast bear motifs which suggest a link between the two styles.

The valleys of the central and south coast, from Huarmey to Nazca, comprise the next general area for discussion.

Little information is available about the ruins in the valleys north of the Rímac, in which Lima lies, with the important exception of the "Fortaleza" at Paramonga (I, PL. 164). This striking structure of mold-made adobe is generally believed to be a southern frontier of the Chimu state at its maximum extent but J. C. Tello regarded it as a ceremonial center. The material and the good state of preservation of the ruin make it certain that it is of late date.

In the Rímac Valley itself, which, from the point of view of site distribution, has been better studied than the others, most of the ceremonial centers are classic or earlier, two of the largest being Maranga, which is a cluster of 12 large and medium-sized pyramids, and Vista Alegre, a large pyramid surrounded by five smaller ones. Farther south, in the Nazca Valley, are the decayed remains of a large classic ceremonial site at Cahuachi, with a pyramid built around a natural hill as its focus, and a number of adobe-walled courts and cemeteries surrounding it. Pyramids of the same age at La Huaca de Alvarado and La Huaca de Santa Rosa, in the Chincha Valley, are very much weathered.

In the Lurín Valley is the major religious center of Pachácamac (I, PL. 162), which retained its importance up to the Spanish Conquest. The site is dominated by the massive Inca terraced structures of the "Temple of the Sun," which was greatly restored in 1939 and damaged again by earthquake in 1940. This overlooks the great terraced block of the Temple of Pachácamac and another much-weathered adobe mass, all within an enclosure wall, outside which is a large area of dwellings and an outlying Inca building said to be a House of the Chosen Women. At least the later stages of the Temple of Pachácamac appear to be late postclassic, but they seem to cover structures that go back to early postclassic or classic times. La Centinela, in the Chincha Valley, resembles the last stage of Pachácamac, since it has a great pyramid, believed to be the local Inca sun temple, with a large Inca "palace" attached to it. There are also remains of dwellings, indicating that there was a sizable population nearby. At La Cumbe, not far away, is a large adobe platform about 50 ft. high belonging to the late postclassic period (slightly pre-Inca); this was regarded by M. Uhle as a daughter sanctuary of the much older pre-Inca one of Pachácamac.

There are two cities in the Rímac Valley, Armatambo and Cajamarquilla. Like those in the north, they have walled compounds and cover a large area. Some of the compound walls have remains of relief decoration, and some are crowned with crenellations. Tambo de Mora, in the Chincha Valley, has remains of a large habitation area round a central structure that has been described as a court, from which stairs lead up to terraces, the whole being flanked by two massive pyramids. Uhle thinks that this was a chief's residence rather than a ceremonial center, but it is likely that it combined the two functions. Like the cities of the Rímac Valley, Tambo de Mora dates from the late postclassic period, and it can probably be classed as a city, though of different type. The well-preserved remains of Tambo Colorado, in the Pisco Valley, are of adobe on a stone foundation. These are long ranges of buildings built on terraces on gently rising ground, which are plastered and painted red, yellow, and white. There are double-recessed niches of contrasting colors, trapezoidal doorways, and openwork balustrades made by setting adobes diagonally in such a way as to form triangular openings. These buildings are probably the residential or administrative center of a small city.

Two other phenomena, which may be related, should also be noted — though they are not, strictly speaking, monuments. Inland from Paracas, near Pisco, is a depiction of a great tree (three branches, leaves, and flowers) formed by cutting trenches down to break the hard pan beneath the sandy surface of the earth. This form is about 420 ft. long and 240 ft. wide and resembles vegetable designs on pre-Inca textiles. It has been interpreted by the present inhabitants as three crosses (whence its modern name, Tres Cruces), and they have scoured it from time to time to preserve its outline. On the desolate pampas near Nazca there are a multitude of very long straight lines, geometrical figures, spirals, zigzags, and occasional representations of birds and fish marked out by clearing away the stones from the surface of the yellow sand. The forms are clearly apparent only from the air. There is some evidence that these may belong to the classic period.

The valleys south of Nazca are small, and no information is available about any monuments there. Moreover, it is unlikely that any notable ones will be found.

c. Stylistic traditions in the minor arts. Some tattered relics of a remarkable art style from the preceramic period have come to light along the coast. Mainly executed in twined weaving, these highly stylized and sophisticated representations of human beings, fish and snakes with heads at either end, birds, animals, and crabs were produced by floating or transposing the warps. Limited color effects were obtained by using natural brown and white cottons, blue dye, and red powdered pigment rubbed into the thread. The origin of this style is unknown, but its influence continued into later periods.

In the preclassic period, three major artistic traditions appeared and continued through the classic period. The first of these begins in the north with the incised or champlevé stone carving of Chavín de Huántar (I, PLS. 166, 167), which shares something of the impressiveness of the art of Mexico, whence indeed the religion it served may have been derived (see MIDDLE AMERICAN PROTOHISTORY). Some of its motifs appear in the mud-plaster modeling of the coastal buildings and in the vigorously modeled monochrome pottery of Cupisnique. Highland sculpture left no direct successor, but the pottery marks the beginnings of a coastal tradition that came to maturity in the well-known classic Mochica style (I, PLS. 176–180, 185–189), in which modeling is combined with a restrained use of color. The modeling is reputed to be highly naturalistic, but in fact the potters succeeded in achieving vivid impressions by skillful representation of only certain features and by stylizing others. The characteristic form of pottery is the stirrup-spouted jar.

The south coast tradition emphasizes color, both in pottery and textiles. Whereas the north tends to show mundane things, the south represents creatures of fantasy, such as demons and masked figures. There is an abundance of textiles, which cannot be entirely accounted for by better conditions of preservation; outstanding among these are the many-colored embroidered garments of the late preclassic Paracas graves (I, PLS. 182, 183). There is far less skill evident in modeling than in the north, and the most characteristic of the many pottery forms is the "double spout and bridge" jar (I, PL. 204). Animal forms are far more stylized than in the north. Starting experimentally with vivid resinous colors applied after firing (preclassic Paracas pottery), the polychrome tradition reached its climax in the mature classic style of Nazca, with pots painted before firing in up to eight colors on a background slip (I, PLS. 190–192, 197).

The third main tradition is that of the southern highlands, which was a great stoneworking district; but figural sculpture amounts to little more than the carving of pillars with designs in low relief, as at Tiahuanaco (see BOLIVIA). The pottery includes some graceful forms, especially a beaker with concave sides; however, the polychrome painting on it tends to heavy angularity in the representation of animal forms. Devices such as the eyes divided vertically into black and white halves and a nostril resembling a ring balanced on the end of the nose are well-nigh universal (I, PL. 203). These features appear in Peru in the preclassic period at Pucára, northwest of Lake Titicaca, in pottery on which the colors are outlined by incising and in a type more closely resembling that of Tiahuanaco found at the classic-period site of Huari, in the Mantaro Basin.

The beginning of the postclassic period on the coast is marked first by the appearance of Tiahuanacan influences in the coastal art styles, and later by their replacement with Tiahuanacoid styles, expressed by angular figures, divided eyes, and other features on pottery and textiles. In its homeland, the Tiahuanacan pottery style degenerated and finally gave way to local styles of no distinction. On the coast, local survivals blended with Tiahuanacan influences to produce various art styles, known chiefly from pottery. In the course of time highland influences died away; the range of pigments employed on pottery was reduced to three (black, white, and red) or fewer; and the local styles finally crystallized into three, corresponding to three coastal states that emerged not earlier than about A.D. 1200. None of these compares in spontaneity, variety, or interest with its classic or earlier predecessor in the same area. The chief of these states, namely, the Chimu in the north, produced a poor revival of the old modeling tradition of the area, including stirrup-spouted jars (I, PLS. 200–202). The modeling, which became hard and mechanical, is generally carried out in black or gray ware. The pots were produced in large quantities for domestic as well as for funerary purposes. Along the central coast, a second state produced the Chancay ware, covered with a crumbly white slip and painted with black details. Poorly modeled figural jars are characteristic of the Chancay output (I, PL. 207). In the south, the pottery style of the Chincha state was chiefly characterized by well-made bowls of angular silhouette, covered with small geometrical elements in black, white, and red that show the influence of textile patterns.

The final episode in the history of the pre-Columbian art of Peru was the spread of the Incas in the 15th century. Over the whole of Peru and beyond, they left traces of their presence in architecture and in pottery. The style known as "Cuzco polychrome" (normally black, white, and red) is expressed in a few rather pleasing standard shapes, chief among which is the graceful so-called "aryballus" (I, PL. 208). Another very common form is a shallow saucer with a handle in the form of a bird's head or loop. Most commonly the pottery is decorated with repeated small geometrical designs, but in rare instances they are charmingly decorated with a sprinkling of small llmas, fish, or other creatures. Some vessels especially the "aryballus," were made in the conquered provinces, and minor differences of outline are found among the various strains. The chief variant is found in the local blackware of the Chimu country, a hybrid form that was spreading throughout the coastlands when the Spaniards arrived.

Geoffrey H. S. Bushnell

Colonial and modern periods. a. Architecture. In the evolution of Peruvian colonial architecture, Renaissance and baroque periods can be distinguished in many works; examples of the Gothic are rare (S. Agustín, in Sana; S. Domingo, in Lima). The Renaissance style (16th and 17th cent.) is expressed either in classicizing tendencies (Ayacucho) or as an assimilation of the Spanish plateresque (church portals, especially those in the Lake Titicaca region and in a few mansions of the viceregal period). The baroque is found in original forms of indigenous derivation (Cathedral of Puno) and as evolved into a Churrigueresque style (S. Agustín, in Lima); and in the 18th century there developed a rococo style that can be traced to French influences (the tower of S. Domingo, in Lima).

The fundamental problem in an analysis of Peruvian colonial architecture is to determine how much of it is of Spanish derivation and how much of it is autochthonous. In architecture especially, the decisive influence is a combination of local factors (apart from social and political considerations) that are peculiarly Peruvian. Thus, plans of churches and houses differ according to local climate, and geological factors often condition the choice of building materials (e.g., in Cuzco, the use of stone from Inca palaces and temples; in Lima, because of frequent earthquakes, the development of earthquake-resistant vaulting known as *quincha*, made of canes and clay; in Arequipa, the use of sillar, which encouraged the development of a decorative mestizo architecture). The essential elements of the Spanish baroque — ardent religious feeling, exuberant ornamentation, amalgamation of forms — found a sympathetic response in Peru; religious fervor, very strong from the earliest years of the Conquest, increased with the first echoes of the Counter Reformation and stimulated the construction of many cathedrals.

From 1821, the year of the proclamation of independence, civil wars, revolutions, and the war with Chile interrupted the development of colonial art and created a genuine crisis in the visual arts. A result of the break with Spain was an anti-Spanish reaction that led Peruvian artists toward French models; but this influence, foreign to the indigenous spirit, remained uncertain and transitory; a return to native sources of inspiration soon followed. Two opposing tendencies were thus developed — one traditional indigenous, the other Europeanizing — and these have survived in the art of modern Peru.

Architecture, more than the other arts, reflected the changing trends of the times. In the first century of the republic, the new spirit of nationalism was responsible for the remodeling of the façades of a number of buildings in the neoclassic idiom. Nevertheless, the old traditional techniques of adobe and *quincha*, as well as certain colonial elements such as the plan with a wide portal, balconies with jalousies, a patio, stables, and garden, remained in favor. European architects spread French neoclassic academicism, which did not develop any distinctly local characteristics and often produced hybrid results. A few houses well adapted to their environment, such as the Iturregui mansion in Trujillo, were exceptions.

At the beginning of the 20th century, the Art Nouveau movement gave rise to a rather vague architectural empiricism that disregarded functional problems and concentrated on decorative elements. Gradually, in the face of practical problems, architectural professionalism took the place of the craftsman's improvisations. Lima was transformed into a garden city; new zoning regulations specified that 45 per cent of the area must be open. A multitude of houses sprang up, creating a stylistic chaos, to which the so-called "Peruvian renaissance" of neocolonial style was a reaction. Rafael Marquina, the Peruvian architect who designed the Plaza San Martín in Lima, and Manuel Piqueras Cotolí, the Spanish sculptor responsible for the Peruvian pavilion at the Seville Exposition, can be regarded as the leaders of the new trend; they were followed by E. Harth-Terré, José Alvarez Calderón, Enrique Seoane, and others. Their model was Spanish indigenous architecture, strongly influenced by the architecture of Arequipa. The results, however, were often hybrid and to a certain extent anachronistic, especially when they disregarded the functional relationship to natural surroundings. A group of young architects led by Luis Miró Quesada Garland reacted against this trend in consideration of more modern social requirements and of an intelligent review of techniques based on international developments. Le Corbusier, Walter Gropius, and Frank Lloyd Wright were the models for this younger generation of architects.

b. Painting and sculpture. Little is known of the origins of the schools of painting of Cuzco and Lima, which were important in Peruvian colonial art. The first painting seems to have been the *Prisoner Atahualpa* by the Spaniard Diego de Mora. Toward the end of the 16th century Juan de Illescas, who later worked also in Mexico and in Quito, was active in Peru. In the second half of the same century there is documentation of paintings executed in Lima for the Hospital of S. Andrés by Francisco Juárez, known as the "Indio of Huarochirí," and also of some altar paintings for the church of the Hospital of S. Ana, by Melchor de Sanabria. The decisive influence in the painting of this period was the school of Seville, and in particular the work of Zurbarán and Murillo, as is shown by many paintings in the monasteries of S. Francisco and of La Buena Morte, in Lima, and in the Church of Santiago, in Pomata. Marqués de Lozoya believes the *Twelve Apostles* in the Monastery of S. Francisco to be an original work of Zurbarán. The portraits of the founders of the religious orders and of the apostles in La Buena Morte seem to be either by Zurbarán or of his school.

Sevillian influence was not exercised solely through the works of great masters, but also by the studios of Sevillian artists active in Lima at the beginning of the 17th century, such as Angelino Medoro, painter of the first portrait of St. Rose of Lima, and Mateo Pérez de Alesio, whose son was the illuminator of the choral books of the Monastery of S. Domingo. Also active in this period were Fray Francisco Bejarano, who produced the first etching in Lima and who did the twelve paintings of the Virgin in the Monastery of S. Agustín, in Lima; Juan Rodríguez Samanez; the Spaniard Cristóbal Daza, who painted *The Flight into Egypt* in the Cathedral of Lima; Miguel Luis Ramales, Juan García, and Francisco Flores, all active in Lima. Noteworthy in the second half of the 17th century was Basilio Pacheco, who painted for the Monastery of S. Agustín a curious *Death of St. Augustine* that shows the Cathedral of Cuzco in the background.

There are also echoes of Flemish painting (especially Rubens) transmitted through the work of José del Pozo, founder of the first school of drawing in Lima (1691).

The Cuzco school flourished in the 17th and 18th centuries. Little is known of its formation, since the majority of the works of this school (more than 10,000 paintings) are anonymous. Isolated from contemporaneous pictorial trends elsewhere, the Cuczo school presents an inexplicable similarity to medieval painting, with Sienese-Byzantine echoes. There is also a link with the paintings of Bermejo and Alejo Fernandez. Depictions of the Virgin have an undeniable similarity to the *Virgen del Buen-aire* in the Casa de Contratación (Exchange) in Seville. The growth of religious mysticism in South America explains the predilection for religious subjects and, in part, other aspects of taste (e.g., gold backgrounds). Indigenous influence is evident in the works of the painter Marcos Uzcamayta (a native of Cuzco), whose *Flight into Egypt* is in the Monastery of La Merced in Cuzco, and to whom is attributed the Corpus Christi series in the parish church of Santa Ana.

Only a few personalities emerge from the anonymous mass of the Cuzco school. Among these are Diego Quispe Tito (17th cent.); Pedro Saldaña, whose work shows Italian influences; Juan Osorio, a painter of realistic portraits; Marcos Zapata, whose works depicting marriage ceremonies of the nobility are of great documentary interest; Cipriano Gutiérrez; Basilio Pacheco; Ignacio Chacón; Antonio Valdez Oyardo, a priest to whom are attributed a number of paintings of St. Anthony in the Seminary of S. Antonio, in Cuzco; Mariano Torres (18th cent.), one of the most significant exponents of Peruvian colonial painting, and interesting technically for his paintings on copper; Juan Espinoza de los Monteros; and Lorenzo Sánchez.

At the beginning of the 19th century, Spanish and Flemish influence was replaced by French influence in painting as well as in architecture. José Gil de Castro (second half of 18th cent.) was the major figure of the period of transition to French romanticism. The mestizo water-colorist Pancho Fierro recorded the life and costumes of colonial and republican Lima. Ignacio Merino, trained in Paris, eschewed the neoclassic influence and adhered to French romanticism; as the director of the Academy of Drawing and Painting in Lima he contributed much to the diffusion of this style. The art of Francisco Laso was more decidedly Peruvian. Other mid-19th-century painters were Luis Montero, who studied at the Academy in Lima, of which he later became director, and Francisco Masías, a painter of still life and landscape.

In the 20th century, painters such as Daniel Hernández followed more modern pictorial concepts. After studying in Europe, Hernández became Director of the Academy of Fine Arts in Lima (1919–32), where he attempted — without notable success in the face of the dominant academicism — to arouse interest in new tendencies in painting. Others who reacted against academicism were Teófilo Castillo, the portrait painter Carlos Baca Flor, and Francisco Domingo Barreda, a landscape artist who lived for a long time in France. José Sabogal, leader of a movement based on the return to indigenous tradition for sources of inspiration, was in the forefront of an antiacademic and anti-French controversy. Among his followers were Enrique Camino Brent, Ricardo Flores, and Domingo Pantigoso.

Another important group of "independents" to return to indigenous sources included such painters as Jorge Reinoso, a pupil of D. Hernández, but mainly self-taught; Mario Urteaga; Alejandro Gonzáles, called "Apurimak," who has attempted to reconcile Peruvian and European trends; César Calvo de Araujo; Teodoro Nuñez Ureta; and Juan Manuel Ugarte Eléspuru, whose work shows Mexican influence.

Another active movement proposes a freely inspired art, neither academic nor traditionally Peruvian; among its followers are Ricardo Grau, proponent of the diffusion of modern European trends in Peru; Macedonio della Torre, trained in Germany and France; Carlos Quispez Asín; Federico Reinoso; Sabino Spingett; Francisco Espinosa; Juan Manuel de la Colina; and Alfredo Ruiz Rosas. All of these artists adhere to a modern, but not necessarily abstract, esthetic. Among the abstract artists are Szyslo (European-trained), Alberto Dávila, Luis Pereira, and Rodriguez Larrain.

European sculpture was introduced into Peru by the religious orders, which imported statues from Spain to adorn their churches (e.g., the crêches of the Franciscans). A decisive influence was exercised by the Sevillian school through Martin Alonso de Mesa, Pedro Gaspar de la Cueva, Luis Espínola y Villavicencio, Pedro Noguera, Luis Ortiz de Vargas, and many others who arrived in Peru in the 17th century. Notable, too, has been the influence of the Far East — quantities of sculpture from China and Japan were imported to Peru (via the Philippines and Mexico). This influence is especially noticeable in certain technical aspects (e.g., metallic polish on some statues) and in the positions of bodies and hands. Mainly anonymous, the prevalent types of early sculpture are polychrome "candlestick" images, Huamanga stone figures, and low-relief decoration on pulpits, choir stalls, retables, altars, confessionals, and window and door frames.

Diverse tendencies, including those inspired by folk art, converged in Lima, where there are many polychrome statues of unmistakable Sevillian derivation (although they are brilliantly polished in the Castilian technique). There are as well many "candlestick" images (*imágenes de candelero*), also of Spanish origin, known by this term because they consist only of head, hands, and feet attached to an easellike frame; they are dressed and adorned with natural hair, fingernails, and eyelashes. The forms of sacred images generally followed the canons of the school of Seville; they were elegantly proportioned, and the most frequent subjects were Christ brought before Pilate and Christ at the Pillar.

The depictions of Christ produced in Cuzco, with natural hair and agonized eyes wet with blood, reflect Asiatic influence. The female nude was proscribed by religious canons; the only nudes allowed were Christ on the Cross, St. Sebastian, and the Child in Nativity scenes. Characteristic works were the crêches of the easily worked Huamanga stone, highly polished and as transparent as alabaster. Among the most important Spanish works are a *St. Apollonia*, possibly of the school of Martínez Montañés, to whom are also attributed a *St. Francis Xavier* in S. Pedro in Cuzco, two busts (the Virgin and St. John) in the Concepción, and a *St. Francis* and a *St. Jerome* in the Jesuit church in Cuzco. It is difficult to determine the actual contribution of this master to Peruvian art. His influence may have been transmitted through members of his school or through his son, Francisco Montañés y Salcedo, who appears to have worked in South America. In the Monastery of S. Francisco, in Cuzco, there is a *Virgin* attributed to Gregorio Hernández.

In Lima the school of the Spaniard Pedro Noguera, who carved the choir stalls in the Cathedral, was influential. Another school was that of Diego de Medina, who worked in the Monastery of S. Agustín. Among the most successful of the 17th-century works is the fountain in the Plaza de Armas in Lima (1650) by the local sculptor Antonio de Rivas, whose original project was revised by Pedro Noguera.

In 18th-century Lima the most prominent sculptor was Baltazar Gavilán, creator of the *Death* in the Monastery of S. Agustín and known for his vividly realistic portraits.

In Cuzco, in the 17th century, Bishop Mollinedo sponsored a school of religious sculpture, with such Spanish artists as Juan Toledano, Martín de Torre, Juan Rodríguez Samanez and Diego Martínez de Oviedo, who decorated the churches of Cuzco. There were also indigenous artists: Juan Tomás Tuyru Tupac carved the beautiful *Virgin of the Almudena* and possibly the pulpit in S. Blas and that in Cuzco Cathedral; Melchor Huamán y Machucama carved a *Last Supper* for the Church of S. Agustín in Cuzco.

The sculpture of altarpieces was particularly noteworthy in Cuzco, Ayacucho, Trujillo, and Lima. In Lima, unfortunately, the architect Matías Maestro replaced many of the baroque altarpieces with cold neoclassic works.

A typical expression of baroque art can be seen in the fine choir stalls with figural carvings. The most celebrated are those in the Cathedral of Lima. Carved in cocobolo wood, they are the work of Pedro Noguera, Luis Ortiz de Vargas, and Martin Alonso de Mesa, artists trained in the Sevillian school. There are also noteworthy choir stalls in the Cathedral and in the Church of La Merced, in Cuzco, and in the Church of S. Francisco, in Lima. The latter are made of Panama cedar, with a double row of seats, and are the work of indigenous 17th-century woodcarvers. Others, equally fine, are in the Church of S. Francisco, in Cuzco, by Fray Luis Montes with the collaboration of Isidoro Fernández Inca and Antonio de Paz, and in the churches of S. Agustín, La Merced, and S. Domingo, in Lima.

The crisis following the Peruvian declaration of independence in 1821 and the subsequent diminution of the religious spirit produced a slow decadence in sculpture; in the 19th century there were no sculptors worthy of mention. In the 20th century, the foundation of the School of Fine Arts (1919) and the teaching of Manuel Piqueras Cotolí led to a revival of the art. Attempting to express the essence of the Peruvian spirit, Piqueras Cotolí helped to form a generation of antiacademic sculptors. Among his pupils were Ismael Pozo, Moises Laimito, and Carmen Saco.

Another movement, which was bound to indigenous tradition and yet open to European influence, included Romano Espinoza Caceda y Ocaña, a sculptor of monumental works; Luis Valdetaro; and Luis Agurto, a pupil of Rodin.

Piqueras (son of Piqueras Cotolí), also a painter, and Joaquín Roca Rey, a follower of Henry Moore, were among the few abstract sculptors.

Alejandro Miró Quesada Garland

Bibliog. *Pre-Columbian period*: *a. General*: W. C. Bennett, ed., A Reappraisal of Peruvian Archaeology (AmAnt, XIII, 4, sup.), Menasha, Wis., 1948; H. Horkheimer, Guía bibliografica de los principales sitios arqueológicos del Perú, B. bibliog. (Lima), XX, 1950, pp. 181–234; H. Ubbelohde Doering, The Art of Ancient Peru, London, New York, 1952; W. C. Bennett, Ancient Arts of the Andes, New York, 1954 (cat.); Antiguo Perú: Espacio y tiempo (Trabajos presentados a la Semana de Arqueología Peruana, 1959), Lima, 1960; W. C. Bennett and J. B. Bird, Andean Culture History (Am. M. of Natural H., Handbook Ser., 15), 2d ed., New York, 1960; H. Leicht, Pre-Inca Art and Culture (trans. M. Savill), New York, 1960; F. Anton, Alt-Peru und seine Kunst, Leipzig, 1962; J. B. Bird, ed., Art and Life in Old Peru, New York, 1962 (cat.); J. Espejo Núñez, Bibliografía de arqueología peruana, 1956–1961, B. bibliog. (Lima), XXXV, 1–2, 1962, pp. 137–86. See also the bibliog. for ANDEAN PROTOHISTORY. *b. Northern highlands*: J. C. Tello, Andean Civilisation: Some Problems of Peruvian Archaeology, 23rd Int. Cong. of Americanists, New York, 1928, pp. 259–90; W. C. Bennett, The North Highlands of Peru (APAmM, XXXIX, 1), New York, 1944. *c. Central highlands*: P. E. Villar Córdova, Las ruinas de la Provincia de Canta, Inca, I, 1923, pp. 1–23; P. E. Villar Córdova, Arqueología peruana: Las culturas prehispánicas del Departamento de Lima, Lima, 1935; W. C. Bennett, The Archaeology of the Central Andes, HSAI, II, 1946, pp. 61–147; W. C. Bennett, Excavations at Wari, Ayacucho (Yale Univ. Pub. in Anthr., 49), New Haven, London, 1953. *d. Southern highlands*: P. Fejos, Archaeological Explorations in the Cordillera Vilcabamba, Southeastern Peru (VFPA, 3), New York, 1944. *e. Northern coast*: A. L. Kroeber, Archaeological Explorations in Peru, II: The Northern Coast (Field Mus. of Natural H., Anthr. Mem., II, 2), Chicago, 1930; J. C. Muelle, Los Valles de Trujillo, Lima, 1937; R. Larco-Hoyle, Los Mochicas, 2 vols., Lima, 1938–39; W. C. Bennett, Archaeology of the North Coast of Peru (APAmM, XXXVII, 1), New York, 1939; R. Larco-Hoyle, Culture of the North Coast of Peru, HSAI, II, 1946, pp. 149–75; R. Schaedel, Major Ceremonial and Population Centers in Northern Peru, 29th Int. Cong. of Americanists, Chicago, 1951, pp. 242–43; G. R. Willey, Prehistoric Settlement Patterns in the Viru Valley, Peru (B. Bur. of Am. Ethn., 155), Washington, D.C., 1953; J. C. Tello, Arqueología del Valle da Casma, Lima, 1956. *f. Central and southern coast*: M. Uhle, Über die Frühkulturen in der Umgebung von Lima, 16th Int. Cong. of Americanists, Vienna, 1908, pp. 347–70; M. Uhle, Explorations at Chincha, Univ. of Calif. Pub. in Archaeol. and Ethn., XXI, 1924, pp. 55–94; E. Yacovleff and J. C. Muelle, Una exploración en Cerro Colorado, Rev. Mus. nacional de Lima, I, 2, 1932, pp. 31–59, 81–89; A. L. Kroeber, Archaeological Explorations in Peru, II: Cañete Valley (Field Mus. of Natural H., Anthr. Mem., II, 4), Chicago, 1937; A. L. Kroeber, Peruvian Archaeology in 1942 (VFPA, 4), New York, 1944; L. M. Stumer, Ancient Centers of Population in the Valley of Rimac, AmAnt, XX, 1954, pp. 130–48.

Colonial and modern periods: Escuela Nacional de Bellas Artes del Perú, Monografía histórica, Lima, 1922; A. Palma, Pancho Fierro acuarelista limeño, Lima, 1935; J. Uriel García, Sintesis de historia del arte en el Perú, Actas II Cong. int. de h. de América (1937), I, Buenos Aires, 1938, pp. 76–81; A. Guido, Redescubrimiento de América en el arte, Rosario, 1940 (3d ed., Buenos Aires, 1944); Marqués de Lozoya, Historia del arte hispánico, III–IV, Barcelona, 1940–45; A. Benavides Rodríguez, La arquitectura en el virreinato del Perú y en la capitanía general de Chile, Santiago de Chile, 1941; E. Harth-Terré, Arquitectura popular peruana, Lima, 1941; M. Buschiazzo, Estudios de arquitectura colonial hispano-americana, Buenos Aires, 1944; A. Miró Quesada Garland, Un siglo de pintura peruana, Lima, 1944; D. Angulo Iñiguez, Historia del arte hispano-americano, 3 vols., Barcelona, Madrid, 1945–56; M. de Castro, La arquitectura barroca del virreinato del Perú, Havana, 1945; E. Harth-Terré, Artífices en el virreinato del Perú, Lima, 1945; A. Miró Quesada Garland, Espacio en el tiempo, Lima, 1945; J. E. Rios, La pintura contemporanea en el Perù, Lima, 1946; H. Velarde, Arquitectura peruana, Mexico City, 1946; M. S.

Noel, Rutas históricas de la arquitectura virreinal altoperuana, Buenos Aires, 1948; H. E. Wethey, Colonial Architecture and Sculpture in Peru, Cambridge, Mass., 1949; P. Kelemen, Baroque and Rococo in Latin America, New York, 1951; E. Marco Dorta, La arquitectura barroca en el Perú, Madrid, 1957; F. Cossío del Pomar, Arte del Perú colonial, Mexico City, 1958 (bibliog.); G. Kubler and M. Soria, Art and Architecture in Spain and Portugal and Their American Dominions, 1500–1800, Harmondsworth, 1959 (bibliog.).

ART CENTERS AND ARCHAEOLOGICAL SITES. Art centers and archaeological sites are listed alphabetically within the general divisions of highlands and coastal areas (northern, central, and southern). Departmental location of major cities is indicated in parentheses following the name of the city.

Highlands. a. Northern. Cajamarca (Cajamarca). Chief city of the department, Cajamarca is situated in the northeast sierra, approximately 9,000 ft. above sea level. It has low clay houses, profusely decorated stone churches, and several patrician country houses of the colonial period, characterized by sober façades enlivened with richly ornamented doors and windows. The city is very Spanish in appearance, and shows no traces of native influence. Wood sculpture, uniformly religious in theme, was a highly developed local art.

The Cathedral, originally a parish church, was elevated to cathedral status in 1682 and consecrated in 1762 but was never completed. It has a basilican plan, and the façade has the rhomboid decoration characteristic of Cajamarca. The Church of S. Francisco has a façade with three portals similar to that of the Cathedral. The adjoining Chapel of La Dolorosa, which can be entered from the Cathedral, has a plateresque portal and magnificent carved-stone wall decorations (recently discovered). The Church of Belén is a mixture of plateresque and baroque elements, with a façade by José Morales (1744). The interior has a noteworthy cupola. Other churches are S. José, La Inmacultada, La Concepción, and La Recoleta.

BIBLIOG. A. Gridilla, Cajamarca y sus monumentos, Cajamarca, 1939; R. Mariátegui Oliva, San Francisco y la Dolorosa de Cajamarca (Doc. de arte peruano, 2), Lima, 1947; H. Villanueva Urteaga, Hacia la ciudad de Cajamarca la grande, Rev. universitaria (Cuzco), XXXVI, 92, 1947, pp. 197–244; H. Villanueva Urteaga, La catedral de Cajamarca, Rev. Inst. Am. Arte, VII, 1954, pp. 103–34.

Archaeological sites (see above, *Pre-Columbian era*): Cajabamba; Callejón de Huaylas: Huaraz (Huarás), Wilkawain; Chavín de Huántar; Kuntur Wasi; Marañon Valley: Kuélape; Marca Huamachuco: Cerro del Castillo; Viracochapampa.

BIBLIOG. W. C. Bennett, Chavín Stone Carving (Yale Anthr. S., 3), New Haven, 1942; A. Soriano Infante, Algo sobre la arqueología de Ancash, 27th Int. Cong. of Americanists, I, Lima, 1942, pp. 473–83; J. C. Tello, Discovery of the Chavín Culture in Peru, AmAnt, IX, 1943, pp. 135–60; T. D. McCown, Pre-Incaic Huamachuco: Survey and Excavations in the Northern Sierra of Peru, Univ. of Calif. Pub. in Am. Archaeol. and Ethn., XXXIX, 1945, pp. 223–400; R. Carrión Cachot, La Cultura Chavín, dos nuevas colonias: Kuntur Wasi y Ancón, Rev. Mus. Nacional de antr. y arqueol., II, 1948, pp. 99–172; J. C. Tello, Chavín: Cultura matriz de la civilización andina, I, Lima, 1960; J. H. Rowe, Chavín Art, New York, 1962.

b. Central. Ayacucho (Ayacucho). The departmental capital, Ayacucho lies between the western and eastern Cordilleras. Somewhat isolated, the city has retained a Spanish colonial character, with many monuments of the 16th century surviving. Founded by Pizarro in 1539 as San Juan de la Frontera, the city has been renamed several times (San Juan de la Victoria, Huamanga, Ayacucho) as a result of military and political events. The present name was given it after the Battle of Ayacucho (1824), during the Peruvian War of Independence.

S. Cristóbal, the city's oldest church, retains its original simple structure, with a single tower and a red tile roof. The Cathedral, however, has been very much restored. The Church of S. Domingo has an unusual façade, with a three-arched portal surmounted by a wooden gallery with a tile roof. (This portion may have been a chapel of the *Indios* or, as other authorities believe, may have been reserved for the tribunal of the Holy Office.) The Church of S. Francisco has a plateresque portal in the style of the period of Charles V. The Church of the Jesuits has a sacristy entrance in rusticated ashlar; the decorations show probable Indo-Asiatic influences, since it contains, for example, a depiction of an elephant, an animal unknown in Peru. Other noteworthy churches are S. Francisco de Paula, S. Francisco de Asís, S. Teresa, S. Ana, and S. Clara, the last of which has a wooden ceiling in the Mudejar style.

Among examples of secular architecture, a number of patrician mansions are noteworthy, such as the Tres Máscaras and the houses of the Marquéses de Cabrera, the García del Barco family, and Cayetano Ruiz de Ochoa. All show influence of the Cuzco and Lima architectural styles. The stone decoration of the façade of the Municipalidad (Town Hall) is extravagant but architecturally harmonious.

BIBLIOG. P. M. Medina, Monumentos coloniales de Huamanga, Ayacucho, 1942; N. Cabrera Bedoya, Guía histórica de los monumentos coloniales de Huamanga, Ayacucho, 1947; C. F. Vivanco, Ayacucho, Lima, 1947.

Archaeological sites (see above, *Pre-Columbian era*): Bonbón (Lake Junín); Canta Province: Canta Marca, Chipprak (Chíprak); Huanuco; Mantaro Basin: Huari, Jauja district.

BIBLIOG. P. E. Villar Córdova, Las ruinas de la Provincia de Canta, Inca, I, 1, 1923, pp. 1–23; P. E. Villar Córdova, Arqueología peruana: Las culturas prehispanicas del Departamento de Lima, Lima, 1935, pp. 292–311; P. E. Villar Córdova, Las ruinas de 'Chipprak' en la Provincia de Canta, Actas Acad. Nacional de Ciencias . . . de Lima, X, 1947, pp. 171–88; J. H. Rowe, D. Collier, and G. R. Willey, Reconaissance Notes on the Site of Huari, AmAnt, XVI, 1950, pp. 120–37; L. G. Lumbreras, La cultura de Wari, Ayacucho, Etn. y arqueol., I, 1960, pp. 130–227.

c. Southern. Arequipa (Arequipa). The departmental capital, Arequipa was founded by the Spanish in 1540 and became an episcopal see in 1577; in 1582 it was destroyed by an earthquake. It was reorganized under the viceroy Francisco de Toledo and served as a place of refuge for the Spanish during the attempted Inca restoration. Certain building materials (especially the local white volcanic stone), ethnic features, and antiseismic building techniques characterize the baroque architecture of the city, which treats Spanish elements in a typically Peruvian idiom of massive edifices with thick walls, small windows, and richly carved portals.

The Church of the Jesuits has exuberant ornamentation of floral and animal motifs in Churrigueresque style; the adjoining monastery (1738) has a cloister with profusely decorated columns and arches. The Church of S. Agustín has a picturesque façade decorated with stylized indigenous and Iberian motifs reminiscent of the Tiahuanacan manner. The Church of S. Domingo is also baroque, but because of the use of semicircular arches, it is Romanesque in feeling; a façade decoration represents the Pantocrator. The Convent of S. Catalina, built with an antiseismic technique, has massive buttresses decorated, in the manner of Andean terraces, with indigenous motifs. The Cathedral is neoclassic in style, having been rebuilt over the remains of a previous colonial edifice destroyed by fire in 1844. Three churches in the surrounding countryside are excellent examples of 18th-century rural architecture: the church at Yanahuara has a fine façade; that at Caima has two elegant towers; and that at Characato has elements of indigenous inspiration. The parish church of Espíritu Santo at Chiguata, also of the 18th century, was built by the Dominicans. It has a single tower and a façade with local motifs carved in stone.

Secular architecture was occasionally influenced by the ecclesiastical; this can be seen in house façades surmounted by richly decorated broken tympanums. In the older houses, a series of modillions and corbels surmount the façades and the windows are low and have decorated frames. The most noteworthy secular buildings are the Moral, Ricketts, Rey de Castro, Quiroz, and Irriberry houses.

The Museo Arqueológico and the Museo Municipal contain noteworthy collections of pre-Columbian and Spanish colonial art.

BIBLIOG. V. M. Barriga, Memorias para la historia de Arequipa, 2 vols., Arequipa, 1941–46; M. S. Noel, En la Arequipa indohispanica, Buenos Aires, 1957.

Cuzco (Cuzco). After destroying the Indian monuments, the Spanish founded a new city — "La muy noble y gran ciudad del Cuzco" ("the most noble and great city of Cuzco") — on the site of the former Inca capital, on Mar. 23, 1534. Thereafter, the two cultures, Spanish and Inca, were superimposed on each other rather than fused. Elegant baroque bell towers, double-arched windows, Iberian heraldic devices set on barren Incaic walls, and the reddish local stone are characteristic features of the city's architecture.

The Cathedral, situated in the Plaza de Armas on the site of the palace and temple of the Inca emperor Viracocha, is an excellent example of Spanish Renaissance architecture. It was begun in 1560 from designs by the Biscayan architect Juan Antonio Veramendi, which were carried out by Juan Correa and Francisco Becerra, the architect of the Cathedral in Lima. The baroque façade has two simple bell towers. The interior, with three aisles and cross vaults, contains paintings by Diego Quispe Tito and the sculptures *Nuestro Señor de los Temblores* (Our Lord of the Earthquakes) and *La Linda*, attributed to a mestizo sculptor. The silver repoussé of the high altar covers an earlier altar of carved cedar; there is also a noteworthy baroque choir. The Church of the Jesuits, built on the ruins of the Inca palace of Huayna Capac and completed in 1668, has a graceful baroque façade of plateresque inspiration by the architect Juan Bautista Egi-

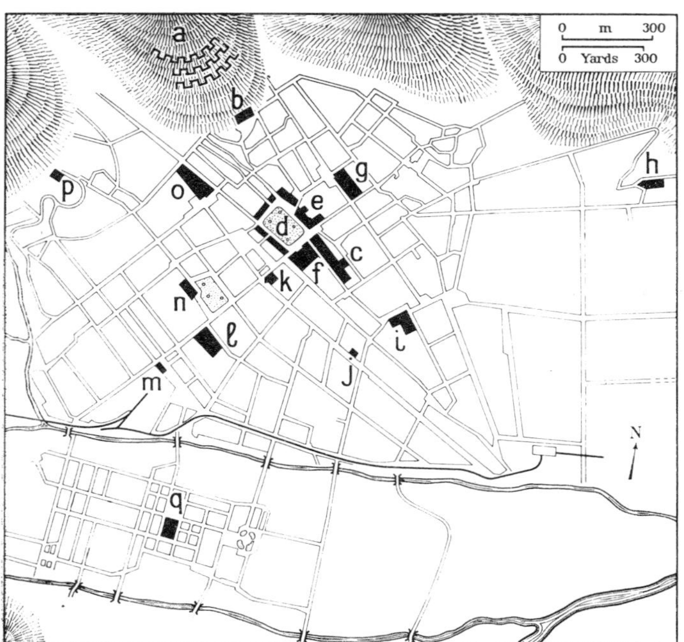

Cuzco, plan of the city. Principal monuments: (*a*) Fortress of Sacsahuaman; (*b*) Colqampata; (*c*) Accla Huasi (now S. Catalina); (*d*) Plaza de Armas; (*e*) Cathedral and Church of El Triunfo; (*f*) Church of the Jesuits, Nuestra Señora de Loreto, and the University; (*g*) Church of S. Antonio Abad and bishop's palace; (*h*) Church of La Recoleta; (*i*) S. Domingo; (*j*) Casa Montes; (*k*) Monastery and Church of La Merced; (*l*) Monastery of S. Clara; (*m*) S. Pedro; (*n*) S. Francisco; (*o*) S. Teresa; (*p*) S. Ana; (*q*) Church of Santiago.

diano. Considered the most beautiful in Peru, this church is distinguished by bull's-eye windows in the towers and a dome covered with polychrome glazed tiles. In the single-aisled interior are works of the Cuzco school. Adjoining the church is the monastery (later the University), with an elegant plateresque façade. The Church of El Triunfo (or the Sagrario), built on a central plan of Byzantine type, adjoins the south side of the Cathedral, with which it communicates. The Churrigueresque high altar is of granite and gold. Another altar holds a life-size wood image of Christ in Agony and a silver-and-wood cross. The church also preserves a painting of the Cuzco school, *Taitacha Temblores* (Great Father of the Earthquakes). The single-aisled church of Jesús y María (known as La Sagrada Familia), which also communicates laterally with the Cathedral, has a restrained baroque façade.

La Merced, founded in 1536 by Fray Sebastian de Trujillo, was rebuilt in the late 17th century from the original plans after its almost total destruction in an earthquake; it is plateresque in style, with a single tower. The interior, with a nave and two side aisles, contains one of the few existing plateresque choir stalls and numerous paintings of the Cuzco school. Inside the adjoining monastery, one of the oldest in Peru (founded 1536), are preserved paintings by Ignacio Chacón and Juan Osorio. S. Domingo, built over the Temple of the Sun (Coricancha), has an imposing Peruvian baroque tower. The cloister contains some excellent examples of 17th-century Cuzco art — paintings by the elder Juan Espinoza and Diego Quispe Tito, as well as wood sculpture by the indigenous artist Melchor Huamán and by the mestizo Juan Rodriguez Samanez. La Recoleta is a simple and severe structure. The sacristy contains a wood figure of Christ and paintings of Franciscan inspiration. The library contains an incunabulum of 1490. S. Francisco, completely remodeled and with few of its original elements, has wooden choir stalls by the Franciscan Luis Montes and a choir lectern and pulpit carved by Fray Pedro Gómez. The monastery has two cloisters, the first of which has a coffered ceiling in Mudejar style and contains paintings by Juan Espinoza de los Monteros. The 17th-century church of the seminary of S. Antonio Abad has a Churrigueresque high altar and a carved cedar pulpit. It contains numerous paintings, among them those attributed to Antonio Valdez Oyardo, known as "Cura Pintor" (Painter-Priest), and fine wood statues. The Monastery of S. Clara survived the earthquakes of 1650 and 1950. The church, with a single aisle, is richly decorated with paintings and fire-gilt mirrors — whence it is popularly known as "Iglesia de los espejos" (Church of the Mirrors). The Monastery of S. Teresa, built over Inca walls, has a single-aisled church, two slender towers, and a socle of glazed tiles that is unique in Cuzco. The high altar is of repoussé silver, and the ceiling is also of silver. The Monastery of S. Catalina and its church are built over the Inca Accla Huasi (House of the Chosen Women). The church, with a single aisle, has two entrances and a tower and contains some Churrigueresque altarpieces, two choirs, gilt grills in Mudejar style, and a series of paintings by Juan Espinoza de los Monteros. The 17th-century church of S. Pedro, built of masonry from Inca ruins, has a subdued red-colored façade with ornamentation recalling the style of Gothic altars. The Churrigueresque high altar has sculptures of the Evangelists and paintings of the Cuzco school.

In the churches of the immediate vicinity of Cuzco, Spanish forms give way to indigenous influences, from whence arose a style that might be termed "Indio-Mestizo." The Church of San Sebastián, approximately 3 miles from Cuzco, is built of adobe and has a granite façade by the *Indio* Manuel de Sahuaraura (1685), with indigenous decorative motifs inspired by textile designs. It contains a wood sculpture attributed to Melchor Huamán and a painting by Diego Quispe Tito. The Church of S. Cristóbal, founded in 1560 by an Inca, Cristóbal Paullu, is attributed to the *Indio* architect Marcos Uzcamayta. Built of adobe, it has a stone tower and contains a number of paintings. The Church of S. Blas has a modest exterior, but it contains a high altar richly decorated in gold and an elaborately carved pulpit. The Church of S. Ana, facing the Inca fortress of Saccsaihuaman, contains twelve paintings (by an unknown painter) of great documentary value, depicting the traditional Corpus Christi procession in Cuzco. The Church of Santiago, founded by the viceroy Toledo in 1571, is of adobe but has been very much remodeled. The atrium contains a stone cross of *Indio* workmanship. The Church of S. Jerónimo, founded by the Dominicans, contains some fine sculpture; and the Church of La Almudena contains the *Virgin of the Almudena*, by the indigenous sculptor Juan Tomás Tuyru Tupac (17th cent.).

Like the churches, residential palaces were often built on Incan foundations, and the stones of Inca palaces and temples were used for their walls. Carved-stone decorations in local idioms, influenced by Spanish styles, were the preferred ornament for façades. The Casa del Almirante (first half of 17th cent.) has a plateresque façade and a balcony with double windows; the coat of arms of Pedro Peralta de los Ríos surmounts the main entrance. Its large patio shows Spanish Renaissance influence, and some of the rooms have magnificent ceilings in the Mudejar style. The Casa de los Cuatro Bustos (said to have belonged to the Pizarro brothers) has a plateresque façade with a historiated coat of arms. The house of the Marquéses de Valleumbroso, with a Renaissance façade and carved wooden balconies, is interesting for the main portal, which retains the indigenous trapezoidal doorway. The episcopal palace, formerly the home of the Marquéses de Buenavista y Rocafuerte, was built on the foundations of the palace of the Inca Roca; it has a simple façade and a corner balcony with double-arched windows. Other noteworthy houses include that of the Marquéses de Casa Jara y Casa Palma, built on Incan walls, with a carved wooden balcony showing indigenous influence; the Casa de San Borja; the Casa de Silva, with an Incan portal; Casa Concha, with a delicately ornamented balcony; the palace of the Marqués de Venero, on Incan foundations; the Casa de los Condes de Peralta; and the Casa Montes.

Museums include the Museo de Antropología, the Museo de la Casa del Senato, and the Museo de la Universidad.

Bibliog. *Pre-Columbian period*: C. R. Markham, Cuzco: A Journey to the Ancient Capital of Peru, London, 1856; J. Jijón y Caamaño, Los orígenes del Cuzco, Anales de la Univ. Central de Ecuador, LII, 1934, pp. 203–36, 285–344, LIII, 1934, pp. 89–130; L. E. Valcárcel, Sajsawaman redescubierto, Rev. Mus. Nacional de Lima, III, 1934, pp. 3–36; L. A. Pardo, Ruinas precolombinas del Cuzco, Cuzco, 1937; J. H. Rowe, An Introduction to the Archaeology of Cuzco (Peabody Mus. Pap., XXVII, 2), Cambridge, Mass., 1944; L. A. Pardo, Historia y arqueológia del Cuzco, 2 vols., Callao, 1957. *Colonial and modern period*: M. E. Cuadros Escobedo, Historia y arquitectura de los templos del Cuzco, Lima, 1946; G. Kubler, Cuzco: Reconstruction of the Town and Restoration of Its Monuments, Paris, 1952; M. Kropp, Cuzco: Window on Peru, New York, 1956; J. M. Covarrubias Pozo, Apuntes para la historia de los monumentos coloniales del Cuzco, Rev. universitaria (Cuzco), XLVI, 1957, pp. 105–407; J. Cornejo Bouroncle, Derroteros de arte cuzqueño, Cuzco, 1960; J. de Mesa and T. Gisbert, Historia de la pintura cuzqueña, Buenos Aires, 1962.

Puno (Puno). Puno, the departmental capital, is situated on the northwest shore of Lake Titicaca, 12,648 ft. above sea level, and is laid out on a regular plan oriented around a central plaza. Missionary activity, the density of population, and the mineral wealth of this zone made it one of the most flourishing centers of the colonial period.

In the church architecture of Puno indigenous influence is blended with Renaissance, plateresque, and Mudejar elements in a "mestizo" baroque style. The Cathedral (18th cent.), built of gray basaltic stone on a Latin-cross plan, has three entrance portals. Designed by Simon de Asto, it is markedly similar to (and was undoubtedly influenced by) the Church of the Jesuits in Arequipa.

In the villages along the western shore of Lake Titicaca, there are several interesting churches. At Zepita, SS. Pedro y Pablo, of local stone and red sandstone, was built in the mid-17th century, with a plateresque façade and flanking walls creating a somewhat classicizing effect. At Pomata, the Dominican church on the main plaza was begun in the 17th century (but not finished until the 18th) from plans by the Dominican priest Juan Moreno. The interior has rich and exuberant decoration with indigenous characteristics. The subjects of the sculptures, inspired by folk art, include a S. Domingo de Guzmán, a Nazarene, a Holy Sepulcher group, and a Virgin of the Rosary. The church at Azángaro, the Asunción, was built of adobe and originally had a single aisle without a crossing. The crossing was added when it was enlarged in the 17th century, and the façade was then remodeled. Of the original church, there remains an ogival arch covered with paintings. The interior contains original colonial paintings and carved confessionals.

Bibliog. R. Mariátegui Oliva, Una joya de arquitectura peruana de los siglos XVII y XVIII, Lima, 1942; R. Mariátegui Oliva, Iglesia de la Ascensión de Azángaro (Doc. de arte peruano, 3), Lima, 1948; R. Mariátegui Oliva, Pomata: Templo de Nuestra Señora del Rosario (Doc. de arte peruano, 4), Lima, 1949; R. Mariátegui Oliva, Zepita: Templo de San Pedro y San Pablo (Doc. de arte peruano, 12), Lima, 1950; W. Cano, La catedral de Puno: Obra arquitéctonica del siglo XVIII, La Plata, 1951; M. S. Noel, La arquitectura mestiza en las riberas del Titicaca, Buenos Aires, 1952; X. Moyssén, La catedral de Puno, Perú, Anales Inst. de investigaciones estéticas, XXXI, 1962, pp. 43–55.

Archaeological sites (see above, *Pre-Columbian era*): Colcampata; Cuzco; Inti Pata; Kenko; Machu Picchu; Maras; Moray; Ollantaytambo; Piquillacta; Pisac; Pucára; Racche; Saccsaihuaman (Sacsahuaman); Sicuani; Sillustani; Tampu Macchai; Titicaca Basin; Titicaca Island; Urubamba Gorge; Wiñay Wayna.

Bibliog. A. F. Bandelier, The Aboriginal Ruins at Sillustani, Peru, AmA, VII, pp. 49–68; M. Uhle, Fortalezas incaicas: Incallacta e Machupicchu, Rev. chilena de h. y geog., XXI, 1917, pp. 154–70; H. Bingham, Machu Picchu: A Citadel of the Incas, New Haven, 1930; L. E. Valcárcel, Las ruinas de Pisaj, Rev. geog. am., IV, 1935, pp. 111–15; V. M. Guillen, El gran templo de Huiraccocha, Rev. universitaria (Cuzco), XXVI, 2, 1937, pp. 82–97; J. M. Franco Inojosa and L. A. Llanos, Trabajos arqueólogicos en el Departamento de Cuzco, Rev. Mus. Nacional de Lima, IX, 1940, pp. 22–32; A. Kidder II, Preliminary Notes on the Archaeology of Pucará, Puno, 27th Int. Cong. of Americanists, I, Lima, 1942, pp. 341–45; M. H. Tschopik, Some Notes on the Archaeology of the Department of Puno, Peru (Peabody Mus. Pap., XXVII, 3), Cambridge, Mass., 1947; L. A. Pardo, Ollantaitampu, Rev. universitaria (Cuzco), XXXV, 2, 1946, pp. 47–73; H. Bingham, Lost City of the Incas, New York, 1948; M. A. Astete Abril, Ollantaytambo, Rev. universitaria (Cuzco), XLI, 103, 1952, pp. 212–28; V. W. von Hagen, The Mystery of Pisac, Archaeology, V, 1952, pp. 33–39; L. A. Pardo, Los monumentos arqueológicos de Ppisacc, Rev. Mus. e Inst. Arqueol. (Cuzco), X, 1957, pp. 4–33; E. Harth-Terré, Piki-Llkta ciudad de Posito y Bastimentos, Rev. Mus. e Inst. Arqueol. (Cuzco), XI, 1959, pp. 41–56; L. A. Pardo, Machupicchu: Una joya arquitectónica de los Incas, Rev. Mus. e Inst. arqueol. (Cuzco), XII, 1961, pp. 223–98.

Coastal area. a. Northern. Trujillo (La Libertad). The chief city of the sparsely populated northern coast is the departmental capital, Trujillo, founded by Francisco Pizarro in 1535. Surrounded by fortifications, it has retained a colonial character, particularly in its domestic architecture, which reflects the prosperity achieved in the 18th century. The baroque in the forms that were popular in the south is not found in the churches of Trujillo, with the exception of Santiago de Huamán, on the outskirts of the city. The baroque style prevalent in Trujillo is exemplified in the Church of Nuestra Señora del Carmen, built in the mid-18th century. Destroyed by an earthquake just before it was to be consecrated, it was immediately rebuilt. The interior is richly decorated with gilt; the Churrigueresque altars, the pulpit, and the tribune are noteworthy. The church preserves numerous pictures and a polychrome wood statue of St. Joseph by the Indian Manuel Chili, known as "Caspicara." Among the reliquaries is a fine ciborium in gold and precious stones. S. Agustín, the only church to survive the earthquake of 1619, has a restored façade with beautifully proportioned 17th-century portals. The baroque church of La Merced has a dome decorated with relief scenes from the life of S. Pedro Nolasco. Also worthy of mention because of their rich interiors are the churches of Belén and S. Clara; and for its lovely tower, the Church of S. Francisco.

In the vicinity, the churches of Mansiche, Huanchaco, and Huamán are particularly interesting for the use of indigenous elements in their decoration.

Bibliog. R. Mariátegui Oliva, Una iglesia relicario: El Carmen de Trujillo (Doc. de arte peruano, 1), Lima, 1945; R. Mariátegui Oliva, Escultura colonial de Trujillo, Lima, 1946; J. Leon Linares, Guía turística de Trujillo, 2d ed., Lima, 1961.

Archaeological sites (see above, *Pre-Columbian era*): Ascope; Batán Grande; El Brujo; El Castillo de Tomaval; Cerro Blanco; Chancaillo; Chan-Chan; Chotuna; Dos Cabezas; Etén; Huaca de la Luna; Huaca del Sol; Moxeke; Pacatnamú (La Barranca); Pallka; Pampa Grande; Pañamarca; Punkurí Alto; Cerro de Sechín.

Bibliog. M. Uhle, Die Ruinen von Moche, JSAm, X, 1913, pp. 341–67; O. Olstein, Chan-Chan: Capital of the Great Chimu, Geog. Rev., XXVII, 1927, pp. 36–61; C. Garcia Rosell, Los monumentos arqueológicos del Perú, Lima, 1942, pp. 35–36, 134–35, 169–71; J. C. Tello, Origen y desarollo de las civilizaciones prehistoricas, 27th Int. Cong. of Americanists, I, Lima, 1942, pp. 589–714 at 702; J. C. Tello, Discovery of the Chavín Culture in Peru, AmAnt, IX, 1943, pp. 135–60; H. Horkheimer, Vistas arqueologicas del Nor-Oeste del Peru, Trujillo, 1944; J. E. Garrido, Descubrimiento de un muro decorado en la "Huaca de la Luna" (Moche), Chimor, IV, 1956, pp. 25–30; D. Bonavia, A Mochica Painting at Pañamarca, Peru, AmAnt, XXVI, 1961, pp. 540–45.

b. Central and southern. Lima (Lima). Capital of the Republic of Peru, Lima was founded in 1535 by Francisco Pizarro, on a wide and fertile plain in the Rímac Valley sloping from the foothills of the Andes to the Pacific Ocean. A bishopric since 1541 and the seat of the University of S. Marcos since 1551, Lima flourished during the colonial period. Markedly Iberian in character from the beginning, the city also showed, always through the agency of Spain, Arabo-Andalusian and even Eastern influences; later, there also developed an independent mestizo and creole style. In spite of the severe earthquakes of 1687 and 1746, much of the original character of the city, with its typical plan — streets intersecting at right angles and interrupted by spacious plazas landscaped with trees and fountains and flanked by low houses — has been preserved. The center of the city is the Plaza de Armas, upon which a monumental complex of building faces.

Both secular and religious architecture display Andalusian baroque characteristics. The Cathedral of La Asunción, erected in 1540, was rebuilt and enlarged in 1552 and again, by Alonso Beltram, in 1564. It was further enlarged after plans by Francisco Becerra in 1582 and was finally consecrated in 1625. Little of the original structure survived the earthquakes; the stone groin vaults were replaced by *quincha* cross vaults, more resistant to seismic tremors. In the interior there are fine carved wooden choir stalls, a silver-covered altar, and a small chapel containing the tomb of Francisco Pizarro. The Church of S. Domingo (consecrated 1547) and its adjoining monastery comprise the oldest complex of religious buildings in Lima. The styles of the 16th, 17th, and 18th centuries are evident in these structures, which also provide one of the few examples of late Gothic style in Peru. The plan of S. Domingo has been attributed to the viceroy Amat, but the architect was Juan de la Roca. The interior of the church has two side aisles and a central nave with eight semicircular arches and a domed crossing. In the monastery are preserved the remains of St. Rose of Lima, as well as the earliest baptismal font in the city.

Little documentation exists for the Monastery of S. Francisco and its church, begun in the mid-16th century and known to be the work of several architects, among them the Galician Costantino Vasconcelos (to whom some scholars attribute the façade, and others the monastery), Manuel Escobar, who built the side portal, and Lucas Melendez, who was responsible for the mestizo-baroque façade of the sacristy. The church has a baroque façade of heavy rusticated ashlar and a three-aisled interior with a vaulted ceiling decorated in Moorish-style motifs that also appear on the arches and the drum of the dome. The monastery cloisters are particularly noteworthy; there are also remains of a ceiling in Mudejar style, which was partially destroyed in the earthquake of 1940. The socles of glazed tiles (*azulejos*) of the colonial period are also noteworthy. The Jesuit church of S. Pedro (formerly known as Templo de San Pablo), finished in 1638, was intended to be a replica of the Church of the Gesù in Rome. Restored in the 19th century, it preserves noteworthy baroque and Churrigueresque altarpieces, a number of gilded wood sculptures, paintings of the colonial period, and glazed-tile socles. The Monastery of La Merced (ca. 1536) and its church are Churrigueresque in style, but the façade has been much altered and consequently has a curiously hybrid appearance, incorporating twisted (Solomonic) columns, shell-shaped niches, and Greek key patterns. The choir stalls are particularly noteworthy. The Church of S. Agustín, very much restored, was founded in 1554 and was the work of Esteban de Amaya. The remodeled stone façade is Churrigueresque; the interior was demolished and rebuilt in the late 18th century by Matías Maestro. In the sacristy, there is a wood sculpture *Death* by Gavilán and numerous paintings of the Cuzco school.

There are numerous smaller, single-aisled churches with sober exteriors that contrast strongly with their sumptuous interiors. Among these are the Church of Jesús María, with Churrigueresque pulpit and fire-gilt altarpieces; the Church of Magdalena Vieja, notable for its fine proportions and its pulpit and six gilt Churrigueresque altars;

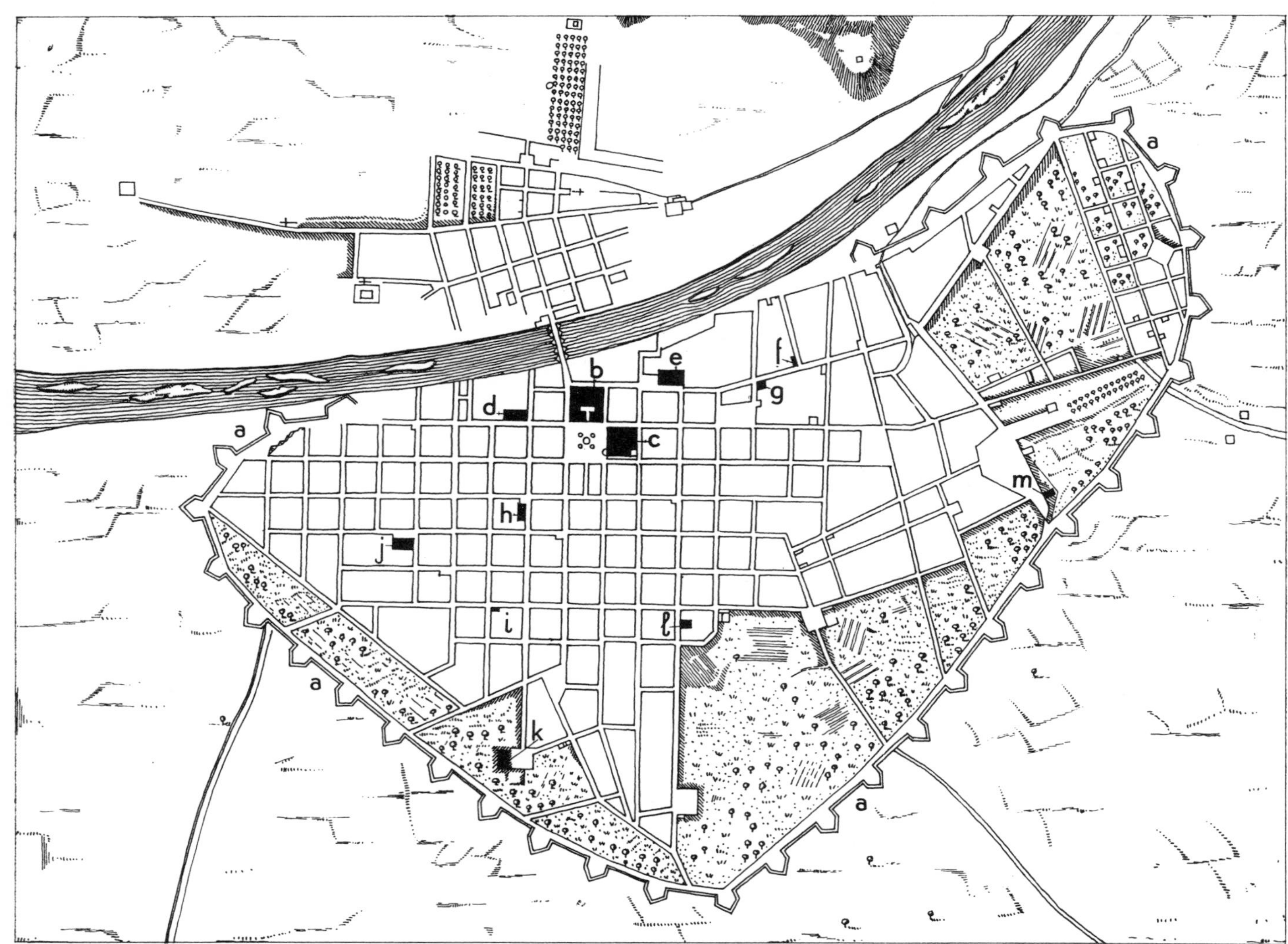

Lima, plan of the city in the 18th century (from a map of 1713). Principal monuments: (*a*) Walls with gates; (*b*) royal plaza and viceroy's palace; (*c*) bishop's palace and Cathedral; (*d*) S. Domingo; (*e*) S. Francisco; (*f*) S. Pedro; (*g*) Monastery of La Merced; (*h*) S. Agustín; (*i*) Church of Jesús María; (*j*) Church of Las Nazarenas; (*k*) Church of Magdalena Vieja; (*l*) Church of El Corazón de Jesús (or de los Huérfanos); (*m*) Nuestra Señora de Cocharcas.

S. Carlos, begun in 1606 but destroyed in the earthquake of 1746 and rebuilt; the Church of El Corazón de Jesús (or de los Huérfanos), elliptical in plan and with a baroque façade attributed to the Jesuit Juan de Rher, who popularized the use of *quincha*. The Church of Las Nazarenas was built around an image regarded as miraculous, which had been painted on a wall in the mid-16th century by a mulatto. This wall was incorporated to form the main altar of the baroque church designed by the viceroy Manuel Amat.

The civil architecture of Lima shows many Andalusian influences. Carved balconies, somewhat resembling cupboards and equipped with jalousies, project from many façades; stone portals with the armorial bearings of the owners, long, narrow balconies of carved wood, and patios recalling those of Seville are other characteristic elements of the important residences. The stone Casa del Pilatos, a typical example of the earliest architectural style introduced by Spanish architects in the 16th century (the so-called "school of the first Viceregal period"), was the only building of its kind to survive the earthquake of 1746. The palace of the Marquéses de Torre Tagle, a jewel of 18th-century colonial domestic architecture, was built of precious materials such as cocobolo from Guayaquil, cedar from Costa Rica, Panamanian stone, and glazed ceramic tiles from Seville, in a mixture of baroque and Churrigueresque styles with strong Eastern and Mudejar influences, especially in the carved wooden balconies and the patio. The rococo Quinta de Presa, built in a style prevalent in Lima in the second half of the 18th century, was the last Spanish residential building to be erected in Lima. It houses the Museo Virreinato (Museum of the Viceroyalty), with its collections of Peruvian colonial decorative arts and crafts, costumes, and furniture. Other museums include the Museo de la Republica (colonial and early-19th-century art); Museo de Antropología y Arqueología (art and history of aboriginal races of Peru); Museo Rafael Larco Herrera (formerly the Chiclin Museum, moved from Trujillo; contains pre-Columbian art, especially of the Mochica period); and the Pinacoteca Municipal (Peruvian painting, colonial to modern).

Bibliog. E. Martín-Pastor, De la vieja casa de Pizarro al nuevo Palacio de gobierno, Lima, 1938; J. M. Peña Prado et al., Lima pre-colombina y virreinal, Lima, 1938; C. Morales Macchiavello, San Francisco de Lima, Lima, 1941; V. M. Barriga, El templo de la Merced de Lima, Arequipa, 1944; M. Buschiazzo, Estudios de arquitectura colonial hispano-americana, Buenos Aires, 1944; M. H. di Doménico Suazo, La fuente de la Plaza mayor de Lima, Lima, 1944; J. Bromley and J. Barbagelata, Evolución urbana en la ciudad de Lima, Lima, 1945; L. A. Eguiguren (Multatuli), Las calles de Lima, Lima, 1945; B. Gento Sanz, San Francisco de Lima, Lima, 1945; A. Santibáñez Salcedo, La sacristia del templo de San Agustín de Lima, Lima, 1945; J. Flóres Aráoz, El local del tribunal del Santo Oficio de Lima, Cultura peruana, VI, 25, Lima, 1946; A. Miró Quesada Sosa, Lima: Ciudad de los reyes, Buenos Aires, 1946; R. Vargas Ugarte, La iglesia de San Pedro de Lima, Lima, 1956; H. Velarde Bergmann, La restauración del palacio de Torre-Tagle, Fanal, XIV, 54, 1958, pp. 9–16.

Archaeological sites (see above, *Pre-Columbian era*): Armatambo; Cahuachi; Cajamarquilla; La Centinela; La Cumbe; La Huaca de Alvarado; La Huaca de Santa Rosa; Lima; Maranga, Nazca; Pachácamac; Paracas; Paramonga; Pisco; Tambo Colorado; Tambo de Mora; Vista Alegre.

Bibliog. M. Uhle, Pachacamac: Report of the W. Pepper Expedition of 1896, Philadelphia, 1903; M. Uhle, Über die Frühkulturen in der Umgebung von Lima, 16th Int. Cong. of Americanists, Vienna, 1908, pp. 347–70; M. Uhle, Explorations at Chincha, Univ. of Calif. Pub. in Archaeol. and Ethn., XXI, 1924, pp. 55–94; S. K. Lothrop, La Centinela, The Independent, CXVI, 1926, pp. 13–16; R. Cúneo-Vidal, Pachacamac, B. Soc. geog. de

Lima, XLV, 1928, pp. 169–83; P. E. Villar Córdova, Arqueología peruana: Las culturas prehispánicas del Departamento de Lima, Lima, 1935, pp. 207–21; L. Langlois, Las ruinas de Paramonga, Rev. Mus. Nacional de Lima, VII, 1938, pp. 21–52, 281–307; A. Casa Vilca, La ciudad muerta de Cajamarquilla, B. Soc. geog. de Lima, LVI, 1939, pp. 101–09; A. A. Giesecke, Las ruinas de Paramonga, B. Soc. geog. de Lima, LVI, 1939, pp. 118–23; A. A. Giesecke, Tambo Colorado, B. Soc. geog. de Lima, LVI, 1939, pp. 110–15; H. A. Urteaga, Tambo Colorado, B. Soc. geog. de Lima, LVI, 1939, pp. 85–94; C. Garcia Rosell, Los monumentos arqueológicos del Perú, Lima, 1942, pp. 34–35, 124–33, 138–43, 181–82; W. D. Strong and J. M. Corbett, A Ceramic Sequence at Pachacamac, Columbia S. in Archaeol. and Ethn., I, 1943, pp. 27–122; F. Engel, Early Sites in the Pisco Valley of Peru: Tambo Colorado, AmAnt, XXIII, 1957, pp. 34–45; H. Buse, Guía arqueológica de Lima: Pachacamac, Lima, 1960.

Illustrations: 5 figs. in text.

(NOTE: Material for sections on art centers provided by Alejandro Miró Quesada Garland.)

PERUGINO (PIETRO DI CRISTOFORO DI VANNUCCI, known as Perugino). Because his biographical data on Perugino have been found generally unreliable, modern art historians tend to disregard the information given by Vasari — according to whom the painter was born in the year 1445 — and instead choose to follow Giovanni Santi's *Cronaca rimata*, in which Perugino and Leonardo are noted as being "par d'etade" (of the same age). Interpreted literally, this phrase would indicate a birth date of 1452 for both artists; taken in a wider sense, to suggest a relative contemporaneity for the two, it would seem reasonable evidence to fix Perugino's birth about 1450.

The earliest document definitely pertaining to Perugino shows that he was in his native town, Città della Pieve, in 1469; in 1472, however, it is known that he was already in Florence and enrolled as a member of the Compagnia di S. Luca. From that date until 1485 he traveled between Florence and Umbria and Rome. It is almost certain that in 1473 he was in Perugia; he returned there in July, 1475, to paint frescoes (lost) in the council chamber of the Palazzo dei Priori (the present Hall of the Notaries in the Palazzo Comunale). In 1478 he was again in Umbria, in Cerqueto. The following year he was probably in Rome, painting in the Cappella della Concezione in Old St. Peter's, and it is known definitely that he was there in October, 1481, when he was commissioned, together with Botticelli, Ghirlandajo (qq.v.), and Cosimo Rosselli, to decorate the Sistine Chapel with ten scenes from the Old and New Testaments. He probably completed this large work — some parts of which, such as the scene of Christ giving the keys to St. Peter, had been approved in January, 1482 — in September of that year, since in October he was asked to go to Florence to execute frescoes in the Sala dei Settanta of the Palazzo Vecchio, a commission that he did not fulfill, however. In November, 1482, he was in Perugia, where the priors commissioned from him the altarpiece for their chapel in the Palazzo dei Priori, a work he did not carry out until 12 years later. Also in 1482, he is believed to have painted the portraits of the Sforzas in Pesaro, which presumably were lost when the Palazzo Ducale burned in 1514. In Rome once again in 1484, he painted standards and coats of arms for the coronation of Pope Innocent VIII; he remained there until July of the following year, when he returned to Perugia for further consultation with the priors concerning their altarpiece, about which time he also received Perugian citizenship.

Between 1486 and 1496, he spent the greater part of his time in Florence. He was there in December, 1486, when it is known that he and Aulista d'Angelo, a painter from Perugia, were involved in a brawl. Two years later Perugino was in Fano, working in S. Maria Nuova. Back in Florence in 1489, he completed an altarpiece for S. Maria Maddalena dei Pazzi. In that year, he also started negotiations with the Operai del Duomo of Orvieto, who entrusted him with the decoration of the Chapel of St. Brizio. Despite his contractual commitment, Perugino never painted the frescoes, even though the Operai sought him out and continued to entreat him to carry out this work as late as 1499. In 1491 he sat on the committee appointed to choose the model for the new façade of the Cathedral of Florence. In the same year he went to Rome in the service of Giuliano della Rovere (later Pope Julius II), and he was there again the next year when he, Antoniazzo Romano, and Pier Matteo di Amelia painted standards for the coronation of Pope Alexander VI. In 1493 he returned to Florence, where he painted many pictures thereafter. During the years 1495 and 1496, though he was still busy in Florence, he was often in Perugia negotiating for important commissions. There he signed the final contract with the priors for their altarpiece and made an agreement with the monks of S. Pietro to paint a large retable. During this period Lodovico il Moro sent for him several times, to work on the altarpiece for the Certosa of Pavia, which was not delivered until 1499. Between May and September, 1497, he was probably once more in Fano, and the next year in Florence, where, along with various other artists, he was called upon to give an opinion about restoring the lantern on the dome of the Cathedral.

By this time, however, the painter's interests were becoming centered in Perugia rather than in Florence (where he engaged a representative to look after his affairs). In Perugia he was at work in S. Pietro and was engaged in painting the extensive decorations in the Sala dell'Udienza of the Collegio del Cambio, which the bankers' guild had decided to commission from him as early as January, 1496. Except for short visits to Florence, he remained in Perugia until 1502, setting up a studio on the Sopramuro and obtaining an appointment as prior (1501). Although fully occupied in painting the frescoes in the Collegio del Cambio, with the help of his many assistants he also managed to paint numerous works for churches in Perugia and other cities.

In 1503 he was in Florence, as a member of the committee that was to decide on the best location for Michelangelo's *David*. From there he carried out his lengthy negotiations with Isabella d'Este, Marchioness of Mantua, for a canvas to be placed in her *studiolo*. This work, the *Combat of Love and Chastity*, was not delivered to Mantua until 1505 (PL. 116).

The "patriarcha perusino," though by then an old man, was still accepting new commitments and working in several different places, especially in Umbria. In 1508 he was in Rome, where he began painting the frescoes on the ceiling of the Stanza dell'Incendio in the Vatican, which were criticized so unfavorably that he was obliged to abandon the work. His last works, in addition to a few panels for churches in Perugia, were frescoes for smaller towns in Umbria, such as Città della Pieve, Montefalco, Spello (XII, PL. 34), and Trevi. While working on a *Nativity* in the country church of Fontignano, Perugino was stricken with the plague and died, either in February or March of 1523. With the epidemic still of grave concern, he was probably buried in haste somewhere in the vicinity of this church, though some scholars believe some bones that were found conserved in the church to be the remains of Perugino.

The traditional opinion set forth by Vasari, that Perugino was a disciple of Piero della Francesca (q.v.), should most likely be construed simply as intending that the young Perugino studied the works of the older master closely, whereas he probably received his training in his craft, on a purely technical level, in his native town. A document concerning Perugino, dated 1472, states that in that year he had already settled in Florence. According to Vasari, he worked in the studio of Verrocchio (q.v.), where he is believed to have met Botticelli. Art historians have made a great effort to trace the early work of Perugino; unfortunately, the frescoes that Vasari says the painter executed in the churches of S. Agostino and S. Caterina in Arezzo, in S. Martino a Porta Prato in Florence, and in the Camaldolite monastery in Arezzo are all lost. Testifying to his activity in the period before he moved to Florence is the *Presentation of Jesus in the Temple* (Rome, Coll. Morandotti; Zeri, 1953). From the time of his connection with Verrocchio are the *Madonna with Rose Garland* in Paris (Mus. Jacquemart-André; Berenson, 1930), the *Crucifixion* in the Church of S. Maria e Angiolo at Argiano (Ragghianti, 1935), *The Archangel Raphael and Tobias* and the *Madonna and Child with Two Angels* in London (Nat. Gall.; Ragghianti, 1935; Zeri, 1953), and the *Madonna and Child* in Berlin (Staat. Mus.; Longhi, 1952). None of this dating, however, is definitely established; consequently the only unquestionable example of his early work is a small panel dated 1473, one of eight representing scenes from the life of St. Bernardino of Siena (VIII, PL. 190) that modern scholars attribute to Perugino (Venturi, 1913; Bombe, 1914; Berenson, 1930; Salmi, 1948).

Supposedly, Perugino painted some of the panels himself and planned the composition of the others, which were then painted by local artists. This hypothesis implies that the painter visited Urbino at some time in his youth, for the architectural backgrounds in these small paintings are typical of that town. On the basis of this and with further evidence provided by the *Pietà* panel in Perugia (PL. 115) — a work that may be attributed to Perugino with some reservations (Gnoli, 1923) — Rotondi (1950) verifies Perugino's visit to Urbino. While no fewer than three painters, including Pintoricchio (q.v.), worked on these panels, placing their figure groups against similar backgrounds to give uniformity to the narrative, in the panel representing St. Bernardino healing Polissena a distinctive artistic personality can be discerned. The symmetrical distribution of the figures and their pronounced isolation, as well as the small but wonderful landscape and its harmonious relationship to the figures, cause this panel to stand out clearly from the others of this series, with a new and extraordinary sense of monumentality and space. Already defined in this work are the characteristic qualities of Perugino's style; apparent, too, are those elements derived from Piero della Francesca and from the teachings of Verrocchio, which are particularly manifest in the figures.

This small masterpiece is the more interesting in that the work which followed it chronologically, the *Adoration of the Magi* (ca. 1475; Perugia, Gall. Naz. dell'Umbria), shows that Perugino returned for a time to the Florentine tradition. In the *Adoration of the Magi* spatial symmetry is broken in the crowded scene, dominated by a large tree, the latter a characteristic so reminiscent of Piero della Francesca that Cavalcaselle and other scholars have questioned Vasari's attribution of the work to Perugino. However, the *St. Sebastian* of the parish church of Cerqueto (PL. 113), a fragment of the fresco decoration of a whole chapel, provides sure evidence of the gradual maturing of his technique. This fragment is also valuable because it is dated (1478) and because it provides the first example of his signature (even though a copy). The iconography of the St. Sebastian in Piero della Francesca's polyptych in Sansepolcro is enlivened with a new conviction of modeling, which Perugino must have developed through his study of Pollaiuolo and Verrocchio; but for the first time there also appears an example of a negative element in Perugino's art, namely, the pathetic attitude.

The painting of the Cerqueto fresco immediately precedes Perugino's Roman period. Adolfo Venturi's hypothesis (1913) that Perugino may have contributed to the work carried out by Luca Signorelli in the Sacristy of S. Giovanni in Loreto between 1479 and 1481 is unacceptable. Even though the influence of Signorelli is apparent both in color and in draftsmanship in many of Perugino's works — particularly in the *Crucifixion* painted for the Compagnia della Calza in 1479, probably in collaboration with Signorelli — Gnoli has pointed out that in those years Perugino must already have been in Rome, working on the frescoes in Old St. Peter's and beginning the decoration of the Sistine Chapel. This would furnish grounds to speculate as to whether he might not then have derived some inspiration from the work of Melozzo da Forlì, who was working in the Vatican and in the Church of SS. Apostoli between 1470 and 1481. This possibility, which was suggested by Schmarsow (1886; 1912) and accepted by Gamba (1949), may account for the new solemnity of Perugino's figures in the Sistine Chapel.

The Sistine Chapel commission, which in allotting most of the decoration to Perugino represented the greatest sign of recognition he received from his contemporaries, had a specific program, namely, to illustrate the parallels between the lives of Christ and Moses. This kind of ideal parallel, very typical of Neoplatonism, is apparent also in the frescoes of the Collegio del Cambio in Perugia and in Raphael's Stanza della Segnatura. In the Moses in Egypt and the Baptism of Christ segments of the Sistine cycle, executed by Pintoricchio, Perugino's sense of composition is only vaguely distinguishable. The three episodes on the altar wall (depicting the Assumption, the finding of Moses, and the Nativity) were destroyed by Michelangelo to make room for his *Last Judgment*; thus, only the fresco representing Christ giving the keys to St. Peter (PL. 113) is left as a basis for appraisal: this was almost certainly painted by Perugino, though Vasari states that it was done in collaboration with Bartolomeo della Gatta. In this work is found one of the more notable examples of Perugino's compositional methods. The row of figures in the foreground, in which each figure is posed at an angle to its neighbor, forms a base for the symmetrical group of buildings in the background. The importance of the architectural background was emphasized by Cavalcaselle and by Venturi (1913); and Salmi (1948) compared it to the view of the ideal city in the Palazzo Ducale of Urbino (PL. 88) and to a perspective painting in the Walters Art Gallery in Baltimore.

It has already been noted that the small panels dated 1473 provide valuable documentation of the development of Perugino's art. The compositional schemes of the Florentine school of Piero della Francesca, as interpreted in the manner of Urbino, endowed Perugino's treatment of space with a rigorous set of perspective formulas. These laws acquired new pictorial validity in the Sistine fresco, in which the strongly lighted, idealized buildings became the real protagonists of the composition. The contemplative isolation of some of the splendid portraits in this fresco recalls the panel representing the healing of Polissena. The scene of Christ giving the keys to St. Peter pointed to new directions for the painters of the Renaissance; perhaps Perugino's highest achievement, this work marks the conclusion of his early period.

The *Crucifixion* (1481–85; Washington, D.C., Nat. Gall.) illustrates a phase of Perugino's activity in which he tended to indulge in a narrative tone, perhaps a consequence of his long association with Pintoricchio. The only other work in which this tone is somewhat in evidence is the *Nativity* (1491) of the Villa Albani, in Rome. The compositional scheme of the *Nativity*, which Gnoli believes was derived from the last Sistine fresco, is repeated in full or in details in later works of Perugino. Although in the period between 1490 and 1496 Perugino's art acquired harmonious and individual qualities, the numerous works done during those years — paintings in which the fundamental motif was monotonously repeated — betray his lack of interest in further development or innovations. He had begun to place his figures inside the architectural frames, and the carefully calculated space of these structures heightened the sense of the infinity in the background landscape; he also deployed his figures in graceful attitudes. This gracefulness became more and more accentuated, eventually to the detriment of Perugino's art. But in this period it was still a valid constituent of the work. In works such as the *Vision of St. Bernard*, formerly in Sto Spirito in Florence (ca. 1491–94; PL. 111), the altarpiece with the Madonna Enthroned, formerly in the Church of S. Domenico, near Fiesole (1493; PL. 115), and the *Pietà* that was once in S. Giusto, near Florence (ca. 1494–95; Uffizi), the precise balance of the poses in relation to one another, as well as their relationship to the architectural contours or to the shape of the panel (e.g., the tondo *The Virgin, St. Rose, and St. Catherine*, ca. 1491–92, Louvre), was the ruling compositional principle; where such a given frame was lacking, the composition would seem to have failed. An effective example of this is furnished by the overcrowded and overpraised *Lamentation over the Dead Christ* formerly in S. Chiara (1495; Florence, Pitti), which Raphael doubtless had in mind when he painted his *Deposition* (IV, PL. 314).

Even at this stage, however, Perugino continued to demonstrate outstanding gifts, as is evidenced by the subtle relationship between figure and setting in the portrait of Francesco delle Opere (1494; PL. 114). The work that brought this period to a close was the *Crucifixion* (shortly before 1496) in S. Maria Maddalena dei Pazzi in Florence, in which the architecture is brought into the foreground, with a very realistic treatment of the masonry, while the figures overlap each other in the background. This was a prelude to the new compositional system that Perugino was to adopt for the altarpieces he did in the last five years of the century: that in the Pinacoteca Nazionale in Bologna (ca. 1495–96; PL. 112) and those for S. Pietro in Perugia (1498; Lyons, Mus. B. A.), the Certosa of Pavia (1499; London, Nat. Gall.), and the abbey at Vallombrosa (1500; Uffizi).

Longhi (1955) suggests that these innovations might have resulted from the study of Luca della Robbia's colored reliefs. The improvement in quality as well as in color — though color was never to be one of the essentials of Perugino's art, notwithstanding its gradual mellowing and the transparency he obtained through superimposing thin layers of paint — is related by some authorities to Perugino's contact with the young Raphael. If the two had met before 1496, no doubt the younger man would have had a stimulating effect upon the master who had become settled in his ways. This theory, suggested by Cavalcaselle and further considered by Longhi (1955), which would seem especially plausible on the evidence of the wonderful triptych for the Certosa of Pavia, is less convincing when one examines a number of other works of this time, in which his customary compositional pattern is monotonously repeated without any real improvement in quality. Among these more or less uninspired examples is the Pala dei Decemviri (Vat. Mus.), once in Perugia, begun in 1495 and completed perhaps in 1496. The *Pietà* for the cyma of this altarpiece (Perugia, Gall. Naz. dell'Umbria), which was perhaps painted earlier, and the altarpiece in S. Maria Nuova in Fano (*Madonna and Saints*, 1497; repeated in the enthroned Madonna of S. Maria delle Grazie, Senigallia) are far superior works. Longhi believes that Raphael's hand is apparent in the Annunciation panel of the Fano predella (1498). This predella panel is related to a small *Annunciation* in Perugia (PL. 113), which is similar in composition to one of the scenes from the life of St. Bernardino (1473), and is dated erroneously both by Van Marle (1933), who thought it to have been executed in 1470–75, and by Venturi (1913), who fixed it at about 1485–90. To this period also belongs the next major work of Perugino executed after the frescoes in the Sistine Chapel, namely, the frescoes in the Collegio del Cambio in Perugia.

The decorations in the Sala dell'Udienza of the Collegio del Cambio (1497–1500; PL. 116) followed the Neoplatonic program, perhaps dictated by the Perugian Humanist Francesco Maturanzio, that perfection is achieved through the combined spiritual experience of the cardinal virtues (which also inspired the heroes and sages of pagan antiquity) and the incarnation of Christ (Nativity and Transfiguration scenes), whose coming into the world was announced by the prophets and the sibyls. Perugino undoubtedly considered it his most important work, for he embellished it with an acute self-portrait (PL. 114) and devoted his full attention (though assisted by a number of minor artists) even to details; adhering harmoniously to the architectural confines, he succeeded in creating one of the most unified decorative schemes of the Renaissance. On the ceiling are the planets, surrounded by elegant grotesques — an early instance of this kind of imitation of the antique (another, perhaps earlier example is found in the Perugino *Annunciation* of the Kress Collection, in the National Gallery, Washington, D.C.). The scholar Giovanni Carandente believes the ceiling to have been painted as early as 1496. The monotonous isolation and the mannered, unbalanced attitudes of the sages, the heroes, and the Virtues [Venturi's attribution (1920) of Fortitude to Raphael must be rejected] were considered by critics as obvious symptoms of decline, though the pure beauty of the prophets and sibyls seems to give further proof of his artistic resurgence during those years. The somewhat affected poses, the undeniably elegant flourish of the scarves and plumes, and even the didactic ranging of the figures in the other groups might well proceed from a continuing somewhat medieval interpretation of a fabulous antiquity, a tone of wonder unchanged by archaeological findings.

Perugino interpreted classic antiquity with a more intimate sense of reality only once, in the *Apollo and Marsyas* (PL. 117), which was formerly attributed to Raphael. Venturi assigned this work to Perugino's youthful period, but it is more likely that it was painted shortly before the frescoes in the Sala dell'Udienza. Thus, it is valuable for a better appreciation of Perugino's capabilities when not bound by the limitations of an elaborate iconographic program.

A faint reflection of the bucolic sweetness of the *Apollo and Marsyas* appears in some of his late works, such as the *Combat of Love and Chastity* (1505; PL. 116), which was condemned by critics and even by Isabella d'Este, who had commissioned it. This picture was painted in various stages, and some of the figures clearly reveal an effort to imitate Venetian models, while others are secular renderings of sacred characters. A weak and slovenly work, its sole interest lies in the completely dominant landscape setting, on which Perugino concentrated what remained of his imaginative powers; the figures, for the most part replicas from other of his works, are devoid of vitality. These characteristics of Perugino's final period are evident especially in the *Adoration of the Magi* (1504) in the Oratory of S. Maria dei Bianchi in Città della Pieve, the *Martyrdom of St. Sebastian* (1505) in S. Sebastiano in Panicale, and the *Marriage of the Virgin* (1506; Caen, Mus. B. A.), painted for the Cathedral of S. Lorenzo in Perugia.

Vasari tells of the sharp criticism directed at Perugino's *Assumption* (1505–07) for the Servite church of SS. Annunziata in Florence by his contemporaries and of his reply: "I have used the figures which you, on other occasions, have praised and infinitely admired; if now you dislike them and do not praise them, what fault is it of mine?" This retort shows that his artistic decline was, to a great extent, mainly a matter of mental attitude. Pictorial imagination could receive no fresh impulse from an essentially unintellectual cast of mind. Still, he found it worth his while to repeat earlier compositions in a routine manner, in order to comply with the demand in the provincial environment to which the aging painter progressively retired and from which he never moved after 1512. Various works to be found in Umbrian churches are nothing but continued and tired reiterations of old themes, especially the Nativity scenes. Only in the *Madonna with Saints* of 1512–23 — a large altarpiece for the Church of S. Agostino in Perugia and his last major work — is it possible to find some substantial value, though it too is but a pale reflection of his best works.

Whatever his later failures, Perugino ranks among the great innovators of Italian art in the transitional period between the early and High Renaissance, for he gave a new dimension to painting. Early in the 15th century, space had been treated by Tuscan artists as an empirically measurable entity, strictly conditioned by the requirements of the figures. Perugino, however, although he started out from the spatial principles of Piero della Francesca, opened this defined space to infinity, as a function of which man has his being. And it is this sense of infinity that he presents which attracts the deep spiritual interest of others. But such a contemplative aura set its own limitations, for the figures, being merely human extensions of this infinite space, could not be permitted to convey any dramatic emotion, lest the serene universal order be disturbed. They might show only tender melancholy or resigned sorrow. The iconographic repertory and style of forms were also conditioned by this attitude, so that only delicate grace, gentle and simple forms, and the tranquil postures of the sacred figures of Christian iconography could be introduced into the calm of this atmospheric, fundamentally spatial vision. The religious emotion deriving from Perugino's works is aroused not by these elements of form, however, but — as Berenson noted (1930) — by the possibility offered to the spectator of identifying with the universe through this "space-composition."

Perugino was appreciated by the "progressive" Florentine painters of the 16th century only as the master of Raphael. He had better critical fortune with the primitivists of the 17th century, particularly Sassoferrato; but only with the advent of neoclassicism and, even more, among the 19th-century neoprimitivists were his genuine values discerned. Nonetheless, it is to Berenson (1930) that the merit of discovering the true inner significance of Perugino's art must be assigned. He identified it in the concept of a "space-composition," held to be a common characteristic of the whole Umbrian school in the second half of the 15th century. However, as Lionello Venturi (1955) rightly observed, it is truly found only in the painting of Perugino and Raphael.

No genuine school was later formed by the artists who were in longest and closest contact with Perugino in Umbria, such as Tiberio d'Assisi, Francesco Melanzio, Giannicola di Paolo, Giovanni di Pietro (Lo Spagna), Francesco Tifernate, Eusebio da San Giorgio, Sinibaldo Ibi, and Berto di Giovanni. They grasped only those elements of his art which were least capable of development (just as proved by Perugino's own late works): the human traits and the pseudoreligious pathos of his figures. Once they had accentuated these elements for obvious devotional purposes, sometimes to a grotesque degree, the younger among them turned elsewhere for examples, especially to Raphael, the true heir to Perugino's art. It was he who provided the exceptional medium through which the new spatial vision of Perugino became part of the most precious heritage of Italian art.

BIBLIOG. A. Mariotti, Lettere pittoriche perugine, Perugia, 1788; B. Orsini, Vita, elogio e memorie dell'egregio pittore Pietro Perugino e degli scolari di esso, Perugia, 1804; K. F. von Rumohr, Italienische Forschungen, 3 vols., Berlin, 1827–31 (new ed., J. von Schlosser, Frankfurt am Main, 1920); A. Mezzanotte, Della vita e delle opere di Pietro Vannucci, Perugia, 1836; J. D. Passavant, Raffael von Urbino und sein Vater Giovanni Santi, 3 vols., Leipzig, 1839–56; G. B. Cavalcaselle and J. A. Crowe, A New History of Painting in Italy, London, 1864; Vasari; G. B. Cavalcaselle and J. A. Crowe, Raphael: His Life and Works, 2 vols., London, 1882–85; A. Schmarsow, Melozzo da Forlì, Berlin, 1886; B. Berenson, Central Italian Painters of the Renaissance, New York, London, 1897; G. Santi, Cronaca (ed. H. Holtzinger), Stuttgart, 1897; J. C. Broussolle, La jeunesse du Pérugin et les origines de l'école ombrienne, Paris, 1901; W. Bombe, Geschichte der peruginer Malerei (Italienische Forschungen, V), Berlin, 1912; A. Schmarsow, Joos van Gent und Melozzo da Forlì in Forlì, Rom und Urbino, Leipzig, 1912; Venturi, VII, 2, 1913, pp. 453–858; W. Bombe, Perugino, Stuttgart, Berlin, 1914; A. Venturi, Raffaello, Rome, 1920; U. Gnoli, Pietro Perugino, Spoleto, 1923; U. Gnoli, I documenti su Pietro Perugino, Perugia, 1924; Van Marle, XIV, pp. 299–411; U. Gnoli, Bibliografia degli scritti su Pietro Perugino, B. R. Istituto di archeol. e storia dell'arte, II, 1928, pp. 14–22; W. Bombe, Urkunden zur Geschichte der peruginer Malerei im 16. Jahrhundert, Leipzig, 1929; B. Berenson, The Italian Painters of the Renaissance, Oxford, 1930 (2d ed., London, 1952); F. Canuti, Il Perugino, 2 vols., Siena, 1931 (bibliog.); C. L. Ragghianti, La giovinezza e lo svolgimento artistico di Domenico Ghirlandaio, L'Arte, N.S., VI, 1935, pp. 174–77, 186–91, 349–51; M. Salmi, Il Palazzo Ducale di Urbino e Francesco di Giorgio, S. artistici urbinati, I, 1948, pp. 9–55; C. Gamba, Pittura umbra del Rinascimento, Novara, 1949; P. Rotondi, Il Palazzo Ducale di Urbino, 2 vols., Urbino, 1950–51; R. Longhi, Quadri italiani a Berlino e Sciaffusa, Paragone, III, 33, 1952, pp. 39–46; F. Zeri, Il Maestro dell'Annunciazione Gardner, BArte, XXXVIII, 1953, pp. 131–39, 233–39; R. Longhi, Percorso di Raffaello giovine, Paragone, VI, 65, 1955, pp. 11–23; L. Venturi et al., Il Perugino, Turin, 1955; E. Camesasca, Tutta la pittura di Raffaello, Milan, 1956; E. Camesasca, Tutta la pittura del Perugino, Milan, 1959 (bibliog.).

Francesco SANTI

Illustrations: PLS. 111–117.

PERUZZI, BALDASSARE. Italian painter and architect (b. Siena, 1481; d. Rome, Jan. 6, 1536). The son of Giovanni di Salvestro Peruzzi of Volterra, Baldassare Tommaso Peruzzi was baptized in Siena on Mar. 7, 1481. His first training was probably as a draftsman and painter.

In July, August, and October of 1501 Peruzzi engaged in painting frescoes (destroyed) on the vaulting of the Chapel of St. John in the Cathedral of Siena. During a short stay in Volterra he became friendly with Pietro d'Andrea da Volterra, with whom he then journeyed to Rome. There remains various testimony of his Roman activity. The frescoes in the apse of S. Onofrio are assigned to Peruzzi. In the Raphael Stanze of the Vatican palace, the decorative framework for the vault of the Stanza d'Eliodoro and the ceiling pictures in the Stanza della Segnatura (the figure of Philosophy, the Apollo and Marsyas, and the putti in the *Contemplation of the Universe*) are attributed to him. Toward 1510, he painted the much-restored *Nativity* in S. Rocco a Ripa.

Between 1509 and 1511 he designed and built, for Agostino Chigi, the villa in Trastevere that is known as the Farnesina. He also painted the exterior decorations in *terretta* and some of the interior fresco decoration, including the scenes from Ovid's *Metamorphoses* in the Sala del Fregio and the mythological-astrological illustrations of Agostino Chigi's horoscope in the Sala di Galatea (PL. 119). Later, in about 1515, he painted the illusionistic architectural scenes with landscapes in the Salone delle Prospettive (PL. 118).

In 1511, he and his brother Pietro took a long lease on two houses in the Rione Ponte, a district of Rome. In 1513, he painted a scene representing the betrayal of the Romans by Julia Tarpea for the temporary festival decorations erected for Giuliano de' Medici. In 1515, he may have supplied a design and a model for the Cathedral of Carpi, and for the façade of the old cathedral ("La Sagra") as well. After 1516 he fulfilled the same tasks for the resumption of work on the nave of S. Niccolò in Carpi.

Peruzzi's *Nativity* (London, P. Pouncey Coll.) was painted between about 1512 and 1516. His frescoes in the Ponzetti Chapel in S. Maria della Pace in Rome were done in 1515–16, and the *Presentation of the Virgin* in the same place in 1516–17. In 1518 Peruzzi took part in the competition for the building of S. Giovanni dei Fiorentini and, in the same year, was a consultant for the building of S. Maria della Consolazione in Todi. He also collaborated on the decorative painting of the Villa Madama, outside Rome, and painted in *terretta* the *St. Bernard* in the garden of S. Silvestro al Quirinale (Vasari; cf. H. Wurm, *AB*, XLIV, 1962, p. 254, fig. 1). On Aug. 1, 1520, he became an associate architect (*coadiutore*), under Antonio da Sangallo the Younger (q.v.), for construction of St. Peter's in Rome. This first appointment lasted until 1527, and in 1531 it was renewed until his death. During this time he worked on the building and prepared designs (IV, PL. 189); he executed the walls around the *confessio* begun by Bramante and also worked on the south transept, in the Chapel of the King of France. In the summer of 1521 he delivered a complete plan (*modello*).

In 1521–22 he was in Bologna, where he prepared designs for the façade and dome of S. Petronio, for the rebuilding of the Palazzo Lambertini (destroyed), and for the main portal of S. Michele in Bosco. He also executed a chiaroscuro *Adoration of the Magi* (pen and ink with wash; London, Nat. Gall.) for Count Bentivoglio. There is no evidence of a stay in Carpi at this time. In 1523 he was again in Rome, living near S. Salvatore in Lauro. He completed the decorations for the coronation of Clement VII in 1524. During the summer and autumn of 1525 he prepared plans for the Cathedral in Siena, and while there he caught a fever.

His activities in Rome during the 1520s include plans for the gardens of Cardinal Trivulzio's villa at Salone, for S. Giacomo in Augusta, and for the rebuilding of the Lateran Baptistery (interior and a dome with lantern). The Palazzo Ossoli and the Palazzetto Spada are attributed to him. When Rome was sacked in 1527, a large ransom was demanded of him, and, being destitute, he went to Siena. In Siena he was appointed architect for the Cathedral; on Aug. 21, 1527, he became architect to the city, and in 1528 he bought a house there. In 1527–28 he repaired the fortifications of Siena and built seven bastions and towers, of which there remains a *fortino* near Porta Pispini. At this time he also painted a fresco of Augustus and the sibyl (1528; PL. 119) in the Church of Fontegiusta.

About 1529, the tomb of Hadrian VI in S. Maria dell'Anima in Rome was completed according to Peruzzi's design. In 1529 he made a tour of inspection of the fortifications of Chiusi, Torrita, and Sarteano and prepared a report on Poggio Imperiale. He finished the plans for the Ghislardi Chapel in S. Domenico in Bologna in 1530. In 1531, he devised a new method for minting coins. His plans for alterations to the Cathedral of Siena can be assigned to these years. On Aug. 1, 1531, during a one-month stay in Rome, he was reappointed for the work on St. Peter's, now becoming one of the two chief architects on the project, together with Antonio da Sangallo the Younger.

In 1531–32 he made a tour of inspection in the Maremma region. After 1532, he prepared designs for S. Domenico in Siena. On Apr. 30, 1534, Clement VII ordered him to Rome, where he fell ill with a high fever in August. The plans for Rocca Sinibalda, outside Rieti, and for the alterations to the Castle of Belcaro, outside Siena, probably belong to that year. Paul III confirmed his appointment as architect for St. Peter's on Dec. 1, 1534. In 1535 he was completely occupied in Rome with works for the Savelli family in the Theater of Marcellus, with the completion and restoration of the Belvedere in the Vatican, with the reconstruction of the Palazzo Massimo alle Colonne (perhaps begun in 1532), and with plans for the Palazzo Ricci in Montepulciano. He died on Jan. 6, 1536, and was buried with solemn ceremonies in the Pantheon.

Among the great Renaissance artists, there are few whose chronological development has been so little studied — and who have received so little critical attention in general — as Baldassare Peruzzi. Moreover, if uncertain traditional testimony and attributions are revised, contemporary studies have little solid evidence left as a point of departure.

The most decisive stimuli for Peruzzi's painting were his contacts with Pintoricchio and Sodoma (qq.v.) in Siena, and his portions of the frescoes in the apse of S. Onofrio in Rome give proof of this. He sought to approximate the classical style of the Roman High Renaissance in his *Nativity* in S. Rocco a Ripa, as well as in his works in the Farnesina and in the Ponzetti Chapel, in which his study of antique sculpture bore fruit. Soon a Venetian richness of color was added to his Roman academicism (e.g., the *Nativity* in the Pouncey Collection). With the *Presentation of the Virgin* of S. Maria della Pace, a genre painting characterized by its refined scenography, he disclosed mannerist tendencies at a notably early date; in this he was in a way resuming Pintoricchio's imaginative vision and reflecting Raphael's late style. An even more decisive and magnificent piece of evidence for this stylistic change is the *Adoration of the Magi* painted for Count Bentivoglio in 1521. His painting of the 1520s and 1530s becomes more and more difficult to comprehend, for the forms are quieter and yet, at the same time, bolder — as in the Fontegiusta fresco (PL. 119). Perhaps a certain decline in the quality of the late works (the Belcaro frescoes) can be explained by his collaboration with a pupil, Daniele da Volterra (Daniele Ricciarelli).

As a painter, Peruzzi seems to have been at his best in decorative work. His relation to other artistic figures in Rome and Siena (e.g., Beccafumi, Sodoma) was more a matter of borrowing and of evolving new variations, rather than greatly transforming their ideas or influencing these artists decisively. Peruzzi's work is of paramount importance in the fields of stage design (see designs in PL. 99; XII, PLS. 422, 431), festival decoration, façade painting, and trompe l'oeil decoration. In the trompe l'oeil wall painting of the Salone delle Prospettive in the Farnesina (PL. 118) he created a work far in advance of his time.

Peruzzi's architecture deserves even more substantial consideration than his painting. It is quite probable that his birthplace, Siena, endowed him with a native gift for fine-lined elegance, but it is not certain that he received any practical architectural training there under Francesco di Giorgio (q.v.). The bases for the design of the Farnesina (PL. 120; FIG. 273) were the older Tuscan and Roman styles, and the Roman idiom before Bramante's and Raphael's (qq.v.) innovations. On the other hand, hints of the approach of mannerism are also evident in its forms; and the Farnesina, despite its lightness and clarity, does not lack certain mannerist tensions. Certainly, Bramante's work and personality were the model for Peruzzi's architectural activity, and the problems posed by the work on St. Peter's and the lengthy discussions of them provided his actual architectural curriculum. The unfinished buildings and the store of ideas left behind by Bramante were inherited by Peruzzi.

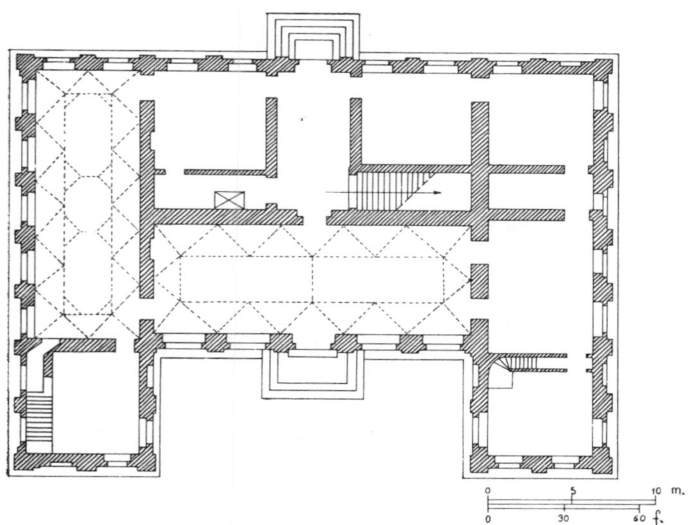

Rome, Farnesina, ground-floor plan.

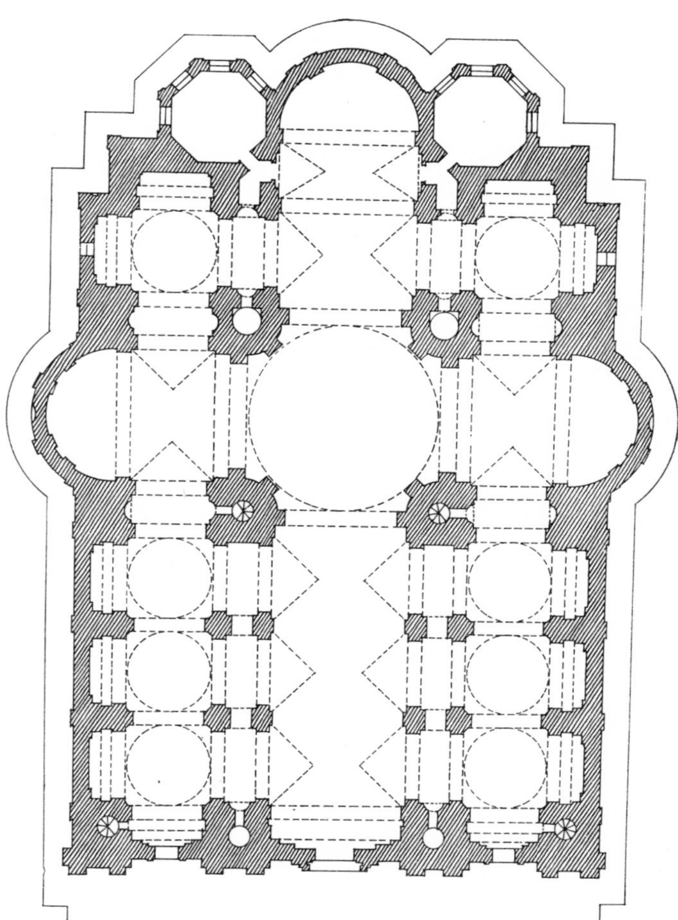
Carpi, the Cathedral, plan (*from. H. Semper, F. O. Schulze, and W. Barth, Carpi: eine Furstensitz der Renaissance, Dresden, 1882*).

The facade for "La Sagra" in Carpi, with its strong formality and intersecting temple fronts, constitutes a notable step in the direction of Palladio (q.v.) in the development of church façades. The plan for the new cathedral in Carpi (FIG. 274) was designed with reference to St. Peter's.

Although during his active career, with the prolific creation of the great classical works of the High Renaissance, Rome became the art center of the world, Peruzzi himself was not fully identified with this prevailing influence. An example of his tendency toward anticlassical conceptions might be seen in the façade of the palace shown in the *Presentation of the Virgin* in S. Maria della Pace. A parallel to this development in his painting is found in his architectural drawings, in which he devised a new mode of handling the representation of space from multiple viewpoints (multiple perspective). He resolved the problem posed by the façade of the Gothic church of S. Petronio in Bologna by going back to pure late Gothic forms, combining the rich late Gothic decorative elements with Renaissance principles of articulation.

After 1520, as has already been noted, Peruzzi was active in the building of St. Peter's in Rome. Work on the great church was still unfinished at his death, and his definitive plan can only be discerned approximately from various existing versions, both original drawings and publications (Uffizi, Gabinetto dei Disegni e Stampe, no. 2A; another formerly in New York, Coll. L. G. White; cf. also Serlio, III, 27). His preoccupation with the problem of St. Peter's over a span of many years is documented in numerous drawings (IV, PL. 189). Among these D. Frey has identified the designs made by Peruzzi before his appointment to the Fabbrica di S. Pietro (Frey, 1915, Group III, B, 1); the reduced ground plans made during his first appointment, 1520–27 (III, B, 2); his further development of the plans left by Raphael (III, C); and the plans from the architect's last creative period (III, D), which were more strongly oriented toward theoretical considerations. In his more tectonic, classical expression of form, Peruzzi differs consistently from Antonio da Sangallo the Younger.

In his work of the 1520s, he succeeded in making the oval into one of the possible monumental architectural ground plans. A drawing in the Uffizi (Gabinetto dei Disegni e Stampe, no. 531A) is the first known attempt in Western architecture to base a construction on a geometrically elliptical plan. With this innovation, where the interior space was not clearly reflected by the exterior form, the classical organic unity of "interior" and "exterior" in architectural form was abandoned.

Peruzzi's designs for S. Domenico and the Cathedral in Siena provide important indications of his late style. The originality of his architectural and spatial forms is characterized by the pillars detached from the wall and freely "disengaged" in space. In the Palazzo Massimo alle Colonne in Rome (PL. 121; FIG. 275; XII, PL. 52), Peruzzi achieved a clearly personal expression, despite the difficulties presented by the site. With

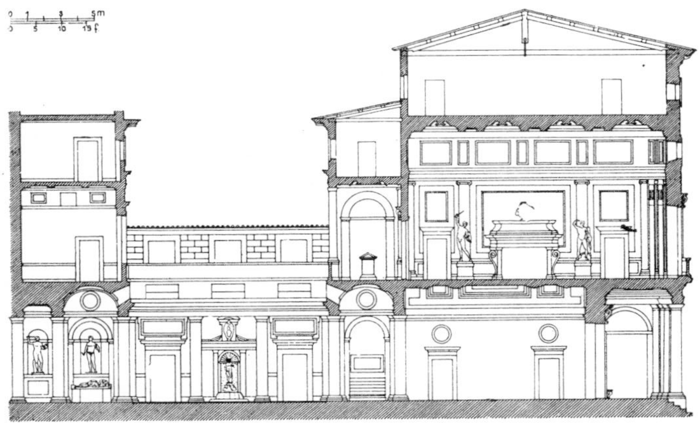

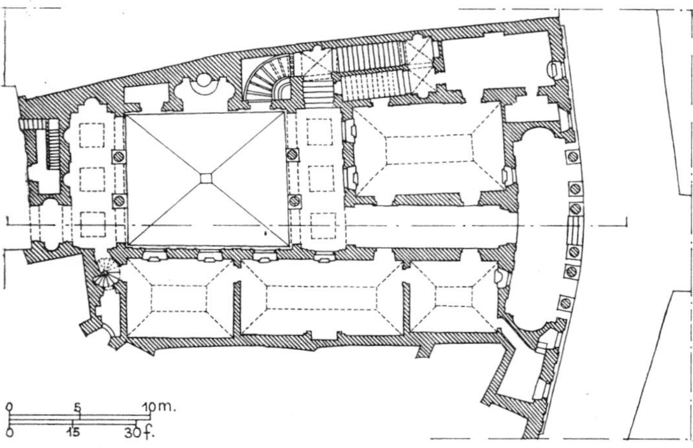

Rome, Palazzo Massimo alle Colonne. *Above*: Section. *Below*: Ground-floor plan (*from P. Letarouilly, Edifices de Rome moderne, Paris, 1860*).

the convex façade and intimate courtyard of this palace, he created one of the major examples in Rome of early mannerist architecture.

Peruzzi's continued concern with the fundamental theoretical premises of his architectural activity was demonstrated by his numerous surveys of ancient structures, by his preparation of a commentary on Vitruvius, and by his own treatise on architecture. His research in these areas went far beyond the prevailing ideas of his time. His theoretical bent is also attested by the fact that many of his projects did not take into consideration the actual architectural possibilities. The small number of his completed buildings and the unfortunate fate of many of his building projects make it difficult to determine whether this scientific and speculative inclination was a dominant trait in his character. A definitive and complete assessment of his artistic personality is not possible before a thorough critical examination of his legacy of drawings has been completed. Nevertheless, Peruzzi — together with Raphael and Antonio da Sangallo the Younger — was one of the most important of the pupils and followers of Bramante and one of the outstanding representatives of early Roman mannerism in its classical phase. Also, he was the only Sienese among the great architects of the Italian Renaissance.

BIBLIOG. Vasari, IV, pp. 589–642; S. Borghesi and L. Bianchi, Nuovi documenti per la storia dell'arte senese, Siena, 1898; H. Egger, Entwürfe B. Peruzzis für den Einzug Karl V in Rom (Inventario critico del "Taccuino senese," Bib. Com. S. IV, 7), JhbKhSammlWien, XXIII, 1902, pp. 1–44; K. Frey, Zur Baugeschichte St. Peters, JhbPreussKSamml, XXXI, 1910, pp. 1–95, XXXIII, 1912, pp. 1–153, XXXVII, 1916, pp. 22–136; A. Bartoli, I monumenti antichi di Roma nei disegni degli Uffizi, 5 vols., Rome, 1914–22; D. Frey, Bramantes St. Peter-Entwurf und seine Apokryphen (Bramante-S., I), Vienna, 1915; W. W. Kent, The Life and the Works of Baldassare Peruzzi of Siena, New York, 1925; C. Ricci, La scenografia italiana, Milan, 1930; P. Metz, ThB, s.v. (bibliog.); F. Saxl, La Fede astrologica di Agostino Chigi: Interpretazione dei dipinti di B. Peruzzi, Rome, 1934; Venturi, VIII, 1, pp. 912–19, 2, pp. 780–83, IX, 5, pp. 375–412, XI, 1, pp. 358–439 (rev. of XI, 1, by G. Giovannoni, Palladio, II, 1938, pp. 107–14); H. Tintelnot, Barocktheater und barocke Kunst, Berlin, 1939; R. Zürcher, Stilprobleme der italienisch Baukunst des Cinquecento, Basel, 1947; J. S. Ackerman, The Cortile del Belvedere, Vatican City, 1954; W. Lotz, Die ovalen Kirchenräume des Cinquecento, Römische Jhb. für Kg., VII, 1955, pp. 1–100; C. L. Frommel, Die Farnesina und Peruzzis architektonisches Frühwerk, Berlin, 1961; H. W. Wurm, Der Palazzo Massimo alle Colonne, Berlin, 1965.

Heinrich Wilhelm WURM

Illustrations: PLS. 118–121; 3 figs. in text.

PEVSNER, ANTOINE. Russian sculptor and painter (b. Orel, Russia, Jan. 18, 1886; d. Paris, Apr. 12, 1962). Antoine Pevsner was the third son of a copper refinery executive. His brothers, including the equally celebrated younger brother who later became a sculptor and took the name Naum Gabo (q.v.), were educated in science and engineering. Antoine, however, spent two years (1908–10) at the Kiev Academy of Fine Arts and a year at that of St. Petersburg. Having become interested in French art by the Ivan Morosov and Sergei Shchukin (Stchoukine) collections in Moscow, he visited Paris in 1912 and was overwhelmed by the art of the cubists. The following year he took a studio there, but at the outbreak of World War I fled to Norway with his brother Naum. Upon the overthrow of the Czar, they returned to Moscow, where Antoine was appointed to the faculty of the Vchutemas (Higher Art and Technical School) along with other modernists such as Kazimir Malevich (q.v.) and Vladimir Tatlin. When the government withdrew its support of the constructivist movement, Pevsner left Russia and, after a brief stay in Berlin, settled in Paris (1923), where he remained for the rest of his life. He was decorated by the French government in 1961, a year before his death.

Pevsner began as a painter of geometric abstractions. While in Russia he made a few masks and reliefs, doubtless following the example of his younger brother. Some of his drawings of this period resemble the architectural inventions of the constructivists. In Paris he continued to paint but devoted himself increasingly to sculpture. The *Torso* (1924–26; New York, Mus. of Mod. Art) is a characteristic work of this period; a semi-abstract contour of head and body, it is composed of intersecting silhouettes cut from metal and celluloid. Volume is thus suggested, but the matter so constituted is itself apparently weightless, dematerialized by the way in which the transparent planes penetrate the surrounding space. A similar work is his relief *Portrait of Marcel Duchamp* (1926; New Haven, Conn., Yale University Art Gall.). At the same time, inspired perhaps by Umberto Boccioni's (q.v.) futurist sculpture, Pevsner was experimenting with a more radical departure from representational imagery in works which suggest dynamic forms — with an effort to suggest the elements of time and motion with material that is actually static. *Projection in Space* (1924; Baltimore, Mus. of Art) is an early example of these aims, and was followed by numerous others, such as *Construction in Space: Project for a Fountain* (1929; Basel, Kunstmus.). Incorporating the void as an essential element, these works move toward an integration of time and space, the extraordinary quality of Pevsner, without precedent in previous sculpture.

During the 1930s Pevsner sought to arrive at a tangible definition of aerial space, as in *Construction for an Airport* (1937; Amsterdam, Stedelijk Mus.), in which the flat planes of diagonals and curves are augmented by a surface web of "lines of force." Instead of continuing with rectlinear planes of transparent plastic and sheet metal, he began to use cast and tooled bronze, worked in a rich variety of spirals, acute angles, and parabolic curves. (It is quite possible that his forms have had some influence on the architecture of Torroja, Eero Saarinen, and others.) His ultimate style was achieved in the 1940s with compositions such as the *Construction in an Egg* (V, PL. 137), *Developable Column* (1942; New York, Mus. of Mod. Art), and especially with the *Developable Column of Victory* (1946; Zurich, Kunsthaus). In these the grooves and embossed rods of the 1930s are refined to an over-all surface of minute striations, which not only enrich the surface but make it seem to vibrate. In the 1950s he produced a few works of a monumental scale, notably a second *Developable Column of Victory*, also called *The Flight of the Bird* (ca. 12 ft. high; 1955; Warren, Mich., General Motors Technical Center).

BIBLIOG. R. Olson and A. Chanin, Naum Gabo–Antoine Pevsner (introd. by H. Read), New York, 1948 (cat.); R. Massat, Antoine Pevsner et le constructivisme (pref. by J. Cassou), Paris, 1956; P. Peissi, Antoine Pevsner, Neuchâtel, 1961.

Henry R. HOPE

PHIDIAS (Φειδίας; PHEIDIAS). Athenian sculptor of the 5th century B.C., the son of Charmides. According to Pliny (*Naturalis historia*, XXXV, 34, 54), the painter Panainos was Phidias's brother, though Strabo (IV, 89) states that he was his nephew. Ancient sources are also at variance concerning the teacher of Phidias, since there is mention of both Hegias of Athens and Ageladas (Hagelaidas) of Argos. However, since Phidias was an Athenian, it would seem likely that he received his training from a master in his own city rather than that he should have gone to study with an artist from Argos who worked mainly at Olympia.

Pliny states that Phidias was originally a painter and that he decorated, for example, a shield in Athens; he states, too, that besides being a sculptor in marble, Phidias also worked in bronze and embossed metal and was highly skilled in the making of statues in gold and ivory. His experience and mastery in diverse arts was reflected in the variety of his recorded output.

Documents quoted by Pliny referred to the 83d Olympiad (448–444 B.C.) as marking the high point of Phidias's activity. It is also known that Perikles commissioned Phidias to carry out work on the Parthenon in 447, and that the consecration of his magnificent statue of Athena Parthenos took place in 438. Plato mentions Phidias as still active in Athens in 433. The friendship and favor of Perikles were of fundamental importance in the career of Phidias, for the sculptor gave brilliant expression to the statesman's ideals of grandeur and beauty through his transformation of the Acropolis into a symbol of the new ascendancy of Athens. Later events in Phidias's life appear to be closely linked with the destiny of his patron and developed out of the political climate of the times. Because he had adorned Athens with temples, statues, and costly marble, Perikles was reproved for drawing on public funds to gild and bedeck the city like a "courtesan" (Thucydides). The comic writers made jokes about the visits that Perikles and his entourage of fashionable women paid to Phidias's workshop. This atmosphere of envy and backbiting reached its height in 432, when Menon, a resident alien, accused Phidias of misappropriating gold and ivory intended for the *Athena Parthenos*. Since the various parts of the statue were detachable, Phidias was able, by means of an actual test, to prove that the charge was groundless, but his enemies resumed the attack and accused the sculptor of sacrilege (*asebeia*) for daring to depict himself and Perikles in the Amazonomachia on the outside of the shield of the *Athena Parthenos*.

Ancient writers give two versions of Phidias's trial. According to the tradition followed by Plutarch (*Perikles*, XXXI, 5), Diodoros Siculos (who draws on Ephoros), and the Pseudo-Aristodemus (XVI, 1), Phidias was condemned and sent to prison in Athens, and it was there that he died; according to a variant tradition that goes back to Philochoros, cited by a scholiast on Aristophanes, Phidias was condemned, but managed to escape to Elis, where he found shelter at Olympia. Details apart, some writers consider this trial to be one of the events leading up to the Peloponnesian War, for Perikles, once more faced with a climate of opposition and accusations — just as he had been in the cases of Damon, Anaxagoras, and Aspasia — adopted a war policy in order to dissipate the general ill-feeling.

The two different accounts of Phidias's later years form the basis for two schools of thought among modern scholars about the dating of his masterpiece, the chryselephantine *Zeus* for the temple at Olympia. This is assigned to the year 448 by those who support the account of his death in prison in Athens, and to about the year 432 by those who believe that he fled to Elis. This second opinion assumes that Phidias did not see his work on the Parthenon through to its conclusion, since the inscriptions indicate that the finishing touches were still being put to the temple in 433–432, in the fifteenth year after the work was begun.

According to Plutarch (op. cit., XXXI, 4), Phidias had depicted himself on the shield of the *Athena Parthenos* as a bald old man, but this information should not be taken as definitive evidence of the sculptor's physical appearance, for baldness might well have been one of the features used to characterize the demiurge as a type. Moreover, as B. Schweitzer has pointed out, following a tradition cited by the Roman writer Ampelius, this figure on the shield was probably intended to represent Daidalos, the first *agalmatopoios* (sculptor), while the other figure, in which contemporaries thought they recognized Perikles, may in fact have celebrated Theseus, traditional hero of Attic unity and symbol of the first Athenian ascendancy.

Aside from the points of disagreement, with the information about the life and work of Phidias provided by written sources, it can reasonably be supposed that the sculptor was born between 500 and 480 B.C. His activity was at its peak about 448 and came to an end toward 432. He is known to have worked at Pellene, Delphi, Thebes, Plataia, and Olympia, but Athens was the main center of his activity, the place where most of his works were to be seen.

The gold and ivory image of Athena for the temple at Pellene in Achaia, mentioned by Pausanias (VII, 27, 2), is no longer extant. The figure of Athena found on Roman coins minted in this city may be a debased representation of this statue, though this is by no means certain. Athena appears on the coins bearing a shield on her arm and holding her helmet and spear aloft in her right hand; her right leg is slightly flexed, with the foot raised sideward. The peplos is adorned with horizontal bands. Pausanias says that this statue was earlier than either the *Athena Promachos* on the Acropolis in Athens or the *Athena Areia* in Plataia; therefore, it is probably an early work.

The bronze *Apollo Parnopios*, which Pausanias (I, 24, 8) mentions as standing near the Parthenon, was erected to commemorate the god's deliverance of the city from an invasion of locusts (*parnopes*). It has been thought that the type of which the *Apollo* of Kassel (III, PL. 353) is the principal replica may be a copy of this statue. This was a well-known type in the Roman world, for about twenty replicas of it are recorded. In the original, the god held a bow in one hand, and in the other probably the purificatory branch of laurel, as would befit his epithet "Alexikakos," he who wards off evil. If this identification of the type is correct, the statue is probably an early work, since certain features are still somewhat severe. It is probably correct to attribute this Apollo type to Phidias, because it stands out sharply from the other images of Apollo produced in the first half of the 5th century. This manifestation of a god, a vigorous and organic whole, is brought to life by its soft coloristic modeling. The rich sculptural quality of the curly hair reveals the temperament of a genius who was able to infuse the iconographic convention with new life, and in doing so, to bring about substantial modifications in the Apollo type. Even in these early works Phidias seems to give proof of his originality.

Pausanias described the Athenian offerings that stood at one end of the Sacred Way at Delphi; there was an inscription on the base stating that the Athenians had used a tenth part of the spoils taken at Marathon to erect this votive monument (X, 10, 1–2). It comprised a series of bronze statues, representing Athena, Apollo, Miltiades (the victor of Marathon), and the heroes after whom the ancient tribes were named (Erechtheus, Kekrops, Pandion, Leos, Antiochos, Aigeus, Akamas), as well as Theseus, Kodros (son of Melanthos), and Phileos. These statues were all the work of Phidias; the statues of Antigonos, his son Demetrios, and the Egyptian Ptolemy were added at a later date. There do not seem to be adequate grounds for supposing that this Athenian group was dedicated immediately after the Battle of Marathon, though it has been suggested that it is to be viewed in connection with Kimon's anti-Persia policy and his desire to honor his father, Miltiades. Rather, it is to be assigned to a later date, either before Kimon was ostracized in 461 or after his recall to Athens in 452. The whole group of statues is no longer extant, and there are no descriptions of how Phidias interpreted these personages. There is a herm of Miltiades from Porto Corsini (Ravenna); however, though its style is classicizing there is no way of verifying whether it is based on the Miltiades in the group at Delphi. The tribal heroes Erechtheus, Kekrops, and Pandion were subjects that Phidias used again, in the sculptures with which he adorned the Parthenon.

Pausanias (XI, 4, 1) also mentions the *Athena Areia* at Plataia, which he called a "xoanon epichryson" (gold-covered statue with unclothed parts of marble); it was only slightly smaller than the bronze *Athena Promachos* on the Acropolis and, like that statue, was erected with a tenth part of the spoils of Marathon. Plutarch (*Aristides*, XX, 3), on the other hand, connected it with the victory of Plataia in 479, when, with Aristides in accord, the allies allocated 80 talents for rebuilding the sanctuary of Athena there. The temple was adorned with paintings by Polygnotos and Onasias (Pausanias, IX, 4, 3). Even if this be the occasion, the work is probably to be dated somewhat later, during the period of undisturbed prosperity after the Peace of Kallias in 449 B.C. It was a national votive offering and, along with the institution of the Panhellenic festivals, was on acknowledgment of divine protection in a great victory. As such, it was an important public work that would justify the summoning from Athens of a top-rank artist such as Phidias, perhaps only shortly before work on the Parthenon was begun. The technique of covering the statue with gold leaf and fashioning the unclothed extremities of marble was perhaps chosen in order to convey an impression of opulence and splendor, as befitted a statue erected from the spoils of a great victory. This work is no longer extant; H. Thiersch ("Die Athena-Areia des Phidias u. der Torso Medicis in Paris," *Nachr. von der Gesellschaft d. Wissenschaft zu Göttingen, Philol.-Klasse*, II, 10, 1938, p. 211 ff.) proposed that it be identified with the type known through the "Athena Medici" (Louvre), a torso of which there are two replicas in Seville (Casa de Pilatos) and fragments in Salonika (Archaeol. Mus.). These would give us a body about 8 ft. high, to which belongs, according to Amelung (*Öjh*, XI, 1908, p. 169 ff.), a head type known through a number of replicas, the best example being the Carpegna head (Rome, Mus. Naz. Rom.). Another, from Hadrian's Villa, is in Vienna (Kunsthist. Mus.), and further examples are in London (Br. Mus.), the Vatican Museums, and Salonika (Archaeol. Mus.); yet another replica, this one from Ephesos, is in a private collection (Rome).

Amelung's reconstruction is borne out by one of the Seville statues, the head of which is preserved, though its appearance has been markedly altered by a modern addition of a helmet. His plaster reconstruction of the statue has the shield on its left arm and a patera in its lowered right hand, but the Salonika replica shows the right hand clasping a spear. Thiersch and F. Chamoux (*BCH*, LXVIII-LXIX, 1944–45, p. 206 ff.) reconstructed the goddess type with a spear rather than with the patera, but G. E. Rizzo preferred the patera in his reconstruction. Both versions are confirmed by Athenian coins of the Roman era. Since this type seems to have been used in libraries during imperial times, it may be conjectured that the original cult statue with a patera in the right hand was later adopted as an emblem of the patroness of knowledge and the arts. The suggested connection between this type and the *Athena Areia* should probably be rejected, since the first-mentioned does not seem to have been a *xoanon epichryson* and its style does not seem to conform to that of the period of Phidias's activity prior to the *Athena Promachos*. It reveals a more mature treatment of the drapery, similar to that of the Parthenon sculpture. A magnificent and monumental work, it displays great richness in its treatment of the complex folds of the soft and crinkly chiton visible under the more substantial peplos; the right leg is flexed, with the foot slightly raised toward the side. There is a subtle but insistent use of contrast, related to the rhythmic draping of the robe over the left shoulder and hip in order to give emphasis to the shield-bearing arm. It is a work presenting similarities to the pediments of the Parthenon, and this author has suggested previously that its original was actually the small-scale *Athena Parthenos*, which Himerios (XXVIII, 8) mentioned as having been produced at the same time as the colossal version.

Pausanias (I, 28, 2) and other writers mention that the *Athena Promachos* was an immense bronze image which stood on the Acropolis and that, as soon as a traveler had rounded Cape Sounion to land at the Piraeus, he could discern the point of her spear and the crest of her helmet glittering. Fragments of an inscription contain accounts that may refer to this statue: expenditures for charcoal, firewood, unminted silver, copper, tin, and horsehair are recorded, and mention seems to be made of a furnace. If the three-column inscription (each column showing the records for three years) is correctly reconstructed, the work must have been spread over nine years, with a total expenditure of 500,000 drachmas, or 83 talents. The statue must date from about 450 B.C.

According to G. P. Stevens ("The Periclean Entrance Court of the Acropolis of Athens," *Hesperia*, V, 1936, p. 497), the statue stood on what are now the remains of a base with two steps and a dado. This was 5 ft. high, had a cornice with ovolo molding, and stood against a wall of Mycenaean terracing between the Propylaia and the Parthenon. This means that without its base the statue must have stood about 25 ft. high, so that the top of the crest reached about 520 ft. above sea level and would have been visible from more than 6 miles away.

A. E. Raubitschek (*Dedication from the Athenian Akropolis*, Cambridge, Mass., 1949, no. 172) has advanced a connection between the statue and two blocks, which he dated about 455 B.C., bearing a dedication by the Athenians of their Persian booty. No longer extant, the statue is reproduced on Athenian coins of the Roman imperial period; these show a north view of the Acropolis from the Agora. Between the Propylaia and the Parthenon towers the huge statue, seemingly holding a Nike in her outstretched right hand. She is turned toward the right, and her shield is not visible; therefore there is no way of knowing whether it rested on the pedestal or was borne on her arm. Pausanias (I, 28, 2) recounts that this shield was decorated with a scene of the struggle between the Lapiths and the Centaurs, executed by the metalworker Mys from a design by Parrhasios. Stevens (op. cit., p. 493; "The Pedestal of the Athena Promachos," *Hesperia*, XV, 1946, pp. 107–14) and Chamoux are of the opinion that the shield was placed low to provide a better view of the reliefs. There have been attempts to relate busts carved on Roman oil lamps to this statue. These depict Athena bearing aloft a Corinthian helmet, and the slanting spear shown on the lamps would then be attributed to the craftsmen's own variation of the composition.

The statue may have been transferred to Constantinople, possibly under Constantine, if it is indeed the *Athena Promachos* that Niketas Choniates is referring to when he describes a 30-ft. bronze statue of Athena standing in the new Eastern capital. This is the statue destroyed by superstitious Crusaders in an excess of zeal after the siege of the city in 1203. The hair bound at the nape of the neck in the figure described corresponds to the type with the Corinthian helmet. Her outstretched right hand, the position of which was formerly interpreted as a gesture of beckoning her enemies, must have once held the Nike; her left hand, lightly touching the drapery, must once have held her shield.

The *Athena Lemnia* is recorded by Pausanias (I, 28, 3) and other writers as standing on the Acropolis. It was given the epithet "Lemnia" because it was dedicated by Lemnians — in reality, perhaps the Athenian settlers who went to Lemnos in about 450 B.C. rather than the actual inhabitants of the island. Lucian admired the outline of the statue's face, the softness of her cheeks and the fine proportions of her nose. Himerios tells us that Phidias did not invariably portray the goddess armed, that he also disguised her aspect by suffusing her cheeks with a blush instead of hiding the face under a helmet. It was A. Furtwängler (*MW*) who identified and reconstructed the type from a torso in Dresden (Albertinum) and a head in Bologna (PL. 122). Other replicas of the head are found in Oxford and in the Vatican Museums, and there is another torso in Kassel (Staat. Kunstsammi.). In her outstretched right hand the goddess held her helmet; in her left hand she held the spear. She wore her aegis across her body, fastened at the waist with the girdle of her peplos. This statue had a note of peace and serenity — absent from the *Athena Promachos* — that was appropriate to the subject. The sculptor infused new life into the inconographic tradition of the Athena without helmet. Perhaps Pliny (op. cit., XXXIV. 19, 54) is referring to this bronze statue when he speaks of an Athena of such supreme loveliness ("eximiae pulchritudinis") that she was called Beauty itself ("formae cognomen acceperit"). The beautiful young head with the hair gathered back into a fillet is also reproduced on gems.

The *Anadoumenos* of Olympia, which Pausanias (VI, 4, 5) credits to "the great sculptor Phidias," is no longer extant, and the name of the youth it portrays is unknown. It is thought, however, that the Farnese *Diadoumenos* (PL. 123) may be a replica of the bronze original. Representing a nude youth binding a fillet about his head as a badge of victory, it is a more subtly conceived figure than the *Diadoumenos* of Polykleitos (PL. 190), a later work. The calm aspect of the naked figure, the soft modeling of the hair, and the restrained rhythm all indicate that the original may have been a work of Phidias dating from about 450 B.C. The facial structure and the effect of the fillet pressing into the hair suggest a parallel with the *Athena Lemnia* head.

Pliny (op. cit., XXXIV, 19, 53) tells us that the *Amazon* of Ephesos was created for a competition with other sculptors, who were then required to judge the works submitted. The *Amazon* of Polykleitos (PL. 193) was unanimously acknowledged the best, ranked by each artist as second only to his own entry, and that of Phidias was chosen as the runner-up. In spite of the great diversity of opinions attributing the types of Amazon figure variously to the different sculptors mentioned by Pliny (see POLYKLEITOS), most scholars recognize the Phidian work in the type that is best represented in the Mattei replica in the Vatican Museums (XII, PL. 138). Other copies are in the Capitoline Museums and at Hadrian's Villa, as well as at Petworth (Sussex, Coll. of Lord Leconfield); there is a torso at Trier (Rheinisches Landesmus.) and one of basalt in Turin. All these replicas are headless; according to Furtwängler the head of one of the competion types is to be identified with the head of the bronze herm from Herculaneum (Naples, Mus. Naz.), the best marble copy of which was found at Hadrian's Villa (Rome, Mus. Naz. Romano). A plaster cast of the marble head was taken by Becatti for his reconstruction of the statue (XII, PL. 138). The arms of the Phidian *Amazon* have survived in the headless copy from the Canopus at Hadrian's Villa (PL. 123): its right hand is raised, clutching a slanted spear, which is gripped lower down by the left hand. The left leg, flexed and with the thigh brought slightly forward, is left uncovered by the chiton to expose the wound, which in other Amazon types appears under the breasts. The statue has an intense, vital rhythm, and the modeling of the chiton is comparable with that of the chitons worn by the maidens in the Panathenaic frieze on the Parthenon, indicating perhaps that the statue was carved at the time when Phidias was working on the Parthenon. Lucian admired the fine modeling of the mouth and neck; seen from the side or rear, the neck takes on an independent prominence, since it is not covered by the drapery as in other Amazon types attributed to Kresilas and Polykleitos.

The *Aphrodite Ourania* of Elis is described by Pausanias (VI, 25, 1) as chryselephantine, who states also that one of her feet was resting on a tortoise. The type of Aphrodite with a tortoise under her left foot is known from Hellenistic statuettes (Dura-Europos; Cyrene, Archaeol. Mus.); very probably it is also to be recognized in the headless Grimani statue (Berlin, Staat. Mus.), in which, however, the tortoise is a restoration. The Grimani statue, which was in Venice as late as the beginning of the 19th century, is a Greek original of Pentelic marble; the delicate, filmy chiton, together with the himation enveloping the legs, is in the style of the Parthenon sculptures, as R. Kekulé observed (*Ueber eine weibliche Gewandstatue aus der Werkstatt der Parthenongiebelfiguren*, Berlin, 1894). H. Schrader, however, attributes it to Paionios, and K. Schefold to Kallimachos (*Orient, Hellas und Rom*, Bern, 1949, p. 121 ff.). C. Blümel is of the opinion that it is later than the Parthenon and akin to the balustrade of the Athena Nike temple. A. Frickenhaus (*JdI*, XXVIII, 1913, p. 363 ff.) refers to it as the *Ourania* of Phidias, but since it is an original and not a replica it cannot be the *Aphrodite Ourania* of Elis, which was made of gold and ivory. It has been suggested that it might be a different *Aphrodite Ourania* by Phidias, that from Melite, which was made of Parian marble and which was mentioned by Pausanias as standing in the temple in that district of Athens, built in the time of Perikles.

Since all the parts left uncovered by the delicate chiton were carved and mounted separately (these are no longer extant), it has been suggested that the more gleaming and luminous Parian marble known to have been used (Pausanias) was purposely chosen for these detachable parts in order to obtain a color contrast with the Pentelic body, since a pronounced three-dimensional contrast was unobtainable with the flimsy texture of the chiton, clinging softly to the shoulders as it does. The full-front pose, impressive and solemn, is entirely appropriate to a cult statue. The style of this statue could well be that of a work made toward the end of Phidias's life, after he had executed the pediments of the Parthenon. Since the statue was once in Venice, it seems quite likely that it had come from Athens.

The *Aphrodite* of the Portico of Octavia in Rome, mentioned by Pliny (op. cit., XXXVI, 4, 15) as being of surpassing beauty, is no longer extant. Identified with a seated type of Aphrodite that exists in nine replicas (e.g., Rome, Mus. Cap., Imp. 84; Uffizi, Amelung nos. 80, 85; Paris, Mus. Rodin, no. 403; Rome, Coll. Torlonia, nos. 77, 79; Ostia, Villa Aldobrandini; and a drawing of De Cavalleriis), the statue has been reconstructed by E. Schmidt (*Corolla Ludwig Curtius*, Stuttgart, 1937, p. 72 ff.) and by Becatti, using the head of the so-called "Sappho" or "Aspasia" type, of which there are fourteen known replicas. The best of these are in Naples (Mus. Naz.), Rome (Gall. Borghese; Mus. Barracco), London (Br. Mus.), Boston (Mus. of Fine Arts), Berlin (Staat. Mus.), Munich (Staat. Antikensamml.), Florence (Palazzo Medici-Riccardi), and in the Vatican Museums. E. Langlotz (*SbHeidelberg*, 1593–54) has reconstructed the work with another head (Oxford, Ashmolean Mus.), which shows a slight variation in the position of the fillet that is wound through the hair. He considers this seated type, with a dog underneath, to be the original and identifies it with the famous *Aphrodite in the Gardens* by Alkamenes (q.v.). This would make the other series of heads replicas of a later variant, possibly of the *Venus Genetrix* by Arkesilaos. Attribution of this seated figure to Alkamenes, however, would seem to conflict with Pausanias' description (I, 18, 2) of the *Aphrodite in the Gar-*

dens as a squared figure, akin to a herm; whereas this seated Aphrodite type is obviously related to the Aphrodite in the east frieze of the Parthenon and, stylistically, might well be one of Phidias's later works. This seated type was very popular in the Roman world for cult statues and was continued in certain representations of St. Helena. No sculptural replicas of the type have survived in Greece, however, though it was soon adopted there in another context: it is frequently reflected in the vase paintings of 430–400 B.C., as Langlotz has pointed out.

The bronze *Athena* in Rome, in the Temple of Fortuna Huiusce Diei, is mentioned by Pliny (op. cit., XXXIV, 19, 54), who says that it was brought there by Aemilius Paulus. The original is lost, and no copies of it have been identified. The *Hermes Pronaos* of Thebes, mentioned by Pausanias (IX, 10, 2), is also lost and has not been identified in any copies. The pallium-clad statues that Pliny mentions as standing in the Temple of Fortuna Huiusce Diei after their dedication by Catullus are no longer extant; nor is the nude colossus that, again according to Pliny, stood in the same temple.

The fame of Phidias was due less to the above-noted statues than to his two gold-and-ivory colossi, both praised by a large number of ancient writers. Of the two, the *Zeus* was considered his masterpiece, one of the Seven Wonders of the ancient world. Pausanias (V, 11, 2–5) describes the *Zeus* of Olympia in great detail, dwelling minutely on the wealth of decoration on the throne. Cicero (*Orator*, II, 9) says that Phidias created it with his mind fixed on an image of supreme beauty formed in his imagination ("in mente insidebat species pulchritudinis eximia quaedam"); and many sources agree in reporting the tradition that Phidias had been inspired by three lines from Homer: "Thus spoke the son of Kronos and bowed his sable brows. The ambrosial-scented locks fall forward from the head of the immortal King, and lofty Olympos shook thereat" (*Iliad*, I, 527–29). Pliny (op. cit., XXXVI, 4, 18) calls it an inimitable masterpiece; Quintilian (XII, 10, 9) declares that its beauty enhanced its religious significance and that the grandeur of the work was worthy of the divinity of its subject ("cuius pulchritudo adiecisse aliquid etiam receptae religioni videtur; adeo maiestas operis deum aequavit"). Dion Chrysostomos (*Orationes*, XII, 51–52) calls it the finest and most godlike of all cult statues, expatiating on the atmosphere the statue creates and the effect it produces on the beholder. In his sixth epigram, Kallimachos gives some idea of its measurements: it was about 40 ft. high and had a base perhaps a yard high, so that the head rose to within a short distance of the ceiling of the cella. As Strabo (*Geography*, VIII, 30) says, had the god stood up, he would have smashed through the roof.

It was perhaps the desire to create something truly monumental that led Phidias to this oversize conception, for he seems to have been more concerned with the statue itself than with its relation to the surrounding architecture and the space it was to fill. According to Pausanias (V, 11, 1–8), the throne was sumptuously decorated with gold, ivory, ebony, and precious stones. On the back were three Charites and three Horai; on the arms, sphinxes and Theban children; on the sides of the seat, reliefs depicting the slaughter of Niobe's children; on the legs, dancing Nikai; on the stretchers of the throne, the struggle of Herakles and the Amazons; and on the front crosspiece, seven statuettes of athletes, one of which resembled Pantarkes, Phidias's paramour. The screen panels between the legs of the throne were covered with paintings by Panainos of various mythological subjects — Atlas and Herakles, Theseus, Peirithoos, Hellas, Salamis, Herakles and the Nemean lion, Ajax and Kassandra, Hippodameia and her mother, Prometheus in chains, the Hesperides with the apples, and Herakles, Achilles, and Penthesileia. The front screen was painted blue. The footrest was adorned with golden lions and a relief of Theseus battling with the Amazons. The base was decorated with a gold relief depicting the birth of Aphrodite from the sea, the goddess being welcomed by Eros, crowned by Peitho, and surrounded with other gods and goddesses, with Helios and Selene placed on the far sides. The *Zeus* was garlanded with leaves and clad in a robe adorned with lilies; he had sandals of gold on his feet, a golden Nike in his right hand, and a scepter with an eagle perched on it in his left.

Among Phidias's assistants were the painter Panainos and Kolotes, an expert craftsman in ivory. Phidias set up a temporary studio on the Altis (VI, FIG. 925) that had the same dimensions as the cella. (This was later converted into a Byzantine church.) In the vicinity of this workshop, excavators have come upon a rubbish pit containing numerous fragments of clay molds, which are thought to have been used to model the gold parts of the drapery of the *Zeus*. Other finds include fragments of vitreous paste decorations with lily motifs, various tools, and lead molds. The inscription "I belong to Phidias" was found scratched on the underside of a broken vessel, made locally during the second half of the 5th century. This was a surprising and fortunate confirmation of the existence of Phidias's ergasterion at the site. The Olympian colossus may have been destroyed during the late Roman imperial period; reproductions of it are found on Hadrianic coins from Elis (PL. 124), either full-length or just the head. Small-sized copies have been identified in a marble head from the Olympieion in Cyrene (PL. 124) and on gems; whereas classicizing elements in the figures portrayed on certain other coins and gems have led some scholars to suppose that these were reproductions of the Hadrianic *Zeus* of the Olympieion in Athens rather than of Phidias's monumental statue.

H. Schrader has identified marble replicas of the Nike figure in the hand of Zeus; these replicas (Louvre; Berlin, Staat. Mus.) represent Nike in descending flight and cross-girdled. B. Schweitzer, however, does not consider this type to be Phidian. F. Eichler has advanced the likely suggestion that the group on the arms of the throne, consisting of a sphinx and a youth, is to be identified with a basalt copy from Ephesos that is presently in Vienna ("Thebanische Sphinx," *ÖJh*, XXX, 1937, p. 75 ff.). There are a number of Neo-Attic copies of the frieze of the slaughter of Niobe's children (Leningrad, The Hermitage; Rome, Villa Albani; Bologna, Palazzo Bevilacqua; Florence, Coll. Milani; Pozzuoli, Antiquarium), including the reliefs on a disk in London (Br. Mus.), all of which show some deviation from the original composition of the two segments of the frieze. According to Schweitzer's and Becatti's reconstructions, the original probably comprised, on one side, the figures of Apollo and Artemis shooting arrows and, on the other, two groups of six children, with possibly a pathetic pair in the middle. The scene on a pyxis with a white ground, from Numana (Ancona, Mus. Naz. delle Marche), may be a distant echo of that on the statue base, representing the birth of Aphrodite. The statue was to be looked after by the Phaidyntai, who were traditionally regarded as the descendants of Phidias. During the 2d century B.C. repairs on the statue were undertaken by Damophon of Messene. Lucian mentions the theft of two gold curls, and in Caesar's time the statue is said to have been struck by a thunderbolt. Memmius Regulus, acting under orders from Caligula, seems to have made a vain attempt to have the statue transported to Rome in A.D. 40. Cedrenus relates that it was taken to Constantinople, but this move is unlikely.

There are two conflicting theories about the dating of the *Zeus*: some scholars hold that it was made in about 448 B.C., while others say 432. The two schools of thought arise from the varied interpretation of what the ancient writers say about Phidias's trial and the events of the last years of his life. Those who date the *Zeus* after the *Athena Parthenos* believe that Phidias took refuge in Elis and that, since his descendants were held in high esteem in Elis, his family must have been settled there. There are three pieces of evidence to show that the relation between Phidias and Pantarkes in Olympia belongs to this later period: first, the fact that Pantarkes was a winner in the Olympiad of 436; second, the likeness of the statuette on the crosspiece of the throne to this youth whom the sculptor loved; and third, the tradition, referred to by Clement of Alexandria, according to which Pantarkes's name was inscribed on the finger of the *Zeus*. Then, according to Schweitzer, the rough state of some parts of the south frieze of the Parthenon would indicate that Phidias discontinued work on them abruptly and went to Olympia; furthermore, this scholar holds that the style of the Niobe frieze is more mature than that of the Parthenon sculpture. E. Kunze maintains that the style of the terra-cotta molds which are linked with the *Zeus* and which were found together with pottery of the last two decades of the 5th century is also later than the style of the Parthenon work.

The *Zeus* can only be dated earlier than the *Athena Parthenos* if it is first conceded that Phidias went to Olympia prior to beginning his work on the Parthenon, and that he eventually died in an Athenian prison. The anecdotes about the portrait of Pantarkes and about the athlete's name being carved on the finger of the statue of the god would also have to be judged merely as popular legend. Moreover, since the peak of an artist's activity is generally marked by his masterpiece, one would have to ascribe the *Zeus* to 448. The Niobe frieze is, indeed, not without stylistic parallels in certain of the Parthenon metopes, some of which were among the first of the temple carvings to have been completed. Also, the disproportionate relation of the colossus to its setting within the cella was remedied in the case of the *Athena Parthenos*. This change may mean that the *Athena* was a considered improvement on the *Zeus* and that Phidias had used his earlier experience at Olympia to good account, organizing the more complicated work on the Parthenon the more smoothly for it.

The oldest reference to the *Athena Parthenos* is found in Plato, who describes the unclothed parts of the body as being of ivory and the pupils of the eyes as being composed of precious stones. Pausanias (I, 24, 5–7) mentions this chryselephantine statue: according to his description, the middle lophos of the helmet was in the shape of a sphinx, and the side lophoi bore reliefs of griffins. Reproductions of the statue show that the headband was decorated with animal protomas. The goddess was bedecked with pendant earrings and a gold necklace; over her peplos she wore her aegis, which had a scalelike pattern and a gorgoneion for decoration. In her outstretched right hand she held a gold Nike, some four cubits in height; in her left, she held a spear and a shield that rested on the base of the statue, with the serpent Erichthonios coiled against it. The soles of her sandals were decorated, Pliny (op. cit., XXXVI, 4, 18–19) notes, with a relief of the battle between the Centaurs and the Lapiths; the base of the statue showed the birth of Pandora in the presence of twenty gods and goddesses. Pliny also records that on the inside of the shield was the Gigiantomachia, and on the outside the Amazonomachia. According to Thucydides and Plutarch, the gross weight of the gold used on the statue was 40 talents. Diodorus Siculus put the figure at 50 talents, and Philochoros at 44. From the inscriptions in the temple, it is learned that the statue was restored a number of times. During the 88th Olympiad repairs were carried out on the gold crown held by the Nike, in the 95th on some of the gold leaves in this crown, and in the 103d on parts of the helmet and shield. The weight of the detachable gold parts was checked every four years; it is known that gold was stolen from the statue in the 4th century and, again, in 296 by Lachares.

The colossal *Athena*, which must have been about 37 ft. high, has survived through a number of small replicas: the Varvakeion statuette (PL. 124), the Lenormant *Athena* from the Pnyx (Athens, Nat. Mus.), a third found at Patras (Archaeol. Mus.), and yet another from Bitolj, Yugoslavia. Larger copies also exist: a classicistic replica by Antiochos (Rome, Mus. Naz. Rom.), the so-called "Minerve au collier" (Louvre), and a replica in Madrid (Prado). The base of the large-scale replica from Pergamon (III, PL. 385) is extremely important, for it shows part of the scene of the birth of Pandora. An Athena torso in Rome (Mus. Cap.) is important for the fragment of the decorated shield preserved with it; other related but less important torsos have been preserved. There are a numer of marble replicas of the head, including those in Copenhagen (Ny Carlsberg Glyptothek), Berlin (Staat. Mus.), Paris (Louvre) and Cyrene (Archaeol. Mus.). The most valuable and most accurate copy of the head, however, is found on the Aspasios gem (PL. 124), which shows the decoration of the helmet and the full oval of the face. Various details are preserved in some gold medallions from Kul Oba (Leningrad, The Hermitage), and echoes of this famed statue are discernible in votive reliefs and in letterheads of official Attic documents.

G. Richter has examined (*Studi in onore di Aristide Calderini e Roberto Paribeni*, III, Milan, 1956, pp. 147–53) the question of whether or not the original statue had a pillar supporting her right hand, as the Varvakeion statuette seems to indicate. Compositional considerations and architectural relationships, as well as technical reasons, indicate that such a pillar was probably present in the original work.

The battle between the Greeks and Amazons depicted on the outside of the shield can be reconstructed by means of the Strangford shield, in London (Br. Mus.), and through others in Patras (Archaeol. Mus.) and Rome (Mus. Cap.; Vat. Mus.). Also useful in this connection are a clay fragment found in the Agora and the rough Lenormant statuette; most significant, however, are the series of Neo-Attic reliefs found near Piraeus (X, PL. 264) and others preserved in Rome (Villa Albani), Berlin (Staat. Mus.), and Chicago (Art. Inst.). Phidias set the struggle between the Greeks and Amazons on the rocky slopes of the Acropolis and filled the whole surface of the shield with groups of adversaries in single combat around the central gorgoneion. E. Buschor (*Medusa Rondanini*, Stuttgart, 1958) recognized the type of this gorgoneion in the so-called "Medusa Rondanini." As the Amazons attempt to scale the Acropolis, the Athenians repel the attack. One Amazon, about to hurl herself into space, is seized by the hair, while another plunges headlong and a third lies dead on the ground, and others fall wounded or are engaged in combat with opponents who are mostly youths. A bearded hero wearing a fillet has sunk to his knees wounded. This figure was mistakenly thought to be Kapaneus until G. Hafner (*JdI*, LXXI, 1956, pp. 1–28) proposed that it depicted Anakreon. The same scholar has identified as Xanthippos another bearded warrior, who wears the felt headdress known as the *pilos*. Schweitzer (*Daidalos*, Leipzig, 1939) put forth the hypothesis that the two chief figures, a bald old man and the warrior with his arm raised who are traditionally supposed to be Phidias himself and Perikles, may in fact be Daidalos and Theseus.

Possible reminiscences of the battle between the gods and the giants, as depicted on the inside of this shield, are found in vase painting; a vase in Naples may relate to the original composition, with its pyramidlike mass of giants battling their way toward the top of the scene as the gods, with Helios on one side and Selene on the other, give battle from above. The theme of Pandora's birth, again with Helios and Selene present at the sides, is suggested in the sketch on the Lenormant statuette. Various other works (which Schweitzer assigns to the circle of Phidias) — a relief in the Palazzo del Drago (Rome), the base of a Neo-Attic candelabrum in St. John Lateran, and a fragment in Corinth — provide further examples of the classical deities in group compositions. In these the gods are arranged in pairs, around a seated Zeus; the type is similar to that of the east pediment of the Parthenon, known from a puteal preserved in Madrid (Prado).

Regarding the Parthenon itself, Plutarch (*Pericles*, XIII, 4) recounts that Phidias was the superintendent (*episkopos*) in charge of the entire commission, which was begun in 447. He was able to set up and run a complex studio organization, with many different teams of craftsmen (*technitai*) who worked together harmoniously under his direction. It is inconceivable that Phidias could have carried out on his own all the decorations envisaged for such an enormously complicated and exacting plan, involving the carving of 92 metopes, the long frieze, and the magnificent sculpture of the pediments. The question therefore arises as to whether there are any traces of Phidias's own hand in the Parthenon marbles and, if so, in how many of them. Some scholars (Schrader, Buschor, Blumel) attach great significance to the fact that a number of different techniques are clearly discernible in the different sections of the work. They have recognized notable stylistic differences and have attributed them to other sculptors of that period such as Alkamenes, Agorakritos, Paionios, Kallimachos, and Kolotes, all of whose styles were said to be different from that of Phidias. Other scholars, however, tend to emphasize the substantial unity of the whole creation, which they attribute to Phidias alone.

Schweitzer's contribution to this problem is especially noteworthy, for he made a detailed analysis of the Parthenon sculpture in which he not only showed that there was a single, unifying Phidian conception for the whole undertaking but also described how the work was actually carried out. He suggested that Phidias first conceived his compositions and sketched cartoons for the marble metopes (PL. 126), and then superintended their execution by sculptors with varied styles — some archaic and severe, others not dissimilar from his own. At times he took a hand in the work himself, carving and retouching some of the marble figures (one definite example of this is seen in the first of the south metopes, where the head of the Centaur is vigorous and extremely lifelike). He designed the Panathenaic frieze (PL. 125) and produced three-dimensional models for the west segments of it. He personally executed some of these in marble as a standard; one such example is the fine central slab with the rearing horse on it. Like the west slabs, those to the east were carved in the studio from Phidias's own models, whereas the long north and south sides (III, PL. 362) were carved *in situ* from Phidias's cartoons by the various sculptors working on the assignment. As time went on, they learned to adjust their styles to one another and showed increasing similarity to the master himself in technique, until finally the difference in techniques and styles became barely perceptible in the sculpture of the pediment, which was the last to be done (PL. 125). The pediment sculpture was certainly based on three-dimensional models made by Phidias himself, who may have participated substantially in the final carving of the marble as well.

Ancient writers also speak of a number of works for which they found it difficult to establish whether they should be ascribed to Phidias or to his pupils. To further confuse the problem of attribution, Phidias was also in the habit of putting the finishing touches on his pupils' works, and even signed them with his own name. In this way, Phidias was credited with the *Athena* in Elis by Kolotes, the *Nemesis* of Rhamnous by Agorakritos, and the *Meter* in the Metroon in Athens — the last-named of which Pliny attributed to Agorakritos, and Pausanias (I, 3, 5) to Phidias. Langlotz believes it to have been by Phidias and he identifies it with a type known from a 2d-century (A.D.) Athenian statue from Lebadeia by Hermias. According to Pliny, the *Aphrodite in the Gardens* of Alkamenes was given its finishing touches by Phidias.

Two works can be attributed to Phidias with greater certainty. One of these, the *Anakreon*, is known through the Borghese replica in Copenhagen (II, PL. 49) and from five heads, the best of which are in Berlin (Staat. Mus.) and Rome (Mus. Cap.). The original was probably a statue seen by Pausanias (I, 25, 1) on the Acropolis: it portrayed the poet as if he were drunkenly singing. If it is compared with the so-called "Kapaneus" or Anakreon figure on the shield of the *Athena Parthenos*, it might be taken for an early work of Phidias, to be dated about 460. But S. Reinach (*RA*, XXI, 1893, p. 63) attributes it to Kolotes, J. Six (*BCH*, XXXVII, 1913, p. 350) to Pythagoras, R. Kekulé (*JdI*, VII, 1892, pp. 119–26) and Hafner to Kresilas, and Furtwängler to Phidias. The other reasonably certain work portrays Pantarkes, victor in an Olympiad, a statue that Pausanias (VII, 10, 6) stated was to be seen at Olympia. It is likely that Phidias carved this work and that it is to be linked with a head type which was popular enough to have been preserved in five replicas (Met. Mus.; Petworth, Lord Leconfeld Coll.; Rome, Coll. Abbati; Trier, Rheinisches Landesmus.; Florence, Palazzo Medici-Riccardi). This head portrays a young boy wearing a fillet, with his hand resting on his softly curling ringlets. The style, tender and coloristically modeled, might well be that of Phidias's mature period.

This long catalogue of works reveals Phidias to have been a sculptor of extraordinary genius, whose career marks an important phase in the development of Greek art. Through his work there occurred the transition from the severe to the classical style; he opened new horizons and with his great creative imagination brought about a transformation in the entire Greek iconographic tradition, giving life to a multitude of visions of heroes and gods. In creating the classical style, he saw the nude with wholly new eyes and produced works of grandeur and power that also displayed a keen sense of the subtle texture of human flesh. He revolutionized the conception of drapery as well, creating effects of thin, transparent, almost veillike and rippling material. He mastered every technical demand and learned how to derive the maximum effect from each technique. With equal nobility of conception he could plan a single statue, a metope, a length of frieze, sculpture for a pediment, or a colossus for a temple cella. With his knowledge of the great achievements in painting, he revolutionized the concepts of three-dimensional composition and attained the unity and grandeur revealed in the shield carvings of the Gigantomachia and the Amazonomachia. His solutions to the problem of pediment composition have never been surpassed for their perfection of spatial sense and for the organic relation of the parts to the whole.

From the very start of his career, Phidias was an innovator and a highly individual artist. The *Apollo Parnopios* and the *Anakreon*, for instance, reveal his formation within the severe style of the early 5th century. By the middle of the century he had reached the high artistic level of the *Athena Lemnia* and the *Athena Promachos*, and had probably begun the onerous task of making the first of his chryselephantine colossi, the Olympian *Zeus*, which was consecrated in 448. In 447 his period of intense activity on the Acropolis began, and there he supervised work on the Parthenon and executed his second colossus, the *Athena Parthenos*, which was consecrated in 438. His creative imagination broadened and his artistic idiom matured into the style that is associated with the Parthenon. He became the greatest interpreter of the ideals of this time: he celebrated the glories of Athens and gave substance to the old Attic myths; he created a school and exerted a strong normative influence over all the art of his time — an influence that brought forth many further developments among his followers and imitators, as well as among the so-called Hellenistic "mannerists." In brief, he laid new foundations on which the whole of subsequent art could build solidly, so that the ancient chroniclers and even his critics justly acknowledged him as the greatest artist of his times and judged his works to be the pinnacle of Greek art.

BIBLIOG. *a. General*: Overbeck, SQ, nos. 618–807; O. Jahn and A. Michaelis, eds., Arx Athenarum a Pausania descripta, Bonn, 1901; A. Hekler, Die Kunst des Phidias, Stuttgart, 1924; H. Léchat, Phidias et la sculpture grecque au Ve siècle, Paris, 1924; H. Schrader, Phidias, Frankfurt am Main, 1924; C. Praschniker, Parthenonstudien, Vienna, 1928; W. H. Schuchhardt, Die Entstehung des Parthenonfrieses, JdI, XLV, 1930, pp. 218–80; G. Lippold, RE, s.v. Pheidias; B. Schweitzer, Prolegomena zur Kunst des Parthenonmeisters, JdI, LIII, 1938, pp. 1–89, LIV, 1939, pp. 1–96; B. Schweitzer, Pheidias der Parthenonmeister, JdI, LV, 1940, pp. 170–241; V. H. Poulsen, Phidias und sein Kreis, From the Coll. of the Ny-Carlsberg-Glyptotek, III, 1942, pp. 33–92; E. Langlotz, Phidiasprobleme, Frankfurt am Main, 1947; E. Buschor, Pferde des Phidias, Munich, 1948; E. Buschor, Phidias der Mensch, Munich, 1948; E. Langlotz, Phidias, Vermächtnis der antiken Kunst, Heidelberg, 1950, pp. 71–79; G. Becatti, Problemi fidiaci, Milan, 1951; J. Liegle, Der Zeus des Phidias, Berlin, 1952; C. Blümel, Phidiasische Reliefs und Parthenonfries, Berlin, 1957; F. Brommer, Athena Parthenos, Bremen, 1957; B. Schweitzer, Neue Wege zu Pheidias, JdI, LXXII, 1957, pp. 1–18; E. Berger, Parthenon Ostgiebel, Bonn, 1959; G. Becatti, EAA, s.v. Fidia. *b. Single works*: C. H. Morgan, Pheidias and Olympia, Hesperia, XXI, 1952, pp. 295–339; F. Brommer, Studien zu der Parthenongiebeln, AM, LXIX–LXX, 1954–55, pp. 49–66; C. Praschniker, Neue Parthenonstudien, ÖJh, XLI, 1954, pp. 5–53; S. Aurigemma, Lavori nel Canopo di Villa Adriana (II), BArte, XL, 1955, pp. 64–78; F. Brommer, Neue Forschungen über die Parthenongiebeln, Atlantis, XXVII, 1955, pp. 368–71; F. Brommer and E. B. Harrison, A New Parthenon Fragment from the Athenian Agora, Hesperia, XXIV, 1955, pp. 85–87; E. Kunze, Die Ausgrabungen in Olympia im Winter 1954/55, Gnomon, XXVII, 1955, pp. 220–24; C. H. Morgan, Footnotes to Pheidias and Olympia, Hesperia, XXIV, 1955, pp. 164–68; E. Berger, Ein neuer Kopf aus den Parthenon-giebel ?, AM, LXXI, 1956, pp. 153–72; F. Brommer, Studien zu den Parthenongiebeln (II–III), AM, LXXI, 1956, pp. 30–50, 232–44; E. Kunze, Die Ausgrabungen in Olympia im Frühjahr 1956, Gnomon, XXVIII, 1956, pp. 317–20; J. Marcadé, Deux fragments méconnus de l'Hélios du fronton est du Parthenon, BCH, LXXX, 1956, pp. 161–82; H. Walter, Der Meister der Nord-Heroen vom Ostfries des Parthénon, AM, LXXI, 1956, pp. 173–79; J. Marcadé, Deux notes parthénoniennes, BCH, LXXXI, 1957, pp. 76–94; W. H. Schuchhardt, Die Eleusinischen Kopien nach Parthenonskulpturen, Festschrift für K. Bauch, Munich, Berlin, 1957, pp. 21–28; E. Berger, Das Urbild des Kriegers aus der Villa Hadriana und die Marathonische Gruppe des Phidias in Delphi, RM, LXV, 1958, pp. 6–32; K. Schunk, Leto in Parthenon-Ostgiebel ?, JdI, LXXIII, 1958, pp. 30–35; E. Kunze, Olympia, Neue deutsche Ausgrabungen im Mittelmeergebiet und im vorderen Orient, Berlin, 1959, pp. 263–310; P. Fehl, The Rocks on the Parthenon Frieze, Warburg, XXIV, 1961, pp. 1–44.

Giovanni BECATTI

Illustrations: PLS. 122–126.

PHILIPPINES. The Republic of the Philippines comprises more than 7000 islands in the western Pacific, which form the northern part of the Malay Archipelago. Only 2773 of these islands are named, and of that number only 462 are greater than a square mile in area. The population of the republic is more than 30 million.

Since prehistoric times the area comprising the Philippines has been a meeting ground for numerous Asiatic peoples. According to R. B. Bean, there have been six principal migratory waves into the Philippines: from southeast Asia (the Indonesian and Tibeto-Burman strata), from southern China, from India, from Micronesia, and from the islands of modern Indonesia and Malaysia. The present inhabitants are generally divided into two groups: the Pygmies (Negritos) and the Indonesians. The latter are further separated into three subgroups, the classifications being based upon origin and culture. The art of the Philippines may be divided into two distinct categories: the art of the indigenous cultures and the Western-influenced art produced after the Spanish colonization in the last half of the 16th century.

Art of the indigenous cultures. *Negritos.* The Negritos, gradually decreasing in number, are related to other Pygmy groups of southeast Asia, the Samang of Malacca and the Andaman islanders (the Öngé, the Járawa, etc.) in the Bay of Bengal. In the Philippines the largest groups of Negritos, known under the Tagalog name of Aëta, are found principally in the northeastern part of Luzon and in the province of Zambales, in the western part of the island. Negritos are also found on Panay and Negros; in Mindanao, in the Camarines region of Luzon, and on Palawan they have become mixed with other groups. Their hunting and food-gathering culture is one of the most primitive on earth, and their artistic production is correspondingly limited, comprising mainly the decoration of various utensils and domestic objects, using black and natural color, and the scratchwork ornament on women's bamboo combs.

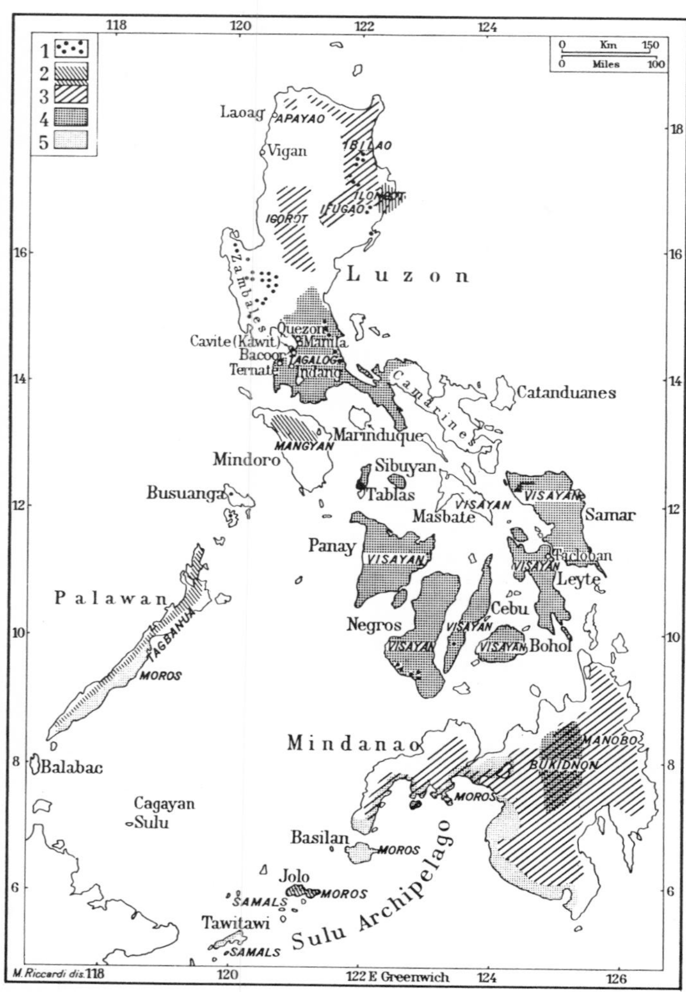

Philippines, principal ethnic components of the indigenous cultures (italic capitals) and significant modern centers. *Key*: (1) Negritos. Indonesian subgroups: (2) Mangyan-Samal; (3) Ibilao-Manobo; (4) Tagalog-Visayan; (5) Moros (with Malayan-Arab admixture).

Indonesians. The oldest of the three Indonesian subgroups established in the Philippines appears to have been mixed with the Aëta and the Malays. Among the tribes composing this first subgroup, the Mangyan occupy Mindoro and other neighboring islands; the Tagbanua occupy Palawan; and the Samals are found in the Sulu Archipelago. The culture of this Indonesian subgroup remains at a relatively undeveloped level, and the only ornamentation appears on the women's girdles, woven of bamboo and painted alternately in black and yellow. Their pottery is undecorated.

The second subgroup is found on the island of Mindanao, in the interior of Luzon, and on smaller islands nearby. The tribes comprising this group are the Manobo, the Bagobo, the Mandaya, and the Ata in eastern Mindanao; the Subanun in western Mindanao; and the Igorot (Bontok, Kankanai, Nabaloi), the Ifugao, the Kalinga, the Apayao, the Ibilao, and the Ilongot in central and northern Luzon. Mixtures with the Aëta and Mangyan are also found. The members of this subgroup are considered to be of Malay origin and of the same stock as the inhabitants of southern Indonesia. They have linguistic affinities with the Malagasy and have been subject to strong Hindu influence. They were at one time head-hunters and, as part of their ritual, preserved trophy heads that were often painted and adorned with such elements as pendants and mother-of-pearl eyes. The aboriginal costume, a simple covering of bark, is on festive occasions exchanged for a kind of long smock, vividly colored and highly decorated and worn with a girdle wound around the waist, its appearance recalling the Malayan sarong or the Burmese skirt. This garment, woven of cotton or other vegetable fiber, is painted with linear and geometric designs that contrast sharply with the background cloth. The costume is completed by numerous necklaces.

Warriors' weapons include the iron lance (*gayang*), the dagger (*talibán*), kept in a wooden scabbard, the battle axe (*aligua*), and a rectangular shield (*acalaba*) with three prongs at the top and two at the bottom. The weapons are decorated with geometric motifs, and the handles are often carved in anthropomorphic or zoomorphic forms.

Pectoral, brachial, and occasionally facial tattooing is widely practiced. The usual designs represent animals (especially the lizard) and flowers, but geometric motifs are most frequent. These last are seen particularly in *chak-lág*, the distinctive tattooing reserved for head-hunters. Still practiced, though with a purely ornamental purpose, *chak-lág* tattooing, which covers the breast, consists of a series of highly stylized curved motifs that originally perhaps represented mythical or totemic beings. Tattoo motifs, which generally vary according to tribe, include realistically drawn lizards, a pattern resembling chain mail, and stylized plants and objects of everyday use, such as hunting weapons and agricultural tools, delineated with very fine parallel lines.

The tribes of this subgroup also produce small wood sculptures, the *anito* figures, which generally represent ancestors who have become guardian spirits. The *anito* carvings portray, crudely but faithfully, standing or seated human figures and are adorned with shell collars, earrings, and hair (human or of vegetable origin); sometimes they wear a girdle at the waist and are covered with tattoos. Though, in comparison with the artistic production of other primitive cultures, the artistic value of the *anito* is slight, they are nonetheless interesting for the spontaneity of their carving. Anthropomorphic forms are also found on carved ritual cups, the bowls of which may rest on bases composed of three *anito* figures or may be supported on the knees of one seated figure or held in its arms.

As with its sculpture, which is linked to the ancestor cult, all other artistic activity of this subgroup has religious significance; and the making of each new object is therefore considered to be, literally, an act of creation and is firmly circumscribed by ritual. Natural materials are used for embellishment: bamboo, stone, wood, bone, tortoise shell, vegetable fibers, dried earth, and various seeds. Incised motifs on wood, bamboo, and bone commonly consist of horizontal and vertical stripes and of schematic and geometric reductions of prescribed conventional forms. Color contrasts consist in opposing red ocher and black or in applying white or yellow earth pigments to blackened objects.

The third and last subgroup of Indonesian origin has been influenced by the Hindu and East Asian cultures, as well as by those of neighboring Indonesian islands. Hindu influence is recognizable in the designs incised in wood on the ancient Panay Island dwellings and on the handles of the betel spatulas of Mindoro. Such influence was particularly strong at the time of the Sumatran empire of Śrīvijaya, which flourished in the 8th century of our era. The Tagalog of the Manila region, the Bukidnon of Mindanao, the Visayans (or Bisayans) inhabiting the islands between Luzon and Mindanao, and the Moros of the Sulu Archipelago and the western regions of Palawan and Mindanao comprise this subgroup. Some ethnologists distinguish as many as seven groups among the Moros, who are a mixture of indigenous elements with the Borneo Malayans and Arab immigrants from Asia and perhaps from eastern Africa.

The Tagalog and related tribes are products of a fusion of invaders from southeast Asia with the autochthonous Negritos; in addition, there has been strong Chinese influence upon them. The coexistence of these diverse elements explains the basic character of Tagalog art, which, in comparison with that of other peoples of the Philippines, is outstanding in craftsmanship, especially in basketry, textiles, and wood carvings. The weaving itself, done on narrow horizontal looms without frames, is rudimentary; but the woven stuffs are richly ornamented with geometric motifs and are embroidered in vivid colors, with red and blue predominant.

Both male and female dress is fashioned of pieces of fabric joined together in a technique called *ginontiñan*. This decoration, linear in its total effect, is composed of vertical or horizontal parallel bands that are especially striking for their contrast of color. Reds and yellows are used for decorative motifs in which stylized human beings, animals, or plants are represented according to the traditional canons found throughout Indonesia. The female headdress is characterized by large combs, Indonesian in type, inlaid with bits of mother-of-pearl, in some examples merely scattered at random and in others forming designs. Armlets, leg bracelets, and metal rings of very fine workmanship are numerous.

Baskets woven of bamboo, rotang, and pandanus leaves are generally left unornamented, but alternate strips of natural color and black create harmonious, decorative effects. Men's and women's hats, woven of rotang or bamboo fiber, recall the large round hats of the Far East. Crudely decorated pots for domestic use are made by the women. Dwellings, erected on pilings, and agricultural implements are simple and undecorated. Wooden chests inlaid with mother-of-pearl in geometric designs are among the furnishings in the dwellings of the wealthy tribesmen.

The Moros are Moslems, and their culture, script, and social structure clearly show the strength and continuity of Islamic influence. Their dress, ample and ornate, recalls Arabic costume, but is sometimes decorated with gold and silver stripes and with designs (flying dragons, stylized clouds, flowers) that evidence Chinese influence. Precious metals worked in motifs reminiscent of Arabic geometric ornament are used in jewelry and frequently in the decoration of weapons such as the *barong*, a short sword with a straight or curved blade, the *campilan*, a two-handed straight sword with a blade that widens and then terminates in a triangular point, and the Malay creese (*kris*), a dagger with a serpentine blade. The handles of these weapons are of wood or of carved and polished ivory inlaid with gold wire. Shields, generally round, have geometric ornamentation that also recalls Arabic motifs. The various types of boats and dugouts in common use (*pabco*, *vinta*, *salisipan*, *cancano*, *pilane*) are often decorated with painting, ivory and mother-of-pearl inlay, and silver filigree of great richness.

Bibliog. A. F. Jagor, Reisen in den Philippinen, Berlin, 1873 (Eng. trans., London, 1875); A. Schadenberg, Über die Negritos der Philippinen, ZfE, XII, 1880, pp. 133–74; A. Schadenberg, Die Bewohner von Süd-Mindanao und der Insel Samal, ZfE, XVII, 1885, pp. 8–37, 45–57; E. Abella y Casariego, Rápida descripción de la isla de Cebú, Madrid, 1886; J. de Lacalle and F. Sánchez, Tierras y razas del archipiélago filipino, Manila, 1886; P. A. Paterno, La natigua civilización tagalog, Madrid, 1887; J. M. Ruiz and F. Sánchez, Pobladores aborígenes, razas existentes y sus variedades . . . de Filipinas, Manila, 1887; F. Blumentritt, Breve diccionario etnográfico de Filipinas, Madrid, 1889; F. Blumentritt, Las razas del archipiélago filippino, 2 vols., Madrid, 1890; A. B. Meyer and A. Schadenberg, Die Philippinen, 2 vols., Dresden, 1890–93; P. A. Paterno, Los Itas, Madrid, 1890; R. Echauz, Apuntes de la isla de Negros, Manila, 1894; P. A. Paterno, Los Tagalog, Madrid, 1894; B. Campa, Etnografia filipina: Les Mayóyaos y la raza Ifugao, Madrid, 1895; R. González Fernández, Filipinas y sus habitantes, Béjar, 1896; B. Low, The Natives of Borneo (ed. H. Ling Roth), 2 vols., London, 1898; F. Blumentritt, List of the Native Tribes of the Philippines and of the Languages Spoken by Them (trans. O. T. Mason), Smithsonian Inst. Rep., 1899, pp. 527–47; A. Pérez, Igorrotes: Estudio geográfico y etnográfico sobre algunos distritos del norte de Luzón, Manila, 1902; A. Pérez, Relaciones augustinianas de las razas del norte de Luzón, Manila, 1904; F. W. Atkinson, The Philippine Islands, New York, 1905; A. E. Jenks, The Bontoc Igorot, Manila, 1905; E. Y. Millar, The Bataks of Palawan, Manila, 1905; W. A. Reed, Negritos of Zambales, Manila, 1905; N. M. Saleeby, Studies in Moro History, Manila, 1905; J. Foreman, The Philippine Islands, 3d ed., London, 1906; N. M. Saleeby, The History of Sulu, Manila, 1908; E. B. Christie, The Subanuns of Sindangan Bay, Manila, 1909; F.-C. Cole, The Wild Tribes of the Davao District, Mindanao, Chicago, 1913; J. R. Arnold, The Philippines, Manila, 1913; D. C. Worcester, The Non-Christian Peoples of the Philippine Islands, Nat. Geog. Mag., XXIV, 1913, pp. 1157–1256; H. O. Beyer, Population of the Philippine Islands in 1916, Manila, 1917; L. R. Sullivan, Racial Types in the Philippine Islands (APAmM, XXIII, 1), New York, 1918; F.-C. Cole, The Tinguian, Chicago, 1922; W. Hough, Synoptic Series of Objects in the United States National Museum, Washington, 1922; S. Y. Orosa, Sulu Archipelago and People, New York, 1923; F. C. Lubach, Peoples of the Philippines, New York, 1925; H. W. Krieger, The Collection of Primitive Weapons and Armors of the Philippine Islands, Washington, 1926; D. C. Worcester, The Philippines Past and Present, New York, 1930; J. M. Garvan, The Manóbos of Mindanao, Washington, 1931; G. F. Nellist, Men of the Philippines, Manila, 1931; V. Hurley, Southeast of Zamboanga, New York, 1935; P. Schebesta, Les pygmées, Paris, 1940; A. L. Kroeber, Peoples of the Philippines, 2d ed., New York, 1943; F.-C. Cole, Central Mindanao: The Country and the People, Far Eastern Q., IV, 1945, pp. 109–18; F.-C. Cole, The Peoples of Malaysia, New York, 1945; H. O. Beyer, Outline Review of Philippine Archaeology by Islands and Provinces, Philippine J. of Sc., LXXVII, 1947, pp. 205–374; Museo Etnológico de Madrid, Guía, Madrid, 1947; R. Lynch, Some Changes in Bukidnon between 1910 and 1950, Anthr. Q., XXVIII, 1951, pp. 95–115; F.-C. Cole, The Bukidnon of Mindanao, Chicago, 1956; Selected Bibliography of the Philippines, Chicago, 1956; J. B. Vaughan, Land and People of the Philippines, Philadelphia, 1956; W. S. Solheim II, Kulanay Pottery Complex in the Philippines, AAs, XX, 1957, pp. 279–88; R. B. Fox, The Calatagan Excavations, Philippine S., VII, 1959, pp. 321–90; J. R. Francisco, A Note on the Golden Image of Agusan, Philippine S., XI, 1963, pp. 390–400; D. Szanton, Art in Sulu: A Survey, Philippine S., XI, 1963, pp. 465–502.

Jean M. Roger Rivière

Art of European derivation. Artistic development in the Philippines subsequent to the Spanish occupation of the archipelago is closely linked with the foundation of Manila (1571), established as the seat of a bishopric in 1580, and with the spread of the Catholic religious orders, which had their bases in that city. The earliest Spanish colonial architecture in the Philippines is characteristically baroque in form and had its center in Manila, which became the capital in 1575. The religious orders were most active in the 16th and 17th centuries. The Augustines began the Church of S. Agustín at the end of the 16th century. It was primarily a work of Fray Antonio de Herrera, son or nephew of the Spanish architect Juan de Herrera (q.v.). In 1587 the Dominicans built the church and monastery of S. Domingo, after the designs of an architect of the order, and subsequently (1615) a building for Santo Tomás University. Of these structures, only S. Agustín escaped destruction during World War II; it is the only important church of the early colonial period surviving in Manila.

Religious and civil architecture of the 18th and 19th centuries was built in heterogenous styles imported for the most part from Europe. A major 19th-century Philippine architect, Felipe Roxas, traveled in England, France, and Spain. He was responsible for the Neo-Gothic reconstructions of the church and monastery of the Dominican order in Manila; for churches in Santo Domingo and in Bacoor, near Cavite; and for various private houses, among them that of Señora Carmen Roxas in Manila, which is of particular interest for its grace and elegance. The Spaniard Hervas, municipal architect of Manila from 1887 to 1893, used forms of Byzantine inspiration, as was evident in the Cathedral (1878–79) and in many other public and private buildings. Considered his masterpiece is the government loan bank in Manila, known as the "Monte de Piedad," built in a classicizing style. The works in Manila of both Roxas and Hervas suffered heavy damage in World War II.

When United States sovereignty replaced that of Spain in 1898, rapid industrialization was accompanied by a notable increase in public building. In the first years of the 20th century the Bureau of Public Works was created to direct city planning and building; under its aegis many schools, markets, town halls, and provincial public buildings were erected. Government offices in Manila and Baguio were built according to plans by the American architect Daniel H. Burnham.

During the 1930s a group of young architects, among them Pablo Antonio, Carlos Arguelles, and Cesar Concio, revolted against traditionalism and began to experiment with new forms in architecture. They and their followers, notably Leandro Locsin and Angel Nakpil, have been responsible for much of the architecture of good contemporary design in the Philippines. The Captain Luis Gonzaga Building by Pablo Antonio in Manila, Concio's Protestant Chapel at the University of the Philippines, and Locsin's Chapel of the Holy Sacrifice, a concrete-shell dome also located at the University, are examples of Philippine solutions to contemporary architectural problems.

During the period of Spanish dominion, sculpture and paintings for churches were usually imported from Mexico and Spain; but European prints sometimes served as models for local artists.

A royal decree of Charles III in 1785 encouraged local artistic production by exempting artists from the payment of taxes.

Philippine sculpture was strongly influenced by Spanish wood sculpture, particularly that of the Sevillian Juan Martínez Montañés (1568–1648). The sculpture was primarily religious in subject matter, and the period of greatest activity in the medium was during the first half of the 19th century, when numerous churches were built. In the last decades of the 19th and the early decades of the 20th century, the sculptors Leoncio Asunción, José Arevalo, and Guillermo Tolentino were active; and later a trend away from the academic was evident in the original and experimental works of José Alcántara, Ildefonso Marcelo, and Abdul Mari Imao.

Philippine painting in Western style began with the Spanish occupation. The first painters were missionaries, and their work, primarily religious in subject matter, was strongly influenced by Spanish painting, especially that of Murillo, Alonso Cano (qq.v.), and Pedro de Moya. Between 1815 and 1820 the first school of fine arts was founded in Manila, largely on the initiative of the painter Damian Domingo. Later it became the Escuela Superior de Pintura, Escultura y Grabado de Manila, and foreign artists, especially Spanish, were invited to teach there. In 1906 the school was incorporated into the University of the Philippines as the Department of Fine Arts and was largely responsible for the formation of a new generation of Philippine artists.

Philippine painters of the late 19th and the early 20th century were almost exclusively followers of contemporary European academic dictums. Two of these painters won recognition in Europe for their work: Juan Luna, for his romantic historical paintings such as the *Spoliarium*, and Felix Resurrección Hidalgo, who painted in much the same representational style, for his *Antigone*. Notable among other painters of the period were Antonio Malantic, whose work is distinctive for its Oriental flavor, and the genre painter Fabian de la Rosa, a colorist of great sensitivity.

Abstract-expressionist painting was relatively unknown in the Philippines until the 1930s, when Victorio Edades, Philippine architect and painter, returned from the United States to direct the College of Fine Arts at Santo Tomás University. Opposed to academicism, he attracted a large group of followers, among them the surrealist Galo Ocampo and the muralist Carlos Francisco, who was particularly concerned with social themes.

Following World War II a group known as the "13 Moderns" (later called the "Neo-Realists") was prominent. It then merged with the "Art Gallery Group," which became the nucleus of experimental painting activity in the Philippines. Anita Magsaysay-Ho, Emilio Lopez, Arturo Luz, Vincente Manansala, Romeo Tabueno, and Fernando Zobel are among the contemporary painters whose work has achieved international recognition.

Art centers. Manila. Site of a Moslem settlement before its establishment as a Spanish colony in 1571, Manila quickly became the center of Spanish commerce with the East and an important naval base. Planned by Miguel López de Legazpi, who is considered its founder, the Spanish port Manila was surrounded by walls and equipped with a fortress (Fort Santiago).

Until World War II there was much evidence of Spanish colonial architecture in Manila, particularly in the old city (Intramuros), but these structures were almost totally destroyed in the fighting. A few reminders of the past survive. Sections of the early walls are still standing, as is Fort Santiago. The oldest existing church is S. Agustín (1599–1614); the Malacañang Palace on the Pasig River, the residence of the president, is another Spanish colonial edifice.

Noteworthy buildings of the late 19th and early 20th century are the Neo-Gothic church of S. Sebastian, built in 1890 with steel parts prefabricated in Belgium; the Philippine General Hospital, the Metropolitan Theater, and the Philippine Normal College. Important structures built since World War II include the Protestant Chapel, the Chapel of the Holy Sacrifice, and the Liberal Arts Building, all at the University of the Philippines; the Children's Memorial Hospital; the Ayala y Cia office building; and the Ever Theater.

The National Museum, founded in 1948, contains permanent displays of archaeological and anthropological material and presents special exhibitions of painting and sculpture, both local and international in scope.

Representative churches of the period of Spanish dominance are found throughout Luzon, particularly in the Ilocos region. Good examples of Spanish colonial architecture exist in Balaoan, Bantay, Laoag, Morong, San Juan, Santa María, San Nicolas, and Vigan. Baclayon, on the island of Bohol, has a pre-Spanish tower erected against the Moros, and on Cebu there are a memorial to Magellan and an early church dedicated to St. Augustine.

Bibliog. D. Aduarte, Historia de la provincia del S. Rosario de la orden de predicadores en Filipinas, Manila, 1640 (continued by B. de Santa Cruz, Manila, 1693); F. Colin, Labor evangelica . . . en las islas filipinas, Madrid, 1663; J. F. de San Antonio, Chronica de la apostolica provincia de S. Gregorio en las islas filipinas, 3 vols., Manila, 1738–44; J. Montano, Voyage aux Philippines et en Malaisie, Paris, 1886; J. Montero y Vidal, Historia general de Filipinas, 3 vols., Madrid, 1887–95; J. de Medina, Historia de los sucesos de la orden de nuestro gran padre Agustín de las islas filipinas, Manila, 1893; E. H. Blair and J. A. Robertson, The Philippine Islands, 55 vols., Cleveland, Ohio, 1903–09; A. P. C. Griffin, A List of Books on the Philippine Islands in the Library of Congress, Washington, 1903; T. H. Pardo de Tavera, Biblioteca filipina, Washington, 1903; D. P. Barrows, A History of the Philippine Islands, New York, 1905 (2d ed., 1924); J. A. Robertson, Bibliography of the Philippine Islands, Cleveland, Ohio, 1908; W. E. Retana y Gamboa, Origines de la imprenta filipina, Madrid, 1911; P. Pastells, Mision de la Compañia de Jesús de Filipinas en el siglo XIX, 3 vols., Barcelona, 1916–17; P. Torres Lanzas and P. Pastells, Catalogo de los documentos relativos a las islas filipinas procedido da una erudita historia general de Filipinas, 3 vols., Barcelona, 1925–28; D. C. Worcester, The Philippines: Past and Present, New York, 1930; F. F. de la Rosa y Cueto, A Brief Sketch of the History of Plastic-Graphic Arts in the Philippines, Manila, 1931; D. Angulo Iñiguez, Planos de monumentos arquitectónicos de América y Filipinas, 4 vols., Seville, 1933–39; Z. M. Galang, ed., Encyclopaedia of the Philippines, IV, Manila, 1935 (2d ed., VII–VIII, Manila, 1953); G. F. Zaide, Catholicism in the Philippines, Manila, 1937, pp. 146–69; W. C. Repetti, Pictorial Records and Traces of the Society of Jesus in the Philippine Islands and Guam prior to 1768, Manila, 1938; P. Kelemen, Baroque and Rococo in Latin America, New York, 1951, pp. 73–74; F. Zóbel de Alaya, Art in the Philippines Today, Liturgical Arts, XXI, 1953, pp. 108–10; A. E. Nakpil, Address on a Review of Philippine Architecture. . . , Manila, 1956; W. S. Smith, Art in the Philippines, Manila, 1958; F. Zóbel de Alaya, Philippine Colonial Sculpture: A Short Survey, Philippine S., VI, 1958, pp. 249–94; J. Camón Aznar, La arquitectura y la orfebrería españolas del siglo XVI (Summa Artis, XVII), Madrid, 1959; M. L. Diaz-Trechuelo Spínola, Arquitectura española en Filipinas (1565–1800), Seville, 1959; D. V. Welsh, A Catalogue of Printed Materials Relating to the Philippine Islands, 1519–1900, in the Newberry Library, Chicago, 1959; B. Legarda y Fernandez, Colonial Churches of Ilocos, Philippine S., VIII, 1960, pp. 121–58; R. Perez, Philippine Architecture, Am. Inst. of Arch. J., XXXIV, 1960, pp. 40–46; E. Torres, The Arts in the Philippines: Painting, Philippine S., X, 1962, pp. 127–33; R. Ahlborn, Spanish Churches of Central Luzon: The Provinces near Manila, Philippine S., XI, 1963, pp. 283–92; F. Zóbel de Alaya, Philippine Religious Imagery, Manila, 1963.

* *

Illustration: 1 fig. in text.

PHILOSOPHIES OF ART. See ARCHITECTURE; ART; CRITICISM; CUBISM AND FUTURISM; ESTHETICS; EUROPEAN MODERN MOVEMENTS; EXPRESSIONISM; FAUVES; HISTORIOGRAPHY; IMPRESSIONISM; MEDIA, COMPOSITE; MIMESIS; MODERNISM; NONOBJECTIVE ART; SURREALISM; TREATISES.

PHOENICIAN-PUNIC ART. The coastal cities of Lebanon were inhabited from the beginning of historical times by a Semitic people, the Phoenicians, who in the 2d and 1st millenniums B.C. developed an art production of notable significance. While closely connected with Near Eastern traditions, especially from Syria and Palestine, Phoenician art presents a number of individual characteristics that arose from the predominantly Mediterranean orientation of the country.

The Phoenicians carried on an intense trade by sea with lands lying to the west; they were present on Cyprus; they contributed to the diffusion of Orientalizing products and motifs in Greek and Italic cultures; they founded trading centers throughout the western Mediterranean, in the islands, in Africa, and on the Iberian Peninsula. In the 6th and 5th centuries B.C. the Phoenician settlement of Carthage assumed the hegemony of the western colonies, including Spain, Morocco, Sicily, and Sardinia, and established a powerful empire that competed with the Greek city-states and with the rising power of Rome. Western Phoenician culture, though linked to the original forms developed by this people, took on a character of its own. Its traditions, which persisted into Roman times, are usually designated by the term "Punic," derived from a Latin name for the Phoenicians (Poeni, Puni). See ASIA, WEST: ANCIENT ART; CYPRIOTE ART, ANCIENT; ORIENTALIZING STYLE; SYRO-PALESTINIAN ART.

Historical background. *Phoenicia.* The Phoenicians were a Semitic people of antiquity celebrated for their skill in commerce and craftsmanship; they spoke a Semitic language of the northern branch, very close to ancient Hebrew. Their true home was the mountainous area of Lebanon between the modern Nahr el-Kelb (Dog River) on the north and Ras-en-nakoura (Ras en-Naqura) on the south, the headland marking the northern limit of Palestine. Here the chief cities were Byblos (Bib., Gebal; mod. Jebeil, Jubayl), Sidon (Tsidon; mod. Saida), and Tyre or Tsur (Gr., Tyros; mod. Es Sur). Tradition assigns to them a very early foundation; Berytus (mod. Beirut) surely existed in the Tell el 'Amarna period. But Botrys (mod. Batrun, between Tripolis and Byblos) was founded in the early 6th century; and the federation of Tripolis, with its capital city of Tripolis, was probably established in the 4th century B.C., to include Tyre, Sidon, and Arvad (Gr., Arados; mod. Ruad).

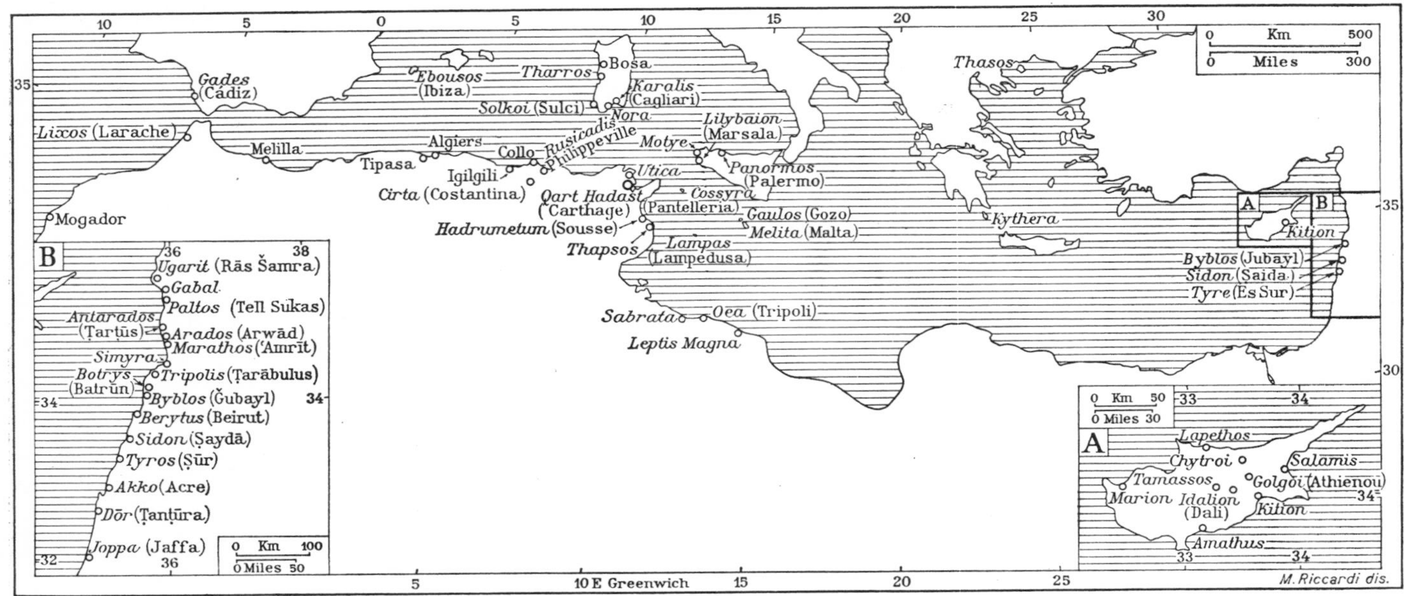

Phoenician-Punic cultural and artistic centers.

Although native tradition, preserved through Philo of Byblos (Philo Byblius), held that the Phoenicians were autochthonous, this claim is unlikely; they were preceded in Lebanon by an apparently different population in the Neolithic period and appear to have established themselves in this area about 3000 B.C. Their place of origin is obscure, and their name itself does not appear before the 16th century B.C. According to reports preserved by classical writers, the Phoenicians came from the Red Sea area (Herodotus, VII, 89). Owing to the paucity of excavation in Phoenicia, very little is yet known of its history before the 2d millennium B.C. The first urban settlement at Byblos, found in Level IV of the early Bronze I period, is dated about 3050–2850 B.C. There is evidence that by 2750 B.C. Byblos had passed through five building levels, and in Level VI, evidently the proto-Phoenician level, it had reached a golden age of prosperity.

Important relations between Egypt and Phoenicia, partly commercial and partly religious, which were most probably maintained by sea, can be traced at Byblos as far back as the 4th dynasty, about the middle of the 3d millennium B.C. (Montet, 1928–29, pp. 128 ff., 271 ff.). The Old Kingdom of Egypt was brought to an end just after 2300 B.C. by an invasion of Asiatic nomadic tribes; and about 2150 B.C. the fifth city of Byblos was destroyed in a violent conflagration evidently caused by the invading Amorites. Under their rule the city was rebuilt (Level VII), and a period of close relations with Egypt began. The Pharaohs of the Middle Kingdom, especially during the 12th dynasty (1991–1786 B.C.), tried to keep hostile Semitic princes in check by means of elaborate curses and to maintain the friendly rulers with costly gifts, some of which have been found in their tombs (K. Sethe, *Die Ächtung feindlicher Fürsten Völker und Dinge auf altägyptischen Tongefäßscherben des mittleren Reiches*, Berlin, 1926; Montet, 1928–29). The same policy of royal gift-giving was pursued by the Pharaohs toward the northern Phoenician colony of Ugarit (mod. Ras Shamra). The remarkable skill and elegance of Egyptian artifacts such as those found at Byblos and Ugarit left a lasting impression on the minds of Phoenician craftsmen. Not only were the Egyptian techniques of glazing and goldsmith's work faithfully imitated, but even Egyptian ideas of representation were adapted as seen in the earliest, somewhat clumsy copper figures of human beings, usually a nude male depicted with right arm raised and striding forward with the left leg (Dunand, 1937–58, II, pls. LXII, LXXII, LXXXI).

The end of Amorite hegemony over Byblos came with the fall of the city of Level II, in an invasion that swept on to Egypt about 1730 B.C. and eventually brought the Middle Kingdom to an end. The new rulers were the Hyksos (Eg. Hiq-khasitu, "nomad people"), about whose origins little is certain except that they came to Egypt from the north and that they included an appreciable Semitic element, for among their gods were the Phoenician deities El, Baal, and Anath. Though the cultural achievements of Egypt under their rule do not appear to have been great and their rule was of brief duration, they furthered the connection between Syria, Phoenicia, Palestine, and Egypt; and when they were driven out from Egypt by Amasis I (Ahmose) in about 1565 B.C., that connection was strengthened by the Egyptian ruler's campaigns in Syria and Phoenicia and by those of his successors, Thothmes I (Thutmose; r. ca. 1525–ca. 1515 B.C.) and Thothmes III (r. ca. 1504–1450 B.C.).

The colonies. By the 2d millennium B.C. the range of influence of the Phoenicians was extended through a string of colonies planted along the Levant coast. Some of these are hardly more than names; others are well known throughout history, such as Joppa (Jaffa), Akko (Acre), and Dor (mod. Tantura, north of Caesarea) to the south. To the north, the island port of Arvad controlled a string of settlements on the coast, beginning with its mainland counterpart Antarados (mod. Tartus), and including Gabal (Gabala), Paltos (mod. Tell Sukas), Balanea, Carnē (mod. Karnum, north of Tartus), Marathus (mod. 'Amrit), and Zimyra. How early these colonies were established is unknown — if indeed they were deliberately founded and were not the natural outgrowth of mere trading stations. It is clear, however, that the Phoenicians made a regular practice of occupying any offshore islets, peninsulas, or vacant natural harbors (including many that have long since silted up) which were easy to approach by sea and to defend from attack by land.

The earliest colony of which much is known is Ugarit, discovered in 1929 at Ras Shamra, about 7 mi. north of Laodicea (Latakia). Ugarit was an important stepping stone to the wealth of Anatolia and Cyprus alike, especially the deposits of metals. The site was already inhabited by the 4th millennium; Phoenician settlement and dominance did not occur, however, till Level II (corresponding to the Egyptian 12th dynasty), when two temples were built to the Phoenician gods Baal and Dagon, although the ruling family appears to have been Hurrian or Mitannian and the population mixed. By the 15th century B.C. an exceedingly strong Mycenaean and Cypriote influence emerged in Ugarit. During the following century the most remarkable feature was the regular use of a script imprinted on clay with a stylus, in principle like the cuneiform in use in other parts of Syria and in Mesopotamia but in fact expressing a true alphabet of 28 sounds. Tablets in this script containing religious, epic, economic, and commercial texts have been recovered and published, providing valuable insights into the early history of the Phoenicians and their neighbors. Rich finds of art objects in stone, gold, silver, bronze, and ivory have also been made (PL. 131; IX, PL. 158). Ugarit, along with other Phoenician cities, was destroyed by an invasion of the Land and Sea Raiders that moved on toward Egypt, finally to be thrown back at its frontiers by Pharaoh Ramses III about 1200 B.C.

By this time, as is clear from the tale of Wenamon (G. Maspero, *Popular Stories of Ancient Egypt*, New York, 1915, p. 209), Egypt no longer claimed any suzerainty over the Levant coast, and for a while the Phoenicians were free. Led by Tyre, they entered a period of prosperity and expansion. The foundation of Tyre's first colony, Utica in Tunisia, is traditionally placed at either 1140 or 1101 B.C. By the 6th century B.C. the Phoenicians appear to have extended to the entire island of Cyprus the name of Kittim (Gr., Kition), from the name of their first and principal colony there. This may imply that Kition did not become important until the 7th century B.C., even if it had been founded as early as the 9th; however, no excavations there have yet verified this. A Phoenician text of the 8th century B.C. (*CIS*, I, 5) found near Amathus and an Assyrian text of Esarhaddon (7th cent. B.C.) mention Kart-hadasht (Qartihadashti, "New Town"), but it is uncertain whether this is the original name of Amathus or of Kition. Other Cypriote sites such as Tamassos, Chytroi (Chytros), Lapithos, and Golgoi (mod. Athienou) have all produced Phoenician inscriptions and were probably Phoenician outposts. Elsewhere on the island, Amathus, Salamis, and Marion were but partly Phoenician settlements. Idalion (mod. Dali) was Phoenician only from the 5th century B.C., when it was captured by Kition.

Several island settlements were also founded by Phoenicia farther west in the Mediterranean; Greek tradition has Greeks placed at Kythera and Thasos (Herodotus, IV, 47). Other island stations — Lampas (Lampedusa), Cossyra (Pantelleria), Gaulos (Gozo), and Melita (Malta) — are known as Phoenician, though only Malta has revealed early remains with finds going back to the 7th century B.C. (A. Mayr, "Aus den phonikischen Nekropolen von Malta," *SbMunchen*, 1905, pp. 467–509; T. Zammit, "The Maltese Rock-cut Tombs of a Late Pre-Christian Type," *B. of the Malta Mus.*, I, 1931, pp. 101-37; Cintas, 1950).

In Sicily the Phoenicians were certainly established by the end of the 8th century B.C. Thucydides (VI, 2) claims that they settled at first over a wide area of the island but, confronted by the Greek colonists, withdrew to the northwest corner, presumably by the 6th century B.C. This statement, however, is doubted by scholars, who believe the Phoenicians arrived there only after the Greeks were installed. The Phoenician settlements were at Motye and, later, at nearby Lilybaeum (Lilybaion; mod. Marsala).

North Africa had a chain of Phoenician-Punic foundations. After Utica came Carthage (according to tradition, settled 287 years after Utica, i.e., 814/813 B.C.). Archaeology has been able to substantiate this only with finds from the second half of the 7th century B.C. Other Phoenician, or more probably Punic, foundations confirmed by archaeological materials are in Tunisia (Hadrumetum; mod. Sousse; 5th–4th cent. B.C. finds), Thapsus (3d–2d cent. B.C. finds), and Hermaeon (Cape Bon; 4th–2d cent. B.C. finds). In Algeria the foundations begin with Djidjelli [Igilgili, 7th cent. B.C.; M. Astruc, "Nouvelles fouilles à Djidjelli (Algérie)," *Rev. Africaine*, XXX, 1937]. Other sites include Rusicadis (Philippeville), Collo, Algiers, Tipasa, Gouray (Gouraya), and Constantine (Cintas, 1950). In the territory of ancient Mauretania, remains have come from Lixos (Larache) and Melilla, in Spanish Morocco, and from Mogador, on the Atlantic coast (Cintas, 1954).

In the Balearic Islands a colony was founded at Ebusus (Ibiza), according to tradition in 654/613 B.C.; the material excavated there, probably Punic, appears to be not earlier than the 6th century B.C. (PLS. 134, 135; A. Vives y Escudero, *La necropoli de Ibiza*, Madrid, 1917; A. Garcia y Bellido, "Archaeological News: Spain and Portugal," *AJA*, LIII, 1949, pp. 150–62). In Sardinia, Phoenician sites are Nora (near Cagliari), Bosa, Tharros (near Oristano), and Sulci (island of Sant'Antioco). The finds in Sardinia, in fact, indicate large-scale settlement in the island only in the late 6th century B.C., presumably after the naval victory of Alalia in Corsica had secured the Carthaginians in their possession of the island.

In Spain the oldest settlement was traditionally that of Gadir, or Gades (Cádiz; Strabo, III, 5, 5), then an islet off the territory of the Tartessians (or Turdetans), with whom the Gaditans struggled to establish themselves (Justin, *Historiarum Philippicarum*, XLIV, 5). The position of Gades at the mouth of the Baetis River (Guadalquivir) gave the settlers access to the wealth in gold, silver, and copper of Tartessus (mod. Andalusia; J. de M. Carriazo, "Gold of Tarshish?" *Illustrated London News*, Feb., 1959). Thus far, the earliest material from Cádiz belongs assuredly to the 5th century B.C. (R. Menendez Pidal, *Historia de España*, I, 2, Madrid, 1952), but it is possible that the tombs from Punta de la Vaca (Cádiz) excavated in 1887 belong to the 6th century B.C. The well-known ivories from Carmona are surely no earlier in date (*Early Engraved Ivories in the Collection of the Hispanic Society of America, from Excavations by George E. Bonsor*, New York, 1928). However, earlier material is known in Spain — for example, the splendid jewelry from La Aliseda, which probably dates from the 7th century B.C. (A. Blanco Freijero, "Orientalia," *AEA*, XXIX, 1956, pp. 3–51; A. Garcia y Bellido, "Materiales de arqueología hispanopúnica: Jarros de bronce, *AEA*, XXIX, 1956, pp. 85–104).

Effect of colonization and commerce. The traditional role of the Phoenician cities in commerce was first developed, it seems, under the Egyptian 18th dynasty (1570–1314 B.C.). An Egyptian tomb painting of this period (destroyed) depicted the arrival at an Egyptian port of seven Phoenician trading ships, bearing a cargo of the distinctive Canaanite wine jars; on shore a market was set up (G. Daressy, "Une flotille phénicienne," *RA*, 3d ser., XXVII, 1895, pp. 286–92). Strategically placed, straddling both sea and land routes, ruled by astute merchant princes, and protected by Egypt, Phoenicia became the center of a network of commerce and the home of thriving crafts, which were well organized into guilds of every kind (C. Virolleaud, "Les villes et les corporations du royaume d'Ugarit, *Syria*, XXI, 1940, pp. 123–51). Her establishment of a network of trading stations and colonies along the Mediterranean seaboard was the outcome of political pressures, commercial needs, skillful organization, and a capacity for meeting supply and demand in the economics of the ancient world, with the aid of business methods evolved much earlier by the merchants of Babylonia and Assyria. The Phoenician aptitude for commerce was undoubtedly stimulated by — if it did not itself stimulate — the invention of the abacus for ready calculation and of the alphabet for keeping records. By the time of Wenamon, about 1100 B.C., there is mention of a Phoenician merchant named Barakat-ilu living at Tanis, Egypt, who owned a fleet of 10,000 ships trading between Tanis and Sidon. In the poems of Homer the Sidonians were already famous as craftsmen and traders. The vast extent of the import-export trade of Tyre in the early 6th century B.C. is described in the Book of Ezekiel (27–28).

The exports of Phoenicia included cedar and pinewood from Lebanon; fine linen from Byblos, Berytus, and Tyre (Lutz,

1932); cloths dyed with the famous purple made from the murex shell at Tyre or Dor; and embroidered cloths, metalwork, glass, and salt from Sidon, as well as wine and glazed faïence. In return, Phoenicia took from Egypt, India, and the less developed peoples of Anatolia, Africa, the Aegean, and the western Mediterranean various goods and raw materials such as papyrus, gold, silver, copper, iron, tin, precious stones and jewels, ivory, ebony, coral, silk, amber, spices and incense, ostrich eggs, slaves, and horses. The network of Phoenician commercial interests extended certainly as far as Spain in the west; and if, as seems likely, Ophir is to be identified with Suppara (Sopara), near Bombay, it reached as far as India in the east (Barnett, 1957). To Ophir, in the 10th century B.C., Hiram of Tyre and Solomon of Jerusalem sent a trading fleet every three years from Ezion-geber (on the Gulf of 'Aqaba) to bring back cargoes of gold, silver, ivory, apes, peacocks, and almug wood (I Kings 9:26–28, 10:11–12, 22). Confirmation of this eastward penetration is found in the fact that the words in the Hebrew text for the last three items are themselves of Indian origin. Whether the Phoenicians, in search of tin, an indispensable constituent of bronze, reached Cornwall in Britain rests only on the testimony of Diodorus Siculus and Strabo, who mentioned the Cassiterides ("Tin Islands") by some thought to refer to the Cornwall coast. (The Cassiterides are now assumed to be the Scilly Islands.) It is only through the Carthaginians that a definite contact between Phoenicia and Britain can be demonstrated, for Punic coins have been discovered in Britain. It is likely, however, that the Carthaginians followed in the footsteps of earlier Phoenician explorers.

Barter methods that the Phoenicians used with backward peoples are described by Herodotus (IV, 196). Our only clue to their business procedure in more advanced regions is the survival in Greek of the Phoenician word *arrabôn* (a pledge). Unfortunately, except for a few small texts from Ugarit, no Phoenician commercial documents have yet been found.

Later history. The tremendous colonizing activity must be viewed as part of the history of the Phoenicians as a whole. It may have owed much of its impetus to the grave incursions into Phoenician territory between the 8th and 6th centuries B.C., which would have crushed a less hardy and adaptable race; instead, these may have led to the seeking of new markets and safer bases as a means of escape. In 732 B.C. Tiglathpileser III invaded Phoenicia. Sargon II, in 715 B.C., even managed to establish some form of occupation over Cyprus. In 700 B.C. Sennacherib pillaged Phoenicia and occupied Sidon. In 677 B.C. Esarhaddon attacked and destroyed Sidon, killing its king and deporting its inhabitants; later he seized Tyre. In about 665 B.C. Ashurbanipal conquered Tyre and Antarados. Next, the founding of Naukratis in Egypt (ca. 650 B.C.) by the Greeks was a blow to Phoenician trade with that country. Finally, Nebuchadnezzar captured Tyre after a siege lasting 13 years (586–573 B.C.). Not surprisingly, the Phoenicians became loyal supporters of the Persians, who overthrew their oppressors and reopened the trade of the Orient to Phoenicia. In the time of Xerxes I, Sidon was reckoned the leading city of Phoenicia, and in the ruler's fleet the ships of Sidon had precedence.

Meanwhile, Carthage had been growing steadily stronger. In 535 B.C. the crushing sea victory of the Carthaginians and Etruscans over the Phocaeans off Alalia (Corsica) confirmed the Carthaginian control of Sardinia and the approaches to the western Mediterranean. Carthaginian material goods, no longer imported from the weakened Phoenician homeland, became mixed with Greek imports in western Phoenician colonies. In the first treaty of Carthage with Rome (508/507 B.C.), Africa, Sardinia, and part of Sicily were recognized as Punic, while it was forbidden to sail past the "fair headlands" (Polybius, III). Treaties were also concluded with the Etruscans, Massaliotes, and Cyrenians (Aristotle, *Politics*, III, 9, p. 1280a; Justin, op. cit., XLIII, 5, 2; Sallust, *History of the Jugurthine War*, 78). There is some reason to think that after the Persians and their allies, the Phoenicians and Carthaginians, were defeated by the Greeks in the Persian Wars and the wars in Sicily, Carthage saw herself cut off from the east by the Greeks; and under pressure of this threat, the Carthaginians made sure of their control of the western Mediterranean. A revolution in the polity of Carthage transferred its conduct from the hands of a monarchy to that of an aristocracy, and the political change was accompanied by a religious reform and a period of severe austerity during which the markets were closed to the import of Greek goods.

After entering upon an unsuccessful war with the Greeks of Sicily in 480 B.C., the Carthaginians retired from the contest for 70 years, during which time they built up an empire in North Africa out of the older Phoenician colonies, which had never been truly united. Whittaker notes that Phoenicia, in contrast to her daughter Carthage, "seems never to have been a united country with a solid basis but was broken up into many townships each independent of the others . . . this failing . . . extended to most of her colonies as well . . . . Although undoubtedly great from a commercial point of view, it never became a really powerful state politically, or maintained an army capable of holding its own against others. . . . To its offspring, the men of the New City [Carthage], was it left to carry out what the older community had failed to do . . . . " (J. I. S. Whittaker, *Motya*, London, 1921, p. 3). Renewed Punic-Greek wars in Sicily (408–368 B.C.) resulted in partition of the island along a line from Himera to the Halycus River. Defeated and disappointed in Sicily by yet another war in 346–339 B.C., Carthage retrieved her fortunes in Spain under the Barca family during the next century. These leaders founded three Spanish settlements: one at Qarthadasht (Cartagena) in 223 B.C.; another known only by the Greek name of Akra Leuke, near Alicante, in 231 B.C.; and a settlement at Baria, near Villaricos (A. Garcia y Bellido, "Materiales de arqueología hispano-púnica: Jarros de bronce, *AEA*, XXIX, 1956). A string of smaller stations round the Baetis Valley, known from coins, are Asido, Oba, Vesci, Lascuta, Arsa, Iptuci, Ituci, Olont, and Turris Regina; these settlements may belong to this period.

Various settlements were also planted by the Carthaginians on the most distant Moroccan shores of western Africa (D. B. Harden, "The Phoenicians on the West Coast of Africa," *Antiquity*, XXII, 1948, pp. 141–50). Great journeys of exploration were undertaken, such as the famous Phoenician Periplus organized by Pharaoh Necho around Africa, an enterprise credibly reported by Herodotus (IV, 42); or that of the Carthaginian Hanno, perhaps undertaken in the late 5th or 4th century B.C.; or that of Pseudo-Scylax (C. Muller, ed., *Geographi Graeci Minores*, I, Paris, 1855). The commercial connections of Carthage reached as far as the Azores, where a pot of Punic and other coins was found in 1740 on the island of Corvo (R. Hennig, *Terrae incognitae*, I, Leiden, 1936, pp. 109–19; cf. Diodorus Siculus, V, 20); as well as to Britain, as has been noted, attested by many finds of Punic coins there (J. G. Milne, *Finds of Greek Coins in the British Isles*, Oxford, 1948).

Carthage grew enormously and is said to have attained a population of 700,000 (Strabo, XVII, 15, p. 833). Greek fashions gained more and more ground. After the Third Punic War with Rome, when the Romans captured Carthage under Scipio Africanus, the city was razed to the ground in 146 B.C. but was rebuilt when a Roman colony was planted there in 44 B.C. During the Roman imperial period seven ancient Phoenician cities in Africa were considered "free": Utica, Theudalis, Hadrumetum, Thapsus, Leptis Magna, Acholla, and Usula. A curious local revival of Punic nationalism then took place, finding expression in the renewal of the Punic language for inscriptions, usually funerary, and in the representation of older Carthaginian deities.

PHOENICIAN ART. *Character and impact.* Many of the splendid *objets d'art*, presumably of precious metal, that New Kingdom (18th–20th dynasties; 1570–1085 B.C.) tomb frescoes show being brought into Egypt from the north were most likely of Phoenician workmanship (Montet, 1937; Shaeffer, 1939–56, II). These illustrations are supplemented by accounts of tribute in the royal annals. Phoenician and Syrian craftsmen and merchants saw an ever-expanding market opening up in Egypt, especially in luxury goods for the nobility, which through the spread of Egyptian culture and fashions during this period were everywhere in increasing demand. The Asiatic imports revealed a new humor and inventiveness. Vases suggesting pools hold ducks' heads peeping out of flowers or kine in papyrus thickets. Sometimes figures of captured Asiatics form the handles. There can be no doubt that the craftsmen of Phoenicia studied the Egyptian taste of the day; some settled in Egypt in craftsman's quarters like those excavated at Deir el-Medineh [B. Bruyère, "Fouilles de Deir el Medineh (1934–1935)," *Mém. de l'Inst. français d'Archéol. orientale du Caire*, XVI, Cairo, 1939].

Meanwhile, Phoenician products made for the homeland acquired an ever-stronger Egyptian tinge. But Egyptian elements and themes were not the only foreign elements to be adopted in Phoenician art: a Mesopotamian love of symmetry, an Aegean taste for galloping animals, the north Syrian skill in fashioning groups of fighting animals, and the sphinxes and griffins of Levantine origin were all grafted onto the basic stock of Phoenician imagery and symbolism (Barnett, 1957, pp. 54–62).

The Phoenicians' special skill lay in using the artistic elements of others for structural solutions. Thus, handles of vases were wedded to the bodies by palmettes, and the necks and bases of the vessels were decorated with lotus flowers. Sacred trees were adapted to make the columns of buildings. "But if for the production of Phoenician art, the Nile spread over all a superficial dressing of its own, the soil below remained purely Semitic . . . . If we judge the Phoenicians fairly, we see that their achievement lay in the sphere of uniting the various forms and devices of their Near Eastern neighbour for the benefit of their own constructive ideas . . . . Unhappily, they failed largely to create from these ingredients a real synthesis, which was only accomplished by a secondary distillation in the alembic of nascent Greek art" (R. D. Barnett, "The Nimrud Ivories and the Art of the Phoenicians," *Iraq*, II, 1935, pp. 179–210).

In the Iron Age, in the period corresponding to the 20th–26th dynasties (1200–525 B.C.), Egyptian influence in Phoenicia remained particularly strong in such classes of imported objects as scarabs, faïence figures, and amulets, which displayed a conservatism in which Egyptian religious prestige played an undoubted part. But by the 5th century B.C. the advances of Greece in the fields of art and thought overwhelmed Phoenicia, especially in the representation of the human figure [e.g., reliefs from Umm el-Aḥmād (Umm-el-Awamid); PL. 133; IX, PL. 352; Sidonian anthropoid sarcophagi; PL. 132] — although some homage is paid to Persia in the "Sarcophagus of the Satrap." A similar process took place in Carthage, where some Greeks certainly settled and worked. In the Hellenistic period this Hellenization continued even more rapidly, although the Ptolemies abandoned the Attic monetary standard in favor of the Phoenician, in an effort to conciliate Carthage and the Phoenicians. That Egyptian influence was still strong at this time is borne out by the fact that, until the destruction of Carthage, the Carthaginians could read the Egyptian hieroglyphics and understood the purposes of their magic; however, they made use of these occult tools only for the living (Vercoutter, 1945).

The Phoenicians did not merely perform the passive role of carriers in art; and their achievement was positive, even when only as intermediaries, for they furnished the newly arising Mediterranean world of the early Iron Age with material comforts and luxuries and with a vocabulary of artistic expression. No more important force than the Phoenicians can be mentioned among the various competing influences from the East that contributed to the Orientalizing period of the 7th century B.C. in Greece and Etruria (Poulsen, 1912; Barnett, 1956; Dunbabin, 1957). This movement had already begun in the previous century. Oriental bronze bowls — some certainly of Phoenician manufacture, others perhaps Syrian — have been found at Delphi, Athens (Kerameikos, or Dipylon, cemetery), and Olympia. Phoenician ivories, probably of the same date, have been found in Rhodes (Ialysos, Lindos, Kameiros), Crete (Mt. Ida Cave), Samos, Perachora (opposite Corinth), and in Italy at Praeneste (mod. Palestrina). In Sparta a school of ivory-workers developed soon after the appearance of these Oriental originals (R. D. Barnett, "Early Greek and Oriental Ivories," *JHS*, LXVIII, 1948, pp. 1–25). In Perachora over 800 scarabs of the 8th–7th century B.C. have been found, apparently the product of some half-Greek, half-Oriental center, perhaps in Corinth itself, which was a great trading center with both East and West. Many more such scarabs have been found elsewhere in Greece — in the Heraion of Argos, for example — and in Etruria (J. D. S. Pendlebury, *Aegyptiaca*, Cambridge, Eng., 1930, p. 58). Greek goldsmiths' work owes much to the Phoenicians, noticeably in the profusion of granulation in jewelry from Rhodes, Ephesus, and Crete (Becatti, 1955; E. Coche de la Ferté, *Les bijoux antiques*, Paris, 1956, pp. 34–39). In Etruscan work also granulation abounds. Alabaster objects, glazed frit beads, faïence figurines, glass vessels, and many other objects found in ancient sites are all, originally at least, imports from or through Phoenicia. The extent of Greece's debt to Phoenicia may be measured by the introduction of the alphabet into Greece and, perhaps as early as the 9th century B.C., by the adoption of the Phoenician standards and names of weights. A quantity of Semitic words in Greek guise testifies to the existence and character of the Phoenician trade, even if many of them may well prove to have been endenizened as early as the late Bronze Age.

The exact routes through which these commercial contacts were made is less certain. Naturally the Phoenician influence upon Greece was felt most strongly in adjacent Rhodes and Cyprus; there the weaker strain of Greek art acquired a hybrid Greco-Phoenician quality. No Phoenician stations or colonies in Greek or Italian territory have as yet been found, despite the literary traditions; nonetheless, the existence of sizable Oriental quarters (e.g., at Corinth) may be conjectured. Vigorous settlements of Greek traders are to be found in the East, however — for example, at Al Mina in northern Syria and at Tell Sukas, which has been under excavation by a Danish expedition [C. L. Wooley, "Excavations near Antioch in 1936," *AntJ*, XVII, 1937, pp. 1–15; E. Forrer, "Eine unbekannte griechische Kolonie des 6. Jahrunderts v. Chr. in Phönikien," *Bericht über den int. Kong. für Archäol.* (1939), Berlin, 1940, pp. 360–65; C. M. Robertson, "The Excavations at Al Mina, Sueidia, IV: The Early Greek Vases," *JHS*, LX, 1940, pp. 2–21].

The sarcophagus of the priestess Shamashtart found at Carthage (PL. 127), a work belonging to the 4th century B.C., has a relief on the lid with a handsome woman wearing a cloth on her head — the Egyptian *klaft*, surmounted by a gilded head of a Horus hawk. Her tunic is surmounted by a red, blue, and gold band, and her lower limbs are enveloped in two birds' wings. This presentation means that the figure was envisaged as impersonating the actual character or role of Isis or Nephthys (the goddess's twin and companion), who are repeatedly shown on Egyptian and Phoenician monuments in an attitude of protecting the babe Horus, with their lower limbs enfolded between the wings of a flying hawk (Barnett, 1957). In her right hand, Shamashtart holds an unguent vase in the form of a duck, which clearly reproduces a traditional type of ivory toilet vessel found in the Near East as far back as the late Bronze Age. Such vases seem to have possessed a religious significance.

Priests are represented on the ivories from Arslan Tash (Haddatu), in the Phoenician homeland, wearing an Egyptian wig and a kilt and holding a ram-headed staff and a jug. On the 5th-century B.C. funerary steles from Sidon and Umm el-Aḥmād, the priests are clean-shaven and wear a Greek chiton and a round hat like a biretta (Bossert, 1951, figs. 507, 511, 1071a); the headgear may be influenced by the crown of the Phoenician king in Persian times (stele of Yeḥawmilk, from Byblos, 5th–4th cent. B.C.; Louvre) and is probably the ancestor of the hat worn by the priests of Palmyra. There is evidence that some Phoenician priests wore a tall, pointed hat; the miter of the Hebrew high priest may have been similar. Punic priests were required to sacrifice in bare feet and with shaven head covered, and without a girdle; their sacrificial garment was distinguished by a *latus clavus* (Silius Italicus, *Punica*, III, 21–30). Carthaginian priests, who may have worn gold earrings and nose rings, are represented as bearded, with their heads unshaven and uncovered (A. L. Delattre, *Les grands sarcophages anthropoides du Musée Lavigerie*, Paris, 1904, figs. 4–7, 34–46). This defiance of rules may explain the presence of the symbolic razors that have been found in tombs. A figure at Umm el-Aḥmād holds an incense spoon that evidently reproduces an ivory original, the handle of which is in the form of a swimming female nude; this is a well-known type of Syrian or Phoenician ivory that, like the duck, seems to have acquired a ritual association.

Architecture. Basic information for study of the architecture is also mainly dependent on the excavations of Byblos and Ugarit. In attempting to follow those at Byblos, however, we are hampered by the inadequacy of publication and by the method of excavation, which rigidly follows a system of *tranches géometriques* rather than one of levels as adopted in other excavations. The first urban settlement at Byblos (Level IV), of the early Bronze I period (ca. 3050–2850 B.C.), contains houses consisting of several rooms which are often attached to an initial cell. The walls of semirigid materials, presumably mud and wattle, stood on a base of well-set sandy stones that were dressed at the corners. The roofs are supported by posts placed axially and along the walls. An early sanctuary has come to light also. In Level V (ca. 2850–2700 B.C.) the houses, while similar, acquire real foundations of split limestone. The first monumental temples — those of Baalat Jubayl (called by Montet the "temple égyptien"), Resheph, and two others — were built in this level, and the town is surrounded by a rampart of double thickness. In Level VII (2150–1800 B.C.) temples are equipped with obelisks, especially that of Resheph. This dedicatory form may be traced from Canaanite religion of the late Bronze Age, in the miniature steles at Hazor, to the steles set up in vast numbers in Carthaginian temples and sanctuaries. A small model of a house or temple was discovered in the foundation deposit of the Resheph temple. It shows a square building with two tall doors in each side; the doorposts appear to be crosshatched (Montet, 1928–29, pl. XLVI, fig. 123; P. Montet, "Les Egyptiens à Byblos," *MPiot*, XXV, 1921–22, pl. XIX). No details of constructional elements from Byblos (e.g., capitals, columns) are available before the Roman period. In the Roman period the precinct of Resheph contained two small sanctuaries and a large one, all oriented toward the east and in conformity with the ancient Canaanite tripartite ground plan, divided into forecourt [all enclosed in a rectangular colonnaded (?) peribolos], pro-cella, and cella. It is clear that this plan influenced early Greek architecture. The entrance has a pair of columns *in antis*, and it is believed that examples of this plan *in antis* at Byblos go back to Level IV (Dunand, 1937–54, II, p. 33). The Canaanite tripartite division has been found in the Temple of the Sun at Hazor, in Galilee, at Nahariya, and in the Temple of Baal at Ugarit. It corresponds exactly to the description of the 10th-century B.C. Temple of Yahweh in Jerusalem, of Phoenician construction and described in the Bible (I Kings 5–7; II Chron. 3-4); these passages constitute the fullest description available of a Phoenician temple. The most striking external feature of the Temple of Yahweh was the pair of freestanding pillars of bronze, Jachin and Boaz, which were set up in the porch by Hiram of Tyre and which bore elaborate lily capitals. In the use of this pair of columns (and perhaps in many other particulars), the temple in Jerusalem was evidently modeled on the Temple of Melkarth in Tyre. The latter was a very ancient foundation, according to tradition originally built about 3000 B.C. (Herodotus, II, 44), a date that corresponds well to the dates of the earliest temples at Byblos. A representation of part of the Melkarth temple survives in a drawing of a lost Assyrian relief from the Palace of Sennacherib at Nineveh (R. D. Barnett, "Phoenicia and the Ivory Trade," *Archaeol.*, IX, 1956, pl. IX; R. D. Barnett, "Early Shipping in the Near East," *Antiquity*, XXXII, 128, 1958, pp. 220–30, pl. XXII). In front of its main doors are twin columns with lily capitals. These are the twin columns of gold and emerald which Herodotus describes as gleaming at night and forming a landmark for sailors. A similar pair of columns with lily capitals recurs in a model shrine from Idalion in Cyprus (Perrot-Chipier, fig. 208; Bossert, 1951, fig. 16), which may represent the Temple of Resheph-Apollo. These freestanding ritual twin columns are reflected in the 6th- and 5th-century Cypriote votive steles (in the form of pillars with elaborate capitals) from Idalion, Larnaca, and Athienou (Bossert, 1951, figs. 23–26). Their analogues on the mainland are the flatter and simpler so-called "proto-Ionic" (sometimes called "proto-Aeolic") capitals found at Megiddo, Samaria, Hazor, and Ramat Rahel (south of Jerusalem). It is still uncertain whether these capitals should be restored as freestanding or with an architectural function, supporting doorways or architraves. It is also uncertain whether these proto-Ionic capitals are the parent of the Greek Ionic capital, which may have derived from farther east (e.g., Persia), but they are definitely the direct parent of the local type of capital (Aeolic) found in such places as Lesbos. It is probable that other types of capitals known to the Greek architects were also of Phoenician derivation — the Corinthian, the Pergamene of palm leaf type, and even the caryatid column (Barnett, 1957, pp. 109–10).

There is evidence that the entrances of some shrines were flanked with figures of sphinxes. Some shrines also were probably built in purely Egyptian style, as represented in the pair of small rock-carved water shrines at Marathus.

Whole blocks of private dwellings have been excavated at Byblos and Ugarit but have produced little more than ground plans. For the appearance of the superstructure of secular architecture in Bronze Age Phoenicia, one must, for want of other material, fall back on Egyptian illustrations. These do not appear to distinguish between Syrian and Phoenician buildings; indeed it is very likely that the difference was not striking. These illustrations are found in narrative reliefs in Egyptian temples; but there also exists at Medinet Habu a full-scale reproduction of a Canaanite *migdol*, or fortress, which Ramses III introduced into his palace there (U. Hölscher, *Das hohe Tor von Medinet Habu*, Leipzig, 1910). What apparently impressed the Egyptians were the lofty gateway buildings, with windows above rising to a second story over the gate and with a flat roof fortified with rounded crenelations. Such crenelations continued in use well into the Iron Age, and examples have been found at Motye (J. I. S. Whittaker, *Motya*, London, 1921). The lofty upper rooms with windows over the gate were a convenient habitation for the women and are so depicted in reliefs carved on the Medinet Habu building. A Canaanite house or palace of about the 13th century B.C. is depicted in an existing ivory (G. Loud, *The Megiddo Ivories*, Chicago, 1939, pl. 32), but the piece is much damaged.

A link between Phoenician architecture of the Bronze Age and that of the Iron Age is furnished by the brief Biblical account of Solomon's palace, a hypostyle hall built of cedarwood, and his harem, built of colossal hewn and sawed stone and cedarwood (I Kings 6–7), probably laid in alternate courses as was done at Tell Atchana and Zincirli (Sam'al; C. L. Woolley, *Alalakh: An Account of the Excavations at Tell Atchana in the Hatay*, 1937–49, Oxford, 1955, pp. 224–25). For illustrations of a palace and houses of the Iron Age, however, one must turn to an Assyrian relief from the Palace of Sennacherib, depicting part of his campaign in Phoenicia in about 700 B.C. (FIG. 305; A. H. Layard, *Monuments of Nineveh*, London, 1849; Barnett, 1957, fig. 53 and p. 147). The relief probably depicts Ushu ("Old Tyre"). Several of the tallest buildings are surmounted by windows that are half-closed by a balustrade supported on small columns. There is Talmudic evidence for calling this type of window "Tyrian." From the roofs of these harems in the Assyrian sculpture and in several carved ivories a female figure looks out of a Tyrian window; the roofs are decorated with fan-shaped objects, presumably trees planted in a roof garden. A building in the foreground very much resembles the house model from Byblos, and another the above-mentioned model from Idalion.

A relief from the Palace of Sargon at Khorsabad (FIG. 306; E. Botta, *Monument de Ninive*, II, Paris, 1848, pl. 114) shows a different type of Phoenician building. Beside a lake among wooded hills is set a kiosk, or hunting lodge, with the roof supported by columns having proto-Ionic capitals. The presence of the horse-headed barge (a Phoenician *hippos*; see R. D. Barnett, "Early Shipping in the Near East," *Antiquity*, XXXII, 128, 1958, pp. 227–28) indicates the location of the building depicted in this relief.

Conventional Assyrian illustrations show the city of Tyre

Eighth-century Phoenician secular architecture, from an Assyrian relief in the Palace of Sennacherib, Nineveh (*from R. D. Barnett, A Catalogue of the Nimrud Ivories, London, 1957*).

perched on its island (L. W. King, ed., *Reliefs from the Bronze Gates of Shalmaneser*, London, 1915, PL. XIII) and probably depict Arvad also (A. Parrot, "La 'scène maritime' de Khorsabad," *Sumer*, VI, 1950, pp. 115–17). From literary references, the height of the buildings at Tyre is determined as at Motye (Arrian, VIII, 128; Diodorus Siculus, XIV). A detailed view of the quay and sea wall of Tyre, found in a relief of Sennacherib, (Barnett, op. cit., pl. XXII) shows that the curtain walls between the towers were hung with shields. This agrees perfectly with the description in Ezekiel (27:11).

A silver bowl from Amathus in Cyprus (PL. 131; see also J. L. Myres, "The Amathus Bowl," *JHS*, LIII, 1933, pp. 25–39) shows a Phoenician (?) city with gate towers undergoing assault by an enemy. Seen within the gateway are a number of beehive-shaped houses. Such houses were built by the Carthaginians and were called *magalia*; a clay model of one was found at Ras Shamra (Bossert, 1951, fig. 649). The beehive house must have been imported into North Africa from the Levant in northern Syria, where such houses are still built (e.g., Neirab, Arslan Tash, and Khan Sheikhun).

The Phoenicians were also fine engineers and stoneworkers. Xerxes entrusted to them the task of cutting the canal through Mt. Athos. Phoenicians also built artificial harbors (*cothones*) at Carthage, Utica, Hadrumetum, Leptis Magna, and Motye.

In funerary architecture the inhabitants of Phoenicia and Palestine, during the middle and late Bronze Age, perhaps under Egyptian inspiration, developed the practice of constructing deep underground tombs cut in the soft rock; these were reached by a vertical shaft and sometimes a short passage. At Byblos the royal tombs of the 2d millennium were of this type, but in the burial chamber was placed a stone sarchophagus clearly modeled on Egyptian prototypes. In Ugarit the subterranean cavern took the form of the square or domical beehive tomb of Aegean type, made of stone blocks carefully shaped and set in place. The tradition of the hypogeum persisted into the Iron Age, the cavern being regarded as more or less an underground room, often approached by steps, with clearly architectural decoration (e.g., at Tamassos in Cyprus, where wooden beams of a gable roof are represented; and Amathus and Pila, where the doorway is surmounted by an apotropaic gorgon's head). Some remarkable funerary monuments, which suggest a mixture of domestic and temple architecture, were also erected aboveground. Such are the strange twin domical monuments called the "Maghazil" in Marathus, one of which is flanked by figures of lions that date perhaps from the 6th century B.C. These structures are possibly derived from the Phoenician circular houses (E. Renan, *Mission de Phénicie*, Paris, 1864–74). Renan (ibid., pl. 17) also recorded a pyramidal tomb, obviously inspired by the Egyptian custom. The well-known mausoleum at Thugga in Tunisia (2d cent. B.C.), represents a fusion of the Egyptian pyramidal funerary monument and a Hellenistic Greek temple.

Sculpture. Of monumental Phoenician sculpture in the round, hardly anything survives. At Byblos, in the forecourt of the Temple of Baalat Jubayl, there are shapeless remains, chiefly the legs, of large statues depicted seated or standing. One large figure is complete, however, showing a stocky form of Egyptian type wearing a kilt, with firmly clenched hands held close to the hips and advancing with the left foot in the Egyptian manner. Its date is uncertain (Dunand, 1937–54, I, pl. 26), but it is possible that the work belongs to the Bronze Age. In Cyprus, there is a tradition of Phoenician limestone statuary. A colossal figure of a deity rending a lion (Louvre), from Amathus, belongs to the 6th century B.C.

Sculpture in relief provides evidence of quite a long tradition. The excavations at Ras Shamra have produced a number of steles belonging to the 14th–13th century B.C. One shows two figures, wearing the wrap-around garment of the period, who are apparently kings swearing a contract together; others show deities (PL. 130) such as Teshub, Resheph, or El (Bossert, 1951, figs. 429, 431–33; the last-cited is of very high quality). From Byblos has come the first major work of Phoenician sculpture, the sarcophagus of Aḥīrām, now dated to the end of the 11th century B.C. (PL. 130; IV, PL. 453; Montet, 1928–29, II, pls. CXXVIII-CXLI; cf. M. Dunand, *Byblia grammata*, Beirut, 1945, pp. 197–200). Bearing a long inscription, this limestone

Eighth-century Phoenician pavilion, from a relief in the Palace of Sargon, Khorsabad (*from E. Botta, Monument de Ninive, Paris, 1848*).

sarcophagus has figures of men and lions carved on the lid and base, and the sides are decorated with funerary scenes showing lamentation over and worship of the dead man. On the dresses of the figures on the lid, there are traces of painting in red and blue. This sarcophagus obviously reproduces in stone a contemporaneous wooden chest decorated with carved panels; it is the earliest example of what later became the established type of Phoenician funerary furniture, continuing with little change through stone sarcophagi such as the late-6th- and early-5th-century pieces from Amathus, Tamassos, and Athienou on to the Satrap and Alexander sarcophagi and that of the "Mourning Women" at Sidon (VII, PL. 151–152; III, PL. 379; Perrot-Chipiez, III, figs. 415–21; Bossert, 1951, figs. 52–54). In the 8th century B.C., architecturally employed relief slabs were used (as at Karatepe). The further influence of the Phoenician decorated sarcophagus on Greek art cannot be traced here, but it should be noted that it was considerable.

The tradition of the relief stele, however, persisted strongly in Phoenician art. An interesting sculpture of a bull(?) from Sidon shows Assyrian influence. From Marathus comes the fine stele showing Shadrafa swinging a lion (ca. 8th cent. B.C.). From nearby Aleppo comes the relief of Melkarth, the tutelary god of Tyre, of like date. In the Persian period there was a return to following Egyptian models more closely, as in the stele of Yeḥawmilk of Byblos (*CIS*, I). From Umm el-Aḥmād, however, comes a graceful series of funerary steles deeply influenced by Greek art of the 5th century (PL. 133).

The Sidonians for a time tried to synthetize Greek and Egyptian artistic currents, bringing the Tabnit sarcophagus up to date by substituting idealized Greek heads as portraits of the dead on the lid of the sarcophagus, made of Greek marble (PL. 132). Several of the artists employed may have been Greeks; it remains debatable as to how far these works are to be regarded as Phoenician. Very popular among wealthy Phoenicians, these sarcophagi were not only found at Sidon but were also exported to Cyprus, Egypt, Antarados, Motye, Malta, Carthage, and Cádiz (Kukahn, 1955; Bühl, 1959). It is clear that by this period Phoenician art had spent itself and was quite derivative and secondary, simply employing the motifs and ideas of others. In Carthage the decadence of the relief, a genuine Phoenician tradition, could not have been more complete. By the 3d century B.C., it was reduced to a few simple subjects coarsely scratched within an aedicula (e.g., a priest worshiping, an upraised hand, the sign of Tanit, a bull, a sun disk or a moon crescent).

Metalwork. The discovery of large numbers of metal objects deposited as foundation gifts in the Temple of Resheph at Byblos (Level VII; ca. 2000 B.C.) demonstrates that the wealth of Phoenicia lay not merely in its control of the cedar, pine, and other timber forests of Lebanon and Anti-Lebanon but also in its exploitation of copper and gold deposits there and in Cyprus. Copper prospecting in Europe may well have been organized from Ugarit (Schaeffer, 1939–56, II, *Porteurs de Torques*). The Phoenicians' skill in goldworking, especially in the ingenious method of decorating a surface with dotted patterns of granulation, is evidenced on many fine votive objects in these deposits (Dunand, 1937–54, II, pls. 121–122, 134–137); this art, later passed on to the Etruscans, may be fairly claimed as a Phoenician invention — its secret afterward lost until modern times. From other finds at Byblos (Montet, 1928–29, pl. 63) it is clear that the Phoenicians had also mastered the technique, highly developed in Egypt, of inlaying stone and glass in *cloisons* of gold. Many statuettes from the foundation deposits at Byblos are covered or partly covered with gold foil (PL. 129). The fame of the Phoenician goldworkers and metalsmiths spread throughout Syria and Palestine (W. M. Flinders Petrie, *Ancient Gaza*, London, 1931–34, IV, pls. XIII-XX). From Ugarit come two splendid gold bowls with decorative scenes (Schaeffer, 1939–56, II, pls. 1–8). In the Iron Age, the influence of Phoenician metalworkers is definitely reflected in archaic Greek jewelry from Rhodes (Becatti, 1955). Fine examples of Phoenician jewelry of the 7th century B.C. have been found at Aliseda (J. R. Mélida y Alinari, *Tesoro de Aliseda*, Madrid, 1921) and on the northwest border of Persia [A. Godard, *Le trésor de Ziwiyè (Kurdistan)*, Haarlem, 1950]; and of the 6th century, from Sardinia, Malta, and Cyprus, (F. H. Marshall, ed., *Catalogue of Greek and Roman Jewellery . . . in the British Museum*, London, 1911, pls. XXIII-IV; Becatti, 1955). Some of these groups, however, may be Carthaginian work (PL. 128; cf. P. Gauckler, *Nécropoles punique de Carthage*, Paris, 1915). A fine gold bowl found near Agrigento, showing passant bulls, appears to be an example of a local Sicilian-Phoenician school of the 7th century B.C. (Marshall, op. cit., pl. LXXIII). Reflections of Egyptian art, unmistakably indicating the presence of strong Phoenician influence, have been found. In the Homeric poems the Sidonians were the most skilled metalworkers. To them are to be attributed at least some of the famous series of bronze bowls found at Nimrud (anc. Calah; A. H. Layard, op. cit.), silver or silver-gilt bowls found hitherto only in Cyprus and Italy (E. Gjerstad, "Decorated Metal Bowls from Cyprus," *OpA*, IV, 1946, pp. 1–18), and bronze bowls found in Greece. Other forms of metallic Punic jugs and bowls have been found in Spain (E. Cuadrado Díaz, "Los recipientes metálicos llamados 'Braserillos punicos,' " *AEA*, XXIX, 1956, pp. 32–84). Hiram, master craftsman of Tyre, was famous (I Kings 7:14; II Chron. 2–3) in the 10th century B.C., especially for his great bronze castings of the paired pillars destined for the Solomonic temple. Of the colossal "brazen sea" and wheeled bases (*mekônôth*) used in the same temple, miniature versions have been recognized in wheeled "braziers" from Cyprus (A. Furtwängler, "Über ein auf Cypern gefundenes Bronzegerät," *Kleine Schr.*, II, Munich, 1913, pp. 298–313; W. Lamb, *Greek and Roman Bronzes*, London, 1929, pl. XII).

Carving in wood and ivory. There is no doubt that the Phoenicians, with the best timber resources of the ancient Near East available to them in Lebanon, were skilled woodworkers and carvers. This is attested by their renowned work on both the Temple and Palace of Solomon (I Kings 7:1–14). The stone sarcophagi reflect wooden prototypes (sarcophagus of Aḥīrām; PL. 130). Only a single piece recognizable as Phoenician wood carving has survived — an 18th-dynasty unguent bowl from Abu Gurob in Egypt (C. Boreux, ed., *Musée du Louvre: Département des antiquités égyptiennes, Guide-catalogue sommaire*, Paris, 1932, pl. LXXVI).

The carving of ivory, closely linked to wood carving, became more and more a Phoenician specialty, perhaps even during the Egyptian New Kingdom (VIII, PL. 237; XIII, PLS. 362, 363). At Ras Shamra Schaeffer found a fine collection of 16 carved panels with religious or ritual scenes forming the decoration of a bed, as well as part of a carved ivory table (C. Schaeffer and G. Chenet, "Les fouilles françaises en Syrie: Nouvelles découvertes importantes à Ras Shamra," *L'illustration*, no. 4631, 1931). In the 9th and 8th centuries B.C. the fame of such furniture decorated with ivory was widespread; collections of ivory panels and other shapes carved in a Phoenician style have been found in Samaria (evidently belonging to the Palace of Ahab), Arslan Tash (presumably made for Hazael, king of Damascus), Zincirli, Khorsabad, and Nimrud, whither the Assyrian kings had carried them off from the west in large quantities, probably as booty. Isolated pieces also found their way to various sites in Greece and Etruria [F. Thureau-Dangin et al., *Arslan Tash*, Paris, 1931; R. D. Barnett, "The Nimrud Ivories and the Art of the Phoenicians," *Iraq*, II, 1935, pp. 179–210 (id., R. D. Barnett, *PEQ*, LXXI, 1939, pp. 169–73); E. Kunze, "Orientalische Schnitzereien aus Kreta," *AM*, LX-LXI, 1935–36, pp. 218–33; J. W. Crowfoot and G. M. Crowfoot, *Early Ivories from Samaria*, London, 1938; G. Loud and C. B. Altman, *Khorsabad*, II, Chicago, 1938; G. Loud, *The Megiddo Ivories*, Chicago, 1931; R. D. Barnett, *Fine Ivory Work*, in C. Singer et al., eds., *A History of Technology*, I, Oxford, 1954, pp. 663–83; Decamps de Mertzenfeld, 1954; H. J. Kantor, "Syro-Palestinian Ivories," *JNES*, XV, 1956, pp. 153–74; Barnett, 1957].

The pillaging of the Phoenician cities by Assyrians and Babylonians and the subsequent inaccessibility of the ivory supplies led to the extinction of the Phoenician school of ivory craftsmen. In the far west, however, their traditions were continued probably into the 5th century B.C. in Spain. The ivories found in the Guadalquivir Valley probably emanated from Carthage (*Early Engraved Ivories in the Collection of the Hispanic Society of America, from Excavations by G. E. Bonsor*, New York, 1928).

Textiles. The Phoenicians were famous for their textiles; the finest linen (*buṣ*) came from Byblos, Tyre, and Berytus (Lutz, 1927). Maresha (Marisa, Tell Sandakanna), in southern Palestine, where a Phoenician settlement and Phoenician influence lasted into Hellenistic times, was also a center for fine linen, a commodity that was extremely popular in Egypt [I Chron. 4:21; J. P. Peters and H. Thiersch, *Painted Tombs in the Necropolis of Marissa* (*Marêshah*), London, 1905]. The name "Motye" itself means "loom" in Phoenician. The famous Tyrian purple dye made cloth from Tyre known all over the Mediterranean. Patterned and decorated textiles — it is not clear whether the designs were woven or embroidered — were a speciality of the Sidonian women (Homer, *Iliad*, VI, 289–94; *Odyssey*, XV, 415–18).

Glass and glazing. Both glass and glazing were known in the ancient Near East from very early times. An alkaline glaze on stone or powdered quartz was used. A lead glaze used on pottery vessels first appeared in the 17th century B.C., at Tell Atchana, on the north Syrian coast. Frit, a partially fused powder not fully vitrified, was made from the same material as glass and was used to fashion amulets, seals, and blue or green vessels, probably in the 18th dynasty (1570–1372 B.C.) in Syria, Mesopotamia, and Egypt. Most vessels were made by the sand-core process, with a decoration of light-colored wavy lines on a dark blue ground. In the Iron Age, from the mid-8th century B.C. to Hellenistic times, glass manufacture was expanded, and the main center of such activity was very probably Phoenicia — the Phoenician traders certainly being active in its distribution (D. B. Harden, 1956; P. Fossing, *Glass Vessels before Glass Blowing*, Copenhagen, 1940).

Many glazed amulets, scarabs, and small figurines of Egyptian or Egyptianizing type have been found in west Phoenician sites of the 8th–5th century B.C.; some of these were probably made at Carthage, where molds have been found (P. Gauckler, *Nécropoles puniques de Carthage*, Paris, 1915, pp. 328, 512; P. Cintas, *Amulettes puniques*, Tunis, 1946, p. 27). Other finds from Greek lands, however, may have been made locally in partly Phoenician workshops.

The main output of the glassmakers of the 8th and 7th centuries B.C. was unguent or perfume flasks of opaque glass, decorated with wavy or zigzag threads; these were a common item in Etruria and Rhodes. In the 6th century the range of shapes was extended. Some believe that such flasks were made in Egypt under the 26th dynasty, whereas others favor Syria or elsewhere as their place of manufacture. The making of mold-pressed or molded glass (i.e., glass vessels molded over a usually positive clay mold) as certainly practiced in Syria in the early Iron Age, though it may have been of Egyptian origin [R. J. Forbes, *Glass* (*Studies in Ancient Technology*, v), Leiden, 1955, pp. 110–231].

The invention of glass is attributed by Pliny to the Phoenicians at the "River Belus" (*Naturalis historia*, XXXVI, 190) — probably as a reflection of the fact that in about 1500 B.C. the manufacture of glass vessels suddenly developed in the Near East and that the Phoenicians played an important part in this. Strabo (XVI, 2.25, cap. 758) localizes the discovery more exactly, as having occurred in the sands between Ace and Tyre, which he says were exploited by the Sidonians. This is not impossible, but as yet there is little evidence to determine the part played by the Phoenicians in glassmaking before Hellenistic times.

Glass blowing was apparently invented in the Phoenician coastal area in the 3d century B.C. or, according to some, in the 1st century B.C. Some of the craftsmen were Syrian, others were Jews. A further development in the 3d–2d century B.C. was mold blowing (though it is possible that it preceded the art of free blowing), with the glass being blown into bipartite or tripartite hollow molds of wood or clay. Many of the artisans who signed these pieces are known to have been "Sidonian," that is, Syrians or Jews.

Richard D. BARNETT

CARTHAGINIAN AND OTHER PUNIC ART. *Carthage.* Unlike the colonies established by the Greeks, the Phoenician colonies for a long time remained simple trading stations with small, transient populations. They had no local industries at first, and manufactured products were imported from Phoenicia. The oldest objects found are terra-cotta vases linked stylistically to the Cycladic sub-Geometric, dating from the second half of the 8th century B.C. These constitute a part of the votive deposit of the Cintas chapel in the Tophet (sacrificial furnace). At the beginning of the 7th century B.C. the influx of immigrants from Tyre to Carthage created a local industry, that can be studied from the grave goods and votive deposits of the Tophet. Thereafter, among the local products were household pottery, including red glazed ware, with very simple decoration derived from the Geometric style; terra-cotta masks (grinning demons, *kouroi*, a smiling goddess; PL. 136), which have no counterpart in Phoenicia itself but are found in various parts of the Aegean world; the very coarse statuettes of the Tophet, which seem to be in the Aegean tradition; shells of ostrich eggs cut in the form of masks or bowls, with painted geometric decoration; and bronze razors in the shape of hatchets, without decoration.

During the 6th century B.C. Carthaginian local industry continued to develop but did not participate in the manufacture of luxury goods (e.g., carved ivories, glassware, gold- and silversmith's work), with the exception of small pieces of jewelry. The first sculptured monuments made their appearance in the Tophet: the cippi in the form of a throne bearing a double-headed ax, and Egyptianizing *naiskoi* enclosing a double-headed ax or Geometric idol. Both of these forms are in the Phoenician tradition. The red pottery had vanished by the end of the 6th century B.C.

Until the 5th century B.C. Carthage remained essentially faithful to Phoenician themes, though it was more subject to Aegean-Hellenic influence than was the Phoenician homeland. Cut off from the East at this time, however, Carthage affirmed its artistic originality. This is revealed particularly in the field of religious sculpture, where a radical reform gave birth to new modes of expression. From the end of the 5th century there appeared in the Tophet, together with the Egyptianizing *naiskoi*, steles in the form of obelisks; the schematic representation of a Greco-Asiatic temple sheltering images of deities (mostly derived from the symbolic representations of Aegean religion); the symbol or sign of the bottle, derived from the violin-shaped idol; and the sign of Tanit (a triangular or trapezoidal shape surmounted by a disk and a bar with raised ends; PL. 133), in imitation of the bell-shaped idol with upraised arms. The same radical reform undoubtedly resulted in the disappearance of the terra-cotta masks. At the same time, razors were covered with decorative depictions of myths whose protagonists were sometimes deities of Egyptian appearance and sometimes gods or heroes of Hellenic appearance. A bronze industry developed at this time (oinochoai with Egyptianizing decoration), and with the return of prosperity jewelry appeared once more.

The 4th and 3d centuries B.C. witnessed a Hellenization of Punic art, which gradually became a provincial Hellenistic style characterized by archaicism and Orientalism. During this period Carthage maintained regular contacts with Sicily and Magna Graecia and subsequently became integrated with the Alexandrian world. Harbors, fortifications, and monuments built with Hellenistic techniques were essential features of the city. The architectural fragments revealed by excavations are mostly in a Hellenistic style with archaicizing tendencies. The houses that have come to light are decorated with small, fanciful columns, mural paintings, and trompe-l'oeil stucco reliefs similar to coeval houses in Delos. The floors are covered with marble inlays (Punic quarter on St. Louis' hill, city of Kerkouane on Cape Bon, Tunisia). Metalwork and terra-cotta sculpture reproduced Sicilian and Italian models in a composite style in which archaic themes survived. Greek artists opened sculpture workshops in Carthage, in which they produced sarcophagi, adorned with reclining figures or in the form of a Greek temple with painted decoration, such as the sarcophagi from the *rab* necropolises. Even the gods assumed the aspect of the Olympian deities, as can be seen in the enthroned Tanit (Carthage, Mus. National). Potters imitated the Campanian black-slip ware; jewelry in general was of Greek type, and only the amulets continued the Egyptian tradition.

The question of the relationship between a colonial art and the art of the mother country is difficult to resolve in the case of Carthage. The earliest salient characteristic of Punic art was its austerity, a tendency accentuated by great material poverty. Later the impact of outside traditions was felt; among those influences soon obliterated, however, was that of distant Mesopotamia. The art of Egypt, a land with which direct contacts were maintained, had a strong and enduring effect on Punic art. Although subject to temporary decline, Egyptian influence demonstrated remarkable periods of revival, for example, at the end of the 5th century B.C. and the beginning of the 4th, and again in the 3d century, with a strong Alexandrian imprint. Aegean and Greek influences were quite consistently paramount, but Carthage ignored Attic classicism both for political reasons and because of spiritual incompatibility. The Hellenic influence appears to have come through Cyprus, the southern Aegean, Asia Minor (e.g., Mausoleum of Thugga, the masterpiece of Punic architecture in its period of decline, directly related to the Anatolian tombs), and, of course, Sicily and Etruria, areas with which Carthage maintained continuous and close political and economic relations. The original quality of Punic art resulted from the severity and austerity demanded by the priestly caste — demands that gave birth to the symbolic abstractions of the 5th century B.C.

It is a curious fact that only after the fall of Carthage did Phoenician art find, among the Numidians, creative agents of great artistic power and individuality (ca. 150 B.C.–A.D. 150). Numidian architecture, like Carthaginian, derived inspiration from various sources: from the Punic mausoleums of Thugga, the Medracen, and the Tomb of the Christian Woman (I, PL. 18), in which Punic influences blended with local traditions, as well as from the Souma of Kroubs (I, PL. 18) and the monumental altar of Kbor Klib, which are purely Hellenistic in character. Nonetheless, Numidian architecture retained a great purity of line and was majestic in effect. The sculptors expressed the Phoenician cosmology with a kind of fantastic surrealism, in which symbols, anthropomorphic personifications, phytomorphic forms, and architectural decorations were all mingled (e.g., stele of Ghorfa, I, PL. 19; stele of Ain Barchouch; see AFRICAN-ROMAN ART).

Spain and Morocco. Although according to tradition the Tyrian settlements of Gades (Cádiz) and Lixos date from the end of the 2d millennium B.C., no finds have been made in Spain or Morocco that are earlier than the 7th century B.C. Between the 7th and the 5th centuries B.C., numerous precious objects were imported from Phoenicia; examples have been found in the native necropolises and in the treasure of La Aliseda (buried ca. 3d cent. B.C., but containing objects from as early as the 6th cent.). There are also jewels and vessels of gold, silver, and bronze; oinochoai of engraved glass; and ivory combs and plaques with incised decoration, found at Osuna and Carmona (6th–5th cent.). In the 5th century there appears to have existed at Gades a local industry that remained under the strong influence of Phoenicia, as exercised across a route passing through Cyprus, Malta, and Sicily. The output of Gades included anthropoid sarcophagi influenced by Greek classicism, glazed red ceramic ware (which persisted until the 4th century), and jewelry in the Egyptian manner.

Until the 3d century B.C. typical Punic objects such as terracotta idols similar to those of the Tophet in Carthage, masks of grinning demons and the smiling goddess, and decorated shells of ostrich eggs were found only in the Balearics, where Phoenician statuettes of a type unknown in Carthage were also found (PL. 135). Carthaginian hegemony, on the other hand, is attested during the Barcide period (237–207 B.C.) by the existence of Punic necropolises in Spain (Gades, Villaricos, La Albufereta, Ebusus, etc.) identical with those of Africa.

Iberian art, which underwent a remarkable development during the 3d century B.C., undoubtedly owed far more to Greece than to Phoenicia. Nevertheless, Eastern influence is clearly revealed in the ornamentation of jewels covered with rosettes and floral designs, pendants, statues and figurines (the "Great Lady" from Cerro de los Santos, Albacete, IX, PL. 408; "The Lady of Elche," IX, PL. 409; statuettes of Puig d'es Molins), and in the embroideries in which plant motifs are intermingled with religious emblems (vestments covering tunics of Puig d'es Molins deities). Iberian architecture may have been partially derived from that of Carthage, although the fragments of capitals or reliefs scattered throughout Spain are too scant to provide substantial proof.

Sicily and Sardinia. Unlike Carthage, Phoenician Sicily never became an independent artistic center but served merely as a link between the Cypro-Phoenician and Carthaginian worlds. At Motye, therefore, the grave goods, pottery, and ex-votos of the Tophet are identical with those from Carthage of the same period (7th–6th cent. B.C.). From Solunto, on the other hand, comes a statue of an enthroned deity flanked by winged sphinxes in an Orientalizing Phoenician style. From the Palermo region (at Cannita) have come two 5th-century anthropoid sarcophagi related to those from Sidon. Moreover, the stamp of Phoenicia was strong enough to survive the fall of Carthage, and even during the Roman period the cippi from Lilybaeum continued to be decorated with Phoenician emblems.

Like the Balearics, Sardinia was subjected to profound and lasting Carthaginian influence from the 6th century B.C. on. As at Carthage, Sardinian tombs have yielded masks of grinning demons and the smiling goddess, razors in the form of hatchets, and the decorated shells of ostrich eggs. The island was also subject to Cypro-Phoenician influence, transmitted through Sicily (steles of Sulcis and Nora in the form of Egyptianizing *naiskoi* or of Greco-Asiatic temples). This influence was particularly strong during the 5th century B.C., when the Carthaginians retreated to Africa, and masks and figurines derived from Phoenician art were influenced by classic Greek art. Although the local Sardinian craftsmen, sculptors, and potters adopted the esthetic principles of the Phoenicians and Carthaginians, they remained faithful to ancestral traditions, which gave an individual character to their works.

Colette and Gilbert-Charles PICARD

BIBLIOG. *a. Historical background*: S. Gsell, Histoire ancienne de l'Afrique du Nord, I–V, Paris, 1913–28; G. Contenau, La civilisation phénicienne, Paris, 1949; P. Cintas, Contributions à l'étude de l'expansion carthaginoise au Maroc, Paris, 1954; G. Picard, Le monde de Carthage, Paris, 1956; P. K. Hitti, History of Syria, 2d ed., London, 1957; R. Carpenter, Phoenicians in the West, AJA, LXII, 1958, pp. 35–53; B. H. Warmington, Carthage, New York, 1960. *b. Nature and development of Phoenician-Punic art*: P. Montet, Byblos et l'Egypte, 2 vols., Paris, 1928–29; P. Montet, Les reliques de l'art syrien dans l'Egypte du nouvel empire, Paris, 1937; C. Schaeffer, Ugaritica, 3 vols., Paris, 1939–56; J. Vercoutter, Les objets égyptiens et égyptisants du mobilier funéraire carthaginois, Pars, 1945; R. Dussaud, L'art phénicien du IIe millénaire, Paris, 1949; H. Bossert, Altsyrien, Tübingen, 1951; H. Frankfort, The Art and Architecture of the Ancient Orient, Harmondsworth, 1954, pp. 133–201; W. Stevenson Smith, Interconnections in the Ancient Near East, New Haven, 1965. *c. Influence of Phoenician art on archaic Greece and Etruria*: F. Poulsen, Der Orient und die frühgriechische Kunst, Leipzig, 1912; R. D. Barnett, Ancient Oriental Influences on Archaic Greece, in The Aegean and the Near East: Essays Presented to Hetty Goldman, Locust Valley, N.Y., 1956, pp. 212–38; T. J. Dunbabin, The Greeks and Their Eastern Neighbours, London, 1957. *d. Architecture*: M. Dunand, Fouilles de Byblos, 2 vols., Paris, 1937–58. *e. Sculpture*: E. Kukahn, Anthropoide Sarkophagen in Beyrouth und die Geschichte dieser sindonischen Sarkophagekunst, Berlin, 1955; M. L. Bühl, The Late Egyptian Anthropoid Stone Sarcophagi, Copenhagen, 1959. *f. Minor arts*: H. F. Lutz, Textiles and Costumes among the Peoples of the Ancient Near East, Leipzig, New York, 1923; P. Cintas, Céramique punique, Paris, 1950; C. Decamps de Mertzenfeld, Inventaire commenté des ivoires phéniciens, Paris, 1954; G. Becatti, Oreficerie antiche, Rome 1955; D. B. Harden, Glass and Glazes, in C. Singer et al., eds., A History of Technology, II, Oxford, 1956, pp. 311–46; R. D. Barnett, A Catalogue of the Nimrud Ivories, London, 1957.

Richard D. BARNETT

Illustrations: PLS. 127–136; 3 figs. in text.

PHOTOGRAPHY. This article will not deal extensively with the mechanical and photochemical processes involved in making photographic images, nor will it discuss specific technical aspects and uses or the notable documentary value of photography. Used by an artist who succeeds in conferring on the image the mark of his own creative imagination, photography can become an art form. In this sense, what is essentially a means of objective reproduction can become a medium of subjective expression that ranks with other accepted artistic techniques.

In its early years photography was, in fact, considered an art; very likely, in that stage no one would have predicted that it was to become a universal means of documentation and communication almost as indispensable as writing. The careful composition and evocative lighting, as well as the use of symbolism, in early experimental photographs clearly indicate that these were considered art works. Moreover, just as the aims and ideals of painting and sculpture were destined to undergo a complete transformation in the course of a century, so also were those of photography, which became a major factor in stylistic change in the other arts.

In the 19th century the highest function of art, especially in academic and realist circles, was thought to be to reproduce reality. Photography, which seemed ideally adapted to that purpose, began to develop just when artists were becoming aware of the extent to which attempts to reproduce reality (viz., the imitative concept in painting) might differ from the true optical image. Photography could thus constitute for them a sort of empirical gauge by which they might measure the accuracy of their own optical perceptions. In so far as it was used in a totally realistic manner, the photography came to be regarded as the antithesis of art. As its techniques were refined, however, photography broadened its range of effects as well as its intent, and there appeared examples of nonrepresentational photographs that, whether or not consciously conceived as an artistic expression, take their place alongside other contemporaneous manifestations of a new kind of imaginative vision.

SUMMARY. Historical development (col. 313). Photography in the study of art (col. 317). The use of photography by artists (col. 318).

HISTORICAL DEVELOPMENT. In the 18th century, artists used the camera obscura to capture images of external objects (X, PL. 387). The light reflected from the objects passed through a hole (with or without a lens) into a darkened box (camera obscura) and formed an image on a glass at the back of the camera. The first to succeed in fixing by chemical processes the image thus obtained, Joseph Nicéphore Niépce (1765–1833) discovered that a chemical substance, bitumen of Judaea, became insoluble under the action of light and could be used to prepare sensitized plates, which were set in the camera obscura instead of the glass. As early as 1826 he was able to fix the camera's image, but this required an exposure time of six to eight hours (PL. 137).

It was not until after Niépce went into partnership with Daguerre in 1829, however, that the first "artistic photograph" was made. Louis Jacques Mandé Daguerre (1787–1851) had already experimented with a similar process, but it was not until some years later that he accidentally discovered a rapid means of developing exposed plates; this new process reduced exposure time to a comparatively brief interval. In 1839 Daguerre's development process was communicated to the Academy of Sciences in Paris, and thereby came into the public domain (X, PL. 389).

Daguerre himself had intended his discovery as an aid to the painter's craft. Many examples of daguerreotypes can still be seen; and in some respects they remain unsurpassed for quality of materials and the tonal clarity of the image. These plates have an artistic quality superior to that of many contemporaneous engravings.

William Fox Talbot (1800–77) made further improvements in photographic technique. In 1833 at Lake Como, after unsuccessful attempts at using the camera lucida to sketch landscapes, Talbot called upon his previous experience with the camera obscura, which he had employed to project images onto paper, and he conceived of the possibility of fixing the projected image permanently. Back in England, he experimented first with fixing the imprint of leaves and laces placed directly on paper prepared with silver chloride. Having achieved some success, he used this process to fix the image captured in a camera obscura. Talbot's productions, which he called "calotypes," were an improvement over daguerreotypes, for, although the image was not as distinct, they could be made without expensive and cumbersome silver or copper plates. His process also made possible multiple reproductions of the same image (i.e., making of many positives from one negative).

These elementary notes on the early history and techniques of photography are essential in following the evolution of its expressive means, which broadened as its mechanical means were refined. For example, Talbot's "photogenic drawings" — that is, impressions obtained by direct contact, without the use of the camera — anticipated the abstract photograms of a century later, which were considered works of art. As has already been noted, in its early phases, photography was not considered important as an instrument of documentation or empirical research but was linked theoretically with drawing and painting, as their complement and auxiliary.

This concept of photography as an artist's tool is illustrated by the portraits made about 1843 by the Scottish painter David Octavius Hill, in collaboration with the chemist Robert Adamson. Commissioned to do a huge painting representing a group of more than 450 Scottish clergymen, Hill enlisted the aid of photography in executing the work. His photographic portraits are superior in quality to the finished painting and are of much greater interest to the modern viewer. Though such portraits, as well as the photographic still lifes, landscapes, and interiors of the period, have gained an aura of 19th-century pathos that is attributable more to fashion than to a deliberate style, they nonetheless compare favorably with contemporaneous oil painting and engraving. In fact, about 1850, an "art" photograph was considered analogous to a landscape or sentimental genre painting. A photograph such as Henry Peach Rosinson's celebrated *Fading Away* (1858), a combination print that joined images from several negatives, is an allegorical composition of clearly symbolic and literary intent. Photographers in the Pre-Raphaelite period sometimes achieved absurd results in their overblown, allegorical combination prints.

Another early photographer, Julia Margaret Cameron, made discerning and evocative portraits of her notable circle of friends; she pioneered in the use of the close-up and carefully controlled lighting to enhance her observations of the personality of the sitter (PL. 138).

In the mid-19th century a realistic school of photography emerged, with the intention of recording the appearance of landscapes (often in exotic locales) and documenting important events of the day. This photographic reportage clearly affirmed man's newly developed power to fix transient moments or images in time and thus to give an enduring reality and significance to something that is seemingly anecdotal. Further noteworthy examples of this journalistic realism are the photographs of the Crimean War by Roger Fenton and the more than 7000 pictures by Mathew Brady and his colleagues (who were the actual source of many of the renowned "Brady" photographs) of every aspect of the American Civil War (PL. 137). After the war, many of these same photographers went on to record the scenery and various events in the history of the expanding American West.

The romantic-symbolic tendency characteristic of the early years of photography was succeeded by a realistic-empirical trend as exemplified in a speech given before a photographic society in London in 1886 by Peter Henry Emerson, who based his photographic theory on scientific postulates. (The optical theories of Hermann von Helmholtz greatly influenced Emerson's conclusions.) Emerson maintained that the aim of art is imitation of the effect of nature on the eye, offering as examples the works of artists such as Constable and Corot. According to Emerson, photography was superior to engraving in accuracy of rendering perspective, and (at that time) inferior to painting only in its lack of color. (This was an era in which the photo-engraving, which has come to be employed primarily as a reproduction technique, was appreciated on the level of the true artistic engraving.) Emerson also advocated using images that were slightly out of focus, as being most like those seen by the human eye, and in the decade 1880–90, among the most disputed matters in photographic practice were "soft focus" and the merits of retouching. Although, at first one of the most impassioned partisans of the artistic photograph, as being a copy of nature that is more faithful than a painting, Emerson later reversed his views, and in 1891 he wrote: "Photography is not an art but solely a means of knowing reality." If he erred, it was in his apparent belief that art was to be equated with painting, instead of granting that the "art" of photography might lie in just those qualities that differentiated it from painting.

During the early years of the 20th century Eugène Atget recorded the monuments of Paris and scenes of Parisian daily life in photographs that greatly influenced the development of documentary photography (PL. 139).

In the first decades of the 20th century Alfred Stieglitz (1864–1946) exercised a strong influence in the development of photography, both as head of the Photo-Secession group (1900) and as director of the magazine *Camera Work* (1902–17). His adherents championed a return to the "pure" — artistic in its intended effect and techniques — as opposed to the "pictorial" photography; Photo-Secession defended the soft-focus technique but admitted no manipulation of the plate after exposure, hence opposing all the alterations possible in developing.

Another American photographer of great repute in the early 20th century, Edward Steichen, is noted for photographs of Rodin and his sculpture (PL. 138) and for vivid portraits such as that of John Pierpont Morgan (1903). He often used soft-focus effects, as did Alvin Langdon Coburn and Clarence H. White. In 1905 Steichen and Stieglitz jointly founded New York's Little Gallery of the Photo-Secession, later known as "291," dedicated to the exhibition of avant-garde photographs and art. A half century later Steichen organized a widely acclaimed exhibition, "The Family of Man," at the Museum of Modern Art in New York (1955).

Improvements in enlarging processes and lens speeds brought enlarged detail and arrested movement within the range of striking and artistic effects possible to photography. Paul Strand, who photographed minute details of plants and natural objects, spoke of "liberating the photograph from the domination of painting." In other works he conveyed the mood and character of his subjects in lyrical studies of people and places (PL. 140).

Many artist-photographers, ruling out chance in their work, hold that in order to achieve true artistry in the medium the final print must be precisely envisioned. Edward Weston was one of these; in photographing details of natural forms and landscapes, he created compositions of abstract beauty and great naturalistic force. Using what might be termed the "straight approach," he insisted that the photographer should meticulously plot his composition in advance, fully visualizing the final print before releasing the shutter. From this attitude followed his abhorrence of enlarging or cropping prints, for he felt that enlargement resulted in a loss of precision in the final print and that cropping or trimming indicated a failure of the photographer's creative vision.

Throughout the first years of the 20th century photography kept pace with avant-garde movements in painting. As early as 1913, Alvin Langdon Coburn presented a series of photographs in which the perspective was so distorted as to achieve abstract effects similar to those found in cubist painting. The photographer himself acknowledged cubism's influence on his work. In 1917 Coburn created his first "vortographs," true abstract photographs, achieved by using mirrors joined to form a triangle with which the images of objects captured by the camera lenses were reflected in entirely new patterns. Christian Schad, creator of the "Schadograph" (1918), participated in the Zurich Dada movement and was among the first to make photograms by placing cut-outs and other objects directly on sensitized paper; such photograms were not unlike cubist collages in effect. Technically, these might be considered repetitions of Talbot's photogenic drawings; but whereas Talbot prided himself on having achieved an identification with reality, the Hungarian artist László Moholy-Nagy (q.v.), another innovator in abstract photographs, proclaimed that the photogram opened vistas hitherto unknown and constituted a completely dematerialized medium. Moholy-Nagy's *Vision in Motion* (1947) and the *New Landscape* (1956), written by his friend and disciple Gyorgy Kepes, are notable treatises in this vein; through their ideas a whole new universe — the universe of the micro- and macroscopic, the telescopic, radioscopic, and stroboscopic — was borrowed from science and given to photographers, as to all other artists, as a rich source of formal motifs.

In 1921, independently of Moholy-Nagy, Man Ray introduced his rayographs and photograms, in which not only the silhouettes but also the shadows of three-dimensional objects were projected onto sensitized paper. The first rayographs were collected in an album, *Champs délicieux* (1922), published with a preface by Tristan Tzara.

From the time of World War I, artistic photography has grown in volume and achievements, accompanied by the emergence of new forms and new personalities. The efforts of photographic "chroniclers" such as Henri Cartier-Bresson, Werner Bischof, Margaret Bourke-White, Dorothea Lange, Walker Evans, Barbara Morgan, Berenice Abbott, and the painter Ben Shahn (q.v.) combined to make photography heir to the historical and anecdotal functions once fulfilled by painting. Among those who have sought imaginative new graphic-pictorial experiences, Harold Edgerton and Gjon Mili have made use of the stroboscopic photograph; Andreas Feininger, author of the *Creative Photographer* (1955), believes that "creative photography" takes on the quality of art according to the photographer's capacity for selection; Otto Steinert maintains that there is an artistic quality inherent in the photographic medium, valid in itself and not in its pictorial content or evocative elements.

Other photographers who have done distinctive work since World War II are Harry Callahan (striking isolation of details of common scenes and objects, imaginative use of multiple exposure; PL. 140), Yousuf Karsh (portraits, with telling use of close-up techniques), Robert Capa (documentary photography), Ernst Haas (expressive color compositions), Aaron Siskind and Paul Caponigro (formal abstraction through close-ups of natural objects; PL. 140), Ken Heyman (realistic straight photography; PL. 140), and Ed van der Elksen (romantic narratives). Worthy of mention also are Robert Frank, Minor White, Richard Avedon, Yuichi Midorikawa, Toshiji Mukai, Bill Brandt, Max Scheler, Herbert List, Franco Grignani, Paolo Monti, Luigi Veronesi, Wayne Miller, Irving Penn, Fulvio Roiter, and Erwin Blumenfeld.

From this brief survey it may seen that the successive technical stages of photography correspond to as many esthetic phases. The principal steps in this evolution may be summed up as follows: the fixing of the static image with the daguerreotype and calotype and other techniques employing long exposures; the invention of the shutter, which made it possible to capture the fugitive scene, whether in the straightforward reportage of events or in the high-speed stroboscopic photograph; discovery of the creative possibilities of detail and enlargement, and the increasing importance of "framing"; the transition from the evocative symbolic-allegorical to the realist photograph, followed by the dispute between the purists and the pictorialists; the use of the camera or of sensitized paper in creating abstract compositions, wholly without narrative or realistic elements; and finally, the emergence of color, which radically altered the black-and-white world of the 19th-century photograph, especially in relation to painting.

The abstract category, which has developed remarkably, may be subdivided in the following manner: (1) the true photogram, in which objects are placed on a sheet of sensitized paper and so illuminated that their silhouettes are fixed on the paper without the intervention of a camera; (2) the so-called "tabletop" system, which also uses the photogram technique but which alters forms by manipulation of mirrors and refracting crystals; (3) the photomontage, which uses multiple exposures fused into a single image; (4) the true abstract photograph, in which the subject may be modified through use of distorting lenses, so that the final effect can be visualized before the exposure is made; (5) finally, all those techniques through which the normal results obtainable with the camera may be altered either before or during the development process — toning, printing with special techniques such as solarization, enlargement of the grain while developing to produce reticulations, relief effects obtained by printing positive and negative together and slightly off-register, and chemical manipulation of photosensitive gelatines.

These and countless other techniques are artistically valid only in so far as the photographer is creative and sensitive in his use of them. Mechanical changes may accidentally result in an interesting or even an excellent picture; but only that which has been preconceived, that is, planned thoroughly and determined by the author, can be considered truly artistic. This touches on a controversial point in what might be called the esthetic of photography: in the creation of photograms and abstract photographs, there undoubtedly figures an element of chance — as in certain types of modern painting, in what has sometimes been called "random art." This author holds that, in the case of photography, this element of chance may be considered an artistic means; the intervention of a factor extrinsic to the will of the photographer, modifying the formal and formative aspects of the final work, is an almost inevitable concomitant of photography. Perhaps this is a concession that man must make to the machine, in exchange for its new expressive possibilities. As was the case with design, advertising, and prefabricated architecture, photography developed and entered into the realm of the arts by merit of technological progress and the intervention of artificial, sometimes mechanical means; however, it has been successfully transformed into a valid medium of artistic expression, presenting new and original creations.

Gillo Dorfles

Photography in the study of art. As an instrument of study — particularly for the study of art — early in its development the photograph began replacing engraved reproductions of art works, which, alone or as text illustrations in catalogues, had become increasingly popular since the 16th century (see reproductions). In one of the first applications of photography to the study of art history, Prince Albert, consort of Queen Victoria of England, made a photographic catalogue of Raphael's works.

An engraving could never be an entirely faithful reproduction of the original art work; a much more exact visual record is obtainable with photography, although tonal values may be altered in the monochrome image and the quality of the original materials may be lost. In response to the demand for accurate visual documentation of art works, firms specializing in photographic reproductions sprang up. Among the pioneers were Alinari (founded in 1854), Anderson, and Brogi in Italy; Giraudon and Bulloz in France; M.A.S. in Spain; Mansell in England; Hanfstaengel and Bruckmann in Germany. Almost all still active, these firms quite early formed important archives of negatives and published catalogues of the reproductions available. Responding to the dominant interests and requirements of scholars, their collections furnished illustrative material for the great scholarly publishing ventures of the time. They also served to propagandize works of art, making photographic reproductions of famous art works widely available, to be framed like original paintings.

Soon, in addition to these private commercial ventures, there were formed official or national photographic archives: for example, Archives Photographiques, Paris; Gabinetto Fotografico Nazionale, Rome; Archives Centrales Iconographiques, Brussels. Photographic archives were also set up in museums and libraries and by organizations devoted to the preservation of art works (q.v.). The list of photographic sources printed at the beginning of each volume of this encyclopedia offers some idea of the extent and variety of such archives. The collections of slides formed by various schools and universities and by private suppliers (e.g., Lichtbildverlag Stoedtner, Düsseldorf) represent another special category of photographic study aids. Art dealers (e.g., Durand-Ruel), connoisseurs, and scholars have also assembled notable private collections of photographic slides. These collections were prototypes of such photographic libraries as the Rijksbureau voor Kunsthistorische Documentatie at The Hague, the Frick Art Reference Library in New York, the Marburg Bildarchiv Foto, and so on. The collection belonging to the Warburg Institute in London has a special character, as it is arranged according to various subject areas (allegory, astrology, magic, etc.) and iconographic themes (liberal arts, virtues and vices, divinities, etc.). Princeton University's Index of Christian Art is an unparalleled example of specialization; it is a reportorium catalogued on index cards classified and cross-referenced according to multiple systems (material, subject, location, etc.).

Photographs of works of art have an essentially documentary value. The only surviving record of many lost works is a photograph; a photograph can also be an invaluable aid in restoration of a damaged work. For example, Mantegna's frescoes in the Church of the Eremitani in Padua, heavily damaged by bombing during World War II, have been reconstructed with the help of photographs made from existing negatives, enlarged to the dimensions of the original paintings (XII, pl. 140). The scientific aspects of photography have made possible investigations that are invaluable in restoring paintings; infrared, ultraviolet and X-ray examinations can reveal a *pentimento*, or a design that has been painted over, and record the metamorphoses of a given work in the course of its creation or subsequent restorations.

The documentary value of the photograph for art history, however, can never be absolute; despite technical progress, the work of art can never be exactly reproduced in it. When a painting is translated into black and white, the tonal relationships — the relations between the lights and darks — are almost inevitably altered, even though the use of filters on the lens may help to guarantee a closer approximation of the original. Nor can the color photograph reproduce the original perfectly, though color technique has been much improved. Even if the hues in the color reproduction were identical with those of the original, the reproduction employs dissimilar materials and does not have the variation in strokes and in thicknesses of the color layers, on which the action of light results in vibrations and optical effects necessarily different

The photograph, finally, may approach or may be less faithful to the original in quality as a result of the subjective response and sensitivity of the photographer, whose own viewpoint and interpretation may prevail over the original. A point in fact is the change in taste apparent in today's architectural photographs when compared with earlier examples: formerly a piece of architecture was isolated and presented as an object with clearly distinguishable component elements, with no sign of human life about it or even of the animated character given by changing natural light and weather conditions; in addition, the viewpoints and details selected were the most fundamental ones. Today it might almost be said that the photographer seeks to question the architectural work with his lens, capturing the effects of light upon it and its vital relationship to its setting, isolating details that might ordinarily pass unnoticed, discovering unexpected angles which the creator himself may not have considered but which can be of great interest, as in the case of aerial photographs. The danger in this more subjective approach, especially with three-dimensional works, is that arbitrary lighting and framing effects may be substituted for those arising most naturally from the expressive capacities of the work itself.

* *

The use of photography by artists. Himself a painter, Daguerre began investigating the camera obscura to improve upon it as a useful tool for his art. From the beginning, therefore, photography was closely connected with painting and was to become a significant influence in the development of 19th- and 20th-century painting. At first its immediate success frightened some painters, and in exaggerated reaction Paul Delaroche proclaimed that painting was dead from that moment on. Photography soon extended its range of possibilities into areas formerly reserved for the artist (e.g., portraiture, landscape), where it found a ready and lucrative market.

The actual effect of photography on the major 19th-century painters, however, was less ruinous than Delaroche had predicted. The artist found in photography a means of facilitating his work, particularly in portraiture, and of bringing to it new, striking, and individual perceptions and viewpoints, yet without damaging its integrity as painting. Eugène Delacroix made several entries in his diary describing how he drew from a model while a friend photographed it, and he even noted that he made sketches from the photographs themselves. Ingres sent his sitters to the photographer Nadar and used the photographs as a guide for his portraits. In his review of the Salon of 1861, Théophile Gautier wrote of the very evident influence of photography in the exhibition and commented on its beneficial effects: "The daguerreotype, which has been given neither credit nor medal, has nevertheless worked very hard for this exhibition. It has yielded much information, spared much posing of the model, furnished many accessories, background, and drapery, which had only to be copied and colored" (*Abécédaire du Salon de 1861*, Paris, 1861). The most revolutionary application of photography by artists in the second half of the 19th century, however, did not involve the officially accepted painters. Courbet's lighting effects in the *Funeral at Ornans* (1848; Louvre) were derived from daguerreotypes, and Corot's misty landscapes (III, PL. 459) were probably greatly influenced by the work of a group of landscape photographers at Arras, where Corot stayed for several years after 1852. Degas (q.v.), an accomplished amateur photographer, in certain of his paintings adopted some of the most unusual features of the candid street-scene stereoscopic photography that was popular after about 1860. His arresting compositions, figures cut at odd angles by the frame, steep perspective, and natural lighting effects can all be related to the influence of photography, although the artist's interest in Japanese prints should also be taken into consideration in this respect (IV, PLS. 134, 135, 137). Another aspect of his art was strongly affected by the sequence of action photographs of horses published by the Englishman Eadweard Muybridge in 1887 (*Animal Locomotion*). Degas made many drawings after the photographic studies of horses, with a clear effort to capture the effect of movement in the animals. In 1872 Muybridge was employed by the American Leland Stanford to investigate, through photographic methods, whether a galloping horse ever had all four hooves off the ground at once. His experiments lasted until 1877, when he finally achieved success by setting up twelve cameras along the track, with strings to trigger the shutters. As the galloping horse touched each string, the camera registered its movement at that particular point. The resulting photographic sequence showed that the horse did have all four hooves off the ground at the same time, but not in the position in which they were traditionally represented. Muybridge continued his motion studies in Philadelphia and worked closely with the American painter Thomas Eakins, who, by using only one camera with a rotating disk, exposed several views of a moving figure on the same plate (X, PL. 390). In his paintings Eakins reproduced the photographic record of movement, sometimes distorting the subject in order to convey better the idea of motion (*Fairman Rogers Four-in-hand*, 1879).

Jules Étienne Marey, a French physiologist, carried Muybridge's experiments even further. He painted a white stripe on the legs, arms, and torsoes of his models and exposed various stages of their movement in a sequence on a single plate. The resulting print showed a series of closely spaced "stick" figures caught in successive phases of motion across a plane. Marey's experiments had marked effects on painting, particularly at the beginning of the 20th century. Although Seurat was perhaps the first to emphasize repeated directional lines suggesting movement, in his painting *Le Chahut* (1889–90; XII, PL. 560), it was Marcel Duchamp in France and the futurists in Italy who developed them to their fullest expression. In his *Nude Descending a Staircase* (1913; IV, PL. 79), Duchamp repeated the figure in a progressive sequence of movement to give a graphic description of her descent. The futurist Giacomo Balla, in his *Abstract Speed* (1913; X, PL. 390), carried these experiments even further and discarded the representational object, painting only the directional lines of movement.

Throughout the latter part of the 19th and into the 20th century, apart from such adaptions of the photography of motion, artists continued to use photographs of static figures as models. Toulouse-Lautrec often had his model photographed while posing, so that he could continue to work in her absence. Cézanne used a photograph in painting the *Bather* (ca. 1885–90; New York, Mus. of Modern Art), and in 1920 Picasso made use of a photograph for his drawing of Renoir. Arshile Gorky, who was to become a leading abstract expressionist in the 1940s, worked for an extended period (ca. 1926–36) on a large canvas of himself and his mother that was based on an old photograph.

The free application of extraneous materials in painting, initiated in cubist collages and *papier collé*, marked a new phase in the artist's use of photography. Photographs themselves or photographic fragments were incorporated in painted compositions and in sculpture, a technique used extensively by the Dadaists and the surrealists in their *objets trouvés* and photomontages (e.g., Max Ernst's *Chinese Nightingale*, ca. 1920, Paris, Coll. Tristan Tzara). Realistic photographs were used within contexts that conveyed an exaggerated and distorted image of reality, thus pointing up the ironic contrast of the real-surreal and the abstract sought by the artists of these tendencies.

The practice of applying cut-up photographs to art works continues to be widespread. In his abstract *Woman* (1961) Willem de Kooning used a photographic cutout for the woman's head. The Venezuelan sculptor Marisol pastes a photograph of her own face or the face of a famous personality to the wooden heads of her sculpture. The "pop" artist Andrew Warhol has gone even further. Using a silk screen process, he transfers repeated images of a face taken from movie stills onto his canvas (e.g., *Silver Liz*), the photograph being the sole matter of the painting. Thus the photograph, which in relation to painting had started mainly as a practical aid to the representation and arrangement of figures on the canvas, has in the 20th century become a major pictorial constituent in itself.

BIBLIOG. G. Potonniée, Histoire de la decouverte de la photographie, Paris, 1925; J. M. Cameron, Victorian Photographs of Famous Men and Fair Women (introd. by V. Woolf and R. Fry), London, 1926; C. Sandburg, Steichen the Photographer, New York, 1929; C. W. Ackerman, George Eastman, Boston, 1930; F. Roh, ed., L. Moholy-Nagy: 60 Fotos, Berlin, 1930; H. Schwarz, David Octavius Hill, Master of Photography, New York, 1931; J. M. Eder, Geschichte der Photographie, 2 vols., Halle, 1932 (History of Photography, Eng. trans. by E. Epstean, New York, 1945); Brassai, Paris de nuit, Paris, 1933; L. Goodrich, Thomas Eakins: His Life and Work, New York, 1933; W. Frank et al., eds., America and Alfred Steiglitz, Garden City, N.Y., 1934; N. Newhall, Paul Strand, New York, 1934; M. Ray, Photographs by Man Ray, 1920–34, Hartford, Conn., 1934; Brassai, Volupté de Paris, Paris, 1935; P. Wolff, My First Ten Years with the Leica, New York, 1935; A. Genthe, As I Remember, New York, 1936; M. Fisher Hammer, History of the Kodak and its Continuations, New York, 1940; B. Abbott, A Guide to Better Photography, New York, 1941; B. Morgan, Martha Graham, Sixteen Games in Photographs, New York, 1941; H. Gernsheim, New Photo Vision, London, 1942; R. Lecuyer, Histoire de la photographie, Paris, 1945; N. Newhall, Paul Strand: Photographs 1915–1945, New York, 1945 (cat.); Brassai, Trente dessins, Paris, 1946; Y. Karsh, Faces of Destiny, New York, Chicago, 1946; R. Meredith, Mr. Lincoln's Camera Man: Mathew B. Brady, New York, 1946; M. Newhall, The Photographs of Edward Weston, New York, 1946 (cat.); J. R. Whiting, Photography is a Language, Chicago, 1946; R. Chapman, The Laurel and the Thorn: A Study of G. F. Watts, London, 1947; L. Moholy-Nagy, Vision in Motion, Chicago, 1947; Mus. of the City of New York, Battle with the Slum: Fifty Photographic Prints, New York, 1947 (cat.); Brassai, Camera in Paris, London, 1949; Brassai, Histoire de Marie, Paris, 1949; A. Liberman, ed., The Art and Technique of Color Photography, New York, 1951; Brassai, Seville en fête, Paris, 1954; A. Feininger, The Face of New York, New York, 1954; H. and A. Gernsheim, Roger Fenton: Photographer of the Crimean War, London, 1954; M. H. Brown and W. R. Felton, The Frontier Years: L. A. Huffman, Photographer of the Plains, New York, 1955; M. H. Brown and W. R. Felton, Before Barbed Wire: L. A. Huffman, Photographer on Horseback, New York, 1955; A. Feininger, The Creative Photographer, New York, 1955; H. and A. Gernsheim, The History of Photography from the Earliest Use of the Camera Obscura in the Eleventh Century up to 1914, London, 1955; J. D. Horan, Mathew Brady: Historian with a Camera, New York, 1955; H. and A. Gernsheim, L. J. M. Daguerre: The History of the Diorama and the Daguerreotype, London, 1955; E. van der Elsken, Love on the Left Bank, London, 1956; The Foral Encyclopedia of Photography, London, New York, 1956; G. Kepes, The New Landscape, Chicago, 1956; J. T. Soby, Four Photographers: Modern Art and the New Past, Norman, Okla., 1957; P. Pollack, The Picture History of Photography from the Earliest Beginnings to the Present Day, New York, 1959; G. Kepes, Language of Vision, Chicago, 1964 (copyright 1944); B. Newhall, The History of Photography from 1839 to the Present Day, rev. ed., New York, 1964 (previous eds. 1937, 1939).

* *

Illustrations: PLS. 137–140.

PIAZZETTA, GIOVANNI BATTISTA VALENTINO. Italian painter of the Venetian school (b. Venice, Feb. 13, 1683; d. Venice, Apr. 29, 1754). After spending a brief period in the workshop of his sculptor father, Giacomo, Piazzetta studied painting under Antonio Molinari in Venice and about 1703 entered the studio of Giuseppe Maria Crespi in Bologna. Listed in the *fraglia* (guild) of Venetian painters in 1711, he apparently never left his native city again. The date of his marriage is unknown, but it is certain that he had seven children, the eldest of whom was baptized on Dec. 6, 1725.

Piazzetta's artistic activity is fairly well documented by contemporaneous sources. In 1717, for example, Antonio Balestra wrote to the Florentine collector Francesco Gabburri about one of Piazzetta's drawings. Antoine Joseph d'Argenville, in his *Abrégé de la vie des plus fameux peintres* (1745–52), states that Piazzetta's first important work was an altarpiece depicting the Virgin and a guardian angel (fragment with the Madonna and Child, Detroit, Inst. of Arts), exhibited at the Scuola di S. Rocco and purchased by Zaccaria Sagredo. The *St. James Led to Martyrdom* (Venice, S. Stae) dates perhaps from 1717. The altarpiece of the Virgin appearing to St. Philip Neri (Venice, S. Maria della Fava) was paid for in 1724. In 1726 the theatrical impresario Owen MacSwiny wrote to the Duke of Richmond about an allegorical tomb (Birmingham, Coll. of Lord Plymouth), with figures by Piazzetta. On Aug. 3, 1727, the artist was paid for four monochrome roundels on the ceiling of the Chapel of St. Dominic in SS. Giovanni e Paolo, Venice, where there is also a *St. Dominic in Glory* executed by him. The Assumption altarpiece formerly in Lille (Mus. B. A.; on loan, Louvre), from about 1735, was commissioned by the Archbishop Elector of Cologne, exhibited with approbation in Venice, and engraved before being sent to Germany. About 1739, Piazzetta painted an altarpiece with SS. Vincent Ferrer, Hyacinth, and Lorenzo Bertrando for the newly built Church of the Gesuati, a work still *in situ*. The later altarpieces were often completed in collaboration with his many pupils. For a patron, Count Johann Matthias von der Schulenburg, he painted the *Idyll on the Sands* (Cologne, Wallraf-Richartz-Mus.) and the *Pastoral Scene* (Chicago, Art Inst.). Piazzetta was commissioned by the publisher Albrizzi, who became a close friend, to illustrate a ten-volume edition of Bossuet (1736–57). Albrizzi also commissioned drawings for an edition of Tasso, published in 1745. About this time the critic Count Francesco Algarotti commissioned a *Caesar and the Pirates* (lost) for the Dresden gallery.

During his lifetime Piazzetta gained artistic recognition both in and beyond Venice. The Accademia Clementina of Bologna made him a member in 1727. He executed several works for German patrons and was urged to fulfill other commissions outside his native city. In 1750 he was appointed director of the newly formed Venetian Academy. Nevertheless, a petition to the doge by his family, dated May 15, 1754, reveals that he died a poor man.

Primarily a painter of religious subjects, Piazzetta also produced a few portraits and genre scenes. He was a prolific draftsman and drew many "semi-caprice" heads in black and white chalk. In an age of rapid, virtuoso painters, there is much contemporaneous testimony to his slow methodical technique. Albrizzi noted his love of solitude and his melancholy disposition. A genius outside the conventions and conditions of 18th-century Venice, he had a poignant grasp of reality and a low-key palette that put him in sharp contrast to Tiepolo in all aspects except essential artistic ability. (Cf. IV, PL. 279; VIII, PL. 222; XII, PL. 183).

BIBLIOG. G. B. Albrizzi, Studj di Pittura già dissegnati da Giambattista Piazzetta, Venice, 1760; R. Pallucchini, L'Arte di Giovanni Battista Piazzetta, Bologna, 1934 (2d ed., Rome, 1943); R. Pallucchini, Piazzetta, Milan, 1956.

Michael LEVEY

PICASSO, PABLO. Spanish painter. Pablo Picasso was born on Oct. 25, 1881, in Málaga, where his father, José Ruiz Blasco, was a professor in the School of Arts and Crafts. His mother was Maria Picasso, from Málaga. From 1901 on, Picasso signed his paintings with the surname of his mother.

Through the influence of his father, Picasso learned to paint at an early age. In 1891 his family moved to La Coruña, where in 1895 Picasso painted one of his earliest works, *The Barefoot Girl* (Picasso Coll.). In 1896 Picasso entered the Barcelona School of Fine Arts and in 1897 was admitted as an advanced student at the Royal Academy of San Fernando in Madrid. He soon, however, tired of this academic training and moved back to Barcelona, where he began to study on his own the works of older and contemporary Spanish and European artists. Their influence is evident in the many drawings and paintings Picasso did between 1898 and 1902, such as Goya's influence in *The Divan* (ca. 1900) and that of Velázquez in *Nana* (ca. 1901; both Barcelona, Mus. de Arte Mod.); the paintings of Gauguin, Van Gogh, and Munch were others that had a profound effect on his early work. At the same time, Picasso participated actively in the intellectual life of Barcelona, forming close friendships with such artists and writers as Jaime Sabartés, Isidre Nonell, and Angel and Mateu Fernandez de Soto. He painted portraits of these friends during his early years in Barcelona (*Portrait of Mateu F. de Soto*, 1901, Picasso Coll.). The young avant garde met at the Café Els Quatro Gats, for which Picasso designed a poster and two menu cards. In 1900 he also published two drawings in the magazine *Joventut* to illustrate two short stories of Joan Oliva Bridgman. In these early years Picasso also turned to sculpture. *The Seated Woman* (ca. 1901; New York, Dr. Peritz Levinson Coll.), a cast bronze, anticipated his works of the Blue Period. Picasso drew his subjects from among his friends, his family (*Portrait of the Artist's Sister*, 1899, Picasso Coll.), from the café he frequented (*Two Women at a Bar*, 1902; New York, W. P. Chrysler, Jr., Coll.), and from the poor in Barcelona (*Mother and Child*, 1901; New York, Mrs. Maurice Wertheim Coll.).

In October, 1900, he left Barcelona for Paris, where he stayed for three months. There he met Pedro Mañach, who helped him to establish himself as an artist in Paris. Mañach introduced Picasso to Berthe Weill, who bought three of his sketches, the first paintings that Picasso sold in Paris. During this trip to Paris, Picasso painted the *Moulin de la Galette* (New York, Justin K. Thannhauser Coll.). In December of the same year, Picasso returned to Madrid, where from January to May, 1901, he joined the writer Francisco de Assis Soler in publishing the magazine *Arte Joven*. As art editor he provided all the illustrations for the magazine. In Barcelona the magazine *Pèl y Plom*, of which Miguel Utrillo was literary editor, sponsored an exhibition of his pastels at the Salón Parés. Picasso returned to Paris in 1901 when Pedro Mañach persuaded Ambroise Vollard to give him a one-man exhibition in his gallery. The exhibition at Vollard's was not a great success, and the influence of Toulouse-Lautrec, Vuillard (qq.v.), and T. A. Steinlen, was very evident in his painting. Picasso, however, aroused the interest of the critic Gustave Coquiot (whose portrait Picasso painted in 1901; Paris, Mus. d'Art Mod.), and the writer Max Jacob, who subsequently became one of his closest friends. Picasso returned to Barcelona at the end of the year. During his absence Berthe Weill exhibited some of his work (*The Blue Room*; Washington, D.C., Phillips Coll.) in her gallery from Apr. 1 to 15, 1902. Picasso went to Paris again briefly in October of 1902, but he returned to Barcelona early in 1903. He left Barcelona to settle permanently in Paris in April, 1904.

In the autumn of 1901, Picasso's painting entered the so-called "Blue Period," which lasted until 1904 (*The Glass of Beer, Jaime Sabartés*; Moscow, Mus. of Mod. Western Art). The artist's figures (such as the old man in the *Blind Man's Meal*, 1902; Met. Mus.) conveyed the pathos and tragedy of the earlier Barcelona figures (*Mother and Son*, 1898; New York, Josef Stransky Coll.), but the draftsmanship, through the influence of the Nabis and the symbolists, became sharply defined and highly stylized. The figures themselves were distorted by the elongation and emaciation of the body. The deep chiaroscuro of the early paintings disappeared and was replaced by a monochromatic blue, which put the figures in a clear yet flat relief. In other paintings, such as the *Life* of 1903 (Cleveland, Mus. of Art), the same blue emphasized the emotional isolation of the figures within their impersonal surroundings. In a review of an exhibition of Picasso's work in October, 1902, at the Berthe Weill gallery (*Mercure de France*, December, 1902), the critic Charles Morice remarked upon "the sadness . . . which pervades all the work of this very young man." Picasso himself stated that art is the "daughter of sadness and pain."

On arriving in Paris in the spring of 1904, Picasso set up his studio at 13 rue Ravignon, in an old building popularly called the "Bateau Lavoir." He soon had a large circle of friends, among whom were Max Jacob, Maurice Raynal, André Salmon, Guillaume Apollinaire (whom he met in 1905), and Juan Gris (who moved into the Bateau Lavoir in 1906). Picasso also gained the admiration of the Americans Leo and Gertrude Stein and the Russian Sergei Shchukin (Stchoukine), who began to buy his paintings.

At the beginning of 1905, circus and harlequin themes became predominant in Picasso's painting. The flat blue of the paintings of the preceding years was replaced by a rose color, which suffused the figures in a softer and more pleasant atmosphere. The circus world penetrated all phases of the artist's work: his drawings (*Family with Monkey*, a 1905 pen drawing for Apollinaire's *Les mammelles de Tiresias*; publ. Paris, 1950); his engravings, of which he did a whole series with circus themes in 1905 (*Two Acrobats*; New York, Mrs. John D. Rockefeller Coll.); his sculpture (*Jester*, cast by Ambroise Vollard; Washington, D.C., Phillips Coll.); and his paintings (the large *Family of Saltimbanques*; Washington, D.C., Nat. Gall.). In *La Plume* of May 15, 1905, Apollinaire, emphasizing the ambiguity of the new works, noted how "the charm and indisputable talent [of Picasso] appear at the service of an imagination that successfully unites the delightful with the horrible, the base with the delicate."

Picasso took a trip to Holland for a month in the summer of 1905 with his friend Tom Schilperoot. His paintings during the trip (*Three Dutch Girls*; Paris, Mus. d'Art Mod.) reveal a solidity and fullness of form that mark a significant change in his art. On his return to Paris, Picasso's concern with the weight and volume of his figures, through his awareness of the work of Cézanne (q.v.), became increasingly evident in the paintings of young boys and horses he did in the last half of 1905 (*Boy Leading a Horse*; New York, Mus. of Mod. Art). In the Salon d'Automne of 1905 a special gallery was set aside for an exhibition of ten of Cézanne's paintings, which Picasso probably saw. The painting done by Picasso at Gosol, where he spent the summer of 1906 with Fernande Olivier, completed the artist's evolution toward a sculptural firmness of form. *La Toilette* (PL. 143) still displayed the softness of expression of Picasso's earlier paintings, but the *Woman with Loaves* (Philadelphia, Mus. of Art) and especially the *Portrait of Gertrude Stein* (Met. Mus.) were more severely modeled. The face of Gertrude Stein, which Picasso repainted without the model on his return to Paris in the autumn of 1906, and *Two Nudes* (Pittsburgh, G. David Thompson Coll.), painted toward the end of the year, clearly show the influence of Iberian sculpture, examples of which Picasso had seen at a Louvre exhibition that spring.

The Iberian sculpture was to have a profound effect on his art. In the numerous studies for *Les Demoiselles d'Avignon* (1906–07) and in the early figures on the left of the painting itself (IV, PL. 75), Picasso sought to adapt the primitive yet expressive shapes of the sculpture to his figures. In the midst of painting *Les Demoiselles d'Avignon*, Picasso's interest veered to another kind of sculpture — the Negro sculpture of Africa, which he saw at the ethnographical museum in Paris in 1907. The two figures on the right of the painting were rendered in a new, incisive, and violent way. The dark striations that accentuated the expressiveness and movement of these figures completely dominated subsequent works such as the *Nude with Drapery* (1907; Moscow, Mus. of Mod. Western Art). *Les Demoiselles d'Avignon*, first called humorously "Le bordel philosophique," was shown privately only to a few friends — among whom were Henri Matisse and Georges Braque, whom Picasso had met in 1906. Under its present title, acquired sometime after World War I, it was published in *La révolution surréaliste* on July 15, 1925, and was first exhibited publicly in the International Exhibition in Paris in 1937.

A more subdued influence of Negro art on Picasso appeared in the *Paysanne* (Moscow, Mus. of Mod. Western Art), painted in the autumn of 1908, in which Picasso eliminated the expressive violence of the *Nude with Drapery* and began to separate the component masses of the figure into roughly geometric volumes. Strongly influenced by Cézanne, for whom a large memorial retrospective exhibition of 56 paintings was held at the Salon d'Automne in 1907, Picasso applied this "analytic" examination of form to several landscapes done at the end of the summer of 1908 at La Rue-des-Bois (*Landscape*; Moscow, Mus. of Mod. Western Art). Picasso's work was also affected at this time by the simple rendering of form which he found in the painting of Le Douanier, Henri Rousseau (q.v.). After having purchased a painting by the artist earlier that year, Picasso gave him a huge banquet at the end of 1908.

In 1909 Picasso spent the summer at Horta de San Juan and executed a series of landscapes (*The Reservoir*; Paris, Coll. of the Heirs of Gertrude Stein) in which, through strong planar shading, the buildings were reduced to their most elementary geometric volumes. On his return to Paris, he applied the same analysis to the human figure. In the *Seated Nude* (V, PL. 124), the background and the face and torso of the woman were divided into numerous strongly shaded and crosshatched facets. In his sculpture of the same period (*Female Head*; V, PL. 128) Picasso also broke up the three-dimensional form into small planes.

In the *Portrait of Wilhelm Uhde* (London, Roland Penrose Coll.), done in the spring of 1910, the broad transparent planes of the figures, painted in a somber monochrome, opened into and blended with those of the background. This fusion of the figure with the surrounding space reached the greatest complexity in the work done at Cadaqués, where Picasso spent the summer of 1910 with Fernande Olivier and André Derain. In paintings such as the *Rower* (New York, Ralph Colin Coll.) and in the four etchings for Max Jacob's *Saint Matorel*, which Kahnweiler published in 1911, Picasso attained his most abstract and hermetic cubist compositions. In the *Portrait of Kahnweiler* (Chicago, Art Inst.), painted after his return from Cadaqués, Picasso turned to a more precise analysis of the figure.

At Céret in the summer of 1911 (the year of his first exhibition in the United States, at the Photo-Secession Gallery, New York), Picasso worked very closely with Georges Braque. It was first Braque, and then Picasso (*Le Torero*; New York, Nelson Rockefeller Coll.), who introduced printed letters into his paintings. These letters served to emphasize the two-dimensional nature of the canvas and acted as symbols through which the spectator could reconstruct the subject represented. In Paris, during the winter of 1911–12, Picasso executed a series of paintings inspired by his love for Marcelle Humbert ("Eva," his companion from 1912 to 1915), incorporating the printed words "Ma jolie" (IV, PL. 76) and "J'aime Eva" (Columbus, Ohio, Gall. of Fine Arts). The inclusion of unorthodox materials in his paintings was continued early in 1912 when Picasso introduced a piece of oilcloth, painted to simulate chair caning, into his *Still Life with Chair Caning* (Picasso Coll.) and framed the painting with rope (see MEDIA, COMPOSITE). He stayed with Braque at Céret and then at Sorgues, near Avignon, during the summer and for several weeks in the autumn of 1912. In the paintings done at Sorgues, Braque was the first to use *papier collé* with the oil medium, and Picasso soon adopted the technique extensively (*Bouteille de Vieux Marc, Verre, Journal*, 1912; Paris, Coll. Cuttoli; see CUBISM AND FUTURISM), integrating the bits of pasted paper into his compositions with linear painting. Picasso also introduced bright color, often in the form of pasted paper, into the paintings of this period.

His *papiers collés* became more complicated in 1913 (*La Bouteille de Suze*; IX, PL. 393), and in the same year they began to be combined with many elements of collage. The *Student with a Pipe* (Paris, Coll. of the Heirs of Gertrude Stein) was executed in oil, pasted paper, and sand. The ideographic representation of the student's features reflects the influence of Wobe African masks. Through the use of *papier collé*, Picasso deliberately exaggerated the thin planes of his figures. He soon translated this tendency into oil painting, in which he imitated the textures of *papier collé* and collage (*Card Player*; IV, PL. 78). At Céret in 1913, at the request of Kahnweiler, Picasso etched three aquafortes for Max Jacob's *Siège de Jérusalem.* In the summer of 1914 at Avignon, Picasso executed a series of still lifes in which he used areas of brightly colored dots to enliven the surface of his canvas (*Vive la . . .*, Chicago, L. B. Block Coll.). During this same period, which was marked by increasing variety in his collages, Picasso also experimented with three-dimensional constructions. Making use of the cubist still-life iconography of musical instruments, in 1912 Picasso created two guitars, one of colored papers (Picasso Coll.) and another of wood and *papier collé.* In 1914 the artist executed a still life with painted wood and upholstery fringe (London, Roland Penrose Coll.). He also had cast in bronze a sculpture of an absinthe glass, which he finished with many bright daubs of paint (New York, Mus. of Mod. Art).

The outbreak of World War I marked the end of the period in which Picasso developed ideas similar to those of his contemporaries Braque and Gris. Working alone during the war and directly after, he became involved with experiments both in a realism that tended toward classicism and in a cubist vein that was resolved in various ways. He considered cubism and classic realism as two different means and two different perspectives for examining reality. Although cubism continued to stimulate him, it was through the classicist tendency that he developed a surrealist esthetic. In his cubist painting during the war (*Harlequin*; PL. 143), Picasso abandoned the use of rough collage materials and concentrated instead on constructing his figures from fewer and larger geometric planes, derived from his *papiers collés.* By the summer of 1920 at Juan-les-Pins, the planes, painted so as to look pleated or pierced, corresponded in arrangement to the form of the figures (*Pierrot and Harlequin at a Café Table*; New York, Gilbert W. Chapman Coll.). In 1919 Picasso painted a series of still lifes at Saint-Raphael (*Table in Front of Open Window*; New York, Daniel Saidenberg Coll.) which were often set in tromp-l'oeil frames and in which cubism and realism tended to blend. Some of Picasso's earliest realist work of this period appeared in 1915 in the fine pencil drawings he made of his friends (e.g., that of Ambroise Vollard; Met. Mus.), in which the attention to line and details of dress reveals the influence of Ingres. This influence was even more apparent in the several portraits (1917–19) of his wife Olga Koklova, a dancer with the Ballets Russes, whom he married in 1918 (*Portrait of Mme Picasso*; Picasso Coll.).

During and after the war, Picasso was frequently occupied in creating designs for the theater. Persuaded by Cocteau in 1917 to design scenery and costumes for the Cocteau-Satie ballet *Parade* (reproduced by Serge Diaghilev's Ballets Russes on May 18, 1917, in Paris), Picasso continued to design sets, costumes, and curtains for the Russian ballet until 1924 (see SCENOGRAPHY). In February, 1917, he went to Rome, where the ballet *Parade* was being rehearsed. The characters of the ballet included three "Managers," whom Picasso arrayed in tall cubist constructions. The curtain portrayed a group of familiar circus and harlequin figures recalling the artist's Blue and Rose periods. In style, however, these figures were heavier and more mannered in gesture; they resembled in some aspects the *Harlequin* painted in Paris in 1917 (PL. 143). During Picasso's short stay in Italy, he visited Naples, Pompeii, and Florence.

In London in the summer of 1919, Picasso designed the costumes for *Le tricorne* of Martinez Sierra and Manuel de Falla and painted a bullfight scene on the curtain drop. In 1920, again in Paris, Picasso sketched a theater interior for the ballet *Pulcinella* of Igor Stravinsky, whom Picasso had met in Rome in 1917 (XII, PL. 443). Owing to confusion and haste in mounting the production, however, Picasso discarded this design and substituted a Neapolitan street background, painted in cubist style. Then, in 1924, he designed the scenery for the Ballets Russes production of de Falla's *Cuadro Flamenco* and the curtain for *Le train bleu*, by Jean Cocteau and Darius Milhaud. Also in 1924, Picasso designed the set and costumes for the Comte Étienne de Beaumont's production of the ballet *Mercure.*

Picasso's paintings of the 1920s, like those of the war and immediate postwar period, juxtaposed various styles: the cubist, a realist vein with surrealist overtones, and the neoclassic. The cubist paintings of 1921 are among Picasso's most significant works. The two versions of the *Three Musicians* (Philadelphia, Mus. of Art; PL. 144), painted at Fontainebleau during the summer, were a synthesis of all Picasso's cubist experiments up to that date. Although Picasso continued to do cubist work throughout the 1920s — for example, *The Red Tablecloth* (1924; New York, private coll.), one of a series of late cubist still lifes done between 1924 and 1926 — his cubist abstractions became more subdued and were adapted to a more objective representation.

One of the first indications of a kind of realism different from that found in the Ingres-type portraits of 1915 occurred in Picasso's gouache *The Sleeping Peasants* (1919; New York, Mus. of Mod. Art). Though realistic in bodily form, the figures were exaggerated in scale and somewhat surrealistic in effect. Their hands and feet were extraordinarily large, and the figures themselves were cramped into a relatively small space. These arbitrary distortions were intensified in *By the Sea* (1920; Pittsburgh, G. David Thompson Coll.), where the exaggerated foreshortening of the female bather in the background, with one very long leg and a minute head, portrays her graphically in the act of running into the sea. Two years later, in *The Race* (New York, Mus. of Mod. Art), a painting that was adapted in 1924 for the curtain of *Le train bleu*, Picasso portrayed the figures with less distortion but with the same concentration on their movement. These paintings were forecasts of the more decidedly surrealistic works done by Picasso in the mid-1920s.

The third type of painting in the 1920s, a neoclassic strain that was also foreshadowed by the large figures of *The Sleeping Peasants* of 1919, appeared between 1920 and 1922 in the series of paintings of monumental nude women. In *Two Seated Women* (1920; London, Douglas Cooper Coll.), the colossal figures are posed in very calm pensive attitudes and completely fill the large canvas. About the same time, Picasso also painted some very small canvases of nude women of similar proportions, such as the *Four Bathers* of 1921 (4 × 6 in.; New York, private coll.), as if to show that the monumental quality of a figure was not determined by its actual size. In 1921, Picasso's son Paolo was born, and the artist celebrated the event with a series of maternal subjects (PL. 143), their monumental figures portraying the mother's love for her child.

This neoclassic phase in Picasso's art lasted until about 1925. The tenderness and classic repose characterizing the mother-and-child groups were evident in many paintings of different subjects executed during the next three years: for example, *Two Women and a Child* (PL. 146), *The Pipes of Pan* (1923), and *The Three Graces* (1924; last two, Picasso Coll.).

Many of Picasso's drawings of the same period used the firm contour line, developed under the influence of Ingres, that had been noted in the *Bathers* (1918; Cambridge, Mass., Fogg Art Mus.). The subject matter was taken from classical mythology (*Centaur and Woman*, 1920; New York, Gilbert Seldes Coll.), thematically in keeping with this neoclassic phase in Picasso's art.

In 1924 André Breton published the *Manifesto* of surrealism. Picasso, interested in the movement, found in the surrealist attitudes a decisive impulse to change the then predominant tendencies of his art. Having experimented previously with the representation of running figures in *By the Sea* (1920) and *The Race* (1922), Picasso began to concern himself with rendering a more violent, frantic action. In the *Three Dancers* (1925; London, Tate Gall.), for instance, Picasso sought to capture the whirling movement of the dance through a cubist fragmentation and distortion of the female figures.

Surrealist elements also began to appear in the still-life compositions done at Juan-les-Pins in the summer of 1925. In the *Studio* (private coll.), truncated segments of sculptured arms lie on a table alongside other, more mundane articles in the artist's studio. The surrealist emphasis increased in Picasso's paintings of the late 1920s. In *Seated Women* (1927; New Canaan, Conn., James Thrall Soby Coll.) he combined a profile view of the face with a frontal view and painted the fingernails like actual nails. The next year, Picasso played ironically with the relation of the real to the abstract creation, in the *Painter and His Model* (1928; New York, Sidney Janis Coll.), in which the profile on the painter's canvas is more realistic than either the painter or his model.

During the summer of 1928, which Picasso spent at Dinard, he interested himself in a series of projects, including constructions and metamorphic sculptures, only some of which were carried through to completion. The *Head* (1928; Picasso Coll.), a construction in painted metal, bears a strong resemblance to the painter in the *Painter and His Model* in its purely abstract shape. Similarly, in the metamorphic *Figure* (Picasso Coll.), an image executed in Paris in 1928, and later in projects for monuments and in his sculpture of the early 1930s, the abstracted and distorted figures of his contemporaneous paintings are clearly carried over into other media.

In 1927 Picasso took a new interest in graphic work. He executed a series of drawings and etchings on the theme of the painter and his model, some of which (e.g., *The Painter with a Model Knitting*) were used as illustrations for an edition of Balzac's *Le chef-d'œuvre inconnu* published by Vollard in 1931. Also reproduced in this edition were 16 pages of abstract designs composed of lines and dots. In the same year, Picasso provided 30 etchings of classical subjects to illustrate the Skira edition of Ovid's *Metamorphoses* (e.g., *Combat of Perseus and Phineas over Andromeda*). Another major illustrative project of Picasso was his series of 33 drawings and 6 etchings for Gilbert Seldes' version of Aristophanes' *Lysistrata* (New York, 1934; PL. 146) About 1937 Picasso provided 31 aquatints for the Comte de Buffon's *Histoire naturelle* (published by Fabiani in 1942). Apart from these book illustrations, Picasso's extensive graphic work of this phase of his career includes a series of etchings he made in 1933 based on the theme of the sculptor's studio (XII, PL. 572). Also during this period, important exhibitions in Paris (Gal. Georges Petit, 1932) and in Zurich (Kunsthaus, 1932) added to an already extensive reputation.

Picasso's work of the 1930s, growing out of his metamorphic experiments in the late 1920s, was dominated by a theme of psychological and physical violence that culminated in *Guernica* (VII, PL. 278), his large canvas commissioned in 1937 for the Spanish government building at the International Exhibition in Paris. In the skeletal figure of the *Seated Bather* (PL. 144) of 1929, Picasso gave a morbid and surrealist interpretation to one of his favorite neoclassic subjects of the early 1920s. The series of still lifes that were painted early in 1931 (*Pitcher and Bowl of Fruit*; New York, Nelson Rockefeller Coll.) and the figures of seated and reclining women that he did in 1932 (*Nude Woman in a Red Armchair*; PL. 142) were relatively restrained in tone. Their strong, fluid contour lines and bright colors, however, were used to brilliant effect in the psychological examination of a girl contrasted with her mirror image in *Girl before a Mirror* of 1932 (New York, Mus. of Mod. Art). In 1933, at Cannes, Picasso did several surrealist drawings that, with their figures partly composed of shutters, windows, and doors (*Two Figures on the Beach*; New York, Mus. of Mod. Art), adhered more than any of his previous works to the current surrealist ideal. The summer of 1933 also marked the date of Picasso's trip to Barcelona and of his renewed and avid interest in the bullfight. The paintings that followed in 1933 and 1934, scenes of bulls attacking the picadors' horses, overturning matadors, or dying under the sword (*Bullfight, 9 September 1934*; Philadelphia, Henry P. McIlhenny Coll.), conveyed a barbaric violence in their angular, distorted draftsmanship and in their tense and anguished expression. At the same time, Picasso continued to paint women in interiors; *Two Women* (PL. 144) is one of two versions of this theme arrived at by Picasso after many preliminary drawings.

The symbol of the minotaur — half man, half bull — begins to appear frequently in Picasso's work in the mid-1930s. In 1933 the artist designed, in pencil and pasted paper, and cloth, a minotaur that was reproduced as the frontispiece of the new magazine *Minotaure*, published in Paris by Albert Skira. Two years later in the famous etching *Minotauromachia* (IV, PL. 443), the minotaur appeared as a huge beast of terror. The scene is one of intense action set in a crowded space, emphasized by the man escaping up a ladder on the left.

The outbreak of the Spanish civil war in July, 1936, signaled the artistic expression of Picasso's strong personal and emotional involvement in the republican cause. The extent of his concern was reflected in a statement he made in 1944 when asked about his motives for joining the Communist party: "I have never considered painting as pure pleasure: I have always wanted to penetrate more deeply into the knowledge of the world and of man. Today I understand that this is not enough . . . that I must fight not only with my art but with my entire being." In 1936 he accepted the position of director of the Prado in Madrid. In January, 1937, he wrote the satirical poem *Sueño y mentira de Franco*, illustrated with two etchings that were divided into nine compartments, each containing a savage scene of bulls, horses, and terrified people. His commitment to the Spanish struggle was fully expressed in the *Guernica*, which, with its many preliminary sketches (IV, PL. 281), was completed in about a month after the bombing of the northern Spanish town of Guernica (late April, 1937). In the painting, with its somber color scheme of black, white, and gray emphasizing the tragic import of the subject, Picasso used the images of violence he had developed over the preceding years. The anguished women, the dead children, and the screaming horse, all overlapping each other, are confined within a room from which there is no escape. Dominating this scene of chaos is the bull, an impassive symbol of brutality and violence.

The theme of violence in Picasso's painting was not abandoned with the *Guernica* but continued to preoccupy him during the following years. From the anguished expressions of the women in *Guernica*, there developed the theme of the weeping women, explored by the artist in numerous drawings, etchings, and paintings of 1937 (*Weeping Women*; London, Roland Penrose Coll.). In the drawing *Seated Woman with Necklace* (PL. 146) and the painting *Man with a Lollipop* (1938; New York, Edward A. Bragaline Coll.), Picasso used a sharp crystalline line in giving a fiercely distorted aspect to the figures. The necklace of the woman is actually made of barbed wire. These distortions continue in the paintings of the World War II period executed in Paris, both in the portraits of Dora Maar (1939; Paris, Coll. Mlle Dora Maar), whom Picasso met in 1936, and in his paintings of seated women (*Woman in Green*, 1943; New York, James Johnson Sweeney Coll.).

In 1939 a large exhibition of Picasso's works held at the Museum of Modern Art in New York confirmed the importance of his work in America (Barr, 1946).

Picasso lived in relative quiet in Paris during the Occupation, although his art was condemned by the Nazis and by Vichy France. In 1941 he wrote a satiric play, *Le désir attrapé par la queue*, which was not published until 1944 in *Messages # 2*, illustrated with four drawings. Not many of his paintings during the war period had the violent tone of the *Woman in Green* (1943). In *Serenade* (1942; Paris, Mus. d'Art Mod.), though the human figures were severely distorted, the atmosphere of the painting remained calm. A painting of his daughter Maia (*First Steps*, 1943; New Haven, Yale Art Gall.) and the series of still lifes with tomato plants (*Tomato Plant and Carafe*, 1944; New York, Samuel Bronfman Coll.), painted just before the liberation of Paris, depict occurrences or elements in his immediate environment seemingly unaffected by the war. During the liberation week itself (Aug. 24–29, 1944), Picasso painted a *Bacchanal* (Picasso Coll.) inspired by Poussin's famous painting. Whereas he remained faithful to Poussin's composition in his preliminary drawing, Picasso painted a final version in which he intensified the exuberance of the original, through his strong distortions and crowding of the figures. At the end of the year, to honor Picasso and his work, the artist was assigned a large gallery at the Salon d'Automne in which to display some seventy-five paintings and five sculptures. In 1944 Picasso also joined the Communist party, a fact that he announced publicly in several interviews, arousing a great deal of controversy.

In 1945, after a long interval, Picasso again became very interested in lithography and began to work regularly in the print shop of Fernand Mourlot. During the following years, he produced many lithographs, among which were portraits of Françoise Gilot, whom he met in 1946, and a series of ten lithographs of David and Bathsheba, a theme adopted from a painting by Lucas Cranach. In 1945, also, Picasso painted a series of Parisian scenes (*Ile de la Cité, Paris*; New York, Mrs. Albert D. Lasker Coll.) in which the buildings, bridges, and the Seine were broken up into brightly colored, heavily outlined geometric shapes.

Picasso's style changed perceptibly at Antibes, where he spent the autumn of 1946. At the invitation of M. Dor de la Souchère, the director of the local museum, Picasso set up a studio on the museum's second floor and produced a large number of paintings on fiberboard (Antibes, Mus. Grimaldi). The subjects were drawn from Greek mythology — fauns, centaurs, and youths with shepherds' pipes — and the paintings had an atmosphere suffused with a lightness and joy, as is evidenced in the title of one of the largest works — *La joie de vivre*.

The following year, at Vallauris in the south of France, Picasso began working avidly in ceramics in the Poterie Madoura. Singly and in series, he created almost 2,000 pieces. His ceramic work also encompassed sculpture (III, PL. 173), with which he had frequently been occupied since the outbreak of the war. During the Occupation, Picasso executed numerous studies for a large sculpture, the *Man with a Lamb* (Philadelphia, Mus. of Art), which he completed in 1944. His sculptural work continued through the late 1940s with several important bronzes of animals: *Goat* (1950; Picasso Coll.); *Baboon with Her Young* (1951; New York, Mus. of Mod. Art); and *Crane* (1951–52; Pittsburgh, G. David Thompson Coll.). His painting during this period included many portraits of Françoise Gilot and of their children — Claude, born in 1947, and Paloma, born in 1949 (*Claude and Paloma with Their Mother Françoise Gilot*, 1951; Coll. Claude Gilot Picasso) — in which the figures were broken up into large, flat, sculptural planes. He also painted large allegories of war and peace (1952; PL. 145) to decorate a chapel at Vallauris, as well as his interpretation of Courbet's *Les demoiselles des bords de la Seine* (1950; Basel, Kunstmus.), which presaged the series of paintings after Delacroix, Velázquez, and Manet that were to follow in the coming years. In the postwar period, Picasso again illustrated many publications with the same enthusiasm he had shown in his book illustrations of the 1930s. He provided etchings and lithographs for Góngora's *Vingt poèmes* in 1945, and in the following year he contributed 38 engravings and 4 aquatints to an edition of Prosper Mérimée's *Carmen*. In 1954, with Kahnweiler, he published an illustrated volume of his own poems, entitled *Poèmes et lithographies*.

After the war and in the early 1950s, Picasso's fame spread to all parts of Europe. In 1948 he participated in the World Peace Congress in Warsaw, and in the following year he designed the widely known dove poster for the Peace Congress held in Paris in April. Major exhibitions of his work were also organized throughout Europe and America at this time. In 1953 alone, for example, important one-man shows were held in Rome, Milan, and São Paulo.

Picasso's work of 1953–54 was deeply affected by his separation from Françoise Gilot. His solitude and depression found expression in his 180 drawings of "The Artist and His Studio," in which the advanced age of the painter is contrasted with the youth and sensuousness of his model. At the same time, Picasso executed several drawings on the theme of the visit to the artist's studio (III, PL. 431), in which he gave full rein to his wit and irony. Picasso's painting of 1954 centered around a large number of portraits, both abstract and realistic, of his new companion Jacqueline Roque, who was to become his wife (*Portrait of Jacqueline*; New York, Allan D. Emil Coll.), and of friends (*Sylvette David*; New York, Dr. Herschell Carey Walker Coll.). In December of that year he began a series of 15 paintings after Delacroix's *Women of Algiers*, in which the figures and background were progressively rearranged and broken up into large geometric planes and then reintegrated into a complex, highly colorful, and exciting composition.

Early in 1955 Picasso bought and moved into a villa at Cannes called "La Californie," where he lived until 1961. His first summer there was occupied with the production of Henri Clouzot's film on the artist, *Le mystère Picasso*. The exotic Moorish influence of his series after Delacroix, together with the influence of Matisse, greatly affected his work of 1955 and 1956. He executed a large number of studio interiors of his Cannes villa (*The Studio*, 1955; New York, Saidenberg Gall.), as well as portraits of Jacqueline Roque in Turkish costume, in which the liberal use of bright color and dark outline created a brilliant and exciting atmosphere.

In 1957 the artist embarked on a new series of 20 paintings after Velázquez' *Las Meninas*, interpreting the original in his own very personal idiom. In the same year he completed a large-scale painting, *Le Baignade*, for the UNESCO building in Paris, and during the next two years he produced many lithographs, linoleum cuts, and paintings on the familiar theme of the bullfight. In his later paintings Picasso has continued to work along the same general lines as in the early 1950s, but introducing some variations. His numerous portraits of Jacqueline, his seated and reclining nudes (*The Divan*, 1960; New York, Kootz Gall.), and his series of paintings after Manet's *Le déjeuner sur l'herbe* in 1962 distort the image in order to give a new expressive, animate, and personal interpretation of the original subject.

WORKS. *a. Paintings*: *Self-portrait*, 1901, Picasso Coll. – *Boulevard de Clichy*, 1901, New York, William S. Farish Coll. – *Harlequin*, 1901, Philadelphia, Henry Clifford Coll. – *The Absinthe Drinker, Portrait of Angel Fernandez de Soto*, 1903, New York, Donald S. Stralem Coll. – *Au Lapin Agile*, 1904, New York, Charles S. Payson Coll. – *The Harlequin's Family*, 1905, Met. Mus. – *Seated Nude*, 1905, Paris, Mus. d'Art Mod. (PL. 141). – *Woman with a Fan*, 1906, New York, W. Averell Harriman Coll. – *Composition (Peasants and Oxen)*, 1906, Merion, Pa., Barnes Foundation. – *The Ballerina of Avignon*, 1907, New York, W. P. Chrysler, Jr., Coll. – *Female Nude*, 1908, Philadelphia, Mus. of Art. – *Woman with Pears*, 1909, New York, W. P. Chrysler, Jr., Coll. – *Portrait of Braque*, 1909, New York, Edward A. Bragaline Coll. – *Girl with Mandolin*, 1910, New York, Nelson Rockefeller Coll. – *The Model*, 1912, New York, Dr. Herschel Carey Walker Coll. – *Violin and Guitar*, 1913, Philadelphia, Mus. of Art. – *Seated Man with Glass*, 1914, Milan, private coll. – *Guitar*, 1916–17, New York, A. Conger Goodyear Coll. – *Nessus and Dejanira with a Satyr*, 1920, New York, private coll. – *Dog and Cock*, 1921, New Haven, Yale Art Gall. – *Three Women at the Fountain*, 1921, New York, Mus. of Mod. Art. – *The Lovers*, 1923, Washington, D.C., Nat. Gall. – *Paul as Harlequin*, 1924, Picasso Coll. – *Atelier de la Modiste*, 1926, Paris, Mus. d'Art Mod. – *The Studio*, 1927–28, New York, Mus. of Mod. Art. – *Woman in an Armchair*, 1929, Picasso Coll. – *Crucifixion*, 1930, Picasso Coll. – *Bather Playing Ball*, 1932, New York, private coll. – *The Sculptor and His Statue*, 1933, New York, private coll. – *Portrait of Marie-Thérèse*, 1936, Picasso Coll. – *The Flower Seller*, 1937, New York, Perls Gall. – *Girl with a Cock*, 1938, New York, Mrs. Meric Callery Coll. – *Portrait of Dora Maar*, 1938, New York, W. P. Chrysler, Jr., Coll. – *Portrait of Jaime Sabartés*, 1939, Paris, Jaime Sabartés Coll. – *Night Fishing at Antibes*, 1939, New York, Mus. of Mod. Art. – *Nude Dressing Her Hair*, 1940, New York, Mrs. Bertram Smith Coll. – *Reclining Nude*, 1941, New York, Nathan L. Halpern Coll. – *Girl with Artichoke*, 1942, New York, W. P. Chrysler, Jr., Coll. – *Seated Woman and Standing Nude in Interior*, 1944, Chicago, Morton G. Newmann Coll. – *Still Life with Skull*, 1945, Picasso Coll. – *The Charnel House*, 1945–48, New York, W. P. Chrysler, Jr., Coll. – *Seated Woman (Françoise Gilot)*, 1946, New York, private coll. – *Pastoral*, 1946, Greenwich, Conn., Richard Deutsch Coll. – *Claude on his Bed*, 1948, Picasso Coll. – *Massacre in Korea*, 1951, Picasso Coll. – *Portrait of Mme H. P.*, 1952, Mrs. Maurice L. Stone Coll. – *Portrait of J. R. with Roses*, 1954, Picasso Coll. – *The Bullfight*, 1956, New York, Daniel Saidenberg Coll. – *Seated Woman in Blue Armchair*, 1960, New York, Saidenberg Gall. – *Pike*, 1960, Los Angeles, Taft B. Schreiber Coll. – *L'enlevement des Sabines*, 1962, Paris, Mus. d'Art Mod. – *The Painter and His Model*, 1963, Paris, Gal. Louise Leiris.

b. Drawings: *Self-portrait*, 1899, New York, Dr. Herschel Carey Walker Coll. – *Seated Woman with Arms Crossed*, 1902, New York, private coll. – *The Embrace*, 1903, Paris, Coll. Mme Jean Walter. – *Peasants from Andorra*, 1906, Chicago, Art Inst. – *African Head*, 1907, New York, G. Seligmann Coll. – *Still Life with Glass*, 1909, New York, private coll. – *Nude*, 1910, Paris, Coll. Pierre Loeb. – *Head*, 1914, London, Roland Penrose Coll. – *Max Jacob*, 1915, Paris, Coll. Mlle Dora Maar. – *Pierrot and Harlequin*, 1918, Chicago, Charles B. Goodspeed Coll. – *Diaghilev and Selisbourg*, 1919, Picasso Coll. – *Three Bathers*, 1920, New York, Thannhauser Foundation. – *Girl in a Yellow Hat*, 1921, New York, W. P. Chrysler, Jr., Coll. – *The Pipes of Pan*, 1923, New York, Walter W. Weismann Coll. – *Bather*, 1927, Picasso Coll. – *Minotaur and Nude*, 1933, Paris, Coll. Pierre Granville. – *Farmer's Wife on Stepladder*, 1933, Paris, Coll. Cuttoli. –

Bull with Woman and Horse, 1935, New York, Lee Ault Coll. – *Weeping Head*, 1937, Picasso Coll. – *Woman Washing her Feet*, 1944, Chicago, Art Inst. – *Faun and Centaur*, 1946, Paris, Coll. H. Berggruen. – *Mother and Child*, 1951, Worcester, Mass., Chapin Riley Coll. – *The Studio*, 1953, Chicago, Morton G. Newmann Coll. – *Bacchanale*, 1955, New York, Daniel Saidenberg Coll. – *Before the Pike*, 1959, New York, Saidenberg Gall. – *Young Girl and Duenna*, 1960, New York, private coll.

c. Graphic works: *Jardin Paris*, 1901–02, design for a poster. – *The Frugal Repast*, 1904, etching on zinc. – *Salome*, 1905, drypoint. – *Figure Turned to the Left*, 1907, woodcut. – *Mlle Léonie*, 1910, etching. – *The Brandy Bottle*, 1912, etching. – *Woman*, 1922–23, etching on zinc. – *Reading*, 1926, lithograph. – *The Picador*, 1934, etching. – *Weeping Woman*, 1937, etching and aquatint. – *Dancer with a Tambourine*, 1938, etching with aquatint. – *The Bull, I–XI*, 1945–46, lithograph. – *Black Pitcher and Death's Head*, 1946, lithograph. – *Dove*, 1949, lithograph. – *Don Quixote and Sancho Panza*, 1951, lithograph. – *Mother and Children*, 1953, lithograph. – *Divinities Visiting a Studio*, 1955, aquatint. – *Bullfighting Game*, 1957, lithograph. – *Bacchanale*, 1957, lithograph. – *Half-length Figure of a Woman after Cranach the Younger*, 1958, linoleum cut. – *Picador and Bull*, 1959, linoleum cut.

d. Sculpture: *Fernande*, 1905, Oberlin, Ohio, Allen Memorial Art Mus. – *Kneeling Woman Combing Her Hair*, 1905, New York, Nelson Rockefeller Coll. – *Head of a Woman*, 1907, New York, Sampson R. Field Coll. – *Mandolin*, 1914, Picasso Coll. – *Decorated Cup*, 1921, New York, private coll. – *Construction*, 1928, Picasso Coll. – *Construction Head*, 1931, Picasso Coll. – *Le Grand Coq*, 1932, Paris, Gal. Louise Leiris (XIV, PL. 473). – *Woman's Head*, 1932, Picasso Coll. – *Tête casquée*, ca. 1932 (I, PL. 306). – *Man with a Bouquet*, 1934, Picasso Coll. – *Death's Head*, 1941, Picasso Coll. – *Reaper in the Big Straw Hat*, 1941–44, Picasso Coll. – *Woman's Head* (*Françoise Gilot*), 1951, New York, Mus. of Mod. Art. – *Flowers in a Vase*, 1953, Milwaukee, Harry Lynde Bradley Coll. – *Six Bathers*, 1956, New York, Otto Gerson Gall. – *Bather Playing*, 1958, New York, Kootz Gall. – *Monument to Guillaume Apollinaire*, 1959, Paris, Place St.-Germaine-des-Prés (X, PL. 146). – *Standing Man*, 1960, New York, private coll.

e. Book illustrations: André Salmon, *Poëmes*, Paris, 1905, drypoint. – Guillaume Apollinaire, *Alcools*, Paris, 1913, portrait drawing of the author. – Max Jacob, *Le cornet à dés*, Paris, 1917, burin engraving. – Guillaume Apollinaire, *Caligrammes*, Paris, 1918, portrait drawing of the author. – Max Jacob, *Le phanérogame*, Paris, 1917, zinc etching. – Jean Cocteau, *Le coq et l'arlequin*, Paris, 1918, portrait drawing of the author. – Igor Stravinsky, *Ragtime*, Paris, 1919, cover design of two musicians. – Max Jacob, *La défense de Tartuffe*, Paris, 1919, burin engraving. – André Salmon, *Le manuscrit trouvé dans un chapeau*, Paris, 1919, 38 drawings. – Paul Valery, *La jeune parque*, Paris, 1921, transfer-lithograph portrait of the author. – Pierre Reverdy, *Cravates de chanvre*, Paris, 1922, 3 etchings. – André Breton, *Clair de terre*, Paris, 1923, drypoint portrait of the author. – José Delgado y Galvez, *La tauromaquia*, Barcelona, 1929–30, 6 etchings. – Gertrude Stein, *Dix portraits*, Paris, 1930, 3 drawings. – Paul Eluard, *La barre d'appui*, Paris, 1936, 3 etchings. – Paul Eluard, *Les yeux fertiles*, Paris, 1936, 5 drawings. – Iliazd (Ilya Zdanevitch), *Afat*, Paris, 1940, 6 etchings, – Georges Hugnet, *La chevre-feuille*, Paris, 1943, 6 wood engravings. – Aimé Césaire, *Corps perdu*, Paris, 1950, 31 engravings. – Adrian de Monluc, *La maigre*, Paris, 1952, 10 drypoints. – Roch Grey, *Chevaux de minuit*, Cannes, Paris, 1956, 13 drypoints and engravings. – José Delgado y Galvez (Pepe Illo or Hillo), *La tauromaquia*, Barcelona, 1959, 26 aquatints and 1 drypoint. – Jacqueline Roque, *Température*, Paris, 1960, 4 etchings.

* *

There are artists whose personalities and work have shaped an entire period in art history, and others who seem to fulfill or embody the developments of their period, almost as if they in themselves represented its natural culmination. Giotto, Masaccio, and Caravaggio belong to the first category; Raphael, Velasquez, and Vermeer to the second. Although Picasso would seem to fall readily into the first category, his position is ambiguous and controversial, for he has forcefully inserted himself into his own era and become obligatory to any consideration of it. Again and again he has succeeded in establishing its artistic laws and models, as the dominating figure over half a century. Earlier in his career the public eye was continually astonished, disoriented, and fascinated by him, but the spirit of the time was not yet identified with him. At the height of his destructive and inventive activity, when Picasso seemed to have made inevitable a final renunciation of the 19th-century development, to have blocked further elaboration of its main tendencies and done away with its remains in a single blow, a residual reaction occurred, and "irrational feeling become libido" overflowed and once again came to the surface even in his own art. Cubism fell, but surrealism was not strengthened by its fall; the abstract avant-garde, moved to react by the abandonment of the very cubism that had created it, eventually came to surpass even Picasso, who was to find himself an isolated titan in the wake of the varied and radically new directions taken by art following World War II.

No one can deny Picasso's place in art and in history: a prophet who proclaimed only himself, who changed his message as he proceeded. The full development of cubism is due most of all to Picasso. Though with Picasso at the start, Braque evolved his cubism from roots that lay in impressionism and therefore he remained closer to the preceding recognized tradition. In reestablishing the supremacy of form and reemphasizing the creative process, as opposed to the ambivalence of art and nature that had been promulgated in the 19th century, Picasso achieved the highest aims that the artist can work toward. But cubism was too cerebral and too inaccessible a concept for the period in which it was created. Whereas, on the one hand, there arose a need to render it even more esoteric — gradually leading to abstract art — on the other, its creative emphasis also nurtured the waves of irrationalism which had already appeared in expressionism, and which Picasso did not resist.

His perturbing personality revealed itself early, a creative imagination teeming with ideas and having an instinctive force that from any point of departure could infallibly arrive at its goal. The provincial milieu of his early youth in La Coruña and Barcelona does not suffice to explain the independence and self-assurance of his artistic beginnings, since once he arrived in Paris and came into contact with the work of the leading painters of the time he chose as exemplars, seemingly at random, Steinlen and Vuillard, Lautrec, and Renoir, and his technical facility permitted him to preserve with equal validity and authority Zuloaga and Goya as a base. Though something of this nature was also observable in Van Gogh, with Picasso it had a very different meaning. For Van Gogh it meant the gradual condensation of a meaningful experience that, like a gas-filled atmosphere, needed only the spark to explode. For Picasso such a preparation served as immunization against impressionism and neoimpressionism. His first brief stay in Paris helped him discard all the artificial trappings he had thus far accumulated, yet without inducing him to take part in any of the avant-garde currents. Taking little notice of Gauguin and perhaps still unaware of Cézanne, Picasso worked out a style of painting sought after by no one else. Monochrome, heavy, and chiaroscuro, it was a mode superficially involved and also inclined to accept, along with Denis, the contribution of Art Nouveau. Notwithstanding such a false start, Picasso was to reach indisputable heights early in his career, above all in the Rose Period, which nonetheless represented anything but a striking innovation in art.

A steady, almost craftsmanlike application has always characterized Picasso's work, even when it appeared to be most revolutionary. The most distinguishing feature of his art, at least until about 1920, was its steady rejection of contemporaneous vogues, yet without inventing an avant-garde trend of its own. His painting has been always and only in the present, and the development or abandonment of a manner has clearly depended upon no one but himself. By decisively placing himself thus between past and future, Picasso has continually disconcerted those who have attempted to foretell the future course of art and has ineluctably forced them to a gradual convergence on his own person. Among painters, it could be said that the only one of his contemporaries who remained quite unaffected by Picasso was Matisse. Picasso's power was that of all great artists, a strong urge to move toward form. This magnetic attraction to form came from various directions. Picasso's art was not the expression of a culture strongly polarized in any particular direction; his opposition to impressionism was not instinctive like that of the Fauves, and clearly he was not one of them. Nor did his work convey an attitude of explicit protest, of categorical rejection, as found in Dada — even though later it was his painting that became the standard bearer of a political consciousness which culminated in *Guernica* (and less forcefully in *War* and *Peace*).

Actually his discovery of the concentration on form in Negro sculpture and in the late work of Cézanne, as well as his recognition of Rousseau, revealed to him that the pictorial image had to evolve in the mind before it could be expressed. The lucidity with which Picasso grasped the two principal phases of the creative process explains his capacity to change completely the art of his time.

If cubism truly represents one of the more notable periods of art, this distinction depends first of all on its decisive negation both of art as imitation of nature and of art as the expression of feeling. The spatial character of the image had never been placed so markedly in antithesis to its ambient space as in cubism; never had the materialization of the inner essence of the object through visual means been accomplished by such a relentless mode of analysis, nor the object been submitted to such a bold spoliation. Analytic cubism demonstrated that the image could be reduced to almost algebraic signs, could lose its commonly accepted description almost completely, and yet could remain an image in so far as it was form. Almost every subsequent tendency in contemporary painting would be unthinkable without the precedent of analytic cubism. There would be little to explain for and comprehend in Kandinsky or Mondrian; and in later phases of nonobjective art, Picasso, if not cubism proper, provided the basis for the experiments of Jackson Pollock and Arshile Gorky.

It cannot be emphasized enough that cubism was not merely an ephemeral episode centered about a formalistic analysis of form. Moreover, it was not an analysis of the object based on preordained canons, but an analysis (involving the space in which the object existed as well as the object itself) that, as it gradually developed, suggested further means of investigation, new means of expression. Of these, the invention of collage was a way of giving form to — and not only a manifestation thereof — the new spatiality of the image, which was at the same time its new mode of representation. In that first brilliant phase Picasso and Braque were able to advance together because it was a work of analysis and decantation, in which there was no preestablished goal; but gradually the way itself supplied the goal. However, in this journey Picasso was something more than a painter, whereas Braque remained a painter by nature bound to a color sense which was still linked with that of impressionism. The color zone as a source of light and means of exact spatial location necessarily tended to recompose the spatial unity that had been broken up by cubist analysis. It is significant that, in the diverse stages of Picasso's career after the Rose Period, what has continued to be absent in his rediscoveries of other periods is some return, some homage to impressionism. Only in the works of the decade 1952–62 has even a slight trace of impressionism appeared, and this emerged through the works of Matisse, which might in one sense be called the ultimate refinement of impressionism.

What has been said about the various impulses that consistently led Picasso toward form also furnishes the only plausible explanation for the disconcerting parallelism with which his neoclassic and cubist styles moved side by side. One of Picasso's distinguishing qualities as an artist is that he has always remained in the present, that is, continuously evolving as he worked: his cubism, for example, did not follow a fixed course and was not aimed toward some goal that he knew in advance. In his greatest period he did not move toward abstraction through cubism but toward a realization of form, of which cubism was only one expression. With much greater maturity than when he had mixed Steinlen and Renoir, he shared in the formal perfection of classic art, which, through Ingres, presented itself as the end point of a fascinating trajectory. On the one hand, there was a course that gradually moved toward the unknown, with abrupt halts, breaks, and changes of direction: this was cubism. And the reconsideration — not a retreat, as it has sometimes been considered — that is designated synthetic cubism disclosed in yet another way Picasso's constant awareness of form (taking into account the participation of Gris as well). On the other hand, there was classic art, a tendency completely opposite in that the end point was established and the trajectory had only to be repeated.

With the phase of his heavy figures, such as *Two Seated Women* (1920), *The Race* (1922), and the *Pipes of Pan* (1923), Picasso introduced an aspiration to the monumental into the neoclassical trend, seeking to re-create from within the essential grandeur and solemnity of classical art. But such an Apollonian aspiration to a new classicism — common, albeit with different means, to both the cubist works of the above-mentioned period and to neoclassic works (almost to the point of being counterfeits, rather than imitations, in the illustrations for the *Metamorphoses* edition of 1931) — was not an aspiration with which the period could indentify strongly. The explosion of surrealism, in which the most varied irrational impulses flowed together, replaced the nihilism of Dada, which had been a harsh and encompassing judgment — like a great X scrawled upon itself by the period — and before which not even cubism had found mercy.

The most obscure aspect relating to the influence exerted by surrealism on Picasso (and it was an unsettling, fragmenting influence) is to find some reason as to why Picasso, who seemed the most prepared to resist, did not resist the force of the new development. There was no overwhelming personality among the surrealist painters, for not even Max Ernst, the most stimulating talent of the movement, could rival the artistic gifts of Picasso. Nor did Picasso himself ever openly agree to participate in the group. The reason for his surrender was undoubtedly a more profound and ineffable one. Picasso could defy the popular mode and could accept an apparent lack of contemporaneity as long as he was sustained by the certainty that he could capture the true and essential face of his time — that is, even if its nature were unknown to the time itself. One such face he had created from zero, with cubism. In addition, he had presented his period with a tragic mask in the form of his resurrection of classicism. The development of surrealism, however, was not so much a new figural mode — though it involved that also — as it was a growing awareness of the deceptive and unreal aspects of daily life, of the barely concealed abyss that is within us. Feelings of profound dissatisfaction and frustration, evasions, and transferrals all had passed from the psychiatric clinics where they were recognized and defined into the typical middle-class family life, where they became the ruling agents.

Picasso's awareness of form vacillated and was no longer his only motive force, his sole internal justification. As early as 1925, in the *Three Dancers*, elements appeared that could no longer be traced back to cubist inspiration: spatial structure grew weaker, the dislocation of the elements of the image became arbitrary, and their disposition was no longer analytically determined by the figural context. Symbolic motifs appeared, and perhaps since Picasso has always been a direct rather than a symbolic painter, the loosening of the spatial rationale led to the introduction of expressionistic elements, into which the surrealist infiltrations were also transmuted, instead of developing into symbolic allusions. Line, which he had manipulated with such mastery in his almost transcendental exercises of interrupted-line drawing and which he had fragmented so thoroughly in cubism, again assumed a decisive role in formulating the image, even outside his exercises in the neoclassic vein. A surprising new series of works was created, in which the image was recomposed with an authoritarian line that "recapitulated" the object in its own way. The stratified and intersecting cubist planes seem to fuse to an ever-increasing degree, but nonetheless remain a distinct substratum in the vortex surrounding the image. In the *Minotaur* (1928) and the *Acrobat* (1930) this recomposing of the image perhaps reached its highest point. Such fabulous figures, which were contemporaneous with the scenes painted at Dinard (1928), clearly indicate that surrealism had not yet exhausted its fascination for Picasso. This was a period of great vigor for the artist; his studies for a monument (1930), for example, which show an awareness of suprematist developments, can be viewed as a point of departure for Henry Moore's vital renewal of sculpture.

If the possibility of Picasso's simultaneous development in almost diametrically opposite directions had to this point in his career been an astonishing and disoriented phenomenon, the two tendencies had maintained a notable degree of independence and purity. A certain osmosis took place at about this time, however, a mutation that was already perceptible in the drawings for the *Lysistrata* (1934), which were no longer as austerely Greek and Apollonian as those for the *Metamorphoses*. The expressionist tendency was strengthened to the point of almost destroying the cubist substructure in the series of still lifes painted in 1937 and 1938.

The recapitulation of all that Picasso had done so far — a recapitulation that was not only formal but ethical and political as well — took place at precisely this same time, in the great monochrome painting *Guernica* (1937). Into this carefully thought out and deeply felt work, with incredible intensity Picasso transferred his entire being, with symbol as with the instinctiveness of *Urschrei*, evoking beauty as well as horror reduced to the level of the grotesque, with the formal grandeur of cubism but also with surrealist *coups de main*. A milestone and one of the great heights of Picasso's career, this work constitutes perhaps the most tragic expression of our epoch.

In the period that followed, roughly that of World War II, the all but Olympian assurance he had gained in everything he touched was qualified only by the fact that he now lacked that urgent straining toward form which had guided him so justly in the most varied directions, and the analytical, structural tension formerly present became noticeably loosened. The inventions he took from life (which he at that time spent on the Riviera) are often surprising in the authoritarian brutality with which they are extracted from their ordinary context. There is no doubt that a painting such as *Night Fishing at Antibes* (1939), among others of this phase, will continue to represent one of the fundamental artistic texts of our time.

Albeit a more attenuated activity and more modest in its aims, Picasso's later interest in ceramics (from about 1949) has been much more than a minor pursuit or mere hobby, particularly in that it reveals his need of activity. In his ceramic work Picasso undeniably redimensioned himself, but at the same time he continued with his painting and sculpture. Indeed, some of the sculpture of this period was to number among his most widely known work, such as the *Goat* (1950), in which a violent new realism, expressionistic in key, displaced the vestiges of Rodin that were still apparent in the *Man with a Lamb* (1944). Although Picasso's activity as a sculptor has been discontinuous, each of his pieces, beginning with those of the early cubist period, has furnished a point of departure for many of his contemporaries.

In his more recent works of large size — the *Massacre in Korea* (1951), *War* and *Peace* (1952), and *Le Baignade* (1957) for UNESCO — the ambitious scale of the composition is not always matched by the intensity of expression. A consciousness of Matisse's color, of his sure and reductive line, which had never before influenced Picasso, forms a sort of mental climate for his work of this period; this explains, in a sense, his sudden shift toward a portraiture that on occasion seeks an adherence to the subject even more precise and meticulous than that evidenced at the time of the *Portrait of Gertrude Stein* (1906). These works might be characterized as the results of a seduction by the lifelike — but as seen through the chromatic filter of Matisse. From such an internal reference may also derive his attraction for the precursors of impressionism, as attested in his series of *Women of Algiers* (1954–55) after Delacroix and *Las meninas* (1957) after Velázquez. In these Picasso transformed his sources, sometimes simplifying, sometimes combining, following an ideal pattern like that of musical variations, but perhaps also modeled after Mondrian's early exercises in decomposition of form, based on the motif of a tree.

Whatever the ultimate judgment on the work of this latest phase of his career, a phase that is still in progress, it remains that its final evaluation cannot impugn what Picasso had achieved and represented up to 1937. Neither a survivor nor an epigone of himself, Picasso continues to evolve wholly from within himself, and the fact that his late work is no longer at the root of contemporary artistic development neither detracts from nor denies his personality and his originality. His vital energy, no less than his preoccupation with laws of form, has kept and will always keep him at the forefront of the artistic history of his epoch.

Cesare Brandi

Bibliog. *General*: A. H. Barr, Jr., Picasso: Fifty Years of His Art, New York, 1946 (full preceding bibliog.); C. Zervos, Pablo Picasso, Milan, 1946 (bibliog.); M. Raynal, Picasso, Geneva, New York, 1953; W. Boeck and J. Sabartès, Picasso, New York, London, 1955 (bibliog.); F. Elgar and R. Maillard, Picasso, Paris, London, 1955; A. Valentin, Picasso, Paris, 1957; R. Penrose, Picasso: His Life and Work, London, 1958; J. Golding, Cubism, London, 1959; P. de Champris, Picasso: Ombre et soleil, Paris, 1960; R. Rosenblum, Cubism and 20th Century Art, New York, London, 1960; H. Jaffé, Picasso, New York, 1964; J. Berger, Success and Failure of Picasso, London, 1965. *Reminiscences, memoirs, and photographic documents*: F. Olivier, Picasso et ses amis, Paris, 1933; G. Stein, Picasso, London, 1938; J. Sabartès, Picasso: Portraits et souvenirs, Paris, 1946 (Eng. trans., Picasso: An Intimate Portrait, New York, 1948); J. Sabartès, Picasso: Documents iconographiques, Geneva, 1954; A. Salmon, Souvenirs sans fin, 2 vols., Paris, 1955–56; R. Penrose, Portrait of Picasso, London, 1956; D. D. Duncan, The Private World of Pablo Picasso, New York, 1958; L.-G. Buchheim, Picasso: A Pictorial Biography, New York, 1959; H. Parmelin, Picasso sur la place, Paris, 1959; G. Apollinaire, Chroniques d'art (ed. L. C. Breunig), Paris, 1960; D. H. Kahnweiler, Entretiens, Paris, 1961.

Works: *a. General*: C. Zervos, Picasso: Œuvre catalogue, 12 vols., Paris 1932–61 (works to 1943). See also CahArt, 1939–54, passim, and Verve, 19–20, 1948; 25–26, 1951; 29–30, 1954. *b. Painting*: G. Apollinaire, Les peintres cubistes, Paris, 1913 (Eng. trans., L. Abel, New Tork, 1944); D. H. Kahnweiler, Der Weg zum Kubismus, Munich, 1920 (Eng. trans., H. Aronson, The Rise of Cubism, New York, 1949); J. J. Sweeney, Picasso and Iberian Sculpture, AB, 1941, pp. 190–98; A. Cirici-Pellicer, Picasso antes de Picasso, Barcelona, 1946 (Fr. trans., rev., Picasso avant Picasso, Geneva, 1950); H. and S. Janis, Picasso: The Recent Years, 1939–46, New York, 1946; W. S. Lieberman, Picasso and the Ballet, 1917–45, Dance Index, V, 1946, pp. 263–308; J. Larrea, ed., Guernica, New York, 1947; J. F. Ráfols, Modernismo y Modernistas, Barcelona, 1949; C. Roy, ed., La guerre et la paix, Paris, 1952; Vercors, D. H. Kahnweiler and H. Parmelin, Picasso: Œuvres des Musées de Leningrad et de Moscou et de quelques collections parisiennes, Paris, 1955; L. C. Breunig, Studies on Picasso 1902–1905, College Art J., 2, 1958, pp. 188–95; J. Sabartès, Les ménines, Paris, London, 1958; J. Runnquist, Minotauros 1900–1937, Stockholm, 1959; J. Padrta, Picasso: The Early Years, London, 1960; D. D. Duncan, Picasso's Picassos, New York, London, 1961; R. Arnheim, Picasso's Guernica, Berkeley, Los Angeles, 1962; A. Blunt and P. Poole, Picasso: The Formative Years, London, Greenwich, Conn., 1962; D. Cooper, Picasso: Les Déjeuners, Paris, 1962. *c. Drawings*: C. Zervos, Dessins de Picasso, 1892–1948, Paris, 1949; D. Cooper, ed., Picasso: Carnet catalan, Paris, 1958; G. Boudaille, ed., Picasso: Carnet de La Californie, Paris, 1959; M. Jardot, Pablo Picasso: Drawings, New York, 1959; K. M. Dominguín and G. Boudaille, eds., Toros y toreros, Paris, 1961 (Eng. trans, Picasso: Bulls and Bullfighters, New York, 1962); A. Miller, The Drawings of Picasso, 1961. *d. Graphic arts and book illustrations*: B. Geiser, Picasso: Peintre-Graveur, Berne, 1933; F. Mourlot, Picasso lithographe, 3 vols., Monte Carlo, 1949–56; B. Geiser, Picasso: 55 Years of His Graphic Work, London, 1955; H. Matarasso, Bibliographie des livres illustrés par Picasso, Nice, 1956; J. K. Foster, Posters of Picasso, New York, 1957. *e. Sculpture*: D. H. Kahnweiler, Les sculptures de Picasso, Paris, 1949; G. C. Argan, Le sculture di Picasso, Venice, 1953. *f. Ceramics*: S. and G. Ramié, Céramiques de Picasso, Lausanne, 1948; C. Zervos et al., Céramiques de Picasso, CahArt, XXIII, 1948, pp. 72–208; J. Sabartès, Picasso ceramista, Milan, 1953; D. H. Kahnweiler, Picasso: Keramik, Hannover, 1957.

Exhibition catalogues: Picasso (Palazzo Reale), Milan, 1953; R. Benet and R. Llates, eds., Els Quatre Gats, Barcelona, 1954; L'œuvre gravé de Picasso (Mus. Rath), Geneva, 1954; Das graphische Werk Picassos (Kunsthaus), Zürich, 1954; M. Jardot, ed., Picasso (Mus. des Arts Décoratifs), Paris, 1955; Arts Council of Great Britain, Picasso: 50 Years of His Graphic Art, London, 1956; A. H. Barr, Jr., ed., Picasso: 75th Anniversary Exhibition (Mus. of Mod. Art), New York, 1957; Picasso: Peintures 1955–56 (Gal. Louise Leiris), Paris, 1957; D. Cooper, ed., Picasso (Mus. Cantini), Marseille, 1959; Picasso: Les Ménines (Gal. Louise Leiris), Paris, 1959; Picasso: Dessins 1959–60 (Gal. Louise Leiris), Paris, 1960; Picasso: 45 gravures sur linoleum 1958–60 (Gal. Louise Leiris), Paris, 1960; Arts Council of Great Britain, Picasso (Tate Gall.), London, 1960; J. S. Boggs, Picasso and Man (Art Gall. of Toronto), Toronto, 1964.

Anna Grelle

Illustrations: pls. 141–146.

PICTURESQUE, THE. The word *pittoresco* appears to have originated in northern Italy early in the 17th century to denote the point of view characteristic of painters — in other words, that which is particular to painting. In England (ca. 1730–1830), "picturesque" came to identify an esthetic concept that had wide practical consequences. This point of view, historically considered, marks a stage in the development of romanticism that is not limited to post-Renaissance culture alone. Common to all its meanings has been an emphasis on certain arresting elements of scenes, and more particularly as presented in "painterly" renderings of their tactile values, so that it is these qualities which affect the beholder as much as or even more than the scene's actual components. *Pittoresco* (and its equivalents *pittoresque* and "the picturesque") is a visual analogue of "poetic."

The interaction of poetic and formal values, with their combined effect of heightening visual appreciation generally, has come about in most cultures at that stage when the society has been educated in the appreciation of the arts to a degree that their values are widely accepted. In *The Romantic Agony* (London, 1933, pp. 13–21), Mario Praz has indicated the affinities of Hellenistic pastoral poetry and genre painting with the 18th-century picturesque, suggesting that Hadrian's Villa at Tivoli revealed "quite as much preoccupation with picturesque views, and as much exoticism, as were shown by owners of English country houses with their 'Gothic ruins' and Chinese pagodas." Passages in Greek romances, which had reflected something of Hellenistic painting, in turn influenced poetry and thus came to be reflected in the courtly and allegorizing painting of the Renaissance. Moreover, Praz points out, it was the Alexandrians who worked out the esthetic formula "ut pictura poesis," which, as elaborated by the 16th-century emblematists and stated in Charles Alphonse Dufresnoy's *De arte graphica* (1665), published posthumously in Paris in 1668, became the polite esthetic of post-Renaissance Europe. As late as 1794 Uvedale Price adopted the remark of Cicero, "quam multis vident pictores in umbris et in eminentia quae nos non videmus," as the motto of his *Essay on the Picturesque*. . . . In ancient China, as in medieval Europe, similar evidence exists of the heightening of visual appreciation by association with diverse arts, particularly landscape painting.

This discussion will be largely concerned with that phase of the artistic associative process in which the term *pittoresco*, along with the conscious formulation of the concept of "the picturesque," came into use. An early occurrence of the word is found in a report by G. A. Costa with reference to architectural design (*Per la facciata del Duomo di Milano*), published about 1654. The word is not, however, given by Filippo Baldinucci in his *Vocabolario toscano dell'arte del disegno* (Florence, 1681), its absence perhaps confirming Price's supposition of its origin among painters of the Venetian school. It is credible that their departures from convention in visual renderings of scenery of the Venetian locale — by means of an over-all tonal unity, accentuated chiaroscuro, and those "sudden variations of form and line" which Price considered to be "the most efficient causes of the Picturesque" — demanded an epithet to distinguish them from other tendencies in painting.

The contemporaneous development of painting and visual appreciation in the Netherlands produced, possibly a little earlier in the 17th century, the apparently synonymous Dutch word *schilderachtig*. W. J. Hipple, Jr., cites its use in a 1618 edition of *Het schilder-boeck*, a work by Karel van Mander published initially in 1604, and its use in German by Joachim von Sandrart, for example, with reference to Rembrandt's painting (*L'academia todesca della . . . pittura*, Nürnberg, Frankfort, 1675–79). *Schilderachtig* is given in no German dictionary, its place, though hardly its connotation, being taken by *malerisch*. Whether or not such Italianate northern painters as Jan Both and Adam Pynacker were *schilderachtig*, their nostalgic landscapes depicted an aspect of scenery that came to be regarded as highly picturesque in England (PL. 148). It may be advanced that many of the visual qualities and psychological reactions subsequently associated with the concept of the picturesque entered into mannerist and, still more, into baroque architecture, most notably in representations of it in stage scenery.

In England W. Aglionby observed of free and natural execution in painting, "This the Italians call working A la pittoresk" (*Painting Illustrated*, London, 1685). At that time, however, his and the ensuing literary use of the new word applied to allegorical painting. Pope regarded it in 1712 as a French term, applying it, for example, to Homeric descriptions as a synonym for "vividly conceived in the Grand Manner." The Académie Française, however, did not admit the word *pittoresque* until 1732, and a century later Stendhal believed that "le pittoresque nous vient d'Angleterre."

Extension of the concept (but not as yet the word "picturesque") to natural scenery is found first in England early in the 18th century, upon increasing familiarity with Italian landscape painting. Making the customary grand tour, many of the British aristocracy acquired landscapes by Claude Lorrain, Salvator Rosa, and the Poussins (qq.v.; PLS. 147, 149; IX, PL. 15), which not only captured vividly the emotional associations of the Italian and Alpine scenery they themselves had experienced but which also imparted to them a fresh appreciation of English scenery — just as they found models for their mansions in the villas of Palladio. At home again, they began to seek or reproduce similar scenic effects in their parks. The castles of the architect-dramatist Sir John Vanbrugh (q.v.) illustrate the influence of these associations; however, the picturesque character of Blenheim Palace (VI, PL. 436), erected for the first Duke of Marlborough, was only recognized initially by Reynolds in 1786. About 1730 the painter and architect William Kent (q.v.) began designing gardens that were to some extent in imitation of landscape paintings (VIII, PL. 444). English poetry after 1725, especially James Thomson's *The Seasons*, abounds in word paintings, the color and composition of which are vivid derivations of landscape paintings. Visually educated tourists in the mid-18th century, notably the poet Thomas Gray and the agricultural reformer Arthur Young, gave critical attention to picturesque aspects of actual scenery. In the 1770s the Reverend William Gilpin commenced his *Observations Relative Chiefly to Picturesque Beauty in Several Parts of Great Britain*, illustrated by characteristic water colors that were reproduced in aquatint when his tours were published between 1782 and 1809. Thus an artistic vogue was transformed into an aspect of esthetic theory.

By suggesting an esthetic basis for the pleasure increasingly excited by abrupt or rude forms and textures in nature, Gilpin met a need then current especially in Great Britain. The heightened sentiment for nature, everywhere an underlying premise of the rococo period, was manifested in the British Isles not only in landscape gardening and in the naturalistic reworking of earlier baroque parks along the "waving" lines recommended by Capability Brown (1715–83) but also, and perhaps more symptomatically, in the general pursuit of romanticism and awareness of natural science. The esthetic thinking of the generation had been formulated by the publication, in 1757, of Burke's *Philosophical Inquiry into the Origin of Our Ideas on the Sublime and Beautiful*. By recognizing the awesome and the infinite as other possible sources of esthetic satisfaction, Burke was the first popular writer to account for a significant element in the taste of the age. Yet, in common with most contemporaries, he continued to trace or attribute emotions ultimately to intrinsic qualities of objects, and his complementary categories failed to account for a wide range of material that gave visual pleasure. Gilpin suggested that such objects displayed "picturesque beauty" and that, ranging from the shaggy animal and withered tree to ruins and mountain crags, they possessed the common quality of roughness. As the distinguishing tactile quality of ideal beauty was smoothness and the illusion of infinity distinguished the sublime, so broken textures and varied shapes in nature were the qualities of the picturesque and presented a chiaroscuro suitable for rendering by painters with a "free bold touch." The artist, selecting from and combining the "variety of parts," formed a broadly conceived — that is, a picturesque — composition. In an "analysis of romantic scenery," based on his tour of the English lakes (1786) and included in the *Observations* . . . , Gilpin gave rules for the selection, treatment, and composition of landscape features according to the conventions of the 17th-century painters. The naïve clergyman's enthusiasm appealed primarily to amateurs like himself, though many of the English water-colorists exemplify the same taste and principles (PL. 153).

Sir Joshua Reynolds (q.v.), with whom Gilpin discussed his thesis, made no explicit reference in his *Discourses on Art* to the picturesque, which, in so far as it denoted anything of visual significance, he regarded as synonymous with beauty. In the twelfth of these discourses (1784), however, he paid tribute to the art of Gainsborough (q.v.), whose landscape paintings and freedom of technique mark him as the outstanding exponent in England (as are Tiepolo and Guardi in Italy) of that visual impressionism which the word "picturesque" had come to denote. Though he held Gainsborough to be blind to the conception of the ideal, Reynolds admitted the advantages of observing nature through the eyes of a painter so deeply perceptive of the particular. He recalled Gainsborough's custom "of continually remarking on peculiarities of countenance, accidental combinations of figures, or happy effects of light and shade," and of assembling found objects as a source of details for his landscapes. Uvedale Price also acknowledged the influence of Gainsborough in his *Essay on the Picturesque . . .*, published in 1794 in conjunction with Richard Payne Knight's didactic poem *The Landscape*.

It is significant that these related manifestoes were produced almost simultaneously by neighboring landowners (in the romantic countryside of the Welsh marshes) who were also dilettanti professing advanced political and esthetic creeds, and that they were primarily intended to amend taste in landscape gardening by forestalling publication of a work by Humphry Repton, the leading successor of Capability Brown. Price and Knight claimed that the rolling, informal effects produced by conventional methods of "improving" landscapes ignored or destroyed the values most significant to eyes sensitized by the study of paintings. The theory they upheld involved a more detailed analysis of these picturesque, or visual, values than that of Gilpin, and it led to critical reappraisals that contributed important elements to neoclassic and romantic esthetics.

Price's *Essay on the Picturesque . . .* is the generally accepted, because the most direct and attractively written, exposition of picturesque theory. It was revolutionary in claiming that much great art and all of nature were to be fully appreciated not according to academic precepts but by means of direct observation, enhanced by study of the principles and technique of great "modern" painting. Retaining Burke's physiological system and objective categories, Price similarly conceived of the picturesque as consisting in physical qualities directly affecting the observer's sensibilities, though he rejected Gilpin's identification of it with the paintable romantic as too vague a formulation. The reaction to the picturesque was the nervous stimulus of "curiosity." The essentially picturesque qualities in nature corresponded to the "leading principles" found in paintings of the Venetian school from Giorgione to Mola, in certain works of 17th-century landscape, notably those of Salvator Rosa (IX, PL. 16), and in the landscape paintings of northern masters such as Rubens and Rembrandt in particular (IX, PL. 19). The first principle of these painters, and therefore of the picturesque, was "connection," a general fusion of all visible qualities similar to that produced by evening light. Of these qualities, the more picturesque consisted in "intricacy," variety of form and disposition, light and shadow, and sensuous coloring. Objects displaying "roughness, sudden variation and irregularity," whether inherent or in consequence of age, and aspects of nature rich in color, such as the season of autumn, had elicited the most "forcible effects" from these painters and were thus to be regarded as the most picturesque, in both their essential attributes and in their consequences.

In the light of these premises, much that had hitherto been considered ugly came to be viewed as picturesque, and formlessness alone remained deserving of that harsh epithet. The traditional concept of an ideal beauty (as Knight subsequently pointed out) was in effect reduced to little more than the insipid, which was to be rendered visually attractive only by an admixture of picturesqueness. The ultimate criterion established by Price in place of the ideal harmonies of classical esthetics was the freedom and energy with which the individuality, whether of the object or of the artist, was expressed. The picturesque could be equated with "character." The application of these principles in the additional essays on garden design and architecture published in the 1810 edition of Price's work had important consequences. In architecture it did not so much foster the revival of the Gothic as it heightened the appreciation of older and irregular buildings in general, from the rustic cottage to the baroque palace; moreover, it encouraged their preservation in ways that their peculiar texture and intricacy might be maintained or reproduced. In actual landscape, the picturesqueness not only of the natural state but also of venerable formal garden plans was preferable to artificial simplification; for example, the terraces and fountains of old Italian gardens, intensely picturesque in themselves, fulfilled the essential quality of "connection" between architecture and natural setting (VIII, PLS. 429–435).

Knight shared Price's aims and evaluations of art in general, but his more acute critical mind saw the concept of the picturesque as part of a strikingly new interpretation of esthetics, founded on physiology and early notions of psychology. The difference of approach is apparent in his poem *The Landscape*, publication of which actually preceded that of Price's essay by a few months, and was expressed in a note to the second edition (1795) in which he denied the objectiveness of the picturesque, since it is "merely that kind of beauty belonging exclusively to the sense of vision, or to the imagination guided by that sense." This distinction is developed as part of Knight's *Analytical Inquiry into the Principles of Taste*. According to the physiological basis for all taste reactions that was established in the first part of this work, visible beauty — that is, that which gives pleasure organically to the visual faculty — consists wholly in broken or graduated light and color, factors which Knight demonstrates to be basic to everything Price had postulated as picturesque. In so far as it is the art of painting that, being devoted to visual pleasure only, separates the visible aspect from all others associated with it in practical experience, the term "picturesque" is valid, the author believes, to signify visual beauty as distinct from that of utility or propriety. But another, more subtle meaning of picturesque derived from an imaginative extension of associated ideas. To a mind conditioned by pictures, not only works of art but "all objects in nature and society" may be significant through association, and though they may give no direct visual pleasure are thus to be termed "picturesque." Like Price, Knight traced the word *pittoresco* to that stage of visual culture reached by the Venetian and Dutch schools when painting presented what the eye sees rather than copying what the mind knows to be from the testimony of other senses (IX, PLS. 14, 17, 20). In a passage that anticipates Wölfflin, he distinguishes the linear and painterly, the clear and unclear. Knight concludes that the picturesque is the beautiful, in the defined sense. Although painting has rendered it visible, the existence of this beauty is independent of both painting and of the mind. This conclusion, so much like Price's view of the picturesque, underlines the similarities between the conceptions of both these men of acute esthetic sensibility and contrasts with Archibald Alison's system, wholly ideal and logically irrefutable, which accounted for all esthetic phenomena entirely by association. His *Essays on the Nature and Principles of Taste*, though published in 1790, appear to have had no influence on Price or Knight, but subsequently went far to shape 19th century romantic sentiment, as distinguished from picturesque discrimination.

In so far as the concept of the picturesque developed into an esthetic system or philosophy, it was originally and almost exclusively a British phenomenon, characteristic of a national empiricism that persistently tends to approach esthetics intuitively from the perception of nature rather than intellectually from formal postulates. In painting itself, the direct influence of formal picturesque theory produced little of worth; but it can be claimed that the tactile values underlying this concept were the same as those which inspired much of the best contemporaneous and 19th-century painting. In many respects, Price's "principles" are expressed in Constable's landscapes (q.v.; III, PLS. 440–444; IX, PL. 23), and Knight's "beauty of pure light and color" in the abstractions of Turner's last phase (PL. 156); thus, at some remove, the theory of the picturesque helped beget French impressionism (q.v.), a trend that at the same time disavowed the 19th-century corruption of what was comprised by the picturesque (see below). In English architecture and garden design, the principles of the picturesque have persisted; further, such theories have affected developments on the Continent, especially in the beginnings of functional architecture, which owe something to the tenets of Price and Knight.

Christopher HUSSEY

In the course of the 19th century, the term "picturesque" acquired other nuances that have remained in the popular usage, where the original significance has been almost entirely lost. The "picturesque" that retains currency in the vocabulary of the amateur artist and the tourist — Webster's "quality or principle which combines what is unusual and charming in scenes, objects, actions, or ideas without attaining beauty or sublimity" — is in reality a corruption of the original meaning. In the sense of the formal theories of the picturesque, the painterly and poetic, it has been substituted by other expressions.

As has been noted above, in its initial usage "picturesque" referred to that which was "truly of painting and painters." Moreover, when used with reference to the creative process, the term came to imply the frenzy of expression, inspiration, the painter's instinct or genius, which is neither subject to rules nor, as a natural gift, to be comprehended by the intellect. Thus, in the 17th century, the expression "lavorare alla pittoresca" (to work in a picturesque manner) was frequently used synonymously with "lavorar di furia" (to work in frenzy).

When extended to style (very likely as a result of works done in such a mode), the designation "picturesque" inferred a pictorial and coloristic perception, translated directly and swiftly onto canvas and more concerned with over-all effect than with careful presentation of details. Characteristic of this bold synthetic, rather than analytic, vision is deemphasis of draftsmanship and linear perspective, with an emphasis instead on color effect and on light and shadow, executed in a vigorous, somewhat nervous manner. (In a letter of Salvator Rosa, for example, the artist mentions a floral painting so rendered "in the picturesque manner.") This conception of the picturesque continues in modern criticism but not in the terminology employed, which favors more explicit terms — "painterly" as opposed to "plastic" effects, "tonal" as opposed to "draftsmanly," and so on.

In the late 18th century, the critic Francesco Milizia (collected works, 1826–28) qualified the definition of "picturesque" with reference to painting: "For picturesque, we mean something unusual which immediately catches the eye and is pleasing." An exponent of neoclassicism, however, Milizia attacked such picturesqueness, which was to be a distinguishing trait of the romantics. Already troubled by the prevalence of these qualities, he lamented, "It is no talent. . . . There is no longer any reflection involved in the placement of the subject, nor any beauty of form, purity of design, character, or expression. The great merit in paintings has become the fine craft of picturesque arrangements, picturesque effects, and picturesque lines. Thus painters have acquired the privilege of no longer thinking; they arrange, maneuver." The conception of the picturesque was developing, thus, into a search for facile surface effects which appealed to the senses or emotions but which could not satisfy the intellect.

In architecture, too, a building filled the requirements of the picturesque when the rational architectural scheme gave way to some decadent state: fragmentary, damaged, or overgrown with weeds and thereby fused with the landscape, reverting variously to the "disorder" of nature. The prevalent concern with ruins in 18th-century painting, for reasons both of *rocaille* taste and of a preromantic exaltation of the "poetry of ruins," was another expression of the picturesque ideal (PLS. 149, 151; IX, PL. 18). Moreover, the search for such effects that had dominated in painting from Giovanni Ghisolfi to Giovanni Pannini, Sebastiano Ricci, Michele Marieschi, Bernardo Bellotto (IX, PLS. 18, 21), and Hubert Robert (PL. 150) also was frequently in evidence in landscape architecture (q.v.) of the era, in the form of fabrications of ruins and classic temples, as well as fragmentary or broken columns and porticoes — in other words, fake ruins with the purpose of suggesting authentic ones to lend a romantic air of antiquity to the landscape. This was an architecture to be reflected in the waters of a lake or pool, to ornament a natural landscape, or to blend with the greenery (PL. 152; VIII, PL. 445).

The varied interests of the romantic movement brought new themes to satisfy the seekers after the picturesque: to the repertory of the classic as a lost golden age, poetic and far removed in time, were added scenes of strange and faraway places, Orientalia, and other exotic motifs (see EXOTICISM), particularly of North Africa and the Near East (IV, PLS. 142–145). Distant lands and cultures seemed, categorically, more picturesque than familiar landscapes and peoples.

The popularity of travel literature and broad curiosity about local traditions, customs, and folklore encouraged what might be termed a search for "sociological" picturesqueness. William Hogarth, Théodore Géricault (qq.v.), and Bartolomeo Pinelli, artists who were different in temperament and style, were all fascinated by the curious and exceptional survivals found in those strata of ordinary life generally considered picturesque because of their ability to retain, apart from the upper and the more progressive levels of society, long-established traditions and customs, and even dress that was often colorful or gaudy. Further, any event that furnished an unusual or colorful spectacle — for example, a ceremony in costumes of other times — became picturesque subject matter and was in itself almost synonymous with the visually stimulating, the scenographic and charming, and therefore was an ideal theme for painting. Interpreted in this way, the picturesque led painting toward a search for particular subjects; and understandably much of the painting of the time took on a literary and historical cast.

In the second half of the 19th century, with impressionism and the affirmation of "pure" vision, there came a reaction against the picturesque that denied any possibility of identifying the term with beauty, even a particular beauty, and of accepting such literary and historical orientation in subject matter as a means to poetical expression. Since that time, the term has acquired a generally negative connotation; or it has been restricted, for the most part, to natural beauty and to architecture in its more indigenous forms, or to instances where the work of man is fused with that of nature. Such corrupted usage, which has become the primary meaning, has made it inappropriate terminology for distinguishing that art which, in resorting to the picturesque, seeks to attract through pseudo-esthetic values intrinsic to the subject matter rather than through carefully considered esthetic values.

Luigi SALERNO

BIBLIOG. E. Burke, A Philosophical Inquiry into the Origin of Our Ideas on the Sublime and Beautiful, London, 1757 (ed. J. T. Boulton, London, New York, 1958); W. Gilpin, Observations Relative Chiefly to Picturesque Beauty in Several Parts of Great Britain, 8 vols., London, 1782–1809; W. Gilpin, Three Essays to Which is Added a Poem on Landscape Painting, London, 1792; U. Price, An Essay on the Picturesque as Compared with the Sublime and the Beautiful, 2 vols., London, 1794–98 (3d ed., 3 vols., London, 1810); R. P. Knight, Analytical Inquiry into the Principles of Taste, 1805; F. Milizia, Opere complete risguardanti le belle arti, 8 vols., Bologna, 1826–28; E. W. Manwaring, Italian Landscape in 18th Century England, New York, London, 1925; C. Hussey, The Picturesque: Studies in a Point of View, London, 1927; J. H. Mayoux, Richard Payne Knight et le Pittoresque, Paris, 1933; W. D. Templeman, The Life and Work of William Gilpin (Univ. of Ill. S. in Language and Lit., XXIV, 3–4), Urbana, 1939; M. Jourdain, The Work of William Kent, London, 1947; D. Stroud, Capability Brown, London, 1950; W. J. Hipple, Jr., The Beautiful, the Sublime, and the Picturesque in 18th Century British Aesthetic Theory, Carbondale, Ill., 1957 (full bibliog. and numerous refs. to ms. sources).

* *

Illustrations: PLS. 147–156.

PIERO DELLA FRANCESCA. Italian painter (b. Sansepolcro, near Arezzo, ca. 1420; d. Oct. 12, 1492). He was the eldest son of Benedetto de' Franceschi, shoemaker and skin dresser, and Romana di Perino da Monterchi. The surname "de' Franceschi" was long ago altered into that of "della Francesca," by which name Piero is traditionally known.

It is not wholly improbable that Piero may have acquired the first rudiments of his art locally in the workshop of that master Antonio di Anghiari who had been entrusted in 1430 with the task of painting the altarpiece for the Church of S. Francesco, a task transferred to Sassetta in 1437. There is no doubt, however, that only in Florence could Piero have had a proper initiation to painting. That he was in that city on Sept. 7, 1439, is proved by a document relating to sums paid to Domenico Veneziano and Piero for the frescoes in the choir of S. Egidio.

In the margin it is stated that "Pietro, son of Benedetto of Borgo a San Sepolcro, is with him." It is not known how long Piero's apprenticeship with Domenico Veneziano lasted or when and where — whether in Umbria or Florence — it began. It is a fact that Piero was again in Sansepolcro in 1442 and was elected to the Town Council. By 1444 Sassetta had finished his resplendent polyptych in S. Francesco; possibly inspired by this, the Confraternita della Misericordia (the Confraternity of Mercy) commissioned Piero, on Jan. 11, 1445, to paint a large polyptych (PL. 157; XII, PLS. 26, 376). The terms of the contract stipulated that the artist must paint the entire work himself and finish it within three years. Piero, however, does not seem to have remained in Sansepolcro long enough for matters to proceed according to the terms of the contract: the polyptych was not completed in the allotted time and part of it was painted by assistants.

It is probable that at this time Piero was already in contact with the Duke of Urbino. An old local tradition maintains that the *Flagellation of Christ* (PL. 158) was intended to commemorate the violent death of Prince Oddantonio da Montefeltro, who is supposedly portrayed with his evil ministers. According to Longhi, it should be inferred therefore that it was painted not long after 1444, the year of the conspiracy of the Serafini. Whatever conclusions are to be drawn on this point, it is quite certain that around the middle of the century — 1448, according to Salmi's calculations based on reflections of the work found in various miniatures — Piero was in Ferrara, and, although the frescoes of the Castello Estense and of the Church of S. Agostino are now lost, they undoubtedly had a decisive influence on the development of Ferrarese art. From Ferrara to Rimini, almost in the very tracks of Leon Battista Alberti, is but a short step, and in 1451 Piero was in Rimini, painting the large fresco in the Tempio Malatestiano showing Sigismondo Malatesta, followed by two greyhounds, kneeling before his patron saint, St. Sigismund.

Closely connected with the Rimini fresco is the series of frescoes in the choir of the Church of S. Francesco in Arezzo, illustrating the legend of the True Cross (PLS. 159–61). Contemporary outside events coincide with the artist's movements: Bicci di Lorenzo, the artist whom the Bacci family in Arezzo had commissioned in 1447 to decorate the entire choir of S. Francesco, died in 1452 and Piero almost immediately took up the work, completing the decoration of the vaulting and entrance arch, most of which had been done by Bicci, and beginning the frescoes on the walls. The frescoes are mentioned as completed in a document dated Dec. 20, 1466. Thus, the undertaking occupied Piero for many years, although it was interrupted on more than one occasion. He was in Sansepolcro on Oct. 4, 1454, and a contract was drawn up for a polyptych ("tabulam que est de tabulis compositam") for the Church of S. Agostino, to be completed within eight years. That he failed to keep to the agreement appears proved by the fact that payment was not made until 1469. Vasari states that Piero was in Rome during the pontificate of Nicholas V, that is, before 1445; a document dated April 12, 1459, mentions a payment to Piero of 150 florins for "paintings made in the room of his Holiness our Lord the Pope" (Pius II). If Vasari's confused indications can be taken as a guide, this would mean that Piero's paintings were in what was to become Raphael's Stanza d'Eliodoro. All this is inferred from documents, but further confirmation of the many interruptions in the decoration of the choir of S. Francesco in Arezzo also comes from works produced elsewhere, for example, the *Madonna del Parto* in the chapel of the cemetery at Monterchi, which it would be rash to date later than the Arezzo frescoes and which should be assigned to the period between the completion of the right wall in Arezzo and the beginning of the opposite wall.

In 1466 Piero, by then at the acme of his fame, was once more in Arezzo. On Dec. 20 the Company of the Annunciation commissioned him to paint their new banner, believing that no painter other than the one who had created the frescoes in the choir of S. Francesco could satisfy their desire for the most beautiful work possible, and they recommended — how very much alive the Gothic tradition still was! — that the faces of the Madonna and the archangel should be as beautiful and gentle as the faces of angels. It can be supposed that the works mentioned by Vasari as being in various churches in Arezzo go back to this period, works of which only the *Magdalen* in the Duomo remains. In 1467 Piero was in Sansepolcro, where he was again appointed to public office. The following year he fled from the plague to Bastia, a small village near Sansepolcro, and there completed the banner for the Company of the Annunciation, which was consigned to two members of the company on Nov. 7.

In 1469 Piero made another brief appearance in Urbino. On this occasion his host was Raphael's father, Giovanni Santi. The Confraternity of Corpus Domini advanced 10 Bolognese florins to Santi "to defray the expenses of master Piero of Borgo who came to see the panel in order to make it." To what panel the document alludes is not clear. Ricci thought the reference was to the large panel of the *Madonna and Child, Angels, Saints, and Federigo da Montefeltro* (PL. 166), but, as the work was commissioned by the Confraternity of Corpus Domini, it seems more reasonable to agree with Longhi that the panel in question is that of the *Communion of the Apostles* (Urbino, Gall. Naz. delle Marche), the predella of which had been indifferently painted by Paolo Uccello in 1467 and which was entrusted in 1474 to Justus of Ghent, who painted it to the satisfaction of neither the Confraternity nor Federigo himself. Piero did not accept the commission and returned to Sansepolcro, where he took up residence, leaving the town occasionally for excursions to Urbino. On Nov. 14, 1469, the friars of S. Agostino in Sansepolcro paid him in money and land for the polyptych commissioned on Oct. 4, 1454. On Feb. 23, 1471, Piero was cited as being in arrears in paying a communal tax. In 1473 a document regarding a power of attorney for his brother shows that he was in Sansepolcro. In 1474 Piero was paid for the frescoes, now lost, in the Chapel of the Madonna of the Badia in Sansepolcro; the frescoes of two saints in the parish church in Sansepolcro, which Vasari mentions as being very beautiful, are also lost.

Piero's last trip to Urbino — when he probably presented Duke Federigo with his treatise on perspective, the *De prospectiva pingendi*, and certainly painted the Sinigallia Madonna (PL. 163) as well as the altarpiece for the Church of S. Bernardino (Milan, Brera) — fell in all likelihood between 1474 and 1478, as the documents of Sansepolcro contain no mention of his name for this period. News of him, though more infrequent, is to be had again starting in 1478, when the Confraternita della Misericordia paid him to fresco a Madonna, now lost, upon a wall "between the Church and the Hospital." Between 1480 and 1482 Piero was in Sansepolcro as head of the priors of the Confraternita di S. Bartolomeo. Not only was he personally esteemed, as shown by the offices he held, but his works were too, as evidenced by the allocation of a sum from the communal funds in 1480 to defray the cost of restoring the wall "where Piero had painted the *Resurrection*."

On April 22, 1482, the artist was again in Rimini, where he leased a house. The reason for his presence in Rimini is not known. After this there are no records except for Piero's last will and testament, which he instructed his notary, Lionardo di ser Mario Fedeli, to draw up on July 5, 1486, in conformity with notes he himself had prepared. This document, discovered by Mancini, is now preserved in the Archivio di Stato in Florence.

Piero, afflicted by blindness, lived six years longer. In book III of the necrology of the Confraternita di S. Bartolomeo in Sansepolcro, the following is recorded: "Piero of Benedetto de' Franceschi, famous painter, on the twelfth day of October, buried in the Badia." In addition to money and some houses, Piero bequeathed to his family the *Nativity* (PL. 165), which belonged to Piero's descendants until the middle of the 19th century, and a small self-portrait, which Vasari used for the engraving of the artist's likeness in the 2d edition of the *Lives*.

WORKS. The works of Piero della Francesca are based on a philosophy so well thought out and on qualities of style so personal that, although there may still be some doubt as to their chronological order, there is very little margin for uncertainty of attribution. The works that are signed and dated, those that can be dated by documents, and those that, though only signed, it is possible to place between works of certain date, are the following: 1445: Polyptych of the Misericordia (PL. 157; XII, PLS. 26, 376). The central panel portrays the Madonna of Mercy with eight worshipers. It is flanked by SS. Sebastian and John the Baptist on the left and by SS. Andrew and Bernardino of Siena on the right. The central panel is surmounted by the Crucifixion with the Virgin and St. John; the lateral panels depict the Angel and the Virgin of the Annunciation and SS. Francis and Benedict. The figures of saints upon the small lateral pilasters of the polyptych and the various compartments of the predella were not painted by the artist, for which Salmi has suggested the name of Giuliano di Amedeo. The difference in styles in the various parts shows that the work extended over many years; it was completed, according to Longhi, as late as 1462, the year when Piero's brother received from the Confraternita della Misericordia a sum of money that was in all probability connected with the completion of the polytych. The earliest parts of the painting are the two saints on the left of the central panel, who still have pronounced Masaccioesque features, and the Crucifixion; then come, in chronological order, the two saints on the right of the central panel (St. Bernardino of Siena was canonized only in 1450), the panels to the left and right of the Crucifixion, and the central panel itself with the towering Madonna of Mercy, of a mature solemnity.–1450: *St. Jerome in Penitence* (Berlin, Staat. Mus.). The panel is signed and dated Petri de Burgo opus MCCCCL. The work is not entirely by Piero's hand; the master painted the lower part, which includes the figure of St. Jerome, the cave, the bench with the books, and at least the outline of the lion. It is probable that the artist left the work unfinished; at any rate, the rest of it is not by him or was subsequently completely repainted.–1450: *St. Jerome with a Devout Suppliant* (Venice, Accademia). The panel is signed (on the left-hand trunk) Petri de Burgo Sancti Sepulcri opus. Under the figure of the suppliant is the inscription Hier. Amadi Aug. F., challenged by Cavalcaselle, but most probably by Piero.–1451: Sigismondo Malatesta kneeling before his patron saint (Rimini, Tempio Malatestiano). The fresco, now transferred to canvas, is signed and dated Sanctus Sigismundus Pandolfus Malatesta Pan. F. Petri de Burgo opus MCCCCLI. In the frame surrounding the oculus with the Rocca Malatestiana is the inscription Castellum Sismundum Ariminense MCCCCXLVI (this being the date of the castle's completion).–1444–51?: *The Flagellation of Christ* (PL. 158). The panel is signed Opus Petri de Burghi Sancti Sepulcri. It is customary to identify the three figures on the right as Oddantonio da Montefeltro (center) and his two evil ministers, Manfredo del Pio and Tommaso dell'Agnello, who started the conspiracy of the Serafini that led to the young prince's murder in 1444. The panel was formerly in the old sacristy of the Duomo, and it is not improbable, in view of the many divergent perspective indications, that it may have been part of a larger composition (a frontal or a threefold predella).– 1452–66: The legend of the True Cross (PLS. 159–61). Piero's frescoes continue the pictorial decoration begun by Bicci di Lorenzo (the vaulting with the figures of the four Evangelists and part of the undersurface of the entrance arch). Piero completed the entrance arch with the figures of SS. Augustine and Ambrose; he also painted the heads of two angels where the ornamentation of the southeast rib of the vaulting begins and St. Peter and an angel on the piers of the entrance arch. The two prophets on the uppermost section of the end wall (in the panels flanking the large window), who were probably chosen from among those prophets who had, in one way or another, some connection with the wood of the cross (Jeremiah, Daniel, Jonah), do not show a uniform intensity and vivacity of style, although both were conceived by Piero. The one on the right seems to dominate the other by his monumental appearance and calm and solemn loftiness. The frescoes on the walls depict episodes from the legend of the True Cross, as narrated essentially in the *Golden Legend* of Jacobus de Voragine. The episodes begin on the right wall. The death of Adam (PL. 160) is followed by the meeting of Solomon and the Queen of Sheba (PL. 159) and, in the middle scene of the end wall to the right of the window, by the removal of the holy wood of the cross. The link between this part, which is a kind of pre-Christian prologue to the legend, and that touching on Constantine is established by an Annunciation scene (which presages Christ's sacrifice), represented in the lowest of the three compartments on the left side of the end wall. It makes a thematic link with Constantine's dream, which has an epilogue in the scene of Constantine's victory over Maxentius (PL. 161), the bottom scene on the right wall. The part of the legend relating to St. Helena begins with the torture of Judas (in the middle compartment on the left of the end wall), continues in the middle compartment of the left wall with the finding and proof of the True Cross, and has an epilogue in the two large scenes at the bottom and top of the wall, which represent Heraclius's victory over Khusrau II and Heraclius bringing the cross back to Jerusalem. That the execution of the work extended over many years and that the intervention of assistants, such as Lorentino d'Andrea and Giovanni da Piamonte, was more extensive as the work progressed explain the discontinuities to be observed in the frescoes, particularly between those on the right wall (the first to be carried out) and those on the left wall (painted last).–1454–69: Parts of the polyptych formerly in S. Agostino in Sansepolcro: St. Augustine (Lisbon, Mus. Nac. de Arte Antiga); St. Michael (London, Nat. Gall.); St. John the Evangelist (New York, Frick Coll.); St. Nicholas of Tolentino (Milan, Mus. Poldi Pezzoli); three small panels originally in the small side pilasters: St. Monica, an Augustinian saint (both, New York, Frick Coll.) and St. Apollonia (Washington, D.C., Nat. Gall.); and a Crucifixion scene (New York, John D. Rockefeller Coll.), probably the central part of the predella. The fragments listed, that is, the four lateral compartments (two on each side), three of the six small panels that formed the side pilasters, and one of the scenes of the predella have been identified by Meiss (1941), Longhi (2d ed., 1946), and Clark (1947).–1465: Diptych of the Duke and Duchess of Urbino. On the front panels are the portraits of Federigo da Montefeltro and of his wife Battista Sforza (PL. 164); the reverse panels depict the allegorical triumphs of Federigo and Battista (IX, PL. 11). The date of the work was established by Cinquini (1906) on the basis of a passage written by the Veronese humanist Ferabò, but it has been observed that this may apply only to Federigo's portrait (which, moreover, is the only one cited) and not to Battista Sforza's, which must have been later, as is also suggested by the verses on the base of her allegorical triumph, which seem to refer to her as already deceased (1472).

The works that are neither signed nor dated but whose authenticity nevertheless is beyond any doubt are listed here in their most probable chronological order: *Baptism of Christ* (PL. 162); ca. 1445.–*Madonna del Parto* (Monterchi, chapel of the cemetery), fresco; ca. 1455.– *The Magdalen* (Arezzo, Duomo), fresco; same period as the frescoes in the choir of S. Francesco in Arezzo. – *Resurrection* (Sansepolcro, Pin. Com.); same period as the Arezzo frescoes. – *Hercules* (Boston, Isabella Stewart Gardner Mus.), detached fresco; of a period later than the Arezzo frescoes. – A saint, perhaps St. Julien, from the choir of the former Church of S. Chiara in Sansepolcro; of a period later than the Arezzo frescoes. – Polyptych (Perugia, Gall. Naz. dell'Umbria); the cusp depicting the Annunciation is shown in PL. 164; described by Vasari as being in the convent of S. Antonio delle Monache in Perugia; around the period of the polyptych of S. Agostino in Sansepolcro. – *Madonna and Child with Angels* (Williamstown, Mass., Clark Art Inst.); ca. 1465. – Sinigallia Madonna (PL. 163); ca. 1475. – *Nativity* (PL. 165); ca. 1475. – *Madonna and Child, Angels, Saints, and Federigo da Montefeltro* (PL. 166), from the Church of S. Bernardino degli Zoccolanti in Urbino; this was a votive work, as is shown by the presence of Federigo da Montefeltro kneeling in the foreground; the realism of the hands, joined in prayer, is incompatible with Piero's stylistic rigor, and they were probably painted, as Longhi has suggested, by Pedro Berruguete, whose presence in Urbino is recorded in 1477.

Works that can probably be attributed to Piero are the following: A panel (Florence, Coll. Contini Bonacossi) that bears on the front a Madonna and Child framed by a window that opens onto a landscape, and, on the reverse, a perspective study, which (as noted by Longhi, who is to be credited with the attribution) reproposes in abstract terms the intense "problematic" quality of the depiction of the front. This seems to be the earliest of Piero's works to have reached us (Longhi dates it around 1440). – A fresco portraying St. Luke the Evangelist on the underside of an arch in the chapel on the right aisle of S. Maria Maggiore in Rome. According to Longhi, this is "the only relic still decipherable of the frescoes of the former chapel of SS. Michael and Peter *ad vincula*." A document confirms the presence of the artist in Rome in 1459: this could be the date of the painting, in relation to the later frescoes of the choir in Arezzo (St. Peter and an angel on the piers of the entrance arch).

As mentioned earlier, Piero set off from Sansepolcro, one of Tuscany's easternmost cities, a city that opens onto the plains and valleys of Umbria. From this geographic datum stems the theory that Piero's artistic formation was not Florentine and that it was predominantly Sienese. It is unnecessary to go into this after Longhi's ample confutation. The datum nevertheless does have a certain bearing on the understanding of the conclusive tie with Domenico Veneziano (q.v.): Piero was with him in 1439, working on the frescoes (now almost entirely ruined) in the Cappella Maggiore of the Church of S. Egidio in Florence. Piero was then in all probability twenty years old. The fame of Masaccio (q.v.) was so great in the more informed and advanced circle of Florentine artists and the significance of his works so irrefutable that Piero could not have failed to be influenced by his work and its message of renewal and to develop this message beyond the limits within which Florentine tradition enclosed itself. Piero was drawn toward Masolino, Angelico, and even Filippo Lippi — that is to say, toward a crystalline sense of light and a tenuous though clear chromatism — by the same preferences of the master, who had renewed the perennial brilliance of these lights and colors in a more open and clear timbre, not unrelated to the naturalistic subtlety of the Flemish painters. The convergence of cultures that enfolded Piero during the years of his apprenticeship was impressive and stimulating, containing as it did diverging trends and motifs derived from conflicting sources and tending toward equally conflicting ends. It was the period of the decline of a culture and the rise of a new one whose complex ramifications were already established in a broad sense. On the one hand, apart from the medieval mysticism and the Gothic fables, there was a newly reconquered dignity of man. On the other, there was the dignity of nature, which, in the mystery of its lights and colors, was released from the Gothic spirit. Piero's great discovery was the luminous revelation of nature, a discovery that did not, however, remain an end in itself but placed man at the center of nature in a splendid synthesis that bypassed the Florentine tradition and nourished European art for about a century.

The mediation between "man" and "nature" or, expressed in figurative terms, between "form" and "color" was realized by perspective, a discovery whose significance Piero appreciated in Florence, either through the study of Masaccio's works or by frequenting Brunelleschi's milieu. And it was this discovery that fired his imagination, made him see in nature what others had not seen, nourished the supreme calm typical of his work, and led him to record in extremely rigorous treatises the date of his geometrical and mathematical experiences. "A perspective synthesis of form and color" is the phrase Longhi used to pinpoint the motionless and spectacular perfection of Piero's art, where the luminous intuition was translated into forms of solemn and pondered greatness.

Indications of these preferences, contacts, and preoccupations are clearly evident in the works deemed to be Piero's earliest. In the Contini Bonacossi Madonna, alongside the genuine derivation from Domenico Veneziano are perspective effects, such as the foreshortened window, the haloes that encircle the heads, and the rapport that binds together the Madonna and the Child and connects the group to the window space. In the brilliant *Baptism of Christ* (PL. 162) the colors of Masolino and Domenico Veneziano seem to be tempered by the high zenithal light that reinforces the perspective opening and confers upon the scene, in its all-pervading silence, the aura of a motionless, diaphanous, and suspended apparition. This work, limpid and terse, already contains the quintessence of Piero's art. The figures are set firmly on the ground and rise up motionless, with the solemnity of human monuments, their forms falling into regular, perfectly balanced patterns, as do the trunks of the trees whose branches, like the clouds that cross the sky, scan in a luminous counterpoint the space that opens on vast distances full of air and light. It is a light without shadows, like the figures, the trees, the leaves; still and silent like the mysterious significance of the sacred event. It is difficult to say whether this painting was done before or after the polyptych of the Misericordia, or at least the earliest parts of the polyptych, such as the two saints on the left (SS. Sebastian and John the Baptist), who are still Masaccioesque. This reference to his earliest and greatest master could be explained by the task entrusted to the artist, which was in a way similar to that undertaken by Masaccio in the polyptych at Pisa: to create a series of figures on previously gilded panels. But if the rapport between figures and background is still plastic in the left-hand panel, despite the perspective layout with which the artist tried to supplant the archaic one, this rapport has already acquired a "luministic" meaning in the Crucifixion scene, owing to the violent counterlight in which the figures are placed. In their turn, the figures on the right-hand panel and the Madonna with her mantle open to receive the devout (PL. 157) already possess, in the complete autonomy of the background, the monumental absoluteness seen in *St. Jerome with a Devout Suppliant* and in the figures of the fresco in the Tempio Malatestiano.

In the Rimini fresco the figures of Sigismondo Malatesta and St. Sigismund are placed in the same pattern as that used for the *St. Jerome* panel, except that in the former the background view of the city is replaced by a solemn and ample Albertiesque architectural frame, enclosing the figure of Malatesta, who thus emerges in full profile from the marble background (a foretaste of this is to be found in the small panel formerly in a private collection in Paris and now in Williamstown, Mass., Clark Art Inst.) with a heraldic arrangement that accentuates the peculiar nature of the painting. The diagonal on which the figures are placed, spaced by the relation of the bright and pure colors that (in the few parts of the fresco still intact) foreshadow those of the earliest paintings in the choir of S. Francesco in Arezzo, has as its vertex St. Sigismond and as its base the two marvelous greyhounds, which balance each other in the *contrapposto* relation of both the pose and the colors. But what determines the effect is the supreme equilibrium linking both the architecture and the figures in the exact relation of a mathematical dream.

Similar considerations, matured between the *Baptism of Christ* and the great fresco in Rimini, but taken to an extreme of optical and chromatic lucidity, are revealed in the *Flagellation of Christ* (PL. 158). Here again the figures are mysteriously and indivisibly bound to the architecture. The flagellation itself takes place within a clearly measureable portico that is marked out on a perspective grid, and the figures are linked in a closed rhythmic sequence. Outside the portico, in the foreground, in front of a square flanked with houses and gardens, stand Prince Oddantonio and his two ministers. One minister is speaking and underlines his words with a slow gesture of the hand; the other two listen, their gaze steadfast. This is the same motif as the three angels in the *Baptism*, but strengthened by an additional charge of restrained energy; however, the space between the figures in the *Flagellation* isolates each of them in a closed world, with the archaic severity of Egyptian statues. The inexorable logic of the perspective layout transforms the two groups of figures into symbols of an inaccessible universe, and the effect is heightened by the limpid inlay work of the pure, clear colors that merge to create an unreal atmosphere, as lucid as it is transparent.

Moving along a path marked by moments of such highly expressive intensity, the painter took over, in 1452, the decoration of the choir of the Church of S. Francesco in Arezzo. The episodes on the three walls of the choir focus on the symbol of man's redemption.

In the death of Adam (the large lunette at the top of the right wall; PL. 160) Piero portrayed the old patriarch at the moment when, having gathered his kinsmen around him, he announces that he is nearing death; the lament over Adam's body; and Adam's son Seth discoursing with the archangel Michael. In the center is a large tree, which prefigures the tree that was to grow above the patriarch's body from the three miraculous seeds placed under his tongue by his son at the suggestion of the archangel. The loftiest, if not the most dramatic, part is that composed of the figures grouped around Adam. The scene is pervaded with an air of renewed classicism, a Greek archaism that has acquired a Christian soul. The effect is impaired by the abrasion of the background, but that absorbed circle of figures, with the Herculean youth seen from behind in the foreground, is one of Piero's most majestic and clearly measured creations. The motif is taken up again in the group gathered around Adam's dead body and the scene culminates in a dramatic cry from one of the figures.

In the large fresco that occupies the middle part of the right wall, the meeting of Solomon and the Queen of Sheba, Piero repeated the basic plan used in the *Flagellation*. The scene is divided into two parts: the first, the Queen of Sheba adoring the holy wood (PL. 159), takes place in a vast space of landscape and sky, against which rises the motionless foliage of two trees.

The nobly attired Queen with the ladies of her train and the equerries with the horses form groups that are spaced and somehow protected by the tall trunks and thick foliage of the trees. The "courtly" attire of the ladies brings to mind the passage in Vasari that describes how Piero used to drape earthen models with soft cloths to study the way the material hung. The rapport between figures and background and the spaces between the groups illustrate what Longhi termed Piero's "spatial foresight," that is, rather than starting from impulse and then proceeding to a stylistic revision of that impulse, the artist proposes a theorem as the framework of the composition and then, "very gently, this theorem is clothed and warmed by a scene or spectacle." The second part, depicting the meeting of Solomon and the Queen of Sheba, occurs within an architectural frame, the vertical perfection of whose Corinthian columns already appeared to Vasari as having been "divinely measured." In the vast space, resplendent with marble and colors, the Queen and her retinue of ladies are portrayed on one side and Solomon and his courtiers on the other. Here also the figures are admirably related to one another and to the architectural structures, and once again confirm Longhi's concept of "spatial foresight" as one of the most personal aspects of Piero's inspiration.

According to the legend, Solomon learns from the Queen of Sheba that the fate of the wood cut from Adam's tree and transformed into a small bridge will be to bear Christ crucified and that this fate is linked to that of the people of Judea. The King then orders the wood to be removed and buried deep in the ground. The depiction of this scene was carried out with the help of an assistant (perhaps, as suggested by Longhi, that Giovanni da Piamonte known for a panel at Città di Castello, dated 1456, that is, at the time of the frescoes in question, also remarked on by Longhi; this is borne out by the gratuitous tangle of hair, more sketched than painted.) Piero endowed the scene with the flavor of an everyday episode, although he did not fail, as pointed out by Longhi, to anticipate here an incident in the Passion of Christ — a slave is depicted toiling under the weight of the wood — thus creating an ideological link with the scene of the Annunciation, an elliptical allusion to Christian history as well as a prologue to the events that were to befall the Emperor Constantine three centuries after the Annunciation. Despite the workaday appearance of the episode, he manages to achieve a surreal effect with the veined wood diagonally hoisted by the slave against a background of bare hills and wintry sky crossed by clouds and streaked with flashes of lightning.

In the Annunciation scene, framed within an asymmetrical spatial construction, diagonally foreshortened to the right, the light-bathed column that divides the scene and supports the upper part of the construction has an emblematic meaning, as do the apparition of the Eternal Father through a rift in the clouds, with the luminous disk of His halo; the supreme equilibrium of the angel; and the noble majesty of the Virgin, who seems to be the courtly counterpart of the *Madonna del Parto*, which was most probably painted during the same period.

On the same wall, in the opposite side to the Annunciation panel, is the scene depicting Constantine's dream, with which Piero continued the legend after the parenthetical reference to the Annunciation. The night sky and the tents form the background against which the field tent where the Emperor sleeps stands out. Two soldiers, one seen from behind, the other facing the spectator, watch over the tent while a servant sits at the foot of the bed. The stillness of the night, punctuated by the somber masses of the tents, is suddenly shattered by the intense light that rains down from the angel and illuminates the scene, bringing into relief the white-clad servant, the red coverlet and white sheet of Constantine's bed, and the Emperor's face, and in the counterlight gives the full measure of the tent's depth between the two soldiers. Constantine's dream, contemporary with the miniatures of *Le livre du Coeur d'Amour épris* (V, PL. 385) written by Master of René of Anjou, is the first night scene of Italian painting, anticipating, by the immediacy of its effect and the rigor of its composition, the work of Caravaggio and of his followers.

The episode of Constantine's victory over Maxentius (PL. 161), the bottom scene on the right wall, shines with the incomparable light of a spring day, so much so that it was termed by Clark "the most perfect morning light in all Renaissance painting." The thick and resolute movement of the lances in the two armies, opposed and separated by a segment of landscape (described by Longhi as "an ineffable glimpse of sunlit country"), immediately establishes who are the victors and who the vanquished: by the cross, reduced to two small strips of light on Constantine's outstretched hand, and the yellow standard that billows above his fearless group of knights. This, although it is in many places ruined, is one of the greatest panels in the series. It is undoubtedly the one in which Piero's hand is revealed at its most luminous.

In the middle fresco of the end wall to the left of the window the story of the empress Helena searching for the spot where the holy wood was buried is depicted. The scene takes place inside a castle surrounded by inaccessible walls, beyond which is the clear sky mottled with clouds. Two jailers strain to draw Judas from a well, for he alone knows where the cross is hidden. As soon as he emerges, the judge seizes him by the hair and awaits his confession. The composition revolves around the pyramid formed by the beams supporting the pulley above the center of the well, and the whole scene, despite the rigor of its setting, acquires a surprising tone of everyday actuality. It cannot be said, however, that in this fresco inspiration and style form a homogeneous whole. Piero's project seems to have been carried out by the same collaborator — perhaps Giovanni da Piamonte — who helped him to fresco the removal of the holy wood of the cross.

At the spot indicated by Judas and in the presence of the Empress and her retinue, three crosses are found (the one on which Christ was crucified and those that bore the two thieves), but Judas is unable to identify the True Cross. In the presence of the Empress and her train, he proceeds with the proof by holding the crosses one by one over a youth about to be buried; the True Cross miraculously brings the youth back to life. The finding and proof of the True Cross are portrayed in the middle panel on the left wall opposite the one comprising the scenes of which the Queen of Sheba is the subject. Again, one of the scenes takes place in the open air, among fields that spread beyond the walled enclosure of the city, its cubical structures shining in the sun. The other, while not actually within a frame of aulic architecture, is set in a square overlooked on one side by the front of a church whose proportions might have been suggested by Alberti (e.g., his churches in Mantua) and, on the other, by a city street flanked with old buildings, towers, and belfries, recalling the background of one of Masaccio's panels in the frescoes of the Brancacci Chapel in S.Maria del Carmine, Florence.

In both scenes the courtly element deriving from the presence of the Empress and her retinue is modified by the rusticity of the characters gathered around the workmen who are digging up the crosses (there is a relevant passage by Vasari on the "readiness" of the peasant who stands by to hear St. Helena speak) and by the presence of the bier from which the youth miraculously brought back to life rises. Through this mixture of rustic and aulic elements, the painting acquires a tone that places it on a level with the one opposite. This tone is accentuated by the fact that the arrangement used for the figures of the Empress and her ladies, as they witness the miraculous event, is similar to the pattern of the fresco in which the Queen of Sheba kneels to adore the holy wood. In the depiction of the proof of the True Cross, however, the light unfolds with a more meridian brilliance and the forms have a more vigorous relief.

The following scene — a contrast to Constantine's victory — represents Heraclius's victory over Khusrau II and depicts Khusrau's exaltation of the cross. The events shown occurred three centuries after the cross was miraculously recovered. Constantine's victory over Maxentius was as bloodless as Heraclius's victory over Khusrau, the heretical Persian king, was fierce. Khusrau had not only stolen the cross from Jerusalem but had set it up, together with other objects related to the

Passion of Christ, on the singular throne — pictured on the extreme right of the composition — profaning the mystery of the Holy Trinity. Vasari described the bloody victory as "an almost unbelievable slaughter of wounded, fallen, and dead," a slaughter that was conceived, although dulled in the execution, on a purely perspective basis, as in the detail where Heraclius calmly stabs Khusrau's son in the throat, or in the one where the pagan king kneels to await decapitation within a circle of amiable gentlemen whom Vasari identifies as members of the Bacci family (who commissioned the work). The painting lacks the clear layout of the others. The left part was started by Piero; it was continued by a collaborator, now customarily identified as Lorentino d'Andrea. The painting is to be dated among the later ones of the series and was conceived after the artist's visit to Rome; this is clearly revealed by certain details, such as motifs drawn from the reliefs on the Arch of Constantine in Rome, noted by Warburg (as cited by Clark).

The picture cycle is concluded by the lunette on the left wall depicting Heraclius bringing the cross back to Jerusalem in triumph. For this scene Piero did not portray the whole story as told in the legend but limited himself to the arrival of the cross, followed by the Emperor and his train, at the walls of Jerusalem. This was obviously done to make the composition essentially a relief of monumental solemnity and to create a balanced complement to the death of Adam in the opposite lunette. Its role as a complement is not only architectural and decorative: it links, in the epilogue, the idea of redemption (the cross) with that of sin (Adam), thus illuminating the meaning of the entire tale. Here, as in the opposite lunette, the scene is developed against a vast landscape dominated by two large trees that are spatial counterpoints as well as intervals between the cross-bearers and the citizens of Jerusalem kneeling in worship before the cross. The Emperor and the courtiers are magnificently attired and wear Oriental hats (it has been suggested that these costumes are an echo of the wonder aroused by the clothing of the members of the eastern missions at the Council of Ferrara-Florence). One of the two citizens who has arrived late also wears a fantastic hat, which he removes before kneeling: an everyday gesture that relates him — apart from the difference in rank — with the rustic Magi of the Romance world. These touches, corresponding perfectly to reality, spread a feeling of veracity, of everyday simplicity, through the solemn and monumental scene in the solitary valley, concluding with majestic accents the spectacular tale. In this combination of grandeur and simplicity the hand of the master is again apparent, indicating that Piero chose to execute in person the episode that closes the fresco cycle with such noble dignity.

The isolated figures that complete the pictorial decoration of the choir of the Church of S. Francesco in Arezzo serve as marginal illustrations to the ampler representations, although some of them stand out for their resplendent beauty (e.g., the figure of the prophet on the top of the end wall next to the window). The earliest single figures are undoubtedly those of SS. Augustine and Ambrose on the undersurface of the entrance arch and the heads of the two angels on one of the piers of the southeast rib. These were probably followed by the two prophets on the top of the end wall (the figure on the left many have been painted by Giovanni da Piamonte). The last figures to be painted must have been the two on the piers of the entrance arch (perhaps the work of Lorentino d'Andrea).

In the years during which the frescoes in the choir at Arezzo were carried out, Piero painted other works: in Arezzo, in Sansepolcro, and in Rome, where the figure of St. Luke on the arch of a chapel in S. Maria Maggiore is proof of his presence there. The *Magdalen* in the Duomo in Arezzo is a twin figure to the *Madonna del Parto* in Monterchi. In the *Resurrection* in Sansepolcro, because of the rigorous perspective arrangement the extreme naturalism of Christ triumphant acquires the arcane significance of a Byzantine Pantocrator. A fragment of a painting has recently been discovered in the former Church of S. Chiara in Sansepolcro. The *Hercules* (Boston, Isabella Stewart Gardner Mus.) belongs to a later period and is probably contemporary with the figurations on the piers of the entrance arch in the Church of S. Francesco in Arezzo.

During the same period Piero also initiated larger undertakings, such as the two polyptychs of Sansepolcro and Perugia. These reveal thoughts and inspirations that are in harmony with those of the choir frescoes, or which are a natural development of the work done in S. Francesco. In this connection it is worth underlining the often insufficiently appreciated, unassuming beauty of the panels composing the predella of the Perugia polyptych, where episodes of extraordinary efficacy and subtlety are to be found. There is in this subtlety more than one foretaste of the works painted at a later date, starting with the two small panels Uffizi (PL. 164) bearing on one side the opposed portraits of the Duke and Duchess of Urbino and, on the reverse, also opposed, their individual allegorical triumphs. Against the vast backgrounds of water, fields, and hills, toned down and blended on the horizon in the blue-gray infinity of the sky, the monumental busts of the Duke and Duchess are placed in the heraldic pose characteristic of ancient medals. Though seen in profile, the perfection of the perspective layout makes them appear as if sculptured in the round, an effect enhanced by the minute precision of the portraits, the extreme care with which the details are handled, and the close-up view contrasting with the vastness of space and light, making the minute elements of the backgrounds indistinct in a view that is hazy in its transparency. Thus Piero blended with the dimension of his personal esthetics the more subtle and filtered naturalism of Flemish painting. What has been said for the portraits also applies to the triumphs, except that in the latter the wagons are more in proportion with the backgrounds, which retain, nevertheless, in the optical lucidity typical of illuminated pages a different span and breadth. In both panels the luminous aspects of nature are pursued with unrelenting acumen, down to infinitesimal details, and are fascinating in their vigor and clarity.

This concern for the luminous naturalism of the Flemish tradition increased during the following years, and Piero's images had a new freedom and gentleness and were more detailed. A sign of the new freedom is already to be found in the off-center relation between architecture and figures in the panel of the *Madonna and Child with Angels* (Williamstown, Mass., Clark Art Inst.). The new gentleness is particularly marked and surprisingly intimate in the Sinigallia Madonna (PL. 163). Here the group is also placed against an asymmetrical background but is tightly bound by the aristocractic fineness of the colors and the precision of the "values" and "rapports." The new interest in detail is remarkably represented in the *Nativity* in which recollections of earlier works — the choir of angels (PL. 165) brings to mind the angels of Luca della Robbia — are joined with a typically northern European layout. The result is a veritable ecstasy of chromatic relationships pursued, as in the Sinigallia Madonna, with a lucidity that anticipates the minute and calm intensity of Vermeer (e.g., the remnant of the dilapidated castle and the rustic roof jutting out from it).

The final point of so much new research and of so many revitalized interests is undoubtedly the great votive altarpiece of the *Madonna and Child, Angels, Saints, and Federico da Montefeltro* (PL. 166), the last known work Piero painted. Old thoughts, new interests, and an exceptional degree of intensity merge in this altarpiece and transform into an assembly of human monuments the 10 figures gathered around the Madonna and Child, and yet leave unimpaired their striking naturalism. In Piero's mind, always fascinated with mathematical problems, architecture became an ideal dimension of his humanity, a way to reveal the nobility of pure existence, and the rapport between architecture and figures is so subtle that, as already noted by Longhi, it is hard to tell "whether the architecture has determined the arrangement of the human structure or whether it is not these latter that are responsible for the architectural ones." In this subtle bond is contained Piero's sublime message, which dominated and stimulated all the most vital centers of painting in his time. The message was heeded, and resulted in the renewal of Italian and European art by, on the one hand, Antonello da Messina and Giovanni Bellini and, on the other, Bramante and Raphael.

In reviewing Piero's activity, reference has often been made to his mathematical mind, not only in connection with his painting but also with his treatises, which must necessarily be mentioned as his fame and fortune appear at one point to have been based on them.

In addition to the treatise entitled *Del abaco* (Florence, Laurentian Lib., Cod. Ashb. 359), those works relating to problems that have a direct bearing on the visual arts are: *De prospectiva pingendi* (Parma, Bib. Palatina, Cod. 1576), with 79 drawings (published in Fasola's edition, 1962), and *Libellus de quinque corporibus regularibus* (Rome, Vatican Lib., Cod. Urb. 632), also with drawings (published by Mancini, 1915). The *De prospectiva pingendi* was offered to Duke Federigo, and was therefore written before 1482; the *De corporibus regularibus*, composed at a later date, was dedicated to Guidobaldo, Federigo's son.

In the *De prospectiva pingendi*, Piero dealt with "measurement, which we call perspective," that is, with one of the three principal parts (the others being "drawing" and "coloring") of which painting consists. He dealt, therefore, with problems previously handled rather empirically by Ghiberti as well as by Brunelleschi and Alberti, treating them on systematic and scientific bases, according to a method that was to be so fruitfully taken up again by Galileo and his followers. The fundamental purpose of the treatise was to demonstrate that the various aspects of nature are more easily grasped by the eye if they are brought back to the simple and regular forms of geometrical figures. This theory was expounded mainly in the first two parts of the treatise, in which Piero moved with extreme precision from the simplest to the most complex perspective problems. In the third part the same problems were clarified by means of the empirical process that was obviously applied in his workshop. Cézanne returned to the very same concept when he stated that "within nature all forms are based on the cylinder, the sphere, the cone." Seurat (who in all probability, as suggested by Longhi, knew the Arezzo frescoes through the copies made by Charles Blanc) was obviously inspired by it in the conception of his works, as his compositions were subjected to a static arrangement and the images unalterably set to achieve effects of a calm and hermetic monumentality.

The *Libellus de quinque corporibus regularibus* acquired wider renown in its time, as well as later, through the literal imitation Luca Pacioli made in his *De divina proportione*, published in Venice in 1509. Pacioli's plagiarism was denounced by Vasari and, Longhi says, mentioned by Danti (1583) in his preface to Vignola's *Le due regole della prospettiva prattica*; but, despite the denunciation of the plagiarism and its subsequent confirmation, the arguments pro and con, though idle and superfluous, dragged on even after the year 1880, when K. Jordan discovered Piero's brief treatise in the Urbino section of the Vatican Library and compared it with Luca Pacioli's.

It is quite certain that, in early times as well as somewhat later, when his fame as a painter was still limited to his native province, Piero's name owed its renown to these treatises and even more to the inspiration drawn by architectural theorists and treatise writers from the work on perspective. On the level of artistic historiography proper, Piero's fame was established not so much through the passages written by Luca Pacioli as through the biography written by Vasari, founded on accurate sources and particularly well-informed as concerns Piero's works. This may be due to Vasari's being from Arezzo (a certain amount of parochialism may therefore have been involved) and having a great-grandfather who had been in contact with Piero. At any rate, Piero was, after Masaccio, the artist whom Vasari praised most lavishly, while also recognizing his role as an innovator. The fact also remains, as Longhi noted in the comparison with the interpretation of Masaccio's work, that Vasari had at least felt by intuition that the characteristic of Piero's art "was in a pictorial tonality distinct from plastic intention." Vasari obviously did not realize the historical significance of Piero's insight, and never foresaw that the choir of S. Francesco in Arezzo would draw as impressive a pilgrimage of artists as the Brancacci Chapel in Florence. However, the critical awareness of what is embryonically contained in his statement — where the eccentric position of Piero's art with respect to the development of Florentine tradition is already basically outlined — is the achievement of modern art history, starting with Longhi's conclusive essays. This is not to imply that between Vasari and Longhi, despite the numerous distortions, the work of Piero had fallen outside the sphere of interest of the scholars, but before Luigi Lanzi's research the contributions made were of an erudite rather than critical nature and were mainly the results of municipal historiography. Lanzi made full use of these contributions and, having restored the artist to his proper historical position (mostly on the basis of the high evaluation of the Arezzo frescoes), realized that Piero's natural precedents were to be sought in the Florentine tradition. He compared Piero's work to that of the Greeks, "who placed geometry at the service of painting," and began to trace Piero's many artistic descendents (Perugino, Raphael, the Ferrarese, and, as classified previously by Sabba di Castiglione, Bramante). Lanzi not only accepted Vasari's views but went beyond them, opening the way for the more recent interpretations, which also used many fundamental documentary contributions, such as the one referring to the collaboration with Domenico Veneziano.

The contribution of Cavalcaselle coincides with the trend mentioned above, for his refined sensitivity as a connoisseur allowed him to strengthen the bases for the catalogue of works and, with his vast experience as a scholar, to put forward the theory of Piero's connections with Flemish painting (Jan van Eyck) and with Alberti. Subsequent studies, on which a detailed report can be found in Longhi's volume, insist on the themes already indicated: the connection of Piero's painting with the Flemish school; Piero's artistic derivation, which included elements of the art of Emilia, the Marches, and central Italy; and the artist's catalogue of works, through which the elusive figure of Fra Carnevale weaves in and out, mostly on the basis of the later paintings. All this applies to the period before the year 1914, when the review *L'Arte* published Longhi's first essay on the subject. On the basis of real critical interests, Longhi outlined a profile of the artist and specified the immense historical significance of his work in the development of Venetian painting: Antonello da Messina and Piero were related and the presentation of relevant data historically clarified "the good perspective foundation" with which Bellini's work had been credited in former times, from Pacioli up to Marco Boschini.

Longhi's essay was followed in 1927 (the most recent revised edition is dates 1946) by the volume in which are reviewed the artist's precedents, the course followed by his work, and its Italian and Mediterranean ramifications. There have been a number of other studies and investigations, especially those by A. Venturi, Focillon, Salmi, Clark, L. Venturi, and Bianconi, which have analyses and clarifications of great refinement and critical value.

Sources. P. della Francesca, De prospectiva pingendi (ed. G. Nicco Fasola), Florence, 1942; L. Pacioli, Summa arithmeticae, Venice, 1494; L. Pacioli, De divina proportione, Venice, 1509; Sabba di Castiglione, Ricordi, Venice, 1554, ch. 109–10; Vasari, Le Vite, Florence, 1550; E. Danti, ed., Le due regole della prospettiva pratica di J. Barozzi da Vignola, Rome, 1583 (preface).

Bibliog. A systematic bibliog. is provided by R. Longhi, Piero della Francesca, Rome, 1927 (2d ed., Milan, 1946; Eng. trans., L. Penlock, London, New York, 1930), and U. Baldini, ed., Mostra di quattro maestri del primo Rinascimento, Florence, 1954 (cat.). See also: L. Pungileoni, Elogio storico di Giovanni Santi, Urbino, 1822; E. Harzen, Über den Maler Piero dei Franceschi und seinen vermeintlichen Plagiarius, des Franziskanermönch Luca Pacioli, Arch. für die zeichnenden Künste, Leipzig, 1856, pp. 231–44; G. Milanesi, Le vite di alcuni artefici fiorentini scritte da G. Vasari: Vita di Piero della Francesca, Giorn. storico degli arch. toscani, VI, 1862, pp. 10–15; F. Corazzini, Appunti storici e filologici su la Valle Tiberina superiore, Sansepolcro, 1874; C. Pini, La scrittura degli artisti italiani (ed. G. Milanesi), 3 vols., Florence, 1876; G. Milanesi, Documenti, Il Buonarroti, 3d Ser., II, 1885, pp. 73–85, III, 1887, pp. 37–44; G. Gronau, Piero della Francesca oder Piero dei Franceschi?, RepfKw, XXIII, 1900, pp. 393–94; W. Weisbach, Ein verschollenes Selbstbildniss des Pietro della Francesca, RepfKw, XXIII, 1900, pp. 388–91; A. Cinquini, Piero della Francesca a Urbino e i ritratti degli Uffizi, L'Arte, IX, 1906, p. 56; C. Grigioni, Un soggiorno ignorato di Piero della Francesca in Rimini, Rass. bibliog. dell'arte it., XII, 1909, pp. 118–21; E. Marini Franceschi, Alcune notizie inedite su Piero della Francesca, L'Arte, XVI, 1913, pp. 471–73; G. Mancini, L'opera "De corporibus regularibus" di Piero Franceschi detto della Francesca, Atti R. Acc. dei Lincei, Mem. della classe di sc. morali, storiche e filologiche, XIV, 1915, pp. 446–87; G. Mancini, La madre di Piero della Francesca, BArte, XII, 1918, pp. 61–63; G. Gronau, ThB, s.v.; G. Zippel, Piero della Francesca a Roma, Rass. d'arte, XIX, 1919, pp. 81–94; Van Marle, XI, 1929, pp. 1–110; J. von Schlosser, Xenia, Bari, 1938; C. Gilbert, New Evidence for the Date of Piero della Francesca's Count and Countess of Urbino, Marsyas, I, 1941, pp. 41–53; M. Meiss, A Documented Altarpiece by Piero della Francesca, AB, XXIII, 1941, pp. 53–70; M. Salmi, Piero della Francesca e Giuliano Amedei, RArte, XXIV, 1942, pp. 26–44; M. Salmi, La Bibbia di Borso d'Este e Piero della Francesca, La Rinascita, VI, 1943, pp. 365–82; M. Salmi, ed., Piero della Francesca: Gli affreschi di San Francesco in Arezzo, 2d ed., Bergamo, 1944; M. Salmi, Piero della Francesca e il Palazzo ducale di Urbino, Florence, 1945; K. Clark, Piero della Francesca's St. Augustine Altarpiece, BM, LXXXIX, 1947, pp. 204–09; P. Rotondi, Linee di livello, fondamento ed altre divagazioni nel Palazzo ducale d'Urbino, Belle arti, I, 1947, pp. 127–31; M. Salmi, Un'ipotesi su Piero della Francesca, ArtiFig. III, 1947, pp. 78–84; J. Alazard, Piero della Francesca, Paris, 1948; A. D. Stokes, Art and Science: A Study of Alberti, Piero della Francesca, and Giorgione, London, 1949; R. Longhi, Piero in Arezzo, Paragone, I, 11, 1950, pp. 3–16; K. Clark, Piero della Francesca, New York, London, 1951; H. Focillon, Piero della Francesca, Paris, 1952; F. Wittgens, La pala urbinate di Piero, Milan, 1952; L. Venturi, Piero della Francesca, G. Seurat, J. Gris, Diogène, III, 1953, pp. 25–30; B. Berenson, Piero della Francesca or the Ineloquent in Art, London, 1954; C. Brandi, Restauri a Piero della Francesca, BArte, XXXIX, 1954, pp. 241–58; M. Meiss, Ovum struthionis: Symbol and Allusion in Piero della Francesca's Montefeltro Altarpiece, S. in Art and Literature for Belle da Costa Greene, Princeton, 1954, pp. 92–101; L. Venturi, Piero della Francesca, Geneva, 1954; R. Longhi, Piero della Francesca: La leggenda della Croce, Milan, 1955; M. Salmi, L'affresco di Sansepolcro, BArte, XL, 1955, pp. 230–36; P. Bianconi, ed., Tutta la pittura di Piero della Francesca, Milan, 1957 (2d ed., 1959); D. Formaggio, Piero della Francesca, Milan, 1957; C. Marinesco, Echos byzantins dans l'oeuvre de Piero della Francesca, BAFr, 1958, pp. 192–203; M. Salmi, Arte e cultura artistica nella pittura del primo Rinascimento a Ferrara, Rinascimento, IX, 1958, pp. 123–40; E. H. Gombrich, The Repentance of Judas in Piero della Francesca's "Flagellation of Christ," Warburg, XXII, 1959, p. 172; W. Kronig, La "Resurrezione" di Piero della Francesca, Arte antica e moderna, II, 1959, pp. 428–32; A. del Vita, La battaglia di Anghiari rappresentata di Piero della Francesca?, Vasari, XVII, 1959, pp. 106–11; A. Parronchi, Paolo o Piero?, Arte antica e moderna, IV, 1961, pp. 138–47; C. de Tolnay, Religious Conceptions in the Painting of Piero della Francesca, New York, 1966.

Stefano Bottari

Illustrations: pls. 157–166.

PIERO DI COSIMO.

PIERO DI COSIMO. Florentine painter (b. 1462; d. 1521). Son of the goldsmith Lorenzo Chimenti, Piero di Cosimo was the pupil of Cosimo Rosselli, from whom he took his name. Vasari's *Vite* is the principal source for reconstructing the life and works of the painter, who was also known for his inventive designs of triumphal processions and pageants. In Vasari's biography Piero's eccentric character and habits, as well as his passion for natural observation, are vividly described. Because of scant documentation it is difficult to establish a precise chronology of Piero's *œuvre*; moreover, this difficulty is heightened by the somewhat eclectic yet individual character of his art, which only partially reflects the transition from the style of the 15th century to that of the High Renaissance.

According to Vasari, Piero assisted Cosimo Rosselli in the execution of frescoes in the Sistine Chapel in 1481. His earliest works, however, reflect the influence of Filippino Lippi and Domenico Ghirlandajo (qq.v.), and a little later that of Luca Signorelli (q.v.); in their wealth of naturalistic detail they also bear affinities with contemporaneous Flemish and German painting (e.g., *The Visitation with Two Saints*; Washington, D.C., Nat. Gall.). There is also an anticlassical quality in Piero's art. Contrasting sharply with the ideals of the Renaissance, this is most apparent in his mythological subjects, which in their unorthodox interpretations convey a mood of poignancy and poetry (*The Fight between the Lapiths and the Centaurs*; VIII, pl. 211). Piero's contribution to portrait painting is no less personal and penetrating. His *Simonetta Vespucci* (XII, pl. 29) is one of the most vivid portraits of the period; those of Giuliano da Sangallo and Francesco Giamberti (Amsterdam, Rijksmus.), in addition to strong characterization, exhibit a pronounced naturalism that is reminiscent of northern painting. This realism is also to be found in his landscapes, which recall Leonardo in their sensitive observation of nature but at the same time possess a highly personal sense of color and tone.

Leonardo's influence is dominant in Piero's paintings that date from the first years of the 16th century. These are more in accord with the new classical ideals in their greater harmony and chromatic unity (the Uffizi *Immaculate Conception*, the Innocenti Altarpiece, the Cini *Madonna*); they are characterized, too, by an intense luminosity and a limpid aura peculiar to Piero. Leonardo's influence is also felt in later mythological paintings, such as *Perseus Freeing Andromeda* (VIII, pl. 194) and the *Legend of Prometheus* (Munich, Alte Pin.; Strasbourg, Mus. B.A.); the style became softer and more fluid but not at the expense of originality.

Piero's last works reveal his unending capacity to absorb new experiences and to move with the times; for instance, his *Madonna Adoring the Christ Child* (Rome, Gall. Borghese) not only finds its place beside the paintings of Fra Bartolommeo but even heralds mannerist works. Piero was the master of Andrea del Sarto (q.v.), and his art was important for subsequent development of 16th-century painting, which found inspiration not only in his use of color but in his anticlassical sense as well.

Major works. *a.* Religious subjects: *Adoration of the Shepherds* (destroyed), formerly, Berlin, Kaiser Friedrich Mus. – *Immaculate Conception*, Uffizi. – *Madonna and Child with Angels and Saints*, Florence, Ospedale degli Innocenti. – *Madonna*, Louvre. – *Madonna Adoring the Christ Child*, Rome, Gall. Borghese. – *Madonna Enthroned with Saints*, St. Louis, Mo., City Art Mus. – *Madonna and Child with Two Angels*, Venice, Coll. Cini. – *The Visitation with Two Saints*, Washington, D.C., Nat. Gall.
b. Mythological subjects: *Venus, Cupid, and Mars*, Berlin, Staat. Mus. – *The Discovery of Wine*, Cambridge, Mass., Fogg Art Mus. – *Perseus Freeing Andromeda*. – *The Fight between the Lapiths and the Centaurs* and *Death of Procris*, both, London, Nat. Gall. – *Legend of Prometheus*, Munich, Alte Pin.; Strasbourg, Mus. B.A. – *The Hunt* and *Return from the Hunt*, both, Met. Mus. – *A Forest Fire*, Oxford, Ashmolean Mus. – *The Discovery of Honey*, Worcester, Mass., Art Mus.

Bibliog. B. Degenhart, Piero di Cosimo, ThB, XXVII, 1933, pp. 15–17; B. Berenson, Drawings of the Florentine Painters, 2d ed., Chicago, 1938, I, pp. 150–54, II, pp. 256–60 (I disegni dei pittori fiorentini, Milan, 1961, I, pp. 219–24, II, pp. 428–35); R. L. Douglas, Piero di Cosimo, Chicago, 1946; F. Zeri, Rivedendo Piero di Cosimo, Paragone, July, 1959, pp. 36–50; S. J. Freedberg, Painting of the High Renaissance in Rome and Florence, Cambridge, Mass., 1961, pp. 24–25, 72–75, 212–14; B. Berenson, Italian Pictures of the Renaissance: Florentine School, London, 1963, I, pp. 175–77; M. Bacci, Piero di Cosimo, Milan, 1965.

Marco Chiarini

PIETRO DA CORTONA. Italian painter and architect (b. Cortona, 1596; d. Rome, 1669). Pietro's father was Giovanni di Luca Berrettini, but the artist is generally known by the name of his birthplace.

Summary. Painting (col. 356). Architecture (col. 360).

Painting. Pietro da Cortona's career is inseparably tied to the history of the baroque movement (see baroque art), and it was not by chance that the phenomenon of "cortonismo" was cited almost two centuries ago as one of the principal reasons for a decadence in Italian painting that lasted for more than a hundred years. This charge is stated explicitly in a well-known passage of Francesco Milizia (*Dizionario delle belle arti*, Bassano, 1, 1797, p. 114): "Borromini in architecture, Bernini in sculpture, Pietro da Cortona in painting, and the Cavalier Marini in poetry represent a diseased taste — one that has infected a great number of artists." Since that harsh judgment Pietro's reputation has risen and fallen in accordance with the critical fortunes of the baroque period as a whole.

There is no doubt that among Italian painters he must be considered the most influential personality of his generation, and this preeminence was recognized by his own contemporaries. His attitudes as well as his forms of expression were in harmony both with the new sense of grandeur and richness of the contemporary Catholic world and with the spirit of absolutism then establishing itself among the monarchs of Europe. It is in this harmony of spirit that one finds not only the reasons for Pietro's attainments but also for his limitations. There was no pronounced break with his society or with its artistic traditions in his superb powers of expression, which, however, did not mask personal defects that were also the defects of contemporary Italian society. Behind his formal eloquence there lies a sense of universal participation, a new vision of nature, that was one of the most positive achievements of the Seicento. Pietro da Cortona changed the course of Italian painting markedly, but his development was part of that general 17th-century cultural and spiritual movement which corresponds precisely to the more limited and valid definition of the baroque formulated later.

Pietro's first teacher was the Florentine painter Andrea Commodi (a pupil of Santi di Tito), who was in Cortona from 1609 to 1611. This early instruction and a general acquaintance with Tuscan art of the first decade of the 17th century, which Pietro had opportunity to study in Cortona in important works by Ludovico Cigoli and others, were not determining experiences in his artistic development. At the beginning of the second decade of the century, while still very young, Pietro moved to Rome. There his early impressions and training were all but erased by the greater artistic force and magnificence of his new ambient. He arrived in Rome in 1612, during the pontificate of Paul V (Borghese), and soon had the good fortune of finding himself in the center of the extraordinary artistic activity that marked the second decade of the century in the papal city. These were restless years, in which cultural tastes changed with great rapidity. It is in this crowded and complex Roman decade that the true sources of Pietro's development must be sought — in the impressions and experiences which for ten years enriched his apprenticeship and which eventually led him, from his first recorded works, to that style of which Giovanni Battista Passeri wrote: "By his originality . . . he changed the face of painting" (1772).

The first endeavors of the youthful Pietro in Rome were conditioned by the circle of Tuscan artists into which Commodi introduced him. He used his new opportunities to perfect himself in the fundamental preliminary of painting — a mastery of drawing. Approaching the study of Roman archaeological remains with extraordinary enthusiasm, he drew vases, sarcophagi, funerary urns, statues, triumphal arches, friezes, and bas-reliefs. This solitary and personal "academy" left an indelible mark on his subsequent development. This profound rapport with the antique was for the young artist a means of escape from the restrictive Tuscan milieu of his former companions from the workshop of Santi di Tito.

He continued to associate with this group even after Commodi left Rome in 1614, having entrusted Pietro to the care of Baccio Ciarpi, another Tuscan, in whose workshop he remained for some time. The limitations of his Tuscan confreres became more and more evident as he observed the great artistic enterprises then under way in Rome, which impelled him toward more modern, grandiose conceptions.

During the decade of Roman apprenticeship that preceded his earliest acceptance by the public, Pietro's intellectual and cultural outlook expanded, principally because of the influence of two quite different circumstances. First, about 1620 he entered a circle of artists, antiquarians, and aristocratic conoisseurs that brought him into contact with members of the Crescenzi family, as well as with Cassiano dal Pozzo, for whom he worked until that collector's departure for France, and with Marcello Sacchetti, who later became his patron. This new milieu into which Pietro moved encouraged classical tendencies of a more practical, active character than the theoretical classicism of Giovanni Pietro Bellori and Giovanni Battista Agucchi. The second formative influence of the artist's career was his early realization of the change of direction then taking place in painting. This new manner of expression is difficult to ascribe to any individual artist, for it constituted instead a general change in the manner of representation. These new tendencies in painting, which became marked in the second decade of the 17th century, reflected the growing establishment of patrons and the influence of their commissions, the adjustment of the artists to their patrons' ideals and wishes, and the consequent development of a mode of expression more appropriate to the long-felt needs of a ruling class that had for some time been making itself heard with increasing authority.

A notable group of works by Pietro in which the beginnings of the new manner are evident can be securely dated before 1624, since Giulio Mancini mentioned them in that year (*Considerazioni sulla pittura*, ed. A. Marucchi and L. Salerno, II, 1957). These were most likely painted, therefore, while Paul V was still pope and during the three-year pontificate of Gregory XV. The group includes the frescoes in the gallery of Palazzo Mattei, as well as *The Sacrifice of Polyxena* and *The Triumph of Bacchus* (PL. 168), both painted for the Sacchetti family. Their brushwork is free, spontaneous, unbound by naturalistic precision, and with little heed for details. There is no emphasis of any one figure over another, or a strong differentiation of foreground against background. The lighting, which pervades the whole picture surface equally, creates that atmospheric verisimilitude which was the true secret of the Venetians, an effect that had been vainly sought after in the early years of the 17th century. These works reveal Pietro's priority in a trend that was later to have fuller and significant development — the so-called "Neo-Venetian" idiom.

After the election of Maffeo Barberini to the pontifical throne as Urban VIII, Pietro, with the help of the Sacchettis, succeeded in establishing a close relationship with the Barberini family through his participation with the circle of artists under the patronage of Cardinal Francesco Barberini (Urban VIII's nephew), thereby putting his reputation on a firmer base. His first important commission under Urban VIII was to fresco a wall in the Church of S. Bibiana, a task on which he worked from 1624 to 1626. Agostino Ciampelli had been commissioned to paint the opposite wall, and when Pietro's fresco was unveiled in the spring of 1626, the confrontation of the two works clearly revealed the distance separating the two generations of artists and, most of all, the great originality of Pietro's style. Episodes from the life of the heroic young Christian maiden St. Bibiana were portrayed by figures with noble gestures and poignant attitudes, in the dress of the Trajanic era, against a background of massive imperial architecture and profuse classical ornament. The relationship of the figures to the setting is so natural and appropriate that there is no hint of artificiality or archaeological preoccupation — an approach that the warm spontaneity of the drama would also have precluded. In comparison to Ciampelli's fresco, Pietro's must have appeared startlingly original.

The success of this work led to other commissions for the artist, and in 1626 Pietro entered upon a period of intense activity. For his first patrons, the Sacchettis, he painted, among other works, *The Rape of the Sabine Women* (PL. 169), a companion piece to *The Sacrifice of Polyxena*, painted some five years earlier. He also supervised the intricate decoration of the Sacchetti villa at Castel Fusano, in which he covered the entire chapel with landscape frescoes.

As a result of these works, the artist gained a preeminent position in the artistic circles of Rome, and a reputation that led the Barberinis to entrust him with the responsibility of what must be considered the major painting commission of the time: the decoration of the immense vaulted ceiling of the main hall (II, PL. 176; XII, PL. 125) of the new palace that the powerful family had constructed at the Quattro Fontane between 1625 and 1633. Pietro worked on this project for seven years, from the beginning of 1633 to the end of 1639. The program of the ceiling decoration was suggested by Francesco Bracciolini, the court poet from Pistoia: "The triumph of Divine Providence and the consummation of her ends by means of the temporal and spiritual power of the papacy." The ceiling of Palazzo Barberini discloses for the first time in painting the true measure of the new baroque expression, which by its very nature demanded complex themes to illustrate and great spaces in which to essay them. Pietro met the challenge in his gigantic fresco, which, notwithstanding the multiplicity of scenes, was conceived as a closely knit whole that acknowledged the unities of time and space. For achieving this unity, the disposition of interconnected groupings was the principal means. Although these painterly groupings may seem to be distributed at random — dense figure groups in some areas, quite thin in others — they actually follow a carefully calculated scheme, revolving about the luminous open space that forms the backdrop for the main group.

Acceptance of this unified conception did not, however, conflict with the exuberant narrative imagination of the artist. Vitality is lent to the various episodes by the use of individual perspectives that are developed within the primary (i.e., the central) perspective focus. These reveal the richness of Pietro's pictorial powers — whether used to create the flickering sylvan light that palpitates over the magnificent scenes of the drunken Silenus, spreads to the enchanted garden of Venus, and fades away on the horizon, beyond the fountain and the farthest reaches of the trees; or in the restless reflections of the flames from Vulcan's forge, which cast flashes of light on the armor scattered about the ground; or in the plunge of the giants, who seem to be hurled downward by the storm whirling about the airborne Pallas in full battle array.

Pietro's artistic activity was not, however, confined to Rome. In the course of three journeys he spent at least seven years in Florence, where he lived in the house of his friend Michelangelo Buonarroti the Younger and entered the service of Grand Duke Ferdinand II de' Medici. Pietro's first trip to Florence in 1637, which interrupted his work on the Barberini frescoes, was of brief duration, but during this stay he did paint the first two frescoes in the Sala della Stufa in Palazzo Pitti — the enchanting *Age of Gold* (PL. 167) and *Age of Silver*. He returned to Florence in 1640 and remained until 1642; finally, he went back toward the end of 1643 and stayed until 1647. The *Age of Bronze* and *Age of Iron*, in the same chamber, were finished in 1640. These are perhaps Pietro's masterpieces, or at least are to be counted among his most inspired works.

The greatest task that Pietro faced in Florence was the responsibility for decorating the grand-ducal apartments in the Palazzo Pitti. The enlargement of the palace, begun by Cosimo II, inspired Ferdinand II and his brothers to seek decoration that involved new and ambitious principles. Pietro found himself again confronting, in 1641, a problem not unlike the one he had found so stimulating eight years earlier in the decoration of Palazzo Barberini. Once again he had to translate into pictorial terms the complex program of a man of letters. Francesco Rondinelli's original scheme called for a unified iconographic program for the ceiling frescoes of seven rooms. The virtues essential to a prince from adolescence to old age, accompanied by the signs of the planets, was the over-all theme. In the Sala di Venere (PL. 170), painted in 1641–42, Pietro put to good use his recent experience on the Barberini ceiling decoration. In his Florentine fresco there is found once more, especially in the central panel, that natural world so atmospheric and spacious and that definition of every object through light which were evident in the *Drunken Silenus* fresco in Rome, but with no effect of a stale formula or mechanical repetition. The Sala di Venere frescoes were successfully completed in less than two years; subsequently, however, work on the remaining decoration of the Palazzo Pitti proceeded with difficulty. The original project entailing the decoration of seven rooms was reduced to five, and of these only two others were painted by Pietro: the Sala di Giove and the Sala di Marte. The Sala di Apollo, which was to have been the second in the series, was barely begun by the artist; for eighteen years it was left encumbered with scaffolding, until finally it was completed by Ciro Ferri, Pietro's most devoted pupil, who used his master's cartoons. Ferri also painted the Sala di Saturno, for which he devised his own composition.

The wealth of stucco decorations used in the Pitti decorative scheme had to be taken into consideration. Their inclusion in the program was a new departure for Pietro, who previously had restricted stuccowork to the simple, classic division of the cornice. Luxuriant stucco decorations now invaded the vault, thereby reducing the pictorial field. These pronounced differences were not the result of any particular influences deriving from the Florentine milieu; rather, they reflected profoundly personal changes in Pietro's artistic aims, for architecture had come to occupy an increasingly important place in his mind from the time he began construction of the Church of SS. Luca e Martina (see below).

Pietro left Florence for the last time in 1647 and returned to Rome. During his absence from Rome, Urban VIII had died, and Innocent X (Pamphili) had succeeded to the papal throne; moreover, Pietro's former patrons, the Barberinis, had fled to France. Still, though the artistic climate and taste had changed notably, Pietro — then just over fifty years old — was quick to regain his previous high standing. After some years of uncertainty, during which he painted (for the Oratorians, who had remained steadfast in their admiration of him) the dome of S. Maria in Vallicella (Chiesa Nuova) in a style recalling, but reworking, that of Giovanni Lanfranco in his S. Andrea della Valle frescoes, Innocent X commissioned him to paint the ceiling of the gallery Francesco Borromini had designed for the Palazzo Pamphili in Piazza Navona. These frescoes, which depict the story of Aeneas (PL. 171), were begun in 1651 and completed by 1654. As well as being the most important works of Pietro's late maturity, they constitute the most important undertaking in painting during the pontificate of Innocent X.

During the pontificate of Alexander VII (Chigi), who came to the papal throne in 1655, Pietro saw a rise in his fortunes and prestige as arbiter of the Roman artistic scene. At the same time, his fame spread beyond the borders of Italy. His interests, for the most part, were dedicated to architecture by this time (see below), and his most significant painting activity was to act solely as director of the commission to decorate the gallery of Palazzo del Quirinale, under the patronage of the new pontiff. In this function, he limited himself to selection of the artists to paint the various scenes. The fresco over the nave in S. Maria in Vallicella (1664–65) and the altarpiece for the high altar of S. Carlo ai Catinari (1667) are his most important late works.

Giuliano BRIGANTI

ARCHITECTURE. As in painting, Pietro da Cortona also made a decisive contribution to the development of high baroque forms in architecture, although few of his many projects were actually realized. His architectural drawings and projects include multiple solutions for the same building, often going beyond the merely decorative elements. The manner of Michelangelo and the classicism of Palladio, as well as his own painterly concepts, all influenced Pietro's development in architecture. Even though Pietro's training was not strictly that of an architect, coming as he did from a family of stonemasons and architects, he had experience in practical construction problems. Moreover, collaboration with his nephew Luca Berrettini, a stonemason of proven ability, assured fine workmanship and precision in carrying out his building programs.

Pietro's first work as an architect appears to date from about 1625, when he undertook the remodeling of Casale Sacchetti at Castel Fusano, a small building with the characteristics of a fortified country house; it was a commission that allowed him little freedom of expression, however. Greater opportunity to display his originality came a few years later (slightly before 1630), with the building of the Villa del Pigneto for the Sacchettis (FIG. 360). This building, now destroyed but known through many representations, stood on a slope of the Valle dell'Inferno,

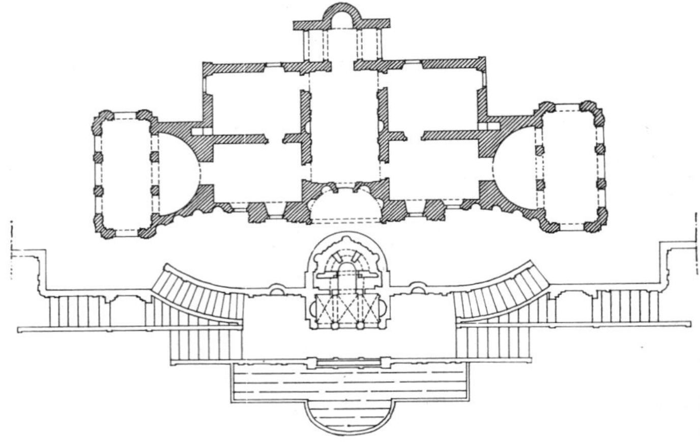

Villa del Pigneto, near Rome, plan according to a drawing by P. L. Ghezzi (*from R. Wittkower, 1958, pl. 80B*).

above a series of terraces with fountains and grottoes. Imposing in character, though the residence itself was small in scale, Cortona's plan called for a central structure with low, concave lateral wings. A monumental niche, inspired by that in the exedra of the Belvedere in the Vatican (II, PL. 346), gave importance to the central block. The arrangement of the various levels of terraces also shows a similarity to the Vatican Belvedere, the Villa d'Este (VIII, PL. 432), and the Sanctuary of Fortuna Primigenia in Palestrina (VII, PL. 199, FIG. 399). Pietro himself later prepared a project for the rebuilding of the Palestrina temple. These earlier models were interpreted by Pietro in a thoroughly baroque manner, however, and with a view to the over-all scenic effect.

As a result of the success of the Villa del Pigneto, the architect was invited to enter the competition for the design of the Palazzo Barberini, to be built at the Quattro Fontane in Rome. Pope Urban VIII rejected Pietro's design — according to Luca Berrettini, because it was too expensive. The ground plan tentatively identified by Rudolf Wittkower (1958, p. 154, pl. 81A; FIG. 361) with the Palazzo Barberini design does not, either in its calligraphy or in its conception, seem to point definitely to Pietro. The *ludus geometricus* characterizing it suggests more the idiom of Borromini than that of Pietro. Only parts of the layout of the garden walls of the Palazzo Barberini were built according to Pietro's plans, namely, the portal and the windows of the section later adapted for use as a theater, the fountain arrangement at the Quattro Fontane corner, and the original portal on the east side of the garden.

Reconstruction of a plan for Palazzo Barberini in Rome (*from Saggi di storia dell'architettura in onore di V. Fasolo, Rome, 1961*).

In the years around 1630 Pietro designed some funerary monuments in which all traces of late mannerism were eliminated in favor of the simple forms of a classicizing baroque. The quiet harmony of the various elements is emphasized by the carefully considered, moderate color scheme of the different marbles used. Examples of such works include the monument dedicated to "Asdrubalis ex comitibus Montis Acuti" in S. Girolamo della Carità, the Landri (Aleandri) monument in S. Lorenzo fuori le Mura, and the project for a funerary chapel known from a drawing in Lille (Mus. B.A.).

In 1633 Cardinal Francesco Barberini commissioned Pietro to design, for the Church of S. Lorenzo in Damaso, a festival decoration to be used in the celebration of the forty hours' devotion (drawing in the Royal Library, Windsor Castle). The Church was transformed into a kind of theater for the occasion, with the interior space articulated by a series of niches with gilt statues between columns and pilasters — a conception directly inspired by Giovanni Battista Aleotti's theater in Parma and Palladio's in Vicenza. Above the altar, framed by an arch, was suspended a tabernacle supported by a glory of life-size angels. Contemporaneous accounts tell of hidden light sources that enhanced the powerful illusionistic effect of the whole. Documentary evidence shows that this decoration, for which Cardinal Barberini paid more than 2,000 scudi, made a tremendous impression on his contemporaries, reflected not only in subsequent festival decorations but also in the architecture of permanent structures.

The use of lighting from hidden sources appeared again in Pietro's design for the high altar of S. Giovanni dei Fiorentini (drawing in Windsor Castle), in which the sculptural group of *The Baptism of Christ* by Francesco Mochi (Mus. di Roma) was to have been illuminated by two concealed side lights and the dove of the Holy Ghost was to appear in an aureole of natural light entering through a window above the altar. This project is dated no later than the beginning of 1634, for in July of that year a full-scale model of the altar in wood and stucco was placed in the church, where it must have remained until Francesco Borromini began work there in 1664. The influence of this ornate, highly theatrical work on the development of baroque altar design cannot be overemphasized. In the use of such hidden light sources within an architectural conception, priority must certainly be given to Pietro da Cortona over Bernini (the latter, for instance, in the De Raymondi Chapel in S. Pietro in Montorio). Moreover, the idea of introducing the pictorial effect created by a glazed window for enhancing a sculptural work is at least contemporary with, if not earlier than, Bernini's memorial to Urban VIII in S. Maria in Aracoeli.

In the same year (1634), in his capacity as *principe* of the Accademia di S. Luca, Pietro proposed the reconstruction of the church of this artists' guild, SS. Luca e Martina. He had already chosen, as various drawings in Munich (Staat. Graphische Samml.) and Milan demonstrate, a central plan in which the four arms of the Greek cross had apsidal ends. At the center was a cylindrical space over which rose the dome, held aloft by eight giant columns without intervening pendentives. (The basic idea for this type of plan was derived from Michelangelo's design for S. Giovanni dei Fiorentini.) In the arms of the cross the architect intended to place the tombs of Urban VIII and his family, apparently in the hope of arousing the Pope's interest in the project. Although the church was not destined to become the burial place for the Barberinis, Urban VIII and, even more, Cardinal Francesco Barberini, protector of the Accademia, developed a keen interest in the project when, in October of 1634, in the course of excavations in the crypt for the foundations of the architect's own tomb, Pietro discovered the body of St. Martina. New plans were then made for rebuilding the entire church, both the upper and the lower levels. The rebuilding was to continue for several decades, but the definitive design for the project was developed from Pietro's conception of 1634–35. In the lower church he linked together a number of catacomblike chapels into a symmetrical scheme. The cult and artistic center is the Chapel of St. Martina, a rectangular space which was only slightly off a square and which had, thus, a centralizing effect. Under the flattened coffered dome, which rests on columns at each corner, is the rich bronze baroque altar of the saint, also designed by Pietro. The flat-domed cruciform vestibule of the principal chapel, which has strongly beveled pilasters at the corners (derived from St. Peter's) and niches containing reliquary statues, is also designed on a central plan. At the eight beveled corners of the pilasters are gray marble columns, which stand out in sharp contrast to the white wall and which serve to define, as related corner elements, the space lying under the dome and between the arms of the cross. This solution was a substantial anticipation of Borromini's for S. Agnese in Agone, in Piazza Navona (II, FIG. 275).

In the upper church, the original plan of a cylindrical central space was modified to one with a dome resting on beveled pilasters and pendentives (PL. 173). As in the vestibule of the lower church, the motif of eight columns placed at the corners of the beveled pilasters is repeated; but here they are painted white like the walls. The interior walls are articulated in a sequence of three planes, one projecting beyond the other in an advancing and retreating movement that is continued in all four arms of the church. A rhythmic effect of light and shadow is created, with the sharply protruding pilasters providing the strongest light values and the set-back wall surfaces the deepest shadows. The cylindrical surfaces of the columns, intervening between these two zones, constitute a transitional medium tone. This optical effect, produced by manipulating light and shadow with architectural elements, became a characteristic of all Pietro's subsequent buildings. From its suggestion of convex movement between projecting lateral pilasters, the façade of SS. Luca e Martina (PL. 172) — the first of its kind in Rome — gives an exterior forecast of the spatial vitality contained within (I, PL. 419).

While he was in Florence painting the frescoes in Palazzo Pitti, Pietro drew up the design for a new church for the congregation of the Oratory of St. Philip Neri (Chiesa Nuova di S. Filippo, the present-day S. Firenze). Construction was started in 1645, based on a wooden model by the architect himself, but it soon became apparent that the project was too large and the cost far beyond the resources of the order. The church was not completed until 1660, according to an extensively revised design by Pier Francesco Silvani and others (PL. 174). Also dating from this period is another project that remained unrealized, namely, Pietro's plan for remodeling of the façade of the Palazzo Pitti (I, PL. 407). The design would have defined a central, portal segment on the façade by adding three orders of rusticated pilasters to frame triple units of apertures on three stories. For the Boboli Gardens he planned a theater, which was to be joined to the two wings of Bartolommeo Ammanati's building by means of a concave arcade over the grotto at the back of the palace courtyard. A pediment was to crown the front of the theater (PL. 174; IV, PL. 192).

Motifs suggested in the Villa del Pigneto and the project for reconstruction of the Sanctuary of Fortuna Primigenia in Palestrina appear in a new, elaborated form in the façade of S. Maria della Pace (1656–57; I, PLS. 381, 419; II, FIG. 307). Here the object was to give a more impressive appearance to the modest façade of the "temple of peace" built by Sixtus IV. Pietro placed a projecting semicircular columned portico in front of the lower part of the narrow façade. The upper part developed the convex motif of a façade framed between two pilasters, a scheme recalling that of SS. Luca e Martina. He also enlarged the façade by adding two lateral wings, which are lower than the central block and which thus give the illusion of a three-aisled interior. The right wing also serves to mask the choir of the adjoining Church of S. Maria dell'Anima, which, had it remained visible, would have interrupted the compositional unity of the whole; the left wing is primarily a balance for this device. These two wings form a concave frame for the central part of the façade, and as a result, the convex central portion appears to project with even greater emphasis from the adjoining recessed areas. Pietro also remodeled the interior of the church with scenographic stucco decoration and later designed a new decoration in marble for the Chigi Chapel.

Because of the alignment with the street and because the church is set between other buildings on the present-day Via del Corso, the façade of S. Maria in Via Lata (1658–62; PL. 172; FIG. 364) could not be given a scenographic treatment. The architect designed a two-storied façade, consisting of Corinthian colonnades separated by a projecting entablature. The colonnades are framed by masonry surfaces decorated with Corinthian pilaster strips and, on the upper level, niches and molding panels. This façade design conveys the strength and dignity of ancient Roman temples. The device of a round arch breaking through the entablature of the upper colonnade and set within a triangular pediment, which is derived from a late-antique model (perhaps the Palace of Diocletian in Split) brings to the restrained classicism of the façade a definitely baroque touch.

Pietro da Cortona also drew up plans for remodeling the in-

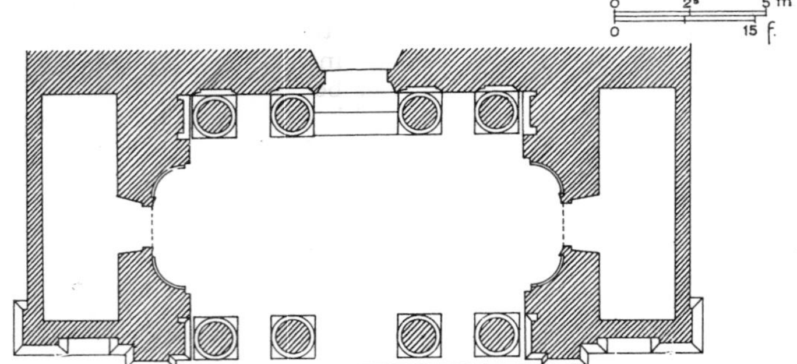

Rome, S. Maria in Via Lata, plan of the entrance porch (*from Via del Corso, Rome, 1961*).

terior of the apse of St. John Lateran (1662; drawing, Vat. Lib.). In his conception, the papal throne was to have been surmounted by a colonnaded aedicula with a narrow tympanum, within which would have been placed the reliquaries of the heads of the princes of the apostles (i.e., SS. Peter and Paul).

In 1664 Pietro da Cortona, Gian Lorenzo Bernini, Carlo Rainaldi (qq.v.), and Candiani (or Landiani) were requested by Louis XIV to submit plans for the completion of the Louvre. It is known that Pietro presented a ground plan (lost) as well as four elevations for façades, of which those for the east and west outside façades and one for the east façade of the courtyard are still in existence (Louvre, Cabinet des Dessins). The architect had been expressly requested to make the fullest possible use of existing structures and to crown the palace with a dome. Pietro chose a colossal order rising from a base for the articulation of the external façade. For the courtyard façade, however, this giant order rises directly from the foundation, as in Michelangelo's Palazzo dei Conservatori. In the design for the west façade Pietro borrowed from the French pavilion system typical of Le Vau, but without abandoning his own parietal forms. This resulted in an extraordinary synthesis of French and Italian styles that calls to mind 18th-century German architecture.

There is a direct stylistic relationship between Pietro's Louvre project (Vat. Lib.) and his plan for a palace for the Chigi family, with a front elevation incorporating a fountain. This residence was to have been erected in Piazza Colonna where the Palazzo Wedekind now stands. For the façade with the fountain, which was to draw its water from the Acqua Vergine, Pietro proposed two plans, both calling for a colossal order rising from a base (as in the Louvre project) and encompassing a full story and a mezzanine. In the first project, the façade is rectilinear; in the second it is divided into three sections, a concave central block and lateral wings. The recessed portion was intended to contain the oval basin of the fountain, set within a kind of rocky landscape with figures of ancient river gods. The juxtaposition of the naturalistic fountain and the severely monumental façade would have created a striking contrast. The Chigi project was the most important precedent for the Trevi Fountain, but the latter admittedly does not equal Pietro's project in grandeur.

During the last years of his life Pietro embarked on three other architectural projects, which he did not live to see completed: the dome of S. Carlo al Corso, the Gavotti Chapel in S. Nicola da Tolentino, and the Chapel of St. Francis Xavier in the Church of the Gesù. Although at one time both were attributed to him, Pietro is the architect neither of the crypt of S. Martino ai Monti nor of that part of the Palazzo Doria-Pamphili facing on Piazza del Collegio Romano. A few minor works in Rome should be noted, however: the restoration of the architect's house (destroyed) in Via della Pedacchia, the Chapel of the Sacrament in S. Marco, and the Ghibbesio tomb (destroyed) in the Pantheon. In Florence, Pietro worked on the restorations of the Ospedale di S. Maria Nuova and the Casa Buonarroti.

Karl NOEHLES

SOURCES. G. Mancini, Considerazioni sulla pittura (ca. 1625), ed. A. Marucchi and L. Salerno, II, Rome, 1957; G. Tetius, Aedes Barberinae, Rome, 1642 (2d ed. 1647); F. Scannelli, Microcosmo della pittura, Cesena, 1656; F. Sansovino, Venezia città nobilissima, Venice, 1663; M. Boschini, La carta del navegar pitoresco, Venice, 1664; Marquis de Seignelay, Voyage en Italie (1671), GBA, XVIII, 1865, pp. 176–85, 357–71, 445–64; I. M. Silos, Pinacotheca sive Romana pictura et sculptura, Rome, 1673; L. Scaramuccia, Le finezze dei pennelli italiani, Pavia, 1674; F.Titi, Studio di pittura, scultura, architettura delle chiese di Roma, Rome, 1674; J. von Sandrart, Teutsche Academie der Edlen Bau-Bild-und Mahlerey-Künste (1675), ed. A. R. Peltzer, Munich, 1925; C. C. Malvasia, Felsina pittrice, Bologna, 1678 (ed. G. P. Zanotti, 2 vols., Bologna, 1841); L. Berrettini, Lettera a Ciro Ferri (1679), in G. Campori, Lettere artistiche, Modena, 1866, pp. 506–15; F. Baldinucci, Notizie de' professori del disegno, I, Florence, 1681; F. S. Baldinucci, Biography of Pietro da Cortona, Ms. Pal. 565, Bib. Naz. of Florence, n.d. (two biographies, of which the first has been published by S. Samek Ludovici, Le Vite di Francesco Saverio Baldinucci, Archivi, XVII, 1950, pp. 77–91); A. Félibien des Avaux, Entretiens sur les vies et les ouvrages, des plus excellens peintres, anciens et modernes, 5 vols., Paris, 1666–88; J. von Sandrart, Academia nobilissimae artis pictoriae, Nürnberg, 1683; L. Pascoli, Vite de' pittori, scultori ed architetti moderni, I, Rome, 1730; A. J. Dézallier d'Argenville, Abrégé de la vie des plus fameux peintres, Paris, 1742; G. B. Passeri, Vite de' pittori, scultori ed architetti, Rome, 1772.

BIBLIOG. *Painting*: F. Milizia, Memorie degli architetti antichi e moderni, 4th ed., II, Bassano, 1785; L. Lanzi, Storia pittorica dell'Italia, I, Bassano, 1795; G. G. Bottari, and S. Ticozzi, Raccolta di lettere, I, V, Milan, 1822–25; M. Gualandi, Nuova raccolta di lettere, II, Bologna, 1845; C. Gurlitt, Barock in Italien, Stuttgart, 1887; N. Fabbrini, Pietro Berrettini da Cortona pittore e architetto, Cortona, 1896; H. Geisenheimer, Pietro da Cortona e gli affreschi di Palazzo Pitti, Florence, 1909; O. Pollak, Neue Regesten zum Leben und Schaffen des Pietro da Cortona, Kchr., N.S., XXIII, 1911–12, cols. 561–67; O. Pollak, ThB, VII, 1911, s.v.; H. Posse, Das Deckenfresko des Pietro da Cortona in Palazzo Barberini und die Deckenmalerei in Rom, JhbPreussKSamml, XL, 1919, pp. 93–118, 126–73; J. A. Orbaan, Documenti sul Barocco in Roma, Rome, 1920; O. H. Giglioli, Disegni inediti nelle R. Gallerie degli Uffizi, BArte, N.S., II, 1922–23, pp. 498–525; H. Voss, Die Malerei des Barock in Rom, Berlin, 1924; O. Pollak, Die Kunsttätigkeit unter Urban VIII, Vienna, 1928; A. Stix, Barockstudien, Belvedere, 1930, 2, pp. 180–85; I. A. Richter, A Rediscovered Painting by Pietro da Cortona, Apollo, XIV, 1931, pp. 285–87; S. von Below, Beiträge zur Kenntnis von Pietro da Cortona, Munich, 1932; E. Lavagnino, Il bozzetto di Pietro da Cortona per la volta della Sala Maggiore del Palazzo Barberini, BArte, XXIX, 1935, pp. 82–89; E. K. Waterhouse, Baroque Painting in Rome, London, 1937; The 17th Century in Europe (cat.), London, 1938; E. K. Waterhouse, Nicolas Poussin's Landscape with the Snake, BM, LXXIV, 1939, p. 103; R. Longhi, Arte italiana e arte tedesca, Florence, 1941; W. G. Constable, Some Unpublished Baroque Paintings, GBA, XXIII, 1943, pp. 219–36; D. Mahon, Nicolas Poussin and Venetian Painting, BM, LXXXVIII, 1946, pp. 15–20, 37–43; U. Barberini, Pietro da Cortona e l'arazzeria Barberini, BArte, XXXV, 1950, pp. 43–51, 145–52; J. Costello, The Twelve Pictures "Ordered by Velasquez" and the Trial of Valguarmera, Warburg, XIII, 1950, pp. 237–84; G. Briganti, 1630 ossia il Barocco, Paragone, II, 13, 1951, pp. 10–17; J. Hess, Tassi, Bonzi e Cortona a Palazzo Mattei, Comm, V, 1954, pp. 303–15; D. Mahon and D. Sutton, Artists in 17th Century Rome (cat.), London, 1955; L. Berti, Nota ai pittori toscani in relazione con Pietro da Cortona: Giovannandrea Commodi, Comm, VII, 1956, pp. 278–82; A. Blunt, The Exhibition of Pietro da Cortona at Cortona, BM, XCVIII, 1956, pp. 415–17; A. Marabottini Marabotti, ed., Pietro da Cortona (cat.), Cortona, Rome, 1956; G. Briganti, Opere inedite o poco note di Pietro da Cortona nella Pinacoteca Capitolina, B. Mus. Comunali di Roma, IV, 1957, pp. 5–14; L. Grassi, Pietro da Cortona e i bozzetti per la Galleria Doria Pamphili, BArte, XLII, 1957, pp. 28–43; R. Wittkower, Art and Architecture in Italy 1600–1750, Harmondsworth, Baltimore, 1958, pp. 162–68; G. Briganti, L'altare di S. Erasmo, Poussin e il Cortona, Paragone, XI, 123, 1960, pp. 12–20; G. Briganti, L'occasione mancata di Andrea Commodi, Paragone, XI, 123, 1960, pp. 33–37; N. Wibiral, Contributi alle ricerche sul cortonismo in Roma: i pittori della Galleria di Alessandro VII nel palazzo del Quirinale, BArte, XLV, 1960, pp. 123–65; C. C. Cuningham, A Modello for Pietro da Cortona's Cupola of the Chiesa Nuova in Rome, B. Wadsworth Atheneum, 5th Ser., IX, 1961, pp. 18–22; W. Vitzthum, A Comment on the Iconography of Pietro da Cortona's Barberini Ceiling, BM, CIII, 1961, pp. 427–33; G. Briganti, Pietro da Cortona o della pittura barocca, Florence, 1962 (bibliog.); E. K. Waterhouse, Italian Baroque Painting, Greenwich, Conn., 1962.

Architecture: G. Giovannoni, Il restauro architettonico di Palazzo Pitti nei disegni di Pietro da Cortona, Rass. d'Arte, VII, 1920, pp. 290–95; V. Moschini, Le architetture di Pietro da Cortona, L'Arte, XIV, 1921, pp. 189–97; A. Muñoz, Pietro da Cortona, Rome, 1921; R. Wittkower, Pietro da Cortona's Ergänzungsprojekt des Tempels in Palestrina, Festschrift für Adolph Goldschmidt, Berlin, 1935, pp. 137–43; G. Incisa della Rocchetta, Il Pigneto Sacchetti e il Mausoleo dell'asino Grillo, L'Urbe, XII, 1949, pp. 9–16; A. Blunt, The Palazzo Barberini; The Contributions of Maderno, Bernini and Pietro da Cortona, Warburg, XXI, 1958, pp. 256–87; R. Wittkower, Art and Architecture in Italy 1600–1750, Harmondsworth, Baltimore, 1958, pp. 153–62; A. Schiavo, Caratteristiche grafiche nei disegni architettonici del Berrettini, Palatino, IV, 1960, pp. 170–74; H. Keller, Ein früher Entwurf des Pietro da Cortona für SS. Martina e Luca in Rom, Misc. Bibliothecae Hertzianae, Vienna, Munich, 1961, pp. 375–84; K. Noehles, Die Louvre-Projekte von Pietro da Cortona und Carlo Rainaldi, ZfKg, XXIV, 1961, pp. 40–74; P. Portoghesi, Gli architetti italiani per il Louvre, Candiani, Rainaldi, Pietro da Cortona, Saggi di storia dell'architettura in onore di V. Fasolo, Rome, 1961, pp. 243–68; A. Schiavo, Il pigneto dei Sacchetti, S. Romani, IX, 1961, pp. 284–87; E. von Hubala, Entwürfe Pietro da Cortona's für SS. Martina e Luca in Rom, ZfKg, XXV, 1962, pp. 125–52; P. Portoghesi, SS. Luca e Martina di Pietro da Cortona, Architettura, IX, 1963–64, pp. 114–27.

Giuliano BRIGANTI and Karl NOEHLES

Illustrations: PLS. 167–174; 3 figs. in text.

PIGALLE, JEAN-BAPTISTE. French sculptor (b. Paris, Jan. 26, 1714; d. Paris, Aug. 21, 1785). Descended from master carpenters, Pigalle began his sculptural training with Robert Le Lorrain at the early age of eight years. About 1734, he continued his training with Jean Baptiste Lemoyne and also studied at the Académie Royale. In 1736, after having competed unsuccessfully for the Prix de Rome, he received permission to attend the French Academy in Rome, but without a stipend. He traveled to Italy on foot, and after suffering illness and want initially, he brought his Roman stay to a successful conclusion by winning a prize in the competition of the Accademia di S. Luca. He returned by way of Lyons, where he worked for local churches. Arriving in Paris in 1741, he presented the model of his *Mercury* to the Académie Royale and was received as *agréé*. At the Salon of 1742, the *Mercury* (possibly the version in the Metropolitan Museum, New York) found great success. It reconciles rococo grace with the idealism of neoclassicism and, at the same time, exhibits vigorous realism in anatomical details. In 1744, the Académie received Pigalle as a full member upon presentation of the small marble version of the *Mercury* (Louvre). As a pendant, Pigalle executed a *Venus* (plaster model executed for the Salon of 1742), markedly more prosaic than the *Mercury*. Life-size versions of the *Mercury* and *Venus*, done in marble at the command of Louis XV, were presented to Frederick II of Prussia in 1749 (Berlin, Staat. Mus.).

Pigalle's rise to fame can be measured by his patrons and academic honors. Appointed professor in 1752, he became chancellor of the Académie Royale in 1785, having in 1769 received the estimable Order of St. Michael. His first important commissions came from the Comte d'Argenson (*Virgin*, 1745; Paris, St-Eustache). Later he was patronized by Mme de Pompadour, and finally by the king himself. In effect, Pigalle after 1750 served, though not so titled officially, as the court sculptor of Louis XV.

He was much in demand as a portraitist and also satisfied the fashionable taste for small-scale work of a decorative or sentimental kind (*Child with Bird Cage*, 1749; Louvre). More important were his royal statues and commemorative monuments. The Revolution brought destruction to his royal portraits, such as the statue of Louis XV (1754) for Mme de Pompadour's Château de Bellevue. Lost, too, was the monument of Louis XV in Reims (1756–63), except for the base, among the figures of which is the seated *Citizen*, a self-portrait of the artist and probably his most powerful work. Between 1770 and 1776, he executed the funeral monument of the Maréchal de Saxe (Strasbourg, St-Thomas), representing the great soldier as he descends into the tomb, mourned by France and received by Strength and Death. This monument was much criticized because of its obscure allegory and disturbingly realistic features. Even more shocking to contemporary sensibilities was the tomb of the Duc d'Harcourt (1769–76; Paris, Notre-Dame), with its representation of a hideous cadaver protruding from the sarcophagus. Pigalle's verism reached its extreme in the *Voltaire* (1770–76; Paris, Palais de l'Inst. de France), which represents the aged philosopher as an emaciated nude. These breaches of conventional taste by such an academic dignitary throw telling light on the liberalism of the age, as well as on Pigalle's independent temperament. Historically, his work stands on the threshold of neoclassicism in its tendency toward strongly defined contour and relieflike arrangements. But its special worth lies in a persisting realism and the power of its characterization, both qualities equally remote from rococo and neoclassical stereotypes.

BIBLIOG. P. Tarbé, La vie et les œuvres de Jean-Baptiste Pigalle, sculpteur, Paris, 1859; S. Rocheblave, Jean-Baptiste Pigalle, Paris, 1919; L. Réau, J.-B. Pigalle, Paris, 1950.

Lorenz EITNER

PILON, Germain. French sculptor (b. Paris, ca. 1535; d. Feb. 3, 1590). The son of a master mason, Pilon was probably a pupil of Pierre Bontemps. His first known works are eight figures (e.g., Paris, Mus. de Cluny) for the base of the tomb of Francis I in the abbey church at Saint-Denis, where Pilon worked with Bontemps. In 1559 Catherine de Médicis commissioned him to execute the monumental receptacle for the heart of Henry II (V, pl. 395). In their extreme elongation and mannered grace, the figures of the Three Graces supporting the urn exemplify Pilon's early debt to the school of Fontainebleau. About 1561 he was active at Fontainebleau itself, where he carved wooden statues of Mars, Mercury, Juno, and Venus for the Queen's garden.

In 1565 he began the tomb of Henry II for the Valois Chapel at Saint-Denis, commissioned by Catherine de Médicis. Started under the direction of Primaticcio, the work continued, with interruptions, almost until Pilon's death. Between 1565 and 1572, he sculptured the kneeling bronze effigies of the King and Queen surmounting the mausoleum, the nearly naked reclining marble figures of the King and Queen placed within the structure, and the reliefs of Faith, Hope, Charity, and Ecclesia on its base. The influence of the Italianate style of Fontainebleau appears much diminished in the royal effigies. Their heavier, markedly corporeal forms and vigor of expression — evident especially in the unsparing presentation of the cadavers — link them with earlier French tradition. In this period Pilon also executed the *Christ Risen* (depicted between two soldiers) for the Valois Chapel. (The figure of Christ is presently located in the Church of SS. Paul et Louis in Paris; the two soldiers, executed by the sculptor's assistants, in the Louvre.)

In 1570 Pilon executed the expressive marble *Virgin and Child* for the altar of the Church of Notre-Dame-de-la-Couture in Le Mans, and during the same year he designed the triumphal arch and other decorations for the marriage of Charles IX. Lodging the artist in the Hôtel de Nesle, the king gave him the title of "sculpteur du roi" and then, in 1573, elevated him further, to the position of "conducteur et contrôleur en l'art de sculpture," which included supervision of the mint. Pilon's principal work of this period was the tomb of Valentine Balbiani, to be placed in the Church of Ste-Catherine du Val-des-Ecoliers (tomb effigy, Louvre). With its recumbent effigy set upon a base on which is carved in relief the half-nude corpse, it echoes the style of the tomb of Henry II. In 1583 Pilon resumed work on the Valois Chapel, adding a seated Madonna (Church of SS. Paul et Louis, Paris; painted terra-cotta model, Louvre) and a *St. Francis in Ecstasy*, a figure of grand dramatic pathos (Church of SS. Jean et François, Paris). In 1584 he began the tomb of Valentine Balbiani's husband, Chancellor René de Birague, of which only the kneeling bronze portrait figure survives (Louvre); it combines a grandiose monumentality, seen in the sweep of the Chancellor's cloak, with an austere realism, particularly apparent in the treatment of the head and hands.

Pilon's work marks the liberation of French Renaissance sculpture from Italianate influences. Although his early career was overshadowed by Primaticcio and the school of Fontainebleau, in his later development elements of a more sober, distinctly French tradition were combined with his personal realistic bent, and he evolved a style of great expressive power and rugged dignity (see also XII, pl. 100).

Bibliog. J. Babelon, Germain Pilon, Paris, 1927; C. Terrasse, Germain Pilon, Paris, n.d. [1930].

Lorenz Eitner

PINTORICCHIO. Italian painter (b. Perugia, ca. 1454, according to Vasari; d. Siena, 1513). Pintoricchio (or Bernardino di Betto di Biagio) is ranked, after Perugino, among the foremost of the Umbrian mural painters. He worked largely in Rome but also, among other places, in Spello, Siena, and Perugia.

The artist's earliest works and his actual teacher are not definitely known, though certain similarities to works by Fiorenzo di Lorenzo have led critics to assume a teacher-pupil elationship. From 1481 to 1484 Pintoricchio assisted Perugino with his frescoes in the Sistine Chapel, among which the *Baptism of Christ* and various figures in the *Journey of Moses*, particularly in the *Circumcision of Gersam*, have often been considered to be by Pintoricchio.

His style becomes mature and identifiable in the frescoes of the Bufalini Chapel in S. Maria in Aracoeli in Rome, and even more so in the Vatican frescoes of the Borgia Apartments of Pope Alexander VI (XII, pl. 34). In these, the figural and compositional treatment seems derived largely from Perugino, but the older artist's unity and clarity of design and his peculiarly lyrical, soft handling of figures are lacking. Instead, Pintoricchio is much more decorative. The renewed and extensive use of gold color, application of gilded stucco, his motley use of color, the abundance of ornamental motifs (he is said to have been the first to make extensive use of ancient decoration), and the more particularized, graphic description of trees, rocks, landscapes, and buildings — all these characteristics contribute to his mundane, festive, and decorative style.

His commissions included the decoration of churches, palaces, and stanze, as well as that of the Piccolomini Library in Siena. The Piccolomini frescoes, commissioned by Francesco Todeschini-Piccolomini (later Pius III), entailed the decoration of a room, annexed to the Cathedral of Siena, that housed the library of his uncle, Pope Pius II (Enea Silvio Piccolomini, known as Aeneas Sylvius). Scenes from the adventurous life of this poet-playwright, historian, diplomat, and churchman are presented on walls ingeniously articulated with illusionistically painted archways that afford vistas into the distant landscapes, piazzas, loggias, and interiors which are the stages for the events depicted. The placement of figures is often confusing in its multiplicity, often repetitively symmetrical; moreover, they fail to create compositional tensions and seem little involved in the scenes portrayed. The stance of the figures, that is, both their attitudes and their significance to the narrative meaning, is indicated by a restricted set of formulas that, rather than individualizing responses according to content, reveal (much like Perugino) an adherence to norms. With dynamic urgency of form and content lacking, the scenes — in spite of the illusionism intended — appear removed from actuality. The liberal use of gold and joyful use of color enhance the decorative effect of the whole.

Thus Pintoricchio, at the beginning of the 16th century, was not able to partake in that synthesis of form and content, that equipoise between expression of feeling and depiction of fact, between the ideal and the real, which characterized the new classic art in Italy. Rather, Pintoricchio remained part of an all-European stylistic situation that — borrowing a term from descriptions of late Gothic sculpture in France — has been dubbed the *détente*.

Works. Rome, S. Maria in Aracoeli, Bufalini Chapel frescoes, soon after 1485. – *Portrait of a Youth* (Washington, D.C., Nat. Gall., Kress Coll.; Dresden, Gemäldegal.), both 1490s. – Vatican, Borgia Apartment frescoes, 1492–94. – *Madonna and Child Enthroned, with the Young St. John* (Perugia, Gall. Naz. dell'Umbria), altarpiece with side and predella panels, painted for S. Maria dei Fossi, Perugia, ca. 1495–98. – Spello, S. Maria Maggiore, Baglioni Chapel frescoes, 1501. – Siena, Cathedral annex, Piccolomini Library frescoes, 1503–08. – Spello, S. Andrea, *Madonna and Child with Saints*, 1509.

Bibliog. A. Schmarsow, Bernardino Pinturicchio in Rom, Stuttgart, 1882; C. Ricci, Pinturicchio, London, 1902; W. Bombe, Geschichte der Peruginer Malerei . . . , Berlin, 1912; E. Carli, Il Pintoricchio, Milan, 1960 (bibliog.).

Curtis Shell

PIRANESI, Giovanni Battista. Italian etcher and architect (b. Mestre, 1720; d. Rome, 1778). Piranesi was educated in Venice, mostly by his father, a stonemason, and his uncle, an engineer and architect. In Rome, during the years 1740–44, he learned etching from Giuseppe Vasi and Felice Polanzani and published his first large volume, *Prima parte di architetture . . .* (1743). After returning briefly to Venice, he settled permanently in Rome in 1745, and had become one of its most famous inhabitants by his death in 1778.

Piranesi's first great prints were four *Grotteschi* and fourteen *Carceri* (IV, PL. 439), which he drew with the ragged brilliance that made Venice the school of etching for the 18th century. Modern-day collectors prefer these early etchings to his later work because of their wit, fantasy, and macabre qualities. In the late 1740s he began the series of *Vedute di Roma*, some 135 in all, which he continued etching until his death. Long considered his most famous achievement because of the manner in which they dramatize antique and baroque Rome, even after more than a century of photography, these works still haunt the viewer with the grandeur of the Eternal City (I, PL. 410; IV, PL. 438). Piranesi himself probably placed a higher value on the series of monumental volumes recording his excavations and reconstructions of famous ruins in and near Rome. During his last summer he measured and drew the temples at Paestum for a publication that his son Francesco finished after his death.

In 1769 he published *Diverse maniere d'adornare i cammini* . . . , a book of original and often eccentric designs for interior decoration, which helped to establish the Empire style. His major work as an architect was the rebuilding of S. Maria del Priorato (X, PL. 267; XII, PL. 308), on the Aventine hill, in 1764–65. (The manuscript daybook for this project is in Avery Architectural Library, Columbia University.) Toward the end of his life he dealt in Roman antiquities. After his death his sons sold the remainder of his ancient marbles to the King of Sweden, and these are still in Stockholm.

Piranesi's prints flooded Europe like travel posters, for he calculated his profits on the basis of 4,000 impressions from each of his 991 copperplates. The poetic Rome of his imagination brought tourists to disappointment upon visiting the crumbling actuality. The 19th century tired of seeing his great, emphatic etchings everywhere; today, however, his volcanic obsessions are no longer consulted as records of fact but are cherished for the grandeur of their intensely personal vision. (See also I, PL. 309; V, PL. 241; XII, PLS. 176, 316.)

WORKS. Prima parte di architetture e prospettive, Rome, 1743; Invenzioni capric. di Carceri all'acquaforte, Rome, 1745 (2d ed., Carceri d'invenzione, Rome, 1760); Antichità romane de' tempi della Repubblica e de' primi imperatori, Rome, 1748 (2d ed., Alcune vedute di archi trionfali ed altri monumenti inalzati da Romani, Rome, after 1765); Varie vedute di Roma antica e moderna, Rome, 1748; Opere varie di architettura, prospettive, grotteschi, antichità sul gusto degli antichi Romani, Rome, 1750; Trofei di Ottaviano Augusto inalzati per la vittoria ad Actium e conquista dell'Egitto con varj altri ornamenti diligentemente ricavati dagli avanzi più preziosi delle fabbriche più antiche di Roma, utili a pittori, scultori e architetti, Rome, 1753; Le antichità romane, 4 vols., Rome, 1756; Lettere di giustificazione scritte a Milord Charlemont e a di lui agenti in Roma . . . intorno la dedica della sua opera delle Antichità Romane, fatta allo stesso Signore ed ultimamente soppressa, Rome, 1757; Della magnificenza et architettura de' Romani, Rome, 1761; Il Campo Marzio dell'antica Roma, Rome, 1762; Osservazioni . . . sopra la lettre di M. Mariette aux auteurs de la Gazette littéraire de l'Europe . . . , Rome, 1765; Diverse maniere d'adornare i cammini [sic] ed ogni altra opera degli edifizi desunte dall'architettura Egizia, Etrusca e Greca con un Ragionamento apologetico in difesa dell'architettura Egizia e Toscana, Rome, 1769; Vasi, candelabri, cippi, sarcophagi, tripodi, lucerne ed ornamenti antichi, 2 vols., Rome, 1778.

BIBLIOG. H. Focillon, Giovanni-Battista Piranesi, 1720–1778, Paris (cat. raisonné); H. Focillon, Giovanni-Battista Piranesi: Essai de catalogue raisonné de son œuvre, Paris, 1918; A. M. Hind, Giovanni Battista Piranesi, London, 1922; A. Hyatt Major, Giovanni Battista Piranesi, New York, 1952; Hylton Thomas, The Drawings of Giovanni Battista Piranesi, London, 1954.

A. Hyatt MAYOR

PISANELLO (ANTONIO PISANO; called "Pisanello"). Painter and medalist. Born in Pisa before 1395 to Puccio di Giovanni da Cereto, of Pisa, and Isabetta di Nicolò, of Verona, the artist identified as Antonio Pisano died presumably between July 14 and Oct. 31, 1455. He was erroneously called Vittore by Vasari, and it was not until 1907 that this name was verified as Antonio (G. Biadego). His native city provided him with his nickname "Pisanello," which is also quoted in documents; he himself signed his work "Pisanus." In 1395, after the death of his father, Pisanello was in Verona, where he received his training. Most probably his first teacher was Stefano da Zevio, whose influence can be clearly traced in his early works. The influence of Altichiero was also appreciable, and persisted even when Pisanello came into the orbit of Gentile da Fabriano, whose pupil and subsequent collaborator he became in Venice. Early drawings also reveal that Pisanello was familiar with Lombard art, especially that of Michelino da Besozzo.

Between 1415 and 1422, Pisanello continued work on the cycle of frescoes begun by Gentile da Fabriano in the Sala del Maggior Consiglio of the Doges' Palace in Venice (repainted as early as 1479). These frescoes are mentioned in the writings of Bartolommeo Facio (d. 1457) and Francesco Sansovino (1581), and a document of 1425 (G. B. Lorenzi, 1868) describes their subject matter as inspired by the struggle between Frederick Barbarossa and Alexander III.

One of Pisanello's drawings (Louvre, no. 2432) is certainly an initial sketch for the fresco representing Otto, Barbarossa's son, at the feet of the pope and the doge; a second drawing (Br. Mus., no. 5226–57) appears to be a copy of the scene of Otto before Frederick (F. Wickhoff, 1883). It is hardly credible, however, that the young Pisanello was already so famous as to be summoned from Verona to Venice on a commission as important as the decoration of the Doges' Palace, previously entrusted to Gentile da Fabriano; it is more likely that Pisanello succeeded his master after having worked directly under his guidance.

This would explain why Gentile's influence on Pisanello, after his period of apprenticeship to the older artists in Venice, was so great as to be paramount in his early works, and why Gentile later asked Pisanello to collaborate with him on the frescoes in St. John Lateran. It could also account for the way in which Gentile's style prevails over Pisanello's personality in certain early works: the frescoes in Treviso (S. Caterina) and in four panels with scenes from the life of St. Benedict (Milan, Mus. Poldi Pezzoli; Uffizi; L. Coletti, 1947). Whatever doubts may persist about attribution of these works to Pisanello, they unquestionably belong to the school of Gentile. This observation has an important bearing on what is known of Pisanello's life, for it supports the thesis that he was one of Gentile's pupils. This author definitely agrees with the attribution of these paintings to Pisanello, and they will be further discussed below.

From 1422 to 1426, Pisanello's name recurs several times in documents both in Mantua and in Verona; on July 4, 1422, he was living in Mantua and bought a piece of land in Verona. In August, 1422, he was confirmed "habitatore Mantue." Toward 1423 he was in Florence, once more in contact with Gentile da Fabriano, with whom he collaborated on (or so this author believes) *The Adoration of the Magi* (VI, PL. 85). About 1424, Pisanello carried out the pictorial decoration of Nanni di Bartolo's monument to Nicolò di Brenzono (d. 1422) in the Church of S. Fermo Maggiore, Verona. The date 1426, discovered on the original tombstone (R. Brenzoni, 1932, 1952), probably refers to the completion of the whole monument.

For the years 1424, 1425, and 1426, there are records of payments made by the Gonzagas for works that are now lost. Most probably, the year 1426 marks the beginning of Pisanello's Roman period, the end of which can be deduced from a safe-conduct granted him by Eugenius IV in July, 1432, for travel throughout Italy. Records of payments in 1431 and 1432 do not necessarily indicate that Pisanello's activity in Rome began at this time, since they refer to the last years of his painting activity in St. John Lateran. Gentile da Fabriano had begun the cycle of frescoes with scenes from the life of St. John in 1425. After his death in 1427, his working material was handed over to Pisanello, who was apparently not only Gentile's continuator and successor but, as in Venice and perhaps in Florence, also his collaborator (from about 1426). Some of Pisanello's drawings, showing scenes from the life of St. John, are probably copies of Gentile's compositions for the Lateran; other drawings that show single figures of saints are copies of, or preparatory sketches for, the chiaroscuro figures of saints painted between the windows.

Evidence of Pisanello's Roman period is also presented by several drawings he made of ancient monuments, particularly sarcophagi; in these, too, the artist appears a faithful follower of Gentile. Later, in Tuscany, he made sketches of paintings and sculpture of the time (by Donatello, the Della Robbia family, Fra Angelico, and Lippi) which indicate that, after leaving Rome, he stopped in Florence and other Tuscan cities before returning to northern Italy. Pisanello's Florentine activity is a controversial matter, and it is mentioned only by Vasari, who speaks of a cycle of frescoes in the Chiesa del Tempio. Before returning to Verona, where his presence was reported in 1433, Pisanello is known to have passed through Ferrara (Jan., 1433); there he offered to Lionello d'Este a painting of a Madonna that he had left for safekeeping with a friend in Rome, which Lionello then requested be sent to him.

Pisanello's meeting with Sigismund can also be assigned to this period, as evidenced in three portraits of the Holy Roman Emperor by the artist: two drawings in profile (Louvre) and a wood panel with a three-quarter view (Vienna, Kunsthist. Mus.). After a stay in Siena (July, 1432–April, 1433), the Emperor made his entry into Rome on Apr. 25, 1433. He left that city in August and stopped at, among other places, Urbino, Rimini, Ferrara, Mantua, and Verona; the preparatory sketches for the panel painting must have been made during some stage of this return journey. Stylistically, as well, the portrait must be dated to the year 1433 or a little later, and thus was produced in approximately the same period as the fresco in the Church of S. Anastasia in Verona. The latter work is all that remains of the Pellegrini Chapel decoration, which can be dated between 1433 and 1438 and which undoubtedly required a long stay in Verona on Pisanello's part (PL. 177; III, PL. 212).

A poem by Guarino da Verona in praise of Pisanello was, in this author's opinion, written after the painting of the emperor Sigismund's portrait, and not after the execution of the medal representing John VIII Palaeologus. On the other hand, Hill (1905) and Venturi (1896) hold that the poem must be dated after 1438, believing that the imperial portraits mentioned in it refer to John, and not (as this author maintains) to Sigismund. Guarino da Verona also alludes to a picture of St. Jerome given him by Pisanello. The *St. Jerome* in the National Gallery, London, cannot be identified with the one mentioned by Guarino in view of its authenticated signature — "Bono da Ferrara." However, it is possible that this Bono, calling himself a pupil of Pisanello, may have copied a St. Jerome by Pisanello, which thus would have been painted before Guarino wrote his poem (after 1425).

In February, 1435, Pisanello sent to Lionello d'Este an "effigy" of Julius Caesar — perhaps an ancient medal, but more likely a small panel painted by himself — as a gift from Verona. His contacts with the Este family were becoming more intimate and more frequent. Margherita Gonzaga (married to Lionello d'Este in 1435; d. 1439) — not Ginevra d'Este, as has been maintained (Hill, Coletti) — is the subject of the portrait of a princess of the House of Este (PL. 175). Ginevra d'Este (1419–40) married Sigismondo Malatesta in 1434. If, judging by her apparent age in the portrait, it is assumed that she was his wife at the time it was painted, there would be no reason for her to wear the Este emblem (the amphora) alone; consequently, the portrait most probably represents Margherita, daughter of Giovanni Francesco Gonzaga. The Gonzaga colors (white, red, and green) appear in the portrait together with the Este emblem. The work may thus be dated between 1435 and 1439.

Two events in which Pisanello was involved, the Council of Ferrara (1438) and the siege of Verona (1439), marked the start of a new period in the life and art of the painter. In Ferrara, Pisanello created the first example of the Renaissance medal — an inspired, wholly original genre. It is known that John VIII Palaeologus, Emperor of Constantinople, stayed in Ferrara from Feb. 29, 1438, to Jan. 10, 1439; this fact is useful in dating his portrait medal. During the siege of Verona, which began on Nov. 17, 1439, Pisanello was a member of the retinue of Giovanni Francesco Gonzaga and Niccolò Piccinino. For this status, Venice retaliated by banishing him from Venetian territory (including Verona, where he was not allowed to return until 1445). In May, 1440, the artist went to Milan to report on the situation at Verona; this journey marked the beginning of a restless period, during which Pisanello wandered from city to city and from court to court, as can be seen from the many medals he designed at the time. Several lost works can be ascribed to this phase of Pisanello's activity: for instance, it is known that in 1441 a competition, for a portrait of Lionello d'Este, took place in Ferrara between Pisanello and Jacopo Bellini. The accolade was won by Bellini.

Furthermore, documents in Mantua mention frescoes decorating a chapel in the Church of S. Paola and a room in the Palazzo Ducale (frescoes already badly damaged in 1480; destroyed) and a canvas representing God the Father (1441–43?). This dating is probably correct, since it is known that during the years 1439–44 Pisanello often traveled between Ferrara and Mantua. Contacts with the Visconti family, as indicated by several medals, also became more frequent at this time; hence his works in the Castello Visconteo in Pavia (works destroyed in 1527; Hill, 1905; Rossi) are most likely to have been painted in this period. A panel (lost) of an unknown subject, executed for the Este residence of Belriguardo, is recorded for the year 1445.

Since so many of his paintings have been lost, the medals still extant are the most important evidence of Pisanello's activity during these years. The medals portraying Giovanni Francesco Gonzaga, Niccolò Piccinino, and Filippo Maria Visconti can be dated about 1441; those of Francesco Sforza and Lionello d'Este can be dated from 1442 on, while the medal commemorating the wedding of Lionello d'Este to Maria of Aragon bears the date 1444. The date on one of the medals representing Sigismondo Pandolfo Malatesta (1445), contemporaneous with that of Malatesta Novello (III, PL. 411), implies a stay in Rimini by the artist. The effigy medals of Cecilia Gonzaga and Belloto Cumano are also dated (1447). Pisanello's last medals before his departure for Naples are those for Vittorino da Feltre (d. 1446), the Humanist teacher of the Gonzaga children in Mantua, and for Pier Candido Decembrio, historiographer to Filippo Maria Visconti. The latter work can be dated to 1448 by the letter of dedication that Lionello d'Este, who had commissioned it from Pisanello, sent to Decembrio with a cast of the medal.

Thus the artist was still in Ferrara in 1447, and possibly stayed there until 1448, in which year he was also in Naples as is testified by the dated sketch (Louvre, no. 2306) for a portrait medal of Alfonso V of Aragon, who had summoned Pisanello to his court. One of the medals of the monarch bears the date 1449, and it was in that year that Pisanello became the privileged intimate of Alfonso. The duration of his stay, and activity, at the court of Aragon has not been established. It has, in fact, proved impossible to find documentary evidence of his works after 1450 or to state exactly the date of his death, which probably occurred between July 14 and Oct. 31, 1455 (Hill, 1905, 1908; Biadego, 1907–08; Zoege von Manteuffel, 1909). Hence Pisanello's career can be traced without interruption only until 1450, and the last five years of his life continue to remain obscure.

After early assimilating Veronese (Stefano da Zevio, Altichiero) and Lombard influences, Pisanello became first the pupil, then the chief collaborator, and ultimately the great continuator of Gentile da Fabriano. As he became aware of the work of Tuscan painters of his time whose creative tendencies transcended those of Gentile, his art evolved from the starting point of those Gothic qualities which still characterized his master's style toward the more monumental, more naturalistic style of the Renaissance. Literary figures such as Leonardo Dati, Guarino da Verona, Ulisse Aleotti, Angelo Galli, Tito Vespasiano Strozzi, Basinio, Decembrio, Porcellio, Flavio Biondo, and Facio all remarked on Pisanello's variety as a significant element in his evolution as a Renaissance artist. Sometimes these writers and poets also set down descriptions of his paintings, which constitute the only available record of many of them. In their own, typically Renaissance fashion they comment on the innovations in form that were Pisanello's original contribution to art.

The artist's 15th-century temperament and his acute observation of reality led him to two genres in which he was to excel: the portrait and the medal. His gift for the first is evidenced not only by the paintings discussed above but also by many drawings; as for the second category, in which his output was also extensive, the Renaissance medal can be regarded as his personal creation. Pisanello's excellence in these two fields made him the favorite artist of princes; his portraits and effigies combine the freshness of a still-evolving style with a mature talent and exquisite refinement that, at least with regard to the portrait medal, has never been surpassed. (Cf. COINS AND MEDALS, col. 739.)

Pisanello's gifts as a draftsman are equally original, and there is great opportunity to appreciate them, for no other artist of the first half of the 15th century left so many drawings (PL. 178; IV, PL. 30). Many of these are preserved in the Codex Vallardi (Louvre), which represents the most comprehensive collection, both in number and in variety of types, of drawings by a European painter on the threshold of modern graphic art. Their methodical composition and excellent state of preservation make the Pisanello drawings an important source for study of a style that bridged medieval and Renaissance art — and perhaps the prime source for tracing the origins of modern European draftsmanship. As a body of work, these drawings provide valuable means of checking existing literary sources, both technical and theoretical, and of learning about working methods the sources do not mention, perhaps because they were then considered to be self-evident.

Even more than his master Gentile, Pisanello reveals in his *œuvre* a universality not to be found again until great Renaissance masters such as Leonardo. In his drawings, especially, Pisanello used every kind of technique and material, confronted various problems of form and content, produced both copies that were medieval in style and realistic studies in Renaissance style, experimented with a new, mathematical approach to perspective and yet retained traditional architectural views, and displayed a mastery of the Lombard (Veronese) Gothic style while demonstrating an understanding of advanced problems of the Tuscan style. It may well be that the Humanistic totality of compass is seen for the first time on such an extended scale in Pisanello, since, moreover, such a large part of his work has been preserved — a unique instance of such a substantial conservation for the entire early Renaissance period.

Because of this rare opportunity for study, but even more because of his great personal gifts and broad scope, this author regards Pisanello as the first universal Humanist among artists, a claim that is certainly well documented by his existing works.

WORKS. Paintings and medals of unreliable attribution have been omitted from this listing, and only selected drawings have been included, those which best reflect the various stylistic phases of Pisanello's career or which serve to identify lost works. (The order of the list is chronological.)

Vienna, Albertina, no. Sc.R.13: This and another early drawing (Paris, Bacri Coll.; Degenhart, 1960, figs. 36, 39) are copies by Pisanello of works by Michelino da Besozzo; they give some indication of the former's early style and confirm his interest in Lombard art. In other early drawings Pisanello copied Altichiero and Stefano da Zevio (Degenhart, 1960, figs. 13, 29). – Louvre, no. 2432: Apart from several studies of animals (cf. PL. 178), the drawings on this sheet include the only surviving sketch by Pisanello of one of the frescoes of the cycle begun by Gentile da Fabriano in the Doges' Palace in Venice (1415–22). It represents Otto, the son of Frederick Barbarossa, at the feet of the pope and the doge. – Four panels with scenes from the life of St. Benedict, *St. Benedict as a Hermit* (Milan, Mus. Poldi Pezzoli), *The Miracle of the Broken Tray*, *St. Benedict Exorcises a Monk*, and *The Miracle of the Glass* (last three, Uffizi) Roberto Longhi (1958) agrees with the author's attribution of this set of panels to Pisanello. Such attribution has been contested no doubt because in these works Gentile's influence is seen at its strongest, so that it is easier to stress the dependence on him than the emergence of Pisanello's own artistic personality. It is not so much their authorship — admittedly very difficult to establish — that matters as the fact that the four panels clearly reveal the style of Gentile's circle in Venice (hence, Pisanello's style as well) but neither derive from nor belong to the Venetian school, as some critics have held. – *Life of St. Eligius* (Treviso, S. Caterina): The above remarks also apply to this fresco, which this author considers the only authentic Pisanello in a group published as his by L. Coletti (1947). The other frescoes mentioned are in the baptistery of S. Caterina, the Museo Civico of Treviso, and the Cathedral of Pordenone and are most likely by minor artists belonging to Gentile's and Pisanello's circle. – *Madonna of the Quail* (ca. 1422; Verona, Mus. di Castelvecchio): Lombard and Veronese elements are here combined with Gentile's sense of composition, resulting in that fusion of northern and central Italian cultures which, from then on, became such a significant quality in Pisanello's art. – Painted decoration of the tomb of Nicolò di Brenzono (Verona, S. Fermo): The architectural and sculptural elements of this monument are by the Florentine Nanni di Bartolo; Pisanello's contributions are an Annunciation and two saints on horseback (St. George and an archangel?). Nicolò di Brenzono died in 1422; the first inscription on the tomb, dated 1426, leads to the assumption that Pisanello completed his decoration in about 1424. These frescoes, with their emphasis on decorative line, still have Gothic echoes, while the later fresco in S. Anastasia in Verona already discloses the first elements of a Renaissance naturalism. – *Baptism of Christ* (Louvre, no. 420) and *St. John the Baptist Brought before King Herod* (Br. Mus., 1947-10-11-20; Degenhart, 1960, figs. 1–3): Both drawings are copies of frescoes from the Lateran cycle painted by Gentile da Fabriano and Pisanello and thus are representative, in style of draftsmanship, of the Roman period of Pisanello. Even if not his own work, they are of value in presenting the late style of Gentile, whom Pisanello succeeded in the decoration of the basilica (ca. 1426–32). A third drawing (Louvre, no. 2594, verso), an autograph sketch by Pisanello for a scene depicting St. John preaching to the animals in the wilderness, completes the fragmentary knowledge of Pisanello's Roman frescoes. – During this period in Rome, Pisanello followed Gentile's example by copying a number of Roman sarcophagi and other art works (e.g., Giotto's *Navicella*). In Tuscany (particularly in Florence), on his return journey from Rome, he made copies of works by Angelico, Filippo Lippi, Donatello, and Luca della Robbia (from 1433 on); these copies document his interest in Tuscan Renaissance art, which after Gentile's death was also manifested in Pisanello's paintings. – *Portrait of the Emperor Sigismund* (Vienna, Kunsthist. Mus.) and two profile sketches of the emperor (Louvre, nos. 2479, 2339): Pisanello made the drawings in 1433, probably after having met Sigismund in Ferrara, Mantua, or Verona after the Emperor's departure from Rome. The painting — which this author considers an authentic Pisanello, though some critics do not include it among his works — must have been done shortly afterward, on the bases of the preparatory sketches. The close affinity between this portrait and the faces of the S. Anastasia fresco (head of St. George, various heads among the king's retinue) supports the dating. – *Legend of St. George* (PL. 177; III, PL. 212): Of the frescoes described by Vasari, those on the interior of the Pellegrini Chapel have been lost; the only segment that has survived is that on the entrance arch of the chapel, with the scene of St. George setting out to slay the dragon, as well as the Pellegrini coat of arms. A large number of preparatory studies of heads, animals, and so on gives a clear picture of the way Pisanello worked (Degenhart, 1951). The S. Anastasia fresco can be dated between the artist's departure from Rome (1433) and his presence at the Council of Ferrara (1438), though more probably it belongs to the last years of this span. – *The Vision of St. Eustace* (XII, PL. 373): This painting can be ascribed to the same period as the S. Anastasia works, or a little later. The many preparatory sketches Pisanello made for it are found among the studies for the *Legend of St. George*. – Portrait of a princess of the house of Este (*Margherita Gonzaga*; PL. 175). – *Portrait of Lionello d'Este* (Bergamo, Acc. Carrara): Though there are no documents to support the conjecture, this may be the portrait painted in Ferrara in 1441, in competition with Jacopo Bellini. It may also be the one described by the Humanist Ludovico Carbone in his study (1461). It is certain that Pisanello painted many different portraits of Lionello; in this one there is evident a stylistic progression with respect to the Louvre portrait of a princess mentioned above. – *Virgin and Child with SS. George and Anthony Abbot* (PL. 176): A mature work, painted in Ferrara, it is difficult to date, but was certainly produced after the S. Anastasia fresco, to which it shows some similarities, and before the Neapolitan period. – Pisanello's works for the court of Naples cover a wide and varied range: they include paintings, medals, architectural and sculptural projects, silver- and bronzework (including cannon barrels), woodworking, and even designs for brocade. Most important of all the Neapolitan works are the projects for the Castel Nuovo triumphal arch. Apart from the medals, which can be dated between 1448 and 1450, a great number of drawings have survived from this period, the most prolific phase in the artist's life. This body of drawings (Louvre) constitutes an extremely important document for Pisanello's late work (Degenhart, 1942, pls. 112, 140; Keller, 1957). – A chronological list of known Pisanello medals is as follows: 1438, John VIII Palaeologus; 1439 or later, Giovanni Francesco Gonzaga; 1441–42, Filippo Maria

Visconti, Niccolò Piccinino, Francesco Sforza; ca. 1443, most of the seven medals portraying Lionello d'Este; 1444 (dated), medal commemorating the wedding of Lionello d'Este to Maria of Aragon; ca. 1445, Sigismondo Pandolfo Malatesta, Malatesta Novello; ca. 1446, Vittorino da Feltre (d. 1446); 1447 (dated), Cecilia Gonzaga, Belloto Cumano; 1447–48, Ludovico III Gonzaga, Pier Candido Decembrio (begun February, 1447, when the subject was staying in Ferrara); 1448, dated sketch for a medal (Louvre); 1449 (dated), "Liberalitas augusta" medal of Alfonso V of Aragon; 1449, "Venator intrepidus" medal of Alfonso V (dated drawing); also ca. 1449, "Fortitudo mea" medal of Alfonso V; 1449–50, Inigo d'Avalos. Other, lost Pisanello medals are known through literary sources (Hill, 1930). He may have made medals of popes Martin V and Nicholas V (Hill, 1930; Degenhart, 1960).

BIBLIOG. F. Sansovino, Venezia città nobilissima, Venice, 1581; B. Facio, De viris illustribus liber (ed. L. Mehus), Florence, 1745; P. Nanin, Disegni di varie dipinture a fresco che sono in Verona, Verona, 1864; G. B. Lorenzi, ed., Monumenti per servire alla storia del Palazzo Ducale, Venice, 1868; F. Wickhoff, Der Saal des grossen Rathes zu Venedig, RepfKw, VI, 1883, pp. 1–37; U. Rossi, Il Pisanello e i Gonzaga, Arch. storico dell'arte, I, 1888, pp. 453–56; G. Gruyer, Vittore Pisano, GBA, XI, 1894, pp. 198–218; G. Vasari, Le Vite di Gentile da Fabriano e Pisanello (ed. A. Venturi), Florence, 1896; F. Malaguzzi-Valeri, Pittori lombardi del Quattrocento, Milan, 1902; G. Zippel, Artisti alla corte degli Estensi, L'arte, V, 1902, pp. 405–07; G. F. Hill, Pisanello, London, 1905 (2d ed., 1911); G. F. Hill, Some Drawings from the Antique Attributed to Pisanello, BSR, III, 1906, pp. 295–303; G. Biadego, Pisanus pictor, Atti R. Ist. veneto di sc., lett. ed arti, LXVII, 1907–08, pp. 837–59, LXVIII, 1908–09, pp. 229–48, LXIX, 1909–10, pp. 183–88, 797–813, 1047–54, LXXII, 1912–13, pp. 1315–29; G. F. Hill, New Light on Pisanello, BM, XIII, 1908, p. 288; K. Zoege von Manteuffel, Die Gemälde und Zeichnungen des A. Pisanello aus Verona, Halle, 1909; L. Testi, Vittore Pisanello o Pisanus pictor, Rass. d'arte, X, 1910, pp. 131–41; J. Guiffrey, ed., Les dessins de Pisanello conservés au Musée du Louvre, 4 vols., Paris, 1911–20; P. Toesca, La pittura e la miniatura nella Lombardia, Milan, 1912; C. Cipolla, Ricerche storiche intorno alla chiesa di S. Anastasia, L'Arte, XVII, 1914, pp. 396–414; A. Venturi, Del quadro attribuito a Bono da Ferrara, L'Arte, XXV, 1922, pp. 105–08; G. Habich, Die Medaillen der italienischen Renaissance, Stuttgart, 1923; Van Marle, VIII, 1927, pp. 54–211; G. F. Hill, Dessins de Pisanello, Paris, Brussels, 1929; G. M. Richter, Pisanello Studies, BM, LV, 1929, pp. 54–66, 128–39, LVIII, 1931, pp. 235–41; G. F. Hill, A Corpus of Italian Medals of the Renaissance, 2 vols., London, 1930; A. H. Martinie, Pisanello, Paris, 1930; A. van de Put, Pisanelliana, Old Master Drawings, VI, 1931–32, pp. 57–60; J. Babelon, ed., Bibliothèque Nationale: Exposition de l'œuvre de Pisanello, Paris, 1932 (cat.); B. Berenson, Italian Pictures of the Renaissance, Oxford, 1932, p. 462; R. Brenzoni, Niccolò de Rangonis de Brenzano e il suo Mausoleo in S. Fermo, Arch. veneto, XII, 1932, pp. 242–70; G. F. Hill, ThB, s.v.; L. Planiscig and O. Fischer, Zwei Beiträge zu Pisanello, JhbPreussKSamml, LIV, 1933, pp. 5–28; A. Venturi, Pisanello, Rome, 1939; B. Degenhart, Pisanello, 3d ed., Vienna, 1942; B. Degenhart, Europäische Handzeichnungen, Berlin, Zurich, 1943; B. Degenhart, Das Wiener Bildnis Kaiser Sigismunds, JhbKhSammlWien, N.S., XIII, 1944, pp. 359–76; A. Da Lisca, Verona, S. Anastasia: La Cappella Maggiore e le sue decorazioni, Atti e Mem. Acc. di agricoltura, sc. e lettere di Verona, XXI, 1944, pp. 33–65; B. Degenhart, Un'opera di Pisanello: il ritratto dell'Imperatore Sigismondo, ArtiFig, II, 1946, pp. 166–85; L. Coletti, Pittura veneta dal Tre- al Quattrocento, Arte veneta, I, 1947, pp. 251–62; A. Hentzen, A. Pisanello: Die Vision des Hl. Eustachius, Berlin, 1948; B. Degenhart, Le quattro tavole della leggenda di S. Benedetto, Arte veneta, III, 1949, pp. 7–22; M. Davies, National Gallery Catalogues: The Earlier Italian Schools, London, 1951, pp. 340–44; B. Degenhart, Zu Pisanellos Wandbild in S. Anastasia, ZfKw, V, 1951, pp. 29–50; R. Brenzoni, Pisanello, 2d ed., Florence, 1952; G. Fiocco, Niccolò di Pietro, Pisanello, Venceslao, BArte, XXXVII, 1952, pp. 13–20; L. Coletti, Pisanello, Milan, 1953 (2d ed., 1958); B. Degenhart, Di una pubblicazione su Pisanello, Arte veneta, VII, 1953, pp. 182–85, VIII, 1954, pp. 96–118; N. Rasmo, Il Pisanello e il ritratto dell'imperatore Sigismondo, Cultura atesina, IX, 1955, pp. 3–7; R. Pallucchini, La pittura veneta del Quattrocento, Bologna, 1956; H. Keller, Bildhauerzeichnungen Pisanellos, Festschrift Kurt Bauch, Berlin, Munich, 1957, pp. 139–52; M. Salmi, La "Divi Julii Caesaris effigies" del Pisanello, Comm, VIII, 1957, pp. 91–95; R. Longhi, Sul catalogo della Mostra di Verona, Paragone, IX, 107, 1958, pp. 74–76; L. Magagnato, ed., Da Altichiero a Pisanello, Venice, 1958 (cat.); C. Marcenaro, Un ritratto del Pisanello ritrovato a Genova, S. in the H. of Art Dedicated to W. H. Suida, London, 1959, pp. 73–79; B. Degenhart, Gentile da Fabriano in Rom und die Anfänge des Antikenstudiums, MJhb, 3d ser., XI, 1960, pp. 59–90; A. Schmitt, Gentile da Fabriano in Rom und der Beginn der Antikennachzeichnung, MJhb, 3d ser., XI, 1960, pp. 91–151; E. Sindona, Gotico e Rinascimento: Pisanello, Paolo Uccello e il Pittore dell'Adorazione, Fede e arte, VIII, 1960, pp. 172–95; G. Marinelli, Il codice Vallardi e i disegni di Pisanello al Louvre, Emporium, CXXXIII, 1961, pp. 251–59; G. L. Mellini, Problemi di archeologia pisanelliana, CrArte, VIII, 47, 1961, pp. 53–56.

Bernhard DEGENHART

Illustrations: PLS. 175–178.

PISSARRO, CAMILLE. French impressionist (b. St. Thomas, Virgin Islands, July 10, 1830; d. Paris, Nov. 12, 1903). Pissarro was the son of Abraham Pizarro, a Portuguese Jew who assumed French nationality, and a Creole mother. Camille was sent to school in Paris, where he showed great talent in drawing, but had to return to St. Thomas to work in the family business. By 1855 his father had yielded to his son's desire to study art and sent him to Paris again. During the next ten years he received a sound academic training and formed close friendships with Monet, Renoir, Sisley, Cézanne, and other future impressionists; in the same period he won and then lost the support of Corot, received the critical backing of Zola, and was both accepted and rejected at the official Salons. Primarily a landscapist, Pissarro also painted many peasant subjects inspired by Millet, but in versions marked by more daring color and brushwork.

With the outbreak of the Franco-Prussian War, Pissarro left his home at Louveciennes (near Paris) and went to London, where, in the company of Monet, he admired the landscapes of Constable and, especially, of Turner. Too much has been claimed, however, for the "English influence" on the two painters; it was more clearly, for them, a case of having discovered like-minded precursors.

In the summer of 1871 Pissarro returned to Louveciennes to discover that the Germans had destroyed all but 40 of the 1,500 paintings he had produced. He immediately began again, soon joined by Cézanne, and for several years they painted in the villages north of Paris, notably Pontoise and Auvers. Pissarro led Cézanne into impressionism, but in return absorbed much of the latter's sense of structure and formal design. In 1874 Pissarro showed in the first exhibition of artists whose work had been rejected by the Salon, the famed group who were dubbed "impressionists" by a hostile critic. He was represented in the next seven of these exhibitions (1876–82), often by as many as 30 works.

Friendly and generous in nature, Pissarro became a kind of patriarch of the impressionists, and with his flowing white beard looked very much the part. He was ten years older than most of the others of the group, and four years the senior of Degas. Among the struggling younger artists who were helped by Pissarro to master the new techniques were Gauguin, Signac, Van Gogh, and Seurat (qq.v.). Pissarro himself experimented with Seurat's pointillist method from 1886 to 1890, but then abandoned it as being foreign to his own more spontaneous technique. He also helped them in other ways when he could; in 1890 he introduced the ailing Van Gogh to Dr. Gachet, the physician and amateur artist of Auvers who cared for Van Gogh until his death.

Pissarro's own life was far from an easy one, but its severe hardships were never mirrored in his serene and beautifully colored landscapes and figure subjects. He was helped by the support of the discriminating art dealer Durand-Ruel, and in 1892 a large retrospective show belatedly established his international reputation.

REPRESENTATIVE WORKS. *The Red Roofs* (VII, PL. 424); *Entrance to the Village of Voisins* (VII, PL. 425); *The Café au Lait* (VII, PL. 426); *The Road to Louveciennes* (VII, PL. 427); *The Boulevard Montmartre*, 1897 (Leningrad, The Hermitage).

BIBLIOG. J. Rewald and L. Pissarro, ed., C. Pissarro: Letters to His Son Lucien, New York, 1943; J. Rewald, The History of Impressionism, New York, 1946 (2d ed., 1961).

S. Lane FAISON, JR.

POLAND. The country, whose inhabitants belong to the group of West Slav peoples, became a kingdom toward the middle of the 10th century and preserved its independence until its partition (1795) between Russia, Prussia, and Austria. Poland was reborn as a republic in 1918. Since 1944 it has been designated a people's republic, Polska Rzeczpospolita Ludowa. In the figural arts, Poland has assimilated and reelaborated French and Italian influences (particularly fruitful from the 16th to the 19th century), as well as German and Bohemian influences.

SUMMARY. Cultural and artistic periods (col. 377): *Antiquity and the early Middle Ages; Pre-Romanesque and Romanesque; Gothic and late Gothic; Renaissance; Baroque and rococo; Neoclassicism; 19th and 20th centuries.* Monumental centers (col. 390): *Warszawa; Bydgoszcz; Poznań; Łódź; Kielce; Lublin; Białystok; Olsztyn; Gdańsk; Szczecin; Wrocław; Opole; Katowice; Kraków; Rzeszów.*

Cultural and artistic periods. *Antiquity and the early Middle Ages.* The oldest artistic manifestations in Polish territory date from the Upper Paleolithic and Mesolithic periods. These objects comprise female figurines and objects of bone with incised decoration. The latter were particularly numerous in the Baltic mesolithic cultures of Maglemose and Kunda, whose settlements have been discovered in northern Poland. In the Neolithic period (3d–2d millennium B.C.) the *Bandkeramik* culture reached Polish territory from the Danube region. In Brześć Kujawski remains of wooden structures that were almost 120 ft. long have been discovered. At this time, in addition to female figurines, small sculptures of men and animals appeared. Sedentary life began to be common in the Aëneolithic period. *Schnurkeramik* pottery and bell beakers have been found in these settlements. Imported decorative objects in stone were followed by copper objects.

Characteristic of the Bronze Age in Poland (1600–700 B.C.) are the pre-Lusatian culture (western Poland) and the Trzciniec culture (central and eastern Poland). Excavations have uncovered remains of wooden houses from this period. Extensive importation across the southern mountain passes brought to Poland from nearby Bohemian lands such bronze objects as necklaces, knives, and small hatchets. Local metalworking developed, particularly in the pre-Lusatian culture. The Lusatian culture (14th–5th cent. B.C.), which seems to have been proto-Slavic, was widespread throughout Polish territory. About the 8th–7th century B.C., iron first appeared, in imported decorative objects. In Biskupin are preserved the remains of an important fortified settlement (ca. 550–400 B.C.) with long wooden houses; similar settlements existed elsewhere as well. Trade relations brought decorative metal objects from Scythia. Later, during the La Tène period, Celtic works were imported into Poland. This was related to the Celtic invasion of Bohemian territories (see EUROPEAN PROTOHISTORY).

In the first four centuries of our era, the Slavic tribes formed distinct ethnic groups. From the provinces of the Roman Empire, mainly luxury objects were imported, as is demonstrated by objects of gold and silver found with Roman coins in treasuries. With the barbarian invasions of the 5th century, there was a decline in interest in artistic production. The next century saw the birth of Polish nationality in the first state organisms established in the regions of

Poland, principal centers of historical and artistic interest. *Key*: (1) National boundaries; (2) voivodeship boundaries; (3) centers with medieval monuments; (4) centers with modern monuments.

modern Kraków and Gniezno. Wooden houses built with various systems indicate a fairly high level of development in building techniques. About the 7th century, there again appeared small fortified settlements, with buildings of wood and occasionally of wood and limestone. Some of these settlements later became administrative centers and seats of secular power. In the 10th and 11th centuries, they assimilated adjacent market areas and became genuine cities. Pagan wooden temples, some adorned with painting and sculpture, are known from descriptions, but only in the area north of modern Poland. Many schematic figurines of deities have been found, executed in stone as well as in wood (e.g., at Janków and Ostrów Lednica). Decorative motifs were chiefly geometric and were widely used not only for jewelry (including necklaces, earrings, and bracelets) executed in silver, bronze, and glass, but for everyday objects and weapons. Local work in gold and silver was noted for its use of filigree and beading. Weapons with incised decoration and smaller objects of worked metal were imported from Asian areas by the Franks and by the Avars.

Pre-Romanesque and Romanesque. The first stone buildings appeared with the coming of Christianity. There is the possibility, though merely hypothetical, of a connection between the remains of small sacred structures in Wiślica and the early spread of the new religion, which probably came from the Great Moravian state. It is certain, however, that the official conversion of Poland in 966 brought about the construction of many churches and marked the beginning of the pre-Romanesque period, which lasted until approximately 1040. The first basilican structures, of which remains are preserved (e.g., in Poznań, after 968) and rectangular- and central-plan churches or chapels (the Rotunda of the Virgin, later known as the Rotunda of SS. Felix and Adauctus, Kraków, ca. 1000) were erected. Excavations have brought to light the remains of a *palatium* and a round chapel (at Lednica and Przemyśl, first half of 11th cent.).

The Romanesque period in Poland extends from about 1040 to the first half of the 13th century. Building activity, fostered by the king, the princes, the Church, and the holy orders, increased about 1100 and reached its greatest development in the second half of the 12th and the first years of the 13th century. Dynastic relations and the provenances of the bishops and the monastic orders, as well as of the various workshops, contributed to the variety of artistic influences (from Saxony, the Rhine-Mosan area, Burgundy, and Italy). The architecture that reflects these influences is generally one of simple forms and little ornamentation. Romanesque forms did not extend beyond the line of the Vistula and San rivers, which marks part of the eastern limit of the diffusion of Romanesque art in Europe. In the provinces, small single-aisled churches were built, which were longitudinal or central-plan structures, and some had square towers. In centers of ecclesiastical or secular power, among which the most important was Kraków, the new capital, basilican-plan churches with three aisles were constructed, sometimes with transepts. Generally the nave had exposed beams, and there were two towers on the west façade. The most important of these structures is the second cathedral of Kraków (ca. 1090–1142), of which the crypt of St. Leonard is preserved. Among the best preserved of the numerous sacred structures built from the 12th century on are the Benedictine church in Kruszwica (1120–40); the collegiate churches in Tum, near Łęczyca (1141–61), and Opatów (second half of 12th cent.); and the church of Kościelec (ca. 1240).

Though there are older examples of architectural sculpture, most of the portals, tympanums, and capitals date from the second half of the 12th century and the beginning of the 13th century (e.g., in Czerwińsk, Tum, Strzelno, and Wrocław). Unmatched in Europe are two columns of the Premonstratensian church of the Holy Trinity (Świętej Trójcy) in Strzelno, the shafts of which are covered with relief figures sculptured within small arches (12th cent.), and two tomb slabs in the Romanesque church in Wiślica, which depict a princely family in prayer and are decorated with symbolic medallions (ca. 1170–80). The most important sculptured monument in Poland and one of the most important in Europe are the bronze doors (XII, PL. 265) of the Cathedral of Gniezno, with scenes of the life and legend of St. Adalbert (Świętego Wojciecha). These doors may have been cast in Gniezno, but they show clear Mosan influences (ca. 1175).

Gothic and late Gothic. The Gothic system of ribbed vaults appeared toward the beginning of the 13th century in the Cistercian abbeys. But the churches, basilican in plan and with rectangular choirs, still preserved Romanesque forms. The most compact group is that of the Małopolska region, which was bound to Burgundian models (Sulejów, Wąchock). This style, which can be considered a transition style, is also manifest in the first Dominican churches [e.g., the Church of St. James (Świętego Jakuba), Sandomierz] and in those of the Franciscans. The building activity of these orders contributed much to the diffusion of brick construction, which was to become typical of Gothic architecture in Poland. Brick was used by itself in northern Poland and in conjunction with stone in southern Poland.

Gothic art spread more extensively after the middle of the 13th century, with the rising prosperity of the country. The royal court in Kraków encouraged artistic activity, especially after the unification of the various Polish provinces in the first quarter of the 14th century, and the artistic ambitions of the cities also increased. The art of southern and central Poland was bound to the central European world, particularly to the Bohemian area, while northern Poland received stimuli from the Baltic area and later from Flemish influences. The German colonization of the Silesian cities from the first half of the 13th century — colonization that spread eastward — reinforced the bonds with central European art. At the same time, the Teutonic Order created an independent state in northern Poland whose capital was Malbork, thereby creating a separate artistic milieu.

In the more important cities, basilican-plan cathedrals were built after western European models. Many of these still exist, for example, those in Wrocław, Kraków, and Gniezno. Noteworthy in Kraków are several large churches with internal buttresses [the churches of St. Catherine (Świętej Katarzyny), the Virgin Mary (Panny Marii), and Corpus Domini (Bożego Ciała)]. Along with the basilican type of church, there was a notable development, from the end of the 13th century, of the hall church (*Hallenkirche*) with nave and side aisles of equal height (e.g., Collegiate Church in Wiślica); this development was most characteristic of the north (e.g., Cathedral of Frombork).

In the cities, which were now built with regular street plans, civil edifices [e.g., town hall (Ratusz) in Toruń] and private houses began to be constructed in Gothic style. On the Baltic coast the Teutonic Order erected castles, including the enormous castle complex in Malbork. Other castles were built in central and southern Poland by King Casimir (Kazimierz) III, the Great (1333–70).

In sculpture and painting, the Gothic style spread more slowly and established itself only toward the end of the 13th century. Most examples of Gothic sculpture and painting come from the regions of Silesia and Malopolska.

Along with architectural decoration (portals, tympanums, capitals, and keystones), funerary sculpture gained importance. Monuments with the figure of the deceased lying on a sarcophagus, surrounded by mourners, were erected for members of the Piast dynasty, for the Silesian princes [the earliest preserved being that of Henry IV in the Church of the Holy Cross (Świętego Krzyża), Wrocław, early 14th cent.], and for the Polish kings (Casimir III, Kraków Cathedral, ca. 1370–80). Simplicity and austerity were the distinctive features of the early sculpture, but from about the middle of the 14th century there was an increasing tendency toward naturalistic forms. In addition to stone, greater use continued to be made of wood in devotional statues (e.g., *Pietà* in Lubiaż, ca. 1370).

Gothic elements appeared in Silesian miniatures in the second half of the 13th century and in wall painting about 1300. Wall painting achieved its greatest development in the 14th century. Bohemian influence, more than any other, contributed to the formation of an idealized linear style; the content reflected the chivalrous and courtly ideals of the time, both in holy images and in the composition of scenes (e.g., Siedlecin, Silesia, 1320–30). After the middle of the 14th century, painting reflected the spatial experiments of the Italian 14th century, principally by way of Bohemia (e.g., church in Niepołomice, ca. 1360–70). Panel painting appeared at the same time, first in Silesia and Pomerania. Stained glass was also produced in the early Gothic period.

The late Gothic period in Poland extended from 1400 to the beginning of the 16th century, but even toward the end of the 16th century the style still survived. In ecclesiastical architecture, building activity was centered in the north. Everywhere, the hall-church type predominated, and there was a tendency to make the choir and the aisles a single unit [e.g., Church of St. Mary (Panny Marii), Gdańsk]. The tendency toward greater decorative effects increased, particularly after the middle of the 15th century. This is especially evident in the upper part of the church and in the elaborate systems of ribbed vaults. The most beautiful example of the flamboyant Gothic style is quite late and in eastern Poland [Church of St. Anne (Świętej Anny), Wilno; second half of 16th cent.]. New town halls (Wrocław) were erected in the cities, and fortified walls were built (Kraków). In Kraków, a group of bourgeois residences were joined by an arcaded courtyard to house the university. Wooden country houses dating from as early as the 15th century survive. Numerous examples dating from the 16th to the 18th century repeat, with a popular imprint, the scheme of the single-aisled church and narrower choir that emerged from the Gothic tradition.

The diffusion, from about 1400, of the altarpiece closed by wings, with painted scenes or scenes sculptured or in relief, contributed to the rapid development of late Gothic sculpture and panel painting. In the first half of the 15th century the so-called "beautiful style" reached Poland from Bohemia. This style was lyrical and idealized, accentuated by rich draperies with soft flowing lines. The most typical expressions of this style in sculpture are the so-called "beautiful Madonnas," which were executed in stone (Toruń, Wrocław) and in wood (Krużlowa; first quarter of 15th cent.). In painting, in contrast to sculpture, Italian influence is more evident and survived until the beginning of the 16th century.

Toward the middle of the 15th century, though there are almost no noteworthy examples of sculpture, there appeared in painting the transition style to realism. Examples include the altar of the Church of St. Barbara (Świętej Barbary) in Wrocław [1447] and, in Małopolska, the productions of the workshops of Kraków and of the region of Nowy Sącz, which bordered Slovakia (e.g., the *Chomranice Deposition* ca. 1450). In a class by themselves are the wall paintings in Russo-Byzantine style, which are rare in Polish art. This production was chiefly sponsored by King Ladislas II Jagello (Władysław Jagiełło), who came from Lithuania, and some of his successors. There are noteworthy examples in Lublin [Chapel of the Holy Cross in the Church of St. Stanislas (Świętego Stanisława) and in the castle, 1418].

The highest expression of late Gothic realism, not only in Poland but in all northern European sculpture, is the wooden altar in the Church of the Virgin Mary (Panny Marii) in Kraków (1477–89). It was carved by Wit Stosz (Veit Stoss; q.v.), who, during his long sojourn in Poland, also created funerary monuments. The decorative exuberance of the broken folds and the realistic expression of faces and hands were characteristic of Stosz's style, which influenced all the figural arts, even beyond the borders of Poland, and endured until the first quarter of the 16th century. Realism in painting, after 1450, owed much more to Flemish influences, and its center was Kraków (polyptychs: Dominican church, ca. 1465; Augustinian church, ca. 1470–80; Olkusz, 1480). Also of high quality were the works of the early 16th century, which reflect the beginning of the influence of the Italian Renaissance (Bodzentyn triptych, ca. 1508). Wall painting declined in the 15th century, while miniature painting achieved its finest expression in the school of Kraków, the most important school in central Europe in the late 15th century to the early years of the 16th century (Behem Codex, 1505, with scenes of bourgeois life).

Renaissance. Because of the dynastic ties between Poland and Hungary — Ladislas II Jagello (Władysław Jagiełło), brother of Sigismund (Zygmunt) I of Poland, was king of Hungary and Bohemia — Hungarian art had considerable influence on the Renaissance in Poland.

During the reign of Sigismund I (1506–48), who married Bona Sforza, there were so many Italian artists in Poland, particularly in Kraków, that one can speak of a genuine "Italian period" in the Renaissance art of Poland. The chief work was the reconstruction of the Romanesque-Gothic castle on the Wawel hill in Kraków (1507–36), first directed by Francesco della Lora, who had come from Hungary in 1502, and then by Bartolomeo Berecci, both of whom were Florentine. In the Cathedral on the Wawel, Berecci built the king's funeral chapel, called the Sigismund Chapel, and there erected the funerary monument of Sigismund I. Among the other Italian artists, architects, and sculptors, mention should be made of Giovanni Cini of Siena and Bernardino de Gianotis Romano, both of whom worked outside Kraków as well, in the cathedral workshops of Płock and Wilno.

A second period of the Polish Renaissance began after the middle of the 16th century, under the reign of Sigismund II Augustus (Zygmunt August). During this period, Renaissance forms spread throughout the country. In Kraków, many middle-class houses were adorned with arcaded courtyards, and the cloth market (Sukiennice), rebuilt with the assistance of Gian Maria Padovano, was enriched with an attic story, a crowning element that, differently conceived and decorated, was to become typical of Polish architecture until the first half of the 17th century. Padovano was the outstanding architect and sculptor of the period. Chief among the Poles was Jan Michałowicz, architect and sculptor, who created chapels and funerary monuments. Numerous castles were built, including the royal castles in Wilno and Niepołomice, while the Castle of Brzeg attests the lively relations between Silesia and the area of Kraków. Middle-class houses and town halls were built. Among the latter, the most impressive is that of Poznań, with a loggia on the main façade, the work of Giovanni Battista Quadro of Lugano. Alongside the predominant Italian influence, Flemish influence increased, nourished by active commercial relations, the center of which was Gdańsk; the dominant personality was Cornelis II Floris de Vriendt. Other artists were Anton van Opbergen, Willem van dem Block from Malines, and Hans Vredeman de Vries from Leeuwarden. The interiors of the royal palaces were adorned with tapestries. More than 100 tapestries have survived that decorated the Wawel Castle; for the most part, they were executed in Brussels after cartoons by Michiel Coxie.

A third Renaissance period extended from about 1575 to the beginning of the 17th century. The city of Zamość, founded by Chancellor Jan Zamoyski and built (beginning in 1579) from plans by the Venetian architect Bernardo Morando, is an outstanding product of Renaissance city planning and architecture. In ecclesiastical architecture, funerary chapels modeled after the Sigismund Chapel were exceedingly popular. (There are about 200 dating from this period.) The most eminent representative of this period in Polish art is Santi Gucci of Florence, who was clearly influenced by mannerism. He designed numerous funerary sculptures, including the monuments of Queen Anna Jagello and King Stephen Báthory in the Wawel Cathedral (1594–97). Gerolamo Canavesi also executed many funerary monuments. Among the castles that were built in these years, particularly noteworthy are those of Pieskowa Skała, Baranów, and Krasiczyn. The Parr family of architects introduced Polish influence in Swedish Renaissance architecture, where, for example, extensive use was made of the attic story.

Baroque and rococo. In the first half of the 17th century, during the reign of the Vasa dynasty — Sigismund (Zygmunt) III, Ladislas (Władysław) IV, and John II Casimir (Jan Kazimierz) — baroque style became established alongside surviving Renaissance and mannerist tendencies. The first manifestations of the baroque style are the Jesuit churches in Nieśwież (1582–99), Kalisz (1588–95), and Kraków (from 1597). At first, the Jesuit churches were modeled after the Church of the Gesù in Rome, but subsequently the Church of S. Andrea della Valle (Rome) became the chief model. The royal architect Giovanni Trevano from Lugano was the prime mover of the new style in secular architecture. He partially transformed the royal castle (Zamek Królewski) on the Wawel in Kraków and the one in Warsaw in the years 1600–19. In 1596, the capital had been transferred from Kraków to Warsaw. At that time, the royal villa, later known as the Kazimierzowski Palace (now the University), and the Castle of Ujazdów (outside Warsaw) were built. Among the many palaces built in Warsaw by the nobility, the most impressive were the Krzyztopor Palace and the Podhorce Palace. In this period, a new and strictly local type of palace gradually was developed, the best example of which is in Kielce. With the Counter Reformation, numerous monasteries were erected, such as the one in Bielany, which has a church built by Andrea Spezza, and the one in Kalwaria Zebrzydowska, which has a group of chapels by Paweł Baudarth. A distinct current produced churches that were still Gothic in tradition but adorned with Renaissance and mannerist stuccoes and some baroque elements (e.g., the parish church in Kazimierz Dolny, near Lublin). A new type of middle-class dwelling with stucco decorations on the façade was developed. Particularly fine examples survive in Lwów and Kazimierz Dolny.

A particularly noteworthy figure in Polish painting was Tomaso Dolabella of Belluno, who had studied in Venice. He was active in Kraków in the first half of the 17th century and attracted many followers. He executed large historical canvases, which decorated the royal castle in Warsaw, and religious paintings, most of which are preserved in the Dominican churches in Kraków and Bielany. Giovanni Battista Falconi was an outstanding stucco-worker of the period. In the middle of the century, Gdańsk was an important center of portrait painting, and the most illustrious portraitist was Daniel Schultz.

Baroque art had a second great flowering during the reign of John III Sobieski (Jan Sobieski; last quarter of the 17th cent.), after a series of wars had caused enormous destruction to works of art. The chief architect of this period was the Italian August Locci. The outstanding painters were the Frenchmen Claude Callot, François Desportes, and Henri Gascar. The outstanding painter of monumental decorations, in Wilanów and elsewhere (Towicz, Wilno, and, in Lithuania, Pożajście), was the Florentine Michelarcangelo (Michelangelo) Palloni. Martino Altamonte painted historical canvases and portraits. Outstanding Polish artists included the portraitist Jan Aleksander Tretko and Rome-trained Jerzy Eleuter Szymonowicz-Siemiginowski. Between 1681 and 1694, the sculptor Andreas Schlüter of Gdańsk, who was later active in Berlin and St. Petersburg, worked in Wilanów and Zotkiew for the king and in the Krasiński Palace in Warsaw.

An outstanding architect and representative of the Palladian style was the Dutchman Tylman a Gameren, who worked chiefly for the nobility and the Church. He designed the Krasiński palaces in Warsaw, Puławy, and Nieborów, the Church of St. Anne (Świętej Anny) in Kraków, the Church of the Sisters of the Holy Sacrament in Warsaw, and the Church of St. Boniface (Świętego Bonifacego) in the suburb of Czerniaków. The Church of St. Anne was decorated with stuccoes by Baldassare Fontana, who worked in Poland in 1695–1704. The Church of S. Maria della Salute in Venice inspired two Polish churches: the Camaldolite church in Pożajście (built after a plan by Lodovico Fredo, 1667–90) and the church in Gostyń. The finest sculptural decoration in stucco is that executed (1677–83) for the Church of St. Peter of Antokol in Wilno by Giovanni Maria Galli and Pietro Peretti of Como.

During the late baroque period, which includes the reign of Augustus II and the first part of the reign of Augustus III, electors of

Saxony, there were very close ties between the courts of Warsaw and Dresden. These ties are reflected, for example, in the reconstruction of the royal castle of Warsaw and in the new royal palace, the so-called "Saxon Palace." Another trend of civil and religious architecture was created by Italians, such as Giuseppe Bellotti [Church of the Holy Cross of the Missionaries (Świętego Krzyża Misjonarzy), Warsaw] and Józef Fontana (Bieliński Palace, Warsaw), as well as by the Rome-trained Pole Kacper Bażanka. Pompeo Ferrari of Rome, a follower of Borromini, was responsible for the building or reconstruction of several churches in the regions of Poznań and Gniezno. That close contacts were also maintained with France is particularly evident in the furnishings of the period. The entire furnishings of palaces were often ordered from Paris, including the wood paneling for the walls (e.g., those planned by Just Aurèle Meissonnier for the Bieliński Palace in Warsaw and the Czartoryski Palace in Puławy).

The period of rococo style lasted from about 1740 until the death of Augustus III in 1763. Among the most noteworthy architects in Warsaw were Jakub Fontana, Józef's son (Collegium Nobilium of the Piarist fathers, Warsaw; Palace of Radzyń), and Jan Zygmunt Deybel (palaces in Warsaw and the reconstruction of the palaces in Puławy and Białystok). The Roman Francesco Placidi was very active in ecclesiastical architecture in Kraków and elsewhere. An important center of the rococo was Lwów, where such architects as Bernardo Merettini (Cathedral; 1748–60) and Jan de Witte (Dominican church, 1749–64). In the Wilno region, there was an extremely decorative tendency in Polish rococo architecture. Representative architects were Jan Krzysztof Glaubitz and Aleksander Osikiewicz. Active in Lwów were the followers of the sculptors Antoni Osiński, Fesinger, and Pinzl, while the school of Jan Jerzy Plersch worked in Warzaw.

Outstanding monumental painters in central Poland included Stanisław Stroiński and Walenty Żebrowski in Lwów. The most popular painter of altarpieces was Szymon Czechowicz, who studied in Rome and worked several years there. Also active in Rome were Tadeusz Kuntze, known as Taddeo Polacco, and the neoclassic artist Franciszek Smuglewicz.

Neoclassicism. During the reign of Stanislas II Augustus (Stanisław II Poniatowski; 1764–95), the period usually described as the Polish Enlightenment occurred. About 1760, because of the patronage of the nobility, which since the end of the 17th century had established close ties with France, influences of French neoclassicism began to penetrate Polish architecture (e.g., C.P. Coustou was in Poland about 1761). At the same time, relationships were established with the Italian "antiquarians."

The most important artistic center was Warsaw, especially the court. In 1765, Victor Louis, summoned from Paris, planned the reconstruction of the royal castle (Zamek Królewski), and J. L. Prieur collaborated on the interiors. In the first years of the reign of Stanislas II Augustus, many noteworthy foreign artists were called to the court. In 1767, Bernardo Bellotto (q.v.), called "Canaletto," arrived in Warsaw, where he remained until his death (1780), painting a series of 26 views of Warsaw. André Lebrun, a student of Jean Baptiste Pigalle (q.v.), was summoned from Paris in 1768 and was named first sculptor to the king, and he lived in Poland until 1811. In 1768, Giacopo Monaldi (d. 1798) came from Italy to become a court sculptor and executed allegorical works, portraits, and funerary monuments. The Swedish portraitist Per Krafft and the painter Jean Pillement, who painted decorative rococo canvases, also sojourned in Poland in the years 1766–68.

The dominant artistic personality at court was Marcello Bacciarelli, first court painter and director of the royal factories. Born in Rome, and active in Dresden and later in Vienna, Bacciarelli settled in Warsaw in 1765 and died there in 1818. The first architect of the king was Domenico Merlini of Valsolda. The patronage of Stanislas II Augustus, who was a great connoisseur and dilettante artist (his painting collection comprised 2,300 pictures), influenced the development of neoclassicism. In fact, the first phase of Polish neoclassicism was called "Stanislas Augustus style," the most notable examples of which were the interiors of the reception halls of the royal castle in Warsaw and the group of buildings in the Łazienki park.

The aristocracy and the upper middle class were also important patrons. Efraim Szreger (Schröger) and Szymon Bogumił (Simon Gottlieb) Zug were two important architects commissioned by members of these classes. Szreger's studies in Italy (1766–67) and his contacts with French architects influenced his work (e.g., the Palace of the Primate of Poland, Skierniewice). On the basis of the models in Neufforge's *Recueil d'architecture* Szreger and Zug elaborated the type of the neoclassic bourgeois house in Warsaw. Zug was responsible for many municipal palaces in the new style and for romantic parks, after French and English models, with neoclassic, Neo-Gothic, and Oriental pavilions, such as the park of Arkadia, near Nieborów, created for the princes Radziwiłł; and that of Jabłonna, near Warsaw, built for the king's brother, the primate Michał Poniatowski. Szreger and Zug also created the forerunners of the modern Polish suburban type of villa.

About 1780, a new generation of artists took over, among them such architects as Stanisław Zawadzki, Piotr Aigner, Jakub Kubicki, and Hilary Szpilowski. Public edifices such as schools and barracks were erected, theaters were designed, and the neoclassic palace with colonnaded portico became established in a form that was to survive until the mid-19th century and thus become a characteristic element in the Polish landscape. In the last twenty years of the 18th century, various Palladian tendencies reappeared (e.g., the round villa in Lubostroń; X, PL. 273). Aigner was a representative of Palladianism, as was his patron, Stanisław Kostka Potocki, who was particularly amenable to Italian style, notwithstanding his contacts with France and England. Potocki, the Polish Winckelmann, is considered the first Polish archaeologist and art historian. The development of Polish neoclassicism was influenced by the travels in Italy and Sicily of Potocki; Jan Michał Borch, author of *Lettres sur la Sicile et l'île de Malthe* (1776–77); August Moszyński (1784–86); and Jan Baptist Kamsetzer, who was in Italy as well as France (1780–82), after a journey to Greece and Constantinople in 1776–77. An outstanding patron and collector was Prince Stanisław Poniatowski, who lived permanently in Rome and Florence.

Between 1780 and 1783, the Italian architect and painter Vincenzo Brenna decorated some interiors with antique-style paintings (e.g., Natolin Palace, near Warsaw). He later worked in St. Petersburg at the court of Peter I. In 1773, Jean Pierre Norblin, a painter of *fêtes galantes* and later an illustrator of Polish life and customs, was brought from France by the princes Czartoryski. His student and successor was Aleksander Orłowski, who worked in St. Petersburg at the beginning of the 19th century.

In the last two decades of the 18th century, Zygmunt Vogel continued the tradition of the architectural view (*veduta*) and Jan Bogumił Plersch that of "ancient style" decoration with arabesques and grotesques. Noteworthy among portraitists were Kazimierz Wojniakowski, Józef Peszka, and the foreigners Giuseppe Grassi and Gian Battista Lampi.

Aside from Warsaw, another important center of Polish neoclassicism was Wilno, where, after 1770, Carlo Spampani worked alongside Polish architects. Under Bishop Massalski the Cathedral (1780–90) and episcopal palace of Werki (near Wilno) were rebuilt from plans by the architect Wawrzyniec Gucewicz, whose work was related to that of Jacques Germain Soufflot and Claude Nicolas Ledoux (qq.v.). Józef (Giuseppe) de Sacco worked in Grodno and its environs. In the last years of the 18th century, Puławy, the residence of the princes Czartoryski, became a genuine art center. Aigner rebuilt the residential complex and erected the round, domed church and, in 1800, the structure known as the Temple of the Sibyl, modeled after the so-called "Temple of the Sibyl" (Temple of Vesta) in Tivoli. The Temple of the Sibyl was to house the famous Czartoryski Collection, which was later transferred to Kraków.

Particularly noteworthy among the minor arts of the second half of the 18th century were ceramics (e.g., the products of Wolff and Bernardi and those of Belweder in Warsaw), glass (e.g., Urzecz, Naliboki), and fabrics.

Toward the end of the century (1791) there was the notable competition for the building of the Church of the Divine Providence in Warsaw. Kubicki won the contest, in which numerous architects competed, most of whom submitted plans for a central-plan domed structure. As a consequence of the political events that led to the partition of Poland, the church was never built.

In the period following Poland's loss of independence (1795), until about 1830, Warsaw remained the country's chief artistic center. Aigner built the round church of St. Alexander (Świętego Aleksandra) and rebuilt the old Radziwiłł Palace (later known as the Pac Palace) for the czar's regent. (The czar became king of Poland after the Congress of Vienna, 1815.) Kubicki erected the municipal barriers and transformed the Belvedere (Belweder) Palace in the classical style. The Florentine Antonio Corazzi, who arrived in Warsaw in 1822, was responsible for the outstanding classical buildings of the city, including the Association of the Friends of Science (originally the Staszic Palace), the Ministry of Finance (formerly the Leszczyński Palace), the Bank of Poland, and the Teatr Wielki. In the same year another Italian architect, Henryk Marconi, settled in Warsaw; he rebuilt the Pac Palace and erected several buildings in eclectic style, chiefly Neo-Renaissance.

In 1817, a faculty of fine arts was established at the University of Warsaw, and in 1819 the first art exhibitions were held. The outstanding painter of the time was Antoni Brodowski, a student of François Gerard and Jacques Louis David (q.v.) in Paris, who was a portraitist and the author of mythological and historical paintings. Antoni Blank and Aleksander Kokular were also important. "View painting" was continued by Marcin Zaleski, Wincenty Kasprzycki, and Jan Sejdlitz (Seydlitz).

The sculptor Paweł Maliński decorated the tympanums of the buildings that Corazzi built as well as those of other monumental palaces in Warsaw. Jakub Tatarkiewicz, a student of Bertel Thorvaldsen (q.v.), created allegorical and religious sculptures as well as portraits (e.g., those of Antonio Canova and Thorvaldsen). Thorvaldsen received many commissions from Poland, including those for the monument to Copernicus and to Prince Joseph Poniatowski.

From the mid-18th century to 1830, Poland acquired the various artistic tendencies of the principal European art centers and to these tendencies added its own traditions.

19th and 20th centuries. Toward the middle of the 19th century, Neo-Gothic and Neo-Renaissance styles appeared in the work of Jan Jakub Gay and Francesco Maria Lanci, as well as in that of Henryk Marconi. After the middle of the century, the historical interest reflected in these styles also produced a neobaroque tendency. Toward the end of the 19th century architects turned to national styles — Józef Dziekoński to the so-called "Vistulian Gothic," and Stefan Szyller to the Renaissance, while Stanisław Witkiewicz revived the autochthonous tradition of popular architecture in the town of Zakopane in the Tatra Mountains.

After 1830, painting became the standard-bearer of Polish patriotism. The greatest painter of the period was Piotr Michałowski, whose style was similar to that of the great French romantics. Henryk Rodakowski's portraiture was in the tradition of French realism. Józef Simmler was a painter of portraits and historical themes. History painting seemed to reawaken patriotic sentiments — particularly after the failure of the national insurrection of 1863, which had inspired the drawings of Artur Grottger — with the appearance of Jan Matejko in Kraków and with the works of Juliusz Kossak and Józef Brandt.

The dominant figure in Warsaw was Wojciech Gerson, a rather academic painter but an active organizer of artistic life. The realistic tendency in landscape and genre painting is represented by Józef Szermentowski, Aleksander Kotsis, Władysław Malecki, Maksymilian Gierymski, and, at the end of the 19th and beginning of the 20th century, by Józef Chełmoński. Another exponent of the realistic tendency was Aleksander Gierymski, who worked in the last quarter of the 19th century. Impressionism appeared briefly in Polish painting about 1890, with the work of Władysław Podkowiński, Józef Pankiewicz, and Leon Wyczółkowski; Olga Boznańska, who settled in Paris, retained an impressionist tendency and settled in Paris. Among the other painters of a realistic stamp who established links with impressionism were Jan Stanisławski, Julian Fałat, and Leon Wyczółkowski, all of whom were active toward the end of the 19th and in the first quarter of the 20th century.

In the sculpture of the second half of the 19th century, Władysław Oleszczyński represented the patriotic and romantic tendency, though his work was based on the neoclassic tradition. This tradition survived until the end of the century in the work of Oskar Sosnowski, Wojciech Brodzki, Pius Weloński, Cyprian Godebski, and Antoni Madeyski. Romantic realism appeared in the work of the portraitist Marceli Guyski and, at the end of the century, in that of Antoni Kurzawa. There were impressionistic tendencies in the work of Antoni Lepla, Józef Biernacki, and Stanisław Ostrowski.

As early as 1897, artists of the most varied tendencies, but agreed in their reaction against history painting, met at the Sztuka (Art) Club in Kraków. In the years at the end of the 19th century and the beginning of the 20th century, currents were formed parallel to Art Nouveau (q.v.). A new decorative sensitivity developed that was often complemented with a symbolist element. Among the artists were Jacek Malczewski, a symbolist; Stanisław Wyspiański, Józef Mehoffer, Jan Stanisławski, and Wojciech Wojtkiewicz, more decidedly expressionistic; and Władysław Ślewiński. The tradition of folk art is quite alive in the work of Wyspiański, Kazimierz Sichulski, Władysław Jarocki, Fryderyk Pautsch, and Teodor Axentowicz. More allied to realism are Ferdynand Ruszczyc and the portraitists Stanisław Lentz and Konrad Krzyżanowski. Sculpture was less interesting, but Bolesław Biegas, Konstanty Laszczka, and Xawery Dunikowski deserve mention. New architectural developments in Kraków were represented by Franciszek Mączyński and Teodor Tałowski.

This multiplicity of artistic orientations characterized the 19th century and much of the first half of the 20th century in Poland. After World War I, when Poland once again became independent, the most current French influences were felt. Functional architecture appeared about 1920, chiefly in Warsaw (Edward Norwerth, Antoni Dygat, and Teodor Tołwiński). In sculpture there prevailed a classicizing tendency, which stemmed particularly from Aristide Maillol (q.v.), for example, in the work of Edward Wittig, Henryk Kuna, Tadeusz Breyer, Zofia Trzcińska-Kaminska, and Alfons Karny. August Zamoyski was influenced by cubism, while Henryk Wiciński and Katarzyna Lobro worked in abstract sculpture. Jan Szczepkowski and Wojciech Jastrzębowski turned to the tradition of popular wood sculpture, modifying it with cubist elements.

In painting, diverse tendencies were current: Figural painting was bound to postimpressionist tradition, to symbolism, and to surrealism, as well as to lines that could be traced to folk tradition. There were realist and expressionist works, as well as abstract tendencies. After the hiatus of World War II, these abstract tendencies were resumed.

In painting, the "formisty" (Leon Chwistek, Stanisław Ignacy Witkiewicz, and Kamil Witkowski) combined the decorative tradition of the 20th century with cubist and expressionist elements. Władysław Strzemiński, Henryk Stażewski, and Maria Jarema introduced abstract painting, while Marek Włodarski was an exponent of surrealism. The program of a modernized "national style," represented by Zofia Stryjeńska, Eugeniusz Zak, and Wojciech Borowski, triumphed at the Exposition des Arts Decoratifs held in Paris in 1925. There was also a realistic movement (Jacek Mierzejewski, Tadeusz Pruszkowski, and Ludomierz Ślendziński). The most outstanding painter of the period, Tadeusz Makowski, after settling in Paris, fell under cubist influence in particular. In 1925, the "Kappist" trend was begun (from KP, the abbreviation of Komitet Pryski, or Parisian Committee, formed by Józef Pankiewicz and his pupils), which was an offshoot of postimpressionism (Jan Cybis, Helena Rudzka-Cybisowa, Zygmunt Waliszewski, Czesław Rzepiński, Tytus Czyżewski, and Zbigniew Pronaszko); the art of Eugeniusz Eibisch was also close to this trend. A monumental pictorial synthesis is to be found in the work of Felicjan Kowarski.

That art could have a social value was asserted in the rigorous antifigurative tendency and consequent abstractionism of the group that centered around the periodical *Blok*. Outstanding exponents were Władysław Strzemiński [who popularized the constructivism and suprematism of Kazimir Malevich (q.v.), had ties with the Cercle et Carré group of Wassily Kandinsky (q.v.), and founded unism] and Henryk Stażewski, both of whom were supported by Katarzyna Kobro and Mieczysław Szczuka. This tendency was substantially continued and largely extended to architecture by the review *Praesens*, which succeeded *Blok*. Alongside the abstract current, there appeared Berlewi's manifesto "Mechanofacture" in 1924.

Avant-garde groups were still very active in 1930 with the "AR" (Revolutionary Artists), promoters of writings and collections of international modern art; the "Kraków Group," with such abstract artists as Maria Jarema and Jonasz Stern; and the "Artes," which was rather closely bound to the surrealism of André Breton and numbered such artists as Marek Włodarski and Ludwick Lille.

During World War II and the years immediately following, socialist realism dominated the official panorama of Polish art. A clandestine group that met at the T. Kantor Theater in Kraków kept the avant-garde tendencies alive during World War II and, after 1945, promoted rich artistic activity, in the work of Tadeusz Brzozowski, Maria Jarema, Tadeusz Kantor, Kazimierz Mikulski, Jerzy Nowosielski. Along with realism, there continued to flourish nonfigurative art, surrealist, and neoprimitivist tendencies. Among the many artists who deserve mention are Maria Teresa Tyszkiewicz, Stefan Gierowski, Jan Lebenstein, Nikifor, Tadeusz Dominik, Wojciech Fangor, Bronisław Kierzkowski, and Aleksander Kobzdei.

The graphic arts, especially xylography (Władysław Skoczylas, Tadeusz Cieślewski, Stefan Mrożewski, and Tadeusz Kulisiewicz), and poster art (Tadeusz Gronowski) achieved about 1930 the high level that has been maintained. Architectural activity has been very important since World War II, in the reconstruction and restoration of destroyed cities and historic monuments, as well as in new building inspired by international architectural styles.

BIBLIOG. *Antiquity*: W. Antoniewicz, Archeologia Polski (Archaeology of Poland), Warsaw, 1928; J. Kostrzewski, Od mezolitu do wędrówek ludów (From the Mesolithic Era to the Migrations of Peoples), Prehistoria ziem polskich (Prehistory of the Polish Lands), Krakow, 1939–48, pp. 147–52; Historia Polski (History of Poland), I, 1, Warsaw, 1955, pp. 15–47.

Various periods and problems: K. Moszyński, Kultura ludowa Słowian (Culture of the Slavs), 2 vols., Krakow, 1925–39; W. Łęga, Kultura Pomorza we wczesnym średniowieczu w świetle wykopalisk (The Culture of Early Medieval Pomerania in the Light of Excavations), Toruń, 1930; Gród prasłowianski w Biskupinie (The Pre-Slavic Stronghold at Biskupin), Poznań, 1938; W. Kowalenko, Grody i osadnictwa grodowe Wielkopolski wczesnohistorycznej (Strongholds and Fortified Settlements of Protohistoric Great Poland), Pcznań, 1938; J. Kostrzewski, Prasłowiańszczyzna (Pre-Slavic Times), Poznań, 1946; R. Jamka, Przeszłość wczesnopiastowskiego Opola w świetle wykopalisk (Opole in the Period of the Early Piasts in the Light of Excavations), Opole, 1949; J. Kostrzewski, Kultura prapolska (Pre-Polish Culture), 2d ed., Poznań, 1949 (Fr. trans., B. Hamel, Paris, 1949); K. Tymieniecki, Ziemie polskie w starożytności (The Polish Lands in Antiquity), Warsaw, 1951; W. Hensel, Słowiańszczyzna wczesnośredniowieczna (Early Medieval Slavdom), Poznań, 1953.

Medieval and modern periods: *a. Sources*: S. Tomkowicz, Przyczynki do historyi kultury Krakowa w pierwszej połowie XVII w. (Contributions to the History of Culture in Krakow during the 1st Half of the 17th Century), Lwów, 1912; A. Chmiel, Wawel, II, Krakow, 1913; J. Ptaśnik, Cracovia artificum, 4 vols., Krakow, 1917–48.

b. Inventories and topographic catalogues: J. Heise and B. Schmid, Die Bau- und Kunstdenkmäler der Provinz Schlesien, 4 vols., Breslau, 1886–94; A. Bötticher, Die Bau- und Kunstdenkmäler der Provinz Ostpreussen, 9 vols., Königsberg, 1891–99; J. Kohte, Verzeichnis der Kunstdenkmale der Provinz Posen, 4 vols., Berlin, 1896–98; Teka Grona Konserwatorów Galicji Zachodniej (Theca of the Group of Western Galician Curators), 2 vols., Krakow, 1900–06; Katalog zabytków sztuki w Polsce (Catalogue of Monuments of Art in Poland), Warsaw, 1951 ff.

c. Dictionaries: E. Rastawiecki, Słownik malarzów polskich . . . (Dictionary of Polish Painters), 3 vols., Warsaw, 1850–57; E. Rastawiecki, Słownik rytowników polskich (Dictionary of Polish Engravers), Warsaw, 1888; E. Swieykowski, Pamiętnik Towarzystwa Przyjaciół Sztuk Pięknych 1854–1904 (Memoir of the Association of Friends of the Fine Arts), Krakow, 1905; Polski słownik biograficzny (Polish Biographical Dictionary), 9 vols., Krakow, 1935–61; S. Łoza, Architekci i budowniczowie w Polsce (Architects and Builders in Poland), Warsaw, 1954.

d. General works: F. Sobieszczański, Wiadomości historyczne o sztukach pięknych w dawnej Polsce (Historical Information on the Fine Arts in Old Poland), 4 vols., Warsaw, 1847–50; J. Topass, L'art et les artistes en Pologne, 3 vols., Paris, 1923–28; Sztuka polska (Polish Art), Warsaw, 1932; J. Starzyński and M. Walicki, Dzieje sztuki polskiej (History of Polish Art), Warsaw, 1936; Sztuka polska czasów nowożytnych (Modern Polish Art), 2 vols., Warsaw, Łódź, 1953–55; Sztuka polska czasów średniowiecznych (Polish Medieval Art), Warsaw, 1953; Trésors d'art polonais: Chefs d'œuvre des musées de Pologne, Bordeaux, 1961 (exhibition cat.); Historia sztuki polskiej (History of Polish Art), 3 vols., Krakow, 1962.

e. Art in various regions: (1) *Central and Eastern Poland*: K. T. Wilgatowie and H. Gawarecki, Województwo lubelskie (The Voivodate of Lublin), Warsaw, 1957; T. Chłudziński and others, Województwo warszawskie (The Voivodate of Warsaw), Warsaw, 1961; Muzeum Narodowe, Sztuka warszawska od średniowiecza do połowy XX wieku (Art in Warsaw from the Middle Ages to the Middle of the 20th Century), Warsaw, 1962. (2) *Great Poland*: H. Ehrenberg, Geschichte der Kunst im Gebiete der Provinz Posen, Posen, 1893; Ziemia lubuska (The Lubusz Territory), Poznań, 1950; A. Dubowski, Zabytkowe kościoły Wielkopolski (Old Churches of Great Poland), Poznań, Warsaw, Lublin, 1956. (3) *Silesia*: H. Lutsch, Bilderwerk schlesischer Kunstdenkmäler, 4 vols., Breslau, 1903; A. Grisebach and others, Die Kunst Schlesiens im Mittelalter, Breslau, 1938; E. Königer, Die Kunst in Schlesien, Berlin, 1927; T. Dobrowolski, Sztuka województwa śląskiego (Art in the Voivodate of Silesia), Katowice, 1933; M. Gębarowicz and others, Historja Śląska od najdawniejszych czasów do roku 1400 (History of Silesia from the Earliest Times to 1400), Krakow, 1936; D. Frey, Die Kunst in Oberschlesien, Breslau, 1938; T. Dobrowolski, Sztuka na Śląsku (Art in Silesia), Katowice, Wrocław, 1948; Dolny Śląsk (Lower Silesia), 2 vols., Poznań, 1948; Górny Śląsk (Upper Silesia), Poznań, 1959; A. Jurkiewicz and S. Ziemba, Województwo katowickie (The Voivodate of Katowice), Warsaw, 1962. (4) *Pomerania and Northern Poland*: F. Adler, C. Friedrich and O. Schmitt, Pommern, aufgenommen von der Staatlichen Bildstelle, Berlin, 1927; B. Makowski, Sztuka na Pomorzu (Art in Pomerania), Toruń, 1932; Pomorze Zachodnie (Western Pomerania), Poznań, 1949; Warmia i Mazury (Warmia and Mazuria), 2 vols., Poznań, 1953; Studia pomorskie (Pomeranian Studies), 2 vols,. Wrocław, Krakow, 1957.

f. Relations with other lands: (1) *Italy*: S. Ciampi, Notizie di medici, maestri di musica . . . , Lucca, 1830; S. Ciampi, Bibliografia critica delle antiche reciproche corrispondenze . . . , 2 vols., Florence, 1839; J. Ptaśnik, Kultura włoska wieków średnich w Polsce (Medieval Italian Culture in Poland), Warsaw, 1922; S. Lorentz, Relazioni artistiche fra l'Italia e la Polonia (Acc. Polacca di Sc. e Lettere, Bib. di Roma, Conferenze, 15), Rome, 1962. (2) *France and Germany*: La France et la Pologne dans leurs relations artistiques, 2 vols., Paris, 1938–39; Art polonais, art français: Études d'influences, Paris, 1939; P. Francastel, Histoire de l'art, instrument de la propagande germanique, Paris, 1946. (3) *The East*: T. Mańkowski, Orient w polskiej kulturze artystycznej (The Orient in Polish Artistic Culture), Wrocław, Krakow, 1959.

g. Architecture: T. Sydłowski, Pomniki architektury epoki piastowskiej (Architectural Monuments of the Piast Period), Krakow, 1928; G. Ciołek, Ogrody polskie (Polish Gardens), Warsaw, 1954; Z. Dmochowski, The Architecture of Poland, London, 1956; J. Zachwatowicz, Architektura polska do połowy w. XIX (Polish Architecture to the Middle of the 19th Century), 2d ed., Warsaw, 1956; B. Guerquin, Zamkie śląskie (Silesian Castles), Warsaw, 1957; M. and K. Piechotkowie, Bóżnice drewniane (Wooden Synagogues), Warsaw, 1957 (Eng. trans., R. Langer, Warsaw, 1959); J. A. Miłobędzki, Zabytki architektury w Polsce (Architectural Monuments in Poland), Warsaw, 1958; W. Krassowski, Architektura drewniana w Polsce (Wooden Architecture in Poland), Warsaw, 1961.

h. Painting: F. Kopera, Dzieje malarstwa w Polsce (History of Painting in Poland), 3 vols., Krakow, 1925–29; W. Drost, Danziger Malerei vom Mittelalter bis zum Ende des Barocks, Berlin, Leipzig, 1938; T. Dobrowolski, Polskie malarstwo portretowe (Polish Portrait Painting), Krakow, 1948; J. Starzyński, Pięć wieków malarstwa polskiego (5 Centuries of Polish Painting), 3d ed., Warsaw, 1952; T. Dobrowolski, Nowoczesne malarstwo polskie (Modern Polish Painting), 2 vols., Wrocław, Krakow, 1957–60; Malarstwo sakralne w Polsce (Religious Painting in Poland), Warsaw, 1958; M. Walicki, Malarstwo w Polsce: Gotyk-renesans-manieryzm (Painting in Poland: Gothic, Renaissance, Mannerism), Warsaw, 1961; Muzeum Narodowe, Katalog galerii malarstwa polskiego (Catalogue of Polish Picture Galleries), Warsaw, 1962.

i. Minor arts: J. Pagaczewski, Gobeliny polskie (Polish Gobelins), Krakow, 1929; L. Lepszy, Przemysł złotniczy w Polsce (Gold- and Silverwork in Poland), Krakow, 1935; T. Mańkowski, Polskie tkaniny i hafty XVI–XVIII wieku (Polish Textiles and Embroidery, 16th–18th Cent.), Wrocław, 1954; B. Kopydłowski, Polskie kowalstwo architektoniczne (Polish Architectural Wrought Iron), Warsaw, 1958; A. Bochnak and J. Pagaczewski, Polskie rzemiosło artystyczne wieków średnich (Polish Artistic Crafts in the Middle Ages), Krakow, 1959.

j. Pre-Romanesque and Romanesque: J. Starzyński and M. Walicki, Rzeźba architektoniczna w Polsce wieków średnich (Architectural Sculpture in Medieval Poland), Warsaw, 1931; A. Goldschmidt, Die Bronzetüren von Nowgorod und Gnesen, Marburg, 1932; Z. Ameisenowa, Les principaux manuscrits à peintures de la Bibliothèque Jagellonienne de Cracovie, B. Soc. fr. des reproductions des manuscrits à peintures, XVII, 1933–34, pp. 1–129; S. Sawicka, Les principaux manuscrits à peintures de Varsovie . . . du Seminaire de Płock et du Capitre de Gniezno, B. Soc. fr. des reproductions des manuscrits à peintures, XIX, 1938, pp. 1–316; E. Kloss, Die schlesische Buchmalerei des Mittelalters, Breslau, 1942; Z. Kępiński, Odkrycia w Strzelnie (Discoveries at Strzelno), B. h. sztuki, VIII, 1946, pp. 202–07; Z. Świechowski, Architektura granitowa Pomorza Zachodniego w XIII wieku (Western Pomeranian Granite Architecture in the 13th Century), Poznań, 1950; Z. Świechowski, Architektura na Śląsku do połowy XIII w. (Silesian Architecture to the Middle of the 13th Century), Warsaw, 1955; Drzwi gnieźnieńskie (The Doors of Gniezno), 3 vols., Wrocław, 1956–59; Z. Świechowski, Les plus anciens monuments de l'architecture religieuse en Pologne d'après les fouilles et les travaux récents, CahA, IX, 1957, pp. 301–17; Z. Świechowski and J. Zachwatowicz, L'architecture cistercienne en Pologne et ses liens avec la France, B. h. sztuki, XX, 1958, pp. 139–73; W. Antoniewicz, Recenti scoperte d'arte preromanica e romanica a Wiślica in Polonia, Rome, 1961.

k. Gothic and late Gothic: M. Lossnitzer, Veit Stoss, Leipzig, 1912; K. H. Clasen, Die mittelalterliche Kunst im Gebiete des Deutschordensstaates Preussen, I: Die Burgbauten, Königsberg, 1927; T. Dobrowolski, Studia nad średniowiecznym malarstwem ściennym w Polsce (Studies on Medieval Mural Painting in Poland), Poznań, 1927; Z. Ameisenowa, Einblattdrücke des XV. Jahrhunderts in Polen, Strasbourg, 1929; H. Braune and E. Wiese, Schlesische Malerei und Plastik des Mittelalters, Breslau, 1929; J. Starzyński and M. Walicki, Malarstwo monumentalne w Polsce średniowiecznej (Monumental Painting in Medieval Poland), Warsaw, 1929; M. Walicki, Malowidła ścienne kościoła św. Trójcy na zamku w Lublinie (Mural Paintings of the Church of the Holy Trinity in Lublin Castle), S. do dziejów sztuki w Polsce, III, 1930, pp. 1–92; K. Estreicher, Minjatury Kodeksu Bema oraz ich treść obyczajowa (The Illuminations of the Behem Codex and Their Documentary Content), Rocznik krakowski, XXIV, 1933, pp. 199–240; M. Walicki, Stilstufen der gotische Tafelmalerei in Polen im XV. Jahrhundert, Warsaw, 1933; T. Szydłowski, Le retable de Notre-Dame à Cracovie, Paris, 1935; M.Walicki, Polska sztuka gotycka (Polish Gothic Art), Warsaw, 1935 (exhibition cat.); M. Walicki, La peinture d'autels et des retables en Pologne aux temps des Jagellons, Paris, 1937; M. Walicki, Malarstwo polskie XV wieku (15th Century Polish Painting), Warsaw, 1938; K. H. Clasen, Die mittelalterliche Bildbauerkunst im Deutschordensland Preussen, 2 vols., Berlin, 1939; J. Dutkiewicz, Małopolska rzeźba średniowieczna 1300–1450 (Medieval Sculpture in Little Poland), Krakow, 1949; T. Dobrowolski and J. Dutkiewicz, Wit Stosz: Ołtarz krakowski (Veit Stoss: The Krakow Altar), Warsaw, 1951; H. Tintelnot, Die Mittelalterliche Baukunst Schlesiens, Kitzingen, 1951; K. Estreicher, Grobowiec Władysława Jagiełły (The Tomb of Ladislas Jagiello), Krakow, 1953; P. Skubiszewski, Rzeźba nagrobkowa Wita Stosza (Sepulchral Sculpture of Veit Stoss), Warsaw, 1957; Z. Ameisenowa, Rękopisy i pierwodruki iluminowane Biblioteki Jagiellońskiej (Illuminated Manuscript and Incunabula of the Jagiellon Library), Wrocław, Krakow, 1958; S. Detloff, Wit Stosz, 2 vols., Wrocław, 1961.

m. Renaissance: S. Odrzywolski, Renesans w Polsce: Zabytki sztuki z wieku XVI–XVII (The Renaissance in Poland: Monuments of 16th and 17th Century Art), Vienna, 1899; A. Hahr, Östeuropeiska stildrag i nordisk renässansarkitektur, Uppsala, 1915; J. Dutkiewicz, Grobowe rodziny Tarnowskich w kościele katedralnym w Tarnowie (The Tombs of the Tarnowski Family in the Cathedral Church at Tarnów), Tarnów, 1922; S. Komornicki, Franciszek Florentczyk i pałac wawelski (Francesco Fiorentino and Wawel Palace), Przegląd h. sztuki, I, 1929, pp. 57–69; S. Komornicki, Kaplica Zygmuntowska w katedrze na Wawelu (The Chapel of Sigismund in Wawel Cathedral), Rocznik krakowski, XXIII, 1932, pp. 47–120; S. Komornicki, Kultura artystyczna w Polsce czasów Ordodzenia: Sztuki plastyczne (Polish Renaissance Culture: Plastic Arts), Kultura staropolska (Culture of Old Poland), Krakow, 1932, pp. 533–605; K. Sinko, Santi Gucci Fiorentino i jego szkoła (Santi Gucci Fiorentino and His School), Krakow, 1933; K. Bimler, Schlesische Burgen und Renaissanceschlösser, II–III, Breslau, 1934– 36; K. Bimler, Die schlesische Renaissanceplastik, Breslau, 1934; W. Husarski, Attyka polska i jej wpływ na kraje sąsiednie (The Polish Attic and Its Influence on Neighboring Lands), Warsaw, 1936; K. Sinko, Hieronim Canavesi, Rocznik krakowski, XXVII, 1936, pp. 129–76; M. Gębarowicz and T. Mańkowski, Arrasy Zygmunta Augusta (Tapestries of Sigismund Augustus), Rocznik krakowski, XXIX, 1937, pp. 1–220; J. Pagaczewski, Jan Michałowicz z Urzędowa, Rocznik krakowski, XXVIII, 1937, pp. 1–84; W. Tatarkiewicz, Typ lubelski i typ kaliski w architekturze kościelnej XVII wieku (Lublin and Kalisz Types in Religious Architecture of the 17th Century), Prace Komisji h. sztuki, VII, 1937–38, pp. 23–60; A. Hahr,

Drottning Katarina Jagellonica och Vasarenässansen, Uppsala, Leipzig, 1940; J. Szablowski, Ze studiów nad związkami artystycznymi polsko-czeskimi w epoce Renesansu (Studies on the Artistic Relationship between Poland and Bohemia in the Renaissance), Prace Komisji h. sztuki, IX, 1948, pp. 27–64; Z. Hornung, Mauzoleum króla Zygmunta I w katedrze krakowskiej (The Mausoleum of King Sigismund I in Krakow Cathedral), Rozprawy Komisji h. kultury i sztuki Towarzystwa naukowego warszawskiego, I, 1949, pp. 69–150; B. Przybyszewski, Stanisław Samostrezelnik, B. h. sztuki, XIII, 2–3, 1951, pp. 47–87; H. and S. Kozakiewiczowie, Polskie nagrobki renesansowe (Polish Renaissance Funerary Monuments), B. h. sztuki, XIV, 4, 1952, pp. 62–132, XV, 1, 1953, pp. 3–57; Krakowskie Odrodzenie (The Renaissance at Krakow), Krakow, 1954; H. Kozakiewiczowa, Renesansowe nagrobki piętrowe w Polsce (Stone Funerary Monuments in Renaissance Poland), B. h. sztuki, XVII, 1955, pp. 3–47; J. Dutkiewicz, Le sculpteur vénitien Gian Maria Padovano . . . , Venezia e l'Europa (Atti XVIII Cong. int. Storia dell'Arte, 1955), Venice, 1956, pp. 273–75; S. Kozakiewicz and M. Zlat, L'attico in Polonia nel periodo del Rinascimento . . , Venezia e l'Europa (Atti XVIII Cong. int. Storia dell'Arte, 1955), Venice, 1956, pp. 275–77; Studia renesansowe (Renaissance Studies), I–III, Wrocław, 1956–63; H. Kozakiewicz, Spółka architektoniczno-rzeźbiarska Bernardina De Gianotis i Jana Cini (The Architectural-sculptural Collaboration of Bernardino De Gianotis and Giovanni Cini), B. h. sztuki, XXI, 2, 1959, pp. 151–74; S. Kozakiewicz, L'attività degli architetti e lapidici comaschi e luganesi in Polonia nel periodo del Rinascimento fino al 1580, Arte e artisti dei laghi lombardi, I, Como, 1959, pp. 393–421; M. Lewicka, Bernardo Morando, Saggi e memorie di storia dell'arte, II, Venice, 1959, pp. 143–55; A. Bochnak, Mecenat Zygmunta Starego w zakresie rzemiosła artystycznego (The Patronage of Sigismund the Old in the Field of Artistic Craftsmanship), Krakow, 1960; L. Kalinowski, Treści artystyczne i ideowe Kaplicy Zygmuntowskiej (Artistic and Ideological Content of the Sigismund Chapel), S. do dziejów Wawelu, II, 1960, pp. 1–129; M. Gębarowicz, Studia nad dziejami kultury artystycznej późnego renesansu w Polsce (Studies in the History of Late Renaissance Artistic Culture in Poland), Toruń, 1962.

n. Baroque and rococo: A. Bochnak, Giovanni Battista Falconi, Krakow, 1925; M. Loret, Gli artisti polacchi a Roma nel Settecento, Milan, Rome, 1929; A. Bochnak, Ze studiów nad rzeźbą lwowską w epoce rokoka (Studies in Lwów Sculpture in the Rococo Period), Krakow, 1931; E. Kloss, Michael Willmann, Breslau, 1934; S. Siennicki, Meble kolbuszowskie (Furniture of Kolbuszowa), Warwaw, 1936; G. Grundmann, Die Baumeisterfamilie Frantz, Breslau, 1937; S. Lorentz, Jan Krzystof Glaubitz, architekt wileński XVIII w. (Jan Krzystof Glaubitz, Vilno Architect of the 18th Century), Warsaw, 1937; T. Mańkowski, Lwowska rzeźba rokokowa (Rococo Sculpture of Lwów), Lwów, 1937; T. Mańkowski, Pasy polskie (Polish Ornamental Bands), Prace Komisji h. sztuki, VII, 1937–38, pp. 101–218; K. Buczkowski and W. Skórzewski, Dawne szkła polskie (Old Polish Stained Glass), Warsaw, 1938; W. Dalbor, Pompeo Ferrari, 1660–1736, Warsaw, 1938; T. Makowiecki, Archiwum planów Tylmana z Gameren, architekta epoki Sobieskiego (Archive of plans by Tylman of Gameren, Architect of the Sobieski Period), Warsaw, 1938; T. Mańkowski, Fabrica ecclesiae, Warsaw, 1946; T. Mańkowski, Genealogia sarmatyzmu (Genealogy of Sarmatism), Warsaw, 1946; S. Kozakiewicz, Valsolda i architekci zniej pochodzący w Polsce (Valsolda and Valsolda Architects in Poland), B. h. sztuki, IX, 1947, pp. 306–21; J. Orańska, Szymon Czechowitz 1689–1775, Poznań, 1948; W. Tomkiewicz, Aktualizm i aktualizacja w malarstwie polskim XVII wieku (Actuality and Actualization in Polish Painting of the 17th Century), B. h. sztuki, XIII, 1, 1951, pp. 53–94, 2–3, 1951, pp. 5–46; E. Iwanoyko, Jeremiasz Falck Polonus, Poznań, 1952; M. Morelowski, Rozkwit boroku śląskiego 1650–1750 (The Flourishing of Silesian Baroque 1650–1750), Wrocław, 1952; W. Tomkiewicz, Z dziejów polskiego mecenatu artystycznego w w. XVII (History of Polish Artistic Patronage in the 17th Century), Wrocław, 1952; A. Więcek, Polscy artyści Wrocława w wieku XVIII (Polish Artists of Wrocław in the 18th Century), Warsaw, 1956; O. Zagórowski, Architekt Kacper Bażanka (The Architect Kacper Bażanka), B. h. sztuki, XVIII, 1956, pp. 84–122; A. Miłobędzki, Krótka nauka budownicza dworów, pałaców . . . (A Brief Architectural Treatise on Country Houses, Palaces . . .), Wrocław, 1957; S. Wiliński, U źródeł portretu staropolskiego (On the Origin of the Old Polish Portrait), Warsaw, 1958; W. Tomkiewicz, Dolabella, Warsaw, 1959; S. Kozakiewicz, Quattro saggi sull'architettura polacca dal XVI al XVIII secolo e sull'influsso palladiano in Polonia, B. Centro int. di s. d'arch. Andrea Palladio, II, 1960, pp. 42–55; E. Hempel, Baroque Art and Architecture in Central Europe, Harmondsworth, 1965.

o. The Enlightenment to 1830: L. Fournier-Sarlovèze, Les peintres de Stanislas Auguste II, roi de Pologne, Paris, 1907; Z. Batowski, Norblin, Lwów, 1911; A. Lauterbach, Styl Stanisława Augusta (The Style of Stanislas Augustus), Warsaw, 1918; W. Tatarkiewicz, Rządy artystyczne Stanisława Augusta (The Artistic Kingdoms of Stanislas Augustus), Warsaw, 1919; W. Tatarkiewicz, Dwa klasycyzmy wileński i warszawski (Two Classicisms: Vilno and Warsaw), Warsaw, 1921; T. Szydłowski and T. Stryjeński, O pałacach wiejskich i dworach z epoki po Stanisławie Auguście . . . (Villas and Country Houses after the Period of Stanislas Augustus), Krakow, 1925; L. Niemojewski, Wnętrza architektoniczne pałaców stanisławowskich (Architectural Interiors of the Palaces of the period of Stanislas Augustus), Warsaw, 1927; T. Sawicki, Warszawa w obrazach Bernarda Belotta-Canaletta (Warsaw in the Paintings of Bernardo Bellotto, called Canaletto), Warsaw, Krakow, 1927; M. Loret, Życie polskie w Rzymie w XVIII wieku (Polish Life at Rome in the 18th Century), Rome, 1930; T. Mańkowski, Galerja Stanisława Augusta (The Gallery of Stanislas Augustus), Lwów, 1932; H. A. Fritzsche, Bernardo Bellotto genannt Canaletto, Burg bei Magdeburg, 1936; J. Sienkiewicz, Malarstwo warszawskie pierwsej połowy XIX wieku (Painting at Warsaw in the 1st Half of the 19th Century), Warsaw, 1936 (exhibition cat.); Z. Batowski, Aleksander Kucharski, Warsaw, 1948; S. Lorentz, Natolin, Warsaw, 1948; T. Mańkowski, Rzeźby zbioru Stanisława Augusta (Sculpture in the Collection of Stanislas Augustus), Krakow, 1948; W. Ostrowski, Świetna karta z dziejów planowania w Polsce, 1815–1830 (A Splendid Page from the History of Town Planning in Poland), Warsaw, 1949; S. Lorentz, O importach zreźb z Włoch do Polski w pierwszej połowie XIX w. i o Thorwaldsenie (On the Importation of Sculpture from Italy into Poland in the 1st Half of the 19th Century and on Thorvaldsen), B. h. sztuki, XII, 1950, pp. 289–309; P. Biegański, Pałac Staszica (Staszic Palace), Warsaw, 1951; S. Lorentz, Architektura wieku Oświecenia w świetle przemian w życiu gospodarczym i umysłowym (Architecture of the Age of the Enlightenment in the Light of Changes in Economic and Intellectual Life), B. h. sztuki, XIII, 4, 1951, pp. 5–48; S. Kozakiewicz, Warszawskie wystawy sztuk pięknych w latach 1819–1845 (Fine Arts Exhibitions at Warsaw in the Years 1819–45), Wrocław, 1952; I. Malinowska, Stanisław Zawadzki, Warsaw, 1953; A. Ryszkiewicz, Początki handlu obrazami w środowisku warszawskim (The Beginnings of Trading in Paintings in Warsaw), Wrocław, 1953; S. Lorentz, Victor Louis à Varsovie, Urbanisme et architecture: Etudes écrites et publiées en l'honneur de Pierre Lavedan, Paris, 1954, pp. 233–38; M. Wallis, Canaletto malarz Warszawy, Warsaw, 1954 (Eng. trans., W. Ostrowski, Canaletto the Painter of Warsaw, Warsaw, 1954); Bernardo Bellotto 1720–1780, Alessandro Gierymski 1850–1901: Opere provenienti dalla Polonia, Venice, Milan, 1955 (exhibition cat.); W. Tatarkiewicz, Dominik Merlini, Warsaw, 1955; B. Maszkowska, Z dziejów polskiego meblarstwa okresu Oświecenia (History of Polish Furniture of the Period of the Enlightenment), Wrocław, 1956; W. Tatarkiewicz and D. Kaczmarzyk, Klasycyzm i romantycyzm w rzeźbie polskiej (Classicism and Romanticism in Polish Sculpture), Sztuka i krytyka, VII, 1–2, 1956, pp. 31–73; Muzeum Narodowe, Aleksander Orłowski (1777–1832), Warsaw, 1957; W. Tatarkiewicz, Łazienki warszawskie (The Łazienki of Warsaw), Warsaw, 1957; R. Pallucchini, Vedute del Bellotto, Milan, 1961.

p. 1830 to the present: S. Witkiewicz, Sztuka i krytyka u nas (Art and Criticism among Us), Lwów, 1893 (repr. Warsaw, 1949); J. Bołoz Antoniewicz, Grottger, Lwów, 1910; Stanisław Wyspiański: Dzieła malarskie (Stanisław Wyspiański: Paintings), Warsaw, Bydgoszcz, 1925; E. Niewiadomski, Malarstwo polskie XIX i XX wieku (Polish Painting in the 19th and 20th Centuries), Warsaw, 1926; A. Kuhn, Die polnische Kunst vom 1800 bis zur Gegenwart, Berlin, 1930; J. Czapski, Józef Pankiewicz, Warsaw, 1936; W. Kozicki, Henryk Rodakowski, Lwów, 1937; M. Treter, Matejko, Lwów, 1939; A. Bartczakowa, Franciszek Maria Lanci, Warsaw, 1951; A. Wojciechowski, Elementy sztuki ludowej w polskim przemyśle artystycznym XIX i XX wieku (Elements of Folk Art in Polish Artistic Production of the 19th and 20th Centuries), Warsaw, 1953; A. Ryszkiewicz, Henryk Rodakowski 1823–1894, Warsaw, 1954; S. Kozakiewicz and A. Ryszkiewicz, Warszawska "cyganeria" artystyczna (The Artistic "Bohemia" of Warsaw), Warsaw, 1955; M. Porębski, Sztuka naszego czasu (Art of Our Time), Warsaw, 1956; A. Banach, Polska książka ilustrowana 1800–1900 (The Polish Illustrated Book 1800–1900), Warsaw, 1959; J. Sienkiewicz and J. Zanoziński, Piotr Michałowski, Warsaw, 1959; Sztuka współczesna: Studia i szkice (Modern Art: Studies and Sketches), Krakow, 1959; M. Wallis, Sztuka polska dwudziestolecia 1919–1939 (Polish Art of "the Twenty Years" 1919–39), Warsaw, 1959; Douze peintres polonais, Paris, 1960 (exhibition cat.); M. Porębski Malowane dzieje (History Painted), Warsaw, 1961; J. Starzyński, Od renesansyzmu do impresjonizmu (From the Renaissance to Impressionism), Rocznik h. sztuki Polska Akad. Nauk, II, 1961; Polskie dzieło plastyczne w XV-lecie PRL (Polish Work in the Plastic Arts on the 15th Anniversary of the Polish People's Republic), 2 vols., Warsaw, 1962 (exhibition cat.).

Stefan Kozakiewicz

Monumental centers. The cities and towns of artistic interest are listed alphabetically within regional divisions.

Warszawa. Warsaw (Warszawa). Situated on the Vistula (Wisła) River, Warsaw was founded in the second half of the 13th century at the crossroads of ancient trade routes. It was the capital of the dukes of Mazovia (Mazowsze) until 1596, when it became the national capital. In the 13th to 15th centuries the city was circled with a double course of brick walls. The walls were freed of later superstructures after 1936. The center of the city, which was built on a geometric plan, is the square Rynek Starego Miasta (Market of the Old City). In the first half of the 17th century, under Sigismund (Zygmunt) III, Ladislas (Władysław) IV, and John II Casimir (Jan Kazimierz) of the Vasa dynasty, the city was much enlarged, and the Gothic castle was transformed into a modern royal residence (Zamek Królewski). Many baroque palaces were erected for the aristocracy, including the Kazanowski, Ossoliński, and Koniecpolski palaces. Outside Warsaw, a royal villa (later the Kazimierzowski Palace and now the University of Warsaw) and the royal castle of Ujazdów were built. During the war of 1655–56 the city was occupied by the Swedes, who destroyed it and dispersed its art collections. It was rebuilt in the second half of the 17th and first half of the 18th century. Several baroque and rococo palaces and churches were erected, and the royal castle was radically altered. A particularly felicitous period was the Age of Enlightenment, during which the city, under Stanislas II Augustus (Stanisław II Poniatowski), became the chief political, scientific, and artistic center of the country and in architecture turned to the neoclassic style. Outside the city, English-style parks with pavilions were built, such as the royal park Łazienki and that of the king's brother, known as the Ogród Księcia Podkomorzego (Garden of the Prince Chamberlain). In the years 1815–30, the large public buildings

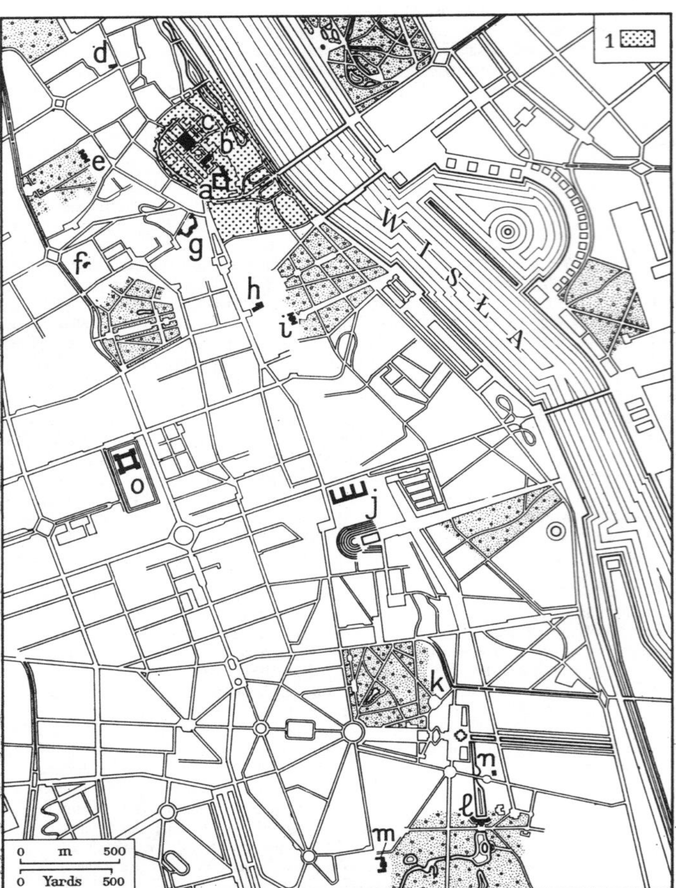

Warsaw, plan. *Key*: (1) Original center. Principal monuments: (*a*) Royal castle (Zamek Królewski) and column of Sigismund (Zygmunt) III; (*b*) Cathedral; (*c*) Rynek Starego Miasta (Market of the Old City); (*d*) Church of the Virgin Mary (Panny Marii); (*e*) Krasiński Palace; (*f*) Radziwiłł (later the Pac) Palace; (*g*) Prymasowski Palace (Palace of the Primate); (*h*) Church of the Visitation; (*i*) Kazimierzowski Palace; (*j*) National Museum (Muzeum Narodowe); (*k*) part of the castle of the dukes of Mazovia (Ujazdów Castle); (*l*) Łazienki Palace; (*m*) Belvedere (Belweder) Palace; (*n*) State Archaeological Museum (Państwowe); (*o*) Association of Friends of Science (formerly the Staszic Palace).

erected by the Florentine Antonio Corazzi and by Piotr Aigner of Warsaw were the most prominent neoclassic structures. Henryk Marconi, of Italian descent, was responsible for several Neo-Renaissance buildings erected in the middle of the 19th century. Until 1939, the oldest quarters bore a Gothic-baroque stamp, while the rest of the city was dominated by the neoclassic tone of the 18th and 19th centuries and by eclectic, primarily Neo-Renaissance, accents.

In the years 1939–44, 90 per cent of the city was destroyed and burned by the Germans. After the war, the old quarters of the city — the main historical streets and buildings — were reconstructed according to their original design, with modern interiors. The Gothic brick churches that have survived are the Cathedral (14th cent.; partially reconstructed), the Church of the Virgin Mary (Panny Marii; 15th cent.), the outer walls and the vaults of the Church of St. Anne (Świętej Anny) and the adjacent Bernardine cloister (15th and early 16th cent.), and the tower of the Church of St. Martin (Świętego Marcina). Most of the baroque and rococo churches destroyed during World War II have been reconstructed in their original form. Entirely preserved were the 17th-century baroque churches of St. Boniface (Świętego Bonifacego) in Czerniaków, the work of the architect Tylman a Gameren, with late-18th-century stuccoes and paintings; the Church of the Visitation of the Virgin Mary (Nawiedzenia Najświęszej Marii Panny Wizytek), with a rococo façade and interior decoration of the mid-18th century; and the Church of St. Joseph of the Discalced Carmelites (Świętego Józefa Karmelitów Bosych), with a baroque-neoclassic façade by Efraim Szreger (Schröger; 1761–80). Partially reconstructed is the baroque church of the Holy Cross of the Missionaries (Świętego Krzyża Misjonarzy), built in the 17th–18th century by the architects Giuseppe Bellotti and Józef and Jakub Fontana.

Civil edifices include the royal castle (Zamek Królewski), which was founded in the late 13th century by the princes of Mazovia, rebuilt in brick during the 14th and 15th centuries in Gothic style, and again rebuilt in Renaissance style in the second half of the 16th century. It was entirely renovated at the beginning of the 17th century (completed 1622). The baroque façade facing the Vistula was completed in 1746, and the series of neoclassic reception rooms was created in the years 1764–90, with the collaboration of such architects as Jakub Fontana, Victor Louis, and Domenico Merlini, the painters Marcello Bacciarelli and Bernardo Bellotto (q.v.; known as "Canaletto"), and the sculptors André Lebrun and Giacopo Monaldi. The structure was severely damaged in World War II, but interior decorations, paintings, and sculptures were saved and the castle is being rebuilt. Ujazdów, the 13th-century castle of the princes of Mazovia, was rebuilt as a royal castle in baroque style in the first half of the 17th century and again rebuilt in the second half of the 18th century; it was completely destroyed in 1944. The Kazimierzowski Palace, built by Ladislas (Władysław) IV in the second quarter of the 17th century as the royal villa, was rebuilt in neoclassic style in 1815–20 by Hilary Szpilowski to house the University. The Łazienki park was laid out in the 17th century, with garden pavilions near the Ujazdów Castle. In the second half of the 18th century, King Stanislas II Augustus (Stanisław II Poniatowski) transformed it into an English-style garden. The villa was reconstructed in neoclassic style, and new residences as well as a theater and an amphitheater were erected. The work was directed by the architect Domenico Merlini (X, PL. 273) and the painter Marcello Bacciarelli, with the collaboration of the architect Jan Krystian Kamsetzer, the painter Jan Bogumił Plersch, and the sculptors André Lebrun and Giacopo Monaldi. The Radziwiłł (later the Pac) Palace, built in the second half of the 17th century by Tylman a Gameren, was rebuilt in the first half of the 19th century by Henryk Marconi. The Krasiński Palace was built in the 17th century by Tylman a Gameren and has sculptures by Andreas Schlüter. The Prymasowski Palace (Palace of the Primate) was rebuilt in the second half of the 18th century in neoclassic style by Efraim Szreger. The late-18th-century Tyszkiewicz Palace was built by Jan Krystian Kamsetzer and Stanisław Zawadzki and was decorated with neoclassic stuccoes. The palace was reconstructed after World War II. The Belvedere (Belweder) Palace, rebuilt in neoclassic style by Jakub Kubicki at the beginning of the 19th century, has been the residence of the chief of state since 1918. The Koniecpolski Palace, built in the second quarter of the 17th century by Constantino Tencalla, became the property of the Radziwiłł family and at the beginning of the 19th century was rebuilt as the regent's palace in neoclassic style by Piotr Aigner; it is now the Presidium of the Council of Ministers. The Pawłowice Palace (X, PL. 272), near Warsaw, dates from about 1786.

Public buildings erected between 1820 and 1833 by Antonio Corazzi include the Association of the Friends of Science (Gmach Towarzystwa Przyjaciół Nauk; formerly the Staszic Palace), the Ministry of Finance (Ministerstwo Skarbu; formerly the Leszczyński Palace), the Bank of Poland (Bank Polski), and the Teatr Wielki (Grand Theater). There are several noteworthy monuments in Warsaw, including the column of King Sigismund (Zygmunt) III (1644), designed by the architect Constantino Tencalla, with a statue by Clemente Molli; the Copernicus monument (1830) by Bertel Thorvaldsen (q.v.); and the Joseph Poniatowski monument (1822–30), also by Thorvaldsen, which was destroyed in World War II and recast after a plaster model in the Thorvaldsens Museum in Copenhagen.

Museums: The National Museum (Muzeum Narodowe), founded in 1862, includes a gallery of ancient art, a gallery of medieval art, a collection of European painting, Polish painting and sculpture, contemporary art, a department of prints and drawings (Polish and European), a collection of ancient, Byzantine, European, and Polish coins and medals. Branches of the National Museum are located in the Łazienki Palace in Warsaw, in the Nieborów Palace and the Arkadia park, and include the Regional Museum of Łowicz (for the last three, see below under Łódz). The Military Museum (Muzeum Wojska) has a collection of arms and weapons, both Polish and foreign, ranging from the Middle Ages to World War II. The State Archaeological Museum (Państwowe Muzeum Archeologiczne) houses prehistoric and protohistoric archaeological material. Other museums are the Adam Mickiewicz Museum, the Historical Museum of the City of Warsaw (Muzeum Historyczne miasta Warszawy), and the Museum of Cultures and Folk Art (Muzeum Kultur i Sztuki Ludowej). The painting gallery of the royal castle (Zamek Królewski) exhibits Flemish, Dutch, Italian, and Spanish works, and the Ujazdów Castle has a collection of ancient sculpture.

BIBLIOG. R. Przezdziecki, Varsovie, Warsaw, 1924; A. Lauterbach, Warszawa, Warsaw, 1925; A. Janowski, Warszawa, Ponzań, 1930; R. Danysz Fleszarowa and J. Kolodziejczyk, Warszawa: Przewodnik krajoznawczy (Warsaw: Tourist Guide), Warsaw, 1938; E. Szwankowski, Warszawa,

Warsaw, 1952; I. Lange and L. Pietrzach, Warszawskie stare miasto (The Old City of Warsaw), Warsaw, 1955; J. Zachwatowicz and P. Biegański, The Old Town of Warsaw (trans. C. Wojewoda), Warsaw, 1956; Vademecum Warszawskie, Warsaw, 1957; Z. Bienicki, Oś barokowa Warszawy (The Baroque Axis of Warsaw), Kwartalnik arch. i urbanistyki, V, 1960, pp. 469–522; E. Kupiecki and Z. Dworakowski, Warszawa, Warsaw, 1960.

Czerwińsk. This town on the Vistula River dates from the foundation of a monastery of the Canons Regular in the second half of the 11th century. The monastery complex dates from the middle of the 12th century. The church preserves its Romanesque basilican appearance despite Gothic, Renaissance, and baroque additions. There are two of the original towers and remains of the original sculptural decoration, including the portal (1148-56), which has been remodeled. In 1951, Romanesque as well as Gothic frescoes were discovered in the Crucifixion Chapel. The Romanesque frescoes, dating from the early 13th century, are the most important in Poland. The monastery has been greatly altered. The Gothic chapel preserves frescoes from the middle of the 15th century. The Gothic tower dates from 1497.

Jabłonna. The neoclassic palace of this town near Warsaw was built in 1775–79 by Domenico Merlini. Burned in 1945, it was rebuilt and is now the seat of the Polish Academy of Sciences. The fine interior, with late-18th-century wall paintings, has been restored. The park in romantic style has pavilions.

Bibliog. S. Lorentz, Jabłonna, Warsaw, 1962.

Płock (Plozk). Situated on the Vistula River, Płock was one of the most important cities of Poland from the Middle Ages until the 16th century. A castle with a settlement existed as early as the 10th century. From 1058 to 1138 it was a residence of the ruling princes, and from 1194 to 1495 it was ruled by the dukes of Mazovia. Since about 1075 it has been the seat of a diocese, and it has been a city since 1237. Its commercial activity was noteworthy in the second half of the 15th century. The city flourished in the 16th century but declined in the 17th century. In 1816 it became the capital of the department of Płock, and in modern times it has become an important city for commerce and culture. The Renaissance cathedral of Płock was built in 1531–34 on the site of a Romanesque cathedral that had been destroyed in the 16th century. The Cathedral was rebuilt in 1784–87 and completely redone in the 19th and 20th centuries. The basilican interior preserves Renaissance and baroque funerary monuments, and the Cathedral treasury is noteworthy. The splendid Romanesque bronze door of the Cathedral was carried off as war booty in the Middle Ages to Great Novgorod, Russia, and is now preserved in the Cathedral of St. Sophia in that city. The bell tower, a Gothic structure enlarged in the 18th century, was originally part of the city walls (15th–19th cent.), as was the Tower of the Nobility. The former collegiate church, a Gothic structure that was later altered, now houses a school. The Gothic parish church of St. Bartholomew (Świętego Bartłomeja) was rebuilt in late baroque style in 1775. The late Gothic Dominican church and its cloister were rebuilt, as was the Benedictine church (1632). The Reformed monastery has a late baroque church (1758–71); the monastery buildings date from the second half of the 17th century. Neoclassic buildings of the first half of the 19th century include the city hall, the bishop's palace, the prison, and several residences. Museums include the Mościcki Museum and the Diocesan Museum (Muzeum Dieczejalne).

Bibliog. A. J. Nowowiejski, Płock, 2d ed., Płock, 1930; M. Sołtysiak, Płock i jego zabytki (Płock and Its Monuments), Płock, 1963.

Wilanów. The Palace of Wilanów (VIII, pl. 439), near Warsaw, was built at the end of the 17th century by the architect August Locci as the suburban villa of King John III Sobieski (Jan Sobieski). Enlarged in the 18th and 19th centuries, the palace is typical of the Polish baroque; restoration was begun in 1955. The fine interior is adorned with sculptures by Andreas Schlüter, among others, and with stuccoes. There are frescoes by Michelarcangelo (Michelangelo) Palloni, and the ceilings were decorated by Jerzy Eleuter Szymonowicz-Siemiginowski, Claude Callot, and others. In the reception rooms of the palace, which serves as a branch of the National Museum of Warsaw, furniture, decorative arts, and paintings of the 17th–19th century are displayed. Part of the beautiful park is in the English style and has pavilions. A permanent exhibit of modern Polish sculpture is housed in the orangery, rebuilt in the neoclassic style by Piotr Aigner, and in the section adjoining the park. Also preserved is the Neo-Gothic mausoleum of Stanisław Kostka Potocki and his wife Aleksandra.

Bibliog. W. Fijałkowski, Wilanów, Warsaw, 1962.

Bydgoszcz. Biskupin. At this site, near Żnin in Wielkopolska, important remains of a fortified settlement of the Lusatian culture (ca. 550–440 b.c.) were discovered in 1933. Partially preserved are wooden houses with remains of furnishings, ceramics, and other everyday objects. There is a small outdoor archaeological museum.

Bibliog. Z. Rajewski, Biskupin: Polish Excavations, Warsaw, 1959; Z. Rajewski, Settlements of the Primitive and Early Feudal Period in Biskupin and Its Surroundings, Archaeol. polona, II, 1959, pp. 85–124.

Chełmno (Culm or Kulm). One of the oldest cities of Poland, Chełmno, situated on the Vistula River, became a bishopric in the 13th century and experienced remarkable economic and cultural development from the 13th to the middle of the 15th century. The Gothic church of St. Mary (Panny Marii), an early-14th-century hall church, preserves its original roof structure. The former Dominican church is a Gothic edifice that was remodeled in the 17th century. The Cistercian church is also Gothic. The town hall, built in the years 1567–97, is one of the most important Renaissance secular structures in Poland. The town walls and tower date from the Gothic period.

Bibliog. F. Schultz, Die Stadt Kulm in Mittelalter, Z. westpreussischer Geschichtsverein, XXIII, 1888, pp. 1–251.

Chełmża (Culmsee or Kulmsee). This ancient city was made a bishopric in 1251. The Cathedral, now the parish church, was built on a hall-church plan (mid-13th–mid-14th cent.). It preserves the Renaissance funerary monument to Bishop Piotr Kostka (late 16th cent.) and some fine baroque furnishings.

Strzelno. The city was founded in the 12th century. The Premonstratensian church of the Trinity (Świętej Trójcy), built about 1175–92 and subsequently remodeled several times, preserves its original Romanesque three-aisled plan. In 1946, four Romanesque columns were discovered beneath the baroque plasterwork: two of the columns have horizontal bands of carvings (12th cent.); there are also two Romanesque tympanums and baroque furnishings. The Church of St. Procopius, near Strzelno, a round Romanesque structure with a ribbed dome, was built in the late 12th century. The tower dates from the 15th century.

Bibliog. Z. Kępiński, Odkrycia w Strzelnie (Discoveries at Strzelno), B. h. sztuki, VIII, 1946, pp. 202–07; T. Kozaczewski, Rotunda w Strzelinie (The Rotunda a Strzelno), Wrocław, 1955.

Toruń (Thorn). There was a fortified settlement on the site of modern Toruń in the 10th to the 11th century. The castle was founded by the Teutonic Knights in 1231. Toruń has been a city since 1233. Situated on the Vistula on important trade routes, it belonged to the Hanseatic League and was particularly prosperous in the 13th to 15th century; most of the monuments in the city date from this period. These Gothic structures are centered around the large market square. Toruń, which was part of Poland from 1466 to 1793, enjoyed a second period of prosperity between 1500 and 1600. Toruń's transformation into a modern city dates from the second half of the 19th century. A Prussian possession until 1919, it became part of Poland in 1920 and has since undergone further transformation. The Gothic church of St. John (Świętego Jana), with three aisles of equal height and a bell tower, was begun about 1250 and completed toward the end of the 15th century. There are Gothic frescoes in the choir (14th cent.), and the late Gothic triptych on the altar dates from 1505. The Gothic, metal tomb slab of Mayor Jan van Soest (d. 1361) and his wife is noteworthy. The church preserves baroque and rococo furnishings. The Gothic church of St. James (Świętego Jakuba; 1309–40) is basilican in plan, with three aisles and two façade towers. The church preserves Gothic sculptures, a painting with scenes of the Passion (second half of 15th cent.), and baroque furnishings. The Gothic church of St. Mary (Panny Marii) has late-14th-century frescoes and Gothic stalls in the choir (early 15th cent.). One of the interesting funerary monuments is the baroque one in honor of Princess Anna Vasa (1636). The tower of the Gothic town hall dates from about 1250. Four wings enclose the 14th-century courtyard, which was enlarged in the 17th century. The building now houses the Pomeranian Museum (Muzeum Pomorskie), with prehistoric collections and 19th- and 20th-century Polish painting. All that survives of the castle of the Teutonic Knights (Zamek Krzyżacki) is an early-14th-century tower. Some of the Gothic towers of the old city walls, including the Bridge Tower (first half of 15th cent.), and private residences in the Gothic, Renaissance, and baroque styles survive in the vicinity of the market square and in adjacent streets of the old and new quarters of the city.

Bibliog. R. Heuer, Thorn, Berlin, 1931; G. Chmarzyński, Sztuka w Toruniu (Art at Toruń), Toruń, 1934; T. Petrykowski, Toruń, Warsaw, 1957; M. Gąsiorowska and E. Gąsiorowski, Toruń, Warsaw, 1963.

Poznań. Gniezno (Gnesen). A castle was founded on this site in the 8th century. Gniezno became the first capital of Poland, under

Prince Miezsko I. In 1000 it was made the first bishopric in Poland, and in the 15th century it became the seat of the primate of Poland. Except for occasional periods of prosperity, it has gradually declined in importance since the 12th century. The Cathedral was built about 980 in pre-Romanesque style. In the second half of the 11th century it was replaced by a Romanesque structure, which was demolished to make room for the Gothic structure erected about 1342–1415. Remodeled several times in later centuries, the Cathedral was partially destroyed in 1945. It preserves a basilican plan, with side aisles and two façade towers, after Western models. There are 12 lateral chapels, of which the baroque chapel of the Potocki family (1727–30), designed by Pompeo Ferrari, is particularly noteworthy. The other chapels date between the 15th and 18th centuries. The magnificent bronze doors (ca. 1175; X, PL. 265) are the most important example of Romanesque sculpture in Poland and among the finest in Europe. Decorated with scenes from the life and martyrdom of St. Adalbert (Świętego Wojciecha), they are the work of local or Mosan craftsmen. The Cathedral preserves the late-15th-century Gothic tomb of St. Adalbert, designed by Hans Brandt. Funerary monuments include that of the primate Zbigniew Oleśnicki, a late-15th-century marble work by Veit Stoss; four Renaissance tomb slabs by Giovanni Fiorentino, which were imported from Hungary by the primate Jan Łąski and dedicated to his family and to his predecessors (1516); the Renaissance monuments of archbishops Jan Krzycki (ca. 1537) and Mikołaj Dzierzgowski (1553), which are related to the work of Gian Maria Padovano; the baroque monuments of archbishops Wójciech Baranowski (ca. 1615) and Wawrzyniec Gębicki (ca. 1625); and the silver sarcophagus with relics of St. Adalbert, a baroque work by Pieter van der Rennen (1662). The Cathedral treasury includes Romanesque, Gothic, and Renaissance illuminated manuscripts and a fine collection of silver and of church furnishings. The single-aisled Gothic church of St. John (Świętego Jana; 14th cent.) preserves 14th-century frescoes (1340–ca. 1360). The Gothic Franciscan church was built in the late 13th century and rebuilt several times. The Gothic church of the Holy Trinity (Świętej Trójey; early 15th cent.) was rebuilt in the 17th century. The town has a Regional Museum (Muzeum Regionalne).

BIBLIOG. I. J. Polkowski, Katedra gnieźnieńska (Gniezno Cathedral), Gniezno, 1874; W. Hensel, Najdawniejsze dzieje Gniezna w świetle wykopalisk (The Oldest History of Gniezno in the Light of Excavations), Gniezno, 1947; T. Ruszczyńska and A. Sławska, Katalog zabytków sztuki w Polsce, V, 3: Powiat gnieźnieński (Catalogue of Monumens of Art in Poland: Gniezno District), Warsaw, 1963.

Gołuchów. The early-17th-century Renaissance castle was rebuilt in 1875–95, with the collaboration of Eugène Emmanuel Viollet-le-Duc, in Neo-Gothic style. In 1880, it received the art collection of the princes Czartoryski, which was transferred from the Hôtel Lambert in Paris. The Greek vases in this collection are now in the National Museum in Warsaw, and the paintings and engravings have been divided between the national museums in Poznań and Warsaw. The branch of the National Museum of Poznań in Gołuchów has some fine interior furnishings and European and Polish paintings.

BIBLIOG. T. Jakimowiczówna, Zamek w Gołuchowie (Gołuchow Castle), S. muzealne, III, 1957, pp. 7–51; A. Kodurowa, Katalog zabytków sztuki w Polsce, V, 19: Powiat pleszewski (Catalogue of Monuments of Art in Poland: Pleszew District), Warsaw, 1959, pp. 11–14.

Kalisz (Kalisch). Situated on the Prosna River, this city seems to be the Calisia that was mentioned toward the middle of the 2d century of our era by Ptolemy. Kalisz was a tribal center and later the capital of the Land of Kalisz. It became the capital of the duchy of Kalisz (until 1279) and then the capital of the voivodate of Kalisz (until 1793). The city enjoyed particular development and prosperity from the middle of the 14th century until the early 17th century. The Gothic church of St. Nicholas (Świętego Mikołaja) has three aisles of equal height. The choir dates in part to the middle of the 13th century, while the aisles (reconstructed) date from the middle of the 14th century. The vaults (1612) are in late Renaissance style. The church preserves fine baroque and rococo furnishings. On the main altar is a *Deposition* painted in the workshop of Rubens. Next to the church is the former monastery of the Lateran canons, a Gothic-style structure (1448; 1538–39) that was subsequently altered. The Franciscan monastery has a Gothic church, erected about 1270 and rebuilt in the 14th and 17th centuries, with three aisles of equal height. The vaults are decorated with stuccoes in late Renaissance style (ca. 1630). The church has baroque furnishings. The Gothic chapel of the Passion of Christ was rebuilt in 1632. The baroque monastery buildings date from the years 1640–80. The Church of the Assumption (Wniebowzięcia) has a Gothic choir (mid-14th cent.) and late baroque aisles and bell tower (1790). The church preserves a late Gothic triptych (ca. 1500) and fine baroque and rococo furnishings. The Bernardine monastery (1594–1622) has a late Renaissance single-aisled church (completed 1607) with baroque and rococo furnishings; the monastery buildings are baroque. The former Jesuit church, built about the end of the 16th century by Gian Maria Bernardoni, was one of the first to be built by this order north of the Alps. The church has baroque and rococo furnishings and houses the baroque funerary monument of Stanisław Karnkowski (1611). Forming a single complex with the church are the former episcopal palace (late 16th cent.) and the collegium (1583–84). These two structures were rebuilt in neoclassic style in 1824–25. The church of the former Reformed monastery has rococo furnishings; the baroque monastery was built in the third quarter of the 17th century.

There are remains of the old city walls in Kalisz. Among the neoclassic secular buildings dating from the first half of the 19th century are the Old Tribunal Court and the Pułaski Palace, as well as private residences. The Museum of the Land of Kalisz is located in the town.

BIBLIOG. K. Dąbrowski, T. Uzdowska and M. Młynarska, Kalisz w starożytności i w średniowieczu (Kalisz in Antiquity and the Middle Ages), Wrocław, 1956; K. Dábrowski, Kalisz prastary (Ancient Kalish), Warsaw, 1960; Osiemnaście wieków Kalisza (18 Centuries of Kalisz), 2 vols., Poznań, 1960; T. Ruszczyńska, A. Sławska, and Z. Winiarz, Katalog zabytków sztuki w Polsce, V, 6: Powiat kaliski (Catalogue of Monuments of Art in Poland: Kalisz District), Warsaw, 1960, pp. 12–45.

Kórnik. The Castle of Kórnik, built in the 16th century and reconstructed in neoclassic style in the middle of the 19th century, now houses a museum and library, with important collections begun after the mid-19th century by the Działyński family. The park is noteworthy.

Ląd. The former Cistercian abbey was founded in 1146, rebuilt in the 14th century, and altered in the 17th and 18th centuries. The abbey church was begun in baroque style between 1651 and 1689 after a plan by Giuseppe Bellotti and continued between 1728 and 1735 after a plan by Pompeo Ferrari. The central-plan single-aisled church has a dome and transept. There are fine stucco decorations (1690–1725), and the frescoes are the work of Adam Swach and G. W. Neunherz (1731–32). The furnishings are baroque. The monastery, partly Gothic (14th–15th cent.), has frescoes from about 1370 in the oratory; this is one of the most important complexes of Gothic wall painting in Poland. Surviving are the chapter room, the vestibule, and the cloister (with Gothic architectural decoration). The upper floor is baroque (17th–18th cent.).

BIBLIOG. M. Kamiński, Dawne opactwo Zalconu Cysterskiego w Lądzie nad Wartą (The Ancient Abbey of the Cistercians at Ląd on the Warta), Ląd, 1936.

Poznań (Posen). Capital of the region of Wielkopolska, the city began to develop in the 10th century with the erection of the ducal castle on the Ostrów Tumski island in the Warta River. Poznań became a bishopric in 966 and was the capital of Old Poland from the third quarter of the 10th century until about 1040 and the capital of Wielkopolska from 1138 until the end of the 13th century. Situated on important commercial routes, Poznań expanded on a regular plan to the west of the island, which remained the seat of ecclesiastical power. In the late Gothic period and in the Renaissance, the city was economically and culturally important. It declined from the mid-17th century until about 1780 and did not begin to flourish again until the end of the 18th century. From 1793 to 1918, it was subject to German rule but remained an active center of Polish culture and developed into a large modern city. Badly damaged in 1945, the old quarter had to be reconstructed and restored.

The Cathedral is the oldest Christian church in Poland. Modern excavations have brought to light remains of a baptismal font (perhaps the one used at the official baptism of Poland in 966), as well as remains of three earlier structures, two pre-Romanesque and one Romanesque (before 1038). The latter may have provided the basis for the Gothic structure on a basilican plan, with three aisles and a choir with radial chapels, the building of which lasted from 1242 to 1262 and from 1344 to 1410; later enlargements have also been made. Destroyed in 1945, the Cathedral has been restored in Gothic style. Some of the ancient furnishings, including a late Gothic polyptych (1512) on the main altar, were saved. The Cathedral also preserves important Renaissance funerary monuments, including those of the bishops J. Lubrański (ca. 1525) and B. Izdbieński (d. 1553), the last work of Jan Michałowicz and those of the Górka family (1574) and that of Bishop Adam Konarski by Gerolamo Canavesi. In the Neo-Gothic Golden Chapel (1836–40), by Francesco Maria Lanci, is a sculpture of the rulers Mieszko I and Boleslav I, the Mighty (Bolesław Chrobry) by Christian Rauch. The late Romanesque brick church of St. John in Jerusalem (Świętego Jana w Jerozolimie), built in about 1200 and later enlarged, preserves a Romanesque portal and a late Gothic triptych. The late Gothic church of the Virgin

Mary (Panny Marii; 1433–44), a three-aisled structure, has wall paintings by W. Taranczewski (1954–55). The late Gothic three-aisled church of Corpus Domini (Bożego Ciała) was begun after 1404 and continued in the years 1465–70; the furnishings are baroque. The Chapel of St. Mary of the Rosary is the work of Pompeo Ferrari (first half of 18th cent.). Late Gothic in style are the Church of St. Margaret (Świętej Małgorzaty) in the Śródka quarter and the Church of St. Mary of Succor (Panny Marii Sukursa; 15th cent.). The late Gothic church of St. Adalbert (Świętego Wojciecha; 13th–16th cent.) preserves Renaissance and baroque furnishings and tombs. The Church of St. Martin (Świętego Marcina) was built in the early 16th century in late Gothic style and has been reconstructed; the main altar is Gothic, and there are murals by Taranczewski (1958). The former Dominican church was rebuilt in baroque style in the late 17th and early 18th centuries. It preserves Gothic elements, including a 13th-century portal, and baroque furnishings; the Chapel of St. Mary of the Rosary and the cloister are Gothic. The parish church (formerly Jesuit) was begun in 1651 and completed in the early 18th century. It is an important example of the baroque in Poland and has fine contemporary furnishings. The main altar is the work of Pompeo Ferrari (1727). There are paintings by Szymon Czechowicz, frescoes by Karol Dankwart, and baroque chapels. The former Jesuit collegium, in baroque style, is now the seat of the department authorities. The Carmelite church (after 1650) is baroque. The baroque Franciscan church (1674–early 18th cent.) has vault paintings by Adam Swach and fine baroque and rococo furnishings. The late baroque Bernardine church (first half of 18th cent.) has a lovely façade and baroque furnishings. The Church of All Saints (1775–86) is neoclassic.

There are remains of the 13th-century Gothic castle of Poznań (in reconstruction). The House of the Psalterists on the Ostrów Tumski island is Gothic (1512). The town hall was rebuilt in Renaissance style (1550–60) under the direction of Giovanni Battista Quadro. Partially destroyed in 1945 and subsequently rebuilt, the town hall preserves Gothic elements. The impressive façade has loggias and an upper story (attic), as well as a tall tower. There is a room with Renaissance stuccoes of allegorical subjects. The town hall now houses the Historical Museum. The Szialński Palace (1773–76) is neoclassic in style, as are the guard barracks (1787). The Raczyński Library (1829) has a façade with a colonnade inspired by that of the Louvre.

The bourgeois houses in the market area, which were in large part destroyed in World War II and subsequently rebuilt, are chiefly baroque and neoclassic in style. The fish vendors' houses in the market square were built in 1530–35 and have been reconstructed.

Museums: The National Museum (Muzeum Narodowe) has collections of ancient art; Polish art; Spanish, Italian, Dutch, and Flemish sculpture and painting; decorative art; prints; and coins. Branches include the City Museum of Poznań (in the town hall) and the Musical Instrument Museum (in a building on the market square). Branches outside the city include those in Rogalin and Gołuchów. Poznań also has a Museum of Prehistoric Archaeology, and an outdoor museum is located at Rydzyna, near Poznań.

Bibliog. N. Pajzderski, Poznań, Lwów, 1922; T. Ruszczyńska and A. Sławska, Poznań, Warsaw, 1953; Sztuki plastyczne, III: Dziesięć wieków Poznania (Plastic Arts: Ten Centuries of Poznań), Poznań, Warsaw, 1956; W. Hensel, Poznań w zaraniu dziejów (Poznań at the Dawn of History), Wrocław, 1958.

Rogalin. The Raczyński family palace in Rogalin (1770–1820) was begun in baroque style and finished in neoclassic style. The interiors are neoclassic and Neo-Gothic. The palace has a fine collection of paintings, fabrics, and decorative art. The beautiful park dates from the end of the 18th and the beginning of the 19th century. The palace and park are a branch of the National Museum of Poznań.

Bibliog. G. Ciołek, Ogsód w Rogalinie (The Garden of Rogalin), Ochrona zabytków, III, 1950, pp. 147–52.

Łódź. Arkadia. The romantic park of Arkadia was founded in 1778 by Princess Helena Radziwiłł. Built after plans by Szymon Bogumił (Simon Gottlieb) Zug, it was completed in 1821. The park has several neoclassic and Neo-Gothic pavilions. Since 1945, it has been, along with the Nieborów Palace, a branch of the National Museum of Warsaw.

Bibliog. J. Wegner, Arkadia, Warsaw, 1948.

Łowicz. This city was subject to the archbishops of Gniezno, primates of Poland, from the 12th to the 18th century. It became a city about 1300 and was particularly prosperous in the 16th century and the first half of the 17th century. The capital of one of the most important regions for folk art, it has preserved many ancient customs. The Collegiate Church was begun before 1464 in Gothic style. It was reconstructed in the style of the transition from Renaissance to baroque (1652–54), probably after a plan by T. Poncino and later (1656–68) by A. Poncino. A basilican structure with three aisles, it has two façade towers and annexed chapels. The main altar is rococo (later than 1761), by Efraim Szreger (Schröger), with sculptures by Jan Jerzy Plersch. The silver altarpiece dates from 1719, and the rococo pulpit from 1754. There are many baroque paintings of sacred subjects as well as portraits. Funerary monuments include those of the archbishops J. Przermbski (by Gerolamo Canavesi) and J. Uchański (altered at the end of the 18th century; by Jan Michałowicz), in the Renaissance style, and that of Archbishop J. Firley and the double Leszczyński tomb, which are baroque works of the 17th century. The church treasury has a splendid collection of furnishings and liturgical objects, including the Reliquary of St. Victoria (1625), a baroque creation decorated with enamels, perhaps the work of Italian goldsmiths; there are also illuminated manuscripts. The Renaissance chapel of St. Victoria (1575–ca. 1580) was built by Michałowicz and was reconstructed (1782–83) in neoclassic style after a plan by Szreger. On the altar is the Early Christian (4th cent.) tomb slab of St. Victoria. The Renaissance chapel of the Holy Trinity (1609–11) houses the funerary monument of Piotr Tarnowski and has an altar and statues by Abraham van dem Block. Other chapels in the church include the late Renaissance chapel of St. Anne (1635–ca. 1640); the baroque chapel of the Holy Sacrament (1640–47), with frescoes by Adam Swach (1718); and the Crucifixion Chapel (1759) by Szreger, a fine example of rococo style, with rococo furnishings and the funerary monument of Archbishop Adam Komorowski (ca. 1759).

The parish church was built in 1404 in Gothic style (later restored) and has a late baroque façade (1778). The church preserves paintings of the 17th and 18th centuries. The Church of St. Bartholomew (Świętego Bartłomeja) is a single-aisled Gothic structure (1582); the former Bernardine monastery, adjacent to the church, is baroque. The Church of SS. Leonard and Margaret is late Renaissance (ca. 1620). The Church of the Virgin Mary and St. Clare and the Bernardine convent, both of which are baroque structures (ca. 1650), were built after plans by T. Poncino. The baroque Piarist church (1672–80), with a late baroque façade (1750), is a three-aisled basilican church containing baroque frescoes (1725) and baroque furnishings.

The Old Seminary of the Missionary Fathers (1689-ca. 1701) may have been built after a plan by Tylman a Gameren. Completely destroyed, except for the chapel, it was rebuilt in 1950–52 in its original baroque form. The Chapel of St. Charles Borromeo, a rectangular structure with vestibule, preserves late baroque frescoes by Michelarcangelo (Michelangelo) Palloni (after 1695); its interior furnishings make the chapel a small museum of Polish baroque art. The seminary building houses the Łowicz Regional Museum (Muzeum Regionalna), a branch of the National Museum of Warsaw, which contains prehistoric objects and a collection of objects relating to the history of the region and the city, as well as one of the most interesting anthropological collections of local folk art. There are ruins of the archbishop's castle, originally a Gothic structure but enlarged in the Renaissance and baroque periods. The castle was almost completely dismantled after 1922. The town hall is neoclassic (1825–28). The neoclassic post office (1819) was rebuilt in 1950, after having been devastated in World War II. A small building erected for General S. Klicki by K. Krauze together with a chapel and a tower form a pseudomedieval complex. On the market square are Renaissance and baroque canons' houses, as well as Renaissance, baroque, and neoclassic private residences, and 18th- and 19th-century wooden buildings.

Bibliog. W. Kwiatkowski, Łowicz prymasowski w świetle źródeł archiwalnych (Łowicz, Seat of Primates, in the Light of Archival Sources), Warsaw, 1939; Katalog zabytków sztuki w Polsce, II : Województwo łodzkie (Catalogue of Monuments of Art in Poland: Voivodate of Łódz), Warsaw, 1954, pp. 124–57.

Nieborów. The Palace of Nieborów (restored), a baroque aristocratic dwelling, was built in the years 1690–96 after a plan by Tylman a Gameren. The staircase has Delft tiles, and the reception rooms are decorated with wood paneling and baroque, rococo, and neoclassic stuccoes. The wall paintings are neoclassic (ca. 1784). The palace is a branch of the National Museum of Warsaw and contains collections of painting, sculpture, and furniture. The part of the beautiful park designed by Tylman a Gameren after 1690 is of the French type; the 18th-century additions are in English style. The various pavilions, including two neoclassic orangeries (ca. 1795), preserve ancient, medieval, Renaissance, and baroque sculptures.

Bibliog. J. Wegner, Nieborów, Warsaw, 1954; J. Wegner, Przewodnik po Nieborowie (Guide to Nieborów), Warsaw, 1960.

Sulejów. The Cistercian abbey was founded in 1176. The Church of St. Thomas (restored), which is a late Romanesque basilican structure with Gothic vaulting, was built before 1232. The façade has Romanesque portals, while the splendid furnishings are baroque and rococo. The convent dates from the 13th century; it was enlarged in the 15th–16th century and later remodeled. The chapter room is Romanesque-Gothic. There is a fine cloister, and the convent walls have towers from various periods.

Bibliog. Z. Świechowski, Opactwo cysterskie w Sulejowie (The Cistercian Abbey of Sulejów), Poznań, 1954.

Kielce. Bodzentyn. Founded in the middle of the 14th century as a residence of the bishops of Kraków, Bodzentyn enjoyed a period of outstanding development in the 16th and first half of the 17th century. The Gothic parish church, with three aisles (1440–52; later reconstructed), has late Renaissance vaults. The church preserves fine furnishings. The main altar (brought from the Wawel Cathedral in Kraków) is Renaissance in style (1545–46). The Gothic triptych (1508–ca. 1515) is an important example of Polish painting. There are remains of the episcopal palace, which was built in the second half of the 14th century and completely reconstructed in Renaissance style in the second half of the 16th century.

Bibliog. Katalog zabytków sztuki w Polsce, III, 4: Powiat kielecki (Catalogue of Monuments of Art in Poland: Kielce District), Warsaw, 1957, pp. 4–8.

Kielce (Keltsy). The capital of the department of Kielce, the city was held in fief by the bishops of Kraków from perhaps the second half of the 11th century until 1789. The city of Kielce developed toward the middle of the 14th century and enjoyed particular prosperity from the 15th to the middle of the 18th century and again in the years 1816–30. The Cathedral was built over a preceding edifice in the 16th century and enlarged in 1632–35 and later. It is a much-remodeled baroque structure built on a three-aisled basilican plan. Fine wooden furnishings in baroque and rococo styles date from 1726–65. The Cathedral contains a late Gothic triptych (ca. 1500) and noteworthy funerary monuments. The Cathedral treasury is also noteworthy. The Church of the Holy Trinity (Świętej Trójcy; 1640–46) is baroque. The former episcopal palace (1637–41), a baroque edifice, may have been built by T. Poncino. Wings were added in the first half of the 18th century. The original plan was square, with loggias on the ground floor and towers at the four corners; within are baroque portals. Some of the rooms retain their original appearance, with baroque stuccoes and historical paintings from the workshop of Tomasso Dolabella (ca. 1641). In Kielce there are also canonical residences and bourgeois dwellings of the 18th and 19th centuries and a museum.

Bibliog. Katalog zabytków sztuki w Polsce, III, 4: Powiat kielecki (Catalogue of Monuments of Art in Poland; Kielce District), Warsaw, 1957, pp. 25–42.

Opatów. A monastery was founded on this site in the 12th century, and in 1282 Opatów became a city. The new city, founded 1317–28, enjoyed a period of great development in the 16th and in the first half of the 17th century. The collegiate church of St. Martin (Świętego Marcina), originally Romanesque, was built in the second half of the 12th century and has been rebuilt many times. The present basilican structure, with three aisles and a transept, preserves some Romanesque elements. There are baroque frescoes (18th cent.) and Renaissance funerary monuments, including that of Great Crown Chancellor Krzysztof Szydłowiecki (ca. 1535), executed with the collaboration of Giovanni Cini, and bearing the bronze relief known as the "Opatów Lamentation." The Bernardine church of the Assumption of the Virgin (1751–65) is late baroque with rococo furnishings. There is also an abbey. Parts of the city walls date from the early 16th century.

Bibliog. Katalog zabytków sztuki w Polsce, III, 7: Powiat opatowski (Catalogue of Monuments of Art in Poland; Opatów District), Warsaw, 1959, pp. 39–51.

Pińczów. Already inhabited in the Middle Ages, Pińczów became a city in 1429. From about 1550 until 1586, it was one of the most important centers of Calvinism and of the Aryan movement known as the Polish Brothers. It enjoyed notable prosperity later as well, particularly in the first half of the 17th century. The presence of stone quarries fostered the activity of workshops that had important effects on the development of art (particularly architecture and sculpture) in the late Renaissance period. In the 18th century the city declined. Partially destroyed in 1944, the city is now being rebuilt. The Church of St. John the Evangelist (formerly of the Paulists) is a late Renaissance reconstruction (ca. 1642) of a Gothic nucleus. A basilican-plan structure, it has baroque stuccoes and furnishings. The baroque monastery (first half of 17th cent.) preserves Gothic elements. The Reformed church of the Visitation (1615–ca. 1620) is late Renaissance and has stucco decorations. The monastery is baroque (1686–1706). The late Renaissance chapel of St. Anne is a central-plan structure (1600). The late Renaissance Old Synagogue (16th–17th cent.) has a Renaissance stone altar. The Castle tower is also Renaissance, and the building known as the Aryan Printing House (ca. 1600) is late Renaissance.

Bibliog. K. Kutrzebianka, J. Z. Łoziński, and B. Wolff, Katalog zabytków sztuki w Polsce, III, 9: Powiat pińczowski (Catalogue of Monuments of Art in Poland: Pińczow District), Warsaw, 1961, pp. 55–71.

Sandomierz (Sandomir). Situated on the Vistula River, Sandomierz was the site of a castle in the 10th century. The city was one of the most important centers in Poland in the early Middle Ages and became the capital of the ducal province in 1138 and later the seat of a voivodate. It officially became a city before 1241. Sandomierz flourished commercially from the 13th to the 15th century and declined from the 17th century on. As the seat of an archdiocese (1818), it attained new importance in the 19th century. The Gothic cathedral has a nave and side aisles of equal height (ca. 1360–82) and in a subsequent alteration acquired a baroque façade. The Russo-Byzantine murals (1430–ca. 1440) in the choir are rare examples in Poland, and there are fine baroque and rococo furnishings. The Cathedral also has a noteworthy treasury and library. The Chapel of the Holy Sacrament is baroque. Near the Cathedral is the so-called "Długosz House." Originally Gothic (1476), it was later rebuilt and now houses the Diocesan Museum (Muzeum Diecezjalne). The canons' houses are baroque. The Dominican church of St. James (Świętego Jakuba; second quarter and mid-13th cent.; later rebuilt) is a late Romanesque edifice with Gothic elements. The brick basilican structure has three aisles, and the portals are particularly noteworthy. The Chapel of the Martyrs of Sandomierz is late Renaissance (first half of 17th cent.). There is a monastery adjacent to the church. The Church of St. Paul (Świętego Pawła; ca. 1430) was built in Gothic style and later reconstructed; it has fine stucco decorations and baroque furnishings. The Church of the Holy Spirit (Świętego Ducha; 15th cent.) was also built in Gothic style and later remodeled, and it preserves baroque furnishings. The Benedictine church and monastery of St. Michael (Świętego Michała) constitute an interesting baroque complex, which preserves fine furnishings, paintings, silverware, and liturgical objects, also chiefly of the baroque period. The church (1686–92) has an external pulpit in baroque style. There is also a library. The Reformed church of St. Joseph (Świętego Józefa) is baroque (1679–89). The former Jesuit college (Gostomianum) is late Renaissance (1605–15). The Old Synagogue dates from 1758. Of the old castle structure, there survives one wing, in part dating from about 1480 and many times reconstructed. There are remains of the 14th- to 16th-century walls, including the Gothic Opatów Tower (after mid-14th cent.), with a Renaissance attic story. The Renaissance town hall, a reconstruction of the Gothic original, was enlarged in the 16th century. On the market square and elsewhere there are many bourgeois dwellings (restored) with Gothic, Renaissance, and baroque elements.

Bibliog. W. Kalinowski and others, Sandomierz, Warsaw, 1956; J. Z. Łoziński and T. Przypkowski, Katalog zabytków sztuki w Polsce, III, 11: Powiat Sandomierski (Catalogue of Monuments of Art in Poland: Sandomierz District), Warsaw, 1962, pp. 49–102.

Wąchock. The Cistercian abbey dates from the first half of the 13th century. The Romanesque church (ca. 1260), of basilican plan with three aisles, contains fine baroque and rococo furnishings and funerary monuments. The cloister is partly Romanesque and partly from the 16th and 17th centuries. The chapter room has Romanesque sculptural decoration, and the refectory (ca. 1250) is Gothic. Tomb slabs, paintings, and sculptures are preserved.

Bibliog. K. Białoskórska, Wąchock: Opactwo Cystersów (Wąchock: Abbey of the Cistercians), Warsaw, 1960.

Wiślica. This site on the Nida River was already settled in the 10th century. Its development was particularly notable from the middle of the 14th century until the 16th century. Important archaeological excavations were undertaken in 1949. There are remains of a small Romanesque church (10th–12th cent.) with what may have been a large baptismal font that is even older. The collegiate church (1350–ca. 1380; later restored) has two aisles of equal height and contains Gothic sculptural decoration, as well as many Gothic, Renaissance, and baroque objects. Under the church have been found remains of a Romanesque church (late 12th cent.) with a crypt; two Romanesque tomb slabs of plaster, dating from about 1165, on which there are engravings filled with a black metallic substance, were discovered in the floor of the crypt. The so-called "Długosz House" (1460) is Gothic.

BIBLIOG. K. Kutrzebianka, J. Z. Łoziński and B. Wolff, Katalog zabytków sztuki w Polsce, III, 9: Powiat pińczowski (Catalogue of Monuments of Art in Poland: Pińczów District), Warsaw, 1961, pp. 95–103.

Lublin. Kazimierz Dolny. This city on the Vistula River, probably founded in the 14th century, had become an important commercial center by the late Middle Ages. From the end of the 16th century to about the middle of the 17th century, many churches and secular structures were built in a regional late Renaissance style. The parish church is a late Renaissance reconstruction (1613), by Jakub Balin, of a 14th-century Gothic edifice. It is a characteristic example of late Renaissance style in Poland. The organ (1620) is baroque, and the furnishings are Renaissance. The beautiful Górski family chapel is Renaissance (1625–29). The baroque church of St. Anne (Świętej Anny) dates from the second half of the 17th century. The Reformed church and monastery are baroque (17th cent.). There are middle-class dwellings that are picturesque examples of Polish late Renaissance style, with fine tall cymae and façades completely decorated with sculptures. Particularly fine examples on the market square include the house "under the sign of St. Nicholas" (pod Świętym Mikołajem) and that "under the sign of St. Christopher" (pod Świętym Krzystofem; ca. 1615). Equally noteworthy is the "Celejowska" (ca. 1635) on Senatorska Street. Of the late Renaissance granaries, many of which survived into the 17th century, only ruins remain.

BIBLIOG. W. Husarski, Kamienke renesansowe w Kazimierzu Dolnym (Renaissance Houses of Kazimierz Dolny), Lublin, 1950; W. Husarski, Kazimierz Dolny, Warsaw, 1953.

Lublin. The capital of the department of Lublin grew up around a castle and settlement that existed as early as the 9th or 10th century and has been documented since the 12th century. Situated at the intersection of important trade routes, Lublin became a city in 1317 and developed rapidly, particularly after the union of Poland and Lithuania in 1386. Lublin became the voivodate capital in 1474 and the seat of a crown tribunal in 1578 and evolved into a political, economic, and cultural center of great importance, where there was

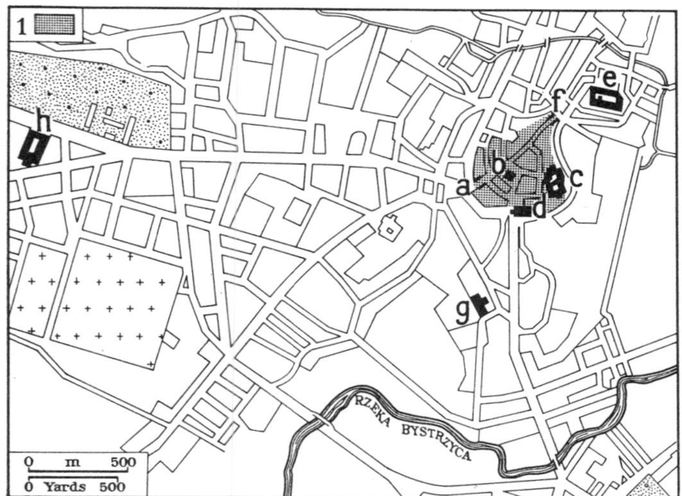

Lublin, plan. *Key*: (1) Area of original center. Principal monuments: (*a*) Fortified Kraków Gate (Brama Krakowska); (*b*) market place with old city hall; (*c*) Church of St. Stanislas (Świętego Stanisława); (*d*) Cathedral; (*e*) castle and Church of the Holy Trinity (Świętej Trójcy); (*f*) Brama Grodzka (Old Gate); (*g*) museum and Church of the Assumption (Wniebowzięcia); (*h*) university.

an architectural florescence under artists, mainly from the Como region of Italy, who initiated the Polish late Renaissance trend. In the middle of the 17th century, the city began to decline. At the end of the 18th and in the first half of the 19th century, the city underwent significant neoclassic transformations in plan as well as in architecture. The former church of the Brigitine nuns is a Gothic structure (1412–26); the choir was rebuilt in the late Renaissance style, and the convent was enlarged in the 17th century. The Dominican church is a late Renaissance enlargement of a Gothic structure. Reconstructed between about 1574 and the beginning of the 17th century, it has three aisles and eleven side chapels and preserves rococo furnishings of the second half of the 18th century. The chapel of the Firlej family (ca. 1625) has late Renaissance stucco decoration and houses the Renaissance funeral monument of Mikołaj and Piotr Firlej. The Tyszkiewicz family chapel (1645–58) has stuccoes by G. B. Falconi, a baroque fresco of the Last Judgement (ca. 1650), and baroque canvases. The Cathedral of Lublin (1586–96), modeled after the Church of the Gesù in Rome, was altered in the second half of the 18th and in the 19th century. The baroque chapel of St. Stanisław Kostka has baroque frescoes by J. Majer (ca. 1750). The Bernardine monastery has a late Renaissance church (1602–07), which replaced a 15th-century Gothic predecessor. The church contains fine stucco decoration and late baroque and rococo furnishings. The 16th-century monastery buildings have been reconstructed. The Orthodox church of the Transfiguration, a late Renaissance structure (1607–33), preserves an iconostasis of the same period. The Church of St. Joseph (Świętego Józefa), of the Discalced Carmelite nuns (1635–44) is late Renaissance and has fine stucco decoration and baroque furnishings; the convent was created in the 17th century from the transformation of a small castle built between 1622 and 1630 by Jakub Balin. The three-aisled late Renaissance church of St. Agnes (Świętej Agnieszki; ca. 1646) has fine stucco decorations.

The Castle (Zamek Lubelski), founded in the 13th century, was later enlarged, rebuilt in Neo-Gothic style in 1826, and restored in 1954. The mid-14th-century Gothic chapel preserves Russo-Byzantine frescoes completed in 1418 with the collaboration of the painter Andrzej. Rooms of the castle have been set aside as a museum and house archaeological, ethnological, and art collections. The remainders of the old fortifications are a Gothic tower and the Kraków Gate (Brama Krakowska), a massive structure, part of which dates from the 14th century (later enlarged). Baroque palaces include the Sieniawski (later the Czartoryski) Palace and the Radziwiłł Palace (now the seat of the university). Both palaces were built at the end of the 17th century, and the Radziwiłł Palace was rebuilt in neoclassic style in the early 19th century. The old town hall (later law courts) was extensively reconstructed in neoclassic style by Domenico Merlini in 1781. The new town hall is also neoclassic (1827–32). There are many Renaissance houses (in part reconstructed) as well as baroque and neoclassic residences. Noteworthy is the Sobieski House (early 17th cent.) on the market square.

BIBLIOG. J. Wadowski, Kościoły lubelskie (Churches of Lublin), Kraków, 1907; K. T. Wilgatowie and H. Gawarecki, Województwo lubelskie (The Voivodate of Lublin), Warsaw, 1957, pp. 50–89; H. Gawarecki and C. Gawdzik, Lublin, Warsaw, 1959.

Puławy. This small locality on the Vistula River has been important since the 16th century. It flourished and became a cultural center, from 1731 to 1831, under the patronage of the princes Czartoryski. The modern development of Puławy dates from the 19th century, when administrative buildings and scientific institutions were erected. The baroque palace was built about 1676 after a plan by Tylman a Gameren. It was reconstructed about 1730 by Jan Zygmunt Deybel and again, from 1788 to about 1800, by Piotr Aigner, and was later remodeled. The ballroom is neoclassic. In the large English-style park (18th cent.), with pavilions, stand the neoclassic Temple of the Sibyl designed by Aigner (after 1798) and the so-called "Gothic House," reconstructed about 1809 by the same architect. Both pavilions originally housed the famous Czartoryski Collection (part of which is now in Kraków). The neoclassic Marysienki Palace (late 18th cent.) and the neoclassic parish church (1803) are also works by Aigner.

BIBLIOG. S. Lorentz, Puławy (Teka Konserwatorska, V), Warsaw, 1962.

Zamość. This fortified city was founded in 1578 by Chancellor Jan Zamoyski after a plan by Bernardo Morando. The construction of the city was not completed until the first half of the 17th century, and it was extensively rebuilt in the 19th century. Zamość is a very rare surviving example of Renaissance urban planning. The city comprises the market square with town hall, two subsidiary markets, and a regular network of streets. Linked to the center by an axial system is the Chancellor's Palace (after 1579; reconstructed several times). Also included in the original city plan are the Collegiate Church (late 16th cent.), the Academy (Akademia) building, founded by the chancellor and later altered, and bourgeois houses. The Collegiate Church, a three-aisled basilican structure, has fine stucco decorations as well as four paintings attributed to Domenico Tintoretto. The bourgeois houses were built according to preestablished plans. Restored and partially reconstructed, with ground-floor loggias opening to the street, some of the houses, which once had attics, preserve fine sculptural decoration. The city was also provided with a modern system of fortification, later dismantled, of which two altered gates survive. The synagogue dates from the late 16th century.

Bibliog. S. Herbst and J. Zachwatowicz, Twierdza Zamości (The Fortress of Zamość), Warsaw, 1936; S. Herbast, Zamość, Warsaw, 1954.

Białystok. Białystok (Belostok, Bielostok). This department capital is documented from the year 1426. It owes its development to the powerful family of the counts Branicki (18th cent.) and subsequently to industrial activity. The Branicki Palace is an impressive baroque structure. Work on the palace continued from 1711 to 1730 but was only completed toward the end of the 18th century. The palace, which is now the seat of the Medical Academy, has a large park. The market is baroque (18th cent.), and there is a museum (Muzeum Podlaskie).

Olsztyn. Frombork (Frauenburg). The first documentation of Frombork dates to 1278. A bishopric at the end of the 13th century, Frombork became a city in 1310. From 1463 to 1772 it belonged to Poland, and from 1772 to 1945 it was part of Prussia; it was returned to Poland in 1945. The city was particularly prosperous in the late Middle Ages. The fortified hill on which the Cathedral is situated preserves Gothic walls and towers, including the so-called "Copernicus Tower." The Gothic brick cathedral (1329–88), a hall church, has a fine façade and contains baroque and rococo furnishings and noteworthy funerary monuments. The Copernicus Museum, inaugurated in 1948, occupies the two canons' houses.

Bibliog. S. Kozakiewicz and J. A. Miączyński, Zabytki Fromborka (Monuments of Frombork), Warsaw, 1948; M. Gajewska and J. Kruppé, Badania archeologiczne w 1958 r. na wzgórzu katedralnym we Fromborku (Excavations on the Hill of the Cathedral at Frombork), Rocznik olsztyński, III, 1960, pp. 81–104.

Lidzbark Warmiński (Heilsberg). A castle was erected here in the 13th century by the Teutonic Knights. The settlement that developed in the 14th century later became a city. It became the seat of the bishops of Warmia (Ermeland) in the middle of the 14th century. Lidzbark Warmiński belonged to Poland from 1466 to 1772 and to Prussia from 1772 to 1945, when it once again became part of Poland. The late Gothic parish church has a nave and side aisles of equal height; the furnishings are late baroque and rococo. The surviving Gothic castle dates from the second half of the 14th century. A square-plan structure, it has an inner courtyard with arcaded loggias and corner towers. The chapel and main hall on the upper floor have late Gothic vaults. There is also a museum in the castle.

Bibliog. A Bötticher, Die Bau- und Kunstdenkmäler der Provinz Ostpreussen, IV, Königsberg, 1894, pp. 134–56.

Olsztyn (Allenstein). The capital of the department of Olsztyn is located on the Łyna River. The castle and city of Olsztyn were founded in 1353. The ecclesiastical center of the diocese of Warmia (Ermeland), Olsztyn belonged to Poland after 1410 and enjoyed particular prosperity in the 16th–17th century. The city became part of Prussia in 1772 and was restored to Poland in 1945. The parish church of St. James (Świętego Jakuba; 15th–16th cent.) is a late Gothic hall church. The Gothic castle (begun in 1353) has a quadrangular plan with an inner courtyard. There is a corner tower of the 16th century. The interior of the castle was remodeled in 1909 and now houses a museum, with prehistoric, ethnographical, historical, and art collections. The late Gothic High Tower survives from the ancient walls.

Bibliog. A. Bötticher, Die Bau- und Kunstdenkmäler der Provinz Ostpreussen, IV, Königsberg, 1894, pp. 7–16.

Gdańsk. Gdańsk (Danzig). The capital of the department of Gdańsk is an important Baltic port near the mouth of the Vistula River. Already known in the 10th century for the castle and settlement of the princes of Pomerania, Gdańsk became a city in 1235. It was the possession of the Teutonic Knights from 1308 to 1454, when it passed to Poland. The city's most important buildings date from this period. From the middle of the 15th century until the early 16th century, late Gothic brick churches with nave and side aisles of equal height were built. (Such structures are typical of the Baltic area.) Many late Renaissance bourgeois houses were built about the end of the 16th and the beginning of the 17th century. At the same time, Dutch architects and sculptors, such as Anton van Opbergen and Willem and Abraham van dem Block, created new fortifications and civic buildings in a style very close to that of Holland. These works were to have great influence on the architecture and sculpture of northern and central Poland. The period also saw great development in the art of furniture. From 1793 until World War I, Gdańsk belonged to Prussia, and from 1919 until 1939 it was a free city. Many of its old buildings were seriously damaged during World War II, after which the city was restored to Poland. The modern reconstruction of the city has followed the old urban plan and architectural style. In the northern quarter is the Old City; in the central quarter is the Śródmieście (Main City), with the Ulica Długa (Long Street) and its prolongation into the Długi Targ (Long Market); in the southern quarter is the Old Town. East of the center are the island between the branches of the Motława (Mottlau) River known by the name of Spichlerze (Granaries) and the suburb known as the Lower City.

The Church of the Virgin Mary (Panny Marii), begun in 1343, was enlarged and altered between 1400 and 1502. A late Gothic structure with nave and side aisles of equal height, the church has chapels set between the external buttresses. The church was severely damaged in 1945 and reconstructed after 1946. The bell tower on the west was begun in 1359–73 and completed in 1452–66. In the floor of the church are numerous tomb slabs. The funerary monument of the Bar family is the work of Abraham van dem Block (1606). There is a Madonna of 1415 and a Pietà dating from the beginning of the 15th century (VI, pl. 363). The *Coronation of the Virgin* on the main altar is the work of the master Michael of Augsburg (1509–16).

The Church of St. Nicholas (Świętego Mikołaja; 1350–late 15th cent.), a late Gothic structure with three aisles, preserves fine baroque furnishings. The Church of St. John (Świętego Jana; 1358–1465) was partially destroyed during World War II. The main altar is the work of Abraham van dem Block (1611). The Church of SS. Peter and Paul (Świętych Piotra i Pawła; late 14th–early 16th cent.) is late Gothic. The church, in which part of the transept was destroyed in World War II, contains noteworthy funerary monuments. The Church of St. Bridget (Świętej Brygidy) was erected in 1392–1402 and reconstructed after 1587. Partially destroyed in World War II, the church was originally late Gothic in style. The 15th-century church of St. Catherine (Świętej Katarzyny), which had vaults dating from about 1500, has undergone reconstruction after being destroyed in World War II. Some of the church's furnishings were preserved, along with a *Crucifixion* by Anton Möller. The late Gothic church of the Holy Trinity (Świętej Trójcy; 1431–1516), part of the former Franciscan monastery, was partially destroyed in World War II. Particularly noteworthy is the west façade (1493–1514); inside the church, funerary monuments and sculptures have been preserved. In the courtyard is the Gothic chapel of St. Anne (1480). The former monastery buildings, dating from the middle of the 15th century and later remodeled, now house the Pomeranian Museum. The Royal Chapel was built in the years 1678–82, probably after a plan by Tylman a Gameren. It is a baroque central-plan edifice with a dome. The Church of St. Ignatius, formerly a Jesuit church, has nave and side aisles of equal height (1722–26).

The town hall of the Main City is a brick Gothic structure (1379–82) that was rebuilt in Renaissance style in the second half of the 16th century. Almost completely demolished in World War II, its reconstruction was begun in 1958. It has an inner courtyard and a tower topped by a gilded statue of King Sigismund II Augustus (Zygmunt August). Some of the furnishings, as well as paintings by Isaak van dem Block, Hans Vredeman de Vries, and Anton Möller, were saved and are now in the Pomeranian Museum. The so-called "Palace of Arthur" (Artushof) was built in its present form in 1476–81. It was burned in 1945, and rebuilding began in 1951. The late Gothic façade was rebuilt in the years 1616–17 with Renaissance additions by Abraham van dem Block. Within the palace there is a three-aisled Gothic hall. The Palace of St. George, built in late Gothic style (1489–94), was burned in 1945 and reconstructed in the postwar period. The town hall of the Old City section was built in 1587–89 in Renaissance style by Anton van Opbergen. Preserved in the town hall are 17th-century frescoes by A. Boy and 16th- and 17th-century paintings, including works by Isaak van dem Block and Anton Möller. The Arsenal (1602–05), designed by Anton van Opbergen, was burned in 1945 and has been reconstructed. The exterior, which has been preserved, has the appearance of four bourgeois houses side by side. The Neptune Fountain (ca. 1620), which stands in front of the Artushof, is Renaissance in style. The Small Arsenal of the Old Town was designed by Jerzy Strakowski (1643–45). Remains of the old fortifications include portions of the Gothic walls with gates and towers of the Main City (14th cent.), the Old City, and the Old Town (late 15th cent.). More modern fortifications, as well as some of the defensive gates, date from the mid-16th century. In the Main City are 15th-century medieval towers (partially reconstructed after 1945), including the Straw Tower, the Bakers' Tower, and the Fish Tower. The Prison Tower was originally built before 1411 and reconstructed in 1586 by Anton van Opbergen; it was rebuilt again in 1948–49. Together with the so-called "House of Torture," the tower forms a kind of barbican. The "House of Torture," built in 1592 by Anton van Opbergen on an older Gothic structure, is of the Renaissance middle-class type. The Green Gate (1568) is Renaissance in style. The High Gate (1576–88), designed by Willem van dem Block, has fine Renaissance sculptural decoration. The Renaissance Gate of Gold (1612) was the work of Abraham van dem Block. The remains of the walls of the Old City include the bastion of St. Elizabeth. The fortifications and towers of the Old Town date from the 16th century. The Renaissance Low Gate (1625) was the work of Jan Strakowski. There are two late Gothic towers (1517–19) in the Lower Town. The Gothic Great Mill dates from the middle of the 14th century. The Crane Gate, a Gothic construction of the mid-15th century, was destroyed in 1945 and subsequently rebuilt. In the 16th century the city boasted some 200 Gothic granaries. Though partially destroyed in World War II, some still survive (reconstructed). Gdańsk boasted many Gothic, Renaissance, and baroque middle-class houses, most of which were destroyed in 1945. Since then many have been faithfully reconstructed, including the old façades which often were elaborately decorated. Generally they were four-story constructions with two or four windows per floor and were generally topped by a fine pediment. Among the most important Renaissance houses are the so-called "Castle of the Lions" (1569), at 35 Ulica Długa; the so-called "Royal House" (1565), at 45 Ulica Długa; the so-called "House of Gold" (1609–19), at 41 Długi Targ; the house (1617) by Willem van dem Block at 3 Ulica Elżbietańska; the house (1569) at 16 Długie Nabrzeże; and the house (early 17th cent.) by Anton van Opbergen on Ulica Mariacka.

The Pomeranian Museum (Muzeum Pomorskie) is housed in the former Franciscan monastery. It contains prehistoric and anthropological collections as well as collections of medieval art, European painting, and Polish 19th- and 20th-century painting.

Bibliog. A. Lindner, Danzig, Leipzig, 1903; G. Cuny, Danzigs Kunst und Kultur, Frankfurt am Main, 1910; T. Kruszyński, Stary Gdańsk i historia jego sztuki (Old Danzig and the History of Its Art), Krakow, 1912; Gdańsk: Przeszłość i teraźniejszość (Danzig: Past and Present), Lwów, 1928; W. Szremowicz, Gdańsk, Warsaw, 1953; G. Brutzer and W. Drost, Kunstdenkmäler der Stadt Danzig, 3 vols., Stuttgart, 1957–59; J. Stankiewicz and B. Szermer, Gdańsk: Rozwój urbanistyczny i architektoniczny (Danzig: Urban and Architectural Development), Warsaw, 1959.

Malbork (Marienburg). This city on the Nogat River developed around a castle built by the Teutonic Knights in 1274 on an earlier site. In 1309, the capital of the Teutonic Order was transferred from Venice to Malbork, and the castle became the residence of the Grand Master of the Order. In the 14th century and the first half of the 15th, the city enjoyed great political and economic prestige. From 1460 to 1772 the city belonged to Poland, and from 1772 to 1945 it was part of Prussia. It was subsequently returned to Poland. In 1945 the old part of the city, except for very few buildings, was completely destroyed. Since then much reconstruction work has been carried out. The castle, which also suffered serious damage, comprises a magnificent complex of fortified structures dominating the Nogat River. Despite the destruction and unfortunate restorations, the castle remains one of the most remarkable Gothic secular structures in Europe. Begun in 1274, it consists of the Upper Castle (completed in the first half of the 14th cent.), the Middle Castle (14th cent.), and other edifices (for the most part dating from 1300–ca. 1450). The Upper Castle was the seat of the Grand Komtur, a high officer of the Teutonic Order, while the Middle Castle was the residence of the Grand Master. The castle was adapted to various uses and was remodeled several times in subsequent centuries. It was rebuilt in Neo-Gothic style in the years 1817–38. The Upper Castle, destroyed in 1945, is being rebuilt. It preserves the "Golden Gate" of the church (ca. 1290). The Middle Castle, a quadrangular-plan structure with a courtyard, contains important interiors, including the Gothic summer refectory and the winter refectory. Among the other edifices of the castle are the arsenal, the foundry, and the Chapel of St. Lawrence, all of which are in part Gothic. The surrounding wall and towers date from the 14th to 15th century. Situated in the Old City are the Gothic church of St. John (Świętego Jana; 1468–ca. 1500), which has a baroque façade, and the town hall (ca. 1380), also Gothic. Parts of the walls and towers also survive.

Bibliog. K. H. Clasen, Der Hochmeisterpalast der Marienburg, Königsberg, 1924; B. Guerquin, Zamek w Malborku (Malbork Castle), Warsaw, 1960 (2d ed., 1962).

Oliwa (Oliva). A Cistercian abbey was built here about 1175, and in the 15th century a town grew up around it. It became a suburb of Gdańsk and since 1925 Oliwa has been a diocesan see. The Cathedral (formerly Cistercian church) was begun in 1224, and work continued later than 1350. It is an impressive construction of basilican type, with transept, and preserves fine furnishings of the 16th–18th century. Noteworthy are the choir stalls (1604), the baroque altar (1688), the organ (after 1760), and the cloister and two-aisled refectory of the Gothic monastery (after 1350).

Bibliog. J. Heise, Die Bau- und Kunstdenkmäler der Provinz Westpreussen, II, Danzig, 1885, pp. 96–122; F. Mamuszka and J. Stankiewicz, Oliwa, Danzig, 1959.

Pelplin. The town developed around the Cistercian abbey that was founded in 1274. In 1824, Pelplin became the seat of the bishopric of Chełmno. The Cathedral (the former abbey church) was begun about 1380 and completed in the first quarter of the 15th century. This noteworthy example of Gothic architecture in Poland was inspired by English models. The furnishings are chiefly 17th-century baroque. The late Gothic choir stalls were restored in 1612. The main altar (1623) is late Renaissance in style. The monastery buildings were begun in 1274 but were drastically altered in Neo-Gothic style in the 19th century; the cloister, chapter hall, and refectory are noteworthy.

Bibliog. P. Skubiszewski, Architektura opactwa cysterskiego Pelplinie (Architecture of the Cistercian Abbey of Pelplin), S. pomorskie, I, Wrocław, Krakow, 1957, pp. 24–102.

Szczecin. Szczecin (Stettin). The city, capital of the department of Szczecin, is situated on the Baltic Sea at the mouth of the Oder (Odra) River. The site was inhabited in Roman times, and there was a Slavic population by the 8th or 9th century. It was a city by 1243, with settlements of German colonists in the same century, and flourished in the Gothic period (13th–15th cent.). The city had lost its commercial importance in the 17th century and had almost completely declined by the 18th century, when it became a possession of Prussia in 1720; its modern development began about the middle of the 19th century. In 1945, it was severely damaged, particularly in the old sections. Restored to Poland, Szczecin became an industrial, commercial, maritime, and cultural center of first importance. The new development of the city has made allowance for the historical monuments, many of which have been rebuilt or restored.

The Gothic church of St. John (Świętego Jana; mid-14th–early 15th cent.) is a noteworthy Baltic example of a church with nave and side aisles of equal height. The beautiful façade has a richly decorated pediment. Partially destroyed in 1945, the church has undergone reconstruction. The late Gothic church of St. James (Świętego Jakuba) has also had to undergo rebuilding after World War II; the Renaissance funerary monument of Duke Barnim III (1543) survives. The Church of SS. Peter and Paul (Świętych Piotra i Pawła), originally Gothic (early 15th cent.), was rebuilt in the baroque period and again in the 20th century. The Castle was partially rebuilt after 1503 and completely rebuilt and enlarged in Renaissance style in 1575–77; its former appearance was lost in the 19th-century rebuilding and in the damage of 1945, though some of the old interiors have been preserved. The Castle has been reconstructed in Renaissance style. The baroque sarcophagi of the last dukes of Pomerania (mid-17th cent.) are noteworthy. Part of the vast edifice houses the Museum of Western Pomerania (medieval art, Polish painting, and historical collections). The Gate of the Virgin (Brama Panny Marii), a Gothic structure, was part of the fortifications. The Gothic town hall has been reconstructed, and interesting middle-class houses have been preserved.

Bibliog. C. Friedrich, Stettin, Berlin, 1927; G. Chmarzyński, Pomorze zachodnie (Western Pomerania), I, Poznań, 1949, pp. 432–49.

Wrocław. Wrocław (Breslau). The capital of the department of Wrocław is situated on the Oder (Odra) River among a group of islands, the principal one being the center of the city. As early as the 10th century, there was a castle and settlement at this site. About 1000 the city became an episcopal see, and in 1264 it became the capital of the duchy of Silesia (under the Piast dynasty). In 1335, it passed with Silesia to Bohemia. An important commercial and cultural center, with an increment of Germanic peoples from the 13th century on, it was particularly prosperous in the 14th and 15th centuries, the period in which the medieval city took on its final form with many Gothic religious and secular monuments. Wrocław was ruled by the Hapsburgs beginning in 1526, and in 1741 it became part of Prussia. The development of the modern city around the old center dates from the 19th and 20th centuries. Wroclaw was severely damaged during World War II; much of its old center was destroyed. Restored to Poland after 1945, the city has been extensively rebuilt in accordance with modern urban criteria.

The Church of St. Aegidius (Świętego Idziego; first half of 13th cent.), a modest Romanesque structure, is the oldest surviving monument. The Cathedral was built over an earlier Romanesque building. The choir was built in the years 1244–72, and the aisles date from the first years of the 14th century. Almost entirely destroyed during World War II, the Cathedral was rebuilt in 1945–51 in Gothic style. The Cathedral is basilican but without transept. The Cathedral preserves Gothic and Renaissance funerary monuments and baroque furnishings. Noteworthy are the baroque chapel of St. Elizabeth (1680) and the Electoral Chapel. The Chapel of St. Elizabeth preserves the statue of Elizabeth of Thuringia, by Ercole Ferrata, and the tomb of Cardinal Frederick, landgrave of Hesse (d. 1682), by Domenico Guidi. The Electoral Chapel was designed by J. B. Fischer von Erlach (1720–27) and has sculptures by Ferdinand Brokof. The Church of St. Adalbert (Świętego Wojciecha) is a Gothic structure, a single-aisled church built on a Latin-cross plan (mid-13th cent.–ca. 1330). The baroque chapel of St. Wenceslaus has sculptures by L. Weber (1719–26). The Gothic church of the Holy Cross (Świętego Krzyża) is a Latin-cross hall church. The choir dates from 1288–95, while the aisles date from about 1350. The crypt is dedicated to St. Bartholomew. The church preserves important medieval Silesian paintings. The Mary Magdalen Church (Marii Magdaleny; second half of 14th cent.) was destroyed in World War II and has been partially reconstructed. A Gothic basilican-plan church, the Magdalen preserves a Romanesque portal brought from the destroyed church of the Romanesque abbey of Olbin, near Wrocław. The Gothic church of St. Dorothy (Świętej Doroty; 1351) is a hall church with baroque furnishings. The hall church of Our Lady of the Sand, built over an earlier Romanesque structure, was reconstructed from the middle of the 14th until the second half of the 15th century. Destroyed during World War II, only the outside walls survived; rebuilding has begun to provide housing for a museum of medieval religious art. The Gothic church of St. Elizabeth (Świętej Elżbiety; late 14th cent.) is a basilican structure without a transept. The Gothic bell tower dates from 1452–56. The church preserves many funerary monuments, including the Renaissance monument of Councilor Rybisch (1534–39). The Church of St. Vincent (Świętego Wincentego; late 14th–15th cent.) is Gothic in style. The Church of Corpus Christi (Bożego Ciała; late 14th–15th cent.) is a three-aisled late Gothic structure. The Church of SS. Peter and Paul (Świętych Piotra i Pawła; 15th cent.) is a single-aisled late Gothic edifice. The Church of St. Barbara (Świętej Barbary; early 15th cent.), a late Gothic hall church, contains noteworthy funerary monuments. The Church of St. Matthias (Świętego Macieja; Jesuit church; 1689–98) is baroque; the interior, with tribunes, preserves frescoes by Johann Michael Rottmayer (completed in 1706) and some of the most beautiful wooden furnishings in all of Poland. The Church of the Ursuline Sisters (1699–1701), by Johann Georg Knoll, is a two-aisled baroque edifice.

The town hall, completed in the years 1470–1505, incorporates earlier elements. The rectangular-plan structure has picturesque sculptural ornamentation and articulated roofs, pediments, and projecting elements. The Jesuit College (now the University) was built in 1728–40 by Johann Blasius Peintner and Joseph Frisch, after a plan by Christoph Tausch. The eastern part was completed in 1895. This fine example of baroque architecture has a long façade overlooking the Oder (Odra) River and splendid interiors. There are many Gothic, Renaissance, and baroque houses in Wrocław, many of which have been rebuilt.

Museums: Museum of Silesia (prehistoric collections, medieval art, Polish painting, and decorative arts); Historical Museum of the City of Wrocław; Department of Prints of the Ossoliński Institute; Archepiscopal Museum.

BIBLIOG. F. Landsberger, Breslau, Leipzig, 1926; L. Burgemeister, Die Kunstdenkmäler der Stadt Breslau, 2 vols., Breslau, 1930–33; W. Güttel, Breslau, 2d ed., Berlin, 1937; T. Dobrowolski, Sztuka na Śląsku (Art in Silesia), Katowice, Wrocław, 1948; K. Maleczyński, M. Morelowski and A. Ptaszycka, Wrocław; Rozwój urbanistyczny (Wroclaw: Its Urban Development), Warsaw, 1956; M. Bukowski, Katedra wrocławska (Wrocław Cathedral), Wrocław, 1962.

Jelenia Góra (Hirschberg). Situated at the juncture of the Bobr and Kamienna rivers, Jelenia Góra is one of the oldest cities of Silesia. It enjoyed particular prosperity in the 15th and 16th centuries, the late 17th century, and the first half of the 18th century. The Gothic parish church (second half of 14th–15th cent.), a basilican structure, has Renaissance and baroque furnishings. The Protestant church of the Holy Cross (Świętego Krzyża; 1709–18), built by the architect Martin Frantz, was modeled after the Church of St. Catherine in Stockholm. The baroque church is built on a Greek-cross plan with a central dome and has vault paintings by the Scheffiers and Neunherz. Parts of the city walls (14th–16th cent.) with partly Gothic towers survive. The town hall is late baroque (1747). There are many interesting private homes, chiefly baroque and rococo. There is also a museum.

BIBLIOG. H. Lutsch, Verzeichnis der Kunstdenkmäler der Provinz Schlesien, III, Breslau, 1891, pp. 455–66; H. Jasieński, Jeleniogórska kamienica mieszczańska (The Bourgeois House at Jelenia Góra), Ochrona zabytków, XI, 1958, pp. 192–205.

Kłodzko (Glatz). A medieval fortress and later a city, Kłodzko is situated on the Nysa Kłodzka (Neisse) River. It was a center of commerce and culture from the late Middle Ages until the beginning of the 18th century. The parish church of the Assumption (Wniebowzięcia; 1364–1468) is a Gothic basilican structure with nave and side aisles and late baroque furnishings. The main altar is noteworthy, as is the Gothic funerary monument of Ernest of Pardubice (second half of 14th cent.). The former Minorite church is baroque (second half of 17th cent.). The stone bridge is a rare example of Gothic structure (1390), and the private houses (17th–18th cent.), chiefly baroque, are richly decorated.

BIBLIOG. H. Lutsch, Verzeichnis der Kunstdenkmäler der Provinz Schlesien, II, Breslau, 1889, pp. 12–26; T. Broniewski, Kłodzko, Wrocław, 1963.

Krzeszów. The Church of St. Mary (Panny Marii; formerly the Cistercian church), built in 1728–35, is the most important baroque monument in Silesia. A single-aisled structure, it has lateral chapels and tribunes. The two-towered façade has sculptures by Ferdinand Brokoff. The late baroque interior preserves paintings by G. W. Neunherz, a splendid organ, and fine choir stalls. The elegant rococo chapel built in 1738 contains the Gothic (14th cent.) tomb slabs of Bolko I and Bolko II, of the Piast dynasty, and their wives. The adjacent monastery is baroque (1662–1768). The baroque church of St. Joseph (Świętego Józefa; 1692) has frescoes by Michael Willmann.

BIBLIOG. H. Lutsch, Verzeichnis der Kunstdenkmäler der Provinz Schlesiens, III, Breslau, 1891, pp. 376–83.

Legnica (Liegnitz). Already a town by the 12th century, Legnica was burned by the Tartars in 1241. It was rebuilt with a regular plan, became the capital of the duchy of Legnica and expanded rapidly in the Gothic and Renaissance periods (14th–16th cent.). It declined after the 17th century but regained importance in the 19th century. The Protestant parish church of the Virgin Mary (Panny Marii), a Gothic structure, was built in the years 1362–80 and 1450–68 and rebuilt between 1824 and 1828. The parish church of SS. Peter and Paul (Świętych Piotra i Pawła; 1333–ca. 1390) has lateral chapels that were built in the 15th century. Reconstructed in the second half of the 19th century, the church preserves a Gothic tympanum (ca. 1400) in the north portal. Within the church there is the Gothic funerary monument of Wenceslaus of Legnica and his wife Anna (second half of 14th cent.). The church also contains Gothic sculpture. The Catholic parish church of St. John (Świętego Jana), built by the Jesuits in the years 1714–27, is a single-aisled baroque edifice. The choir of the old Gothic church was transformed in 1677 into the Mausoleum of the Piasts of Legnica. This structure, with a dome and four baroque statues of princes, was designed by Matthias Rauchmiller (Rauchmüller). Next to the church is the Jesuit College, which is also baroque (1700–06). There are remains of medieval walls and towers. The castle, rebuilt after 1835 after plans by F. Schinkel, retains structural elements from the 13th–15th century as well as a Renaissance portal (1533). The fine sculptural decoration is the work of George of Hamburg. Both the Academy of the Nobles (1726–35) and the town hall (1737–41) are baroque. The city also preserves noteworthy Renaissance and baroque houses.

BIBLIOG. H. Lutsch, Verzeichnis der Kunstdenkmäler der Provinz Schlesien, III, Breslau, 1891, pp. 201–57.

Trzebnica (Trebnitz). The Cistercian abbey church, which was built between 1202 and 1240, was the first brick building in Poland. Despite the rebuilding in the mid-18th century, it is the best surviving example of a medieval Polish abbey church. Though it is an early Gothic structure, it incorporates Romanesque elements. Built on a basilican plan, with side aisles and transept, the church preserves excellent late Romanesque sculptural decoration (first half of 13th cent.). The magnificent interior furnishings are baroque. The Chapel of St. Jadwiga (after 1267) is the finest example of early Gothic art in Silesia and preserves Gothic funerary monuments (e.g., that of Duke Konrad II, 1409) as well as baroque tombs (e.g., that of St. Jadwiga, 1680).

BIBLIOG. D. Frey, G. Grundmann and A. Zinkler, Die Klosterkirche in Trebnitz, Breslau, 1940; T. Broniewski, Trzebnica, Wrocław, 1959; T. Broniewski, Trzebnica, Warsaw, 1961.

Opole. Brzeg (Brieg). One of the most important monumental centers of Silesia, Brzeg was the capital of a duchy from 1311 until the 17th century, except for certain intervals. It was particularly prosperous in the second half of the 14th century and again in the 16th century. The Gothic church of St. Nicholas (Świętego Mikołaja; 1370–1420) is of the basilican type. Reconstruction was undertaken after most of the church was destroyed during World War II. The Church of the Holy Cross (Świętego Krzyża; 1735–39), built by the Jesuits, is a three-aisled baroque structure with tribunes. There are frescoes by Johannes Kube (ca. 1745) and fine furnishings. The ducal palace was erected in the 14th century and completely rebuilt in Renaissance style in 1538–60. From 1544, a group of artists from Ticino, including Jacopo and Francesco Parr, worked on the palace. The palace originally comprised four wings and an inner courtyard with loggias. After the destructions and rebuildings of the 18th and 19th centuries, all that remain are some rooms of the east wing and the entryway to the courtyard. The entryway is decorated with fine sculptures, including a series of portraits of princes of the Piast dynasty. Next to the castle is the Chapel of St. Jadwiga (later the collegiate church). Built in Gothic style after 1342, it was remodeled in the 18th century. The Gymnasium was built by Jacopo Parr in 1564–69 and was later radically altered. The town hall (1569–72), which incorporates a Gothic structure, was built by Bernhard Niuron and Jacopo Parr in Renaissance style. There are Renaissance and baroque houses.

BIBLIOG. M. Zlat, Brzeg, Wrocław, 1960.

Nysa (Neisse). This town on the Nysa Kłodzka (Neisse) River was an ecclesiastical property and capital of the duchy of the diocese of Wrocław from 1138 to the 18th century. Nysa was probably founded in the second half of the 12th century and was much enlarged by archbishops and the middle class in the 13th–15th century and in the 16th century. It became an important center of Humanist culture. Occupied by Prussia in 1740, Nysa served as a border fortress of that state. The city, greatly damaged in the war, was restored to Poland in 1945 and has been extensively rebuilt. The three-aisled Gothic collegiate church of St. James (Świętego Jakuba; 1401–30), a hall church, was partially destroyed in 1945 and rebuilt. It preserves Renaissance and baroque altars. Particularly noteworthy funerary monuments in the church include the Gothic monument to Bishop Wenceslaus (1491) and the Renaissance monuments of the bishops Jakub of Salza (d. 1539), Baldasar of Promnik (d. 1562), Gaspar of Lagow (d. 1574), and Jan Sitsch (d. 1608). The Gothic church of St. Barbara (Świętej Barbary; 1341) has a late-15th-century pediment. The Church of the Assumption (Wniebowzięcia; 1688–92), formerly Jesuit, is baroque and contains frescoes by Karol Dankwart. The Church of the Holy Cross (Świętego Krzyża; 1715–30), of the Friars of the Holy Sepulcher, is late baroque, with fine baroque furnishings and frescoes by Thomas Christian and Felix Anton Scheffler; the adjacent monastery is also baroque. Parts of the Gothic town hall (late 15th cent.) have been preserved. The Public Weighhouse, built in 1604 (destroyed in 1945; rebuilt), is an example of late Renaissance secular architecture. The former Jesuit College (late 17th cent.), a late baroque structure, preserves a beautifully decorated hall. The baroque episcopal palace (ca. 1729) was built by Christoph Tausch. Many of the interesting private houses in Nysa, chiefly Renaissance and baroque, and many of the city gates were destroyed in 1945 but were later rebuilt. There is a baroque well in wrought iron (1686). Among the late Gothic towers of the old fortifications are the Ziebice Tower (early 16th cent.) and the Wrocław Tower (ca. 1600). Nysa also has a museum.

BIBLIOG. H. Lutsch, Verzeichnis der Kunstdenkmäler der Provinz Schlesien, IV, Breslau, 1894, pp. 78–126; T. Chrzanowski and M. Kornecki, ed., Katalog zabytków sztuki w Polsce, VII, 9: Powiat nyski (Catalogue of Monuments of Art in Poland: Nysa District), Warsaw, 1963.

Katowice. Częstochowa (Chenstokhov, Czenstochau). Situated on the Warta River, Częstochowa was recorded as early as 1220. It became a city in 1377 and an important pilgrimage site after the foundation of the monastery of the Pauline Fathers in 1382. The Gothic church of St. Sigismund (Świętego Zygmunt; 15th cent.) was rebuilt in the Renaissance period. The late baroque church (after 1690) of the Pauline monastery incorporates remains of an earlier Gothic edifice. There are frescoes by Karol Dankwart (1694–95), and the main altar has sculptures by Baldassarre Fontana. The Chapel of the Virgin Mary preserves the miraculous image of the Black Madonna, a work of the Sienese school executed before 1383. The monastery buildings are baroque (17th–18th cent.), and additions were made in the 20th century. The 17th-century refectory has frescoes from about 1670. The library and treasury are also noteworthy.

BIBLIOG. J. Braun, Częstochowa, Warsaw, 1959.

Kraków. Kraków (Cracow, Krakau). Situated on the Vistula (Wisła) River, Kraków may have been the capital of the Vistulian (Wislanie) tribe in the 9th century. It was the capital of Poland from about 1040 until 1596. The original settlement of wooden structures stood on the Wawel hill, where toward the end of the 10th century the first stone church was built; it was a round edifice with four apses, dedicated to the Virgin Mary. Early in the 11th century, under King Boleslav I, the Mighty (Bolesław Chrobry), the Cathedral and castle were erected nearby. On the nearby Skałka

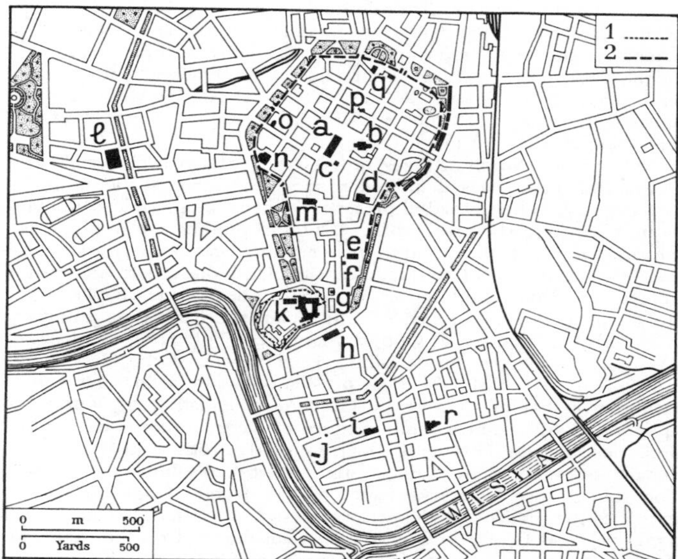

Kraków, plan. *Key*: (1) Line of the fortified walls of the old citadel; (2) line of the city walls of the 13th to the 15th century. Principal monuments: (*a*) Sukiennice (cloth market); (*b*) parish church of the Virgin Mary (Panny Marii); (*c*) Church of St. Adalbert (Świętego Wojciecha); (*d*) Church of the Holy Trinity (Świętej Trójcy) and Dominican monastery; (*e*) Church of SS. Peter and Paul (Świętych Piotra i Pawła) and Jesuit monastery; (*f*) Church of St. Andrew (Świętego Andrzeja); (*g*) Church of St. Aegidius (Świętego Idziego); (*h*) Bernardine church and monastery; (*i*) Church of St. Catherine (Świętej Katarzyny) and Augustinian monastery; (*j*) Church of St. Michael (Świętego Michała) and Pauline monastery; (*k*) Wawel Castle and Cathedral; (*l*) National Museum (Muzeum Narodowe); (*m*) Franciscan monastery; (*n*) university; (*o*) Church of St. Anne (Świętej Anny); (*p*) Church of St. John (Świętego Jana); (*q*) Piarist church and monastery; (*r*) Church of Corpus Domini (Bożego Ciała) and monastery of the Canons Regular.

hill, the round Romanesque church of St. Michael (Świętego Michała) was built in the 11th century. In the Romanesque period the city developed around the settlement and market square at the foot of the hill. Many stone churches were built at this time, and some still survive. After the Tartar devastation in 1241, the city was rebuilt on a geometric plan, with a square market at the center, in 1257. About 1285 the city was surrounded by brick and stone walls and fortifications. In the 14th and 15th centuries, gates were opened in the double ring of walls, and 40 towers were erected (in part preserved). Late in the 15th century barbicans were also added. In the second half of the 13th century and in the 14th and 15th centuries, religious and secular buildings were constructed in brick and stone. Among these were the town hall, the cloth market on the main market square, and the buildings of the University, which was founded by Casimir (Kazimierz) III, the Great, in 1364 and enlarged by Queen Jadwiga and King Ladislas II Jagello (Władysław Jagiełło) in the 15th century. Under kings Sigismund (Zygmunt) I and Sigismund II Augustus (Zygmunt August), that is, in the years 1506–72, Kraków was transformed into a Renaissance city. The royal castle on the Wawel, the cloth market, and the town hall were rebuilt. Numerous Renaissance palaces and middle-class houses were erected, and many funerary monuments were installed in the various churches. Contacts with Italian artists were very close in the Renaissance (the architect Francesco della Lora; the architect and sculptor Bartolomeo Berecci of Florence; and Gian Maria Padovano) and in the baroque period (the architect Giovanni Trevano from Lugano, Switzerland; the painter Tomaso Dolabella from Belluno; and the sculptor Giovanni Battista Falconi). About the end of the 17th century and the beginning of the 18th, the sculptor and stuccoworker Baldassare Fontana of Chiasso, whose style resembled that of Gian Lorenzo Bernini (q.v.), worked in Kraków. One of the most important architects active here in the first half of the 18th century was Francesco Placidi. In the 17th and 18th centuries, after the capital had been removed from Kraków to Warsaw, religious architecture prevailed over secular architecture. By the end of the 18th century, the city had 64 churches, most of which were abbey churches. The general aspect of the city was dominated by the Gothic and baroque styles. Toward the end of the 19th century, owing to the liberal policies of the Austrian Empire, the city developed into an important artistic and cultural center, particularly when it became the headquarters of the Polish Academy of Sciences (Polska Akademia Nauk). The architecture of that period was eclectic, with several stylistic revivals: Romanesque (churches); Gothic (e.g., Collegium Novum of the University); and Renaissance [Academy of Science (Akademia Umiejętności) and Academy of Fine Arts (Akademia Sztuch Pięknych)]. About 1900, there emerged international tendencies, locally known as the "Young Poland" (Młoda Polska) movement. After 1948 the industrial twin city of Nowa Huta was built near Kraków (XIV, PL. 150).

The round church of the Virgin Mary on the Wawel (second half of the 10th cent.) was dedicated to SS. Felix and Adauctus in the 14th century. The Cathedral preserves remains of a basilican Romanesque church (early 11th cent.). The surrounding walls date from the end of the 11th century, the crypt dedicated to St. Leonard from the beginning of the 12th century, and the pillars and vaults from the Gothic rebuilding of 1330–64. The Gothic chapel of the Holy Cross preserves Russian Byzantine paintings (ca. 1470). Funerary monuments include the Gothic monuments of King Ladislas I (Władysław Tokietek) and Casimir (Kazimierz) III, the Great (14th cent.); that of Ladislas II Jagello (Władysław Jagiełło; first half of 15th cent.); that of Casimir IV (Kazimierz Jagiełło), by Veit Stoss (q.v.; 1492); the bronze tomb plaques of Piotr Kmita and Cardinal Frederick Jagello (Fryderyk Jagiełło; early 16th cent.); and the Renaissance monument of King John Albert (Jan Olbracht; 1502–05). The Renaissance funerary chapel of the Jagello (Jagiełło) family, known as the Sigismund Chapel, was designed by Bartolomeo Berecci (1519–31), and Giovanni Cini of Siena collaborated on the execution of the sculptural decoration. The church also preserves the funerary monuments of Sigismund (Zygmunt) I, by Berecci; of Sigismund II Augustus (Zygmunt August), by Padovano; and of Queen Anna Jagello, by Santi Gucci. Other Renaissance funerary monuments in the Cathedral are those of the bishops Zebrzydowski (1563) and Padniewski (1575), both by Jan Michałowicz of Urzędów, and the late Renaissance monument to King Stephen Báthory is the work of Santi Gucci (late 16th cent.). The Chapel of St. Stanislas (1626–29) is the work of Giovanni Trevano, and the silver coffin of the saint was made by Pieter van der Rennen (1671). The Vasa Chapel is baroque (1667). There are sculptures by Bertel Thorvaldsen (q.v.) in the Cathedral. Among the precious objects in the treasury are the so-called "lance of St. Maurice," a 12th-century Saracen-Sicilian silver chest, a 13th-century cross, a 13th-century bishop's miter, a Gothic chalice with an Egyptian glass cup that dates from the time of the Fatimids (11th–12th cent.), a 14th-century French ivory chest, the Reliquary of St. Florian (ca. 1420), the Gothic reliquary of St. Stanislas (1504) designed by the goldsmith Marcin Marcinek (Marczinecz), the chasuble of Piotr Kmita (1504), and the sword of Sigismund II Augustus (1540).

Of the Romanesque churches erected toward the end of the 11th century and in the first half of the 12th, only those of St. Adalbert (Świętego Wojciecha), St. Andrew (Świętego Andrzeja), and the Holy Saviour (Świętego Salwatora) survive. The parish church of the Virgin Mary (Panny Marii), on the market square, was founded in the first quarter of the 13th century. It was reconstructed of brick in Gothic style in the second half of the 14th century. There are stained-glass windows dating from the years 1370–1400. The Gothic main altar, with the Dormition and Assumption in the center, is Stoss's masterpiece. There is also a crucifix by Stoss (ca. 1491) as well as a marble ciborium (1551–52) by Padovano and mural paintings by Jan Matejko (1889–92). The Church of the Holy Trinity (Świętej Trójcy) and the Dominican monastery were founded in 1222. There are Romanesque stone and Gothic brick remains of the original structures. The Gothic brick church dates from the 13th–14th century; the stained glass is Gothic (1430–40), as is the polyptych (1460–70; preserved in the Kraków National Museum). The church also preserves the bronze tomb plaque of the Humanist Filip Kallimach (late 15th cent.). There are four late Renaissance chapels (1608–29) in the church and paintings by Tomasso Dolabella in the Chapel of St. Hyacinth (1619–25) and in the monastery. The Franciscan church and monastery were founded in 1237. In the cloister are 15th–century frescoes, as well as portraits of the bishops of Kraków (16th–18th cent.). In the church are murals and stained glass by Stanisław Wyspiański (late 19th cent.). The Church of St. Catherine (Świętej Katarzyny) and the Augustinian monastery (1363–1426) are brick constructions; a polyptych (ca. 1470–80) from there is now preserved in the National Museum in Kraków. The Church of Corpus Domini (Bożego Ciała) and the monastery of the Canons Regular (1340–1405) have stained-glass windows from about 1420. The Church of St. Mark (Świętego Marka; 13th–15th cent.) is a brick structure that was partially rebuilt in the 17th century. The Church of the Holy Cross (Świętego Krzyża; 14th–15th cent.) is a stone and brick edifice. There are frescoes from approximately the years 1420–1530. The brick church of St. Aegidius (Świętego Idziego; 14th cent.) has Renaissance stalls of marble, part of which came from the funerary monument of St. Hyacinth, formerly in the Dominican church (1583). The Church of St. Barbara (Świętej Barbary; late 14th cent.), another brick structure, has a late-15th-century sepulchral chapel. The Church of SS. Peter and Paul (Świętych Piotra i Pawła), adjacent to the Jesuit monastery, was begun in 1596 and continued in the years 1605–19 by the architect Giovanni Trevano. The church structure was influenced by that of the Church of the Gesù and the Church of S. Andrea della Valle in Rome. The church has stuccoes by Giovanni Battista Falconi (1619–33). The Bernardine church and monastery, rebuilt by Krzysztof Mieroszewski (1670–80), has canvases by Franciszek Leksycki. The Church of St. Anne (Świętej Anny), by Tylman a Gameren (1689–1703), has stuccoes by Baldassare Fontana (1695–1700) and murals by Karol Dankwart (ca. 1700). The Church of St. John (Świętego Jana) was rebuilt in the 17th century. The Piarist church (1759) has a façade by Francesco Placidi. The Church of St. Michael (Świętego Michała) and the Pauline monastery were built on the Skałka hill, over Romanesque edifices, by Jan Münz (1733–51). The Church of the Missionary Fathers was the work of Placidi (1719–28).

The Sukiennice cloth market, on the market square, is a covered market hall begun about 1300 in stone. It was enlarged in brick about the middle of the 14th century and was rebuilt in Renaissance style in the years 1557–59, with the collaboration of Padovano.

There is a stone and brick bell tower (1383) that was once part of the town hall (destroyed). The house at number 17 on the market square, known as the Hetman's House, has Gothic vaults and sculptured keystones (1390). The university, which has a Gothic arcaded courtyard 1492–97), was founded in 1364 by King Casimir (Kazimierz) III, the Great, and it has been known as the Jagellonian University (Uniwersytet Jagielloński) since the 15th century. There are private houses of the Renaissance and early baroque period (late 16th and early 17th cent.), with porticoed courtyards, at numbers 18 and 21 Ulica Kanoniczna. There are similar courtyards in the Zbaraski Palace, the "Pod Krzysztofory" house (in the market square), in the Collegium Nowodworski in Ulica Świętej Anny, and in the Collegium Juridicum in Ulica Grodzka. Remains of the Gothic city walls include several towers: the Wieża Lubranka, Wieża Sandomierska, and the Baszta Zlodziejska (the last was begun in 1390). The original Gothic nucleus of the castle on the Wawel was under construction from the first half of the 14th century until the end of the 15th. The present structure is Renaissance, begun in 1502 and completed in 1536. The large courtyard with three stories of arches dates from this period. The work was directed by Francesco della Lora and later by Bartolomeo Berecci. The wooden beamed ceiling of the Chamber of Deputies (Sala Poselska) is decorated with heads sculptured inside small coffers. Only 30 of the original 195 heads still survive. Of the interior furnishings of the castle, more than 100 Flemish tapestries (mid-16th cent.) have been preserved. Commissioned by King Sigismund II Augustus, these tapestries include scenes of Biblical subjects, some of which were woven from cartoons by Michiel Coxie.

The National Museum (Muzeum Narodowe), founded in 1879, has Polish and foreign art, medieval works (including local works), and a collection of decorative art. The museum also houses part of the Czartoryski Collection. The House of Jan Matejko has a collection of works commemorating that artist. The State Art Collections (Państwowe Zbiory Sztuki) of the Wawel Castle include 16th-century tapestries, paintings of various European and Polish schools, furniture of the 15th–17th century, decorative art and arms, including the coronation sword of the Polish kings (Szczerbiec; 12th cent.). The Museum of the History of Polish Science (Muzeum Historii Nauki Polskiej) in the Jagellonian University has a collection of scientific instruments (15th cent.) and an art collection. In the university treasury there are three scepters of the rectors of the university (from the 15th cent. on).

BIBLIOG. Rocznik krakowski, Krakow, 1898 ff.; L. Lepszy, Krakau, Leipzig, 1906 (Eng. trans., R. Dyboski, London 1912); M. A. de Bovet, Cracovie, Paris, 1910; L. Réau, L'art du moyen âge et de la Renaissance à Cracovie, GBA, III, 1910, pp. 397–423, 499–507; A. Lauterbach, Die Renaissance in Krakau, Munich, 1911; K. Estreicher, Kraków, 2d ed., Krakow, 1938; T. Dobrowolski, Sztuka Krakowa (Art of Krakow), Kraków, 1950; Studia do dziejów Wawelu (Studies in the History of the Wawel), I–II, Krakow, 1955–60; H. Münch, Kraków do roku 1257 wtacznie (Krakow's Development up to 1257), Kwartalnik arch. urban., III, 1958, pp. 1–40.

Kalwaria Zebrzydowska. The site developed as a pilgrimage place in the 17th century. The sanctuary includes the Bernardine church and monastery as well as small churches, chapels, and other religious edifices in the neighboring hills and valleys. A typical expression of the Counter Reformation period, the complex was founded by the Zebrzydowski family and developed chiefly in the first half of the 17th century. The church and monastery are the work of Gian Maria Bernardoni and Paweł Baudarth, who was also responsible for most of the other structures of the sanctuary. The church was built in 1603–09 and enlarged in the 17th and 18th centuries. Notable among the baroque chapels is that of St. Mary of the Miracles (1658–1667). The monastery (1603–09 and 1654–66) has a fine collection of paintings, silver, and sacred furnishings. There is great variety of style in the buildings and in the chapels, which are dedicated to the Virgin and the *Via Crucis*.

BIBLIOG. J. Szablowski, Architektura Kalwarji Zebrzydowskiej (Architecture of Kalwaria Zebrzydowska), Rocznik krakowski, XXIV, 1933, pp. 1–118; Katalog zabytków sztuki w Polsce, I: Województwo krakowskie (Catalogue of Monuments of Art in Poland, I: The Voivodate of Krakow), Warsaw, 1953, pp. 469–85.

Pieskowa Skała. This castle on a mount overlooks the Prądnik River. Originally a Gothic structure, the castle was begun in the 14th century. The work continued in the 15th and 16th centuries, and the castle was enlarged and rebuilt with Renaissance forms in the years 1542–44 and was later remodeled. It is built on an irregular plan with two courtyards, one of which has beautiful Renaissance arched loggias that were exposed during restoration work in 1947. The interiors have been remodeled. There is a mid-17th-century baroque chapel.

BIBLIOG. A. Majewski, Zamek w Pierskowej Skale (The Castle at Pieskowa Skała) (Teka Konserwatorska, II), Warsaw, 1953.

Stary Sącz. Documented from the year 1186, Stary Sącz became a city in 1357 and was particularly prosperous until the middle of the 17th century. The Gothic parish church of St. Elizabeth (Świętej Elżbiety; 14th–15th cent.) was reconstructed and preserves baroque and rococo furnishings. The Gothic church of the Holy Trinity (Świętej Trójcy; ca. 1332) has a single aisle. The main altar and two others, with stucco decoration, were the work of Baldassare Fontana (1696). The furnishings are baroque. The adjacent convent dates from 1601–05. Noteworthy are the Chapel of the Blessed Cunegonde, the chapter room, and the Loreto Chapel with its rich furnishings. The church houses relics of the Blessed Cunegonde (13th cent.) and jewels. The defensive walls have a tower. There are also middle-class residences of art-historical interest.

BIBLIOG. Katalog zabytków sztuki w Polsce, I: Województwo krakowskie (Catalogue of Monuments of Art in Poland, I: The Voivodate of Krakow), Warsaw, 1953, pp. 326–36.

Tarnów. Having existed as early as the beginning of the 12th century, Tarnów became a city in 1332. It was long the possession of powerful aristocratic families such as the Tarnowski (until 1567) and the Ostrogski. It enjoyed great prosperity in the late Gothic and Renaissance periods (15th–16th cent.). It became a see in 1785. The Cathedral was begun about 1400 and was enlarged and rebuilt in the 15th to 19th century. Within the church are Gothic stalls and extremely important Renaissance funerary monuments, including those of the Tarnowski family by Gian Maria Padovano and that of Janusz Ostrogski and his wife by Jan Pfister (1612–20). The church also preserves Gothic chalices and fine furnishings. The Bernardine monastery is baroque. The wooden churches of St. Mary (Panny Marii), built in the 16th–17th century, the Holy Trinity (Świętej Trójcy) are noteworthy. There are remains of Tarnów's old fortifications (14th and 16th cent.). The nucleus of the town hall is Gothic but the structure was enlarged in the second half of the 16th century. It has a tower and typical Polish attics and is a characteristic example of Polish Renaissance art. Now housing the state museum, it has collections of painting, sculpture, decorative arts, and so on. The Gothic-Renaissance house known as the Mikołajowska (1524) houses the Diocesan Museum (Muzeum Diecezjalne), with an important collection of Polish medieval painting (e.g., the *Chomranice Deposition*, ca. 1450), sculpture, and decorative art. There are Renaissance, baroque, and neoclassic houses in Tarnów.

BIBLIOG. Katalog zabytków sztuki w Polsce, I: Województwo krakowskie (Catalogue of Monuments of Art in Poland, I: The Voivodate of Krakow), Warsaw, 1953, pp. 436–46; J. E. Dutkiewicz, Tarnów, Warsaw, 1954.

Rzeszów. Biecz. The center of this small city has preserved the appearance it acquired in the 15th and 16th centuries, the period of its greatest prosperity. The Gothic parish church has a choir dating from about 1470, aisles from 1516–18, and a late Renaissance main altar. The church preserves many Gothic, Renaissance, and baroque sculptures and paintings. A Gothic tower of the 15th-century fortifications has been preserved. The Gothic market square dates from the 15th and 16th centuries. There are Gothic and Renaissance houses, one of which is occupied by the Regional Museum (Muzeum Regionalne).

Jarosłau (Jaroslav, Yaroslav). This city on the San River was founded in the 11th century and was an extremely important center of east–west trade in the late Middle Ages and in the 16th and 17th centuries. The parish church of St. John (Świętego Jana), one of the first Jesuit churches in Poland, was built about the end of the 16th century and was later reconstructed. The Church of St. Anne (Świętej Anny; 1614–24) is late Renaissance. The Dominican church (first half of 17th cent.) has a late baroque façade, and the Reformed church is late baroque. Among the middle-class houses of the 16th to 18th century is the Renaissance Orsetti House, which is now the Regional Museum (Muzeum Regionalne).

BIBLIOG. J. Sas-Zubrzycki, Miasto Jarosława i jego zabytki (The City of Jarosław and Its Monuments), Sprawozdania Komisji h. sztuki, VII, 1906, pp. 357–95; M. Dayczak, Kamienica mieszczańska w Jarosławiu (The Bourgeois House at Jarosław), B. h. sztuki, XVIII, 1956, pp. 24–54.

Krasiczyn. This castle, one of the most important examples of Polish Renaissance architecture, was built between 1592 and 1614 and has a courtyard, corner towers, and a characteristic crowning.

BIBLIOG. F. Kruc, Zamek w Krasiczynie (Krasiczyn Castle), Ochrona zabytków, II, 1949, pp. 61–62.

Łańcut. This small city was founded in the 14th century. In the first half of the 17th century, the impressive castle was built. Originally the seat of the Lubomirski family, it later passed to the Potocki family and was rebuilt at the end of the 19th century. A rectangular-plan structure, it has an inner courtyard and corner towers. There is a room decorated with baroque stuccoes by Giovanni Battista Falconi. The Chinese room dates from the second half of the 18th century, and there are late-18th-century rooms decorated with frescoes by Vincenzo Brenna. Now a museum, it has an important collection of furniture, along with European paintings, sculptures, and old vehicles. The park has pavilions.

BIBLIOG. J. Piotrowski, Zamek w Łańcucie (Łańcut Castle), Lwów, 1923; K. Kieszkowska-Kotzowa, Prace remontowokonserwatorskie w zamku Łańcuckim (Preservation Work at Łańcut Castle), Ochrona zabytków, XII, 1959, pp. 269–79.

Przemyśl (Peremyshl). This city on the San River was a commercial center as early as the 9th century. From the 10th to 14th century it was alternately a Polish and Ruthenian possession, but in 1344 it passed to Poland. It flourished in the 14th to 16th century, declined at the end of the 17th century, and gained new importance in the 19th century. There are remains of a *palatium* and a round chapel, pre-Romanesque structures of the early 11th century, which were recently discovered. The Cathedral, originally Gothic (15th cent.), has been rebuilt several times. It has Renaissance funerary monuments, and some of its furnishings are baroque. The castle, part of which is Gothic (14th cent.), was rebuilt in Renaissance style in the early 17th century and subsequently remodeled. Przemyśl has Renaissance, baroque, and neoclassic middle-class houses. There is a Regional Museum (Muzeum Regionalne) and a Diocesan Museum (Muzeum Diecezjalne) in the city.

BIBLIOG. K. Arlamowski, Château de Przemysl au XVIe siècle, B. h. sztuki, XI, 1949, pp. 43–58.

Rzeszów. The capital of the department of Rzeszów is situated on the Wisłoka River. It was probably a small town in the 12th century but achieved the status of a city in 1354. It was prosperous from the 15th century to the middle of the 17th century and declined until the 19th century. The choir and nave of the parish church of St. Stanislas (Świętego Stanisława) date from about 1434, while the side aisles and additions were built in the 18th century. The structure of the church combines Gothic and baroque elements, while the interior furnishings are baroque. The late Renaissance Bernardine church (1627–29) houses the baroque funerary monument of the Ligęza family (early 17th cent.), and the furnishings are also baroque. The Piarist church is baroque (mid-17th cent.). There are two baroque synagogues in the city (17th and 18th cent.). The castle, which has been rebuilt, was erected in the early 17th century. There is a Regional Museum (Muzeum Regionalne) in Rzeszów.

BIBLIOG. L. Rubach, Województwo rzeszowskie (The Voivodate of Rzeszów), Warsaw, 1954, pp. 40–54.

Stefan KOZAKIEWICZ, Stanisław LORENTZ

Illustrations: 4 figs. in text.

POLLAIUOLO, Antonio and Piero. Italian painters, sculptors, and goldsmiths. The proper family name of the brothers was Benci, their father being Jacopo di Antonio Benci, a poulterer. From his trade derived the name Pollaioulo, which was adopted by Antonio and Piero in accordance with a Florentine custom. The exact dates of their births are unknown. According to the inscription on his tomb in S. Pietro in Vincoli, in Rome, which states that Antonio was seventy-two years old when he died in 1498, he would have been born in 1425/26. But in a tax return of May 31, 1433, his father gave Antonio's age as a year and a half, and in another, dated 1457, listed Antonio's age as twenty-four. Despite the slight discrepancy between the two, more reliance should be placed on these documents, and therefore Antonio must have been born in 1431 or 1432. Similarly, his brother's age was given as fourteen in 1457 and thirty-three in 1480. One can only assume that he was born between 1443 and 1447, with the earlier date being the more likely.

Antonio Pollaiuolo was trained as a goldsmith and metalworker, possibly by Vittorio Ghiberti. Vasari (q.v.), following L'Anonimo Magliabechiano, makes him a pupil of Lorenzo Ghiberti (q.v.) and attributes to him a delicately wrought quail on one of the jambs of the east door (Porta del Paradiso) of the Baptistery in Florence; but Antonio's name does not occur in the lists of Lorenzo's assistants. His first recorded commission dates from 1457, when Miliano Dei and Betto Betti were asked to collaborate on a reliquary crucifix for the Baptistery. The final payment for this work was made in 1459. During the following years Antonio was occupied with a number of important pieces of goldwork, all of which have been lost. In 1460 he furnished the Church of S. Pancrazio in Florence with a reliquary tabernacle (lost) for the arm of its patron saint; in 1462 he made jewelry for Cino Rinuccini, and in 1465 he was paid for silver candlesticks for the Baptistery. By then he must have been well established in Florence, for in 1467 and 1468 he was a member of the commission for deciding about the *palla* to be placed on the lantern of the dome of the Cathedral — an honor that he shared with, among others, such highly esteemed masters as Luca della Robbia and Andrea del Verrocchio (qq.v.).

At about the same time, Antonio apparently extended the field of his activities. Beginning in 1466 he designed embroideries for the vestments of the clergy of the Baptistery (Florence, Mus. dell'Opera del Duomo), for which he is known to have been paid in 1469 and again in 1473; one design is recorded as late as 1480. In 1469 he decorated a suit of tournament armor for Benedetto Salutati, and in 1472 a helmet for the Count of Urbino. In the same year he made a silver crucifix for S. Maria del Carmine in Florence. All these works are lost, as is a silver basin commissioned by the Florentine Signoria in 1473. Between 1476 and 1483 he was engaged on a reliquary cross for the Monastery of S. Gaggio near Florence, fragments of which survive as part of a 16th-century crucifix. The first indication that Antonio was also a painter is given in Vasari's statement that in 1475 he painted, in conjunction with his brother Piero, *The Martyrdom of St. Sebastian* as an altarpiece for the Pucci Chapel in the SS. Annunziata. In 1477 Antonio received the commission for the *Birth of John the Baptist* (pl. 184), a relief for the silver altar frontal of the Baptistery. He was paid for the model in the same year, and the last payment for the completed relief was made in 1483. In 1484 he moved to Rome, where he worked on the bronze tomb of Sixtus IV until 1493. After the completion of this task, he worked on the tomb of Innocent VIII in St. Peter's. He made his last will in 1496 and died two years later.

Piero is reported by Vasari to have been a pupil of Andrea del Castagno (q.v.). Though not impossible, this association is doubtful. Castagno died in 1457, and thus, even if Piero was born in 1443, he would have been only fourteen at the time of his master's death. His earliest recorded work dates from 1466/67, when he painted wall decorations and the altarpiece with SS. Vincent, James, and Eustache for the Chapel of the Cardinal of Portugal in S. Miniato al'Monte (pl. 182). Possibly the two brothers collaborated on this task. In 1469 Piero was commissioned to paint the series of Virtues for the council chamber of the Mercatanzia (pl. 183). Of these he executed only six; Botticelli painted the *Fortitude* in 1470 (II, pl. 327). In 1477 Piero competed with Verrocchio for the Forteguerri tomb in Pistoia, but did not receive the commission. A year later he was again unsuccessful in competition for the altarpiece for the Chapel of St. Bernard in the Palazzo Vecchio, a work that was ordered from Leonardo. Piero's only signed and dated work is the *Coronation of the Virgin* (1483) in the Church of S. Agostino at San Gimignano. He accompanied his brother to Rome and assisted him on the papal tombs, but he seems to have died sometime before Antonio, probably in 1496.

It is worth stressing the fact that all the well documented works of Antonio are in precious metals or bronze, while all those of Piero are paintings. Sources prior to Vasari describe the output of the brothers in the same terms, calling Antonio a goldsmith (Filarete, Ugolino Verino, L'Anonimo Magliabechiano, Antonio Billi) and Piero a painter (Verino, L'Anonimo Magliabechiano, Billi, Francesco Albertini). Collaboration in some instances is admitted: the *Libro di Antonio Billi* lists a *St. Christopher* (lost) in S. Miniato fra le Torri as by Piero according to the design of Antonio, and there is general agreement about Piero's assistance with the Roman bronze tombs. On the other hand, the 1472 edition of the *Libro Rosso* of the Compagnia di S. Luca, the artists' guild, listed Antonio as a member without indicating his trade; but in the entry for the following year it listed him as a goldsmith and painter. When Antonio signed the tomb of Sixtus IV, on the *cartellino* placed behind the head of the Pope, he wrote: "OPUS ANTONI POLAIOLI FLORENTINI ARG.[ENTO] AURO PICT.[URA] AERE CLARI." In doing so, he referred to all his artistic activities. On his own tomb in S. Pietro in Vincoli he is called a famous painter, though at the same time the bronze monuments are mentioned.

After giving a detailed account of Antonio's career as a goldsmith, Vasari wrote that comparatively late in life the artist, realizing the hazards to which works of metal were exposed in times of war, decided to take up painting for the sake of his lasting fame. According to the same source, Antonio turned to his younger brother to learn the handling of colors and became a proficient painter within a few months. Although this account is usually discredited by modern critics, there is no strong reason to do so. A number of Florentine Quattrocento painters had received their initial training in goldsmiths' workshops; furthermore, Vasari states quite clearly that Antonio only sought information about the technique of painting. The fact that the brothers seem to have collaborated on a number of paintings lends credence to Vasari's claim, and there is no evidence that Piero was merely an assistant to his brother, as has sometimes been said. In fact, it is obvious from the tax return Piero made in 1480 that he had his own workshop where he did his painting.

The silver reliquary from the Baptistery (VI, pl. 265) is the earliest extant work of Antonio. It was meant to house a precious relic, a splinter from the True Cross. According to the terms of the commission, Antonio and Miliano Dei were charged to make the lower part, and Betto Betti the upper. No payments to Dei are recorded, however, and Antonio and Betti must have been responsible for the design and execution. The reliquary consists of three parts, held together by wooden dowels: the foot, with inlaid decorations; the *tempietto*, or small tabernacle for the relic, with small statues; and the Cross on a calvary. Because the whole was heavily restored in the 18th century, the original appearance is uncertain. It is not known which artist designed the reliquary, but many features, particularly in the little inlaid reliefs and the statues, point to Antonio's authorship. The conception of the structure is still Gothic, as is easily recognized not only in the splayed cuspidal foot and in such details as the quatrefoil decorations but also in the verticality and slender proportions. Some very obvious Quattrocento elements have been introduced as well: for example, the *tempietto* is a small-scale imitation of Brunelleschi's lantern on the dome of the Cathedral, the wooden model of which Antonio could have known. The praying angels, although now set on what are obviously 18th-century bases, seem also to come from Antonio's hand. They are clearly derived from the figures in the frame of Ghiberti's Porta del Paradiso, with which they share slender proportions, square but narrow shoulders, and the curve of their bodies. The relationship is so close that comparison seems to lend weight to Vasari's assertion that Antonio learned his trade from Lorenzo Ghiberti. It is difficult to do justice to the plaques around the foot, because they have lost their enamel and only the delicately engraved design remains. Nevertheless, the Baptism scene and the Moses reveal the hand of a master keenly interested in the appearance and the anatomy of the human body — although at this early stage Antonio also delighted in the display of rich patterns of folds and heavy draperies.

Proceeding chronologically with Antonio's authenticated works, one comes next to the embroideries for liturgical vestments. The designs themselves are lost, but 27 well-preserved embroideries, detached from the vestments, still remain (Florence, Mus. dell'Opera del Duomo). These aid in gaining a clearer idea of the sources of Antonio's style. In the embroidery depicting the feast of Herod, for instance, thin-limbed, elegantly moving figures in their slender proportions bring to mind the Burgundian International Gothic style, which enjoyed such popularity in Florence in that very period. The artist was also interested in perspective, however, and in the rational construction of pictorial space. This interest is quite pronounced, for example, in the embroidery showing Zacharias leaving the temple. Accompanying this interest is the introduction of classical details, with a more natural gesticulation and firmer modeling of the figures than were to be found on the plaques of the reliquary cross. There are affinities between the embroideries and Donatello's Paduan reliefs, *The Miracles of St. Anthony*: in the scene of St. John's sermon before Herod, Antonio used a background that repeats the three arches from Donatello's *Miracle of the Mule*, and a number of details seem borrowed from the *Miracle of the Miser's Heart*. The arrangement of the figures to suggest depth owes something to the relief with the story of the irascible son, in which St. Anthony heals the young man's foot (IV, PL. 251). Donatello's Paduan works must have been known in Florence; still, the fact that Pollaiuolo drawings are known to have been in Francesco Squarcione's workshop in 1474 may indicate that Antonio himself had visited Padua.

Vasari places the altarpiece for the Chapel of the Cardinal of Portugal (PL. 182) at the head of the list of works jointly executed by the brothers, but earlier sources do not mention Antonio's share. The value of this panel is to be found more in a certain charming, naïve awkwardness than in maturity or perfection. There is no real cohesiveness between the precariously balanced saints, and their gestures are rather wooden. The artist's handling of perspective is insecure, and there is no convincing sense of space. Most noticeable of all are the overly slender proportions and the comparatively small heads. The same characteristics are found in the panel *Tobias and the Angel* (PL. 180) — which can be identified with a work of Piero mentioned by Vasari — and again in the six Mercatanzia Virtues. These qualities are also true of the design for the Caritas figure (on the back of the actual panel), which has sometimes been mistakenly attributed to Antonio. All the paintings just mentioned lack the features that are typical of the embroidery designs and must therefore be wholly from the hand of Piero. His documented *Coronation of the Virgin* (PL. 183) fits easily into this group.

Antonio and Piero seem to have worked together in painting *The Martyrdom of St. Sebastian* (PL. 182), though it is impossible to define the exact degree of their collaboration. Certain features indicate that Antonio must have taken the leading part: the clear design, the firm modeling of the human figures, the interest in anatomy and movement, and above all the understanding of the problems of pictorial space. On the other hand, certain weaknesses in execution may be laid to Piero's share. The picture was highly praised by Vasari, and one can understand why he should have singled it out. Antonio has managed to combine in this painting most of the features characteristic of Florentine Quattrocento art and has followed faithfully the advice given by Alberti in his *Treatise on Painting*, written some forty years earlier. Realizing what it meant to be truthful to nature, at the same time Antonio understood the demands imposed by dignity and restraint of subject matter.

The same characteristics can be found in Antonio's one and only signed engraving, *Battle of the Nudes* (IV, PL. 424). Various attempts have been made to identify the subject, but none of these are convincing. However, one need not assume that the work illustrates a specific incident from myth or history. Vasari, who knew this engraving, described it in terms which allow an interpretation of it as simply an expression of Antonio's interest in anatomy and the play of the human body: "Antonio was more advanced in the study of nudes than the masters who preceded him. He dissected many corpses in order to find out about their anatomy, and he was the first man to demonstrate how one can find the muscles . . . and from these investigations he made a battle scene. . . ."

Antonio could treat the same artistic problems in the round. A bronze statuette that must certainly be ascribed to him is the *Hercules and Antaeus* (PL. 179) from the Medici collection, where its presence was recorded in 1495. The dynamic subject allowed the artist to display all his talents — that is to say, his knowledge of the human body in action, his interest in expression, and his gift of combining these details in a dramatic composition. The choice of subject may well have been suggested to Antonio by Alberti's short treatise on sculpture, and the statuette proved very popular and was widely imitated during the Renaissance. There is only one other bronze statuette that can be attributed to Antonio with any degree of certainty, a *Hercules* standing on a triangular base adorned with Harpies (Berlin, Staat. Mus.).

The date of the *Hercules and Antaeus* is not known, but the treatment of the figures strongly resembles that of *The Martyrdom of St. Sebastian*. It is characteristic of Antonio that he should have used a composition previously employed for a painting of the same subject, done for the Palazzo Medici in about 1465. The original version is lost, but its appearance — and also that of a companion piece, *Hercules and the Hydra* — is known through a small replica (PL. 181; X, PL. 249; since restored to the Uffizi). Such an economical mode of design must have been the result of Antonio's anatomical studies, which allowed him to assemble what approached a graphic vocabulary of the human figure. Similar repetitions of patterns that have once been mastered can be observed in *The Martyrdom of St. Sebastian* and in the figures of the Liberal Arts surrounding the tomb of Sixtus IV (PL. 184).

The tomb was begun immediately after the death of Sixtus in 1484. By 1490 the bronze effigy of the Pope was completed, and Antonio signed the finished tomb in 1493. The monument consists of two parts: the effigy, surrounded by reliefs depicting the Virtues (PL. 186), and the high, chamfered base, with the ten Liberal Arts in deep relief and framed by acanthus leaves. The figures of the Virtues must be Piero's part of the work, since in design they are in close accord with the Mercatanzia Virtues. The recumbent figure of the Pope, lying with hands folded, may well have been suggested by the bronze slab of Martin V in St. John Lateran; and it is worth noting that Pollaiuolo concentrated his powers on the monumentality of the naturalistic head, which is in much higher relief than any other part of the tomb. Nevertheless, the artistry and sophistication of his mature style can best be studied in the ten figures of the Liberal Arts, which exhibit a subtlety that is not confined to the individual figure and its complex posture but is applied to the composition as a whole. The graceful relationship between the figures establishes a rhythmic sequence of movement and mass all round the tomb.

The tomb of Innocent VIII was somewhat changed when it was reerected in its present position in 1621, but the original composition is known from two drawings made before the old basilica of St. Peter's was demolished. Antonio had placed the seated image of the Pope, with the reliefs of the Virtues surrounding it, beneath the sarcophagus with the recumbent effigy, and the whole tomb was placed under a richly decorated triumphal arch. For the Virtues, Piero's old designs were used once again, but the most striking and historically important feature of the tomb is the seated figure of the Pope with his right hand raised in blessing. The immediate reason for this innovation was no doubt a historical event which occurred shortly before the death of Innocent, for the Pope is depicted holding in his left hand the relic of the Holy Lance, which had been given to him in 1492 by the Ottoman Sultan.

Antonio's originality can also be discerned in two studies for an equestrian monument (Munich, Staat. Graphische Samml.; New York, R. Lehman Coll.), which are of uncertain date. Vasari claims that Antonio hoped to be commissioned by Lodovico Sforza to make the memorial to his father, Francesco. Antonio seems to have intended an equestrian statue very different from the *Marcus Aurelius* or from Donatello's *Gattamelata*. His prancing horse riding over a fallen warrior must have been inspired by reliefs on Roman battle sarcophagi. It is typical of his interests that he chose the technically difficult task of representing horse and rider in violent motion and that he was inspired by a motif requiring close attention to expression.

But Antonio has left no completed monumental bronze statues, and it may well be that temperamentally he was unsuited for such a task. First and foremost, he was a master of the delicate bronze statuette and subtle reliefs. His knowledge of anatomy and his innate sense of form make him an important forerunner of the great 16th-century masters. On the other hand, knowledge of how Antonio posed and reposed his models, drawing them from all sides, also suggests that he was well on the way toward the later practices of fine arts academies.

In conclusion, reference must be made to a number of works traditionally attributed to Antonio, but not always with sufficient reason. Two female portraits (Berlin, Staat. Mus.; XII, PL. 29) appear in several publications on the Pollaiuolo brothers but have no real stylistic affinities with the work of either. A fresco with dancing nudes in the Villa della Gallina, Florence, has a number of figures in energetic movement reminiscent of Antonio's figure types; but their heads and, particularly, the illusionistic architectural details below them make his authorship unlikely. On the other hand, the *Apollo and Daphne* (PL. 185) and *The Rape of Deianira* (also known as *Hercules and Nessus*; PL. 181) must be works of Antonio, though it is possible that Piero collaborated on the former painting.

The most important of the drawings attributed to Antonio are *Hercules and the Hydra* (IV, PL. 270), possibly done in preparation for the canvas once in the Palazzo Medici; *Prisoner Led before a Judge* (Br. Mus.), which furnishes another striking example of Antonio's interest in the nude figure in violent action; *Eve Spinning* (PL. 181) and *Adam*; and a design for a reliquary inscribed "Antonio del pollaiuolo horafo" (last three, Uffizi, Gabinetto dei Disegni e Stampe).

BIBLIOG. M. Cruttwell, Antonio Pollaiolo, London, 1907 (bibliog.); A. Sabatini, Antonio e Piero Pollaiolo, Florence, 1944 (prior sources and bibliog.); S. Ortolani, Il Pollaiuolo, Milan, 1948; Piero e Antonio Pollajolo: La Comunione mistica di S. Maria Egiziaca, B. Ist. centrale di restauro, 9–10, 1952, pp. 79–85; L. D. Ettlinger, Pollajuolo's Tomb of Sixtus IV, Warburg, XVI, 1953, pp. 239–71; E. Steingräber, Studien zur Florentiner Goldschmiedekunst, I: Ein unbekannte Arbeit des Antonio del Pollaiuolo für das Kloster S. Gaggio bei Florenz, Mitt. des kunsthist. Inst. in Florenz, VII, 1953–56, pp. 87–92; E. H. Gombrich, Antonio di Giovanni, Warburg, XVIII, 1955, pp. 29–32; E. Panofsky, The Life and Art of Albrecht Dürer, 4th ed., Princeton, 1955, passim; J. G. Phillips, Early Florentine Designers and Engravers, Cambridge, Mass., 1955; M. Levi d'Ancona, Un'opera giovanile del Pollaiuolo già ascritta ad Andrea del Castagno, RArte, XXXI, 1956, pp. 73–91; J. Pope-Hennessy, Italian Renaissance Sculpture, London, 1958; J. Pope-Hennessy, Italian Bronze Statuettes, I, BM, CV, 1963, pp. 14–23.

Leopold D. ETTLINGER

Illustrations: PLS. 179–186.

POLLOCK, JACKSON. American painter (b. Cody, Wyo., Jan. 28, 1912; d. East Hampton, L.I., Aug. 12, 1956). One of the most powerful of the so-called American "action painters," Jackson Pollock was raised in Arizona and California and then went to New York in 1929. There he studied with Thomas Hart Benton at the Art Students League and became interested in the Mexican muralists, especially José Orozco (q.v.). From 1938 to 1942 he worked on the Federal Art Project. In an interview in 1944, Pollock spoke of his work with Benton as important because it gave him "something against which to react very strongly," acknowledged an influence from American Indian art, named Ryder (q.v.) as "the only American master who interests me," and listed Picasso and Miró (qq.v.) as "the two artists I most admire."

Pollock's first New York one-man show was held in 1943. The next year his *She-Wolf* (1943) was purchased by the Museum of Modern Art. In 1950 he had one-man shows in Venice and Milan, and a documentary film about his work was made by Hans Namuth. From this time until his violent death in an automobile accident, he was probably the best-known and most influential painter of the abstract-expressionist group, both in America and Europe.

In the early 1940s Pollock created powerful totemistic images, sometimes with classical themes, as in *Pasiphaë* (1943). By 1945 a more completely rhythmic movement, freed from descriptive imagery, filled the entire canvas, as in *Shimmering Substance* (1946; Ridgefield, Conn., Coll. Mrs. Emily Walker). By 1947 he had entered a new phase, in paintings such as *Full Fathom Five* (New York, Mus. of Mod. Art). These enormous canvases were nailed to the floor, and over them Pollock dripped and flung different kinds of paint (oil, aluminum, Duco), molding his works through the action of his own movements, in a method related to the dance and to the work of the potter (I, PL. 125). These paintings suggest deep involvement in the very act of creation and seem to signify the infinite cosmic proliferation of life itself. In 1951 the human image began to emerge within the galactic movement of his materials, confined to austere black on unsized canvas, as in *Black and White No. 5* (1952; New York, Coll. Mrs. Lee Krasner Pollock). Pollock then returned to the dripped, swirling paintings (X, PL. 342) and also revealed a somewhat unexpected reminiscence of Matisse, as seen in *Easter and the Totem* (1953; New York, Coll. Mrs. Lee Krasner Pollock). In his final canvases, such as *Scent* (1955; New York, Coll. Mrs. Leo Castelli), he returned also to the use of brush and palette knife, with much overpainting, glazing, and textural variation.

BIBLIOG. J. Pollock, Jackson Pollock, Arts and Arch., February, 1944; P. Tyler, Jackson Pollock: The Infinite Labyrinth, Mag. of Art, XLIII, 1950, pp. 92–93; R. Goodnough, Pollock Paints a Picture, AN, L, 1951, pp. 38–41, 60–61; B. H. Friedman, Profile: Jackson Pollock, Art in America, XLIII, 1955, pp. 49, 58–59; P. Tyler, Hopper/Pollock, AN, XXVI, 1957, pp. 87–107; F. O'Hara, Jackson Pollock, New York, 1959; B. Robertson, Jackson Pollock, New York, 1960.

Allen S. WELLER

POLYGNOTOS (Πολύγνωτος). Perhaps the most famous Greek painter, Polygnotos is, nonetheless, one of the least susceptible of biographical reconstruction, for information about his life and artistic personality is scant. The son and pupil of Aglaophon the Elder, he was born in Thasos; later he became an Athenian citizen, and in recognition of his work he was decreed a public guest of Delphi. Polygnotos was a bronze caster as well as a painter, but nothing of a more precise nature is known because, though he is widely mentioned in ancient sources (more often perhaps than any other artist), it is in terms that are very general and conventional, almost always without any real critical judgment.

More recent scholarship places earliest among the known works of Polygnotos the painting of *Odysseus after the Slaying of the Suitors* executed in Plataia, datable after 479 B.C.; next are the Athenian paintings, assigned to the time of Kimon (470–460) on the basis of the friendship between Polygnotos and Elpinike, the sister of Kimon, which include the painter's latest known works — those which Polygnotos painted for the Lesche of the Knidians in Delphi, datable about 468, since they were executed during the period when Phocis was in control of the sanctuary, and before the death of Simonides of Keos, whose epigram was inscribed on them (Pausanias, X, 27, 1). The works that were in the pinakotheke of the Propylaia in Athens do not establish a fixed point in the chronology, since they could have been transferred to the gallery (built between 437 and 432) after the death of the artist. Polygnotos was certainly living in the year 444, when he was included in a list of the Theoriai of Thasos (*IG*, XII, 277, 44–45). Pliny's statement that Polygnotos was already a celebrated painter sometime after the period when Panainos flourished (ca. 450 B.C.) — when, also, painting competitions were held at Corinth, Delphi, and the Pythian games — and before the 90th Olympiad (*Naturalis historia*, XXXV, 58) would appear to conflict with another passage (ibid., 60), in which he noted that Aglaophon also lived at the time of this same Olympiad, unless he is referring to the younger Aglaophon, Polygnotos's nephew, rather than to Aglaophon the Elder, his father. Polygnotos might, then, have been born quite early in the 5th century B.C. and have witnessed in his youth the epic Greek resistance to the Persian invasions.

Notwithstanding the detailed account of Pausanias (X, 25–31), little can be said of the quality of the imposing paintings in Delphi, with the *Ilioupersis* (Troy Sacked) on the one side and the *Nekyia* (the Lower World) on the other, and even less about the paintings in the Stoa Poikile in Athens (Pausanias, I, 15, 2) — among them an *Ilioupersis* — which, as indicated by Synesius (*Epistolae*, 136; *PG*, LXVI, 135), were taken to Constantinople about A.D. 400. Nothing whatsoever is known about the *Rape of the Leukippidai* painted for the Anakeion (Temple of the Dioskouroi) or about a painting of an unknown subject executed for a thesauros. Only the titles of the paintings in the pinakotheke in Athens have come down to us: *Odysseus Stealing the Bow of Philoktetes, Diomedes with the Palladion, Pylades and Orestes Killing Aegisthos, The Sacrifice of Polyxena, Achilles at Skyros, Odysseus and Nausicaa.* In Thespiai, paintings by Polygnotos in an encaustic technique were restored by Pausias (Pliny, op. cit., XXXV, 123). A painted panel with Kapaneos as its subject was transported to Rome and placed in the Portico of Pompey (ibid., 59). A painting by Salmoneos and many drawings on papyrus and parchment, used as models by other artists, are also mentioned in the sources.

Polygnotos lived in that most glorious period of Greek — in particular, Athenian — history which extended from the victories over the Persians through the time of Perikles and which saw Phidias active in Athens. Aware of the grandeur of the epoch, in his themes he celebrated its glory in ideal terms, though he himself was now too old to assimilate the Phidian contributions in his own creations.

In attempting, at least theoretically, to reconstruct the art of Polygnotos, one must bear in mind two elements that tradition indicates were distinguishing features of his art: the representation of the transparently draped female figure, and his ability to convey the emotional state of the personages portrayed. Of these characteristic elements, it should be noted that the first accorded perfectly with the study of the human form — the central concern of all classical art — and constituted an indispensable factor for it, since the drapery-anatomy relationship was of primary importance to exploration of the problems presented by the representation of the human body. That this problem should now capture the attention of artists was the logical result of the change in viewpoint of the artist himself regarding the nature of the principal axes of the human body. During the archaic period in Greek art the artist viewed the body in terms of certain fixed axes, whereas the artist of the period under discussion moved around the body and observed the variety of its aspects — besides the basic frontal, side, and back views — and in turn took note of the various planes on which the body is organized. The draping of the figures served to punctuate these planes, to afford points of reference between them and the surrounding space; hence arose the necessity of making the relationship visible, that is, of revealing the anatomy underneath the garments. This was probably achieved by linear rather than painterly means, though there are contrary opinions in this matter (Borrelli Vlad, 1950). Since the artists of the classical age, especially Polygnotos, were concerned with the problems of perspective, and since they conceived of spatial representations as essentially a perspective problem, it followed that their formal solutions could only be achieved by draftsmanship, that is, obtained through line rather than through color, by applying principles of foreshortening and not by shading. The transparency of the drapery was therefore probably suggested by the outlines of the body under it rather than by means of flesh tones (Ferri, 1946, p. 146, note).

As for the other distinguishing feature of Polygnotos, his portrayal of the ethos of his personages (in recognition of which he was termed "ethographos" by Aristotle), it is apparent that the painter used the techniques of his art not merely for narrative purposes but to suggest to the spectator the psychological state of his figures as well. For him the study of man transcended the depiction of his body, to present an analysis of his soul: man was the measure not only of the physical space around him but also of spiritual elements of the world beyond the senses. From this point it was an easy step to reducing the gods to human scale. Polygnotos abandoned the purely narrative style, fettered to the existential aspect, in order to create in his paintings a synthesis, from the human point of view, of the great historicomythological episodes. In this lay his originality, that which made him so great in the eyes of his contemporaries and so different from the other artists of his time.

As to the definition of his style and the description of individual works, there is little basis for discussion in view of the fact that his entire output has perished and cannot even be reconstructed reliably through vase painting. The endeavor of Löwy (1929) in this direction, it must be said, goes too far; surely the many vases he catalogues cannot all repeat motifs and decorative schemes created by Polygnotos but must be compositions that reflect a cultural phase common to many artists. The vase paintings do, however, suggest certain tentative remarks. Polygnotos placed the figures in his paintings on a smooth monochrome background, arranging them in echelons on several levels, thus giving the impression of depth without really adopting a perspective system. Wavy lines indicated, or rather suggested, landscape but did not constitute a determinant motif of the composition. The figures themselves were linked not by any formal relationship to one another but solely by their psychological response to the action — or, more properly, to their common situation. To this end, Polygnotos used means which Pliny (op. cit., XXXV, 58) credits him with developing: "instituit os adaperire, dentes ostendere, voltum ab antiquo rigore variare" ("he introduced showing the mouth wide open and displaying the teeth and giving expression to the countenance in place of the archaic rigidity"). These innovations had, in substance, appeared before Polygnotos, even in vase painting; what Pliny means is that he was the first to make use of them to convey an emotion or a state of mind.

It has been suggested that in the representation of transparent draperies Polygnotos might have used some kind of shading, painting lighter colors over darker colors in such a way as to permit the color beneath to show through. It has already been noted, however, that the effect of transparency was achieved by linear rather than painterly means; this is clear from the comments of Lucian and Aelian, who speak of thin, rather than transparent, fabrics such as would reveal the form and not the color of the flesh. Moreover, such virtuoso use of color would have been somewhat alien to the severity of the art of this period in general, as well as to the work of Polygnotos in particular.

To attempt to visualize the imagery of Polygnotos, one must turn to sculpture, in the work of the Olympia master (III, PLS. 352, 368), and to ceramics, in the work of those vase painters who sometimes even signed their paintings with the name of Polygnotos, perhaps because they repeated his models or, in any case, because they looked upon him as a master (PLS. 187, 188; II, PL. 47). Finally, for a comparative appraisal, one may turn to the paintings of certain Etruscan tombs, such as the Tomb of the Triclinium or the Tomb of the Funeral Couch in Tarquinia.

SOURCES. For citations of Pliny, Naturalis historia, Loeb Classical Library ed. (Eng. trans., H. Rackham), 10 vols., London, Cambridge (Mass.), 1952; Overbeck, SQ, 1012–79; A. Reinach, Recueil Milliet, Paris, 1921, nos. 100–34.

BIBLIOG. C. Robert, Die Nekyia des Polygnot, Halle, 1892; C. Robert, Die Marathonschlacht in der Poikile und weiteres über Polygnot, Halle,

1895; E. Löwy, Polygnot, 2 vols., Vienna, 1929; A. De Capitani d'Arzago, La "grande pittura" greca, Milan, 1945, pp. 34–40; S. Ferri, Plinio il Vecchio, Rome, 1946, p. 146; L. Borrelli Vlad, Qualche scheda sulla tecnica della pittura greca, B. Ist. centrale del restauro, 2, 1950, pp. 55–57; M. Cagiano de Azevedo, Il distacco della tomba del Triclinio, B. Ist. centrale del restauro, 3–4, 1950, pp. 85–93; C. Weickert, Studien zur Kunstgeschichte des V. Jahrhunderts v. Chr., I: Polygnot, Berlin, 1950; G. Gullini, La coppa di Taranto del Maestro di Pentesilea: Una nota sul problema cromatico nella pittura di Polignoto, AC, III, 1951, pp. 1–12; G. Lippold, Antike Gemäldekopien, AbhAkMünchen, XXXIII, 1951, pp. 10–21; G. Lippold, RE, s.v.; Rumpf, MZ, pp. 91–103.

Michelangelo Cagiano de Azevedo

Illustrations: pls. 187–188.

POLYKLEITOS (Πολύκλειτος). The exact life dates for Polykleitos (I) the Elder, sculptor of Argos, are unknown. According to Plato (*Protagoras*, 328c), the sons of Polykleitos were contemporaries of the sons of Perikles; and from Plato's statement scholars have tended to fix the period of Polykleitos's activity as concurrent with the career of Perikles, that is, in the third quarter of the 5th century B.C. This view is supported by other literary references to Polykleitos's works. It is recounted, for instance, that Polykleitos, Phidias (q.v.), Kresilas, and Phradmon each presented a sculpture of an Amazon in a competition, and tradition also holds the lives of these four artists to be roughly contemporaneous, within a period that coincides with Plato's statement. Another work by Polykleitos, the portrait of Artemon, Perikles's engineer, dates from the same period. Pliny, on the other hand, has dated the peak of Polykleitos's activity as late as the 90th Olympiad (420–417 B.C.; *Naturalis historia*, XXXIV, 49). The cult statue of Hera in Argos, made for the new temple after 423 B.C., when the old temple burned, has been cited as proof of this later date. Still, Pliny's statement alone should not discredit the above-noted dates for Polykleitos's life, since in the same passage Pliny's dates for Myron and Pythagoras are also too late.

As for the beginning of Polykleitos's activity, his earliest works are statues of winners of athletic contests. The victors and the dates of their victories are recorded on papyri; but, since statues of winners were often made years later, the date of the victory affords only a *terminus ante quem*. For the same reason, the dates of the victories of Kyniskos (in the Olympic year 464 or 460 B.C., or 448 B.C. or after), Pythokles (452 B.C.), and Aristion (Ariston; 452 B.C.) do not prove that Polykleitos was active before the middle of the century. The bases and inscriptions of these statues testify to the uncertainty of these dates (Amandry, 1957). The character of the inscription on the Kyniskos base seems to indicate the second quarter of the 5th century, but the traces of the position of the statue suggest a motif that could not have been done earlier than the third quarter of the 5th century. The Pythokles inscription points to the end of the 5th century, and that of Aristion to an even later date, in the early 4th century. For the dates of the victories to have corresponded to the dates of execution of the statues, it must be assumed that the original inscriptions were defaced and then renewed soon afterward.

The end of Polykleitos's activity is even more difficult to establish, because there apparently existed another sculptor of the same name, only slightly younger, who in ancient tradition is not always clearly distinguished from the more famous Polykleitos the Elder. It is this younger sculptor, Polykleitos II, who is believed responsible for the *Zeus Philios* in Megalopolis (Pausanias, VIII, 31, 4), which, for historical reasons, could not have been made before 367 B.C. The relationship in time of these two sculptors can, to a certain point, be established from the following three facts: (1) Pausanias (VI, 6, 2), in mentioning the statue of the wrestler Agenor, distinguishes its creator, the Argive Polykleitos who was the pupil of Naukydes of Argos, from the more celebrated Polykleitos who made the statue of Hera. (2) In another passage Pausanias (II, 27, 7; i.e., if one follows the principal tradition of the manuscripts) designates Naukydes as a brother of a Polykleitos. (3) Naukydes and Daidalos, in the inscriptions on the bases of their statues, cite Patrokles as their father, and Pausanias (VI, 3, 4) also states that Daidalos was the son of Patrokles. There are, then, two possibilities: (A) If Naukydes was a brother of Polykleitos I, then Patrokles was also the father of Daidalos and Polykleitos I. (B) If, however, Naukydes was a brother of Daidalos and Polykleitos II, then Polykleitos I would have to be placed at least a generation earlier and Naukydes's pupil was either Polykleitos II (a younger brother) or some Polykleitos III. Regarding the first possibility, it could be allowed that Patrokles was the father of Polykleitos I only if two Patrokleses were known to have existed, for Pausanias writes that Patrokles worked on the votive offering for the victory in the battle of Aigospotamoi (405 B.C.). This Patrokles cannot possibly be identified with the father of Polykleitos I, and the existence of a second Patrokles is not attested by any written sources. Moreover, such a supposition becomes unnecessary if the second, more probable, alternative is accepted. If Patrokles was the father of Polykleitos II, Naukydes, and Daidalos, then Patrokles could still have been working after 405 B.C. The chronological data that can be deduced from the works of Naukydes and Daidalos also allow for the possibility of a father still active at the end of the century. The relationship between Patrokles and Polykleitos I cannot be determined more precisely than this. Furtwängler has suggested that Patrokles be designated the younger brother of Polykleitos I, but the hypothesis that Polykleitos I was the father of Patrokles remains more probable. The suggested conclusion is that Polykleitos I (active ca. 450–420 B.C.) was the father of Patrokles (active ca. 420–390 B.C.) and that Patrokles was the father of Naukydes, Daidalos, and Polykleitos II (active ca. 390–360 B.C.). This succession of generations coincides with the historical facts connected with the works of these masters. If it is correct, Polykleitos II could hardly have been active before 410 B.C., an important factor to be considered in the attribution of the various works.

The more celebrated elder Polykleitos is designated as Argive as well as Sikyonian. The theory that Polykleitos was from Sikyon is supported both by a direct reference of Pliny (op. cit., XXXIV, 55) and by the direct evidence that Polykleitos's school is known to have emigrated from Argos to Sikyon. The remaining related sources speak of Polykleitos as Argive, which is probably his correct identification. Plato designated Polykleitos as Argive (*Protagoras*, 311c). Similarly, it can be proved, in an indirect way, that Pausanias (VI, 6, 2) treated both Polykleitos I and II as Argives. In addition, there is the inscription (renewed) on the base of the statue of Pythokles, where "Argeios" (Argive) is expressly designated. The emigration of Polykleitos's school from Argos to Sikyon must have taken place by Daidalos's generation, since the Sikyonian origin of Daidalos, brother of Naukydes and of the younger Polykleitos is attested in an inscription as well as by Pausanias (VI, 2, 8).

The statement of Pliny (ibid.) that Hageladas (Ageladas, Hagelaidas) was the teacher of Polykleitos is untenable. Their names were linked by writers of the Roman imperial period, just as those of Myron and Phidias were, but actually, Hageladas and Polykleitos must have been more than a generation apart.

No work by Polykleitos is preserved in the original, but it has been possible to identify definitely at least two works by this master among the copies from imperial times. (A few words should be said here about the ancient concept of a copy, which was not a cast in the present-day sense. It was a more or less exact translation, and often also a transformation, from bronze into marble; both deliberate and unconscious deviations from the original were frequent. The testimony of the copies is often only piecemeal, since reconstruction of the original must be based on a series of clues, the reliability of which is not certain.)

The nude male statue was Polykleitos's principal theme. The *Doryphoros* (spear bearer), attributed by both Pliny and Cicero to Polykleitos, is his best-known work. Even though the identifying accessory of the spear has not been retained in any of the two-dimensional reproductions, the copies of this work show that it existed. The most complete copies of this athlete type are in Naples (III, PL. 359), in Florence (Uffizi, inv. 114), and in Rome (Vat. Mus., no. 126). From these, the original appearance of the statue can be fairly well reconstructed; some doubt remains only as to the length of the spear. Other representations of the work are carved on a stele found at Argos (Athens, Nat. Mus.) and on a gem in Berlin (Staat. Mus.). The theme is of a heroic youth, carrying a spear in a relaxed fashion — conveying the impression that he is both modest and yet assured. The weight of the body rests mainly on the right leg; the left leg is drawn back, with the toes scarcely touching the ground. The new freedom of pose encompasses the whole body, with the limbs adapted to a *contrapposto* arrangement as well. The head no longer looks straight ahead as in archaic figures, but is inclined toward the right shoulder. The right leg, which carries the body weight, is balanced by the left arm, which bears the spear; the relaxed left leg is balanced by the right arm, which swings freely. The four nearly complete copies reproduce only the general motif of the statue, and even in that they do not agree entirely. The head of the copy in Naples, for example, is more erect than the heads of the two full copies in Florence, which have a more pronounced rightward tilt.

In order to investigate the new idiom of forms of this masterpiece, one must look for the best elements in each of the isolated torso and head copies. How much the full copies differ from the original can be shown, for example, by comparing them with the fine green basalt torso in Florence (Uffizi, no. 208). The dark stone of this copy allows for more effective visualization of the original bronze, though the delicate, rich modeling possible in bronze is difficult to achieve in hard stone. This torso is less bulky and heavy in its proportions than the full copies in marble; articulation between body parts is less hard, with the transitions between them thus being smoother. A torso in Berlin (Staat. Mus., K151; formerly Pourtalès Coll.) is also one of the better copies. The transitions between body details are accomplished in a more flowing, pictorial manner, as would be understandable in the softer material of marble. One particular distinguishing this torso from all the others is that the veins are rendered with marked emphasis, not only in the arms but also on the abdomen. It is difficult to tell which of the two torsos is closer to the original. The classic structure of a Polycletan body, as shown in such copies, is characterized by the predominance of the chest articulation. The abdominal muscles are not delineated horizontally but are accommodated to the oblique contour of the diaphragm. At the height of the navel there does not appear the horizontal groove characteristic of most Attic works. The clavicles are almost horizontal, and the muscles of the neck relating to the inclination of the head are strongly marked. On the rear side of the figure, the most striking feature is the strong contraction of the muscles of the lower back. The juncture with the buttocks and their lateral hollows are shatply modeled.

Among the numerous copies of the head, only a few convey more than the general outline of the original; the bronze herm from Herculaneum (PL. 189) and the green basalt head in Leningrad (The Hermitage) are outstanding among these, particularly since the choice of materials enables a closer approximation of the original effect. Moreover, the agreement of the two copies in detail is remarkable. There are significant differences only in the treatment of the back of the head, where the bronze copy is more meticulous and richer in detail. In both copies the concise basic structure of the head is rendered in all its metallic sharpness; in the basalt copy there is less surface modeling, and the transitions are softer. The contour line of the cheeks, the delineation of the eyelids and brows, and the curves of the mouth are strongly and clearly related. The thick hair hugs the head like a cap, arranged in somewhat schematic layers and protruding in small tufts. There is great clarity of design, yet the components are surprisingly varied.

In Munich a valuable contribution has been made to reconstructing this masterpiece by projecting a combination of the Pourtalès torso and the bronze head from Herculaneum. Even more convincing should be a proposed reconstruction (to be realized in Basel) incorporating both of the green basalt fragments, derived from the same copy.

That *Doryphoros* is but a conventional identification is confirmed by Pliny (op. cit., XXXIV, 55). The statue could not actually have represented an ancient athlete. The figure bears a spear, not the short javelin of the palaestra. The size, larger than life, also seems incompatible with the customary scale of a victor statue. Since, according to Pliny (op. cit., XXXIV, 10), statues of naked spearmen were called "effigies Achilleae" by the Romans, the prototype for the *Doryphoros* can be interpreted as Achilles. It might even be said that only from this presumed identity does the motif gain full significance.

The attribution of this "Achilleus Doryphoros" to Polykleitos is based both on the motif and on the presumed identification with the Achillean type — arguments that, by themselves, are not conclusive. The many copies prove how famous the work was; nevertheless, a *Doryphoros* is also, for example, attributed by tradition to Kresilas. However, another statuary type that, on the evidence of similarities in style, must have been made by the same master as the *Doryphoros* can be proved to coincide with a work of Polykleitos, the theme of which is also transmitted by tradition: namely, the *Diadoumenos*. This stylistic correspondence cannot be due merely to chance; thus, the identification of each work as such would serve as mutual verification.

As with the *Doryphoros*, the general aspect of the *Diadoumenos* can be reconstructed quite fully from the available copies. A late Hellenistic copy from Delos (PL. 190) is rather freely executed, aside from minor details; only the hands are lacking. In an excellent copy in the Prado the left hand is almost intact. A replica in the British Museum supplies the right wrist. The head, arms, and the lower legs are to be found in yet another, accurately executed replica in the Metropolitan Museum (no. 38).

The young man represented, putting on the victor's fillet, is about to knot it at the back of his head. Except for the band segments spanning the distance from the back of the head to the hands, this basic motif is preserved in all copies of the statue. The affinities with the *Doryphoros* in style and type are evident. Even the disposition of the arms corresponds in some degree to the *contrapposto* of the *Doryphoros*. Subtle differences in the over-all aspect of this most likely later creation are apparent, but only upon very close observation: the head is slightly more turned and is more inclined forward; the heel of the relaxed leg is more strongly turned toward the supporting leg. Because of these differences, the weight has a more unsteady distribution and the effect is more rhythmic and active.

The quality of the individual copies varies. The complete replicas so far considered are all rather mediocre; as with the *Doryphoros*, the best copies are found among the detached torsos and heads. By far the best is a torso in Basel (Antikenmus., A3), a version as yet unpublished. Compared with this, the other replicas seem bulky and heavy. In a Polycletan work it is not enough merely to indicate dividing lines on the surface of the body; much more depends on the subtlety of modeling, on the degree of the depths and curves. The Basel torso is the only *Diadoumenos* copy that renders the veins on the abdomen. In its rich modeling and its taut, and at the same time pliant, expression of skin texture, it is very like the *Doryphoros* torso from the Pourtalès Collection. A small *Diadoumenos* torso in Berlin (Staat. Mus.) is also outstanding. Among the replicas of the head, there are only a few trustworthy copies as well. Of notable quality is a fragmentary head in Rome (Mus. Barracco, no. 107). Much of the face is destroyed, but the exceptional detailing of the hair shows what a great master was at work here and suggests how much the heads of the full copies lag behind their model. Even the beautiful heads of Dresden (Albertinum, Skulpturensamml., no. 71) and Kassel (Staat. Kunstsamml., no. 6) are not so faithful, though notable feeling and comprehension is evident in these copies.

Like the *Doryphoros*, the youthful *Diadoumenos* cannot represent a mortal athlete. Even if its dimensions are smaller than those of the *Doryphoros* (slightly over 6 ft. and ca. 6½ ft., respectively), it is more than life-size. The copyist of the version from Delos added a quiver to the youth's support — an attribute often used to characterize Apollo. Although the statue type seems unusual for an image of this god, it is expressly described for Apollo in an ancient source (Pausanias, I, 8, 4). Hafner (1961) had proposed interpreting it as Theseus, and he explained the quiver of the Delian copy as an indication of place. In itself a plausible explanation, it does not, however, suffice for another presumed "copy" in Portugal, in which the quiver and bow also appear on the support, for in this version they cannot be connected with the "founder" of the Delian games. One must therefore hold with the interpretation of the statue as an image of the god Apollo represented as the embodiment of victory and the prototype of youth, an identification that makes his self-assured, almost theatrical attitude more understandable. With these subtlest of nuances the master conjured up a completely new vision of the gods. Among the copies are two other statue types, with a similar general attitude, that can be accepted as independent creations of Polykleitos. One of these, which is better accounted for by tradition, can be interpreted as Herakles; the other, as Hermes. Statues of these two deities were explicitly attributed to Polykleitos — though without further description of their appearance — by Pliny (op. cit., XXXIV, 56) and Cicero (*De oratore*, II, 16, 70).

The Herakles type is preserved most completely in a Roman statuette (PL. 192), of which only the lower legs, the feet, and part of the right arm are missing. Two significant traits, constituting departures from the *Doryphoros*, allow recognition of this as an independent type. The body has greater bulk and is more compact. The left arm is drawn back, with the back of the hand resting on the left buttock; because of this, the upper arm is demarcated from the shoulder more sharply than in the *Doryphoros*. The right arm, most of which is lost, was originally brought forward and probably bent upward and back, as far as can be judged from the marks on the right breast. The *contrapposto* is on the whole more pronounced than in the *Doryphoros*, which would thus infer a later date for the *Herakles*. The rhythm of the contours is almost at the stage of the *Diadoumenos*.

Head and body combined can be traced with certainty only through this replica and from another statuette, of inferior quality, in the Vatican Museums (no. 269A). The head for another especially notable statuette, a torso in the Capitoline Museum (Palazzo dei Conservatori, Mus. Nuovo, inv. 1874), is perhaps to be identified, as suggested by Berger, in an example preserved in Copenhagen (Thorvaldsens Mus., no. 35). With the help of these statuettes, it has been possible to trace the principal full-size elements belonging to copies of the *Herakles*, which convey a more exact idea of the original. The most important of these parts are a torso in Copenhagen (Ny Carlsberg Glyptotek, no. 352) and another in Rome (Mus. Naz. Romano, no. 106184), which because of its provenance has been joined to a head that perhaps belongs to it, though the breaks do not correspond exactly. Among the isolated heads, a bronze head in Naples (Mus. Naz.) and a head on the Capitoline (Palazzo dei Conservatori, Mus. Nuovo) deserve particular notice also. The copies in the original scale differ from the statuettes in one essential: in the full-scale versions the right arm was not brought forward but was directed downward, alongside the body (as proved by the *puntelli* on the right side). Against the prevailing opinion, the author finds it obligatory to accept the convincing evidence presented by the statuettes. Furthermore, a forward-stretched right arm would correspond more satisfactorily to the retracted left arm; indeed, it would seem required in order to achieve a *chiasmos*, or counterpoise.

What attribute was held in the right hand? Its identification would be decisive for interpretation of the subject. Unfortunately, it is not preserved in any replica, and it is difficult to establish without some more precise indication of the original attitude of the right arm. The problem is also complicated by a passage in Cicero (*De oratore*, II, 16, 70) that seems to refer to two different statues of Herakles by Polykleitos: a Herakles with the Hydra and a Herakles with the lion pelt. A statement by Pliny (op. cit., XXXIV, 56) can also be read in different ways. But in view of additional reproductions of this statuary type in relief art, its interpretation as Herakles remains assured. A certain air of self-assurance and defiance would not be compatible with a representation of a mortal athlete, but is in perfect accord with the heroic behavior of the legendary Herakles. The body proportions also exceed the ideal of Polycletan youths: the figure is sturdier and bulkier — precisely Herculean. The fact that some detached replicas of the head wear a fillet, an addition of the copyists, is further support for the identification of this type as Herakles.

The second statuary type related to the *Doryphoros*, but differing enough to merit an independent identity, is more difficult to reconstruct. Only the head is recorded in exact copies, which are true to size and unquestionably identified. The best examples of this head type are one in Boston (Mus. of Fine Arts) and two in Rome (Palazzo Valentini; private coll.). A fourth head, a replica in Leningrad (The Hermitage, no. 97), is of value chiefly because traces of wings can still be discerned at the sides: by this means the copyist sought to characterize the statue as Hermes.

A statue of Hermes in the Boboli Gardens in Florence, though not a very felicitous example, does furnish some information as to what the original body type was like. The head, which is most likely the original one belonging to this statue, is a replica of the Boston head, but is provided with additional wings. Other additions by the copyists, such as the cloak or chlamys and the figure of a small boy (Dionysos as a child?), as well as its frontality, make the sure judgment of this statue very difficult. Nonetheless, it constitutes an important link to an excellent small bronze from Annecy (Paris, Mus. du Petit-Palais), with which the position of the arms, the general attitude, and the head type seem (in so far as the marked reduction in size permits an opinion) largely to correspond. Hermes Logios is represented here in his characteristic pose, with the right hand raised as if calling for attention and with the *kerykeion* (caduceus) held in his lowered left hand.

On the basis of the position of this right arm, with the forearm upraised and bent back close to the right breast, a torso from Gortyna in Crete (Heraklion, Archaeol. Mus.) has been connected with this statuary type. Final judgment is as yet impossible because of inadequate publication of this torso. A torso in Rome (Mus. Naz. Romano, inv. 58638) shows the *puntello* as well as a similar position for the left arm, but the left shoulder appears more strongly drawn in and the body is heavier and closer to the Herakles ideal. A torso from the former Somzée Collection (Morlanwelz, Belgium, Mus. de Mariemont), connected by Furtwängler with the Hermes type, is most closely related to the bronze statuette in gesture; but the un-Polycletan body form lacks any strong character. However, an excellent torso in the Palazzo Mattei in Rome (Matz-Duhn, no. 1005; *EA*, no. 2066), by some wrongly designated as a replica of the *Doryphoros*, and a statue in Karlsruhe (Landesmus.; *EA*, nos. 3375–79) could well belong to this type. The contour of the shoulders and of the upper part of the breast differs from the *Doryphoros*; because of its richer modeling and stronger incurvature, it is reminiscent of the bronze statuette in Annecy.

Another statuary type, the so-called "Diskophoros," can be identified on the basis of numerous copies clearly derived from the same original. This type can be recognized as a work of the same master, of an earlier date than the *Doryphoros*. The main characteristic distinguishing it from the other works of this type mentioned so far is the probable position of the legs, which are not completely preserved in any of the replicas. But the position of the feet is known with certainty through a statue in the Vatican Museums (no. 392), of which the right, supporting leg is almost entirely an ancient original, as well as the left foot and part of the left thigh. It follows that this youth was balanced firmly on both feet; the relaxed leg, unlike those in the other works by Polykleitos, was not drawn backward, and the heel touched the ground. The connection between head and body is not directly confirmed by any complete replica; nonetheless, twice in the 19th century, working independently, restorers joined the same head type with the same body type (Vat. Mus., no. 251; Wellesley, Mass., Jewett Arts Center), and the reassembling of both types appears quite certain. The head of the original was, according to these, turned toward the side of the supporting leg and was probably bent forward and to the side somewhat more than in the *Doryphoros*. For the original position of the arms, another Roman replica (Rome, Coll. Torlonia, no. 76) gives us the most complete evidence, since the right hand, a small part of the right forearm, and the left hand, together with the wrist and the discus, are probably the only segments restored (Blümel, 1930). The left arm is alongside the body, the right arm slightly bent at the elbow, and the forearm slightly raised in a forward direction. According to general opinion, this replica proves the interpretation of this type as a diskophoros. Although the discus in the left hand is restored, the original break seems to sustain such a restoration. Other copies do not support this conclusion, however, and they might even be said to render it doubtful. The masterly statuette in the Louvre (PL. 191), which is clearly dependent on the same original, in spite of the varying position of the feet, does not help to solve the problem, since in this version the right hand seems to have held a cup.

Of some help, perhaps, is a much-copied Herakles type, thus far found only in mediocre copies, which is very closely related to the so-called "Diskophoros." A statuette in Paris (Bib. Nat., no. 549), which as yet has been little considered, best represents this collateral line of copies, even though it reveals certain later features in the formation of the head. Herakles is represented holding the apples of the Hesperides in one hand, and in the other hand probably the club. Moreover, several unmistakable replicas of the head of the "Diskophoros" by Roman copyists have been expressly characterized as Herakles through inclusion of the the symposiac band. A fine head in Berlin (Staat. Mus., K146) — in which, however, the back portion is inaccurately rendered — is of the same type, but newly interpreted as Hermes by the addition of wings.

Notwithstanding these related Herakles and Hermes examples, the "Diskophoros" remains a unique type. Its *contrapposto* is even more accentuated, with the relaxed leg resting on the heel — Pliny's "nudum talo incessentem" perhaps refers directly to this work (Berger). It should thus most likely be dated even before the *Doryphoros*. In contrast to the statues already discussed, this type presents a more youthful conception of the body, a fact well illustrated by another torso from the former Somzée Collection. Because the relaxed leg also rests on the whole sole, the figure has a somewhat heavy laborious aspect. The treatment of the head, more inclined than elsewhere, and the tense forehead muscles contribute to this effect. A mood almost of sadness, which is exaggerated in several replicas, is the particular distinction and great attraction of this work; its impressive immediacy can be sensed most strongly in several other copies, such as the beautiful heads in Copenhagen (Ny Carlsberg Glyptotek, no. 114) and Zurich (Archäologische Samml. der Univ.).

The statues of young men so far considered, for the most part, follow the same fundamental scheme: the right leg as the supporting member, the head turned toward the right side. Differences in the basic *contrapposto* movement are small but decisive, because they are so appropriate to the theme at that moment. Each Polycletan creation is an independent, deeply felt work, as is further proved by Polykleitos's statues of boys, in which the basic compositional principle of the *Doryphoros* is apparent, but modified to conform to the new theme.

The so-called "Westmacott Athlete" (or "Ephebe") represents the statue of a boy athlete by Polykleitos that was very frequently copied. Lacking a more definite interpretation, this statue type is named after a nearly complete copy formerly in the Westmacott Collection (PL. 191), from which only the right, upraised arm is missing. The remainder is essentially of ancient origin, although certain scholars have expressed doubt (erroneously) that the left hand really belongs to this statue. In this type a boy is represented in a more active, asymmetrical pose. Unlike the limbs of the other statues thus far considered, here the left leg supports the weight, and the relaxed leg swings out widely. Another difference is readily apparent: the head is inclined toward the side of the relaxed leg, not toward the supporting leg, thereby establishing a very different *contrapposto* composition more fitted to the character of the boy. The raised arm on the side of the relaxed leg is preserved in a carelessly executed replica in Rome (Mus. Barracco, no. 99), but only down to the wrist. The upper arm stands out from the body sidewise and almost horizontally; the arm is bent back at the elbow so that the hand (or its attribute) lay fairly close to the right temple. For this reason, in some replicas *puntelli* can be seen at the front of the hair, above the right temple (e.g., replica in Villa Barberini, Castelgandolfo; the Van Branteghem, or Vincent, head of the Br. Mus.), or on unfinished parts of heads preserved in Dresden (Albertinum), in Kassel (Staat. Kunstsamml.), and in the Lateran collections in Rome. On the front of the raised shoulder in the excellent torso in Baltimore (Walters Art Gall.), traces of a support that must have served to hold the lifted forearm are evident. On the basis of this copy and from certain other evidence reflecting the same original (cf. sarcophagus relief, Vat. Mus., 393a), it has generally been agreed, for more than a half century, that the boy is in the act of placing a crown on his head. Of the crown itself, however, no trace has been preserved on any of the numerous copies of the head. It is possible that the crown did not rest on the head; if it was suspended over the head, it would be necessary to modify, accordingly, plaster reconstructions of the statue in Dresden and Munich. A reconstruction in Stettin is also unsatisfactory, since here the hand does not appear bent downward, as in the Museo Barracco replica. If it is not possible to reconstruct the statue with a crown in the hand, as Hafner (1955) has asserted, it would be more natural to relate this type to the *astragalizontes* (dice players; Berger) than to the *apoxyomenos* type (cf. IX, PL. 220). A reconstruction with an astragal held in readiness for the throw could explain the pronounced and measured movement of the arm and the bent head; a strigil held very high, a pose that is recorded in other cases, would here appear affected.

The identification of the subject with Kyniskos, youthful victor in an Olympic boxing match, is hardly likely, even though the traces of the stance on the preserved base of the victor statue correspond to the position of the feet of the "Westmacott Athlete." The identification of the two is impossible, since epigraphists have attributed the base inscription to the second quarter of the 5th century B.C. Moreover, the statue is, in its over-all rhythm as well as in detail (e.g., the rendering of the hair), clearly executed in a style of later than the victory date, at least of the time of the *Doryphoros*.

The head of the *Doryphoros*, bent toward the side of the supporting leg, accords with the subject's heroic character; whereas the simpler attitude of the boy called for a simple composition. This subtle comprehension of what is appropriate to the age depicted can be traced in the smallest details: for instance, the hair is less carefully arranged, falling in shorter, more irregular locks. The contours of the cheeks and the lips are less decided in their modeling; the body is more slender, the indications of the musculature less pronounced.

A closely related statue of a boy by the same master is recognizable in a statue that appears in its best and most complete form in a well-known copy in Dresden (Albertinum), the so-called "Dresden Boy." Since the right hand and the left forearm are missing, it is again difficult to give a sure interpretation. The boy looks like the Westmacott ephebe, but reversed as in a mirror image. The heel of the relaxed leg is slightly lower, and the head is less inclined and turned; the distribution of the weight therefore seems more unsettled. The left shoulder is slightly raised and thrust forward. Some tomb reliefs reproduce this type in mirror-image versions, with a strigil in the hand. Perhaps the original, too, should be visualized with the same attribute, and be identified with the *apoxyomenos* mentioned by Pliny (*Naturalis historia*, XXXIV, 55). Here, too, the boy, is characterized less by means of the attribute than by means of his attitude. Compared with the Westmacott ephebe, his body is depicted at a somewhat more mature stage, and the bone structure, particularly of the chest, has become more pronounced.

There are few exact replicas of the torso of the "Dresden Boy." A beautiful example in Rome (Mus. Naz. Romano, inv. 653) deserves greater critical attention. Among other torsos designated as replicas of the "Dresden Boy" (e.g., that in the municipal building of Limni; Mus. Naz. Romano, inv. 52400; Leningrad, The Hermitage, no. 90), there is recognizable a somewhat later, perhaps independent creation which, except for deviations in the position of the arms and parts of the shoulders, seems to represent a slightly younger figure and which might even be a counterpart to the Westmacott ephebe.

Another work that was previously cited as deriving from the "Dresden Boy" and as a work by Polykleitos is no longer accepted as such: namely, the *Pan*, known through many replicas. The difference in size — the replicas are only two-thirds of life-size — reveals a fundamental new attitude, one less determined by an ideal canon. The independence of this motif is shown above all in the more emphatic turning and inclination of the head, as well as in the position of the shoulders and arms. The left hand held the syrinx; the right hand may have held a pedum resting on the shoulder. [A different way of bearing this attribute, that seen in a copy in Leiden (Rijksmus. van Oudheden, no. 1.62), seems affected.] The god's animal nature is expressed outwardly by animal ears and horns. The statue is characterized by an unstable attitude, particularly in the flexion of the head, which registers emotion strongly. The *contrapposto* is more refined than in the accepted works of Polykleitos, and the body forms flow into one another even more smoothly than in the Westmacott ephebe and the first variant of the "Dresden Boy." The articulation between parts of the body structure is kept to a minimum. The outline of the chest is no longer noticeably raised. The modeling of the front of the statue is largely limited to defining lateral grooves below the edge of the ribs. These suggestions of an even more youthful subject are in curious contrast to the fundamentally more complex emotional tone, which would appear to go beyond boyhood. On this level, as well as on the stylistic level, the Polycletan consistency seems thus to be broken.

The number of female statues attributed to Polykleitos is relatively small. Literary sources mention the following statues of women in connection with Polykleitos's name: the cult image of Hera in Argos; an Amazon figure for Ephesos; two *kanephoroi* (canephorae), illegally seized by Verres; a statue of Hekate for Argos; an image of Aphrodite, as a tripod support for Amyklai; and figures of Leto and Artemis, as part of a group with Apollo that was perhaps on Mount Lykone (Pausanias, II, 24, 5). Not all these works are to be connected with the celebrated Polykleitos I: the *Aphrodite* executed for Amyklai was dedicated in honor of the Spartan victory at Aigospotamoi in 405 B.C. Since, according to tradition and sources (op. cit., X, 9, 7), Polykleitos I had not participated in creating the far more important memorial to this battle in Delphi, the work dedicated at Amyklai should be ascribed to his namesake Polykleitos II. For the *Hekate* and the group with Apollo, there are no reliable chronological clues, so that in principle they might be assigned either to Polykleitos I or to Polykleitos II. The statues of Leto and Artemis were, according to Pausanias, executed in marble — a fact that perhaps favors their attribution to Polykleitos II, since Polykleitos I was especially noted as a sculptor in bronze.

Amandry has tried to assign the chryselephantine cult statue of Hera to the younger Polykleitos, but literary evidence seems to indicate that Polykleitos I was responsible for this image, since he is greatly praised for it by ancient authors, who compare it with the great cult statues of Phidias. Unfortunately, however, it is not known when the temple of Argos (burned in 423 B.C.) was rebuilt and, therefore, when the erection of the cult image was undertaken. If the attribution to Polykleitos I is accepted, the date of its creation cannot be much later than the date of the fire, since Polykleitos, as a contemporary of Perikles and Phidias, must already have been of advanced age in 420 B.C. There are no certain copies of this statue, but it can be visualized to some extent through the descriptions of ancient authors, especially Pausanias, as well as through coin effigies. The statue was undoubtedly smaller than the colossal figures of Phidias. In one hand the seated Hera held a pomegranate; in the other was a scepter surmounted by a cuckoo (op. cit., II, 17, 4). On her diadem, the Charites and Horai were represented. The suggestion of Waldstein (1901), who associates a head in the British Museum with this cult statue, is not wholly convincing, though it remains the most credible of many attempts to visualize the original head through copies.

For the identification and reconstruction of Polykleitos's *Amazon*, there is much better evidence. Pliny (*Naturalis historia*, XXXIV, 53) tells of a contest among Polykleitos, Phidias, Kresilas, and Phradmon, each of whom produced an Amazon figure for the Artemision in Ephesos; Polykleitos was said to have obtained the first prize, Phidias the second, Kresilas the third, and Phradmon the fifth. (The assignment of the fourth prize to a certain "Kydon" is clearly in error, as noted by Richter.) It is difficult to determine how closely this traditional account corresponds to the facts and how much of it is pure anecdote. It is certain, however, that three (perhaps even four) statue types of Amazons have been preserved in copies which agree in proportions and theme and, in general, in date. The fundamental traits, which were perhaps determined in the rules for the contest, are the following: the wounded Amazon is a standing figure dressed in a chiton. Among the copies, it is possible to exclude with certainty assignment of the "Mattei" type to Polykleitos (PL. 123; XII, PL. 138), and it is generally attributed to Phidias. Here the composition is completely determined by the drama of the subject: the Amazon, wounded in the left thigh, holds herself upright by supporting herself with her spear. There is no attempt at a *contrapposto* effect, such as Polykleitos regularly, and successfully, achieved. The attribution of this type of Amazon to Phidias is further supported by the fact that Lucian draws special attention to the spear motif in the *Amazon* by Phidias. The accessory spear is also attested for one of the two other Amazon types, a version in which, however, the motif is not so central and in which other details of the conception are quite different from the characteristic Phidian work.

The choice for the Polycletan work lies then between the "Sciarra" type in Copenhagen (Ny Carlsberg Glyptotek) and Berlin (Staat. Mus.) and a statue signed by Sosikles, referred to as the "Capitoline" type (PL. 193). It is difficult to decide between the two. Both types show the so-called Polycletan "stepping motif" (see below). Because of the *contrapposto* posture of both statues, too, it is possible to discern a special connection with the works of Polykleitos. In the Sosikles version (right arm improperly restored) everything draws attention to the wound. The Amazon, who lifts the garment from her wounded breast with her left hand, once supported herself with her right arm, leaning wearily on the spear. Her glance is toward that same side. The treatment is emphatically realistic, and the Amazon is not a pathetic portrayal. Quite different is the "Sciarra" type in which the Amazon leans on a pillar (cf. the "Lansdowne statue"; PL. 193); in this attitude the depiction is already separated from any immediate relationship with the combat and is fixed in a self-contained atmosphere. The right arm, by the way in which it is raised over the head, has no reference to the wound on the right side, and the spear motif is absent. In contrast to the "Capitoline" Amazon of Sosikles, the arrangement and stylization of the drapery is completely subordinated to the body rhythms. Furtwängler's view (*MW*, p. 247 ff.), which has generally been abandoned, should perhaps be seriously reexamined and the "Sciarra" type in Berlin recognized as the Polycletan type. Admittedly, the heads considered in themselves could also be attributed differently; the Sosikles head, for instance, by far surpasses the "Sciarra" type in appropriateness to the situation portrayed. But such partial attribution would not answer the main problem. Moreover, it should perhaps be remembered, in this connection, that Kresilas must have been a superior portraitist, since he received the commission for the portrait of Perikles (cf. III, PL. 372).

As has already been indicated, there are but few certain data or clues for Polykleitos's biography and for the identification of his works. From the literary evidence it has not been possible to draw any definite proof as to his life dates or the precise limits of his career. Nor can one, with any degree of certainty, succeed in identifying from among the copies any of the statues of victors made by Polykleitos, though the dates of the victories are known and the inscribed bases are preserved (which is not surprising, since it is not probable that there was any demand for copies of statues portraying comparatively unimportant historical individuals). The activity of the copyists in imperial times was concentrated, for the works of Polykleitos as well as for those of other sculptors, on those for which it was possible to find some new topical application — to throw a "mythological bridge." This is why it has been possible to retrace the mythological figures of Polykleitos from among the copies. For other types whose mythological content, if any, cannot be precisely interpreted (the so-called "Diskophoros," the "Dresden Boy," and the "Westmacott Athlete"), besides judgments from a purely stylistic basis, it has been possible to support connections with Polycletan works on the basis of literary mention according to motif: for example, the "nudum talo incessentem," "destringentem se," and "astragalizontes" of Pliny. Of the cult statue of Hera at Argos and the *kanephoroi* figures, which are known from literary evidence, it has been impossible to find any sure copies, since the descriptions of these works are so brief. The portrait of Artemon and the *Hekate* should most likely be attributed to the younger Polykleitos.

Although no exact dates have been recorded for any of those statues of Polykleitos identified through copies, a relative chronology can be established on the basis of style considerations. In relation to the *Doryphoros*, the *Diadoumenos* has been clearly recognized as a later creation. On the basis of subtle comparisons of details, the other accepted works can, with a single exception, be organized between these two set points. The *Hermes* can be only slightly later than the *Doryphoros*, the second *Herakles* only slightly earlier than the *Diadoumenos*. The two statues of boys are to be placed between these. The *Amazon* should be put as late as possible, yet hardly after 430 B.C., unless one disregards the traditional account of the artistic competition. The so-called "Diskophoros" is to be placed before the *Doryphoros*, because of the pose; comparison of the heads shows, however, that it can only be a question of a few years. For the *Doryphoros* a date of 440 B.C., or shortly before, is almost generally admitted, on the basis of a stylistic comparison with such securely dated monuments as the sculpture of the Parthenon. The development leading to the *Diadoumenos* must have entailed about 30 years. With these guidelines, Polykleitos's career can be established in about the years 450–420 B.C., a dating that agrees with biographical considerations as well. This time range might be extended even later, since the reconstruction of the temple of Argos after the fire of 423 B.C. quite probably took a long time. Since it is said that he did not participate in the creation of the Delphic votive offering of the Spartans for the victory of Aigospotamoi (405 B.C.), it would seem reasonable to place the end of his activity — and the boundary between his works and those of the younger Polykleitos — in the next to the last decade of the 5th century B.C.

An evaluation of Polykleitos's distinctive qualities and of his importance in the history of art can be illustrated by some significant remarks of ancient authors. Pliny (op. cit., XXXIV, 55) notes as peculiar to Polykleitos the innovation of statues standing on one leg ("proprium eius est uno crure ut insisterent signa excogitasse"). What he meant by this expression we can better understand now because of our knowledge of his work. This description must not be interpreted literally. Single statues standing on one leg not only are difficult to imagine in the 5th century B.C. but also could hardly be called peculiar to a single master. In all the Polycletan statues discussed (except the so-called "Diskophoros"), the relaxed leg touches the ground only on tiptoe, and Pliny's phrase must have referred to this characteristic. This attitude is often erroneously interpreted as a stepping pose, but it cannot be explained in such realistic terms. It is, instead, a general principle, or an original esthetic formula, that allows the movements of the human body to be developed in the most balanced way, with the perfect equilibrium of relaxed and moving parts creating a classical harmony.

Polykleitos explained his archetypal solution theoretically in his *Canon*. According to the preserved fragments of this work, two principles were decisive for Polykleitos: (1) the proportional relationship to the whole; (2) the nuances that cannot be grasped by the compass, that is, the irrationality of the ultimate creative moment. Whereas the first premise largely derived from empirical considerations, the second — to which little attention has so far been paid — includes, besides the artistic modeling, the incalculable, or the properly creative, element. It is astonishing to find such a notion of limits already consciously formulated by an artist of the 5th century B.C. And it is only by starting from this premise that one can understand how Lysippos, with such a different style of his own, could nonetheless designate precisely the *Doryphoros* — Polykleitos's sculptured "Canon" — as his inspiration and master (Cicero, *Brutus*, 86).

The statues of Polykleitos are called by Varro "square-built" (*quadrata*) and "uniform" (*et paene ad unum exemplum*), both expressions based on discerning observation. Compared with the preceding as well as the following works of others, the Polycletan creations give the impression of being broad and heavy. Specifically, the effect is due to the following reasons: in the works of the severe style, such as Stephanos's *Athlete* (cf. X, PL. 262), the hips are narrower in relation to the shoulders; in the works of the 4th century B.C., such as the *Apoxyomenos* of Lysippos (III, PL. 382), the legs are longer in relation to the body, and the head is smaller. The "uniformity" of his statues arises from the fact that Polykleitos creates them almost always on the same fundamental principle — without, however, falling into any mechanical repetition. The statues of Achilles, Apollo, Herakles, and Hermes show how varied motifs have been devised through slight nuances of the basic principle. Furthermore, in the representation of the immature physique of the boys, for example, he has deviated quite deliberately from the ideal body structure of the *Canon*. His imagination, though strongly tinged with intellect, never conflicts with genuineness and simplicity of feeling in his theme. Even in that most curious expression of his canon, the *Amazon*, Polykleitos remains free from classic exaggeration.

A characterization that is more concerned with the content of the Polycletan works is given by Quintilian (*Institutio oratoria*, XII, 10, 8), when he says: "Polykleitos seems to have endowed the human figure with a beauty (*decorum*) that takes us beyond reality, but he appears not to have grasped quite adequately the majesty and authority of the gods (*non explevisse deorum auctoritatem*); indeed, he seems to have avoided the representation of advanced age and to have dared nothing except smooth cheeks." Although it was a general tendency of the Greeks to deify men and humanize gods, Polykleitos went especially far on this road, accomplishing the merger of both worlds with a particular consistency. For this reason it was possible for the original significance of the *Doryphoros* and the *Diadoumenos*, among others, to remain obscured for so long, to have been forgotten and contested to the present day. One can affirm that Polykleitos has not dared anything beyond "smooth cheeks" from a review of the existing copies. But this statement does not at all assert that Polykleitos always represented the same type of young man. On the contrary, there was probably no other sculptor who demonstrated such variation and sensitivity as Polykleitos in discriminating subtly within a limited scale of age, from the boy to the fully developed young man.

SOURCES. Overbeck, SQ, nos. 932–77.

BIBLIOG. *General*: Furtwängler, MW, pp. 279, 413–509; C. Anti, Monumenti policletei, MALinc, XXVI, 1920, cols. 501–792; R. Bianchi Bandinelli, Policleto, Florence, 1938; Lippold, GP, pp. 162–72; G. M. A. Richter, The Sculpture and Sculptors of the Greeks, 3d ed., New Haven, 1950, passim· E. Paribeni, ed., Museo Nazionale Romano, Sculture greche del V secolo, Rome, 1953, pp. 35–38; D. Schulz, Zum Kanon Polyklets, Hermes, LXXXIII, 1955, pp. 200–20. *Single works*: F. Studniczka, Eine neue Athletenstatue Polyklets?, ÖJh, II, 1899, pp. 192–98; C. Waldstein, The Argive Hera of Polycleitus, JHS, XXI, 1901, pp. 30–44; G. Lippold, Zu Polyklet, JdI, XXIII, 1908, pp. 203–08; J. Sieveking, Hermes des Polyklet, JdI, XXIV, 1909, pp. 1–7; C. D. Caskey, On a Polyclitan Head in

Boston, AJA, XV, 1911, pp. 215–16; A. Brückner, Polyklets Knöchelwerfer (77. Wpr), Berlin, 1920; C. Blümel, Der Diskusträger Polyklets (90. Wpr), Berlin, Leipzig, 1930 (rev. G. Lippold, Gnomon, VIII, 1932, pp. 49–51); P. Wolters, Polyklets Doryphoros, MJhb, N.S., XI, 1934, pp. 4–25; S. Ferri, Una statuetta romana di Ercole e un passo di Plinio, BArte, XXIX, 1935, pp. 437–41; J. Marcadé, A propos des statuettes hellénistiques en aragonite du Musée de Délos, BCH, LXXVI, 1952, pp. 96–135; G. Hafner, Zum Epheben Westmacott, SbHeidelberg, 1955, 1; P. Amandry, A propos de Polyclète: statues d'olympioniques, Charites, Bonn, 1957, pp. 63–87; G. M. A. Richter, Pliny's Five Amazons, Archaeology, XII, 1959, pp. 111–15; G. Hafner, Geschichte der Griechischen Kunst, Zurich, 1961, pp. 205–06.

Ernst Berger

Illustrations: PLS. 189–193.

POLYKLEITOS THE ARCHITECT. So designated is the Polykleitos (Πολύκλειτος) from Argos to whom is attributed the building, about the middle of the 4th century B.C., of the two most famous structures in the sanctuary of Asklepios in Epidauros, the Tholos (or Thymele) and the Theater (PL. 194, Fig. 435; III, PL. 340): "What architect could compete with Polykleitos in harmony and beauty? For it is Polykleitos who is the author of the theater and of the circular building" (Pausanias, II, 27, 5). However, precise identification of this Polykleitos raises many problems. Before the excavations of the Asklepieion of Epidauros by the Archaeological Society (Archaiologiki Hetairia) of Greece, the Tholos and the Theater were unhesitatingly attributed to the celebrated Argive sculptor of the same name, the Polykleitos (q.v.) of the *Doryphoros* and the *Diadoumenos*, whose period of activity falls in the middle of the 5th century B.C., coming to an end about 420. The discovery of the buildings themselves and of the inscription concerning the construction of the Tholos (*IG*, IV, 1485) demolished this hypothesis, since the text of the inscription cannot be earlier than 360 B.C.; it indicates a lengthy interruption in the process of construction, which must have lasted at least one generation; and, finally, this evidence is confirmed by the architectural and decorative style of the Tholos. The architect of Epidauros was, after this, identified with an Argive sculptor usually referred to as Polykleitos the Younger, whom Pausanias (VI, 6, 2) distinguishes from the master of the *Hera* of Argos, citing this one as a pupil of Naukydes. While this identification is possible, it has not gone uncontested (C. Weickert), nor is it without its difficulties. It is, in effect, accepted that this Polykleitos (the Younger) was born about 435/430 B.C., and that one of his latest works was the *Zeus Philios* in Megalopolis, which postdates the year of the founding of the city (ca. 367) but predates 360, which is approximately when the artist is thought to have died. Thus it would have to have been toward the end of his life that, having turned to architecture, he drew up the plans for the Tholos; he would not have been able to bring it to completion, and the decoration would not have been executed by him. There are even greater difficulties in connection with the Theater, for the building can hardly be earlier than 350 B.C., inasmuch as certain features, for example, the decoration of the door of the parodos, appear to be copied from the Tholos. A study by A. von Gerkan (1961) would bring the date of the building to as late as the 3d century B.C.

What, then, becomes of the testimony of Pausanias? It is very difficult to reconcile it with the above-indicated period of activity of Polykleitos the Younger, whose earliest work appears to have been the *Aphrodite* of Amyklai, erected out of the spoils of the battle of Aigospotamoi (405 B.C.), and whose statues of Olympic victors are datable to the early 4th century. Must one concede the existence of an Argive architect of the same name, but a little younger (C. Weickert)? This is not impossible, since the name was widespread in Argos; Lippold (*GP*) mentions four sculptors in succession, from the 5th to the 3d century, named Polykleitos. Or could there have been some confusion in the tradition transmitted by Pausanias, such as would have made it possible for him to misapply the first portion of the statement, on harmony and beauty, which formed part of the ideas traditionally handed down by ancient authors in connection with Polykleitos the Elder? This is the conclusion arrived at by G. Roux (1961) in a study of the architecture of Argolis. It is all the more regrettable that the architect's name was not preserved in the inscription on the Tholos, as was the case with the accounts for the Temple of Asklepios. The hypothesis by which H. Pomtow (1912) attributes the tholoi both of Epidauros and of Delphi to Theodotos, the architect of the temple, seems unacceptable (M. Schede, 1913). If Pausanias was repeating a well-attested tradition about Epidauros, it is more plausible that the builder, in the second half of the 4th century B.C., of the Theater and Tholos in the sanctuary of Asklepios was a different member of the illustrious Argive family, rather than Polykleitos the Younger.

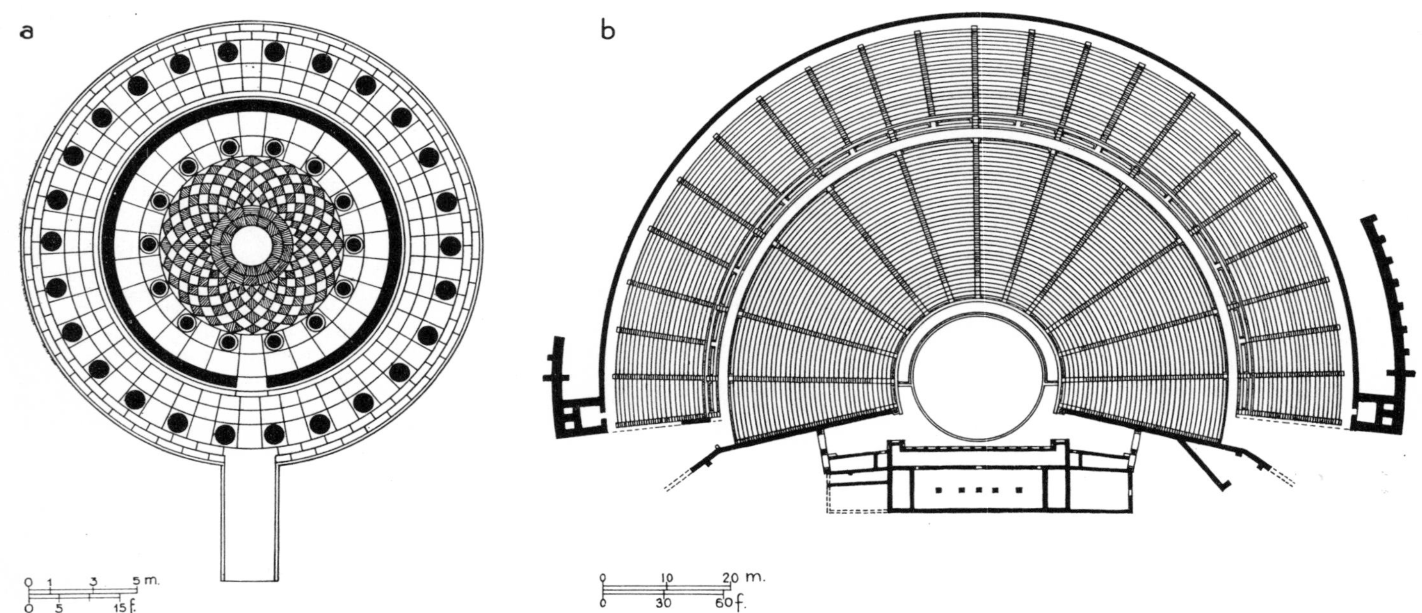

Epidauros: (*a*) Tholos, plan (*from A. W. Lawrence, Greek Architecture, London, 1957*); (*b*) Theater, reconstructed plan (*from EAA, s.v. Epidauro*).

The artistic personality of this architect, as indicated by his works, is most arresting, and the esthetic judgment expressed by Pausanias remains valid. The Tholos, on a series of three concentric circular foundations, supported a triple crown consisting of an exterior peristyle of 26 Doric columns, the cella walls, and an inner circle of 14 Corinthian columns of marble (diam., respectively, ca. 72 ft., 44 ft., 30 ft.). The interior pavement, composed of lozenges of black and white marble, covered a kind of subterranean crypt divided into corridors by three concentric walls of blocks of poros, each wall interrupted by a door leading to the center, where doubtless there was once an altar. This crypt did not, as was long believed (F. Noack, 1927), constitute an earlier structure (F. Robert, 1939). While there are many theories about the original purpose of the building (sacred well, hall for music, tomb of Asklepios, chthonian altar, prytaneion), agreement as to its architectural preeminence is unanimous. The design was striking for its geometric, slender, airy structure, devoid of figural decoration; the proportions of the columns and the entablature and the width of the bays combined to create the effect of a linear geometric composition yet without rigidity. A taste for floral decoration — apparent on the exterior in the rosettes that ornament the metopes, in the cyma richly decorated with voluted rinceaux and framed with acanthus leaves, and in the flowered acroterium crowning the roof (of radially arranged tiles of marble) — was given free rein in the interior. The door of the cella, Ionic in style, is flanked by pilasters richly adorned with astragals enclosing rosettes. The wall of the cella was divided in two by a band of palmettes and lotus flowers in low relief, surmounted by an ovolo; the lower part of the wall was composed of marble orthostats, white for the exterior, black for the interior; the ceilings of the peristyle and of the cella, made up of marble coffers (III, PL. 341), were enriched with a luxuriance of motifs, including lilies, rosettes, and acanthus leaves. "The entire decoration of the tholos of Epidauros, up to the very ceiling, is a hymn to the flower" (Picard). The same efflorescence continues in the interior, with the Corinthian capital framed by a double crown of crisp and vigorously carved acanthus leaves. The inner ceiling may also have been decorated with paintings by Pausias (Pausanias, II, 27, 3; Pliny, *Naturalis historia*, XXXV, 124). Delicacy of line, polychromy of materials, exuberance of decoration made this building a splendid realization of the architectural style of the 4th century, even as it foreshadowed the degeneration of the classical orders.

The theater, with its circular orchestra (diam., 32 ft.), its cavea of 55 tiers (34 below the diazoma, 21 above), is justly termed, as it was by Pausanias, a work of the first rank. Its greatness consists in the clarity, simplicity, and harmony of its lines and in the pure relation of the various parts to one another. Whether it was Polykleitos the Younger or another member of the same family who designed these two monuments, the architect endowed them with those qualities of harmony which had been the glory of the previous century.

SOURCES. Pausanias, Description of Greece, II, 27, VI, 6, 2; IG. IV, 1485; Pliny, Naturalis Historia, XXXV, 124.

BIBLIOG. *a. General*: M. Fränkel, Zur Zeitbestimmung der Tholos von Epidauros, RM, XVII, 1902, pp. 336–37; H. Möbius, Die Ornamente der griechischen Grabstelen, Berlin, 1929, pp. 39–46; C. Praschniker, Zur Geschichte des Akroters, Brünn, 1929, pp. 36–39; J. Jongkees, Zur Chronologie der Münzen von Olympia, JdI, LIV, 1939, pp. 219–29; C. Weickert, ThB, s.v.; Picard, III, 1, pp. 315–22; E. Fabricius, RE, s.v. Polykleitos, no. 15; M. Bieber, The Sculpture of the Hellenistic Age, 2d ed., New York, 1961. *b. Tholos*: P. Kabbadias, Fouilles d'Epidaure, I, Athens, 1893; A. Defrasse and H. Lechat, Epidaure, Paris, 1895, pp. 95–128; P. Kabbadias, Die Tholos von Epidauros, SbBerlin, 1909, pp. 536–41; M. Schede, Antikes Traufleisten-ornament, Strasbourg, 1909, pp. 66–68; E. R. Fiechter, Die alte Tholos und das sikyonische Schatzhaus in Delphi, AAnz, XXVII, 1912, cols. 17–20; H. Pomtow, Die grosse Tholos zu Delphi, Klio, XII, 1912, pp. 171–218, 281–307; M. Schede, Zur grossen Tholos in Delphi, Klio, XIII, 1913, pp. 131–33; F. Noack, Der Kernbau der Tholos von Epidauros, JdI, XLII, 1927, pp. 75–79; F. Robert, La destination cultuelle de la tholos d'Epidaure, REG, XLVI, 1933, pp. 181–96; F. Robert, Thymélè, Paris, 1939, pp. 338–58; G. Roux, L'architecture de l'Argolie aux IVe et IIIe siècles avant J.-C., Paris, 1961, pp. 131–200, pls. 37–52. *c. Theater*: W. Dorpfeld and E. Reisch, Das griechische Theater, Athens, 1896, pp. 130–33; A. Fossum, Harmony in the Theatre at Epidauros, AJA, XXX, 1926, pp. 70–75; H. Bulle, Untersuchungen an griechischen Theatern, AbhAkMünchen, XXXIII, 1928; M. Bieber, History of the Greek and Roman Theater, Princeton, 1939; C. Anti, Teatri greci arcaici, Padua, 1947, pp. 21–23, 320–22; O. A. W. Dilke, Details and Chronology of Greek Theatre Caveas, BSA, XLV, 1950, pp. 21–62 at 42; J. Bousquet, Harmonie au théâtre d'Epidaure, RA, XLI, 1953, pp. 41-49; A. von Gerkan and W. Müller-Wiener, Das Theater von Epidauros, Stuttgart, 1961 (bibliog.).

Roland MARTIN

Illustrations: PL. 194; 1 fig. in text.

POLYNESIA. Comprising the vast scattering of islands and archipelagoes of the central Pacific Ocean between 30° north and 47° south latitude, Polynesia includes the island groups of Hawaii, Marquesas, Society, Austral, and Cook; Samoa and Tonga in western Polynesia; various groups of atolls; Easter Island; and the islands of New Zealand (see NEW ZEALAND). Its indigenous peoples are mainly of the brown race, although there are local differences; they speak dialects of one language. (See POLYNESIAN CULTURES.)

SUMMARY. General characteristics (col. 438). Decorative arts (col. 438). Local styles (col. 441) : *Hawaiian Islands; Society Islands; Marquesas Islands; Austral Islands; Cook Islands; Tonga; Samoa; Atolls; Easter Island.*

GENERAL CHARACTERISTICS. The arts of Polynesia present an array of similarities and differences that reflects the isolation of the various areas as well as the cultural diffusion among them. The relationship between physiographic features and the arts is readily apparent in this area; the art potential of the luxuriantly verdured so-called "high" islands contrasts clearly with that of the dry, surf-beaten coral atolls. The existence of such variations and similarities in styles and forms raises the question of whether the ocean has been a means of or a barrier to communication.

In Polynesia, as elsewhere where art is ancillary to religion, environment also effects certain marked differences in emphasis, from high to low islands, of this or that aspect of ceremony, of ceremonial appurtenances, and of decoration.

DECORATIVE ARTS. Tattooing, widely practiced throughout Polynesia, was most highly developed in the Marquesas, where the patterns used were involved versions of carved reliefs, and where full body covering was common — so encompassing that the result approached a complete change of skin color.

Tahitian tattooing was much less complicated; and in the Society Islands and the nearby Cook and Austral groups, natural forms such as animals, trees, and plants were sometimes rendered pictorially. In this area the allover pattern was seldom used, and then only on men. Sometimes the limited tattooing of women fulfilled a social purpose; in certain areas, for example, only tattooed hands were permitted to serve food.

The angular wood-carving patterns of the western (Samoa-Tonga) area accounted for almost the whole of its tattooing repertory. In Samoa the tattoo designs surpassed wood carving in ingenuity and complexity, and the full waist-to-knee covering still marks the man of rank in those islands. In Tonga, however, missionary influence very quickly brought about the discontinuance of tattooing. In the low islands, tattooing was negligible.

To some degree, the decoration of tapa cloth also uses the designs of carving and tattooing. It seems strange to find such restricted motifs used in this potentially freer medium, until it is remembered that the designs were often printed from carved wooden stencils. Tapa decoration also illustrates the specialization that occurs in the arts of Polynesia. In the Marquesas, where elaborate relief carving and tattooing were produced, little tapa decoration exists; Hawaii, where the first two techniques were neglected, excelled in tapa designs and in the technique of applying them.

The Cook Islands patterns were more prosaic, being for the most part rectilinear, with only occasional curved lines and loops and an infrequent turtle or bird motif. Samoan designs were of the same simple order, and in both regions patterns were painted and stamped. The tiny animal forms introduced here and there among Tongan rectilinear wood-relief patterns also appeared sporadically among its tapa designs. In nearly every island group fine white tapa cloth, used in voluminous windings, had ceremonial significance and no doubt afforded a measure of esthetic satisfaction.

In the low islands the art of tapa decoration is, predictably, absent; little clothing was worn there, fibers available to make the cloth were few, and the paper mulberry could not be cultivated.

The making of textiles as an artistic activity occupied a place secondary to tapa painting in Polynesia, which was without the loom. The plaiting technique by which mats and baskets were also

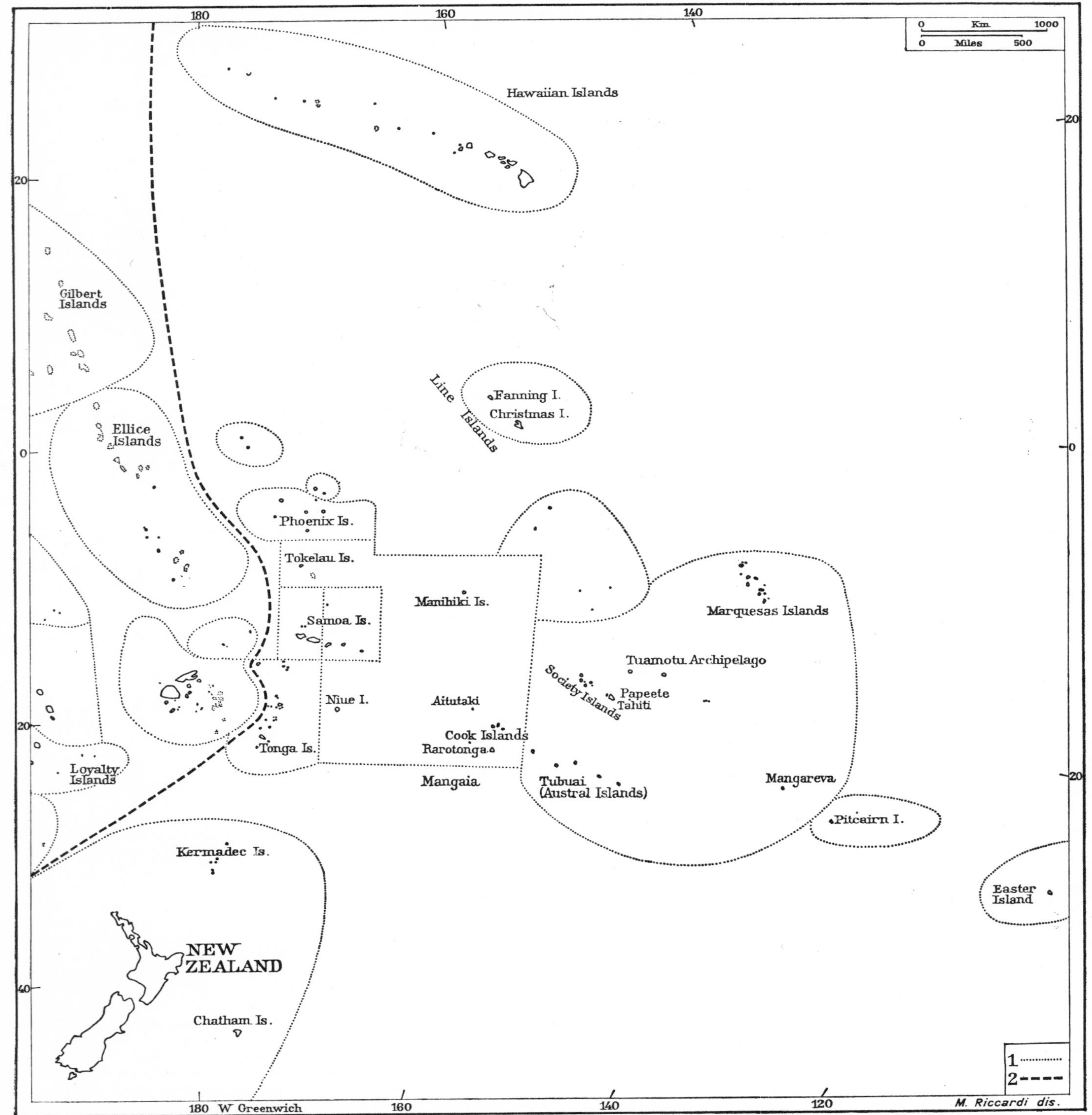

Polynesia. *Key*: (1) Modern national boundaries; (2) ethnic and geographical boundary between Polynesia and Melanesia.

made produced only simple checkered patterns. A notable exception was the decoration of plaited textiles with feathers. Hawaii produced magnificent, velvety royal cloaks made with close-covering red, gold, and green feathers arranged in bold contrasting patterns. Associated with these as part of a chief's raiment were feather-covered helmets with high curved crests.

Ceremonial regalia in central and western Polynesia took the form of tall headdresses of feathers mounted on cane and of elaborate gorgets and breastplates patterned in pearl, shell, and ivory — elaborations, no doubt, of simpler personal ornaments.

Because Polynesia lacked the material to make pottery, it was deprived of further surfaces for decoration. Gourds were substitute vessels, and occasionally simple designs were scratched on their surfaces. Plaited-cord carriers in pleasing designs sometimes appeared, notably in Hawaii, as a compensating form of applied decoration.

The making of such utilitarian vessels and ceremonial appurtenances hardly ranks as an art in Polynesia; it is, rather, an example of the same fine workmanship evident in the making of combs and ornaments, in the careful fashioning of adzes and other tools, and in the making of fishhooks and agricultural implements. Houseposts were carefully dressed and polished, and the lashing often went beyond strict functionalism. Adz lashing is, indeed, one of the most common and successful forms of ornamentation in Polynesia. Tonga and Samoa excel in ornamental lashing, as if to compensate for their more modest achievements in the other arts. However, it is in the low islands, where nature is niggardly and conditions are hard, that what has to be done is done well, that what has to be made is made with care.

Although high standards of workmanship and finish frequently are evident in these utilitarian crafts, it is in sculpture — in wood carving and in the fashioning of ornaments — that what may properly be called works of art are encountered. Standards of achievement in wood sculpture vary, and each area has developed a distinctive manner of stylization and a correspondingly distinct form of ornamental pattern that proceeds from the characteristic regional stylization.

BIBLIOG. See bibliog. for POLYNESIAN CULTURES.

LOCAL STYLES. *Hawaiian Islands.* A chain of volcanic and coral islands, 20 in all, Hawaii is a state of the United States. Named the Sandwich Islands by Captain Cook when he discovered them in 1778, the islands were visited frequently by American missionaries from 1820 onward.

Although Tahiti in the Society Islands seems to be the historical center of Polynesia, it is more convenient in discussing wood sculpture to commence with Hawaii — not only because Hawaii stands far apart to the north, but also because it was undoubtedly the most vital center for sculpture. Cleanly hewn wooden figures, 6 ft. or more in height, were vigorous in posture and emotionally expressive. These often had decorative extensions, such as elaborate crests representing hair and beard or a royal headdress. Although some of the figures were moderately stylized or used in groups as supports for bowls or drums, their forms were never reduced to small allover patterns. The Hawaiian artist possessed, besides a strong sculptural awareness, a sense of pure form and an awareness of surface quality, as exemplified in the simple shapes and softly flowing smooth surfaces of unadorned wooden bowls.

BIBLIOG. W. T. Brigham, Hawaiian Feather Work, Mem. B. P. Bishop Mus., I, 1899–1903; pp. 1–81, 437–53; W. T. Brigham, Old Hawaiian Carvings, Mem. B. P. Bishop Mus., II, 1906, pp. 165–82; W. T. Brigham, Ka hana kapa: The Making cf Bark Cloth in Hawaii (Mem. B. P. Bishop Mus., III), Honolulu, 1911; W. T. Brigham, Additional Notes (II) on Hawaiian Feather Work, Mem. B. P. Bishop Mus., VII, 1918, pp. 1–69; K. P. Emory, The Island of Lanai: A Survey of Native Culture (Mem. B. P. Bishop Mus., XII), Honolulu, 1924; H. M. Luquiens, Hawaiian Art (B. P. Bishop Mus. Special Pub., XVIII), Honolulu, 1931; S. Cammann, Notes on Ivory in Hawaii, JPS, LXIII, 1954, pp. 133–40; Te Rangi Hiroa (P. H. Buck), Arts and Crafts of Hawaii (B. P. Bishop Special Pub., XLV), Honolulu, 1957 (bibliog.).

Society Islands. Volcanic in origin and quite mountainous, the Society Islands group is 650 sq. mi. in area. Its chief islands are Tahiti and Papeete. Figures of near-primitive crudeness in bodily form, posture, and expression are characteristic of the naturalistic sculpture of the Society Islands; but when the figures became stylized, as in decorative fly-whisk handles, a clear sense of form and of arrangement of plane and surface emerges. As in Hawaii, however, no complex patterns were used. A few small ceremonial staves that have survived present stylized human figures grouped in designs of high quality. It is interesting, even curious, to find such accomplished sculptural art in objects that were destined to be hidden by a covering of red feathers, which was believed to be the abiding place of the god. Later, the "gods" became mere shapeless bundles of sennit covered with feathers, and the motifs and sculptural technique of the early products were all but lost.

Efforts to determine the distribution or localization of types of Polynesian art quite often end in uncertainty. Early explorers, missionaries, and traders did not always record the localities in which they had collected objects; and of the records made many have been lost. Not infrequently, when a recorded locality seemed to the ethnographer to be impossible to accept for attribution, recourse was made to typology in order to ascertain locality, in some cases thereby adding to the uncertainty. This complication, however, applies more to figures than to patterns, which have a readily recognizable local character.

BIBLIOG. K. P. Emory, L'art tahitien, B. Soc. des ét. océaniennes, XIX, 1927, pp. 236–39; E. S. C. Handy, History and Culture in the Society Islands (B. P. Bishop Mus. B., LXXIX), Honolulu, 1930; K. P. Emory, Stone Remains in the Society Islands (B. P. Bishop Mus. B., CXVI), Honolulu, 1933; D. S. Marshall, A Working Bibliography of the Society Islands, Particularly Tahiti, Auckland, 1951; R. Teissier, Ile Tapuae Manu ou Maiao Iti, B. Soc. des ét. océaniennes, X, 1956, pp. 517–32; B. Danielsson, A Unique Tahitian Stone Figure, JPS, LXVI, 1957, pp. 396–97.

Marquesas Islands. The rocky and mountainous Marquesas Islands, of which there are 13, have a population of less than 3,000. Set well out in the ocean, away from any land masses and apart from the rest of Polynesia, the islands have produced one of the most distinctive and versatile of the regional decorative patterns. A certain degree of formalism appeared in almost every type of Marquesan wood sculpture; the abundant carved images, of wood or stone, had stylized features and a well-ordered design in the disposition of the limbs. Stronger stylization of such figures was used in highly decorative fan handles and also inspired surface patterns of rich decorative quality. Stylized insects and centipedes, as well as almost every part of the human form, are found in ingeniously complicated designs. Marquesan patterns demonstrate a particular appreciation of the form of the object to be decorated; major and subsidiary areas were clearly organized, and although spaces were always filled, even crowded, they seldom were inappropriately decorated. Curvilinear ornament appeared nowhere else in Polynesia except in New Zealand, where figure stylization developed into even richer, more abundantly flowing designs.

BIBLIOG. H. U. Hall, Art of the Marquesas Islanders, J. Penna. Univ. Mus., XII, 1921, pp. 253–92; W. C. Handy, Tattooing in the Marquesas (B. P. Bishop Mus. B., I), Honolulu, 1922; E. S. C. Handy, The Native Culture in the Marquesas (B. P. Bishop Mus. B., IX), Honolulu, 1923; R. Linton, The Material Culture of the Marquesas Islands, B. P. Bishop Mus. B., VIII, 1923, pp. 263–471; R. Linton, Archaeology of the Marquesas Islands (B. P. Bishop Mus. B., XXIII), Honolulu, 1925; K. von den Steinen, Die Marquesaner und ihre Kunst, 3 vols., Berlin, 1925–28; A. Level, Iles Marquises, CahArt, II-III, 1929, pp. 105–07; W. C. Handy, L'art des îles Marquesas, Paris, 1938; R. C. Suggs, The Archaeology of Nuku Hiva, Marquesas Islands, New York, 1961.

Austral Islands. About 300 miles south of the Society Islands and southwest of the Tuamatu Archipelago, the group of small volcanic islands known as the Austral Islands forms a chain 850 miles long. The inhabited islands of the group are Rimatara, Rurutu, Tubuai, Raivavaé, and Rapa. The local pattern of this group is distinctly angular and much less varied than that of the Marquesas, yet it is of undoubted decorative value.

The figure sculpture from which the patterns were evolved is at its best a clear statement of the human form in precisely related planes and their margins, a style that could have been the inspiration of the formally designed Tahitian fan handles mentioned above. Unlike Tahitian designs, however, the Austral decorative compositions alternated rows of small, highly stylized versions of the sculptures with rows of loops and of angular elements clearly derived from the figural models.

BIBLIOG. W. S. and K. Routledge, Notes on Some Archaeological Remains in the Society and Austral Islands, JRAI, LI, 1921, pp. 438–55; H. Bodin, Note sur les statues de Raivavai (Vavitu), B. Soc. des ét. océaniennes, V, 1933, pp. 275–78; M. Urbain-Faublée, Note sur les pagaies sculptées des îles Tubuai, JSO, III, 1947, pp. 113–18; D. S. Marshall, Ra'ivavae, Garden City, N.Y., 1961.

Cook Islands. Southwest of the Society Islands, the Cook Islands, a dependency of New Zealand, include the Mangaia, Rarotonga, Aitutaki, Atiu, Mauke, Mitiaro, and Hervey islands. Mangaia, 300 miles west of the Australs, has a characteristic angular decoration, the so-called "K" pattern, that could be derived from the human figure. Unfortunately the images reported by missionaries from Mangaia have all been lost. Nearby Rarotonga had figure sculptors of undoubted competence; their small so-called "fishermens' gods" are a notable expression of volume, both in the squat torso and in the strongly formalized head and features. In symbolic compositions that comprise the essential part of elongated staff gods, Rarotongan wood carving approached decorative design. By using figures in the local manner, but more strongly stylized, and setting them along the staff alternately in fullface and in profile, the Mangaians achieved a rhythm that verged on, but did not quite become, an abstract pattern.

Images in a local style but of indifferent quality appear also in the high-island middle chain of the Cook group, especially in Aitutaki; here again the stylized figure became the basis of angular decorative patterns, in this case on flat or cylindrical staff gods. The style is related to that of Rarotonga, but occasional examples of Tahitian influence — or perhaps of Society Island work — appear in the eastern Cook Islands.

These three groups — the Society, Cook, and Austral islands — are manifestly related in their general culture. In their similarities and differences they present a small-scale version of the somewhat erratic chain of artistic relationships across the greater area of Polynesia. These relationships become clearer in the light of centuries of voyaging in frail canoes across the seas separating the islands; contacts certainly existed, but these were intermittent and probably infrequent.

The northern Cook Islands are low atolls; they have neither sculpture nor carved patterns. Art here is, for the most part, a matter of skilled craftsmanship in the production of utilitarian objects.

BIBLIOG. Te Rangi Hiroa (P. H. Buck), Arts and Crafts of the Cook Islands (B. P. Bishop Mus. B., CLXXXIX), Honolulu, 1944; D. S. Marshall, A Working Bibliography of the Cook Islands, Auckland, 1951.

Tonga. An archipelago 150 miles long, Tonga is divided into three major groups, Tongabu, Vavau, and Haabai, in addition to Hiaufor and Niuatobutabu further to the north. Along with Samoa, it is on the western boundary of Polynesia.

Small squat images, naturalistic and expressive, in polished hardwood were characteristic of the sculpture of Tonga. Similar images attempted in whale ivory were seemingly too small-scaled for satisfactory sculptural expression. Ornament on paddles, clubs, and bowls consisted of shallow, angular gouged patterns, with tiny human and animal figures and the crescent moon interspersed. These patterns, which do not suggest derivation from natural forms, resemble Fiji patterns, and both are possibly skeuomorphs of lashing and binding.

Samoa. Located north of the Tonga Islands on the western boundary of Polynesia, the islands of Samoa are well populated and extensive in area. They are divided into American Samoa and the Territory of Western Samoa (a trust territory of New Zealand).

Only one sculptured figure is known from Samoa, and this is possibly of Tongan manufacture. Wood carving, though regularly practiced, constituted but spare arrangements of shallow, angular incisions. Except for textiles and tattooing, Samoa is poor in art forms.

BIBLIOG. A. F. Kramer, Die Samoa-Inseln, 2 vols., Stuttgart, 1902; W. C. McKern, Archaeology of Tonga (B. P. Bishop Mus. B., LX), Honolulu, 1929; Te Rangi Hiroa (P. H. Buck), Samoan Material Culture (B. P. Bishop Mus. B., LXXV), Honolulu, 1930; Te Rangi Hiroa (P. H. Buck), Material Representatives of Tongan and Samoan Gods, JPS, XLIV, 1935, pp. 48–53.

Atolls. Five groups of atolls exist within the area of Polynesia: the Gilbert Islands (including Tarawa, Makin, Albaiang, Abemama, Tabitenea, Nonouti, and Beru), a densely populated group on the equator; the Ellice (or Lagoon) Islands, a group of nine coral atolls; the Phoenix Islands; the Line Islands, south of Hawaii (including Kingman Reef and Palmyra, under United States control, and Washington, Fanning, and Christmas islands, under British control; and the Jarvis, Malden, Starbuck, Caroline, Vostok, and Flint islands claimed by both the United States and Great Britain); and the Tuamotu Archipelago, an extensive group of 80 small islands that are mostly coral atolls, the chief of which are Makatea, Fakarava, Rangiroa, Anoa, Hao, and Réoao.

An absence of sculpture and a paucity of art generally characterizes the atolls of Polynesia; but here, also, religion was served with material appurtenances. In many of the low islands, there have been found large rectangular blocks of coral — a very unsatisfactory sculptural medium — marking the boundaries of open, flat temple courtyards, as well as upright slabs placed seemingly at random but no doubt intended to give place and order to the ceremonies.

In Mangareva, an isolated high-island outpost at the southeastern limit of the Tuamotu group, wooden figural sculpture appears. All but one example of these are naturalistic — the most decidedly so in all of Polynesia. The exceptional piece is strikingly formalized. Mangareva is also exceptional in that it produced no patterns.

BIBLIOG. Te Rangi Hiroa (P. H. Buck), Ethnology of Mangareva (B. P. Bishop Mus. B., CLVII), Honolulu, 1938; K. P. Emory, Tuamotuan Religious Structures and Ceremonies (B. P. Bishop Mus. B., CXCI), Honolulu, 1947; B. Danielsson, Work and Life on Raroia, London. 1956.

Easter Island. At the extreme eastern boundary of Polynesia, more than a thousand miles east of the nearest inhabited island, is Easter Island, discovered on Easter Day in 1722 by Jacob Roggeveen. It is removed from the rest of Polynesia not only in geographic location but in its art as well. The great stone statues that rose above the burial platform have no counterparts elsewhere in Polynesia, though the larger than life-size stone figures in Raïvavaé and the Marquesas and the wooden ones in New Zealand indicate that a taste for the heroic scale did exist elsewhere. The Easter Island achievement was, of course, conditioned by the available material, a soft volcanic tufa. For the opposite reason — the scarcity of a material — wood carvings had to be small; no ornamental patterns were evolved, only quaint but finely fashioned images. The curious Easter Island inscriptions should not be overlooked; though perhaps merely mnemonic pictographs, they are certainly decorative in effect.

At the opposite pole of sculptural achievement from the grandeur of the Easter Island sculpture there occur in several islands figure portrayals that are little more than low reliefs cut into rock faces. Of this same order are the petroglyphs found almost everywhere in Oceania. The motifs are various: natural forms (human, animal, and plant), accompanied by circles, dots, loops, and crescents. Figures and faces surmounted by ceremonial headdresses, which appear to be narrative or symbolic in intent, can scarcely be rated as a high achievement in sculpture.

Petroglyphs occur much less frequently in western Polynesia; quite rare in Tonga, they are unknown in Samoa or the nearby Loyalty Islands. Dendroglyphs like those that are abundant in the Chatham Islands, near New Zealand, do not seem to have occurred in Polynesia; or if they did exist, the trees bearing them have not survived.

BIBLIOG. T. Jaussen, L'île de Pâques, CahArt, II–III, 1929, pp. 108–15; A. Métraux, Ethnology of Easter Island (B. P. Bishop Mus. B., CLX), Honolulu, 1940; J. Röder, Das Bustrophedon der Osterinselschrift, Ethn. Anz., IV, 1, 1944, pp. 475–80; C. Rusconi, Objetos arqueológicos de la Isla de Pascua, Anales Soc. cientifica argentina, CXLI, 1946, pp. 213–19; C. Schuster, Some Artistic Parallels between Tanimber, the Solomon Islands, and Easter Island, Cultureel Indië, VIII, 1946, pp. 1–8; M. D. Sahlins, Esoteric Efflorescence in Easter Island, AmA, N.S., LVII, 1955, pp. 1045–52; A. Métraux, Easter Island: A Stone-age Civilization of the Pacific (trans. M. Bullock), London, New York, 1957; T. Heyerdahl, Aku-Aku, Chicago, 1958; Norwegian Archaeological Expedition to Easter Island and the East Pacific, 1955–1956, Reports, I, Stockholm, Chicago, 1961.

Gilbert ARCHEY

Illustration: 1 fig. in text.

POLYNESIAN CULTURES. Although the arts of Polynesia are the arts of not one but of fifteen or more peoples, they may be identified as comprising a single, distinctive art. The differences between the products of the various areas of Polynesia are broad, being the outcome of individual development in widely separated groups of islands; the unity among them is to be seen in a common, but by no means uniformly presented, basic form. Developments from that form, though similar in procedure, led to vastly different results in each region.

Although the influences of geography, history, and religious beliefs and customs inevitably influence the artistic forms, it would be a mistake to consider such influences as the only forces at work, and to regard the tribal craftsmen (as ethnologists are sometimes inclined to do) as merely reacting to material and social conditions. The works to be treated in this article are genuine works of art, lacking neither in individuality nor in enterprise; and although figure sculpture may be mainly in wood where great trees grew, or in stone where they were lacking or were scarce, there is recognizable everywhere some measure of command over material and an undoubted sense of design and the uses of symbolism.

GENERAL CONSIDERATIONS. The Polynesian artistic character stems from two factors: the immense distances between its archipelagoes, and the tropical abundance with which most of them were favored. The first made for individuality in the arts; the second gave each island group the potential for its own particular artistic development. Because of various limiting factors, this development was not exceptionally rich; and the island groups of this enormous area had been long isolated in time and space. They lacked certain basic materials — for instance, metals and pottery clay — and their peoples had not developed the loom or the wheel. In the absence of these materials and technical advances, the arts of Polynesia remained the products of individual handicraft: sculpture in wood and stone, finger-woven textiles, decorative plaiting and lashing, the decoration of tapa cloth, tattooing, the elaboration of ceremonial regalia, and the fashioning of personal ornaments in bone, ivory, shell, and horn. Among these various forms, the sculpture stands out as the important artistic achievement.

In Polynesia a common core of religious belief and observance is manifested in the commemoration of outstanding leaders, who were often considered more to be feared in death than in life; in the veneration of forebears; and in the worship of remote ancestors to whose reputation, or mana, legend and tradition had added miraculous and supernatural attributes, making of them the gods of the tribe. It is to be expected, therefore, that figure sculpture should be the common basic element of Polynesian art. Although figure sculpture is present throughout Polynesia, the images carved in each archipelago are quite distinctive, and in each area the sculptors became aware of form itself, not merely to the extent of developing their own identifiable style of sculpture, but also proceeding further to group figures in rhythmic compositions and to stylize. They developed complex patterns through local stylizations, and in one remarkable instance (Easter Island) figure sculpture was expanded to heroic scale.

Hawaiian Islands. A discussion of figure sculpture must begin with Hawaii, not as a suggested place of origin but as the region of its richest development. The great achievement of Hawaiian sculpture lies in its actively rendered naturalism and in its unusual technique in Polynesian representations of the gods. Occasionally, natural objects might be considered as symbols of the gods, but these were only for personal or vocational purposes (e.g., house-builders' familiars or fishermen's amulets); for public worship the representations used were human images, commonly carved in wood, but sometimes in stone or occasionally made of wickerwork covered with feathers.

Where there were multitudes of gods, both personal and community, the art of carving flourished, providing opportunity for individual expression. In Hawaii, nevertheless, a general local style of anatomical representation was firmly established, and characteristic Hawaiian images are easily recognizable. The Hawaiian figures are generally squat, but they are more accurately represented than in other areas and are often shown in action (pl. 196). The figure characteristically has a well-developed chest, defined by a clean curved edge from a normally rounded abdomen; the arms hang free from the body, usually straight down but sometimes in a dramatic gesture; the short, stout legs are flexed, occasionally in a kneeling attitude, but expressive movement is not attempted. Although features may be naturalistically rendered to the point of portraiture, strong emotional expression, sometimes to the extent of demonic ferociousness, is the common rule.

Large shell eyes and natural hair attached with small pegs were sometimes used in attempts at further realism. Carved hair and beards are more interesting as decorative features; these, rendered in deep-cut rectangles, soon became a stylization that in turn developed into an elaborate cresting (pl. 196). Another exaggerated form, the broad-lipped dumbbell mouth, was also adapted for decorative extensions.

The feeling for ornament and pattern was thus satisfied by an emphasis on and a decorative elaboration of anatomical detail; that it possibly could have been directed toward figure groupings can be seen in the orderly arrangement of actively posed figures used as supports of bowls or drums (pl. 197). The rhythm achieved in these group compositions in itself suggests a pattern. Artistic efforts went no further in this direction, however, for the Hawaiian sculptor's attention and interest seem to have found satisfaction in the lively and varied naturalism of the above-mentioned type of figure, in the decorative extensions of parts of the head, and in certain rhythmic groupings of the figures. These features and the distinction attained within them formed a large measure of the individuality of Hawaiian art, and perhaps some of this quality would have been lost had a tendency to pattern prevailed. In any case, the Hawaiian artist found ample scope for ornament in tapa-cloth decoration, which will be discussed below.

The coarsely vesicular volcanic rock of Hawaii, a clumsy medium for sculpture, was seldom employed, and never with notable success. Nevertheless, through its very economy of statement — no doubt demanded by the material — a double figure in the Musée de l'Homme, Paris, presents an effective aspect of realism; a single figure in the same museum is similar in form. Nine stone gods in the Bernice P. Bishop Museum, Honolulu, are merely heads carved on a piece of natural rock of a selected shape. The features are not much more than chipped-out depressions, though some attempt at decoration appears in a simple crest on two of them. Two fish gods (Br. Mus.) are done in the same rather crude fashion. A more positive achievement is apparent in some half-dozen squat figurines from long-uninhabited Necker Island; all are in the same style and of simple outline. Although the coarse rock all but forbade a convincing naturalism, in two instances the carver presented his work in a form that was both a perceptive and a positive expression of sculptural potential.

According to Te Rangi Hiroa (1957), the chief authority on Polynesian textiles, representation of the gods received the "highest technical recognition" when Hawaiian craftsmen turned to fashioning images in wickerwork, with effects that were sometimes startling. The clear imitation of the fierce aspect of the wooden statues argues both intention and success in a difficult medium. The most striking of these works (Paris, Mus. de l'Homme) lacks, however, the contrasting red and yellow feather covering (but not the pearl-shell eyes and dog teeth) that gives the examples in Berlin, Göttingen (Inst. für Völkerkunde), and the British Museum (pl. 199; V, pl. 265) such a vivid and dramatic effect.

Even more ambitious examples of featherwork were executed in Hawaii; the art approached the level of Western esthetic in its softly textured royal cloaks, which revealed the full range of color, form, and design that lay within its competence. Technically the feather cloak is entirely Hawaiian in invention and doubtless also in design. It was certainly developed through royal patronage, for royal prerogative claimed red and certain other colors of feathers because of their rarity. Only men were allowed to make these cloaks.

It is possibly true, as Te Rangi Hiroa explains, that the natural sagging of a rectangular length of fabric initiated both the shape of royal capes and cloaks and the curved elements comprising their designs; but the cloaks were handled imaginatively and individually, and in their great variety no two are alike among the 137 known examples. Red, yellow, black, and green (in order of preference or availability) were the colors chosen; crescents, lozenges, triangles, and circles were the design shapes employed. The patterns were usually balanced, though one or two have a pleasing asymmetry. The most notable cloak, that of King Kamehameha I (1737–1819), was made almost entirely of the rare mamo-bird feathers, a deep yellow in color. A feather-covered helmet completed the royal regalia. Produced by the wickerwork technique, it was surmounted by a crest, either in continuous openwork or as a row of stalked disks, the same type of cresting that appeared in a variety of forms on the carved wooden images.

The decoration of tapa cloth, the universal Polynesian fabric, had a special technique in Hawaii, where mechanical means of decoration were employed instead of freehand brush painting. In making the tapa cloth, the women liked to obtain a "watermarked" effect by grooving the surface of the wooden mallets, or tapa beaters. On discovering that the marks could be made more effectively by first wiping the mallet with pigment, they soon learned to make light printing stamps — some adapting the beater designs, others more freely inventive. With such stamps and a range of vegetable dyes (black, gray, red, yellow, green, lavender), they achieved an inexhaustible variety of geometric patterns.

The tapa-beating process itself was used to produce patterns, and also to make a fabric white on one side and tinted on the other. This overlaying consisted of applying a sheet of colored tapa to one of white and beating the two together.

"Snapping," the releasing of a stretched cord charged with dye, resulted in lines that had an added textural quality from the twisted fibers of the cord. However, by far the greater number of Hawaiian tapa-cloth patterns were produced basically by carving, that is, with printing blocks; similar patterns incised on gourd surfaces were darkened — literally inked — by soaking first in bark extract (tannic acid), then in black mud containing iron.

Apparently the abundance of mechanical means developed for tapa-cloth printing satisfied interest in surface ornament, for the common Polynesian custom of tattooing was only occasionally practised in these islands, and then as a mark of mourning. The designs were less enterprising than in the other island groups such as the Marquesas. Surface patterning was also absent from the wood sculpture; here again artistic expression was directed to active, realistic figure portrayal and rhythmic compositions, as well as to the creation of the splendid feather cloaks that are the special artistic achievements of the Hawaiian Islands.

MARQUESAS ISLANDS. The Marquesas Islands are an archipelago 2000 miles southeast of Hawaii and 750 miles northeast of Tahiti, isolated by the intervening atolls of the Tuamotu Archipelago, which are wholly lacking in art forms. In the Marquesas, more completely and uniformly than anywhere else in Polynesia, except perhaps in New Zealand, there was evident a progression from naturalistic figure sculpture through stylization to abstraction.

Marquesan figure sculpture was vigorous, but it displayed less freedom than that of Hawaii. A degree of restraint is evident: images are stiffly erect in attitude, the heads are disproportionately large, and the limbs are usually confined within the natural bulk of the selected wood or stone. (Stone was especially characteristic of Hiva Oa, where a fine-grained tufa was available.) Generally, the figures in wood are more slender, as though contained within the limits of a tree trunk intended for house support; figures in stone were broader, as if appropriate to a unit in a solidly built wall (PL. 198). This is not to say that no naturally proportioned, freestanding figures exist; such figures are to be found, but not frequently, and in general a restrictive stylization prevails. It is interesting, moreover, to observe figures rendered in this conventional manner not only varying in proportions in accordance with the medium used, but also ranging in size from miniature to massive — from the diminutive tiki, carved on small cylinders of human bone and used as suspension cord toggles, to solid monoliths several feet high.

In the head, particularly, a characteristic Marquesan manner or style of presenting the features became almost a universal convention; the eyes are large and round, the nose is broadly spread, and the mouth is formed from three long parallel ridges. The face is a shallow relief mask, rendered with a great economy of carving, and a simple grooving serves to distinguish arms and legs and to outline their strong flexures (PL. 198).

When the Marquesan carver turned to smaller tasks, he used the same sculptural conventions but carried them forward into more thoroughgoing stylizations; these enabled him to produce, for example, closely joined pairs of figures to decorate chieftains' official staves or the handles of fans (FIG. 449, b) used in funeral ceremonies. Even finer stylizations of almost purely abstract quality (although the human figures are always recognizable) were essayed in harder but more evenly grained materials such as bone and ivory, which enabled craftsmen to create, for example, the intricate fretwork for women's ear pendants.

Nearly every article made was covered with shallow surface patterns, all variations of the same motifs of stylized figures, with some additions such as stylized insects, foliage, and flowers. Here, also, designs varied in their degree of delicacy as well as in theme, for as the application of ornament progressed from sizable wooden bowls (PL. 197), to the thin turtle-shell segments of eye-shades, and finally, in the form of tattooing, to the human skin (PL. 208), it became finer and more involved. In Marquesan decorative relief, the human figure as a whole is stylized differently from the face: the figure tends more toward angular simplicity, whereas the face is presented in a curvilinear manner. Hands and feet, as well as eyes and ears, are drawn out freely into loops or curved into spirals, with repetitions that could be axial, radial, or consecutive, and there was apparently no end to the variation possible either in general theme or in details. One especially characteristic design should be mentioned, however, for the interesting simplicity of its arrangement. This appears on the expanded head of a club (I, PL. 436), on each side of which there is a face with eyes and nose in the form of tiny stylized heads. Wherever simple decoration is required — as on the butt of a pestle, spear, or paddle — one of these small heads of faces is found. Departures from this characteristic stylization are few, so firmly was it established throughout the Marquesas.

As tapa-cloth decoration was the primary graphic art of Hawaii, so tattooing constituted that of the Marquesas. In Hawaii the decoration of tapa cloth was a printing art, having developed from the incised or carved designs on the mallets; similarly, in the Marquesas the low-relief patterns developed in wood carving provided the details and designs for tattooing — a development that is not surprising, since tattooing also employs a mallet-and-chisel technique. The result was often an amazingly rich pattern, usually handled with great effectiveness. Apparently it sometimes went too far, however, partly through the virtuosity of the tohunga (tattooer) and partly through the subject's pride in his own fortitude. The man's ambition was to be tattooed all over, and "all over" in this case extended beyond a decorative enhancement of the perfect masculine figure, to become instead such a minutely detailed covering of the skin that the pattern itself almost disappeared into a uniform blue. In view of the painfulness of the process, it is likely that the customer determined the extent of the tattooing and that the artist, of necessity, surrendered his feeling for design to pride in technical finesse.

The art of tapa-cloth decoration seems to have been almost completely absent in the Marquesas, for it is rarely mentioned. Dr. Willowdean C. Handy refers to the tapa-covered head of a club (on which a typical face pattern is painted) as the only example of Marquesan tapa-cloth decoration known to her, and this author knows of only one other, that in the Dominion Museum in Wellington, New Zealand. Similarly, the notable feather art of Hawaii was little practiced in the Marquesas.

CENTRAL POLYNESIA. The nature and condition of the arts of Hawaii and the Marquesas are quite clear and unconfused, as their geographical isolation made almost inevitable. The same cannot be said of Central Polynesia, however, that area comprising the Society, Cook, and Austral islands. The first uncertainty arises from the uneven distribution of the islands and the difficulty of establishing whether the area should be considered as a single extensive entity, as five small archipelagoes, or as two definite groups and twenty separate islands. To complicate the matter further, the first missionaries quickly caused the religious symbols — that is, the art objects — to be destroyed, and the places of origin recorded for the few that have survived are often doubtful. It remains clear, however, that there existed in this area, as in the Marquesas, figure sculpture, stylized figure compositions, and patterns. Whether the typical stylization derived from the sculpture, or whether the patterns and stylizations are from but one locality, cannot always be decided with certainty. There is no certain evidence that planned and regular communication existed between these islands, but only of chance contact from time to time. This is reflected in the arts, for although certain anatomical details are repeated in all images in the area, faces, figural ornaments, and decorative patterns are distinct in each locality. Thus, in Central Polynesia, as in Polynesia generally, there is a basic unity underlying the local development, but here on a lesser scale.

Society Islands. This island group, of which Tahiti is the major island, is an archipelago sufficiently separated from the rest of Polynesia to be regarded as a subarea. It undoubtedly had its own sculpture style, but evidence points to a relationship with other areas. That figure sculpture formerly flourished in the group is known from the observations of the early explorers, who unfortunately left no descriptions, and only about twenty surviving images are available from which to try to fix upon the type. This may be described as short, heavily proportioned, with a large, forward-projecting, triangular-faced head; the upper arms are narrowly separated from the body, the forearms and hands are affixed to the abdomen. A flat, horizontal, shoulder ridge, a hollowed back, projecting buttocks, and strongly flexed legs are consistent characteristics. Shallow carving gives only sufficient representation of features with no particular expression (PL. 198).

These figures are small and fashioned with a rather artless realism. Their character was not marked strongly enough to establish a convention, and there is a moderate range of variation among them; but with so few examples localized it would be hazardous to assign a particular style to any one island. There are a few images in stone, clumsy versions of the wooden prototype. Working with stone tools on the available coarse vesicular rock must have been difficult; it is surprising that it was attempted, except on behalf of some religious observance.

All in all, Tahitian figure sculpture is of very modest achievement, possibly because of mild interest in the carved figure itself. In Tahitian belief the image itself was not always, or essentially, the god; the god might abide within it from time to time, but deity was more surely immanent in the red feathers fastened to the image by sennit bands. As the fiber binding increased in order to provide more feather attachment, the image itself became a mere misshapen billet, completely enclosed. But even here the human form still seemed to have significance, for when the sennit covering had itself become a shapeless sac, it sometimes had worked upon it a crude human outline in fiber.

Images in wickerwork were not unknown in Tahiti. Cook in his first voyage reported meeting with "a very extraordinary curiosity called Mahuwe [the Polynesian demigod Maui]; it was the figure of a man made in basket work 7½ feet high and [in] every way large in proportion; the head was ornamented with four nobs (sic) resembling stumps or horns, three stood in front and one behind, the whole of this figure was covered with feathers, white for the ground upon which [was] black imitating hair and the marks of tattou." Tahitian applied sculpture was more successful. Fan and fly-whisk handles bore back-to-back figures like those of the Marquesas (though less elaborately stylized), sometimes angular, sometimes curved, sometimes the two together. In a Huahine canoe ornament, for instance, a pair of back-to-back images stand above a completely angular figure convention. Fan handles can be arranged in a series starting with an elegantly stylized pair of figures, continuing in their steady reduction to become nothing more than a perforated oval end-tab, but always with sufficient repetition of the decorative detail of the original figures to establish the derivation.

It is in ceremonial fly-whisk handles, however, that Tahiti reveals its most interesting and successful art form. Once again a series begins with simplified paired figures, but instead of showing the diminution of the fan handles, there is a development of formal statement that, through curved extension of the heads and angular modification of the limbs, achieves design of abstract quality (FIG. 449, *c*). It seems curious indeed that the naïve naturalism of Tahitian figures should have given rise to such clearly achieved formalism.

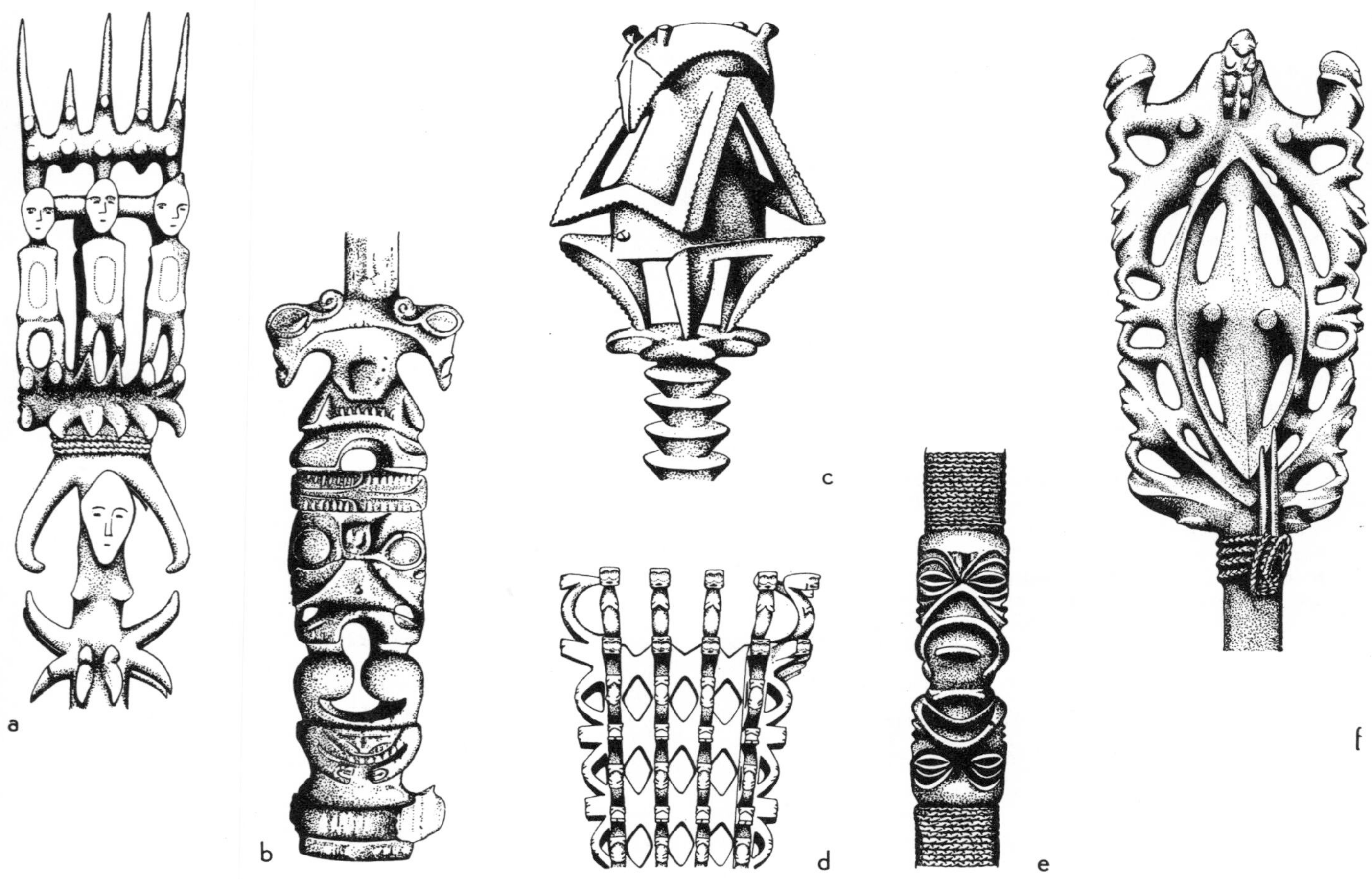

Decorative motifs of the Polynesian archipelagoes. (*a*) Spiked god, from Mangaia (Cook Islands), detail; (*b*) fan handle, from Marquesas Islands, detail; (*c*) fly-whisk handle, from Tahiti, detail; (*d*) staff god, from Mangaia (Cook Islands), detail; (*e*) decoration of ceremonial staff, from Rarotonga (Cook Islands); (*f*) staff god, from Society Islands, detail.

Finally, there is one small object that bespeaks undoubted design competence somewhere in the Society Islands. It is a "staff god" (FIG. 449, *f*) from an unrecorded island; the manner of stylization suggests Tahiti or a nearby locale and reveals that someone envisaged a design, sensed the precise degree of stylization required, and carved with sure skill. Possibly it was the fate of this charming creation to have been smothered under sennit and feathers. This piece is much too typical of its area to be an introduction; its existence is clear evidence of an active and competent school of wood carving — yet it illuminates the fragmentary and uncertain nature of the records of the condition of art in Central Polynesia. It is evidence, too, of versatility, for it is in quite a different mood from the assured abstraction of the fly-whisk handle designs.

Austral Islands. Primitive art is often thought to be bound or limited by its medium, as in the Marquesan figures that seem to be little more than shallow relief cut into the tree trunk, but certain Polynesian craftsmen, such as the Hawaiians, escaped this limitation to render freedom of both pose and action. Raïvavaé, at the southeastern extremity of the Austral Islands, also exercised freedom of design, but to exploit another interest — that of formal sculptural expression; and this gives rise to the speculation that the interrelation of planes so successfully achieved in this sculpture had inspired the angular figure convention of the Tahitian fly-whisk handle. Despite the 400 miles of open ocean between these islands, the suggestion is not farfetched, for Polynesian canoes are known to have traveled greater distances; moreover, ethnographic evidence points to Tahitian-Austral contact.

The formalism of the Austral sculptures is evident in a two-foot-high stone figure (Oxford, Pitt Rivers Mus.) and, in more robust proportions, in the massive 10- and 12-ft. statues now in the grounds of the Papeete Museum, Tahiti. Austral surface pattern evolved a stylization that did not follow the sculptural manner closely, but included such details as the wide collar and the triangular breasts that leave no doubt as to its origin. These patterns comprise bands of stylized human forms, alternating with curved and angular elements taken from the details customarily ornamenting the figures.

To judge from the large number of Austral Island decorated paddles (PL. 201) and bowls in existence there must have been an extensive souvenir trade in them in early shipping and whaling days, as there was in the equally popular Mangaia Island ceremonial adzes.

A three-foot wooden image, somewhat Tahitian in its general aspect (IV, PL. 226) was found on the Austral island of Rurutu. Spread over the image's surface and grouped on the face to indicate features are some two dozen or more high relief or semidetached figurines. These, however, are carved differently from the large figure; they are clumsily semi-stylized and have a crude, rather angular rendering of the face found elsewhere only on three other small carvings, one definitely from Rurutu and the other two supposedly so. They are very different from Raïvavaé work and altogether inferior, but they have the interest of revealing a clear local manner or school in an island at no great distance from this other distinctive style center. Communication between islands undoubtedly existed, but, it would seem, sporadically and not always with influence on well-established local art practice.

Cook Islands. Mangaia, the southeastern island of the Cook group, is known for its characteristic angular surface decoration. There is a mild controversy as to whether or not this so-called "K" pattern (PL. 208) is a human figure derivation. Hundreds of elaborate adz handles covered with this pattern are known; for the rest, Mangaian art comprises only four "staff gods," and one of these is of doubtful origin. There were also two large stone images, but shortly after they were found both were broken up and the fragments built into the walls of a church.

If the unlocalized staff god (FIG. 449, *a*), or "spiked" god, as it is sometimes called, could be certainly attributed to Mangaia, a definite human figure stylization could be traced as the source of the "K" pattern. Bearing on this point is the existence, on examples of staff gods definitely known to be Mangaian (FIG. 449, *d*), of arched shapes that may be human figures, in the place corresponding to where figures stand on the spiked god. Moreover, staff gods with similarly placed arched shapes appear elsewhere in the Cook Islands.

The art of Mangaia, therefore, comprises stone statues about which we know nothing except that they were very large; much-stylized figures or decorated arches, depending upon interpretation; and the purely decorative, characteristically Mangaian "K" pattern. The great adzes, or rather the adzes with great handles that carry this pattern, have been called "peace axes," but one of the first missionaries described them as a kind of god. Like the Austral Island paddles they were made extensively for the souvenir trade; in the early 1800s they were in strong supply in Rarotonga, 120 miles away.

In Rarotonga the same three elements — sculpture, stylization, and pattern — again appear. Pattern is not used lavishly or even extensively, though with pleasant decorative effect. The sculptural form is distinct, despite the inclusion in it of the flat horizontal shoulder ridge first noted in Tahiti and that also was found in Raïvavaé, as well as the typically Tahitian disposal of arms and hands. Rarotongan sculpture is more finely finished, however, and, at least in one figure, much more naturalistic in proportion. This particular image carries on the breast three small images, almost full figure, and two in relief on each arm, a feature seen in Rurutu and that also occurs in New Zealand. The five other known images, small "fishermen's gods," are short and stout, though definitely in the same style as the first mentioned.

The treatment of head and face especially typify Rarotongan sculpture; a high domed forehead marked medianly by a ridge continuing downward as indication of the nose, and clearly defined curved grooves and ridges forming enlarged eyes and mouth are present in every figure, freestanding or relief. This formalized face, reduced or compressed, became a decorative motif, but only on cylindrical shafts (FIG. 449, *e*) or terminal bosses. The large staff god, Rarotonga's most common religious symbol, became its most characteristic art form. It consists of a long thin staff wrapped in tapa, with an exposed part at either end expanded into a figure composition. The upper expansion carries figures in two ranks, those of the lower rank in shallow relief in an alternation of full figures and faces, both strongly stylized; those of the upper rank project freely and appear alternately in fullface and profile. This expansion terminates in the large formalized head, dominant both as art and symbol. The figure succession achieves a certain rhythm in its alternation of attitudes, but it is crowded and gives the impression that it is primarily a gathering of ancestors with care taken to omit no one. It is interesting, however, to note how the accepted manner of indicating anatomical details in more naturalistic sculpture can be traced in these stylized figures. The staff god composition is almost certainly a symbol of tribal genealogy, and the statue with supplementary figures represents parents and children. The same two symbolic groupings appear in New Zealand, interpreted by Maori wood carvers.

Consideration must also be given to the chain of small high islands, from Aitutaki to Mauke, standing some 120 miles to the north of the Cook Islands. A certain lessening of achievement is apparent, as if the endeavor were affected by the proximity to the aridity of the atolls still farther north, the most northern of the Cooks.

The only figures known to be of this group are a few from Aitutaki. They, too, are reminiscent of Tahiti in the attitude of body and head, in the line of the flat shoulder ridge, and in the position of arms and hands; but they are poorly worked, roughly carved and expressionless, the features being little more than simple incisions. The legs particularly are thin and ineffective. Paired and superimposed figures occur, recalling similar examples in Tahiti.

The presence of staff gods in these four islands, in Aitutaki, Atiu, Mitiaro, and Mauke, completes a chain of them around Central Polynesia, excluding the Austral group. All are of one general form: a thin shaft with an expanded upper portion, very like a mace, but rendered on each island in an individual way. The elegant design of the highly stylized figures in the Society Islands specimen has been previously noted; Rarotonga has one with almost the same grouping, though with figures unmistakably in the somewhat stolid local style. In the neighboring island of Mangaia — some 120 miles distant — the side arches and their decoration can hardly be described as being of human form, but it is not unreasonable to interpret them as having derived therefrom as the end result of simplification in general form with angular modification (the so-called "K" pattern) of features and limbs. The indication of human form in specimens from Atiu and Mitiaro, and in one from Mangaia, would constitute the intermediate stages of development.

Aitutaki staff gods have two decorative conventions, the arches previously described and an angular serration of reduced human figures along the edges. A Rarotongan canoe ornament has a somewhat similar, though probably unrelated, edge design, with the figures more naturalistic and standing differently.

Although general similarities are discernible in the style and design and even in certain details of these emblems they do not warrant an attempt to define a pattern of distribution, particularly because a grouping on the basis of the details (especially the arches) would not be consistent with a grouping according to general form. At most there might be a connection between Rarotonga and Aitutaki and perhaps between Atiu and Mitiaro, but Tahiti, Mangaia, and Mauke each stands apart.

The general picture presented by Central Polynesian wood sculpture, then, would be one of common elements in carved images and a common derivation for the form of the staff gods; however, the local renderings of these differ to a degree that indicates long periods of separation, during which there was infrequent and irregular contact among the various areas.

It is not inappropriate to refer to tapa-cloth decorations and tattooing as minor arts, for, although they might be more highly developed than the other arts in a particular locale, they do not have the universal impetus of religion to stimulate them, but only domestic or personal satisfaction. Although they were not universally found, they may have been in more regular use in Central Polynesia than is thought. The first explorers of Polynesia often mentioned, but seldom described, these arts; the few positive remarks recorded may therefore be worth attention.

Captain Samuel Wallis stated in the 18th century that the women of Tahiti were "universally tattooed" about the thighs and loins; and Captain James Cook observed that both men and women were proud to display the prominent arches tattooed on their loins, adding that their buttocks also were covered with a deep black. The art seems nevertheless to have been somewhat free in choice of motifs, using, as Cook said, a variety of ill-designed figures of men, birds, and dogs, circles or crescents, "just as fancy leads them." Ellis reported seeing convolvulus wreaths, coconut palm, human figures in action, any kind of animal, and, a modern touch, even muskets used as design elements.

Tapa-cloth designs in the Cook Islands were angular or curved, and bold, but very simple. With dyes obtained from leaves, berries, bark, and flowers, as well as from swamp mud, the whole piece would sometimes be dyed, perhaps a light brown, and the pattern brush-painted in red, yellow, or black. Dyeing frames with leaf midrib stretched across were colored and pressed on to the cloth, introducing a printing process.

In one example of the Tahitian royal funeral costume (Br. Mus.) a pearl shell crown is surmounted by a magnificent radiating crest of long red feathers. The face mask of two shells with a small perforated view hole lies within a great vertical crescent of five pearl plates. Below it is an apron evenly studded with pieces of shell. Worn over a folded poncho of white tapa cloth is a long cloak of blue-black feathers; long streamers fall from its side and from the horns of the crescent. The wearer bore a long feather-plumed staff.

New Zealand. Although located in the southwest of the Polynesian area, New Zealand seems most related artistically to Central Polynesia. This affinity is seen not so much in forms and style, which are unique, but in origin, that is, in the figure sculpture that is part of every Polynesian art, and also in the primary figure grouping that led the way to designs and patterns. Wood carving, after a period of deplorable degeneration, was revived around the turn of the century. One must distinguish therefore, between steel-tool and stone-tool carving, the latter being native Maori work untouched by Western cultural influences (see NEW ZEALAND).

Few of the large, freestanding statues of the stone-tool era are fully in the round; the majority are broad, lacking depth, their limbs being contained within the frontal plane of the body. The arms, moreover, although naturally rendered, are usually arranged in a pose of rhythmic order.

The few images that are fashioned in natural proportions are small, only inches in height; their somewhat rigid stance is sometimes relieved by more freely rendered arms, and occasionally by a softer, more pliant body modeling.

Each post and plank in a Maori tribal meetinghouse carried the relief figure of an ancestor of the tribe. These structural units, being necessarily of different sizes and proportions, were at once an opportunity and an incentive for sculptural versatility. Figures in this "applied sculpture" were nearly always appropriately proportioned — tall, slender, and in high relief on posts (PL. 205), but in low relief on wide upright wall planks (PL. 204). Where a post was long and the extension of a single figure throughout its length would make it appear functionally weak, two or three figures were superimposed. The three figures might change in proportion and attitude, from the stout upstanding figure at the base to a lighter "ancestor" in the middle, and a still lighter, actively disposed figure above. So frequently does this arrangement occur that it is not unreasonable to suppose that the order was the result of positive awareness of an artistically appropriate succession.

The lively attitudes of the relief figures illustrate the interest in design that pervades Maori art. Freestanding figures are never in motion, only in an ordered pose, and naturalistic figures portray activity only when they are part of a composition. In relief figures action is frequently presented, but here again design takes command; rarely does a single figure fail to express the form and order of the plank that contains it.

The detailed surface decoration usually applied to figures was similarly appropriate in texture to the major volumes, being angular, even rugged, when these were massive, and appearing as light arabesques or softly turning spirals when they were smoothly modeled. Although the spirals were occasionally over-assertive, they introduced liveliness into an otherwise static form.

In the treatment of the face the carver also revealed his interest in exploiting form. Freestanding figures often bore a wide-mouthed, glare-eyed expression of defiance. In Hawaii the same defiant aspect remained in undiminished fierceness notwithstanding the ornamental detail added to it (PL. 196). In Maori carving, however, the decorative form bestowed on this originally vigorous grimace frequently made of it no more than a placid mask. The same mask turned sideways gave the carver another decorative idea; sometimes the profile seemed to develop, as in a series of faces turned successively more and more to one side; sometimes it was secured directly by dividing the fullface mask in half (FIG. 454, *a*). Although this half mask

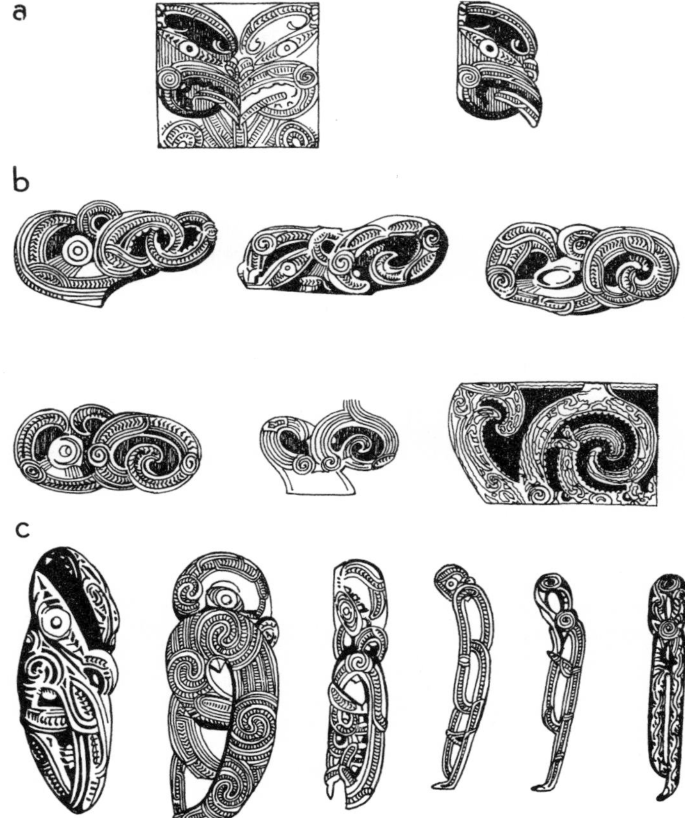

Decorative motifs from New Zealand. (*a*) Mask, full face and profile; (*b*) faces in profile with spiral mouths; (*c*) faces in profile of the "manaia" type.

or profile became conventional, it was not stereotyped, but rather was the starting point for further development. For example, it was often drawn out into a decorative band consistent with a narrow space in a composition. Its flexibility also allowed the lips to bend, to curl into a spiral; when two of these manaia heads were opposed and their curled lips interlocked, a double spiral resulted (FIG. 454, *a*). Thus the Maori achieved his two most frequently used decorative elements — pitau, the double spiral, and manaia, the human face and figure in profile. The foregoing account presents them as natural developments within Maori art; another view, discussed below, is that they are introductions.

There was a further progress of decorative invention, first through compositions of naturalistic figures, then through works of increasing degrees of stylization, until finally patterns of rich complexity were achieved. Numerous carvings exhibiting all these phases have been preserved, amply documenting the development or transition from one to the other.

To demonstrate the full range and variety of stylistic moods in Maori art would require an abundance of illustrations, but examples from house decoration can provide a bare outline of styles. The graven ancestor figures standing along the walls and forming part of the essential structure of the house avoided monotonous uniformity through variations in size, though never in style in any one house, and through their placement in alternation with brightly colored reed panels (VIII, PL. 91). But when, as sometimes happened, figures stood directly alongside one another, invariably a rhythm of alternate attitudes, full-face and profile, tiki and manaia, was established. This is the theme of the verandah threshold, and it appears also on canoe wash strakes and on other carvings.

Smaller structural units, for example doorways, called for a lighter theme, and here again unity is maintained both in architecture and art. Superimposed upright figures provide the doorjambs, and the lintel becomes a medium for infinite variety of design. The lintel panel may contain naturalistic figures in active posture, perhaps with stylized figures intertwined, tiki alternating with manaia, or either or both of them alternating with lively spinning spirals. In lintel decoration alone no less than five styles or "schools" of design have been developed; they are to some extent localized, but each is clearly an expression of the common theme of human figure stylization (PL. 203). From this basic motif of alternating the stance of figure and figure derivatives has developed an intriguing variety of decorative patterns, not only in lintels but in articles of such different structural forms and sizes as canoe prows, stern posts, treasure boxes (PL. 208), and musical instruments. Where only a touch of decoration is called for, as on a comb or a small instrument, there may be found a single element, a stylized eye or mouth, or perhaps a hand, always aptly applied. The basic concept and the standard theme remain, however; nowhere is an art with a basic motif underlying locally developed variations more clearly seen than in Maori wood carving.

The discovery of the semiprecious stone, nephrite, enabled New Zealand alone among Polynesian islands to have an art in jade. Again, the human form was its theme, and the squat figure amulet, the hei tiki (PL. 200; VI, PL. 52), was developed. Contorted attitudes contrived in the green stone are reminiscent of Chinese dragon motifs, but they are human, some in normal profile stylization, others (the pekapeka) rather more involved, with a head at either end. Either the stone was too hard or its smooth surface too much appreciated for surface decoration to be attempted. Useful articles sometimes became personal ornaments, such as the fish hook amulet (hei matu) or the leg ring (kaka poria) for a tame bird. Maori carving in jade is relatively simple and nowhere matches the expressive forms achieved, for example, in the Americas. Unlike the Americas, however, it had no accompanying art of pottery modeling to inform and stimulate its sculpture.

Textiles as well as houses were more substantial in New Zealand then elsewhere in Polynesia. Tapa, the beaten bark that provided Oceania with clothing, and incidentally a surface for decoration, was too flimsy for New Zealand conditions, and a woven fabric was called for. The fiber of the long-leaved "flax," really a kind of lily, provided the material, and a basketry technique called finger-weaving produced a close-knit fabric. This could be ornamented with thrums of colored cord or with feathers, but the range of geometric patterns was limited. The taniko cloak, which was decorated only with a plaited border design of close-set lozenge elements, was especially pleasing for the lightness and sheen of its material made from the softest fiber plaited as finely as the technique would allow.

Sharing the common Polynesian lack of the art of pottery as well as of weaving, the Maori were deprived of another opportunity for decoration offered by a clear surface. The gourds used as water containers were too hard-surfaced for paint, but somewhat constricted loops were pricked or scratched upon them and sometimes rubbed with color. Painting was a small part of Maori art, used more in pattern than in representation. The uniform house color, red, was occasionally relieved by painting white, or leaving uncolored, an uncarved band on the front verandah barge-board; the inside rafter decoration was a bright, flowing design of loops and scrolls in black, white, and red; the same designs in rather more freely undulant rendering made a fitting decoration for the bow and stern of the canoe.

Rock-shelter painting or drawing was more ambitious. A few scattered sites exist on the North Island, but in certain districts of the South Island there are many limestone shelters with drawings of men, birds, and reptiles. Some are rather schematic attempts at realism; others, although influenced by wood-carving stylizations, are as free and extended as the use of brush or charred stick allowed. Drawings of houses and ships bring this art forward to relatively recent times; but it is not possible to date the earliest efforts, even by the remarkably detailed drawings of the moa, because it is uncertain when this giant wingless bird became extinct. The moa drawings could, of course, be perpetuations of a lingering memory. In any case, the inclusion of elements of well-established stylization indicates that these drawings belong more to the developed Maori period than to the period of the earlier "moa-hunters," whose history and art are becoming better known through archaeological exploration.

In speaking of the moa-hunters, ethnographical typology and art must be discussed together. Coastal archaeological sites in the South Island have hitherto been the chief source of knowledge of a culture earlier than that of the traditional fleet-migrants, the creators of the arts already described. Only recently have North Island sites, also coastal, been clearly identified. Possibly the finest example of the art of this culture — and it undoubtedly is a work of art — is an implement, a stone adz; it is also the prime typological criterion, marking the culture as Polynesian in character. Adzes of this type are rectangular, with the backs curving gracefully into the tanged grips; their large size, finegrained stone, and beautiful finish mark them as works of heirloom quality; their inclusion among grave goods of important burials also attests to their status. The occurrence of the same type of adz, finely wrought, in 10 other island groups indicates their widespread Polynesian usage.

Also of marginal Polynesian occurrence are ridged cylindrical "reels" in stone and bone found in the moa-hunter levels, and whale-tooth pendants, so called from the shape as well as from the material, though they were also made in moa bone. These ornaments are representative of the general culture and art that the earliest Polynesian arrivals brought with them to New Zealand, probably 10 centuries ago. Unfortunately no woodwork of this period can be definitely identified. Some ivory pendants of more ambitious design open up the question of whether the lintel chevron ornament is early or late in Maori art development. These "tooth pendants" are bordered or fringed with two or three rows of close-set chevrons, which are clearly limb conventions related to those of the lintel. From their wide distribution, particularly along the South Island coast, an early or even moa-hunter age might be inferred, but they have not been found in any moa-hunter horizon; indeed, they are all surface occurrences except for one in a burial (not an occupation level), and this was accompanied by a bone comb of recent type. As part of an essay in art the pendants could be no more than instances of stylization by simplification, the so-called degeneration of a motif, and this alone would hardly be evidence of age. The lintel, however, presents a finely realized abstract form which, if it were proved to be old, would constitute evidence of undoubted, perhaps even unexpected, assurance and maturity in the earliest endeavor of Maori art.

Mention was made earlier of possible art relationships with Central Polynesia. The first relationship is in content. The appearance of human images carrying smaller figures, presumably children, is fairly widespread. Some examples are a stone image recently discovered in Tahiti; the Rurutu "box god" with its two dozen surface godlets; an image from Rarotonga; and many such images in New Zealand. This figure composition, plainly narrative, is a family record, handled, at least in Central Polynesia, with no particular feeling for art expression, except perhaps in the well-carved Rarotonga figure. The Maori in New Zealand incorporated this theme in his decorative art. Naturalistic renderings similar to the Rarotongan are known; but as the main figure became stylized in low relief, the others, wife or children, followed suit to form a decorative design on the parent's body. Sometimes they appear between his legs or in the narrow space between the head and the edge of the plank.

The second relationship, or at least resemblance, is in composition; it is the alternate succession of relief figures in fullface and profile attitudes. The stylization of the staff gods of Rarotonga is somewhat labored, which is not surprising, for the wood is very hard. In New Zealand, where relatively soft, but fortunately durable, totara is available, the same alternation of figures was extended and developed into decorative designs, including figure-derived abstract forms (the chevron and spiral) of undoubted accomplishment.

Western Polynesia. The archipelagoes of Tonga and Samoa together comprise Western Polynesia. The region had frequent contact with the nearby Melanesian group of Fiji, and certain art resemblances are predictable. Tongan art comprises a few sculptured figures and an abundance of shallow surface patterns, but with nothing in the way of stylization.

The images, carved in hardwood, are small and stout, usually about a foot in height, with every member rendered in clearly stated terms. They were obviously produced with a minimum of adzing or cutting; nevertheless the surface is finely finished and volumes are fully expressed. A figure in Captain A. W. F. Fuller's collection (Chicago, Mus. of Natural History) reveals a degree of modeling of features and of bodily form that contrasts with the more angular carving of arms and torso and the more simply indicated features of the Auckland Museum specimen (PL. 198). The latter, by the way, was the only one brought away of a group of five seen in a temple in Haabai. A wooden image in the British Museum, very doubtfully assigned to Samoa, is so unmistakably Tongan in style, particularly resembling in features Captain Fuller's specimen, that Tongan workmanship is to be strongly suspected.

In addition to this small group of sculptures there is a wealth of paddles, some long spears or ceremonial staves, but chiefly clubs, the latter closely covered with small angular patterns in shallow relief. These articles rival Austral and Mangaian carvings in number, possibly again in response to the desire for easily carried souvenirs.

The patterns have a general, but by no means exact, resemblance to those of Fiji; both may be skeuomorphs of lashing patterns or simply the carving of the hand grip extended as an allover decoration. Some small part of the similarity between these patterns of the two areas could have been the outcome of early European contact when natives traveled widely in ships; there are, for example, undoubtedly Fijian paddles beautifully carved in the style of Maori double spirals. A distinctively Tongan feature is the invariable inclusion here and there in the pattern of small reliefs of men, the moon, stars, and animals. They are like the hallmarks of silver, and no doubt were of symbolic significance. Tongan skill in fine handwork appears in combs bound in finely plaited designs in hair and in baskets plaited so closely as to be able to hold water (II, PL. 232).

It is noteworthy that there formerly existed in Tonga stone-walled mounds, intended for the denoting of rank and for ceremonial observances, that were architectural at least in concept. The most important were "langi," stepped pyramids over 100 ft. square at the base, bordered by carefully fitted masonry of large limestone slabs sometimes decorated with simple geometric devices. These edifices were royal burial places with the tomb vaults paved or lined with limestone flags. An outstanding monument is the Haamonga, a trilithon, on Tongatabu. It stands 17 ft. high, 14 ft. wide at the base, and 12 ft. wide across the top (i.e., the lintel), the uprights being $4\,{}^1/_2$ ft. and the lintel 3 ft. thick. These massive limestone blocks were set up by an 11th- or 12th-century Tui Tonga (royal chief) to symbolize his hope for unity between his two sons. While Tongan wall masonry is considered to have affinity with Fijian stonework, the latter islands have no trilithons or great monoliths.

If it is appropriate to include among the art achievements of Tonga its technical skills, it is even more so in Samoa. No carving is extant except the doubtfully Samoan figure already mentioned and sparse shallow surface patterns of quite modest accomplishment. But, as in most of Polynesia there are minor arts of high quality. In Samoa tattooing is an important form of expression for creative design. Until very recently Samoan young men submitted to tattooing in an allover pattern from loins to knee. Unlike the conventionalized creature designs of Marquesas and the natural plant and animal forms seen in early Tahiti, the Samoan patterns are, or were, closely angular and by no means unbeautiful. The Tongans at one time were profusely tattooed, but missionary influence quickly discouraged the practice. In both regions ceremonial headdresses were made, but they were inferior to those of Tahiti.

In judging Samoan art it should be noted that house builders went far beyond practical necessity in the elaborate and finely wrought lashing patterns they produced. Tapa-cloth decoration also had merit; floral and angular designs were applied through carved stencils. Modern printed fabrics worn in Samoa show adaptations of these fine patterns. Samoan interest in social formality, in political power and precedence, in feasting and graceful dancing, were art expressions of such significance that they could well have displaced the more material arts.

Atolls. Northwest from Samoa, near Melanesia, are isolated islands with Polynesian inhabitants. The islands and communities are too small for much more than economic subsistence, however, and they are of interest more to ethnography than to art. Similarly, eastward at about the level of the equator lie the coral atolls, the southern fringe of the great diffuse skein of Micronesian islets.

The art of these atolls lies in craft well done, but it does not extend to figure sculpture of any merit or even to pattern. In the Ellis and Phoenix islands, the Tokelau and Northern Cook atolls, the area north of Tahiti, and the great scattered archipelago of Tuamotu (Bougainville's "Iles Dangereux"), contain few articles of interest. This does not mean that these were people without esthetic feeling, for they loved ornaments, albeit only of shell or seeds. They made the finest and softest plaited fabrics with varied knot or ply to produce patterns that were either in self-color or polychrome. Small pieces of salvaged driftwood were put together with exquisite precision to create outrigger canoes, the slender lines of which made them the fastest on the seas. Temple areas, of which abundant archaeological artifacts remain, were laid out with courtyards, platforms, and walls of monumental size.

To find examples of what might be considered works of art, it is necessary to find a high island, such as Mangareva, an upraised volcanic mass with attendant hilly islets lying within a lagoon. Unfortunately Mangareva's only surviving art contribution is a group of five wooden figures, each about 3 ft. in height. Four of them are in features, anatomy, and proportions the most naturalistically formed images in the Pacific. It is curious, therefore, to find the fifth (PL. 198) stylized and with lines simplified and organized in quite an abstract manner. The absence of pattern from this group and the assurance evident in two diverse sculptural styles lead to speculation as to what accomplishment in Mangarevan art may have been lost. The many wooden figures known to have existed were burned at the insistence of Père Laval.

Pitcairn Island, another isolated outpost, was uninhabited when it was discovered in the 18th century. Unfortunately three small stone figures standing on a platform were toppled over the cliff by the Bounty mutineers, and the headless fragment recovered reveals very little. The great wealth of beautifully formed and finished adzes and other stone implements found buried speaks of the expert craftsmen who formerly lived there.

EASTER ISLAND. In remote Easter Island (FIG. 459) the task, in popular expectation, is to explain the inexplicable — the ranks of huge stone images that have conjured up a race of giants to fashion them, or a despotic ruler driving the slave population of a great land, now below the sea, to transport

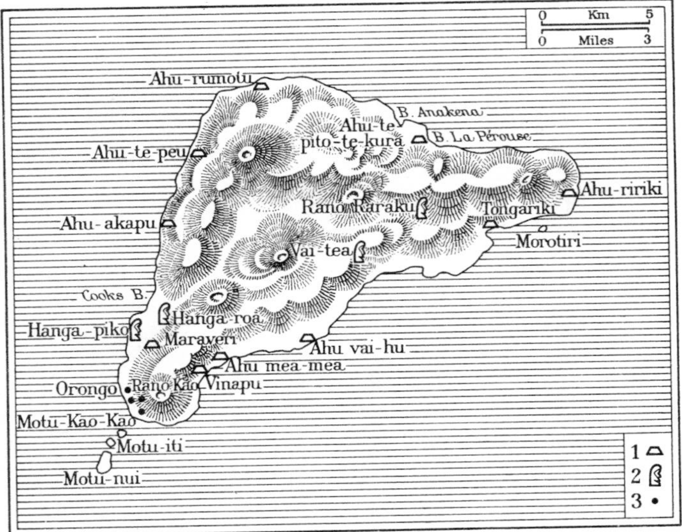

Easter Island, principal areas of archaeological interest. *Key*: (1) Stone platforms; (2) monolithic statues; (3) petroglyphs.

and erect them. So deeply ingrained is the idea of the mystery of Easter Island that a deprecatory attitude toward it could seem almost impertinent. Nevetheless the facts, well reviewed by ethnologists, seem to support the conclusion that Easter Island is still within Polynesia, though at its farthest boundary.

It is possible to be drawn into a lengthy discussion of Thor Heyerdahl's theory that Easter Island is, with respect to its outstanding art, an outlying island of Peru, but the theory is without conclusive proof at present. Heyerdahl's book, *Aku-Aku* (1958), presents three epochs: a first, with small, primitive-looking stone images and highly skilled wall masonry of putatively Peruvian origin; a second, of local sculptural expansion (the great statues) and deteriorated wall masonry; and a third, commencing hardly more than a century before Roggeveen's discovery in 1722, of Polynesian invasion followed by the overthrow of the statues. Later (*Scientific American*, Sept., 1958), however, Heyerdahl accepted the Polynesian occupation as beginning in the 14th–15th century, and it is not clear what contribution, if any, the Polynesians are thought to have made to the sculpture during those two centuries. To discuss Heyerdahl's views in detail would require a longer and more precise review of uncertain and contradictory legends and of questions of race and language than is appropriate here. One would only comment that it seems anomalous for a non-Polynesian art and presumably a non-Polynesian people to have been in the eastern Pacific islands for 600 years and yet to have left no substratum in the Polynesian languages of the region.

It might seem unusual to see sculptural assurance as marked as in Easter Island, not only in two materials, wood and stone, but also in such vastly different dimensions. Nevertheless, parallel cases exist in Polynesia itself, in the challenge met by the Maori of New Zealand in fashioning ornaments in fine jade and in hewing out great statues in wood, or in the Marquesan's equal facility in major sculpture and intricate ivory pendants. In each instance, opportunity and challenge became one, a condition true also in Easter Island. Curiously, the measure of one opportunity on Easter was abundance, of the other, scarcity. The crater wall of Rano-raraku, composed within and without of relatively soft volcanic tufa, was opened up in great quarries; these also became sculptors' studios where statues were hewn out (PL. 195). Wood was scarce, however, and therefore a precious material in itself, reason enough for devoting care and skill to the fashioning of small wooden figures and ceremonial staves. Thus two arts developed, one superheroic and the other small in scale but apt in detailed modeling. There is little in common between them save only that in both carved human figures, the "long-ears" that typified the ascendant element in the population during the period of strong social cohesion, are present. Another similarity, which may be merely incidental, is a certain economy of outline in frontal views, with freer modeling both in features and in attitude evident when viewed from the side. Excluded from this comparison are certain small, quaint animal figures derived from twisted branches.

Roggeveen, in 1722, and Cook, in 1774, saw most, or at least many, of the great statues that were still standing — those on the coastal ahu, or great stone platforms, placed with their backs to the sea and others ranging over the inland slopes. It was left to Mrs. Scoresby Routledge (1914) to discern that the latter had not been set up haphazardly, but had marked, or stood along, "three magnificent avenues on each of which the pilgrim was greeted at intervals by a stone giant facing him and guarding the way to the sacred quarry volcano, Rano-raraku."

In this ordered arrangement, religious symbols are turned to the service of art, as the image itself had already become a work of art in its own right. But these giant images were the sculptural concept of a community, rather than of individual artists, for they are all manifestly of the same type. What then was this type? It was not figure sculpture suffused with rhythmic pose as in New Zealand; neither was it actively postured naturalism as in Hawaii nor the sculptured relief favored in the Marquesas. Instead, it is the same formalism of apprehended sculptural volume that was noted in Raïvavaé, but realized here with manifold weight and power. Two kinds of bases occur, flat for the ahu images and tapering for the inland sculptures; the eyes were deep circular sockets for the former and ill-defined depressions for the latter. It is not known whether all were intended for ultimate erection on an ahu or whether they were carved in one form for the field and another for the platform. Heyerdahl states that all the images still in the quarry were without eye sockets. Only the ahu images wore the great cylindrical crowns of red rock quarried on Punapu.

Some of the field statues still retain the median ridge along the back by which they were attached to the mother rock, others have the back fully modeled. The finishing work was necessarily done on the erect statues, but it seems unlikely that the eye sockets would have been left for secondary excavation, a task infinitely more difficult than finishing off the back. The interesting girdle rings in relief on the small of the back no doubt held symbolic meaning; they were repeated on some of the small wooden figures and were painted on the back and buttocks of children chosen for the "bird-child" ceremony. Captain Cook observed in 1774 that the inhabitants were already indifferent to the statues. Intertribal fighting apparently was the cause of their being overthrown, and it is now over 70 years since the last was hurled from its ahu. Many of the field images, however, still stand erect, protected by the rock detritus and windblown sand that has accumulated about them.

Easter Island possessed but one domestic animal, the fowl, almost certainly introduced from the Marquesas Islands. The fowl was undoubtedly important; it acquired social as well as culinary status, and was the offering when "bird-children" were initiated; it was consumed as a placatory tribute for a dangerous enterprise; and it was acceptable as a conciliatory social gift.

The sooty tern, a sea bird, was given an even higher status, at least for the brief period each year when it began to occupy its breeding site on a small off-shore island. A semi-stylized narrative art arose in association with an annual ceremony, partly competitive and partly religious. There was a competition to swim out to this island, to secure the first egg of manutara, and thus to become the "bird-man" of the year; or at least his backer was so named, for he, together with the other "competitors," had been waiting safely if not altogether comfortably on the nearby cliff top.

This "place of waiting" — Orongo, on the seaward rim of the crater of Rano Kao — was a maze of caves and rock-built huts; their walls and ceilings and the rocks around became a gallery of murals, relief carvings, and paintings, sometimes of human faces or birds, but most commonly of bird-headed men. No doubt the latter formed a kind of calendar record of successive winners. Mrs. Routledge counted 111 "bird-men," and she determined that many reliefs must have disappeared. The British Museum image, which came from Taura-renga, the "house of the statue" at Orongo, has three "bird-men" and other devices proper to the cult carved on its back. The winner received a new name, the bird name by which the year itself then became known; for a year he received food and tribute while living under strict tapu in a special house on the slope of the other sacred mountain, Rano-raraku. Nearby was an ahu reserved for the burial of "bird-men."

The carvings at Orongo were purely symbolic, naïvely realistic in form and arranged in no particular order but only fitted in wherever space was available. The "bird-men," however, did appear in a decorative design on small wood carvings. The lack of trees that turned the Easter Island sculptors' attention to stone was not absolute; there were a few stoutly growing shrubs whose branches produced the toro miro wood, hard and clear-grained. Moai are the small carved images (20 to 30 in. high) through which Easter Island art is best known in museums and collections (PL. 207). It is a continuing art, or craft, supplying the ever-avid souvenir trade with works of good, bad, and indifferent quality and with shapes that grow increasingly curious. It is not untimely to quote Mrs. Routledge who wrote in 1919: " There is, of course, no sentiment connected with the figures of today; they are roughly done and merely for sale. The trade is extended to copies of stone images which are bought by unsuspecting visitors, with circumstantial tales as to their history that would deceive the very elect."

Apart from the slight resemblances to the stone figures already noted, the wooden images are distinctive. They are erect with arms free but aligned with the side and hands usually joined to the thigh. The "long-ears" are treated with decorative emphasis. Anatomical details are clearly carved, and the surface well polished. The face is sculptured with interesting feature emphasis; whether this was intended as emotional expression or was the outcome of direct interest in form is a matter of question, though it should be noted that in the expression of form the manner is stereotyped. The overly clear rendering of ribs and backbone suggests emaciation, but again it could be form emphasis, perhaps with decorative intent. It is of particular interest, and perhaps not without significance, that almost exactly the same rib exaggeration is found in two Chatham Island (New Zealand) figures (the only ones known), one in wood (Auckland, Mus.), the other in stone (Oxford, Pitt Rivers Mus.). These small images carry surface ornament, the girdle and rings on the back previously mentioned, and a pattern of opposed birds on the top of the head, which links them with the Orongo rock reliefs. This detail is always in softly finished low relief such as that appearing in small carvings in New Zealand. It is altogether too practical to attribute this to the keen obsidian knives and scrapers available in both places.

Discussion of the well-known Easter Island script could be interminable, as well as inconclusive. Perhaps for the immediate purpose it little matters what the precise ideographic meaning may have been. What can be accepted, however, is the clearly representational derivation of many of the symbols, as there can also be recognized in them a manifestation of surprising graphic competence. The graver patently followed the traditional boustrophedon arrangement expected by the reader; whether he also was aware of the decorative aspect of what he was doing is a matter for speculation, but there is at least one object, a breast ornament known as a rei miro (Br. Mus.), to which a band of script has been effectively applied.

The girdles and rings carved on statues and tattooed or painted on "bird-children" do not appear to have been elements in secular body decoration. Tattooing, as seen by the early explorers, was extensively practiced; it included close geometrical patterns, a "complete" pair of breeches, and designs of all kinds of natural and artificial articles set in agreeable arrangement on arms and body. Facial tattoo was also in well-balanced designs that kept elaboration well within bounds.

Tapa-cloth manufacture presented difficulties. The necessary trees, the paper-mulberry and the hibiscus, grew poorly and a weak fiber resulted; it was strengthened, and to some extent ornamented, however, by a form of needle quilting. The cloth does not appear to have been otherwise decorated. Both tattooing and tapa-cloth decoration are in a way combined in a curious little tapa-covered figure in the Peabody Museum (PL. 206), in which typical tattoo design has been painted on the face and neck. Feather headdresses were made for social or ceremonial occasions; they were of higher quality than those of Tonga or Samoa, but fell short of the magnificence provided for Tahitian royalty. One can do no more than guess what ceremonial style was affected in the greatest days of the ahu and their images. Perhaps we have a hint of it in the 11-ft.-high Ko-peka, a basket image clothed in white tapa cloth seen by the Comte de La Pérouse in 1786.

PETROGLYPHS. Petroglyphs or drawings of one kind or another have been noted in all parts of Polynesia except in Samoa, Wallis, and Futuna. They are rare in Fiji. The art, however, is somewhat desultory everywhere, as if it were an occasional occupation. There is nothing systematic about it in either form or arrangement. Human figures with schematic outlines, such as children draw, are the most common images used. Sometimes a face or part of a figure is set above an outline that suggests a canoe. There are fowl, pigs, dogs (the last two are scarcely distinguishable), and turtles, as well as semigeometric devices.

The petroglyphs would appear to be of simple narrative intent, perhaps to recount some event or tell of some practical need, or they may have been no more than the passing fancy of some dallying traveler.

EVALUATION. The common core of Polynesian art is figure sculpture, from which each area pursued an independent course. Most frequently this was by way of figure stylization that often, though not always, proceeded into decorative pattern; where neither convention nor pattern appeared, sculpture itself flourished into a high achievement. Sculptured images, however, were by no means uniform throughout the region. In every group of islands they present local characteristics that bespeak the islands' long separation from one another.

The rate of differentiation and development would have been slow; tradition is strong among primitive folk and the apprenticeship tradition in arts and crafts would be conservative. One agency for differentiation would doubtless have been the material available, the density and grain of wood here, the texture of stone there. But beyond this one observes, in the figures themselves, sculptural possibilities so clearly realized that no one can doubt that awareness of and interest in emergent form were part of the carver-craftsmen's outlook and competence.

This manifests itself in the opposing claims of emotional expression and decoration as for example in Maori images where the defiant mask, notwithstanding its bared teeth and glaring eyes, is frequently presented in a decorative fashion and emerges as only a mild symbol of hostility. In Hawaiian images, however, where decoration is restricted to the borders — that is, to the hair and feather headdress incorporated in the sculpture — the defiance remains. A more thoroughgoing surrender to the formal is seen in Marquesan house-post images, which are in effect renderings in relief of body and face in a firmly established local convention.

In the Society Islands naïvely naturalistic figures seemed lacking in vital sculptural incentive; yet formalized figure design achieved such success in other forms of Tahitian art that one senses a diminished interest in images themselves in Tahiti, possibly because of the high importance accorded feather attachments as the abiding place of the deity.

The interrelation of planes and volumes in purely sculptural, nondecorative terms is the special contribution of Raïvavaé, or High Island, to Polynesian art. This approach pervaded nearly all this island's sculpture, small in wood or massive in stone, in contrast with that in the Marquesas, where the known monoliths appear more as reliefs than as statues.

Surprising contrasts can occur even in one small island; in Mangareva, for instance, a figure that is abstract in almost the contemporary manner stands among the most naturalistic images in Polynesia. In explanation one can only credit an individual sculptor with artistic awareness of the forms he was handling and an urge to develop them.

The major influences on the art of Easter Island are not difficult to ascertain; the island was treeless and contained great beds of reasonably soft rock that invited sculpture on the grand scale. The blunt tools and the need to work in a trench would themselves be conducive to creating works of large dimensions, and the spirit arising from working in groups would encourage competitiveness for size. But whatever the opportunities, difficulties, or adventitious aids, no one need search for words or phrases to extol Easter Island sculpture; the statues themselves speak volumes.

One readily understands why Easter Island wooden images were small, for its timber was stunted. It is possible, too, that the lack of decorative pattern here was owing to the same scarcity. The first stages of a stylization need to be of fair size, and space is necessary to draft a pattern. Pattern was not entirely absent from Easter Island — the script would have been a very satisfying substitute.

If, in comparing Easter Island in its grandeur with Tonga on the western boundary of Polynesia, Tongan sculpture seems meager, a moment's reflection reveals that it, too, is expressive, particularly so since the materials used were the hardest of wood and still harder ivory. Neither did Tonga lack the monumental; not only were there the great temple terraces common to the whole of Polynesia but also the massive triblithon in Haabai. In Tonga, then, is another of the surprising antitheses that are so often present in Polynesian art. Antithesis is hardly the right word for the situation in New Zealand, however. The diversity in size is there, as are the extremes of simplicity and complexity, but they do not stand in opposition to each other; there is an even range between them, complete in every stage of progression and development. It is significant that these several stages are not found separately in historical succession or in archaeological horizons. Every manner and variation was practiced contemporaneously: free sculpture and applied; tiki and pou wore the structural elements of architecture, and from them emerged the single figure in heraldic form; groupings and rhythmic progressions of these evolved a real art of decorative design that could express the liveliness of dancing figures alone or the involved complexity of tracery or of arabesques in high relief. Or again, in an entirely different key, it could curve and extend limb conventions into simple but elegant abstractions.

Here, however, we embark upon a point of controversy that bears on the question of the origin and course of development of Polynesian art. To the view adopted here, that Maori art is basically an art of figure sculpture from which all its patterns have been derived by local development, H. D. Skinner of Otago Museum, Wellington, poses the alternative view that important elements of Maori art have affinity with similar elements found elsewhere in Oceania. Skinner relates the manaia of Maori art to the bird-headed men of the Solomon Islands and of Easter Island, and the New Zealand double spirals are held to be genetically connected with spirals in Marquesan patterns and in the Massin area (Louisiade Archipelago and Trobriand Islands east of New Guinea).

In this view, therefore, the curvilinear character of New Zealand art is an introduction, and as there is considerable evidence in tradition and language that the region from which New Zealand was colonized was Central Polynesia (as defined in this essay), this must also be the area from which the curvilinear art was derived.

Skinner recognized the difficulty of deriving a particular art form from a region where it does not exist, or was not found at the time of European discovery. His explanation (1951) is that Polynesian ancestors, possessing Indonesian culture elements and art, moved out into the Pacific in about the 7th or 8th century of our era. Traveling via Micronesia they reached Samoa and Tahiti "not as cultureless fishermen but as well-equipped explorers and colonists taking with them all domesticated animals and plants their vessels could carry." Notwithstanding technical losses on the way (the loom, pottery, metalwork) "the first culture established in the Tahitian group was characterized by elaborate arts and crafts. When groups of Tahitians hived off between A.D. 1000 and 1300 to the marginal areas of Hawaii, the Marquesas, the Tuamotus, Easter Island, the Australs, the Cooks, and New Zealand, they took with them a well-developed decorative art as well as the purely utilitarian arts. In these new marginal settlements representational and decorative art either declined, as in Hawaii, the Tuamotus, and Mangareva, or else continued to flourish, as in the Marquesas and New Zealand."

"This hypothesis also requires that in the Tahitian Islands as generations passed representational and decorative art dwindled almost to nothing." Skinner suggests this was due to the immense elaboration of socioreligious ritual; he continues, "Whether or not this is the correct explanation, the European discoverers found in Tahiti an elaboration of socio-religious ceremonial unknown in such marginal areas as New Zealand, coupled with almost complete absence in Tahiti of representational and decorative art."

"The absence in recent times of decorative and representational art in Tahiti has been treated as the trump card in the hands of those who argue that Maori decorative and representational art is a purely local product. For, say they, if this side of Maori art is not local but ancient, why is it not present in Tahiti? The reply is that it was there anciently but has since been lost. It is admitted that the New Zealand Maori developed a characteristic local style, especially on the decorative side. So also did the Polynesians of the Marquesas. But the more closely the art of these areas is studied the more numerous are to be found the motives which they have in common."

There are three elements to be dealt with in Skinner's thesis: first, the bird-headed man, second, the double spiral, and third, the supposed former existence in Tahiti of an elaborate curvilinear art.

First, the bird-headed man in Easter Island is the undoubted symbol, or almost the narrative presentation, of an important annual event. There was no such event in New Zealand nor even any mention of one in tradition. On the other hand the manaia can be seen readily, even inevitably, in Maori art, either in development in all stages of the gradual turning of a characteristic mask face to one side, or directly by cutting the fullface mask in two. In brief, it is the decorative rendering of the mask face in profile.

Second, the double spiral can similarly be seen in New Zealand art in every possible stage from the curling of the manaia lip to the interlocking of two such curved-lip manaia faces. The Marquesan spirals appear as the normal extension of the outlines of the ears of the conventionalized face mask, or of the legs or antennae of insects. The double spiral of the Louisiades is obviously a motif formed by extension from the interlocking beaks of two frigate birds, a type of bird with special significance for a fishing community. In each of these three widely separated localities, therefore, every stage of development of the spiral from a natural form is present; each is patently locally developed.

Third, there is no evidence in Tahiti that an elaborate decorative art formerly existed and has since disappeared, such a theory is purely hypothetical, postulated to provide a supposed central place of origin for the elaborate arts of marginal Polynesia. Moreover, it is an unnecessary postulation, for, far from being without representational and decorative arts as Skinner thinks, Tahiti at the time of European discovery had both an active representational art in human figures and a highly competent decorative art in three forms: in simplification (fan handles), in formal stylization (fly-whisk handles), and in designs involving stylization (staff gods). It will be seen that the forms in which the decorative or stylized art of Tahiti presents itself are peculiar to this one locality and that none of them is ancestral to the forms seen in New Zealand. They therefore give evidence not in conflict with, but in support of, the main thesis that Polynesian arts are locally developed stylizations and patterns derived from a common basic form, the commemorative human figure.

The question of the sources of Polynesian art as a whole is part of the wider problem of Polynesian racial and cultural origins. The differences observed between the forms in central Oceania and those in marginal Pacific areas point to a long period of separation; they indicate, too, that by whatever migration route the Polynesians traveled, eastern or western, the distances were long, the hazards great, and the movements doubtless attended by a considerable loss of material culture and art. What a people loses least is its physical form and its language; in this connection, despite the present controversies, one cannot fail to recognize in the Caucasoid racial characteristics of the Polynesians and their undoubted Austronesian speech the dominance of Indonesian or Southeast Asian influences. The prototype is not so much modern Indonesian as the Indonesian of at least a thousand years ago, or even earlier; radiocarbon datings show that central Polynesia was already occupied by A.D. 900, and the travels of the canoe voyagers must have been going on for centuries before that, and cost them many casualities of both men and material, to reach their several Hawaiki, or homelands.

It is better, then, to look to the primitive Indonesia of about A.D. 500, rather than to Indian and Southeast Asian civilizations for source relationships. The arts of Polynesia themselves suggest this in their simple anthropomorphism, in their local awareness of form, and in their apprehension of decorative potential reflecting the opportunities, and sometimes the poverty, of resources available.

Time and space, history and geography, are ever-present influences in art. The arts to be examined were those of not one but of about fifteen peoples; not to be ignored were their 1500 years of sojourn in the Pacific and their thousands of miles of hazardous sea voyaging in any direction to or from the center of Oceania.

Polynesian art should be evaluated in relation to its natural environment and its social milieu. There have been criticisms of the achievement of this or that Polynesian area. For example, Willowdean Handy's comment that the art of the Marquesans "offers little true sculpture, for the natives have never known how to model although they could design," is not without validity. There is truth too in Metreaux's suggestion that Easter Island accomplished less than it might: "In our view, the Easter Island accomplished less than it might; the Easter Island sculptors worked in a material so soft that it made for a certain laziness. They too readily accepted a single formula that they could reproduce without effort. Their first conception was bold; their mistake was to remain satisfied with it. They did not always avoid the weakness of mass production." It is also possible that Maori wood carving lost sculptural vitality through its early absorption into the decorative potential of the forms it had created. One could follow a similar line of comment in other areas, observing that Hawaii failed to develop a decorative art in either relief or tattooing, and that Western Polynesia had sculpture only, with a limited decorative art of perhaps only mechanical or geometrical derivation. Such comment is not without some justification, but it ignores the fact that the Oceanic environment is less stimulating than, say, the continental. The arts of Europe and Asia experienced constant crosscurrents of influences; they enjoyed the repeated stimulus of new ideas as well as the use of materials and methods in great variety. The resources of Oceanic islands, even of high islands, were restricted; their contacts with an outside world rare, and these only with a world of similarly limited opportunity and experience.

The most impressive fact is that each area, notwithstanding its isolation, has developed so successfully and with such enterprise and achievement the resources available to it. Moreover, in New Zealand, the one area that may be regarded as a land area rather than a small island, the arts display a surprisingly wide range in their exploitation of material and in the scope and variety of the forms they created. The use of wood, stone, bone, ivory, and jade, and the development of sculpture, stylization, pattern, and abstractions (involved or austere) are not a meager record for stone-age craftsmen. Furthermore, their religious ceremonies and social observances, suffused with poetry and romance, and their literature, though unwritten, can stand alongside the sagas of Scandinavia. Stone-age "primitives" though they be, the Polynesians of Oceania were possessed of both poetry and art.

SOURCES. L. A. Bougainville, Voyage autour du monde, par la frigate du roi La Boudesse et la flûte l'Etoile en 1766, 1767, 1768 et 1769, Paris, 1771 (Eng. trans., J. R. Forster, London, 1772); J. Hawkesworth, An Account of the Voyages ... by Commodore Byron, Captain Wallis, Captain Carteret and Captain Cook, 3 vols., London, 1773; J. Cook, A Voyage towards the South Pole and round the World in "Resolution" and "Adventure," 3 vols., London, 1777, J. G. A. Forster, A Voyage round the World in His Britannic Majesty's Sloop "Resolution," 2 vols., London, 1777; J. Cook, A Voyage to the Pacific Ocean ... in "Resolution" and "Discovery," 4 vols., London, 1784; S. Parkinson, A Journal of a Voyage to the South Seas in His Majesty's Ship the Endeavour, 2d ed., London, 1784; J. F. Galaup, C. de La Pérouse, Voyage de La Pérouse autour du monde redigé par M. L. A. Milet-Mureau, 4 vols., Paris, 1797; G. H. von Langsdorff, Voyages and Travels in Various Parts of the World during 1803 to 1807, 2 vols., London, 1813–14; L. I. Duperrey, Voyage autour du monde ..., I: Histoire du voyage, Paris, 1825; L. C. Desaulses de Freycinet, Voyage autour du monde ... sur les corvettes ... L'Uranie et La Physicienne: Histoire du voyage, 3 vols., Paris, 1825–39; J. S. C. Dumont d'Urville, Voyage de la corvette L'Astrolabe: Histoire du voyage, 6 vols., Paris, 1830–33; J. S. C. Dumont d'Urville, Voyage au Pole sud et dans l'Océanie sur les corvettes L'Astrolabe et La Zelée: Histoire du voyage, 11 vols., Paris, 1841–46.

BIBLIOG. R. H. Greiner, Polynesian Decorative Designs (B. P. Bishop Mus. B., VII), Honolulu, 1923; E. von Sydow, Polynésie et Mélanesie: L'art régional des mers du sud, CahArt, II–III, 1929, pp. 61–64; T. Tzara, L'art et l'Océanie, CahArt, II–III, 1929, pp. 59–60; W. O. Oldman, The Oldman Collection of Polynesian Artifacts (Mem. Polynesian Soc., XV), New Plymouth, N.Z., 1943; R. Linton, P. S. Wingert, and R. d'Harnoncourt, Arts of the South Seas, New York, 1946; M. Leenhardt, Arts de l'Océanie, Paris, 1947 (Eng. trans., M. Heron, London, New York, 1950); P. S. Wingert, Art of the South Pacific Islands, London, New York, 1953; L. Adam, Primitive Art, 3d ed., London, 1954; F. Hewicker and H. Tischner, Oceanic Art, London, 1954; M. Heydrich and W. Frölich, Plastik der Primitiven, Stuttgart, 1954; V. L. Grottanelli, L'image de l'homme dans les arts primitifs, Brussels, 1955 (cat.); W. Münsterberger, Sculpture of Primitive Man, London, 1955; T. Bodrogi, Oceanian Art, Budapest, 1959; B. W. Smith, European Vision and the South Pacific, 1768–1850, Oxford, 1960; A. Bühler, T. Barrow, and C. P. Mountford, The Art of the South Sea Islands, New York, 1962. For single localities, see NEW ZEALAND; POLYNESIA.

Gilbert ARCHEY

Illustrations: PLS. 195–208; 3 figs. in text.

PONTORMO (JACOPO CARUCCI). Italian painter of the Florentine school (b. Pontorme, near Empoli, 1494; d. Florence, late Dec., 1556, or Jan. 1, 1557). Pontormo was a pupil of Piero di Cosimo and of Mariotto Albertinelli; according to Vasari's report alone, he also studied under Leonardo. About 1512 he was associated with Andrea del Sarto, as an assistant rather than as a pupil.

Pontormo's first independent works date from about 1513. He sought in his early works to assimilate the reigning classical idiom of his older Florentine contemporaries, in particular Fra Bartolommeo and Andrea del Sarto. But in 1518 Pontormo broke with the classical style: in his altarpiece for S. Michele Visdomini he created a radically experimental work in which certain of the major premises necessary to the mannerist style were set forth. The following seven or eight years were spent in restless, daringly original inquiries in directions that had been intimated in the Visdomini altarpiece. In his fresco lunette for the Medici villa at Poggio a Caiano (1520–21; IV, PL. 201) he tested, with remarkable success, an unclassical union of abstract, eccentrically ordered pattern with sharply observed elements of reality. In the fresco series depicting the Passion of Christ in the Certosa of Galluzzo (1523–24) he exploited, in his search for new artistic means, not only non-classical but non-Italian sources, such as the prints of Albrecht Dürer. By 1526–28 Pontormo had achieved a solution in his quest for a new style. It would appear that the influence of Michelangelo, then in Florence working on the sculpture of the Medici Chapel, rewon Pontormo to at least some of the tenets of classicism. In the Evangelist tondos, the Annunciation fresco, and the Deposition altarpiece (IX, PL. 288) in the Capponi (Barbadori) Chapel in S. Felicita, Florence, and the *Visitation* in the parish church at Carmignano, Pontormo fused an ideal of harmony and ornamental beauty derived from classicism with the liberty of form and expression he had gained in the course of his experiments, achieving a style that is discernibly a mature mannerism. The classicistic style of these works provided the basis for much of the subsequent development of the Florentine *maniera*.

From about 1530 Pontormo came increasingly under the sway of Michelangelo. The obvious elements of the Michelangelesque vocabulary entered Pontormo's style, but they were in almost every instance remarkably transformed. Michelangelo's grandiose demonstrations of spiritual and physical power were transmuted, in Pontormo's subtler idiom, into evocations of singular, haunting states of mind and personality. Pontormo's figures, so profound in emotion, are almost precious in their artistic fabric; his ornamental will is at once more refined and more libertarian than Michelangelo's.

The last quarter-century of Pontormo's activity is recorded mostly through his drawings (IV, PL. 274). None of the important works on a large scale that he executed in this period have been preserved: loggia decorations in the Medici villas of Careggi (1535–36) and Castello (1537–43) had perished by the end of the 16th century; the crowning work of his late career, the frescoes of the choir of the Medici church of S. Lorenzo (1546–56; finished by Bronzino), were destroyed in the 18th century. The interest and merit of Pontormo's drawings for these works, however, are almost a compensation for the loss of the originals, since throughout his career Pontormo's drawings remained at a remarkable level of accomplishment. Among the Florentines, with the exception only of Leonardo and Michelangelo, Pontormo created the highest expression of the great Florentine tradition in draftsmanship.

Pontormo is of the greatest importance for the history of postclassical Renaissance painting in Florence. He was a principal innovator in Florence of mannerist style. Among his exact contemporaries only Il Rosso (q.v.) — equally original but more eccentric — is in some respects the rival of Pontormo in inventive gifts and in quality of achievement. In a final analysis, Pontormo is the only Florentine painter of his time whose genius can support the historically inevitable comparison with that of Michelangelo.

WORKS. Faith and Charity fresco (badly deteriorated), 1513–14, SS. Annunziata, Florence. – Madonna fresco (from church of San Ruffillo), ca. 1514, SS. Annunziata, Florence. – St. Veronica frescoes, 1515, S. Maria Novella, Florence. – Visitation fresco, 1514–16, SS. Annunziata, Florence. – Three panels with scenes from the life of Joseph, 1515–17, Henfield (Sussex), Coll. Lady Salmond. – Visdomini altarpiece, 1518, S. Michele Visdomini, Florence. – *Joseph in Egypt*, 1518–19, London, Nat. Gall. – *SS. Michael and John the Evangelist*, ca. 1519, Empoli, Gall. della Collegiata. – Vertumnus and Pomona lunette fresco, 1520–21 (IV, PL. 201). – *Portrait of Two Youths*, ca. 1522, Venice, Coll. Cini. – Passion fresco cycle, 1523–24, Certosa of Galluzzo, near Florence. – *The Supper at Emmaus*, 1525 (XII, PL. 77). – *Portrait of a Youth* (Alessandro de' Medici), ca. 1525, Lucca, Pin. Naz. – *Maria Salviati and Cosimo de' Medici*, ca. 1526–27, Baltimore, Walters Art Gall. – Capponi Chapel decorations, including Deposition altarpiece, 1525–28, S. Felicita, Florence (IX, PL. 288). – *St. Jerome*, ca. 1527–28, Hannover, Landesmus. – *Portrait of a Halberdier*, ca. 1527–28, New York, Chauncey D. Stillman Coll. – *Madonna and Child with Infant St. John*, ca. 1527–28 (IX, PL. 289). – *Visitation*, ca. 1528, parish church at Carmignano. – Wayside tabernacle with Crucifixion, ca. 1528–29, Florence, Forte di Belvedere (formerly at Boldrone, near Florence). – St. Anne altarpiece, ca. 1529, Louvre. – *Legend of the Ten Thousand Martyrs*, 1529–30, Florence, Pitti. – *Portrait of Alessandro de' Medici*, 1534–35, Philadelphia, Mus. of Art. – *Portrait of Giovanni della Casa*, ca. 1540–44, Washington, D.C., Nat. Gall. – *Portrait of a Woman*, ca. 1543–45 (PL. 218). – Tapestries from Pontormo designs (lamentations of Jacob, Benjamin at the court of Pharaoh), ca. 1545–49, Rome, Palazzo Quirinale.

WRITINGS. Letter to Varchi (1548), B. Varchi, Due lezioni, Florence, 1549, pp. 132–35; Diario "fatto nel tempo che dipingeva il coro di San Lorenzo" (1554–56), ed E. Cecchi, Florence, 1956.

BIBLIOG. Vasari, VI, pp. 345–95; F. Goldschmidt, Pontormo, Rosso und Bronzino, Leipzig, 1911; F. M. Clapp, Les dessins de Pontormo, Paris, 1914; F. M. Clapp, Pontormo: His Life and Works, New Haven, 1916; W. Friedländer, Die Entstehung des antiklassischen Stiles in der italienischen Malerei um 1520, RepfKw, XLVI, 1925, pp. 49–86 (Eng. trans., Mannerism and Anti-Mannerism in Italian Painting, New York, 1957, pp. 1–43); B. Berenson, The Drawings of the Florentine Painters, 2d ed., Chicago, 1938, I, pp. 310–21, II, pp. 273–307; E. Toesca, Il Pontormo, Rome, 1943; L. Becherucci, Manieristi toscani, Bergamo, 1944; G. Nicco-Fasola, Pontormo o del Cinquecento, Florence, 1947; C. de Tolnay, Les fresques de Pontormo dans le choeur de S. Lorenzo, CrArte, IX, 1950, pp. 38–52; R. Oertel, Pontormos büssender Hieronymus, Mitt. der Kunsthist. Inst. in Florenz, VII, 1953–56, pp. 111–200; L. Berti, L. Marcucci, and U. Baldini, Mostra del Pontormo e del primo manierismo fiorentino, Florence, 1956 (cat.); J. Cox, Pontormo's Drawings for the Destroyed Vault of the Capponi Chapel, BM, XCVIII, 1956, pp. 17–18; C. Gamba, Contributo alla conoscenza del Pontormo, Florence, 1956; J. Shearman, Rosso, Pontormo, Bandinelli, and others at SS. Annunziata, BM, CII, 1960, pp. 152–56; A. Videtta, Attualità del Pontormo, L'Arte, LIX, 1960, pp. 223–31; S. J. Freedberg and J. C. Rearick, Pontormo's Predella for the S. Michele Visdomini Altar, BM, CIII, 1961, pp. 7–10; S. J. Freedberg, Painting of the High Renaissance in Rome and Florence, Cambridge (Mass.), 1961; D. Wild, Le sembianze di Jacopo da Pontormo nel ritratto e nell'autoritratto, Riv. d'Arte, XXXVI, 1961–62, pp. 53–64; J. Shearman, Pontormo and Andrea del Sarto, 1515, BM, CIV, 1962, pp. 478–83; C. de Tolnay, Un disegno sconosciuto del Pontormo a Bergamo, CrArte, X, 56, 1963, pp. 43–45; H. S. Merritt, The Legend of St. Achatius: Bachiacca, Perino, Pontormo, AB, XLV, 1963, pp. 258–63; J. C. Rearick, The Drawings of Pontormo, Cambridge (Mass.), 1964.

Sydney J. FREEDBERG

PORCELAIN. See CERAMICS.

PORTA, GIACOMO DELLA. Italian architect (d. Rome, 1602). The earliest biographer of Giacomo della Porta, Giovanni Baglione says that the architect was about 65 at the time of his death, so that he was born presumably about 1537. Baglione also stresses that Della Porta was a Roman, and it is now generally accepted that reports of his residence in Genoa and his relationship to the family of the sculptor Guglielmo della Porta are fictitious. His entire known career was spent in Rome, where he became "architetto del popolo Romano," and where he was engaged in all the major building enterprises of the late 16th century. He is first mentioned as a direct follower of Michelangelo, working on the Capitoline hill complex after Michelangelo's death. In 1568 he executed the central window of the Palazzo dei Conservatori (IX, PL. 542) and in 1598 a door in the Palazzo dei Senatori, in both of which he altered Michelangelo's designs. With Domenico Fontana (q.v.), he completed the dome of St. Peter's (1588–90; IX, PL. 541). Here again he changed Michelangelo's design by raising the curve of the cupola, but this modification may have resulted from structural requirements.

Notwithstanding his stylistic dependence on the highly personal mannerism of Michelangelo, Della Porta was early influenced by the current of academic classicism represented by Giacomo da Vignola (q.v.). This style had triumphed before Vignola's death in 1573, and Della Porta continued it up to his own death. He was commissioned to complete Vignola's façade of the Church of the Gesù, mother church of the Jesuits and one of the most important of the numerous 16th-century Roman churches because it was reproduced all over the world by Jesuit missionaries (IX, PL. 312). However, the taut elegance of Vignola's façade (known from an engraving) was lost in Della Porta's version, with its great scroll buttresses. These conspicuous and easily imitated elements proved to be one of the most popular features of his design.

Among the many other churches in Rome by Della Porta are Madonna dei Monti (1580–81), S. Maria Scala Coeli (1581–84), S. Atanasio (1580–83), and S. Luigi dei Francesi (partly by D. Fontana, consecrated in 1589; XII, PL. 90). The last two seem to have been new departures in façade desing (although obviously inspired by Michelangelo's design for S. Lorenzo in Florence; cf. I, PL. 406), for in these churches the façade is treated as a rectangular screen with a central pediment and flanking towers (not executed at S. Luigi), closing off the ends of the nave and aisles.

Among his secular buildings, the most notable are the loggia (1589) at the rear of Palazzo Farnese, Palazzo della Sapienza (the present-day Archivio di Stato), and several other structures distinguished by their elegance of fenestration — all in Rome. His last work, the Villa Aldobrandini in Frascati (1598–1604), gives prominence to a huge broken pediment, effective from a distance but uncharacteristically clumsy in a close view.

BIBLIOG. G. Baglione, Le vite de' pittori . . . , Rome, 1642; W. Arslan, Forme architettoniche civili di Giacomo della Porta, B. d'Arte, N.S., VI, 1926–27, pp. 508–28; W. Körte, ThB, XXVII, 1933, s.v.; G. Giovannoni, Saggi sulla architettura del Rinascimento, Milan, 1935; A. Venturi, Storia dell'arte italiana, XI, 2, 1939, pp. 784–845.

Peter MURRAY

PORTRAITURE. Portraiture in its broadest sense is the representation of an individual, living or dead, real or imagined, in drawing, painting, or sculpture, by a rendering of his physical or moral traits, or both. In modern times portraiture has been extensively considered in its historiographical and sociological aspects: the evolution of types and conventions of representation, the relation between portraiture and society, and the moral or spiritual attitudes that portraits reflect and record.

CONCEPTS OF PORTRAITURE. From the Latin verb *protraho*, used in the Middle Ages to mean "reproduce" or "copy," come the French *po(u)rtraire* which dropped out of use in the 18th cent.), *portrait* (in use since the 12th cent.), *portrai(c)ture* (in use since the 12th–13th cent.), and consequently the English forms "portray," "portrait," and "portraiture," the German *Porträt*, and the Russian *portret*. From the related Latin verb *retraho* are derived the Italian *ritratto* and *ritrarre* in their meaning of "copy" (in use since the early 16th cent.); *ritratto*, however, in its commonly accepted meaning of "portrait" derives from the Spanish *retrato* (in use since the 17th cent.).

In art terminology all these forms, with their derivatives, have undergone changes of meaning with the gradual development of the concept, so that such terms as the Greek εἰκών and the Latin *imago*, *effigies*, and *simulacrum* can only broadly be used as substitutes, although all imply an association with the concepts of image and imitation, as Erwin Panofsky has noted. The painted or sculptured image of a person, because it demands a degree of likeness to the model, raises the problem of imitation (see MIMESIS), here understood as the relation of the lifelike to the real. Shortly after the middle of the 16th century, moreover, writers on art (particularly Vincenzo Danti) began to make a more subtle distinction between imitating (i.e., representing according to particular principles) and portraying (i.e., representing things as they are and not as the artist believes they should be). The conceptual life of portraiture, its underestimation and overestimation, and its limitations and fashions have always depended on the varying attitudes toward imitation and portrayal at different historical periods, involving the questions: To what extent can or must the artist pursue the likeness? In what manner and how far is he allowed to depart from it? How much should he idealize his model, mend its defects, or exaggerate them?

Every portrait is, in a sense, a unique solution to these problems. The inherent limitations of the medium (whether stone, marble, or oil with a varnished surface) reduce the supposed verisimilitude to an illusory artifice and hence a rather deceptive reconstruction of the model; and the static quality of the image forces the artist to resort to a whole series of narrative and stylistic conventions. The legends of magical transformations of statues into living beings are themselves evidence of the disparity that artists and viewers sense between the lifelike and the real.

There are, then, in the history of figural art, two conflicting tendencies arising from these opposite poles, namely, the tendency to seek resemblance to the subject and the tendency to pursue its idealization or transfiguration. The former tendency is perhaps associated with the origin of portraiture in its narrow sense, in which portraits were substitutes for actual preserved bodies of the deceased. The various procedures of mummification and preservation of bodies, whether of enemies or of ancestors, indicate both a desire to retain the personal likeness and an assumption that it contains a magical power. This power was long attributed to portraits, especially to those of kings: during the Byzantine age, for instance, the portrait of the emperor was hung in the law courts to symbolize his presence and thus the impartiality of judgment. A faithful likeness seems to have been appreciated especially by those civilizations in which this magical quality, particularly in funerary art, had been traditionally important; in which the fear of the transcendent was diminished (for when it was intense, abstract stylization prevailed); and in which, it would seem, material prosperity, and religious indifference triumphed over religious and political theocracy, as in Hellenistic art (including the prodigious portraits of the Fayum; PL. 214), the Italic and Roman worlds, the Gothic period, the northern Renaissance, Spain at the time of Velázquez, and the bourgeois 19th century. R. Bianchi Bandinelli (*Archeologia e cultura*, Milan, Naples, 1961) has rightly observed that the Roman realistic portraits, which brought the charge that the artists were more concerned with truth than with beauty, were a cultural and artistic manifestation rather than a consequence of the process by which they were made (e.g., death masks). This accounts for the lack of a steady tradition of physiognomical portraiture and its appearance only in response to cultural conditions.

The Italian Renaissance offers examples of the clear conceptualization of both tendencies. Leonardo da Vinci, although his canvases show the augmenting of light and shadow that were believed to add grace and beauty to the subject, also developed a scientific analysis of physiognomy intended to capture the moral character of the model from the facial planes and the lines dividing them. The opposite tendency toward idealization was enunciated by Alberti (q.v.), who reiterated the concept of Plutarch: "When they were painting kings, the painters of antiquity, if there was some defect in their models, tried to mend it as best they could, keeping the likeness" (*Della Pittura*, III, 38). To Alberti, portraying a likeness did not mean merely rendering the natural appearance of the model but ennobling it by hiding the physical defects. The same principle was expounded in the 16th century by Benedetto Varchi, who wrote of the freedom befitting poets and painters to improve on nature, citing Pliny, who told how Apelles painted the image of Antigonos so as to hide the fact that he was blind in one eye. In Venice, at the time of Titian, Paolo Pino and Lodovico Dolce held that great portraits were the result of the "perfection of art" — in other words, that they must go beyond mere imitation, though both authors accepted the validity of *similitudine* and believed that portraits had to be sufficiently lifelike to be identifiable with the subjects. Lomazzo attempted to reconcile the different viewpoints in his *Trattato dell'Arte della Pittura, Scultura ed Architettura* (1584). In his view, although a portrait is primarily a likeness, allowing the viewer to recognize immediately the person represented, the painter must take other elements into consideration, such as the rank of the subject and the "sign" of his dignity (e.g., the crown, for an emperor); in addition, he must use discretion ("to mend what is ugly in body and costume") and must magnify "the nobility and majesty of the features." Consequently portraits that reflected the concept of the executor were to be considered superior to those that did not, and the educational value of a portrait was increased as the rank and moral character of the model were better represented.

In summary, then, it may be said that the portrait must, first of all, indicate the personality of the subject, who inevitably poses before the painter no matter how simple and spontaneous the representation seems. Portraits offer a veritable anthology of the ways of conceiving of man, and it is difficult to determine how much in the differing portrait styles of, say, a Van Eyck and a Titian is the result of differing moral attitudes and how much is, instead, to be attributed to differences of artistic personality. Moreover, conventions of costume and gesture may loom as large as — or larger than — physiognomical fidelity. The attributes or signs used in a portrait must always be considered in their historical and social context, since the significance of an attribute (a small room filled with identifying objects for an antiquary, a manuscript for a scholar), a symbol (a withered flower, a chain of asps around the neck, or a red berry), the pose, or even the length and aspect of the portrait (full-length or bust, fullface or profile) can vary with different epochs and cultural traditions. The new and more complex classifications of portraiture that are needed must be based as much on the varying functions of portraiture as on the changing fashions in iconography and style.

Tullio DE MAURO, Luigi GRASSI, Eugenio BATTISTI

PORTRAITURE AMONG PRIMITIVE PEOPLES. From the previous discussion it is clear that one can consider portraiture a well-defined artistic type, with a chronology and a historical development, only in connection with those epochs or civilizations to which the strictly modern concept of portraiture is applicable, such as the Western cycle of Greco-Roman art, modern European art, and its immediate antecedents. Outside these civilizations, so-called "portraiture" may be a matter of the identification of a figure with a specific individual (human or divine) by the use of conventional means such as the pose, an attribute, or an inscription; or it may refer to isolated phenomena of realistic characterization or pronounced somatic imitation that can only rarely be described as conscious attempts to render the individual human personality through its physical actuality.

The suggestion that portraiture has universally and necessarily evolved from the indistinct and generic to the specific and characteristic cannot be proved. In fact, the opposite may be true, for some of the oldest images of prehistory (q.v.) and the simple ones made by savages, uneducated people, and children (see PRIMITIVISM; PSYCHOLOGY OF ART) seem to be representations of specific individuals rather than of collective, abstract types or symbols of men or gods. Apart from those cases in which the intention of the "primitive" artist is known, there exist, even in the oldest human iconography, representations apparently intended as caricatures (PL. 332). Even the prehistoric or ethnological "aniconism," when it reflects taboos (see IMAGES AND ICONOCLASM), refers to recognizable human figures. One must not, however, altogether rule out the possibility that the representation of "typical" human figures may have become customary even at a very early time among prehistoric artists.

* *

The figural art of "primitive" peoples, whether prehistoric or modern, commonly shows marked conventionalization or stylization in the representation of the physical features and little or no interest in the expression of the moral or spiritual traits that distinguish the personality of the model. In plastic art (including masks), such as painting, carving, and ceramics the human figure is represented in styles varying greatly from one culture to another, from one district to another, and between spheres of art within the same artistic culture, yet there is no evidence of that precise and specifically individualizing characterization which, in modern cultures, is typical of the art of portraiture.

A significant example of what was intended as portraiture of actual historical figures by primitive artists is the group of royal statuettes of the Bakuba tribe of the Congo (see BANTU CULTURES; II, PL. 113). Each of the 18 extant statuettes represents a king and each one supposedly reproduces the royal personage in his exact individuality. Despite some stylistic distinctions (two different styles have been identified), they are so much alike, however, that it has been suggested that these so-called "portraits" cannot have been made from life — and, indeed, all seem to belong to the same artistic school, which flourished for a limited time, perhaps at the beginning of the 19th century. The different identities are expressed symbolically by the addition of distinguishing attributes; for example, a king who was famous for having permitted his courtiers to marry slave women is identified by a small slave figure placed before his knees.

Among primitive peoples the concept of the portrait is interwoven with magicoreligious associations that are basically alien to European culture. The Mpongwe natives of Gabon (II, PL. 100) believe, for instance, that their masks are the actual portraits of women whose faces the artists have "stolen." Despite the geometric stylization of the lines, these masks are elegant and refined in their representation of feminine and physiognomic traits; but they are "portraits" only in the opinion of the natives — in other words, in a magicoreligious sense and by no means in an artistic one, despite their realistic and naturalistic elements.

Among conventional and stylized representations of ancestors that are portraits only symbolically, are the wooden statues and masks of the totemic tribes of northwestern Canada. These occur as parts of totem poles representing mythical genealogies, or sometimes as autonomous works made for potlatch ceremonies (see NORTH AMERICAN CULTURES). The representation of ancestors is generally one of the most widespread figural themes of art at the ethnological level, though in most cases these images can scarcely be considered portraits in the modern sense of the word. They are almost always symbolic portraits or pseudo portraits, more allusive than expressive, as in some of the figures on the carved wooden panels of the Maori, the *malanggan* anthropomorphous figures of New Ireland (IX, PLS. 447, 448), and especially the *uli*, also of New Ireland (IX, PL. 448; see MELANESIAN CULTURES).

Generally speaking, artistic representation in cultures at the ethnological level only rarely rises to the level of characterization (q.v.), as in occasional instances in Africa (see GUINEAN CULTURES; VII, PLS. 107–112, 115) and in the late Mexican-Andean cultures. Even most of these can scarcely be considered authentic portraits, for the artists seem more concerned with the subject's social function, calling, and human type than with the representation of an individual person with his somatic and psychic characteristics. Among the Mexican-Andean productions only the Mochica fictile art (see ANDEAN PROTOHISTORY; I, PLS. 176, 177, 180, 185–187) and the Olmec and Mayan sculptures (see MIDDLE AMERICAN PROTOHISTORY; X, PLS. 2, 4, 17) offer figures of such physiognomic precision and such differentiated and vivid personal traits as to show clearly the makers' interest in portraits as expressive portrayals of individual personalities.

Vittorio LANTERNARI

THE ANCIENT WORLD. *Mesopotamia and Egypt*. In ancient Middle Eastern art, representation of the human figure presents itself in terms of the identification of a specific person. There were considerable differences, however, in the development of portraiture between Mesopotamia (q.v.) and Egypt, the two main centers of artistic culture. In Mesopotamia (and in the areas within reach of its cultural influence) the characterization of physical peculiarities developed early (sculptures of Khafaje; IX, PLS. 468, 469, 473), along with an expressionistic intent even in works of such formal character as the head from Erech (Uruk), perhaps a goddess, in which the eyes, eyebrows, and hair are indicated (IX, PL. 464). These tendencies lasted for centuries, without, however, maturing into a real stylistic interpretation of the features as those of an individual rather than a type, even in the remarkable group of images of King Gudea (I, PL. 511; IX, PL. 472).

In ancient Egypt portraits were plentiful, extending over a period from the beginning of figural art to the Greco-Roman age (see EGYPTIAN ART). Particular attention must therefore be given to this group of works, even though some of the general problems are common to both Egypt and Mesopotamia. First of all, every Egyptian statue that is neither ritualistic nor a representation of a god is potentially a portrait, for nothing could have been more alien to the Egyptian mentality than a figure for its own sake, an abstract type or ἄγαλμα, created for the artist's delight and with no function other than impersonal ornament. Egyptian statues represented actual persons and had proper names but were seldom commemorative; they were not a memento of some prominent personage for posterity or a sign of recognition for special merit, as with the statuary from classic art onward (see MONUMENTS). In the great majority of cases they were meant for the tomb or the temple. If meant for the tomb, they were located in the places for ritual observances, or more frequently, especially in the early age, they were put in chambers, apart, known by the Arab term *sirdab*, connected with the ritual chambers by narrow apertures that shut the statues out from the visitors' view but set the visitors under the eyes of the statues. If meant for the temple, they were located within the sanctuaries, perpetuating the departed believer as worshiper or, in a much later period, as intercessor before the gods.

In both cases the statue was simply a substitute for the person and by no means a memorial. The existence of the *sirdab* shows that the spectator was taken into no account, whereas the statue was regarded as being endowed with a life of its own. This special significance of the statue, therefore, involved concepts different from those arising from merely figural problems. The term used for "sculpturing" was "bringing forth," as in birth; a magic was worked on the statue ("the opening of the mouth") like that performed on the dying person to ensure him or her the mysterious afterlife. There was a quality of interchangeability between the person and its image, ultimately magical or religious in origin, which is understandable in view of the Egyptian belief that a thing, a person, or an entity could reappear in, and almost take the shape of, another one. Portraits to the Egyptians were magical, and this explains why there was no need for the resemblance, or rather the identification of the model on the basis of its features, that is fundamental to Western art. Though the model could be identified by its name, by details of dress and pose, or by physiognomical elements, name and image were interchangeable data of identification: a statue represented someone else if the name was changed, though the statue remained the same.

This archaic general attitude underlay all Egyptian portraiture, which, however, materialized in monuments different from one another for both stylistic and historical reasons. The first great portraits date from the beginning of the Memphite period (4th dynasty). Belonging to this time are, in particular, a number of heads (the so-called "reserve" heads) that were situated in the burial chambers, rather than in the places of worship, probably for the guidance of the souls in their travels back into the bodies. In these heads the characteristic trait (a certain nose or mouth, certain eyes or jaws) appears in the geometric simplicity of the art of the time, in which there is no trace of archaic expressionism. The subject was indicated by an unmistakable identifying element, which itself had undergone the process of geometric stylization.

The richest production of portraits comprised representations of the King. Although a single trait was chosen to identify the model, the tradition of an "ethos," which relates all these statues to one another, was established. In the late Memphite period the figures generally became more markedly characterized, though it is doubtful that they were actual portraits; for instance, in the *Seated Scribe*, the head (IV, PL. 342) is so stylistically consistent with the body in every detail that there is good ground to suspect that the artist was mainly concerned with making a unitary work. On the other hand, in popular art the typical was too overstressed to be anything but the expression of a general taste and of the reaction to the aristocratic stylizations of the Memphite court art.

With the coming of age of the Middle Kingdom (2400–1580 B.C.), a new way of conceiving of reality originated precisely from this descriptive and narrative interest, as is also evident from the literature of the time, in which novels and moralizing speculations began to appear. The earlier purely geometric shapes and numerical proportions had brought about a rational concept of volume; with the Middle Kingdom, the placement of figures in space was no longer prompted by calculations and set laws but was determined instead by a concrete sense of reality and the ability to perceive the figure organically. Thus a richer and more personal sense of life was engendered, and portraits became the ideal form of expression, since they involved study of the subject's personality.

It must be noted, however, that at this time portraits had a set theme in the person of the sovereign (PL. 209; IV, PLS. 348–350, 353, 354), and the characteristics of this art are completely exemplified in the official images of the Pharaoh. His gradual physical decay was represented with frankness and even with greater psychological interest. Attempts to render the psyche of the sovereign were not new, but whereas in the Memphite period the stress had been on his sublime peace, and the same aura had spread to the portraits of his subjects, in the Middle Kingdom he was characterized by a disillusioned, melancholy forcefulness. He had become the "Good Shepherd," mindful of the welfare of his people yet aware of the ingratitude that would attend his efforts. It might seem a modern interpretation to attribute to these portraits such a fine play of expressions, but this kind of psychological insight is found also in the literature of the time in connection with the sovereign. An unusual inscription from this period describes the emotions as they are expressed in set facial expressions; such matters, then, must have played a role in this Egyptian culture. The abundance of statues of the Pharaohs in this period also makes it possible to identify trends of taste — something exceptional in Egyptian art, which had always been unitary in its manifestations. A more classicist idealization is evident in the cultural areas of Memphis and the Fayum, with a more expressive and dramatic tone appearing in the works of the peripheral areas (the Thebaid and the Delta). In Memphis and the Fayum there was probably a heritage of older traditions, and in the final analysis the Egyptian statuary of the Middle Kingdom reflects the style of the Old Kingdom.

Though many examples of the art of the 18th dynasty (the beginning of the New Kingdom) have survived, portraiture, strictly speaking, can scarcely be considered in this connection. The rise of an urban middle class, the gradual refinement of taste, and the wide diffusion of art objects in daily use reduced the art of the time to the level almost of craft work, pleasing, graceful, and conventionally elegant. Personal traits can be detected in the statues of Queen Hatshepsut and of the great conqueror-Pharaohs such as Thutmosis III (IV, PL. 370), but these traits are submerged in a sort of pervasive sweetness. This lack of incisiveness lasted even beyond the 19th dynasty, despite the changes in the concepts behind figural art. Though some details vaguely characterized the individual kings, the readiness with which old statues were renamed and reemployed shows how indifferent this culture was to resemblance in portraiture.

During these centuries of impersonal generic statues the art of Tell el-'Amarna stands out as unique. Outside the large centers of traditional culture (Memphis and Thebes), in an atmosphere of intense concern with moral and religious problems, a taste revived in Tell el-'Amarna for the meticulous reproduction of reality, a stressing of personal traits, and even an interest, completely new to Egyptian art, in deformities and disproportions (IV, PLS. 365, 374). The 'Amarna portraits, most of which were found in a sculptor's atelier, were based on largely unretouched plaster masks taken from living models (IV, PL. 375) — evidence of a great concern, for which even mechanical means were acceptable, to identify individual traits and fix them in durable form. It must be noted, however, that generally speaking both the sculptures and the paintings of the time represented royal personages. The sickly, abnormal appearance of the king and the slender grace of the queen and princesses were themes repeated in admirable portraits and were the model also for the features of statues of commoners.

There is reason to suspect that the importance attributed to the likeness of the king served an official purpose also, as had been true in the art of the Middle Kingdom, though in a different context and with a different significance. In Tell el-'Amarna the king, who was trying to regain the divine prerogatives of the ancient sovereigns, was imagined as being so remarkable and exceptional that his "mask" consisted precisely of the physical peculiarities by which he could be recognized. The fidelity to the model in this case, then, is best interpreted as having a mythological character, and the same interpretation must also be given to the domestic scenes, whether gay or sad, that began to appear at that time in connection with the royal family.

This essentially archaistic reaction toward restoration of a divine monarchy, which had by then become unacceptable, was regarded with suspicion by the people, and with its eventual downfall came the end of this phase of portraiture. After the conclusion of the 'Amarna cycle no evidence of any characteristic way of rendering physiognomy can be found until the Ethiopian period (IV, PL. 386). The strictly traditional attitudes of these southern sovereigns, who were proud of their old Egyptian culture and hostile to the novelties appearing in Egypt proper, brought about a revival of the Memphite and Middle Kingdom manner. Structure was simplified and bodily proportions clarified, and the kind of art that was felt to be the most essentially and classically Egyptian came into favor. The Memphite period and that of the Middle Kingdom, which in the historiography of the Egypt of the Pharaohs go together, were then visualized as being a model of unity, and the portrait, which had been their typical form of expression and which had later dropped out of fashion, again grew in importance. The portraiture became fundamentally intellectualistic — a trend that influenced all subsequent Egyptian art and appeared clearly in the neoclassical works of the Saite dynasty, when portraits reached such formal perfection that the contemplation of the scheme to which the object could be reduced was in itself a source of inspiration to the artist (IV, PL. 388).

This intellectualism was the source of the more complex type of portraiture that first appeared in the Persian period and flourished in the Greek age, when the rigorous modeling that dated back to the Middle Kingdom was counterbalanced by the attention given to details and by a technical skill inherited from the Saite period and perhaps revitalized by the influence of Greek art. It may seem farfetched to connect this late Egyptian artistic taste with the early portraiture of republican Rome, but the possibility should at least be borne in mind as an indication of the far-reaching influence of the figural culture of Egypt.

Sergio DONADONI

Iran. A careful pursuit of realism, not limited to the outer aspect but striving to express also the spiritual values of the model, characterizes some terra-cotta and bronze works of the Elamite and approximately contemporaneous periods. Besides the bronze statue of Queen Napirasu, the wife of Untash-Huban (see METALWORK; VIII, PL. 121), which must have been a portrait though this is hard to detect because of its mutilations, the Elamite terra-cotta man's head from Susa (Louvre) was undoubtedly intended as a portrait, especially since it is a funerary image. Other notable works, such as the bronze heads from Hamadan, have all the characteristics of genuine portraits or at least of refined attempts at the characterization, with keen insight, of specific, though perhaps imaginary, persons.

No tendency toward portraiture is apparent in the protohistorical works of art of the northern centers and Luristan (see ASIATIC PROTOHISTORY; IRANIAN PRE-SASSANIAN ART CULTURES), which were dominated by a symbolic and religious approach and a concern with decoration. The official art of the Achaemenid period relied largely on peculiarities of race and costume, along with symbolic expression of the *majestas* and magico-religious power attributed to the king. Portraiture, however, was not completely neglected, as can be seen from the small lapis-lazuli paste head found at Persepolis (VIII, PL. 129), which some scholars believe is an idealized portrait of Xerxes I (r. 485–464 B.C.); from the damaged image of the Egyptian minister Ptahhotep in Iranian costume; and from some coins issued by the satraps, whose heads (the images are restricted to the heads) are obviously portraits after the Greek fashion. The image of the king, bow in hand, on the obverse of the golden daric is without portrait interest, though it was the first royal image to appear on the obverse of a coin.

The Seleucid mintings influenced the Parthian, Greco-Bactrian, and Indo-Greek styles of coinage (see COINS AND MEDALS), and the best Hellenistic coin portraits in fact appeared on the Greco-Bactrian and Indo-Greek issues. The Parthian images, though at first under the Greek influence, were different in costume and sometimes also in style; only the drachmas and tetradrachmas of Mithradates I (171–138 B.C.), with their idealized images, are unquestionably of the Greek type. The images on Parthian coins soon developed into two different types: stylized though still rather personalized images, and naturalistic images, which might even be said to be crudely realistic as in the coins of Mithradates II (ca. 123–90 B.C.). Later still, the king was portrayed fullface — a view characteristic of Parthian coins — and the stylizing tendency became more marked; for a time realistic portraits continued to appear, but eventually these were replaced by conventionalized figures that varied little from one king to another. This departure from individualizing portraiture might be considered a natural consequence of Greek conventions, but the more likely explanation is that it was simply the result of a revival of a taste wholly alien to classicism.

The Parthians' interest in the human face and in the study of physiognomy is apparent from the several extant figural works that suggest real portraits, such as the limestone man's head from Susa (Louvre); the head of the colossal bronze statue from Shami (VIII, PL. 141); the fragments of the Hellenistic head from Shami (Teheran, Archaeol. Mus.); and the Hellenistic head bearing the signature of the artist Antiochus, the son of Dryas, which perhaps came from Susa and which some authorities believe to be the portrait of Musa, the wife of Phraates IV (Baghdad, Iraq Mus.). If one considers, moreover, the number and the accuracy of the Palmyrene funerary portraits and the Parthian elements that enliven them, the image of King Antiochus I of Commagene on Nimrud Dagh, the presumed portrait of Vologases III, and other stone or bronze works, one may unhesitatingly ascribe to the Arsacid period a remarkable development of likeness in portraiture, despite the various trends found on coins.

During the Sassanian period the interest in portraiture was more limited. The human face was still used in decoration, either represented classically, as in the isolated ornamental heads of the Bishapur mosaics, or stylized, as in the fullface images adorning some silver vessels. The images of the kings, however, on the coins, on the large silver bowls, and on the rock reliefs were identified by symbolic elements, such as the shape and structure of the crown, rather than by physiognomical aspects. Though variations of fashion can be detected, for instance, in the hairdos, and though the colossal statues (e.g., the statue of Shāpur I at Bishapur) were obviously meant as portraits and must have had some value as representations of particular individuals, the figures still cannot be called genuine portraits since they lack all psychological analysis, being rich instead in symbolic and religious elements. An anonymous Islamic chronicle mentions, however, a work containing miniatures of all the members of the Sassanian dynasty. From this information one may infer, with some caution, that portraiture was not altogether a lost art during the Sassanian period but was most probably pursued in miniatures. This seems to be indirectly confirmed by the Manichaean miniatures and paintings of central Asia (see MANICHAEAN ART), in which portraits or images of people often show that the artist had no intention of reproducing the model's likeness. In central Asia many figures of donors carried scrolls identifying the subject. In such cases there is no evidence of interest in physical resemblance, even though the work is meant as a portrait.

Mario BUSSAGLI

Greece. The conception of an individualized portrait was not born in Greece until the first half of the 5th century B.C. Before that, during the whole of the archaic period, the statues erected to commemorate Greek men and women in sanctuaries and on tombs were generalized representations. Kleobis and Biton, Chares, and Aristidikos are indistinguishable from the other statues of their time, except for the inscribed name that accompanies them. These inscriptions, however, show that specific people were intended and supply valuable evidence of the early Greek idea of a portrait. The adoption of such generalized representations was natural, of course, when one considers that in Egypt and Mesopotamia, as well as throughout the ancient world, the representation of a type rather than of an individual was the norm and that during the 7th and 6th centuries B.C. Oriental influence was potent in Greece. When, at the end of the archaic period, the Greek artist had mastered the anatomical structure of the human body, the time was ripe for a new concept in portraiture. In characteristic fashion he applied himself to this task systematically, evolving in a gradual, steady progression a novel attitude toward portraiture.

Before tracing this remarkable development and citing the most conspicuous examples, a word must be said about the extant examples in general. As is well known, the great majority of Greek portraits are preserved only in Roman copies, the originals having long ago disappeared; moreover, these Roman copies consist mostly of herms and busts, whereas the Greek originals commonly were statues. Much, therefore, of the original renderings has been lost, for the Greeks sought to convey the personality of the man not only in the face but in the stance of the body. Furthermore, the painted portraits, which played as important a role in the Greek periods as did the sculptured ones, are almost entirely lost. It is only from occasional descriptions of them in ancient literature that one can derive some idea of their importance.

Fortunately, however, the Roman sculptured copies of Greek portraits provide a fair visualization of this great art, that is, of the evolution from a generic to an individualized likeness, as well as of the various personalities represented: first, because most of the copies were made mechanically by the pointing process and so are, on the whole, accurate; second, because the copies exist in such quantity. The Romans greatly admired Greek culture; they had a profound interest in Greek philosophy and poetry and liked to have portraits of the outstanding personalities of Greek history in their houses, gardens, and public places. As a consequence, of an especially popular individual such as Demosthenes (VII, PL. 160), there exist over forty representations found throughout the Roman Empire; and by a comparison between them one can judge their relative accuracy. There are also still extant a few original Greek portraits (not, however, any of those for which Roman copies have survived), and these help in evaluating the lost examples. One might mention, for instance, the magnificent bronze head of a bearded philosopher that was found in the sea off Antikythera (VII, PL. 156).

Another characteristic of Greek portraits distinguishing them from those of later date is that they represent not ordinary individuals but persons of eminence — the well-known statesmen, generals, poets, philosophers, and orators (only exceptionally artists) who contributed to the greatness of Greece. And these portraits were erected, at least at first, not in private houses but in sanctuaries and public places, some of them being dedicated by friends and relatives, others by the state.

It is true that tomb sculptures with representations of obscure individuals were commonly set up on public or private burial plots, especially in Athens; but, for a long time these figures, inscribed with their names, remained generalized, even during periods when realistic portraiture was current. The heroization of the dead was clearly here a determining factor.

One may ask how the various individuals portrayed can be recognized. The chief evidence is supplied by the inscribed name, which was occasionally — alas! much too rarely — added on the Roman copy. Other useful clues are furnished by the finding places, by double herms (when one of the individuals is known), or by descriptions in ancient literature. When such clues are missing, the identification is difficult; and many a time preconceived ideas of the appearance of well-known individuals have been proved erroneous.

The first signs of an individualized representation are to be seen in some heads of helmeted generals of the first quarter of the 5th century B.C. and in the statue of Aristogeiton, which was erected in 477/476 B.C. in the market place of Athens and which has been preserved in several Roman copies. From the second quarter of the century is the striking portrait of Themistokles (PL. 210) found in 1938 at Ostia, in which the personality of the man is successfully conveyed, despite vestiges of archaism in the modeling. But this and similar attempts at realism that appear in some sculptures and vase paintings of the time were submerged during the second half of the 5th century by the idealizing trend introduced by Phidias, Polykleitos (qq.v.), and other leading artists. Nevertheless, the portraits that can be assigned to this period on the basis of style — those of Anakreon (II, PL. 49), Miltiades, and Pericles (III, PL. 372), for instance, and others found on coins and gems (Pythagoras, Tissaphernes, Pharnabazos) — show increasing interest in depicting the specific character of the individual, though the generalized type is still dominant. Indeed, it is this mingling of the type with individual traits which gives these early portraits their peculiar interest.

From the 4th century on, individualization in portraiture became increasingly marked. The heads of the historians Herodotos and Thucydides (III, PL. 372) and of the orator Lysias evince a more subtle perception, and this is deepened in the portrait of Plato and especially in that of Socrates (III, PL. 372), the original of which was perhaps created by Lysippos.

From literary evidence, it is known that the orator and statesman Lykourgos set up, in the 110th Olympiad (ca. 340–330 B.C.), bronze statues of Aeschylus, Sophocles, and Euripides in the Athenian theater. All three portraits have been more or less persuasively identified in Roman copies: the Sophocles in the well-known Lateran statue (II, PL. 57); the Euripides in a herm inscribed with his name, in Naples (Mus. Naz.); and the Aeschylus through appearing in a double herm with Sophocles.

For the portraits of these three eminent poets, as in many other cases, the question arises as to how faithful the likenesses are. By about 335 B.C., Sophocles and Euripides had been dead about seventy years, and Aeschylus a good deal longer. No person who knew the poets and who might have assisted the sculptor in achieving a faithful likeness could have been alive when the statues were created. Nevetheless it is probable that, at least in the case of Euripides and Sophocles, the portraits were to some degree real rather than invented, for while these poets were alive and famous, informal sketches of them could have been made by artists who had become interested in the new art of individualized portraiture. This is specifically suggested by the fact that masks worn by actors in the theater were often made to resemble the actual people. On the other hand, portraits of people who lived before individualized portraiture was current in Greece — those of Homer, Hesiod, and Periandros, for instance — must have been invented.

The portrait of Aristotle (d. 322 B.C.) initiates a new era. It shows a new intimacy of perception and gives a convincing likeness of a sensitive man with a powerful, analytical mind. Some of the many portraits of Alexander the Great that were erected during his short lifetime and after his death are powerful creations but usually have strongly idealized features, due to his early heroization (VII, PLS. 151–55; IX, PL. 224).

With the 3d century B.C., the great era of realistic Greek portraiture is entered. The dated portrait of Demosthenes, said to have been erected 42 years after his death (i.e., 280 B.C.), and the portrait of Epikouros (d. 270) serve as chronological landmarks, bringing out vividly the striking personalities of the two men: the diffident orator and courageous patriot, and the philosopher whose teachings remained popular for centuries. They were succeeded in the late 3d century and during the 2d by other vivid likenesses, for instance those of the philosophers Zeno, Chrysippos (VII, PL. 170), and Diogenes, of Aesop, of the poet-astronomer Aratos, of the physician Hippocrates and the philosopher Karneades. The heads of the blind Homer (III, PL. 209) and the so-called "Pseudo-Seneca" (Hesiod?; VII, PL. 170) are outstanding achievements of the "baroque" Hellenistic age. Both are invented and thus resemble great works of fiction, having their origin in man's inspired imagination and yet being essentially real in effect.

A special place in late Greek portraiture is occupied by the representations of the Hellenistic rulers who inherited Alexander's vast empire and then carried on ceaseless warfare with one another, preparing the way for the conquest of Rome. Their likenesses exist primarily on the obverse of the coins that were issued by the kings themselves, and from these a few sculptured heads have been identified. The great value of these coin types is that they are datable Greek originals that supply a continuous series, from the beginning of the 3d to the 1st century B.C. and into the 1st century of our era. They are small, however, and give only profile views; so the identification of life-size sculptured portraits by comparison with them is often precarious. Moreover, such sculptured portraits are few, consisting mostly in examples found in the countries of the respective rulers, for the Roman collectors naturally were less interested in the various kings of the Hellenistic provinces they had conquered than in Greek intellectuals.

Nevertheless, the importance of these Hellenistic ruler portraits is great, both esthetically and historically. Some of them, especially the Bactrian, Pontine, and Cappadocian, are artistic creations of the first order, evoking personalities of the most varied types — some (e.g., the Ptolemies and the Seleucids) historically famous, others known only through their coins. Such portraits as those of the Bactrian Antimachos I (ca. 185 B.C.; II, PL. 84) and Heliokles (ca. 150–135 B.C.), of the Pontine Pharnakes I (ca. 190–169 B.C.), of the Indian Archebios, and even the remarkable head of Cleopatra (51–30 B.C.), shown with an aquiline nose and somewhat forbidding features, indicate the degree of realism attained by Greek portraitists of the 2d and 1st centuries B.C. They make one realize that the Greeks initiated realistic portraiture, preparing the way for that of republican Rome. If such late Hellenistic portrait achievements were available in a greater number of life-size sculptures, such as the Euthydemos of the Torlonia Collection (PL. 210), the connection between Hellenistic and Roman portraiture would be better understood.

Etruria. Etruscan portraiture, though retaining throughout a character of its own, was closely allied to the Greek, and its development runs parallel to it. During the archaic period the features of the men and women on the sarcophagi and urns are still types rather than individuals. Those of the 5th century B.C. are generalized, though in attitude and gestures they often reflect the Italian liveliness and grace. In the 4th century, an increasing realism makes itself felt, and this becomes heightened in the Hellenistic period. The remarkable portraits sometimes encountered in the figures on late Etruscan sarcophagi, urns, and wall paintings evince the physiognomies of these Etruscans, so different from both the Greek and the Roman — less intellectual than the former and less forceful than the latter, but with a charm of their own. Their often uncompromising realism is admirably shown, for instance, in the famous urn in Volterra, the sarcophagus in Chiusi (V, PL. 49), and the painted portrait of Vel Saties from the François tomb at Vulci (V, PL. 52). Some of the portrait heads in bronze, marble, and terra cotta that have occasionally survived enlarge the knowledge of Etruscan achievement in portraiture. The charming bronze head of a boy in Florence (Mus. Archeol.; Giglioli, 1935, pl. 366, no. 1), assigned to the 3d century B.C., is still quite Greek in style, whereas the bronze statue of a magistrate in the stance of an orator, the "Arringatore" (2d cent. B.C.?; V, PL. 53), with its dry realism presages the Roman. The Etruscan inscription on this statue contains the name of the subject, Aule Metele (Aulus Metellus); it is an excellent example of the mixture of Greek, native Italic, and Roman elements that gives late Etruscan art its special character. The so-called "Portrait of L. Junius Brutus" (PL. 213) has been claimed to be an Etruscan work of the 4th century B.C. or of the Hellenistic period, but it is more probably a product of the 1st century B.C. in the Greek tradition.

Rome. As has been noted, realistic portraiture was achieved by the Greeks during the Hellenistic period. With the Romans, however, it entered a new phase. In the late Greek renderings there had always remained a trace of the former idealizing, generalizing tendency, but in the Roman portraits the realism became more uncompromising. The portraitists of the republican period — many of whom no doubt were artists who had come from Greece — had inherited the Greek knowledge of naturalistic form but were confronted by entirely different racial types. Their new clients had different physiognomies from the Greek, and most of them were practical men of affairs rather than philosophers and poets. Moreover, there had always been inherent in Italy a feeling for "verism," as evinced in late Etruscan portraits, and this natural tendency had been stimulated and fostered by the old Roman custom of making and displaying wax images of the dead. Furthermore, portraiture became more widespread. Not only eminent men but also obscure private individuals had their likenesses made. This greatly enlarged the scope of the artist, who could now study the physiognomies of both the great and the humble and could include women and children, thereby enlivening his art. Indeed, the whole history of the Roman Republic and Empire is reflected in the varied personalities that succeeded one another generation after generation, from the 1st century B.C. to the 3d and 4th centuries of our era. They reveal the Roman character as it developed and changed and help one to understand the great achievements of Rome, as well as its limitations and ultimate downfall.

In contrast to the Greek portraits that have been preserved in Roman copies (which, as previously mentioned, consist mostly of busts and herms), the portraits of Roman men and women have been preserved in many full statues. They show emperors and generals in military attire and civilians wearing the toga, as well as men represented in the nude, in heroized form. For both men and women, the head is sometimes combined with a body copied from a Greek type, and the two elements do not always form a harmonious whole.

For identification of the individuals, the portraits of the emperors and their families appearing on coins inscribed with their names have proved especially useful — though here, as on the coins of the Hellenistic rulers, the heads regularly appear only in profile views. There are also other clues that help in the dating of sculptured portraits: for example, the form of the bust changes from period to period. In the Augustan age, only a small part of the chest adjoining the neck was included, then more of the chest was added, until finally the bust comprised the whole upper part of the body. The coiffures of women also changed from generation to generation, reflecting current fashions. First came the simple, parted hairdo with central roll typical of the Augustan and Julio-Claudian periods, then the high, honeycombed fronts of the Flavians and the hair coiled on top of the head worn by the Antonine women, and in the 3d century the waved hair descending to the nape of the neck, as in the portraits of Julia Domna.

The fundamental consideration, however, in any classification of the portraits is their general style and the type of the physiognomy. First one encounters the portraits of the stern republicans, who were responsible for Rome's rise to world power. One can see in their virile faces the force, the determination, and the practical sense of these men. If one compares them with some of the later Ptolemies and Seleucids, one can understand the inevitable course of history. As conspicuous examples one may cite the statue of a general in Rome (Mus. Naz. Romano, inv. 106513) and a head in New York (Met. Mus., no. 12.233), whose rugged countenance has often been compared with those of the leaders of the American business world.

With the reign of Augustus (31 B.C.–A.D. 14), a new style appeared in the portraits of the ruling family (VII, PL. 370). The restrained "classicism" of these heads and statues is in striking contrast to the realism of those of the republican age. Rome had by then become a cosmopolitan center and was more and more open to Greek influence in art; moreover, the temperate character of Augustus, who dominated the history of the time, no doubt also influenced artistic expression. The *Ara Pacis* is a conspicuous example of this tendency, and the statue of Augustus from Prima Porta, in its commanding stance and with its reliefs combining representations of Roman victories and prancing Greek chariots, sums up the new mentality. The portraits of the Julio-Claudian family have been recognized in a number of life-size heads and in smaller versions on engraved gems. They show a strong family likeness to one another, most of them having the same clear-cut, aristocratic features as Augustus.

That, however, the veristic Roman sense was not wholly submerged by a tendency toward classicism is apparent from the portraits appearing on sepulchral monuments (PL. 211), which represent ordinary men of the time. Their sturdy countenances are depicted in a thoroughly realistic style. Nor were the classical works of Greece the only ones admired during the Augustan and Julio-Claudian periods; the many Hellenistic works that were copied show a catholic taste, as would indeed be natural at a time when works of art of all periods were being continually imported.

With the bourgeois, intelligent Flavians (A.D. 69–96), a new style asserted itself. There was more stress on individuality, but it was less harshly and more subtly expressed than in the republican heads, the features being less strongly accentuated. Characteristic examples of this style are the head of Vespasian in the Museo Nazionale Romano, with its almost fleeting expression, and the famous relief of a shoemaker in the Palazzo dei Conservatori, as well as the bronze portrait of the sober banker Lucius Caecilius Jucundus from Pompeii (PL. 211). To this period may also be assigned the painted portraits of a married couple, also from Pompeii, both with expressive, typically Italian faces. These are among the few painted specimens to have survived.

After the conquests of Trajan (r. 98–117) and the consolidation of the Roman Empire came the Hadrianic and Antonine ages (117–38, 138–61), which brought an era of peace and prosperity. The many extant portraits of this period, both of members of the reigning families (PL. 215) and of private individuals, show a new suaveness and sensitivity. A pictorial quality is imparted to the heads by the often deeply drilled locks that create strong shadows and by the high polish sometimes given to the faces; and this quality is accentuated by rendering the eye with pupils indicated by two adjoining dots, giving direction to the glance and animation to the expression. In some of the portraits of Marcus Aurelius (r. 161–80) his philosophical outlook is suggested in the wide-open, dreamy eyes, whereas in the majestic equestrian statue of the Capitoline his unquestioned rule over the vast Roman Empire found expression.

Then came the 3d century, with the constant menace of barbarian invasions and with the army virtually in control. Many of the portraits, in their sad and troubled expressions, reflect the restlessness of the times (PL. 215). Furthermore, the foreign blood which was introduced into the Latin stock and which gradually neutralized the Roman physiognomy is increasingly evident in the types of countenances. Especially interesting are several portraits of women, contemporaries of Julia Domna, the Syrian wife of Septimius Severus (r. 193–211), with their long, wavy locks descending to the nape of the neck and brought up in a knot at the back of the head. The hair of the men was no longer indicated by curly locks in high relief, as in the preceding period, but was short — indicated by a series of cuts — or was arranged in closely adhering ringlets. The portraits of the sinister Caracalla (r. 211–17), of Maximinus Thrax (r. 235–37), of Gallienus (r. 253–68), and of Philip the Arabian (r. 244–49) rank among the great achievements of this century.

To the 2d and 3d centuries also belong the remarkable portraits in encaustic that were found in the Egyptian Fayum, embedded in mummy cases (PL. 214). The physiognomies are mostly Greek but also include Roman types.

The story ends with the portraits of Constantine the Great (r. 306–37; IX, PL. 53), the first Christian emperor and founder of Constantinople. In contrast to the 2d and 3d century portraits with their sense of movement and life, the heads now assume a more static character. They are carved in few planes, with clear-cut features, in a two-dimensional design. They thereby acquire a monumental character that harbingers the Byzantine era.

Gisela M. A. RICHTER

Late-antique and Early Christian portraiture. From the time of Constantine onward, portraiture became an ever more independent art form, sometimes almost completely isolated from the rest of world art, as indicated by the persistence from the 5th century on of certain iconographic types for apostles and saints.

New elements appeared in portraiture as courts began to multiply and move from one spot to another, so that court art ceased to be centered in a single place. Moreover, differences developed between royal portraiture and that of lesser dignitaries. In the Missorium of Theodosius (II, PL. 487) and in the reliefs on the base of the Column of Theodosius (II, PL. 467; VII, PL. 265; IX, PL. 57), the face of the Emperor is frozen in a hieratic impassivity that makes him seem detached from the work, and this is also true of the figures of Justinian and Theodora in the mosaics of S. Vitale in Ravenna (II, PLS. 440, 446; IV, PL. 18). In contrast, the portraits in the diptych of Stilicho and Serena (IX, PL. 84) and the faces of the two ladies accompanying Theodora in the S. Vitale mosaic are full of vivacity; these people were not portrayed as symbols of majesty but simply as themselves.

At about the same time, portraits of private citizens appeared, in which fashions, although they exercised some influence, were less important than cultural and spiritual factors, which even brought about a return to historical and artistic precedents, especially in funerary monuments; the relief in Villa Albani (end of the 4th cent.) representing a husband and wife, in which the wife is portrayed as Venus emerging from the water, is a noteworthy example. There were, however, other kinds of private portraits, such as those in chrysographies, in which the rendering of the model's pose or physiognomical traits shows that the artisan tradition, which had persisted in Roman art since the late republican period, had not yet died out (e.g., the family group in Brescia, Mus. Romano; and the man's portrait in Arezzo, Mus. Archeol.).

Another new element that influenced portraiture was the Christian religion. Though on the one hand the Church disapproved of portraits for fear of idolatry, on the other hand it appreciated the evocative possibilities of figures of the saints used for teaching purposes. This attitude had been expressed as early as the apocryphal *Acta ioannis* of the 2d century (R. A. Lipsius and M. Bonnet, ed., *Acta apostolorum apocrypha*, II, 165 ff., 25–27, Hildesheim, 1959), in which the apostle admonishes the portraitist, "You have painted a lifeless picture of a dead man," and adds that a clever painter sees the soul and not does merely reproduce the actuality of the model. Porphyry and all the Neoplatonists took a similar attitude toward the portrait by Carterius of Plotinus (*De vita Plotini*, 1). Thus both Christianity and antiquity were less concerned with the reproduction of physiognomy than with the interpretation of character. This essential factor in the art of portraiture after Constantine is a cause of difficulty in the identification of the surviving images: the so-called *Colossus* of Barletta (IX, PL. 60), for instance, has been identified as both Marcianus and Heraclius.

At this point artists were only a step away from portrayal that "meant" the model instead of rendering its physical actuality. A product of this trend is the porphyry group of the Tetrarchs in Venice (IX, PL. 56), which can represent neither the first nor the second group of these rulers, because their beards are a later addition; they must have been executed outside the great art centers about the second half of the 4th century, near the time of the last Fayum portraits. On the same principle, the iconography of some coins could represent more than one emperor, merely by changing the inscription, just as with the royal effigies of ancient Egypt.

During the same period paganism was replacing a myriad of minor gods with personifications related to time, the seasons, qualities, and virtues, creating images completely lacking in physiognomical details and recognizable only by their attributes; for instance, the personifications in the mosaics of Antioch or those in the mosaic floor of Argos. Christian art appropriated this kind of image and developed it for the concepts of its own teaching (e.g., the *ktisis* or *ananeosis* on the mosaic floors of various churches such as that of Kurion in Cyprus). But Christian art also went further, applying the symbology of attributes and qualifications to the images of saints, apostles, and prophets and to the scenes in sacred books, in keeping with its tendency to identify the image with the sign. When carried to an extreme, this process resulted in the annulment of the portrait as an expression of a physical reality. But this happened only with the iconography of saints, for the portraits of living persons were true portraits made by specialized artists; the portrait of Pope Zacharias (741–52) in the fresco in S. Maria Antiqua (Rome), for instance, was executed by a different painter from the one who made the rest of the fresco. This specialization is also confirmed by the detailed description by John the Deacon (*Vita S. Gregorii Papae*, IV, 83–84; *Acta sanctorum, XII martii*, VIII, 136) of the portraits of Gregory the Great and his parents that were hung, at the time of his papacy (590–604), in the refectory of his monastery on the Caelian hill.

Michelangelo Cagiano de Azevedo

The European West. *The Middle Ages.* The principal theme in the art of the Middle Ages was the portrayal of Christ, though not as an actual man but as a hallowed figure. The images of the saints were also posthumous representations derived from and designed to serve their memory; these might be without relation to the actual appearance, since they were created solely to honor and glorify the subjects.

In keeping with the principle of characterizing, rather than depicting, in an image, the typical medieval portrait was that made after the subject's life was completed: the sepulchral effigy sculptured on his tomb. Already present in Etruscan art, this type reappeared in Ottonian Salic art (bronze tomb relief of Duke Rudolf of Swabia, 1081, Merseburg), possibly as a development from the mosaic tomb image (e.g., that of Optimus in the necropolis of San Fructuoso, Tarragona, 4th cent.). The deceased, represented as eternally youthful yet impersonal, was characterized only by an inscription and the insignia of his rank. By the second half of the 13th century a different concept prevailed in Italy, with the subject depicted as a corpse on his deathbed (tomb of Pope Clement IV, by Pietro d'Oderisio, Church of S. Francesco, Viterbo, 1271–74) following the French precedent (royal tombs, Fontevrault; tomb of Dagobert I, St-Denis, Paris). Above this the deceased was occasionally depicted again as he had been in life, realistically — on a throne, or standing, or even on horseback. In the tomb art of the Germanic lands, toward the middle of the 14th century, the deceased was often shown kneeling before a saint or other important personage in an independent mural relief that had more of the character of a true portrait. In this same period there were instances, particularly in England, of the use, for public display, of a wooden effigy of a deceased ruler or lord dressed in his robes of office. These effigies, however, seem to have made no pretense to portraiture.

In life, a man of the Middle Ages might be represented as the holder of high office, the tradition regarding images of rulers having been established in antiquity. From the times of Alexander the Great and Caesar the ruler's head appeared on coins and on the state seal as a warranty of their worth and authenticity. Charlemagne still used the head of Constantine the Great and, as a seal, a Roman portrait cameo with a new inscription, but he also used his own face (with a moustache; III, pl. 409). Gradually it became commonplace to represent secular or religious princes on coins, usually characterized by inscription or costume but sometimes also by their features or a beard. Frederick II, like Caesar, had coins struck bearing his own features as emperor (III, pl. 407).

The monument in honor of the ruler goes back to the Greco-Roman traditions of the deified emperor. The first Christian statue of an emperor represents Constantine with the cross, and the Byzantine rulers who followed him also had statues dedicated to them. Except for the statuette of Charlemagne (9th cent.; I, pl. 303), there is only one freestanding medieval equestrian monument, that of Emperor Otto I in the market place of Magdeburg (mid-13th cent.), set under a baldachin. Occasionally the ruler's image was placed on the city gates, as with Emperor Henry V (1106; Speyer Mus.) and Frederick II, seated on a throne with his ministers shown beside him in portrait busts in the Roman style (ca. 1235; I, pl. 303). Other similar monuments of seated dignitaries are those of Charles I of Anjou (I, pl. 458) and of Pope Boniface VIII (1296–1301; Florence, Mus. dell'Opera del Duomo), both by Arnolfo di Cambio, and that of Charles IV at the bridge gate in Prague (after 1378), by Peter Parler. In Italy local rulers were sometimes portrayed on the city halls (Oldrado da Tresseno in Milan, 1233) or on public fountains, e.g., Matteo da Correggio, probably by Nicola Pisano (q.v.), and Ermanno da Sassoferrato, probably by Giovanni Pisano (q.v.), in Perugia (1278). Statues of the royal family were placed by the great winding stairs of the Louvre in Paris (ca. 1365), and in the great hall of the palace at Poitiers there are four life-size statues of the royal family (ca. 1390).

In religious communities the images of Church authorities were often presented in long series showing the succession of princes of the Church or, in the case of the popes, the apostolic succession. In the years 440–55 Leo the Great had the portraits of his predecessors painted in medallions (*imagines clipeatae*) in S. Paolo fuori le Mura. There are two similar series of popes in St. Peter's and one in St. John Lateran (Rome). Calixtus II, sometime after 1122, had represented in the Lateran the full-length figures of the popes who had been concerned in the investiture dispute, up to and including himself. Later, portraits of bishops and abbots began to appear in their churches, along with portraits of their predecessors, as at Oberzell, Reichenau (10th cent.), and in the stained glass of the cathedrals of Cologne and Reims (13th cent.; XIII, pl. 158). On the choir stalls of the Cathedral of Cologne, 57 archbishops and a statue of Pope Sylvester I on one side face 69 Roman emperors and a statue of Constantine on the other — the whole of the *sacrum imperium* (14th cent.). The Strasbourg stained glass presents the sequence of the emperors (12th–14th cent.), and at Reims portraits of the archbishops confront in stained glass the images of the kings crowned by them (13th cent.). The portraits in such series lack personal traits and are identifiable only through costume or inscription. The statues of Pope Boniface VIII which he had commissioned in his own honor (and which led to an accusation of idolatry against him) were, however, of contemporary significance. His opponent Philip the Fair followed his example in Paris in 1313 (Notre-Dame), as did later popes from time to time. A unique example is the gilt-bronze head of Frederick Barbarossa in the Stiftskirche of Cappenberg (Westphalia), which, though in the style of a head reliquary, has his own facial features and personal characteristics, as can be confirmed by other portraits of him executed between 1155 and 1171.

Even outside the ruling circles, a man of the Middle Ages might also be considered worthy of a portrait as a founder of a church or monastery or as a donor or benefactor. Statues were erected to such founders as King Alfonso X and Doña Violante, in the Cathedral of Burgos (13th cent.; popularly identified as Ferdinand III and his wife, Beatrice of Swabia); Count Ezzo, or Ehrenfried, and his wife in the abbey church of Brauweiler (12th cent.); and King Dagobert I, in the monastery of St-Denis, and King Childebert (seated), originally in the refectory of the Abbey of St-Germain-des-Prés, in Paris (13th cent.; now in St-Denis). The most impressive monument of this type comprises the 12 life-size statues of the benefactors of the Cathedral of Naumburg (mid-13th cent.; VI, pl. 360).

A founder's portrait in relief on the façade of a church is found as early as the 7th century in Mtskhet, Georgia (VI, pl. 112). In many medieval churches the founder is depicted kneeling along with figures of the saints on the tympanum. Founders were also represented among the rows of saints on the outside of churches (even when they were not, like Emperor

Henry II and King Louis IX of France, canonized as saints themselves). Among others, on the façades of churches in southwestern France there were statues of two abbots (Moissac, 12th cent.) and two bishops (Moreaux, 12th cent.; now in Oberlin College, Ohio, Allen Mem. Art Mus.). Emperor Otto I and Empress Edith are included among the wall figures at Meissen (13th cent.); and the French king Charles V and Queen Jeanne de Bourbon were portrayed on the chapel portal of the Hospices des Quinze-Vingts (after 1380; Louvre), although they represent not themselves but the true founders, St. Louis of France and his queen, Margaret of Provence. In 1399 Claus Sluter of Haarlem produced the monumental founder portraits of Philip the Bold, Duke of Burgundy, and his wife on the portals of the Chartreuse of Champmol at Dijon (XIII, PLS. 77, 78). Princely founders and other patrons of the Church have also been presented in other sculptural details of the architecture such as the pulpit (Frederick II, Cathedral of Bitonto, Italy) and on the capitals of columns (Clermont-Ferrand, Notre-Dame-du-Port; Volvic, Puy-de-Dôme). Even more often representations of founders are found on examples of the minor arts, for instance, Henry II and his empress on the golden altarpiece of Basel (1002–19; VI, PL. 263); Otto I with a model of the church on the 10th-century ivory antependium of Magdeburg Cathedral (Seitenstetten Monastery, Austria); Otto III among the saints and clergy on the fountain of S. Bartolomeo all'Isola in Rome.

In painting, donor portraits appeared at an early date, as in the Turtura Madonna of the Catacomb of Commodilla in Rome (7th–8th cent.). Shortly after came the votive painting, an image of a saint with the donor (St. Demetrios with the healed children, Church of St. Demetrios, Salonika; Chapel of SS. Quiricus and Julita, S. Maria Antiqua, Rome). The finest founder portraits of these early times are the two mosaics in S. Vitale in Ravenna, showing Justinian (II, PL. 440) and Theodora (II, PL. 446) as patrons. In the early Christian mosaics from John VII (705–08) onward the popes are pictured as builders of the church next to the Christ and are identified as living persons by a square blue or green nimbus.

In book illumination the princely or ecclesiastical sponsor is often shown, generally depicted enthroned among his followers, accompanied by saints or personifications of the virtues, or invested with authority by the hand of God (III, PL. 64; X, PL. 461); or, in the case of emperors, sometimes with a halo. Occasionally the sponsor is pictured, according to Byzantine precedent, below Christ on the Cross, but most often at the dedication, with the scribe offering the book to the person who commissioned it, or that person offering it to the abbot or the bishop. These portraits of those who commissioned books usually represented living persons. Rulers are shown on their thrones, seen from a distance, and are customarily not personally characterized, but these sponsors are differentiated in illuminations by costume and by hair color (X, PL. 464). A unique example of Ottonian painting is the portrait of the sponsor Gundold; fat, with a bald pate and bushy eyebrows (Stuttgart, Landesbib., Bibl. 4°2: Evangelistary, fol. 9v).

Following antique traditions, the author is often pictured in the frontispiece, as a sort of pictorial title without real portrait characteristics: the Evangelists (X, PL. 462), St. Augustine, Rabanus Maurius, Uandalgarius (who compiled the *Leges barbarorum*, a collection of Salic law; depicted in a manuscript of 793 from Saint Gallen), as well as the poets of the Manessa Codex (ca. 1300; Heidelberg, Universitatsbib., Cod. pal. ger. 848). It is probably on this basis that Charlemagne is represented in manuscripts of the *Lex Saxonum* collected by him, and Barbarossa with his sons in the *Chronicle of the Guelphs* of about 1180 (Fulda, Landesbib., Cod. D. 11). Frequently princes appeared as guarantors of documents, for example, Lothair II (or III) in the Book of Traditions of Formbach Monastery (ca. 1145; Cod. München Reichsarchiv-Formbach Lit. 1f.3a) and Alfonso II of Spain handing over his testament to the bishop of Oviedo (12th cent.). The copyist might also be portrayed, and the illuminator sometimes included a self-portrait. Up to Romanesque times these were usually monks, members of the large monastic scriptoriums, who were shown at their work or kneeling before a saint, or, more often, in the act of delivering their work to the person who had commissioned it or to their abbot (X, PL. 464). Occasionally lay artists are distinguished from monastic copyists by their dress and appearance (e.g., Hildebertus and the illuminator Everwinus, in a 12th-cent. Ms. in Stockholm, Royal Lib., Ms. theol. A.144; the illuminator Alanus and the copyist or writer Petrus Cantor with Abbot Berthold I of Ottobeuren, who commissioned the work, Br. Mus., Ms. add. 19767). Examples of self-portraits are those of the monk Eadwine in a Ms. of about 1150 (Trinity Coll., Cambridge, England) and the artist Herrad von Landsberg (copy of *Hortus deliciarum*, ca. 1200, original destroyed 1870).

Self-portraits appear also, though less often, in other types of art. Thus the artist of the mosaic floor in St-Denis shows himself kneeling in a medallion (ca. 1140, Louvre); and glass painters frequently portray themselves in their work: Gerlachus with paint pot and brush, ca. 1170 (XIII, PL. 160), and several others in Chartres. Self-portraits are found, too, in tapestries of this period, as in a 13th-century one at Wessobrunn. The metalworkers of the Middle Ages, too, often depicted themselves: among others, Wolvinius, the *magister faber*, or chief artist, of the golden altar of S. Ambrogio in Milan (III, PL. 63); the monk Airardus on the door panels of the Carolingian St-Denis; the unknown Magdeburg master of the bronze portals of St. Sophia, in Novgorod; and the younger of the masters of S. Zeno in Verona (12th cent.). This practice extended to bell casters (1281, Mülhausen, Thüringen) and goldsmiths, such as Fridericus (St. Maurinus shrine, Church of St. Pantaleon, Cologne, after 1191) and the eminent Hugo d'Oignies (book cover, Namur). Sculptors rendered themselves in the act of dedication (the "Heimo" capital, with the kneeling figure so inscribed, Church of Our Lady, Maastricht, 12th cent.) or at work (choir stalls of Pöhlde, 1284; XIV, PL. 278). On occasion the stonemasons, who might also be the architects, are identified by their instruments or tools: Gerlannus, St-Philibert, Tournus, France, ca. 1000; Master Henri, St. Patroklus, Soest, 12th cent.; Master Humbert, Colmar, ca. 1263; Master Matthew (dates recorded as 1161–92) in Santiago de Compostela; and others in Vendôme, Limburg an der Lahn, Magdeburg, Reims (labyrinth), and elsewhere. In the Cathedral of Prague, the bust of sculptor-architect Peter Parler appears on an equal footing with the 20 busts of rulers and patrons (1375–85).

In spite of the concern for and skill in the reproduction of physiognomy in the Middle Ages (Bamberg Cathedral, sculptures of the prophets, ca. 1230; heads on the tomb of Hadrian V in Viterbo, d. 1276; masks in Reims Cathedral, VI, PL. 350), the individual rarely emerges in the image, because the preoccupation was with the universal rather than with specific individual characteristics. As a result, the features are more lifelike in figures of the saints than in portraits, and toward 1300 the features in portraits began to be stylized into an ideal noble type (tombs of St-Denis; the Manessa Codex, Heidelberg, Universitatsbib.). It was at this time that Giotto (q.v.) introduced new possibilities in the field of portraiture.

Giotto is known to have painted the fresco portrait of Pope Boniface VIII with his entourage, blessing the people (St. John Lateran, Rome; in poor condition), and that of Enrico degli Scrovegni as founder of the Scrovegni Chapel in Padua. The Scrovegni portrait, unlike traditional founder portraits, appears in the same proportions, on the same level, and in the same light as the Mother of God, before whom Scrovegni kneels to present the chapel, and the subject is identified not by a coat of arms or by an inscription but by his personal likeness. The figure of the priest accompanying him also seems to be a true likeness. That "true-to-life" quality which had previously been found only in rare examples was extended in Giotto's art to all human portrayal.

The effects of this revolutionary innovation were immediately apparent: both the profile portrait of Guglielmo di Castelbarco at S. Fermo Maggiore in Verona (1314) — like Giotto's prototype, showing the church model and the priest — and Simone Martini's portrait of Robert of Anjou, King of Naples (IX, PL. 336), are true likenesses verifiable from contemporaneous sculptural representations. Portraits in profile of individual founders or groups of founders became common and soon spread to northern art, and portraiture in action scenes was incorporated into the tradition of mural, glass, and book painting. The northern artists went beyond the religious altar and fresco painting of Giotto to the painting of independent portraits, the representation of a likeness for its own sake. The striking portrait on gold ground of King John the Good (PL. 216), definitely a descendant of Italian painting, dates from about 1365 and is contemporaneous with the portrait of Duke Rodolf IV (Vienna, Erzbischöfliches Dom- und Diözesan-Mus.) in three-quarter profile on a dark ground, a product of the progressive Bohemian art. Profile portraiture was to continue for some time in the south, especially in the Florentine Quattrocento, while the art of Van Eyck (PL. 217) did much to bring the oblique view to preeminence in modern portraiture.

Kurt BAUCH

Renaissance to 20th century. With the rise of Humanism, portraiture acquired a new conceptual dignity as a result of various theoretical and cultural factors. The study and imitation of old coins, for instance, and especially of medals (see COINS AND MEDALS), helped to propagate the idea of the official portrait and to limit it to the bust. Unlike 14th-century portraiture, which was still hindered by religious beliefs, the use of the portrait in the 15th century was accepted and developed, especially for didactic purposes, within the framework of the series of "illustrious men" (a theme that can be traced back to Petrarch). With the Humanistic cult of individuality, even the legendary heroes and heroines came to be characterized as much as possible by physiognomical traits, as in the works of Andrea del Castagno (q.v.) and the frescoes in the Castello della Manta, Piedmont (VIII, PL. 94). Despite the qualities of *gravitas, auctoritas,* and *majestas* that Humanism attributed to its heroes and the fact that, especially in sculpture, the portrait was consciously based on late-antique plastic art, Renaissance portraiture is above all characterized by an intense, even pitiless, insight into the model. One could say that the distinction of the first generation of the century, both in southern and northern Europe, lay in its exceptionally acute analytical studies, especially of the human face. The motto of this portraiture could well be that under the *Portrait of a Young Man* by Jan van Eyck (PL. 217), dated Oct. 10, 1432: *Leal Souvenir*. This approach to portraiture was probably based on the Aristotelian theory, well known during the Renaissance, contrasting history (embodying the presentation of truth) and poetry (the idealization of truth). The theoretical difficulty of reconciling the creative freedom of the artist with the accuracy of the representation was also resolved by Aristotle (see MIMESIS).

Jan van Eyck's concept of portraiture can be inferred too, from his masterpiece, perhaps the major work of its genre, the portrait of Giovanni Arnolfini and his wife Jeanne de Cename (V, PL. 223) at their espousals, in the presence of the painter, who acted as a witness and testified in writing that he had attended the ceremony. The analysis of this picture by Erwin Panofsky (*BM*, LXIV, 1934, pp. 117–27) shows that it had the value of a legal document, its verisimilitude being a decisive element from the legal point of view. From a somewhat later reference (16th cent.) by Cardinal Gabriele Paleotti, it can be deduced that the use of portraits as legal means to prove one's membership in a family through physical resemblance was fairly common. Van Eyck's thorough psychological study of his subjects, his keen eye for every facial imperfection or peculiarity, and his abundance of detail (V, PLS. 218, 219, 224) were part of a general pursuit of realism.

Jan van Eyck's portraits make a clear-cut distinction between the sacred, which was idealized, and the secular, which was treated with mimetic realism; and this approach was eventually imitated throughout Europe. By the middle of the 15th century Van Eyck's influence dominated in Flanders and had reached Germany and France, where without the Flemish influence the sharply realistic portrait, in the Pietà at Villeneuve-lès-Avignon (V, PL. 383), of the man who commissioned the picture would be inexplicable; it had affected Spain (portraits of Councilors Ramon Vall and Antonio de Vilatora in the altarpiece that they commissioned from Luis Dalmau; XIII, PL. 128), Portugal (portrait of Dom Henrique in the polyptych of St. Vincent by Nuno Gonçalves; XIII, PL. 126), and of course Italy, where it became even stronger in the second half of the century with such painters as Piero della Francesca (portrait of Federigo da Montefeltro; PL. 164), Andrea Mantegna (portrait of the Gonzaga family; X, PL. 487), Antonello da Messina (*Portrait of a Man*; XII, PL. 43), and Domenico Ghirlandajo (*Old Man and His Grandson*; VI, PL. 178). In Catalonia, about the end of the century, the work of Bartolomé Bermejo (portrait of the deacon in the *Pietà* of the Cathedral of Barcelona) is evidence of a strong revival of interest in physiognomical studies. The last wave of this influence made itself felt at the beginning of the 16th century, with Lorenzo Lotto (q.v.) in Venice and others. Naturally, portraiture was especially cultivated in Flanders (see FLEMISH AND DUTCH ART), in a direct continuation of the Van Eyck manner; but none of Van Eyck's followers came close to his forcefulness, the feature that most interested his European imitators. One might even say that the Van Eyck realism was the only manifestation in the sphere of 15th-century portraiture to have a European, rather than a purely national, significance, although the 15th century is nevertheless characterized by isolated great masters or local schools with trends of their own.

Portraits, especially painted portraits, were rare in the first decades of the 15th century, having been an almost exclusive prerogative of the courts, the nobility, and the clergy; the few portraits from the first half of the century were the product of exceptional and limited cultural circles. In the Duc de Berry's castle of Bicêtre (destroyed 1412) there was, in addition to a collection of iconographic coins (perhaps the first of its kind), a gallery of portraits of popes, kings, and great lords, both living and dead. In addition, between about 1411 and 1415, Jean Malouel was active as a portraitist at the court of Burgundy, where he painted a portrait of John the Fearless. Perhaps it was in the Burgundy court, too, that the elements characteristic of French portraiture were established. Sterling (1959) rightly defined a typical northern portrait, with its lyric simplicity, as "a definition of man based simply on the acute observation of his familiar appearance." At the court of Bohemia, in contrast, there developed a more sumptuous style of portraiture, typically late Gothic, with the subject usually in profile against a dark background, but sometimes in a three-quarter pose on a light background; the features of the models, though fairly well defined, are almost eclipsed by the rich clothing.

Rather similar are the portraits of Pisanello (q.v.), also in profile, with carefully detailed but idealized dress (PL. 175). Unlike Jan van Eyck and the other Flemish painters, in whose portraits the setting as well as the model is represented as realistically as possible, so as to indicate to the viewer the profession, rank, and circumstances of the person portrayed, Pisanello pictured Lionello d'Este (PL. 216) with scarcely any background (only the richness of the work itself conveys the idea of the subject's aristocratic status), at an age between youth and maturity, with little modeling (even less than in the profile portraits of the court of Burgundy) and in a rigid pose (whereas the Burgundy subjects are all portrayed in action, however slight). The Pisanello type of profile portrait, elegant and almost abstract but showing to advantage the elaborate ladies' hairdos of the time, became, understandably enough, a favorite for feminine portraiture, as in the work of Domenico Veneziano, Sandro Botticelli (PL. 223), and Antonio Pollaiuolo (XII, PL. 29), but eventually acquired a more three-dimensional quality. Pisanello's portrait of Lionello d'Este, done in competition with Jacopo Bellini, gave rise, according to Angelo Decembrio (*Politia litteraria*, VI, pp. 503–10, Padua, 1562) to discussions at the Ferrara court as to the appropriateness of a greater or lesser degree of physiognomical realism and of the use of realistic clothing or of idealized clothing that would not later appear outdated — factors that may have influenced the outcome of the competition.

A more sober and rather effective realism characterizes the profile portraits of the Lombard school, not only those of

Bonifacio Bembo but also those of Andrea Mantegna (PL. 217), Baldassare d'Este, and others. With their decreased emphasis on dress and more objective representation of reality, these works bear comparison with the portraiture of Jean Fouquet (PL. 217; V, PL. 362).

With Masaccio and Donatello (qq.v.) Tuscany had already provided instances of realistic physiognomical characterization comparable to that of Jan van Eyck, though tending toward a more ideal *majestas*. Besides the wonderful portraits of the donors in the *Trinità* of S. Maria Novella in Florence (IX, PL. 345), Masaccio painted a series of figures in the great fresco of the *Sagra del Carmine* in the cloister of S. Maria del Carmine in Florence, about which Vasari (I, p. 267) wrote : " . . . the master painted the portraits of a great number of the citizens who make part of the procession, clothed in hoods and mantles; among these figures were those of Filippo di Ser Brunellesco in 'zoccoli' [clogs], Donatello, Masolino da Panicale, who had been his master, Antonio Brancacci, . . . Niccolò da Uzzano, Giovanni di Bicci de' Medici, and Bartolommeo Valori, all of whose portraits, painted by the same artist, are also in the house of Simon Corsi, a Florentine gentleman. Masaccio likewise placed the portrait of Lorenzo Ridolfi, who was then ambassador from the Florentine republic to the republic of Venice, among those of the picture of the consecration; and not only did he therein depict the above-named personages from the life, but the door of the convent is also portrayed as it stood, with the porter holding the keys in his hand." This fresco, unfortunately lost, like its copies (though echoes of it no doubt survive both in the etchings that accompany Vasari's *Lives* and the other known iconography of those personages, and presumably even in parts of the portraits by Filippino Lippi in the Brancacci Chapel of S. Maria del Carmine), is a turning point in the history of portraiture, anticipating Jan van Eyck in quantity and quality and in the class of persons portrayed. Portraiture thus became one of the main themes of Renaissance art and ceased to be a courtly prerogative. The cult of the individual and, even more, the moralism and political fervor typical of Florentine Humanism all contributed to the raising of contemporary portraiture to a level with that of the famous men of antiquity.

Renaissance artists made use of all artistic media for portraiture, including the medal, which Pisanello (followed by his imitators) began using in 1438, and the coin, after 1450 (III, PLS. 410, 411). The first coin portraits appeared on Francesco Sforza's ducats from the mint of Milan, followed by others from Naples, Mantua, Ferrara, Modena, Reggio Emilia, Carmagnola, Asti, Bologna, Casale, Urbino, Mirandola, and occasionally Rome. The custom spread everywhere except to Genoa, Venice, and Florence, where it was delayed for political reasons. Only with the end of the republic in 1530 did the portraits of the dukes make their first appearance on Florentine coins (Rosati, 1961).

In sculpture, as on medals and coins, the classical bust-length portrait prevailed, and realism was pursued even more than in painting, perhaps because of the custom of faithfully reproducing death masks in marble. Some of the portraits obtained by this method are as keenly revealing of personality as the best of northern European art; for instance, the works of Mino da Fiesole, Antonio Rossellino, Benedetto da Maiano (XII, PL. 17), and Antonio Pollaiuolo (qq.v.). With Florentine artists, however, realism — sometimes allusive realism only — was not irreconcilable with the heroic or laudatory exaltation of the model; this is borne out by the equestrian monuments of classical inspiration, whether cast in bronze, like Donatello's in Padua (IV, PL. 249) and Andrea del Verrocchio's in Venice (XIV, PL. 359), or painted, like Andrea del Castagno's (I, PL 246) and Paolo Uccello's in the Cathedral of Florence (XII, PL. 27). A tendency toward idealization prevailed only in portraits of women, where it showed even in the marbles from death masks, for which Francesco Laurana was especially famous (PL. 221; XII, PL. 17).

Though it is difficult to give a short summary of the typical qualities of early Tuscan Renaissance portraiture, one may cite the opinion of Zervos (1950) that it evinces an equilibrium between knowledge acquired by sensory perception and knowledge acquired through the feelings — that is, a balance between the natural and the supernatural, achieved by arriving at the reality of the model after having visualized the whole image in the mind's eye. The fact that 15th-century portraiture had its roots in the classical world helped in setting the physiognomic peculiarities of the subject within the framework of a literary and ideal order or "decorum."

In the second half of the 15th century, despite recurrent waves of Flemish influence, portraiture began to acquire greater monumentality and majesty, even sacrificing realism to the representation of the ideal, precisely because of the Italian contribution, a contribution highly varied and abundant and in some cases (as in the Veneto) aimed at preserving late Gothic elegance. The great portraiture of the 16th century carried this tendency toward monumentality even further and opposed the faithful reproduction of details and the portrait document, which in the hands of such painters as Antonello da Messina and Domenico Ghirlandajo had won public favor at the end of the 15th century. This reaction reached one of its climaxes with Michelangelo (q.v.), who detested "making resemblances of life" and instead made wholly imaginary sculptures of Brutus (III, PL. 389) and of the two Medicis, Giuliano di Lorenzo (IX, PL. 530) and Lorenzo di Piero, convinced (perhaps through the influence of Alberti) that what really mattered in portraiture was not so much the physical likeness as the faithfulness to the personality and expression. When his critics argued that the statues on the Medici tombs were not accurate portraits, Michelangelo is supposed to have answered: "In a thousand years no one will know the difference."

At the opposite extreme were the Venetian portraitists, who were praised for their ability to depict not only the features but also the flesh tones of their sitters. They formed one of the greatest and most homogeneous schools of portraiture in the world, led by Titian (who painted all the most important political and literary personages of his time), Tintoretto, and Paolo Veronese (qq.v.), and including also Lorenzo Lotto (PL. 224) and Palma Vecchio (qq.v.), Paris Bordone, and the Bassanos (II, PL. 243). Even in the canvases of the Venetians, however, there is a marked taste for idealization which was obtained through a kind of impressionistic execution in which the physiognomical and identifying elements are reduced to a minimum. These portraits are far more allusive than descriptive; their lifelike quality, especially in the works of Titian, arises from the heightened flesh tones of the faces in contrast with the surrounding color scheme of deep values of the same tones and from the choice of a pose that seems to be spontaneous and momentary (PL. 218). In addition, the subject's gaze is directed straight ahead, establishing a kind of direct contact with the viewer. This quality of vitality was an extraordinary achievement, especially in the case of Titian, since most of his portraits were based on other artists' sketches or on preexisting portraits such as medals. In his *Discorso intorno alle immagini sacre e profane* (Bologna, 1582), Gabriele Paleotti wrote that some painters ". . . from the mere verbal report on the face, coloring, demeanor, features, and other characteristics of a person whom they have never seen, . . . apprehend the whole so wonderfully in their art that they would seem to have had it before them a long time."

In their pursuit of idealization, 16th-century artists made use of various schemes whose development and variations can be followed from generation to generation. In Venice, Giorgione introduced an idea derived from the Flemish portrait painters: the almost waist-length portrait, with the subject leaning over a parapet, his head bent to one side or looking sideways. This innovation was a reaction to the classical bust-length portrait with a single-color background, which Antonello da Messina, Giovanni Bellini (II, PL. 265), and their school favored. The portrait thus acquired greater directness and, by its broader background settings, a touch of mystery. The half-spontaneous, half-affected pose of the sitter still satisfied the requirement of "noble, majestic, dignified demeanor," the decorum and grandeur that the aristocracy expected of art and literature (Weise, 1961). Florentine portraiture was characterized by a grandiose style; the study of physiognomy was even more prominent than in Venice, and color often had an expressionistic function. A typical example is the illustrious *Portrait of a Young Man*, which perhaps represents Alessandro de' Medici, by Pontormo (q.v.; Lucca, Pin. Naz.).

In the 16th century, portraits executed for biographical purposes began to be preserved. The Museo Gioviano, near the ruins of the villa of Pliny the Younger on Lake Como — for which a catalogue with biographical sketches of the personages depicted in the collection was published in 1546 by Paolo Giovo, its creator — achieved wide renown. It included copies of canvases and frescoes, as well as original paintings, many of them by Titian. Etchings made from them were subsequently reproduced in Lyon (1549), Paris (1552), and Basel (1577), and some of the originals were copied for the similar portrait collections of Cosimo I de' Medici (in the passage joining the Uffizi to the Pitti Palace), of Archduke Ferdinand II of Tirol in the Castle of Ambras (collection now partly in Vienna), and of Cardinal Federigo Borromeo in Milan. At the Château of Blois in France, Catherine de Médicis built up a vast collection including 341 portraits, of which an inventory was made after her death in 1589. Another enormous iconographic collection is the group of etchings — generally reliable portraits of famous artists — with which Vasari prefaced each of his *Lives* in the 1568 edition.

Along with the idealizing tendency, the allegorical portrait was renewed and developed during the Renaissance period. In the 14th and 15th centuries there had already been many instances of donors disguised as saints, and a fashion had later developed for portraits that alluded through symbols — flowers, fruits, and various other objects — to death and to the transience of life, almost as if to belie the accusations of vanity that some reformers (e.g., Girolamo Savonarola) made against the aristocratic models; one example is the distorted skull in the double portrait by Holbein the Younger (see HOLBEIN), *The Ambassadors* (VII, PL. 304) that transforms it into an allegory on vanity. The Italians, too, especially the Venetians, had made use of withered flowers, skulls, and other symbols. But in the Renaissance period the image of the model itself, often disguised in mythological dress, became allegorical. Among the earliest examples, from the court of the Medicis, was the portrayal of Simonetta Vespucci, both by Botticelli in his *Primavera* (Uffizi) and by Piero di Cosimo (probably as Proserpina; XII, PL. 29). Leonardo, who carried these devices to an extreme, portrayed Ginevra Benci (Washington, D.C., Nat. Gall.) on a background of juniper (It., *ginevro* or *ginepro*); and Giorgione, who painted a woman presumed to be Laura (VI, PL. 194) on a background of laurels, introduced an iconographic scheme that became very popular by portraying himself as David conquering Goliath (Vienna, Kunsthist. Mus.). Among other artists influenced by this example were Michelangelo, whose self-portrait appears on the skin held by St. Bartholomew in the *Last Judgment*, and Caravaggio, who for the severed head of Goliath depicted his own face as he thought he might appear when old (Rome, Gall. Borghese). The fashion for portraying models in the guise of the ancient gods, however, became widespread only in the second half of the 16th century. Among the significant examples of this vogue are the portrait of the military leader Andrea Doria in the guise of Neptune by Bronzino (q.v.) in the Brera; the cycle of portraits in honor of Diana, relating to Diane de Poitiers (V, PL. 392; XII, PL. 103), from the French school of Fontainebleu; and the "portraits" in honor of Minerva or Flora in which the allegorical aspect is so marked that it is difficult to tell where the imagination ends and the actual features of the model begin.

The bust-length portrait by Leonardo of the nude Gioconda (known from the cartoon in the Musée Condé, Chantilly, and from an old copy in The Hermitage, Leningrad) must have been famous in its time. The allegorical flavor arising from a vast and mysterious background landscape can also be found in the *Mona Lisa* (IX, PL. 126); and there are innumerable other works in which the composition, even apart from the obscure symbolic attributes, transcends the function of the portrait. One of these is the wonderful *Portrait of a Woman* by Bartolomeo Veneto (PL. 217), a near-personification of Flora, which is related to the many half-nude portraits of courtesans.

Influenced by the Italian, especially the Venetian, Renaissance, Germany developed a brilliant manner of its own. Instead of subordinating the actual physical characteristics of the model, in the idealizing process these aspects tended to be expressively, or rather expressionistically, heightened while preserving some elements of late Gothic grandeur. This combination resulted in intensely spiritual works, bordering at times on the devotional, as in Dürer's famous portraits of Erasmus (IV, PL. 301). In 15th-century Germany there had existed a considerable tradition in painted portraits of locally developed types, such as the double and the full-length portrait, which were later imitated in Italy; but the creation of German Renaissance portraiture must be specifically credited to Dürer (q.v.) and the masters of his school. In 1498 Dürer painted his self-portrait (IV, PL. 292) after the manner of the Italian and Flemish artists, such as Rogier van der Weyden (q.v.), depicting himself as a thinker and a scholar but with no attributes of his profession; in fact, he declared his determination to rid the local art of its "wild tree" quality, so devoid of conceptual roots. In the four decades during which Dürer devoted himself to portraiture the variations of his style were remarkable, but he helped to develop the full-face and three-quarter-face half-length figure, whereas because of northern influence the Venetian painters more often employed the full-length figure, which eventually took on a definitely aristocratic character. Above all, Dürer achieved a unique balance among the elements of dignity conferred on the model, the forceful presentation of the image, and the perceptive rendering of expression — a combination obtained by drawing the outlines of the face in a tense, often harsh and brutal fashion. In the best of Dürer there is an objectivity in the portrayal of his sitters that carries with it a highly dramatic atmosphere, as in the pencil portrait of his mother (IV, PL. 302). Holbein the Younger later carried this narrative expressionism to a more lyrical, but no less intense, naturalism.

The height of expressionism in portraiture was reached, however, outside any national school, by El Greco, who furthermore tended to restore the spirituality of medieval portraiture. Instead of emphasizing the model's features, El Greco elongated and schematized them, concentrating on the eyes, which in his work have an extraordinary range of expression and are sometimes — as in the famous group painting, *The Entombment of the Count of Orgaz* (VI, PLS. 454, 455) — the strongest characterizing element of the face. El Greco, as compared with Titian, was more perceptive in his observation and more unconventional in his choice of natural features to serve as signs of character (VI, PL. 460).

Religious controversies and the ensuing austerity, both north and south of the Alps, influenced the adoption of somber and modest dress for the subjects in portraits and, more important, the return to unembellished realism. Paleotti, who was first bishop and from 1556 archibishop of Bologna, explains in his *Discorso* that since the desire to have one's own portrait made was naturally connected with "a certain wish to display one's own excellence, which indicates considerable weakness of character" a portrait was justifiable only as a souvenir of a faraway relative, as a document, or in negotiations for betrothals. Moreover, after restricting its appropriateness for the living to people of exceptional political or moral merits (e.g., kings or "other prominent men in various professions"), Paleotti commented: "Since they are called portraits from life, the artist must take care that the face or any other part of the body not be made more beautiful or more dignified than it is, nor changed in any way from what nature has made it at that age, and even if there are any natural or accidental defects that greatly disfigure it, these must not be overlooked." Anticipating the attacks that were to come in the second half of the 18th century, he also rebuked women who wanted their portraits painted "with their faces colored and pretty," in other words, made-up.

What particular artists can be associated with these religious precepts, which remained valid also during the Reformation? A keen and unembellished objectivity in portraiture is characteristic, for instance, of Anthonis Mor (V, PL. 293; IX, PL. 297), Frans Pourbus the Younger, and Justus Sustermans, all of whom were active in the main Catholic courts; of Alonso Sánchez Coello (IX, PL. 297), who worked at the Spanish court; and of the elegant, yet austere, portraits of Jean and François Clouet (q.v.; V, PL. 403), especially in their few colored pencil drawings. But this vein of realism had direct precedents in the portraits of Holbein the Younger (VII, PL. 303), an artist who was personally involved in the religious wars and whose fidelity to fact is affirmed in the portrait epigraphs in which he diligently recorded the age of the model and the year of the painting's execution. Apart from any religious reasons, objectivity seems to have been a spontaneous vocation of provincial portraitists bent on satisfying the requirements of the country nobility and the rich middle classes, as in Bergamo, some of whose works were in striking contrast with the stately Venetian production; for example, the portraits by the painters Cariani (Giovanni de' Busi) and particularly Moroni (q.v.; PL. 218). But even in Venice, the family portraits by Bernardino Licinio stood apart from the rest. Realism was typical not only of portraits of high Church dignitaries but also of group portraits of officials and magistrates, especially in Paris and in the south of France, and of portraits for iconographic series (e.g., the portraits of donors in the 1555 collection at the Pio Albergo Trivulzio in Milan).

As a result of the fame of Titian's portraits and the importance they acquired among the rulers of the age as an indication of the models' political and cultural adherence to the ideals of the Renaissance (e.g., at the Spanish court), 17th-century portraiture gained precedence over the historical and allegorical genres. Diego Velázquez' portraits of kings were so famous that innumerable replicas were made of them, especially those of Philip IV. This monarch's change in spiritual attitudes and moods, coinciding perfectly with the political events of his reign, and his gradual aging are recorded in Velázquez' portraits. In the royal equestrian portraits for the Salón de Reinos in the Buen Retiro, Velázquez, both alone and with assistants, painted Prince Baltasar Carlos (XIV, PL. 326), Philip III, Margarita of Austria, Philip IV, and Isabel of Spain (all, Prado) in a glorifying tone after the manner of Titian's portrait of Charles V. He also modeled this series on the allegorical etchings by Jan van der Straet (Giovanni Stradano) illustrating Suetonius's *Lives of the Caesars*. Later, however, his style became independent of any baroque or ideal formula and he executed a number of extremely realistic canvases, like the two of Philip IV and his son in hunting dress (both, Prado) for the Torre de la Parada in the hunting lodge of El Pardo and the even more astonishingly realistic, almost repellent, portrait of a court buffoon (PL. 219), meant for a nursery or playroom of the lodge. The irreverent, caricaturelike portraits of Aesop and Menippus (both, Prado), which were obviously made from living models, were perhaps hung in the same room. In addition, Velázquez gave a portraitlike character to mythological scenes, for example, *The Drunkards* (XIV, PL. 322), and replaced the allegorical-celebrative compositions fashionable at other courts with family portraits (XIV, PLS. 330, 331) or professional or trade groups. (The same thing happened in Flanders, to which Spain remained linked in its artistic traditions.) The quantity of portraits Velázquez executed for the court (II, PL. 190; XIII, PL. 150; XIV, PL. 328) was extraordinary; such a range in this genre was unprecedented except in Venice.

Velázquez represented his models with an objectivity and physiognomical insight never found in the 16th century (XIV, PL. 324). In his library were scientific treatises on physiognomy, and his interest not only in the typical but even in the deformed is clearly manifested, especially in his portraits of buffoons and dwarfs (XIV, PL. 327). That such a searching and rigorous study was possible at all is explained by the deep infiltration into the Spanish court of the ideas of the early Counter Reformation, which considered portraits justifiable only if they were completely realistic. Velázquez reproduced the model faithfully and preferably full length. He did not even try to hide the fact that the model was posing in his studio, and he conscientiously reproduced the subject's expression of fatigue, boredom, or curiosity.

At the opposite pole are the portraits by Gian Lorenzo Bernini, Peter Paul Rubens, Frans Hals, and, above all, Anton van Dyck (qq.v.), whom Giovanni Pietro Bellori admired so much for his tendency toward idealization. Bernini was perhaps the one who most clearly expressed the baroque principles of the art of portraiture. He preferred to concentrate the expression in the face and therefore chose the bust-length rather than the full-length portraits favored by the late-16th-century artists and Velázquez. He also believed, even more firmly than the mannerists, that the resemblance must indeed be there but only in what is "noble and grand." Rather than idealizing the model, he strove to heighten his character and dignity by an act or movement that generally corresponded to the rank or office of the person portrayed. He did what a director does with an actor: he made his model perform the role for which he was suited. Bernini's execution was meticulous enough to allow for a close-up, almost eye-to-eye, view, but it was nevertheless allusive. Paul de Fréart de Chantelou says in his *Journal de voyage du chev. Bernin en France* (ed. L. Lalanne, Paris, 1885) that once Bernini had posed his sitters to his full satisfaction, he resorted to a whole series of illusionistic tricks in sculpturing them, convinced that "sometimes in a portrait in marble it is necessary, in order to imitate nature well, to create what is not present in nature . . . . To represent the dark circles some people have around their eyes, one must hollow out the place in the marble where the dark circles would be to portray the effect of this color and by this device compensate, so to speak, for the defect of the art of sculpture that cannot give things color. However, nature is not, as has been said, the same thing as imitation." The portrait gallery that Bernini offers is, therefore, fictitious but incomparably lifelike (PL. 221; II, PLS. 274, 276–278).

Rubens carried physiognomical characterization further than Bernini did, but with his habitual tendency to generic compositions he did not try to establish a relationship between the persons portrayed and their setting. He was content to dramatize and ennoble his models by his pictorial ardor (PL. 220; XII, PL. 325). He attributed to them, especially if they were in large compositions, marked and strong passions, which command the attention of the viewer. Paraphrasing a contemporaneous preacher, one could say that their authority "lies in their vehement, forceful utterance, which by its sheer power subjugates the spirits of others." Rubens's portraits, with their rhetorical expressiveness, foreshadowed, more than any others, romantic portraiture and served as models for study, at least until the time of the expressionists, while the sensuous air that surrounds his women's bright and rosy faces seems to anticipate the tenderness of Renoir (q.v.).

Van Dyck devoted himself almost exclusively to portrait painting. He profited by all the previous examples but generally tended to idealize his models. He successfully adopted the 16th-century convention of the standing pose, with the models placed on a step platform (IV, PL. 304) to enhance their dignity. Sometimes they were shown in romantic landscapes (IV, PLS. 309, 310), giving to the compositions a tone that is both Arcadian and lyrical. Van Dyck worked on one of the main enterprises in the field of portraiture in Europe: the collection, known as the *Iconography* (printed in Antwerp in 1645), of nearly 200 etchings representing the most outstanding personages of the time, for which Van Dyck etched 18 plates (IV, PL. 433). Though he deprived baroque portraiture of none of its stateliness, Van Dyck made it more restrained and faithful, and also brought to it much dramatic and human feeling (PL. 219; IV, PLS. 28, 308).

In Europe at this time there were also various tendencies ranging between the opposite poles of realism and idealism. In the Low Countries portraits had a more humble function, commemorative rather than glorifying, and their production was therefore abundant. The best and most widely imitated painter in this vein was Frans Hals (II, PL. 213). In his canvases he gradually passed from an early emphatic style of composition (VII, PL. 144) to an increasing introspection, culminating in the severe and relentless physiognomical fidelity of the *Governors of the Old Men's Home at Haarlem* (PL. 220), which he executed at the age of eighty-four in exchange for a subsidy. Even more than Rubens, Frans Hals caught his models in the midst of an action and did not harmonize the various activities of the people in the group portraits, often overstressing their individuality.

Rembrandt, in his intense group portraits, took the opposite approach: he set all his figures in a context of related group activity, harmonizing them by means of a single color tone, which gave the portraits a psychological unity. The *Anatomy Lesson of Dr. Nicolaas Tulp* (The Hague, Mauritshuis), the *Militia Company of Captain Frans Banning Cocq* (or the *Night Watch*, PL. 459), and the sensuous portraits of bridal couples (PL. 463) and families are typical. Jacob Jordaens (q.v.) painted in the same manner although on quite another level (II, PL. 206). Samuel van Hoogstraten, a pupil of Rembrandt, praised his master's skill at achieving unity, saying that, by comparison, the older group compositions seemed like card games. The rigorous quality of Rembrandt's painting confers an allegorical, even dramatic character on his canvases, which thus take on the aspect of historical scenes. Physiognomical study of the individual face is very through (PL. 456; II, PL. 217; III, PL. 212), but its function is mainly expressive. Rembrandt preferred to show his models in meditation, repose, or reading (PL. 461).

Realism, an approach that was perhaps religious in origin, prevailed in the Low Countries, with these great masters as well as with the lesser ones (see BAROQUE ART; FLEMISH AND DUTCH ART), and even Jan Vermeer's ability as a portraitist is noteworthy, despite his entirely different conceptual outlook (XIV, PL. 340). Among many names, that of Cornelis de Vos should be mentioned here; his painting was simple and restrained but for this very reason intensely human (V, PL. 303). Many Flemish and Dutch artists were active abroad, among them Justus Sustermans, the official court painter in Florence.

Realism also was prevalent in France, partly because of the scenes of popular life by the brothers Le Nain (q.v.), but mainly because of the influence of Marie de Médicis's court painter, Philippe de Champaigne (q.v.), a Fleming who respected and explored the personality of his models, both in children's portraits and in those aristocratic portraits that were soon to be called *d'apparat*, or ceremonial (PL. 219; V, PL. 402). His paintings of nuns, outgrowths of his connections with Port Royal, are famous (II, PL. 213). In the work of François Girardon and Antoine Coysevox (qq.v.) academic classicism and Berninian baroque are combined. Girardon's masterpiece is Cardinal Richelieu's tomb (II, PL. 174). Coysevox's spirited style is evident in his portrait busts (II, PL. 178) and in the interior decorations at Versailles (V, PL. 408).

Generally speaking, European portraiture of the second half of the 17th century was dominated by the styles of Bernini and Van Dyck. This influence, especially in funerary monuments and portraits of heroes, led to a full, though delayed, acceptance of the glorifying approach of baroque portrait art.

In the 18th century, before the advent of neoclassicism, three trends prevailed: the typically baroque ceremonial portrait, whose greatest exponent at the French court was Hyacinthe Rigaud (q.v.; V, PL. 406) and which was extremely widespread in funerary sculpture; the allegorical-Arcadian portrait reminiscent of Fontainebleau mannerism, which transformed the style of Van Dyck into decoration; and a realistic type of portraiture, mostly bourgeois in taste, which, unlike that of the early Counter Reformation, no longer rejected a certain splendor. As often happens, the three trends affected each other and developed along parallel lines, each prevailing in a social class rather than in a particular artistic center. The portrait production of the century increased rapidly. The inventories made by Georges Wildenstein (1950, 1956) show that in Paris, where in the 17th century portraits in homes had almost invariably represented royal personages and members of the court (the king, Marie de Médicis, the ministers Richelieu and Mazarin) and were hung as indications of political loyalty and allegiance, by about 1700 even the homes of the artisan class contained portraits of members of their own families. The influence exerted by the general taste, which was prompted by the example of the Low Countries, is shown by the fact that at the 1699 Salon 60 portraits and 150 historical canvases were exhibited, but a few years later the situation was so completely reversed that financial and legislative measures had to be taken to aid the historical painters for whose works there was no longer a market, despite the high opinion they still enjoyed with the critics. In the second half of the century the demand for portraits was so great that it gave rise to patents and inventions for executing them mechanically (mostly by the use of the camera obscura); and both hand- and machine-made profile "silhouettes" became immensely popular.

The typically political ceremonial portraits with their triumphant emphasis had their direct precedents in the magnificent monuments that Francesco Mochi erected in Piacenza in the 17th century to Ranuccio Farnese and Alessandro Farnese (X, PL. 149); they can also be traced back to the portraits by Bernini, especially his equestrian monument (II, PL. 276) and bust of Louis XIV (PL. 221). In France, in the ceremonial portraits along the lines developed by Philippe de Champaigne, despite the respect for physical likeness, the dress was already more prominent than the personality of the model. With the expansion of the French monarchy the ceremonial type became established and enormously successful through the works of such painters as Rigaud, Nicolas de Largillière (q.v.; V, PL. 406), and François de Troy. Because of the attention it gave to dress it was also used for women's portraits and therefore became an instrument for the detailed documentation of fabrics, clothing styles, and society life, as with Louis Tocqué, Maurice Quentin de Latour (q.v.; V, PL. 411), Jean Baptiste Gautier d'Agoty, Jean Baptiste Perroneau (V, PL. 411), and François Boucher (q.v.; II, PL. 334).

When it came to portraying women, however, the mythological or, better still, the pastoral style was more fashionable. Women were portrayed half-nude or in simple clothing disguised as Diana if homage was being paid to their chastity, as Minerva if they were fond of literature, or more generically as nymphs. Some of these canvases are extraordinarily uninhibited. This mythological and pastoral type originated in mannerism, but its return to favor at the beginning of the 18th century was connected with the *fêtes galantes*. Jean Marc Nattier (PL. 222; V, PL. 415) started a fashion for the allegory of Hebe, the goddess of youth. The works of these painters show not only the pursuit of idealization but also a tendency to beautify their models. As a rule, in this kind of portraiture only the face had a resemblance to the model and not the body (which was nevertheless generously exhibited), although at this time the nude portrait from life also became popular (VI, PL. 394), especially with the canvases of Boucher. The figures often melted beautifully into the landscapes in a decidedly preromantic way. By analogy, many men's portraits were also painted with an Arcadian flavor, justified, although not always sufficiently, by the subjects' hunting suits (V, PL. 405). In such cases Arcadia was more than ever set in contrast to heroism. As time passed the ceremonial portrait became limited to victorious generals and navigators, giving rise to an independent class of portraits (PL. 226).

The tendency to beautify the model was more than ever present in the allegorical portraits of this century. According to Antoine d'Argenville, women loved De Troy most of all because, as he said, "They knew that he had a talent for making them beautiful even when they were not. In painting them as pagan goddesses, he gave them a poetic quality and his flattering brush lent them new graces without altering their traits." Bazin (1957) reported some of the contemporary debates on the limits of resemblance. The opposing points of view were more or less those which Diderot expressed at the 1765 Salon: " 'Well,' some said, 'who cares whether the works of Van Dyck are real likenesses or not? . . . The merit of likeness is short-lived; it is that of the brush that gives delight at the time and that makes the work immortal . . . .' 'It is nice,' others replied, 'to see once again the exact image of our forefathers on the canvas . . . .' " There were scores of legal complaints, documented by "reports of experts," charging a lack of resemblance (when they were not protests against excessive and disrespectful realism). Rigaud confessed: "If I paint women as they are, they do not find themselves beautiful enough; if I flatter them too much, their portraits do not resemble them." Actually, rather than flattering, the portraiture of this century expressed a serene, often frankly humane, ideal of life.

The atmosphere was particularly tender in family portraits, which by this time were almost all Arcadian, as in the works of Jean Baptiste Oudry, De Largillière (XII, PL. 160), François Hubert Drouais, and Louis Michel van Loo. The characteristic femininity of the 18th century enabled women painters such as Rosalba Carriera (X, PL. 497), Elisabeth Vigée-Lebrun, and Angelica Kauffmann (VI, PL. 155) to make a name for themselves. These painters and their models suited each other perfectly. In the same atmosphere the great English portraiture of Sir Joshua Reynolds (PL. 222; XII, PLS. 145–48) and Thomas Gainsborough (V, PLS. 465–67, 469) developed, which Hogarth considered the only flourishing branch of art in his country, recalling baroque painting and Van Dyck in particular. In their portraits the models, rather than being faithfully portrayed, are dramatized through gesture, ornament, and clothes.

Still bearing traces of the 16th century, especially of Titian and Veronese, Venetian portraiture flourished once again, but with an additional element of realism, which was particularly pronounced in the works of the provincial painters. Despite the fame of this school, however, only a few artists' names are remembered today: Bartolommeo Nazari, Jacopo Amigoni, Lodovico Gallina, Pietro and Alessandro Longhi (q.v.; IX, PL. 186), and Johann Baptist Lampi I, who was active in Austria and Russia.

The third trend in France about the middle of the 18th century was a decidedly realistic type of portraiture; it served as an introduction to 19th-century naturalism and even, it might be said, to photography. This was not merely a matter of simpler dress and plainer compositional settings in accordance with the general tendency to neoclassicism, whose peak in portraiture was reached by Anton Raphael Mengs (VI, PL. 156), or of the development on a large scale of the realism (Bazin, 1962) that had already been a particular feature of French drawings and pencil portraits. It was rather that this kind of portraiture actually satisfied the desire of the middle classes for exact reproductions of themselves and their surroundings. Its representatives in France were Jean Baptiste Siméon Chardin (III, PLS. 216, 218) and Jean B. Greuze, with whose work this trend acquired a conscious theoretical basis, Nicolas Bernard Lépicié, and Maurice Quentin de Latour. In Spain, Goya (V, PL. 120; VI, PLS. 393, 398, 400) was the exponent of this tendency; in his works there is also an element of caricature. In Italy, there were Giacomo Ceruti and Vittore Ghislandi (known as Fra Galgario; PL. 222); the latter's work was far more baroque than Ceruti's. In Switzerland, Jean Etienne Liotard was exceptionally sensitive to his models and their settings. In England, William Hogarth popularized the conversation piece, or group portrait containing a touch of anecdote (VII, PL. 291); his later single and group portraits are outstanding for their visual and psychological realism (VII, PL. 292). This tendency to portray the subjects in faithfully reproduced interiors almost came to be the only kind of painting. Another aspect of this type of portraiture was the depiction of artists, poets, and the like at work.

Sculpture, however, partly because of its commemorative purpose, which prevented it from following the fashion for Arcadian portraits, did not feel this influence so strongly. Nevertheless, before it was swept by the wave of academic neoclassicism, sculpture also showed an occasional inclination toward naturalism. This appears in the political portraits of Jean Antoine Houdon (PL. 221; VII, PLS. 307, 310), where the exact reproduction of physical traits, which Houdon obtained by first making plaster molds directly on the model, is charged with heroic nobility comparable to that found in historical painting. Houdon said that one of the most rewarding aspects of the difficult art of sculpture was the opportunity it afforded to preserve and make imperishable the likenesses of the men who contributed to the glory and happiness of their country.

The comparative equilibrium between faithful reproduction and idealization, between history and poetry, which the portraitists of the 18th century had reached and maintained and which had sprung less from theory than from taste, broke down in the almost anarchic proliferation of contrasting tendencies after the French Revolution and throughout the 19th century. These tendencies may be classified as follows: the idealizing tendency, which was archaeological and allegorical in the case of the neoclassicists and dramatic and melancholy in the case of the romanticists; the naturalistic tendency, which achieved some especially effective and dramatic exemplars in caricature; the bourgeois-realistic tendency, which became increasingly related to photography and impressionism; and the expressionist tendency (e.g., Vincent van Gogh), which was the last turning point in portraiture before its naturalistic elements were broken down in cubism and abstract art.

The neoclassicists and romantics did not agree on the manner of idealizing their models but both demanded that they be idealized. The words that Napoleon Bonaparte addressed to his court painters — Baron Gros, Jacques Louis David (qq.v.), and Pierre Narcisse Guérin — are typical of the neoclassic attitude (and strangely reminiscent of ideas expressed by Michelangelo). He said: "What do you want a model for? Do you think that the great men of antiquity sat for their portraits? Who cares whether the busts of Alexander the Great look like him so long as the image we have of him is in keeping with the greatness of his genius? That is how great men should be portrayed." There can be no doubt that in this pronouncement, which corresponded to the official taste of the time, Napoleon was affirming that he would not be content even with the generic idealization then in fashion, which was achieved by associating the model with an ancient image or traditional iconography (though it happened that Napoleon was portrayed as a nude Apollo by Canova and placed in the middle of the courtyard of the Palazzo di Brera in Milan). His opinions were very similar to those expressed by Lessing several decades earlier. According to Lessing (*Laokoon*, 1766) the representation of the human figure must exclude not only realistic imitations — that is, portraits proper — and caricaturelike deformations but also expressions of agitation and passion, which contort the face and put the body in unnatural postures, depriving both of the beautiful contours that circumscribe them when they are in repose. To Lessing the task of art is to suggest and not to describe; therefore the artist ought always to portray the least fitful and most abiding emotions. Neoclassic portraiture, intended as it was for eternity, was almost by definition cold and funereal, and it is no wonder that its truest manifestations are sculptures made for cemeteries and commemorative statues, such as the powerful bust of Immanuel Kant in clothing of antiquity by Emanuel Bardou (1744–1818), formerly in Berlin.

In passing from neoclassic portraiture to romantic portraiture, one finds that the general intention appears to have been very much the same, though the manner is different. Eugène Delacroix (q.v.), who argued with equal force against both the realist Courbet (q.v.) and the neoclassicists, wrote in his *Journal* (trans. W. Pach, New York, 1937, p. 234): "Cold exactitude is not art; ingenious artifice, when it *pleases* or when it *expresses*, is art itself . . . . It would be interesting to write a treatise on all the falsehoods which can add up to the truth." Baudelaire (1955) must have had Delacroix no less in mind than Rembrandt and Reynolds, among others, when he wrote: "The . . . method, which is the special province of the colourists, is to transform the portrait into a picture — a poem with all its accessories, a poem full of space and reveries . . . . Here the imagination has a greater part to play, and yet, just as it often happens that fiction is truer than history, so it can happen that a model is more clearly realized by the abundant and flowing brush of the colourist than by the draughtsman's pencil."

Apart from the contrast between drawing and color — evidenced, for example, by the extremely sharp, clear drawings of Ingres (IV, PL. 280) and the rapid sketches of Delacroix, especially for his self-portraits — and that between the statuesque isolation of the figures and their cosmic exaltation obtained by steeping them in darkness and infinity, the traits shared by neoclassic and romantic portraiture are more numerous than the differences between them. The models were intellectualistically idealized in keeping with that scheme of "melancholy" of which the 16th-century artists had been so fond; they were set in an unnatural relationship to the surrounding space and were always given a fourth dimension that was more dramatic than poetic. In his *Journaux Intimes* Baudelaire says that the ideal portrait contains something of passion and sadness, of spiritual longing and frustrated ambitions, a suggestion of restless, unapplied energy, and sometimes the hint of hard vindictiveness and perhaps even something of mystery and misfortune. The two masterpieces of the opposed tendencies, David's *Death of Marat* (IV, PL. 123) and Delacroix's *Self-portrait* of 1838 (IV, PL. 146), are equally artificial, allusive, and bent on the pursuit of the sublime. When idealization prevails, however, portraiture tends to fade. In fact, despite the theoretical importance of the neoclassic and romantic schools, their production of portraits, though highly effective, was small and intermittent.

The naturalistic tendency was far more widespread. Starting with Goya (q.v.) it constitutes the kind of 19th-century portraiture that is regarded as the most modern and significant. At the time there was a lively new interest in scientific treatises on physiognomy, such as Johann Kasper Lavater's *Physiognomische Fragmente zur Beförderung der Menschenkenntnis und Menschenliebe* (1775–78), of which in France alone there were no less than 15 reprints in complete or abridged editions between 1781 and 1845. Balzac himself resorted to these treatises for descriptions of his characters in *La Comédie Humaine*. And the artists extended such studies to pathological cases and the insane, of which they gave an increasingly moving and dramatic interpretation. Such examinations of insanity started with Hogarth and Goya and reached down to Géricault (q.v.), who was the first to make a systematic study of it. Physiognomical portraiture covers a broad range of works, from the merciless descriptive harshness — especially in portraits of living models — of Goya (who, moreover, could take advantage of the earlier tradition of Velázquez) to the harsh political caricatures of Daumier (q.v.; PL. 228; IV, PL. 120) and down to realism (q.v.). (D. Durbé has suggested that this medicopsychiatric knowledge formed a link between these artists whose styles were so different, a view that is not only acceptable but should be explored further.)

Ingres (q.v.), the greatest portraitist of the century, may be placed within this movement, though his very greatness sets him apart. By his own admission, Ingres pursued neither neoclassic nor romantic idealizations (PL. 227; VIII, PL. 73). It was his view that the masterpieces of antiquity were based on models such as one would find in contemporaneous Paris, that the secret of beauty was found through the real, and that art was never so perfect as when it resembled nature so closely that it could be taken for nature itself. He was the last artist of world portraiture to combine faithful reproduction and intuition in his works. Ingres used suggestive effects in his settings, though (as Baudelaire noted) he disdained extravagant dramatic elements such as stormy skies, airy backgrounds, poetical "furniture," and languid postures. In posing his models Ingres proved himself an incomparable stage director and "layout man." In the 1846 *Salon* Baudelaire (1955) wrote that historical — that is, naturalistic — portraiture consisted in rendering "the contours and the modelling of the models faithfully, severely, and minutely; this does not, however, exclude idealization, which, for enlightened naturalists, will consist in choosing the sitter's most characteristic pose — the attitude which best expresses his habits of mind. Further, one must know how to give a reasonable exaggeration to each important detail — to lay stress on everything which is naturally salient, marked and essential, and to disregard (or merge with the whole) everything which is insignificant or which is the effect of some accidental blemish."

At the opposite pole, that of realism, which was soon to be linked with photography (Delacroix had used photographs), there was bourgeois-realistic portraiture, which, despite the opposition of certain factions and styles, extended from Edouard Manet and Gustave Courbet (qq.v.) to late impressionism. This trend was characterized at first by a favorably disposed generalized study of the model, together with an increasingly marked tendency to include it on an equal footing with other elements in the urban or landscape setting. This spontaneous manner of representation, which must be distinguished from the selectiveness of a highly instinctive taste, marked the end of that heroic conception of individuality which Ingres had preserved, although he had adapted it to the middle classes. In fact, while the geometric and overstudied face of his *Mademoiselle Rivière* (VIII, PL. 73) stands out with commanding plasticity against the far more stylized and highly detailed background, the clothing in Courbet's *Young Ladies by the Seine* (IV, PL. 35) or the sheets of the bed on which lies Manet's psychologically inert *Olympia* (IX, PL. 278) are pictorially almost more effective than the faces. With few exceptions, the faces of Courbet are markedly impersonal, like masks from a genre scene. It is no mere accident that the best 19th-century portraits are those which show the greatest concern with physiognomy, either through the influence of such masters as Goya and Rembrandt or as a consequence of the artist's contact with the above-mentioned scientific learning. This is seen in many works of Manet, such as *Lola of Valencia* (V, PL. 120), *The Fifer* (Louvre), *Luncheon in the Studio* (Munich, Neue Staatsgalerie), and *Bar at the Folies Bergères* (London, Courtauld Inst. Galls.; sketch of same subject, IX, PL. 279). It is also the case with Monet (X, PLS. 126, 127), Degas (IV, PL. 134; VII, PL. 420), and Renoir (PL. 228; VII, PL. 432; XII, PLS. 113, 114, 117), despite their variations in expression and the fact that their attention to their models was motivated mainly by personal sentimental attitudes.

Almost as a reaction against photography, to which it had come too close, the portraiture of the late 19th century gradually accentuated the tendency to evoke rather than show. Thus investigation turned once again from physical to moral character as in the early romantic period. But the insight of the romantic portraits (with the exception of Géricault's) had been limited to the theme of the visionary artist, the poet, and the hero, whereas now no human category was to be excluded, as Courbet had advocated earlier (e.g., *The Artist's Studio*; IV, PL. 38). The world of Toulouse-Lautrec (q.v.), even more than that of Daumier, was crowded with people of all social classes (IV, PL. 281; VII, PL. 435; XIV, PLS. 119–121, 123, 124). Deformations, whether graphic, plastic, or chromatic, became instrumental to the rendering of expression. In sculpture, especially in the works of Auguste Rodin (q.v.; XII, PL. 185), this effect was obtained by leaving the works unfinished — that is, by singling out only some of the peculiarities of the model's face or attitude.

Sometimes, as in the works of Vincent van Gogh (q.v.) and even more emphatically in those of James Ensor (q.v.), the model was transformed into a deformed and allusive mask. Painters as well as sculptors no longer reproduced the image of the model but instead freely exposed the effects of their reactions to it. Van Gogh's approach was explained by him in a letter (No. 520) to his brother Theo : " . . . instead of trying to render exactly what I have before my eyes, I use color more arbitrarily to express myself forcefully . . . . I should like to make the portrait of an artist friend of mine who dreams lofty dreams and works as the nightingale sings because such is his nature. This man will have to be blond. I should like to put into the canvas my admiration and love for him. So, to start with, I shall paint him just as he is, as faithfully as I can. But the picture will not be finished yet. To finish it I shall now become an arbitrary colorist. I shall exaggerate the blond of his hair and reach tones of orange, chrome yellow, and pale lemon. Behind his head, instead of the usual shabby apartment wall, I shall paint infinite space, a simple background of the richest, deepest blue that I can mix, and, through this simple combination the blond head glowing on the rich blue background will achieve a mysterious effect like a star in the deep blue sky." These words, as well as the masterpieces Van Gogh painted from 1888 onward, serve to confirm that his works were formulated on a plane of abstraction rather than idealization. Though the human image was by this time no more than a pretext, it continued to function as a highly intense stimulus; its character determined not only the setting but also the style of the picture to a degree unparalleled even by the most emphatic baroque art.

The next step, taken by the Fauves (q.v.) and the German expressionists (see EXPRESSIONISM), brought further psychological intensity to the pictorial image, thus leading to the disintegration of the portrait as a motif in its own right. (At the same time there began to appear the conventional oleographic society portrait.) In Matisse's "Notes d'un peintre," published in December, 1908, in *La Grande Revue*, he wrote: "Expression to my way of thinking does not consist in the passion mirrored upon a human face or betrayed by a violent gesture. The whole arrangement of my picture is expressive. The space occupied by figures or objects, the empty spaces around them, the proportions, everything plays a part." (Cf. V, PLS. 123, 202, 258; IX, PLS. 387, 388.) Baudelaire is said to have spoken of "divinatory" portraits; and one may legitimately wonder whether this breaking up into fragments of the personality while it was under investigation was connected with psychoanalysis and mysticizing irrationalism. However, down to cubism, which fragmented the image but did not completely destroy it, portraiture continued to offer an impressive display of human personalities. After that the genre was all but abandoned: the individual's spiritual world, that which is distinct from and independent of his physical aspect, came to matter most to the painter. In his classes, Paul Klee (q.v.) no longer taught how to reproduce the human face; he declared that a completely faithful self-portrait would be but a "shell," inside of which there would be himself, like a kernel in a husk. On the other hand, for over a century it has been the function of photography (q.v.) to document the human image; and photography has transformed a royal privilege into a common and daily gesture of affection among people. In turn, caricature has inherited the place once held by romantic and expressive portraiture. Art historians who organize exhibitions of portraiture or attempt to compile catalogues of portraiture on a national or international basis find themselves at grips with one of the most varied and elusive fields of artistic representation — one in which it is all but impossible to distinguish between what the artist represents passively and what the model, through pose and projection of the image, creates inventively.

Eugenio BATTISTI

THE ORIENT. *Islam.* Literary sources mention the existence in Sassanian Iran (see SASSANIAN ART) of albums of royal portraits that survived the Islamic conquest, but the fashion for portraits spread only in the Timurid period (see TIMURID ART), and there are no instances of this genre from either Iran or the rest of the Islamic world that can be dated before the end of the 15th century — about the time when Iranian artists might have felt the influence of the Western artists who lived at the court of Meḥmet II Fatih (Mehmet the Conqueror; see OTTOMAN SCHOOLS).

The early Islamic Persian portraits are nothing but the calligraphic reproduction of some characteristic traits of the persons portrayed. In fact, the faces are always lifeless, the features have hardly any relief, and the absence of shadows makes them look like masks rather than human figures. This is the case, for instance, with the portrait of a Timurid prince (Teheran, Gulistan Lib.) belonging to the school of Herat of the early 16th century.

During the Safavid period (see SAFAVID ART) portraits became far more popular and the literature of the time records the names of several excellent painters of the genre. In the best exemplars of the time there is a decided improvement in the pursuit of likeness and of facial expression, as, for instance, in the portraits attributed to Bihzād (q.v.) and to Sulṭān Muḥammad — for example, the portrait of Sām Mīrzā, the younger son of Shah Ismā'īl in a miniature of the *Dīvān* by Ḥāfiẓ (Paris, Coll. Louis Cartier). The portraits by Qāsim 'Alī, a student of Bihzād, and by Shāh Muḥammad, a student of Sulṭān Muḥammad, are colder and more impersonal; these artists, however, left some graceful images of princes and of girls. The series of young men's portraits by Ustād Muḥammadī, who was active during the second half of the reign of Shah Ṭahmāsp I, is extremely interesting. This artist seems to have favored natural expressions, and traits, as is also apparent from a self-portrait (Boston, Mus. of Fine Arts) executed when he was about fifty years old.

It was at this time that the Safavid traditions in portraiture spread to the Moghul court of India, when Mīr Sayyid 'Alī and 'Abdu 's-Ṣamad (qq.v.) followed Humāyūn, the second Moghul emperor, there upon his return from exile at Tabriz in 1555. The earliest Moghul portrait is the large canvas representing the Timurid ancestors of the imperial family (X, PLS. 95, 96), usually attributed to Mīr Sayyid 'Alī. The influence of the Persian school is clearly seen in 'Abdu 's-Ṣamad's portrait of the young Akbar giving his father a miniature (I, PL. 16).

Among the ablest painters of the school that Shah 'Abbās I founded in Isfahan was Āqā Riẓā, who started a style that was soon to degenerate into a tedious obsession: this consisted in drawing the figure with a slight forward inclination and slightly bent knees. He gave great attention to elegance of line and to detail. During the first half of the 17th century, the most interesting personality of the Safavid school was Riẓā-i-'Abbāsī (q.v.), who was active for about thirty years and had so many imitators that it is sometimes difficult to identify his works. Shah 'Abbās held European painters in great esteem (some of his palaces were decorated by Dutch artists) and sent some of his court painters to Italy. Among them was Muḥammad Zamān, who according to some accounts was converted to Catholicism and took the name of Paul (Paolo Zamān). His sketch of Ḥusayn (1694–1722), the last Safavid shah, which he drew on a lacquer inkpot (Cairo, Haloun Coll.), is in a hybrid and decadent style, a mixture of European characterization and Persian traditions.

The Western influence was more successful during the Qajar period (see QAJAR SCHOOL) from the end of the 18th century through the 19th century. It helped to make portraiture more realistic and finally led to a complete Westernization of the Islamic painting of Persia (see ORIENTAL MODERN MOVEMENTS). Among the best-known painters of the Qajar court were Muḥammad Ḥasan Ghaffārī, a favorite of Fatḥ 'Alī Shāh, and his son Abū'l-Ḥasan, who left many portraits of richly clothed courtiers with strongly characterized faces. Abū'l-Ḥasan also worked in Italy, where he learned the technique of oil painting. Upon his return home he started, among the family of the shah and the courtiers, a fashion for compositions consisting of several panels. Among the foreign dignitaries represented is Comte Joseph Arthur de Gobineau, who was chargé d'affaires for his country.

In Turkey, portraiture spread in the second half of the 15th century from Persian and European sources. Among the Persian painters who worked at the court of Süleyman the Magnificent the most famous were Shāh Qulī and Wālī Jān from Tabriz. The former was placed at the head of a group of Turkish painters who were working at the Topkapi Saray. Of the production of Wālī Jān there are extant a few portraits of dervishes and of youths and girls in Turkish costumes, all in stiff, unnatural poses. The presence at the court of Istanbul of the Italian painters Costanzo da Ferrara and Gentile Bellini, however, served to give new life to the local art by the introduction of Western realistic elements. Their influence also helped to bring out the realistic spirit of the Turks in contrast to the more imaginative spirit of the Persians. The wonderful portraits by Gentile Bellini (e.g., *Portrait of Mohammed II* and *Portrait of a Turkish Artist*; see BELLINI, JACOPO, GENTILE, GIOVANNI) were unsurpassed models and a source of inspiration for local artists, who sometimes achieved remarkable results. Unfortunately, few works have survived because of the frequent fires that devastated Istanbul and because of the iconoclasm of Bāyazīd. Upon his election, Bāyazīd ordered all the pictures and objects of art that his father Meḥmet II had collected to be sold in the bazaar of the capital. This explains how Bellini's *Portrait of Mohammed II* came to be in Venice, where Sir Austen Henry Layard bought it.

Many portraits, however, are still preserved in the Topkapi Saray Museum. Among them is a miniature of Meḥmet II, which has been variously attributed but was certainly executed prior to the arrival of the Italian artists in Turkey. It portrays Meḥmet II seated, with a rose in one hand and a handkerchief in the other. Some experts believe this old portrait to be the work of Sinān Bey, the first known Turkish painter, who supposedly studied with Master Paoli of Venice. From the hand of Haydar, called Nigāri, the favorite painter of Selīm II, two portraits are known: one representing the sultan in the act of shooting an arrow (Istanbul, Topkapi Saray Mus.), the other representing the famous corsair Barbarossa II, better known as Khair ed-Dīn. An interesting portrait, though a difficult one to attribute to a particular artist, is that of the professional storyteller La'līn Kabā (Boston, Mus. of Fine Arts), who enjoyed a great reputation under the reign of Murād III, at the beginning of the 17th century. The old man is shown with a fly-whisk on his shoulder, like professional storytellers in Turkey to this day. This realistic detail suggests that the picture is by a Turkish rather than an Iranian artist.

Portraits of women were more in keeping with the esthetic laws of the contemporary Persian Qajar culture, as may be seen in a portrait of the daughter of Shah Rukh by an anonymous artist of the 18th century and in that of a Turkish beauty by 'Abdullāh Bukhārī, who was active in Istanbul about the early 1760s. The latter, together with the Armenian Raphaël, was among the important artists of the century, and the many surviving pictures by him are valuable documents of the costume of the time. Besides portraits of men and women in Oriental clothing, 'Abdullāh Bukhārī also left those of a barber, of some *chibūq* smokers, and of a Christian woman, who was perhaps a Greek from the islands. In a signed work dated 1743 he represented a bathing girl nude except for a towel resting on her knees. The figure is rather awkward, but the realistic details show that it was taken from life. As early as the end of the 17th century Mu'īn Muṣawvir, who declared himself a disciple of Riẓā-i-'Abbāsī though he had never seen that master, had represented licentious subjects such as the intimacies of lovers; 'Abdullāh Bukhārī most probably derived his inspiration from Mu'īn, though his palette is more lively and original. Three Armenian miniaturists belonging to the Manasse family became famous in the 19th century. A delicate miniature on ivory representing Fāṭima Sulṭāne, the daughter of 'Abd-al-Majīd, is signed in French, Rubens Menassé, and is dated 1850. There is nothing Oriental about this portrait, for both the technique and the costume are completely European in taste, and the same is true of the anonymous portrait of a lady of the Persian *bourgeoisie* dated 1843. As in Persia, the imitation of Western art in Turkey had by then degenerated into a kind of mannerism that, though not altogether lacking charm, is somewhat heterogeneous and academic.

Bianca Maria ALFIERI

India. Portraiture in India has a long and interesting tradition, comprising funerary and donor statues, historical and votive reliefs, funeral steles and masks, effigies on coins and seals, murals, and miniatures. The subjects of many indubitable portraits, however, cannot be identified for lack of epigraphic evidence or because the likeness has been idealized or conventionalized beyond recognition.

The images of the great religious and political leaders, or chakravartins, were required to conform to certain concepts of perfection. Representations of the Buddha (see BUDDHISM) or of Mahāvīra (see JAINISM), of their disciples, and even of the kings who were their contemporaries have, therefore, no more historical authenticity than medieval European representations of Christ, the apostles, or Charlemagne. Many figures of kings in legendary scenes or of *dvārapālas* (shrine guardians), river goddesses, and the like that are found at the entrances of temples are actually portraits of princely donors. This practice was merely a variation of the practice of representing deceased princes and princesses as the gods Śiva, Viṣnu, or Brahmā, for example, and the Buddha and Dhyani Buddhas as Umā Durgā, Lakshmī, or Prajñāpāramitā — a practice based on the belief that the deceased had become identified with some aspect of the primordial deity. Though more common in Indonesia and the Khmer and Cham kingdoms, where such funeral images can always be recognized by certain deviations from the accepted iconography, the custom was known in India also (see ESCHATOLOGY). But innumerable representations of praying donors on the socles of idols and votive reliefs are so conventionalized that they have no portrait value, and the same is true of representations on funeral steles.

Possibly the earliest portrait figures may be the yaksha statues found in the neighborhood of Patna, which the historian K. P. Jayaswal (1919) has tentatively identified as several of the Śaiśunāga emperors of Magadha (5th–4th cent. B.C.). From at least the 1st century of our era onward there are indisputable donor statues and relief figures. The most interesting set would have been the inscribed but almost wholly destroyed figures of the Śātavāhana rulers of the Deccan (1st cent. B.C.-early 3d cent.; see ANDHRA; DECCAN ART; DRAVIDIAN ART) in the Nana Ghat Pass east of Bombay. There are life-size donor reliefs (without inscriptions) at the entrance of the Buddhist cave-temples of Karli and Kanheri, possibly representing King Gautamīputra Śātakarni, Pulumāyi Vasishthīputra, and Queen Bālaśrī Gautamī. Of the same type are the figures of the satrap Bhūmaka (ca. A.D. 100) and his wife in Nagpur (Central Mus.) — he a clumsy Central Asian barbarian, she an elegant Indian lady.

Of about the same time are the headless statues from the *devakūla* at Mat, near Mathura (q.v.), representing in the Parthian manner (see PARTHIAN ART) the Kushan emperors Wima Kadphises, Kaniṣka, and Huviṣka (see KUSHAN ART) and the satrap Chashtana. A number of portrait heads, either in high Scythian bonnets or with round caps, sometimes decorated with rams' horns (as on some Sassanian silver dishes; see SASSANIAN ART) have been found at Mathura and Bhilsa (Malwa); similar statues in the Indo-Greek style of Gandhara (q.v.) are either disguised as bodhisattvas or lokapalas or are frankly donors, like the fine princess from Sahri Bahlol. A Gandahran statue of the yakshi Hāritī found at Mathura has been tentatively identified as Kambojā, wife of Chashtana. A Jain donor relief (*āyāgapaṭa*) from Mathura shows the rich courtesan Āmohinī parading under an umbrella.

Both the Kushan emperors (1st–3d cent.) and the Indo-Greek kings before them (2d–1st cent. B.C.) left behind portrait coins, probably all engraved by Hellenistic artists, in the best Greek tradition (heads of both male and female rulers). The earliest, modeled on imperial Roman coins, show us the heavily built Central Asian nomad chieftains in their barbarian riding dress and high caps, either standing or sitting cross-legged on the floor or on a bed (VIII, PL. 409). The silver coins of the satraps of western India show profile heads modeled after those on Parthian coins.

The Gupta emperors (4th–early 6th cent.; see GUPTA, SCHOOL OF) originally drew the inspiration for their coins from Roman prototypes but soon replaced the designs with a purely Indian imagery. With but one or two exceptions they show the full-length figure of the emperor, sitting or standing, very rarely also with the empress. The likenesses are indifferent, however, and after about 470 the coins degenerated rapidly in quality. There has also survived a portrait of the emperor Kumāragupta I (ca. 415–55) on a carnelian seal (Br. Mus.). The Ephthalite (White Hun) invaders simply copied the Sassanian coinage, and on the medieval Hindu coins this conventional royal bust degenerated into a meaningless set of dots.

That portrait painting was common in the Gupta period is known from literary sources — in fact, every member of upper-class society, including courtesans, was expected to be able to draw a recognizable likeness; but nothing has survived. Whether any of the figures of kings and queens in the caves at Ajanta (q.v.) represent members of the Vākāṭaka or Kalacūrī dynasties (contemporaries of the Guptas) is unknown. A royal audience scene in Vihara II has been variously explained as either the Chalukya emperor Pulakeśin II (609/10–42) receiving an embassy from the Sassanian Khusrau II of Persia (more probably, from some Central Asian people) or as Kubera, the divine guardian of the north; but these interpretations need not be mutually exclusive. In Cave XXVI, one of the group of mourners at the feet of the dying Buddha may represent Nannarājā Rashtrakuta (ca. 600), mentioned in an inscription there. In the Pallava temples of Mamallapuram and Sittanavasal there are documented portraits of Siṃhaviṣṇu (ca. 574–600) and Mahendravarman I (ca. 600–30) with their queens; in the Shore Temple at Mamallapuram and the Vaikuṇṭha Perumāl at Kanchīpuram there are inscribed very conventional likenesses of most of the Pallava rulers; and there are donor groups in the Chalukyan temples of Pattadakal, probably representing Vikramāditya II (ca. 743–47) with his queens, Lokamahādevī and

Trailokyamahādevī. In northern India the evidence is less clear, but there are strong reasons to surmise that the royal couples at the entrance of the Deogarh temple represent some late Gupta emperors, that the river goddesses on the Telī-kā Mandir at Gwalior represent the queens of Yaśovarman of Kanauj (reigned until 753?), and that the couples on the Osia temples are portraits of Abhira princes.

From medieval times in northern India and the Deccan there are donor statues and reliefs in many temples: at Martand, that of Lalitāditya (ca. 725–56); at Avantipur (mod. Vantipur), the figure of Avantivarman (856–83); on a Buddha frame from Divsar in Srinagar (Sri Pratas Siṅgh Mus.), that of Śaṅkaravarman (883–902/5) of Kashmir; at Dilwara and Kumbharia, representations of the Gujarat ministers Vimala Shā and Vastupāla (11th and 13th cent.); at Ellora (Tīn Thal) and Jalor, reliefs of Paramāra rulers; at Sandur, of Indra III Rashtrakuta (915–27); at Konarak, representations of Narasiṃha II Ganga (1278–1305) and his court; at Belur, the figure of the Hoysala Viṣṇuvardhana (1100–52). Many figures of heavenly nymphs, too, seem to be portraits of queens, court ladies, and courtesans; Viṣṇuvardhana's queen Śāntale is known from a portrait figure as well as in such a heavenly disguise. From the time of the Chola dynasty (10th–13th cent.), votive statues of kings, queens, and temple dancers, executed in stone as well as bronze, have been common in southern India; the most famous is the fine group of Kṛṣṇa Deva Rāya of Vijayanagar (1509–30) with two queens, in the Tirupati temple. There are also many statues of famous Saiva (Nāyanār) and Vaishnava (Ālvār) saints, but their portrait value is doubtful. After the 7th century but more commonly after the 11th, funeral steles for chieftains and warriors killed in battle or duels, as well as for their wives who had become *satī*, became customary; but these were generally purely conventional (the *pāliyā* in northern India, the *vīrakkal* or *satīkkal* in the Deccan). The earliest example is a stele at Sangsi (Kolhapur) depicting the burning of the Kadamba queen Halīdevī (early 6th cent.). The only known funerary statue is that of Bappā Rāval, the founder of the Guhilot dynasty of Mewar, near Eklinji; but this may be a 15th-century work.

Wherever Saivism or Mahayana Buddhism were dominant there are funerary statues. In ancient Java and Cambodia practically all temples were originally funerary shrines, the ashes of the deceased person being deposited beneath the cult images; hence these images were commonly idealized portraits of the dead [Prajñāpāramitā and Durgā statues at Leiden, several images in Berlin, the Belahan group of king Erlaṅga (1037–42) as Viṣṇu on Garuḍa, many East Javanese statues]. In Cambodia the daily life, official duties, and military achievements of the kings also were depicted on the reliefs of the royal palace (the "Leper King," etc.) and of the funeral temples (e.g., Angkor Wat, the Bayon), often with a decided measure of individual characterization. In Champa (Annam) the Śiva linga on the yoni was often replaced by a bust (*kut*; influence of the Chinese ancestor tablets) of the deceased king, sometimes depicted in the arms of Śiva; full-length funeral statues are also known. Where Hinayana Buddhism prevailed, portrait statues are rare and controversial, such as that of an unidentified royal figure found near the Ruanweli Dagoba at Anuradhapura and the so-called "Parākramabāhu" at Polonnaruva (a rishi), in Ceylon; another example is a 15th-century statue of a princess, from Ayuthia, Siam.

Under Moslem rule the Hindu princes and nobles continued to erect funeral steles (*pāliyās*), but from the 17th century on, these were decorated with relief figures in the style of Rajput paintings (see RAJPUT SCHOOL). They are found by the thousands all over Rajasthan, are less numerous in central India, and are rare in the Himalaya. Ajīt Siṅgh of Jodhpur (1686–1724) built a rather conventional "hero gallery" at Mandor; and the Moghul emperor Akbar erected statues, in a similar conventional Rajput style, in memory of Paṭṭa and Jaimal, the Rajput defenders of Chitorgarh in 1568 (now at the Red Fort in Delhi). In the Marāṭhā states (17th–19th cent.) it was the custom to set up portrait "dolls" of the deceased princes behind the linga of the memorial temple (Gwalior, Indore, Baroda, Satara, Kolhapur).

Moslem portraiture flourished even earlier in the Yamīnid palaces of Ghazni (see GHAZNEVID ART) and Lashkari Bazaar, the ruins of which have yielded reliefs and murals with strongly individualized figures. Fīrūz Shah III Tughlaq (1351–88), in a fit of bigotry, destroyed the many royal portraits in the Delhi palaces. The representations of the Śaka ruler in the Jain *Kālakācāryakathā* manuscripts were probably inspired by similar portraits of the sultans of Gujarat (see JAINISM). The earliest extant portraits of Indian sultans are those of Ghiyās-ud-dīn of Malwa (1469–1500; *Ni'mat-nāma*, London, India Office Lib.) and of Fīrūz Shah of Bengal (1533; Br. Mus.); there are later copies of the portraits of some of their contemporaries, such as the first sultans of Bijapur and Golconda.

Portraiture reached a high standard in the late 16th century and especially in the 17th at the courts of the Moghul emperors Akbar, Jahāngīr, and Shah Jahān, Ibrāhīm II and Muḥammad 'Ādilshāh, and 'Abdullah and Abūl-Hasan Quṭbshāh. The earlier Indo-Moslem portraits are usually full-face or almost so, but from the 17th century on, the profile portrait became the rule. The shah or sultan was always depicted with a halo round his head. Portrait heads are rare, but there are some exquisite sketches for this form in the British Museum. Portrait busts were generally framed by a *jharokhā* window. Full-length portraits were shown either against a green or chocolate-brown background or in a garden setting. Portraits on horseback were probably inspired by European paintings. Group portraits were in the form of either an imperial audience or a procession, with the name written alongside each figure.

Female portraits are not rare, but the individuality of the subject was subordinated to the ideal of beauty then current; normally the subjects are anonymous, the names inscribed (often those of famous heroines of romance) being later additions intended to divert attention from the real identity of the lady hidden in the harem. Portraits of Moslem and Hindu saints came into fashion in the late 17th century. There are a few self-portraits of painters of the late 16th and early 17th centuries. In the late 18th century the Moghul portrait type was replaced by European oil paintings; the first foreign portraitist to achieve success was Zoffany, who worked for the nawabs Shujā'-ud-Daula and Āṣaf-ud-Daula of Oudh. An intaglio showing the emperor Jahāngīr killing a lion is in Paris (Cabinet des Médailles), and a small marble relief of Shah Jahān is in Amsterdam (Rijksmus.). Jahāngīr also had his portrait (head, or full figure with wine cup) and that of Akbar placed on his gold coins (intended mainly for honorific presentation).

From the Moghuls and the sultans of the Deccan the Hindu Rajputs (see RAJPUT SCHOOL) and Marāṭhās took over the portrait miniature, but their portraits were much cruder and tended to become wholly stylized. During the 18th century, portraits of the nobles became common, and in the last half of that century the old custom of representing a ruler and his wives or concubines in a religious guise (especially as Kṛṣṇa, Rādhā, and the *gopīs*) was revived. By the middle of the 19th century, growing European influence had brought Rajput and Marāṭhā portraiture to an end.

Hermann GOETZ

China. Figure motifs surviving on pictorial tomb slabs and on tomb furniture make it possible to draw conclusions as to the quality of Chinese portraiture during the first three centuries of our era. They disclose a sense of characterization based on life and not on legend. The Hui-hsien type of terra cottas, if they are genuine, would place the beginning of free movement in the 5th century B.C.; but it was only during the Han dynasty (206 B.C.–A.D. 220), and probably in its later part (after A.D. 25), that Chinese figure design attained full freedom as well as realistic truth. Psychology could then be said to have replaced animism, and the independence of a liberated eye to have broken the spell of mythic bondage. On slabs of Chu Wei's tomb (middle on the 1st century of our era) there are engraved figures that, despite later recarving, still attest to the designer's skill in character portrayal spiced with mockery. Groups on the Lo-lang (Korea) "Painted Basket" Tomb (VIII, PL. 413), of the 2d or 3d century, recede diagonally into space, turning and gesticulating with histrionic ardor.

The outstanding example, however, is found in some of the figures on the Boston Museum of Fine Arts tiles, probably to be dated about the 3d century. The group of five men in conversation (III, PL. 234) rivals in its casual realism Degas's painting of the Lepic family in the Place de la Concorde (formerly Berlin, Coll. O. Gerstenberg). Perhaps only a craftsman's whim, the Boston group is nevertheless a master drawing of the first order. The heads are unique in the history of Chinese painting: turned, inclined, full of attention and surprise, they have been jotted down with the controlled and meaningful accents of the Chinese brush, never again to be used in figure design with such bite. Isolated, minute, and graffiti-like as these brush sketches are, they remain to give at least an idea of the nervous elegance alive among the Chinese intelligentsia at the end of the Han dynasty and during the later part of the 3d century. With these decades came both the conclusions and the aftermath of a period of groping liberalism and spirited decadence, just before the apocalyptic invasions from the Steppes and the advent of an austere new transcendentalism. It was a Chinese *fin-de-siècle* that may well have had, in the originator of the Boston group, its own Toulouse-Lautrec.

Traits of Han realism and formal elements of Han figure design lingered on into the following centuries and into the T'ang dynasty (618–906). The Han motif of the human form sitting low, legs pulled under, serves as a prototype for the early monastic image, where it is isolated and shown in three-quarter view. Modified to the cross-legged yogi position, it was used for depiction of monks and other Buddhist-inspired images down into the Ming (1368–1644) and Ch'ing (1644–1912) dynasties. In secular art, continuing the Han tradition, portraiture, not landscape painting, still occupied first place. In the period of the Six Dynasties (220–589), the portraits of such masters as Ku K'ai-chih and Lu T'an-wei were considered their highest achievements. Yet no portraits of these centuries survive, and the figures in the "Admonitions" scroll (III, PL. 235; VIII, PLS. 406, 407) attributed to Ku K'ai-chih (344–406) and in the still-archaic Scroll of the Emperors (VII, PLS. 256, 394; X, PL. 492; XIV, PL. 451) of Yen Li-pên (d. 673) show only faint traces of true portraiture. Their realism is stereotyped rather than endowed with life. S. Elisseev (1932) has discussed this and related questions from the historian's point of view.

To do justice to what is art, and not only representation, in Chinese portraiture one must go far beyond the scope of the ceremonial image. One must include fictional as well as historical figures, and even genre ones, provided they transcend barren realism and possess that quality of *ch'uan-shên* ("transmitting the soul") demanded by Ku K'ai-chih (Waley, 1933, p. 159) — but of course, this is the *raison d'être* of a good portrait anywhere. In this connection the bas-reliefs representing *śrāvaka* carved in the style of the Wei dynasty (386–557) from brush-drawn originals deserve mention (Ecke, 1957, pl. 1). The pronounced verisimilitude of their features is of a glowing unworldly kind and yet of intense, lifelike actuality, like that of the Elders of Revelation on the portal of the church at Moissac. It is a realism of the soul, revealing the labors of samadhi, those endless years of concentration that change a human face into a scarred, incorporeal visage.

From the T'ang period two portraits survive. One is Li Chên's likeness of Amoghavajra, painted about 800 (thirty years after the patriarch's death), which shows him seated the archaic way, on a simple platform in classic T'ang style (Sirén, 1956, vol. III, pl. 113). He is represented in three-quarter view, with hands folded in hieratic pose, the body cloaked and compact; the head is drawn in fine firm lines that capture the austere physiognomy with penetrating characterization (Elisseev, 1932, p. 180).

The other T'ang portrait represents the Han scholar Fu Shêng (XIV, PL. 398). An imaginary creation, but intensely realistic and alive, it is attributed to Wang Wei (699–759). Although this attribution has been questioned, the 8th-century date of the painting seems certain: the layout of the picture, the even flow of the contours, the placement of the emaciated figure bending over the table, even the structural details of the table itself support a mid-T'ang-dynasty date. The figure is again seen in three-quarter view but dressed only in an archaic apron, so that the skeletonlike frame with its wasted members appears in all its nervous, torsional strain. A complex of contrasting axes, converging diagonals, and undulating curves is fenced off from the void in rhythmic silhouette and adjusted to the picture plane. To throw this configuration toward the beholder it has been placed, in accordance with the surviving Han tradition, at a 45° incline, with the table drawn in inverted perspective. As Cohn (1951, p. 51) has said: "The work can only be credited to a revolutionary master of the first rank, such as Wang Wei was known to be." Here form and subject interact to visualize a grand idea, that of the Confucian gentleman.

The change in the mode of domestic living that characterized the transition from the T'ang to the Sung dynasty (960–1279) meant for Chinese official portraiture a break with a great formal tradition. While the Yamato-e (q.v.) painters and their later followers retain the archaic mode of low sitting, to the advantage of their composition, the Chinese sitter is raised to the height of the ceremonial armchair then in fashion. The 13th-century portrait of Amoghavajra attributed to Chang Ssŭ-kung, as compared with the T'ang portrait by Li Chên, reveals this fundamental innovation. In the Sung image the thronelike chair predominates, and with its brocade cover it seems almost more interesting than the personage placed upon it. It is a civilized and beautiful portrayal, but it has little of the *ch'uan-shên* quality. The same holds true of the anonymous portrait of the priest Wu-chun dated 1238 (Sirén, 1956, pls. 209, 210). Both portraits are important historical and cultural documents, but as paintings they are merely descriptive.

While this was a decline, it was a decline with taste, and what can be assumed from surviving copies of the portraits of the Sung and Ming emperors confirms this view. However, the mortuary portraits of the Ming and Ch'ing official class cannot be even so moderately commended; Waley (1933, p. 161) finds them among "the most repulsive paintings in existence." Some of the Ming ancestral portraits, when done directly from the corpse, are at least impressive through a certain macabre quality, combining accuracy of features with a blank stare of the eyes that makes them worthy to be hung at Madame Tussaud's.

Imaginary portraiture, often not to be distinguished from figure painting, came into its own in the Sung period. Masters like Li Kung-lin (q.v., also known as Li Lung-mien) and, in the Yüan period (1280–1368), Ch'ien Hsüan attest to the fact that humanism in figure painting had not died out; their figures and images are of a high order. The so-called "Ch'an" painters (IV, PL. 282) of the Southern Sung dynasty (1127–1279) created a kind of intuitive portraiture that has remained unmatched. Representative of this type of portraiture is Liang K'ai's lost *Drunken Man* (formerly Tokyo, Coll. Kishichiro Okura; reproduced in *Sogen Meigashu*, vol. III, Tokyo, 1930). Its composition is uncommon for the 13th century, as Liang K'ai has retained the outmoded low seat to accent the blissful lethargy of his *clochard* and for compositional reasons. But the body volume, in Li Chên's *Patriarch* so powerfully compact, appear almost in a state of decomposition: the painter has succeeded in suggesting a living corpse with the help of his shorthand technique, employed to charge his vision with shocklike impact.

With the advent of the Ming period in 1368 and its new aspirations, interest in portraiture took a new turn. Chou Ch'ên, who worked in the early 16th century, excelled in figure painting (his landscapes were slightly academic). If naturalism can be defined as an interpretation of life through its gross facts, then Chou Ch'ên certainly belongs in this category — one unknown in China until then. In an album series of beggar types in the collection of the Honolulu Academy of Arts there is a monk's head (PL. 230) that is the incarnation of evil. It is an impression taken directly from life and realized with the greatest economy in firm, massive, modulated lines, not with the nervous accents employed by the master of the Boston tiles. T'ang Yin, Chou Ch'ên's follower, was reputed for his figure design, of which, however, very little has survived. His *Gentleman Hermit* in the Honolulu Academy of Arts may be a self-portrait; its inspiration is elegance, *tournure* as actual as the depravity of Chou Ch'ên's beggar monk.

There were other excellent figure painters with portrait

interests during the last hundreds years of the Ming dynasty, among them Ting Yün-p'êng and Ch'en Hung-shou. Portraiture from life, however, came to new significance with an innovator of a high order, Tseng Ch'ing (1568–1650), the author of a veritable portrait gallery of gentlemen scholars and painters. Examples of his fascinating images are perhaps hidden away in private collections, but only one, so far, has come to the knowledge of the West, that of the noble young Wang Shih-min (Omura Seigai, *Bunjin Gasen*, Tokyo, 1921, part 1, fasc. XI, pl. 8). Sirén (1938, vol. II, p. 42) quotes a Chinese critic as saying that Tseng "painted portraits which looked exactly like reflections of the models in a mirror, and grasped the spirit and emotions of the people." In the portrait of Ch'ên Chi-ju (1558–1639), a friend and *confrère* of Tung Ch'i-ch'ang, the wisdom and cultured dignity of the subject are transmitted to the beholder in the best *ch'uan-shên* manner. As with all portraits by Tseng Ch'ing, the head is seen full-face. His innovation, the plastic modeling of the face through shading, is said to derive from the lost art of the Jesuit Matteo Ricci.

Distinguished among the early Ch'ing portraitists was Yü Chih-ting (1647–1706), a *wên-jên* and a fairly good landscapist, known in the West for his portrait of a young lady with attendant, in the British Museum. His *pai miao* ("contour") manner is pure Chinese. It possesses an ethereal quality foreign to the somewhat earthy realism of Tseng Ch'ing. Among his other works is the likeness of Wang Yün-ch'i, the grandson of Wang Shih-min, one of China's greatest landscapists (*Shen Chou Ta Kuan*, Shanghai, 1912, fasc. XIV, pl. 19). Outstanding in charm and gentle derision is the portrait of the aged Sung Li-ch'ang, seen reading on a mat, with piles of books as backrest and arm rests and with two slender girls at his side who have turned away — one is seen from the back — and are gossiping while guarding the old man's fishing gear and hat (*Shen Chou Ta Kuan*, Shangai, 1912, fasc. IX, pl. 18). It is a composition of intimate Chinese flavor, a genre scene that Watteau, so influenced by *chinoiserie*, would have loved.

The renewed wave of Western influence in the 18th century, represented mainly by Father Giuseppe Castiglione, gradually led to a new approach in portraiture. Little is known about Wu Chün, to whom the portrait of Juan Yüan (1764–1849; PL. 230) is attributed. The treatment of the face, seen in half-profile, is Western; its relief is achieved through shading and tonal values, very different from Tseng Ch'ing's grave strokes and with a greater interest in subtlety of line.

Gustav ECKE

Japan. The oldest known portrait by a Japanese artist is an 8th-century one of the Chinese monk Chien Chên (Jap., Ganjin; Nara, Toshodaiji). This is a dry-lacquer statue internally supported by a simple wooden skeleton — a prototype of that series of portraits by the great Buddhist masters which was to be particularly appreciated in the Heian (794–1185) and Kamakura (1185–1333) periods. The figure, painted in pink, red, and green (the colors are still visible), is modeled in a simple but intensely expressive way. The hands and the face are completely relaxed, and the eyes are closed (not as an invitation to meditation, however, but because the monk had lost his sight on a trip to China). Another portrait of the Nara period (645–793) is a dry-lacquer statue of the monk Gyōshin (in the Hōryūji, near Nara). In these works the different natures of the two monks are clearly expressed: while the statue of Ganjin shows him as noble and kindly, a man of wisdom and knowledge, that of Gyōshin reveals a powerful and ambitious personality.

The only painted portrait of the time is of Shōtoku Taishi and his children (PL. 230), which was probably made from life and is the earliest Japanese painted portrait. It is composed according to the then-current Buddhist pattern; that is, the main figure (the prince) is in the middle, with the two minor figures (the two young attendants) on each side. The face certainly does not fully reveal the prince's character, and in this first Japanese attempt at painted portraiture the influence of Chinese precedents can be clearly seen.

During the Heian period that followed (late 8th–late 12th cent.) there developed a fashion for portraying the patriarchs of the various Buddhist sects. The most famous work of this type is that representing the Chinese monk Tz'ŭ-ên Tai-shih (Jap., Jion Taishi or Daishi; Nara, Yakushiji), though the portraits of Ryūzo and Zemmui, patriarchs of the Tendai sect, better express the spirit of the time. Meanwhile sculptured portraiture continued to develop, and the first attempts at realism appeared. Notable is the statue of the monk Rōben, the first abbot of Todaiji monastery (II, PL. 392).

In the Kamakura period (late 12th–early 14th cent.) portraiture underwent profound stylistic changes, mostly due to the marked taste for realism at the time. Portraits of laymen took the place of those of priests. A famous work is the portrait of Minamoto Yoritomo, the founder of the Kamakura shogunate (PL. 229), which is attributed to Fujiwara-no-Takanobu (1143–1205) of the Yamato-e (q.v.) school, to whom is also attributed the portrait of Taira-no-Shigemori (Kyoto, Jingoji). In the portrait of Yoritomo the artist succeeded in giving the face remarkable psychological distinction. Although the setting and dress are abstract, the pinched and delicate features reveal the painter's pursuit of realism: the closely set eyes, the big curved nose, and the long face that grows broader around the jaws are all strongly identifying physiognomical traits, although simply and flatly rendered. In spite of its markedly decorative quality, this is a realistic portrait. Since Takanobu was a contemporary of Yoritomo, it is almost certain that it is a portrait from life and not the representation of an idealized type. Fujiwara-no-Nobuzane (1176–ca. 1268), the son of Fujiwara-no-Takanobu, painted less realistic though no less artistic portraits, as can be seen from his paintings of the Thirty-six Poetic Immortals (VIII, PL. 304).

The tendency to realism made itself felt also in sculpture, although in this medium priests continued to be portrayed along with laymen. There are famous statues of Asaṅga (Jap., Muchaku) and Vasubandhu (Jap., Seshin), two great masters of Indian Buddhism, in the Kōfukuji in Nara; according to the inscription on the pedestal, they were made by Unkei (q.v.) and his son in 1208. They are idealized images, but the faces are extremely expressive, as in other works of Unkei (VIII, PLS. 291, 292). Realism, however, is more decided in the portraits of laymen, for example, the statue of Uesugi Shigefusa (VIII, PL. 294), done in a style similar to the contemporaneous Yamato-e painting but with a markedly realistic face that vividly expresses the personality of the subject. Seen from the front the figure is perfectly symmetrical, with the loose trousers counterbalanced by the solid upper part of the body and the high headgear.

In the Muromachi period (1334–1573) the doctrines of Zen Buddhism were reflected in portraiture. This was the period of the portraits of famous monks, a theme, originally Chinese, that was continued by the Japanese in an entirely new way. Among these portraits is the interesting one of Musō Kokushi, painted by Mutō Shūi in 1349 (Kyoto, Myōchiin). The style here, too, is realistic; it is a portrait from life, and the artist gave it a simple directness by eliminating all superfluous elements. It is less formal than the Yoritomo portrait discussed above, especially in such details as the clothes, which are given a simple linear treatment; and the physiognomy is represented more realistically and with greater sensitivity.

The Muromachi period practically marks the end of development in Japanese portraiture. In the following centuries there were no noteworthy advances, and with the restoration of the Meiji (1868), the Japanese traditions of portraiture were replaced by those of the West.

Antonio PRIORI

BIBLIOG. *General*: H. E. A. Furst, Portrait Painting: Its Nature and Function, London, 1927; M. J. Friedländer, Landscape, Portrait, Still-life: Their Origin and Development (trans. R. F. C. Hull), Oxford, 1949; A. Gwynne-Jones, Portrait Painters, London, 1950; J. Roubier and J. Babelon, Dauernder als Erz: Das Menschenbild auf Münzen und Medaillen von der Antike bis zur Renaissance, Vienna, Munich, 1958; E. Buschor, Das Porträt, Munich, 1960. See also bibliogs. for PAINTING; SCULPTURE.

Ancient world: a. Mesopotamia and Egypt: G. Steindorff, Der Ka und die Grabstatuen, ZäS, XLVIII, 1910, pp. 152–59; G. Maspero, Essais sur l'art égyptien, Paris, 1912; H. Schäfer, Das Bildnis im alten Ägypten, Leipzig, 1921; V. V. Pavlov, Skulpturnyi portret v drevnem Egipte (Portrait Sculpture in Ancient Egypt), Moscow, Leningrad, 1937; A. Scharff, Typus und Persönlichkeit in der ägyptischen Kunst, Arch. fur Kulturgeschichte, XXIX, 1939, pp. 1–24; J. H. C. Kern, Antike Portretkoppen, The Hague, 1947; A. Shoukry, Die Privatgrabstatue im Alten Reich, Cairo, 1951; H. Frankfort, The Art and Architecture of the Ancient Orient, Harmondsworth, 1954; J. Vandier, Manuel d'archéologie égiptienne, III: Les grandes époques, La statuaire, Paris, 1958; J. A. H. Potratz, Die menschliche Rundskulptur in der sumero-akkadischen Kunst, Istanbul, 1960; H. Zaloscer, Porträts aus dem Wüstensand: Die Mumienbildnisse aus der Oase Fayum, Vienna, 1961.

b. Iran: I. Stchoukine, La peinture indienne à l'époque des Grands Moghols, Paris, 1929; E. Babelon, Traité des monnaies grecques et romaines, III, 1: Numismatique de la Perse antique, Paris, 1933; L. Binyon, J. V. S. Wilkinson, and B. Gray, Persian Miniature Painting, London, 1933; A. Godard, Les statues parthes de Shami, Āthār-e-Īrān, II, 1937, pp. 285–305; SPA, passim, L. Binyon, Arte orientale, Rome, 1939; H. Seyrig, La grande statue parthe de Shami et la sculpture palmyrénienne, Syrian, XX, 1939, pp. 177–82; E. Diez, Iranische Kunst, Vienna, 1944; M. Rutten, Arts et styles du Moyen-Orient ancien, Paris, 1950; R. Ghirshman, L'Iran dès origines à l'Islam, Paris, 1951 (Eng. trans., Harmondsworth, 1954); H. Frankfort, The Art and Architecture of the Ancient Orient, Harmondsworth, 1954; E. Kühnel, Persische Miniaturmalerei, Berlin, 1959; I. Stchoukine, Les peintures des manuscrits safavis de 1502 à 1587, Paris, 1959; L. Vanden Berghe, Archéologie de l'Iran ancien, Leiden, 1959; R. Ghirshman, 7000 ans d'art en Iran, Paris, 1961 (cat.); B. Gray, Persian Painting, Geneva, 1961.

c. Greece: E. Q. Visconti, Iconographie grecque, 3 vols., Milan, 1824–26; D. Comparetti and G. de Petra, La villa ercolanese dei Pisoni, Turin, 1883; F. Imhoof-Blumer, Porträtköpfe auf antiken Münzen hellenischer und hellenisierter Völker, Leipzig, 1885; ArndtBr; J. J. Bernoulli, Griechische Ikonographie, 2 vols., Munich, 1901; J. Sieveking, Porträt-Darstellungen aus der griechischen Literaturgeschichte, in W. Christ, Geschichte der griechischen Literatur, 5th ed., Munich, 1905, pp. 983–96; H. Oehmann, Porträttet den grekeska Plastiken, Helsingfors, 1910; R. Delbrück, Antike Porträts, Bonn, 1912; A. Hekler, Die Bildniskunst der Griechen und Römer, Stuttgart, 1912; G. Lippold, Griechische Porträtstudien, Munich, 1912; F. Poulsen, Ikonographische Miscellen, Copenhagen, 1921; F. Poulsen, Greek and Roman Portraits in English Country Houses (trans. F. G. Richards), Oxford, 1923; E. Pfuhl, Die Anfänge griechischer Bildniskunst, Munich, 1927; E. G. Suhr, Sculptured Portraits of Greek Statesmen, Baltimore, 1931; R. Paribeni, Il ritratto nell'arte antica, Milan, 1934; R. P. Hinks, Greek and Roman Portrait Sculpture in the British Museum, London, 1935; E. T. Newell, Royal Greek Portrait Coins, New York, 1937; A. Hekler, Bildnisse berühmter Griechen, Berlin, 1940; L. Laurenzi, Ritratti greci, Florence, 1941; J. Babelon, Le portrait dans l'antiquité d'après les monnaies, Paris, 1942; K. Schefold, Die Bildnisse der antiken Dichter, Redner und Denker, Basel, 1943; E. Buschor, Bildnisstufen, Munich, 1947; Picard, III–IV; E. Buschor, Das hellenistische Bildnis, Munich, 1949; B. M. Felletti Maj, Museo Nazionale Romano: I ritratti, Rome, 1953; E. B. Harrison, Portrait Sculpture, Princeton, 1953; V. Poulsen, Les portraits grecs, Copenhagen, 1954; G. M. A. Richter, Greek Portraits, 4 vols., Brussels, 1955–62; Boston Museum of Fine Arts, Greek and Roman Portraits, Boston, 1959; G. S. Dontas, Εἰκόνες . . . (Monuments . . .), Athens, 1960; M. Bieber, The Sculpture of the Hellenistic Age, 2d ed., New York, 1961; G. A. Mansuelli, Galleria degli Uffizi, Le sculture, II: Ritratti greci ed ellenistici, Rome, 1961; G. M. A. Richter, The Portraits of the Greeks, 3 vols., London, 1965.

d. Etruria and Rome: E. Q. Visconti, Iconographie romaine, 4 vols., Paris, 1817–26; J. J. Bernoulli, Römische Ikonographie, 3 vols., Stuttgart, 1882–94; G. von Kaschnitz-Weinberg, Studien zur etruskischen und frührömischen Porträtkunst, RM, XLI, 1926, pp. 133–211; L. Curtius, Ikonographische Beiträge zum Porträt der römischen Republik und der julish-claudischen Familie, RM, XLVII, 1932, pp. 202–68; XLVIII, 1933, pp. 182–243, L, 1935, pp. 260–340, LIV, 1939, pp. 112–44; A. N. Zadoks, Ancestral Portraiture in Rome and the Art of the Last Century of the Republic, Amsterdam, 1932; C. Blümel, Römische Bildnisse, Staatliche Museen, Berlin, 1933; H. P. L'Orange, Studien zur Geschichte des spätantiken Porträts, Oslo, 1933; R. West, Römischen Porträtplastik, 2 vols., Munich, 1933–41; J. M. C. Toynbee, The Hadrianic School, Cambridge, 1934; G. Q. Giglioli, Arte etrusca, Milan, 1935; M. Wegner, Das römische Herrscherbild, 3 vols., Berlin, 1939–56; L. Goldscheider, Roman Portraits, New York, 1940; O. Vessberg, Studien zur Kunstgeschichte der römischen Republik, Lund, Leipzig, 1941; A. Boëthius, On the Ancestral Masks of the Romans, ActaA, XIII, 1942, pp. 226–34; H. P. L'Orange, Apotheosis in Ancient Portraiture, Cambridge, Mass., 1947; G. M. A. Richter, Roman Portraits in the Metropolitan Museum of Art, New York, 1948; B. Schweitzer, Die Bildniskunst der römischen Republik, Leipzig, Weimar, 1948; P. H. von Blankenhagen, Das Bild der Menschen in der römischen Kunst, Marburger Jhb. für Kw., XV, 1949–50, pp. 115–34; A. de Franciscis, Il ritratto romano a Pompei, Naples, 1951; G. M. A. Richter, Who Made the Roman Portrait Statues — Greeks or Romans?, Proc. Am. Philosophical Soc., XCV, 1951, pp. 184–208; G. M. A. Hanfmann, Observations on Roman Portraiture, Brussels, 1953; G. Hafner, Späthellenistische Bildnisplastik, Berlin, 1954; G. M. A. Hanfmann, Etruskische Plastik, Stuttgart, 1956; G. Daltrop, Die stadtrömischen männlichen Privatbildnisse trajanischer und hadrianischer Zeit, Münster, 1958; C. C. Vermeule, Hellenistic and Roman Cuirassed Statues, Berytus, XIII, 1959, pp. 1–82; V. Poulsen, Claudische Prinzen, Baden-Baden, 1960; E. Rosenbaum, A Catalogue of Cyrenaican Portrait Sculpture, London, 1960; R. Bianchi Bandinelli, Archeologia e Cultura, p. 172 ff., Milan, Naples, 1961; P. R. Franke, Römische Kaiserporträts im Münzbild, Munich, 1961; H. von Heintze, Römische Porträt-plastik aus 7. Jahrhunderten, Stuttgart, 1961; V. Poulsen, Les portraits romains, Copenhagen, 1962; A. Boncasa, Ritratti greci e romani della Sicilia, Palermo, 1964 (cat.); R. Calze, Scavi di Ostia V, I Ritratti, I, Rome, 1965; H. P. L'Orange, Art Forms and Civic Life in the Late Roman Empire, Princeton, 1965.

The West: *a. Middle Ages*: F. Burger, Geschichte des florentinischen Grabmals, Strasbourg, 1940; M. Kemmerich, Die frühmittelalterliche Porträtmalerei in Deutschland, Munich, 1907; M. Kemmerich, Frühmittelalterliche Porträtplastik in Deutschland, Leipzig, Berlin, 1909; C. Ricci, I ritratti di Bonifacio VIII, Santi ed artisti, 2d ed., Bologna, 1910, pp. 26–27; R. Delbrück, Porträts byzantinischer Kaiserinnen, RM, XXVIII, 1913, pp. 310–52; C. Sommer, Die Anklage der Idolatrie gegen Papst Bonifaz VIII. and seine Porträtstatuen, Freiburg im Breisgau, 1920; J. Banko, Ein Porträtkopf aus merovingischer Zeit, Festschrift J. von Schlosser, Vienna, Leipzig, 1927, pp. 68–69; S. Steinberg, Grundlagen und Entwicklung des Porträts im deutschen Mittelalter, Kultur und Universal-geschichte: Festschrift W. Goetz, Leipzig, 1927, pp. 21–34; L. Bréhier, L'art chrétien, 2d ed., Paris, 1928; P. E. Schramm, Die deutschen Kaiser und Könige in Bildern ihrer Zeit, 2 vols., Leipzig, Berlin, 1928; P. E. Schramm, Umstrittene Kaiserbilder, N. Arch. der Gesellschaft für ältere d. Geschichtskunde, XLVII, 1928, pp. 469–94; E. Stollreither, Bildnisse der IX.–XVIII. Jahrhunderts aus Handschriften der Bayerischen Staatsbibliothek, I, Munich, 1928; W. Hager, Die Ehrenstatuen der Päpste, Leipzig, 1929; H. K. Mann, Tombs and Portraits of the Popes of the Middle Ages, London, 1929; J. Prochno, Das Schreiber- und Dedikationsbild in der deutschen Buchmalerei, I: 800–1100, Leipzig, Berlin, 1929; S. Steinberg, Die Bildnisse geistlicher und weltlicher Fürsten und Herren, I: 950–1200, Leipzig, Berlin, 1931; H. Focillon, Les origines monumentales du portrait français, Mél. offerts à N. Iorga, Paris, 1933, pp. 259–85; L. Olschki, Il ritratto di Ottone II sul Cod. Vat. Lat. 4939, Bibliofilia, XXXVI, 1934, pp. 213–24; J. Bolten, Die imago clipeata, Paderborn, 1937; S. Steinberg, I ritratti dei re normanni di Sicilia, Bibliofilia, XXXIX, 1937, pp. 29–57; H. Borgwardt, Die Typen des mittelalterlichen Grabmals in Deutschland, Schramberg, Schwarzwald, 1939; H. Keller, Die Entstehung des Bildnisses am Ende des Hochmittelalters, Römisches Jhb. für Kg., III, 1939, pp. 227–256; J. J. Rorimer, 13th Century Statues of King Clovis and Clothar at the Cloisters, BMMA, XXXV, 1940, pp. 121-26; G. Ladner, I ritratti dei papi nell'antichità e nel Medioevo, I, Vatican City, 1941; J. Déer, Ein Doppelbildnis Karls der Grossen, Wandlunger christlicher Kunst im Mittelalter, Baden-Baden, 1953, pp. 103–56; H. Keller, RIDKg, s.v. Denkmal; J. Kollwitz, Rl. Antike und Christentum, s.v. Bild (III); T. Rensing, Der Kappenberger Barbarossa-Kopf, Westfalen, XXXII, 1954, pp. 165–83; W. Franzius, Das mittelalterliche Grabmal in Frankreich, Tübingen, 1955; J. Kollwitz, Rl. Antike und Christentum, s. v. Christusbild; S. Schrade, Zur Frühgeschichte der mittelalterlichen Monumentalplastik, Westfalen, XXXV, 1957, pp. 33–64; H. Keller, RIDKg, s.v. Effigie; N. M. Chegodaeva, The Portraits of Jan van Eyck (Rus.), Iz istorii russkogo i zapadno-evropeiskogo iskusstva: Sbornik . . . V. N. Lazareva (On the History of Russian and Western European Art: Collection . . . V. N. Lazarev), Moscow, 1960, pp. 197–231.

b. Renaissance to 20th century: A. Marquet de Vasselot, Histoire du portrait en France, Paris, 1880; J. d'Auriac and R. Pinset, Histoire du portrait en France, Paris, 1884; H. Bouchot, Les portraits au crayon du XVIe et XVIIe siècle conservés à la Bibliothèque Nationale, 1525–1646, Paris, 1884; F. Kenner, Die Porträtsammlung des Erzherzogs Ferdinand von Tirol: Die italienische Bildnisse, JhbKhSammlWien, XVII, 1896, pp. 101–274, XVIII, 1897, pp. 135–261; E. Müntz, Le Musée des portraits de Paul Jove, Mém. Acad. des Inscriptions et Belles Lettres, XXXVI, 1900, pp. 249–343; A. Hagelstange, Eine Folge von Holzschnittporträts der Visconti von Mailand, Mitt. aus dem Ger. Nationalmuseum, 1904, pp. 85–100; E. Schaeffer, Das florentiner Bildnis, Munich, 1904; G. C. Williamson, The History of Portrait Miniatures, 2 vols., London, 1904; Bibliothèque Nationale, Portraits peints et dessinés du XIIIe au XVIIe siècle, Paris, 1907 (cat.); W. Waetzoldt, Die Kunst des Porträts, Leipzig, 1908; L. Dumont-Wilden, Le portrait en France, Brussels, 1909; J. Schlosser, Geschichte der Porträtbildnerei in Wachs, JhbKhSammlWien, XXIX, 1911, pp. 171–258; P. Bautier, Juste Suttermans, Brussels, 1912; J. Gürtler, Die Bildnisse der Erzbischofe und Kurfürsten von Köln, Strasbourg, 1912; H. F. Secker, Die Skulpturen des Strassburger Münster seit der französischen Revolution, Strasbourg, 1912; G. Ring, Beiträge zur Geschichte der niederländische Bildnismalerei, Leipzig, 1913; L. Roblot-Delondre, Portraits d'enfants, XVIe siècle, Paris, Brussels, 1913; A. de Beruete, El Greco: Pintor de retratos, Madrid, 1914; Junta de iconografía nacional, Retratos de personajes españoles: Indice illustrado, 2 vols., Madrid, 1914–19; G. Wildenstein, Etudes et documents pour servir à l'histoire de l'art français du XVIIIe siècle: Rapports d'experts, Paris, 1921; E. F. Kossmann, Giovios Porträtsammlung und Tobias Stimmer, Anz. für schweiz. Altertumskunde, N.S., XXIV, 1922, pp. 49–54; B. Croce, Problemi di estetica, 2d ed., Bari, 1923, pp. 258–63; B. Viallet, Gli autoritratti femminili delle R. R. Gallerie degli Uffizi in Firenze, Florence, 1923; J. Alazard, Le portrait florentin de Botticelli à Bronzino, Paris, 1924; L. Dimier, Histoire de la peinture de portraits en France au XVIe siècle, 3 vols., Paris, 1924–26; G. Marlier, Anthonis Mor van Dashorst (Antonio Moro), Brussels, 1924; M. von Boehn, Miniaturen und Silhouetten, 4th ed., Munich, 1925 (Eng. trans., E. K. Walker, London, New York, 1926); W. Beetz, Österreichische Porträtausstellung, 1815–1914, Vienna, 1927 (cat.); E. Benkard, Das Selbstbildnis von 15. bis zum Beginn des 18. Jahrhunderts, Berlin, 1927; P. Pecchiai, I ritratti dei benefattori dell'Ospedale Maggiore di Milano, Milan, 1927; J. Schlosser, Präludien, Vorträge und Aufsätze, Gespräch von der Bildniskunst, Berlin, 1927, pp. 227–46; L. Rovelli, L'opera storica ed artistica di Paolo Giovio comasco: Il Museo dei ritratti, Como, 1928; H. Deckert, Zum Begriff des Porträts, Marburger Jhb. für Kw., V, 1929, pp. 261–82; Fédération française des artistes, Quatre portraitistes français du XVIIIe siècle: Aved, Danloux, Drouais, Tocque, Paris, 1930 (cat.); N. von Holst, Die deutsche Bildnismalerei zur Zeit des Manierismus, Strasbourg, 1930; H. W. Singer, Allgemeiner Bildnis Katalog, 2 vols., Leipzig, 1930; C. Terlinde, L'utilité historique de l'étude des portraits anciens, Actes XIIe Cong. int. d'h. de l'art, Brussels, 1930, pp. 247–81; E. Wind, Humanitätsidee und heroisiertes Porträt in der englischen Kultur des 18. Jahrhunderts, Vorträge der Bib. Warburg, IX, 1930–31, pp. 156-229; A. Riegl, Das holländische Gruppenporträt, 2 vols., Vienna,

1931; K. Gläser, Das Bildnis im Berliner Biedermeier, Berlin, 1932; D. Grafly, Self-portraits in Prints, Am. Mag. of Art, XXV, 1932, pp. 167–72; Die Porträtsammlung des Erzherzogs Ferdinand von Tirol, Vienna, 1932; M. Belzoni, Ritratti polacchi di artisti trentini, Trentino, 11, 1933, pp. 451–59; L. Chamson, Das Porträt in der Avignoner Schule des 15. Jahrhunderts, Pantheon, XI, 1933, pp. 118–34; L. Foscari, Autoritratti di maestri della scuola veneziana, Riv. di Venezia, XII, 1933, pp. 247–62; J. Muls, Cornelis de Vos; Schilder van Hulst, Antwerp, 1933; P. de Nolhac, Portraits du XVIIIe siècle, Paris, 1933; U. Christoffel, Das deutsche Bildnis der Dürerzeit, Pantheon, XIII, 1934, pp. 1–9; A. W. Shaw, The Early English School of Portraiture, BM, LXV, 1934, pp. 171–84; S. H. Steinberg, Bibliographie zur Geschichte des deutschen Porträts, Hamburg, 1934; F. M. Godfrey, Male Portraits of the Venetian School, Apollo, LVIII, 1935, pp. 96–97; Knoedler Gallery, 15th Century Portraits, New York, 1935 (cat.); G. Ladner, Zur Porträtssammlung des Erzherzogs Ferdinand von Tirol, Mitt. ö. Inst. für Geschichteforschung, XLIX, 1935, pp. 367–91; G. Pudelko, Florentiner Porträts der Frührenaissance, Pantheon, XV, 1935, pp. 92–98; T. Bodkin, Some Problems of National Portraiture, BM, LXIX, 1936, pp. 246–51; L. Goldscheider, 500 Selbstporträts, Vienna, 1936 (Eng. trans., L. B. Shaw, Vienna, 1937); W. Boeck, Die Fürstenbusten in gotische Hause zu Wörlitz, Z. d. Vereins für Kw., LV, 1937, pp. 31–50; G. Brière, M. Dumolin and P. Jarry, Les tableaux de l'Hôtel-de-Ville de Paris, Paris, 1937; G. Gombosi, Über venezianische Bildnisse, Pantheon, XIX, 1937, pp. 102–10; H. W. Singer, Neuer Bildniskatalog, 5 vols., Leipzig, 1937–38; E. Wind, Studies in Allegorical Portraiture, Warburg, I, 1937, pp. 138–62; L. Douglas, 19th Century French Portraiture, BM, LXXII, 1938, pp. 253–63; P. Wescher, Das französische Bildnis von Karl VII. bis zu Franz I., Pantheon, XXI, 1938, pp. 1–10; A. Greifenhagen, Antiken als Beiwerk auf Porträts des 18. Jahrhunderts, Pantheon, XXVI, 1940, pp. 292–93; A. C. Sewter, Kneller and the English Augustan Portrait, BM, LXXVII, 1940, pp. 104–11; V. N. Whitehill, Consider the Portrait, Parnassus, XII, 4, 1940, pp. 5–9; C. Björkbom, Svensk portrattliteratur, Gävle, 1941; New-York Historical Society, Catalogue of American Portraits (ed. D. A. Shelly et al.), New York, 1941; C. King, Naval Portraits, Studio, CXXIV, 1942, pp. 1–11; L. Ortigão Burnay, Retratos em gravura de portugueses de seculo XVII, B. Acad. nacional de belas artes, X, 1942, pp. 70–76; D. Redig de Campos, Il "Pensieroso" della Segnatura, Michelangelo Buonarroti nel IV Centenario del "Giudizio Universale" (1541–1941): Studi e saggi, Florence, 1942, pp. 205–19; W. Pinder, Rembrandt's Selbstbildnissen, Leipzig, 1943; E. Greindl, Corneille de Vos, Brussels, 1944; L. Haunah, Retratos coloniales, Rev. Servicio de Patrimonio Historico e Artistico Nacional, 1945, pp. 251–90; M. Florisoone, Portraits français, Paris, 1946; G. Fogolari, Scritti d'arte, Milan, 1946; L. Keil, Os retratos de personages portuguesas, B. Acad. nacional de belas artes, XV, 1946, pp. 18–23; E. Buschor, Bildnisstufen, Munich, 1947; B. Denvir, The King's Pictures, Studio, CXXXIII, 1947, pp. 160–66; L. Frerichs, Antonio Moro, Amsterdam, 1947; O. Deubner, Der Dichter auf Wanderschaft Marburger, Wpr, 1948, pp. 20–26; T. Dobrowolski, Polskie malarstwo portretowe (Polish Portrait Painting), Cracow, 1948; P. O. Rave, RlDKg, s.v. Bildnis; K. Röthel, Le portrait allemand à la Renaissance, Les arts plastiques, 3–4, 1948, pp. 127–40; F. Winkler, Augsburger Malerbildnisse der Dürerzeit, Berlin, 1948; C. Winter, The British School of Miniature Portrait Painters, Br. Acad. Proc., XXXIV, 1948, pp. 119–37; H. Adhémar, Double portrait allégorique, Mus. de France, I, 1950, pp. 11–14; H. Adhémar, Portraits français du XIVe, XVe et XVIe siècles, Musée du Louvre, Paris, 1950; J. Bouchot-Saupique, Le portrait dessiné français de Fouquet à Cézanne, Mus. de France, I, 1950, sup., pp. 11–18; R. L. Douglas, Some Portraits of Ceremony of the Jacobean School, The Connoisseur, CXXVI, 1950, pp. 162–66; Galerie Charpentier, Cent portraits de femmes, Paris, 1950 (cat.); A. Gerlo, Erasme et ses portraitistes, Brussels, 1950; S. Moholy-Nagy, Retreat from the Model, College Art J., X, 1950–51, pp. 370–76; T. Pignatti, I ritratti settecenteschi della Quirini Stampalia, BArte, XXXV, 1950, pp. 216–18; G. Wildenstein, Le goût pour la peinture dans la bourgeoisie parisienne au début du règne de Louis XIII, GBA, XXXVII, 1950, pp. 153–63; C. Zervos, Notes sur les portraits et les figures de la Renaissance italienne, CahArt, XXV, 1950, pp. 97–114, 365–82; R. A. d'Hulst, Pieter Pourbus, Ghent, 1951; D. Piper, Eton Leaving Portraits, BM, XCIII, 1951, pp. 201–02; F. A. Sweet, Mezzotint Sources of American Colonial Portraits, AQ, XIV, 1951, pp. 148–57; V. M. Zimenko, Sovetskaia portretnaia zhivopis' (Soviet Portrait Painting), Moscow, 1951; P. Fierens, Le portrait flamand de Memling à van Dyck, Paris, 1952 (cat.); P. Fierens, Le portrait flamand de Memling à van Dyck, Art et style, 25, 1952, pp. 1-4; Galerie Charpentier, Cent portraits d'hommes du XIVe siècle à nos jours, Paris, 1952 (cat.); H. Geller, Die Bildnisse der deutschen Künstler in Rom, 1800–1830, Berlin, 1952; O. Millar, The Brunswick Art Treasures at the Victoria and Albert Museum: The Pictures, BM, XCIV, 1952, pp. 267–68; O retrato ha Franza do Renascimento ao neoclassicismo, São Paulo, 1952 (cat.); C. Seymour, A Group of Royal Portrait-busts from the Reign of Louis XIV, AB, XXXIV, 1952, pp. 285–96; E. Buchner, Das deutsche Bildnis der Spätgotik und der frühen Dürerzeit, Berlin, 1953; Musée Communal, Le portrait dans les anciens Pays Bas, Bruges, 1953 (cat.); E. R. Peixoto, Exposição de retratos femininos, Anuario Mus. nacional de belas artes, XII, 1953, pp. 69–90; D. R. Reilly, Portrait Waxes, London, 1953; R. dos Santos, Portratos do século XV, Belas Artes, V, 1953, pp. 33–37; M. Valsecchi, Il ritratto nella pittura lombarda dell'Ottocento, Milan, 1953 (cat.); E. Auerbach, Tudor Artists, London, 1954; E. Bernareggi, Le monete d'oro con ritratto del Rinascimento italiano, Milan, 1954; P. du Colombier, Portrait d' "Illustres," La rev. fr., 53, 1954, pp. 19–26; G. Doria and F. Bologna, Ritratto storico napoletano, Naples, 1954 (cat.); N. Ivanoff, I ritratti dell'Avogaria, Arte veneta, VIII, 1954, pp. 272–83; C. Baudelaire, The Mirror of Art (ed. and trans. with notes by J. Mayne), London, New York, 1955, pp. 93–98; J. Desneaux, Jean van Eyck et le portrait de ses amis Arnolfini, Misc. E. Panofsky, B. Mus. royaux des beaux-arts, IV, 1955, pp. 128–44; F. H. Dowley, French Portraits of Ladies as Minerva, GBA, XLV, 1955, pp. 261–86; J. Dupont, Le portrait de Zaccaria Contarini Ambassadeur de Venise à la cour de Charles VIII, Misc. E. Panofsky, B Mus. royaux des beaux-arts, IV, 1955, pp. 121–27; M. Masciotta, ed., Autoritratti dal XV al XX secolo, Milan, 1955; E. S. Ovchinnikova, Portret v russkom iskusstve XVII veka (Portraiture in 17th Century Russian Art), Moscow, 1955; E. Panofsky, Facies illa Rogeri Maximi Pictoris, Late Classical and Mediaeval S. in honor of A. M. Friend, Jr., Princeton, 1955, pp. 392–400; M. Picone, La mostra del ritratto storico napoletano, Emporium, CXXII, 1955, pp. 70–77; N. Rasmo, Il Pisanello e il ritratto dell'imperatore Sigismondo a Vienna, Cultura atesina, IX, 1955, pp. 11–16; S. Sulzberger, Autoportraits de Gerard David, Misc. E. Panofosky, B. Mus. royaux des beaux-arts, IV, 1955, pp. 176–84; O. Teixeira, G. Ribeiro and M. Costantino Exposição de retratos masculinos, Anuario Mus. nacional de belas artes, XIII, 1955–56, pp. 70–77; J. C. Woodiwiss, 19th Century Profiles, Apollo, LXI, 1955, pp. 12–14; C. de Azevedo, Masters of Portuguese Portraiture, The Connoisseur, CXXXVII, 1956, pp. 3–9; P. Huisman, Une école oubliée, L'Oeil, 16, 1956, pp. 18–24; C. Mauri, I ritratti dei benefattori dal XVI secolo ai contemporanei, Milan, 1956; Royal Academy, British Portraits Exposition, London, 1956 (cat.); P. Striede, Zur Nürnberger Bildniskunst des 16. Jahrhunderts, MJhb, 3d Ser., IX, 1956, pp. 120–37; F. Valcanover, Il ritratto veneto da Tiziano al Tiepolo, Warsaw, 1956 (cat.); F. Valcanover, Il ritratto veneto da Tiziano al Tiepolo a Varsavia, Arte veneta, X, 1956, pp. 240–44; G. Wildenstein, Le goût pour la peinture dans le cercle de la bourgeoisie parisienne autour de 1700, GBA, XLVII, 1956, pp. 113–94; J. Wilhelm, Les tableaux de l'Hôtel-de-Ville de Paris et de l'Abbaye Sainte-Geneviève, B. Soc. de l'h. de l'art fr., 1956, pp. 21–32; J. Adhémar, British Portraits: Exposition à la Royal Academy, GBA, XLIX, 1957, pp. 249–52; G. Bazin, Le portrait français de Watteau à David, Paris, 1957 (cat.); W. Boeck, Ritratti di manieristi italiani a Tübingen, CrArte, IV, 1957, pp. 57–61; J. W. Goodison, Cambridge Portraits: 16th and 17th Centuries, The Connoisseur, CXXXIX, 1957, pp. 213–18; P. Grotemeyer, Da ich het die Gestalt: Deutsche Bildnismedaillen des 16. Jahrhunderts, Munich, 1957; M. Serullaz, Le portrait français de Watteau à David, RArts, VII, 1957, pp. 283–90; O. Strettiovà, Das Barockporträt in Böhmen, Prague, 1957; P. du Colombier, Portraits du XVIIIe, La rev. fr., 98, 1958, pp. 46–48; H. Keutner, Zu einigen Bildnissen des frühen Florentiner Manierismus, Mitt. des kunsthist. Inst. in Florenz, VIII, 1958–59, pp. 139–54; Le portrait française au Musée d'Orléans, La rev. fr., 97, 1958, pp. 25–32; A. W. Rutledge, Portraits of American Interest in British Collections, The Connoisseur, CXLI, 1958, pp. 266–70; F. J. Sanchez Cantón, Retratos ingleses en el Museo del Prado, Goya, XXIV, 1958, pp. 342–47; O. Warner, Wax Portraits and Reliefs, Apollo, LXVIII, 1958, pp. 116–17; G. Wildenstein, A propos des portraits peints par François Hubert Drouais, GBA, LI, 1958, pp. 97–104; W. P. Belknap, American Colonial Painting: Materials for a History, Cambridge, Mass., 1959; F. Fournier, Portraits anonymes et "portrait parlé," GBA, LIII, 1959, pp. 345–56; F. H. Huebner, Sitten- und Gesellschaftsmaler in Belgien, Weltkunst, XXIX, 20, 1959, p. 9; F. Neugass, Das Bildnis des Menschen in unserer Zeit, Weltkunst, XXIX, 22, 1959, pp. 7–8; Portraits français de Largillière à Manet, Copenhagen, 1959 (cat.); J. Siegfried, The Romantic Artist as a Portrait Painter, Marsyas, VIII, 1959, pp. 30–42; C. Sterling, La peinture de portrait à la cour de Bourgogne au début du XVe siècle, CrArte, VI, 1959, pp. 289–312; E. E. Veevers, The Source of Inigo Jones' Masquing Designs, Warburg, XXII, 1959, pp. 373–74; I. Bergström, On Religious Symbols in European Portraiture of the 15th and 16th Centuries, Umanesimo e esoterismo: Atti V Conv. int. di s. umanistici, Padua, 1960, pp. 335–43; A. M. Crinò, Documents Relating to Some Portraits in the Uffizi and to a Portrait at Knole, BM, CII, 1960, pp. 257–60; H. von Einem, Karl V. und Tizian, in P. Rassow and F. Schalk, ed., Karl V. der Kaiser und seine Zeit, Cologne, Graz, 1960, pp. 67–93; A. Emiliani, Ritratti in cera del Settecento bolognese, Arte figurativa antica e moderna, VIII, 44, 1960, pp. 28–35; P. Meller, Ritratti "bucolici" di artisti del Quattrocento, Emporium, CXXXII, 1960, pp. 3–10; M. U. Smith, Russian Imperial Portraits at Luton Hoo, The Connoisseur, CXLV, 1960, pp. 7–10; C. Canter, A Gallery of Artist's Portraits, Apollo, LXXIII, 1961, pp. 3–7; R. J. Clements, Michelangelo's Theory of Art, Zürich, New York, 1961; Galerie Cailleux, Âmes et visages de France au XVIIIe siècle, Paris, 1961 (cat.); L. Grassi, Lineamenti per una storia del concetto di ritratto, Arte antica e moderna, IV, 1961, pp. 477–94; O. Millar, The Restoration Portrait, J. Royal Soc. of Arts, CIX, 1961, pp. 410–33; Musée Cantini, Le portrait en Provence de Puget à Cézanne, Marseille, 1961 (cat.); Museo Poldi Pezzoli, Daumier scultore, Milan, 1961 (cat.); F. Panvini Rosati, Monete italiane del Rinascimento, Rome, 1961 (cat.); J. R. F. Thompson and F. G. Roe, Some Oil Portraits in the British Museum, The Connoisseur, CXLVII, 1961, pp. 114–18; G. Weise, L'ideale eroico del Rinascimento e le sue premesse umanistiche, Naples, 1961; G. Bazin, Il ritratto francese da Clouet a Degas, Rome, 1962 (cat.; annotations by O. Dutilh and P. Rosenberg); C. Brandi, Carmine o della pittura, Turin, 1962; M. Levy, Public Image or Private Face? The Dilemma of a Portrait Painter, Studio, CLXIII, 1962, pp. 64–67; D. Piper, ed., Catalogue of 17th Century Portraits in the National Portrait Gallery, 1625–1714, Cambridge, 1963.

The Orient: a. Islam: G. Migeon, Manuel d'art musulman, II: Les arts plastiques et industriels, Paris, 1907; E. Blochet, Peintures de manuscrits arabes, persans et turcs de la Bibliothèque Nationale, Paris, 1910; T. W. Arnold, Painting in Islam, London, 1928; I. Stchoukine, Miniatures indiennes de l'époque des Grands Moghols au Musée du Louvre, Paris, 1929; I. Stchoukine, La peinture indienne à l'époque des Grands Moghols, Paris, 1929; L. Binyon, J. V. S. Wilkinson, and B. Gray, Persian Miniature Painting, London, 1933; F. Edhem and I. Stchoukine, Les manuscrits orientaux illustrés de la Bibliothèque de l'Université de Stamboul, Paris, 1933; A. Sakisian, Contribution à l'iconographie de la Turquie et de la Perse: XVe–XIXe siècles, Ars Islamica, III, 1936, pp. 7–22; Y. A. Godard, An Historical Survey: Post-Safawid Period, SPA, III, pp. 1898–1900; E. Kühnel, History of Miniature and Drawing: Book Painting, SPA, III, pp. 1829–97; A. Sakisian, The Portraits of Mehmet II, BM, LXXIV, 1939, pp. 172–81; G. Wiet, Miniatures persanes, turques et iraniennes, Cairo, 1943; M. S. Dimand, Turkish Art of the Muhammadan Period, BMMA, N.S., II, 1944, pp. 211–17;

J. V. S. Wilkinson, Mughal Painting, London, 1948; A. A. Pallis, Islamic Portraiture, Islamic Rev., XXXIX, 3–4, 1951, pp. 14–17; V. Kubicková, Persian Miniatures, London, 1959; E. Esin, Turkish Miniature Painting, London, 1960; B. Gray, Persian Painting, Geneva, 1961; R. Ettinghausen, Arab Painting, Geneva, 1962. *b. India*: J. P. Vogel, A Statue of King Kaniśka, J. Pañjab Historical Soc., II, 1913, pp. 39–49; K. P. Jayaswal, Statues of Two Śaiśunāka Emperors, JBORS, V, 1919, pp. 88–106, 550–51 (rev. by V. A. Smith et al., JBORS, V, 1919, pp. 512–49); J. L. Moens, Hindoe-Javaansche Portretbeelden, Tijdschrifte Bataviaasch Genootschap, LVIII, 1919, pp. 493–526; H. Goetz, Indische historische Porträts, Asia Major, II, 1925, pp. 227–50; H. Goetz and H. Kühnel, Indian Book Painting, London, 1925; H. K. Sastri, Two Statues of Pallava Kings at Mahābalipuram, Calcutta, 1926; O. C. Gangoly, On the Authenticity of Feminine Portraits of the Moghul School, Rūpam, XXXIII–XXXIV, 1928, pp. 11–15; I. Stchoukine, Les portraits Moghols, RAA, VI, 1929–30, pp. 212–41, VII, 1931–32, pp. 163–76, 233–43, IX, 1935, pp. 190–208; J. P. Vogel, La sculpture de Mathurā, Paris, 1930; T. G. Aravamuthan, Portrait Sculpture in South India, London, 1931; P. Mus, Cultes indiens et indigènes au Champa, BEFEO, XXXIII, 1933, pp. 367–410; H. Heras, The Royal Portraits of Mahabalipuram, ActaO, XIII, 1934, pp. 163–72; B. Karimi, Un album de portraits des princes tîmûrides de l'Inde, Āthār-e-Īrān, II, 1937, pp. 179–281; W. Stutterheim, Note on Śāktism in Java, ActaO, XVII, 1938, pp. 144–52; H. D. Sankalia and M. G. Dikshit, A Unique 6th Century Inscribed Stela from Sāngśi, Kolhāpur State, B. Deccan Coll. Research Inst., IX, 1948–49, pp. 161–66; D. Barrett and B. Gray, Painting of India, Geneva, 1963. *c. China*: O. Fischer, Die chinesische Malerei der Han-Dynastie, Berlin, 1931, pp. 113–14, pls. 51–52; S. Elisseev, Notes sur le portrait en Extrême-Orient, Et. d'orientalisme publiées par le Musée Guimet à la mémoire de Raymonde Linossier, I, Paris, 1932, pp. 169–202; K. Tomita, Portfolio of Chinese Paintings in the Museum of Fine Arts, Boston, 1933, pl. 8; A. Waley, An Introduction to the Study of Chinese Painting, London, 1933; Y. Harada, The Tomb of Painted Basket of Lo-Lang, Seoul, 1934, pl. XLVIII; O. Sirén, A History of Later Chinese Painting, 2 vols., London, 1938; A. Priest, Portraits of the Court of China, New York, 1942; W. Franke, Two Yüan Treatises on the Technique of Portrait Painting, O. Art, III, 1950–51, pp. 27–32; W. Cohn, Chinese Painting, London, 1951; G. Ecke, Hui Hsien Ware, Honolulu, 1954, pls. XII, XIV; G. Ecke, Notes on a Portrait by Hasegawa Nobuharu, Arts asiatiques, III, 1956, pp. 125–29; O. Sirén, Chinese Painting: Leading Masters and Principles, 3 vols., London, New York, 1956; Tsêng Yu-ho, Notes on T'ang Yin, O. Art, N.S., II, 1956, pp. 103–08; G. Ecke, Ananda and Vakula in Early Chinese Sources, Sino-Indian Studies, V, 3–4, 1957, pp. 40–46. *d. Japan*: A. Morrison, The Painters of Japan, 2 vols., London, 1913; L. Binyon, Painting in the Far East, 4th ed., London, New York, 1934; Hōshu Minamoto, An Illustrated History of Japanese Art, Kyoto, 1935; Pageant of Japanese Art, I–II, Tokyo, 1952; H. Munsterburg, The Arts of Japan, Rutland, Vt., Tokyo, 1957; Y. Yashiro, Art Treasures of Japan, I–II, Tokyo, 1960.

* *

Illustrations: PLS. 209–230.

PORTUGAL. The modern Republic of Portugal, which occupies the Atlantic coastal region of the Iberian Peninsula and includes the archipelagoes of Madeira and the Azores, has essentially the same continental boundaries that defined the Portuguese community in the Middle Ages. Its name is derived from the Latin toponym Portus Cale. In addition, the African territories of Angola and Mozambique (qq.v.) remain under Portugal's jurisdiction.

The area of Portugal is known to have been inhabited in the Neolithic era. From the Roman period on, the history of the country became closely tied to that of Spain (q.v.). A more independent artistic development characterized the medieval period, especially after the reconquest from the Moors, when more distinct Portuguese styles began to evolve. In the 15th century, as a result of discoveries and conquests, Portugal entered a period of particular splendor, leading to the spread of Portuguese culture overseas, especially in South America (see AMERICAS: ART SINCE COLUMBUS; BRAZIL). Later, Spanish influence once again became prevalent, and there were periods in which French and Italian influences were also appreciable. Such artistic influences, especially apparent in architecture and sculpture, were variously introduced, often through political and commercial relations as well as by the presence of foreign artists (see SPANISH AND PORTUGUESE ART).

SUMMARY. Prehistory and antiquity (col. 516). The Middle Ages (col. 518): *Pre-Romanesque art; Romanesque art; Gothic art.* Modern period (col. 522): *The late 15th and the 16th century; The 17th and 18th centuries; The 19th and 20th centuries.* Artistic and monumental centers (col. 532): *Lisbon; Minho; Trás-os-Montes and Alto Douro; Douro Litoral; Beira Alta; Beira Baixa; Beira Litoral; Estre madura; Ribatejo; Alto Alentejo; Baixo Alentejo; Algarve.*

PREHISTORY AND ANTIQUITY. The oldest known remains of human habitation in Portugal have been dated to the Lower Paleolithic era. Such archaeological material has been found in the coastal regions of the provinces of Minho, Douro Litoral, Estremadura (Nazaré, Caldas da Rainha, and Peniche), Baixo Alentejo (Cabo de Sines, Vila Nova de Milfontes, and Cabo de São Vicente), and Algarve; in the Guarda district of eastern Beira Alta; along the banks, in the valley, and around the bay of the Tejo (Tagus) River (Alpiarça, Muge, Lisbon and environs), including the peninsula of Setúbal; near the Mira and Guadiana rivers and their tributaries (Baixo Alentejo and Algarve), and elsewhere. From the Upper Paleolithic era, the most extensive evidence of artistic activity has been provided by various dwellings and caves in the vicinity of Lisbon and of Rio Maior (Estremadura) and in Alentejo.

Remains from periods after the Upper Paleolithic era, which are more abundant, have been found along the Atlantic coast, in Minho, Douro Litoral, in the vicinity of Muge, and in neolithic settlements from the north to the south of Portugal, where villages, cemeteries, crypts, and caves have been discovered. There are many megalithic funerary and religious monuments (dolmens and menhirs), and there

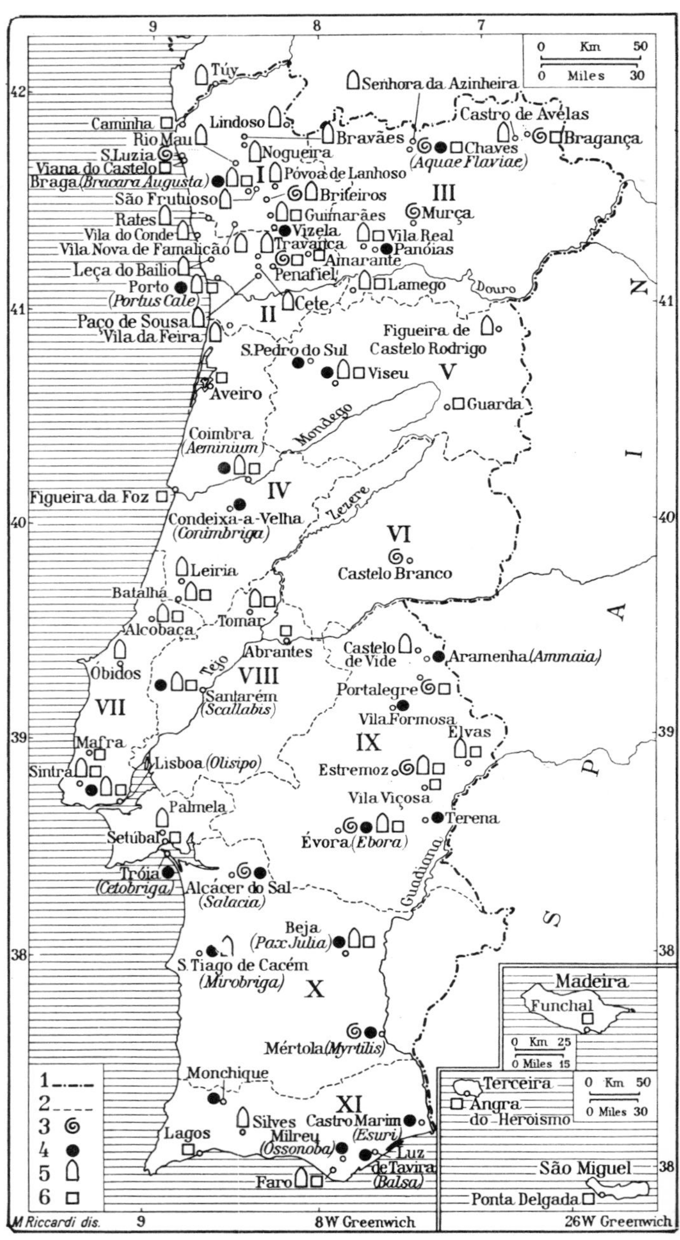

Portugal, major artistic and archaeological centers. *Key*: (1) National boundaries; (2) provincial boundaries. Provinces: (I) Minho; (II) Douro Litoral; (III) Trás-os-Montes e Alto Douro; (IV) Beira Litoral; (V) Beira Alta; (VI) Beira Baixa; (VII) Estremadura; (VIII) Ribatejo; (IX) Alto Alentejo; (X) Baixo Alentejo; (XI) Algarve. Archaeological sites and monumental centers: (3) Prehistoric; (4) Roman period; (5) Middle Ages; (6) modern period.

have been important discoveries of anthropomorphic funerary steles, metal objects (of iron, lead, tin, bronze, copper, silver, and gold), pottery, wheels, cloth fibers, and so on. Some of the more remarkable finds date from the early Bronze Age, when the goldsmith's art appeared in the area of Portugal (see EUROPEAN PROTOHISTORY).

From the late Bronze Age come rock engravings and dolmen paintings, as well as numerous sculptures, ornaments, and tools and weapons (swords, daggers, lance tips, spears and arrows, axes, scythes, spikes, etc.), found in Beira (north, south, and coastal region), Estremadura, Alentejo, and Algarve.

From the occupation of the area of Portugal by the Celts (see CELTIC ART), who are thought to have invaded the Iberian Peninsula in the 8th century B.C., by the Phoenicians, and by the Greeks, there remains a wide variety of pottery and metal relics. Outstanding examples are a Celtic helmet in the Museu Regional, Lagos; a wine vessel from Alcácer do Sal in the Museu Etnológico, Lisbon; and a gold bracteate with the signature of Euai(n)etos, from Bragança, in the Museu Nacional de Soares dos Reis, Oporto.

The most valuable archaeological remains date from the time of the Iberians and the Romans. Outstanding among the Iberian settlements of Minho, Trás-os-Montes, Alto Douro, and Beira, with their characteristic circular houses, are those of Santa Luzia (near Viana do Castelo), Sanfins (Paços de Ferreira), and Briteiros and Sabroso (near Guimarães). Interesting artistic documents of Iberian castrensian culture include the sculptures of warriors and boars (Castelo Mendo) and the statue of the "Porca" in the public garden at Murça; the petroglyph of Penafiel; the engraved "pedras formosas" ("beautiful stones") from Briteiros (e.g., Guimarães, Mus. de Martins Sarmento); fragments of crude pottery; and bronze goats from Évora and Mértola. There also exist examples of the goldsmith's art, worked on a lathe and punched, among which there are some exceptionally fine examples of jewelry — torques (V, PL. 182), bracelets (from Chaves, Castelo Branco, Portalegre, Estremoz), brooches, rings, and earrings (e.g., Lisbon, Mus. Etnológico; a pendant earring in Guimarães, Mus. Regional de Alberto Sampaio; and one from Sabrosa, private coll.

Among the archaeological sites from the Iberian period and the Roman occupation (see HISPANO-ROMAN ART) are those of Conimbriga (Condeixa-a-Velha, near Coimbra), a stipendiary city mentioned by Pliny, that belonged to the *Conventus Scalabitanus*, Cetobriga (Tróia, near Setúbal), Merobriga (Santiago de Cacém), and Ossonoba (Milreu, near Faro), Ammaia (Aramenha), and Balsa (Luz de Tavira).

Of the extensive and complex network of roads built by theRomans, there are remains, for example, of those which went from ancient Olisipo (Lisbon) to Bracara Augusta (Braga), through Scallabis (Santarém), Conimbriga (Condeixa-a-Velha), and Aeminium (Coimbra); from Esuri (Castro Marim) to Ebora (Évora) and Pax Julia (Beja); from Bracara Augusta to Asturica Augusta (Astorga, Spain); and from Olisipo to Emerita Augusta (Mérida), with branches at Salacia (Alcácer do Sal) and Ossonoba (Faro). Various bridges which served these roads still retain their original appearance and construction or have been reconstructed; the one at Chaves (anc. Aquae Flaviae), built in the reign of Trajan (98–117), and that of Vila Formosa, between Ponte de Sor and Alter do Chao, on the road from Lisbon to Mérida, are outstanding examples.

Military architecture is principally represented by earthworks at Viseau (Cava de Viriato), Antanhol (Cidade da Mata), and Alpompé (Chões), as well as by a stretch of roughly constructed wall at Conimbriga.

Noteworthy examples of religious architecture are the so-called "Temple of Diana" in Évora; the temple integrated into the parish church of Sant'Ana do Campo, in the environs of Arraiolos, which according to Vergílio Correia was probably dedicated to Apollo Carneus; and the sanctuary at Panóias, near Vila Real, with inscriptions referring to the cult of Serapis.

The largest and most important monument of civic architecture is a cryptoporticus of ancient Aeminium, lying under the building of the Museu de Machado de Castro in Coimbra, which resembles constructions at Aosta, Arles, Bavai, and Ferentino (J. M. Bairrão Oleiro). Other typical examples of Roman building in Portugal are the ruins of the so-called "Tower of Centum Coeli," a granite structure with a rectangular plan situated about 1 1/4 miles from the Castle of Belmonte; the ruins of the vast "domus" at Conimbriga, lying outside the walls (which were built later), and those of the "villae" of the landing places Santa Maria do Ameixial and Torre da Palma; the aqueduct of Alcabideque, about 2 miles from Conimbriga; the tanks for salting fish at Cetobriga (Tróia), Balsa (Luz de Tavira), and Bôca do Rio; the thermae at Vizela (in Minho), Chaves (in Trás-os-Montes), São Pedro do Sul (in Beira Alta), Conimbriga (in Beira Litoral), Lisbon and Tróia (in Estremadura), Santiago de Cacém, Monchique, and Milreu (in Algarve); and the arch at Bobadela (in Beira Litoral). In addition to the monuments that are partially extant, ancient documents and various inscriptions refer to others that have disappeared, such as a theater and a circus, baths, arches, walls, fountains, and lighthouses.

Excavations have yielded sculpture of stone and of bronze, as well as Roman reliefs, representing male and female figures, divinities, and other subjects (examples in museums in Lisbon, Coimbra, Abrantes, Évora, Beja, etc.). Among the finest examples of sculpture are the statue from Quinta de Baeta in Abrantes (Mus. Regional de Dom Lopo de Almeida), in which the lack of detail prevents its identification as a god, a supplicant, a priestess, a muse, or an allegorical figure (J. M. Bairrão Oleiro); portraits of the Roman emperors Gallienus, found in Milreu, and Claudius, one of which was found at Bodabela and two others in the cryptoporticus of Aeminium (Coimbra); and female portraits in Balsa and Conimbriga (Mus. Monográfico). The statue of Apollo found at Alcoutim and the presumed images of Endovélico from the sanctuary at Terena also present evidence of Roman stone sculpture.

There are also sarcophagi decorated in relief: that of Vila Franca, which is an exceptional find in the Iberian Peninsula; one from Reguengos showing the seasons of the year (Oporto, Mus. Nacional de Soares dos Reis); one from Alfeizerão decorated with figures of muses (Lisbon, Mus. Arqueol. do Carmo); a sarcophagus lid from Chelas depicting writers and muses; and a bas-relief fragment with Mithraic representations found in the Mithraeum of Cetobriga (Lisbon, Mus. Etnológico). Among the small bronze sculptures are statuettes of Minerva (from Conimbriga), Mercury (from Casal Comba and Monte Molão), Fortuna (from Torres Novas and Pombalinho), a legionary (from Loulé), and other subjects.

The remains of mural paintings at Conimbriga, Cetobriga, Marim, Balsa and Bôca do Rio, and Salacia and Ossonoba are of a simple and vulgar type, with architectonic and geometric designs. There are also many Roman mosaics of varied design, both polychromatic and in slate and white combinations, notable examples of which have been found at Conimbriga, Torre da Palma (IX, PL. 76), and Santa Maria do Ameixial (Estremoz).

Numerous pieces of functional pottery (lucernes, vessels, tiles, and clay bricks), as well as complete examples and fragments of terra sigillata and of glass, continue to come to light at various sites under study. One of the finest specimens of glassware, a vessel with a representation of the port of Puteoli (mod. Pozzuoli), was discovered at Odemira, but its present location is unknown.

There are interesting pieces of protohistoric goldwork and jewelry, of which the Museu Etnológico in Lisbon has a rich and representative collection, including a fine silver patera from Lameira Larga (Penamacor) showing Perseus slaying Medusa in the presence of Hermes and Athena.

THE MIDDLE AGES. The medieval period in Portugal is considered to have extended from the 4th century, when conversion to Christianity began, to the last decade of the 15th century, when Vasco da Gama discovered the sea route to India. The Portuguese Middle Ages thus includes the national renascence of the 13th and 14th centuries, overlaps the period of expansion begun under King João II (r. 1481–95), and culminates with the auspicious reign of Manuel I (1495–1521).

The history of Portuguese medieval art is divided into three periods: the pre-Romanesque, including Visigothic art, from the 4th to the beginning of the 8th century; the Romanesque, from about 711–12 to the 11th, 12th, or mid-13th century, depending upon the date of Christian reconquest of the different regions; and a third, or post-Romanesque (mainly Gothic), period that lasted from the 13th century to the final decade of the 15th, when the "Manueline" style was developed. (See GOTHIC ART; PRE-ROMANESQUE ART; ROMANESQUE ART.)

Pre-Romanesque art. Because of the battles of the reconquest and the devastation carried out by al-Manṣūr (939–1002), little Visigothic art remains in Portugal: there are some damaged and altered buildings, the architectural outlines of which can be reconstructed in part, and remains of destroyed buildings that were integrated into later structures. Apart from inscriptions of the period, which are somewhat scarce, mention must be made of the beautiful Church of S. Frutuoso, near Braga. Initially Visigothic, but with clear Mozarabic features, it is without doubt the most Byzantine church of the Iberian Peninsula (H. Schlunk). Built on the Greek-cross plan, it originally consisted of four equal arms terminating in perfect semicircular apses. Marble columns and pilasters with capitals of Corinthian type support the dividing arches of the apses, which were probably covered by brick domes like the surviving central dome.

The chief remains of Visigothic architecture are the basilican church of S. Pedro de Balsemão (near Lamego); the vestiges of a basilican cathedral at Idanha-a-Vélha; the ruins of the basilicas of São Torcato (near Guimarães), Arnal (environs of Leiria), and Tôrre da Palma (between Monforte and Vaiamonte), as well as S. Miguel de Odrinhas (between Sintra and Ericeira) and S. Miguel da Mota (vicinity of

Terena and Alandroal); the remains of a baptistery at Tróia (near Setúbal); and ruins of two polygonal towers in Évora.

Apart from these remains should be noted two funerary plaques from the church at Dume, one of which shows the four Evangelists with the heads of their respective symbols; the sarcophagus of Évora, of white marble and decorated with strigils, cantharus, and lion heads; and the Early Christian sarcophagus of Braga, ornamented on one side with a crater and chrismon surrounded by a crown of laurels. There are also fragments with ornamental reliefs from Lisbon, Chelas, Mértola, and Beja.

There are few remains from the period of Moslem domination in Portugal. Of the military architecture — both the Roman constructions that fell into the hands of the Moors and the castles they built themselves — little survived after successive rebuildings by the Portuguese kings of the first two dynasties. The finest example of Moorish military architecture is the Castle of the Moors in Sintra, although it was rebuilt during the reigns of Sancho I (1185–1211) and Fernando I (1367–83) and again after the earthquake of 1755, which destroyed much of the walls.

As for Moorish religious architecture, apart from the small dwellings of the Moslem hermits found all over the south, the outstanding example is the old mosque of Mértola, now part of the parish church.

The church at Lourosa, uniting Arabic elements with the original Visigothic structure, is an example of the Mozarabic style (see MOZARABIC ART). Another example of this style is the Church of S. Amaro in Beja, a Visigothic construction with Mozarabic additions.

The gilded silver chalice in the museum of the Cathedral of Braga is the most beautiful example of pre-Romanesque goldwork in Portugal; dating from the beginning of the 11th century, its inscription with the names of the donors, Count Menendo Gonzales and Countess Doña Tude, gives it exceptional historical value.

Romanesque art. The peninsular Romanesque style developed under French influence, especially Burgundian, with the southward advance of the reconquest. From the Kingdom of Navarre it spread through the territories of the Portuguese earldom. On the one hand, the Benedictine monastery of Cluny directly and indirectly influenced the style and techniques of constructing those religious edifices which in turn exercised a profound and durable influence on Portuguese architecture. On the other hand, during the wars against the Moors in the 12th century, the military orders applied new methods in the construction of castles and other fortifications in the western part of the Iberian Peninsula.

Romanesque architecture, which faithfully mirrored the religious, political, and military activities of Count Henrique and the first kings of Portugal, lasted in the Galega region until the 14th and 15th centuries in a relatively pure form and persisted in Portugal for a long time, particularly in the northern coastal provinces; it was closely allied to the style of the monuments of neighboring Galicia, a province with which the new kingdom of Portugal had strong cultural links until long after its separation (Joaquim de Vasconcelos).

The Romanesque religious architecture of Portugal comprises cathedrals (e.g., Braga, Oporto, Lamego, Coimbra, Lisbon, and Évora), single-aisled churches and churches with a nave and side aisles, and small sanctuaries and chapels.

The foundations of the Cathedral of Braga were laid late in the 11th century, and the original structure was completed in the following century. Certainly the most representative Romanesque church building in northern Portugal, it greatly influenced contemporaneous architecture in Minho and Douro, where there were 111 monasteries of the powerful order of Cluny. The Cathedral of Oporto was built in the 12th century by order of Bishop Hugo (who had been archdeacon of the Cathedral of Santiago de Compostela), to resemble a medieval fortress, but it was altered in the ensuing century and more extensively in the 16th and 17th centuries. Between 1162 and 1176, the Old Cathedral in Coimbra was rebuilt according to a plan presumably elaborated by the architect Roberto and executed under the direction of Master Bernardo and, later, Master Soeiro. A pilgrimage basilica, it appears to have followed a tradition stemming from the region of Auvergne — particularly from Clermont-Ferrand — which reached the Iberian Peninsula by way of such churches as St-Léger in Royat, St-Julien in Brioude, St-Sernin in Toulouse, and the Cathedral of Santiago de Compostela.

The Cathedral of Lisbon, dating from the 12th century and damaged by several earthquakes (especially those of 1344 and 1755), was also disfigured by reconstructions made during the reign of Afonso IV (1325–57) and by mutilations and additions dating mainly from the 17th and 18th centuries. Had it not undergone such profound alterations, the Cathedral of Lisbon would probably be the most remarkable church in the Auvergne-Peninsular style of the region between the Douro and Tejo rivers; it was probably built by the architects Roberto and Bernardo, who worked on the Old Cathedral of Coimbra. The Cathedral of Évora, the most unusual and original example of early Romanesque construction in Portugal, reflects the architecture of Auvergne and Burgundy, by way of Santiago de Compostela and Poblet. Dozens of Romanesque churches, sanctuaries, and chapels — scattered chiefly throughout the provinces of Minho, Trá-os-Montes and Alto Douro, and Douro and Beira Litoral — show peculiarly Portuguese features alongside direct or indirect Spanish and French influences. A great many of these structures do not have vaulted ceilings.

The polygonal church of the Templars in Tomar, inspired by the Mosque of 'Omar (Dome of the Rock), differs from other Romanesque religious structures in Portugal and recalls, instead, the Church of the Holy Sepulcher in Jerusalem and the Church of Vera Cruz in Segovia. Probably built in the third quarter of the 12th century and owing its unusual design to the demands of military defense, the church consists of a small octagonal sanctuary in the middle, surrounded by a broad ambulatory with radiating chapels.

The fierce conflicts with the Moslems and the struggles with peninsular neighbors explain the existence of numerous castles built in the 12th and 13th centuries or reconstructed after being captured from the Moors. The Castle of Guimarães, birthplace of Afonso I and thus regarded as a symbol of the founding of the Portuguese nation, is an outstanding example of Romanesque military architecture. In the second half of the 12th century the Templars rebuilt and erected many castles, typical of which are those of Tomar (1160), Idanha-a-Nova (1187), and Pombal and Almourol (both, ca. 1175). A unique representative of Portuguese civil architecture in Romanesque style is the Domus Municipalis in Bragança (ca. 1200; XIII, PL. 120).

Romanesque sculpture is profusely in evidence, in architectural embellishments, images, and tombs. The motifs of architectural ornamentation — found chiefly on porticoes, shafts and capitals, cornices and corbels, friezes and modillions — are abundant. They are not so much representational as geometrical, showing stylized flora and fauna in a manner greatly influenced by Near Eastern art. An exception is the historiated decoration of the type found on the archivolts of the monastery at Vilar de Frades, on the capitals of the church at Rio Mau, and on the shafts of the main portico of the church at Bravães. In the Romanesque style of Coimbra, the design is more delicate because of the soft and smooth calcareous stone of the region. An angel from the Cathedral of Oporto and a relief that is thought to represent the Paschal Lamb (*Cordeiro Pascal*), both in Coimbra (Mus. de Machado de Castro), are especially noteworthy examples of Romanesque statuary and relief. Also notable is the funerary sculpture, such as the tombs at Alcobaça, the sarcophagus of Egas Moniz in the church at Paço de Sousa, the supposed tomb of S. Ovìdio in Guimarães, and that of a bishop in the Church of S. Pedro in Rates (XIII, PL. 118).

There are some fine and unusual specimens of Portuguese metalwork from the Romanesque period. Especially remarkable for their sobriety, balance, and delicate filigreework and chiseling are the gold cross of Sancho I from the Monastery of Sta Cruz in Coimbra (Lisbon, Mus. Nacional de Arte Antiga); the gilded silver chalice presented by Geda Mendes (Menendiz) to the Monastery of S. Miguel de Refoios (Coimbra, Mus. de Machado de Castro); and a gilded copper crosier also in the last-named museum. Other chalices (e.g., from Guimarães and Alcobaça), crosses, and reliquaries (e.g., a diptych reliquary in the monastery church of Arouca) provide further evidence of Romanesque goldsmith's work in Portugal.

Gothic art. Just as the Abbey of Cluny provided the spiritual and artistic inspiration from which the Romanesque style in Portugal developed, so the Cistercian abbeys of Pontigny and Clairvaux inspired the beginnings of Portuguese Gothic, as seen in the Monastery of S. Maria at Alcobaça, begun in the second half of the 12th century and completed in the first quarter of the 13th (VI, PLS. 308, 309, FIG. 496). This monastery, erected on lands in the fertile Estremadura Valley donated by Afonso I, became the main religious foundation of the kings of the first dynasty and a most important center of artistic activity.

During the 13th and 14th centuries, as the construction of Romanesque monuments continued, the Gothic style — sometimes incorporating ogival forms into buildings of a more ancient character, sometimes incorporating older, traditional designs into buildings in the new style — evolved from the first efforts at Alcobaça to the triumph of a national architectural style in the Monastery of S. Maria da Vitória at Batalha.

In the second half of the 13th century, S. Maria dos Olivais at Tomar was the model for Portuguese churches with nave and side aisles. Built in a very pure, unpretentious, and noble Gothic style, it was the type most adopted in Portugal until the end of the 15th century, with four or five bays, illuminated by lofty windows, with portals, apse, and apsidioles covered by ogival vaulting, and with great lacy rose windows opening on the façade. The Church of

S. Clara-a-Velha (consecrated 1330) in Coimbra, in spite of the Gothic manner of certain structural and ornamental features, follows the traditional evolutionary line in conception and execution. The Cathedral of Silves and the churches of Graça in Santarém, S. Francisco in Oporto, and S. Domingos in Guimarães herald the advent of the flamboyant Gothic style.

The Monastery of S. Maria da Vitória at Batalha, begun in 1388 in fulfillment of a vow made by King João I before the battle of Aljubarrota, represents the fullest realization of the Portuguese late Gothic style (cf. XIII, PL. 120). It was conceived by Afonso Domingues, the initial master architect (1388–ca. 1402), in an eclectic style, in which traditional forms inspired by northern architecture and adopted in Portugal are fused with techniques of English and French origin that were probably introduced by Afonso Domingues's successor at Batalha, the master Huguet (d. ca. 1438), and with designs that are purely ornamental and typical of the exuberance of the "Manueline" style. In the combination of all these elements Batalha shows definite originality and a truly Portuguese character.

This monastery was the outstanding goldsmith's workshop in the country and the school in which were trained the builders and architects of many of the later monuments in the flamboyant Gothic style and of the first in what may be termed the "peninsular" style, such as Nossa Senhora do Carmo in Lisbon (by the brothers Afonso, Gonçalo, and Rodrigo Anes), the Church of S. Francisco in Guimarães, the Church of S. Domingos in Vila Real, the Cathedral of Guarda, the Convent of the Conceição in Beja, the Church of S. João Baptista in Tomar, and the Church of S. Francisco and the Monastery of the Lóios in Évora.

The most representative examples of regional Gothic in the south are sanctuaries of the type of S. André in Beja and such churches as that of the Monastery of S. António at Serpa; S. Maria, also in Beja; S. Sebastião at Alvito; S. Diniz at Pavia; and, especially, S. Brás in Évora (all in the province of Baixo Alentejo).

Among the military and civic monuments remaining from this period the most notable are the castles of Vila da Feira (Beira Litoral) and Pôrto de Mós (Estremadura) and the palace of the dukes of Bragança at Guimarães.

The medieval renascence in the plastic arts reached its most powerful expression in the sculpture of the 14th century and in the painting of the 15th (see GOTHIC ART). The most representative categories of sculpture are tombs and religious images. In the 14th and 15th centuries Coimbra, Évora, and Lisbon were the main centers for Gothic funerary sculpture (while in the north such sculpture — mainly executed in granite and therefore coarse and austere — continued to be related to the Romanesque of those provinces). Coimbra became the most famous center for works executed in fine and easily worked stone from Ançã and Portunhos. The creative power and nobility of style that characterize the Coimbra school are evidenced by various funerary monuments of the period in the Old Cathedral and S. Clara-a-Nova (i.e., the stone tomb commissioned by St. Isabel, Queen of Portugal, for herself); by the tombs of Domingos Joanes and his wife in the Capela dos Ferreiros of the parish church of Oliveira do Hospital; and most notably by the tomb of Gonçalo Pereira, by Master Pero de Coimbra in association with Telo Garcia of Lisbon, in a chapel in the Cathedral of Braga. A masterpiece of this school, with its balanced proportions, rhythm, and harmonious design, is the sepulcher with the recumbent Christ in the Museu de Machado de Castro in Coimbra.

Among the greatest works of tomb sculpture of all periods, however, are the tombs of Pedro I (1320–67) and Inês de Castro (XIII, PL. 124) in the monastery church at Alcobaça; apart from motifs of Aragonese and French inspiration, these works are characteristically Portuguese in conception and execution.

Batalha and Santarém were, along with Coimbra, the main centers of funerary sculpture in the 15th century. In the monastery at Batalha, directed in the second stage by the northern master Huguet, the double tomb of João I (1357–1433) and Philippa of Lancaster, the tomb of King Duarte (1391–1438) and Queen Leonor, and the monument of the infante Henrique (all later than the first third of the 15th century) show stylistic features that deviate from traditional lines. At Santarém, where the funerary sculpture of the time adheres to traditional lines and reflects the influence of Coimbra, the most noteworthy sarcophagi are those of João Martins Docem and Duarte de Menezes (both in the Mus. Municipal) and that of Pedro de Menezes (in the Church of Graça).

The continuity of style and technique of Coimbra's traditional school is also affirmed by the works of the famous 15th-century masters and their associates and followers. The great sculptors of tomb statuary in Coimbra in the 14th and 15th centuries — for example, Pero de Coimbra, João Afonso, the Master of the Albadas, the Master of Maiorca, and Diogo Pires the Elder — were also the creators of numerous statues of the Virgin and Child and saints and of isolated ornamental images for tomb niches, such as the gravid Virgins of the Annunciations by Master Pero and the enthroned Virgins by the Master of Maiorca. Other fine sculptures identified with this group of artists are a wooden Christ in the Museu de Machado de Castro, Coimbra, and another in the Monastery of Almoster, a bas-relief of the Nativity in the Church of S. Leonardo at Atouguia da Baleia (14th cent.), the numerous images on the portico of the Monastery of Batalha, and the *Corpo de Deus* retable (15th cent.), also in the Museu de Machado de Castro.

Despite various references in manuscripts to painters and paintings in Portugal in the 12th, 13th, and 14th centuries, little remains from these periods, apart from a few miniatures and murals. Among the existing frescoes from the 14th and 15th centuries attributed to Portuguese artists are those in the Cathedral of Braga, ascribed to Gonçalo Anes of Lisbon (1325 and 1348), and those in the Abbey of Florença, attributed to João Gonçalves (1436–39).

The oldest paintings identified as the work of Portuguese artists are by painters who worked in Italy and were influenced by the spirit and technique of the Tuscan and Florentine masters of the 15th century. Among these are Alvaro Pires of Évora, active in the first half of the 15th century, whose works reflect Sienese traditions and training (under Taddeo di Bartolo) and Luís de Portugal, who was directly influenced by the Quattrocento Florentines, especially Benozzo Gozzoli.

A supreme artistic achievement of profoundly Portuguese character is the *Veneration of St. Vincent*, the famous panels attributed to Nuno Gonçalves (known 1450–71), from the Monastery of S. Vicente de Fora and now in the Museu Nacional de Arte Antiga in Lisbon (XIII, PL. 126). Stylistically these panels derive from the esthetic movement that was the forerunner of the Italian Renaissance and from the works of the great Florentine fresco painters — the Lippis, Gozzoli, and Ghirlandajo. With incomparable vigor and an extraordinary mastery of composition Gonçalves presents, in a suggestive atmosphere, a large assembly of persons of all the social classes, impressively delineating the character of the people in the second half of the 15th century. Another characteristic example of Portuguese painting of the 15th century, in the same museum, is the *Ecce Homo* by an unknown artist.

The production of gold- and silverwork (q.v.) increased notably during the national renaissance in the reign of King Diniz (1279–1325). Outstanding pieces from the treasure of his queen, St. Isabel, are a gilded silver and enamel reliquary statuette of the Virgin and Child (the *Virgem do Pilar*), an agate and silver cross, a gold necklace, and a silver statuette of St. Nicholas (all in Coimbra, Mus. de Machado de Castro). Also among the most beautiful pieces of Portuguese medieval art is a valuable group of rock-crystal and silver crosses in the museums of Coimbra, Lisbon, Mafra, and elsewhere. The treasury of the monastery church of Arouca contains a fine reliquary diptych of gilded silver.

MODERN PERIOD. *The late 15th and the 16th century.* Portuguese Gothic, which during the 15th century evolved from sober works of Cistercian inspiration to the characteristic creations in the flamboyant style, entered at the end of the 15th and beginning of the 16th century a highly ornamental phase known as the "Manueline style."

This phase extended through the 16th century, descending from a naturalistic trend which spread through the Iberian peninsula. It is most clearly defined in works of national character, in which are combined motifs of Portuguese and Moorish inspiration. The Manueline style terminated in a disorderly conglomeration of Gothic and Renaissance features, the latter incorporated reluctantly and late.

Architectural influences probably emanated from Batalha, where, among others, Diogo (?) Boytac and João de Castilho worked; these architects were the masters of some of the most representative Manueline monuments of Setúbal, Lisbon, and Tomar. Typical Manueline works created for the monastery at Batalha are the Unfinished Chapels (Capelas Imperfeitas) and the Royal Cloister. The remarkable portal of the Unfinished Chapels, eclectic (i.e., northern and peninsular) in inspiration, is by the Portuguese architect Mateus Fernandes I (the Elder), who is known to have worked at Batalha from about 1480 to 1515. The Church of Jesus in Setúbal, begun in 1494 by Boytac, has been judged the best example of a late Gothic church with the addition of certain Manueline features (Watson).

On the water front of Belém (Lisbon), from which Vasco da Gama sailed to discover the sea route to India, King Manuel I (r. 1495–1521) ordered the construction of the Monastery of the Hieronymites (Jerónimos), which remains the most representative monument of Manueline art, reflecting the singular Portuguese attitude toward the Renaissance. For this reason, perhaps, it is the finest expression of Portuguese architectural genius, complementary to Batalha. In these two monuments is synthesized the evolution of Portuguese Gothic from the severe and aloof inspiration of the Cistercians to the creations of Renaissance classicism, by way of the exuberant flamboyant Gothic and Manueline styles.

In conception the work at Belém falls into two distinct phases: the first (1502–16), with Boytac as architect and master of works, in a naturalistic and flamboyant Gothic style, is closely related in structure and ornamentation (like Batalha) to the same architect's work in the Church of Jesus in Setúbal and the Monastery of Sta Cruz in Coimbra; the second (from 1517), under the direction of João de Castilho, in collaboration with the French sculptor Nicolas Chanterene (one of the initiators of the French Renaissance ideas in Portugal), is Manueline with plateresque and Renaissance features.

After the Monastery of the Hieronymites the most notable architectural work is that of João de Castilho, the Biscayan (?) master, in the Monastery of Christ at Tomar, where there are two richly decorated architectural details that are typical of Manueline art: the main portal, by Castilho, and the window of the chapter house, by Diogo de Arruda (active 1508–31). The Monastery of Sta Cruz in Coimbra, another important center of Manueline art where the first influences of the French Renaissance appeared, was rebuilt, beginning in the first quarter of the 16th century, by Boytac and Marcos Pires, with the collaboration of the French sculptors Chanterene, Filipe Udarte (Hodart; from 1530), Jacques Loguin (from ca. 1520), and João de Ruão (Jean de Rouen; 1530–80).

While Manueline monuments such as those discussed above were being executed in the provinces of Beira Litoral, Estremadura, and Ribatejo, monuments associated with the final stage of the peninsular naturalistic Gothic style were being built in the northernmost provinces, especially Minho, by Biscayan masters.

Other typical examples of the religious architecture of this period are the parish churches of Caminha and Vila do Conde; the 16th-century works in the cathedrals of Braga and Viseu; the churches of Nossa Senhora do Pópulo in Caldas de Rainha, S. João Baptista in Moura, and Nossa Senhora da Misericórdia in Lisbon; and the chapels of the Immaculate Conception in the Convent of S. Clara at Vila do Conde and S. Jerónimo in the monastery at Belém.

In the first decades of the 16th century, when Manueline art was widespread, the Renaissance style was imported, either directly from Italy or through French and Spanish artists, and flourished in Portugal. The most beautiful Renaissance monuments, which date from the first half of the 16th century, are of three types: (1) the Norman style, brought to Portugal by Nicolas Chanterene and João de Ruão; (2) the peninsular style, the creation of João de Castilho and of the artists whose work in Minho and Trás-os-Montes reflected aspects of the plateresque style; and (3) the style of French and Spanish masters as realized by their associates and followers, such as João Braz, Pero de la Gorreta, Afonso Pires, João Luís, and Gaspar Denis.

Outstanding Renaissance monuments are the Church of S. Maria de Conceição (mid-16th cent.) in Tomar and the Church of Graça (rebuilt second quarter of 16th cent.) in Évora.

Although certain Gothic formulas persisted in Portugal until the 17th century (Mário Chicó), the Renaissance style was already established there by the last quarter of the 16th in the classicism of the architects Diogo de Torralva (d. ca. 1566), Miguel de Arruda (d. ca. 1563), Filippo Terzi (1520–97), and Jerónimo de Ruão (d. after 1593). The finest work of this period is the main cloister of the Monastery of Christ in Tomar, which was begun by Torralva, master of works from 1554, and completed by Terzi.

Examples of 16th-century secular architecture are the Paços de Concelho (Town Hall) of Viana do Castelo and the royal palace at Sintra, representative of the Manueline-Moorish style and the inspiration for similar structures in Estremadura and Alentejo, such as the manors of Sempre Noiva, near Évora, Água de Peixes, near Viana do Alentejo, and the Castle of Alvito; the palaces of Quinta da Bacalhoa, near Setúbal, and the counts of Carreira in Viana do Castelo; the manors of Sub-Ripas in Coimbra, Pitas in Caminha, and Gisteira in Évora; and the houses known as the Casa Soure and the Casa Cordovil, also in Évora.

An outstanding example of Manueline military architecture from the period of conquest and the voyages of discovery is the Tower of Belém (begun ca. 1515), designed by Francisco de Arruda (active 1510–47).

During the reign of Manuel I sculptural decoration, in both northern and southern Portugal, was influenced by Galician and Basque artists such as Juan de Riaño, Sancho Goya, and Pero Gallego, who contributed to the mother church of Vila do Conde (1500), the Cathedral of Braga (1509), and the parish churches of Caminha (1511) and Viana do Castelo (1512). In Tomar and Lisbon, from 1510 onward, the Basque architects João de Castilho and his younger brother Diogo (d. ca. 1575) exercised strong influence. At the Monastery of the Hieronymites Pero de Trilho, Fernando de le Formosa, Rodrigo da Pontesilha, Francisco de Benavente, and the statue maker Horta were contracted to work on the south portal under the direction of João de Castilho.

In the 16th century Coimbra remained the chief center for sculpture, reflecting medieval tradition through the work of Diogo Pires the Elder. Traditional influence is seen mainly in the works of Diogo Pires the Younger, the brothers Pedro and João Álvares (e.g., a Manueline window at Tentúgal, signed and dated 1507; the *Cristo das Maleitas*, in the Mus. de Machado de Castro), and Pedro and Francisco Henriques, or Anriquez (baptismal fonts in the New Cathedral of Coimbra and the Church of Nossa Senhora do Pópulo in Caldas da Rainha), and in the tomb sculpture in the Convent of S. Clara at Vila do Conde.

Works by or influenced by the greatest of the Manueline sculptors, Diogo Pires the Younger (active 1511–35), include a baptismal font in the monastery church at Leça do Bailio and the calvary nearby (both, 1514–15); the angels and garland for Sta Cruz in Coimbra (ca. 1517); and the tombs of Friar João Coelho in Leça do Bailio and Diogo de Azambuja (ca. 1515) in the Church of Nossa Senhora dos Anjos at Montemor-o-Velho and those in the Monastery of S. Marcos, near Tentúgal, in Beira Litoral (1522).

In wood carving and imagery the work of the Flemish artist Olivier de Gand (van Ghent; d. 1512) and his followers Fernando de Muñoz, Master Machim, João Alemão, Francisco Loreto, and Diogo de Sarça is preeminent for its stylistic dignity and perfect execution. Apart from isolated designs for churches in the Beira Litoral region (Coimbra, Montemor-o-Velho and Botão), Olivier de Gand executed the most beautiful carved retable to survive in Portugal, that of the Old Cathedral in Coimbra (1498–1503), and, between 1503 and 1508, various works for the Church of S. Francisco in Évora and, in 1511–12, choir stalls of the Monastery of Christ in Tomar, which were destroyed when the monastery was devastated by Napoleonic troops in 1810. Also attributed to him on stylistic grounds are the carvings of the great retable and of the choir stalls in the Cathedral of Funchal (Madeira). His followers later created choir stalls for the Monastery of Sta Cruz in Coimbra, the Monastery of Christ in Tomar, and the Monastery of the Hieronymites at Belém (Lisbon).

In statuary and stone decoration, four French sculptors worked at Coimbra, together with traditional and archaizing sculptors. These men, Filipe Udarte, Nicolas Chanterene, Jacques Loguin (Longuin), and Jean de Rouen, introduced the Renaissance style into Portugal in the first half of the 16th century and revitalized sculpture by giving it a new direction.

Udarte, who worked at Toledo before he came to Portugal, made a vigorously modeled terra-cotta *Last Supper* for the refectory of the Monastery of Sta Cruz in Coimbra, a work of mannerist and monumental expression, now in the Museu de Machado de Castro.

Chanterene, perhaps the most cultured of the four, was commissioned to create the tomb effigies of the first two kings of Portugal (1515), also in Sta Cruz, aud the main portal (1517) of the church of the Hieronymite monastery at Belém. He also worked at Coimbra, the Monastery of S. Marcos, Braga, Alcobaça, Obidos, and Sintra (until about 1532), and in Alentejo, particularly at Évora, from 1533.

Jacques Loguin, who produced the most delicate and refined sculpture of this group, was probably the creator of the masterpiece of Renaissance sculpture at Coimbra, the famous pulpit of Sta Cruz. Among other works for the same monastery attributed to him are the reliefs of the Claustro do Silêncio and the small retables from the Claustro (or Jardim), da Manga, now in the Museu de Machado de Castro.

Of these four sculptors, the work of Jean de Rouen (João de Ruão) was perhaps the most extensive and elaborated, and also the most Portuguese in sensibility. As sculptor, architect, contractor, and creator he was an associate of João de Castilho at Coimbra, at Buarcos and Bouças, at Guarda, Pedrogão Grande, and Pombal, and elsewhere. His pupil and collaborator Tomé Velho (fl. 1561–1621), in an age when mannerism predominated, remained free of this influence and retained the taste for harmonious proportions and nobility of style that characterized traditional sculpture.

In the first two decades of the 16th century — that is, in the last twenty years of the reign of Manuel I — at least three great centers of pictorial activity flourished in Portugal: Lisbon, Évora, and Viseu. Contemporary manuscripts relate that, at that time, there were in Lisbon the studios of two celebrated painters — Jorge Afonso, the court painter, who examined and supervised all royal painting under Manuel I and João III, and his brother-in-law Francisco Henriques. Among the works attributed to Jorge Afonso are the monumental panels, of eclectic character, in the ambulatory of the Monastery of Christ in Tomar; the altar panels in the Church of Madre de Deus in the Xabregas quarter of Lisbon; and some works for the Church of Jesus at Setúbal (e.g., parts of the panels from the dismantled main altar now in the Mus. Municipal). Francisco Henriques, who probably died in Lisbon during the plague of 1518, created the retables for the high altar and lateral chapels of the Church of S. Francisco in Évora, which are now in the Museu Nacional de Arte Antiga, Lisbon, and in the Casa dos Patudos, at

Alpiarça (XIII, PL. 130). In 1517, Frei Carlos, a painter of Flemish origin, took his vows in the Hieronymite monastery at Espinheiro, near Évora, and painted there a series of panels (Lisbon, Mus. Nac. de Arte Antiga), the spirit and technique of which seem to be directly descended from the artists of Bruges and Antwerp. Another painter who also worked in the monasteries of the Hieronymite brothers in the first quarter of the 16th century was the Master of Lourinhã; his work — characterized by harmony of composition and framing, vigor and delicacy of design, and rich, transparent color — is outstanding among contemporary peninsular painting.

The character of Portuguese painting during this period can be seen most clearly in the work of the artists of Beira Alta and Litoral: at Viseu, where the genius of Vasco Fernandes (d. between 1541–43), often called "Grão Vasco," predominated; and in Coimbra and surroundings, where the blunt, intensely strong personality of the monogrammist MN stood out. Vasco Fernandes marked all his work with a distinctly regional character — realistic and penetrating in portraiture, idyllic in the rendering of landscape, and precise in the presentation of the details of fabrics, accessories, and jewelry.

The second quarter of the 16th century is dominated by the extensive and varied output of the three painters of the period whose works are found in the greatest number throughout Portugal. These are Gregório Lopes, Cristóvão de Figueiredo (XIII, PL. 130), and Garcia Fernandes, the so-called "Masters of Ferreirim." Lopes was painter to kings Manuel and João III, and Figueiredo, by appointment an examiner of royal paintings, worked for the infante Cardinal Afonso. These artists and their third associate, Fernandes, were most representative of the cosmopolitan trend and of the profound and far-flung artistic communication that Portugal had established with the various intellectual centers of Western Europe.

In Portuguese painting of the second half of the 16th century the direct and indirect influences of the mannerists and the art of the Italianizers, particularly the Romanists, resulted in the work of such artists as Gaspar Dias (1560–90), Simão Rodrigues (1562–1612), and Diogo Teixeira (1565–98). The art of portraiture (q.v.), which was developing in Portugal during this period, was certainly influenced by Anthonis Mor van Dashorst, Alonso Sánchez Coello, and Joris van der Straeten. In Portugal, Cristóvão Lopes (1516–ca. 1594), Cristóvão de Morais (fl. 1551–71), and, especially, the unknown painter of the portrait *Lady with a Rosary* (Lisbon, Mus. Nac. de Arte Antiga) raised the genre to a level that makes them worthy successors to the art of Nuno Gonçalves.

During the 16th century various Portuguese painters worked in Spain, where many of their works are known to have integrated with the currents of Catalonia and Andalusia. Among them are Henrique Fernandes, Pedro Nunes (active in Spain 1518–46), João Henriques, and Vasco Pereira (active at Seville 1561–1609).

Portuguese decorative arts in this century were inspired by traditional motifs; by the flamboyant Gothic, Manueline, and Italian Renaissance styles — sometimes in hybrid association but with balanced and pleasing results; and by subjects of African, Brazilian, and Oriental origin, as a result of the discoveries and conquests overseas. Lisbon was at this time the center of activities and a great warehouse in Western Europe for the new products and minor arts from the East, such as cloth, chinaware, lacquer, carpets, and ivories from Persia, India, China and Japan.

The Renaissance and mannerist altars, the archaic façades and forerunners of the baroque style, and the *mudéjar* roofs of the royal palace at Sintra, the cathedrals of Coimbra and Funchal, and the Church of Madre de Deus in Lisbon and that of Caminha all show heterogeneous aspects.

In ceramics, formal and ornamental influences from Oriental porcelain were first introduced into Europe in profusion through Portugal; Chinese pieces with Portuguese motifs and the increase in the use of glazed tile brought to this genre of decorative art completely original expression. The tiles in the palace of Quinta da Bacalhoa (Azeitão) date from 1565, and those of the Church of S. Roque (Lisbon), signed by Francisco de Matos, from 1584.

There are also interesting and original collections of embroidery, counterpanes, furnishings for personal and religious use, and rugs originating from the countries that were discovered and conquered at this time.

Of all the decorative arts, the richest and most varied is that of the goldsmiths, inspired by traditional Portuguese and foreign motifs. Its development can be studied from the establishment of the Portuguese monarchy in the 12th century to the present. From the 16th century there comes a wide range of religious and secular objects which are remarkable for their originality, including chalices, custodials, lamps, reliquaries, jugs, trays, salvers, and pitchers. Among the most notable of these are the custodial from the Monastery of the Hieronymites, which was made by Gil Vicente in 1506 from enamel and gold from the Quilon region of India, and the Renaissance reliquary, also of gold and enamel and decorated with pearls and precious stones, made by Master João for Queen Leonor. There are many examples in national and foreign collections, of Indo-Portuguese goldsmith's art, with Christian motifs, personages, and incidents and with African, American, and Asian flora and fauna.

The 17th and 18th centuries. The removal of the court to Madrid during the period of Spanish domination (1580–1640) and the discovery of the mines of Brazil in 1693 contributed greatly to the decline of national art: until the late 1640s, part of the royal orders were issued from the Spanish capital, and the Brazilian gold attracted to Portugal a large numer of foreign artists — mainly Italian, German, and French.

Various artists who were inspired by the Italian schools greatly influenced Portuguese architecture in the 18th century, notably the Prussian João Frederico Ludovice (Johann Friedrich Ludwig; ca. 1670–1752) and the Tuscan Niccolò Nasoni (Nazzoni; d. 1773). Equally important to monumental art of the 18th century were the influences of works by foreigners that were conceived outside Portugal, such as the precious Chapel of S. João Baptista by Nicola Salvi and Luigi Vanvitelli, begun in Rome in 1742 and incorporated into the Church of S. Roque in Lisbon in 1749, and the magnificent Hospital of S. António in Oporto, designed by the architect John Carr of York and begun in 1770.

The triumph of Portuguese monumental art in the 18th century lies, moreover, in the adaptation of the baroque, the rococo, and the neoclassicism of Ludovice, Nasoni, and Alessandro Giusti (1715–99), of Carlos Mardel (d. 1763) and Giacomo Azzolini (d. 1786/87), and of Jean Baptiste Robillon (d. 1748) and Francisco Xavier Fabri (d. 1807) by such Portuguese architects as Mateus Vicente de Oliveira (1706–86), Reinaldo Manuel dos Santos (d. 1789/90), José de Figueiredo Seixas, and Carlos Luil Ferreira da Cruz Amarante (1748–1815) in the north and Manuel Caetano de Sousa (d. 1802) and José da Costa e Silva (1747–1819) in the south.

The monastery at Mafra, by Ludovice, with its magnificent classical church and elegant rococo library (the latter created by Manuel Caetano de Sousa), is the most representative building of the age and the grandest architectural creation in the country; during the era of the Marquês de Pombal it became a school for Portuguese architects and sculptors, under the direction of Ludovice (1717–30) and, later, Giusti (1753–70).

Just as the outstanding German architects exercised a strong influence on 18th-century architecture in the south, so Nasoni influenced the architecture of the north. Nasoni's masterpieces in Portugal are the Church and Tower of S. Pedro dos Clérigos (the former begun in 1732 and consecrated in 1779; the latter built between 1748 and 1763; cf. II, FIG. 281); the house of Quinta da Prelada, and the palace and garden of Quinta do Freixo, all in Oporto and environs. After Mafra, the most extensive project to be commissioned by King João V (r. 1706–50) is the aqueduct of Áquas Livres (1729–48) in Lisbon, designed by Manuel de Maia (1688–1768); begun by the designer and Antonio Canevari (b. 1681), Custódio Vieira (active in first half of 18th cent.), and José da Silva Pais; and finished by Carlos Mardel, Reinaldo Manuel dos Santos, Francisco António Ferreira, Malaquias Ferreira, Rodrigo Franco, and Miguel Angelo Blasco. The reconstruction of Lisbon after the earthquake of 1755, for which the chief designer was Eugénio dos Santos de Carvalho (d. 1760), incorporating industrial installations, buildings in series, and workers' quarters, attests the prodigious endeavors of de Pombal.

In line with the spirit and taste of the era, and stimulated by the splendor of the court of João V, Portuguese palace and manorial architecture entered a golden period in the 18th century and reflected various trends. Gardens, carvings, tilework, statues, paintings, and stuccoes enriched the palaces and manors of the 18th century, which came to have a spaciousness, flexibility, and graciousness unknown in the seignorial residences of the 17th century.

The palace at Queluz (XIII, PL. 137), with its highly refined neoclassical and rococo character of French inspiration, mainly the work of Jean Baptiste Robillon and Mateus Vicente de Oliveira, is "the most perfect expression of aristocratic society in the second half of the 18th century" in Portugal (Júlio Dantas).

Portuguese sculpture of the 17th century is represented mainly by portraits and by carvings for retables, polychrome and gilded, principally for the churches of Braga, Tibães, Vilar de Frades, Vila do Conde, Santo Tirso, Oporto, Coimbra, Torres Vedras, Évora, and so on. Apart from Italian inspiration, the works show Spanish influences from Granada and Valladolid.

Typical of Portuguese baroque is the conjunction of terra cotta and wood in the monastery at Alcobaça. The monumentality and noble style of the figures of the high altar and of the Chapel of S. Bernardo, dating from the last quarter of the 17th century, make them the highest achievements of baroque sculpture in Portugal. Manuel Pereira (1588–1683), who was born in Oporto but worked in Spain, is the foremost exponent of the classically inclined movement of the 17th century.

The sculpture of the 18th century, like the architecture, reflects various influences: there were the traditional artists of Braga, Arouca, Aveiro, Évora, Elvas, Lisbon, and Alcobaça; Claude de Laprade, who was active in Portugal from 1699 to about 1730, and, above all, Alessandro Giusti (1715–99), who founded a school of sculpture at Mafra, were the most important sculptors active in Minho, Douro, Beira Litoral, and Estremadura during this splendid and adventurous era for Portugal.

Among the notable artists of the first three quarters of the 18th century are José de Almeida (1700–69), who had studied with Carlo Monaldi in Rome; João José de Aguiar, a student of Antonio Canova (q.v.) in Rome (1785); and Joaquim Machado de Castro (d. 1822), a disciple of Giusti and creator of the famous equestrian statue of King José Manuel, which stands in the Terreiro do Paço in Lisbon, and of Nativity scenes. The latter was a popular art, in which Joaquim José de Barros Laborão (1762–1820), Frey Manuel Teixeira, and António Ferreira (b. before 1731, d. ca. 1795) also excelled. Current artistic trends in Portugal also appeared in her overseas territories, especially in Brazil, where the art of sculpture was dominated by the extraordinary figure of António Francisco Lisbôa, called "Aleijadinho" (1738–1814; I, PL. 143; XIII, PL. 154).

In this rich and prosperous century, the art of coachbuilding achieved unprecedented splendor. The Museu Nacional dos Coches in Lisbon has one of the world's finest collections, in which the coach of the Portuguese ambassador to Pope Clement XI (XIV, PL. 301) and that of King João V, attributed to José de Almeida, are most remarkable.

The most important 17th-century Portuguese painter whose work has been identified is Domingos Vieira (1627–52), the greatest artist to follow in the tradition of Nuno Gonçalves. Vieira's work is psychologically profound and executed in a sober style, reflecting the new tendencies that, since the mid-16th century, had been impressed on Portuguese painting by Cristóvão de Morais, by the anonymous master of the *Lady with a Rosary* (Lisbon, Mus. Nac. de Arte Antiga), and by Spanish and Dutch painting.

In the monastic centers, such as Aveiro, national painting, with its curious attributes of clothing and accessories, acquired a very distinctive character because of its graceful, devout, and decorative ingenuousness. The walls of monasteries, the ceilings of churches and chapels, and altars were covered with panels of cloth, copper, and wood on which were piously portrayed mystical images of saints and miracles, frequently surrounded by garlands of flowers in the Flemish manner. Josefa d'Ayala, known as Josefa d'Óbidos (ca. 1630–84), who was born in Seville, daughter of the Portuguese painter Baltazar Gomes Figueira, was a foremost practitioner of this genre of painting; it also brought fame to other painters of the time, whose work must be admired more for its expression of religious devotion than for its creative power, technique, or style.

Independent of the peculiarly national characteristics of 18th-century Portuguese art, painting, at least until the third quarter of the century, shows the predominant influence of the Italians — especially Carlo Maratta (or Maratti), Pompeo Girolamo Batoni, Vincenzo Baccherelli (or Baccarelli), Ferdinando Galli Bibiena, and Giacomo Azzolini — in religious, decorative, and theatrical subjects; in portraiture, genre painting, and landscape painting, the preponderant influence is German and French, from João Glama (Strebel or Ströberle; 1708–92), Jean Ranc (1674–1735), Pierre Antoine Quillard (1701–33), Jean Pillement (1728–1808), and others.

Among Portuguese artists, the dominant figure of the first half of the 18th century is the brilliantly inspired Francisco Vieira de Matos (1699–1783), better known as "Vieira Lusitano," the designer, engraver, and painter who studied in Rome under Benedetto Lutti and Francesco Trevisani (called "Romano"). After Vieira Lusitano, the foremost painters of the century are André Gonçalves (1692–1762), Pedro Alexandrino de Carvalho (1730–1810), and Cirilo Volkmar Machado (1748–1823). Many Portuguese artists of the 18th century, mainly influenced by the Italians, specialized in perspective ceiling paintings.

At the end of the 18th and beginning of the 19th centuries, the three most prominent figures in Portuguese painting are Morgado de Setúbal (1752–1809); Francisco Vieira, known as "Vieira Portuense" (1765–1805); and Domingos António de Sequeira (1768–1837; XIII, PL. 149). The canvases of Morgado de Setúbal show a vigorous realism, with portrait figures, costumes, and still lifes of animals, vegetables, and household utensils. Vieira Portuense, like Sequeira, is one of the immediate precursors of early-19th-century romanticism in Portuguese art. His paintings, very delicate in feeling and extremely pure in form, are executed in sensitive and gently harmonious colors; he was educated in Italy and assimilated in his work the artistic trends then prevalent in Germany and England.

The most representative figure of this movement, however, is Domingos António de Sequeira, justly considered the Portuguese Goya. A genius, Sequeira left a vast quantity of work, in which he showed a rare gift of imagination, sensitivity, and technical skill. He was a prolific draftsman, lithographer, and designer of coins, medals, and jewelry; also a remarkable painter, he expressed the anxieties and enthusiasms of a revolutionary artist in his portraits and in his historical and allegorical paintings.

During these centuries, the decorative arts showed great originality in style, ornament, and design. Furniture (q.v.) of national and Indo-Portuguese type, carving, embroidered rugs, ivories, china, tiles, and goldwork all reflect a mixture of European and overseas influences, particularly Spanish, Italian, Dutch, English, French, African, Asiatic, and South American. In furniture of the so-called "Philippine" style, dating from the reigns of João V (1706–50), José Manuel (1750–77), and Maria I (1777–1816), predominantly peninsular and Dutch influences are seen in the rendering of such imported styles as Louis XV, Louis XVI, Queen Anne, Chippendale, Hepplewhite, and Adam. Some of these pieces are made of exotic woods (e.g., *pau santo* and *jacarandá*), with spiral columns and carvings and inlays of bone, mother-of-pearl, and ivory. Inlay, lacquer, and *chinoiserie* are occasionally used to decorate oratories, cabinets, chests of drawers, tables, coffers, and desks.

The same mixture of styles is found in ceramics; the most unusual Oriental motifs to decorate European china of the 17th century appear on Portuguese faïence, along with ideas inspired by Italian majolica and by pieces from Talavera, Rouen, Moustiers, and so on. In the second half of the 18th century, as a result of official measures taken by the Marquês de Pombal to stimulate and protect industry, there was a considerable development in Portuguese ceramics with the establishment of faïence factories. The most notable were those of Rato (Lisbon, 1767), Darque (Viana do Castelo, 1774), Rocha Soares (Oporto, 1775), Massarelos (Oporto, 1783), Brioso e Vandelli (Coimbra, 1784), Cavaquinho (Oporto, 1789), and Bica do Sapato (Lisbon, 1796). Numerous examples of the production of these centers may be seen in the museums of Lisbon, Coimbra, and Oporto.

Tilework — blue-and-white or polychrome — entered a most beautiful decorative period in the second half of the 16th century, as seen in the palace of Quinta da Bacalhoa (Azeitão; 1565) and in the panels by Francisco de Matos (1584) in the Church of S. Roque in Lisbon; this art reached monumental proportions in the facing of entire walls in both religious and civic buildings.

The goldwork from the workshops of Lisbon, Oporto, Braga, Guimarães, Aveiro, Coimbra, Santarém, Setúbal, and Évora comprises one of the most original decorative arts of the period. Baroque, rococo, and neoclassical designs, with exuberant relief and repoussé work and gracefully executed engraving and scalloping, are found on tabernacles, custodials, chalices, crosses, reliquaries, missals, bookbindings, candelabra, chandeliers, salvers, trays, and jugs; there is a most varied repertory of forms and ornamental motifs that enhance columns, balusters, volutes, modillions, the representations of birds, flowers, and leaves, and work in precious stones (diamonds, emeralds, sapphires, and rubies). Outstanding examples of silverwork are the tomb of St. Isabel (1612), in S. Clara-a-Nova, in Coimbra; the altar (1639–51) of the Cathedral of Oporto; a tabernacle (1663) in the Church of the Hieronymites at Belém; the altar and tabernacle of the *comendadoras* of Avis (early 18th century); and a tabernacle (1742) in the Cathedral of Oporto.

These two centuries also produced some magnificent works adorned with precious stones, such as the cross in the Palace of Vila Viçosa (17th century) and the monstrance from Bemposta (18th century; Lisbon, Mus. Nac. de Arte Antiga).

The 19th and 20th centuries. Portuguese architecture of the first quarter of the 20th century reflected the confusion that was the legacy of the 19th century, which had seen the decline of national institutions, conscience, and activity arising from the nation's social disorganization.

Vain attempts were made to reestablish traditions in the midst of disorderly activity, which, at one moment, displayed a conventional cult of classicism and, at another, copied the surface forms of consecrated styles from the great periods of the nation's history. The balance found in such works in Lisbon as Fortunato Lodi's Teatro Nacional (1842–46), Domingos Parente's Câmara Municipal (1867–75), G. Davioud's and Elias Robert's monument to Pedro IV (1870), and António Tomás da Fonseca's monument to the Restauradores (1866) is opposed by the romantic fantasy expressed by the German Baron von Eschwege, in the Castelo da Pena in Sintra (1840), with its medieval suggestions; by Italians such as Cinati in his Manueline imitation (1859) for the Monastery of the Hieronymites at Belém, and Luigi Manini and Nicola Bigaglia, in the Neo-Manueline Palácio Hotel at Buçaco (1888–1907); and by the English architect Bermett (James Knowles?) in the Oriental-style Palácio Monserrate in Sintra (1865). The tendency toward imitation of past styles was also fostered by Portuguese artists, as shown by the Renaissance-style Fountain of El Rei (second quarter of 19th cent.) in Lisbon; the Neo-Gothic Capela dos Pestanas (1878–88) in Oporto, by Albano Cordeiro Cascão; the Neo-Gothic tomb of the poet Alexandre Herculano de Carvalho e Araújo (1886), by the sculptor Manuel Raimundo Valadas, in the Monastery of the Hieronymites, Belém; the Neo-Manueline Rossio Railroad Station in Lisbon, by José Luís Monteiro; the Byzantine-style building of the Sociedade Martins Sarmento (1881) in Guimarães, by José Marques da Silva; the Arabian salon in the Palácio da Bolsa (bourse) of Oporto; and the tombs of Vasco da Gama and Luís de Camões in the Belém monastery by the sculptor Costa Mota.

In a society that was artistically, intellectually and politically weakened, the work of the architect Miguel Ventura Terra (1866–1919) stands out as exemplary because of its rationality, discipline and control. His Palácio do Congresso, adapted from the Convent of S. Bento da Saúde (1599–1615), "is fashioned with the greatest purity of line, without any subjection to mere canons or servile conventionalism" (Figueiredo, 1908).

The architect Raul Lino (b. 1879), designer of the Casa dos Patudos in Alpiarça, collaborating in Afonso Lopes Vieira's plan to reestablish the national spirit of Portugal and make it European, was the forerunner of the nationalistic current. His followers include the brothers Carlos and Guilherme Rebelo de Andrade (b. 1887 and 1891, respectively), architects of the Museu das Janelas Verdes in Lisbon, and Vasco Palmeiro-Regaleira (b. 1897). Tertuliano Marques and Rogério de Azevedo (b. 1898) were also talented exponents of the transition toward the new trends.

The architects José Luís Monteiro, in Lisbon, and José Marques da Silva, in Oporto, laid the foundations, based on solid academic principals, for the coming generations, which were to transform completely the nature and expression of contemporary Portuguese art "not by imitating or stylizing elements or designs from past eras, but by the logical solution of synthesizing new plans that would embody the possibilities and aspirations of the present" (Pardal Monteiro). The idea of a new order was supported by the architects Pardal Monteiro, Carlos Ramos (b. 1897), Luiz Cristino da Silva (b. 1896), José Angelo Cottinelli Talmo (1897–1948), Jorge Segurado (b. 1898), Gonçalo de Mello Breyner, Paulino Montez (b. 1897) and Couto Martins; António Veloso Reis (b. 1899), Jacobety Rosa, Luís Benavente, Raul Tojal, Adelino Nunes, Raul Rodrigues Lima (b. 1909), Francisco Keil do Amaral (b. 1910), and others. As the 20th century advanced monumental architecture came to reflect the aspirations and reforms of a period of national renascence and progress. Artists of the current generation, representing the most advanced movement in Portuguese architecture, come primarily from the schools of fine arts in Oporto and Lisbon.

Although part of a generation that lent itself servilely to a cult of decadent conventionalism, João José de Aguiar (1769–1840) is one of the few memorable sculptors of his time. After him, Portuguese sculpture entered an uncharacteristic phase in the weak romanticism of the first half of the 19th century, which was followed by the naturalism of Simões d'Almeida (1844–1926) and Costa Mota (1847–1930).

António Soares dos Reis (1847–89) was the dominant sculptor of the second half of the 19th century. His portraits of the Count of Ferreira (1876; Oporto, Mus. Nac. de Soares dos Reis), the Viscountess of Moser (1884), Avelar Brotero (1887; in the Botanical Garden, Coimbra), his busts of Flor Agreste (1876) and an English lady (1888), and his statues *The Exile* (1872; Oporto, Mus. Nac. de Soares dos Reis) and *Dom Afonso Henriques* (1887) show strength and discipline, evocative power and psychological insight, and dignity and nobility of style, all of which reveal the vigor of the sculptor's personality and his extraordinary technique.

During the first quarter of the 20th century, when the persistence of the realists and the verists was leading Portuguese art to weakness and decadence, there emerged an irrepressible movement in reaction to failing academism. The foremost exponent of this movement was the sculptor Francisco Franco (1885–1955). Among the sculptors of Franco's and the following generation, the most important are João da Silva (b. 1880), António Alves de Sousa (1884–1922), Diogo de Macedo, and Ernesto Canto da Maia (b. 1890), as well as Leopoldo de Almeida (b. 1898), Álvaro De Brée (b. 1903), and Rui Roque Gameiro (1907–35), Salvador Barata Feyo (b. 1902), António Duarte (b. 1912), and Joaquim Martins Correia (b. 1910); the last six, with their varied figural styles — classic and romantic, realist and mannerist, essentially spiritual — represent the most balanced and constructive trends of the contemporary movement. An abstract school, without previous roots in Portuguese art and based on ideology, esthetics, and existentialist writings, appeared after World War II, especially among those artists who attended the fine arts schools of Lisbon and Oporto.

Among the painters of the 18th century who lived into the 19th century Francisco Vieira ("Vieira Portuense"; d. 1805) and Domingos António de Sequeira (d. 1837) are the most notable. Of the same period as Sequeira, but showing a weak academicism or incipient romanticism without any particular personality, are José da Cunha Taborda (1766–1836), Joaquim Rafael (1783–1864), João Baptista Ribeiro (1790–1868), and the painter-sculptor António Manuel da Fonseca (1796–1890).

Romanticism, whose chief literary exponent in Portugal was the poet and playright Almeida-Garrett (1799–1854), was fostered by painters of the second and third quarters of the 19th century who were attracted to this new movement of philosophical, political, social, and artistic ideas, which had begun in England and Germany. Naturalism, which arose throughout Europe in reaction to the romantic movement, also found expression in Portuguese painting. The artists most representative of these tendencies were Tomás José da Annunciação (1818–79), Luís de Miranda Pereira Henriques de Menezes (1820–78), Franz Xaver Winterhalter (1805–73), Francisco Augusto Metrass (1825–61), João António Correia (1825–97), and João Cristino da Silva (1829–77).

The two figures of this period who were the forerunners of the naturalistic movement of the last quarter of the century were Tomás José da Annunciação and Miguel Angelo Lupi (1826–83). Of the latter, Diogo de Macedo says that "no Portuguese romantic painter surpassed him in the interpretation of the subjects which gave him inspiration, nor had any a more delicate, vigorous and sober style or greater technique."

In the last quarter of the 19th century, there emerged a number of remarkable painters who added luster to the history of Portuguese art: among them were António Carvalho da Silva Porto (1850–93) and João Marques da Silva Oliveira (1853–1927), the main landscape painters; Alfredo Keil (1850–1907); José Júlio de Sousa Pinto (1856–1939) and António Carneiro (1872–1930), outstanding for their poetic vision and subtle rendering of line and color; Columbano Bordalo Pinheiro (1857–1929), a major portrait painter; António Monteiro Ramalho (1858–1916); Henrique César de Araújo Pousão (1859–94) and José Vital Branco Malhôa (1855–1933), noted for their impressionistic treatment of light; João Vaz (1859–1931); Carlos Reis (1863–1940); José Veloso Salgado (1864–1945); Aurélia de Sousa (1865–1922); Zoé Wauthelet Batalha Reis (1867–1949); and António Ezequiel Pereira (1868–1943). Alfredo Roque Gameiro (1864–1935) was an outstanding water-colorist of the same period; in this medium João Alves de Sá (b. 1878), Alberto de Sousa Leitão de Barros, Helena Roque Gameiro (b. 1895), and Paulino Montez (b. 1897) also became notable.

The modern movement in Portuguese painting began in Paris with Amadeu de Sousa Cardoso (1887–1918) and was continued by Guilherme Augusto Santa Rita (1889–1918), Eduardo Viana (b. 1881), and Simão Dórdio Gomes (b. 1890). The constellation of Portuguese painters that emerged in the first half of the 20th century also includes Francisco Semith, Henrique Franco (b. 1883), Alfredo Miguéis (1883–1943), Manuel Jardim, Luíz Eduardo de Ortigão Burnay (b. 1884), Milly Possoz (b. 1889), Armando de Lucena (b. 1886), Abel Manta (b. 1888), Armando Basto (1890–1923), António Soares (b. 1894), Jorge Barradas (b. 1894), Luiz Varela Aldemira (b. 1895), Mário Augusto, Lino António, Clementina de Moura, Carlos Botelho (b. 1899), Mário Eloy (1900–51), Agostino Salgado (b. 1907), José Dominguez Alvarez (1906–42), Carlos Carneiro (b. 1900), Maria Keil (b. 1914), Luciano Santos (b. 1911), Celestino de Sousa Alves (b. 1913), and João Hogan (b. 1914).

The following contemporary artists should be noted in connection with their styles or fields of artistic endeavor: for the most advanced tendencies, Simão Dórdio Gomes, Helena Vieira da Silva (b. 1908), and Júlio Resende (b. 1917); for true-to-life realism in portraiture, Eduardo Malta (b. 1900) and Henrique de Medina (b. 1901); in landscape painting, Armando de Lucena, Abel Manta, Agostinho Salgado, Luciano Santos, Celestino de Sousa Alves, and João Hogan; in mural and fresco painting, Jaime Martins Barata (b. 1899) and José de Almada Negreiros (b. 1893); in drawing and water color, Eduardo Malta, Henrique de Medina, Lima Freitas, João Carlos Celestino Gomes, José Tagarro (1902–31), Milly Possoz, and Bernardo Marques (b. 1900); and in engraving, Júlio Pomar, Teresa de Sousa, and Bartolomeu dos Santos. In the work of artists who have emerged more recently, such as Santiago Arcal e Lanzner, there is an eclecticism and a sincerity that indicate a search for creative freedom.

With the resurgence of interest in the decorative arts, new life was given to the artistic industries and the applied arts, which have developed individual character and fresh ideas and techniques and have been given dignity through the work of talented technicians, artists, and artisans. Distinguished among them are Almada Negreiros, Carlos Botelho, and Tomaz de Mello in various fields; Jorge Barradas in ceramics, Guilherme Camarinha in tapestry; and Bernardo Marques and Luís de Jesus Moita in graphic arts.

Luís Reis-Santos

BIBLIOG. *General*: C. Volkmar Machado, Collecção de memórias relativas ás vidas dos pintores e esculptores, architectos e gravadores portuguezes, Lisbon, 1823 (2d ed., Coimbra, 1922); A. Raczynski, Les arts en Portugal, Paris, 1846; A. Raczynski, Dictionnaire historique-artistique du Portugal, Paris, 1847; J. de Vasconcelos, Historia da arte em Portugal, 6 vols., Oporto, Coimbra, 1881–85; I. Vilhena Barbosa, Monumentos de Portugal, Lisbon, 1886; F. de Sousa Viterbo, Diccionário historico e documental do architectos, engenheiros e constructores portuguezes e a servicio de Portugal, 3 vols., Lisbon, 1899–1922; A. Belino, Archeologia Christã, Lisbon, 1900; A arte e a natureza em Portugal, 8 vols., Oporto, 1902–08; J. de Figueiredo, Algumas palavras sobre a evolução da arte em Portugal, Lisbon, 1908; W. C. Watson, Portuguese Architecture, London, 1908; Notas sobre Portugal, II, Lisbon, 1909; M. Dieulafoy, Art in Spain and Portugal, London, New York, 1913; J. de Vasconcelos, Arte religiosa em Portugal, 2 vols., Oporto, 1914–15; E. Bertaux, Les arts en Portugal, in Guides Bleus, Espagne et Portugal, Paris, 1916, pp. 451–59; P. M. Aguiar Barreiros, Elementos de archeologia e belas artes, Braga, 1917; A. Pereira de Almeida, Portugal artístico e monumental, Lisbon, 1923; Guia de Portugal, 3 vols., Lisbon, 1924–44; R. dos Santos, A arquitectura em Portugal, Lisbon, 1929; Marqués de Lozoya, El arte hispanico, 5 vols., Madrid, 1931–49; B. Direcção Geral dos Edificios e Monumentos Nacionais, Oporto, 1935 ff. (107 vols. pub. by 1962); R. dos Santos, L'art portugais, Paris, 1938; L. Reis Santos and C. Queiroz, Paisagem e monumentos de Portugal, Lisbon, 1940 (Eng. trans., Monuments of Portugal, Lisbon, n.d.); R. dos Santos, Conferências de arte, 3 vols., Lisbon, 1941–49; A. de Lacerda et al., Historia da arte em Portugal, 3 vols., Oporto, 1942–53; E. Lambert, L'art en Espagne et au Portugal,Paris, 1945; V.Correia, Obras, 3 vols., Coimbra, 1946–53; D. de Macedo, Sumário histórico das artes plásticas em Portugal, Oporto, 1946; R. dos Santos, L'art portugais, Lisbon, Oporto, 1949; R. dos Santos, L'art portugais, Paris, 1953; S. Sitwell, Portugal and Madeira, London, 1954; Influências do Oriente na arte portuguesa continental a arte nas províncias portuguesas do ultramar, Lisbon, 1957 (cat.); E. Lambert, Art musulman et art chrétien dans la péninsule ibérique, Paris, 1958; A. de Amorim Girão, Atlas de Portugal, 2d ed., Coimbra, 1959; R. dos Santos, Historia del arte portugués, Barcelona, Madrid, 1960; Sindicato nacional dos Arquitectos, Arquitectura popular em Portugal, 2 vols., Lisbon, 1961; J. Barreira, ed., História da arte em Portugal, 4 vols., Lisbon, n.d.

Antiquity: a. Pre-Roman Lusitanian art: A. de Paço, Páleo e Mesolítico português, Rev. de Guimarães, XLVI, 1936, pp. 221–30, XLVII, 1937, pp. 8–24; A. de Paço and E. Jalhay, Páleo e mesolítico português, Anais Acad. Portuguesa de H., IV, 1941, pp. 9–98; P. Montez, História da arquitectura primitiva em Portugal: Monumentos dolménicos, Lisbon, 1943; J. Olivier, Le paléolithique supérieur au Portugal, B. Et. portugaises et de l'Inst. fr., XI, 1947, pp. 103–12; M. Cardoso, Algunas observaciónes sobre el arte ornamental de los "castros" del nordeste de la Peninsula Ibérica, Cartagena, 1949; A. García y Bellido, Tartessos y los comienzos de nuestra historia, in R. Menendez Pidal, ed., Historia de España, I, 2, Madrid, 1952, pp. 281–308; A. García y Bellido, La Peninsula Ibérica en los comienzos de su historia, Madrid, 1953; A. Schulten, ed., Fontes Hispaniae Antiquae, 2d ed., Barcelona, 1955 ff.; A. Schulten, Iberische Landeskunde, Strasbourg, 1955; G. Zbyszewsky, Le Quaternaire du Portugal, B. Soc. Geológica de Portugal, XIII, 1–2, 1958, pp. 1–227; M. Cardoso, Die "Castros" in norden Portugal, Rev. de Guimarães, LXIX, 1959, pp. 417–36; A. Souto, O Ocidente ibérico, Oporto, 1960; A. Vasco Rodrigues, Arqueologia de Peninsula Hispânica, Oporto, n.d. *b. Roman art*: B. Taracena, Arte romano, Ars Hispaniae, II, Madrid, 1947, pp. 11–178; A. García y Bellido, Esculturas romanas de España y Portugal, 2 vols., Madrid, 1949; J. M. Bairrão Oleiro, A escultura romana em Portugal, Lisbon, 1950; J. M. Bairrão Oleiro, Elementos para o estudo da "Terra Sigillata" em Portugal, 2 vols., Guimarães, Coimbra, 1951–54; G. Battelli, Una città romana in Portogallo, Rome, 1952; J. A. Ferreira de Almeida, Introdução ao estudo das lucernas romanas em Portugal, Lisbon, 1957; A. García y Bellido, Del caracter militar activo de las colonias romanas de la Lusitania y regiones immediatas, Trabalhos de antr. e etn., XVII, 1959, pp. 299–304.

Middle Ages: J. de Vasconcelos, Da architectura manuelina, Coimbra, 1885; A. Fuschini, A architectura religiosa na Edade Media, Lisbon, 1904; J. de Vasconcelos, Arte românica em Portugal, Oporto, 1918; D. J. de Pessanha, Arquitectura pre-românica em Portugal: S. Pedro de Balsemão e S. Pedro de Lourosa, Coimbra, 1927; V. Correia, A arquitectura em Portugal no século XVI, Coimbra, 1929; J. das Neves Larcher, Castelos de Portugal, 2 vols., Lisbon, Coimbra, 1933–35; R. da Costa Tôrres, A arquitectura dos descobrimentos e o rinascimento ibérico, Braga, 1943; F. Perez Embid, El mudejarismo en la arquitectura portuguesa de la época manuelina, Seville, 1944 (2d ed., Madrid, 1955); H. Schlunk, Arte visigodo, arte asturiano, Ars Hispaniae, II, Madrid, 1947, pp. 227–306; E. Lambert, L'art manuélin, Lisbon, 1949; R. dos Santos, O estilo manuelino, Lisbon, 1952; M. Tavares Chicó and M. Novais, A arquitectura gótica em Portugal, Lisbon, 1954; R. dos Santos, O românico em Portugal, Lisbon, 1955; P. A. Evin, Etude sur le style manuélin, Paris, 1956; A. de Gusmão, A expansão da arquitectura borgonhesa e os mosteiros de Cister em Portugal, Lisbon, 1956; E. Lambert, Art portugais, Et. médiévales, III, 1957, pp. 249–97; M. M. Ribeiro Martins, Elementos para a cronologia da arquitectura românica em Portugal, Coimbra, 1957; M. Cocheril, Abadias Cistercienses portuguesas, Lisbon, 1959; M. Cocheril, Recherches sur l'Ordre de Cîteaux au Portugal, B. Et. portugaises et de l'Inst. fr., N.S., XXII, 1959–60, pp. 30–102; M. A. Soares de Azevedo et al., A arquitectura portuguesa da época dos descobrimentos, Lisbon, 1960; D. F. de Almeida, Arte visigótica em Portugal, Lisbon, 1962; J. Grave, Castelos de Portugal, Oporto, n.d.

Modern and contemporary periods: J. de Vasconcelos, A pintura portuguesa nos séculos XV e XVI, Oporto, 1881 (2d ed., Coimbra, 1929); J. de Vasconcelos, Da architectura manuelina, Coimbra, 1885; A. Haupt, Die Baukunst der Renaissance in Portugal, 2 vols., Frankfurt am Main, 1890–95; A. Haupt, Portuguesische Frührenaissance, Berlin, 1898; F. de Sousa Viterbo, Notícia de alguns esculptores portuguezes ou que exerceram a sua arte em Portugal, Lisbon, 1900; F. De Sousa Viterbo, Notícia de alguns pintores portuguezes, 3 vols., Lisbon, 1903–11; E. Bertaux, La Renaissance en Espagne et en Portugal, Michel, IV, 2, 1911, pp. 817–991; V. Correia, A pintura a fresco em Portugal nos séculos XV e XVI, Lisbon, 1921; V. Correia, Pintores portugueses dos séculos XV e XVI, Coimbra, 1928; V. Correia, A arquitectura em Portugal no século XVI, Coimbra, 1929; T. Vitry, Essai sur l'œuvre des sculpteurs français au Portugal pendant la première moitié du XVIe siècle, B. Et. portugaises, II, 1932, pp. 1–33; J. das Neves Larcher, Castelos de Portugal, 2 vols., Lisbon, Coimbra, 1933–35; L. X. da Costa, As belas artes plásticas em Portugal durante o século XVIII, Lisbon, 1934; J. Couto, Pinturas quinhenistas do Sardoal, Lisbon, 1939; Comemorações Nacionais de 1940, Exposicão de os primitivos portugueses (1450–1550), Lisbon, 1940 (cat.); E. Lavagnino, Gli artisti in Portogallo, Rome, 1940; R. dos Santos, Os primitivos portugueses, Lisbon, 1940 (2d ed., 1957); R. dos Santos, As relações artisticas entre a Itália e Portugal (Estudos it. em Portugal, 2), Coimbra, 1940; R. da Costa Tôrres, A arquitectura dos descobrimentos e o rinascimento ibérico, Braga, 1943; A. de Lucena, Pintores portugueses do Romantismo, Lisbon, 1943; F. de Pamplona, Um século de pintura e escultura em Portugal (1830–1930), 2d ed., Oporto, 1943; F. Perez Embid, El mudejarismo en la arquitectura portuguesa de la época manuelina, Seville, 1944 (2d ed., Madrid, 1955); D. de Macedo, A escultura Portuguesa dos séculos XVII e XVIII, Lisbon, 1945; R. dos Santos, A escultura em Portugal, 2 vols., Lisbon, 1948–50; P. Fierens, Les primitifis portugais, Brussels, 1949; E. Lambert, L'art manuélin, Lisbon, 1949; L. van Puyvelde, Les primitifs portugais et la peinture flamande, Lisbon, 1949; A .de Gusmão, Os Portugueses e a Renascença, Historia da Arte: Pintura, Lisbon, 1950; D. de Macedo, Académicos e Românticos, Lisbon, 1950; P. Monteiro, Eugénio dos Santos: Precursor do urbanismo e da arquitectura moderna, Lisbon, 1950; R. dos Santos, O estilo manuelino, Lisbon, 1952; L. Reis Santos, Gregorio Lopes, Lisbon, 1954; R. Huyghe, Nuno Gonçalves dans la peinture européenne du XVe siècle, Lisbon, 1955; R. dos Santos, Nuno Gonçalves, London, 1955; P. A. Evin, Etude sur le style manuélin, Paris, 1956; G. Kubler and M. Soria, Art and Architecture in Spain and Portugal and Their American Dominions 1500–1800, Harmondsworth, 1959; J. A. França, Da pintura portuguesa, Lisbon, 1960; J. A. França, Pintura portuguesa abstracta, Lisbon, 1960; J. Lees-Milne, Baroque in Spain and Portugal, London, 1960; J. J. Martin González, La huella española en la escultura portuguesa (Renacimiento y barocco), Valladolid, 1961; A. de Carvalho, D. João V e a arte do seu tempo, Lisbon, 1962; J. Grave, Castelos de Portugal, Oporto, n.d.

Minor arts: *a. General*: Exposição retrospectiva da arte ornamental portuguesa e hespanhola, 2 vols., Lisbon, 1882 (cat.); J. de Vasconcelos, Industrias portuguesas: Resumo histórico, 2 vols., Oporto, 1886–87; F. de Sousa Viterbo, Artes e artistas em Portugal, Lisbon, 1892 (2d ed., 1920); A. Guimarães and A. Sardoeira, Mobiliário artístico português, 2 vols., Oporto, Gaia, 1924–35; R. dos Santos, ed., Exposição de arte portuguesa, London, 1955 (cat.). *b. Metalwork*: J. de Vasconcelos, Historia de ourivesaria e joalharia portuguesa, Oporto, n.d.; J. de Vasconcelos, Toreutica, Oporto, 1904; J. Couto, Ourivesaria portuguesa, Lisbon, 1929; R. dos Santos and I. Quilhó, Ourivesaria portuguesa nas colecções particulares, 2 vols., Lisbon, 1959–60; J. Couto and A. M. Gonçalves, A ourivesaria em Portugal, 2 vols., Lisbon, 1960–63. *c. Ceramics*: J. de Vasconcelos, Ceramica portuguesa, Oporto, 1884; F. de Sousa Viterbo, Artes industriais e industrias portuguesas: O vidro e o papel, Coimbra, 1903; J. Queiroz, Cerâmica portuguesa, Lisbon, 1907 (2d ed., 2 vols., Lisbon, 1948); V. Correia, Azulejos datados, Lisbon, 1916; D. J. de Pessanha, A porcelana em Portugal, Coimbra, 1923; P. Vitorino, Azulejos datados, Oporto, 1925; L. de Castro e Solla, Cerâmica brazonada, 2 vols., Lisbon, 1928–30; V. Valente, Porcelana artística portuguesa, Oporto, 1949; V. Valente, O vidro em Portugal, Oporto, 1950; R. dos Santos, Faiança portuguesa: Séculos XVI e XVII, Oporto, 1960; J. M. dos Santos Simões, Iconografia Olissiponense em azulejos, Lisbon, 1961; J. de Campos e Sousa, Loiça brasonada, Oporto, 1962; V. Valente, Cerâmica artística portuense dos séculos XVIII e XIX, Oporto, n.d. *d. Furniture and textiles*: D. S. de Pessanha, Um núcleo de tecidos, 2 vols., Lisbon, 1918–19; C. Bastos, Subsídios para o estudo das origens e evolução da industria textil em Portugal, Oporto, 1950; A. F. da Silva Nascimento, Leitos e camilhas portugueses, Lisbon, 1950; A. Cardoso Pinto, Cadeiras portuguesas, Lisbon, 1952. *e. Miniatures and illumination*: R. dos Santos, Les principaux manuscrits à peintures conservés au Portugal, Paris, 1933; R. dos Santos, A tomada de Lisboa nas iluminaras manuelinas, Lisbon, 1939; J. Faro, O Livro de Horas de Dom Manuel, Lisbon, 1957; J. Brandão, Miniaturistas portuguesas, Oporto, n.d. *f. Engraving*: F. de Sousa Viterbo, A gravura em Portugal, Lisbon, 1909; E. Soares, Dicionário de gravadores portugueses e dos estrangeiros que trabalharam para Portugal, Lisbon, 1937; E. Soares, Historia da gravura artistica em Portugal, 2 vols., Lisbon, 1940–41; E. Soares and H. C. Ferreira Lima, Dicionário de iconografia portuguesa, 3 vols., Lisbon, 1947–50; E. Soares, Evolução da gravura de Madeira em Portugal: Séculos XV a XIX, Lisbon, 1951; E. Soares, Dicionário de iconografia portuguesa: Suplemento, 2 vols., Lisbon, 1954–60. *g. Arms*: F. de Sousa Viterbo, A armaria em Portugal, 2 vols., Lisbon, 1907–08.

ARTISTIC AND MONUMENTAL CENTERS. The following survey is organized by provinces, from north to south in a general west-to-east direction, with the exception of the nation's capital, Lisbon, which will be discussed first.

Lisbon (Port., Lisboa; anc. Olisipo). The ancient settlement was probably founded by Phoenicians, who appeared to have occupied the summit and southern slope of the hill on which the Castelo de S. Jorge now stands. They called their colony Alis Ubbo, meaning "fair haven." As a result of the search for an etymological explanation for the name, the founding of the city was attributed in ancient times to various legendary figures — most commonly to Ulysses.

Occupied by the Romans from 205 B.C., Olisipo was elevated to the status of municipium during the reign of Augustus (27 B.C.–A.D. 14) and given the name Felicitas Julia. After the decline of Roman power and an ensuing period of Visigoth domination, the city fell to the Moors in 714.

Although kings of León and Asturias temporarily recaptured the city from the Moors three times from the early 9th to the mid-12th century, it was not until 1147 that Lisbon was finally conquered by the first king of Portugal, Afonso Henriques (Afonso I), with an army of crusaders recruited mainly from England, Germany, and Flanders. Olissibone, or Lissibona, as it was called, then spread beyond the walls of the Moorish castle to the banks of the Tejo (Tagus) River affluent (which later silted up), where the lower part of the city is now situated.

Capital of the kingdom from the reign of Afonso III (1248–79) onwards, it was extended under Fernando I (1367–83) by a new wall, which eventually enclosed an area of about 255 acres. Lisbon was enlarged several times during the Middle Ages and became the principal commercial center of Portugal during the era of discovery and conquest; it was even further enriched and adorned with buildings and monuments during the reign of João V (1706–50). However, a great part of the city's artistic heritage was damaged and destroyed by earthquakes in the 14th, 16th, and 18th centuries, especially by those of 1531 and 1755. It was rebuilt by the Marquês de Pombal, the plan of the lower city having been delineated by Carlos Mardel (d. 1763), Manuel da Maia (1688–1768), and Eugénio dos Santos de Carvalho (1711–60). From that period on, especially from the second quarter of the 18th century, the area of the capital has increased considerably, with the construction of major streets and the development of new zones, most notably the large districts of Arieiro and Alvalade.

One of Lisbon's most prominent landmarks is the Castelo de S. Jorge, the oldest parts of which date back to the 9th–12th centuries. Built on a foothill which was certainly the site of a Roman castrum, the present structure incorporates part of the original Moorish castle and another built during the Middle Ages. Badly damaged by the earthquakes of 1531 and 1755, it was extensively restored during the second quarter of the 20th century. Preserved within the enclosure are the castle proper (Castelejo), with ten rectangular towers, the ramparts, and the doors of St. George (southeast), Martim Moniz (northeast), and the Betrayal (north).

The Cathedral (Sé) of Lisbon retains little of the 12th-century Romanesque structure, having been restored at the end of the 19th and in the first half of the 20th century. Built in a style that is commonly termed "Auvergne-Peninsular," the Cathedral has a cruciform plan, with nave and side aisles spanned by six wide vaulted arches, a triforium above the side aisles opening onto both the nave and transept, a lantern tower of Clermont type above the crossing, and a narthex and the main portico between two towers. At present, apart from the remains of Romanesque construction, the most important architectural and sculptural features of the Cathedral are in the Gothic style, mainly of the 14th century, and include tombs, the cloister, the chancel (known as the *capela-mor*), the ambulatory (XIII, PL. 120), and the apsidal chapels and that of Bartolomeu Joanes.

Among the works of historical and artistic note in the Cathedral are the font at which, according to tradition, St. Anthony of Padua (b. 1195) was baptized; the retable in the style of Garcia Fernandes, painted in 1537, and the 18th-century crib by the sculptor Joaquim Machado de Castro in the Chapel of Bartolomeu Joanes; canvases by Pedro Alexandrino (1730–1810) in the body of the church and high altar; and the beautiful medieval wrought-iron latticework in the cloister and in one of the apsidal chapels. The Cathedral treasure includes rich furnishings and plate, some set with precious stones, dating chiefly from the 16th to the 18th century.

The monastery church of Nossa Senhora do Carmo, founded in 1389, was designed by the architects Afonso, Gonçalo, and Rodrigo Anes (or Eanes). Badly damaged by the earthquake of 1755, all that remain are the façade, with its Gothic portal, the south face, and the main and four apsidal chapels. The Museu Arqueológico is housed in the remains (see below).

In the history of Portuguese architecture, the Monastery of the

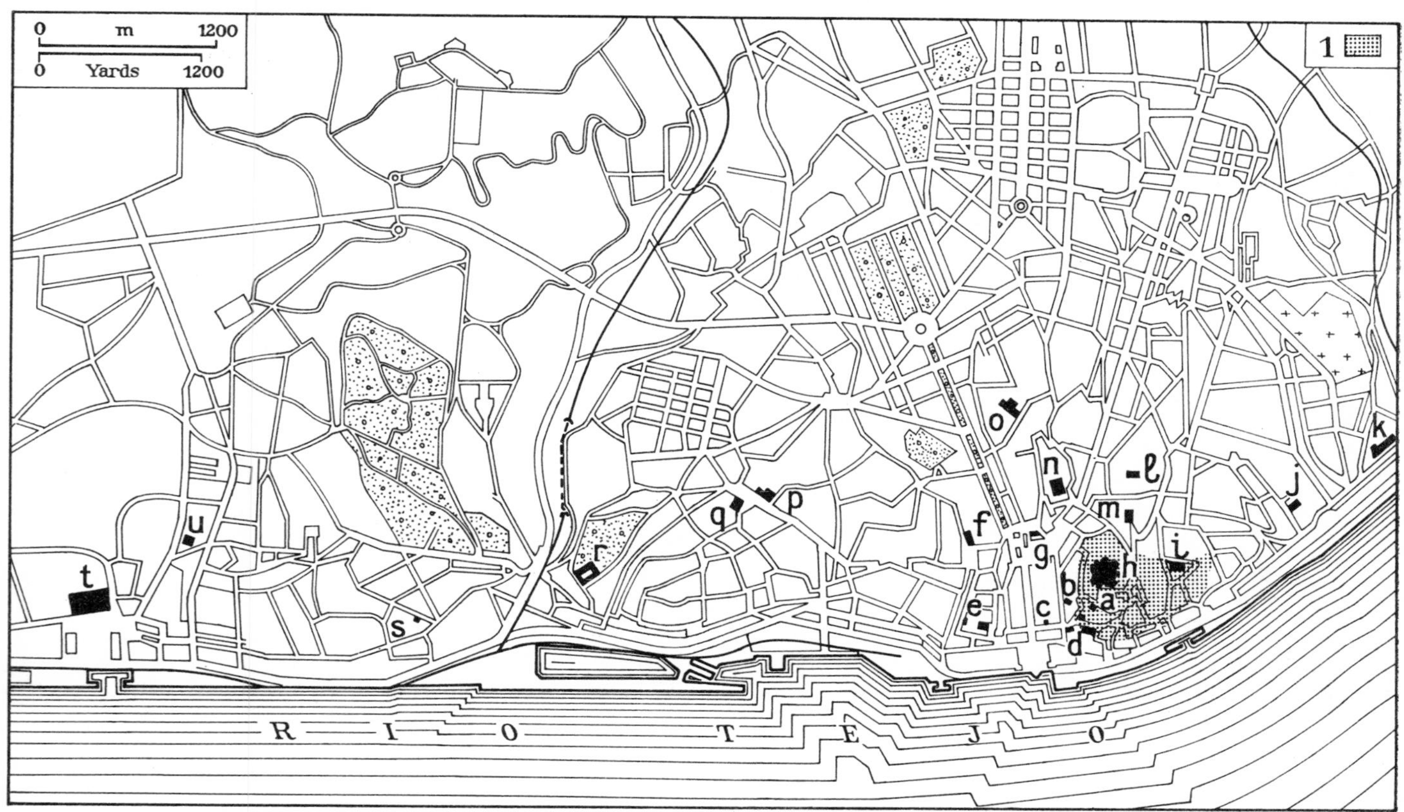

Lisbon, plan. *Key*: (1) Zone of the old Alfama quarter. Principal monuments: (*a*) Remains of Roman theater dedicated to Nero; (*b*) remains of baths of the Cassii; (*c*) remains of Baths of the Augusti; (*d*) Cathedral (Sé), Church of the Magdalen, and S. António da Sé; (*e*) Teatro Nacional de S. Carlos and Museu de Arte Contemporânea; (*f*) Church of S. Roque and Museu de Arte Sacre; (*g*) Church of S. Domingos, near the Praça Rossio (Praça D. Pedro IV); (*h*) Castelo de S. Jorge; (*i*) S. Vicente de Fora; (*j*) Sta Engrácia; (*k*) Maria Pia Asylum and Monastery of Madre de Deus; (*l*) Church of Nossa Senhora del Monte; (*m*) Church of Graça; (*n*) Hospital of S. José (former monastery of S. Antão); (*o*) Hospital of S. António dos Capuchos; (*p*) Military Hospital (former convent of S. João de Deus); (*q*) Basilica of Estrêla; (*r*) Palácio das Necessidades; (*s*) Church of S. Amaro; (*t*) Monastery of the Hieronymites (Jerónimos), at Belém; (*u*) Church of Memória.

Hieronymites (Jerónimos), in the Belém quarter of Lisbon, is an expression of the great maritime achievements that marked the climax of the era of conquest and discovery. The interior of the monastery church consists of three aisles of equal height (i.e., of the *Hallenkirche* type) separated by two rows of pillars carved with reliefs. On the west end is the choir, situated above two chapels, with a vault of Manueline tracery. The arms of the transept terminate in baroque sanctuaries. The apsidal chapel was rebuilt by Jerónimo de Ruão and completed in the third quarter of the 16th century. Apart from the over-all structure of the church, the most important features are: the vault of the transept by João de Castilho; the unusually harmonious cloisters by Boytac, João de Castilho, and Diogo de Torralva; the south and west portals by João de Castilho and Nicolas Chanterene, respectively; the upper choir, with choir stalls attributed to Diogo de Carça; the sacristy, which is decorated with 17th-century paneling; and the 17th-century silver tabernacle of the high altar.

The Monastery of Madre de Deus was founded in 1509, during the reign of Manuel I. The original construction commissioned by Queen Leonor, widow of João II, was enlarged during the reigns of kings João III (16th century) and José (18th century). The large classical cloister is perhaps the work of Diogo de Torralva (d. 1566). A national monument, the monastery contains interesting 16th- and 17th-century paintings, carvings, and blue and polychrome tilework.

A gem of Manueline architecture, the Tower of Belém is admired for its tracery and stonework. It was built under the direction of Francisco de Arruda, designated "Master of the Fortress of Restelo" in 1516, who was one of the best stone masons of a great family which over four generations built notable monuments at Batalha, Tomar, and Belém and in Morocco.

The Church of Conceição Velha occupies the site of an earlier church, Nossa Senhora de Misericórdia, which had been built by order of Manuel I and completed under his son, João III. After the Manueline church was almost entirely destroyed by the earthquake of 1755, its fine portico and the Chapel of the Most Holy Sacrament were incorporated into the newer structure, which was built during the reign of King José (1750–77).

Of the 16th-century Casa dos Bicos (meaning "points"), or "House of Diamonds" as it was called in the 17th century, only two stories remain. Built for Braz son of the famous Afonso de Albuquerque, the faceted stonework of the façade resembles that of the Casa de los Picos in Segóvia and the Palazzo dei Diamanti in Ferrara (cf. XII, PL. 12).

The Church of S. Roque, designed by Filippo Terzi, was built about 1570, and its façade was reconstructed after the 1755 earthquake. The single-aisle interior is richly decorated with marble, mosaics, gilded carvings, and *azulejos* (glazed tiles). The Chapel of S. João Baptista, by Salvi and Vanvitelli, was commissioned from Rome in 1742 by João V and arrived in Lisbon in 1747; it was installed in its present position two years later.

Apart from the rich collection of mid-18th-century Italian art (of marble, mosaic, silver, fabric, etc.) in the chapel, the outstanding decorative features of the church and sacristy include 16th- 17th- and 18th-century paintings (by Gaspar Dias, André Reinoso, José de Avelar Rebelo, Bento Coelho de Silveira, Vieira Lusitano, and André Gonçalves), 16th- and 17th-century tombs, gilded carvings, marble mosaics, silver, and polychrome tiles by Francisco de Matos (dated 1584) and Sevillian tiles (late 16th century).

Other notable churches of the 16th and 17th centuries are: S. Vicente de Fora (1582–1627), a typical work of Terzi, which was completed after the architect's death under the direction of João Nunes Tinoco (or Tinouco) and Leonardo Turriano (XIII, PL. 131); and Sta Engrácia, begun in 1682 and never completed, which was designed by João Antunes (see BAROQUE ART, cols., 307–08).

In the 18th century, and especially after the earthquake of 1755, important secular construction was undertaken in Lisbon. The Aqueduct of Águas Livres (1729–48), designed by Manuel da Maia and carried to completion by other engineer-architects, is, after Mafra (see below), the most ambitious of the works commissioned by João V. Praça do Comércio (Commerce Square) — still commonly called by its old name, Terreiro do Paço — was laid out on the site of the square before the old palace, which, along with adjacent buildings, was destroyed by the earthquake. The square is surrounded by public buildings and has in its center an equestrian statue of King José Manuel by Machado de Castro and in the middle of its north side a triumphal arch, begun in 1755 and completed in 1873, which is surmounted by statues by Vítor Bastos (1830–94).

Other prominent buildings from the 18th century are Palácio Foz (begun 1755) by the Italian architect Francisco Xavier Fabri, which now houses the National Secretariat for Information, Popular Culture and Tourism; and the Teatro Nacional de S. Carlos (1792), by José da Costa e Silva, whose plans were inspired by the original Teatro San Carlo in Naples.

The Basilica da Estrêla, founded by Queen Maria I, was built between 1779 and 1790 after designs by Mateus Vicente de Oliveira and Reinaldo Manuel dos Santos, disciples of Ludovice at Mafra. The interior and the façade are decorated with works by the sculptor Joaquim Machado de Castro, and the chapels and high altar with panels by Pompeo Girolamo Batoni.

According to tradition, the Church of S. António da Sé stands on the site of the house in which St. Anthony of Padua was born. The present building, which replaced a 15th-century chapel destroyed by the 1755 earthquake, was designed by the architect Mateus Vicente de Oliveira and finished in 1812. Of special interest are the main chapel and the panel paintings by Pedro Alexandrino Carvalho.

The Palácio Nacional da Ajuda, the last residence of Portuguese monarchs before the proclamation of the republic in 1910, was initially designed by Fabri, whose plan was modified by the Portuguese architects who succeeded him. The building, which was begun in 1802 but never finished, houses sculpture and paintings by famous Portuguese and foreign artists of the 19th century as well as a valuable collection of furniture, tapestries, china, crystal, bronzes, miniatures, and so on.

The Teatro Nacional Dona Maria II was built between 1842 and 1846 from plans by the Italian architect Fortunato Lodi. Most of the statues and reliefs on the exterior were designed by António Manuel de Fonseca; the ceiling of the auditorium is the work of the painter Columbano Bordalo Pinheiro (1857–1929). The Câmera Municipal was built between 1867 and 1875 to replace the old municipal building, destroyed by fire in 1863; it was designed by Domingos Parente da Silva (1836–1901). Notable features of the interior include the staircase, with a gallery painted in grisaille by the French decorator Pierre Bordes, and the various halls containing works by 19th-century Portuguese painters and sculptors.

A prominent 20th-century landmark is the Monument of the Discoveries, designed by the architect José Angelo Cottinelli Telmo on the occasion of the Historical Exposition of the Portuguese World (1940), with sculptures by Leopoldo de Almeida (b. 1898); it is situated on the Tejo River, in front of the Monastery of the Jerónimos, from where the ships of Vasco da Gama sailed to discover the sea route to India in 1498.

Museums. The Museu Etnológico Leite de Vasconcelos has departments of archaeology, anthropology, and ethnography relating to Portugal, overseas territories, and other areas. The Museu Arqueológico do Carmo, located in the remains of the monastery church of the same name, is a repository of art works dating from Iberian prehistoric times into the Middle Ages. The Museu Nacional de Arte Antiga, housed in the 17th-century Palácio das Janelas Verdes and in annexes built in the second quarter of the 20th century, contains valuable collections of Portuguese and overseas art (African, Indian, Chinese, Japanese) with Portuguese themes (sculpture, paintings, drawings, illuminations, furniture, goldwork, ceramics, ivories, tapestries, carpets, etc.), as well as valuable works of the medieval, Renaissance, mannerist, baroque, and neoclassical periods from other European countries (most notably paintings of Italian, Spanish, Flemish, Dutch, German, French, English, and eastern European schools).

The Museu de Arte Sacre, adjoining the Church of S. Roque, has an outstanding collection of Italian sacred art of the mid-18th century, which consists largely of the treasure from the Chapel of S. João Baptista, along with paintings by important Portuguese artists of the 16th century and later. The Museu Nacional dos Coches exhibits, in the riding arena and dependencies of the Palace of Belém (designed by the Italian artist Giacomo Azzolini), an outstanding collection of coaches, as well as paintings, engravings, livery of the Portuguese royal houses, harnesses, and so on.

Other important museums in Lisbon are the Museu de Artes Decorativas (founded in 1953 to house examples of such art from the 16th through the 19th century from the collection of Ricardo Espirito Santo Silva); the Museu de Arte Contemporânea (sculpture and paintings by Portuguese artists from about 1850 to the present); the Museu de Arte Popular (exhibits and materials pertaining to Portuguese folk arts and folkways); the Arquivo Nacional da Tôrre do Tombo (illuminated manuscripts and miniatures); the Museu da Cidade (art works and iconographic and bibliographical materials that document the history of the city); and the Museu de Azulejaria (tiles and tilework of various styles, eras, and places of origin, especially Portuguese *azulejos* of the 16th through the 19th century).

BIBLIOG. E. Freire de Oliveira, Elementos para a historia do Municipio de Lisboa, 16 vols., Lisbon, 1885–1910; F. N. de Faria e Silva, A egreja da Conceição Velha e várias notícias de Lisboa, Lisbon, 1900; F. de Sousa Viterbo and R. Vicente d'Almeida, A Capella de S. João Baptista erecta na egreja de S. Reque, Lisbon, 1900; A. Vidal, Lisboa antiga e Lisboa moderna, Lisbon, 1901; J. Castilho, Lisboa antiga: Bairro Alto, 2d ed., 2 vols., Lisbon, 1902–04; M. J. da Cunha Brandão, As ruinas do Carmo, Lisbon, 1908; F. W. Feilchenfeld, Die Meisterwerke der Baukunst in Portugal, I: Das Kloster "Dos Jeronymos" zu Belem, Vienna, 1908; F. L. Pereira de Sousa, Effeitos do Terremoto de 1755 nas construções de Lisboa, Lisbon, 1909; V. Correia, Notas de arqueologia: Lisboa prehistorico, Lisbon, 1912; V. Ribeiro, A velha Lisboa e os estudos de arqueologia na Capital, Lisbon, 1914; R. dos Santos, A torre de Belem, Lisbon, 1922; A. Coelho Gasco, Das antiguidades da muy nobre cidade de Lisboa, Coimbra, 1924; L. Gonzaga Pereira, Monumentos sacros de Lisboa em 1833 (ed. A. Vieira da Silva), Lisbon, 1927; A. Vieira da Silva, A velha Lisboa, Lisbon, 1927; S. Matos Sequeira and F. Nogueira de Brito, Sé de Lisboa (Monumentos de Portugal, 8), Oporto, 1930; M. Ribeiro, A Sé de Lisboa (A arte em Portugal, 13), Oporto, 1931; R. dos Santos, O mosteiro de Belém (Jerónimos), Oporto, 1931; R. de Sousa Lobo Ramalho, Guia de Portugal artistico: Lisboa, 8 vols., Lisbon, 1933–41; J. Castilho, Lisboa antiga: Bairros Orientais, 2d ed., 12 vols., Lisbon, 1935–38; N. de Araújo, Inventário de Lisboa, Lisbon, 1939; G. de Matos Sequeira, O Carmo e a Trinidade, 3 vols., Lisbon, 1939–41; J. Castilho, A Ribeira de Lisboa, 2d ed., 2 vols., Lisbon, 1941; A. Vieira da Silva, A Cerca Fernandina da Lisboa, 2 vols., Lisbon, 1948–49; M. V. Ferreira de Andrade, Palácios reais de Lisboa, Lisbon, 1949; F. Cancio, Coisas e loisas de Lisboa antiga, Lisbon, 1951; L. Pastor de Macedo and N. de Araújo, Casas da câmara de Lisboa, Lisbon, 1951; J. de Sousa Nunes, Torre de Belem, Lisbon, 1959; R. dos Santos, Historia del arte portugués, Barcelona, Madrid, 1960; I. Baldassarre, EAA, s.v.; N. de Araújo, Pequena monografia de São Vicente, Lisbon, n.d.; A. de Lacerda, Madre de Deus, Barcelos, n.d.

Minho. Barcelos. Notable among the town's predominantly Gothic architectural monuments are the ruins of the palace of the counts of Barcelos (15th–16th cent.), now a museum, and the parish church (13th cent., with 14th and 18th cent. alterations). The central-plan, octagonal church of Bom Jesus, attributed to João Antunes, dates from 1701.

In the environs there are important examples of Romanesque architecture, such as the Church of S. Maria de Abade and that of Manhente, the latter flanked by a crenellated tower, and the Benedictine monastery of Vilar de Frades.

Bibliog. A. M. do Amaral Ribeiro, Noticia descritiva da muito nobre e antiga Vila de Barcelos, 2d ed., Barcelos, 1867; J. Leitão, Guia ilustrado de Barcelos, Oporto, 1908; M. d'Aguiar Barreiros, A egreja de Villar de Frades no concelho de Barcelos, Oporto, 1919; M. d'Aguiar Barreiros, A portada românica de Villar de Frades e o seu symbolismo, Oporto, 1920; F. de Azevedo, O paço dos Condes-duques de Barcelos, Oporto, 1954.

Braga. (anc. Bracara Augusta). Founded by the Romans in the 2d century A.D., it immediately became an important communications center, from which radiated five military roads. Occupied in 583 by the Visigoths and in 710 by the Moors, who destroyed it almost completely, it returned to Christian rule in 1040. In the 16th century Braga experienced a period of great building activity under Archbishop Diogo de Sousa, after whom the Museu Regional is named. Minor remains from the Roman period, including many inscriptions and milestones, have come to light; the Quintal do Idolo, a rock shrine with a spring, two inscriptions, a standing male figure, and a niche containing a bust carved from the rock, is interesting. The Cathedral, begun in the late 11th century, retains part of the original Romanesque structure, though it has been greatly transformed by Gothic and baroque additions; it contains notable examples of funerary sculpture, among which are the 15th-century gilded bronze monument of the Infante Afonso, son of João I, and the tombs of Henry of Burgundy and his wife Teresa, and those of the archbishops Gonçalo Pereira and Diogo de Sousa. The Chapel of Nossa Senhora da Conceição (dos Coimbras), built in 1525, in the form of a tower, has remarkable flamboyant Gothic windows and finely caved granitic statuary on its façade.

Located in the environs of Braga is the important church of S. Frutuoso (see above: *Pre-Romanesque art*), built in the 7th century by the Visigoths and the only Byzantine church in Portugal.

Bibliog. M. d'Aguiar Barreiros, A Cathedral de Santa Maria de Braga, Oporto, 1922; A de Lacerda, A Capela de Nossa Senhora da Conceição (em Braga), Oporto, 1922; M. d'Aguiar Barreiros, Egrejas e capelas românicas da Ribiera Lima, Oporto, 1926; M. d'Aguiar Barrieros, Braga monumental (A arte em Portugal, 2), Oporto, 1927; A. Feio, Bom Jesus do Monte, Braga, 1930; M. Monteiro, S. Fructuoso, Braga, 1939; Guia de Braga: Arte e turismo, Braga, 1959; J. M. Bairrão, EAA, s.v. Bracara.

Guimarães. The birthplace of Afonso I, first king of Portugal, Guimarães contains notable medieval monuments. Outstanding among these are: the Romanesque basilica of Nossa Senhora da Oliveira, founded in the 10th century and rebuilt at the end of the 14th century, which still retains the original tower and the chapter house (now the Museu Regional de Alberto Sampaio); the imposing castle and adjacent Romanesque church of S. Miguel do Castelo (12th cent.), a small single-aisle church with chancel; the Gothic palace of the dukes of Brangança (15th cent.); and the Monastery of S. Domingos (founded 1271), in the cloisters and annexes of which are housed the collections of the Museu de Martins Sarmento, consisting of finds from the nearby Iberian sites (citânias) of Briteiros and Sabroso (see above: *Prehistory and antiquity*).

In the environs of Guimarães there are two interesting examples of Romanesque ecclesiastical architecture: the chapel of the old monastery at São Torcato, now the parish church, and the church of Serzedo, built for the Templars.

Bibliog. A. Guimarães, Guimarães monumental (A arte em Portugal, 11), Oporto, 1930; L. de Pina, O castelo de Guimarães, Gaia, 1933; A. Guimarães, Guimarães, 2d ed., Guimarães, 1953.

Viana do Castelo. Among the town's important monuments are the Church of the Misericórdia, built in Renaissance style by the 16th-century architect João Lopes, and the palace of the Counts of Carreira. The Museu Municipal contains a library, paintings, ceramics of various regional types, archaeological finds, and other objects of art-historical interest.

Bibliog. L. de Figueiredo da Guerra, Guia de Viana do Castelo, Viana do Castelo, 1923; L. de Figueiredo da Guerra, Viana e caminha (A arte em Portugal, 7), Oporto, 1929.

Trás-os-Montes and Alto Douro. Bragança (Braganza). Of medieval origin, it was the seat of the ducal house that ruled Portugal from 1640 to 1910. The Castelo (late 12th cent.), in ruins, is enclosed by the walls of the old fortress and has a very tall keep. The Domus Municipalis (Town Hall; XIII, pl. 120), dating from about 1200 and resting on pillars over a reservoir, is one of the few surviving examples of civil architecture from the Romanesque period in Portugal. The Museu Regional Abade de Baçal contains a collection pertaining to local history and archaeology.

Bibliog. F. M. Alves, Memórias arqueológico-historicas do distrito de Bragança, 7 vols., Oporto, 1909–31.

Chaves (anc. Aquae Flaviae). The town was founded as a thermal spa by the Romans, who in the time of Trajan built the 460-ft. arched bridge which still spans the Tamega River. Much archaeological material has been recovered, including numerous inscriptions. Important examples of military architecture include the Fort of S. Francisco and the 15th-century castle of the dukes of Bragança.

The Romanesque church of Nossa Senhora da Azinheira, in the environs of Chaves, is perfectly preserved and contains noteworthy Renaissance frescoes (1533). Near Abobeleira there is a rock covered with more than 350 prehistoric engravings of stylized anthropomorphic figures, axes, and other designs.

Bibliog. M. A. Rodrigues, Guia-album de Chaves e seu concelho, Oporto, 1915; J. M. Bairrão Oleiro, EAA, s.v.

Lamego. The remains of a massive 13th-century castle overlook the town. The Cathedral, one of the most important in Portugal from the art-historical point of view, has a Gothic façade and a 12th-century Romanesque tower; the Renaissance cloister dates from 1557. The Museu Regional, in the old episcopal palace, contains the five surviving panels painted by Vasco Fernandes ("Grão Vasco"; d. between 1541–43) for the high altar of the Cathedral.

The small pre-Romanesque church of S. Pedro de Balsemão, situated in the environs of Lamego, retains its original basilican plan and a Visigothic chapel of the 7th century.

Bibliog. J. de Azevedo, Historia eclesiastica da cidade e bispado de Lamego, Oporto, 1877; V. Correia, Vasco Fernandes: Mestre do retabulo da Sé de Lamego, Coimbra, 1924.

Vila Real. The Gothic church of S. Domingos, formerly part of a Dominican monastery, was begun in the late 14th or early 15th century. The form of the capitals and the proportions of the columns reveal late Romanesque influences. In the cemetery stand the ruins of the Romanesque chapel of S. Brás.

At Mateus, about 2 miles away, there is a fine 18th-century palace in the Louis XV style.

Bibliog. J. A. Aires de Azevedo, Origens de Vila Real, Coimbra, 1899.

Douro Litoral. Amarante. In this old town, located on Tamega River, many picturesque buildings of the 16th-18th centuries are preserved. The former convent church of S. Gonçalo (founded 1540), with a Renaissance cloister, contains important 16th- and 18th-century paintings and sculpture.

In the vicinity of Amarante there are fine Romanesque churches at Freixo de Baixo, Gâtão, and Marco de Canaveses.

Bibliog. F. de Alpoim e Menezes, Historia antiga e moderna da sempre leal e antiquissima Villa de Amarante, London, 1814.

Anreade. The Church of S. Maria de Cârquere was once part of a convent built in 1099, but only the funerary chapel dates back to the time of its foundation. The nave is in Manueline style, and the choir is 13th-century Gothic.

Arouca. The 12th-century monastery was partly rebuilt in the 17th and 18th centuries. The monastery church houses the crystal and silver tomb (1734) of Queen Mafalda, wife of Henry I of Castile (r. 1214–17).

BIBLIOG. B. de Brito, História de fundação e dedicação do mosteiro de S. Paulo de Arouca, in F. de São Boaventura, Memórias para a vida da Beata Mafalda, Coimbra, 1814, pp. 213–345.

Oporto (Port., Pôrto; anc. Portus Cale). Of pre-Roman origin, the city was a flourishing Roman port under the Empire. After a period of occupation (540–997) by the Visigoths and then the Moors, it became the seat of the counts of Portugal for most of the period of the Christian reconquest of southern Portugal. Now the nation's second largest city, Oporto spreads in a series of concentric rings, but with an irregular street plan, along the broad granite bluffs on the right bank of the Douro River. The so-called "walls of Dom Fernando," which circumscribe the area of the medieval city, were actually begun under Afonso IV in 1336 and completed in 1374. Tall, narrow granite houses with gardens between them are characteristic of the older sections.

The Cathedral (Sé) was built in the 12th and 13th centuries, but the original Romanesque structure has undergone many alterations, especially in the 17th and 18th centuries. The lower part of the façade towers and the *rosace* above the main portal remain from the initial construction. On the north side there is a loggia attributed to the Italian architect Niccolò Nasoni (Nazzoni; d. 1773), who was active and influential in the city. He was also responsible for the rebuilding and enlargement of the former episcopal palace, adjacent to the Cathedral, in 18th-century baroque style.

The typical, small, single-aisle Romanesque church named the Cedofeita (recently restored) is known to have existed in 1120, though it may have been founded much earlier. The Church of S. Francisco, founded in 1233 and rebuilt in the years 1383–1410, is mainly Gothic with Romanesque traces. The carved and gilded woodwork of the interior (in the vaults and around the columns) are baroque additions of the 17th and 18th centuries.

Other notable churches are: S. Clara, founded in 1416 and rebuilt (early 16th cent.) in the late Gothic style, blending Manueline and Renaissance motifs; S. Lourenço, more commonly known as the Grilos, a fine work of the late 16th and early 17th century, with a sumptuously ornamented interior; S. Bento da Vitória, built between 1597 and 1646 after plans by Diogo Marques Lucas, a student of Terzi; the Congregados, a baroque church of the late 17th and early 18th century, containing fine examples of goldwork; S. Pedro dos Clérigos (cf. II, FIG. 281) and its high tower (Tôrre dos Clérigos), the city's most prominent landmark, built 1732–48 according to designs by the above-mentioned Nasoni, who was the chief exponent of the baroque in northern Portugal; and the Misericórdia, built in 1750, with a façade that is also by Nasoni.

Two important secular buildings from the architectural standpoint are the Hospital of S. António (1770–95), designed by the English architect John Carr, and the Bolsa (stock exchange), a neoclassic edifice built in the mid-19th century from plans by Thomas A. Soller.

The Museu Nacional de Soares dos Reis has important and extensive collections of archaeological material, goldwork, faïence and Portuguese and foreign paintings and sculpture. The Casa-Museu de Guerra Junqueiro, housed in a building erected between 1730 and 1776 and probably designed by Nasoni, contains the valuable collection of art works (mainly 15th–18th cent.) that belonged to the poet for whom the museum is named, including stone and wood sculpture, furniture, metalwork, tapestries, faïence, glass and crystal, ivories, and engravings. The Museu de Etnografia e História has a rich and varied collection of local material, including handicrafts.

The palace at Quinta do Freixo, near Oporto, is also attributed to Nasoni and remains a most beautiful example of 18th-century secular architecture. Also in the vicinity of the city are the fortresslike former monastery church at Leça do Bailio, dating from 1336, and the circular-plan monastery church of Nossa Senhora do Pilar at Vila Nova de Gaia, with its unusual round cloister, which was begun in 1598.

BIBLIOG. C. de Passos, Porto: Noticia historico-archeologica e artistica da Cathedral e das egrejas de Santa Clara, S. Francisco e Cedofeita, Oporto, 1926; J. A. Ferreira, Porto (A arte em Portugal, 1), Oporto, 1928; C. de Passos, Guia histórica e artística do Porto, Oporto, 1935; A. de Magalhães Basto, Silva de história e arte (notícias portucalenses), Oporto, 1945; G. B. Barreiros, Aspectos arqueológicos e artísticos da cidade do Porto, Oporto, 1949; M. J. Rodrigues Monteiro, Igrejas medievais do Porto, Oporto, 1954.

Rates. The Church of S. Pedro (XIII, PL. 118), built in the 12th century by order of Henry of Burgundy, is one of Portugal's most significant monuments from the Romanesque period. It has sculptural decoration of particular note.

BIBLIOG. M. Monteiro, S. Pedro de Rates, Oporto, 1908.

Santo Tirso. In the former Monastery of S. Bento (now a school) there is a Romanesque church with a single aisle nave and rectangular apse. The Gothic arches of the 14th-century cloister are supported by double columns crowned by capitals carved with fantastic animal figures.

About 7 miles distant, at Roriz, there stands one of the most beautiful Romanesque churches in Portugal, S. Pedro, with a perfectly preserved 12th-century choir.

BIBLIOG. A. Pimentel, História de Santo Tirso da Riba d'Ave, Santo Tirso, 1902.

Vila da Feira. The Castle, standing in the midst of luxuriant vegetation, is regarded as a notable example of medieval fortification. The oldest part dates from the 11th century, but it has been restored at various times, particularly in the 15th and 16th centuries. Access to the rectangular donjon, flanked by four turrets, is provided by an iron ramp.

BIBLIOG. F. de Tavares e Tavora, O Castello da Feira, Oporto, 1917.

Vila do Conde. The Church of S. Clara, dating from the 14th century, has a fine Manueline chapel (1526) and contains some exceptional 15th-century tombs.

BIBLIOG. J. A. Ferreira, Villa do Conde a seu alfoz, Oporto, 1923; J. A. Ferreira, Os tumulos do mosteiro de Santa Clara de Villa do Conde, Oporto, 1925; J. A. Ferreira, Villa do Conde (A arte em Portugal, 3), Oporto, 1928.

Beira Alta. Figueira de Castelo Rodrigo. The church of the nearby Cistercian Monastery of S. Maria de Aguiar, founded prior to the 13th century, is of particular art-historical interest as an early illustration of the use of Gothic forms in Portugal. It is a spacious and well-proportioned structure, with nave and side aisles. The chapter house dates from the same period.

BIBLIOG. J. Couto, O convento de Santa Maria de Aguiar em Riba-Côa (Termo de Castelo Rodrigo), Oporto, 1927.

Guarda. Situated over 3,000 feet above sea level, Guarda is the highest city in Portugal. The Cathedral, erected between the late 14th and mid-16th centuries, is a majestic edifice in late Gothic style. Two three-tiered towers flank the façade. On the west side there is a fine Manueline portal. The Cathedral contains an important retable, which was begun by João de Ruão about 1552. The remains of a 14th-century castle are also interesting. The Museu Regional occupies part of the former episcopal palace.

BIBLIOG. J. de Almeida, A Guarda, capital da Beira, Lisbon, 1937; J. de Almeida, Roteiro dos monumentos de arquitectura militar do concelho da Guarda, 2d ed., Lisbon, 1940.

Oliveira do Hospital. The Capela dos Ferreiros in the parish church houses the notable 13th-century tombs of the nobleman Domingos Joanes and his wife, as well as an interesting polychrome retable of about the same period.

At Lourosa, approximately 6 miles away, the pre-Romanesque church, with a basilican plan, dates from the early 10th century, though it was completely restored in 1931–32. The nave and side aisles are divided by horseshoe arches, showing Moorish influence.

BIBLIOG. A. de Abreu, Oliveira do Hospital: traços historico-críticos, Coimbra, 1893.

Trancoso. A large medieval castle rises above this historic village. The massive structure, surrounded by a double wall, is in an excellent state of preservation.

BIBLIOG. D. B. Soares Moreira, Terras de Trancoso, Oporto, 1932.

Viseu. The Cathedral, dating from the 12th century, is one of the most imposing in Portugal in terms of its size and its position in the surrounding city. The interior, comprised of nave and side aisles, has the original 12th-century grouped columns and an extraordinary Manueline vaulted ceiling, which was completed in 1513. The Museu de Grão Vasco in the old seminary building next to the Cathedral contains an important collection of Portuguese paintings from the 14th to the present century.

Bibliog. F. de Almeida Moreira, Os quadros de Sé de Viseu, 2d ed., Oporto, 1925; M. de Aragão, Viseu, 2 vols., Oporto, 1928; Guia da cidade de Viseu, Oporto, 1932; F. de Almeida Moreira, Imagens de Viseu, Oporto, 1937; L. Reis Santos, Vasco Fernandes e os pintores de Viseu do século XVI, Lisbon, 1946; A. de Lucena e Vale, Viseu monumental e artistico, Viseu, 1949.

Beira Baixa. Castelo Branco. The old quarter, which is remarkably unchanged, surrounds the remains of a castle built by the Templars. The Museu Regional exhibits an important collection of archaeological materials (prehistoric, Roman, and medieval), as well as paintings and tapestries.

In the environs of the city, there is a 15th-century Gothic chapel, Nossa Senhora de Mércoles. Excavations at the nearby site of a fortified settlement, Castro de S. Martinho, have yielded archaeological finds from both prehistoric and Roman times.

Bibliog. J. A. P. da Silva, Memorial chronologico e descriptivo da cidade de Castelo Branco, Lisbon, 1853; A. Roxo, Monographia de Castelo Branco, Elvas, 1891; F. Tavares Proença, Jr., Archeologos do districto de Castelo Branco, Leiria, 1910.

Beira Litoral. Aveiro. The church of the Monastery of Jesus (begun 15th cent.) is notable for its carved decorations of gilded wood — especially that of the high altar — installed in the 17th and 18th centuries. The magnificent marble-mosaic tomb of the beatified infanta Joana, daughter of Afonso V, in the lower choir is an early-18th-century work. Influences from the monastery church can be seen in the sumptuous baroque decoration (18th cent.) of the Carmelite church.

The Museu Municipal of Ilhavo (about 4½ mi. away) has an extensive collection of porcelain and crystal ware from the famous factories of neighboring Vista Alegre. The tomb of Manuel de Moura Manuel, Bishop of Miranda, in the chapel at Vista Alegre, is a renowned work by the French sculptor Claude de Laprade (active 1699–1730).

Bibliog. J. A. Marques Gomes, O districto de Aveiro, Coimbra, 1877.

Batalha. The Monastery of S. Maria da Vitória, begun in 1388, is an exemplary work in the Portuguese late Gothic style, the evolution of which can be seen in the various stages of enlargement and embellishment under successive monarchs and masters (see discussion above, *Gothic art*). The complex consists of six main architectural units: (1) The church, with a façade (XIII, pl. 120) suggestive of the English Perpendicular style and a deep sculptured portal of seemingly Burgundian inspiration, has a lofty central nave and side aisles, with ogival vaulting and clerestory. (2) Also belonging to the first stage of construction (1388–1438), the Chapel of the Founder (which contains, among other funerary monuments, the remarkable tomb of João I and Philippa of Lancaster) has at its center an octagonal area circumscribed by pillars, above which there is an octagonal lantern. (3) The Chapter House is especially remarkable for its star vault, which spans, without supports, an area over 60 ft. square. (4) The Unfinished Chapels (Capelas Imperfeitas), begun about 1435 as a mausoleum for King Duarte and his family, consist of seven radial chapels around a rotunda; work was resumed under Manuel I (r. 1495–1521), who commissioned the never-completed decoration of the walls and portals — most notably, the main portal (1509), over 50 ft. in height, designed by Mateus Fernandes I — in the style that now bears this monarch's name. (5) The Royal Cloister begun in the second stage of building (1438–81), was subsequently completed in the highly ornamental Manueline style. (6) The Cloister of Afonso V, completed in the second stage, is a pleasing work in the unadorned Gothic style, in striking contrast to the remainder of the monastery.

Bibliog. W. Beckford, Recollections of an Excursion to the Monasteries of Alcobaça, and Batalha, London, 1835; Visconde de Condeixa, O mosteiro da Batalha em Portugal, Paris, 1892; L. da Silva Mousinho de Albuquerque, Memoria inedita acerca do edificio monumental da Batalha, Lisbon, 1897; V. Correia, Batalha (Monumentos de Portugal, 1, 9), 2 vols., Oporto, 1929–31; P. Vitorino, Mosteiro da Batalha (A arte em Portugal, 12), Oporto, 1930; J. das Neves Larcher, Monumentos de Portugal, Alcobaça e Batalha, 3d ed., Oporto, 1932.

Coimbra (anc. Aeminium). Of Roman origin, the city acquired its present name from ancient Conimbriga (see below), which was situated at nearby Condeixa-a-Velha. During the Middle Ages, it became an important center under the Visigoths, and under the first Portuguese rulers it was the capital of the kingdom. Portugal's oldest university has been permanently located at Coimbra since 1537.

The Sé Velha (Old Cathedral) is one of Portugal's finest churches of the pilgrimage basilica type in the Auvergne-Peninsular style. The present building was erected between 1162 and 1176. Aside from the over-all plan and architectural detail (XII, pl. 215), among the especially noteworthy features are the medieval tombs; a Manueline baptismal font, perhaps the work of Diogo Pires the Younger; the Gothic retable of the high altar, in gilded and polychrome wood, an early-16th-century work by Olivier de Gand; and the Chapel of St. Peter and Chapel of the Sacrament (1566), which are fine works in the Portuguese-French style of the designers and executors, Jacques Loguin (or Loquin), João de Ruão (Jean de Rouen), and Tomé Velho. On the north face there is a beautiful portal in the Renassaince style. A late-13th-century Cistercian cloister, a work of great merit, is situated on the south side.

The Monastery of Sta Cruz was founded by the Augustinians in the reign of Afonso I (1139–85), who is buried there. The present church and the cloister were designed and built under the direction of Boytac and Marcos Pires during the reign of Manuel I (1495–1521). Under his successor, João III, it was further enriched and improved by Diogo de Castilho and João de Ruão. Also notable are the tombs of the founders of the kingdom, Afonso Henriques and Sancho I, flanking the high altar; the pulpit in the nave; the Claustro (or Jardim) da Manga; the chapels and reliefs at the angles of the Claustro de Silêncio; the chapels of Christ, St. Michael, and St. Teotónio in the Portuguese-French style of Chanterene, Loguin, João de Ruão, and Tomé Velho; and the sacristy (1662), attributed to the architect Pedro Nunes Tinoco and decorated with panels by the 16th-century painters Vasco Fernandes (Grão Vasco), Cristóvão de Figueiredo, and Garcia Fernandes and 18th-century paintings by André Gonçalves. In the upper choir there are Manueline choir stalls, attributed to the master João Alemão and extended in 1531 by the French carver Francisco Loreto, and a valuable collection of Persian rugs. The 18th-century sanctuary contains interesting statuary, painting, goldwork, relics, and other objects. Throughout the monastery there are *azulejos* dating from the 16th to the 18th century.

The Church of S. Salvador was most likely erected in the 11th century and was rebuilt in the following century. There are funerary chapels and a retable in Renaissance style. The Church of S. Tiago was built in the 12th century, consecrated in 1206, and restored in the second quarter of the 20th century. The portals are ornamented with Romanesque sculpture of the Coimbra school. The interior, consisting of nave and side aisles, is very subdued. A lateral chapel, of the 15th century, is in the flamboyant Gothic style. The Monastery of Celas was founded at the beginning of the 13th century and was enlarged and restored during the reign of João III (1521–57). In the cloister there are remarkable historiated capitals of the late 13th and the early 14th century. The Church of S. Clara-a-Velha, completed in 1330, is an example of the beginnings of the Gothic style in Portugal. After having been inundated by the waters of the Mondego River, the church was abandoned in the last quarter of the 17th century and remains empty. The *arcosolium* in the south aisle was built to shelter the tomb of St. Isabel, Queen of Portugal (d. 1336).

Other ecclesiastical monuments of particular note are the Sé Nova (New Cathedral; 16th–18th cent.), a spacious and severe Jesuit church with an imposing windowed façade; the baroque convent church of S. Isabel (1649–96), popularly known as S. Clara-a-Nova, designed by João Turriano, and the neoclassic cloister adjacent to it; the collegiate churches of Graça and Carmo (founded in 1541 and 1548, respectively); the Colégio Novo (begun 1593), with a cloister by Terzi; and the Seminário (ca. 1745–65).

Of the palaces (Paços das Escolas) that constituted the original nucleus of the University of Coimbra the principal surviving works are the Manueline university chapel (1517–21), by Boytac and Marcos Pires, with a beautiful double entrance portal and pulpit, paintings, *azulejos*, and goldsmith's work of the 16th–18th centuries; the Sala dos Capelos (the hall in which degrees are conferred), restored and richly decorated between 1654 and 1658; and the Claustro dos Gerais, with sculptural decoration (1699–1702) by Claude de Laprade and rare *azulejos* of the 17th-century type. Among the 18th-century works are the magnificent old general library, a neoclassical building commissioned by João V with interior decoration by Laprade (XIII, pl. 137); the tower, known as Tôrre da Cabra, completed in 1733 and attributed to J. F. Ludovice, and the colonnaded gallery called the Via Latina. Beyond the Porta Férrea (1634) lies the New University City, including a general library and the buildings of the faculty of letters and medicine, which were completed in the 1950s.

The Aqueduct of S. Sebastião, which probably dates to Roman times, was restored under the direction of Terzi between 1568 and 1570. The Palace of Sub-Ripas, with its Manueline doorway, remains a fine example of 16th-century secular architecture.

The Museu de Machado de Castro, which occupies the former episcopal palace, contains one of Portugal's finest and most extensive collections of paintings, sculpture, gold- and silverwork, and tapestries.

Bibliog. A. C. Borges de Figueiredo, Coimbra antiga e moderna, Lisbon, 1886; T. Braga, Historia da Universidade de Coimbra, 4 vols.,

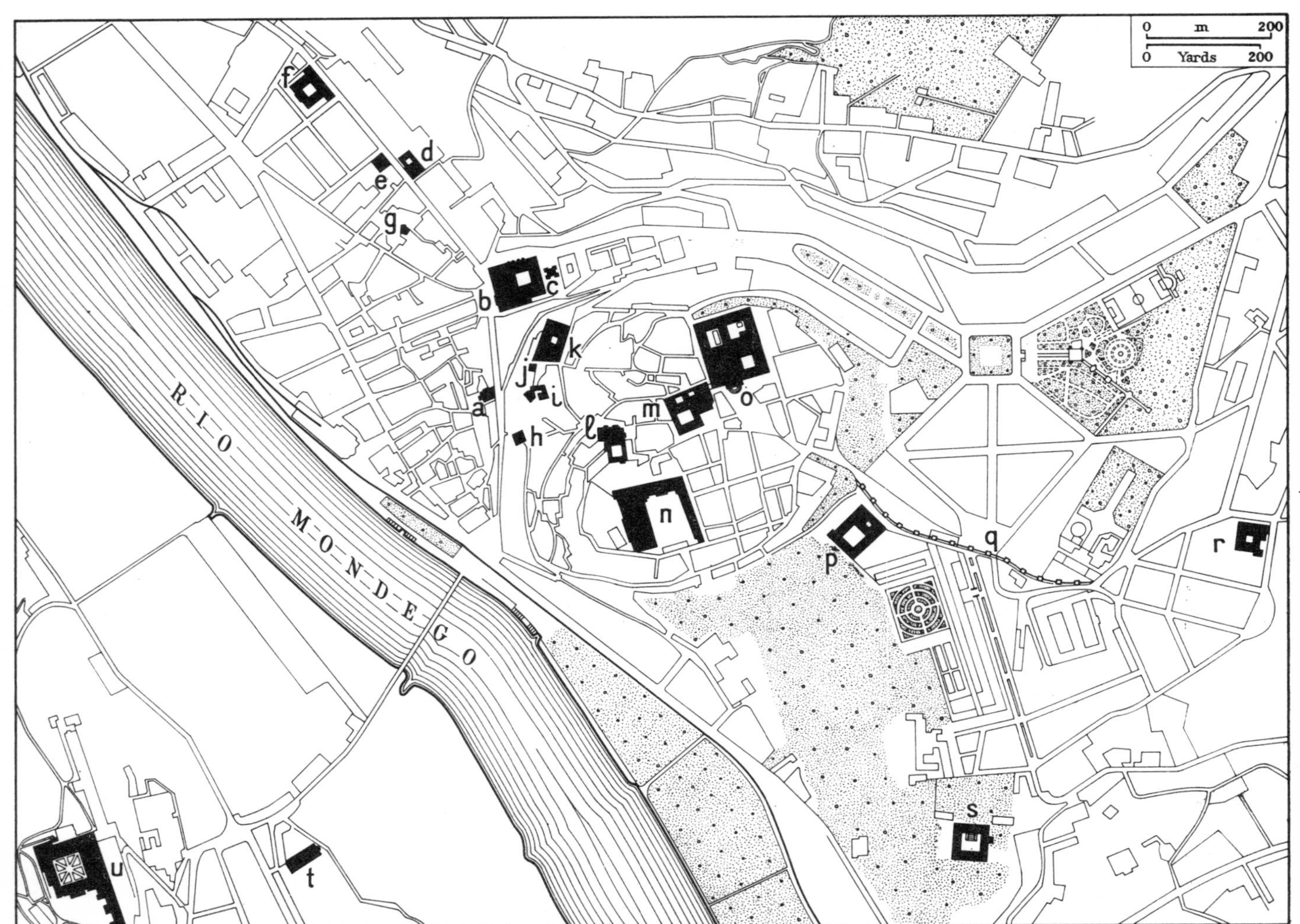

Coimbra, plan, showing principal monuments. *Key*: (*a*) Church of S. Tiago; (*b*) Monastery of Sta Cruz; (*c*) Claustro (Jardim) da Manga; (*d*) Church of Carmo; (*e*) former church of S. Domingos (with Capela do Tesoureiro); (*f*) College of S. Tomás (Palace of Justice); (*g*) remains of the church of S. Justa-a-Velha; (*h*) Arch of Almedina and Tower of Rolaçom; (*i*) Palace of Sub-Ripas; (*j*) Tower of Anto; (*k*) Colégio Novo; (*l*) Sé Velha (Old Cathedral); (*m*) Museu de Machado de Castro; (*n*) University; (*o*) Sé Nova (New Cathedral); (*p*) College of S. Bento; (*q*) Aqueduct of S. Sebastião; (*r*) Convento das Teresinhas; (*s*) Seminário; (*t*) Church of S. Clara-a-Velha; (*u*) Convent of S. Clara-a-Nova.

Lisbon, 1892–1902; F. de Sousa Viterbo, O mosteiro de Santa-Cruz de Coimbra, 2d ed., Coimbra, 1914; A. A. Gonçalves, Coimbra (A arte em Portugal, 5), Oporto, 1929; V. Correia, Coimbra romana, Coimbra, 1930; A. Garcia Ribeiro de Vasconcelos, Sé velha de Coimbra, 2 vols., Coimbra, 1930–35; A. Nogueira Gonçalves, Arquitectura românica de Coimbra, Coimbra, 1939; P. Mereia, Sôbre as origens do concelho de Coimbra, Coimbra, 1940; A. Nogueira Gonçalves, A frontaria românica da igreja de Santa Cruz de Coimbra, Coimbra, 1940; A. de Amorim Girão et al., Coimbra, 2d ed., Coimbra, 1942; F. Martins, A porta do sol; Contribução para o estudo da cerca médieval Coimbrã, Biblos, XXVII, 1951, pp. 321–59.

Conimbriga. This site at Condeixa-a-Velha, inhabited at least since the Bronze Age, was occupied by the Romans, probably in the mid-2d century B.C., and became a flourishing city under the Empire. The city wall, which has survived intact, is almost triangular in form, and a north-south sector (probably built in the second half of the 3d century of our era) passes through the center of the site. Within the walls, thermae and a complex of commercial character (*tabernae*) have been excavated. Outside the older defensive line, there are remains of an aqueduct and a *domus*. Fine mosaics have also come to light. The greater part of the recovered material is preserved in the Museu de Machado de Castro in Coimbra.

BIBLIOG. V. Correia, Conimbriga: Noticia do oppidum e das escavações nele realizadas, Coimbra, 1935 (new ed., 1952); V. Correia, Las mas recientes excavaciones romanas de interes en Portugal: la ciudad de Conimbriga, AEA, XIV, 1940–41, pp. 257–67; J. M. Bairrão Oleiro, Conimbriga e alguns dos seus problemas, Humanitas, IV, 1952, pp. 32–41; J. M. Bairrão Oleiro, EAA, s.v., Conimbriga.

Góis. The parish church contains the tomb of Luís da Silveira (1531) and other notable Renaissance sculptures.

Located near the neighboring village of Arganil, the Church of S. Pedro is an interesting structure of the transition period from the Romanesque to Gothic style.

BIBLIOG. J. Afonso Baeta Neves, Noticia historica e topografica da vila de Goes e seu termo, Lisbon, 1897; V. Correia, Um tumulo renascença: A sepultura de Dom Luis da Silveira em Goes, Coimbra, 1921.

Montemor-o-Velho. Within the fortifications of the 11th-century castle stands the Church of S. Maria de Alcáçova, an early Gothic structure with nave and side aisles (the Moresque *azulejos* were added in the 16th century). The Church of Nossa Senhora dos Anjos, built in the early 16th century, is particularly noteworthy for the tomb of Diogo de Azambuja and for its altar with a sculptured Pietà (ca. 1542). All the above-mentioned monuments have been greatly restored.

BIBLIOG. Guia de Portugal, III, Lisbon, 1944, pp. 141–51.

Estremadura. Alcobaça. The Monastery of S. Maria, one of the principal Cistercian abbeys of the Iberian Peninsula, was begun about 1152 and essentially completed in the 13th century (VI, PLS. 308, 309, FIG. 496). The present façade, added in the 18th century, has to a considerable extent altered the external appearance of the medieval edifice, though the Gothic main portal has been retained. The inte-

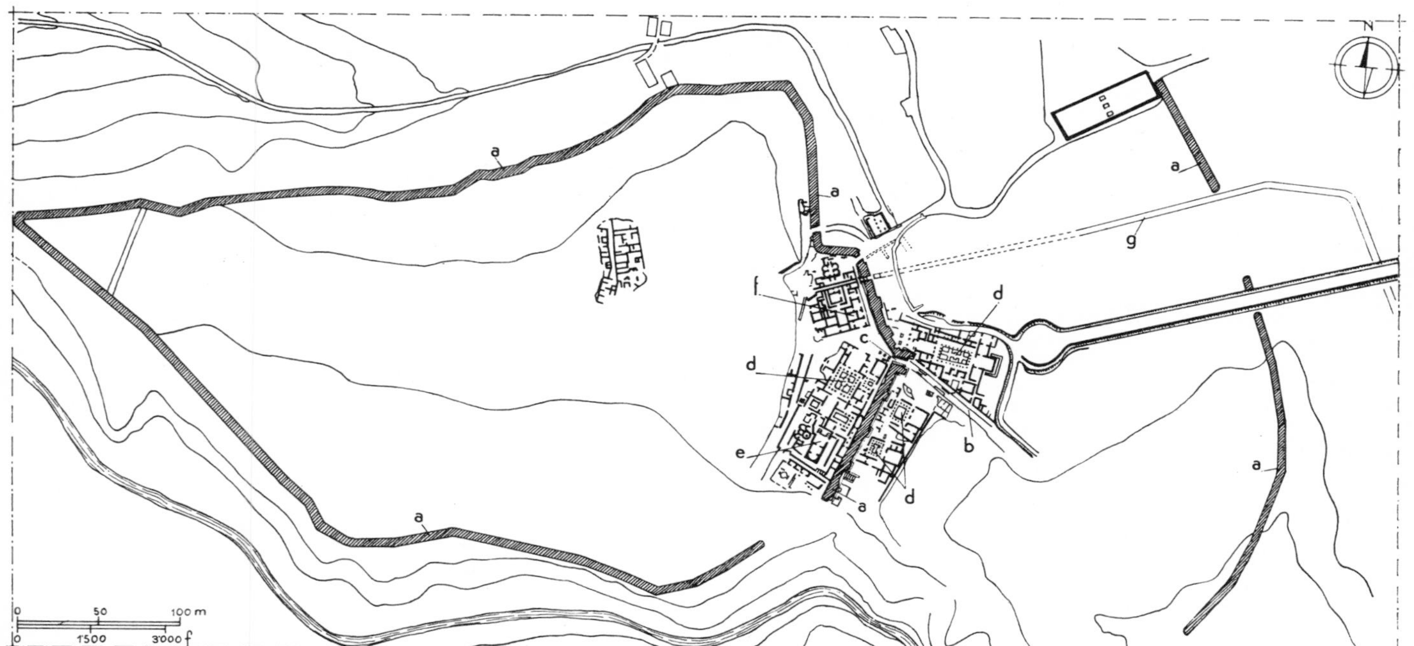

Conimbriga (Condeixa-a-Velha), plan of site. *Key*: (*a*) Wall; (*b*) road to east gate; (*c*) east gate ;(*d*) *domus*; (*e*) thermae; (*f*) *tabernae*; (*g*) aqueduct.

rior of the monastery church — of the *Hallenkirche* type, with nave and side aisles of equal height — is about 350 ft. long, 75 ft. wide, and 70 ft. high. The Claustro do Silêncio (Cloister of Silence) was built in 1308–11 and raised to a second level at the end of the 15th century. The chapter house, in transitional Romanesque style (i.e., with ogival arches), is exceptionally beautiful. The monastery is also famous for its polychrome terra-cotta sculptures dating from the 13th century onward.

Bibliog. W. Beckford, Recollections of an Excursion to the Monasteries of Alcobaça and Batalha, London, 1835; V. Correia, Alcobaça, 5 vols., Coimbra, 1929–31; E. Korrodi, Alcobaça (Monumentos de Portugal, 4), Oporto, 1929; J. Vieira Natividade, O mosteiro de Alcobaça (A arte em Portugal, 9), Oporto, 1929; J. das Neves Larcher, Monumentos de Portugal: Alcobaça e Batalha, 3d ed., Oporto, 1932.

Azeitão. The palace of Quinta da Bacalhoa dates from the last quarter of the 15th century and, in 1528, was passed on to Afonso de Albuquerque, by whom major alterations were made. It is a remarkable example of secular architecture, reflecting Florentine and Moorish influence, and its 16th-century *azulejos* rival those of the former royal residence at Sintra (see below).

Bibliog. J. Rasteiro, Quinta e palacio da Bacalhôa em Azeitão, 2 vols., Lisbon, 1895–98.

Mafra. The monastery-palace at Mafra was built by Ludovice (Johann Friedrich Ludwig) for João V in the years 1717–35. The monumental complex consists of royal apartments, sacristy, refectory, chapter house, chapels, and living units for the monks. The most important Portuguese architects and artists of the era collaborated in its construction and decoration.

Bibliog. J. da Conceição Gomes, O monumento de Mafra, 4th ed., Lisbon, 1887; J. Ivo, O monumento de Mafra, Lisbon, 1906; J. P. Freire, Mafra, Lisbon, 1925; J. Ivo, O monumento de Mafra (A arte em Portugal, 9), Oporto, 1930; J. P. Freire and C. de Passos, Mafra (Monumentos de Portugal, 2d ser., 1), Oporto, 1933; P. Montez, Estudos de urbanismo em Portugal: Mafra, Lisbon, 1933.

Oeiras. The Quinta dos Marqueses de Pombal is a characteristic 18th-century manor, which was built under the direction of the architect Carlos Mardel.

Queluz. The former royal residence (18th cent.; now the Palácio Nacional), designed by Jean Baptiste Robillon and Mateus Vicente de Oliveira, shows marked French influence (xiii, pl. 137). The surrounding gardens and park are laid out in the French and Italian manners of the epoch.

Bibliog. A. Caldeira Pires, História do Palácio Nacional de Queluz, 2 vols., Coimbra, 1924–26.

Setúbal. The Church of Jesus was built in 1494 from designs by Boytac, one of the creators of the Manueline style. Founded by fishermen, the Church of S. Julião was rebuilt in 1513; of particular interest are the Manueline portals on the main façade and north side and the 18th-century *azulejos* decoration inside the church. The Museu Municipal contains gold- and silverwork, 16th-century paintings, and a series of retables from the above-mentioned Church of Jesus.

Opposite Setúbal, on the south bank of the Rio Sado estuary, is the site of ancient Tróia, where the ruins of a Roman city have been excavated. Dwellings of several stories, thermae, a system of salting tanks (*cetariae*) and wells, remains of what was probably a mithraeum, a necropolis, and mosaics and paintings have been brought to light.

Bibliog. M. M. Portela, Noticia dos monumentos nacionaes e edificios e logares notaveis do concelho de Setúbal, Lisbon, 1882; F. Pacheco, Setúbal e as suas celebridades, Lisbon, 1930; J. M. Bairrão Oleiro, EAA, s.v. Cetobrige; F. B. Ferreira, O problema da localização de Cetóbriga, seu estado actual, Conimbriga, I, 1959, pp. 41–70.

Sintra. The Royal Palace of Sintra is a reconstruction and enlargement of an older palace (probably of Moorish origin), in the late Gothic, Manueline, and Moresque styles of the 16th century. The interior decoration includes Mudejar motifs and 16th–18th century accessories, such as goldsmith's work, textiles, and tapestries. The Castle of the Moors, one of Portugal's best extant examples of Moorish military architecture, was rebuilt under the kings Sancho I and Fernando I and has been restored many times since the earthquake of 1755.

The picturesque Castelo da Pena, in the environs of Sintra, was begun in 1840 from plans by the German architect Baron von Eschwege. Nearby there is a small former monastery church with a notable Renaissance retable sculptured by Chanterene for João III.

Bibliog. T. Lino de Assuncão, Cintra, Collares e seus arredos, Lisbon, 1888; A. C. Inchbold, Lisbon and Cintra, London, 1907; A. Haupt, Lissabon und Cintra, Leipzig, 1913; N. C. Cardoso, Cintra (Monumentos de Portugal, 7), Oporto, 1930; J. Pessanha, Sintra (A arte em Portugal, 15), Oporto, 1932; J. de Sousa Nunes, O Palacio da Pêna em Sintra, Lisbon, 1933.

Ribatejo. Almourol. The Castelo de Almourol, situated on a small island in the Tejo River, was erected in 1171 over the ruins of an ancient Roman fortress. Some of the battlements were restored in the 19th century.

Santarém. (anc. Scalabis). An important center in Roman times, on the road from Olisipo (Lisbon) to Bracara Augusta (Braga), it was elevated to the status of colonia under the name Praesidium Julium. In the former convent of S. Francisco (partly used as military barracks) there is a Romanesque-Gothic church of the 13th century, with a

14th-century choir. The Church of Graça (1380) is in the flamboyant Gothic style, with a nave and two lateral aisles. The former basilica of S. João d'Alporão (13th cent.) is now the Museu Municipal.

BIBLIOG. Z. N. G. Brandão, Monumentos e lendas de Santarém, Lisbon, 1883; J. Ozorio, Guia de Santarém, Santarém, 1923; F. Nogueira de Brito, Santarém (Monumentos de Portugal, 5), Oporto, 1929; Z. Sarmento, Santarém (A arte em Portugal, 14), Oporto, 1931; A. Schulten, EAA, s.v. Scallabis.

Tomar. The historical importance of the city stems principally from the Knights Templars, to whom Afonso I donated the site of the town in 1159. The Monastery of Christ, founded by the Templars and passed on to the Hospitalers after the suppression of the former order in 1314, is the largest in Portugal; it constitutes a museum of Portuguese architecture from the 12th to the 17th century. The Church of the Templars, with an octagonal sanctuary surrounded by a 16-sided ambulatory, based on Moslem and Byzantine prototypes, was begun in the years 1150–62. It is preceded by a 16th-century Manueline nave by Diogo de Arruda, with a south façade designed by João de Castilho (1515). The main cloister was begun after 1557 by Diogo de Torralva and completed by Filippo Terzi. Of the six other cloisters, the 15th-century Gothic cemetery cloister built by Fernão Gonçalves for Henry the Navigator is outstanding. The chapter house has an exceptionally fine Manueline window, originally one of three (of which one now serves as a door, the third having being walled up during later construction), designed by Diogo de Arruda.

Nearby, the Church of the Conceição (begun before 1551) is built in the Renaissance style and has a nave and two side aisles divided by Corinthian columns. The Church of S. João Baptista (15th cent.) is flanked by a fine Manueline tower.

BIBLIOG. J. A. dos Santos, Monumentos das ordens militares do Templo e de Cristo em Tomar, Lisbon, 1879; J. M. de Sousa, Noticia descriptiva e historica da cidade de Tomar, Tomar, 1903; F. A. Garcez Teixeira, Tomar (A arte em Portugal, 6), Oporto, 1929; J. Vieira Guimarães, Tomar (Monumentos de Portugal, 2), Oporto, 1929.

Alto Alentejo. Castelo de Vide. The small town is noted for its well-preserved 15th- and 16th-century quarters and for its 13th-century castle with excellent Gothic portals.

BIBLIOG. J. A. Gordo, Castelo de Vide, Portalegre, 1903.

Elvas. The parish church, formerly the cathedral, is an essentially Manueline structure begun about 1517 after plans by Francisco de Arruda. The Renaissance-style church of the Dominican nuns has an octagonal plan; the interior of the dome, decorated with 17th-century *azulejo*, rests on columns that were painted and gilded in the 18th century. The Moresque-Romanesque castle is known to have existed in 1226.

BIBLIOG. V. de Almada, Elementos para um diccionario de geographia e historia portuguesa: concelho de Elvas e extinctos de Barbacena, Villa Boim e Villa Fernando, 3 vols., Elvas, 1889–90; A. Varella, Theatro das antiguidades de Elvas, Elvas, 1915.

Estremoz. The castle keep, built in 1258, rises to a height of 88 ft. and retains its machicolated battlements. The remains of a royal palace, which was rebuilt in the reign of João V (1706–50), adjoin the castle.

BIBLIOG. S. J. Beçam, A vila de Estremoz, Estremoz, 1913.

Évora. (anc. Ebora). A fortified town of indigenous origin, it became a colonia, Liberalitas Julia, under the Romans and, later, a bishopric under the Visigoths. After the city was taken from the Moors in 1165, the kings of Portugal sometimes resided there, and it developed into an important cultural center.

From the Roman period there survive sections of the city walls (partly rebuilt by the Visigoths and in the 14th and 17th centuries), including a rectangular tower and a gate known as the Arch of Dona Isabel; the substructure of the Aqueduct of Prata (rebuilt ca. 1531–38); and the well-preserved ruins of a Corinthian temple, the so-called "Temple of Diana."

The Cathedral, begun in 1186, was built in the transitional style of the end of the Romanesque and beginning of the Gothic periods, with ribbed vaulting above the aisles and rose windows in the transept. The Church of Graça (16th cent.), attributed to Diogo de Torralva, is an interesting example of the Portuguese adaptation of the Renaissance style, showing a blending of 15th-century Italian elements with plateresque idioms. The baroque church of the former Jesuit College of Espírito Santo (1567–74), with tribunes, is a prototype of Jesuit architecture by Manuel Pires.

There are also some fine mansions of the 15th to 17th centuries (e.g., Casa Soure and Casa Cordovil), some of which are decorated with painted friezes.

The Hermitage of S. Brás (late 15th cent.), outside the city walls, is a curious mixture of Gothic and Mudejar styles, a blend that is frequently found in Alentejo.

The Museu Regional, in the former archbishop's palace (16th–18th cent.), contains collections of Roman and medieval sculpture and carving and Portuguese and Flemish painting and decorative arts.

BIBLIOG. M. Cardoso de Azevedo, Historia das antiquidades de Evora, Évora, 1739; C. da Camara Manoel, Atravez e cidade de Evora, Évora, 1900; A. F. Barata, Memoria historica sobre a fundação da Sé de Evora e suas antiguidades, Évora, 1903; A. F. Barata, Evora antiga, Évora, 1909; J. Rosa, Iconografia artística eborense, Lisbon, 1926; C. David, Evora (A arte em Portugal, 8), Oporto, 1930; T. Espanca, Cuadernos de historia e arte eborense, 9 vols., Évora, 1944–49; M. Tavares Chico, A Cathedral de Evora na Idade Média, Évora, 1946; F. J. Wiseman, Roman Spain, London, 1956, pp. 181–82; P. Romanelli, EAA, s.v.

Baixo Alentejo. Alcácer do Sal (anc. Salacia). The Romanesque church of S. Maria do Castelo (begun 12th–13th cent.), inside the walls of the ruined castle, has a nave with side aisles, with timbered ceilings, and a rectangular apse. In the Gothic church of Senhor dos Martíres (13th–14th cent.) there is the interesting octagonal chapel of S. Bartolomeu (1333; now the sacristy). The monastery church of S. António (founded 1524), with a Renaissance portal, is divided into two aisles which are strikingly dissimilar in design. The Museu Arqueológico contains a small collection of archaeological finds from Roman times.

BIBLIOG. V. Correia, Monumentos e esculturas (seculos III–XVI), 2d ed., Lisbon, 1924, pp. 137–58; V. Correia, Excavações realizadas na nécropole pré-romana de Alcacer do Sal em 1926 o 1927, Coimbra, 1928; A. Schulten, RE, s.v. Salacia, no. 2.

Alvito. The castle of the marquises of Alvito, begun in 1494 and completed in the first quarter of the following century, has a rectangular plan, with rounded towers at the corners, and shows pronounced Moorish characteristics in the architectural details.

BIBLIOG. J. Fialho de Ameida, O castelo de Alvito, Lisbon, 1946.

Beja. Originally a Celtic settlement, Beja became the Roman colonia of Pax Julia and was subsequently held by the Visigoths and then the Moors. Remains of pavements, buildings, and the foundation of a temple from the Roman period have come to light. The Castle, originally a Roman construction, was rebuilt under Afonso III in the 13th century, and the massive rectangular keep was built by King Diniz in 1310 (restored 1940). Near the Porta de Évora, Beja's sole surviving Roman gate, stands the pre-Romanesque church of S. Amaro; the nave and side aisles are divided by rounded arches supported by columns with Corinthian-type capitals. The former convent of Conceição (15th cent.) is in the flamboyant Gothic style with Moresque details and has a 16th-century cloister decorated with *azulejos*.

BIBLIOG. A. Viana, Mosteiro da Conceição e Palácio des Infantes, Arquivo de Beja, III, 1946, pp. 161–226, 278–305; A. Viana, Pax Julia: Arte romano-visigotico, AEA, XIX, 1946, pp. 93–109; A. Viana, Restos de um templo romano em Beja, Arquivo de Beja, IV, 1947, pp. 77–88; F. C. da Silva, Historia das antiguidades da cidade de Beja (1792; ed. A. Viana), Arquivo de Beja, V, 1948, pp. 225–42, VI, 1949, pp. 3–36, 292–324; A. Viana, Estelas discoides do Museu de Beja, Arquivo de Beja, VI, 1949, pp. 37–85; Guia turistica de Beja, 1950; J. Fragoso de Lima, Aspectos de romanização no território portugês da Bética, Arqueol. português, N.S., I, 1951, pp. 171–211; A. Viana, Notas históricas, arqueológicas e etnográficas do Baixa Alentejo, Arquivo de Beja, XI, 1954, pp. 3–31, XIII, 1957, pp. 110–67, XIV, 1957, pp. 3–57; EAA, s.v.

Mértola (anc. Myrtilis). Located on the Guadiana River, the town is still surrounded by its ancient walls. Nossa Senhora da Assunção (late 13th-century), the parish church, has a very unusual mosquelike interior divided into five aisles. The fortified castle is Moorish, though the keep was added in 1292. On the river there are remains of a quay also built by the Moors.

Moura. The Church of S. João Baptista was rebuilt in the Manueline style during the reign of Manuel I. The elaborate main portal is an excellent example of that style.

BIBLIOG. J. Segurado, A igreja de S. João de Moura, da sua arquitectura e de sua historia, Lisbon, 1929.

Serpa. Within the walls of the ruined castle there is the Gothic church of S. Maria, built in the reign of King Diniz (1279–1325), with 17th-century decoration of polychrome *azulejos.*

BIBLIOG. J. M. da Graça Afreixo, Memorias historico-economicas do concelho de Serpa, Coimbra, 1884.

Algarve. Faro. Most of the city was destroyed by the earthquakes of 1722 and 1755. The Cathedral, which has a Renaissance interior, retains vestiges of the original Gothic structure, most notably the base of the tower that now forms the main entrance. The ruined convent of Nossa Senhora de Annunção (founded between 1518 and 1523) has a portal dated 1539 and a Renaissance cloister (1543) by Afonso Pires. The Museu Arqueológico Infante D. Henrique, in the old church of S. António dos Capuchos, contains prehistoric finds and Roman antiquities from Ossonoba, an important archaeological site at nearby Milreu.

BIBLIOG. O Algarve illustrado, Faro, 1880; S. J. Baçam, A cidade de Faro, Faro, 1912.

Silves. The town was a very important center under the Moors. The Moorish castle (restored in 1940), one of the most important in Portugal, is built of red sandstone; beneath it there are several enormous cisterns and cellars, which are well-preserved. The 13th-century Gothic cathedral is noteworthy, though badly marred by later renovations and additions.

BIBLIOG. P. Mascarenhas Júdice, Atraves de Silves, Silves, 1911.

* *

Illustrations: 4 figs. in text.

The geographical section includes the contributions of Luís REIS-SANTOS and additional information provided by Giorgio STACUL.

POSTER ART. See PUBLICITY AND ADVERTISING.

POSTIMPRESSIONISM. See EUROPEAN MODERN MOVEMENTS; EXPRESSIONISM.

POUSSIN, NICOLAS. French painter (b. Les Andelys, Normandy, June, 1594; d. Rome, Nov. 19, 1665). Considered the greatest of living painters in his own time, Poussin has come to be generally regarded as that artist who best expresses the French genius, in his explorations of the resources of formal expression and in his constant need to meditate and rationalize (Bernini, striking his forehead, is said to have remarked, "He is a painter who works from there"; cited by Chantelou, 1665).

Three factors explain his rather special place in 17th-century painting: an acquaintance with literature and philosophy rare among artists of the time; training in France, followed by a long career in Rome (whereas most artists received their training in Rome before returning to their native countries); and a total creative freedom secured comparatively early in his career (after the age of forty he accepted hardly any specific commissions).

His father, who came from Soissons, served in the royal army during the religious wars, married the widow of an attorney, Marie Delaisement, and cultivated a small property in Les Andelys. Despite these modest circumstances, however, he could look back on more or less noble origins and seems to have destined his only son for the law — hence, a sound education for Nicolas and, doubtless from childhood, a knowledge of Latin. His artistic vocation, already awakened in his school days, was sharply opposed by his parents. The sojourn of the painter Quentin Varin in Les Andelys in 1612 (paintings by him, some signed and dated, are preserved in the church of the locality) seems to have spurred the youth to decision and led to his clandestine flight from home. A stay in Rouen with the painter Noël Jouvenet, an early member of the famous dynasty, may be placed about this period. Once in Paris, he obtained the protection of a young nobleman from Poitou, and for several months he probably frequented the studios of Ferdinand Elle, a popular portraitist, and Georges Lallemand, a fashionable mannerist from Lorraine. About 1614 there must have occurred the Poitou episode confided to Giovanni Bellori: the departure of Poussin's patron, who took him along to his castle in Poitou, where the artist was rebuffed by the young nobleman's mother; Poussin's return on foot, sustained along the way by meager commissions (no certain evidence of these is preserved); his arrival in Paris, worn out and ill, followed by a year of rest with his parents in Les Andelys.

His health restored, Poussin returned to Paris and doubtless made his living through commissions, from the provinces as well as from the capital. No authenticated work from that period is preserved. Of the few paintings suggested as possibilities, the most interesting is the *St. Denis Crowned by an Angel* formerly in St-Germain-l'Auxerrois in Paris (Rouen, Mus. B.A.). Two attempts to reach Rome proved abortive. The first time (ca. 1620–21) he stopped in Florence, where he was to have his first direct contact with Italian art and where he became acquainted with Jean Mosnier, Jacques Stella, who remained his most faithful friend, and perhaps Jacques Callot; the second time he got no farther than Lyons, another important center of painting. In Paris he seems to have been attracted particularly by the "classical" example offered at that time by Toussaint Dubreuil and, above all, by Pourbus the Younger (*The Last Supper*, 1618; Louvre). He must also have studied the mannerist cycles of Fontainebleau, especially the frescoes of Primaticcio (q.v.); and he knew the works of antiquity and the Italian paintings (Raphael, Leonardo, etc.) of the royal collection. The connoisseur Alexandre Courtois, a member of the royal household, introduced him to Italian engraving, in particular to examples after Raphael and Giulio Romano, which made a profound impression on him.

During this time Poussin also acquired a sound theoretical knowledge of architecture, perspective, and anatomy, which he studied in a hospital. There are many indications of his frequenting intellectual circles and of his finding patrons there, while diligently broadening his cultural outlook. He lived for a time at the Collège de Laon in Paris (ca. 1622) .His first substantial commission came in the summer of 1622 from the Jesuit College (Collège de Clermont) in Paris, which was celebrating the canonization of St. Ignatius and St. Francis Xavier. The six temperas he executed on this occasion (all trace of them has been lost) won him the notice of the celebrated Giambattista Marino, who was then in Paris (1615–23), feted by the whole literary world. The poet befriended Poussin and had him execute a series of drawings — those known as the "Massimi drawings," after Cardinal Camillo Massimi, who later owned them (15 in Windsor, Royal Lib.; another in Budapest, Mus. of Fine Arts). Then the archbishop of Paris, Jean-François de Gondi, ordered a painting for Notre-Dame, a *Dormition of the Virgin* (1623) which has disappeared but which may be rendered in a drawing still extant (Hovingham Hall, Yorkshire, Coll. Sir William Worsley). Poussin had by now achieved appreciable success; indeed, at thirty he appeared a fully developed artist, the master of a highly personal style, who liked to organize complex and thoroughly deliberated compositions according to noble and clearly marked cadences. It was at this point in his career that he gave up these promising circumstances to leave for Rome.

His presence in Rome in March, 1624, is documented. A brief halt in Venice, noted by Giulio Mancini, must have taken place in the course of the journey. Although the ensuing period was rich in encounters and artistic experiments, the first years in Rome (1624–25) were a time of personal misfortunes. Marino left for Naples, where he died, and Cardinal Francesco Barberini, to whom Poussin was recommended, and Cavaliere Cassiano dal Pozzo, who was to become one of his most important clients, both left for France. The painter eked out a wretched living, studied Raphael and ancient works of art, and took up the study of anatomy again; he continued to paint in a style close to that of his Parisian years, however, little affected by the Roman fashions, Caravaggism not excepted. At this time he produced the two battle pieces with Joshua (Moscow, A. S. Pushkin Mus. of Fine Arts; Leningrad, The Hermitage). In 1626–27, because of his literary cast of mind — "his literary erudition enables him to evoke any story, fable, or poem whatsoever" (Mancini, ed. 1956–57, I, p. 261) — Poussin once more began to win the esteem of the intelligentsia, gaining favor in the circle of the Barberinis and Dal Pozzo. In 1626, also, he lived with the Flemish sculptor François (Frans) Duquesnoy, and with him and Alessandro Algardi (q.v.) he studied ancient sculptures and works of Titian, whose *Bacchanals* (then in the Vigna Ludovisi) particularly attracted him.

In the wake of the Joshua battle pieces, and following the first version of *The Capture of Jerusalem by Titus* (1625–26; only the engraving of a drawing is preserved), Poussin painted the *Death of Germanicus* (1627), greatly admired for its historical truth and emotional expression — its *costume* and *affetti* — and the *Triumph of Flora* (1626–27?), with greens and blues directly borrowed from Titian. Seeking to win recognition with virtuoso pieces, such as the great *Massacre of the Innocents* for the Marchese Vincenzo Giustiniani, he finally obtained, in February, 1628, an important official commission for an altar of St. Peter's (finished Sept., 1629). This brilliant work, *The Martyrdom of St. Erasmus*, was an attempt to combine naturalistic effects with "baroque" movement and bright color, and it securely established his reputation with the public. Another major commission, *The Virgin Appearing to St. James the Greater* (1629–30), executed for the city of Valenciennes in Flanders, reveals a return to a darker color scale but the same effort to give the masses a vibrant aspect through manipulation of lighting, color, and brush stroke. Important paintings of about the same period, such as the *Childhood of Bacchus* in Chantilly, mark the success of this "baroque" vein, which placed Poussin in direct rivalry with Pietro da Cortona, who was about the same age, had the same patrons, and employed the same themes.

Then, in 1630, there occurred in Poussin's art another decisive change of direction, which seems to have coincided with a grave crisis in his life. An illness, no doubt contracted several years earlier, threatened his life; he owed his recovery to Jacques Dughet, a French pastry cook, who took him in and nursed him. There were other sources of anxiety: the increasingly fierce competition of Cortona; the jealousy of Lanfranco; the challenge presented to the *Martyrdom of St. Erasmus* by its pendant, Valentin de Boullogne's *Martyrdom of SS. Processus and Martinianus* (1629; Vat. Mus.) — "a big dispute," according to Joachim von Sandrart; and the preference accorded Charles Mellin over Poussin for the decoration of a chapel in S. Luigi dei Francesi (July, 1630). At this point he seems to have made a definitive choice regarding the direction of his career. On Sept. 1, 1630, he married Anne Marie Dughet, the daughter of his benefactor: a humble marriage, hardly suited to furthering a mundane career, and one that fixed his residence in Rome. He declined any official commissions, altarpieces and decorations, thus leaving Cortona a clear field; he now intended to devote his art to "cabinet pictures" for a cultivated audience — works of limited format, painstaking workmanship, and carefully deliberated composition, which from this time on were much sought after, highly remunerative (110 *écus* for the *Plague at Ashdod* in 1631), and often copied. His art emerged from this personal crisis purified, and more and more, after the intervening "baroque" experiments, he returned to and renewed his earlier formulas, and was seemingly affected by the example of Andrea Sacchi and his blond tonality as well. The *Plague at Ashdod* (PL. 234) and the *Realm of Flora* (1630–31) present him in full mastery of his formula and esthetic.

The ensuing decade (1631–40), especially, marked a deepening commitment to his new course. In close contact with erudite circles (Dal Pozzo, French travelers and *libertins*), he enlarged upon his Humanistic and philosophical knowledge; he studied theoretical treatises such as that of Matteo Zaccolini on perspective; he made drawings for Leonardo's *Trattato della pittura*, from which the demonstrations on movement were applied in the two *Rape of the Sabines* and the *Rescue of the Young Pyrrhus*; and he measured and drew ancient sculptures. The presence of his young brother-in-law Gaspard Dughet, the landscapist, in his house (1631–35) may have sharpened his awareness of the countryside, and he made direct notations of landscape (cf. the studies of trees in the Louvre, nos. 32466 and 32467) during his walks in the Roman campagna with Claude Lorrain (q.v.), Sandrart, and "il Bamboccio" (see BAMBOCCIANTI). He elaborated various types of compositions, in which he sometimes made use of a dramatic linear linkage (*Rescue of the Young Pyrrhus, Armida Carrying the Sleeping Rinaldo*) that in some instances approaches bas-relief (*Venus Arming Aeneas*) or he dispersed small groups in depth — but without ever sacrificing the analytic composition — each of these representing a separate phase of the action (the two *Rape of the Sabines, The Israelites Gathering Manna*).

In a production for which reliable points of reference are few (the correspondence preceding 1639 has almost completely disappeared), the variations in technique and palette seem to succeed one another rapidly. From the luminous "blond group" of about 1631 (the Louvre *Narcissus*, the Prado *David Victorious*, etc.), to the "copper group," with its smooth texture and "stony" flesh (ca. 1637–38?; *Rescue of the Young Pyrrhus*, the Louvre *St. John Baptizing the People*), it is perhaps unwise to seek to establish too rigid a pattern of evolution. A single example should serve to confirm this cautionary note: the tender *Pan and Syrinx* is contemporaneous (ca. 1637–38) with the *Schoolmaster of Falerii*, a work severe to the point of ponderousness. Problems of archaeological exactitude (the second *Capture of Jerusalem by Titus, The Schoolmaster of Falerii*), of linear perspective and accurate construction of the scene in space (the *Rape of the Sabines* in the Louvre; the *Seven Sacraments* for Dal Pozzo, begun ca. 1636), and of aerial perspective, or rendering the transparency of the atmosphere outdoors (*Venus Arming Aeneas*) or in an interior (the *Confirmation* for Dal Pozzo; PL. 235), became the object of Poussin's careful experimentation. His artistic reputation, by then solidly established among Roman connoisseurs, reached Paris again, especially after the return to France of his friend Jacques Stella (1634) and the shipment of the *Bacchanals* commissioned by Cardinal Richelieu (1636). A number of important works were henceforth painted directly for French collectors, such as Melchior de Gillier, Louis de La Vrillière, and Paul Fréart de Chantelou.

A pressing invitation from Richelieu and Louis XIII, transmitted as early as 1638, as well as the title of First Painter and a 3000-livre pension, finally persuaded Poussin to return to Paris. He arrived at the end of 1640 and was to remain until September, 1642. Received like a prince, he was feted by the king, the court, and especially the intellectuals and was lodged in "a little palace," a pavilion in the Tuileries. But before long, on the part of both the King and the painter, hopes were disappointed. Simon Vouet (q.v.), until then in first place and beloved by his pupils, was vigorously defended by a whole coterie and was supported by others who felt slighted through the prerogatives of the First Painter — the landscapist Jacques Fouquier (Fouquières), the architect Jacques Lemercier. The official circle (Richelieu, Building Superintendent Sublet de Noyers) had hoped for a great "master of works," such as Vouet and later Charles Lebrun (q.v.), one capable of creating a "style," even in the applied arts, and of directing teams of artists and artisans while remaining active as a courtier. Poussin had to execute mantel paintings (*Moses Kneeling before the Burning Bush* for Richelieu), altarpieces (*The Institution of the Eucharist* for the chapel of the Château of Saint-Germain-en-Laye; *The Miracle of St. Francis Xavier* for the Jesuit Novitiate in Paris), and decorative compositions (*Time and Truth*, a ceiling composition for Richelieu); he had to plan over-all decorative schemes (for the Orangery of the Luxembourg Palace and, most notable of all, that for the Grande Galerie in the Louvre) and provide designs for books (frontispieces for a Bible, a Vergil, and a Horace printed by the Typographie Royale), bindings, tapestries, and so forth. But the art he had formulated during the preceding ten years was fundamentally opposed to the notion of "style" and a decorative function. Incapable of improvising to order, he required a contemplative atmosphere that was not available to him at the French court. The obligation to take up once again modes of painting he had long since abandoned, under the eyes of adversaries always ready to criticize, demanded an exhausting effort of him. His works of this time clearly reveal an anxious attempt to adapt himself to these demands and to astonish every time. They reveal, in turn, an emphasis on austerity and a rigorous combination of volumes in the *Institution of the Eucharist*; on dramatic and psychological values — greatly admired and for a long time afterward a model for the young Parisian painters — in the *Miracle of St. Francis Xavier*; and on a light color range and illusionistic foreshortening in *Time and Truth*. The results were uneven, and the poetic quality was often submerged under the official aims. The change in climate, moreover, was affecting Poussin's health. He finally obtained permission to leave, on the pretext of bringing his wife from Rome. Richelieu's death in December, 1642, and that of Louis XIII in May, 1643, and the serious political disturbances that followed, made his return to Rome permanent. After futile negotiations the decoration of the Grande Galerie was interrupted, and despite his protests, his lodging in the Tuileries was withdrawn (1644).

Nearing his fifties (PL. 233), he was never again to leave Rome and his true course, his former path of artistic inquiry, was now rediscovered. He led a modest but comfortable life in his house on the Via Paolina, with his wife and nieces, avoiding encumbrances and honors (in 1657 he turned down the principal office of the Accademia di S. Luca). Thereafter, the only eventful occurrences in his life were the minor shifts in his household (his brother-in-law Jean Dughet, the sculptor Giovanni Perraca, his nieces and nephews all came and went), local political incidents (the disgrace of the Barberinis, 1644–46, and other disturbances) the visits of notable and engrossing figures such as Abbé Louis Fouquet (1655–56), brother of the powerful Nicolas Fouquet, and the artist's more important commissions (the *Seven Sacraments* for Chantelou; the models for terms for Nicolas Fouquet's château Vaux-le-Vicomte, 1655–56; the *Four Seasons* for the Duc de Richelieu, 1660–64). His interior life continues to be elusive, for almost nothing is known of his sentimental relationships, which even after 1630 were more important perhaps than his biographers would lead one to believe. It is known, however, that illness, and especially a severe bladder ailment, tormented him continuously and that he was sometimes immobilized for long periods, as in August-September, 1646. His hand began to tremble at a quite early age — the first symptoms appeared no later than 1641 — and he was compelled to draw more and more schematically, thus achieving ever-bolder effects (the wash *Carrying of the Cross*, 1646; Dijon, Mus. B. A.). Thus compelled to paint with a stroke that became less and less precise, in compensation he became increasingly subtle in rendering color values (*Landscape with Diogenes*). The admirable serenity found in his last works was attained only through the most cruel physical suffering ("I do not spend a day without pain"; *Correspondance*, 1911, Aug. 2, 1660).

The sojourn in France greatly enhanced not only the renown of "Monsù Possino" among Roman art collectors such as Dal Pozzo and Monsignor (later Cardinal) Massimi and among men of letters such as Carlo Dati, but also that of "Monsieur le Poussin" among French collectors such as Chantelou, Pointel, Michel Passart, Mauroy, and the Duc de Richelieu, grandnephew of the Cardinal, and poets such as Georges de Scudéry, François (Tristan) L'Hermite, and Jean Desmarets de Saint-Sorlin. The collections of Cassiano dal Pozzo in Rome and of Chantelou in Paris, which served as sanctuaries open to all devotees of painting, fostered Poussin's reputation. He had assumed the status of master, and in Rome itself his eminence was unchallenged except for Cortona, who was valued for his large decorative projects and altarpieces. Young painters and intellectuals clustered around him during his walks; his remarks were carefully noted by writers such as André Félibien and Giovanni Bellori; French artists and travelers came to visit him (Nicolas Loir and Félibien, 1647–49; Abbé Arnauld de Pomponne, the poet François de Maucroix, canon of Reims, 1661; Balthazar de Monconys and the young Duc de Chevreuse, C. H. de Luynes, 1664); numerous tributes to his genius were printed (the Jesuit Giovan Battista Ferrari in his *Hesperides*, 1646; Scudéry in his *Cabinet* . . . , 1646; the engraver Abraham Bosse in his *Sentimens* . . . , 1649; Roland Fréart de Chambray in the *Traité de la peinture de Léonard*, dedicated to Poussin, 1651; the painter Hilaire Pader in *La peinture parlante*, 1653; etc.). His paintings were contested for at very high prices (about 1660 the *Plague at Ashdod* was bought for 1,000 *écus*); it was taken as a signal favor if he would consent to begin a commission, and one hardly dared specify "religious subject" or "secular subject," leaving the creator complete freedom.

It is readily apparent that, from quite early in his career, Poussin pursued a wholly personal course in evolving his art, one determined solely by inner demands. On his return from Paris his art passed through a tense and dogmatic phase, in which the composition was increasingly deliberated, down to the smallest detail (the *Seven Sacraments* for Chantelou; PL. 236), and in which his efforts at psychological interpretation of the subject were intensified (*Judgment of Solomon*, 1649; II, PL. 295) and the forms became ever more monumental (*Eliezer and Rebecca*, 1648). The result is always severe (*The Finding of Moses*, 1647, for Pointel), sometimes grandiose (*The Crucifixion*, 1646), and often forbidding (*The Infant Moses Trampling Pharaoh's Crown*; Louvre).

Then, in about the 1650s, it is as if this tension broke and allowed the emergence of some forgotten accents: violent pathetic effects (*The Dead Christ before the Sepulcher*), visionary architectural backgrounds or "cubistic" constructions (*The Woman Taken in Adultery*, 1653; PL. 232), dramatic austerity (the London *Annunciation*, 1657), and a mood of poetic meditation in which myths are strangely intermingled (*The Birth of Bacchus*, 1657). From 1647–48 particularly, landscape assumed an unforeseen importance in Poussin's work, bringing a new dimension to his world. His landscape depictions, at first studied and dramatic (*The Funeral of Phocion*, 1648, PL. 235; *The Woman of Megara Gathering the Ashes of Phocion*, 1648; *Pyramus and Thisbe*, 1651), evolved toward a pure contemplation of nature (*Landscape with Two Nymphs*, ca. 1659), in which pebbles, blades of grass, the slightest nuances of light and water, fondly painted for their own sake, draw attention from the human drama and convey a soothing quality. The last works (*Landscape with Hercules and Cacus*, *The Four Seasons*, *Apollo and Daphne* or *The Misfortunes of Apollo*) — quite paradoxical in the production of the period — mark that final moment, known to Michelangelo and Titian at the end of their lives, when technical mastery is forgotten and emotion transcends reality. On Nov. 19, 1665, death carried off an infirm old man, who had been cruelly stricken by the passing of his wife (1664) and who had put aside his brushes to wait for the end without trembling: "God grant that it be soon, for life weighs too heavy upon me" (July 26, 1665; cited in *B. de la Société Poussin*, I, 1947, p. 50).

Poussin's *œuvre* is relatively well preserved — of approximately two hundred compositions engraved in the 17th century, only about a third of the corresponding paintings have disappeared — and it forms, even though the youthful works are missing, a balanced whole that is exceptional for the 17th century for its being elucidated by an abundant correspondence and contemporaneous testimony. These works, in their order of creation, convey a series of principles whose theoretical elaboration seems to have taken place about 1630–35, principles set forth by Poussin himself in his letters (notably that to Chantelou of Nov. 24, 1647, and that to Chambray of Mar. 1, 1665) and in some fragments preserved by Bellori. These guidelines can be summarized as follows: (1) A subject, or the "noble matter," ought to be chosen that is capable of sustaining a poetic contemplation which is both complex and profound; hence, genre subjects, still life, current events, and other such are to be excluded. (2) The chosen theme should be handled by juxtaposition of the "peripeteia," analogous to either the strophes of an ode (lyrical breakdown) or the scenes of a tragedy (dramatic breakdown), without ever offending reason (law of verisimilitude) but also without submitting to the prosaic verity of time (in *The Israelites Gathering Manna*, for instance, the Israelites are famished on the left, sated on the right, and pass through various intermediate stages in the area between) or place (in the *Birth of Bacchus* the solitary drama of Echo and Narcissus intervenes in counterpoint) or action (in the *Moses Exposed*, at Oxford, a pagan divinity mingles with the protagonists). (3) The effect of the painting should depend not only on representation (principle of the expression of the passions through the aspect, gestures, and mime of the personages; cf. the *Judgment of Solomon*, II, PL. 295) but also on the organization of the formal elements. In accordance with the application to painting of the ancient musical system of modes, as set forth in the *Istituzioni harmoniche* of Gioseffo Zarlino, the basic harmony, the color effects, the dimensions and distribution of volumes, and the play of arabesques should all be consciously calculated in relation to the subject ("in this consists the whole artifice of painting," *Correspondance*, 1911, Nov. 24, 1647). For example, the sacrament of Confirmation, a "grave and severe" subject, requires the Doric mode (stable volumes, continuous lines, uniform colors); war subjects such as the capture of Jerusalem or the rape of the Sabines require the Phrygian mode (multiple small elements, acute angles, linear patterns that are continually broken), to obtain an effect of "vehemence" and "fury," while preserving a severe aspect; bacchanals (PL. 231), scenes of joyous movement, demand the Ionic mode; the Holy Family or the story of Rebecca should be treated in the hypolydian mode, which expresses "sweetness" and "gentleness"; the Entombment and other scenes of sadness and lamentation require the Lydian mode.

Such a theory could apply only to a creative process of a very special type, in which the artist does not — like Veronese, Cortona, or Rembrandt — choose his subject as a pretext for illustrating a poetic realm congenial to him. He gives himself up entirely to his themes, relives them within himself (Poussin, speaking of a projected *Carrying of the Cross*: "I should not be able to resist the distressing and serious thoughts with which one must fill mind and heart in order to succeed with these subjects in themselves so sad and lugubrious"; 1646; cited by Brienne, 1692–95), and seeks to create for each a particular world. The universality of art is thus sought midway between pure objectivity ("the subject matter or argument" that assures the bond with the spectator) and subjective interpretation ("the concept . . . pure offspring of the mind, which sets about grappling with things"; Bellori, 1672, p. 461). The role of the painter is conceived along the lines not of the elegiac but of the epic poet — there are numerous references to Vergil — and especially of the dramatist.

Thus arises the apparent paradox in Poussin's work: visible discontinuity in his *œuvre*, yet a personality that cannot be denied. Although his painting does not, like that of Rubens or Rembrandt, take one into a self-contained world, the part played by inspiration and the expression of a personal spirit remains essential: "It is the Golden Bough of Vergil that none can find or pluck, be he not led by Fate" (*Correspondance*, 1911, Mar. 1, 1655). Attempts have often been made to define this quality of inspiration through the creation of some "noble" world. This is a complete misconception, however, for what is characteristic of Poussin — what places his work on a level with the most exalted productions of the human spirit — is precisely that he does not offer some fixed, ideal universe but a continuing meditation on the relations between man and the natural world. In his early works there appears a kind of breach between, on the one hand, the free flowering of the senses in harmony with the life of nature (*Triumph of Flora*, *Bacchanal with a Lute Player*) and, on the other, the cruel fate of man, marked by conflict, blood, and death (the first battle pieces, of a "romantic" brutality; *The Massacre of the Innocents*). The two themes, treated separately in the battle pieces and the bacchanalian scenes, are interwoven in subjects taken from Ovid's *Metamorphoses*. Man is always directly confronted with his destiny, be the hero Pan or Moses; Bible story and myth are treated in the same spirit. At first entirely at the mercy of impulsiveness and the tumultuous play of the passions, the protagonists are increasingly portrayed, especially in the works from 1637–40 onward, as resisting the vicissitudes of chance and attempting "to remain firm and immovable before the efforts of that blind madwoman" (*Correspondance*, 1911, June 22, 1648).

From about 1647–48, however, with the artist having entered his fifties and the masterly cycle of the *Sacraments* for Chantelou completed, the two general themes were treated more and more profoundly, in a new manner. The human drama is henceforth placed in the midst of nature, and landscape gains increasing importance in his work. Man's struggle is contrasted with the joyful fecundity of the universe, in which it tends to become secondary (*The Funeral of Phocion*; PL. 235) or even to be harmonized with it (*Landscape with Polyphemus*, *Landscape with Hercules and Cacus*). The flight of time ceases to be perceived as a threat. Love appears senseless unless fulfilled in fruition, life in creation: the birth of Bacchus is henceforth an answer to the painful and sterile death of Narcissus. The measured and peaceful flow of nature seems equally to invade all living beings, objects, and events (*Ruth and Boaz* or *Summer*, part of the "Four Seasons" series), and the earlier stoic attitude gives way to a kind of pantheistic wisdom, through which Poussin, laden with years, returns to his peasant origins. *Apollo and Daphne*, given unfinished in 1664–65 to Cardinal Massimi, is truly a poetic testament, which shows the poet-god Apollo in his timeless realm, serenely enthroned amid a fecund universe.

Complex and difficult to put into formulas as Poussin's work is — Wölfflin, failing to integrate it into his "classical-baroque" framework, excluded it with the epithet "reactionary" — it has never yet known disaffection, and each period in turn has drawn on it according to its needs. Lebrun and the Academy derived from it a taste and precepts (the theory of the expression of the passions) that long continued to influence French painting. Artists of the late 17th and the 18th century rejected Poussin's color but continued to take him as a model in the realm of "invention," the interpretation of subject matter. David and Ingres subsequently placed their education under his inspiration. His example has seldom ceased to have some effect on the most varied tendencies. Those who admire his rational, reflective approach — Degas, Puvis de Chavannes, Seurat, Lhote, Cézanne ("every time I come away from Poussin I know better who I am") — balance those who admire his creative freedom and his sensuous poetry, such as Delacroix (*Essai sur le Poussin*, 1853) and Picasso, who in 1944 copied, or rather interpreted, the *Triumph of Pan* in the Louvre.

WORKS. This "basic catalogue" rests exclusively on explicit and trustworthy testimony of the 17th century. The most valid reference for each item — literary source, notices in archives, faithful engravings from before 1700 — is given in parentheses. A question mark indicates that the value of the reference or the identification of the work is open to doubt. See SOURCES for full entries on the main literary sources: Bellori (abbr. B); Brienne (Br); Chantelou (Ch); *Correspondance de Poussin* (C); Félibien, 1685 (F); Passeri (P); Sandrart (S). Thuillier, 1960, reproduces other 17th-century sources: Robert de Cotte; Félibien, 1647, 1679, 1681; Louis Fouquet; Pompeo Frangipani; Hilaire Pader; Balthazar de Monconys; Sublet de Noyers; as well as pertinent passages from Brienne, Chantelou, and Sandrart. In addition: for the Barberini inventory, see J. A. F. Orbaan, *Documenti sul Barocco in Roma*, Rome, 1920; for the Borghese archives, P. Della Pergola, "Appunti: Due documenti . . . ," *Paragone*, no. 83, 1956, pp. 66–68; for the Giustiniani inventory, L. Salerno, "The Picture Gallery of Vincenzo Giustiniani," *BM*, CII, 1960, pp. 21, 93 ff.; for the inventory of Louis XIV, N. Bailly, *Inventaire des tableaux du Roy rédigé en 1709 et 1710*, ed. F. Engerand, Paris, 1899; for the Massimi inventory, Orbaan, op. cit.; for the Mazarin inventory, Henri d'Orléans, Duc d'Aumale, *Inventaire de tous les meubles du Cardinal Mazarin dressé en 1653*, 1861, and G.-J. de Cosnac, *Les Richesses du Palais Mazarin*, Paris, 1884; for the inventory of the Duke of Richelieu, C. Ferraton, "La Collection du Duc de Richelieu au Musée du Louvre," *GBA*, 1949, p. 437 ff.; for the Ruffo inventory, V. Ruffo, "Galleria Ruffo nel secolo XVII in Messina," *BArte*, 1916, p. 167; for the archives of the Valguarnera trial, J. Costello, "The Twelve Pictures 'ordered by Velasquez' and the Trial of Valguarnera," *Warburg*, XIII, 1950, p. 237 ff.

a. Self-portraits: Executed for Pointel, 1649 (C), Berlin, Staat. Mus. – For Chantelou, 1650 (C), Louvre (PL. 233). – For Dal Pozzo (?) (De Cotte, 1689), lost.

b. Old Testament: *The Sacrifice of Noah*, with God upon clouds (F?) (engr. pub. by E. Gantrel), lost. – *The Sacrifice of Noah*, with God carried by angels (?) (engr. by L. Cossin), lost. – *Eliezer and Rebecca*, 1648 (Ch), Louvre. – *Rebecca*, for Dal Pozzo (De Cotte, 1689), lost or London, Coll. A. Blunt (?). – *Jacob Complaining to Laban* (?) (anon. engr., dubious), lost. – *Joshua's Victory over the Amalekites* and *Joshua's Victory over the Amorrheans*, pendants, 1624–25 (B),

Leningrad, The Hermitage, and Moscow, A. S. Pushkin Mus. of Fine Arts. – *The Plague at Ashdod*, 1630–31 (Valguarnera trial, 1631), Louvre (PL. 234). – *David Victorious*, ca. 1630–31? (B), Prado. – *The Judgment of Solomon*, 1649 (Ch), Louvre (II, PL. 295). – *The Judgment of Solomon* (?) (Giustiniani invt., 1638), lost. – *Samson Seated* (Barberini invt., 1631), lost. – *The Fainting of Esther* (Ch), Leningrad, The Hermitage. – *A Prophet*, 1624–25 (F), lost.

Moses Exposed, for Stella, 1654 (Ch), Oxford, Ashmolean Mus. – *Moses Exposed, with a Great River*, ca. 1626–28? (Br?), Dresden, Gemäldegal. – *The Finding of Moses*, 1638, owned in 17th cent. by Le Nôtre (F), Louvre (IV, PL. 315). – *The Finding of Moses*, for Pointel, 1647 (C), Louvre. – *The Finding of Moses*, for Reynon, 1651 (Br), Dorking (Surrey), Bellasis House, Coll. Mrs. Derek Schreiber. – *The Infant Moses Trampling Pharaoh's Crown*, ca. 1643–45 (Ch), Woburn Abbey (Bedfordshire), Coll. Duke of Bedford. – *The Infant Moses Trampling Pharaoh's Crown* and *Moses and Aaron before Pharaoh*, pendants, ca. 1645? (B), Louvre. – *Moses Driving Away the Shepherds* (engrs. by A. Bouzonnet-Stella and A. Trouvain), lost. – *Moses Kneeling before the Burning Bush*, 1641 (B), Copenhagen, Statens Mus. for Kunst (?). – *The Crossing of the Red Sea* and *The Adoration of the Golden Calf*, pendants (Monconys, 1664), Melbourne, Nat. Gall. of Victoria, and London, Nat. Gall. – *The Adoration of the Golden Calf* (F), destroyed ca. 1647–48, fragment in London, Liddell Coll. (?). – *The Israelites Gathering Manna*, 1637–39 (C), Louvre. – *Moses Striking the Rock* (engr. by Jean Lepautre; B), lost. – *Moses Striking the Rock*, for Gillier, ca. 1636–37? (Paris, Arch. Nat., invt. after death, 1650; see Wildenstein, 1957, no. 20), Mertoun St. Boswells (Scotland), Coll. Earl of Ellesmere. – *Moses Striking the Rock*, for Stella, 1649 (C), Leningrad, The Hermitage.

c. New Testament: *Virgin and Child*, three-quarter length (Ch?; engr. by Jean Pesne), lost. – *Little Virgin under a Roof*, 3 figs. (Br), lost. – *Virgin and Child with St. John* (engr. by Jean Pesne), lost. – *Virgin and Child with St. Joseph*, for Roccatagliata, 1641–42 (C), Detroit, Inst. of Arts. – *Virgin and Child with St. Joseph and St. John*, life-size, known as the "Grande Sainte Famille," 1655 (C), lost, engr. by A. Voet II, copies. – *Virgin and Child with St. Joseph and St. John*, who holds a cross (engr. by S. Vouillemont), lost. – *Virgin and Child with St. Joseph, St. John, and the Lamb* (engr. by G. Chasteau), formerly Lugano, Thyssen-Bornemisza Coll. – *Holy Family*, 5 life-size figs., the so-called "Chantelou Virgin," 1655 (C), Leningrad, The Hermitage. – *Holy Family*, 5 figs. including Joseph with hands joined, 1656 (F), Louvre. – *Holy Family*, 5 figs. before a group of trees (Br), Louvre. – *Holy Family*, 5 figs. on steps, 1648 (F), Paris, private coll. (?). – *Holy Family with Basin* (engr. by Jean Pesense), Cambridge (Mass.), Fogg Art Mus., perhaps a poorly preserved original. – *Holy Family with Fruit Basket*, 9 figs. (engr. by T. Roger), Winterthur, Coll. O. Reinhart. – *Holy Family with Flower Basket*, 10 figs., 1649 (Ch), lost or Dublin, Nat. Gall. of Ireland (?), engr. by C. Bouzonnet-Stella. – *Holy Family with Flower Basket*, 11 figs., 1651 (engr. by E. Baudet), Derbyshire, Trustees of the Chatsworth Settlement.

The Adoration of the Shepherds, with 5 angels (Ch?), London, Nat. Gall. – *The Adoration of the Shepherds*, 7 figs. (engr. by J. B. Nolin), Munich, Alte Pin. – *The Adoration of the Shepherds*, 8 figs.(engr. by T. Roger, dubious), private coll. (?). – *The Adoration of the Shepherds*, 9 figs. in closed stable (engr. by Jean Pesne), lost. – *The Adoration of the Shepherds*, 9 figs. with landscape (engr. by J. Hainzelmann), lost. – *The Adoration of the Magi*, 1633 (Ch), Dresden, Gemäldegal., or Louvre (?). – *The Massacre of the Innocents*(?), ca. 1625(?), in Altieri Palace, Rome, in 17th cent. (Tessin, 1686–87; see O. Sirén, *Nicodemus Tessin*, Stockholm, 1914, p. 187), Paris, Mus. du Petit Palais. – *The Massacre of the Innocents*, for Giustiniani, ca. 1626–28? (Giustiniani invt., 1638), Chantilly, Mus. Condé. – *The Flight into Egypt*, with the boatman (Pader, 1653), Cleve. Mus. (?). – *The Flight into Egypt*, with a recumbent traveler, 1657? (Ch), lost, engr. by P. del Pò. – *The Flight into Egypt*, no particulars known (Br), lost. – *The Rest on the Flight into Egypt*, with an elephant (anon. engr.), lost. – *The Rest on the Flight into Egypt*, with a procession of Osiris, 1655–57 (C), Leningrad, The Hermitage (PL. 236). – *St. John Baptizing the People*, owned in 17th cent. by Le Nôtre (F), Louvre. – *St. John Baptizing the People*, for Dal Pozzo (B), Zurich, Coll. of the late E. Bührle. – *St. John Baptizing Christ*, with God the Father, 1648 (C), New York, Wildenstein Coll. – *St. John Baptizing Christ*, with the dove, ca. 1655–58? (engr. by P. del Pò), Philadelphia, Mus. of Art (Johnson Coll.). – *Christ and the Woman of Samaria*, for Dal Pozzo (De Cotte, 1689), lost. – *Christ and the Woman of Samaria*, for Chantelou, 1661–62, (C), lost, engr. by Jean Pesne. – *The Blind of Jericho*, 1650 (Richelieu invt., 1665), Louvre. – *The Woman Taken in Adultery*, 1653 (Ch), Louvre (PL. 232). – *The Institution of the Eucharist*, 1641 (C), Louvre. – First series of the *Seven Sacraments*, for Dal Pozzo, 1636–42 (C), Leicestershire, Belvoir Castle, Coll. Duke of Rutland (PL. 235), except for *Baptism*, Washington, D.C., Nat. Gall., and *Penance*, destroyed. – Second series of the *Seven Sacraments*, for Chantelou, 1644–48 (C), Mertoun St. Boswells (Scotland), Coll. Earl of Ellesmere, on loan to Edinburgh, Nat. Gall. of Scotland (cf. PL. 236). – *Christ in the Garden of Olives* (S), lost, perhaps drawing preserved at Windsor, Royal Lib. – *The Crucifixion*, 1646 (C), Hartford (Conn.), Wadsworth Atheneum (?). – *The Descent from the Cross* (engr. by F. Chauveau), Leningrad, The Hermitage. – *The Dead Christ in the Lap of the Virgin* (engr. by R. Vuibert), Munich, Alte Pin. – *The Dead Christ before the Sepulcher* (engr. by Jean Pesne), Dublin, Nat. Gall. of Ireland. – *Christ Appearing to Mary Magdalen*, 1653 (F), lost. – *Christ Appearing to Thomas* (anon. engr., dubious), lost.

The Conception of the Virgin, 1628 (Borghese arch., 1628), lost or to be identified with the following work (?). – *The Annunciation*, with the Virgin kneeling (engr. by Gérard Edelinck), Chantilly, Mus. Condé. – *The Annunciation*, with the Virgin seated (not documented, but signed and dated 1657), London, Nat. Gall. – *The Dormition of the Virgin*, 1623 (B), lost. – *The Assumption of the Virgin*, with a choir of angels (Giustiniani invt., 1638), lost (?). – *The Assumption of the Virgin*, with 4 angels, 1650 (C), Louvre.

d. Saints: *The Martyrdom of St. Erasmus*, 1628–29 (Vatican arch., 1628–29; see O. Pollak, *Die Kunsttätigkeit unter Urban VIII*, II : Die Peterskirche in Rom, Vienna, 1931, pp. 79–83, 87, 540 ff.), Rome, Vat. Mus.; Barberini sketch (De Cotte, 1689), Rome, Coll. Barberini; Passart sketch (Br), lost. – Six temperas for the canonization of St. Ignatius of Loyola and St. Francis Xavier, 1622, including *St. Ignatius in Ecstasy*, *St. Ignatius Writing His Meditations*, *Christ and the Virgin Appearing to SS. Ignatius and Francis Xavier*, and *St. Francis Xavier Persecuted by Demons* (B), lost. – *The Miracle of St. Francis Xavier*, 1641 (C), Louvre. – *The Virgin Appearing to St. James the Greater* or the *Madonna del Pilar*, 1629–30 (Ch), Louvre. – *St. John the Baptist* and *St. John the Evangelist*, half-length, pendants, 1628 (Borghese arch., 1628), lost. – *St. Paul*, half-length, before 1624? (engr. by J. Lenfant), lost. – *SS. Paul and Barnabas before the Proconsul* (anon. engr., dubious), lost. – *The Scourging of SS. Paul and Silas* (engr. by Jean Le Pautre), lost. – *The Ecstasy of St. Paul*, for Chantelou, 1643 (C), Sarasota (Fla.), Ringling Mus. – *The Ecstasy of St. Paul*, for Scarron, 1650 (C), Louvre. – *SS. Peter and John Healing the Lame Man*, 1655 (F), New York, Met. Mus., or private coll. (?). – *The Death of Sapphira* (B), Louvre. – *The Marriage of St. Catherine*, ca. 1629–30? (De Cotte, 1689), formerly Richmond (Surrey), Cook Coll. (?). – *St. Mary Magdalen in the Desert* (S), lost. – *S. Francesca Romana* (engr. by P. del Pò), lost. – *St. Margaret* (engr. by F. Chauveau), Turin, Gall. Sabauda.

e. Mythology: (1) *Apollo*: *Apollo and Vergil* or *The Inspiration of the Poet* (Mazarin invt., 1661), Louvre. – *The Inspiration of the Poet* (invt. of Johann Friedrich von Hannover, 1679; see G. Parthey, *Deutscher Bildersaal*, II, Berlin, 1864, p. 294, no. 47), Hannover, Landesmus. – *Apollo Presiding over Parnassus* (Félibien, 1647), Prado. – *Apollo and Daphne*, "first manner" (F?; engr. by F. Chauveau), Munich, Alte Pin. – *Apollo and Daphne* (B), lost, drawing preserved, Derbyshire, Trustees of the Chatsworth Settlement. – *Apollo and Daphne* or *The Misfortunes of Apollo*, for Cardinal Massimi, unfinished, 1664 (B), Louvre. – *Apollo Besought by Phaethon* (arch. of the Académie, 1674; see A. de Montaiglon, *Procès-verbaux de l'Académie royale de peinture et de sculpture*, II, Paris, 1875–1909, p. 21), Berlin, Staat. Mus. – (2) *Bacchus*: *The Birth of Bacchus*, 1657 (B), Cambridge (Mass.), Fogg Art Mus. (?). – *The Birth of Bacchus* (engr. by G. Verini), lost. – *The Childhood of Bacchus* (Louis XIV invt., 1683), Louvre (II, PL. 199). – *The Childhood of Bacchus*, ca. 1629–30 (anon. engr.), Chantilly, Mus. Condé. – *Bacchus and Erigone* (detail engr. by Jean Pesne), Stockholm, Nationalmus., repainted and modified by another artist. – *Bacchus and Ariadne* (S), lost, copy in Leningrad, The Hermitage (?). – *Bacchus and Midas* (Valguarnera trial, 1631), Munich, Alte Pin. – *Midas and a Kneeling Figure*, ca. 1629–30 (Valguarnera trial, 1631), Ajaccio (Corsica), Mus. Fesch. – *Midas Washing in the Pactolus*, ca. 1629–30 (Massimi invt., 1677), New York, Met. Mus. – *Bacchanal with a Lute Player* (Richelieu invt., 1665), Louvre. – *Bacchanalian Revel before a Term of Pan* (anon. engr.), London, Nat. Gall. (PL. 231). – *Bacchanal before a Temple* (F), lost, engr. by Jean Mariette, copies. – *Bacchanal with a Nymph, a Bacchante, a Little Satyr, and Two Putti* (Ruffo arch., 1647–49), Leningrad, The Hermitage (?). – Three bacchanals for Cardinal Richelieu, 1635–37 (B; letter of Frangipani to Richelieu, 1636): *The Triumph of Bacchus*, Kansas City, Nelson Gall. of Art and Atkins Mus. (?); *The Triumph of Pan*, Sudeley Castle (Gloucestershire), Morrison Coll., or Louvre (?); *The Triumph of Silenus*, lost, copy in London, Nat. Gall. – (3) *Diana*: *Diana* (or *Selene*) *and Endymion* (Mazarin invt., 1653), Detroit, Inst. of Arts (PL. 234). – (4) *Flora and the Metamorphoses*: *The Loves of Flora and Zephyrus*, ca. 1630 (F), lost. – *The Triumph of Flora*, ca. 1626–27? (B), Louvre. – *The Realm of Flora*, 1630–31 (Valguarnera trial, 1631), Dresden, Gemäldegal. – *The Coloring of the Rose* (B), lost, drawing preserved at Windsor Castle, Royal Lib. – *The Coloring* (or *Origin*) *of Coral* (B), lost, drawing preserved at Windsor Castle, Royal

Lib. – *Narcissus*, owned in 17th cent. by Le Nôtre (F), lost (?). – *Narcissus*, ca. 1630 (Louis XIV invt., 1683), Louvre. – *Pan and Syrinx*, ca. 1637–38 (C), Dresden, Gemäldegal. – (5) *Hercules*: *Hercules between Vice and Virtue* (F), Stourhead (Wiltshire), The Nat. Trust (?). – *Hercules Carrying Off Deianira*, ca. 1637–38 (Ch), lost, drawing preserved in the Louvre. – (6) *Jupiter*: *Jupiter Nourished by Nymphs* (engr. by G. Chasteau), Berlin, Staat. Mus. – *Jupiter Carrying Off Europa* (C), lost. or not executed (?), drawing preserved in Stockholm, (engr. by L. de Châtillon), lost. – (7) *Neptune*: *The Triumph of Neptune* Nationalmus. – *Jupiter and Antiope*, often called *Hermaphrodite* (?) (engr. by B. Picart), lost. – *Danaë Reclining* (F), lost. – *Leda and the Swan* (engr. by L. Châtillon), lost. – (7) *Neptune: The Triumph of Neptune and Amphitrite*, ca. 1636 (B), Philadelphia, Mus. of Art. – (8) *Venus*: *Venus after the Bath* (engr. by E. Baudet), lost. – *Venus Guarded by Three Loves* (Br), disfigured in the 17th cent., lost. – *Venus and Adonis with Four Putti* (Ruffo invt., 1649), lost. – *Venus Mourning Adonis* (Louis XIV invt., 1683), Caen, Mus. B.A. – *Venus and Mars* (engr. by F. Chiari), lost. – *Venus and Mercury* (engr. by F. Chiari), cut up in the 18th cent., poorly preserved fragments in the Louvre and London, Dulwich College Picture Gall. (?). – *Venus Arming Aeneas*, 1639 (B), Rouen, Mus. B.A. – (9) *Miscellaneous: Acis and Galatea* (engr. by A. Garnier; Br), Dublin, Nat. Gall. of Ireland (?). – *Aurora and Cephalus* (B), Hovingham Hall (Yorkshire), Coll. Sir William Worsley.

f. Legend and ancient history: Achilles and the Daughters of Lycomedes, with Achilles unsheathing his sword (B), lost or Boston, Mus. of Fine Arts (?). – *Achilles and the Daughters of Lycomedes*, with Achilles holding a mirror, 1656 (B), Richmond, Virginia Mus. of Fine Arts, perhaps a much overpainted original. – *Medea* (B), lost, drawing preserved at Windsor Castle, Royal Lib. – *Theseus Discovering His Father's Sword* (anon. engr.), Chantilly, Mus. Condé (?). – *The Testament of Eudamidas* (B), Copenhagen, Statens Mus. for Kunst (?). – *Battle Scene with Porus on an Elephant* (De Cotte, 1689), lost (?). – *The Rescue of the Young Pyrrhus* (Richelieu invt., 1665), Louvre. – *The Rape of the Sabines*, ca. 1635, owned in the 17th cent. by Duchesse d'Aiguillon (F), New York, Met. Mus. (VII, PL. 268). – *The Rape of the Sabines*, for Cardinal Omodei (B), Louvre. – *Coriolanus* (B), Les Andelys (Eure), Mus. Poussin. – *The Schoolmaster of Falerii*, for Passart, ca. 1635 (F), private coll. – *The Schoolmaster of Falerii*, for La Vrillière, 1637 (C), Louvre. – *The Capture of Jerusalem by Titus*, 1625–26 (Barberini arch., 1626; see *Actes du Colloque . . . Poussin*, I, 1960, pp. 3–4, n. 12), lost. – *The Capture of Jerusalem by Titus*, second version (B), Vienna, Kunsthist. Mus. – *The Death of Germanicus*, 1627 (Barberini invt., 1631), Minneapolis, Inst. of Arts.

g. Jerusalem Delivered: Armida Coming upon Rinaldo Asleep (engr. by G. Audran), London, Dulwich College Picture Gall. (?). – *Armida Carrying the Sleeping Rinaldo*, 1637–38 (C), lost, engr. by G. Chasteau, copies. – *Armida Carrying the Sleeping Rinaldo* (B), lost, drawing preserved at Windsor Castle, Royal Lib. – *Two Knights on Their Way to Deliver Rinaldo* or *The Companions of Rinaldo* (De Cotte, 1689), Vienna, Coll. Harrach.

h. Allegorical subjects: Et in Arcadia ego, ca. 1629–30 (B), Derbyshire, Trustees of the Chatsworth Settlement. – *Et in Arcadia ego*, second version (F), Louvre (I, PL. 307). – *The Dance of Human Life*, symbolized by four children (Mazarin invt., 1661), lost (?). – *The Dance of Human Life*, symbolized by the Four Seasons in the guise of women (B), London, Wallace Coll. – *Time and Truth*, ceiling, 1641 (B), Louvre. – *Time and Truth*, painting (B), lost, engr. pub. by Jean Dughet. – *Victory above the World* (Massimi invt., 1677), lost.

i. Miscellaneous subjects: Women Bathing, ca. 1635 (F), lost, engr. by B. Picart. – Two renditions of *Bacchanal with Children* (Chigi invt., 1693; see G. Incisa della Rocchetta, "I 'Baccanali Chigi' di Nicolas Poussin," *Paragone*, II, 15, 1951, p. 38 ff.), Rome, Coll. Incisa della Rocchetta. – *Five Children Playing* (anon engr.), formerly London, Grosvenor House, Coll. Duke of Westminster (?).

j. Landscapes: Four landscapes with scenes from the Old Testament, known as the *Four Seasons*, 1660–64 (Richelieu invt., 1665), Louvre. – *Landscape with St. John on Patmos* (engr. by L. de Châtillon), Chicago, Art Inst. – *Landscape with Mercury and Argus* (Giustiniani invt., 1638), Berlin, Staat. Mus. – *Landscape with Polyphemus* (F) and *Landscape with Hercules and Cacus* (not documented, but pendant), Leningrad, The Hermitage, and Moscow, A.S. Pushkin Mus. of Fine Arts. – *Landscape with Orion*, 1658 (B), New York, Met. Mus. – *Landscape with Orpheus and Eurydice* (arch. of 1685; see J. Guiffrey, *Les Comptes des bâtiments du roi sous le règne de Louis XIV*, II, Paris, 1887, pp. 586, 664), Louvre (IX, PL. 15). – *Landscape with Diogenes* (Richelieu invt., 1665), Louvre. – *The Funeral of Phocion* (PL. 235) and *The Woman of Megara Gathering the Ashes of Phocion*, pendants, 1648 (Ch), Oakly Park, Ludlow, Coll. Earl of Plymouth, and Knowsley Hall (Lancastershire), Coll. Earl of Derby (?). – *Landscape with Pyramus and Thisbe*, 1651 (C), Frankfort on the Main, Städelsches Kunstinst. – *Landscape with a Snake*, 1648 (Félibien, 1679), London, Nat. Gall. – *Landscape with a Large Paved Road*, 1648 (engr. by E. Baudet), lost. – *Landscape with a Large Sandy Road*, 1648? (engr. by E. Baudet), lost. – *Landscape with Three Monks* (F), Belgrade, Palace of the President of the Yugoslav Republic. – *Landscape with Three Men* (F?), Prado. – *Landscape with a Woman Washing Her Feet*, 1650 (F), Ottawa, Nat. Gall. of Canada (?). – *Landscape with a Tree Blasted by Lightning* (F?), lost, engr. by L. de Châtillon. – *Landscape with Two Nymphs*, ca. 1659? (engr. by L. de Châtillon), Chantilly, Mus. Condé.

k. Decorations: Loggia in the Château of Cheverny, before 1624? (Félibien, 1681), lost. – Oratory of Father Caravita at the Collegio Romano in Rome, ca. 1633 (P), lost. – Grande Galerie in the Louvre, 1641–44 (C), lost, engravings and drawings preserved. – Orangery in the Luxembourg, Paris, 1642 (Sublet de Noyers correspondence, 1642), planned but not executed. – Terms for Vaux-le-Vicomte, 1655–56 (correspondence of Abbé Louis Fouquet, 1655–56), and gardens of the Château of Versailles. – Two vases (correspondance of Abbé Louis Fouquet, 1655–56), lost.

l. Illustrations: Plate for the *Hesperides* (1646) of G. B. Ferrari, before 1640, engr. by C. Bloemaert. – Frontispieces for the Bible (1641), a Vergil (1641), and a Horace (1642) issued by the Typographie Royale, engr. by C. Mellan.

SOURCES. Correspondance de Nicolas Poussin, ed. C. Jouanny, Paris, 1911; G. Mancini, Considerazioni sulla pittura (1614–28), ed. A. Marucchi and L. Salerno, 2 vols., Rome, 1956–57; Fréart de Chantelou, Journal du voyage du cavalier Bernin en France (1665), ed. L. Lalanne, GBA, 1877–84, new ed., Paris, 1885; G. P. Bellori, Le Vite de' pittori, scultori et architetti moderni, Rome, 1672; J. von Sandrart, Teutsche Academie . . . , Nürnberg, 1675, II, 3, xxvi; G. B. Passeri, Vite de' pittori, scultori ed architetti che anno lavorato in Roma, morti dal 1641 fino al 1673 (written before 1678), Rome, 1772, ed. J. Hess, Leipzig, Vienna, 1934; A. Félibien des Avaux, Entretiens sur les vies et sur les ouvrages des plus excellens peintres, 4e Partie, 8e Entretien (on Poussin), Paris, 1685; L.-H. Loménie de Brienne, Discours sur les ouvrages des plus excellens peintres anciens et nouveaux (ca. 1692–95), Paris, Bib. Nat., Ms. fr. 16986, pub. in Thuillier, 1960, pp. 210–24. For other 17th-century sources see J. Thuillier, Pour un "Corpus Pussinianum," Actes du Colloque international Nicolas Poussin, II, Paris, 1960, pp. 49–238 (ca. 150 texts reproduced).

BIBLIOG. For a comprehensive bibliog. to 1960 see Exposition Nicolas Poussin, ed. A. Blunt, C. Sterling, et al., 2d corr. ed., Paris, 1960, pp. 285–328. *Monographs and essays*: E. Delacroix, Essai sur le Poussin, Moniteur Universel, June 26–30, 1853 (reprinted in E. Delacroix, Œuvres littéraires, II, Paris, 1923); P. Desjardins, Poussin, Paris, 1906; W. Friedlaender, Poussin, Munich, 1914; O. Grautoff, Poussin, 2 vols., Munich, 1914; E. Magne, Poussin, Brussels, Paris, 1914; G. de la Tourette, Poussin, Paris, 1929; P. du Colombier, Poussin, Paris, 1931; W. Friedlaender, ThB, s.v., 1933; A. Gide, Poussin, Paris, 1945; B. de la Société Poussin, ed. T. Bertin-Mourot, I–III, Paris, 1947–50; A. Blunt, Poussin Studies, BM, I–II, 1947, III, 1948, IV–V, 1950, VI, 1951, VII, 1958, VIII, 1959, IX–XI, 1960, XII, 1961, XIII, 1962, XIV, 1964; P. Jamot, Connaissance de Poussin, Paris, 1948; Actes du Colloque international Nicolas Poussin, ed. A. Chastel, 2 vols., Paris, 1960; G. Kauffmann, Poussin Studien, Berlin, 1960; D. Mahon, Poussiniana, Paris, New York, 1962; L'ideale classico del Seicento in Italia e la pittura di paesaggio (cat.; incl. article on Poussin by D. Mahon), Bologna, 1962; W. Friedlaender, Poussin, New York, 1965.

Catalogues: *a. Paintings*: J. Smith, A Catalogue Raisonné of the Works of the Most Eminent Dutch, Flemish and French Painters, VIII, Poussin, London, 1837; W. Friedlaender, Poussin, Munich, 1914; O. Grautoff, Poussin, II, Munich, 1914 (with reproductions of all paintings); E. Magne, Poussin, Brussels, Paris, 1914; T. Bertin-Mourot, Addenda au catalogue de Grautoff, B. de la Société Poussin, II, 1948, pp. 43–87; J. Thuillier, Tableaux attribués à Poussin dans les archives révolutionnaires, Actes du Colloque international Nicolas Poussin, II, 1960, pp. 27–44; J. Thuillier, Tableaux attribués à Poussin dans les galeries italiennes, loc. cit., pp. 263–78; J. Thuillier, Tableaux "poussinesques" dans les musées de province français, loc. cit., pp. 285–99; Exposition Nicolas Poussin (ed. A. Blunt), 2d ed., Paris, 1960, pp. 39–185. *b. Drawings*: The Drawings of Nicolas Poussin: Catalogue Raisonné, ed. W. Friedlaender, A. Blunt, et al., I–IV, London, 1939, 1949, 1953, 1963. *c. Engravings*: F. Le Comte, Cabinet des singularites, II, 2, Paris, 1699, pp. 129–39 (reproduced in Wildenstein, 1957); A. Andresen, Poussin: Verzeichniss der nach seinen Gemälden gefertigten . . . Kupferstiche, Leipzig, 1863; G. Wildenstein, Les Graveurs de Poussin au XVIIe siècle, Paris, 1957 (with reproductions of all engravings); M. Davies and A. Blunt, Some Corrections and Additions to M. Wildenstein's "Graveurs de Poussin au XVIIe siècle," GBA, 1962, pp. 205–22.

Jacques THUILLIER

Illustrations: PLS. 231–236.

POZZO, FRA ANDREA. Italian perspective painter and architect (b. Trent, Nov. 30, 1642; d. Vienna, Aug. 31, 1709). After being apprenticed to an obscure painter in Trent, Pozzo entered the order of Discalced Carmelites in 1661, as a novice in the Convento delle Laste, near Trent. After a year, because of illness, he left the monastery and Trent for Milan, where he worked for one year. It is known that afterward he was in Como. No painting from this period of his career has survived.

On Dec. 23, 1665, he became a lay brother of the Society of Jesus in Milan, in which his task was to be "cook and dishwasher." His gifts as a painter were discovered when he executed a *machina* for the forty hours' devotion.

In 1671 he did altar paintings for chapels in the Chiesa del Gesù (SS. Ambrogio e Andrea) in Genoa. His other activity in Liguria includes works for S. Maria Maggiore in Novi Ligure and for S. Stefano in San Remo. In 1675 he was sent to Mondovì to correct with "perspective frescoes" the interior proportions of the Jesuit church, S. Francesco Saverio; his work on this project lasted until 1678. Thereafter he was in Turin (where he had already visited briefly in 1675 to see, as he himself says, "le cose belle") on a commission for the Duke of Savoy. But in local Jesuit files he is still recorded as "socius coqui." In 1681 he was back in Milan, and from there he was probably sent to Modena to decorate the Church of S. Bartolomeo, on which he did architectural work as well, in conjunction with Fra Luigi Barbery.

On the advice of Carlo Maratti, early in 1682, he was summoned to Rome by the Jesuit general Padre Oliva, who died that same year. Pozzo continued to be "cocus" in the Jesuit Casa Professa. When, in 1685, he offered to paint a forty hours' *machina* (engraved in his *Tractatus perspectivae pictorum et architectorum*, Rome, 1692–97; Eng. ed., 1707), the Jesuits commissioned him to decorate a corridor of the Casa Professa, near the Gesù, and to correct with decoration the proportions of the apse of the Church of S. Ignazio. About the same time, on a large canvas he painted the false dome (now restored) for the same church, a work that aroused the admiration of the whole city. Three years later, in 1688, he began the large-scale frescoes of the ceiling of the nave, the first segment being the *Triumph of St. Ignatius* (II, PL. 172); these grandiose paintings were all based on a subjective, but very carefully calculated *prospettiva* (see PERSPECTIVISTS). This ambitious project was finished in 1694. Earlier, in 1687, he had done the festive decoration in the Jesuit Collegio Romano for the visit of the English ambassador.

Pozzo did several projects for the great St. Ignatius altar in the Church of the Gesù in 1695, and the chosen design was finished in 1699. In 1697 he did the altar of St. Louis in the transept of the Church of S. Ignazio. A lay brother, Carlo Maria Bonacina, assisted him in the work on both altars. With the help of the Jesuit Antonio Colli, in 1697 he decorated with frescoes the Jesuit church in Frascati. Active in Montepulciano in 1702, Pozzo executed frescoes in the Palazzo Contucci as well as other works for several churches, and about the same time he decorated the dome of the Badia in Arezzo.

Pozzo left Rome for Vienna in the same year, having been called there by the Prince of Liechtenstein. Enroute he stopped in Trent, where he drew up plans for the local Jesuit church. Previously he had prepared a project for the Jesuit church in Dubrovnik (1699) and the plan for the Ljubljana Cathedral (1701). Upon his arrival in Vienna in 1703, he painted the ceiling of the imperial villa "La Favorita" (destroyed). In only one year (1703–04), he rebuilt the Universitätskirche and decorated it with perspective frescoes (repainted) and altar paintings; though he had many assistants in this undertaking, Pozzo himself furnished all the drawings. His masterpiece in Vienna is the ceiling fresco in the main reception hall of the Liechtenstein Palace, depicting the Labors of Hercules. In the Marmoorsaal of the same palace he painted some profane subjects, and he also did the paintings (sold in Paris during the 19th cent.) for the small chapel. Afterward he made several altars (almost all lost) for various Jesuit churches in Vienna and furnished a project for the Doroteakirche (destroyed). After being back in Milan briefly in 1709, he returned to Vienna, where he died in the same year.

Pozzo had a very decided influence on Austrian and German painting and decoration of the 18th century. His treatise was translated into many languages (even into Chinese) and became a widely used method in many countries ("un vero e proprio strumento di lavoro"). According to sources, he did many portraits and other easel paintings, but, except for two self-portraits (1685, Uffizi; 1703, Trent, Mus. Naz.), none of these has as yet been identified. (See also, XII, PL. 431.)

BIBLIOG. R. Marini, Andrea Pozzo pittore, Trent, 1959; N. Carboneri, A. Pozzo architetto, Trent, 1961 (bibliog.).

(The dates cited in this article have been taken from documents preserved in the archives of the Jesuit mother house in Rome).

Fernanda de' MAFFEI

PRAXITELES. The most eminent Athenian sculptor of the 4th century B.C., Praxiteles worked both in bronze and in marble. He was probably the son of the sculptor Kephisodotos, since he named one of his own sons Kephisodotos, and it was customary to give a son the name of his grandfather. The dates of his birth and death are unknown, but his working life seems to have extended from about 375 to about 330 B.C. Pliny (*Naturalis historia*, XXXIV, 50) places him in the 104th Olympiad (364–361 B.C.), which may be the date of his most famous statue, the *Aphrodite of Knidos*. Two signatures survive, but they cannot be dated with precision. Most of his works were in Athens (where, in addition to statues of gods, he is known to have made at least one grave monument: Pausanias, I, 2, 3) and in Greece proper. When he became more widely known, he made statues for other cities, especially in eastern Greece, but except for an *Eros* at Messene, said to be a replica of another at Thespiai, he seems not to have worked for the Greeks in Italy or Sicily. There is no reason to assume that he traveled abroad in order to execute commissions, although he may have done so.

Some idea of his artistic development can be formed by identifying copies from Roman times with works attributed to him by ancient authors and arranging them in order of stylistic evolution. Assistance is given by reproductions of some works on coins of the Roman period, but many mentioned by ancient authors have not been recognized in copies. Six surviving sculptures have been claimed as originals: namely, the *Hermes* from Olympia (PL. 239), the Petworth head (Sussex, Petworth House), the Aberdeen head (Br. Mus.), the so-called "Eubouleus" from Eleusis (Athens, Nat. Mus.), the bronze youth from the sea near Marathon (II, PL. 56), and a head from Chios (Boston, Mus. of Fine Arts).

Of the acknowledged copies, that of a young satyr pouring wine (PL. 237) reproduces a bronze, perhaps the one seen by Pausanias in the Street of the Tripods in Athens (I, 20, 1). The style shows that it must be an early work (ca. 375 B.C.) and that it owes much to the statues of young athletes by Polykleitos.

It is known from Pausanias that a satyr and a marble Eros were together in the studio of Praxiteles and that his mistress Phryne dedicated the *Eros* at Thespiai, her native city, where it became exceedingly famous and was twice carried off to Rome (Pausanias, I, 20, 1–2; IX, 27, 3). An Eros closely related to the young satyr is known in copies (e.g., one from the Palatine; Louvre), and this possibly represents the type; but other claims have been made.

Not identifiable with any statue mentioned by ancient authors but assigned to this period on grounds of style is the youthful *Artemis* in Dresden (PL. 241), a marble copy of a bronze. It resembles in some respects the *Eirene* of Kephisodotos but bears the marks of a new personality, and forms a link between the satyr and another bronze appreciably later in development (ca. 370 B.C.), the *Apollo Sauroktonos*.

The *Apollo Sauroktonos* has been identified, with the aid of a description by Pliny (*N.H.*, XXXIV, 70), in a number of copies (one in bronze, PL. 237; cf. also III, PL. 374) and also on Roman coins of Nikopolis on the Danube and Philippopolis in Thrace: a boy waiting with an arrow to kill a lizard leans forward with the weapon poised in his right hand and rests his left elbow on the tree trunk up which the lizard crawls.

The most famous of all the works of Praxiteles was the marble statue of Aphrodite dedicated at Knidos (Pliny, *N.H.*, XXXVI, 20; III, PL. 374). Reproductions on Roman coins of the city have served to identify copies, but these differ from each other so seriously that it is possible that the original was never cast; most are from Roman times, but fragments from Tralles in Berlin (head and middle of body; PL. 238) may be part of a Hellenistic version. The goddess was naked — a bold innovation at this period, although common in primitive times — and was shown about to step forward, laying aside her drapery over a waterpot, these two elements forming the support necessary for a marble figure in such free movement. The model was said to have been Phryne at the ritual bathing before an Eleusinian festival (Athenaeus, XIII, p. 590). Ancient eulogies show that the beauty of the statue depended equally on its main design and on the quality of its modeling — of the latter, surviving copies tell us little.

According to Pliny (*N.H.*, XXXVI, 20), Praxiteles had offered the people of Kos a choice between two types of statues of Aphrodite, and they selected a draped figure. The type chosen has not been identified, but the *Aphrodite* found at Arles, which is draped from the waist downward, seems to reproduce a Praxitelean work of the period 360–350 B.C.

Also close to the Knidian work in date, though sometimes attributed to Kephisodotos, was a Dionysos, bearded and heavily draped (known as the "Dionysos Sardanapallos," from an inscription on the replica in the Vatican Museums), of interest as a 4th-century version of a type that had originated in the 5th century.

A statue of Hermes carrying the child Dionysos (PL. 239), excavated in the Heraion at Olympia in 1877, was (after initial hesitation) accepted as an original of Praxiteles, seen there by Pausanias (V, 17, 3) in the 2d century of our era — though he does not say that it was signed or how he knew its authorship. Doubts have been raised on technical grounds as to whether this could be a work of the 4th century B.C. (C. Blümel, 1927), and since these initial reservations there has been a prolonged and still undecided discussion as to whether it is an original or a copy (perhaps from a bronze, in which case the drapery might be a copyist's addition) substituted when the original was carried off to Rome. The marble *Eros* by Praxiteles at Thespiai, where a copy by Menodoros of Athens was substituted, is cited as analogous, but Pausanias recognized the copy as such. An alternative suggestion is that the *Hermes* is a work of a Hellenistic sculptor of the same name. Whatever the truth, the statue, apart from certain mainly technical details (e.g., the very high polish, the form of the *puntello* between the tree trunk and figure, the unfinished and uncertain state of the back, and perhaps the drapery), fits well into the work of Praxiteles between 350–340 B.C. Thus, the *Hermes* can be accepted with the reservation that it may be a Hellenistic or Roman version and therefore not trustworthy in details. It is known that Kephisodotos made a group with the same subject (Pliny, *N.H.*, XXXIV, 87); moreover, the subjects and composition strongly recall the bronze group of Eirene with the child Ploutos on her arm, also by Kephisodotos, dedicated on the Areopagos at Athens in the years after 375 B.C. (III, PL. 373).

Two heads stand very close to the *Hermes* in both style and technique: one, presumably of Aphrodite (Sussex, Petworth House), the model for which was a woman older than the one for the Knidian work; the other, the Aberdeen head (Br. Mus.), which may be a young Herakles.

Less close to the *Hermes* in style is the head found at Eleusis near an inscription to Eubouleus, the swineherd of Eleusinian myth. That Praxiteles made a statue of Eubouleus is proved by an inscribed (but headless) herm in the Vatican Museums. The head from Eleusis, of which there are other replicas, is sometimes thought not to be an original, and the subject has lately even been identified as Alexander the Great.

The *Eros* of Parion in Mysia (Pliny, *N.H.*, XXXVI, 22) seems from its composition to have belonged to this period; it is represented on local coins of Roman date, and one copy in marble has been found on the island of Kos.

The statue of *Artemis Brauronia* on the Acropolis at Athens was by Praxiteles, according to Pausanias (I, 23, 7). A statue from Gabii (PL. 241) may be a copy, since the action, that of fastening the cloak, would have been appropriate in a shrine where offerings of garments are known to have been made. An inscription of 345 B.C. mentions a cult statue in this shrine, but this date seems somewhat early for the style of the statue from Gabii.

A young satyr (PL. 240), not identified with any mentioned by ancient authors (though it may have been that in the Street of the Tripods in Athens, where the account of Pausanias is confused), from its composition and style was almost certainly by Praxiteles in his middle period. It was immensely popular in Roman times, more than sixty copies being known: the satyr rests with his left elbow on a tree trunk (as in the *Apollo Sauroktonos* an integral part of the composition), and this allows the free leg to be moved backward; but the view intended is still frontal. In the features there is an attempt to show the blend of animal and human in mind as well as in body.

A late work, probably an original, is the bronze youth, perhaps Hermes, from the sea off Marathon (II, PL. 56). The features show the fully developed Praxitelean style, and the spiral composition, related to but more advanced than the *Apollo Sauroktonos*, begins to be truly three-dimensional: the right arm is stretched out but seems not to have been supported; a large object was held on the left forearm.

Pausanias records a group of Leto, Apollo, and Artemis by Praxiteles in a temple at Mantineia, on a base with Apollo, Marsyas, and the Muses (VIII, 9, 1); three of the four slabs of this base have survived (PL. 241). Probably the work of pupils, in late Praxitelean style, they show various types of draped figures current in this period and school.

Close in date and style to these is the exquisite head from Chios in Boston, an original, but perhaps of the next generation, in which the main design is strongly geometric but the lips are closely studied from nature and the surface treatment is extremely soft.

Praxiteles was a pupil, even if not a son, of Kephisodotos and with him reacted against the excessive refinement of the Attic school of the late 5th century. They studied the grand and simple statues of a century earlier and on this basis developed a new classical style that was sensitive and even sweet, without at first any loss of strength. Praxiteles experimented with new poses and with the subtleties of subsidiary modeling, softening the sharp divisions of 5th-century musculature to produce a flowing surface, a delicate play of light and shade, and contrasts of texture with hair and drapery. He depicted emotion, but it was of mood rather than passion; and in subject he had a predilection for youth and gentleness: Eros and Aphrodite were favorites, and athletic subjects were rare. His treatment of myth was apt to be idyllic or even whimsical: the *Apollo Sauroktonos*, for example, is an echo of the solemn theme of the killing of the dragon at Delphi. But the gods were brought down to human level: Phryne frankly served as the model for statues of Aphrodite, and her dedication of a statue of herself at Delphi caused a scandal even at the time.

In technique and in choice and treatment of subject, he foreshadowed many of the developments of the Hellenistic age, which, along with Skopas, he influenced profoundly — and thus ultimately the sculpture of the Renaissance and of modern times. His fame, especially as a marble sculptor, was great in his lifetime and even greater after his death throughout antiquity.

BIBLIOG. G. E. Rizzo, Prassitele, Milan, Rome, 1932; M. Bieber, ThB, s.v. Praxiteles, II; T. L. Shear, The Campaign of 1936, Hesperia, VI, 1937, p. 339 (signature); T. L. Shear, The Campaign of 1937, Hesperia, VII, 1938, p. 329 (same); H. K. Süsserott, Griechische Plastik des 4. Jahrhunderts v. Chr., Frankfurt am Main, 1938, pp. 129–95; G. Becatti, Un dodekatheon ostiense e l'arte di Prassitele, ASAtene, N.S., I-II, 1939–40, pp.85–137; I. V. Zeest, Praksitel', Moscow, 1941; Picard, III, pp. 406–632, IV, pp. 237–410; Lippold, GP, pp. 234–43; G. M. A. Richter, The Sculpture and Sculptors of the Greeks, 3d. ed., New Haven, 1950, pp. 259–67; G. Lippold, RE, s.v. Praxiteles, no. 5; L. Alscher, Griechische Plastik, III, Berlin, 1956, pp. 147–52. *Aphrodite of Knidos*: C. Blinkenberg, Knidia, Copenhagen, 1933; T. Kraus, Die Aphrodite von Knidos, Bremen, 1957. *Hermes of Olympia*: C. Blümel, Griechische Bildhauerarbeit (JdI, sup., XI), Berlin, 1927, pp. 37–48; R. Carpenter et al., Symposium on the Hermes of Praxiteles, AJA, XXXV, 1931, pp. 249–97 (bibliog.); D. A. Antonsson, The Praxiteles Marble Group in Olympia, Stockholm, 1937; C. H. Morgan, The Drapery of the Hermes of Praxiteles, *'Εφημ*, 1937, pp. 61–68; C. Blümel, Hermes eines Praxiteles, Baden-Baden, 1948; R. Carpenter, Two Postscripts to the Hermes Controversy, AJA, LVIII, 1954, pp. 1–12; L. Laurenzi, Il prassitelico Eros di Parion, RIASA, N.S., V–VI, 1956–57, pp. 111–18. *Eubouleus*: E. B. Harrison, New Sculpture from the Athenian Agora, Hesperia, XXIX, 1960, pp. 369–92 at 382. *Eros of Parion*: L. Laurenzi, Sculture inedite del Museo di Coo, Ann. della Scuola Archeol. di Atene N.S., XVII–XVIII, 1955–56, pp. 66–69.

Bernard Ashmole

Illustrations: PLS. 237–241.

PREHISTORY. The concept and the term "prehistory" originated in the 19th century to designate, broadly, the epochs preceding the civilizations of the past of which there remains "historical" evidence — that is, either written records or oral traditions subsequently committed to writing. This definition, obviously, is in the first place a relative one, in that it is based on the nature of our sources of knowledge, which are strictly limited, for the cultures we call "prehistoric," to physical remains. Yet the distinction between "prehistory" and "history" does have an absolute value, both chronologically and culturally, prehistory referring to older and inferior stages of human development. It goes without saying that these stages, however remote, nonetheless belong to the history of human progress; and for this reason the term "prehistory" is less appropriate than that of "primitive history" or "origins" (an idea better expressed by the German *Urgeschichte*). Moreover, it is clear that no precise demarcation can be established between prehistoric and historic times, if one excepts the purely relative one of the first use of writing, which in any case does not correspond to a general chronological delimitation and does not represent (or represents only one-sidedly) an authentic and decisive evolutionary stage of human civilization. Consequently there is great latitude and diversity in the use and application of the term "prehistory" by scholars and modern scientists.

Since the concern here is primarily with cultural and artistic phenomena, "prehistoric" is taken to designate the whole complex of human activities belonging to the geological epoch anterior to the present one — that is, the Pleistocene (with which the cultural periods of the Paleolithic, or Old Stone Age, are coeval) — as well as the cultures of the hunters of paleolithic tradition flourishing in the Mesolithic cultural period at the beginning of the present geological epoch, before the establishment of the sedentary cultures of the farmers in the Neolithic period and the metal ages. These sedentary cultures, which heralded the formation of the earliest historical civilizations, have been termed "protohistoric." Such a restrictive concept of prehistory finds its justification in the character and in the modern critical evaluation of "prehistoric art," recognized essentially in the great Paleolithic figural manifestations and their derivatives.

This article, therefore, deals primarily with the Paleolithic, whose productions have a fundamental world-wide importance as the oldest artistic relics of humanity. After Paleolithic art, however, the article considers the art of the Mesolithic hunters and of the conservative cultures of more recent times that have adopted these traditions, especially rock painting and engraving. The article also touches on the artistic achievements of primitive peoples living in the present period or one not far removed from it. Those areas whose prehistoric art is given detailed consideration in other articles are omitted or are given only summary treatment with cross references to the relevant information. Not specifically treated here are the artistic phenomena dependent on the cultural revolution of the Neolithic period (see ASIATIC PROTOHISTORY; MEDITERRANEAN PROTOHISTORY and, for the Americas, ANDEAN PROTOHISTORY and MIDDLE AMERICAN PROTOHISTORY).

Framework and chronology. The accidental discovery in 1879 of the cave paintings of Altamira in the Cantabrian Mountains of Spain first revealed to the modern world a pictorial art dating from the Paleolithic and belonging therefore to the geological epoch preceding the present one. Some years earlier the paleontologist E. Lartet had subdivided the Lower Quaternary, or Pleistocene, period into three phases, named (after the dominant fauna) the Age of the Hippopotamus, the Age of the Mammoth and the Great Bear, and the Age of the Reindeer. Today the earliest prehistory is subdivided, on the basis of climatic variations and cultures, into the following approximate periods:

1. The Lower Pleistocene, embracing the Günz and Mindel glaciations and the intermediate warm interglacial period. It probably endured for several hundred thousand years (estimates place the beginning of this period about a million years ago) and witnessed the emergence of very primitive human races and the making of crude tools, roughly shaped from pebbles (cores) or splinters struck from these (flakes). The best-defined and most widespread of these tools is the bifacial pointed tool, the *coup de poing*, or hand ax, peculiar to the Abbevillian culture (named from the terraces of the Somme at Abbeville, in France).

2. The Middle Pleistocene, covering the long Mindel-Riss interglacial period and the climaxes of the Riss glaciation, a span of two or three hundred thousand years. It is characterized by the Acheulean culture (named from Saint-Acheul, near Amiens), with its hand axes of more advanced workmanship, slender, flattened, and regular in form (PL. 242).

3. The period including the Riss-Würm interglacial and the first climax of the Würm glaciation. Beginning about 180,000 years ago and lasting for roughly 140,000 years, it saw the emergence of Neanderthal man and the development of the Levalloisian and Mousterian cultures, characterized by specialized tools made by retouching stone flakes.

4. The period including the interstadials of the Würm glaciation and its last two climaxes, beginning about 40,000 years ago and lasting some 30,000 years. It was marked by the European appearance of contemporary man (*Homo sapiens*) and the fashioning of specialized tools from stone blades (longer, more regular flakes) and from bone. Its cultural subdivisions, because of their importance, are treated below in greater detail.

The general, broadly inclusive term "Paleolithic" has been adopted for the entire development of the Pleistocene cultures. The first three periods, of long duration and difficult to circumscribe, are referred to as the Lower Paleolithic (or, by some authors, as the Lower and Middle Paleolithic, the latter term generally designating the third period). The fourth period, of more limited duration, is commonly referred to as the Upper Paleolithic.

A fundamental fact in the history of world art is the absence of any evidence of figural representation in the phases of human development prior to the Upper Paleolithic. Production seems to have been limited to the manufacture of utilitarian objects, in which, however, some well-defined and constant traditions of form reflect functional and technical requirements. The hand ax, in particular, assumed in its Acheulean version (PL. 242) the character of a carefully worked tool, not without a certain elegance of line, in which the refining of execution had already become a "gratuitous" act, a gratification of the esthetic sense. To a far greater extent regularity of form, symmetry, geometrizing tendencies, and minute retouches and indentations were to give a decorative value to the stone and bone tools produced in the later Paleolithic phases.

Art in the sense with which this article is concerned, however, can be said to appear only with the occurence of signs intentionally incised or painted on objects or on rock walls and graphic or sculptural reproduction of the forms of living beings. Such manifestations began in the Upper Paleolithic and continued, practically without interruption, in succeeding prehistoric epochs. The cultures of the Upper Paleolithic evidently mark a decisive "leap" in the development of humanity.

The traditional division of the European Upper Paleolithic into the cultures of the Aurignacian (from the cave of Aurignac, Haute-Garonne), Perigordian (named after the region of Perigord and identified by D. Peyrony in 1933), Solutrean (from Solutré, Saône-et-Loire), and Magdalenian (from the cave of La Madeleine at Tursac, Dordogne), with relatively minor subdivisions, offers a rough scheme, substantially valid for France, allowing a relative chronological classification of artistic phenomena. The method of dating by radiocarbon (carbon 14), applied since the late 1940s, gives more precise information on the relations between these cultural horizons and on their absolute chronology. Indeed, notwithstanding the reservations that may legitimately be entertained on technical grounds as to the exactitude of this method (reservations other than those concerning the statistical uncertainty affecting the measurements, which must be expressed with a standard deviation of a few hundred years), it must be recognized that it is far superior to even the best-informed subjective estimate. The dates that follow, obtained by carbon 14 measurements, are to be understood as the number of years before Christ (not from the present); they are medial values and are merely indicative.

The first artistic manifestations are attributed to the Aurignacian. The typical Aurignacian of the shelter of La Quina in Charente has been dated 29,200, and that of Willendorf in Moravia — the deposit that contained a celebrated Venus — has yielded dates of 29,880 for level IV and 28,350 for level VII. It is at about 30,000 B.C., then, according to the present state of knowledge, that the origins of prehistoric cave art may be placed. The Perigordian IV of Abri Pataud (at Les Eyzies, Dordogne), which yielded a delicate female figure deeply engraved in a block of limestone, is dated about 21,600. An archaeological stratum of the Lascaux cave in which conifer charcoal was associated with a large bone spear and some flints (among them a blunt-backed blade of Gravettian tendency) is dated 13,550. To be sure, this archaeological stratum is not necessarily contemporaneous with the paintings of Lascaux; but Abbé Henri Breuil, by a subtle stylistic analysis, places Lascaux in the Aurignacio-Perigordian cycle, of which it would mark the apogee. The engraved pebbles of La Colombière (PL. 246; IV, PL. 261) were found in an archaeological setting assessed as Perigordian. Dating of the levels of this site by carbon 14 gave the figures 11,650, 13,400, and 15,500, and later analyses gave 14,500 and 14,950, hence 14,000 may be taken as a rough average. The comparative lateness of the dates is surprising, but they must be accepted as scientifically grounded.

For the chronology of the Magdalenian the dates obtained by the carbon 14 method still offer only a loose network, but they provide reassuring concordances. The archaeological strata of the middle Magdalenian of Saint-Marcel, brought to light by J. Allain, give dates of 12,986, 11,550, and 11,109. The middle Magdalenian of El Juyo cave in the Cantabrians (Magdalenian III in a series established from Magdalenian I to VI) is dated 13,050. The German Magdalenian of Meiendorf (called "Hamburgian") goes back to 13,750, and at Altamira the Magdalenian level marked by delicate shells has been dated 11,950. The excavations conducted by L.-R. Nougier and R. Robert in the cave of La Vache, in Alliat (Ariège), uncovered a significant stratum of the final phase of the Magdalenian — the Magdalenian VI of Abbé Breuil — containing harpoons with double rows of barbs. This stratum, which yielded excellent examples of *art mobilier* (i.e., portable art objects such as engraved fragments of bone; see PL. 251), corresponds to a considerable human settlement. These men of La Vache were certainly also the ones who frequented the sanctuary of Niaux, on the other side of the deep valley of the Vicdessos: the habitation was on the left side of the torrent, the "temple" on the right side. The excavators subdivided this stratum, of an average thickness of some 30 in. and sealed by a thick layer of pulverulent stalagmites, into four horizons, numbered, beginning at the top, from I to IV. Charcoal from horizon II, treated in a laboratory of Columbia University in New York, established a date of 9850. Charcoal from another sector of this same horizon, examined in a laboratory of the University of Groningen, indicated 10,500. The difference, which may well be due to variations in the human settlement (the position of the hearths in the cave changed), is infinitesimal. As to horizon IV, the deepest Magdalenian layer at La Vache, Groningen dated it 10,900. In this Pyrenean valley, then, the last phase of the Magdalenian seems to have lasted from about 11,000 to about 9500 B.C.

The chronology of Upper Paleolithic cave and rock art has thus been established along its fundamental lines. The first tentative artistic manifestations go back to about 30,000. They belong to what is called the Aurignacio-Perigordian style, whose apogee, represented by Lascaux, occurred about 14,000–13,500. That style was followed closely, with some overlapping, by the Magdalenian style, the beginnings of which can be placed about 14,000 and the end about 9500. In view of the modest and obscure beginnings of artistic activity, amounting to no more than stumbling, rudimentary efforts, one cannot actually speak of an art before 20,000–15,000. Instead of the 40,000 years proposed by Breuil, prehistoric art lasted at most 10,000 years, and all its masterpieces belong to the final 5000 years. Its evolution, then, considered on a prehistoric scale, appears singularly contracted and accelerated.

The problems of prehistoric art, no longer falsified and weighed down by too vast a chronology, thus reenter the sphere of modern historical conceptions and proportions: henceforth they seem more nearly on a "human" scale. The clarification is fundamental, even though discussion may continue as to whether, for example, the great bovids of Lascaux are Perigordian or Magdalenian. Lascaux represents a transition of style and of school, and it may be called Perigordian if it is posited that the Perigordian school remained attached to certain traditional modes of representation such as that of showing bovids in profile with their horns fullface. But this mode of representation is indicative of a school; it has little chronological value.

This so-called "twisted perspective" is found also in the art of the Spanish Levant, and, although there was no doubt of the connection of this art with Perigordian art (compare the animal figures of Alpera with certain Perigordian ones), the long chronology gave rise to some uneasiness. Scientific dating has put everything in order and has shown the art of the Spanish Levant to be the direct and temporally logical heir of Perigordian art. Twisted perspective is frequent also in the finest engravings of the Maghrib. The great bison of Gouiret bent Saloul shows the body in profile and the large ringed horns fullface, as do the antelopes and ruminants of Arréchin; and at Kreloua Sidi Cheikh the rock face exhibits two facing bison under the same pair of frontally depicted horns. The art of the Spanish Levant and of the Maghrib must be placed between 10,000 and 5000 B.C. The shelter of Sefar in the Tassili-n-Ajjer (or Azger), with paintings believed to be much more recent, is dated 3070 by carbon 14. Everything falls into place, even the distant paintings of southwestern Africa, which in Philipp Cave do not date back further than 1468 B.C. The unity of prehistoric art is world-wide.

The creators of prehistoric art are not so very remote. They belonged to the species *Homo sapiens*; to appropriate Abbé Breuil's remark, "they are men like ourselves, neither handsomer nor uglier, neither more stupid nor more intelligent, than we are." The same races will be encountered again in the Mesolithic cultures; they were to undergo an admixture of outside influences, but these were perhaps less important than had been thought.

Excavations in the Rouffignac cave, conducted by C. Barrière, have demonstrated the continuity of its prehistoric occupation: the stratigraphy revealed an uninterrupted sequence from the Magdalenian level until the ages of metal and historical times. Burials in the outer portion yielded 10 skulls, dolichocephalic and brachycephalic, in a post-Tardenoisian stratum. A deep layer (C 4c), dated 7005 by carbon 14, as well as the two layers above and below it, contained flint knives showing the typical gloss given by cereals (undoubtedly cultivation was not yet known, but cereal grasses were gathered). Until the scientific dating is completed, it may therefore be postulated that the gathering of cereals dates back at least to the 8th millennium in the West. The dates of the agricultural village of Jarmo (Iraq) lie between 9000 and 6700. The peasant level of the Belt Cave (Ghar-i-Kamarband, Iran), at which pottery was as yet unknown, is dated 7790. These dates, marking the origins of plant gathering and cultivation in the East, are synchronous with the gathering of cereals at Rouffignac.

It was long the fashion to interpret the manifestations of prehistoric art by multiple references to ethnological data. But such connections between continents and across the millenniums, regardless of chronology, are hazardous: the resemblances may be fortuitous, mere products of convergence. Caution requires, at the very least, a chain of ethnological facts confronting a chain of prehistoric facts. The latest discoveries, the tighter chronological sequence revealed by carbon 14, invite even greater restraint in the realm of ethnological parallelism, for they seem to indicate the existence of a tradition that connects the world of the Upper Paleolithic with our own historical experience.

THE ORIGINS OF ART IN EUROPE. *The natural environment and its effects.* About forty thousand years ago a thick blanket of ice was weighing down Europe, gnawing away its northern edges and overrunning the limits of the future Baltic Sea. In the west a rigorous climate reigned over the hills and plateaus and the old Hercynian massif, all gripped in a vise between the great glaciers of the north, covering part of Germany, and those of the Alps, overflowing onto the Bavarian plateau or petering out upon the slopes of Fourvière, above Lyons. Cold winds blew fiercely from the heart of Europe, carrying with them the rigors of the Arctic. The cold grassy expanse of the tundra, with its meager vegetation, its islands of pine, birch, and juniper, and its frozen swamps spread to the west. The violence of the winds, more than the severity of the temperatures, was hostile to the growth of trees, which could find a hold only in sheltered zones or milder regions. As a result of the westward displacement of the glacial front, the central massif came to form a decisive screen, a precious barrier for the protection of humanity. The northeastern sections of the massif received the cold in full force and were deserted, except during the more temperate climatic phases, which permitted some hunters' settlements in France at Châtelperron (Allier), about 35,000–30,000, and at Arcy-sur-Cure (Yonne), where an industry of Aurignacian type flourished about 29,000–28,370 (the so-called "Arcy interstadial" period). Protected by the massif itself, the southwestern sections, with their broad valleys oriented toward the southwest, facing the Atlantic — a source of temperate influences — became privileged regions. During the great Würm glaciation a favorable human environment began in the Charente region, covered Aquitaine, and continued toward the south. In the Cantabrians the snow line descended to 4,600 ft. on the Picos de Europa, and in the French Pyrenees to 5,600–6,000 ft. But even though the mammoth, the woolly rhinoceros (*Rhinoceros tichorhinus*), and the reindeer pushed as far as the Cantabrians, the rigors of the climate, rather like that of present-day Scotland, were episodic and largely attenuated.

This sheltered region, from Charente to the Cantabrians, is the cradle of the prehistoric art commonly called "Franco-Cantabrian," although the term is no doubt too limited geographically: since the ornamented caves extend farther toward the center of Spain (e.g., Los Casares) and even southward (La Pileta in Andalusia), it would be more appropriate to speak of a "Franco-Hispanic" distribution. In southeastern France a multitude of caves with paintings or engravings dot Languedoc and the valley of the Rhone; then this art seems to make a leap to southern Italy and Sicily; more accurately, then, one might speak of an Aquitanio-Cantabrian cradle of art and a Franco-Mediterranean extension, whose southern tips are marked by La Pileta in Andalusia and Levanzo and Addaura in Sicily (see FIG. 571).

Climate "suggests" a certain kind of life to man but imposes it on animals. This biological rule accounts for the birth of art in the Aquitanio-Cantabrian region. The basic animal species of the cold periglacial zones descended to lower latitudes as these zones themselves descended before the extension of the ice, but the species remained attached to their biological environment and stopped their advance to preserve the most favorable conditions of existence. Aquitaine was, and for a long time remained, their place of refuge, and man found it a rich and bountiful hunting preserve.

To the powerful species just evoked — the mammoth, the woolly rhinoceros, the great cave bear, all species that played their role in prehistoric imagery — must be added the swarms of more modest and accessible edible species, among which the most typical and widespread was the reindeer. This animal rightly lends its name to the period and to the cultures of the great hunters of the ornamented caves. The economy of the reindeer was fundamental. In this "universal" animal, as in the bison, everything was useful to man: fresh meat, fat, skin, antlers, bones, tendons. Perhaps man even drank the milk of the female and kept lost calves near his rock shelter or hut. In this way domestication, or at least an attempt at domestication, may have originated in the West, although it may never be possible to prove this.

The reindeer dominates in the food remains of La Ferrassie, La Quina, and Abri Castanet at Sergeac (lower Aurignacian), Badegoule and Le Roc de Sers (Solutrean), La Madeleine (Magdalenian IV), the Mège shelter at Teyjat (Magdalenian V), the Villepin shelter (Magdalenian VI), and elsewhere. In the Aurignacian strata of La Ferrassie the three fundamental species, the reindeer, the bison, and the horse, are in the lead. These examples indicate that the "culinary" preferences of human beings remained constant over 20 millenniums, from 30,000 to 10,000 B.C., despite differences of culture, which, indeed, were rather secondary. But some original "menus" were composed, which can be attributed to climatic alterations affecting the temporary distribution of fauna, to local influences, or to some other cause. Periods of cold and dry weather in the Solutrean (Laugerie Haute, lower stratum, 18,690) and in the lower and middle phases of the Magdalenian created landscapes of cold steppes, often enlivened by the saiga antelope. This animal is preponderant in certain strata of Saint-Germain-la-Rivière; in second place, after the reindeer, at Le Roc (Cave of Les Fées) at Marcamps (Magdalenian III and IV); and second after the horse at Isturitz in the Pyrenees.

The appearance of the red deer, the boar, the roe deer, and the beaver at the end of the Magdalenian indicates the climatic change that was gradually taking place. These "Atlantic" species, adapted to milder and more humid environments, slowly but inexorably replaced the tundra and steppe species that had occupied Aquitaine for over 200 centuries. The food remains of the cave of La Vache (Ariège) have made it possible to analyze with precision the daily diet of the last Magdalenians of the mountains, flourishing about 10,000 B.C. Of the bones of mammals excavated there in 1958 and 1959, the Pyrenean ibex accounts for 85 and 91 per cent respectively, the reindeer for 5 and 3 per cent. These proportions reflect, on the one hand, the lateness of the deposit (the reindeer was in the course of being replaced, even totally eliminated, by the ibex) and, on the other, the mountainous character of the site (reindeer were always scarce in the region of Tarascon-sur-Ariège, because the structure of their hoofs was suited to the plains rather than the mountains). It might be supposed that the Magdalenians' interest in the ibex would be confirmed by the animal representations in the cave of Niaux, the sanctuary opposite La Vache; and indeed ibexes are numerous on the walls of Niaux's "Salon Noir" (PL. 269), whose specimens, with those of the Great Ceiling of Rouffignac (PL. 273), are among the finest known. Yet it is not the ibex that is in the majority at Niaux but the bison (although it accounts for but 0.4 per cent of the bones in the excavations of La Vache of 1959), followed by the steppe horse, which is nearly absent from La Vache's bony remains (1 bone out of 10,044). The reindeer does not appear at all on the painted walls. In its numerical proportions, then, the fauna represented at Niaux seems to reflect neither the diet of Magdalenians nor its actual distribution. The implications for the iconography of cave art are fundamental.

Although the reindeer, as already mentioned, played a

Principal European caves and shelters with Paleolithic rock art. *Key*: (1) Modern political boundaries.

primary economic role during the whole period of the last glacial recession, there is no clearly identified representation of it in the cave of Lascaux (Perigordian cycle or, better, school) or at Cabrerets, Rouffignac, or Niaux, the three caves exemplifying the different artistic aspects of the Magdalenian school. Abbé Breuil noted that even in the caves where it is encountered the representations are few: six at Font-de-Gaume, among hundreds of figures; a dozen at Les Combarelles, among over three hundred images. He concludes that the reindeer played a role of scant importance in Magdalenian mural art, in spite of its abundance, because it was stupid and not very alert, hence easy to capture without drawing on the resources of magic. But is cave art in fact a magical art, or at least, an exclusively magical art? The problem merits further analysis.

The harsh climatic conditions in the West that created the severe environments of taiga, cold steppe, and even tundra led to a systematic search for possible human refuges. From mere hollows in the rock to monumental halls carved out of high limestone cliffs, from slight, exposed overhangs to deep caverns extending several miles underground, every shelter was investigated as a permanent or temporary dwelling or a mysterious repository of art, a sanctuary.

The Aquitanio-Cantabrian triangle was the most densely populated area in Europe and the percentage of inhabited and ornamented caves in it far higher than in the rest of the continent. This concentration of humanity was decisive for the development and expansion of artistic activity. The first experiments, the first finger tracings in the clay of the caves, the first interpretations, found a "public"; they were not immediately consigned to oblivion. Transmission from neighbor to neighbor, from generation to generation, became possible through the increasing concentration of people. The phenomenon developed slowly at first, then speeded up about the 15th millennium in accordance with the ineluctable law of "human acceleration." Thus it is not surprising, in the last artistic phases, to find much overlapping of schools. There was a veritable artistic effervescence, and the works of Lascaux, Rouffignac, Niaux, Altamira — to name them in their presumptive chronological order — span hardly 5000 years. Concentration in space and concentration in time are two fundamental characteristics of prehistoric art.

Function of prehistoric art. A direct consequence of the climate during the Age of the Reindeer was that man was obliged to seek his food almost exclusively among the fauna, rather than the flora, which was relatively poor. Life in the Aurignacian, Solutrean, and Magdalenian turned on game, and hunting was the basic occupation. An indication of its crucial importance is the very full equipment of the hunter, a remarkable enrichment of which dates from the cultures characterized by flakes of Mousterian type, at the time of the great cave bear. The Châtelperron blade, with its tapered cutting edge and thicker back, is the prototype of the modern knife. Elongated into a Gravette blade or point and solidly fastened to a wooden haft, it became a redoubtable lance. Flint weapons multiplied; the Solutrean repertory included tanged arrowheads with barbs, shouldered points, and large "laurel-leaf" lance heads. Bone was used for spears, points hafted directly or in grooved supports, harpoons, fishhooks, and darts for birds. The missiles were either thrown directly or projected by means of hooked spear throwers. Deposits such as those of Fontalès in the Aveyron Valley, La Madeleine in the Vézère Valley, and La Vache in the Vicdessos Valley have yielded astonishingly complete hunting and fishing arsenals. The great majority of the instruments — even the delicate bone needle with eye, symbol of leatherwork, which made its appearance in the Solutrean and was widely disseminated in the Magdalenian — evoke the problems of the chase. The themes of prehistoric art are hunting themes, and representations of the animal world hold an overwhelming place among them.

In 1837 the notary Brouillet discovered in the cave of Le Chauffaud at Savigné (Vienne) a reindeer leg bone ornamented with a delicate engraving of two hinds, one following the other, the shoulder of each pierced by an arrow; and in 1864 the lower cave of Massat, excavated by F. Garrigou, revealed the first engraving on stone found in the Pyrenees: an excellent bear carved on a flat river pebble. When Altamira was discovered in 1879 innumerable examples of *art mobilier* (portable art objects, including statuettes and decorated implements, also known as "chattel art" or "home art") were already known; in 1864 E. Lartet and H. Christy had published in the *Revue Archéologique* an important memorandum on this art. All these objects provided remarkable proofs of man's activity, the varied manifestations of his creative genius; yet one can scarcely, in this connection, say with E. Piette, the great discoverer of the Pyrenean caves, that "the leisure of an easy life engenders the arts."

Life was certainly anything but easy for the Aurignacian or the Magdalenian man, although it must not be judged according to modern civilized standards determined by a temperate climate. The hardships of Magdalenian life doubtless stemmed from the fact that it was almost exclusively a hunting culture; for the more limited the base of an economy, the more difficult and hazardous it becomes. The Neolithic period marked a revolutionary turning point in the human economy because it brought about a multiplication of resources: to the ancestral food gathering, hunting, and fishing were added the new resources of agriculture and animal husbandry. Such variety was the best guarantee against famine. The Magdalenian dependence on a single resource was dangerous and fertile in crises. Periods of famine, caused by lack of game, might be followed by periods of plenty, but the thought of tomorrow must have been a constant source of anxious fears. Man lived from day to day, from week to week, without solving the problem of reserves for the future. Even in a climate of cold steppes, dried meat could not be kept indefinitely; the marvel of husbandry was to consist in the preservation of live meat on the hoof. Piette's phrase might well be revised to explain prehistoric art by the leisure of a *difficult* life.

Depicting the environment of the artist, representational art necessarily reflects his fundamental and vital preoccupations. The Aurignacian or Magdalenian hunter engraved or painted the animal species that surrounded him: reindeer and bison, mammoth and rhinoceros. But why did he carve these fine profiles, trace with his finger these enormous mammoths; why did he paint these friezes? To fulfill an esthetic desire? In response to some mysterious intellectual call, an urge to translate his thoughts into images? Or for more materialistic, utilitarian reasons — to satisfy an imperious magic? There is the dilemma that confronts one again and again: art or magic?

Obsession with the coveted prey, a prey scarce and hard to capture, must have reigned uncontested. A civilization devoted exclusively to hunting is one of anxiety; and the idea of art for art's sake seems incongruous in such an environment. Such a conclusion, however, can be only tentative; distributions must be made. The understanding of prehistoric cultures has been considerably delayed by the fact that they were too long considered cultures of "savages," with the disdain inherent in that term. Perhaps they should not even be qualified as "primitive," this term, too, having suffered abuse and deterioration. They are, rather, "primary" cultures.

The very circumstances of the discoveries — the finding of engraved bones or pebbles in kitchen middens and of mural works deep in the most remote sections of the caves — suggest the function of the art. The prehistoric hunters seem to have lived in almost permanent contact with the representations, which were thus part of their daily lives, constituting more than a simple pleasure; they were a utility, rather, an aid to better living. In truth these engravings, these drawings, were indispensable. The obsession with game seems to have led to an obsession with the image; the image became the reality.

For the prehistoric hunter, to *represent* an object or an animal by an image must have seemed senseless, at least in the beginning. But by his hand to *create* an object, an animal, to make it present — in that lay a meaning; to draw, paint, or engrave the desired animal must have been the equivalent for him of true creation. What was important was not the image itself but the *act* of representation, the act of sketching, painting, or modeling. Prehistoric art is essentially action. And in the tumultuous lives of the great hunters, which can hardly have favored pure contemplation, creative activity, in the strict, forceful sense of the term, seems to find a natural place.

The dynamic conception of this art explains most of the "anomalies" that characterize it. There has been, for example, a somewhat exaggerated insistence on the almost systematic superimposition of figures. True, in the deepest part of the cave of Les Trois Frères, under the protection of the "Horned God," the delicate engravings are crowded upon one another; and the vault over the well at Lascaux, the black stone at Gargas, certain wall surfaces at Les Combarelles, and the Great Ceiling at Rouffignac reveal the same frequent superimpositions. But far more numerous are figures simply placed side by side or in direct relation to one another. The vast Salon Noir of Niaux offers more examples of juxtaposition and composition than of the confused piling up of images.

The engraved pebbles of La Colombière (PL. 246; IV, PL. 261), among other examples, show an accumulation of these "reality-pictures." A fine-grained limestone pebble, 4 $^{3}/_{4}$ in. long and 3 $^{1}/_{4}$ in. wide, discovered there by H. L. Movius in 1948, bears on one face a thick-maned horse, a reindeer with true modeling obtained by fine incisions, an ibex, a headless bison, the hindquarters of a cervid, and two incomplete rough sketches of carnivores, undoubtedly bears. On the opposite face are another horse with a handsome mane, two unfinished heads of woolly rhinoceroses, a complete rhinoceros, and a rough sketch of a reindeer. This pebble, a veritable palimpsest, is a talisman rather than a work of art; the esthetic aspect is present, but it is not fundamental. The superimpositions themselves are deliberate and in no way accidental. It would surely have been easy to find fresh pebbles, one for each animal, in the fluvioglacial gravel of the Ain, which flows at the foot of the vast overhang of La Colombière. The only reason for the artist to have disregarded this supply was that the original pebble had in some way been sanctified by the first figure traced. The reality had come up to the image. That being so, why reject the "good" pebble?

Just as there were auspicious rock surfaces, or portions of the rock face, such as the Great Ceiling of Rouffignac, so also there were certain auspicious objects. That is why prehistoric works of art must be studied and analyzed in relation to their placement and surroundings. To study these examples of a functional art out of their environmental context is to deprive them of their underlying significance.

Genesis of representational art. a. Natural imprints and finger tracings. Man's imprints in the snow of his rigorous environment and in the soft clay of the caves played an indirect but decisive role in the obscure genesis of representational art. The imprints were signs with universal meaning, and man's hunting experience had certainly shown him the value and interest of such marks. The bilobate imprint of an ibex in the snow or on fresh earth in a certain sense *was* the ibex. From trace to trace, it was possible to overtake the animal, to seize its reality. Still indistinct glimmered the notion of the part for the whole, already a symbolic device.

There are records of human prints belonging to phases prior to the Upper Paleolithic. In the Basura Cave, or Cave of the Witch, near Toirano in Liguria, imprints have been found that date back to the Mousterian and can be attributed to Neanderthal man. More are encountered, however, in the caves frequented during the Age of the Reindeer. In a gallery beyond the Salon Noir of Niaux the Magdalenian soil has preserved the imprints of some twenty adolescents (4 ft., 7 in.–5 ft., 3 in. in height). At Cabrerets a highly stalagmitic clay preserves the traces of a woman and child, the child about ten years old. The galleries of Rouffignac bear witness to the passage of bears and to the advance of men, sometimes on their knees in the very low passages, with imprints of hands, wrists, and other parts of the body.

The cave of Gargas, which opens on the western flank of the hill at Aventignan (Hautes-Pyrénées), is important not only for its archaeological deposit, which ranges from Mousterian levels, with remains of the great cave bear, to the various horizons of the Perigordian, but also for decorations essential for an insight into the origins of art. Accompanying the numerous handprints in the cave are finger tracings made on the clay of certain sections of the ceiling; they are among the oldest known and can be considered contemporaneous with the early Aurignacio-Perigordian style (ca. 30,000). To be sure, these "macaronis," as some scholars call them, by no means deserve the name of art, but they constitute a step toward it.

Particularly suggestive is a panel of finger tracings at Rouffignac, on the left-hand wall of the Henri Breuil Gallery (FIG. 576), the gallery of the great friezes:

1. On the left can be seen the widely spaced claw marks of a cave bear that must have drawn itself up against the wall. This print covers some transversal ones in which the claw marks are likewise widely spaced but less deep. These marks also belong to cave bears and are necessarily older than the first — how much older, it is impossible to say.

2. On the lower register of the panel the claw marks are thick-strewn and deep. They are shorter than those noted above, and the spaces between the claws are less wide. They belong to the brown bear (a fine, very characteristic brown bear is incised on the left-hand wall of the gallery a few dozen yards beyond).

3. Immediately to the right of the great claw marks of the cave bear we find the human replica. Under a rounded flint nodule, at a convenient human height, the hand was placed slightly under and to the right of the bear's paw and the three middle fingers were swept down the wall, impressing their triple vertical track in the marly limestone. Again to the right, a complete left hand was pressed hard against the wall. Examples can be multiplied. The parallel finger tracings constitute

Rouffignac Cave (Dordogne), Henri Breuil Gallery, panel with bear claw marks, finger tracings, and a mammoth drawn with the finger.

the first human "engravings" of the panel: man had discovered and experienced the creative effect of the first engraving tool, the finger.

4. The next stage is the creation of "expressive" tracks. The essential one is the one-fingered outline of a mammoth. With a sure finger, man, an artist henceforth, traced the dorsal profile of the animal: the emphatic curve of the head, the typical cervical depression, the sloping line of the back, vanishing with the lifting of the light creative finger. This vigorous profile cuts across prior multifingered tracks as well as some distinct claw marks of bears. A later, slanting, four-fingered tracing clearly suggests the trunk of the mammoth.

5. The last phase is one of destruction. Two fingers swept across the mammoth. A four-fingered hand was placed on top of this cancellation, forming a cross.

This modest panel, then, of which Rouffignac offers hundreds of similar examples, exposes a logical sequence of events that hints at a possible origin of art. It testifies to an "experiment" of the Magdalenian period. At Gargas the experiment is older: that is, Aurignacio-Perigordian. It should be pointed out, however, that "macaronis" do not in themselves have any precise chronological value. Finger tracings are encountered in all periods, styles, and schools.

The cave of La Clotilde de Santa Isabel, discovered in 1906, contains a long gallery, to which a narrow passage gives access about 100 ft. from the entrance. Farther on, about 560 ft. from the entrance, the rather low ceiling is covered with a clayey substance on which animal figures, in a very archaic style, are traced with the finger: bulls with horns in twisted perspective, the body sometimes crossed by short finger tracks. Breuil, no doubt correctly, considered these finger tracings among the oldest known. At Altamira a ceiling bearing red clay formed by decalcification is striated with tracings, among which can be discerned the head of a bovid turned to the left; the twisted perspective places this profile, too, in the earliest school. The ceiling of the Hall of the Hieroglyphs at Pech-Merle (Cabrerets), which measures 33 × 13 ft., belongs to the same artistic phase. The tracings are innumerable and must be scanned attentively before a mammoth or a human figure emerges. One wall bears a magnificent archaic Megaceros stag.

A vast ceiling at Rouffignac (FIG. 577, 1*b*), 3,000–4,000 sq. ft. in area, bears some highly original and captivating tracings. Relatively low (max., 6 ft., 7 in.), it vaults a large chamber surrounded by a number of subsidiary rooms that give the effect of apsidioles. Water invaded this large subterranean complex, and the red clays of decalcification deposited a thin crimson pellicle on the ceilings. Countless finger tracings, made by scraping this thin clay skin, show up light against the red background. Discernible in the tangle of interlaced and superposed lines are meanders traced with one or two fingers and

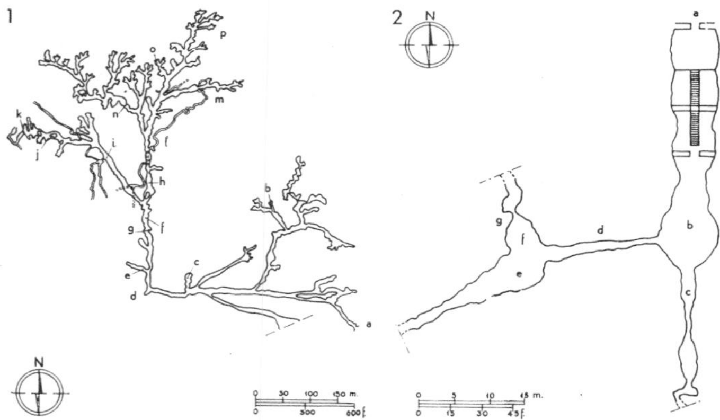

Plans of French caves with Paleolithic representations. (1) Rouffignac (Dordogne): (*a*) Entrance; (*b*) Ceiling of the Meanders; (*c*) Red Ceiling; (*d*) well; (*e*) Gallery of the Two Mammoths; (*f*) "Sacred Way"; (*g*) "Discovery Mammoths"; (*h*) Henri Breuil Gallery; (*i*) Great Ceiling; (*j*) "Mammoth with the Roguish Eye"; (*k*) saiga antelope; (*l*) "Little Lascaux"; (*m*) "Great Confessional"; (*n*) "Adam and Eve"; (*o*) painted mammoths; (*p*) Gallery of the Concretions (*from Nougier and Robert, Rouffignac, 1957*). (2) Lascaux (Dordogne): (*a*) Entrance; (*b*) Great Hall of the Bulls; (*c*) axial gallery; (*d*) gallery with engravings, or lateral passage; (*e*) "Nave", or main gallery; (*f*) "Apse"; (*g*) well, or Shaft of the Dead Man (*from Graziosi, 1956*).

comprising six to eight volutes, some of which draw three-quarter loops around the flint nodules in their path. They frequently attain 28 in., a length that corresponds to the reach of a man standing in one place. In the midst of these meanders appear representations of snakes with large, well-formed heads and outthrust bifid tongues. So far no large animal figures have been discovered on this vast vault, but this does not mean that none exist. The only figures that have been deciphered are human ones, finger-traced, often caricatural and grotesque.

In the Mediterranean region the huge cave of La Baume-Latrone offers tracings made with one, two, or three fingers dipped in red clay. Outlines of animals (elephants and a large snake) are projected with vigor, the red ocher vivid against the naturally pale background on the rock. Finger drawing here seems to pass insensibly into painting.

b. "Positive" and "negative" hands, mutilations. Characteristic for the early schools — the Aurignacian of Gargas and the early Magdalenian of Cabrerets — are the imprints of hands in black or red, the basic colors, on the walls of the galleries. The red was obtained from ocher, a clayey earth colored by iron oxides; the black from manganese oxides, deposits of which can be scraped off the base of the walls. The impressions were made by firmly placing the hand smeared with color flat against the wall. Certain well-preserved hands show the thumb slightly at a slant, in three-quarter view, and the shadow, as it were, of the Aurignacian thumbnail.

In the "negative" hands, paint delimits the contours of the fingers, the palm, and the wrist, forming a colored halo around the hand, which retains the natural tone of the wall. The felicitous graduation of color justifies the process. The paint was blown against the wall where the hand was placed, either from the mouth or through a hollow bone. When the hand was removed, a negative impression remained. This astute technique reveals two artistic innovations: spray-gun painting and stenciling. In the absence of art in the classic sense, here is at least the basis of industrial art and of decorative art, some thirty thousand years before the Christian era!

Sometimes, as in the Sanctuary of the Hands at Gargas, the prints are repeated, aligned, in friezes. The repetition is a form of art, expressing a feeling for motif, the notion of the decorative band; it marks a sense of the horizontal and of alignment, the consciousness of a rhythm between images and intervening spaces. At Gargas, too, are found friezes of stenciled thumbs, and the elevated portions of the cave bear imprints of hands pressed into the clay and friezes of fingers, also aligned and pressed into the clay. There is a strong link at Gargas between imprints and finger tracings and the paintings of hands.

The significance of the hands is a more delicate problem. The magical and ritual value of the hand is undeniable. It is often a symbol of possession, especially in an art where the image is taken as the reality. In the cave of Pech-Merle at Cabrerets two admirable spotted horses, one of which gains in impressiveness by utilizing a natural rocky projection for the head, are surrounded by five negative hands with graduated black halos (PL. 262). They are later than the horses, which they frame with precision. Do they not suggest a mark of possession?

Only three caves are known, so far, to bear imprints of mutilated hands on their walls: Gargas (FIG. 578), its annex Tibiran, and Maltravieso, in the Spanish province of Estremadura. Gargas and Tibiran show a great variety of mutilated hands and an astonishing variety of mutilations. The count once made by Breuil has since been raised to more than 200. The hands are those of men, women, and children; some recur several times. The most frequent mutilation is the amputation of the last two phalanges of the last four fingers; next are the amputations of the two terminal phalanges of the third finger, and of the top

Gargas Cave (Hautes-Pyrénées), negative imprints of mutilated hands.

phalanx of the last four fingers. The most frequently mutilated is the middle finger; next, in decreasing order, the ring finger, the little finger, and the index finger. There seems to be only one instance of a mutilated thumb.

The problem is being reexamined by A. Sahly, who has kindly permitted some of his conclusions to be cited here. Two explanations for the mutilations have been advanced: ritual amputation and disease. The extreme severity and variety of the mutilations incline Sahly to attribute them to disease, specifically, vascular disorders of the type of Raynaud's disease, with gangrene of the extremities after frostbite of the last degree. Raynaud's disease characteristically attacks the terminal phalanges of the last four fingers, leaving the thumb intact, and strikes both adults and children. The cold and humid climate at the time of the occupation of the caves, the humidity of the caves themselves, the precarious living conditions in the Aurignacian, the permanent shortage of food and the vitamin deficiency — all may have favored an intense local development of Raynaud's disease, with the resulting mutilations recorded at Gargas. The mutilations of Maltravieso, on the other hand, all affecting the two distal phalanges of the little finger — the least indispensable — and showing a clean, straight cut, seem to validate the hypothesis of ritual sacrifices.

c. Utilization of natural features. Halfway into the narrow Breuil Gallery at Le Mas d'Azil, on the right, one comes upon a strange rocky projection that recalls the muzzle of an animal, perhaps a feline. The impression is reinforced, on further examination, by a few black strokes marking the eye, the line of the neck encircling a hollow in the rock, and the profile of the forequarters. For prehistoric man the resemblance was doubtless increased by the play of light and shadow produced by a feeble lamp that made the "magical" creation mysteriously appear or disappear.

To this day natural features in the caves attract attention and excite the imagination by their resemblance to familiar objects. In a cave with concretions visitors exclaim over the "bell tower" or the "pagoda." In the same way prehistoric man was led to recognize objects familiar to him, above all the animals he hunted. In a secondary gallery at Altamira the angle of a rock suggests a nasal ridge. Two little black circles on either side represent the eyes, and a natural fissure completes the lower part of the nose. At El Castillo a stalagmitic column evokes an upright bison. The illusion is reinforced by engraved lines and large black patches on the hump, the belly, the inner portion of the thigh, the breast, the dewlap, and the head; the loins, tail, and exterior contour of the thigh benefit from the strong relief of the pillar. One of the spotted horses of Pech-Merle (PL. 262) is equally suggestive. The head is inscribed on a natural projection of the rock. The intermingling is so complete, the shadow cast gives so much relief to the rocky outline, that one may wonder where the representation actually lies, in the painting or in the natural relief. The Magdalenian artist made use, too, of natural "hollow reliefs." In the Salon Noir at Niaux an oval cavity suggests the head of an animal, and black antlers are painted on either side to complete the resemblance.

A rather low ceiling in Le Combel, the section of Pech-Merle discovered in 1949, is covered with stalagmitic protuberances, of globular and pyriform shapes, some of them perfect evocations of female breasts. What heightens the illusion is that the tips are delicately tinted with crimson. A magico-sexual explanation, questionable here, can be offered with more reason, perhaps, for the ithyphallic figures of Le Portel. A natural stalagmitic relief, which by its form and its position suggested a phallus, was taken by the Aurignacian artist as the starting point for a human silhouette. The body of this figure is presented frontally, while the long head with protruding jaw and bulging forehead appears in profile. A few feet farther on some slender stalactites seem to represent on the left a leg and in the center a phallus with distinct meatus; on the right a limestone swelling rather imperfectly suggests the left leg. The figure, defined in red, is presented fullface. During humid climatic phases a thin stream would run from the central stalactite. Closely related is a figure at Cougnac that vividly brings to mind the *Manneken-Pis* of Brussels. Its functioning, again, depends on the humidity of the cave. It is a masterpiece in the utilization of natural resources. In this connection may be mentioned the delightful engraved reindeer of Les Combarelles, whose long neck and head are bent toward a crack in the rock from which, when rain falls for a long time and feeds the subterranean streams, water slowly trickles into the animal's open mouth.

Natural features served as points of departure for the most varied representations; the picture evolved around the "motif." Sometimes the figure even took its start from a dot. The process is best exemplified by the bison traced in clay, at Niaux, known as the "Bison with the Natural Cupules" (PL. 270), perhaps the most affecting work, in the creative psychology it implies, in all the art of the Paleolithic period. In the clay ground on which the bison is drawn are innumerable little cupules, or hollows, formed by the intermittent dripping of water from the ceiling. The animal, which looks toward the right, precise, neat, and realitistic, incorporates, within the outlines of its body, three of these cupules, skillfully utilized toward magical ends; three arrows directed toward them show that they are intended to suggest wounds. A more obvious procedure would have consisted in first tracing the bison in the clay and then piercing it with three ritual wounds, as was done in the clay sculpture at Bédeilhac. Here, on the contrary, three cupules, promoted to the magical significance of wounds, were selected and the outline of the bison was drawn around them, while others in the vicinity were neglected. An attentive examination reveals that the eye of the animal also utilizes a small, regular cupule. This, then, is the actual point of departure for the whole work. That there is not the slightest error in proportion, that the eye is in the correct spot, that the body and the members, too, are impeccably in place indicates that the creator had an exact preliminary vision of the figure he was going to engrave.

A similar procedure was followed in all schools of the Upper Paleolithic. In the "Throne Room" at La Pasiega a very fine head of a hind, of majestic carriage, was drawn starting from a natural round protuberance envisioned as the eye. The animal's pricked ears are not in profile but turned slightly frontward, in accordance with the Perigordian style. A little red horse at Le Portel was engraved starting from a calcite concretion, again interpreted as an eye. An outline filled in with red paint, it recalls the painting of the Spanish Levant and seems to belong to the same school as the horses of Alpera, discussed later.

A cupule could also be the point of departure for a sculpture. The shelter of Le Roc aux Sorcier at Angles-sur-Anglin revealed, *in situ*, an admirable sculptured frieze buried by archaeological strata of the late Magdalenian. The center is occupied by three "Venuses" modeled from the waist to the bottom of the legs; head and feet are not depicted. It has been conjectured that the central figure represents a woman four months pregnant, since the abdomen, rendered by a skillfully adapted natural protuberance of the rock, shows some swelling (which, however, is more likely an effect of obesity, especially in view of the heaviness of the thighs and the scant marking of the waist). The two other figures, which are lower, underline the preeminence of the central one. For an understanding of the genesis of the three figures, one fact is essential. The navel of the middle woman is an adapted natural cupule, carefully evened out and softened. The sculptor evidently chose it among many others as the most suitable point of departure for the entire group of "Three Graces."

Sometimes the utilization of natural features is even more subtle. In the Breuil Gallery at Rouffignac a few strokes of black on an oddly shaped flint nodule bring to life an admirable horse's head. The delicate hollow of the cheek and the slight prominence of the cheekbone are accented by the natural relief.

CHARACTERISTICS OF PALEOLITHIC ART. *Hunting magic.* A large number of animal representations are "reality-pictures," a direct and living echo of the hunter's way of life. About 10 per cent of all animals represented are depicted as game, struck by arrows or other weapons. Certain caves are particularly rich in depictions of wounded animals, Niaux for example (PLS. 269, 271). In the Salon Noir of that cave a magnificent bison is shown struck in the flank by a double-barbed red arrow. The figure reveals the artist as a virtuoso of line — the isolated line or lines tightly grouped together to form descriptive hatchings, in a technique reminiscent of etching. The two elegantly curved horns, presented in felicitously exact perspective, are delineated with a very fine brush. Nearby, a large bison, whose expressive eye inevitably recalls that of the great mammoth situated in the most remote section of Rouffignac, is pierced by three black arrows. In the last panel an ibex drops its head under the blows that assail it; it has two javelins in its breast and a long lance in its flank. Nearly half the animals at Niaux are wounded in this manner.

Lascaux, too, harbors many representations of wounded animals. At the entrance of the "Nave," on the side nearest the "Apse" and the well (see FIG. 577), is a frieze of three horses: a pregnant mare followed by a stallion pierced by seven arrows, then by a second pregnant mare. At a short distance is a large black and ocher bison, painted after the horse, which occupies a niche in the left-hand wall. The horns are in twisted perspective; the hoofs, very black, are strongly emphasized. Seven parallel arrows, engraved after the painting, strike the animal obliquely from behind. The simplest interpretation of the scene is that it is one of destruction by sympathetic magic.

It is at Montespan that the most striking illustrations of the pictorial annihilation of animals are to be found (IX, PL. 241): witness the headless bear lacerated by 30 spear thrusts or, in the frieze of three horses modeled and incised in the clay of the wall, the great horse in the middle, striated by heavy vertical finger tracks in which some have seen a palisade but which, more probably, are intended to erase, to annihilate, the animal. The horse that precedes shows, besides finger tracks, spear thrusts dealt both before and after the finger tracks. The superimpositions bespeak a very frenzy of destruction. Contributing to the magic-laden atmosphere are the human footprints, imprints of heels of reduced dimensions, as if adolescents had walked in a crouch, bearing down on their heels with all their might. This must have been a ritual walk like the one recognized around the two coupled bison modeled in clay at Le Tuc d'Audoubert (PL. 259). At Le Portel a black horse of the Aurignacian school retains a lump of mud thrown with such violence that it adhered and calcified (the calcification is a proof of authenticity). Though more primitive, is not this the same movement of destruction, following one of creation, that occurred at Montespan?

All schools give examples of headless animals. They occur even among archaic representations of typically Aurignacian style, such as the horses of Las Monedas, 3 out of 13 of which are headless. Drawn with a heavy hard line, these figures cannot be considered works of art, though they occasionally evidence a good sense of observation; what importance they have resides in the light they may shed on the thoughts and motives of Paleolithic man. Also headless is a horse of Magdalenian style at Le Portel, distinguished by a happy firmness of line, an excellent rendering of the hoofs, and even an attempt at modeling along the line of the belly and the mane. At Altamira, toward the near edge of the great painted ceiling, is an excellently preserved polychrome bison (PL. 268), 5 ft. long, whose head is missing, even though the artist had enough room for it between a bison on the left and a galloping boar (in all likelihood earlier) on the right. Is this to be considered a systematic rite, connected with hunting magic, or simply unfinished work?

The incompleteness of all these figures seems deliberate and systematic. In a hind at Las Chimeneas, for example, marked by genuine qualities of draftsmanship, the omission of the dorsal line, of the muzzle, and especially of such a vital element as the eye seems to point to a magical significance rather than want of industry. Or consider an engraved horse of the Breuil Gallery at Rouffignac, a figure of astonishing clarity and incisive power drawn with long vigorous strokes that proclaim the perfect mastery of its creator. When the eye — the detail that would have given it the greatest intensity of life — was omitted in such a figure, the reason could not have been incapacity or chance. It is conceivable that these animals were depicted without head, without muzzle, without eyes to symbolically deprive them of the scent and the sight that would have permitted them to smell, detect, and avoid the hunter; in short, they may have been thus mutilated to make them defenseless against capture.

The opposite phenomenon, heads without bodies, is also frequent (PL. 260). In the gallery of the headless bear at Montespan, rising out of the ground, appears a fine horse's head, clear, precise, with a well-marked eye and a small, sensitively pricked ear. It is easier to account for the isolated heads than for headless bodies: the head obviously stands for the entire animal ("200 head of sheep," the shepherd says). Oddly enough, the number of headless animals in prehistoric art is approximately equivalent to the number of isolated heads.

Human representations. a. Realistic figures. Although Paleolithic art has bequeathed many masterpieces of animal art in all mediums of expression, it offers very little, by comparison, in the realm of human representation, either quantitatively or, more important, qualitatively. Among the few works that can be regarded as having any esthetic merit may be cited the little ivory female head of Brassempouy (PL. 243; VIII, PL. 235); the steatite head, closely related to it, from the Grimaldi caves of the Balzi Rossi; a female bust from the Magdalenian of Le Mas d'Azil, carved in the incisor of a horse; possibly the "Lady with the Hood" of Bédeilhac, of the same period and material; and, finally, more to the east, a female head of ivory found at Dolní Věstonice in Moravia. Interesting from the points of view of conception and typology are the prehistoric "Venuses," of which a few dozen have been discovered in various European localities. Among the best-known are those of Laussel (PL. 255), Lespugue (VII, PL. 351), Sireuil, Willendorf (IX, PL. 247), and Savignano (PL. 257). It should be noted that these figurines represent the earliest examples of small-scale human statuary.

Chronologically the statuettes can be divided into two groups, one distinctly Perigordian, the other Magdalenian. The figures of the second group have often lost their specifically female characteristics and assume a more generalized anthropomorphous aspect. With their generous volumes, often unequally and badly distributed, these statuettes can hardly be taken as the canon of female form of the period, but, according to the most widely accepted hypothesis, are to be considered magical representations, images of female fecundity. The discovery of the Perigordian "Venus" of Tursac in 1959 confirms certain observations based on earlier finds. The upper portion of the body is not represented; the sculpture begins at the waist. The abdomen is perfectly defined. The nether members are folded under the body in a squatting position; the knees are thick, and the lower legs, almost horizontal, are joined at the back; the feet are not depicted. One unusual detail gives this figure special importance: the presence in the lower portion of a peg, which could be inserted in a holding stick or directly into the ground. Thus the figurine was perhaps intended to be planted in the earth like a sort of magical icon, a guardian of the hearth. This technical contrivance lends support to the hypothesis that the "Venuses" were symbols of fecundity, or mother goddesses.

Aside from the *Manneken-Pis* already mentioned, the cave of Cougnac contains other human representations that are interesting on an intellectual or a religious plane but artistically quite secondary. Notable is a human figure advancing almost prone, as though on all fours, and pierced by seven arrows, while to the right a second personage falls, head first and feet in the air, pierced by three arrows. These are the only known human figures that are transfixed like vulgar prey, and they may have the same significance as the horses and bison of Lascaux and Niaux.

The part for the whole is a principle that found application also in anthropomorphous representations. Some rocky fragments from the Aurignacian of the Blanchard shelter at Sergeac bear deep incisions representing vulvas; the same theme recurs in the Magdalenian phase, in the form of a clay modeling at Bédeilhac, not far from a small modeled bison. Thus the female symbol reappears, with fine chronological consistency, in the course of the two cycles. But such representations, although important for the magical thought they disclose, are infrequent, and this infrequency should moderate the elaboration of explanations that would give prehistoric sex a disproportionate role.

b. Composite figures. Of the representations combining human and animal features the most famous is the "Sorcerer" (also called the "Horned God"; X, PL. 133) in the cave of Les Trois Frères. In the "Sanctuary" in the depths of that cave, it is the only one among 89 Perigordian and 193 Magdalenian engravings that is not only engraved but also painted, in black. It dominates the dense crowd of animals below from a height of about 13 ft., in a position seemingly inaccessible but which can be reached by a hidden corridor leading up in a spiral. It has round eyes with pupils. The ears, erect, are those of a stag, and the forehead bears antlers. The feet mark a dance movement, and a long, broad tail, that of a wolf or a horse, completes the personage. As Breuil has pointed out, the Magdalenians evidently considered this figure the most important in the cave, probably as a representation of the spirit governing multiplication of game and hunting expeditions. It should be noted that in the tangle of animal figures two other composite representations can be deciphered: one, bison above and man below, plays a little musical bow while dancing; the other, also half bison and half man, is equally enigmatic. The interpretation of these singular figures is necessarily hypothetical, but tentatively one may see in them, under religious or magical disguises, "dancers" whose movements had a bearing on the chase and its success.

The Sanctuary at Les Trois Frères must have been the scene of actual ceremonies with processions and dances. The very layout of the place lends itself to rites designed to strike the imagination. An aspect of prehistoric art too seldom considered is that of performance, of singing and dancing, an aspect sometimes evoked far underground, as in Les Trois Frères and other large caves such as Niaux and Rouffignac, but even more vividly in such open shelters as Le Roc aux Sorciers (Angles-sur-Anglin) and Cap Blanc.

c. Caricatures and masked figures. Various painted and engraved human caricatures, often crude sketches, are found on rock walls, as at Marsoulas, Altamira, Font-de-Gaume, and Rouffignac (on the Ceiling of the Meanders), as well as on engraved stone plaquettes, such as those of Limeuil and of La Marche at Lussac-les-Châteaux. These images present an enigma in the contrast between the masterly realism of the animal figures and the grotesque absurdity of the human ones. The explanation cannot be inexperience or lack of skill: the draftsman capable of sketching the drinking reindeer of Les Combarelles would have been well able to sketch a human silhouette. But the talent of the prehistoric artist seems to have been paralyzed before the human face. Profound psychological or social causes must be adduced, perhaps some mysterious taboo on representing the human face, permitting only caricatures or masked figures. Man was not to be recognizable, whereas animals might shout their truth. Only the existence of such a taboo can explain this strange paradox (see also IMAGES AND ICONOCLASM).

Nevertheless these wretched treatments of the human countenance sometimes convey real feeling. In an extension of the Breuil Gallery at Rouffignac, on the ceiling, can be seen two facing personages. The larger, whose head alone is more than 2 $^1/_2$ ft. high, has an oversized nose and mouth: it is a veritable carnival face, yet the felicitously suggested upper lip is not devoid of a certain humanity. The face opposite is a more modest one, all curves, with a mocking nose, a long slender neck, and an eye at once tender and ironic, rendered by a dent in the wall and a fine incision. This pair has been named "Adam and Eve."

Composition and placement. It is too often held that the prehistoric animal representations were conceived singly, executed one after the other, without concern for the context. At times the piling up is undeniable. The figures obliterate one another, constituting those "reality-pictures," already mentioned, which accumulated on surfaces that had proved auspicious, such as the wall under the "Horned God" at Les Trois Frères or the Great Ceiling of Rouffignac; the privileged surface might also be that of an object. But very often new paintings and engravings were added with a patent respect for prior ones. The concern for the earlier work was an esthetic preoccupation, a beginning of the sense of composition.

It is hard to say when this new tendency emerged, but it was already present at Lascaux. In the Great Hall, wedged between two immense bulls in black line, over 13 ft. long, appear the head and shoulders of a horse with a fluffy mane. This figure can be taken as later than the horns of the great bull on the right, on the basis not of superimposition but of juxtaposition: its muzzle fits exactly between the horns of the bull, and the motive for leaving it unfinished was respect for the bull on the left.

The Salon Noir of Niaux multiplies examples of respect shown for previous work. The large figures, the bison, are often superimposed, because their volumes would have made it difficult to paint the new figures while respecting the old. But the small figures could more easily be avoided and could more readily find place among the big ones. Notice, for example, between the legs of a large black bison a transpierced little black horse (PL. 271) and, emerging from the mane of another large bison with five black arrows, a little ibex's head whose muzzle falls precisely between the horns of the bison. At Altamira can be seen a little black bison under the neck of a large polychrome hind. It is painted in an older technique, and its lower portion is very faded. The author of the hind evidently made a deliberate effort to avoid any obliteration of it.

The delineation of an animal with respect for its neighbor marks an esthetic concern that foreshadows the sense of composition. Sometimes two animals are fully associated and form a couple. A complete survey would reveal a considerable number of such couples, of which the Salon Noir at Niaux supplies a good example in a female bison followed by its partner. Often, too, the partners face each other, sometimes in the same position, sometimes in a complementary one. A pair of bison at Rouffignac and a pair of reindeer at Font-de-Gaume are depicted in the same relation: the male standing, the female kneeling before him. At Rouffignac the male bison, executed before the female, was first sketched in light black line, then the hindquarters were engraved with a thick, heavy flint burin, and the animal finished with sweeping strokes of black paint.

Facing pairs in the same position appear frequently in most of the caves. This scheme, illustrated by the masterpiece of the Breuil Gallery at Le Portel, a Magdalenian panel of two black bison portrayed muzzle to muzzle and horn to horn, can be taken as the compositional leitmotiv of Rouffignac. The first figures discovered there, the famous "Discovery Mammoths," very realistic, and vigorously rendered by finger tracing, face each other; so do the two painted mammoths of the little gallery named after them (FIG. 577 *1e*) and some large mammoths engraved in white on the clay of the Red Ceiling (*1c*). But of all the face-to-face encounters, that of the leaders of the mammoth herds in the great frieze of the Breuil Gallery is no doubt the most spectacular.

Even in the earlier cycle an attempt was made to solve, by means of overlapping, the delicate perspective problem of representing animals advancing side by side or slightly staggered. The celebrated early Magdalenian prancing horse at Le Portel, with its raised foreleg, is preceded by the forequarters of a black horse, probably of later execution, which — less elaborately drawn, more schematic — seems more remote, as if some details were blurred by distance. At Rouffignac both engraved and painted examples of animals overlapping in this manner are numerous; they occur notably in the Breuil Gallery, in the monumental frieze of painted mammoths on the right-hand wall and in the series of engraved mammoths on the left-hand wall. The technique of overlapping is also often encountered in the rock art of the Maghrib, where the long columns of mammoths are replaced by long caravans of elephants. Although, as is to be expected, Aquitanian art and the art of the Maghrib are strongly individualized by the geographic conditions and the ways of life peculiar to each region, they are identical in essence. The themes of both are animals; they utilize the same compositional schemes (animals in pairs, face to face, mirror images, in procession); they are amenable to the same explanations, subject to the same esthetic, and capable, sometimes, of the same errors, such as representation in "twisted perspective." This close resemblance is hardly surprising, for the art of the Maghrib is the direct heir of Franco-Cantabrian art; only 5000 years lie between their extremes.

The concentration on single figures has often hindered recognition of the importance of the composition, which, perhaps more than any other factor, testifies to the quality of the artist's vision. A balanced arrangement of figures was achieved very early. At Covalanas, for instance, there is a panel of red hinds that demands to be seen as a whole. In the center is a complete figure, neck outstretched, nostrils anxiously sniffing the air, about to flee, that is brought to life by the simplest means: the use of a vivid red laid on in successive, sometimes connected, dashes. Flanking it are the forequarters of two hinds, both executed later. This indisputable triangular composition (one of the oldest known: early Perigordian) was doubtless suggested by two convergent fissures that frame the scene, giving it the aspect of a gable. At Rouffignac, again, on the right-hand wall of the "Sacred Way," a panel of five engraved mammoths offers a most suggestive example of pyramidal composition. Two groups of two animals face each other. The admirably preserved leader on the left, with its high rounded head, sloping back, and long stiff hair fringing trunk and body, is one of the finest mammoths in the cave. Trunk coiled, curved tusks pointing toward its forehead, eyebrows puckered, it stands firm-footed with a melancholy gaze, awaiting its adversary. Between this animal and the facing one, in the triangle left free by their great masses, appears a baby mammoth.

Many figures take on their true esthetic significance only in relation to their original surroundings; often their meaning does not emerge except in the light of these surroundings. A case in point is a negative black-encircled hand at Gargas, placed, as though in a jewel casket, in a natural niche and thus endowed with an esthetic value, a magical significance perhaps, far superior to those of the two hundred-odd other hands in the cave, of which, in a sense, it is the archetype or symbol. A vault in the axial gallery of Lascaux displays, at a height of 13 ft., a large red cow that bestrides the void, its rear legs supported by the right-hand wall, its front legs by the left-hand wall, its body adopting the bend of the vaulting. An apsidiole of the Chapel of the Mammoths at Pech-Merle shows a mammoth rearing up on its hind legs, its trunk following the curve of the rock. These figures would be meaningless divorced from the surroundings that presided over their conception.

In a large sketch at El Buxu a cervid straddles a vault, seeming to emerge victoriously from the shadows of the gallery to dominate the extrados of the vault. This figure is an excellent introduction to the very numerous animal figures unmistakably associated with shadowy galleries, fissures, openings of narrow passages — all kinds of "pockets of darkness." It has been suggested that these dark recesses were envisioned as traps or pits, into which the animal would fall, and some examples can indeed be given of horses and hinds appearing to fall. But far more numerous are the animals that issue from pits, springing from dark recesses. It is thus necessary to formulate an inverse explanation: these animals rise from the bowels of the earth, the Earth Mother, creator of life and of game.

At La Pasiega a large horse, dating from the Perigordian, emerges from a narrow fissure as though escaping from the mysterious blackness. In the nearby cave of Las Chimeneas a large stag with tossing antlers rises out of the depths into the open air and joins the hind galloping before him. A little farther, the head of a stag and, farther still, the head and neck of a horse come out of the shadows. At Rouffignac a frieze of mammoths engraved on a ceiling emerges from the great pit situated halfway between the Great Ceiling and the saiga antelope. The last mammoth in the monumental frieze of the Breuil Gallery, the "Solitary," also comes out of a pit, and an engraved mammoth, opposite, does not fall into the nearby hole but seems to fairly issue from it.

The figures of the Great Ceiling at Rouffignac, as stirring, perhaps, as the great friezes and the admirable engravings of the Breuil Gallery, form an indescribable accumulation of "reality-pictures" heaped upon one another in apocalyptic excess and disorder (PL. 273). An inventory has totaled 20 mammoths, 11 ibexes, 9 bison, 7 horses, and 3 rhinoceroses, all running in different directions. Why is this ceiling, this section of the cave, so wealthy in figures? It is some 870 yd. from the entrance and is preceded and followed by hundreds, perhaps thousands, of square yards of surface as regular or more so, and often more conveniently placed for painting. Conscious esthetic judgment, so manifest in the friezes of rhinoceroses and mammoths in the Breuil Gallery, is noticeably absent here (that many of the figures are masterpieces must be taken as a sheer bonus). The choice of the ceiling is actually easy to explain; it is a choice, moreover, that sheds light on one of the major meanings of prehistoric art, one of the underlying reasons, indeed, for its existence.

The smooth surfaces of the Great Ceiling were chosen because they spread over the largest subterranean chasm of Rouffignac, a vast funnel that gives access to the second subterranean level. The mammoths and horses, then, were conceived as escaping from the depths. The sides of the chasm became steeper after the time of the Magdalenian painters; later falls of earth would have precluded the drawing of certain mammoths now suspended above the void. To reach the second level the dangerous slopes of the chasm must be negotiated until, 35–50 ft. below, a diaclase 7–10 ft. wide is reached that winds between limestone walls of astonishing whiteness. The Magdalenians may have seen a new rite in the act of following the serpentine diaclase, in view of the numerous serpentiforms of the cave, the snakes, and the red Ceiling of the Meanders with the hissing snake, tongue thrust out, opposite the entrance. On ground level, in an angle of the diaclase, a narrow fissure opens out; two rocky lips give access to a kind of chimney, just wide enough to allow the passage of a man, leading down to the third and lowest level, where a stream flows.

Evidence that the Magdalenian artist followed this path is seen in the column, with three faces projecting from the wall, over the fissure that leads to the lowest level. The first face, parallel to the fissure, shows two painted mammoths. The second, perpendicular to the fissure, displays a little engraved mammoth, a painted mammoth, and, curiously aligned along the angle of the column, four figures all looking toward the left: from top to bottom, a black bison, a black anthropomorphous head, a second black bison, and the forehead of a red horse. On the third face, parallel to the other lip of the fissure, appear two black bison and a red bison, more widely spaced. This large assortment of images thus includes two red figures; the only other red figure known in the cave is a mammoth, a little before the saiga antelope (FIG. 577, 1*k*), the most remote figure at Rouffignac. The richness of the column draws attention to the chimney that opens underneath. This passage into the bowels of the earth is, par excellence, the sacred place of the cave; it may be said that Rouffignac had its origin in a subterranean track coiled in the depths of the earth, the Earth Mother, creator of men and beasts, creator of life.

As further evidence that the column indeed identified a sacred spot, it may be pointed out that, except there, no figures appear on over 50 yd. of wall. The second-level gallery extends along increasingly narrow corridors through which passage becomes impossible. About 7–10 ft. before the fissure leading to the third level is another fissure, one not marked, however, because it does not permit access to the subterranean stream. Only the usable fissure is indicated. The painted and engraved animals on the column that surmounts it are, in effect, signposts. Special importance attaches to the anthropomorphous figure at the angle of the column; may it not be interpreted as the guardian of the depths?

The great ceiling of Altamira (PL. 282) marks the apogee of late Magdalenian painting, the swan song of prehistoric art. The ceiling is 59 ft. long and 26–30 ft. wide. Its original height ranged from about 4 to 6 1/2 ft.; its average height was the same as that of the Great Ceiling at Rouffignac. Here, as at Rouffignac, the original ground level was considerably lowered in order to safeguard the paintings, a measure that profoundly altered the visual effect. Even though the whole ceiling, or at least large portions of it, can now be taken in at a glance, something of its power, something of its impact, has been lost. The surface is hummocky, bristling with protuberances, which, sometimes reaching a length of nearly 20 in., reduced clearance to 2 ft. in the most constricted places and to 5 ft. in the roomier spots. There were enough obstacles, it would seem, to dishearten any artist. Yet the ceiling became one of the choicest gems of prehistoric mural art. The bosses proved of great suggestive value and imposed the theme of Altamira: every boss carries a bison. From what at first appears a disorderly mass of animal figures there emanates, after analysis, an impression of order. In the section farthest from the entrance, and facing it, a figure of flawless perfection, a large polychrome hind, 8 ft., 8 in., long, seems to occupy the vanishing point of a seething bison-dominated pyramid of animals, a cascade of bodies that suggests a "Last Judgment" of bison. They roll in the dust, twisting about; eight run toward the back, six return; four whirl in pirouettes, backs to the entrance. It is as though a tumultuous sea bore the animals on waves of rock, in a living and dynamic ebb and flow.

Schools of art. In the cave of Hornos de la Peña, both dwelling and sanctuary in the Aurignacian phase (in the Magdalenian these functions were to be rigorously separated), excavations brought to light a bone engraved with the hindquarters of a horse; and the left-hand wall of the vestibule of the cave bears the same horse (length, 16 in.), deeply engraved. In an early Magdalenian stratum of Altamira a hind delicately engraved on a shoulder blade was found; it appears again, almost line for line, engraved in larger format on a wall of the cave of El Castillo. Such links, instances of which need to be multiplied, are precious, for they offer a basis for dating mural works of unknown archaeological context by analogy with other works of established date.

Prehistoric artists, then, apparently made preparatory sketches and kept "notebooks of jottings," in the form of engravings on plaquettes, for example, as at Limeuil. One fact remains rather surprising, however: the high ratio of mural works of quality to what may validly be considered sketches (on plaquettes or bone or even on the walls themselves). The same is true of *art mobilier*: successful representations outnumber awkward and misshapen ones. No doubt the act of engraving or painting was considered a serious one, religious in some sort, to be performed only by the worthy — a supposition compatible with the underlying character of prehistoric works and with their atmosphere. A purely secular art might well have adorned the walls of the caves right from the entrance. A religious art alone can explain the choice of remote sites (the Salon Noir of Niaux, 870 yd. from the entrance; the Breuil Gallery of Rouffignac at the same distance) and the singular placement of the figures, mentioned in the preceding section. It is likely that many sketches were made on perishable materials — engravings on wood, paintings on skins — and therefore have not survived.

The unity of animal art, its consistencies, presuppose the existence of schools. It is revealing to compare two inventories made by Breuil, in 1902 and 1924, of Gallery I at Les Combarelles. In 1902 he reported 14 mammoths and 2 bison; in 1924, 13 mammoths and 37 bison. From the first, then, he recognized all mammoths (and one extra), but a minute study was necessary before he could decipher the bison. May it not be concluded that mammoths and bison were engraved by different hands, that Les Combarelles had a sorcerer-artist of remarkable talent for the mammoths, whose engravings are recognizable at first glance, and for the bison a sorcerer-artist of lesser resources, whose works cannot be recognized without close scrutiny? It may be pointed out that the apparent specialization in the representation of certain animals and the preponderance of given species in specific caves suggest some kind of totemism in the social structure of the Upper Paleolithic — a possibility on which Breuil speculated.

The Pyrenean region, rich in Magdalenian deposits often containing engraved objects, offers evidence of the existence of schools of art. In 1941 the Péquarts discovered at Le Mas d'Azil the famous spear thrower of reindeer antler known as the "Fawn with the Birds." In this engaging and humorous work a young fawn turns its head with a charming movement of its neck to watch a tenderly billing pair of birds perched on a protrusion from its anus. A few years later R. Robert discovered in the Magdalenian level of a corridor of Bédeilhac, less than 20 miles as the crow flies from Le Mas d'Azil, a second spear thrower, likewise of reindeer antler, on the same theme, but with only one bird. Most likely it is the older work, the prototype. In the more evolved piece from Le Mas, in which the feet are treated in the round and not merely in relief, the theme has been enriched and has assumed its definitive flavor. Other ornamented objects proceed from the same inspiration: three spear throwers from Arudy and one from Isturitz mark the spread of this Pyrenean school.

Isturitz has proved rich in spears and above all in bones carved in an original style: deeply incised interlocking or juxtaposed curvilinear motifs, combined here and there with small angular or denticulate motifs (FIG. 599, 1–4). These motifs, of rather ponderous "baroque" tendency, are encountered in only five Pyrenean deposits: those of Arudy, Lourdes, Lespugue, Isturitz, and Massat. Other analyses would underline relationships between the works of Le Mas d'Azil, Massat, and the cave of La Vache and, beyond the Pyrenean radius, the close links between the Magdalenian of La Vache and that of Fontalès, in the Aveyron Valley, not far from Bruniquel.

There is no "miracle" of prehistoric art. It originated in indistinct gropings: finger tracings made in idleness, from which one day a form emerged; then a multitude of experiments, which, fortunately, could be transmitted, because the life of the hunter required collective action and common dwellings. Thus schools of art arose. The migrations of the hunters explain the migration of artistic themes, the transmission of techniques, the spread of fashions; they explain the remarkable unity of that art which blossomed forth throughout the favored regions of France and Spain.

PALEOLITHIC ART IN EUROPE. *Western Europe.* Abbé Breuil (1952) ordered the available artistic data in two vast cycles: the first, the Aurignacio-Perigordian, of which Lascaux was the apogee, and the second, the Solutreo-Magdalenian, of which Altamira was the apogee. These two cycles, according to him, were characterized by different "styles" that succeeded one another chronologically. Analysis of carbon 14 datings, as already mentioned, has demonstrated a lesser duration of cave art and a probable overlapping of the two cycles. The apogee of the first cycle occurred about 14,000–13,500 B.C.; the second cycle evolved from 14,000 to 9500.

A certain confusion has arisen because such terms as "Perigordian" and "Magdalenian" can assume two meanings, one stylistic, the other chronological, and in given contexts priority may be given to one or the other of the two senses. If "Perigordian" is used in its stylistic sense, implying a specific artistic school or specific artistic traditions, La Pasiega, Lascaux, and Alpera (in the Spanish Levant) can be grouped together, with indications of successive stages. If, on the other hand, "Perigordian" is used in its chronological sense, it must be referred to a period corresponding to the industrial levels sometimes denominated "Chatelperronian" (after Châtelperron) and "Gravettian" (after La Gravette), and La Pasiega would then be placed largely in the Aurignacian (chronological sense); Lascaux would be debatable, because its dating inclines toward the early Magdalenian (chronological sense); and Alpera would be placed beyond the bounds of the Upper Paleolithic, in the Mesolithic.

The same ambiguity impairs all archaeological terms in this area. A certain type of harpoon is called "Magdalenian" in the Cantabrians, in the Pyrenees, and in the plain of Hamburg, although south to north its age will differ by 5000 years. Since absolute dates are becoming increasingly available, it would be well to leave such terms their cultural meaning and empty them of their chronological content. Thus to say: Saint-Marcel, Magdalenian III, 11,550, and El Juyo, Magdalenian III, 13,050 implies that the Cantabrian deposit presents the same cultural characteristics as that of the Creuse Valley but is 1500 years earlier. In that period the Magdalenian populations must already have begun their northward push, following the displacements of the reindeer herds.

In the analysis of the great periods of prehistoric art, the various names are here given an essentially cultural sense, that of artistic schools. It is convenient to slightly adapt Breuil's fundamental division and recognize a vast Aurignacio-Perigordian school and a no less vast Solutreo-Magdalenian school.

a. The Aurignacio-Perigordian school. The primitive phase of the Aurignacio-Perigordian school is the hardest of all phases to identify, as its recognition can be based only on stratigraphic proofs. This primitive phase can be characterized by the majority of the early works at Gargas: the finger tracings on clay, often confusingly intermingled, among which it is difficult to recognize any intelligible designs, although these do exist, representing, for instance, rhinoceroses and bison. The same phase is assuredly illustrated at La Clotilde de Santa Isabel in the finger tracings of much-simplified bulls with bodies often striated by vertical lines. The horns are shown frontally, although the animal is in profile. The legs are simply two or four sticks drawn without any attention to perspective. The finger tracings on clay at Pech-Merle — the mammoths and the women with hanging breasts on the ceiling of the Hall of the Hieroglyphs, and the panel of the Megaceros stag — may also belong to this phase, but there is no way of proving it; they may well be more recent and take their place among the archaic works of the Magdalenian. The innumerable finger tracings at Rouffignac, on the white walls and ceilings, as well as the white tracings on the red ceilings, belong to the early Magdalenian, the same artistic phase as the great engravings and paintings. Some engravings on limestone found at occasional sites such as Laussel and La Ferrassie correspond to the finger tracings on clay. They represent, for the most part, female sexual symbols and, more rarely, crude animal heads.

Painting in this primitive phase makes its appearance with the hands, negative and positive, imprinted on rock walls. Gargas, with its annex, Tibiran, again provides the type site. The heads of El Castillo, Altamira, and Pech-Merle are later and must be synchronized with the animal figures with which they are associated. The fine negative hands that frame the horses at Pech-Merle, for example, belong to the Magdalenian cycle (PL. 262).

As the primitive phase of the Aurignacio-Perigordian is characterized essentially by finger tracings — the simplest and most archaic form of engraving — the classical phase is distinguished primarily by painting. The lines are generally dotted, with the dots sometimes widely spaced (as in an ibex at Pech-Merle), sometimes close together and joined (as in some hinds at Covalanas; PL. 275). The most frequently used colors are yellow and red; black seems to have come later. Often, in a complex panel, the red figures are the oldest (see PL. 267). In the "Chapel of the Mammoths" at Pech-Merle groups of red dots clearly underlie the black outline of an early Magdalenian mammoth. Typical of this phase are the fine red horses, with dots close together, of La Pasiega, in the Cantabrians, and the figures of Le Portel, in the Pyrenees. At Le Portel the development can be followed from the simple black line of some archaic horses to the full modeling of other horses' heads. Some early figures at Lascaux are in the relatively evolved style.

The mural engravings — of varying depth, depending on the nature of the limestone — show the same stylistic characteristics as the paintings: stiffness, twisted perspective, and a generalized rendering without refinement of detail. Breuil classified as Perigordian the shallow engravings of El Castillo, La Pasiega, Altamira, Gargas, and Les Trois Frères, as well as the deeper engravings of Altamira, Hornos de la Peña, Pair-non-Pair, and La Grèze (PL. 260). Most of the engravings of Lascaux belong to this phase.

The sculptures of the Laussel shelter (PLS. 254, 255), in the Beune Valley, deserve special mention, not only because of their intrinsic interest but more particularly because they present a new original aspect of prehistoric art: outdoor sculpture. The site is a vast ledge, over 370 ft. long, facing south. About 20 ft. from the back wall an enormous block, the remainder of a vault that caved in, delimits a rectangular space about 30 ft. long. On a second monumental block the Perigordians sculptured the "Venus" of Laussel (PL. 255), which was detached after its discovery. It is a female nude with massive hips and voluminous breasts, whose right hand holds a bison's horn decorated with incisions. No less interesting are some smaller blocks, of which the two best display female figures with the same characteristics. The left hand of the one and the right hand of the other are outstretched and hold some enigmatic object in the same gesture of offering as the "Venus." The open-air esplanade might easily have lent itself to dancing and ceremonies of magic, under the benevolent gesture of the "Ladies of Laussel." The importance of subterranean art and rituals must not be allowed to overshadow outdoor art and ceremonies.

It is evident that the symbolic and ritual and, by and large, the representational concepts expressed in the "Venus" of Laussel also underline the small-scale sculpture in the round — those female statuettes of emphatic shape already mentioned in connection with human iconography. Yet it is difficult to fit their origin, function, and technical and stylistic characteristics into the framework of the great Aurignacio-Perigordian school, particularly in view of their formidable geographic dissemination, from Aquitaine to Italy, through central Europe, to Russia and Siberia. The ivory figures from Brassempouy (PL. 243; VIII, PL. 235) — the famous little head, the small torso of markedly sexual tendency, and lesser pieces — are southern French; so are the "Venus" of Lespugue (VII, PL. 351), also of ivory, a Pyrenean example of the stylization of body masses characteristic of the reliefs of Laussel.

It is in the realm of painting, at Lascaux, that Perigordian art reached its peak. Breuil noted intimations of the evolution toward bichromy at La Pasiega, El Pindal, and Pech-Merle, as well as on blocks, detached from the wall and painted, from La Ferrassie and the shelters of Sergeac; and he recognized the climax of this evolution at Lascaux, in the large and beautiful figures of the Great Hall of Bulls, for example.

The cave of Lascaux (FIG. 577; PLS. 258, 263–265, 267), discovered by some boys in 1940, has survived in a perfect state of preservation. Closed, doubtless by some geological accident, shortly after its decoration, it remained obstructed for 15,000 years. The original entrance is still unknown; the present passage is the result of a cave-in that occurred in the first decades of this century. Breuil, taking into account the respective positions of the figures, the superimpositions, the evolutions of style and process, recognized 13 successive series, 13 stages in the pictorial formation of Lascaux. These make it possible to break down the art of Lascaux, to note its successive states, just as the sketches of a painter make it possible to follow his thought, his changing conceptions of his future work. But these stages cannot convey any useful chronological information, for some may have been separated by centuries, others by a few brief moments. The oldest work is probably what Breuil identified as the arm of a child (or a woman), surrounded by a faded red, at the end of the axial corridor, on the left. If, as is possible, it belongs to the primitive phase of the Aurignacio-Perigordian school, Lascaux contains within itself the entire evolution of that school. At any rate, it is certain that the atmosphere of Lascaux is profoundly and typically Perigordian. Its characteristics are rooted in a more remote past, whereas Magdalenian art, newer, more evolved, less bound by archaisms and traditions, fully assimilated previous contributions.

Some of the horses — e.g., the frieze of ponies (PL. 258) — are executed with a "spray gun," like the hands of the primitive phase at Gargas. Perspective effects are still rendered by distortions, but with tendencies toward a more realistic presentation. The big black bulls (the biggest attains 18 ft.), the cervids, and the cows with small horns all show horns or antlers in twisted, or at least semitwisted, perspective. It is worth-while to cite the estimate of Abbé Breuil (1952): "The art of Lascaux, notwithstanding subsisting vestiges of primitive conceptions of draftsmanship, already testifies to a remarkable mastery in the execution of sometimes immense figures, of skillful, sure, and varied technique. The works attain elegance and power; sometimes they attain the level of masterpieces, as in certain big bulls, despite means of great simplicity. They testify to a culminating point in the ultimate blossoming of the first phase of Upper Paleolithic art; and their primitive lines help give them a freshness of expression, sometimes a little rugged and naïve, reminiscent in its way of the early Renaissance. The multiplicity of techniques, succeeding one another in a relatively short space of time, is the sign of a kind of artistic fever, rich in inspirations and experiments. Nothing could have made one foresee, in that remote period, . . . such an explosion of truly great art, perfect of its kind."

The Aurignacio-Perigordian school spreads over the whole classic Franco-Cantabrian area. Take, for example, the representations of bears belonging to this school. Mural representation of them occur in the caves of Atapuerca, Venta de la Perra, Las Monedas, Gargas, Pair-non-Pair, Cabrerets, Barabao (Bara Bahau), Font-de-Gaume, Lascaux, Villars, and, in addition, Aldène, in the Mediterranean orbit. Bears pictured on pebbles or plaquettes have been found at three sites: Isturitz and Péchialet, in the same geographic area as the caves named above, and particularly La Colombière (Ain), which markedly displaces the zone eastward (the late absolute chronological position of this deposit has already been noted).

The art of the Spanish Levant can be considered an extension of the Aurignacio-Perigordian tradition; some of its characteristics are adumbrated by certain figures in the Franco-Cantabrian domain. Thus the treatment of the little red horse of Le Portel, first sketched with a very fine engraved line, then filled in with vivid red paint, occurs again in a shelter of Valltorta; and a small bright-red horse in the Breuil Gallery at Le Mas d'Azil is strongly reminiscent of the style of Alpera. Breuil noticed the same applications of twisted perspective in the figures of Minateda as in those of Lascaux. For anyone who studies the extensions of the Aurignacio-Perigordian school, it is henceforth essential to formulate clearly the problem, already indicated, of the derivation from it of the art of the Spanish Levant (see below: *Mesolithic art in Europe and later survivals*).

Other extensions of the Aurignacio-Perigordian style can be observed along the shores of the Mediterranean, where they contributed to the formation of an original artistic province (see below: *The Mediterranean province*). Some horses and bovids engraved on plaquettes from the upper Aurignacian levels of the cave of El Parpalló (Valencia) exhibit the semitwisted perspective of Lascaux, a feature also observable in the bovids of the cave of La Pileta (Málaga), which, in addition, presents long serpentine finger tracings in red, very difficult to date. Farther to the east, the serpentiforms and big elephants of La Baume-Latrone, above Nîmes, and the engravings of Ardèche, of which the cave of Ebbou is typical, are so many landmarks of that Mediterranean art whose farthest known points are Romanelli, Levanzo, and Addaura. At Romanelli, stratum B, with its engravings, is dated 9980; and the lower level of the deposit with engravings at Levanzo is dated about 7750. These data are compatible with the development and apogee of the Aurignacio-Perigordian style in the Aquitanian basin and the Cantabrians. The geographical displacement explains the time lag, as it does the distinctive features of this Mediterranean art.

b. The Solutreo-Magdalenian school. No mural painting, so far, can be attributed to an initial phase archaeologically and chronologically identifiable with the Solutrean. Possibly future discoveries will modify the present complexion of the problem, for some painted slabs, depicting, for example, a horse and a hind, have been discovered in the Solutrean strata of El Parpalló (in Valencia, a Mediterranean province). Artistically this initial phase is of an undeniable originality, an originality reflected throughout its geographic distribution.

It may be useful to recall that the whole Solutrean culture constitutes an intrusion, as it were, in the archaeological complex of the Upper Paleolithic, edging itself between the Perigordian industries of blades and knives, scrapers and gravers, and the Magdalenian industries of blades, needles, and other bone objects. It produced outstanding examples of "industrial art" and indisputably brought the cutting of flint to its apogee. Pieces of remarkable fineness testify to an esthetic delight in good workmanship, which was manifested as early as the Acheulean, was later apparently lost, and was then revived for a few millenniums in the Solutrean. The forms have originality and elegance. The large regular "laurel-leaf" points of the middle horizons exhibit a perfect aerodynamic line, and the shouldered points of the upper horizons are of a cut and functional shape not without genuine esthetic qualities. The problem of the genesis and origin of the Solutrean has not been resolved. Its solution must be the first objective, if the formation of the Solutrean school of outdoor sculptures in southwestern France is to be understood.

Three remarkable open-air complexes belong to the Solutrean: the sculptured friezes of Le Roc de Sers and La Chaire-à-Calvin, in the department of Charente, and that of Isturitz, in the western Pyrenees. The sculptures of Le Roc de Sers, 24 miles from Angoulême, were discovered in 1927–29 by H. Martin. Their sculptured sides against the ground, the blocks lay buried under Solutrean archaeological strata. Excavations in 1950 revealed some sculptures *in situ*. Originally there must have been a continuous frieze occupying the base of the shelter and framing the wide, slightly sloping esplanade. The vigorously modeled representations comprise half a dozen horses, three or four bison, a composite animal with the body of a bison and the head of a boar, and two splendid ibexes facing each other in an admirably balanced composition of pyramidal type (PL. 256); one block shows a musk ox pursuing a small man with a stick on his shoulder. La Chaire-à-Calvin near Mouthiers-sur-Boëme is an important archaeological shelter, some 65 ft. wide and about 16–20 ft. deep, forming a little semicircular esplanade, continued in a slight incline on the exterior. The frieze, discovered by P. David in 1927, extends over 8 ft. and includes one or two bovids and two or three horses, among them a stallion in action. The figures of Isturitz, in low relief, include two reliefs of deer carved on a reindeer over 3 ft. long, a bear, and some horses.

Whether sculptured on powerful blocks or on the wall itself (at Le Roc the blocks may have become detached through the dissolution of the limestone and through temperature changes), these friezes doubtless share some antecedents with Laussel. But the very fact that the figures are ordered and composed into friezes differentiates them from those of Laussel, which are merely juxtaposed. The setting, however, is the same, and the question arises whether this outdoor art is susceptible of the same explanations as the underground paintings and engravings. The subject matter — animal life — is the same; the presence of man is still exceptional. The scene of the musk ox pursuing the little man at Le Roc is analogous to that of the well, or Shaft of the Dead Man, at Lascaux, in which an eviscerated bison stands over a bird-headed man (PL. 264). The esthetic organization is the same; the feeling for the frieze existed earlier (the ponies and deer of Lascaux, PLS. 258, 265) and is found again later (the mammoths and rhinoceroses of Rouffignac, PL. 272). Certain motifs, such as that of facing figures (the ibexes of Le Roc), are likewise known both before and after.

What actually sets this open-air art apart is its form: sculpture. Sculptures were not executed in the heart of the caves, although in the middle Magdalenian clay figures were modeled at Le Tuc d'Audoubert (PL. 259), Montespan, and Bédeilhac. The reason may be a purely technical one. Stone is more easily worked in the open, by daylight, whereas clay, abundant in the caves and lending itself to rapid handling, could be utilized even under the difficult conditions prevailing there. The feeling of this open-air art does not differ essentially from that of the caves. As already noted, at Montespan and Le Tuc the imprints of human heels suggest processions and ceremonies. Before the friezes of Le Roc, La Chaire-à-Calvin, Cap Blanc, and Angles-sur-Anglin (described below) identical actions must have taken place: processions, songs, dances — true manifestations of a scenic art.

A certain break seems to have occurred between Perigordian art, with its fruitful experiments in modeling, the rendering of volumes, and even a skillful bichromy leading to polychromy, and the art of the early, or preclassical, Magdalenian, which marks a return to black drawings and very simple linear sketches. To the same period, no doubt, belong what Breuil describes as "rather broad, slurred, generally coarse black tracings."

The difference is essentially one of draftsmanship, notably a correct treatment of paws or hoofs, horns, antlers, and, in general, of all perspective effects, without much concern for detail and finish. Numbers of horses and bison in the cave of Le Portel (PL. 266), mammoths and bovids in the Chapel of the Mammoths at Pech-Merle, and deeply engraved figures, sometimes in high relief, belong to this early phase of the Magdalenian.

Shortly after the discovery of Rouffignac its figures, too, were attributed to the very beginning of the second cycle. Breuil had connected the monumental works of the cave, whose leitmotiv is the combat of facing mammoths, with a pierced "bâton," also depicting facing mammoths, found by D. and E. Peyrony at Laugerie Haute in a stratum qualified as "pre-Magdalenian," a stratum superimposed on Perigordian but anterior to all Solutrean levels. (The stratigraphy of Laugerie Haute seems to confirm the previously indicated "intrusive" character of the Solutrean). A more complete analysis of Rouffignac, although it confirms the homogeneity of the works, suggests attribution to a less primitive phase, in fact, to the classical phase of the Magdalenian. Indeed these works are a far cry from the harsh profiles, the schematic contours, of the early phase; on the contrary, they display excellently rendered volumes and great expressive power in the poses, the heads, the glances — values achieved though a subtle play of painted strokes or fine incisions. If the great frieze of mammoths gives an impression of lifeless drawing, it must be taken into account that alterations have occurred, that internal calcification has embedded, drowned in the calcite, many secondary lines, the very ones that give a work expression and life. The better-preserved frieze of rhinoceroses (PL. 272), the Great Ceiling, and above all the whole ensemble of engravings demonstrate more convincingly the "classicism" of Rouffignac.

This school of open-air sculpture, located mainly in Charente and profoundly Solutrean in its characteristics, seems to have had considerable power of expansion. A wider geographic area and a later epoch, the early Magdalenian, are highlighted by outdoor works of the first rank. One of the most impressive is at Cap Blanc. This shelter, about 50 ft. long and 10–13 ft. deep in the center, carries along its base the most admirable frieze of horses known to prehistoric art. It is indeed a frieze: the animals are not simply juxtaposed but are ordered in a rhythmic composition. One horse is over 6 1/2 ft. long; the highest relief attains almost 12 in. At the time of discovery traces of red ocher were detected, an indication that the frieze was painted in red, the magical color (the other friezes, too, were painted). The vast shelters with southern exposure at Angles-sur-Anglin have yielded sculptures on big broken-off blocks; and other sculptures were found *in situ*, on the rock wall itself: a mare, a bison, and horses preceding the "Three Graces," or "Venuses," previously described, and five ibexes (among them a female following a kid) after them. Here are the traditional animal themes, notably the family theme, related to fertility. Another indication of continuity in open-air sculpture is the presence of numerous rings in the wall; these are also found at Laussel and may have been used to fasten captured prey during religious ceremonies. Both Cap Blanc and Angles-sur-Anglin, where the frieze was buried under deposits of Magdalenian VI, belong to the early Magdalenian.

In the classical Magdalenian, works multiply, contacts accelerate and sharpen, the time scale shrinks, and a rigorous classification becomes impossible. The greatest masterpieces of animal art known to the prehistoric world must be placed within the limits of three millenniums, perhaps only two. Some day classification may be feasible, but there are as yet no exhaustive inventories of the manifold works of *art mobilier*, which are often attributable to specific archaeological strata.

Small-scale sculpture in the round, not unrelated to the large reliefs in stone (Cap Blanc, Angles) and clay (Le Tuc), is illustrated by statuettes of horses, bison, mammoths, and other animals, generally carved in bone or ivory and apparently the first venture of southern French art into animal representation in this reduced format (PL. 248). The same trend found expression in the sculptured decoration of objects such as spear throwers (PL. 249) and "bâtons," whose purpose, perhaps a ceremonial one, remains obscure (PL. 247). In these objects the variety of animal themes and of couplings, the adaptation of the figures to shafts and butts, the boldness of the poses (e.g., the kids on the spear throwers from Le Mas d'Azil and Bédeilhac), and the close rhythms (the reindeer following each other, from Bruniquel; PL. 247) attest a keen sense of ornamentation and an inventiveness almost without limits. Many carvings show, instead of figures, decorative motifs consisting of meanders, spirals, stars (Isturitz, Lespugue, and elsewhere; FIG. 599). Small-scale human figures reappear in more modest forms than in the Perigordian, with marked stylizations and less particularized sexual characteristics.

Once the immense fund of *art mobilier* — a wonderfully rich and varied one — has been adequately studied, it will be possible to define, geographically and chronologically, particular traditions and artistic schools and to relate them to mural works. Thus a better knowledge of the objects found at, for example, Le Mas d'Azil (PL. 249), Massat, Lorthet (PL. 252), and La Vache (PL. 251) will permit the definition of a late-Magdalenian Pyrenean art; Bruniquel (PL. 247) and Fontalès will describe a closely related Aveyronian art. At present, however, only the type sites can be defined, the sites that are assumed, in the present state of knowledge, to be the artistic centers of the various schools.

Rouffignac is first in the field of engraving. At every step, along mile after mile of gallery, the cave reveals engravings of the greatest purity and realism: the long friezes of the left-hand wall of the Breuil Gallery, the panel of the five mammoths, the figures of the ceilings, the "Mammoth with the Roguish Eye" of the depths. The quality of these engravings may be partly explained by the grain of the rock, and their outstanding legibility by the absence of superimpositions. The style of Les Combarelles is no less excellent but much harder to appreciate, because of the superimpositions and the unfavorable conditions of viewing. Les Trois Frères contains numerous engravings of the same school. Rouffignac deserves to be considered the chief center in the field of drawing (PLS. 272, 273). If the Great Ceiling were cleared of the scribblings that have marred it since the 18th century, it could rival the Salon Noir of Niaux and the ceiling of Altamira. The style of Le Portel is perhaps less evolved, more "rustic," but some attempts at modeling (PL. 266) herald interesting late techniques. The facing bison are a fine example of the classical Magdalenian. In modeling, primacy must doubtless be shared by Le Tuc d'Audoubert and Montespan.

Shortly before its disappearance Magdalenian art gave the world its most astonishing masterpieces. Mural engraving reached its apogee in the marvelous sketches delicately traced on a stalagmite formation in the cave of La Mairie at Teyjat, engravings of the same purity as those found on late Magdalenian objects. The peak of draftsmanship is seen in the Salon Noir of Niaux, in works of peculiar expressive force. The elegant curve of the bison's horns, the slenderness of the horses' hocks, the quivering of the manes still fanned by the icy winds of the great steppes, the intense life in the eyes of the bison, all these details, orchestrated within splendid forms, remain inimitable (PLS. 269, 271). Some echo of this purity of line is perhaps found in the elegant silhouettes of the Arctic, in the bear of Finnhågen (V, PL. 151) and the reindeer of Sagelva, a few thousand years later. Magdalenian painting at its zenith decked itself in the glamour of polychromy and chiaroscuro: in the facing reindeer and the cohorts of bounding bison at Font-de-Gaume; in the horses and the bison of the Vidal Gallery at Bédeilhac, which time is effacing and changing into phantoms; in the impressionistic, pointillistic bison of Marsoulas; in the bison of El Castillo, so close to those of Altamira; and, finally, in the bison of the mighty bossed ceiling of Altamira itself, the savage and impetuous herd casting itself at the feet of the great hind (PLS. 268, 277, 278, 281, 282). It is Altamira, indeed, that remains the triumph of prehistoric art and one of the peaks of all artistic achievement.

Louis-René NOUGIER

Principal sites of discovery of Paleolithic *art mobilier* in Europe. *Key*: (1) Modern political boundaries.

The Franco-Cantabrian region, whose art spread over the European continent, passing beyond the boundaries of its western focal area with its *art mobilier*, was not the only Paleolithic artistic province of Europe. The discoveries made after World War II, especially, demonstrate that during the vigorous development of realism in the Franco-Cantabrian province other kinds of art flourished outside this area, characterized now by a representational idiom not far removed from the Franco-Cantabrian, now by a contrary tendency toward the creation of a world of decidedly nonrepresentational graphic expression. Of these various artistic provinces, one, particularly extensive and important — the Mediterranean — can be identified and roughly defined, in its characteristics and in its zones of dissemination, by an analysis of the postwar discoveries (especially in Italy) and the application of new selective criteria to the longer-known works.

One peculiarity that differentiates Mediterranean and eastern European from Franco-Cantabrian art is the existence of nonrepresentational works, of inorganic geometric designs, in some instances actual abstractions. In some places, particularly in eastern Europe, the geometric form appears as the sole pictorial expression, even though it is sometimes recognizable as the product of extreme stylization of organic subjects. Elsewhere, in the Mediterranean province, the geometric and the nonrepresentational mingle with a realism of high level that reveals points of contact, and sometimes a close bond, with the best products of Franco-Cantabrian art, even while retaining its own well-defined stylistic peculiarities.

Eastern Europe. Judging by the data so far available, the pure Paleolithic geometric form seems to present itself as the appanage of a rather restricted geographic area and of a relatively small group of works, all of them examples of *art mobilier*, from deposits in eastern Europe, specifically Moravia and the Ukraine.

The first discoveries go back to the beginning of the century. Between 1907 and 1909 T. Volkov and P. P. Ephimenko found in the Upper Paleolithic deposit of Mezine in the Ukraine (in strata whose industries the authors do not clearly link with any specific cultural "facies" but which can be roughly synchronized with the Magdalenian) curious objects of rounded form, sometimes bottle-shaped, fashioned in mammoth ivory and decorated with incisions in V forms, zigzags, and complicated rhomboidal motifs. Their significance is a matter of dispute. They have been interpreted as phallic representations, but the most commonly accepted hypothesis describes them as anthropomorphous figures. Accompanying them were what is probably a bracelet, also of ivory, covered with a very regular rhomboid decoration, and ivory plaquettes incised with Greek-fret and zigzag patterns. The station of Mezine, then, shows the establishment of geometric ornamentation with well-defined motifs, based on the broken line, that were developed in meander and rhomboid designs — designs very different in taste from the dynamic ornamental themes displayed by such objects (not very numerous, by the way) attributed to the French Magdalenian.

Considerably to the west, in Moravia, strata of the Perigordian period (or Solutrean, according to Czechoslovakian authors) of the deposit of Předmostí have yielded fragments of ivory and bone decorated with series of V motifs and with braid patterns, concentric loops, and "fingerprints" or whorls. In addition they have yielded a curious figurine, interpreted as the schematization of a female body, with patches of concentric ellipsoids representing the breasts, the abdomen, and the pelvis. The Moravian station marks the westernmost limit of the geometric-schematic art whose motifs are illustrated with particular distinctness at Mezine and were, much later, to find wide application in the Neolithic of the Carpatho-Danubian region.

The representational idiom in eastern Europe finds expression in the "Venus" of Willendorf (IX, PL. 247), the figurines of Dolní Věstonice (PLS. 243) and Brno in Moravia, of Kostenki and Gagarino in Russia, and of Malta near Lake Baikal (VIII, PL. 235), some of these offering striking conceptual analogies with the statuettes of Lespugue, the Balzi Rossi, and Savignano. Others reveal more abstract and geometrizing forms; for example, an ivory statuette from Dolní Věstonice in the shape of a simple cylinder with swollen pendulous breasts (PL. 244). Animal sculpture in the round, on the other hand, in ivory and bone, appears to be limited to east-central Europe (PL. 245); connections with the French Perigordian are lacking.

The Mediterranean province. This province covers a large area nearly equal to the Franco-Cantabrian (which, however, offers an incomparably more abundant and varied assortment of works). The art centers so far known are, compared to the Franco-Cantabrian, few and far between. Mural works and *art mobilier* may be found in the same station, but in general one or the other form either exists alone or is noticeably dominant. The province unfolds along a semicircular band that takes its start in the region of Málaga, touches that of Valencia, and, advancing northward, penetrates the valley of the Rhone, then runs along the Ligurian coast, descends the Italian peninsula across Latium, Apulia, and Calabria, and finally joins Palermo and the Egadi Islands. But Mediterranean Africa, too, especially the Maghrib, contributes some examples of decorated objects — notably the ostrich eggs of the Capsian, incised with schematic, purely ornamental shapes — and consequently extends even farther the zone of dissemination of this Paleolithic or epipaleolithic art, which thus encloses, in a broad perimeter, the whole western basin of the Mediterranean.

The identification of a Mediterranean province, distinct from the Franco-Cantabrian notwithstanding the indubitable and fundamental ties between them, began to appear feasible after the discovery in 1950 of a large number of admirable mural engravings in the cave of Cala Genovese on the little island of Levanzo (Egadi Islands). These engravings, for the most part naturalistic animal figures, could immediately be linked to the prepottery strata of the Anthropozoic deposit in the same cave, strata containing industries of Paleolithic type and wild fauna common in levels of the late Sicilian Pleistocene. The link was amply confirmed by subsequent excavations.

The engravings (PL. 283) are expressions of a lively naturalism that compares favorably with its Franco-Cantabrian counterpart but offers stylistic characteristics of its own, which, essentially, can be defined as the rendering of forms, movements, and perspective by means of outline without internal details (except for some rare suggestions of the manes of equids) or any graphic indication of light and shade. This primacy of line is accompanied by certain distinctive stylistic devices, such as the mode of representing the horns of bovids: always a single horn, turned forward, drawn in outline, and broken off at the apex. The existence of these fundamental and constant features, clearly identifiable in such a large and important series of works, underlined the significance of similar characteristics in previously known Paleolithic mural works and *art mobilier* in the Mediterranean orbit (e.g., in the caves of Ebbou in Ardèche and of El Parpalló near Valencia). It thus became possible to recognize a distinctive character common to all these works, one separating them from the great Franco-Cantabrian traditions and allowing their regrouping and geographical isolation. So far as their relations with Franco-Cantabrian art were concerned, it was apparent that they did not belong in the sequence of the triumphant experiments of the Magdalenian school but followed, rather, in the wake of Perigordian artistic mastery (see above: *Western Europe*).

In *art mobilier* the geometric (very different from the eastern European just described) and the schematic mingle with the naturalistic; indeed, the simultaneous existence of these opposite artistic tendencies constitutes one of the earmarks of Mediterranean art. The abstract designs themselves, moreover, exhibit a certain individuality. Whereas in the Franco-Cantabrian region the geometric or purely ornamental motifs seem to derive from an actual stylization and disintegration of naturalistic prototypes — a process that can be traced through its various stages (FIG. 599, 9–11) — in the Mediterranean province the origin and meaning of the inorganic motifs are generally unrecognizable. Sometimes they were evidently inspired by concrete subjects (e.g., ribbons and loops); yet their hidden significance does not seem to be what their form would at first sight suggest. As to the zigzags, plait motifs, rectangles, and the like, they do not lend themselves to any explanation at all. An obscure symbolism seems to dominate these nonrepresentational themes, which appear for the most part on portable objects, although they are not absent from rock art (Romanelli, La Pileta).

The cave of El Parpalló near Valencia, excavated and studied by L. Pericot García (1942), has yielded about 5000 painted and engraved fragments of rock from Gravettian, Solutrean, and Magdalenian levels, which compose an impressive series crowned by industries of Magdalenian IV. The cultures are comparable, from a typological viewpoint, to the classic French horizons; but the engraved figures (and less numerous painted ones that accompany them) exhibit the stylistic traits described above as peculiar to the Mediterranean province. In the representations of animals (bovids, deer, equids, ibexes) an evolution is noticeable from the lower to the higher levels of the deposit. In the older ones the figures are crude and incorrectly drawn; they improve in the Solutrean levels, showing tendencies toward grouping and composition; in the Magdalenian levels, finally, the drawing becomes more refined. The geometric and schematic motifs, present in all levels, attain

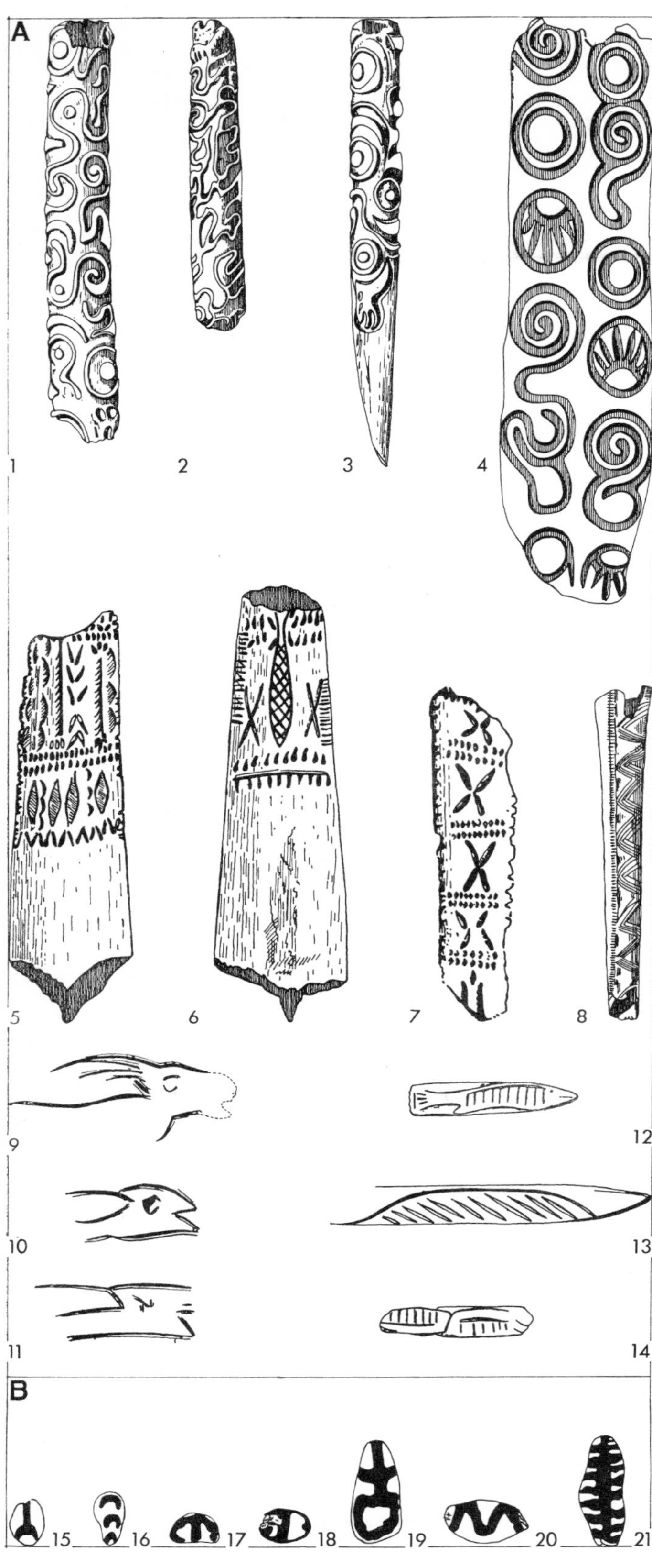

Decorative motifs of the Upper Paleolithic and the Mesolithic. (*A*) Geometric motifs and schematizations in Franco-Cantabrian *art mobilier* of the Magdalenian: (1–4) Spiral motifs on bone from Isturitz; (5–7) geometric motifs on bone from Marsoulas; (8) geometric motifs on bone from Le Placard; (9–11) increasingly stylized ibex heads on the same fragment of bone from Massat; (12–14) stylized figures on horn, derived from representations of fish, from La Madeleine. (*B*) Schematic motifs of the Mesolithic: (15–21) Pebbles from Le Mas d'Azil.

a special development in the Magdalenian ones; they include bands of parallel lines — wavy, straight, serpentine, or zigzag — crisscrossing lines, radiating lines, rectangles, and alberiforms, as well as ribbons sinuously unrolling or seeming to form knots and sometimes ending in a kind of loop. There are close affinities with the cave of Romanelli in Apulia.

Geometric motifs also appear on a few pebbles from the Grimaldi caves of the Balzi Rossi near Ventimiglia. Unfortunately no trustworthy stratigraphic information about them is available; it is known only that they were picked up in the Barma Grande cave in strata of the Upper Paleolithic. The pebbles are engraved with zigzag and reticulated designs; two of them bear rudimentary and fragmentary engravings of a horse's head and the head of a bovid. Three perforated "bâtons" of elk antler were found in the cave of Arene Candide near Finale Ligure; they are decorated with groups of parallel lines radiating from the hole at one end, while other groups of lines are incised on the shaft.

The cave of Polesini near Tivoli in Latium has yielded, in Gravettian levels, a considerable series of bones and pebbles delicately engraved with naturalistic and geometric figures in nearly equal quantities. The animal representations include heads of bovids with forward-pointing horns drawn in open or closed outline, the head of an equid and that of a deer, the hindquarters of a cervid, and a full-length wolf of somewhat rigid profile. The deer and the horse have a Franco-Cantabrian flavor; the other figures are readily connected with the Mediterranean style. Schematism and geometrism are illustrated by a bifid figure resembling a fish tail, spindle shapes crossed by parallel lines, zigzag patterns, small rectangles disposed regularly along the edges of fragments of bone, leaf-shaped figures that suggest feathered arrows, and, finally, a curious and complex meander or Greek-fret design, which seems to anticipate ornamental motifs of much later date.

In the famous Romanelli cave in Apulia, whose mural works are discussed below, A. C. Blanc has on various occasions drawn attention to a large collection of limestone fragments bearing delicately incised designs: some few of seminaturalistic tendency (a feline, two presumed cervids, and a boar) and others, more numerous, of schematic and geometric character, whose significance is obscure. Also noteworthy is a limestone block with comb shapes painted in red. The age of these pieces is certain: they were found in a deposit with Gravettian (Romanellian) industries, whose upper level (B) has been dated by carbon 14 to about 10,000 B.C. Like El Parpalló, Romanelli has revealed such motifs as clusters of lines, tree shapes, and sinuously winding ribbons. One pebble is encircled by two such ribbons, which cross as if they were real strings tied around it; and this impression of reality is enhanced by a kind of knot halfway along one of the ribbons and a loop at the end of the other, as well as by the tasseled and fringed extremities. One of the surfaces of another fragment of stone is engraved with a motif of interwoven lines. Of particular interest are the painted comb shapes, which evoke similar designs in the Neolithic and later rock art of Spain and other countries and anticipate the more or less similar signs on Azilian pebbles (see below: ***Mesolithic art in Europe and later survivals***).

Some isolated examples of *art mobilier* from two Italian caves and two French ones may also be cited. From the Mura cave at Monopoli in Apulia come two pebbles with naturalistic figures, one of them the head of a bovid with a single horn bent forward, in the Mediterranean style; and from the prepottery levels of the cave of Levanzo, dated about 7750, comes a bovid engraved on stone. From the caves of La Salpêtrière (at the foot of the Pont-du-Gard) and Nicolas (Gard) come two painted pebbles, one of the Aurignacian period and the other chronologically less certain, the first with a motif of parallel transverse lines, the second with a complex geometric decoration somewhat akin to the meander decoration of the fragment from Polesini discussed above.

To the Mediterranean orbit belong the steatite statuettes of the Balzi Rossi, the "Venus" of Savignano (PL. 257), of green serpentine, with an unformed hooded or conical head, and a few examples of uncertain period.

In rock art, too, nonrepresentational designs occasionally appear alongside naturalistic ones; but the naturalistic ones predominate and are generally found alone. In fact, so far the walls of only two caves, La Pileta and Romanelli, have shown clearly nonrepresentational designs. La Baume-Latrone near Nîmes contains certain figures that are indeed highly stylized but basically realistic.

A survey of the chief monuments of rock art of this period in Spain, southern France, and Italy may well begin with the cave of La Pileta near Málaga, known since 1911 through the discoveries of W. Verner, which contains paintings in red, yellow, and black, assigned by those who have studied them to four phases, three of them Paleolithic; a more precise classification is not possible. They include naturalistic and seminaturalistic animal figures, as well as motifs of obscure significance, apparently nonrepresentational. Some animals, such as an ibex and some horses and bovids, present more or less marked affinities with Franco-Cantabrian art, but in general the paintings of La Pileta are in a more denuded style. The outline is at times rather rigid; there is no suggestion of light and shade, and, except for some schematic indications of wounds on the bodies of certain animals, internal details within the outline are lacking. These features place La Pileta outside the Franco-Cantabrian panorama and relate it rather to El Parpalló and Romanelli. The seemingly nonrepresentational designs consist of serpentine tracings, radiating lines, rectangles, and other forms.

Multidigital tracings produced by streaking the walls with fingers dipped in reddish clay are found at La Baume-Latrone, which, as already mentioned, also contains markedly stylized realistic works such as curiously distorted mammoths and a feline reduced to a head with wide-open jaws and a long wavy line to represent the animal's body.

The cave of Ebbou near Vallon, in the valley of the Ardèche, is famous for its engraved figures of oxen, horses, deer, and ibexes. They are simple outlines omitting bodily details; the legs are summarily drawn and end in a point, without any indication of feet; sometimes the lines of the outline cross at the extremities. Despite the simplicity and even crudeness of these figures, they are skillfully drawn, with a firm, sure hand. The bovids have a single outlined horn turned frontward, in the usual Mediterranean fashion.

The Romanelli cave in the vicinity of Otranto, at the southeastern extremity of the Italian peninsula, offers, besides the *art mobilier* already mentioned, some rough mural engravings, in large part incomprehensible. The most realistic figure is an ox that recalls La Pileta. Also interesting is a series of spindle shapes, some of which can be interpreted as synthetic representations of the female body, others probably as fish, while still others seem to suggest the gradual transformation of the first form into the second. Finally, there is a geometric design made up of clusters of vertical and slanting lines, alternating with a certain regularity.

In the 1960s new rock pictures came to light in southern Italy, in the Romito shelter near Papasidero in Calabria and in the cave of Paglicci on Mt. Gargano. These findings suggest a far wider extension of Mediterranean art than was previously imagined. The Romito shelter revealed a superb bull of notable dimensions (nearly 4 ft. long), whose style — the general structure of the engraved outline and such peculiarities as the forward-tilted horns and the form of the extremities — places it among the works of the Mediterranean province; but unlike other Mediterranean engravings, it shows various details besides the outline, such as the eye (though this is only summarily indicated), the ear, the nostrils, and the broad folds of skin at the neck. Nothing can be said as yet about the chronology or a possibile connection with the archaeological horizons of the cave. A brief test excavation yielded a bone awl delicately incised with a geometric design of rectangles and groups of parallel zigzag lines, a design that fits in with the known schemes of geometric ornamentation.

The cave of Cala Genovese, on the west coast of the island of Levanzo, is composed of an inner and an outer chamber. The deposit of the outer chamber disclosed industries of Gravettian type in the lower levels and of the Neolithic in the upper ones. The inner chamber is completely dark. Its walls bear engravings of numerous animal figures and a few human ones (PL. 283), which, as already indicated, can be related to the prepottery deposit of the outer chamber. Besides the engravings there are numerous paintings (PL. 287), all of the Neolithic period or later except for one or two that may be contemporaneous with the engravings, consisting of some thirty-odd figures, only four of them human. The fauna includes *Cervus elaphus*, *Bos primigenius*, *Equus hydruntinus*, probably another type of equid, and perhaps a feline. The representations of animals, of decidedly naturalistic style, reveal a great command of draftsmanship as well as the lively sense of reality that animates the best Franco-Cantabrian works. As already indicated, no detail appears inside or outside the outline of the animal, except in some rare instances where the eye is sketched or a few dashes suggest the equids' manes. A few figures are outstanding: a vivid fawn turning its head in a lively manner (PL. 283); a vigorous running bull with massive head, robust chest, and sinewy back, which seems in its expressive force almost to anticipate the bulls of Minoan art; a bull with protruding tongue (a detail repeated in another bull's head) following a cow; a little horse (*E. hydruntinus*) propped on thin legs, and another walking, with its voluminous head thrust forward; and a pair of equids, mare and foal, whose simple profiles are barely sketched. Three human figures, more or less complete, seem grouped together, perhaps for a dance (PL. 283). The central personage, larger than the others, has a wedge-shaped head and a long beard. His face, as well as that of the little figure on the right with curving arms turned downward, lacks features. The head of the third personage, who holds his arms raised, is provided with a kind of beak. Undeniably these human forms are less realistic than the animal figures and are drawn far less skillfully. As will be seen, the reverse is true of the contemporaneous Sicilian cave of Addaura. It is probable, although not certain, that Levanzo's wall art should be assigned to the same phase as the previously mentioned fragment engraved with a bovid found in the outer chamber. This phase is geologically and chronologically post-Paleolithic, a fact that demonstrates the notable duration, persistence, and survivals of the Mediterranean style, which can be traced back as far as the Aurignacian industries of El Parpalló.

Discovered two years after the engravings of Levanzo, those of Addaura on Monte Pellegrino, near Palermo (PLS. 284, 285), which consist in large part of human figures, form a complex that differs substantially from previously known works of Paleolithic art (or works of direct paleolithic tradition). Their novelty lies in the dynamism that pervades the human figures, the scenic aspect of the main group, and the radically different style of treatment, testifying to unusual anatomical knowledge. Notwithstanding its extraordinary originality, the art of Addaura must be considered contemporaneous with that of Levanzo because of the obvious iconographic and stylistic affinities between the animal figures of the two caves. At Addaura the equids and cervids particularly are excellently drawn, showing great purity of line. The few bovids are artistically and technically inferior by reason of the stiffness of the outlines, the lack of proportion, and the crudeness of the drawing (one of the specimens in the main cave has only two legs, which end in a point, as at Ebbou); it is assumed that some of these figures are later than the others. But it is the human figures that have the greatest claim to attention, especially the animated principal scene (PL. 284), the interpretation of which is still much debated. Some naked men are disposed in a circle around two others lying on the ground. All these personages lack features, but some seem to wear a kind of beaked mask; some have substantial heads of hair, while others are hairless. Two of the standing men raise their arms, as though to express wonder or a strong emotion, and they and the others seem to watch the two personages on the ground, who lie prone, their legs so sharply bent at the knees that the heels touch the buttocks. One of them bends his arms at the elbows, carrying his hands to his neck, and seems to grasp and pull at a kind of rope that reaches from his shoulders to his feet; the other stretches his arms forward, having abandoned the rope stretched from his shoulders. Both are ithyphallic or, more probably, are supplied with a phallic sheath; a kind of belt crosses their waists. The ithyphallism of these two figures and the stretched rope have suggested a ritual scene of human sacrifice by strangling; but the hypothesis of exercises or a gymnastic contest seems more probable. Beyond this main group some personages move briskly in various directions. They too have thick heads of hair, and many wear the curious mask with a beak. Some carry sticks or bundles; one woman is laden with some kind of large sack or receptacle.

Two years after the discoveries of Addaura, the small cave of Niscemi, on the slope of Monte Pellegrino facing Palermo, revealed some engraved figures of bulls and small equids, in the style of Levanzo and Addaura. As at Levanzo, the horns of the bulls are drawn in open outline and point forward.

Several conclusions may be drawn from this survey. The art of the Mediterranean associates — polyphonically, as it were — geometric and representational themes. These are often contemporaneous and present in the same deposit (El Parpalló, Polesini), but schematic designs (Romanelli) and naturalistic figures (Levanzo, Addaura) may also appear separately. It is clear that this art endured, preserving the same characteristics, beyond the end of the glacial epoch and flourished in a phase that, chronologically speaking, is definitely Mesolithic (or post-Paleolithic), as is proved by the dating of Levanzo; so much cannot be said for an analogous prolongation of Franco-Cantabrian art. Also to be considered is the presence of schematic and geometric elements characteristic of decidedly more recent periods. There is no doubt, for example, that the comb-shaped painted signs of Romanelli, although they belong to an age and to cultures in a chronological sense unquestionably Paleolithic, are the same as those of the Azilian pebbles and of the rock art (Neolithic and later) of the Iberian peninsula; the meander design incised on the fragment of bone from a Gravettian Paleolithic level at Polesini heralds motifs that appeared in the Apennine Bronze Age.

It should be kept in mind, too, that the expressions of naturalistic art of the Mediterranean province seem to have found continuation and undergone dissemination even after the definitive ending of the Paleolithic and Mesolithic civilizations. Indeed, turning to North Africa, to the world of Saharan and North African rock figures, that immense reservoir in which graphic and pictorial records of the most diverse cultures and ages have accumulated through the millenniums, one realizes that the products of certain artistic phases of the pastoral civilizations repeat, often in a surprising way, some of the stylistic formulas of Paleolithic Mediterranean art. Even certain human figures (Fezzan, Tassili-n-Ajjer), by their form and their liveliness, recall those of Addaura (see below: *Prehistoric rock art in other areas*). Apropos of human representation, a final comparison may be offered, relating to another extraordinary flowering of Mesolithic rock art (treated in the next section), namely, the painting of the Spanish Levant. The human creations of this art are pervaded by a vital and dynamic spirit in many ways akin to that of the scenes of Addaura, although the figures of that cave appear more concrete and realistic and, in that sense, more paleolithic. It is not inconceivable that future discoveries will reveal — possibly by means of findings testifying to intermediate phases — still closer bonds between the Sicilian pictorial world and that of the Spanish Levant.

Paolo Graziosi

Mesolithic art in Europe and later survivals. *Mesolithic art provinces.* When the Pleistocene came to an end after the last cold wave of the Würm glaciation, about 10,000 B.C., the European ice age was gradually succeeded by a more temperate climate. As a result, the fauna and flora throughout the continent underwent changes. The men of the Upper Paleolithic adapted their hunting culture to these new natural conditions, which, originating in the Mediterranean and Atlantic regions, spread from there to the east and north of Europe. The outlook of these men was in turn transformed, and new forms of artistic expression originated during the Mesolithic period (an era that might more accurately be termed "epipaleolithic," since the peoples then inhabiting Europe were the same as those of the Upper Paleolithic, in spite of the profound changes in their environment). Above all, new methods of hunting were adopted, appropriate to the species of prey found after the change in fauna. Various large animals such as the mammoth, the horned rhinoceros, and the cave bear had become extinct; others — the bison, the reindeer, the elk, the aurochs, and many other animals that had nourished the Magdalenian and other cultures of the Upper Paleolithic in Europe — had emigrated toward the east and north. The tribes of hunters had to gather shellfish, snails, and such vegetation as nuts and berries; they hunted birds, rodents, goats, deer, wild boar, and other small animals. In the face of mankind's new environment, the magic rites and cults of the hunt — rites and cults that had inspired those great works of Paleolithic art, the painted sanctuaries hidden in deep caves — were forgotten. The cultures developed by the Mesolithic hunters are only a weak offshoot of the great Aurignacio-Perigordian, Solutrean, and Magdalenian art.

In only three of the Mesolithic cultures derived from the Upper Paleolithic can the evolution of distinct artistic cycles be seen. These correspond to the three basic cultural areas of Europe during the transition from the climatic conditions of the last Würm glaciation to the present geological age.

The first of these artistic provinces of Mesolithic Europe is that in which the great Paleolithic art of western Europe had flourished. This art was developed by the peoples of the Azilian culture (named from Le Mas d'Azil in Ariège) and continued in the Tardenoisian culture (named from Fère-en-Tardenois, Aisne); it is not possible as yet to distinguish between the works of these two peoples. There are few examples of this art, and they are a poor and increasingly faint echo of the great Paleolithic art. They are strongly oriented toward imaginative conceptions and consistently remote from the vigorously naturalistic figural art of the Pleistocene.

As examples of *art mobilier* there are only some very simple engravings on bone and certain pebbles or plaques painted with dots, parallel stripes, simple lines, and zigzag motifs or cruciform or circular marks that can be interpreted as schematic representations of the human form or as symbolic representations of animals. The fame of the painted pebbles of Le Mas d'Azil (FIG. 599) found by Piette (1895–96) has not helped to solve the question of their precise stratigraphic position and their significance, although they have been interpreted as symbolic representations of ancestors. Certainly the tendency toward an imaginative, symbolic, conceptual art had already begun to be discernible on some of the bones and "bâtons" of the late Magdalenian era, especially in the Cantabrians. But it is difficult to find at Azilio-Tardenoisian levels anything other than simple linear motifs, in no respect similar to the schematic figures of the late Magdalenian; nor are figures such as those painted on the pebbles of Le Mas d'Azil found in a clear stratigraphic position anywhere else. Occasionally in some caves small boulders or stone plaques have been discovered on which there are red dots or at most single or parallel lines. Such remains have been found at the Azilian levels of the caves or shelters of La Crouzade and Bize, or Bise (Aude); Bobache (Drôme); Montfort, Montardit, and Le Mas d'Azil (Ariège); Marsoulas, La Tourasse, and Gourdan (Haute-Garonne); Sordes (Landes); La Salpêtrière (Gard); Pagès (Lot); Rochereil (Dordogne); Birseck in Switzerland; and Ofnet in Bavaria. There are also some remains in the Cantabrian region of Spain, but they are rare and not very expressive, and their interpretation is uncertain.

No rock art can be conclusively attributed to this era. It is possible, though by no means certain, that the Azilian culture may have produced the dots and the branching and zigzag marks found in the cave of Marsoulas (FIG. 599) and the series of dots found in the caves of Niaux (Ariège), Yssat (Cantal), and elsewhere in France and also in Spain, especially those of El Castillo (PL. 274). Breuil and Obermaier (1935) have hesitated to attribute any rock art to the Azilian age. The only thing that appears evident is that the great Paleolithic rock art and *art mobilier* were not continued by the men of the Mesolithic cultures of southwestern Europe. However, the tradition of ornamenting the walls of caves must have persisted among the indigenous poeples of Europe until it merged with the rock art of the early Bronze Age, an era that produced paintings and engravings in various caves, especially in the regions of Ariège and the south of France.

A second artistic province of the Mesolithic is that of northern Europe, a region gradually inhabited by Mesolithic peoples as the polar cap of the Würm glaciation receded. These peoples, hunters of Arctic fauna, were descended from the Ahrensburg culture of northern Germany. They spread toward the northern tundras which were forming in the wake of the constantly receding northern polar cap and which were in turn succeeded by woodlands. From approximately 8000 to 1000 B.C., during the evolution of a moderate climate in those regions, cultures arose there that have left behind interesting works of both chattel art and rock art.

The little schematic human figurine from Hophenbach (Ahrensburg), sculptured in wood, is the most ancient example of artistic concern left by those early hunters. Its style coincides with that of certain schematic works from the end of the late Magdalenian culture and from the Azilian culture, which flourished simultaneously in the regions of the Rhine and southern Germany. From such influences originated the cultural groups of Tjorger, Rissen, and Wehlen; farther to the east, those of Probstfels and Dobritz; and westward, that of Creswell Crags in England.

There are no true works of art from this period. Shortly afterward, from about 6000 B.C., these cultures became united in the extensive Klosterlund-Maglemose culture, which offers several seminaturalistic figures engraved on bone (V, PL. 152), such as those on a perforated club from the Stettin-Grabow region, a distant echo of the Magdalenian "bâtons." This Maglemosean art, oriented toward schematization and showing clear similarities to Azilian art, contrasts with a fine series of vigorously naturalistic engravings created in Scandinavia, especially

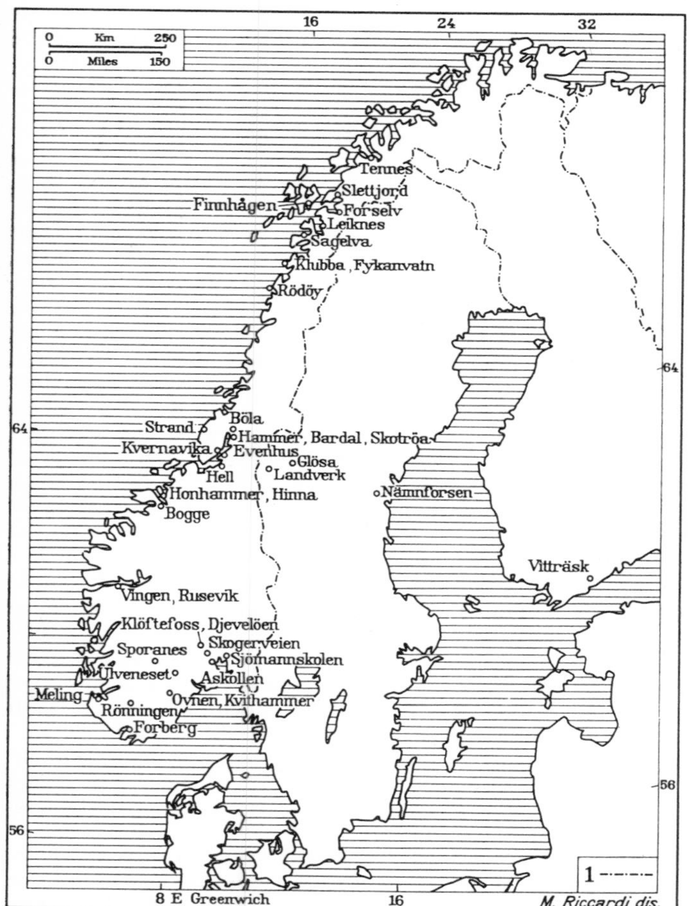

Principal sites of naturalistic rock art in Scandinavia. *Key*: (1) Modern political boundaries (*prepared by A. Vigliardi Micheli*).

in Norway, by the last Mesolithic hunters (PLS. 290, 291; V, PL. 151), who spread along the Norwegian coast as far as North Cape, as well as inland toward Sweden. It is clear that after a naturalistic stage of art that recalls the Paleolithic creations of southwestern Europe a seminaturalistic phase developed that ultimately produced highly imaginative schematic works, always, however, remaining within the limits of figural art, as distinct from the complete abstraction of Azilian, Neolithic, and early Bronze Age symbolic art. A cycle derived from these same Scandinavian artistic conceptions developed in the regions between Lake Onega and the White Sea and even farther east, toward Siberia, where works of art not unlike those of northern Russia are found, although they are for the most part of a later period.

Frequent superimpositions make it possible to follow the technical and stylistic evolution of the European Nordic cycle from the early naturalistic figures to the increasingly stylized later ones. In Norway, for example, on the rock of Bardal (Nord-Trøndelag) are found the first images of boats, dating from the height of the Bronze Age. Beneath these are engraved seminaturalistic animal figures, and below both appear beautiful naturalistic forms of elks, deeply engraved in a style of great realism. On the rock of Sletfjord in Ofoten (Nordland) there are engraving of birds and snares dating from the era of the Scandinavian hunting peoples; superimposed on these figures are drawings characteristic of an agricultural era. In Meling (Rogaland), superimposed on a drawing of a fish, are stylized shapes of boats. Some of these stylizations, such as the schematized human forms at Tennes (Balsfjord), are the result of early Bronze Age influences brought from Iberia along the Atlantic coast of Europe. Similar superimpositions are found in Hammer and Skotrøa (Nord-Trøndelag), in Bogge (Møre og Romsdal), in Rusevik (Sogn og Fjordane), in Sporanes (Telemark), among other places.

Three stages or phases have been identified in the development of this Mesolithic art cycle in northern Europe. The first phase, beginning between about 6000 and 5000 B.C., is naturalistic, and its forms are deeply engraved; they reveal a strong sense of realism and a marked taste for expressing movement. The second phase, dating from about 3000 B.C., is characterized by more angular and less realistic forms and by the use of grooved lines, as in the figures of Vingen (Nordfjord, Norway); animals, fish, and birds are almost the only motif of this phase. The third phase, beginning about 2000 B.C., offers stylized men and animals; the conceptual art created during this phase (within which must be placed the rock art of northern Russia) belongs entirely to the Bronze Age and is not discussed here. It has been possible to establish with reasonable precision the absolute chronology of these works of art because the time it took for the Scandinavian glacier to recede is known, and the stages by which man reached these regions can be quite closely dated.

Although most of the works of Scandinavian rock art are engraved, there are also paintings. To judge from their appearance, they were executed with iron oxides diluted in a fatty substance, perhaps whale or seal oil. There are 11 sites with paintings in Norway and 6 in Sweden. In both countries these paintings follow the same phases of stylistic development as do the engravings, the two earliest phases being the product of hunting peoples. The most ancient style is represented at Forberg and Furodden (Vest-Agder); examples of the second, or less naturalistic, phase are the animal and fish paintings at Honhammer (Møre og Romsdal). The paintings of the last phase, at Ovnen, Kvithammer, Rønningen, and Ulveneset (all in Telemark), belong to the Neolithic and the Bronze Age and are not considered here.

All these works have a marked magical character. Success in hunting and the fertility of the economically important animals — elk, reindeer, seals — preoccupied the artists who created these figures. Even today, among the Lapps and the Eskimos, the practice persists of drawing those animals which are objects of the hunt; and in 1910, the Lapps were seen making votive offerings before a rock painting of Seitjaur on the Kila peninsula.

A third artistic province developed during the Mesolithic in the Mediterranean; its character is different from both the

Azilian art and that of the Russo-Scandinavian province. It is interesting both for the number and completeness of its works and for the amount of discussion it has aroused.

Even during the Upper Paleolithic, the great pictorial art of this province revealed, in its less naturalistic and more dynamic qualities and its use of the human figure, a character distinct from that of the great Paleolithic art of the Dordogne, the Pyrenees, and the Cantabrians, despite marked affinities. Thus the anthropomorphous figures shown swimming and fishing on the walls of the cave of Los Casares in Riba de Saelices (Guadalajara) are reminiscent of the dancers of the cave of Addaura in Sicily (PL. 284); and the painted and engraved plaques from Romanelli have similarities to those of El Parpalló and La Cocina (the last clearly Mesolithic). The tendencies of that art lead to the most famous Mesolithic art of the Mediterranean, that of eastern Spain, or the Spanish Levant. It is unlikely that these Paleolithic and post-Paleolithic artistic complexes were very distant in time from the art of the provinces already discussed; but it would seem that the vigorous figural realism of the Mediterranean province preceded that which developed elsewhere in Europe. This sequence somewhat parallels the development of the industries of the primitive hunters, whose microlithic tendency, originating in the Mediterranean regions, spread from there (particularly after the Magdalenian era) to the Azilio-Tardenoisian stone industries found more or less well developed in all European Mesolithic cultures.

The geographic extent of this rock art, at least in its most characteristic naturalistic phase, is limited to the mountainous zone extending along the Mediterranean coast. The most northerly examples are found at Cogul in Lérida and in the mountains of Tarragona (PL. 289); examples appear also in the Castilian and Aragonese Maestrazgo, in the Teruel district of Baja Aragon, and along the mountain ridges of Albarracín, Cuenca, and Valencia, ending at the boundaries of Albacete Province, north of Murcia. Along the Jaén-Almería mountain ridges and others still farther south and in the group of painted rock shelters near Laguna de la Janda (Cádiz) there are only small and crude works (PL. 292), the last distant echo of the healthy and vigorous realism of Levantine art that is occasionally reflected in the central area of Spain, as at Las Batuecas (Salamanca).

The works of this area are always found in simple rock shelters, almost in the open air, and never in the deeply hidden caves that house the works of Paleolithic art. Since scholars of prehistory began studying them, an immense number of paintings have been found on the protected walls of these rock shelters. As a rule numerous groups of figures in the same shelter are superimposed one upon the other, making it possible to establish the sequence of stylistic and technical developments.

Principal sites of Mesolithic rock art in eastern Spain. *Key*: (1) Modern political boundaries.

Now that many groups of paintings have been published, the character of this Levantine rock art can be more clearly seen, as the technical, stylistic, and thematic differences that set it apart fom Franco-Cantabrian Paleolithic art become more distinct. Studied as a whole, the artistic legacy bequeathed by these prehistoric painters of the Spanish Levant reveals a mentality much more complex than that of their predecessors, the Paleolithic artist-magicians. The Levantine paintings reveal a new human spirit, richer in nuances, keener, more sensitive, and able to express itself in richer forms and colors. While the Paleolithic artist represented only animals on the walls of his sanctuaries, rendering the human figure only in crude and ambiguous forms, the prehistoric Levantine artist took a giant step forward in presenting a complete panorama of the human life of his time, with man appearing always as the chief protagonist. It was in the Spanish Levant that the human figure first became the principal inspiration for the artist. In addition, the artist now managed to construct scenes. His more developed mind no longer saw merely isolated animals or individual men but vivid events; he composed artistic narratives of the simple and hazardous life of these primitive hunters, with man shown in action and playing the chief role. Group views of the hunt, the pursuit of wounded animals, and the gathering up of slaughtered prey are among the dominant themes of these artists. They also reproduced scenes of fighting between tribes, dance scenes, realistic scenes of the execution of enemy chiefs and tribesmen, images of men and women in various postures, masked magicians, and even harvest scenes. Thus the prehistoric Levantine artist succeeded in creating an art in which beauty and grace reside in the total composition. The individual figures are sometimes carelessly executed, but taken together they often create an enchanting and expressive work. There was no feeling for landscape; the natural setting surrounding a figure or scene was ignored. Plants are seldom shown together with human or animal forms: a pine in the "Cave of Doña Clotilde" in Albarracìn (PL. 289) and what are apparently pines and other plants in the rock shelter of Alacón, which depicts harvesters, are the only nonanimal forms.

This art continues the convention of "twisted perspective," in which animal figures were rendered in profile but with their horns and hoofs in front view; one no longer finds the marvelous perspectives and foreshortening seen in the best Magdalenian work. In this aspect, too, what Levantine art gained in spontaneity and lively expressionism, it lost in the perfection of its technique and its drawing. This spontaneity and simplicity of technique do not rule out the existence of regional schools, each with its special way of painting. The conception of figures and scenes, the prevalence of certain themes, and the use of colors were not homogeneous throughout this geographic area. Purely evolutionary criteria can lead to errors in the chronological evaluation of styles; and it must be remembered that figures have sometimes been repainted within previously drawn outlines, so that a figure that is recent in its coloring may nevertheless be more ancient in style. Such repainting has often been confused with true polychromy. The figures in these works, however, were painted in a single color, generally red, ranging from bright red to maroon; black and white were also used. The colors were obtained from hematites and kaolins mixed with resin and other substances impossible to identify today because of the mineralization and fossilization of the paintings. Generally animals and men were painted by simply filling in their outlines with an even color, always of a uniform shade. Sometimes they were only silhouetted with wide or narrow lines, and on occasion these silhouettes were filled in with simple stripes, crude and irregular.

There was no true engraving, although at Cogul there are fine lines that were apparently preliminary sketches for the paintings of bulls found there. These outlines are simple and not very skillful. Thus it may be said that Spanish Levantine rock art was exclusively the work of painters.

In spite of the variety of themes found in Levantine rock art, there is no doubt that all the scenes and figures had the same magico-religious value as the earlier Paleolithic animal figures. Evidence of this is the fact that the chosen rock shelter acquired the value of a sanctuary, with scenes of the most diverse activities often superimposed one over the other, even when there were other nearby shelters equally suitable for such works. Thus at Minateda (Albacete) 13 different styles have been distinguished, and at Tormón (Albarracín) 8 series of paintings in different techniques can be seen in the same rock shelter, although there are innumerable well-protected shelters suitable for painting near both sites. Sometimes the figures became worn away or faded, and, since the figure as well as the site was regarded as an indispensable source of magical power, they are frequently found repainted in obviously later techniques and colors but with no changes in the animal scene or human figure. Examples of such repainting are the bulls of Tormón and the bulls and some figures in the phallic dance at Cogul.

Certain scholars have maintained that some of these works may be merely commemorations of battles or of important events that the artists wanted to perpetuate; or even that some of the figures and scenes have no transcendent purpose and sprang only from the artistic satisfaction of reproducing scenes of domesticity, war, or the hunt. Such opinions are difficult to reject dogmatically; however, history offers much evidence of the religious origins of art, and the author believes that it was exclusively in the service of such magico-religious beliefs that this Levantine art continued to live. This view is confirmed by the discovery and transcription (Almagro Basch, 1952) of votive inscriptions at Cogul dating from the Iberian and Roman periods; they show that the site continued to be held sacred even far into the historical era.

There are obvious difficulties in establishing with certainty even a relative chronology of the development of this art, much less an absolute chronology. Breuil (1912) regarded the paintings preserved in the rock shelters of the Spanish Levant as works of Paleolithic art, and this view has been upheld by Obermaier (1938) and is accepted by most archaeologists, on the basis chiefly of Breuil's identification (questionable, according to the investigations of the present author) of Paleolithic fauna in Levantine art and of certain stylistic and technical analogies with Franco-Cantabrian Paleolithic rock art. This Paleolithic dating has, however, been rejected by Hernández Pacheco (1924) and by Kühn (1941–42), among others; and the marked differences between Franco-Cantabrian rock art and that of the Spanish Levant offer grounds for assigning the latter to the Mesolithic period and in some instances even later — that is, approximately 7000–3000 B.C.

The most emphatic difference between Paleolithic art and that of the Spanish Levant is found in their themes. Paleolithic art offers only isolated animals and, very rarely, scenes, and the human figure was never represented except in crude and ambiguous form; whereas, as already indicated, the Levantine artist was a composer of complicated scenes and discovered man as the chief protagonist of art. There are vivid and attractive scenes of hunting and warfare, dancing, and even harvesting. An agricultural period with a developed textile art, rather than a truly Pleistocene age and climate, is reflected in numerous examples of Spanish Levantine art: among others, the scene of honey gathering in the Cave of La Araña (PL. 288) and other scenes at La Gasulla, Alpera, and Alacón, showing baskets and rope ladders; the long-skirted women of Cogul and Alpera and especially the woman wearing a comb in her hair and in a long dress, at Dos Aguas (Valencia); the harvest scenes in the Alacón shelter, which seem to depict women gathering fruit along a row or furrow; the woman gathering crops with a basket, in Dos Aguas; and the scenes of women planters in Alacón and Dos Aguas. It is not impossible that such scenes could have occurred during the Pleistocene, in spite of the cold climate that must then have prevailed in the high mountains of eastern Spain; but no such scenes are known anywhere in all of Franco-Cantabrian Paleolithic art. In addition, it may be pointed out that some of these rock shelters contain rather competent naturalistic representations of domesticated animals: for example, the horse led by reins in Villar del Humo, the plumed rider in La Gasulla, and the apparently domesticated asses in Alacón.

The size of the figures in the rock shelters of the Spanish Levant is another important and interesting difference. The tendency there was to draw "micro-figures," always full of vigor and grace but averaging only about 6–8 in. and sometimes as small as 1 or 1½ in. In the whole of Levantine painting, only a few rare exceptions show figures as large as those in the Paleolithic cave paintings and engravings.

The fact, already referred to, that some painted shelters have continued to be places of worship in historical times lends further credence to the view that this rock art is of relatively modern origin. The same conclusion is at least suggested by recent investigations (Almagro Basch) that have repeatedly found, at the foot of the painted shelters of the Spanish Levant, consistent evidence of a stone industry dating entirely from a post-Paleolithic period.

Schematic rock art. Rock art, so deeply rooted in the Paleolithic, did not die out with the introduction of new techniques and the changed way of life that characterized the Neolithic. Its forms may have changed — and these changes were already foreshadowed in the Mesolithic — but it remained basically prehistoric in concept. The dominant artistic trend everywhere was schematism.

The development of naturalistic and expressionistic Spanish Levantine art had come to an end with the pursuit of increasing abstraction and the complete loss of expressiveness and clarity in the presentation of figures. Scenes no longer represented actual events but became purely symbolic. All spontaneity disappeared; the sense of movement was lost; the range of themes narrowed to include only schematizations of animals and of the human figure and certain enigmatic signs (FIG. 611). What makes this schematic rock art remarkable is that its images reveal, more or less plainly, the intellectual process by which the figures were abstracted from natural forms. It is almost possible to detect here the same process that in the Near East, at nearly the same time, led to the birth of pictographic writing.

Works reflecting these new artistic currents are occasionally found in the naturalistic art of the Levantine rock shelters, and the technique and distribution of such works indicate that they were for a time contemporaneous with the lively naturalistic and figural art of the last Mesolithic hunters. Originating probably in the eastern Mediterranean area, this schematic art traveled across the Mediterranean to reach the Iberian Peninsula, where it gave rise to a profusion of painted rock shelters containing an unsurpassed wealth of images, found most abundantly in the mountainous areas of southern and southwestern Spain.

Portugal, the central Iberian mountain ridges, and northern Spain also offer numerous examples of this art, and it is certain that it persisted in some interior regions, until protohistoric times, as evidenced by several works of schematic art in Estremadura showing representations of central European four-wheeled carts, which reached Spain only with the so-called "Celtic invasion" beginning about 800 B.C. Some of these works must be dated even later, although the peak of this artistic cycle coincides with the early Bronze Age, or Los Millares culture, at which time there appeared sculptured idols, engraved plaques, and especially pottery decorated with schematic motifs similar to those of the rock art.

Martín ALMAGRO BASCH

Similar schematic and geometric designs are found engraved on the stones of the great dolmens of Andalusia and Portugal, as well as on those of certain megaliths in France, especially in Brittany, and in other regions of western Europe. Most of the supporting slabs of the monumental dolmen on the island of Gavr'inis in Brittany are deeply engraved with closely grouped concentric tracings, semicircular, elliptic, or horseshoe-shaped; and on a stone of the dolmen of Luffang (in Crach, Brittany) can be seen a schematized human face, which perhaps evokes the Mother Goddess. The finest dolmen engravings are found in Scania on the supporting slabs of the Bronze Age dolmen of Kivik (V, PL. 180). The subjects include schematic "filiform" human figures, ships, and two-wheeled carts drawn by two horses. Designs of the same style are found in some caves of southern France.

Motifs from Iberian schematic painting. Seminaturalistic animals and derived comb motifs: (1) Seminaturalistic group of stag and hind with fawn from Rabanero; (2) seminaturalistic goat and kid from Rabanero; geometrized comblike quadrupeds from (3) Rabanero, (4) Las Viñas, (5) Malas Cabras, and (6, 7) Las Viñas. Schematic human figures: (8–13) Anchor motifs from Malas Cabras; (14) superhuman figure from La Batanera; (15–17) superhuman figures or "fir-men" from Las Moriscas; (18–21) inscribed "fir-men" from Las Moriscas; (22, 23) W motifs from Murrón del Pino; (24) sequence of zigzag human figures, a further schematization of the W type, from Estrecho de Santonge; Φ motifs from (25) Cueva Ahumada, (26) El Muro de Helechosa (Hoz de la Guadiana), and (27) Reboso del Chorillo; variants on Φ motif from (28, 29) Covatilla de San Juan and (30) Las Vacas del Retamoso; (31) dolmenic idol motif from Cueva Ahumada; mixture of Φ and dolmenic idol motifs from (32, 33) Peñon del Águila and (34) Reboso del Chorillo; bar motifs from (35) Nuestra Señora del Castillo and (36) Los Buitres; (37) variant of bar motif from San Blas; double-triangle (or hourglass) motifs from (38–40) Las Viñas, (41) Callejones del Río Frío, (42) El Escorialejo, and (43) Las Viñas; (44) double-triangle group from Puerto Palacios; (45) double-triangle pair from Las Viñas; (46) variant of double-triangle motif from Las Viñas; dumbell motifs from (47) Cueva de los Arcos, (48) Malas Cabras, and (49) Covatilla de San Juan; (50–52) plantlike figures from El Navajo. Star motifs from (53) Garganta de la Hoz, (54) Las Vacas del Retamoso, and (55) Sierra Grajera ("la Chica"). Bounded areas (huts?) from (56, 57) Nuestra Señora del Castillo and (58) Los Buitres (*after Breuil, 1933–35, and Acanfora, 1960*).

In remote places high in the Pyrenees or the Alps the traditions of rock art in the Neolithic and the Bronze Age reflected a pastoral way of life. Two very different styles can be distinguished. A technique of finely incised lines, often hard to detect, contrasts with a technique of broad stippled lines and whole stippled surfaces (PL. 294). The first linear work is perhaps of Neolithic, or in some instances of Mesolithic, date, thus affording a valuable link with the rock art of the ornamented caves through the art of the Spanish Levant and schematic rock art.

At a spot called La Peyra Escrita, at Formiguères, Abbé Abélanet discovered a large schist slab covered with linear engravings (tree shapes, geometric figures, circles and rowels, pentacles); schematic representations of birds, cervids, warriors, and female "dancers"; and some curious crossbow shapes, extreme stylizations of the human silhouette. Shortly after the first discovery, investigations of an engraved stone revealed a scene of a stag hunt in which the animal is pursued by a pack of dogs — a theme that recurs in southern Spain, in the Val Camonica in the Alps, and in Sweden. In the region of Olargues (Hérault), R. Guiraud found the same themes again finely engraved on schist. The similiarity to the Pyrenean finds is striking; the Olargues region thus provides a precise and valuable link between Pyrenean and Alpine linear engravings.

C. Conti (1946) was the first to call attention to the linear engravings of Mont-Bégo. The world of Mont-Bégo is one of high valleys, difficult of access, and of vast, high plateaus studded with glacial lakes, dominated by the peak of the Bégo. The creators of the tens of thousands of engravings found there can only have been shepherds who for some three months brought their flocks to the high summer pastures. The themes (PL. 294), which repeat themselves monotonously, can be reduced to four classes: horned figures (comprising almost 50 per cent of all the engravings); weapons and tools (ca. 20 per cent); "enclosures" and their derivatives (20 per cent); and indeterminable designs (10 per cent). The horned figures are always schematic, and the pair of horns is very clear and precise. Some of these horned figures yoked in pairs to a plow, a harrow, or a sledge are indisputably oxen, indicating a population dedicated to agriculture. Other animals have been tentatively identified: aurochs, hinds, antelopes, stags, elk, she-goats, mouflons, rams, chamois, wild goats; the stags, mouflons (now found only in Corsica and Sardinia), goats, and sheep are unmistakable. Weapons and tools are often shown full size; the object was probably placed against the stone and carefully traced, then the outline was filled in with stippling. Thus the engraving was a "double" of the real object and possessed all the ritual value of a substitute. Among the objects represented are daggers, hunting weapons, shepherds' knives, plowshares, and scythes. The "enclosures" are planlike representations of dwellings — often round, sometimes rectangular — and of fields and pastures, drawn with a wealth of detail.

Very different from the austere, closed, in some ways impenetrable world of the Bégo is that of the Val Camonica. It is dominated by high mountains — with the Concarena rising to 8361 ft. — but the engravings (PL. 294) occupy the lower-lying cultivated or wooded terraces of the Oglio. The great engraved rock of Naquane is located at an elevation hardly over 1470 ft., and the engravings of Cemmo are in the midst of vineyards. The peasant of the Val Camonica remained close to his cabin, returning to it every evening. His art, less harsh and more human than that of Mont-Bégo, is alive and joyful. The subjects are houses and fields, teams and four-wheeled wagons. The finest as well as the oldest engravings are found on the great rock of Cemmo, with its dagger-knives and its admirable stags arranged in friezes, which carry the viewer back to the Bronze Age. But the tradition of rock engraving was a durable one: the last of the twenty-thousand-odd engravings of the Val Camonica are as recent as the Roman era.

Louis-René NOUGIER

PREHISTORIC ROCK ART IN OTHER AREAS. The recognition and study of evidences of prehistoric art or that of prehistoric traditions, especially rock art, outside of Europe is barely beginning (except with reference to some areas of Africa). The available data, unlike those for Europe, are too scanty to afford a basis for a chronology; but the material so far available (again, except the African) has been recognized as post-Paleolithic. Hence the basic problem concerning the area or areas of origin of human visual art remains unresolved: whether there were ever, in any part of the world outside the Franco-Cantabrian area, centers that gave rise to engravings and rock paintings, or, more generally centers outside the Eurasian area that originated the small sculptures of the Upper Paleolithic. Present knowledge of Paleolithic art is largely limited to the production of the European area.

But the term "prehistoric" may, as already indicated, be applied to art and culture in a broader sense than the chronological. Various peoples have either passed through a level comparable to the Stone Age in quite recent times (Eskimos) or have remained on that level into the modern era (e.g., Australia). Their art is prehistoric in its cultural context, and even its recent manifestations are so much part of a long, unbroken tradition that there is no practical difference between new and old works. Rock art in particular may seem unrelated to specific cultures (the Sahara, India) and without a finite chronology; even when produced by known historical peoples or in a historical period it is, like the European rock art considered in the preceding section, a survival or continuation of a prehistoric tradition and is included here as such.

Africa. The African continent unquestionably provides the richest documentation of rock art (see PALEO-AFRICAN CULTURES), although much of it belongs to the native tradition of historical, and often even quite recent, times. Particularly oustanding are the art complexes of North Africa, especially in the Saharan areas: from the Atlantic to the Red Sea there stretches a chain of centers of engraving and painting distributed over vast areas, with more or less broad intervals between them. In this area, too, the features of an art that is authentically prehistoric and is connected with that of Europe can be distinguished.

The rock art of the Sahara, seen in its entirety and especially with respect to its engravings, can be subdivided into a few basic phases that reflect the climatic conditions, fauna, and cultural characteristics of the peoples who created it; but it offers little basis as yet for a sure chronology either relative or absolute.

What may be considered, in a very general way, a first, most ancient phase was that of hunting, or chiefly hunting, peoples who have left, especially in the Fezzan region of southern Algeria, some superb monumental works representing large wild fauna of the tropical type and hunting scenes in which human beings participate, armed with arrows and generally nude except for a loincloth (PLS. 10, 11, 295). This is the so-called "hunting art," which testifies to climatic conditions in the Sahara far different from modern ones, with sufficient water resources for animals that have long since migrated toward the equatorial forests. It seems certain, however, that even at that time there were domesticated animals, although not very many of them.

The phase of the hunting peoples was followed by the so-called "pastoral" one. In this period the rocks of North Africa, from the Fezzan to Algeria and Morocco, became covered with countless representations of domestic animals, chiefly bovids, although representations of wild animals also persisted. It has been established, in any case, that the economy of the human groups of the period was definitely based on the raising and grazing of livestock.

Unexpectedly and suddenly a new domestic species, the horse, appeared, and eventually spread through most of North Africa. At first it is seen harnessed to chariots (PL. 15), in the Garamantic phase whose graphic and pictorial works can be connected with the Garamantic peoples who lived in the Fezzan and are mentioned by the historians of antiquity, beginning with Herodotus. The arrival of the wheeled vehicle in the Sahara can be traced back to the second half of the 2d millennium B.C., shortly after its introduction in Egypt. Subsequently the horse appeared no longer harnessed to a cart but with a rider, and lastly came the dromedary, which spread rapidly, indicating the period when the Sahara became a wholly desert area, after the gradual drying out that had occurred over the preceding eras. The dromedary's arrival in the desert can be dated to approximately the first centuries of the Christian Era.

The presence of representations of this ruminant on the rocks of North Africa has provided a point of reference for the subdivision of all Saharan rock art into two major phases: "pre-cameline" and "cameline."

Extremely complex problems are raised by the chronological and cultural interpretation of the paintings discovered in great numbers, especially in recent years, in southern Algeria (Tassili-n-Ajjer), the Fezzan (Acacus massif), the Libyan desert (Auenat massif) and the massif of Tibesti (still little known in this respect but probably rich in visual documents). North African rock painting seems to have spread exclusively in a southerly direction; so far there has been no trace of it to the north of the Sahara. Here too, as with the engravings, an older phase depicting hunting scenes (although not so well-defined as in the engravings) can be distinguished, followed by a pastoral phase represented by the most impressive and varied complex of Saharan paintings (PLS. 12–15, 17, 18). The pastoral phase, again, is continued in a Garamantic or "bitriangular" phase (the latter term derived from the hourglass shape given the human figures), with horses harnessed to war chariots.

The cultural world depicted by Saharan painting is the most complex and mysterious so far revealed by the rock art of the African continent. There are scenes of daily life, war, dancing, and strange ceremonies of unknown significance. In some phases there is an obvious relation to dynastic Egypt, but it is still an open question which of the two artistic areas influenced the other.

East of Libya, in the Nile Valley and on the rocks of the Arabian Desert, a vast production of engraved figures is found, ranging from an older phase (represented by some authors as comparable to the hunting art of the Fezzan but actually decidedly inferior in quality and with less variety of subject) up to the dynastic and cameline period.

No satisfactory outline of the development and chronology of Nile Valley rock art has yet been established; indeed, apart from a few specific areas, the great body of these works is still largely undefined. So far, Winkler is the only authority who has offered a more or less acceptable classification system for the rock engravings of an area of Upper Egypt roughly covering the zone between Luxor and Aswan. He classes among the most ancient the group of paintings in the "wedge-shaped style" (*Keilstil*), so called from the shape of the anthropomorphous figures, which show affinities with the bitriangular figures of the Fezzan. This phase is followed by the figures called the "phallic sheath men" (*Penistaschen-leute*), the "men with feathered ornaments," and others. It would appear that, as a whole, the figures so far known in the Nile Valley are considerably later than those of the hunting-art phase and of the first part of the pastoral phase of the Fezzan and Tassili.

In Nubia, too, there are large numbers of figures engraved on rocks, although all of them appear to be of much later date. Engravings have also been found farther south, in the Upper Nile region (PL. 295).

To the west, specifically in the Ennedi massif, are paintings similar to those of the Auenat, Fezzan, and southern Algeria — that is, in a style resembling that found much farther north, in the Mesolithic art of the Spanish Levant, and much farther south, in South Africa.

Rock painting is also widespread in the Ethiopian highlands. The paintings of Genda Biftu and other shelters between Dirre Dawa and Harar have been known since 1934, when H. Breuil reported them as the work of pastoral peoples and executed in a naturalistic style, far removed, however, from that prevailing in the Saharan and South African paintings. An extensive series of rock shelters has been reported in Ethiopian Eritrea at Acchelé Guzai, containing a wealth of painted figures and somewhat fewer incised ones. Eritrean painting, which is still being studied, presents many phases as yet impossible to date even approximately. One of the oldest paintings shows figures that fall stylistically within the great tradition of the Saharan paintings, although rougher and less dynamic. In this phase there are representations of warriors armed with shields and lances that in some instances (Libanos, Zeban Kabessa) closely recall similar human figures in the art of the Auenat massif, the Spanish Levant, and South Africa. Some of the bovids (Sollum Ba'ăt; PL. 296) of this Eritrean phase are also related to those of the Auenat. Other, later phases show characteristics peculiar to this region; in some of them schematic human and bovine figures seem to represent a development similar to that of European schematic art, especially that of the Iberian Peninsula. At present, however, it is impossible to explain this similarity.

It is clear that Ethiopia, in its most ancient artistic phase, represents a stage in the route by which the great Mediterranean art currents spread, passing through the Sahara, the Ennedi, and the more southerly regions of East Africa to the southernmost part of Africa. (For the art of the latter area see PALEO-AFRICAN CULTURES).

Paolo GRAZIOSI

The Near East. Works of prehistoric chattel art (*art mobilier*) originated in various areas, especially Palestine and Jordan; they are very ancient, and their stratigraphic dating is certain. They are products of the Natufian culture developed by Mesolithic hunting peoples at the beginning of the present geological era and are subdivided into four periods. The figure of a gazelle found at Umm el-Zuweytina in Palestine, in an early Natufian deposit, sculptured in limestone and measuring only about 6 in. in length (the head in missing), is an example of a vigorous and sensitive realism that is no way suffers from comparison with the finest European Upper Paleolithic sculptures; its stiff feet are modeled with great delicacy. Other figures that may be cited include gazelles in bone (from Mugharat el-Wad), chamois (from Mugharat el-Kebara), a limestone sculpture, representing two embracing human figures (from the cave of Wadi Khareytun), and a few small human heads, also in limestone. Carbon 14 analysis has established dates from 7000 B.C. for the Natufian culture. Most of the rock engravings widely spread over the entire Near East, however, are works of a period after the spread of agricultural cultures and are often, in fact, the products of comparatively recent historical times.

The most numerous and recent discoveries of prehistoric art have been made in the Negev desert between Egypt and Jordan. Anati (1956) has established a sequence of seven styles distinguished by marked technical and stylistic differences. Style I, which is the most naturalistic, presents realistic and expressive animal figures, traced by a burin with an incised and continuous line; this style appears at Wadi er-Ramliya and shows some similarities to the more ancient art of the Kilwa region in Jordan, discussed below. Other styles follow, with a pecked technique (as at Mont-Bégo, see above), containing animals and human figures and even a scene with dancers and harpers. The chronological sequence probably extends from the Palestinian Eneolithic era (ca. 4000 B.C.) into Arab times. Beginning with style II, groups of engravings representing animals and, in some cases, open human hands are found together with Thamudic and Nabataean inscriptions; style V has Greek inscriptions and style VI ancient Arabic signs, while style VII corresponds to that of the Bedouins of modern times, who continue to incise similar inscriptions in southern Arabia.

Special mention must be made of the fine series of engravings of the Kilwa region in Jordan (PL. 298), which are certainly more homegeneous in style and technique than the Palestinian ones of the Negev. The subjects are predominantly animals, sharply engraved with a burin: bulls, goats, and gazelles, some wounded with a javelin, and even some crude scenes such as that of two people embracing, which recalls the above-mentioned Palestinian sculpture of the Natufian era. Rhotert (1938) has maintained that these figures were made by a Mesolithic hunting people contemporaneous with the Natufians of the Palestinian caves, although granting that they survived into the Bronze Age. Undeniably, some of these engravings, such as the wild goats shown with darts thrust into their bodies, reveal an ancient conception and a highly vigorous and realistic art. Most of the representational repertory, however, seems to belong to a livestock-raising and herding culture.

Evidently in various mountainous regions of the Near East, peoples of a hunting and primitive herding culture must have survived into an age of well-developed agricultural and protohistorical cultures. Przeworski (1935) has reported rock engravings of wild goats in Demir Qapu, on the road from Nusaybin to Mosul in Upper Mesopotamia. Similar engravings in shelters have also been found in southern Kurdistan and near Lake Gökcha in Armenia. The appearance of the goats and other animals recalls the figures of the painted vases of Iran and Mesopotamia of the 4th and 3rd milleniums, but they may be much later. Some groups of similar engravings, of uncertain period, have also been found in northern Caucasia.

A recent discovery made at Çatal Hüyük in the Anatolian highlands, under the auspices of the British Institute of Archaeology at Ankara, adds remarkable new data on Near Eastern prehistoric painting and reopens the entire question of the origin and spread of post-Paleolithic visual art of naturalistic tradition in the Mediterranean. In an excavation of a settlement of the early Neolithic, archaeologists brought to light a building made of dried unbaked bricks with the walls of the rooms faced with plaster and covered with fresco paintings. These frescoes consisted of enormous compositions showing scenes of hunting, dancing, and the like, with animals (deer, bulls) and human figures painted in a highly realistic and lively manner (PL. 297). The human figures are shown in a wide variety of poses, skillfully foreshortened and with a fluidity of outline giving a strong sense of movement; they have headdresses, clothing, and other personal attributes. Single colors (red and brown) fill in the outlines of the animal and human figures, but polychromy also is used to distinguish the figures from one another or to emphasize details of dress (by black dots). Taken as a whole, both in the freedom of composition and in the style of the human and animal figures these paintings reveal many more similarities with the rock art of Asia, the Mediterranean (Addaura, Levanzo, the Spanish Levant), and North Africa than with the subsequent drawings and painted works of the protohistoric and historic civilizations of the Near East. This find was probably a sanctuary, even containing tombs: in the same building stone statuettes of a naturalistic style were found, one of them depicting a nude Mother Goddess. The agricultural form of culture, represented here by the entire comparatively advanced complex (architecture, obsidian tools, dressed stone, and pintaderas, or stamped patterns, of terra cotta incised with meander patterns), does not rule out a tradition of hunting rites. On the basis of carbon 14 dating of the most recent culture of Hacilar, these documents can be dated to about 6000 B.C., an era not much later than that of the Mesolithic works of the central Mediterranean and contemporaneous with or preceding nearly all the Mesolithic and post-Mesolithic rock art of Spain, Scandinavia, North Africa, and Western Asia Minor.

The survivals of rock art into advanced historical times are considerable in the entire Middle East and especially in the desert areas with pastoral cultures. Engravings in the intermediate styles of the Negev desert have been reported in various sites of the Sinai peninsula, especially at Serabit and Quseir. In the mountain region of Safa and Jebel el-Druz in Syria, Dussaud and Macler (1901) discovered a group of engravings most of which belong to the 1st and 2d centuries of the Christian Era. Some of them, however, are certainly of an earlier period and all are related to the tradition of the pastoral peoples of the Near East. In southern Arabia too, especially in the Hejaz area, groups of rock engravings have been reported that are of the same style and technique as those of the Palestinian and Syrian steppes.

* *

The Indian subcontinent. The first published mention of Indian rock paintings appeared in 1899, drawing attention to some paintings discovered in the Kaimur and Mahadeo ranges near Pachmarhi in Madhya Pradesh, the mountainous region of the southern Ganges in central India. These Indian rock paintings were compared with those discovered in European Paleolithic caves and were at first judged more ancient; and the rock art found at Singanpur in the Mahanadi Valley (Anderson, 1918) reinforced the hypothesis that India, like Africa and Europe, had been a center of Paleolithic art. Later discoveries and more thorough studies of these rock shelters, however, led to a modification of the earlier impressions and a different view of the origin and development of this rich and varied art. Findings to date have been reviewed and analyzed by Gordon (1958), and the author has drawn freely on his presentation.

In the rock shelters of the Pachmarhi area of the Mahadeo Hills are a great number of paintings showing myriads of small figures of men on foot and on horseback, sometimes grouped in rows, three by three. The figures that chiefly attract attention are the archers, which at first glance look as though they might have been found in the painted shelters of southeastern Spain or of southern Africa. Closer analysis, however, shows, intermingled with the archers, figures of men with swords and shields, riding richly caparisoned horses and in postures of skilled horsemanship or parade positions. There are also figures of warriors without metal weapons, some of them in scenes that appear to represent battles between organized bands of archers fighting on the open plains. On the basis of the abundant superimpositions, Gordon describes a sequence of four series of styles, each with an ancient and a later phase, that has served as a key to the relative chronology of the extensive art of the Indian province.

The two most recent of the Mahadeo style series are the work of agricultural and pastoral peoples of historical times. Series IV, the most recent, comprises rather poorly drawn figures of animals and men painted in white ocher and, in the earlier phase, outlined in red. The paintings of Series III are in red or rose. The human figures in both these series are shown in the most diverse activities: the men hunting, dancing, playing the harp, and gathering wild honey, the women weaving, tending children, carrying water, and milling grain; and there are representations of their dwellings, pottery, clothing, baskets, and benches. The people who produced these paintings had extensive military equipment: spears, swords, shields, battle-axes, daggers, and bows, as well as drums and war trumpets or horns; and they rode elaborately caparisoned horses and sometimes elephants. They had oxen, donkeys, goats, and dogs and also pictured monkeys and bears.

The two older Mahadeo series are the production of hunting peoples. Series I, the most ancient of all, is composed of rather schematic paintings with stylized figures of animals and men in red and cream. These works are scarcer than examples of the later styles. The human figures have triangular heads and square bodies and sometimes wear what appears to be ornamented clothing. Over the paintings of this series is superimposed a later style, Series II, in which the triangular heads have disappeared and bodies have elongated. These figures are rose-colored, and their development continues into the first phase of the red and rose Series III, a fact that makes untenable the supposition that many milleniums intervened between the two styles and rules out a Paleolithic dating for any of this art.

Most of the paintings discovered by Anderson at Singanpur are in dark red, but some are in reddish orange. There are some abstract or symbolic signs, impossible to interpret, as well as many human and animal figures. A scene of hunters is very similar to one at Kabra Pahar, some 18 miles to the southeast. Gordon points out that the human figures with square bodies filled in with parallel wavy lines seem to represent a transition from the early to the later phase of Series I, and that at least one of the Kabra Pahar figures also closely resembles the early Series I of the Mahadeo Hills; there are also similarities between the animal figures and those of the first and especially the second Mahadeo. The hunters carry bows and arrows and, in the most ancient phases, javelins; they frequently wear loincloths. Arrowheads, apparently of metal, appear as early as the last phase of the most ancient series here, and metal spearheads are common in the paintings of Series II. Thus, as in the Mahadeo Hills shelters, an early schematic phase can be distinguished followed by a more naturalistic one, the production of a primitive hunting people of the central Indian plains (there are animated scenes of struggles with tigers, crocodiles, and the great porcupine). Although they painted horsemen and elephants, these must have belonged to other, more evolved, peoples of the plains, for the painters themselves were evidently simple hunters.

In the Likhunia shelter of the Mirzapur region there appear, exceptionally, several horsemen that can be attributed to Series II. Other paintings of this series are found in shelters in the environs of Manikpur in Banda District (Uttar Pradesh) and at Sarhat, Karpatia, and Malwa; they contain figures of horsemen and archers, and even a cart of an obviously historical period. Also assigned to Series II is the interesting painting at Ghormangur of a rhinoceros hunt, in which the spearheads used by the hunters are of the same type as a Ganges Valley bronze harpoon that was first used about 800 B.C. and continued in use for a long time.

No secure dating seems possible except for the more recent Mahadeo style series. The military equipment depicted in Series III can be related to the arms used by the army of Porus in 326 B.C., which continued in use in India for some centuries. Gordon has remarked an exact correspondence between the late-third–early-fourth series warriors and those represented on a sculpture in the Mahadeva Temple at Harasnath (Rajasthan) dating from the mid-10th century of the Christian Era; and figures of these series found in the Jhalai rock shelter and in several shelters at Adamgarh in the Mahadeo Hills can be related to the Ajanta sculptures dating from the 6th century. Gordon therefore believes that the paintings of the third and fourth series can be dated between the 5th and 10th centuries, and that the whole cycle of prehistoric Indian rock art is comprised between 700 B.C. at the earliest and the 10th century of the Christian Era.

It is of interest that this art is still practiced among the Saora (or Sawara) tribes in the districts of Ganjam and Korapur in the large province of Orissa, some 350 miles south of Mahadeo and about 150 miles from the painted shelters of Singanpur. Verrier Elwin (1951) has studied these paintings and has pointed out parallels between them and the rock paintings of the Mahadeo Hills and other places discussed here.

There is no doubt that a number of Indian rock paintings have religious significance; in the later ones especially, the figures of magicians can be recognized. There are figures with rats' heads, carrying huge rats; giants holding a tiger as though it were a dog; and other masked personages. But most of the paintings represent simple, actual events of the hunt and of war. In such scenes there are sometimes representations of "foreigners," men of a more evolved culture than the Indian rock painters. This phenomenon is not unique in rock art: the rock art of the Sierra de Córdoba in Argentina, in its last cycle, includes representations of the Spanish conquerors with their flags and weapons; and in the shelter of La Gasulla (Castellón), one of the most beautiful of the Spanish Levant, there is a mounted figure wearing a helmet of the Spanish late Bronze Age, perhaps seen by the last Mesolithic hunters of those mountains as he passed through the plains along the Mediterranean coast of Spain.

In addition to the rock paintings, central and southern India offer several groups of rock engravings that have a clear cultural affinity with the paintings. Occasionally, though not often, paintings and engravings occur together in the same shelter. The engravings, however, seem to be distributed over a wider geographic area. Among the more notable examples of rock engravings are those of Vikramkhole in the Sambalpur region (Orissa); Bihar (naturalistic representations of eagles with open wings); Ghatsila; and the Mahadeo Hills shelters, although engravings are not numerous there (a man on horseback, a bull, and several abstract figures).

Rocks bearing both paintings and engravings have been found in a rock shelter among the Gombigudda slopes near Jamkhandi (Maharashtra) and in a wide area extending eastward as far as Lingsugur in the Raichur District of Mysore. The engravings are schematic and very crude, with several representations of oxen, horses, and elephants bearing howdahs. They seem to date from the first centuries of the Christian Era. Gordon credits F. R. Allchin with calling attention to other such figures in the region — at Koppal, Piklihal, Maski, and Billarayan Gudda — and also in the Benkal forest; to which group can be added the engravings and paintings of Kallur, Togal Gudda, and Chik Hesrur. Allchin has divided them into three groups according to their style and character, the latest group including drawings of temples and of objects familiar in early Indian iconography, such as the trisula. The more ancient works represent horsemen and one elephant; the horsemen carry swords and battle-axes. These older works are somewhat comparable (although with reservations) to the late Series II Mahadeo paintings, especially to the scene of an elephant hunt found in the Likhunia shelter; thus they must probably be dated to the first centuries of the Christian Era.

The representations of bulls found in these shelters and in Kupgallu are probably somewhat more ancient. They have archery bows tied to their horns and objects placed vertically on their heads; bulls with the same attributes have been discovered on painted pots in Nagda, not far from Ujjain (Madhya Pradesh). Gordon believes that this more naturalistic style, characteristic of a hunting people, dates from the final pre-Christian centuries. Farther south in the Raichur region, in Bellary District, is another group suggesting the same cultural background but of more ancient appearance. A group of engravings in Kupgallu represents men, animals, birds, and scenes of cattle herding. They employ a pecking technique that produces something resembling a crude bas-relief. There are no figures of horses among these engravings; there are, however, phallic images, which are not found in other Indian rock shelters, with the single exception of a shelter in the Benkal forest. There have been attempts to date these figures tentatively between 900 B.C. and 200 B.C.; however, it is not logical to suppose that any of these engravings are older than the most ancient paintings, which must date from about 700 B.C.

Some 185 miles farther south of Bellary are two other groups of engravings. One of these, at Gotgiri Betta, contains five different patinas, on the basis of which a relative chronology has been suggested: the most ancient works are stylized figures and are lightest in color; subsequent works, in which figures of scorpions appear, are more yellow. Most of the engravings are schematic figures of animals and men in a bright shade of gray, almost like the rock; the most recent, however, are blackish. Absolute dating of these is uncertain, but the other group in the same style (located just north of Dod Kannelli, on the road from Bangalore to Sarjapur), considered later in date, contains figures of horsemen that are probably of about the 3d or 2d century B.C.

Also worth mention are the engravings discovered by Fawcett (1901) in the Edakal cave, among the Edakal Hills near the Sultan's Battery in Wynaad, some 55 miles northwest of Ootacamund. The walls of this rock shelter are covered with engravings of human and animal figures and abstract symbols. The great number of superimpositions makes the group extremely difficult to interpret. The human figures are highly stylized; some seem to be carrying bows. The animal figures are even more schematic. There are various abstract signs, most of them basically crosses.

Important rock engravings have also been found in Pakistan. In 1882 a fine group was discovered at Chargul, in the Mardan District. This group dates from the same period as other groups of engravings from the central Indus Valley, especially near Attock Bridge, where the Mandori deposits were found; from Gandab on the western bank; and from Ghariala on the eastern bank, near where the Haro flows into the Indus. In Mandori these crude engravings appear together with two Kharoshthi inscriptions, one of which is accompanied by a mythological figure mounted on an elephant and carrying a man in one hand and a woman in the other. Elsewhere in these deposits are engravings representing animals, as well as men mounted on horses, camels, and elephants and armed with swords, lances, and occasionally bows and arrows and battle-axes. There is one representation of a cart drawn by oxen, similar to engravings of carts in the Bellary region of India. Human and animal abstractions abound, but many of them cannot be interpreted. To these engravings of the Indus must be added others discovered by

Stein (1929) in northern Baluchistan: at Andarbes near the source of the Zhob and in the Bashor Valley northeast of Pishin. In the latter place the figures of an elephant and of two men fighting were found, as well as three Kharoshthi letters. All these works are of the same cultural and chronological cycle and can be dated between 200 B.C. and A.D. 200.

In Afghanistan an expedition from the American Museum of Natural History in 1950–51 collected several hundred petroglyphs of the same type. Other groups have been located near Dilaram and Farah on the border between Pakistan and Afghanistan. Among these are figures of men on horseback carrying round shields, scenes depicting the hunting of an ibex with bow and arrow, and a male figure brandishing two torches.

It is not easy to connect these Pakistan and Afghanistan examples of apparently ancient rock art with specific peoples or ethnocultural phenomena. They seem to express the spirit of warriors and of hunting and pastoral peoples of historical times, but it is impossible to establish anything more definite concerning their chronology or to relate them to such groups of rock engravings as those of the Mahadeo Hills or Singanpur in India.

Southeast Asia, Indonesia, and Australia. A group of petroglyphs discovered by V. Gouloubew in Chapa, North Vietnam, consists of a series of human or semihuman figures with spiral-shaped additions that have sometimes been interpreted as phallic allusions. The engravings are isolated and show no clear relationship to the prehistoric cultures of India or of Indonesia, perhaps because the prehistoric art of southeast Asia has so far been little known or studied.

There have been reports of deposits of prehistoric art on the islands of Ceram, Kai (Kei), and Celebes, those in McCluer Gulf, and those of Kamrau Bay, off Kaimana in northwestern New Guinea. In all these areas there seem to be relatively abundant deposits of rock art, and it may be assumed that this type of art spread to other oceanic islands as well. Rock paintings have, in fact, been found in the Fiji Islands that are obviously related to the rock art of New Guinea and Celebes, but they have been little studied; they must be quite ancient, since there is no evidence of modern migrations in the cultural sequence of the Fijis.

It has been postulated that this art, especially the frequently found representations of hands and feet, was the creation of proto-Australian peoples. However, the painting of a running wild boar discovered in the "Cave of the Paintings" at Patta on the island of Maros in southern Celebes (Van Heekeren, 1957), for instance, recalls, in its movement and vitality, the best paintings of the Indian subcontinent.

The motif of negative handprints in Indonesia is of great interest because of its resemblance to the negative handprints of western European Paleolithic rock art. Van Heekeren has found a fine series of these prints in the above-mentioned "Cave of the Paintings," situated in the Turicale District of Maros, near Liang-Liang, on the east coast. A short distance away, in the cave of Burung, he found representations of two left and two right hands, as well as other prints that were badly deteriorated. Previously C. J. H. Franssen had discovered many representations of hands in the so-called "Cave of the Fingers" (Leong Djari) on the same island. These are all negative handprints against a red background, similar to those found in the Paleolithic caves of southwestern Europe. One of them has only four fingers, another only three; some lack the little finger and ring finger; and in one group the thumb is missing.

The period to which these paintings can be attributed is indicated by the relics of hunting, traces of hearths, and plant carbon found in the "Cave of the Paintings," together with thousands of Brolia and Thiara mollusk shells, at various levels. (In all strata red hematite was gathered, no doubt connected with the presence of rock paintings in the cave.) These findings identify an industry corresponding to the Toleanian culture, a type of Upper Paleolithic regional culture that has been found in 19 different caves. The people who originated this culture apparently lived chiefly on fresh-water mollusks, large and small game, fish, and edible forest plants. Since the fauna associated with this culture is modern, these prehistoric representations of hands and other motifs cannot be of a very early period. Van Heekeren believes that all of the handprints found in Celebes belong to the Toleanian III period, an era corresponding in Celebes to the late Mesolithic. Sarasin (1938) believes this culture to be related to that of the Vedic peoples of India, a theory rejected by Van Heekeren, although the relationship between the rock art of this culture and that of India seems rather obvious.

From the islands of Maros, Kai, and Ceram this artistic province extends to an area of more symbolic and less naturalistic art on the islands in and around MacCluer Gulf, from the north shore of which an important group of prehistoric rock-art complexes reaches southward to Kamrau Bay. Its discoverer, Röder (1959), has brought to light a large number of rock shelters containing an abundance of figures in red, black, and yellow, with occasional superimpositions that have made possible the identification of three closely related styles. What appears to be the most ancient group comprises negative prints of hands and feet similar to those already described. Over these, other more modern figures have been painted; some tend toward the naturalistic, but most of them are geometric or schematic. They include paintings of anthropomorphs and of animals and fish (birds, tortoises, starfish, etc.). Boats and symbolic motifs are also found. These figural schematic motifs eventually gave way to a final style of marked abstraction and simple geometric motifs, which can still be seen today in the chattel art and tattoos of the natives of New Guinea.

Some of the objects represented in these paintings — certain hatchets, for instance — can be related to an Indonesian Bronze Age culture of Indo-Javanese influence. The representations of boats, such as those found in the Sosorra rock shelter of New Guinea, seem to suggest the same or a somewhat later period. It is clear, in any case, that when the first European colonizers arrived in New Guinea in the second half of the 19th century this art was no longer a living part of the local cultural heritage, although the Papuan peoples of the island retained legends concerning the handprints in caves and rock shelters (they attributed the prints to their ancestors, who, they believed, were blind when they arrived on the island and therefore had to feel their way along the walls). Röder considers that the blackish figures are the most ancient and the red paintings later, although the two styles were not very distant from each other in time. It is evident that this entire art is the work of a hunting people who survived into the period when the higher cultures that put and end to such rock art reached Oceania.

This art has affinities to the art of Australia. Its remote relation to Indian rock art, however, ought not to be ruled out.

In the extensive Australian subcontinent there are several series of prehistoric painted and engraved shelters of uncertain date (see AUSTRALIA). The artistic practices of aboriginal peoples still surviving in some regions of Australia make it possible to interpret at least some of these prehistoric works (see AUSTRALIAN CULTURES).

The Australian works fall into two provinces. The first is that of the northern and northwestern coastal regions, the chief characteristic of which is the naturalism of the figures. This art has certain similarities to some of the art of southwestern Australia but manifests a more vigorous naturalism, more grace and originality, and a richer, more suggestive, and more complex variety of motifs. Instead of the uniform background common elsewhere, the backgrounds here are shaded with polychrome crosshatching.

In the northwest, especially in the Kimberleys, are found many examples of the curious *wondjina* (II, PLS. 64, 65) — more or less realistic representations of anthropomorphous beings surrounded by a kind of aureole. Many tribes believe that they were made by Ungud, the creator of the world, and that the tribes are descended from them. The oldest person of the clan must repaint them in order to preserve their magical, protective, and representative values. These *wondjina* figures are surrounded by zoomorphous and anthropomorphous figures without the halo or aureole, usually painted in light red. The various techniques and styles sometimes found superimposed indicate that the practice of painting these figures continued for a long time.

The *wondjina* style in the Kimberleys existed side by side with the so-called "elegant" (or "filiform") style; the *wondjina* style, which continued longer and was more widespread (its influence reached farther south and west, as well as into central Australia), may be thought to derive from the filiform style, examples of which are rarer. The monochrome anthropomorphous "elegant" figures (II, PLS. 60, 62) are graceful and animated filiform drawings that seem quite foreign to the spirit of the rest of Australian art. The best groups of this style, in which the artists composed scenes of a symbolic character, are found in the northwest, particularly in Arnhemland. In and around that area are also found so-called "X-ray" paintings (i.e., revealing the internal organs) of fish turtles and serpents (II, PL. 61). The natives regard these as pictures of "spirits" and attribute far greater significance to them than to any of the other animal figures in the shelters.

The second artistic province extends across the whole interior of southern Australia from east to west, coming into contact with the naturalistic art at both its eastern extremity (in the western part of New South Wales) and its western extremity (in Victoria and western Australia). The art of this area is usually geometric and symbolic, consisting chiefly of simple labyrinthine and circular motifs. The Devon Downs deposit contains a number of these figures, which are difficult to interpret. The archaeological level of the site has been dated by carbon 14 analysis to 2200 B.C. The same styles are found in Burra (in the Flinders Range) and toward central Australia. Occasionally some crude anthropomorphous sketches are found, as in shelters near the South Para River and in Malkaia. This schematic style and the labyrinthine figures seem to have influenced the naturalistic art of the north and northwest: reptiles are represented in labyrinthine fashion even as far as the Yule River and Ngungunda in the Kimberleys.

A detailed description of the sites of prehistoric art is given in AUSTRALIA and a discussion of the styles of various areas and their development is found in AUSTRALIAN CULTURES. It is impossible to offer a detailed relative chronology for most regions, and no absolute chronologies have been established. Investigations to date indicate, however, that the healthy, vigorous, and expressive figural works of the north and northwest are the oldest examples of Australian art, as is shown also by their rarity and their limited geographic distribution.

Martín ALMAGRO BASCH

The Americas. Rock art in North America is found particularly in the northwest and southwest. The Eskimos of southwestern Alaska have produced interesting rock art that includes both paintings and engravings. Most of it cannot be dated with any certainty, but even when it appears to be of the historical era it is unquestionably part of a tradition unchanged for centuries or perhaps even millenniums. On the shores of Prince William Sound and of Cook Inlet, immediately to the north, cave paintings of stylized human and animal figures and of men in boats are found, as well as engravings — mostly geometric designs — on pebbles and on curious triangular stone plaques. On Kodiak Island there are, in addition to geometric designs, engraved figures of whales and other animals and human figures and faces, some of them on granite cliffs along the shore. The style of some of the apparently oldest rock paintings and engravings appears to be reflected in the woodcarving of the historical period in the northwest (see AMERICAN CULTURES; ESKIMO CULTURES).

In the southwestern portion of the United States comprising the states of Utah, Arizona, Colorado, and New Mexico there are innumerable petroglyphs, often concentrated along the banks of rivers, especially the Colorado and its tributaries. These works, probably produced for the most part during the second half of the first millennium of the Christian Era by peoples of the Anasazi culture sequence, include a great variety of geometric designs — spirals, Greek frets, concentric circles (often considered a sun symbol), and many others. The numerous naturalistic animal and human figures, often quite stylized, range in size from tiny ones like those on the famous "Newspaper Rock" in Petrified Forest National Monument (Arizona) to towering ones visible from a considerable distance, like those painted on an 80-ft. expanse of sandstone cliff in Horseshoe Canyon (Utah). Similar or related motifs and styles seem to extend into Texas and Middle America, especially Mexico, but present knowledge has not been able to trace a clear relationship.

It is uncertain whether the crudely carved Malakoff heads found in Texas are part of this artistic cycle; they are believed to be much earlier. Some small stone carvings of animals discovered in Ventura County, California, seem to represent a different style, of greater elegance and maturity; they have been tentatively dated to the 4th millenium B.C. (see NORTH AMERICAN CULTURES).

South American rock art, distributed throughout the continent and known by various regional names (*letreiros, piedras lavradas, piedras escritas, piedras garabateadas, pintados, riscos*), can be assigned to four main stylistic groups: Patagonian, Peruvio-Andean, Colombio-Venezuelan, and Brazilian. With some exceptions, as among the Taino of the Greater Antilles, it does not seem to have had a religious character.

The engravings, on great boulders and in caves or shelters, are numerous along rivers or near rapids and waterfalls (see GUIANA). Figures on dressed stones are rare; they occur in Colombia on slabs that delimit the temples and tombs of the pre-Columbian epoch, in northwestern Argentina and Puerto Rico on stone pillars, and on the islands of Hispaniola and Puerto Rico on slabs that delimit the sphericity. Most of the engravings, produced by pecking or incision, are of small dimensions; only in some places in Guiana and in the district of Tarapacá in Chile are enormous engravings encountered, visible even from a great distance. The paintings are generally red and white, but black and other colors also appear. The styles are sometimes naturalistic, sometimes schematic and geometric. In general the paintings are more elaborate than the engravings. The fauna represented belongs to recent epochs; as in North America, it includes animals introduced by Europeans, such as horses and swine.

Patagonian rock art, which embraces the manifestations of Tierra del Fuego and the southern part of central Chile, is in general somewhat crude. The repertory includes animal and human prints, geometric figures, and outlines of hands and of human beings, which are drawn schematically. The figures, in caves or rock shelters, are generally painted, sometimes after being engraved.

The Peruvio-Andean group extends over the central and southern Andes of Peru, Bolivia, northwestern Argentina, and northern Chile. The works are found both in the open and in caves and shelters. They are either painted or engraved; the two techniques are rarely found together. The motifs are the most complex in all of South America. The style may be realistic or strongly conventional, but geometric motifs are clearly in a minority by comparison with representational ones and seem to derive from textile patterns. Group scenes are rather frequent, especially in northwestern Argentina; some reveal an attempt at perspective representation.

Colombio-Venezuelan rock art is limited to the Andean regions of the two countries and is rather crude. The outlines of the figures are usually traced and then filled in, most often with straight lines. For Colombia alone, Pérez de Barradas (1941) has distinguished eight stylistic areas; the style of the department of Huila seems to present affinities with Chibcha gold- and silverwork (see ANDEAN PROTOHISTORY; GOLD- AND SILVERWORK).

The rock art of Brazil (q.v.), concentrated near the sources of various tributaries of the Amazon, forms a single stylistic complex with that of Guiana (q.v.) and the Antilles. On the continent, engraving predominates in the north and painting in the south. Representations of animals, fishes, and human beings are characteristically traced in a schematic manner; sometimes these representations are partial, being reduced to a head, for example. Circles and spirals are also common.

In general, South American rock art continued into historical times and in many instances appears to be the work of Indian tribes now living in the areas where it is found, as in the Chaco and the Guianas. It shows no genetic connections with the prehistoric mural production of the Old World.

* *

BIBLIOG. *Prehistoric art in general and the European Upper Paleolithic*: H. Breuil and R. Lantier, Les hommes de la pierre ancienne, Paris, 1951 (2d ed., 1959); H. G. Bandi and J. Maringer, L'art préhistorique, Basel, 1952 (Eng. trans., R. Allen, Art in the Ice Age: Spanish Levant Art: Arctic Art, London, New York, 1953); H. Breuil, 400 siècles d'art pariétal, Montignac, 1952 (Eng. trans., M. E. Boyle, Montignac, 1952); H. Kühn, Die Felsbilder Europas, Stuttgart, 1952 (Eng. trans., A. H. Brodrick, The Rock Pictures of Europe, Fair Lawn, N.J., 1957); L.-R. Nougier, La préhistoire, in L.-H. Parias, ed., Histoire universelle des explorations, I, Paris, 1955, pp. 21–110; H. Breuil, L.-R. Nougier, and R. Robert, Le "lissoir aux ours" de la grotte de la Vache à Alliat et l'ours dans l'art franco-cantabrique occidental, Préhistoire et spéléologie ariègeoises, XI, 1956, pp. 15–78; P. Graziosi, L'arte dell'antica età della pietra, Florence, 1956 (Eng. trans., Palaeolithic Art, New York, London, 1960); E. O. James, Prehistoric Religion, New York, London, 1957; L.-R. Nougier and R. Robert, Le rhinocéros dans l'art franco-cantabrique occidental, Préhistoire et spéléologie ariègeoises, XII, 1957, pp. 15–52; L.-R. Nougier and R. Robert, Rouffignac, Paris, 1957 (Eng. trans., London, 1958); H. Bégouen and H. Breuil, Les cavernes du Volp, Paris, 1958; J. Maringer, L'homme préhistorique et ses dieux, Grenoble, 1958 (Eng. trans., M. Ilford, The Gods of Prehistoric Man, New York, 1960); L.-R. Nougier and R. Robert, Le "lissoir aux Saïgas" de la grotte de la Vache à Alliat et l'antilope Saïga dans l'art franco-cantabrique, Préhistoire et spéléologie ariègeoises, XIII, 1958, pp. 13–28; L.-R. Nougier, Réflexions sur l'origine de l'art, Trav. Inst. d'art préhistorique, I, 1958; L.-R. Nougier, Géographie humaine préhistorique, Paris, 1959; L.-R. Nougier, Unité de l'art préhistorique, Trav. Inst. d'art préhistorique, II, 1959; A. Varagnac, ed., L'homme avant l'écriture, Paris, 1959; C. Zervos, L'art de l'époque du renne en France, Paris, 1959; M. Almagro Basch, Manual de historia universal, I: Prehistoria, Madrid, 1960; H. G. Bandi et al., L'âge de la pierre, Paris, 1960 (Eng. trans., The Art of the Stone Age, London, New York, 1961); H. Breuil, La caverne ornée de Rouffignac, Mém. Acad. des Inscriptions et Belles Lettres, XLIV, 1960, pp. 147–67; A. Leroi-Gourhan, Art et religion au paléolithique supérieur, 2 vols., Paris, 1960–61; L.-R. Nougier and R. Robert, Les "loups affrontés" de la grotte de la Vache à Alliat et les canidés dans l'art franco-cantabrique, Steinzeitfragen der alten und neuen Welt: Festschrift für L. Zotz, Bonn, 1960, pp. 399–420; L.-R. Nougier, Signification de Rouffignac, Trav. Inst. d'art préhistorique, III, 1960; E. Patte, Les hommes préhistoriques et la réligion, Paris, 1960; S. Giedion, The Eternal Present: The Beginnings of Art, New York, London, 1962; A. Laming-Emperaire, La signification de l'art rupestre paléolithique, Paris, 1962.

Mediterranean province: H. Breuil, H. Obermaier, and W. Verner, La Pileta a Benaojan (Malaga), Monaco, 1915; G. A. Blanc, Grotta Romanelli, II, Arch. per l'antr. e l'etn., LVIIII, 1928, pp. 365–411; P. Graziosi, Les gravures de la grotte Romanelli (Puglia, Italie): Essai comparatif, IPEK, 1932–33, pp. 26–46; L. Pericot García, La Cueva del Parpalló, Barcelona, 1942; A. Glory, Les gravures préhistoriques de la Grotte d'Ebbou à Vallon (Ardèche), La Nature, LXXV, 1947, pp. 257–62, 283–85; P. Graziosi, Le pitture e i graffiti preistorici dell'Isola di Levanzo nell'Arcipelago delle Egadi (Sicilia), RScPr, V, 1950, pp. 1–43; J. Marconi Bovio, Incisioni rupestri dell'Addaura (Palermo), BPI, N.S., VIII, 5, 1952–53, pp. 5–22; F. C. E. Octobon, Contribution à l'étude des couches supérieures de la Barma-Grande, Cah. de préhistoire et d'archéol., I, 1952, pp. 3–28; J. Marconi Bovio, Sulle forme schematizzate dei graffiti dell'Addaura (Palermo), Actes IVe Cong. int. du Quaternaire, II, Rome, Pisa, 1953, pp. 769–75; E. Drouot, L'art paléolithique à la Baume Latrone, Cah. ligures de préhistoire et d'archéol., II, 1953, pp. 11–46; J. Marconi Bovio, Nuovi graffiti preistorici nelle grotte del Monte Pellegrino (Palermo), BPI, N.S., IX, 1954–55, pp. 57–72; A. M. Radmilli, La produzione d'arte mobiliare nella Grotta Polesini presso Roma, Quartär, IX, 1957, pp. 41–59; P. Graziosi, L'art paléolithique de la "Province Méditerranéenne" et ses influences dans les temps post-paléolithiques, Österreichisches Symposion, IV, Burg Wartenstein, 1960; P. Graziosi, Ciolotti dipinti del Gard: Il disegno schematico paleo- e postpaleolitico nella "Provincia Mediterranea," Steinzeitfragen der alten und neuen Welt: Festschrift für L. Zotz, Bonn, 1960, pp. 171–78; P. Graziosi, A New Masterpiece of Palaeolithic Art Discovered in Italy: A Superb Bull of 10,000 B.C. Carved in a Calabrian Rockshelter, ILN, CCXXXVI, 1961, pp. 578–79; P. Graziosi, Levanzo: Pitture e incisioni, Florence, 1962; P. Graziosi, La scoperta di incisioni rupestri di tipo paleolitico nella grotta del Romito presso Papasidero in Calabria, Klearchos, IV, 13–14, 1962, pp. 12–20.

Azilian-Tardenoisian art: E. Piette, Hiatus et lacune: Vestiges de la période de transition dans la grotte du Mas-d'Azil, B. Soc. d'Anthr. de Paris, 4th ser., VI, 1895, pp. 235–67; E. Piette, Les galets coloriés du Mas-d'Azil, L'Anthropologie, VII, 1896, pp. 383–427; H. Breuil and H. Obermaier, La cueva de Altamira en Santillana del Mar, Madrid, 1935 (Eng. trans., M. E. Boyle, Madrid, 1935).

Northern European rock art: K. Lossius, Helleristninger på Bardal i Beitstaden, Årsberetning fra Forengen til norske førtidmindesmerkers bevaring, 1896, pp. 145–49; K. Lossius, Arkaeologiske Undersøgelser i 1897, Kgl. norske videnskabernes selskab, Skrifter, 5, 1897, pp. 1–10; A. W. Brøgger, Den Arktiske stenalder i Norge, Oslo, 1909; A. W. Brøgger, Kulturgeschichte des norwegischen Altertums, Oslo, Leipzig, 1926; J. Bøe, Felszeichnungen im westlichen Norwegen (Bergens Mus. Skrifter, 15), Bergen, 1932; G. Gjessing, Artiske Helleristninger i Nord-Norge, Oslo, Cambridge, Mass., 1932; E. S. Engelstad, Østnorske ristninger og malinger av den arktiske gruppe, Oslo, 1934; G. Gjessing, Die Chronologie der Schiffsdarstellungen auf den Felsenzeichnungen zu Bardal, Trøndelag, ActaA, VI, 1935, pp. 125–39; G. Gjessing, Veideristningen på Stein i Ringsaker, Univ. oldsaksamlings Arbok, 1935, pp. 52–68; G. Gjessing, Nordenfjelske ristninger og malinger av den arktiske gruppe, Oslo, 1936; V. Ravdonikas, Naskal'nye izobrazheniia Onezkogo ozera i Belogo Moria (Rock Paintings of the Lake Onega Shores and the White Sea), 2 vols., Leningrad, Paris, 1936–38; G. Hallström, Monumental Art from the Stone Age of Northern Europe, I: The Norwegian Localities, Stockholm, 1938; O. N. Bader, Drevnie izobrazheniia na pobolkakh grotov v Priazov'e (Ancient Paintings on the Ceilings of the Grottoes in the Area of the Sea of Azov), MIA, II, 1941, pp. 126–29; G. Gjessing, Yngre steinalder i Nord-Norge, Oslo, 1942; G. Gjessing, Norges Steinalder, Oslo, 1945; P. Simonsen, Arktiske Helleristningerr Nord-Norge, II, Oslo, 1958; G. Hallström, Monumental Art of Northern Sweden from the Stone Age, Stockholm, 1960.

Rock Art of the Spanish Levant: H. Breuil, L'âge des cavernes et roches ornées de France et d'Espagne, RA, XIX, 1912, pp. 193–234; J. Cabré Aguiló, El arte rupestre en España, Madrid, 1915; P. Wernert, Nuevos datos etnográficos para la cronología del arte rupestre de oriente de España, B. Real Soc. Esp. de H. Natural, XVIII, 1917, pp. 139–42; E. Hernández Pacheco, Las pinturas prehistóricas de la Cueva de la Araña (Valencia), Madrid, 1924; H. Obermaier and P. Wernert, La edad cuaternaria de las pinturas rupestres del levante español, Mem. Real Soc. Esp. de H. Natural, XV, 1929, pp. 527–37; H. Obermaier, Nouvelles études sur l'art rupestre du levant espagnol, L'Anthropologie, XLVII, 1937, pp. 477–98; H. Obermaier, Probleme der paläolithischen Malerei Ostspaniens, Quartär, I, 1938, pp. 111–19; H. Kühn, Die Frage des Alters der ostspanischen Felsbildern, IPEK, XV–XVI, 1941–42, pp. 258–60; M. Almagro Basch, Los problemas del epipaleolítico y mesolítico en España, Ampurias, VI, 1944, pp. 1–38; M. Almagro Basch, El covacho con pinturas rupestres de Cogul (Lérida), Lérida, 1952; M. Almagro Basch, La cronología del arte levantino de España, Actes IIIe Cong. int. de sc. préhistoriques et protohistoriques, Zürich, 1953, pp. 142–49; M. Almagro Basch, Las pinturas rupestres levantinas, Madrid, 1954; J. Sánchez Carillero, Avance al estudio de las pinturas rupestres de "Llana de las Covachas," pedania de Rio-Moral (Nerbio-Albacete), Noticiario arqueol. hispanico, V, 1956–61, pp. 1–12; E. Ripoll Perelló, Los abrigos pintados de los abrededores de Santolea (Teruel) (Monografias de arte rupestre: arte levantino, 1), Barcelona, 1961.

Schematic rock art: *a. General*: M. O. Acanfora, Pittura dell'età preistorica, Milan, 1960. *b. Spain*: J. Cabré Aguiló and E. Hernández Pacheco, Avance al estudio de las pinturas prehistóricas del extremo Sud de España (Laguna de la Landa), Madrid, 1914; E. Hernández Pacheco and J. Cabré Aguiló, Las pinturas prehistóricas de Peña Tú, Madrid, 1914; H. Breuil and M. C. Burkitt, Rock Paintings of Southern Andalusia, Oxford, 1929; H. Breuil, Les peintures rupestres schématiques de la Péninsule Ibérique, 4 vols., Lagny, 1933–35. *c. France*: G. Vézian, Gravures rupestres dans l'Ariège, Rev. anthr., XXXIV, 1924, pp. 357–60; A. Glory, Gravures rupestres schématiques dans l'Ariège, Gallia, V, 1947, pp. 1–45; A. Glory et al., Les peintures de l'âge du métal en France méridionale, Préhistoire, X, 1948, pp. 7–135; J.-L. Baudet, Note préliminaire sur les peintures, gravures et enceintes du Sud de l'Ile de France, B. Soc. préhistorique fr., XLVII, 1950, pp. 326–36; J.-L. Baudet, Les peintures préhistoriques de l'Ile de France, B. Soc. préhistoriques fr., LVII, 1960, pp. 210–13; P. R. Giot, Brittany, London, New York, 1960.

Southern Alpine areas: C. Bicknell, The Prehistoric Rock Engravings in the Italian Maritime Alps, Bordighera, 1902; M. and St. J. Pequart, Téviec: Station nécropole mésolithique du Morbihan (Mem. Arch. Inst. de paléontologie humaine, 18), Paris, 1937; C. Conti, Nuove figurazioni rupestri di Monte Bego, Atti Acc. dei Lincei, 8th ser., I, 1946, pp. 145–66; E. Süss, Le incisioni rupestri della Valcamonica, Milan, 1958 (Eng. trans., L. Krasnik, Milan, 1959); E. Anati, La civilisation du Val Camonica, Paris, Grenoble, 1960 (Eng. trans., L. Asher, New York, 1961); E. Anati, La grande roche de Naquane (Mem. Arch. Inst. de paléontologie humaine, 31), Paris, 1961; L.-R. Nougier, Archéologie préhistorique du Mont Bego, Trav. Inst. d'art préhistorique, IV, 1961, pp. 133–49.

Near East: R. Dussaud and F. Macler, Voyage archéologique au Safâ et dans le Djebel ed-Druz, Paris, 1901; J. Jaussen and R. Savignac, Mission archéologique en Arabie (II), 2 vols., Paris, 1914; R. Neuville, Statuette animale du mésolithique palestinien, L'Anthropologie, XLII, 1932, pp. 546–47; G. Horsfield, Prehistoric Rock-drawings in Transjordania, AJA, XXXVII, 1933, pp. 381–86; R. Neuville, Statuette érotique du désert de Judea, L'Anthropologie, XLIII, 1933, pp. 558–60; S. Przeworski, Prähistorische Felszeichnungen aus Vorderasien, Arch. Orientalni, VII, 1935, pp. 9–15; H. Grimme, Altsinaitische Forschungen, Paderborn, 1937, pp. 15–22; H. Rhotert, Transjordanien: Vorgeschichtliche Forschungen, Stuttgart, 1938, pp. 161–240; B. Howe, Two Groups of Rock Engravings from the Hejas, JNES, IX, 1950, pp. 8–17; E. Anati, Les gravures rupestres de Neguev Central, B. Soc. préhistorique fr., LII, 1956, pp. 722–28; E. Anati, Rock Engravings from the Jebel Ideid (Southern Negev), PEQ, LXXXVIII, 1, 1956, pp. 5-13.

Africa: H. Breuil, Peintures rupestres préhistoriques du Harar (Abyssinie), L'Anthropologie, XLIV, 1934, pp. 473–83; H. A. Winkler, Prehistoric Rock-drawings of Southern Upper Egypt, 2 vols., London, 1938–39; P. Graziosi, Le pitture rupestre dell'Amba Focadà (Eritrea), Rass. di s. etiopici, I, 1941, pp. 61–70; A. Mordini, Un riparo sotto roccia con pitture rupestri dell'Amba Focadà, Rass. di s. etiopici, I, 1941, pp. 55–60; C. Conti Rossini, Incisioni rupestri all'Hagghèr, Rass. di s. etiopici, III, 1943, pp. 102–06. For North Africa and sub-Saharan Africa, see bibliog. for PALEO-AFRICAN CULTURES.

The Indian Subcontinent: F. Fawcett, Prehistoric Rock Pictures near Bellary, Asiatic Q. Rev., 2d Ser., III, 1891, pp. 147–57; J. Cockburn, Cave Drawings in the Kaimūr Range, JRAS, 1899, pp. 89–97; F. Fawcett, Notes on Rock Carvings in the Edakal Cave, Wynaad, Indian Ant., XXX, 1901, pp. 409–21; C. A. Silberrand, Rock Drawings of the Banda District,

J. Asiatic Soc. of Bengal, N.S., III, 1907, pp. 567–70; C. W. Anderson, Singapur Rock Paintings, JBORS, IV, 1918, pp. 298–306; A. Stein, Archaeological Tour in Waziristan and N. Baluchistan (Mem. Archaeol. Survey of India, XXXVII), Calcutta, 1929, pp. 79–89; H. C. Das Gupta, Bibliography of Prehistoric Indian Antiquities, J. Asiatic Soc. of Bengal, N.S., XXVII, 1931, pp. 1–96; R. S. M. Ghosh, Rock Paintings and Other Antiquities of Prehistoric and Late Times (Mem. Archaeol. Survey of India, XXIV), Calcutta, 1932; D. H. Gordon, Indian Rock Paintings, Science and Culture, V, 1939, pp. 142–47, 322–27; C. King, Drawings on the Indus, Man, XL, 1940, pp. 65–68; D. H. Gordon, The Rock Engravings of the Middle Indus, J. and Proc. Royal Asiatic Soc. of Bengal (Letters), VII, 1941, pp. 197–202; B. De Cardi, On the Borders of Pakistan, Art and Letters, XXIV, 1950, pp. 52–57; V. Elwin, The Tribal Art of Middle India, London, 1951; D. H. Gordon, The Rock Engravings of Kupgallu Hill, Bellary, Madras, Man, LI, 1951, pp. 117–19; W. A. Fairservis, Future Archaeological Research in Afghanistan, Southwestern J. of Anthr., IX, 1953, pp. 139–46; D. H. Gordon and F. R. Allchin, The Rock Paintings and Engravings in Raichur, Hyderabad, Man, LV, 1955, pp. 97–99; D. H. Gordon, The Prehistoric Background of Indian Culture, Bombay, 1958, pp. 98–117; A. H Dani, Prehistory and Protohistory of Eastern India, Calcutta, 1960.

Southeast Asia and Indonesia: M. Colani, L'âge de la pierre dans la province de Poabinh (Tonkin), Mém. Service Géologique de Indo-Chine, XXX, 1927, pp. 299–422; M. Colani, Le protonéolithique, Praehistorica Asiae Orientalis, I, 1932, pp. 93–95; J. Röder, Felsbilder im Flussgebiet des Tola (Süd West Ceram), Paideuma, I, 1938, pp. 19–28; J. Röder, Felsbildforschung auf West Neu Guinea, Paideuma, I, 1938, pp. 75–88; F. Sarasin, Über Spuren einer früheren weddide Bevölkerung auf der Insel Roti oder Rote bei Timor, Z. für Rassenkunde, VII, 1938, pp. 251–54; F. D. McCarthy, A Comparison of the Prehistory of Australia with That of Indochina, the Malay Peninsula and the Netherlands East Indies, Proc. 3d Cong. of Prehistorians of the Far East (1938), Singapore, 1940, pp. 30–50; P. H. Buck, Les migrations des Polynésiens, Paris, 1952 (bibliog.); H. R. van Heekeren, The Stone Age of Indonesia, The Hague, 1957; J. Röder, Felsbilder und Vorgeschichte des MacCluer-Golfes, West Neuguinea, Darmstadt, 1959, pp. 38–161.

Australia: N. B. Tindale, Natives of Groote Eylandt and of the West Coast of the Gulf of Carpentaria, Records of the South Australian Mus., II, 1925, pp. 61–102, 103–34; D. S. Davidson, Aboriginal Australian and Tasmanian Rock Carvings and Paintings (Mem. Am. Philosophical Soc., V), Philadelphia, 1936; L. Black, Aboriginal Art Galleries of Western New South Wales (Aboriginal Customs of the Darling Valley and Central New South Wales, 3), Melbourne, 1943; A. P. Elkin, Grey's Northern Kimberley Cave Painting Re-found, Oceania, XIX, 1948, pp. 1–15; F. J. Hall, R. G. Gowan, and G. F. Guleksen, Aboriginal Rock Carvings: A Locality near Bimba, Records of the South Australian Mus., IX, 1951, pp. 375–80; A. Leonhard, Rock Paintings near Glen Island, Victoria, Mankind, IV, 1952–53, pp. 343–45; D. J. Tugby, Coning Range Rockshelter: A Preliminary Account of Rock Paintings in Northeast Victoria, Mankind, IV, 1952–53, pp. 446–50; H. Read and C. P. Mountford, Australia: Aboriginal Paintings — Arnhem Land, New York, 1954; E. A. Worms, Contemporary and Prehistoric Rock Paintings in Central and Northern Kimberley, Anthropos, L, 1955, pp. 546–66; A. S. Schulz, North-west Australian Rock Paintings, Mem. Nat. Mus. of Victoria, XX, 1956, pp. 7–57; F. D. McCarthy, The Cave Paintings of Groote and Chasm Island (Records of the American-Australian Scientific Expedition to Arnhem-Land, 1948, II), Melbourne, 1959. See also the bibliogs. for AUSTRALIA; AUSTRALIAN CULTURES.

Americas: C. Bruch, La piedra pintada del arroyo Vaca Mala y las esculturas de la cueva de Junín de los Andes (Territorio de Nequén), Rev. Mus. de la Plata, X, 1902, pp. 173–76; J. W. Fewkes, The Aborigines of Porto Rico and Neighboring Islands, Bur. of Am. Ethn. Ann. Rep., XXV, 1903–04, pp. 3–220; T. Koch-Grünberg, Südamerikanische Felszeichnungen, Berlin, 1907; E. Boman, Antiquités de la région andine de la République Argentine et du désert d'Atacama, 2 vols., Paris, 1908; M. Uhle, Explorations at Chincha, Univ. of Calif. Pub. in Am. Archaeol. and Ethn., XXI, 1924, pp. 55–94; G. A. Gardner, Rock-painting of North-west Córdoba, Oxford, 1931; A. Costa, Introdução à arqueología brasileira (Etn. e h. brasiliana, 5th Ser., XXXIV), São Paulo, 1934; M. A. Vignati, Resultados de una excursión por la margen sur del Río Santa Cruz, Notas preliminares del Mus. de la Plata, II, 1934, pp. 77–151; A. T. Jackson, Picture-writing of Texas Indians, Austin, 1938; R. R. Latcham, Arqueología de la region Atacameña, Santiago de Chile, 1938; A. Mattos, Prehistoria brasileira: Varios estudos, São Paulo, 1938; R. Herrera Fritot, Informe sobre una exploración arqueológica a Punta del Este, Isla de Pinos, realizada por el Museo Antropológico Montané de la Universidad de La Habana (Rev. Univ. de La Habana, XX–XXI), Havana, 1939; F. H. Douglas and R. d'Harnoncourt, Indian Art of the United States, New York, 1941; J. Pérez de Barradas, El arte rupestre en Colombia, Madrid, 1941; I. Rouse, Petroglyphs, HSAI, V, 1949, pp. 493–502; I. Rouse, Porto Rican Prehistory, in New York Academy of Sciences, Scientific Survey of Porto Rico and the Virgin Islands, XVIII, 1949, pp. 307–460, 463–578.

* *

Illustrations: PLS. 242–298; 9 figs. in text.

PRE-RAPHAELITISM AND RELATED MOVEMENTS. Among the artistic movements of the 19th century, Pre-Raphaelitism and purism are parallel phenomena which share not only their ideological origin — a mysticizing romantic esthetic — but also their purpose, to protest academizing neoclassicism and to uplift by faithful illustration of the Gospels. One also finds in Pre-Raphaelitism and purism the same reference to tradition and the same programmatic position aimed at a direct and naïve study of nature, a common desire to return to simple, pure technique without recourse to the expedients and virtuosities of the craftsmen-painters of the past, as well as a tendency of the artists to form groups or small communities and to identify their artistic ideal with an ideal of religious life. Historically, the religious background of these movements — as opposed to the pagan and secular spirit of neoclassicism — corresponded to the inclination of contemporaneous literature toward its heritage of language and letters and toward local and national traditions. Although Pre-Raphaelitism and purism did not produce outstanding esthetic results, they had a considerable effect on the arts, which in the preceding period had been almost completely conditioned by imitation of antiquity. Also, English Pre-Raphaelitism was particularly important because of its influence, through Ruskin and the Morrises, on the formation of the English Arts and Crafts movement and Art Nouveau (q.v.). See also FOLK ART; NEOCLASSIC STYLES; PRIMITIVISM; ROMANTICISM.

PRE-RAPHAELITISM. Pre-Raphaelitism, or Praeraphaelitism (the latter form being used by the Rossetti family), was the name selected by a group of young English artists in the late 1840s to describe their theories about necessary reforms in the visual arts. The leaders of the movement were William Holman Hunt (1827–1910), John Everett Millais (1829–96), and Dante Gabriel Rossetti (1828–82). Hunt was the son of a warehouse manager, Millais of well-to-do upper-middle-class parents, and Rossetti of an emigrant Italian poet. The other members of "the Brotherhood" were Frederic George Stephens (1828–1907) and William Michael Rossetti (1829–1919), who were to be better known as critics than painters; Thomas Woolner, a sculptor (1823–92); and James Collinson (1825?–81), a young painter of some promise who abandoned painting on his conversion to the Roman Church. The aims of Pre-Raphaelitism were described by Holman Hunt: (1) To have genuine ideas to express; (2) to study directly from nature, disregarding all academic rules of lighting and grouping; and (3) to envisage events as they must have happened rather than as the rules of design required. Put into practice, these rules resulted in a selection of literary or historic subjects — Keats, Shakespeare, and Tennyson being largely drawn on — where the incident chosen was a serious and significant one; a prolonged study of the setting out of doors and its rendering with a higher range of tones than was usual in studio pieces and with a minutely detailed study of the foreground, the greatest care being expended on the accuracy of leaves, flowers, and rocks; and the posing of the figures — for whom they demanded faces of character rather than conventional good looks — in attitudes that bore out the action of the story, however awkward and ungainly they might appear. Much of this was a revolt against the fashionable painting of the time, in which trivial, sentimental subjects treated within an accepted range of coloring and lit according to the rules of academic chiaroscuro held the field. In particular it was a protest against the "monkeyana" of Edwin Landseer (1802–73), whose animals posing as classical heroes, faithful friends, or sorrowing widows were — however technically accomplished — an unfortunate aberration of early Victorian taste. In 1848, when the Brotherhood was founded, Hunt and Rossetti were only twenty and Millais nineteen: young men full of ardor and self-importance are apt to incur the rebukes of their elders, and there is no doubt that in 1848, the year of revolutions, the name Brotherhood, with its suggestion of Continental secret societies, was highly suspect. In the Academy Exhibition of 1849 the initials "P.R.B." (Pre-Raphaelite Brotherhood) on Millais's and Hunt's paintings and on Rossetti's *The Girlhood of Mary Virgin* (London, Tate Gall.), shown at the

Hyde Park Gallery in the same year, were not understood, and the paintings were on the whole well received. Hunt's *Rienzi* (Coll. Mrs. E. M. Clarke), where Rossetti posed for the Roman tribune and Millais for the young knight, is a good revolutionary subject handled with vigorous gestures and considerable invention in the grouping. Millais's *Lorenzo and Isabella* (1849; Liverpool, Walker Art Gall.) from Keats's poem *Isabella; or, the Pot of Basil* is a remarkable achievement for so young a man. Painted on a white ground in bright, clear colors, it is today, owing to the Pre-Raphaelites' conscientiousness in technical matters, wonderfully well preserved; and Millais had imposed a rhythm of bending heads on the difficult theme of a group at table, seen in depth, and linked the whole with the daring gesture of the wicked brother kicking the hound. Little else could have constituted a more direct challenge to accepted canons of art, but the critics, while somewhat amused, were prepared to be friendly and encouraging. All this was changed in the following year, when through an unguarded remark by Rossetti the meaning of the initials was discovered and published in the press. Hunt, the chronicler of the movement, always asserted that they meant by the term no disrespect to Raphael himself but had in mind his influence as the source of academic painting. It was easy, however, to twist the word into an arrogant insult to the most revered name in visual art — the painter whose cartoons, then exhibited at Hampton Court, were the greatest works of High Renaissance art outside Italy and the particular pride of English connoisseurs. In fact, the famous attack on the Brotherhood by Charles Dickens in his periodical *Household Words* had this supposed arrogant assumption of superiority as its main contention. Millais's painting, however, *Christ in the House of His Parents* (PL. 304), generally known as *The Carpenter's Shop*, laid itself open to other complaints. In design and luminosity of paint it is as fine as *Lorenzo and Isabella*, but, whereas the latter applied a realistic sense of gesture and characterization to a poem by Keats, here the same process was applied to Holy Writ. Millais, in the search for seriousness of subject, was using both realism in the worn faces, rough hands, and poor setting of the family at Nazareth, and symbolism in the Virgin looking at her Child's hand accidentally pierced by a nail, in the young Baptist gazing at the bowl of water in his hand, and in the sheep in the background looking up to be fed. To Dickens the weary, pallid face of Mary seemed to be that of a woman "so horrible in her ugliness . . . that she would stand out as a Monster in the vilest cabaret in France." The general opinion was that the painting was shocking and blasphemous. Rossetti's white and gleaming *Annunciation* (1850; London, Tate Gall.) and Hunt's *Christians Escaping from Persecuting Druids* (1850; Oxford, Ashmolean Mus.) were deemed less clearly obnoxious but shared in the general condemnation. It looked as though the new movement might be crushed, and it did in fact break up under the public outcry. Rossetti took alarm and for a time abandoned painting in oils, retreating from the actualities advocated by the Brotherhood into a dreamlike medieval world for which he found water color a more facile expressive medium. In 1851 he first met Elizabeth Siddal, whose unusual beauty inspired all the Pre-Raphaelite group. She had some talent, painting water colors that reflect Rossetti's influence but with a distinct character of their own. In 1860 she and Rossetti were married, but her health had always been precarious, and in 1862 she died of an overdose of laudanum. It was for Rossetti a period of emotional stress ending in tragedy. Hunt and Millais meanwhile had held to their course despite adversity. Some older artists such as Augustus Egg (1816–63) encouraged them, and despite all the outcry the Academy Exhibition of 1851 did not refuse to hang Hunt's and Millais's pictures, as well as one by Charles Collins (1828–73) — *Convent Thoughts* (1850–51; Oxford, Ashmolean Mus.), painted with brilliant colors and exact delineation of all the details. The mocking once more began, but help was at hand. A letter appeared in *The Times* by John Ruskin, signed "The Author of *Modern Painters*," claiming that these young men were laying "in our England the foundations of a school of art nobler than the world has seen for three hundred years." Ruskin's praise marks the turning point. Public opinion was reassured and began to find the Pre-Raphaelite paintings not only respectable but enjoyable. The following year Millais's *The Huguenot* (1852; New York, Gall. of Mod. Art, Huntington Hartford Coll.) and *Ophelia* (PL. 302) were more readily appreciated than some of their predecessors, and two years later the Academy Exhibition of 1854 included Hunt's *The Light of the World* (Oxford, Keble College; replica by Hunt in Manchester, City Art Gall.), a didactic painting even more full of symbolism than *The Carpenter's Shop*, but one that was much less aggressively contemporary in approach and had, moreover, Ruskin to expound its meaning. In the visual imagination of the Protestant world the painting has proved a very powerful image, widely reproduced and copied; and when a replica of it later toured the English-speaking world, the crowds that visited it in North America and Australia bore witness to its authority and power, however much we feel repelled today by a sentimentality that is not now to our taste.

In 1854, the year *The Light of the World* was exhibited, Hunt departed on a journey to Palestine, where he spent two years. His search for accuracy in the representation of his subjects led him to the inevitable conclusion that Biblical scenes required a Palestinian background. From that time onward, throughout his long life, he produced a series of paintings in which scenery, architecture, costume, and facial types were all studied in Palestine: *The Scapegoat* (Port Sunlight, Cheshire, Eng., Lady Lever Art Gall.), *The Finding of Christ in the Temple* (1860; Birmingham, Eng., City Art Gall.), *The Shadow of Death* (Manchester, City Art Gall.), *Triumph of the Innocents* (1884; London, Tate Gall.) — paintings which did much, at least in the English-speaking world, to break down older conventions of Biblical representation. *The Scapegoat* revealed the strange brilliant coloring of the Dead Sea and the mountains of Moab to a Western public which had never before realized the reds and purples of the Palestinian landscape. In its uncompromising design — a series of horizontal lines hardly broken by any vertical emphasis — it illustrates the Pre-Raphaelite adherence to actuality and rejection of the artifices of picture making. The subject also — the lonely, dying beast — could hardly have been a less inviting one. Hunt painted it on the spot, a remote part of the Dead Sea coast, threatened by hostile Arabs and exposed to a blazing sun; and something of his intense sincerity survives in the canvas. Exhibited in the Royal Academy Exhibition of 1856, it puzzled the critics, and Ruskin severely condemned the composition and many of the details, though in sum he held it a "truly honorable" picture. It certainly has a lasting quality and has become one of the memorable images of English art. *The Finding of Christ in the Temple*, begun in Palestine but not completed until 1860, had a very different reception. It was purchased by the art dealer Gambart for the very considerable price of 5,500 guineas and exhibited by him in his own gallery, where crowds gathered to see it. The authenticity of its costumes and settings were recognized as being of great interest, and the figure of the youthful Christ had a taut intensity that held attention. From then on, Hunt's position as one of England's leading artists was established.

Meanwhile, in 1856, Millais had produced his two most notable paintings, *The Blind Girl* (Birmingham, City Art Gall.) and *Autumn Leaves* (Manchester, City Art Gall.). These are scenes of mild, sentimental anecdote but painted with such a complete fusion of mood and landscape, such boldness in design and color, that they achieve genuine poetic quality. Millais never again equaled them. They came at a period of great disturbance in his life: Mrs. Ruskin had left her husband and secured an annulment of marriage, and in the summer of 1855 she married Millais. His two greatest paintings date from this first year of married life — a year in which he also finished, in silent sittings, that portrait of John Ruskin against the background of a Scottish burn, which in the vivid use of detail is one of the most convincing Pre-Raphaelite documents.

With their seriousness of purpose the members of the Pre-Raphaelite group were bound to be drawn to topics more urgent than those suitably treated in historical form. Ford Madox Brown (1821–93) had never been a member of the Brotherhood but had been much affected by its aims; between 1852 and 1865 he completed his large painting *Work* (Manchester, City

Art Gall.), depicting a group of workmen excavating a stretch of the main street in Hampstead, a road-making theme that Courbet had recently been treating very differently in France in his painting of 1850, *The Stonebreakers* (destroyed; IV, PL. 36). Brown studied the movements of his laborers as carefully as Courbet, but he crowned the scene with details, leaving out nothing that could make it a tract for the times. His *Last of England* (1855; Birmingham, City Art Gall.) is a much more satisfactory work, vigorous in color and pose and admirably fitted to the roundel shape. It was inspired by the departure of the sculptor member of the Brotherhood, Thomas Woolner, for the gold fields of Australia.

Though the original group was thus scattering, some younger painters had enthusiastically adopted their methods and doctrines: William Burton (1824–1916), Arthur Hughes (1832–1915), Walter Deverell (1827–54), Robert Braithwaite Martineau (1826–69), John Brett (1830–1902), Henry Wallis (1830–1916), William Lindsay Windus (1822–1907), and James Smetham (1821–89). They were producing work of merit and originality, and the Royal Academy Exhibition of 1856, the year of Millais's *Blind Girl* and *Autumn Leaves*, also had Burton's *A Wounded Cavalier* (London, Guildhall Art Gall.), Windus's *Burd Helen* (Liverpool, Walker Art Gall.), and Wallis's *Death of Chatterton* (London, Tate Gall.). It was in many ways the high point of the movement. From then on, Millais's art passed into a strange decline. No longer in close contact with Hunt, he abandoned Pre-Raphaelite outdoor painting, exactness of detail, and seriousness of subject. He began to paint more broadly, with great facility, and to supply the public with trivial or sensational subjects much to their liking. His popularity, furthered by his handsome presence and genial hospitality, brought him in the end to the presidency of the Royal Academy, but never has an artistic career that opened with such promise so sadly deteriorated. His decline seemed to rob his disciples of their impetus. Long-lived as several of them were, they did their best work in the fifties and sixties and seldom afterwards recaptured their earlier success.

Before the movement waned, however, it achieved a considerable triumph in another field. With their emphasis on content, the Pre-Raphaelites had a natural bent for illustration. It was a great period of the English novel, and in monthly or quarterly publications such as the *Cornhill Magazine* or *Good Words* the Pre-Raphaelite group matched the writers with series of woodcuts that set a new standard in interpretation and execution. The weakness that was gaining on Millais's brush did not affect his pencil, and his woodcut designs in the sixties are fresh and vital. Moxon's edition of the works of Tennyson (1857) stands as a striking example of the Pre-Raphaelite achievement in this field (PL. 304).

Rossetti (PL. 301), too, had a group of pupils, of whom William Morris (1834–96; q.v.) and Edward Burne-Jones (1833–98) were to become the most celebrated. Burne-Jones's elegant, wan figures, classical subjects in medieval poses (PLS. 305, 306), were very far removed from all that the Brotherhood had stood for; and Rossetti himself always made clear that the real protagonist of the movement was Holman Hunt and that his own connection with it had been one of personal friendship rather than artistic principle. Ruskin, however, and Rossetti's brother, the writer William, continued to include Rossetti under the Pre-Raphaelite heading; and despite protests from Holman Hunt the general public gradually began to apply the term to the new estheticism, the wallpapers and fabrics of William Morris, Burne-Jones's paintings, tapestries, and stained glass, Rossetti's long-necked ladies, and the poems of Swinburne. It is little wonder that this later phase is hard to define, for the term was now being employed, owing to personal associations, to theories that were almost antitheses of those for which the word had been coined. In its early days, the Brotherhood had been confused by the public with the "Nazarenes" (see below), the German revival of Early Christian art, or rather of Quattrocento mannerisms and a flat, simplified style of painting. Hunt had always strongly disowned any relationship with the German school, and William Dyce (1806–64), the most Nazarene of English painters, in fact completely changed his style under Pre-Raphaelite influence. Now the name gained lasting currency as describing the languid affectations of a fashionable esthetic cult whose solid achievements were in the decorative arts rather than in imaginative painting.

To the 20th century looking back at Pre-Raphaelitism over an interval of many developments in the arts, its most striking feature seemed to be the minute examination of the whole field of vision. The paintings that first made the name famous were pondered works, where the whole foreground was sharply focused and precisely rendered. Hunt and Millais could and did paint backgrounds broadly, as in *The Hireling Shepherd* (1851; Manchester, City Art Gall.) and *The Blind Girl*, where the row of poplars or the sloping meadow shows considerable grasp of visual effects of distance and recession; but their study of nature was prolonged and exacting, while the future was to lie with the rendering of the immediate impression. While the landscapes were conscientiously painted out of doors, the figures, dressed in carefully chosen costumes, had to be painted in the studio, and the two stages of composition did not always blend satisfactorily. The contribution of the movement to the development of English art lay in their scrupulous insistence on quality in the materials of their craft, in their line drawings for woodcuts, and in the decorative arts. Hunt alone remained faithful to the early principles of the Brotherhood: in his *May Morning on Magdalen Tower* (1890; Birmingham, City Art Gall.; larger version, Port Sunlight, Cheshire, Eng., Lady Lever Art Gall.) and *The Miracle of the Sacred Fire* (1899) he was still painting with the uncompromising precision of his early days, though the lines had grown harder and the new vision had become a formula. In his *Pre-Raphaelitism and the Pre-Raphaelite Brotherhood* (1905) he left a literary monument as striking as the paintings of the movement's prime and one that has done much to keep alive interest in and appreciation of the movement's achievement.

T. S. R. BOASE

PURISM. *Friedrich Overbeck and the "Nazarenes" in Rome.* The term "purism" (*purismo*) was invented by Antonio Bianchini in 1833 to indicate a movement that caused many artists, particularly after 1810, to seek out the more austere representational elements in pre-Renaissance art, specifically in the art that preceded the maturity of Raphael. The term is applied to a wide range of work carried out in the transition period between neoclassicism and romanticism. These works can be reduced to a common denominator only with difficulty, as the period abounds in contradictions, affecting even the purist productions of individual artists. Typical features of the movement are a return to primitivism (q.v.) in art and the desire to add emotional content that is often linked with a religious attitude inspired by the Catholic revival that followed the French Revolution and the Empire.

As early as the end of the 18th century, the art of the period before Raphael had begun to attract the attention of historians and scholars such as Giovanni Lami, Guglielmo della Valle, Leopoldo Cicognara, and the brothers Sulpiz and Melchior Boisserée. These men were already moving along the path to romanticism under the influence of the writers Ludwig Tieck and the brothers Friedrich and August Wilhelm von Schlegel; at the same time they were stimulated by the artist Peter Cornelius (1783–1867) to study the works of the early German painters. In 1812 J. N. Paillot de Montabert (1771–1849), a pupil of David (q.v.), published his *Peintures du moyen âge*, while Séroux d'Agincourt, in his *Histoire de l'art par les monumens . . .* (1810–23), tried to show "what artists must avoid."

At the end of the 18th century, the leader of these primitives ("Penseurs" or "Barbus") was the painter Maurice Quaï (or Quay; 1779–1804), who accused neoclassical artists of not "idealizing" nature, maintaining that ideal beauty can only be achieved by accentuating certain aspects of reality. The Danish artist Asmus Jakob Carstens (1752–98), who worked in Rome after 1792, took classical art as his ideal but admired the primitive and rejected Mengs and David in the name of "feeling." Under Carstens's influence the sculptor Bertel Thorvaldsen (q.v.;

X, PL. 252), who worked in Italy from 1797 to 1841, rejected certain fundamental neoclassical ideas and tried to combine the theories of J. J. Winckelmann with the demands of religious feeling.

In *Herzensergiessungen eines Kunstliebenden Klosterbruders* (*Outpourings of the Heart of an Art-loving Monk*), published in Berlin in 1797, W. H. Wackenroder put forward the view that the artist should abandon himself entirely to irrational inspiration. A similar position was taken by Friedrich von Schlegel. Of some importance, too, is the attitude of Goethe, who criticized Wackenroder's book with cutting irony and expressed similar disapproval of the "new German religious and patriotic art" that drew inspiration from Wackenroder's work.

In July, 1809, a group of six artists known as the "Nazarenes" (because of their beards and long hair) founded in Vienna an artistic association that came to be called the Lukasbund (Guild of St. Luke). The members were Friedrich Overbeck (1789–1869; PL. 299), Franz Pforr (1788–1812), Georg Ludwig Vogel (1788–1879), Johann Hottinger (1788–1828), Joseph Wintergerst (1783–1867), and Joseph Sutter (1781–1866). In that same year Overbeck and Pforr moved to Rome, first residing at Villa Malta and then in the deserted Monastery of S. Isidoro a Capo le Case. In 1810 Vogel and Hottinger followed, to be joined in 1811 by Wintergerst. Thus was formed the nucleus of a group that was, at one time or another, to include or influence numerous other artists. The group was distinguished by religious orthodoxy and a belief in the virtue of communal activity.

Overbeck had learned much from J. H. W. Tischbein (1751–1829; cf. XII, PL. 311), Philipp Otto Runge (1777–1810; cf. XII, PL. 318), and the Italian frescoes studied in the drawings of the brothers Franz and Johannes Riepenhausen (1786–1831, 1788–1860, respectively), who had come to Italy in 1805 with the historians K. F. Rumohr and L. Tieck. In a letter to his father in 1808, Overbeck mentions as the main grounds of disagreement with the Vienna Academy of Fine Arts (which he had entered in 1806) their attitude of eclecticism that tried to combine Michelangelo's mastery of design with Titian's sense of color, neglecting "the heart, the soul, the feelings." In that same year (1808) Overbeck painted his well-known *Supper at Emmaus* (Lübeck, Behnhaus). Vogel generally supported the ideals of the Nazarenes, but he was closer in spirit to Cornelius, Joseph Anton Koch (1768–1839), and Thorvaldsen, who themselves were influenced by the purist movement. By 1813 Vogel had already returned to Zurich. Hottinger, in Rome with Vogel, never became seriously committed to the ideals of the Lukasbund, and he too left the city in 1813. Of the few works Hottinger produced, his sketches for a *Judgment of Solomon* and a *David before Saul* are worthy of mention. Wintergerst joined the group at S. Isidoro, but in 1813 he returned home in the company of Johann Christian Xeller (1784–1872). Sutter left the Viennese Academy in 1811; in 1816 he came to live in Rome and remained there until 1828.

Other artists — such as Cornelius, Philipp Veit (1793–1877; PL. 300), Julius Schnorr von Carolsfeld (1795–1872; VII, PL. 270), Johann Evangelist Scheffer von Leonhardshof (1795–1822; also known as Raffaellino) — became at various times part of the group (Veit in 1815, Schnorr von Carolsfeld in 1818). For most of them, however, their association with the Nazarenes and with Overbeck, the dominant personality, was merely a phase of youthful romanticism. Cornelius maintained an independent position. Schnorr von Carolsfeld, who was associated in Vienna with friends of Overbeck — particularly the Von Olivier brothers, Ferdinand and Friedrich — was influenced at first by both Overbeck and Koch. In 1817 he painted *St. Roch Distributing Alms* (Leipzig, Mus. der bildenden K.) and *The Family of St. John the Baptist with the Family of Christ* (Dresden, Gemäldegal.), which showed the influence of 15th-century art. Of his Roman period the most notable works are an *Annunciation* (1820; Berlin, Staat. Mus.) and *The Marriage at Cana* (Hamburg, Kunsthalle). Veit shows some independence from Overbeck in the freer use of color. Worthy of mention are his *Madonna with the Angels* (Rome, Church of Sta Trinità dei Monti), *The Three Marys at the Tomb* (Frankfort on the Main, Städelsches Kunstinst.), and his work on the vault of the Galleria dei Candelabri in the Vatican Museum. Scheffer von Leonhardshof, in Rome between 1814 and 1816, was mainly influenced by the art of Michelangelo, Raphael, and Perugino. His portrait of Pope Pius VII shows the influence of Overbeck. Also worth mentioning are his paintings of St. Cecilia and St. Ludovica in the Viennese Church of St. Anne (both 1820).

One of the greatest achievements in the artistic production of the Nazarenes is the painting of St. Elizabeth in the Cathedral in Naumburg by Gustav Heinrich Naecke (1786?–1835), a pupil of Josef Grassi, who came to Rome in 1816–17. Also significant is a series of frescoes depicting the adventures of Joseph in Egypt painted by the Nazarenes between 1815 and 1818 in the Palazzo Zuccari (later Bartholdy), home of the Prussian consul-general J. S. Bartholdy at Sta Trinità dei Monti (transferred 1887 to Berlin, Staat. Mus.); those engaged on the project included Cornelius, Wilhelm Schadow (1788–1862), Veit (who painted *The Seven Fat Years*), and Overbeck (*The Seven Lean Years*). Another important joint enterprise was the Casino of the former Villa Giustiniani Massimo, also in Rome, where the Nazarenes decorated three rooms with frescoes depicting episodes from classics of Italian literature: The *Divine Comedy* (Veit; PL. 300), *Orlando Furioso* (Schnorr von Carolsfeld, much influenced by Raphael's Vatican frescoes), and *Gerusalemme Liberata* (Overbeck; PL. 299). Veit took over the *Divine Comedy* frescoes from Cornelius, who had prepared the sketches but returned to Germany (1819) before the fresco was finished. Veit's work was completed by Koch, a landscape painter who shared the ideals of the Nazarenes and whose painting showed the influence of both Umbrian art and Giorgione. Overbeck worked slowly, and when in 1827 he left to fulfill a commission at the Church of S. Maria degli Angeli at Assisi, he turned his unfinished fresco over to Josef von Führich (1800–76), who came to Italy in 1827 and was to become the greatest Austrian Nazarene.

Among Overbeck's paintings are *Italia and Germania* (1811–28; Munich, Neue Pin.; cartoon in Lübeck) and *The Triumph of Religion in the Arts* (1840; Frankfort on the Main, Städelsches Kunstinst.). In later life he sketched many works but could not bring himself to complete any of them for fear that he would lose his intensity of emotional expression. Among the great masters mentioned by critics in discussing influences on the evolution of Overbeck's purism are Fra Angelico, Michelangelo, Pontormo, Luca Signorelli, Perugino, Pintoricchio, Francesco Francia, and the young Raphael. The same uncertain eclecticism is reflected in the works of other members of the group. Characteristic features of their painting are an elementary conception of color, an almost calligraphic attention to line, a tendency toward the elimination of chiaroscuro, with a resulting weakness in the definition of contrasts, and a preoccupation with linear values. As for subject matter, the Christian spirit of the Nazarenes inclined them to avoid the nude and mythological themes and to pay little attention to subjects from daily life or contemporaneous history.

Purism in France and Italy. The studio of Ingres (q.v.) was the formative center for the French equivalent of the Nazarene movement. Ingres had certain purist tendencies, which in some of his works — his portraits, for example — reached an independent expression; elsewhere, however (e.g., *Christ among the Doctors*; Montauban, Mus. Ingres), his refined archaism was closer to the Nazarenes. One of Ingres's pupils, Eugène Emmanuel Amaury-Duval (1808–85), reveals the direct influence of Overbeck (cf. the paintings at St-Germain-l'Auxerrois), though Amaury-Duval tended to restrict himself to certain formulas derived from study of the Italian masters. Another pupil of Ingres to feel the influence of the Nazarenes was Alphonse Henri Perin (1798–1874), who lived in Rome between 1817 and 1826; his mural paintings (1836–52) can be seen in Paris at the Church of Notre-Dame-de-Lorette (*Glorification of the Eucharist*; 1836–52).

Also on view at Notre-Dame-de-Lorette are works by Victor Orsel (1795–1850), who studied the primitives and was influenced by Cornelius and Overbeck. Another painter whose work showed certain of these influences was Victor Louis Mottez, who traveled in Italy (1837–47) and England (1851–56) and did some work at St-Germain-l'Auxerrois on a gold background in the manner of the Italian primitives; in 1858 he translated the *Libro dell'Arte* of Cennino Cennini. Hippolyte Flandrin (1809–64) became a pupil of Ingres in 1829; later he moved closer to the German Nazarenes and became strongly influenced by Giotto, probably attempting to preserve a certain primitivistic tone without sacrificing the technical resources of contemporaneous art. Among Flandrin's works are *St. Clare Healing the Blind* (1837; Nantes, Cathedral of St-Pierre) and frescoes in Paris (1840; St-Séverin; 1842–44, 1855–61, St-Germain-des-Prés; 1849–53, St-Vincent-de-Paul), Nîmes (1846–48, St-Paul), and Lyons (St-Martin-d'Ainay). Louis Janmot (1814–92), a pupil of Orsel in Lyons and later of Ingres in Paris, was in Italy between 1841 and 1845 and was drawn to the early 15th-century Italian painters, particularly Masaccio; he was also attracted by the theories of Overbeck. Janmot's principal works are the frescoes of the *Eucharist* (1844–46; Lyons, Hospice de l'Antiquaille) and the *Poem of the Soul* (34 compositions of a mystical and symbolic nature). Another French visitor was Paul Delaroche (1797–1856), who in 1834 began his study of Early Christian painting in Italy. The artist Paul Chenavard (1807–95), who was in Paris in 1825, met both Cornelius and Overbeck in Italy; in 1848 he designed a series of paintings for the Paris Panthéon entitled *Palingénésie universelle*.

Even among students of the French Academy in Rome there was a tendency toward "the primitive taste of the art of Giotto and Fra Angelico" (see letter by H. Vernet, 1834). In 1836 the first volume of a significant publication, *De l'art chrétien* by A. F. Rio, appeared in Paris. Rio was a friend of the French priest and philosopher F. Robert de Lamennais and made no secret of his sympathies for German mysticism and for Montalembert, cofounder (with De Lamennais) of a journal; he championed the cause of the purists against various attacks, including that of Chateaubriand.

At the opposite pole to the idealism and religious zeal of the Nazarenes is the spirit of "Jacobin purism" (Maltese, 1960) that animated the Italian sculptor Lorenzo Bartolini (1777–1850), a one-time pupil of David in Paris. A friend of Ingres, Bartolini shared with him the desire to reform neoclassicism, stressing at the same time the need to imitate reality; an inclination toward certain motifs of Tuscan 15th-century art is evident in his *Carità educatrice* (1824; Florence, Gall. Pitti) and in the Tomb of Sofia Zamoyska (1837; Florence, Sta Croce). Bartolini's pupil in Carrara, the sculptor Pietro Tenerani (1798–1869; later a pupil of Thorvaldsen in Rome, 1813), shows purist tendencies in some of his works, notably the Severini Monument (1822; Rome, S. Lorenzo in Lucina), the relief of *The Deposition* (1844; Rome, St. John Lateran), and *St. Benedict* (PL. 303). Tommaso Minardi (1787–1871), in an early painting, *Apparition of the Virgin* (PL. 303), already reveals an interest in certain motifs of the early Renaissance. Other Roman works by Minardi in this vein are *The Madonna of the Rosary* (1840; Gall. d'Arte Mod.) and paintings for the Palazzo Doria, the Church of Verano (1861), and the Palazzo del Quirinale (1864).

The foremost exponent of "Raphaelism" and of the Roman academic style (Maltese, 1960) in the second half of the 19th century was Francesco Podesti (1800–95), who based the compositional scheme and models for his figures in the Sala dell'Immacolata of the Vatican (PL. 303) on Raphael's *Disputa del Santo Sacramento* (*Dispute on the Holy Sacrament*) in the Stanza della Segnatura.

Among the purist painters who are considered followers of Overbeck (despite other influences such as Cornelius) are Wilhelm von Kaulbach (1805–74), who first came to Italy in 1835; Edward Jakob von Steinle (1810–86), who was in Rome in 1828–29 and worked with Overbeck at Assisi; Alexander Maximilian Seitz (1811–88), pupil of Cornelius, who came to Rome in 1833 and who in 1848 decorated the Villa Torlonia at Castel Gandolfo with paintings of evangelists and apostles based on sketches by Overbeck; and Alberto de Rohden (b. 1850), son and pupil of Franz.

Various purist sculptors were also interested in the Nazarenes. Theodor Wilhelm Achtermann (1799–1884), attracted by the Nazarenes' religious sincerity and by their interest in primitive art, went to Rome in 1840. An original plaster model of his *Deposition* can still be seen in Rome in Sta Trinità dei Monti. Giovanni Dupré (1817–82) is also connected with religious *purismo*. His statue *Abel* (Florence, Gall. d'Arte Mod.) was executed in 1842; a Pietà (1863–65) for the Misericordia Cemetery in Siena shows the influence of early 16th-century Tuscan and Ferrarese art. The painter Luigi Mussini (1813–88) was another who shared these same interests. About 1835 he painted the avowedly purist *Samuel Anointing Saul*. His *Sacred Music* (1840; Florence, Gall. d'Arte Mod.), praised by Bartolini, is remarkably reminiscent of Overbeck, although Mussini also tried to adapt his art to the manner of Ingres, whom he met in Rome between 1840 and 1844. In 1849 Antonio Bianchini, a pupil of Minardi, published *Del purismo nelle arti*, a work which became known as the "Manifesto of the Purists"; Bianchini's views were endorsed by Overbeck, Minardi, and Tenerani. The Manifesto, which states that "purists are concerned only with the essence of the expression of emotion," was a defense of the style elaborated by the Nazarenes and by those already known as purists. The sculptor Ignazio Jacometti (1819–83; created the *Moses* at base of Piazza Mignanelli column, Rome) was another to join the ranks of the *puristi*. The early works of Vincenzo Vela (1820–91) also show the influence of Bartolini's formal themes.

In architecture, Lavagnino (1956) includes under *purismo* such 19th-century architects as Giuseppe Valadier (1762–1839), Luigi Poletti (1792–1869), Antonio Sarti (1797–1880), Carlo Amati (1776–1852), and Lorenzo Nottolini (1787–1851), whose ideas were similar to those expressed by Francesco Milizia (1725–98) and who looked for inspiration to Giacomo da Vignola and Palladio rather than to Vitruvius. In French architecture, a parallel movement includes the names of Guy de Gisors (1762–1835), Etienne-Hippolyte Godde (1781–1869), and L. H. Lebas (1782–1867); with some architects, such as Jacques Félix Duban (1797–1870; IV, PL. 316), the influence of the Italian Renaissance is obvious.

A Spanish artist whose views on painting were similar to those of the Nazarenes was Federico de Madrazo y Kuntz (1815–94). He had been a pupil of Ingres in Paris and had met Overbeck and his circle about 1841. Another Spanish painter, Claudio Lorenzale y Sugranes (1814–89), was a pupil of Overbeck and the master of José Maria Fortuny y Carbo (1838–74); before developing the style for which he became famous, Fortuny y Carbo was in Rome (1858) and came under the influence of the Nazarenes. Paul Melchior von Deschwanden (1811–81), also connected with Overbeck in Rome, had an important influence on religious painting in Switzerland. The Russian painter Aleksander Andreevich Ivanov (1806–58) was another artist associated with the Nazarenes in Rome. His style shows their influence, particularly in his use of the linearity of the Primitives (*Christ before the People*, 1835–55; Leningrad, Russian Mus.).

An artistic movement in Russia with both religious and romantic overtones was the Society for Traveling Art Exhibitions (Tovarishchestvo Peredvizhnykh Khudozhestvennyk Vystavok; 1870), which aimed at correcting the vices of society and awakening the conscience of the people. Among the more important "peredvizhniki" (travelers, or wanderers) were I. N. Kramskoi (1837–87), N. N. Gay (1831–94), and Vasili Polenov (1844–1927). M. V. Nesterov (1862–1942), who began as a member of this group, later turned for inspiration to the Byzantine tradition, as did V. M. Vasnetsov (1848–1927).

In France, Pierre Puvis de Chavannes (q.v.; PL. 305) showed some leanings toward primitivism, acquired from Ingres by way of Théodore Chassériau (q.v.). Another French painter, Maurice Denis (1870–1943), found in the imitation of the primitives the justification of his own linearism and predilection for religious subjects; in 1890 he published *Art et critique*, the manifesto of a movement called Néotraditionisme. In Beuron,

Germany, in the Benedictine Monastery of St. Maur, some painter monks founded, after 1868, a movement that took the art of Fra Angelico as its moral and esthetic ideal. Their work had a hieratic quality and pretentions to monumentality. Peter (Desiderius) Lenz (1832–1928), painter and architect, and fellow Benedictines Jakob Wüger (1829–92) and Fridolin Steiner (1849–1906) worked at Monte Cassino (1874–78; 1900–13) in mosaic and fresco. In 1894 the Dutch painter Willibrord Verkade, attracted by the theories of Lenz, entered the monastery at Beuron.

In the last decade of the 19th century, some Italian painters passed through a Pre-Raphaelite phase in which a leaning toward the English Pre-Raphaelites became confused with an interest in a more authentically Renaissance taste, despite attempts of Aristide Sartorio (1860–1932) to maintain a distinction. *Purismo* became increasingly subordinate to new concerns, even among artists associated with the purist position. Giovanni (Nino) Costa (1827–1903), for example, had tried to graft a certain "idealism" onto Renaissance realism; his painting *La Lettrice* (*The Reader*) is explicitly Pre-Raphaelite. The drawings of Francesco Paulo Michetti (1851–1929) for the Amsterdam Bible belong to a *purismo* attenuated by the demands of realism. Suggestions of Pre-Raphaelite taste are present in works such as *Love at the Fountain of Life* (Leningrad, Russian Mus.) by Giovanni Segantini (1858–99; cf. PL. 442), *Woman of the Fountain* (1898; Bologna, private coll.) by Adolfo de Cardis (1874–1928), and *The Kiss* by Gaetano Previati (1852–1920). Sartorio, who for a decade or so was particularly interested in this movement, initiated in Italy the study of the English Pre-Raphaelites, on whom he wrote an essay in the periodical *Il Convito* (1895–98) based largely on photographs supplied by Gabriele d'Annunzio. Typical works by Sartorio during this period are *The Wise Virgins and the Foolish Virgins* (1891–94; Frankfort on the Main, private coll.) and *The Madonna of the Angels* (1895; Rome, private coll.).

Twentieth-century art has also manifested purist currents, though these are unrelated to historical purism. The term is often applied to movements of simplification, such as "analytical cubism," De Stijl, and "hard-edge abstraction," which attempt to isolate the elements of form and color and object-space relationships and to "purify" art of emotional content. For discussion of the development of these modern trends, see CUBISM AND FUTURISM, EUROPEAN MODERN MOVEMENTS, MODERNISM, and NONOBJECTIVE ART.

Gaetano BRUNDU

BIBLIOG. *a. Pre-Raphaelitism:* J. Ruskin, Pre-Raphaelitism, 1851, Lectures on Architecture and Painting, 1854, The Three Colours of Pre-Raphaelitism, 1878, The Art of England, 1884 (Complete Works, ed. E. T. Cook and A. Wedderburn, London, 1903–12, XII, XXXIII, XXXIV); E. Chesneau, La peinture anglaise, Paris, 1882; F. G. Stephens, Dante Gabriel Rossetti, London, New York, 1894; R. Motier de la Sizeranne, La peinture anglaise contemporaine, Paris, 1895 (Eng. trans., H. M. Poynter, Westminster, 1898); W. M. Rossetti, D. G. Rossetti's Family Letters: With A Memoir, 2 vols., London, New York, 1895; F. M. Hueffer, Ford Madox Brown, London, 1896; J. W. Mackail, The Life of William Morris, 2 vols., London, New York, 1899; H. C. Marillier, Dante Gabriel Rossetti, London, 1899; J. G. Millais, The Life and Letters of Sir John Everett Millais, 2 vols., London, 1899; W. M. Rossetti, Ruskin, Rossetti, Pre-Raphaelitism, London, New York, 1899; P. Bate, The English Pre-Raphaelite Painters, Their Associates and Successors, 2d ed., London, 1901; R. Muther, Geschichte der englischen Malerei, Berlin, 1903; G. Burne-Jones, Memorials of Edward Burne-Jones, 2 vols., London, New York, 1904; W. H. Hunt, Pre-Raphaelitism and the Pre-Raphaelite Brotherhood, 2 vols., London, New York, 1905–06; A. Woolner, Thomas Woolner, R.A.: Sculptor and Poet, London, 1917; W. Gaunt, The Pre-Raphaelite Tragedy, London, New York, 1942; R. Ironside and J. Gere, Pre-Raphaelite Painters, London, 1948; O. Doughty, Dante Gabriel Rossetti: A Victorian Romantic, London, 1949; Dante Gabriel Rossetti and His Circle, Univ. of Kansas Mus. of Art (cat.), Lawrence, Kansas, 1958; T. S. R. Boase, English Art 1800–1870, Oxford, 1959; The Pre-Raphaelites, Herron Mus. of Art (cat.), Indianapolis, New York, 1964.

b. Related movements: A. F. Rio, De l'art chrétien, Vol. I, De la poésie chrétienne dans son principe, Paris, 1836; C. Laderchi, Sulla vita e sulle opere di F. Overbeck, Rome, 1848; A. Bianchini, Del purismo nelle arti, Rome, 1849; P. Selvatico, Del purismo, Venice, 1851; C. Guasti, Del purismo, Florence, 1852; A. F. Rio, De l'art chrétienne, II, Paris, 1855; H. Delaborde, La peinture religieuse en France: H. Flandrin, Rev. de deux mondes, 2d ser., XXIV, 1859, pp. 862–92; H. Delaborde, Lettres et pensées d'H ppolyte Flandrin, 2 vols., Paris, 1865; A. F. Rio, Epilogue à l'Art chrétien, 2 vols., Freiburg im Breisgau, 1870; E. E. Amaury-Duval, L'Atelier d'Ingres, Paris, 1878; M. Howitt, F. Overbeck: sein Leben und Schaffen, 2 vols., Freiburg im Breisgau, 1886; J. Schnorr von Carolsfeld, Briefe aus Italien, Gotha, 1886, L. von Donop, Die Wandgemälde der Casa Bartholdy in der Nationalgalerie, Berlin, 1889; L. Mussini, Epistolario artistico (biog. by L. Anzoletti), Siena, 1893; P. Lenz, Zur Asthetik der Beuroner Schule, Vienna, 1898; A. R. Willard, History of Modern Italian Art, 1898; A. Germain, Les artistes lyonnais, Lyon, 1910; L. Hautecoeur, Rome et la renaissance de l'antiquité à la fin du XVIIIe siècle, Paris, 1912; P. Ettinger, A. Ivanoff und die Nazarener in Rom, ZfbK, N.S., XXIV, 1913, pp. 145–46; R. Jean, Puvis de Chavannes, Paris, 1914; W. Verkade, Die Unruhe zu Gott, Freiburg im Breisgau, 1919; L. Réau, L'art russe de Pierre le Grand à nos jours, Paris, 1922; F. H. Lehr, Die Blütezeit, romantischer Bildkunst: Franz Pforr der Meister des Lukasbundes, Marburg, 1924; L. Venturi, Il gusto dei primitivi, Bologna, 1926; C. G. Brandis, Briefe von P. Cornelius und F. Overbeck, RepfKw, XLIX, 1928, pp. 124–30; C. G. Heise, Overbeck und sein Kreis, Munich, 1928; M. L. Gengaro, Della polemica Rio-Rumohr sul valore dell'arte cristiana, L'Arte, N. S., II, 1931, pp. 351–54; A. M. Brizio, Ottocento e novecento, Turin, 1939; A. De Rinaldis, L'arte in Roma dal Seicento al Novecento, Bologna, 1948; E. Lavagnino, L'arte moderna, I, Turin, 1956 (2d ed., 1961); C. Maltese, Storia dell'arte in Italia (1785–1943), Turin, 1960; F. Novotny, Painting and Sculpture in Europe, 1780–1880, Harmondsworth, 1960; T. Talbot Rice, A Concise History of Russian Art, London, 1963; K. Andrews, The Nazarenes, Oxford, 1964.

* *

Illustrations: PLS. 299–306.

PRE-ROMANESQUE ART. Pre-Romanesque art is considered to include the artistic manifestation of the western European countries (excluding the areas under direct Byzantine influence) during the period between the late antique (see LATE-ANTIQUE AND EARLY CHRISTIAN ART) and the beginning of the Romanesque (see ROMANESQUE ART), that is, approximately from the 6th to the 11th century A.D. The arts of this period, which is also known as "early medieval," did not arise from a common base, nor were they expressed in a single style. They present, instead, as is natural in a period of such ethnic, political, and economic upheavals, a diversity of expressions, the manifestations of which in the different regions should be examined separately. The detailed study of the individual artistic areas (see ANGLO-SAXON AND IRISH ART; CAROLINGIAN PERIOD; EUROPE, BARBARIAN; MOZARABIC ART; OTTONIAN PERIOD) is here supplemented by a general view that is all the more significant because each of the pre-Romanesque schools and movements made its own contribution to the formation of the Romanesque style, in which western Europe once again found a fundamentally unified artistic expression.

INTRODUCTION. Pre-Romanesque, in the broad sense of the term, comprises the history of art in the West during the centuries between the decline of the late-antique period and the rise of the Romanesque. However, although almost all of the European countries participated with common fervor in the formation and development of Romanesque art and, hence, of Gothic art (France, in particular, distinguishing herself in the Gothic period), in the pre-Romanesque period the accents were distributed in various ways in a succession of artistic cultures that had their centers in different regions and countries and spheres of influence of well-defined limits. In a general view, Romanesque and Gothic art embraced the artistic developments of all the European countries. Notwithstanding differences from area to area, there was substantial community of style. This was not the case during the pre-Romanesque period, which lacked an intrinsic unity, not only from a European point of view but also in the artistic activities in the individual countries. Consequently, links between artistic trends that came into being contemporaneously in different parts of Europe were often very tenous, as, for example, those between Mozarabic and Ottonian art. In Italy completely

Pre-Romanesque architecture, chief centers of development and influence.

different artistic expressions, such as the 9th-century mosaics of the Roman basilicas, which were born of the fusion of local and Byzantine elements, correspond in time to Carolingian art, of northern European origin. Similarly, the decorative qualities of sculptured interlace, which was widespread in the central and northern regions of Italy, had nothing in common with contemporaneous mosaics of the Roman basilicas.

The original pre-Romanesque achievements were the contributions of the barbarian and northern European peoples (see EUROPE, BARBARIAN); these, nurtured by eastern influences that created a new abstract and expressionistic artistic outlook, were in marked opposition to Greco-Roman classicism. In particular, the decorative boldness of Irish and Anglo-Saxon art (see ANGLO-SAXON AND IRISH ART), which in part descended from Celtic tradition (see CELTIC ART), is one of the most original manifestations of incipient medieval European art. The purest artistic expression of the barbarian populations can be found in their gold- and silverwork; in the countries they invaded — Italy, France, Spain — an architectural and sculptural activity developed from local indigenous elements that were characterized by contributions from the Mediterranean tradition permeated by late-antique and Byzantine elements. Evidence of this can be found in Aquitanian sarcophagi of the Visigothic period in France, in the architecture and sculpture of the Visigothic period in Spain, and in the ornamental interlaced sculpture of Italy, all of which reflect the taste and tendencies of the Latin populations of these countries in the 5th century and in the period from the 6th to the 10th century. Carolingian art (see CAROLINGIAN PERIOD), in which the expressive intensity of the northern European artistic genius was allied with the deep-felt reevocation of classical forms, was to result from a dialectical confrontation with Mediterranean artistic cultures; and it is precisely this return in Carolingian art to the representation of the human figure in classical terms that checked the spread of the abstract stylizations of Irish art which threatened to destroy the structure of the human figure as it had been elaborated by the Mediterranean tradition. Carolingian art — which had a notable influence on English art but only a slight effect in Italy and Spain — sprang from the union of Frankish and Germanic creative forces; but Ottonian art (q.v.), the subsequent artistic flourishing of a very high qualitative level that preceded the birth of Romanesque art (q.v.) by a short time, is a typical creation of Germanic artistic genius, which, beyond the borders of Germany, had noteworthy developments only in Italy. Mozarabic art (q.v.), which is imbued with Islamic and Persian reminiscences, and Asturian art (see below) developed in Spain and occupy a place of their own in the history of pre-Romanesque art, some of the artistic achievements of which are among the finest creations of the medieval period.

* *

FRANCE. *Art in Gaul from the 5th to 8th century.* (*a*) *Architecture.* During the period from the 5th to the 8th century the formula of the church built in a private dwelling (Reims, Bourges, Auxerre, Narbonne) was progressively abandoned in favor of construction of the group of episcopal churches. This group generally comprised three churches, dedicated respectively to the Blessed Virgin, St. Stephen, and St. John the Baptist, although in a few cases churches were dedicated to the Holy Cross (Lyons), the Maccabees (Vienne), SS. Sergius et Bacchus (Chartres), and SS. Gervase et Protase (Le Mans). These episcopal groups are mentioned in several cities: Arles, Vienne, Lyons, Tours, Le Mans, Chartres, Paris, and Toul. In the 5th century, among the bishops who were builders, there were Patient (Lyons), Rustique (Narbonne), Niçaise (Reims), Namace (Clermont); in the 6th century, Gregory (Tours) and Delmas (Rodez); in the 7th, Didier (Cahors). Little is known of the structure of the episcopal churches. The basilican form is given for St-Jean-Baptiste (Lyons), with a columned nave and flat ceiling, a double portal, and an atrium. The churches of Clermont, Nantes, and Orléans probably had transepts. The triapsidal form (*forma triformis*) is found in Nantes, Chartres, and Vaison-la-Romaine; in the latter edifice, the apses are horseshoe-shaped. The excavations at Nîmes brought to light an apse turned to the west. Didier's buildings at Cahors recalled the ancient manner: *more antiquorum . . . quadris ac dedolatis lapidibus . . . .*

The baptisteries were placed near the episcopal churches; to the north (as in Angers, Marseilles, Nevers, Mélas), to the south (Aix, Valence, Poitiers), or to the west (Fréjus; PL. 307). Others existed near monastery churches and even in isolated places (Port-Bail, in Normandy). The need of a water supply, of a font, and of adjoining rooms account for the similarity between the plans of these buildings and those of the ancient baths. The reuse of ancient columns and capitals in the baptisteries in Provence and in Poitiers, the type of masonry, and the stone arches alternating with brick (Fréjus) accentuate the resemblance to Gallo-Roman architecture. The octagonal plan occurred frequently; sometimes it was inscribed within a square, the angles of which were recessed by niches (Marseilles, Saint-Rémy, Riez, Fréjus); at Mélas small niches were also opened in the middle of each side of the square; the baptisteries of Aix and Riez (FIG. 643) have a circular colonnade, as did, formerly, the one in Marseilles. At Nevers, square apsidioles alternated with horseshoe or pentagonal apses. The baptistery of Port-Bail was hexagonal and that of Valence in the shape of a cross, as was the one at Venasque, where the arms of the cross were terminated by rounded apsidioles. The baptistery in Poitiers is rectangular, with three small apses; it is built on older, differing substructures; its pedimented façades relate it to the Roman tradition, in spite of the barbarian ornamentation of the exterior. The windows of the Fréjus baptistery were provided with openwork marble *claustra*; the font was protected by a ciborium.

The number of funerary churches, basilicas, and chapels, isolated or in groups, grew as a consequence of the increasing veneration of relics. Monastic communities often assumed the religious custody of services associated with the holy remains. In such cases the funerary church engendered the foundation of a monastery, especially in the cities whose bishops of the first centuries had been canonized (e.g., Denis of Paris, Hilary of Poitiers, Martin of Tours, Saturninus of Toulouse, Genès of Arles, Germain of Auxerre, Aignan of Orléans). At Angers and in Vienne the first bishops were venerated in churches dedicated to St. Martin and St. Peter. In other places where churches were dedicated to pioneers of the monastic life, their disciples venerated their relics; this gave a funerary character to one of the churches of the monastery (e.g., St. Cesarius of Arles, Ste-Croix in Poitiers, St-Léger in Saint-Maixent, St-Philbert in Noirmoutier, St. Maurus in Glanfeuil, Jouarre; PL. 307). At the same time, various other foundations arose (e.g., Ligugé; St-Pierre, Metz; St-Paul, Besançon; Jumièges; St-Quentin, Solignac; Fleury-sur-Laire). Others (SS. Apôtres, Ste-Geneviève, Ste-Croix et St-Vincent, in Paris; St-Martin at Autun; St-Denis) were founded by kings Clovis, Childebert I, Brunehaut, and Dagobert.

The funerary character of the majority of these churches is shown by the importance of the martyriums, oratories, and crypts which were sometimes complex or multiple; they communicated with the sanctuaries by means of stairs, corridors, or fenestellas on a level with the floor of the church. The martyrium of St-Sernin in Toulouse was undoubtedly a rotunda. The crypt of St. Victor in Marseilles was hewn from the rock and provided with arcosolia; it was connected with a basilican church with an atrium. The crypt of St-Laurent in Grenoble has a double apse and a trefoil choir. At St-Léger in Saint-Maixent the inverted T-plan and the two staircases made circulation easier. On the other hand, the hypogeum in Poitiers is a closed hall, access to which is by a single staircase. The groin vaults, barrel vaults, and semicircular vaults are supported by antique columns and by rectangular pilasters.

The churches varied in plan as well as elevation: there was a basilica with prothesis and with diaconicon in St-Laurent in Lyons and in St-Pierre in Metz; one with galleries in St-Irénée in Lyons and possibly in Tours; a ground plan in the form of a cross with apsidioles in Ligugé; a decagonal plan in La Daurade in Toulouse; and one with three apses in Lérins.

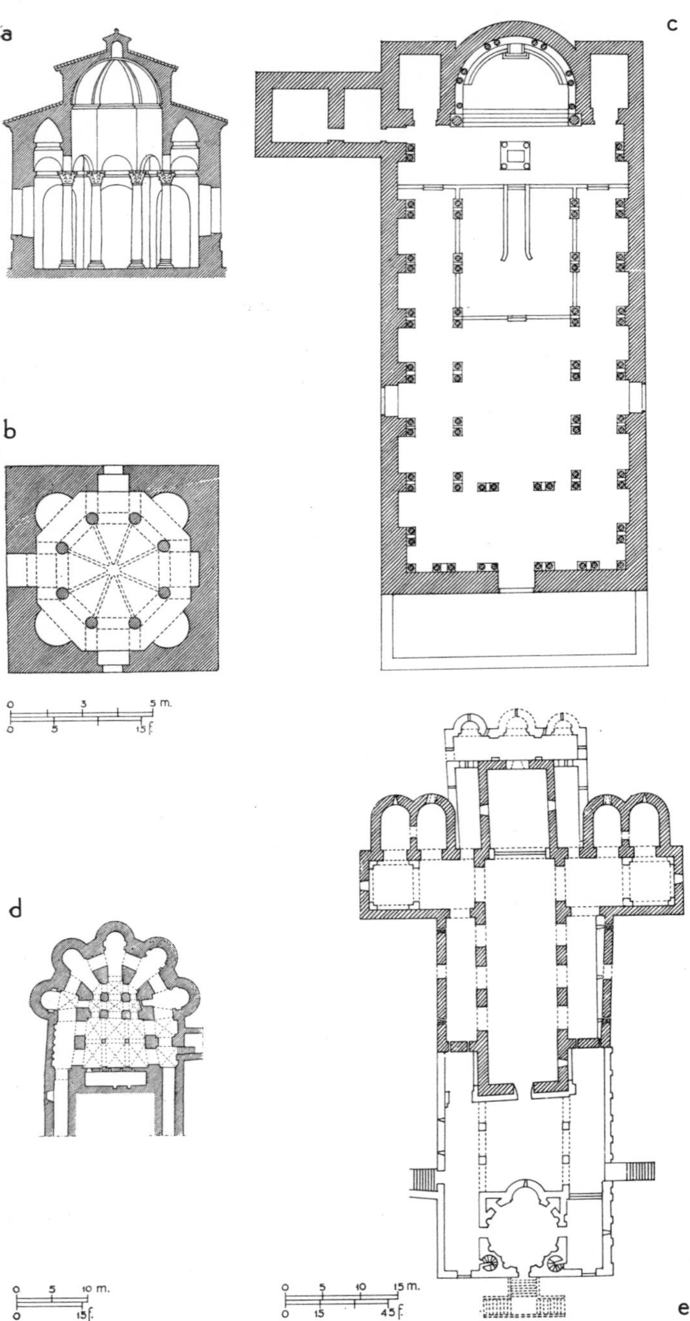

France. Riez (Basses Alpes), baptistery, (*a*) section, (*b*) plan (*from G. Bandmann, Die Bauformen des Mittelalters, Bonn, 1949*); (*c*) Vienne (Isère), St-Pierre, reconstructed plan (*from E. Lundberg, Arkitekturens formspråk, Stockholm, 1949, Vol. VIII*); (*d*) Orléans, St-Aignan, crypt, plan (*from G. Bandmann, op. cit.*); (*e*) Saint-Michel de Cuxá, plan of main church with shading to indicate 10th-century structures (*from K. J. Conant, 1959*).

Two towers flanked the facade of St-Martin in Autun, St-Julien in Broude, and in Jumièges; there is a single tower in Tours.

The Gallo-Roman heritage survived in the use of large-stone masonry (Cahors, Ligugé, St-Léger in Saint-Maixent) and of small-stone masonry combined with brick (St-Bertrand in Comminges, St-Pierre in Metz, and in Méris) or the use of brick in regions where stone was scarce (La Daurade in Toulouse). Ancient material was reused in both structural parts and in decoration.

(*b*) *Architectural decoration.* Decoration during this period was far from being a subordinate technique. On the contrary, various methods of architectural decoration were used, and the decoration (crypt of Jouarre, Civaux; PL. 307) contributed to the embellishment of the masonry. Panels with rose designs in slight relief were inserted in the walls, as were geometric designs in baked clay (Poitiers baptistery) or stamped brick representations of figures or scenes (St-Martin in Vertou; St-Similien in Nantes). These processes, derived from the art of the barbarian gold-and silversmiths, were used on a scale to fit a monumental application. The antique heritage is seen in stucco facing (St-Laurent in Grenoble; St. Victor in Marseilles; Lérins; the last two with a vine-branch design). Marble facing was also used. It was sometimes incised with Christian figures (the Virgin, Abraham's sacrificial offering, Daniel, at Saint-Maximin). The mosaics and mural painting were derived from the antique and early Christian traditions. Literary sources describe the mosaics of St-Jean-Baptiste in Lyons (Jerusalem, the Lamb, the Four Rivers of Paradise); of Nantes Cathedral with its realistic figures described by Fortunat; and of La Daurade in Toulouse, where depictions from the Old and the New Testament appeared together. Mosaic floors have been found in St-Jean-Baptiste in Lyons, in the Aix baptistery, and in St-Genès in Thiers, where a bestiary shows strong Oriental influence. Ligugé has a polychrome pavement with various designs. From texts mural painting is known in St-Maur de Glanfeuil (Christ as a Child, the Suffering Christ, and Christ Triumphant) and in Tours with scenes from the Life of Christ and the miracles of St. Martin as well as a second version of the miracles in the cathedral. In Poitiers the steps of the hypogeum are encrusted with colored glass.

The capitals bear witness to the vigor of the survival of classic traditions, in the use of marble as well as by their faithfulness to Corinthian, Composite or (seldom) Ionic models. Nevertheless, the stylized acanthus leaf developed into the palmette; volutes grew smaller or, on the contrary, turned into the detached handles of the bell of the capital (St-Brice de Chartres; Jouarre); spiraliform molding was of barbarian origin; the cushion bearing Christian emblems came from Byzantium or Ravenna; the cushion was superimposed on the bell of the capital (Jouarre; St-Laurent at Grenoble). The dolphins on a capital in the Poitiers baptistery are in the early Christian tradition.

(*c*) *Liturgical objects and sarcophagi.* Altars in the form of cippi or tables and decorated with Christian motifs were fairly numerous in Provence (Tarascon, Cavaillon). In the Poitiers hypogeum there is a cippus altar adorned with two roughly carved crucified martyrs. Some chancels (Arles, Vaison-la-Romaine, Lyons) are covered with intricate geometric traceries, foliage issuing from a vase, or clumsy human figures (St-Pierre in Metz); the antique strigil motif was employed with the cross and the tree of life (St-Denis); the Visigothic slabs of Narbonne and Septimania, carved with symbolic animals, studded crosses, and interlacings, resemble those of Catalonia (see SPAIN).

From the standpoint of art history the production of sarcophagi in the pre-Romanesque period is more interesting, especially in the regions rich in marble (Pyrenees) or hard limestone (Poitou, Burgundy). Other workshops have been identified in Champagne and in the Paris region, where softer limestone and even plaster were used. In Aquitaine the sarcophagi were made of marble from the Pyrenees. They were used in Narbonne (PL. 311), Toulouse, Agen, Auch, and Bordeaux; some were exported to the Massif Central (Rodez, Clermont) and to the Paris region. Usually they were rectangular, hollowed out, and with a gable-shaped lid with four sloping sides. From early Christian art they retained the subdivision into panels by means of small twisted columns or of pilasters, at times by round or mitre-shaped arcatures; these panels frame figures and religious or secular scenes (Daniel, Christ and the Apostles, miracles from both Testaments, or scenes symbolically representing a hunt or a trade, such as blacksmiths or peasants at their work). Another type of sarcophagus was adorned with foliage, grapevines, strigils, oblique fluting, and Christian symbols (the chrismon, doves, etc.). In Burgundy, sarcophagi with rounded ends and convex covers were more frequent. They too were decorated with antique and early Christian designs. Many trapezoid sarcophagi were produced in the Poitou workshops; the lids were decorated by one band lengthwise and three bands across; this type is found as far as Brittany; certain lids show geometric designs and crosses like those found in the northeast. The sarcophagus from Charenton in Berry and the sarcophagus of St-Léonien in Vienne, which are decorated with carved figures (Daniel, griffins, peacocks), are very similar, in spite of the distance that lies between them. The sarcophagi of Jouarre are of exceptionally fine quality, not only for the beauty of their inscriptions but also for the variety of their decoration (intersecting circles, scalloped frieze, Christ Triumphant between the symbols of the Evangelists). The use of incrustations, as on the sarcophagus of Bishop Boece in Venasque and in the hypogeum of Poitiers, with the coarsely engraved symbols of the evangelists Matthew and John and the archangels Raphael and Gabriel, was derived from the gold- and silversmith's art. Its influence on sculpture shows how important this technique had become; it combined the elegance of design with the richness of the material by the use of embossing and cloisonné, inserting between the fillets colored and precious substances. This period also had a taste for rich fabrics, used in religious ceremonies and for clothing; a considerable amount of these fabrics came from the East; by the variety and style of the woven or embroidered ornaments these objects enriched the repertory of patterns from which designers in other media were able to draw (see EUROPE, BARBARIAN).

Development of Romanesque art. The Carolingian period falls between the period discussed above and the period of the 10th and 11th centuries, which saw the development of Romanesque art in France. Earlier artistic experiences were developed under the more vigorous stimulus of the emperor and his court, which included many ecclesiastics. Both had a sense of grandeur. The vastness of the empire favored the synthesis of the various cultures existing from the Mediterranean to Ireland; the memory of the grandeur of Rome and the example of the Byzantine empire inspired what is known as the "Carolingian renaissance." Architecture and decorative techniques (mosaic, gold- and silverwork, ivories) attained a high degree of development. The manuscripts from the imperial and monastic scriptoria were brilliant examples of the art of illumination (see CAROLINGIAN ART).

10th and 11th centuries. (a) Architecture. The end of the Carolingian period was followed immediately by a period of prolific building activity. This reflected an intense religious life, sponsored by bishops and abbots who were also builders: Etienne II in Clermont and Conques, Gauzlin at Saint-Benoît-sur-Loire and at Bourges, Mayeul at Cluny, Fulbert at Chartres, among others. It also reflected the division of France into large feudal units, the masters of which — the dukes of Normandy and Burgundy, the counts of Poitiers and Angers — contributed to the enrichment and embellishment of the abbeys under their protection (Caen, Fécamp, Cluny, Maillezais, Vendôme, etc.). The Capetian kings did the same in their own territories — at Orléans, Melun, and Stampes. This, however, did not signify a partition into watertight compartments; the freedom of movement of the clergy was unlimited: the Lombard William of Volpiano divided his activity between Dijon, Fécamp, and Bernay; Abbot Oliva was at the same time abbot of Ripoll and of St-Michel de Cuxá and owing to him, and to many other historical factors, artistic interchange overcame the barrier of the Pyrenees. At the same time, through the Rhone valley, across the Alpine passes, and along the Mediterranean coast links were established with northern Italy and Catalonia. On the other hand, the territories of the Meuse, Saône, and Rhone remained within the Germanic world of the Ottonian empire, where great architectural formulas were being evolved (see OTTONIAN ART). This, however, did not prevent the survival of ancient traditions, nor the constant influx of Eastern elements, nor the further development of previous experiences that the invasions of the 9th and the beginning of the 10th century had by no means entirely destroyed. The year 1000 did not, however, mark the begining of a new era. Nearly fifty large workshops (cathedrals and abbeys) were active in the second half of the 10th century, and at least as many could be counted in the first half of the 11th century.

Creative effort was especially concentrated on the development of the sanctuary, the crypts, and the transept. The addition of an ambulatory with radiating chapels was an innovation of the greatest consequence and a formula full of promise for the future of Romanesque and Gothic art. Its place of origin is hard to determine. It appeared in central France (Clermont) as well as in the middle region (Tours, Orléans, Auxerre), and in the west (Chartres, Rouen, Saint-Gildas de Ruis, Landévennec). The simpler formula of the so-called Benedectine plan with parallel apsidioles, either partitioned or laterally connected, is found in Auvergne (Thiers) as well as in the north (Saint-Bertin). It was also used in Burgundy (Cluny II, the Charité), in Berry (Dèvre, Méobec), in Anjou (le Ronceray d'Angers), and in Normandy (Cerisy-la-Forêt) and was later developed considerably during the 11th and 12th centuries. Even simpler was the combination of transept with the apse between two small apsidioles, or transept and the apse between two groups of two small apses of equal depth (St-Michel de Cuxá) with a nave with or without side aisles. In some cases this structure had small annexes, survivals of the prothesis and the diaconicon (*secretaria* at Nouaillé). In other cases a monumental development was given to the transept by the lateral aisles built around it (Cathedral and St-Aignan in Orléans; St-Rémi at Reims; Saint-Bertin); but sometimes there is no transept at all (Elne).

The increasing veneration of relics explains the shape and size given to the crypts, to the point of making their different parts — confessional, corridors, ambulatories with radiating chapels — govern the arrangement and the dimensions of the church above (Fulbert's Cathedral at Chartres). The modes of vaulting remained the same (barrel vaults, groin vaults with or without transverse ribs). Often sculptures were reused (columns, capitals, and tracery panels at St-Brice de Chartres, Auxerre, Saint-Geosmes).

The early Christian tradition of the basilican plan (nave with side aisles) was all the better preserved as builders still hesitated to vault a wide span and preferred to remain faithful to the formula of galleries built over the side aisles or along the interior of the façades. From St-Michel de Cuxá to Notre-Dame-de-la-Basse-Oeuvre in Beauvais, from Strasbourg to St-Hilaire in Poitiers, a long list could be made of the large churches with nave covered by a timbered ceiling, even when the lateral aisles are vaulted. Galleries were frequently used in Normandy (St-Pierre, Jumièges), in Maine (La Couture, at Le Mans), in Touraine (St-Martin, Tours), in the Orléanais (Orléans Cathedral), in Poitou (Maillezais), and in Champagne (St-Rémi in Reims, Montier-en-Der). In Champagne this usage was established to the extent that the master-mason of Vignory (PL. 309) opened, between the arches below and the windows above, a series of openings that are not the equivalent of a gallery but still give the impression of a three-story structure. After the beginning of the 11th century, however, large churches, such as Cluny II and Tournus (PL. 308), St-Rémi in Rheims, Ronceray in Angers, Maillezais, and others, were vaulted throughout or in part by longitudinal or transverse barrel vaults. The churches with sanctuaries at the western end, with or without transept (the cathedrals of Verdun and Nevers), are reminiscent of certain distant Carolingian or Ottonian models; the same may be said for those which have a forestructure with an upper story forming a gallery flanked by two staircase towers; these are found in Alsace (Strasbourg) as well as in Lorraine (Toul), Champagne (Montier-en-Der), Burgundy (Tournus), Normandy (Jumièges), Touraine (St-Mexme at Chinon, St-Melaine at Preuilly), and Poitou (Maillezais); a simpler formula is that of a gate tower at the front of the nave (St-Germain-des-Prés in Paris; St-Julien in Tours; the Cathedral and St-Martial in Limoges; Souillac); in other places lantern turrets continued to be built over the crossing (Châtillon-sur-Seine, Bayeux Cathedral, St-Etienne at Caen) or an isolated tower was erected near the church (St-Hilaire in Poitiers, St-Florent-lès-Saumur).

Funerary traditions, the Carolingian heritage, and relations with the East, which were intensified through the pilgrimages to the Holy Land, all contributed to the continued use of circular or octagonal plans. They were used for funerary structures and structures built to preserve the remains of saints [the crypts of St-Pierre-le-Vif in Sens, St-Béningne in Dijon (PL. 308), St-Pierre-de-Lémenc, St-Michel de Cuxá], for smaller scale imitations of the Palatine Chapel of Aachen (Ottmarsheim), and also for monuments inspired by the Holy Sepulcher in Jerusalem (Ste-Croix in Quimperlé; Neuvy-Saint-Sépulcre; St-Léonard). They are fairly frequent in Provence (Grasse); the use of octagonal or triconch baptisteries continued in Corsica.

On the basis of certain peculiarities of the masonry (flat stones), ornament (pilaster strips, dentils), or types of support (cylindrical piers without capitals), Puig y Cadafalch has identified "early Romanesque" art which, starting from the Mediterranean coasts of Provence, Languedoc, and Catalonia, moved up the Rhone and Saone valleys into Burgundy; from Elne and Saint-Martin-du-Canigou (PL. 308), from Corsica, from Vence, and from Saint-Dalmas it spread as far as Tournus, Dijon, Lons-le-Saulnier, and Châtillon-sur-Seine. On the other hand, the horseshoe arches of St-Michel de Cuxá, St-Martin-des-Puitz (Aude), and the monastery at Thiers reveal Mozarabic influences (see MOZARABIC ART).

Another form of early Romanesque art prevails from Anjou, northern Poitou, and Touraine, through the Orléanais and Nivernais to Chalon-sur-Saône; it is distinguished by the small limestone masonry with ornamental effects (Distré, Saint-Généroux, Cravant, Monthou-sur-Cher, etc.), the insertion of sculptured panels, timber ceilings, and the simplicity of the ground plans.

Meanwhile, as in the early Middle Ages, important monuments were built of large masonry, *ex quadris lapidibus* (St-Rémi in Reims, Jumièges, Maillezais), while others, some of which were later (St-Hilaire at Poitiers), followed the tradition of the small masonry construction or *opus reticulatum* that continued to exist during the Romanesque period.

(*b*) *Sculpture*. The part played by sculpture in certain monuments, which are impressive mainly by the organization of their comparatively unadorned architectural masses, is far from negligible. Preserved in the design of capitals was the ancient Corinthian tradition, which had such excellent results as Bernay, St-Hilaire in Poitiers, Maillezais (PL. 310), and Saint-Benoît-sur-Loire. The distant examples of Byzantium and Ravenna determined the shape of the strictly geometric capitals surmounted by a block that is the equivalent of the cushion and covered by a surface decoration composed of linear, plant, and animal elements; nowhere do they alter the volume of the block (Bernay, Brantôme, Chamalières, Issoudun, Vignory, Orléans; PL. 310). Interlace still appeared on inserted panels (Bessuéjouls, PL. 311), pilasters (Saint-Oustrille), and crosses carved into the outer walls of churches in Touraine (Esves-le-Moûtier) and Berry. The carved panels in the nave of Bayeux suggest Scandinavian influences. Promising innovations can be seen in the increasingly frequent use of figured capitals, as in the churches in Paris (St-Germain-des-Prés, Ste-Geneviève), Orléans (St-Aignan), and Saint-Benoît-sur-Loire, which are early examples full of inventiveness in the adaptation of iconographical themes to the mass of the bell.

On the exteriors of the churches, the dominant sculptural feature is the carved panels arranged either in a relative order in the form of a frieze (Graville-Sainte-Honorine, Selles-sur-Cher, Marcilhac, etc.) or placed at random (Cormery, Chabris, etc.). The subjects were taken from the Bible, from the local hagiography, from daily life (war, hunting, rural occupations), or from real or imaginary animals of Eastern origin. In certain cases there is a noticeable effort to achieve a composition related to the architectural elements of the façade (St-Mexme in Chinon) or of the gable of a transept (Beaulieu-lès-Loches). The sculptured lintels of Saint-Genis-des-Fontaines (PL. 311) and of Saint-André-de-Sorède and the tympanum of St-Ursin at Bourges (which is signed "Girauldus") heralded the future arrangement of sculpture on Romanesque portals.

The dominant technique was flat carving and the strict confinement of the design within a frame from which it does not project either in surface or in depth. The human figure was often represented within the compass of a small arch; the semicircular arch matches the outline of the head or, better still, of the halo; the small columns give a hieratical rigidity and frontality to the human figure, which is seldom changed (Azay-le-Rideau, St-Hilaire at Poitiers). This method recalls the techniques used on steles, sarcophagi, and antependia made of precious metals. Funerary sculpture sometimes followed the same practice (the tomb of Abbot Isarn in Marseilles, ca. 1048). The hypothesis that the sculptors of this period were incapable of reproducing the human figure in the round is contradicted by the presence of smaller works in the round, such as the early Virgins from Auvergne and the gold or silver reliquary statues (Ste-Foy in Conques), both of the 10th century.

(*c*) *Painting and minor arts*. From numerous texts it is known that mural painting was practised in the 10th century and at the beginning of the 11th century in several abbeys (St-Florent at Saumur, Saint-Benoît-sur-Loire, St-Martial in Limoges) and in the cathedrals of Beauvais and Auxerre, which reflected the Carolingian traditions. But this art can be studied at first hand only in smaller churches. At Ternand, in the Lyons region, the crypt contains paintings representing the childhood of Christ and His triumph. In Saint-Pierre-les-Eglises there are a moving Crucifixion (PL. 311) and animated scenes from the Apocalypse. There are notable Byzantine elements (e.g., in the Bathing of the Infant Jesus). St-Michel l'Aiguille at Le Puy (ca. 962) shows a similar inspiration, if not the same style. A definite connection existed between the paintings in St-Julien, Tours, and the local illuminations. Mosaics were still made in the 10th century, in St-André-le-Bas in Vienne and in the Cathedral of Reims, which also has windows with human figures.

The invasions of the 9th century undoubtedly slowed the development of the art of illumination, but a vigorous revival set in toward the end of the 10th century; at Saint-Bertin, under Abbot Odbert, the scriptorium combined Franco-British Carolingian traditions with Anglo-Saxon contributions from Winchester; St-Waast at Arras, Saint-Amand, and Marchiennes were also among the centers of reproduction that were destined to have a brilliant future. There was some activity in Chartres under Fulbert; at Saint-Germain d'Auxerre owing to Abbot Heldric (989–1010) and at Saint-Béningne, Dijon, at the time of Guillaume de Volpiano. At St-Martial in Limoges there began a remarkable flowering with a famous Bible (Bib. Nat., Ms. lat. 3) and the works of the monk Adhémar de Chabannes, who was both a chronicler and an artist, reveal southern and eastern influences. These are also seen in the Beatus Apocalypse of St-Sever (XII, PL. 207), which, by the violence of its colors and the boldness of its compositions, bears witness to the strong creative power common to all the art forms of that period.

René CROZET

GERMANY. *Merovingians*. After the middle of the 3d century A.D., the Roman boundary (limes Germanicus) on the Danube and the Rhine gave way under the attacks of the Germanic tribes, who were then able to advance deep into Gaul. From that time on the Rhine formed the border of the Gaulish empire, and Trier, under Postumus (258–73), became the capital. The Roman rulers erected in Trier the great "Basilica" begun (with the south church) in 326 and completed in 348.

During the 4th century other important buildings were erected in the sees of the Rhenish bishops. In Cologne, under the Church of St. Severin, an early Christian martyr's cella memoria (ca. 390) was found with an apse facing west; in St. Gereon, an early centrally planned structure with a narthex was preserved; and in Xanten a cella memoria (ca. 390) was found. Important early Christian structures exist in Trier (St. Maximin, St. Paulinus), Koblenz (Liebfrauenkirche), Speyer (St. German), Regensburg (St. Emmeran), and Kempten.

In Austria a series of important early churches were built during the same century, as in Klosterneuburg, near Vienna, and Lavant. Although soon after the year 400 and toward the end of the 5th century much was destroyed in the north, late-antique culture survived until about 600 in the south of Germany where architectural forms remained close to their Mediterranean models. The churches of the Noricum, especially, are closely patterned, even if in a provincial form, after their counterparts in northeast Italy, such as those of Parenzo (Poreč, now Yugoslavia) and Aquileia.

Even after the occupation of the areas of the south by the Germanic tribes and the ensuing almost total destruction of early Christian religious edifices, there was, nevertheless, a continuity in building; just as the religious buildings of the 4th and 5th centuries were mostly erected on the sites formerly sacred to pagan cults, so were succeeding generations inclined to use them for their religious edifices; as these were mostly modest structures, some made of wood, no new stylistic forms in architecture such as appear in decorative arts are found in them. The depopulation and impoverishment of the lands around the Rhine and the Danube precluded larger architectural undertakings. It was not until the consolidation of power under the Merovingian kings and the centralization of the newly occupied (central-western) areas that architecture received a new stimulus — this in close connection with the art of the richer regions of the northern and southern parts of the Frankish empire; the uniform style of the eastern and western areas was determined above all by the Benedictine monasteries and the cathedrals. An additional influence was the growing activity of the great missionaries who began to penetrate into unchristianized areas, where they built churches. St. Boniface (d. 754) stimulated the artistic activities of the diocese of Mainz, and the new structures in Fulda, Bremen (789), Reichenau (724), Regensburg, Eichstätt, Chur, and Paderborn are associated with the names of great missionaries such as St. Emmeram, St. Rupert, St. Corbinian, and St. Pirmin. Thus the artistic centers spread toward the Elbe together with the missions and the ensuing political expansion. At the same time, the old episcopal sees of Cologne, Mainz, Augsburg, Regensburg, and Passau continued to provide definitive points of reference.

The cultural center of the area, however, was the region between the Loire and the Rhine, where the old Roman tradition mixed with Germanic-Frankish elements and developed new forms. Southern Gaul exercised a particularly strong influence, which was disseminated by bishops, such as Sidonius in Mainz, who came from old senatorial families. There are also records of direct connections with Italy; Nicetius, for example, brought Italian builders to Trier. Oriental and Byzantine influences penetrated across the Alpine regions from Italy, and through Spain and Gaul, from Egypt and North Africa. The artistic accomplishments of this epoch were modest, but they laid the foundation for the brilliant rise that took place in the Carolingian period.

Surviving examples of Merovingian handicrafts and a few sculptures permit survey of this development. Even though an ever stronger local character was expressed in the individual forms, the heritage of classical antiquity remained so strong and vivid that it survived latently during the next centuries. In Gaul this classicizing tradition was cultivated — undoubtedly consciously — by the families of the senatorial nobility who continued under the Frankish Merovingian kings, to provide high ecclesiastical dignitaries. The "imperial art" of the western world never entirely disappeared, not even during the time of the Frankish Merovingian Empire; thus, in the classical treatment of the robes of SS. Peter and Paul in the ivory diptych from Kranenburg (New York, Met. Mus.) the relationship to Rome's early Christian sculpture and the Marseilles sarcophagi is still visible. In the same conservative spirit classical ornamentation was retained in the acanthus and the cymatium, which frequently appeared alone or combined with new motifs such as the simple braided band or Germanic zoomorphic designs.

The rise of a feudal ruling class was accompanied by the development of local styles that differed sharply from the classicizing works of a restricted aristocratic class. Sculpture almost entirely disappeared, and in decoration plant motifs gave way to geometric patterns, particularly in metalwork, in which the effect was achieved primarily through the contrast of latticed gold settings with inlays of red and other colored glass ("verroterie cloisonné"). This technique, which had already appeared at an early date among Eastern peoples, particularly in Iran and the Pontus region, was brought to the West with the beginning of the migration from the East, the invasion of the Huns, and also partly through direct importation.

The invasion of the Huns (375) gave the first impetus to the migration of the peoples. The Germanic tribes of the Alans and the Vandals, the former probably in Hunnish array, crossed the Rhine (ca. 406–07) in the vicinity of Worms and Mainz to invade Gaul. Some objects of Hunnish workmanship have come to light in a grave on the Rhine near Altlussheim in the vicinity of Mannheim (Baden), among them a sword from about 450 embellished with a gold cloisonné mount set with almandines (V, PL. 59). The rich find in a grave at Wölfsheim on the middle Rhine includes a gold ring with a pendant (Wiesbaden, Städt. Mus.) also embellished with almandine inlay and engraved with the name of the Persian king Ardachir I (224–41). Undoubtedly craftsmen from the Black Sea region accompanied the invaders from the East and directly transmitted the technique to German artisans. This first phase of the Pontic art of inlay ended with the defeat of the Huns in 451.

In the following epoch the settings of the gold objects became narrower and richer. This stage of the style is called the Childeric stage, because the best examples of it were found in the middle of the 19th century at Tournai, in the grave of Childeric (d. 481), first king of the Franks. The wavy gold bands recall the technique used in the small fibula of Wölfsheim (Budapest, Nat. Mus.) and in the related fibulae of Apahida and Rüdern. In Frankish territory this richly embellished type was also found in the graves of Rhenish princes such as those at Flonheim, Planig, and in somewhat later works in Morken, from the end of the 6th or the beginning of the 7th century.

After the 6th century Mediterranean elements again made their appearance northward across the Alps. The great influence exerted by these elements is revealed in metal seals of southern origin and in the earrings and gold-filigree brooches found in the rich Frauengrab (grave of a woman) in the Cologne Cathedral. The presence of coins make it possible to date these articles to the middle of the 6th century. This combination of red almandine and green stones was already present in the fishes from the burial ground of Bülach in the first half of the 6th century (Zürich, Mus.) which recall Ostrogothic workmanship. This technique was also widespread among the Lombards of upper Italy, where it was combined with Byzantine and classical elements, and in the course of importation by the Lombards arrived north of the Alps and influenced Germanic art. Pressed gold-leaf crosses, some imported, some imitated, which were found almost exclusively in southern Germany, testify to this influence. Lombard trends were also evident in the fibulas of the Princess of Wittislingen (V, PL. 79). Similarly the fibulas found in Soest, Westphalia, grave 106 (Münster, Landesheim), and those from Pfullingen and Schretzheim (V, PL. 62) must have been patterned after Lombard pieces, even though the animal heads of the Schretzheim piece, laid out as they are in a swastika pattern, betray Germanic origin.

With the rise of Islam in the 6th and 7th centuries the importation of almandines was stopped; and these stones were at first replaced by glass inlays and then in the course of the 7th century the "polychrome style" vanished entirely. The cloisonné or cell technique lived on for a long time in the arts of metal plating and inlay (also stemming from an antique tradition) which imitated it on brooches, buckles, and the like; it was generally replaced in the course of the 7th century by stones in high settings. The final stage of the style can be seen in the reliquary in St-Maurice d'Agaune, commissioned from Undiho and Ello by the presbyter Teuderique in the second half of the 7th century. In pieces that combine the cloisonné technique with crosses of encased stones there can be seen the transition to the Carolingian technique of setting large stones in the shape of a cross, as is found, for example, in the 8th-century Eugern reliquary (Berlin; Staatliche Mus.).

The flow of Eastern art that penetrated to the West and the consequent turning away from a realistic style is noticeable from the end of the 5th century in the representation of human figures and animals. It is not always clear to what extent the invading Germanic peoples were responsible for this stylistic change (see EUROPE, BARBARIAN). According to Aberg (1945), the Mediterranean element predominated; sculpture in the round disappeared completely and reliefs became increasingly flatter and more linear. All spatial effects and lively movements were lost. Figures became abstract, often geometric, and as symmetrical as possible. Toward the end of the 8th century the attempt at illusionistic representation was completely abandoned, the preference for toreutic work was striking, and stone and marble were used less and less frequently. At the same time artistic imagery became increasingly subject to the requiments of the growing power of the Church. Here the use of stone was generally limited to structural sculpture such as capitals, pulpits, choir screens, ciboria, and sarcophagi. Metal was used for the decoration of ciboria, antependia, altar crosses, candelabra, chalices, and patens. At the same time the increasing veneration of relics required the production of costly reliquaries. The history of early German medieval metalwork must be reconstructed from written sources, inasmuch as most works of this type have been lost, either through vandalism or through attempts at modernization in later centuries. Fortunately a mass of smaller objects have survived, such as brooches and buckles which, from the 6th century onward, were deposited in the so-called "line graves." Moreover the finds from the graves, which are partly datable by coins, illustrate the differences among the Germanic tribes. It is clear that the greater part of these objects were more or less mass-produced for popular consumption. This "folk art" is quite different from the art produced for the ruling classes, but it also reveals Germanic elements to a far greater degree. In producing it a stamping technique was commonly used. Thus by the 5th century medallions of the Roman emperors were copied in northern lands; there are the peculiar so-called "gold bracteates" (*Goldbrakteaten*), which were stamped pendants in which the classic models were restyled in an increasingly radical manner (V, PLS. 71–73). The amuletic character of these pieces is undeniable, particularly where they depict Christian motifs such as the adoration of the Magi or the knight-saint, but Germanic reinterpretations are not to be excluded, as in the metal plates from Strasbourg, Plieshausen, and Gültingen. These in turn must be related to such pieces as those from Braünlingen (V, PL. 95) decorated by means of perforated bronze plates. Beginning in the 8th century increasingly strict regulations of the Church almost brought an end to the practice of placing objects in tombs. Only a few pieces from highly qualified workshops survive.

The few pieces of large sculpture, such as the tombstones from the Frankish cemeteries of Niederdollendorf (V, PL. 77), Hornhausen (V, PL. 98), and Leutesdorf (Bonn, Rheinisches Landesmus.), are similar in the flat quality of the relief to the carved bone reliquary in Werden (ca. 750); they suggest Mediterranean, possibly Egyptian, prototypes. The primitive technique recalls wood carvings. Connections with the art of the British Isles may also be seen in such pieces as the stele with a Christ on the cross from Meselkern, near Kochem (Bonn, Rheinisches Landesmus.).

Thus, at about the beginning of the 8th century, in all areas of Merovingian art there are indications of a long development drawing to a close; the last influences of late Roman art together with currents of Eastern culture and the adaptation of new motifs from both the north and south of the continent as well as from the British Isles produced an art in central Europe with a unifying character found in the conscious surface distribution of the design and the will to a tectonic order whereby figural art is almost completely repressed.

Carolingians. In the year 751 the last of the Merovingian kings was banished to a monastery by his mayor of the palace, Pepin, who had himself anointed king of the Franks by the pope, thereby founding the Carolingian dynasty which was to become a world power under his son, Charlemagne. The latter, in close cooperation with the papacy, opened the way to the establishment of the Holy Roman Empire. On Dec. 25, 800, Charlemagne was crowned emperor in Rome by Pope Leo III. This act established his equality with the Byzantine emperor, and European unity and a Christian empire were created.

The political tendencies of the emperor were clearly reflected in the art that he sponsored and cultivated. Although in the independent monasteries the Merovingian style was still practiced, Charlemagne's desire for a renaissance of the classical greatness of the Roman Empire was asserted in the "New Palace School" and in the buildings and art works that he commissioned (see CAROLINGIAN PERIOD). This was expressed above all in his castles, such as the one at Ingelheim; in his churches, such as the Palatine Chapel in Aachen (III, PL. 52) and the Salvatorkirche in Paderborn; and in the miniatures and the ivory and gold objects produced in the "New Palace School."

Typical of the revival of early Christian elements was the preference for the undivided "Roman" transept before the apse, which, primarily derived from St. Peter's in Rome, occurred in as early a church as St-Denis (consecrated 775) and is found in the abbey church of Fulda (consecrated 819; III, PL. 44), where this feature led to the characteristic facing chancels of Carolingian architecture. It is this arrangement that appears clearly developed in the abbey churches of St. Gall (830–35), the Salvatorkirche (ca. 799) in Paderborn, St-Maurice d'Agaune, and Echternach. The connection between the large churches of eastern parts of the Frankish empire is very close, as a comparison of Corvey and Saint-Riquier (anc. Centula) will indicate (III, PL. 44).

Basilicas with three apses, such as those in Müstair (Münster) and Disentis, can be traced back to stronger Oriental influences. Foreign influences, however, are most apparent in the centrally planned Carolingian structures commonly used as baptisteries and memorial chapels and which are reminiscent of Eastern circular buildings such as those in Jerusalem. St. Michael (ca 822) in Fulda is typical of this sort of structure. Even the first large structures, such as the Palatine Chapel in Aachen (795–805), were centrally planned, as was the building (790–99) in Saint-Riquier. There is a clear relationship between the Cathedral in Aachen and such structures as S. Vitale in Ravenna. Charlemagne's veneration for Theodoric, whom he regarded as one of his political models, may possibly have played a part in this. The characteristically heavy *Westwerke* (westworks) of this period were also centrally planned. Like the Syrian churches of the early Christian era, Carolingian structures had massive towers over the entrance that enclosed a kind of ante-church. These westworks, which contained a special sanctuary above the entrance, exist in Lorsch, in the cathedrals of Halberstadt (consecrated 859), Hildesheim (consecrated 872), Corvey (873–85), and Minden (consecrated 952). Since in Aachen this floor served as the emperor's tribune, it is possibile that it had a similar function in the other related structures. The gateway of the monastery at Lorsch is clearly derived from St. Peter's in Rome. It is also reminiscent of a drawing preserved from the lost reliquary of Einhard (Paris, Bib. Nat.).

The increasing veneration of saints and relics also exerted an influence on the development of new structural forms, requiring as it did a greater number of altars. This led to the development of double and lateral apses, as well as to the building of crypts modeled on Roman prototypes. These include the ones in Regensburg (before 771), in the Abbey of Werden (before 830), in the Cathedral of Hildesheim (consecrated 872), and in Corvey (mid-9th cent.).

The degree to which the art of the Carolingian court strove to imitate the magnificence of pre-Christian architecture was evidenced by the mosaic decorations, now unfortunately lost, for the dome of the Cathedral in Aachen, which depicted Christ Triumphant and the 24 elders of the Apocalypse before their thrones. Technically the mosaic was probably related to the one in Germigny-des-Prés (III, PL. 53). Unfortunately very little of the German wall painting of this period has survived; the best surviving examples are the wall paintings in the monastery church of St. Johann, in Müstair (Münster), Switzerland (III, PL. 57). The paintings in this triple-apsed hall church (ca. 780–90) are reminiscent of those in Oberzell, Germany (X, PL. 463). The few remains of the vestibule of the monastery at Lorsch show even more clearly the influence of this classicizing court style. Important remains of the old painted decoration were discovered in the crypt of St. Maximin in Trier portraying a crucifixion, the evangelists, and saints (III, PL. 54). These frescoes, from the second half of the 9th century, reveal a close connection with the contemporary illuminations painted in the court style. Secular painting was also revived under the Carolingian rulers; the loss of the frescoes in the Castle of Ingelheim is therefore all the more regrettable.

To a certain extent the many illuminated manuscripts which have survived offer a substitute for the lost large-scale paintings, in that they also reveal the imprint of Charlemagne's taste and political aims. The manuscripts of the preceding Merovingian epoch, only sparingly decorated and then primarily with beautiful initials, were created in various monasteries such as Fleury, Corbie, Luxeuil-les-Bains, Laon, and Limoges and bore the stylistic marks of their respective schools. Although the schools independent of the court were able to maintain their characteristic styles until the Carolingian period, in the imperial workshops the styles underwent a basic change. The works commissioned by Charlemagne influenced the techniques of book decorations under his successors, each of whom chose according to his favor a different monastery for their production.

The clearest evidence of Charlemagne's humanistic tendencies is found in the elaborate, richly decorated manuscripts partly produced by the Rhenish schools of Mainz, Trier, and Aachen. Among these is the Godescalc Gospels (III, PL. 155), which was written in 781–83, possibly for Charlemagne's sister Ada, whose name has become associated with the entire group of manuscripts (the "Ada Group"). The Majestas Domini and the Evangelists, as well as similar manuscripts such as the so-called "Psalter of Dagulf" (Vienna, Nationalbib., Cod. 1861) written for Pope Adrian I or the Gospel Book of St. Médard of Soissons (Paris, Bib. Nat., Lat. 8850), clearly reveal their derivation from Greco-Italian models. A second group shows a stronger derivation from the illusionistic, painterly style of classical antiquity. This group, previously known as the "New Palace School," comprises such manuscripts as the Gospel Book in Aachen (PL. 314) and the one in the Vienna Schatzkammer, as well as the famous Utrecht Psalter (III, PL. 58), which may have been produced in Reims. Works like these gradually shifted the center of influence increasingly toward the imperial monasteries in the west such as Metz, Tours, and Corbie. There was close contact between them and the schools of southern Germany and the Rhine which began to flourish at that time. Here strong traces of the influence of, for instance, St-Denis and of Tours are visible, even though there was at the same time a strong backward pull to the art of Charlemagne's time. In these Rhineland scriptoria, particularly in Cologne, western influences met with those from the British Isles in the early period of the Franco-Saxon group. These monasteries — Prüm, the family monastery of the Carolingians, its neighbor, Echternach, and Cologne, where the Monastery of S. Pantaleon became an influential art center — were closely associated with the imperial house.

After the division of the empire in 843 (Treaty of Verdun), these schools all became increasingly autonomous. From Cologne and Mainz the church organization extended its rule to conquered Saxony, and shortly established new schools of art, particularly in the newly founded dioceses of Paderborn, Verden, and Minden, and later in Hildesheim. The important school of Fulda developed almost independently of Cologne. Several of its excellent manuscripts are preserved in Würzburg (Up. theol. lat. 66) and in Erlangen (Univ. Lib. 9). Here in the second quarter of the 9th century Rhabanus Maurus, the great humanist and pedagogue, was active. A pupil of Alcuin, he was also a great patron of the arts, whose strong humanistic leanings are clearly revealed in the manuscript *De laudibus s. Crucis* (Vienna, Cod. Vindob. 652), which he presented to Archbishop Otgar of Mainz and to the pope, and which can therefore be dated between 831 and 840.

Even though in the art of this period the sense of space gradually began to vanish, the influence of the imperial "Ada Group" was still felt. Würzburg was closely connected with Fulda, whereas the other south German dioceses formed a separate historically and culturally unified group of their own. St. Gall, however, remained independent; because of its geographical position the monastery school was somewhat influenced by the British Isles, but under the rule of Abbot Grimald (d. 872) it came under Frankish influence, as can be seen in one of its most precious manuscripts, the Psalterium Aureum (St. Gall, Stiftsbib. Cod. 22). The diocese of Salzburg also shows signs of having been in close contact with the Anglo-Saxon world; artistic creation there, particularly under the Irish monk Virgilius (743–84), shows an exceptionally strong British character in works such as the Codex Millenarius of Kremsmünster, as well as in related goldsmith's work, particularly the Tassilo Chalice (I, PL. 287).

The political chaos which was result of the Norman and Hungarian invasions, together with the weakening political power of the eastern Frankish kingdom, led to a regression in artistic creation as well. With the almost contemporaneous extinction of the eastern Frankish Carolingian dynasty (911), this epoch came to its political and artistic end. With the spread of Christianity the custom of depositing objects in graves fell into disuse; hence it is difficult to form a picture of folk art in Carolingian times; knowledge is limited to the few religious sculptures and handicrafts that have survived in church treasuries. Compared to the original richness mentioned everywhere in the written sources only a few remains of these magnificent works of art are known. Almost all the large cathedrals possessed huge treasuries of gifts from sovereigns and prince-bishops. These were increased by the victorious campaigns of Charlemagne (e.g., against the Avars). Parts of this booty were presented by the emperor to the pope and to ecclesiastical institutions. In addition, friendly relations with the powerful rulers of Islam and the Byzantine court also brought numerous works of art to the Frankish empire and the Carolingian court, partly by the sea route, partly by way of the "Baltic Bridge" in the north. These foreign works in gold, glass, and textiles in their turn influenced the Frankish workshops, just as Scottish missionary activities had brought artistic currents from the British Isles not only to the Salzburg area but also up the Rhine as far as St. Gall and into Frankish territory. In comparison, the work of the secular workshops, often mass-produced, remained more bound to the Germanic tradition. Some of these pieces — swords, spurs, strap buckles, and spangles — have been preserved in Scandinavian countries from Norman booty (e.g. the gold buckle from Hon in Norway; XII, PL. 412).

During the Carolingian period there was, moreover, a new stylistic development of forms and objects. The evolution of the liturgy, the exigencies of ritual, and the wealth of the patrons and founders influenced the designs of diverse utensils, whose decoration became richer. Chalices became larger and more magnificent; the covers of the sacred scriptures were richly encrusted with gold, ivory, and precious stones. Although the ivory thrones disappeared they were replaced, in imitation of classical antiquity, with metal thrones (III, PL. 68); holy water stoups were made of metal and ivory and there were richly decorated antependia and costly ornaments of tombs, cabinets, and ciboria. Rich altar crosses are often mentioned. Sculptures in the round, however, were rare, appearing again at the end of the 10th century, as a result of the worship of relics, to serve as reliquaries. Enamel technique was in widespread use and the old techniques of metal filigree and jewel-setting remained in use and were perfected; but silver overlay became rare, and inlay techniques disappeared completely. Simultaneously the currents in Germanic art mentioned above continued to develop the Merovingian style until the end of the 8th century, even though they were strongly influenced by the works of the Hiberno-Saxon provinces, as shown by the Tassilo Chalice

(I, PL. 287), the purse reliquary of Enger (PL. 313), and the older cover of the book of Lindau (I, PL. 287). In a further development of the Nordic zoomorphic style, the full form of the animal appears in this later stage, as in the purse reliquary in the Cathedral Treasury of Chur, which again reveals its derivation from British models.

Concurrently with the products of the group working in the Merovingian tradition there appeared, at the end of the 8th century, artistic trends that consciously aimed at a revival of the art of classical antiquity, even attempting to copy it. Particularly important in this category are the ivory reliefs, among them two pyxides (Vienna, Kunsthist. Mus.; Br. Mus.) and a diptych from Palermo (London, Vict. and Alb.). As in illumination, early Christian examples were taken as models but were freely adapted. The beautiful panel in Oxford (Bodleian Lib.), also derived from early Christian models, was copied from the lateral panels of an early-5th-century five-part diptych (Berlin, Staat. Mus.; Louvre). The covers of the so-called "Psalter of Dagulf" (PL. 313) were also derived from western works of the 5th century. On the other hand, the St. Michael in Leipzig and the two panels from Lorsch (III, PL.66; London, Vict. and Alb.) probably were done after 6th-century Byzantine models. This would certainly indicate that at that time the church treasuries in France and on the Rhine still contained a great number of late-classical and early Christian wood carvings that could be used as models. It has not yet been possible to determine from which workshops these Carolingian ivory carvings came, although it seems certain that they were created at the same time as the corresponding miniatures. Trier, Aachen, and Lorsch may have possessed such ivory carvings; these workshops, in their turn, were closely connected with the western court and such monastery schools as Metz, Tours, Reims, Corbie, and St-Denis.

The so-called "Carolingian renaissance" by no means ended with the death of Charlemagne in 814; it is possible to follow the attempt to achieve a balance between classical elements and a contemporary style until the 9th century. Moreover, the great humanists who had formed part of the emperor's court tried even after his death not only to keep these tendencies alive but to further their development. Thus under Louis the Pious, Louis II, called "the German," and Lothair I, artistic activity continued essentially unaltered and on the same level. It is improbable that a workshop of any significance was maintained at the imperial court under these rulers, but the style of Charlemagne's "New Palace School" spread increasingly among the smaller art centers, with the powerful bishops and the rich Benedictine monasteries exerting increasing influence on the arts. With the waning of strong central power these small centers again began to become artistically active, as for example in Metz, under the powerful Bishop Drogo (826–55), who was a noble closely related to the imperial house. In Metz it is possible to observe the close stylistic relationship between ivory work and miniatures, as shown by comparing the Drogo Sacramentary (III, PL. 62) with the miniatures of this school.

Ottonians. By the close of the Carolingian epoch the German tribes had been fused into an empire. After 911 and the end of the terrible wars, the empire experienced, under the Saxon emperors of the Ottonian dynasty, a renaissance that reached one of the high points of medieval art around the year 1000. In contrast to Charlemagne's concept of empire, which still remained in the minds of the new rulers as an unachieved ideal, the princes and the bishops (who were generally drawn from the palatine chapel) occupied a dominating position alongside that of the emperor in both political life and its related artistic activity. This led to continual tensions; however, during this period Germany received its basic ecclesiastical structure, which also determined the course of its artistic activity. Under a number of princes of the church, the court style assumed strong individualistic traits. Thus Egbert in Trier (977–93), Willigis in Mainz (975), Bruno and Gero in Cologne, Meinwerk in Paderborn, Bernward in Hildesheim, Notker in Liège, and Ulrich in Augsburg were active not only as great builders but also as founders of workshops that produced magnificent examples of sculpture, painting, and the goldsmith's art. Even though a close relationship existed between the different successful schools, the individual character of each of these prince-bishops left special imprints on the products of their workshops, in which local regional character played no small part. The center of the empire, now in the process of closer unification, shifted toward Magdeburg and Quedlinburg, in the dynasty's native territory in the Harz district. The strong political bonds to Byzantium already indicated by the marriage of Otto II to Theophano, daughter of the Eastern emperor Romanus II, also found expression in art. An example of this is the ivory relief in the Musée de Cluny, Paris. It portrays Christ crowning Otto II and his wife and is an imitation of a Byzantine panel showing the coronation of the Emperor Romanus and Eudokia (Paris, Cabinet des Médailles). At the same time a great quantity of precious Byzantine work arrived in Germany: valuable silks such as the "elephant cloth" from the tomb of Charlemagne in Aachen, the so-called "Cloth of Günther" in Bamberg, and the *Willigiskasel* in Mainz, ivory carvings such as the death of the Virgin Mary on the cover of the Gospel Book of Otto III (Munich, Bayerische Staatsbib., Clm. 4453) and the diptych from the Prayer Book of Henry II in Bamberg (Staat. Bib.); and enamel work such as that on the cover of the Book of Pericopes of Henry II (Munich, Bayerische Staatsbib.). Admiration for Byzantine culture, in spite of any political antagonism, is evident in many works — in such Franconian ivory carvings as the Martyrdom of St. Kilian on a book cover in Würzburg (Universitätsbib.); in the cast-bronze seven-armed candelabrum (PL. 312) and the enamel work in the Cathedral Treasury in Essen, and in the various crosses of Duke Otto (d. 982) and his sister, the Abbess Mathilde of Essen (973–1011; X, PL. 470).

The important works of large painting such as the Pantokrator in the Church of Frauenchiemsee (Bavaria) are difficult to explain without taking into account Byzantine influence. The style, however, of the largest preserved cycle, with its scenes depicting Christ's miracles (Church of St. Georg, Oberzell; X, PL. 463), can more readily be understood as deriving from the Carolingian schools of St. Gall and Reichenau Island, from whence its obvious classical influences would have emanated. The same is true of the development of the superb Ottonian book illumination, which overcame the eclecticism of the imperial Carolingian workshops and developed an independent style despite the fact that they both derived from the same models. This truly new "chivalrous" style can be found in a Gospel Book from Quedlinburg (New York, Pierpont Morgan Lib., Ms. 775). It is characteristic of the manuscripts in this group that they were produced in the vicinity of the Monastery of Corvey, in the native territory of the new imperial house. Illumination in Fulda developed within an even more intense relationship with Carolingian painting. Early works of this school retain a connection to book painting from the Weser district, as the illumination from the legends of St. Kilian and St. Margaret (Hannover, M 189), dating from the last quarter of the 10th century. The illumination from the Wittechindeus Codex of about 970–80 (Berlin, Staatsbib. Ms. theol. lat.) represent the high point of this school. Most of these scriptoria were sponsored by prince-bishops such as Egbert of Trier, Bernward of Hildesheim, and Willigis of Mainz. But even at this period the large, elaborated manuscripts owe their origin to the munificence of the emperor, which made possible the superb Reichenau Codices; the Gospel Book of Otto III (X, PL. 462; XIII, PL. 347), the Book of Pericopes of Henry II, (PL. 314; VIII, PL. 97; X, PL. 72), and the Bamberg Apocalypse (IV, PL. 467). Closely related to the Reichenau works are the illuminations of the Codex Egberti (Trier, Stadtbib., Cod. 24), the dedicatory page of which shows the masters Keraldus and Heribertus, "Augigenses," alongside Bishop Egbert. The master of this codex has much in common with one of the most significant painters of the epoch, namely the Master of the Registrum Gregorii, of which a detached leaf shows various lands paying homage to Otto III (X, PL. 461). Also in his style is the Ste-Chapelle Gospel Book (ca. 1014) with its portrayal of Henry II (Paris, Bib. Nat., Ms. lat. 8851). Here the Ottonian chivalrous style is displayed at its peak of perfection. This great master was certainly inspired by the Carolingian "New Palace School," but he also received a stimulus from classical and Byzantine art. He probably was active in Trier. Yet to be investigated are certain ivory carvings of the same period, such as the Baptism of Christ on a Gospel Book from Bamberg (Munich, Bayerische Staatsbib., Clm. 4451) and a Madonna Enthroned (Mainz, Altertumsmus.) and to what degree they may be related to the schools of Trier or Mainz.

The precious manuscripts of the Upper German School, particularly those from Regensburg and Salzburg, also owe their existence to imperial patronage. Thus the Sacramentary of Henry II (VII, PL. 266) was created in Regensburg as was the Uta Codex (X, PL. 472), but the Salzburg school first received its great new impetus under Bishop Friedrich (954–99).

Artistically, the most significant creations in the field of gold- and silverwork were the imperial gifts, such as the golden altarpiece commissioned by Henry II for the Cathedral of Basel (VI, PL. 263) and the golden antependium in Aachen (X, PL. 473), which recalls the Carolingian antependium of S. Ambrogio, in Milan (III, PL. 63). The art of Bernward of Hildesheim's court also has strongly Italian aspects, as seen in the reliefs of the bronze door of the Cathedral (X, PL. 467) and in the Paschal candlestick in the form of a column in the Cathedral. Without models from early Christian Rome these works would not have been possible. The bronze doors of Augsburg Cathedral (IX, PL. 508), although very different in theme and arrangement from the one in Hildesheim, and the wooden door of St. Maria im Kapitol in Cologne recall 5th-century Roman models such as the wooden door of S. Sabina in Rome. The large wooden crucifix in Cologne, commissioned by Archbishop Gero (969–76; X, PL. 466), was very influential on works of the 11th century, when Ottonian art achieved a unity and individual style in works of great beauty.

The development of the arts under the Ottos is also clearly visible in architecture and can be traced from its origins in the Carolingian epoch to the monumental creations of the Salic-Romanesque period. A change in style became clearly evident after 919. With the eastward advance of the German emperors and the missionary work undertaken in these newly conquered territories, Ottonian architecture spread as far as Bohemia and southward to the Alps. At the same time, however, the border line between the Ottonian style and early Romanesque art is often undefined. A clear change is recognizable in the sculpture of about 1050. In architecture as well as in painting and sculpture, the will of the princely patrons was predominant and the artist was of secondary importance. The development of new structural forms was primarily determined by liturgical requirements. That this art was still using Carolingian models can be seen in the centrally-planned structure of the Münster in Essen (PL. 312), which is clearly derived from the one in Aachen. The columned basilica, such as the one in Magdeburg (begun ca. 955) with its three aisles, was also derived from the preceding epoch; but soon, as in the Cathedral of Mainz (975–1009), transepts began to be wider. A variety of reasons can be given for the ever-increasing number of elaborate and important structures; primary among them must be the needs of the powerful princes of the church to display their material might in order to counterbalance the central power of the emperor. At the same time the reformist ideas emanating from the Monastery of Cluny since the beginning of the 10th century began to exert a strong influence; the church began to demonstrate its power in opposition to the secular imperium in a prelude to the great struggle for the right of lay investiture that was to shake the empire.

W. F. VOLBACH

ENGLAND. English art of the period between 900 and 1100 must be examined against its historical background. The 30 years following the death of Alfred the Great (899) were filled with great political and military struggles with the viking settlers of north and west Britain. About the middle of the 10th century peace came to England and, with the accession of Edgar in 959, "things," in the words of the Anglo-Saxon Chronicle, "improved greatly." The turn of the 10th century found the viking raids renewed and, in 1017, a viking king, Canute, came to the throne of England. By 1066, when England was finally conquered by the Normans, the whole administrative organization was on a sound footing and French influence was very strong. The greatest achievements of English art in this period were accomplished during the reign of Edgar. The monastic reform of St. Dunstan, himself a great craftsman, was associated with a flowering of English art unparalleled since the early 8th century.

Opus Anglicanum. In the south of England in the first half of the 10th century various influences were at work in the art of the painter and embroiderer. English embroidery of this period was famous throughout Europe; William the Conqueror, in the middle of the 11th century, returned to France from England and it is recorded that the Normans were astonished by the richness and the quality of the robes he had acquired in England. Later, in the 12th century, opus Anglicanum was to become famous throughout Europe. Unfortunately, few pieces of embroidery survive from this period. A fragment in S. Ambrogio in Milan may possibly be of English workmanship, but there is one series of embroideries, of undoubted English workmanship — the embroidered stole and maniple ordered by Queen Aethelfloed and placed in the tomb of St. Cuthbert at Durham (I, PL. 277). An inscription on the stole demonstrates that it was executed, probably at Winchester, between 909 and 916. The stole of St. Cuthbert bore embroidered representations of the standing figures of 16 prophets set on either side of a central Agnus Dei and with the busts of St. Thomas and St. James at either end. A similar arrangement, with a variation of identity of the portrayed subjects, occurs on the maniple. The figures are of elongated proportions and wear loosely draped garments. The heads are nimbed, and both bearded and unbearded faces occur, executed with remarkable sensitivity. The feet are set at an angle to each other and the stance is well-balanced and naturalistic. The figures are separated from each other by stiffly formal acanthus leaves. The whole effect of the embroidery is extremely rich, with the figures, in (now faded) blues, greens, and browns set against a background of gold.

An analysis of the ornament of these textiles shows that its roots must be in the 9th-century Carolingian school of art; the figural style has been traced back to Byzantine sources. Whatever their origin, the ornamental details of this group of textiles demonstrate the beginnings of an art that apparently reached its zenith in the middle of the 10th century, with the manuscript known as the Benedictional of St. Aethelwold. The art of this school takes its name from the town, Winchester, in which St. Cuthbert's stole was made.

Miniatures and illumination. Few manuscripts survive that can be dated with any certainty to the early 10th century. The embroideries found in the coffin of St. Cuthbert hint at a richness that is not found in surviving manuscripts. One of the early 10th-century manuscripts is the Durham Ritual (Durham, Cathedral Lib., A.IV,19), which combines, with its confidently drawn animal heads of southern English origin, an acanthus leaf pattern that has Carolingian roots. Most of the manuscripts that survive from the early 10th century, however, are poverty-stricken imitations of 9th-century English manuscripts. In 937 King Athelstan prsented to Durham Cathedral a copy of Bede's *Vita Cuthberti* (Cambridge, Corpus Christi College, Ms. 183) that was probably painted in the south of England, at Winchester or Glastonbury, under Frankish influence. The paneled border of the frontispiece of this book contains acanthus leaf scrolls confined, in the traditional Anglo-Saxon manner, within a series of panels. The scrolls, of Continental inspiration, contain animals and birds which, without being in any way stylized, have a strained, taut appearance that might suggest a continuity from the Anglo-Saxon tradition. The figure style of this manuscript is stiff and formal, lacking in vitality and dull in color; it must be seen as a poor attempt to imitate rich Continental prototypes. The initials, however, are more lively and foreshadow more nearly the final Winchester style.

A manuscript combining both English and Continental influences is the Junius Psalter (Oxford, Bodleian Lib., Ms. Junius 27): it contains 152 initials of finely drawn and boldly colored acanthus scrolls combined with winged dragons and ribbon interlace in a style entirely different from that of the Bede. The acanthus, freed from its confining borders, is used in a much freer manner and is at times given animal heads instead of leaves, the heads being derived from such 9th-century manuscripts as the Rome Gospels (I, PL. 275) and the Canterbury Gospels (I, PLS. 276, 279). Another manuscript, probably by the same hand as the Junius Psalter, is the Helmingham Hall *Orosius*, which is dated by Wormald to the second half of the 10th century. Such manuscripts as these demonstrate in their illumination and decoration the beginnings of the style known as Winchester.

The first surviving manuscript showing the new style in its fully developed form is the charter given to the New Minster at Winchester on its foundation by King Edgar in 966. The text of this manuscript is in gold, and there are three ornamental pages at the beginning as well as a principal illuminated page (on purple vellum) showing the King, between the Virgin Mary and St. Peter, offering the charter to Christ, who sits in a mandorla attended by four angels (PL. 319). The well-proportioned figures are drawn with incisive freedom. The faces are fairly coarse, but there is a grace and movement in the design of the page that shows a complete mastery of the Carolingian idiom of, for instance, the Metz group of ivories. An important feature of this page is the heavy formal border consisting of a double, plain, bandlike frame entwined by a rather ponderous acanthus ornament that seems at odds with the floating quality of the figures it encloses.

The heavy border occurs again, with even greater richness, in the highest achievement of Anglo-Saxon art of the 10th century — the Benedictional of St. Aethelwold (Br. Mus., Ms. Add. 49598), written and illuminated at the command of the saint between 975 and 980; here the importance of the borders is revealed by the mention of *circos multos* in the dedicatory inscription. These borders are so important that the pages must be seen, as Kitzinger has pointed out, not as individual portraits but as large and sumptuous openwork ornament. And yet, although the borders dominate each page of the manuscript, they do not overbalance the constructional equilibrium of the whole. The artist achieved a high vitality in his human portrayal; the Annunciation scene, for instance, shows an angel with swirling draperies in a light and naturalistic stance, and the faces of the figures in the crowd scenes are full of expression and have lost the coarseness noted in the New Minster Charter. Where the artist had to portray a single figure, however, the result was sometimes stiff and occasionally lifeless; in his drawing of St. Benedict he has lost all movement and is, interestingly enough, nearer in style to the figure portrayal of the 8th century than to that of his own era. The coloring of the manuscript is extravagant in the extreme; pastel tints are blended and contrasted with rich purples, gold, greens, and blues. The corners of the borders of many of the pages are filled with thin, rather complicated acanthus scrolls, somewhat heavy and muddled in effect.

This sumptuous, fleshy painting of the Winchester school continues with little development into the 11th century. The Missal of Robert of Jumièges (Rouen, Bib. Municipale, Ms. Y. 6), which is dated between 1013 and 1017, and the Grimbald Gospels (Br. Mus., Ms. Add. 34890), although less florid, are directly in succession from the rich Winchester manuscripts. The tradition of the heavy English border can still be seen in such late-11th-century manuscripts as the Arundel Psalter (PL. 315; VI, PL. 429).

The decorated initials that form so important a part of the ornament of the manuscripts are found in two forms. The first is typified in the Vercelli Codex (Oxford, Bodleian Library), in which the initials are made up of creatures with complete bodies; the second is typified in another manuscript in the Bodleian Library (Digby 146) in which only the heads of the animals occur. In the 11th century these two styles were producing a result that might be recognized as early Romanesque (e.g., Cambridge Univ. Lib., Ms. Ff. 1, 23). The second type can still be seen, however, at the end of the 11th century (Durham, Cathedral Lib., Ms. B.II,13).

The term "Winchester School," which is applied to the art of those manuscripts of the 10th and early 11th centuries produced under the influence of the monastic revival inspired by St. Oswald of Worcester, St. Dunstan of Glastonbury, St. Aethelwold of Winchester, and others, is but a convenient label. Certain of the manuscripts of this school, such as the New Minster Charter and the Benedictional of Robert of Jumièges (Rouen, Bib. Municipale, Ms. Y.7), were certainly produced at Winchester; but other books were produced at Canterbury and in the great monasteries of East Anglia, and the Winchester style must be seen as a style common to the Reformed English monasteries of the 10th and 11th centuries.

The figure drawing of the period is very much influenced by the art of the Utrecht Psalter and the Reims School. The 9th-century Utrecht Psalter (III, PL. 58; IV, PL. 231) has a figure style that is Byzantine in origin; it is one of the few survivors of a group of contemporary manuscripts written on the Continent (in this case at Hautevillers, near Reims) which are of the utmost importance for the history of later Anglo-Saxon drawing. That it is difficult to underestimate the importance of the influence of this style on 10th- and 11th-century English art is demonstrated by the copy (Br. Mus., Harley Ms. 603) that was made in southern England (probably at St. Augustine's, Canterbury) about 1000, which reproduces all the breathless activity of the original. The style of illustration is impressionistic, the figures have a tendency to be spindly, have protruding eyes, and are hunchbacked and slightly twisted; the clothes are elaborately swirled, with simple strokes of the pen; rocks, hills, and clouds are sketched in quickly and lightly. The impression of movement and life is accurately translated, in a slightly more linear style, from the original. This sense of movement is lacking in the richer costly books, such as the Benedictional of St. Aethelwold; it is completely absent, for example, in the portrait of St. Benedict. Hints of the style do, however, occur in the rather lifeless portrait of St. Etheldrythe and in some of the crowd scenes. The pure Utrecht style can be seen in three late-10th-century manuscripts: a Psalter from one of the Fen monasteries (Br. Mus., Harley Ms. 2904), with a crucifixion scene in which Mary and John are shown standing on hills of a sketchy, cloudlike texture that very definitely belong to this tradition, and two others (Br. Mus., Harley Ms. 2506; Orléans, Bib. Municipale, Ms. 45). The same qualities are to be seen within the typical heavy Winchester border of the Benedictional of Robert of Jumièges, and are even more evident in a Gospel Book in New York (Pierpont Morgan Lib., Ms. 709), although in this the figures are more graceful and controlled in their sense of movement. One of the most noticeable features of this style of drawing is the treatment of the draperies, an airy rendition that occurs throughout the corpus of manuscripts after the middle of the 10th century. Toward the end of the 11th century the Utrecht style, while still influencing English drawing, is developed into something specifically English and more accomplished. The figures become larger and more angular and the flickering, quick strokes of the pen are reduced to more successful, stylized pattern forms (as in the secondary hands in Br. Mus., Harley Ms. 603).

That the Anglo-Saxon artist of the period still loved the grotesque can be seen from portrayals such as the "Mouth of Hell" (Br. Mus., Cotton Ms. Tiberius B.V, and Stowe Ms. 944). Extreme insularity, which had its roots in Northumbrian and Celtic art of a much earlier period, is seen alongside a formalized Utrecht style in a manuscript from about 1050 (Br. Mus., Cotton Ms. Caligula A.XIV). The Arundel Psalter (PL. 315; VI, PL. 429), which was written about 1060, combines the conventional Winchester-style borders and initials with an emaciated tautness and formality of figure drawing completely foreign to the Anglo-Saxon school of the period. The influences in this manuscript are Continental in origin; both German and French parallels have been drawn with their style. These influences are to be seen again in the ornament of the manuscripts written in Durham under Bishop William Carileph (d. 1096).

Sculpture. The pre-Romanesque stone sculpture of 10th- and 11th-century England is dependent on two traditions: the old Anglo-Saxon sculptural tradition, which can be compared with the Winchester style, and the Northumbrian tradition of formalized patterns of interlace and vine scroll. Examples of the former are few and far between, but examples of the latter are numerous and often seem to be, in the words of Kendrick, a vast and dreary assemblage of carvings that are either of indifferent quality or downright bad. No thorough examination has been undertaken of the sculptural style of England at this period and it is impossible to give more than a rough review of the subject within the present critical apparatus.

In the West Saxon area only about 20 fragments of stone sculpture survive. The angels from Bradford-on-Avon, the Bristol Cathedral "Harrowing of Hell," and the Inglesham Virgin and Child are the most famous examples. The clothes are stiffly draped about the bodies in the manner of the manuscripts, and the faces, though generally rather coarsely modeled, are not without distinction; there is little attempt at modeling the body, the details being executed in finely drawn lines. Where crucifixes occur, the figure of Christ is elongated with thin, outstretched, slightly arched arms and straight legs in low relief and linear style.

There are a large number of stone carvings in the north of England. The sculptures mainly comprise memorial and preaching crosses and fragments of such objects. They are crude, but not without interest in that they reflect the survival of the Hiberno-Saxon styles of the previous centuries. The paneled style (in which each panel contains a figure subject, scroll, or interlace) and the developed vine scroll, which are the chief elements of this ornament, are but pale imitations of the sculpture of the Northumbrian golden age. The viking settlement of this area was responsible for the decline in standards of execution; this is reflected in certain aspects of figural representation — as, for example, those on the cross from Halton, Lancashire. This cross bears a representation of certain passages of the Scandinavian Sigurd Saga (scenes from the same cycle also occur on the Isle of Man, then a Norse kingdom, at the same period). The style of the carvings is strictly linear and there is never any attempt at carving in the round. The figures are representational and naïve and rarely have any great charm. In the carvings of these crosses can be seen the beginnings of the Jelling style and the major styles of Scandinavia. It is agreed that England took an important part in the development of this art, and the area of 5 miles round Normanby, Yorkshire, has produced a number of sculptured stones that are representative of the earliest period of development of this Anglo-Scandinavian style. The stones of this area bear representations of animals with a double contour and interlaced limbs and tails and must be dated to the first half of the 10th century. The ornament on monuments such as these, with the art of the more florid portions of the Winchester manuscripts, must be one of the chief elements in the developments of the Jelling style. It is from these sources also, with the Scandinavian developments of the style, that the English expressions of the Scandinavian Ringerike and Urnes styles are developed (but see below).

The crosses of Mercia and eastern England continue in the same dull tradition as that of the north. In Mercia the square-sectioned cross shafts are replaced in certain places by round shafts. This series is centered on Macclesfield, but the most famous example is the cross from Gosforth, Cumberland, which bears the familiar Northumbrian interlace together with scenes from both Norse and Christian mythology. These tall, round shafts are not without grace, but the other crosses of the same area, as well as of the Midlands and East Anglia, are dull and poverty-stricken in their ornament. Occasionally the competence of the carver reaches a higher plane, as at Norbury, but these instances are rare. In the extreme southeast of England few examples of sculpture survive, but fragments of the shaft of a cross from All Hallows, Barking, London, demonstrate that certain sculptors were using a paneled style akin to that employed in the north of England.

The advent of Scandinavian influence in southern England is demonstrated by a number of stones, of which the two best examples come from the old churchyard of St. Paul's Cathedral, London. The first example (London, Guildhall Mus.), although in many ways comparable with the Jelling style, is in effect a lighter treatment of a form that was popular during the period when the Jelling style flourished. The style is a modified version of the Ringerike style of Scandinavia with certain features following in the tradition of such initials as those portrayed in a manuscript in Cambridge (Univ. Lib., Ff 1, 23). The animal has great clawlike paws, with foliated tail and ears, and is surrounded by typical Ringerike motifs. The foliate characteristics can also be seen on the other stone from the same place (Br. Mus.). Similar designs occur on metalwork and in a manuscript in Oxford (Bodleian Lib., Ms. Junius 11).

Minor Arts. The remains of the minor arts, metalwork, pottery, and ivory carving of the period are scarce. The great period of Anglo-Saxon ornamental metalwork was past. The artist at the beginning of the century was struggling along familiar lines to produce finished products of the kind made by his forebears. The struggle is reflected in the metalwork on such pieces as the pair of plates from a house-shaped casket in the British Museum and on a disk brooch found at the King's School, Canterbury. The casket plates are of silver, carved and inlaid with niello, in the technique of the 9th-century Trewhiddle style; as in the Trewhiddle style, one of the plates is divided by billeted lines into triangular fields filled with speckled animals. The clumsily drawn animals are an attempt to develop or imitate the Jelling and Ringerike styles in such features as the double contour of the body, the florid aspect of the head, and the encircling and interlacing rings that are standard features of these two styles. On these pieces the northern art styles and the traditional Anglo-Saxon theme meet. On the Canterbury brooch gold plates let into the silver base are decorated with Ottonian-style filigree ornament, adding yet another influence to the assemblage noted on the casket plates. Meanwhile, in northern England and Ireland metalworkers were producing silver penannular brooches which show a continuity in form from the Hiberno-Saxon brooches of the 9th century. The same continuity is to be seen in their ornament and in the method of division of the fields; these brooches are not, as has been stated by Coffey and Shetelig, of Norse origin in their decoration. The brooches are firmly dated to the early part of the 10th century by their occurrence in closely dated coin hoards.

In the minor arts the classic Winchester style is found on a small group of objects which includes two ivories. The first is a triangular plate from Winchester (Winchester, City Mus.) bearing two angels with censers in high relief. The bodies are well modeled and the flowing garments are executed most competently, in a successful translation of the art of the manuscript to that of carving. The second is an ivory seal of Godwin and Golgytha (Br. Mus.) which can be dated on parallels with coins to about 1000. The first two Persons of the Trinity, in seated position, are carved on the handle in a graceful, but more sculptural, manner than is displayed by the Winchester piece. Other pieces of ivory of the same period, and of the same accomplished style, also exist (they include a crucifix in London, Vict. and Alb.), but are not very common.

A number of minor antiquities serve to demonstrate the popularity of the 10th-century inhabited vine scroll. The finest object, is a pen case of ivory (Br. Mus.) and a group of small bronzes, including some strap ends, two censer covers, and a cruet, have similar ornament based on animals and acanthus leaves. Few metal objects survive from the 11th century, but among them are a group of engraved bronze plates in the Ringerike style and one or two strap ends and openwork mounts in the Scandinavian Urnes style. In this latter style is the crozier head of Bishop Rannulf Flambard of Durham (d. 1128). The ornament of this object, although in the direct tradition of the Viking Urnes ornament, contains a number of details (e.g., a lightly drawn terminal knot) that are not present in the Scandinavian varieties of the ornament.

Architecture. Although many of the ornamental details of Anglo-Saxon architecture are to be seen in Romanesque contexts, there is still a wide gulf between vernacular and imported architecture in England. The appearance of a great church of the 10th century can be drawn from the description of the Minster at Ripon, dedicated in 980. From this it appears that the church had an arcaded atrium and a great tower, resembling the staged towers of the Continental churches (e.g., Saint-Riquier; III, PL. 44). "The lofty peaks of the tower," says the poet, "are capped with pointed roofs and are adorned with various and sinuous vaults . . . above these stands a rod with golden balls and at the top a mighty cock." Unfortunately, no complete example of a great church of this period has survived, but from descriptions such as this and the remaining evidence it can be seen that the ecclesiastical architecture of England at this period was a direct offshoot from the Continental stem.

The complete ground plan of the Cathedral at North Elmham survives and shows a T-shaped church with nave, a long transept, a shallow, semicircular apse, and a western tower; this was, however, a small church, being only about 40 yards in length. The plan of this church shows a reversion to the plan of the Constantinian age, which also occurred on the Continent (e.g., Saint-Denis, Fulda); the main tradition of the period, however, was the church of double-cross plan.

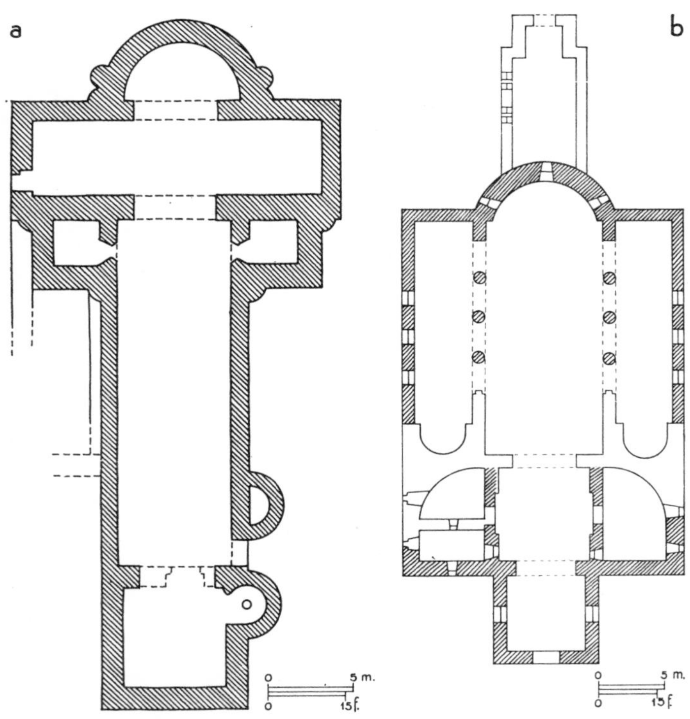

England and Germany. (*a*) North Elmham, church, plan (*from G. Webb, 1956*); (*b*) Reichenau-Oberzell, St. Georg, plan (*from E. Gall, Dome und Klosterkirchen am Rhein, Munich, 1956*).

In the small church of the period the cruciform plan began to come into fashion: the side porticoes developed into transepts, which reached their full size in the 11th century. Apses were rare, and the chancel commonly had a squared end. The towers (as at Barnack and Earl's Barton in Northants.), placed at the western end, were massive and often decorated with pilaster strips. Notable architectural features of the period are the "long-and-short" work on the corners of the building (upright stones alternating with thin stones laid flat on their natural bed) and the megalithic method of building quoins from large stones set lengthwise, alternately to one side and another of the angles of the building. Although many doorheads are round, triangular arches are often found (e.g., Colchester, Essex, Barton-on-Humber, Lincolnshire); window openings are narrow, and the single splay has given way to the double splay, where the narrowest part of the opening is in the middle of the wall. In the 11th century a central tower was a common feature and the tympanum made its appearance. The general architectural trend was much closer to the contemporary Norman Romanesque of France.

David M. Wilson

SPAIN. *Visigothic period.* During the Visigothic domination of the Iberian Peninsula a series of artistic expressions of Roman-Byzantine derivation developed. In studying the art of this period, it should be noted that there already existed in Spain (particularly in the south and in the Levant) a flourishing early Christian culture that provided the foundation for that Hispano-Roman art known as Visigothic. The Visigoths consolidated their dominion of the Iberian Peninsula near the beginning of the 6th century, particularly after the Battle of Vouillé (507); but it was not until the Third Council of Toledo (589), in the time of Recared, that the Hispano-Roman and Visigothic populations coalesced. By this time the style known as Visigothic had existed for almost a century.

Among the basilicas preceding or contemporaneous with the Visigothic invasion are those in the Balearic Islands — Son Peretó, Puerto de Manacor, and Santa María, in Majorca; Son Bou, Es Fornás de Torelló, and Fornells, in Minorca — which, with the cella memoriae in Ampurias (Gerona), form a group of structures of African derivation, with apse enclosed within a straight wall, a prothesis, and a diaconicon, and structures that are accompanied by sepulchral mosaics, baptisteries, and other elements of the same origin. This complex must be dated from the second half of the 4th through the 5th century.

Examples from the same period are found all over the eastern part of the peninsula, at Tarragona, Tarrasa (Barcelona), Játiva (Valencia), Elche (Alicante), La Vega del Mar, and San Pedro de Alcántara (Málaga). The important Constantinian mausoleum of Centcelles (Tarragona), with a dome of mosaics, and the martyrium of La Alberca (Murcia), both dating from the 4th century, have parallels with structures in Italy and Dalmatia respectively.

A series of structures from the 5th and 6th centuries provided the source for Hispano-Visigothic building. Of particular interest are the basilicas of Cabeza del Griego (Cuenca) and Aljezares (Murcia). The former, dated by the epitaph of the bishops Higinius and Sefronius, is earlier than 550; it consists of a long nave with a crypt preserved at the sanctuary, a long transept, and a horseshoe-shaped apse, as in the basilicas of San Fructuoso de Tarragona and San Cugat del Vallés (Barcelona). Cabeza del Griego repeats African models, as does the shrine with three aisles at Aljezares, with baptistery attached at the back, containing a circular baptismal font with a double diametrical stair for access. Of less interest are the basilicas of San Pedro de Alcántara (Murcia), Alcarecejos (Córdoba), Casa Herrera (Mérida), as well as in the eastern Tarraconensis, in San Cugat del Vallés (Barcelona), in the Villa Fortunatus in Fraga (Huesca), and in Zorita de los Canes, the supposed Recópolis (Guadalajara).

The greatest number of churches of cruciform plan from the 7th century are situated outside Baetica (mod. Andalusia), cradle of Visigothic art. The evolutionary line of this group can be followed through the decorative styles from known creative centers. A distinction must be made between ornamental sculpture in stone or marble and that which was made of brick impressed by molds. Perhaps it is in these ceramic works that the relation between the Roman Christian style of African origin and the ornamental forms of the 7th century can best be found. From the examples imported from Béja (Tunisia), found in Son Peretó and in Seville, up to the complex pieces with repeated motifs among the interlace of S. Juan Bautista at Baños de Cerrato of the 7th century, there exists a wide range of pictorial motifs. It is interesting to note the Andalusian origin of this repertory.

The most important centers of decorative art were Córdoba after the 6th century, with its reutilized Visigothic capitals in the Mosque (X, PL. 162); Seville, which is but scantily known;

Mérida, very important for its pilasters, screens, and niches and opulent ornamentation of Hispano-Roman origin; Toledo, which was the capital of the kingdom (with an aulic art, now totally lost, strongly influenced by Mérida) and was artistically very active from the 6th century to the end of the Visigothic reign; and Lisbon, where a clear Byzantine strain (in the theme of animals in borders) in the manner of the Sassanian and Byzantine silks developed fully in the 7th century.

The 7th-century churches of basilican plan and of short and wide dimensions are S. Juan Bautista at Baños de Cerrato (PL. 316) dedicated to St. John by King Receswinth in 661; São Pedro at Balsemão, near Lamego, Portugal; and Quintanilla de las Viñas, the latest of all. Of cruciform plan are S. Comba at Bande, San Pedro de la Mata, of the period of Wamba, and San Pedro de Nave (PL. 316) from the end of the 7th century, which have, to a certain degree, a central structure. Of cruciform plan and martyrium structure are the Church of São Frutuoso, near Braga, Portugal, dating from 665, and, possibly, the remains of the crypt of S. Antolín in the Cathedral of Palencia, from the time of Wamba in 672.

Among the characteristics common to all these churches are protruding apses, sometimes rectangular, with lateral recesses as in Mata and Baños de Cerrato; excellent ashlar masonry with no mortar; the typical horseshoe arch, which is generally lacking in the façades of the churches in Tarraconensis; and coverings of vaulting and cupolas. The main apse is flanked by chambers, except in São Frutuoso, near Braga, S. Juan at Baños de Cerrato, and São Pedro at Balsemão. In Nave and Viñas such chambers are located at the ends of the arms of the transept.

The ornamental sculpture of San Pedro de Nave and of Quintanilla de las Viñas is of great interest. Two masters decorated San Pedro de Nave; the work by one of them is particularly noteworthy on four of the capitals in the center of the transept; two of these depict Daniel in the Lions' Den and the Sacrifice of Abraham (PL. 321); these are flanked by the apostles Peter, Paul, Philip, and Thomas; on the capital nearest the apse are carved birds among vines, repeated on the imposts. In Quintanilla de las Viñas the friezes on the exterior of the walls of the apse are composed of birds and tendrils, together with anagrams of Byzantine style; in the interior, the imposts are richly decorated with representations of the sun and the moon within circles held by angels (PL. 321); the Pantocrator, flanked by apostles, is on the springer of the vaulting of the dome of the ciborium, and a figure with a cross, perhaps Christ, and a female figure identified as the Virgin is portrayed as the wall of the sanctuary.

In the minor arts a wide cleavage is found until the close of the 6th century between Germanic group and the Mediterranean types of Coptic derivation. Examples of the latter are the censers of Aubenya (Majorca) and Lladó (Gerona); the series of openwork disks; the harnesses for horses with *crismones*, and the liturgical jars and patens which, bought in Coptic or Italian centers, were imitated in the workshops in the region of León after the middle of the 7th century.

Asturian Art. While the Arabs of Córdoba were developing their important caliphate art, a small group of Christian refugees established themselves in the north and soon made Oviedo their capital. They provided a continuity with the ealier Christian civilization and created a superbly rich group of pre-Romanesque works of great importance in which non-Hispanic elements are also present. Compared to the contemporary art of the rest of Christian Europe, this Asturian art was strongly individual. Most of these monuments have been preserved, and their development can be traced through three important sources: the Chronicle of Albelda, the Chronicle of Alfonso III, and the *Liber Testamentorum* of the Cathedral of Oviedo.

The first constructions of interest were erected by order of Alfonso II, the Chaste (812–40). In a second, more original and creative stage a mature art appears under Ramiro I (842–50); and, in a third period, under Alfonso III (866–910), Córdoban Mozarabic influences are obvious and explain the later development of Leonese Christian art. Before Alfonso II there are few records. The Church of Santa Eulalia de Velamio, of which little is known, is attributed to Don Pelayo; it is said that in the year 737 Favila and his wife, Froliuba, consecrated in Cangas de Onís the Church of Sta Cruz, which follows the plan and Visigothic tradition of San Pedro de Nave. From the primitive court of Pravia there are records mentioning a church at Santanes and a royal pantheon, built by Silo (774–83), the plan of which broke with Visigothic tradition.

The aforementioned chronicles and that of Silo give concrete data on the structures of Alfonso II. In the year 812 he consecrated the Cathedral of S. Salvador in Oviedo (constructed by the architect Tioda) together with the Church of S. María, the tribune, and royal pantheon to the north, and the chapel of SS. Miguel y S. Leocadia to the south; of these only the latter, called the Cámara Santa, remains. The basilica of S. Tirso, of which the head of the apse remains, as well as a palace with baths, triclinia, and other chambers in the Roman style, and its church, dedicated to S. Julián de los Prados (PL. 318), are extant.

The Cámara Santa, set against a small rectangular tower, is formed by two rectangular, superimposed naves. The lower one is covered by a low vaulting of brick; the upper one, with a beautiful Romanesque apostolate, was probably covered in wood. It is related in style to the Adriatic martyriums.

S. Julián de los Prados (PL. 318; III, PL. 46; FIG. 95) is of basilican plan, and has three aisles, each with three bays separated by pilasters and round arches, a triumphal arch, a wide transept, and a chevet with three protruding rectangular apses of varying heights, the only distinguishing feature that can be observed from the exterior. The height is achieved by a high chamber with a three-light window atop the central apse (III, PL. 46). The system of support is by means of buttresses. S. Julián is richly decorated (see III, cols. 106–07); the paintings covering the walls are examples of the court art of the time. The molded lattice work of the windows has parallels with work done in Italy after 800. Under Ramiro I this architecture reached its major fruition in the group of structures at Naranco (S. María and S. Miguel de Lillo) and Pola de Lena (S. Cristina).

S. María (III, PL. 51) was originally a royal annex — a belvedere or reception hall. Of rectangular plan, it has two stories, the lower with vaulted baths. The upper story has a tribune on one of the long sides and a double stairway at the opposite side with two loggias, one at each end, separated by columns leaving ample space for windows. The vault has transverse arches supported by buttresses richly decorated with strap-and-medallion motifs. S. María resumes the traditions of the epoch of Alfonso II (Cámara Santa) and others, perhaps Roman, but strongly influenced by the East; for this reason it is considered to be the work of a non-Hispanic architect.

S. Miguel de Lillo (PL. 318), incomplete but of extraordinary beauty, is formed by a vestibule and two side chambers. According to excavation data the plan of the church must have been of rectangular shape with a triple apse terminating in a continuous wall. The interior is divided into four transverse sections, of which only the first is preserved. An important feature is the tribune over the entry door to the main aisle. The structural and decorative unity with S. María is clear, and, according to the Albedlan chronicler, S. Miguel should form part of the same royal group.

Of the same epoch is S. Cristina at Pola de Lena (PL. 317), of rectangular plan and vaulted. On each of the four sides of the rectangle another chamber was added so that the plan seems to be that of a cross (FIG. 667). These added chambers comprise a vestibule, a chapel, and two side wings. The tribune above the porch, which is like the one in S. Miguel de Lillo, is particularly interesting.

With Alfonso III appears the first contact with Mozarabic art. In the year 893 S. Salvador de Val de Diós in Oviedo (III, PLS. 46, 47) was consecrated. Its prototype was S. Miguel de Lillo, but it lacks the latter's boldness.

From this time and later are other churches of popular tendency, such as Santiago de Gobiendes, S. Adriano at Tuñón (891), which is important for its paintings, and S. Pedro at Nora and S. Salvador at Priesca (912), also with painted decoration.

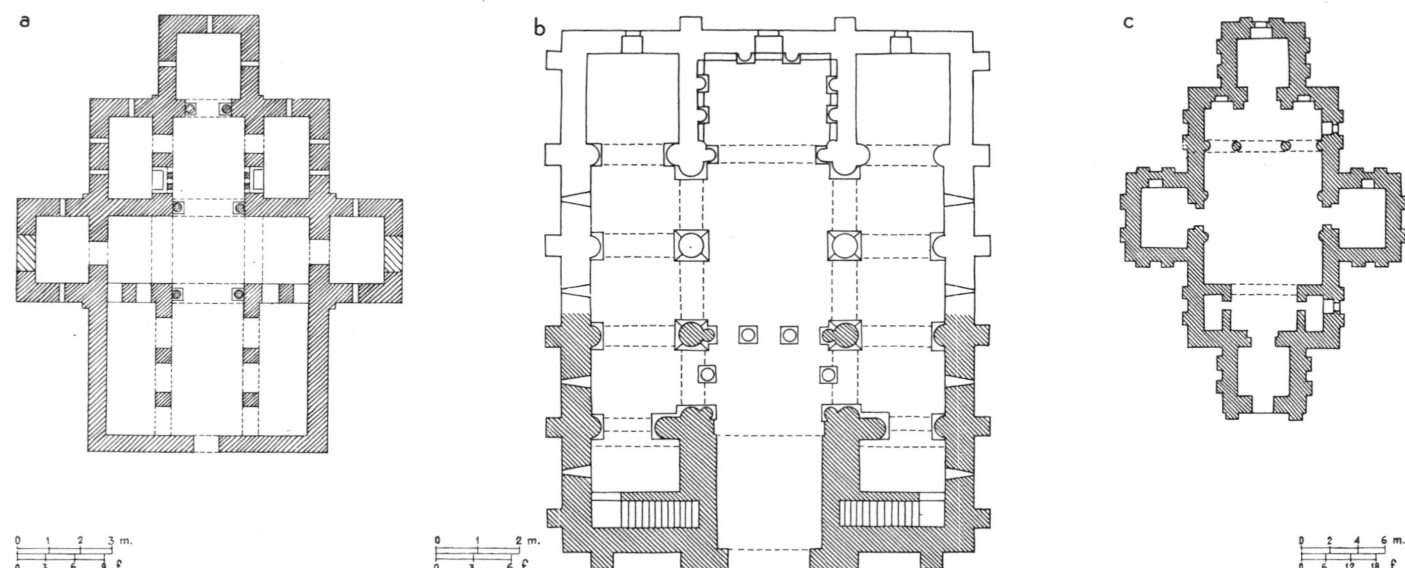

Spain. (*a*) San Pedro de la Nave, plan, 7th century (*from J. Puig y Cadafalch, 1961*); (*b*) S. Miguel de Lillo, Naranco, plan, 9th century (*from G. Bandmann, 1951*); (*c*) S. Cristina, Pola de Lena, plan, 9th century (*from E. Lundberg, Arkitekturens formspräk, Stockholm, 1949, Vol. VIII*).

The pictorial decoration of the Asturian churches is among the most important examples of pre-Romanesque painting. Modern studies have reevaluated this art. There are paintings in the churches of S. Julián de los Prados (812–42), S. María at Bendones (before 905), S. Miguel de Lillo (842–50), S. Adriano at Tuñón (891), S. Salvador de Val de Diós (893), and S. Salvador at Priesca (912). S. Julián de los Prados and S. Salvador at Priesca provide the best examples of classical decoration employing architectural motifs, in the Pompeian manner or, rather, as in the mosaics of St. George in Salonika and Ravenna. Do they perhaps reflect Toledan aulic prototypes of the Visigothic era that have disappeared? Efforts have been made to relate the painted human figures in S. Miguel de Lillo with the Visigothic sculpture of San Pedro de Nave and Quintanilla de las Viñas, although nothing is known of Visigothic painting. The third group of structures, represented by S. Adriano at Tuñón, shows unmistakable Eastern influences absorbed from Spanish Arabic art. This, the first reflection of the presence of Mozarabs in the Asturian domain, was the result of a personal desire of Alfonso III, who brought foreign artists to his court. This Mozarabic tendency is again found in S. Salvador de Val de Diós, which furthermore presents abundant architectural elements in the manner of S. Julián de los Prados and figures in the style of S. Miguel de Lillo. There are no Carolingian influences in this group, but the spirit of the Byzantine east, perhaps introduced by way of Italy, is certainly evident. Some elements are from the Roman epoch.

Pedro de Palol

Miniatures and illumination. The history of Spanish illumination from the 5th to the 8th century is difficult to reconstruct because of the destruction of manuscripts during the Moslem invasion. The various surviving examples of the Visigothic era were written with uncial or semiuncial letters and had few illustrations. Some initial letters were illuminated, but in a simple color. Concentric circles were used into the Mozarabic period; and in a codex in Verona (Bib. Capitolare, Cod. 89) there are letters in several colors. The Ashburnham Pentateuch (Paris, Bib. Nat., Nouv. acq. lat. 2334), which has been the subject of much scholarly discussion (see X, cols. 132–133), is an exception. It is written in semiuncial letters typical of the 5th and 6th centuries with full-page illustrations icononographically related to North Africa; but, judging by the writing, it was probably executed in Spain. Some scholars maintain that it is a local copy of a Syrian text; others (e.g., Hempel), that it is a copy of an ancient Hebrew Bible. In the 8th and 9th centuries concentric and geometric circle decorations are found, for example, in a fragment of the Historia Evangelica (Cambridge, Corpus Christi College, Cod. 304) and in a manuscript in Paris (Bib. Nat., Ms. 4667). Beatus of Liebana wrote a commentary on the Apocalypse, of which a number of illuminated manuscripts still exist; the most important are in Mozarabic style (X, pls. 73, 199; XII, pl. 279). The León Bible (pl. 322), written in 920, is in traditional Spanish style; it has crude but expressive pictures in which Visigothic elements are set off by Andalusian elements. Among the oldest Andalusian manuscripts are the manuscripts of the Dialogues of St. Gregory of 938 (Seo de Urgel, Cathedral Arch.) and the Biblia Hispalense, variously dated between the 8th and 10th centuries (Madrid, Bib. Nacional, Ms. Vitrina 13.1). Their rough, figured illustrations are similar to the sculpture of the church in Quintanilla de las Viñas. But the ornamental elements are in a different style, one in which there can be seen the first Mozarabic elements, which were to achieve full expression in the 10th century. (See mozarabic art.)

* *

Gold- and silverwork. The gold- and silverwork of this period, the richness and profusion of which can be observed on the cover of the *Liber Testamentorum* in the Cathedral in Oviedo, is noteworthy. The few objects that remain are of exceptional artistic interest and stylistically are entirely different from the Hispano-Visigothic examples. The Cross of the Angels dates from 808 and the great Cross of Victory from 908 (both Oviedo, Cathedral, Cámara Santa). The Cross of Santiago de Compostela (874) was a copy of the Cross of the Angels; it had a medallion of perhaps earlier date in the center; this cross disappeared in 1906. In the Cross of the Angels the arms are decorated with filigree on the obverse and an inscription on the reverse around a central circular medallion. It is reminiscent of Carolingian or Italian works, and has been compared to the Cross of Desiderius of Brescia. It is, supposedly, the work of foreigners, perhaps itinerant artists, which gave birth to the legend of the angel-goldsmiths as its creators. The Cross of Victory is, perhaps, the one of greatest interest; it was executed on order of Alfonso III at the castle of Gauzón. The richness of its enamels and encrusted stones has suggested Carolingian parallels — the relations of Alfonso the Chaste with Charlemagne are known. The work has been compared with the golden altar of S. Ambrogio (III, pl. 63, and cols. 120–21) and with the cover of the Codex Aureus in Munich (Bayerische Staatsbib.), among other works.

The Caja de las Reliquias (PL. 321) of the 9th century carries an inscription that bears reference to Alfonso III, the Great, and it presents marked stylistic similarities with the Cross of Victory. The Caja de las Ágatas (PL. 321), made in 910, was donated by Fruela II to the Cathedral of Oviedo ; it incorporates a plaque, with almandines in enamel, of an earlier epoch and of foreign origin. It has Anglo-Saxon or Carolingian parallels, and it has been suggested that it was from the same workshop as the cover of the Book of Lindau (I, PL. 287), although there are notable differences. It is not certain whether an enamel workshop existed in Asturias that may, among other things, have produced the circular medallion of the Cross of Santiago de Compostela and this plaque; it appears more probable that the latter was produced beyond the Pyrenees.

Caliphate art. From the beginning of the 8th century to the beginning of the 11th century there developed in the Spanish al-Andalus, under the Córdoban caliphs and emirs, a new art form that, even within the general sphere of Arab art, possessed strongly individual characteristics. This caliphate art was destined to exercise a considerable influence upon future artistic developments, Arab and Christian in Barbary, in Sicily and in Egypt, as well as in Christian Europe, from the so-called Mozarabic period to the true Carolingian and later.

Of this new style there exist very few examples antedating 'Abd-ar-Raḥmān I, who in the year 786 constructed the first mosque in western Europe (X, PLS. 161-63). It was built on the foundation of the Visigothic Cathedral of S. Vicente in Córdoba. This work, with its subsequent enlargements, was to be the most important source of prototypes and influences in all caliphate art. Its Hispanic past is evident, not only in the presence of a great number of Roman and Visigothic capitals, bases, and columns but also because the basilican tradition was maintained even more firmly than in Damascus itself, and for having absorbed such elements as the horseshoe arch, essentially a Visigothic feature (leaving aside discussions of that feature's Eastern origin in late Roman times), and round arches superimposed on others of horseshoe form, as in the Roman aqueduct of Los Milagros in Mérida. The first mosque of Córdoba was composed of 11 aisles, the center one the widest.

The mosque was enlarged toward the south by 'Abd-ar-Raḥmān II (833-48). From this same epoch are the mosque in Seville and the one in Tudela. 'Abd-ar-Raḥmān II constructed the Alcazaba of Mérida (835), called "El Conventual." This edifice, of square plan with towers at the projections and the corners, was constructed with Visigothic and Roman materials brought from the capital of Lusitania. Its decorated Visigothic pilasters on doors and cisterns are famous.

Following the ending of the dependent emirate there are clear differences between the elements of the Moslem architecture of Spain and that of the East, specifically in the arches of horseshoe form and intersecting arches, in contrast to those of the East, where the arches are pointed, the columns at an angle, and towers circular. (See MOORISH STYLE.)

Under al-Manṣūr another artistic center emerged — Toledo — with more intense relations with the East, perhaps through the builder Ahmed el-Omarī, who, before 1012, had erected two mosques in Toledo. Among the most notable monuments in this city, besides the old city gate of La Visagra, is the mosque (later the Church of Cristo de la Luz) dated 1000. It is of square plan, with four columns that sustain arches dividing eight vaulted sections of which the central one is the highest; brick is used abundantly. Musā ibn-'Alī is referred to as its architect. In Toledo the mosque which is the present Church of S. Salvador has retained two arches resting on Visigothic piers.

With the destruction of the Córdoban caliphate and the appearance of the so-called *taifa* ("faction") kingdoms, Toledo became increasingly independent, and its majestic styles, together with those of Córdoba, powerfully influenced the famous Aljafería of Saragossa, built at the beginning of the 11th century. Saragossa, Granada, Málaga, and Almería showed the continuation of Córdoban art, which also influenced Egypt, as in the mosque of Ibn-Tūlūn.

Within the group of minor arts, of which a wealth of literary descriptions exist, special mention should be made of the ivory boxes from the atelier of Medina az-Zahra, where two masters were at work: an anonymous master from Zamora, who carved the box (Madrid, Mus. Arqueológico Nacional) given by al-Ḥakam II to his wife in the year 964; and Ḥalaf, who signed two pieces. The latest work is from 1005, and was made for 'Abd-al-Mālik, son of al-Manṣūr (Pamplona, Cathedral). This production was continued by another atelier in the 11th century in Cuenca, particularly by ibn-Zayān and his family, who worked for the court at Toledo.

There exist a large collection of bronzes of animal representations, such as a bronze doe (Córdoba, Mus. Arqueológico Provincial), an aquamanile in the shape of a peacock (Louvre), and the lion of Monzón de Campos (Louvre). Various caskets and reliquary chests were made of chased silver, for example, the Wafer Chest of Roda and the Reliquary Casket of al-Ḥakam II, in the Cathedral of Gerona. Among the objects for personal use still extant are a variety of perfume burners, braziers, jars, mortars, and little niches or shrines of unknown use, which reveal Coptic influences in caliphate art.

Rich textiles and ceramics complete this group, the styles of which exerted such a powerful influence on the medieval Christian world.

Pedro de PALOL

ITALY. *Architecture.* In the 7th century in Italy the last traces of early Christian art had all but vanished. At this time Rome was the center of Greek culture, as reflected in the basilicas of Honorius I (625-38), S. Pancrazio, with its tripartite transept (as in Greek basilicas), and S. Agnese (7th cent.), a noble echo of Byzantium, with its women's gallery. The early Christian influence survived more noticeably in Sicily (in buildings difficult to date, but from about the early 6th and 7th cent.), such as in the so-called "Trigona" church at Citadella dei Maccari, near Noto, in the so-called "Torre Cuba di S. Teresa" (near Syracuse), in the small Byzantine church near Malvagna, and in the small basilica at Nésima (near Catania) of the 7th century.

On the other hand, new ideas appeared in that part of Italy dominated by Germanic peoples. From 572 Pavia was the capital of Lombardy; and it was in the 7th century that, alongside buildings of the traditional type (Baptistery of Lomello), others were built, such as the Pavian churches of S. Pietro Apostolo, S. Bartolomeo "de Strata," S. Giovanni Domnarum (erected 636-52 by King Rothari of the Lombards), of which some traces remain, the Oraculum Domini Salvatoris of Aribert I, S. Ambrogio, and S. Michele in the so-called "Faramania" (i.e., the Foro Magno). These were lost when the city was destroyed by the Magyars in 924; little is known of their appearance except that they were departures from the traditional. At the beginning of the 7th century, the Basilica (rectangular, of the 5th cent.) of S. Giovanni Evangelista at Castelseprio was given an apse divided by pillars. From the beginning of the 7th century there appeared all over western Europe the type of church with simple rectangular hall with a square presbytery of narrower width. This type was to be even more common in the 8th and 9th centuries. In addition, the remains at Pavia of a triconch church and transept with apsed arms, what is known of the Church of SS. Apostoli, founded in Rome by Nerses, and the probably cruciform plan of the Cathedral of Monza (adjacent to a royal palace, both erected by Theodolinda) lead to the assumption that in the 6th and 7th centuries churches in the form of Latin and Greek crosses were not unusual. The Church of S. Eufemia in Isola Comacina, the Church of S. Maria del Monte in Varese, and others (in Dalmatia) were also cruciform. Early Christian examples (Church of S. Simpliciano) inspired the use of buttresses on the sides of some early medieval churches, such as St. Martin in Cazis (Graubünden), possibly of the 7th century, and 8th-century edifices. In southern Italy, Arechi I erected a palace at Benevento; and about 671, Romualdus I founded churches in Canosa, on the Gargano promontory, and at Volturno (S. Vincenzo).

In conclusion, what remains of 7th-century architecture in Italy does not show a unified character and appears to be bound in part to the 6th century; but it also seems marked by elements derived from the Christian East and from the peripheral areas of the Empire.

In the last quarter of the 7th century the reinvigorated Lombard society intensified its building activity. The *magistri comacini* built houses of stone and brick roofed with tegulae. Among the churches of Pavia there is precise information about the Church of S. Maria in Pertica, of about 677. It was a rotunda in which the niche motif of the early Milanese-type baptisteries (*ambrosiani*) was multiplied with six columns supporting domical vaults, not unlike (except for the niches) the plan of the Baptistery of Torcello (640). For other churches of Pavia there are only descriptions available (Church of S. Agata del Monte; Church of S. Maria Teodote; Church of S. Michele Maggiore, in which, alongside the work of the Lombard stonecutters, there was, perhaps, southern influence, brought by St. Damian 680–711).

The 8th century was marked by important structural innovations in church architecture. The hall with three nonprotruding apses (or three parallel altars) was probably common in northern Italy and Rhaetia, in Milan (Church of S. Benedict), in Rome (the basilicas of S. Angelo in Pescheria and S. Maria in Cosmedin), and in Swiss churches. The plan of the Church of S. Maria of Castelseprio (PL. 324), which is from the middle of the 8th century and is unusual in that it has three apses and "keyhole" windows, is clearly of Eastern importation, as is that of the Church of S. Domenica, near Castiglione Etneo di Sicilia (different in arrangement but not dissimilar in the forms used), which is possibly of the 8th century. Very few surviving structures render the idea of the splendor and taste peculiar to these last decades of Lombard domination. At Benevento an inspired architect gave a stellar shape to the perimeter of the Church of S. Sofia (founded by Arechi II, 762) around a double circle of columns and pillars supporting various types of vaults (FIG. 672). The three-aisled Church of S. Salvatore in Brescia (end of 8th cent.; PL. 323; FIG. 672), illuminated by large windows and surrounded on the outside by small arches similar to those in Ravenna (and to those of the Church of S. Maria delle Cacce in Pavia), was decorated with stuccowork (XIII, PL. 275) and contemporary frescoes and is related to the Church of S. Maria in Valle at Cividale del Friuli, another remarkable structure, with even more refined stuccoes (PL. 324, XIII, PL. 274) than those of Brescia and with three altars bounded by a rectilinear wall and surmounted by barrel vaults. A mixed character (Roman and Eastern-basilican) is recognizable in the surviving 9th- and 10th-century churches of Capua (Church of S. Michele in Corte, Church of S. Salvatore Maggiore, Church of S. Giovanni in Corte).

The last quarter of the 8th century marks a radical change. In Rome under Pope Leo III (795–815) a reaction set in against Eastern influence and there was a return to the styles of the 4th century (such as in the no-longer-existing triclinium of Leo III of the Lateran, which had 12 exedras, the Church of S. Anastasia; the Church of S. Stefano degli Abissini) that was to continue under Pope Paschal I (817–24) with the churches of S. Prassede (III, FIG. 95), S. Cecilia in Trastevere, and S. Maria in Domnica. Toward the middle of the 9th century the churches of S. Marco (restored 792), S. Martino ai Monti, and SS. Quattro Coronati were also inspired by early Christian prototypes, and a similar revival of early Christian modes can be noted north of the Alps (Saint-Denis, Fulda, Seligenstadt, Hersfeld). In Aquileia it is likely that already at the beginning of the 9th century the patriarch Maxentius planned the basilica (later renovated by Poppo), which recalls examples from north of the Alps. The Monastery of Farfa, an imposing 9th-century complex which had a church (S. Maria) with two opposing apses, recalls northern European examples. The foundations of the large church with a nave and four aisles under the Cathedral of Vicenza of uncertain date, recall the plans of 9th- and 10th-century French churches. In the rest of Italy, from Montecassino (Basilica of Gisulf, 798–817) to the Venetian lagoons (the three-apsed church of S. Ilario, 827–28), the usual basilican plan was widely used, but simpler early Christian forms, also survived, as, for example, the free cruciform with barrel vaults that has

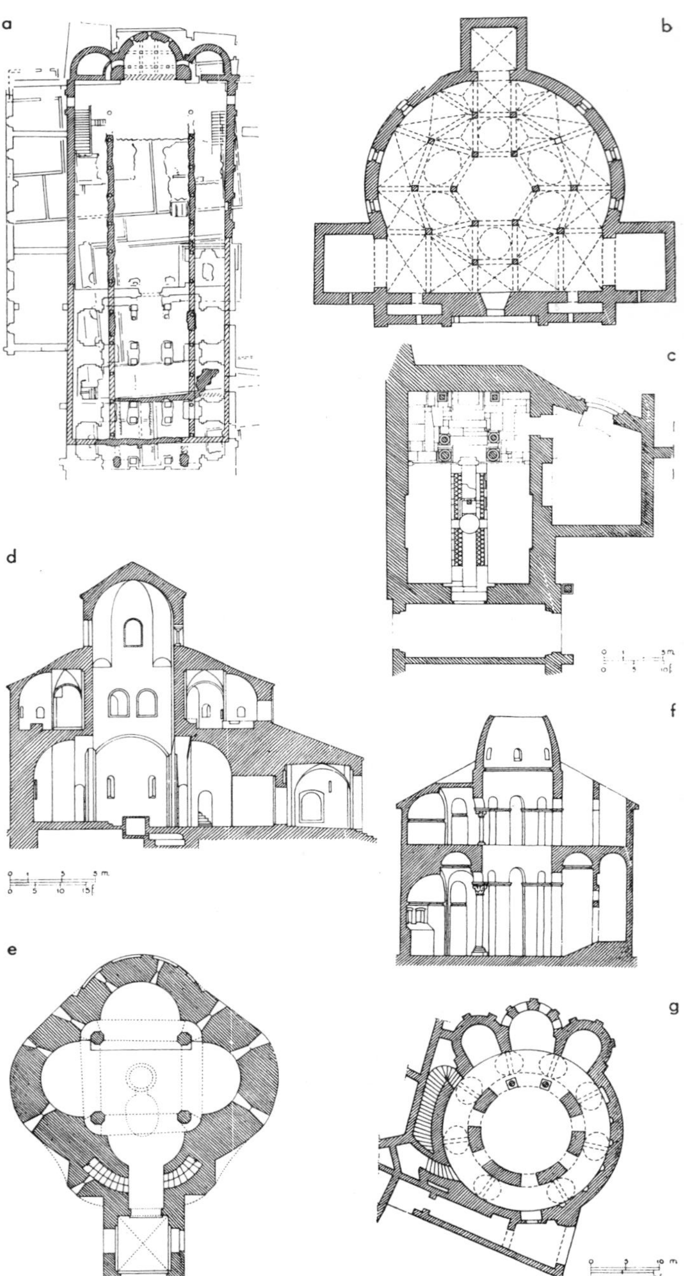

Italy and Yugoslavia. (*a*) Brescia, S. Salvatore, plan of upper church, 8th century (*from Arte lombarda, 1960*); (*b*) Benevento, S. Sofia, plan, 8th century (*after F. Benoit, L'architecture, L'occident médiéval du Romain au Roman, Paris, 1933*); (*c*) Cividale, "Tempietto," 8th century (*from L'Architettura, no. 42, 1959*); (*d, e*) Galliano, S. Giovanni, section and plan, 10th century (*from L'Architettura, II, no. 16, 1957*); (*f, g*) Zadar, S. Donat, section and plan, 10th century (*after F. Benoit, op. cit.*).

its precedents in the two cruciform constructions (one in the Mezzagnone zone and the other in the Vigna di Mare zone) near Santa Croce Camerina in Sicily, and is also found in the Church of S. Zeno in Bardolino (ca. 875, but according to some scholars, considerably later; III, FIG. 102), in the Church of S. Salvatore in Iglesias, and in the Church of S. Giovanni di Assemini in Sardinia; or in the form of a cross with domical vaults (Shrine of S. Zenone in S. Prassede, Rome).

In the 9th century, Byzantine art exerted a direct influence in various regions. The inscribed Greek-cross plan is first exemplified by the Church of S. Salvatore (8th–9th cent.), in Rometta (Sicily), with its dome on an octagonal base and its barrel vaults on the arms of the cross, and by the Church of S.

Maria delle Cinque Torri in Cassino (778–87; destroyed). The most impressive structure of this type must, however, have been the first Church of S. Marco in Venice (ca. 830), a Greek-cross structure with a nave and two aisles and perhaps surmounted by a single dome. In Milan the Cappella della Pietà in the Church of S. Maria presso S. Satiro (III, FIG. 102), built by Archbishop Ansperto (876), still exists. A central-plan structure with four supports, smaller vaults in the corners, and three apses, it has a definite link with 9th-century Byzantine architecture. Similar, but later in date, are the churches of S. Giuseppe in Gaeta, S. Costanzo in Capri, and S. Pietro in Otranto. The last-named is reminiscent of the Cappella della Pietà in S. Maria presso S. Satiro, Milan, in the breadth of its central vault in relation to the corner ones.

Another obvious connection with Byzantine architecture is recognizable in the Calabrian churches, of inscribed-cross plan with drum domes and equal cylindrical apses. Examples of these are La Cattolica in Stilo (PL. 323) and S. Marco in Rossano, which are not later than the 10th century, and also St. Nicholas in Split, Yugoslavia. In La Cattolica, the accentuated polychromy of the wall facing is derived from Byzantine styles. (It should be remembered that in about 732 Leo III the Isaurian placed the Calabrian churches under the patriarchate of Byzantium.) A resemblance to the Palatine Chapel of Aachen (III, PL. 52, FIG. 102) can be seen in the Church of St. Donat at Zadar (PL. 323; FIG. 672). From the 8th to the 10th century six-lobe plans were also common in Dalmatia.

Crypts were common in the 8th century (Luni, Ventimiglia) and at first were semicircular in structure; but in the 9th and 10th centuries their structure became more complex. They became, in effect, subterranean oratories (Church of S. Stefano, Bologna; Church of S. Apollinare Nuovo, Ravenna; Church of S. Maria delle Cacce, Pavia's S. Michele, near Santarcangelo di Romagna). In the second half of the 10th century, together with the first attempts to organize the structure according to a rhythmical principle applied to the distribution of space (which was to develop into the Romanesque style), there appeared the multiplication of pilasters and columns supporting vaults and intrados, but without a regular framework (crypts of the Basilica Ursiana, Ravenna; crypt and apse of the Church of S. Stefano, Verona; crypt of the Cathedral of Ivrea; crypt of S. Giovanni Domnarum, Pavia; and crypt of S. Zaccaria, Venice, 888–912).

The 10th century marks the passage to the Romanesque style. The Church of S. Quintino di Spigno, Alessandria (ca. 991), and the Church of S. Pietro in Valle, Gazzo Veronese, demonstrate a distribution of space that was common in northern Europe. The few remaining examples of this period ascertain the perpetuation of archaic schemes (the Chapel of S. Lino in the Church of S. Nazaro Maggiore, Milan, 10th cent.), or the definite abandonment of that paratactic style characteristic of the basilican plan, giving preference, even if somewhat tentatively, to rough pilasters set in rows to divide rooms, as in the three older aisles of the Basilica of S. Vincenzo, Galliano (PL. 324), or with more regularity in those of the Church of S. Stefano, Verona. Still, at the very outset of the Romanesque period, the Baptistery in Galliano (PL. 323) repeats, with the addition of women's galleries, the central-plan principle of the shrine of the Church of S. Maria presso S. Satiro in Milan and, in the thickness of its unusual walls, announces the Romanesque.

Sculpture. The 7th century in Italy was a period of serious decline in sculpture. In Rome the decline was marked by the continuation of 6th-century motifs, with the addition of rosettes, lilies, and quatrefoils. Early Christian elements, reduced to uncertain shapes, survived, but were subject to barbarian infiltrations in the exarchate of Ravenna, and the rhythm of form of the 6th century disappeared (the sarcophagus in the Church of S. Decenzio, Pesaro, and that of a young boy, no. 410, in Ravenna, Mus. Naz.). The Felix sarcophagus (XIII, PL. 342) in Ravenna, which shows crude early Christian formulas but demonstrates, at least, a certain rhythmic order, dates from 723. The same survivals appear in Venice (plutei in S. Marco; the Tiepolo sarcophagus, in the Church of SS. Giovanni and Paolo, of the 6th–7th cent.).

In comparison, the small group of reliefs existing in Lombardy shows more force even in its poor quality. Those *magistri comacini* who were explicitly mentioned in the codex of Rothari must already have been at work in Lombardy (the group of slabs in Pavia, Mus. Civ.; the pluteus on the outside of the Cathedral of Monza, where the clearly hollowed-out surfaces indicate a taste completely different from anything that had been done before). Obviously, works like the cathedra of St. Mark, in the Treasury of S. Marco, Venice or, in the same church, the so-called "Carmagnola" head, which perhaps represents Justinian II Rhinotmetos (705–11), do not enter into this development, for they are not the work of Italian craftsmen. The consular diptych with King David and St. Gregory (Monza, Cathedral Treasury), however, because of its similarity to preceding works, may be of the 7th century.

The first decade of the 8th century saw a new style of relief sculpture which appears to have been the work of Lombards (almost certainly from the region of the lakes). The repertory combined purely classical elements with elements of Eastern and barbarian origin, and there are even neolithic echoes and reworkings (in the barbarian idiom) of ancient motifs such as interlace decoration. It was this latter element that increasingly gave an unmistakable character to this decorative sculpture, which was applied to ciboria, plutei, capitals, small pilasters, and well-copings; it appeared from Provence to lower Austria, in Carinthia, Croatia, the Balkans, and Greece, as well as, of course, in the whole of Italy. In Italy, however, to avoid confusion, it is always necessary to distinguish between the productions of the stone carvers and stuccoworkers of the Lombard lakes, who, beginning in the early Middle Ages, migrated all over the West, and their local imitators. The works that mark the advent of this new sculptural concept date from the end of the 7th century (Ancona, ambo of S. Maria della Misericordia, now destroyed, 687–701; fragments in Albenga and Luni); but it may have already existed at the time of Rothari (636–52). A documented work such as the ciborium of S. Giorgio (712), near San Ambrogio di Valpolicella, shows in its robust design a definite use of classicizing motifs and a marked naturalistic style. The assumption must be made that this orientation continued to characterize sculpture under Liutprand. Subsequently it gradually lost these characteristics and in the 9th century greater regularity, tighter interlace motifs, and a dryer, more mechanical execution became usual. The better marble carvings of the Verona ciborium are in keeping with the practice and taste that distinguished the two marble slabs of the sarcophagus of Theodota (or Theodote, d. 720; Pavia, Mus. del Castello), the slab of S. Cumiano in Bobbio, and the peacock relief of Brescia; later works are the archivolts (736–44) of the Baptistery of Callisto in Cividale del Friuli. A notably different feeling characterizes the marble carvings that were formerly in the Church of S. Maria d'Aurona (now in the Castello Sforzesco) in Milan, the contemporary, ornate inscriptions of Pavia of the first half of the 8th century (tomb slabs of Cuningperga and of Ragintruda; both, Castello Visconteo, Mus. Civ.), certain sculptures in the chapter house of Garda, a fragment in the Church of S. Giovanni in Assemini, and other reliefs; all of these would seem to indicate a contact with the decorative sculptures of Christian Egypt that supposedly came about through the agency of the Arabs.

The variety of styles that must have existed by the first half of the century is demonstrated by the Altar of Ratchis (PL. 326; VII, PL. 373), in which the proportions of the figures (which are similar to Visigothic sculpture, e.g., Quintanilla de las Viñas; PL. 321) recall the provincial sculpture of North Africa, Mesopotamia, Phrygia, and above all, of the Balkan Peninsula (Francovich, 1962). An adequate explanation of such an extensive phenomenon is difficult, and the difficulty increases outside the confines of the Po valley. The Magister Ursus who signed a pluteus at Ferentillo in 739 was undoubtedly a local imitator of the stone carvers who, especially at the beginning of the second half of the 8th and throughout the 9th century, worked throughout the Italian peninsula and crossed the Alps and the sea. But the ciborium of S. Prospero in Perugia (perhaps of the 8th cent.) is a remarkable work; and the outstanding altar frontal of Sigwald (XIII, PL. 342) provides proof of a definite movement away from the naturalism of Liutprand. These forms finally became codified in the direction of a more dense and regular distribution of the usual decorations and in the 9th century acquired a sharply defined perfection that attests to the long practice and ample diffusion of these skilled and industrious workers. It should not be forgotten that even in this later period there was considerable variation in quality, from city to city and according to the taste of the individual stone carver. In the second half of the 8th century many motifs that had been in use until then began to disappear (fusaroles, lilied crosses, paired S's, etc.) and more strictly symmetrical interlaces predominated.

In the 9th century the human figure and the large animal shapes disappeared and small figurations, highly abstract and stylized, became more prevalent. These are evident in the relief fragment in the Castello Scaligero, Sirmione, upon which the name of Desiderius (756–74) is inscribed; in the arch of a ciborium (Mus. Vat.), of 795–816; in the carvings in Ventimiglia and Albenga (tracery windows), of 775–800; in the beautiful ciborium of S. Apollinare in Classe (PL. 325); and in the plutei of the Church of S. Sabina in Rome (PL. 326). At the end of the 8th and beginning of the 9th century the Roman churches (S. Maria in Cosmedin, S. Agnese, S. Maria in Trastevere, S. Cecilia in Trastevere) were rich in similar sculptures, as were the churches of Latium, Umbria, and the Adriatic coast. The casket of carved cypress wood from the Treasury of the Sancta Sanctorum (795–816; Mus. Vat.) is unique. This type of sculpture lasted beyond the 9th century, and well into the 10th, at which time the modeling lost its strength and the regularity of the design slackened. In the 11th century the centuries-old taste for interlace carvings reappeared (ca. 1080) in the earliest Lombard Romanesque churches. The spread of these sculptures to territories dominated (or formerly dominated) by the Lombards and to neighboring territories, the presence in northern Italy at the very beginning of the 8th century of those examples that were of high quality, and the tendency (which was to continue for centuries) of the Lombard stone carvers to emigrate confirm the theory that the artists of the north Italian lake region were responsible for all these sculptures. The Lombard artists, by the way, had always — in the late Middle Ages, in the Renaissance, and in the baroque period — shown a willingness to assimilate the most varied stylistic influences. This explains the conflicting theories of scholars who attempted to tie this production to the Germanic world or to the East. There is further proof of this in the fact that this sculpture is lacking in southern Italy, where Byzantine influence was stronger; the large-figured marbles of the Cathedral in Calvi and of Capua (Mus. Campano) undoubtedly are linked more closely to the East and to Byzantium than to the barbarian world. The 10th- and 11th-century figured plutei of the basilica at Aquileia, in Venice, in Sorrento, in Greece, and in Constantinople, even though they are related to the art of the Lombards, seem more refined because of the contact with Byzantine and Moslem art.

Stuccowork was very common. The most important examples of this art are represented by the very lovely decorations of the Church of S. Maria in Valle in Cividale del Friuli (PL. 324; XIII, PL. 274) and by those, inferior in quality, discovered in the Church of S. Salvatore in Brescia (XIII, PL. 275). Their dating is disputed, but it has been maintained that they are a classicizing Italian production of the 8th–9th century, comparable to the stuccoes of the Church of St-Laurent in Grenoble and marking one of the aspects of the Carolingian "renaissance" (Peroni).

These light and quickly produced stuccoes (and many others of less importance) are remarkable evidence of a very unstable style (as contrasted with that of the stone carvers), and this makes the dating of some undocumented works extremely uncertain and controversial. The most important of these is the ciborium of the Church of S. Ambrogio in Milan (PL. 325; XIII, PL. 278). It has been variously assigned to the 9th, 10th, and also (as this author believes) to the 12th century. There are other stuccoes in Malles (III, PL. 60), Pomposa, Testona, and Fontaneto d'Agogno. Polychrome decorations in brickwork can be seen in Grado, Venice, and Pomposa; undoubtedly they are derived from Ravenna and from provincial Roman art (Römerturm, the so-called "Roman Tower" of Cologne). Those of La Cattolica, Stilo, however, should be considered of Byzantine derivation.

Among strictly Italian ivories of the 8th–10th century there are no constant characteristics that reveal a single school. The bone pax of Duke Orso in Cividale del Friuli (Mus. Arch.) and the diptych from Rambona (PL. 326) are difficult works to date. The first, which could be dated any time between the 8th and 10th centuries, is similar to the ivories of the St. Gall school; the second, of about 900, shows elements of diverse sources (Syrian, barbarian, and from contemporaneous illumination). A more consistent group of 9th- and 10th-century ivories can now be attributed (Francovich, 1942–44; Volbach, 1952) to the Milanese school, which can take its place beside the various Carolingian schools. This Milanese school has certain characteristics of its own that can be seen in the diptychs of the Cathedral Treasuries in Milan and Aachen and in London (Vict. and Alb.) and in a cover in Oxford (Bodleian Lib.). The aspersorium of Bishop Gotfredus of Milan (ca. 974–80) in the Cathedral Treasury in Milan, an ivory cover in the Castello Sforzesco in Milan (with Otto I and his family standing before the Virgin), an aspersorium in the Victoria and Albert Museum (Basilewsky Coll.), and the parts of an altar frontal scattered in various Museums throughout Europe are all later, that is, from the end of the 10th century. For a long time these were believed variously to be the products of the workshops of Reichenau or Milan, but now the tendency is to attribute a strictly Milanese origin to them. The balance and rhythmic details of these works give weight to the belief that they are the work of Italian craftsmen. Three ivory reliefs in Munich (Bayerisches Nationalmus.) seem to herald the style of Wiligelmo da Modena (q.v.). (The comb in the Treasury of Monza is, on the other hand, believed by Volbach to be of the 11th–12th cent.)

Painting. The echoes of the classical idiom, of Hellenistic elegance, of Syrian expressionism, of Byzantine abstractions, and, in general, of the complex world of forms that marked all Mediterranean art from the 4th century on were retained in the 7th century in Italy. The very few examples that remain from this period give the impression of a variety of styles that did not develop into a single well-defined trend. It is therefore convenient to limit the discussion to works presumably created in Italy itself. The illumination in the so-called "Gospels of St. Augustine" (ca. 600, Cambridge, Eng., Corpus Christi College, Ms. 286), which were perhaps executed in Rome and are characterized by vivid coloring within fine black outlines, seem to have nothing in common with other pictorial works, but their anticipation of certain monumental arrangements in Carolingian illumination (Ada Group) is significant. The madonnas of the Pantheon and of the Church of S. Francesca Romana, both in Rome, are possibly of this period. Also in Rome, the votive fresco of the matron Turtura, in the Catacomb of Commodilla, still shows a strongly linear style. On the other hand, the mosaic panels of the "privileges and sacrifices" in S. Apollinare in Classe in Ravenna still reflect forms of the 6th century.

The fact that there were 13 Greek and Syrian popes from 606 to 741 is still the best explanation for what happened in painting after the end of the 6th century (frescoes in the Church of S. Maria Antiqua; PL. 327; II, PL. 445) in Rome, where, from the first decade of the 7th century, a linear and rigid style alternated with others that were more fluid and softly Hellenizing. But a noticeable rigidity in taste is evident in the triumphal arch in the Church of S. Lorenzo fuori le Mura (Rome); and despite the Hellenistic revivals it is likely that the more severe and abstract style tended to prevail. (Remarkable surviving evidence of this is the apse mosaic of the Church of S. Agnese, 625–38; however, in the mosaics of the Oratory of S. Venanzio, 640–42, which is annexed to the Lateran Baptistery in Rome, there is evident a revival of plastic sense.)

The pontificate of John VII (705–09) gives further evidence of disparate tendencies, and confirms the absence of a single style in Rome and the prevalence in the city of Eastern influences at this time. There are echoes of the idiom of Constantinople in the mosaics for the Oratory of John VII in Old St. Peter's (X, PLS. 99, 180); in the St. Sebastian mosaic in the Church of S. Pietro in Vincoli; and in the painting of the Madonna della Clemenza of the Church of S. Maria in Trastevere (completely revealed during a restoration). During the same years there was, however, a last solemn declaration of Eastern taste in the magnificent Adoration of the Cross in the Church of S. Maria Antiqua. In the Maccabees and Their Mother Salome (II, PL. 445), also in S. Maria Antiqua, there are "flashes of light upon dark complectioned faces, as in classical frescoes" (Toesca, 1912).

Lombardy was undoubtedly an important center of painting. There, Theodolinda had already commissioned secular frescoes for her royal palace. These works were most certainly by Roman artists. Frescoes adorned the churches of Pavia of the second half of the 7th century (e.g., S. Michele ad Palatium) and the Church of S. Giovanni Evangelista at Castelseprio. All that has survived of such painting is the fresco found in a tomb in S. Giovanni in Conca, Milan, possibly of the first decades of the Lombard conquest, and the beautiful illuminations of a codex in Vienna (Nationalbib., B.N. 847, theol. 682), formerly in Bobbio, patterned after the Ravenna plutei (6th–7th cent.). On the other hand, the decorations of the baptismal font of Lomello (6th–7th cent.) are in a strictly linear style. The motifs of Celtic and Merovingian illumination also appear in Lombardy where they became known with the founding (612) of the Monastery of Bobbio. Throughout Italy the illuminations of the 7th and 8th centuries seem to have attempted to imitate, though with difficulty, northern European models (cf. the Dialogues of St. Gregory, ca. 757, Milan, Bib. Ambrosiana, B. 159 Sup). In Northumberland, however, St. Benedict Biscop, Abbot of Jarrow, had the Codex Grandior of Cassiodorus copied and sent a copy, the Codex Amiatinus (I, PL. 282), to the pope (beginning of 8th cent.).

After various attempts, ranging from the 6th to the 10th century, at dating the cycle of paintings of Christ in the Church of S. Maria, Castelseprio (II, PL. 445), it now seems reasonable to assign them to the 8th century. The cycle was undoubtedly executed by a Greek who drew heavily (contrary to Carolingian custom) on the apocryphal gospels; it is one of the most skillful and successful examples of early medieval painting. It is the work of a remarkable artist whose fresh Eastern style has perplexed scholars because of its unique character within the realm of Italian art. Justifiably, contacts have been pointed out between this style — in which the agility of forms is accompanied by very free and rapid brush strokes — and the 10th-century Joshua Roll (II, PL. 447; X, cols. 134–35), which repeats the styles of earlier illuminations (the disagreement among scholars regarding the date of the first edition is well known). The discovery of the frescoes originally decorating the structure of the Church of S. Salvatore in Brescia, of the beginning of the 9th century, in which there appear undeniable similarities to the Castelseprio cycle, will, perhaps — more than the excursions variously attempted into the realm of Carolingian and Byzantine illumination — facilitate the dating and critical understanding of such an important work.

The frescoes of S. Procolo, Naturno (III, PL. 57), fall within the sphere of the St. Gall illuminations, and the Verona illuminations of this period reflect, through the influence of Salzburg (Codex Millenarius of Kremsmünster, etc.), characteristics of Celtic art.

It seems probable that toward the end of the 8th century (and there are many examples of this in Roman churches) a more continuous and coherent idiom was finally becoming established, and not only in Rome. The frescoes of the time of Paul I in the Church of S. Maria Antiqua (757–67) show contacts with the Verona illuminations of the Codex of Eginus (790–99; Berlin Staatsbib.), and the frescoes also show anticipations of and similarities to the great Carolingian illuminations (the Ada Group). The Crucifixion in the Church of S. Maria Antiqua (PL. 327) is perhaps the first surviving work in which strong Western characteristics outweigh the various Eastern and local components. In a very broad sense, and far ahead of its time, this work is genuinely pre-Romanesque.

National and regional characteristics slowly and almost imperceptibly began to be defined. In northern Italy important fresco cycles, datable around 800, made their appearance within the sphere of the Milanese ecclesiastical jurisdiction; they are marked by a very robust style of their own, a style of purely Western character that also announced the later establishment of the Romanesque idiom. These cycles are the important complex of works in the Church of St. Johann at Müstair (Münster, Graubünden; II, PL. 287; III, PL. 57) and some frescoes in the Church of S. Benedetto, Malles, in Val Venosta (III, PL. 57). The similarities to Carolingian painting seem obvious (St-Germain, Auxerre; in Lorsch; in Trier); but it should be noted that the artists of those works, undoubtedly Lombards, approached artistic representation with a sense of harmony and fullness of expression unknown to northern European painting. Their relationship to the illuminations of the Codices Conciliorum (Ms. CLXV), of the Etymologiae of St. Isidore of Seville (Ms. CCII), and of the Homiliae in Evangelia of St. Gregory (Ms. CXLVIII; all, Vercelli, Cathedral Treasury) has been pointed out (Francovich, 1956).

There are examples in Italy of even the most vividly expressionistic currents of Carolingian painting. It suffices to mention the frescoes in the crypt of S. Lorenzo in San Vincenzo al Volturno (826–43; PL. 320) and various Roman frescoes (e.g., the fragment of the Ascension, in S. Clemente, 847–55). Nevertheless, this "rapid" style virtually ended in the 9th century. Henceforth colors were no longer "fused in a continuous gradation, but conventionally divided by shading, almost always greenish, by half-tones, and by bright white highlighting" (Toesca, 1912). This was later to become the most abstract idiom taken up by Romanesque art until the end of the 13th century.

A Roman school of mosaicists, active under Leo III (795–816) and succeeding popes, had its own particular characteristics that seemed to perpetuate the "Syrian" taste that still survived in Rome during the 7th and 8th centuries. A color sense enriched by subtle relationships of flat tones, which are bound by a thin outline, distinguishes the mosaics of the churches of SS. Nereo e Achilleo, S. Maria in Domnica, S. Prassede, S. Cecilia in Trastevere (PL. 327), and S. Marco. The mosaic (M. Salmi, *Un problema storico-historico medioevale*, BArte, July-Sept., 1958, pp. 213 ff.) in the Cathedral of Narni (ca. 800) proves that these very fine colorists did not work only in Rome. Painting activity must also have been outstanding in Rome at the end of the 9th century, when Pope Formosus redid the decorations of St. Peter's and Sergius III redid those of St. John Lateran. The remains of the frescoes in the temple of Fortuna Virilis, which are perhaps of the 10th century, reecho Carolingian illuminations.

The illumination of the 8th and 9th centuries (in addition to the above-mentioned Vercelli manuscripts) showed varied characteristics. The Collectio Canonum in Rome (Bib. Vallicelliana, Ms. A. 5), which may have been written and illuminated in northern Italy, clearly shows contacts with the Carolingian school of Reims. Other illuminations also have Eastern characteristics, but are similar to work produced in Reichenau; a missal from Bobbio (Milan, Bib. Ambrosiana, Cod. D. 84 inf.) from the end of the 10th century likewise brings to mind the violent forms of the Reichenau school. In southern Italy (mss. preserved in Rome, Vatican Lib., in Patmos, Athens, Leningrad, from Reggio, Salerno, etc.) illumination was dominated by Byzantine characteristics, which sometimes (e.g., Theophylact fresco of 959 at S. Maria delle Grazie, near Carpignano) are evidence of an irreversible decline.

In this period codices were often decorated with pen drawings (Rome, Bib. Vallicelliana, Cod. 25), in which the forms within the pen-drawn outlines are suggested by summary coloring. This tendency of the form to dissolve, the disappearance of classical proportions, became particularly noticeable in the 10th century (Benedictio fontis, Rome, Bib. Casanatense, Ms. 724), even though the outlines were relatively firm. This disintegration of form was gradually to give rise to the Romanesque idiom, based upon new rhythms, while infiltrations of Northern painting, Ottonian frescoes, and the St. Gall illuminations persisted although in an ever-dwindling stream (as in the frescoes, of 996, of the sacellum of the Church of SS. Nazaro e Celso, Verona; fragments, also of the 10th century, of a chapel in the Church of S. Nazaro, Milan; numerous illuminations). The Christ in Limbo, in the Church of S. Clemente, of the 10th century, already showed decidedly Romanesque linear formulations.

There is an almost polemical, very free attitude (and one that was full of promise for the future) in the drawings in the codices ordered by Varmondo d'Ivrea (969–1002); and in certain aspects the illuminations in a Milanese psalter (Vatican Lib., Lat. 83) are already Romanesque in style. At last the time was ripe for the birth of a style that would be common to all of the West.

Minor Arts. In the 5th century, the void caused by the disintegration of political and social structures owing to the disastrous effects of the war in Italy was particularly noticeable in the minor arts (see EUROPE, BARBARIAN). But the Italian peninsula was still subject to classical and Mediterranean influences, and thus, as early as the time of Theodolinda, works of a very different type contrasted with the mature style of the Lombard goldsmiths, who were in direct contact with northern Europe. Coptic pottery has been found in the necropolises of northern Italy; circular fibulas indicate contact with the Byzantine world; Mediterranean and Eastern elements contributed to the formation of Nordic style II, which was elaborated by the Lombards in the 7th century; and the well-known copper plaque with the figure of Agilulf (XII, PL. 501) and the ornamentation on the shields preserved in Lucca (Mus. Naz. di Villa Guinigi) and Bern (Historisches Mus.) are not barbarian, but classicizing. The cover of Theodolinda's evangelistary (VI, PL. 258) is a work of great balance; and the well-known Hen with Seven Chicks (VI, PL. 259) is of a naturalism that is entirely Italian, not barbarian. (See also IX, PL. 163.)

The first Italian goldsmith's work using cloisonné is of the 7th century. Examples of this techniques are the "Castellani Fibula" (IV, PL. 407) and the fibula from Senise (Naples, Mus. Naz.). The thickness of the dividing strips and the crudeness of the enamelwork create a style that is obviously undeveloped and summary. The Cross of Agilulf (Monza, Cathedral Treasury) is even more decidedly influenced by the East.

A later, more barbarian taste is evident in the tooth reliquary (III, PL. 68), which is thickly covered with precious stones and is similar to the Enger purse reliquary (PL. 313) and to the Altheus reliquary in Sion (Switzerland). All of these are of the 8th or 9th century. A spirit equally opulent, albeit undisciplined, imbues the Cross of Desiderius (Brescia, Mus. Cristiano).

Concomitant with the Carolingian renaissance, Italy too in the 9th century witnessed the perfection of the ancient techniques, which achieved new stylistic heights. Under Paschal I, Rome was a center of primary importance. The Lateran Cross (817–24) and the Beresford-Hope cross (PL. 315) were executed in a cloisonné technique; but two silver cross-shaped caskets (Mus. Vat.) of 817–24 are adorned with vividly designed scenes in light embossed work.

Angilbertus II (824–59) commissioned the golden altar of S. Ambrogio in Milan, in which various sacred scenes in repoussé work are framed by ornamentation of enamel and precious stones (III, PL. 63, and cols. 120–21). The harmonious division of the piece attests in itself to an Italian origin. Undoubtedly the "iron crown" of Monza (VI, PL. 260) is of the same workmanship; and the Cross of Berengar I has similar incrustations of gems (III, PL. 68). The openwork part of a 9th- or 10th-century sacramentary cover (Monza, Cathedral Treasury) is similar to the decorative sculpture of the same period. The four enamels of the apostles in the upper part of the crown given to St. Stephen of Hungary (995–1038) by Sylvester II (Budapest, Fortress Mus.) are possibly from northern Italy, but it is more likely they were made in some workshop north of the Alps.

Nevertheless, in the works of these centuries it is possible to note the difference between the superb creations of the Byzantine goldsmiths, of which examples exist in Italy, and the labored imitations of their art. A 10th-century silver pyx in the Cathedral of Cividale del Friuli has flattened-out figures that are imitative of the Byzantine manner but lack its fine technique.

With the exception of the cover of the Evangelistary of Aribertus (1018–45; Milan, Cathedral Treasury) and the plaque in Chiavenna (Collegiate Church), which some would credit to German art, there are no noticeable traces in the works of Italian artists of the great schools of gold- and silversmiths that were centered during the Ottonian and Salian periods in Echternach, Trier, Prüm, and Metz. In German works of the same period, there is a quieter, smoother treatment of the surface in handling the contrast between enamels (even if not of remarkable quality), gems, filigree, and metal reliefs. But here there is already a spirit that no longer belongs to the early Middle Ages. The Cross of the Field in the Rotonda, or Old Cathedral, of Brescia is close in feeling to the Aribertus cover.

As far as tesselated paving is concerned, the early Middle Ages saw the passage from *opus sectile* flooring (S. Maria in Castelseprio) common in early Christian architecture to the tile flooring of the 9th century (Gazzo Veronese, S. Maria), which in turn was reminiscent of early Christian mosaic floors. Other new types of flooring are represented by the work in the Church of S. Ilario near Fusina of 827–28 (now in Venice, Mus. Archeologico del Palazzo Reale), imitating Sassanian cloth; that in the Church of S. Zaccaria in Venice (829–30); and that in the Church of SS. Nazaro and Celso in Verona (10th cent.). A floor in Ivrea that is similar to the designs of the period of Varmondo belongs to the end of the 10th century.

Among the textiles that were not imported should be mentioned the Lombard altar cloth of the Sancta Sanctorum (Mus. Vat.), the Veil of Classe (Ravenna, Mus. Naz.), an 8th–9th-century Veronese work echoing contemporary Roman painting; and the fabrics woven in the Lombard monasteries (e.g., the so-called "Apostle Tapestry" in Monza, Cathedral Treasury).

Edoardo ARSLAN

SCANDINAVIA. It is not customary to speak of pre-Romanesque art in the Scandinavian countries; general Christianization took place there at such a late date — during the 11th century — that the term has lost its validity, at least from the point of view of Christian art. Art forms of the era prior to the 11th century are pagan or prehistoric, and those that appear to coincide with pre-Romanesque art in the European sense would, generally speaking, be that of the Scandinavian viking era (VI, PL. 259).

Nevertheless, it seems possible to distinguish certain categories within Scandinavian art that deserve to be included in the category of pre-Romanesque art and that have a Christian rather than pagan bias. Of course, much of this art is strongly provincial, but in some respects it shows great variety, and the very fact that certain tendencies evident in it survived into the Romanesque period deserves attention.

A small silver crucifix from Birka, Sweden, is perhaps the oldest of its kind in Scandinavia (PL. 328). It is a hanging cross with a simple and primitive image of Christ in filigree and granulation, and probably dates from the second half of the 10th century. Akin to this one, but somewhat later in date, are two filigree crucifixes from Trondheim, Norway (Trondheim, Archeologisk Mus.). Both were made in the same workshop, but differ somewhat. What seems to be a Celtic influence is noticeable in the delineation of the hair, turned-up moustaches, thoracic vertebra, and in the design of the arms. This influence is even more evident in a crucifix encolpion from Gåtebo, Öland, Sweden (PL. 328). Here the basic form of the cross is Byzantine, but the figure of the Christ very clearly shows that it is Scandinavian work. It is an engraved and niello representation with a Christ provided with a halo and a moustache, clad in a long garment that falls from curious figure-eightlike pleats into deep drapes. On the figure's breast are two spiral coils. The upper part of the cross shows an interlace pattern that looks very much like the snakes and dragons customarily found on Swedish runic stones and the side arms have similar intertwining bands. Several of these details might derive from local viking art, but this influence is not clearly so striking as the Western influence. In an illuminated Celtic manuscript from the 10th century, the so-called "Southampton Psalter" (Cambridge, Eng., St. John's College), is found a crucifix group in which the Christ wears practically the same curiously tied mantle as that of the encolpion from Gåtebo. On each arm of an Irish metal cross from Cloyne the same standing figure is repeated (the Christ?), with head bent toward the middle of the cross and legs entwined by snakes that bear great resemblance to the snakes on the Norse runic stones. In other Celtic works similar details are found (e.g., in a Christ crucified represented on the so-called "Lismore crosier"). It seems obvious that all these instances cannot be merely coincidental, but must be admitted to prove direct contact — one that manifests itself in many other ways. The well-known crosiers from Aghadoe (Stockholm, Statens Historiska Mus.) and Clonmacnoise (Dublin, Nat. Mus.) are outstanding examples of this cultural interchange, and in the Scandinavian countries there are countless other examples. For a long time it has been asserted that the stylistic natures that met in this context reflect Scandinavian influence on the art of the British Isles; however, the fact has been overlooked that, although the material in question is for the most part later than the Scandinavian occupation, it coincides with the epoch of the Anglo-Irish missions in the Scandinavian countries. During the entire 11th century and far into the 12th examples of this peculiar art are found. They seem to be composed of equal portions of the British Christian and Scandinavian pagan characteristics that gradually developed into Christian Romanesque art in Scandinavia. (See SCANDINAVIAN ART.)

A great many small, modest, hanging crucifixes of Byzantine origin from the second half of the 11th century and from the 12th century reveal their Scandinavian origin, above all in clothing and other details, such as the drapes of the garment falling down obliquely on the sides, the peculiar border of the tunic, and the intertwining bands around the arms. In certain cases these details may be of Western origin, thus ascertaining that the Scandinavian hanging crosses, although they are typically Scandinavian work, reflect both Byzantine and Western (British) characteristics.

Still another group of hanging crosses, most generally represented by the cross from Bonderup, Sjælland (Zeeland), Denmark (Copenhagen, Nationalmus.), have arms shaped like palm leaves; their Scandinavian origin is abundantly proved by the fact that the same palm-leaf cross recurs in a great many runic stones and on coffins, especially from central Sweden (Södermanland). Generally speaking, these crosses can be dated to the second half of the 11th century, but it is not impossible that they recur somewhat later. In any case, a certain number of Scandinavian round silver brooches have similar palm-leaf crosses, and these no doubt belong to the 12th century. It is not difficult to trace the inspiration for these palm-leaf crosses; the name usually given in Scandinavian literature to these palm leaves is "Irish loops." Indeed, it seems to have been fully ascertained that these plant-motif crosses have their origin in English and Celtic art. A prime example to be used as reference is the sculptured stone from St. Paul's Churchyard in London (London, Guildhall Mus.), not to mention the many English illuminated manuscripts with corner loops in the same style (I, PL. 280).

The "Irish loops" and the palm-leaf crosses constitute, however, only a small portion of the plant ornamentation which, owing to inspiration from the West, appears in the earliest Scandinavian Christian monuments, such as the sarcophage of the Eskilstuna type (PL. 328), the runic stones marked with crosses, the sculptured stones of the Ringerike type, the Christian weather vanes, the earliest stave churches (Stavkyrkorna), such as those at Urnes in Norway and Hemse in Gotland, and, naturally, in such important monuments as the Jelling stone in Denmark with its representation of Christ (XII, PL. 414), as well as in the well-known reliquaries in the cathedrals of Cammin, Poland, and Bamberg, Germany.

Agreement has, for the most part, been reached among scholars that the plant elements in the decoration of these monuments were borrowed from British art, but one finds at the same time an astonishing tendency to label similar features in England and Ireland as "Scandinavian influences." One of the reasons for this tendency is that a too-early dating has been given to the Scandinavian monuments, whereas English and Irish monuments often had to be given fairly late dates, because of the presence of inscriptions. When examining the existing material as impartially as possible one arrives at the conclusion that by far the greatest part of the Scandinavian material must be dated somewhat later than has been customary; at the same time it must be acknowledged that the artistic influences of the British Isles in Scandinavia during the 10th and 11th centuries remains considerably more firmly established than those in the opposite direction.

Scandinavian pre-Romanesque art during the later part of the 11th century is very well represented by the above-mentioned two reliquaries from Cammin and Bamberg; they are cut in horn and abound in plant and animal ornamentation. Much of the cheerfulness and recklessness that characterize the Scandinavian spirit persists here, along with the more peaceful Christian characteristics from the West.

The weather vanes of bronze from Heggen (Oslo, Universitetets Oldsaksamling) represent essentially the same artistic conception. They show elegant and concise lines, but their style does not strive for realism; on the contrary, both the twining plants and the animal figures are idealized in a Norse spirit of very ancient origin (see XIV, PL. 463).

The old stave church at Urnes (X, PL. 438; XII, PL. 233), from the second half of the 11th century, is rightly considered one of the most perfect examples of Scandinavian art from the transitional period preceding the Romanesque era. A special artistic style (the Urnes style), which is also widely represented in the British Isles, was named after it. Most Scandinavian scholars classify this style as having derived from eastern Sweden, and most British scholars define all monuments on the British Isles showing these elements as of Scandinavian influence. The opposite point of view, however, might very well be maintained; in the British Isles this style is used in all feasible materials, including religious stone architecture, and it is often of a high order of workmanship. Even though it is mainly to be found in areas formerly occupied by Scandinavians, this special style developed only several decades after the Norse disappeared from Britain. It reached its peak during the second half of the 11th century, that is, during a period in which Norway was thoroughly subjected to cultural and Christian influence from the British Isles. One might well imagine that the new style took root in Norwegian villages at this stage, and that, among others, the superbly decorated Urnes stave church was built under this influence. The style might subsequently have spread to eastern Sweden, among other areas, in connection with the veneration of St. Olaf (King Olaf II, d. 1030), Norway's patron saint. Many facts that now seem unclear in the history of Scandinavian art might be explained in the light of a development that followed the diffusion of the cult of Olaf.

In the religious art of the 12th century, many of the stylistic elements dealt with above recur, so that they come within the definition of pre-Romanesque. The cornices, transverse beams on the walls, baptismal fonts, and porch sculptures of the churches often cling strikingly to the old traditions. This should not, however, be interpreted as a tenacious adherence to the past, but rather as a direct and natural continuation of an art that was at its peak during the later part of the 11th century and still lived on in the following century (see XII, PL. 415).

A good example of this tradition continuing through the 12th century is the curious crucifix from Bru, Norway, in which the figure of Christ is a 9th-century work done in continental Europe, whereas the cross and the base are from the 12th century and by a local goldsmith. Other examples showing the same stylistic elements are the beautiful crucifix from Halikko, Finland (Helsinki, Nat. Mus.), with its animal-head decorations, and a reliquary.

In secular artistic production there is a wealth of material, especially in metalwork. During the 11th and part of the 12th century, exactly as in religious art, certain tendencies appear that do not belong in the Romanesque scheme but are of different origin. As a rule this is a minor art of provincial nature having little to do with the evolution of international art, and it should serve as a warning to all those who believe that Northern art during the 11th century proved a source of inspiration far beyond the Scandinavian borders.

The round brooches or pendants of gold and silver deserve special attention. During the 11th and 12th centuries there continued to be a flourishing local production, which from the latter part of the 9th century had received strong influences from the British Isles, but nevertheless succeeded in maintaining a style of its own. The composition of the ground, usually divided by a cross or into three portions, is of particular interest. Loops, rings, and palm knots are customary, as well as animal motifs. In all these details these products reveal a clear inspiration from British sources, whereas continental European influences seem much less evident.

A group of round gold pendants from Gotland, Sweden (Stockholm, Statens Historiska Mus.), are among the most interesting of this group. Characteristic of them is a surface richly ornamented with filigree and divided into three fields separated from each other by luxuriant triquetras. Among these is an animal ornament strongly reminiscent of those animals that in the 10th century were shown coiled up into a loose knot of loops; others resemble birds with extended wings. In some of the pendants the composition has been transformed into a loose palm knot in which the animal motifs can scarcely be distinguished. Probably most of these pendants belong to the 11th century, but there are many indications that they continued to be made even later; several of them were found as part of treasures which were unlikely to have been buried until the second half of the 14th century.

Another rich and decorative group consists of round filigree brooches of silver from Bredsätra and Kumla (Stockholm, Statens Historiska Mus.), the surfaces of which are often divided by a cross and have animal figures in the fields. The figures are birds alternating with snakes or dragons. They are clearly derived from earlier Norse art, specifically from the 10th century, but they were made more or less according to the same pattern throughout the 11th and 12th centuries. Whether these brooches show any noteworthy Romanesque influence is debatable; the ground composition is, in fact, not Romanesque, nor are the animal figures. The latter are definitely local, even though they originally came from a British environment. It might be possible to interpret the often quite exuberant pampres found between the animal figures as a Romanesque element, or perhaps a Carolingian or Ottonian influence.

From the latter part of the 11th century there is a group of brooches akin to those previously discussed, and common mainly in southern Scandinavia. The brooches are of gold and have a wide border with embossed bird figures similar to those mentioned above. One of them shows a reniform snake coiled among the birds. Both the animal figures and spaces between are richly ornamented with filigree. Two magnificent gold brooches with filigree from Hornelung, Jutland (Copenhagen, Nationalmus.), are characteristic examples of the stylistic eclecticism that dominated the Scandinavian goldsmith's art on the threshold of the Romanesque era. These brooches, certainly local, are based on the principles of composition that had been adopted from British sources by the end of the 9th century and that were to distinguish Scandinavian gold- and silverwork during the 10th and 11th centuries. In one of them the composition is cruciform with four animal shapes arranged as palmettes forming the cross arms and intertwining loops and rings in the intervening spaces. The other brooch shows a tripartite composition with a profuse ornamentation of rings, loops, and palms. Local Scandinavian as well as British traits are quite obvious here, but there is some question as to whether a strain of continental European origin can be traced in the vine and palm borders of the outer fields.

In pre-Romanesque Scandinavian metalwork the silver bowls — with or without fluting and as a rule decorated with mixed festoons below the upper rim — form a group with a character of its own. They are found in treasures from the second half of the 11th century and from the 12th century. Earlier it was assumed that these bowls were objects imported from the East, but such an assumption is untenable for many reasons. The pampres engraved below the rims had long since been incorporated in Scandinavian art by the time these bowls were made; furthermore, part of the pampres show specific Scandinavian characteristics, as, for instance, in a bowl from Älvkarleby, Uppland (PL. 328). In no case has it been possible to prove direct oriental influence. The animal figures all belong to the fauna that, from the 11th century, occupied an increasingly important place in Scandinavian art and were then absorbed into the religious art of the Romanesque period. There are many signs that these silver bowls, as well as other contemporary works in precious metals, originated in a group of workshops in eastern Sweden.

A particular type of metal incrustation that is especially well-represented in Finland and Iceland is also pre-Romanesque in origin. The earliest examples appear during the final stage of the viking era, but the phenomenon can subsequently be traced through the 13th century. Ornamentation chiefly consists of festoons, palmettes, and arbor vitae motifs, which cannot be defined as belonging to ordinary Scandinavian ornamental flora, although they are not Romanesque in the proper sense of the word. The richest material comes from Finland (sword hilts, axes, spear heads) and from Iceland (porch rings of iron). A few examples have also been found in the other Scandinavian countries and the British Isles. The question of origin is still not solved, but there are several reasons for assuming that this ornamentation penetrated the North along Eastern cultural routes during the late viking era and the early Middle Ages, and that its true origin is to be found in Byzantine art. For the time being, the most significant objects of comparison are some iron camp chairs; and at least a few of these have an ascertained Italian origin (one from Nocera Umbra and one from Pavia). Here, roughly the same ornamental flora as in the Scandinavian products is found. The dating of these pieces oscillates between the 6th and 12th centuries, and is probably somewhere in between. Since it can hardly be doubted that the said chairs belong to a Byzantine environment, or one influenced by Byzantine art, the same conclusion can probably be drawn with regard to the Scandinavian finds. The local character of the material (axes, swords, porch rings) nevertheless seems to indicate that it comes from local workshops.

The above-mentioned several groups in Scandinavian religious and secular art are often neglected in descriptions of the major periods in the history of art, although they present features of interest. The stylistic characteristics which they present are markedly pre-Romanesque, yet they survive as a sort of national undercurrent during almost the entire Romanesque period up to the Gothic era. These characteristics must be regarded as national, or even provincial, since on the whole they preserve the Norse traditions of the viking era; but at the same time they are the bearers of artistic borrowing from the East, the West, and the South and thus contribute to enriching the history of world art.

Wilhelm HOLMQVIST

BIBLIOG. *General*: E. H. Zimmermann, Vorkarolingische Miniaturen, 5 vols., Berlin, 1916; P. Frankl, Die frühmittelalterliche und romanische Baukunst, Wildpark-Potsdam, 1926; J. Baum, Die Malerei und Plastik des Mittelalters, II: Deutschland, Frankreich und Britannien, Wildpark-Potsdam, 1930; H. T. Bossert, Geschichte des Kunstgewerbes aller Zeiten und Völker, V, Berlin, 1932; R. Hinks, Carolingian Art, London, 1935 (repr. Ann Arbor, Mich., 1962); E. Knögel, Schriftquellen zur Kunstgeschichte der Merowingerzeit, BJ, CXL–CXLI, 1936, pp. 1–258; H. Focillon, Art d'occident, Paris, 1938 (Eng. trans., D. King, 2 vols., New York, 1963); J. Hubert, L'art préroman, Paris, 1938; C. R. Morey, Mediaeval Art, New York, 1942; N. Åberg, The Occident and the Orient in the Art of the 7th Century, 3 vols., Stockholm, 1943–47; G. Bandmann, Mittelalterliche Architektur als Bedeutungsträger, Berlin, 1951; H. Focillon, L'an mil, Paris, 1952; W. F. Volbach, Elfenbeinarbeiten der Spätantike und des frühen Mittelalters, Mainz, 1952; Settimane di studio del Centro italiano di studi sull'alto medioevo, Spoleto, 1953 ff.; A. Grabar and C. Nordenfalk, Early Medieval Painting, Geneva, New York, 1957; Karolingische und ottonische Kunst: Werden, Wesen, Wirkung (Forsch. zur Kg. und christlichen Archäol., III), Wiesbaden, 1957; L. Grodecki, Au seuil de l'art roman: L'architecture ottonienne, Paris, 1958; K. J. Conant, Carolingian and Romanesque Architecture, 800–1200, Harmondsworth, 1959; H. Jantzen, Ottonische Kunst, 2d ed., Hamburg, 1959. See also bibliogs. for CAROLINGIAN PERIOD; OTTONIAN PERIOD.

France: E. Le Blant, Les sarcophages chrétiens de la Gaule, Paris, 1886; E. Michon, Les sarcophages de Saint-Drausin de Soissons . . . et les sarcophages chrétiens dits de l'école d'Aquitaine ou du Sud-Ouest, Mél. Schlumberger, II, Paris, 1924, pp. 376–85; L. Bréhier, L'art en France dès invasions barbares à l'époque romane, Paris, 1930; J. Baum, La sculpture figurale en

Europe à l'époque mérovingienne, Paris, 1937; J. B. Ward Perkins, The Sculpture of Visigothic France, Archaeologia, LXXXVII, 1938, pp. 79–128; L. Gischia and L. Mazenod, Les arts primitifs français, Paris, 1939; A. Grabar, Martyrium, 2 vols., Paris, 1943–46; R. Rey, L'art roman et ses origines, Toulouse, 1945; D. Fossard, Les chapiteaux de marbre du VIIe siècle en Gaule: style et évolution, CahA, II, 1947, pp. 69–85; R. Lantier and J. Hubert, Les origines de l'art français, Paris, 1947; E. Salin, La civilisation mérovingienne, 4 vols., Paris, 1949–59; E. Mâle, La fin du paganisme en Gaule et les plus anciennes basiliques chrétiennes, Paris, 1950; P. Deschamps and M. Thibout, La peinture murale en France: Le haut moyen âge et l'époque romane, Paris, 1951; J. Hubert, L'architecture religieuse du haut moyen âge en France, Paris, 1952; S. McK. Crosby, L'Abbaye Royale de Saint-Denis, Paris, 1953; G. H. Forsyth, Jr., and W. A. Campbell, The Church of St. Martin at Angers, 2 vols., Princeton, 1953; D. Fossard, Répartition des sarcophages mérovingiens à décor en France, Et. mérovingiennes, Paris, 1953, pp. 117–24; Y. Christ, Les cryptes mérovingiennes de l'Abbaye de Jouarre, Paris, 1955; A. Sassier, L'évolution de la sculpture paléochrétienne et préromane en Septimanie, Et. roussillonnaises, VI, 1957, pp. 167–214; D. Costa, Le décor architectonique à l'époque mérovingienne dans le pays nantais, B. Soc. archéol. et h. de Nantes et de Loire-Atlantique, XCVIII, 1959, pp. 173–93; E. Salin, Les tombes gallo-romaines et mérovingiennes de la basilique de Saint-Denis, Mém. Acad. des inscriptions et belles-lettres, XLIV, 1, 1960, pp. 169–263; J. Puig y Cadafalch, L'art wisigothique et ses survivances, Paris, 1961.

Germany: B. Salin, Die altgermanische Tierornamentik, Stockholm, 1904; A. Goldschmidt, Die Elfenbeinskulpturen aus der Zeit der karolingischen und sächsischen Kaiser, 4 vols., Berlin, 1914–26; A. Goldschmidt, Die deutsche Buchmalerei, 2 vols., Leipzig, Munich, 1928 (Eng. trans., German Illumination, 2 vols., Florence, New York, 1928); E. Gall, Karolingische und ottonische Kirchen, Burg bei Magdeburg, 1930; W. Köhler, Die karolingischen Miniaturen: Die Schule von Tours, 2 vols., Berlin, 1930–33; A. W. von Jenny and W. F. Volbach, Germanischer Schmuck, Berlin, 1933; R. Hamann-MacLean and J. Verrier, Frühe Kunst im westfränkischen Reich, Leipzig, 1939; W. Holmqvist, Kunstprobleme der Merovingerzeit, Stockholm, 1939; G. L. Micheli, L'enluminure du haut moyen âge et les influences irlandaises, Brussels, 1939; A. W. von Jenny, Die Kunst der Germanen im frühen Mittelalter, Berlin, 1940; G. Behrens, Merowingerzeit, Mainz, 1947; E. Lehmann, Der frühe deutsche Kirchenbau, 2d ed., Berlin, 1949; G. Haseloff, Der Tassilokelch, Munich, 1951; H. E. Kubach and A. Verbeek, Die vorromanische und romanische Baukunst in Mitteleuropa: Literaturbericht, ZfKg, XVI, 1951, pp. 124–48; R. Noll, Frühes Christentum in Österreich, Vienna, 1954; W. Holmqvist, Germanic Art during the First Millennium A.D., Stockholm, 1955; R. Wesenberg, Bernwardinische Plastik, Berlin, 1955; J. Werner, Beiträge zur Archäologie des Attila Reiches, Munich, 1956; H. Schnitzler, Rheinische Schatzkammer, 2 vols., Düsseldorf, 1957–59; W. Köhler, Die karolingischen Miniaturen: Die Hofschule Karls des Grossen, 2 vols., Berlin, 1958; E. Steingräber, Deutsche Plastik der Frühzeit, Königstein im Taunus, 1961.

England: A. W. Clapham, English Romanesque Architecture before the Conquest, I, Oxford, 1930; F. Wormald, Decorated Initials in English Manuscripts from A.D. 900–1100, Archaeologia, XCI, 1945, pp. 107–35; T. D. Kendrick, Late Saxon and Viking Art, London, 1949; D. Talbot Rice, English Art, 871–1100 (Oxford History of English Art, II), Oxford, 1952; F. Wormald, English Drawings of the 10th and 11th Centuries, London, 1952; C. F. Battiscomb, The Relics of St. Cuthbert, Oxford, 1956; A. L. Binns, 10th Century Carvings from Yorkshire and the Jellinge Style, Bergen, 1956; G. Webb, Architecture in Britain: The Middle Ages, London, 1956; D. M. Wilson, Two Plates from a Late Saxon Casket, AntJ, XXXVI, 1956, pp. 31–39; H. M. Taylor, Some Little-known Aspects of English Pre-Conquest Churches, The Anglo-Saxons: Studies . . . presented to Bruce Dickins, London, 1959, pp. 137–58; M. and L. de Paor, Early Christian Art in Ireland, 2d ed., London, 1960; F. Henry, L'art irlandais, I, Paris, 1963. See also bibliog. for ANGLO-SAXON AND IRISH ART.

Spain: *a. Visigothic period*: E. Camps Cazorla and J. Ferrandis, Arte visigodo, in R. Menéndez Pidal, ed., Historia de España, III, Madrid, 1940, pp. 437–666; H. Schlunk, Arte visigodo, Ars Hispaniae, II, Madrid, 1947, pp. 227–416; P. de Palol Salellas, Arqueología paleocristiana y visigoda, Atti IV Cong. int. de ciencias prehistóricas y protohistóricas, II, Madrid, 1954, pp. 1–38; P. de Palol Salellas, Esencia del arte hispánico de época visigoda: romanismo y germanismo, Settimane di studio del Centro italiano di studi sull'alto medioevo, III, 1955, pp. 65–126; P. de Palol Salellas, Los monumentos paleocristianos y visigodos estudiados en España desde el año 1939 a 1954, Actes Ve Cong. int. d'archéol. chrétienne (Vatican City, 1954), Paris, 1957, pp. 87–95 (full bibliog.); J. Fontaine, Isidore de Seville et la culture classique dans l'Espagne wisigothique, Paris, 1959; J. Puig y Cadafalch, L'art wisigothique et ses survivances, Paris, 1961. *b. Asturian art*: F. de Selgas, Monumentos ovetenses, Madrid, 1910; M. Gómez Moreno, Iglesias mozarabes, 2 vols., Madrid, 1919; M. Gómez Moreno, El arte románico español, Madrid, 1934; H. Schlunk, Arte asturiano, Ars Hispaniae, II, Madrid, 1947, pp. 327–44; H. Schlunk, The Crosses of Oviedo: A Contribution to the History of Jewelry in Northern Spain in the 9th and 10th Centuries, AB, XXXII, 1950, pp. 91–114; D. Andújar Polo, Repertorio bibliográfico de arte y arqueología asturiana, B. Inst. de est. asturianos, IX, 1955, pp. 1–33; H. Schlunk and M. Berenguer, La pintura mural asturiana de los siglos IX y X, Madrid, 1957; L. Menéndez Pidal, Influencia y expansión de la arquitectura pre-románica asturiana, en alguna de sus manifestaciones, B. Inst. de est. asturianos, XV, 1961, pp. 417–30. *Caliphate Art*: See bibliogs. for MOORISH STYLE; MOZARABIC ART.

Italy: *a. Architecture*: R. Cattaneo, L'architettura in Italia dal secolo VI al Mille circa, Venice, 1889 (Eng. trans., I. Curtis-Cholmeley, London, 1896); A. K. Porter, Lombard Architecture, 4 vols., New Haven, 1915–17; C. Cecchelli, Il Tempietto longobardo di Cividale, Dedalo, III, 1922–23, pp. 735–60; C. Cecchelli, Miscellanea cividalese, Mem. storiche forogiuliesi, XXIII, 1927, pp. 57–82; Toesca, Md; G. Galassi, L'architettura protoromanico dell'Esarcato, Ravenna, 1928; T. Horia, Les églises à cinq coupoles en Calabre, EphDr, IV, 1930, pp. 149–89; T. Horia, Eglises cruciformes dans l'Italie méridionale, EphDr, V, 1932, pp. 22–34; G. Chierici, Note sull'architettura della contea longobarda di Capua, BArte, XXVII, 1933–34, pp. 543–53; J. Puig y Cadafalch, La géographie et les origines du premier art roman, Paris, 1935; A. Santangelo, Catalogo delle cose d'arte e di antichità d'Italia: Cividale, Rome, 1936; S. Steinmann-Brodtbeck, Herkunft und Verbreitung des Dreiapsidenchores, ZfSAKg, I, 1939, pp. 65–95; P. Verzone, L'architettura dell'XI secolo nell'Esarcato, Palladio, IV, 1940, pp. 97–112; G. Chierici, La Chiesa di S. Satiro a Milano, Milan, 1942; R. Krautheimer, The Carolingian Revival of Early Christian Architecture, AB, XXIV, 1942, pp. 1–38; G. Panazza, L'arte medievale nel territorio bresciano, Bergamo, 1942; P. Verzone, L'architettura religiosa dell'alto Medioevo nell'Italia settentrionale, Milan, 1942; C. Cecchelli, Monumenti del Friuli dal secolo IV all'XI, I: Cividale, Milan, 1943; C. Cecchelli, Rassegna di storia, archeologia ed arte del mondo barbarico, Mem. storiche forogiuliesi, XXXIX, 1943–51, pp. 122–98; G. P. Bognetti et al., S. Maria di Castelseprio, Milan, 1948; A. de Capitani d'Arzago, Le recenti scoperte di Castelseprio, BArte, XXXIII, 1948, pp. 7–23; M. Salmi, Influssi degli edifici antichi di culto sulle chiese dell'alto Medioevo in Italia, Atti IV Cong. int. di archeol. cristiana (1938), II, Rome, 1948, pp. 231–70; B. Pace, Arte e civiltà della Sicilia antica, IV, Milan, 1949, pp. 340–45; Atti I Cong. int. di studi longobardi (1951), Spoleto, 1952; L. Coletti and V. Piazzo, Il tempietto di Cividale, Rome, 1952; E. Dyggve, Il tempietto di Cividale, Atti II Cong. int. di studi sull'alto medioevo (1952), Spoleto, 1953, pp. 75–80; M. Salmi, L'architettura in Italia durante il periodo carolingio, Settimane di studio del Centro italiano di studi sull'alto medioevo, I, 1953, pp. 227–40; P. Verzone, I capitelli del tipo corinzio dal IV all'VIII secolo, Wandlungen christlicher Kunst im Mittelalter (Forsch. zur Kg., II), Baden-Baden, 1953, pp. 87–97; E. Arslan, L'architettura dal 568 al Mille, Storia di Milano, II, Milan, 1954, pp. 501–608; E. Arslan, Remarques sur l'architecture lombarde du VIe siècle, CahA, VII, 1954, pp. 129–37; Frühmittelalterliche Kunst in den Alpenländern (Akten III. int. Kong. für Frühmittelalterlicherforschung, 1951), Olten, Lausanne, 1954; E. Lehmann, Vom neuen Bild frühmittelalterlichen Kirchenbaus, Wissenschaftliche Z. der Martin-Luther-Univ., Halle-Wittenberg, VI, 1956–57, pp. 213–34; C. Cecchelli, Modi orientali . . . nell'arte del VII secolo in Italia, Settimane di studio del Centro italiano di studi sull'alto medioevo, V, 1957, pp. 371–426; M. Mazzotti, Cripte ravennati, Felix Ravenna, 3d Ser., XXIII, 1957, pp. 28–63; E. Franco, Il tempietto di Cividale, L'Architettura, IV, 1958–59, pp. 842–49; G. Chierici, Cimitile, Atti III Cong. int. di studi sull'alto medioevo (1956), Spoleto, 1959, pp. 125–37; Misc. di S. bresciani sull'alto medioevo, Brescia, 1959; G. Panazza, Reliquie di due monasteri longobardi nel Bresciano, Arte Lombarda, IV, 1959, pp. 17–28; G. Panazza, Le scoperte in S. Salvatore a Brescia, Arte Lombarda, V, 1960, pp. 13–21.

b. Sculpture: M. G. Zimmermann, Oberitalische Plastik in frühen und hohem Mittelalter, Leipzig, 1897; E. A. Stückelberg, Langobardische Plastik, Munich, 1909; C. Albizzati, Il ciborio carolingio nella basilica ambrosiana di Milano, RendPontAcc, II, 1924, pp. 197–265; C. Cecchelli, Reliquie trentine dell'età barbarica, S. trentini di scienze storiche, IX, 1928, pp. 193–210; A. Haseloff, Pre-Romanesque Sculpture in Italy (trans. R. Boothroyd), Florence, 1930; L. Karaman, Notes sur l'art byzantin et les Slaves catholiques, Orient et Byzance, V, 2, Paris, 1932, pp. 332–80; R. Kautzsch, Die langobardische Schmuckkunst in Oberitalien, Römisches Jhb. für Kg., V, 1941, pp. 1–48; P. Verzone, Note sui rilievi in stucco dell'alto medio evo nell'Italia settentrionale, Le Arti, IV, 1941–42, pp. 121–28; W. F. Volbach, Oriental Influences in the Animal Sculpture of Campania, AB, XXIV, 1942, pp. 172–80; P. Verzone, L'arte preromanica in Liguria ed i rilievi decorativi dei secoli barbari, Turin, 1945; G. de' Francovich, Il problema delle origini della scultura cosiddetta "longobarda," Atti I Cong. int. di studi longobardi (1951), Spoleto, 1952, pp. 255–74; G. Matthiae, La iconostasi della chiesa di S. Leone a Capena, BArte, XXXVII, 1952, pp. 293–99; G. Panazza, Lapidi e sculture paleocristiane e pre-romaniche di Pavia, Arte del primo millennio (1950), Turin, 1952, pp. 211–96; G. Rosa, La scultura decorativa preromanica a Milano, Storia di Milano, II, Milan, 1954, pp. 609–21; Corpus della scultura altomedievale, I–II, Spoleto, 1959–62.

c. Painting: P. Toesca, La pittura e la miniatura in Lombardia, Milan, 1912; J. Wilpert, Die römischen Mosaiken und Malereien der kirchlichen Bauten vom IV.–XIII. Jahrhundert, 4 vols., Freiburg im Breisgau, 1916; Toesca, Md; E. Arslan, La pittura e la scultura veronese dal secolo VIII al XIII, Milan, 1943; P. Cellini, Una Madonna molto antica, Proporzioni, III, 1950, pp. 1–6; G. Matthiae, Tradizione e reazione nei mosaici romani dei secoli VI e VII, Proporzioni, III, 1950, pp. 10–15; K. Weitzmann, The Fresco Cycle of S. Maria di Castelseprio, Princeton, 1951; N. Gabrielli, Le miniature delle Omelie di San Gregorio della Biblioteca capitolina vercellese, Arte del primo millennio (1950), Turin, 1952, pp. 301–11; C. Cecchelli, Pittura e scultura carolingie in Italia, Settimane di studio del Centro italiano di studi sull'alto medioevo, I, 1953, pp. 181–214; A. Grabar, La peinture byzantine, Geneva, 1953 (Eng. trans., S. Gilbert, Geneva, 1953); E. Arslan, La pittura dalla conquista longobarda al Mille, Storia di Milano, II, Milan, 1954, pp. 625–61; G. de' Francovich, Problemi della pittura e della scultura preromanica, Settimane di studio del Centro italiano di studi sull'alto medioevo, II, 1954, pp. 355–519; G. de' Francovich, Il ciclo pittorico della chiesa di S. Giovanni a Münster nei Grigioni, Arte Lombarda, II, 1956, pp. 28–50; M. Salmi, Un problema storico-artistico medievale, BArte, XLIII, 1958, pp. 213–31; G. P. Bognetti, Castelseprio, Milan, 1961; E. Kitzinger, Römische Malerei vom Beginn des VII. bis zur Mitte des VIII. Jahrhunderts Munich, n.d. (diss.).

d. Minor arts: M. Salmi, Il tesoro del duomo di Milano, Dedalo, V, 1924–25, pp. 267–88, 358–82; C. Cecchelli, Il tesoro del Laterano, Dedalo, VII, 1926–27, pp. 139–66, 231–56; Toesca, Md; W. F. Volbach, Mittelalterliche Bildwerke aus Italien und Byzanz, Berlin, 1930; W. F. Volbach, Reliquie e reliquari orientali in Roma, BArte, XXXI, 1937, pp. 337–50; Y. Hackenbroch, Italienische Emails des frühen Mittelalters, Basel, Leipzig, 1938; A. de Capitani d'Arzago, Antichi tessuti della basilica ambrosiana, Milan, 1941; G. de' Francovich, Arte carolingia ed ottoniana in Lombardia, Römisches Jhb. für Kg., VI, 1942–44, pp. 113–255; G. B. Tatum, The Paliotto of Sant'Ambrogio in Milan, AB, XXVI, 1944, pp. 25–45; C. Cecchelli, La decorazione paleocristiana e dell'alto Medio Evo nelle chiese d'Italia, Atti IV Cong. int. di archeol. cristiana (1938), II, Rome, 1948, pp. 127–212; C. Cecchelli, La vita di Roma nel Medio Evo, I: Arti Minori, Rome, 1952; V. H. Elbern, Der karolingische Goldaltar in Mailand, Bonn, 1952; V. H. Elbern, Der Ambrosiuszyklus am karolingischen Goldaltar zu Mailand, Mitt. des kunsthist. Inst. in Florenz, VII, 1953–56, pp. 1–8; G. Matthiae, La cultura artistica in Roma nel secolo IX, RIASA, N.S., III, 1954, pp. 257–74; G. Rosa, Le arti minori dalla conquista longobarda al Mille, Storia di Milano, II, Milan, 1954, pp. 665–716; P. L. Zovatto, Decorazioni musive pavimentali del secolo IX in abbazie benedettine del Veneto, Settimane di studio del Centro italiano di studi sull'alto medioevo, IV, 1956, pp. 417–22; G. Panazza, Sculture preromaniche ... della riviera occidentale del Garda, Mem. Ateneo di Salò, XVIII, 1957–59, pp. 137–59; P. J. Nordhagen, Nuove constatazioni sui rapporti artistici tra Roma e Bisanzio sotto il pontificato di Giovanni VII, Atti III Cong. int. di studi sull'alto medioevo (1956), Spoleto, 1959, pp. 445–52.

Scandinavia: B. Salin, Nagra krucifix och kors i Statens Historiska Museum, Svenska fornminnesföreningens tidskrift, VIII, 3, 1893, pp. 277–312; H. Shetelig, Urnesgruppen, Foreningen til norske fortidsmindesmerkers bevaring, Aarsberetning, LXV, 1909, pp. 75–107; S. Lindqvist, Den helige Eskils biskopsdöme, Stockholm, 1915; H. Shetelig, To danske pragtskrin fra vikingetiden, Kunst og haandverk, Oslo, 1918, pp. 193–99; B. Salin, Förgylld flöjel från Söderala kyrka, Fornvännen, 1921, pp. 1–22; J. Brøndsted, Early English Ornament, Copenhagen, 1924; C. A. Nordman, Karelska järnåldersstudier, Helsinki, 1924; A. W. Brøgger, Bronsefliøene fra Heggen og Tingelstad kirker (Norske Oldfunn, V), Oslo, 1925; S. S. Lindqvist, Yngre vikingastilar, in H. Shetelig, ed., Konst (Nordisk Kultur, XXVII), Stockholm, 1931, pp. 144–79; J. Brøndsted, Inedita aus dem dänischen Nationalmuseum: Eisenzeit, ActaA, V, 1934, pp. 167–83; H. Shetelig, Specimens of the Urnes Style in English Art of the Late 11th Century, AntJ, XV, 1935, pp. 22–25; M. Rydbeck, Skånes stenmästare före 1200, Lund, 1936; C. R. af Ugglas, Gotländska silverskatter fran Valdemarstågets tid, Stockholm, 1936; E. Kivikoski, Die Eisenzeit im Auraflussgebiet, Helsinki, 1939; N. Åberg, Keltiska och orientaliska stilinflytelser i vikingatidens nordiska konst, Stockholm, 1941; R. Skovmand, De danske skattefund, Aarbøger for nordisk oldkyndighed historie, 1942, pp. 1–275; C. A. Nordman, Gotländisch oder deutsch: ein Silberkruzifix von Halikko im eigentlichen Finnland, ActaA, XV, 1944, pp. 29–62; W. Holmqvist, Sigtunamästaren och hans krets, Situne Dei, VII, Sigtuna, 1948; T. D. Kendrick, Late Saxon and Viking Art, London, 1949; W. Holmqvist, Viking Art in the 11th Century, ActaA, XXII, 1951, pp. 1–56; E. Kivikoski, Die Eisenzeit Finnlands, II, Helsinki, 1951; H. Shetelig, The Norse Style of Ornamentation in the Viking Settlements, Viking Antiquities, VI, Oslo, 1954, pp. 113–50; M. MacDermott, The Kells Crozier, Archaeologia, XCVI, 1955, pp. 59–113; M. Blindheim, Brukrusifikset — et tidlig-middelaldersk klenodium, Årbok Univ. oldsaksamling, Oslo, 1956–57, pp. 151–93; D. M. Wilson, An Inlaid Folding Stool in the British Museum, Medieval Archaeol., I, 1957, pp. 39–56; J. Brøndsted, Vikingerne, Copenhagen, 1960.

• •

Illustrations: PL. 307–328; 5 figs. in text.

PRESERVATION OF ART WORKS. By preservation of art works is meant the sum of all the measures taken by a community to guarantee the survival and public enjoyment of art works and other objects that constitute the cultural heritage of a community. An essential prerequisite to preservation is, therefore, the recognition of a collective interest in certain works of art regardless of their actual ownership. Even when such objects are not the property of the community at large but of an individual, it is felt that the possession of such works entails certain moral and objective responsibilities toward the community interest, and the legal aspects differ from those arising from ownership of any other type of property. The determination of what works are to be subject to preservation is closely related to the appreciation of their historicocultural and esthetic value; therefore, preservation is of importance in the history of criticism (q.v.), of style, and of all that can be considered as manifestations of criticism "in action" (see MUSEUMS AND COLLECTIONS; PATRONAGE; TOWN PLANNING). Preservation also includes preventive measures exercised through the application of technical and legal regulations restricting the use and disposal of art works and measures for safeguarding them. The recognition and care of art works (see RESTORATION AND CONSERVATION), both of which are regarded as an active part of preservation, are carried out by special technical organizations (see INSTITUTES AND ASSOCIATIONS).

SUMMARY. Principles (col. 688): *Reasons for preservation; Determination of works to be preserved; General implementation.* Historical precedents (col. 691): *Antiquity; Middle Ages to modern times.* Modern legislation (col. 694). International preservation (col. 701).

PRINCIPLES. The preservation of works of art consists of the following fundamental aspects: (1) the reasons for preservation; (2) the determination of which objects are to be preserved; and (3) the general technical and legal means by which preservation is effected.

Once the esthetic, anthropological, sociological, and historical values of art works have been recognized, the desire to retain them for the enjoyment of future generations gives rise to various measures for their preservation and tutelage, in view of their possible destruction by natural causes and by unintentional or willful human actions. Natural depreciation is brought about by the physicochemical properties of the atmosphere, climatic conditions, soil erosion, microorganisms, animal and vegetable parasites, combustibility, etc. In dealing with these causes, preventive measures can be taken to forestall their effects, as well as restorative measures to repair damage and to render the material less subject to natural depreciation (see RESTORATION AND CONSERVATION).

Many of the human actions that, unintentionally or otherwise, jeopardize the integrity of the historical and artistic heritage are connected with the predominance of the practical necessities of community life over the sense of respect for such a heritage. Great damage has been caused (and is still being caused despite greater awareness of the problem and more specific preservative measures) by population increase and economic expansion. The constant growth of cities has radically changed the character of whole areas as well as the individual structures that formed an integral part of their appearance. There have been many cases in which new forms of community life and new city planning measures have led to the demolition of ancient structures, parks, and gardens in order to make room for other buildings, streets, or other public areas. In many cases the inappropriate use of monuments has disfigured them or subjected them to undue wear and tear. The commercial development of land and waterways has often resulted in disfigurement of the landscape and sometimes even in the devastation of areas of archaeological interest.

Other human actions, while not intentionally destructive or injurious, nevertheless result in destruction or alteration such as the removal of objects or rendering them permanently inaccessible to public view, the failure to carry out such conservation measures as circumstances require, or the improper handling of works of art. These kinds of actions derive from the predominance of other interests, ignorance or indifference to the artistic value of the objects (as when paintings and sculptures on sacred themes have been altered for purposes of worship), or a lack of awareness of the need for preservation; thus it has become necessary to place preservation on a legal basis to ensure the priority of the reasons that make an object subject to protection over all other interests.

Acts of deliberate vandalism are a source of injury to the historical and artistic heritage, but fortunately these are becoming increasingly rare. In many cases, however, the knowledge of the artistic — and consequently the commercial — value of art objects has led to what might be termed "hoarding," that is, removing them from public view, or to their dispersal, thereby changing their reference to a certain place or to their original function, which often forms a large part of an object's historical interest. The enormous list of such cases ranges from the theft of art works belonging to public institutions, to archaeological excavations conducted for the sole purpose of appropriating any valuable objects, and to the clandestine export of art works of special significance. Even the collecting of art, while it has frequently ensured the physical survival of art works, has, precisely because it encourages their dispersion, had negative effects upon the completeness of the historical and artistic heritage (see MUSEUMS AND COLLECTIONS).

Reasons for preservation. A society that specifies certain art works for preservation makes a choice whereby the esthetic appreciation of those works is explicitly expressed (see CRITICISM) or a special significance is attributed to them. The protection of certain objects of an artistic nature was undertaken in primitive cultures primarily in connection with the social function of such objects, inasmuch as they had a religious significance, or symbolized the societal structure, social institutions, or ruling power, or because their material value warranted their inclusion in the community treasury. Preservation, therefore, was originally motivated not by the appreciation of the object's artistic qualities but by the importance attributed to *monumenta,* in the etymological sense of the word, that is, to symbols of the collective ethos. Consequently, objects appreciated for their artistic qualities but not characterized by special ethicosocial values were at most subject only to general legislation relating to ownership of public or private property. The progressive refinement of esthetic sensitivity led to a collective concern with artistic production and came to include within the concept of *monumenta* art objects that did not fulfill any specific religious or commemorative function but rather constituted an ornament to cities or other public places and were therefore intended for the public's enjoyment (see MONUMENTS).

Certainly there existed, if only in embryonic form, an awareness that certain works of art reflected the spiritual life of the society and were in an ideal sense the property of the community at large. The purpose underlying the plundering of a region by its conquerors was not only to appropriate material wealth or symbols of religious or temporal power but also to deprive the conquered community of symbolic objects of which it might be proud. Polybius, who in his *Histories* pleaded not to have the statues of the Achaean hero, Philopoemen, pulled down, expressed a sorrow commonly shared for the looting that had taken place in his homeland. In the hands of the victors the looted art objects acquired a new symbolic value as material signs of victory (*monumenta victoriae*); for this reason they were preserved, together with a record of the circumstances under which they had been acquired, in temples, churches, and public buildings and became objects placed under special protection. Just as in ancient times, modern wars have caused the most severe damage to the historical and artistic heritage. The effects of war are unpredictable and in regard to preservation raise special technical problems and require specific international agreements.

Determination of works to be preserved. In advanced societies the concept of the artistic heritage as including art objects that the society means to preserve together with the concept of monuments as historical documents of common interest (in the broadest sense of the term) provide a basis for determining what is to be preserved by the society as a whole. While the work of art has an intrinsic value per se, the historical document has meaning only in its context. Only modern historically oriented cultures have asserted the permanent and universal character of the concept of the *monumentum.* This historical approach has also made possible the definition of artistic phenomena in such a way that all works of art are by their very nature covered by it (regardless of their authors' intentions) and are therefore to be regarded as historical documents (monuments). Inherent in this concept is the idea of uniqueness and typicality, which is amplified to include phenomena of nature considered part of man's cultural and natural heritage. Monuments to be protected therefore include such a variety of things as sacred places, belongings of famous people, curios, battle trophies, relics of saints, and wildlife preserves as well as pure works of art. This general theoretical principle has given rise to the definition of "the historical and artistic heritage," which comprises things that help to explain the remote past and are specifically archaeological. Under the more historical aspects of man's culture fall those objects connected with civic and military history, customs and mores, political and legal codes, systems of religion, economics, town planning, and technology. Finally, this category includes everything of a biographical historical interest, such as manuscripts, bygones of various kinds, the residences of distinguished personalities, and sites that have become famous through literary references.

Included as specifically artistic are not only works of fine art of every kind but also those works which, though not unique or particularly important qualitatively, are of interest as documents of particular aspects of artistic culture such as handicraft products, coins and medals, even copies of lost works, as well as products of industrial design. In short, for the purposes of preservation the definition of a work of art is taken in its broadest historicocultural meaning, although its purely esthetic context continues to play an important role. Landscape architecture consisting of modifications done with specifically esthetic intentions belongs by all rights to the category of art works (see LANDSCAPE ARCHITECTURE). When the landscape itself has acquired traditional significance or typical forms through modifications resulting from man's occupation of the site, or when unique natural elements need to be preserved, these also form part of man's heritage.

General implementation. Measures for the preservation of man's historical and artistic heritage logically take the form of legislation but also depend on common sense when the law makes no specific provisions. Each country has generally dealt with its own problems, although beginnings have been made at international controls. The laws reflect a country's attitude toward its monuments and works of art and are conditioned by its general concept of individual rights and private property, variously delegating the necessary powers to the state or entrusting them to private organizations. The criteria underlying protective action fall into the general categories of preventive and restrictive measures and those that are primarily conservational (see RESTORATION AND CONSERVATION).

When the state itself assumes the responsibility of preserving monuments of various kinds, the first step is usually the formation of an administrative organ that decides what comprises the national heritage, the general regulations necessary, and who is to implement them. These recommendations are, if possible, translated into laws. The greater part of protective legislation is based on restrictions generally applied to privately owned sites and works and regulating their use. The nature of the lien is generally proportionate to the importance or the condition of the object: it may be purely interdictive, prohibiting any act that might change the *status quo* of the object, or it may be a relative restriction reconciling the requirements of preservation with other interests and with property rights. In the first case the restriction prohibits building on a particular panoramic or archaeological site, altering the appearance or structure of a building or putting it to uses incompatible with its dignity, or causing eventual damage. This type of restriction also prohibits private owners from withdrawing, removing, or exporting certain objects from their historical place of origin or from breaking up collections of particular interest. The prohibitive restriction is applied especially to archaeological excavations and even to the possession of chance archaeological finds. In many countries archaeological excavation by private individuals is illegal, and other types of excavation must be reported in advance to the proper authorities. Objects of archaeological interest are generally the property of the state, although the landowner or the discoverer may be compensated.

In the case of a relative restriction, statutory regulations limit the use of the protected object; certain building standards are established for a given zone (zoning regulation), or else new constructions must conserve the character of the landscape. While permits may be given for the export of some works of art or for the removal or withdrawal of some art collections, the community has an option on their purchase; moreover, it is obligatory to declare publicly the location of movable art works of particular interest.

Restriction, however, not only limits future action but also exacts positive protective measures. When the community delegates trusteeship of an object in its possession to a private owner, he is obligated to provide for proper maintenance of the buildings, to guarantee suitable settings for sites, and to permit the public to view private art collections.

The most effective way of implementing preservation is that of active trusteeship by the community or state, which provides directly, through appropriate conservation measures, for care and upkeep of its own historical and artistic heritage, removed from all private interests and acquired for the public domain by expropriation, legacy, or purchase. Having first declared all objects in the public domain inalienable and unalterable, the community then can provide for their custody and maintenance through appropriate organizations, house and display them to public view and enjoyment (see MUSEUMS AND COLLECTIONS), undertake their restoration (see RESTORATION AND CONSERVATION), and assure the proper setting for monuments and maintenance of their grounds.

Finally, monuments and the reasons underlying their custody and preservation must be publicized and understood. It is self-evident that the application of even the most rigorous legal measures can be of little value if there is no corresponding sense of responsibility among the individual members of society toward their heritage and if such a sense of responsibility is not deeply rooted in the civic awareness of the entire community.

HISTORICAL PRECEDENTS. *Antiquity.* While prescriptions for the custody and even the restoration of monuments and works of art can be encountered in both the ancient Near East and in Greece, the motives were primarily religious, practical, and esthetic. The concept of the *monumentum* was, in itself, a relative one; an object that was a *monumentum* for a given community at a given time was not one for another community and might even cease to be one for the same community at a later period in its history. Nor was a purely esthetic consideration, which is even more susceptible to change with time, sufficient to uphold the privilege of preservation for art works to which the definition of *monumentum* had ceased to apply. Because of the relativity of the concept of the *monumentum* and of esthetic judgement, preservation in an organized form has until recently been transitory and circumscribed.

As time went on works of art gradually came to be identified with their cities in addition to having a religious or decorative significance. Cicero, refuting Verres who claimed that he had purchased and not stolen the works of art he had collected when governor of Sicily, stated (*In C. Verrem*, II, iv, 59) that no city in Greece had ever consented to the sale of the statues or other art works that adorned them.

As Rome became filled with works of art brought from the conquered provinces, the ancient indifference gradually abated, though Fabius Maximus and even Cicero himself (*op. cit.*, II, iv, 14; and II, iv, 59–60) still regarded these works as superfluous and a sign of moral laxity. But the new interest was only in part due to an appreciation of the esthetic qualities of the works themselves; they were considered primarily as commemorative symbols of victories over subject peoples (*monumenta victoriae*) and were explicity declared to be the inalienable property of the Roman people. Cicero himself recalls that the consul Publius Servilius "brought to the Roman people" the works of art seized from the conquered cities and "had them entered in full in the official catalog of the public treasury" (Cicero, *op. cit.*, II, i, 21). Soon not only the works of art taken as spoils of war but also the statues and monuments that the Senate and Curia had erected in honor of illustrious persons, as well as the statues that private citizens in their wills ordered to be raised to their memory in public places, began to be considered as belonging to the Roman people. The latter were considered to be private property but for public use, on the assumption that they were intended principally for the adornment of the city.

The Romans were always jealously aware of this collective ownership of art works. Pliny records (*Naturalis historiae* XXXIV, xix, 61) that when Tiberius transported to his palace a marble statue by Lysippos (q.v.) that Agrippa had placed near his baths, the protest of the citizenry was so strong that the Emperor was obliged to return the work to its original location. Sometimes this sense of public ownership reached such a pitch that the citizens felt they could assert their right not only over the works that were, in fact, for the public's enjoyment but also over those collected in private homes and villas, on the pretext that they had been acquired as a result of conquests by the Roman people.

This already existing awareness of a common heritage was an important factor in the promulgation of laws providing for the protection as well as the appearance of the city. These laws also envisaged the imposition of severe restrictions and constituted one of the exceptional limitations on the full right of disposal of private property that was one of the basic principles of Roman law.

Special magistrates were appointed who were called *comites nitentium rerum* during the Augustan period. It was their duty to see to it that private owners did not neglect to conserve their buildings and repaired them when necessary. Later an edict of Vespasian, ratified by the *senatus consultum* in Hadrian's reign and then entered into the *Codex de aedificatis privatis*, forbade the demolition of buildings for the purpose of disposing of their ornaments on pain of the annulment of the sale and a fine amounting to twice the value of the object sold. At the same time the sale and export of any art object that decorated a building was forbidden even if such objects were not an integral part of the edifice (e.g. movable statues, paintings, and vases). The only course open to the owners was to donate or leave them as a legacy to the community.

All these measures for the supervision of public and private buildings formed part of the various city planning laws that were extended to all the territories of the Empire, such as the *Lex Municipalis* of Julius Caesar, the *Lex coloniae genetivae Iuliae* and the *Lex Malacitana.*

The practical enforcement of the principles underlying these laws were to the credit of enlightened emperors such as Vespasian, Hadrian, Caracalla, Diocletian, Theodosius (who ordered the demolition of constructions adjacent to monuments that disfigured their appearance), and Honorius.

During the first centuries of Christianity, when the respect for ancient civilization and ancient religions was opposed by Christian beliefs, the civil authorities were nevertheless aware of the need for reconfirming the traditional protective measures and for making them even more severe; in 450, the emperors Leo and Majorian passed an edict prescribing severe monetary penalties against the magistrates who had issued permits for the removal of artistic ornaments from ancient monuments for reuse in new public or private buildings and ordered very severe corporal punishment (amputation of a hand) to those who had benefited from these permits.

Middle Ages to modern times. During the Middle Ages respect for and interest in ancient monuments waned. The attitude of the Church to paganism and everything related to it, the instability of the times, the crusades, and the consequent lack of economic means resulted in the destruction, gradual decay, or dispersal of much of the artistic heritage of Europe and the Near East.

In the Carolingian period a historical and esthetic justification was found for the despoliation of buildings according to which the reuse of ancient parts meant an appreciation of them and had a symbolic value that attested to the cultural (and hence political) continuity between the Roman Empire and that of Charlemagne. In the same way the use of architectural elements from pagan structures in St. John Lateran (Rome) was intended to "pour into the new constructions the strength and glory of the old" (H. Lützler, *Vom Sinn der Bauformen*, I, Freiburg, 1953).

There were some noteworthy exceptions, particularly in Italy, where Theodoric, prompted by his minister Cassiodorus, wished not only to increase the beauties of the cities with art works that would perpetuate his memory, but also to restore the ancient monuments: "It is our intention not only to build new things, but also to preserve the old, because it is no less praiseworthy to discover as many things as possible than it is to acquire those which have been preserved" (Cassiodorus, *Variae*, II, 35). He ordered the architect Aloysius (Aloisius) to restore the baths of Abano and granted privileges to private individuals who provided at their own expense for conservation. Cassiodorus himself blamed the authorities in Ostuni for the poor state of their monuments and told them to send to Ravenna for craftsmen capable of restoring them.

Several popes followed Theodoric's example, granting the use of buildings to patrician families on condition that they assume the responsibility for maintaining them. That the ancient Roman city planning ordinances must still have been in vigor in Charlemagne's time can be seen from his appointment of Einhard as curator of the monuments in his kingdom.

Finally all conservation measures disappeared completely other than those for the simple maintenance of buildings and artistic accessories of religious worship or for the custody of private family possessions that might be of artistic value. The very concept of a common artistic heritage died out. Throughout Europe the Church was the custodian of learning, and the cultural heritage was preserved in cathedrals and abbeys. With Renaissance humanism the legislative tutelage of works of art was again taken up.

The legislation in the Papal States was particularly important in the historical development because of its direct basis in the Roman past and its cultural continuity. In Rome the destructions and looting through the centuries, particularly during the popes' residence in Avignon, reduced the appearance of the city to such a miserable condition that after the papacy's return to Rome (1377) the popes decided to repair the damages and prevent the perpetration of further ones.

Martin V, in the bull *Etsi de cunctarum* (1425), defined the destruction of public or private buildings as sacrilegious, ordered the demolition of all structures illicitly attached to monuments, exhorted the citizens to restore their own houses, and reestablished the ancient Roman office of Maestri delle Strade, (masters of the streets), whose duties included supervision of the city's artistic appearance. More detailed measures for the legal preservation of monuments were established later by Pius II (who had in verse decried the vandalism perpetrated in Rome) with the bull *Cum alman nostram urbem* (1462). Sixtus IV, as part of his grandiose scheme for the city's embellishment and the restoration of its monuments (Temple of Vesta and Arch of Titus), devised new measures for the preservation of the state's artistic heritage, and in another bull, *Etsi de cunctarum* (1474), reasserted the authority of the papal chamberlain and masters of the streets and forbade the removal of art works located in churches.

These measures had only a limited effect; the custom of taking marble from ancient monuments to use in new buildings or, worse, to make into lime continued unabated. A moving and dramatic testimony of this is found in a report that Raphael addressed to Leo X: "I would make bold to say that all this new Rome that we see today, however great, however beautiful, however adorned with palaces and churches and other buildings, has been built with lime made from ancient marbles" (J. D. Passavant, *Raffaello D'Urbino*, Florence, 1882, p. 376).

With the increasing appreciation of art, the need for preservation also became more apparent and was extended to paintings, sculpture, and other art works. When Raphael was made a commissioner of antiquities by Leo X, official authority was for the first time associated with technical competence. Paul III subsequently appointed the Latin scholar Giovenale Mannetti Commissioner of Antiquities and made him responsible for the custody of all the monuments in the city. Among other things he was directed "to do what is possible to conserve and free [antiquities] from thorn bushes, brush, trees, and especially ivy and wild figs; and not allow new houses or walls to be added to them, so that they are not allowed to fall into ruins, be reduced in size, destroyed, burnt for lime, or transported outside the city." At the same time certain restrictions began to be imposed to remove privately owned art works from completely arbitrary actions of their owners.

Other papal edicts limited excavations within the Papal States and prevented the export of archaeological objects without a permit approved by the pope himself. The objects to be preserved were better defined, and a distinction was made between works that were of particular artistic interest and those that were not. One reason advanced for preservation was the international prestige that works of art conferred upon the city and state, and the consequent revenue that would accrue to the economy from tourism. These measures were given a more unified form with an edict (1750) by Cardinal Valenti and another prepared by Cardinal Doria on the basis of a chirograph (1802) of Pius VII that reestablished the office of Inspector General of Fine Arts and appointed Antonio Canova to it.

The pontifical edicts of 1750 and 1802 served as the basis for the edict (1820) of Cardinal Pacca that is regarded as the most complete juridical text on the subject at that time. It established the first administrative structure in this field, with a central commission in Rome and a number of commissions of technicians located in the principal cities of the Papal States. It provided for the cataloguing of art works kept in public buildings; it even extended supervision to objects connected with folk art; it established a special customs duty on the export of art works; and it laid down specific norms for imposing restrictions, for the restoration of public and private monuments, and for the financing of such restorations. Though this edict was only partially applied in practice (it did not prevent the removal and export of sizable collections such as that of Cardinal Fesch in 1845 and the Campana collection in 1871), the importance of the modern principles that inspired it and made it the basis of much future legislation is undeniable.

In no other Italian state was the problem of the preservation and supervision of the artistic heritage so clearly defined as in the Papal States. Various provisions existed elsewhere, indicating an awareness of the problem, but they do not reveal a genuine legal tradition of preservation of art works.

Modern legislation. The current laws on preservation are based on the legal groundwork and the practical experiences of the various European countries. Although the laws have a common aim, their characteristics differ mainly with respect to the interpretation of property rights and to the organization of the services responsible for the application of these laws.

A thorough comparative study of legislation in this sphere does not exist, although such a study has been proposed many times, especially by the Secretariat of UNESCO, with a view to the coordination of the fundamental principles of such legislation on an international scale.

In France, more scrupulous care of the historical and artistic heritage was stimulated not only by national ideals but also by regret for the destruction and damage caused by the Revolution (even perpetrated intentionally: in 1792 a decree had been issued ordering the destruction of all monuments that recalled the "feudal age"). Demolition continued later as well, for in 1811 it had been impossible to prevent the destruction of the church of Cluny (see monuments).

Outstanding people in cultural fields were of great importance in creating a greater awareness of preservation problems. The writings of Chateaubriand and Guizot were a case in point; in 1825 Victor Hugo published an essay *Guerre aux Démolisseurs* and in 1832 an article of the same title (in the *Revue des Deux Mondes*); in 1833 Montalembert published an article entitled "Du vandalisme en France." Also noteworthy were the activities of the Société Française d'Archéologie pour la Conservation et Description des Monuments Historiques (and its *Bulletin Monumental*), which gave rise to many local associations inspired by the same ideals. After the July revolution (1830) the Ministry of Education, evidently inspired by the Pacca Edict, established the office of Inspector General of Historical Monuments, appointing first Louis Vitet and then the writer Prosper Merimée to it. The functions of the office included general supervision, and it had a special budget for restorations. Later the Inspector General was replaced by a Commission des Monuments Historiques (1837), which acted as a central administrative body, assisted by a Comité Historique des Arts et Monuments with special technical functions; both these bodies were responsible for the compilation of an inventory of the principal historical and artistic monuments whose maintenance was to be provided for by the state. These decrees and the technical and administrative organization they established were given full legal sanction by the law of 1887. As a result France possesses some of the most complete legislation on preservation. Fundamental is the concept of *classement* (classifying monuments by their historical significance), which dates back to 1837. The law of 1887 limited *classement* to monuments and objects belonging to the state that were declared inalienable and to those owned by other public entities that were alienable only by authorization of the state. In 1913 *classement* was extended to privately owned immovable and movable monuments (in the broadest definition) of particular public interest. When private individuals did not consent to classification, a decision of the Conseil d'Etat was necessary. A change of ownership of classified objects in private possession had to be made public, and the sale of such objects was limited by the law that prohibits exportation of classified objects. The law of 1929 provided for the classifying of all privately owned objects of exceptional interest that could be placed in national collections. It also gave the state the option of purchase in all public sales of works of art.

Archaeological excavations in France have to be authorized by the state; the latter has the right to conduct excavations on privately-owned land upon payment of an indemnity. A law also provides for the expropriation of privately owned land by communal or departmental administrations for such excavations. The principle of public ownership of archaeological finds does not exist in France; only half the objects found in excavations sponsored by the state on privately owned land belong to the state, and all movable objects found by private individuals in the course of authorized excavations or by chance are retained by the finder, subject only to their being reported to the communal authorities. Even immovable finds remain private property, although they must be classified.

The preservation of natural monuments and sites of artistic, historical, scientific, legendary, or picturesque character is subject to various laws (1930, 1945, 1947) that provide for the classification of natural scenic beauties and extend to them the provisions covering immovable monuments. Of particular interest in French legislation is a law (1962) that provides for the protection and use of buildings and building complexes of special esthetic and environmental interest and facilitates their restoration, whether undertaken voluntarily by the owners or by order of communal authorities.

In Italy, the measures in the various Italian states for the preservation of art works remained in effect even after the unification of the country. The first national law in this field was promulgated on June 12, 1902 (No. 185) and modified by the law of June 20, 1909 (No. 364). The formulation of a new general law had long been delayed because it was necessary to find a legal solution to the problems of fideicommissary restrictions and rights of primogeniture and entail to which private art collections were traditionally subject. This problem had already been raised during the last years of the papal government and had reached a critical point, inasmuch as those restrictions were contradictory to the Italian Constitution. (A temporary solution had been found in 1871 that to some extent guaranteed the integrity of the main private collections.) Other special measures of importance had been passed: in 1865 the principle of the declaration of public utility was extended to monuments, so that if their integrity was visibly endangered, the state had the right to expropriate them; in 1866, when the religious orders were suppressed, their properties were claimed by eminent domain and the art works and archives that had been preserved in such buildings were entrusted to the surveillance of the Ministry of Public Education; in 1881 the first technical-administrative body was formed on a national basis, the Direzione Generale delle Antichità e Belle Arti which was part of the Ministry of Public Education.

Modern Italian legislation has inherited the longest and most complex legal tradition in the field of preservation. The law of 1939 concerns both movable and immovable objects that are of artistic, historical, archaeological, or ethnographic interest. It does not, however, cover the works of living artists or works less than fifty years old. This control is extended to all objects regardless of ownership. Those belonging to the state are directly under its care through its technical-scientific organs; objects belonging to public bodies (communes, provinces, etc.), to legally recognized organizations, or to ecclesiastical bodies must be catalogued at the expense of these organizations or of the bodies that are responsible for their preservation. Objects belonging to private persons are subject to tutelage by means of a notification (*notifica*) — a precise and detailed declaration of public interest by the Ministry of Public Education.

The purpose of the law is, first of all, to assure respect for the rights of use and enjoyment that the public has acquired over things defined as objects under tutelage; and secondly, to guarantee the conservation, integrity, and security of the objects themselves. To this end it specifies that their status quo may not be changed without special authorization and that they may not be put to uses incompatible with their character or dignity. Private owners of movable "notified" objects are also under obligation to report any change in the location of such objects. The law also entitles the state to impose conservation measures on objects under tutelage at the expense of their owners; the state itself can carry out preservative work, carrying all or part of the cost if the objects belong to public bodies, or demanding repayment if the objects are privately owned.

While objects subject to supervision or those belonging to the state are, in general, inalienable, removal of objects belonging to public bodies may be authorized provided this does not compromise the national historical and artistic heritage. The alienation of privately owned "notified" objects must be reported by the owners to the supervisory bodies of the state, which may forbid it whenever it is prejudicial to the national heritage or in conflict with the dispositions that confirm the former fideicommissary liens on particularly important art collections. When such alienation is of a commercial nature the state has the right of preemption.

The export of objects under supervision is also forbidden if it injures the national heritage; otherwise it is permitted on payment of a tax proportionate to the value of the objects. Nevertheless the state has the right of preemption at the declared value.

The problem of archaeological excavation is particularly important in Italian legislation. The subsoil is state property, and the right to conduct excavations is reserved exclusively to the state, even on privately owned property, granting the owners an indemnity equivalent to one-fourth of the objects found or of their monetary value. The state also grants permission to cultural institutes but retains ownership of objects found in the course of these specially authorized excavations. The state can also expropriate objects under its supervision, and such action is regarded as extremely important to the conservation or increase of the national heritage. Strict penalties are prescribed for violators of the law.

The 1939 legislation also placed under supervision scenic beauties including villas, parks, and buildings not covered by the previously mentioned law, as well as panoramas, landscapes, etc. For obvious historical reasons, the law on the preservation of the landscape is not limited to protecting what is specifically defined as natural scenic beauty but also includes the traditional aspects of towns and cities. The state takes measures with its laws and regulations on town planning (see TOWN PLANNING).

The application of these fundamental laws and their corollaries is entrusted to the Direzione Generale delle Antichità e Belle Arti in the Ministry of Public Education, to the subsidiary departments of the Ministry, or to the Sopraintendenze alle Antichità, alle Gallerie, ai Monumenti. These offices are responsible not only for the administration, technical supervision, and improvement of that part of the heritage which is state property (the direction of state museums and monuments, the carrying out of archaeological excavations, the restoration of art works, etc.) but also for the periodic revision of catalogues, the supervision of conservation of objects belonging to public organs, the notification of those which are privately owned, the supervision of exports and the granting of permits for them, and the imposition of architectural, landscape, and archaeological restrictions.

It is evident that in Italy the preservation and supervision of the historical and artistic heritage is almost exclusively assumed by the state, which, moreover, includes this as one of its institutional responsibilities mentioned in the fundamental principles of the Constitution. Under the Italian regulations "delegated supervision" is granted to certain legally recognized institutions such as the Ente per le Ville Venete (Organization for Venetian Villas), a joint body of state and provincial administrations and tourist organizations for the maintenance and enhancement of historical villas of the three Venetian regions.

In England, the problem of the preservation of historical and artistic monuments had been debated at length in the *Gentlemen's Magazine* (founded in 1731) and, after 1770, in the periodical *Archaeologia* published by the Society of Antiquaries. The renewed devotion to national art history was visibly expressed in the romantically suggestive forms of the Gothic revival and produced also an enthusiastic fervor for the restoration of medieval monuments, although the results were hardly what had originally been intended. The establishment of proper legal measures for preservation of monuments, similar to those being promulgated in other European countries, met with a formidable obstacle in the preeminence that the English legal system and, above all, public opinion assigned to the rights of private property. In 1882, during the debate on a proposed Ancient Monuments Protection Act that was to force a private owner who intended to destroy a historic or artistic monument to offer it first for sale to the state at a price to be agreed upon with the Treasury, some members of Parliament protested that this would constitute "an invasion of the right of property . . . in order to gratify the antiquarian taste of the few at the public expense."

Under the Ancient Monuments Consolidation and Amendment Act (1913) boards were set up in England, Scotland, and Wales with the task of scheduling the monuments. The commissioners of works or any local authority had the power to acquire ancient monuments by purchase, accept ownership under a deed, or assume guardianship of any ancient monument offered them by the owners. This law was amplified by the Ancient Monuments Act (1931) and the Historic Building and Ancient Monuments Act (1953). These laws protect all monumental buildings (with the exception of ecclesiastical buildings and those being used for religious purposes) which are of "historic, architectural, traditional, artistic and archaeological" interest and which are listed by the Historical Building Council of the Ministry of Works, on the advice of the Ancient Monuments Boards. The owners and occupants of such monuments are notified of the inclusion of their property in this list and have the right to oppose such listing. If, however, they consent, or if the scheduling is imposed by authority, the owners cannot alter the status quo of the monument without giving prior notice to the Ministry of Works.

Whenever the Ancient Monuments Boards (which also have inspectorial functions) have reason to believe that a monument is in danger of damage either from neglect or improper maintenance, the Ministry of Works may issue a Preservation Order (which may eventually be confirmed by Parliament) to restrain the owners or the occupants from carrying out any alterations without official authorization and proper technical supervision. The Ministry of Works can place the monument, if it is not inhabited, under its own direct custody and assume the responsibility for its conservation. It is authorized to contribute to the costs of restoration of privately owned monuments; it can also acquire monuments by deed of gift or by purchase and can assume guardianship when it is offered. The export of art works is under the control of the Board of Trade, which avails itself of the advice of museum and university administrations to establish the taxable value of the works themselves and, in certain cases, to refuse an export license. Archaeological excavations must be authorized by the state, and the law provides that the objects recovered are the property of the finder or of the owner of the land where they have been found; usually such objects are then donated to a museum or a university. An exception is made for coins and gold and silver objects found by chance; these are considered to be treasure-trove, the possession of which is a prerogative of the Crown.

In England there is no law specifically regarding movable art objects, nor are there any laws concerning the preservation of landscape. The latter is covered by city planning ordinances, which are generally very strict in this regard.

Conservation activities provided by English law cover only a part of the historical and artistic heritage, and they are exercised by the state primarily in the form of supervision and technical consultation; direct action by the state, although provided for in the Act of 1953 (Part I, article 4), is rare. However, state action is supplemented to a great extent by a special institution, the National Trust for Places of Historic Interest and Natural Beauty.

Founded in 1894 by Hardwicke Drummond Rawnsley (1851–1920), Sir Robert Hunter (1844–1913) and Octavia Hill and incorporated in 1907, the National Trust is a nonprofit (charity) private association, supported by membership fees and other contributions from members, by donations and legacies, by state contributions (tax exemptions on property, occasional donations of money or of immovable property of historical and artistic interest, etc.), and by the income from its own properties. The state has no voice in its administration.

The National Trust is concerned with the conservation of buildings and their artistic accessories, of movable art objects, landscape sites, gardens, parks, natural scenic beauties, characteristic sites, and natural and zoological reservations in England, Wales, and Northern Ireland. (The Scottish National Trust takes care of such places in Scotland.) Most of these properties — three entire villages (Lacock, West Wycombe, and Chiddingstone), numerous farmhouses, the ruins of Hadrian's Wall at Housesteads (Northumberland), etc., amounting to a total of over a quarter million acres — are directly owned and administered by the National Trust, which places almost all of them at the disposal of the public. The National Trust also manages properties entrusted to it under special conditions. The owner of a country house or other historical building may offer it as a gift to the National Trust together with an income to defray part of the expenses of its maintenance, reserving to himself the right to continue to live on the premises, and agreeing to open it to the public at stated times.

The British National Trust is undoubtedly the most remarkable example of a collective awareness by society of the importance of preserving the national heritage and is developed to the point of freely taking initiatives that not only supplement the forms of supervision exercised by the state but also supplant them when they are restricted by respect for private property rights. In effect, the National Trust safeguards these rights, inasmuch as it is itself legally a private proprietary body as well as an administrator of properties voluntarily committed to its care.

In no other country is the preservation of the national heritage based on criteria similar to the ones in England. There are, of course, many private associations that exist for the purpose of preserving this heritage (see INSTITUTIONS AND ASSOCIATIONS), but they carry on supervisory functions mainly by virtue of a formal delegation of such powers on the part of state authorities.

In reviewing briefly other current legislation, it is evident that artistic supervision is most often the exclusive responsibility of the state (sometimes so assigned in the constitution). This is the case in the U.S.S.R., Spain, Yugoslavia, and Austria. The laws in force in Austria are very similar to the Italian ones, the main difference being a more rigorous discipline over exports and the concession of private ownership of objects recovered during archaeological excavations, provided they are reported to the central authority. A technical organization, connected with the Zentralstelle für Denkmalschutz, has *Landeskonservatoren* (provincial curators); in addition there are the administrations of the federal art collections.

In Greece, the laws of 1834 and 1899 on the national ownership of the archaeological heritage were confirmed and extended in 1902 to cover all historical monuments of every age. In Hungary, in Czechoslovakia, and particularly in the Scandinavian countries, by ancient tradition artistic preservation is left to the state. Public ownership of the subsoil and therefore of archaeological finds is recognized. (In Belgium, on the contrary, there are no regulations whatever concerning archaeological excavations.)

In the Scandinavian countries the earliest laws go back to Gustavus Adolphus in Sweden (r. 1611–32), who appointed official custodians of national treasures and began the office of Antiquary of the Kingdom (*Riksantikvar*), and to Christian IV (r. 1610–48) in Denmark. The Danish custom, according to which anything found in the subsoil is the property of the state if there is no legitimate owner, is much older. In 1666 Charles XI of Sweden issued detailed instructions on preservation and included castles, fortifications, mounds, gravestones, and burial grounds among historical monuments. The supervisory measures passed by Christian IV were supplemented by decrees in 1737 and 1752 that regulated the possession of archaeological finds by private persons. In 1807 a permanent state commission was appointed to collect information regarding monuments of all kinds, to see that they were not destroyed, and eventually to provide for their restoration. In all the Scandinavian countries interest centered mainly on prehistoric remains, and edicts concerning them had been passed by the end of the 18th century.

In the 19th century a sense of national unity developed in Germany, stemming from romantic idealism and from the release from Napoleonic domination. In a report made to the Prussian government in 1815, the architect Karl Friedrich Schinckel addressed a stirring appeal to his countrymen to respect the medieval monuments, which he described as the first manifestations of the collective German ethos; he aroused wide response and agreement. In 1852, the Gesamtverein der deutschen Geschichts-und Altertumsvereine was founded, organizing individual preservation societies into one association. Generally, all objects characteristic of their time and not private property come under governmental protection. Supervision is the prerogative of the individual *Länder* making up the federation (cantons in Switzerland); these draw up lists of their monuments and control archaeological excavations. Private property is dealt with under the general building laws. Only Bavaria seems to have laws (1949) in any way comparable to those of Italy and France. It is interesting to note that as early as 1780 an edict of the Margrave of Bayreuth declared all his subjects fully responsible for the preservation of the medieval monuments existing in his territories.

In the Netherlands the law of 1918, modified in 1933, applies to immovable historical and artistic properties. In 1950 a national commission was set up to schedule monuments and act as a consultant in their preservation.

In non-European legislation, the principle of total state jurisdiction in the supervision of the historical and artistic heritage prevails in the legislation of Iraq, Egypt, Mali, and Argentina. In Brazil (where the Serviço do Patrimônio Historico e Artistico Nacional also authenticates and regulates the sale of antiques), works of art and natural amenities fall under the law of 1937, which provides for an inventory and for the inalienability of monuments listed therein. Legislation in Argentina is specified in the Constitution of 1949. Mexico has protected its archaeological monuments since 1897; the protection of natural monuments and amenities was added by a presidential decree of 1930. In Turkey and Morocco the preservation of immovable objects, archaeological excavations, and exports are subject to regulations.

In India, in 1861, Lord Canning established the Archaeological Survey of Northern India. Later, there were other archaeological surveys of various areas and commissions occupied with preservation. Besides establishing and placing on a permanent basis the seven archaeological areas into which British India was divided, the Ancient Monuments Preservation Act (1904) empowered the local government of any province to declare any monument to be "protected."

In Japan, the governmental protection of cultural properties began about 1868. It was followed by various legislative steps: the law for the Preservation of Ancient Shrines and Temples of 1897 was expanded in 1919 with the law for the Preservation of Historic Sites, Places of Scenic Beauty, and Natural Monuments. The 1897 law was replaced in 1929 by the National Treasures Preservation law, which included other properties valuable from historical and artistic points of view. In 1933 the law for the Preservation of Important Objects of Art prescribed permits from the Minister of Education for the exportation of valuable objects of art if they were not registered as national treasures — in which case they were inalienable except in special cases. In 1950 a new law for the Protection of Cultural Properties was passed that included movable historical or art works and buildings, folk art, monuments, historical sites, and nature reserves. The important items are designated by a commission that exercises independent authority. Custody and repair is done by owners or custodians who can be subsidized and are expected to keep the commission informed of the status of the monument. The designated cultural properties must be open to the public, and the expenses are born by the national treasury. Both archaeological excavation and excavation for public works of grounds known to contain cultural properties must be announced in advance. The number of designated items totaled over 10,000 in 1962. Should an owner want to sell, the commission holds priority of purchase. Emphasis is placed on the utilization of works of art and on restoration for purposes of conservation.

In the United States, federal preservation falls under the National Park system. The first national park was set aside in 1872, instituting a form of land use that eventually spread to other continents. The Antiquities Act (1906) empowered the president to name areas containing historical, prehistoric, or scientific objects as national monuments. The National Park Service was established in 1916 and was directed to "promote and regulate the use of the federal areas known as national parks, monuments, and reservations." With the Historic Sites Act (1935) the investigation, selection, and protection of nationally significant sites was begun. This Act also authorized the secretary of the interior to designate outstanding areas of whatever ownership as historical sites and to enter into cooperative agreements with nonfederal agencies for their conservation. Permits to excavate on archaeological sites are issued only to specially qualified cultural institutions on condition that the work is done in the interest of these institutions and that the collections are preserved in a public museum. On the whole, the National Park Service program stresses preservation rather than restoration or reconstruction. There are more than 60 historical national monuments and sites that are federally controlled.

Beside federal legislation, the states themselves have set up legislation protecting individual remains. The principle of expropriation has been accepted by some states. The Constitution of the Commonwealth of Massachusetts states: "The preservation and maintenance of ancient landmarks and other property of historical or antiquarian interest is a public use, and the Commonwealth and the cities and towns therein may, upon payment of just compensation, take such property or any interest under such regulations as the General Court may prescribe." In 1921 the Kansas legislature validated an act that extended the power of eminent domain to any parcel of land that had exceptional historical interest. In the case of the Shawnee Mission, which the state of Kansas wanted to purchase, the court of appeals upheld the legislature, asserting that the utilization by the state of places of exceptional historical interest is a matter of public interest. The Supreme Court again confirmed this right regarding the National Military Park in Gettysburg. The various town-planning laws also take into consideration the protection of landscapes and historical sites.

Much is done by private archaeological and historical societies partly maintained by the states. Historic houses are often administered by states or nonprofit organisations. However, private initiative has not always succeeded in saving architectural monuments from destruction. Frank Lloyd Wright's Larkin Building (Buffalo) was torn down in 1950, but dismantling of his Frederick C. Robie house (Chicago) was stopped by a stay of execution; when its present occupants leave it will be turned over to a national preservation group. Some of the private associations are the Committee for the Recovery of Archaeological Remains, the Society for the Preservation of New England Antiquities (Boston), the Daughters of the American Revolution, and the Colonial Dames; the Mount Vernon Ladies Association of the Union maintains the George Washington estate as a national shrine. The files of the Historic American Buildings Survey contain much valuable material that is used for conservation and restoration.

The National Trust for Historic Preservation (Washington, D. C.) is another private organization concerned with America's cultural heritage. The Trust itself administers few properties (Woodlawn Plantation in Virginia, the Wayside Inn at Sudbury, Mass., a house occupied by Commodore Stephen Decatur in Washington, etc.). Like the British National Trust, it does not accept gifts of buildings unless their maintenance is assured. The Trust's chief function is offering advisory, educational, and technical assistance to local organizations. Its affiliated societies number in the hundreds.

International preservation. Even though the supervision of the historical and artistic heritage is almost entirely within the competence of the national laws and administrative organizations of the individual countries, an awareness that this heriage transcends the limits of nationality and belongs to all human societies has come to be increasingly felt in all cultural fields. This has led to a recognition of the need for universal acceptance of the fundamental principles of supervision and preservation and for a practical formulation of this common aim. The international organizations have expressed an interest in establishing homogeneous criteria in international law that will not conflict with the constitutions and laws of the different countries; they are also anxious to promote cooperation for the preservation, conservation, and diffusion of knowledge of the various countries' cultural heritages.

The problem of international artistic supervision had already emerged in the attempt to settle the damages in certain countries as a result of wars, particularly with respect to the restitution of art works taken away by invaders. The Treaty of Paris (1815) provided for the restitution to the Papal States of the art works removed by the Napoleonic armies. Even after Canova and Luigi Gaetano Marini had recovered part of the looted works the influence of the Treaty was felt, especially an important clause that ordered the return of the art works removed from Italy "because they are inseparable from the country to which they belong"; this clause had been included in the Treaty by the victorious powers as a result of the interest that the Duke of Wellington and Marshall Blücher (the respective representatives of the English and Prussian governments) had shown in the matter (see the note of Viscount Castlereagh in Mariotti, 1892, p. 62). In this clause the principles of preservation and custody of an artistic heritage became for the first time part of international law.

The Hague conferences of 1899 and 1907 formulated the principle that buildings dedicated to arts and sciences should be saved wherever possibile in the case of war. In 1923 this was extended to include protection of monuments of great historical value, on condition that they were not used for military purposes. In 1935 an international accord was signed between the United States and Latin America, known as the Roerich Pact, in which the neutralization of historical monuments and of institutions consecrated to the sciences, arts, education, and culture was affirmed. The Office International des Musées was established by the League of Nations to prepare two projects for international covenants, one for the "repatriation of objects of artistic, historic, or scientific interest lost, stolen, or alienated and exported illegally"; the other for the protection of monuments and art works in the event of war.

During World War II the problem of preservation and safeguarding of the historical and artistic heritage became dramatically acute, and the principles established by the League of Nations were of no avail. Each of the belligerent countries was obliged to devise its own protective measures without being able to rely on respect for the agreements that were supposed to guarantee the integrity of its cultural heritage. The only international organization of a supervisory nature was a special office, the Fine Arts, Monuments and Archives Branch attached to the Allied Forces Headquarters (with field representatives at army level), whose mission was to take charge in those cases where the technical organizations of the occupied nations were no longer able to operate, to brief the air forces in an attempt to avoid bomb damages, and to try to recover lost or looted art objects. After the war the search continued. The recovery policy was initiated by the American Commission for the Protection and Salvage of Artistic and Historic Monuments. Special groups to trace and protect lost or looted works were formed in many countries.

Since World War II and the signing of the peace treaties, the need for reaching some form of international agreement for the preservation of monuments and art objects has again been felt. Starting from the premise that such agreements cannot be limited to action in the exceptional event of armed conflict but must be of a permanent nature, UNESCO, in its 5th session (Florence, 1950), examined the proposals put forth in 1949 by a committee of experts under the chairmanship of the jurist G. Berlia and published its conclusions in the form of a report signed by the representatives of all the members of the United Nations; the report covered measures for cooperation among the states interested in the protection, conservation, and restoration of antiquities, monuments, and archaeological sites and for establishing an international fund to subsidize conservation and restoration. This report established the general lines for international action on a permanent basis on the following points: (1) moral and educational activities for adults and children; (2) legislative and administrative action; (3) technical action; (4) international legislative action; (5) financial assistance; (6) action against damage caused by armed conflicts.

At the same session UNESCO also adopted proposals for an international covenant for prevention of and protection against war damage. Considered further in the course of the 7th session (1952), these proposals were approved as a supplement to the Convention for the Protection of Cultural Possessions in case of Armed Conflict (The Hague, 1954). Another important milestone in the program of international supervisory action has been reached by UNESCO with its recommendations concerning the safeguarding of the beauty and character of landscapes and sites and the proposal to launch an international campaign for monuments of historic and artistic value in all the member countries.

UNESCO also conducts continuous activities to stimulate the governments of member nations with its recommendations aimed at increased efficiency of the various technical structures and at more intense educational campaigns to stimulate public interest. UNESCO has also aided many of its member states through the Program of Participation in the Activities of Member States. The International Campaign to Safeguard the Monuments of Nubia was an outstanding example. It began in 1955, when the governments of the United Arab Republic and the Sudan turned to UNESCO for help in saving the temples and monuments of Abu Simbel, threatened by the construction of the Aswan High Dam. In 1960 an international campaign was launched by UNESCO and three plans for preservation were studied. The Italian plan ran into difficulty because of its high cost, and a Swedish proposal that called for cutting the temples into blocks, dismantling them, and reeerecting them at a new site on a plateau was considered more feasible. Contributions were pledged by 46 member and associate member states of UNESCO as well as by the U.A.R., and work began in 1964, with the actual reconstruction to begin in 1966.

Other UNESCO programs include the establishment, debated at the 9th session (New Delhi, 1956), of the International Center for the Study of the Preservation and Restoration of Cultural Property, with headquarters in Rome.

In addition to UNESCO there are other international organizations concerned with the problems of artistic preservation and supervision. The International Council of Museums (ICOM), with headquarters in Paris, works on a purely technical basis; it is a private organization of official technical organs, museum directors, curators, and scholars. ICOM is particularly interested in museological problems and organizes propaganda activities such as the Annual International Museum Week.

Outstanding for facilitating international collaboration in art preservation and supervision is the Council of Europe, which in its 5th session (1950) adopted preservation of art works as one of its institutional functions. The Council of Europe has organized a number of important international exhibitions and has taken a particular interest in the preservation of urban centers of historical and artistic interest. In June, 1963, the Council published a report on the defense and enhancement of historical and artistic sites and centers that also takes into account the results of the conference held in Gubbio (1960) by the Associazione Nazionale Italiana per i Centri Storici, as well as the studies of the Civic Trust (London, 1960); the report examines the problem of giving a unified direction to the reorganization of urban centers of particular interest. The Council of Europe has also sponsored a pilot project for the conservation and enhancement of a historic section of Venice.

BIBLIOG. G. Azzurri, Il vero proprietario dei monumenti antichi, Rome, 1865; J. J. A. Worsaae, La conservation des antiquités et des monuments nationaux en Danemark, Mém. Soc. royale des ant. du Nord, 1872–77, pp. 343–60; A. von Wussow, Die Erhaltung der Denkmäler in den Kulturstaaten der Gegenwart, 2 vols., Berlin, 1884; Cong. int. pour la protection des œuvres d'art et des monuments, Procès-verbaux sommaires, Paris, 1889; G. Ducrocq, La loi du 30 mars 1887 et les décrets du 3 janvier 1889 sur la conservation des monuments et objects mobiliers, Paris, 1889; G. Giacomini, L'editto Pacca del 7 aprile 1820, Rome, 1891; A. Bonetti, Le Belle Arti e il loro commercio, Rome, 1892; P. Mariotti, La legislazione delle belle arti, Rome, 1892; G. Nobili, Sul progetto di legge per la conservazione dei monumenti, oggetti d'arte e d'antichità, Florence, 1892; R. Sabfiles, La législation italienne relative à la conservation des monuments et objects d'art, Dijon, 1894; R. Hunter, Memorandum as to the Steps Taken in Various Countries for the Preservation of Historic Monuments and Places of Beauty, London, 1896; L. Tétreau, Législation relative aux monuments et objects d'art, Paris, 1896; Per la libertà delle Belle Arti in Italia: Brevi osservazioni sull'annunziata riforma dell'editto Pacca . . . , Bologna, 1897; Reports from Her Majesty's Representatives Abroad as to the Statutory Provisions Existing in Foreign Countries for the Preservation of Historical Buildings, London, 1897; G. B. Brown, The Care of Ancient Monuments, Cambridge, 1905; G. De Montemayor, Diritto d'arte, Naples, 1909; J. Metman, La législation française relative à la protection des monuments historiques et des objets d'art, Dijon, 1911; J. W. Frederiks, Monumentenrecht, Leiden, 1912; R. Biamonti, Natura del diritto dei privati sulle cose di pregio storico o artistico, Foro it., 1913, I, cols., 1010–23; N. A. Falcone, Il codice delle Belle Arti e delle antichità, 2d ed., Piacenza, Prato, 1914; W. Gödel, Riksantikvarieämbetet, Stockholm, 1930; J. Locquin, Rapport fait au nom de la Commission des Finances chargée d'éxaminer le projet de loi des beaux-arts, Paris, 1930; L. Parpagliolo, ed., Codice delle antichità e degli oggetti d'arte, 2d ed., 2 vols., Rome, 1932–34; International Museum Office, La conservation des monuments d'art et d'histoire, Paris, 1933; H. d'Arnoux de Fleury de l'Hermite, Objects et monuments d'art devant le droit des gens, Paris, 1934; U. Aloisi, Protezione internazionale delle cose di pregio storico e artistico, Città di Castello, 1935; A. Bertini-Calosso, Conoscenza e difesa del patrimonio artistico, Rome, 1937; H. G. J. Mass Geesteranus, La protection des monuments historiques aux Pays-Bas, Mouseion, XI, 37–38, 1937, pp. 171–81; Wichtigste Gesetze und Verordnung über Denkmalpflege in Österreich, Vienna, 1937; I. Zemp, Lois et reglements suisses en matière d'antiquités et de fouilles, Mouseion, XII, 43–44, 1938, pp. 143–50; A. Bentivoglio, Ritrovamenti archeologici ed extraterritorialità, Capitolium, XIV, 1939, pp. 485–87; C. de Visscher, ed., Art et archéol.: Rec. de législation comparée et de droit international, 2 vols., Paris, 1939–40; M. Lazzari, L'azione per l'arte, Florence, 1940; Les mésures de précaution pris dans divers pays pour protéger les monuments et les œuvres d'art au cours de la guerre actuelle, Mouseion, XIV, 49–50, 1940, pp. 9–27; U. Costa and L. Matarazzo, L'amministrazione e la tutela delle arti in Italia, 2d ed., Rome, 1942; La protezione del patrimonio artistico nazionale dalle offese delle guerra aerea, Florence, 1942; G. Mariani, La legislazione ecclesiastica in materia d'arte sacra, Rome, 1945; G. C. Argan, Autonomie regionali e difesa del patrimonio artistico, Ulisse, I. 1947, pp. 343–49; The National Trust: A Record of Fifty Years' Achievement, 3d ed., London, 1948; R. B. Sugden, Safeguarding Works of Art; Storage, Packing, Transportation and Insurance, New York, 1948; C. de Visscher, International Protection of Works of Art and Historic Monuments, U. S. Dept, of State Doc. and State Pap., I, 1949, pp. 821–71; R. Burrows, ed., Halsbury's Statutes of England, 2d ed., XVII, 2, London, 1950, pp. 372–437; M. Grisolia, La tutela delle cose d'arte, Rome, 1952; M. Salmi, Per la difesa dell'arte e delle bellezze naturali, Comm., III, 1952, pp. 167–72; The UNESCO Courier, 1952 ff.; Kingdom of Jordan, Antiquities Law No. 33 of 1953, Amman, 1953; T. Mirabella, Considerazioni su alcune particolari forme di protezione giuridica in materia d'arti figurative, La Giara, II, 1953, pp. 79–81; F. Ulivi, La tutela delle opere d'arte, La Giara, II, 1953, pp. 112–15; W. G. Constable, Curators and Conservation, S. in Conservation, I, 1954, pp. 97–102; P. O. Geraci, La tutela del patrimonio d'antichità e d'arte, Naples, 1956; G. Cecchini, La tutela delle opere d'arte a Siena nel secolo XVIII, B. senese di storia patria, LXIV, 1957, pp. 182–83; A. Venditti, Il patrimonio artistico e lo Stato, Nord e Sud, IV, 1957, pp. 81–83; Z. Wirth, Vývoj zásad a prakse ochrany památek v obdobi 1800–1950 (Development of Principles and Practice for the Protection of Monuments in the Czech Countries and in Slovakia 1800–1950), Umění, V, 1957, pp. 105–16; M. Gianturco, La protezione dei beni culturali in caso di guerra, Ann. Ministero della Pubblica Istruzione, IV, 1958, pp. 656–64; H. Noblecourt, Protection of Cultural Property in the Event of Armed Conflict, Paris, 1958; K. H. Buhse, Der Schutz von Kulturgut im Krieg, Hamburg, 1959; M. Gianturco, La protezione dei beni culturali e artistici, Nuova antologia, CCCCLXXVII, 1959, pp. 543–46; A. Khater, La régime juridique des fouilles et des antiquités en Egypte, Cairo, 1960; Pámátková péce, Prague, 1960 ff.; J. F. W. Rathbone, The National Trust: Its Development and Problems, London, 1960; W. Treue, Art Plunder: The Fate of Works of Art in War, Revolution and Peace, London, 1960; Administration for Protection of Cultural Properties in Japan, Tokyo, 1962; R. Assunto, Introduzione alla critica del paesaggio, De Homine, V–VI, 1963, pp. 252–78; A. Cantone, Ordinamento dell'Amministrazione delle Antichità e Belle Arti, Rome, 1963; Council of Europe, Rapport sur la défense et mise en valeur des sites et ensembles historiques ou artistiques, Strasbourg, 1963; M. Pallottino, Che cos'è l'archeologia, Florence, 1963.

Oreste FERRARI

PRIMATICCIO, FRANCESCO. Italian mannerist painter (b. Bologna, Apr. 30, 1504; d. Paris, between May 15 and Sept. 14, 1570). For six years, beginning in 1525 or 1526, he assisted Giulio Romano (q.v.) in the Palazzo del Te in Mantua (VIII, PL. 428). At the beginning of 1532 he was called by Francis I to work on the decorations of Fontainebleau. Il Rosso (q.v.), Pellegrino Tibaldi, and other Italian artists were already there, but from 1541 (the date of the suicide of Il Rosso) Primaticcio became the principal figure in the work on the Château. He was indeed the founder of the school of Fontainebleau and became the dictator of French artistic taste. It was through his work that the French first truly felt Italian influence. Primaticcio's style goes back essentially to Parmigianino, the first, and perhaps the greatest, mannerist painter of north Italy. Parmigianino's assistant Antonio da Trento (who may or may not have been the same person as the Antonio Fantuzzi who assisted Primaticcio in Fontainebleau) is supposed, according to Vasari, to have stolen his master's drawings and copperplates and to have made his way to Fontainebleau, where he sought and found employment. If this story is true, it establishes a direct link between Parmigianino and Primaticcio.

Like many Italian mannerists, Primaticcio was also a decorator, architect, and accomplished *stuccatore*. As the works at Fontainebleau reveal, he had a fine sense of the enframement of the single elements in a scheme of decoration, as well as of the total effect. Primaticcio finished the Gallery of Francis I (XII, PL. 100; XIII, PL. 281), for which Il Rosso had been mainly responsible, and collaborated with that painter elsewhere in the Château. The decoration of the Chambre de la Duchesse d'Etampes (V, PL. 395) was his alone, however; it includes stucco feminine figures of great elegance that support panels (repainted in the time of Louis Philippe) depicting the story of Campaspe and Alexander the Great (with allusions to the relationship between Francis I and the Duchess).

Primaticcio continued to work at Fontainebleau after the death of Francis I (1547) and was responsible for the Gallery of Henry II (IV, PLS. 165, 402). In 1563 he revisited his birthplace, Bologna, and met Vasari there — that is, in the interval between the publication of the first and second editions of the biographer's work, in which Primaticcio occupies a place. He was back in Paris the following year, and on May 15, 1570, made his will. Since his successor, Tristan de Rostaing, was appointed on September 14 of that year, it can be deduced that Primaticcio's death occurred between these two dates.

No doubt Primaticcio was a painter of the second rank, but undeniably he had a highly developed inclination to elegance and a particular sensitivity to one kind of feminine beauty, which he helped to make fashionable. These qualities, together with his superior Italian training, brought him exceptional influence as an artistic personality.

There is a self-portrait of Primaticcio in the Uffizi and another signed in red chalk in the Albertina in Vienna. (See also XII, PL. 81).

BIBLIOG. L. Dimier, Le Primatice, peintre, sculpteur et architecte des rois de France, Paris, 1900 (cat. of works and bibliog.); P. Barrochi. Precisazioni sul Primaticcio, Comm, II, 1951, pp. 203–32.

Arthur MCCOMB

PRIMITIVISM. A conscious return to the art of an undeveloped state, whether in subject, technique, or form, is known as primitivism. Historically the term "primitive" has been accredited with three distinct meanings. Art critics of the 19th and 20th centuries referred to painters and sculptors (rarely to architects) of the late Middle Ages and early Renaissance (not the Byzantine or pre-Romanesque) as primitives. Exhibits and monographs dedicated, for example, to the Sienese or French "primitives" refer to artists of the period between the 13th century and the end of the 15th. The term acquired this meaning during the "humanistic" epoch, beginning in the 16th century and ending in the late 19th, an epoch in which the classical Renaissance was considered the insuperable high point of art, whose prelude or immediate antecedent was the "primitive" art of the 13th to 15th century.

At the beginning of our century, the term "primitive" was used to refer to peoples who are at present the objects of ethnological studies, many of whom still live on a Stone Age level. By analogy to these contemporary "Stone Age men," cut off from progress and civilization, the term "primitive" has been applied to those artists who remain isolated from the culture of their immediate environment. Synonymous with this definition of primitive are "naïve," "instinctive," "childlike."

The present commonly accepted meaning of primitive implies on the one hand primordiality and on the other ingenuousness, simplicity, and inexperience. Primitivism under this definition has been exhaustively considered by modern studies in the psychology of art (q.v.). Primitive art must take its meaning from its context in the general process of human historical development.

A BRIEF VIEW OF PRIMITIVISM IN ART UP TO THE 19TH CENTURY. Ever since the late Middle Ages, when classic art was rediscovered and exalted as "the art of the wise," while that of the 4th to the 13th century was condemned as "the art of the ignorant" (Boccaccio), the ideal of an art based on ancient models, on the imitation of nature, and on the intellect constituted the very foundation of Humanism (q.v.) and dominated the theory of art until the 19th century. In Giorgio Vasari's (q.v.) conception of history as a cyclic succession of periods of progress and decadence, the Byzantine and pre-Giottesque period represented old age and death, and the period of Cimabue and Giotto a rebirth; thus the first masters of the late 13th and early 14th centuries were the first artists (hence primitives), the forerunners of modern art, who hesitantly and imperfectly initiated the Renaissance. Vasari could discern no art in the primitive, the barbarian, and the savage. Medieval art to him was the art of barbarians, and Gothic, or German, art was unnatural, lacking true proportion, order, and thought.

After this sweeping condemnation, the history of the taste for primitive art in the first meaning coincided with the reevaluation of the Middle Ages, with a widespread anti-Humanistic or anticlassic movement that triumphed in the neo-Gothic of the 18th century and became generally accepted in the 19th. An initial tendency in this direction is to be found in mannerism (q.v.). Furthermore the development of archaeology and sacred scholarship, though rooted in church history rather than in the history of art, led to the study and evaluation of medieval monuments. Still another factor stimulating the study of medieval remains was the provincial chauvinism of historians, impassioned by research into past local glories. This movement did not fail to influence art, and such artists as Scipione Pulzone and Siciolante da Sermoneta represent a pre-Raphaelite tendency destined, in turn, to have a following in the 17th century in Sassoferrato. This art was inspired by the primitive masters to the extent that it exhibited a perfect spirit of devotion, reviving now a Giottesque simplicity, now a composite simplicity in the graceful manner of Perugino.

Toward the end of the 17th century the concept of ancient Roman or classical art as perfect and exemplary was overthrown, and the whole ancient world began to appear primitive to certain thinkers. This view represented a liberation from dogmatism, the uprooting of the authority of the classics, and gave new faith to the idea of modern progress. Furthermore, it made the ancient world appear in a new perspective, as but a stage in mankind's infancy, almost a golden age, now forever lost.

The 18th century went even further, rediscovering the origins of the classic world, fascinated by remote antiquity. Giovanni Battista Vico (1725) did not include art in his concept of the history of mankind as a succession of stages of perfection; instead, it belongs to an initial phase: The first peoples, who were the children of humanity, suddenly founded the world of the arts; then, much later, philosophers founded the sciences. Vico was the first thinker to consider art as creative imagination, as the primordial form of human consciousness, so distinct from the activities of intellect and reason that Vico found imagination correspondingly stronger as intellect was weaker. This the first genuine esthetic theory of the primitive.

Vico's thesis that poetry precedes thought, that the senses precede reason, was further developed by Johann Georg Hamann, Johann Gottfried Herder, Jean Jacques Rousseau, and Denis Diderot. The discovery of America had brought with it the myth of the noble savage, the mythical creature who seemed to represent the survival of the earthly paradise, and who aroused the aspiration toward the state of nature that exists in every individual. While the beautiful was still attributed to the classic, a different kind of beauty was discovered in the preclassic, the "sublime" (see TRAGEDY AND THE SUBLIME). Blake and others evoked the world of the progenitors, with their gigantic forms and the visionary character of their images. They created a style designed to exalt the primordial world, a style that in Asmus Jakob Carstens was already that of a purist *ante litteram,* based wholly on line without chiaroscuro effects. It was the very self-conscious cultural fullness of the modern world, the very refinement of the rococo, that fostered a search for mankind's origins and the myth of a nostalgic return to nature, to the delightful imagined golden age of the primitives.

Yet in the field of the visual arts the concept of ideal beauty still predominated. Johann Joachim Winckelmann had recognized the art of the Greeks as superior to that of the Romans, but he still considered Greek art before 480 B.C. as an initial imperfect stage in the evolution toward classicism. The concept of the archaic is one owing to Winckelmann; moreover, he provided the first description of the archaic in terms of antinaturalistic elements: stylization, abstraction, frontality, rigidity, and so forth. After Winckelmann the archaic (or primitive) came to signify an initial moment in the evolutionary progress toward the classic; and in the field of archaeology until the beginning of the 20th century, historical criticism remained bound to the Vasarian biological cycle as applied to ancient art: youth, archaic; maturity, classic; senility, Hellenistic; death, late-antique.

THE DISCOVERY OF THE PRIMITIVES IN MODERN TIMES AND PRIMITIVISM IN MODERN ART. The evolutionist criterion that prevailed in the biological sciences characterized positivist criticism. According to Alfred Haddon, Gottfried Semper, and Alexander Conze (see HISTORIOGRAPHY), an expression of cultural and artistic inferiority is evident in the primitives, "primitive" being synonymous with the infantile and the inferior. All the archaic stages of art, characterized by stylization and geometric ornamentation, were considered the oldest because they were closer to the purely technical necessities: the nature of the material, the working techniques, the purpose of the object, these were the elements that determined the form. The evolution from the primitive (or archaic) to the cultivated was equated with the evolution from the stylized (or abstract) to the naturalistic.

Positivism thus confused technique with the ability to imitate. Indeed, in their employment of technique, the primitives were very often superior to the artists of more evolved civilizations. One need only consider the engineering perfection of the Romanesque and Gothic architects and the poor constructions of the 17th and 18th centuries; or compare the careful preparation involved in the paintings of the 14th century, which have survived the passing of the centuries, with that of the 17th and 18th centuries. In fact, presumed artistic progress was often accompanied by a veritable regression in technique, as was well understood by William Morris (q.v.), who wished for a return to the quality of primitive craftsmanship, to the dedicated handicraft characteristic of the 14th- and 15th-century workshops.

When speaking of technical inferiority, however, the positivists referred not to the actual technique of painting but to the ability to represent nature according to certain already determined schemes, that is, to the technique of successful imitation. Thus only when one has freed himself from the preconceptions of imitation and naturalism can he recognize the esthetic validity of abstract or geometric art or of archaic, primitive art. This, however, raised the questions of esthetics and interpretation. While classic art was defined according to certain formal categories that appeared instrumental (line, volume, and color were the means of imitation), primitive art presented formal values that were ends in themselves. Although it was possible to classify classic art historically as a manifestation of progress toward a perfect imitation of nature, primitive art lacked an external reference point. Classic art could be considered historically in terms of its content; primitive art appeared to be without any clear content. Classic art, the expression of individual personality, could be considered as a history of individual artists and, furthermore, appeared to be related to general cultural development. Primitive art, however, posed the problem of an art without any clear relationship to general cultural development, and in certain cultures it appeared to be without any development at all.

Alois Riegl was the first to recognize the esthetic qualities of the geometric style, thus redeeming the Middle Ages from comparison with the classic and from the prejudice that placed it on a level with barbarian craftsmanship. Riegl found the criterion of imitation completely extraneous to the medieval mentality: what guided the "barbarian" artists (like all artists) was *Kunstwollen*, that is, not the will to imitate but the desire to create and express. According to Riegl, artistic intention does not derive from the materials — as the positivists had maintained — but rather from a need to struggle against those materials in order to dominate them and give them form. Another fundamental intuition of Riegl's was that geometric stylization did not precede but often followed the naturalistic stage.

The theory of "pure visibility," to which Riegl adhered, became a premise for the understanding of primitive art and developed parallel with the crisis of the old naturalistic theory of art. Out of this crisis emerged the possibility of considering less cultivated art as exemplary. The attraction of the primitive (which soon included not only the art of the "barbarian" Middle Ages and of archaic civilization but also that of prehistory and of tribal Africa) was a flight toward purity, virginity, a need to go back to the origins of creative possibilities. It was not a romantic, sentimental flight toward an evocative and picturesque world, as with exoticism (q.v.), but a fully conscious, rigorous, and dedicated search for an understanding of form and its creative processes. There were two aspects to the problem, one objective, one subjective: one concentrating on the value of formal, stylistic elements and the other on subjective expression.

The need to make a *tabula rasa*, to escape from the dogmatism and rationalism of modern civilization, led Paul Gauguin (q.v.) to abandon Europe for the West Indies and the South Sea Islands (in 1887–88 he lived in Martinique and from 1891 to 1893 in Tahiti, where he returned in 1895 to remain until his death, in 1901). Gauguin was attracted by the myth of a people who lived in a state of natural innocence, in close contact with nature, where the response to life was more intense. Gauguin's exoticism was not that of an Alexandre Gabriel Decamps or a Eugène Fromentin; he did not seek the spectacle of the picturesque but rather a life that was still primitive and uncorrupted. Barbarism for him was the innocence of childhood. He was attracted not by the art of the primitives but by their human condition. He did not aspire to an art dominated by nature but rather one that dominated nature. Van Gogh (q.v.) also anticipated this psychological aspect of modern primitivism by establishing a new relationship, free from every convention, between feeling and object, a relationship through which it was possible for him to translate even his own hallucinations into images. Thus with Gauguin and Van Gogh an esthetic attitude was reached that preferred the barbaric, the childlike, even the irrational, to civilization.

This primitive world suggested a wealth of fascinating forms to the creative imagination, images that were, in turn, completely foreign to naturalistic civilization. Impressionism had already accepted an optical (scientific) truth different from that which had been generally accepted until then, thus opening the way to formal research into the abstract. Yet the relationship between art and criticism, like that between art and science (see the discussion of the influence of physics and psychoanalysis below) was very close: the evolution of modern art from impressionism on was closely tied to the evolution of thought in general and of the criticism and history of art in particular. For example, the direct correspondence sustained by Gauguin and the Nabis between form and emotion and between form and ideas anticipated the theory of the *Einfühlung*. These painters departed from classical tradition by liberating line from volume and developed the autonomous vitality of these elements. And when, following the opening of trade relations with Japan in the middle of the 19th century, Japanese prints first reached Paris, not only Gauguin and the Nabis but also Edouard Manet, Claude Monet, Edgar Degas, Vincent Van Gogh, Toulouse-Lautrec (qq.v.), and Emile Bernard discovered a stylistic value in them: the arabesque, a linear harmony that coordinated perfectly with the ideal of pure color spread over flat areas. The abandoning of the three-dimensional (a premise of classical and naturalistic vision) in favor of the symbolic and the expressive occurred in the name of a formal element discovered in a civilization whose art had until then been considered primitive; and an analogous sympathy for forms foreign to the Occidental tradition accounted for the sudden success of African art.

The phenomenon of primitivism was not limited to a search for an innocence and liberty of vision or to a sympathy for certain formal achievements of primitive art. It represented a much broader revolution than that. Scientific progress had enlarged the range of human consciousness to such an extent that the immensity of the unknown was now revealed in all its vastness. Following 19th-century optimism, with its faith in scientific progress, came a wave of pessimism during which the very idea of progress was shaken to its roots. As the ideal of the Renaissance man had already disappeared, so now the entire universe seemed to dissolve. Classical physics was suddenly superseded. Space and time were no longer considered realities but mere conventions (Albert Einstein), and man discovered the unknown even within himself, in the abyss of his unconscious (Sigmund Freud). More than ever reality appeared to be obscure and extraneous to the relationships that man had established between himself and the objective world and between himself and his own inner world, relationships that turned out to be mere conventions. Art, attracted to the unknown because of its faith in the perspective value of the imagination, sought the prelogical, the state of nature, the primordial, the original, an idiom still free from reason, determined only by the pure creative and expressive necessity.

The epithet "Fauves" that was given to the painters who exhibited in the famous Salon d'Automne in 1905 emphasized the qualities these painters had in common: they were not only beasts, they were "wild beasts" — mainly because of their polemic refusal to identify with any civilization, with any academic tradition (see FAUVES). From this moment on it became a common tendency for avant-garde artists to cultivate spontaneity, barbarism, innocence. The Italian primitives, which art history was reevaluating, appeared cultivated and refined compared with the authentic primitives, the Negroes of the African tribes and the cavemen. In 1904 Maurice Vlaminck bought two Negro sculptures, and André Derain, Henri Matisse, and Pablo Picasso (qq.v.) followed his example. Matisse began studying Egyptian, archaic, and Oriental forms. Georges Rouault (q.v.) went through a Negro period. The influence of Negro sculpture on Picasso is apparent in his great picture *Les Demoiselles d'Avignon* (IV, PL. 75) of 1907, the painting that marked the beginning of cubism.

Cubism sought to formulate a system founded on laws extraneous to those of conventional vision. It felt the fascination of Negro art (PL. 332) because of its geometric character, its forms cut in simple planes, its simplified, squared-off volumes whose high relief gave an impression of great vitality. Roger Fry recognized that the primitive conceived of form itself in three dimensions, so that a given work was not so much the image of an external reality as a reality in itself, expressive of a life of its own. The cubists emphasized the value of materials that maintained their own physical individuality, their own arbitrary liberty from any possible model; and this constituted a stimulus for the development of modern techniques of mixed media (see MEDIA, COMPOSITE). What was later called "intellectual realism," recognized as being typical of the savage and the child, was the essential element of Picasso's esthetic. "I make objects the way I imagine them, not the way I see them," he declared. The decomposition of forms, the simultaneity of points of view, and the free construction of reality independent of sense perception are the basis of cubism (see CUBISM and FUTURISM).

Modigliani (q.v.; PL. 332), too, responded to the fascination of African sculpture, in particular that of the Guro peoples of the Ivory Coast. It suggested to him schemes for the deformation and simplification of volume, and between 1910 and 1915 it was the subject of his sculpture and painting. In Paris, the struggle against subject and descriptive intention occupied the Romanian sculptor Constantin Brancusi (q.v.; PL. 332), whose works revealed a constant search for the geometrically essential, for pure volume, for forms free from all accidental elements and temporal references, eternal in their abstraction. He studied the art of Africa and the intimately decorative art of Oceania and he revived the ritual, magical function of imagery. Picasso also studied Aztec, Mayan, and Inca art. Always considering himself a classic artist, he thought of the work of these peoples as classic because it expressed genuine human values. Classicism and the Renaissance now appeared to be nothing more than limited, circumscribed moments in the history of human expression.

In these years primitive art became identified with the geometric and thus stimulated stylistic research in this direction. Paul Cézanne propagated the idea of form as volume. This was the time when the critics insisted on distinguishing between the decorative and the illustrative (Bernard Berenson) and tended to identify the former with "tactile values." It was the moment in which the autonomous and objective values of form were emphasized and in which much research was done in the evolution of form in the abstract (Roger Fry).

That primitivism was even then a widespread movement, including virtually every form of "primitive" art, is confirmed by the fact that cubism discovered the painting of Marie Laurencin, as well as the work of popular, ingenuous painters. Guillaume Apollinaire, a key personality in the cubist group, discovered in Henri Rousseau (q.v.) an authentic primitive who operated in the midst of modern society with total detachment, in complete isolation, painting what his imagination conjured up without reference to the cultural environment of his time. Both the cubists and Rousseau shared a tendency toward objectivity; Rousseau, however, was essentially indifferent to the geometric. This objectivity was reached from a condition of subjective liberty by substituting free creation for conventional imitation: the very bohemianism of the *peintres maudits*, such as Modigliani, and the quest for narcotic states had the purpose of liberating the spirit from all conventions, so that a purely instinctual state could be recovered.

In the same years in Germany other artists were touched by the fascination of primitive art. In 1904 the Brücke (Bridge) painters (see EUROPEAN MODERN MOVEMENTS), including Ernst Ludwig Kirchner, Emil Nolde (qq.v.; V, PLS. 203, 206, 212), and Erich Heckel (V, PLS. 209, 211), discovered the art of Africa and Oceania in the ethnological museum of Dresden and assimilated its essential forms. Heckel was also interested in Etruscan art. In 1913 Nolde traveled in Russia, Japan, and Oceania studying the primitives of those regions. Franz Marc and Heinrich Campendonk were inspired by popular Bavarian art. German expressionism (q.v.) was thus stimulated by primitive art at least as much as, if not more than, cubism was, and it was particularly concerned with form as expression.

In Munich, which became a vital center of modern art and was the seat of the Blaue Reiter (Blue Rider) group (1911–12), Wilhelm Worringer published *Abstraktion und Einfühlung* in 1908 and *Formproblem der Gothik* in 1912. He based his work on the *Kunstwollen* of Riegl: empathy was the presupposition of the will to art when it tended toward the vitally organic (naturalism), while an opposite impulse led toward abstractionism. Thus Worringer rediscovered the creative spirit of medieval Germany. The tendency of contemporary art toward the nonobjective (see NONOBJECTIVE ART) called for the justification and understanding of the abstract art of the past. Wassily Kandinsky (q.v.), in his essay *Ueber das Geistige in der Kunst* (1910), founded art on expression, formulating his theory of the "great reality." Later Paul Klee (q.v.; PL. 333) distinguished between form and gestalt. Gestalt was a form founded on vital functions of a spiritual nature, the basis of which was the need to express. Klee sought expression in archaic visualization, in children's art, and in the art of the insane; he sought not so much the form but the "sign."

It was, then, in Germany, above all, that the problem was posed not of abstract form but of the creative process that led to it. In the field of art criticism Max Dvořák and Erwin Rosenthal analyzed the two opposing processes, from the abstract to the concrete and from the concrete to the abstract, involved in the creative act. Emanuel Löwy recognized that the art of the primitives was the result not of technical inexperience but rather of antinaturalistic mental schemata and, further, that art often has its origins in magic and that its development shows characteristics as constant and universal as psychophysical laws. Luquet (1930) tried to determine the characteristics of primitive art through the drawings of children. The child does not create a geometric, abstract, ideoplastic art, but a representative, concrete, physioplastic, sensory art. He does not draw what he sees, but what he thinks the object is, a practice common to the primitives also. This thesis would confirm that of Riegl — that sensory art preceded geometric. Bearing this in mind, Luquet went on to trace all the expressions of primitive vision throughout history. Cesare Lombroso and Freud discovered close affinities between the symbolism of the insane and that of the primitives. Thus the sphere of primitivism grew to include both children's art and that of psychopaths, that is, the products of the unconscious, the purely instinctual, common to children, madmen, and savages. The old preromantic theory of genius and poetic madness had only timidly hinted at all this.

Once the plane of the unconscious had been reached, the doors were opened to the metaphysical and the surreal. It was not without reason that a certain ingenuousness of conception, an elementariness of vision, was revealed in dream painting — especially that of Marc Chagall (who, however, was inspired chiefly by folklore) — and in Italian metaphysical painting. Surrealism (q.v.) attempted an integral primitivism, if such an expression may be used. The surrealists, attracted not so much by the formal values of the primitives as by the conditions in which their works were produced, wished to return to the consciousness of the first man: to a life of unanalyzed sensations without logical categories of space and time, of objective truth and falsity, of normality and abnormality. For these men art was psychic automatism, pure magic. All the surrealists felt the myth of the primitive; Joan Miró (q.v.; PL. 333) took an evident interest in archaic artistic forms, particularly the art of the Neolithic period.

Among moderns who shared the fascination of the primitive was Henry Moore (q.v.). Moore's sculpture is primordial in form (compare his *Family Group* with sculptures from New Guinea, for example); he looks toward ancient American sculpture and the paleolithic paintings of Altamira for inspiration.

The artists who followed the myth of the primitive, from Picasso to Moore, were highly cultivated men who raised to the same level as classic art the art of civilizations hitherto condemned as inferior; and in so doing they completely destroyed the traditional idea of the artistic inferiority of the primitives. It was the artists who revealed the esthetic quality of primitive art to the ethnologists and historians, pointing out its essentiality, its expressive force, its symbolic immediacy; thus Negro art, in the works of Frobenius and of Kühn (1923), was finally recognized as the result of man's irrepressible esthetic need to create. Since then museums have begun to collect the works of primitive peoples not merely as ethnological documents but as authentic works of art.

Children's art. It was inevitable that primitivism should culminate in the analysis of children's art (PLS. 331, 368, 369, 373) because the child is the primitive par excellence. Vasari's theory, which compared the individual life cycle with the historical cycle of humanity as a whole, anticipated the theory of recapitulation, according to which the evolution of the child's mind corresponds to the evolution of the human race. Interpreting children's art as the original art, historians and scientists have tried to deduce from its analysis the characteristics of the historically primitive. Modern psychology has clarified the various evolutionary stages of infantile expression. The first stage, which occurs when the child is between two and three years old, is characterized by so-called "fortuitous realism" or "manipulation," that is, by the use of flourishes, blots, and movements that trace lines and circles. The second stage, when the child is between three and six years, is often referred to as the period of "unfulfilled realism" (Luquet) or "descriptive symbolism": that is, the child shows a desire actually to depict (mostly houses, landscapes, members of his family), and in his images his unconscious appears. This primitive infantile production is often fascinating because of its independence from reality: already, in the image selected, a principle of choice is manifested, a dawning of knowledge is apparent, but the creation is essentially mysterious in its origins. The third stage — a transition between "visual realism" and "intellectual realism" or between schematism and syncretism — is characterized by the decreased use of symbolic constants; the representation of things not according to sight but to reason and therefore with optical incongruencies, transparent forms, a multiplicity of viewpoints, and an enumeration of single objects without connections between them; the use of conventional instead of imitative elements; and by lacunosity and two-dimensionality. In this stage children produce noteworthy and charming images that express an ingenuous world with its own peculiar characteristics, but one where individual talent is allowed greater liberty and the possibilities of expression have already reached an artistic level. This last stage, which is followed at puberty by repression, is the most interesting of all and the one usually compared with the historical development of primitive civilization.

These researches present many difficulties, however. Psychology has shown that all peoples manifest the same psychic laws and the same tendencies toward graphic representations in their infancy but great differences exist because of individual personality and the cultural influence of environment. Thus children's art teaches us that although the primitive, in an absolute sense, is reducible to the merely psychophysical, still, even in children, artistic creation is not merely the expression of certain laws of nature but also of individuality and culture. How is it possible, then, to compare the creative processes of the child, who is the primitive operating in a cultivated adult society, with the savage who represents an adult in a civilization that is still infantile because not yet developed? Every attempt at such a comparison on the basis of statistics must be considered arbitrary, especially in regard to civilizations known chiefly from ethnological studies. There are no visual schemata peculiar to the primitive, schemata that can distinguish and characterize that world that historically has come to be considered primitive. The lack of geometric perspective could be considered a constant if one then excluded almost all the *naïfs* considered below. On the contrary, establishing the geometric, long considered a characteristic of the primitive, as a constant would exclude children's art and prehistoric art; in fact, every attempt to establish a concept about a certain way of seeing would automatically exclude a large part of the primitive world.

Modern primitives. Artists who may earn this appellation are the ingenuous, the spontaneous, the "neoprimitive." That it is possible for a man without any education to create a work of art starting from an absolute ingenuousness and lack of experience, from a total isolation, from a total lack of rapport with official art was demonstrated by the painter regarded today as a pillar of modern art, Henri Rousseau (q.v.), called "Le Douanier." Although Picasso and other moderns already mentioned made an intellectual ideal of the primitive, Rousseau and others of his category were not "primitivists" but genuine primitives, artists who flowered unexpectedly in an ultracultivated and ultrarefined society. For this man of bourgeois origins, this humble clerk, art was an escape, the amateur pastime of a dreamer, and he was roundly jeered for his aspirations.

From 1890 on Rousseau painted exotic pictures, in many compositions fusing daily life and fantasy to produce surreal effects; among his works are *The Snake Charmer* (PL. 330), *The Sleeping Gypsy* (New York, Mus. of Mod. Art), *The Dream of Yadwigha* (New York, Sidney Janis Coll.), and *Storm in the Jungle* (New York, Henry Clifford Coll.). The influence of the novelist Pierre Loti, whose exoticism stimulated Rousseau's imagination, is evident in these pictures. In the *Dream of Yadwigha* a nude woman is lying on a divan in a tropical forest amid exuberant vegetation, phosphorescent light, bright and intense colors, and ingenuous, almost symbolic, forms. In his repertory Rousseau combined real, familiar objects with the stuff of dreams — the household divan with the jungle he had never seen. Lions and tigers, unusual plants with fantastic fruits, fire balloons, scenes from daily life such as *The Country Wedding* (Paris, Coll. Mme Jean Walter) and *The Soccer Players* (Providence, Mrs. Henry O. Sharpe Coll.) — he observed everything with the same curiosity and painted everything in the same enchanted way. In *War* (PL. 335) he expressed the horrible and the terrible in a childlike ingenuous way. It is difficult to forget the paintings of Rousseau. Because of the absence of conscious striving for stylistic effects it is irrelevant to trace their chronology or study their development. Each painting is self-sufficient, and each one captivates with its gripping reality.

Characteristically, Rousseau and most of the other primitivists did not employ preconceived or conventional forms and ignored primitive art itself in favor of looking at reality, whether perceptible or dreamed, with an ingenuous heart. These painters gave equal weight to imagination and reality in the same way as children and savages do, operating on the level of "great reality" (Kandinsky) or "magic realism" (Roh, 1925). This world of theirs, which was supposedly the result of a perfect objectivity, was instead completely subjective; precisely because of a lack of conscious stylistic considerations they did not set out to paint a tree but rather each strip of bark, breaking down objective detail in a way that paradoxically led to the antithesis of objectivity.

Perhaps only Rousseau's work can be considered genuine fine art. The other *naïfs* of our century, sought out and exalted by the critics, represent a phenomenon of folk art. There is, however, a profound difference. Folk art, which has produced artistic objects that are particularly gratifying to our taste today, has always been rooted in tradition or collective taste. In the sphere of folk art (q.v.) the individual, in creating, attempts to express not so much his own individuality as his adherence to the social community and its customs and beliefs. Today, however, with the dissolution of community life and the weakening of tradition, folk art is also vanishing. The anonymous creative forces no longer have a tradition to channel them and form them. The modern primitives, lacking a tradition, are self-taught men who invent and create according to elementary psychological laws, like uneducated children not yet corrupted by instruction and culture. It is thus evident that the primitives do not constitute a current of modern art but are instead a phenomenon of modern times, the object of interest to artists and painters whose taste is oriented toward primitives.

An admirer of Rousseau, Wilhelm Uhde, spent years discovering and collecting the works of ingenuous painters. He presented the works of five of these painters in an exhibition in Paris (Galerie des Quatre Chemins, 1928), designating them "the painters of the *Coeur Sacré*," alluding to their ingenuous expressivity. Among them was a woman, a shepherdess and later a kitchen helper, by the name of Séraphine Louis (1864–1934), who painted fantastic flowers, expressing a powerful passion through her creations. Another of the painters, André Bauchant (b. 1873; PL. 335), a gardener, fused myth and daily life, nobility and banality, with ironical fairy-tale effects quite similar to those of metaphysical painting. The other painters represented were Louis Vivin (1861–1936), a postal inspector; Camille Bombois (b. 1883); and Dominque Peyronnet (1872–1943).

"Primitives" of every nation were then sought out, discovered, and exhibited in various shows and publications in every country. In Yugoslavia (PL. 337) spontaneous art flowered in the rural villages, particularly Hlebine, where a noteworthy representative was the painter Ivan Generalić, and Kovačica, where the painting is still rooted in old folk and peasant art. In Germany, the work of Oluf Braren, born in 1787 on Föhr, an island in the North Sea, was discovered, as well as that of many other artists, all of them simple folk. In Italy the paintings of the shoemaker of Terni, Orneore Metelli (PL. 337) and those of a Neapolitan, Rosina Viva, came to public attention.

It is not difficult to distinguish the authentic primitives from those artists who are deliberately ingenuous and simple. In Germany, Felix Muche (1865–1947), called Ramholz, was an ingenuous painter but a highly cultivated man: original works by Chagall, Picasso, Franz Marc, and Paul Klee hung on his studio walls. A conscious ingenuousness of vision appeared in metaphysical painting, which approaches the banal and puerile. Perhaps Maurice Utrillo (q.v.) is the best known of the artists whose work is consciously ingenuous, but F. Gentilini, Antonio Donghi, Ottone Rosai, and the American, Ben Shahn — to cite but a few — made ingenuousness an ideal of taste and a moral program.

Luigi SALERNO

AMERICAN PRIMITIVES. The most prolific period of primitive and "folk" painting in the United States began early in the 1800s and continued up to the time of the Civil War (PL. 334; I, PLS. 96, 97, 99). During colonial days and the first years of the young republic — up to perhaps the second decade of the 19th century — most American painters were conscious of the academic tradition of Europe, and many went to Europe to study. Benjamin West, John Singleton Copley, Gilbert Stuart, and Charles Willson Peale, familiar names of the 18th century, cannot be considered primitive. The next generation of native artists, however, many of them born during the first years of the nation's independence, not only were untaught but worked in cities and towns that no longer aped English styles. Ralph Earl was a painter who bridged the two cultures; before studying in England he painted (probably in the summer of 1775) a *View of the Town of Concord* (Mrs. Stedman Buttrick, Sr., Coll.) in which the strong sense of design characterizing primitive work is apparent. Houses and troops are placed squarely on the fields in almost geometric patterns, and no single part of the picture can claim interest over another. After Earl's return to the United States, in 1785, he handled his backgrounds with more sophistication, subordinating one element to another through the use of shadow and perspective. Only his portraits retain the early simplicity that makes them so appealing (I, PL. 100).

The primitive American painter or "limner" often grew into his art by way of a craft. The men might begin as house painters, drum or sign painters, carvers of ship figureheads, or cabinetmakers. The women might be weavers or needleworkers, although it was not uncommon for them to begin to paint or cut portraits at an early age, perhaps because it was part of a girl's education to sketch and "take likenesses." Eunice Pinney (1770–1849) was an unusually bold painter whose first water colors date from roughly 1810, when she was forty years old; they are vigorous and solid, the figures are stocky, and there is much contrast and strong design. A favorite subject was the memorial, usually dedicated to a specific person, in which she showed the family mourning at the grave; she sometimes composed appropriate verse and lettered it in the margins. Mary Ann Willson also worked in the early 19th century, from about 1810 to 1825, painting narrative scenes ranging from events in the life of George Washington to the story of the Prodigal Son. The artist made many of her own pigments from crushed berries, bricks, and vegetable dyes, the unusual quality of which heightens the interest of the arabesques and geometric patterns adorning subjects such as *Pelican* (1820; Providence, Rhode Island School of Design, Mus. of Art) and *Mermaid* (1820; Jean and Howard Lipman Coll.).

Edward Hicks (1780–1849) is now known for his many paintings of *The Peaceable Kingdom* (I, PL. 107). He was a Quaker preacher who began to paint and letter signs while an apprentice to a coachmaker; indeed, he reconciled his later creative painting with his religion by considering his pictures not so much as art for its own sake but as a useful craft. He delighted in painting over and over again William Penn's treaty with the Indians, which at first was the subject of his signs but later was incorporated in every one of the many *Peaceable Kingdom* versions. Nobody knows how many of these Hicks painted, but close to 100 have survived. They illustrate the prophecy of Isaiah: "The wolf also shall dwell with the lamb and the leopard shall lie down with the kid; and the calf, and the young lion, and the fatling together; and a little child shall lead them."

The work of Erastus Field (1805–1900) was more diversified than that of Hicks but in some ways recalls the older man. As a young man Field had 3 months of formal training with Samuel F. B. Morse (q.v.) in the latter's New York studio — training that modified to some extent the technique of his early portraits. Field's most striking works are his allegorical and mythological scenes, which relate him to Hicks and to Rousseau. The *Garden of Eden* (1850; Boston, Mus. of Fine Arts), is highly original. The *Historical Monument of the American Republic* (1875; Springfield, Mass., Mus. of Fine Arts) guesses at the future rather than the past; it is a huge canvas, 13 × 9 ft., showing fantastic towers, stepped like wedding cakes and connected at their summits by narrow bridges. Even in his early portraits painted before his formal training with Morse, Field had shown considerable competence in drawing and brushwork, and some critics have supposed that he learned these techniques directly from an itinerant portrait painter.

A considerable number of these itinerant painters were at work in the first half of the 19th century; they found a market among the more prosperous citizens of New England, who in the days before photography were willing to pay substantial fees to make a record of their families and furniture for generations to come. The portraitist might spend a week or more with the family while completing his work — often an agreeable arrangement for the host, who saw a way to lower the fee in exchange for board. Rufus Porter (1792–1884), founder of the *Scientific American*, was a confirmed wanderer who spent the years between 1815 and 1825 as an itinerant portrait painter, traveling from New England to as far south as Virginia. Although he is now known especially for his murals, which were a substitute for expensive foreign wallpaper in many New England homes, he was first a mass-producer of portraits, which he could make in 15 minutes with the help of a camera obscura to focus the sitter's silhouette on paper. It was even proper for young ladies to visit other homes to paint family portraits; there is an instance of a young lady, Deborah Goldsmith (1808–36), who met her future husband, George Throop, at such a sitting.

It is a mistake to think that because primitive painters were untaught they paid little attention to technique. On the contrary, the individual painter worked hard to perfect his own technique, which was often rather elaborate among the portraitists. James Ellsworth developed a style in his miniatures that he apparently never varied; his subjects are always posed in profile, and the white faces almost always are dramatized by being placed against a dark cloud. Joseph Davis, an itinerant portraitist of approximately the same period (the 1830s), developed a compromise between the blank faces of the silhouette cutter and the more elaborate expensive oil portrait; Davis drew the features in pencil, then filled in with water color and added a body and a stylized setting (in water color plus pen and ink) that often had nothing to do with what he saw. The result was a picture in which the subject saw himself in whatever dress he chose, sitting at a handsome table with an assortment of books on top and an elaborately patterned carpet beneath. William Prior (1806–73) worked in both tempera and oil and perfected two entirely different styles according to the price the sitter was willing to pay, advertising portraits "without shade or shadow" to be had at one-quarter the price of an oil. Apparently both kinds were a great success, for the whole Prior family was engaged in preparing canvases, grinding paints, and making frames. In the 1840s Prior set off on foot with as many canvases as he could carry, stayed in various homes while painting, then returned to his home for more canvases and frames, which his two sons had been making in the interim.

Although portrait painting was an especially lucrative form of art it was not the only one: several of the portraitists, including Prior, also tried landscapes; Rufus Porter, as was noted, painted murals; and finally there developed a group of genre painters whose work became more popular as daguerreotypes — readily obtainable after about 1870 — made portraits less so. Two painters of the Middle West farm scene were Olaf Krans and Paul Seifert. Krans, with his parents and other emigrants from Sweden, came to Illinois in 1850 to work on the Bishop Hill communal farm, about 100 miles southwest of Chicago. Many scenes of colony life, showing bobsledding, haying, and pile driving, date from the early 1900s. At about the same time Seifert was painting water colors of Wisconsin farm scenes, adding to the water-color technique such novel additions as the use of colored paper to set the dominant tone of the painting.

The most recent primitive genre painter of note is Anna Mary Robertson Moses, known as Grandma Moses, who like many other American primitives was "discovered" in the 1930s after the Museum of Modern Art and some other New York galleries had shown the work of self-taught artists. Painting was relatively new to her then; she began at the age of seventy-five, when she proclaimed herself "too old for farm work, too young to retire." Before her death at one hundred and one, she painted more than 1000 scenes of farm life in which children play recognizable games and high-booted men drag Christmas trees through thick snow. She, too, perfected an individual technique, which consisted in coating the canvas first with three separate coats of flat white paint, giving luminosity to the finished picture, which had first been drawn sketchily with pencil, then finished with a fine brush. Like many other primitives, Grandma Moses painted what she knew was there, not what might be seen by a trained eye. To friends who urged her to vary the snow scenes with blue shadows she would reply, "I have looked at the snow and looked at the snow and I can see no blue."

Notable among other primitive and self-taught artists of the late 19th and the 20th century whose works may be found in major American collections are Edwin R. Elmer (PL. 334), Joseph Becker (PL. 336), Horace Pippin (PL. 338), Joseph Pickett, John Kane, Morris Hirshfield (PL. 336), Thorvald A. Hoyer (PL. 336), and Patrick J. Sullivan.

Noteworthy collections of American primitive painters are found in the M. and M. Karolik Collection in the Boston Museum of Fine Arts; the Museum of the New-York Historical Society; the New York State Historical Association in Cooperstown; the Morgan Wesson Memorial Collection in the Museum of Fine Arts, Springfield, Mass.; the Jean and Howard Lipman Collection; the Abby Aldrich Rockefeller Folk Art Collection in Williamsburg, Va.; and the Gilcrease Institute of American History and Art in Tulsa, Okla.

* *

BIBLIOG. G. B. Vico, La scienza nuova, Naples, 1725 (3d ed., 2 vols., Naples, 1744; Eng. trans., T. G. Bergin and M. H. Fisch, Ithaca, N. Y., 1948); E. B. Tylor, Primitive Culture, 2 vols., London, 1871; P. Gauguin, Avant et après (1902–03), Paris, 1923 (Eng. trans., Van W. Brooks, The Intimate Journals of P. Gauguin, London, 1930); W. Uhde, Henri Rousseau le douanier, Paris, 1911; G. Apollinaire, Les peintres cubistes, Paris, 1913 (Eng. trans., L. Abel, New York, 1949); H. Kühn, Die Kunst der Primitiven, Munich, 1923; F. Roh, Nachexpressionismus: magischer Realismus, Leipzig, 1925; L. Venturi, Il gusto dei primitivi, Bologna, 1926; E. von Sydow, Primitive Kunst und Psychoanalyse, Vienna, 1927; Galerie des Quatre Chemins, Les peintres du Coeur Sacré: Exposition organisée par W. Uhde, Paris, 1928 (cat.); G. H. Luquet, L'art primitif, Paris, 1930; E. Panofsky, Das erste Blatt aus dem "Libro" Giorgio Vasaris, Städel Jhb., VI, 1930, pp. 25–72; F. Toor et al., eds., Las obras de José Guadalupe Posada, grabador mexicano, Mexico City, 1930; L. Venturi, Art populaire et art primitif, Actes XIIIe Cong. int. d'h. de l'art, I, Brussels, 1930, pp. 333–36; Art populaire: Travaux artistiques et scientifiques du Ier Cong. int. des arts populaires (Prague, 1928), 2 vols., Paris, 1931; H. Cahill, ed., American Folk Art, 1750–1900, New York, 1932; N. Pevsner, Pioneers of the Modern Movement, London, 1936 (3d ed., Harmondsworth, 1960); J. Cassou, Les maîtres populaires de la réalité, Art vivant, 1937, pp. 201–03; R. J. Goldwater, Primitivism in Modern Painting, New York, London, 1938 (bibliog.); Museum of Modern Art, Masters of Popular Painting: Modern Primitives of Europe and America, New York, 1938 (cat.); R. Huyghe, La peinture française: Les contemporains, Paris, 1939 (2d ed., 1949); S. Janis, They Taught Themselves: American Primitive Painters of the 20th Century, New York, 1942; J. Lipman, American Primitive Painting, New York, London, 1942; M. Gauthier, André Bauchant, Paris, 1943; H. Read, Education through Art, London, 1943 (3d ed., New York, 1958); P. Courthion, Henri Rousseau, Geneva, 1944; R. R. Tomlinson, Children as Artists, London, New York, 1944; E. O. Christensen, Popular Art in the United States, London, 1948; Ouverture en juin de la salle W. Uhde, consacrée aux primitifs du XXe siècle au Musée d'Art Moderne, Paris, 1948; A. E. Ford, Pictorial Folk Art: New England to California, New York, London, 1949; M. Gauthier, Henri Rousseau, Paris, 1949; A. Jakovsky, La peinture naïve, Paris, 1949; W. Uhde, Cinq maîtres primitifs, Paris, 1949 (Eng. trans., R. Thompson, New York, 1949); M. Gauthier, Jean Eve, Paris, 1950; J. Lipman and A. Winchester, eds., Primitive Painters in America, 1750–1950, New York, 1950; H. Read, Art and Society, 2d ed., London, 1950; H. Bing-Bodmer, Camille Bombois, Paris, 1951; P. Francastel, Peinture et société, Lyons, 1951; H. Bing-Bodmer, Lettre sur Louis Vivin, Du, XII, 2, 1952, pp. 21–22; A. Jakovsky, Louis Vivin: Peintre de Paris, Paris, 1952; F. Meyer, Primitive im 20. Jarhundert, Du, XII, 2, 1952, pp. 6–18; H. Read, The Philosophy of Modern Art, New York, London, 1952; M. Valsecchi, La pittura di Rosina Viva, Milan, 1952; E. Ambron, Rapporti fra l'arte primitiva e gli artisti europei, Realtà nuova, 4, 1953, pp. 301–11; G. C. Argan, Dell'idea dei "primitivi" nella storia dell'arte, Arch. di filosofia, 1953, 1, pp. 91–97; L. Adam, Primitive Art, 3d ed., Harmondsworth, 1954; R. Arnheim, Art and Visual Perception, Berkeley, Calif., 1954; M. Basicevic, ed., Les primitifs yougoslaves, Dubrovnik, 1956 (cat.); A. Chastel, Le goût des préraphaélites en France, preface to M. Laclotte, De Giotto à Bellini, Paris, 1956 (cat.); A. Jakovsky, Les peintres naïfs, Paris, 1956; H. Perruchot, Le douanier Rousseau, Paris, 1957; H. Read, The Significance of Children's Art: Art as Symbolic Language, Vancouver, 1957; A. Werner, Henri Rousseau, New York, 1957; A. Buttitta, Individualismo e sincretismo dell'arte infantile, Atti Acc. di Sc., Lettere e Arti di Palermo, 4th ser., XIX, 2, 1958–59, pp. 291–99; La peinture naïve: du douanier Rousseau à nos jours, Knokke-Zoute, 1958 (cat; introd. by A. Jakovsky, Ces peintres de la semaine de sept dimanches); O. Bihalji-Merin, Das naive Bild der Welt, Cologne, 1959 (Eng. trans., N. Guterman, Modern Primitives: Masters of Naive Painting, London, New York, 1961); O. Bihalji-Merin, Yugoslavian Primitive Art, Belgrade, 1959; G. Previtali, Bottari, Maffei, Muratori e la riscoperta del medioevo artistico italiano, Paragone, X, 115, 1959, pp. 3-18; G. Previtali, Collezionisti di primitivi nel Settecento, Paragone, X, 113, 1959, pp. 3–32; G. Previtali, La controversia secentesca sui "primitivi," Paragone, X, 119, 1959, pp. 3–28; C. Roy, Le arti selvagge, Milan, 1959; G. Luzzatto, Una curiosa polemica contro la moda dei primitivi nel 1824, Comm., XI, 1960, pp. 87–90; G. Previtali, Le prime interpretazioni figurate dai "primitivi," Paragone, XI, 121, 1960, pp. 15–23; G. Cocchiara, L'eterno selvaggio: Presenza e influsso del mondo primitivo nella cultura moderna, Milan, 1961; 101 Masterpieces of American Primitive Painting from the Collection of E. W. and B. C. Garbisch, New York, 1961 (cat.); A. M. Cirese, Aspetti della ricerca folklorica, Ann. Mus. Pitrè, XI, Palermo, 1962.

Luigi SALERNO

Illustrations: PLS. 329–338.

PRINTING. See GRAPHIC ARTS.

PROPORTION. The concept of proportion is a mathematical concept that has acquired great importance in the visual arts, having supplied a set of norms corresponding to those of meter in music and poetry. The search for balance, for unity, for "style," the need to fix art within a technical rationalization, in an intelligible language of forms and the psychophysiological structure of vision itself have induced the artist to base himself largely on geometric forms and their proportional relationships, be it either intuitively or consciously, even on a theoretical level. Since proportions exist in objective reality and are the subject of a separate science, it is, or at least should be, evident that a historical study of proportions in works of art can have meaning only if one intends to go beyond apparent facts, that is to say, beyond mere statement of those relationships which exist objectively in form (and which are very often unknown even to the artist who created them); such a study should attempt to reconstruct the intention of the artist — that is, his theory of proportions, be it implicit or openly declared. In such a historical survey of proportion, it will be seen that proportion has always been a subjective element in artistic creation, and it is therefore variable through the different civilizations and even through the different artists who have made of it a personal ideal and an instrument of their own creation.

SUMMARY. Theories of proportion (col. 717). Proportion in the human figure (col. 722). Proportion in architecture (col. 731). Proportion in pictorial composition (col. 738). Proportion in the Orient (col. 738).

THEORIES OF PROPORTION. By a theory of proportion "we mean a system of establishing the mathematical relations between the various members of a living creature, in particular of human beings, in so far as these beings are thought of as subjects of an artistic representation" (E. Panofsky, *Meaning in the Visual Arts*, New York, 1955, p. 56). Such relationships obtain expression by proceeding from the whole and dividing it into parts, or by the multiplication of a unit in order to obtain the whole.

The object of the artist's research on proportions can be traced back to three distinct requirements: the desire for beauty, the need for a norm and the necessity of establishing a convention. Furthermore, it may be taken for granted that there exist objective proportions and subjective or technical proportions (i.e., those by which the artist translates the former into his works). The art historian must limit himself to the study of subjective or technical proportions, which are based on three possible theoretical categories: theories that aim at establishing objective proportions without relating them to techniques; theories that aim at establishing technical proportions without reference to their connection with objective proportions; and theories which aim at the fusion of technical and objective proportions.

No sooner had the "geometric" style succeeded prehistoric naturalism than rationalism replaced sensuality; and art thus became the language of ideas (that is, of symbols), and fixed forms made their appearance, inspired by the figures and laws of geometry. The actual mathematical conception of proportion was first elaborated by the Babylonians, then by the Egyptians, to whom the Pythagorean and Euclidian science of Greece was closely related. Obvious examples of the purest geometric style are the pyramids of Egypt; but also in sculpture and in painting the Egyptians constructed the figure by reducing its height, width, and depth to measurable size (as they did in architecture for the frontal and lateral elevations and the ground plan). For this conception of construction they could make use of the division by squares, subdividing the figure into equal parts according to a conventional criterion; in the same way, they were able to build with blocks of equal size which assumed the value of a standard form (FIG. 717).

The Egyptians constructed but did not imitate the human figure; their realism and their technique produced an art that was not naturalistic (that was not, in other words, an imitation of reality); they were not concerned with transient and ever-changing life but with timeless and static eternity. Egyptian sculpture represents bodies that await resurrection; it creates a reality that is magical. Thus, form has a real and an ideal value, both concrete and abstract, live and symbolic.

The Greeks of the classical age considered proportion in a very different manner. They made of it an ideal esthetic reality, an instrument of mimesis (q.v.) capable of representing the appearances of life, the image of the organic function of the living man. In this way, technical proportion was separated from that of reality, because the problem was the representation of the latter in a verisimilar manner by means of optical devices and illusions of perspective. Thus, the laws arose out of experience and not, as in Egyptian art, out of preestablished rules. The canon of Polykleitos (q.v.) (which has not come down to us), was certainly of an anthropometrical character and a perfect example of this subjective investigation (FIG. 717).

In order to grasp the value that proportional theory was to assume in the medieval and modern worlds, it is indispensable to recall the thought of the great Greek thinkers Pythagoras, Euclid, and Plato. While Euclid fixed the rules of geometry, the Pythagorean school distinguished three distinct types of proportion: arithmetic, geometric, and harmonic. Pythagoras discovered that tones could be measured by spatial standards. Musical consonances could be determined by numerical relationships. If two strings were made to vibrate under identical conditions, one being half as long as the other, the note of the shorter one would differ by an octave (diapason) from that of the longer one. If the ratio of the two strings were 2 : 3, the difference would be a fifth (diapente), and if the ratio were 3 : 4, it would be a fourth (diatessaron). Thus the consonances on which the Greek musical system is based can be expressed in the progression 1 : 2 : 3 : 4.

This discovery seemed to reveal the mystery of the order of the universe. Plato, in his *Timaeus*, explained that the har-

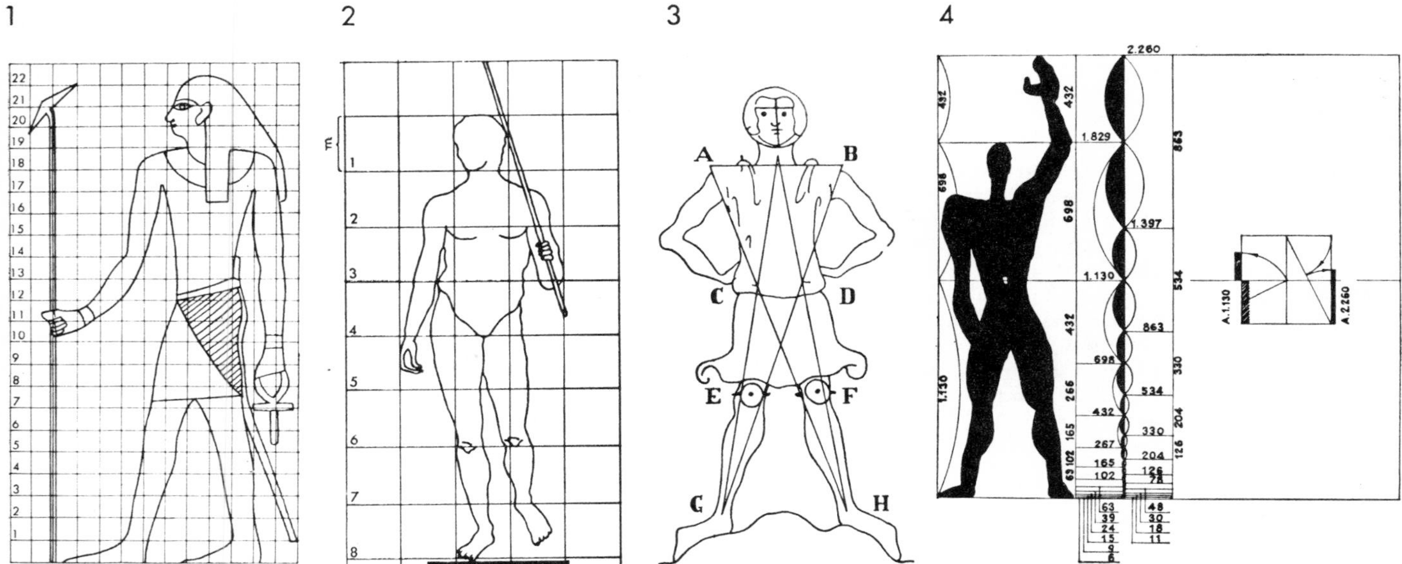

Canons for the proportions of the human figure. (1) The "Later Canon" of Egyptian art (*from E. Panofsky, 1957*); (2) graphic scheme of the Polycletan canon; (3) construction of the frontal figure, on the basis of Villard de Honnecourt (*Paris, Bib. Nat., MS. fr. 19093, fol. 19*); (4) Le Corbusier, *modulor*.

mony and the order of the cosmos are based on certain numbers and their multiples, that is, on the progressions 1, 2, 4, 8 and 1, 3, 9, 27, in the following form:

```
      1
    2   3
  4       9
8           27
```

The musical consonances of the universe — in other words, the music of the spheres, which cannot be perceived by man because he is too accustomed to it — would therefore be contained within the relationships between these figures. Thus originated the symbolism and mysticism of numbers that were to influence human thought for 2000 years.

According to Pythagoras, 3 was a perfect number because it has a beginning, a middle, and an end. From then on, 3 was considered the essential and divine number, symbol of the Trinity; the square and the cube of 3 were also considered perfect.

In the same way, certain forms appeared to possess absolute beauty. In *Philebus*, Plato writes: "I do not mean by the beauty of form such beauty as that of animals or pictures, which the many would suppose to be my meaning; but, says the argument, understand me to mean straight lines and circles, and the plane and solid figures which are formed out of them by turning — lathes and rulers and measures of angles; for these I affirm to be not only relatively beautiful, like other things, but they are eternally and absolutely beautiful...." (trans., in B. Bosanquet, *A History of Aesthetics*, London, 1892, p. 33).

This is the origin of the standards for the preference of certain forms and geometric bodies above others. Modern psychology (see PSYCHOLOGY OF ART) has undertaken research on the question of the more pleasing geometric figures, for there is no doubt that the sense of balance and of harmony is an innate psychophysiological factor. But the preference for some abstract forms rather than others has its origin above all in their link with cosmologic conception.

The only source from the classical age for the theory of proportion is Vitruvius, who gives the names of various theoreticians whose works are lost (Nexaris, Theocydes, Demophilus, Pollis, Leonidas, Silanion, Melampus of Samaca, Euphranor). For Vitruvius, beauty consisted in eurythmy and symmetry; he defined it as proportional harmony of the various parts to each other and to the whole, so that nothing could be modified in an organism, which is an end in itself and perfect. This harmony is a reflection of celestial harmony and perfection. Plato had already asserted that man was the measure and center of all things, a microcosm reflecting the macrocosm. Anthropomorphic references are obvious in Vitruvius, who furthermore established that a man is six times as tall as the size of his foot and a woman eight times as tall as the size of her foot. The numbers 6 and 8 are therefore the module of the relationship between height and diameter for columns of the Doric and Ionic orders; 6 is thus the symbol of masculine force, and 8 of feminine gentleness. In this way, Vitruvius established relationships between all the single architectural elements and, on the basis of relationship modules, fixed the elements of the architectonic orders. (See STRUCTURAL TYPES AND METHODS.)

In the Middle Ages, the proportional concept continued to be applied to philosophical and theological speculation on cosmology. Elements of an esthetic of proportions and numbers are to be found in St. Augustine (*De vera religione*), in Boethius, who also insisted on the relationship to music, and in Cassiodorus, who insisted on the value of certain numbers. The number 4 was made to represent the four winds, the four cardinal points, the phases of the moon, and the four seasons, and it was applied also to man, *homo ad quadratum*, a tetragon also in a moral sense. The square is, in fact, a perfect form because it can be inscribed in the circle (Vincent of Beauvais, *Speculum doctrinale*, XXVIII, c2).

But medieval cosmology, the metaphysical interpretation of the structure of the human body, the correspondences between man and the universe, and the esthetics of proportion and numbers had no influence on the arts (see COSMOLOGY AND CARTOGRAPHY). In actual fact, proportions in art were reduced to simple practical schemes, to mere rules of the profession. Medieval Byzantium replaced the "constructive" proportional concept of the Egyptians, and the "anthropometric" concept of classical antiquity, with a concept that Panofsky has described as "schematic." The image is conceived as a surface and is not studied from nature; in this manner, the different relationships arise from the simple application of a module. For example, the *Painter's Manual of Mount Athos* fixes as a module that the height of the forehead is equal to those of the nose and the chin, thus permitting the human face in a frontal view to be inscribed in three concentric circles and to be executed without any study of anatomy.

In the late Middle Ages, during the Gothic period, the withdrawal from objective proportion persisted. The process was not one of gradual ascent from experience to an ideal canon of perfection (a process of idealization) but one of movement from a preconceived and abstract ideal toward the bringing out of concrete form — that is, from geometric forms and bodies toward architectonic human and animal forms. A French architect of the 13th century, Villard de Honnecourt, left a treatise, a kind of manual, *Livre de Portraiture*, in which his intention was to establish a practical method of drawing. He inscribed human figures (FIG. 717), horses, faces, and limbs within pure geometric forms (in triangles, circles, squares, etc.; PL. 341). In the process, natural images were deformed, but from the structural point of view the result was the concreteness of the image peculiar to Gothic sculpture and, later, to Giotto's painting, both of which follow the same movement from the abstract to the concrete.

Especially in architecture, the Middle Ages was characterized as much by abstract geometric forms and by geometric processes as by a sense of symmetry; but there are many exceptions to this, especially in the pre-Romanesque and Romanesque periods. At the end of the 14th century, various experts were called to Milan to give their advice on what proportions to give the new cathedral. Opinion was divided into two currents: the French maintained that circles should be used, while the Germans preferred triangles. C. Cesariano's edition of Vitruvius's *De architectura* (1521) contains an engraving of the cathedral with a text explaining how the proportions were finally established according to the German method, using equilateral triangles.

With the Renaissance, the theory of proportion ceased to be an expedient or a method of construction reserved to architects. Once again, a direct link was established between technique and nature, between man (the microcosm) and the universe (the macrocosm), and, most important of all, the concept of beauty itself was again conceived of on the basis of proportion.

At the beginning of the 15th century, Brunelleschi and Donatello (qq.v.) were in Rome measuring the ancient monuments and were called "the treasure hunters"; but the treasure they sought among the ruins was the secret of the perfect proportions attained by the ancients.

Leon Battista Alberti (q.v.), the father of the artistic theory of Humanism (q.v.), like Leonardo da Vinci (q.v.) after him, wished to establish a science that would not limit itself to reproducing the Vitruvian rules but would study nature and by study and experience extract the "ideally beautiful" identified with perfect physical beauty, which could serve as a criterion for artists. This was the beginning of the theory later developed by Raphael (q.v.) and revived by Baldassare Castiglione, according to which it is necessary to study the most beautiful parts of many different models in order to obtain absolute beauty, which cannot be found in nature because of the incidence of recurring defects. Thus nature bows to art, and objective proportions give way to technical ones. In fact, Alberti studied proportions not only on the surface but also in depth, that is to say, in relation to perspective. In architecture he recommended nine geometric figures for churches, all with a central plan; that is, they could be inscribed within a circle, which he considered the most perfect form. This preference was a characteristic of Renaissance architecture up to the time of Bramante and Raphael.

Research was thus directed toward perspective, optics, and anatomy, and the figural arts were presented as being the science of the visible. The purist theoreticians of this period did not hesitate to consider mathematics the formal foundation of all art, and there was perfect concordance between the artistic theory of Piero della Francesca (q.v.) and his painting. Precisely because mathematics proposes an esthetic ideal an almost metaphysical purity appears in art, an idealization which rationalizes and transcends all.

Francesco di Giorgio (q.v.) wished to be the modern Vitruvius, and in his treatise he attempted to combine the centralized part of the edifice with the longitudinal one, creating a direct relationship between architecture and the human figure. Man is in fact the measure of all things, asserted another great theoretician, Luca Pacioli, in his *De divina proportione* (written in 1479, but published in 1509 in Venice; PL. 339). He remarks that the ancient temples were built according to the human figure (the circle, the square, and the rectangle), because man, created in the image of God, is the mirror of the universe. According to Pacioli, divine proportion is the golden section, in which the whole is to the largest part as the largest part is to the smallest ($1 : x = x : 1 - x$). In a rectangle, constructed according to Euclid (11 . 11), the width should be $\frac{1}{2}(\sqrt{5} - 1)$, or .618 of the length. This proportion, according to other later, theoreticians as well, is the perfect proportion, that which, in other words, best represents unity in diversity.

Pacioli was a product of Milanese culture, of the same cultural world as Leonardo da Vinci, Foppa, and Bramante (qq.v.). Leonardo associated proportion with human movement and with perspective; in other words, he affirmed technique as a subjective aspect in respect to objectivity. In fact, he studied movement and foreshortening and again based his work on analogies with music, which is the "sister" of painting because both are based on harmony. In this way he established the law of perspective diminution of distances, by which objects of equal size whose distance from the eye increases in mathematical progression (1 : 2 : 3 : 4) diminish in inverse proportion, that is, in harmonic proportion (1/2 : 1/3 : 1/4).

The study of human proportion culminated in the work of Dürer (q.v.), whose treatise *Vier Bücher von menschlicher Proportion* appeared posthumously in 1528. He made a basic innovation by denying the existence of a single canon of beauty and by studying proportions of different types, both masculine and feminine (PLS. 346, 347) as well as those of children. In this manner he included in the sphere of art those forms that were merely characteristic and even ugly. With Dürer's theory arose the first crisis in the Renaissance proportional system.

The treatises of Serlio, Palladio, and Scamozzi (qq.v.; see also TREATISES) establish a true system of rules in architecture that remained valid into our era, until the modern reaction against architectural "orders" set in (see STRUCTURAL TYPES AND METHODS). During the baroque period the tendency was to neglect the certainties of proportional norms in favor of effects of perspective, reality in favor of illusion, and real proportion in favor of intuitive proportion. Actually, the myth of abstract proportional perfection was already condemned, even in architecture, and had become philosophically and theoretically untenable.

With the coming of the Gothic revival (see NEO-GOTHIC STYLES) the proportions of medieval edifices were studied in order to discover basic rules and, in a more general way — particularly in the 19th century with its theory of "styles" and its eclecticism — historians became more and more interested in the search for the basic schemes of the buildings of the past, not only those of the classical and medieval periods but of all civilizations. From A. Aurès to Viollet-le-Duc (q.v.), to Auguste Choisy, A. Zeising, J. Hambidge, A. Thiersch, E. Mössel, M. C. Ghyka, and Charles Texier, this kind of investigation has proceeded, very often producing esthetic or metaphysical deductions that are decidedly nonhistorical and unhistorical and not justified by modern esthetics. For example, Ghyka has insisted on the role of the pentagon and of numbers in living nature, and he has asked whether certain proportions do not arise in the mind of the architect out of some obscure instinct. Other writers have maintained the absolute validity of some proportional formulas or have turned results into causes, but such research makes the mistake not only of being unhistorical, but also of projecting into the past our own modern knowledge. As Hautecoeur (1937) observes, the fact that an architect should make use of a pentagon or any other geometrical figure in his work implies that certain numerical properties of which even the artist may not be aware are involved; in actual fact, certain objective mathematical properties were discovered later and could not have been known to the artists. In recent times the study of proportion in art has been undertaken from a historical point of view, as an analysis of the development of theories of proportion, by scholars such as E. Panofsky and R. Wittkower, with extremely interesting results for art history and for artistic culture.

Regarding the reactions to the theories of proportion, none of the subjective or pictorial styles (e.g., 17th-century Dutch painting, 18th-century Venetian painting, French impressionism, or any of the expressionist movements) ever had a preconceived theory of proportions. Though proportions exist in reality or are implicit in perspective, they have not interested artists of these schools.

Apart from this indifference, however, there did occur a definite historical reaction that was already visible in mannerism. In particular, the school of Michelangelo (q.v.), with its deformation of the human figure by foreshortening, its lengthenings and contortions (the famous Michelangelo *contrapposto*), produced an ideal of the undulating line, which is not a fixed module but an arbitrary device of the artist. In the treatise *Il primo libro del trattato delle perfette proporzioni* of Michelangelo's pupil Vincenzo Danti, and in spite of its title, one finds beauty defined no longer on the basis of a proportional canon but on that of the organic adaptability of the body to its object (i.e., on the concept of "decorum"), and so the proportions are based on anatomy, imitation, and movement.

It is true that, in the 17th and 18th centuries, the theory of the mathematical universe subject to harmonic laws, with all its derivations, still dominated the thought of Johannes Kepler (*Harmonice mundi*, 1619) and Galileo, as it did, by reflection or tradition, that of many theoreticians of architecture. But contrary principles were beginning to be vigorously asserted; they, too, were of Neoplatonic origin, that is, centered on the concept of genius, of inspiration (poetic "furore"), and of grace. Giordano Bruno wrote that there were no rules and that the artist was the creator of his own rules — a principle that has become essential since the foundation of modern esthetics (Immanuel Kant). The intuitive, uncontrollable element in creation was attached to a certain type of beauty and gracefulness, which Ludovico Dolce has called a "je ne sais quoi" and which Franciscus Junius was to declare a "harmony which does not arise from proportions." Thus, the battle against rules and in favor of the relativity of taste was continued by theoreticians such as Claude Perrault, by artists such as William Hogarth (q.v.), and by essayists such as David Hume, Edmund Burke, and Archibald Alison.

Since mannerism, proportions have been a matter of individual sensitivity, no longer aimed at imposing a universal law. It has been demonstrated that the possibilities in proportion are infinite.

The formalistic currents of modern art have again attempted a study of ideal proportion and have once more felt the magic of numerical relationships and abstract laws of geometric forms. In cubism, especially, there has been an attempt to discover universally valid laws of formal relationships. Le Corbusier has proposed his theory of the "modulor" (FIG. 717), and Mondrian has given theoretical support to his language of abstract forms. Even "informal art" (see EUROPEAN MODERN MOVEMENTS; NONOBJECTIVE ART) is a field of investigation into harmony and hidden relationships between the visual elements, on the level of a constant search for form that should be important for man and that should have universal value and validity.

Luigi SALERNO

PROPORTION IN THE HUMAN FIGURE. Like all norms that are imposed upon the practice of art, proportions are at once an impediment — giving rise to fixed classes, to artisanal derivations, and to exaggerated simplifications — and an invitation to the imagination. For thousands of years they have inspired the search for new treatments of the human figure, so that it might be varied and adapted to the transformations of the various civilizations (Panofsky). This is due also to the fact that a proportional canon, however complex, always leaves a large margin of ambiguities, as in the case of the commentaries on Vitruvius. Moreover, the movement of the depicted figure, which unbalances and often deforms it, or else the position of the viewer, who has to submit to distortions produced by perspective, by distance, or by vantage point, all these are enough to bring out the insufficiencies of objective canons and to cause their modification time and again. But precisely because of its complexity, the matter of proportions is one of the most fascinating in the history of art, evidently arising from a profoundly felt psychological and esthetic necessity, since one finds that the proportions established in the Far East in the 18th century by Masayoshi Kitao (pseudonym, Keisa Kuwagata) are in many respects similar to those of the Vitruvian tradition.

It is impossible, however, to undertake a history of proportional concepts, and of their concrete application in particular, either in the East or the West, because of the lack of specialized studies even on vast cultural zones (the proportions of baroque figures, for instance, have never been analyzed); in too many other cases, studies have been limited to arbitrary measurement that takes no account of the ancient systems, though these are well known and accredited from various existing sources. Also regrettable is the tendency to complicate, through pointlessly artful geometric and algebraic constructions, proportional systems that must in general have been simple and practical, based mostly on the use of the compass and the square, either during the drawing of the cartoon or model or on the actual surface of the stone or marble blocks.

The oldest direct documentation of the use of regular and geometric canons come from Egypt, where the practice appears to have become general at the end of the 2d dynasty. Indeed, the tablet commemorating King Narmer (IV, PL. 324) seems to have been constructed according to an objective canon. Various unfinished tomb decorations clearly show the grid, or proportion squares, upon which they were executed; roughed-out sculpture shows that similar grids were traced on the four vertical sides of the block and sometimes on the top horizontal plane.

In the Staatliche Museen, Berlin, there is a papyrus showing a very accurately proportioned prospect and plan of a sphinx. The grid originally consisted of 18 squares from top to bottom and was then changed to 22. The head, excluding the hair, usually consisted of 3 units, so that, in relation to the canon later prevalent in the West, the complete height was about 6 or 7 times that of the head. Because of the complete absence of perspective foreshortening and the absolute flatness of the figure, similar conventions permitted the establishment of the proportions of the principal gestures and movements (e.g., the length of a step was fixed at $10^1/_2$ units from the point of one foot to the other). It was also possible to use different canons in the same picture; for instance, the Berlin drawing mentioned above displays three canons: one for the lion's body of the sphinx, one for the anthropomorphic head, and a third for the little goddess who stands between the paws.

Similar systems of proportioning must have been elaborated for the monumental sculpture of Mesopotamia, if only to make it possible to work separately on the various parts of the figures; probably, there too, the most practical system was the division into squares. In Mesopotamia, as in Egypt, the placing of the figures in procession, the alternation of movements, and even the repetition take on a rhythmic value precisely because they allow recurrences of the module; actually, the distance between the figures is fixed and is in geometric proportion to the height of the entire sculptured fascia.

The system of division by squares is also well adapted to cases in which the human figure is not seen alone but is projected onto a continuous space or into a narrative context. Frescoes, bas-reliefs, and especially mosaics offer such conditions. Much more difficult — in fact, almost impossible to solve — is the problem of establishing a valid canon for sculpture in the round, which is meant to be seen from all sides and should therefore be composed of harmonic correlations between the different members. This necessity appears to be connected, at least in the beginning, with an attempt at greater naturalism, with real progress in anatomical knowledge, and, of course, with the great value attributed to sculpture as such. The canons developed little by little; the first convention that seems to have been applied in Greek art, together with the most rigorous symmetry of the members (as in the archaic kouroi), is the repetition of the measurements of the head, which is used as a module, seven and a half times. L. D. Caskey (1924), commenting on the Apollo of Tenea, says that from the study of proportions it is possible to understand the psychology of the archaic sculptor, who, faced with the problem of reducing a rectangular block of marble into a figure, would soon discover that the height of the head is repeated approximately in the distance from the chin to the pectoral muscles and from there again to the navel, and in the width of the waist and of the legs at the knees. He found the width of the neck to be about one half of this unit; the breadth of the shoulders was originally equal to two units, as is shown by the statues from Thera and Delphi. Later, as in the Apollo of Tenea, he reduced this excessive breadth of the shoulders, but by freeing the arms from the sides, he was still able to conform to this scheme.

It was precisely with the increase in the freedom of attitude, and therefore with the impossibility of adapting to an elementary system of proportion like the Egyptian one of designing a grid on the faces of the marble block, that the problem of proportion became the central theme of Greek theory. So far, however, the measuring of Polykleitos's works has not revealed any facts that would permit us to reconstruct his famous canons with any degree of certainty. The fragments of his writings that have survived are quite insignificant. However, Galen's *Placita Hippocratis et Platonis* (v, 3), describes the anthropometric process adopted by the sculptor: "Chrysippus . . . holds that beauty does not consist in single elements but in the harmonious proportion of the parts, the proportions of one finger to the other, of all the fingers to the rest of the hand, of the rest of the hand to the wrist, of these to the forearm, of the forearm to the whole arm, in short of all the parts to all the others . . . ." It is difficult to explain how and in what manner Greek canons evolved into those of late antiquity and of the early Middle Ages. The fundamental flatness of Byzantine painting could, it is true, facilitate the checking of the proportions adopted by the artists, but, on the other hand, theoretical writings are extremely scarce. The Mount Athos canon, besides being of late date, is somewhat invalidated by a profound infiltration of Western ideas, specifically Italian ideas, and so its canon is occasionally close to those of the Renaissance. Nevertheless, considering that the first Byzantine period bases itself on an anatomy that is fairly precise from a scientific point of view, one may deduce that artists also availed themselves of anatomic distinctions and that these anatomic data were lost in the high Middle Ages (e.g., the use of a third of the face or the head for the neck and of a similar proportion for the height of the foot). Judging by the imperial portraits in S. Vitale, Ravenna, even if the deformations inherent in the use of mosaics are taken into account, it seems possible to assert that the head was divided into three equal parts including the whole forehead, even when this part is covered by hair or ornaments. These values, however, are not related in any way to those of the breadth of the face, at least not in these portraits: the elongated and emaciated face of Theodora (IV, PL. 18) belies any basic canon. The habit of dividing the face into concentric circles with the length of the nose as radius and module, which later became so widespread, seems to have taken root only from the 11th century on (mosaic with the Emperor Constantine Monomachus and the Empress Zöe in S. Sophia; II, PL. 448) and culminates in the Pantocrator of the cupola of Daphni. Also at Daphni, and with perfect smoothness, considering that the compass method tends to cause the neck to be absorbed within the measurements of the whole face, the size of the body is exactly eight times the module, at least judging by the nude figure of Christ in the baptismal scene.

The feet and the neck, in fact, appear to be smoothly assimilated into the legs and the head respectively. It must be added that the use of concentric circles has a strongly hieratic and symbolic function, causing the structure of the face to coincide very closely with the cross within the halo. The complication of the measurements and a progressive slackening in anatomical knowledge characterize stylistic evolution in the West, which seems to have aimed mainly at the adaptation of the picture to the story or to the architectural context. The reading of Vitruvius and explicit cosmologic reference to the four parts of the universe led to a revival of the motif of the *homo ad quadratum*, probably at the Carolingian court, as magnificently exemplified in the crucifix with outstretched arms and erect head whose width is equal to its height. (A crucifix of this sort was given to the Basilica of St. Peter, and the famous "Holy Face" of Lucca is of the same type.) For the Benedictines and the Dominicans, this was to remain a normative example; it allowed the arms and knees to bend dramatically without sacrificing symbolic schematicism or emotional references to the Passion. This is not the only case of structuring the human body on the basis of symbolic numbers. Just as frequent was the motif of man numerically assimilated to Noah's ark, that is, with the length of his body being ten times its thickness and six times its width; this was done even in architectural projection. Misunderstandings were possible: the breadth of the body included the arms. Almost certainly, the crucifixes with arms stretched upwards in the shape of a tau and those with arms inclined conformed to this type.

In complete contrast with the relative stability of the Byzantine canon, a great variety of canons were used by artists of the Romanesque period. In the 224 sculptures measured by J. Laran, the relationship between the total height and that of the head ranges between 3.8 (Notre-Dame du Port) and 10.5 (Chartres), although the average norm was of $7^1/_2$ times the head; between the total height and the width of the shoulders, the proportion varied from 3 to 8.2. Moreover, "the greater the height of the whole statue, the smaller the proportional size of the head. When the dimensions of a figure are modified, the height of the head varies much less than the canon for the total height of the figure." A parallel theory of great interest also because it is almost certainly reflected in the geometrical schemata used by Villard de Honnecourt, is Robert Grosseteste's concept of a universe constituted like a scaffolding of spheres and cones, lines and planes, curves and angles, which have their origin in light and vary according to the resistance of matter. The return to order, at the beginning of the 13th century, besides being inspired by classical sculpture, implies a less fanciful interpretation of figural schemes and of the proportions of the human body; the Chartres *Annunciation* is based on 8 heads and that of Reims on $8^1/_2$ or 9, judging by the photographs.

The majority of crucifixes, with eyes either open or closed, seem on the whole to respect the Byzantine proportional system based on 9 heads. The peculiar iconography that abnormally elongated the feet, which were seen from the front as though the soles were flat on the wood of the cross, and also limited, or even eliminated, the neck with the sagging of the head caused the artist to resort to the subterfuge of dividing the figure into nine parts: one for the head (not only the face), three for the torso (one reaching to the beginning of the stomach), two for the legs down to the knees, two for the legs below the knees, and one for the feet. The hands are counted as one unit of measurement, while the arms appear to adhere to no rules whatever. On the whole, however, the extension of the figure in breadth is double the height of the body from the top of the head to the pubis.

One also finds certain subtleties that help to make the scheme of the crucifix tally with the anthropometry of the 9-faces standard: thus, in a crucifix (no. 20) in the Museo Nazionale, Pisa, the reclining head has the same value as the face, if calculated on a vertical axis, so that the whole perfectly retains the measures of the 9 faces. (It seems to be inscribed in a circle.)

In Giunta Pisano's work (particularly the crucifix in S. Domenico, Bologna) a tetragonal structure is prevalent, which causes the figure to be as wide as it is high. The *hanchement*, or drooping, of the arms is a valuable expedient for avoiding the risk of making the image look too squat. More or less the same system can be found in the works of later imitators of this particular type. For example, the crucifix by the St. Francis Master in the Galleria Nazionale dell'Umbria, Perugia, conforms to the same symmetrical law, except for the halo, which is included in the total height of the image (and which, on the contrary, is omitted in Giunta's crucifix).

A very important example of the return to Vitruvius's ideas, also because it occurs precisely at the beginning of the Tuscan Renaissance, is the crucifix by Cimabue (q.v.) in Arezzo, which follows the 10-head module, as is testified by a contemporaneous literary source that has so far been ignored, Ristoro d'Arezzo's *Della composizione del mondo* (1282): "And the learned designers, to whom nature has given the power to divide and to design the things of this world, when it came to drawing the human figure, divided the space into ten equal parts; and of the top part they made the face, and from that downwards there remained nine parts; and according to the face, they proportioned the hands, and the feet, and the torso, and all the body; and below the face there remained nine equal parts, so that the whole was made up of ten equal parts. They perceived and knew the form of a person well-proportioned and perfect, and this was the result of their nobility of imagination and of intellect."

It would be interesting to calculate the proportions used by Giotto on this basis, at least in the case of his youthful crucifix for S. Maria Novella in Florence, which marks the transition from Byzantinesque stylization to naturalism. Though it is difficult to deduce the possible constructional schemes from photographs, it is possible to perceive the use of a canon of 8 faces, which adapts itself to the iconography of the dead Christ leaning to the left, as in Cimabue's work in Arezzo, with slight modifications. Because the head is sunken, the point of metric reference is the neck and not the chin; the arms are in four parts, and the stretching of the left arm is taken into account (this, however, is much more evident in Cimabue). But instead of adopting stylizations — and this is partly what produces the remarkable reality — the work retains as its own measure the width of two faces, at the risk of appearing stiff like a parallelepiped. The same measurements apply to the Giotto crucifix in the Museo Civico of Padua. Unfortunately, drawings and sinopias, which never give proportional grids, are of little help in this kind of study; this phase of the work was evidently executed and entirely finished at the studio table before the work was transferred to the wall.

Among the changes from Gothic to Renaissance were, presumably, changes in the canons of proportion; it is common knowledge that in the Gothic there was a preference for elongated figures and that it was therefore based on a smaller and more repetitive canon. Considering the contemporary progress in anatomy (and if one remembers that a similar reduction of canon was brought about in Greece by the growth of this science), it may be assumed that the determining factor was essentially a more subtle analysis of the human body and a more differentiated way of measuring its parts. In other words, instead of proceeding by arbitrary stylistic breaks from the chin to the top of the stomach to the navel and accentuating the work with unnatural anatomic markings, the divisions of the body were respected and studied more rigorously. A famous example of this is given by Cennino Cennini in chapter 70 of his *Libro dell'arte*: "Take note that, before going any farther, I will give you the exact proportions of a man. Those of woman I will disregard for she does not have any set proportion. Firstly, as I have said above, the face is divided into three parts, namely: the forehead, one; the nose, another; and from the nose to the chin, another. From the side of the nose through the whole eye up to the ear, one of these measures. From one ear to the other, a face lengthwise, one face. From the chin under the jaw to the base of the throat, one of the three measures. The throat to the top of the shoulder, one face; and so for the other shoulder. From the shoulder to the elbow, one face. From the elbow to the joint of the hand, one face and one of the three measures. The whole hand, lengthwise, one face. From the pit of the throat to that of the chest, or stomach, one face. From the stomach to the navel, one face. From the navel to the thigh, one face. From the thigh to the knee, two faces. From the knee to the heel of the leg, two faces. From the heel to the sole of the foot, one of the three measures. The foot, one face long. A man is as long as his arms crosswise. The arms, including the hands, reach to the middle of the thigh. The whole man is eight faces and two of the three measures in length" (trans. D. V. Thompson, Jr., New Haven, 1933, pp. 48–49). Cennini wrote this in 1437, and while it is true that he had transmitted recollections of the school of Giotto, it is also true that his is a much more evolved conception of proportion. In fact, one can verify it even with Donatello and Mantegna (qq.v.), the latter having respected it scrupulously in his early works and especially in his *St. Sebastian* (IX, PL. 325), the anatomy of which has an almost geometric partitioning because of it. As for the height of the figure, the basic novelty stated by Cennini is the inclusion of 1/3 of a measure for the neck and 1/3 for the width of the foot (regarding this detail, it is possible that similar notions in the Mount Athos canon were derived from the West, and not the reverse). The inclusion of a measure for the neck solved a problem that artists had always remedied rather ambiguously, most often by adding a beard or by shortening the length of the head. Cennini also established for the first time the proportions of the width of the figure in such a satisfactory manner that they remained unchanged and unchallenged until the 16th century. Michelangelo's *David* (VII, PL. 383), for instance, follows these canons, except for the addition of an extra measure between the shoulder and the elbow.

Cennini also recorded invaluable information regarding the proportioning of the entire painting and of the architectural background surrounding the human figure. In chapter 30, he advises, "as the prime measurement which you adopt for drawing, adopt one of the three which the face has, for it has three altogether: the forehead, the nose, and the chin, including the mouth. And if you adopt one of these, it serves you as a standard for the whole figure, for the buildings, and from one figure to the other; and it is a perfect standard for you provided you use your mind in estimating how to apply these measurements. And the reason for doing this is that the scene or figure will be too high up for you to reach it with your hand to measure if off" (Ibid., p. 17). This is found, for example, in Cesariano's famous commentaries on Vitruvius (PLS. 342, 343), which seem too laborious and perhaps superfluous for a fresco composition, the canon for which have been more simply based on 1 head or 1 face. In the panel with Eve at Villa Pandolfini (formerly Villa Legnaia) by Andrea del Castagno, even the space around the figure follows the canon of 1 face above, 1 face to the left and 2 to the right. The dark space that limits it is 10 faces by 4. The criticism of Cennini's method by the artists who came to maturity in the second half of the 15th century was directed at the generalization implicit in the proposed canons and concerned the attempt to replace them by a more elastic anthropometry. This denotes a more naturalistic inclination, even though in many cases the man who served as point of departure was but a hypothesis of eclectic character. Three artist-theoreticians, Leon Battista Alberti, Piero della Francesca and Albrecht Dürer, were mostly concerned with the improvement of the proportioning of man standing and motionless, at most as seen in foreshortening. In his *De statua* Alberti proposed canons of 1/6 of the height (*exempeda*), subdivided into 60 *unceolae* (thumbs) and 600 *minuta*: this very differentiated system was to permit, at least in theory, the registration of even the smallest personal variations; the table of measurements given by Alberti aimed at establishing a universally valid standard, though it was the fruit of actual experimentation. The extremely precise mensuration of the different parts of the face and of two different types of skull by Piero della Francesca, who made his calculations according to accurate geometric systems, were also aimed at finding an ideal standard. Dürer took Vitruvius as the starting point of his investigation (probably through contact with Jacopo de' Barbari), adopted a canon whereby the head was 1/8 of the total height, and used geometric figures (rectangles, squares, etc.) to schematize the different parts of the body and make them symmetrical (PLS. 346, 347). Later, when Leonardo's ideas became known, he undertook extremely precise investigation toward the discovery of more forms of relative beauty in nature and discarded the myth of unique and ideal beauty. In his *Four Books on Human Proportions* he concluded "that no man lives who can grasp the ultimate in beauty in the meanest living creature, let alone of man who is an extraordinary creation of God, and unto whom other creatures are subject.... But beauty is so much hidden in men and so uncertain is our judgment of it, that we may perhaps find two men both beautiful and fair to look upon, and yet neither resembles the other in any single point or part, whether in measure or kind; we do not even understand whether of the two is the more beautiful, so blind is our knowledge" (from E. G. Holt, *Literary Sources of Art*, Princeton, N.J., 1947, pp. 287, 291). Evidently, the brink of characterization without rules had arrived. Dürer does, however, limit himself to the designation first of 5, then of 13 types of men and women, using Alberti's detailed method of 6 parts and their subdivisions. He also studied movement but was far behind Leonardo, who, as early as 1498, had compiled a treatise on human movement, thus adopting what might be called a phenomenological attitude similar to that of Dürer and with as little, or even less, confidence in measures: "The same action will appear to be infinitely different because it can be seen from an infinite number of positions which have continuous quantity and continuous quantity is divisible to infinity. Therefore, infinitely different points show that every human action is in itself infinite" (Leonardo da Vinci, *Treatise on Painting*, trans. A. Philip McMahon, Princeton, 1956, p. 137, no. 361); and, "Beauty of face may be equally fine in different persons, but it is never the same in form, and should be made as different as the number of those to whom such beauty belongs" (Ibid., p. 113, no. 278). The taste for variety underlies the similar ideas of Francis Bacon who, in several passages, asserted that there is no beauty that does not have some strangeness in its proportions. Leonardo, however, used modules, at least in the establishment of the essential articulation points of the members and the maximum outstretch of the different movements (PLS. 344, 345). There is an ample record of these, certainly based on the master's models, by the anonymous Lombard painter who compiled the Codex Huygens about 1570; here we see the body rotating around the ideal central point of the navel, bending and rising from a prostrate to an erect position but always retaining its proportional balance (PL. 348). It is known that Michelangelo also undertook similar studies. In his *Trattato delle perfette proporzioni* Vincenzo Danti says that Michelangelo, "seeing with excellent judgment that the modern painters and sculptors and, as far as we can see, also those of antiquity, had looked for and found some perfection in all things, but that that none had as yet entirely seen nor known the most perfect proportion of man, realized that this could have no other cause than the complexity of its composition." The path followed by the great Tuscan seems, however, more related to Aristotelianism; in other words, its object was to make beauty, and hence proportion, coincide with the actual length of the members in the exercise of their functions. "And for this reason," says Danti, "those will be closest to human perfection whose parts, both external and internal, are made so as to best operate in keeping with that which nature has ordained." Thus, ultimate proportional harmony is a composite of different, even contradictory, beauties, whose functions should be investigated anatomically. Moreover, the human body is "from beginning to end, mobile, that is to say, it has no stable proportions." Since the members also develop with age, it is impossible to measure them from definite points of departure. Danti also made a series of important observations on the proportions of animals, the anatomic study of which he advised, thus overcoming a prejudice that had endured throughout the 15th century, according to which only the horse could be treated proportionally (studies to this effect were made by Alberti, Leonardo, etc.). Discussions of proportion in Italy and in Germany aroused the greatest interest and violent disputes; an echo of them in Spain has been transmitted by Juan de Arphe y Villafañe in his *De Varia Commensuracione para la Esculptura y Architectura* (1585), where an amplified canon consisting of Felipe de Borgoña's $9^1/_3$ heads is quoted and in which Berruguete (q.v.) is praised for having first introduced the practice of "perfezionamento" and Gaspar Becerra for having improved this practice, with "figuras compuesta de màs carne que las de Berruguete." But research on the proportions of the body in movement was little developed. Among the few pertinent texts, Teofilo Gallaccini's short and still unpublished treatise based on the Vitruvian canon of 8 heads (which it in some ways completes, especially in regard to length of the arms, also according to the propositions of Pomponius Gauricus and Gironimo Cardano), attempts to fix centers of movement, considering them as the apexes of triangles whose sides represent the limbs in movement. Thus, a walking figure bending forward from the waist forms a triangle composed of the legs with the pubis as apex. This system, which superficially resembles the medieval one of Villard de Honnecourt's and that of Dürer, led to some arbitrary conclusions but was well suited to the exaggerated stylization of late mannerism.

The need, expressed by Danti, for a proportional system based not so much on extrinsic canons as on the internal construction of the parts of the body is probably reflected in certain drawings like those of Luca Cambiaso (Uffizi, no. 13736; Städelsches Kunstinst., Frankfurt am Main), which, being based on flexible models, divide the members into differentiated geometric volumes. Nevertheless, respect for the more traditional canons is visible in these drawings (e.g., the thorax and chest are three times the size of the head including the neck). Apart from the necessity of reconciling the proportional canons, which are always connected with the idea of a simple module with multiples and subdivisions that may be easily measured, with the deformations in the human body caused by the movement, the crisis in the proportional theories of mannerism and, later, the baroque was accentuated by the difficulties created by foreshortening, that is, by a deformation due to the position of the viewer in relation to the picture and to the specific conditions of the wall against which it is placed. Foreshortening is a consequence of naturalism (even though as it evolved it became mostly fanciful or decorative); it is, in fact, based on the presupposition that the picture, in spite of being removed from the normal optical field, must nevertheless retain all its characteristics and present an absolute verisimilitude. From the proportional point of view interest lies not in the expedients resorted to for stressing the optical aberrations of reality (such as models, transparent frameworks for the recording of images, etc.) but only in the calculations of the limits of the aberrations from a strictly pictorial point of view, such as that of curving upward the part of the wall on the onlooker's side (which was suggested by Dürer and soon after adopted by Michelangelo in the *Last Judgment*; IX, PLS. 533, 536) or the numerous distorted-image paintings in which the subjects were recognizable only through lateral vision. In these studies, of which J. Baltrusaitis has given a full account, one perceives an elongation of the modules according to very precise progressions, with the result that the visible proportional harmony appears to be derived from a much more complex and sophisticated type of proportioning. The theoretical interest of such investigation goes far beyond the practical use that can be made of it and is perhaps centered in the recognition, which in our era is quite general, that effect is not connected with reality, even though the laws of both may be the same. We are therefore fairly close to the baroque conception of perspective, which is not recognition of universal structures but deception of the senses.

Proportional canons ceased to have real artistic importance as soon as anatomic plates became common property (even though they contain very rigorous proportions). The consultation of such studies was also habitually practiced with the object of discovering the presumed meeting point of natural beauty and artistic beauty, which has been a constant aspiration in the West. But at this point artistic beauty becomes identified quite simply with ancient statuary, and proportions as studied in the sculptures become stereotyped and of secondary importance in comparison with the rendering of muscles or of the vague and ductile general outline of the image.

Proportion, in other words, becomes a latent element, if not altogether absent. To encounter it, one may turn to anatomic treatises, which demonstrate how traditional the proportional modules remain (the head multiplied by a number between 8 and 9) underneath the scientific objectivity and how their constant application is accentuated, sometimes almost too blatantly. Thus the plate on the external morphology of the male and female bodies in Andreas Vesalius's *Epitome* shows regular application (except for certain alterations for reasons of perspective) of the 8-head module, with the back strictly divided into 3 parts from top to bottom; and since the bodies are shown at a slightly slanting angle because they are facing sideways, it may be deduced that, for a clearer relationship of the points of measurement to the nipples, the navel, the pubis, and the knees, the metric scale according to which the figures were constructed was imagined to be at their left (Vesalius does not take the hair into account). An identical 8-head module (the head including the neck) is used by Pietro da Cortona in his anatomic treatise and, among others, by Godfried Bidloo; variations (e.g., the fact that in the 17th century the figures have larger thoraxes and therefore appear more thickset and balanced) are due to the persisting uncertainty as to the module that should be applied to the width of the human body. The trend toward animating all figures increasingly impelled artists to find particular solutions. Even the open-armed crucifix figure, which was the valid canon in the Middle Ages and which the Renaissance again associated with Vitruvius, lost its importance as a standard. In his famous *Christ on the Cross* (XIV, PL. 325), Velázquez uses either a very complicated and not easily recognizable proportional system or simply limits himself to general principles of symmetry. As had happened in the 16th century with the Venetians, the mannerists, and the scene painters, the determining factor came to be the grid system of drawing, which was no longer used only for practical reasons (i.e., to facilitate the transfer of the model to the real dimensions of the cartoon), but actually to condition the picture from without and not from within; thus, the picture was schematized and included within a geometric pattern, as it was in the times of Villard de Honnecourt, but with greater flexibility. The proportion squares only dictated the general layout, and the use of a great variety of measurements proves that the grid system had lost its value as a module; images were indifferently divided into 6, 8, or 10 parts. The same measurements might produce either very slender figures or clumsy imitations of statuary. The series of studies for the monochromes with dancers and nymphs by Antonio Canova, now in the Museo Civico of Bassano, are typical because they are by a sculptor. In them the 8-head canon is observed for the main figures, but the lines of the grid do not join the canonic points of the body; immediately next to these figures, other dancers are divided into 7-head figures and so forth.

Artists who were quite unconventional in their choice of subjects and their pictorial methods — for instance, Gauguin and Degas — also used proportion squares, which helped to give their pictures a pleasing balance; they were free of schemata, but the result was, so to speak, always static. (In 1886, Félix Fenéon accused Degas of making his figures excessively geometric.)

With cubism, proportional problems became more acute, or at least those concerning the rational relationship between the geometric forms it had introduced into painting, became more acute. Attempts were made to place these geometric exercises on a mathematical basis, even to the point of conducting studies in algebra. But the human figure was rarely involved, excluded as it was from the representational interest of the time, at any rate in the major works (though its proportions were discussed in reviews like *De Stijl* and the *Section d'Or*). The schemes, which were mostly geometric, depended on an allusive value of dynamic interplay between empty and full spaces: thus, according to Georges Vantongerloo ("Reflection," in *De Stijl*, July 9, 1917), circles that can be inscribed within a figure express the vibrations caused by external and internal volumes and even by the empty spaces, the triangle symbolizes infinity, and so on. Within the scope of these ideas, of which the principle theoreticians were Amédée Ozenfant and André Lhote, one finds a fairly widespread attempt to extract from the human figure, through only partial deformation, geometric forms that would give it a metaphysical quality, or to find within it dynamic tensions that could expand within its sphere. Oskar Schlemmer's scenographic and plastic experiments are especially revealing in this light. At least for the present, they conclude, in an uncertain and anachronistic accord between geometry and the human figure, the long history of vicissitudes in the field of human proportions.

PROPORTION IN ARCHITECTURE. An embryonic conception of proportion must have taken shape as early as the Neolithic age, that is, at the very beginning of monumental architecture. Examination of groups like Stonehenge (V, PLS. 164, 165; XIII, PL. 252) reveals that the structure of the sanctuary creates a rhythmic sequence between full and empty spaces which anticipates that of the colonnades of the archaic Greek temples. The geometric schematizations resorted to are explained by the cosmologic allusion to solar symbols; with the same careful symmetry, the sanctuary was cut by transversal lines that "squared" (in the classical use of the word) its surface and was surrounded by regular circular cinctures. Almost all the civilizations in which cosmologic symbolism prevailed are extremely important in the history of proportion. It has been observed that the Egyptian temples are strictly oriented in terms of direction and are divided into two symmetrical parts, north and south. The ornamentation is also closely related to this symbolism. The same applies to the tombs. The use of geometric forms, such as the square and the pyramid, is connected with an equally symmetrical concept of the parts of the universe; it seems that the slope of the sides of the pyramids was calculated according to the stars. Similarly, in the diminishing steps of the Babylonian ziggurat (III, PL. 485) the heights of the horizontal divisions were the presumed reciprocal orbits of related planets. This abstract symbolism, which has given rise more to imaginative interpretations than to exact reconstructions of the way in which the symbolism of the cosmos and the heavens was actually materially expressed, is reflected in the emphasis on proportion in the Bible, which, especially in the cases of Noah's ark and the Temple of Solomon, gives measurements and a simple proportional key based on divisions by 3 and ratios of $1:2$, $1\,{}^1/_2 : 2\,{}^1/_2$, and so on It was from God himself that the Hebrews adopted the proportional standards that were to render their buildings suitable for sacrifices and other rites. Moreover, the numerology seems to be connected, at least originally, with the planets known to man (e.g., 2 would represent the polarity of the sun and the moon; 3, that of the sun, the moon, and Venus). It is therefore possible that ancient proportions should be read symbolically, as it is true in large part of medieval religious architecture (12 columns stand for the 12 apostles; 3 apses for the Trinity). Proceeding in time, it is found that several of the buildings that have been most studied by artists and theoreticians since the Renaissance because of their proportions belong to the class of cosmologically significant edifices, such as the Septizonium, which alludes to the 7 planets and is related to the sun cult of the emperor, and the Pantheon (I, PL. 401), which is proportionally built as a sphere in reference to the cosmos and is based on complicated repetitions of the number 12.

Proportion, even in Pythagoras's time, had already acquired true esthetic value apart from its allusive function, so much so that its mere presence sufficed to sanction the quality of the work, for the same reasons that a well-proportioned statue was considered suitable for purposes of worship; this was also true of correctly designed churches during the Renaissance.

Much of the ancients' research on proportion was connected with simpler practical needs. An echo of these considerations is found in Vitruvius: speaking of temples, he observes that columns too large or too close to one other would hinder a processional advance by twos to the inner sanctuary and that columns placed too far apart would cause the collapse of the architrave. The ideal measures which he gives for the orders (XIV, PL. 378; see STRUCTURAL TYPES AND METHODS) are a compromise between the demands of statics and those of practical function. What was new in Greece with respect to the proportioning on a monumental scale in Egypt and Mesopotamia was, on the one hand, the organic conception of the edifice in relaion to its ideal resemblance to man and his physical type and, on the other, the conviction that proportioning should be applied not only to certain parts but uniformly to the whole construction. The Greek edifice was thus conceived as a module, conditioned in turn by the materials employed, by its relationship to man and perhaps to landscape, or to the divinity to be worhiped and so on. The column, deemed the synthesis of all these requirements, thus became the moving force behind all constructional development, on the basis of a rather complex and rigid series of interrelations between it and the other adjacent parts (FIG. 733). As is well known, moreover, the Greeks took into account optical and perspective adjustments. Proportion thus seems to have been considered both as an objective and a subjective fact.

Without entering into a matter that will be examined later, it should be mentioned that this same flexibility between norm and nature occurs in the attribution of typical characteristics to the various orders: the 6-ft. Doric column is thought to correspond to the proportions of the male body, the 8-ft. Ionic column to those of feminine beauty (PLS. 351–353). Proportional relationships and canons therefore vary according to the character of the architectonic type.

The Greek temple has been the object of endless examination beginning with Vitruvius, but late-antique edifices still in existence would probably yield even more fruitful information. In certain cases, their proportions resulted from external factors, such as the ready-made columns sent from the imperial quarries (this was presumably the case for the Ravenna basilicas). But perhaps never as in these times of political decline and intense spiritualization, has the musical value of proportions, and their power to suggest another world without resorting to allegoric or schematic interpretations, been realised. It would be interesting to compare the praise lavished on the "Mausoleum of Galla Placida" (II, PL. 425; IV, PL. 458) with what Boethius wrote at the same period on the value of musical harmony: that in itself it could give the most complete spiritual satisfaction. The whole evolution of rhythm in late-antique architecture, which aimed at alternations of pilasters and columns, at a more complex spatial character, and even at extreme proportional contrasts (which were to influence the Romanesque and the Gothic), could be considered in its relation to the concept, also proclaimed by Boethius, of harmony as a synthesis of contrasts, of squares and rectangles, of stability and movement. The relationships given by the philosopher are, in fact, far more complex than the simple ones, mostly of 1:2, that had been the basis for the construction of squares, basilicas, and temples in the classical and imperial ages.

Ravenna, Byzantium, and Vitruvius were, also in the matter of proportion, at the focus of high medieval Palatine culture, which resumed, in particular, the search for order, symmetry, compatibility, and harmony between the different parts and for the ideal reflection of the human body in the edifice, sometimes with curious practical applications (e.g., the following, from a 7th-century formulary published in *Bulletin Monumental*, 1907, pp. 5–6: "Si in altitudine quattuor staturis fuerit fabrica, unius statuarie altitudo, usque ad bifurcum erit fundamentum; si cuvrem unius statuarae altitudo, usque ad geniculum erit fundamentum"; also an important prescription because it is related to the proportioning of the *homo ad quatratum*). In the Palatine Chapel at Aachen there is an inscription that praises its proportions; recently, it has been demonstrated that the edifice is based on a module of 25 Carolingian feet, which recurs in the diameter

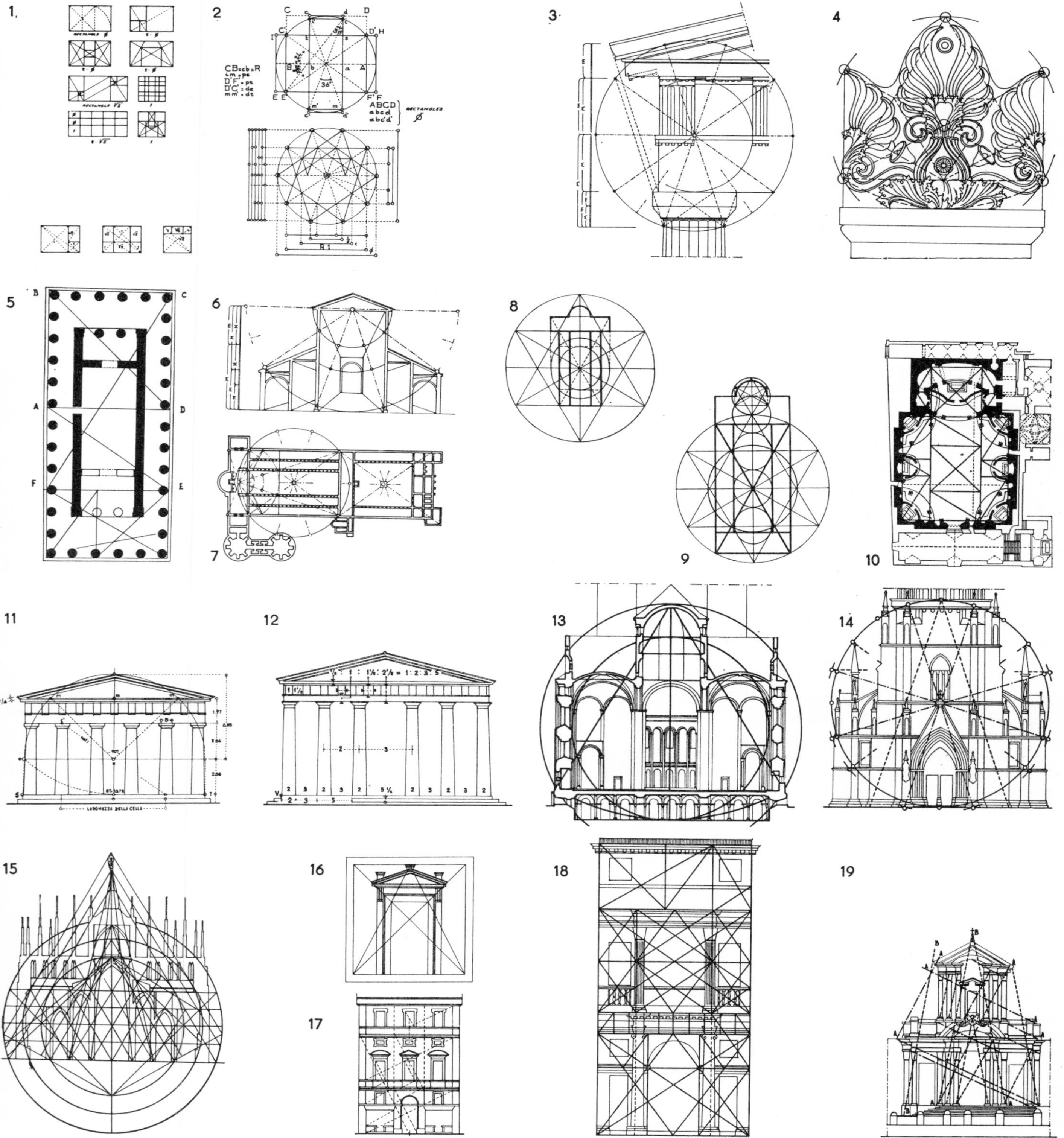

(1, 2) Geometric proportions: (1) harmonic rectangles (Φ and $\sqrt{5}$), after Hambidge; harmonic sectioning of square; modulation $\sqrt{2}$; (2) two systems of proportion obtained by the polar segmentation of the circle according to Moessel (*from Ghyka, 1931*). (3, 4) Proportional interpretation of architectural elements: (3) Athens, trabeation of the propylaea of the Acropolis, proportional relationships of the individual members; (4) acroterion of an Attic funerary stele, with system of triangulation indicated (*from E. Mössel*). (5–19) Proportional studies of building plans and elevations: (5) Athens, so-called "Theseion," plan, with scheme of proportions of the rectangle, formed by the bases of the columns (*from C. Bairati*); (6, 7) Rome, St. Peter's, section and plan; (8) Rome, S. Maria in Aracoeli; (9) Rome, S. Maria in Trastevere (*from O. Wolff, Tempelmaasse, Vienna, 1912*); (10) Turin, S. Lorenzo, proportional scheme of the plan, based on the square broken up into squares and golden rectangles (*from C. Bairati*); (11) Athens, so-called "Theseion," front with Vitruvian proportions indicated; (12) front view of a Doric hexastyle temple, ideal construction based on Vitruvian proportions (*from Bairati*); (13) Speyer, Cathedral, section (*from F. Klimm. Der Kaiserdom zu Speyer, Speyer, 1953*); (14) Freiburg, Cathedral, façade (*from E. Mössel*); (15) Milan, Cathedral, schematic drawing of the elevation (*from Cesariano's edition of Vitruvius, 1521*); (16) analysis of a door (*after Serlio, I*); (17) Rome, Palazzo Gaddi Nicolini, front, according to Thiersch's theory (*from G. Giovannoni, L'architettura del Rinascimento, Milan, 1953*); (18) Vicenza, Palladio's house, front; (19) Rome, SS. Vincenzo e Anastasio, double-rhythm scheme of proportions (*from G. Giovannoni*).

of the octagonal room, in the height and the width of the ambulatory, in the subdivisions, and in the opening, between the forecourt and the chapel proper. Simple proportional relations, closely linked with those of music and with a graphic construction based on rectangles and squares, characterize Ottonian and, later, monastic architecture. Pertinent in this regard is E. de Bruyne's analysis (1946) of the church *ad quadratum* planned for the Cistercians, and reproduced by Villard de Honnecourt (see also H. R. Hahnloser's edition of *Livre de Portraiture*, Vienna, 1935): "This church is designed within a rectangle of 3/2, that is, a double squared third, which corresponds to the fifth; the length is of 12 bays and the width of 8. The choir is a projection of the fourth, 4/3, and thus the two transepts represent the ratio of the octave 4/2; the transept as a whole follows the same rule of 8/4; the crossing of the nave and the transept forms a perfect square, 4/4 — that is, the unit, the origin of all harmony — and the square clearly has a fundamental role in the succession of the bays; the nave repeats the third, 5/4. The choir and the nave together, without the central square, have the same value as the entire transept, and their ratio to the nave and the central square is 9/8. All the fundamental intervals of music are to be found here: all those proportions which are included in the perfect harmony of 6/8 : 9/12."

Compared to the disorder of an architecture that had turned frequently vague, picturesque, and impulsive, the proportional theorizing of the Cistercians had the effect of a classicistic revival. H. Hahn (*Die Frühe Kirchenbaukunst der Zisterzener*, Berlin 1957), has demonstrated how its theoretical origins are directly linked with the thought of St. Bernard of Clairvaux. Decreed in 1134, it was rigorously applied from 1139–40 on. In the same period, geometry was notably developed in the great Romanesque cathedrals, first and foremost in Chartres (VI, PLS. 291, 293, 296); this was probably due to a knowledge of texts translated from Arabic. It was in this technical and cultural climate, according to J. Gimpel [*Les batisseurs de cathédrales*, Paris, 1958 (Eng. trans., C. F. Barnes, Jr., New York, 1961)] that the most important building secrets originated, such as the one of building an elevation from the plan itself, which in 1459 was still rigorously protected from any indiscretion. The same problem was raised in Villard de Honnecourt, and it was revealed publicly in 1486 by the German architect Roriczer: given a square, construct geometrically a square which is its half. The system consists in using for the side of this second square the semidiagonal of the first one, as Plato had already shown in *Meno*. In this case too, the basic proportion is the simple one of 2:1, but it is reflected in a great number of geometric figures, a good example of which are the thousands of pinnacles of the cathedral. The rise of the Gothic, at least at St-Denis, is also marked by a fairly complex problem of geometrical construction, that of dividing a square into 9 equal squares by means of parallel lines. The module adopted was the "foot" (*pied de roi*), which was prevalent in Paris and was in common use elsewhere. According to S. Mck. Crosby, (*L'Abbaye Royale de Saint-Denis*, Paris, 1953, pp. 61–62), Pierre de Montreuil, architect of St-Louis, determined all the more important measurements of his edifice according to this module (the span of the interaxes of the transept and the nave, measured from a center at the top of the pilasters, is 6.50 meters, or 20 modules; the sides of the transept measure 13 meters, or 40 modules, like the width of the nave; etc.). Another important dimension that appears to have been calculated on modules and in harmony with the general proportions of the transept is the height of the nave up to the vault. In fact, the proportion of width of the interaxes to the height is 1:2. The plans of the more complex Gothic cathedrals show that the multiplication of such relationships, and not their dissolution, enriches the pictorial quality of the edifice, causing it to attain an almost encyclopedic symbolism. In the same way that Eastern and Western monsters and mythologies were accepted in the carved ornamentation, a great deal of the symbolism of numbers was adapted in the form adumbrated by the theologists, who were also imbued with mythical syncretism (cf. V. F. Hopper, *Mediaeval Number Symbolism*, New York, 1938).

The complicated Gothic geometric proportioning (FIG. 733) appeared also in Italy, both in more evolved forms as in the Cathedral of Milan (VI, PL. 357), for which many consultations, even with eminent mathematicians, were necessary in order to decide how to harmonize the width with the height, and the simpler, more crystalline forms related to Cistercian architecture. Unfortunately there is a lack of studies on the successive transitions from the abbeys of central Italy to the churches of the mendicant orders, which were equally rigorously built on a rapport of 3:2, and finally to Brunelleschi, whose return to the antique only resumes this trend towards symmetry and musicality in a more congruous manner. The mysticism of proportion as a value in itself beyond given circumstances (for which reason, plans could be drawn up hundreds of miles away or merely thought out with no reference whatsoever to actual building practices) and as harmony made visible continued to mark the thought of Alberti and, a century later, that of Palladio (PLS. 351–353) and countless other architects around them (PLS. 349–350).

In the Tempio Malatestiano (formerly Church of S. Francesco; I, PL. 52) at Rimini the calculations of the façade can be schematized thus: $1+1\frac{1}{2}+1+2+1+1\frac{1}{2}+1$, in all, 9 times the width of the pilaster. The niches in the sides are disposed in the ratio $1:1\frac{1}{2}:2 = 2:3:4$. The complete dimensions of the sides are of 9 modules by 12. Renewed attempts to rediscover a golden section as a formal theme must have been considered ineffectual. Relationships established with such simplicity suffice to create precise musical resonances. In his drawings, Francesco di Giorgio carried even further the Vitruvian link between man and edifice and attempted hypothetical proportional correlations. A modular plan of an entire city that would be both symbolic and practical, is proposed by Filarete (q.v.) in the Sforzinda (he, however, developed a medieval theme, and even that under the influence of Vitruvius). The first well-known example of a mechanical squared plan is one of St. Peter's in Rome, which may be dated 1506–16 and is linked with Bramante. Henceforth, geometry became an instrument of compostion rather than of proportioning; in this sense, Leonardo was ahead of his time, for he knew the very elaborate system of late German Gothic planning and used the works of Euclid, Archimedes, and others. The greatest problem of the 16th century became, as is illustrated by Serlio, that of consolidating a great many geometric forms within a dynamic structure. This is true above all of the plan. In the prospect itself, an effect of remarkable simplicity is found, as in the Farnesina by Peruzzi (q.v.; Frommel, 1961), in which the dominant proportion is 2:1. Thus, in spite of the spread of knowledge of the golden section, owing to Pacioli, proportions remained elementary in buildings with a simpler plan, like the Venetian basilican structures: Francesco Giorgi, Palladio's adviser for the church of S. Francesco della Vigna, recommended ratios of 1:2:3:4 (R. Wittkower, *Architectural Principles in the Age of Humanism*, London, 1949). Not even Palladio seems to have used irrational proportions, even though it was precisely during the second half of the 16th century that they came to be associated with a tendency toward mystical interpretation (e.g., in Philibert Delorme; q.v.). In architectonic theory and practice, something very much akin to the pictorial studies on foreshortening and anamorphosis was attempted. Very properly, G. Soergel (1958) opens his treatise on Italian Renaissance proportion with an anthology of extracts from Serlio, Vasari (q.v.), Barbaro, Martino Bassi and Giovanni Paolo Lomazzo, all of whom seem to be commenting on Michelangelo's statement that "modern painters and sculptors should have proportion and the proper measurements in their own eye." For instance, Lomazzo says that he who does not know that by placing the orders on top of each other in the proper order and according to the proportions that they show without taking into account the requirements of perspective would cause them to appear inordinately low and to be receding from the facade over the horizon, notwithstanding their having their own proper proportions, because of their lacking that other proportion which exists through the manner of seeing things according to rules of a given distance. In this manner, the protrusions, the reliefs of the architraves, the pedestals and the cornices of the top part would appear too obtrusive, and the columns would have none of the beauty that is seemly due to the eye (trans.

from *Trattato dell'arte*, I, 1844, pp. 123–24). The multiple risks of such proportional empiricism were well understood by Giacomo da Vignola (q.v.), who, though he was quite aware of the difference between concrete proportional harmony and apparent harmony, and though he chose the latter, nevertheless asserts that the appearance must successfully follow objective rules and standards.

In Scamozzi, one finds a proportional query very similar to the one originated by Michelangelo: that of knowing whether proportional harmony is above all the result of adherence to function. He is certainly anti-Palladian when, for example, he makes the problem prosaic by declaring that the proportions of the parts should be in harmony with the whole and that they should serve their function even though they are not properly placed as they should be and that they should result in harmony. His real tendency, however, was towards rationalism, and his idea of beauty, though functional (as may be observed in his plans for Utopian cities), is of great geometric simplicity, so much so that it proves him to have been an adherent of that widespread, almost rigorist movement which arose in southern Europe after the Reformation, the masterpiece of which is the Escorial (a complex of buildings built on strict proportions of 2:3). Scamozzi's attitude is also interesting in respect to Gothic architecture, wherein he praised above all else the symmetry, even at the cost of falsifying the factual data. This is all the more important since certain churches reproduced by him with geometric exactitude, such as St-Denis or St-Nicholas-du-Port, contain such deviations of the main nave or such obvious anomalies in the distribution of the lateral chapels that one cannot help but think that in drawing them the draftsman deliberately transposed them into "regular" forms (F. Barbieri, ed., *Taccuino di Viaggio da Parigi a Venezia*, Venice, 1959). On the contrary, in the North there was much resistance against the spreading of Italian classicistic ideas precisely for proportional reasons, in spite of the fact that books like those of Palladio, Vignola, and Scamozzi had sanctioned the system of the five orders; Vignola's manual *Regole delle cinque ordini* (1562), especially, had become a standard text for students and had been translated many times. It is significant that Claude Perrault (q.v.), the author of one of France's most classical structures, the Louvre colonnade, in his argument against the liberties taken by the baroque, asserted that architectonic proportions were to be considered more elastic and subjective than those of music and more dependent on public tastes. As is well-known (see TREATISES), the problem of proportions and orders was put aside and even banished by the modernists, be it in a decorative (see ROCOCO) or in a functional sense, allowing their application to remain immature and mechanical (A. Capra, *La nuova architettura familiare*, Bologna, 1678, p. 110). While the thematics of proportion were disintegrating, however, the literature on the subject was constantly nourished by the nostalgic conviction that there existed an essentially mathematical and geometrical root to beauty, and its aim was to track it down as a proof of objective esthetic values in the works of all periods and civilizations. (e.g., Viollet-le-Duc; A. Zeising; A. Thiersch, 1904). Esthetic philosophy (first Schopenhauer) promptly furnished the means of distinguishing the authentic efficacity of proportion within a sensitive-objective framework, more as an element of balance than as a calculation, as an open problem, rather than as a rigid end in itself. In this manner, architecture appeared to Wölfflin as "the representation of the noble feelings of our existence," as the mirror and reflection of the soul. Architectonic rationalism (see EUROPEAN MODERN MOVEMENTS) promoted a return to an interplay of geometric volumes that was concise and essential, but such structuring was thought of in terms of *Einfühlung* (i.e., feeling one's way). Le Corbusier and his school are probably alone in remaining faithful to the idea of a module as both a metaphysical and an objective point of reference. In his research on the "modulor," which was sometimes also decorative, Le Corbusier (PL. 354) tried to reconcile the irrational proportions made fashionable by the effects of cubism in France and classical anthropomorphism (FIG. 718). Prefabrication, which gives the module a practical value and which in a certain sense returns architecture to the pure technicians, makes Le Corbusier's invention seem somewhat remote. It also runs the risk of losing all esthetic significance, not because it is too technical but because it is too obscure. Proportion must not only exist but must reveal itself.

PROPORTION IN PICTORIAL COMPOSITION. Apart from the pictorial or plastic construction of the human figure, the problematic nature of proportion is just as decisive a factor in the construction of the painted image, independently of its subject. First and foremost, any painting (and, in the same way, any sculptural composition) is conditioned by its reciprocal geometric relations due to height and width and by the carefully distributed optical effects resulting from the frame or the friezes that surround it. Secondly, any figurative or chromatic insertion within the painting creates disturbances that must be readjusted and balanced. In this wider sense, the history of proportion almost coincides with the history of art itself and may be regarded as an indication of cultural motivations operative in the various periods. It has already been observed (see HUMAN FIGURE) how the dimensions of figures with respect to landscape give positive evidence of the prevalence of an active and optimistic conception of man, who, sometimes with exceptional boldness, dominates the world around him (the giants and colossi are a good example of this) or, on the other hand, of pantheistic pessimism that gradually eliminates the boundary between the rational and the irrational [as in the paintings of Turner (q.v.; XII, PL. 317), in which figures and objects are absorbed by mist], or the reduction of the figures to minute dimensions against an immanent background. It is precisely in landscape painting that complex proportional calculations were used and created latent schematicism. So far, the most extensive study has been that of 17th-century Dutch painting (cf. Åke Bengtsson, "Studies on the Rise of Realistic Landscape Painting in Holland, 1610–1625," *Figura*, 3, 1952), in which the straight line of the horizon actually cuts the painting in two at a quarter of its height from the bottom and acquires a great intensity; nevertheless, the vast area of the sky is thus perfectly balanced. The attraction of proportion in Dutch painting is also expressed in the depiction of interiors, especially in the works of Pieter Saenredam (V, PL. 307) and Emanuel de Witte of Delft (q.v.; V, PL. 317) and in the architectonic views, in which concern with orthogonal perspective is relegated to the background in favor of the exclusively proportional balance of masses, the arrangement of planes of color, and so on — in all the better works at least. Since such balance is not based on exact numerical relationships, the result is an intangible spatial quality, which, in a sense, lacks continuity as opposed to the precise monumental compositional elements of 15th-century Florentine paintings. This sensitivity was rediscovered, as is well known, by Mondrian, nonfiguratively and without any pedantic return to structures prevalent in preceding centuries, but with perhaps a hint of the rustic architecture of the Dutch islands; his poetics became more and more clearly centered on the problem of proportional balance, which once again took on more of a cosmologic value than a symbolic one.

Eugenio BATTISTI

PROPORTION IN THE ORIENT. Indian technical treatises develop organically and in great detail, a whole series of studies on the proportions of the human figure, which comprise a specific branch of the technical knowledge required by artists that is generally called *tālamāna*, which may be translated as "iconometry." The word *tāla*, which also indicates musical tempo, refers to a precise unit of iconometric measurement based on the width of the hand. The *aṅgula*, another unit of measurement equivalent to 634.6 mm., is also widely used in the designing of pictures and statues. The proportional basic measure for all human figures is the head, which in real nature was assumed to measure 12 *aṅgulas* from the tip of the chin to the top of the skull. Texts like the *Viṣṇudharmottara*, for example, describe with great precision the proportions between the head and each of the limbs. It is a very interesting and curious fact that, as prescribed by the *Citrasūtra*, the proportions of the limbs vary not only according to their movements (*aṅgahāra*) and positions (which can be connected with research on foreshortening and muscular tension but cannot be expressed anatomically owing to the peculiar qualities of Indian esthetics) but also with the various divinities. For these divinities, diverse dimensions are established that, within a proportional scale which varies from one text to another but on the whole is fairly homogeneous, correspond — in certain conditions at least — to a proportional symbolism which is most effective as a reflection of the hierarchy of metaphysical values. Such theoretical directions obviously apply above all to the sacred images, which were meant either for veneration or for the expression of symbolic or mythical values for the purpose of edification. A similar but rather less homogeneous and precise group of directions also exists for the technique of architecture. While it is easy, however, to discern a close relationship between the indication of the texts and the actual cult images, be they carved or painted, architectural work retained greater freedom with respect to the elements of proportion, even though within the limits of a specific style the relations between the various parts of an edifice remained more or less invariable. With time, on the contrary, proportions changed, in the same way as the forms of certain parts of the edifice itself changed; this phenomenon is most evident in the evolution of the stupa, which, from a flat form with a semispherical cupola, came to have an extremely elongated one that was made to seem even slenderer by its wooden superstructure. The plans of the great medieval temples frequently reveal constant use of proportional canons, but these were only valid and coherent for the specific group of buildings in which they were applied, with the result that the proportions of similar groups sometimes differed substantially, even though these contained similar structural forms. Little is known of Chinese research on proportion. Apart from the great aulic paintings, which during the period of the supremacy of Ch'an Buddhism (see BUDDHISM; CHINESE ART), for symbolic reasons, established unbalanced proportions between the human figure (q.v.) and the surrounding landscape (see LANDSCAPE IN ART), proportioning was usually left to the inspiration of the artist. In the evolution of Chinese sculpture, it is easily perceived that the stylistic variations depend in part on proportional modifications, which are particularly evident in Buddhist works and above all in the faces, so much so that they provide a fairly certain indication of chronology even for periods very close together. In a late period Japanese texts refer to the structure of Buddhist images with schemes of a proportional type. From the architectonic point of view, the presence of proportional canons in pagodas and other constructions, including bridges, is obvious, but this has not yet been studied sufficiently. The problem of proportion in the Asiatic world has yet to be gone into, though there do exist a few tentative studies, mostly Japanese. With the exception of the Indian world, which presents difficulties in the interpretation of the various texts of the *tālamāna*, it might be possibile to extract from the mass of more important data indications that would not only clarify the stylistic evolution of the different civilizations but also establish considerable similarities with the Western world.

Mario BUSSAGLI

BIBLIOG. *Ancient, medieval and modern West*: See generally H. Graf, Bibliographie zum Problem der Proportion, Speyer, 1958 (1800–1957). See also: A. Zeising, Neue Lehre von den Proportionen des menschlichen Körpers, Leipzig, 1854; E. Henszlmann, Théorie des proportions appliquées dans l'architecture depuis la XIIe dynastie des rois égyptiens jusqu'au XVIe siècle, I, Paris, 1860; F. Dieterici, Die Propädeutik der Araber im 10. Jahrhundert, Leipzig, 1865; G. T. Fechner, Vorschule der Aesthetik, 2 vols., Leipzig, 1876; F. X. Pfeifer, Der goldener Schnitt und dessen Erscheinungsformen in Mathematik, Natur und Kunst, Augsburg, 1885; A. Kalkmann, Die Proportionen des Gesichts in der griechischen Kunst (Berliner Wpr. 53), Berlin, 1893; G. Dehio, Untersuchungen über das gleichseitige Dreieck als Norm gotischer Bauproblemen, Stuttgart, 1894; G. L, Raymond, Proportion and Harmony of Line and Color in Painting, Sculpture and Architecture, New York, London, 1899; W. Cantor, Vorlesungen über die Geschichte der Mathematik, II–III, Leipzig, 1900; T. Lipps, Aesthetik; Psychologie des Schonen und der Kunst, I, Hamburg, Leipzig, 1903, pp. 66–67; P. Fauré, Architecture-musique, système rationnel des proportions, L'architecture, XVII, 1904, pp. 418–20; A. Thiersch, Proportionen in der Architektur, Handbuch der Architektur, 3d ed., IV, 1, Stuttgart, 1904, pp. 37–90; C. C. Edgar, Remarks on Egyptian "Sculptor's Models," Rec. de trav. relatifs à la philologie égyptienne, XXVII, 1905, pp. 137–50; J. Laran, Recherches sur les proportions dans la statuaire française du XIIe siècle, RA, IX, 1907, pp. 436–59, XI, 1908, pp. 331–59, XIV, 1909, pp. 74–93, 216–49; F. Burger, Vitruv und die Renaissance, RepfKw, XXXII, 1909, pp. 199–218; I. Herwegen, Ein mittelalterliche Kanon des menschlichen Körpers, RepfKw, XXXII, 1909, pp. 445–46; V. Mortet, Recherches critiques sur Vitruve et son oeuvre, VI: Le canon des proportions du corps humain RA, XIII, 1909, pp. 46–78; V. Mortet, La mesure de la figure humaine et le canon des proportions d'après les dessins de Villard de Honnecourt, d'Albert Dürer et de Léonard de Vinci, Mél. E. Chatelain, Paris, 1910, pp. 367–82; S.Colman, Nature's Harmonic Unity: A Treatise on Its Relation to Proportional Form, New York, London, 1912; O. Wolff, Tempelmasse; Das Gesetz der Proportion in den antiken und altchristlichen Sakralbauten, Vienna, 1912; P. Frankl, Die Entwicklungsphasen der neueren Baukunst, Berlin, 1914; O. Stein, Die Architekturtheoretiker der italienischen Renaissance, Berlin, 1914; E. Panofsky, Dürers Kunsttheorie vornehmlich in ihrem Verhältnis zur Kunsttheorie der Italiener, Berlin, 1915; E. Mackay, Proportion Squares on Tomb Walls in the Theban Necropolis, JEA, IV, 1917, pp. 74–85; M. Theuer, Der griechisch-dorische Peripteraltempel; ein Beitrag zur antiken Proportionlehre, Berlin, 1918; F. M. Lund, Ad quadratum, 2 vols., London, 1921; E. Panofsky, Die Entwicklung der Proportionslehre als Abbild der Stilentwicklung, Mnh. für Kw., XIV, 1921, pp. 188–219; L. D. Caskey, The Geometry of Greek Vases, Boston, 1922; J. Meder, Die Handzeichnung: Ihre Technik und Entwicklung, 2d ed., Vienna, 1923; L. D. Caskey, The Proportions of the Apollo of Tenaea, AJA, XXVIII, 1924, pp. 358–67; E. Berti, Un manoscritto di Pietro Cattaneo agli Uffizi e un codice di Francesco di Giorgio Martini, Belvedere, VII, 1925, pp. 100–03; M. Borissavliévitch, La science de l'harmonie architecturale, Paris, 1925; B. Kossmann, Einstens massgebende Gesetze bei der Grundrissgestaltung von Kirchengebäuden, Strasbourg, 1925; E. Mössel, Die Proportion in Antike und Mittelalter, 2 vols., Munich, 1926–31; K. Rathe, Ein Architektur-Musterbuch der Spätgotik mit graphischen Einklebungen, Festschrift der Nationalbibliothek Wien, Vienna, 1926, pp. 667–92; M. C; Ghyka, Esthétiques des proportions dans la nature et dans les arts, Paris, 1927; W. Überwasser, Spätgotische Baugeometrie: Untersuchungen an den "Basler Goldschmiederissen", Jahresbericht der öffentl. Kunstsamml. Basel, 1928–30, pp. 79–122; F. Durach, Mittelalterliche Bauhütte und Geometrie, Stuttgart, 1929; H. Tietze, Aus der Bauhütte von St. Stephan, JhbKhSamml-Wien, N.S., IV, 1930, pp. 1–46, V, 1931, pp. 161–87; M. C. Ghyka, Le Nombre d'Or, Paris, 1931; C. Linfert, Die Grundlagen der Architekturzeichnung, mit einem Versuch über französische Architekturzeichnungen des 18. Jahrhunderts, Kunstwissenschaftliche Forsch., I, 1931, pp. 133–246; K. F. Discher, Deutsche Bauhütten und ihre Geheimnisse, Vienna, 1932; I. A. Richter, Rhythmic Form in Art, London, 1932; G. D. Birkhoff, Aesthetic Measure, Cambridge, Mass., 1933; P. Fontana, Osservazioni intorno ai rapporti di Vitruvio colla teoria dell'Architettura nel Rinascimento, Misc. di Storia dell'Arte in onore di I. B. Supino, Florence, 1933, pp. 305–22; M. C. Ghyka, Influence de la mystique pythagorienne des nombres sur le développement de l'architecture occidentale, Actes XIIIe Cong. int. d'h. de l'art, Stockholm, 1933, pp. 315–29; W. Thomae, Das Proportionwesen in der Geschichte der gotischen Baukunst und die Frage der Triangulation, Heidelberg, 1933 (rev. O. Kletzl, ZfKg, IV, 1935, pp. 56–63); W. Überwasser, Nach rechten Mass: Aussagen über den Begriff des Masses in der Kunst des XIII. bis XIV. Jahrhunderts, JhbPreussKSamml, LVI, 1935, pp. 250–72; L. Hautecoeur, Les proportions mathématiques et l'architecture, GBA, XVIII, 1937, pp. 265–74; E. Oertel, Wandmalerei und Zeichnung in Italien, Mitt. kunsthist. Inst. in Florenz, V, 1937–40, pp. 217–314; O. Kleitzl, Ein Werkriss des Frauenhauses von Strassburg, Marburger Jhb. für Kw., XI–XII, 1938–39, pp. 103–58; O. Kletzl, Plan-fragmente aus der deutschen Dombauhütte von Prag in Stuttgart und Ulm, Stuttgart, 1939; W. Überwasser, Der Freiburger Münsterturm im "rechten Mass", Oberrheinische Kunst, VIII, 1939, pp. 25–36; L. Wolfer-Sulzer, Das geometrische Prinzip der griechisch-dorischen Tempel, Winterthur, 1939; E. Panofsky, The Codex Huygens and Leonardo da Vinci's Art Theory, London, 1940; M. Weinberger, The First Façade on the Cathedral of Florence, Warburg, IV, 1940–41, pp. 67–79; N. Pevsner, Terms of Architectural Planning in the Middle Ages, Warburg, V, 1942, pp. 232–37; E. Fiechter, Raumgeometrie und Flächenproportion, Concinnitas, Basel, 1944, pp. 59–81; O. Kletzl, Die Kressberger Fragmente: Zwei Werkrisse deutscher Hüttengotik, Marburger Jhb. für Kw., XIII, 1944, pp. 129–70; H. Siebenhüner, Deutsche Künstler am Mailänder Dom, Munich, 1944; R. Wittkower, Principles of Palladio's Architecture, Warburg, VII, 1944, pp. 102–22, VIII, 1945, pp. 68–106; P. Frankl and E. Panofsky, The Secret of the Mediaeval Masons, AB, XXVII, 1945, pp. 46–64; C. J. Moe, Numeri di Vitruvio, Milan, 1945; E. de Bruyne, Etudes d'esthétique médiévale, 3 vols., Bruges, 1946; W. M. Ivins, Art and Geometry, Cambridge, Mass., 1946; J. S. Ackermann, "Ars sine scientia nihil est": Gorhic Theory of Architecture at the Cathedral of Milan, AB, XXXI, 1949, pp. 84–111; C. Funck-Hellet, L'équerre des maîtres d'oeuvre et la proportion, Cah. techniques de l'art, II, 1949, pp. 37–82; Le Corbusier, Le Modulor, Boulogne, 1949 (Eng. trans., P. de Francia and A. Bostock, London, Cambridge, Mass., 1954); P. H. Michel, De Pythagore à Euclide, Paris, 1950; K. T. Steinitz, A Pageant of Proportion in Illustrated Books of the 15th and 16th century in the Elmer Belt Library of Vinciana, Centaurus, I, 1950–51, pp. 309–33; M. Borissavliévitch, Les théories de l'architecture, 2d ed., Paris, 1951; C. Funck-Hellet, De la proportion: L'équerre des maîtres d'oeuvre, Paris, 1951; H. A. Groenewegen-Frankfort, Arrest and Movement: An Essay on Space and Time in the Representational Art of the Ancient Near East, Chicago, 1951; C. Jouven, Rythme et architecture: Les tracés harmoniques, Paris, 1951; H. Koch, Vom Nachleben des Vitruv, Baden-Baden, 1951; M. Velte, Die Anwendung der Quadratur und Triangulatur bei der Grund- und Aufrissgestaltung der gotischen Kirchen, Basel, 1951; H. Weyl, Symmetry, Princeton, 1951; C. Bairati, La simmetria dinamica: Scienza ed arte nell'architettura classica, Milan, 1952; Cong. int. sulle proporzioni nelle arti (Milan, 1951), Atti e Rass. Tecnica della Soc. degli Ingegneri e degli Architetti in Torino, VI, 1952, pp. 119–35; R.

Wittkower, Architectural Principles in the Age of Humanism, 2d ed., London, 1952; P. du Colombier, Les chantiers des cathédrales, Paris, 1953; B. Grimschitz, Die Risse von Anton Pilgram, Wiener Jhb. für Kg., XV, 1953, pp. 100–18; K. M. Swoboda, Geometrische Vorzeichnungen romanischer Wandgemälde, Alte und Neue K., II, 1953, pp. 81–100; R. Wittkower, Systems of Proportion, Architects' Y. B., V, 1953, pp. 9–18; J. S. Ackermann, Architectural Practice in the Italian Renaissance, J. Soc. of Arch. Historians, XIII, 3, 1954, pp. 3–11; R. Billig, Die Kirchenpläne "al modo antico" von Sebastiano Serlio, Op. Romana, I, 1954, pp. 21–38; L. Hautecoeur, Mystique et architecture: Symbolisme du cercle ed de la coupole, 3d ed., Paris, 1954; E. Iversen, Canon and Proportion in Egyptian Art, London, 1955; Le Corbusier, Le Modulor 2, Boulogne, 1955 (Eng. trans., P. de Francia and A. Bostock, London, Cambridge, Mass., 1958); W. Lotz, Die ovalen Kirchenräume des Cinquecento, Römisches Jhb. für Kg., VII, 1955, pp. 7–99; P. Booz, Der Baumeister der Gotik, Munich, 1956; K. Clark, The Nude, New York, London, 1956; E. D. Ehrenkrantz, The Modular Number Pattern: Flexibility through Standardization, London, 1956; E. Forssman, Säule und Ornament, Uppsala, 1956; G. Hellmann, Studien zur Terminologie der kunsttheoretischen Schriften Leon Battista Alberti, Cologne, 1956; W. Lotz, Das Raumbild in der italienischen Architekturzeichnung der Renaissance, Mitt. kunsthist. Inst. in Florenz, VII, 1956, pp. 193–226; Modulor Coordination in Building: Project 174, European Productivity Agency of O.E.E.C., Paris, 1956; Schlosser; W. Boeckelmann, Von der Ursprüngen der Aachener Pfalzkapelle, Wallraf-Richartz Jhb., XIX, 1957, pp. 9–38; J. Bousquet, Le trésor de Cyrène, AJA, LXI, 1957, pp. 402–11; A. Fournier des Corats, La proportion égyptienne et les rapports de divine harmonie, Paris, 1957; N. Pevsner and others, A Report of a Debate on the Motion "that Systems of Proportion make good design easier and bad design more difficult", J. Royal Inst. of Br. Arch., LXIV, 1957, pp. 456–63; N. Speich, Die Proportionslehre des menschlichen Körpers, Andelfingen, 1957; J. Summerson, The Case for a Theory of Modern Architecture, J. Royal Inst. of Br. Arch., LXIV, 1957, pp. 307–13; J. White, The Birth and Rebirth of Pictorial Space, London, 1957; G. de Angelis d'Ossat, Enunciati euclidei e "divina proporzione" nell'architettura del primo Rinascimento, Il mondo antico nel Rinascimento, Florence, 1958, pp. 253–64; M. Borissavliévitch, The Golden Number and the Scientific Aesthetic of Architecture, London, New York, 1958; S. Ferri, Figure "quadrate" nel Rinascimento, Il mondo antico nel Rinascimento, Florence, 1958, pp. 249–52; P. H. Scholfield, The Theory of Proportion in Architecture, London, 1958; G. Soergel, Untersuchungen über den theoretischen Architekturenwurf von 1450–1550 in Italien, Munich, 1958; H. Schenck, Der goldene Schnitt, 4th ed., Augsburg, 1959; R. Wittkowever, The Changing Concept of Proportion, Daedalus, LXXXIX, 1960, pp. 199–215; C. L. Frommel, Die Farnesina und Peruzzis architektonisches Frühwerk, Berlin, 1961; D. E. Gordon and F. de L. Cunningham jr., Polykleitos' "Diadoumenosi": Measurement and Animation, AQ, XXV, 1962, pp. 128–50; C. L. Ragghianti, Mondrian e l'arte del XX secolo, Milan, 1962. *The East*: A. Foucher, L'iconographie buddhique de l'Inde, 2 vols., Paris, 1900–05; T. A. Gopinatha Rao, Tālamana or Iconometry (Mem. Archaeol. Survey of India, III), Calcutta, 1920; S. Kramrisch, The Viṣṇudharmottara, 2d ed., Calcutta, 1928; Liang Ssŭ-ch'en, Ch'ing shih ying-tsao tsê-li (Rules for Building in Ching Dynasty Style), Peking, 1934 S. N. Das Gupta, L'intimo aspetto dell'antica arte indiana, Rome, 1940; Seiichi Mizuno, Chinese Stone Sculpture, Tokyo, 1950; A. K. Coomaraswamy, The Transformation of Nature in Art, 2d ed., New York, 1956; W. Willetts, Chinese Art, 2 vols., Harmondsworth, 1958; P. Brown, Indian Architecture, I: Buddhist and Hindu Periods, 4th ed., Bombay, 1959.

* *

Illustrations: PLS. 339–354; 2 figs. in text.

PROVINCIAL STYLES. In the history of art, as in that of literature and, more generally, of civilization and customs, the terms "provincial" and "provincialism" are frequently employed to denote local styles that reveal dependence on, imitation of, or differentiation from the art produced in primary centers of creative activity. The term "provincial" is often used in a somewhat derogatory sense, which is not necessarily pertinent to a consideration of provincial art from the historical point of view.

CONCEPT OF PROVINCIAL ART. Provincialism in art has been subjected to adequate critical investigation only in certain areas. One of these is the art of the Roman provinces — and in this case the term "provincial" is precisely related to its derivation from the Latin *provincia*, with its administrative, territorial, and historical delimitation. Outside this specific frame of reference, "provincial" customarily denotes the cultural milieu of territories and small or more remote centers as opposed to that of the capital or other major centers.

The correlation between center and periphery in any particular period or area is an essential element in the study of artistic development. It can be analyzed on diverse levels: (1) within the narrow limits of a single aggregation such as a city, as the relationship between socially and culturally distinct groups or between the central and the suburban or other sections; (2) in an autonomous city with its own territories (e.g., the Greek *polis*, the medieval commune), as the relationship between the countryside and the city; (3) within a state, as the relationship between districts and the capital; (4) in the case of expansion or conquest, as the relationship between colonies and mother country, or between subjugated countries and the dominating power; and (5) even more generally (without regard to political unity), as any relationship between areas that are subject to influences and the great centers from which religious or cultural influences stem.

In art the distinguishing marks of "provinciality" may be summed up in the following characteristics and tendencies: (1) variations of elements in the dominant artistic style, which may take the form of a reduction and repetition of motifs; a tendency to schematize and geometricize forms for primarily decorative purposes; overworking of single, often obvious motifs; alterations in design, proportions, range of color, and materials used; a tendency to increase the areas of ornamentation; an emphasis on pure technique, sometimes at the expense of artistic creativity; (2) retarded acceptance and prolonged continuance of elements of the dominant art, which produces an effect of archaism or conservatism; (3) retention of elements of the local culture that existed prior to the introduction of the dominant style, and also, in some cases, a freer acceptance of elements from extraneous sources; (4) a certain tendency toward eclecticism (q.v.), evidenced by a mixture of elements that, in the development of the dominant art, had emerged successively or by a fusion of elements typical of different cultures; (5) a tendency toward spontaneous and instinctive expression that, as in the areas of primitivism (q.v.) and folk art (q.v.), results in the creation of forms not typical of the prevailing art of the time.

Obviously, in many cases these categories overlap. For example, some forms of expressive ingenuousness might be construed as evidence either of artistic impoverishment or of archaism. On a technical level, the tendency of the artisan to facilitate his work by repeating selected models inevitably concurs with a certain geometricization of the forms.

The category of provincial art cannot be isolated from other tendencies involved in the concept of secondary expressions of a dominant art — particularly that of "peripheral art," which, as defined by A. Boëthius (*Atti del I Congresso Internazionale di Preistoria e Protostoria Mediterranea*, Florence, 1952), includes the Mediterranean and European artistic developments influenced by the art of Greece before the Roman Empire (see ETRUSCO-ITALIC ART; GREEK ART, WESTERN; MEDITERRANEAN, ANCIENT WESTERN; PHOENICIAN-PUNIC ART). "Colonial art," such as the Spanish and Portuguese styles of the Western Hemisphere, and what is more strictly called "folk art" are also related forms of expression. Still, certain useful distinctions can be made. Undoubtedly "provinciality" commonly refers to style relationships within a single organic unit, whereas the term "peripheral art" is applied to manifestations influencing broad and diversified geographical areas and societies that are ethnically and culturally different from their centers of influence. More closely related to provincialism than peripheral art is the concept of colonial art. As far as folk art is concerned, it clearly presents many analogous qualities, and there are superpositions of folk influence on the provincial, and vice versa. However, the concept of "folk" implies the existence of social and cultural strata that may be totally lacking in an artistic "province." A further distinction is that provincial art finds expression within the highest traditions of the fine arts as well as in the forms of folk art.

In its historical development provincial art has followed one of two opposite courses: an ingrown, terminal process that is devoid of inspiration and ultimately without creative issue (giving rise to the derogatory connotation of "provincialism"); or a vigorous development leading to new forms of creation and even to the formation of a new style or art period, as evidenced, in the gradual evolution of Western medieval art from within the diversity of influences that entered into the cultural milieu of the Roman provinces.

* *

Historical examples. *Egypt and the Near East.* Manifestations of provincialism are to be observed only with the rise of organized states with vast expanses of territory or with the development of substantial cultural unity. In ancient Egypt provincial taste is easily traced, for its rise was constantly connected with the weakening of central power and the strengthening of local autonomy. In the Old Kingdom, as the Memphite tradition consolidated the stylistic innovations of the Thinite period into a formal, court manner, the provinces during the 5th and 6th dynasties continued to express archaic modes with liveliness and freshness, laying the foundations for a new tradition, that of the Middle Kingdom. The First Intermediate Period witnessed the full fruition of these earlier developments (pl. 355); the minor statuary of this period revealed a crude expressionism, sometimes intermingled with realistic tendencies, a new sense of mass that was far removed from the geometric intellectualism of Memphis, and a heightened chromatic sensitivity, evident in such famous pieces as the statue of Mentuhotep (IV, pl. 348). These new qualities, which found expression in the necropolises of the provincial cities from the First Intermediate Period to the Middle Kingdom (Hierakonpolis, Hamamiyeh, Beni Hasan, Meir, Deir el-Gebrawi, Gebelein, El-Kab, Aswān), were to be absorbed by the official art of the 12th dynasty and were to contribute to the formation of the Theban tradition, at a time when the quickened provincial activity was marked by adoption of the Memphite modules, a detailed grid system for copying the Memphite prototypes then in vogue.

Comparable examples, though less clearly defined, can be noted in Mesopotamian regions. During the Old Babylonian period, schools of engraving flourished in Assyria and Cappadocia; these produced a very particular type of glyptics, in which the iconographic heritage of Neo-Sumerian tradition was greatly modified by the inclusion of Anatolian elements and in which there prevailed a taste for strong emphasis of the positive values of shadow, a preference far different from contemporaneous tendencies in southern Mesopotamia. Also within the Old Babylonian artistic culture, other local schools showed typical provincial tendencies. Thus, in the large-scale statuary of Mari, there were examples that faithfully expressed the most severe aspects of the Neo-Sumerian official tradition — for instance, the goddess with flowing vase (IX, pl. 485) — whereas others revealed a crude expressionism that aspired to a closer contact with reality, such as the statue of Ishtup-ilum. Certain provincial modes of Old Babylonian glyptics later had a profound influence on Mesopotamian taste in the Kassite period.

Finally, the renewal that took place in Mesopotamian art in Assyria during the 1st millennium B.C. should be regarded as part of a complex of formal and compositional developments that is known only through the Middle Assyrian and Mitannian glyptics. Such divergent trends, breaking the continuity of the Neo-Sumerian and Old Babylonian traditions, are to be interpreted as the outcome of northern provincial tendencies, even though the facts of their rise and original ties with Babylonian centers are difficult to determine. An example of late Assyrian provincial art is found in the wall reliefs of Arslan Tash, where the alteration of traditional modes is revealed in a new compositional conception, in which the figures are isolated from any narrative context in the background (I, pl. 516; III, pl. 301). The same approach, with the addition, however, of the weight of the late Hittite volumetric conception of form, can be found in the reliefs of Sakçegözü; the reliefs in Zincirli, of a later date, maintained a greater autonomy of style. In late Hittite sculpture, in itself a marginal and composite art (which, however, taken as a whole, cannot be considered provincial), there are manifestations of provincialism in relation to the major centers (Zincirli, Carchemish); for example, the reliefs of Karatepe are closer in style to archaic works in Zincirli (I, pl. 253) than to the coeval ones.

Also lying within the sphere of ancient Near Eastern culture, the art of Nubian centers, which was dependent on that of Egypt and was characterized by the realism of its portraiture and by its "baroque" tendencies (see nubian art), can be considered, for the most part, as another typical manifestation of provincialism.

Paolo Matthiae

Greece. Provincial traits were discernible in the relation of the dominant Greek cultural centers (Athens, Corinth) to poorer and less vigorous centers or territories such as Boeotia and Euboea. A significant circumstance, the plentiful supply of fine materials, is also confirmed by the relatively slight production of sculpture in regions that did not possess the incomparable marble found in the islands and Attica. The formal deterioration, the simpler and more awkward rhythms that are to be seen in the later production of Boeotian steles cannot, however, be considered true examples of provincialism. A truly provincial style is evident in Thessaly, especially after the emergence of the first classical tendencies. There, fine prototypes such as a stele from Pharsalos or the head of a young man from Skiathos (pl. 356) degenerated into a stiff, mechanical production displaying crude workmanship, poverty of invention, and increasing aridity. At a later period, as a consequence of Macedonian hegemony and the economic and political ascendancy of northern Greece, the extensive artistic production centered about the palaces in Pella at times was characterized by an exuberance and excess that bears the traces of provincialism; this is particularly evident in the sumptuous, heavy goldwork of Thessaly, Thrace, and Macedonia from the Hellenistic age (VII, pl. 49; see greek art, northern).

In Asia Minor, instead, there was a confluence of diverse artistic traditions that were comparable in their levels of excellence and authority. Ekrem Akurgal (1961) has attempted to group as many of these as possible roughly according to a national scheme: thus there is a Phrygian and a Lydian art, either opposed to or paralleling the Hellenistic traditions of the coastal regions (see asia minor, western). In many instances, however, the fusion of varied currents produced merely a hybrid art rather than works of a decidedly national character or a genuinely provincial strain. The frescoes of Gordion (I, pl. 366) or the ceramics in Hellenistic style of Sardis should be judged as such hybrids; moreover, there is no trace of provincialism in the refined sculpture of Xanthos (IV, pl. 167), which Akurgal places among the monuments of Asian art, or in the production of sarcophagi at Sidon (I, pl. 535; VII, pl. 151), which continued for centuries. In Cyprus, continuing Hellenistic influence, along with Near Eastern influence from Mesopotamia, Egypt, and Syria, was reduced to formulas so consistently inferior and with such indifferent workmanship that designating this Cypriote production as provincial is fully justified (see cypriote art, ancient).

Contrary to what might be imagined, Greek colonial enterprises, which introduced Hellenic nuclei into distant countries that often had different and well-established artistic traditions of their own, seldom produced a genuine provincial style. This observation, though also valid for the West Greek colonies, is especially true of the eastern centers of colonization (see greek art, eastern; greek art, western). A certain degree of provincialism may be implicit in the backwardness of the sculptural forms of Temple C at Selinunte (VII, pl. 58), where, together with vestiges of a crude Doric archaism, there appears a highly developed technique of rendering drapery that clearly suggests a later date for these works. The bastardized copies of Tarentine terra cotta that are found in the interior of Italy, as far north as Fratte, near Salerno (pl. 356), also present provincial aspects. In contrast, in isolated localities such as Serra di Vaglio, terra cottas made with good molds in pure Greek style existed, without transition, alongside rude, uncultivated artistic products of local character. Crude manifestations of provincialism can be found, too, in the painted tombs of Lucania. These mark the limits of diffusion of Greek artistic influence on territories with a non-Hellenic population, for whose artistic expressions it is better to adopt the less restrictive designation "peripheral art."

In the Hellenistic age, offshoots of Greek art had increasingly broad and complex influence, moving across entire continents and modifying the aspect of their art as a result of Alexander's conquests. These far-ranging political and cultural upheavals were among the circumstances that gave birth to Parthian and Greco-Indian art (see BACTRIAN ART; BUDDHIST PRIMITIVE SCHOOLS; INDIAN ART; PARTHIAN ART); and with this large-scale development, minor local islands of provincialism were inexorably swept away.

Enrico PARIBENI

Roman provincial art. Roman provincial art (PLS. 357, 358) is generally understood to be the production from cultural centers outside Rome and the Italian peninsula during the period from the administrative establishment of provinces by Augustus in 27 B.C. to A.D. 212, when Caracalla granted Roman citizenship to all the inhabitants of the Empire — or, according to others, to 285, the beginning date of the Tetrarchy. Such a definition of limits becomes too imprecise if, for example, it is recognized that rural artistic production in sizable areas of the Italian peninsula was, in the time of Augustus, already quite different from that of Rome and was more closely related to what is generally considered popular or folk art (see ITALO-ROMAN FOLK ART). From the 1st century of our era the artistic development in the Greek-speaking provinces or those where Oriental languages were spoken demonstrated distinct differences from, and at times was even in anticipation of, the course of the visual arts in Rome (see ROMAN ART OF THE EASTERN EMPIRE).

Difficulties arise from a lack of discerning investigation into the general character and history of the Roman Empire in relation to its art. This lacuna, sometimes the result of personal predilections or bias, was already recognizable in some methodological approaches developed during the second half of the 19th century and became accentuated during the period between the two World Wars in studies that overvalued the significance of national and racial unity in regard to the ancient world. On the one hand, some historians attempted to contrast with Rome, as the focal point of the Empire, certain highly autonomous national cultures; others concentrated their efforts on emphasizing its absolutist character, the idea of Rome as colonizer of the world and the radix of artistic creation. These two attitudes, apparently contradictory, arise from the same basic premise: namely, viewing the Roman Empire as a unified, rigidly centralized structure in which Rome, as the capital, was supposed to have promulgated a culture valid for the entire Empire. As an extension of this interpretation, also, autonomous cultural and artistic forms are supposed to have arisen in some provinces as local reactions to centralized authority. The attempt to resolve any apparent contradiction by considering Roman art to be formed of two diverse components, the imperial (see ROMAN IMPERIAL ART) and the folk (especially diffused by Roman legions in less-developed cultural milieus) is only partly valid, for at best it explains only certain manifestations limited to the urban environment and the western provinces.

The unity of the Roman Empire should be understood, rather, as consisting in the relationships established through juridical and administrative conventions between the varied cultures of individual provinces, each one having a more or less autonomous identity as well, often deriving from its pre-Roman traditions. The role of Rome was to sustain this administrative elasticity: it was the primary urban force which was receptive to and which achieved a balance among oftentimes diverging interests. This position was understood even by the ancients themselves: Roman sources, in particular the oration by Aristides in honor of Rome, written in A.D. 144, insist on this equilibrating power of Rome.

A more comprehensive, and yet more accurate, definition of Roman provincial art would include, therefore, the art and the craftsman production of the provinces of the Empire from their inception, taking note that in the period from Augustus to the Tetrarchs such production was variously influenced by urban activity (to a minor degree in the eastern and Greek-speaking provinces, more intensely elsewhere) and in turn influenced the urban centers. Moreover, individual provincial strains sometimes interacted with each other directly, without the mediation of Rome. Thus, the study of the art of any given province should consider all aspects of its history (administrative, commercial, etc.), in order to discern the premises on which to base a reconstruction of its artistic character.

To try to define a general artistic continuity for the Roman provinces is an impossible aim, but it is possible to perceive, from certain more general tendencies, unified artistic trends that were restricted to one or to several provinces. The study of the art of the imperial provinces must begin with the assumption that the works of art and monuments found in a given geographical and cultural locale are not necessarily related to the stylistic tradition of that region and, in fact, may have been imported from other cultures, thus testifying to historical vicissitudes, commercial exchanges, or cultural superimpositions. In studying the production of individual provinces, one should be aware of the possibility that it may differ according to social class as well, for works of art often assume characteristics based on the use for which they are intended.

Such extended stylistic areas can be described on the basis of the artistic development that occurred during the Hellenistic era: namely, as Attic, Pergamene, Seleucid, and Alexandrian schools. To these recognizable strains should be added as well stylistic patterns that evolved during the same period in the Celtic, Etrusco-Italic, Judaic, Parthian, and Arabic culture areas, for even within the broad sphere of Hellenistic influence such cultures were already responsible for specific artistic trends, such as the Iberian and the eastern Anatolian, which evolved primarily through varying combinations of new elements and older local traditions. Moreover, in many areas, usually those quite remote from the creative centers of the period, there continued older cultures that were fundamentally of Punic, Egyptian, or Achaemenid type.

As a consequence of reciprocal exchanges and the direct and indirect influences arising from various historical events, the period between the reigns of Augustus and Diocletian witnessed the emergence of a number of definite cultural and artistic ambients in various parts of the Empire; apart from minor subdivisions, such distinct developments may be identified in Greece, Asia Minor, Syria, Egypt, and the provinces of the Rhine and Danube limites. The importance of these provinces, as far as their influence on Rome itself was concerned, varied with time; whereas the culture of the western provinces was significant principally as a base for the succeeding culture of the Middle Ages, Attica in the 1st century of our era, and again in the 2d, Asia Minor in the 2d century, and Egypt and particularly Syria in the 3d century had a marked and growing artistic influence on Rome, on the provinces of northwest Africa (PL. 359), and on the western European ones (through the Danubian provinces). (See AFRICAN-ROMAN ART; DANUBIAN-ROMAN ART; GALLO-ROMAN ART; ROMAN ART OF THE EASTERN EMPIRE.)

The most highly developed and creative civilization of the entire Hellenistic world, that of the Hellenic East, a culture that during the Roman era flourished in areas which were within some of the oldest geographical confines of Roman dominion, became from the 3d century the primary source for ensuing elaborations of imperial art, which can be considered truly the koine of the vast imperial world only from the time of the Tetrarchy on. It is only from then, perhaps, that one can justly speak of provincial art — with all the limitations implicit in the present interpretation of the designation "provincial" — in referring to the artistic output of particular regions of the Empire.

Antonio GIULIANO

Byzantine to modern times. In the Byzantine world there existed a consciousness of the different, and inferior, value of the art of the various provinces — from Syria and Cappadocia to the Balkan Peninsula — as compared with the art of the capital of the Empire, Constantinople. This was a reflection of the ideological and administrative structure of the Empire, focused on the capital, whence radiated artistic and cultural splendor as well as the political power. Provincial styles, with a diversity of local inflections, can be distinguished within the general context of Byzantine art. In some instances, these emerged in distant territories or long after the height of Byzantine political sway; but, nonetheless, they disclosed a clear relationship to the dominant Byzantine tradition (PLS. 360, 361).

In the Middle Ages, artistic centers became as numerous as the monasteries in which the artists worked. Among these there were, naturally, centers of major importance and others of minor influence. The array of "dialects" and stylistic exchanges, even across great distances, reached a remarkable level, for the existence of many of the artists was an essentially nomadic one, encouraged by the far-reaching enterprises of the monastic orders and by the impetus of religious pilgrimages. Since artistic centers no longer coincided with political centers and since a centralized, exemplary art was lacking, a dependent (i.e., truly provincial) art was also lacking. It is therefore more usual to treat lesser or peripheral artistic output of this period, in comparison with art emanating from the more important centers, as manifestations of minor or folk art, rather than as genuine provincial strains.

In the art of the 14th and 15th centuries in Italy, the age of the commune and then of the seigniory, when the identification of artistic schools again coincided with the new urban centers (see ITALIAN ART) — favoring the development of some cities, generally capitals or noble seats, over others — there again arose the possibility of a basis of comparison between the culture of the leading centers and that of the provinces. Still the level of art in many minor centers was extremely high, and the secondary art produced in such places, manifesting a cultural retardation (generally typical of folk art), is not readily defined as provincial, except for certain instances of artists who were slow to absorb new influences or of persisting local traditions (PL. 362).

Many Italian centers or artistic schools had phases in which they emerged from or fell back into provincial status. For example, Umbria had been a receptive artistic center, characterized by eclectic adoption of outside influences, until it developed, with the advent of Perugino, a school of far-reaching significance; then, in the 16th century, having reached the height of its importance, it again closed itself off within the limits of a minor province. Sicily had consistently been provincial with respect to the main course of Italian art, and not even an artist of the stature of Antonello da Messina succeeded in stimulating a school of painting of permanent consequence (his art stirred the greatest reaction in Venice); Sardinia and the Abruzzi (PL. 363) have also occupied the same position in the Italian development. In general, all the regions that presently might be called "depressed" remained cut off from the creative development of Europe: that is, they had no significant contact with the principal centers in which this development was determined. Toward the end of the 16th century Florence, which had been an artistic center of fundamental importance from Giotto to the end of the Renaissance, virtually lost this preeminent position. The late mannerism and baroque of Florence were but provincial variants of an evolution that had its main centers elsewhere — in Rome, Emilia, Paris, and other locales.

A typical example of provincial style was the baroque architecture and decoration of Lecce, a minor and local manifestation that was favored by tradition and the materials found in the area (a soft and easily worked stone), but one nevertheless superior to the qualifications implied in the designation "folk art," because it was an original, expressive, and in no way *retardataire* development (PL. 365; II, PL. 146).

Even when, in the 16th century, Rome became a prime artistic center for all of Europe, other Italian and European schools could not be defined as truly provincial, for the characteristic dependency of provinces on an administrative capital was lacking between Rome and the centers it dominated culturally. Moreover, in Rome no single canon of taste triumphed over all others; on the contrary, Rome had become the meeting ground and the crucible of the most significant and disparate artistic currents.

It is not until the 17th century in France, when the political and artistic capitals again coincided, that a clear distinction can again be made between the art of Versailles and Paris, which centered around the monarchy, and the art of the various provinces, which was distinct from and opposed to this centralized dictatorship of taste. Thus critics often write of "provincial painters," in referring to masters from the various regions of France who, during the 16th century, exhibited tendencies quite different from those prevailing in the capital and at Versailles — for instance, a movement toward an extremely vivid realism.

The activity of princely patrons and collectors, intensifying throughout Europe from the 16th century on, proved a major antidote to provincialism (see MUSEUMS AND COLLECTIONS; PATRONAGE). In order to ally themselves with those courts already noted for their artistic opulence, many sovereigns called foreign artists into their service, formed collections of art, and founded academies. Appreciation of art, and the prestige accruing from it, became such a widely based ideal that the small German courts of the 18th century, for example, earnestly sought to imitate the splendor of Versailles.

A common failing of provincialism is its tendency to imitate the dominant art of the place or the epoch, with a consequent incapacity for original creation. Elements such as obvious or even vulgar imitation, cruder workmanship, and the use of more economical or local materials are frequent corollaries of the concept of provincial style. For such reasons, perhaps, country houses frequently have an outmoded aspect, solid but less fine than that of city houses of the same owners. That this is not necessarily the rule, however, is evident in Renaissance and baroque villas built by great artists in surprising and seemingly inappropriate locations. In such cases, the exceptional presence of an exemplary monument often had a prolonged influence on local art and craftsmanship, sometimes encouraging the formation of particular local traditions that continued the style of the celebrated master who had worked in the province.

After the great age of explorations, a provincial art eventually developed in the European colonies overseas. In the Americas, in the beginning, art was imported from Europe; soon, however, differences in craftsmanship and materials resulted in the creation of monuments and objects characterized by distinctive local variations (PL. 364), often even richer in imagination and flavor than their prototypes. The styles evolved from such variants are commonly called "colonial" rather than provincial (see AMERICAS: ART SINCE COLUMBUS; FURNITURE; SPANISH AND PORTUGUESE ART).

By the 19th century the traditional patronage of the aristocracy, which for centuries had promoted international artistic exchanges, had to a great degree declined, but the diffusion and leveling of culture had begun to diminish the force of local tradition as well. Artists now felt the danger of remaining closed within a restrictive local tradition, cut off from the general cultural progress of their day, so that from this time, on an ever-widening scale, they sought to establish themselves in what were considered the more active and avant-garde centers (with little regard for patronage possibilities or commissions), leaving the confines of the provinces — even the borders of their native countries — and becoming more and more governed by an attitude toward art that transcended such geographical limits and regional styles.

Nevertheless, certain styles of the 19th century are commonly referred to as "provincial," and with reason (PL. 366). An example is the American impressionism exemplified by T. W. Dewing, which reflected a wholly Europeanized taste in contrast to the original contemporaneous work of such American masters as Homer and Eakins. Similarly R. Fontana, a Milanese painter of the 19th-century Lombard school, found his sources of inspiration primarily in the outstanding painters of Florence and Naples. Notable examples of provincial or "peripheral" art in the 19th century, extending in some instances to the mid-20th, are to be found in the Oriental countries, where a process of Westernization was under way and artists were at first dependent on European and American trends (see ORIENTAL MODERN MOVEMENTS). Ravi Varma, for example, working in India in the late 19th century, adopted Western techniques and at the same time adapted Western compositions to Indian themes in what may properly be called a "provincial" style (PL. 366).

BIBLIOG. *Ancient Near East: a. Egypt:* H. Kees, Studien zur ägyptischen Provinzialkunst, Leipzig, 1921. *b. Western Asia*: H. Frankfort, The Art and Architecture of the Ancient Orient, Harmondsworth, 1954; J. A. H. Potratz, Die Kunst des Alten Orient, Stuttgart, 1961. *c. Late Hittite art*: M. Vieyra, Hittite Art 2300–750 B.C., London, 1955; E. Akurgal, The Art of the Hittites, London, New York, 1962. *Greek world*: E. Langlotz, Frühgriechische Bildhauerschulen, 2 vols., Nürnberg, 1927; L. Quarles van Ufford, Les terrecuites siciliennes, Assen, 1941; E. Langlotz, Wesenzüge der bildenden Kunst Grossgriechenlands, Antike und Abendland, II, 1946, pp. 114–39; T. J. Dunbabin, The Western Greeks, Oxford, 1948; P. Zancani Montuoro and U. Zanotti Bianco, ed., Heraion alla foce del Sele, 2 vols., Rome, 1951–54; T. J. Dunbabin, The Greeks and Their Eastern Neighbours, London, 1957; E. Akurgal, Die Kunst Anatoliens von Homer bis Alexander, Berlin, 1961. *Roman world*: T. Mommsen, Römisches Staatsrecht, 3 vols., Leipzig, 1871–88; T. Mommsen, Die römischen Provinzen von Caesar bis Diocletian, Berlin, 1885 (Eng. trans., W. P. Dickson, 2 vols., London, 1886); M. I. Rostovtzeff, Social and Economic History of the Roman Empire, Oxford, 1926; Cambridge Ancient History, IX–XII, Cambridge, 1932–39; T. Frank, ed., An Economic Survey of Ancient Rome, 5 vols., Baltimore, 1933–40; M. I. Rostovtzeff, Social and Economic History of the Hellenistic World, 3 vols., Oxford, 1941.

* *

Illustrations: PLS. 355–366.

PRUD'HON, PIERRE-PAUL. French painter (b. Cluny, Apr. 4, 1758; d. Paris, Feb. 16, 1823). The son of a stonecutter, Prud'hon was sent to Dijon in 1774 to study at the Ecole de Dessin under François Devosge. In 1776, he won first prize in painting; and in 1780, under the patronage of Baron de Joursanvault, he was sent to Paris, where he studied at the Académie Royale until 1783. Returning to Dijon, he won the Prix de Rome in 1784, and thereafter lived in Italy until 1788. During this sojourn, in 1786, he was commissioned to paint a free copy of Pietro da Cortona's ceiling in the Palazzo Barberini for the Salle des Statues of the Palais des Etats in Dijon. He returned to Paris via Dijon and Cluny, and began exhibiting at the Paris Salon in 1791. After the fall of Robespierre, he returned to his native Burgundy for a two-year sojourn near Gray. There he did illustrations for the editions of Firmin Didot, including Ovid's *L'Art d'aimer*, Rousseau's *La Nouvelle Héloïse*, and Bernardin de Saint-Pierre's *Paul et Virginie*. In 1796, he returned permanently to Paris, where, despite his isolation from such leading artists as Gros and Guérin, he was highly favored in Napoleonic society — even becoming the drawing master of Joséphine and, later, of Marie Louise. He also received many commissions through the Minister of Arts, Vivant Denon. In 1803 his legal separation from his wife, Jeanne Pennet, ended an unfortunate domestic situation, and was followed by a period of happiness with a student of his, Constance Mayer La Martinière. After 1815, however, his professional life was clouded by the loss of earlier patronage, and in 1821 his private life was saddened by the suicide of Mlle Mayer.

In contrast to most of his contemporaries, Prud'hon avoided the orthodox neoclassicism of David. In its place, he substituted a highly personal style in which the sleek mannered figures of his friend Canova are transformed by the chiaroscuro effects, murky atmosphere, and fluent rhythms that he admired especially in the work of Leonardo and Correggio. Despite his efforts at classical and political allegory, his strongest work pursues erotic and sentimental directions established in the mid-18th century. To these tendencies are added a tone of private melancholy and an obscure *ambiance* that foreshadow much of the languor and mystery of later French romantic painting and, in particular, the paintings of Delacroix, who admired, copied, and wrote about Prud'hon's work. (See also X, PL. 280.)

MAJOR WORKS. *Portrait of Mme Anthony* (1796; Lyons, Mus. des B. A.). — *Portrait of the Empress Joséphine* (1805; X, PL. 269). — *The Rape of Psyche* (1808; Louvre). — *Justice and Divine Vengeance Pursuing Crime* (1808; Louvre). — *Crucifixion* (1825; Louvre).

BIBLIOG. C. Clément, Prud'hon: Sa vie, ses œuvres, et sa correspondance, Paris, 1872; E. de Goncourt, Catalogue raisonné de l'œuvre peint, dessiné et gravé de P.-P. Prud'hon, Paris, 1876; J. Guiffrey, L'Œuvre de P.-P. Prud'hon (Collection de la Société de l'Art Française, XIII), Paris, 1924; G. Grappe, P.-P. Prud'hon, Paris, 1958; Pierre-Paul Prud'hon, 1758–1823, les premières étapes de sa carrière, Dijon, 1959 (cat.).

Robert ROSENBLUM

PSYCHOLOGY OF ART. The need to consider art from a psychological point of view has distant origins, but the establishment of a definite avenue of approach based on experimental methods is relatively recent. This article attempts to give not only a historical outline of the various methods of investigating the field but also a definition of the characteristics of the new science and a description of its scope of research. The strong influence of artistic and literary styles has been stressed, as well as new prospects opened up by psychology in the study of philosophies of art and of historical and critical research. For those art forms which may express an elementary and spontaneous idiom and hence reflect more directly a natural psychic mechanism (such as children's art, the art of aboriginal peoples, and the so-called "primitives") see also FOLK ART and PRIMITIVISM.

GENERAL CONSIDERATIONS. *Definition.* As the historical survey given below will indicate, it has proved difficult to delimit the scope of a psychology of art, and in particular to separate it from esthetics (q.v.). Esthetics may be divided into two main branches: the philosophy of art and the psychology of art. But, in general, it has been felt that a philosophy of art must be based on empirical evidence; and the emergence of an esthetic philosophy from a priori considerations, such as is found in Schiller, Schopenhauer, or Nietzsche, has not led to any generally acceptable conclusions. In short, it is now almost impossible to distinguish a psychology of art from esthetics. According to the comprehensive definition of Edward Bullough (1957), "Aesthetics is the systematic study of aesthetic consciousness," and its study comprises: "I. *A.* That form of aesthetic consciousness which is directed to, or rather results in, the creation of aesthetic objects, i.e. *artistic production. B.* The consciousness which enjoys or contemplates works thus produced — and other objects of Nature and Reality, in so far as they are susceptible of being aesthetically contemplated, i.e. *aesthetic contemplation.* II. The *objective* products, resulting from artistic creation, in relation to the productive and receptive consciousness, i.e. *the world of Art, fine and applied.* III. The aesthetic consciousness extended to other spheres, and applied to Life in general, i.e. *aesthetic culture.*" It is difficult to see how a psychology of art could be given any more comprehensive or less exclusive definition.

It follows that a psychology of art should be aware of two points of view: the psychology of the artist and the psychology of the percipient; and that it should draw its evidence from all the arts — from music and poetry as well as from the plastic arts. It is more concerned with establishing distinctions between the arts than with defining the concept of art itself, which is a semantic problem. Problems of social or ethical evaluation may be left to the philosophy of art, properly so called, though it must be assumed that the esthetic validity of the evidence has first been established by a psychology of art.

Scope. In view of the confused boundaries that separate the various disciplines concerned with art, it is necessary to make a preliminary attempt at a definition of the scope of the psychology of art. Before this can be done, art itself must be defined, at least to the extent of describing its function. No subject has been discussed so often and so inconsistently by philosophers, from Plato and Aristotle in the ancient world to Leibniz and Kant at the beginning of the modern period. Until comparatively recent times these discussions centered on the concept of beauty, with the feelings of pleasure and displeasure as the accompanying psychological reactions; and on this basis Kant could affirm the objective validity of esthetic judgments. The nature of such judgments might be classified as idealist, formalist, or sensualist, but in each case it was assumed that the purpose of artistic activity was to create forms that caused pleasure in a percipient. The pleasure might be derived from the content of the work of art, which was in general the idealist point of view, or from the form of the work of art, which was the sensualist or empirical point of view.

The fallacy of this assumption was perhaps first exposed by Friedrich Nietzsche, who pointed out (1872) the dialectical opposition of two contrary elements in Greek tragedy which he called the Apollonian and the Dionysian. Psychology was subsequently to find more scientific terms for these opposed directions of artistic purpose, but for the moment they can be simply described as the contrary principles of beauty and vitality. Nietzsche's intuition was to be amply confirmed on a wider historical basis by the discovery, toward the end of the 19th century, of various types of primitive or tribal art in which qualities were gradually recognized and appreciated that could not be reconciled with the ideal of beauty. Once vitality had been admitted as an esthetic principle, then a revaluation of art could take place which accomodated various phases of the historical evolution of art hitherto rejected as primitive or immature. In brief, it was discovered (and this is the essential fact for the psychology of art) that behind the work of art, determining its specific form, was a "will to art": the work was an expression of an immediate purpose (*Kunstwollen*) and could be judged or appreciated by its fulfillment of the purpose.

The term *Kunstwollen* was invented by Alois Riegl (1910) and first formulated in connection with postclassic Roman art (see HISTORIOGRAPHY). Riegl wished to show that the condemnation of such a style as "decadent" was valid only in relation to the esthetic ideals of another age (i.e., the classical age, whose ideals the 19th century had adopted) and that in relation to the aims of the late Roman artist the art might be adequate and admirable. In other words, the development of new social conditions and new ideals required new modes of expression, new forms of representation.

This teleological relativism has remained basic to the psychology of art, providing common ground to both formalists and sensualists. Granted the existence of a universal will to art, all styles can be studied as modifications due to a particular environment or a particular psychological type. The main task of the psychology of art, on such a supposition, would be to define the nature of this will to art and to trace its working in society and in the individual artist.

Though instinctual theories of art are found in Plato and in even earlier Greek philosophers, the ideal of will in its modern sense of an activating force probably originates with Schopenhauer and was first applied to esthetics by Robert Vischer. Vischer's vast system of esthetics, Hegelian in structure but anti-Hegelian in intention, need not concern us here. But in one of his other works (1873) he introduced the notion of empathy (*Einfühlung*), which in its most general sense may be briefly defined as the projection of human feelings, emotions, and attitudes into inanimate objects. In this form, empathy may mean nothing more precise than a feeling of identity with the forms of nature, an objectified self-enjoyment that can attach itself to trees, rocks, landscape, the external world as a whole. But Vischer made the concept much more precise and subtle. In artistic empathy a process of transformation takes place, for the forms into which the feelings are projected define those feelings and may, if the forms are appropriate, induce a general state of well-being. Two later definitions of emphathy will show how the concept has been developed and refined since Vischer's time. The first is from Martin Buber (1920): "Empathy means to glide with one's own feeling into the dynamic structure of an object, a pillar or a crystal or the branch of a tree, or even of an animal or a man, and as it were to trace it from within, understanding the formation and motoriality (*Bewegtheit*) of the object with perceptions of one's own muscles; it means to 'transpose' oneself over there and in there." More recently David A. Stewart (1956, p. 12) has given a much broader definition of empathy as "deliberate identification with another, promoting one's knowledge of the other as well as of oneself in striving to understand what is now foreign but which one may imagine... to be something similar to one's own experience.... It is felt to be ethical because it is grounded in feeling, presupposes goodwill, and strives for mutual understanding. It is seen as a sound psychological concept, because the process it stands for produces our most authentic and genuine personal experiences. It is esthetic in its creative and selective activities. These three aspects of empathy, the psychological, the ethical, and the esthetic, are inseparable in practice. Empathy as here defined will be subsequently represented . . . as the ground at once of ethics and of personality theory and as an act of first importance in all art."

It will be seen from these definitions that, apart from any other merits the theory may have, it offers a common ground for both the psychology of the artist ("its creative and selective activities") and the psychology of art appreciation ("mutual understanding," "genuine personal experiences"). It leaves us free to analyze the formal aspects of art ("dynamic structure," "formation," "motoriality," "perceptions of one's own muscles") and yet at the same time effects transposition of feeling, promotes knowledge and understanding, and relates esthetics to ethics (moral evaluation). In view of the long standing of the theory and its present currency, an important part of any psychology of art must necessarily be concerned with the experimental verification of the theory. This was begun by Theodor Lipps (1851–1914), whose two volumes of *Aesthetik* first appeared in 1903 and 1906. Many other psychologists of art have either developed or criticized the theory, but subsequent developments in psychology, above all the theory of projection in Freud's psychoanalytical therapy, have all tended to bring support to the doctrine of empathy. Even those theories of art which are confined to the analysis of modes of visual or aural perception in the artist and in the layman do not necessarily conflict with the concept of empathy, which presupposes such perceptual activities; whereas those theories which give art a biological function (the play theories of Herbert Spencer, Grant Allen, and Karl Groos) can be construed as an analysis of the physiological symptoms of empathy. That is to say, the doctrine of empathy is wide enough to accomodate the extremes of idealism and materialism and is, in the words of Vernon Lee (1856–1935) "part and parcel of our thinking." Lee, one of the earliest and best exponents of the doctrine, described its universality thus (1913): "Although nowhere so fostered as in the contemplation of shapes, empathy exists or tends to exist throughout our mental life. It is, indeed, one of our simpler, though far from absolutely elementary, psychological processes, entering into what is called imagination, sympathy, and also into that inference from our own inner experience which has shaped all our conceptions of an outer world, and giving to the intermittent and heterogeneous sensations received from without the framework of our constant and highly unified inner experience, that is to say, of our own activities and aims." The whole psychology of art is necessarily a discussion of the relation of form to feeling and of feeling to form; empathy is the science of that relationship. We shall proceed to a discussion of these two terms, which denote the objective and subjective aspects of the psychology of art.

Form and content. Empathy does not necessarily imply a pleasurable response. Indeed, the notion that the function of art is to give pleasure must be abandoned. According to Baensch (1923–24) the function of art is "to raise the emotional content of the world to the level of an objectively valid cognition," in other words, to create concrete and sensuously apprehensible symbols for any kind of feeling. This theory would seem to owe something to earlier but little publicized theories of Konrad Fiedler (1841–95), which in turn were based on observation of and conversation with such artists as Hans von Marées and Adolf von Hildebrand. (The latter was to make his own important contribution to the discussion in *Formprobleme der Kunst*, an essay first published in 1893.)

Baensch further states that "the function of art is not to give the percipient any kind of pleasure, however noble, but to acquaint him with something he has not known before. Art, like science, aims primarily to be 'understood'. Whether that understanding which art transmits then pleases the feeling percipient, whether it leaves him indifferent or elicits repugnance, is of no significance to art as such." This definition is not essentially different from the earlier and better-known definition of Benedetto Croce (1866–1952), according to which art "does not classify objects, nor pronounce them real or imaginary, nor qualify them, nor define them. Art feels and represents them. Nothing more. Art therefore is *intuition,* in so far as it is a mode of knowledge, not abstract, but concrete, and in so far as it uses the real, without changing or falsifying it. In so far as it apprehends it immediately, before it is modified and made clear by the concept, it must be called *pure intuition*."

It would not be difficult to correlate Croce's "pure intuition" with Baensch's "mental activity," whereby we bring the world's emotional content into objectively valid cognition; and indeed the main line of development in esthetics and the psychology of art leads from the first formulations of the doctrine of empathy to the theories of symbolic discourse advanced by Ernst Cassirer (1874–1945) and Susanne Langer (b. 1895). Psychology of art is now inseparable from a science of symbolic discourse; art is not imitation of nature but a discovery of reality, it embodies an original, formative power, a particular image world which does not merely reflect the given world but is constituted in accordance with an independent principle, a system of symbolic forms equivalent to but independent of a system of intellectual symbols. "In every linguistic sign, in every mythical or artistic image, a spiritual content which intrinsically points beyond the whole sensory sphere is translated into the form of the sensuous, into something visible, audible, or tangible. An independent mode of configuration appears, a specific activity of consciousness, which is differentiated from any datum of immediate sensation or perception but makes use of these data as vehicles, as means of expression" (Cassirer, 1923).

It will be seen how impossible it is, in such a theory, to distinguish the elements of form and content. The vague feeling, the fluid impression, assumes form and duration for us only when it is molded by symbolic action: content cannot be defined, or even imagined, until it has form. The contents of works of art are dim feelings brought to expressive shape; art, as Baensch would say, takes the fluid emotional content of our consciousness, objectifies it, clarifies it, and gives it a heightened and enduring realization. Art can have no higher function than this, and this has been its social, even its biological, function throughout history.

We must still ask by what modes feeling takes shape — what is the principle of formation in art? Baensch found it in rhythm, which he defined as "the alternation between heavy (stressed) and light (unstressed or less stressed) parts, in so far as it follows certain rules . . . . In the construction of such rhythmic sensuous unities, next to each other, in each other, and one above another, the form of the work of art originates: it is nothing else but its total rhythm. Only inasmuch as a work has form, i.e., rhythm, is it a work of art."

This is a dogmatic and limited definition of form, with obvious application to the arts of dance, poetry, music, and architecture. Rhythm is also characteristic of organic life as well as of mathematical (spatial and temporal) unities; it therefore serves as a link between the two primary elements of art — beauty and vitality. Langer has convincingly demonstrated its presence in drama and comedy (1953), though there is a difference of precision, amounting almost to a difference of kind, between the measurable rhythm of poetry and music and what Mrs. Langer calls the rhythm of animal existence — "the strain of maintaining a vital balance amid the alien and impartial chances of the world." She states: "The pure sense of life springs from that basic rhythm and varies from the composed well-being of sleep to the intensity of spasm, rage or ecstasy." Comedy "expresses the elementary strains and resolutions of animate nature, the animal drives that persist even in human nature, the delight man takes in his special mental gifts that make him the lord of creation; it is an image of human vitality holding its own amid the surprises of unplanned coincidence." Tragedy has a different basic feeling and therefore a different form: a "cadential" form; a rhythm of action, springing from the poet's original conception of the fable and dictating the major divisions of the work, "the light or heavy style of its presentation, the intensity of the highest feeling and most violent act, the great or small number of characters, and the degrees of their development. The total action is a cumulative form, and because it is constructed by a rhythmic treatment of its elements, it appears to *grow* from its beginning. That is the playwright's creation of 'organic form.'"

Admittedly rhythm in such a context is a question-begging word borrowed from the realm of physiology, but Langer is no doubt justified in applying to dramatic form, "without distortion or strain," a concept of rhythmic structure usually considered more appropriate to musical forms. We may conclude that the vitality of any form, organic or artistic, is based on its rhythmical structure.

Form, however, is more than rhythmical structure. It is quite evident that certain works of art — Shakespeare's *Hamlet* may be given as an example — are essentially "formless" and do not exhibit any demonstrable rhythmic unity. The 20th century has seen the rise of a style in painting which is described by its practitioners as "informal," and though not all examples of such art are lacking in rhythm, rhythm is not their distinctive quality. Apart from such particular exceptions, there have always existed works of art to whose content Cassirer has given the name "mythic consciousness," and for these a different conception of form may be appropriate. Such a conception is the archetype, a hypothesis first suggested by Jung (1943) and later elaborated in relation to works of art by Neumann (1949). According to this hypothesis the human brain inherits a disposition to specific pattern formations ("thought-forms") which are "psychic residua of numberless experiences of the same type" endured by the human race, and these inherited patterns of response act as a priori determinants of the form in which individual experience will be represented.

This hypothesis cannot be dismissed (as Langer dismisses it) on the grounds that it is concerned with the motif and not with the form of art, for Jung's hypothesis relates essentially to form and is an attempt to explain why the forms of such dramas as those of Orestes and Oedipus, Hamlet and King Lear, in spite of their separate origins, possess a common structure. The fact that some of the dramas in question can be characterized as rhythmical does not explain either the origin or the universality of the basic form. The psychic mold into which these dramas are cast is (according to another hypothesis of great relevance to the psychology of art) an element of the collective unconscious, and its nature is not to be defined nor its dimensions measured by the rational intellect. Archetypes are grounded in the peculiarities of the living organism and are therefore, according to Jung, direct expressions of that organism's being.

An irrational element extends to both the form and the content of works of art, and its justification lies in that process of *enantiodromia* (Gr., *enantio*, opposite; *dromos*, running), discovered by Heraclitus, which Jung calls "the most marvelous of all psychological laws." According to this law, the excessive rationality of our culture would require for equilibrium an art of excessive irrationality, and that is precisely what it has got. To attempt to accomodate such irrational forms and images within an architectonic process involving the construction and ordering of forms in space or time according to laws of rhythm and proportion is vain: the most we can demand is that form as such is perceptible and meaningful. Rhythmics, Baensch contends, is the logic of art; it may be, but to subsume art under rhythmics is to repeat the fallacy of identifying art with beauty and thus to make pleasure the only criterion of art. But Baensch admits that when a form is felt to fit the content inevitably, we may be aware of such inevitability without being able to explain it; we assume that its inevitability is due to its rhythmic-formal nature, though the artist does not know why he gives the content, which he shapes, just such a form: "There is no bridge between the two which could be constructed by the intellect. The 'unconscious' factor in artistic creation is nothing but this impossibility of comprehending the necessity which connects the feeling to be captured with the form in which it will be bound. And since this necessity remains incomprehensible even after the completion of the work of art, this unconsciousness never really becomes consciousness in the artist, nor the beholder." The value of a work of art, Baensch concludes, lies in the degree of necessity that obtains in it, and this is a definition that would account for the artist's recourse to forms which embody the highest degree of necessity — those primordial images which we call archetypes.

Confirmation of this hypothesis is found in children's art, mainly in connection with the psychology of education. The origin of basic forms (e.g., the circle, the cross, the square, and the rectangle) may be attributed to the gradual coordination of random muscular activity (cf. Arnheim), but the child's concentration on certain motifs (houses, trees, trains, etc.) can be significantly correlated according to sex as the determining psychological type; certain features characteristic of works of art in general, such as haptic emphasis (exaggerations due to internal somatic sensations) appear most clearly in children's work (cf. Löwenfeld, 1939, 1957; Read, 1958). Similar observations have been made in primitive art, where formal distinctions between ideoplastic and physioplastic types have been worked out by Verworn (1914), the first being determined by conceptual preoccupations, the second by somatic sensations. Kühn's distinction between imaginative and sensorial types (1923) is almost identical. There can be little doubt that a systematic correlation between modes of expression in art and psychological types, as differentiated by psychologists, can be made; a considerable literature on the subject already exists.

Herbert Read

Origins and history. *Major changes from ancient to modern times.* In tracing the history of a present science or branch of scholarship, it is well to start with some description of the subject as it is today. Early examples of discussion along similar lines can then be found by looking backward through the history of thought. One must not expect to find a single clear-cut path or logical sequence from ancient to modern. At the start there is likely to be a scattering of vague and fragmentary comments, occasional brilliant but undeveloped insights, dogmatic assertions, and poetic metaphors. Often these occur in the midst of rambling discussions of other topics by men whose main interests lie elsewhere. The history of the subject then appears as a gradual convergence of these scattered lines of thought, a clarification of basic problems and possible solutions detached from the irrelevant and nonessential. Little by little the subject takes on more definite form as distinct from neighboring subjects, though perhaps overlapping them at certain points.

The basic conception of many separate sciences dividing up the total field of knowledge and cooperating in its growth (as in Francis Bacon's *Advancement of Learning*) was only vaguely achieved in antiquity. There was no such concept as "psychology" or "art" in the modern sense of these terms. But many anticipations of what is now called the psychology of art are to be found in Greek and Roman philosophy and poetry, combined with legend, myth, fantasy, metaphysical speculation, and moral precept.

The psychology of art is still not fully purged of irrelevancies or drawn together into a single clearly organized science like mathematics or biology. It is precise and quantitative in a few limited areas, but it falls far short of exact scientific status as a whole; it is still mixed with philosophical speculation and literary, personal expression. Only a few of its present exponents attempt, or think it possible and desirable, to develop it much further in the direction of science. Contributions come from different fields: from philosophical esthetics, where the psychology of art had its principal source; from the writings of artists; and more recently from the young science of general psychology. The contributions often arrive in the form of abstract hypotheses, with little controlled, systematic observation and experiment. They tend to supplement and correct each other, so that a body of knowledge and theory is developing; but little has been done in the way of comprehensive synthesis or logical proof.

The psychology of art can best be characterized by its long emphasis upon a certain set of problems. In brief, what is the nature of human behavior and experience in relation to works of art? The term "art" in this connection is meant in its esthetic sense, as including the visual, musical, literary, and theater arts — all the skills and products which commonly have the function of arousing some kind of satisfactory esthetic experience. From the standpoint of the artist they involve the expression and communication of his own emotional experience. Also included are the so-called "fine" arts and the "useful" arts such as architecture. How to describe more specifically the kinds of experiences now vaguely indicated by such terms as "esthetic satisfaction," "the sense of beauty," "creative inspiration," and the like is a perennial task of the psychology of art.

Two basic types of phenomena in art have been distinguished under various names since ancient times: those which characterize the artist (as creator, performer, or craftsman) and those which characterize the observer (the viewer, listener, reader, user, appreciator, or evaluator). Economists have referred to the "producing" and "consuming" phases of art. What psychophysical functions, abilities, and processes are involved in each of these phases? How do they vary from culture to culture, period to period, individual to individual, and from childhood to maturity in the same individual? What factors within the individual himself and in outside objects such as works of art combine to cause these types of experience and behavior? How are they related to the various types and styles of art and to the general cultural climate of an age and a people? Around these central problems there is a broad margin of others whose importance in the psychology of art is more questionable.

In the shifting, controversial fabric of modern science the psychology of art has an uncertain status. Some would place it in the broader subject of esthetics, which has itself been regarded in the past as a branch of philosophy — the philosophy of beauty — but is changing in the direction of descriptive, empirical science. Other branches or specialized interests in modern esthetics deal with the forms and styles of art (esthetic morphology), esthetic value (axiology), and the philosophy of art history. Whereas some of these fields focus on works of art, the psychology of art looks more at the human beings who make and enjoy them, seeking to understand what happens within the mind of the artist and appreciator. But neither artist nor appreciator can be adequately understood in isolation: a sympathetic study of the artist's works may throw light on his personality and motives, and vice versa.

In relation to psychology as a whole, the psychology of art is sometimes regarded as a branch of applied psychology, since it studies how the basic psychophysical functions and abilities are applied in a special, cultural realm — the situations and products of the arts. From this point of view the psychology of art is analogous to the psychology of religion, of education, and of politics. But it has a far wider, deeper significance than many of the other current applications of psychology, such as advertising and propaganda. Some theorists have held that a special ability operates in artistic creation and nowhere else. This theory has declined in favor of the belief that art involves only a special way of developing the same basic powers and processes that operate in all human experience. In any case, some of these powers and processes appear in art more clearly and fully than elsewhere; hence a study of their artistic applications throws light upon their nature in general. The psychology of art deals with well-nigh universal processes of symbolic thinking, creative imagination, visual perception, hedonic tone, and emotional preference. A fully developed system of general psychology might be expected to deal with the more complex and subtle manifestations of thought and feeling, but in fact none has done so adequately; general psychology contains large blank, or almost blank, areas where these higher developments are concerned. The metaphysical accounts of them in the old "philosophy of mind" are largely rejected, but empirical science has not yet filled the gaps with verifiable knowledge. Thus relevant information about these more complex manifestations from the psychology of art and other branches of applied psychology can help illuminate some of the dark corners of general psychology. But it is always hard to distinguish in such information what is comparatively basic and universal (and hence material for general psychology) from what is true only of certain cultures, periods, or social groups.

It was mentioned above that art criticism, including all comments on art by discerning observers, has been a source of insights for the psychology of art. Whereas specialists in esthetics and psychology have often been content with generalizations remote from art as such, those who deal directly with art through designing, performing, admiring, criticizing, or even buying and selling have usually had more concrete experience. When the ability to think and write is added to this experience, the results have provided valuable data and hypotheses for psychology. Of course, these need to be checked and interrelated with other information. Since the late 19th century art critics and historians have been stressing the fact that a style in art is also a "mode of seeing" — of looking at the world and representing it in accord with a total, though local and temporary, world view. Biographies of individual artists today are almost all pervaded by amateur psychologizing, especially in the application of current psychoanalytic and sociological hypotheses. Fallacious as the author's conclusions may be, he can at the same time present useful data for the scientific psychologist to reconsider. Thus criticism, history, and biography in the arts tend to become, when well thought out, both an application of the psychology of art and a source of material for it.

Early discussion often combined theorizing on the psychology of art with theorizing on metaphysics and ethics. This is especially notable in Plato, Plotinus, and St. Thomas Aquinas. They were concerned with the part played by an immortal soul or spirit in artistic creation, with the divine origin of inspiration, and with the good or bad effects of art on moral character. Other philosophers such as Lucretius tried to explain perception, imagination, and art in terms of material atoms without the intervention of an incorporeal spirit. One great change in the psychology of art has been from this metaphysical and moral approach to a more purely psychological one. A considerable shift in that direction occurred in the 18th century in English, German, and French criticism. It appeared in more detailed description of how people actually look at art and listen to music; of how they feel beauty as distinct from sublimity; of the difference between genius and mere technical skill. But the psychological element had also been present in Plato and Aristotle, for example, in discussions of how tragedy, comedy, and simple and complex music affect the superior and the vulgar mind. The historian of psychology must try to distinguish the psychological element at each point from the context of religion and metaphysics. A step toward the psychology of art is made when any theory is advanced as to how music affects the emotions or how the artist gets his inspirations, even if the effects are credited to magic and the sources to Apollo.

Psychological theorizing on art was still mixed with epistemology and metaphysics in the philosophy of Kant and with metaphysics and philosophy of history in Hegel. There was no persistent effort to detach it from these other lines of thought until well along in the 19th century. The metaphysical traditions of idealism and dualism are still strong in European esthetics. In other parts of the world, and in other branches of psychology, naturalism has gained more ground.

The desire of empiricists to eliminate all metaphysical speculation from scientific psychology often hides a tacit naturalism. In ceasing to rely on the concept of an incorporeal soul or spirit as the active agent in artistic imagination it tends to imply the belief that no such agency exists and that all the phenomena of art have a physical basis. It also assumes that the physical and sensuous aspects of art and esthetic experience are worthy of respectful study. These ideas are in sharp conflict with traditional supernaturalism. However, psychologists of different schools — naturalistic, idealistic, and dualistic — have been able to cooperate to some extent in attacking the central problems of esthetic psychology, notably in developing the theories of empathy (*Einfühlung*) and of unconscious symbolism in art; they can agree substantially on the nature of the phenomena while disagreeing on the deeper explanations.

The psychology of art today still has many interests in common with philosophy. For example, psychologists and philosophers of the analytic or logical positivist school share an interest in semantics, in the nature of symbols, signs, and meanings in art, and in the process of critical judgment.

Before the 20th century most discussions of the psychology of art were combined with moral and evaluative considerations and were often dominated by them. The conception of a descriptive, rigorously objective, and factual science such as mathematics, physics, or biology had not yet spread through the social and psychological fields. Comments on art and the artist were made not merely to explain their nature and causation but also to state what kinds were best, what "good taste" implied, rather than the actual varieties of taste. The basic concepts of art and the artist were evaluative, implying high merit. Not all the poems and pictures were considered art, nor were all painters and poets artists — only the best or "true" ones. By the same token, the psychology of art was not an unprejudiced study of the processes involved in certain activities but was restricted to those examples which measured up to accepted standards. Since people disagreed on the standards, no scientific study of art and artists could proceed very far. For example, the psychology of children's drawings and primitive music was generally ignored, since these were not "true art." Tolstoi went so far as to exclude from art those symphonies and paintings which appealed to the more sensual and decadent tastes, including some of the most respected works of Wagner and Michelangelo. If there is any one change that distinguishes the psychology of art as a modern science, it is the determined effort to describe without evaluative prejudice the relevant facts as they appear within a particular area of human experience. This does not mean that evaluation is unimportant or that any kind of art and esthetic experience is as good as any other. It means, first, that such evaluation can best be done as a separate undertaking in ethics and esthetic axiology; second, that much more psychological information is needed as a basis for it.

One of the main turning points in the history of the psychology of art, and indeed of all psychology and esthetics, was the publication in 1876 of Gustav Theodor Fechner's *Vorschule der Ästhetik*. It called for a new approach "from below" (i.e., from empirical observation and experiment) instead of the old one "from above" (i.e., by deduction from metaphysics). It did much to change psychology as a whole from a branch of speculative philosophy ("the philosophy of mind") into an empirical science based on laboratory experiment and statistical measuring. The new method (also called "psychometric") yielded significant information on many problems of limited scope, such as the extent of individual preference for rectangles of different proportions (FIG. 771). But in esthetics and the psychology of art as a whole, the method proved disappointingly superficial, incapable of throwing much light on the central problems of creation and taste. After a few decades of experimental esthetics in Europe and America, its use declined in favor of other approaches with less emphasis on quantitative exactness.

Behaviorism in a broader sense, as applied to esthetics (i.e., the observation of all human activities in art without restriction to quantitative methods or laboratory situations), has not declined along with it. The varieties of behavior, both individual and social, including the external phenomena of taste and art production, are inexhaustible, and hardly a beginning has been made in describing them. But certain fundamental obstacles have so far thwarted the attempt to make the psychology of art into an empirical science, with or without exact measurement. These limitations arise from the fact that so large and important a part of esthetic experience is internal, subjective, and inaccessible to outside observation. Esthetic experience can be directly felt and its objects observed only by the individual involved. This is true of a great deal of human experience and hence applies to all psychology, but it applies with special force to the creation and appreciation of art. Each of these has its inner or subjective phase and its overt or objective phase, only the latter of which is directly observable from without. The inner phase is, by common consent, crucially important — much more so in art than in many other areas of life. Behind the impassive face and quiet body of an artist the decisive steps in creative imagination may proceed. Equally unobservable from without may be the esthetic response of others to his product. What happens in the mind or brain of the artist and the appreciator, often with no discernible sign at the moment, is the central problem of esthetic psychology.

Since Hume, no one hopes to be able to look within and see his own soul or mind at work or to observe that which in an ultimate sense performs the creative or appreciative act. But it is easy to pay attention to one's own thoughts, mental images, sensations, and feelings. These are inner or subjective phenomena. Epistemologists have pointed out that there is a subjective element in our perception of all apparently external objects, and some go on to insist that all such objects are really and completely mental, products of a divine mind if not of individual minds. That philosophical issue is avoided, so far as possible, in modern psychology. The data which the psychology of art investigates can be roughly divided into those which are obviously inner and subjective (including hallucinations, which falsely appear to be external) and those which are commonly regarded as external after careful, sane, wakeful, collective observation. Both types of data can be observed or perceived. The idea of an empirical approach to science can be broadly conceived so as to include introspection or autoperception as well as external observation through the senses. The psychology of art has always relied on both approaches to some extent, except in the case of extreme behaviorism, which rejects introspection as completely unreliable. Since the advent of depth psychology everyone recognizes the difficulty of observing and describing one's own inner mental life with any accuracy. Much of it is unconscious. The would-be observer is always emotionally biased in what he thinks he sees within himself, and still more in how he reports it to the public. Nevertheless there is reason to hope that the process of observing, reporting, and comparing the subjective experiences of individuals can be made increasingly reliable through scientific safeguards. In this hope a variety of techniques for interpreting and correcting introspective data have been elaborated. They are used to supplement the data collected through external observation of behavior and expression in situations related to the arts. The primary aim of the psychologist in interpreting both sorts of data is to discern in them significant recurrences and patterns. Generalizations are sought on the correlation between inner and outer events to serve as principles of causal explanation, prediction, and control. As in other behavioral sciences, however, it is recognized that the variables involved are too numerous and irregular to be reduced to absolute universal laws.

The difficulty of using introspection as a scientific method is increased by the fact that what appears to be inner and subjective is intimately blended in experience itself with what appears to be outer and objective. This is especially true in the experience of art, where there is a frequent shifting of the focus of attention between the sensory object and the inner play of imagination. Inner feelings, beliefs, and attitudes are constantly projected upon a sense object such as a work of art and regarded as among its integral parts or qualities. In analyzing a work of art it is hard to distinguish between (*a*) what is really in and of the object and (*b*) what is contributed by the observer. Furthermore, that very confusion, that lack of self-conscious awareness of the differences between oneself and the object, is prized as an invaluable element in creative inspiration and esthetic enjoyment. On the other hand, the development of such clear distinctions is indispensable to philosophical or scientific understanding. Persons of a scientific temperament tend to seek them, while those of a predominantly artistic or esthetic temperament tend to avoid them.

Before the 19th century philosophical interest in the epistemological problem usually involved an emphasis on the individual aspects of art and esthetic experience: on the ego as a perceiving, knowing, feeling, desiring mind in relation to the external object and the world. The individual beholder and judge of beauty was then conceived, as by Kant, in somewhat abstract and absolutistic terms — not with attention to the actual variations of individual taste and personality but in terms of an ideal, universal type which stood for all humanity. In the 19th century, however, the social and evolutionary approach developed in the psychology of art as in general esthetics, art history, and related fields. Anthropology and cultural history turned away from the lonely individual and his inner life to interpret the objective data of primitive artifacts and the role of art in various cultural settings. This brought a great increase of empirical knowledge and theory about the overt, collective aspects of art. Much of it was regarded as marginal or extraneous to the psychology of art by those who still emphasized the individual, subjective element. But much was relevant to psychology in a broader sense, for example, Darwin's theory that man's emotional life and expression develop out of animal origins; Taine's theory of the mental and emotional climate that, along with physical climate, helps to determine artistic styles; and the Marx-Engels theory that socioeconomic factors determine styles and tastes in every epoch. In the early 20th century greater attention was paid to the actual varieties of individual personality in relation to art: by the sociologists, to the kinds of individuals fostered by different cultural patterns; and by the psychoanalysts, to specific types of individuals resulting from unconscious motivation and conflict in the growth of personality. From the attempt to conceive a single type of ideal individual such as "the artist" or "the connoisseur" and a single type of ideal experience such as "the sense of beauty" the emphasis shifted to the analysis of many varieties of individuals and processes, no one of which was taken as necessarily best or universal.

All modern subjects, especially in the humanities, are now so permeated by the psychological point of view that it is often hard to say where psychology ends and the others begin. There are books on the psychology of philosophers, of literary criticism, of mathematics, of law, and of military strategy. The psychological element in such studies is often unscientific but far-reaching, especially in trying to show how different types of persons in different cultures tend to think and how their thoughts and acts are influenced by unconscious motivation. Most contemporary discussion of the arts is psychological in this broad sense. But it would be confusing and inexpedient to enlarge the definition of the psychology of art to include all its ramifications in esthetics, art history, and social science. Hence it is well to emphasize the central area both in defining it and in tracing its origins — that is, the study of the creative and appreciative processes in art, with some attention to both inner and outer, individual and social phases. The "outer" includes consideration of the techniques, materials, and forms of artistic production, while the "social" includes the study of art as an evolving institution. But the psychology of art, in a strict sense, still tends to focus mainly on the inner, the individual, and the comparatively basic universal aspects of the process and to organize its materials accordingly.

Psychological speculations in ancient poetry and philosophy. The beginning of philosophic inquiry about man and his works is customarily dated from the time of Socrates. There are only fragmentary records of the earlier philosophers, most of whom seem to have been more interested in physical and mathematical problems than in those of human nature. Nevertheless, the arts were important enough in prehistoric and early historic times to make it certain people wondered about them. The origin of psychological theorizing about the nature of the arts can perhaps be traced to the widespread primitive belief in the supernatural power and inspiration of the artist. It was thought that some artists, like shamans, priests, and oracles, could be divinely inspired; their artifacts, chants, and dances could have magical power for good or ill. The right kind of verse or amulet could avert evil spirits. The right kind of statue, temple, or ritual drama could please the gods, who apparently preferred some to others; this assumed some esthetic taste on their part.

Said Hesiod, "By grace of the Muses and archer Apollo are men minstrels upon the earth and players of the lyre" (*Theogony*, 85). This refers not only to the origins of art as a divine gift to man but also to the source of present inspiration. Both Homeric epics begin with an invocation to the Muse, and this became common practice. Implicit in the religious explanation is a psychological observation: artists often feel and act as if their best creative work were directed by some outside power over which they have little control; it is not consciously willed and planned from within. To explain this is still a scientific problem, on which some light has recently been thrown by depth psychology. Lucretius, on behalf of the Epicurean naturalists, followed poetic convention by invoking a goddess as the creative power of the universe but later made it clear that he did not seriously believe in the divine gift of language and the arts. Man had originated them himself by accident, reason, skill, and industry (*De rerum natura*, v).

Besides the nature of artistic creation, the other basic problem in the psychology of art was that of appreciation — the effects of different kinds of art on the observer. Here again ancient poetry foreshadowed science in pointing out a psychological fact: the power of music over the emotions. In his *Ode to King Hiero of Etna*, Pindar sang: "O lyre of gold, Apollo's treasure, shared with the violet-wreathed Muse . . . , you quench the warrior Thunderbolt's everlasting flame; on god's scepter the eagle sleeps; . . . his eyelids are shut with a sweet seal . . . . Your quivering song has conquered him. Even Ares the violent, leaving aside his harsh and pointed spears, comforts his heart in drowsiness."

In the Greek maxim "Know thyself" lies the germ of all psychology. Socrates and Plato developed it along partly rationalistic lines, though not without poetic figures. In his *Republic* Plato declares, from the standpoint of his idealistic metaphysics; "Musical training is a more potent instrument than any other because rhythm and harmony find their way into the inward places of the soul" (III, 401 D.). Simple, stately, old-fashioned music helps to produce a self-controlled and spirited youth; soft Lydian airs and complex rhythms tend to soften and disintegrate character. In explaining the creative side of art, Plato's rationalism is counterbalanced by a mystical theory of divine inspiration. "A third kind of possession and madness," he says in the *Phaedrus*, "comes from the Muses. This takes hold upon a gentle and pure soul, arouses it and inspires it to songs and other poetry, and thus by adorning countless deeds of the ancients educates later generations. But he who without the divine madness comes to the doors of the Muses, confident that he will be a good poet by artistic skill, meets with no success, and the poetry of the sane man vanishes into nothingness before that of the inspired madmen." Technical skill alone, we still agree, is not enough to make a creative artist; but how can the concept of divine madness be translated into modern psychological terms? Modern criticism would also agree with Plato that exact imitation or representation of the sensory appearances of things, including mimicry of characters on the stage, is not the most important kind of art. Some would favor instead abstract, geometric designs, at which he seems to hint in the *Philebus*. Naturalists reject his idealistic theory of true artistic creation as based on awareness of the eternal universal forms of perfect beauty through sudden inspiration or remembrance of having known them in heaven before birth.

Aristotle describes appreciation in a more empirical way by pointing out that each type of art (e.g., tragedy and comedy) can arouse a special type of pleasure. A work of art can succeed or fail as a means to that end, and Aristotle shows what in his opinion are the requirements for success — for example, the types of characters and plot in a tragedy. He conceives of art, on the whole, as a device for producing certain psychological effects on the beholder, not as a mode of expression having intrinsic value. His conception has since been disputed by those who deny that a work of art is essentially a means to an end or that its value depends on how it affects observers. As to the effects of tragedy, Aristotle further specifies that it has "incidents arousing pity and fear, wherewith to accomplish the purgation of such emotions" (*Poetics*, 6, 1449b). Scholars still debate the exact meaning of "purgation," but the philosopher probably had in mind the power of poetry to bring a sense of peace and rational self-control through beholding the tragic events as necessary parts of an all-inclusive moral and artistic order.

Lucretius (1st cent. B.C.) is close to modern psychology in reporting a specific phenomenon (visual perspective) in a direct, empirical way. He is not much concerned with esthetics, but what he has to say of sensation in general is relevant to the perception of visual art: "Though a colonnade runs on straight-set lines all the way and stands resting on equal columns from end to end, yet when its whole length is seen from the top end, little by little it contracts to the pointed head of a narrow cone, joining roof with floor, and all the right hand with the left, until it has brought all together into the point of a cone that passes out of sight" (op. cit., IV, 25).

In Plotinus (3d cent. of our era) the empirical and rationalistic elements are overwhelmed by a mystical religiosity. There is less description of the psychological phenomena of art and more metaphysical theorizing about the cosmic role of art as transmitting the divine radiance and elevating the mind to a knowledge and love of spiritual beauty [see especially *Enneads*, I, 3 (Dialectic), 6 (Beauty)]. St. Augustine, also much influenced by Platonism, similarly emphasizes the moral and religious aspects of art, especially the danger of temptation from sensuous beauty as manifested in music and the visual arts. His *Confessions* include acute self-observation on his experiences with art, toward which he was sensitive and scholarly, especially as to the extent to which music tended to reinforce or distract from the meaning of religious texts (X, 33:50). The influence of St. Augustine on theories of art remained strong throughout the Middle Ages. Through it the mystical, idealistic, ascetic side of Platonism was transmitted. Toward the end of the period the influence of Aristotle gradually exerted itself through St. Thomas Aquinas and others, leading in time to a more rationalistic approach. Meanwhile Gothic art became more naturalistic in daring to represent physical nature with greater sympathy and accuracy; but naturalistic theory was much later in reviving.

Renaissance, baroque, Enlightenment, and early romantic theories. Neoplatonism was a powerful influence on esthetic theory during the Renaissance and on into the 18th century. In Italy, Girolamo Fracastoro (*Naugerius*, or *De poetica dialogus*) and Marsilio Ficino (*Opera*, Basel, 1561, 1756; especially his commentary on Plato's *Symposium*) labored to reconcile the ever-growing naturalism and humanism of art and life with the Platonic ideal of perfect unchanging spiritual beauty which could raise men's minds from sensuous nature to the world above. Where lay the artist's task between these competing values? Rationalists looked to Aristotle and Horace (*Ars poetica*) for a set of rules to guide the artist: his work, they reasoned, need not await the inspiration of divine madness, for classical art and philosophy offered models and rules. Writers differed on how binding these were upon the modern artist; how much freedom he had to follow his own genius. In this controversy psychological conceptions of the creative process were still mixed with moral and esthetic evaluation. In England, Lord Shaftesbury (1671–1713; *Characteristics of Men, Manners, Opinions, Times*; *Letters to a Young Man*) and Francis Hutcheson (1694–1746; *An Inquiry into the Original of Our Ideas of Beauty and Virtue*) carried on the Platonic approach to esthetics. Shaftesbury's concept of an "inner sense" of beauty and goodness implied a spiritual harmony between the superior mind and the divine harmony of the universe. Descartes, Pascal, Spinoza, Leibniz, Bacon, Hobbes, and Locke all discussed the imagination in relation to reason, sense, and passion — some from the standpoint of metaphysical dualism, some from pantheism, and some from materialism. They referred occasionally to the use of these faculties in art; but on the whole the arts and the entire esthetic of life were a minor concern to the great baroque philosophers. Many were inclined to fear and disparage fantasy and emotion unless these were thoroughly controlled by reason.

John Locke paved the way for revolutionary thinking in the 18th century; the full impact of his liberalism and empiricism was not felt until long after his death. His empiricist theory of knowledge undermined the Platonic and Cartesian belief in innate ideas, including the "recollection" of heavenly perfection. It tended to raise sensation in esteem, not only as the sole source of true knowledge but also as the chief source of pleasure and pain, hence of moral and esthetic value. David Hume developed this empiricist, hedonist approach in epistemology, ethics, and esthetics (especially in his essay "The Standard of Taste"). In psychology it became the "associationist" theory, which sought to explain imagination and knowledge by showing how small units of sensation (impression, images, or ideas) rearranged themselves into complex forms in the mind through similarity, contiguity, and other "laws" of association. The individual mind and its processes, including those of art, were subjected to intensive analysis. Transcendental metaphysics almost faded from the picture for a time in England; except for the idealism of Bishop Berkeley there was little tendency to project mind, art, and beauty into the surrounding universe as realities of cosmic importance.

English, French, and German theories of criticism in the 18th century became strongly psychological in tone. While still concerned with value, they also introduced a great deal of factual observation, partly in characterizing various types and styles of art by their psychological effects (e.g., Winckelmann's description of Greek art as nobly simple, reserved, and quietly dignified). Lessing commented on the effects of emotional expression and of represented motion in painting and sculpture (*Laokoön*, 1766). In part the psychological approach was directed into detailed observation of the creative and appreciative processes without an attempt to explain their ultimate, metaphysical basis. Joseph Addison (1672–1719) made a notable contribution along this line with his papers *The Pleasures of the Imagination*, in which he explored the difference between primary (directly sensed) and secondary (remembered and rearranged) images. Pleasant imagining, he pointed out, results from three sets of qualities: greatness or sublimity; novelty, surprise, or uncommonness; and beauty, from the harmonious adaptation of parts to wholes. Beauty was thus not the only value in art. The mental process of comparing an original object with artistic imitations is itself pleasant, especially when the imitation is suggestive rather than exact. Addison's work stimulated further inquiries in England and on the Continent, notably Edmund Burke's essay *On the Sublime and Beautiful* (1756). Burke emphasized passion and the social instinct rather than reason, thus moving toward romanticism. Feelings of sublimity he traced to the instinct of self-preservation in confronting pain or danger; feelings of sympathy help explain our pleasure in tragedy; ambition or emulation leads us to admire the sublime; the pleasure of imitation makes painting and poetry agreeable. Blake, Shelley, Wordsworth, Coleridge, and Hazlitt all glorified the imagination over cold, calculating reason as a source of vital richness in the arts. But all of them, especially Coleridge in his later years, felt drawn toward intellectual analysis in the effort to understand as well as to create in art. Coleridge made a careful distinction between imagination and fancy in poetry. The one achieves a joint effect by stressing and integrating as many as possible of the characteristics of each part; the other selects only a few of them or admits irrelevant discrepancies for a mixed effect such as burlesque. Imagination is more capable of expressing a complex, integrated view of life and the world (see PL. 370).

Among those stimulated by English psychological criticism was Kant. His discussion of sublimity was indebted to Addison and Burke, while his concept of esthetic pleasure as "disinterested" had been anticipated by Shaftesbury, Hutcheson, and Lord Kames (1696–1782). But Kant criticized the psychological approach to esthetics as inconclusive. He had little respect for a merely empirical description of the varieties of taste and imagination. His own experience of the arts was very limited and his theoretical classification of them was empirical rather than transcendental. His main concern was to discover an a priori condition or sanction for esthetic judgments — not how men actually judge taste, but how they *should* judge it. A judgment of taste sets itself up as universally valid, he says; we demand of all men that they agree with us when we esteem something beautiful. Psychologically this generalization is doubtful if not obviously false; yet from it Kant proceeded to set up an a priori standard of taste. His vast influence on subsequent philosophy, including esthetics, can hardly be overestimated. Its immediate effect was to discourage the growing interest in a descriptive, empirical psychology of art as applied to creation and taste and to the interpretation of different kinds of art. It helped to turn esthetic theory once more along metaphysical and moral lines, thus separating it throughout most of the 19th century from psychological, sociological, and morphological studies of art and artists. This deflection is not due to Kant alone, however. The new German variety Platonic idealism formulated by Kant and Schelling found theoretical support in Coleridge and was akin to Wordsworth's lyrical pantheism. Early romantic thought throughout Europe was turning away from science and neoclassical rationalism and was ready to be impressed, for a time, by the monumental structures of German romantic idealism. Toward these the 18th century had already contributed another essential ingredient: the concepts of historical development — of progress and evolution — which were to reach fruition in the next century.

The 19th century. Strictly speaking, neither the criticism of Addison nor the romantic metaphysics of Schiller and Hegel belongs to the science of psychology. But there is an ingredient of prescientific psychology in both. British criticism, while literary in character, is closer to 20th-century psychology because of its prevailing empiricism, German metaphysics is farther away because of its heavy reliance upon hypothetical entities such as the "world mind" and "cosmic will," which can not be experimentally verified; the same is true of Bergson's vitalism, with its incorporeal *élan vital*. From the standpoint of naturalistic science, transcendental idealism is an imaginary projection of the human mind from its only real habitat, in living human bodies, throughout the universe at large, there to receive its apotheosis, equally fictitious, as the directive force of cosmic history. From this divine status the mind is made to return, through the radiance of Plotinus's sun and equivalent modern devices, to inspire the creative human artist. So conceived, this is again a hypothesis for the explanation of psychological phenomena: how the artist imagines and how the sensitive beholder responds to his expressed imaginings.

For Schiller (1759–1805) the essence of art is a kind of play characterized by freedom, as distinguished from the coercions of physical fact and law which operate in necessary work: art helps to reconcile the material and spiritual phases of life; the artist plays in creating new verbal, musical, or visual forms. In Hegel's system the "world mind" evolves toward ever-greater clarity and detail in rational understanding; successive styles of art express successive stages in world civilization and in the development of the world mind; an individual artist or thinker is formed and limited by the general stage of advancement in each age; progress is mainly rational, and philosophy may replace art. Schopenhauer, on the other hand, conceived the world spirit as essentially will, not reason; man is the slave of his irrational desires, but art (especially music) helps him to a partial escape at times. Friedrich Nietzsche (1844–1900), through his studies of classical drama, advanced the polar concepts of Apollonian and Dionysian art: the one is characterized by clarity, distinctness, measure, and restraint; the other, more primitive, by a passionate melting away of distinctions in orgiastic excitement, as in drunkenness. These concepts were partly reconciled in Greek tragedy but recur in every age; one or the other tends to dominate in different artists, styles, and works. While oversimplified, this antithesis has been suggestive in characterizing individual personalities and historic styles. The emphasis placed by late romantic writers on the irrational and unconscious in life and on its productive role in art was one of the lines of thought which eventually led to Freud, Jung, and depth psychology in the early 20th century.

The Platonic tradition had given art a status of power and dignity in the universe and in the social life of man. In the more restrained analyses of 18th-century criticism it had sometimes been reduced to a leisure-time amusement for the gentleman of means. German philosophy stressed again the moral and social aspects of art as a major factor in the history of civilization. This last had also been done in a more naturalistic way by the French encyclopedists and social philosophers from Diderot to Condorcet, who saw in art a means and manifestation of unaided human progress. Auguste Comte carried on in early 19th-century France the empiricist or positivist approach to social philosophy, rejecting supernaturalist religion and metaphysics as prescientific. His recognition of the need for art to satisfy the emotional and imaginative side of man, even in an age of scientific humanism, reserved a place for esthetic theory in the new science of sociology. Though overshadowed by idealism in the 1800s, the naturalistic approach to esthetics was gradually reinforced by new developments in other fields. It led Herbert Spencer in England to a vast philosophy of history, conceived as the working out of a universal natural law of evolution, in which the evolution of the arts had a place. Spencer showed that more complex and definite types of art, along with increasing differentiation of the arts and forms of art, was an integral part of cosmic and social evolution. Across the channel, Taine (*Philosophie de L'Art*, 1865) advanced his theory of a psychological climate in each culture epoch that determines national and period styles in art; this climate results from the interaction of hereditary race, environment, and the momentary configuration of historical circumstances. Meanwhile Karl Marx, Friedrich Engels, and their followers proposed another theory of art as socially determined; the artist's personality and style were seen chiefly as expressions of his socioeconomic class and situation.

During the mid-19th century anthropology was also accumulating masses of concrete evidence on the origins and early stages of cultural history (including the arts), on the relations between prehistoric and modern primitives, and on the role of art in cultural evolution. Anthropologists Lewis Henry Morgan and Edmund Tylor elaborated their somewhat oversimplified unilinear theories of cultural evolution, and James Frazer advanced a naturalistic account of the origins of art and religion in primitive society. All these lines of investigation, while extending far beyond the strict limits of psychology, had bearings on the psychology of art. They helped to weaken the Platonic-romantic conception of the artist and his work as purely spiritual, above and apart from nature and the physical world, and strengthened the view of art as intimately bound up with the struggle for existence and the animal, physical basis of life. This conception, like evolutionism in general, seemed to some to demean and devaluate art, as did psychoanalysis to others in the following century. But still others chose to emphasize the opposite aspect: that man had arisen from primitive misery and ignorance, had achieved some measure of freedom and enlightenment with the aid of art, and might go on to more.

During the mid-19th century, still broader foundations for a scientific psychology of art were laid in the fields of physiology and general psychology. Hermann Helmholtz, Wilhelm Wundt, Oswald Külpe, and others examined the physical basis of visual and auditory perception and also experimented with the active processes of sense perception. Like Grant Allen in England, they sought to discover how the rudimentary sensation and hedonic tones of animal and infantile experience could develop into the subtle, complex perceptions and esthetic emotions of civilized art. Following Fechner, they emphasized controlled experiment and quantitative measurement.

Historians of the arts in the late 19th century, some naturalistic or positivistic and some idealistic or vitalistic, began approaching their subject matter from a more psychological point of view. Jakob Burckhardt, Alois Riegl, Konrad Fiedler, and, after 1900, Heinrich Wölfflin built up the new school of historic thought which came to be described as *Kunstwissenschaft*, or "science of art." As distinct from earlier art history and criticism, it sought to eliminate judgments of value on the whole: that is, to cease argument over the respective merits of Byzantine and Gothic art and to understand them as different modes of seeing. (Taine had already written along a similar line.) Historians now sought to discern a diversified but self-consistent way of perceiving, feeling, and interpreting the worlds of nature and art in each culture epoch. This would help them understand "the spirit of the styles." Some historians explained this spirit in terms of a supernatural *Zeitgeist* or of a "will to form," and other in a more naturalistic way.

Likewise open to different metaphysical interpretation was the 19th-century conception of *Entfühlung* (empathy), which helped to throw much-needed light on the processes involved in esthetic experience. In brief, it is the theory that an observer, when in the "esthetic attitude" of contemplation, tends to project his feelings into a work of art or another suitable object and to find a peculiar kind of satisfaction therein (see PL. 371). If the work is complex, such as a cathedral or a tragedy, he can imagine himself as living and moving through it. This constitutes a sort of mental play. It was not a new discovery that the mind tends to project some of its experiences into perceived objects, regarding these experiences as inherent attributes of the latter; something of this is involved, as Locke pointed out in the attribution of secondary qualities such as color, to a material thing. It is not a new idea that beauty is in the eye of the beholder and that people tend to attribute their own feelings to other humans or even to inanimate objects such as a "cheerful fire." But only in the 19th century was the extensive, important role of empathy in esthetic responses to art and nature described in detail. We tend to "feel ourselves" kinesthetically into the flight and thrust of Gothic vaulting and the onward rush of music. In imagination we live and move through the scenes of a drama, sometimes forgetting ourselves entirely. The term *Einfühlung* was applied to this process by Vischer (1873), and the phenomena were further described by Theodor Lipps (1891), Johannes Volkelt, Vernon Lee, and others.

The early 20th century. In the decades before the World War I and in the interval before World War II significant work was done along all the lines just reviewed. In addition, psychoanalysis or depth psychology contributed a revolutionary scientific breakthrough into unconscious and preconscious levels of creative and appreciative experience. The atomistic approach of associationist psychology and the overreliance on innate instincts as a principle of explanation were corrected by the Gestalt approach (see below). This was productive in esthetics since it stressed the complex total response to a complex situation, as in experiencing art. A great deal of original experimental work was done in the psychology of children's art, especially drawing. The development of a number of artists from infancy to professional maturity was traced with sympathetic patience, not only for the sake of knowledge but also as a guide to fostering creativeness through the right kind of education. Some theorists held to the "recapitulation" view, that the stages in a child's artistic development correspond roughly to those of the development of art in cultural history; but this was not widely accepted. Toward the middle of the 20th century more was learned about Oriental esthetics, especially those of India, including such psychological principles as *rasa*, or esthetic quality. Centuries ago, Abhinavagupta and other philosophers of art had subtly distinguished a variety of moods and feelings in response to the arts, along with corresponding qualities in art.

Two world wars and the continuing threat of a third, together with social and economic maladjustments, have seriously retarded work in such "nonessential" fields as esthetic psychology. Germany, formerly in the advance, has contributed little. However, several important synthetic works appeared in the twenties and thirties. In the United States Chandler (1934) emphasized the laboratory, or psychometric, approach. Several came from Germany, most notably those by Dessoir (1923), Müller-Freienfels (1923), Plaut (1935), and Sterzinger (1938); the last of these emphazises the psychological aspects of form in the arts, whereas Müller-Freienfels and Plaut emphasize the processes of creation and appreciation and the personality of artists.

Thomas MUNRO

EXPERIMENTAL PSYCHOLOGY. Art and science are generally regarded as being quite distinct, and the notion that esthetic appreciation could itself become the subject of scientific investigation is of relatively recent origin, having first been put forward by Fechner (1876), as we have seen. The purpose of experimental esthetics has been well described by Sander (1913): "Experimental esthetics, because of the twofold dependence of the esthetic feeling on subjective and objective conditions, can pursue a twofold aim. On the one hand it can put its main emphasis on the investigation of the objective side of the esthetic experience. Its task will then be to investigate all kinds of esthetically effective *Gestalten,* singly or in combination, with respect to their quantitative characteristics, by presenting these objects to a very large number of subjects in extended quantitative differentiations and thus determine their esthetic effectiveness.

"The aim would be to discover, from the infinite manifold of possible combinations, certain quantitative peculiarities and proportions which are marked out from all others through their particular effectiveness. These proportions should then, if possible, be expressed mathematically, and opposed, as objectively and universally valid factors of beauty, to the infinitely extended field of subjectively varying factors.

"The other aim would be to regard the objects as stimuli whose task it is to produce that esthetic experience whose analysis would then be the main problem. The definition of the particular feeling called 'esthetic experience' . . . would then be the main task of this part of experimental esthetics."

The logic of experimental esthetics. Most experimental work has been related to the first of these two projects, and so little has been done with respect to the second that there would be little point in reviewing such studies as have been carried out in the latter category. Before turning to an actual survey of the evidence, we must anticipate one frequent objection to studies of this kind, namely, that esthetic judgments are entirely subjective and consequently are not to be made the basis of any rule or law. The experimentalist will answer by going back to Kant (1790), who claimed that the judgment "this is beautiful" was universal and necessary, inasmuch as "it implies that every normal spectator must acknowledge its validity"; in other words, Kant held that esthetic judgments must show agreement because beauty is objective. The experimentalist would accept this but turn it around and say that esthetic judgments are objective to the degree that they reflect agreement between judges. (It is assumed, of course, that nonesthetic bases of judgment have been carefully excluded, either experimentally or statistically.) Thus Kant's a priori judgment, from which agreement among spectators is logically derived, becomes an experimental problem; we now ask about the degree of agreement between judges and make no a priori assumptions. Philosophical subjectivists and objectivists assume without experiment that there does or does not exist agreement among judges with respect to esthetic merit; the experimental psychologist asks, instead, precisely how much agreement can be found in actual fact, what properties of the stimulus produce favorable and unfavorable responses, what part training, heredity, sex, intelligence, and other factors play in determining judgments, and other questions of a similar kind.

Apart from asking questions which can be answered experimentally, the experimental psychologist has made two major contributions: In the first place, he has abandoned the usual habit among philosophers of regarding agreement as either perfect or nonexistent. By recognizing that there are intermediate degrees of agreement and by developing mathematical formulas to express these degrees of correlations in terms of indices ranging from +1 (perfect agreement) through 0 (no agreement at all) to −1 (perfect disagreement), he has refined the measurement of agreement to an extent which makes appropriate mathematical treatment of experimental results possible.

The second contribution has been that of developing experimental methods for the study of preference judgments which enable the psychologist to control many factors that everyday observation leaves uncontrolled. There are three main methods in this field, the first being that of "absolute judgment." Here the object to be judged is shown to the subject, or judge, who must rate it for esthetic merit on a predetermined scale having three, five, seven, or more steps. (The meaning of these steps has, of course, been carefully explained to him previously.) This method is most like that used in ordinary life and is probably the least valuable. The second method is that of "relative judgment," which appears in two forms: The first is the so-called "ranking method," in which all objects to be judged are presented simultaneously and ranked by the subject in order from the most pleasant to the least pleasant (or most displeasing). The other form is that of "paired comparison," in which all the objects to be judged are presented two at a time and the subject is required in each case to say which of the two he prefers. (It is, of course, essential that all the possible combinations should be judged.) These two methods have certain advantages and disadvantages which are discussed by Woodworth (1939); they usually give very similar results, however.

The last method to be described is that of "adjustment." In this the subject physically changes the stimulus until he has found a setting which gives him the greatest esthetic satisfaction. Thus in judging the esthetic merits of different rectangles the subject may adjust the length of the rectangle while keeping its width constant until he reaches the proportion he likes best. This method is in many ways preferable to the others, but of course it can only be used with very simple stimuli and would be useless for more complex works of art. It has therefore been used less extensively than the other methods mentioned.

In developing our discussion of the use of experimental methods in esthetics, we shall first deal with simple shapes and colors and later proceed to more complex esthetic objects. The question sometimes raised as to whether simple judgments of single colors or proportions can justifiably be called esthetic cannot be answered on a priori grounds; however, experimental evidence appears to favor an affirmative conclusion, and the point will be discussed later on.

The esthetics of simple forms. Beginning with Pythagoras and his followers there has been much argument about the most pleasing proportions of rectangles. This prompted Fechner (1876) to start his experimental work by placing rectangles of different proportions before subjects, asking them to indicate the most and the least pleasing. He was particularly interested in the claims made for the golden section, in which the whole is to the larger part as the larger is to the smaller ($1 : x = x : 1 - x = .618$). In a rectangle the golden section is obtained by making the width .618 times the length, and it had been maintained that this gave the ideal proportion by best presenting "unity in diversity." Results obtained by Fechner in 1876 and by the French psychologist Lalo in 1908 are shown in Table 1. It will be seen that the golden section is the most preferred ratio; but even so, only about 1/3 of the subjects preferred it to other ratios. Indeed, there appears to be a whole band of values from .57 to .67 which are almost equally attractive, and there also appears to be a second center of preference for squares (ratio of 1). Others who obtained similar results were Witmer (1894), Thorndike (1917), Weber (1931), Haines and Davies (1904), and Davis (1933).

Some of these results are presented in FIG. 771. Number I shows data from Thorndike's study, in which each of 200 subjects ranked the rectangles shown at the top of the graph; these were presented all at once but with better spacing than is shown here. Number II shows data from Weber's study, in which the rectangles were projected on a screen, the long side vertical. Sixty-eight female college students took part, and the method of paired comparison was used. Weber was especially interested in testing a theory advanced by Hambidge (1921), according to which rectangles would have the most pleasing shapes when the width was to the length as 1 is to the square root of 2, the square root of 3, and so forth. Weber was also interested in the effects of practice, and it will be seen that on the second occasion of judging there was a definite shift of preferences toward lower ratios of width to length. Number III in the same figure shows results obtained by Haines and Davies using the adjust-

TABLE 1. *Preference for Rectangles*

Ratio of width to length	*Best rectangle, per cent* Fechner	Lalo	*Worst rectangle, per cent* Fechner	Lalo
1.00	3.0	11.7	27.8	22.5
.83	.2	1.0	19.7	16.6
.80	2.0	1.3	9.4	9.1
.75	2.5	9.5	2.5	9.1
.69	7.7	5.6	1.2	2.5
.67	20.6	11.0	.4	.6
.62	35.0	30.3	0	0
.57	20.0	6.3	.8	.6
.50	7.5	8.0	2.5	12.5
.40	1.5	15.3	35.7	26.6
	100.0	100.0	100.0	100.1

ment method. The experiment was carried out on 23 students, each having four or more trials; each dot records one trial.

In testing esthetic preference Fechner also used other figures beside rectangles, in particular ellipses, and he was followed in this by some of the other writers mentioned. Still other experimenters were interested in the concept of balance and asked their subjects to divide a horizontal line into two parts in such a way that the result would be most pleasant (Angier, 1903; Pierce, 1894, 1896; Puffer, 1903; Legowski, 1908). Most of these workers found unequal divisions preferred to equal divisions, and quite generally their work as well as that already described shows a certain amount of agreement, while falling far short of perfect coincidence of judgments.

The esthetics of simple colors. A great deal has been done since the appearance of a paper by Cohn in 1894 on preferences for simple colors; references will be found in the book by Chandler (1934) and the bibliography by Chandler and Barnhart (1938). A review of this work, together with some independent experiments, has been made by Eysenck (1941), who summarized results from previous workers in many countries. Table 2 and Fig. 771, No. IV, indicate the main conclusions of this work, namely, that there is considerable agreement among people of all races and colors and both sexes. Two further points emerge: First, it is found that preference for any color varies inversely with the luminosity factor of that color. (Brightness at any wave length relative to that at 550 A.U. along the equal energy spectrum is known as the luminosity factor; cf. Murray and Spencer, 1939, for the meaning of these terms.) Secondly, there is a direct relation between preference for a color and its differentiation from white, as shown by the minimum amount of spectral color that must be added to the test field before it is seen to differ from white.

TABLE 2. *Average Rankings of Color Preferences Obtained in the Various Experiments*

Color	*White subjects* (12,175)	*Colored subjects* (8885)	*Weighted total* (21,060)
blue	1.12	1.83	1.42
red	2.32	2.03	2.20
green	3.32	2.98	3.18
violet	3.66	4.28	3.92
orange	5.30	4.76	5.07
yellow	5.28	5.12	5.21

Much of the work reviewed by Eysenck was technically imperfect. In particular, little care had been taken to separate out and control for the different dimensions into which color sensations could be divided. As is well known, surface colors have three main attributes: hue (the quality that distinguishes one color from another), brightness (the amount of lightness or darkness), and saturation (strength or weakness). A. H. Munsell (cf. Nickerson and Lentall, 1943) has produced relatively pure samples of surface colors showing various values of these three principal attributes, which he prefers to call hue, value, and chroma. They are represented in his system by the coordinate axes of a cylindrical diagram in which value is indicated by altitude on a vertical axis, chroma by horizontal distance from the axis, and hue by angle relative to the axis (see FIG. 771, Nos. V and VI). Each dimension is intended to form a scale of perceptually equidistant space. The Munsell color chips were used by Granger (1955) in a study in which he carefully controlled any two of these factors while investigating the third. Working with 50 subjects of normal color vision, and also controlling very carefully the illumination in which the color chips were inspected as well as the background against which they were seen, he found: (*a*) There is a general order of preference for each of the three attributes of color (in other words, people agree in their judgments on the esthetic value of all the different properties of color). (*b*) A person who on one of the tests agrees strongly with the mean order derived from the whole group also tends to agree strongly on any of the other tests with the mean order of the whole group. If we can regard such agreement as evidence of "good taste," then the experiment presents evidence that a subject's "good taste" is shown in relation to all the properties of color stimuli. (*c*) There are no sex differences in color preferences.

There is much agreement on the fact that with respect to colors judges tend to agree with each other on esthetic properties. If this is so, then the mean judgments of groups of observers have the interesting property of becoming a standard against which individual judgments can be evaluated. A first attempt to prepare such standards was made by Guilford (1934, 1939, 1940, 1949), who obtained affective judgments of colors at different chroma and tint levels. He then prepared charts (see FIG. 771, No. VII) containing lines which he called "isohedons" (lines of equal pleasure) by analogy with weather maps; each chart, representing a constant hue, shows lines of equal affective value in steps of one-half unit. Guilford's work reaches degrees of complexity which make it unsuitable for fuller description here, but the underlying idea will be readily grasped by anyone studying the figure.

The esthetics of color combinations. Color combinations also have been frequently studied, as is shown in the bibliography of Chandler and Barnhart (1938). The most recent work, as well as the best, is that of Granger (1955), whose experiments are outstanding for the careful control of all relevant variables. He found that preferences for color combinations for any given person are highly predictable in terms of (*a*) that person's relative liking for the individual colors making up the combinations and (*b*) certain general laws relating to the properties of the colors making up the combination. Thus preference tends to increase for an increasing size of hue interval and to decrease with increasing size of saturation interval. He also found that people who showed "good taste" as here defined in one test would also show it in other tests; he found as well that this "good taste" extended beyond the field of judgment of color combinations to that of single colors. (In other words, a person who agrees with the average in his judgment of single colors will also agree with the average in his judgment of color combinations.)

These findings suggested the possibility of deriving an empirical formula for the prediction of preference for binary combinations of colors. This formula included two sets of factors: (*a*) the preference for the component colors making up the combination and (*b*) the relational term, dependent on the combination as such. When this formula was applied by Granger to new groups of subjects it was found to give excellent prediction, suggesting that the possibility of predicting reactions to more complex stimuli from knowledge about the esthetic properties of simple stimuli cannot be ruled out.

An esthetic formula. It has been seen that experimental work supports the notion that it may be possible to express the esthetic value of objects in terms of a formula. Many such attempts have been made by Rashevsky (1939), Emch (1900), and particularly by Birkhoff (1933), a mathematician who in

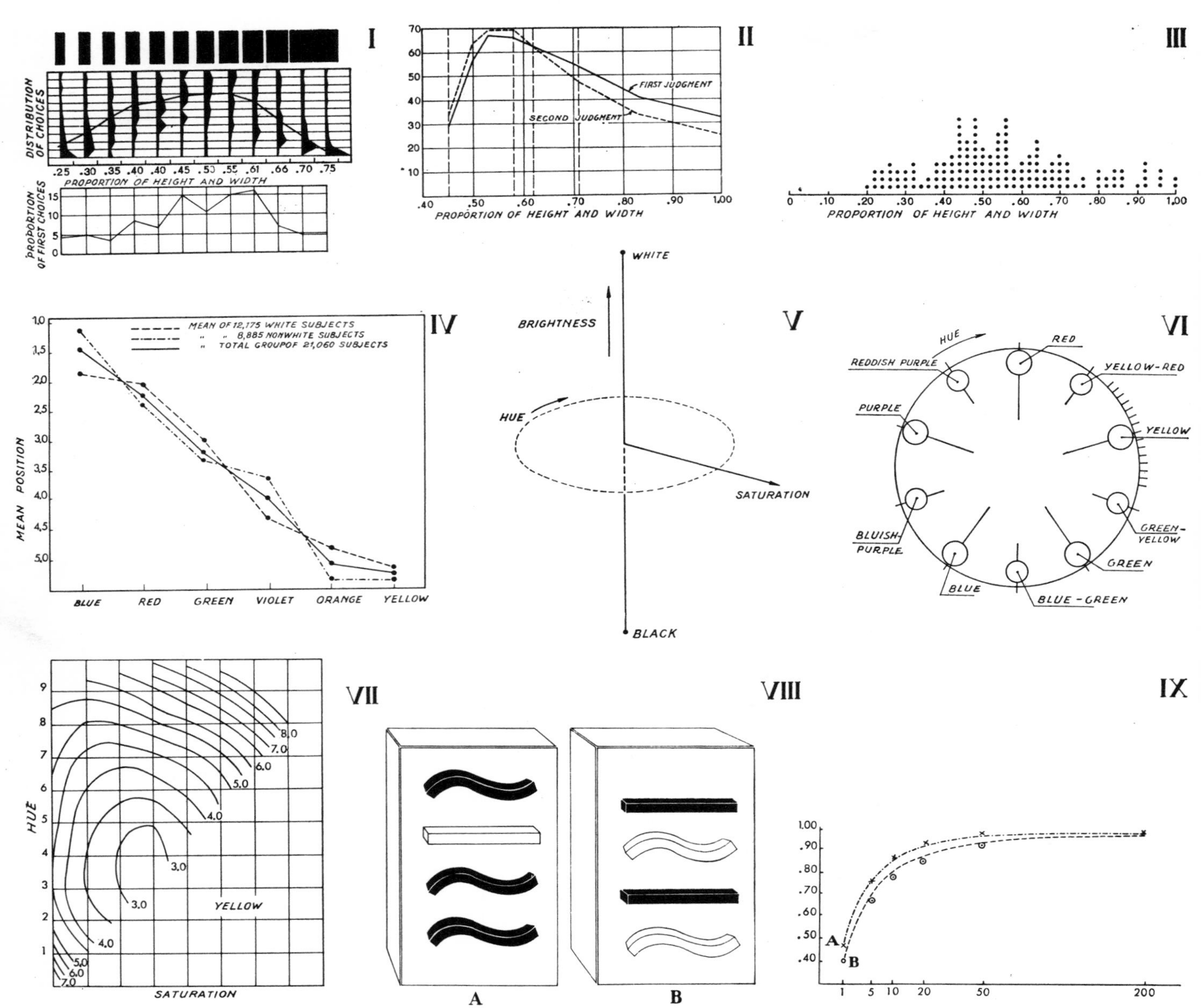

Graphs illustrating psychological tests based on responses expressing preference in shapes and colors: (I) Distribution of preference judgments by 200 subjects who ranked the rectangles shown above the graphs (*from Woodworth, 1939*). (II) Preference for rectangles as affected by practice (*from Woodworth, 1939*). (III) Preference for rectangles as determined by means of the method of "adjustment" (*from Woodworth, 1939*). (IV) Average rankings of color preferences. (V) System coordinated by Munsell. (VI) Hue scale of the Munsell system. (VII) An "isohedon" chart showing on the Munsell plane for yellow the lines of equal preference as judged by the average man. (VIII) Sample item from the Maitland Graves Design Judgment Test requiring subject to select preference for *A* or *B*. (IX) Relation between number of judges and correlation of their pooled judgments with independent criterion: (*A*) Esthetics experiment; (*B*) weights experiment.

his book *Aesthetic Measure* maintained that our pleasure in any work of art depends upon two variables: the amount of order (*O*) and of complexity (*C*) in the object. These are measured in various ways for different classes of objects, but all classes obey the general formula *M* (measure of pleasure derived) $= O/C$. Birkhoff worked out in detail the objective properties of esthetic objects which could be measured and which would be subsumed under either order or complexity elements, and his book is largely devoted to a discussion of methods of doing this. Like most other writers in this field, his work has one fatal weakness in that it is entirely a priori and relies on hypotheses which are not in line with the evidence. He assumes, for instance, that the eye in looking at a polygonal figure will follow the lines of the polygon, whereas in actual fact the eye makes a number of saccadic jumps which are quite different from the movements postulated (Woodworth, 1939). As many of Birkhoff's arguments are derived from this assumption, it can be seen that the resulting detailed formula can have little value. This has been shown by R. C. Davis, C. M. Harsh, H. W. Miller, J. B. Parry, A. Schnittkind, E. Fischer, H. J. Eysenck, and others who have tried to correlate values of the esthetic measure calculated from Birkhoff's formula with actual preference judgments. Table 3 gives some of the results: it will be seen that the correlations are low throughout but nevertheless tend to be positive.

It is possible to make use of the notion of an esthetic formula, however, without relying on a priori hypotheses. Eysenck (1941) determined certain objective features of a number of polygons taken from the series published by Birkhoff, had these polygons ranked in order of preference, and then determined the correlations between each of the measured bases of judgment and the final ranking. The results are shown in Table 4. He then combined these 12 bases in a formula, using the correlations as weights. He tested the accuracy of the formula by applying it to a new set of subjects and found near-perfect agree-

TABLE 3. *Correlation of Preference Judgments with Values of the Esthetic Measure Calculated from Birkhoff's Formula*

Investigator	*Material*	*Subjects*	*Correlation*
R. C. Davis	10 polygons	162 students	.11
R. C. Davis	10 polygons	55 art students	.05
F. W. Swift*	45 polygons	6 laymen	.53
F. W. Swift*	45 polygons	3 art students	.22
F. W. Swift*	15 polygons	6 students	.16
C. M. Harsh et al.	26 polygons	30 students	.34
H. J. Eysenck	32 polygons	14 observers	.53
H. J. Eysenck	32 polygons	14 observers	.62
H. J. Eysenck	15 polygons	12 observers	.48
R. C. Davis	10 poems	63 English students	.55
H. W. Miller*	7 Lines of poetry	16 English students	.11
H. W. Miller*	7 lines of poetry	5 students	.08
J. B. Parry	19 lines of poetry	14 observers	.34
H. W. Miller	7 nonsense lines	5 students	.68
A. Schnittkind*	15 vases	5 students	.23
A. Schnittkind*	15 vases	8 observers	.16
A. Schnittkind*	15 vases	7 art students	.09
E. Fischer*	10 chordal seq.	15 observers	.05
E. Fischer*	5 unharm. melodies	145 observers	.50
E. Fischer*	5 unharm. melodies	59 observers	.50
E. Fischer*	5 unharm. melodies	19 observers	.70

* Reprinted in Beebe-Center and Pratt (1937).

ment. It would seem, therefore, that empirical methods are greatly superior to those used by Birkhoff and the others mentioned if the aim is to predict preference judgments. Evaluating Eysenck's results together with the results of Granger on preferences for color combinations, it would appear that the search for an esthetic formula is by no means doomed to failure. Eysenck (1942) also found that Birkhoff's general formula $M = O/C$ was wrong; he showed that a multiplicative function works much better, for example, $M = OC$.

TABLE 4. *Correlation of Measured Bases of Judgment of Polygons with Preference Rankings*

Basis of judgment	*Coefficient of contingency*
(x_1) vertical of horizontal symmetry	.71
(x_2) rotational symmetry	.69
(x_3) equilibrium	.51
(x_4) repetition	.45
(x_5) compact figure	.37
(x_6) complexity six or more	.33
(x_7) both vertical and horizontal symmetry	.31
(x_8) pointed top and/or base	.20
(x_9) complexity three or more	.10
(x_{10}) complexity two	-.17
(x_{11}) reentrant angles	-.52
(x_{12}) angles close to 90° or 180°	-.63

The esthetics of complex stimuli. So far we have been dealing with groups of stimuli which many people might not regard primarily as examples of esthetic value at all. We must now turn to more complex materials and discuss the work of Dewar (1938), Burt (1933), and Eysenck (1940, 1941). Burt and Dewar found that both laymen and experts tended to put pictures of widely different merit in much the same order of liking; they concluded that a positive answer could be returned to the question originally raised by Burt: "If we could brush aside all irrelevant associations, and take a completely detached view . . . would there be any solid ground for preference left?" This, however, is a questionable conclusion. The series of pictures ranked in these studies included "reproductions from classical masters, second rate pictures by second rate painters, every variety and type down to the crudest and most flashy birthday cards." Now it is quite possible that the subjects of these experiments might have demonstrated not so much their esthetic ability as their cognitive knowledge that certain painters and their work are preferable to certain others. The experiment leaves far too many factors of knowledge, judgment, and so forth uncontrolled to derive any definite conclusions.

Eysenck (1940, 1941) attempted to overcome this difficulty by designing an experiment in which the objects within each of 18 tests were so homogeneous that irrelevant outside influences of this kind could not affect the issue. One test, for instance, contained 20 reproductions of pencil drawings by Claude Lorrain; by thus restricting himself to a single artist and to drawings all of which were unknown to any of his subjects he made it impossible for knowledge of cultural judgments to take the place of genuine esthetic ones. Another test contained 15 color reproductions of bookbindings from the British Museum; a third, 12 curves of mathematical functions; a fourth, 15 reproductions of pieces of silver plate; and so forth. Having thus experimentally controlled for the main sources of error, Eysenck then proceeded to have his subjects rank the objects in each of the sets in order of esthetic merit. Analyzing their responses he found that (*a*) in each of the tests there was substantial agreement between the judges, and (*b*) those who agreed most with the rest in one test also tended to agree most with the rest in the other tests. In other words, "good taste," defined as agreement with the average, extended over a great variety of different tests and different esthetic objects.

It will undoubtedly be objected by most readers that "good taste" surely cannot be so defined. The man in the street, it will be pointed out, prefers popular songs to Beethoven and Marilyn Monroe to the *Mona Lisa*. This is perfectly true but quite irrelevant. Such comparisons are between different *classes* of objects and are, therefore, not esthetic judgments at all; they reflect, if anything, a liking for rhythm and for women liberally endowed with secondary sex characteristics. Objections of this sort illustrate the necessity of controlling external sources of error; they do not indicate that when such control has been successfully applied the results are still subject to criticism of this kind.

Granted that the results obtained can best be explained in terms of a general property of individuals with respect to which they differ quantitatively (in the sense of some having a considerable amount of it and others having very little, with the majority in between), we may attempt to measure this "good taste" in a quantitative manner. Attempts to do so have been made by many psychologists, the most successful probably being Maitland Graves (1946) in his Design Judgment Test. This test consists of figures in which the subject has to choose one of two alternative versions he prefers (see FIG. 771, No. VIII). The correct choice is one which agrees with three criteria: (*a*) agreement among art teachers as to the better design; (*b*) greater preference for the design by art students than by nonart students; and (*c*) greater preference for the design by those who have achieved high scores on the entire test than by those achieving low scores.

If this test were valid, one would expect advanced students of art to score better than, say, students of engineering or dental technology. The figures seem to bear this out. The mean score of several hundred art students was about 75, that of engineering students 48, that of dental technology students 41. Altogether it has been found that students with good scores tend to do better at art schools than do students with poor scores.

It is possible to link this discussion of judgment of complex stimuli with those of simple stimuli by quoting one particularly interesting experiment reported by Granger (1955) in which subjects who showed "good taste" on experiments in which simple colors and simple color combinations had to be ranked also tended to have high scores on the Maitland Graves Design Judgment Test. He demonstrated that this relationship could not be due to intelligence, and it is difficult to find an alternative hypothesis to the one that suggests itself naturally, namely, that there is some underlying biological property of the nervous system which determines our esthetic judgments, both of simple colors and color combinations and of complex objects such as those used in the Maitland Graves test.

The validity of esthetic judgments. It will be seen from what has been said so far that to the psychologist esthetic judgments are simply one subset of the general set of psychological experiments called judgments. We would expect esthetic judgments to have certain mathematical properties which are known to belong to other sets of judgments. One of the most important of these (which is related to what was said in the above discussion on the derivation of a "standard of good taste") is the relationship between the validity of a judgment and the number of judges. If under appropriate experimental conditions the average judgment of a group of people is indeed the expression of some underlying biological feature characterizing the whole human race, and if, therefore, the mean is to be regarded as a "true" judgment, then it would be desirable to know something about the degree to which a "true" judgment can be approached by increasing the number of judges. An experiment along these lines has been carried out by Eysenck (1939), who obtained 900 ranks in order of preference for 12 uncolored pictures. Of these 900 judgments, 700 were used as a standard or an independent criterion; the remaining 200 constituted the experimental group. Correlations were calculated between the criterion and single judgments, pools of 5 judges, pools of 10 judges, pools of 20 judges, and pools of 50 judges. As the number of judges increased so did the correlation of their mean ranks with the criterion (as indicated in FIG. 771, No. IX). Also included in this figure is the result of a similar experiment carried out with weights that had to be ranked in order of heaviness. (The weights were so similar that the task was very difficult.) It will be seen that the shapes of the curves are almost identical, and both follow a general formula which reads

$$r_{\bar{k}g} = \sqrt{\frac{n\bar{r}_{kk'}}{1 + (n-1)\,\bar{r}_{kk'}}}$$

where $r_{\bar{k}g}$ stands for the correlation of the average ranking order derived from the judges and $\bar{r}_{kk'}$ stands for the average intercorrelation of all the various rankings.

It is possible, therefore, to achieve a reasonable estimate of the "true" value of the esthetic measure of objects with something like 200 subjects. It will also be seen that esthetic judgments are subject to the same laws as are other types of psychological judgments and, lastly, that experimental esthetics can gain considerably from mathematical and statistical treatment. The extent to which this is true, of course, depends entirely on the value of the psychological theories being tested and the rigor with which controls are applied to the experimental situation. Unfortunately, very few psychologists are interested in this field, and consequently progress has been far less rapid than it might have been; nevertheless, certain fairly well substantiated facts are already known, and methods have been worked out in sufficient detail to make further advance possible.

H. J. EYSENCK

GESTALT PSYCHOLOGY. The explicit contributions of gestalt theory to the psychology of art are few. There is an extensive essay by Koffka (1940) on general problems of esthetics; and there is Arnheim's book (1954) on visual perception as applied to painting, sculpture, and architecture. No written record is available of M. Wertheimer's academic lectures on the psychology of art and music, in which he dealt mainly with the expressive characteristics of visual and auditory patterns. Yet from the very first statement of the gestalt principles by Von Ehrenfels (1890), writing on the subject has been replete with references to the arts. No product or activity of the human mind had been more resistant than the arts to the approach of the "atomistic" school of science, which for good reasons of its own describes organic and inorganic entities by adding up descriptions of their parts. Therefore the unbreakable wholeness of a painting or musical composition was often called upon to demonstrate the basic gestalt tenet that physical and psychical objects and events tend to be organized by an over-all structure that determines the size, shape, character, place, and function of the parts and is in turn, determined by the parts and their interrelations.

Art was quick to benefit by serving as an attractive exhibit in psychology: it found in gestalt theory the scientific rigor to which philosophical speculations on esthetics hardly aspired. At the same time, scientific discipline as applied by the new school of thought did not require the chopping up of the artistic phenomenon into isolated part problems — a procedure that had made Fechner's and Helmholtz's earlier explorations of proportion, brightness constancy and other aspects of the artistic media look like interesting but somewhat distracting marginalia. Thus gestalt became a household concept of esthetic theory (see Kepes, 1944; Read, 1958).

Motivation. A principle underlying much modern theorizing on the mainsprings of behavior can contribute an answer to the question why art exists. According to this principle the organism strives constantly toward a minimum of tension. This hypothesis was not introduced into our thinking by gestalt theory but is one formulation of the fundamental gestalt statement on the direction of processes: any organized whole tends to assume the simplest, most stable, most regular form available to it. In his writings W. Köhler has repeatedly traced this principle, best known as Wertheimer's law of the "good gestalt," to the thinking of physicists in the 18th and 19th centuries. It became influential as Boltzmann's second law of thermodynamics, according to which the physical universe tends toward an equilibrium eliminating all "asymmetries." In the life sciences its realm extends all the way from Cannon's homeostasis (1932) to Freud's motivational theory (1921), according to which pleasure coincides with decrease of tension. Today's textbooks of psychology assert that needs aim at reestablishing a condition of psychological equilibrium.

From this point of view, artistic activity appears as a response to the stimuli of experience which upset the balance of the mind. To create a work of art means to clarify the obscure, to find order in confusion and simplicity in complexity, to derive the effect from its causes, to display the forces underlying action, to set things straight or, psychoanalytically speaking, to escape the pressures of reality. The principle of equilibrium thus provides a theory of artistic motivation. At the same time this principle can serve as a criterion of esthetic excellence in that the better the work of art the less energy it requires from the recipient who endeavors to understand its content or message. Economy as a gauge of artistic value goes back to Fechner (1876) and Allen (1877) and has been conveniently discussed by Eysenck (1942). The idea reverberates also in the esthetics of Freud, who asserts that the mechanism of jokes is a "saving in psychical expenditure" (1905). In gestalt terms, the artist strives toward a structural pattern of maximum simplicity, stability, and regularity as a means of bringing about a similar condition in his own mind.

The principle of equilibrium, although of basic importance to modern thinking, leads to a one-sidedly static conception if it is made to sustain a theory of organic process all by itself. Critics of homeostasis in its original biological as well as derived psychological version have pointed out that while the organism covets the stationary state of optimal functioning it is more essentially characterized by vital aims, ranging from the satisfaction of animal instincts to the spiritual aspirations of man. Freud's "conservative drives" are open to the same objection, and the law of the "good gestalt" has been said to lead to the absurd consequence that a mere circle, being the simplest and stablest structure, would make the greatest work of art.

This misunderstanding of the gestalt position arose because the early experimental gestalt work in perception was limited by necessity to situations in which the stimulus object provided practically the only dynamic impulse. The principle of the "good gestalt" was indeed capable of dealing singlehandedly with the ways in which a quasi-closed mental system tackled the impinging perceptual stimuli. But Köhler in particular has made it quite clear that in its tendency toward a minimum of tension the organism participates in the behavior of all physical matter, while the unique property of the organic open system is that of constantly drawing in energy in order to ward off the state of deadly rest and pursue its own vital aims (1938). No preference for equilibrium is inherent in gestalt theory, which is impartially concerned with the various processes occurring in organized wholes. Consequently, gestalt psychology is in a position to point out that "just as the emphasis of living is on directed activity and not on empty repose, so the emphasis of the work of art is not on balance, harmony, unity, but on a pattern of directed forces that are being balanced, ordered, unified" (Arnheim, 1954).

Perception. With this proviso, the analysis of visual and auditory patterns profits greatly from the rules of perceptual organization, which are derived from much painstaking experimental work. Very little practical help can be gained from the traditional view that the mind interprets the perceptual stimulus material according to previously acquired information, connections, images, or assumptions. Describing the mind rather sadly as a mere administrator of its own past, such an approach could hardly support, systematize, and explain the experiences of the artist, who finds that certain properties of shape, size, distance, spatial orientation, intensity, or pitch regularly produce certain perceptual effects. Wertheimer's analysis of these inherent stimulus properties started with a set of rules on how elements such as dots or lines combine in larger wholes (1923). Attempts to reduce his factors of proximity, good continuation, similarity, and so forth to one were made by Musatti ("omogeneità"; 1931) and Arnheim ("similarity"; 1957).

The procedure of building up perceptual compounds "from below" by tracing the relations between elements was considered transitional by Wertheimer. It applied an approach, partially anticipated by the older laws of the association of ideas, to the new analysis of inherent perceptual organization. Characteristically, these rules of grouping gained wide popularity, whereas the true gestalt procedure "from above," which Wertheimer applied in the further pursuit of his classical investigation (1923), has continued to leave even psychologists somewhat uncomfortable. Here he showed concretely how visual patterns will subdivide into parts in such a way as to produce the simplest available organization. Later, the principle of "good gestalt" was applied to the third dimension: depth was shown to be perceived whenever the third dimension yielded a simpler, more stable, more regular pattern than the second (Koffka, 1930; Arnheim, 1954).

Most gestalt studies on perceptual structure dealt with immobile visual patterns. And since the findings demonstrated that much of what had been attributed to past experience by the empiricists was due to the spatial stimulus configuration itself, gestalt psychologists were accused of denying the importance of the time dimension. Actually the principles of organization derived from spatial figures apply to temporal patterns as well, so that any perceptual experience can be said to result from the four-dimensional interplay of space and time factors. Michotte's book (1946) on the perception of causality may be cited (see also Arnheim, 1954) as a gestalt investigation of temporal processes. The musicologist L. B. Meyer (1956) has applied Wertheimer's principles to melodic and harmonic structure.

In most of the gestalt processes of which we possess concrete experimental accounts the tendency to simplest structure — that is, to tension reduction — is at work. Under certain conditions, however, the mind can be shown to heighten tension by intensifying the dynamics of perceptual patterns. Asymmetries are increased; angles become more pointed; and intervals are stretched. Wulf (1922) demonstrated this phenomenon in his study on the effect of memory on visual patterns and named it "sharpening," as opposed to "leveling." Musical examples can be found in Abraham's experiment on intonation in singing (1923). In the arts, expressionism relies heavily on the devices of sharpening.

Reference should be made to the so-called "law of *prägnanz*" (clarification), by which Wertheimer (1923, p. 318; 1945, p. 199) described a tendency of *gestalten* to assume the most clear-cut organization. The term has created much trouble. It has been confused with the English word "pregnant" and has also been identified with the principle of "good gestalt," to which it seems to bear no simple relation, since changes that make a pattern more clear-cut may decrease or increase tension. However, "clarification" is an essential aspect of art, and the concept of *prägnanz*, properly translated and defined, is likely to be of help in esthetic theory.

Expression. In keeping with its preference for facts that can be measured by yardstick, protractor, and clock, the psychology of perception limits itself commonly to the quantitative properties of shape, size, distance, intensity, speed, and so forth. It thereby neglects the primary perceptual fact (primary certainly as far as esthetic perception is concerned) of the expressiveness of sensory phenomena. Even before an observer ascertains the geometric-technical properties of what he sees or hears he is impressed with the "physiognomic" character of objects and actions (Warner, 1948). Most clearly in pictorial and musical compositions "shapes" are perceived as patterns of directed forces, not as static extensions, locations, or pitches. According to Arnheim (1949), these dynamic qualities of percepts are the conscious counterpart of the pulls and stresses accompanying the formation of perceptual patterns in the brain.

Gestalt psychology, then, does not consider expressiveness a subjective attribution projected upon the object through empathy by the observer but rather a property determined by the stimulus object and, in fact, constituting it, like shape and color. In contradiction to the traditional assumption that one learns to interpret expressive qualities by associations formed in the past, gestalt psychology asserts their immediate meaningfulness (Köhler, 1947, pp. 216–47; Koffka, 1935, pp. 654–61). The perceptual patterns of forces displayed by the muscles of a face, the curve of a wave, or the wailing of a siren are said to be "isomorphic" (i.e., structurally identical) with the dynamic patterns of mental strivings and thereby acquire human significance. Expressiveness is in no way limited to the human body. Since it is nothing more nor less than the dynamic aspect of any perceptual quality, it appears in the images of inorganic as well as organic things, in immobile shape as well as in action. The fundamental importance of this approach for the theory of art is evident. In particular, the long-debated problem of the semantics of music finds a tangible solution (Langer, 1942).

The creative process. While gestalt psychologists have shown how effectively the external stimulus controls perceptual experience they have also paid attention to the mind's capacity for restructuring a given state of affairs in the interest of a particular goal. Problem solving always involves conceiving of the problem situation in a way not spontaneously suggested by it and often requires a reshaping of relations, shift of focus, change of range, and other structural modifications. The gestalt psychology of productive thinking is directly relevant to art in that the creative process can be described as problem-solving behavior. Artistic creation means shaping the raw material of experience in order to make it a vehicle of the theme or content to be embodied. It also means adapting the material to the stylistic form, which reflects the artist's attitude toward reality. Since the material is not amorphous but possesses an articulation of its own, there results the dramatic wrestling with the problem often described by artists and concretely analyzed in its formal mechanisms by Wertheimer. Some of these mechanisms closely resemble those observed by Freud (1900) as the devices of "dreamwork": displacement, condensation, inversions, transformation, change of stress, and so forth.

Very little concrete material is available to the psychologist who wishes to follow the successive transformations of a gestalt as it grows from the first conception into the completed work of art (Abbott, 1948). Much fuller sets of data exist, naturally, for the related attempts to consider the entire *œuvre* of an artist as a continuous process of problem solving, culminating in the late works in the realization of his personal world. In an even broader sense the stylistic development of an entire period of the history of art often appears as the stepwise solution of a particular esthetic task. In the field of art education, the school of Britsch and Schaefer-Simmern has described the development from structurally simple forms in the early stages of the art work of children and primitives to increasingly more complex conceptions.

Objectivism and value. It is characteristic of gestalt psychologists not only to describe creative efforts sympathetically but also to point out that in order to be productive the conception must be truthful, that is, meet the intrinsic requirements of the problem situation. They have found themselves in opposition to the relativism inherent in the empiricist approach. If one assumes that the stimuli are a neutral substratum upon which the observer projects whatever patterns derive from his personal memories and wishes, then the nature of the resulting percept depends entirely upon who is looking and listening. If, on the other hand, the stimuli are not an assortment of elements but an organized whole, they will do their share in determining the nature of the resulting experience. Thus gestalt psychology assures the artist that his work is not entirely at the mercy of the recipient. Instead, the more clear-cut (*prägnant*) the organization of his composition, the more effectively it will steer the beholder's or listener's reception. The gestalt view does not deny the influence of individual differences, memories, and biases but holds that it is the interaction of the forces issuing from the stimulus situation and the forces deriving from the personality of the recipient that determines the outcome of the encounter. This respect for the objectively definable aspects of the work of art can be said to be shared by most artists.

Going beyond the description of psychological facts, gestalt theory emphasizes the duty of the individual to act and react according to the requirements of the given circumstances, not in the sense of conformist adjustment but out of devotion for the demands of the task. This position is supported by the testimony of many artists, who have indicated that when the work goes well they proceed according to the dictates of the composition itself. Similarly the recipient can expect to do justice to the work of art only if he submits to its demands and guidance rather than follows his own fancies.

This attitude leads finally to a theory of value according to which "good" and "bad" do not derive from what a person happens to like or to prefer, or from some preordained standards, but rather from the requirements of the situation (Köhler, 1938). A certain brush may be good for painting a certain picture; but the picture in turn may not be so good in terms of what — according to the best human knowledge — art is suited to do. Certainly the functions of art, derived from its inherent virtues, are broad enough to satisfy a variety of needs by a corresponding variety of styles; but the claim is made that use and misuse of art can be distinguished from each other on factual grounds and that artistic value can be argued objectively in terms of what the work attempts, and should attempt, to accomplish.

Rudolf Arnheim

Children's art. It has been demonstrated that the impulse to create artistically, especially in the visual arts, is spontaneous in every normal individual. Until a few decades ago children were directed to draw and paint according to adult conceptual schemes rather than left free to express their own personalities. New methods have been introduced, however, and the results both esthetically and psychologically have been astounding. Through the free production of forms and colors children have been able to project some of their feelings in such a way as to reveal many repressed as well as conscious psychic elements. Among the first to study children's painting were some of the great educators — Pestalozzi, Herbart, Salzmann, Froebel, and, above all, Franz Cizek, the real pioneer of freedom of expression for children.

At present there are several different evaluations of the artistic activity of children. Numerous attempts have been made to distinguish certain psychological types already defined by psychologists, such as integrated and disintegrated, cycloid and schizoid, Jung's classification of extroverts, introverts, and so forth, and Jaensch's subdivision into T type, B type, and S type (1930). Undoubtedly children's drawing and painting are an important diagnostic means and may be considered the equivalents of the well-known projective tests such as the Rorschach Ink Blot Test, the Thematic Apperception Test, the Pyramid, the Tree Test, and so on.

It is possible to classify in a general way certain principal stages in the development of the child's artistic activity. The first is "manipulation" (age 2–3 years), in which the more or less incoordinate miokinetic movements predominate. According to H. Read's classification, this stage corresponds to that of the scribble; according to R. Arnheim, it is characterized by the more elementary patterns based on the line and the circle (PL. 368). The second stage is that of "descriptive symbolism" (3–6 years), in which the unconscious material brought to light in painting is richer and offers a greater esthetic interest (PLS. 369, 373). This is followed by a transitional stage (8–10 years, corresponding to Read's "visual realism"), in which realistic and even detailed motifs appear more frequently, while the symbolic constants weaken. Finally a fourth stage (which some authorities term the stage of "repression") coincides with the crisis of puberty and marks the end of child art. According to the theory of recapitulation, the evolution of the child's mind follows the same course as that of the whole human race; if the theory is correct, it should be possible in children's art to find significant analogies with elements to be found in the art works of primitive and barbarian peoples.

The rapid extinguishing of children's creative faculties is attributed by several scholars (e.g., E. R. Jaensch) to the disappearance of the eidetic faculty, that peculiar quality of infantile perception which allows the child to give life to his visual images with an extreme vividness of detail; it is a quality almost always lost in the adult.

Children's art has been the subject of a vast amount of research, among which are the highly interesting studies of Tomlinson (1944), Arnheim (1954), and Read (1958).

Art and psychopathology. Much in the visual arts has particular psychological and psychopathological significance. Interest in the art of psychotics originated during the 19th century, when directors of psychiatric institutions began to take an interest in the spontaneously produced art of their patients. But only since the first decades of the 20th century has it been possible to speak of study dedicated especially to this problem (Prinzhorn, 1922; PL. 367). Before Prinzhorn, Cesare Lombroso in Italy and Max Simon in France had analyzed with some accuracy the works of the mentally deranged. The relationship between certain forms of a phallic character appearing in the painting of psychotic patients and the art forms of various barbarian peoples and primitive and archaic cultures was noted by Lombroso (1894), but only Freud was able to give a systematic interpretation of this relationship. Other psychiatrists such as Kraepelin and Jaspers also had occasion to study the artistic expressions of the mentally deranged; they, however, were concerned with diagnostic indications rather than esthetic values and importance. Not only did the Freudian interpretation of the art of psychotics, neurotics, and children have a considerable following, but also the work of Jung gained many adherents particularly interested in collecting in a symbolic-expressive lexicon the great amount of symbolic material found in psychopathological images; this approach was based on Jung's theory of archetypes and the collective unconscious.

It is not possible to arrive at a judgment of the "value" of the art of the psychologically disturbed any more than it is possible to arrive at such an evaluation of children's art; judgments must be based on the creative and technical qualities of the individual artist. What is of undoubted importance is an analysis of the use which numerous surrealist painters have made of psychopathological images, especially sexual images and those suggesting a delirious or dreamlike state (PL. 374). There is great similarity between certain surrealist productions and the fantasies of not only neurotics and psychotics but also drugs addicts and persons in a state of spontaneous oneirism (mediums, spiritualists, etc.). Surrealism, as is well known, drew heavily on the most recondite structures of the unconscious and made use of all the art forms aroused in oneiric or oneiroid states, either spontaneous or induced. It is not a coincidence that this movement was defined as "pure psychic automatism" by André Bréton (*Manifeste du Surréalisme*) and that later Salvador Dali actually spoke of "critical paranoia," which he defined as a "spontaneous method of irrational knowledge based on the interpretative-critical association of deliriant phenomena." Bréton in his manifesto referred to the experiences of scientists and psychologists and emphasized the esthetic values of hallucination, even when clearly pathological. In this way he gave new values to all the psychopathological, hallucinatory, and dissociated elements hitherto not considered pertinent to art. In the work of many artists of this tendency (Magritte, Max Ernst, Tanguy, Dali; PLS. 372, 374; V, PLS. 133, 134, 144) there is a predilection for erotic, sadistic, and sadomasochist elements, as well as coprophagy and putrefaction.

It should also be noted that all modern art has been accused of presenting a marked similarity to the art produced by children and demented persons. This observation is not so startling and may easily be justified if one bears in mind how certain dissociative manifestations in the thought and creativity of the modern world suggest a parallel between modern art and conditions of alienated communication (cf. Dorfles, 1962, chap. 3).

Another interesting aspect of the relationship of art to psychology and psychopathology is the use of art for therapeutic purposes. This has been studied particularly by the American psychologist Margaret Naumburg (1953), who demonstrated how often works of great vividness and originality can be created by both child and adult patients. Through these works it has been possible to reveal the oneiric, iconic, and symbolic material latent in the patient's unconscious and thus to arrive at successful therapy.

Gillo DORFLES

BIBLIOG. I. Kant, Kritik der Urtheilskraft, Berlin, 1790 (Eng. trans., J. C. Meredith, Oxford, 1928; 2d ed., 1952); F. Nietzsche, Die Geburt der Tragoedie aus dem Geist der Musik, Leipzig, 1872 (Eng. trans., F. Golffing, Garden City, N. Y., 1956); R. Vischer, Über das optische Formegefühl, Leipzig, 1873; G. T. Fechner, Vorschule der Ästhetik, Leipzig, 1876; G. Allen, Physiological Aesthetics, New York, 1877; C. von Ehrenfels, Über "Gestaltqualitäten," Vierteljahrsschrift für wissenschaftliche Philosophie, XIV, 1890, pp. 249–92; T. Lipps, Ästhetische Faktoren der Raumanschauung, in A. König, ed., Beiträge zur Psychologie und Physiologie der Sinnesorgane, Hamburg, 1891, pp. 219–307; A. Hildebrand, Das Problem der Form in der bildenden Kunst, Strasbourg, 1893 (Eng. trans., M. Meyer and R. M. Ogden, New York, 1907); J. Cohn, Experimentelle Untersuchungen über die Gefühlsbetonung der Farben, Helligkeiten und ihre Combinationen, Philosophische S., X, 1894, pp. 562–603; C. Lombroso, L'uomo di genio, 6th ed., Turin, 1894; E. Pierce, Aesthetic of Simple Forms, II. Functions of the Elements, Psychol. Rev., 1, 1894, pp. 487–95, I. Symmetry, 3, 1896, pp. 220–82; L. Witmer, Zur experimentellen Aesthetik einfacher räm ulicher Formverhiltnisse, Philos. Stud., 9, 1894, pp. 96–144, 209–63; A. Emch, Mathematical Principles of Aesthetic forms, The Monist, XI, 1900–01, pp. 50–64; S. Freud, Die Traumdeutung, Leipzig, Vienna, 1900 (Eng. trans., J. Strachey, New York, 1959); A. Riegl, Die spätromische Kunstindustrie, 2 vols., Vienna, 1901–23; R. P. Angier, The Aesthetics of Unequal Divisions, Psychol. Monographs, IV, 1903, pp. 541–61; A. Binet, Etude expérimentale de l'intelligence, Paris, 1903; T. Lipps, Aesthetik, 2 vols., Hamburg, Leipzig, 1903–06; E. P. Puffer, Studies in Symmetry, Psychol. Monographs, IV, 1903, pp. 407–539; T. N. Haines and A. G. Davies, Psychological Aesthetic Reactions to Rectangular Forms, Psychol. Rev., XI, 1904, pp. 249–81; S. Freud, Der Witz und seine Beziehung zum Unbewussten, Leipzig, Vienna, 1905 (Eng. trans., A. A. Brill, New York, 1917); E. Bullough, The Perceptive Problem in the Aesthetic Appreciation of Single Colours, Br. J. of Psychol., XI, 1906–08, pp. 406–63; C. Lalo, L'ésthétique expérimentale contemporaine, Paris, 1908; L. W. Legowski, Beiträge zur experimentellen Aesthetik, Arch. für das gesamte Psychologie, XVII, 1908, pp. 236–311; V. Lee (V. Paget), The Beautiful, Cambridge, Eng., 1913; F. Sander, Elementarästhetische Wirbungen zusammengeskter geometrischer Figuren, Psychol. Stud., 9, 1913, pp. 1–34; M. Verworn, Ideoplastische Kunst, Jena, 1914, G. L. Thorndike, Individual Differences in Judgement of the Beauty of Simple Forms, Psychol. Rev., XXIV, 1917, pp. 147–53; M. Buber, Die Rede, die Lehre, und das Lied, Leipzig, 1920; J. Hambidge, Dynamic Symmetry, New York, 1920; S. Freud, Jenseits des Lustprinzips, 2d ed., Leipzig, 1921 (Eng. trans., J. Strachey, New York, 1950); C. G. Jung, Psychologische Typen, Zurich, 1921 (Eng. trans., H. C. Baynes, London, New York, 1923); H. Prinzhorn, Bildnerei der Geisteskranken, Berlin, 1922; F. Wulf, Über die Veränderung von Vorstellungen, Psychol. Forsch., I, 1922, pp. 333–73 (Eng. summary in W. D. Ellis, ed., A Source Book of Gestalt Psychology, New York, 1938, pp. 136–48); O. Abraham, Tonometrische Untersuchungen an einem deutschen Volkslied, Psychol. Forsch., IV, 1923, pp. 1–22; O. Baensch, Kunst und Gefühl, Logos, XII, 1923–24, pp. 1–28; E. Cassirer, Philosophie der symbolischen Formen, I, Berlin, 1923 (Eng. trans., R. Manheim, New Haven, Conn., 1953); M. Dessoir, Ästhetik und allgemeine Kunstwissenschaft, 2d ed., Stuttgart, 1923; H. Kühn, Die Kunst der Primitiven, Munich, 1923; R. Müller-Freienfels, Psychologie der Kunst, 3d ed., Leipzig, 1923; M. Wertheimer, Untersuchungen zur Lehre von der Gestalt, Psychol. Forsch., IV, 1923, pp. 301–51 (Eng. summary in W. D. Ellis, ed., A Source Book of Gestalt Psychology, New York, 1938, pp. 12–16); G. Britsch, Theorie der bildenden Kunst, Munich, 1926; E. Jaensch, Eidetic Imagery and Typological Methods of Investigation (Eng. trans., O. A. Oeser, London, New York, 1930, 2d ed., 1955); K. Koffka, Some Problems of Space Perception, in C. Murchison, Psychologies of 1930, Worcester, Mass., 1930, pp. 161–87; C. L. Musatti, Forma e assimilazione, Arch. it. di psicologia, IX, 1931, pp. 61–156; C. O. Weber, Aesthetics of Rectangles and Theories of Affect, J. of Applied Psychol., XV, 1931, pp. 310–18; W. B. Cannon, The Wisdom of the Body, New York, 1932; G. D. Birkhoff, Aesthetic Measure, Cambridge, Mass., 1933; C. Burt, The Psychology of Art, in How the Mind Works, London, 1933, pp. 267–310; R. C. Davis, Aesthetic Proportion, Am. J. of Psychol., XIV, 1933, pp. 298–302; N. C. Meier, ed., Studies in the Psychology of Art, 3 vols., Princeton, Albany, N.Y., 1933–39; A. R. Chandler, Beauty and Human Nature, New York, 1934; J. P. Guilford, The Affective Value of Color as a Function of Hue, Tint and Chroma, J. of Experimental Psychol., XVII, 1934, pp. 342–70; K. Koffka, Principles of Gestalt Psychology, New York, 1935; P. Plaut, Prinzipien und Methoden der Kunstpsychologie, Handbuch der Biologischen Arbeitsmethoden, VI, C, 2, Berlin, 1935, pp. 745–966; R. C. Davis, An Evaluation and Test of Birkhoff's Aesthetic Measure and Formula, J. of General Psychol., XV, 1936, pp. 231–40; E. Kretschmer, Physique and Character, London, 1936; J. G. Beebe-Center and G. C. Pratt, A Test of Birkhoff's Aesthetic Measure, J. of General Psychol., XVII, 1937, pp. 339–53; A. R. Chandler and G. N. Barnhart, A Bibliography of Psychological and Experimental Aesthetics, Berkeley, Calif., 1938; H. Dewar, A Comparison of Tests of Artistic Appreciation, Br. J. of Educational Psychol., CIII, 1938, pp. 29–49; W. Köhler, The Place of Value in a World of Facts, New York, 1938; O. H. Sterzinger, Grundlinien der Kunstpsychologie, Graz, 1938; J. Evans, Taste and Temperament, London, 1939; H. J. Eysenck. The Validity of Judgements as a Function of the Number of Judges, J. of Experimental Psychol., XXV, 1939, pp. 650–54; J. P. Guilford, A Study in Psychodynamics, Psychometrika, 4, 1939, pp. 1–23; C. M. Harsh, J. G. and R. Beebe-Center, Further Evidence Regarding Preferential Judgements of Polygonal Forms, J. of Psychol., VII, 1939, pp. 343–50; V. Löwenfeld, The Nature of Creative Activity, London, 1939; H. D. Murray and D. A. Spencer, Colour in Theory and Practice, London, Boston, 1939 (new ed., R. Donaldson and others, London, 1952); N. Rashevsky, Contributions to the Mathematical Biophysics of Visual Perception with Special Reference to the Theory of Aesthetic Value of Geometric Patterns, Psychometrika, 3, 1939, pp. 253–71; R. S. Woodworth, Experimental Psychology, New York, 1939; H. J. Eysenck, The General Factor in Aesthetic Judgments, Br. J. of Psychol., XXXI, 1940, pp. 94–102; J. P. Guilford, There Is System in Color Preferences, J. of the Optical Soc. of Am., 30, 1940, pp. 355–59; K. Koffka, Problems in the Psychology of Art, Bryn Mawr Notes and Monographs, IX, 1940, pp. 180–273; H. J. Eysenck, A Critical and Experimental Study of Colour Preferences, Am. J. of Psychol., LIV, 1941, pp. 385–94; H. J. Eysenck, The Empirical Determination of an Aesthetic Formula, Psychol. Rev., XLVIII, 1941, pp. 83–92; H. J. Eysenck, 'Type' Factor in Aesthetic Judgments, Br. J. of Psychol., XXXI, 1941, pp. 202–70; H. J. Eysenck, The Experimental Study of the "Good Gestalt": A New Approach, Psychol. Rev., XLIX, 1942, pp. 344–64; S. K. Langer, Philosophy in a New Key, Cambridge, Mass., 1942 (3d ed., 1957); C. G. Jung, Über die Psychologie des Unbewussten, 5th ed., Zurich, 1943; D. Nickerson and S. M. Lentall, A Psychological Color Solid, J. of the Optical Soc. of Am., 33, 1943, pp. 419–22; G. Kepes, The Language of Vision, Chicago, 1944; R. R. Tomlinson, Children as Artists, London, New York, 1944; M. Wertheimer, Productive Thinking, New York, 1945; M. Graves, Design Judgment Test, New York, 1946; A. Michotte, La Perception de la Casualité, Louvain, 1946 (Eng. trans., T. R. and E. Miles, London, 1963); A. H. Osório, Psychologie de l'art, Lisbon, 1946; W. Köhler, Gestalt Psychology, 2d ed., New York, 1947; A. Malraux, Psychologie de l'art, 3 vols., Geneva, 1947–54; A. D. Stokes, Inside and Out, London, 1947; C. D. Abbott, ed., Poets at Work, New York, 1948; H. Schaefer-Simmern, The Unfolding of Artistic Activity, Berkeley, Los Angeles, 1948; H. Werner, Comparative Psychology of Mental Development, 2d ed., Chicago, 1948; R. Arnheim, The Gestalt Theory of Expression, Psychol. Rev., LVI, 1949, pp. 156–71; J. P. Guilford, System of Color Preferences, J. Soc. of Motion Picture and TV Engineers, LII, 1949, pp. 197–210; S. Honkavaara, On the Psychology of Artistic Enjoyment, Helsinki, 1949; E. Neumann, Ursprungsgeschichte des Bewusstseins, Zurich, 1949 (Eng. trans., R. F. C. Hull, The Origins and History of Consciousness, New York, 1954); E. Kris, Psychoanalytic Explorations in Art, New York, 1952; S. K. Langer, Feeling and Form, New York, 1953; M. Naumburg, Psychoneurotic Art: Its Function in Psychotherapy, New York, 1953; R. Arnheim, Art and Visual Perception, A Psychology of the Creative Eye, Berkeley, Los Angeles, 1954; G. W. Granger, Aesthetic Mesaure Applied to Colour Harmony: An Experimental Test, J. of General Psychol., LII, 1955, pp. 205–12; G. W. Granger, An Experimental Study in Colour Harmony, J. of General Psychol., LII, 1955, pp. 21–35; G. W. Granger, An Experimental Study of Color Preferences, J. of General Psychol., LII, 1955, pp. 3–20; G. W. Granger, The Prediction of Preference for Color Combination, J. of General Psychol., LII, 1955, pp. 213–22; L. B. Meyer, Emotion and Meaning in Music, Chicago, 1956; D. A. Stewart, Preface to Empathy, New York, 1956; W. Abell, The Collective Dream in Art, Cambridge, Mass., 1957; E. Bullough, Aesthetics: Lectures and Essays (ed. E. M. Wilkinson), London, Stanford, Calif., 1957; V. Löwenfeld, Creative and Mental Growth, 3d ed., New York, 1957; H. Read, Education Through Art, 3d ed., London, New York, 1958; E. H. Gombrich, Art and Illusion, New York, 1960; F. Heidsieck, L'inspiration, Paris, 1961; J.-P. Weber, La psychologie de l'art, 2d ed., Paris, 1961; G. Dorfles, Simbolo, comunicazione, consumo, Turin, 1962; C. W. Valentine, The Experimental Psychology of Beauty, London, 1962; R. and M. Wittkower, Born under Saturn, London, 1963.

* *

Illustrations: PLS. 367–374; 1 fig. in text.

PUBLICITY AND ADVERTISING. Publicity in the broad sense of making information public, and by extension of giving impact to the ideas and facts publicized, has existed since antiquity. The Persepolis reliefs showing the king enthroned with processions of tribute bearers, the Egyptian carvings depicting the slaying of enemies, the Roman aureus commemorating the imperial distribution of food to the poor, and the use of art in the service of religion during the Middle Ages are examples of those forms of publicity that might today be called political or religious "propaganda."

Advertising art, in the sense of art used to stimulate buying and selling, was clearly established in the shop signs and painted wall announcements of the commercially active Roman Empire. The association of labels or symbols with particular products and crafts was elaborated in the era of the guilds. With the invention of printing the insertion of announcements in publications and the use of printed labels and handbills became possible, and, in fact, the relation of advertising art to the printed word has remained close ever since.

However, it was the industrial revolution, followed by mass production, that gave birth to advertising art as it is known today, just as the evolution of mass communications has given new dimensions to the meaning of publicity. One of the earlier modern manifestations was the lithographed poster, which maintained a high degree of rapport between the so-called "commercial" and fine arts and became in the late-19th and early-20th century a distinguished art form. Following World War II a tremendous expansion in the products advertised, accompanied by increasing variety in media and in the techniques of both creation and reproduction, served to shape one of the most complex and extensive categories of 20th-century art.

THE ROMAN PERIOD. The figured signboards of commercial enterprises and emblems on dwellings are mentioned by ancient authors [Martial, x, xx (xix), 10 ff.; Quintilian, *Institutionis oratoriae*, IV, iii, 38; Cicero, *De oratore*, II, lxvi, 266; Pliny, *Naturalis historia*, XXXV, viii, 25; Livy, IX, xl, 16]. Allusions to figured signboards (painted, in mosaic, or in relief) on shops or workshops and to emblems on houses may perhaps be recognized in certain place names, especially when the name is formed by a topographic addition to the name of a tradesman.

Most of the figured signboards come from Pompeii, where signs were painted on the exteriors of many of the shops, either above the architrave of the door or, more often, to the right or left of the entrance. Pompeii has been divided into nine regions (abbreviated here as reg.), with each region subdivided into insulae (abbreviated ins.), and with each house having an entrance number (abbreviated no.). The signs include a ship filled with rowers (reg. VI, ins. 10, no. 10); varied crockery for a tavern (reg. IX, ins. 11, no. 4); a carpenter sawing a beam with the help of his apprentice (reg. VI, ins. 14, no. 37; Sogliano, 1879, 654); a man carrying recently dyed cloths (reg. VII, ins. 2, no 11; Helbig, 1868, 1502b); a porter (?; *ibid.*, 1502c); the manufacture of terra-cotta pottery (reg. II, ins. 3, no. 7); the sale of cloth (reg. IX, ins. 7, no. 7; VI, PL. 61); an elephant with a pygmy (the sign for an inn, reg. VII, ins. 1, no. 44–45; *ibid.*, 1601; *CIL*, IV, 806–07); and an anchor (reg. III, ins. 1, no. 3). The sign almost always contained an indication of the type of trade or industry carried on in the shop. In all these signs — generally not very well executed — there seems, in contrast with the greater part of the Pompeian wall paintings, to be little or no direct influence of Hellenistic art. Two of these signs, one depicting the sale of pottery and the other a miming scene (reg. I, ins. 8, no. 10), are of pre-Roman date (2d or 1st cent. B.C.); the others can be assigned to the 1st century of our era, the greater part from the years 63 to 79.

Many of the sacred paintings (depictions of divinities, the phoenix, in some cases a procession, found either singly or in groups) on the exteriors of Pompeian shops or workshops may have been the emblems of the shop and served to inform the tutelary deities as to the nature of the trade or industry (VIII, PL. 233). This is known to have been the case with the painting of a procession of carpenters (reg. VI, ins. 7, no. 8–12; *ibid.*, 1480) and one with Bacchus pressing grapes (*ibid.*, 25). Similarly a personification of Alexandria, painted on the exterior of a shop, must be classed as a shop sign and appears to have indicated that imported Egyptian wares were sold there (reg. I, ins. 12, no. 3).

In many Pompeian shops, and especially in taverns and *thermopolia*, the outer wall of the counter was decorated with paintings, colored marbles, or reliefs, which sometimes had the function (rarely recognizable, e.g., reg. III, ins. 3, no. 1 and reg. VI, ins. 16, no. 33) of a signboard or served to attract the attention of the public. A similar advertising function can be ascribed to certain paintings found in the interiors of taverns (Pompeii, reg. VI, ins. 10, no. 1; *ibid.*, 1504; reg. VI, ins. 14, no. 36; Sogliano, 1879, 657; Ostia Antica, reg. I, ins. 2, no. 5; PL. 376) and fulling works (Pompeii, reg. VI, ins. 8, no. 20; Helbig, *op. cit.*, 1502; reg. VI, ins. 14, no. 21–22; Sogliano, *op. cit.*, 653).

In Pompeii some emblems are executed in relief. They are usually carved from tufa, but some are of terra cotta and even of marble. Generally they were placed on the corners of edifices. Some of them seem to have a direct connection with the premises where they are placed. One, for example, contains the tools of the mason's craft and a mason's name (*CIL*, X, 868; reg. VII, ins. 15, no. 15); another depicts the tools of a smith (reg. IX, ins. 1, no. 5); another, located near the entrance of a brothel and gambling house (reg. VI, ins. 14, no. 28), has a *fritillus* (dicebox) between two phalli. An apotropaic function can be attributed to the numerous phallic emblems, nearly all of terra cotta or tufa.

Among the Roman reliefs, in addition to the famous frieze from the funerary monument of the baker M. Vergilius Eurysaces (VIII, PL. 229), are one with an architect's instruments (Rome, Mus. Cap.), one with a bricklayer's hammer (Met. Mus.), two with scenes showing the sale of clothing and of pillows (Uffizi) — all from Rome — one with a scene that may depict a restaurant (Naples, Mus. Naz.), and some from Neumagen, Germany (Trier, Rheinisches Landes.).

Surviving epigraphs, which must have been as widely diffused as the figural signboards, include a few stone signs of private baths open to the public (*CIL*, VI, 29764–29770, Rome; X, 1063, Pompeii, but possibly originating in some other locality; XI, 721, Bologna; XIII, 1926, Lyons; XIV, 4015, Rome), signs of inns (*CIL*, XIII, 2301, Lyons), signs of shops (*CIL*, VII, 265, perhaps from Isurium), the sign of the "Horrea Epagathiana et Epaphroditiana" in Ostia (*CIL*, XIV, 4709), one, in both Greek and Latin, of a stonemason (*CIL*, X, 7296, Palermo), some painted inscriptions on the exteriors of inns (*CIL*, IV, 806–807, 3779, Pompeii), a few very rare painted inscriptions indicating the type of merchandise on sale in a shop or the conditions of sale (*CIL*, IV, 7124, 7678, Pompeii; *NSc*, 1958, 81–82, no. 23 and others), and a few of doubtful function (*CIL*, II, 4284, Tarraco; IV, 7384; X, 4104, Capua). The majority, in addition to indicating the type of activity, also draw attention to the excellence of the wares sold or of the services rendered. The artistic taste of the advertiser must have found expression in the execution of the writing and in the placement of the inscription.

Some of the funerary inscriptions (e.g., *CIL*, VI, 9556) have every appearance of being imitations of the inscriptions placed on the exteriors of shops. Funerary inscriptions made during a man's lifetime could also serve as an advertisement if the owner of the tomb was a manufacturer or a merchant (e.g., *CIL*, VI, 9526).

In addition to the figural and written shop signs, another advertising medium was in the form of mosaic floors. They were common in Ostia Antica, particularly in the 2d and 3d centuries; their artistic quality is rather low. They are found in the great stands of the guilds (*CIL*, XIV, 4549), in shops (e.g., reg. II, ins. 6, no. 1; *CIL*, XIV, 4756; reg. IV, ins. 5, no. 1; *CIL*, XIV, 4757; reg. IV, ins. 7, no. 4; reg. V, ins. 5, no. 1), and in the premises of the professional associations of Ostia Antica (e.g., reg. I, ins. 19, no. 3; PL. 375; VIII, PL. 232). In Pompeii there are no certain examples of this type; a few mosaics on the sidewalks and in the vestibules of houses — for example, the boar in the House of the Wild Boar (reg. VIII, ins. 3, no. 8), the anchor in the House of the Anchor (reg. VI, ins. 10, no. 7), and the ships in a house (reg. VII, ins. 15, no. 1–2) — seem to have been family emblems; some also indicate the family profession.

The offers of contracts or of premiums to the public, being of an occasional and temporary nature, were generally made by means of a crier, except when they were explicitly included in the signs of the enterprise. There are, however, some written "for rent" signs in Pompeii painted on the exteriors of the buildings (*CIL*, IV, 138, 1136, and possibly also 807). The services offered by the Roman horrea (warehouses) were written on stone; they also specified the conditions of business (*CIL*, VI, 33747, 33860).

In Pompeii notices regarding the loss of animals or objects were painted on the exteriors of buildings and tombs; they included a promise of a reward for the finder (*CIL*, IV, 64, 3864, 8938; *NSc*, 1958, 140, no. 330, Pompeii). Humorous "lost and found" notices with similar texts are found in Petronius (*Satirae*, XCVII, 1–2) and in Apuleius (*Madaurensis Metamorphoseon*, VI, 8), from which it is clear that it was also the custom for a crier to make such announcements orally.

Trademarks, either impressed or in relief (for the most part bearing only the name of the manufacturer but occasionally that of the craftsman), are frequently found on bricks, tiles, and various utensils of terra cotta and metal. Among the oldest certain examples are the gold fibula of Manios (7th cent. B.C.) and the bronze Ficoroni cist (V, PL. 50). Numerous Hellenistic and Roman fictile vases with relief ornament bearing an indication of the name of the maker have been found in various parts of the Mediterranean world, in the Crimea, and in Asia Minor. In Italy, particularly in the 3d to the 1st century B.C., the most famous were the vases of C. Popilius, the so-called "Calene" vases, and the Aretine ware. In some cases the mark was a symbol or figure (appearing either separately or in combination) or even a scene that perhaps served primarily as an ornament rather than as a trademark. Such identifying marks were even used on bread and cakes.

Amphoras, casks, jars, and other types of containers also had indications of the contents and the manufacturer painted on them. Many inscribed amphoras (sometimes with other information) have been found in Pompeii (*CIL*, IV, 2551–2880, 5510–6600, 6911–7007, 7109–7115, 9313–9821; *NSc*, 1958, 158–70, nos. 406–519, Pompeii); a few have come to light in other places. The later amphoras, particularly those dating from the 3d century on, sometimes have a figural sign in addition to, or in place of, a written inscription. Possibly its function was purely ornamental. In some cases it is difficult to ascertain whether a name is that of the manufacturer, the producer, or the owner.

In Rome the programs of the entertainments (circus, theater, and amphitheater) were announced by means of handbills that were also on sale (Cicero, *Philippicus*, II, xxxviii, 97; Seneca, *Epistulae morales*, CXVII, 30; *Scriptores Historiae Augustae*, V, 5). In Pompeii and other Campanian cities, paintings (always without figures) containing announcements of gladiatorial fights have been found on the exterior walls of buildings and tombs.

Approximately 3000 inscriptions of electoral propaganda in favor of candidates for the civic magistrature have been found in Pompeii and Herculaneum (PL. 375). The majority can be ascribed to the years 50–79 (a few examples are older); they are all painted, often in characters of studied elegance, on the exterior walls of buildings and occasionally in the vestibule. These announcements are never accompanied by figural scenes; the few graffito figures that are sometimes found near them have no connection with the written propaganda. These inscriptions could also serve as a publicity medium when the name of the supporter of the candidate was included. The written inscriptions painted on the piers of the portico of the so-called "Little Market" in Ostia Antica (reg. I, ins. 8, no. 1) were probably electoral propaganda, but they may have been commercial publicity or advertisements.

Paintings, sculptures, and emblems (sometimes alone, sometimes in conjunction with written posters) made to be carried in the triumph of a Roman general belong to the category of political propaganda (Ovid, *Tristia*, IV, ii, 20; Suetonius, *Divus Julius*, I, xxxvii; Appian, *De bello civico*, II, xv, 101). A considerable number of references regarding the use of such propaganda exist, some of them fragmentary. Depending on the case, the images and symbols were those of the cities and provinces over which the conqueror had triumphed (Cicero, *De officiis*, II, viii, 28; Cicero, *Philippicus*, VIII, vi, 18; Livy, XXVI, xxi, 6–7, XXXVII, lix, 3; Ovid, *op. cit.*, 37 ff.; Ovid, *Epistulae ex Ponto*, II, i, 37 ff., III, iv, 105 ff.; Propertius, II, i, 31 ff.; Florus, III, xiii, 88; Valerius Maximus, II, viii, 7; Pliny, *Naturalis historia*, V, v, 36; Tacitus, *Annales*, II, xli, 2), or of the slain enemy king or general (Plutarch, *Vitae*, XXI, lxxxvi; Dio Cassius, XLIII, xxiv, 1, LI, xxi, 8; Appian, *op. cit.*), or even the portrait of the triumphant general himself (Pliny, *op. cit.*, XXXVII, vi, 14, which mentions the portrait of Pompey done in pearls). For the triumph of Aemilius Paulus, after his victory over Perseus (167 B.C.), the philosopher-painter Metrodoros of Athens was chosen to execute the necessary paintings (Pliny, *op. cit.*, XXXV, 135). There are references to many other paintings and sculptures made for the purpose of political propaganda, especially to celebrate military victories.

Pio CIPROTTI

THE MIDDLE AGES TO THE 19TH CENTURY. In contrast with the religions of the ancient world, which were based on mysteries and whose proselytization was directed to the individual, Christianity engaged in the most widespread propaganda in order to conquer the masses. The principal media of this propaganda were the figural arts: the mosaics, frescoes, and reliefs served to illustrate the Bible and the Gospels to the illiterate. The decorative and narrative cycles were commented on by the preachers from the pulpit; they pointed out the scenes depicted in the works of art and explained their symbols and contents. The most successful example of recruiting and war propaganda was the Church's proselytization for the Crusades over the course of two centuries.

Later, the Counter Reformation made use of the arts to illustrate the sacred texts with greater realism (see REFORMATION AND COUNTER REFORMATION). The Sacred Congregation of Propaganda was founded in 1622 to carry on foreign missionary propaganda. The baroque (see BAROQUE ART) was to be the last great expression of a religious art directed toward the masses, a typical example of the art of persuasion, of the *delectando docere*. In the 19th century the Church realized that other and more efficacious forms of propaganda could take the place of the arts, and abandoned her patronage of the arts which had lasted for more than a millenium.

Publicity connected with various secular activities, such as sports, theatrical entertainments, and trade, continued to make use of various media; these can be divided into three main categories: signboards, trademarks, and wall posters. The written word, the image, and the symbol continued to be the primary means of communication.

In the Middle Ages emblematic publicity came to occupy a more important place than it had in the ancient world. Symbols increasingly replaced the written word, penetrating the mentality of the individual as an essential form of thought and communication. The emblem became diffused as the badge of the different strata of the social hierarchy, distinguishing individual families and guilds from the collective mass. The widespread use of emblems also served as propaganda for chivalry at the time of the Crusades. Heraldry originated in the 12th century (see EMBLEMS AND INSIGNIA). The signs of ships and artisans' workshops were fashioned from the emblems of the craft. Taverns, public houses, and inns sported wrought iron or wooden signs (PL. 376), some of considerable artistic merit, depicting the popular name of the premises, such as the Bear or Cock tavern, indicated by a bear or a cock in profile over the door or on the roof. Examples survive all over Europe, especially in central Italy, Austria, and Bavaria. This medieval tradition explains the widely diffused existence of emblems in many modern trademarks and is the prototype of modern signboards and shop windows.

With the era of urban civilization and mercantilism, which began in the 15th century, there was a progressive increase in population, and consequently in trade. Markets and fairs were developed and the manufacturing centers grew wealthy through their exports and their trade with distant places. This required that goods be advertised, which was generally done by word of mouth. Criers and canvassers announced the arrival of merchants in the villages and praised the quality of their wares. This type of publicity still survives in some countries. The public scribes prepared copies of proclamations and notices of public interest; this was an embryonic form of journalism.

The invention and diffusion of printing (see GRAPHIC ARTS) opened up many new possibilities, which were to have far-reaching developments. Printing became the most suitable and widely used medium for publicity and advertising. The satirist Pietro Aretino was one of the first to realize the publicity value of the printed word and book. It was the invention of printing that gave rise to the saying: "verba volant et scripta manent" (words fly and writing endures). People soon began to believe that what was printed must undoubtedly be true. The printing press could exalt a person or destroy him with satire. It was difficult for someone who had been publicly libeled to redeem himself. Ideological propaganda by means of the printed word became so effective that the Church had recourse to the Index.

Books bore printers' or publishers' devices and title pages to make the contents known and at the same time to advertise themselves (VI, PLS. 408–09, FIG. 699). Books were illustrated with woodcuts (VI, PLS. 405–07, 415). Printers also used woodcut illustrations for publicity purposes, which must have been very widely diffused since they inspired a satire on publicity.

Between the 16th and 18th centuries there was a fairly widespread production of printed matter of an occasional nature that served practical or informative purposes, rather than fulfilling cultural and esthetic exigencies. Included in this category are handbills (PL. 378), announcements with or without illustrations, such as the announcements for a lottery (PL. 377), proclamations, permits, and invitations to public ceremonies. Printed sheets, pasted on walls, invited young men to enroll in the army (PL. 378) and gave information regarding entertainments and other items of interest to the public. They were also sent by mail or delivered by hand. Their circulation increased with the development of social life in the 18th century.

During the baroque period shops continued to use signboards. In many places beautiful shop signs (mostly of the 19th century) still survive (PL. 376). Sometimes these signs were paintings, often by a good artist, such as the signboard for a barbershop by Vittore Ghislandi (Fra Galgario; Bergamo, Acc. Carrara) and for the antiquarian Gersaint by Watteau (IV, PL. 29; XIV, PL. 404).

The 17th century saw the publication of the first newspapers and periodicals in the form of multiple-paged pamphlets. Newspaper advertising began with Théophraste Renaudot, who founded the *Gazette* (later the *Gazette de France*) in Paris in 1630. In its sixth issue the first paid advertisement, placed by a physician, appeared. The most famous journal devoted to commercial publicity appeared in 1751: *La Petite Affiche*, published by Abbé J. L. Aubert. Such novelties, however, were generally regarded with suspicion, and for a long time recourse to these media was considered indecorous. In England weekly papers carried a few advertisements, for example, in 1658 the *Mercurius Politicus* published the first advertisement for Chinese tea. The outstanding daily paper in the mid-18th century was the *General* (later *Public*) *Advertiser* of London, which contained many advertisements. These were the origins of advertising, which was to develop enormously in the 19th century.

Luigi SALERNO

POSTER ART IN THE 19TH AND 20TH CENTURIES. Advertising as it is known today had its origin in the European poster art of the late 19th century. The increased use of posters both as an art form and as a selling device resulted largely from the development of printing techniques throughout the 19th century that accelerated the progress and improved the quality of color reproduction. In 1795 a Bavarian named Alois Senefelder invented lithography. This process of drawing directly on limestone with wax crayon or other water-repellent media permitted the artist a freedom not attainable in engraving, where he had to rely on a carefully graduated intensity of line for tone and contour. In 1811 Friedrich König used steam power for the first time to run a printing press, and he and Andreas Friedrich Bauer produced a flat-bed press with revolving cylinder which began its career in the service of the London *Times* in 1814. The American lithographers, Nathaniel Currier and James Merritt Ives, in a constant effort to polish their pictorial chronicles of contemporary life and to satisfy their customers by issuing accounts of newsworthy events as soon as possible after they took place, progressed from printing in black and coloring by hand to exact color register and rapid full-color lithography in consecutive runs through the press.

In 1866 one of the early French impressionists, Jules Chéret, established a printing shop in Paris. By drawing directly on the lithographic stone and printing in three colors, Chéret introduced to French poster design a spontaneity and appeal that Henri de Toulouse-Lautrec was to perfect and modernize. Chéret, Toulouse-Lautrec, and other impressionists, inspired by Japanese prints, established certain basic principles of design that are still used in good advertising. Background elements were subdued or deleted altogether and replaced by a solid block of color. Details and shapes were radically simplified to focus attention on the subject, and with the elimination of a frame the viewer was drawn into the activity of the illustration. Toulouse-Lautrec's penetrating depictions of Jane Avril (PL. 383), Yvette Guilbert, and Aristide Bruant sold not only the entertainer's performance but also the sensual atmosphere of the music hall itself (XIV, PL. 120). Other poster artists of the late 19th century included Jules Alexandre Grün; Pierre Bonnard; Théophile Steinlen (PL. 384) from Switzerland; Alphonse Mucha, a Czech working in Paris; Aubrey Beardsley in England (I, PL. 470); James Pryde and his brother-in-law William Nicholson (known as the Beggarstaff Brothers) in England; and Jan Toorop, one of the foremost practitioners of Art Nouveau, in the Netherlands (PL. 381).

The undulating lines, simplicity of composition, stylized forms, and purely decorative backgrounds of Art Nouveau (q.v.) were particularly appropriate to poster art. The decorative forms of Art Nouveau invaded every area of applied art, influencing artists to abandon simple illustration in favor of design.

In the United States at the turn of the century poster art compensated in quantity and exuberance for what it lacked in taste and style. Baroque layouts and wildly inventive 19th-century wooden type faces recalled the more primitive forms of European advertising; the narrative element of illustration was prominent. Countless posters for theater and circus performances were lithographed in black and the primary colors, but speed and applicability discouraged original thought. Spaces were left for names and dates so that one poster could be printed in quantity and sold to several different companies. The same was true in the production of handbills distributed before the arrival of a patent medicine man to persuade the public to watch his demonstrations. Elaborate and often impossible claims were made, as there was no Federal Trade Commission to institute controls for truth in advertising.

Carol Stevens KNER

An exhibition of posters was held in New York in 1890. The first exhibition of posters in England took place in London in 1894. Collectors were extremely interested in poster art. In January, 1896, it was estimated that there were more than 6000 collections of posters in the United States alone. The poster not only stimulated the graphic and calligraphic styles of the Art Nouveau movement but it also influenced the pictorial styles based on the decorative effects of surface color: the Fauves (q.v.) and the Brücke and Blaue Reiter groups (see EXPRESSIONISM).

In France the influence of suprematism and other abstract trends, which stressed the primacy of pure color and geometric shapes, was strong. About 1920, posters appeared with more brilliant and contrasting colors, which clearly showed the influence of cubism and then of Delaunay and of Orphism. The breaking down of masses into planes was a technique particularly suited to poster art, and so was dynamic vision, brought about by the fact that the poster was seen by a spectator in motion, for example, in the Paris Métro stations. An outstanding example of this type of poster was the one that A. M. Cassandre did for Dubonnet (PL. 393).

In Germany the two main centers of poster art in the early 20th century were Munich, where the leading poster artists remained tied to the Art Nouveau style and were interested in color and movement, and Berlin, where some of the artists attempted to create the "objective" poster. Although the Bauhaus favored the abandonment of traditional academicism, political conditions prevented developments in this direction. After World War II the German graphic artists reorganized the Bund Deutscher Gebrauchsgraphiker (founded in 1919), which has held annual exhibitions and has furthered the development of avant-garde trends.

In England poster art from 1900 to 1914 was dominated by Fred Taylor and Tom Purvis. As in the other countries mentioned above, a direct relationship was established in England in the years between World War I and II between poster art and modern art movements, a relationship that was encouraged by perceptive, farsighted industrialists. The Shell Oil Company commissioned work from Paul Nash and Graham Sutherland. Frank Pick, head of advertising for the London underground, commissioned a series of posters inviting people to ride the subway, which influenced not only travel posters but international advertising in general. Among the artists who worked for Pick were Gregory Brown, Dame Laura Knight, Austin Cooper, and an American, E. McKnight Kauffer.

In Switzerland a leading figure in the graphic arts was Ernst Keller, professor at the Kunstgewerbeschule in Zurich, who sponsored a movement to reform the graphic elements of advertising. Modern Swiss poster art is extremely effective. Swiss designers reduce copy to a minimum and rely on strong visual symbols for impact and clarity.

In Italy the first artistic posters appeared in 1902; they were in the Stile Floreale (Art Nouveau) idiom. Later, posters appeared that showed the influence of futurism (always sensitive to industrial phenomena). After 1927 the rationalist movement, founded by a group of architects who were followers of the futurist Sant'Elia, exerted a decisive influence on the graphic arts. The illustrative element was progressively discarded in favor of visual unity and of a union between image and content, form and function. Since World War II abstract concepts have prevailed, and large Italian companies, such as Olivetti (PL. 390), Einaudi, Rinascente, Pirelli, and Fiat, have commissioned many works in this vein.

In Belgium, the Netherlands, Denmark, Norway, and Sweden poster art attained a high artistic level. The film and theater posters of the Polish designer Jan Lenica show the influence of the Fauves and the Blaue Reiter group in exploiting the psychological and symbolic meanings of color.

Poster design has become a specialized activity, almost an autonomous profession. Posters are used to advertise plays (PL. 389), ballets (III, PL. 302), operas (PL. 386), and motion pictures (PLS. 386–87), for commercial and tourist publicity, and for political propaganda. The use of posters to advertise art exhibitions began in 1927, and prominent artists like Dufy, Léger, Matisse, Braque, Chagall, and Miró have executed such posters. In 1946 Picasso designed posters for an exhibition of ceramics held in Vallauris and later he executed posters for exhibitions of his own works. Many artists now design the posters for their own exhibitions (PL. 390).

Poster design has its own functional and formal requirements. Most important are clarity and visibility. The poster must not contain too much text or too small type. At the same time the image must lend itself to a hasty and comprehensive "reading." Clarity, synthesis, and impact are the fundamental characteristics of an effective poster.

A poster, moreover, tends to remain a one-dimensional picture, even if it attempts, or achieves, effects of relief or depth. It never is a "set piece" in the sense that painting may be, because the image, typography, figures, and letters compose themselves into an indivisible unit like the illustrated page of a book. The chromatic element is not always achieved only by brilliant or vivid colors, or even, as in some more recent examples, by the use of luminous paints; it can also be achieved by the simplicity of the tonal harmony. While many poster artists have relied on realism and attracted the attention of the public by their technical virtuosity, others have discovered that it is possible to capture attention by the abstract or almost abstract beauty of an image. They have created masterpieces of graphic art worthy of the label of "fine art," concentrating on the composition of the lettering and its relation to the forms and colors as a part of an indivisible whole, obtaining effects that sometimes recall cubist collages. With the remarkable development of photographic reproduction (PL. 382), the poster artist, working in a wide size range, has availed himself of every means that photography and technical printing processes can offer.

Together with this qualitative development there has been an increase in the number of new collections and of exhibitions of posters, and books devoted to this medium have been published. Specialization has also given rise to the present type of professional organization and to the founding of advertising agencies, of periodicals, and of specialized editorial and industrial bodies. There has been an astounding increase in the number of advertising and publicity organizations in the United States, including the American Institute of Graphic Arts, the Art Directors Club of New York, and the National Society of Art Directors.

Luigi SALERNO

THE 20TH CENTURY. Advertising agencies existed before 1900, but only to place advertisements, not create them. N. W. Ayer, in 1900, was the first agency to assist clients in preparing copy. Art departments originated about 1905 when two- and three-color magazine advertisements became popular. Agencies became increasingly important as new publications proliferated and their circulation increased.

Advertisers eager to reach the public sought newspapers and periodicals with wide distribution. In America numerous magazines vied for attention on the newsstands. The covers were often printed in larger sizes to be used as advertisements in places other than those where the periodicals were sold. Traditional, standard cover designs were abandoned and American artists such as Edward Penfield and Maxfield Parrish, inspired by European poster design, created original and highly competitive cover designs. In 1895 *Century Magazine* sponsored the first contest for a magazine cover. In the late 19th century two German magazines, *Die Jugend* and *Simplicissimus*, began publishing, and German artists evinced a growing interest in publications from other parts of the world. Influenced by the candor of American magazine covers and posters, Lucian Bernhard, with his poster for Priester Matches, and Ludwig Hohlwein, in his work for Marco Polo Tea and Munich Löwenbräu, reduced their poster designs to a simplicity unknown before.

In the United States publication advertising was far more prevalent than it was in Europe, where poster advertising was more common. An important factor in the growth of publication advertising was the development of photoengraving. Halftones could be reproduced in any size, and photographs were widely used, which at first presented the merchandise in a straightforward manner. During the 1920s and 1930s photography came to be recognized more and more as an art form. A photographer's choice of focus, lighting, size, film, and emulsion allowed him great freedom in interpreting the subject. The magazines *Vogue*, *Harper's Bazaar*, and *Vanity Fair* used avant-garde photography in their fashion illustrations. The influence of abstract art gave rise to experimentation with new techniques such as double exposure, blurred focus, grainy printing, distortion, retouching, and silhouetting. Type too could be distorted by the camera, and dramatic and surrealist effects were produced.

Color photography came into being with the development of color film in the 1930s and 1940s, and four-color process reproduction was gradually perfected. The choice of location, time, and subject matter was increased by the development of stroboscopic lighting and improved photographic equipment. With better printing processes photographs could be reproduced on an enormous scale and used for murals and billboards.

The spread of automobiles resulted in an increased use of advertising on the highways. Between 1900 and the 1920s a group of companies concerned with the appearance of the countryside and the standards of advertising design in general organized the Outdoor Advertising Association, later called Outdoor Advertising, Inc. This organization influenced the placing of billboards and brought about a reasonable amount of conformity in their size and construction, which facilitated the printing and designing of advertisements on a national scale.

During these years both business and government became increasingly aware of the power of advertising. Throughout World War I governments all over the world solicited the moral and emotional support of their citizens with posters, circulars, and various other types of propagandistic material. After the war American advertising agencies established European branches, but until the business boom following World War II these foreign offices had little success. The Europeans regarded as vulgar the American predilection for high-powered, hard-sell advertising, preferring publicity to persuasion. The depression in 1929 cut advertising budgets drastically both in the United States and Europe.

In the United States, with the example of government-sponsored graphic and architectural projects during the depression and big business's growing patronage of the artist, the gap between strictly commercial and fine art began to narrow. Large corporations created for themselves a kind of double publicity by buying fine art for offices, private collections, and small museums open to the public and by employing artists for their advertising and publicity. The International Business Machines Corporation was one of the first companies to collect art for the benefit both of its employees and the artist. In 1936 Thomas Watson, president of the company, paid the costs for 150 American paintings exhibited at the Venice Biennale. In 1939 the Connecticut Mutual Life Insurance Company published the first calendar reproducing paintings by American artists. The Pepsi-Cola Company in 1941, 1942, and 1943 issued calendars featuring American works of art, and in 1944 it sponsored a competition for artists to paint a "Portrait of America." The Abbott Laboratories in 1942, as a contribution to the national war effort, commissioned Thomas Hart Benton to paint *The Year of Peril*, a series of powerful dramatic illustrations depicting the horrors of war. The Standard Oil Company commissioned a series of paintings illustrating the history of the oil industry during the war years.

One of the companies internationally known for consistent excellence in advertising is the Container Corporation of America. Realizing that customers would be more likely to accept suggestions for improved package design if the company practiced what it preached in its own publicity, it instituted campaigns such as the "United Nations" series, and, more recently, the "Great Ideas of Western Man" series, employing such outstanding artists as Willem de Kooning (PL. 396), Fernand Léger, Miguel Covarrubias, Henry Moore, and Ben Shahn. DeBeers Diamonds Ltd., in introducing an advertising policy that would distinguish the company from slick credit jewelers, engaged the services of Picasso, Dufy, and Covarrubias.

Between World Wars I and II the dynamic concepts of the early 20th-century art movements began to be felt in every visual sphere. Whether the appearance of new graphic principles in advertising influenced the public's acceptance of modern art or whether the public accepted the application of abstract techniques in advertising design because it had already become accustomed to abstract painting is impossible to say. The Dada and De Stijl movements (see EUROPEAN MODERN MOVEMENTS; NONOBJECTIVE ART) contributed substantially in creating a climate for change and in overcoming the fear of radical innovation. More of an attitude than an artistic movement, Dada was a rebellion against bourgeois conventions in art, life, and politics, and was especially influential in shattering traditional concepts in typographic design. In the bold, unconventional work of the Italian typographer Franco Grignani and in Klaus Oberer's advertisements for Bols, the continuing stimulus of Dada principles is apparent.

The designer gained a new awareness of structure, shape, and texture from the cubist painters (see CUBISM AND FUTURISM). Theo van Doesburg and Mondrian explored asymmetrical balance, the interdependence of rectangles, and the relationship of primary colors in their work, which was reflected in the simplicity, clarity, and organization of book design, packaging, and other aspects of applied art. The French designer A. M. Cassandre absorbed from Le Corbusier and Léger a feeling for simplified volumes and symbolic forms, which influenced practically all subsequent trademark design (PL. 393). The collages of Bracque, Picasso, and Gris have had an effect on modern design. The movement and spatial freedom in the work of Balla, Severini, Duchamp, and other futurists and the mobiles of Alexander Calder resulted in a new originality in window decoration and store displays (PL. 394). Recently *Life* magazine ran several large-scale mobile advertisements that were modeled on kinetic art (see MODERNISM).

Expressionist concepts (see EXPRESSIONISM) affected the use of color in advertising design and, in many cases, a designer's entire attitude toward the psychological implications of his subject matter. The influence of the brilliant cut-paper forms of Matisse and the mischievous drawings of Klee are apparent in the work of the designer Paul Rand. The dynamic surrealist techniques of Dali, Chirico, and Tanguy have been used with dramatic results in advertising design. Dali created a series of whimsical stocking advertisements for Bryan (PL. 396).

The Bauhaus, founded in Germany in 1919 (see EUROPEAN MODERN MOVEMENTS; GROPIUS), encouraged experimentation with materials and investigated the relationship of colors and the plastic properties of space. The impact of Bauhaus principles on typography and page layout was revolutionary. New concepts were evolved for organizing space and type. Mondrianesque rules, bleed photographs, austere type faces, and the freedom of large areas of empty space created vibrant oppositions of horizontal and vertical. Headlines, columns, and blocks of type were arranged with regard to their textural qualities and operated as design elements. Sans-serif type faces and lower-case letters were used extensively, resulting in streamlined design.

When World War II started many European artists and designers sought refuge in the United States, some of them from the Bauhaus, which was closed by the Nazis in 1933. Trained in the tradition of good typography, book production, and illustration, they raised the standards of American design and exerted a tremendous influence on American designers. This influence created the distinction between hack "commercial art" — gaudy, commonplace window or road signs and hastily compiled copy for trite publication ads — and designs as conscientiously considered and composed as a Dürer painting.

After the war, improved communications and increased travel brought about an internationalization of advertising styles and media. Cartoon characters had always been popular in American advertisements. The influence of American comic strips and motion-picture cartoons, which had inspired Cassandre's famous three-stage Dubonnet poster (PL. 393), was evident in the animation and humor that appeared in European advertising.

Animation, humor, and pictorial rather than written symbols were used in television advertisements, which during the 1940s and 1950s became the American advertisers' most popular medium. Manufacturers spent more and more of their budgets on television commercials, and magazines, to meet the competition, were forced into drastic redesign programs to stimulate new interest. The redesign of *McCall's* magazine had an enormous effect on the entire magazine industry both in the United States and Europe, where *Paris Match* and *Twen* followed *McCall's* lead.

Business cards (PL. 377; VI, PL. 424), which were an early form of advertising in Europe, were the antecedents of all types of direct mail — pamphlets, folders, and brochures. Gadgets, gimmicks, and gift items are sometimes included with this kind of publicity material.

Packaging has become the most direct form of advertising (PL. 394). The first packages concerned with informing the consumer of the manufacturer's identity consisted simply of business cards or labels attached to or printed on wrappers and containers (PLS. 377–79). The importance of a package design, which, like a book jacket or record cover (PL. 392), is the last advertisement the consumer sees before he buys the product, was recognized in the United States in the 1930s with the growth of self-service markets. With the increase of supermarkets in Europe the importance of attractive packaging has been recognized. New materials such as cellophane, plastics, and light metals and improved color-printing processes have opened up new possibilities in package design. The package has become important as an ambassador for the manufacturer (VIII, PL. 61).

Manufacturers have become increasingly interested in establishing corporate identity. Some trademarks, like those for Uneeda Biscuit and His Master's Voice, designed in 1900, continue to denote experience and quality. Corporations such as International Business Machines, Olivetti, Geigy, and Braun have developed comprehensive programs to help promote their names and the quality of their services. Brochures, letterheads, promotion material, and packages are all designed in relation to one another. Specially designed luxury items unavailable to the general public, such as the Container Corporation Atlas, playing cards from Time Inc., and Olivetti calendars, are produced as gifts for clients and suppliers. Even an office building can promote a corporate image, either by virtue of elements in its structure that indicate the company's line of work or simply because it was designed by an accomplished architect and improves the appearance or adds to the prestige of the city in which it was built (I, PLS. 94, 137; V, PLS. 103, 105, 110, 114). The graceful projections and streamlined contours of Eero Saarinen's Trans World Airlines Flight Center in New York (PL. 398) suggest flight; the Alcoa Building in Pittsburgh (Harrison & Abramovitz) is sheathed with aluminum panels; the Maritime Union Building in New York (Albert Ledner) has portholes and tiers like the decks of a ship; and Saarinen's building for the John Deere Company in Moline, Ill., brings to mind the color, strength, and function of farm machinery. In New York City Lever House (Skidmore, Owings & Merrill; I, PL. 95) and the Seagram Building (Mies van der Rohe and P. C. Johnson; I, PL. 95), with its spacious plaza, brought a modern elegance to Park Avenue; Saarinen's building for the Columbia Broadcasting System is probably the most original of all the recent architecture on the Avenue of the Americas; and the Pan American Building at 45th Street by Walter Gropius has completely changed the midtown skyline.

Other forms of commercial publicity are the buildings erected by large corporations at world fairs, the display of new products at industrial exhibitions (q.v.), and outdoor advertising (billboards and illuminated signs; PL. 395).

Advertising has had an enormous influence on other creative work. Typographic forms and patterns as well as some of the gaudier aspects of full-color photographs used in publication advertising attracted the interest of artists as compositional devices. In certain portions of Picasso's *Guernica* (VII, PL. 278) gray and white dot patterns simulate rows of newspaper type and add a suggestion of journalistic sensationalism to the drama of the composition. Picasso, Braque, and Gris, fascinated by the textures of newstype, labels, and other commercial printing, incorporated these items in their collages and even sometimes used whole sheets of newspaper as a background for a painting or drawing (IV, PL. 80; IX, PL. 393). Signs and lettering painted in his own style appear as abstract, graphic elements in many paintings by Stuart Davis (I, PL. 126). Tom Wesselmann has surrounded his paintings of *Great American Nudes* with magazine cutouts of sundaes, chocolate pudding, and travel posters.

The influence of advertising on fine art is evident in "pop" art. Robert Indiana paints literary quotations in his canvases in lettering like the "fragile" and "handle with care" lettering on shipping crates. Andy Warhol (PL. 397) reproduces Brillo boxes and cartoon characters. Claes Oldenburg concocts hot dogs, hamburgers, pies, and cakes in plaster of Paris. In many of Robert Rauschenberg's compositions flashing lights evoke the garish rhythm of neon signs. These and other artists have adopted the gaudy commercial media of the 20th century as a means of spotlighting commercialism and identifying it as the culture of our time.

Carol Stevens KNER

BIBLIOG. *The Roman period. a. Signboards*: W. Helbig, Wandgemälde der vom Vesuv verschütteten Städte Campaniens, Leipzig, 1868; A. Sogliano, Le pitture murali campane scoperte negli anni 1867–1879, Naples, 1879; F. Matz and F. von Duhn, Antike Bildwerke in Rom, Leipzig, 1881, passim (particularly nos. 2858–74 and 3548–51); A. Mau, RE, s.v. Aushängeschilder; P. Gusman, Pompei, Paris, 1900, pp. 217–22 (Eng. trans., F. Simmonds and M. Jourdain, London, 1900); V. Chapot, DA, s.v. Signum (bibliog.); H. Blümner, Technologie und Terminologie des Gewerbe und Künste bei Greichen und Römern, 2d ed., I, Leipzig, Berlin, 1912, passim; S. Reinach, Répertoire de reliefs grecs et romains, II–III, Paris, 1912, passim; H. Gummerus, Darstellungen aus dem Handwerk auf römischen Grab- und Votivsteinen in Italien, JdI, XXVIII, 1913, pp. 63–126; K. Miller, Itineraria romana, Stuttgart, 1916, p. XLVIII; J. W. Kubitschek, RE, 2d ser., s.v. Signum; Mostra augustea della romanità, 4th ed., Rome, 1938, pp. 648–64, 887–91, 898–904, app. bibliog., 293–302, 410–12, 415–19 (cat.); G. Calza, La necropoli del Porto di Roma nell'Isola Sacra, Rome, 1940, pp. 203–04, 248–57; R. Calza, Museo Ostiense, Rome, 1947, pp. 26–28; A. Levi and D. Faccenna, Insignia, in V. Spinazzola, Pompei alla luce degli scavi nuovi della via dell'Abbondanza, I, Rome, 1953, p. 635 (incomplete list of Pompeian examples); M. Della Corte, Case ed abitanti di Pompei, II, Rome, 1954, passim. *b. Makers' marks*: CIL, X, 8041–74, 8330–39, 8422, XI, 6672–6731, 8105–35, XIII, 1001–10036, 12091–13099, XIV, 5308–19; H. Dessau, Inscriptiones latinae selectae, II, 2, Berlin, 1906, pp. 951–85. *c. Brick marks*: H. Bloch, I bolli laterizi e la storia edilizia romana, 3 pts., Rome, 1939–47 (bibliog.). *d. Marks indicating property in immovables*: H. Dessau, Inscriptiones latinae selectae, II, 1, Berlin, 1902, pp. 475–78; V. Arangio-Ruiz, Fontes iuris romani anteiustiniani, III (negotia), Florence, 1943, pp. 343–48. *e. Political propaganda*: CIL, IV, passim (inscriptiones pictae); S. Willems, Les élections municipales à Pompei, Brussels, 1886; E. Pais, Fasti triumphales populi Romani, Rome, 1920, passim; M. Della Corte, Case ed abitati di Pompei, Rome, 1954, passim.

Modern period: *a. Posters*: E. Maindron, Les affiches illustrées, GBA, XXX, 1884, pp. 415–23, 533–47; O. Uzanne, Les collectionneurs d'affiches illustrées, Livre moderne, III, 1891, pp. 193–206, 257–67; C. Hiatt, The Collection of Posters: A New Field for Connoisseurs, The Studio, I, 1893, pp. 61–64; C. Hiatt, Picture Posters: A Short History of the Illustrated Placard, London, 1895; M. Bauwens et al., Les affiches étrangères illustrées, Paris, 1897; V. Pica, Attraverso gli albi e le cartelle, VII: I cartelloni illustrati in Francia, America, Inghilterra, ecc., Bergamo, 1921; E. McK. Kauffer, The Art of the Poster, New York, London, 1925; G. Mauclair, Jules Chéret, Paris, 1930; R. J. Goldwater, L'affiche moderne, GBA, XXII, 1942, pp. 173–82; G. Lo Duca, L'affiche, Paris, 1947; C. Périsseau, Les peintres et l'art de l'affiche, Rev. fr., V, 47, 1953, pp. 21–25; R. Gandilhon, Pour un inventaire des affiches de théâtre: Affiches conservées aux Archives de la Marne (XVIIe–XVIIIe siècles), Rev. d'h. du théâtre, IV, 1954, pp. 295–99; R. Koch, The Poster Movement and Art Nouveau, GBA, L, 1957, pp. 285–96; W. Rotzeler, Il manifesto pubblicitario svizzero d'oggi, La Biennale di Venezia, 1958, pp. 23–30; F. Mourlot, Les affiches originales des maîtres de l'école de Paris, 1959; F. Filippini, Arte, industria e costume nei cartelloni svizzeri, Civiltà delle macchine, X, 1, 1962, pp. 43–48; E. Metzl, The Poster, New York, 1963.

b. General and special studies. A. Lancellotti, Storia aneddotica della réclame, Milan, 1913; L. Ramo, L'arte in réclame, Milan, 1917; H. Tipper et al., Advertising: Its Principles and Practices, 2d ed., New York, 1919; E. Roggero, Come si riesce con la pubblicità, Milan, 1920; S. E. Hall et al., Advertising Handbook, New York, 1921; D. C. A. Hemet, Traité de publicité, 2 vols., Paris, 1922; K. Lauterer, Lehrbuch der Reklame, Vienna, 1923; P. V. Bradshaw, Art in Advertising, London, 1925; G. French, 20th Century Advertising, New York, 1926; S. R. Hall, Theory and Practice of Advertising, New York, 1926; A. Halbert, Praktische Reklame, Hamburg, 1927; J. H. Picken, Advertising, 10 vols., New York, 1927; G. Russel, Advertisement Writing, London, 1927; R. E. Ramsay, Effective Direct Advertising, 2d ed., New York, London, 1928; K. Lauterer, Die Reklame von Morgen, Zurich, Frankfurt am Main, 1929; R. Seyffert, Allgemeine Werbelehre, Stuttgart, 1929; M. Rosenberg and E. W. Hartley, Art of Advertising, New York, 1930; H. W. Hess, Advertising: Its Economics, Philosophy and Technique, Philadelphia, London, 1931; K. M. Goode, Manual of Modern Advertising, New York, 1932; H. Bayer, W. Gropius, and I. Gropius, eds., Bauhaus 1919–28, New York, 1938 (2d ed., Boston, 1952); E. McK. Kauffer, Advertising Art Now, AD, VIII, 2, 1941–42; H. Sternberg, Silk Screen Color Printing, AD, VII, 1, 1941; G. Kepes, Language of Vision, Chicago, 1945; P. Theobald, Modern Art in Advertising, Chicago, 1946; E. M. Ettenberg, Type for Books and Advertising, New York, 1947; E. Mc Causland, Work for Artists, What, Where, How, New York, 1947; L. Moholy-Nagy, Vision in Motion, Chicago, 1947; Graphis Press, Trade Marks and Symbols, Zurich, 1948; R. Goldwater and R. d'Harnoncourt, Modern Art in Your Life, New York, 1949; L. Hogben, From Cave Painter to Comic Strip, New York, 1949; C. Coiner, Advertising Art in 1900, Portfolio, I, 2, 1950; C. P. Hornung, Handbook of Early American Advertising Art, 2d ed., 2 vols., New York, 1953; W. Verkauf, ed., Dada, New York, 1961; W. Amstutz, ed., Who's Who in Graphic Art, Zurich, 1962; G. Dorfles, Simbolo, comunicazione, consumo, Turin, 1962; R. Schmutzler, Art Nouveau, New York, 1962.

c. Annuals and periodicals: Art Directors Club, Ann. of Advertising Art in the U.S., New York, 1927 ff.; Posters & Publicity, London, New

York, 1927; Publicité, Geneva, 1943 ff.; Int. Poster Ann., Teufen AR, 1948–49 ff.; Graphis Ann. 1965/66, New York, 1965. The following periodicals in general offer valuable coverage of advertising and the graphic arts: *American*: Modern Packaging, 1927 ff.; Print, 1939 ff. (esp. XVII, 3, 1963, XVIII, 1, 1964, XIX, 4, 1965); Art Direction, 1950 ff.; Industrial Design, 1954 ff.; CA, 1959 ff.; Packaging Design, 1960 ff. *English*: Modern Publicity, 1924 ff. (esp. 1930, 1931, 1934–35, 1935–36, 1951, 1951–52, 1954–55). *German*: Gebrauchsgraphik, 1942 ff. *Swiss*: Du, 1941 ff.; Graphis, 1944 ff. (esp. 1952–53 ann. and II, 60, 1955); Neue Graphik, 1958 ff. *Japanese*: Idea, 1953 ff.; Graphic Design, 1959 ff.

* *

Illustrations: PLS. 375–398.

PUCELLE, Jean. Although no certain dates outside a limited known working span — from before 1324 until at least 1327 — can be deduced from the five extant documents that mention the name of Jean Pucelle, it is probable that this gifted miniaturist worked for one or two decades longer than this brief period. It is not known where he was born or in whose shop he was trained, but his art was presumably founded on the French tradition of illumination established by the artist called Maître Honoré. In any case, Pucelle was a recognized artist in Paris when he was asked to design the great seal for the Confraternity of St-Jacques-aux-Pèlerins. His name appears in the account books of the confraternity for the years 1319–24, at a place in the manuscript suggesting that he probably received payment for his work at the end of 1323. Although the seal itself has been lost, it is known from an engraving and the reproduction of an impression, both made in the 19th century.

Pucelle's name appears four other times, in connection with books that he illuminated. The Belleville Breviary (VI, PL. 315), a two-volume Dominican service book which was made for Jeanne de Belleville and which later belonged to King Charles V of France, bears marginal notes about payments for the work done by Pucelle and three assistants. This book is convincingly dated after 1323, but before 1326. Robert of Billyng, an Englishman who transcribed a Latin Bible (Paris, Bib. Nat., Ms. lat. 11935), carefully noted along with his signature that the book had been illuminated by Jean Pucelle and two assistants and completed on Thursday, Apr. 30, 1327. Two other mentions of the name of Pucelle are thought to be references to the same book. One of these appears in a codicil to the will of Jeanne d'Evreux (d. 1371), who bequeathed to the reigning monarch, Charles V, a very small prayer book that Pucelle had illuminated for her at the order of her late husband Charles IV (le Bel), whom she had married three years before his death in 1328. Finally, the name of the miniaturist occurs in the inventories (1401, 1413, and 1416) of the Duc de Berry, who owned a book known as the *Heures de Pucelle*, with illuminations in black and white. The predominantly grisaille technique, of which Pucelle was especially fond, the small size, and above all the unmistakable evidence of Pucelle's authorship have led scholars to believe that a fine Book of Hours in New York (Met. Mus., The Cloisters) was the book made between 1325 and 1328 for Jeanne d'Evreux and to assume that this book went from its original owner to Charles V and then to his brother, the Duc de Berry. The Cloisters book is a precious example of the work of Jean Pucelle, for its miniatures are remarkable not only for their uniformly high quality but also for an extraordinary stylistic consistency, which seems to indicate that in carrying out this important royal commission the artist worked alone; whereas a number of other manuscripts reveal Pucelle as the chief illuminator, the director of a large and influential workshop manned by capable assistants.

The work of Jean Pucelle breaks with medieval tradition and marks the beginning of a new era in French painting. It is generally assumed that at the beginning of his career he visited Italy, where he found in the paintings of Duccio and Giotto inspiration and guidance in solving the artistic problems that arose from his eagerness to achieve greater naturalism, more convincing representation of space, and more vivid and moving expression of emotion. His style is characterized by extraordinary vitality and freshness, with a graceful and elegant technique that was perfectly suited to giving form to the myriad creations of his imagination.

Bibliog. E. Panofsky, Early Netherlandish Painting, I, Cambridge, Mass., 1953, p. 27 ff.; Metropolitan Museum of Art, The Hours of Jeanne d'Evreux, New York, 1957; K. Morand, Jean Pucelle, Oxford, 1962.

Margaretta M. Salinger

PUGET, Pierre. French sculptor, painter, architect, decorator, ship designer, and military engineer (b. Marseilles, Oct., 1620; d. Marseilles, Dec. 2, 1694). Pierre Puget was the son of a master mason. After a brief apprenticeship to a ship carpenter in Marseilles, in 1638, he worked in Florence. From about 1640 until 1644 he was a pupil of Pietro da Cortona (q.v.), assisting in the decoration of the Palazzo Barberini in Rome and the Palazzo Pitti in Florence. Through the auspices of the Duc de Bézé, Admiral of France, he was active as a decorator of warships at the Arsenal of Toulon. During the time of the Fronde he devoted his talents to religious painting (1649–54), preferring the Italian academic vein of the Carracci followers (notably Guercino), but as a colorist being closer to Rubens.

Churches in Toulon, Marseilles, and Aix-en-Provence have examples of his painting; many of these have been confused with works of his son François (1651–1707). In 1656, as part of his design for the new south portal of the Hôtel de Ville, in Toulon, he executed the telamones supporting the balcony, his initial sculptural effort. Though these figures are Roman high baroque in inspiration — synthesizing the evolution of Michelangelo, the Florentine mannerists, Bernini, and Pietro da Cortona — Puget infused them with a rude vitality and tense expressionism peculiarly his own, in a style that was incompatible with the strict classical taste of official French art at the time.

Their success, nevertheless, was immediate in nonofficial circles, for Claude Girardin invited Puget to Normandy to make statues for his Château de Vaudreuil (e.g., Rouen, Mus. B. A.). Nicolas Fouquet, the Minister of Finance, subsequently ordered the *Hercule Gaulois* (Louvre), a terra-cotta model of which (Paris, Ecole B. A.) demonstrates Puget's masterful amalgamation of 16th-century mannerist and 17th-century baroque formulas, which is heightened by Hellenistic attitudes with respect to the importance of coherent design. This coarsely virile Hercules, more like a hunted convict than suggestive of the calm self-containment of a strictly classical god, brings to mind an effective fusion of the Belvedere torso and the pugilist of the Museo Nazionale Romano. Puget's association with Fouquet ruined his hopes for court recognition under Fouquet's successor, Jean Baptiste Colbert. The aversion of the latter official for Puget was probably more political than esthetic, for Colbert did take the finished *Hercule Gaulois* for his own gardens at Sceaux.

From 1661 to 1667, during what might be termed his middle period, Puget remained in Genoa, where he produced what many consider to be his best works, those esteemed for their greater academic restraint. The *St. Sebastian* (II, PL. 170) affords a rewarding comparison with Bernini's *Daniel* for the Chigi Chapel in S. Maria del Popolo, Rome. It testifies to Puget's tendency to respect, in a truly classical fashion, the geometric confines of the marble block, as opposed to Bernini's typical space-breaking approach to it. This innately French attitude toward the validity of geometric structure may have been the basis for the ultimate acceptance of Puget's late works at Versailles. After directing various tasks at the Arsenal of Toulon (1667–79), recorded in over a hundred fine ink drawings on vellum, he secured permission from Colbert to execute three sculptural works for Versailles (1671–93): two for the gardens, *Milo of Crotona* and *Perseus and Andromeda* (both, Louvre); and one for the vestibule of the chapel, the *Alexander and Diogenes* (II, PL. 173).

The arrival at Versailles of the *Milo*, the most sensuous and exuberant baroque work by a French 17th-century sculptor to grace the gardens of Versailles — in strident contrast to the mild mythology of Michel Anguier, François Girardon, and Antoine Coysevox (q.v.). — marks the beginning of a period in which the austere tenets of the severely classical French "grand manner" were gradually mitigated. Though this period is sometimes characterized as the late baroque in France (ca. 1680–1700), constituting the maximum concession of the French to Italian high baroque canons of the period 1620–45, for Puget the dates range from 1683, the year of Colbert's death, to 1691, the date of the death of Louvois, Puget's supporter and Colbert's successor. The *Milo* was given a prominent place at the entrance to the Tapis Vert, and Charles Lebrun (q.v.), the rigorously academic dictator of the arts, wrote Puget to express his admiration. Despite its tragic message of man's overweening self-confidence destroyed by nature, in spite of the violence of movement and excessive naturalism of both the human anatomy and accessory elements, the over-all design is cautiously, even academically controlled and presents a synthesis of the principles of Myron's *Discobolus*, the *Laocoön*, and Michelangelo's *Slaves*. Viewed directly from the side, the essential geometric conception of the design is apparent in the parallelogram scheme defined, first, by the right arm and the tree trunk and, second, by the definition of the left arm in relation to the legs. Within this self-contained composition, there is a maximum elaboration of the realistic adjuncts, approaching the descriptive excesses of Bernini's *Apollo and Daphne*.

During the height of his official recognition, Puget's naïve sincerity often led him to make proposals marked by a brash enthusiasm. Anxious to contribute something to the royal glory, he proposed a gilded colossus of Apollo to bestride the Grand Canal at Versailles. When, as a result of intrigue, his *Alexander and Diogenes* failed to reach Versailles and payment for his accepted works was nearly impossible to secure, he retired permanently to Marseilles. His last work, *St. Carlo Borromeo Arresting the Plague of Milan* (Marseilles, Mus. B. A.) is, through Puget's respect for the Greek metope tradition, more classically conceived than comparable high-relief works even of Alessandro Algardi.

Puget's fusion of pictorial effects with sculptural forms has posed a persistent problem for admirers, however, and it was conspicuous even to a romantic individualist such as Delacroix. Though the scope of Puget's activities gives him many claims to a universal attitude, one might still regret his being both sculptor and painter, perhaps somewhat to the detriment of each. His principal followers were Christophe Veyrier, Antoine Duparc, and Carlo Simonetta, and his works contributed to the inauguration of the 18th-century rococo style that culminated in the fragile restlessness of Clodion and Fragonard.

BIBLIOG. J. Bougerel, in Mémoires pour servir à l'histoire de plusieurs hommes illustres de Provence, Paris, 1752; L. Lagrange, Pierre Puget, Paris, 1868; P. Auquier, Pierre Puget, décorateur naval et mariniste, Paris, n. d. [1903]; F. P. Alibert, Pierre Puget, Paris, 1930.

George Van Derveer GALLENKAMP

PUGIN, AUGUSTUS W. N. English architect, author, and designer (b. London, Mar. 1, 1812; d. Ramsgate, Kent, Sept. 14, 1852). Son of Augustus Charles Pugin, a talented architectural draftsman and illustrator frequently employed by Rudolph Ackermann and John Nash, the younger Pugin received his initial training from his father. His first independent professional undertaking was the continuation of *Examples of Gothic Architecture*, which A. C. Pugin had begun to publish before his death in 1832. In 1833–34 Pugin traveled widely to study Gothic art and architecture for the preparation of this book as well as to further his own education. He became a convert to Roman Catholicism in June, 1835. In 1835 and the first six months of 1836 he worked for Sir Charles Barry (q.v.) on the preparation of that architect's entry in the Houses of Parliament competition and on revisions of the design after Barry had received the commission. In 1836 he built and took up residence in St. Marie's Grange, near Salisbury, the first building constructed entirely from his designs; about the same time he completed his series of four volumes illustrating medieval decorative arts, published by Ackermann, and published independently the first edition of *Contrasts*. This book, a letterpress edition of 50 pages and 15 trenchant illustrations, contrasting the art of the 19th century with that of the late Middle Ages, brought him immediate celebrity as a controversialist in religious and architectural matters. By his own wish and because of the tense situation of his church in England, Pugin worked in the main for Catholic clients, the most generous of whom was John Talbot, Earl of Shrewsbury, who paid entirely or in part for many of the buildings designed by Pugin. The association of Pugin and Barry was renewed in 1844, and thereafter Pugin was much involved with designing the decoration and details for the Houses of Parliament.

Although about seventy buildings were constructed from his designs and he published more than twenty works during his lifetime (including books, pamphlets, and important groups of illustrations of architectural designs and decorative details), Pugin's major period of creativity lay between 1838 and 1844, when he wrote the books propounding his architectural theories and his interpretation of the history of ecclesiastical architecture; this was also the period in which he produced the buildings that established his personal style. The publications of his later years demonstrate an increasing preoccupation with matters of religious rather than architectural import.

By 1844 he had become closely associated with craftsmen and manufacturers who worked from his designs. With their assistance he contributed to the interior decoration of the Houses of Parliament and carried out his ecclesiastical and domestic architectural commissions. In pursuing his craft interests, Pugin created designs in many media, including metalwork, stained glass, furniture, tiles, fabrics, wallpaper, bookbindings, and even type faces. Pugin and his craftsmen associates installed examples of their work in the medieval court of the Crystal Palace for the Great Exhibition of 1851. By that date he was acknowledged to be the most notable figure of the Gothic revival, though many others had now entered the field he once dominated so completely.

Afflicted by ill health from 1836 onward, Pugin suffered from the chronic demands made upon his strength and emotions. In 1852 he became mentally ill and died at his home in Ramsgate.

Pugin's importance lies in his perception that from the study of fine architecture and decorative art general principles of quality could be discerned without reference to the particular characteristics of any historic style. In constructing this argument, he employed methods of reasoning derived from French theorists of the 18th century, and his originality lay in the application of this method to Gothic. Within his lifetime Lord Lindsay and Ruskin accepted his premises, but without his Roman Catholic bias. In his Utopian view of the Middle Ages, best seen in *Contrasts*, Pugin introduced into architectural theory concepts that had been current in English political and literary circles since 1800. His close partnership with his builder George Myers, his metal and stained-glass specialist John Hardman, and the firm of John Crace, who manufactured his furniture, foreshadows the arts and crafts movement later headed by William Morris (q.v.). In his acceptance of the Decorated style, Pugin was in advance of the Camden Society of Cambridge. Bibliophile, art collector, and authority on the history of medieval art and its adaptation in the Gothic revival, Pugin set a high standard for his contemporaries, both in his own expert knowledge and in his gifts as an artist.

WORKS. St. Wilfrid's Church (1839–42), Hulme, Manchester. – Alton Castle (ca. 1840), Alton, Staffordshire. – St. Oswald's parish church (1840–42), Old Swan, Liverpool. – St. John's Hospital (1840–44), Alton, Staffordshire. – The Grange (Pugin's home; 1841–43), Ramsgate, Kent. – St. Giles's Church (X, PL. 292), Cheadle, Staffordshire. – Bilton Grange (1841–46), Rugby, Warwickshire. – Rolle family chapel and tomb, Bicton Grange, Devonshire. – St. Barnabas's Cathedral, bishop's house, and convent (1842–44), Nottingham. – St. Mary's Church (1843), Brewood, Staffordshire. – St. Augustine's Church (1847–51), Ramsgate, Kent. – Chapel of St. Edmund's College (1853), Ware, Hertfordshire.

WRITINGS. Contrasts, London, 1836 (2d ed., 1841); The True Principles of Pointed or Christian Architecture, London, 1841; An Apology for the Revival of Christian Architecture in England, London, 1843; The Present State of Ecclesiastical Architecture in England, London, 1843; A Glossary of Ecclesiastical Ornament and Costume . . . , London, 1844.

BIBLIOG. M. Troppes-Lomax, Pugin, A Mediaeval Victorian, London, 1932; H.-R. Hitchock, Early Victorian Architecture in Britain, 2 vols., New Haven, 1954.

Phoebe B. STANTON

PUNIC ART. See PHOENICIAN-PUNIC ART.

PUPPETS, MARIONETTES, AND DOLLS. Numerous references can be found in the writings of Apuleius, Aristotle, Horace, Plato, and Xenophon to figures moved by the pulling of strings. Since figures with a rod projecting vertically from the head in the manner of the Sicilian puppet and terra-cotta dolls with articulated limbs have been found in ancient Greek tombs, puppets are believed to be as old as civilization itself. According to one authority (C. Magnin, 1852), the mobile sculpture known to have existed with the Greeks may have given rise to puppets through the priests, who adopted the use of mobile figures from a belief in the propitious effect of having the idols enact what was hoped for from the gods rather than from a desire to trick the faithful or a mere inclination toward theatricality; as the rituals became more complex, the effigies were further developed, until, in the later formal representations, human actors took the place of the puppets.

The concept of puppets as substitutes for human actors is evidently of more recent origin. According to one account of an incident dating from the 17th century, the success of the string-puppet performances of La Grille in the Marais in Paris provoked a protest by actors that drove the puppet theater back to the fairs. Though dolls — perhaps the earliest evidences of the human imitative process — both with and without movable members have been extensively studied, the history of puppets has failed to interest many scholars because of the common use of the puppet for puerile entertainment and because examples from ancient and primitive societies are rare and in poor condition.

The early conceptions and production of miniature theaters and conventional personages were less directed toward a detailed naturalistic representation of beings acting in a realistic dramatic situation than toward the creation of typifying gestures and simplified expressions that were a magical or symbolic evocation of some aspect of reality; in our era, however, realistic trends have predominated, especially in the production of dolls. The modern European puppet evolved in Italy, where puppets had been favored for religious drama, and was introduced in neighboring countries by traveling showmen. In 1573, an Italian puppeteer had established himself in London, and evidently puppet theaters were familiar to the Elizabethans. References to puppets appear in the works of Shakespeare and Ben Jonson, and a puppet version of the story of Adam and Eve is believed to have inspired Milton in his *Paradise Lost.*

Though the classic marionette was most in evidence in Europe during the 16th century, the hand puppet being favored in the 17th century, a folk branch of marionettes existed that can be traced, largely through literary sources, from the jigging puppets and marionettes *à la planchette* used by wandering musicians. This type of puppet is similar in principle to the type described by Geronimo (Girolamo) Cardano in *De rerum varietate* (Nürnberg, 1557). According to this source, two rudimentary puppets were joined together at the chest with a cord and were moved by two men (PL. 403).

By the 17th century, tales of chivalry and satirical plays, in addition to the former religious subjects, were presented with puppets. Later, masques were born with the appearance of comic puppet characters based on local types and characters borrowed from the *commedia dell'arte,* and puppet theaters became well patronized in Rome, Milan, Genoa, Naples, Munich, Basel, Vienna, and Paris, all of which were famed for their puppet theaters. The improved puppet performances began to attract and hold the attention of sophisticated writers and painters, as well as the skilled craftsmen and showmen, who began to discover the immense range of dramatic expression of which the puppet is capable; these artists themselves began carving or fashioning the figures, costuming them, designing the settings, and writing the plays. In the 18th century, when puppet plays had become more elaborate in treatment and presentation, many aristocrats established private theaters; at his palace in Eisenstadt, Prince Miklós József Esterházy presented five marionette operettas for which Haydn composed the music.

The trend toward realism, which began in the 19th century, especially in Germany with the rise of nationalistic spirit, predominated in most Western-influenced countries. In Italy, entire ballets, plays, and operas presented with marionettes were not uncommon. The famous Viennese showman and craftsman Geisselbrecht created highly realistic marionettes that could cough, choke, spit, and effect mysterious transformations. In the 20th century, though puppetry was further developed through experimentation during the first decades, when popular interest was renewed, the advent of animated cartoons, movies, and television has brought about a decline of general interest in puppetry, and the art is practiced by relatively few individual artists and puppet companies. (See AUTOMATA; CHOREOGRAPHY; CINEMATOGRAPHY; FOLK ART; GAMES AND TOYS; HUMAN FIGURE; MIMESIS; MYTH AND FABLE.)

HAND PUPPETS. The hand puppet is constructed to fit over the puppeteer's hand and to be manipulated by the puppeteer's fingers; for this reason it is also called a glove or fist puppet. Legs are sometimes added to the figure but are not directly controlled. The hand-puppet theater is constructed without a stage floor, permitting the operator to manipulate the puppet from below and to remain hidden from the audience. A widely popular type of puppet theater developed in many parts of the world, the repertory of which usually consisted of lively burlesques involving frequent beatings. The faces of the puppets were therefore not delicately carved but were rather crude and primitive. In Italy, hand puppets are called *burattini,* which was probably derived from *buratto,* a kind of rag for bolting grain, and which referred to their crude garment; they were called *bagatelli* in the 16th century (since they were displayed with the trifles and bric-a-brac of the street vendors), *magatelli* (referring to the art of magic), *fraccuradi* (rudimentary hand puppets made from pea pods), and *romanitt* (from the nickname "Il Romanino," by which the Milanese puppeteer Massimo Bertelli was popularly known). They were called *títeres* in Spain, *Handpuppen* in Germany, and *marionettes à main* in France. Traditional puppet characters, having their origins in obscure folk traditions or perhaps from local characters, developed in many countries. One of the oldest puppet characters is the Neapolitan Pulcinella, a clown type probably derived from the *commedia dell'arte,* created about the beginning of the 17th century. Similar protagonists, usually a small, deformed character with a large hooked nose and squeaky voice produced by a small ivory whistle held in the puppeteer's mouth, appeared in France as Polichinelle (ca. 1630), under the direction of Giovanni Briocci; in England as Punch (end of 17th cent.); in Austria and Germany as Hanswurst and Kasperl; and in Russia as Petruška. The guignol theater in France, the English parallel of which is the Punch-and-Judy show, evolved from the Lyonnais puppet character named Guignol, created during the Revolution by Laurent Mourguet (PL. 404) and continued by Louis Josserand.

Because hand puppets must be worked from below, the stage *décor* of the puppet theater was naturally elementary, the changes of scenery being limited to the backdrops of the stage and, rarely, to the two wings. The puppeteer's house, or castle, was a colorful element that has been depicted in many famous prints. For example, in several prints of Roman piazzas by Bartolomeo Pinelli (1781–1835), there are puppet theaters designed after those of Ghetanaccio (Gaetano Santangelo, 1782–1834); many French prints depicting the gate of Nesle show the booth of the Briochés (Giovanni and Francesco Briocci), famous puppeteers in Paris at the time of Jules Mazarin and Cyrano de Bergerac; and Pietro Longhi (1702–85; q.v.) drew the puppet house of Borgogna before the Doge's Palace in Venice. Though the manner of presentation varied among puppet theaters (e.g., in ancient China, the theater was simply a large sheet held above the puppeteer's head), the repertory was consistently popular. A plot that originated in 1587 was even credited with being the inspirational source for Goethe's *Faust,* and Goethe is said to have composed several plays in 1753 for a puppet theater that had been presented to him. From 1831 to 1903, the Bolognese puppeteers Filippo and Angelo Cuccoli were famous for their representations of the puppets Fagiolino and Sandrone made by Francesco Campogalliani, to whom a statue was erected at Mantua. In France in 1861, the writer-painter Louis Emile Edmond Duranty tried to ennoble the repertory of popular puppets in his castelet in the Tuileries gardens. With the collaboration of the sculptor Leboeuf, he attempted to produce puppet plays of high literary quality, distinguished by a certain philosophic flavor; in an attempt to achieve success, however, he resorted to the usual finale of canings. Among the famous nonroving puppet theaters was the Théâtre des Amis (1847–89) of George and Maurice Sand. At the peak of its popularity, the troupe of puppets comprised over 800 characters, with costumes fashioned by George Sand and the puppets and plays by her son Maurice and his friend, the painter Eugène Lambert. The Théâtre des Pupazzi (1843) of Lemercier de Neuville began as a gallery of caricatures of famous people and eventually became a performed show that traveled to all parts of Europe. At first it was a little theater of *pantins,* which were flat figures with movable arms, but later, on the advice of Gustave Doré, he modeled and dressed his puppets in the round after types borrowed from contemporary life. His puppet Pierrot playing a guitar was his masterpiece of characterization.

Puppet theaters of the 20th century include the theater of Gaston Baty, who reconstructed traditional guignol puppet plays of Lyon; the Opera dei Burattini of Maria Signorelli, who organized the International Festival of Puppets and Marionettes in Rome in 1961; and the Burattini all'Italiana of Maria Perego, who, in addition to directing experimental television presentations, for example, interpreting the opera *El retablo de Maese Pedro* (Master Peter's Puppet Show) by Manuel de Falla with an all-puppet cast, adapted the tradition of the ancient Japanese marionette theater for the animation of the large puppet figure of Don Quixote; the controllers were three, a main puppeteer and two assistants dressed in black.

The puppets of Sergei Vladimirovich Obraztzov (PL. 405) director of the Central Puppet Theater created in 1931 in Moscow, are inspired by tradition but have been modernized and revised for cultural purposes. Though almost all types of puppets are used, the hand puppet has been preferred. The Moscow puppeteer Nina Efimova, who has specialized in hand puppets, was an innovator in the use of new materials for costuming and lighting and in the technique of manipulation. Particularly original among 20th-century companies of hand puppets in Paris were those of Yves Joly (1945), Jacques Chesnais, who produced many Polichinelle plays, and Marcel Temporal; in the United States, those of Perry Dilley, whose repertory ranged from simple clowning to philosophical tragedies, and Elena Mitcoff; in Germany, those of Carl Schröder, Hermann Rulff, Max Radestock, and Max Jacob; in England, those of Bruce Macloud, Roy Newlands, and Walter Wilkinson; and in Switzerland, that of Fred Schneckenburger, whose surrealist glove puppets were controversial.

Alessandro BRISSONI

ROD PUPPETS. The traditional rod puppet, which, unlike the Sicilian type, is worked from below, is a complete, jointed figure mounted on a sturdy rod; in its simplest form the arms, attached to thin rods, are controlled and the legs hang free. The rounded wooden figures of the *wayang golek* play in Java are perhaps the most refined examples of this type. The figures have thin waists and long, slender arms jointed at the shoulder and elbow. The slow, graceful gestures that characterize rod puppets in general compliment the ornamental representation of the human form in the *wayang golek* figures, which are noteworthy also for their fantasy and expressiveness.

Another traditional type of rod puppet, characteristic of the Rhineland puppet theaters, evolved in Europe with the founding of the Hänneschen Theater by C. Winter in 1802. Apparently influenced by a variation of the Sicilian rod puppet that appeared in Flanders, Winter created a rod puppet, the right arm of which was controlled by a rod manipulated from below. The principal character, a fun-loving peasant youth named Johann Knoll, was popularly known as Hänneschen. Both the Alte Hänneschen Theater (Cologne, directed by Jakob Dingarten) and the Landestheater Saarpfalz (Kaiserlautern) continued the traditions of the Hänneschen Theater, and the type of puppet used became popular in smaller German puppet theaters. The same type was traditional in Liège, Antwerp, and Ghent for nativities, romances of chivalry, and extempore plays in local dialects. A variant of this type of Rhineland puppet, developed in the Aachen Puppet Theater (directed by Will Hermanns) was designed with the traditional thick rod holding the puppet and thin rod actuating the right arm and, in addition, with a special mechanism for animating the head.

Richard Teschner, a Viennese painter and sculptor, created a unique puppet theater, some works of which reflected influences from the Javanese round figures and Indo-Chinese art (*Der Drachentöter* and *Prinzessin und Wassermann*), with elegant, fragile figures of the rod-puppet type. Behind a gold-framed concave lens encircled with zodiacal symbols, Teschner manipulated his figures in mime to an accompaniment of slow, delicate music, like that of a music box, and harmoniously blended lighting effects of the setting. His *Die Orchidée* is said to have possessed a Vermeer-like quality. Teschner's theater was seen in London at the Austrian Exhibition in 1934, and his collection belongs to the Austrian nation.

Though not restricted to the presentation of rod puppets, the theater of the Detroit puppeteer Paul McPharlin included them in the many technical experiments for which it was well known. McPharlin was largely responsible for the formation of The Puppeteers of America, in 1937, and he published the annual *Puppetry*, one of his several works directed toward furthering the art. The Frenchman Jacques Chesnais, who was known for his glove puppets, produced his ballet *Boxers* with modernist rod puppets designed by Fernand Léger.

A special type of rod puppet known as the pedal, or piano, puppet, was conceived by Edmé Arnaud and elaborated by M. Bellée for the poet Henri Signoret, whose Petit Théâtre in the Galerie Vivienne was highly praised by artists and writers. During the several years of his theater's existence, Signoret revived plays that, though not well known, were of high literary quality. The figures themselves were mounted on an iron rod fitted into a base with pedals, to which the controlling strings were attached, and the arms and legs were animated by depressing the pedals. A variant of this type of puppet has since been used for special characters, a famous example being the puppet called Misery, designed by S. de Walles for the Théâtre de l'Arc-en-ciel.

* *

MARIONETTES. The term *marionette,* which came into use in the 17th century, is usually applied to the full-length jointed puppet manipulated by strings overhead. The term is probably of Italian origin, perhaps derived from the Festival of Mary during the Venetian republic when strolling vendors hawked *Marione*, miniature copies of huge processional statues, made to be moved by strings; but it may be derived from *marotte*, which refers to the wooden head carved on the scepter of jesters. In India, the director of a theatrical company is called *sūtradhāra* (the one who holds the strings), and the meaning seems to substantiate the theory that performances of marionettes were earlier than those of actors there. The term, however, is probably used in a broad sense to refer to the one who holds the thread of the dramatic action. In the *Satyricon* attributed to Petronius, a rather sophisticated marionette, a small silver skeleton, is mentioned; it was moved by strings and danced, and it provoked a discourse by Trimalchio on life and death.

The classic marionette, worked by many strings, is the most difficult type of puppet both to construct and to manipulate (PL. 403). The famous conventional marionette characters, such as Gianduia, Famiola, and Gerolamo, had as many as thirteen strings, which were maneuvered by one person. The body of the marionette, which in early times was rigid or semirigid, became highly flexible; the Englishman Thomas Holden, who traveled throughout Europe about 1875, was an expert innovator in this regard. The height of marionettes ranges from about 30 to 35 in. and the weight, complete with costume, from 6 to 16 lb.

The marionette stage is a miniature replica of any stage, adapted to conceal the marionettists and to allow the passage of strings, puppets, and controlling devices through the wings. The dialogue of the puppet theater, which at first was entrusted to only one storyteller assisted by a small orchestra, usually consisting of a drum and a mandolin, came to be read by the puppeteers themselves. The puppeteer controlled the strings from about 10 ft. above the stage on a scaffolding with a parapet capable of holding as many as six puppeteers, and in front of the parapet was the lectern holding the script, written in large letters. In our era, the custom was established of employing actors and singers having nothing to do with the handling of the marionette or of taping a sound track in advance.

Though throughout its history the hand puppet almost always played to popular audiences, the marionette acted in both the popular squares and in the palaces of the aristocracy and upper middle class. In the Middle Ages in Italy, marionettes were used in sacred mystery plays presented in the Church, and eventually they were carried to street or palace by means of portable booths, or *castelli*. In the 18th century, the Polichinelle marionette of the Fair of St-Laurent and St-Germain in Paris proudly proclaimed, after having attacked the privileges of the Opéra and the Comédie Française, "J'en valons biens d'autres" and was received in the salons of both the duchess of Maine and Voltaire. During the same period in Italy, Pier Jacopo Martelli, originator of the *martelliano* verse form, wrote various *bambocciate*, or marionette plays, for the little theater of Angelo Maria Labia in Venice. Original musical scores for marionettes were written for other private theaters in Venice, Bologna, Florence, Rome, and Naples. In the 19th century, Filippo Teoli wrote a series of comedies for his theater in the Fiano (now the Almagià) Palace in Rome, known as *The Masque of Cassandrino*. This skillful satire of the Roman clerical circles was described by Stendhal and was highly successful. In Munich, Count Franz von Pocci became the permanent "poet" of Josef Schmidt's famous marionette theater. Even when it was not original and was addressed to the common people, the marionette repertory always had a plot development and stage production of high theatrical effectiveness, the performances generally concluding with scenes grouping the entire cast in an apotheosis, as it was called in the 19th century, in imitation of the dances of the major theaters.

Marionette theaters were so successful in Europe in the 19th century (in France the theaters of Henri Signoret produced texts from Aristophanes and Shakespeare with marionettes) that they gained the attention of and were reviewed by official critics, and in many cities nontouring theaters were established that lasted for a century, even in Italy, which had had a tradition of touring theaters. In 1818, the Teatro Gianduia was established in Turin by the Lupi family marionette masters, and the Teatro Gerolamo was founded in 1815 in Milan by the Colla family marionettists. The technique of the puppet masters was handed down from father to son for many generations, but in our era it is inherited by very few. Vittorio Podrecca (1883–1959), creator of the Teatro dei Piccoli of Milan (1919), with the collaboration of well-known painters for the stage settings and with a repertory that drew from both the comic opera and the review, successfully presented traditional Italian marionettes all over the world. Among their varied presentations, the troupe illustrated the puppet's ability to imitate real people in its interpretations of famous theatrical personalities and types, and it featured productions with the traditional Sicilian puppet as well as with the classic marionette.

In Russia, the marionette theater has been officially recognized by the state, and there are many permanent theaters. In Czechoslovakia, homeland of the famous touring puppet master Matěj Kopecký (1775–1847), there are thousands of marionette and puppet theaters supported by magazines and specialized periodicals, by national and international conventions and groups, and by studies at the College of Marionettology of the Academy of Musical Arts in Prague. Josef Skupa's theater, established at Pilsen, frequently toured through Central Europe, and the productions of Jan Malík were well known for their fine composition and unusual, stylized quality.

Alessandro Brissoni

In Belgium, the painter Carlo Speder, who founded the Théâtre du Perruchet at Brussels, is noted for his marionette productions. Though in France, as in Russia, puppeteers in general seemed to prefer the glove puppet to the string puppet, the marionette productions of Robert Desarthis, Les Pajot-Walton, and the Théâtre de l'Arc-en-ciel were esteemed. For the latter theater, Gera Blattner conceived many original marionette figures.

In Germany artistic puppet plays were developed in Hilmar Binter's Künstlerische Puppenspiele, with well-characterized puppets designed by Walter Oberholzer, in Bendel and Flach's Würzburg theater, and in Georg Deininger's theater in Stuttgart, noteworthy for its artistry and fantasy. Hermann Aicher's marionette theater in Salzburg, opened in 1913 by his father Anton, was known for the high artistic quality of its period pieces and for the refinement of the productions. Harro Siegel, of Berlin, an authority and instructor in all phases of puppetry, formed a company noted for its puppet presentation of Mozart's opera *Bastien et Bastienne*. The marionettes of Fritz Gerhard, whose theater was outstanding for its carefully executed and well-conceived fantasies, were finely carved in the best traditions of German medieval art, with some figures of baroque influence. Peter Kastner, who founded his theater in Dortmund in 1934, attempted situations and themes, both serious and humorous, that were beyond the scope of human players. Zangerle's aim in founding a Rhineland marionette theater (1925) was to revive the local traditional puppetry, but using the string puppet instead of the rod puppet. His figures, some of them controlled by as many as 40 strings, were capable of an extraordinary range of expression.

A sculptor and wood carver who was a pioneer of modern English puppetry, William Simmonds designed, executed, and operated the marionettes in his own productions in London from 1916 to 1934. To an accompaniment of a muted fiddle and virginals, he operated a puppet in each hand while speaking the dialogue or singing. Kenneth Wood, who was perhaps best known as a painter, was also associated as a puppeteer with the London Marionette Theater, which was cofounded by Waldo Lanchester and H. W. Whanslaw. The two men later established separate theaters, and Lanchester's marionette theater in Malvern, England, was formally opened by George Bernard Shaw. Other companies were created by The Gair Wilkinsons, whose stage setting for *St. George and the Dragon* possessed the quality of an early Italian painting; John Carr's Jacquard Puppets; R. W. Blanchard; Jan Bussell and Ann Hogarth; John Bickerdike, a wood sculptor whose puppets showed an influence of 18th-century painting; and The Roel Puppets, a group formed in 1927 with the aim of exploring the possibilities of the string puppet as an art form. Other well-known designers and makers in England were L. J. Bywaters, Victor Hotchkiss, Helen Binyon, and Margaret Hoyland, who developed unusual marionettes of heavy paper and string.

During the early 20th century in America, the troupe known as the Tatterman Marionettes produced an elaborate version of Ibsen's *Peer Gynt* that arose from the belief of William Duncan and Edward Mabley that the realm of fantasy could be more faithfully conveyed by puppets than by human actors. Ralph Chessé, known for his designs of puppets and scenery, was especially interested in the presentation of literary drama, and the slender, refined figures of his Hamlet, Macbeth, and Lady Macbeth are noteworthy. The marionettes of Jerome Magon, Gustave Baumann, Roger Hayward, and those of the popular theater of Tony Sarg are representative of the American revival of puppetry during the years after 1915. Rufus Rose and his troupe not only produced dramatized versions of classical children's stories, but featured ensembles of both classical and popular dance. (In this regard, the Canadian Puppeteers of Toronto specialized in the interpretation of classical and modern music with string puppets and appeared with the Toronto

Symphony Orchestra.) Of experimental interest were the terracotta string puppets by Yolande Gregory and the surrealist marionettes designed by R. Bruce Inverarity, who helped devise the play *Z-739*, intended for production with puppets contrived from household utensils. The extraordinary flat marionettes designed by Robert Edmond Jones and executed by Remo Bufano for special effects in a staging of Stravinsky's *Oedipus Rex* directed by Stokowski (Philadelphia; 1931), were 10 ft. in height and were controlled partly from overhead by operators concealed high above the stage and partly from below by operators dressed in black standing behind the figures.

Puppetry in Burma was a popular pastime during the reign of the Burmese kings and was a means used by the court ministers to control the king's actions. Marionettes originated there before 1782, yet the troupe of Dagon Saya Lin from Rangoon has preserved in our era the traditional plays from the 550 birth stories of the Buddha and scenes and plays from the nation's great historical events; featured in the troupe was the first Burmese woman to enter the difficult profession of puppetry, traditionally restricted to men.

* *

The Sicilian rod puppet is commonly referred to as a marionette, since unlike other rod puppets it is worked from above and requires staging similar to that of string puppets. An early popular marionette character of this type was the heroic knight, the first example of which was the wandering puppet master of Cervantes. The literary evidence from Spain indicates that marionettes of the same type and with the same repertory survive today in such places as Amiens, Ghent, Flanders, Liège, Antwerp, Brussels, and Naples, which were once occupied by Spain and from which they were then carried to Sicily in the 19th century, thus giving birth to the Sicilian *opra di'pupi*, or puppet opera (PL. 400). Compared with the usual marionette, the Sicilian puppet is much heavier (a puppet with armor may weigh up to 78 lb.), and the puppet must be supported by a strong iron bar projecting vertically from the head instead of by thin strings. Naturally, the technical handling of the classic Sicilian puppet differs from that of the classic marionette, and the theater is particularly bound by a strict tradition of dueling "laws," which at one time was severely controlled by the public. In Czechoslovakia, puppets of a type similar to the Sicilian marionette developed, in which the knees were fitted with strings so that the legs could be controlled. In our era, Alessandro and Ermenegildo Greco introduced innovations in the repertory and stage production of the Sicilian puppet theater, which they made known even abroad. It is curious that the popular puppet repertory, consisting of imaginatively reworked story cycles of knights errant, was presented in installments for hundreds of consecutive evenings, as in Japan, where the cyclical performance with marionettes had flourished from the end of the 17th to the end of the 18th century.

Alessandro Brissoni

The traditional Japanese round puppet, used at the Bunraku-za Theater in Osaka, is controlled by a principal operator, dressed in ceremonial costume, who holds the puppet and animates the head and right hand; a second operator, dressed in veiled black, who manipulates its left arm; and a third operator, dressed similarly as the second, who manipulates its legs and arranges the folds of the costume from behind. The operators, puppets, readers, and musicians are specially placed in full view of the audiece. The head of the figure is carved from cypress wood and painted; eyes, eyebrows, nose, and mouth can be actuated by interior strings threaded through the neck to pegs in the head stick. The head is characteristically small, owing to its weight in relation to the light frame of the body; the figure as a whole generally weighs from 22 to 33 lb. Puppets in minor roles are controlled by one man, who sometimes is seated on a stool with three wheels so that the puppet's legs can be controlled by the operator's feet. The popular drama that originated in the 17th century in Japan was shared by both the marionette and the Kabuki theaters; Chikamatsu (1653–1724?) composed plays for both types of theater. Rivalry grew between the Kabuki and marionette actors and culminated in the 18th century with the human actors learning to act by studying the marionettes. A related phenomenon also occurred in the Sicilian puppet opera when at certain times the puppeteers in the company of Giovanni Grasso actually took the place of their puppets.

* *

Analogies between marionettes and actors or men have been a recurrent theme of writers and students of the theater. They have been historically studied by Charles Emmanuel Nodier (1780–1844) and Charles Magnin, and Goethe and Heinrich von Kleist have made them the subject of reflections and comparisons in the art of the theater. Edward Gordon Craig, the English scene designer, extolled the superiority of the marionette over the actor, and Gaston Baty preferred companies of marionettes to those of human actors. The marionette as a symbol for man is a recurring motif in literature. Many plays not intended for an actual marionette theater have been written "for marionettes" by such authors as Paul Louis Charles Claudel, Gerhart Hauptmann, Maurice Maeterlinck, and Luigi Antonelli.

Shadow figures. Like hand and rod puppets, the shadow figure is worked from below. The figure is a flat cutout attached to a slender stave by which it is held and moved behind or in front of a cloth. A light is placed behind the cloth, which may be painted to suggest scenery, in order to throw the shadow plainly into view. The arms or jaw of the figure can be made movable by wires or threads attached to the back of the figure, which is seen only in profile. Originally associated with religious rites that exist even in our era in certain countries of the Far East, the shadow theater can be considered to have originated certain techniques that have often been employed in moving pictures, for example, the fade-out or the fondu (obtained in the shadow theater by the movement of the figure illuminated from behind by a small source of light away from or toward the screen). Shadow figures form part of the history of puppets because of their style of movement and folk-art quality. They entered the history of art when Gustave Doré and Caran d'Ache took an interest in them. Like puppets and marionettes, they have been enjoyed by children as toys and have delighted and inspired adults as well as children in formal stage presentations.

The French name for shadow figures, *ombres chinoises*, places the origin of these figures in China (PL. 406), and there is evidence of shadow performances in China as early as the 11th century. In North China, the preferred material for the shadow figure was donkey hide; in South China it was sheepskin. The translucent hide was colored and elaborately decorated, and the figure was supported by a thin wire either hooked or sewn to the neck of the puppet and set in a wooden rod. The limbs were articulated with pieces of thread as pivots and were moved by rods from below against a background of picturesque landscapes. Since the head could be made detachable at the neck, a new character could be instantly created by substituting a different head, and by this device ceremonies, processions, and retinues composed of many characters were presented, parading in a mysterious atmosphere to the rhythm of a drum.

Outside of China, the best-known shadow figures are those of Java used in the *wayang purwa* and *wayang gedog* puppet plays. The figure, mounted on a central rod and with the arms jointed at the elbows and shoulders, is manipulated by control rods attached to the hands; some figures have movable jaws, controlled by an additional string. The height of the figures varies from 12 to 32 in., depending upon the importance of the character represented, the largest being demoniacal spirits and the smallest, heroes and gods. This kind of hierarchy, a reversal of the usual representation of hero or god as the largest, is a unique element, not only in the history of puppets, but in that of art itself. The wayang shadow figures, cut from buffalo hide with a remarkable feeling for line, are colored, gilded, and decorated with a pierced design that gives them an iridescent quality; when illuminated, the shadows range from pitch black to light gray. In ancient times, the wayang performances were given on the veranda of the Javanese houses. The women watched from the outside and thus saw an authentic shadow performance, while the men, seated behind the chief operator (*dalang*) and his assistants, saw a double performance: the shadows on the screen and the colored figures themselves. These performances, which probably came from India and were also widespread in Bali, Malaysia, Indo-China (PL. 406), and in other countries of the Far East, open and conclude with propitiatory religious ceremonies and songs. The *dalang* was not only a master storyteller and puppeteer, but a kind of priest as well. The repertory consists of story cycles from the *Rāmayāna* and the *Mahābhārata*, the great epics of the Hindus; for the most part it is of Indian derivation.

Turkish shadows, though similar in plan to the Chinese figures, are heavier and cruder in effect. They are cut from camel skin or other nontransparent material and pierced in decorative designs filled with colored transparent materials; the subdued colors of these translucent shadow puppets suggest figures from early stained-glass windows. They, too, have jointed limbs maneuvered by rods from below, and normally they do not exceed a height of 12 in. The hero of this theater, the jester Karagöz (meaning "black eye" and implying "gypsy"), is noted for his large turban and his rather heavy-handed amatory-farcical adventures that are accepted with equal hilarity by an audience of both adults and children. In the last century, theaters of this type were scattered throughout almost all of North Africa; and in Greece there exist shadows of the same simplicity but with a heroic-chivalrous repertory. In the 18th century, the fashion for shadow theaters, especially in France and Germany, drew inspiration from these Eastern examples, though European shadow figures generally were not pierced and were worked by strings or threads in preference to rods.

Ombres chinoises or silhouettes (after Etienne de Silhouette, a minister of finance who was caricatured by a ridiculous profile cut from black paper), were the names given the miniature protagonists of the 18th-century French shadow theaters. Johann Kaspar Lavater (1741–1801), philosopher, poet, and student of physiognomy, built the first device for exactly reproducing a silhouette. Baron Melchior von Grimm (1723–1807) wrote that the *Schattenspiel*, or shadow play, had changed from a child's game to a performance for adults. In 1767, the script of a performance for *les ombres à scènes changeantes*, the *Happy Fishing Party*, was published. In 1755, a certain Ambroise achieved great success in Paris with his Théâtre des Récréations de la Chine. Probably the most famous shadow theater was that of Dominique Séraphin, whose silhouette cardboard figures were remarkable for their artistry. The theater was established first in Versailles (ca. 1774) and then a few years later at the Palais Royal in Paris. The most representative work of his repertory was the *Pont cassé*, which was inspired by a simple popular yarn. The performances of Séraphin were reproduced and popularized by the prints of Nancy and Epinal and had innumerable imitators. At the Exposition Universelle (Paris, 1889), the collector Arthur Maury offered the public the remains of the famous Théâtre Séraphin and its *petits bonhommes noirs* in a revival.

In the 19th century, characters and scenes in silhouette became a recurrent theme in many book and newspaper illustrations of the Western countries. Delarne in *Le Charivari*, Caran d'Ache in *L'Illustration*, and Karl Hermann Fröhlich and Rodolphe Töpffer drew shadow figures for lithography. The great English illustrator, Arthur Rackham, illustrated several famous stories by Charles Perrault (1628–1703) exclusively with silhouettes, creating a theatrical production of shadow figures for each book. In 1886, Paul Eudel, in his book *Les ombres chinoises de mon père*, gathered together the backgrounds, characters, and repertory of his own household theater. One of the last examples of printed theater toys was the *Théâtre d'ombres du Père Castor*, published by Camille Flammarion. With the development of cinema technology, animated shadow figures appeared in shadow movies; two images alternating through grillwork gave the illusion of movement. In 1899, an album using this principle was published under the title *Le motographe*, and the cover was drawn by Henri de Toulouse-Lautrec.

Shadow figures produced only by the hands, occasionally aided by using small accessories (PL. 406), were probably accidentally originated by prehistoric man in caves lit by fires; certainly they were favored in ancient times by the Chinese, the Indians, and the Japanese. They not only became the fashion in 19th-century France, but attained the honor of an almost scientific consideration. Thus, there appeared such books as those of Théodore Révei, *Les silhouettes à la main*, and of Victor Effendi Bertrand, *Les silhouettes animées*.

The heritage and fame of Séraphin were recognized anew in the 1880s with the revival of his *ombres chinoises* in a more modern form of shadow theater at the Chat Noir in Paris. Originally a café that catered to an artistic clientele, the Chat Noir became perhaps the most famous shadow theater in history. Under the direction of its proprietor, Rodolphe Salis, the shadow figure of uniform density, as used by Séraphin, became a figure with many new qualities and values obtained by varying the distance of the reflected object from the screen so that the shadows ranged from pale grey to deep black and by the clever use of perspective. Multicolored and controlled sources of light were substituted for the candles of ancient times, and interesting affects were achieved through the use of figures cut from sheets of zinc rather than of cardboard. The Théâtre du Chat Noir was officially opened in 1887 with the presentation of *L'éléphant* by Henri Somm, and among the successful productions were *L'épopée* by Caran d'Ache, *La tentation de Saint Antoine* and *Le marche à l'étoile* by Henri Rivière, *La conquête de l'Algerie* by L. Bombled, and *L'age d'or* by A. Willette. The most refined and expert creator in this theater was the painter Henri Rivière, who, by projecting colored glass in an oxyhydrogen light machine of his own invention created ingenious background effects to frame the shadows as they passed. The performances of the Théâtre du Chat Noir virtually ended with the death of its director, Salis, in 1897. L'Ecole Polytechnique, where performances in which professors and students were caricatured were annually given, and the Théâtre Noir et Blanc, created in 1910 by Paul Vieillard, a former student of L'Ecole Polytechnique who used suitably controlled electrical means for illumination, can both be considered descendants of the 19th-century shadow theaters.

DOLLS. The doll, which in present-day society has come to be regarded largely as a child's plaything, with a great variety of sizes and types and with varying degrees of detail (from the common, generalized infant doll to elaborately costumed regional and historical figures), has had various magical or religious functions in other epochs and societies (see GAMES AND TOYS). While human figure carvings and manikins generally conceived of as ancestor figures or spirits are found in South Pacific art, their use as marionettes, puppets, or dolls is rare. In the New Hebrides, faces, sometimes with the addition of arms (*temes nevinbür*), were modeled and attached to a stick to be exhibited over the top of a fence in the Nevinbimbaau play, given for the benefit of women, children, and the uninitiated. Among the Maori there is found a sort of jumping jack (*karetao*; PL. 406). In Africa, too, dolls generally have a magical function, such as the Ashanti fertility figurines (*akua'ba*; VI, PL. 1) carried by women who desire children or the Bambara and Yoruba twin figures used as a replacement for a dead twin. Cloth dolls decorated with seeds and beads (PL. 401) found in various parts of Africa also probably had a magical intent. In the Hopi Indian culture, cult images were customarily given to the children after the completion of the religious ceremony; but the colorful fertility doll of Rhodesia, even though made with small glass beads of Western manufacture, retained its magical purpose throughout its existence. A clear distinction between the simple figurine and the supernatural image is, however, generally nonexistent in primitive cultures.

The ancient dolls of India, carved from wood and from buffalo horn and ivory, were highly esteemed; according to Indian mythology, the god Śiva fell in love with his wife Pārvatī's doll. The Greeks called an image or doll moved by strings *neurospaston.* Mobile wooden and ivory dolls have been found in tombs at Thebes and Memphis (mod. Mit Rahina); the separate parts of the jointed figures were connected with metal wires. Dolls discovered in ancient Roman tombs, both pagan and Christian, are of ivory, bone, and terra cotta, with the components joined by copper wires (PL. 401; VI, PL. 1). An indication of the variety of Roman dolls and animated figures is given by the various Latin terms: *pupae*, *imagunculae animatae*, *sigillae*, and *homunculi*.

Miniature figures used in scenes of daily life have also been found in Egyptian tombs (IV, PL. 460; X, PL. 100); throughout history, such compositions of small figures in groups were often created for recreational or religious purposes. Artistic crèches, which flourished in Naples in the 18th and 19th centuries, were initially the privilege of royal and aristocratic families and later spread to the upper middle class. The heads and hands of the figures were frequently carved by well-known sculptors or by specialized artists who became famous for their skill (e.g., Lorenzo Vaccaro, Matteo Bottiglieri, Giuseppe Sanmartino, Giuseppe Gori, Nicola Somma, Francesco Gallo, Ferdinando Mosca, Francesco Celebrano, and Antonio Maria Vassallo; PL. 399). An unusual *Betième* (Bethlehem) in Verviers, Belgium, consists of a series of tableaux with dolls and settings in glazed cases. Some crèches were animated by mechanisms like those used for mechanical theaters and museums which presented historical events or events of the day. Belonging to the family of Cecil C. Brinton since 1800, the Heathrington Theatre Royal has small dolls fitted with an iron base that are activated by an electromagnet manipulated beneath the stage.

Accessories in the large-scale group compositions such as the artistic Neapolitan crèches (clothing, food, tools, furniture, etc.), though miniature in scale, were executed with exacting realism. Dollhouses have long been and are still made with the same realistic emphasis, and for this reason they have become valuable sources of data in the history of furniture and costume. Elaborate hand-made examples of such dollhouses, which were most popular from 1600 to 1800 are preserved in various museums (PL. 402). Though in our era mass production has caused a loss in originality, devices such as talking dolls that could stammer one or two syllables, depending upon the position given the doll, have been improved upon; by means of interior sound devices, dolls capable of reproducing phrases have been created.

Besides the larger-than-life-size animated figures on festival floats, the category of mobile dolls includes "stop motion" puppets employed in films; these achieve an effect of animation through the successive shifting of a part or all of the doll between one shot and the next. This technique, which has been experimented with in the United States by George Pal, has been especially successful in Czechoslovakia, where the most famous maker of such films is Jiří Trnka (PL. 405; see CINEMATOGRAPHY).

Alessandro BRISSONI

BIBLIOG. C. Magnin, Histoire des marionnettes en Europe, Paris, 1852; A. Avril, Saltimbanques et marionnettes, Paris, 1867; M. Monnier, Théâtre de marionnettes, Geneva, 1871; L. Lemercier de Neuville, Théâtre des Pupazzi, Lyon, 1876; P. Eudel, Les ombres chinoises de mon père, Paris, 1886; E. Maindron, Marionnettes et guignols, Paris, 1900; Yorick (P. C. Ferrigni), La storia dei burattini, 2d ed., Florence, 1902; H. S. Rehm, Das Buch der Marionetten, Berlin, 1904; H. H. Joseph, A Book of Marionettes, New York, 1920; E. G. H. Craig, Puppets and Poets, London, 1921; A. Altherr, Schatten- und Marionettenspiele, Zürich, 1923; M. Anderson, The Heroes of the Puppet Stage, New York, 1923; A. Pandolfini Barberi, Burattini e burattinai bolognesi, Bologna, 1923; P. Jeanne, Bibliographie des marionnettes, Paris, 1926; S. Lo Presti, I pupi, Catania, 1927; W. H. Mills and L. M. Dunn, Marionettes, Masks and Shadows, New York, 1928; Tsunao Mujajima, Contribution à l'étude du théâtre de Poupées, 2d ed., Paris, 1928; M. von Boehn, Puppen und Puppenspiele, 2 vols., Munich, 1929 (Eng. trans., J. Nicoll, London, 1932; 2d ed., Boston, 1956); D. S. Green, Puppet Making, London, 1935; C. W. Beaumont, Puppets and the Puppet Stage, London, 1938; G. Baty, Trois p'tits tours et puis s'en vont, Paris, 1942; J. Bussell, The Puppet Theatre, London, 1947; J. Chesnais, Histoire générale des marionnettes, Paris, 1947; A. C. Gervais, Marionnettes et marionnettistes de France, Paris, 1947; P. McPharlin, The Puppet Theatre in America: A History, New York, 1949; D. Keene, The Battles of Coxinga: Chikamatsu's Puppet Play, London, 1951; G. H. Rivière and P. Soulier, Théâtres populaires des marionnettes, Paris, 1952 (cat.); C. Roy, La Chine dans un miroir, Lausanne, 1953; R. L. Mellema, Wayang Puppets: Carving, Colouring and Symbolism (trans. M. Hood), Amsterdam, 1954; E. K. Early, English Dolls, Effigies, and Puppets, London, 1955; G. Speaight, The History of the English Puppet Theatre, New York, London, 1955; D. Bordat and F. Boucrot, Les théâtres d'ombres, Paris, 1956; L. von Wilckens, Zageslauf in Puppenhaus, Munich, 1956; W. Alberti, Il cinema d'animazione, Turin, 1957; J. Cuisinier, Le théâtre d'ombres à Kélatan, 2d ed., Paris, 1957; E. Li Gotti, Il teatro dei pupi, Florence, 1957; S. V. Obraztzov, My Profession (trans. R. Parker and V. Scott), Moscow, 1957; Il Presepe Napoletano (ed. fuori commercio), Naples, 1957; J. E. Varey, Historia de los títeres en España, Madrid, 1957; R. and R. M. Leydi, Marionette e burattini, Milan, 1958; P. Martin and M. Vaillant, Le monde merveilleux des soldats de plomb, Paris, 1958; Saito Seijiro et al., eds., Masterpieces of Japanese Puppetry, Tokyo, Rutland, Vt., 1958; G. Baty and R. Chavance, Histoire des marionnettes, Paris, 1959; O. Spies, Türkisches Puppentheater, Emsdetten, 1959; E. G. Laura, Il film cecoslovacco, Rome, 1960; M. Signorelli, Ghetanaccio, Padua, 1960; R. Benayoun, Le dessin animé après Walt Disney, Paris, 1961; P. Lorenzoni, Storia del teatro giapponese, Florence, 1961; H. Harris, Soldatini, Milan, 1962; E. Maingot, Les Automates, Paris, 1962; M. M. Rabeq-Maillard, Histoire du jouet, Paris, 1962; I. Schaarschmidt-Richter and Takabayashi Yasushi, Japanische Puppen, Munich, 1962; D. Keene, Bunraku: The Art of the Japanese Puppet Theater, Rutland, Vt., 1965 (photographs, K. Hiroshi).

* *

Illustrations: PLS. 399–406.

PUVIS DE CHAVANNES, PIERRE. French painter (b. Lyons, Dec. 14, 1824; d. Paris, Oct. 10, 1898). As a youth, Puvis intended to become an engineer, but a trip to Italy convinced him to take up painting as a career. Upon his return to Paris, he studied briefly with Henri Scheffer and Thomas Couture, but the strongest influences on his style came from Delacroix and Théodore Chassériau (qq.v.). In the 1850s the works he submitted to the Salons were almost invariably rejected, but gradually he won official acceptance, partly through the efforts of the writer Théophile Gautier, who championed his work. By the 1880s he was an established figure in the Salons, and in the 1890s, as Robert Goldwater (1946) wrote, "Puvis's triumph was complete."

The reputation of Puvis rests largely on his mural decorations — chiefly mythological and religious in subject — for numerous public buildings in Paris and other major French cities. These were executed in a distinctive style which makes use of rather flat, slightly modeled figures inspired by classical sources and which is characterized by simple contours, a pale, chalky color range, and ample, uncluttered spaces. Because Puvis respected the integrity of the wall surface to be decorated, he flattened and simplified his forms radically. While he employed a classicizing style that was acceptable to academic taste, his unprecedented simplification of pictorial means made him a hero of the Symbolist painters and poets of the late 1880s and 1890s. They saw him not as a dull classicist but as one who pointed the way to an antirealistic art — an art that Paul Adam wrote was "capable of translating a thought by means of a symbol."

Other Symbolists praised the dreamlike, poetic qualities of his work, despite Puvis's claim that he was not an avant-garde painter. Nonetheless, he traveled in Symbolist as well as academic circles, and his art influenced that of Gauguin, Seurat (qq.v.), and Maurice Denis, among others. Puvis was a cofounder and also served as president of the Société Nationale des Beaux-Arts (the "new" Salon), thus exercising substantial influence on official, academic art. A celebrated figure by the time of his death, Puvis, who had been decorated by the French government, was also considered a prophet by artists of a more progressive tendency.

WORKS. *Ludus pro patria* and *Work . . . War . . . Peace*, murals, 1863, 1869, 1879, Amiens, Mus. de Picardie. – *Hope*, 1872, Louvre. – *Life of St. Genevieve*, mural, 1874–78, 1898, Paris, Panthéon. – *The Prodigal Son*, 1879, Paris, private coll. – *The Poor Fisherman*, 1881, Louvre. – *The Happy Land*, 1882, Bayonne, Mus. Bonnat. – *The Sacred Grove*, *The Antique Vision*, and *Christian Inspiration*, murals, 1884–87, Lyons, Mus. B. A. – *Science, Art, and Letters*, mural, 1887, Paris, Sorbonne. – *Summer* and *Winter*, murals, 1889–93, Paris, Hôtel de Ville. – *Inter artes et naturem*, 1890–92, Rouen, Mus. B. A. – *Pastoral Poetry*, mural, 1895–98, Boston, Public Lib.

BIBLIOG. M. Vachon, Puvis de Chavannes, Paris, 1895 (2d ed., 1900); M. Goldberg, Puvis de Chavannes, Paris, 1901; A. Michel and J. Laran, Puvis de Chavannes, Paris, n.d. [1911]; R. Jean, Puvis de Chavannes, Paris,

1914 (new ed., 1933); L. Werth, Puvis de Chavannes, Paris, 1926; C. Mauclair, Puvis de Chavannes, Paris, 1928; A. Declairieux, Puvis de Chavannes et ses œuvres, Lyons, 1928; R. Goldwater, Puvis de Chavannes, AB, I, 1946, p. 33 ff.; W. Seitz, Some Studies by Puvis de Chavannes, Record of the Art Mus., Princeton Univ., X, 2, 1951, pp. 10–22; R. L. Herbert, Seurat and Puvis de Chavannes, Yale Univ. Art Gall. B., XXV, 2, 1959.

William I. Homer

QAJAR SCHOOL. The arts under the Qajar (Per. Qājār) dynasty, which ruled Persia (see IRAN) from 1794 to 1925, were not a direct continuation of those of the immediately preceding period (Nādir Shāh Afshār, 1736–47; Zand sovereigns, 1750–94). Qajar art shows three fundamental characteristics: a progressive isolation of Persian culture from the great Islamic tradition, owing to the triumph of the Shiite sect and the rivalry of the Ottoman Empire; an increasing intrusion of elements of folk art; and a growing subjection to Western influences. Although qualitatively inferior and not comparable to the glories of earlier epochs, the art of this period developed a clearly defined individual character.

Eighteenth-century antecedents of Qajar style: Zand period. During the reign of Shah ʿAbbās I (1587–1629; see SAFAVID ART) European merchants introduced Western products and manufactured articles from the Far East to Isfahan. In the second half of the 17th century Muḥammad Zamān was sent on long journeys by ʿAbbās II (1642–67). His travels took him to Rome, where he was converted to Christianity. He returned to Persia completely under the influence of Italian painting techniques. Muḥammad Zamān can be regarded as the symbol of a restless artistic consciousness that was to become increasingly provincial in its isolation. Persian artists sought new means of expression, but their works tended to become merely imitative; and the feudal conservatism of court circles was not able to find an alternative to this progressive provincialization, since the feudal ruling class was politically weak and rent by internal strife. When the Safavid dynasty was overthrown by the Afghan invaders, the Persian people were spiritually isolated and economically subjected to the worst form of exploitation. The insurrection against foreign rule was led by Nādir, who belonged to the feudal class; the solidarity of the nation was only apparent, for Nādir's policies were in sharp opposition to the interests of the people and aggravated the economic ills of the country, which labored under the illusion that Persia was still a great power.

In the arts, the popular uprising led to an attempt to create a more realistic style, although bound by lifeless academic conventions. This is clearly seen in the few surviving works of the period and is particularly evident in pictorial works. Certain new elements of popular realism are evident in the numerous portraits of Nādir Shāh. In portraits of other sovereigns, too, there is a primitive immediacy that is in marked contrast to the stereotyped conventional models. The two depictions of Nādir in the paintings of the battles of Karnal and Chaldiran (in the Chihīl Sutūn of Isfahan) also represent a considerable departure from the traditional miniaturist style. Together with this "nationalistic" type of painting, which enjoyed considerable favor from the time of Nādir on, many decorative paintings of flowers and birds were produced, idyllic creations that contrasted sharply with the realistic style of the period.

The architecture of the 18th century lacks originality. This is clearly seen in the Masjid-i-Ganj of Kerman (from which comes a good marble mihrab; Teheran, Archaeol. Mus.), in the Masjid-i-Jāmiʿ in Qum, in the mosque in Resht, and in the sanctuary of ʿAlī ibn-Mihryār in Ahvaz (the last two built toward the end of the 18th century). Popular taste is apparent in the restoration of the sanctuary of Mashhad (Meshed) by Nādir, who is said to have embellished some of the edifices with golden bricks.

These restorations still survive and lend to the whole complex of buildings a note of rustic splendor, unusual in the Persia of the delicate glazed blue faïence tiles known as kashi. The Tālār-i-Nādirī, in Qazvin, which apparently was the most interesting architectural work of Nādir's reign, no longer exists. The strong popular current, which politically took the form of an attempted restoration of the Safavid dynasty by means of uprisings in favor of false pretenders, by no means exhausted itself in the time of Nādir but increasingly influenced the arts during the reigns of his successors. After the collapse of Nādir's short-lived empire and the difficult reconstitution of the political unities that spontaneously came again into being within the various natural economic divisions of the country, Persia proper became a mere fragment that survived the general ruin of the great Iranian world of former times. The course of artistic development was unsettled by the myth of a return to ancient traditions. The newly awakened national consciousness regarded this tradition with the same ingenuous detachment with which it regarded Europe. The history of the visual arts in modern Persia is marked by simultaneous but contradictory attempts to return to aulic traditions and at the same time to follow Western models. The results were neither an autonomous return to traditional forms nor an adaptation of European academicism but an archaizing Westernism in a popular vein.

Starting from these premises, the subsequent evolution of Persian art can be divided into distinct phases, essentially represented by three separate epochs corresponding with the reigns of Karīm Khān Zand (1750–79), Fatḥ ʿAlī Shāh (1797–1834), and Nāṣir ad-Dīn (1848–96). In the great public works in Shiraz, which was intended to become a second Isfahan, Karīm's attempt to emulate Shah ʿAbbās is very clear. In fact, the whole reign of Karīm was dominated by the myth of a political and cultural restoration of the Safavid civilization. Karīm himself modestly assumed the title of *Vakīl* (regent) to an obscure Safavid prince. Of the fortifications of Shiraz, only the Arg, an inner citadel with four towers, survives; the outer walls were destroyed in 1792–93 by Āqā Muḥammad, who carried to Teheran all the recoverable fragments of any value. The great bazaar, however, still stands; it is remarkable for its cruciform plan and for the excellent arrangements for ventilation and lighting. The Masjid-i-Vakīl, completed in 1773–74, retains typical Safavid structural elements but differs sharply from the classical design of the old Iranian mosques: the 48 stone columns that divide it into three aisles reflect, instead, the style of the ancient Arab mosques, and the floral decoration in marble is also somewhat archaic in manner. The Masjid-i-Vakīl also has kashi decoration, as well as a notable carved mimbar hewn from a single block of marble. Other noteworthy structures in Shiraz are the Museum of Fars and the Haft Tan.

The octagonal pavilion occupied by the Museum was originally the mausoleum of the sovereign. Its sober exterior decoration consists of decorative kashi, sometimes with figural depictions, including a hunting scene and a representation of Solomon on his throne derived from Safavid prototypes. The interior of the pavilion is divided into eight rooms surrounding the *hauz* (central pool).

The Haft Tan is a pavilion with a flat-roofed liwan supported by two columns. Variants of this design already existed in 18th-century Shiraz. Surviving structures that give some idea of this design are the Takht-khāna liwan in Teheran, built by Āqā Muḥammad, and, in Shiraz, some buildings built a century later, including the Dīvān-khāna of Qavām al-Mulk, the Bagh-i-Irām villa, and the villa of Dilgushā. Since the other two surviving Zand buildings in Iran, the so-called "Ṣofā" and "Khalvat-i-Karīm-Khanī" in the Bāgh-i-Fīn near Kashan, are very similar in style, it may be assumed that this design constituted the dominant type of 18th-century Persian civil architecture.

Both the Museum and the Haft Tan still contain the paintings originally destined for them. Many of the subjects in the seven large paintings of the Museum and the five surviving frescoes in the Haft Tan — including the sacrifice of Abraham, Moses the shepherd, and the love of Shaykh Sanʿān for a Christian maiden — are similar and seem to indicate that court iconography drew its inspiration from sacred writings as well as from epic and genre subjects. Although the paintings are stereotyped, their execution reveals a tendency toward the anecdotal that is not found in the official art of earlier periods. Similar characteristics are evident in the surviving Zand paintings in other collections, the most interesting of which is that of the Georgian National Museum of Art in Tbilisi (Tiflis). In all Zand painting the rather provincial execution (though not the interpretation) gives the vaguely idyllic subjects a modest and familiar character, and certain pseudorealistic touches give them a banal quality that detracts from their value as genre sketches.

A tendency toward three-dimensional composition led, tentatively at first, to an interest in a type of portraiture unknown in previous periods, although certain court portraitists in Isfahan had made attempts to characterize their subjects, depending on the individual tastes of the various sovereigns. The relationship between late-Safavid Isfahan and 18th-century Shiraz is not very clear; Safavid models may have survived in the Qajar period, since during the Safavid period Shiraz was an important center of portrait painting. During the Zand period Shiraz witnessed a growing tendency toward combining Safavid models, the essentially European style of court portraiture, and popular taste. The results were often rigid and clumsy, since the Zand artists in order to correct what they considered an excessive emphasis on three-dimensionality, attempted to lighten the composition by the introduction of extrinsic decorative elements, such as a profusion of dainty multicolored flowers on robes, tapestries, frames, and floors. Sometimes pearls and various jewels were painted on the headgear and clothing of the subjects. A striking contrast to the rigid figures, whose conception repeats the canons of the lacquers of Āqā Ṣādiq and Āqā an-Najaf (eyes depicted frontally, each one twice as wide as the mouth, and curls hanging on the cheeks), is provided by the naturalistic capes and saddlecloths. The best examples of this type of painting, such as the small Luṭf ʿAli Khān (1779) of Bahrām Shīrāzī, recall the work of Riẓa-i-ʿAbbāsī (q.v.). Other examples make one wish that the artist's influence had been more extensive. All the examples suggest that the contemporary production of textiles must have been quite notable.

First phase of Qajar style: reign of Fatḥ ʿAlī Shāh. Only with the reign of Fatḥ ʿAlī Shāh, the second of the seven rulers of the Qajar dynasty, did Persian art become consciously neo-Safavid. At the same time it was archaizing and receptive to European influences. The moat and the clay walls that surrounded the Arg (citadel) of the new capital city of Teheran were probably built in the reign of Āqā Muḥammad (the first Qajar sovereign), as well as the moat outside the city walls. The surviving Takht-khāna (throne room), within the Arg (now the Gulistan Palace complex), can definitely be dated to Āqā Muḥammad's reign. The Takht-khāna is a pavilion in pure Zand style; stones, columns, and mirrors were taken from an adjacent pavilion built by Karīm. Behind tall twisted columns decorated with painted foliage and surrounded by portraits of sovereigns in gilded frames and by faceted mirrors stands the famous Takht-i-Marmar (marble throne), supported 3 feet off the ground by caryatids and twisted columns, one of which has two seated lions at the base. The throne is completely surrounded by stonework carved as delicately as lace. The rich decoration of the throne gives this fine work a barbaric grandeur.

Under Fatḥ ʿAlī Shāh there was a clear return to tradition. Together with minor architecture, which continued, without any fixed program, the Safavid concern for public works was maintained in such projects as the bridges of Maragha on the Sufi River, the Sar Chashma underground conduits by means of which Hājī Mīrzā ʿAlī Riẓā assured the water supply of Teheran, and the restoration of the great national sanctuaries of Mashhad, Qum, and Rayy. In Isfahan the governor, Muḥammad Ḥusayn Khān Ṣadr, created, in the Chahār-bāgh-i-Khājū, a new version of the Chahār bāgh of Shah ʿAbbās I. In Yezd (Yazd) the Bazaar Gate was built, which repeats the vertical style of the nearby Jāmiʿ Gate (erected under the Muẓaffarid dynasty). At the same time certain reflections, not always unpleasant, of late-18th-century European court style appeared

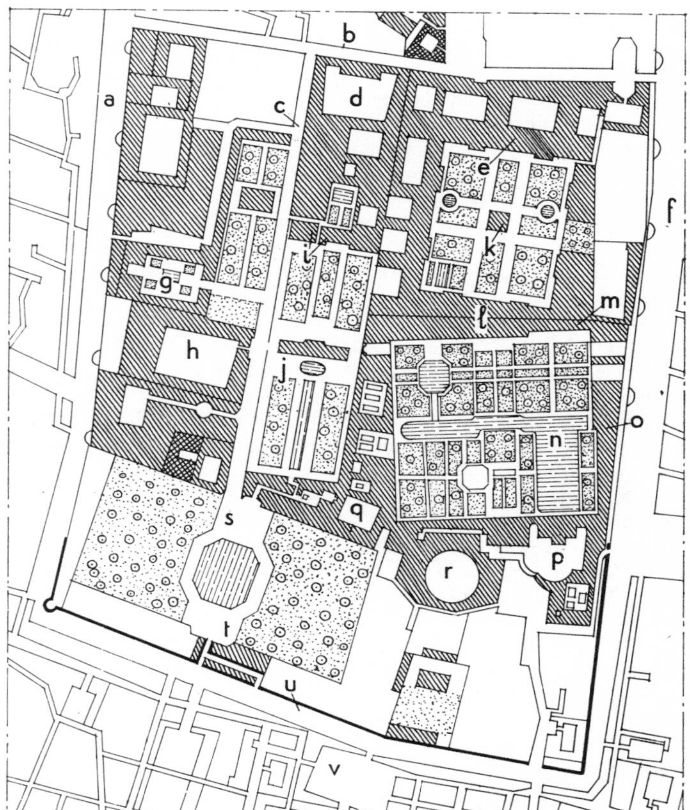

Gulistan complex, Teheran, plan: (*a*) Galilabad (now Hayyam) Street; (*b*) Street of the Andarūn (now Azar); (*c*) Naʿib al-salṭana Street; (*d*) palace of Naʿib al-salṭana; (*e*) *andarūn* (private royal residences); (*f*) Nasiriya street (now Nāṣir-i-Khusrau); (*g*) guards' barracks; (*h*) royal stables; (*i*) Ministry of War; (*j*) Takht-Khāna (throne room); (*k*) Kwabgah (pavilion of repose); (*l*) Museum; (*m*) Nāranjistān; (*n*) Gulistan (court); (*o*) Shams al-ʿImāra; (*p*) Bādgīr Palace; (*q*) Daftar-khāna (administrative offices); (*r*) Takita-i-Daulat (theater for religious drama); (*s*) Maidan-i-Arg; (*t*) Naqqara-khāna (music pavilion); (*u*) Giba-khāna (now Buzurgmitir) Street; (*v*) Sabza Maidan (*from D. Feuvrier, Trois années à la cour de Perse, Paris, n.d.*).

in the early attempts to create a unitary organization of the royal palaces in Teheran. In the decorations, white stucco — the so-called "Italian style" *gach* — was used extensively, either gilded or painted the color of lapis lazuli. European influence appears in the Negārestān, with its bearded family portraits; in the Sulaymāniya of Karaj; in the Qaṣr-i-Qājār, with its ingenious rows of pylons to support the acqueduct that carried water uphill; and in the attempts to create furniture in a kind of Empire style. Even in carelessly executed traditional types there is a certain imagination and variety.

In architecture, European influence is evident in the columned nave of the Mosque of Ibrāhīm Khān Zahīr in Kerman (1816–17) and in the highly accentuated bulb of the dome of the Imāmzāda Shāh Chirāgh in Shiraz (1834). New colors appeared in decoration, with an abundant use of white and yellow, as well as new forms, such as clusters, flower vases, and tiles decorated with human figures. In some cases popular influences predominated in architecture, and academic concepts of interior space were rejected in favor of functional considerations of climatic suitability. In contrast, the mosques built at the order of Fatḥ ʿAlī (all known as "Shah mosques") in Teheran, Burujird, Qazvin, Simnan, and Zinjan are traditional in design. The finest are those in Simnan (1826–27) and Qazvin (1786–1808). The former, which is the work of Ustād Muḥāmmad Bāqir, Ustād Zayn-ul-ʿĀbidīn of Simman, and Ṣafar ʿAlī of Isfahan, with inscriptions by Ḥabibullāh ibn-ʿAlī Akbar and modest kashi by Muḥammad ʿAlī, contains the best example of the principal architectural innovation of the Qajar style — a system of small satellite domes and windows opening from the vaults of the building above the liwan, which revolutionized the interior lighting of mosques and is a dominant feature of civil architecture in central Iran. This innovation was accompanied by ventilation installations whose towers and brick gratings are a characteristic element of the architecture of the region. One of the most advanced examples of this architecture is the Burūjirdi house in Kashan.

Fatḥ ʿAlī Shāh was particularly receptive to ancient Iranian influences, and numerous rock reliefs were carved in a neo-Sassanian style (see SASSANIAN ART) depicting this Qajar sovereign in the guise of Khursrau (PL. 407). The best-known reliefs are at Chashma-i-ʿAlī, at Tāq-i-Bustān, and in the vicinity of the Koran Gate in Shiraz. The various governors also followed the example of their sovereign. Among these was Ḥusayn ʿAlī Mīrzā Shujaʿ al-salṭane, who built himself a Kākh-i-Khorshīdī ("palace of the sun") in the Bāgh-i-Naẓar of Shiraz, next to the mausoleum of Karīm Khān; it was adorned with statues depicting the heroes of Firdausī's *Shāh-nāma*. Although this palace was later destroyed, the statues still remain, scattered among pools and flower beds.

The interaction of popular style, archaizing taste, and European influence is even more evident in painting, and above all in the lacquer work used for the characteristic *qalamdān* (pen box). The style is particularly cosmopolitan and characteristic of a court that attempted to combine the styles of Persepolis, Isfahan, and Versailles. The use of shading and the introduction of the art of printing (1816–17) brought about a revolution in the esthetics of miniature art. The work of Mīrzā Bābā Naqqāsh Ḥusayni-al-Imāmī, the chief court miniaturist, is characterized by a certain eclecticism, particularly in the illustrations (Teheran, Gulistan Lib.) of the poems of Khāqān (the sovereign himself), the masterpiece of the art of miniature of this period. The illustrations include many portraits of court personalities, including one of Fatḥ ʿAlī Shāh (1801–02). In 1847 the same artist painted a portrait of Dūst ʿAlī Khān, in which he introduced landscape elements of European inspiration. In the portrait of a youth in the Mu-ʿayyir al-Mamālik Collection, there is, to the right, a half-open tent set in a green landscape, a motif that attracted even the most archaizing artists.

Flemish and Florentine elements, which reached Iran by way of India, appear in the paintings of the dancer Mazdā by Madhī Shīrāzī (1819–20), while in the portrait of a young aristocrat (Tbilisi, Georgian Nat. Mus. of Art) there is attention to the spirit of the subject and a certain attenuation of concern for the decorative values of fabrics and costume. This tendency was to lead to undeniable, and sometimes successful, attempts to achieve a form of psychological portraiture. In other works the pressure of popular taste coupled with a decided preference for courtly subjects achieved odd results. When the popular style was combined with archaizing aspirations, as was most often the case in the portraits of royal personages (typical are the depictions of Fatḥ ʿAlī Shāh with an Assyrian beard), the result was a mannered exoticism both obvious and facile. Far better results were achieved when the artists depicted clearly popular subjects, such as a juggler with his monkey or a huntsman with his rifle. As a result of closer ties with Europe, the Western academic idiom, combined with the reborn consciousness of local tradition, influenced the late Qajar genre scenes, which did not entirely lose their original popular inspiration. The tendency to psychological portraiture degenerated into the copying of photographs and resulted in a host of veristic productions.

The reign of Fatḥ ʿAlī Shah also saw the production of precious textiles (e.g., those of ʿAlī Naqī Qājār in the Museum of Mashhad). In the Gulistan in Teheran is the carved Takht-i-Ṭāʾūs ("throne of Ṭāʾūs"), named after one of the wives of the Shah. This throne, by the way, should not be confused with any of the so-called "peacock thrones" (from *ṭaʾus*, "peacock"), such as the Moghul one carried off by Nādir Shāh, of which no trace remains. The throne in the Gulistan was once known as the Takht-i-Khorshīdī ("throne of the sun"); it was commissioned in 1799–1800 by Fatḥ ʿAlī Shāh, who turned to the best craftsmen in Isfahan, which was renowned as a center of wooden mosaic work and metal objects. Important centers of chased silver work were Tabriz, Zinjan, and Shiraz, whose craftsmen drew inspiration from the Achaemenid past. A remarkable copper piece decorated with miniatures from Varamin is now preserved in the Gulistan.

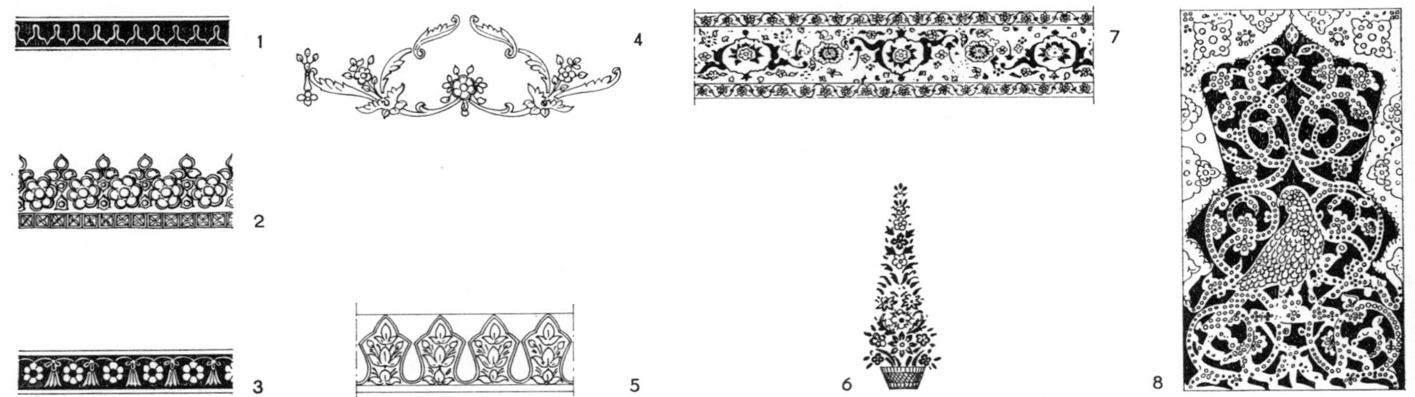

Persian decorative motifs of the Qajar period: (1–6) Floral ornamentation taken from gold and enamel objects belonging to the treasury of the Iranian royal house; (7) Herat carpet, detail; (8) side panel of the Takht-i Ṭāʾūs (throne of Ṭāʾūs).

Reigns of Muḥammad Shāh and Nāṣir ad-Dīn. The reign of Muḥammad Shāh, the successor of Fatḥ ʿAlī, was one of consolidation rather than evolution. The most significant architectural work — more important than the Muḥammadiya, on the road to Avin, or the garden of Abbasabad — is the Ṣafā bath in Qazvin, commissioned in 1843–44 by Hājī-Ḥasan ibn-Hājī ʿAbdullāh of Tabriz. The painting of the period — represented by Mīrzā Shāfiʿ of Shiraz (1778–1847), poet and calligrapher, and by the works of Ṣādiq Naqqāsh in the Chihīl Sutūn (if these are not to be attributed to Rajab ʿAlī) — is far more interesting. Particularly noteworthy are the works of Lutf ʿAlī Shīrāzī, perhaps the greatest painter of the period, whose stylistic coherence distinguishes him among his contemporaries. His flower paintings show a notable balance between the Persian tradition and the Oriental and Western idioms.

During the long reign of Nāṣir ad-Dīn there were no substantial novelities either in religious architecture or in the traditional architecture of bazaars, baths, and caravansaries. Religious structures of the period include the madrasah of Sipāhsālʿār in Teheran (1878–90) and the Shāhzāda Ḥusayn of Qazvin. What was new was the civil architecture, which was highly imaginative even in the design of military emplacements, gates, and barracks. Noteworthy is the Darwāza-i-Arg of Simnan (1884–85) with arches and turrets, kashi, and painted decoration. More interesting, however, is the architecture of the royal and government palaces, especially in Teheran. In Isfahan, Ẓill al-Sulṭān (the local governor and son of the Shah) commissioned the building of the Qaṣr-i-jihān-nimā and the three-storied octagonal tower of his government palace.

During the reign of Nāṣir ad-Dīn the Gulistan complex in Teheran took on its present appearance. After it had been enlarged it assumed the form of a large rectangle, the interior of which has several buildings of various types to the north and west of the palace of Āqā Muḥāmmad, which continues to be the center of this complex. To the south of the palace is the only public square, the Maidan-i-Arg, which communicates with the city by means of the gate of the music pavilion. To the east, a new appearance was given to the royal palaces themselves. This eastern zone was clearly divided into three sectors, northern, central, and southern, occupied, respectively, by the *andarūn* (private residence quarters), the *bīrūn* (quarters for public functions, grouped around a court planted with trees and flowers known as the Gulistan), and the court theater for the performance of sacred plays. The *andarūn* was rebuilt in 1882–83 (after the buildings of Fatḥ ʿAlī Shāh had been torn down) around a small palace inspired by the Dolmabahçe in Istanbul. From here a gate leads across the greenhouses of the Nāranjestān into the northern portion of the group around the Gulistan garden. On the north side the Museum, built to house a part of the royal treasures, contains a famous hall of mirrors, a library, and collection of porcelain. The eastern side of the court contains the masterpiece of Shir Ghaʿfar — the delicate but solid Shams al-ʿImāra, a Europeanizing pastiche, of which only the fragility is Oriental. On the south side is the Bādgīr Palace, known for its kashi, golden ceilings, colored crystal chandeliers, and famous paintings. On the western side is a French-style palace. All four groups of buildings are joined together, on the sides facing the Gulistan garden, by a system of blind arches.

There are many buildings of this period in the vicinity of Teheran, including the Bāgh-i-Shāh, built around a circular park, with watercourses and pavilions; and the complex of Arak (formerly Sultanabad), with towers, guard posts, and four palaces. At Niyavaran the Shah erected between 40 and 50 pavilions that together comprised the Shāhib-i-Qirāni, of which the principal palace still stands, although it has been remodeled. It contains a rich collection of paintings and the finest hall of mirrors of the period. Other palaces and pavilions are those of Surkh-hisar, ʿIshratabad, and Shikar-gah. It is difficult to tell, in these edifices, where Iran ends and Europe begins; the determining factor of their style is a progressively more European taste, particularly oriented toward Russia. One could easily reconstruct the path by which the interest in landscape and urban styles, as well as specific architectural elements (e.g., window decoration, steep roofs, and certain neoclassic elements) filtered from eastern Europe across the Caucasus and Central Asia to Persia. The exotic variations did not attempt to modify or adapt these European elements but tended merely to give them a provincial character. The Bādgīr Palace, with its ventilating towers and *sardāb* (underground rooms), is a rare example of an aulic version of a simpler and functional local architectural style, which failed to penetrate the court, being overcome by a decorative style that was foreign to it.

Apart from the imperial palaces, the following buildings are worth noting: in Teheran, the Niẓāmiya (1853–54), whose walls are covered with paintings; in Shiraz, the villas of Irām, Dilgushā, ʿAfīfābād, and Jihān-nimā, as well as the houses of the Qavām al-Mulk, including the Nāranjistān and the adjacent house of Ẓinat al-Mulk Qavāmī. In Kashan there are some extreme examples of a Persian Art Nouveau style, with decorative elements derived from Europe and from Achaemenid and Sassanian traditions. These elements tended to ruin the genial architectural concepts represented by the Burūjirdi house. In Qazvin, the Taqavi house and the three halls separated by glass partitions in the Ḥusayniya of the Amīnī are the only surviving portions of the group of 16 buildings for religious meetings that Ḥājj Muhammad Riẓā Amīnī built in 1858–59.

A similar juxtaposition of elements is found in the painting of this period. Two members of the Ghaffārī family — Mīrzā Abū Ṭurāb Naqqāsh-bāshī and Mīrzā Abūʾl Ḥasan Muḥammad Ḥasan Khān Naqqāsh-bāshī (known as Sani al-Mulk after 1860–61) — appear as the first artists who consciously attemped to fuse Persian tradition with academic Western idioms. The former was the illustrator of the *Ruznāma-i Vugāyiʿ-i-ittifāqiya*, one of the first Persian newspapers. The formative experience of the latter artist was his journey to Rome and Paris, where he spent about six years copying European masters, especially Raphael. In his portrait of Khorshīd Khānum, the landscape is reduced to a few conventional arabesques and the interest of the artist is concentrated on the person of the woman. Sani al-Mulk was charged with supervising the drawing of portraits of princes and notables in the second important Persian weekly, *Ruznāma-i daulat-i ʿāliya-i-Irān*. Nāṣir ad-Dīn also commissioned him to illustrate the *Thousand and One Nights* now in the Imperial Library. Among Sani al-Mulk's best-known works are a portrait of his father on horseback, a portrait of the minister Mīrzā Āqāsī (Leipzig, Schultz Coll.,) a *qalamdān* (Vict. and Alb.), and, above all, a painting of the court of Nāṣir ad-Dīn (Teheran, Nat. Mus.), which has been considered his masterpiece. A number of water-color studies for this painting are divided among the Museum, the Gulistan, and the Imperial Library. Other works by Sani al-Mulk include a portrait of Nāṣir ad-Dīn (1855–56; Teheran, Najmābādi Coll.); a portrait of ʿAbbās Qulī Khān Muʿtamid al-daula (1856–57); a painting of a court personage (Muḥsin Maqaddam Coll.); a portrait of Farrukh Khān ʿAmin al-daula, a cousin of the artist; a portrait of the artist's three sons; and another painting of Āqāsī (1866; formerly Moscow, Shchukin Coll.). Apart from being the founder of the first official school of fine arts in the country, Sani al-Mulk, in his unreserved acceptance of the academic canons of the 19th century, marked the end of the Persian miniature.

The court of Nāṣir ad-Dīn (who considered himself a connoisseur of the arts) included several other artists. Mīrzā ʿAlī Akbar Muʿayyir al-daula, who lived in Paris at the time of Napoleon III, was a musician as well as a painter. Mīrzā ʿAbdal-Muṭṭalīb, the last father-in-law of the Shah, achieved remarkable success in the European colony of Teheran with a painting depicting a beggar. Āqā Mīrzā Ismāʿil Jalāyir, calligrapher, portraitist, and landscape painter, painted in brilliant colors as well as in black and white. Mīrzā Bābā was known for the distant views that he used in the backgrounds of his portraits. Prince Mubārak Mīrzā, a painter of *qalamdāns*, specialized in battle scenes. Ustād Bahrām decorated the small *andarūn* of Nāṣir ad-Dīn with a famous fresco depicting a banquet scene. Sayyid Naqqāsh painted the portrait of Nāṣir ad-Dīn on the sovereign's grave. Mīrzā Mahdī Khān Muṣavir al-Mulk, a portraitist and painter of landscapes, faithfully copied a number of works by Austrian artists and lived long enough to paint Muḥammad ʿAlī on the day of his coronation in 1907. Masʿūd Khān Ghaffārī was another important painter. A winter landscape by him, crowded with horsemen and nomad tents (formerly in a private house in Teheran), represents, within the limits of the then-prevailing academism, a search for more local motifs.

The leading artists of this last phase of the Qajar period was another member of the Ghaffārī family, Muḥammad Kamāl al-Mulk (a nephew of Sani al-Mulk), considered in Iran the country's greatest modern painter. Born in Kashan in the middle of the 19th century, Muḥammad Kamāl al-Mulk entered at the age of fifteen, the Dār al-Funūn, the university that Nāṣir founded in Teheran in 1852. The artist attracted the attention of the Shah with a portrait painted from a photograph and thereafter enjoyed his patronage and later that of his successor, Muẓaffar ʿAlī; his *Hall of Mirrors*, which depicts the Shah in the famous hall of the Gulistan, is generally regarded as his greatest work. In 1903–04, his desire to create a new idiom led him to Mesopotamia in search of exotic Oriental color. Ten years later, under the influence of Rousseau, he abandoned this exoticism in favor of nature. He turned to teaching and founded the Ṣanāyiʿ-i Mustaẓrifa Madrasah. All the Persian academic artists belonged to his school. Toward the end of the 19th century there had been a revival of interest in miniature art, represented by the Persian scholar Bihzād (not to be confused with the great Kamāl al-Dīn Bihzād) and Ḥājj Mīrzā Aqā Imāmī of Isfahan, who was a restorer of ancient miniatures and the painstaking painter of a miniature of a game of polo and one of a nomad.

In the field of applied arts, only weaving (see TAPESTRY AND CARPETS) continues to have an importance that extends beyond the borders of Iran. In the 20th century this craft has begun to flourish again in Kashan, in its first revival after the splendors of the Safavid period. In Arak four types are produced: the Sarouk, with orange-red and brown motifs; the Maḥall, with dark leaves and flowers in a red-and-brown frame; the Mashirabad; and the thick Armenian Lilahan. The Farahan carpets use motifs of branches, flower beds, and fish on a dark ground, while those made in Kerman depict animals and the tree of life against a light ground. Many other cities, including Shiraz, Hamadan, Mashhad, and Tabriz, are important centers for the production of this esteemed and splendid product of Iranian craftsmanship.

BIBLIOG. See IRAN, s.v. Teheran, Shiraz, Qazvin, Mahan, Mashhad. See also: Mīrzā Muḥammad Sādiq Mūsavī Isfahānī Nāmī, Ta'rikh-i Gī-tigushā (History of the Conqueror of the World), Teheran, 1317 A.H. (1938–39), pp. 153–58; M. Bahrāmī, Shabīhsāzī dar fann-i naqqāshī-i Īrān (Persian Miniature Portraits), Īrān-i imrūz, III, 1320 A.H. (1940–41), pp. 9–10, 29–33; Tasvīr-i Luṭf 'Alī Khān Zand (Portrait of Luṭf 'Alī Khān Zand), Yadgar, I, 5, 1323–24 A.H. (1944–45), pp. 63–66; 'Abdullāh Musraufī, Sarh-i zindigānī-i man (Autobiography), I, Teheran, 1324 A.H. (1945); A. Tāhirī, Daura-yi tahavvul-i naqqāshī-i Īrān (Period of Evolution in Persian Painting), Ruzigār-i nau, V, 2, 1945–46, pp. 13–27; A. Mazdā, Nufūz-i salk-i Urapā'ī dar naqqāshī-i Īrān (The Influence of European Style on Persian Painting), Payām-i nau, II, 10, 1324–25 A.H. (1946–47), pp. 59–72; Yaka sūratsāzì dar naqqāshī-i qurūn-i ahīr (Scarcity of Portraits in Painting of the Last Centuries), Payām-i nau, II, 10, 1324–25 A.H. (1946–47), pp. 46–58; M. Qārābāgiyān, Muḥammad Ghaffārī Kamāl al-Mulk, Payām-i nau, II, 10, 1324–25 A.H. (1946–47), pp. 81–92; A. A. Hikmat, Tasvir-tāza-i az Nādir-Shāh (A New Portrait of Nādir Shāh), Yadgar, IV, 1–2, 1326–27 A.H. (1947–48), pp. 35–36; Buzurgtarīn naqqash-i asr-i Qājāriya (Sanī al-Mulk) (The Greatest Painter of the Qajar Period: Sanī al-Mulk), Ittilā'āt māhāna, I, 2, 1327 A.H. (1948–49), pp. 29–32; I. Afsār, 'Akshā-ya qadīmī va ma'rūf-i Nādir Shāh (Ancient Images and Notes on Nādir Shāh), Jahān-i nau, IV, 1328 A.H. (1949–50), pp. 590–91; A. H. Navā'ī, Kamāl al-Mulk, āfiranda-yi zībā'ī (Kamāl al-Mulk, Creator of Beauty), Ittilā'āt māhāna, III, 1329 A.H. (1950–51), 4, -pp. 9–13, 48, 5, pp. 17–20, 43–47; Q. Ghanī, Kamāl al-Mulk, Yaghma, III, 1329, A.H. (1950–51), pp. 338–42, 361–65; I. Bohnām, Tābluhāy az shāhkarhā-yi naqqāshvī-i ma'ruf-i Īrān (Some Notable Masterpieces of Persian Painting), Ittilā'āt māhāna, IV, 12, 1330 A.H. (1951–52), pp. 41–43; Naqqāshān-i 'asr-i Nāṣirī (Painters of the Time of Nāṣir), Yaghma, X, 1336 A.H. (1957-58), pp. 168–75, 216–17; Chand nash-i hunarmand dar yak jahān-i chand sāla der Kirman, Naqsh u nigār, 6, 1339 A.H. (1960–61), pp. 30–44; M. T. Mustafavī, Mi 'mārī-i Īrān ba 'dar daurān-i Zandīya (Persian Architecture after the Zand Period), Bānk-i sakhtimani, 1340 A.H. (1961), pp. 8–9, 44–45; G. R. Scarcia, La Casa Borūğerdī di Kāšān: Materiali figurativi per la storia culturale della Persia Qāğār, Ann. Ist. univ. orientale di Napoli, N.S., XII, 1962, pp. 83–95; Dūst 'Alī Han Mu'ayyir al-mamalik yāddāshthā ez zindigānī-i husūsī-i Nāsir al-dīn Shāh (Memoirs of the Private Life of Nāsir al-Dīn Shāh), Teheran, n.d.

Gian Roberto SCARCIA

Illustrations: PLS. 407–410; 2 figs. in text.

RAEBURN, SIR HENRY. Scottish portrait painter (b. Edinburgh, Mar. 4, 1756; d. Edinburgh, July 8, 1823). Although he was trained as a jeweler, he began early to paint in oils, making copies of portraits by David Martin (1737–98); but he received no instruction from him, and it seems probable that as an oil painter he was largely self-taught. He had a natural gift for bold and vigorous modeling, and he very rapidly established a personal style and a local reputation. His first important commission, *George Chalmers of Pittencrieff* (Dunfermline, Scotland, Town Hall), painted for the corporation of Dunfermline in 1776, shows all the characteristics of his later work and, except for the inclusion of a view of Dunfermline Abbey as seen through an open window, it might have been painted at any time during the succeeding 20 years. Raeburn occasionally modeled heads in clay, and the "square touch" with which he blocked out the features of his sitters sometimes gives the impression of painting by a sculptor.

In 1780 Raeburn married Anne Leslie, a widow of independent means. In 1785 he went to Rome, returning to Edinburgh in 1787; but Italy appears to have made no impact whatsoever on his art. The portrait of Francis, Lord Elcho, painted in Rome in 1786, certainly shows the influence of P. G. Batoni, but it could just as well have been painted in Scotland. On his way to Italy Raeburn spent some time in London. Though it is not certain that he had been there before, it is difficult to believe that he had not already seen the work of George Romney (q.v.), with whom he had in common a bold, assured, and easy manner of painting; like Romney, Raeburn worked directly on the canvas without making preliminary drawings or studies.

Raeburn first exhibited in London in 1793, when *Sir John and Lady Clerk* was shown at the Shakespeare Gallery in Pall Mall. This is an excellent example of the overly emphatic lighting which he tended to use at this stage. The shadows are so deep as to appear artificial, and it was remarked at the time that it was not clear whether the scene was in moonlight or sunlight. At the same time the composition is natural and informal. Raeburn had the gift of making a portrait forceful and spontaneous, so that it seems as if his sitter has just looked up and been caught unexpectedly by a flash-bulb camera. This nonchalance, which sometimes degenerates into carelessness, accounts for occasional weaknesses and perhaps for the fact that Raeburn's style changes so little. The most noticeable change in his later work is that the lighting becomes more diffused and less artificial and the general tonality lighter.

Raeburn established a steady practice at his studio in York Place, Edinburgh. In 1810 he considered moving to London but decided against it. In 1812 he became president of the Society of Artists of Edinburgh and was subsequently elected to the Royal Academy. In 1815 he became a full academician, and on the occasion of the visit of George IV to Edinburgh in 1822 he was knighted. He has sometimes been called the "Scottish Reynolds" or the "Reynolds of the North," which is misleading as his work has neither the intellectual content nor the variety of Reynolds; but it has virility, purpose, and a very considerable sympathy with and understanding of character.

MAJOR WORKS. *George Chalmers* (1776; Dunfermline, Scotland, Town Hall). – *Sir John and Lady Clerk* (1790; Coll. Sir Alfred Beit). – *Lady Raeburn* (1790; Romsey, Hampshire, Broadlands, Countess Mountbatten of Burma). – *Dr. Nathaniel Spens* (ca. 1794; Edinburgh, Archers' Hall). – *The Macnab* (ca. 1805–10; London, Messrs. John Dewar). – *Lord Newton* (ca. 1806–11; Earl of Rosebery). – *Sir Walter Scott* (1808; Duke of Buccleuch). The best single collection of Raeburn's work is in the National Gallery of Scotland, Edinburgh.

BIBLIOG. W. Raeburn Andrew, Life of Sir Henry Raeburn, Edinburgh, 1886; Sir Walter Armstrong, Sir Henry Raeburn, London, 1901; Catalogue of Raeburn Bi-Centenery Exhibition, National Gallery of Scotland, 1956.

Kenneth GARLICK

RAIMONDI, MARCANTONIO. Italian engraver (b. near Bologna, ca. 1480; d. ca. 1534). As an apprentice to Francesco Francia (q.v.), the Bolognese painter and medalist, Marcantonio learned to decorate silver belt buckles and other ornaments in niello and also to make engravings with the same technique. The dark, close crosshatching of niello shows in his earliest engravings (ca. 1500). In 1505, possibly in Venice, he simplified his line when he imitated the open effect of the woodcut in copies of scenes from Dürer's *Life of the Virgin* (V, PL. 226). In Rome, from about 1510 to 1527, Marcantonio and his growing number of assistants produced engravings of works of Raphael (IV, PL. 425). After Raphael's death in 1520 he made engravings after designs by Giulio Romano and Baccio Bandinelli. In 1524 the Pope jailed him for engraving Giulio's lascivious drawings of the *Modi*. In 1527, when the Sack of Rome scattered the Roman art world, Marcantonio is said to have gone home to Bologna, and it is known that he was dead by 1534.

By simplifying the technique of Dürer and Lucas van Leyden (qq.v.) Marcantonio perfected an intelligent and teachable system of lines that curve around forms and suggest sculptural bulk. [Gian Giacomo Caraglio, Nicolaus Beatrizet (Beatricetto; XII, PL. 307), Giulio Bonasone, Enea Vico, Giovanni Battista Scultori ("Il Mantovano"; IV, PL. 426), and Giorgio Ghisi were later greatly influenced by his work.] His clear style laid the basis for all subsequent engraving, because its impersonality enabled several engravers to collaborate on one plate. The hundreds of engravings by Marcantonio's group of artists conveyed the High Renaissance style to the north of Europe.

A. Hyatt MAYOR

SELECTED WORKS. *Pyramus and Thisbe*, 1505 (first dated work). – *The Baptism of Christ*, ca. 1505 (after Francesco Francia). – Allegorical subject, called "Raphael's Dream," ca. 1506 (after Giulio Campagnola, and possibly Giorgione). – *The Life of the Virgin*, ca. 1506 (V, PL. 226; 17 engravings after Dürer). – *The Climbers*, ca. 1509 (after Michelangelo and Lucas van Leyden). – *The Death of Lucretia* (IV, PL. 425; after Raphael). – *The Massacre of the Innocents* (after Raphael). – *The Judgment of Paris* (after Raphael). – *The Phrygian Plague* (after Raphael). – *I modi*, 1524 (after Giulio Romano). – *The Martyrdom of St. Laurence* (after Baccio Bandinelli). – *Portrait of Pietro Aretino* (after a Venetian model).

BIBLIOG. Vasari; A. Bartsch, Le Peintre-Graveur, XIV, Leipzig, 1867; H. Delaborde, Marc-Antonine Raimondi, Paris (n.d.); A. M. Hind, Marcantonio and Italian Engravers and Etchers of the Sixteenth Century, London 1912; A. Petrucci, Disegni e stampe di Marcantonio, BArte, XXX, 1936–37, pp. 392–406; A. Petrucci, Il Mondo di Marcantonio, BArte, XXXI, 1937–38, pp. 31–44; A. Petrucci, Linguaggio di Marcantonio, BArte, XXXI, 1937–38, pp. 403–18.

* *

RAINALDI, CARLO. Roman baroque architect (b. Rome, May 4, 1611; d. Rome, Feb. 8, 1691). He was the son of Girolamo Rainaldi (1570–1655), "Architetto del Popolo Romano," and received a solid education before studying under his father in Rome and northern Italy. After Bernini, Borromini, and Pietro da Cortona, Carlo Rainaldi was the leading architect in 17th-century Rome and was often employed by Pope Innocent X (see BAROQUE ART). Both Rainaldis were commissioned to design S. Agnese in Agone, Piazza Navona (1652), but the work was probably done by Carlo (Girolamo then being eighty-two). In any case, the design was profoundly modified by Borromini, who took over the project on Aug. 7, 1653 (II, PL. 306, FIG. 275). Borromini's transformation was in turn altered when Carlo Rainaldi was reappointed in 1657.

Rainaldi's first independent work — and his masterpiece — is the Church of S. Maria in Campitelli (I, PL. 405; II, FIG. 269), founded by Alexander VII in 1660 (to replace the existing church in Piazza Campitelli) but not begun until 1663. The final design combines a central plan (here a Greek cross) with a long nave formed by extending the eastern arm of the cross into a domed sanctuary. R. Wittkower has shown this type of plan to be both mannerist in its implications and northern Italian in inspiration, as the axial directions are ambiguous and the sources are to be found in Rainaldi's early training under his father. The façade, with stressed vertical elements, also derives from northern prototypes. S. Maria in Campitelli was finished in 1667, and its façade is approximately contemporary with the architect's completion of Carlo Maderno's façade for S. Andrea della Valle.

At this time Rainaldi was also working on S. Maria in Montesanto and S. Maria dei Miracoli (both founded March 15, 1662), the twin churches in Piazza del Popolo (IV, PL. 188). However, S. Maria in Montesanto was taken over by Bernini and completed by him and Carlo Fontana. Fontana also helped Rainaldi to complete S. Maria dei Miracoli (ca. 1675–79). The major importance of these two churches lies in the skill with which they are positioned at the ends of three main streets, all emerging into the piazza and focusing on the obelisk. In fact, the two sites formed by the convergence of the three streets are not equal, and this was compensated for by giving one of the churches an oval plan and the other a circular plan; the oval dome of S. Maria in Montesanto is so placed that it appears, when viewed from the piazza, to be identical with that of the other church. (See also II, PL. 132.)

BIBLIOG. F. Baldinucci, Notizie de' Professori del disegno . . . , XIII, Milan, 1812, p. 353 ff.; E. Hempel, Carlo Rainaldi, Munich, 1919; R. Wittkower, Carlo Rainaldi and the Roman Architecture of the Full Baroque, AB, XIX, 1937, p. 242 ff.; R. Wittkower, Art and Architecture in Italy, 1600–1750, Harmondsworth, 1958; F. Fasolo, L'Opera di Hieronimo e Carlo Rainaldi, Rome, 1962.

Peter MURRAY

RAJPUT SCHOOL. The Rajput school represents the most important stylistic phase of the art of northern India during the Islamic period (see INDIAN ART; ISLAM). Formally it was somewhat eclectic, borrowing freely from medieval Hindu art (see HINDUISM) and from all stages of Indo-Moslem art (see INDO-MOSLEM SCHOOLS); but in its esthetic approach and in its subject matter it achieved a pronounced character of its own, distinguishing it from all other forms of Indian art. Its chief characteristics were conservatism, simplicity of apperception combined with passionate emotion, geometric composition, strong rhythm, and chaste romanticism.

HISTORICAL AND SOCIOPOLITICAL BACKGROUND. The Rajput school formed part of a general recovery of Hindu art after the first destructive shock of the Moslem conquest had spent itself. Other, provincial styles flourished contemporaneously in Assam and Bengal, Orissa, and especially in the Vijayanagar empire of the south. The Rajput style predominated first in Rajasthan and soon spread to Malwa, Bundelkhand, and Baghelkhand in central India, then to the Himalaya (especially the valleys adjoining the Punjab but also Kumaon and Nepal). It exercised a strong influence on Moghul art during the reigns of Akbar and Jahāngīr and on the Maratha states and the Punjab under the Sikhs in the 18th and early 19th centuries, eventually becoming the principal source of the revival of Indian architecture inspired by the British in the late 19th century and one of the sources of the Bengal school of modern Indian painting.

The reasons for the special character of Rajput art must be sought in the geography and sociopolitical structure of Rajasthan. It is a very poor country of deserts and steppes interrupted by oases and low mountain ranges, in the east slowly changing into the jungles of central India. Primitive tribes (Bhils, Mīnas) are still numerous in the hills; in the deserts and plains, nomads form a considerable portion of the population. The ancestors of most of these nomads came to Rajasthan during the invasions of the Greeks, Parthians, and Scythians (2d cent. B.C.–3d cent. of our era), the Huns (middle of the 5th–early 7th cent.), or the Moslems (11th–13th cent.). Most of the aristocracy (since the 10th century bearing the title of *rājaputra* or "king's son") is probably also descended from these invaders, despite pedigrees (spurious before the 7th century) going back to the heroes of the *Mahābhārata* and *Rāmāyana* and the families of the sun and moon gods (Sūryavaṃśīs and Sōmavaṃśīs).

During the Middle Ages the population of Rajasthan was regarded as uncouth and barbarian; but at the imperial, royal, and minor courts a highly refined civilization based on the Gupta heritage (see GUPTA, SCHOOL OF) had developed, a culture that was later wiped out by the Moslem conquest. By the late 14th century a feudal society very similar to that of contemporaneous Europe had emerged, comprising among its clans and tribes the Guhilōt of Partabgarh and Mewar, the Rāṭhōr of Marwar and Bikaner, the Yādava of Jaisalmer, the Kachhwāha of Amber and Jaipur, the Hāṛa (Chauhān) of Bundi, Kotah, and Jhalawar, and the Bundēla of Bundelkhand. In the 15th century these formed a great coalition under the leadership of the Sisōdiā maharanas of Chitorgarh (later of Udaipur), which disintegrated after its defeat in 1527 by the Moghul emperor Bābur. In the following years Maldēō of Marwar (Jodhpur) tried to build up another feudal empire. By the middle of the century the Sūr sultans Shēr Shāh and Islām Shāh controlled most of

Rajasthan, and between 1561 and 1570 the Moghul Akbar (r. 1556–1605) forced the Rajputs into vassalage.

Between 1570 and 1720 the Rajputs, led by the Kachhwāha clan of Amber, were among the leading nobles of the Moghul court and provided its best troops, were governors of many provinces and commanders of many garrisons, and brought home enormous booty. Only Udaipur retained a sort of semi-independence, though encircled by Rajput fiefs loyal to the Moghuls (especially in Malwa). The Rajput states proper were threatened by disintegration due to constant family quarrels and partitions. The attempts of the emperor Aurangzēb to annex these weakened states, combined with his religious persecution of the Hindus, led to a series of revolts. In the reign of Farrukhsiyar (1713–18) the Rajput states regained their independence, but their energies continued to be dissipated in incessant wars among Jaipur, Jodhpur, Bikaner, and Udaipur, during which they became increasingly demoralized. First the Marāṭhā dynasty, and then the British, gained control. Under British suzerainty (from 1818) the Rajput courts degenerated, though superficially their old civilization was preserved. In 1948 the states were merged and the Rajput nobility partly dispossessed.

Farther east the Tomār rulers of Gwalior (originally from Delhi) played a considerable role from the 14th to the early 16th century, and the Bundēla tribes were in revolt from 1530 to 1628 and again in the 18th century. There were other small Rajput states in Kathiawar (Saurashtra), Cutch, and Gujarat in the south and Malwa in the southeast. In the Himalaya the Rajputs, expelled from the plains by the Moslems, founded numerous small kingdoms in the fertile belt of the outer valleys: Rajauri, Poonch, Jammu, Basohli, Kashtwar, and Bhadrawah west of the Ravi River; Nurpur, Guler, and Kangra in the Kangra Valley and Chamba to the north; Suket, Mandi, Kulu, Bilaspur, and Sirmur in the east; Garhwa and Almora in Kumaon. They were politically and culturally unimportant, however, until the late 18th century, when the chaos in the Punjab deflected international trade into the protected mountain valleys and engendered a short-lived but refined court culture.

Rajput society was strictly feudal. The state was parceled out into fiefs and subfiefs held mainly by various branches of the ruling clan. The peasantry and nomads were serfs, treated with patriarchal benevolence. The nobles, proud and brave to foolhardiness, spent most of their time in the field or at court and left the civil administration in the hands of rich Jain or Vaishnava merchants (see JAINISM), who farmed taxes and customs, advanced loans against the security of whole districts and provinces, and had practically a monopoly of imports and exports. The nobles lived in their innumerable strong castles, and the merchants in rich town houses. The ladies, no less proud, lived apart in the *rāwala* (a sort of harem, but without eunuchs) but went unveiled, wore arms, and could ride, hunt, and fight; some were famous heroines and even fell in battle. Suttee was practiced, however: a wife was expected to immolate herself on the pyre of her dead husband, or with his turban if he had perished in battle. At the death of a prince who maintained many wives and concubines, it was not unusual for more than one hundred ladies to follow him as *satīs* and in desperate times even thousands thus committed suicide together.

Kings and queens, nobles, and merchants spent great sums on temples. The older ones were generally dedicated to Śiva or the Great Mother in some form (most commonly as Durgā, Mahiṣamardinī, Cāmuṇḍā, or Kālī). Kālī (as Rājrājēśvarī) and Viṣṇu were the patrons of royalty. The most popular deity, however, especially with women, was the cowherd god Kṛṣṇa (the most important avatar of Viṣṇu, venerated under many names), whose love for Rādhā and the milkmaids (*gopīs*; PL. 414) was interpreted as the mystery of the relations between God and the human soul. Jainism (q.v.), on the other hand, was a rather rationalistic doctrine of ethics and salvation, in practice consisting in the veneration of the great world teachers (tirthankaras) and the support of monks and nuns.

ARCHITECTURE. The most common and most important monuments of Rajput architecture are the castle and the fortress. Throughout Rajasthan and central India almost every hill is crowned by a castle, usually consisting of an exterior fortification that served as refuge for the tenants in time of war and an inner castle comprising the audience halls, the ladies' quarters and one or more temples. In the desert the castle afforded control over wells or ponds, without which besiegers had to retreat. In central India, eastern Rajasthan, the Rajasthan-Gujarat border country, and the Himalaya, the difficult hill country and the jungles surrounding the castle or fortress made it possible to harass the attackers from the rear. Waterside castles such as Orchha in Bundelkhand and Nagarkot (Kangra) in the Himalaya are rare, no doubt because huge substructures and platforms had to be constructed for their fortifications. Most castles are perched on hilltops or plateaus, where only a line of battlements and lookouts had to be constructed along the top of the cliffs and occasional gullies closed (such gaps either were used as fortified entrances or cisterns or were filled with storage vaults). Palaces also might be built on such exposed but favorable sites, and if the plateau was large enough a town might even be included (Gwalior, Narwar, Chitorgarh, Jaisalmer). If a town was in a valley, it never covered the valley floor completely, so that armies and caravans could pass outside but remain under observation; often an outer line of fortifications spread over the adjoining hills protected the whole valley, and in some cases the passes also were closed by walls and gates flanked by forts, as in the Girwa and Mukunddarra passes.

The palaces normally had three courts (stable and kitchen, durbar, and ladies' quarters), separated by strong walls and gates. The durbar court comprised an audience and banqueting hall and luxurious smaller reception rooms and private rooms. The ladies' quarters (*rāwala*), protected by double doors connected by corridors, consisted of the richly decorated common room (*rang mahal*), small "flats" for the princesses of rank, and many small rooms for the concubines and artistes. Besides the palace temple or temples, there were many small shrines at the entrances or in the courts for various (mainly protective) deities. Staircases and corridors were very narrow so that they could be defended by a single person.

Mansions and private houses could not expand within the confines of the fortifications and thus tended to rise in several stories on narrow plots. The ground floor is often erected on a stone platform, with an entrance provided with stone benches inside and often outside also; the private rooms, and especially the ladies' quarters, occupy the upper stories, with grilled balconies projecting over the narrow road. Balconies as well as windows have projecting roofs for protection from the sun. Interiors are plain except for many wall niches, but public rooms may have finely carved doors and, occasionally, cupolas. The façades of homes of the wealthy are frequently covered with rich murals or reliefs (decorative or religious scenes). Enclosed gardens and garden houses are common and are generally of the Moghul type, with watercourses, tanks, fountains, open pavilions, and so on. Often they are situated at the foot of an irrigation dam, the principal pleasure house opening on one side toward the lake, on the other to the garden.

The most common type of temple is the small chapel, generally on a platform, beside the roadway or under a tree or along the embankment of a pond. Larger temples sometimes imitate the medieval type, though their sculptural decoration is much simpler and often clumsy or reduced to such ornamental motifs as appear on Islamic buildings. Others consist in late Moghul pavilions with *bangaldār* roofs and domes, or mansions in which a chamber at the back of the principal reception room has been transformed into a chapel; in the latter type extensive murals and, quite often, sculptures emphasize the religious character of the building. These types may be combined, a small medieval type of temple being placed in the center of an open court on the upper floor of a mansion.

The monument for the dead (erected on the site of cremation) may be either a stone pillar (*deval*) or a stone slab (*pāliyā*) covered with reliefs, often protected by a *chattrī* (dome or canopy resting on 4 to 12 pillars; PL. 420). Funeral temples were sometimes erected for sovereigns, and chapels for religious leaders such as abbots of Hindu and Jain monasteries and for suicides who had cursed those who had wronged them.

Irrigation ponds and lakes were constructed in great number as charitable works offered for the salvation of the soul of the donor or his deceased relatives, the donation being commemorated by a stone column (*kīrtistambha*). An earthen dam would be constructed across a depression or a valley and strengthened by a stone embankment, serving as a bathing ghat and therefore dotted with *chattrīs* for protection against the sun, as well as temples, pleasure houses, and so forth. Some of these embankments, such as the Raj Samand at Kankroli and the Jai Samand (or Dhebar Lake), compare favorably with the best works of modern hydraulic engineering.

Fortification walls and substructures were constructed of rubble or roughly dressed stones embedded in mortar or clay, often revetted with sandstone slabs. Here Moslem arches and vaults were common, but in the upper stories Hindu pillars and trabeate lintels and balconies predominated, important rooms often being covered with domes. A special type of construction was a sort of joinery of red sandstone slabs set into grooves of sandstone beams (very similar to modern steel construction), for which the numerous quarries on both sides of the Aravalli Range provided unlimited material; this type of construction was common for the roofing above colonnades, balconies, and windows and later for the balconies themselves and even for whole houses. Other materials used included yellow limestone (from Jaisalmer), white as well as black marble (especially from Makran), and greenish schist (from Partabgarh); a tough red clay for wall coating, often modeled into gilded and painted ornaments; encaustic tiles during the 16th century; and, from the late 17th century on, marble stucco. In the Himalaya, of course, wood predominated: houses, even palaces, were constructed with large "posts" (consisting of rubble-filled "boxes" of short planks superimposed crosswise) connected by horizontal beams, having clay-bound rubble walls set between them, the whole being covered with mortar or whitewash and adorned with wood-carved decorations.

Sculpture. Sculpture in the round is rare and consists mainly of small cult bronzes, and imitations of medieval icons. Reliefs, in a folk style or in imitation of Moghul Rajput paintings, were the rule for funeral monuments as well as for temple and palace walls. On the *deval* and the *kīrtistambha* the principal deities, often with the deceased person or the donor kneeling before the Śiva linga, are represented on four small panels; but when used in association with an actual linga, one side is treated like the *pāliyā*, on which customarily the deceased person is depicted, on foot or on horseback, with sword and shield or in full armor, often in the company of the wives and concubines who died on his funeral pyre. On the *satī* stones erected for the latter is shown either the raised hand of the *satī* or her figure seated on the lap of her husband or of the deity. The accompanying inscription, always dated, is placed between the symbols of the sun and the moon. Below donor inscriptions there is occasionally the figure of an ass copulating with a nude woman, symbolizing the special torment in hell awaiting anyone desecrating the donation. The *gōvardhan* was a similar stele with the relief of Śrī Kṛṣṇa lifting Mount Govardhana; but in many areas the term simply stands for a *deval* or *pāliyā*. Reliefs in the Moghul Rajput style depict the Kṛṣṇa myth, especially the Rās Līlā (the divine round dance in the Vṛndā forest; PL. 411), as well as other deities and court scenes. Under the influence of these reliefs, religious sculpture in the round (in Vaishnava as well as in Jain temples) also began to represent contemporaneous costumes; thus the *surasundarī* figures (apsarases, or heavenly nymphs) came to be depicted as court dancing girls. In the late 18th and early 19th centuries similar figures were introduced into the decoration of *chattrī* and garden palaces.

Painting and miniatures. Murals were common in temples and palaces from at least the 16th century, as niche fillings, wall panels, cornice friezes, and ceiling decoration. The normal technique was fresco secco worked over with tempera, but along wall dadoes may be found sgraffito on white or brick-red marble stucco, as well as glass mosaic; on walls, mosaic of encaustic tiles; and on skylights and windows, figures and ornaments cut out of perforated limestone, alabaster, or stucco slabs, sometimes filled in with colored glass.

Miniatures were generally painted on individual sheets of thick paper. First the outline was sketched in black or red and the colors fixed by means of brief notations or sample patches; then a coating of finely powedered *chunam* (chalk) and glue was applied and polished; and finally the outline was given its definitive form and the colors filled in.

These miniatures, which formed the house library of every princely or aristocratic family, were normally kept in the ladies' quarters. They comprised portraits of the prince and his wife or wives, concubines, and children; scenes of durbars, weddings, celebrations of Holī and other festivals, temple ceremonies, hunting parties, and war cavalcades, as well as city views (PL. 412); and small genre pictures presented by the household painter as a *nazar* (gift) on special occasions (for instance, a nude lady making her ablutions after she had given birth to a child; a married couple either at their wedding or upon their reunion after a voyage or campaign; or a lady kneeling before a yogi on the occasion of a vow for the health or safety of her husband). When these miniatures left the ladies' quarters, the identity of the subjects was concealed by the fanciful addition of the names of famous romantic heroines such as Padmāvatī, Chānd Bībī, Nūr Jahān, or Zēb-un-Nisā Begam; such miniatures are always portraits (within the limits of fashionable idealization), but not of the persons whose names are inscribed on them. Even in illustrations the faces of the heroes and heroines are fashionably beautiful, the expression being achieved by means of attitudes and poses; only low-class persons and demons were given a characteristic physiognomy.

Other miniatures formed loose illustration sets kept between wooden covers, carved, painted, or lacquered and held together by a string. Each sheet has its corresponding text on the back. The books usually illustrated fall into five classes: (1) religious books: the tenth *skandha* of the *Bhāgavata Purāṇa* (the youth of Śrī Kṛṣṇa; PL. 413), the *Devīmāhātmya* (Cāṇḍī), the *Bhāgavad Gītā*, and occasionally other texts such as the devotional songs of Mīrā Bāī or Sūrdās, the *Saundarya Laharī*, the *Gaṇēśa-Stōtra*, or extracts from other puranas; (2) classic literature: the *Rāmāyaṇa* and *Mahābhārata*, the *Mēghadūta*, *Kumārasaṃbhava*, and other works of Kālidāsa; (3) bardic poetry: *Prithvīrāj Rāsau*, *Hāmir Hāth*, *Padumāvatī*, *Dhōla Rāī and Marōnī*, *Sasūī and Puṇhūn*; (4) erotic poetry: the *Kavipriyā* and *Rasikapriyā* of Keśava Dās Sanādhya Miśra (PL. 414), the *Satsāī* of Bihārī Lāl Chaubē, the *Bhāshā Bhushana* of Jaswant Siṅgh, the *Rāsamañjarī* of Bhānudatta, the *Nāyikā Bheda*; (5) musical texts: the *Rāgmālā*. The favorite sets were the *Bhāgavata Purāṇa*, the *Rāmāyaṇa*, and the *Rāgmālā*; next came the *Devīmāhātmya*, the *Mahābhārata*, the *Rasikapriyā*, and the *Nāyikā Bhēdā*.

The erotic and musical motifs are closely interlinked, in so far as all *rāgas* and *rāgiṇīs* (or *putras*), the various musical modes (PL. 415), are connected with some *nāyaka* (hero, swain) or *nāyikā* (heroine, mistress). Each of these poems is a psychological characterization of a man or woman — the shy, coy one or the brazen, audacious one, the lonely and yearning or the happily united, the jealous or the domineering, whether the legitimate spouse or a concubine, an adulteress or a dancing girl (courtesan) admitted as entertainer to the palace. The whole approach to love, notwithstanding its delicacy, is understandable only in the context of polygamy, the closed world of the *rāwala*, with all its exterior splendor and secret tragedies. This poetry is further related, on the one hand, to the aspects and moods of nature (spring, summer, the rains, autumn, winter, the dawn and dusk, noontime and night) and, on the other, to religion — the god Kṛṣṇa (PLS. 411, 413, 414) as the ideal lover and the consoling substitute when the princely lover forsook his beloved.

Minor arts. Furniture has always been sparingly used in India (see FURNITURE). Wall niches normally serve as cupboards, wardrobes, lamp stands, and so forth. Door jambs and lintels are often richly carved with knobs, bells, lotus rosettes, lambrequins, and reliefs of the Hindu gods, especially Gaṇeśa and Śrī Kṛṣṇa. The doors themselves are inlaid with ivory or sandalwood, studded with wooden or metal rosettes or with heraldic figures cut out of brass sheets; or they are covered with painted and lacquered panels, occasionally even with reliefs. Especially in Marwar, small wooden horse heads are often mounted on both sides of the door. Beds and seats are of the light charpoy type used everywhere in India; their legs may be plain, painted and lacquered, carved in ivory, or chased in silver (sometimes in the shape of heavenly nymphs). Small taborets often have brass feet and are covered with chased copper or silver. The brass chains and supporting wooden frames of swings (*hindōlas*) are decorated with figures of maidens (*gopīs*). Palanquins were made of painted and lacquered wood, with small silver and gold umbrellas on top as a symbol of rank for maharanis. The ceiling fan (*pankhā*), introduced in the 19th century, is embroidered with heraldic peacocks; hand fans often have pictures (painted or embroidered) of Śrī Kṛṣṇa and the *gopīs*. Until the late 17th century convex metal mirrors were used; thereafter, looking glasses set in lacquered wood frames were prevalent. From the end of the 18th century on, sleeping rooms were often completely lined with mirrors. Wooden or leather chests and toilet boxes were treated in appliqué work or painted and lacquered. Small stands for water vessels often consist of three or four intersecting horse heads. Carpets were generally of the Persian or Moghul type, but enriched by Indian animal motifs.

Beautiful wall hangings (*pichhwāī*), painted or embroidered with cows and *gopīs*, were used especially in the temples of the Vallabhacharya sect of Kṛṣṇa devotees (PL. 414). Bedcovers and embroideries were often decorated in appliqué work set with small mirrors or in phulkari (flower) designs. Petit-point embroidery was used in Rajasthan for prayer gloves, knuckle pads of shields, wall hangings, and other decorative purposes. Embroidered rumals (kerchiefs) were placed, as covers, over trays of wedding presents; those of Chamba, Kangra, and Basohli especially have become famous for their figural embroideries (partly done in gold thread). Cross-stitch embroidery is found in India only in Rajasthan (among the Gujar and Mēr nomads) and in a corner of the Deccan (the Ahirs). For turbans, very fine muslin treated in *bandhnī* technique (dyeing in various designs by tying off certain parts) was much in fashion; Rajasthan brocades (Moghul technique) have become famous.

The Rajput costume, characterized by intense, harmonious colors, is of three types: the pointed male coat and the female wheel skirt, which belong to a very old (Central Asian?) tradition; the early Islamic (Seljuk) costume, especially the small tight turban (*pagri*); and the late Moghul costume. Jewelry (especially rich for women, but used also by men), most often of silver, is of immense variety, though few genuine early examples are known. A characteristic technique, especially of Jaipur, is enamel (taken over from the Moghuls). Generally the weapons are of Persian type, though some archaic forms (e.g., the stick shield and the katar, or dagger with H-shaped hilt) have survived. They are beautifully engraved with benedictions, charms, reliefs of the family deities (especially Durgā), or hunting and battle scenes. The sword hilts repeat the late Moghul decoration. Armor consists of chain mail with four-piece cuirass, armlets, and greaves. The pointed helmet has a visor and chain collar. The round shields, of buffalo or rhinoceros hide or of steel, have no armorial bearings. The armor of the war horses was of leather on which hand-size steel scales were sewn.

EVOLUTION OF RAJPUT ART. The medieval temple and palace art of northern India represented a continuation and elaboration of late Gupta art (see GUPTA, SCHOOL OF) in the same sense in which Romanesque art evolved from the art of the Roman Empire. As in Romanesque art, however, primitive motifs or reinterpretations of classical motifs broke through: plait-band ornaments and knots (often disguised as snakes), dissolution of architectural or floral motifs into such bands, replacement of figures by chessboard or rhomboidal designs, costumes like those of the Hindu Kush or the Pamirs, and sculptures suggestive of Visigothic, Lombard, or Frankish figures. The *pāliyā* occurs first in Osia (near Jodhpur) and around Gwalior in the 7th and 8th centuries; stele reliefs of horsemen, first found on provincial examples of the 11th century, can be linked with Caucasia, Turkistan, and the Altai. However, this undercurrent could rise to the surface only after the dissolution of medieval Hindu art, and even then not immediately.

Transitional period. With the collapse of the northern Indian kingdoms, the art cultivated by the courts and the aristocracy was all but wiped out. The Hindu principalities surviving in the desert and the mountains were unable even to maintain existing buildings that had escaped destruction. The tradition was, however, kept alive by the Jain bankers and merchants, who, accepted as government contractors by the Moslems, built many temples on mountain tops (e.g., Mount Satrunjaya, Mount Girnar, Taranga Hill, Pawagarh, Mount Abu) and in the desert (Jaisalmer, Lodurva, Pallu, Bhadreshwar, etc.) and equipped them with stone and bronze figures and illustrated manuscripts (see JAINISM), although these monuments had to be rebuilt with endless patience after each desecration. Thus the tradition was kept alive, though ossified. From the late 14th century on, there was intensive reconstruction, but the artistic tradition had been so badly weakened in the meantime that even the Jains had to summon sculptors from the Vijayanagar kingdom of the Deccan, and in erecting new Hindu temples architects and sculptors had to fall back on the study of the Śilpaśāstras (see TREATISES) and on imitation of ancient ruins. At first many mistakes were committed, but by the middle of the 15th century, in the reign of Rana Kumbha (1433–68), a fairly successful "neomedieval" art had developed, and in the 16th and early 17th centuries the pastiche was often perfectly executed.

This very renaissance of the medieval style, however, was its undoing, for it depended on a simultaneous growth of all intellectual activity and on direct artistic inspiration. Two factors contributed to the development of a new, original style: in secular life, a far-reaching adaptation of the contemporaneous Indo-Moslem civilization, especially in military and palace architecture, costume, and armaments; and in religious life, the reawakening of a fervent mysticism after a long period of sterile retreat into pedantic ritualism. This religious revival centered around the cult of Śrī Kṛṣṇa, especially the contemplation of his love for Rādhā, as described in Jayadeva's passionate poem, the *Gīta Govinda*. Great poet-saints such as Chaitanya, Narsiṅgh Mēhta, Mīrā Bāī, Vallabhāchārya, Haridās, and Sūrdās sang his praise, following their own spiritual experiences; and with little regard to orthodox teachings, they evolved a new theology, in which Śrī Kṛṣṇa practically became the one and only godhead. The cult of Durgā Mahiṣamardinī (the new royal patroness in place of Viṣṇu) experienced a lesser revival. Later, in the 17th century, came a revival of the cult of Śrī Rāma (another avatar of Viṣṇu), owing to the popularization of the *Rāmāyaṇa*, in a more emotional form, by the poet Tūlsī Dās; Śrī Rāma became the patron of the strengthening of the monarchy at the expense of the feudal lords.

These religious trends had a double effect. The cosmic symbolism of the medieval temple lost its meaning and would soon have ceased to have effect if the destruction of existing temples, occasionally by Shah Jahān and on a grand scale by Aurangzēb, had not inspired a political revival of this architecture in defiance of the Moghuls [best example: the Jagannāth Temple at Udaipur (1652); others at Jaisalmer (1681), Kulu, Chamba, etc.]. The second effect was the creation of a whole new iconography for which the ancient texts and monuments could offer no prototypes, and eventually also another type of sanctuary, the mansion temple.

Early Rajput style. The Rajput style first made its appearance in the royal palace of Chitorgarh, which after the capture of the fortress by the Moslems in 1303 was rebuilt by the ranas Lākhā (1382–97), Mōkal (1397/8–1433), and Kumbha. Its main front, on the principal access to the town from the Rām Pōl, or main gate (toward the north), and the adjoining administrative buildings ("Kumbha's Treasury") and Jain temples, is several stories high, rather plain except for a number of balconies and roof *chattrīs*. On the east side are the chief entrance, guard-rooms, and stables; next comes the durbar hall, then the private rooms and the *rāwala* overlooking the Gambheri Valley; farther south are gardens, ponds, and the palace of the heir apparent. The fortifications (except for the old gates) are of the contemporaneous Islamic type. Within the palace are found Islamic vaults and domes (on a high drum), plastered walls, highly simplified Hindu columns and lintels resting on Śrīdhara-style brackets, projecting eaves (*chajjas*), balcony parapets sloping slightly outward (as in medieval temples), and Islamic battlements and Hindu lotus roundels as decoration.

This style was widely imitated until the middle of the 17th century and occasionally even later. The next example of it is the Rām Mandir (the oldest palace, later transformed into the royal family temple) in the Bundēla capital, Orchha, founded in 1530 on an island in the Betwa River. It is a uniform building block around a square court, with its lower stories resting on Islamic arches and vaults, its upper ones organized into wall panels enclosing windows with fantastic cusped Lōdī arches, crowned by pavilions with fluted domes of the Lōdī type (see INDO-MOSLEM SCHOOLS). Similar are the oldest palace remains at Bundi. The Hindōla Mahal, on the west side of the Bundi Range, is perhaps one of the best extant examples of this style, with an outer court, the palace proper rising on high substructures above the slopes, and the living rooms arranged in picturesque asymmetry. What can still be traced of early Rajput architecture in Marwar is also similar: the "medieval" gates of Malkot (Merta), the Lōdī-type gateways of Maldēō in Jodhpur Fort, the castle of Phalodi, the earliest gateways and palaces in Bikaner (1582), and Jaisalmer Fort. The oldest palaces at Amber (the early-16th century Narsiṅgh Mandir, or Purāṇa Mahal, and the *rāwala* erected by Mān Siṅgh) — as well as the Mān Mandir at Benares, Rothas Palace (Mān Siṅgh's residence as governor of eastern India, 1589–1604), and the Hāḍī-Rāṇī Mahal at Toda Raisiṅgh — represent a more ornate style. Next come the buildings of Bhōj Siṅgh (1585–1607) in Bundi Fort: namely, the Hathīśālā, Badal Mahal, and Phūl Mahal; and then, those of Ratan Siṅgh (1607–31): Ratan-Daulat Hall, a series of private apartments, and the Hathiā Gate. Thereafter the style was taken up for the palaces of Rana Amar Siṅgh (1597–1620) and Rana Karan Siṅgh (1620–28) at Udaipur, where the grilled balconies of the *rāwala* form a series of half towers projecting from the *chattrī*-crowned façade (PL. 416). The last representatives of this style are the Talēti Mahal (Lower Palace) in Jodhpur town and the Fateh Mahal in Jodhpur Fort (early 18th cent.), the latter with slim "Hindu" columns (inspired rather by the high pillars of Indo-Moslem mosques, formed of several Hindu columns).

Soon a more ornate style developed, the first example of which is the Mān Mandir, or palace of Mān Siṅgh Tomār (1586–1616) of Gwalior, at whose court new styles of painting and music also developed. The walls and towers on the valley side of this palace are decorated with a frieze of traditional Hindu toranas combined with floral motifs, animals, and human figures in encaustic tiles, and there are roof *chattrīs* with high domes in the Malwa style; the interior has southern Indian (Vijayanagar) brackets, rich columns freely interpreting many long-forgotten Hindu motifs, huge tympanum rosettes, and friezes of dancers and musicians cut in silhouette in the perforated stone slabs lighting the dance hall from above (FIG. 829). Some tile decoration is found in the rather cramped palace of Rana Ratna Siṅgh II (1527–30) at Chitorgarh (north of the Rām Pōl); however, since Gwalior Fort was taken by the Moslems in 1618, the Mān Mandir has left no other mark on subsequent Rajput architecture.

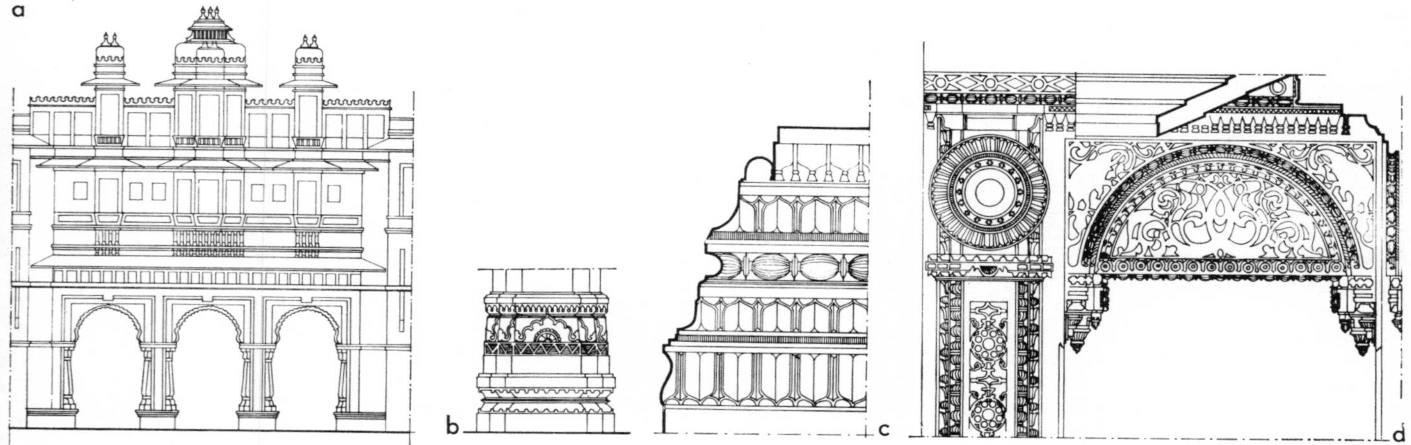

Rajput architectural elements and decorative motifs: (*a*) Udaipur, Gangor Ghat, portal of an inner court; (*b–d*) Gwalior, Mān Mandir, details of architectural elements and ornament (*from C. Batley, The Design Development of Indian Architecture, Bombay, 1954*).

Instead, there was new inspiration from the eclectic Moghul style of the time of Akbar and Jahāngīr, which had itself been the product of a synthesis of Indo-Moslem and Rajput art (see MOGHUL SCHOOL). The most important representatives of this style are the fine palaces of Raja Bīr Siṅgh Deo (1605–28) at Orchha (Jahāngīrī Mahal) and Datia (Govind Mandir; PL. 416; FIG. 831), the latter never occupied because of a curse. Both were *rāwalas*, consisting (like Mān Siṅgh's palace at Amber) of a court surrounded by the individual quarters of the ranis and concubines, rising in receding stories, four units in the corners and four on the chief axes, with adjoining terraces around the court. There is a single pavilion on the roof terrace. At Datia the individual quarters are connected by bridges, with a central tower surrounded by balconies rising above the whole block. Outside, the walls are covered by galleries with perforated stone windows; only the monumental gate and the balconies supported by it project from the simple square of the walls. The lower stories rest on heavy Moslem vaults; the upper ones appear to be made up of light columns, brackets, and grills, and end in fluted domes of the Malwa type. The decoration is very rich: relief, sgraffito, murals, and encaustic tiles combining Hindu and Moslem features, lotus ceilings, cornices with *Rāgmālā* illustrations, niches filled with female figures, *jālīs* (perforated-stone grills) decorated with *haṃsas* and every type of contemporaneous ornament, arabesques as well as geometric designs. The palaces of the Bundēla rajas of Chanderi (the Rāj Mahal and Rām Mahal), which are of the mid-17th century, and the old palace (built by Bhīm Siṅgh, 1707–20) in Kotah Fort reveal the decline of the tradition. In Mewar this style is represented by the dams of the Raj Samand (Kankroli), with its beautifully sculptured *chattrīs*, and the Jai Samand.

The last phase, directly inspired by the Moghul architecture of Fatehpur Sikri, is found chiefly in the Dīwān-i-Am of Amber Palace (ca. 1620); the Sūr Mandir, the Karan Mahal Chowk, and the Dēvī Dvāra of Bikaner Fort; the *chattrī* of Karan Siṅgh (1674) and that of Anūp Siṅgh (PL. 418) at Devikund, increas-

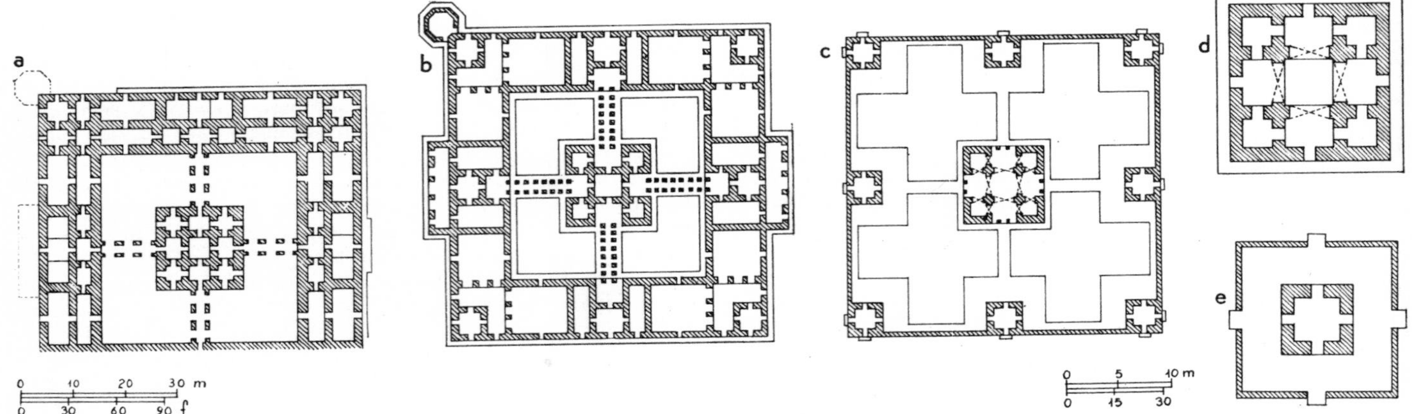

Datia, Govind Mandir (built by Raja Bīr Siṅgh Deo), plans: (*a*) First story; (*b*) second story; (*c*) third story; (*d*) fourth story; (*e*) fifth story (*from P. Brown, Indian Architecture, Islamic Period, 3d ed., pl. C, Bombay, n.d.*).

ingly permeated with classic Moghul ornamental motifs; several other *chattrīs* outside Devikund Fort (1740) and at Kolayat (PL. 420); the original Karnījī Temple at Deshnokhe; and the Sūraj Pōl, as well as towers and balcony brackets under the Fateh Mahal, Ajīt Vilās, and Daulat Khāna of Jodhpur Fort.

Temple architecture thus far had been an imitation of the medieval style. The rajas of Amber, however, were the supporters of the most liberal trend in Hinduism, the Kṛṣṇa mysticism of Chaitanya and Mīrā Bāī; and under their rule arose a new temple style which combined the traditional with Moslem forms and popular sculpture. Its principal representatives are the Jagat Śiromani Temple at Amber (PL. 417), erected by Mān Siṅgh; the Govind Deo (1590), Jagat Kishōr (1627), Gōpīnāth (ca. 1630), and Madan Mohan at Brindaban; and the Hardēōjī at Gobardhan, near Mathura. Raja Bīr Siṅgh Deo of Orchha adopted this style in the Caturbhuja temple at Orchha (PL. 417) and in the gigantic Keśava temple at Mathura and the Viśveśvara at Benares (both destroyed by Aurangzēb in 1669 and replaced by mosques). Finally, Raja Bāsu Deo of Nurpur (Kangra Valley) built there a large Kṛṣṇa temple (PL. 418) in the same style, probably with the help of masons from Amber (the power and wealth of which had declined since 1605); this was destroyed by Jahāngīr after the rebellion of 1613.

Though differing greatly in detail, all these temples have many characteristics in common. The mandapa is a high hall covered with a Moghul vault (as in the Great Mosque of Fatehpur Sikri) or even a dome; the walls are broken up into galleries of more or less pure Hindu type, and the sanctuary (*garbhagṛha*) is joined to the end of the hall. The height of the hall required a raising of the cella, so that the *śikhara* could overtop the mandapa. Where the mandapa had a cruciform plan, the temple might have several sanctuaries (like many Chalukya and Hoysala temples). Inevitably in such buildings Moslem features (e.g., wall panels and blind niches) and ornament were freely mixed with Hindu motifs adapted to Moslem art and with pure, though much simplified, Hindu forms. The sculptures are free adaptations of traditional models or are derived from folk styles. This temple style may have its antecedents in certain Jain temples; in these the sanctuaries, because of their great number (one for each of the 24 tirthankaras), tended to be reduced to mere adjuncts to a gallery or hall. Thus, as early as the beginning of the 16th century, the temple with an ordinary house plan was known, though it might be decorated in an imitation of medieval sculpture (e.g., Sanganer, Bairat, Bikaner).

Under these circumstances the funeral shrine (samadh) also changed. The normal *chattrī* type (i.e., the canopy on pillars) probably had been the result of a fusion between the primitive Bhil and Mīna type of funeral shrine and a small mandapa roof (still apparent in the earliest Amber *chattrīs* and in those of Jaisalmer and of Arh near Udaipur). But its domical cover developed under Moslem influence, as did the complete mausoleum (of Lōdī type, but generally of circular ground plan, as at Bairat, Chanderi, and Mandor). A unique monument is the Staīburj at Mathura, a two-story tower erected in 1570 in memory of the mother of Raja Bhagwant Dās.

The sculpture of this period vacillated between an imitation of medieval prototypes, carefully copied but without their sensuousness and their understanding of anatomy, and a primitive folk style that can be traced as far back as the donor figures of the 10th century, which dominated the decoration of the *pāliyā* and which in the course of the 16th century invaded temple architecture also. As in ancient Egyptian art, scenes are arranged in tiers separated by distinct lines and are grouped side by side without overlapping. Individual figures are built up so that the parts of the body are given fullest visibility; heads, however, are usually shown frontally. The movements are vivid but angular and naïve; facial expression is entirely absent. In the temples and *chattrīs* of the 16th and early 17th centuries the styles fused, sometimes with odd results. In the Jagat Śiromani at Amber this new sculpture is still very clumsy and tentative; in the Govind Deo at Brindaban a harmonious and expressive, though still archaic, style has developed; and in the Kṛṣṇa Temple at Nurpur (PL. 418) the sculpture reveals a great wealth of invention and expression. Thereafter, under Moghul pressure, the style bifurcated, one form returning to the traditional line, though enriched by new types, and the other developing a three-dimensional adaptation of the pure Rajput style of painting. The finest examples are the Rās Līlā ceilings at Datia, Bikaner, Devikund, and Kolayat (PL. 420); the ceilings of the Cāmuṇḍā temple and the reliefs of the Brahmōr Kōthī (17th cent.) in Chamba; other later reliefs at Brahmor, Chamba, Bilaspur, and elsewhere and innumerable *pāliyā* reliefs all over Rajasthan; and finally, the beautiful reliefs in later *chattrīs* (e.g., Gethor near Jaipur), Jain temples (Merta), and palaces (Kotah).

The origins of Rajput painting are still controversial, since the premises on which earlier research tried to build a classification and chronology have proved untenable. There were neither regional styles that were characteristic over a long period nor a uniformity of style covering the entire area or even a large part of it. Rather, certain styles evolved at the court of one or another art-loving ruler, moved to vassal courts and then to noble households, and at last merged with middle-class and folk art as they were superseded by more modern fashions. During the 16th through 18th centuries the Amber-Jaipur court was the most modern and that of Udaipur the most conservative. Certain prominent centers and the styles successively evolved can be identified; but as yet it seems quite impossible to trace the later migrations and changes of the individual styles. Moreover, the wars with the Moghuls led to the destruction of much early material; and the few known dates refer partly to court and partly to provincial material, whose interrelation is uncertain. Only a tentative reconstruction can be attempted, within the context of all other aspects of Rajput art, especially the reliefs.

The first development toward a Rajput painting style is reflected in the illustrations for the late-15th-century Vaishnava manuscripts in the Jain style of western India, especially the *Vasantavilāsa* and the *Bālagopālastūti*. As in the contemporaneous sculpture of Rana Kumbha's reign, the whole formal typology of the Jain *Kalpasūtras* of the 13th to early 15th centuries (an ossification of the preceding Solanki-Vāghelā style of Gujarat and southern Rajasthan) was taken over, including the eyes projecting from the outline of the face, the beaklike noses, and the rich hairdo of the ladies. But, like marionettes set in motion, the figures assumed new postures and new groupings, wore new emblems, and were seen against new settings, and there was an intensity of feeling never known to the *Kalpasūtras*.

Early in the 16th century a revolution in artistic style became evident all over India. At Vijayanagar (Deccan) the folk style dominated the reliefs of the throne platforms (1518) of Kṛṣṇa Deva Rāya; a *Bhāgavata* manuscript from Assam, dated 1539, is illustrated in a style very similar to Rajput painting; in the Jain manuscripts the old style degenerated and absorbed many Indo-Moslem elements; and all over Rajasthan (except Mewar) *pāliyās* in the folk style became common. In the Mān Mandir at Gwalior the animal friezes in encaustic tiles reveal the early Rajput style, and the human figures there and on the *jālīs* of the dance hall occupy an intermediate position between the Jain and the Rajput (and Moghul) styles. Many vestiges found in later works point to significant development throughout the 16th century, but no original works have survived, because the important Rajput capitals were repeatedly stormed or occupied at least temporarily and looted.

The first center seems to have been Amber in the time of Mān Siṅgh (1592–1614), under whom the style appears in the Hindu figures in the otherwise Moghul *Rasm-nāma* of 1584 at Jaipur (Royal Lib.), the murals in the royal *chattrīs* at Amber (ca. 1578–84), Mān Siṅgh's garden house at Bairat (1587), the bas-reliefs in the Govind Deo Temple at Brindaban (1590), and the earliest leaves of the Laud *Rāgmālā* [Oxford, Bodleian Lib., Or. 149 (Laud)]. This style became one of the chief components of later Moghul painting. Echoes can be traced also in West Bengal and Orissa, of which Mān Siṅgh had been governor. The later development of the school after 1605 is unknown, although miniatures in this style continued to be produced until the middle of the 17th century (Baroda and Bombay *Rasm-nāmas*, 1598; later leaves of the Laud *Rāgmālā*, ca. 1620; *Rasm-nāma*, Br. Mus., ca. 1630; Bikaner *Kṛṣṇa-līlā*).

Next, probably, came the Bundēla school. Its earliest representative is the *Rāgmālā* in Boston (Mus. of Fine Arts), probably done under Madhukar Shāh (1554–92); then came the *Rasikapriyā* made for Indrajīt Siṅgh of Indargarh (ca. 1600); and last the murals in the Datia and Orchha palaces. This school, probably dispersed during the revolt of Jhujhār Siṅgh, survived at Chanderi and merged into the so-called "Malwa style" (PL. 413). In Marwar a related style (with round instead of square heads) can be traced up to the middle of the 17th century. Another style, influenced by the Moslem art of Gujarat, seems to have flourished in Jalor.

Mewar had probably been one of the most important early centers of Rajput painting, but the long guerrilla war with the Moghuls, with its scorched-earth policy, wiped out all vestiges of its production. A recovery set in under Amar Siṅgh (1597–1620) and Karan Siṅgh II (1620–28), as seen in the Chawand *Rāgmālā* (containing vestiges also of a lost genuine Malwa style of the 16th century) and some portraits. The manuscripts of the last years of Jagat Siṅgh (1628–52) represent a degenerate late style, evidently imported (though it is difficult to determine from where — Marwar? Bundelkhand?). Under Rāj Siṅgh II (1652–80) this archaic style reached its zenith; under Jai Siṅgh (1680–98) it degenerated, to survive partially in provincial Mewar and in Malwa.

The Basohli school (PL. 413; VII, PL. 234), which dominated the Punjab Himalaya until about 1760 and was the precursor of the mature Pahari painting, must go back at least to the reign of Bhūpat Pal (1598–1625); otherwise the many survivals of the Akbar style it evidences would be unaccountable. However it can be traced in that region only from the reign of Kīrpāl Pāl (1678–95), in certain *Rāsamañjarī* sets, despite somewhat earlier vestiges at Mandi (mid-17th cent.) and Brahmōr (1670).

Period of Moghul domination. Beginning with the reign of Shah Jahān (1628–58), the influence of the imperial style made itself felt increasingly in Rajput art, owing to the long stay (often for decades) of the Rajput princes at the imperial court or with the imperial armies, as hostages, generals, fortress commanders, and governors. Aurangzēb's persecution of the Hindus after 1668 brought about not a return to the old traditions but artistic competition with the Moghul court. The emperor's puritanism and the bankruptcy of the empire after about 1680 induced many of the best Moghul artists to seek service with the Hindu rajas. As the empire disintegrated after Aurangzēb's death, more and more artists migrated to the rising Rajput courts; the movement continued as artists sought refuge from the invasions of Nādir Shāh (1739) and Aḥmad Shāh Durrānī (1747–61) and from the Marāṭhā depredations, until eventually, with the decline of Delhi, Moghul artistic influence was absorbed.

The Kachhwāhas of Amber (Mīrzā Raja Jai Siṅgh I, 1625–67) and Marwar were the first to take over the Shah Jahān style. Jai Siṅgh I built the Jai Mandir (with its Shīsh Mahal and Dīwān-i-Khās) and the Sohāg Mandir in the Old Palace at Amber in the best Shah Jahān style. Under him also the *Satsāī* of his court poet Bihārī Lāl (1648–50, Bikaner, Coll. Maharaja of Bikaner; another set in Baroda Mus.), as well as several *Rāgmālā* sets (Baroda and elsewhere), were illustrated in a style deriving its technical details and decorative motifs from the Moghul tradition, although the more sweeping line, the strict geometric pattern, the lack of spatial sense, and the romantic mood are characteristically Rajput. After Jai Siṅgh's death, Amber again declined but recovered under Sawai Jai Siṅgh II (1693–1743), who added the Jalāib Court and the Gaṇeśa Gate to Amber Palace and in 1728 founded the new capital at Jaipur. Most of this architecture is an adaptation of the Moghul style under Aurangzēb, Bahādur Shāh I, and Farrukhsiyar. The same can be said of painting; and the Moghul style in the minor arts (especially textiles, metalwork, and enamel) was taken over to such an extent that even genuine creations of the imperial workshops have been classified in museums as "Jaipur work." From the later years of the reign of Jai Siṅgh II on, especially under Iśrī Siṅgh (1743–51) and Mādhō Siṅgh (1751–68), the "baroque" Moghul style of Muḥammad Shāh's and Aḥmad Shāh's reigns was introduced; its best-known monuments at Jaipur are the Chandra Mahal and the Hawā Mahal (XIII, PL. 236). The larger compositions and romantic atmosphere of late Moghul painting were also brought to Jaipur; their best transposition into Rajput paintings is found in the Kishangarh miniatures painted for the deposed Sawant Siṅgh (1748–64; known under the pen name Nāgarī Dās).

At Jodhpur most of the 17th-century buildings were destroyed during the Moghul occupation of 1678–1705; but between the Lōha and Sūraj gates and beyond them there are still a number of galleries composed of cusped arches in the Shah Jahān style (beneath the Phūl Mahal, Chaukalās Mahal, and Takht Mahal). A few miniatures of Sūr Siṅgh's time (1585–1620) reveal the penetration of the Jahāngīr taste; others, of Jaswant Siṅgh's time (1638–78), show an adapted Moghul style, technically better but emotionally harsher than the contemporaneous Amber paintings.

At Bikaner classic Moghul architecture was introduced by Anūp Siṅgh (1674–98), the builder of the Karan Mahal (PL. 420), Rai Mahal, and Anūp Mahal (durbar and conference rooms); the later, excessively ornate taste was promoted by Gaj Siṅgh (1745–87), who erected the Gaj Mandir, a beautiful Shīsh Mahal, and richly decorated rooms for his two chief consorts, Chand Kaur and Phūl Kaur. As early as the time of Karan Siṅgh, governor of Daulatabad, a charming school of painting (in a mixed Moghul-Deccan-Rajput style) had evolved; Anūp Siṅgh and Sujān Siṅgh (1700–36), as governors of Adoni (Deccan), engaged several accomplished Moghul painters (e.g., Rashīd, Rukn-ud-Dīn), who soon adopted the Rajput style when they had recourse to earlier miniatures and bronzes in illustrating the *Rasikapriyā* and other books. Under Gaj Siṅgh there was a final influx of Moghul painters, who executed mainly portraits.

At Udaipur the Moghul style made its first, sporadic appearance in the Jai Mandir, erected on the island of Jagmandir by Karan Siṅgh for Shah Jahān in 1627; it became more pronounced in the Sabrat Vilās (under Rāj Siṅgh, ca. 1660) and then became general in the new rooms built on top of the Maharana's Palace (PL. 416) by Amar Siṅgh II (1698–1710) and Sangrām Siṅgh II (1710–34). Under the latter two rulers, Moghul painting also was introduced. At Bundi, Rao Chattarsāl (1652–58) and Rao Bhāō Siṅgh (1658–78) fostered the Moghul style; and in Bundelkhand the Rāj Mahal of Orchha, planned like the Jahāngīrī Mahal but finished with halls and murals in the Moghul taste of Aurangzēb's time, and the Phūl Bāgh (a garden palace) represent the same development. The round mausoleum of Chattarsāl of Panna (1662–1732) at Mau, finally, is an elaboration in the late Moghul style of the earlier round samadhs at Bairat and Chanderi.

Later Rajput costume in Bikaner, Jaisalmer, and Marwar was based on the Moghul fashion under Farrukhsiyar, and in Jaipur and Udaipur on that under Muḥammad Shāh and Aḥmad Shāh. According to the miniatures, Rajput princesses also wore Moghul costume in this period (ca. 1670–1730/40), except on religious occasions.

In the Himalaya Moghul architecture was first adopted at Taragarh (early Shah Jahān style), Nurpur (the palace and Ṭhākurdvāra, in Aurangzēb style), and Rajnagar, or Kankroli (also Aurangzēb style). Later it developed under Sansār Chand II (1775–1823), at Alampur, Tira Sujanpur (palace; Sansārchandeśvara Temple; Narbadeśvara Temple, PL. 420), and Nadaun (Moghul gardens), and under Govardhan (ca. 1735–60) and Prakāsh Chand (ca. 1760–90) at Guler (Oudh style). Moghul painting was first imported from Lahore by Pandit Sēu, who had fled to Jasrota from the Persians, and by his sons Manik and Nainsukh, who brought it to Guler and Jammu.

Late Rajput style. The transition from the Moghul to the late Rajput style was so gradual as to be almost imperceptible, yet by the last quarter of the 18th century the two styles were totally different. Rajput conservatism faithfully preserved the techniques, structural forms, and iconography of Moghul art; but without the imperial prestige and model, these traditions were submerged in the Rajput esthetic and emotional attitudes and underwent a subtle reinterpretation. Meanwhile the early Rajput tradition, which had been displaced and relegated to minor provincial centers, returned and fused with the Moghul trend. But this early Rajput style, too, was changed: the old simplicity and naïveté gave way to a more sophisticated mentality, chivalry to rakish gallantry, personal religious feeling to ritualism, love to sensuousness, naturalness to artificiality.

The style of the many extant architectural monuments of this time represents the peak of the baroque and rococo trends in late Moghul art: paired or repetitive and involuted forms; increasing and decreasing accent and rhythm; curving or even circular plans; *bangaldār* roofs carried downward into a semicircle, or combined with horizontal roofs into an outline like a Turkish bow, or combined with domes and their flutings continued into cusped eaves; disintegration of pillars, arches, railings, and other architectural forms into a jungle of floral and plant motifs; introduction of figural sculpture and murals. In the 19th century the process was reversed: the multitude of forms and ornaments was again organized into vertical or horizontal units, simple in their general effect though rich in detail. An important new type was the mansion temple, developed from the house chapels, which were the only places of worship to survive during the high tide of Moghul intolerance.

In the various Rajput states, this style developed with differences determined by local conditions and cultural contacts. The most conservative area was Mewar, where the Moghul forms of the early 18th century were more or less retained into the 20th century, though they became increasingly slim and elegant. The most important buildings are the upper stories of the Maharana's Palace in Udaipur (PL. 416; the Karan Mahal, Baṛē Mahal, Chīnī Chitra Mahal, and Choṭā Chitraśālā — with Portuguese and Delft tile — and later the Mānak Mahal and Mōtī Mahal), the round pavilion on Jagmandir Island, the palace on Jagnivas Island, and the Sahēliōn-kī Bāṛī. In Jaipur, too, the architecture and decoration became rather conservative after it had absorbed the Muḥammad Shāh style. There are more *baradari* roofs than in Udaipur, but they are of the classic Moghul type (e.g., the Purāṇa Ghat and Galta outside Jaipur). In the famous Hawā Mahal (XIII, PL. 236) the bays of the upper stories, grouped in an irregular rhythm, rise in a gradually steepening curve; the entrance (inside the palace) is richly decorated with astronomical reliefs.

In Marwar the outstanding features were round bulidings or pavilions (e.g., the Fatteh Mahal, Phūl Mahal, and Daulat Mahal in the fort, built by Mān Siṅgh, 1803–43; late Jodhpur and Mandor *devals*); the mansion temples (the Kunjbihārījī, built by Gulābrāijī, a concubine of Bijai Siṅgh, 1753–93; the Ganghāmjī Mandir; the Tījī Mājī Mandir; the Bāghēlījī Mandir), with toranas and elephant statues at the entrance; and houses and mansions with semicircular sun roofs (Zenāna Mahal in the fort; Bal Samand Palace). Above the *garbhagṛha* the Kunjbihārījī has an odd shikara; in the Mahā Mandir (built for the Kanphata yogi Dēōnāthjī, assassinated 1808) this spire was transformed into a pyramid of bulbous miniature shikaras. In the palaces of Takhat Siṅgh (1843–73), a return to simpler forms is evident.

Bikaner and Jaisalmer (PL. 419) occupy an intermediate position between Jaipur and Marwar. The mansion temple (e.g., Jagannāth, Dhunīnāth, Rājratan Binārījī, Rasikśiromani) and the semicircular sun roof (Anūp Mahal and Sūrat Nivās) occurred, but less frequently. Predominant, especially in the time of Sūrat Siṅgh (1787–1828), was a mixture of Oudh-Moghul and older indigenous forms; the redecoration of the Anūp Mahal with red, green, and gold clay moldings is similar to the work in the Golden Temple at Amritsar and Rañjīt Siṅgh's tomb at Lahore. Under the later rulers, especially Duṅgar Siṅgh, the aspect of the palaces of Bikaner Fort was reduced to a play of simple rhythms by the construction of new façades of uniform galleries.

Another distinct style developed in Bundelkhand, from Chanderi and Datia in the west to Panna in the east, characterized by semicircular or nearly semicircular *bangaldār* pavilions (often little more than façades) with many pinnacles. For this type of *bangaldār* roof structure the large domes (often paired), as well as the small domes of the roof *chattrīs*, were adapted. Sometimes the *bangaldār* roof was also swung out horizontally to end in elephant heads. The temple spires are sometimes simple cones ending in a *kalaśa* (vase-shaped top); or they may consist of a tower of several stories, ending in a cupola surrounded by *bangaldār*-roofed half *chattrīs*. Eventually a complicated ground plan was evolved, as in the Lakṣmī-Nārāyaṇa Temple at Orchha (triangular colonnaded court, entrance rising in front of a circular projection, sanctuary under a tower at the back of the court).

In the western Himalaya only a few late Rajput buildings are found: the Sansārchandeśvara and Narbadeśvara at Tira Sujanpur, the Rang Mahal at Chamba, the Basohli Palace in its present (Sikh) form, and the palaces and temples of Jammu, especially the Raghunāthjī, Jwālamukhī, and Vajrēśvarī (Bhavan) temples. The style of the older ones forms a transition from the Oudh to the Sikh type.

The sculpture of this period includes cult images (generally very simple because they were clothed), decorative figures on pillars (representing dancing or servant girls), and occasional reliefs (including some beautiful ones in the *chattrī* of Sawai Jai Siṅgh II at Gethor, near Jaipur, and very curious ones at the entrance to the Hawā Mahal). The Jaipur idols, of white and black marble, continue the late Jain style even when representing Hindu deities; these were exported all over India. At Tira Sujanpur and Alampur in the Kangra Valley a local imitation, not without a quaint "medieval" charm, was attempted.

Painting of this period is more easily classified because the 18th century brought a clear political grouping, and hence distinct cultural affiliations; but in minor provincial centers the early Rajput style continued to be cultivated beside the court styles into the first half of the 19th century. The miniatures of this period, however, can be distinguished from genuinely old ones by their careless workmanship and lack of expression and by the introduction of contemporaneous costumes and, occasionally, of late architectural forms and color schemes. Some of the miniatures in the late Rajput style, especially portraits of minor nobles, were rather crude; but paintings (miniatures as well as murals) by the court artists were usually of high quality.

The general trend was for Moghul naturalism (as far as it ever matured) to be replaced by a pronounced stylization, with rhythmically flowing lines, simple color contrasts, shading reduced to a minimum or entirely lacking, a flat decorative pattern replacing composition in depth, size of figures and other objects determined by their importance, and expression through vivid gesture or movement. Occasional portraits of the same person executed by a Moghul and by a Rajput artist attest the amazing divergence of interpretation.

As with architecture and costume, the development of painting showed variations in the different kingdoms. In Mewar, devastated by the Marāṭhās and the feud between the Chondāwat and Saktāwat clans in the last half of the 18th century, pictorial art began to recover only in the early 19th century, producing a prolific but crude derivative of the Moghul style of about 1700–20, though with later costumes.

In Marwar (Jodhpur), Jaisalmer, and Saurashtra, the Moghul style of about 1710–40 formed the basis for a much more lively style, with fantastic high turbans, skirts and coats swinging out in a quarter-circle, and female figures whose leaf-shaped eyes extend in an elegant curve over the temples up to the hairline. This style reached its acme in the last years of Bijai Siṅgh and under Mān Siṅgh (ca. 1780–1830). Mān Siṅgh had many Saivite texts illustrated with exceptionally large miniatures. Under Takhat Siṅgh the miniatures became a crude but effective chronicle of endless court festivals, with thousands of alluring damsels. A rich variety of the Marwar style was that cultivated for the Kṛṣṇa temples of the Vallabhacharya sect, from Nathadvara and Kankroli in Mewar to Jodhpur and Jaipur. In the Kishangarh paintings (PL. 414) the elaborate late Moghul composition was reinterpreted in the spirit of a genuine Vaishnava mysticism, with wasted faces and the elongated eyes of Jodhpur.

Jaipur evolved its finest style, based on the Moghul tradition of Muḥammad Shāh, under Sawai Pratāp Siṅgh (1778–1803), Sawai Jagat Siṅgh (1803–18), and Sawai Jai Siṅgh (1818–35) — three young profligates for whom the love of Kṛṣṇa for Rādhā and the *gopīs* became a pretext for debauchery with concubines and prostitutes. The paintings depicting these festivals, miniatures as well as large cartoons (PLS. 411, 414), are among the best creations of Rajasthani art, with their grandiose and heavily flowing rhythms, perfect proportions, and an artificiality that heightens their erotic charm. Under Sawai Rām Siṅgh (1835–80) painting declined, at first to correct but lifeless portraits and court scenes, then to cruder work, until at last it was reduced to the fabrication of stenciled and gaudily colored *Rāgmālā*s for export.

After Gaj Siṅgh's death, painting in Bikaner fell first under the spell of Jaipur, then of Jodhpur. The architectural features of many *Rāgmālā* miniatures found in museums, with small, slim figures, point to Bundelkhand; but they are often attributed to Jaipur as well. The only well-known Bundēla "school," that of Datia, reveals a succession of widely differing styles; the so-called "Bundēla style" is characteristic only of the reign of Shatrujīt Siṅgh (1762–1801), and the late murals of Orchha reveal a different style, distinguished by heavy figures.

In contrast, the small western Himalayan states, so poor in architecture, in the late 18th century produced, in the Pahari style, the finest Indian paintings of the time, worthy of comparison with the best works of Indian art of any era. Here the ideals of Rajput chivalry and religious mysticism had been less corrupted than elsewhere; moreover, the chaos in the Punjab brought prosperity to these states by deflecting international trade into their safe valleys. Sansār Chand I (1775–1823) of Kangra, the most successful of the western Himalayan princes, employed Guler artists but also engaged others from outside; he kept an active studio and took a personal interest in its activities.

Kangra paintings (see MINIATURES AND ILLUMINATION) have larger figures than those in the Guler and Jammu styles but the same flowing rhythm; they have a sweetness in the depiction of female life and a spiritual intensity that the early Rajasthani and Basohli artists had been unable to express. In many respects they are reminiscent of the work of Fra Angelico and of the Sienese school of the Trecento, especially Giovanni di Paolo. The Kangra style soon spread from Tira Sujanpur to Garhwal in the east and Kashmir in the west, but by the early 19th century local political intrigues and the Gurkha invasion (1806) and Sikh conquest (1797–1809) had brought about its decline, through an overly ornate phase to static dullness. Only in isolated Garhwal, where this style had superseded an older tradition represented chiefly by the poet-painter Mōlārām (ca. 1750–1833), was the higher standard retained somewhat longer in a local variation. Under Gulāb Siṅgh (1846–57) and Ranbīr Siṅgh (1857–85) of Jammu, the Kangra style was introduced also in Kashmir and became the starting point for a local school of Hindu (Śiva-Śākti) book illumination, represented in many libraries (PL. 415). Other Kangra painters were employed by the Sikhs; their art, combined with Moghul realism, became coarsened, though it was not without a rather bawdy humor.

BIBLIOG. *Historical background*: J. Tod, Annals and Antiquities of Rajasthan, 2 vols., London, 1829–32 (new ed. by W. Crooke, 3 vols., London, 1920; 2d ed., 2 vols., London, 1957); E. C. Bayley, History of Gujarat, London, 1886; A. Adams, The Western Rajputana States, London, 1900; C. E. Luard, Chhatarpur State Gazetteer, Lucknow, 1907; C. E. Luard, Datia State Gazetteer, Lucknow, 1907; C. E. Luard, Eastern States (Bundelkhand) Gazetteer, Lucknow, 1907; C. E. Luard, Panna State Gazetteer, Lucknow, 1907; A. K. Forbes, Ras Mālā, or the Hindoo Annals of Goozerat (new ed. by H. G. Rawlinson), 2 vols., London, 1924; G. H. Ojha, Rājputānā-kā itihās (History of Rajputana), 7 vols., Ajmer, 1927–41; H. C. Ray, Dynastic History of Northern India, 2 vols., Calcutta, 1931–36; J. Hutchison and J. P. Vogel, History of the Punjab Hill States, 2 vols., Lahore, 1933; Bhupendranath Datta, The Rise of the Rājputs, JBORS, XXVII, 1941, pp. 34–49.

Cultural setting: F. S. Growse, Mathurā: A District Memoir, 2d ed., Allahabad, 1880; G. A. Grierson, The Satsāiya of Bihari Lal, Calcutta, 1896; Dinesh Chandra Sen, History of Bengali Language and Literature, Calcutta, 1911; F. E. Keay, History of Hindi Literature, Calcutta, 1920; J. C. French, Himalayan Art, London, 1931; Lala Kannū Mal, Kāma-Kalā: A Comprehensive Survey of Erotics, Rhetorics and Science of Music, Lahore, 1931; N. A. Thooty, The Vaishnavas of Gujarat, Bombay, 1935; Kshiti Mohana Sen, Mediaeval Mysticism of India, London, 1936; H. Goetz, The Background of Pahārī-Rājput Painting, Roopa-lehkā, XXII, 1951, pp. 1–16; W. Eidlitz, Die indische Gottesliebe, Olten, Freiburg am Breisgau, 1955; H. Goetz, Mīrā Bai: Her Life and Times, J. Gujarat Research Soc., XVIII, 1956, pp. 87–113; W. G. Archer, The Loves of Krishna in Indian Painting and Poetry, London, 1957.

Art in general: B. L. Dhama, A Guide to Amber, Bombay, 1931; H. Goetz, Bundela Art, JISOA, VI, 1938, pp. 181–94; H. Goetz, The Coming of Muslim Cultural Influence in the Panjab Himālaya, India antiqua: A Volume of Oriental Studies Presented . . . to J. P. Vogel, Leiden, 1947, pp. 156–66; H. Goetz, Rajput Art: Its Problems, Art and Thought, Issued in Honour of Dr. A. K. Coomaraswamy, London, 1947, pp. 87–94; H. Goetz, Art and Architecture of Bikaner State, Oxford, 1950; D. R. Patil, Descriptive and Classified List of Archaeological Monuments in Madhya Bharat, Gwalior, 1952; H. Goetz, The First Golden Age of Udaipur, Ars Orientalis, II, 1957, pp. 427–37.

Medieval antecedents and the period of transition: A. Cunningham, Ann. Rep. Archaeol. Survey of India, II–XX, 1871–83; D. R. Bhandarkar, Chaumukh Temple at Ranpur, Ann. Rep. Archaeol. Survey of India, 1907–08, pp. 205–17; W. N. Brown, Early Vaishnava Miniature Paintings from Western India, EArt, II, 1930, pp. 167–206; H. Cousens, Somanatha and Other Mediaeval Temples in Kathiawad, Calcutta, 1931; N. C. Mehta, Gujarati Painting in the 15th Century, London, 1931; D. R. Bhandarkar, The Kirtistambha of Rana Kumbha, JISOA, I, 1933, pp. 52–56; R. C. Kak, Antiquities of Basohli and Ramnagar, Indian Art and Letters, VII, 1933, pp. 65–91; Muniraj Shri Jayanlavijayaji, Abu, 3 vols., Ujjain, 1934–41; M. R. Majumdar, Earliest Devī Mahatmya Miniatures in Gujarat, JISOA, VI, 1938, pp. 118–36; W. N. Brown, Manuscript Illustrations of the Uttarādhyayana Sūtra, New Haven, 1941; Sarabhai Manilal Nawab, Jaina Tirthas in India and Their Architecture, Ahmedabad, 1944; H. Goetz, The Post-Medieval Sculpture of Gujarat, B. Baroda Mus., V, 1947–48, pp. 29–42; Moti Chandra, Jain Miniature Paintings from Western India, Ahmedabad, 1949; H. Goetz, Decline and Rebirth of Mediaeval Indian Art, Mārg, IV, 1950, pp. 36–48; R. C. Agrawal, Some Famous Sculptors and Architects of Mewar, Indian H. Q., XXXIII, 1957, pp. 321–34.

Architecture and sculpture: S. S. Jacob, Jeypore Portfolio of Architectural Details, 10 vols., London, 1890–98; Hirananda Sastri, Ruined Temple in the Nūrpur Fort, Ann. Rep. Archaeol. Survey of India, 1904–05, pp. 110–20; J. Fergusson, History of Indian and Eastern Architecture (new ed. by J. Burgess), 2 vols., London, 1910; G. Sanderson and J. Begg, Types of Modern Indian Buildings, Allahabad, 1913; O. Reuther, Indische Paläste und Wohnhäuser, Berlin, 1925; M. B. Garde, Guide to Chanderi, Gwalior, 1928; M. H. Kuraishi, List of Ancient Monuments Protected under Act VII of 1904 in the Province of Bihar and Orissa, Calcutta, 1931; H. Goetz, Late Indian Architecture, ActaO, XVIII, 1940, pp. 81–102; H. Goetz, The "Basohli" Reliefs of the Brahmor Kothi, 1670 c., Roopa-lekhā, XXV, 1, 1954, pp. 1–12; H. Goetz, Rajput Sculpture and Painting under Raja Umēd Singh of Chambā, Mārg, VII, 4, 1954, pp. 23–34.

Rajasthāni painting: A. K. Coomaraswamy, Rajput Painting, London, 1916; A. K. Coomaraswamy, The Rasikpriyā of Kesava Dās, BMFA, XVIII, 1920, pp. 50–52; A. K. Coomaraswamy, Catalogue of the Indian Collections in the Boston Museum of Fine Arts, V–VI, Boston, 1926–30; N. C. Mehta, Studies in Indian Painting, Bombay, 1926; O. C. Gangoly, Masterpieces of Rajput Painting, Calcutta, 1927; O. C. Gangoly, Rāgas and Rāginīs, Calcutta, 1935; Hirananda Sastri, Ancient Vijñaptipatras, Baroda, 1942; H. Goetz, The Kachhwāha School of Rajput Painting, B. Baroda Mus., IV, 1946–47, pp. 33–47; H. Goetz, The Mārwār School of Painting, B. Baroda Mus., V, 1947–48, pp. 43–54; B. Gray, Rajput Painting, London, 1948; B. Gray, Western Indian Painting in the 16th Century, BM, XC, 1948, pp. 41–45; E. Dickinson, The Way of Pleasure: Kishangarh Paintings, Mārg, III, 4, 1949, pp. 29–35; H. Goetz, Notes on Indian Miniature Painting, B. Baroda Mus., VII, 1949–50, pp. 53–66; Moti Chandra, An Illustrated Set of the Amaru-Sataka, B. Prince of Wales Mus., II, 1951–52, pp. 1–63; H. Goetz, A New Key to Early Rajput and Indo-Muslim Painting, Roopa-lekhā, XXIII, 1952, pp. 1–16; G. K. Kanoria, An Early Dated Rajasthāni Rāgamālā, JISOA, XIX, 1952–53, pp. 1–5; K. Khandalavala, A Gītā Govinda Series in the Prince of Wales Museum, B. Prince of Wales Mus., IV, 1953–54, pp. 1–18; N. H. Stooke and K. Khandalavala, The "Laud" Rāgmālā Miniatures, London, 1953; H. Goetz, The Early Rajput Murals of Bairat, Ars Orientalis, I, 1954, pp. 113–18; H. Goetz, The Laud Rāgmālā Album and Early Rajput Painting, JRAS, 1954, pp. 63–74; Moti Chandra, Mewar Painting in the 17th Century, New Delhi, 1957; W. G. Archer, Central Indian Painting, London, 1958; Pramod Chandra, Bundi Painting, New Delhi, 1959; E. Dickinson and K. Khandalavala, Kishangarh Painting, New Delhi, 1959; R. Reiff, ed., Indian Miniatures: The Rajput Painters, Tokyo, Rutland, Vt., 1959; S. E. Lee, Rajput Painting, New York, 1960 (cat.).

Pahari painting: J. P. Vogel, Catalogue of the Bhuri Singh Museum, Chamba, Calcutta, 1909; A. K. Coomaraswamy, Indian Drawings, 2 vols., London, 1911–12; A. K. Coomaraswamy, The Eight Nāyikās, JIAI, XVI, 1914, pp. 99–112; Hirananda Sastri, Hamīr Hath, or the Obstinacy of Hamīr, JIAI, XVII, 1915–16, pp. 35–40; A. K. Coomaraswamy, Rajput Painting, London, 1916; Mukandilal, Some Notes on Molā Rām, Rūpam, 8, 1921, pp. 22–30; O. C. Gangoly, An Editio Princeps of Sundara-Śrigāra, Rūpam, 30, 1927, pp. 47–51; Keshava Prasad Misra, An Illustrated Manuscript of Madhu-Mālatī, Rūpam, 33–34, 1928, pp. 9–11; B. N. Treasuryvala, A New Variety of Pahari Paintings, JISOA, XI, 1943, pp. 133–35; J. C. French, Kāngrā Frescoes, Indian Art and Letters, XXII, 1948, pp. 57–59; W. G. Archer, Indian Painting in the Punjab Hills, London, 1952; W. G. Archer, Kangra Painting, London, 1952; J. Mittal, Mural Painting in Chamba, JISOA, XIX, 1952–53, pp. 11–18; W. G. Archer, Garhwal Painting, London, 1954; M. S. Randhawa, The Krishna Legend in Pahārī Painting, Delhi, 1956; K. Khandalavala, Pahārī Miniature Painting, Bombay, 1958; M. S. Randhawa, Basohli Painting, Delhi, 1959; M. S. Randhawa, A Note on Rasamañjarī Paintings from Basohli, Roopa-lekhā, XXXI, 1, 1960, pp. 16–26.

Minor arts: S. S. Jacob and T. H. Hendley, Jeypore Enamels, London, 1886; T. H. Hendley, Decorative Art in Rajputana, JIAI, II, 21, 1888, pp. 43–49; T. H. Hendley, Ulwar and Its Treasures, London, 1888; T. H. Hendley, Industrial Art in Bikaner, JIAI, IV, 33–34, 1892, pp. 1–6; T. H. Hendley, Memorials of the Jeypore Exhibition 1883, 4 vols., London, 1893; G. Watt and P. Brown, Indian Art at Delhi, Calcutta, 1904; T. H. Hendley, Asian Carpets, 16th and 17th Century, from the Jaipur Palaces, London, 1905; T. H. Hendley, Indian Jewellery (JIAI, XII), London, 1906–09; B. Chaubay, Jaipur Pottery, JIAI, XVII, 1916, pp. 27–28; H. Goetz, Kostüm und Moden an den indischen Fürstenhöfen der Grossmoghul-Zeit, Jhb. der asiatischen K., I, 1924, pp. 67–101; Pūran Singh, Some Rumals from Chambā, Rūpam, 32, 1927, pp. 133–34; K. K. Ganguli, Chambā Rumāl, JISOA, XI, 1943, pp. 69–70; H. Goetz, An Early Basohli-Chambā Rumāl, B. Baroda Mus., III, 1945–46, pp. 35–42; Mulk Raj Anand, Chambā Rumāls, Mārg, VII, 4, 1954, pp. 35–40; R. N. Mehta, Picchavāīs: Temple Hangings of the Vallabhācārya Sect, J. Indian Textile H., III, 1957, pp. 4–14.

Allied artistic movements: S. N. Gupta, The Sikh School of Painting, Rūpam, 12, 1922, pp. 125–28; A. Ghose, Bemalte Buchdeckel aus Alt-Bengalen, OAZ, N.S., V, 1929, pp. 119–21; H. Goetz, The Art of the Marāthas and Its Problems, B. C. Law Commemoration Volume, II, Bombay, 1946, pp. 433–44; H. Goetz, Two Illustrations to the Sundar-Sringār from Gujarat, B. Baroda Mus., VII, 1949–50, pp. 62–64; Harinarayan Dattabaruva, Chitra-Bhāgavata, Nalbādi, 1950; Mulk Raj Anand, Painting under the Sikhs, Mārg, VII, 2, 1954, pp. 23–31; Rai Krishnadasa, An Illustrated Avadhi Manuscript of Laur-Chandā, Lalit-Kalā, I–II, 1955–56, pp. 66–71.

Hermann Goetz

Illustrations: PLS. 411–420; 2 figs. in text.

RAPHAEL (Raffaello Sanzio). Painter and architect (b. Urbino, Apr. 6, 1483; d. Rome, Good Friday, Apr. 6, 1520). He probably received his first instruction in painting from his father, Giovanni Santi. A painter of the school of Melozzo da Forlì (q.v.), Giovanni Santi was an artist of no great merit, but he was a man of culture, the author of the well-known *Cronaca rimata*, who found favor at the court of Guidobaldo da Montefeltro after Melozzo's departure from Urbino. But the earliest artistic influences to which the young Raphael was subjected were not so much those of his father (of which few traces are noticeable in his style) as those of the artistic climate of Urbino, where the spirit of Luciano Laurana, Piero della Francesca, and Leon Battista Alberti (qq.v.) was still preeminent. Vasari (q.v.) writes that Giovanni Santi, unable to find in Urbino masters worthy of the remarkable genius of his son, took him to the studio of Perugino (q.v.) in Perugia. This would have occurred before Aug. 1, 1494, the date of Giovanni's death (his mother, Magia Ciarla, had died in 1491). There exists no definite confirmation for Vasari's statement, but a considerable mass of evidence indicates that Raphael's apprenticeship with Perugino could not have begun much later. If not actually in 1494, it must have begun shortly after, because Raphael's youthful style recalls that phase of Perugino's art which can be dated approximately between 1494 (with reference to his *Crucifixion* in S. Maria Maddalena dei Pazzi, in Florence) and the first years of the 16th century. Crowe and Cavalcaselle (I, 1882) have, in fact, correlated the visible rise in the artistic level of Perugino's work at that time — from the altarpiece of Fano to the *Ascension* of Lyons, from the *Madonna* of S. Pietro Martire (Perugia) to the *Madonna* of Sinigaglia, and then to the *Assumption* of Vallombrosa and the *Virgin* of the Certosa of Pavia — with the presence in his studio of the young Raphael. Later, Venturi also considered the problem of Raphael's collaboration with Perugino and identified work by Raphael's hand in the frescoes in the Hall of the Collegio del Cambio in Perugia, particularly the figure of Fortitude, which stands out from the others for its quality. It is easier to identify the many influences of Perugino in the works of Raphael from before 1504, such as the St. Jerome of the Mond *Crucifixion* (London, Nat. Gall.), of which Vasari wrote, "if his name had not been written on it, no one would believe it to be by Raphael, but by Pietro," and the musician angels of the Vatican *Coronation of the Virgin* and the predella of the same altarpiece. The problem becomes more complicated when one investigates the possible inserts by Raphael in the Fano predella (S. Maria Nuova). In the panel depicting the birth of the Virgin, the figure of the woman with the newborn girl reechoes the fresco with the Madonna and Child in the house of Raphael in Urbino, a work formerly attributed to Giovanni Santi but now regarded as a very early work of Raphael.

Although Raphael absorbed and made Perugino's idioms his own, in his hands they underwent a subtle conceptual transformation which rendered them more poetic and harmonious. With his acute intelligence he sought out their original sources, particularly in the works of Piero della Francesca. Among Raphael's youthful experiences, his contacts with Pintoricchio (q.v.) should not be overlooked, although the old tradition of Raphael's collaboration in the frescoes of the Piccolomini Library in the Cathedral of Siena may now be discounted. The *Resurrection*, formerly in the Kinnaird Collection and now in São Paulo (Mus. de Arte), is derived from a Peruginesque prototype, either the predella of the *St. Peter* polyptych, now in Rouen (Mus. B.A.), or the *Resurrection* altarpiece, formerly in S. Francesco al Prato and now in the Vatican Museums, in which the possibility of Raphael's collaboration cannot be excluded. But in the ornate flourishes of the cover of the sarcophagus, the cuirasses, the scrolls, and the pictorial details of the landscape there are definite suggestions of Pintoricchio. The hypothesis that Raphael studied under Timoteo Viti, who apparently came to Urbino in 1495, seems less tenable.

The first documented work of Raphael is the *Coronation of St. Nicholas of Tolentino*, commissioned by the Church of S. Agostino in Città di Castello on Dec. 10, 1500, to Raphael and Evangelista di Pian di Meleto, formerly "famulus Johannis Sanctis, pictoris de Urbino," an indication that Raphael, while in Perugia, maintained his relations with Urbino and whatever remained of his father's workshop. Only three fragments of this altarpiece survive: the bust of an angel, signed by Raphael (Brescia, Pin. Civica Tosio Martinengo), the head of the Virgin, and the Eternal Father (both, Naples, Mus. di Capodimonte); the last two were painted in collaboration with Evangelista. A sheet of very fine drawings for this altarpiece has also survived (Lille, Mus. B.A.); of an admirable delicacy and vigor, they are much freer, more personal, and independent of Perugino's influences than the paintings. The pictorial quality of the rapid and light stroke is very striking, particularly in the trampled devil, which suggests the influence of Leonardo da Vinci (q.v.). It is probable that Raphael become acquainted with Florentine art circles several years before 1504, having accompanied Perugino on one of his frequent journeys to Florence about the turn of the century.

In addition to the fresco with the Madonna in the house in Urbino, the *Coronation of St. Nicholas of Tolentino*, and the São Paulo *Resurrection*, the following works belong to the early years of Raphael's activity: the badly preserved Trinity standard, with the creation of Eve depicted on the reverse, from the Church of the Trinità in Città di Castello (now in the Pin. Com., Città di Castello); earlier than this, but of questionable attribution, because of its very poor state of preservation, another standard in the same museum depicting the Crucifixion on one side and the Madonna della Misericordia on the other; a bust of St. Sebastian (Bergamo, Acc. Carrara); and the Solly *Madonna*, the *Madonna between St. Jerome and St. Francis*, and the somewhat later Diotallevi *Madonna* (all three, Berlin, Staat. Mus.). Of the two altarpieces, the Mond *Crucifixion*, painted for the Church of S. Domenico in Città di Castello, and the *Coronation of the Virgin* (Vat. Mus.) commissioned by Maddalena degli Oddi for the Church of S. Francesco in Perugia, it is not known which is earlier. In the *Crucifixion* the figures are fewer and the composition is simpler, although the wide landscape background is more open. The *Coronation of the Virgin* is more complex and richer in ideas, but the relationship between the two parts, celestial and terrestrial, is not completely resolved. The predellas of both are very beautiful. That of the *Coronation* is still attached to the altarpiece; it recalls the Fano predella but has a wider spatial concept and greater luminosity. The predella of the Mond *Crucifixion*, which has been separated (New York, J. Weitzner Coll.; Lisbon, Mus. Nac. de Arte Antiga), is later in relation to the Peruginesque idiom; in the secular portions, however, certain elements can be found that relate it to the *Combat of Love and Chastity* (PL. 116), the large panel which Perugino was commissioned to paint in 1503 for the *studiolo* of Isabella d'Este in Mantua, into which, possibly because of its destination, a number of Emilian-Venetian elements have been introduced.

His *Marriage of the Virgin* (PL. 423), signed and dated 1504, formerly in the Church of S. Francesco in Città di Castello, represents the furthest limit of the Peruginesque phase, the moment at which Raphael was about to enter a new one. The original idea of this composition has generally been ascribed to Perugino's fresco of Christ giving the keys to St. Peter (PL. 113) in the Sistine Chapel. But another almost contemporaneous and much more similar work by Perugino must be taken into account, namely, the *Marriage of the Virgin* in Caen (Mus. B.A.), which some historians have wrongly omitted from the catalogue of Perugino's works (although it must be admitted that this painting shows a fairly large amount of collaboration by assistants). In this altarpiece the massive octagonal temple presses down heavily on the figures, while Raphael placed his temple much farther in the background and inserted between the high steps of the temple and the figures in the foreground the parallel rows of the paving stones of the piazza, which are rendered in perspective. He also increased the number of sides of the temple from 8 to 16, thus bringing it closer to a circle, and decorated it with an open portico and a series of volutes joining the portico to the dome, which is completely visible within the arc of the frame; in Raphael's picture the dome follows the curve of the frame, whereas it is cut off abruptly in the Perugino work. Critics have unanimously praised not only the pure spatial values of the Raphael background but also the light, which seems to spring from the clarity of the colors; critical judgment differs, however, in regard to the figures. Nevertheless, even these, when compared with the monotonous and often static grouping of Perugino's figures, reveal better proportion in relation to the depth and spacing. In the drawings for some of the heads (Oxford, Ashmolean Mus.; Br. Mus.) the characterization is carried further than in the painting; they even reflect the influence of Leonardo. The small portrait of a young man in the Galleria Borghese, Rome, can be dated to approximately the same time as the Brera *Marriage*.

A school of criticism that is still romantically inclined to regard the miracle of spontaneous inspiration as the source of pure poetry until recently assigned to early in the first period of Raphael's activity three small panels: *The Vision of a Knight* (London, Nat. Gall.), *The Three Graces* (Chantilly, Mus. Condé), and the Conestabile *Madonna* (Leningrad, The Hermitage). It is true that these paintings bear the mark of Perugino's influence less than the other works hitherto mentioned, not however because they precede this influence, but, rather, because they already anticipate the surpassing of it. The lightness of the forms is an indication of contact with Florentine art. The softness of the modeling suggests that the surrounding atmosphere is incorporated into the form, relating it to the landscape. *The Vision of a Knight* and *The Three Graces* are two small "poems" conceived in the idiom of the period; they are inspired by that taste for undetermined naturalistic allegories which found more favor in the Venetian-Emilian-Urbinate ambient than in the Tuscan-Florentine milieu.

To the Urbino period, or possibly a little later (ca. 1505), belong also the two panels in the Louvre, the *St. George and the Dragon* (PL. 421) and the *St. Michael*, which is more strongly marked by Florentine traits. Müntz's hypothesis (1886) that they were painted for Guidobaldo was opposed by Cavalcaselle but has been revived by Lynch (1962) on the basis of old inventories; moreover, the somewhat emblematic character of the panels, which almost suggests that they — especially the *St. George* — had been composed for a medal or to commemorate a feat, makes them appear suited to an aristocratic court.

In the background of the *St. Michael*, the writhing mass of small figures and monsters reminiscent of the temptations of St. Anthony or the "infernos" of Bosch (q.v.) suggests obvious Flemish influences, which were not alien at the court of Urbino. On the other hand, the very fine drawing for the *St. George* (Uffizi, Gabinetto dei Disegni e Stampe) brings out much more clearly than the painting Raphael's study of Leonardo's work, as revealed by the idiom and the intrinsic value of the line and the chiaroscuro. The other *St. George and the Dragon* (Washington, D.C., Nat. Gall.) — less successful but more mature — may be assigned to the year 1505–06. The same date can also be assigned to the fine portrait *Young Man with an Apple* in the Uffizi; it stands out from the rest of Raphael's works because of its clear reflections of Flemish portraiture.

By the end of 1504 Raphael had settled in Florence, though he occasionally returned to Perugia and Urbino. From that moment he began an intense study of the idiom of Florentine art, which for four years he pursued without pause and with remarkable insight. He studied the works of Leonardo and Michelangelo in particular, but he did not fail to go further back, to Antonio Pollaiuolo and Donatello (qq.v.), following the whole stream of this tradition to its most authentic sources. The drawings, more than the paintings, bear witness to the profundity of his studies. In a drawing in Oxford (Ashmolean Mus., no. 523) the clothed figure of Donatello's statue of St. George (cf. IV, PL. 241) for Orsanmichele in Florence stands out between three nude youths. On the other hand, the drawing of Hercules struggling with three centaurs, in the Uffizi (Gabinetto dei Disegni e Stampe), recalls Pollaiuolo, even though it is interpreted with a far more modern pictorial freedom. But, naturally, Michelangelo and Leonardo formed the focal center of Raphael's interest. Among Michelangelo's works Raphael studied not only the Doni Tondo [(IX, PL. 527), as evidenced by the kneeling woman in the Borghese *Deposition* (IV, PL. 314), which is apparently derived from the Doni Madonna] but also his sculptures. A drawing in the British Museum reproduces the *David* from the rear. On the back of the

British Museum study for the Borghese *Deposition* there is a drawing of Michelangelo's great unfinished statue of St. Matthew. In *La Belle Jardinière* (XII, PL. 65) the infant Jesus strongly resembles Michelangelo's figure in the Bargello Tondo (*Madonna and Child with the Infant St. John*; IX, PL. 525).

The derivations from Leonardo are far more numerous. The *Leda and the Swan* appears in a drawing in Windsor (Royal Lib.). A group from Leonardo's *Battle of Anghiari*, together with another typically Leonardesque head, is sketched in one corner of the Oxford study for the head of St. Placidus (or possibly St. Benedict) in the S. Severo fresco *Trinity with Saints* in Perugia. But the importance of Leonardo goes far beyond mere textual derivations; his influence permeates the entire canon of Raphael. In the drawings of the Madonna and Child in the Albertina in Vienna, the British Museum, and the Uffizi, the animated and vivid Leonardesque "scrawl" (*scarabocchio*) — the pen strokes tangled into an inextricable dynamic mass, catching in motion the changing poses and movements of the figures, the clusters of shadows, and the snatches of lights — is interpreted by Raphael in its truest significance as the opening up of forms, the suggestion of movement and of atmospheric envelopment. Even the fundamental subjects on which Raphael worked tirelessly during his Florentine period have their roots in Leonardo. The models for the Madonna and Child were the so-called "Benois Madonna" and the Madonna in the *Adoration of the Magi* (IX, PL. 117); those for the group of the Madonna and Child with the Infant St. John in a pyramidal composition were the *Virgin of the Rocks* (IX, PL. 121) and related drawings, as well as the various versions of St. Anne (IX, PLS. 122, 123); and, finally, that for the new manner of portraiture was the *Mona Lisa*.

In only four years (1504–08) Raphael's stylistic development was remarkable — from the Terranuova *Madonna* (Berlin, Staat. Mus.) to the Canigiani *Holy Family* (PL. 425), from the Ansidei *Madonna* (London, Nat. Gall.) to the Borghese *Deposition*. At the beginning of his Florentine period he was engaged in painting two altarpieces for Perugia: the above-mentioned altarpiece, commissioned by the Ansidei family for the St. Nicholas Chapel in the Church of S. Fiorenzo, and the Colonna *Madonna* (Met. Mus.), executed for the Convent of S. Antonio. The predellas of both of these altarpieces have been dismembered. Only one panel of the Ansidei predella is known: the *John the Baptist Preaching* (Calne, Wiltshire, Bowood Park, Marquess of Lansdowne Coll.). The three predella panels of the Colonna altarpiece are divided between the Metropolitan Museum, New York (*The Agony in the Garden*), the National Gallery, London (*The Procession to Calvary*), and the Isabella Stewart Gardner Museum, Boston (*Pietà*); the end figures (SS. Francis and Anthony) are in the Dulwich Picture Gallery, London. Stylistic analysis leaves open the question as to which altarpiece is the earlier. The edge of the robe of the Ansidei *Madonna* bears a date that is of no real value since, owing to abrasions, it has been read as MDV, MDVI, and even MDVII (which can be ruled out on stylistic grounds). The composition of the Ansidei altarpiece, the most traditional of these works, is based on Perugino's Pala dei Decemviri (Vat. Mus.), but the freer and more spacious execution, the beauty of the great light arch, which takes the place of Perugino's fragmentation into multiple piers and arches, and the sculptural quality of the two saints all indicate that the work proceeded slowly and that the style of the artist was changing as the painting progressed. In the predella, the arrangement of the putti close to the seated youth listening to the Baptist begins to suggest the movement of the Carità group in the predella of the Borghese *Deposition*. The Colonna altarpiece is more original in organization: the great canopied throne and the saints, in the absence of any architectural framework, stand directly in the landscape. The execution is, however, more uneven. These three beautiful predellas are filled with impulses which reveal the results of Raphael's new Florentine studies and the continual sprouting of new ideas.

The theme of the Madonna and Child, in which he frequently included the Infant St. John, sometimes St. Joseph, and also St. Anne (Canigiani *Holy Family*), stands at the center of Raphael's finished works and compositional studies of his Florentine period. The variety of his inventions and accents is remarkable, and even more so when the numerous drawings that accompany the paintings are considered. Leonardo is the principal source from whom Raphael draws, but the derivation is one of the spirit and not of the letter. The motif of Jesus and St. John, over whom hovers the protective figure of the Madonna, while the two children seem to be adoring or blessing each other, or sometimes even playing together, is derived from Leonardo (cf. IX, PLS. 121–23). But, devoid of the *terribilità* (awesomeness) and the sum of the Leonardesque meditations and cosmic allusions that make those supreme rarefied exemplars of Leonardo so disturbing and remote, Raphael's spontaneous, though stylistically complex and elaborate, compositions result in a smiling and affectionate expression of the natural tender links that unite woman and child and one child with another. From this quality in his painting results the unique position of Raphael in the history of art: he is the painter most highly esteemed by academicians for his compositional ability and harmony; at the same time he is popular with laymen who find beautified in his Madonnas the expression of their most cherished and natural sentiments.

Between the Terranuova *Madonna*, an early work of the Florentine period, and the Canigiani *Holy Family*, the span of Raphael's development is wide. It is not a question of quality. The Terranuova *Madonna* is a creation of the highest clarity, with the undulating suppleness of the composition carried out by the three children, giving the impression of a delightful garland strung on the knees of the Virgin; the Canigiani *Holy Family*, on the other hand, so complex and learned, is not without a faint trace of academicism, although of a very high level. The problem is to trace the stylistic evolution of the artist during his years of intense study in Florence. It is not always easy to establish the chronological sequence of his work with any degree of certainty. Few of the paintings are dated, and those dates that do exist are often difficult to read. The problem of the Ansidei *Madonna* has already been discussed. The date of 1505 for the S. Severo fresco in Perugia cannot be accepted: either it was added later or it is actually spurious; the style would indicate a date of about 1507. The reading of the date of the *Holy Family with the Lamb* (Prado) is uncertain, though probably 1507. Dates remain for the *Madonna del Prato* (PL. 425), 1506; *La Belle Jardinière*, 1507; the Borghese *Deposition*, 1507; and the large Cowper *Madonna* (Washington, D.C., Nat. Gall., 1508).

The following chronological order would seem to be acceptable: the Conestabile *Madonna*, the Terranuova *Madonna*, the *Madonna del Granduca* (PL. 424), the small Cowper *Madonna* (Washington, D.C., Nat. Gall.), the Orléans *Madonna* (Chantilly, Mus. Condé), the *Madonna with the Beardless St. Joseph* (Leningrad, The Hermitage), the *Madonna del Prato*, the *Madonna of the Goldfinch* (Uffizi), the *Holy Family with the Palm* (London, Earl of Ellesmere Coll.), *La Belle Jardinière*, the *Holy Family with the Lamb*, the Colonna *Madonna* (Berlin, Staat. Mus.), the Canigiani *Holy Family*, the Bridgewater *Madonna* (London, Eare of Ellesmere Coll.), the large Cowper *Madonna*, and the Tempi *Madonna* (PL. 422). However, the dates of the Northbrook *Madonna* (London, Northbrook Coll.) and the Esterhazy *Madonna* (Budapest, Mus. of Fine Arts) are controversial. The conflicting opinions of various critics range from as early as 1505–06 to as late as 1508–15. Neither seems to belong to the early Florentine period, but rather to a more advanced stage: the Esterhazy *Madonna* would seem to belong practically to the threshold of the Roman period (even disregarding the Roman ruins in the background). The latest dated painting is the large Cowper *Madonna* (1508). But the finest, efflorescing from the summit of the long examination of his Florentine years, is the Tempi *Madonna*.

In portraiture, too, the paintings of Leonardo provided a model for Raphael. The influence of Leonardo is not yet evident in the above-mentioned portrait, *Young Man with an Apple*, which is strongly marked by Flemish influences, but it can be read like a watermark behind the pose in the portrait of Maddalena Doni, the companion piece to the portrait of

Angelo Doni (both, Florence, Pitti), which can be dated around 1506. Here the Raphaelesque image in the foreground emerges with quite different accents: objective, calm, convincing, dominating the quiet landscape, and forming a line with the horizon at its elevation. Although the form resembles that of Leonardo, Raphael's Humanism is distinguished by a superior fullness of feeling. As soon as he had absorbed all that the models of Leonardo could offer him, Raphael began to follow him less. The *Pregnant Woman* (Florence, Pitti), superbly blocked out in its form, is more freely positioned in space; its fullness of color makes the spectator realize the inadequacy of limiting the "problem of color" in the art of Raphael to his Roman period and of attributing it to the influence of Sebastiano del Piombo. An image like this comes alive in its perfect color relationship, achieved in a very different manner from that of a Venetian painter but with equal sensitivity to the tones. It is manifested not as a surrounding environment but as a part of the synthesis of the planes in the image, which thus acquires a supreme naturalness and authority. In a more cautious manner, because of the greater complexity and the aristocratic quality of the model, this supreme ability of Raphael to penetrate and entirely recreate a "person" is revealed in the portrait known as "La Muta" (PL. 424), a painting of unequaled quality. It is stylistically very close to the *Pregnant Woman*, particularly for a certain impalpable Nordic subtlety in the clear gradations of light and shadow on the facial planes. The attribution to Raphael of either the so-called "Munich Self-portrait" (Alte Pin.) or the portraits of the *Veiled Woman* in Hannover (Landesgal.), Elizabeth Gonzaga (Uffizi), and Emilia Pia di Montefeltro (Baltimore, Epstein Coll.) cannot be sustained. Of fairly dubious attribution (because of its very bad state of preservation, which eliminates the possibility of resolving this question) is the *Portrait of a Young Man* (Budapest, Mus. of Fine Arts.); also of dubious attribution are the *Portrait of Guidobaldo* and the *Self-portrait* (both, Uffizi). The *Portrait of a Young Man* at Hampton Court (Royal Colls.) is less doubtful.

To the end of the Florentine period belong certain works in which a monumental character begins to appear: the Borghese *Deposition*, dated 1507, commissioned for the Church of S. Francesco al Prato in Perugia by Atalanta Baglioni, in memory of her son Grifonetto, who was killed in the course of the violent fratricidal feud among the members of the Baglioni family; the *St. Catherine* in the National Gallery in London; the above-mentioned fresco of the Trinity in S. Severo in Perugia, which, although it bears a false date (1505), on stylistic grounds can be considered more or less contemporary with the Borghese *Deposition*. The Dei altarpiece, or *Madonna of the Baldachino*, commissioned by the Dei family for Sto Spirito (Florence, Pitti) was left unfinished at the time of his departure for Rome.

A series of drawings shows that the composition of the Borghese altarpiece underwent a long evolution and that the original design was considerably modified. The scene was first conceived as a lamentation for the Dead Christ. In the very beautiful drawing in Oxford (Ashmolean Mus.) the grouping of men and women is the reverse of that in the painting. On the left of the Ashmolean drawing the holy women form a pathetic group of mourners around the body of Christ, which lies in the foreground. The group of four men, on the other hand, stands vertically on the right but slightly farther back.

In a later drawing in the British Museum the mourning motif is transformed into an Entombment, the attitudes and the movements being strongly accentuated. The positions of the men and women are reversed; Christ is being lifted by the shoulders and knees by two robust bearers, whose positions, owing to the effort, are divergent. Three standing figures are leaning over the body and echo the descending rhythm that in the preceding drawing was represented by the group of holy women. The Madonna, accompanied by two women, takes the place of the men in the Oxford drawing and moves pathetically to accompany the procession. But Raphael must have felt that this group was too elegiac a contrast with the more dynamic group of the carriers; in successive phases he reintroduced the motif of the fainting Virgin (which was widely imitated in the 16th century), thus heightening the dramatic and dynamic qualities of the group of women as well. The kneeling figure, twisted around with arms raised to support the Virgin, is clearly derived from the Madonna of the Doni Tondo.

The fact that the composition was originally conceived in two parts is still apparent in the painting; in order to bind the two together, Raphael was obliged to exaggerate the proportions of the handsome young bearer, who throws himself backward so that the upper part of his body is carried over into the right sector and forms part of both groups. Every detail has been carefully worked out, but such intelligent thought and studied composition notwithstanding, the final composition is lacking in dramatic effect. The highly sculptural integrity of the individual figures, their formal beauty barely touched by the otherwise clever and coordinated expressions of pain, the hard intarsia of the color areas, which, instead of providing a unifying element, results in an enamellike polychromy — all these together freeze the dramatic affective relationships in a sort of sculptural living tableau.

The composition did not spring from a single inspiration, as it did in so many of Raphael's preceding works. Formal virtuosity took the upper hand over a real and total commitment of the artist's personality. It reveals how the Florentine experience, a period in which he was strenuously engaged in an attempt to master a modern idiom, had begun to confront him with the dangers of intellectual formalism. His ability now needed other horizons, not of stylistic refinement but of new wealth of content, of human and historical problems to solve, in order to leaven his new formal creations. Raphael was now ready for the great Roman endeavors. The predella of the *Deposition* is very fine; now separated from the altarpiece, it is preserved in the Vatican Pinacoteca. It represents the three Virtues — Faith, Hope, and Charity — each flanked by two putti; they are painted in a delicate and luminous grisaille that emphasizes but does not attenuate the sculptural quality of the modeling. The Charity group, with its impeccable rhythmic involution that binds the figures together, is worthy of the *Madonna of the Chair*; the other two virtues are younger sisters of the London *St. Catherine of Alexandria*, which must belong to this same period.

The Trinity fresco of S. Severo in Perugia contains reminders of the *Last Judgment* (Florence, Mus. di S. Marco) by Fra Bartolomeo, but it also contains the germ of the composition of the *Disputa* (detail, IV, PL. 230) in its semicircular positioning of the saints, seated on clouds that form lateral wings and lead the eye toward the central figure of Christ. The fine drawing for the head of St. Placidus (previously noted) is close in time and style to the other very beautiful drawing for the head of St. Catherine, also in Oxford (Ashmolean Mus.).

The *Madonna of the Baldachino* is also composed with an ample sonority of semicircular rhythms, repeating those of the apse, the classicizing architecture of which is already fully representative of the 16th century. The canopied throne is taken from the Colonna altarpiece but, by dividing the composition, gives it a semicircular movement. The draperies of the canopy swell out into a whirlpool of shadows, from which the Madonna and Child emerge in a manner similar to both the Colonna and the large Cowper Madonnas. The two angels soaring overhead are a prelude to mannerist forms; they must have made a profound impression on Lorenzo Lotto (q.v.), who introduced them into the Bergamo altarpieces. The derivation from Fra Bartolomeo, which at one time was believed to be evident in this altarpiece, must in the light of a more exact chronology be reversed; it was Raphael who created a model which inspired the monk and not vice versa.

At the end of 1508, Raphael moved to Rome. A payment dated Jan. 13, 1509, apparently connected with the Vatican frescoes, resolves in favor of the earlier date the controversy as to whether the artist arrived in Rome in 1508 or 1509. Vasari states that he was called to Rome by the Pope at the suggestion of Bramante (q.v.). There is no proof of this, but it is certain that Bramante, who was the architect of the papal palaces, introduced Raphael to the papal court; he held Raphael in great esteem and went as far as to designate him as his successor in the building of St. Peter's.

Pope Julius II had only just begun the decoration of new rooms on the floor above the Borgia Apartments, in which he did not care to live. He had engaged a host of artists, including Sodoma, Peruzzi, Bramantino, and Lotto (qq.v.), to decorate them with frescoes. But when he saw samples of the work of Raphael, a young man "nec adhuc stabili autoritate" according to the words of Giovio (1781), Julius II, who had a divinatory intuition for genius, did not hesitate to dismiss all the other artists — and even to remove all the paintings already completed — in order to leave the field open to Raphael.

The subjects for the frescoes of the first room, originally destined to be the pope's library but later turned into the seat of the ecclesiastical tribunal — hence the name Stanza della Segnatura — were, according to Giovio, determined "ad praescriptum Julii Pontificis." They derive their inspiration from Neoplatonic doctrine and form a single cycle intended to glorify the Idea of Truth, of Goodness, and of Beauty. The first fresco to be completed, the *Disputa* (IV, PL. 230) celebrates the triumph of supreme Truth: God is seen in a celestial vision of the Church Triumphant, with the prophets, the apostles, and the saints grouped in a semicircle around the Trinity in heaven. Truth is proclaimed on earth by the Church Militant: the fathers and doctors of the Church, saints, popes, and the simple faithful — to many of whom Raphael gave the physiognomy of historical or living persons — are grouped in a semicircle around the altar, which stands isolated, with the Host rising on an axis with the Holy Ghost, Christ, and the Eternal Father. "The Triumph of the Church" would therefore be a more appropriate title than that assigned by tradition. Behind the terrestrial assembly there rises a large structural block, an allusion to the beginning of the building of St. Peter's. In the landscape at the left another edifice under construction can been seen in the distance.

This is the first of the frescoes painted by Raphael in the Stanza della Segnatura. The drawings indicate a long evolution of the original idea before the large composition was outlined in two superimposed concentric semicircles, pivoting on the axis between the Trinity and the consecrated Host, which revolve within the large spaces of the final version of the fresco. But in the upper portion a few elements from the old ornate Umbrian manner remain in the raised gilded stucco rays that radiate from the great arch of the lunette and open out downward in the form of a fan, surrounded by a host of small angel heads. Moreover, if certain details in the grandiose assembly of figures are studied separately, they reveal a number of reminiscences, not yet completely controlled by a unitary style, such as the gentle and obviously Leonardesque figure in the left foreground, turned toward the viewer and pointing to the altar. In the older figure next to him, leaning on a balustrade, some critics have seen the portrait of Bramante — another allusion to the construction of St. Peter's.

Every hesitation vanishes in the *School of Athens* (PL. 426), which symbolizes the rational search for truth through philosophy and science. In the *Disputa* the host of the saints and of the faithful are placed under the vault of heaven, constituting in themselves a living architecture; in the *School of Athens* the figures stand solidly on the ground within a magnificent walled architecture of a very new style that had, in the first years of the 16th century, no counterpart in reality. It anticipates, in an idealized image, the as yet unrealized building of St. Peter's to the extent that tradition attributes the drawing to Bramante himself. Undoubtedly the idea was derived from Bramante, but Raphael was perfectly able to draw it himself, taking only the inspiration from Bramante. At the vanishing point the great arch of the vault is open to the sky, and in front of it two figures engaged in conversation are striding forward: they represent Plato and Aristotle. A single gesture summarizes the essence of their philosophy: Plato raises his arm to point to the sky; Aristotle stretches his arm before him, with the palm turned downward toward the earth. Raphael's extraordinary ability to express the most complex ideas and to sum up a situation in a most convincing concrete image or attitude is also manifested here with the greatest simplicity and authority. And if, beginning with the figures of the two protagonists (Raphael has given Plato the semblance of Leonardo), all the figures are studied individually, each of them reveals the same pertinence of gestures, features, and disposition in an ever-new and always appropriate variety of accents (III, PL. 209). The features of the figures in the foreground assume the guise of portraits; the relationship between the person of antiquity and the contemporary personage to whom he lends his features gives an unusual concreteness to the whole composition. Past and present are combined in a single superior concept, opening up a more universal view of history and giving a particular vividness to the whole image. The painter of the beautiful Florentine Madonnas, faced with a new task, was revealed as a great historical painter, and it is easy to understand the extraordinary and immediate celebrity of his work among the literati and Humanists of the papal court. Cardinals, men of letters, princes, and prelates vied to obtain Raphael's works. Although he created the images of the themes they proposed, he resolved these themes in the most modern and universal manner, at an incomparably higher and more intellectual level than these patrons could ever have conceived or foreseen.

Many of the figures in the foreground of the *School of Athens* have been identified as portraits. Michelangelo has been recognized in the figure of the thoughtful Heraclitus seated on the lowest step leaning on a great square stone block. Redig de Campos (1938) has clearly demonstrated that this figure was inserted at a later date and that its more monumental idiom is a direct allusion to Michelangelo's style and renders homage to the subject of the portrait. The analysis of the joints in the plaster surface have confirmed the fact that this is an addition. On the left, the young man with a white toga who turns around with a smile has the features of Francesco Maria of Urbino, while the head of the child behind Epicurus, crowned with vine leaves, is that of the young Federigo Gonzaga. At the right, the figure of Euclid, who is bending over to draw the terms of a theorem with compasses, is a portrait of Bramante; a little further to the rear, the two young men together are Raphael himself and Sodoma.

The two minor lunettes contain depictions of Parnassus (I, PL. 307) as a celebration of Beauty and of the Virtues as an exaltation of Good. Although the curvature of Mount Parnassus, where the muses are grouped around Apollo, is actually very slight, it is accentuated by the disposition of the figures: the poets ascend on the left toward the central group and descend to the right. Here too portraits abound; and if some, like Dante and Petrarch, are reproduced with their own actual features as known from recent traditions, Raphael has given to the Latin poets the features of living Humanists. Thus, with supreme naturalness, he has made incarnate the linguistic program proclaimed by the Humanists themselves, namely, to give richness and dignity to the vernacular by means of quotations from the Latin tongue. Despite some very beautiful details, the Parnassus lunette is somewhat more rhetorical than the other frescoes, perhaps because of the particular nature of Raphael's meditations. The first strong accentuation of classical and even archaeological motifs appears in this fresco, evidently in relation to the tendencies and interests of the Humanists. In the two preceding frescoes the overwhelming impression of the antiquities of Rome on the young Raphael when he arrived from Florence resulted in the rapid transformation of his style into a monumental idiom. In the *Parnassus* lunette there appear the signs of a more detailed study and the beginnings of those archaeological interests that were to increase year by year. Winternitz (1952–54) has found that the musical instruments in the hands of Sappho and the Muses were copied exactly from a sarcophagus "of the Muses," now in the Museo Nazionale Romano. In the lunette on the opposite side, the large figures of the virtues Fortitude, Prudence, and Temperance are placed with a beautiful rhythm. Small putti in animated movement link them together. According to an unusual iconography, three of them symbolize the "theological" virtues: the small putto with a raised finger on the right represents faith; the one in the center holding a torch, hope; and the one on the left picking fruit from a branch, charity. The size and the monumental placing of the three virtues, particularly that of fortitude, indicate that Raphael knew and had assimilated Michelangelo's Sistine frescoes; together, the figures compose a sort of festooned frieze of inimitable grace and harmony.

Below, flanking the windows, the two scenes of the presentation of the decretals and of the pandects are an allusion to the institution of law, derived from the concept of justice, the supreme virtue (in fact, Justice dominates the others from the vaulted ceiling above). These two scenes are probably of somewhat later date. While the first, which is a fine collection of portraits, was largely painted by Raphael himself, of the second only the composition, but not the execution, can be attributed to him.

On the ceiling there remain traces of the work belonging to the phase before Raphael took over. It is divided by frames painted with grotesques into nine major panels: a central octagon, four tondos corresponding to the top of the lunettes, and four rectangles in the pendentives. There are four minor compartments between the octagon and the rectangles. The octagon, the minor compartments, and the frames of grotesques are the work of Raphael's predecessors. He, on the other hand, inserted the four tondos and the four large rectangles, illustrating in them, with an even more pronounced symbolism, the Neoplatonic principles. In the tondo above the *Disputa* is the personification of theology; above the *School of Athens*, that of philosophy; above the *Parnassus*, that of poetry; and above the *Virtues*, that of the supreme virtue, justice. The scenes depicted in the four rectangles are also related to the themes developed on the walls below them. The *Sin of Adam and Eve* corresponds to the *Disputa*; the so-called "First Motion" (*Primo Moto*), the figure of a woman bent over the great transparent globe of the cosmos, with the small sphere of the earth at the center, corresponds to the *School of Athens*; the *Judgment of Solomon* corresponds to the *Virtues*; *Apollo and Marsyas* to the *Parnassus* (this last panel was not executed by Raphael, although the cartoon is undoubtedly his). As in the *Parnassus*, the ceiling frescoes reveal precise classical references; the executioner in the *Judgment of Solomon*, for example, is taken directly from one of the Dioscuri on the Quirinal.

The frescoes in the Stanza della Segnatura were completed in 1511. The ones in the adjoining room, known as the Stanza d'Eliodoro, were painted between 1511 and 1514. From the allegorical rendition of a spiritual, Christian Neoplatonism in the Stanza della Segnatura Raphael turned to dramatic historical subjects, depicting four events which exalt the divine intervention in aid of the church: the *Expulsion of Heliodorus from the Temple* (PL. 426), drawn from the Bible; the *Mass of Bolsena*, from medieval chronicles (PL. 427); the *Liberation of St. Peter*, from the Acts of the Apostles; and *The Repulse of Attila* (PL. 430), also from medieval chronicles. The program of the decorations for this room must have also been proposed by Julius II, as the choice of episodes clearly alludes to the tumultuous events of his papacy or contains other references to himself. The choice of so unusual a subject as the Mass of Bolsena, with the bleeding of the Host that dispels the doubts of the officiating priest, must be considered in relation to the Pope's devotion to the feast of Corpus Domini. The choice of the liberation of St. Peter is to be associated with S. Pietro in Vincoli, Julius's titular church when he was a cardinal. The last to be executed, the episode of the repulse of Attila (PL. 430), was painted, or at least completed, after the death of Julius II, because the figure of the Pope, who solemnly advances toward the tumultuous hordes of Attila, bears the features of Leo X. The date, 1514, written below the *Liberation of St. Peter*, alludes to the completion of the whole cycle and not to this single fresco.

The *Expulsion of Heliodorus*, the first fresco to be painted, takes place, as does the *School of Athens*, in a magnificent temple. But the character of the architecture and the placing of the figures could not be more different. Everything assumes a vertiginous centrifugal rhythm of an intensely dramatic quality. The mass of the faithful on the left, pressing one against another and crowding against the columns, and the avenging angels on the right, hurling themselves upon Heliodorus, who falls backward against his fleeing accomplices, diverge violently in an inverted V, leaving a large void in the center. The architecture, in contrast with the compact curtain of walls of the *School of Athens*, opens laterally in massive arcades; the circular rhythm of the series of drums of the domes and extradoses in the center seems to give the primary impetus to the vertiginous movement in the foreground. They emphasize, as does the rotundity of the columns inserted in the rectangular mass of the piers, the shafts of light and the deep masses of the shadows. Remote and alone in the background, in an interplay of light and shade, the high priest Onias is seen praying, invoking heavenly assistance; his prayer is answered by the unleashing of the divine punishment in the foreground.

In the group of fugitives and in the figures of the horseman and Heliodorus some critics claim to see already the hand of Giulio Romano. This may be true of two or three figures crowding against the arch of the lunette, but the figure of Heliodorus is so new in its formal concept and its foreshortening that only Raphael at this phase of his artistic development could have carried it out. It became a prototype which Michelangelo remembered, perhaps unconsciously, in his frescoes for the Pauline Chapel. In a way this was a kind of repayment for the even deeper influence that from this time onward the frescoes of the Sistine Chapel began to exert on Raphael. The ceiling of the Stanza d'Eliodoro itself is proof of this: Raphael retained the original decorations of the ribs of the vaulted ceiling, and he painted four trompe l'œil tapestries, which appear to be hung like a tent, depicting four Biblical scenes, executed by one of his assistants, probably Guglielmo di Marcillat.

The only static group in the Heliodorus fresco is that of Julius II carried by chair bearers, which has just crossed the arch of the lunette. The significance of this group's appearance has always been puzzling to critics. A very persuasive explanation is given by D. Redig de Campos, who suggests that this represents a real portrait of the Pope, who has been carried into the room to survey the work of his painter. With remarkable compositional ability, and at the same time giving proof of his great intellectual freedom, Raphael successfully reconciled this insertion with a scene whose development is completely independent of it.

Powerful light effects and dramatic expressiveness also characterize the *Mass of Bolsena*, but the drama is elevated to and contained on a plane of purely spiritual conflicts and emotions that finally find calm reassurance in the certitude of the faith. Raphael's feeling for architecture is demonstrated in his solution to the great problem posed not only by the presence of the window but also by its asymmetric position in respect to the axis of the lunette, leaving the space on the left smaller than that on the right. The figures of the faithful are massed in the smaller space on the left, pressing against the steps and the platform of the altar, spatially and psychologically gravitating in their movement toward the point where the miracle occurs. On the right, however, Raphael painted an extension of the upper edge of the window, making it into a spacious platform for the central scene. On it kneels the Pope (in the guise of Julius II), who watches the miracle. Arranged on the altar steps behind him are the cardinals, each of whom is thus articulated individually; at the base, between the jamb of the window and the arch of the lunette, the Swiss guards, firm and solid, are the immovable custodians. The painting here is extraordinarily vivid and strong: in the grouping at the left the colors are broken up into gradating shades in the large soft forms of the women and children; on this side the images form large areas of flat color, standing out and assuming an immediacy of apparition and extraordinary individualization. This is undoubtedly one of Raphael's most famous "pieces." There has been so much insistence on the theory attributing the chromatic strength of this fresco to the influence of Sebastiano del Piombo (q.v.) that it is now high time to point out the inadequacy of such an explanation. To attribute such a novel and original coloristic solution to the influence of Sebastiano does not help us to understand its magnitude and depth. Nor is it sufficient to add the influence of Lotto, as has been suggested recently, after it was discovered that Lotto had also worked in the Stanze immediately before Raphael. The roots of Raphael's astounding use of color go both wider and deeper than this; they begin in his native Marches, where Piero della Francesca had been active and where Venetian art had always been widely diffused and had exerted a deep influence on local artistic development. From his early youth Raphael had found a link with Venetian culture, not to mention the Flemish art with which he had become acquainted in Urbino. In his Florentine period, color solutions such as those in the *Pregnant Woman* had already taken him far beyond the Perugian and Tuscan idiom, to the extent that some art historians have turned to the theory that Fra Bartolomeo, after his return from Venice, had transmitted to his younger companion in art some of the knowledge gained in his Venetian experience. This may have contributed to Raphael's color techniques, but the fundamental fact remains that he came from Urbino and from circles where contacts with Venice, and with Ferrara and Mantua were continuous and profound. Only this fact can explain the constant pictorial quality of his style, the softness of his modeling, the links between figure and landscape, and the harmony which subdues the vivacity of the local colors to the effect of the whole.

Another completely new solution, quite different from the preceding ones, was devised by Raphael for the *Liberation of St. Peter*. His sensitivity to the effects of light which in his youth had caused him to search for a style beyond those of his father and Perugino and to go back to Piero della Francesca, and, in his Florentine years, had attracted him to Leonardo, emerges in this fresco as the generating element of the scene. The miracle is made manifest in terms of light. As the scale of significance grew, so in Raphael's imagination did the scale of his light values. Here the sacred event, which in the previous subjects had taken place in the presence of chorus-like masses of figures, is simplified and reduced to two personages, St. Peter and the angel of liberation. Raphael had the happy idea of returning to the archaic form of continuous narration: each character appears twice in two successive phases of the story. In the center of the lunette above the window, the entire wall of the prison falls away and becomes a grate, which with its black bars, increases in a remarkable way the splendor of the aura of light diffused about the angel. Incandescent and ethereal, the angel leans over the sleeping St. Peter, "and he smote Peter on the side and raised him up, saying, Arise up quickly.... And he [Peter] went out and followed him . . . but thought he saw a vision" (Acts 12: 7–9). And in fact, on the right beyond the door, the figure of the angel, suffused with light, reappears leading the still-dreaming St. Peter by the hand. "When they were past the first and second ward, they came unto the iron gate that leadeth unto the city; which opened unto them of its own accord: and they went out, . . . and forthwith the angel departed from him" (Acts 12:10). Even the gate and the guards are shown. Raphael's fresco, which adheres faithfully to the Scriptures, is magnificently visionary. One of the guards is asleep on his feet, leaning against the wall and on his spear, which is inside the cell. His cuirass reflects the light of the angel. The other guards are asleep in the shadows, barely touched by the few flickers of light on the steps of the stairway leading down to the right foreground. Descending to the left are other stairs, symmetrical with those to the right, on which soldiers excitedly raise the alarm. Above them is a patch of light from a torch, and from on high the light of the moon is seen among the clouds. This is the first work in the whole of Italian art based entirely and exclusively on light effects.

The three frescoes in the Stanza d'Eliodoro, although they contain a variety of solutions, show a close unity of development that makes them into a single conceptual cycle. A departure from this unity is evident, however, in the *Repulse of Attila*. This is one reason, and the most valid one (aside from the presence of a portrait of Leo X), for dating it as the last of the frescoes in the room, painted after the death of Julius II. Up to that point, although his compositions had become increasingly dynamic and dramatic, Raphael had organized them on a central axis that coincided with the highpoint of the scene, effecting a fundamental symmetry between the two sides. In the *Repulse of Attila* both the central axis and symmetry are abandoned. The barbarian hordes overflow the midpoint of the lunette and are brought to a halt at the edge of an imaginary demarcation line. Cutting across the composition diagonally from the foreground to the background, this line traces a furrow between the tumultuous soldiers of Attila and the solemn advance of the Pope accompanied by his attendants, while SS. Peter and Paul, brandishing their swords from on high, give divine aid. From this point onward Raphael developed this compositional scheme, whose novelty and far-reaching consequences have hitherto been too often ignored by art historians. Fascinated as they were by the supreme harmony of the earlier compositions, critics have been far too prone to consider every departure from it as "decadent."

Raphael, ever sensitive to changes of artistic climate, which under Leo X was to differ greatly from that of the pontificate of Julius II, became acutely aware of the new problems with which he would now be faced; the way in which he faced them was to have incalculable consequences in his artistic development. It is therefore necessary at this point to interrupt analysis of the Vatican frescoes in order to consider Raphael's entire activity under Julius II, which ranged over a very wide field. As in Florence, he continued to paint beautiful Madonnas, although this subject no longer had the same significance and importance in relation to his total output. He continued to paint superb portraits and executed two large altarpieces: the *Madonna of Foligno* (PL. 434; IV, PL. 314) and, after the death of Julius II but still in the spiritual clime of that great creative phase, the *Sistine Madonna* (PL. 433). He also worked for Agostino Chigi (his second patron) in the Farnesina, in S. Maria del Popolo, and in S. Maria della Pace. Finally, during his first Roman years, he also turned to architecture — an important but often neglected aspect of his art.

The Alba *Madonna* (Washington, D.C., Nat. Gall.), "the true sister of the Virtues" (Ortolani, 1942) belongs, therefore, to the period of the Stanza della Segnatura. It is close in time to the Mackintosh *Madonna*, or "Madonna of the Tower" (badly damaged but erroneously questioned), and the Aldobrandini *Madonna*, also known as the "Garvagh Raphael" (both, London, Nat. Gall.). Within this same period could also be assigned the unpublished small *St. Sebastian* in the J. de Baranowicz Collection, Paris, which is known from literary sources. The famous *Madonna of the Diadem* (Louvre), which shows traces of collaboration, appears to be a little later; the landscape background contains certain light effects which are related to those in the *Madonna of Foligno*.

The *Madonna of Foligno*, commissioned by Sigismondo de' Conti in fulfillment of a vow made on occasion of his escaping uninjured when his house had been struck by lightning, can be dated between the end of 1511 and the beginning of 1512. It differs considerably from Raphael's previous altarpieces. The group of the Virgin and Child raised up into the sky is not, however, separated from the saints and the donor, who from the ground look up at them in contemplation. Every element of the composition, the sequence of rhythms, the accentuated gestures and, above all, the circular effects of the atmosphere and of the lights between the landscape and the sky, are all intended to bind the two parts together. The great orange-gold disk behind the sacred group becomes the amplifying aureole and the visual center to the extent that the Virgin and Child, although retaining a natural proportional relationship with the other figures, are raised to a superior rank by the intensity of the light. The Child, in the dark folds of the mother's cloak, retains the posture of Michelangelo's Doni Tondo, but the torsion of the figure loses every feeling of strain by drawing toward itself all the compositional lines through which flow, with a pleasing naturalism, the sentiments emanating from all the characters in the picture. The background landscape (PL. 434) is lit up with shafts of light from the flaming trajectory of the bolt of lightning (or fireball). Crowe and Cavalcaselle (1882–85) saw in these light effects the signs of collaboration with Battista Dossi. Longhi (1944) proposed that they reveal the hand of Dosso Dossi. Ortolani (1942) is perplexed by this feature and suggests that, on the contrary, the new inventions in the altarpiece were taken up by the school of Ferrara, thus accounting for seeming resemblances, and this would appear to be the case. In other works of this period, Raphael animated his landscapes with a richer luminosity by means of individual sources of light; this is seen in the *Madonna of the Diadem*, in the background of the *Repulse of Attila*, and in the light values of the great gold disk behind the Virgin and the prancing host of cherubs among the clouds in the *Madonna of Foligno*. A short while later, the *Liberation of St. Peter* was entirely interpreted in terms of light.

The portrait of the donor, Sigismondo de' Conti, is very fine. Moreover, all the frescoes in the Stanze are filled with excellent portraits. During this period Raphael executed a number of independent portraits, but of those painted in the years before the death of Julius II only one original remains, namely, the *Portrait of a Cardinal* (Prado). Raphael found in his subject (not further identified) an air of detachment which gives the image a sense of importance as the very symbol of high rank and culture. The great arm that lies with natural elegance on the unseen armrest of the chair makes the figure recede slightly and creates around it a new spatial breadth. The finely chiseled face above the pyramid of the red robes and under the Cardinal's biretta, rendered paler by the great masses of red, acquires a sharp aristocratic delicacy. The portrait of Julius II, which is known to have been painted for S. Maria del Popolo, has only come down to us through copies, two of which are in Florence. One, in the Uffizi, is more static and was probably executed by a painter from Raphael's circle; the other, in the Pitti, is reinterpreted with a greater pictorial freedom so Venetian in manner that Ortolani has proposed Titian as the copyist (this has been accepted in the catalogue of the gallery).

The effigy of Julius II transfigured into St. Sixtus, a name beloved by the Della Roveres, reappears in the *Sistine Madonna* (PL. 433), painted for the Church of S. Sisto in Piacenza. Two curtains are drawn apart, and the Virgin, holding the Child, appears on luminous clouds as if advancing with a miraculous lightness. The mantle descending from her head swells out like a sail and acts as a counterbalance to the florid figure of the Child. From the inspired figure of the pontiff, on the left, expressions of adoration rise toward her and, after they have enveloped her, descend filled with the divine image through the lowered eyes of the kneeling St. Barbara. This magnificent concept, carried out with sovereign mastery, humanizes the divine images fully and serenely, bringing them closer to mankind, yet elevating them to the highest sphere by means of light, beauty, and the feeling of ecstatic adoration and supreme yearning that rises up from the hearts of the faithful. Raphael created, by means of his imagery, a powerful weapon "de propaganda fidei." The dating of this very celebrated Madonna has been widely discussed; in fact, the gap between it and the *Madonna of Foligno* is immense, yet the interval cannot have been more than two years. Taking into consideration the turning point in Raphael's art which occurred in 1514, the most suitable chronological place for the *Sistine Madonna* is close to the *Liberation of St. Peter*.

Stylistically, chronologically, and even typologically, the *Sistine Madonna* is close to the *Madonna of the Chair* (Florence, Pitti). While Raphael's mind was taken up with an idea, evolving and resolving a flow of images and investigating all their possible developments, he nearly always created collateral works connected with those of major importance. These works are "minor" only in so far as they are of smaller size and lesser complexity; often they are of greater purity and inspiration, as exemplified by the *Madonna of the Chair*. The concept had its origins in the group of women and children gathered at the foot of the stair in the *Mass of Bolsena*; it was developed further in the *Madonna of the Curtain* (Munich, Alte Pin.), which actually precedes the *Madonna of the Chair*.

The *Galatea* (III, PL. 392), painted in fresco for Agostino Chigi in the loggia of his villa on the Tiber (now the Farnesina), belongs to the same category of meditations that gave birth to the *Parnassus* but with more satisfactory results. It is undoubtedly a little later in date; his interest in archaeology, which appeared openly for the first time in the *Parnassus*, is here expressed with greater freedom by means of a metaphoric transposition in which antiquity is evoked and turned into a myth by a Humanist. Raphael, entranced by the wealth of motifs that the antiquities of Rome disclosed to him, gave himself over to depicting them with uninhibited abandon. Both the face and the *contrapposto* of the *Galatea* are reminiscent of the London *St. Catherine*, but the impulse of her motion is freer, and the cloak that covers her so scantily flutters, together with her hair, behind her in the wind. A shell and two dolphins serve as an aquatic chariot — an image undoubtedly suggested by a grotesque motif, but which here acquires the joyous flavor of a marine sport. The surface of the sea has the polish of the marble of *opus sectile*. To the Humanists, who at that time dearly loved to write poetry in Latin and sought inspiration from classical models, Raphael's fresco must have appeared as a revelation, the very image of the world of their dreams. This is borne out by the enthusiastic chorus of eulogies that greeted this work on its completion.

To the time of the Stanza d'Eliodoro also belong the frescoes with sibyls and angels in the extrados of the Chigi Chapel in S. Maria della Pace in Rome (PL. 432). Raphael also drew the prophets (Uffizi, Gabinetto dei Disegni e Stampe), but the fresco was not executed by him. At the center of the sibyls and angels a small winged genius holding up a torch turns on himself; with one knee bent he leans on a dado that nearly forms the keystone of the arch. This figure is actually the key to the whole composition, which opens in both directions and falls along the curve of the arch like a magnificent festoon. The basic idea is still similar to that of the Virtues in the Stanza della Segnatura, but it is carried out with a less lyrical rhythm and a more complex and monumental counterpoint of movements, revealing the impression made on Raphael by the figures in the lunette of the Sistine Chapel.

More literal reminiscences of Michelangelo are found in the frescoes of the prophet Isaiah in the Church of S. Agostino, painted for the apostolic prothonotary Jan Goritz, presumably in 1512. This work actually has the feeling of an exercise, almost as if Raphael had been seized once more by his old desire to investigate to the very roots the meaning of another artist's style, even to the extent of actually repeating him.

Raphael was also active as an architect for Agostino Chigi; but not in his villa on the Tiber (the modern Farnesina), which through vindicatory enthusiasm Geymüller (1870) tried to attribute to Raphael; stylistic character, documentary evidence, and contemporary witnesses, including Vasari, indicate that Peruzzi was the architect. However, a few years later, in 1512, Raphael was commissioned to build the Chigi stables on the side of the villa facing on the street, away from the Tiber.

Previously, he had been engaged on another architectural project, the building of the Church of S. Eligio degli Orefici in Rome. In spite of alterations, the interior has an airy and luminous quality, a clarity of surfaces, and an elasticity in the design of the arches that are characteristic of Raphael. His architectural elements are clearly based on Bramante, as evidenced by the relation between the diameter and the height of the dome, the manner in which the pendentives are set in the spaces between the arches, and the linking together of pilaster strips, trabeations and arches. Bramante's choir in S. Maria del Popolo (II, PL. 344) immediately comes to mind. Two drawings in the Uffizi are only apparently contradictory regarding the names of the architects. One, by Sallustio Peruzzi, bears the note, "S. Alò degli Orefici . . . work of Raffaello da Urbino." It reproduces the ground plan and the exterior and interior elevations of the church, which differ in part from the present form of the building. Geymüller points out that the variants in the drawing are more typical of Raphael than the present building. The other drawing, by Bastiano da Sangallo, reproduces the exterior of the cupola and the lantern as they appear today. Bearing the notation "by Maestro Baldassari da Siena," it seems probable that Peruzzi completed Raphael's unfinished work. The direct testimony of Baldassare Peruzzi's son Sallustio and the characteristics of the work provide a sound basis for the attribution of the original design and of the earlier phases of the construction to Raphael himself. The documentary evidence fixes the beginning of the work in June, 1509, the date of the papal bull authorizing the construction of the church.

A comparison of S. Eligio with the Chigi stables at the Farnesina, as they appear in Frommel's reconstruction (1961), partially modified from the previous ones by Pontani (1845) and Geymüller (1884) — actually only a fragment with plinths and the bases of columns remain — seems to indicate that they are more or less contemporaneous. Documents show that the project for the stables dates from the year 1512. In spite of its functional purpose, the building was primarily designed to form a monumental façade to Chigi's dwelling, on the side facing the Via della Lungara. Raphael's design adhered to that of the building already erected by Peruzzi, but he widened it and proportionately increased its height by the addition of another half story. On the other hand, he reduced the vertical partitions by reducing the number of paired columns from nine to seven, thus obtaining an articulation of the building far more monumental than Peruzzi's, using a strictly Bramantesque idiom.

From the very beginning of his Roman period Raphael moved forward, basing his architecture on Bramante's. He fully understood the latter's spirit and style, even at a time when other architects, including Peruzzi, still expressed themselves in 15th-century forms. That Bramante designated Raphael as his successor in the building of St. Peter's shows the older architect's high regard for Raphael. Until he assumed this heavy burden, Raphael's architectural activity had been limited to S. Eligio, the Chigi stables, and the beginnings of the Chigi Chapel.

The Chigi Chapel in S. Maria del Popolo (PL. 436; FIG. 855), the funerary chapel of Agostino Chigi, springs from an idea similar to that of S. Eligio, although it is more complex and executed with greater power. In the Chigi Chapel the emphasis is on the interior, since the external structures are incorporated into the fabric of the church. Only the dome can be seen from the outside; it is simple and unadorned, rising from a high drum. Raphael concentrated all his interest on the interior space and its continuous development. The surfaces extend rather than swell, allowing the light to spread over the whole like a luminous skin. At the four corners of the square plan are four sturdy piers with niches; they cut the corners diagonally and support the thrust of the four great arches on which rests the windowed drum and the cup-shaped dome. This is a solution inspired by Bramante's design for the dome of St. Peter's, but Raphael developed it in a different manner. Within the chapel itself the four piers are hidden, and only a single face is visible; they have the appearance of the shorter walls of an octagon obtained by cutting the corners of a square. Their mass is discernible only from under the great arch of the entrance of the chapel, of which they form the abutments; from within the chapel the mass is hardly apparent. The spans of the three great arches (the fourth is open, forming the entrance) are closed by walls having a purely superficial value. The walls are engaged to the piers by means of pilaster strips set at an angle like a half-open book, as in Bramante's sacristy for S. Maria presso S. Satiro in Milan. The niches in the piers contain statues; the slight shadows within them provide a contrast to the whiteness of the pilaster strips. All the members that indicate the structural lines — arches, pilasters, cornices, capitals, and garlands — are white, while the surface planes are covered with polychrome marble. Despite the alterations to which it has been subjected, the chapel still substantially reflects Raphael's original project, which included the pyramidal tombs, even though the present ones are partial modifications executed much later by Bernini. It is remarkable how Raphael's design manages to withstand, without loss of harmony, Bernini's intrusions, which differ so greatly in their dimensions. [It is only necessary to compare Bernini's *Habakkuk* (I, PL. 381) with the *Jonah* carved by Lorenzetto from a design by Raphael (XII, PL. 59) to see how the latter is proportionally more harmonious with its niche.] Raphael consciously continued this coloristic animation of his architecture, culminating in the dome covered by mosaics with a blue ground, with the figures enclosed within a gold framework — an extremely original motif for the Rome of the 16th century. The mosaics were executed by the Venetian Luigi da Pace (X, PL. 192), who signed them with the date 1516. The subject must have been suggested by Chigi himself, for he had commissioned a similar one for the Farnesina. It represents the signs of the Zodiac and the influences of the astral conjunctions on the destinies of man. In the central eye is the great foreshortened figure of the Eternal Father, his arms stretched upward in a grandiloquent and impetuous gesture, in the manner of Michelangelo but with a more circular flowing movement.

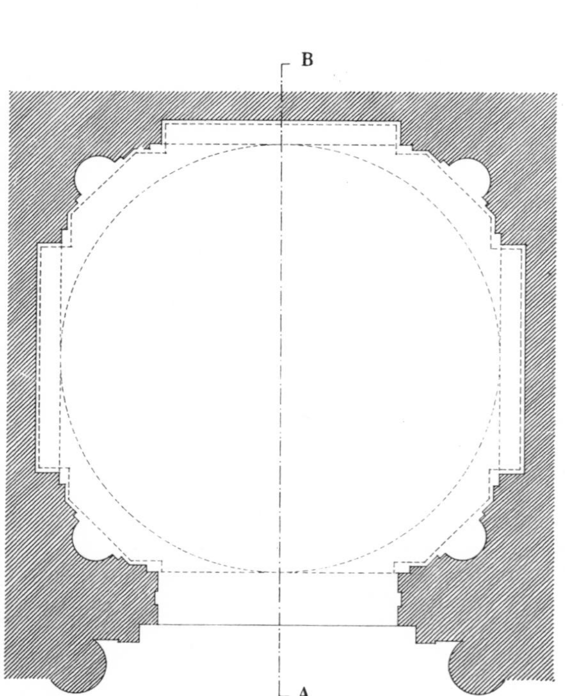

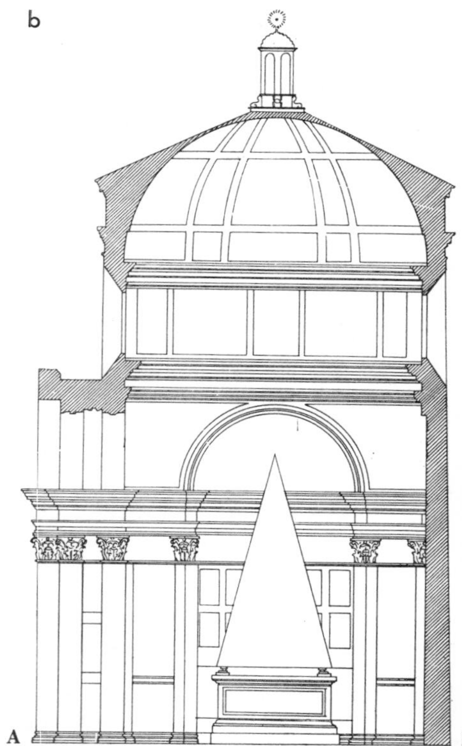

Rome, S. Maria del Popolo, Chigi Chapel: (*a*) ground plan; (*b*) section.

The beginning of the Chigi Chapel is generally dated about 1512–13. It may perhaps be necessary to propose the slightly later date 1513–14 because of the surehandedness of the solutions revealed by the construction, the continuity with which the interior space is developed without the slightest break, and by the study and knowledge of ancient architecture, especially of the Pantheon, which the building reveals.

The raising of Leo X to the pontifical throne changed the whole tone of the papal court. A pleasure-loving Humanist and patron of the arts, Leo X surrounded himself with men of letters, favored the researches into antiquity and sponsored the diffusion of a culture based on Latin civilization, well understanding how much the prestige of ancient Rome would help to raise the prestige of papal Rome. This was the moment when men of letters concerned themselves with linguistic matters and attempted to enrich the vernacular by the introduction of Latin elements in order to arrive at a unified Italian idiom. They also wrote Latin poetry and set out to imitate classical comedy and revive the ancient drama.

Raphael's art is remarkably synchronous with this cultural movement. His incomparably open mind absorbed its terms, and he used them with the greatest mastery. Raphael has been mentioned in connection with eclecticism, but this judgment does not take into account his remarkable critical intelligence and his exhaustive study of the major artists and schools of his own time. In fact, while the men of letters discussed the vernacular, he, with a far deeper insight into the future, made use of his rich artistic experiences in Urbino, Umbria, Florence, Venice, and Rome to create a new artistic language which in the span of a decade became truly Italian and gave a new direction to the course of Italian and European art. His style, originally lyric and epic, became more dramatic: at times, it was even eloquent. These aspects of Raphael's art may be regarded as decadent by those who are subject to the fascination of his more lyric youthful works. Nevertheless they have a new grandeur and correspond to the new currents of the century that are found in all other fields. It is true, of course, that in his last years, overburdened as he was with commissions, he was obliged to leave the execution of a number of his works to his assistants. However, his creations continued to pour forth in such an unceasing flow that, while he left to his assistants the execution of the frescoes in the third Vatican Stanza and of the innumerable Madonnas which his clients never ceased to commission, he himself created the compositions for the tapestries (the cartoons are almost all by Raphael's own hand and signed by him) and devised a new type of altarpiece that was a prototype for this art form for over a century.

The frescoes in the third Stanza were painted almost entirely by Raphael's collaborators, but the original ideas were his own. They represent four incidents in which the leading figures are Leonine popes. The most important of the four frescoes, *The Fire in the Borgo* (PLS. 429, 430) must be interpreted scenographically. The architecture in the foreground has the spectacular character of the pseudoarchitectural elements used as settings for theatrical events of festivities: they are wings and backdrops and do not constitute an articulated structural organism. The perspective is entirely illusionistic — that is, it ceases to be rational. The great fluted columns on the left are not symmetrical with the center of the lunette but are placed according to the golden section. The figures in the foreground are enlarged disproportionately in relation to the architectural elements and thus acquire the heroic proportions of characters in classical tragedy. The inclusion of the episode from the burning of Troy, wherein Aeneas saves his father, Anchises, and his son Ascanius, was undoubtedly suggested by some Humanist to give a mythical and heroic quality to the fire in the Borgo, in the same way that the writers turned to classical themes for their tragedies. The principal event in the fresco (drawn from the *Liber pontificalis romanus*) is placed in the background. Pope Leo IV appears on the loggia of the Vatican (with the old St. Peter's in back) and extinguishes the fire by making the Sign of the Cross. The telescopic effects created by the void in the center and by the gestures of the clamoring women in the foreground irresistibly draws the spectator's attention to the Pope. The idea of placing the principal motif in the background like a vision was completely new: it was to be adopted and used throughout the 17th century. Unfortunately, the execution, left for the most part to assistants — to Giulio Romano in particular — makes the picture somewhat prosaic and heavy in spite of the beauty of the original concept.

The results are even worse in the other frescoes: *The Battle of Ostia*, *The Coronation of Charlemagne*, and *The Oath of Leo III*. Moreover, they are in a very bad state of preservation. Nevertheless, careful scrutiny reveals a number of new motifs that were to be widely echoed in Italian art for a long time.

Raphael was also directly engaged in designing stage scenery. An envoy of the Estes in Rome wrote to the Duke of Ferrara on Mar. 2, 1519, to tell of a visit with Pietro Bembo, during which they talked of nothing but plans for masques and of the scenery created by Raphael for Ariosto's play *Gli Suppositi*, which was to be performed the following Sunday.

The Stanza dell'Incendio was painted between 1514 and 1517. At the end of 1517 the ceiling of the loggia of the Farnesina (VIII, PL. 101) was also completed, with frescoes depicting the myth of Psyche, "a despicable thing for a great master; far worse than the last Stanza of the palace," wrote Leonardo Sellaio to Michelangelo with venomous malice on Jan. 1, 1518. Undoubtedly the work is heavy, and little or none of it was executed by Raphael himself. But once again Raphael's inventive imagination is a delight: he transformed the loggia into an arbor composed of great festoon of fruits and flowers, which outline the empty spaces of intense sky blue, against which stand out the white and luminous nude figures of the tale of Apuleius. In the center there are two trompe l'oeil tapestries depicting the council of the gods (X, PL. 250) and the marriage of Cupid and Psyche.

But the high point of Raphael's creative process in these years moved in other directions, namely, in executing the cartoons (London, Vict. and Alb.) for the tapestries (Pin. Vaticana; XIII, PL. 400) intended for the Sistine Chapel; in designing a new type of altarpiece; in architectural endeavors; and in the study of the antiquities of Rome.

The cartoons for the tapestry mark a return to the deep meditations of the Stanza della Segnatura and the Stanza d'Eliodoro but are stronger and more eloquent in concept. The majority of the scenes are not developed symmetrically along a central axis but rather from left to right (the scenes are reversed in the cartoons; Raphael bore in mind the fact that in the weaving the design would be reversed). This is the normal direction in reading and marks a rising of the tension of the action towards the final climax. In the *Miraculous Draught of Fishes* only the apostle who stretches out his joined hands to Christ is actually Peter, but in their attitudes also the other figures seem to impersonate three successive moments of Peter's approach to Christ. Amid the rising tumult of movements and gesture Christ is purposely the least dramatic and monumental figure, but He is instead the clearest and most luminous in the light of the sky and water, separated from the others by an infinitesimal and yet significant hiatus. In the *Charge to St. Peter* the figure of Christ is more static but also more luminous than the others, and His gesture imperceptibly determines the distance that isolates Him. In both these scenes in which Christ appears the background is an open luminous landscape. In the other scenes, which depict the stories of SS. Peter and Paul, the surrounding space is strongly characterized as the "theater" of the event, whether it be a temple, a square, or a palace. St. Paul, preaching, stands on a platform before the temple, three steps above the crowd that surrounds him at a distance measured by the steps and by the eloquent gesture of his outstretched arms. In the same way, St. Peter, in the *Death of Ananias* (PL. 432), stands with his followers on a platform that is really a stage; his accusing gesture causes a movement among the crowd of bystanders which is so intense that they break apart, while Ananias falls backwards to one side. The artists of the 17th century — a theatrical century — from the Carraccis to Poussin held this late Raphael in great esteem. The *Death of Ananias* anticipated the work of Poussin in a most remarkable way. The great distorted figures carrying heavy bundles who lean over Ananias became the prototypes of the gigantic secondary figures of mannerist and baroque art. The payment on June 15, 1515, of "three hundred ducats ... in partial payment for the cartoons or drawings to be sent to Flanders to be made into tapestries" (Golzio, 1936, p. 38) is an indication that the cartoons had been finished by that date (cf. XIII, PL. 400).

At about the same time Raphael was working on a new type of altarpiece, inspired by new human and dramatic concepts and by the renewed affirmations of the universality of the Roman Church. A new link between the human and the divine is established by visual means in Raphael's three last altarpieces, the *Ecstasy of St. Cecilia* (XII, PL. 367), the *Sistine Madonna*, and the *Transfiguration* (PL. 435). The exact dates of these works are not known. The date of the *St. Cecilia* seems to be indicated by a document from the Monastery of S. Giovanni, in Monte in Bologna: "In the year 1514 the Blessed Elena ... Dell'Oglio ... caused to be erected the chapel of St. Cecilia and commissioned from Raphael of Urbino the painting of St. Cecilia ..." (Golzio, 1936, p. 29). This date, which immediately precedes the tapestries, seems appropriate for the painting, in which the figure of St. Paul is of the virile, mature, and powerful type found in the tapestries, although the other four figures seem to indicate a slightly earlier formal concept. This altarpiece, which was renowned in the 16th century, for a time ceased to be popular. The critics in the romantic period declared it a frigid piece of academicism. According to them, Raphael's composition was too vertical, too balanced and calculated: the placing of the two figures in the middle ground was regarded only as a means of filling the spaces left empty by the other three. The still life of musical instruments in the foreground, while beautiful in itself, was said to detract from the main subject. Despite the passing of many years, Raphael's approach in this work is remarkably similar to that of the Borghese *Deposition*. In the *St. Cecilia* in Bologna there appears a very new idea, however: no longer is the divine figure presented to the faithful for adoration; rather, the feeling of adoration itself, with all its implications of devotional inspiration, becomes the entire theme of the composition. Raphael anticipated — in fact, in a sense he proposed — a position that the Roman Catholic Church was to adopt, fully conscious of its programmatic significance, just a half century later during the period of the Counter Reformation. In that period the fervor for the saints, represented on the altar in full view of the faithful, acquired the greatest importance in sacred iconography. In Raphael's altarpiece, the image is naturally filled with classical serenity, even though it is touched by a slight tinge of mannerism (evident in the draperies). Although it may exude an excess of intellectualism it shows an extraordinary historical intuition. Perhaps this was a flash of intelligence which as yet had not taken hold of the entire personality of the artist. The absence of the other pole of the dialectic of the devotional relationship, namely the divine image, also results in a lack of emotion in the representation.

Overburdened as he was with appointments and commissions, Raphael during these years increasingly turned the actual execution of his works over to his pupils. The greater part of the Madonnas and Holy Families, which became more numerous again during the pontificate of Leo X, after a decline under Julius II, were painted in his studio, although Raphael never neglected the actual designing and supervision of the work, sometimes painting certain sections himself. Within the limits of this article it is not possible to examine for chronology and greater or lesser degree of collaboration such works as the *Madonna of the Fish* (Prado), the *Madonna of the Linen Window* (*dell'Impannata*) (Florence, Pitti), the *Madonna of the Divine Love* (Naples, Mus. di Capodimonte), the *Madonna of the Rose*, *The Holy Family under the Oak Tree*, and the *Holy Family* known as "La Perla" (last three, Prado), the small *Holy Family* and the large *Holy Family* painted for Francis I of France, signed and dated 1518 (Louvre), and other pictures of various sacred subjects: the *Visitation* (Prado), the *John the Baptist in the Desert*, and the *St. Michael* (Louvre), also painted for Francis I, and the so-called "Spasimo di Sicilia" (*Christ Falling on the Way to Calvary*; Prado), painted for the Church of S. Maria dello Spasimo in Palermo and completed in 1517. This painting is important for the novelty of the composition, often imitated in the 16th century; the procession of the climb to Calvary unfolds and turns almost in the middle of the picture and forms the center of the scene: Christ falls under the weight of the Cross, whose arms together with the position of the figures form the pattern of a St. Andrew's cross.

Beautiful, but in a different way, is the *Vision of Ezekiel* (Florence, Pitti). Berenson (1930) commented on it: "Is it thus that Jehovah revealed himself to his prophets? Is it not rather Zeus appearing to a Sophocles?". Gamba (1932) conceived of the possibility of projecting it into the curvature of an apse. In fact, this painting of small dimensions could bear enlargement to a monumental scale. It can be regarded as largely the work of Raphael himself, since the slightest defect in the foreshortening and spacing, however small, would have disorganized the composition of the divine group, which is raised above the ethereal and distant vision of a delicately shaded landscape of water, clouds, and snow-clad mountains, filled with shafts of light and airy masses. From this relation between the upper regions and the distant landscape, reduced to a dim line under the swelling and luminous cloud mass, the divine group is projected forward and enlarged with the effect of "prophetic vehemence."

In his late period, light increasingly assumed a generating function in Raphael's compositions. The *Transfiguration* (PL. 435) commissioned by Cardinal Giulio de' Medici early in 1517 in competition with Sebastiano del Piombo's *Raising of Lazarus*, was originally conceived to present only the Transfiguration of Christ on Mount Tabor, surrounded by a circle of apostles and prophets, in a single great halo of light. But after his Humanist statement in *St. Cecilia* and the *Sistine Madonna*, the reduction of his subject to the divine nucleus must have seemed to Raphael to constitute a regression from the new ideas he had been pondering. He added the miracle of the healing of the possessed boy, so that the altarpiece became the dramatic antithesis between the Transfiguration, which takes place on the mountain in a miracle of light, and the agitation of the men on earth in the shadow of the mountain. Though the light from above falls upon them too, it is not diffused; hence they have a greater massiveness and objectivity. The contrast of opposites is perhaps too dialectically obvious, but it enabled Raphael to arrive at remarkable solutions that were to have great influence on future painting. The 16th-century artists, as well as the Carraccis, Guido Reni (q.v.), and Poussin, took freely from Raphael. Even Caravaggio undoubtedly studied the *Transfiguration* with great perception; the huge book and the tree stump on which is seated St. Andrew, who is reflected in the dark water, and the light beating down on the series of gestures, heads, and shoulders rising from the left-hand corner toward the mountain and, above all, the syntactical frontal view of the large foreshortened foot must have impressed themselves on Caravaggio's mind as superb formal models.

Of a quite different nature were the idealistic motives that proceeded from Raphael's inborn, traditional "Christianity" mingled with a highly civilized sense of humanity and moral responsibility. This led him to take part, with an acute and concrete understanding of the historical situation, in the events of his time, which, though seemingly glorious, were filled with tension and potential drama. Evidence of this is the series of splendid portraits, each one of which is not only the image of a person in the most individual sense of the term but also of a society and of an entire culture. The best among these are the *Portrait of Baldassare Castiglione* (Louvre), the *Lady with a Veil* (Florence, Pitti), and *Leo X with Cardinals Giulio de' Medici and Luigi de' Rossi* (PL. 428). In the latter the introduction of additional dramatis personae increases the intensity of the evocation of a human and historic situation. Among the other portraits are those of Bindo Altoviti (Washington, D.C., Nat. Gall.) and Cardinal Fedra Inghirami, in two versions (Florence, Pitti, and, perhaps the better of the two, Boston, Isabella Stewart Gardner Mus.), the *Portrait of a Young Man* (Kráków, Czartoryski Mus.) and the powerful *Portrait of Two Men* (Louvre), as well as a number which are lost but mentioned in the sources (e.g., those of Pietro Bembo, Tebaldeo, and Federico Gonzaga. Still others are known through replicas or copies: those of Julius II (Florence, Uffizi, and Pitti), Cardinal Bibbiena (Florence, Pitti), Joanna of Aragon (Louvre), Lorenzo de' Medici (Montpellier, Mus. Fabre), and Giuliano de' Medici (New York, Met. Mus.). There is some doubt about the authenticity of the double portrait of Andrea Navagero and Agostino Beazzano (Rome, Gall. Doria Pamphili). On the other hand, the so-called "Fornarina" (Rome, Gall. Naz.) is believed to be mainly Raphael's own work.

In the last four years of his life Raphael created a new type of fresco decoration intimately connected with his new archaeological and architectural interests. He followed and emphasized the architectural divisions, decorating them with a dense pattern of grotesques (V, PL. 239), cupids, small mythological scenes, and festoons, alternating with stuccoes of a similar nature treated in the manner of cameos. At other times he filled smooth, bare walls with an illusionistic complexity of joinings. This second type of decoration was, however, used more by his imitators and followers than by Raphael himself.

Raphael's multiple activities in the fields of painting, architecture, and archaeology were so closely interrelated, and the influences of each penetrated the others to such an extent that it would be necessary to study these aspects along parallel lines.

After the death of Bramante on Apr. 11, 1514, Raphael was called to direct the building of St. Peter's, and a papal brief dated Aug. 1, 1514, confirmed his appointment as *magister operis.* The same brief named Giuliano da Sangallo *operis administer et coadiutor*. A year later, a brief dated Aug. 27, 1515, appointed him *praefectum marmorum et lapidum omnium,* which gave him the right to obtain building material for St. Peters, with its needs taking precedence over all others; the brief also charged him with the supervision of the preservation of marbles which contained inscriptions "for the benefit of the literature and of the eloquence of the Latin language." Interest in antiquity had been increasing in Rome since the beginning of the 16th century and particularly after the election of Leo X.

This appointment brought Raphael into permanent contact with the antiquities of Rome, but his interest went far beyond the prevailing literary and erudite interest in inscriptions, being directed especially to the study of ancient monuments. The consequences, as was always the case with him, were remarkable. Another brief, dated November, 1517, increased his responsibilities and made him a commissioner of antiquities. From this was born the idea of "illustrating ancient Rome," using the methods explained in the famous letter, formerly believed to have been written by Castiglione but later attributed to Raphael and even to Bramante (q.v.). The Bramante attribution has been reproposed by Förster (1956), but there are many reasons for believing that the letter was written by Raphael. The most cogent of these is the fact that such an undertaking could not have been conceived and proposed under Julius II but only at the height of the pontificate of Leo X. In a little-known letter from Celio Calcagnini to Jacob Ziegler (ca. 1519), Raphael is extolled not only as the Prince of Painters but as "architectus uero tantae industriae, ut ea inveniat ac perficiat, quae solertissima ingenia fieri posse desperarunt." The letter also speaks of his knowledge of Vitruvius, of his familiarity with Fabio Calvo, and of his remarkable undertaking to show Rome "in antiquam faciem et amplitudinem ac symmetriam instauratam . . . fundamentis profundissimis excavatis, reque ad scriptorum veterum descriptionem ac rationem revocata." These, in embryo, formed the premises of future archaeological methods.

The sources say that when Raphael died he had already completed the First Region. Marcantonio Michiel wrote in his diary that Raphael's death was especially mourned by men of letters because he was unable to finish the description and paintings of ancient Rome. "Such was the greatness of this man," wrote Vasari, "that he maintained draughtsmen all over Italy, at Pozzuolo, and even in Greece." It is indeed a great and almost incredible loss that of the numerous archaeological drawings by himself and his collaborators not one survives.

The reflections of this intense archaeological activity, so significant and valuable in itself, are seen in his paintings, especially in the new type of fresco and stucco decoration derived from the Domus Aurea, which Raphael used in the Stufetta of Cardinal Bibbiena, in the Vatican (completed June, 1506); in the Vatican Logge (PL. 431; II, PL. 291); and in the loggia of Villa Madama (VIII, PL. 206). They are also found in his architectural works, which increased after Bramante's death.

Raphael was nominated architect of St. Peter's in April, 1514, and, as was mentioned above, the appointment was confirmed in the brief of August 1; his collaborators, at an equal salary, were Fra Giocondo and Giuliano da Sangallo, who had been Bramante's assistant since January 1. During this period the sequence of events in the building of the church is confused. In the beginning work continued with no less fervor than it had under Julius II and Bramante, but soon activity slackened. It is known that the transformation from a central plan to a basilican plan occurred when Raphael was *magister* of the Fabbrica di S. Pietro. For the time being, this change appeared only on paper (which can be seen in the plan reproduced in Serlio's treatise *Regole generali di architettura sopra le cinque maniere degli edifici*, IV, 1537), since actual construction continued on the central nucleus, planned and partly built by Bramante.

After the death of Fra Giocondo on July 1, 1515, and the departure of Giuliano da Sangallo for Florence a short time later, Raphael asked for a new collaborator. On Dec. 1, 1516, Antonio da Sangallo the Younger was nominated to this post, at a salary which was half that of Raphael. Antonio worked with Raphael until the latter's death, not only in St. Peter's but also at Villa Madama, for which Antonio's large drawing survives. The detractors of Raphael as architect lean heavily on Antonio, attributing to him not only collaboration on Villa Madama but the actual design for it. Giovannoni (1959) goes farthest in this direction; nevertheless, reading between the lines of his book, it appears that Antonio, at least in his earlier years, was above all an able builder and the head of a well-organized firm, in which several members of the Sangallo family collaborated, including his cousins Giovanni Francesco and Bastiano, known as Aristotile.

Designs for Villa Madama by all three Sangallos exist, but these can be regarded only as evidence of their position as collaborators, which was much the same as that of Raphael's assistants in his efficiently organized painting studio. He made use of his assistants to draw up the detailed plans from the preparatory designs and then to execute the work, but the concepts and original plans were his own. When, after the death of Bramante, he acquired heavy obligations also in architecture, not having a contracting (to use a modern term) organization of his own, he had recourse to the most efficient building firm then in existence in Rome — that of the Sangallos, headed by Antonio the Younger. In this period architecture and construction were not yet separate activities; probably the process of disassociation actually began with Raphael. Having had Antonio at his side in St. Peter's, Raphael also engaged him for other undertakings as his growing fame brought him an increasing number of commissions, and after the death of Raphael, Antonio succeeded him as the director of works at St. Peter's.

The building of St. Peter's soon ground to a halt, for Leo X did not bring to it the passionate interest of Julius II. The gigantic project involved enormous expenditures, and funds began to grow scarce. Disagreements also arose regarding the Greek-cross plan; it was probably as a result of pressures from the clergy that Raphael altered the plan to a basilican one. Leo X was more interested in the building of the Vatican palaces, and Raphael carried forward Bramante's work on this project as well. He finished the second level of the Logge, following and refining Bramante's model and designing the beautiful stucco and fresco decoration (1517–19).

Villa Madama, commissioned by Cardinal Giulio de' Medici on the slopes of Monte Mario, fits well with the Vatican Logge into Raphael's intellectual development and architectural evolution after 1516, which was stimulated by his increasingly profound studies of ancient monuments. The loggia motif, one of the essential elements in the design of Villa Madama, is developed in a new and more magnificent manner, structurally it is clearly inspired by the motifs of the Roman thermae. In Villa Madama Raphael moved away from Bramante's manner, which had been his point of departure, in a new direction that was entirely his own. This is attested by the attention directed to the interior and to achieving greater continuity of the perimeter; by a noticeable tendency to heighten the coloristic effects by using pictorial decoration of the same type as that of the Vatican Logge — but in this case the work was executed by Raphael's pupils (VIII, PL. 206); by the tendency to open the interior toward the exterior and to achieve unity with the surrounding landscape by means of large loggias; and, finally, by direct, or one might say, archaeological derivations from ancient art. Raphael was able to make full use of the natural setting. In order to ensure against the possibility of landslides, he provided the villa with a magnificent retaining structure in the form of large niches, which "make a beautiful ornamentation for the building." Serlio's testimony is very clear; the attribution to Raphael is repeated in every one of the sources, all of which praise the beauty and nobility of the edifice. Palladio, during one of his stays in Rome, made drawings of the niches in Villa Madama and, like Serlio, in his studies of ancient art made two exceptions in favor of modern architects: Bramante and Raphael. The original project envisaged the landscaping of large grounds around the villa in the form of an amphitheater, with terraces and gardens at various levels, towards the summit of the mountain and in a larger area toward the valley. It was a grandiose project, of which only a small part was ever actually carried out. The villa itself is also only a small portion of what was originally planned; moreover, it was built at various periods under different supervision and has been damaged, abandoned, and remodeled. Hence it falls far short of reproducing faithfully Raphael's original project; but even as it stands today Villa Madama has remained a model for architects. The culmination of a long tradition originating in the Tuscan villas of Michelozzo (e.g., Villa Medici in Fiesole) and of Giuliano da Maiano (at Poggioreale in Naples), Villa Madama marks the beginning of another series of villas tending towards the Venetian type. (To some avail did Palladio study Raphael so thoroughly.) Furthermore, the Mantuan architecture of Giulio Romano, which exerted so strong an influence on the development of northern Italian art, has its roots in the last stage of Raphael's architecture, particularly as is represented by Villa Madama and also, as far as can be deduced from extant graphic records [the well-known drawing by Parmigianino (q.v.) in the Uffizi and the engraving by Pietro Ferrerio] of the Palazzo Branconio dell'Aquila in Rome, which was demolished to make room for Bernini's colonnade in St. Peter's Square. Five great arches supported by six columns formed the ground floor. Above this ran a large cornice with multiple moldings which touched the keystones of the arches and strongly defined the division between the ground floor and the second story. This had the effect of making the ground floor into a base, albeit elastic, a substructure rather than a mass. On the second floor there was a series of large windows, each one flanked by a column and surmounted by a pediment, alternately triangular and curvilinear, and by niches. Between the second and third stories ran a mezzanine, and to judge especially from the Parmigianino drawing it must have appeared like a single large decorative frieze consisting of shields with heads in relief, a large emblem in the center, and stucco festoons. Vasari describes the Palazzo Branconio as being "decorated with stuccoes on the facade" and adds that it was regarded as "a singular edifice." At the top of the structure there was another simpler story surmounted by a balustrade, which eliminated its harsh straight lines and established an aerial relationship with the atmosphere. It was an architectural work rich in pictorial and mannerist details.

The attribution of the Palazzo Cafarelli-Vidoni to Raphael is not very persuasive. It repeats too literally the forms used by Bramante in the palace he built for the Caprini, which Raphael purchased from them in October 1517 (since then it has been known as the House of Raphael). A phrase in the contract, "in ea forma quae nunc est," led A. Venturi to suppose that at the time of purchase the latter palace was not yet finished and that Raphael finished it in the same form used in the Palazzo Cafarelli-Vidoni, which he had built previously. But the order of events and the derivation are actually the reverse. The first of these houses to be built was the one by Bramante, with a massive ground floor of rusticated construction that Vasari clearly alludes to as a new invention — that is, a rusticated wall which is not made with cut stone masonry but of lime and pozzuolana poured into wooden molds and then modeled into the form of rusticated ashlar. It is better to return to Vasari's attribution of the Palazzo Cafarelli to Lorenzetto. This places the authorship of building within the circle of Bramante and Raphael without ascribing to it any original inventive imprint. Raphael's later development in the field of architecture followed completely different formal lines, more open, elegant and airy, and also more intimately linked with his new and ever-increasing archaeological interests.

According to tradition Raphael prepared the design for the Palazzo Pandolfini in Florence. The frieze surmounting it bears the date MDXX and the name of Giannozzo Pandolfini, Bishop of Troia, for whom the palace was built by Antonio da Sangallo the Younger (here again, as in Rome, the names of Raphael and one of the Sangallos are found together). Certain characteristics peculiar to Raphael are found in this palace, such as the horizontal link between the cornices of the second-floor windows and the garden loggia, but on the whole the building bears the imprint of Sangallo.

In addition, Raphael took part in the competitions for the façade of S. Lorenzo in Florence and for the Church of S. Giovanni dei Fiorentini in Rome. He also shared the post of "Master of the Streets" (*magister stratarum* with Antonio da Sangallo the Younger. From a *motu proprio* of Leo X, which on good evidence can be dated approximately 1518, it appears that the two architects had prepared a design for a square on the location of the modern Piazza del Popolo, envisaging two entrances to it, from the Via Lata (the present-day Via del Corso) and from the Via Leonina (the modern Via di Ripetta). The Apostolic Camera had already adopted another project of the *magistri* Bartolomeo del Valle and Raimondo Capodiferro, but the Pope ordered the adoption of Raphael and Antonio's plan, placing the latter in charge of the actual construction. The designs reveal that Raphael was also interested in city planning, and even in this field his ideas were vital and progressive.

Apart from the works that Raphael actually executed, the letters of cardinals and Humanists and the reports of the envoys of princes tell of the abundance of commissions he was constantly being offered, from which he defended himself with promises and procrastination since he could not satisfy them all; this was true despite the fact that he could count on the assistance of a large number of pupils and collaborators, of whom the principal ones were Giulio Romano, Giovanni da Udine and Francesco Penni, and for architectural work he could avail himself of the Sangallos. The collateral activity of the studio of Marcantonio Raimondi should also be remembered. Raimondi maintained a working relationship with Raphael from the time of his arrival in Rome in 1509, and not only did he diffuse the paintings of Raphael by means of engravings but he also executed engravings for which Raphael had prepared the drawings.

One should also bear in mind the rapid spread of Raphael's inventions and the mannerist deviations that soon appeared even in the work of immediate followers. However, a discussion of this matter would involve a number of problems that lie beyond the limits of a monographic article.

It would be better to glance, though fleetingly, at the interpretations given to the art of Raphael in the course of the centuries, from his Roman years down to the present time, placing them in their proper historical perspectives in order to take the measure of the richness of his art and the vitality of its motifs. Few artists have enjoyed such early and enthusiastic recognition in their own lifetimes. It was perhaps even more enthusiastic and unanimous among the circle of literati and Humanists surrounding the papal court than it was among the artists. His rivalry with Michelangelo had the almost immediate effect of dividing artists into two opposing factions. The contraposition of Raphael and Michelangelo, beginning on a practical and day to day basis and easily discernible in their works, soon tended to crystallize into a series of critical formulations. Founded as these were on Michelangelesque precepts, they did not always take into account the themes and problems peculiar to Raphael. Apart from the contemporary literary texts — and a greater importance than has been hitherto accorded it must be attributed to Baldassare Castiglione's *Il Cortegiano* (1528), not only in those passages concerning Raphael directly but also in its entirety — the sources of the traditional critical studies of Raphael were written in the middle of or after the second half of the 16th century, when Michelangelo was still alive (but after Raphael's death). When Vasari and Dolce were writing, Raphael already belonged to the past; he formed part of an era before the overwhelming catastrophe of the Sack of Rome and before those other less spectacular but more fundamental disturbances, the Reformation and the Counter Reformation. Michelangelo, however, was still a leading actor in the events in which Vasari and Dolce, his contemporaries, were participants. By the force of events and circumstances, their points of view were conditioned by the living, active, and overwhelming presence of Michelangelo. When they discuss Raphael it is almost inevitable that their points of view should be based on Michelangelo's antithesis to him, even when (as in the case of Dolce), the critical judgment is favorable to Raphael. The *Vite* of Vasari, published (1550) before Dolce's *Dialogo della pittura* (1557), followed the line of Florentine art, which inevitably led to Michelangelo as apex and summing up rather than to Raphael, even though he too was imbued with Tuscan culture. Vasari's misunderstanding of Raphael is evident in spite of the acumen and perception of so many of his passages and critical judgments (e.g., on the lunette of the *Liberation of St. Peter*, on the portraits, and on other works). Dolce's and Aretino's debates induced Vasari to reexamine his critical opinions of Michelangelo and Raphael. The fruit of these meditations is found in the long passage that he added to his life of Raphael in 1568. Here Vasari recognizes the possibility that other artists besides Michelangelo might achieve perfection: "Raphael knew, nevertheless, that he could never equal Michelangelo [in the field of the nude] but, since he was a man of great intelligence, he knew that painting does not consist only in the delineation of the nude form; rather it embraces a much wider field, so that he who can express and compose his ideas clearly and easily may be numbered among the great painters." This long passage is well known: it constitutes a great improvement on the edition of 1550, where the discussion is still carried out from Michelangelo's point of view. Not only the criticism but the Italian *maniera* was at that time under the influence of Michelangelo rather than of Raphael. Certain Raphaelesque elements had wide circulation but were distorted in a mannerist spirit. Nor did the purist classicism of Dolce understand the vastly different stature and essence of Raphael's classicism; on the contrary, it carried the seed of the misunderstandings of the formalist and academic interpretations of the future.

The most authentic interpretation of Raphael's spirit seems then to have flourished in the Veneto, not in the writings of a critic nor in the works of a painter but in the works of Palladio (q.v.). For this reason too, Raphael's architectural activity is of capital importance in the field of Italian 16th-century art, even though it was limited to a few years and resulted in a small number of buildings that either remained unfinished or were soon destroyed.

Raphael's hour came with the reaction against mannerism, which began in Bologna but soon encompassed Rome and resulted in a long and continued dialectical controversy with the baroque throughout the 17th century. The Carraccis first concentrated mainly on the Venetians, but they soon turned to Raphael. They did not consider him from the purist point of view of Dolce, nor from a formalist one, but in the light of that classical ideal of adherence to nature and of the objectivation of universal values which Raphael, with his increasing progression of new horizons, had brought into being from the Humanistic foundation of the Italian Renaissance. From this moment on, Raphael's art became the basis of the classicist movement of the 17th century, including Poussin, although the latter expressed certain rigid and severe reservations regarding the art of Raphael in relation to the art of antiquity. Bellori, in his *Discorso sull'idea* (1664), which serves as an introduction to his *Lives* (1672), quotes Raphael directly: "Raphael of Urbino, the great master of thinking men, wrote to Castiglione about his *Galatea*: 'To paint a beautiful woman I would need to look upon many. But since there is a famine of beautiful women, I make use of a certain Idea that springs from my mind.'" The new, more profound, and less academic interpretation of the "classical ideal of the 17th century in Italy," accompanied by Longhi's studies of the Carraccis, has stimulated biennial exhibitions of ancient art in Bologna as well as new studies of the art of Raphael, especially, of the last phase, beginning 1513–14. This was necessary, since for the classicist artists, and for the critics and historians of the 17th and 18th centuries, the works of Raphael most generally studied and appreciated (Félibien, 1666–88; De Piles, 1699; De Brosses, 1739–40; Algarotti, 1762; Winckelmann, 1764; Reynolds, 1769; Mengs, 1780–83; Goethe, 1786–88) were those that belong to the period between the Stanze and the *Transfiguration*. In the 19th century, however, this point of view underwent a complete change.

The beginnings of modern scholarly studies of Raphael (Quatremère de Quincy, 1824; Pungileoni, 1829; Von Rumhor, 1831; Passavant, 1839 ff) coincided with the period when the art and ideals of Raphael were farthest removed from the intellectual and spiritual directions of the times. The exaltation of subjectivity in art, which was one of the essentials of the romantic movement, and the rediscovery of the primitives could only act as obstacles to a true understanding of Raphael. Nor could the Pre-Raphaelitism of a large group of the romantics be countered by the Raphaelism in Ingres, which was in itself so out of tune with the true substance of Raphael's art. (It was, however, Delacroix who wrote in several passages of the *Journal* the most telling arguments in rebuttal of Poussin's limitative criticism of Raphael vis-a-vis the ancients.) Passavant had links with the Nazarenes, and they are linked with the Purists, who, over the signature of Antonio Bianchini, proclaimed in their manifesto of 1849 that "from the *Disputa* onward, all the works of Raphael" were to be deprecated. Moreover, the very name Pre-Raphaelite is in itself a program. While, on the one hand, scholarly research contributed decisively to the documentation and attribution of Raphael's life and works (the fundamental monograph of Crowe and Cavalcaselle, 1882–85, should be remembered), on the other, critical opinion of his art declined because of a lack of a spiritually motivated understanding and interest in the creative development of the artist. The effects of the romantic and Pre-Raphaelite climate that was current at the time of the revival of studies of Raphael have made themselves felt until very recent times; they are evidenced in the greater interest and appreciation expressed by modern writers for the works of Raphael's youth, and the summary condemnation of the later works (after the second Stanza) as decadent and as the fruit of collaboration and "studio" production.

To what extent ideological premises can help or hinder even the most intelligent and sincere critic is clearly demonstrated by Ortolani's *Raffaello* (1942), the latest book on Raphael to appear in Italy. This work is well thought out on a critical level and clarified through a continuous direct analysis and, one might even say, almost through a direct dialogue and debate with the work of the artist. The passages on the *Sistine Madonna* or the *Transfiguration* are striking for their profound understanding of the artistic creative process, of the results, and of the values that they express. Nevertheless, the author hesitates to arrive at a fully positive judgment. He is prevented from doing so by certain romantic reservations, reinforced by the distinctions between poetry and literature, between lyricism and rhetoric, derived from Croce (who also extols the pure lyricism of Raphael's early works, to the detriment of the later works, because of his romantic idealization of the pure lyrical effusion, of the spontaneity of genius and of poetic inspiration). He continues to accept the *Vision of a Knight* and the *Three Graces* as early works (these are youthful works, but they are already the expressions of a profound intellectual meditation) but hesitates to draw conclusions from his own sharp intuition when he passes from the serene harmony of the first Stanza to the increasingly dramatic, monumental concepts of the subsequent works.

On the other hand, C. Gnudi, in his introduction to the catalogue of the Bologna exhibition of 1962, expresses a completely different opinion with determination and clarity although in abbreviated form and barely outlined.

Moreover, the employment of a less dogmatic and polemic approach than in the past and at the same time a fuller and more comprehensive kind of historical research in conjunction with the most current orientations of art criticism — which is abandoning the formal approach of pure visibility and turning with increasing interest to the investigation of the process from which a work of art springs, in all its constituent elements, media, and significance — has resulted in a revival of the study of Raphael. In this sense, few subjects can offer so many cues to interesting investigation.

SOURCES AND GENERAL ART HISTORY: P. Giovio, Raphaelis Urbinatis Vita, in D. Tiraboschi, Storia della letteratura italiana, IX, Modena, 1781, pp. 290–93; Vasari, IV, pp. 315–46; L. Dolce, Dialogo della pittura, Venice, 1557 (ed. P. Barocchi, in Trattati d'arte del Cinquecento, I, Bari, 1960, pp. 141–206; Eng. trans., London, 1770); G. P. Lomazzo, Trattato dell'arte della pittura, Milan, 1584 (Eng. trans., R. Haydock, Oxford, 1598); G. P. Lomazzo, Idea del tempio della pittura, Milan, 1590; V. Golzio, ed., Raffaello nei documenti, nelle testimonianze dei contemporanei e nella letteratura del suo secolo, Vatican City, 1936 (16th-century documents, biographical accounts, and comments); F. Zuccaro, Discorso all'Accademia di S. Luca, March 20, 1594; N. Poussin, Correspondence (ed. C. Jouanny), Paris, 1911; G. P. Bellori, Le vite de' pittori, scultori, et architetti moderni, Rome, 1672; G. P. Bellori, Descrizione delle immagini dipinte da Raffaello da Urbino nelle camere del Palazzo Apostolico Vaticano, Rome, 1695; A. Félibien des Avaux, Entretiens . . . sur des plus excellens peintres anciens et modernes, 5 vols., Paris, 1666–88; R. de Piles, Abrégé de la vie des peintres, Paris, 1699 (Eng. trans., London, 1706); C. de Brosses, Lettres familières écrites d'Italie en 1739 et 1740 (ed. R. Colomb), Paris, 1836; J. J. Winckelmann, Geschichte der Kunst des Altertums, Dresden, 1764 (Eng. trans., G. H. Lodge, 2 vols., Boston, 1872–73); F. Algarotti, Saggio sopra la pittura, Bologna, 1762 (Eng. trans., London, 1764); J. Reynolds, Discourses (1769–91), ed. Royal Academy, London, 1924; A. R. Mengs, Opere, 2 vols., Parma, Bassano, 1780–83 (Eng. trans., 2 vols., London, 1796); F. Milizia, Dizionario delle belle arti del disegno, 2 vols., Bassano, 1787; L. Lanzi, Storia pittorica dell'Italia, Bassano, 1789; W. Goethe, Italienische Reise, 1786–88 (Eng. trans., W. H. Auden and E. Mayer, London, 1962); J. P. Eckermann, Gespräche mit Goethe, 3 vols., Leipzig, 1836–48 (Eng. trans., R. O. Moon, London, 1951).

BIBLIOG. *a. General*: A. C. Quatremère de Quincy, Histoire de la vie et des ouvrages de Raphaël, Paris, 1824 (Eng. trans., W. Hazlitt, London, 1849); R. Pungileoni, Elogio storico di Raffaello Santi da Urbino, Urbino, 1829; K. F. von Rumohr, Italienische Forschungen, III, Berlin, 1831; J. D. Passavant, Raffael von Urbino und sein Vater Giovanni Santi, 3 vols., Leipzig, 1839–58; J. Burckhardt, Der Cicerone, Basel, 1853 (Eng. trans., Mrs. A. H. Clough, London, 1908); H. Grimm, Das Leben Raffaels, Berlin, 1872 (2d ed., 1886; Eng. trans., S. H. Adams, London, Boston, 1889); J. Lermolieff [G. Morelli], Die Werke italienischer Meister in den Galerien von München, Dresden und Berlin, Leipzig, 1880 (Eng. trans., L. M. Richter, London, 1883); J. A. Crowe and G. B. Cavalcaselle, Raphael: Life and Works, 2 vols., London, 1882–85; E. Müntz, Raphaël: Sa vie, son œuvre et son temps, Paris, 1886 (2d ed., 1900); R. Vischer, Studien für Kunstgeschichte, Stuttgart, 1886; B. Berenson, Central Italian Painters, New York, London, 1897; B. Berenson, The Study and Criticism of Italian Art, I–II, London, 1901–02; A. Rosenberg and G. Gronau, Raffael, Stuttgart, Leipzig, 1919; A. Venturi, Raffaello, Urbino, 1920; V. Wanscher, Raffaello Santi da Urbino: His Life and Works, London, 1926; Venturi, IX, 2, pp. 1–336; B. Berenson, The Italian Painters of the Renaissance, Oxford, 1930; B. Berenson, Italian Pictures of the Renaissance, Oxford, 1932; C. Gamba, Raphaël, Paris, 1932; O. Fischel, ThB, s.v. (bibliog. prior to 1935); S. Ortolani, Raffaello, Bergamo, 1942 (3d ed., 1948); W. E. Suida, ed., Raphael, Oxford, 1942 (2d ed., London, 1948); O Fishel, Raphael, 2 vols., London, 1948; A. Venturi, Raffaello (rev. ed. by L. Venturi), Milan, 1952; E. Camesasca, Tutta la pittura di Raffaello, 2 vols., Milan, 1956 (Eng. trans., L. Grosso, 2 vols., New York, 1963); W. Schöne, Raphael, Berlin, Darmstadt, 1958; C. G. Stridbeck, Raphael Studies, I, Stockholm, 1960.

b. Monographs and studies of individual works: W. Suida, Beiträge zu Raphael, Belvedere, XII, 1934–36, pp. 161–67; K. Lanckorońska, Zu Raffaels Loggien, JhbKhSammlWien, N.S., IX, 1935, pp. 111–20; P. Cellini, Il S. Luca di Raffaello, BArte, XXX, 1936, pp. 282–88; G. Glück, Ein wenig beachtetes Werk Raphaels, JhbKhSammlWien, N.S., X, 1936, pp. 97–104; O. Fischel, Le gerarchie degli angeli di Raffaello nelle Logge del Vaticano, Ill. vaticana, VIII, 1937, pp. 161–64; O. Fischel, Raphael's Auxiliary Cartoons, BM, LXXI, 1937, pp. 167–68; G. Gronau, Some Portraits by Titian and Raphael, Art in Am., XXV, 1937, pp. 93–104; E. Wind, Platonic Justice Designed by Raphael, Warburg, I, 1937–38, pp. 69–70; O. Fischel, Ritratti poco conosciuti di Leone X, Ill. vaticana, IX, 1938, pp. 361–63; D. Redig de Campos, Il concetto neo-platonico cristiano nella Stanza della Segnatura, Ill. vaticana, IX, 1938, pp. 101–05; P. Sanpaolesi, Due esami radiografici di dipinti, BArte, XXXI, 1938, pp. 495–505; H. Weitzsäcker, Raffaels Galatea in Lichte der antiken Überlieferung, Die Antike, XIV, 1938, pp. 231–42; E. Wind, The Four Elements in Raphael's Stanza della Segnatura, Warburg, II, 1938–39, pp. 75–79; O. Fischel, A Motive of Bellinesque Derivation in Raphael, Old Master Drawings, XIII, 1939, pp. 50–51; O. Fischel, Raphael's Pink Sketchbook, BM, LXXIV, 1939, pp. 181–87; H. B. Gutman, The Medieval Content of Raphael's "School of Athens," J. H. of Ideas, II, 1941, pp. 420–29; P. Rotondi, Il Palazzo Ducale d'Urbino nella formazione di Raffaello, S. urbinati, ser. B, XV, 1941, pp. 420–29; C. Cecchelli, Di un'ignorata fonte letteraria della Galatea di Raffaello, Roma, XX, 1942, p. 426; E. Lavagnino, Influssi della pittura antica nella evoluzione dell'arte di Raffaello, Atti V Cong. naz. di s. romani, III, Rome, 1942, pp. 365–68; F. Hartt, Raphael and Giulio Romano, AB, XXVI, 1944, pp. 67–94; P. Oppé, Right and Left in Raphael's Cartoons, Warburg, VII, 1944, pp. 83–94; W. Suida, Marcantonio Raimondi: His Portrait Painted by Raphael: His connection with Venetian Painters, AQ, VII, 1944, pp. 239–48; G. J. Hoogewerff, Documenti in parte inediti che riguardano Raffaello ed altri artisti contemporanei, AttiPontAcc, XXI, 1945–46, pp. 253–68; G. J. Hoogewerff, Raffaello nella Villa Farnesina, Capitolium, XX, 1945, pp. 9–15; J. A. Richter, The Drawings for the Entombment, GBA, XXVIII, 1945, pp. 335–36; C. L. Sommer, A New Interpretation of Raphael's Disputa, GBA, XXVIII, 1945, pp. 289–96; D. Redig de Campos, Raffaello e Michelangelo, Rome, 1946; J. Hess, On Raphael and Giulio Romano, GBA, XXXII, 1947, pp. 73–106; G. J. Hoogewerff, La Stanza della Segnatura, RendPont Acc, XXIII, 1947–49, pp. 317–56; C. L. Ragghianti, La Deposizione di Raffaello Sanzio, Milan, 1947; C. L. Racchianti, Il percorso della Deposizione di Raffaello, Belfagor, III, 1948, pp. 159–72; F. Zeri, Raffaello Arcangelo e Raffaello Sanzio, Proporzioni, II, 1948, pp. 178–80; H. Friedmann, The Plant Symbolism of Raphael's Alba Madonna, GBA, XXXVI, 1949, pp. 213–30; F. Hartt, Lignum Vitae in medio Paradisi: The Stanza di Eliodoro and the Sixtine Ceiling, AB, XXXII, 1950, pp. 115–45; J. Pope-Hennessy, ed., The Raphael Cartoons, Vict. and Alb. Mus., London, 1950 (2d ed., 1958); D. Redig de Campos, Le stanze di Raffaello, Florence, 1950 (2d ed., Rome, 1957); W. Schöne, Raffaels Krönung des Heiligen Nikolaus von Tolentino, Eine Gabe der Freunde für K. C. Heise, Berlin, 1950, pp. 113–36; O. Bock von Wülfingen, Ein unbekanntes Bild aus der Jugendzeit Raffaels, MJhb, 3d ser., II, 1951, pp. 105–17; G. J. Hoogewerff, Leonardo e Raffaello, Comm, III, 1952, pp. 173–83; S. Liberti, Relazione sulle analisi dei colori degli stendardi di Raffaello, B. Ist. Centrale del Restauro, 9–10, 1952, p. 197; D. Redig de Campos, Dei ritratti di Antonio Tebaldeo e d'altri nel Parnasso di Raffaello, Arch. Soc. Rom. di Storia Patria, LXXV, 1952, pp. 51–58; G. Urbani, Raffaello: SS. Trinità con i SS. Rocco e Sebastiano, Creazione di Eva, Stendardo processionale, Città di Castello, Pinacoteca Comunale, B. Ist. Centrale del Restauro, 9–10, 1952, pp. 94–95; E. Tietze-Conrat, A Sheet of Raphael Drawings for the Judgment of Paris, AB, XXXV, 1953, pp. 300–02; V. Zubov, Zur Komposition von Raphael Sposalizio, ZfKg, XVI, 1953, pp. 145–53; G. Fiocco, Fra' Bartolomeo e Raffaello, RArte, XXIX, 1954, pp. 42–53; P. D'Ancona, Gli affreschi della Farnesina, Milan, 1955; R. Longhi, Percorso di Raffaello giovane, Paragone, VI, 65, 1955, pp. 8–23; M. Putscher, Raphael's Sixtinische Madonna: Das Werk und seine Wirkung, Tübingen, 1955 (full bibliog.); W. Suida, Raphael's Painting the Resurrection of Christ, AQ, XVIII, 1955, pp. 3–10; E. H. Gombrich, Raphael's Madonna della sedia, London, 1956; C. Volpe, Due questioni raffaellesche, Paragone, VII, 75, 1956, pp. 3–18; M. Alpatov, La Madonna di S. Sixto, L'Arte, LVI, 1957, pp. 25–72; H. Biermann, Die Stanzen Raphaels, Munich, 1957; E. Nasalli-Rocca, Nota sulle vicende della Madonna di S. Sisto, L'Arte, LVI, 1957, pp. 72–76; A. Blunt, The Legend of Raphael in Italy and France, It. S., XIII, 1958, pp. 1–20; P. Cellini, Il restauro del S. Luca di Raffaello, BArte, XLIII, 1958, pp. 250–62; H. B. Gutman, Zur Ikonologie der Fresken Raffaels in der Stanza della Segnatura, ZfKg, XXI, 1958, pp. 27–39; F. Hartt, Giulio Romano, 2 vols., New Haven, 1958 (bibliog.); M. Rivosecchi, Variazioni sul tema della Galatea, Capitolium, XXXIII, 10, 1958, pp. 14–15; J. White and J. Shearman, Raphael's Tapestries and Their Cartoons, AB, XLI, 1958, pp. 193–221, 229–323; K. Badt, Raphael's Incendio del Borgo, Warburg, XXII, 1959, pp. 35–59; A. Chastel, Art et humanisme à Florence, Paris, 1959; E. Larsen, A New Interpretation of a Passage from Vasari's Life of Raphael, L'Arte, LVIII, 1959, pp. 303–08; F. de' Maffei, Il ritratto di Giuliano fratello di Leone X, dipinto da Raffaello, L'Arte, LVIII, 1959, pp. 309–36; P. A. Riedl, Raphaels Madonna del Baldacchino, Mitt. des Kunsthist. Inst. in Florenz, VIII, 1959, pp. 233–46; S. Sulzberger, Dürer a-t'il vu à Bruxelles les cartons de Raphaël?, GBA, LIV, 1959, pp. 177–84; K. Dettlinger, Zu Raphaels Kardinaltugend im Vatikan, Mouseion: Studien für O. H. Förster, Cologne, 1960, pp. 101–02; G. E. Mâle, La transfiguration de Raphaël, RendPontAcc, XXXIII, 1960–61, pp. 225–36; A. Bertini, La Transfigurazione e l'ultima evoluzione della pittura di Raffaello, CrArte, VIII, 1961, pp. 1–19; C. Brandi, Raffaello e Bramante, L'approdo letterario, VII, 1961, pp. 129–32; L. D. Ettinger, A Note on Raphael's Sibyls in S. Maria della Pace, Warburg, XXIV, 1961, pp. 322–23; M. Hirst, The Chigi Chapel in S. Maria della Pace, Warburg, XXIV, 1961, pp. 161–85; D. Redig de Campos, La Madonna di Foligno di Raffaello, Misc. Bib. Hertziana zu Ehren von L. Bruhns, Munich, 1961, pp. 184–97; C. Gnudi, ed., L'ideale classico del Seicento in Italia, Bologna, 1962 (cat.), pp. 21–24; M. Imdahl, Raphaels Castiglione Bildniss im Louvre, Pantheon, XX, 1962, pp. 38–45; M. Levey, Raphael Revisited, Apollo, LXXVII, 1962, pp. 678–83; J. B. Lynch, The History of Raphael's Saint George in the Louvre, GBA, LIX, 1962, pp. 203–12; K. Oberhuber, Vorzeichnungen zu Raphaels Transfiguration, Jhb. Berliner Mus., N.S., IV, 1962, pp. 116–49; C. de Tolnay, Poussin, Michel-Ange et Raphaël, Art de France, II, 1962, pp. 260–62; R. Wittkower, Young Raphael, Oberlin College B., XX, 1963, pp. 150–68.

c. Drawings: 1. *Basic works*: O. Fischel, Raphaels Zeichnungen, Strasbourg, 1896 (2d ed., 1898; prior bibliog.); O. Fischel, Raphaels Zeichnungen, 8 vols., Berlin, 1913–41 (to 1512 or shortly after). 2. *Principal catalogues after 1898 of important collections*: S. Colvin, Drawings of the Old Masters . . . , 3 vols., Oxford, 1903–07; A. Stix and L. Fröhlich-Bum, Die Zeichnungen der tuskanische, umbrische und römische Schulen, Vienna, 1932; G. Rouchès, Musée du Louvre: Les dessins de Raphaël, Paris, 1938; A. E. Popham and J. Wilde, The Italian Drawings of the 15th and 16th Centuries at Windsor Castle, London, 1949; K. T. Parker, Catalogue of the Collection of Drawings in the Ashmolean Museum, II, Oxford, 1956; Collection J. B. Wicar du Musée de Lille, Les dessins de Raphael, Lille, 1961; P. Pouncey, Italian Drawings in the British Museum: Raphael and His Circle, 2 vols., London. 1962 (cat.). 3. *Collections of selected drawings*: U. Middeldorf, Raphael's Drawings, New York, 1945; A. E. Popham, Raphael and Michelangelo: Selected Drawings from . . . Windsor Castle, London, 1954; G. Castelfranco, Raffaello, Milan, 1962.

d. Archaeology: A. Fulvio, Antiquitates Urbis, Rome, 1527; M. F. Calvo, Antiquae urbis Romae cum regionibus simulacrum, Rome, 1532; J. B. Marliano, Antiquae Romae Topographia libri V, Rome, 1534; C. Calcagnini, Opera aliquot, Basel, 1544, pp. 100–01; D. Francesconi, Congettura che una lettera creduta di Baldassare Castiglione sia di Raffaello d'Urbino, Florence, 1799; P. E. Visconti, Lettera di Raffaello a Papa Leone X, Rome, 1840; G. Gruyer, Raphaël et l'antiquité, 2 vols., Paris, 1864; K. von Pulszky, Beiträge zu Raphael im Studium der Antike, Leipzig, 1877; E. Müntz, Raphaël archéologue et historien d'art, GBA, XXII, 1880, pp. 307–18; H. Thode, Die Antiken in den Stichen Marcanton's, Leipzig, 1881; R. Lanciani, La pianta di Roma antica e i disegni archeologici di Raffaello Sanzio, RendLinc, Classe di sc. morali . . . , III, 1894, pp. 791–804; E. Loewy, Di alcune composizioni di Raffaello ispirate ai monumenti antichi, Arch. storico dell'art, 2d ser., II, 1896, pp. 241–51; R. Lanciani, Storia degli scavi di Roma, I, Rome, 1902; J. Vogel, Bramante und Raphael, Leipzig, 1910; A. Bartoli, I monumenti antichi di Roma nei disegni agli Uffizi di Firenze, 5 vols., Florence, 1914–22; A. Venturi, La lettera di Raffaello a Leone X sulla pianta di Roma antica, L'Arte, XXI, 1918, pp. 57–65; A. Lazzari, Un enciclopedico del secolo XVI: Celio Calcagnini, Atti e mem. Deputazione ferrarese di storia patria, XXX, 1936, pp. 83–164; E. Winternitz, Archeologia musicale del Rinascimento nel Parnasso di Raffaello, RendPontAcc, XXVII, 1952–54, pp. 359–88; V. Mariani, Raffaello e il mondo classico, S. romani, VII, 1959, pp. 162–72; R. Weiss, Andrea Fulvio antiquario romano, Ann. Scuola Normale Superiore di Pisa, XXVIII, 1959, pp. 1–44; R. Weiss, Il primo Rinascimento e gli studi archeologici, Lettere it., XI, 1959, pp. 89–94; C. Pedretti, A Chronology of Leonardo Da Vinci's Architectural Studies after 1500, Appendix I: A Letter to Pope Leo X on the Architecture of Ancient Rome, Geneva, 1962.

e. Architecture and scene-painting: F. Biondo, De Roma instaurata, Rome, 1471 (It. trans., L. Fanno, Venice, 1558); Il terzo libro di Sebastiano Serlio Bolognese, Venice, 1540; F. Milizia, Memorie degli architetti antichi e moderni, 3d ed., I, Parma, 1781, pp. 202–04; C. Percier and P.-F.-L. Fontaine, Choix des plus célèbres maisons de plaisance de Rome . . . , Paris, 1809; M. Missirini, Dell'eccellenza di Raffaello Sanzio nell'architettura . . . , Bib. it., XCVIII, 1840; C. Pontani, Opere architettoniche di Raffaello Sanzio, Florence, 1845; G. Campori, Documents inédits sur Raphaël, GBA, XIV, 1863, pp. 442–56; H. von Geymüller, Trois dessins d'architecture inédits de Raphaël, GBA, XXVIII, 1870, pp. 79–91; A. Bertolotti, La casa di Michelangelo e quella di Raffaello, Arch. storico, artistico, archeol. e lett. della città e provincia di Roma, I, 1875, pp. 163–64; E. Müntz, Les maisons de Raphaël à Rome, GBA, XXI, 1880, pp. 353–59; H. von Geymüller, Raffaello Sanzio studiato come architetto, Milan, 1884; D. Gnoli, La casa di Raffaello, Nuova ant., 3d ser., IX, 1887, pp. 401–23; H. Stegmann and H. von Geymüller, Die Architektur der Renaissance in der Toskana, VII, Munich, 1887–1908 (sup., H. von Geymüller, Raphael von Urbino: Der Palazzo Pandolfini in Florenz und Raphael Stellung zur Hochrenaissance in Toskana, Munich, 1908); D. Gnoli, Documenti relativi a Raffaello d'Urbino, Arch. storico dell'arte, II, 1889, pp. 248–51; M. Ermers, Die Architekturen Raphaels in seinen Fresken, Tafelbildern und Teppichen, Strasbourg, 1909; M. Rosenthal, Die Architekturen in Raphaels Gemälden, Strasbourg, 1909; J. Vogel, Bramante und Raphael, Leipzig, 1910; A. Mercati, Raffaello d'Urbino e Antonio da Sangallo maestri delle Strade di Roma sotto Leone X, RendPontAcc, I, 1923, pp. 121–27; F. Hermanin, La Farnesina, Bergamo, 1927; F. Ehrle, Dalle carte e dai disegni di Virgilio Spada, MPontAcc, II, 1928, pp. 1–98; D. Gnoli, La Roma di Leone X, Milan, 1938; Venturi, XI, 1, pp. 177–266; M. Bafile, Il giardino di Villa Madama, Rome, 1942; H. Leclerc, Les origines italiennes de l'architecture théâtrale moderne, Paris, 1946; G. Zorzi, Due schizzi archeologici di Raffaello fra i disegni palladiani di Londra, Palladio, N.S., 1952, pp. 171–75; O. H. Förster, Bramante, Vienna, Munich, 1956 (bibliog.); G. Giovannoni, Antonio da Sangallo il Giovane, 2 vols., Rome, 1959; C. L. Frommel, Die Farnesina und Peruzzis architektonisches Frühwerk, Berlin, 1961 (bibliog.).

Anna Maria Brizio

Illustrations: PLS. 421–436; 1 fig. in text.

REALISM. In criticism and art history, the term "realism" is used in three different ways. It can have a generic sense (to designate the artist's specific attitude to reality), a historical sense, and, finally, an esthetic-philosophic sense. From the historical point of view, realism generally indicates the philosophy of art dominant during the second half of the 19th century in Western art and certain artistic movements of the 20th century, which, with various inflections and nuances, and frequently with altogether different intentions, took up some of the elements of the 19th-century movement. This article considers realism from the historical point of view, with particular emphasis on 19th-century realism.

Use of the term and its history. The name "realism" was adopted by an artistic movement that, having appeared in France with the revolution of 1848, had its first major representative in the painter Gustave Courbet (q.v.), its first theorist in Champfleury (J. F. Husson), and its first group manifestation at the Exposition Universelle in Paris (1855). Having arisen in opposition to the idealism of the classicists and romantics, the proponents of realism saw objective reality as a representational theme valid in itself, without recourse to any embellishment, correction, or preconceived choice; they maintained, with great polemic vigor, the necessity of treating themes of contemporary life, introducing the most humble social classes as protagonists of the work of art. Notwithstanding its clear position concerning romanticism, realism did not ignore the importance of subjective feeling, which was esteemed as artistic sincerity. Taken as a whole, realism must be considered not so much as a reaction to romanticism but as a continuation and development of trends that were more valid than romanticism and that were strongly rooted in the culture of the Age of Enlightenment. The realism of Courbet and Champfleury was therefore only one of the most obvious aspects of a vast process that — not only in art, but in philosophy, political ideals, and customs — emerged in the second half of the 19th century and involved artists of typically romantic formation such as Corot, Daumier, Millet, and Daubigny (qq.v.), leaving its mark, with various inflections and nuances, on the entire period, from approximately 1850 to 1890, that is usually called the "period of realism." Favored by the naturalism widely propagated by romanticism and even more by the democratic passion that had been reawakened in many of the countries of Europe by the Revolution of 1848, realism had a rapid and wide diffusion. At the same time, in France, the death of many social illusions and the establishment of positivism gave rise to a more detached and pessimistic view of reality and to a new philosophy of art, which Castagnary was the first to call "naturalism." Naturalism, a typically literary phenomenon that prevailed in France during the Third Republic, thanks principally to Emile Zola, combined forces with realism (as, for instance, in Italian *verismo*) and had important repercussions in the history of art, chiefly because it prepared the way for impressionism and expressionism.

If it is used in its precise 19th-century application, the definition of realism has a particular historical and descriptive value and also helps the understanding of, by comparison or contrast, certain periods in the history of art, which, under social or cultural conditions that may be profoundly different, nevertheless present an attitude toward reality in some way similar to that of the 19th century. Less valid, however, are the attempts that have been made to attribute to realism (as to other historical definitions, such as Gothic, baroque, romantic, etc.) the value of a recurring and broadly constant attitude in human history, or to institute, regardless of all sane historical principles, a metachronistic category of realism.

Historical precedents for such an attitude can be seen in 19th-century theories of realism, such as that of Champfleury, who was the first to extend the term to include the works of the Le Nain brothers (in an essay of 1850, which, in 1862, was expanded into a book entitled *Les peintres de la réalité sous Louis XIII: Les frères Le Nain*). The writings of T. Thoré (W. Bürger) on Dutch and Flemish painters, Champfleury's *Histoire de la caricature* (1865–80), H. Taine's *Philosophie de l'art dans les Pays-Bas* (1869), and E. Fromentin's exemplary *Maîtres d'autrefois* (1876), helped to render this extension of the term increasingly justified and articulate. The progress of this use of the term is clearly evident in the monumental *Histoire des peintres de toutes les écoles* (1847–75), edited by Charles Blanc in collaboration with various authors. From the French painters "of reality," the term began to be applied to Dutch, Spanish, and Venetian artists "of reality" and gradually to the Flemish and Florentine 15th century, to certain aspects of the German 16th century, and to other, very different aspects of the French, English, and Venetian 18th century. With regard to the 18th century, a significant work was *L'Art au dix-huitième siècle*, by the Goncourt brothers, which appeared between 1859 and 1875, in which the entire century was subjected to very close analysis.

In all this literature 19th-century bourgeois culture sought historical precedents in the realm of art for its own mode of thinking and feeling. It is understandable that, while in the field of literary criticism such precedents were found in the *Satyricon* and the *Golden Ass*, in the medieval romances and in the *Decameron*, in the *Divine Comedy* and in Shakespeare's plays, the presence, to a greater or lesser degree, of a realistic spirit should also have become a criterion of judgment in art criticism. This tendency (which may certainly seem today to be debatable on the esthetic level) was extremely important, since it permitted the rediscovery of artists, some of them very great, who had fallen into oblivion because of the prejudices of classicism. Moreover, many acute observations made in those years, particularly those of a historic-sociological nature (e.g., on the association between certain phases of the development of the modern bourgeoisie and the growth of realism in art; the indubitably stimulating observations of Taine on the typical character of artistic creations, later explored by Engels, and extensively used by Marxist esthetics), have been felicitously used by 20th-century art historians.

Long before its introduction into art criticism, the term realism had a precise significance in philosophy. Here, beginning with the dualistic basis that Descartes gave to the problem of reality, it indicated an attitude of mind that asserted, in opposition to idealism the reality of the external world, this being a phenomenon that, in the words of Kant, "need not be deduced, but is immediately perceived." The habit of presenting the gnosiological problem as an alternative between idealism and realism and the frequent use of the term realism itself go back to Kant and Hegel; and it is not surprising that this expression was introduced into artistic literature by such a critic as Gustave Planche, who was formed in the school of Victor Cousin and therefore accustomed to Hegelian terminology. In his forceful polemic against the Academy, eclecticism, and the capricious eccentricities of the romantics, Planche, from 1833 on, used "realism" to indicate an art which was not the fruit of the imagination or of the "Idea" but was born of the careful observation of reality — of "a hand-to-hand combat with nature and with truth." Planche was virtually the only user of the term realism until about 1838; but given his position as a moderate critic, as much an admirer of Ingres and Corot as a resolute adversary of the romantics, the real importance of his esthetic ideas, which he liked to sum up in the formula "inventer dans le cercle de la nature et de la tradition," should not be misunderstood. Notwithstanding his modern tone, realism was, for Planche, almost synonymous with naturalism, in the traditional and Carracciesque sense, which is to be found as early as the 17th century. in the Academy lectures, to indicate a studied imitation of truth sustained by a high plastic ideal. In this sense Charles Baudelaire himself could, in complete agreement with his contemporaries, consider Ingres "the most illustrious representative of the naturalistic school of drawing" and accompany such an assertion with the observation that "the purest draughtsmen are always naturalists."

In those same years, however, the term realism was being used in artistic circles to indicate the recently diffused tendency to reproduce nature with a minute and almost photographic fidelity. In the spirit of an ingenuous scientism (or rather, a virtually superstitious faith in the positive sciences) that was a typical trait of the culture of the time, the objectivity of photographic reproduction (the first diffusion of which took place in the decade between 1830 and 1840) met the artist's requirements for truth as a model of total realism, fostering the taste for trompe l'œil (see PHOTOGRAPHY). That this sense of the word realism was rapidly and widely diffused is demonstrated by the fact that Planche soon refuted the term that he had coined, and realism for him became, many years before Courbet and Champfleury invested it with the meaning generally understood today, a true polemic idol. Furthermore (and this is worth noting) Théophile Thoré himself, quite rightly considered the dean of realist criticism, for many years used the word realism in a deprecatory sense, giving it precisely the significance that Planche had rejected. Believing that a purely objective vision does not exist, there being always present in every sensation "a greater or lesser part of the individual who perceives it," Thoré strongly opposed the mechanical and coldly analytical imitation of nature; he considered the belief in the existence of a "reality that is manifest to all in the same form" as a capital error for a painter. Only in 1855 did Thoré confer a positive meaning on the word, that of total mastery of reality — that is to say, the meaning that the term had meanwhile acquired in the controversy aroused by the presentation to the public of the first works of Courbet and the writings of Max Buchon and Champfleury. This meaning of the term realism was established by Courbet's pamphlet on the occasion of the Exposition Universelle of 1855 and by Champfleury's book entitled *Le Réalisme* (1857). Here reality was no longer fixed once and for all by the daguerreotype or by the analyses of the natural sciences. Neither was reality the common and the plebeian aspects of life that were sought out by the artist for their very vulgarity (another sense of the word, which appeared as early as 1844 in Théophile Gautier, Arsène Goussaye, and others, and which derived from the romantic acceptance of ugliness as something characteristic). Reality was now the world in all its natural and social complexity, a dynamic synthesis of nature and history, by means of which the artist is formed and, becoming a living expression of his own time, displays his creative activity. In this significance the term was extended for the first time to art and artists of the past (in Champfleury's *Les peintres de la réalité sous Louis XIII*).

Meanwhile, even the expression "naturalism" lost the usual meaning it had in artistic and academic writings and became synonymous with realism, assimilating both the philosophical sense of the word and the sense derived from the natural sciences. Philosophically, in the 18th century, naturalism was equivalent to Spinozan materialism ("the opinion," as Diderot wrote, "of those who attribute everything to nature as first principle," "the doctrine of atheists and epicurists, who do not admit the existence of God"). Naturalism kept this meaning in the 19th century. In 1839 Sainte-Beuve considered materialism, pantheism, and naturalism perfectly synonymous; the *Revue des deux mondes*, in 1852, criticized the poetry of Gautier and his school as being "toute naturaliste" and "professing the cult of created things, without ever going back from these to God." At the same time, the noun "naturalist" became more and more frequently used to indicate the scholar concerned with the natural sciences. Balzac, in presenting *La comédie humaine* to the public (1842) declared that he had studied the social species as if he were studying zoology; even Victor Hugo, in the preface to *La Légende des Siécles* (1859) asserted his wish to "experiment with social phenomena in the same way that the naturalist experiments with zoological ones." Accepted in its former philosophical sense, but pregnant with its more current scientific connotations and the effort made by modern painting to represent reality with frankness, the word naturalism, introduced into art criticism by Castagnary (1862), slowly came to replace the term realism. For the term naturalism seemed more in keeping with that ideal of a regeneration of art through science and democracy that was peculiar to the radical bourgeoisie of the Third Republic. "The naturalists," wrote Sainte-Beuve in 1863, "tend to introduce and make prevalent in everything the procedures and results of science; being themselves emancipated, they attempt to emancipate humanity from its illusions, vague disputes, vain solutions, idols, and delusive powers." With the work of Emile Zola, beginning, roughly with the preface to *Thérèse Raquin* (1867), the expression naturalism triumphed rapidly over the term realism, firmly established itself about 1880, and was used with increasing frequency to indicate the entire complex of esthetic, political, and social ideals of the age of Positivism. In the subsequent period of reaction against Positivism, the word naturalism continued to be used with this precise and well-defined sense, but it also reacquired — in literary and artistic criticism of more or less openly philosophical tendencies — its former and more genuine sense of Renaissance hylozoism or pantheistic naturalism. In this sense it was frequently applied to certain aspects of Renaissance art and 17th-century art in Italy, Holland, and Spain.

REALISM IN FRANCE. A twofold process took place in the second half of the 18th century: the establishment, on the one hand, of the principles of rationality, clarity, and civic commitment of the Age of Enlightenment, and on the other hand the development, often in a single individual, of a moral uneasiness that was all the more pronounced the stronger was the manifestation of the contrast between the principles of rationality and historical reality. This duality had its own clear expression in art, where the dominant neoclassic style was challenged by those phenomena defined as "preromantic." The most conspicuous examples of this preromanticism are found in certain aspects of English painting, in Piranesi, and above all in Goya (qq.v.). Although the Bourbon Restoration in France greatly facilitated the acceptance of the ideas of romanticism and shifted the accent of artistic interest from objective to individual reality, the rationalistic foundations of French culture were not seriously shaken. On the contrary, rationalism retained its influence, not only in philosophy and science but also among men of letters and artists (from Stendhal to Balzac and Géricault) and even, in many aspects, among some of the major representatives of romantic spiritualism, such as Chateaubriand and Delacroix. The frequently repeated and substantially acceptable assertion that realism was nothing but the following through the ideals of romanticism is therefore valid, bearing in mind the concrete historical character of French romanticism; but another observation (one not frequently made despite its obviousness) must be added, namely that, like Positivism in philosophy and socialism in politics, realism in art constituted a powerful revival of certain motifs of the Enlightenment. This was so both with regard to the relationship of man and nature and with regard to the concept of man in society. Only these considerations explain how, when the romantic experience had run its course, realism could establish itself in France on the wave of the Revolutions of 1830 and 1848. Realism reached its height during the Second Empire, that is, in the period of the greatest economic and technical development of the bourgeoisie and of the social optimism this development generated. Here, too, is the explanation of the profoundly unified character of French art of the 19th century. It is no accident that particular studies have been made on the classicism and realism of the romantics and, vice versa, on the romanticism and the realism of the neoclassic artists and that the most serious scholars added a whole series of cautions and qualifying statements to the traditional subdivisions between neoclassicism, romanticism, and realism. F. Antal's studies on the art of the 18th and 19th centuries (1935–41), those by K. Berger on Géricault (1946, 1952), by Zeitler on classicism and Utopia (1954), and others by Hauser, P. Francastel (1951, 1955, 1956–57), K. Scheffler (1942, 1947, 1952), and Hans Hoffman (all of which set out to clarify the relationship between the ideals contained in the major artistic phenomena of the 19th century and their stylistic form) contributed considerably toward putting the problem of 19th-century realism in historical terms, even though, for lack of a profound general study, many questions, some of them basic, still await a satisfactory explanation. Only a critical effort aimed at viewing clearly the complexity of the artistic and ethical-social interests of the major personalities of the century can give an articulate and organized vision of this substantial unity. Its thread of continuity can be traced in the history of the effort made by the bourgeoisie in the 18th and 19th centuries to impose its own outlook on modern culture. Taking romanticism as the "crucial phase of mind of the Enlightenment" (see ROMANTICISM), 19th-century realism may be defined as the culmination of that phase, in the determination to integrate the basic principles of the Enlightenment with historical reality.

The figure who best expresses this development in French art is undoubtedly Théodore Géricault. In his intellectual preparation, social conscience, and ethical ideals, Géricault was a true son of the Enlightenment. Few other artists recognized as acutely as he did the necessity (fundamental to realism) to be of one's own time, to give artistic dignity to contemporary reality — not to its heroic aspect, already celebrated by such artists as David and Gros, but to its everyday aspect — with its problems of a common humanity no less worthy of engaging the feelings and imagination of an artist. Géricault's depictions of Napoleon's retreat from Russia, the drawings of the African slave trade, full of fervid humanitarian feeling, and even his *Raft of the Medusa* (VI, PL. 118; XII, PL. 314) bear witness to this. Although until recently Géricault was considered chiefly as a precursor of romanticism, recent studies have shown that his personality was deeply rooted in an order of human and cultural interests very close to those of nascent realism. Géricault had a strong influence on Courbet and Millet, by his capacity for achieving effects of an austere and near-heroic solemnity in subjects inspired by the life of the people (PLS. 437–39).

Besides evident similarities in political formation and the strong social conscience common to Géricault and Daumier (which in such a work as *Rue Transnonain, le 15 avril, 1834,* for example, find expression even in a similarity of forms), Géricault attracted the attention of Daumier, the scholar of realism, by his position vis-à-vis English art. (English art exerted a formative influence in France throughout the first half of the 19th century.) Géricault was the first to adapt to the French idiom certain aspects of English art (aspects that were clearly of the Enlightenment), such as prints of everyday life with a social flavor and sporting prints. At the same time he managed to confer on works of this inspiration the characteristics of great art. Not, of course, that Géricault was the only, or the major, disseminator in France of the English style in genre painting, social prints, and caricature. A whole group of the public, while revering, on the critics' authority, the great canvases of the neoclassic and romantic masters, remained substantially insensible to their courtly and solemn idiom, which was better suited to museums and churches than to middle-class homes. The genre scene, in the English rather than in the Dutch sense, presented a means of adhering closely to immediate, everyday reality. (This was true even when subjects were drawn from historical novels, theatrical dramas, and comic operas, for novels and the theater, just as much as sporting events, formed a part of daily life.) In almost every case the adherence to reality was only external and, following J. B. Greuze's example (q.v.), tended to bring out chiefly the pathetic, sentimental, and moralistic side of the genre picture (VI, PL. 77); but Géricault caught the essence of reality with his refined "dandy" spirit, with his innate attraction to the ephemeral sense of modernity, what might be called the metaphysical value of the transitory. This constitutes a central point as much for the art of Daumier (who found inspiration precisely in the passing events chronicled in the newspapers) as it did later for impressionism. Géricault did not limit himself to repeating, in lively drawings and lithographs (an eminently democratic and popular medium, on which he was the first to confer a high artistic importance), the themes and figures of genre painting (in *The Fish Seller*, Louvre, for example, already so close to Daumier); with his designs of boxers and trapezists he showed that the forms of a fashionable print and a sporting subject could constitute an extremely valid subject for a truly great painting; this he achieved with his *Racing at Epsom* (VI, PL. 119), which suggests Degas and in which the Baudelairian theme of "the heroism of modern life" is in some ways already present.

One could make the same observation, albeit for other reasons, about other pictures by Géricault — those that also bring to mind Daumier's social awareness. Such pictures are *The Plaster Kiln* (Louvre) and the "Charrettes" series, some lithographs representing dray horses, which shed a ray of light on a whole tendency in French painting between 1830 and 1850, a tendency inspired by life on the outskirts of the city and so profoundly bound up with the mores of the provinces as to confer a new immediacy on the intimate and severe vein of the Le Nain brothers (q.v.). It was an austere tendency frequently stirred by a vehement breath of working-class passion, which was promoted not only by Daumier, but also in various ways by such artists as A. G. Decamps, G. Michel (PL. 440), P. A. Jeanron, and Adolphe and Armand Leleux; and during the course of the 19th century it more and more frequently assumed a humanitarian and socialistic aura. Géricault had some of this aura himself, and it is the democratic character of this trend (which culminated in the painting of Millet and Courbet) that helps to bring out the immediacy of other cultural components, very important in the forming of 19th-century realism, such as the influence of Dutch and Spanish painting. A direct line connects such a painting as Courbet's *Funeral at Ornans* (Louvre) with the large group portraits of the Dutch 17th century, in which a very real sense of collectivity shows the importance that the Dutch gave to community and corporate life. The direct connection between these two periods may be explained, in fact, by the rebirth in French provincial society of a precise consciousness of its own historical function. At the same time the semianarchic populism that pushed artists toward the Le Nains could, through the example of some Spanish painting, succeed in conceiving a scene of peasant life with a breadth of vision unusual in those old French painters who were still confined in a certain sense, within the limits of *bambocciata* painting (see BAMBOCCIANTI). The new vision could also try to recapture in a scene of peasant life some of the heroic grandeur of Caravaggio.

It could be said that, as a tendency of style, interest in Dutch art never diminished in France, beginning with the 18th century; and this tendency, both in genre painting and in landscape, was nourished by the predilections of the bourgeois public. But in such tireless artisans as Martin Drolling, whose gifts for scrupulous imitation of reality and minute objectivity were developed by contact with the Dutch, and as in the works of A. Trimolet of Lyons, who followed Drolling's example, this interest in Dutch art generated only a mastery of technique. It cannot be said of Trimolet, or of A. F. Cals, who was also a painter of some sensitivity, or of Philippe Rousseau, that there was a true assimilation of the Dutch spirit. Such a picture as Philippe Rousseau's *City Rat and Country Rat*, greeted in 1845 as a small masterpiece, is basically nothing but a bravura piece, although it has a certain finesse and a sobriety unusual in genre painting of romanticism. Continuing this tendency, such work as E. Meissonier's would inevitably be produced, as well as all the inferior anecdotal painting of the second half of the 19th century. (Meissonier's cloying taste for truth of expression was complicated by the pedantic requirement for historical exactness. Baudelaire considered him a "Fleming without a Fleming's imagination, charm, color, and naiveté.") The first great artist of the century who knew how to assimilate completely the teaching of an art that, like Dutch art, saw every aspiration to the absolute and to the divine satisfied by the faithful and intimate representation of domestic life was [after J. B. Chardin's (q.v.) example] Corot (q.v.) whose similarities with Jan Vermeer, however problematical for the critics, are not any less real. The process of profound humanization, which (in contrast to the substantial mysticism of the Dutch and to the "high silence" of their interiors and landscapes) was achieved in the art of Chardin and Corot and which gives the painting of the latter an unmistakable sweetness, would have been unthinkable had there not appeared a new religiosity, completely laic and earthy in family feeling — a new sentiment that, frequently taking on the colors of the idyllic and utopian, acconpanied the formation of the democratic consciousness of the *petit bourgeois*. Along with the steady intensity of contemplation there is, in Chardin, the exceptional emphasis that, by means of the glow of the material, his brush succeeds in conferring on each object — not rich and precious objects, like those of Vermeer, but, on the contrary, such humble and modest ones as an unadorned table or the most insignificant crockery — and his prodigious capacity to express the inner moral significance in them through eminently tactile values. And thus it is with such painters as Corot, Chardin, and the Dutch that was born the idea of pure painting (the central idea of 19th-century realism), that is, of an art of painting that, by passing directly from object to representation and nourished exclusively by the concrete visual patrimony of the artist, would secure itself against any literary or rhetorical imposture.

Such a valuation of the means of painting as an element fully capable of translating the sensible world that almost palpably surrounds the artist not only conditioned the great success of still life (q.v.) in the 19th century but went on to become a central fact of the work of Courbet, in which the solidity, the thickness, and the tangible presence of the paint almost become an existential demonstration of truth and beauty.

A different order of cultural pressures is implicit in the 19th-century French interest in Spanish painting, which was known in France from the end of the 18th century — Fragonard himself was inspired by Murillo (qq.v.) — and reached its maximum popularity and currency with romanticism. The love of Spanish art and folklore was an aspect of a more general tendency of the time, a tendency to seek in civilizations distant and different from that of France a new dimension for man. From the Middle Ages, which were admired by the purists, to the Morocco that inspired Delacroix, to the Spain loved by Baudelaire and Prosper Mérimée, and even to the land of classicism, Italy, to which Stendhal turned, there was a search for experiences — not exclusively esthetic — which helped the artists to free themselves from a too-intellectually based vision of the world. These experiences stimulated new approaches, brought neglected attitudes back to life, rediscovered a lost "naturalness." The French painters found expressed in Spanish painting not only the sense of daily life in all its former ardor — an ardor that in Spain seemed to survive in the customs, in the physical types, in dress — but they saw great vitality and physical exuberance represented in paintings in which the classical foundation (albeit veiled) was organically present. In this direction exoticism and the taste for folklore (which so strongly engage the modern imagination) became, in Courbet and Edouard Manet (q.v.) a genuine recovery of naturalness, truth, and sincerity of expression in works that have a classical breadth of conception.

The complexity of social and civil life in the 19th century controlled, even in art, the formation and interrelations of the cultural currents and styles of the middle classes. Proof of this is the great attraction toward science, which does not affect as much the ethical and sentimental life (like those attitudes that have been considered above) as it does the intellectual attitude of the century; the curiosity aroused by themes tied to physiology and the interest in natural sciences. There was the reflection in the esthetic creation, of the scientist's world of ideas and feeling. Precise motifs of the Enlightenment are involved here, too, motifs that were quickened with new meanings by the romantic sensitivity. That craze for physiognomy, standing at the border between physiology, psychology, and magic, was a characteristic expression of this new attitude and lasted from Goethe to Balzac, from the neoclassicists to Géricault, from Thoré to Daumier. The importance of this phenomenon has not yet been sufficiently appreciated, but it becomes evident in the celebrated series of portraits of insane people that Géricault executed between 1822 and 1823 to illustrate the work of a notable mental specialist; and in the same way, in the other, equally memorable, series of portraits by Daumier, dating from 1830 to 1835, in which the artist represented, first modeling them in clay, then reproducing them lithographically, the portraits of many political figures of the July Monarchy. These two series, which mark two dates in the history of modern realism, take their places (precisely through their interest in physiognomy) within a much larger phenomenon, one that explains the extraordinary richness, vivacity, and continuity of realistic motifs in French portraiture of the 19th century, from David to Géricault, from Ingres to Degas (qq.v.). Friedrich Antal has very effectively underlined the realistic significance of such a portrait as that of Lepeletier de Saint-Fargeau (Paris, Bib. Nat.), sketched by David with great physiognomic immediacy in 1780. One cannot miss the Balzacian "physiological" treatment evident in the subtle effects of many of the portraits of Ingres; unequaled models of penetrating observation and impeccable decision in the selection of indications of character. With the establishment of the bourgeoisie, the portrait had a very wide popularity during the July Monarchy, and not only turned systematic minds away from their prejudices but also confirmed that contemporary custom had a right to be recognized in painting, since there was no figure so despised, so irregular, or so ugly that it did not have its own interest for art. There is, in this honest objectivity, much of the dispassionate objectivity of the natural sciences. The impression that the way of feeling and thinking of the scientist made on French culture accounts for many aspects of the personality of Ingres that have at times, and not without justification, brought him under consideration as a precursor of realism (Duval, 1936). Such aspects were his intransigent sense of loyalty and his artistic honesty, which he transmitted to his students; his intrepidly conducted war against romanticism in the name of one of the most important principles later held by the realists — sincerity; and his condemnation of the romantic exaggeration of color and of the exaggerated rendering of characterization as being against nature.

The principal field in which it is possible to grasp the affinity of the esthetic interests with the scientific-naturalistic ones, and which expresses better than any other the continuity of the Enlightenment through the romantic crisis into realism, is, without doubt, landscape painting (see LANDSCAPE IN ART). Here the artificial schemes coined to distinguish various currents — classic, romantic, and realistic — have demonstrated their weakness more than anywhere else. From the middle of the 18th century there was an increasingly extensive diffusion of the taste for landscape painting, with its epicenter initially in England, then, after 1830, in France; it was an aspect of the establishment of that 18th-century concept of nature which, at the beginning of the century, in the very homeland of sensationism and empiricism, had found a champion in the Earl of Shaftesbury and had become popular in Europe through a renewal of Spinozan themes that took various forms in different countries. Dutch landscape painting, from Rembrandt to the Ruysdaels (qq.v.) was in some ways the artistic equivalent of the philosophy of Spinoza; and Dutch landscape painting constitutes the foundation for the painting of John Constable (q.v.) and for the Barbizon school.

After the grandiose vision of the Dutch, with their deeply rooted sense of the objectivity of reality, one is struck by the clear dominance in these painters of *sensiblerie*, an effusive and lyrical sense of nature, although this is offset by a minute realism that is sometimes pushed to the point of pedantry. Whereas Constable had considered painting as a branch of natural philosophy, Théodore Rousseau approached the landscape with a curiosity every bit as intense as that of the naturalist, the geologist, or the ethnologist; he wished to penetrate structure, light, habits of life (PL. 440; IX, PL. 22; XII, PL. 321). But the requirement of truth and analytical rigor rarely succeeded in integrating itself with Rousseau's sentimental inclinations, which seem not to effect the pictorial fabric but almost to gleam directly out of the colored artifice of the dusks he painted. These contradictory traits, which also constitute the prime esthetic problem of the great personality that was Constable, were characteristic symptoms of a philosophical crisis in which the cosmic sense of nature challenged the individuating clarity of the scientific concept. The fact is that for these artists, in a way different from that of the Dutch, nature was not a reality to be discovered but rather a projection of sentiment, a shelter to which one went to find consolidation for the disillusionments of society. This attitude, already manifest in the English school, emerged very clearly in the Barbizon painters, who, like Géricault, belonged to a generation frustrated in its social and civil ideals by the political involutions that followed the French Revolution and Napoleon's Empire. All more or less influenced by the utopianism of Saint-Simon and Charles Fourier (a phalanstery designed by Daubigny has been preserved), their own way of life at Barbizon had much of those early communistic colonies in it, of those Icarias of which Étienne Cabet dreamed and later tried to achieve in America. From the social point of view, the Barbizons took up a position of having broken with the bourgeoisie, which had been dominant since the revolution of 1830; but the Barbizons were unlike Daumier, who, compelled to abandon the political fight, continued his battle with a satire on customs. The Barbizons, instead, found in isolation and union in the forest the only way in which they could come to terms with themselves.

This sentimental limitation had been predicted by Thoré, in the significant letter he wrote to Théodore Rousseau for the Salon of 1844. Fervently republican, Thoré sought to make his friend understand the dangerous consequences of an excessive withdrawal from commerce with men, for this sentimental withdrawal generated an abstract and pedantic faith in intellect. The utopians and the social visionaries of those years had, along with their dreamy and tender style, a love of the most minute details of the organization of their phalansteries and Icarias, details they delighted in describing with a pseudo-scientific seriousness. Something similar to this happened in the representation of nature by the landscape painters of the 1830s. Only Courbet (who, as a landscape painter, directly derived from Barbizon) succeeded in solving the problem with his capacity to identify completely with the pictorial event. Later some of the impressionists, strongly attracted by the anarchistic life of their Barbizon precedessors and by the fine sense of light of Daubigny (PL. 441), in particular, took up the question again, on Courbet's example, placing the accent in landscape painting decidedly on individual autonomy. Thus, although the Barbizon school was clearly a product of romanticism in its inquietude in the face of nature, its longing for the infinte, its conception of the landscape as an expression of melancholy sentiments, the problem it posed of the relation between scientific objectivity and individual sensitivity (which marked one of Thoré's greatest moments of impassioned esthetic investigation) opened the road to realism.

PROGRAMMATIC REALISM AND ITS THEORISTS. As a programmatic movement, realism developed in the conversations that occurred, about 1846, between the painter Gustave Courbet, the writer Champfleury, and the poet Max Buchon. It found its best terrain for development in the fervid spiritual climate preceding the Revolution of 1848, and had as initial points of reference the story *Chien-Caillou* by Champfleury (1847) and, even more, the first masterpieces of Courbet, which were shown publicly in 1849 and 1850. The Andler Keller, the Parisian *brasserie* that immediately after the Revolution became an animated meeting place for discussion (the physical meeting point of new ideas), expressed almost symbolically the democratic exuberance that came to characterize the artistic debate, as compared with the aristocratic refinement which had marked the controversy between the two major representatives of classicism and romanticism, Ingres and Delacroix. Numerous French artists, including Corot, Daumier, Aimé Chenavard, A. L. Barye (q.v.) and A. A. Preault, passed through this celebrated meeting place, this "temple of realism," as it was called, where for several years men of letters, artists, theater people, social preachers, economists, and politicians gathered around Champfleury, Max Buchon, and above all, Courbet. Among the regular frequenters were the painter François Bonvin, Courbert's first Paris friend and his guide in discovering the museums and collections of the city, the poet Pierre Dupont, author of *Le Chant des ouvriers*, and a varied, numerous representation from the French provinces, including Etienne Baudry and Castagnary, from Saintes (Charente-Maritime); the painter Jean Gigoux, who came from Besançon, as did Courbet; and Alfred Bruyas, the celebrated collector from Montpellier, to whom Courbet owed much valuable and positive encouragement. Théophile Silvestre (who, sincere admirer of Courbet that he was, nevertheless always maintained great reservations about Courbet's ideas), Gustave Planche (who was a confirmed opponent of Courbet's ideas), and Louis Duranty, the future director of the review *Le Réalisme* and later a supporter of the impressionists, were among the habitués of the Andler Keller, as was Baudelaire, who in those years was very close to Courbet and occupied a special position. In this lively atmosphere the ideas matured that were later developed in the writing of Courbet, and even more in those of Champfleury and Castagnary. Courbet's ideas were elementary in the extreme; he stated that he had studied outside of any system and without prejudice both ancient art and modern art, not to imitate the one any more than to copy the other; nor to achieve the useless goal of art for art's sake. "To know in order to be able, this was always my idea," he said; "to make a living art; this is my aim." Notwithstanding the painter's polemic spirit and his exuberance, which frequently found him wielding the banner of realism as if it were that of independence and truth, Courbet never gave much importance to theories: "The title of realist has been imposed on me as that of romantics was imposed on the men of 1830. In all times, titles have never given a correct idea of things; otherwise, works would be superfluous." Nor can it be said that in the works of Champfleury and Castagnary the ideas on realism have ever received organic exposition. However, there are, essentially, two points upon which the ideas rest: on the one hand, a clear conviction of the objectivity of the external world, of the possibility of understanding it and representing it plastically; on the other, strong exploitation of subjective feeling as a means through which this knowledge of the world can be realized. A third point, which derives directly from this last and is of decisive importance, is sincerity, which was considered an absolutely necessary condition in order to insure the validity of the artistic creation. Since, in fact, the inner man and his psychological structure are no more or less than the historical result of the meeting within him of nature and society, then only through the genuine expression of himself, of his own individuality, can any artist succeed in making a real contribution to the knowledge of the contemporary world. From these fundamental points derive various consequences. For one thing, there is the absolute necessity to be of one's own time, to live intensely the life of the society of which the artist is an integral part, in order to be able to express its usages, ideas, and appearance. The polemic against the historical painting of the neoclassicists and the romantics (judged to be subject to the pure reign of fantasy, a dream for archaeologists) seemed the only one worth representing. Also deriving from the realistic attitude was the appreciation and utilization of all that could in any way constitute a direct experience of the artist. Primary among these was nature, in all of its aspects, even the most modest, even the less pleasant and edifying, as long as they expressed (aside from all values of convention or preconceived choice) a genuine relation to the artist's world of ideas and feelings (whence the utilization of the humble landscape, without pretensions of the ideal, and of the simplest, uncultivated nature); and alongside nature, man in his social function, whatever it might be, even in the humblest and most derelict strata of the population. For each fragment of nature was granted its own value and its own beauty, and in the same way a degrading condition of life could be a worthy object of representation. Thus the requirement of direct experience of reality met with the social convictions of the century, which loved to find in the poorest classes the highest human dignity.

The principle of sincerity as the first quality of the artist naturally had incalculable consequences. Such an idea, in fact, undermined a series of deeply rooted convictions about the study of the classics and the literal respect for classical forms; the tendency to repeat without variation the worn-out schemes that (as was then said) had made an entire period forget its own personality and even its past came to be considered fanaticism, a "fever of imitation." Moreover, since genuineness (in the sense of independence and immediacy in the works) was considered the only solid value (even in ancient artists), therefore, to paint in the manner of Raphael, Veronese, or Rembrandt began to seem a pretentious impotence, and it appeared clear that if there were any sense in mastering the discoveries of these great masters it was in order to be able to achieve the most efficacious expression of the artist's own world. "I have traversed tradition as a good swimmer would cross a river," Courbet wrote; "the academicians drown in it."

Of particular note in the development and investigation of these ideas were the critical writings of Théophile Thoré and Champfleury, and, from 1857 on, those of Castagnary. His writings followed the clear assertion of positivistic ideas in esthetics and even, in some ways, the passage from the impassioned realism of the middle of the century to the academic realism of its last decades. Théophile Thoré is a personality of the greatest interest for the understanding of how certain romantic ferments came out of French materialistic culture, as it had been popularized through the concept of the *idéologues*. Undoubtedly Thoré's revolutionary political activity (which only the *coup d'etat* of 1851 succeeded in interrupting with an exile protracted until 1860) constituted a very significant aspect of his personality; but a no less important and much less well known trait of the writer was, undoubtedly, his philosophical formation, which can be observed in a short work produced in his youth, *Dictionnaire de Phrenologie* (1832), in which he collected the philosophical results of some of his scientific experience at the side of the celebrated anatomist and phrenologist Dumousier, and in which he fused the ideas of Pierre Cabanis and Destutt de Tracy, of Saint-Simon and Pierre Leroux, the physiognomics of J. K. Lavater, and Friedrich Wilhelm von Schelling's idea of the *symbolisation de la vie humaine* as popularized by Théodore Jouffroy in France in his philosophical lectures. What emerges in Thoré's book is a sort of universal pantheism, founded on the idea of a fundamental unity, indissolubly joining, through the affinity of being and the infinite vibrations or correspondences of nature, the present with the past, the body with the soul, and man with all that surrounds him. This concept was to inspire all of Thorés' critical writings; those that appeared before 1849 in *L'Artiste*, in *Le Siècle*, in *Constitutionnel*, in *L'Art moderne* (the latter a review Thoré himself directed); and the important works that he wrote in exile (under the pseudonym of W. Bürger) on the Flemish and Dutch painters, in particular the memorable monograph on Vermeer (1866), which revealed for the first time the physiognomy of that painter; and finally, the new *Salons* appearing after his return to France, from 1861 to 1868. His fine understanding of Delacroix's art, his friendship with Théodore Rousseau, and his extremely original ideas on landscape painting made Thoré a link between the romantic and realistic generations.

The writer whose name is most strictly connected with the programmatic realism of the middle of the century, and who merits, for this, a particular mention, was Jules Fleury-Husson, called Champfleury, born in Laon (the home of the Le Nains) in 1821. When still very young, Champfleury ran away from school to go to Paris, where he found work as a salesman in a bookstore. His first story, *Chien-Caillou*, was, it might be said, the manifesto of realism in literature; a knowledge of it is helpful in reconstructing the environment in which the movement developed. Many characters in the story adumbrate real people, the same who, along with others described in the narrative prose of the years that followed (especially in *Les adventures de Mlle Mariette* and *Les confessions de Sylvius*), reappeared in the *Souvenirs et portraits de jeunesse*. Champfleury concentrated on writing on the history of art chiefly during the second half of his life, but his vocation as a militant critic began extremely early; in criticizing a Salon of 1846 he revealed his intuition, backing Delacroix and the landscapists against the eclectics, the sentimentalists, and the neopagans. In 1848 he was already securely oriented, being the first to consider Daumier the equal of the greatest living French artists — according to Champfleury — Delacroix, Ingres, and Corot. After a brief period of contributing to Victor Hugo's *Événement*, Champfleury (still in the heat of the Revolution of 1848 and still fervidly taking part in the encounters and discussions at the Andler Keller) switched to writing for the *Messager de l'Assemblée*. His first writings on Courbet appeared there, as did those on the Le Nain brothers. These varied interests — of lived experience and study of the past — were synthesized in 1857 in the small volume *Le Réalisme*, which was followed in 1859 by *Les amis de la nature* (with a frontispiece by Courbet and a piece by Louis Duranty). Champfleury is also known for his studies of folk literature, songs, and folklore, and for his major work, *L'Histoire de la caricature*, which appeared in several volumes from 1865 to 1880.

ARTISTS. Honoré Daumier (q.v.; PL. 439) and the sole declared champion of the realistic movement, Gustave Courbet (q.v.; PL. 438), profoundly reveal the contradictory complexity of the entire period that produced realism in art. They did not resolve, in their work, the spiritual anxieties of the romantic age, which, on the contrary, continued to be manifest during the entire second half of the century, both in literature (it is enough to think of Baudelaire) and in art (from Rodin to Gustave Moreau to Odilon Redon). But their exceptional moral temper was a reflection of the period of maximum united effort exerted by an entire society around the democratic ideals that had been the spirit of the century. Such a richness and complexity of ideas cannot be found in any other representative of strict realism; not even in Jean François Millet (q.v.), who was, after Daumier and Courbet, the most significant exponent of the movement. In Millet, as in Théodore Rousseau and Charles Daubigny (notwithstanding the undeniably higher level of Millet's personality) a vein dominated that was idyllic and idealizing, but incapable of a successful commitment to reality notwithstanding the touching realism of his subjects.

Although the heritage of Daumier and Courbet as authentic creators of realism was directly assimilated by the chief artistic sequel to 19th-century realism, namely, impressionism, realism's contemporary popularity in Europe was due to its sentimental content. It would not be possible to imagine the painting of Monet, Renoir, or Cezanne without Courbet; without his work it would be difficult to explain the art of Manet, even if Manet had drawn his primary sustenance from direct contact with Venetian, Spanish, and Dutch sources. An entire aspect of impressionism, that which uses themes of the city and modern life, has its precedents in the work of Daumier. Degas also owed a great deal to Daumier and, in another way, to the penetrating realism of Ingres. The contribution made by Corot and the Barbizon painters (especially Daubigny) was of major importance to impressionist landscape painting, into which, by way of Pissarro and, later, Van Gogh (qq.v.), even the rustic and sentimental vein of Millet was assimilated.

Millet, however, was of prime importance (perhaps more important even than Courbet) in the popularization of realism. Born in Normandy in 1814 to a family of farmers, Millet was, by definition, the painter of peasants. From his class origin he derived an austerity and a religiosity that he transferred into his artistic vision. In his work the daily toil of the life of the fields was amplified into a solemn eloquence of warm human sympathy and austere Biblical transfiguration. These literary and idyllic tendencies were already manifest in Millet's first (so-called "flowery") production, inspired by the painters of *scènes galantes* of the 18th century, but his ethical commitment appeared in the strong portraits he painted in his youth, paintings in the tradition of Géricault, which remain among his most straightforward works. After the *Winnower* (1848) Millet presented at the following year's Salon (the one in which Courbet's *Stonebreakers* was displayed) his *Man with a Hoe*, then, in 1850, his celebrated *Sower*, which inspired a poem by Victor Hugo (collected in *Chansons des rues et des bois*, 1866) lauding the eloquent gesture of the sower. With these works Millet arrived at a simple broad style that remained substantially unchanged at least until about 1870, when he moved to Barbizon, where he passed the rest of his life, living like a peasant in a house at the edge of the forest with his family of 14 children. The rustic paintings of Millet include single figures of peasants, groups of peasants absorbed, with solemn dignity, in the work of the fields, but never pure landscapes. "When you paint a painting," he wrote, "should it concern a house, a wood, the ocean, or the sky, think always of the presence of man." In the best of these paintings and in Millet's many admirable drawings there is a sense of resignation to the hard necessities of destiny. He said, "I would like the creatures I represent to seem consecrated to their place and that it would be impossible to imagine that it could enter their minds to be another thing" (PLS. 437, 439).

The first paintings by Courbet and Millet, which appear today, because of their realistic accent, to break with the entire production that preceded them, are not actually isolated phenomena. At least in the choice of subject and in inspiration, they participated in a movement of which a first broad panorama appeared at the Salon of 1848. Thanks to the revolutionary government, the Salon of 1848 was open to all artists without exception. With a total number of 5500 works (an unprecedented number), the Salon displayed for the first time a large number of the works of artists who were in direct contact with the life of the provinces. Such were I. F. Bonhommé, with his interiors of smithies; Emile Loubon, from Marseilles, with his *Mineurs du tunnel de la Nerthe*; C. J. Chaplin, with his *Montagnards* and *Fileuses d'Auvergne*; A. A. E. Hébert, with his *Femme battant au beurre*; E. V. Luminais and the Leleux brothers, with their paintings of Breton peasants. These last, who first aroused talk of realism, now clearly reveal their literary limitations. (In 1844 Adolphe Leleux's *Roadworkers* appeared to Theophile Gautier as "Nature itself, in the act, without interpretation or embellishment.") Luminé and the Leleux brothers made a trip to Spain in 1843. This trip inspired the painting that created a sensation, Adolphe Leleux's *La Posada*, in which a depth of characterization unusual for that time and a pure feeling for the exact and typical gesture inspired Charles Blanc to appraise its creator, in 1846, as the direct descendant of "that robust family of painters which, in France, dates back to Le Nain, and which knows how to find a character for every object." It is of particular interest to note in all of this production the accent of provincial exuberance, which had already found its way into landscape painting.

In England the landscapes of Constable (III, PLS. 440, 441, 443, 444; IX, PL. 23) — rejecting the Roman countryside, the Dutch beaches, and the German forests as preferred subject matter — had become truly English landscapes; French landscape painters traversed every region of France, recording the national countryside there, too. A statistic of 1835 states that of 480 landscapes exhibited at the Salon, 336 were inspired by France and only 44 by Italy.

Théodore Rousseau (q.v.; PL. 440; IX, PL. 22) traveled in the Auvergne, in Normandy, the Loire Valley, the Pyrenees, Gascony, Picardy, the Franche-Comté, and in Switzerland, forming numerous friendships among the local painters and exerting a wide influence. Charles Daubigny, the painter who undoubtedly was most influential in the second half of the century and who constituted the bridge between the romantic landscape and the impressionistic one, gained his fame through a series of landscapes with strongly regional accents, painted between 1850 and 1860 (PL. 441). Typical is his *Lock at Optevoz* (1855; Rouen, Mus. B. A.; a replica painted in 1859 is in the Louvre), a delicate view of the Dauphiné. Daubigny was at Morestel on the Isère with Ravier, Corot, and Courbet, and later at Trouville with Courbet, Menot, and E. Boudin (1865). Daubigny discovered Auvers-sur-Oise, which became a famous spot for painters, and his *atelier-bateau*, from which Monet took an example, traveled the full length of the delightful and sweet valley of the Oise. The same tone also marked the work of two landscape painters who, together with Daubigny, reflect more the example of Corot: Antoine Chintreuil, painter of the astonishing *Space* (PL. 441); and, at least in his earliest (and best) efforts, Henri Harpignies.

The interest in the local character of landscape took root much earlier, however, as shown by the statistics of 1835 that most of the landscapes exhibited at the Salon were inspired by France, and by the fact that those were precisely the years in which the various regional schools were being formed. In addition to the school of Normandy, there were important ones in Marseilles, where E. Loubon was the first of a large group of local landscape painters, and in Lyons. Loubon went to Italy with Granet in 1829, spent two years there, and then moved to Paris, where he frequented the studios of Delacroix, Corot, and the Barbizon painters. In 1849 Loubon established his residence permanently in Marseilles. Artists influenced by him there were A. Aiguier (1819–65), M. Engalière (1824–57), P. Grésy (1804–74) and, above all, P. Guigou (1834–71), a notable talent who, sensitive as he was to Courbet's art, knew better than the others how to express strongly the character of his region. On the other hand, the landscapes of the painters of Lyons — Ravier (1814–95), Carrand (1821–99), and Miel, known as Vernay (1821–96) — had a conspicuously romantic accent, with their melancholy inspiration and their murky tones.

This elegiac painting, seemingly immersed in an autumnal atmosphere and morning mists, profoundly influenced the painting of the Piedmontese Antonio Fontanesi, certainly one of the major Italian painters of the 19th century.

Fontanesi met the Fontainebleau painters at the Exposition Universelle of Paris in 1855, which marked the beginning of the spread of realism throughout Europe. Through the Romanians Gregorescu and Andreescu, the Hungarian László de Paál, the Swiss Bodmer and Menn, the Belgian painters of the Tervueren school, who turned the forest of Soignes into their own Fontainebleau, and the Dutch painters of The Hague school, French painting of the 1830s became known throughout Europe in the second half of the 19th century. Realism reached America, too, in the work of George Inness (q.v.) and J. La Farge and became very popular there (see AMERICAS, ART SINCE COLUMBUS; I, PL. 109). The Belgian school drew its energy from Hippolyte Boulanger, and, in Louis Artan and Guillaume Vogels, had two painters of a sensitivity very close to that of Daubigny and who were called "delicate bards of the rain, the sea, and the Flemish landscape." The Dutch realist school, certainly the most notable artistic phenomenon in Holland since the end of the 17th century, applied the teachings of Rousseau and Millet to national tradition, drawing from them a minute, precise, and literal realism that nevertheless is saved by a melancholy sensitivity sustained by traditional Dutch technique. Anton Mauve and the Maris brothers (Jacob, Matthys, and Willem), who were particularly close to Daubigny, constituted, along with H. Mesdag (a painter of marine scenes and collector of French painting), the circle in which the young Van Gogh developed. Even Johan Barthold Jongkind, notably close to the impressionists in feeling, started out in The Hague school. In France, in 1846, Jongkind frequented the Antwerp group and met Daubigny, Lépine, and Boudin; he exhibited at the Exposition Universelle of 1855 in the French section. He was among the first to succeed in creating form through the use of light, and, although he did not discover the laws of atmospheric and chromatic vibration, he nevertheless expressed the unexpected quality of sensation, capturing fleeting movements and atmospheres. Together with Boudin, who maintained that the romantics had their day and that one must "seek out the simple beauties of nature," he represented the transition to the impressionist landscape.

In Italy, Naples had early relationships with the Barbizon school, facilitated by particular cultural conditions and a traditional affinity with Dutch painting, then reinforced by the presence in Naples of a Dutch artist, Anton Pitloo. One of the Palizzi brothers, Giuseppe, who moved to Paris (1844–45), maintained a close relationship with the landscape and animal painters of Barbizon, exhibiting repeatedly with them and encouraging his brother Filippo toward a minute *verismo*. Serafino De Tivoli (IX, PL. 240) of Leghorn also shared a close relationship with the French; upon his return from the Exposition of 1855 he brought news of the developments in French painting to the group of artists who shortly after were to take the name "Macchiaioli" (q.v.). A tradition of landscape painting (still little studied) with an evident relationship to the teachings of Ingres and Purism must have had a greater influence on the character of the Macchiaioli than did the Barbizon school. In the early 19th century, Italy was the enduring goal of innumerable landscape painters. P. H. de Valenciennes (IX, PL. 22), Michallon, Bertin, Aligny, and Granet (the intimate friend of Ingres) all painted in Italy. Corot was in Italy three times: between 1825 and 1828 (ending in Naples), in 1834, and again in 1843; Ingres himself drew and painted Roman landscapes. After the chaste luminosity of Purism's network of drawing, there appeared the painting of the abbreviated impression, and then the paintings that seemed to be composed of patches of color (*macchie*) and which, particularly through Nino Costa (*Women Loading Wood at Anzio*, 1852; Rome, Gall. Naz. d'Arte Moderna), was transmitted to the painting of Giovanni Fattori (IX, PLS. 229, 232, 233, 236) and the Macchiaioli about 1860. It was, in some ways, a preimpressionism; there were similar developments in France, the Netherlands, and other countries before 1870.

The spread of genre painting had a much less unitary nature than that of landscape painting because of the multiplicity of genre's themes and the influence of the various centers of local diffusion. Genre painting was very much alive in the France of the July Monarchy (in the work of a host of minor masters, water-colorists, draftsmen, and lithographers, such as E. Lami, Thassaert, Monnier, and Paul Gavarni) and was very popular, in those same years, in the German-speaking countries: in Prussia with Drüger, in Bavaria with Spitzweg, and Austria with Jacob Alt and Moritz von Schwind. In these countries, the anecdotal and photographically narrative landscape painting practiced in Vienna by Ferdinand Georg Waldmüller, or in Berlin by Gärtner, recalls genre painting. Probably as a consequence of the long Austrian domination, genre painting found its way easily into the Lombard-Veneto region. In Milan, Domenico Induno painted in a Mürgerian vein, and not without a certain charm. He painted little patriotic and sentimental scenes (*School for Seamstresses*, 1858), opening the way to a kind of romantic anecdotal quality in painting that was also to characterize the artists of the so-called "Lombard scapigliatura" — Cremona, Ranzoni, and Conconi; the sketches of Mosè Bianchi; and, in Venice, Giacomo Favretto showed, in a straightforwardly local idiom, the measure of his notable qualities as a painter in *The Anatomy Lesson* (1873; Milan, Gall. d'Arte Mod.).

The vivacity of the romantic vignette, its charm, and its implicit tendency to cherish the fleeting moment gave rise to an entire production of *macchia* impressions in which the palette was extremely simplified, with strong contrasts of light and dark. This style quickly became generally popular, and it was out of the accentuation of chiaroscuro effects through a *macchia* technique that there sprang the first realist works of the Tuscan painters. The German Adolf von Menzel attempted something similar in his *Memories of the Théâtre Gymnase* (1856; Berlin, Staat. Mus.). Only in the years between 1860 and 1870 was there a more marked tendency toward characterization of settings and the development of the element of light. There is clearly a similarity of attitude and outlook in a water color by the French painter Boudin, *On the Beach* (St. Louis, City Art Mus.), the *Palmieri Rotunda* by the Italian painter Fattori (1866; IX, PL. 229), and *The Croquet Game* by Winslow Homer (1866; Chicago, Art Inst.), which is reflected in the similarity of their formal solutions. No less singular, given the absolute improbability of direct contact, are similarities between the small portrait of T. Duret, painted by Manet in 1868, and certain small figures painted in the same years by Fattori (*Portrait of the Lawyer Bongiovanni*; Coll. Mario Galli), Giuseppe Abbati (IX, PLS. 228, 231), and Giovanni Boldini (IX, PL. 239). The *Piazza S. Marco in Venice* (1869; Rome, Gall. Naz. d'Arte Mod.) by Michele Cammarano, a jewel of Italian 19th-century painting, also brings to mind the intense and luminous blacks of Manet. Cammarano, a painter of enormous military canvases and ostentatiously realistic paintings of brigands, achieved his only worthwhile work in this painting. At the same time, *Picnic in May* (PL. 443) by the Hungarian P. Szinyei-Merse, was a subject in clear harmony with those of the impressionists. Contrasting with this lively vein of genre painting, which found its most effective means of expression in rapidity of execution, was another tendency, especially in Germany and Italy, that was also a development from romantic premises — a tendency toward a tranquil representation of domestic themes and the intimacy of the family. In Germany, certain works by Menzel are significant examples of this (*Room with Balcony*, PL. 445), as are, in Austria, the paintings of Moritz von Schwind (*Interior in the Morning*, 1858), and, in Italy, certain small works by Dalbono (*From the Terrace*) and a large part of the production of the Macchiaioli, from the silent cloisters of Abbati and Puccinelli to the domestic scenes of Lega (PL. 444; IX, PLS. 227, 230, 238) and the tranquil little family scenes of Adriano Cecioni (IX, PL. 239). Gioacchino Toma was capable of giving life to a somber setting through the evocative and subdued opalescence of his color (*The Orphan's Journey*; *The Wheel*; *Luisa Sanfelice in Prison*, PL. 444). Very different from this, and quite close to the art of Courbet, was the painting of Wilhelm Leibl in Germany, from 1869; later, in certain vigorously painted portraits (e.g., *Old Parisienne*, 1870, Cologne, Wallraf-Richartz-Mus.), Leibl was influenced by Manet. Leibl's best work is *Cocotte* (Cologne, Wallraf-Richartz-Mus.), painted in 1869 in Paris. In Belgium Hendrik Leys (about whom Delacroix and Baudelaire wrote) was particularly influenced by the lesser Dutch masters; and the same may be said of his nephew, Henry de Braekeleer, who, however, showed greater finesse, a more modern sensitivity, and a notable richness of color (*Man at the Window*; Brussels, Mus. Royaux des B. A.). The most superficial heir to the romantic vignette in France was J. L. Meissonier, who turned this genre to the minute and photographic representation of the past, obtaining (as the Catalan, Mariano Fortuny, did later) great admiration throughout Europe, and sowing everywhere the bad seed of that cloying and melodramatic realism that represents the worst aspect of 19th-century painting. Artists like Menzel, in his paintings inspired by the court of Frederick the Great, like Favretto in Venice, E. Dalbono and V. Gemito in Naples, and G. De Nittis and Boldini in Paris suffered the consequences of this. From Fortuny, who lived out his last days in Italy, came the inspiration for the superficialities of Francesco Paolo Michetti and the insipid local color of Antonio Mancini.

If the Exposition Universelle of 1855 marked the high point of success for the painting of the 1830s, the Salon des Refusés of 1863, in which Manet, Whistler, Henri Fantin-Latour (qq.v.), and Alphonse Legros participated, was a clear index of the ground gained by Courbet — if not by his theory, certainly by his manner of painting. Fantin-Latour was a complex personality, whose admiration for Courbet was complicated by a romantic inquietude that welled up in many artists of his generation, inspiring a realism of an intimate quality particularly successful in his portraits and still lifes, some of which were already very close to impressionism. Of particular interest are his *Hommage à Delacroix* (1864; Louvre) and his *Studio at Batignolles* (1870; Louvre), two large canvases in which he depicted several of the major representatives of French art and culture of the time. The paintings of Legros have an academic character; he attempted to give a modern accent, through Courbet, to the lessons of Holbein, an artist for whom he had a predilection in his youth. The *Ex-voto* (1861; Dijon, Mus. B. A.), *Les Demoiselles du mois de Marie* (1875), and *Le pélerinage* demonstrate, nevertheless, painting of a certain forcefulness. The Salon of 1863 was also a measure of the debt French realism owed to Spanish art. Along with Manet, who exhibited numerous paintings of direct Spanish influence in the one-man show that was open contemporarily with the Salon, and with Legros, whose *Amende honorable* brings to mind Francisco de Zurbarán (q.v.), the influence of Spain was also noticeable in Léon Bonnat, the future official portrait painter of the Third Republic, in Carolus-Duran (C. E. A. Duran), and, above all, in Théodule Ribot. Through these painters, realism came rapidly to form a new academy: Jean Paul Laurens, in France, and Hans Thome in Germany were two typical examples.

The art of Courbet and, even more, that of Millet came to kindle an entire production with a social and humanitarian basis, of which Jules Breton was the most characteristic representative in France. Jules Bastien-Lepage, Alfred Roll, and Henri Gervex worked within this orientation; but the trend probably bore its best fruits outside of France. Artists worthy of mention are, in Belgium, Charles de Groux (PL. 443) and Constantin Meunier, painter and sculptor, who depicted the ambient of the working class with a simplicity that is not without grandeur (PL. 442); in Italy, Giovanni Segantini (PL. 442) and Giuseppe Pellizza da Volpedo, both divisionist painters; Max Liebermann in Germany; and Jozef Israëls in the Netherlands.

Social and humanitarian realism had a particular success in Russia and the Slavic countries, but was a pure realism of subject. Notwithstanding the decisively antiacademic position of many artists from 1863 on, and notwithstanding the notable writings by N. G. Chernyshevski concerning the relationship between the "beautiful" and the "real" (1865), and the so-called "Society for Traveling Art Exhibitions," intended to educate the people of Russia to art (1870), there were no true and proper artistic results; neither with V. G. Perov and his positivistic anticlericalism (XIV, PL. 247), nor with the social painting with religious overtones of I. N. Kramskoi and N. N. Gay, nor with the pacifism of the military paintings of V. V. Vereshchagin, nor the antiauthoritarian protests of V. I. Surikov (XIV, PL. 247). Not even Ilya Repin (XIV, PLS. 245, 247), most famous of the Slavic painters, managed to shake off the yoke of academicism; this is substantially true also for the work of the portrayer of the virtues and unhappiness of the Magyar people, Mihály Munkácsy.

Dario Durbé

Realism in the 20th century. The socially and humanitarianly based divisionism that was very popular in the heart of Europe at the end of the 19th century was the starting point for a group of works done in the formative period of Italian futurism — works by Umberto Boccioni (*The City Rises*; V, PL. 131), by Giacomo Balla (*The Day of the Worker*, 1904, Rome, Coll. Balla), and by Carlo Carrà (*The Funeral of Galli the Anarchist*; IV, PL. 81). For their evident social commitment, notwithstanding the near anarchistic ambiguity of the ideals that stimulated them, these works represent the first original voice of the 20th century that had anything at all to do with realism. It is worth noting that Boccioni wrote about *The City Rises*, in a letter to Barbantini (ca. 1911), that "its only fault is a slight persistence of veristic details in a work that is a complete mental vision born out of reality"; but he went on to say that a painting that intends to exalt modern life is "infinitely superior to any more or less objective reproduction of real life."

In the years immediately following World War I, in the revolutionary ferment of the young German republic, a clearly committed artistic current appeared in which political satire and veristic reportage found themselves in marked antithesis to the most abstract tendencies of expressionism. The sculpture and painting of Käthe Kollwitz (q.v.) opted for realism, but with Die Neue Sachlichkeit (the new objectivity) group — which, in 1925, organized its first show at Mannheim — the movement manifested its intentions and all of its notable ethical-social commitment. Otto Dix and Georg Grosz were its most important exponents, and if the portraits by the former anticipate surrealism with their hallucinating form, the painting and caricature of the latter, who originally had followed the Dadaist movement, are cold and sarcastic indictiments of militarism and the German bourgeoisie, and at the same time, depictions of an extraordinary veracity (PL. 446; III, PL. 431; V, PL. 214; XII, PL. 572).

Other members of the group were Max Beckmann and Alexander Kanoldt. Beckmann, who had taken part in the Sezession, liked large-scale compositions, at times triptych in form, in which he strove for a violent and dramatic realism. Kanoldt, with his still lifes and petrified landscapes, arrived at that particular type of figurative painting of sharps focus and precise representation of objects, real or imagined, which is approximatively defined as "magic realism." Grosz's most vital production ceased with his move to the United States (1933) and his progressive departure from the influence of Otto Dix toward the more poetic and less committed American School.

The realist tradition in the United States — supported by the group called "The Eight" — had among its numerous representatives John Marin (q.v.), Ivan Albright (q.v.; I, PL. 123), John Sloan (q.v.; I, PL. 113), and Edward Hopper (q.v.; I, PL. 122), who was perhaps its most notable and interesting personality. Even though seeking the most objective aspects of surrounding reality, theirs was a spontaneous and ingenuous vision in which they attempted to analyze the multiform life of great cities. Ben Shahn (q.v.), raised in the poor sections of New York, established himself as a designer, poster painter, and colorist, who with a vivid and sarcastic representation of contemporary events achieved in his work a vibrant social expressionism (PL. 446; I, PL. 120).

In Mexico, during almost the same years, there came into being one of the most interesting developments of the realistic movement. After the revolution of 1911–17 Mexican artists, restored esthetically and impelled by a strong progressive charge, began to produce, about 1922, their great *murales* for public buildings of the large cities. These were to exalt the recaptured liberty, the emancipation of the popular classes, and the political entrance into history of Mexico. Each with his own sensitivity, but all starting out from the assumption of the validity of the popular and pre-Columbian tradition, such artists as Diego Rivera (q.v.; PL. 446; I, PL. 145), José Clemente Orozco (q.v.; I, PL. 119), and David Alfaro Siqueiros (q.v.; I, PL. 145) revived the art of fresco in scenes of large dimensions having contemporary social themes. The political and propagandistic aims of these artists directed them towards woodcuts and lithography as means of direct and efficacious communication with the masses. Such aims led, in fact, to the artists' associations called "El Taller de Gráfica popular," around which the foremost Mexican artists gathered, vigorously participating in the new artistic idioms that had developed in the world, from expressionism to futurism and Fauvism.

Things happened differently in the Soviet Union, where, in 1932, on the occasion of the 15th anniversary of the October Revolution, a large group show was organized, in which the established deans of art as well as groups of young artists participated. The official theme of the exhibit was the exaltation of the Revolution. While it was declared that "realism is first of all the whole truth of the subject and of the artist's characterization of the phenomena of reality," it was also affirmed that Soviet artists "want to interpret and reflect the beautiful, reject evil, through comprehensible forms, as the voice of their people." Out of this came a monotonous repetition of populist themes and a return to 19th-century portraiture, by then out of date in the rest of the world. Few works were of the quality of the sculptures of Vera Mukhina or of the paintings of A. Deineka, in which there is a greater maturity and pictorial sensitivity than was commonly found in socialist realism.

Very different from Soviet socialist realism, however, was magic realism, in which painters attempted to recreate the "lucid wonder" of certain "magic spells" in day-to-day life. This technique was absorbed by surrealism, which, using trompe-l'œil, strange juxtapositions of familiar objects, and various methods of "primitive" painting, achieves similar results.

In Italy, the most interesting aspect of the cultural and political opposition to the official esthetic of the 20th century developed in Rome in the years 1930–35, with the Roman school. Formed principally around the names of Scipione, Mario Mafai, and Corrado Cagli, the Roman school brought together such artists as Renato Guttuso (PL. 446; VII, PL. 278), who, as spokesman for the group, said: "In order that a work live, the man who produces it must be angry and express his anger in the way that best suits that man .... he must act in painting as one who makes war or a revolution would act. As he, in short, who dies for something" (Guttuso, 1952).

In 1942, at the Bergamo Prize competition Guttuso exhibited his famous *Crucifixion* (Velate di Varese, Coll. Guttuso), an event of notable importance from the esthetic point of view, but chiefly as a clamorous break with precedent and a cry of rebellion.

The term "realism" (which later, by analogy with the usage in the Italian cinema, became "neorealism") appeared for the first time in reference to modern Italian art in *Corrente*, and had its first critical definition in an article by M. De Michelis in 1945. In it he spoke of "dialectic realism," from which "must, however, be removed any interpretation in the veristic and naturalistic sense." Another notable manifestation of this attitude was seen at the Venice Biennale of 1948 in the *Fronte Nuovo delle Arti*, a movement of nine artists, among whom was Guttuso, who demonstrated great thematic and stylistic differences but who were united by the need for a profound renovation of Italian art, for liberation from a too-rigid interpretation of the realist credo, and for regaining the intimate human dimension through formal and thematic research.

Bianca SALETTI ASOR ROSA

In the late 1950s and early 1960s a new approach to realism made its appearance in the movement known as popular realism, or more familiarly, "pop" art. This art of the commonplace faithfully reproduced was called variously "junk art," "nonart," and "anti-art" but it nevertheless quickly won the acceptance of collectors and critics, an example winning (albeit over the angry protests of many) the first prize at the Venice Biennale in 1964. Pop art was based upon the use within a fine-arts framework of images, symbols, and actual objects drawn from the media of mass production — advertising art, packaging, and comic strips. Its impact upon the viewer depended in large part upon the shock value of seeing objects such as were normally found on supermarket shelves, in hardware stores, or even in junk heaps dispayed in galleries as serious works of art. (See MODERNISM.)

• •

SOURCES. Of primary importance for the entire period are articles in newspapers and periodicals; these are collected in E. Hatin, Bibliographie historique et critique de la presse periodique française, Paris, 1866, while a selection of articles more directly pertaining to the arts may be found in M. Tourneux, Salons et expositions d'art à Paris: Essai bibliographique, IV–V, Paris, 1910–11. Also fundamental are the writings of G. Courbet: see P. Courthion, ed., Courbet raconté par lui-même et par ses amis, ses écrits, ses contemporains, sa postérité, 2 vols., Geneva, 1948–50; M. De Micheli and E. Treccani, ed., Il Realismo: Lettere e scritti di G. Courbet, Milan, 1954 (a complete collection has not yet appeared).

See also: C. Blanc, Histoire des peintres français au XIXe siècle, Paris, 1845; P. Mantz, Salon de 1847, Paris, 1847; Champfleury (J. F. Husson), Essai sur la vie et l'oeuvre des Le Nain, peintres laonnais, Laon, 1850; G. Planche, Portraits d'artistes, peintres et sculpteurs, 2 vols., Paris, 1853; E. De Mirecourt, ed., Les Contemporains, 99 vols., Paris, 1854–65; Champfleury, Les peintres de Laon et de Saint Quentin, Paris, 1855; T. Gautier, Les beaux-arts en Europe, 2 vols., Paris, 1855–56; G. Planche, Etudes sur l'Ecole française (1831–52), 2 vols., Paris, 1855; T. Silvestre, Histoires des artistes vivants: Les artistes français, Etudes d'après nature, Paris, 1856 (2d ed., Paris, 1878, under the title Les artistes français; 3d ed., Paris, 1926, under this same title, with numerous additions by E. Faure); Champfleury, Le Réalisme, Paris, 1857; W. Bürger (T. Thoré), Les musées de Hollande, 2 vols., Paris, 1858–60; W. Bürger, La galerie d'Aremberg, Brussels, 1859; Champfleury. Les amis de la nature, Paris, 1859; E. and J. de Goncourt, L'art du XVIIIe siècle, 12 vols., in 1, Paris, 1859–75; W. Bürger and G. F. Waagen, Catalogue de la Galerie Suermondt (Aachen), Brussels, 1861; J. A. Castagnary, Les artistes du XIXe siècle: Salon de 1861, Paris, 1861; Champfleury, Grandes figures d'hier et d'aujourd'hui, Paris, 1861; W. Bürger, Musée d'Anvers, Brussels, Paris, 1862; Champfleury, Nouvelles recherches sur la vie et l'oeuvre des frères Le Nain, Paris, 1862; Champfleury, Les peintres de la réalité sous Louis XIII: Les frères Le Nain, Paris, 1862; E. Chesneau, La peinture française au XIXe siècle: Les chefs d'école, Paris, 1862; J. A. Castagnary, Grand Album des expositions de peintures et de sculpture, Paris, 1863; J. A. Castagnary, Les libres propos, Paris, 1864; E. Chesneau, L'art et les artistes modernes en France et en Angleterre, Paris, 1864; Champfleury, Documents positifs sur la vie des frères Le Nain, Paris, 1865; Champfleury, Histoire de la caricature antique, Paris, 1865 (3d ed., 1879); Champfleury Histoire de la caricature moderne, Paris, 1865 (3d ed., 1885); P. J. Proudhon, Du principe de l'art et de sa destination sociale, Paris, 1865 (2d ed., 1875); H. Taine, Philosophie de l'art, Paris, 1865 (Eng. trans., J. Durand, London, New York, 1865; numerous later ed.); W. Bürger, Van der Meer de Delft, GBA, XXI, 1866, pp. 297–330, 458–70, 542–75; H. Taine, Philosophie de l'art en Italie, Paris, 1866 (2d ed., 1876); T. Couture, Méthodes et entretiens d'atelier, Paris, 1867; T. Duret, Les peintres français en 1867, Paris, 1867; E. and J. de Goncourt, Manette Salomon, Paris, 1867 (3d ed., 1876); H. Taine, De l'Idéal dans l'art, Paris, 1867 (2d ed., 1879); C. Baudelaire, Oeuvres complètes, 7 vols., Paris, 1868–92 (rev. ed., Y. Le Dantec and C. Pichois, Paris, 1961); E. Baudry, Les camps des Bourgeois, Paris, 1868 (with 12 drawings by Courbet and a colloquy on art by him); J. A. Castagnary et al., Le bilan de l'année 1868, Paris, 1868; Champfleury, Histoire de l'imagerie populaire, Paris, 1869 (2d ed., 1886); A. Moreau, Decamps et son oeuvre, Paris, 1869; H. Taine, Philosophie de l'art dans les Pays-Bas, Paris, 1869; W. Bürger, Salons de 1861 à 1868, 2 vols., Paris, 1870; Champfleury, Histoire de la caricature au moyen âge, Paris, 1870 (2d ed., 1875); Champfleury, Souvenirs et portraits de jeunesse, Paris, 1872; A. Sensier, Souvenirs sur T. Rousseau, Paris, 1872; A. Sensier, Etude sur Georges Michel, Paris, 1873; Champfleury, Histoire de la caricature sous la République, l'Empire et la Restauration, Paris, 1874 (2d ed., 1877); A. de la Fizelière, Champfleury, and F. Henriet, La vie et l'oeuvre de Chintreuil, Paris, 1874; C. Tillot, ed., Catalogue de la vente qui aura lieu par suite du décès de J.-F. Millet, peintre, Paris, 1875; C. Blanc, Les artistes de mon temps, Paris, 1876; E. Fromentin, Les maîtres d'autrefois: Belgique, Hollande, Paris, 1876 (Eng., trans., A. Boyle, New York, 1948); A. Theuriet, La poésie populaire en France et la vie rustique, Rev. des deux mondes, 3d Ser., XXI, 1877, pp. 43–74; H. d'Ideville, Vieilles maisons et jeunes souvenirs, 3 vols., Paris, 1878; P. Burty, Charles Meryon: A Memoir and Complete Catalogue of His Works (trans. M. B. Huish), London, 1879; Champfleury, Henry Monnier: sa vie, son oeuvre, Paris, 1879 (2d ed., Paris, 1889; E. Zola, Mes haines: Causeries littéraires et artistiques, 2d ed., Paris, 1879; Champfleury, Histoire de la caricature sous la Réforme et la Ligue, Paris, 1880; L. Gonse, E. Fromentin: Peintre et écrivain, Paris, 1881; A. Sensier, P. Mantz, and A. Lebrun, La vie et l'oeuvre de J.-F. Millet, Paris, 1881; J. Claretie, Peintres et sculpteurs contemporains, 2 vols., Paris, 1882–84; Champfleury, Les vignettes romantiques: Histoire de la littérature et de l'art, 1825–40, Paris, 1883; V. Fournel, Les artistes français contemporains: peintres, sculpteurs, Tours, 1883 (2d ed., 1885); A. Gill, Vingt années de Paris (pref. by A. Daudet), Paris, 1883; J. K. Huysmans, L'art moderne, Paris, 1883; A. Chevassus, Max Buchon: Sa vie, son oeuvre, Paris, 1884; T. Duret, Critique d'avant-garde, Paris, 1885, J. Gigoux, Causeries sur les artistes de mon temps, Paris, 1885; A. Piédagnel, J.-F. Millet, Paris, 1885; C. Yriarte, J.-F. Millet, Paris, 1885; J. Brabey d'Aurevilly, Sensations d'art (XIXe siècle: Les oeuvres et les hommes, VII), Paris, 1886; Champfleury; Les artistes célèbres: La Tour, Paris, 1886; C. Clément, Decamps, Paris, 1886; A. Schanne, Souvenirs de Schaunard, Paris, 1886; Catalogue descriptif des peintures, aquarelles, pastels, dessins rehaussés, croquis et eaux-fortes de J.-F. Millet, réunis à l'Ecole de Beaux Arts . . . , Paris, 1887; A. Lebrun, Catalogue of the Etchings, Heliographs, Lithographs, and Woodcuts Done by Jean François Millet (trans. F. Keppel), New York, 1887; C. Bigot, Peintres français contemporains, Paris, 1888; Champfleury, Le Musée secret de la caricature, Paris, 1888; J. K. Huysmans, Certains, Paris, 1889; P. Mantz, Salon de 1889, Paris, 1889; J. Breton, La vie d'un artiste: Art et nature, Paris, 1890; P. Hippeau, Les fédérations artistiques sous la Commune, Paris, 1890; J. Proust, ed., L'art français, 1789–1889, Paris, 1890; J. Veth, J.-F. Millet, Haarlem, 1890; J. A. Castagnary, Salons, 2 vols., Paris, 1892; W. Bürger, Les salons (1844–68), 3 vols., Paris, 1893; E. and J. de Goncourt, Etudes d'art (pref. by Roger-Marx), Paris, 1893; Champfleury, Salons (1846–51), Paris, 1894; M. André, Notes sur l'art moderne . . . à travers les Salons, Paris, 1896; J. Breton, Un peintre paysan: souvenirs et impressions, Paris, 1896; M. Pittaluga, ed., La critica dei Salons, Florence, 1948; E. Fromentin, Salon de 1845 (ed. M. Pittaluga), Comm, I, 1950, pp. 50–57, 114–20; D. d'Angers, Les Carnets (introd. by A. Bruel), Paris, 1958.

BIBLIOG. *Theories of realism and naturalism*: A. David-Sauveaugeot, Le réalisme et le naturalisme dans la littérature et dans l'art, Paris, 1889; P. Lenoir, Histoire du réalisme et du naturalisme dans la poésie et dans l'art, Paris, 1889 (2d ed., 1899); L. Berg, Der Naturalismus: Zur Psychologie der modernen Kunst, Munich, 1892; G. Larroumet, L'art réaliste et la critique, Rev. des deux mondes, 3d ser., CXIV, 1892, pp. 802–42, CXVI, 1893, pp. 100–36; G. Larroumet, Etudes de littérature et de l'art, Paris, 1893; S. Rocheblave, L'art français dans ses rapports avec la littérature au XIXe siècle, in L. Petit de Julleville, ed., Histoire de la langue et de la littérature française des origines à 1900, VII, Paris, 1899, pp. 742–94 (5th ed., 1924); A. Cassagne, La théorie de l'art pour l'art en France chez les derniers romantiques et les premiers parnassiens, Paris, 1906 (repr. 1959); W. Balzer, Gustave Planche: Eine Untersuchung zur Geschichte der französischen Kunstkritik im 19. Jahrhundert, Leipzig, 1908; T. Troubat, Sainte-Beuve et Champfleury, Paris, 1908; A. Fontaine, Les doctrines d'art en France de Poussin à Diderot, Paris, 1909; G. Beaume, Fromentin, Paris, 1911; A. Schmarsow and B. Klemm, ed., W. Bürger's Kunstkritik, III, Leipzig, 1911; G. Pellissier, Le réalisme du romantisme, Paris, 1912; P. Martino, Le roman réaliste sous le Second Empire, Paris, 1913; E. Bouvier, La bataille réaliste (1844–57), Paris, 1914; L. Rosenthal, Du romantisme au réalisme, Paris, 1914; A. Dresdner, Die Kunstkritik, Munich, 1915; M. Pittaluga, Eugenio Fromentin e le origini della moderna critica d'arte, L'Arte, XX, 1917, pp. 1–18, 115–39, 240–58, 337–49, XXI, 1918, pp. 5–25, 66–83, 145–89; T. M. Mustoxidi, Histoire de l'esthétique française (1700–1900), Paris, 1920; P. Sabatier, L'esthétique des Goncourt, Paris, 1920; F. Doucet, L'esthétique d'Emile Zola et son application à la critique, The Hague, 1923; P. Martino, Le naturalisme français (1870–95), Paris, 1923; H. Marguery, Un pionnier de l'histoire de l'art: Thoré-Bürger, GBA, XI, 1925, pp. 229–45, 295–311, 367–80; E. Kris, Der Stil rustique, JhbKh-SammlWien, N.S., I, 1926, pp. 137–208; H. A. Needham, Le développement de l'esthétique sociologique en France et en Angleterre au XIXe siècle, Paris, 1926; L. Venturi, Il gusto dei primitivi, Bologna, 1926; P. d'Ancona and F. Wittgens, Antologia della moderna critica d'arte, Milan, 1927; H. U. Forest, "Réalisme": Journal de Duranty, Mod. Philology, XXIV, 1927, pp. 463–79; L. Manoliu, Aperçu sur Champfleury comme éclaireur et critique d'art (Mél. Ecole roumaine en France, IX), Paris, 1930; E. Maynial, L'époque réaliste, Paris, 1931; P. Moreau, Le classicisme des romantiques, Paris, 1932; A. Ferran, L'esthétique de Baudelaire, Paris, 1933; E. Starkie, Baudelaire, London, New York, 1933 (new ed., 1957); F. Antal, Reflections on Classicism and Romanticism, BM, LXVI, 1935, pp. 159–68, LXVIII, 1936, pp. 130–39, LXXVII, 1940, pp. 72–80, 188–92, LXXVIII, 1941, pp. 14–22; H. Händel, Champfleury: Sein Leben und Werk, Borna-Leipzig, 1935; R. Janssens, Les maîtres de la critique d'art, Brussels, 1935; M. I. Jehle, Das deutsche Kunstmärchen von der Romantik zum Naturalismus, Urbana, Ill., 1935; M. B. Bras, Gustave Planche (1808–57), Paris, 1936; R. Dumesnil, Le réalisme, Paris, 1936 (2d ed., 1945; rev. ed., Le réalisme et le naturalisme, Paris, 1955); T. E. Duval, The Subject of Realism in the Revue des deux mondes (1831–65), Philadelphia, 1936; L. Venturi, History of Art Criticism (trans. C. Marriott), New York, 1936; B. Weinberg, French Realism; The Critical Reaction 1830–70, New York, London, 1937; G. Boas, ed., Courbet and the Naturalistic Movement, Baltimore, 1938; E. B. O. Borgerhoff, "Réalisme" and Kindred Words: Their Use as Terms of Literary Criticism in the 1st Half of the 19th Century, Pub. Mod. Language Assoc. of Am., LIII, 1938, pp. 837–43; A. Heppner, Thoré-Bürger en Holland, de ontdekker van Vermeer en zijn liedfe voor Nederland's kunst, Oud-Holland, LV, 1938, pp. 17–34, 67–82, 129–44; G. Macchia, Baudelaire critico, Florence, 1939; H. Frey, Max Buchon et son oeuvre, Besançon, 1940; F. Fosca, Edmond et Jules de Goncourt, Paris, 1941; S. Meltzoff, Rediscovery of Vermeer, Marsyas, II, 1942, pp. 145–66; S. Meltzoff, Revival of the Le Nain, AB, XXIV, 1942, pp. 259–86; A. Tabarant, La vie artistique au temps de Baudelaire, Paris, 1942; M. Gilman, Baudelaire, the Critic, New York, 1943; A. Blum, Vermeer et Bürger-Thoré, Geneva, 1945; V. Giraud, Eugène Fromentin, Niort, 1945; M. Arland, Les échanges, Paris, 1946; S. Damarion, Une grande revue d'art, L'Artiste: Son rôle dans le mouvement artistique du XIXe siècle, Paris, 1946; E. Dolléans, Proudhon, Paris, 1948; E. Gardner, Thoré's Sphinx, BMMA, N.S., VII, 1948, pp. 73–78; J. C. Sloane, Tradition of Figure Painting and Concepts of Modern Art in France from 1845 to 1870, J. Aesthetics and Art Criticism, VII, 1948, pp. 1–29; C. E. Gauss, The Aesthetic Theories of French Artists: 1855 to the Present, Baltimore, 1949; M. Shapiro, Fromentin as a Critic, Partisan Rev., XVI, 1949, pp. 25–51; V. Bloch, Fromentin e i suoi "Maîtres d'autrefois," Paragone, I, 1, 1950, pp. 24–28; A. Billy, L'époque 1900 (1885–1905), Paris, 1951; J. Wilcox, The Beginnings of L'art pour l'art, J. Aesthetics and Art Criticism, XI, 1952–53, pp. 360–77; N. Barbantini, Baudelaire critico d'arte, Scritti d'arte, Venice, 1953, pp. 171–212; R. Julian, Delacroix et Baudelaire, GBA, XLII, 1953, pp. 311–26; L. E. Tabary, Duranty (1885–1905): Etude biographique et critique, Paris, 1954; C. Baudelaire, The Mirror of Art (ed. and trans. J. Mayne), London, New York, 1955; R. Huyghe, L'esthétique de l'individualisme à travers Delacroix et Baudelaire, Oxford, 1955; J. Roos, Théophile Gautier et les Beaux-Arts, Atti V Cong. int. di lingue e letterature moderne (1951), Florence, 1955, pp. 391–97; G. Bazin, Le Salon de 1830 à 1900, Scritti di storia dell'arte in onore di Lionello Venturi, II, Rome, 1956, pp. 117–23; G. May, Diderot et Baudelaire, critiques d'art, Geneva, Paris, 1957; G. Morpurgo Tagliabue, L'estetica francese fra il Positivismo e la Fenomenologia, Riv. di estetica, II, 1957, pp. 374–404; T. H. Bornecque and P. Cogny, Réalisme et naturalisme: L'histoire, la doctrine, les oeuvres, Paris, 1958; P. Grate, Deux critiques d'art de l'époque romantique: Gustave Planche et Théophile Thoré, Stockholm, 1959; A. Ottino della Chiesa, Natura ed estensione del Neoclassicismo, Arte figurativa antica e moderna, VII, 5, 1959, pp. 22–29; D. Sutton, A proposito del movimento romantico, Arte figurativa antica e moderna, VII, 5, 1959, pp. 44–51; F. Fosca, De Diderot à Valéry: Les écrivains et les arts visuels, Paris, 1960; J. Thuiller, L'homme qui retrouva Vermeer, L'Oeil, 65, 1960, pp. 51–57 (bibliog.).

19th-century French realism: J. W. Mollett, The Painters of Barbizon, London, 1890; A. Dayot, Charlet et son oeuvre, Paris, 1893; L. Roger-Miles, Le paysan dans l'oeuvre de J.-F. Millet, Paris, 1895; J. Cartwright, J.-F. Millet: His Life and Letters, London, New York, 1896 (3d ed., 1910); H. Naegely (H. Gäelyn), J.-F. Millet and Rustic Art, London, 1898 (2d ed., 1902); L. Bourges, Daubigny: Souvenirs et croquis, Paris, 1900; G. Cahen, Eugène Boudin: Sa vie et son oeuvre, Paris, 1900; L. Souillié, Peintures . . . de Millet relevées dans les catalogues de ventes 1849–1900, Paris, 1900; G. Lanoë and T. Brice, Histoire de l'école française du paysage depuis Poussin jusqu'à Millet, Paris, 1901; R. Muther, Ein Jahrhundert französischer Malerei, Berlin, 1901; L. Bénédite, Fantin-Latour, Paris, 1902; G. Geffroy and A. Alexandre, ed., Corot and Millet, London, New York, 1902; W. Gensel, Millet und Rousseau, Bielefeld, Leipzig, 1902; R. Rolland, Millet, London, New York, 1902; H. Marcel, J.-F. Millet: Biographie critique, Paris, 1903; A. Tomson, Millet and the Barbizon School, London, 1903; Les dessins de J.-F. Millet illustrés de cinquante reproductions en fac-simile . . . , Paris, 1906 (Eng. trans., London, Philadelphia, 1906); J. Guiffrey and P. Marcel, Inventaire général illustré des dessins du Musée du Louvre et du Musée de Versailles: Ecole française, 11 vols., Paris, 1907–38; C. G. Laurin, J.-F. Millet, Stockholm, 1907; V. Fantin-Latour, Catalogue de l'oeuvre complet (1849–1904) de Fantin-Latour, Paris, 1911; E. Diez, J.-.F. Millet, Bielefeld, 1912; P. Leprieur and J. Cain, J.-F. Millet, Paris, 1913; P. Marmottan, Le peintre Louis Boilly (1761–1845), Paris, 1913; L. Dimier, Histoire de la peinture française au XIXe siècle, Paris, 1914; E. Moreau-Nélaton, Millet raconté par lui-même, 3 vols., Paris, 1921; A. Kuhn, Die neuere Plastik von 1800 bis zur Gegenwart, 2d ed., Munich, 1922; A. Fontainas and L. Vaucelles, Histoire générale de l'art français de la Révolution à nos jours, I, Paris, 1923; H. Hildebrandt, Die Kunst des 19. und 20. Jahrhunderts, Wildpark-Potsdam, 1924; L. Meier-Graefe, Entwicklungsgeschichte der modernen Kunst, 4th ed., 3 vols., Munich, 1924; P. Dorbec, L'art du paysage en France: Essai sur son évolution de la fin du XVIIIe siècle à la fin du Second Empire, Paris, 1925; E. Moreau-Nélaton, Daubigny raconté par lui-même, Paris, 1925; C. Saunier, Barye, Paris, 1925; H. Focillon, La peinture au XIXe siècle, 2 vols., Paris, 1927–28; E. Waldmann, Die Kunst des Realismus und des Impressionismus im 19. Jahrhundert, Berlin, 1927; P. Gsell, J.-F. Millet, Paris, 1928; L. Hautecoeur and P. Jamot, La peinture au Musée du Louvre: Ecole française, XIXe siècle, 2 vols., Paris, 1929; W. Friedländer, Hauptströmungen der französischen Malerei von David bis Cézanne, Bielefeld, 1930; A. Rümann, Das illustrierte Buch des 19. Jahrhunderts in England, Frankreich und der Schweiz, Leipzig, 1930; L. Bénédite, Théodore Chassériau: Sa vie et son oeuvre, 2 vols., Paris, 1931; P. Jamot, Théodore Chassériau, Paris, 1933; G. Vollmer, Millet, Bergamo, 1938; E. Wind, The Revolution of History Painting, Warburg, II, 1938–39, pp. 116–27; E. P. Richardson, The Way of Western Art 1776–1914, Cambridge, Mass., 1939; N. Pevsner, Academies of Art, Past and Present, Cambridge, 1940; R. Escholier, La peinture française: XIXe siècle, 2 vols., Paris, 1941–44; L. Venturi, Peintres modernes, Paris, 1941 (Eng. trans., New York, 1947); E. Sacchetti, Millet e il socialismo, Nuova Antologia, CCCCXXI, 1942, pp. 28–33; K. Scheffler, Die grossen französischen Maler des 19. Jahrhunderts, Munich, 1942 (2d ed., 1949); J. J. Seznec, The "Romans of the Decadence" and Their Historical Significance, GBA, XXIV, 1943, pp. 221–32; O. Benesch, An Altar Project for a Boston Church by J.-F. Millet, AQ, IX, 1946, pp. 300–05; R. Huyghe, Millet et Théodore Rousseau, Geneva, 1946; K. Scheffler, Verwandlungen des Barocks in der Kunst des 19. Jahrhunderts, Vienna, 1947; K. Clark, Landscape into Art, London, 1949; P. Gay, ed., J.-F. Millet (1814–75), Paris 1950; M. Selluzac, Van Gogh et Millet, Etudes d'art, V, 1950, pp. 87–92; P. Francastel, Peinture et société, Lyon, 1951; A. Hauser. The Social History of Art, 2 vols., London. 1951–52; M. Raynal, Le XIXe siècle: De Goya à Gauguin, Geneva, 1951 (Eng. trans., J. Emmons, New York, 1951); J. C. Sloane, French Painting between the Past and the Present (1848–70), Princeton, 1951; C. Roger-Marx, Le paysage français de Corot à nos jours, Paris, 1952; K. Scheffler, Das Phänomen der Kunst: Grundsätzliche Betrachtungen zum 19. Jahrhundert, Munich, 1952; F. H. S. Bauman, 150 Years of Artists' Lithographs (1803–1953), London, 1953; C. Roger-Marx, Maîtres du XIXe et du XXe siècle, Geneva, 1954; R. Zeitler, Klassizismus und

Utopia: Interpretationen zu Werken von David, Canova, Carstens, Thorvaldsen, Koch, Stockholm, 1954; C. O. Zieseniss, Les aquarelles de Barye, Paris, 1954; K. Berger, Poussin's Style and the 19th Century, GBA, XLV, 1955, pp. 161–70; P. Francastel, Histoire de la peinture française, II: Du Classicisme au Cubisme, Paris, 1955; F. Fosca, La peinture française au XIXe siècle, Paris, 1956; P. Francastel, Les grande tendances de l'art européen au XIXe siècle, Cah. d'h. mondial, III, 1956–57, pp. 909–40; F. Baumgart, Geschichte der abendländischen Plastik, Cologne, 1957; P. J. Jouve, Tombeau de Baudelaire, Paris, 1958; C. Sterling and H. Adhémar, Musée National du Louvre, Peintures: Ecole française, XIXe siècle, 4 vols., Paris, 1958–61; L. Hautecoeur, Histoire de l'art, III: De la Nature à l'abstraction, Paris, 1959; C. G. Heise, Grosse Zeichner des 19. Jahrhunderts, Berlin, 1959; J. Laran, J. Adhémar, and J. Prinet, L'estampe, 2 vols., Paris, 1959; G. Charensol, J.-F. Millet, Rev. des deux mondes, Nov. 15, 1960, pp. 346–51; K. Clark, Looking at Pictures, London, New York, 1960; F. Novotny, Painting and Sculpture in Europe, 1780–1880, Harmondsworth, 1960; A. M. Brizio, Ottocento e Novecento, 3d ed., Turin, 1962; R. L. Herbert, Barbizon Revisited, Boston, 1962 (cat.); R. L. Herbert, Millet Revisited, BM, CIV, 1962, pp. 294–305, 377–83.

Realism in other countries: *a. Austria*: L. von Hevesi, Österreichische Kunst im 19. Jahrundert, 2 vols., Leipzig, 1903; L. von Hevesi, Rudolf Alt: sein Leben und sein Werk, Vienna, 1911; E. H. Zimmermann, Das Alt-Wiener Sittenbild, Vienna, 1923; H. Tietze, Das vormärzliche Wien in Wort und Bild, Vienna, 1925; B. Grimschitz, Die österreichische Zeichnung im 19. Jahrhundert, Vienna, 1927; G. Probszt, F. von Amerling: Der Altmeister der Wiener Porträtmalerei, Vienna, 1927; B. Grimschitz, Die österreichische Zeichnung im 19. Jahrhundert, Vienna, 1940; K. Ginhart, Die bildende Kunst in Österreich von Ausgang des 18. Jarhunderts bis zur Gegenwart, Vienna, 1943; K. Ginhart, Wiener Kunstgeschichte, Vienna, 1948; O. Benesch, Kleine Geschichte der Kunst in Osterreich, Vienna, 1950; L. Münz, Rudolf von Alt: 24 Aquarelle, Vienna, 1954; W. Buchowiecki, Geschichte der Malerei in Wien (Geschichte der Stadt Wien, Neue Reihe, VII, 2), Vienna, 1955; H. Schwarz, Salzburg und das Salzkammergut: Die künstlerische Entdeckung der Stadt und der Landschaft im 19. Jahrhundert, 3d ed., Vienna, 1958. *b. Low Countries*: M. Liebermann, Jozef Israëls, Berlin, 1901; G. H. Marias and W. Martin, Johannes Bosboom, The Hague, 1907; E. Hancke, Anton Mauve, Kunst und Künstler, XIII, 1914–15, pp. 356–68; G. Vanzype, L'art belge du XIXe siècle, 2 vols., Brussels, Paris, 1923–25; P. Fierens et al., L'art en Belgique, Brussels, 1938; P. Haesaerts, Henri de Braekeleer, Brussels, 1943; P. Fierens, L'art flamand, Paris, 1946; L. Christophe, Constantin Meunier, Antwerp, 1947; H. E. van Gelder, Jozef Israëls, Amsterdam, 1947; J. Knoef, Van romantiek tot realisme, The Hague, 1947; H. E. van Gelder and J. Duverger, ed., Kunstgeschiedenis der Nederlanden, 3d ed., 3 vols., Utrecht, 1954–56; F. van den Wijngaert et al., Album gewijd aan Henri de Braekeleer en zijn werk, Antwerp, 1956 (cat.); P. K. van Daalen, Nederlandse beeldhouwers in de Negentiende eeuw, The Hague, 1957; H. J. Aubert, The Etched Work of Josef Israëls, Amsterdam, n.d. *c. Germany*: A. Rosenberg, Franz von Lenbach, Bielefeld, Leipzig, 1898; H. Thode, Hans Thoma: Des Meisters Gemälde im 874 Bildern, Stuttgart, 1909; M. Osborn, Franz Krüger, Bielefeld, Leipzig, 1910; R. Hamann, Die deutsche Malerei im 19. Jahrhundert, Leipzig, Berlin, 1914; E. Waldmann, Wilhelm Leibl: Eine Darstellung seiner Kunst, Berlin, 1914; H. Wolff, ed., Adolph von Menzels Briefe, Berlin, 1914; J. A. Beringer, Hans Thoma: Griffelkunst, Frankfurt, 1916; J, A. Beringer, Hans Thoma: Radeirungen, Munich, 1920; L. Justi, Deutsche Malkunst im 19. Jahrhundert: Ein Führer durch die Nationalgalerie, Berlin, 1921; H. Uhde-Bernays, Münchener Landschafter im 19. Jahrhundert, Munich, 1921; A. Heilmeyer, Adolf von Hildebrand, Munich, 1922; R, Oldenbourg and H. Uhde-Bernays, Die Münchener Malerei im 19. Jahrhundert, 2 vols., Munich, 1922–27; E. Bock, Adolph Menzel: Verzeichnis seines graphischen Werkes, Berlin, 1923; C. Gurlitt, Die deutsche Kunst seit 1800. 4th ed., Berlin, 1924; R. Hamann, Die deutsche Malerei vom Rokoko bis zum Expressionismus, Leipzig, Berlin, 1925; A. Rümann, Die illustrierten deutschen Bücher des 19. Jahrhunderts, Stuttgart, 1926; A. Feulner, Skulptur und Malerei des 18. Jahrhunderts in Deutschland, Wildpark-Potsdam, 1929; L. Justi, Von Runge bis Thoma: Deutsche Malkunst im 19. und 20. Jahrhundert (Ein Gang durch die Nationalgalerie), Berlin, 1932; H. Karlinger, München und die deutsche Kunst des 19. Jahrhunderts, Munich, 1933; E. Waldmann, Der Maler Adolph Menzel, Vienna, 1941; E. Waldmann, Wilhelm Leibl als Zeichner, Munich, 1942; P. O. Rave, Die Malerei des 19. Jahrhunderts, Berlin, 1945; R. Neuhaus, Die Bildnismalerei des Leibl-Kreises, Marburg, 1953; L. Justi, Zeichnungen deutschen Meister von Klassizismus bis zum Impressionismus (Nationalgalerie), Berlin, 1954; E. Bülau, Der englische Einfluss aid die deutsche Landschaftmalerei im frühen 19. Jahrhundert, Freiburg im Breisgau, 1955 (diss.). *d. Italy*: A. R. Willard, History of Modern Italian Art, 2d ed., New York, London, 1900; C. Carrà, Ranzoni, Rome, 1924; E. Guardascione, G. Toma: Il colore in pittura, Bari, 1924; E. Cecchi, Pittura italiana dell'Ottocento, Rome, 1926 (3d ed., Milan, 1946); S. Di Giacomo, V. Gemito: Vita e opere, Naples, 1928; U. Ojetti, La pittura italiana dell'Ottocento, Milan, Rome, 1929; A. Soffici, Medardo Rosso, Florence, 1929; T. Silani, F. P. Michetti, Milan, Rome, 1932; M. Bernardi, A. Fontanesi, Milan, 1933; M. Biancale, G. Toma, Rome, 1933; G. Nicodemi, Tranquillo Cremona, Milan, 1933; A. De Rinaldis, G. Toma, Milan, 1934; L. Vitali, Incisione italiana moderna, Milan, 1934; M. Sarfatti, D. Ranzoni, Rome, 1935; O. Morisani, Vita di Gemito, Naples, 1936; A. Savinio, Seconda vita di Gemito, Rome, 1938; Tranquillo Cremona e gli artisti lombardi del suo tempo, Pavia, 1938 (cat.); N. Barbantini, G. Segantini, Venice, 1945; A. Dragone and J. Dragone Conti, I paesisti piemontesi del'Ottocento, Turin, 1947; G. Briganti, A. Fontanesi, Reggio Emilia, 1949 (cat.); M. Borghi, Medardo Rosso, Milan, 1950; D. Valeri, G. Favretto, Venice, 1950; F. Bellonzi and R. Frattarolo, Appunti sull'arte di Gemito, Rome, 1952; R. Longhi, Paesisti piemontesi dell'Ottocento, Cat. XXVI Esposizione biennale int. d'arte, Venice, 1952, pp. 33–42; R. Longhi, Ricordo di Enrico Reycend, Paragone, III, 27, 1952; pp. 43–55; A. Mezzetti and E. Zocca, Pittori italiani del secondo Ottocento, Rome, 1952 (cat.); F. Bellonzi and C. Lorenzetti, A. Mancini, Rome, 1953; N. Ciarletta, Teofilo Patini, Rome, 1954 (cat.); E. Lavagnino and L. Salerno, G. Toma, Rome, 1954 (cat.); A. Mensi, Mostra del pittore Giuseppe Pellizza da Volpedo, Alessandria, 1954 (cat.); C. A. Petrucci, La caricatura italiana dell'Ottocento, Rome, 1954; G. Nicodemi, G. Segantini, Milan, 1956 (bibliog.); G. De Carli, ed., Segantini, Arco, 1958 (cat.); C. Maltese, Storia dell'arte in Italia, 1785–1943; Turin, 1960; P. Ricci, I fratelli Palizzi, Busto Arsizio, 1960. *e. Czechoslovakia*: J. Neumann, Die tschechische klassische Malerei des 19. Jahrhunderts, Prague, 1955 (Eng. trans., I. Urwin, Prague, 1955). *f. Hungary*: B. Lazar, Ladislaus de Páal: Un peintre hongrois de l'école de Barbizon, Paris, 1904; I. Genthon, Az új magyar festömüvészet története (History of the New Hungarian Painting), Budapest, 1934; L. von Balás-Piry, Die ungarische Malerei des 19. und 20. Jahrhunderts, Berlin, 1940; B. Biro, A magyar müvészettörténeti irodalom bibliográfiája (Bibliography of Hungarian Art History), Budapest, 1955; Ö. G. Pogány, 19th Century Hungarian Painting, 2d ed., Budapest, 1956; L. Végváre, Katalog der Gemälde und Zeichnungen Mihály Munkácsys, Budapest, 1959. *g. Romania*: G. Opresco, Roumanian Art from 1800 to Our Days, Malmö, 1935; J. Tianu, Civigovescu, Bucharest, n.d.; R. Bogdan, Tendinte şi orientări în pictura romînească din a doua jumătate a secolului al XIX-lea (Tendencies and Currents in Rumanian Painting of the 2d Half of the 19th Century), S. cercetări de istoria artei, VII, 1960, pp. 111–42. *h. Serbia*: Dwa weka srpskoga slikarstwa (Two Centuries of Serbian Painting), Belgrade, 1943. *i. Scandinavian countries*: C. A. Been, Danmarks Malerkunst: Billeder og biografier, 2 vols., Copenhagen, 1902–03; G. Nordensvan, Schwedische Kunst des 19. Jahrhunderts, Leipzig, 1904; A. Aubert, Die norwegische Malerei in 19. Jahrhundert, Leipzig, 1910; C. Hintze, Kopenhagen und die deutsche Malerei um 1800, Würzburg, 1937; O. Okkonen, Die finnische Kunst. Berlin, 1943; H. Cornell, Den Svenska Konsten Historia, II, Stockholm, 1946; A. Lindblom, Sveriges Konsthistoria från Forntid till Nutid, III, Stockholm, 1946; E. Zahle, ed., Danmarks Malerkunst fra Middelalder til Nutid, 3d ed., Copenhagen, 1947; V. Thorlacius-Ussing, Danmarks Billedhuggerkunst fra Oldtid til Nutid, Copenhagen, 1950; O. Barregard and H. Rosenthal, ed., Svenskt Konstnärslexikon, 2 vols., Malmö, 1952–53; Tecknade och målade av svenska 1800 — talskonstnärer, Stockholm, 1956 (cat.). *j. Switzerland*: D. Baud-Bovy, Les maîtres de la gravure suisse, Geneva, 1935; W. Hugelshofer, Schweizer Kleinmeister, Zürich, 1943; L. Fromer, Die Entwicklung der schweizerischen Landschaftsmalerei im 18. und 19. Jahrhundert, Basel, 1945.

20th-century realism: E. M. Benson, Forms of Art: Phases of Naturalism, Am. Mag. of Art, XXVIII, 1935, pp. 222–29; R. Gavelle, La réalité et son imitation suggérée: aspects du trompe-l'oeil, L'amour de l'art, XIX, 1938, pp. 231–40; F. Whiting, Realism Today and Tomorrow, Am. Mag. of Art, XXXII, 1939, p. 261; R. Guttuso, Pensieri sulla pittura, Primato, II, 16, 1941, p. 20; R. B. Winn, Beauty of Nature and Art, J. Aesthetics and Art Criticism, I, 5, 1942, pp. 3–13; D. Brian, Is the Sharp Focus Clear? Realists and Magic Realists Exhibition, Museum of Modern Art, AN, XLII, March 1943, pp. 18–20; M. W. Brown, Cubist Realism: An American Style, Marsyas, III, 1943–45, pp. 139–60; M. Riley, Americans 1943: Realism and Magic Realism, Art Digest, XVII, Feb. 15, 1943, p. 6; L'arte contro la barbarie, Rome, 1944 (cat.); W. Born, American Still Life Painting from Naturalism to Impressionism, GBA, XXIX, 1946, pp. 303–18; M. De Micheli, Realismo e poesia, Il '45, I, 1, 1946, pp. 35–44; Manifesto "Oltre Guernica," Milan, 1946; H. Sawkins, Realism and Impressionism, Artist, XXXII, 1946, pp. 70–72; M. W. Brown, Early Realism of Hopper and Burchfield, College Art J., VII, 1, 1947, pp. 3–11; Galleria della Spiga, Prima mostra del Fronte Nuovo delle Arti, Milan, 1947 (cat.); R. Guttuso, Letter to E. N. Rogers, Domus, 223–25, 1947, pp. 42–46; E. Mastrolonardo, Il Fronte Nuovo delle Arti, Alfabeto, III, 13–14, 1947, p. 4; A. Frankenstein, Haberle: Or the Illusion of the Real, Am. Mag. of Art, XLI, 1948, pp. 222–27; U. Apollonio, Pittura moderna italiana, Venice, 1950; N. Ciarletta, Realismo e astrattismo, Alfabeto, VII, 1–2, 1951, p. 3; C. H. Morgan and M. C. Toole, Notes on the Early Hudson River School, Art in Am., XXXIX, 1951, pp. 161–65; M. Valsecchi, Astrattismo e neorealismo, Panorama dell'arte it., II, 1951, pp. 416–18; B. Biral, Condena y defensa del "realismo," Histonium, 162, 1952, pp. 41–42; R. De Grada, Il movimento di Corrente, Milan, 1952; L. Ferrante, Arte e realtà, Venice, 1952; R. Guttuso, Sulla via del realismo, Alfabeto, VIII, 3–4, 1952, pp. 1–4; R. Guttuso, Sulla via del realismo, Società, VIII, 1, 1952, pp. 80–88; Realismo: Mensile di arti figurative, Milan, 1952, ff.; P. Tyler, Magic Realism in American Painting, Am. Artist, XVI, 1952, pp. 40–43; G. C. Argan et al., Perchè l'arte non è popolare, Ulisse, IV, 19, 1953, pp. 1–125; M. G. Biovi, Noi "farisei" e il realismo, Il punto nelle lettere e nelle arti, II, 1, 1953, pp. 3–6; F. Clerici et al., Grand Illusion: Some Considerations of Perspective, Illusionism and Trompe-l'Oeil, AN Annual, XXIII, 1953, pp. 98–178; R. Guttuso, L'insegnamento di Picasso, La Biennale, 13–14, 1953, pp. 58–59; R. Guttuso, Lettera a Picasso, Realismo, II, 9–10, 1953, pp. 1–2; Il nuovo corriere, Dec., 1953 (research into contemporary art); B. Shahn, What is Realism in Art?, Look, Jan. 13, 1953, pp. 44–45; J. I. H. Baur, American Luminism: A Neglected Aspect of the Realist Movement in 19th Century American Painting, Perspectives USA, IX, 1954, pp. 90–98; Il Contemporaneo, Rome, 1954 ff. (18–19, 1959, reports a discussion on "Avanguardia e decadentismo"); A. Garrett, The New Realism in English Art, The Studio, CXLVII, 1954, pp. 161–69; L. Reidemeister, ed., Espressionismo e arte tedesca del XX secolo, Turin, 1954 (cat.); CahArt, XXX, 1955 (special no. on modern art, with articles by C. Zervos, etc.); M. Maccari, Il Selvaggio (ed. C. L. Ragghianti), Venice, 1955; J. P. Hodin, Quattro giovani pittori inglesi, XXVIII Esposizione biennale int. d'arte, Venice, 1956, pp. 412–16 (cat.); K. Kuh, American Artists Paint the City, Chicago, 1956; C. Maltese, Materialismo e critica d'arte, Rome, 1956; W. Grohmann, ed., Arte tedesca dal 1905 ad oggi, Rome, 1957 (cat.); M. Rieser, Aesthetic Theory of Social Realism, J. Aesthetics and Art Criticism, XVI, 1957, pp. 237–48; B. Shahn, Realism Reconsidered, Perspecta, 4, 1957, pp. 28–35; V. Barker, From Realism to Reality in Recent American Painting, Lincoln, Neb., 1959; C. Cagli and others, Astratto o figurativo?, Ulisse, VI, 33, 1959, pp. 9–149; G. Castelfranco and D. Durbé, La scuola romana dal

1930 al 1945, Rome, 1959 (cat.); W. George et al., Gibt es einen modernen Realismus?, Kunstwerk, XIII, 4, 1959, pp. 3–29; A. Bertin, La nozione di realismo nella critica contemporanea sovietica, Riv. di estetica, V, 1960, pp. 273–80; J. Giono, Peintres de la réalité du XXe siècle, Paris, 1960; E. Riccomini, ed., Rinnovamento delle arti in Italia, Ferrara, Bologna, 1960 (cat.; bibliog.); La Biennale, XII, 46–47, 1962 (articles on realism by R. Assunto, etc.); A. Moravia, Renato Guttuso (F. Grasso, La vita e l'opera di Guttuso), Palermo, 1962 (bibliog.); M. Paoli and R. Sitti, ed., Il dopoguerra: La pittura in Italia dal 1945 al 1955, Ferrara, 1962 (cat.; full bibliog.).

* *

Illustrations: PLS. 437–446.

REDON, Odilon. French painter and graphic artist (b. Bordeaux, Apr. 22, 1840; d. Paris, July 6, 1916). Redon's artistic training began in 1855 under Stanislas Gorin at Bordeaux and then, in 1864, continued more briefly and conventionally under Gérôme at the Ecole des Beaux-Arts in Paris. The most decisive influence on his work, however, was that of the print maker Rodolphe Bresdin, with whom he was in close contact (Bordeaux, 1863–65) and about whom he wrote an article (1869). After the Franco-Prussian War Redon settled in Paris, with frequent visits to his childhood home in Peyrelebade (Gironde). Although he worked with oils and pastels throughout his life, much of his energy was devoted to series of charcoal drawings and albums of lithographs, whose titles suggest their imaginative and often literary contents [e.g., *Dans le Rêve*, 1879; *A Edgar Poë*, 1882; *Les Origines* (PL. 371), 1883; *Hommage à Goya*, 1885; *La Nuit*, 1886; *La Tentation de St. Antoine*, 1888; *Les Fleurs du Mal*, 1890; *Les Songes*, 1891; *L'Apocalypse*, 1899]. In 1881 and 1882 Redon held small exhibitions of his charcoal drawings, and he later participated in the first Salon des Indépendants (1884) as well as in the first exhibition of painter-engravers at Durand-Ruel (1889). In 1894, he had his first comprehensive one-man exhibition. During these years Redon's art attracted the attention of many symbolist artists, writers, and critics, including J. K. Huysmans (who discussed him in *A Rebours*, Paris, 1894), Gauguin, G. A. Aurier, Emile Bernard, and Maurice Denis. Occasionally Redon executed decorative panels and screens for the homes of friends such as Mme. Ernest Chausson and Robert de Domecy, but the last decades of his life remained outwardly uneventful. His international fame increased just before his death as a result of his being represented more copiously than any other artist in the Armory Show of 1913.

Though Redon was an exact contemporary of Monet, his art nevertheless contradicted the realist premises of the impressionist generation. Instead, Redon, like Gustave Moreau, extended the romantics' sensibility for the mysterious into a realm that coincided with the visionary goals of a younger generation of symbolists. From such sources as the magically tiny landscape etchings of Bresdin, the exotic coloration of Delacroix, and the muted mood of Corot, Redon evolved a personal style that was directed to the creation of an unreal *ambiance* for his fantastic subjects. Like the iridescent and jewellike colors of his oils, pastels, and water colors, the resonant and often ominous shadows of his lithographs and charcoal drawings evoke a filmy, gravityless world of apparitional character, in which even still lifes can float. Redon, like Moreau, often transformed the classical subjects of the academies into strange and precious dreams, but his iconographical range was broader and more inventive. Frequently inspired by such macabre literary and visual sources as Baudelaire, Poe, and Goya, Redon developed a weirdly personal imagery, in which Cyclopean and humanoid creatures loom large. Elsewhere, his exploration of lower forms of plant and sea life, seen as if through an imagery microscope, suggests his early friendship with the botanist Armand Clavaud and an awareness of Darwinian theories of evolution. In this, Redon's art partakes of the biological primitivism recurrent in Art Nouveau. From the viewpoint of the 20th century, Redon may be seen as a vital link of continuity between the romantics' exploration of the bizarre and the hallucinatory and the surrealists' preoccupation with the subconscious.

Writings. A soi-même (1867–1915): Notes sur la vie, l'art et les artistes, Paris, 1922; Lettres d'Odilon Redon (1878–1916), Paris, Brussels, 1923.

Bibliog. André Mellerio, Odilon Redon, peintre, dessinateur et graveur, Paris, 1923; Sven Sandström, Le Monde imaginaire d'Odilon Redon, Lund, 1955; Roseline Bacou, Odilon Redon, 2 vols., Geneva, 1956; John Rewald, Odilon Redon, in Redon, Moreau, Bresdin (cat.), Museum of Modern Art, New York, 1961; K. Berger, Odilon Redon, New York, 1965.

Robert Rosenblum

REFORMATION AND COUNTER REFORMATION. The far-reaching religious, political, and cultural movement known as the Reformation not only deprived the Church of its power and authority over vast territories, which in turn were further broken up among the various new religious denominations, but also, in so far as it counteracted the Renaissance artistic tradition, laid the foundations for an artistic renewal. At the same time, within the Catholic Church itself there occurred a deep-rooted reaction known as the Counter Reformation, which was largely the outcome of the Council of Trent (1545–63) and which affected the arts, as essential tools in spreading the prestige and teachings of the Church. The relationship between these ideological conflicts and art has increasingly come to be regarded as a phenomenon that had a vital historical development, assumed various aspects at different periods, and profoundly affected all stages of the Renaissance (q.v.), mannerism (q.v.), and, to a lesser extent, baroque art (q.v.). This influence was not restricted to esthetics, artistic theories, criticism, style, and iconography; by its effect on the attitudes of society as a whole and of individuals in particular, it also created an atmosphere of spiritual conflict that was manifested in all the arts from the middle of the 15th century to the middle of the 17th and beyond.

Introduction. Erasmus lamented: "Ubicumque regnat Lutheranismus ibi literarum est interitus"; but history has belied the fears that the Reformation would bring about not only the undermining but the complete disappearance of the artistic tradition and that the Catholic Counter Reformation would result in the production of works of art of a merely exterior and practical character (B. Croce, *Storia dell'Età Barocca in Italia*, Bari, 1929). Matthias Grünewald, El Greco, and Caravaggio (qq.v.), originally regarded as men born out of their time and in conflict with the religious culture of the day, now appear, on the contrary, as the protagonists of an artistic renewal that matched the religious revival all along the line. Even Bernini's (q.v.) religious sense (PL. 454), as recorded by Filippo Baldinucci, is seen to be both authentic and deeply felt (R. Wittkower, *Gian Lorenzo Bernini, the Sculptor of the Roman Baroque*, London, 1955). Moreover, the modern tendency in religious historiography is to highlight the points of convergence rather that those of conflict between the Protestant and the Catholic worlds, and this calls for a far-reaching reappraisal of the art of the 16th and 17th centuries, a period particularly rich in stylistic and theoretical contrasts.

In all civilizations there are relationships between art and religion that give rise to iconographic, stylistic, and esthetic concomitants. During the period of the Reformation — roughly from the second half of the 15th century to the end of the 17th — the traditional alliance between artists and the Church (which had been their chief and most faithful patron for over a thousand years) was disrupted, and there was heated debate over such perennial religious themes as images, devotions, and the cult of the saints (see SAINTS, ICONOGRAPHY OF). Furthermore, there had been a great social change. To begin with, such discussion — for instance, between the advocates of vast architectural projects and those who, espousing poverty, opposed all representational or ceremonial pomp — instead of being conducted within the Church and even within a particular religious order, produced vast rival camps capable of airing their views on a large scale and imposing them on whole nations. Artists discovered that they had a lay market, with patrons affording them many opportunities for employing their abilities in secular art. They were therefore in a position to choose their own personal attitude, although often at the risk of having to go into exile or to conduct their activities clandestinely. The Inquisition and censorship (in both the Catholic and the Protestant camps) brought about dramatic situations and provoked violent reactions.

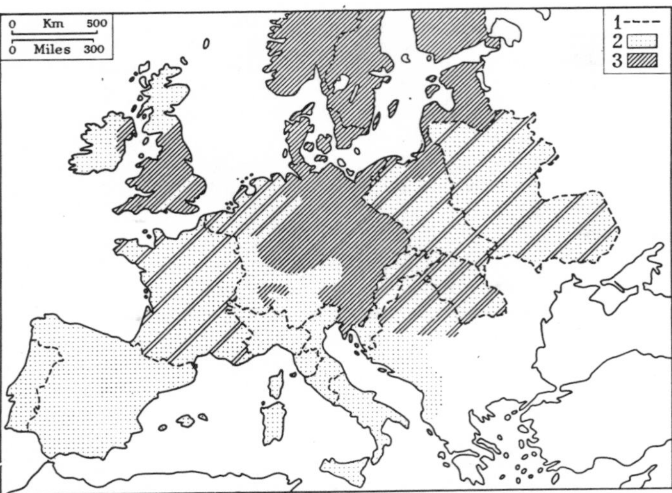

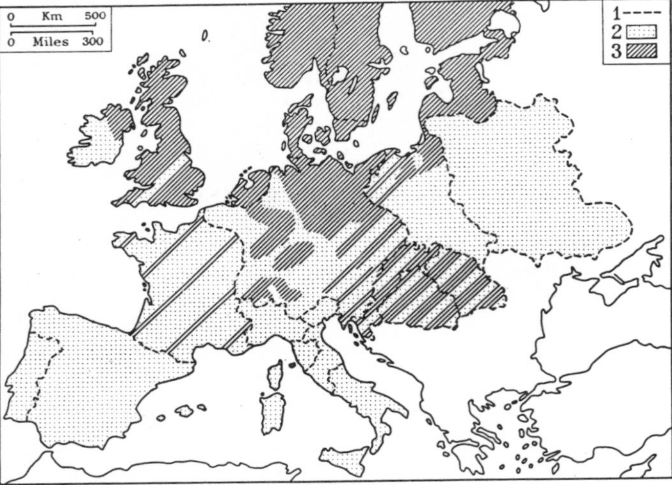

Distribution of the religious confessions in Europe. *Above*: In 1546, at the beginning of the Reformation. *Below*: In 1650, after the Counter Reformation. *Key* (1) Political boundaries; (2) Catholic areas; (3) Protestant areas.

The varying degree of importance attached to Humanistic culture had the effect of dividing Europe in two parts, roughly separated by the Alps and the Rhine: in the north aniconism and austerity prevailed, and in the south, the baroque and the cult of saints. Nevertheless it would be an oversimplification to consider the north of Europe as diametrically opposed to the south in those strife-torn decades of wars, disputes, and spiritual explorations; at least so far as the arts were concerned, there were as many parallels as differences. Generally speaking, men of the same generation reacted more or less in the same way, and it was by no means unusual for ideas that first appeared in the south to be followed through to their logical conclusion in the north and vice versa.

ARCHITECTURE. Tradition has it that it was the expenditure incurred at the beginning of the 16th century for the monumental reconstruction of St. Peter's and the Vatican Palace that was responsible for the intolerable extension of the practice of selling indulgences to replenish the papal coffers. The facts of the matter, whether economic or religious, were, of course, more complex. It is, nonetheless, certain that the bitterest diatribes of the reformers, both in Italy and beyond the Alps, returned repeatedly to the theme of the pomp of religious ceremonial and the attempt to turn Rome into an unrivaled political and religious center. There had already been a prelude to this dispute when, at the end of the 13th century, the popes first attempted to wield a power and authority equal to those of the emperor. The resulting differences of opinion and divisions between the religious orders and the violent altercations occasioned by the reconstruction on a grand scale of S. Francesco in Assisi and Sta Croce in Florence are well known; even at that early stage, architectural splendor with its colossal cost was the focal point of schisms and extremist heresies. In the 15th century the idea of Rome as the earthly Jerusalem was upheld by Pope Nicholas V (1447–55), and the repeated condemnation of his project by the popes who succeeded him is significant as a prelude to the great "protest." It is worth recalling that Pius II, in opposition to the Renaissance attempt to reconcile the classic temple and the Christian basilica, wanted the inside of Pienza Cathedral to be in the Gothic style, considering it better suited to arousing devotion in the faithful; to him the Tempio Malatestiano was a temple of infidels worshiping the devil.

Luther's trip to Italy in 1510–11 makes it possible to follow, almost as in a day-to-day account, the growth of this rebellion against classicist ideals. The first thing that scandalized Luther on this trip was the great wealth that was evident in the holdings of monasteries such as that at S. Benedetto Po, outside Mantua. In Rome he was both fascinated and repelled by the imposing ruins, which disturbed him deeply: "How could pagans build such great things without the help of God?" Later he came to the conclusion that the Pantheon, at least, was the work of the devil. The architectural splendor and the magnificent churches had their counterpart in ostentatious ceremonies for which the Pope was clothed in vestments that no German prince would have dared to wear. In this as in his other reactions, it is significant that Luther was an Augustinian; that is, he belonged to the order that above all others, both in Rome and Florence, had stood for sincere adherence to Humanist ideals and to Renaissance art theories.

About the same time a satirical dialogue by Andrea Guarna (*Scimia*, Bologna, 1517) was going the rounds; it denigrated the memory of Bramante (q.v.), destroyer of Old St. Peter's, accusing him of having abandoned himself unreservedly to caprice and of having "raised up thousands of fallen statues of the ancient gods." He is accosted by St. Peter: "Why have you ruined my temple in Rome, when its very antiquity seemed to turn even the most irreligious souls toward God?" In the end Bramante suggests remaking paradise itself on modern lines; failing to get permission, he wants to go down to "make a new hell, overturning the old one, decrepit and worn out as it is by the old flames." The satire is an indication of the controversies once again provoked by the papal building policy, which prudence had prevented from becoming as widely known in Italy as beyond the Alps. It should be recalled, in this connection, that during the Sack of Rome in 1527, bonfires were lit in the Sistine Chapel in an attempt to destroy the frescoes.

In this period a famous writer from Padua, "Il Ruzzante" (Angelo Beolco), who in his lifetime was accused of heresy, wrote a speech addressed to Cardinal Pisani, who was planning to reconstruct the local cathedral, containing turns of phrase faintly reminiscent of Cicero and apparently inspired by the condemnation by Swiss cities of those who were anxious to restore works of art destroyed by the iconoclasts: "If you have so much money to throw about, give bread to those who are hungry, a husband to those who want and cannot have one because they have no dowry, freedom to those who are in prison, have clothes and shirts made for those who are dying of cold, and many other things of this nature, so that all may speak well of you: in brief, try to restore the spiritual church and not the one made of stone, because I have never heard it said . . . that anyone has gone to heaven because of constructions of that sort. And if you are absolutely determined to build, then build a hospital for the poor and . . . see to it that our workers all receive just wages, in short, provide water for our dried-up plants since it is you who hold the keys to the fountains." This program of Il Ruzzante (ed. A. Mortier, Paris, 1925–26), who was one of the most interesting supporters of Erasmus in Italy, highlights a change in stylistic emphasis toward the functional, which became very noticeable towards the mid-16th century, particularly in architecture.

A whole chapter could be devoted, on the one hand, to the regulations issued by the bishops concerning church architecture, which were to culminate in the famous provisions laid down by St. Charles Borromeo in his *Instructiones fabricae et suppellectilis ecclesiasticae*, issued shortly after 1572; and, on the other, to the Utopian community theories propounded by writers attached to the smaller courts. Such theories had occasionally been implemented on a large scale, as in the model city district of no less than 53 one-story houses erected between 1519 and 1523 by the great Fugger bankers in Augsburg: "Because they were born for the common good and owe their great fortune in the first place to the Most High and Merciful Lord and for this very reason are bound to give it back again in charity and generosity, which may be taken as an example to all, they have given, donated and consecrated this foundation to those of their fellow citizens who are poor but upright" (W. Winker, *Fugger il ricco*, It. trans., Turin, 1943, pp. 244–46). The houses comprise two rooms, a kitchen, a water closet, and a woodshed; they are still inhabited and are made available, free of charge, to those in need. There were a number of such foundations, particularly in the feudal agricultural domains, inspired by this mixture of moral and religious sentiment. In Abruzzi, for instance, Pescocostanzo was founded by Vittoria Colonna in 1535; this settlement's development was to be supervised by the members of a council elected by the inhabitants themselves, as if to confirm the concept of a community created for the common good. The town of Sabbioneta was built by Vespasiano Gonzaga, beloved nephew of the Giulia Gonzaga to whom Juan de Valdés dedicated his *Christian Alphabet*, in an atmosphere of religious crisis. It goes without saying that where function prevailed over decoration and the monumental, the very concept of style changed, becoming restricted to a proportioning of unadorned surfaces with a uniform and repetitive rhythm. These were the features that gave the Escorial (with the exception of the solemn cathedral) its wonderful harmony and lent an air of extraordinary composure to the walls of the Collegio Romano in Rome. They suggest a rigorous conceptual life, which was something new that pervaded all Europe until the 17th century.

Not content with austerity, the Reformation only just failed to do away with religious architecture altogether, in the conviction (Acts 7, 48–49) that "the Most High dwelleth not in temples made with hands, as saith the prophet: Heaven is my throne and the earth my footstool. What house will ye build me? saith the Lord, or what is the place of my rest?" Moreover, there is some evidence of sermons preached in the open air in far-off Catholic mission lands and of ceremonies performed in open squares. Nevertheless, an even stronger Biblical influence prevailed: the theme (centuries old even in its symbolic or concrete Jewish manifestations) of the Temple of Solomon, which was at this time interpreted as a place suitably designed (without any divisions or hierarchical barriers) to enable the faithful to take a direct part in religious worship and ceremonial and, in particular, in the reading of the Bible.

One of the earliest and most significant examples of the new reformed churches, from the point of view of architectural and decorative quality, was the chapel (consecrated by Luther in 1544) in Schloss Hartenfels in Torgau. Originally it was on two floors (according to a custom intended at first to separate the people from the rulers but serving during this period to accommodate larger numbers of the faithful). Its furnishings were the simplest, consisting of altar, baptismal font, pulpit, and organ. The richest decorations were the carvings (by Simon Schröther) around the pulpit, illustrating Simeon meeting the Infant Christ, Child Jesus preaching to the doctors in the temple, and Christ driving the merchants out of the temple — three episodes that also refer to Luther's religious life. The altar, no longer containing relics and no longer the center of attention, is no more than a stone slab resting on four pillars or, rather, angels used as caryatids. Later a painted altarpiece illustrating the Last Supper was added.

Even larger places of worship, capable of holding thousands of people, were either rectangular or, later, elliptical in shape and so retained the appearance of a lecture hall, mainly due to the care taken to eliminate all pillars and columns and any hierarchical division between one portion of the building and another, thus ensuring maximum visibility for all. This was at the opposite pole from the large Gothic (and some Renaissance) churches, which, because of the cult of saints and relics, were made up of dozens of side chapels fronted by grilles and gratings, partially cut off and designed for a select few of the faithful. The introduction of galleries was an excellent solution, which probably spread from churches to theaters with boxes.

In "unreformed" Europe the new type of church could gain ground only in the areas occupied by the Huguenots and similar groups. In Italy almost the only examples are the unassuming Waldensian edifices in Angrogna and Ciabas, both built in 1555 and both having three aisles linked by wide arches, without bell tower or apse; the walls, evenly lit by windows, are painted white and innocent of all decoration. But similar trends, designed to emphasize the unity of the faithful during worship and to give the clearest possible view, appeared everywhere simultaneously. In this context may be mentioned the oratories, originally meeting places or chapels devoted largely to hymn singing and preaching; the idea originated with Gaetano Tiene (1480–1547) and was later developed by the Fathers of the Oratory. Some of the late-16th-century or baroque churches that now have side chapels were initially designed with unbroken walls and were altered only at a later date. The Jesuits widened the nave, making it more imposing and roomy enough to accommodate large numbers of the faithful.

The basic problem with which both Catholics and reformers were almost always faced was that of adapting existing churches. Generally speaking the procedure was the same: to destroy the presbytery, or choir, even when this was of value as a monument (in Italy Vasari, in particular, devoted himself to this type of remodeling). In the second half of the 16th century even Catholic churches, above all those owned by the stricter orders such as the Jesuits, were very simply decorated and whitewashed all over (one example still intact is S. Giovanni dei Fiorentini in Rome). The return to ornamentation came only later, when prodigal use was made of gold and paintings. The Gesù in Rome (PL. 450; FIG. 899) constituted a prototype that was later copied all over the world, developing the Renaissance conception of the central-plan church. In this connection Dorfles (1954) has remarked that "in the circular or elliptical plan single-aisle churches of the baroque period, the faithful are placed in a completely new position . . . , in more direct and intimate contact with the scene of the religious service . . . and with the altar, instead of being cut off from it." In the Romanesque basilica the faithful were still partly cut off from the scene of the sacred action.

Even though, as once before at the end of the age of antiquity, the Catholic world recognized these demands, it was extremely reluctant to give up the architectural repertory of orders, columns, and tympanums built up largely from the study of pre-Christian antiquity, which had come to symbolize the rebirth of the empire and the religious power that for centuries had dominated the political plans of the Roman Curia and guided its art patronage. It is perhaps this component of post-Reformation Catholic art that explains why the spread of the baroque was largely limited to the major political capitals of Europe and was fostered by those religious orders most directly subordinate to the Curia and therefore more or less officially representing it.

Probably only the collapse of the papacy could have achieved a wholesale simplification of architectural decoration such as had taken place in northern Europe, where religious edifices came to have bare whitewashed walls and nothing more; but

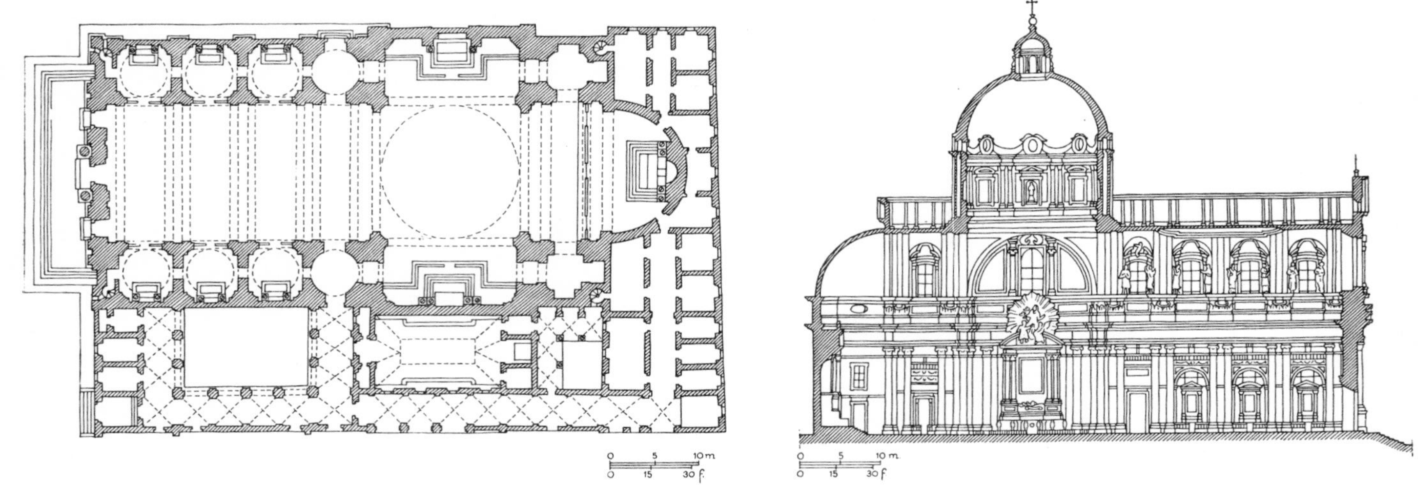

Rome, the Gesù, ground plan and side elevation.

it is interesting to follow some of the attempts made in the south to find a new basis for reconciliation between the pagan and Christian worlds — in other words, between culture and faith — based on more fitting symbolic interpretations. S. Serlio (*Regole generali di architettura* . . . Venice, 1584, IV, 6, 7, 8) furnishes a significant early example: The Doric order, formerly used for temples of Jupiter, Mars, and Hercules, now came to be dedicated to "Jesus Christ our Redeemer, or to Saint Paul, or to Saint Peter, or to Saint George, or to other such saints who were not soldiers but were manly and courageous in endangering their lives for the faith of Christ." The Ionic order, often seen in temples of Apollo, Diana, and Bacchus, was now recommended "for those saints whose lives were a combination of strength and tenderness, hence for those who were also wives and mothers." The Corinthian order, linked with the theme of virginity, was suitable for "temples dedicated to the Virgin Mary and to all those saints, both men and women, who lived as virgins," and for "monasteries and cloisters that house virgins dedicated to divine worship." One of the architectural masterpieces of the baroque era is the Church of S. Maria in Campitelli in Rome; from the point of view of its religious significance, it was been shown that here the Corinthian order is employed in its most monumental form in order to exalt Marian themes to combat the Lutherans.

PAINTING AND SCULPTURE. *The reaction against the Renaissance.* The visual arts in general and some specific artistic manifestations and themes in particular were the subjects of violent controversies directed against the use of images inspired by ancient art and associated — particularly the nude figure, both heroic and mythological — with representations of an erotic nature (certainly a much more widespread phenomenon than has yet been documented), with legendary hagiography, and the like. Even before the famous bonfire sponsored by Savonarola there were a number of condemnations of vanity, even in such a Humanist stronghold as Tuscany. St. Bernardino of Siena, as well as Nicolaus Cusanus, criticized some hagiographic works as remnants of paganism. St. Antoninus, Bishop of Florence, condemned the errors of artists not only *contra fidem* (as, for instance, the representation of the Trinity with three heads) but also those against morality: "quando formant imagines provocativas ad libidinem, non ex pulchritudine, sed ex dispositione earum, ut mulieres nudas et hujusmodi." Thus Botticini's *Assumption of the Virgin* (London, Nat. Gall.), which had been inspired by the writings of Matteo Palmieri, was covered between 1485 and 1500 (Blunt, 1956, p. 109).

In his *Treatise on Painting*, Leonardo, in what might be regarded as a comment on one of Savonarola's strict injunctions not to include portraits of real people in altarpieces, cited the case of someone who fell in love with one of his Madonnas: "The lover bought the picure and wanted to remove the emblems of divinity in order to be able to kiss the picture without scruples. But finally conscience overcame his sighs of lust and he was obliged to remove the painting from his house."

Devotional opposition to the Renaissance was accompanied by a parallel exaltation of the painting of the northern countries, which, less open to charges of immorality and neopaganism, was much favored by a number of popes, who used such works for their private devotions and summoned Flemish artists to work for them. During the Sack of Rome in 1527, a Flemish triptych, now in Cagliari Cathedral, was stolen from the bedroom of Clement VII. Adrian VI gave the post of keeper of the Vatican collections in the Belvedere to Jan van Scorel, who thus became Raphael's successor in this position.

The reassessment of medieval architecture and painting (unadorned churches were praised by Ghiberti and others), though sporadic, was slowly spreading. This trend culminated in high appreciation of the hagiographic treatment of Fra Angelico, whose *Last Judgment* was copied by Bartholomaeus Spranger toward the end of the Renaissance. Francisco de Hollanda praised the religious severity of the medieval images; in France, the Jesuits stoutly supported Gothic architecture.

Iconoclasm and Protestant sacred art. There is a vast difference, of course, between these manifold but localized instances of criticism and condemnation and the campaigns of destruction organized by citizens from various social classes, including even artists who had been converted to the Reformation; these campaigns were conducted in Switzerland, Germany, France, and England. The reformer Carlstadt (Karlstadt) began by tearing down and burning images in Wittenburg with the enthusiastic assistance of the crowd. The iconoclasm spreading throughout northern Europe would seem to have sanctioned the revolt against the ecclesiastical hierarchy. In the beginning this movement was directed against images and especially against those to which there was the greatest devotion. There were instances of aniconic outbreaks recorded in Weininger as early as 1523; in Paris, on May 31, 1528, a band of Huguenots destroyed the statue of the Madonna at the corner of Rue du Roi de Sicile (L. Réau, *Histoire du vandalisme*, Paris, 1959); in Switzerland the movement was fostered by Zwingli, who was an extremist on this point. Eventually this iconoclasm extended to the decapitation of the statues on porches and façades and the destruction of the screens and choirs that cut off the faithful from the altar, culminating in the firing and destruction of entire churches. The destruction wreaked in 1562, in particular, blotted out centuries of religious sculpture in many parts of France. Even John Calvin, although much opposed to images, was forced to protest: "Dieu n'a commandé d'abattre les idoles si non a chacun en sa maison," and he ordered the pastors in Lyons to put a stop to all violence, as it was injurious to religion. The procedure adopted in this wave of destruction has been vividly described in various chronicles and court records and is represented in such works as an engraving by Frans Hogenberg, executed in 1566. Paintings were burned and stained-glass windows smashed; the heads of statues were broken off and generally thrown into wells, from which they might still be recovered. However, altars and reliquaries were entrusted to private individuals for protection or were consigned to the local commune or sent to other cities, and a remarkably high proportion of images and similar objects were saved.

Although iconoclasm burst out in Italy at the time of the Sack of Rome, the movement never became so bitter as it did across the Alps, though there were some significant incidents. In 1558 the vicar of Farra (Aquileia) removed all images from the churches; ten years later, the record kept by the Venetian ambassadors relates that in Amandola "the exiles, who are said to have been accompanied by a number of unfrocked clergy, got in and cruelly burnt the churches and tore down and broke the images, displaying great contempt for all things sacred." In 1569 the bishop of Cortona (Florence, Archivio di Stato, Cosimo Records, No. 198) informed the grand duke of a widespread rumor in the city that it was forbidden to possess crucifixes or images, so that these were being either hidden or destroyed. It would seem that Rome itself was not immune to this movement of revolt. Late in the 17th century there were reports in Venice of sacrilegious acts being perpetrated collectively but in private against an unclothed statue of Christ.

Documentation of the opposition to the use of sacred images by private individuals in private and family chapels and particularly in monumental tombs would, if amply collected and studied, constitute an important chapter in this history. So far there are a number of indications of a specific and deliberate trend in this direction: for example, Renata di Francia (Renée of France) left a chapel decorated with nothing but geometric designs in the Castello Estense in Ferrara; and especially in Emilia and in those areas most exposed to Protestant influence, there was widespread use of tombs devoid of all images or with, at the very most, a bust of the deceased such as those that Michelangelo had helped to make popular. Evidence that the Inquisition was aware of this trend is demonstrated by the distrust and even prohibition of the use of private chapels and the tombs in them.

The attitude to images among the reformers themselves differed considerably, rising steadily in intensity from Erasmus's mild criticism to the complete rejection expressed by Calvin and Zwingli. In Calvin's *Christianae religionis institutio* (1st edition, 1536) there occurs the violent denunciation: "It was a Father of the Church who declared that it was a horrible abomination to see an image of Christ or of a saint in a Christian church. Moreover, this was not the voice of just one man; a decree to this effect was promulgated by a Church council stipulating that the things that are venerated must not be painted on the walls. They are far from respecting these limits, considering that they do not leave a single corner bare of images."

Fearful of all excess, Luther, in his *Wider die himmlischen Propheten* (against Carlstadt), reiterated his defense of at least the main images: "If it is not a sin, but indeed a good thing, for me to bear the image of Christ in my heart, why should it be a sin to have it before my eyes?" Luther was fond of Biblical stories, especially that of the Last Supper; however, he was convinced that revelation came by hearing and not by sight. There was a relationship between Luther and Lucas Cranach the Elder similar to that between Erasmus and Hans Holbein the Younger. In his religious paintings Cranach depicted the Biblical episodes exactly as they were recounted in the text and drew inspiration from such themes as original sin and the Redemption, which, particularly about 1529, were the very basis of the evangelical "protest." These paintings are based on accurate Biblical quotations, even in the smallest details, which therefore often strike an unusual note (as, for instance, the unicorn in the earthly Paradise, inspired by Num. 23: 22 and Ps. 92: 10). The altarpiece of the Stadtkirche of Wittenburg (the first iconoclastic city), executed in 1547 by Lucas Cranach, represents the cult of the Reformed Church (communion, baptism, confession, and preaching) and portrays its principal spiritual leaders: Luther, Melancthon, and Johann Bugenhagen.

It is quite possible that the popularity of prints beyond the Alps during the first half of the 16th century was related to the hatred of images. Il Ruzzante, addressing himself to Cardinal Cornaro, pointed out that the Lutherans in Padua did not want to go to confession, did not fast, no longer went to church, and would not look at "any images except those printed on paper." This attitude is easily understandable: the iconoclastic quarrel was with idols, that is, sculptured images (with constant reference to the Biblical golden calf); the print, an illustration of events in the Gospel, was saved by the fact that it was a narrative and not an anthropomorphization of the divine. On the contrary, it acquired great importance for its function in the educational activities so vigorously carried on by the reformers. From the esthetic point of view, even in the Catholic camp there were some interesting distinctions between the image or representation and its physical substance (anonymous manuscript, Rome, Bib. Casanatense, Ms. 2116, fol. 170–71: "Imago non est lignum, aurum, aut argentum, sed est in ligno, auro, et argento").

The importance attached to the arts in northern Europe is attested by the fact that antipapal propaganda was conducted largely by means of prints and illustrated books, ranging from Holbein's drawings for Erasmus of Rotterdam's *In Praise of Folly* to the astronomical and apocalyptic forecasts distributed in iconographic sheets and the collections of illustrated sacred texts. A vast quantity of material of high quality was published in this way and constitutes a high point in Renaissance graphic art, with a rich vein of themes ranging over the most widely differing fields, from the demoniacal to the magical, from caricature to realism, and from history to everyday life (PLS. 447, 448; Saxl, 1957). Even in Italy, as early as the 15th century Jacopo Sannazzaro in one of his poems (A. Altamura, *Jacopo Sannazzaro*, Naples, 1951) described a painting depicting the tricking of Ferdinand II of Aragon by three friars.

However, a Protestant type of sacred art developed only in the north and east of Germany, which were the only areas to remain Lutheran. This art, like that of the Catholic world, became gradually richer. In addition, in the field of the arts there was a certain tolerance for the religion professed by the artist: Cranach, for example, was able to work for Catholic sovereigns, although they were aware of his religious convictions. Nevertheless, a serious artistic consequence of the Reformation was the breaking off of all contacts between Germany and Italy (which had been very close at the beginning of the 16th century), even at the level of the employment of Italian masons and plasterers.

As already pointed out, Luther had taken a reasonably broad view of the representational complement to preaching, and Lutheran churches often had outstanding furnishings, with sculptures and paintings of high esthetic quality. The altar also performed a complex iconographic function, and its development was to a large extent furthered by the parallel narrative enrichment of tombstones, which obviously satisfied the need for prestige of the local dignitaries. Statistics prepared by Von Haebler (1957) concerning the subjects recurring in altarpieces between 1560 and 1660 show that the most popular subjects were the Last Supper and the Crucifixion and scenes connected with them, the Resurrection, the Evangelists, Moses, and allegories of the Virtues. Tombstones, on the other hand, bore scenes of the apostles, Christ meeting with sinners, and allegories of death. Confessionals, too, began to bear Biblical scenes. Occasionally (especially in Styria) there appeared whole cycles inspired by events in the Bible, accompanied by allegories and at times even making use of Christianized grotesques (such manifestations had occurred as early as 1530 in D. Hopfer's etchings). The most important of these cycles (which have been studied by E. Guldan and U. Riedinger, 1960) is in the chapel of Schloss Strechau and dates from 1579. In Switzerland, the Rhineland, and other areas there were examples of very delicate stuccowork in ceilings, devoid of anthropomorphic motifs and characterized by a certain aridity of design.

The Protestant and Reformed symbolism is of interest because of the absence of anthropomorphism. There are a number of 15th-century precedents for this: for instance, St. Bernardino of Siena's sun symbol and woodcuts such as that of Master E.S. (ca. 1470) depicting the Child Jesus in a heart surmounted by a cross. Luther's seal (used from 1516 onward and intended as the emblem of his theology) conveyed the following complex message: in the center a heart, surmounted by a black cross, was to recall the fact "that faith in Him Who was crucified makes us blessed; because if one believes with all one's heart, one is justified. And although this cross is black and mortifies and must also cause pain, it nevertheless leaves untouched the color of the heart, it does not destroy nature, in other words it does not kill but gives life, because the just man shall live by his faith, that is, his faith in the cross. However, this cross must be set in a white rose to indicate that faith gives joy, consolation, and peace and, in brief, places us in a pure and joyous rose; not the joy and peace that the world gives, and for this reason the rose must be white and not red; because white is the color of spirits and all the angels. Then the rose is set against a background of heavenly blue, because that joy in spirit and faith is the beginning of future heavenly joy which is already understood and hoped for here and now but which has not yet become manifest. And around the field there is a golden ring as a sign that the bliss of heaven lasts for eternity and is endless and also that it is more precious than any jewel, than any good, just as gold is the loveliest and most precious of metals" (letter dated July 8, 1530, to Lazarus Spengler). The Counter Reformation, too, made use of such abstract symbolic emblems: typical examples that have become widespread are the Heart of Jesus, the Heart of Mary pierced by seven swords, and the highly stylized Jesuit emblem.

Everywhere, moreover, "the number of emblematists who advertised this function of their books by including the word 'moral' or 'morality' in the title is almost distressing.... Baudouin, Bonomi, Borjia, Bruck, Bruno, Covaruvias y Horozco, Espinosa, Farley, La Perrière, Menestrier, Ricci, Schoonhovius, to mention a few. Probably, there were more 'moral emblems' in Spain than elsewhere" (R. J. Clements, *Picta Poesis, Literary and Humanistic Theory in Renaissance Emblem Books*, Rome, 1960, p. 103). It was very common for these works to begin with an invocation to God, and there was a great deal of specialization in the moralizing and religious sense (see *RlDKg*, s.v. Emblematica).

Developments in secular art. At this time there was considerable development in what had hitherto been regarded as the minor forms of art, such as portraiture, still life, and landscape painting, particularly in the richer and more peaceable areas of northern Europe. This trend had the effect of offsetting the serious economic and social predicament of many artists when commissions from ecclesiastical dignitaries were no longer forthcoming and they found themselves compelled to flee or, at least, to move from one place to another (as happened to Holbein). It may well be that the development of some of these art forms as independent specialties was due to the organization of specialized art shops following the politico-religious upheavals, as well as to the desire to transfer the pleasure derived from painting and sculpture to permissible fields far removed from theological troubles. Furniture, silverware, and objects wrought in precious stones or quartz also acquired a new magnificence: in other words, the profane took on the luxury of the sacred. It did not succeed, however, in getting away from the sacred altogether but continued to be at least allusive and symbolic.

It was in portraiture, which, in spite of the reformers' scruples, became very popular in the north owing mainly to the patronage of Erasmus, that religious and profane themes were most inextricably mingled. Massys (q.v.), Dürer (q.v.; who in a letter to Erasmus dated May 17, 1521, avowed his devotion to Luther), and Holbein (q.v.) painted dozens of portraits of him, which were distributed in the form of replicas, reproductions, engravings, coins, and seals. The popularity of portraiture would almost seem to be a form of psychological compensation for the condemnation of all types of hagiographic representation. In fact, the portraits of the Reformation leaders were objects of veneration. Next to Holbein, it was Cranach who produced the greatest number of such portraits (portraits of Luther dated 1519–29, 1532–33, 1537, 1543, 1546–48, 1552–53); many of these portraits were to be found even in the collections of Catholic sovereigns. A similar phenomenon is reported to have occurred in Florence toward the end of the 16th century, when, in spite of opposition from the Church, portraits of Savonarola were eagerly acquired and prized by people of all social classes.

In still life (q.v.), which had acquired an independent status by about 1550, an important theme, from the point of view of both style (e.g., the lifeless and dark coloring) and subjects (which frequently dealt with the passage of time and life: pipes smoking, flowers wilting, dead animals) was that of *memento mori* and *vanitas vanitatum* (mirrors, musical instruments, jewels, weapons). It thus took on a character of piety, but its message was more intimate and related more directly to the fate of the observer. The mingling of sacred and secular (encouraged by the 15th-century Flemish tradition, which had developed the theme of the *theatrum mundi* as a divine manifestation) is equally apparent in scenes of everyday life and allegorical portraiture: portrayals of skulls, either alone or with other objects, were added in pursuance of the *vanitas vanitatum* theme or perhaps, as Pariset has remarked, in observance of St. Ignatius Loyola's suggestion that one should pray with a skull in hand and eyes closed (see, for example, Georges de La Tour's *Magdalen*, IX, PL. 91).

The urge to ensure the moral purpose of all art made itself felt also in private dwellings in Catholic Europe, where personifications and allegories had for some time taken on an eminently didactic character. Perhaps the most outstanding example of this trend is the château of Diane de Poitiers at Anet (designed by Philibert Delorme): funeral motifs (palm trees and black marbles) symbolizing inconsolable widowhood (in spite of the fact that the lady in question was the King's favorite) mingle with others symbolizing the flight of time (on the clock set above the entrance, 1554, was inscribed the legend: "Cur Diana oculis labentes subjicit horas? Ut sapere adversis moneat, felicibus uti") and referring to the contrast between the earthly and the divine ("Splendida mereris magni palatia, Coeli. Non haec humana saxa polita manu"), not to mention the symbolism of the chapel itself, which was not intended to have altarpieces (perhaps an indication that Delorme, at least, may have had Calvinist tendencies). In Italy a similar, though isolated, example was the palace at Bomarzo (dating from 1552), commissioned by Vicino Orsino, and its richly adorned and highly original park. Such edifying texts as "Sperne terrestris; post mortem vera voluptas," "Medium tenuere beati," "Quid ergo?" and "Dirige gressus meos Domine" accompanied the path through the wild grounds, laid out in a symbolic-allegorical arrangement, along which were encountered monstrous and imaginary figures (X, PL. 137) and architectural structures that provided a kind of moral orientation emphasized by epigraphs, culminating in the little temple guarded by a three-headed Cerberus and surrounded by triangular columns bearing skull and crossbones.

Apart from the extreme cases cited here, the intent of "moralizing" profane art was manifested widely, as illustrated by the strange interpretations or explanations provided by commentators and at times even by the artists themselves for the mythological themes used in decorating homes and palaces. The task of defending the Castello del Buonconsiglio in Trent fell to Andrea Mattioli (*Il Magno Palazzo del Cardinale di Trento descritto in ottava rima*, Venice, 1539). In his reply to a caustic critic who had remarked that he would have preferred to see "a St. Peter, a Paul, a John, or Old Testament scenes" in the chapel forecourt, Mattioli, instead of emphasizing their allegorical significance, explained that since the ancient divinities depicted had been defeated by Christ, they were represented there "as a sign of victory." A few decades later the moral problem arose even in connection with the decoration of ordinary private galleries; thus Jacopo Zucchi, the painter, who was at work in the Palazzo Ruspoli in Rome, wrote a treatise about it that constitutes a commentary and defense of his work (*Discorso sopra li dei de' gentili e loro imprese*, Rome, 1602, reprinted in Saxl, 1927). However, classical culture and mythological themes occupied too prominent a position to be done away with altogether in Catholic countries and particularly in Italy. In any case, as is made abundantly clear in the various treatises of the period concerning emblems, a didactic or allegorical justification, even if a little farfetched, could always be found for them. Thus it was that, according to the sculptor himself, as recorded by Chantelou, Bernini's sensual *Apollo and Daphne* (II, PL. 272) escaped the censure of a scandalized prelate by acquiring at the last moment its still extant Latin distich, which transformed it from a highly erotic myth into a morality fable: "Quisquis amans sequitur fugitivae gaudia formas. Fronde manus implet, baccas seu carpit amaras."

The representation of everyday life, already deeply rooted in the Flemish painting tradition, was given new impetus by the Reformation in two directions: as a representation of country life, presumably coincident with the peasant revolts, and as a hidden allegory of Biblical episodes. The first type is represented by the series of engravings of peasants by Dürer (q.v.) and the monument he designed and included as a woodcut in his *Underweysung der Messung mit dem Zirckel und Richtscheyt in Linien* (IV, col. 525). Among the earliest treatments of the theme is a Lucas van Leyden engraving of 1510, which probably inspired the pastorals of Titian and later those of the Bassanos. The peasant theme culminated with Pieter Bruegel (q.v.), who may have been connected with a heretical sect or a religious group with unorthodox tendencies (Auner, 1956). Hidden allegories and even outright transposition of Biblical episodes appear in representations of the parable of the prodigal son, which served as a pretext for the inn scenes, scenes of seduction, and the like, so popular with the Caravaggio school.

After the great wave of expression in northern Europe in the early part of the 16th century, the greatest artistic preoccupation was the immeasurable distance between the divine creation and man, expressed with remarkable originality in landscapes, which ceased to be a mere background to a group of people and became rather the context within which people moved and against which they were measured until they were absorbed and lost in it. Such landscapes ranged from those of Joachim Patinir (q.v.) to those of the great 17th-century Dutch landscape painter Jacob van Ruisdael (q.v.; whose Mennonite education has been recorded by W. R. Valentiner, 1957, who drew attention also to the fact that Ruisdael's figures appear to be detached from the world at large but have still not found the peace and serenity they had hoped to find in solitude). Moreover, the insistence on the horizontal, on the flat line of the horizon (usually placed low), in marked contrast to the optimistic verticality of Gothic architecture and *imagerie*, can be taken as a hint of an unresolved fear of transcendence or of dramatic isolation.

With the suppression of the hagiographic and devotional tasks traditionally allotted to the arts came a compensatory growth of drawing and scientific illustration. Here again the origins are to be found in Italy, and in Leonardo in particular, for it was he who brought these arts to a peak of accuracy and skill; but there can be no doubt that in the Reformation areas the study of nature eventually acquired a powerful moral purpose.

The naturalists Gesner and Belon and the skilled illustrators who worked for them were convinced that it was only by investigating the world that it was "possible to have a more certain science . . . a surer knowledge" (E. Battisti, *L'antirinascimento*, Milan, 1962, p. 257). In this way drawing came to assume the function of a scrupulous document confirming the rationality of the universe. It is difficult to tell whether the realistic illustrations acquired, south of the Alps or the Pyrenees, the same character of heresy as the research that gave rise to them (Gesner's work was placed on the Index). This seems likely, however, in view of the vicissitudes of the celebrated ceramicist Bernard Palissy and of his *Recepte véritable par la quelle tous les hommes de la France pourront apprendre a multiplier e augmenter leurs trésors*, in which, with a multitude of Biblical and Gospel texts, he proposes "to build a palace or an amphitheater as a refuge for the Christians exiled during the periods of persecutions" (a reference to the Calvinists, whom he resolutely defended). In this refuge there were to be rustic grottoes with bays and niches in which were to be reproduced faithfully (in casts from life) all living things, for he was convinced that "sans Sapience, est impossible de plaire a Dieu." This idea of the study and reproduction of nature as a means of moral edification in the Catholic world was seized upon by Ulisse Aldrovandi in debate with Gabriele Paleotti, one of the most authoritative supporters of the value of devotional images. After referring to the collections of zoological and botanical drawings of the ancient Mexican emperors, Aldrovandi comments: "So much the more should Christian princes, who know St. Paul's saying that the invisible things of God are known through those which are visible, cause to have painted all those things in their kingdoms that are continually produced by nature."

The new spirituality in art. Without plodding through all the links between the new religious movement and painting and sculpture north of the Alps, it may be emphasized that the religious fervor and the trend toward mysticism and the interior life that were the glory of the second and third decades of the 16th century were reflected in the arts throughout Europe. Dvořák (1928) established a link (which later criticism was to examine still further) between El Greco's stylization and the renewal of spirituality all over Europe arising from Luther's determined effort to move religion into the sphere of the subjective and the personal. There is an undeniable connection between a more intense participation in the faith and the dramatic interiorization of artistic vision, even though the beginnings of this change may be dated back to the end of the 15th century, coinciding with the apocalyptic type of preaching that occurred all over Europe, the first great religious trials (such as that of Savonarola), and the first artistic expressions of unrest. The religious crisis in the arts is merely anticipated, or rather, summed up prophetically, in Dürer's terrifying Apocalypse series of 1498 (IV, PL. 467); it stands fully revealed, strengthened by mysticism, in Grünewald's Isenheim Altar of 1515 (IV, PL. 180; VII, PLS. 85–89, 91, 92), and in the south, in Botticelli's Savonarolian *Mystic Nativity* of 1501.

It is highly significant that Dürer became an enthusiastic admirer of Luther and that three of his Nürnberg pupils, G. Pencz, H. Sebald, and B. Beham, as well as others of Dürer's collaborators, were accused of no longer performing their traditional religious duties; that Grünewald appears to have been officially charged with inciting the peasants to revolt; that the intense and highly spiritual Tilman Riemenschneider (q.v.) was tortured for having taken part in the Peasants' War; that Niklaus Manuel Deutsch, the famed author of moral allegories, became a Calvinist and enrolled in the militia, giving up painting to do so; that the painter Hans Leu the Younger died in battle alongside Zwingli; that Urs Graf reproduced in a self-portrait an antiphon of Luther's; and that the most intense graphic art and the most biting sculpture in France was the work of Protestants. It would be possible, and probably accurate, to attribute even the principle of artistic genius, in the modern sense of inner inspiration, to the new Reformation spirituality. Such a concept of genius is, of course, a development of Neoplatonic theses, but it also represents the emergence of a new dimension of awareness, which, as compared with the simplicity and optimism of 15th-century art, is the common factor in the art of the 16th century.

No image better represents the new religious depth brought about by the Reformation, both in the north and in the south, than the pathetic figure of the suffering Christ (*ecce homo*). This derived from 14th-century prototypes (such as the famous one in Nürnberg), widely distributed in the form of prints at the end of the 15th century; but the suffering, and with it the humanity, were now accentuated and the sublime aspects of power and potential vengeance emphasized. Around this figure could be compiled an anthology of devotional texts of prime importance, beginning with St. Theresa herself, whose ecstasies took place in front of this very image.

A little later Giordano Bruno was to write: "We can love the gods only through their images" (*Il Candelaio*, VI, i), in a passage indicating also that the effectiveness of the image is dependent on the realism with which it is painted (visual realism, of course, rather than portraiture, allusive rather than descriptive, providing a starting point for the experiences of a St. Theresa, a St. Ignatius, a Vittoria Colonna, or a Giulia Gonzaga). It was at this time, too, that the question of imitation was discussed by theologians, who opposed the imaginative interpretations of the mannerists and the late Gothic artists and favored a scrupulous veracity in painting appropriate to the teachings of sacred Scripture, emended and expunged of arbitrary medieval interpolations. The result, after various instructions and orders issued by bishops, was an intense and dramatic narrative type of painting, at once sublime and tangible, in which the limited stylization is offset by close adherence to existential reality. This style was evidenced in Bologna, Milan, and Tuscany in the work of Lodovico Carracci, in the gloomy and devotional circle of Lodovico Cigoli, and in Caravaggio. It has been suggested (Battisti, 1960) that this unified style be termed, provisionally, the "Tridentine," a designation now accepted by Church historians.

The Reformation concept of grace, with its inherent pessimism concerning man's salvation, is dramatically reflected in the arts of each country — in the "majestic splendor, grandiose nobility, impressiveness, and sublime quality" imparted to images of the divine and of saints (Weise, 1957). The full daylight that, in religious works of the 15th century and the classicist 16th century, indicated the presence of the Divinity and his beneficent influence on the life of the world was gradually dimmed; now the miracle became an instantaneous act emphasized by a blinding light, striking the protagonists like a fiery sword. In churches, too, the sometimes dazzling radiance of the Gothic tradition and of the Tuscan Renaissance was replaced by an indirect and filtered lighting, so that the gilding, the marble veneers, and the frescoes with their deep colors all contributed to a shadowy mystical atmosphere. In painting, Venice and Lombardy tended especially to dark colors; in architecture it would seem to have been Rome that gave to the whole of Europe a new type of church designed for preaching.

Heretical artists in the Catholic world. It is difficult to identify artists aligned with the Reformation in Catholic countries because of the extreme caution necessitated by the stringent condemnations of the Inquisition and the reluctance in official circles and even among private individuals to give any hint of problems of faith. Moreover, the part played by artists in religious struggles was frequently ambiguous, since their adherence to Reformation theories might be apparent only in the easily saleable woodcuts and prints for which there was no rigorous Index like that for heretical books (although an imprimatur for these, too, had been considered, as in Spain in 1562).

The greater the cause for suspicion, the more the biographers of an artist skimmed over or camouflaged the facts of his private life. But it is known that at the beginning of the 16th century, within the framework of the fairly widespread libertinism linked with the religious rebellion, both Il Riccio (Andrea Briosco, d. 1532) and Piero di Cosimo (after 1515) died without the Last Sacraments; and violent accusations of heresy were made in writing (and apparently with some justification, in view of the iconography, which alludes to salvation by faith alone) against Pontormo's *Last Judgment* (1546–56) in the Church of S. Lorenzo in Florence (De Tolnay, "Les Fresques de Pontormo dans le chœur de S. Lorenzo a Florence, Essai de reconstitution," *CrArte*, 1950, pp. 38–51). It is well known, and in fact constitutes a starting point for any interpretation of Michelangelo's personality and work, that there was a close relationship between him and Vittoria Colonna and that he was on terms of friendship with Valdés's circle and the publishers of Lutheran texts. Both Palladio and the Sansovinos, in Venice, appear to have been in dangerous contact with Francesco di Giorgio Martini, who left a mystical interpretation of proportions and was tried as a heretic. Sebastiano del Piombo gave up painting just when iconoclasm began to spread, and in Venice Pier Paolo Vergerio ordered the removal of images of SS. Christopher and George, ex-voto offerings, and the like. Moreover, even in the liberal atmosphere of Venice, Veronese was brought to trial in 1573 merely because he had included some superfluous figures in his *Feast in the House of Levi* (VI, PL. 67) for the refectory of SS. Giovanni e Paolo, although he was not compelled to remake the painting; and Marescalco (Giovanni Buonconsiglio) was brought to trial in 1572 because he did not want to depict purgatory. Elsewhere the Inquisition was even more severe: Pietro Torrigiani was sent to prison in Spain (and died there) for having destroyed a sacred statue he himself had made (ca. 1528); and Passignano (Domenico Cresti) was sentenced to exile in 1581, in which year he moved to Venice. In Genoa Cesare Corte, a beloved pupil of Luca Cambiaso, was arrested in 1613. It was the extraordinary iconography of some masters that betrayed them, as with Lelio Orsi, the creator of visionary devotional pictures, about whom, however, contemporary sources are singularly silent.

In France, where the religious wars brought about a fearful decline in the visual arts (with the possible exception of architecture), it is easier to identify major personalities found openly even in the Calvinist camp, which, as has already been pointed out, was the most hostile to the use of images. These leading figures appear alongside such minor artists as François Main, Girard Lenet, and Jean Lefèvre, who have now practically disappeared from the history of art. Jean Duvet was very active both in Dijon and in Switzerland as a coinmaker, military engineer, enameler, designer of windows and tapestries, and above all print engraver (his violently antipapal Apocalypse series published in 1561 is particularly famous). The home of Jacques Ducerceau (q.v.), who was in touch with Renée of France, was sacked because he was a Protestant, and political upheavals compelled him to break off his series of illustrations of French buildings; in 1585 he left the king, who had summoned him to his service, declaring that he would rather leave than return to the Mass. It is known that at least one of his relatives, Salomon de Brosse, was a Protestant and built, among other things, the Temple of Charenton. In the field of painting, Sébastien Bourdon limited himself, for religious reasons, to landscapes, bambocciades, and portraiture; and violent religious unrest is at the root of Jacques Bellange's extraordinary stylization. In sculpture, Pierre Bontemps saw his career broken off by the religious persecutions in 1562. Ligier Richier, one of the most dramatic sculptors of tombs, fled from France in 1567 and died shortly afterward.

Art of the Counter Reformation. After prolonged theoretical discussions between the north and the south, the Council of Trent (PL. 449) promulgated an official ruling concerning liturgical art. It represented a moderate position similar to that adopted by Erasmus and Luther except for the marked predominance of hagiographic art in the Catholic world and the acceptance, in the name of tradition, of even the most superstitious and popular forms of devotion, with only an iconographic control. Thus Catholic Europe declared its support for an allusive idiom of images, one alien to excessive rationalization and almost unconcerned with the problem of illiteracy — the visual approach, with all its dangers (lack of knowledge, superficiality, excessive emotionalism, etc.), could take the place of the written word.

The decree, promulgated in Trent Dec. 3, 1563, during the 25th session of the Council, read as follows: "Representations of Christ, of the Virgin Mother of God, and of the other Saints are to be placed and retained especially in churches. . . . The cult given to them refers to the prototypes which these representations portray. . . . The bishops should attentively consider that the people are instructed and strengthened in reliving and continually calling to mind the truths of faith by the story of the mysteries of our redemption as portrayed in paintings and other representations . . . . The people should be taught, moreover, that the deity is not a material body such as can be seen with bodily eyes or represented in colors and shapes.... Finally, everything of a suggestive nature is to be avoided.... statues are not to be painted or decorated with excessive ornamentation.... Nothing that is purely secular, nothing that is unseemly should be seen. No one has permission to erect or cause to be erected in any place or church, however exempt, any unusual statue without the approval of the bishop...."

The theoretical defense of images was reinforced by legends as in the very early days of iconoclasm. As early as 1550 Fra Aleandro Bolognese, had, with the approval of Pope Julius III, sung the glories of the Holy House of Loreto, the miracle of Bolsena, and the apparition of St. Michael the Archangel on Mt. Gargano; and St. Charles Borromeo had restored the 15th-century images to their honorable position in the shrines of Rho and Saronno. But as the triumph of the images spread, the dispute between the rigorists and the laxists broke out afresh in Italy itself, where it was becoming urgent to follow up the general decree of the Council with specific regulations. Attempts to do so brought to light interesting diversities of taste and artistic criticism from one diocese to another, which have been treated by Paolo Prodi (1962) in the first systematic treatise of its kind.

St. Charles Borromeo came to the fore in this field in Lombardy because of his interest in art; his prohibitions were accompanied by positive instructions (meetings of artists to discuss iconographic problems with the bishops, as well as instructions governing the construction of churches, 1573; commitment to the use of images to instruct the faithful in connection with preaching, 1576; etc.). Themes based on the Apocryphal Gospels and on hagiographic legends were prohibited, and strict adherence to the facts recorded in the sacred texts was required, the addition of decorative and imaginative detail being discouraged. This severity of interpretation had already been advocated by Jan Vermuelen Molanus in Louvain (1570 and 1594) on theological and moral grounds.

In Bologna, Gabriel Paleotti was equally interested in the arts and more than ever convinced not only of their didactic but also of their expressive value; in 1582, in collaboration with, among others, the painter Prospero Fontana and the architects Domenico Tibaldi and Pirro Ligorio, he advanced a plea for greater pathos and an emotional and rhetorical style, always, however, within the context of a sober decorum and respect for historical truth. The difference in atmosphere between the diocese of Milan and that of Bologna corresponds fairly closely to the contrast between the sober, dramatic, and restrained naturalism of a Caravaggio and the cultural and iconographic eclecticism of the Carracci family, who had a much more modern approach in their acceptance of sensuality and and expressive eloquence. Paleotti was violently opposed, however, to any revival or continuation of classical iconography, although he was prepared to allow, in the details of pictures, the inclusion of sketchily suggested decorative or imaginative elements. He was equally suspicious of the allegories that had a considerable development in those areas more directly influenced by classical culture. Paleotti must therefore be regarded as very close in his rigorism to the sober religious sense of Lodovico Carracci (PL. 452). In 1583 Paleotti was appealed to for precise instructions as to how Scipione Pulzone should paint the *Assumption* in the Bandini Chapel in S. Silvestro al Quirinale in Rome — a work in which, in accordance with his advice, the Apostles are not looking at the Virgin assumed into heaven. The relative austerity and harshness in the condemnation of iconographic errors (which almost certainly had their effect in later condemnations of the work of Rubens, Caravaggio, and others) aroused significant counterattacks in Rome, where Paleotti moved in 1590. These attacks originated with the traditionalists — for example, the objections of Silvio Antoniano (P. Prodi, 1962), who pointed out that excessively severe criticism of images merely played into Protestant hands. In Florence, too, where the situation with regard to the arts was less complex, there was a sober rigorism, which led to the creation of works distinguished by an intensely dramatic atmosphere and remarkably controlled decoration against dark, simplified backgrounds.

Though he does not indicate the lines of cultural development, Zeri (1957) shows that in Rome this austerity appears to have been restricted to certain orders and patrons and particularly to the Jesuits until as late as the last decade of the 16th century; but there were various admixtures of a superficial, academic, illusionistic type of devotional art devoted entirely to moralistic and utilitarian purposes and excluding all poetry and esthetic emotion. Great religious painters like Caravaggio were regarded as incomprehensible exceptions. Caravaggio's realism, almost a heritage from Erasmian philosophy, was interiorized to a suspicious extent. In any case it is revealing to note that in his *Calling of St. Matthew* (III, PL. 41) Caravaggio seems to have had in mind Urs Graf's drawing of the enrollment of a mercenary, the real theme of which was Death lying in wait.

However, the great issue over which Church and artists were at loggerheads was undoubtedly the wide-scale use of the nude, resulting from the Humanistic and Neoplatonic exaltation of man and the influence of archaeology. Vasari has noted the scandal caused even by Fra Bartolommeo's *St. Sebastian*. In 1549 the discovery of Baccio Bandinelli's *Adam and Eve* in S. Maria del Fiore in Florence provoked enraged correspondence, and Michelangelo, who was blamed for having provided the model for this particular work, was denounced as "the inventor of filth." In a public confession in 1582, Bartolommeo Ammanati rejected one of his own nude carvings as sinful and urged members of the Accademia del Disegno in Florence to depict only clothed figures. An erotic legend surrounded Guglielmo della Porta's beautiful allegorical figures for the tomb of Paul III (VIII, PL. 208; IX, PL. 365); and a number of popes even proposed to dismiss all artists from their service. Michelangelo's *Last Judgment* (IX, PL. 536) aroused fiery discussion, and his nudes, having been condemned at Trent in 1564, were clothed by papal order; Clement VIII had to be prevented from doing away with them altogether by the Accademia di S. Luca, whose own activities were supervised by the Church. Other similar episodes occurred in the course of the 17th century; Innocent X and Innocent XI were leaders in the campaign to cover nudes. Michelangelo's *Leda* and other works were burned at the French court on moral grounds. There was violent opposition to the nude in Spain. In 1601 the friar José de Jesus Maria, a Carmelite, writing on the "excellence of the virtue of chastity" (*Excellentia de la virtud de la chastidad*, Alcalá, 1601), commented: "Angry at being dispossessed of his kingdom of the Indies, Satan has wrought his revenge by bringing this plague, this poison, into Spain — I refer to the scandalous images of nude women." Nevertheless Velázquez' sublime Venus (XIV, PL. 329) is an outstanding masterpiece of this century. Female fashions, too, underwent a change; extreme décolletage disappeared; dull, dark colors predominated; and skirts were worn longer. However, both tradition and the wide diffusion of archaeological culture made it impossible to suppress the nude completely in Catholic countries.

The great care taken to ensure that even the Child Jesus should not be depicted nude contrasts oddly with Ottonelli's observations (*Treatise on painting and sculpture, their use and abuse*, 1652, p. 38): "... another thoughtful theologian ... told me that in his opinion it was possible to paint the image of either a man or a woman completely nude, with the various parts of the body so placed and with such skillful arrangement of attitudes and positions and overlapping of the parts as to show no undesirable shamelessness." It was along this road, that of purity become prudery, that religious baroque art was inevitably to tread.

Viewed in historical perspective, it is obvious that the iconoclastic controversy acted as a sort of catalyst, and when the cult of images was restored and even extended, new shrines to the Virgin and the saints and in honor of the principal relics were erected on the very frontiers between the two faiths [e.g., the Chapel of the Holy Shroud (Sta Sindone) in Turin, in open reference to the Holy Sepulcher in Jerusalem so scorned in Reformation circles]. At first this trend was chiefly a new flowering of devotion to Mary; later, however, there developed specific iconographies such as those of the martyrs, in allusion to the risks run in preaching the faith among the infidels (cycles in S. Vitale and S. Stefano Rotondo in Rome, PL. 452), and of the apotheosis of the new saints and the triumph of the Catholic orders (see SAINTS, ICONOGRAPHY OF).

Once devotion and emotionalism had been accepted, and, above all, once Rome's political role and hegemony and her cultural mission had been confirmed, it was no longer possible to call a halt, at least in regard to the major public works designed to hymn the glories of the Church and in particular of the saints, and these became more than ever trophies and symbols of victory. The predominant theme in painting, in sculpture, and in architecture was that of triumph, of the Heavenly Jerusalem, of universal apotheosis. It is a theme that was unfolded with remarkable consistency from the baldachino to the throne of St. Peter in St. Peter's, from the emphatic "sacre conversazioni" by Rubens to the mystical ecstasies of St. Theresa (II, PL. 273) and the Blessed Lodovica Albertoni (II, PL. 279), a theme of great intimacy exposed to the full light of day as a proof of faith.

Devotional art. Devotional painting and sculpture in Catholic Europe developed, almost as an independent form of art, out of the need to express the pathetic, to make it possible through pictorial imitation for the onlooker to share in the feelings expressed by the devotional image. In this sense the painter became the rival of the preacher, for he was using similar rhetorical, symbolic, and analogical expedients, though he renounced all violent effects. To begin with, therefore, it was only a question of developing the emotional appeal already characteristic of Flemish painting and of that influenced from the north (such as the Venetian school) and abandoning the great Humanistic conceptual themes and the problems of tragedy and the sublime.

The violent reaction of Michelangelo's circle, representing the Humanist intellectuals, to this tendency is recorded by Francisco de Hollanda in Buonarotti's cutting comment: "Generally speaking, Flemish painting will be much more popular than Italian painting with those of a devout turn of mind. Italian painting will never cause the shedding of a single tear, whereas the Flemish variety will have to answer for the shedding of many. This is due to no vigor or goodness in the painting itself, but rather to the goodness of the devout individual in question. It will be liked by women, particularly the oldest and youngest among them, as also by nuns, monks, and those amiable souls who have no musical appreciation of true harmony." Vasari, in the *Lives* of 1550 (p. 809), contemptuously dismissed Giovannantonio Sogliani: "He was popularly acclaimed because he depicted pious and devotional scenes as is the practice of hypocrites." In 1568, however, (II, 519), obviously for censorship reasons, Sogliani was praised through clenched teeth: "His manner was popular because he depicted religious scenes as such scenes are liked by those who, without appreciating the fine points of art and of certain mannerisms, like things that are forthright, easy, gentle, and graceful."

In any case, devotional art had by then become established as a specialized branch in its own right. Girolamo Muziano, for example (as Peroni has pointed out) was highly praised in this field, his work being "free from all worldly lust and vanity but filled instead with all due modesty and gravity." Edifying legends came into being, the most outstanding of them concerning Fra Angelico. Later Carlo Dolci made a vow to paint only"sacred images, or sacred histories, represented in such a way as to bring forth fruits of Christian piety in those who look at them," and in Holy Week he "would never paint anything that was not in some way connected with the Passion of Our Lord" (F. Baldinucci, *Notizie de' professori...*, 2d ed., 1767–74, X, 379).

Affinities between certain devotional movements, particularly those which constituted cultural groupings or elites, and certain styles of painting may be observed. Thus, quietism, which was not formulated doctrinally until the latter half of the 17th century, had roots going much further back ("The true disposition of a soul wishing to contemplate God is a true desire to listen to God by silencing all its own thoughts, all affections of the will, all discourse.... When it has become passive, the soul loses itself in God, so much so that it is no longer interested even in its own salvation"; M. Petrocchi, 1948). This doctrine corresponds, in the field of painting, to a series of great masterpieces dating from the end of the 16th century onward and representing meditation, the reliving of the life of God in the soul, with perfect indifference to everything, in an atmosphere of night, untroubled by earthly affections or memories and therefore in complete and motionless solitude. This is true of the nocturnal Madonnas by Luca Cambiaso, in which are represented also the symbols of household tasks, as if to reinforce the theme of domestic life as a continual act of devotion. (It may be recalled that one of Cambiaso's pupils, Cesare Corte, was sent to prison for possessing heretical books, and he was himself the author of a commentary on the Apocalypses based on the writings of Luther and Calvin.) Cambiaso's works were collected by Rembrandt, whose religious sensibility is recognized though perhaps not fully explained; moreover, it has been assumed that there was a link between Cambiaso and Georges de La Tour (q.v.), whom F. G. Pariset suspects of occult practices.

The link with quietism can easily be confirmed in Spanish art: "If Zurbarán preferred to paint saints in mystic contemplation rather than cruel martyrdoms, one reads in Molinos that 'the soul gains more in ... prayer in complete withdrawal of the senses and mental powers ... than in penitent exercises, disciplines, etc. All this punishes only the *body*, but by withdrawal the *soul* is purified.' Molinos [founder of quietism] seems to contrast the religious approach of Murillo and that of Zurbarán by asserting: 'there are two ways of praying — tenderly, fondly, lovingly, and full of feeling — or humbly, dryly, without comfort and darkly' " (George Kubler and Martin Soria, *Art and Architecture in Spain and Portugal and Their American Dominions, 1500 to 1800*, Harmondsworth, 1959, p. 247). This relationship can be further traced not only in a great number of anonymous Spanish images (such as the Palazzo Spinola *Christ* in Genoa) but also in Georges de La Tour's silent and pathetic figures, with a candle symbolizing the inevitable passage of human life, equally futile in its earthly actions.

In discussing the various aspects of devotional painting, mention should be made of the admittedly exceptional tendency apparent in the *Seven Sacraments* series painted by Poussin (q.v.) for Cassiano dal Pozzo between 1640 and 1642 (PLS. 235, 236). His work provides a clear anticipation of the neoclassicism of a David; sacred scenes are peopled with figures garbed in ancient costume or idealized attire, set against an architectural background of extreme monumental simplicity, so much so that the buildings almost give the impression of being archaeological reconstructions, in spite of the contemporaneous details included. The action unfolds deliberately, with statuesque severity. It would be a mistake to regard this cycle merely as an expression of Poussin's classicism; it reflects a specific moral standpoint, which would appear to be connected with that vast current, so widespread in the Low Countries and in France, based on the works of Joest Lips (Justus Lipsius) and aiming at a synthesis of stoic morality (inspired by Seneca in particular) and Christianity, founded on the control and restraint of the passions. This moral standpoint later encountered violent opposition as being a "phantasm of virtue and constancy"; it would appear that passion was to be neither suppressed nor excessively moderated, since in the sphere of the arts Catholicism found it an indispensable basis.

Judging by a perusal of contemporaneous criticism, Guido Reni would appear to represent the extreme of moderation allowed at the time; Reni was undoubtedly the sacred painter par excellence in the 17th century, moderate in his habits (though generous), delicate in his imagination (Bernini described his work as heavenly), tender of sentiment, readily moved, with a deep devotion to the Blessed Virgin, and modest; buried in his Capuchin habit, his body remained incorrupt in the tomb ("not like one dead but rather like one who sleeps"). Even he, however, encountered difficulties over the addition of an illicit iconographic detail that he refused to remove and was forced to flee Rome.

Anton van Dyck (q.v.) was a fairly close counterpart of Guido Reni in Flanders, as was Murillo (q.v.) in Spain. Even as late as the 18th century, no one could look at Van Dyck's works, said Fierens (1956), "sans être touché et ému de compassion . . . par l'expression de douleur qu'on remarque sur les visages." And, he continued, using an expression applicable to all paintings of this type: "C'est donc parce qu'il fait appel à la pitié du spectateur, parce qu'il l'invite à pleurer sur la mort du Christ, a verser les larmes avec la Vierge et les saints, que le peintre 'inspire la devotion.' " Murillo, too, used the pathetic sweetness of his tenderly idealized Madonnas (X, PL. 203) to establish an emotional rather than a rational relationship, with great and lasting success.

It is significant that devotional art rapidly became established as a popular art serving the purpose of group participation in identical sentiments and activities such as pilgrimages, ex-voto offerings, and the acquisition of religious prints at local country markets. On the folk level it is apparent that the Catholic cult was accompanied by a most extraordinary and spontaneous representational, or at least ornamental, vitality. In the Alpine zone the popularizing of the baroque flourished to such an extent and was of such a quality as to lay dramatic emphasis on the division between the two religious and cultural areas for anyone traveling from north to south or vice versa, far more than the presence or absence of altarpieces in the churches. Those traveling from the German-speaking part of Switzerland into the canton of Ticino as early as the second half of the 16th century were fascinated by the crowds of faithful in the cemeteries, which were filled with wooden crosses and flowers; by the churches adorned with silver reliquaries held in great reverence; by the lamps burning continually before the altars and by the great wax candles; by the chapels and pillars bearing representations of the Crucifixion and by the images of the saints along the streets; by the great crosses erected in memory of the Passion at the entrance to each village (*Die Nuntiatur von Giovanni Francesco Bonhomini, 1579–1581*, Documenten, bearbeitet von Franz Steffens und Heinrich Reinhardt, Solothurn, 1906).

Nevertheless, only rarely (in Andrea Sacchi, Giovanni Paolo Pannini, etc.) did the magnificent spectacle of Roman Catholic ceremonies continue to provide material for pictorial inspiration and sincere religious feeling. The austere Gothic churches of the north, on the other hand, whitewashed like tombs, aroused the intimate but intense poetry of a Saenredam (q.v.; PL. 451) and prompted artists to the humility required to approach everyday life with conviction — in other words, they stimulated the transformation of profane art into something sacred.

BIBLIOG. C. Cantù, Gli eretici d'Italia: Discorsi storici, 3 vols., Turin, 1865–66; C. Dejob, De l'influence du concile de Trente sur la littérature et les beaux-arts chez les peuples catholiques, Paris, 1884; C. Gurlitt, Kunst und Künstler am Vorabend der Reformation, Halle, 1890; A. Ratti, ed., Acta ecclesiae mediolanensis, 54 fasc., Milan, 1890–99; P. Lehfeldt, Luthers Verhältnis zur Kunst und Kunstlehre, Berlin, 1892; A. Peltzer, Deutsche Mystik und deutsche Kunst, Strasbourg, 1899; J. Braun, Die belgischen Jesuitenkirchen, Freiburg im Breisgau, 1907; J. Braun, Die Kirchenbauten der deutschen Jesuiten: Ein Beitrag zur Kultur- und Kunstgeschichte des 17. und 18. Jahrhunderts, 4 vols., Freiburg im Breisgau, 1908–10; W. Weibel, Jesuitismus und Barockskulptur in Rom, Strasbourg, 1909; M. Reymond, L'école bolonaise, Rev. des deux mondes, LV, 1910, pp. 109–27; M. Reymond, L'art de la Contre-Reforme: Ses caractères généraux, Rev. des deux mondes, LVI, 1911, pp. 301–413, LVII, 1911, pp. 37–61; H. Tietze, Programme und Entwürfe zu den grossen österreichischen Barockfresken, Jhb-KhSammlWien, XXX, 1911, pp. 1–28; C. Rogge, Luther und die Kirchenbilder seiner Zeit (Schriftens des Vereins für Reformationsgeschichte, 108), Leipzig, 1912; J. Braun, Spaniens alte Jesuitenkirchen, Freiburg im Breisgau, 1913; G. Dehio, Die Krisis der deutschen Kunst im 16. Jahrhundert, Kunsthistorischen Aufsätze, Munich, Berlin, 1914, pp. 145–62; A. Foratti, La Controriforma bolognese e i Carracci, L'Archiginnasio, IX, 1914, pp. 15–28; H. Lehmann, Bildnisse auf Glasgemälden, Zwingliana, I, 1917, pp. 273–77; W. R. Valentiner, Zeiten der Kunst und der Religion, Berlin, 1919; H. Voss, Die Malerei der Spätrenaissance in Rom und Florenz, 2 vols., Berlin, 1920; A. Warburg, Heidnische-antike Weissagung in Wort und Bild zu Luthers Zeiten, Heidelberg, 1920; H. Grisar and F. Heege, Luthers Kampfbilder, 4 vols., Freiburg im Breisgau, 1921–23; W. Weisbach, Der Barock als Kunst der Gegenreformation, Berlin, 1921; Dictionnaire de théologie catholique, s.v. Images, VII, 1922, cols. 766–843 (bibliog.); L. Fendt, Der Lutherische Gottesdienst des 16. Jahrhunderts, Munich, 1923; A. Schramm, Die Illustration der Lutherbibel (Luther und die Bibel, I), Leipzig, 1923; E. Strong, La Chiesa Nuova, Rome, 1923; J. Braun, Der christliche Altar in seiner geschichtlichen Entwicklung, 2 vols., Munich, 1924; Concilium Tridentinum: Actorum pars sexta, IX, Freiburg im Breisgau, 1924, pp. 77–79; H. Feurstein, Zur Deutung der Bildgehaltes bei Grünewald, in Oberdeutsche Kunst der Spätgotik und Reformationszeit, Augsburg, 1924, pp. 137–63; R. Günther, Die Brautmystik im Mittelbild des Isenheimer Altars, Leipzig, 1924; N. Pevsner, Gegenreformation und Manierismus, RepfKw, XLVI, 1925, pp. 243–62; H. Preuss, Die deutsche Frömmigkeit im Spiegel der bildenden Kunst, Berlin, 1926; F. P. Barnard, Satirical and Controversial Medals of the Reformation, Oxford, 1927; F. Saxl, Antike Götter in der Spätrenaissance, Berlin, 1927; F. Buchholz, Protestantismus und Kunst im 16. Jahrhundert, Leipzig, 1928; M. Dvořák, Geschichte der italienischen Kunst im Zeitalter der Renaissance, II, Munich, 1928; D. Frey, Architettura barocca, Rome, Milan, 1928; W. Friedländer, Der manieristische Stil um 1580 und sein Verhältnis zum Übersinnlichen, Vorträge der Bib. Warburg, XIII, 1928–29, pp. 214–43 (Eng. trans., Mannerism and Anti-Mannerism in Italian Painting, New York, 1957); N. Pevsner, Beiträge zur Stilgeschichte der Früh- und Hochbarock, RepfKw, XLIX, 1928, pp. 225–46; N. Pevsner, Die italienische Malerei vom Ende der Renaissance bis zum ausgehenden Rokoko (Barockmalerei in den romanischen Ländern, I), Wildpark-Potsdam, 1928; M. Wackernagel, Apologetische Kirchenkunst, Schweizer Rundschau, XXIX, 1928, pp. 961–72, XXX, 1928, pp. 161–69; W. Weisbach, Gegenreformation, Manierismus, Barock, RepfKw, XLIX, 1928, pp. 16–28; M. Wackernagel, Die kirchliche Kunstbewegung, Schönere Zukunft, V, 1929, pp. 1014–15, 1036–37; O. Thulin, Johannes der Täufer in geistlichen Schauspiel der Mittelalters und der Reformationzeit, Leipzig, 1930; H. Bethe, Die Torgauer Dedikationstafel: Ein Beitrag zur sächsischen Bronzeplastik der Renaissance, Jhb-PreussKSamml, LII, 1931, pp. 170–75; P. Tacchi-Venturi, Storia della Compagnia di Gesù in Italia: La vita religiosa in Italia durante la prima età della Compagnia di Gesù, 2d ed., 2 vols., Rome, 1931; E. Mâle, L'art religieux après le Concile de Trente, Paris, 1932 (2d ed., Paris, 1951); P. Pirri, S.J., Chi fu l'architetto del Collegio Romano? Nota sopra il Padre Giuseppe Valeriani, La civiltà cattolica, LXXXIII, 3, 1932, pp. 251–64; G. Gabrieli, Federico Borromeo a Roma, Arch. Soc. romana di storia patria, LVI–LVII, 1933–34, pp. 157–217; C. L. Ragghianti, Cultura artistica e arte barocca, La Cultura, XII, 1933, pp. 196–209; F. Diethelm, Die Zürcher Geistlichkeit bekämpf Kirchenstuhlwappen als verwerfliches Scheinwerk, Zwingliana, VI, 1934–38, pp. 173–88; K. Simon, Die Züricher Täufer und der Hofgoldschmied Kardinal Albrechts, Zwingliana, VI, 1934–38, pp. 50–54; M. Vorenkamp, Bijdrage tot de Geschiedenis van het hollandsch Stilleven in de 17. eeuw, Leiden, 1934; W. Wiesbach, Zum Problem des Manierismus, Strasbourg, 1934, pp. 15–20; R. Longhi, Momenti della pittura bolognese, L'archiginnasio, XXX, 1935, pp. 111–35; H. Swarzenski, Quellen zur deutschen Andachtsbild, ZfKg, IV, 1935, pp. 141–44; W. Ohle, Die protestantischen Schlosskappellen der Renaissance in Deutschland, Stettin, 1936; M. Pest, Die Finanzierung des Süddeutschen Kirchen- und Klosterbaues in der Barockzeit, Munich, 1937; G. Schnürer, Katholische Kirche und Kultur in der Barockzeit, Paderborn, 1937; K. L. Schwarz, Zum ästhetischen Problem des "Programmes" und der Symbolik und Allegorik in der barocken Malerei, Wiener Jhb. für Kg., XI, 1937, pp. 79–88; G. Stuhlfauth, Künstlerstimmen und Künstlernot aus der Reformationsbewegung, Z. für Kirchengeschichte, LVI, 1937, pp. 498–514; G. Weise, Titians als religiöser Maler, Die christliche Kunst, XXXIII, 1937, pp. 161–86; E. Wind, Studies in Allegorical Portraiture, Warburg, I, 1937–38, pp. 138–62; P. Francastel, Le réalisme du Caravage, GBA, XX, 1938, pp. 45–62; G. Weise and W. Otto, Die religiösen Ausdruckgebärden des Barock und ihre Vorbereitung durch die italienische Kunst der Renaissance, Stuttgart, 1938; U. Weymann, Die Seussesche Mystik und ihre Wirkung auf die bildende Kunste, Berlin, 1938; A. Graziani, Bartolomeo Cesi, CrArte, IV, 1939, pp. 54–95; B. Knipping, Die iconographie van de Contra-Reformatie in de Nederlanden, 2 vols., Hilversum, 1939–40; G. Rehbein, Malerei und Skulptur des deutschen Frühmanierismus, Würzburg, 1939; F. Saxl, Rembrandt's Sacrifice of Manoah, London, 1939; A. L. Mayer, Notas sobre la iconografia sagrada en las obras del Greco, AEArte, XIV, 1940–41, pp. 164–68; H. Rosenau, The Synagogue and Protestant Church Architecture, Warburg, IV, 1940–41, pp. 80–84; G. Schreiber, Prämonstratensische Frömmigkeit und die Anfänge des Herz-Jesu-Gedankens, Z. für katholische Theologie, LXIV, 1940, pp. 181–201; G. Weise, Il rinnovamento dell'arte religiosa nella Rinascita, La Rinascita, IV, 1941, pp. 659–80; G. Weise, Das Transzendente als Darstellungsvorwurf der Kunst des Abendlandes, Festschrift für W. Waetzoldt, Berlin, 1941, pp. 65–80; W. Pinder, Rembrandts Selbstbildnisse, Leipzig, 1943; J. Ehrmann, Massacre and Persecution Pictures in 16th Century France, Warburg, VII, 1944, pp. 195–99; P. Lavedan, L'architecture française, Paris, 1944 (Eng. trans., London, 1956); T. P., A proposito di iconografia trinitaria, Ephemerides Liturgicae, LVIII, 1944, pp. 181–84; O. Benesch, The Art of the Renaissance in Northern Europe, Cambridge, Mass., 1945; R. M. de Hornedo, El arte en Trento, El Concilio de Trento: Exposiciones e investigaciones pro colaboradores de "Razón y Fe," Madrid, 1945, pp. 333–62; E. Kirschbaum, L'influsso del Concilio di Trento nell'arte, Gregorianum, XXVI, 1945, pp. 100–06; O. Prjevalinsky Ferrer and F. S. Escribano, Lutero y Calvino en el Palacio Real á fines de siglo XVIII, B. Soc. española de excursiones, LVIII, 1945, pp. 163–66; D. Cantinori, Riforma cattolica, Società, II, 1946, pp. 820–34; H. Jedin, Katholische Reformation oder Gegenreformation?, Lucerne, 1946; C. L. Ragghianti, Miscellanea minore di critica d'arte, Bari, 1946, pp. 178–81; M. Petrocchi, Il quietismo italiano del seicento, Rome, 1948; P. Boesch, Der Zürcher Apelles: Neues zur den Reformatorenbildnissen von Hans Aspern, Zwingliana, IX, 1949–53, pp. 16–50; E. Mâle, L'art religieux du XIIe au XVIIIe siècle, Paris, 1949 (Eng. trans., New York, 1949, repr. 1958); E. Massa, Egidio da Viterbo, Machiavelli, Lutero e il pessimismo cristiano, Arch. di filosofia, "Umanesimo e Machiavellismo," 1949, pp. 75–123; F. de Dainville, S.J., Lieux de théâtre et salles des action dans les collèges de Jésuites de l'ancienne France, Rev. d'h. du théâtre, II, 1950, pp. 185–90; P. Francastel, Baroque et classicisme: Histoire ou typologie des civilisations, Annales, XIV, 1950, pp. 142–51, A. Gerlo, Erasme et ses portraitistes, Brussels, 1950 (bibliog.); L. Hiesberger;

Die österreichische Reformation und Gegenreformation im Spiegel der Geschicht-Schreibung, Vienna, 1950; A. Huxley, Art and Religion, AN, XLIX, April, 1950, pp. 21–23; C.-A. Isermeyer, Il Vasari e il restauro delle chiese medievali, Studi vasariani, Florence, 1950, pp. 229–36; G. Spini, Ricerca dei libertini, Rome, 1950; C. Galassi Paluzzi, Storia segreta dello stile dei Gesuiti, Rome, 1951, pp. 30–31; J. A. Jungmann, Das Konzil von Trient und die Erneuerung der Liturgie, Das Weltekonzil von Trient: Sein Werden und Wirken, I, Freiburg im Breisgau, 1951, pp. 325–36; P. Roques and R. P. Regamey, La signification du Baroque (La Maison Dieu, XXVI, 2), Paris, 1951; G. Schreiber, Der Barock und das Tridentinum: Geistgeschichtliche und kultische Zusammenhängen, Das Weltekonzil von Trient: Sein Werden und Wirken, I, Freiburg im Breisgau, 1951, pp. 381–425; C. Angeleri, Il problema religioso del Rinascimento: Storia della critica e bibliografia, Florence, 1952; St. Charles Borromeo, Arte sacra (De fabrica ecclesiae, ed. C. Castiglioni and C. Marcora), Milan, 1952; A. Chastel, L'Apocalypse en 1500, Bib. d'humanisme et renaissance, XIV, 1952, pp. 124–40; J. A. Jungmann, Missarum solemnia, 3d ed., 2 vols., Vienna, 1952 (Eng. trans., F. A. Brunner, The Mass of the Roman Rite, 2 vols., New York, 1951–55); K. Lankheit, Dürers "Vier Apostel," Z. für Theologie und Kirche, XLIX, 1952, pp. 238–54; P. Pecchiai, Il Gesù di Roma descritto ed illustrato, Rome, 1952; H. M. Rotermund, The Motif of Radiance in Rembrandt's Biblical Drawings, Warburg, XV, 1952, pp. 101–21; H. M. Rotermund, Rembrandt und die religiösen Laienbewegungen in der Niederländer seiner Zeit, NedKhJb, IV, 1952–53, pp. 104–92; C. de Tolnay, The Autobiographic Aspect of Fra Filippo Lippi's Virgins, GBA, XXXIX, 1952, pp. 253–64; C. A. Beerli, Le peintre poète Nicolas Manuel et l'évolution sociale de son temps (Tr. d'Humanisme et Renaissance, II), Geneva, 1953; E. Castelli, ed., L'umanesimo e il demoniaco nell'arte (Atti II Cong. int. di s. umanistici), Rome, Milan, 1953; G. G. Coulton, Art and the Reformation, 2d ed., Cambridge, 1953; C. Galassi Paluzzi, Essenza e funzionalità dell'arte della Riforma Cattolica, Fede ed arte, I, 1953, pp. 42–51; L. Grassi, Barocco o no, Milan, 1953; Lukas Cranach der Altere: Der Künstler und seine Zeit, Berlin, 1953; W. Mrazek, Ikonologie der barocken Deckenmalerei, SbWien, CCXXVIII, 3, 1953, pp. 1–88; M. Petrocchi, Il problema del Lassismo nel secolo XVII, Rome, 1953; R. Robres Lluch, El Beato Ribera y el Greco, AEArte, XXVII, 1953, pp. 245–55; J. Seznec, The Survival of the Pagan Gods: The Mythological Tradition and Its Place in Renaissance Humanism and Art, New York, 1953; P. Chaix, Recherches sur l'imprimerie à Genève de 1550–1564 (Tr. d'Humanisme et Renaissance, XVI), Geneva, 1954; R. Crozet, La vie artistique en France au XVIIe siècle, 1598–1661: Les artistes et la société, Paris, 1954; G. Dorfles, Antiformalismo nell'architettura barocca della Controriforma, Rettorica e Barocco (Atti III Cong. int. di s. umanistici), Venice, 1954, pp. 47–52; P. Giovanni da Locarno, Saggio sullo stile dell'oratoria sacra nel Seicento, esemplificata sul P. Emmanuele Orchi, Rome, 1954; H. E.'s Jacob, Idealism and Realism, Leiden, 1954; J. Pinkerfeld, The Synagogues of Italy: Their Architectural Development since the Renaissance, Jerusalem, 1954 (in Hebrew); U. Spirito, Barocco e Controriforma, Retorica e Barocco (Atti III Cong. int. di s. umanistici), Venice, 1954, pp. 209–16; W. Drost, Tizian und Tintoretto: ein Epochenwende, Universitas, X, 1955, pp. 935–48; J. Ehrmann, Antoine Caron: Peintre à la cour des Valois (Tr. d'Humanisme et Renaissance XXXVII), Geneva, 1955; W. F. Friedländer, Caravaggio Studies, Princeton, 1955; F.-G. Pariset, Mise-au-point provisoire sur Georges de la Tour, Cah. de Bordeaux, II, 1955, pp. 79–89; P. Pirri, G. Tristano e i primordi della architettura gesuitica, Rome, 1955; L. Réau, Iconographie de l'art chrétien, 3 vols. in 6, Paris, 1955–59; U. Spirito, Barocco e Controriforma, Medioevo e Rinascimento, Misc. in onore di Bruno Nardi, Florence, 1955, pp. 701–14; O. Thulin, Cranach-Altäre der Reformation, Berlin, 1955; J. Weerda, Holbein und Calvin: Ein Bildfund, Neukirchen, 1955; F. Arcangeli, Sugli inizi del Carracci, Paragone, VII, 79, 1956, pp. 17–48; M. Auner, Pieter Breugel: Umrisse eines Lebensbildes, JhbKhSamml-Wien, LII, 1956, pp. 51–122; A. Blunt, Artistic Theory in Italy, 1450–1600, Oxford, 1956; A. Chastel, Le problème de Caravage, Critique, XII, 1956, pp. 949–67; N. Z. Davis, Holbein's Pictures of Death and the Reformation at Lyons, S. in the Renaissance, III, 1956, pp. 97–130; E. du Gué Trapier, Valdés Leal: The Baroque Concept of Death and Suffering in His Paintings, New York, 1956; P. Fierens, Le sentiment religieux chez Van Dyck, Scritti di storia dell'arte in onore di Lionello Venturi, II, Rome, 1956, pp. 43–58; C. Gnudi, ed., Mostra dei Carracci (introd.), Bologna, 1956 (cat.); P. Michaelis, Die Dornerkrönung als Triumph Christi, Festschrift W. Sas-Zaloziecky, Graz, 1956, pp. 119–24; W. J. Müller, Der Maler Georg Flegel und die Anfänge des Stillebens, Frankfurt am Main, 1956; L. van Puyveld, Bernardo Passeri, Marten de Vos and Hieronimus Wierix, Scritti di storia dell'arte in onore di Lionello Venturi, II, Rome, 1956, pp. 59–64; G. Radetti, Umanesimo e riforma nella prima metà del secolo XVI, Giorn. critico della filosofia italiana, XXXV, 1956, pp. 210–25; Schlosser; S. Slive, Notes on the Relationship of Protestantism to 17th Century Dutch Painting, AQ, XIX, 1956, pp. 3–15; E. Tietze-Conrat, Il "Cristo portacroce" di Sebastiano del Piombo, Emporium, LXII, 1956, pp. 99–104; L. A. Veit and L. Lenhart, Kirche und Volksfrömmigkeit in Zeitalter des Barock, Freiburg im Breisgau, 1956; I. Bergström, Ikonologiska studier (Symbolister, I), Stockholm, 1957; H. von Campenhausen, Die Bilderfrage in der Reformation, Z. für Kirchengeschichte, LXII, 1957, pp. 96–128; D. Cantinori, Note su alcuni aspetti della propaganda religiosa nell'Europa del Cinquecento, Aspects de la propagande religieuse (Tr. d'Humanisme et Renaissance, XXVIII), Geneva, 1957, pp. 348–50; A. Chastel, L'art et le sentiment de la mort au XVIIe siècle, B. Soc. d'ét. du XVIIe siècle, 36–37, 1957, pp. 287–93; M. Florisoone, La mystique plastique du Greco et les antécedents de son style, GBA, XLIX, 1957, pp. 19–44; P. Francastel, Baroque et classicisme: Une civilisation, Annales, XII, 1957, pp. 207–22; E. Guldan, Ein protestantischen Bildzyklus in der Steiermark, Ö. Z. für K. und Denkmalpflege, XI, 1957, pp. 8–12; H. K. von Haebler, Das Bild in der evangelischen Kirche, Berlin, 1957; G. Nicodemi, L'accademia di pittura, scultura ed architettura fondata dal cardinale Federigo Borromeo all'Ambrosiana, S. in onore di C. Castiglioni, Milan, 1957, pp. 653–96; A. Plebe, La sacralità della musica in Platone, negli Stoici e nello Pseudo. Plutarco, Arch. di filosofia, 1957, no. 2, pp. 185–94; E. Przywara, S.J., Schön, Sakral, Christlich, Arch. di filosofia, 1957, no. 2, pp. 11–29; F. Saxl, Lectures, 2 vols., London, 1957; V. L. Tapié, Baroque et classicisme, Paris, 1957 (Eng. trans., A. R. Williamson, The Age of Elegance, London, New York, 1960); A. Tenenti, Il senso della morte e l'amore della vita nel Rinascimento, Turin, 1957; W. R. Valentiner, Rembrandt and Spinoza: A Study of the Spiritual Conflicts in 17th Century Holland, London, 1957; G. Weise, Manieristische und frühbarocke Elemente in den religiösen Schriften des Pietro Aretino, Bib. d'humanisme et renaissance, XIX, 1957, pp. 170–207; F. Zeri, Pittura e Controriforma: L'arte senza tempo di Scipione da Gaeta, Turin, 1957; G. Alberigo, Studi e problemi relativi alla applicazione del Concilio di Trento in Italia (1545–1558), Riv. storica it., LXX, 1958, pp. 239–98; E. Castelli, Le démoniaque dans l'art, Paris, 1958; H. Gouhier, Le refus du symbolisme dans l'humanisme cartésien, Umanesimo e simbolismo (Atti IV Cong. int. di s. umanistici), Padua, 1958, pp. 65–74; H. B. Gutman, Zur Ikonologie der Fresken Raffaels in der Stanza della Segnatura, ZfKg, XXI, 1958, pp. 27–39; L. Ponnelle and L. Bordet, S. Philippe Neri et la société romaine de son temps, 1515–1595, 2d ed., Paris, 1958; E. Schaffran, Der Inquisitionprozess gegen Paolo Veronese, Das Münster, XI, 1958, pp. 209–12; Umanesimo e simbolismo (Atti IV Cong. int. di s. umanistici), Padua, 1958; G. Alberigo, I vescovi italiani al Concilio di Trento (1545–1547), Florence, 1959; A. Bea et al., Cor Jesu: Commentationes in litteras evangelicas Pii P. P. XII "Haurietis aquas," Rome, 1959; S. Bottari, Situazione del Vignola, Atti e mem. della Deputazione di storia patria per le antiche prov. modenesi, 6th ser., XI, 1959, pp. 3–8; M. C. Donnelly, Calvinism in the Work of Jacob Jordens, AQ, XXII, 1959, pp. 356–66; F. Fasolo, Moralità del "cantiere" barocco: in margine a note d'archivio, Arch. sacra, VII, 1959, pp. 48–59; P. Francastel, Baroque et classicisme: Histoire ou typologie des civilisation ?, Annales, XIV, 1959, pp. 142–51, 719–31; C. Gilbert, The Archbishop on the Painters of Florence, AB, XLI, 1959, pp. 75–87; A. Krücke, Der Protestantismus und die bildliche Darstellung Gottes, ZfKw, XIII, 1959, pp. 59–90; K. Lankheit, Das Triptychon als Pathosformal (Abh. Heidelberger Akad. der Wissenschaft, 3), Heidelberg, 1959; P. Prodi, Il cardinale Gabriele Paleotti, Rome, 1959; A. Rüstow, Lutherana Tragoedia Artis, Schweizer Mnh., XXXIX, 1959, pp. 891–906; G. Sinibaldi, ed., Mostra del Cigoli, San Miniato, 1959 (cat.); P. Barocchi, ed., Trattati d'arte del Cinquecento, 3 vols., Bari, 1960–62; E. Battisti, Rinascimento e Barocco, Turin, 1960; J. E. D'Angers, Sénèque et le stoicisme dans le "De bono senectutis" (1595) du cardinal G. Paleotti, B. de lit. ecclésiastique, LXI, 1960, pp. 39–49; K. Goldammer, Kultsymbolik des Protestantismus, Stuttgart, 1960; E. Guldan and U. Riedinger, Die protestantischen Deckenmalereien der Burgkapelle auf Strechau, Wiener Jhb. für Kg., XVIII, 1960, pp. 28–86 (bibliog.); T. Müller, Zur Augsburger Goldschmiedekunst der Reformation, Pantheon, XVIII, 1960, pp. 16–19; E. Raimondi, Trattatisti e narratori del Seicento, Milan, Naples, 1960 (bibliog.); G. Cozzi, Intorno al cardinale Ottavio Paravicino, a Monsignor Paolo Gualdo e a Michelangelo da Caravaggio, Riv. storica it., LXXIII, 1961, pp. 36–68; C. J. Friedrich, The Age of the Baroque, New York, 1961; A. Peroni, Il Collegio Borromeo di Pavia: Architettura e decorazione, IV Centenario del Collegio Borromeo di Pavia, 1561–1961, Pavia, 1961, pp. 111–61; G. C. Argan, S. Maria in Campitelli, Barocco Europeo e Barocco Veneziano, Venice, 1962, pp. 63–76; P. Barocchi, Un "Discorso sopra l'onestà delle imagini" di Rinaldo Corso, Misc. di s. di storia dell'arte in onore di Mario Salmi, III, Rome, 1962, pp. 173–91; P. Prodi, Ricerche sulla teoria delle arti figurative nella riforma cattolica, Arch. it. per la storia della pietà, IV, 1962, pp. 123–212; G. Weise, L'ideale eroico del Rinascimento e le sue premesse umanistiche, Naples, 1962.

Eugenio BATTISTI

Illustrations: PLS. 447–454; 2 figs. in text.

REMBRANDT. Dutch painter, etcher, and draftsman [b. Leiden, July 15 (?), 1606; d. Amsterdam, Oct. 4, 1669]. From 1632 onward he signed himself Rembrandt on paintings and etchings and occasionally on drawings; in documents his full name, Rembrandt Harmensz. van Rijn or Rhijn, is used.

Although esthetic measures differ, Rembrandt is considered the greatest artist of the northern Netherlands. For most 20th-century observers Rembrandt's greatness lies in his understanding of and sympathy for the complex world of feelings and emotions, in his ability to represent these in all types of subjects (portraits as well as scenes from real life, the Bible, history, and mythology), and in his use of the media, exhausting their possibilities for expressing these emotions. Since these characteristics are fully developed in Rembrandt's late art, these works are today the most appreciated.

Rembrandt's work and the documentary evidence confirm that he was a deeply religious man. He was a member of the Reformed Church as a young man and perhaps throughout his life. In his later years he came into close contact with the Mennonite community. He probably was not dogmatic in religious questions. Hardships, such as the deaths of relatives or a deterioration in social circumstances, did not hamper his creativity. He collected art and art objects eagerly (as Rubens did) and dealt in art (as Vermeer did). His collection in 1656 was large, varied, and of high quality. Probably less a sound businessman than an art lover, he acquired articles for their beauty rather than their value. Because of his collecting, his comparative wealth, his manifold ties with the circles of scholars, preachers, and writers, and his success as a teacher, Rembrandt represented in the 1630s and 1640s the type of the worldly, humanist artist (not that of the artist-craftsman or the artist who combined another profession with painting, types common among his contemporaries). His personal interpretation of that type was criticized by those who favored a traditional continuation of the artist as humanist and academician (Hoogstraten, Sandrart, Houbraken). Thus Rembrandt's use of the teaching methods practiced in academies, especially that of having classes draw after the nude or from plaster casts, was ridiculed by Sandrart, because for him the results made Rembrandt's instruction a mockery of academic teaching. Rembrandt had a great interest in Italian art, greater than was often assumed. He borrowed principles of composition and color schemes, the latter especially from Venetian 16th-century art, but interpreted them in a highly personal way. His art, especially after 1642, occupied a singular position in the 17th century. It has neither the "classical" characteristics of Nicolas Poussin or Claude Lorrain nor the "baroque" of Rubens or Bernini; it lacks the mathematical clarity of Vermeer, the religious devotion of Francisco de Zurbarán, and the dramatic directness of Caravaggio. Nor are the essential characteristics of his art found in that of other artists, not even of his pupils. Although probably not a "heretic in art" (Andries Pels in *Gebruik en Misbruik des Tooneels*, 1681), Rembrandt was, in life as well as in art, a nonconformist.

He was equally creative as a painter, etcher, and draftsman. Approximately 600 paintings, 290 etchings, and 1400 drawings are generally attributed to him. He did not subordinate drawing to painting or etching, nor did he consider the graphic technique a reproductive medium. With few exceptions the three were practiced independently and in their own right. The three categories are therefore treated separately: first, before 1631 during his life in Leiden when rapid changes took place each year; then, after his move to Amsterdam, in each of the periods in which Rembrandt's life and work can be divided.

The three basic catalogues for Rembrandt's works are: A. Bredius's *The Paintings of Rembrandt* (Vienna, London, 1935), abbreviated as Br. throughout; A. M. Hind's *Catalogue of Rembrandt's Etchings* (2 vols., 2d ed., London, 1923), abbreviated as H.; and O. Benesch's *The Drawings of Rembrandt, First Complete Edition in Six Volumes* (London, 1954–57), abbreviated as B.

LEIDEN, 1606–31. Rembrandt was the fifth of six children of a miller, Harmen Gerritsz. van Rijn and his wife, Neeltgen Willemsdr. van Zuytbroeck, daughter of a baker. Rembrandt's father owned half a mill and some houses in Leiden and can be considered as rather well-to-do (Groot, 1906, pp. 5–8, esp. 82). At the age of thirteen, in May, 1620, he was enrolled as a student at the University of Leiden (Groot, 1906, p. 11). It is likely that his parents wanted him to receive a humanist training (Orlers, 1641) and that he received a thorough education primarily in theological, but also in classical and historical, matters. It is not known at what age he left the university. Orlers's statement that Rembrandt "had not the slightest inclination or desire" for such an education and that he followed his natural predilection for art, which forced his parents to take him out of school, is probably exaggerated and perhaps borrowed from the legendary lives of great artists (cf. Vasari on Giotto). After leaving the university, Rembrandt became a pupil of Jacob Isaacsz. Swanenburgh (ca. 1571–1638), with whom he stayed, according to Orlers, for three years (1621–23?; the time may actually have been shorter, for Orlers may have wanted to overemphasize the teaching of a Leiden artist), then probably of Jan Pynas in Amsterdam (according to Houbraken), and certainly of Pieter Lastman in the same city (1624/25?). Swanenburgh's teaching was probably limited to that of the fundamentals of painting. Of the three, Swanenburgh's influence is not noticeable, that of Pynas can be established, and the impact of Lastman was very strong and lasting. All three artists were Italianate painters of historical scenes; none was among the avant-garde artists who created a new naturalistic style, especially in landscape and portraits. Soon after Rembrandt had established himself as a painter he started teaching, an activity to which he devoted much of his time and attention throughout his life. In this period, more than later, some of Rembrandt's pupils and assistants worked with him on a painting. Rembrandt seems to have established a kind of association with Jan Lievens (1607–74). Lievens was influenced by Rembrandt, although he differed in style and character, a difference noted by Constantijn Huygens. Gerard Dou (1613–75) became a pupil of Rembrandt in February, 1628. A painting by Dou with important additions by Rembrandt is in the National Gallery, London; paintings by Rembrandt in collaboration with Lievens are in museums in Schwerin and Amsterdam. Other pupils between 1628 and 1631 were Isaac de Jouderville, Jacques de Rousseau, and Jacob van Spreeuwen (see Bauch, 1933 and 1960).

The earliest known paintings by Rembrandt are dated 1625 (*Stoning of St. Stephen*, Lyons, Mus. B. A.) and 1626 (*Consul Cerealis*, also called *The Justice of Brutus*, Br. 460, Leiden, Stedelijk Mus. De Lakenhal; *Balaam*, Br. 487, Paris, Mus. Cognacq Jay; *Expulsion from the Temple*, Br. 532, Moscow, A. S. Pushkin Mus. of Fine Arts; *Tobit and Anna*, Br. 486, Paris, Coll. Baroness Bentinck; and *Musical Party*, Bauch, 1960, fig. 101, U.S.A., private coll.). The earliest etchings can be dated in the same year (*Circumcision*, H. 388; *Rest on the Flight into Egypt*, H. 307). These paintings and etchings depict scenes from the Bible or historical subjects, and are traditional in style. The composition and color of these paintings and etchings are mainly influenced by Lastman; the technique of the etchings, by Gerrit Pietersz. (Sweelinck). Only one painting, undated but probably executed in 1626, deviates from this style (*Head of a Soldier*, Br. 132, Lugano, Thyssen-Bornemisza Coll.) in showing strong contrasts in light and dark on a face that is not embellished or idealized. Here Rembrandt for the first time portrayed a person in historical costume. The first drawings known, probably executed in or around 1627, are in red and black chalk, a technique borrowed from Lastman, and represent standing soldiers (B. 3–5) similar to those in his paintings of this time.

By 1627 or 1628 certain features that would be characteristic of the artist's later work were already apparent: the *Gold Weigher* (1627, Br. 420, Berlin, Staat Mus.) is dominated by chiaroscuro effects; *St. Paul in Prison* (1627, Br. 601, Stuttgart, Staatsgalerie) is primarily a representation of such feelings as doubt, surprise, and bewilderment; the *Presentation in the Temple* (1628/29, Br. 535, Hamburg, Kunsthalle) conveys the mystery of the presence of God; *Two Scholars* (1628 — perhaps representing Elisha and Elijah; see print by Lorenzo Monaco and Bauch, 1960, p. 143 — Br. 423, Melbourne, Nat. Gall. of Victoria) is Rembrandt's humanized interpretation of the old theme of the Dispute. The *Triumph of David* (1627, Br. 488, Basel, Offentl. Kunstsamml.) is, in execution and composition, still very much indebted to Lastman.

In the etchings certainly made in 1628 (*Rembrandt's Mother*, H. 1–2), Rembrandt exploited the possibilities of the medium for swift, expressive lines and heavy contrasts between light and dark. In drawings his use of chalk in some studies of beggars (B. 30–32), still close to Lastman, became entirely personal in the *Elevation of the Cross* (B. 6, Rotterdam, Mus. Boymans-Van Beuningen) and in the study of a *Seated Old Man* in red chalk (B. 7, Berlin, Kupferstichkabinett) for the painting in Melbourne. Here too the dual purpose is the creation of meaningful contrast and expressive physiognomies.

Paintings. From 1629 to 1631 Rembrandt began to formulate his ideas with greater conviction and precision. His paintings of historical subjects, as well as those from the Bible and mythology, decreased in number, while his studies of heads and portraits gradually increased. His style moved rapidly toward a concentration on one event or statement, while scenes, figures, and objects of secondary importance were subordinated to the main point: from *Judas Returning the Thirty Pieces of Silver* (1629, Whitby, Mulgrave Castle, Marquis of Normandy Coll.) to the *Supper at Emmaus* (ca. 1630, Br. 539, Paris, Mus. Jacquemart-André) and the *Raising of Lazarus* (Br. 538, Solothurn, Dübi-Müller Coll.). In single figures the representation of mood and character became more sophisticated: from the abovementioned *St. Paul* to *St. Paul* (ca. 1630, Br. 602, Nürnberg, Germanisches Nat.-Mus.) and *Jeremiah Lamenting the Destruction of Jerusalem* (ca. 1630, Br. 604, Amsterdam, Rijksmus.). In the *Flight into Egypt* (Bauch, 1960, fig. 86, Tours, Mus. B.A.; painted 1628/29), dominated by chiaroscuro effects, he introduced a new element, that of everyday simplicity and intimacy. In the *Presentation in the Temple* (1631, Br. 543, The Hague, Mauritshuis), Rembrandt summed up the achievements of his previous four or five years in historical painting by reintroducing a large number of people (in the earlier *Presentation*, Br. 535, there were no onlookers), but subordinating them to the main theme by the distribution of light and dark, thus heightening the dramatic effect of the scene.

Studies of heads and portrait studies increased in number every year after 1629: *Rembrandt's Father* (1630, Br. 76, Innsbruck, Tiroler Landesmus. Ferdinandeum; Br. 78, Kassel, Germany, Staat. Kunstsamml.; Br. 81, Chicago, Art Inst.); *Portrait of an Old Man* (1630, Br. 141, Kassel, Staat. Kunstsamml.); *Portrait of a Young Man* (1631, Br. 143, Toledo, Ohio, Mus. of Art). Self-portraits followed at regular intervals, but with great variety, ranging from quick character studies (1629, Br. 4, Cambridge, Mass., Fogg Art Mus., P. M. Warburg loan coll.; Br. 5, Amsterdam, Rijksmus.) to finished portrait studies of the artist in fancy costumes (1629, Br. 8, Boston, Isabella Stewart Gardner Mus.; 1631, Br. 16, Paris, Mus. du Petit-Palais) to penetrating self-analyses that were, in addition, attempts to convey mood and character through form and posture as well as through light and shadow (ca. 1628, Br. 1, Kassel, Staat. Kunstsamml.; also 1629, Br. 2, Munich, Alte Pin.) to the first "official" carefully finished self-portrait (ca. 1629, Br. 6; PL. 456).

Etchings. Between 1629 and 1631 Rembrandt used etching only occasionally to depict Biblical scenes, first in an effort to give the print the character of a pen sketch (ca. 1629, *Peter and John at the Gate of the Temple*, H. 5, and *Flight into Egypt*, H. 17), and later to create more finished and detailed representations with strong contrasts in light and dark (1630, *Christ Disputing with the Doctors*, H. 20, and the *Presentation in the Temple*, H. 18), the last two similar in composition and spirit to the painted *Presentation* (1629, Br. 543). Apparently Rembrandt tried without success to achieve a deep and uniform black (the plate of H. 17 was cut and reused after very few impressions). The *Raising of Lazarus* (ca. 1631, H. 96; PL. 464), although based on a composition by Lievens, shows that Rembrandt had mastered the technique of etching to obtain a most dramatic effect in composition and expression.

In self-portraits Rembrandt used etching at this time primarily to investigate the expressive possibilities of the line; his etched self-portraits are studies in facial expressions (1629, H. 4, the largest and boldest of the early etched self-portraits; 1630 and 1631, H. 3, 29–34, 36, 55–58, probably also H. 2*). In some studies of an old man, probably his father, Rembrandt tried in addition to capture the essence of old age and to study what contrasts in light and dark could be obtained in this medium (e.g., H. 23–24, 26–28, 47). A number of studies of beggars, picturesque in their shabbiness and rendered with sensitivity, originated in these years (e.g., H. 7–16). As in painting, Rembrandt completed before the end of 1631 the first etched portraits that can be called finished, those of himself, his father, and his mother (H. 54, 92, 51–52) of which H. 51, 92, 54 (in that order, from left to right) probably were meant to form a triptych.

Drawings. In his drawings as in his etchings of this period, Rembrandt sought to create meaningful chiaroscuro in well-defined forms in studies after nature, mainly of old people (B. 37–46), using black or red chalk or both and occasionally the brush (self-portraits B. 53–54). During this period he sometimes used the brush in drawings to approximate the effect of oil paint when the drawings were intended as studies for paintings, such as *Judas Returning the Thirty Pieces of Silver* (B. 8, B. 9 recto; the most elaborate brush drawing, *The Reading*, B. 52, Bayonne, Mus. Bonnat, was probably also intended as a study for a painting), or, more often, to increase the contrast between light and dark, either in substitution for the pen (beggars, B. 24–25) or together with the pen (self-portraits, B. 53–54; beggar, B. 36). One of the most remarkable and touching drawings from this time is the portrait of his father in the year of his death (1630), drawn with red and black chalk and brown wash (B. 56, Oxford, Ashmolean Mus.); this is the first complete portrait drawing Rembrandt made.

The only drawings of religious scenes done before Rembrandt moved to Amsterdam that are extant, apart from the sketches for the *Judas* (B. 8, B. 9 recto), are the *Elevation of the Cross* (ca. 1628, B. 6, Rotterdam, Mus. Boymans-Van Beuningen) and *Christ at Emmaus* (1629, B. 11, formerly Berlin, W. Bode Coll.). Since these as well as the *Judas* sketches are studies for paintings, it seems that Rembrandt had not yet developed the habit of making drawings of Biblical scenes whenever the need to record a passage struck him.

AMSTERDAM, 1631–36. Rembrandt was firmly established in Leiden and had a flourishing studio there (Aernout van Buchell and Constantijn Huygens reported how famous he was). Exactly when he moved to Amsterdam is not known, but the event was of profound significance for him. He went from a town where the university set the tone to a world trading center.

Rembrandt is last mentioned as living in Leiden on June 30, 1631 (Groot, 1906, p. 20). In July, 1632, he was staying in the house of Hendrick Uylenburgh, art dealer, painter, and publisher, in Amsterdam. Whether he moved to Amsterdam in 1631 or early in 1632 is of little importance. The *Anatomy Lesson of Dr. Nicolaas Tulp* (Br. 403, The Hague, Mauritshuis) must have been commissioned shortly before or after Jan. 31, 1632, when the anatomical demonstration that provided the subject for the painting took place. Furthermore, the portraits of two Amsterdam merchants, that of Nicolaas Ruts, who headed a firm trading with Moscovia, dated 1631 (Br. 145, New York, Frick Coll.) and that of Maerten Looten, dated Jan. 11, 1632 (Br. 166, Los Angeles, County Mus.) must have been painted in Amsterdam. Rembrandt's most successful years were those between 1631 and 1636. He became the leading portrait painter of Holland, received important commissions for paintings of religious subjects, led the life of a well-to-do citizen, and married Saskia, the wealthy daughter of Hendrick Uylenburgh, mayor of Leeuwarden (they were betrothed June 6, 1633, and married June 22, 1635). Rembrandt's art during these years represents the phase in his work — often called "baroque" — when movement, the outward expression of emotions, and strong lighting effects prevailed.

Soon after he arrived in Amsterdam he had a large number of pupils, some of them talented and gifted. Jacob Adriaensz. Backer came to Rembrandt's studio in 1631 or 1632, Ferdinand Bol, Govert Flinck, and Jan Victors about 1633, and Gerbrand van den Eeckhout about 1636. Other artists were strongly influenced by Rembrandt's style, above all Salomon Koninck by the works of 1630–32.

Paintings. The most important painting and perhaps the commission that decided the move to Amsterdam is the *Anatomy Lesson of Dr. Nicolaas Tulp.* It is surprising that the commission was given to Rembrandt instead of to one of the well-established Amsterdam portrait painters such as Nicolaes Eliasz. Pickenoy or Thomas De Keyser. He had never painted a group portrait and was much younger than his colleagues. The anatomical demonstration in painting had a long and well-established tradition (see VII, cols. 663–64), from which Rembrandt radically deviated. Instead of retaining a symmetrical, isocephalic, and planar scheme of the persons around the corpse, he placed the figures in an asymmetric, triangular, and spatial arrangement. The painting, in addition to being a group portrait of Tulp and the members of the Amsterdam Guild of Surgeons, is a tribute to Vesalius, a *memento mori*, and a reminder of the punishment for sin (since only cadavers of capital offenders could be used for dissection; see Heckscher, 1958). The colors are subdued, the painting surface smooth, the brush stroke fine.

Undoubtedly Rembrandt was influenced by De Keyser in establishing his portrait style in 1631–32. It is already apparent in the afore-mentioned portraits of Nicolaas Ruts and Maerten Looten and even more noticeable in the knee-length, life-size ones such as those of the poet Jan Hermansz. Krul (1633, Br. 171, Kassel, Staat. Kunstsamml.) and of an *Elegant Gentleman* and his wife (1633, Br. 172, Cincinnati, Art Mus., and Br. 341, Met. Mus.), or portrait busts such as *Portrait of a Gentleman* and its companion piece (1632, Br. 159, 338, both in Brunswick, Herzog-Anton-Ulrich Mus.), or the *Portrait of a Man* (1632, Br. 163, Vienna, Kunsthist. Mus.). Also close to De Keyser in composition is the *Portrait of a Couple* (1633, Br. 405, Boston, Isabella Stewart Gardner Mus.). Completely different from the Gardner portrait is *The Ship-builder and His Wife* (1633, Br. 408, London, Buckingham Palace, Royal Colls.), in which Rembrandt introduced a daring new concept of action and even an element of genre. He transformed the double portrait tradition, of which the Gardner painting is part, just as he had modified the tradition of anatomical demonstrations. He did not, however, take up this trend in double portraits — except in the "disguised" self-portrait with Saskia (1635/36, Br. 30, Dresden, Gemäldegal.) — and in 1641 he even reverted to the De Keyser type (portrait of the Mennonite preacher Cornelis Claesz. Anslo and his wife, Br. 409, Berlin, Staat. Mus.). The subdued colors, the careful, often meticulous rendering of details, the evenly distributed light, the excellent likenesses in these and the other portraits of 1632–33 (approximately 50) apparently met a demand. The "official" style reaches its climax in the portraits of Marten Soolmans and Oopjen Coppit, formerly thought to be of Maerten Day and Machteld van Doorn, of 1634 (Br. 199, 342, Paris, Coll. Baron Robert de Rothschild). In a few portraits of perhaps less official character, those of Jacques de Gheyn III (1632, Br. 162, London, Dulwich College Picture Gall.), Maurits Huygens (1632, Br. 161, Hamburg, Kunsthalle), and Saskia (1633, Br. 97, Dresden, Gemäldegal.), a less smooth brush stroke is noticeable.

Rembrandt's numerous self-portraits between 1631 and 1936 show the same characteristics as his portraits of sitters of these years. He portrayed himself in carefully posed attitudes (Br. 12, 16–17, 20–23, etc.), sometimes in fancy dress (Br. 16, 20, 22–25), and his facial expression seems self-imposed and self-conscious (especially Br. 16–17, 20, 22). The self-portrait that sums up Rembrandt's mood and his approach to portraiture is the *Self-Portrait with Saskia on His Knee* (Br. 30), which contains an allusion either to the story of the Prodigal Son (Neumann, 1922; Sumowski, 1958) or to that of Samson and Delilah (Goldscheider, 1960).

Although Rembrandt was very busy as a portrait painter he still had time to paint historical scenes. He had started out as a painter of historical subjects and would always remain more inclined toward the historical than the present, even when painting landscapes. His mythological and religious subjects of these years are full of movement and space; the feelings and emotions are expressed with pathos; the action of the figures is theatrical. Mythological scenes, often with many figures, reflect the influence of Rubens as well as of Antonio Tempesta, whereas in the nudes the emphasis is on realism rather than on classical beauty: *Rape of Proserpina* (1631/32, Br. 463, Berlin, Staat. Mus.); *Rape of Europa* (1631/32, Br. 464, New York, Paul Klotz Coll.); *Landscape with the Bath of Diana and the Discovery of Callisto's Fault* (1635, Br. 472, Rheda, Germany, Coll. Fürst zu Salm-Salm). The format of these paintings is considerably larger than the one Rembrandt used in Leiden. Single figures were either personifications of a general kind, such as of Wisdom (*Minerva*, Br. 465, formerly New York, Wildenstein Coll.; Br. 466, Berlin, Staat. Mus.), studies of the nude (*Bathsheba*, 1632, Br. 492, Rennes, Mus. B. A., Br. 494; Ottawa, Nat. Gall. of Canada), or portraits in disguise (Bellona, 1633, Br. 467, Met. Mus.; *Sophonisba*, also known as *Artemisia*, 1634, Br. 468, Prado; *Minerva*, 1635, Br. 469, formerly Coll. Wennergren). To the latter category belongs the *Portrait of a Boy as Amor* (1634, Paris, Coll. Baroness Bentinck). The last of his paintings with mythological subjects during these years was the *Rape of Ganymede* (1635, Br. 471, Dresden, Gemäldegal.), a most anticlassical representation of the theme; its purpose and meaning remain enigmatic (a connection with Aquarius has been suggested by J. G. van Gelder, 1948).

Rembrandt rarely treated religious subjects in Amsterdam until 1633. One of the few was the remarkably advanced *Holy Family* of 1631 (Br. 544, Munich., Alte Pin.), which, because of the large size of the figures and the air of tender care of parents for their child, foreshadows his later paintings of the same subject. Probably by 1633 Rembrandt also finished two paintings for Prince Frederick Henry of Orange, an *Elevation of the Cross* and a *Descent from the Cross* (Br. 548, 550, Munich, Alte Pin.); perhaps a recently found *Christ on the Cross* (1631, Le Mas d'Agenais, near Agen, France, parish church) is connected with these (cf. Bauch, 1962). Shortly afterward Rembrandt took up the subject of the *Descent* again (1634, Br. 551, Leningrad, The Hermitage), now using some of the effects obtained in the reversal of the composition in the etching he had made of the earlier painting. In 1636 Rembrandt added an *Ascension* to the series for Prince Frederick Henry (still later, in 1639, an *Entombment* and a *Resurrection*, and ca. 1646 two more; see below). In this series, Rembrandt's most distinguished commission, it is evident that he tried to live up to "official" standards by borrowing heavily from Rubens.

The agglomeration of small, moving figures to a group that forms the expressive center of a composition, already established in the *Rape of Proserpina* (Br. 463), was fully developed in the dramatic *Storm on the Sea of Galilee* (Br. 547, Boston, Isabella Steward Gardner Mus.) where the colorful figures are tossed on a dark sea under an ominous sky by waves illuminated by a golden light. Soon the drama would be expressed by fewer people, and by individuals rather than crowds. This important point in Rembrandt's development is marked by two paintings, *Abraham's Sacrifice* (1635, Br. 498, Leningrad, The Hermitage) and *The Blinding of Samson* or *The Triumph of Delilah* (1636, Br. 501; PL. 460). Movement remained an important element: in the former, the angel is rushing to grab Abraham's hand at the very moment he is going to cut his son's throat; in the latter, one of the soldiers is piercing the eye of the struggling Samson while Delilah is leaving the tent swiftly and triumphantly, holding Samson's hair in one hand and the scissors in the other. The colors in both paintings are vivid; in *The Blinding of Samson* they are dominated by the luminescent blue of the light streaming into the tent.

In contrast with these rather large, colorful paintings with few figures, Rembrandt executed two smaller paintings in grisaille, both with large crowds of people. Both works are oil studies for etchings, one (1633, Br. 546, London, Nat. Gall.) for *Christ before Pilate* (H. 143), the other, *St. John the Baptist Preaching* (Br. 555, Berlin, Staat. Mus.), for an etching that was never executed. In contrast with Rubens, Rembrandt did not make sketches in oil for larger paintings but only for etchings, apparently to establish in grisaille the values of light and dark. In *St. John the Baptist Preaching* he gathered a great number of people in a spacious landscape setting for the last time. The painting presents a broad spectrum of reactions from different types of people to one trying to make his message understood.

Etchings. Rembrandt etched few portraits after his move to Amsterdam; apparently he preferred to paint portraits and to etch religious scenes. The self-portraits, such as that of 1633 (H. 108), remained primarily character studies in light and dark, although less sketchy than before, or studies in fancy dress, such as the one of 1634 (H. 109) in which he represented himself in Oriental costume, holding a saber. The portrait of the preacher Jan Cornelis Sylvius (1634, H. 111), seated, and that of another preacher, Jan Uytenbogaert (1635, H. 128), in pose and of oval format, are reminiscent of earlier engraved portraits, such as those by Jan van de Velde. As in the paintings, a separate category consists of "disguised" portraits. To this group belong, in addition to the self-portraits, the *Four Oriental Heads* (1635, H. 131–34), partly based on etchings by Lievens and executed with the assistance of pupils (it is not clear whether historical figures are meant) and the *Great Jewish Bride* (1635, H. 127), a carefully and thoroughly finished etching, the subject and meaning of which are not known. In its manner of presenting the figure, with grandeur in the pose and costume, as well as in the strong light and intense blacks, the print is characteristic for this period.

The religious subjects of these years are full of movement and pathos, particularly his two most ambitious undertakings, the *Descent from the Cross* (1633, H. 103), executed with the assistance of pupils, and more especially the *Angel Appearing to the Shepherds* (1634, H. 120). In its subordination of man to nature, in the darkness of the setting of figures and animals, and in the vehemence of motion, the latter etching is comparable to the mythological paintings mentioned above. The same element of pathos is present in *Christ before Pilate* (H. 143), which in size, finish, and ambitiousness is comparable to prints made after Rubens's paintings around the same time. A similar pathos is present in *The Tribute Money* (H. 124), *Christ at Emmaus* (1634, H. 121), *Christ Driving the Money Changers from the Temple* (1635, H. 126) and *The Stoning of St. Stephen* (1635, H. 125). *The Return of the Prodigal Son* (1636, H. 147; IV, PL. 434) marks the beginning of a new approach to the religious scene in etchings (see below).

Drawings. Very few studies for portraits and even fewer finished portrait drawings by Rembrandt are known. Two portrait drawings from this period are the delicate, touching *Portrait of Saskia*, delineated with silverpoint on vellum right after she and Rembrandt became engaged (June 8, 1633; B. 427, Berlin, Kupferstichkabinett), and the large *Portrait of a Man in an Armchair Seen through a Window Frame*, executed in colored chalks and ink on vellum (1634, B. 433, New York, Mrs. Charles S. Payson Coll.), in which Rembrandt for once followed the tradition of Federico Zuccari and Hendrik Goltzius in drawing colored portraits.

The most numerous of the drawings are the figure studies, only a few of which can be related to known paintings and etchings, such as the study for the *Great Jewish Bride* of 1635 (B. 292, Stockholm, Nationalmus.). The distribution of light and dark in the etching is studied in this drawing, which for that reason differs stylistically from other contemporaneous ones. The sketching of figures from life, a practice Rembrandt had started in Leiden, was intensified in Amsterdam. Beggars, women, and children were studied in the streets and in homes. The number of drawings Rembrandt made of women and children increased considerably toward 1636.

In Rembrandt's treatment of religious subjects (see II, cols. 513–14) his drawings have characteristics similar to those found in his paintings and etchings, although sometimes the drawings are more advanced than the paintings. Similar to paintings in expressing strong emotionality are such drawings as *Jacob Lamenting* (ca. 1635, B. 95), *Christ Bearing the Cross* (B. 97, probably a study for a painting), and the *Calvary* (B. 108) — all three in Berlin, Kupferstichkabinett — and the *Annunciation* (B. 99, Besançon, Mus. B. A.). *The Raising of the Daughter of Jairus* (ca. 1632/33, B. 61, formerly Haarlem, Coll. F. Koenigs) is further advanced than the etching of the *Raising of Lazarus* (ca. 1631, H. 96), with which it is comparable. Most sophisticated are the expressions on the faces in the studies for a *Disciple at Emmaus* (ca. 1633/34, B. 87, Paris, Inst. Néerlandais, Lugt Coll.). The first "layout" drawing for a painting dates from this period: *Rape of Ganymede* (B. 92, Dresden, Kupferstichkabinett), a study for the painting (Br. 471).

AMSTERDAM, 1636–42. In 1636 Rembrandt turned away from the boisterous and the dramatic, introduced more quiet in the scenes and more warmth in the colors, and gradually substituted the suggestion for the display of emotions. In painting his brush stroke became freer, the colors more tonal. The change, a gradual one, was apparent first in drawings and in paintings of religious subjects, and by 1642 was extended to all media and subjects.

The most important events in Rembrandt's life during these years were the births and subsequent deaths of four children, the birth of Titus in 1641 (baptized Sept. 22), the death of Saskia (June 14, 1642), and the acquisition of a large house in St. Anthonie Breestraat (the present Jodenbreestraat 4–6, the Rembrandthuis). Among the most important pupils at this time were Lambert Doomer (ca. 1640), Samuel van Hoogstraten (ca. 1642), and Carel Fabritius (ca. 1641–43).

Paintings. The stylistic change is apparent in *Tobias Healing Tobit's Blindness* (1636, Br. 502, Stuttgart, Staatsgal.), where a grayish-brown tonality prevails, although the figures are still depicted in strong movement. The large *Danae* or *Semele* of the same year (Br. 474; PL. 457) is remarkably advanced: with a minimum of gesture the reclining figure welcomes the warm, golden light that is coming toward her (the vexing problem of the subject remains unsolved). In the *Parable of the Laborers in the Vineyard* (1637, Br. 558, Leningrad, The Hermitage) the almost monochromatic coloring is combined with a free brush stroke and a greater restraint in movements. This development continued through the *Susanna Bathing* (1637, Br. 505, The Hague, Mauritshuis) and the *Noli me tangere* (1638, Br. 559, London, Buckingham Palace, Royal Colls.), where Mary Magdalene's gesture is still highly dramatic and the colors have assumed a warmth hitherto unknown in Rembrandt's paintings, to the *Holy Family* (1640, Br. 563, Louvre), where quiet intimacy prevails. In some paintings Rembrandt harked back to earlier concepts, not only in *Tobias and the Angel* (1637, Br. 503; II, PL. 293) but more especially in the large *Wedding of Samson* (1638, Br. 507, Dresden, Gemäldegal.), praised by Philips Angel in 1641 for Rembrandt's knowledge of Roman customs. The change was complete in 1642 in *The Parting of David and Jonathan* (1642, Br. 511; PL. 460). The brush stroke has become broad, the figures monumental in their expressive silence, the composition restful.

Two paintings with historical subjects date from this period: the two grisaille oil sketches for etchings, *Joseph Telling His Dreams* (ca. 1637, Br. 504, Amsterdam, Rijksmus.) and *The Concord of the State* (1641, Br. 476, Rotterdam, Mus. Boymans-Van Beuningen). The latter is unusual for Rembrandt, being a political allegory; it would seem to have been commissioned for an etching that was never executed.

In portraits the change becomes apparent when the portrait entitled *A Polish Nobleman* of 1637 (Br. 211, Washington, D.C., Nat. Gall., Mellon Coll.) is compared with the so-called *Portrait of an Oriental* of 1635 (Br. 206, Amsterdam, Rijksmus.). The colors have merged and become warmer, the expression has gained in directness, and the brush stroke has become rugged without detail being sacrificed. This can also be observed in the portrait of Eleazar Swalmius (Br. 213, Antwerp, Mus. Royal B. A.) and *Portrait of a Preacher* (Br. 214, London, Earl of Ellesmere Coll.). One of the few full-length portraits, the *Portrait of a Gentleman* (1639, Br. 216, Kassel, Staat. Kunstsamml.) invites comparison with that of *Marten Soolmans* (Br. 199) painted five years earlier: the sitter is equally stylish but has come to rest, the light is more even, and the detail is subordinated to the whole. Such elegant and precise, yet restful, portraits were painted by Rembrandt through 1643. Already in 1640 Rembrandt painted two portraits in which pose and ostentation have vanished and which represent the sitters with unprecedented immediacy: the frame maker Herman Doomer (1640, Br. 217, Met. Mus.) and his wife Baertjen Martens (Br. 357, Leningrad, The Hermitage). Saskia is not represented as often as in earlier years. In *Saskia with a Red Flower* (Br. 108, Dresden, Gemäldegal.), painted in the year of Titus's birth, a smile is only faintly visible; the attitude and the light are Venetian (cf. Titian's *Flora*, Uffizi).

Rembrandt's self-portraits of these years (Br. 26–29, 31-34) generally show him with a serious expression, occasionally in action, as in *The Bittern-Shooter* (1639, Br. 31, Dresden, Gemäldegal.); painted in warm colors, they are sometimes close to Titianesque prototypes in colors and in pose [1640, Br. 34, London, Nat. Gall., based on Titian's portrait of Ludovico Ariosto (Indianapolis, John Herron Art Inst.) rather than on Raphael's Baldassare Castiglione (Louvre)].

It was at this time that Rembrandt took up a new subject in painting — landscape. He painted fantastic scenes with threatening skies and sudden shafts of light (1636, Br. 439, Hannover, Landesgal.; ca. 1638, Br. 441, PL. 458; 1638, Br. 442, Kraków, Czartoryski Mus.; and Br. 443, Boston, Isabella Stewart Gardner Mus.) and occasionally representations of the countryside in Holland, either under dramatic light (*Landscape with Stone Bridge*, ca. 1636–40, Br. 440, Amsterdam, Rijksmus.) or under a rather even light (*Evening Landscape*, 1639, Br. 448, Oslo, Nazjonalgall., overpainted). In between these two groups stands the *Landscape with a Drawbridge* (ca. 1640, Br. 446, Madrid, Duke of Alba Coll.) where naturalistically represented motifs are combined with a contrasting light and dark and fantastic mountains. Especially in this landscape, but also in the other "dramatic" ones, the influence of Hercules Seghers is unmistakable.

The painting that incorporates all the changes in Rembrandt's style from 1636 to 1642 is *The Militia Company of Captain Frans Banning Cocq*, which is traditionally called *The Night Watch* (Br. 410; PL. 459). The painting, finished in 1642, was one of seven commissioned to various artists for the Kloveniersdoelen in Amsterdam, probably to commemorate Maria de Médicis' triumphal entry into Amsterdam in 1639 and the part the militia companies played in the processions. Rembrandt changed the Dutch tradition of the group portrait by representing the company as if it were in the process of starting a march rather than in the customary stationary position. The presence of movement, so pronounced in Rembrandt's many-figured religious paintings executed between 1632 and 1636, has been reduced in intensity but retained in principle in the transfer to the group portrait; the colors have assumed a warmth and the distribution of light a richness in contrasts that are influenced by Venetian art; the individual portraits have been subordinated to the over-all arrangement of the figures; the brush strokes have become broad without any loss of detail.

Etchings. Rembrandt's style also changed about the year 1636 in portrait etching, as shown by *Rembrandt and Saskia; Busts* (1636, H. 144). In his own portrait as well as in Saskia's, the emphasis on pose has given way to an effort to represent the personality. (The two portraits are not combined into a double portrait but are etched separately on one plate.) Apart from a few earlier self-portrait studies, this is the first etched portrait sketch. Rembrandt had never used the medium of etching as if it were drawing in connection with portraits; many were to follow in this period, such as *Studies of the Head of Saskia and Others* (1636, H. 145) and *Three Heads of Women, One Lightly Etched* (1637, H. 153). Rembrandt now created the printed sketch or *griffonnement*; characteristic is the *Sheet of Studies, with a Woman Lying Ill in Bed* (ca. 1639, H. 163). Finished portrait etchings from these years are few; the most important ones are, in their completeness, similar to portrait engravings of Goltzius and his school and are part of that tradition (the portraits of Jan Uytenbogaert, or *The Gold Weigher*, 1639, H. 167, and of Cornelis Claesz. Anslo, 1641, H. 187).

In etching as in painting, a change in Rembrandt's style of treating Biblical subjects became apparent around 1636. In *The Return of the Prodigal Son* (1636, H. 147), although its composition is based on a print after Maerten van Heemskerck, the etched line has become freer, and the emotion of the figures is expressed with less movement but greater understanding and intensity than before. The same applies to *Abraham Caressing Isaac* (H. 148), *Abraham Casting Out Hagar and Ishmael* (1637, H. 149), and especially to *The Raising of Lazarus* (1642, H. 198); here the etched line is used for its linear quality, and Christ's miracle has been rendered without the pathos of the earlier representation of the same subject (H. 96). The most ambitious etching of this period is *The Death of The Virgin* (1639, H. 161; PL. 464), where a great freedom of line is combined with dense etching, which was frequently used to obtain dark areas in the preceding period. This combination helped to create a scene in which monumentality in space and actions is balanced with sensitivity in the representation of the Biblical text and where richness of color is obtained by contrasts of black and white. A similar combination is often found in etchings of these years (as in *The Triumph of Mordecai*, ca. 1640, H. 172).

During this period Rembrandt etched some landscapes, which, in contrast to most of the painted landscapes of this time, are not fantastic visions but views in the neighborhood of Amsterdam (*The Windmill*, 1641, H. 179; *Landscapes with Cottages*, 1641, H. 177–78; *View of Amsterdam*, ca. 1641, H. 176). All the properties of the etched line were exploited to suggest atmosphere and light.

Drawings. In drawings the change around 1636 is less clear and more gradual than in paintings or etchings. The change was toward a greater autonomy of line, a simplification of pose and composition, and a greater expressiveness. Studies from nature became numerous, as well as those of single figures and of two or more figures (B. 299–389; 654–83); especially beautiful is the *Woman Carrying a Child Downstairs* (ca. 1636, B. 313, New York, Pierpont Morgan Lib.). Rembrandt depicted many mothers and children; he also portrayed actors and scenes on the stage (B. 94, 293–97, 321, 316–68). His increased interest in drawing from nature is apparent in the studies of animals (*Elephants*, B. 457–58, Vienna, Albertina, B. 459, Br. Mus.; *Horses*, B. 461, Amsterdam, Rijksmus.; *Birds of Paradise*, B. 456, Louvre) and over 30 landscapes.

In composition the landscapes at first were based on those of Jan van de Velde and similar printmakers (B. 471–72, Vienna, Akad. der Bildenden Künste); then they became freer. Most are studies from nature, some executed with black chalk, and they are still detailed. A dramatic effect, as in *Cottages before a Stormy Sky* (ca. 1641, B. 800, Vienna, Albertina), was an exception. The drawings of English sites (London, Windsor, and St. Albans, B. 785–88) are puzzling, since in all probability Rembrandt never visited England, so they may have been based on prints or drawings by other artists.

Many of the drawings of religious subjects of these years must have been made for their own sake, even those that are summary sketches. Rembrandt must have drawn subjects from the Old and New Testaments, as well as the Apocrypha, immediately upon reading the text, probably as a visual comment to the text. This supposition is based on the fact that most subjects are represented with the exact wording of the text in mind, and that so many representations are not based on an existing pictorial tradition but are without precedent. An outstanding example of the expressive, yet detailed and moving — although nearly motionless — representation of a Biblical subject, is *The Return of the Prodigal Son* (ca. 1640–42, B. 519, Haarlem, Teylers Mus.). One of the most remarkable finished drawings of this category is *The Brethren of Joseph Requesting Benjamin from Their Father* (B. 541, Amsterdam, Rijksmus., Prentenkabinet). In the arrangement of the figures Rembrandt was influenced by contemporary theater. The pen-line drawings of figures with their expressions of tension and expectation are done with the utmost care. Those drawings that were made for etchings or paintings often have characteristics directly related to their function; thus *The Triumph of Mordecai* (B. 487, Lvov, Staat. Gemäldegal.) is a quick study for the layout of the etching (H. 172), while B. 527, 526, and 528 are studies for both the composition and the individual expressions of the figures in the etching *Joseph Telling His Dreams* (H. 160); the study for the group of the sick of the so-called *Hundred Guilder Print* (B. 188, Berlin, Kupferstichkabinett) anticipates the composition of part of the etching and at the same time, in the heavily drawn lines and shaded areas, the distribution of light and dark in the etching.

During this period Rembrandt made few portrait drawings (*Self-portrait*, B. 437, Washington, D.C., Nat. Gal., Rosenwald Coll.; *Portrait of a Woman*, 1639, B. 441, Stockholm, Nationalmus.).

Amsterdam, 1642–55. In *The Night Watch* Rembrandt gave new form to a theme that until then had been defined by tradition. It marks the point of greatest independence in his career. From then on Rembrandt's art departed in style further and further from that of his contemporaries. This development was expressed by a greater simplification of form, a reduction of movement, a suppression of detail, an intensification of the expression of sentiments, and, in painting, the appearance of darker and warmer colors. In brush work, colors, and sometimes composition, a Venetian influence, already noticeable in the previous period, became stronger. All the changes were meant to subordinate the representation of the outer appearance of man to that of his emotions. This development was not interrupted in 1655 but continued into Rembrandt's last years. A few milestones mark an intensification in this development, although there were occasional regressions. One important milestone, coming at a time when etching seems to have attracted Rembrandt's attention more than painting, is the late state of *Christ Presented to the People* (H. 271) where the crowd in the foreground has been replaced by a blind wall with arches. Executed in 1655, it can be assigned as a point of demarcation between two periods, a position which is, to an extent, arbitrary.

Rembrandt's personal life in the years between 1642 and 1655 was marked, at first, by difficulties centering around the nurse of the young Titus, Geertghe Dircx, who stayed with the artist until June, 1649. She ended up in a madhouse in Gouda. In 1649 or perhaps as early as 1645, Hendrickje Stoffels became Rembrandt's common-law wife. She was a kind-hearted and mentally strong person from a simple background. A marriage probably could not be contracted because of Saskia's will. Shortly after 1650 Rembrandt's financial situation became very difficult, reaching a particularly low point in 1653. In 1654 Hendrickje gave birth to a daughter, Cornelia, who was the only one of Rembrandt's children to survive him.

In the first half of this period Rembrandt had a considerable number of pupils, some of whom were to become prominent, but in the second half the number decreased. His most important pupils were Nicolaes Maes, who worked with him between 1648 and 1650, and Barent Fabritius, who probably studied with him in 1652.

Paintings. After 1642 the commissions for portraits diminished in number and came mainly from people in Rembrandt's own circle. Rembrandt also portrayed members of his family, especially Hendrickje (ca. 1650, Br. 113, London, S. Morrison Coll.) and Titus. The delicate and sensitive face of Titus was Rembrandt's preferred model in the early 1650s. He represented him gay (1650/51, Br. 119, Los Angeles, County Mus.), pensive (1655, Br. 120; pl. 461), reading (ca. 1655, Br. 122, Vienna, Kunsthist. Mus.), and serious and a little sad (1655, Br. 123, London, Wallace Coll.). Many of his other sitters belonged to learned circles, the ministry, and the art world: the preacher Jan Cornelis Sylvius, a posthumous portrait of 1645, and his wife (Br. 237, Cologne, Wallraf-Richartz Mus., Br. 369), the painter Hendrick Martensz. Sorgh and his wife (1647, Br. 251; Br. 370, London, Trustees of the late Duke of Westminster), the physician and writer Ephraim Buëno (Br. 252, Amsterdam, Rijksmus., a study for the 1647 etching H. 226), the art dealer Clement de Jonghe (Br. 265, Berkshire, Eng., Buscot Park), and Jan Six (1654?, Br. 276, Amsterdam, Verzameling der Six-Stichting). Most of the sitters whose identity is unknown are elderly people (e.g., Br. 236, 239–40, 266–67, 269–70; a portrait of an old woman seated, Br. 381, Leningrad, The Hermitage, which is a companion piece to Br. 270, is one of the most remarkable representations of old age); so are those whose faces Rembrandt sketched on a smaller scale, such as Br. 249 (1647, Rotterdam, Mus. Boymans-Van Beuningen) and Br. 271 (1654, Groningen, Mus. van Oudheden voor Provincie en Stad Groningen). Rembrandt was equally capable of depicting the *élan* of young people, as in the dynamic portrait of Nicolaes Bruyningh (1652, Br. 268, Kassel, Staat. Kunstsamml.), and their lack of sophistication, as in *Young Girl at a Window* (1645, Br. 368, London, Dulwich College Picture Gall.). Reminiscent of an earlier period are some portraits, carefully and smoothly executed with a more dramatic light, such as the *Man with a Falcon* and *Lady with a Fan* (1643, Br. 224, 363, London, Trustees of the late Duke of Westminster), both with strong Venetian influences, and even the *Portrait of a Man in Military Costume*, done as late as 1650 (Br. 256, Cambridge, Eng., Fitzwilliam Mus.). Other portraits in disguise are the *Woman in Arcadian Costume* (not Hendrickje Stoffels; Br. 114, Met. Mus.) and the *Man in a Golden Helmet* (1646/47, Br. 128; pl. 461).

In the few self-portraits of these years, the pose became less defiant; attention was concentrated on the face, especially the eyes, and the expression was always serious (Br. 35–39, 42, etc.). One of the most remarkable self-portraits dates from 1652, where Rembrandt painted himself standing, hands at his sides, with a searching look in his eyes (Br. 42, Vienna, Kunsthist. Mus.).

Scenes from the Old and New Testaments became rare in paintings, especially after 1650. Here, too, an intensification and simplification similar to that in portraits is noticeable. Scenes with few figures were preferred, often representing an idea, a condition of man, or a mood rather than an event. Sometimes Rembrandt harks back to an earlier style, as in the *Woman Taken in Adultery* (1644, Br. 566, London, Nat. Gall.), which has many small figures and is rather smoothly painted. A quiet, simple home that is a refuge in old age is the central theme of *Tobit and Anna*, not only in the work executed in 1645 (Br. 514, Berlin, Staat. Mus.) but also, more pronounced, in the one executed in 1650 (Br. 520, Rotterdam, W. Van der Vorm Coll.). The care of a young mother for her child is the theme of *The Holy Family* (1645, Br. 570, XIII, pl. 113; 1646, Br. 572, Kassel, Staat. Kunstsamml.). The youth of Christ is repeatedly the subject of paintings between 1645 and 1650. In 1646 Rembrandt twice painted an *Adoration of the Shepherds*, once for Prince Frederick Henry of Orange (Br. 574, Munich, Alte Pin., with a *Circumcision*, now lost, the last of the series of New Testament scenes painted for the Prince), and once for an unknown patron (Br. 575, London, Nat. Gall.). In both, but especially in the latter, the execution is broad and attention is focused on the Holy Family and on the mysterious light emanating from the Child in the midst of the surprised, simple, and humble shepherds. The contrast between Rembrandt's last work for the series for the Prince and his earlier ones is significant: instead of a composition in the manner of Rubens a personal interpretation of the scene is the guiding force. The *Rest on the Flight into Egypt* (1647, Br. 576, Dublin, Nat. Gall.), based on Adam Elsheimer's composition of the same subject, is a night scene where darkness and solitude prevail. The greatest simplicity is achieved in *Christ at Emmaus* (1648, Br. 578; pl. 455), so much more expressive than the earlier, more pathetic version (ca. 1630, Br. 539), and in *Christ Appearing to the Magdalen* (1651, Br. 583, Brunswick, Herzog-Anton-Ulrich Mus.), where the figures are enveloped in darkness. The mystery of Christ's appearing to man is likewise the subject of *Christ and the Woman of Samaria* (1655, Br. 589, Met. Mus.; Br. 588, pl. 465).

Rembrandt depicted the nude in *Bathsheba*, musing upon the message she receives from King David, Rembrandt's largest, most monumental, and most Italianate nude (1654, Br. 521, Louvre), and in *Susanna and the Elders* (1647, Br. 516, Berlin, Staat. Mus.), with remarkably warm reds and browns. Similar tones prevail in the two representations of *Joseph and Potiphar's Wife* (1655, Br. 523, Washington, D.C., Nat. Gall.; Br. 524, Berlin, Staat. Mus.). The "seminude" *Woman Bathing in a Stream* (1655, Br. 437; pl. 460) is perhaps connected with a representation of Susanna; the *Woman in Bed* (1649?, Br. 110, Edinburgh, Nat. Gall. of Scotland, probably not Hendrickje) with another representation of *Joseph and Potiphar's Wife.*

The landscapes Rembrandt painted in these years continue and clarify the two trends present before 1642: the imaginary landscapes have become wider and the forces of nature stronger (ca. 1650, Br. 450, Louvre; Br. 454, Kassel, Staat Kunstsamml.; Br. 451, London, Wallace Coll.); and the "Dutch" landscapes are more naturalistic: *Winter Landscape* (1646, Br. 452, Kassel, Staat. Kunstsamml.) and *Landscape at Sunset* (1654, Br. 453, Montreal, Van Horne Coll.) were probably painted in plein-air. It remains to be established whether the boldly painted

Butchered Ox (1655, Br. 457; PL. 466) should be considered a gruesome still-life painting or a genre scene with an allegorical meaning.

One painting stands by itself in subject and in Rembrandt's interpretation of it. When the Sicilian nobleman Antonio Ruffo commissioned Rembrandt to paint a "philosopher," the artist represented *Aristotle Contemplating the Bust of Homer* (1653, Br. 478, Met. Mus.); here the philosopher has become a seer, and the magic light seems to indicate the bond between poetry and philosophy.

Etchings. This is the great, classical period of Rembrandt as an etcher. The characteristics previously established were developed further and became more pronounced. Rembrandt produced many etchings, especially after 1650, when etching together with drawing seems to have received more of his attention than painting. He preferred landscapes and New Testament subjects.

The *Three Trees*, Rembrandt's most dramatic etched landscape (1643, H. 205), is his largest etched landscape and the only one with the kind of tension the painted imaginary landscapes have. The greatest purity in the etched line was achieved in the etchings of 1645, above all in *Six's Bridge* (H. 209), but also in *The Boathouse* (H. 211) and in the background of *The Omval* (H. 210). After an interval of about five years Rembrandt took up landscape etching again and executed a long series of works using the dry-point technique (H. 238–47, 263–65). Most of the landscapes, although simple in motifs, present true (albeit reversed) views of the farms, meadows, and dikes, and were often done on the spot, as in the case of the pure dry-point etching *Clump of Trees with a Vista* (H. 263). Other landscapes were probably sketched on the plate in the open air and finished in the studio. With the combined effects of etched line, dry point, white paper, and partial wiping, Rembrandt created the effect of space, atmosphere, and color, especially in such landscapes as *Landscape with Trees, Farm Buildings, and a Tower* (H. 244) and *Landscape with Three Gabled Cottages* (1650, H. 246). Because of the great difference between early and late impressions, only the former fully represent the artist's intention. In one work, *The Gold Weigher's Field* (1651, H. 249), Rembrandt obtained a distance and light reminiscent of the Campagna painters and Claude Lorrain.

Rembrandt also made etchings of subjects other than landscapes by drawing from nature on the prepared plate. To this category belong some nudes and genre scenes. The etching *Studies from the Nude: One Man Seated and Another Standing* (H. 222) and probably also the related etchings H. 220 and 221 must have been made in this way. While he sketched the standing nude of H. 222 with the needle, two pupils sketched the same figure on paper (B. 709 and 710, according to Benesch drawn by Rembrandt, Vienna, Albertina, and Br. Mus.). Perhaps *The Hog* (1643, H. 204) also should be considered a "printed sketch."

During this period Rembrandt made three etchings with religious subjects, in which the print assumed characteristics of paintings in size, completeness, and monumentality. Together with some of the landscapes of Seghers these are the largest Dutch etchings of the first half of the 17th century, and they represent a counterpart to the large prints that various printmakers executed for Rubens after his own paintings. In contrast, Rembrandt's three etchings were independently made etchings that in size, monumentality, and texture resembled paintings. The so-called *Hundred Guilder Print* (H. 236; PL. 464), illustrating various verses of Matthew 19, was finished around the middle of the 1640s after Rembrandt had worked on it for some years. The combination of etching and dry point which produced colorful, painterly effects and the varied expressions of the many figures arranged around the figure of Christ are factors that have made this Rembrandt's best known print. *The Three Crosses* (1653, H. 270; IV, PL. 435), Rembrandt's most monumental representation of Christ crucified, is larger than the *Hundred Guilder Print* and is entirely executed in dry point. The third print, *Christ Presented to the People* (H. 271), approximately the same size as *The Three Crosses*, marks, in its fifth state (1655), a departure from the idea of confronting Christ with a multitude by deleting the crowd in front of the estrade. The substitution of a void for man, resulting in an intensification of the drama, subsequently was applied to *The Three Crosses*; Rembrandt completely reworked the plate with vehemently applied dry-point lines, replacing the original groups of figures with a rocky terrain, and then narrowing the light streaming down to earth. Segher's idea of "printing paintings" had found another expression.

In other Biblical etchings from this period, Rembrandt first developed the line to a high degree of expressiveness, contrasting the whiteness of the paper with the line more than before (*Rest on the Flight into Egypt*, 1645, H. 216; *The Doubting of St. Thomas*, 1650, H. 237; *Christ Disputing with the Doctors*, 1652, H. 257). Subsequently he intensified the effects of darkness and mystery by an increasing use of parallel lines and crosshatching (*David in Prayer*, 1652, H. 258; *The Virgin and Child with the Cat*, 1654, H. 275; *The Adoration of the Shepherds*, 1654, H. 273; *The Circumcision*, 1654, H. 274) or by adding dry-point lines to those already etched (*St. Jerome Reading, in a Landscape*, 1651/52, H. 267; *The Agony in the Garden*, ca. 1653, H. 293; *Christ between His Parents*, 1654, H. 278). Both techniques are present in a series of four etchings, probably executed in 1654 (and the undated ones perhaps slightly later): *The Presentation, The Descent from the Cross, The Entombment*, and *Christ at Emmaus* (H. 279–82).

Only a few portraits — in addition to three self-portraits — were made in this period, all of friends and acquaintances, all great achievements, and each a unique work: the posthumous portrait of Jan Cornelis Sylvius (1646, H. 225), in pose and execution reminiscent of an earlier period; Ephraim Buëno and Jan Six (both 1647, H. 226, 228), highly finished, almost full-length portraits; Jan Asselyn (1648, H. 227); Clement de Jonghe (1651, H. 251), remarkable for the colorful effects obtained with the etched lines; Jacob Thomasz. Haaringh (1655, H. 288); Thomas Jacobsz. Haaringh (ca. 1655, H. 287); and Arnold Tholinx (1655/56, H. 289), with much dry point and painterly effects, and an intensity of expression on the same level as *The Three Crosses* and *Christ Presented to the People*. Reminiscent of Rembrandt's portraits of men in an interior (Lieven Willemsz. van Coppenol, H. 269; Jan Uytenbogaert, H. 128) and also of his representations of philosophers (*Old Man in Meditation*, the so-called *Archimedes*, H. 218, and the 1653 painting of *Aristotle*) is the etching called *Faust in his Study* (1652/53, H. 260), which remains problematical as to the subject, although recent investigations (Bojanowski, 1956; Rotermund, 1957; Wegner, 1962) have suggested various interpretations. The etching probably represents a scholar practicing the magic art of crystal gazing with the use of a round mirror. In identifying the magic apparition of the words, reflected from the mirror and arranged as on an amulet, with the light coming through the window, Rembrandt used a device he also employed for the mysterious revelation of Christian sainthood (*The Virgin and Child with the Cat*, H. 275).

Drawings. Rembrandt's activity as a draftsman increased greatly in this period, especially from 1650 on. He had already established the habit of drawing religious subjects as independent works of art. Now he made such drawings in great quantities (approximately 200 have been preserved). The main element of these drawings is the line. The brush is used only sparingly and seldom extensively as, for example, in *St. Jerome Reading* (ca. 1652, B. 886, Hamburg, Kunsthalle); characteristically, this is a study for an etching in which Rembrandt tried to establish the contrasts between light and dark. He used black chalk rarely. At the end of this period he often used the reed pen, achieving the greatest boldness in two drawings related to etchings — the *Presentation in the Temple* (B. 1032, Rotterdam, Mus. Boymans-Van Beuningen) and the *Supper at Emmaus* (B. A66, Amsterdam, Rijksmus., Prentenkabinet) — if these can be considered as studies for the etchings H. 279 and 282. The development is toward a further economy of line. If in such drawings as *Jacob's Dream* (B. 557, Louvre; B. 558, Rotterdam, Mus. Boymans-Van Beuningen) and *Hagar by the*

Fountain (B. 560, Louvre) the line is still descriptive and details are still represented, soon this descriptive function of the line would be reduced, as it was in such drawings as the *Deposition* (B. 587, Berlin, Kupferstichkabinett; according to Benesch ca. 1647, according to Lugt, 1931, ca. 1650) where forms are suggested rather than defined. Previously Rembrandt had allowed himself such economy only in quick layout studies for compositions, such as the *Holy Family in the Carpenter's Workshop* (B. 567, Bayonne, Mus. Bonnat, a study for the 1645 painting, Br. 570, Leningrad, The Hermitage). The development toward an ever more expressive economy is a gradual one; such drawings as *Christ Comforted by the Angel* (B. 626, Cambridge, Eng., Fitzwilliam Mus.; according to Benesch ca. 1648–50) and *Satan Showing Christ the Kingdoms of the World* (ca. 1649, B. 635, Berlin, Kupferstichkabinett) show this economy combined with detail. Occasionally Rembrandt reverted to an earlier method and made drawings with greater stress on particulars, such as *Christ in the House of Martha and Mary* (B. 631, Munich, Staat. Graphische Samml.; B. 632, Br. Mus.) and *Tobias and His Wife Sarah Praying* (ca. 1648–50, B. 633, Met. Mus.). At the same time the line gradually lost much of its roundness and, especially at the end of this period, became angular. This is clear particularly in the impressive series of drawings representing scenes from the life of Christ (B. 921, Rokeby, Yorkshire, H. E. Morritt Coll.; B. 922, Berlin, Kupferstichkabinett; B. 923, Haarlem, Teylers Mus.; B. 927, Amsterdam, Rijksmus.), in the *Susanna Led to Judgment* (B. 942; IV, PL. 278), and in the unusual representation of *Joseph Supporting Mary when Departing for Egypt* (B. 902, Berlin, Kupferstichkabinett).

As for composition, the symmetry and rest which are noticeable in the previous period, still present in the *Calvary* (B. 586, Frankfurt am Main, Städelsches Kunstinst., according to Benesch ca. 1647), gradually gave way to an asymmetrical composition, such as in the *Calvary* (B. 652, Louvre, according to Benesch ca. 1649/50). As in his work of the 1630s the asymmetry was introduced to stress the emotional content, but the various parts of the asymmetrical composition are at rest instead of in motion (e.g., B. 652 or *David Receiving the News of Uriah's Death*, B. 890, Amsterdam, Rijksmus., Prentenkabinet).

Second in number to drawings with religious subjects are the landscapes. Approximately 180 such drawings are preserved, almost all drawn with the pen; often, shadows were added with the brush. These farms, polders, and dikes in the neighborhood of Amsterdam were sketched directly from nature in Rembrandt's sketchbooks or on loose sheets of paper. A trip or trips to Rhenen and Arnhem and others to Muiden and Muiderberg also provided motifs. Light and atmosphere were the prime objectives in the rendering of trees, farms, and farmlands. Never before in art had the immediate surroundings of the artist been sketched with such attention to the simple motif. The most remarkable collection of these landscapes is in the Library of the Duke of Devonshire, Chatsworth, England — a collection once owned by Nicolaes Flinck, son of Rembrandt's pupil Govert Flinck.

During these years Rembrandt made most of his studies from life in black chalk (cf. B. 673, 693–701, 705, 714–21, etc., also B. 751, 769, 1074–76). The pen sketches of figures, especially the late ones such as B. 1071, 1079, and 1080, as well as of genre scenes such as B. 1153, show the same economy of line as other drawings of these years. Most of the portrait drawings are studies for paintings (the posthumous portrait of Jan Cornelis Sylvius, 1644/45, B. 762, Washington, D.C., Nat. Gall., Rosenwald Coll., a sketch for painting Br. 237 rather than a portrait in the conventional sense of the word) or for etchings (another portrait of Sylvius, B. 762a, 763, a study for etching H. 225 of 1646; the broadly sketched portrait of Jan Six, B. 767, for etching H. 228 of 1647, and the careful preparatory drawing for the same, B. 768, both in Amsterdam, Verzameling der Six-Stichting). Among the few other portrait drawings is the most impressive self-portrait Rembrandt ever sketched, the remarkable full-length *Self-portrait in Studio Attire* (1655/56, B. 1171, Amsterdam, Rembrandthuis).

Probably at the very end of this period Rembrandt made a series of drawings after Moghul miniatures, of which 20 have been preserved. With a fine sharp line and carefully applied washes, to which sometimes a few colors were added, he recreated, generally on Chinese paper, the delicate, exotic quality of the miniatures which at that time must have been in an Amsterdam collection, perhaps his own.

Amsterdam, 1655–69. The financial difficulties that beset Rembrandt from 1647 on, partly owing to misfortunes in the art trade in which he participated, became so serious that in July, 1656, he asked to be granted a *cessio bonorum*, that is, the formal surrender of his goods and chattels to his creditors. This was granted and meant that his bankruptcy was not a dishonorable one but that the administration of his debts and the distribution of his possessions among the creditors would be dealt with by the department of the city government called the "chamber of bankruptcy." On July 25, 1656, the chamber drew up an inventory listing all the goods in Rembrandt's house: the works of art, curiosities, and other objects (furniture was almost entirely absent). The chamber began to liquidate the goods. In December, 1657, the first sale took place, under the direction of Thomas Haaringh; in February, 1658, there was a second one (the house and furniture); and in September of the same year a third one (prints and drawings). In this same year (probably in December) his son Titus and Hendrickje Stoffels, as partners, set up as art dealers. This probably meant that Rembrandt's activities were officially carried on by them. It seems that most creditors, of whom Titus was represented with the largest sum, were satisfied by Dec. 10, 1660, and that the remaining creditors agreed to be paid in paintings to be done by Rembrandt. Rembrandt then joined Hendrickje and Titus's business in the capacity of adviser. In 1660, when the large house on the Jodenbreestraat was turned over to its new owner, Rembrandt moved to a smaller, although comfortable, house on the Rozengracht. From that time on his financial circumstances, although not nearly so favorable as in the 1630s and the early 1640s, were such that Rembrandt could not be called poor. He continued to receive commissions, although not in large numbers. At the time of his death he left 9,000 guilders to his daughter and granddaughter.

For a time his family life was not disturbed by such events as those which had previously cast a shadow over his life. The artist seems to have been somewhat withdrawn, although he remained active as a collector and dealer. On July 24, 1663, Hendrickje died. Titus married Magdalena van Loo in February, 1668, and died half a year later, leaving Rembrandt alone with his daughter Cornelia, at that time fourteen years old. Rembrandt's only grandchild, a girl called Titia, was born in March, 1669. Rembrandt died on Oct. 4, 1669. His daughter-in-law, Magdalena, died on Oct. 21 of the same year.

After 1655 Rembrandt had very few pupils. One was his son Titus; another was Aert de Gelder (1645–1727), who, in contrast to earlier pupils, did not switch to a French-inspired classicism but transformed Rembrandt's late style into an 18th-century idiom.

Paintings. Between 1648 and 1655 a new Town Hall (now the Royal Palace) was built in Amsterdam. It is very likely that Rembrandt's painting representing Quintus Fabius Maximus requesting his father to pay him homage (1655, Br. 477, formerly Belgrade, King of Yugoslavia) was painted for a chimney piece in the Town Hall, although a painting of the same subject by Govert Flinck was used instead. Also painted for the Town Hall were *Moses Showing the Tables of the Law* and *Jacob Wrestling with the Angel*, although neither was placed (Br. 527, 528, Berlin, Staat. Mus.); the *Woman in Armor* (Athena or Bellona?; 1655, Br. 479, Paris, Gulbenkian Coll.) and the *Man in Armor* (Mars?; 1655, Br. 480, Glasgow, Art Gall. and Mus.) were probably, as Van de Waal (1952, p. 218) supposed, done for the same building. Another painting made for the Town Hall, the *Conspiracy of the Batavians* (1661, Br. 482; PL. 463), was placed but soon removed. It was a completely novel interpretation of Tacitus's text: instead of extolling the merits of the Batavians as ancestors of the Dutch, Rembrandt has depicted a conspiracy by night with a number of people in various walks of life swearing allegiance to their one-eyed leader. Like the earlier *Night*

Watch, these paintings were commissioned from the leading artists of the time. Whatever the reasons for the refusal or the removal of Rembrandt's paintings, it is proof of the divergence of Rembrandt's artistic concepts from those of the other artists. This does not imply that Rembrandt was unappreciated or did not receive commissions.

More and more in his portraits of this time he concentrated on the face, sometimes also on the hands. The rest of the portrait merely served to indicate posture and costume and to provide a background or an accent to the expression of the character of the sitter. Most of these sitters were people from Rembrandt's immediate surroundings and from a circle of friends and acquaintances: *Hendrickje Stoffels* — at least if the identification is correct — in simple costume resting her arms on a table (ca. 1658, Br. 115, Munich, on loan to Alte Pin.), informally posed in a door opening (1655–58, Br. 116; II, PL. 217), or as Venus accompanied by Amor (Cornelia?, Br. 117, Louvre); the physician Arnold Tholinx (1656, Br. 281, Paris, Mus. Jacquemart-André); the painter Gerard de Lairesse (1665, Br. 321, New York, R. Lehman Coll.); and the poet Jeremias de Decker (1666, Br. 320, Leningrad, The Hermitage). The commission for a third group portrait, the second anatomical demonstration (*The Anatomy Lesson of Dr. Deyman,* 1656, Br. 414, only a part of which is preserved; II, PL. 209), also came from the circle of learned people. The composition is symmetrical and thus entirely different from the earlier representation of a similar subject; the corpse is foreshortened, the feet stretching toward the viewer. The second important portrait commission of these years was that for *The Syndics of the Drapers' Hall* (finished 1662, Br. 415, Amsterdam, Rijksmus.). Portrayed to be seen from below (the painting was made for a chimney piece), the five syndics (and one servant) are depicted as if they were gathered for one of their meetings. Rembrandt did not represent the meeting during a real or imaginary moment, nor the syndics as if they were addressing an audience. He chose a composition that allowed him to depict the features and posture of each syndic, one of whom is depicted as the secretary or treasurer; their attitudes were partly defined by the composition, partly by the characteristics of the sitters as Rembrandt understood them. The sitters, clearly differentiated and defined, have in common an air of authority and self-assurance. This composition, like that of the *Conspiracy of the Batavians* (Br. 482), contains reminiscences of Leonardo's *Last Supper.*

At the same time Rembrandt painted two of his most noble portraits of old people, monumental in their human dignity, reminiscent of Tintoretto in composition and distribution of light: Jacob Trip, one of the leading merchants of his time (ca. 1661, Br. 314), and his wife Margaretha de Geer (ca. 1661, Br. 394; she is also portrayed in Br. 395; all three, London, Nat. Gall.). The identity of many sitters is unknown, for example, the masterful *Old Man* in the Uffizi (Br. 285), the *Man Holding a Manuscript* (1658, Br. 294, Met. Mus.), and a number of late, impressive portraits (Br. 312–13, 315, 323, 327, 398, which is very "Venetian," and 402). Also unknown is the identity of the man portrayed in the *Equestrian Portrait,* probably painted late in the 1650s (Br. 255, London, Nat. Gall.). More than in other portraits of this time Rembrandt introduced here an element of ostentatious grandeur, traditionally present in equestrian portraits. However, by placing the man off-center and against a dark background, he gave a novel interpretation to the theme.

As before, Rembrandt painted a few "disguised" portraits (*Old Man as St. Paul,* 1659, Br. 297, London, Nat. Gall.) and portrayed some sitters in historical or exotic costumes, the meaning of which is not always clear (*Bearded Man in a Cap,* Br. 293, London, Nat. Gall., the same figure used in 1653 for the *Aristotle,* Br. 478, and in 1657 for the so-called *Portrait of a Rabbi,* San Francisco, Calif. Palace of the Legion of Honor; and *Old Man in a Beret,* Br. 324, Dresden, Gemäldegal.). Hendrickje is depicted as Venus, probably with her daughter Cornelia as Amor (ca. 1656, Br. 117, Louvre). In the *Bridal Couple* or the *Jewish Bride* (Br. 416; PL. 463) the figures are painted in warm red, brown, and golden tones, applied in broad strokes with brush and palette knife. The man and woman are portrayed in exotic costumes, as a bridal pair in Biblical guise. Although Rembrandt probably intended the pair to represent a specific couple from the Bible (Valentiner, Noch einmal Die Judenbraut, 1957), such as Tobias and Sarah, Isaac and Rebecca, or Jacob and Rachel, he emphasized the temporal and individual nature of the bond that united this particular man and woman rather than striving for the durability and general validity of a historic type. The names of the sitters remain unknown, even though both were also portrayed separately (so-called *Portrait of a Man with a Magnifying Glass,* Br. 326, and *Portrait of a Lady with a Carnation,* Br. 401, both Met. Mus.). The attractive supposition that the *Bridal Couple* represents Titus and his wife (Valentiner, *ibid.*) needs further evidence. The lady probably recurs in the *Family Portrait* (Br. 417, Brunswick, Herzog-Anton-Ulrich Mus.), which in warmth of tone, breadth of painting, and depth of sentiment ranks equally with the *Bridal Couple.* Both paintings were done sometime after 1660. The *Two Negroes* (Br. 310; III, PL. 212) is both a disguised portrait and a genre painting; one of the negroes is wearing a Roman costume.

The dividing line between portraits and mere representations of saintly figures had become very thin by 1660 or 1661. Apparently Rembrandt set out to paint a number of apostles and evangelists, perhaps as part of a series. Some of these are undoubtedly portraits, or at least can be presumed to bear a great likeness to the model used. All have pronounced individual features; *St. Matthew* (Br. 614, Louvre), *St. Bartholomew* (Br. 615, Paul Getty Coll.), *Praying Apostle* (Br. 616, Cleveland, Mus. of Art), *St. James* (Br. 617, Cambridge, Eng., L. C. C. Clark Coll.), *Writing Evangelist* (Br. 618, Rotterdam, Mus. Boymans-Van Beuningen; Br. 619, Boston, Mus. of Fine Arts), *Simon* (Zurich, Kunsthaus), *Self-portrait as St. Paul* (Br. 59, Amsterdam, Rijksmus.), as well as the afore-mentioned Br. 297 and others. The so-called *Portrait of a Nun* (representing the Virgin Mary, 1661, Br. 397, Epinal, Mus. Départemental des Vosges) belongs to this group.

A number of self-portraits from these years have been preserved. Generally his look is serious and penetrating, especially in portraits showing only the head and shoulders (1657, Br. 48, London, Earl of Ellesmere Coll.; Br. 49, Vienna, Kunsthist. Mus.). Later his expression is mellow and wise, tired and resigned, the face itself rounder and fuller (ca. 1664, Br. 60, Uffizi; ca. 1667, Br. 55, London, Nat. Gall.; 1669, Br. 62; PL. 456). The life-size, knee-length self-portraits of these years are some of the most beautiful ever painted, especially the one where Rembrandt represented himself seated (1658, Br. 50, New York, Frick Coll.), showing dignity and wisdom colored with melancholy and a touch of amusement about the folly of this world, and the mysterious one where Rembrandt, holding palette and brushes, is standing in front of what seems to be a canvas on which two large circles are drawn (ca. 1664, Br. 52, London, Kenwood, Iveagh Bequest); does this, perhaps, contain an allusion to Jupiter, Painter of the World? Revealing of Rembrandt's view of the world is his *Self-portrait as Democritus* (1665/67, Br. 61, Cologne, Wallraf-Richartz Mus.).

Religious scenes from these years are few in number, but have become almost without exception universally known and acclaimed interpretations of the subject: *Jacob Blessing the Sons of Joseph* (1656, Br. 525; PL. 462), with the unforgettable characterization of the old Jacob (his face modeled upon the bust of Homer), the expression of expectation and tension in Joseph and Agnath, his wife (her presence unusual); *Saul and David* (Br. 526; PL. 461), a subject Rembrandt had first treated long before (ca. 1631, Br. 490, Frankfurt am Main, Städelsches Kunstinst.); *The Adoration of the Magi* (1657, Br. 592, London, Buckingham Palace, Royal Colls., in contrast to the other paintings of religious scenes of these years, containing a large number of comparatively small figures); *The Denial of Peter* (1660, Br. 594; PL. 462); *The Fall of Haman* or *David and Uriah* (1665, Br. 531, Leningrad, The Hermitage; Valentiner, *ibid.,* has suggested that this represents *Jonathan's Anger*); and *The Return of the Prodigal Son* (after 1660, Br. 598, Leningrad, The Hermitage). The greatness of these paintings lies ultimately in their humanity. The large *Prodigal Son,* one of Rembrandt's last paintings, is in many respects his greatest in its simplicity and grandeur, depth of understanding of man, and expression of the ineffable.

During these years Rembrandt also painted a few subjects of an allegorical, historical, and mythological nature: the so-called *Polish Rider* (ca. 1655, Br. 279, New York, Frick Coll., perhaps his interpretation of the *miles Christianus*), and *The Falconer* (ca. 1661, Br. 319, Göteborg, Konstmus., which Valentiner, 1948, thinks may represent Count Floris V). For Antonio Ruffo, Rembrandt painted as a companion piece to the above-mentioned *Aristotle*, a *Homer Dictating His Verses* (1663, Br. 483, The Hague, Mauritshuis). In the manner of painting it is close to the *Bridal Couple* (the painting is cut down; of the scribe only part of a hand remains; its composition can be reconstructed to some extent on the basis of Rembrandt's study for the painting, the drawing in Stockholm, Nationalmus., B. 1066). For the *Conspiracy of the Batavians* (Br. 482) see above. The only mythological painting of these years is *Philemon and Baucis* (Br. 481; X, PL. 251).

Rembrandt's last painting, which is not finished, forms a significant close to his career: *Simeon in the Temple* (Br. 600, Stockholm, Nationalmus.). Simeon is represented as the man who knowns that, after seeing the Light of Christ, he need see no more.

Etchings. Rembrandt's activity as an etcher declined sharply after 1655. A few portraits were executed: the silversmith Jan Lutma the Elder (1656, H. 290), who is portrayed against the light falling in through the window behind him, something Rembrandt had never done in painting; the writing master Lieven Willemsz. van Coppenol (1656, H. 300); and the art dealer Abraham Francen (ca. 1656, H. 291). The last commission for a portrait came to Rembrandt through his son Titus. The rather lifeless etching of Jan Antonides van der Linden (1665, H. 268), done after a painting by Abraham van den Tempel (The Hague, Mauritshuis), for an edition of Hippocrates by Van der Linden, was refused. It was Rembrandt's last etching.

A few etchings of religious subjects, in addition to a series of four illustrations done for the book *Piedra gloriosa* by Manasseh ben Israel (1655, H. 284), close his fruitful production in this field: *Abraham's Sacrifice* (1655, H. 283), a subtle recasting of the painting of 1635 (Br. 498, Leningrad, The Hermitage); *Abraham Entertaining the Angels* (1656, H. 286), based on one of the Moghul miniatures he copied in drawings; *Christ and the Woman of Samaria* (1657–58, H. 294); and *St. Francis Beneath a Tree, Praying* (1657, H. 292), done almost entirely in dry point, in which Rembrandt obtained a color and atmosphere reminiscent of Titian and Giorgione.

In 1658 Rembrandt etched a number of nudes as well as sketches of them. In 1661 he executed the *Woman with an Arrow* (H. 303), probably representing Venus, not Cleopatra, taking an arrow from Cupid; in his last etching of a nude Rembrandt returned to the antique.

The last of the self-portraits was done in 1658 (H. 300A). Toward the end of his activity as a graphic artist Rembrandt etched a subject that, whatever its exact meaning, must have had autobiographical overtones: *The Phoenix* (1658, H. 295).

Drawings. After 1655 Rembrandt's activity as a draftsman gradually decreased. From 1661 or 1662 on he seems to have made drawings only occasionally. In this last period he used the reed pen and the brush more than ever before, while his style developed further along the same lines as in the previous period: detail was suppressed more and more in favor of a representation of light and of the basic characteristics of a scene, person, or idea. Most drawings are, as in previous periods, done for their own sake. Exceptions are the study for the *Bridal Couple* (see below) and the layout studies for the *Conspiracy of the Batavians* (B. 1061, Munich, Staat. Graphische Samml.) and the *Denial of St. Peter* (B. 1050, Madrid, Bib. Nacional).

Only 10 to 15 landscapes from this period exist. Some are large, boldly sketched views of farms, done with a reed pen and brush (B. 1366, Paris, Inst. Néerlandais, Lugt Coll.; B. 1367, Berlin, Kupferstichkabinett; both, according to Benesch, 1657/58), but most of them became more and more abbreviated renderings of nature, such as the *View of Diemen* (B. 1360, Washington, D.C., Nat. Gall., Rosenwald Coll.) and the *View of the Pesthuis near Amsterdam* (B. 1359, formerly Amsterdam, Dreesman Coll.; both, according to Benesch, 1655/56). In these drawings light has absorbed the forms to the extent of making them almost unrecognizable and giving the lines and dots an abstract quality.

From this period only two self-portraits (B. 1176, Rotterdam, Mus. Boymans-Van Beuningen; B. 1177, Vienna, Albertina) and a few portraits were done. A sequence of drawings for *The Syndics of the Drapers' Hall* is a rare example of Rembrandt's preparing the composition of a group portrait (B. 1178, Berlin, Kupferstichkabinett) and trying to find the attitudes of two syndics that would best fit the composition and be in accord with the subjects' mentality (B. 1179, Amsterdam, Rijksmus., Prentenkabinet; B. 1180, Rotterdam, Mus. Boymans-Van Beuningen). The purpose of three other broadly sketched portraits of men is not known, although it is likely that they were done independently, sometime between 1655 and 1660 (B. 1181, Amsterdam, Verzameling der Six-Stichting; B. 1182, Louvre; B. A80a, Cambridge, Mass., Fogg Art Mus.). Large in size (especially the two mentioned first) and done with a broad reed pen, they have many characteristics in common with the late painted portraits. The difference between the two media has been reduced to a minimum.

During this period Rembrandt seems to have sketched figures from life only rarely, and these life studies assumed a more finite character, such as a portrait of a figure in a certain attitude or action. Thus the difference between genre and figure studies disappeared. Some of the most remarkable studies of figures belong to this category, for example, the two sketches of a *Girl Seated at a Window* (B. 1101–02, Stockholm, Nationalmus.) and a *Girl Sleeping* (B. 1103, Br. Mus.). The first two, unusual in composition, are executed with pen and brush, the third entirely with broad brush strokes. In all three, but especially in the *Girl Sleeping*, Rembrandt "painted" the forms as lighting up amid the dark.

Between 1658 and 1661, at the same time that he was etching similar subjects, Rembrandt made many studies of nudes, mainly seated. A sketch of Rembrandt's studio with a nude seated in front of a stove (B. 1161, Oxford, Ashmolean Mus.) shows that only a limited amount of rather diffuse light illuminated the nude from above; a drawing by a pupil of Rembrandt, Constantijn Renesse (Weimar, Schlossmus.) represents a class drawing from a model, this time reclining. Rembrandt had adopted the academic practice of drawing from the nude, but changed the purpose and method by substituting broad, painterly sketched light and dark for the traditional linear rendering of three-dimensional forms (cf., above all, B. 1107, Munich, Staat. Graphische Samml.; B. 1116, Oxford, Christ Church Lib.; B. 1117–18, 1142, Amsterdam, Rijksmus., Prentenkabinet; B. 1121, Rotterdam, Mus. Boymans-Van Beuningen; B. 1122, 1127, Chicago, Art Inst.).

Most numerous from this period are drawings of religious subjects. Rembrandt preferred a friezelike arrangement of the figures, eliminating depth. This is apparent even when he still defined the figure with care, such as in *Christ and the Woman Taken in Adultery* (ca. 1655, B. 964, Rotterdam, Mus. Boymans-Van Beuningen, where the background consists in a pattern of walls, a technique used in the previous period). It becomes more pronounced in such drawings as *The Doubting of St. Thomas* (ca. 1656, B. 1010, Louvre) and in almost all later drawings of religious subjects, such as *Christ Walking on the Waves* (1659/60, B. 1043, Br. Mus.). While eliminating depth and detail Rembrandt simultaneously stressed the main point of each scene, such as the recognition of Christ in *The Doubting of St. Thomas* (B. 1010) and *Noli me tangere* (B. 993, Paris. Inst. Néerlandais, Lugt. Coll.); the grandeur of Christ in his death, as opposed to the lack of insight of the masses crucifying him, in *The Elevation of the Cross* (B. 1036, Berlin, Kupferstichkabinett); and the sorrow of departure in *The Dismissal of Hagar* (ca. 1656, B. 1008, New York, Pierpont Morgan Lib.). In two of his last drawings, which are among the most impressive of his works, Rembrandt took up the fine pen again, now incorporating in the delicate lines all the power of expression he had learned to convey in various media: *The Presentation in the Temple* (1661, B. 1057, The Hague, Koninklijke Bib.) and *Isaac and Rebecca Spied Upon by Abimelech* (after 1660, B. 988, New York, S. Kramarsky Coll., a study for the *Bridal Couple*). The two drawings, one a representation of Christ's first sacrifice, the other of the love tendered one human being by another, can be understood as a summation of a major theme in Rembrandt's life: Christian love.

SOURCES. A. H. Kan and G. Kamphuis, ed., De jeugd van Constantijn Huygens door hemzelven beschreven (autobiographical fragments, 1629–31), Rotterdam, Antwerp, 1946; J. Orlers, Berchrijving der Stadt Leyden, 2d ed., Leiden, 1641, p. 375; J. von Sandrart, Teutsche Academie . . . , 2 vols., Nürnberg, Frankfort, 1675–79 (ed. A. R. Peltzer, Munich, 1925); S. van Hoogstraten, Inleyding tot de Hooge Schoole der Schilderkonst, Rotterdam, 1678; F. Baldinucci, Cominciamento e progresso dell'arte dell'intagliare in rame . . . , Florence, 1686; C. Hofstede de Groot, Die Urkunden über Rembrandt (Quellenstudien zur holländischen Kg., III), The Hague, 1906; S. Slive, Rembrandt and His Critics, 1630–1730, The Hague, 1953; I. H. van Eeghen, De kinderen van Rembrandt, Maandblad Amstelodamum, XLIII, 1956, pp. 144–46; H. E. van Gelder, Rembrandt's finantiele moeilijkheden, De Groene Amsterdammer, Aug. 25, 1956, p. 10; J. Goudswaard and I. H. van Eeghen, Hendrickje Stoffels, jeugd en sterven, Maandblad Amstelodamum, XLIII, 1956, pp. 114–16; H. E. van Gelder, De jonge Huygensen en Rembrandt, Meded. van het Rijksbureau voor Kunsthist. Documentatie, XIII (Oud-Holland, LXXIII), 1958, pp. 238–41; H. F. Wijnman, Uit de kring van Rembrandt en Vondel: Verzamelde studies over hun leven en omgeving, Amsterdam, 1959; H. Gerson, Seven Letters by Rembrandt, The Hague, 1961 (contributions by I. H. van Eeghen and Y. D. Ovink).

BIBLIOG. The literature on Rembrandt is vast. Since complete and systematic bibliographies covering all publications on Rembrandt through 1943 are easily available (see below), the following bibliography includes only a few basic publications from before 1943 and is mainly concerned with later publications. An effort has been made to list all those books and articles published between 1943 and 1961 that considerably increase our knowledge or enhance our understanding of the artist. For a complete bibliography of material published since 1943 see the *Rijksbureau voor kunsthist. en ikonografische documentatie, Bibliog.*, to which the following bibliography is heavily indebted (consulted through 1958). O. Benesch, ThB, s.v. Rembrandt Harmensz van Rijn; O. Benesch, Rembrandt: Werk und Forschung, Vienna, 1935; H. van Hall, Repertorium voor de geschiedenis der Nederlandsche schilder- en graveerkunst, I, The Hague, 1936, pp. 528–91, II, The Hague, 1949, pp. 310–31; Rijksbureau voor kunsthist. en ikonografische documentatie, Bibliog., The Hague, I, 1943–45, pp. 6, 43–44, 68–71, II, 1946, pp. 31–32, 89–91, III, 1947, pp. 41-47, IV, 1948, pp. 72–76, V, 1949–50, pp. 159–71, VI, 1951–52, pp. 164–74, VII, 1953–54, pp. 164–72, VIII, 1955–56, pp. 161–89, IX, 1957–58, pp. 188–202 (in progress); J. Białostocki, Najnowsze rembrandtiana, B. historii sztuki i kultury, X, 1948, pp. 328–45; H. Roosen-Runge, Zur Rembrandt-Forschung, Literaturbericht und Bibliographie, ZfKg, XIII, 1950, pp. 140–51; H. van de Waal, Rembrandt, 1956, Museum, Tijdschrift voor filologie en geschiedenis, LXI, 1956, pp. 193–209.

General: E. Kolloff, Rembrandt, Raumers historisches Taschenbusch, 1854, pp. 401–58; C. Vosmaer, Rembrandt, The Hague, 1877; E. Michel, Rembrandt: Sa vie, son œuvre et son temps, Paris, 1893 (Eng. trans.; F. Simmonds, 2 vols., London, 1894); K. Neumann, Rembrandt, 2 vols., Munich, 1922; A. M. Hind, Rembrandt, Cambridge, Mass., 1932; K. Bauch, Die Kunst des jugen Rembrandt, Heidelberg, 1933; O. Benesch, Rembrandt: Werk und Forschung, Vienna, 1935; T. Borenius, Rembrandt: Selected Paintings, London, 1942 (3d ed., 1952); R. Hamann, Rembrandt, Berlin, 1948 (rev. H.-M. Rotermund, Die Samml. Göttingen, VI, 1951, pp. 189–91); J. Rosenberg, Rembrandt, 2 vols., Cambridge, Mass., 1948 (rev. H. E. van Gelder, De Nieuwe Stem, 1949, pp. 291–95; W. Stechow, AB, XXXII, 1950, pp. 252–55); H. E. van Gelder, Rembrandt, Amsterdam, 1948; J. G. van Gelder, Rembrandt's vroegste ontwikkeling, Meded. K. Akad. van Wetenschappen, Afdeeling letterkunde, N.S., XVI, 1953, pp. 273–300 (see G. Knuttel, BM, XCVII, 1955, pp. 44–49); L. Münz, Rembrandt Harmenszoon van Rijn, London, New York, 1954; G. Knuttel, Rembrandt: De meester en zijn werk, Amsterdam, 1956; V. F. Levison-Lessing, ed., Rembrandt van Rijn 1606–1669, Moscow, 1956 (in Rus.); Rembrandt, Stockholm, 1956 (cat.; rev. by C. Müller-Hofstede, Kchr., IX, 1956, pp. 89–96); D. C. Roëll et al., ed., Rembrandt: Tentoonstelling, 3 vols., Amsterdam, Rotterdam, 1956 (cat.; rev. by J. Rosenberg, AQ, XIX, 1956, pp. 380–90; J. Rosenberg, Kchr., IX, 1956, pp. 345–52; W. R. Valentiner, AQ, XIX, 1956, pp. 380–90; F. Winkler, Kchr., X, 1957, pp. 141–47); A. B. de Vries, Rembrandt, Baarn, 1956; O. Benesch, Rembrandt: Etude biographique et critique, Geneva, 1957; J. Hulsker, Wie was Who was Qui était Quién fué Rembrandt, The Hague, 1957; J. Jahn, Rembrandt, Leipzig 1958; E. R. Meijer, Rembrandt, Novara, 1958; K. Bauch, Der frühe Rembrandt und seine Zeit, Berlin, 1960; L. Goldscheider, Rembrandt: Paintings, Drawings and Etchings, London, 1960.

Paintings. a. Critical general catalogues: W. R. Valentiner, Rembrandt: Des Meisters Gemälde in 643 Abbildungen, 3d ed., Stuttgart, 1908; C. Hofstede de Groot, Beschreibendes und kritisches Verzeichnis . . . holländische Maler . . . , VI, Stuttgart, 1915; W. R. Valentiner, Rembrandt: Wiedergefundene Gemälde, 2d ed., Stuttgart, 1923; A. Bredius, The Paintings of Rembrandt, Vienna, London, 1935 (2d ed., London, 1942); O. Benesch, The Rembrandt Paintings in the National Gallery, AQ, VI, 1943, pp. 20–33; N. Maclaren, National Gallery Catalogues: The Dutch School, London, 1960, pp. 302–52. *b. Individual works*: J. S. Held, Rembrandt's Polish Rider, AB, XXVI, 1944, pp. 246–65; J. Rosenberg, Rembrandt and Guercino, AQ, VII, 1944, pp. 129–34; W. Stechow, Rembrandt-Democritus, AQ, VII, 1944, pp. 233–38; J. A. van Hamel, De eendracht van het land, 1641, Amsterdam, 1945, T. Koot, Rembrandt's Nachtwacht in nieuwen luister, Amsterdam, 1947 (Eng. trans., Rembrandt's Night Watch: Its History and Adventures, Amsterdam, London, 1949); W. Martin, Van Nachtwacht tot feestoet . . . , Amsterdam, Antwerp, 1947; L. Münz, The Original Shape of Rembrandt's Shipbuilder and His Wife, BM, LXXXIX, 1947, pp. 251–54; A. C. van Schendel and H. H. Mertens, De restauratie van Rembrandt's Nachtwacht, Oud-Holland, LXII, 1947, pp. 1–52; AN, XLVII, 6, 1948, p. 37 ("Rabbi", 1657); J. G. van Gelder, An Unknown Landscape by Rembrandt, BM, XC, 1948, pp. 18–21; L. Münz, A Newly Discovered Late Rembrandt, BM, XC, 1948, pp. 62–67; San Francisco Acquires a Rembrandt, Art Digest, XXII, 20, 1948, p. 14; W. R. Valentiner, Rembrandt's Conception of Historical Portraiture, AQ, XI, 1948, pp. 116–35; C. Brière-Misme, Autour de Rembrandt, Mus. de France, 1949, pp. 122–28; H. von Einem, Der Segen Jacobs, Bonn, 1950; R. van Luttervelt, Rembrandt's Pallas Athene in the Gulbenkian Collection, GBA, XXXVII, 1950, pp. 99–106; W. Martin, Nachtwacht overdenkingen, Oud-Holland, LXVI, 1951, pp. 1–9; W. R. Valentiner, Rembrandt's Landscape with a Countryhouse, AQ, XIV, 1951, pp. 341–47; C. Brière-Misme, La Danaë de Rembrandt et son veritable sujet, GBA, XXXIX, 1952, pp. 305–18, XLI, 1953, pp. 27–36, XLII, 1953, pp. 291–304, XLIV, 1954, pp. 67–76; C. Müller-Hofstede, Rembrandt's Familienbilde und seine Restaurierung, Brunswick, 1952; J. Q. van Regteren Altena, Retouches aan ons Rembrandt-beeld, III: Het genetisch probleem van de Heendracht van het Land, Oud-Holland, LXVII, 1952, pp. 30–50, 59–67; H. E. van Gelder, Rembrandt's portretjes van M. Huygens en J. de Gheyn, Oud-Holland, LXVIII, 1953, p. 107; O. Benesch, An Unknown Rembrandt Painting of the Leiden Period, BM, XCVI, 1954, pp. 134–35; J. Q. van Regteren Altena, Retouches aan ons Rembrandt-beeld, II: Het landschap van den Goudweger, Oud-Holland, LXIX, 1954, pp. 1–17; G. Knuttel, De Nachtwacht en de Gysbrecht, Nederlandsch Kunsthist. Jb., VI, 1955, pp. 151–56; J. Q. van Regteren Altena, Quelques remarques sur Rembrandt et la Ronde de Nuit, Actes XVIIe Cong. int. d'h. de l'art (Amsterdam, 1952), The Hague, 1955, pp. 405–20; C. Bille, Rembrandt's Eendracht van het land en Starters Wt-Treckinge van de Borgerij van Amsterdam, Oud-Holland, LXXXI, 1956, pp. 24–35; I. H. van Eeghen, De echtgenoot van Cornelia Proock, Maandblad Amstelodamum, XLIII, 1956, pp. 111-12; I. H. van Eeghen, Maria Trip of een anonym vrouwsportret van Rembrandt, Maandblad Amstelodamum, XLIII, 1956, pp. 166–69; I. H. van Eeghen, Marten Soolmans en Oopjen Coppit, Maandblad Amstelodamum, XLIII, 1956, pp. 85–90; W. G. Hellinga, Rembrandt fecit 1642, Amsterdam, 1956; Konsthistorisk Tidskrift, XXV, 1956, pp. 1–93 (articles by C. Bille and others on the Claudius Civilis painting); I. Linnik, The Hermitage Rembrandt and Its Subject (in Rus.), Iskusstvo, XIX, 1956, pp. 46–50; A. C. van Schendel, De schimmen van de Staalmeesters, Oud-Holland, LXXI, 1956, pp. 1–23; H. van de Waal, De Staalmeesters en hun legende, Oud-Holland, LXXI, 1956, pp. 61–107; K. Bauch, Rembrandt van Rijn: Die Nachtwachte, Stuttgart, 1957; P. van Eeghen, Eenssem was mij Amsterdam, Maandblad Amstelodamum, XLIV, 1957, pp. 150–54; W. G. Hellinga, De bewogenheid der 'Staalmeesters' . . . , Nederlandsch Kunsthist. Jb., VIII, 1957, pp. 151–84; V. F. Levinson-Lessing, History of Rembrandt's Painting "David and Jonathan" (in Rus.), Soobshcheniia Gosudarstvennogo Ermitazha, XI, 1957, pp. 5–8; W. R. Valentiner, Noch einmal Die Judenbraut, Festschrift Kurt Bauch, Munich, 1957, pp. 227–37; M. F. Wynman, Rembrandt's portretten van Joannes Elison en zijn vrouw Maria Bockenolle naar Amerika verkoocht, Maandblad Amstelodamum, XLIV, 1957, pp. 65–77; I. H. van Eeghen, Een Amsterdamse burgemeestersdichter van Rembrandt in Buckingham Palace, Amsterdam, 1958; I. H. van Eeghen, Frederick Rihel: een 17de eeuwse zakenman en paardenliefhebber, Maandblad Amstelodamum, XLIV, 1958, pp. 73–81; I. H. van Eeghen, De Staalmeesters, Oud-Holland, LXXIII, 1958, pp. 80–84; W. S. Heckscher, Rembrandt's Anatomy of Dr. Nicolaas Tulp: An Iconological Study, New York, 1958 (rev. by C. E. Hellett, BM, CI, 1959, pp. 150–52; J. R. Judson, AB, XLII, 1960, pp. 305–10); C. de Tolnay, A Note on the Staalmeesters, Oud-Holland, LXXIII, 1958, pp. 85–86 (reply by H. van de Waal, pp. 86–89); P. Brachin, Les Syndics des drapiers vus par W. Hellinga, L'information d'h. de l'art, IV, 1959, pp. 87–90; J. Gantner, Rembrandt's Falkenier in Göteborg: ein letztes Echo aus dem Abendmahl des Leonardo, Festschrift K. M. Swoboda, Vienna, 1959, pp. 97–102; J. G. van Gelder, Een Rembrandt van 1633, Oud-Holland, LXXV, 1960, pp. 73–78; I. Manke, Zu Rembrandt's Jakobsegen in der Kasseler Galerie, ZfKg, XXIII, 1960, pp. 252–60; J. I. Kuznetsov, Rembrandt's Painting The Parable of the Laborers in the Vineyard (in Rus.), Trudy Gosudarstvennogo Ermitazha, VI, 1961, pp. 60–88; K. Bauch, Rembrandt's Christus am Kreuz, Pantheon, XX, 1962, pp. 137–44; Jeugdwerk van Rembrandt, Nieuwe Rotterdamsche Courant (Overzeese Weekeditie), Nov. 6, 1962.

Etchings. a. Catalogues and general works: A. Bartsch, Catalogue raisonné de toutes les estampes qui forment l'œuvre de Rembrandt, Vienna, 1797; A. M. Hind, Catalogue of Rembrandt's Etchings, 2d ed., 2 vols., London, 1923; J. Q. Regteren Altena, ed., Rembrandt Harmenszoon van Rijn (1601–1669): Mostra di incisioni e disegni, Rome, Florence, 1951 (cat.); L. Münz, Rembrandt's Etchings, 2 vols., London, 1952 (rev. by W. Stechow, AB, XL, 1958, pp. 164–67); W. Boeck, Rechts und Links in Rembrandt's Druckgraphik, Wallraf-Richartz Jhb., XV, 1953, pp. 179–219; H. van de Waal, Rembrandt's Radierungen zur Piedra Gloriosa des Menasseh ben Israel, Imprimatur, XII, 1954–55, pp. 52-61; G. Biörklund and O. H. Barnard, Rembrandt's Etchings True and False, Stockholm, 1955 (rev. by E. Trautscholt, Wallraf-Richartz Jhb., XVIII, 1956, pp. 238–43); H. de la Fontaine Verwey, Rembrandt als illustrator, Maandblad Amstelodamum, XLIII, 1956, pp. 104–10; L. Münz, Chronologie der späten Rembrandt-Radierungen, Kchr., X, 1957, pp. 150–52. *b. Individual works*: M. Bojanowski, Rembrandt's Faust: Radierung von 1648. D. Vierteljahresschrift, XXX, 1956, pp. 526–32; D. Frey, Die Pietà-Rondanni und Rembrandt's Drei-Kreuze,

Kunstgeschichtliche, S. für Hans Kauffmann, Berlin, 1956, pp. 208–32; W. Wolthuis, Rembrandt's Faust, Jb. Amstelodamum, XLVIII, 1956, pp. 91–112; H.-M. Rotermund, Untersuchungen zu Rembrandt's Faustradierung, Oud-Holland, LXXII, 1957, pp. 151–68; O. H. Lehmann and E. Ettlinger, Contributions to the Interpretation of Rembrandt's Etching Known as Faust in His Study, The Connoisseur, CXLI, 1958, pp. 111–19; H. van de Waal, Rembrandt's Faust ets, Meded. Departement van Onderwijs, Kunsten en Wetenschappen, XXII, 1958, p. 549; W. Wegner, Die Faustdarstellung vom 16. Jahrhundert bis zur Gegenwart, Amsterdam, 1962, pp. 18–26, 126.

Drawings: F. Lippmann and C. Hofstede de Groot, Zeichnungen von Rembrandt in Lichtdruck nachgebildet, Ser. I-IV, Berlin, The Hague, 1888–1911; C. Hofstede de Groot, Die Handzeichnungen Rembrandt's, Haarlem, 1906; J. Kruse and K. Neumann, Die Zeichnungen Rembrandt's und seiner Schule im National-Museum zu Stockholm, 2 vols., The Hague, 1920; W. R. Valentiner, Rembrandt: Des Meisters Handzeichnungen, 2 vols., Stuttgart, 1928–34; F. Lugt, Musée du Louvre, Inventaire Général, Ecole Hollandaise, III: Rembrandt, ses élèves, Paris, 1933; F. Lugt, JhbPreussK-Samml, 1931, p. 58; F. Lugt, Bibliothèque Nationale: Inventaire général des dessins des écoles du Nord, Paris, 1936; M. D. Henkel, Catalogus van de Nederlandsche teekeningen in het Rijksmuseum te Amsterdam, I: Teekeningen van Rembrandt en zijn school, The Hague, 1943; O. Benesch, Rembrandt: Selected Drawings, Oxford, London, 1947 (New York, 1948; rev. by J. G. van Gelder, BM, XCI, 1949, pp. 206–07; J. S. Held, College Art J., VIII, 1949, pp. 234–36); F. Lugt, Inventaire général des dessins: Ecole des Beaux-Arts, I, Paris, 1950; O. Benesch, The Drawings of Rembrandt, 6 vols., London, 1954–57 (rev. by J. G. van Gelder BM, XCVII, 1955, pp. 395–96, CIII, 1961, pp. 149–51; J. Rosenberg, AB, XXXVIII, 1956, pp. 63–70, XLI, 1959, pp. 108–19; W. Sumowski, Wissenschaftliche Z. der Humboldt-Univ. Berlin, Gesellschafts- und sprachwissensch. Reihe, VI, 1956–57, pp. 255–81; E. Haverkamp-Begemann, Kchr., XIV, 1961, pp. 10–28, 50–57, 85–91; W. Sumowski, Bemerkungen zu Otto Beneschs Corpus der Rembrandtzeichnungen, II, Bad Pyrmont, 1961); J. Rosenberg, Rembrandt, the Draughtsman: With Consideration of the Problem of Authenticity, Daedalus, LXXXVI, 1956, pp. 122–36; W. Wegner, Rembrandtzeichnungen, Munich, 1957 (cat.); J. Rosenberg, Great Draughtsmen from Pisanello to Picasso, Cambridge, Mass., 1959, pp. 69–84; O. Benesch, Rembrandt as a Draughtsman, London, 1960; C. White, The Drawings of Rembrandt, London, 1962.

Formal influence on Rembrandt: J. L. A. A. M. van Ryckevorsel, Rembrandt en de traditie, Rotterdam, 1932; O. Benesch, Rembrandt and the Gothic Tradition, GBA, XXVI, 1944, pp. 285–304; E. M. Bloch, Rembrandt and the Lopez Collection, GBA, XXIX, 1946, pp. 175–86; J. Rosenberg, Rembrandt and Mantegna, AQ, XIX, 1956, pp. 153–61; J. Q. van Regteren Altena, Rembrandt en Wenzel Hollar, De kroniek van de vriendenkring van het Rembrandthuis, XIII, 5, 1959.

Iconography and iconology. a. General: J. Białostocki, Ikonographische Forschungen zu Rembrandt's Werk, MJhb, 3d Ser., VIII, 1957, pp. 195–210. *b. Biblical subjects*: W. Stechow, Jacob Blessing the Sons of Joseph from Early Christian Times to Rembrandt, GBA, XXIII, 1943, pp. 193–208; F. Landsberger, Rembrandt, the Jews and the Bible, Philadelphia, 1946 (rev. by J. S. Held, College Art J., VII, 1947–48, pp. 143–45); W. A. Visser 't Hooft, Rembrandt et la Bible, Neuchâtel, Paris, 1947 (Eng. trans., London, 1957); H. van de Waal, Hagar in de woestijn, Nederlandsch Kunsthist. Jb., I, 1947, pp. 143–69; J. Q. van Regteren Altena, Rembrandt's Way to Emmaus, Kunstmuseets Aarsskrift, XXXV–XXXVI, 1948–49, pp. 1–26; J. Kalff, Rembrandt en de Bijbel, Amsterdam, Antwerp, 1949; H.-M. Rotermund, The Motif of Radiance in Rembrandt's Biblical Drawings, Warburg, XV, 1952, pp. 101–21; H.-M. Rotermund, Rembrandt und die religiösen Laienbewegungen in den Niederlanden seiner Zeit, Nederlandsch Kunsthist. Jb., IV, 1952–53, pp. 104–92; Bibeln eller den Heliga Skrift med bilder av Rembrandt, Stockholm, 1954; H.-M. Rotermund, Wandlungen des Christus-typus bei Rembrandt, Wallraf-Richartz Jhb., XVIII, 1956, pp. 197–237; W. Sumowski, Rembrandt erzählt das Leben Jesu, Berlin, 1958; J. Bruyn, Rembrandt's Keuze van Bijbelse onderwerpen, Utrecht, 1959; H.-M. Rotermund, Unidentifizierte beziehungweise misseverstandene Zeichnungen Rembrandt's zu biblischen Szenen, Wallraf-Richartz Jhb., XXI, 1959, pp. 173–208; H.-M. Rotermund, Handzeichnungen und Radierungen Rembrandt's zur Bibel (in press.). *c. Rembrandt and the past*: R. Kieser, Über Rembrandts Verhältnis zur Antike, ZfKg, X, 1941–42, pp. 129–62; W. R. Valentiner, Rembrandt's Conception of Historical Portraiture, AQ, XI, 1948, pp. 116–35; H. von Einem, Rembrandt und Homer, Wallraf-Richartz Jhb., XIV, 1952, pp. 182–205; H. van de Waal, Drie eeuwen vaderlandsche geschied-uitbeelding, 1500–1800, The Hague, 1952; H. Schmidt, Rembrandt, der islamische Orient und die Antike, Aus der Welt der islamischen Kunst; Festschrift für Ernst Kühnel, Berlin, 1959, pp. 336–49. *d. Rembrandt and the culture of his time*: J. A. Emmens, Ay Rembrandt maal Cornelius stem, Nederlands Kunsthist. Jb., VII, 1956, pp. 133–65; J. G. van Gelder, Rembrandt en de 17de eeuw, De Gids, CXIX, 1956, pp. 397–413; J. Q. van Regteren Altena, Rembrandt und die Amsterdamer Bühne, Kchr., X, 1957, pp. 135–37; W. R. Valentiner, Rembrandt and Spinoza: A study of the Spiritual Conflicts in 17th Century Holland, London, 1957 (rev. by H. Gerson, BM, CI, 1959, p. 464). *e. Varia*: F. Lugt, Mit Rembrandt in Amsterdam, Berlin, 1920; L. Münz, Rembrandt's Bild vom Mutter und Vater, JhbKhSammlWien, N.S., XIV, 1953, pp. 141–90; H. Gerson, Aktdarstellungen bei Rembrandt und seinen Schülern, Kchr., X, 1957, pp. 148–50.

The school of Rembrandt: O. Benesch, Rembrandt's Artistic Heritage, I: From Rembrandt to Goya, GBA, XXXIII, 1948, pp. 281–300; Matthiesen Gallery, Rembrandt's Influence in the 17th Century, London, 1953 (cat.); Rembrandt als Leermeester, Leiden, 1956 (cat.); W. R. Valentiner, Rembrandt and His Pupils, Raleigh, N.C., 1956 (cat.).

Criticism of Rembrandt: J. Gantner, Jacob Burckhardts Urteil über Rembrandt und seine Konzeption des Klassischen, Concinnitas, Basel, 1944; pp. 83–114; J. Burckhardt, Rembrandt, und Van Dyck, Bern, 1947 (repr. of 1877 lecture); J. S. Held, Debunking Rembrandt's Legend, AN, XLVIII, 10, 1948–49, pp. 20–24, 60–62; S. Slive, Rembrandt and His Critics, 1630–1730, The Hague, 1953; J. A. Emmens, Rembrandt als genie, Tirade, I, 1957, pp. 49–51; R. W. Scheller, Rembrandt's reputatie van Houbraken tot Scheltema, Nederlands Kunsthist. Jb., XII, 1961, pp. 81–118.

[This article was submitted for publication early in 1963. Since that time many new ideas have been propounded concerning Rembrandt, which of necessity could not be incorporated into the text and bibliography].

Egbert HAVERKAMP-BEGEMANN

Illustrations: PLS. 455–466.

PLATES

Pl. 1. The Great Extinction (*mahāparinirvāṇa*) of the Buddha, 9th–10th cent. Stone, ht., 26 in. London, British Museum.

Pl. 2. *Left*: The goddess Tārā, from Magadha, 9th–10th cent. Sandstone, ht., 11 3/4 in. Paris, Musée Guimet. *Right*: The Buddha Śākyamuni, from Kurkihar, Bihar, 10th–11th cent. Stone,

Pl. 3. *Left*: The goddess Mārīcī, Nalanda, Bihar, 11th cent. Stone. *Right*: Nativity scene (Birth of Śiva or of Kṛṣṇa?), detail, ca. 11th cent. Stone. Calcutta, Indian Museum.

Pl. 4. The Bodhisattva Avalokiteśvara, 9th–10th cent. Stone, ht., 46 in. London, British Museum.

Pl. 5. The goddess Tārā, 10th–11th cent. Stone, ht., 28 in. London, British Museum.

Pl. 6. The sun god Sūrya, 10th–11th cent. Stone, ht., 51 in. London, British Museum.

Pl. 7. The goddess Tārā, Bihar, late 11th cent. Stone. Calcutta, Indian Museum.

Pl. 8. *Left*: Jambhala, from Nalanda, Bihar, 9th–10th cent. Copper, ht., 7 3/4 in. Seattle, Art Museum. *Right*: The goddess Tārā, from Kurkihar, Bihar, 10th–11th cent. Bronze. Patna, India Museum.

Pl. 9. *Above*: Leaf from the *Aṣṭasāhasrikā Prajñāpāramitā*, Nepal, 1015. Palm leaf. Cambridge, England, University Library (Ms. Add. 1643, fol. 164). *Below, left*: Śiva and Pārvatī, Bengal, ca. 12th cent. Bronze, ht., ca. 6 in. Boston, Museum of Fine Arts. *Right*: Model stupa with Dhyani Buddhas, bodhisattvas, and attendants, Bengal, 10th–11th cent. Ivory, ht., 4 1/2 in. Seattle, Art Museum.

Pl. 10. *Above*: Elephant, hunting phase, In Habeter II, Wadi Bergiug, Fezzan. Rock engraving. *Below*: Two giraffes and an elephant superimposed, hunting phase, In Habeter III*d*. Rock engraving.

Pl. 11. Giraffes, hunting phase, Matendusc, Wadi Bergiug, near Murzuq, Fezzan. Rock engraving, polished.

Pl. 12. Hippopotamus, early pastoral phase, Shelter of Uan Muhuggiag V, Acacus massif, southwestern Libya. Rock painting (reproduction).

Pl. 13. Cattle, pastoral phase, Cave of Uan Amil I, Acacus massif, southwestern Libya. Rock painting (water-color copy).

Pl. 14. *Above*: Family scene, pastoral phase, Karkur Talah, Auenat massif (Jebel Uweinat), southeastern Cirenaica. Rock painting. *Below*: Archers, pastoral phase, Cave of Uan Amil I, Acacus massif, southwestern Libya. Rock painting. (Both, reproductions.)

Pl. 15. *Above*: Cow, giraffe, antelope, and three hunters, superimposed, pastoral phase, Shelter of Wadi Kessan II, Acacus massif, southwestern Libya. Rock painting. *Below*: Horse-drawn cart, animals, and men adorned with feathers, period of the horse, Shelter of Wadi Teshuinat VI, Acacus massif. Rock painting. (Both, reproductions.)

Pl. 16. Eland (*above*) and giraffe, southwestern Transvaal. Coarse-pecked petroglyphs. Pretoria, Old Museum.

Pl. 17. Giraffe, pastoral phase, Cave of Uan Amil I, Acacus massif, southwestern Libya. Rock painting.

Pl. 18. Battle over a cow, pastoral phase, Karkur Talah, Auenat massif (Jebel Uweinat), southeastern Cirenaica. Rock painting (water-color copy).

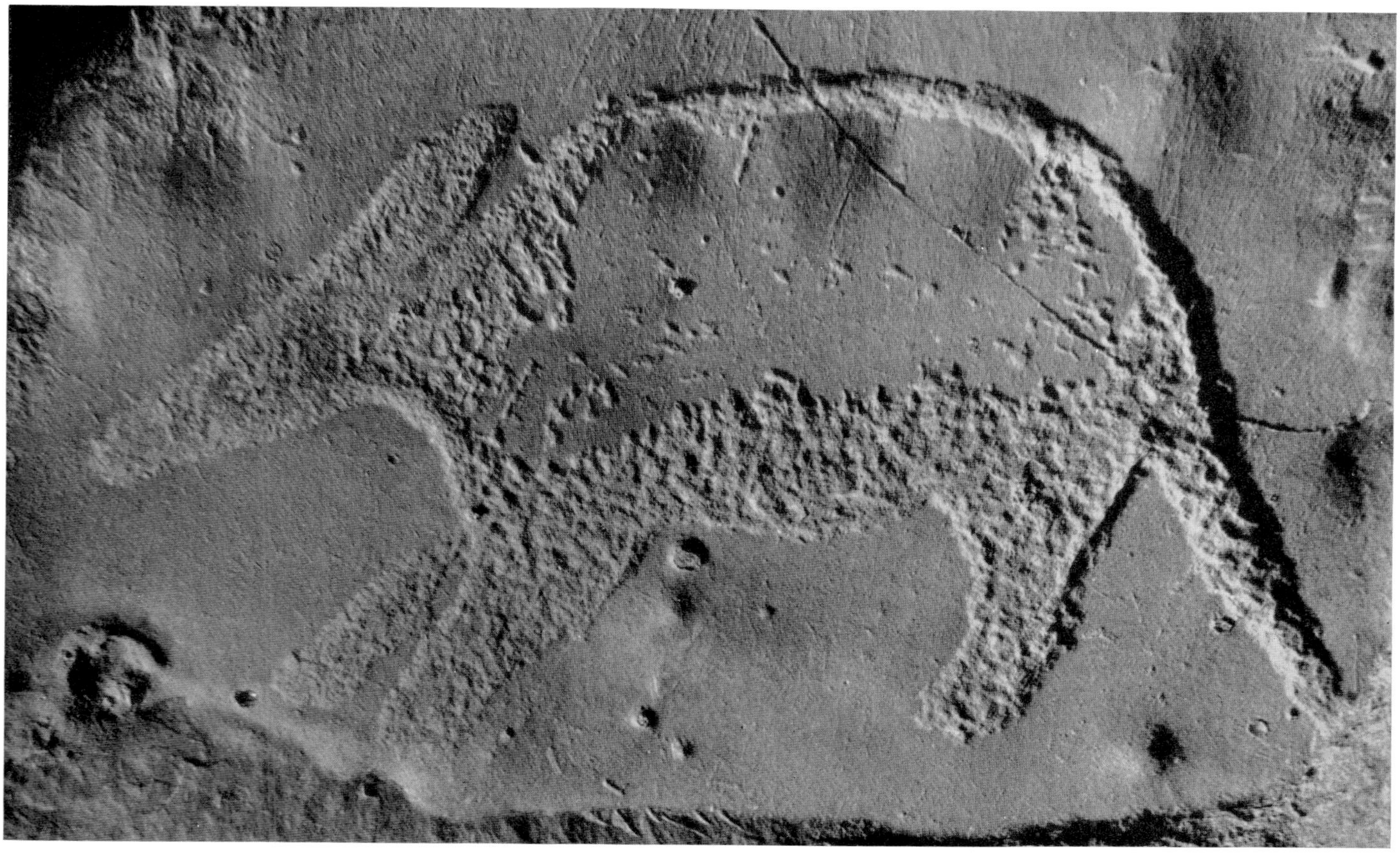

Pl. 19. Anteater (*above*) and eland, southwestern Transvaal. Coarse-pecked petroglyphs. Pretoria, Old Museum.

Pl. 20. *Above*: Eland, Kamberg district, Natal. Rock painting. *Below*: Dead eland, one being quartered, with mantis figure, Herschel district, Cape Province. Rock painting.

Pl. 21. Running hunters with bow and arrows, Cathkin Peak, Drakensberg, Natal. Rock painting.

Pl. 22. Bushman cattle raid, detail with Bantu giving chase, cave on Ventershoek farm, near Wepener, Orange Free State. Rock painting (water-color copy).

Pl. 23. *Above*: Ostriches and hunter disguised as an ostrich, Witteberg, Herschel district, Cape Province. Rock painting, 14×26 in. *Below*: Eland and two hunters, cave on Boomplaats farm, near Kornet Spruit, Zastron district, Orange Free State. Rock painting, 22 1/2×33 3/4 in. (Both, water-color copies.)

Pl. 24. *Above*: Dancing figures, seated woman, and animals, cave on Upper Longreach farm, Cathcart district, Cape Province. Rock paintings (from six different slabs). *Below*: Dance of masked women, Tiffin Kloof, Madeira Hill, Queenstown district, Cape Province. Rock painting, 18×22 in. (Both, water-color copies.)

Pl. 25. Zimbabwe, Rhodesia. *Above*: Wall of the western enclosure of the so-called "Acropolis." *Below*: Exterior wall of the elliptical building or "Temple."

Pl. 26. Zimbabwe, Rhodesia, the great conical tower.

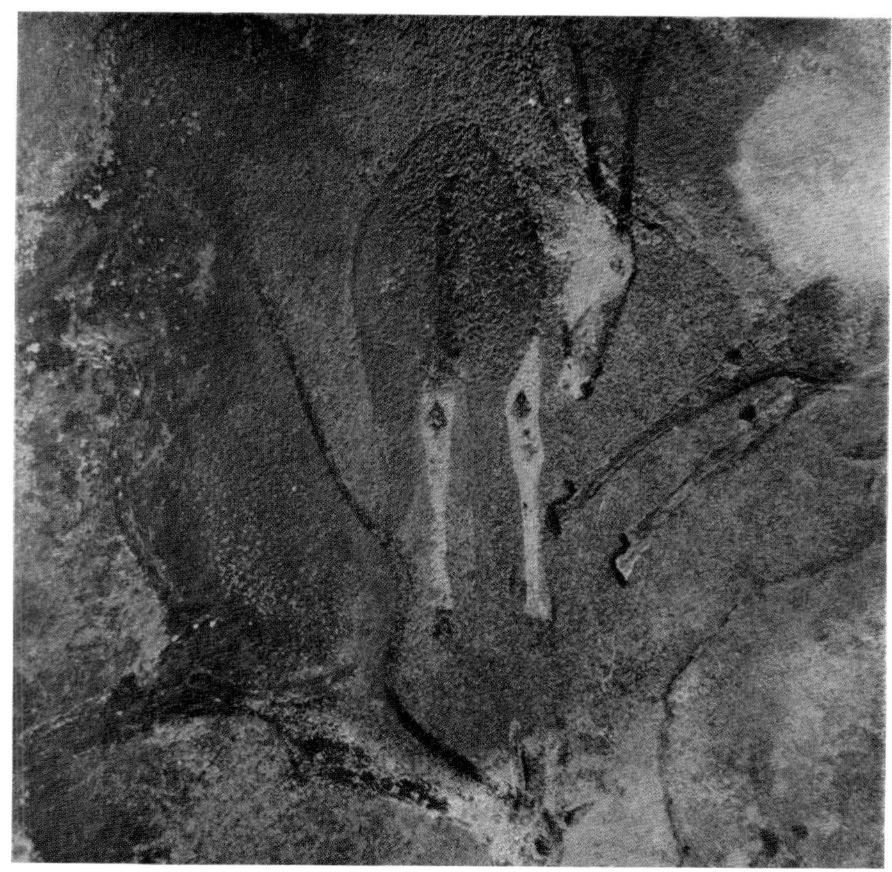

Pl. 27. *Above*: Eland, foreshortened view, Matatiele, Griqualand East, Cape Province. Shaded polychrome rock painting. *Below*: Eland and human figures, Underberg district, Natal. Shaded polychrome rock painting.

Pl. 28. Battle between two groups of Basuto warriors, cave on La Belle France farm, Brakfontein Ridge, Rouxville district, Orange Free State. Rock painting (water-color copy).

Pl. 29. Vicenza, Palazzo della Ragione, called the "Basilica," general view and detail of the upper loggia.

Pl. 30. Vicenza. *Above*: Palazzo Chiericati (now Museo Civico), façade. *Below, left*: Palazzo Porto-Barbaran. *Right*: Palazzo Valmarana, façade.

Pl. 31. *Above*: Vicenza, villa called "La Rotonda." *Below*: Fusina, Veneto, Villa Foscari, called "La Malcontenta," rear view.

Pl. 32. *Left*: Fusina, Veneto, Villa Foscari, called "La Malcontenta," side view of the portico. *Right*: Santa Sofia di Pedemonte, near Verona, Villa Sarego (now Boccoli), main front, right wing.

Pl. 33. Maser, near Treviso, Villa Barbaro (now Volpi), façade (cf. VIII, pl. 434).

Pl. 34. Venice. *Above*: S. Giorgio Maggiore, interior. *Below*: Chiesa del Redentore, interior.

Pl. 35. *Above, left*: Venice, Chiesa del Redentore, façade. *Right*: Maser, near Treviso, Villa Barbaro (now Volpi), the *tempietto* (cf. VIII, PL. 435). *Below*: Vicenza. *Left*: Casa Cogollo, detail of façade. *Right*: Loggia del Capitanio.

Pl. 36. Vicenza, Teatro Olimpico, detail of the stage (IX, PL. 312), completed by V. Scamozzi.

Pl. 37. The Flood, detail of PL. 41.

Pl. 38. Creation of the Animals and Creation of Adam, lunette in a cycle of scenes from Genesis. Fresco. Florence. S. Maria Novella, Chiostro Verde.

Pl. 39. The Battle of San Romano, showing the unhorsing of the Sienese commander, detail. Panel; full size, ca. $6 \times 10\frac{1}{2}$ ft. Florence, Uffizi.

Pl. 40. St. George and the Dragon. Canvas, $22\,{}^{1}/{}_{4} \times 29\,{}^{1}/{}_{4}$ in. London, National Gallery.

Pl. 41. The Flood and portion of Noah's Sacrifice and Drunkenness (*below*), in a cycle of scenes from Genesis. Fresco. Florence, S. Maria Novella, Chiostro Verde.

Pl. 42. The Battle of San Romano, showing Niccolò Mauruzi da Tolentino leading the attack against the Sienese. Panel, ca. $6 \times 10\frac{1}{2}$ ft. London, National Gallery.

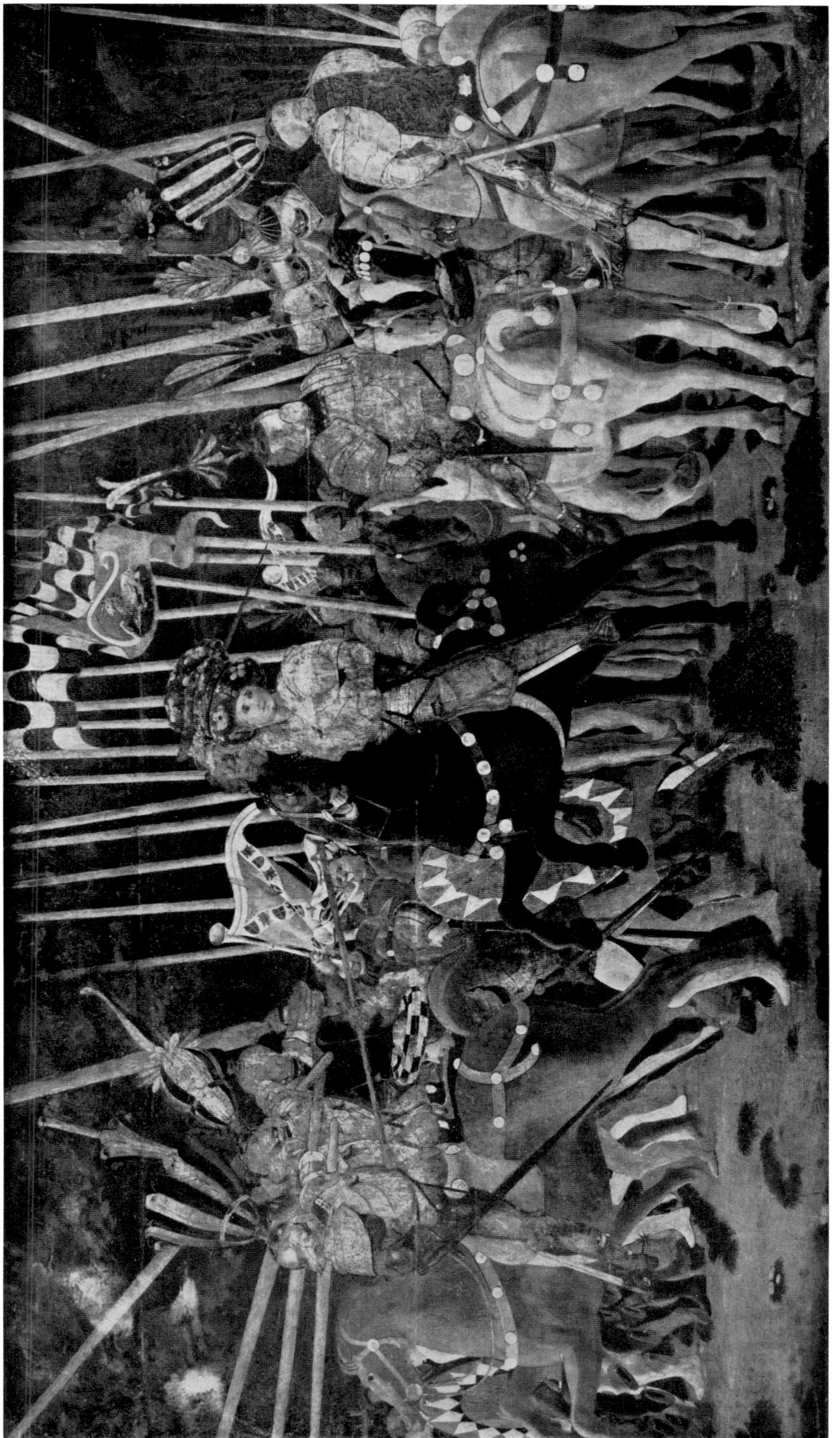

Pl. 43. The Battle of San Romano, showing the Florentines in counterattack. Panel, ca. 6×10½ ft. Paris, Louvre.

Pl. 44. *Above*: Adoration of the Magi (attrib.?), central scene on a predella. Panel, ht., 8 1/4 in. Bagno a Ripoli, near Florence, S. Bartolomeo a Quarata. *Center*: Night Hunt. Panel, 2 ft., 1 5/8 in. × 5 ft., 5 in. Oxford, England, Ashmolean Museum. *Below*: Legend of the Profanation of the Host, two of six scenes on a predella. Panel, ht., 16 1/2 in. Urbino, Italy, Galleria Nazionale delle Marche.

Pl. 45. The Vision of St. Jerome. Panel, 11 ft., 5 3/4 in. × 5 ft., 1/4 in. London, National Gallery.

Pl. 46. *Left*: Holy Family. Panel, $43\frac{1}{4} \times 35$ in. Madrid, Prado. *Right*: St. Roch and a Donor. Panel, 8 ft., $10\frac{1}{4}$ in. $\times$ 6 ft., $5\frac{1}{2}$ in. Bologna, S. Petronio.

Pl. 47. *Left*: Madonna of the Rose. Panel, $42^7/_8 \times 34^5/_8$ in. Dresden, Gemäldegalerie. *Right*: Portrait of Gian Galeazzo Sanvitale. Panel, $42^7/_8 \times 31^7/_8$ in. Naples, Museo di Capodimonte.

Pl. 48. Madonna and Child with St. Stephen, John the Baptist, and a Donor. Panel, 8 ft., 3 1/2 in. × 5 ft., 3 1/4 in. Dresden, Gemäldegalerie.

Pl. 49. Madonna and Child, the Infant St. John, the Magdalen, and St. Zacharias. Panel, $28^{3}/_{4} \times 24^{3}/_{8}$ in. Florence, Uffizi.

Pl. 50. Portrait of a young woman, called "Antea." Canvas, $54^3/_4 \times 34^5/_8$ in. Naples, Museo di Capodimonte.

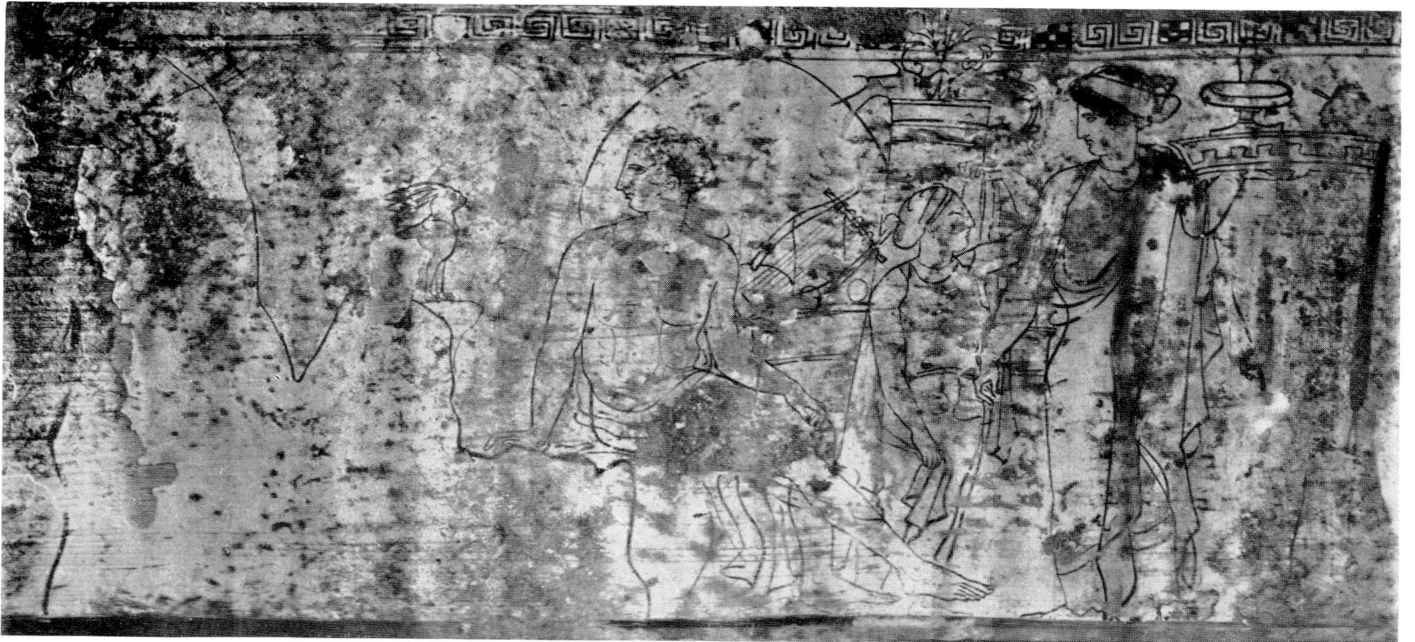

Pl. 51. *Above*: Funerary scene, in the manner of Parrhasios, decoration around an Attic white-ground lekythos, ca. 440 B.C. Athens, National Museum. *Below*: Relief probably reflecting the Parrhasian Philoktetes, on a cup signed by Cheirisophos, Augustan age, found at Hoby, Denmark. Silver, ht., $4^{3}/_{8}$ in. Copenhagen, Nationalmuseet.

Pl. 52. Warrior near tomb, in the manner of Parrhasios, on a white-ground lekythos from Eretria, ca. 430 B.C. Ht., 19 in. Athens, National Museum.

Pl. 53. *Above*: Khurha, Iran, ruins of the temple, ca. 3d cent. B.C. *Below*: Hatra, Iraq, main palace, details of the south liwan (*left*) and an archivolt, 2d cent. B.C.

Pl. 54. *Above*: The gods Agli-bol, Baal-shamin, and Malak-bel, relief from Palmyra, Syria, 1st cent. Stone, 22×27 in. Paris, Louvre. *Below*: Limestone statues from Hatra, Iraq. *Left*: Ubal, daughter of Jabal, from Temple IV, 2d cent. Ht., 5 ft., 7 in. Baghdad, Iraq Museum. *Right*: Warrior, from Temple III, 1st–3d cent. Ht., 6 ft., 6 in. Mosul, Iraq, Museum.

Pl. 55. Head of a Parthian king, fragment of a portrait statue, from Hatra, Iraq, 2d cent. Stone. Baghdad, Iraq Museum.

Pl. 56. *Left*: Veiled women, 1st cent. Marble, ht., 51 in. Palmyra, Syria, Temple of Bel. *Right*: Head, from Susa, Iran, 1st–3d cent. Limestone, ht., 10¼ in. Paris, Louvre.

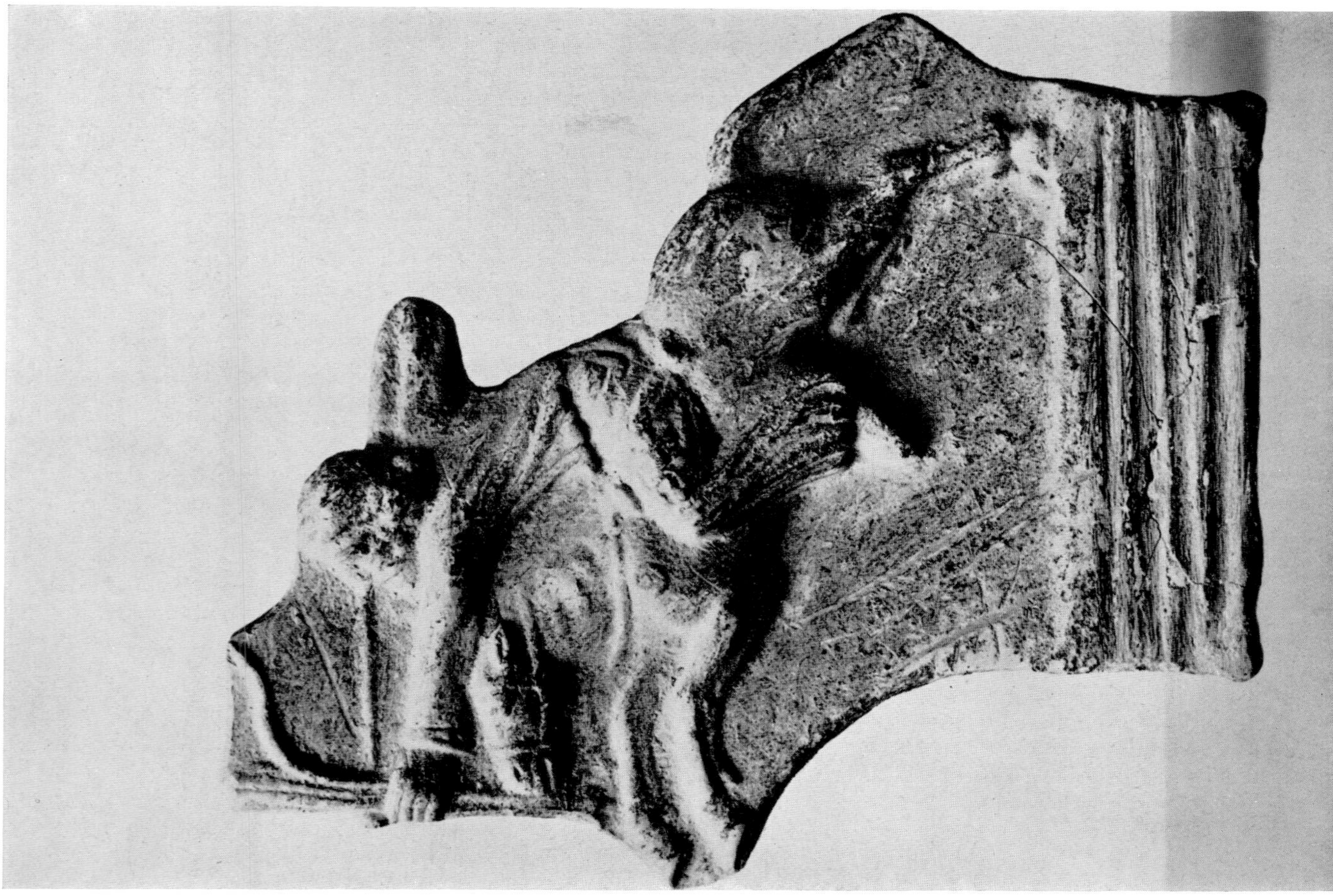

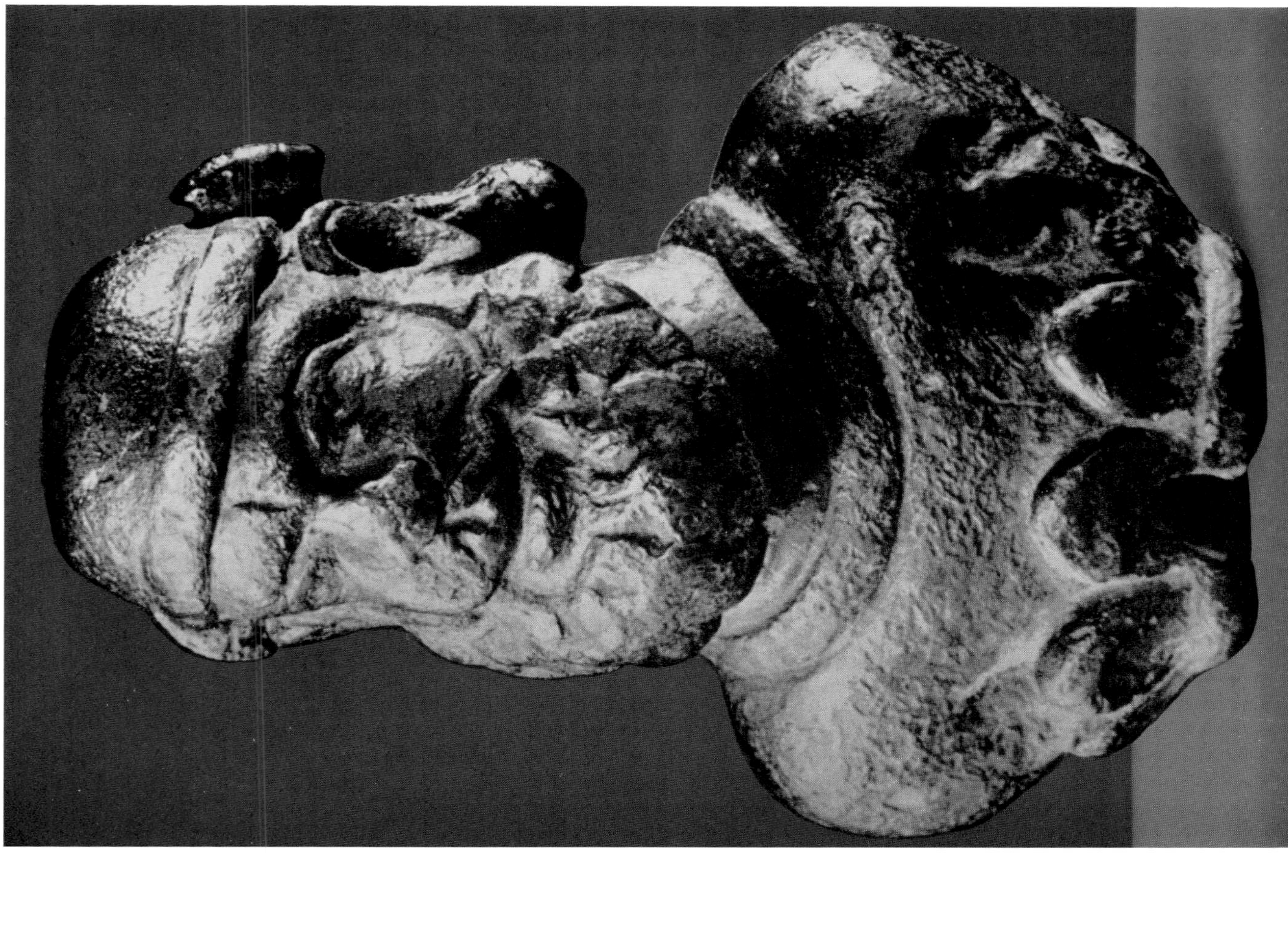

Pl. 57. *Left*: Finial with portrait of Orodes (?), early 1st cent. (?). Bronze. *Right*: Mounted archer, Syria, 1st–3d cent. Terra cotta, ht., $6^3/_4$ in. Both, Berlin, Staatliche Museen.

Pl. 58. Nihavand hoard, Iran, found with coins of 1st–2d cent. *Left*: Repoussé and engraved bowl, with radiating pattern of lotus leaves and acanthus foliation. Gilded silver; diam., 5 3/4 in. Berlin, Staatliche Museen. *Right*: Clasp with symbolic eagle and deer. Gold set with turquoise. New York, Metropolitan Museum.

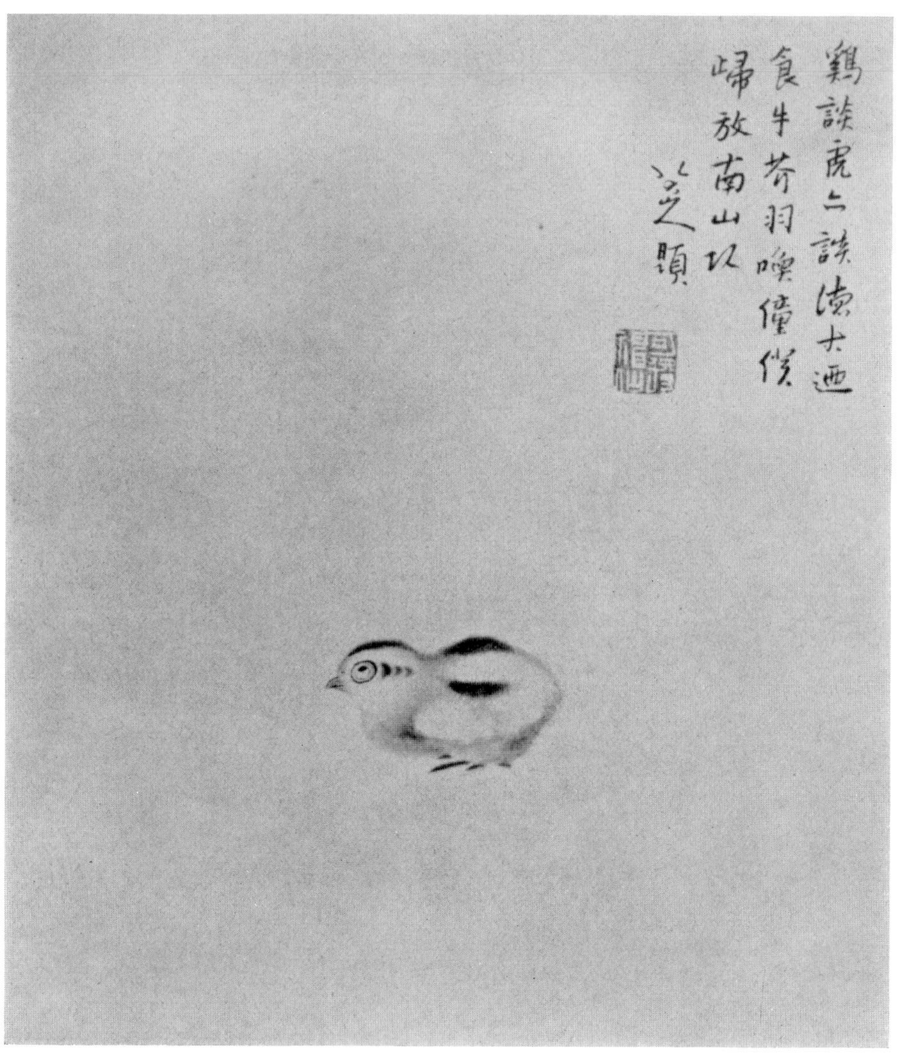

Pl. 59. Album leaves. *Left*: Chick. Ink on paper. China, private coll. *Right*: Fish. Ink on paper. Oiso, Japan, K. Sumitomo Coll.

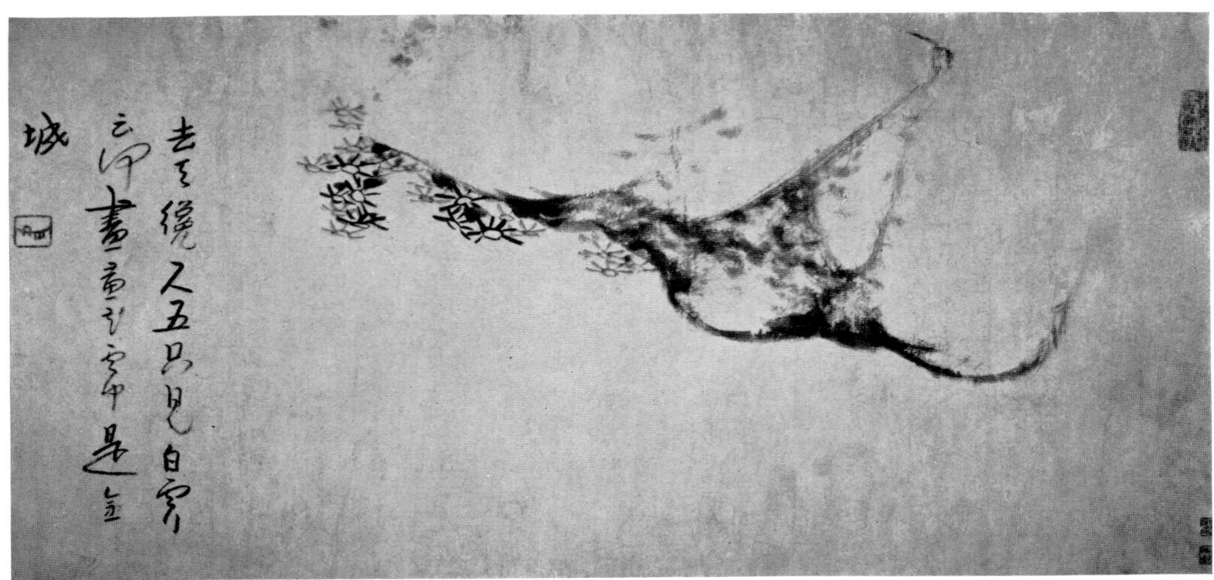

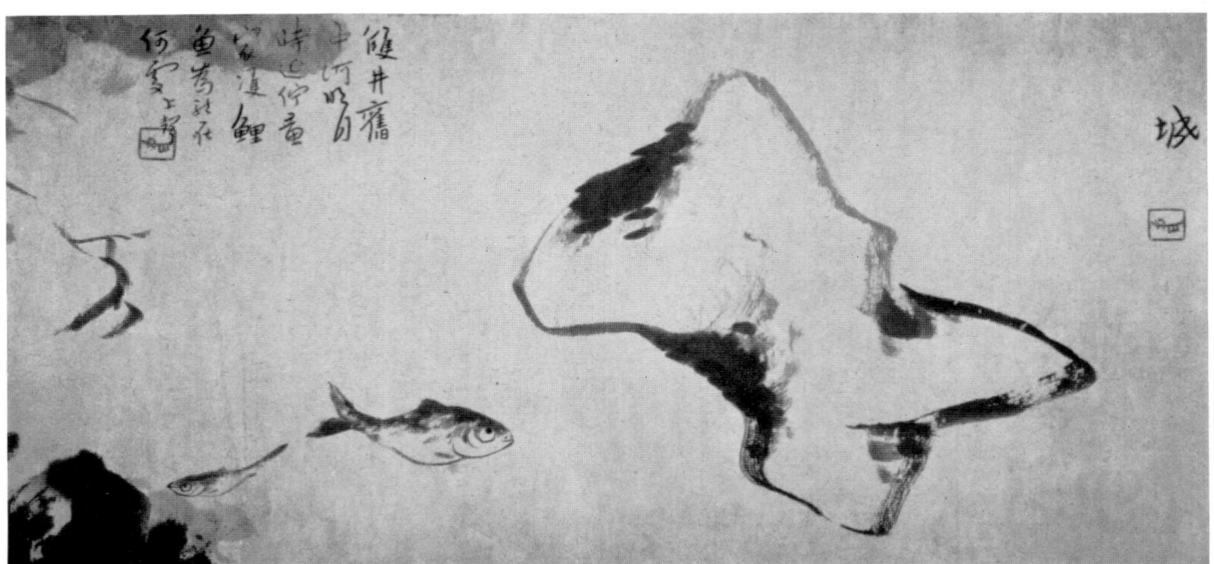

Pl. 60. Fish and Rocks, details of a hand scroll. Ink on paper; full size, 11½ in. × 5 ft., 2 in. Cleveland, Museum of Art.

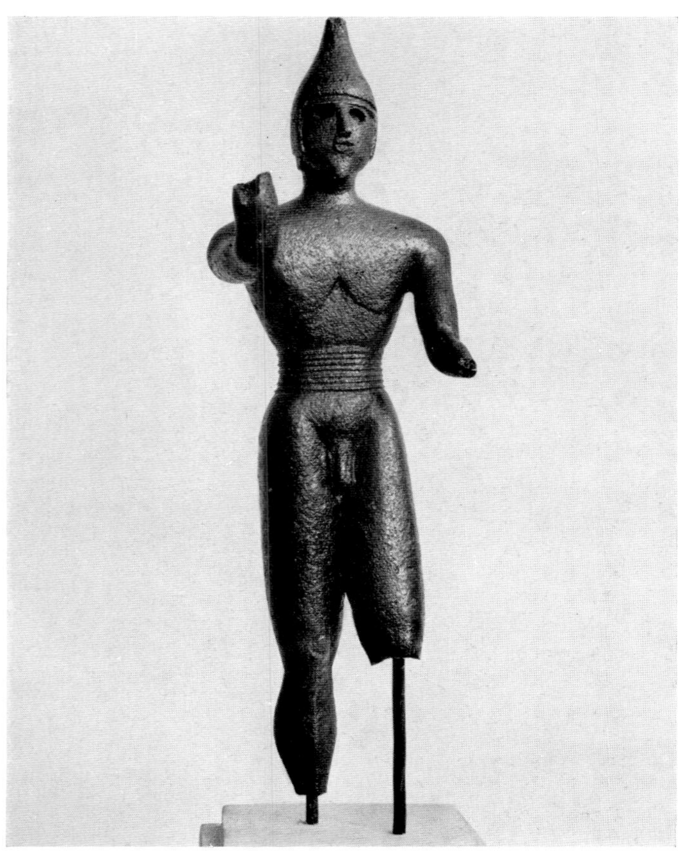

Pl. 61. Bronzes, from Olympia. *Above, left*: Mare and foal, second half of 8th cent. B.C. Ht., ca. 4 in. Athens, National Museum. *Right*: Griffin protoma, from a caldron, second half of 7th cent. B.C. Ht., 14 in. *Below, left*: Warrior, first half of 7th cent. B.C. Ht., ca. 7 in. *Right*: Laconian statuette, old man with staff (now missing), second half of 6th cent. B.C. Ht., ca. $5^{1}/_{2}$ in. Last three, Olympia, Archaeological Museum.

Pl. 62. *Above, left*: Proto-Corinthian oinochoe, from Syracuse, Sicily, mid-7th cent. B.C. Ht., 6 1/4 in. Syracuse, Museo Archeologico. *Right*: Proto-Corinthian aryballos, from Thebes, ca. mid-7th cent. B.C. Ht., 2 3/4 in. Paris, Louvre. *Below*: Corinthian kylix, ca. 600 B.C. Ht., 5 1/8 in. Athens, National Museum.

Pl. 63. Corinthian vases, first half of 6th cent. B.C. *Above*: Low-footed cup, found in Etruria. Diam., $8\frac{5}{8}$ in. *Below, left*: Amphora with Tydeus and Ismene. Ht., $12\frac{5}{8}$ in. Both, Paris, Louvre. *Right*: Column crater with the departure of Amphiaraos. Berlin, Staatliche Museen.

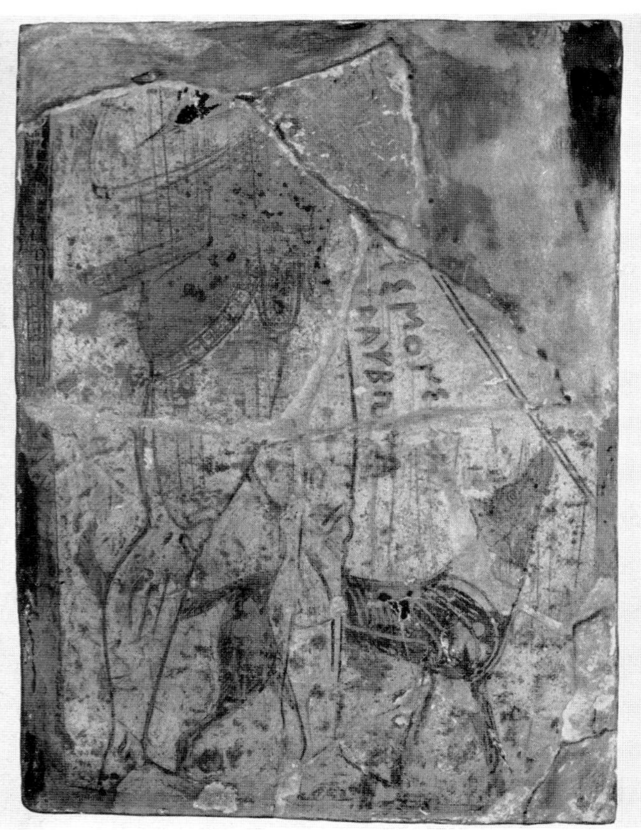

Pl. 64. *Above*: Corinthian terra cottas. *Left*: Demeter and Persephone on a cart, found in Thebes, late 7th cent. B.C. Ht., 6 1/2 in. London, British Museum. *Right*: Pinax fragment, signed by Timonidas, second quarter of 6th cent. B.C. Berlin, Staatliche Museen. *Below, left*: Bronze mirror, from Corinth, ca. late 6th cent. B.C. Athens, National Museum. *Right*: Corinthian silver stater, obv., ca. 530 B.C. Zurich, private coll.

Pl. 65. Fragmentary metopes found under the so-called "Treasury of the Sikyonians" at Delphi, first half of 6th cent. B.C. Limestone, ht., 23 1/4 in. *Above*: The Argo with Orpheus and the Dioskouroi. *Below*: Rape of Europa. Both, Delphi, Archaeological Museum.

Pl. 66. *Left*: Fragment of relief with female figure, from Mycenae, ca. 630 B.C. Limestone, ht., 16 in. Athens, National Museum. *Right, above*: Fragment of crater with the blinding of Polyphemos, from Argos, mid-7th cent. B.C. Terra cotta. Argos, Museum. *Below*: Mask, from Tiryns, early 7th cent. B.C. Terra cotta. Nauplia, Museum.

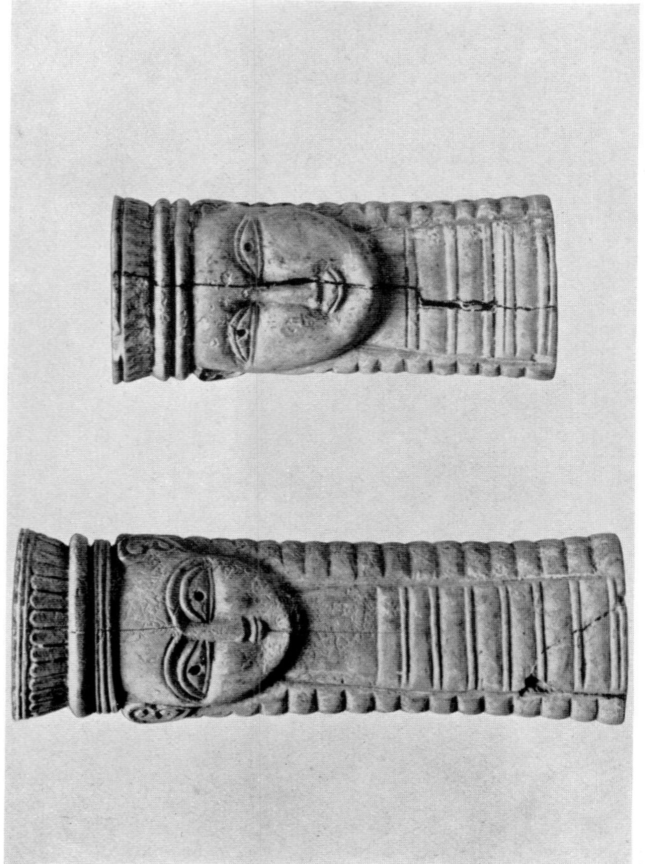

Pl. 67. *Left, above*: Votive sculptures, from the sanctuary of Artemis Orthia, Sparta, late 7th cent. B.C. Bone; max. ht., 4 in. Athens, National Museum. *Below*: Laconian kylix with Atlas and Prometheus, mid-6th cent. B.C. Rome, Vatican Museums. *Right*: Head of a youth, from Sparta, ca. 550–530 B.C. Bronze, ht., ca. 6 in. Boston, Museum of Fine Arts.

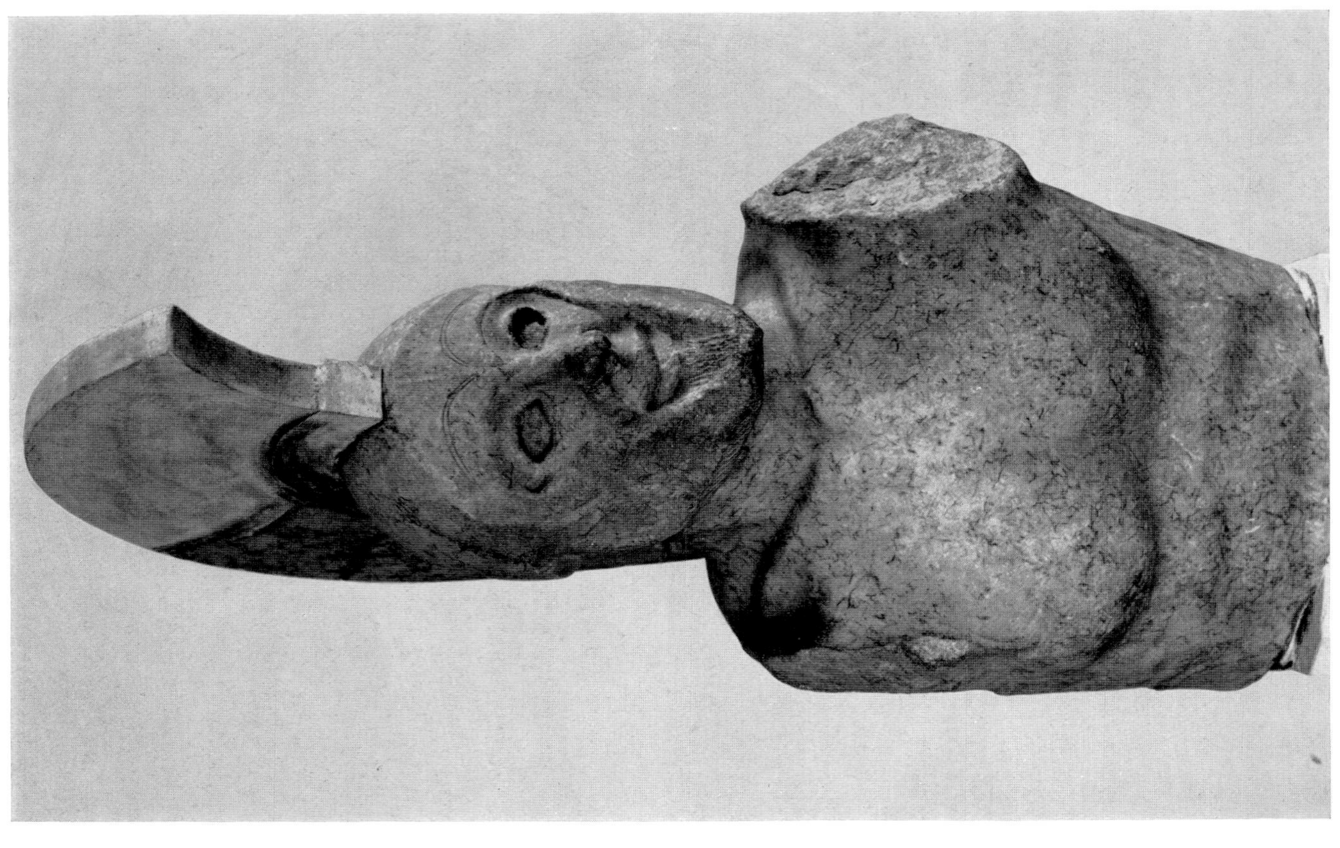

Pl. 68. *Left*: Relief with so-called "Menelaos and Helen" (Orestes and Elektra?), from Sparta, late 7th–6th cent. B.C. Marble, ht., 26 3/4 in. *Right*: Warrior, so-called "Leonidas," from Sparta, early 5th cent. B.C. Marble, ht., 30 3/4 in. Both, Sparta, Museum.

Pl. 69. Laconian cup with dolphins and other fish, found in Taranto, Italy (anc. Taras), late 7th–early 6th cent. B.C. Diam. without handles, $8^1/_4$ in. Taranto, Museo Nazionale.

Pl. 70. Proto-Corinthian vase, called the "Chigi oinochoe," from a tomb at Veio, Formello, Italy (anc. Veii), details with trumpeter and warriors (*above*) and horsemen (*below*), ca. 640–630 B.C. Full ht., 10¼ in. Rome, Museo di Villa Giulia.

Pl. 71. Head of Hera (?), from the Heraion of Olympia, ca. 600 B.C. Limestone, ht., 20 1/2 in. Olympia, Archaeological Museum.

Pl. 72. Head of Athena, from the east pediment of the Temple of Aphaia, Aegina, early 5th cent. B.C. Marble, ht., ca. 12 in. Munich, Staatliche Antikensammlungen.

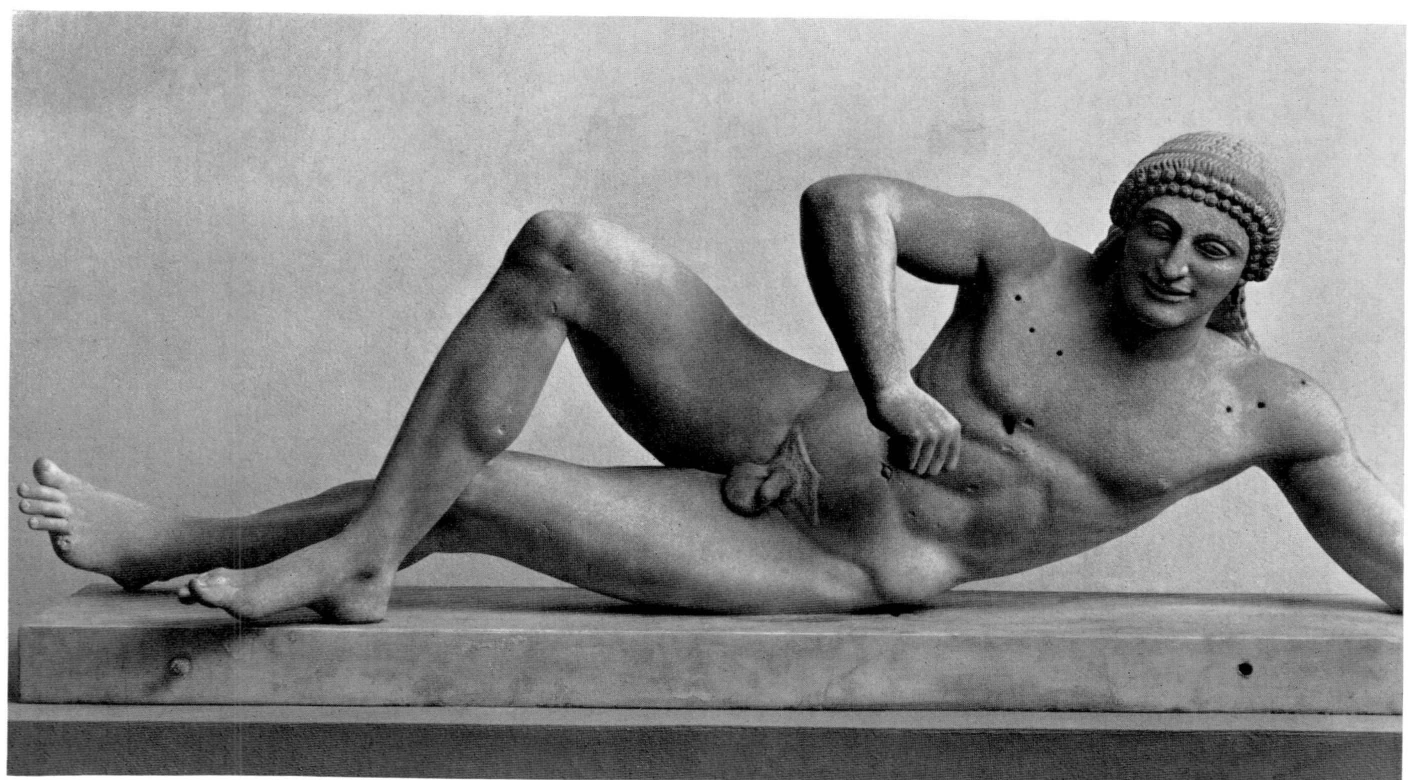

Pl. 73. *Above, and below, left*: Details of the west pediment of the Temple of Aphaia, Aegina, early 5th cent. B.C. Marble. *Above*: Fallen warrior. L., 5 ft., 2 1/2 in. *Below, left*: Head of warrior. Full ht., 4 ft., 8 in. Both, Munich, Staatliche Antikensammlungen. *Right*: Sphinx, from Aegina, ca. 460 B.C. Marble, ht., 37 3/4 in. Aegina, Museum.

Pl. 74. Statues, probably Oinomaos and Sterope, from the east pediment of the Temple of Zeus, Olympia, ca. 460 B.C. Marble; max. ht., 9 ft., 8 in. Olympia, Archaeological Museum.

Pl. 75. Marble sculptures, from the Temple of Zeus, Olympia, ca. 460 B.C. *Above*: Lapith woman and centaur, from the west pediment. *Below, left*: Head of Theseus, from the west pediment. Ht., $13^{3}/_{4}$ in. Both, Olympia, Archaeological Museum. *Right*: Metope with Herakles and the Cretan bull. Ht., 5 ft., 3 in. Paris, Louvre.

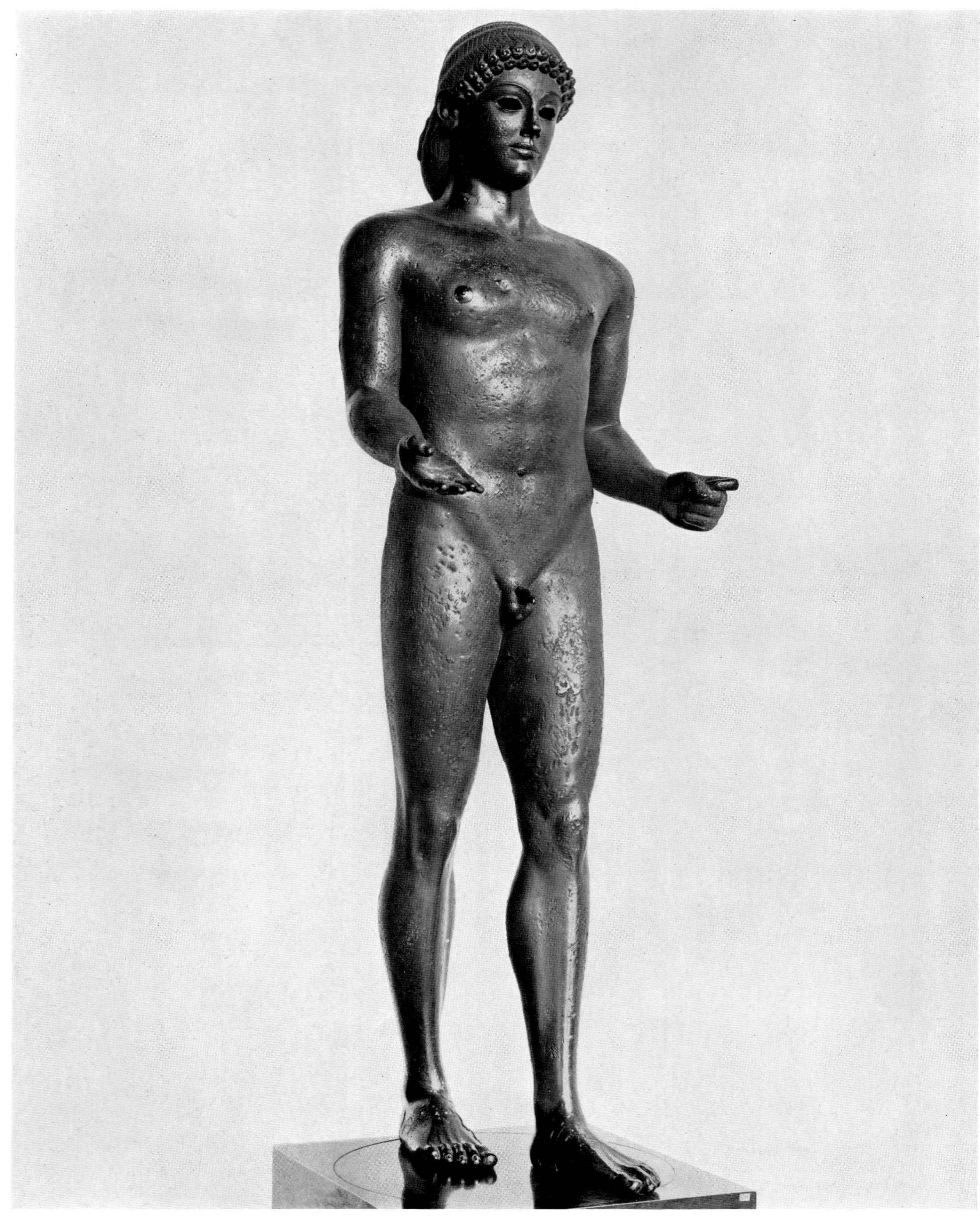

Pl. 76. Kouros, found in the sea at Piombino. Italy, perhaps from Aegina, first half of 5th cent. B.C. Bronze, ht., 45 1/4 in. Paris, Louvre.

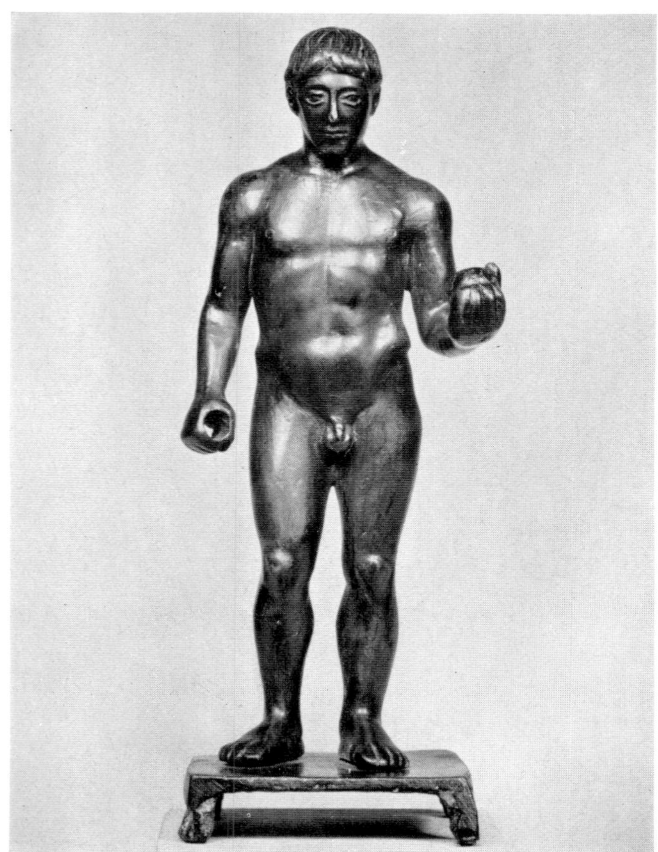

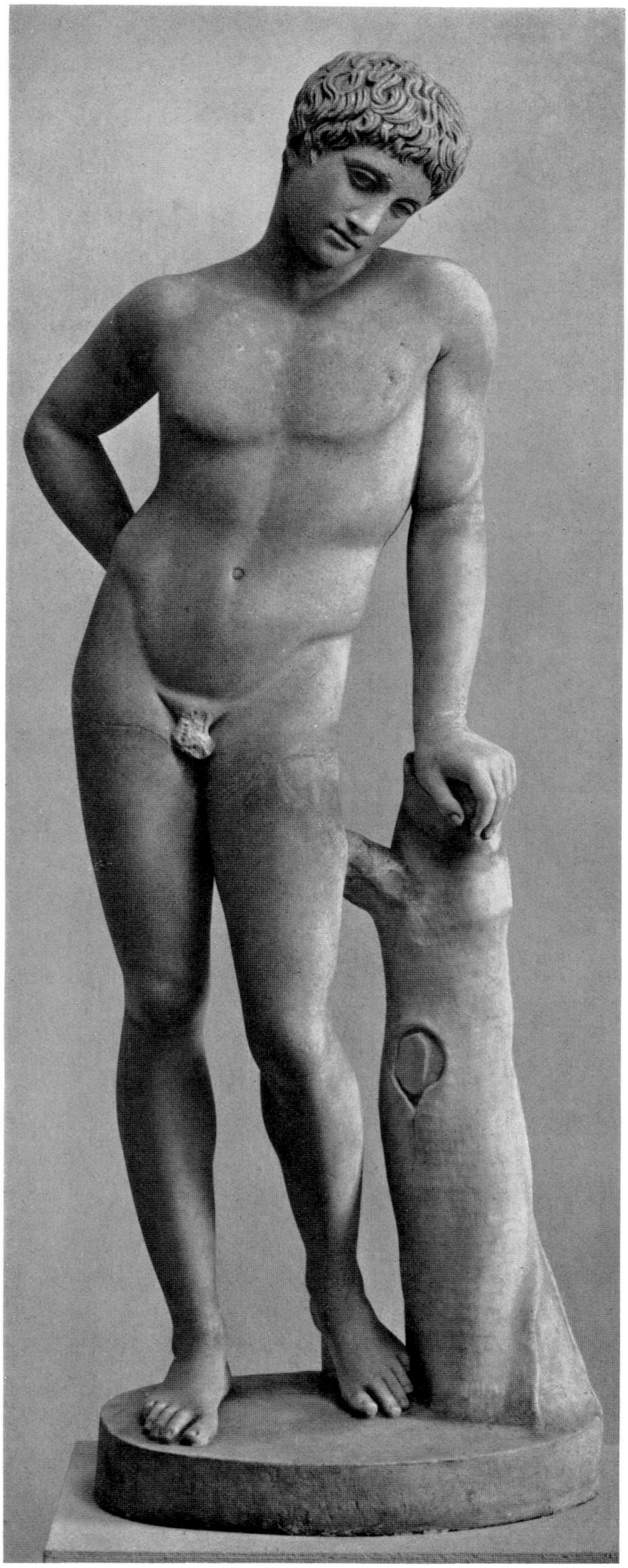

Pl. 77. *Left*: Bronzes, from Argolis, ca. mid-5th cent. B.C. *Above*: Head of a youth, fragment of a victory statue. Munich, Staatliche Antikensammlungen. *Below*: Athlete, from Lygourion. Ht., ca. $4^{1}/_{2}$ ft. Berlin, Staatliche Museen. *Right*: So-called "Narcissus," Roman statue of Polykleitan derivation. Marble, ht., $42^{1}/_{8}$ in. Munich, Staatliche Antikensammlungen.

Pl. 78. Female head, from Tegea, third quarter of 4th cent. B.C. Marble, ht., 11 1/4 in. Athens, National Museum.

Pl. 79. Indications of the third dimension in prehistoric art. Paintings of horse and bison, Lascaux Cave, Dordogne, France.

Pl. 80. *Above*: Conventionalized representation of elements in space, ancient art. *Left*: The Battle of Kadesh, detail, relief in the rock-cut temple of Ramses II at Abu Simbel, 13th cent. B.C. *Right*: Interior of a fortification, detail of a war relief of Ashurnasirpal II, from Nimrud, 9th cent. B.C. London, British Museum. *Below*: Foreshortening and space in Greek vase painting. *Left*: Scene in a palaestra, detail of a kylix by Euphronios, from Capua, ca. 500 B.C. Full ht., 13 3/4 in. Berlin, Staatliche Museen. *Right*: The bride Alkestis with her friends, detail of an onos (epinetron) by the Eretria Painter, ca. 420 B.C. Full l., 11 3/8 in. Athens, National Museum.

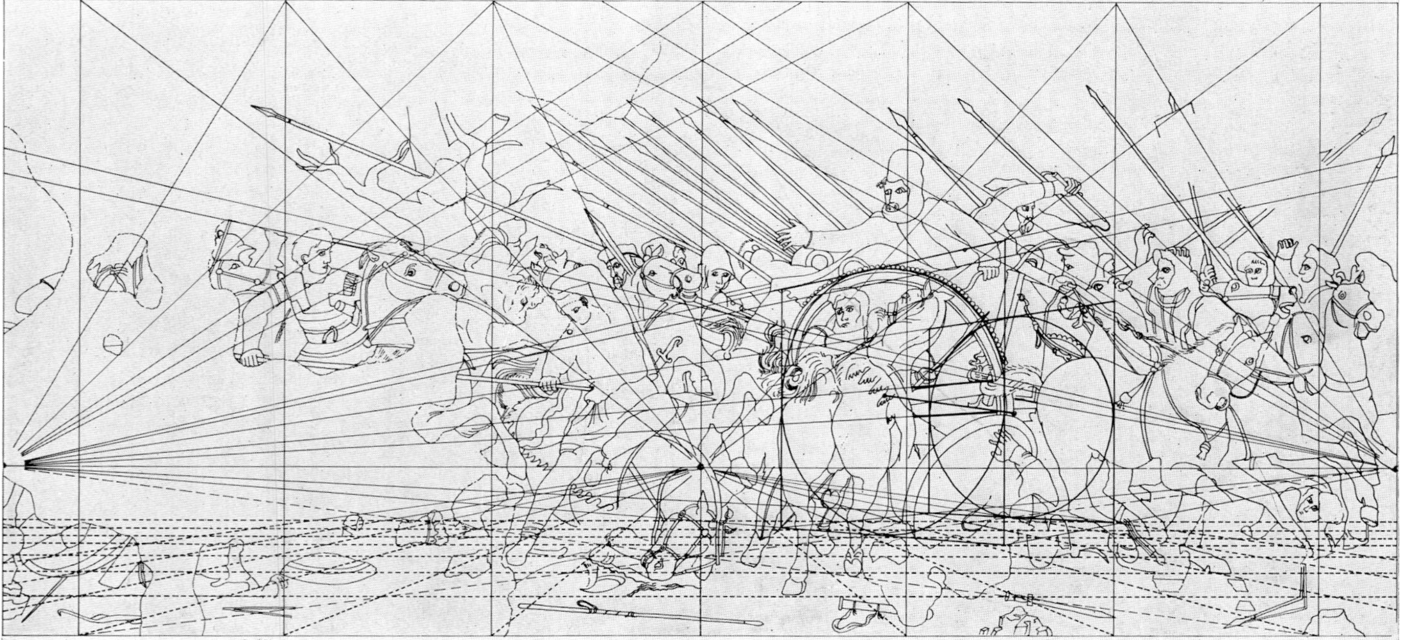

Pl. 81. *Above*: The Battle of Alexander, from the House of the Faun, Pompeii, 3d–2d cent. B.C., after an original painting of the late 4th–early 3d cent. B.C. Mosaic, 8 ft., 10 3/4 in. × 16 ft., 9 5/8 in. Naples, Museo Nazionale. *Below*: Diagram of a hypothetical perspective scheme.

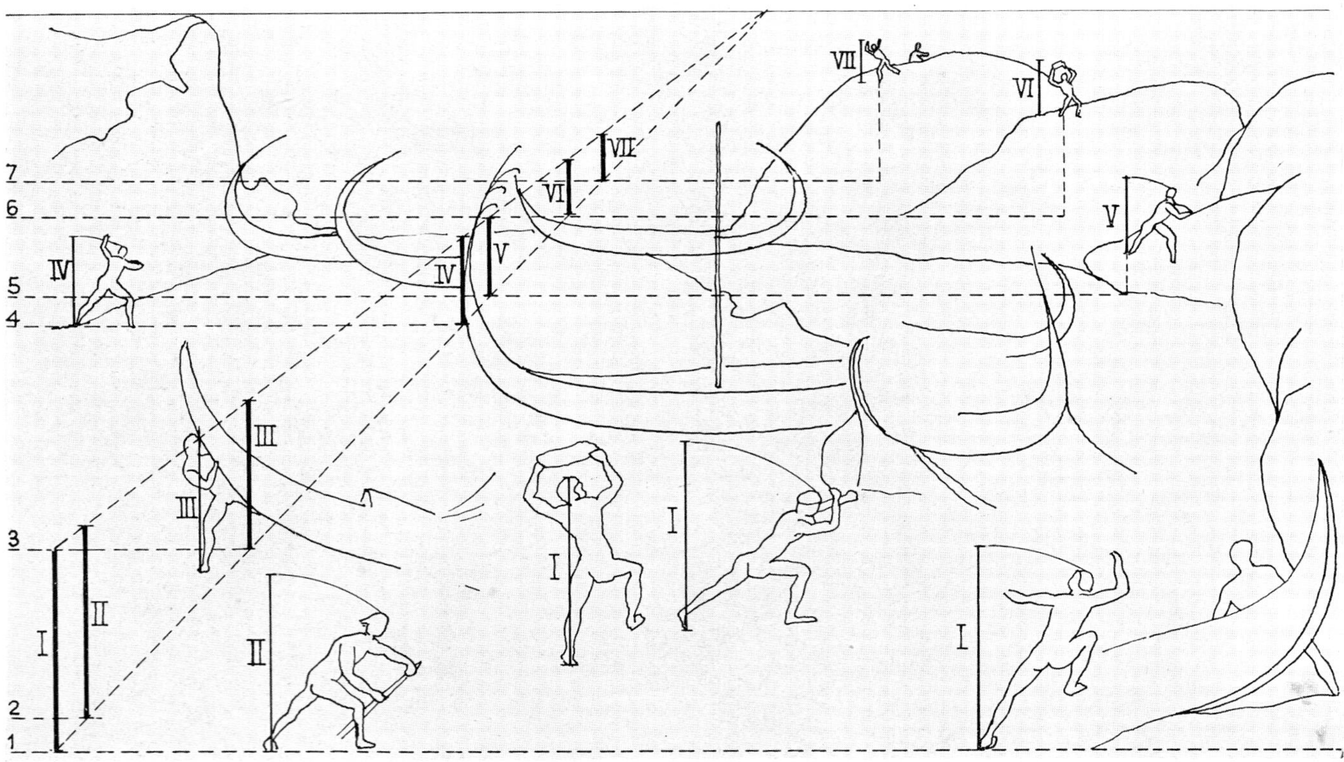

Pl. 82. *Above*: The Laestrygones destroying the fleet of Odysseus, wall painting, Rome, ca. 50 B.C., after a Greek original of ca. mid-2d cent. B.C. Rome, Vatican Library. *Below*: Diagram indicating the relative scale of the figures in recession.

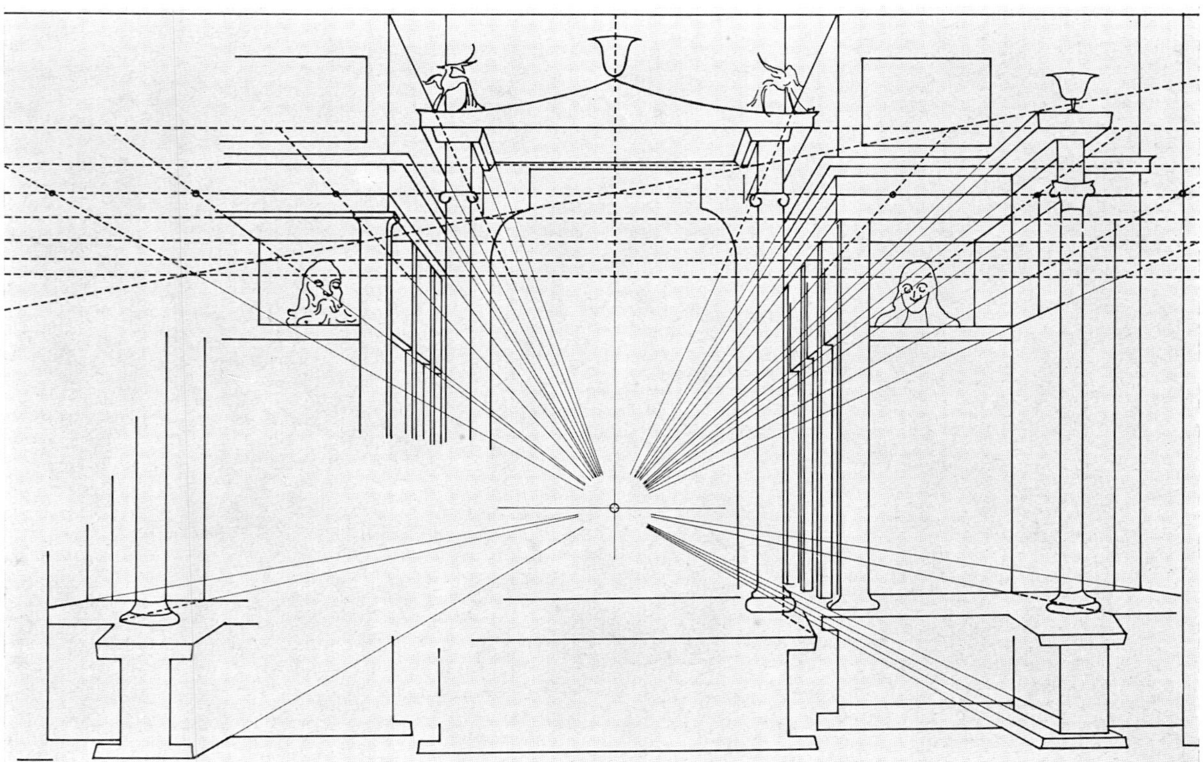

Pl. 83. *Above*: Painting with architectural elements, east wall of the Room of the Masks in a house of the Augustan period on the Palatine hill, Rome. *Below*: Diagram postulating a vanishing point.

Pl. 84. Conventionalized representation of elements in space. *Above*: Military scene on the base of the Column of Antoninus Pius (d. 161), erected by his sons in Campo Marzio, Rome. Marble, ht., 8 ft., 1 1/4 in. Rome, Vatican, Giardino della Pigna. *Below*: Māyā's dream, relief from the railing of the stupa at Amaravati, India, ca. 2d cent. Stone. Calcutta, Indian Museum.

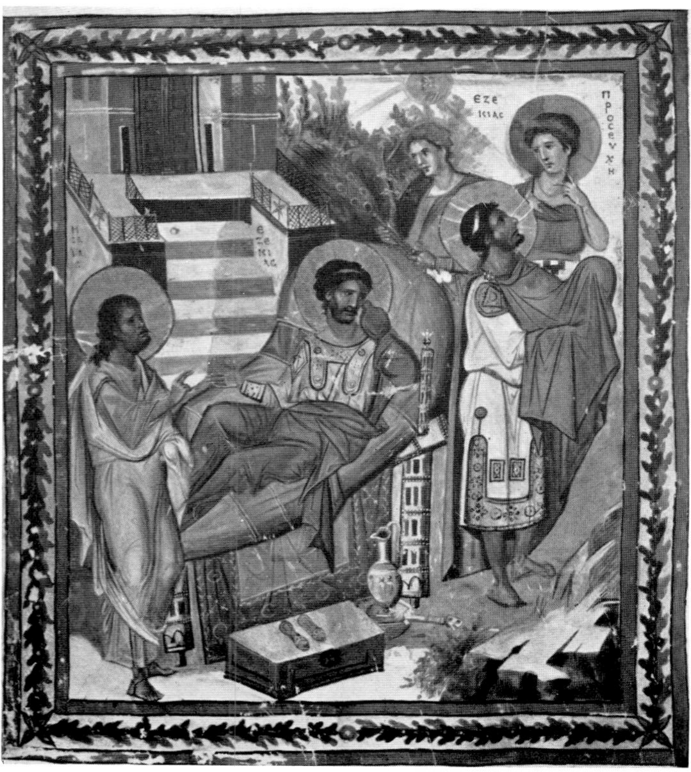

Pl. 85. Architectural and landscape elements. *Left, above*: Mosaic in the precincts of the Antonine baths, Carthage, ca. mid-4th cent. *Below*: Psalter. The Prayer of Hezekiah, 10th cent. (?), after an Alexandrian prototype. Illumination. Paris, Bibliothèque Nationale (Ms. gr. 139, fol. 446v). *Right*: Hsia Kuei, Landscape in a Rainstorm, China, Sung period. Ink on paper, $35^{7}/_{8} \times 15^{3}/_{4}$ in. Kobe, formerly Kawasaki Coll.

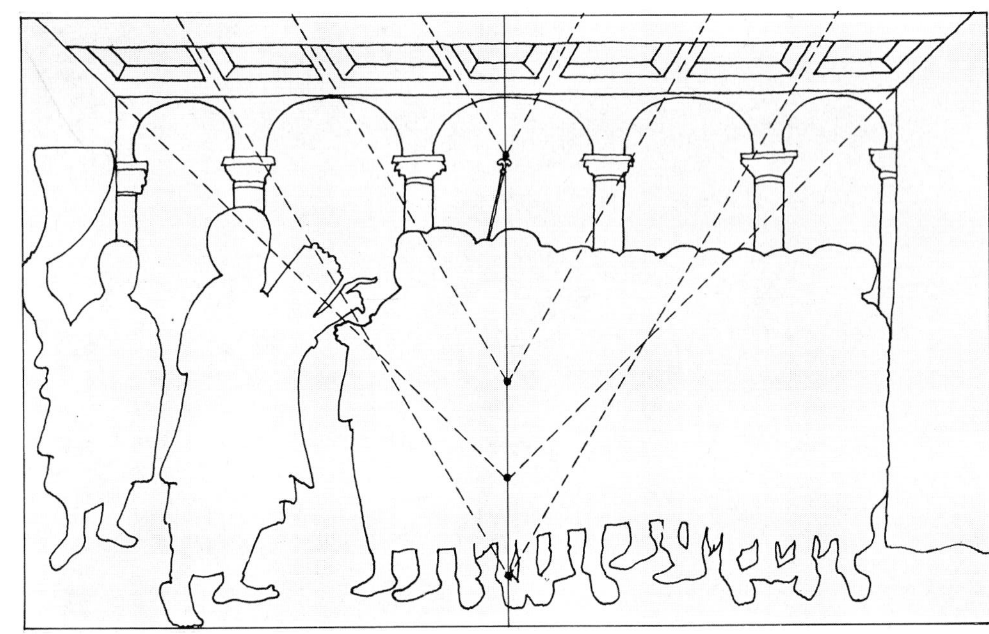

Pl. 86. *Left, above*: Offerings of Abel and Melchizedek, 6th cent. Mosaic. Ravenna, Italy, S. Vitale. *Below*: Diagram (simplified) indicating major spatial areas. *Right, above*: Moutier-Grandval (so-called "Alcuin") Bible. Moses reading the law to the Israelites, Tours, France, 9th cent. Illumination. London, British Museum (Add. 10546). *Below*: Hypothetical analysis of the coffered ceiling.

Pl. 87. Painting of the west wall, Room of the Masks in a house of the Augustan period on the Palatine hill, Rome. (Cf. PL. 83.)

Pl. 88. Ideal city (variously attrib.), 15th cent. Panel. Urbino, Italy, Galleria Nazionale delle Marche.

Pl. 89. Giotto and assistants, St. Francis cycle. Fresco. Assisi, Italy, S. Francesco, Upper Church. *Left*: The Approval of the Rule. *Right*: Hypothetical analysis of the triple-arch motif shown at left and in two other scenes: The Sermon before Honorius (*center*) and The Pentecost (*below*).

Pl. 90. J. van Eyck, The Madonna with Chancellor Rolin, ca. 1433. Panel, 26×24 3/8 in. Paris, Louvre.

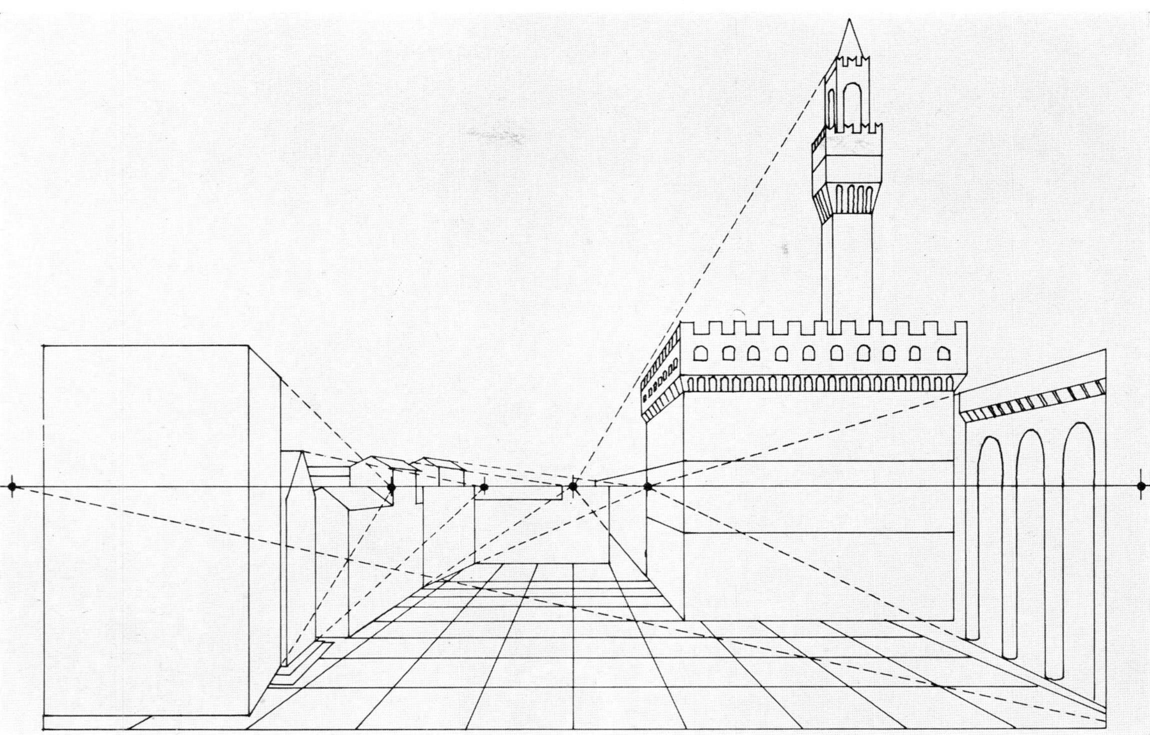

Pl. 91. *Above*: Schematic reconstruction of Brunelleschi's perspective panel of Piazza della Signoria in Florence, after D. Gioseffi. *Below*: Florentine painting of Piazza della Signoria, depicting the death of Savonarola, 15th–16th cent. Panel. Florence, Convento di S. Marco.

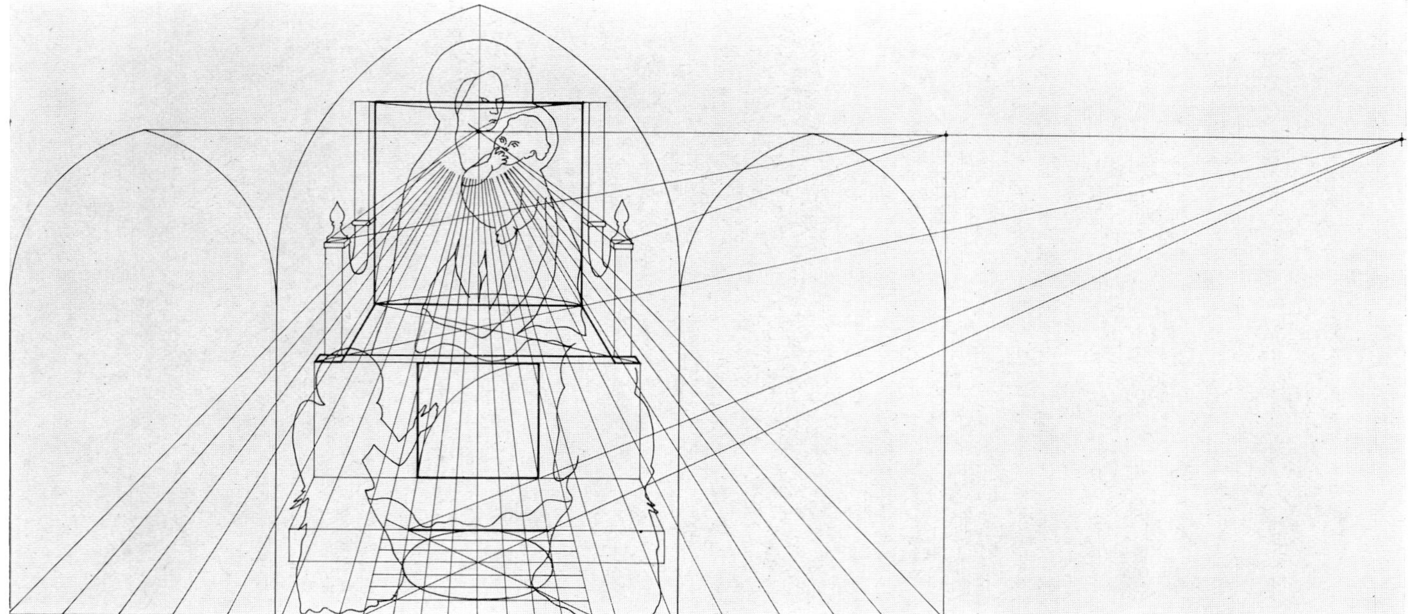

Pl. 92. *Above*: Masaccio, Madonna enthroned with saints and two adoring angels, 1422. Panel. Cascia di Reggello, near Florence, S. Giovenale. *Below*: Hypothetical analysis of the central panel.

Pl. 93. *Above*: Paolo Uccello, The Nativity and Annunciation to the Shepherds, mid-15th cent. Detached fresco. Florence, S. Martino alla Scala, cloister. *Below*: The sinopia, recovered after detachment, showing the preparatory perspective drawing.

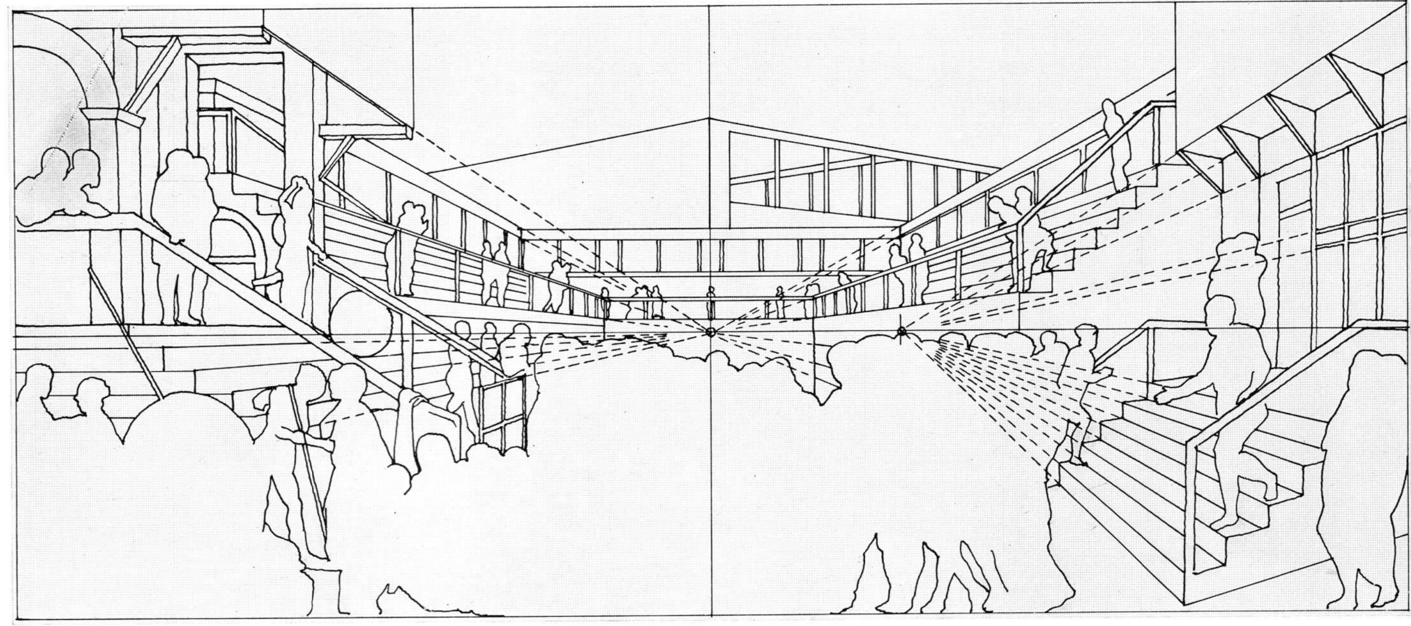

Pl. 94. *Above*: Donatello, St. Anthony healing the young man's foot. Bronze with gold and silver, 22 1/2×48 1/2 in. Padua, Italy, S. Antonio, high altar. *Below*: Diagram of the perspective scheme.

Pl. 95. F. Galli Bibiena, method of designing a stage setting with a room seen at an angle, illustration from ... *l'Architettura civile* . . ., Part III, Bologna, 1731–32.

Pl. 96. P. Cézanne, The Kitchen Table, ca. 1888–90, showing multiple points of view. Canvas, $25^5/_8 \times 31^7/_8$ in. Paris, Louvre.

Pl. 97. C. and G. Alberti, vault fresco. Rome, S. Silvestro al Quirinale.

Pl. 98. G. Curti (called Dentone, ca. 1570–1632), fresco decoration, detail. Rome, Palazzo Odescalchi.

Pl. 99. *Above, left*: M. Palmezzano, fresco in the dome of Feo Chapel, before 1495 (destroyed, 1944). Forlì, Italy, S. Biagio. *Right*: Rinaldo Mantovano, fresco in chapel dome, 1534. Mantua, Italy, S. Andrea. *Below*: B. Peruzzi, drawing of the ancient monuments of Rome. Florence, Uffizi, Gabinetto dei Disegni e Stampe.

Pl. 100. *Above*: P. Tibaldi, Ulysses cycle, detail of ceiling frescoes (cf. IX, PL. 298), after 1549. Bologna, Palazzo dell'Università (formerly Palazzo Poggi). *Below*: G. and C. Alberti, vault frescoes of the Sagrestia dei Canonici, ca. 1600. Rome, St. John Lateran.

Pl. 101. *Above*: A. M. Colonna (1600–87), ceiling fresco. Trebbo, near Bologna, Casino Malvasia. *Below*: A. Tassi, fresco with painted architecture, ca. 1613. Bagnaia, near Viterbo, Italy, Villa Lante, Loggia della Palazzina Montalto.

Pl. 102. *Above*: G. A. Fumiani, ceiling frescoes, detail, 1680–1704. Venice, S. Pantalon. *Below*: D. Baroni and M. Aldobrandini, vault frescoes, detail. Bologna, Oratory of S. Giovanni de' Fiorentini.

Pl. 103. C. Bononi, Coronation of the Virgin, ceiling fresco, ca. 1620. Ferrara, Italy, S. Maria in Vado.

Pl. 104. A. and G. Rolli, vault frescoes with the Triumph of St. Paul, central section, 1695. Bologna, S. Paolo.

Pl. 105. V. M. Bigari (1692–1776), Belshazzar's Feast. Canvas. Bologna, Pinacoteca Nazionale.

Pl. 106. S. Ricci, Hercules Received on Olympus, ceiling fresco, 1706–07. Florence, Palazzo Marucelli (later, Palazzo Fenzi).

Pl. 107. V. M. Bigari and S. Orlandi, fresco with allegory of Painting. Bologna, Casa Tacconi.

Pl. 108. *Above*: P. Scotti and G. Baroffio, frescoes in the Ordensbausall, 1730. Ludwigsburg, Germany, Palace. *Below*: Frescoes of central hall, with figures by J. Amigoni, 1720. Schleissheim, near Munich, Schlösschen Lustheim.

Pl. 109. G. B. Tiepolo, allegorical fresco with the four parts of the world, ceiling of grand staircase, 1752–53. Würzburg, Germany, Residenz.

Pl. 110. G. B. Crosato, frescoes in the Antechapel of St. Hubert, 1732. Stupinigi, near Turin, royal hunting lodge.

Pl. 111. Vision of St. Bernard. Panel, 5 ft., 8 in. × 5 ft., 7 in. Munich, Alte Pinakothek.

Pl. 112. Madonna in Glory and Saints. Panel, 5 ft. × 4 ft., 1 in. Bologna, Pinacoteca Nazionale.

Pl. 113. *Above, left*: St. Sebastian, fragment. Fresco, ca. 5 ft., 7 in. × 3 ft., 4 in. Cerqueto, Italy, parish church. *Right*: Annunciation. Panel, 21 5/8 × 16 1/2 in. Perugia, Italy, Coll. Ranieri. *Below*: Christ giving the keys to St. Peter. Fresco, ca. 11 × 18 ft. Rome, Vatican, Sistine Chapel.

Pl. 114. *Left*: Portrait of Francesco delle Opere. Panel, $20^{1}/_{2} \times 17^{3}/_{8}$ in. Florence, Uffizi. *Right*: Self-portrait. Fresco, $15^{3}/_{4} \times 12$ in. Perugia, Italy, Collegio del Cambio, Sala dell'Udienza.

Pl. 115. *Left*: Madonna Enthroned, John the Baptist, and St. Sebastian. Panel, 5 ft., 10 in. × 5 ft., 4 1/2 in. Florence, Uffizi. *Right, above*: Pietà. Panel, 4 ft., 8 3/4 in. × 5 ft. Perugia, Italy, S. Pietro. *Below*: The Marriage at Cana. Panel, 15 3/8 × 33 1/8 in. Perugia, Italy, Galleria Nazionale dell'Umbria.

Pl. 116. *Above*: Fortitude and Temperance with Heroes of Antiquity. Fresco, 9 ft., 6 1/2 in. × 13 ft., 1 1/2 in. Perugia, Italy, Collegio del Cambio, Sala dell'Udienza. *Below*: Combat of Love and Chastity. Canvas, 5 ft., 1 1/2 in. × 6 ft., 3 1/2 in. Paris, Louvre.

Pl. 117. Apollo and Marsyas. Panel, 15 3/8 × 11 3/8 in. Paris, Louvre.

Pl. 118. Trompe l'oeil colonnade with view of Rome. Fresco. Rome, Farnesina, Salone delle Prospettive.

Pl. 119. *Above*: The constellation Callisto. Vault fresco. Rome, Farnesina, Sala di Galatea. *Below, left*: Augustus and the sibyl. Fresco. Siena, Church of Fontegiusta. *Right*: Adoration of the Magi and scenes from Genesis. Vault fresco. Rome, S. Maria della Pace, Ponzetti Chapel.

Pl. 120. Rome, Farnesina. *Above*: North façade. *Below*: Detail of exterior frieze.

Pl. 121. Rome, Palazzo Massimo alle Colonne. *Above*: Façade. *Below, left*: East end of portico. *Right*: Courtyard.

Pl. 122. Head of Athena Lemnia, Roman copy. Marble, ht., 15 3/4 in. Bologna, Museo Civico.

Pl. 123. *Left*: Anadoumenos (Farnese Diadoumenos). Roman copy. Marble, ht., 4 ft., 10¹/₄ in. London, British Museum. *Right*: Amazon, Roman copy. Marble, ht., 7 ft., 2 in. Hadrian's Villa, near Tivoli, Italy, Antiquarium.

Pl. 124. *Left, above*: The Zeus of Olympia, copy on a Hadrianic coin of Elis. Florence, Museo Archeologico. *Below*: Head of Zeus from the Olympieion of Cyrene, Libya, an example of work showing Phidian influence. Marble, ht., 15 in. Cyrene, Archaeological Museum. *Right*: Copies of Athena Parthenos. *Above*: Engraved gem, signed by Aspasios, 1st cent. B.C. Red jasper, ht., ca. 1 1/8 in. Rome, Museo Nazionale Romano. *Below*: The Varvakeion statuette, Roman copy. Marble, ht., 41 3/8 in. Athens, National Museum.

Pl. 125. *Above*: Fragment of the east frieze of the Parthenon, showing maidens in the Panathenaic Procession. Marble; ht. of frieze, 3 ft., 7 in. Paris, Louvre. *Below*: Dionysos or Herakles (?), so-called "Theseus," with Demeter and Persephone, from the east pediment of the Parthenon. Marble, ht., 3 ft., 11 1/4 in. London, British Museum.

Pl. 126. Centaur carrying off a Lapith, detail of a metope from the south side of the Parthenon. Marble; ht. of metopes, 4 ft., $4^{3}/_{4}$ in. London, British Museum.

Pl. 127. Lid of the so-called "sarcophagus of priestess," from Carthage, ca. 300 B.C. Carthage, Tunisia, Musée National.

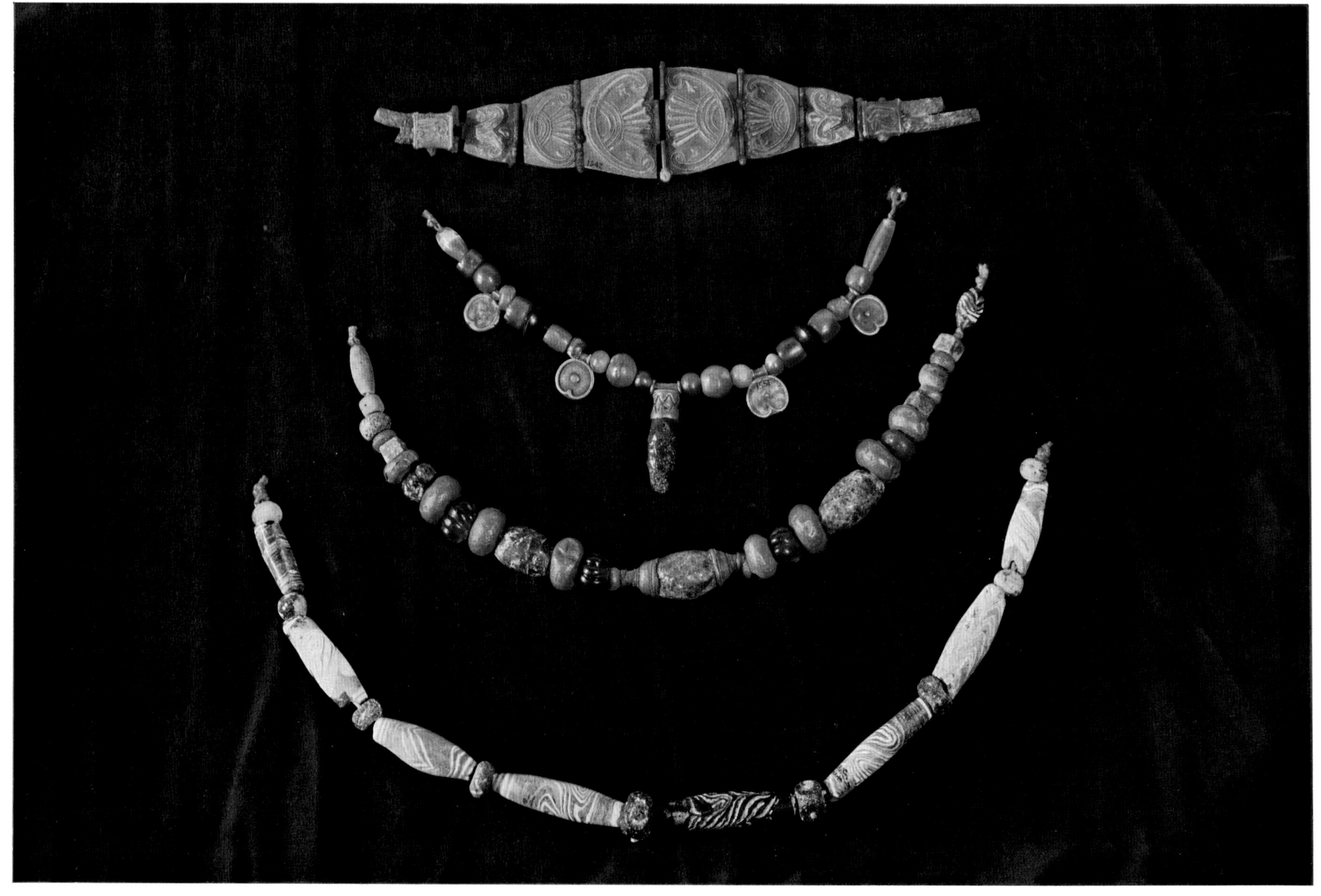

Pl. 128. Punic jewelry, from Tharros, Sardinia, 7th–6th cent. B.C. *Top to bottom*: Gold connected with silver; carnelian, agate, and black stones, with amber and gold pendants; gold and glass beads; glass beads. L. of bracelet at top, 7 7/8 in. All, London, British Museum.

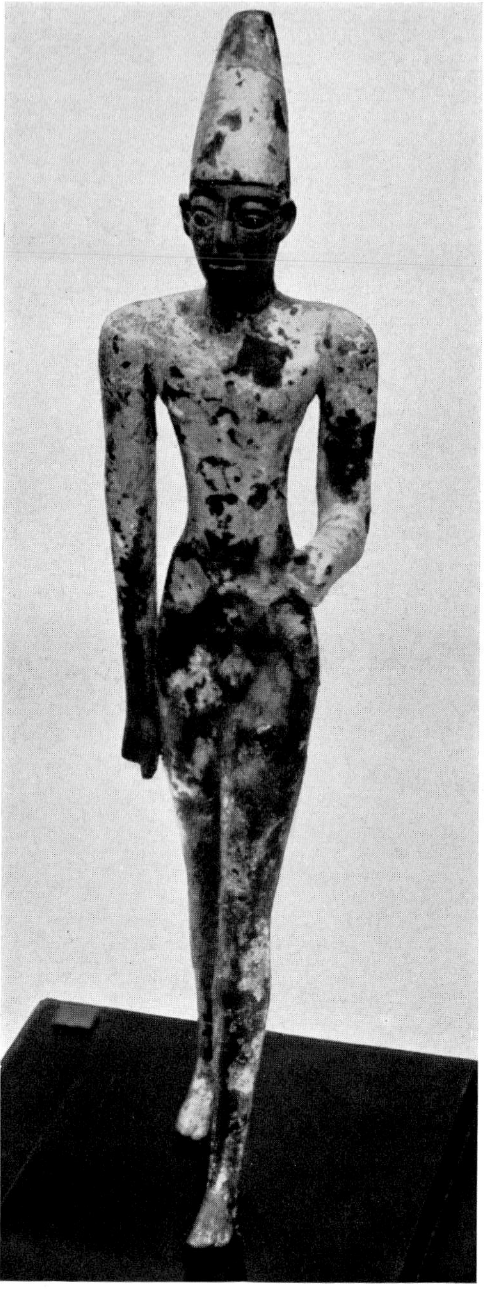

Pl. 129. Phoenician figurines, 2d millennium B.C. *Left*: From Byblos. Bronze covered with gold, ht., 13 3/8 in. Beirut, National Museum. *Center*: From Sidon (?). Copper, ht., 10 in. *Right*: From Ras Shamra (anc. Ugarit). Bronze, ht., ca. 10 in. Last two, Paris, Louvre.

Pl. 130. *Left*: Stele showing Baal, from Ras Shamra (anc. Ugarit), ca. 13th cent. B.C. Limestone, ht., 4 ft., 7 7/8 in. Paris, Louvre. *Right, above*: Stele with seated divinity and offering bearer, from Ras Shamra, ca. 13th cent. B.C. Serpentine, ht., 18 1/2 in. Aleppo, Syria, Musée National. *Below*: Sarcophagus of Aḥīrām, from Byblos, end view of IV, PL. 453.

Pl. 131. *Above, left*: Bowl, from Ras Shamra (anc. Ugarit), ca. 1450–1350 B.C. Gold, diam., 7 1/2 in. Paris, Louvre. *Right*: Phoenician plate, 8th–7th cent. B.C. Silver, diam., ca. 10 1/4 in. Leiden, Rijksmuseum van Oudheden. *Below*: Fragmentary bowl, from Amathus, Cyprus, 7th cent. B.C. Silver. London, British Museum.

Pl. 132. *Above*: Sarcophagus of King Tabnit, a reused Egyptian piece with Phoenician inscription added on ledge, from Sidon, 6th cent. B.C. (?). Black basalt. *Below*: Sarcophagus, from Beirut, 4th cent. B.C. Marble. Both, Istanbul, Archaeological Museums.

Pl. 133. *Left*: Stele of Baalyaton, from Umm el-Aḥmād, Lebanon, ca. 3d cent. B.C. Limestone, ht., 5 ft., 11¼ in. Copenhagen, Ny Carlsberg Glyptotek. *Right*: Stele in the shape of a symbol of Tanit, from Carthage, ca. 5th cent. B.C. Limestone, ht., 24⅜ in. Tunis, Musée du Bardo.

Pl. 134. Punic pinax, from Ibiza, Balearic Islands, ca. 5th cent. B.C. Terra cotta, $7^{1}/_{8} \times 5$ in. Madrid, Museo Arqueológico Nacional.

Pl. 135. Punic female statuette, from Ibiza, Balearic Islands, 5th–4th cent. B.C. Terra cotta. Madrid, Museo Arqueológico Nacional.

Pl. 136. Masks from Carthage. *Left*: 6th cent. B.C. Terra cotta; ht. (*below*), 7 3/4 in. Paris, Louvre, and Carthage, Tunisia, Musée National. *Right*: 4th cent. B.C. Glass, ht., ca. 1 3/4 in. Tunis, Musée du Bardo.

Pl. 137. *Above*: The world's first photograph (8-hour exposure). J. Nicéphore Niépce, courtyard of the photographer's estate, Gras, near Chalon-sur-Saône, France, 1826. London, Gernsheim Coll. *Below*: Photographic journalism. M. B. Brady (attrib.), Ruins of Richmond, 1865. New York, Museum of Modern Art.

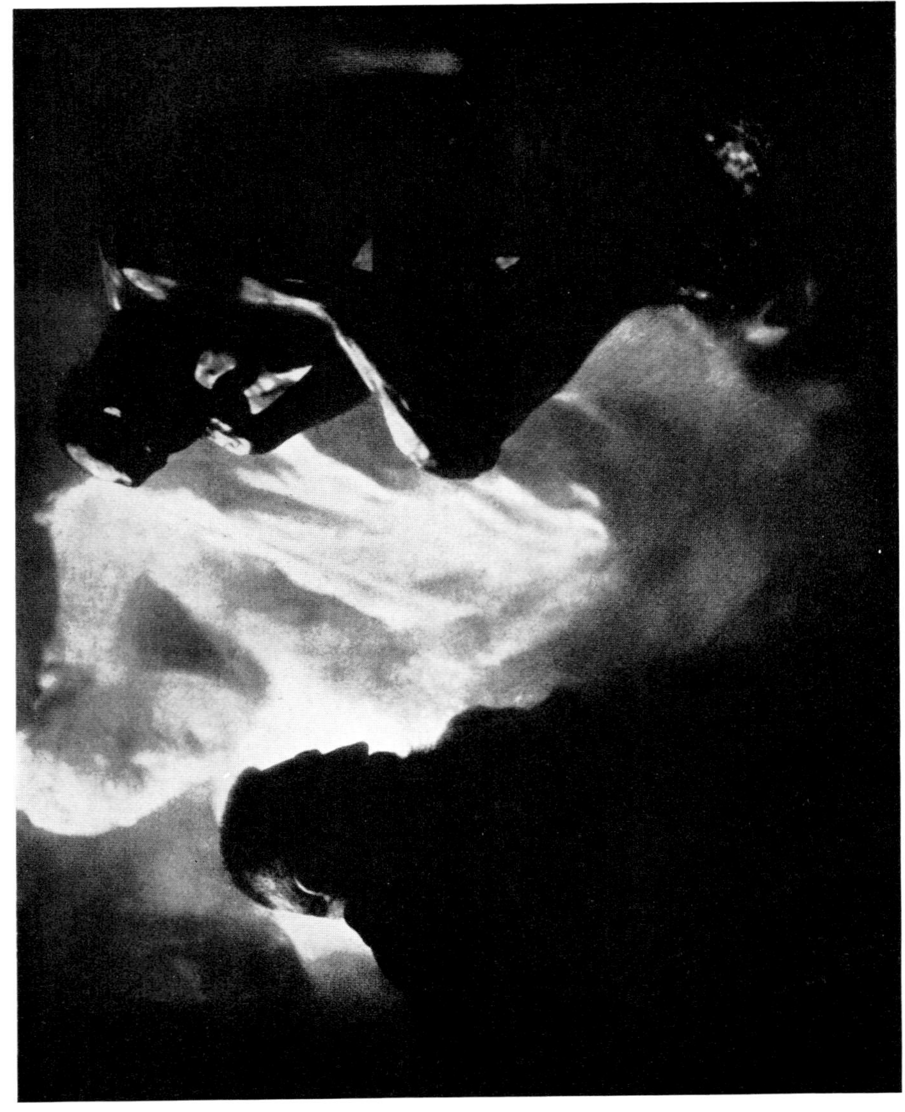

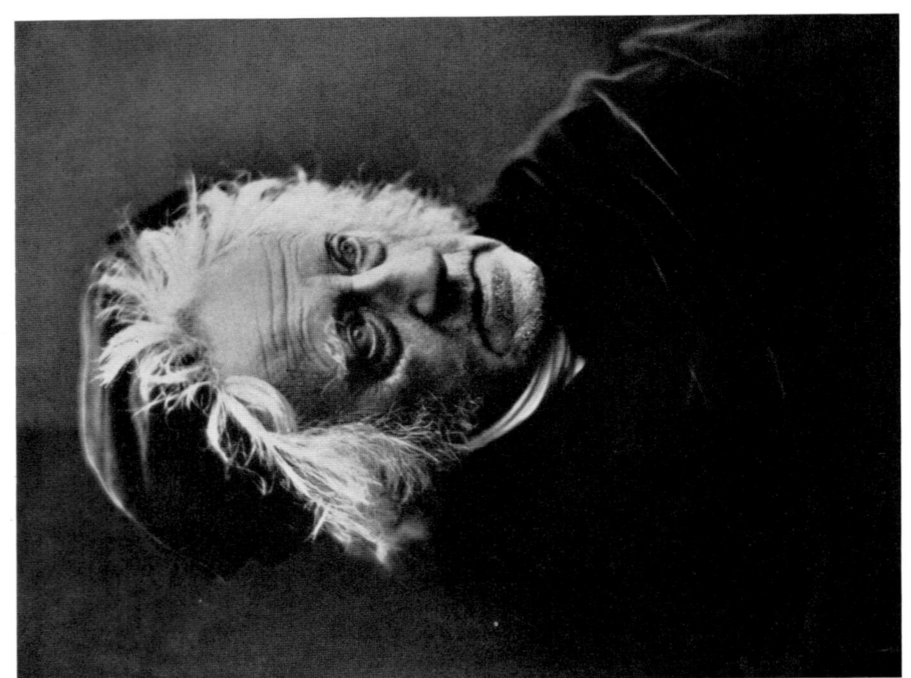

Pl. 138. Portraiture. *Left*: J. M. Cameron, Sir John F. W. Herschel, 1867. New York, Museum of Modern Art. *Right*: E. Steichen, Rodin—Le Penseur, 1902. Stockholm, Mr. and Mrs. H. Hammarskjöld Coll.

Pl. 139. The genre scene. *Left*: A. Stieglitz, The Steerage, 1907. New York, Museum of Modern Art. *Right*: E. Atget, Candy Seller, ca. 1910 (*copyright, Berenice Abbott*).

Pl. 140. Modern realism and abstraction. *Above, left*: P. Strand, Woman and Boy, Tenancingo, Mexico, 1932. *Right*: H. Callahan, Alley, Chicago, 1948 (multiple exposure). *Below, left*: P. Caponigro, Rock Wall No. 2, 1959. *Right*: K. Heyman, Nigerian Rail Splitter, 1962 (*copyright, Magnum Photos*). All, courtesy of the photographers.

Pl. 141. Seated Nude, 1905. Oil on cardboard, $41\,^3/_4 \times 30$ in. Paris, Musée d'Art Moderne.

Pl. 142. Nude Woman in a Red Armchair, 1932. Canvas, $51^{1}/_{8} \times 38^{1}/_{4}$ in. London, Tate Gallery.

Pl. 143. *Above, left*: La Toilette, 1906. Canvas, $59^{1}/_{2} \times 39$ in. Buffalo, N.Y., Albright-Knox Art Gallery. *Right*: Harlequin, 1917. Canvas, 45×34 in. Barcelona, Museo de Arte Moderno. *Below, left*: Harlequin, 1915. Canvas, 6 ft., $^{1}/_{4}$ in. × 3 ft., $5^{3}/_{8}$ in. New York, Museum of Modern Art. *Right*: Maternity, 1921. Canvas, 5 ft., 4 in. × 3 ft., 2 in. Property of the artist.

Pl. 144. *Left*: Seated Bather, 1930. Canvas, 5 ft., 4¼ in. × 4 ft., 3 in. *Right, above*: Three Musicians, 1921. Canvas, 6 ft., 7 in. × 7 ft., 3¾ in. Both, New York, Museum of Modern Art. *Below*: Two Women (La Muse, or L'Atelier), 1935. Canvas, 4 ft., 3 in. × 5 ft., 5 in. Paris, Musée d'Art Moderne.

Pl. 145. Peace, one of two panels with allegories of war and peace, 1952. Oil on fiberboard, 15 ft., 5 in. × 33 ft., 5 1/2 in. Vallauris, France, Chapel, vault of the nave.

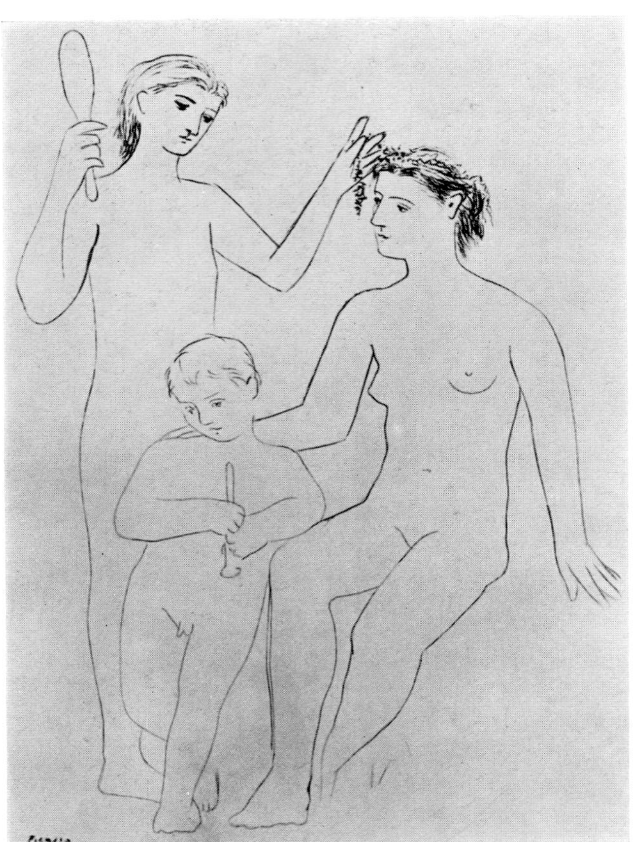

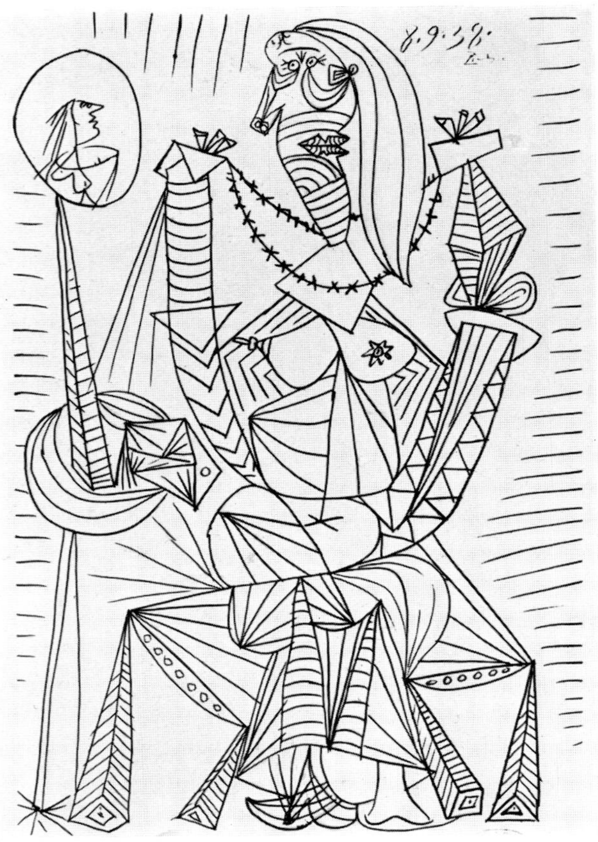

Pl. 146. *Above, left*: Two women and a child, 1922. Drawing in oil on canvas, 6 ft., 1 in. × 4 ft., 1 in. *Right*: Seated woman with necklace, 1938. Pen and ink, $26^{1}/_{2} \times 17^{1}/_{2}$ in. *Below*: Study, 1934, for the illustrations of Aristophanes' *Lysistrata.* Pen and ink, $14^{1}/_{8} \times 19^{3}/_{4}$ in. All, private colls.

Pl. 147. *Above*: N. Poussin, Winter (The Flood), 1660–64. Canvas, 3 ft., 10 1/2 in. × 5 ft., 3 in. Paris, Louvre. *Below*: C. Lorrain, Landscape with the Nymph Egeria, 1669. Canvas, 5 ft., 1 in. × 6 ft., 6 1/2 in. Naples, Museo di Capodimonte.

Pl. 148. A. Pynacker (1622–73), Lake Shore in Italy. Canvas, $38^{1}/_{2} \times 33^{3}/_{4}$ in. Amsterdam, Rijksmuseum.

Pl. 149. *Above*: S. Rosa (1615–73), Arch and Ruins by the Sea. Canvas, 26×59 in. Rome, Galleria Doria Pamphili. *Below*: L. Coccorante, Harbor with Ruins, first half of 18th cent. Canvas, $28\frac{3}{4} \times 38\frac{1}{4}$ in. Naples, Museo di Capodimonte.

Pl. 150. *Above*: H. Robert, Transformation of the Bains d'Apollon at Versailles in 1775. Canvas. Versailles, Palace. *Below*: P. Patel the Elder, Landscape, 1652. Canvas, $30 \times 44^1/_2$ in. Leningrad, The Hermitage.

Pl. 151. *Above*: M. Marieschi (1696–1743), Imaginary View. Canvas. Milan, formerly Coll. Chiesa. *Below*: B. Bellotto (1720–80), Roman *capriccio*. Canvas, $17^{1}/_{4}\times27$ in. Rome, Coll. L. Albertini.

Pl. 152. Artificial ruins. *Above*: L. C. Carmontelle, La Naumachie, ca. 1778, Parc Monceau, Paris. *Below*: C. Unterberger, the "Temple of Antoninus and Faustina," ca. 1787, Villa Borghese, Rome.

Pl. 153. *Above*: P. Sandby, An Ancient Beech Tree, 1794. Water color, $23^1/_4 \times 41^1/_2$ in. London, Victoria and Albert Museum. *Below*: P. F. Poole, The Vision of Ezekiel, ca. 1875. Canvas, 4 ft., $5^1/_4$ in. $\times$ 6 ft., $1^1/_4$ in. London, Tate Gallery.

Pl. 154. J.-B.-C. Corot, Souvenir de Mortefontaine, 1864. Canvas, $25^{3}/_{4} \times 35$ in. Paris, Louvre.

Pl. 155. J. C. Ibbetson, Conway Castle, 1794. Canvas, $13\frac{1}{2} \times 17\frac{3}{4}$ in. London, Victoria and Albert Museum.

Pl. 156. J. M. W. Turner (1775–1851), Waves Breaking against the Wind, late period. Canvas, 23×35 in. London, Tate Gallery.

Pl. 157. The Madonna of Mercy, central panel of a polyptych. Panel, 4 ft., 4 3/4 in. × 2 ft., 11 3/4 in. Sansepolcro, Italy, Pinacoteca Comunale.

Pl. 158. The Flagellation of Christ. Panel, $23^{1}/_{4} \times 32$ in. Urbino, Italy, Galleria Nazionale delle Marche.

Pl. 159. The Queen of Sheba adoring the Holy Wood, detail, scene from The Legend of the True Cross. Fresco. Arezzo, Italy, S. Francesco.

Pl. 160. The Death of Adam. Fresco. Arezzo, Italy, S. Francesco.

Pl. 161. Victory of Constantine over Maxentius. Fresco. Arezzo, Italy, S. Francesco.

Pl. 162. Baptism of Christ. Panel, 5 ft., 6 in. × 3 ft., 9 3/4 in. London, National Gallery.

Pl. 163. Madonna and Child with angels, from Sinigallia. Panel, 24×21 in. Urbino, Italy, Galleria Nazionale delle Marche.

Pl. 164. *Above*: The Annunciation, cusp of a polyptych. Panel, 4 ft. × 6 ft., 4 1/2 in. Perugia, Italy, Galleria Nazionale dell'Umbria. *Below*: Portraits of Battista Sforza and Federigo da Montefeltro. Panel, each, 18 1/2 × 13 in. Florence, Uffizi.

Pl. 165. The Nativity, detail. Panel; full size, $49 \times 48^{1}/_{4}$ in. London, National Gallery.

Pl. 166. Madonna and Child, Angels, Saints, and Federigo da Montefeltro. Panel, 8 ft., 1 3/4 in. × 5 ft., 7 in. Milan, Brera.

Pl. 167. The Age of Gold, detail, from the Allegories of the Four Ages of Man. Fresco. Florence, Pitti, Sala della Stufa.

Pl. 168. The Triumph of Bacchus. Canvas, 4 ft., 9 in. × 6 ft., 9 in. Rome, Palazzo dei Conservatori, Pinacoteca Capitolina.

Pl. 169. The Rape of the Sabine Women. Canvas, 9 ft., 1 in. × 13 ft., 10 in. Rome, Palazzo dei Conservatori, Pinacoteca Capitolina.

Pl. 170. Allegory representing a youth (Cosimo I de' Medici?) snatched by Minerva from the arms of Venus and being led toward Hercules. Fresco. Florence, Pitti, ceiling of the Sala di Venere.

Pl. 171. Scenes from the story of Aeneas. Fresco. Rome, Palazzo Pamphili in Piazza Navona, ceiling of the *salone.*

Pl. 172. Churches in Rome. *Left*: SS. Luca e Martina. *Right*: S. Maria in Via Lata.

Pl. 173. Rome, SS. Luca e Martina. *Left*: View into dome. *Right*: Interior.

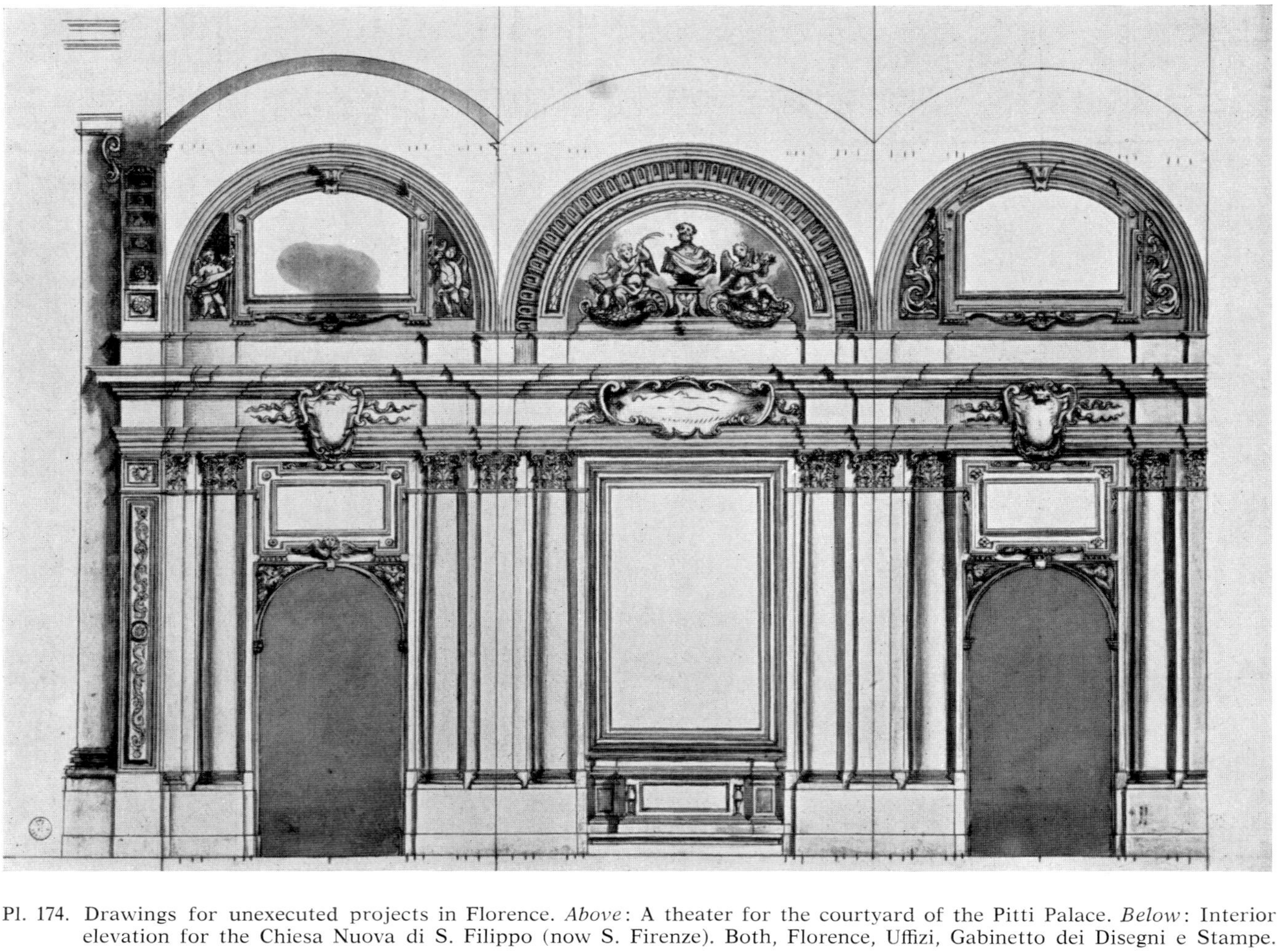

Pl. 174. Drawings for unexecuted projects in Florence. *Above*: A theater for the courtyard of the Pitti Palace. *Below*: Interior elevation for the Chiesa Nuova di S. Filippo (now S. Firenze). Both, Florence, Uffizi, Gabinetto dei Disegni e Stampe.

Pl. 175. Portrait of Margherita Gonzaga. Panel, $16^{7}/_{8} \times 11^{3}/_{4}$ in. Paris, Louvre.

Pl. 176. The Virgin and Child with SS. George and Anthony Abbot. Panel, 18 1/2×11 1/2 in. London, National Gallery.

Pl. 177. St. George and the Princess, detail of the Legend of St. George. Fresco. Verona, S. Anastasia, Pellegrini Chapel.

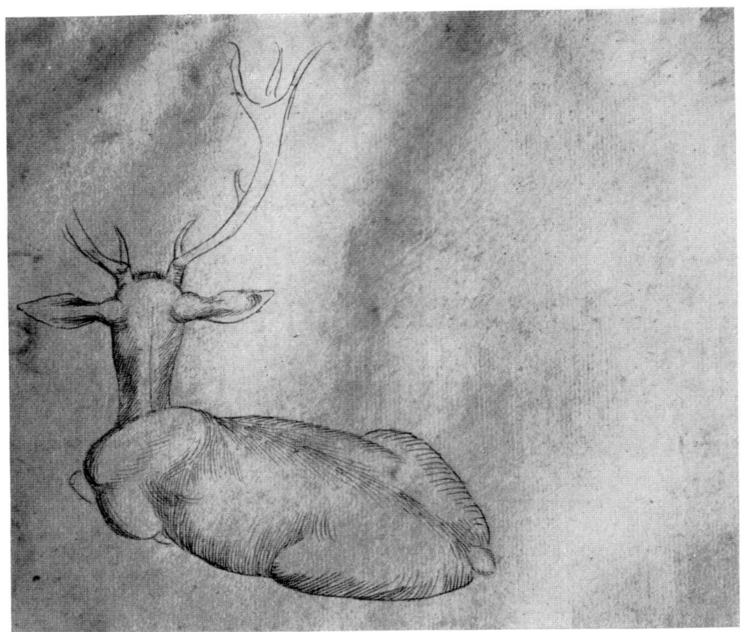

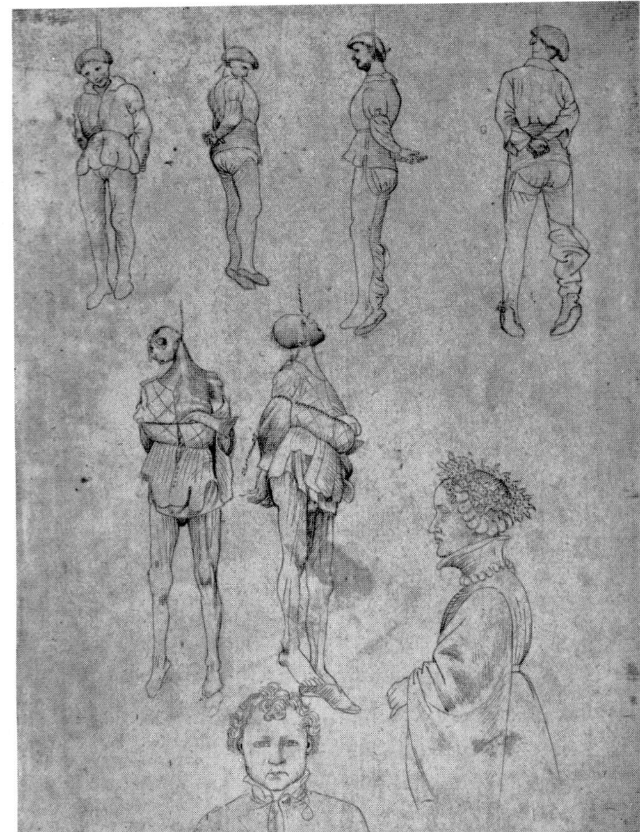

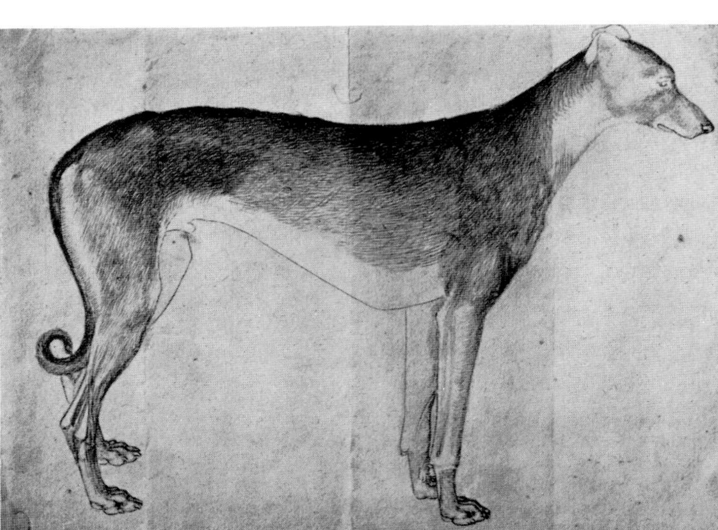

Pl. 178. Drawings. *Left, above*: Deer. Pen, $10^{7}/_{8}\times11^{1}/_{2}$ in. *Center*: Herons. Pen, $6^{1}/_{2}\times9^{3}/_{4}$ in. *Below*: Greyhound. Pen and water color, $7^{1}/_{4}\times9^{5}/_{8}$ in. All three, Paris, Louvre. *Right, above*: Studies with hanged men. Pen over metal-point drawing, $11^{1}/_{8}\times7^{5}/_{8}$ in. London, British Museum. *Below*: Studies with female nudes. Pen on parchment, $8^{5}/_{8}\times6^{1}/_{2}$ in. Rotterdam, Museum Boymans-Van Beuningen.

Pl. 179. A. Pollaiuolo, Hercules and Antaeus. Bronze, ht., 17 3/4 in. Florence, Museo Nazionale.

Pl. 180. P. Pollaiuolo, Tobias and the Angel. Panel, 6 ft., 1 1/2 in. × 3 ft., 10 1/2 in. Turin, Galleria Sabauda.

Pl. 181. A. Pollaiuolo. *Above, left*: Hercules and Antaeus. Panel, $6^1/_4 \times 3^3/_4$ in. Florence, Uffizi. *Right*: Eve Spinning. Pen and bistre wash over black chalk, $10^7/_8 \times 7^1/_4$ in. Florence, Uffizi, Gabinetto dei Disegni e Stampe. *Below*: The Rape of Deianira. Panel painting transferred to canvas, $21^3/_8 \times 31^3/_4$ in. New Haven, Conn., Yale University Art Gallery.

Pl. 182. *Left*: P. Pollaiuolo, SS. Vincent, James, and Eustace, altarpiece from the Chapel of the Cardinal of Portugal, S. Miniato al Monte, Florence. Panel, 5 ft., 7 3/4 in. × 5 ft., 10 1/2 in. Florence, Uffizi. *Right*: A. and P. Pollaiuolo, The Martyrdom of St. Sebastian, altarpiece. Panel, 9 ft., 6 3/4 in. × 6 ft., 7 3/4 in. London, National Gallery.

Pl. 183. P. Pollaiuolo. *Left*: The Coronation of the Virgin. Panel, 10 ft., 3 in. × 8 ft., 2 1/2 in. San Gimignano, S. Agostino. *Right*: Prudence. Panel, 5 ft., 1 3/4 in. × 2 ft., 6 1/4 in. Florence, Uffizi.

Pl. 184. A. Pollaiuolo. *Above*: The Birth of John the Baptist, relief from the Altar of St. John, formerly in the Baptistery, Florence. Silver. Florence, Museo dell'Opera del Duomo. *Below*: Allegorical figure of Arithmetic, bas-relief on the Tomb of Sixtus IV. Bronze. Rome, Vatican Grottoes.

Pl. 185. A. (and P.?) Pollaiuolo, Apollo and Daphne. Panel, 11 5/8 × 7 7/8 in. London, National Gallery.

Pl. 186. A. Pollaiuolo, Tomb of Sixtus IV, view from above. Bronze. Rome, Vatican Grottoes.

Pl. 187. Scene probably derived from a painting by Polygnotos, Herakles with heroes, detail of a calyx-crater by the Niobid Painter, from Orvieto, Italy, ca. 460 B.C. Full ht., 21 1/4 in. Paris, Louvre.

Pl. 188. Scenes of the Gigantomachy, probably derived from paintings by Polygnotos, details of two calyx-craters from Spina, Italy, ca. 440 B.C., attributed to the vase painter also named Polygnotos, and (*below*) to his workshop. Full ht., 23 in. and 21 5/8 in. Ferrara, Italy, Museo Archeologico.

Pl. 189. Herm of the Doryphoros, copy signed by Apollonios, from Herculaneum. Bronze, ht., 21 1/4 in. Naples, Museo Nazionale.

Pl. 190. Diadoumenos, late Hellenistic copy, from Delos. Marble, ht., ca. 6 ft. Athens, National Museum.

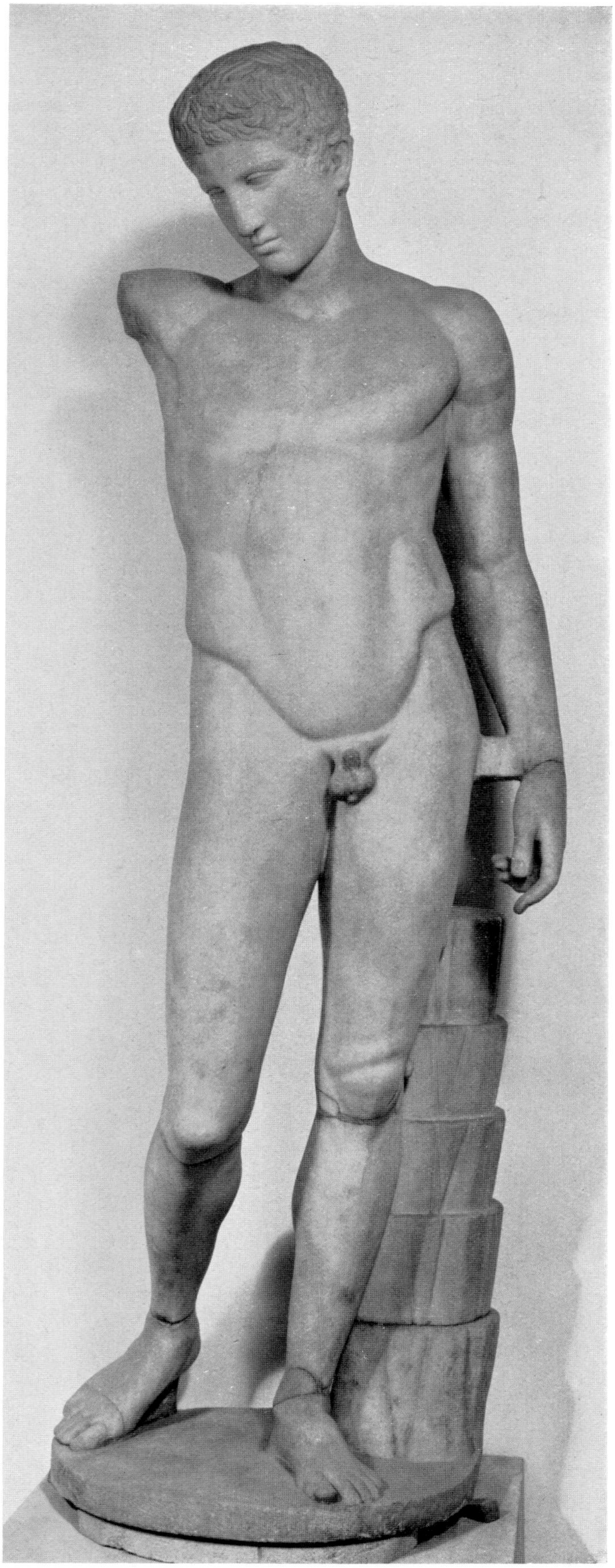

Pl. 191. *Left*: So-called "Diskophoros," Roman copy. Bronze, ht., 8 1/4 in. Paris, Louvre. *Right*: The "Westmacott Athlete," Roman, possible copy of Kyniskos. Marble, ht., 4 ft., 10 5/8 in. London, British Museum.

Pl. 192. Herakles, Roman copy. Marble, ht., 19 1/4 in. Rome, Museo Barracco.

Pl. 193. Wounded Amazon, possible copies. *Left*: Statue signed by Sosikles (original by Kresilas?). Marble, ht., 6 ft., 7 in. Rome, Palazzo dei Conservatori. *Right*: The "Lansdowne statue." Marble; ht. as restored, 6 ft., 8 1/4 in. New York, Metropolitan Museum.

Pl. 194. Epidauros. *Above*: The Theater, view from topmost row of seats. (Cf. III, PL. 340.) *Below*: The Tholos, showing the concentric foundations.

Pl. 195. *Above*: Colossal monolithic images on the slope of Rano-raraku, Easter Island. Tufa. *Below*: Remains of a marae (Marae Faretou), Huahiné, Society Islands.

Pl. 196. Figure with "defiant face" and carved cresting, from a marae, Hawaii. Wood, ht., 30 in. London, British Museum.

Pl. 197. *Above*: Food bowl supported by two human figures, Hawaii. Wood with shell and bone; w., $19^{5}/_{8}$ in. London, British Museum. *Below*: Bowl with shallow carving, Marquesas Islands. Wood., diam., $12^{3}/_{8}$ in. Auckland, Museum.

Pl. 198. *Above, left*: Tiki statuette, Marquesas Islands. Volcanic rock, ht., 6 in. Paris, Musée de l'Homme. *Right*: Double-headed image, Tahiti. Wood, ht., 22 3/8 in. *Below, left*: Male figure, Mangareva, Tuamotu Archipelago. Wood, ht., 46 3/8 in. Last two, London, British Museum. *Right*: Female figure, Tonga Islands. Wood, ht., 13 1/8 in. Auckland, Museum.

Pl. 199. The war god Kukailimoku, Hawaii. Feathers over wickerwork, pearl shell, and dog teeth; ht., 40 1/4 in. London, British Museum.

Pl. 200. Maori hei tiki, New Zealand. Nephrite, ht., 8 1/2 in. London, British Museum.

Pl. 201. *Above*: Pommel of ceremonial paddle with forked handle, Austral Islands. Wood, w., 9 1/8 in. Salem, Mass., Peabody Museum. *Below*: Paddles of Austral Islands type, details. Wood; ht. of portion shown, 8 1/2 in. (*left*) and 4 1/2 in. Auckland, Museum.

Pl. 202. Staff god, detail, Rarotonga, Cook Islands. Wood; full ht., 2 ft., $^{3}/_{4}$ in.; ht. of head, 7$^{3}/_{4}$ in. Auckland, Museum.

Pl. 203. Maori door lintels, from (*top to bottom*) Kaitaia, Waitara (Taranaki), Patetonga, and Rotorua, New Zealand. Wood; w., 7 ft., 8 1/2 in.; 4 ft., 10 1/2 in.; 7 ft., 8 in.; 4 ft., 2 in. Auckland, Museum; (*second from top*) Wellington, Dominion Museum.

Pl. 204. Carved wall plank of Maori assembly house, with ancestor figure and two "children," South Auckland, New Zealand. Ht., 4 ft., 5 1/2 in. Auckland, Museum.

Pl. 205. *Left and right*: Maori wooden figures from veranda posts of large assembly houses, East Coast area, New Zealand. *Left*: Figure in high relief on the front of a post (at the base). Wellington, Dominion Museum. *Right*: Freestanding figure forming the base of a post. Ht., 3 ft., 9 1/4 in. *Center*: House post with applied sculpture in low relief. Ht., 6 ft., 1 1/2 in. Last two, Auckland, Museum.

Pl. 206. Tapa-covered female figure with painted tattoo designs, Easter Island. Ht., ca. 15 in. Cambridge, Mass., Peabody Museum.

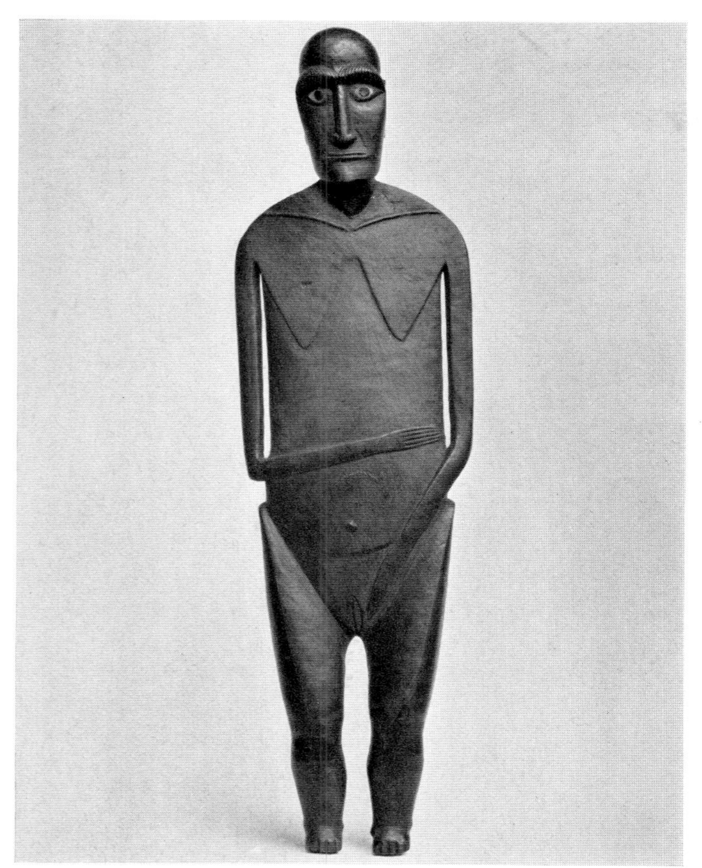

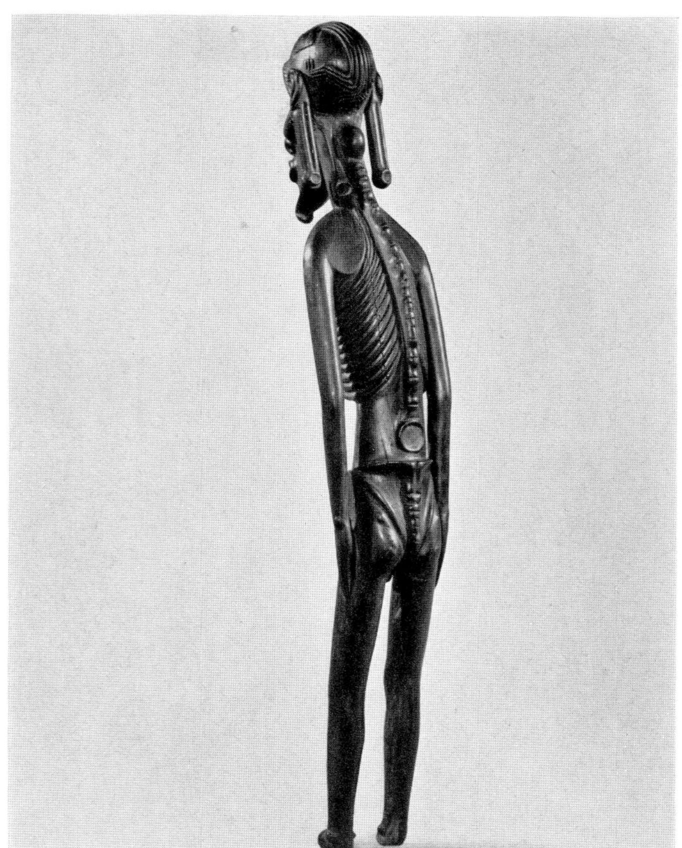

Pl. 207. Easter Island sculpture. Wood. *Above, left*: Flat-bodied female figure. Ht., 21 1/2 in. London, British Museum. *Right*: Male figure in "skeletal" style (cf. IV, PL. 451). Ht., 17 3/4 in. Auckland, Museum. *Below*: Lizard figure. L., 12 3/4 in. London, British Museum.

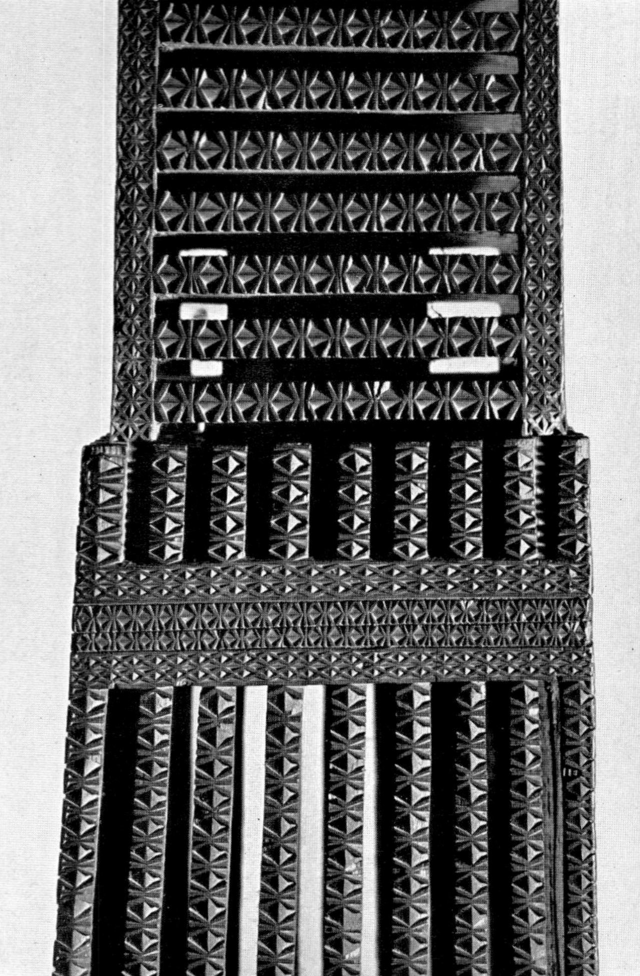

Pl. 208. *Above, left*: Tattooer's pattern sample, Marquesas Islands. Wood, $34^{1}/_{4} \times 10^{1}/_{4}$ in. Paris, Musée de l'Homme. *Right*: Handle of ceremonial adz, detail, decorated with K pattern, Mangaia, Cook Islands. Wood. Rome, Museo Pigorini. *Below*: Maori waka huia (treasure or feather box), New Zealand. Wood, l., $22^{1}/_{4}$ in. Chicago, Natural History Museum.

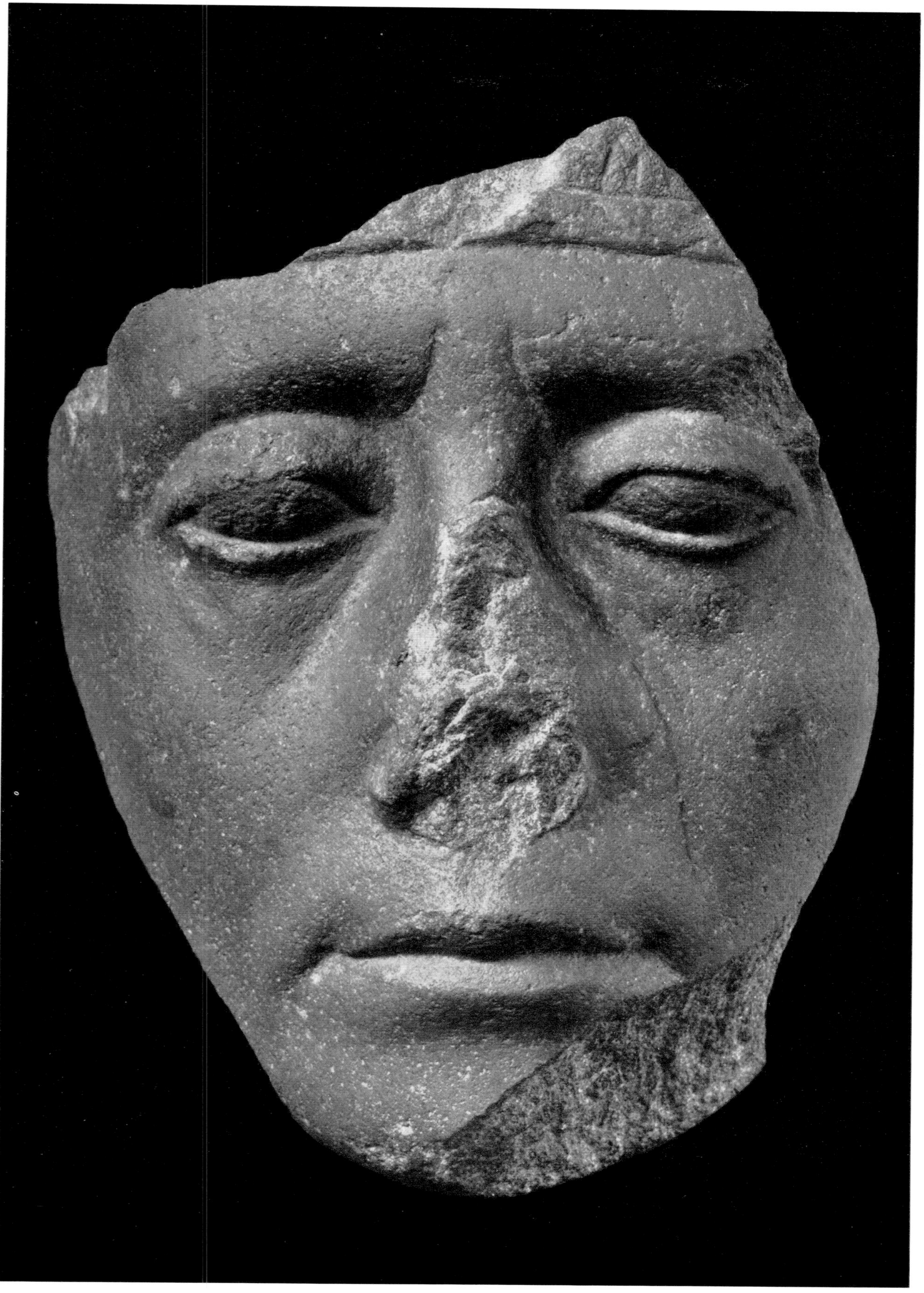

Pl. 209. Fragment of a head of Sesostris III, 12th dynasty, Egypt. Quartzite, ht., 6 1/2 in. New York, Metropolitan Museum.

Pl. 210. *Above, left*: Themistokles, copy of a Greek original of 5th cent. B.C., found in Ostia. Marble, ht., $15\frac{3}{4}$ in. Ostia Antica, Italy, Museo Ostiense. *Right*: Menander, Roman copy of a Greek original of ca. 200 B.C. Venice, Seminario Patriarcale. *Below, left*: Euthydemos of Bactria, ca. 200 B.C. Marble, ht., $13\frac{3}{4}$ in. Rome, Torlonia Coll. *Right*: Harsinebef, from Sais, Egypt, 1st cent. B.C. Black granite. Berlin, Staatliche Museen.

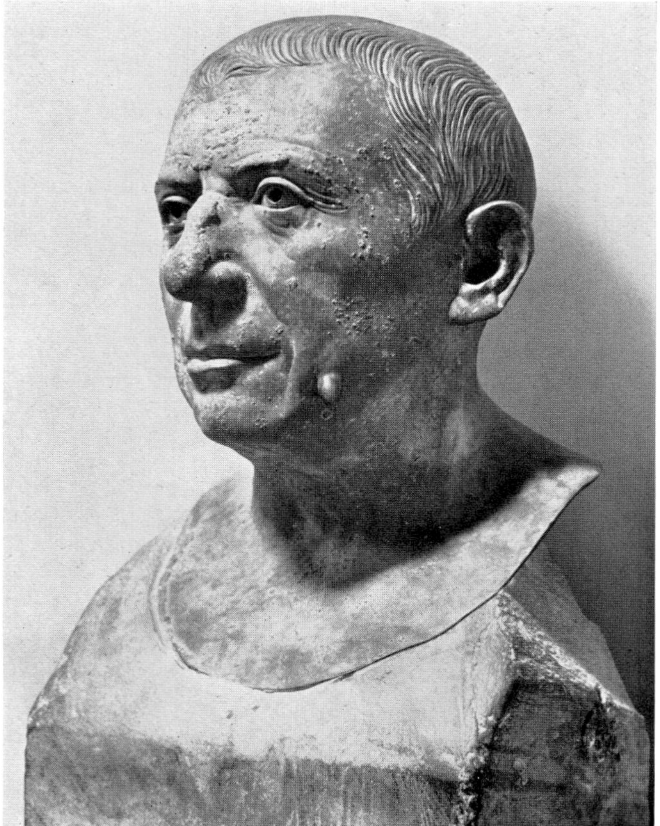

Pl. 211. Roman. *Above*: Funerary stele of L. Vibius and family, Augustan period. Marble, ht., $29^{1}/_{2}$ in. Rome, Vatican Museums. *Below, left*: Minatia Polla (?), Julio-Claudian period. Marble, ht., $12^{1}/_{4}$ in. Rome, Museo Nazionale Romano. *Right*: The banker Lucius Caecilius Jucundus, Pompeii, Flavian period. Bronze, ht., $13^{3}/_{8}$ in. Naples, Museo Nazionale.

Pl. 212. Claudius as Jupiter, mid-1st cent. Marble; ht., excluding base, 8 ft., 4 in. Rome, Vatican Museums.

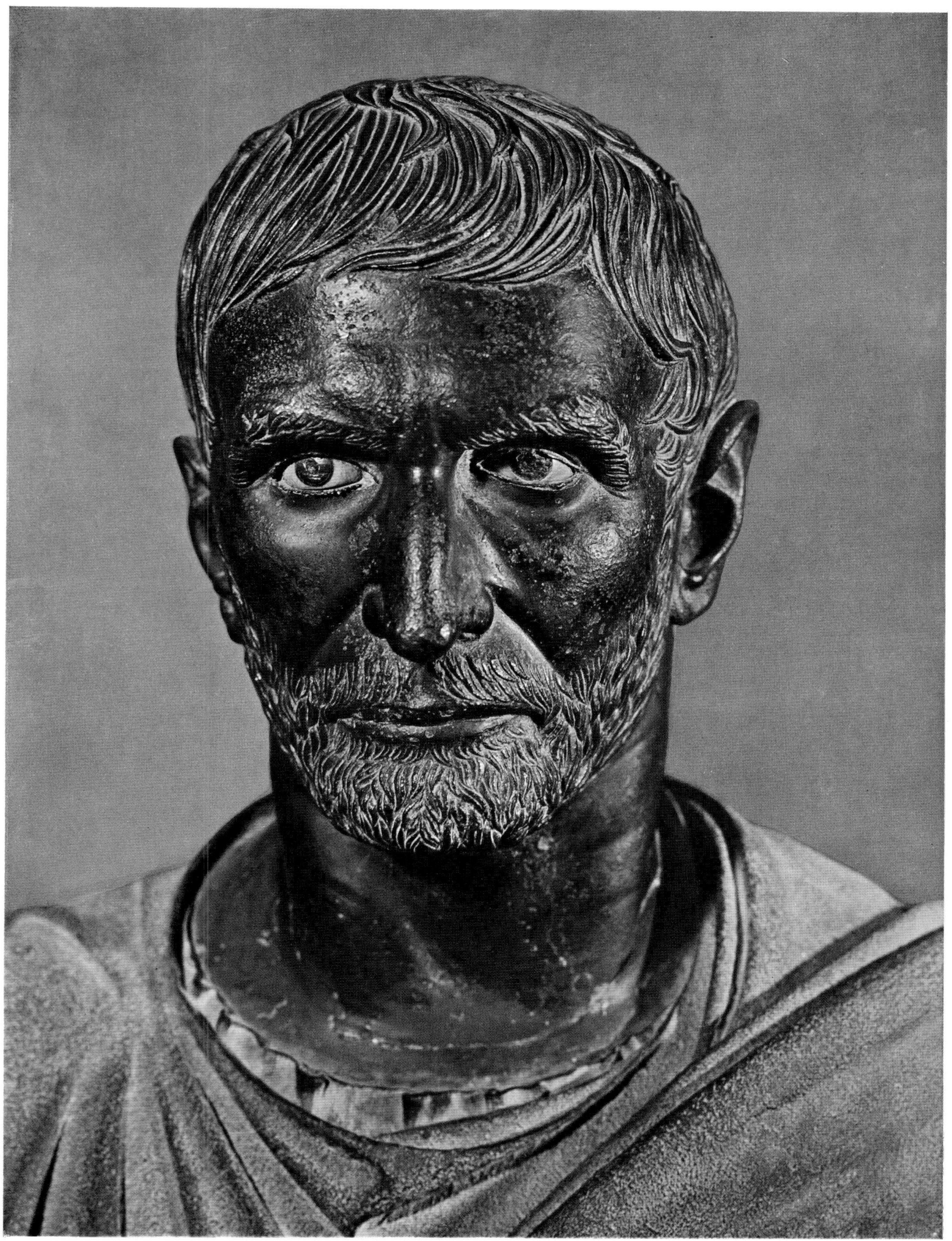

Pl. 213. Portrait known as L. Junius Brutus, of controversial date (4th–1st cent. B.C.). Bronze head (adjoined to bust of 16th cent.), ht., 12 5/8 in. Rome, Palazzo dei Conservatori.

Pl. 214. Portrait of a man named Perseus, from a mummy case, Fayum, Egypt, 2d cent. Encaustic on panel, 17 5/8×9 in. Berlin, Staatliche Museen.

Pl. 215. Roman. *Above, left*: Hadrian, 117–38. Marble, ht., $13^{3}/_{4}$ in. Rome, Museo Nazionale Romano. *Right*: Probus, 276–82. Marble, ht., $17^{3}/_{4}$ in. Rome, Capitoline Museum. *Below, left*: Licinius I, obverse of aureus, 313–14. Gold, diam., 20 mm. Rome, Museo Nazionale Romano. *Right*: So-called "Eutropius," from Ephesus, 5th cent. Marble, ht., $12^{5}/_{8}$ in. Vienna, Kunsthistorisches Museum.

Pl. 216. *Above, left*: Abbot Epiphanius, detail of Crucifixion, 9th cent. Fresco. Abbey of San Vincenzo al Volturno, Abruzzi, Italy, crypt of S. Lorenzo. *Right*: Genghis Khan, from a manuscript of the *Jāmi' at-Tavārīkh* by Rashīd al-Dīn, ca. 1310. Miniature. Paris, Bibliothèque Nationale (Ms. sup. pers. 1113, fol. 116v). *Below, left*: King John the Good, 1360–64. Panel, $22^7/_8 \times 14^1/_8$ in. Paris, Louvre. *Right*: Pisanello, portrait of Lionello d'Este, ca. 1440. Panel, $11 \times 7^1/_2$ in. Bergamo, Italy, Accademia Carrara.

Pl. 217. *Above, left*: J. van Eyck, Portrait of a Young Man, 1432. Panel, $13^{1}/_{8} \times 7^{1}/_{2}$ in. London, National Gallery. *Right*: J. Fouquet, Charles VII, after 1444. Panel, $33^{7}/_{8} \times 28^{3}/_{8}$ in. Paris, Louvre. *Below, left*: Bartolomeo Veneto, Portrait of a Woman, first half of 16th cent. Panel, $16^{7}/_{8} \times 13^{3}/_{8}$ in. Frankfort on the Main, Städelsches Kunstinstitut. *Right*: A. Mantegna, Cardinal Francesco Gonzaga, ca. 1462. Panel, $9^{7}/_{8} \times 7^{1}/_{8}$ in. Naples, Museo di Capodimonte.

Pl. 218. *Above, left*: Pontormo, Portrait of a Woman (Maria Salviati?), ca. 1543–45. Panel, $34^{1}/_{4}\times28$ in. Florence, Uffizi. *Right*: G. Moroni (d. 1578), so-called "Titian's Schoolmaster." Canvas, $38^{1}/_{8}\times29^{1}/_{4}$ in. Washington, D.C., National Gallery. *Below, left*: Titian, Self-portrait, 1550 or later. Canvas, $37^{3}/_{4}\times29^{1}/_{2}$ in. Berlin, Staatliche Museen. *Right*: A. Mor (d. 1576), Self-portrait. Panel, $44^{1}/_{2}\times34^{1}/_{4}$ in. Florence, Uffizi.

Pl. 219. *Above*: P. de Champaigne, Triple Portrait of Cardinal Richelieu, 1640. Canvas, $23 \times 28^{1}/_{2}$ in. London, National Gallery. *Below, left*: D. Velázquez, The Fool of Coria (the court jester Juan de Calabacillas), detail, ca. 1636. Canvas; full size, $41^{3}/_{4} \times 32^{5}/_{8}$ in. Madrid, Prado. *Right*: A. van Dyck, The Virgin and Child Adored by the Abbé Scaglia, detail, ca. 1634. Canvas; full size, $42 \times 47^{1}/_{2}$ in. London, National Gallery.

Pl. 220. *Above*: P. P. Rubens, The Countess of Arundel and Her Attendants, 1620. Canvas, 8 ft., 8 1/2 in. × 8 ft., 9 in. Munich, Alte Pinakothek. *Below*: F. Hals, Governors of the Old Men's Home at Haarlem, 1664. Canvas, 5 ft., 8 in. × 8 ft., 4 3/4 in. Haarlem, Netherlands, Frans-Hals-Museum.

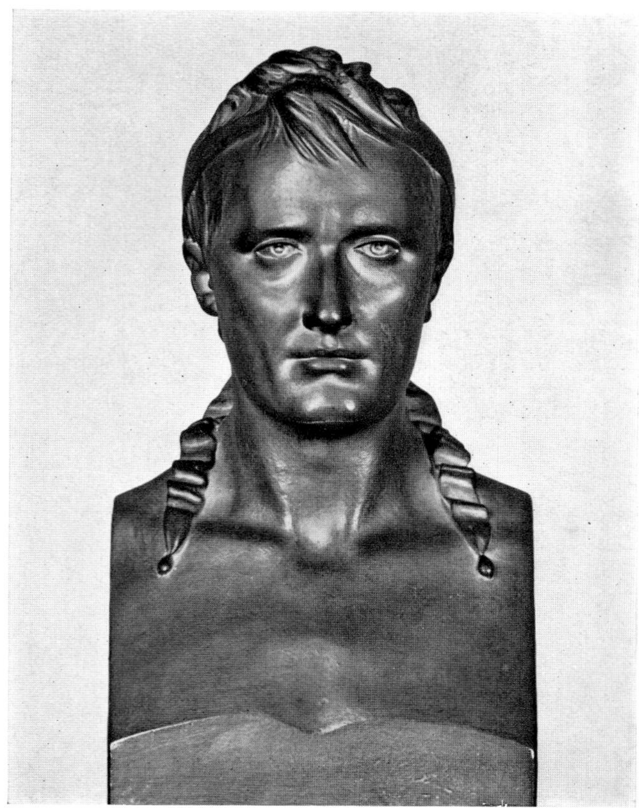

Pl. 221. *Above, left*: F. Laurana, Eleanor of Aragon (d. 1405), ca. 1468. Marble. Palermo, Galleria Nazionale della Sicilia. *Right*: D. Cattaneo, Lazzaro Bonamico, ca. 1552. Bronze, ht., $29^{1}/_{2}$ in. Bassano, Italy, Museo Civico. *Below, left*: G. L. Bernini, Louis XIV, 1665. Marble, ht., $31^{1}/_{2}$ in. Versailles, Palace. *Right*: J. A. Houdon, Napoleon I, 1806. Terra cotta, ht., $20^{1}/_{8}$ in. Dijon, France, Musée des Beaux-Arts.

Pl. 222. *Above, left*: V. Ghislandi (1655–1743), Portrait of a Nobleman. Canvas, $42\,{}^{7}/_{8} \times 34\,{}^{1}/_{4}$ in. Milan, Museo Poldi Pezzoli. *Right*: J. Reynolds, Lord Heathfield, Governor of Gibraltar, 1787. Canvas, $56 \times 44\,{}^{3}/_{4}$ in. London, National Gallery. *Below, left*: J. M. Nattier (1685–1766), Marie Leczinska. Canvas, $53\,{}^{1}/_{8} \times 40\,{}^{1}/_{2}$ in. Dijon, France, Musée des Beaux-Arts. *Right*: J. L. David, Marquise d'Orvilliers, 1790. Canvas, $51\,{}^{5}/_{8} \times 38\,{}^{5}/_{8}$ in. Paris, Louvre.

Pl. 223. S. Botticelli, Giuliano de' Medici, ca. 1478. Panel, 29 3/4 × 20 5/8 in. Washington, D.C., National Gallery.

Pl. 224. L. Lotto, Young Man in His Study, 1516–27. Canvas, $38^5/_8 \times 43^3/_4$ in. Venice, Accademia.

Pl. 225. *Left*: Bronzino, Cosimo I de' Medici in armor, 1545. Panel, $29\frac{1}{8} \times 22\frac{7}{8}$ in. Florence, Uffizi. *Right*: F. Clouet, Elisabeth of Austria, ca. 1571. Panel, $14\frac{1}{8} \times 10\frac{5}{8}$ in. Paris, Louvre.

Pl. 226. P. G. Batoni, Marshal Cyril Grigoriewicz Razumowski, 1766. Canvas, 9 ft., 9 1/2 in. × 6 ft., 5 in. Vienna, A. Razumowski Coll.

Pl. 227. J.-A.-D. Ingres, Louis-François Bertin, 1832. Canvas, $45\frac{5}{8} \times 37\frac{3}{8}$ in. Paris, Louvre.

Pl. 228. *Above, left*: H. Daumier, F. Guizot, ca. 1833. Bronze, ht., $8^1/_2$ in. Milan, Coll. A. Borletti. *Right*: A. Modigliani, Renée, 1917. Canvas, 24×15 in. São Paulo, Brazil, Museu de Arte. *Below, left*: P. A. Renoir, By the Seashore, 1883. Canvas, $36^1/_2 \times 28^1/_2$ in. New York, Metropolitan Museum. *Right*: O. Kokoschka, Self-portrait, 1917. Canvas, $30^3/_4 \times 24^3/_8$ in. Wuppertal, Germany, Städtisches Museum.

Pl. 229. Fujiwara-no-Takanobu (attrib.), Minamoto Yoritomo, Japan, Kamakura period. Color on silk, $53^{1}/_{8} \times 43^{1}/_{4}$ in. Kyoto, Jingoji.

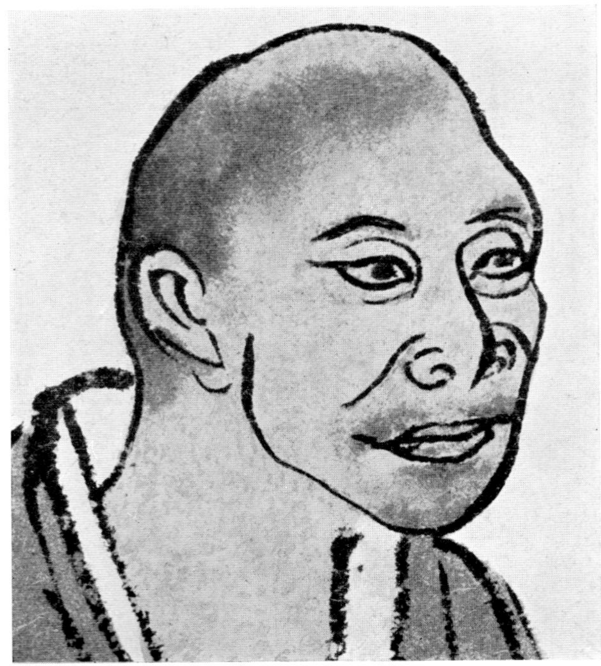

Pl. 230. *Above left*: Shōtoku Taishi and his sons, Japan, Nara period. Color on paper, 49 1/4 × 20 1/8 in. Nara, Hōryūji. *Right*: Ming princess, China, 15th cent. Color on silk. China, Li Family archives. *Below, left*: Chou Ch'ên, study of a monk, detail, China, early 16th cent. Honolulu, Academy of Arts. *Right*: Wu Chün (attrib.), portrait of Juan Yüan, detail, China, early 19th cent. Formerly, Peking, Ecke Coll.

Pl. 231. Bacchanalian Revel before a Term of Pan. Canvas, $39^1/_4 \times 56^1/_4$ in. London, National Gallery.

Pl. 232. The Woman Taken in Adultery, detail. Canvas; full size, ca. 4 ft. × 6 ft., 5 in. Paris, Louvre.

Pl. 233. Self-portrait. Canvas, 38 1/2×29 in. Paris, Louvre.

Pl. 234. *Above*: The Plague at Ashdod. Canvas, 4 ft., 10 in. × 6 ft., 6 in. Paris, Louvre. *Below*: Selene and Endymion. Canvas, 4 ft. × 5 ft., 2 1/2 in. Detroit, Institute of Arts.

Pl. 235. *Above*: Confirmation, from the first series of the Seven Sacraments. Canvas, $37^{1}/_{2} \times 47$ in. Belvoir Castle, Leicestershire, England, Duke of Rutland Coll. *Below*: The Funeral of Phocion. Canvas, 3 ft., 9 in. × 5 ft., 9 in. Oakly Park, Ludlow, England, Earl of Plymouth Coll.

Pl. 236. *Above*: Penance, drawing for the second series of the Seven Sacraments. Pen and bistre wash, $8^1/_4 \times 12^1/_4$ in. Montpellier, France, Musée Fabre. *Below*: The Rest on the Flight into Egypt. Canvas, $41^1/_4 \times 57$ in. Leningrad, The Hermitage.

Pl. 237. Roman copies. *Left*: Young satyr pouring wine, from Torre del Greco, Italy. Marble, ht., ca. 5 ft. Palermo, Museo Archeologico. *Right*: Apollo Sauroktonos. Bronze, ht., 37 1/2 in. Rome, Museo di Villa Albani.

Pl. 238. Aphrodite of Knidos, fragment of a Hellenistic copy (?), from Tralles (mod. Aydın), Turkey. Marble, ht., 11 3/4 in. Berlin, Staatliche Museen.

Pl. 239. Hermes with the child Dionysos (original?). Marble, ht., ca. 7 ft. Olympia, Greece, Archaeological Museum.

Pl. 240. Resting satyr, Roman copy. Marble, ht., 5 ft., 7 in. Rome, Capitoline Museum.

Pl. 241. *Above*: Three of the Muses, relief from the Mantineia base, probably by pupils. Marble, ht., 3 ft., 1 3/4 in. Athens, National Museum. *Below, left*: Artemis, Roman copy. Marble, ht., 4 ft., 11 1/2 in. Dresden, Albertinum, Skulpturensammlung. *Right*: Artemis of Gabii, Roman copy of Artemis Brauronia (?). Marble, ht., 5 ft., 5 in. Paris, Louvre.

Pl. 242. Acheulean hand ax, from Coq-Gaulois, Somme, France. Flint. Abbeville, France, Musée Boucher de Perthes (on loan).

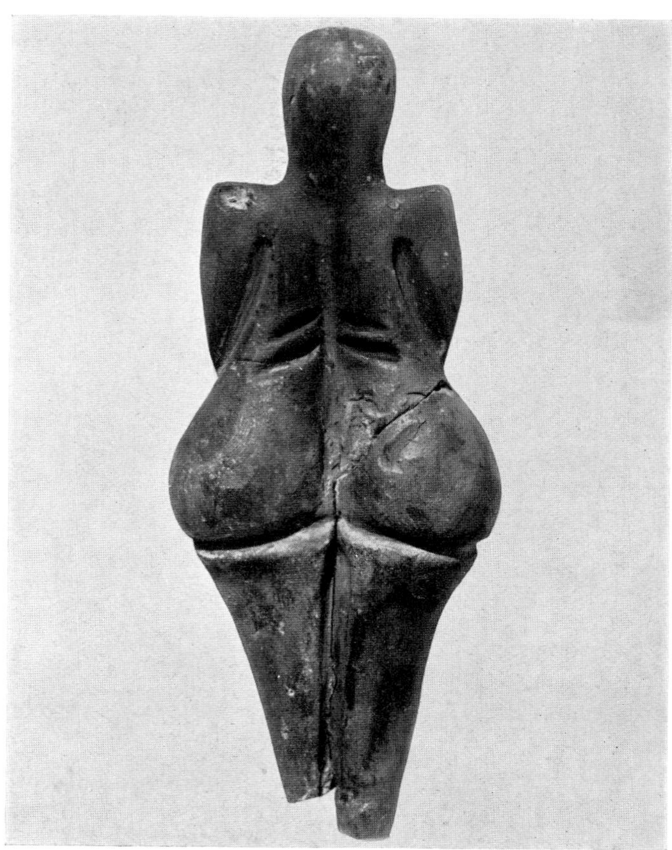

Pl. 243. Sculpture, Aurignacian-Perigordian period. *Above*: Ivory carvings from the Grotte du Pape, Brassempouy, Landes, France. *Left*: Female head. Ht., $1^3/_8$ in. (See VIII, PL. 235.) *Right*: Female torso. Ht., $3^1/_8$ in. Both, Saint-Germain-en-Laye, France, Musée des Antiquités Nationales. *Below*: So-called "Venus of Věstonice" (two views), from Dolní Věstonice, Moravia. Clay with bone ash, ht., $4^1/_2$ in. Brno, Czechoslovakia, Moravian Museum.

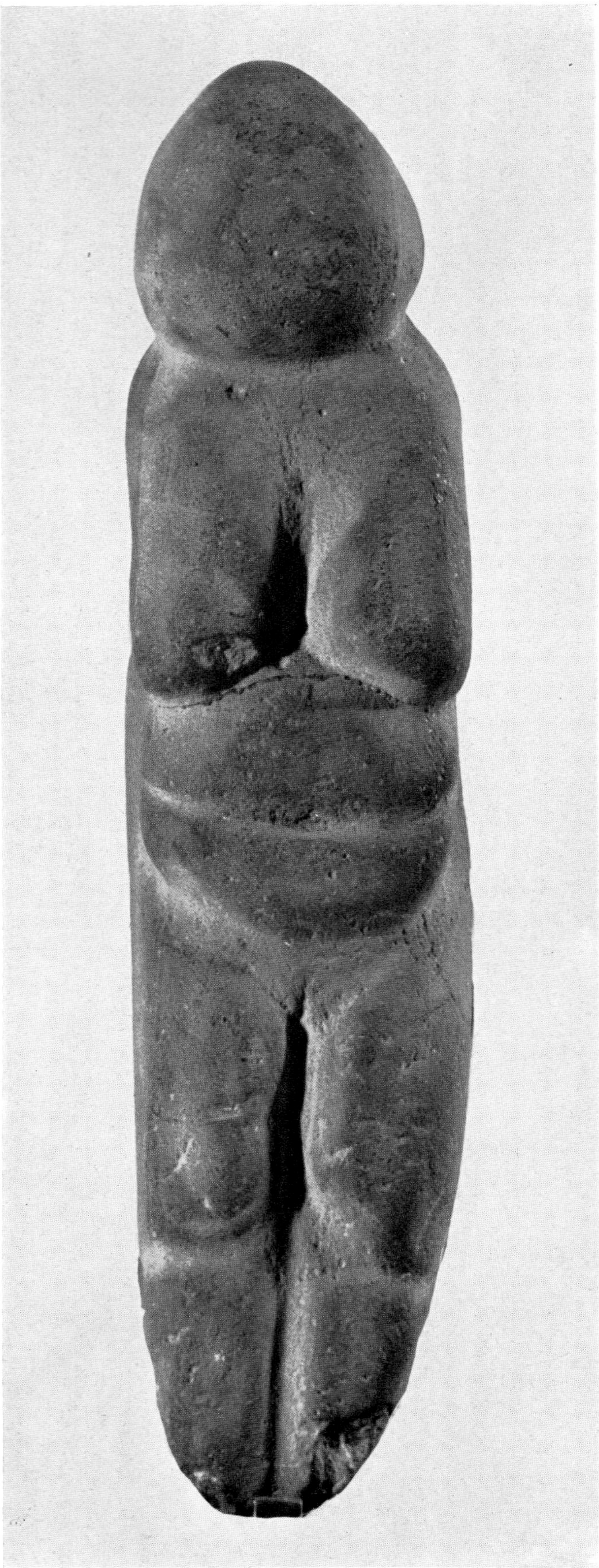

Pl. 244. *Left*: Stylized figurine, from Dolní Věstonice, Moravia, Aurignacian-Perigordian period. Ivory, ht., 3 1/2 in. Brno, Czechoslovakia, Moravian Museum. *Right*: So-called "Venus of Chiozza," found near Scandiano, Emilia, Italy, Aurignacian-Perigordian type. Sandstone, ht., ca. 8 in. Reggio Emilia, Italy, Musei Civici.

Pl. 245. Animal sculpture, Aurignacian-Perigordian period. *Above*: Feline, from Vogelherd, Württemberg, Germany. Ivory, l., 3 1/2 in. Tübingen, Germany, Vorgeschichtliche Sammlung der Universität. *Below*: Bear, from Dolní Věstonice, Moravia. Clay, l., 3 in. Brno, Czechoslovakia, Moravian Museum.

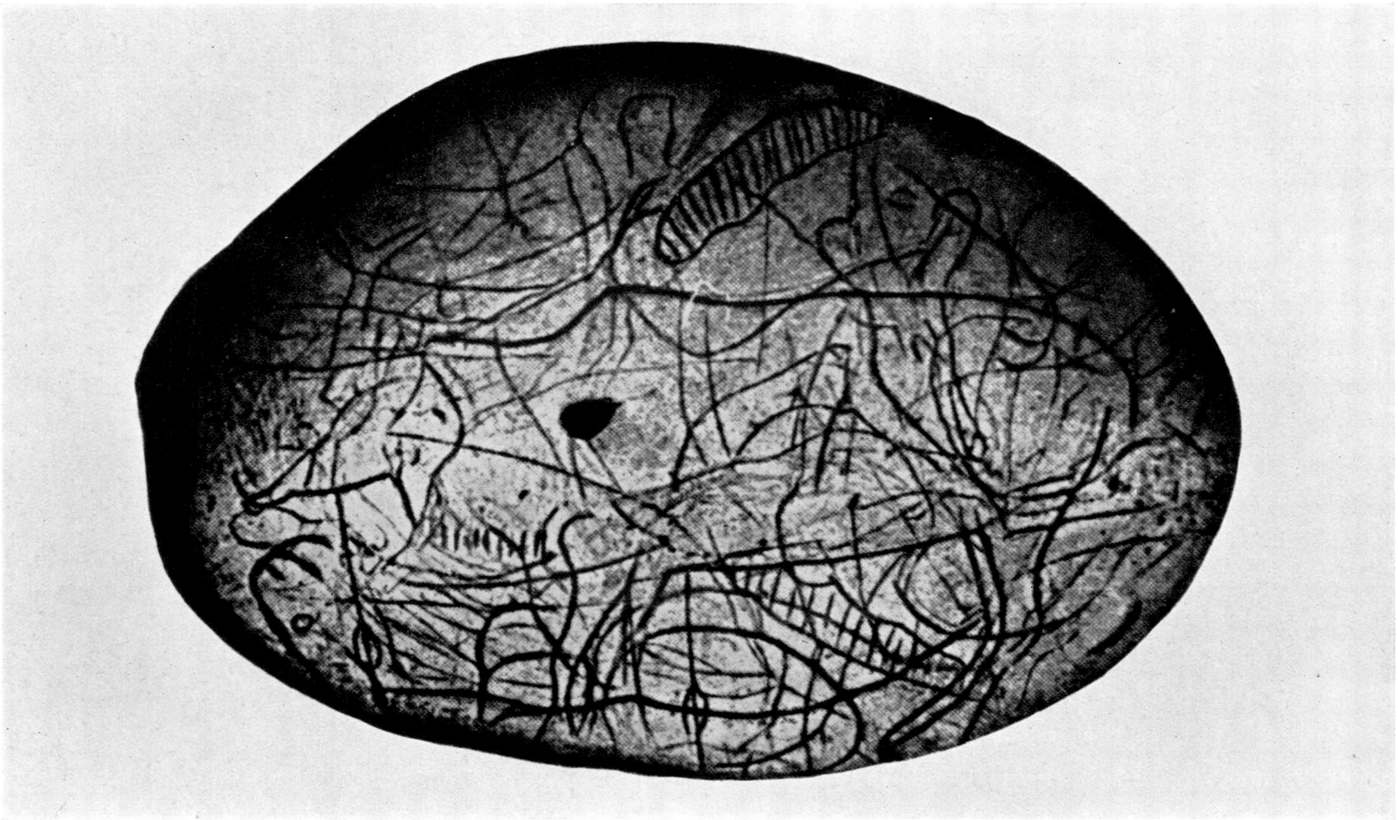

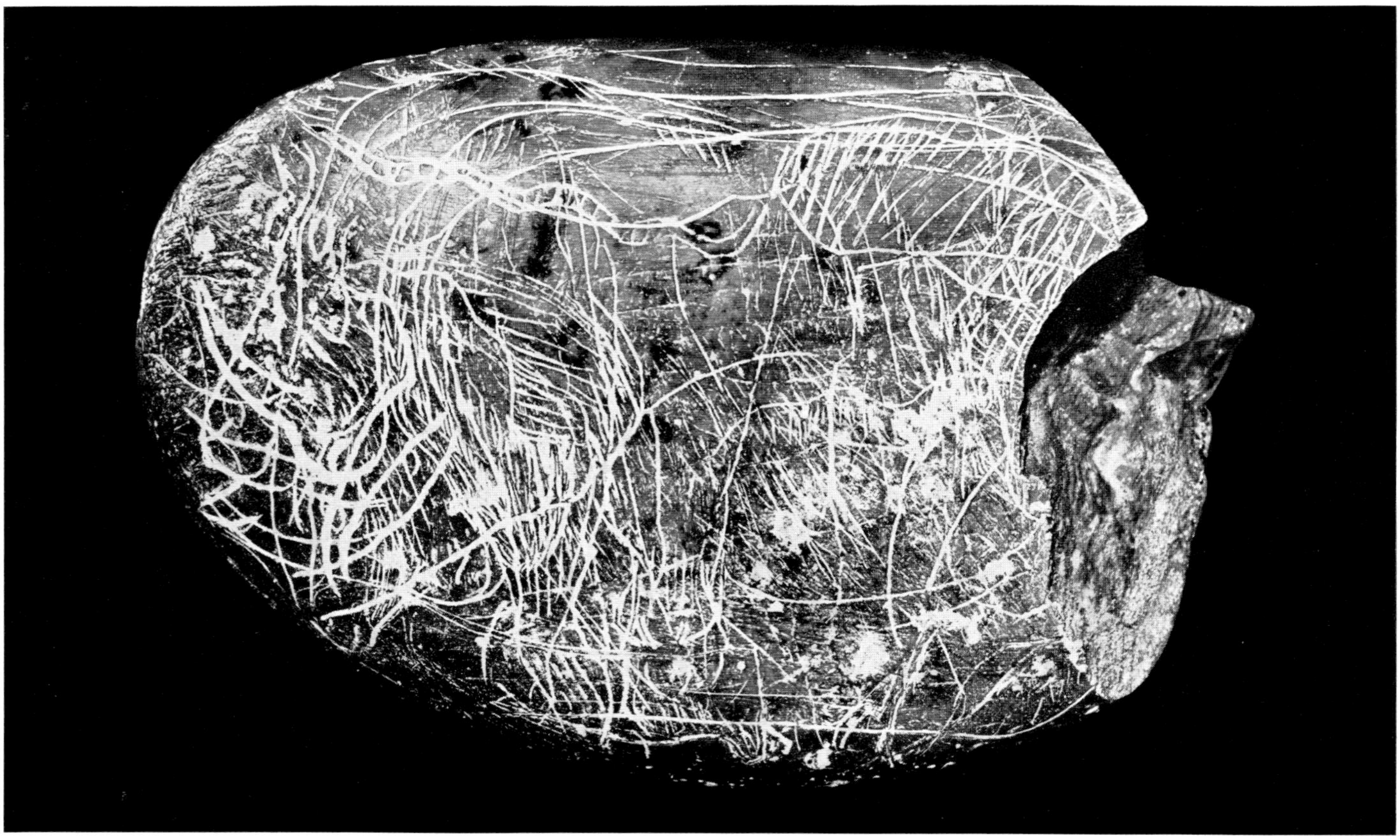

Pl. 246. Engraved pebbles with superimposed animal figures, from La Colombière, Ain, France, Aurignacian-Perigordian period. *Above*: L., ca. 4 in. Lyons, University, Laboratoire de Géologie. *Below*: L., 4 3/4 in. (Harvard University excavations, 1948.)

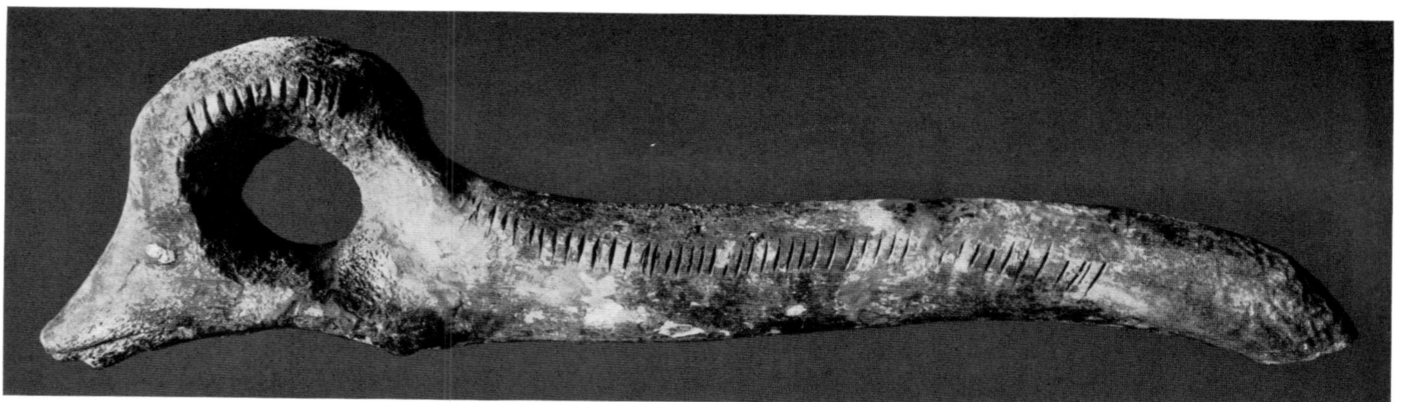

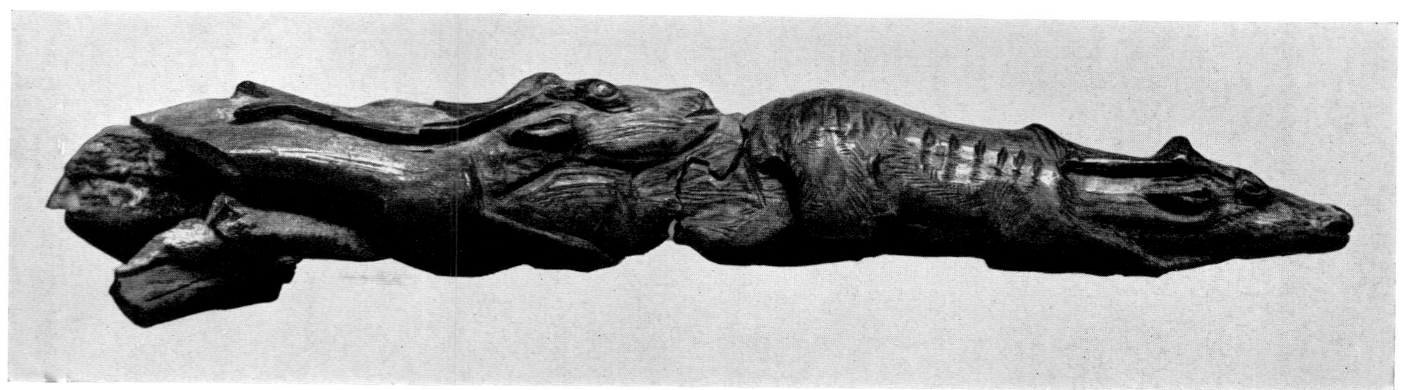

Pl. 247. Sculpture, Magdalenian period, France. *Above*: Pierced " bâton " with animal head, from Le Placard, Charente. Bone, l., ca. 13 in. Saint-Germain-en-Laye, France, Musée des Antiquités Nationales. *Center*: Two reindeer, from Bruniquel, Tarn-et-Garonne. Ivory, l., ca. $8^{1}/_{2}$ in. *Below*: Mammoth, from Bruniquel. Antler, l., ca. 5 in. Last two, London, British Museum.

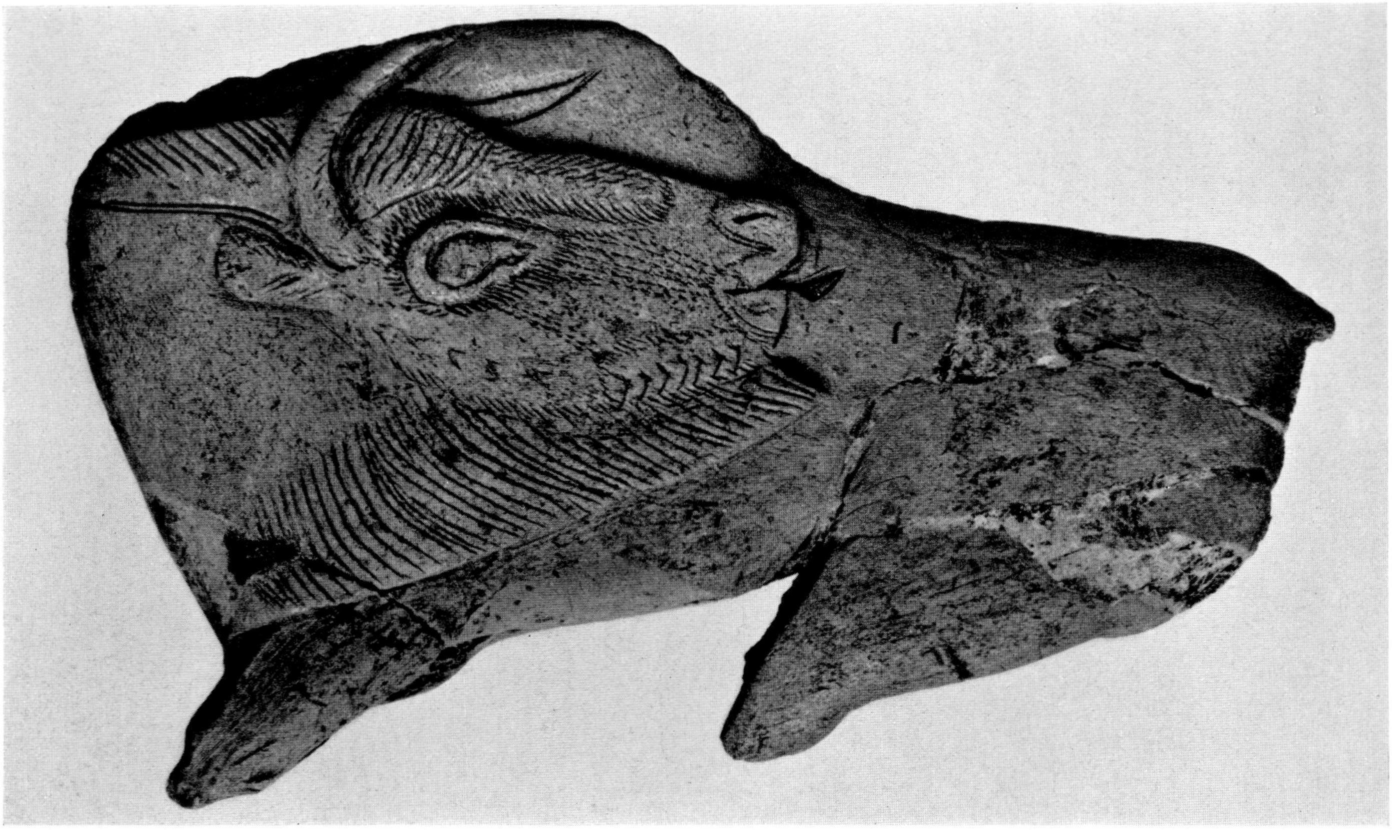

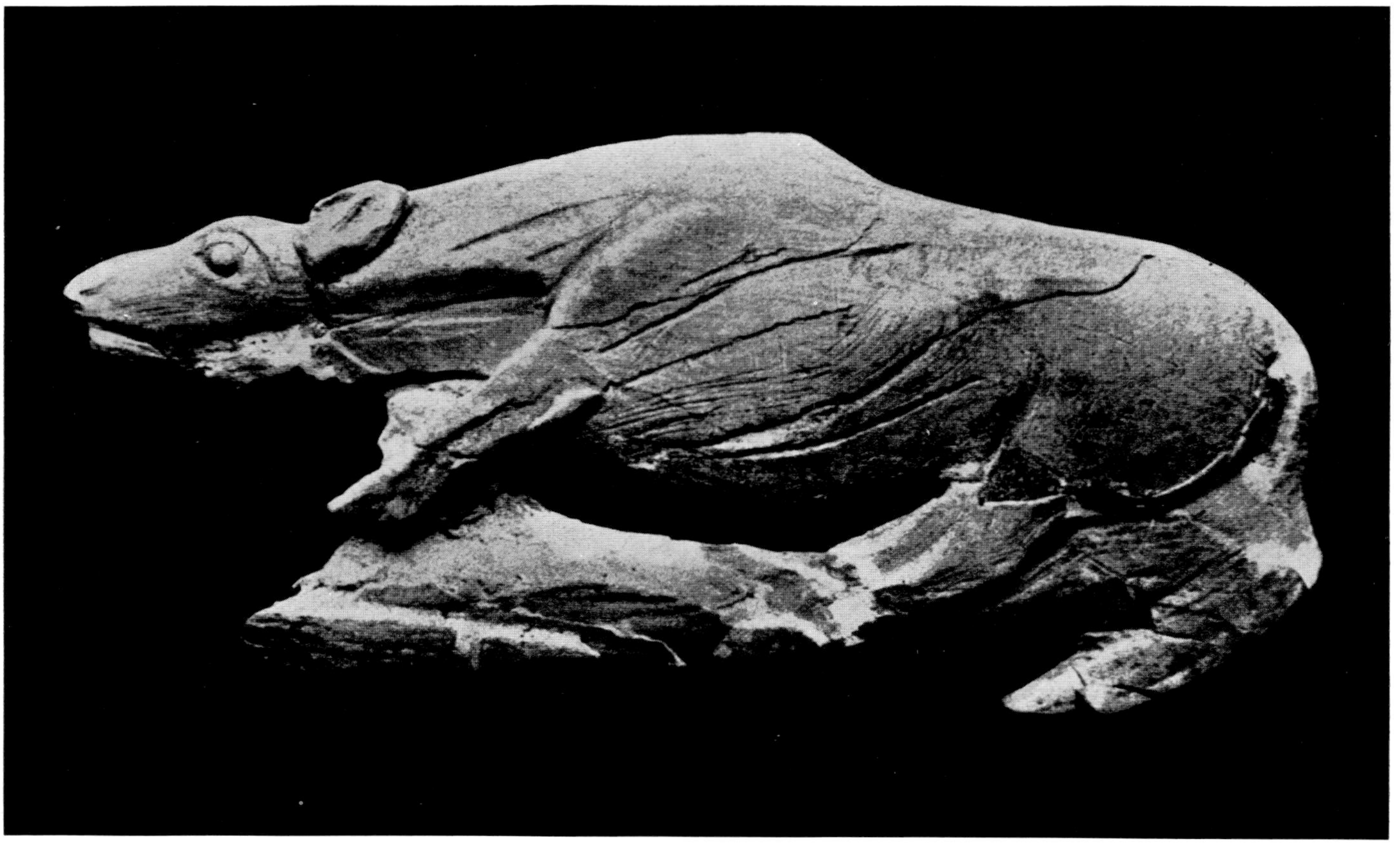

Pl. 248. Animal sculpture, Magdalenian period, from La Madeleine, Dordogne, France. *Above*: Bison. Antler, l., ca. 4 in. *Below*: Hyena (?). Ivory, l., 4 3/8 in. Both, Saint-Germain-en-Laye, France, Musée des Antiquités Nationales.

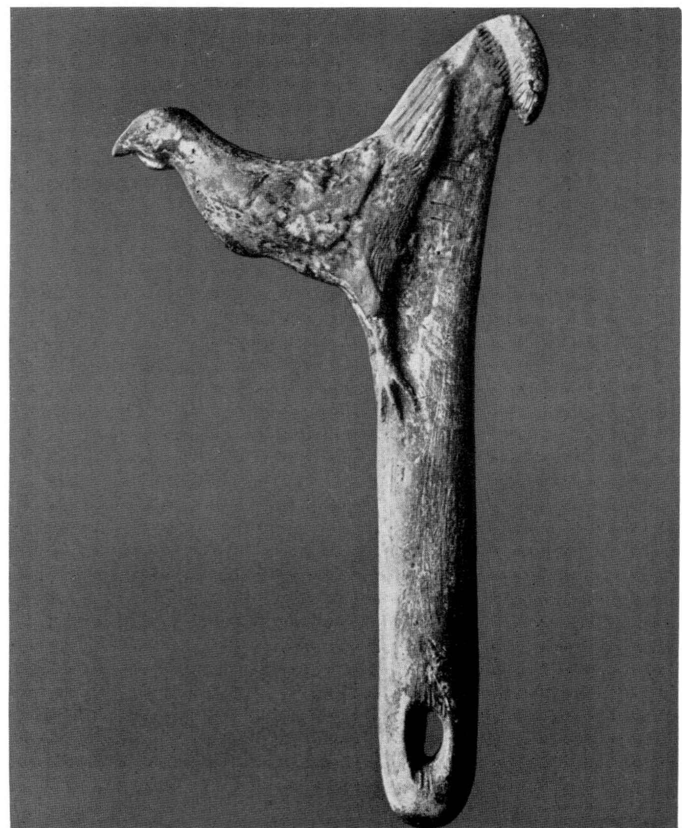

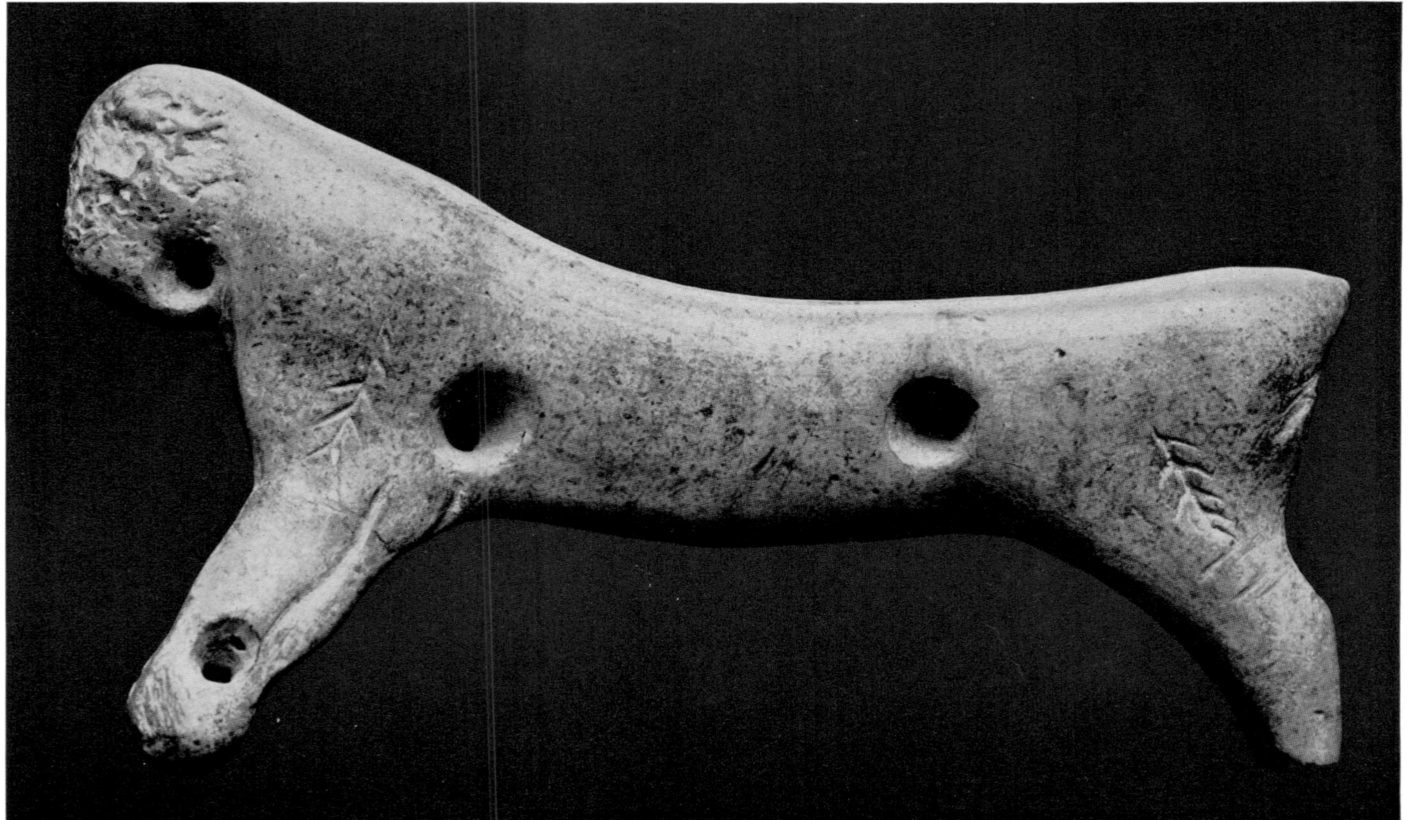

Pl. 249. Carved antler, Magdalenian period, France. *Above, left*: Two horses, from Le Mas d'Azil, Ariège. Ht., ca. $6^1/_2$ in. *Right*: Bird fragment (in Breuil's reconstruction of a spear thrower, head and foot added), from Le Mas d'Azil. L. of fragment, $2^3/_4$ in. *Below*: Feline, with perforations and engraved motif, from Isturitz, Basses-Pyrénées. L., ca. 4 in. All, Saint-Germain-en-Laye, France, Musée des Antiquités Nationales.

Pl. 250. Fragmentary reliefs with reindeer, from Isturitz, Basses-Pyrénées, France, Magdalenian period. Stone, l. of each, ca. 6 in. Saint-Germain-en-Laye, France, Musée des Antiquités Nationales.

Pl. 251. Engravings, Magdalenian period, France. *Above*: Wolves, from La Vache, Ariège. Bone, l., 3 1/2 in. Tarascon-sur-Ariège, France, Coll. R. Robert. *Below*: Human and animal figures, from La Madeleine, Dordogne. Antler, l., ca. 6 in. Saint-Germain-en-Laye, France, Musée des Antiquités Nationales.

Pl. 252. Impression of engraving around an antler, deer and fish with lozenges, from Lorthet, Hautes-Pyrénées, France, Magdalenian period. L. of design, 5 1/2 in. Saint-Germain-en-Laye, France, Musée des Antiquités Nationales.

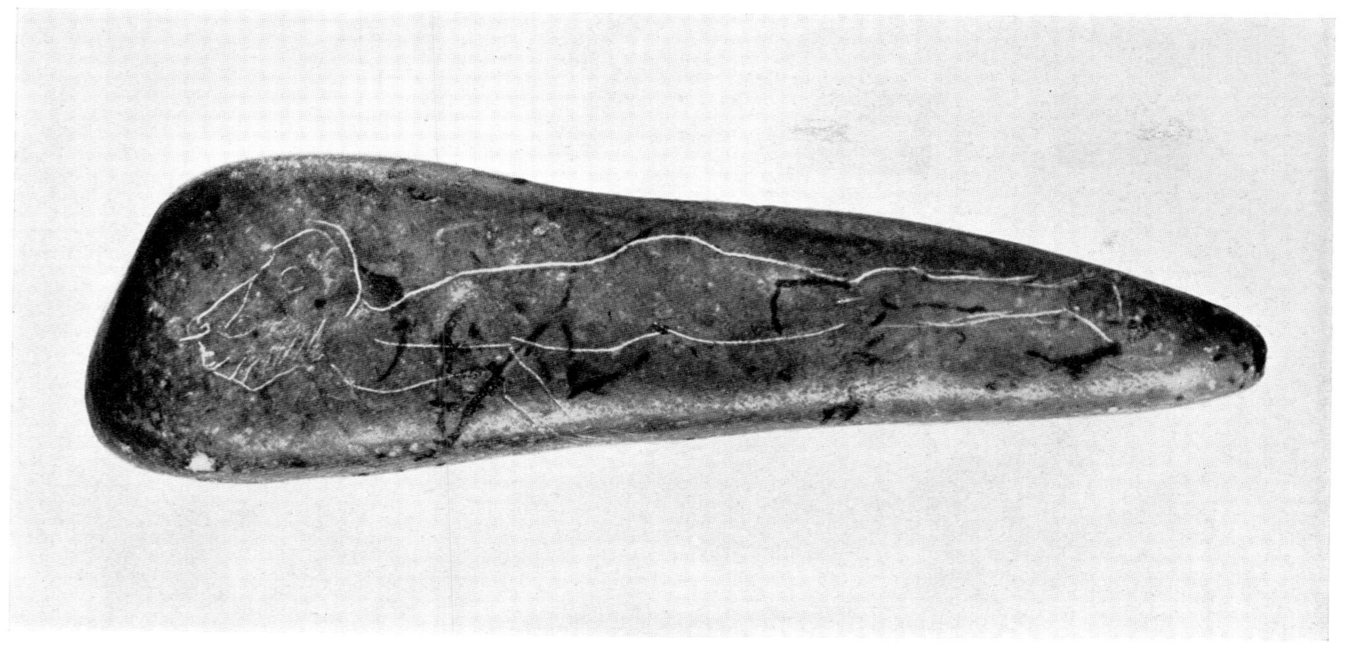

Pl. 253. Engravings, Magdalenian period, France. *Left*: Stone fragment with hare, from Isturitz, Basses-Pyrénées. Ht., $5^3/_8$ in. *Right*: Pebble with (masked?) human figure, from La Madeleine, Dordogne. Ht. of figure, 3 in. Both, Saint-Germain-en-Laye, France, Musée des Antiquités Nationales.

Pl. 254. Stone slab with male figure (partially flaked), from the Laussel Shelter, Dordogne, France, Aurignacian-Perigordian period. Ht. of figure, 15 in. (Photograph of a cast.)

Pl. 255. So-called "Venus of Laussel," stone slab from the Laussel Shelter, Dordogne, France, Aurignacian-Perigordian period. Ht. of figure, 18 in. (Photograph of a cast.)

Pl. 256. Animal reliefs, from Le Roc de Sers, Charente, France, Solutrean period. Stone. *Above*: Mare and bison with boar's head. L. of slab, 5 ft., $4^{1}/_{2}$ in. *Below*: Ibexes. L. of ibex at right, $21^{1}/_{2}$ in. Both, Saint-Germain-en-Laye, Musée des Antiquités Nationales.

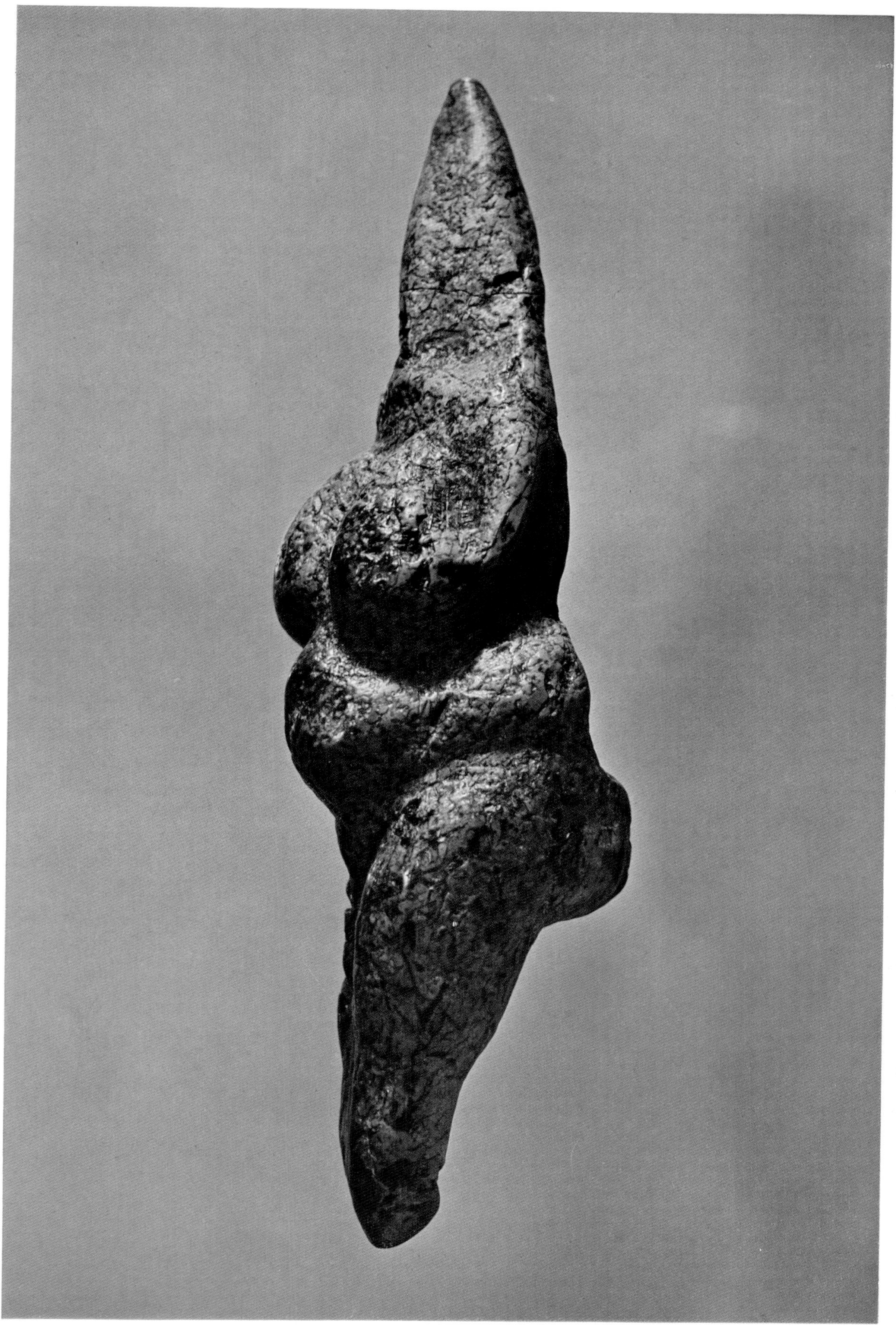

Pl. 257. "Venus" from Savignano sul Panaro, near Modena, Italy, Aurignacian-Perigordian type. Serpentine, ht., 8$^{5}/_{8}$ in. Rome, Museo Pigorini.

Pl. 258. Lascaux Cave, Dordogne, France, painting of leaping cow and small horses, superposed on red painting with lattice sign. L. of cow, 5 ft., 8 in.

Pl. 259. Cave of Le Tuc d'Audoubert, Ariège, France, two bison modeled in clay. L. of bison at right, 24 in.

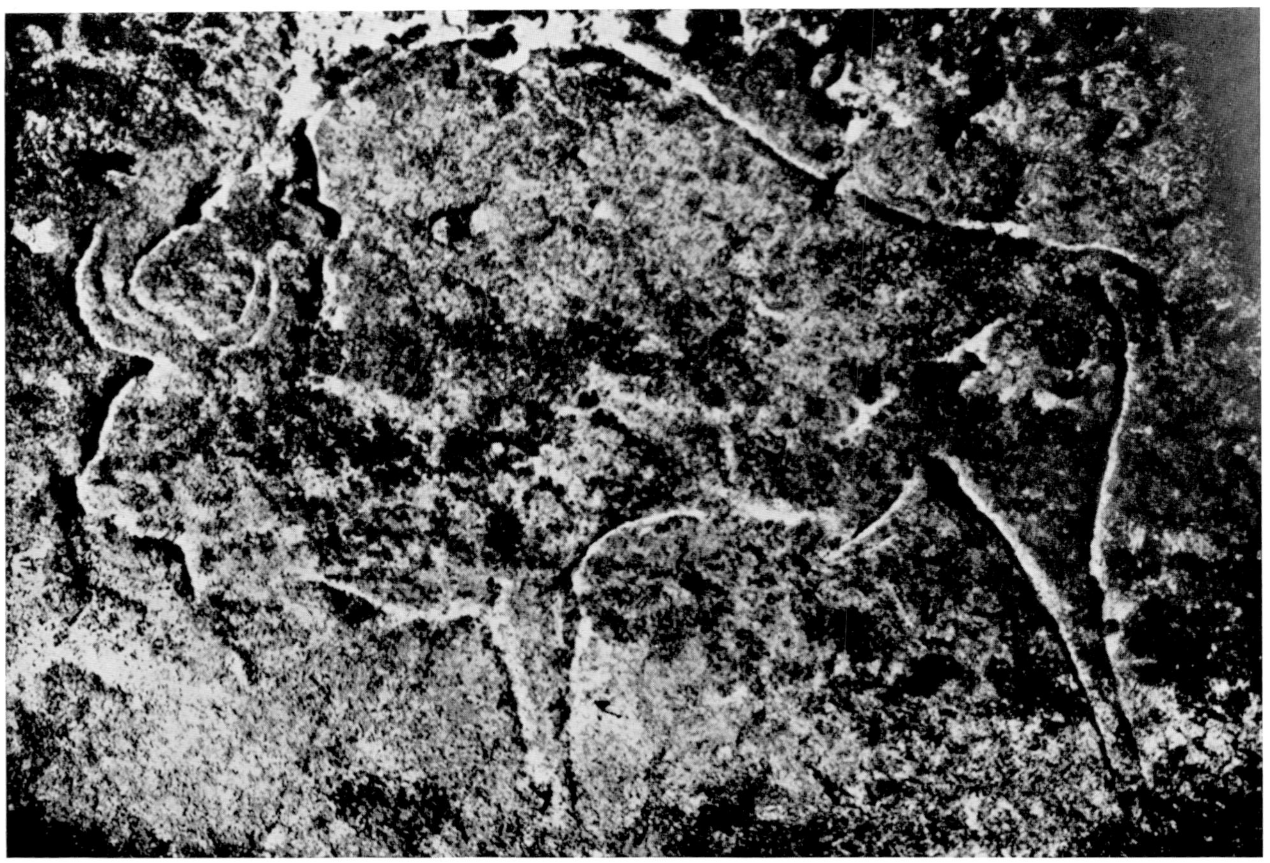

Pl. 260. *Above*: Cave of La Grèze, Dordogne, France, engraved bison. L., $23^{1}/_{2}$ in. *Below*: Bédeilhac Cave, near Tarascon-sur-Ariège, France, engraved horse's head. L., nostrils to poll, $43^{3}/_{8}$ in.

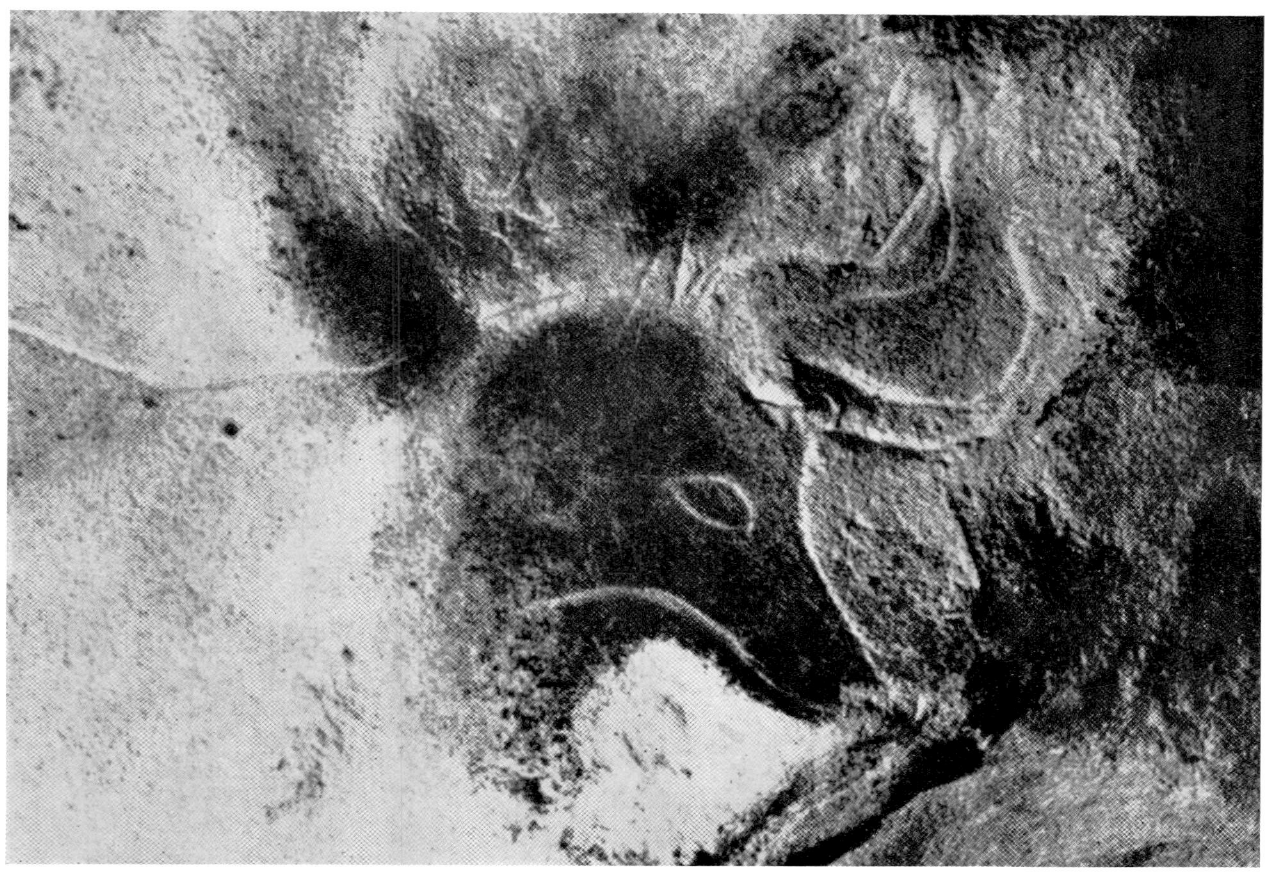

Pl. 261. Cave of Le Gabillou, Dordogne, France, engravings. *Above*: Hare. L. ca. 10 in. *Below*: Bull's head, engraved and painted. Ht., $8^{5}/_{8}$ in.

Pl. 262. Pech-Merle Cave, Lot, France, painting of horses and stenciled hands. W., 11 ft., 2 in.

Pl. 263. Lascaux Cave, Dordogne, France, painting of a bull in the main hall ("Hall of the Bulls"), superimposed in black over animals in red. W., ca. 18 ft.

Pl. 264. Lascaux Cave, Dordogne, France, painting of a wounded bison and fallen man; rhinoceros (at left) in different technique. L. of bison, 43 in.

Pl. 265. Lascaux Cave, Dordogne, France, painting of stag heads, called the frieze of "swimming deer," in the "Grande Nef." W., ca. 16½ ft.

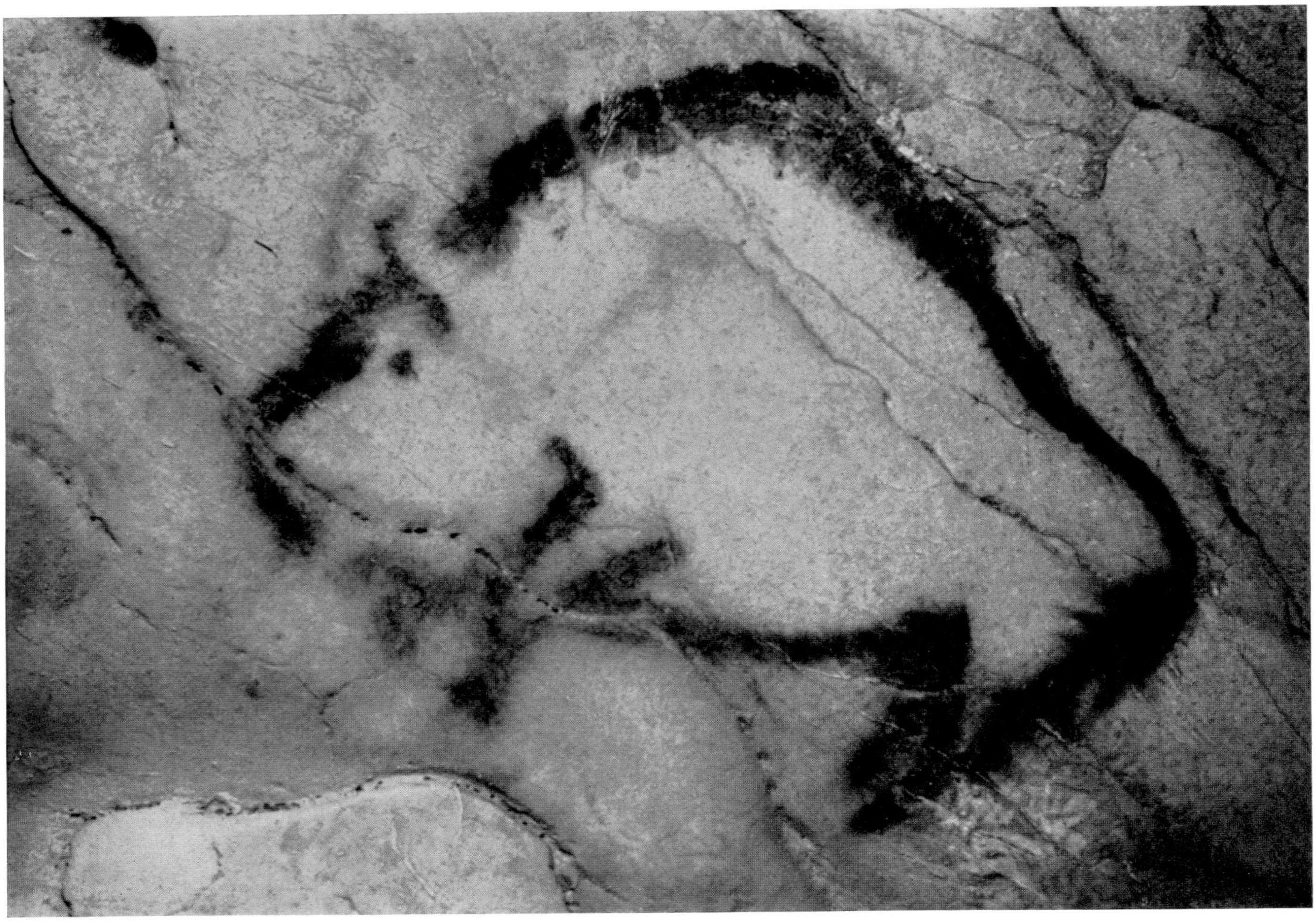

Pl. 266. Cave of Le Portel, Ariège, France. *Above*: Painted bison. L., 17 1/2 in. *Below*: Bison, painted with engraved outline. L., ca. 28 in.

Pl. 267. Lascaux Cave, Dordogne, France, painting of bull, detail with triple-branched sign, superposed on animal paintings in ocher. Full l., ca. 10 ft.

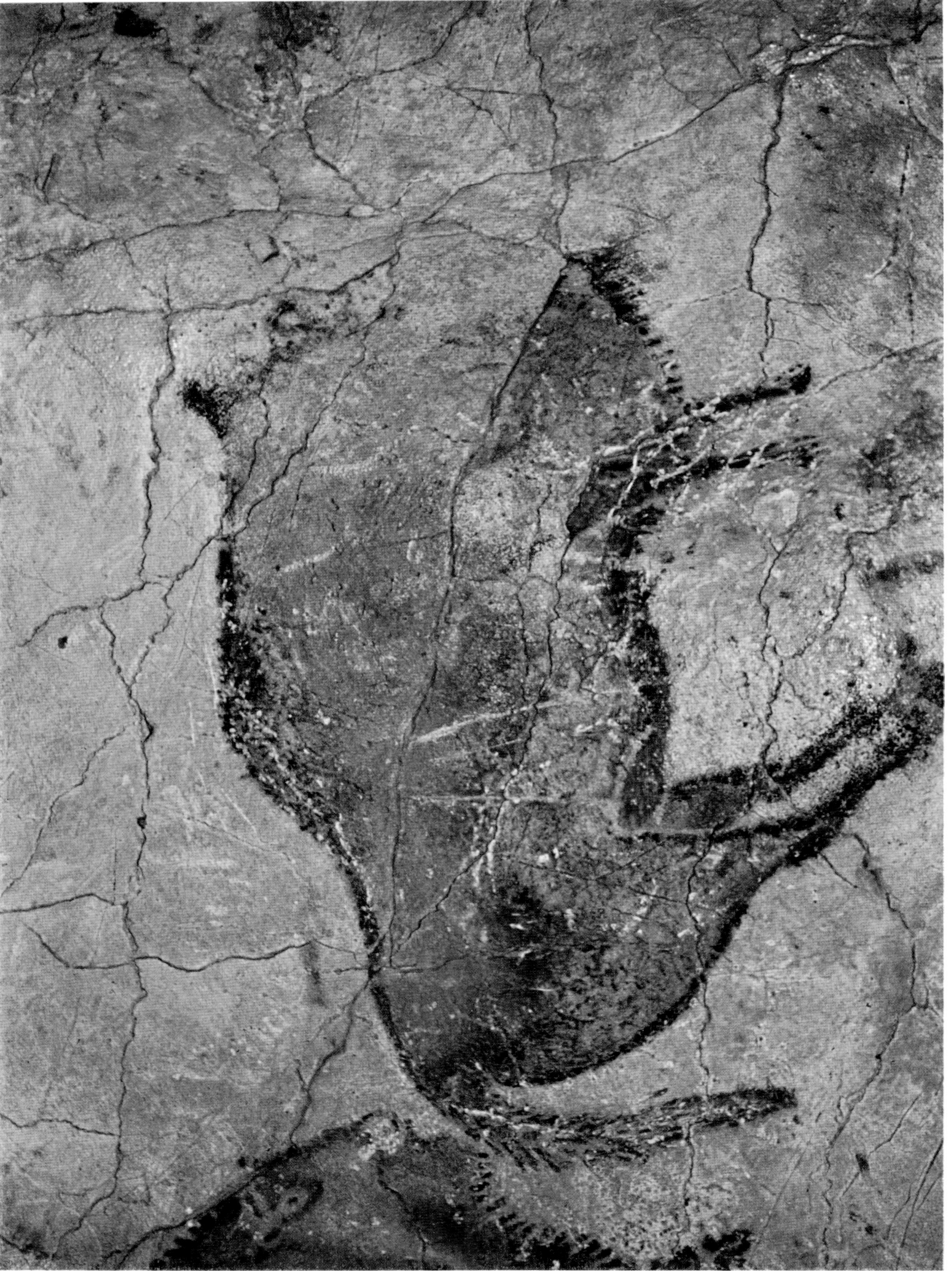

Pl. 268. Altamira Cave, Santander, Spain, polychrome painting of bison, Magdalenian period.

Pl. 269. Niaux Cave, Ariège, France, painting in the " Salon Noir " of a bison pierced by arrows, with an incomplete bison and ibex at right. W., ca. 40 in.

Pl. 270. Niaux Cave, Ariège, France, bison incised in clay, detail. Full l., 23 in.

Pl. 271. Niaux Cave, Ariège, France, animal paintings in the " Salon Noir." *Above*: L. of deer, 32 in. *Below*: L. of bison at right, 50 in.

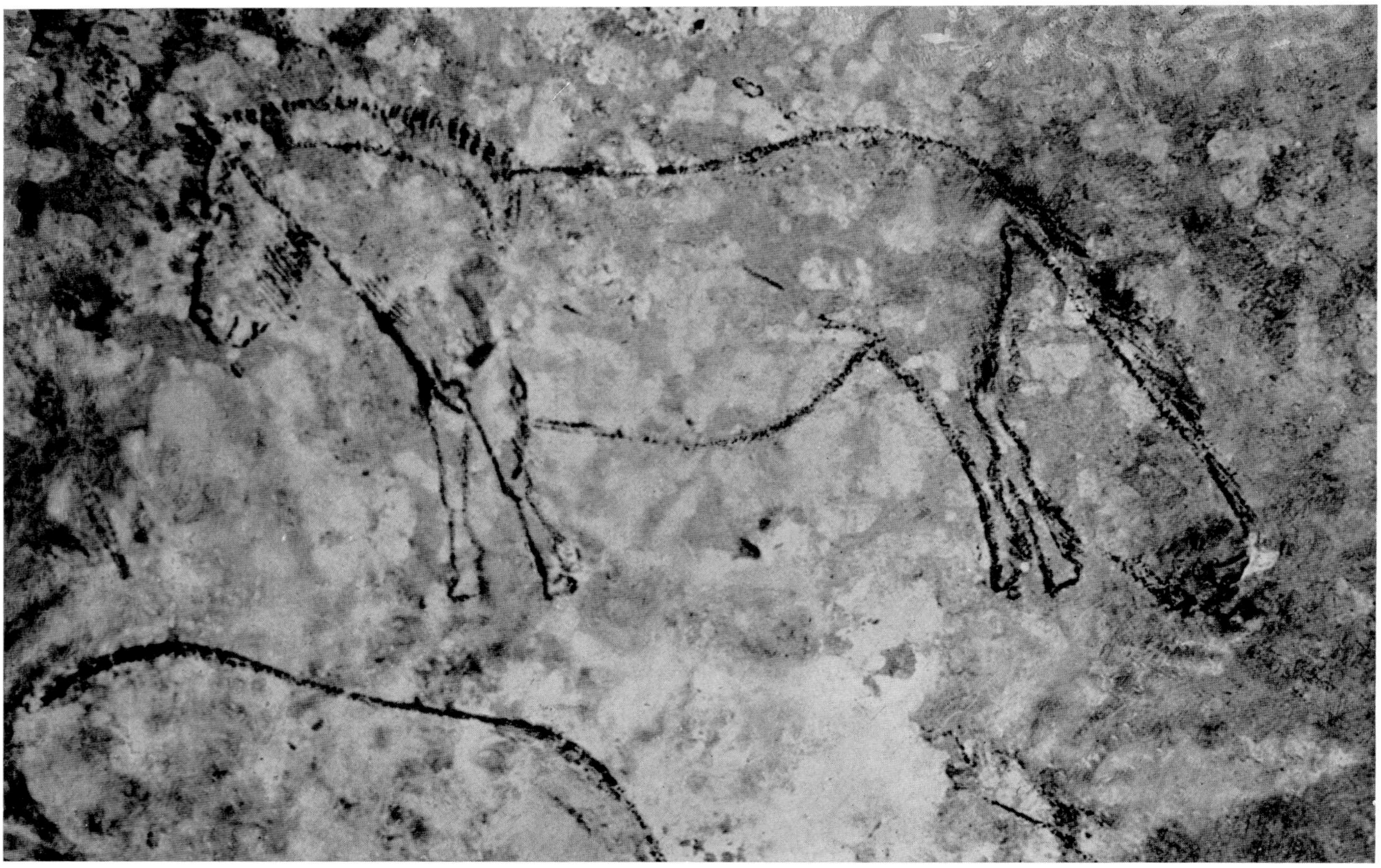

Pl. 272. Rouffignac Cave, Dordogne, France, paintings. *Above*: Detail of rhinoceros frieze in the "Galerie Breuil." *Below*: Horse at the center of the "Grand Plafond."

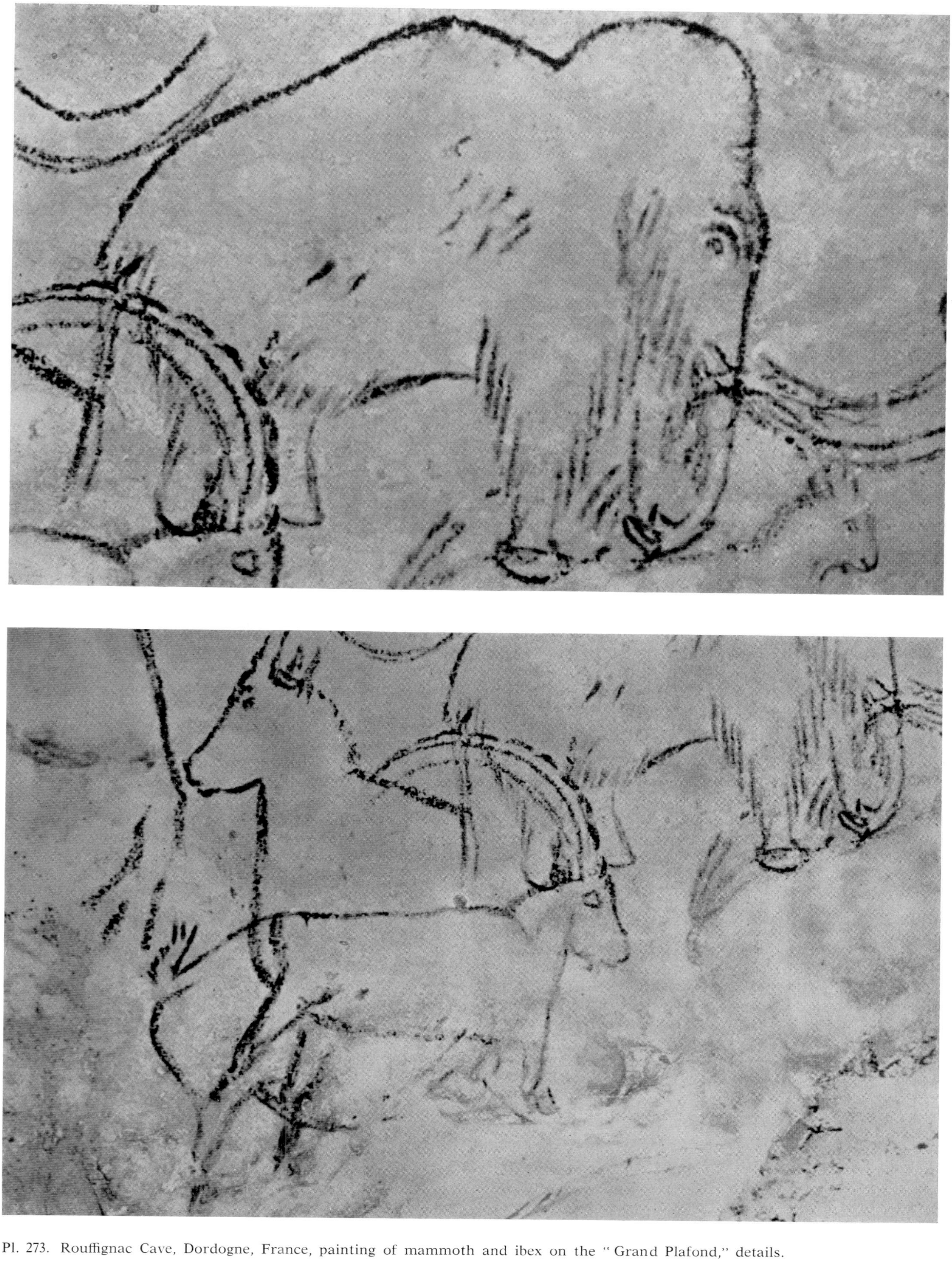

Pl. 273. Rouffignac Cave, Dordogne, France, painting of mammoth and ibex on the "Grand Plafond," details.

Pl. 274. Cave of El Castillo, Santander, Spain, painted tectiforms and dotted lines. Ht. of larger figure, ca. 25 in.

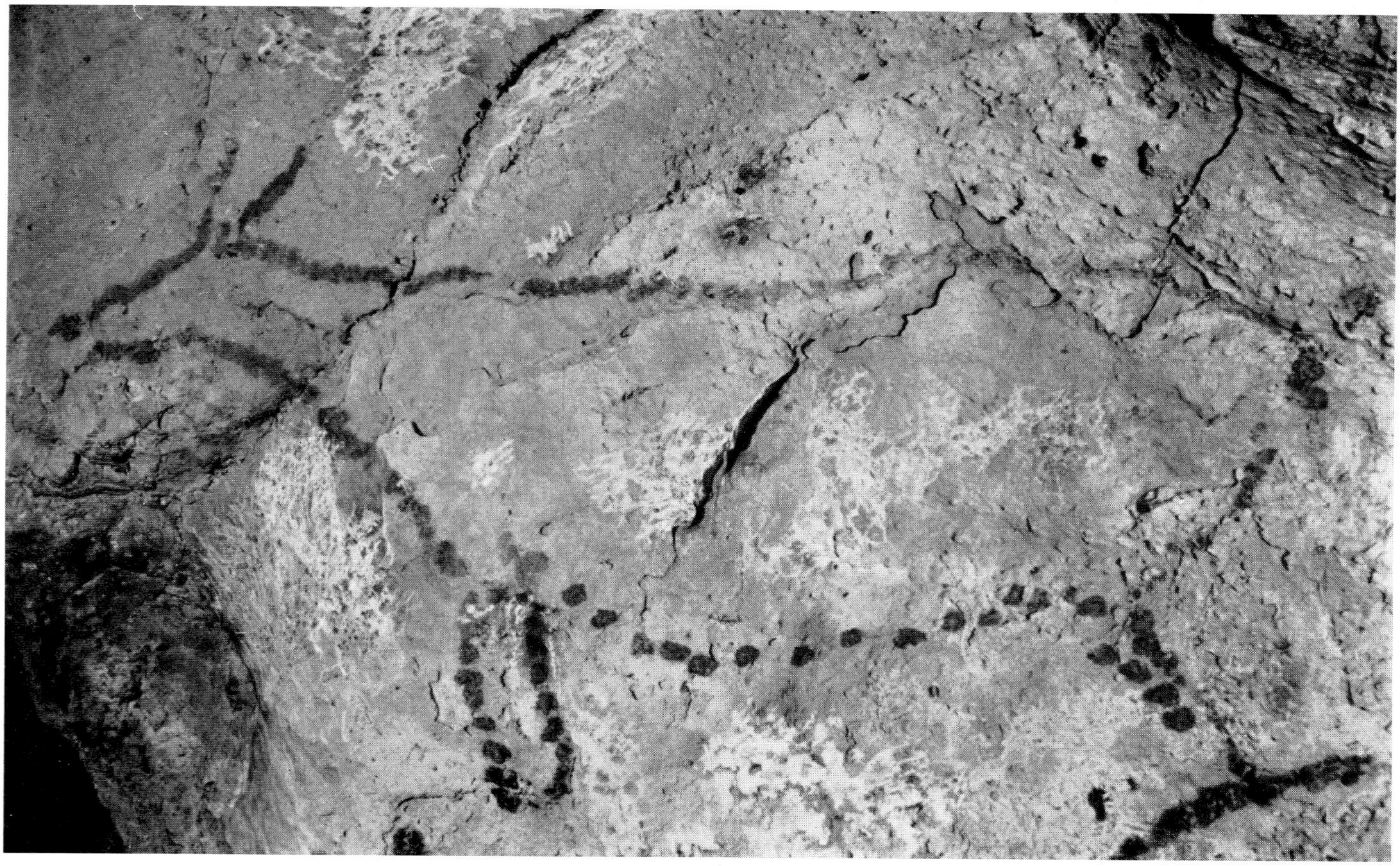

Pl. 275. *Above*: Cave of La Pasiega, Santander, Spain, painted horse. L., ca. 19 in. *Below*: Covalanas Cave, Santander, deer, combining continuous and dotted line. L., 31 1/2 in.

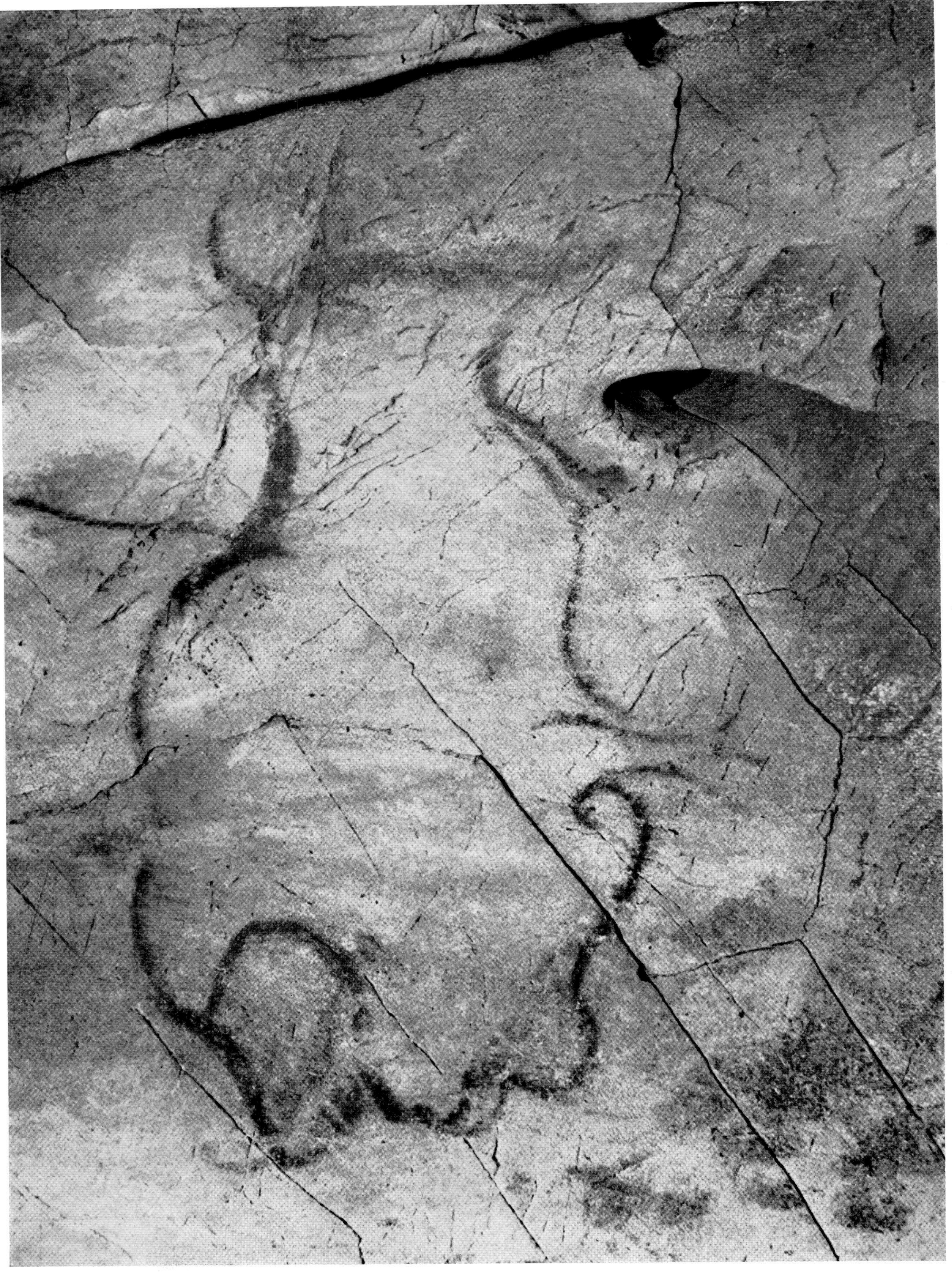

Pl. 276. Cave of La Pasiega, Santander, Spain, painted bison. L., 26 in.

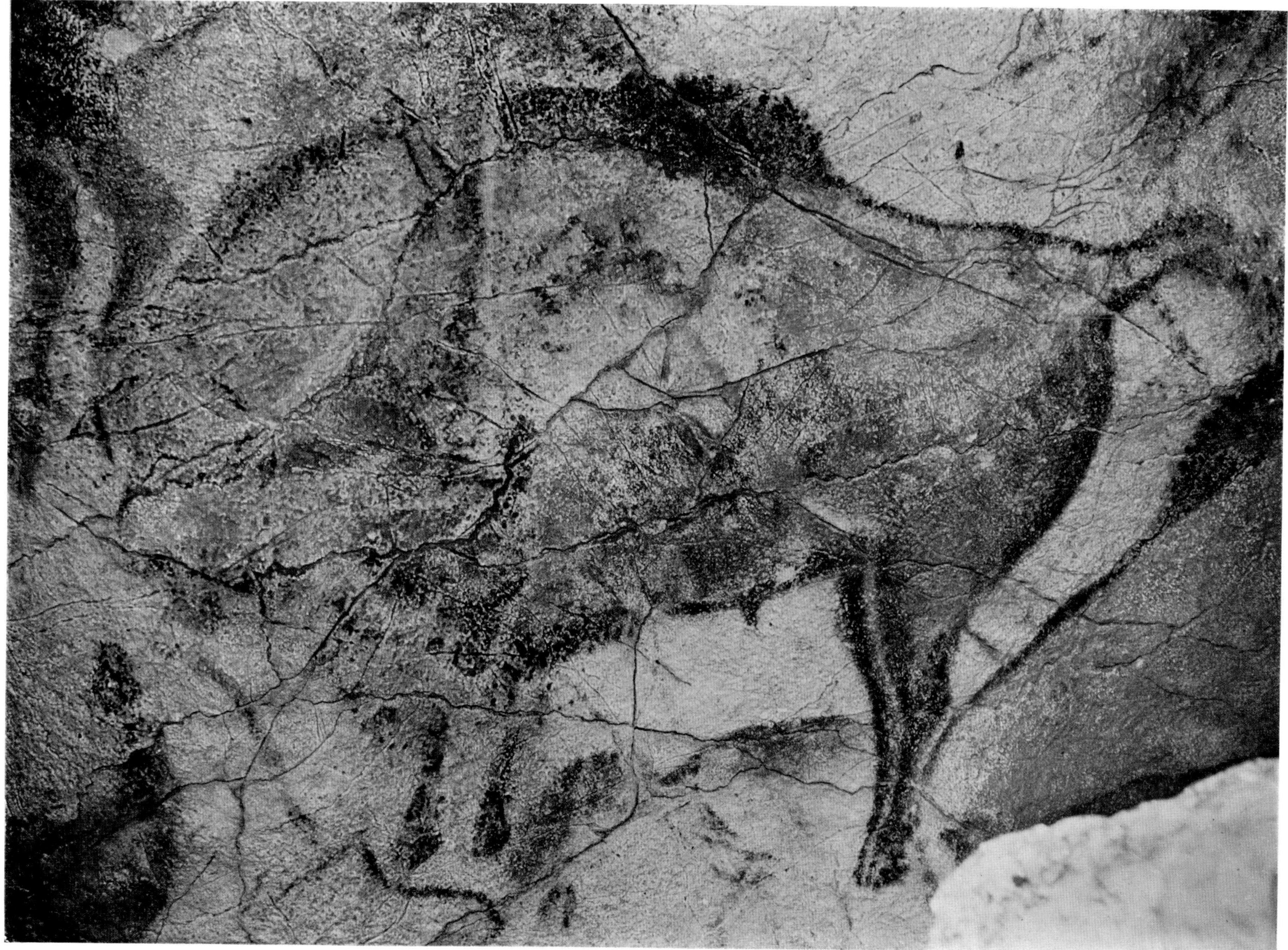

Pl. 277. Altamira Cave, Santander, Spain, polychrome painting of bison, Magdalenian period.

Pl. 278. Altamira Cave, Santander, Spain, polychrome painting of bison, Magdalenian period. L., ca. 6½ ft.

Pl. 279. Santimamiñe Cave, Vizcaya, Spain, painted bison. L., $27\frac{1}{2}$ in.

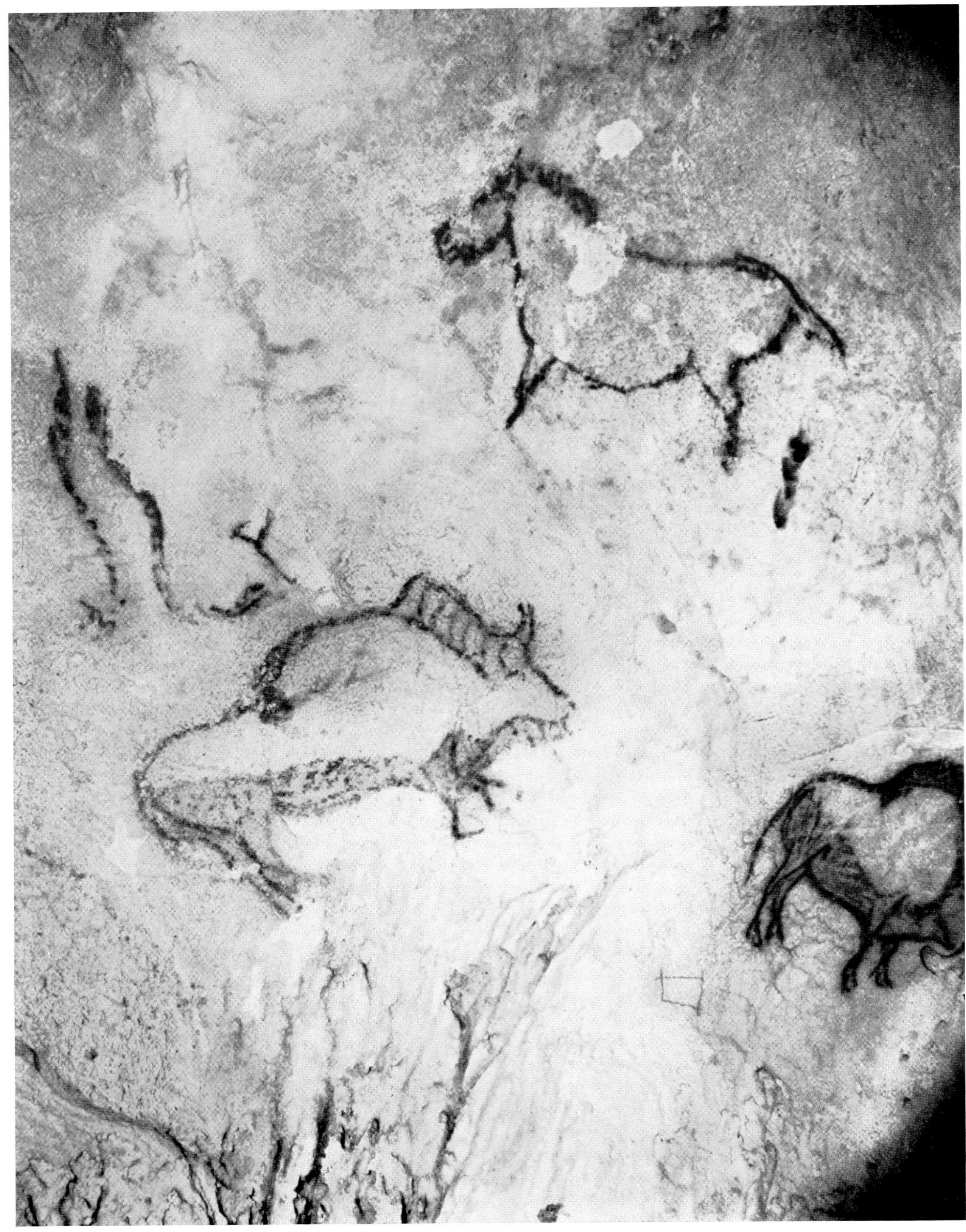

Pl. 280. Santimamiñe Cave, Vizcaya, Spain, painting with horse and bison. L. of horse, ca. 18 in.

Pl. 281. Altamira Cave, Santander, Spain, painting of a bovine head.

Pl. 282. Altamira Cave, Santander, Spain, detail of ceiling with painted bison.

Pl. 283. Cave of Cala Genovese, Levanzo, Egadi Islands, Sicily, engravings. *Left*: Deer. L., ca. 8 in. *Right*: Human figures. Ht. of central figure, ca. 12 in.

Pl. 284. Addaura Cave, Monte Pellegrino, near Palermo, Sicily, engraved human figures. Ht. of top figure, 9 1/2 in.

Pl. 285. Addaura Cave, Monte Pellegrino, near Palermo, Sicily, engraving with human and equine figures. Ht. of figure at top right, ca. 6 in.

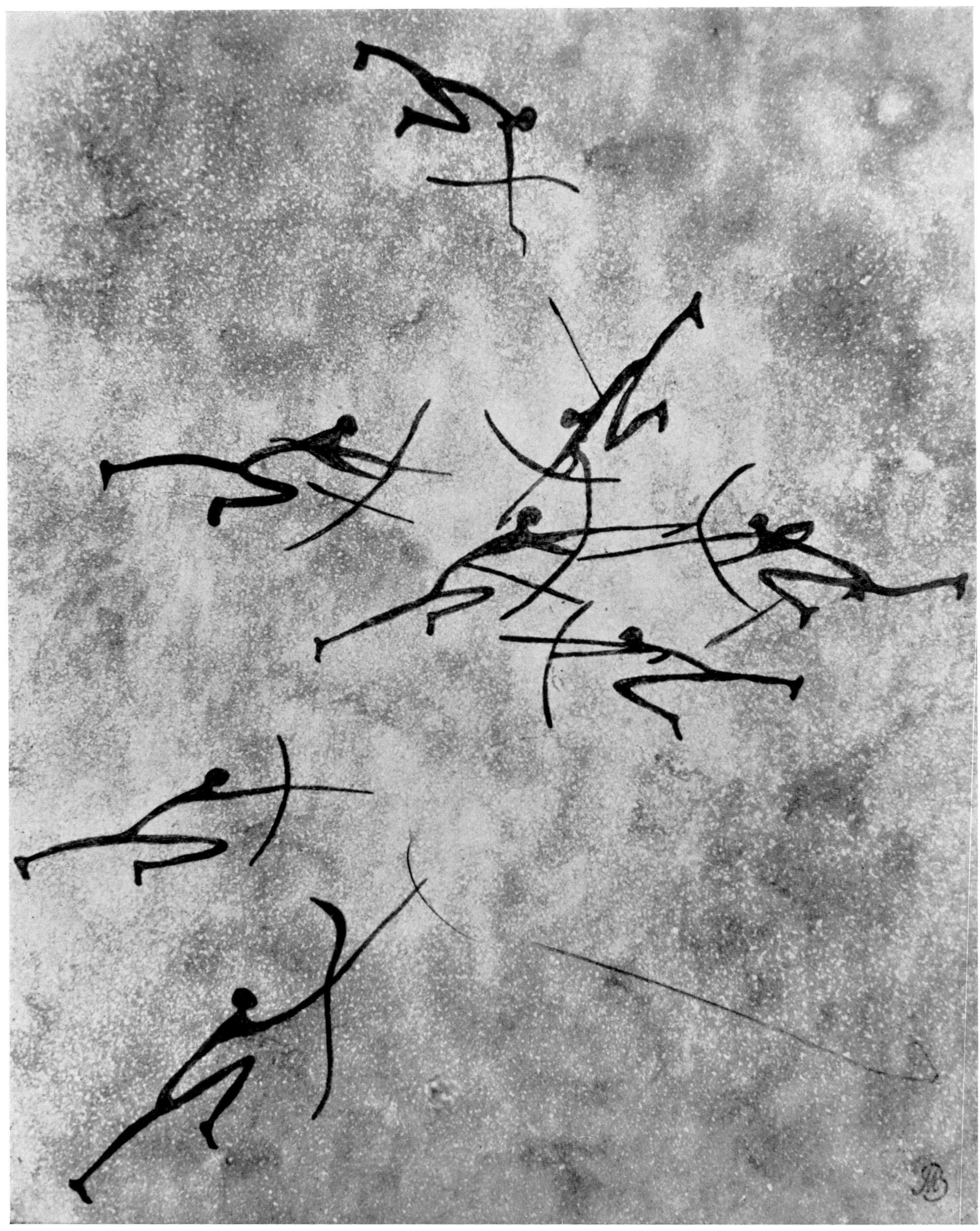

Pl. 286. Morella la Vieja, Castellón de la Plana, Spain, rock painting depicting archers in combat. Ht. of figures, ca. 4–6 in. (Watercolor copy.)

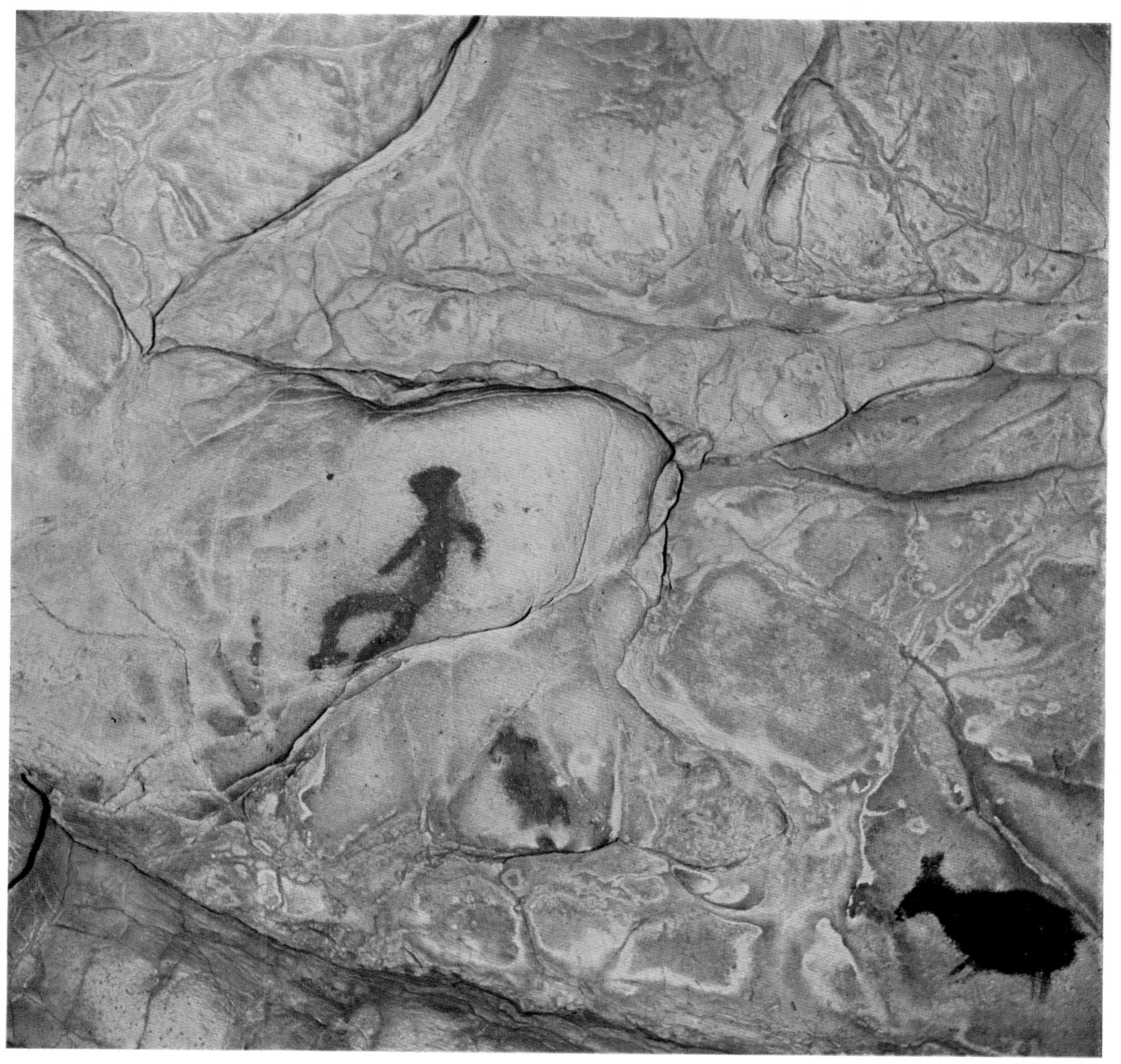

Pl. 287. Cave of Cala Genovese, Levanzo, Egadi Islands, Sicily, paintings of human and animal figures. Ht. of human figure, ca. 1 ft.

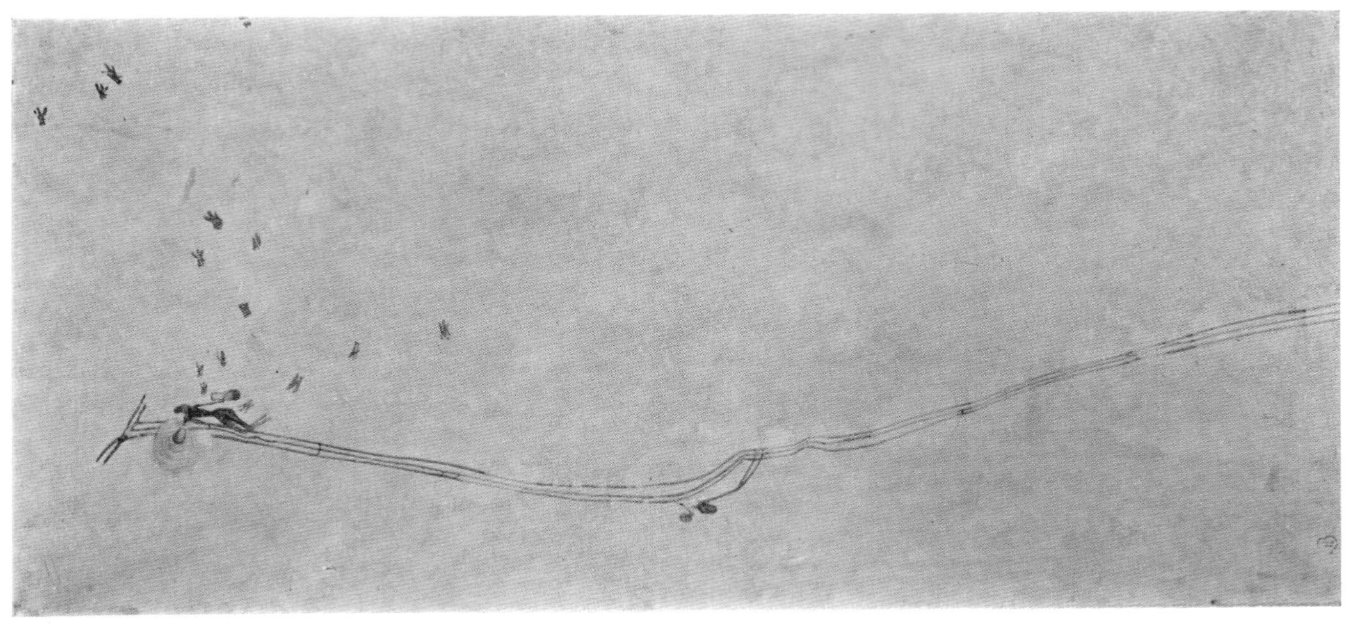

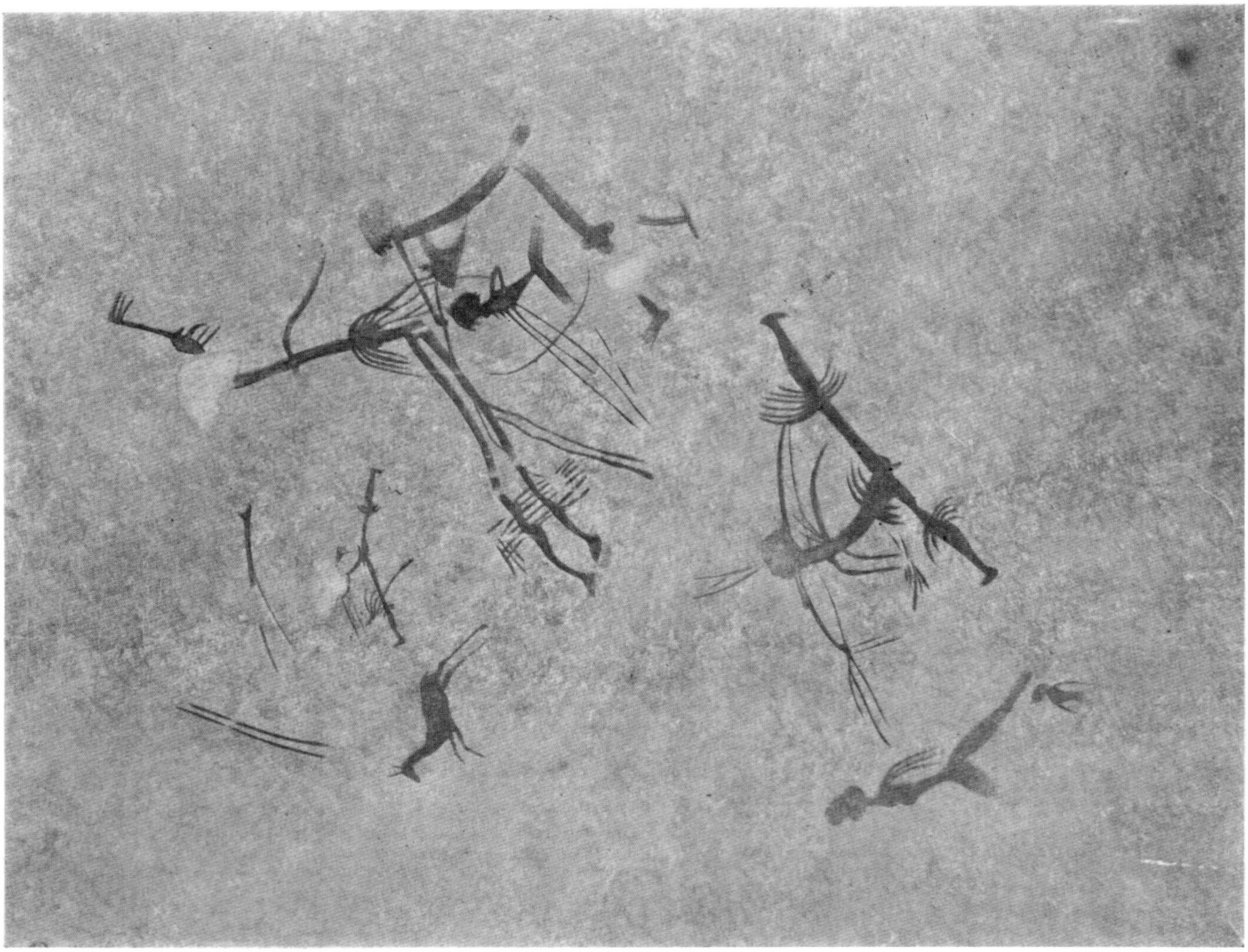

Pl. 288. Rock paintings of eastern Spain. *Left:* Cerro Felío, near Alacón, Teruel, hunting scene. *Right:* Cueva de la Araña, near Bicorp, Valencia, honey gatherers on ropes, surrounded by bees. (Both, water-color copies.)

Pl. 289. *Left*: Peixet, Perelló, Tarragona, Spain, rock painting with archer and animals. *Right*: Cueva de Doña Clotilde, near Albarracín, Teruel, Spain, rock painting, detail with a man leading an animal, stylized figures, and tree. (Water-color copies.)

Pl. 290. Norwegian petroglyphs, at Skjomen (*left*) and Leiknes, Nordland.

Pl. 291. Skjomen, Nordland, Norway, rock engravings.

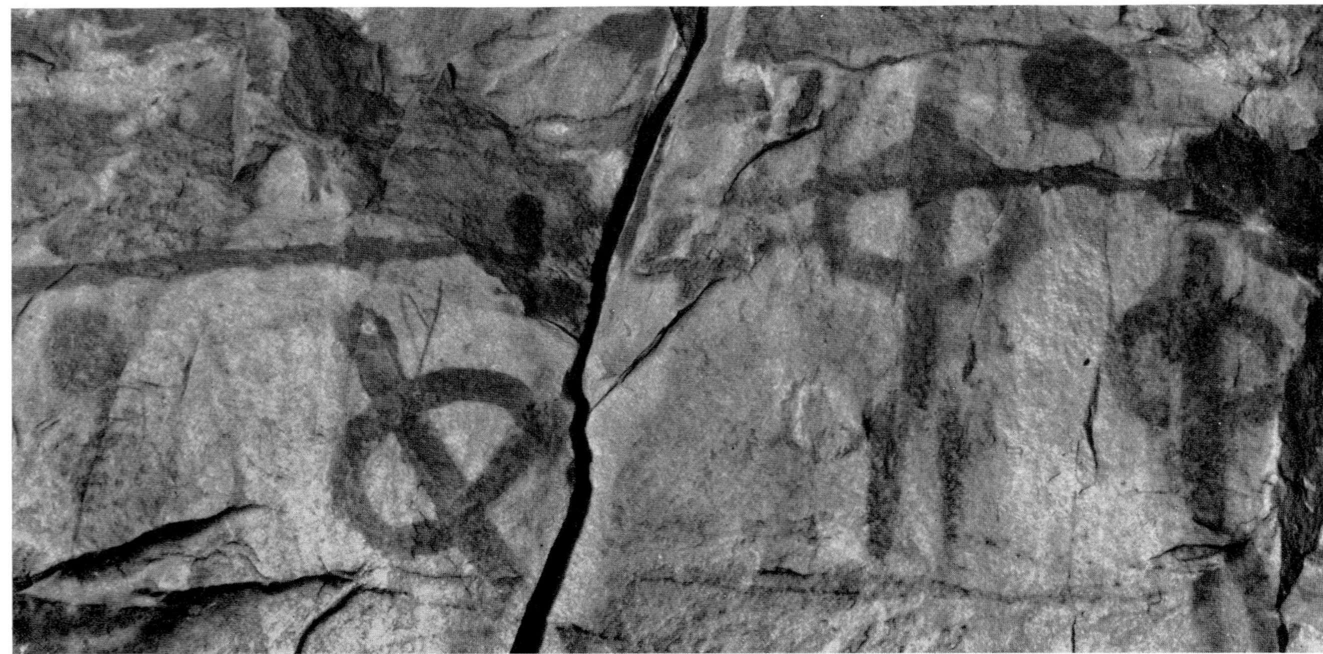

Pl. 292. *Above*: Tajo de las Figuras, near the Laguna de la Janda, Cádiz, Spain, rock painting of hunters and animals with superimposed schematic figures of later epochs. (Water-color reproduction.) *Below*: Detached rock painting with stylized human figures, from Aldeaquemada, Jaén, Spain. Madrid, Museo Nacional de Ciencias Naturales.

Pl. 293. Cave of Cala Genovese, Levanzo, Egadi Islands, Sicily, two details of paintings in the inner chamber.

Pl. 294. Alpine rock engravings. *Above*: Mont-Bégo, near Tende, Alpes-Maritimes, France. *Below*: Capo di Ponte, Valcamonica, Lombardy, Italy.

Pl. 295. North African rock engravings. *Above*: Sabagura, Nubia, Egypt. *Below, left*: In Habeter III*d*, Wadi Bergiug, Fezzan, Libya. *Right*: Matendusc, Wadi Bergiug, near Murzuq, Fezzan.

Pl. 296. Sollum Ba'āt, Acchelé Guzai, Eritrea, Ethiopia, rock painting of bovine animals.

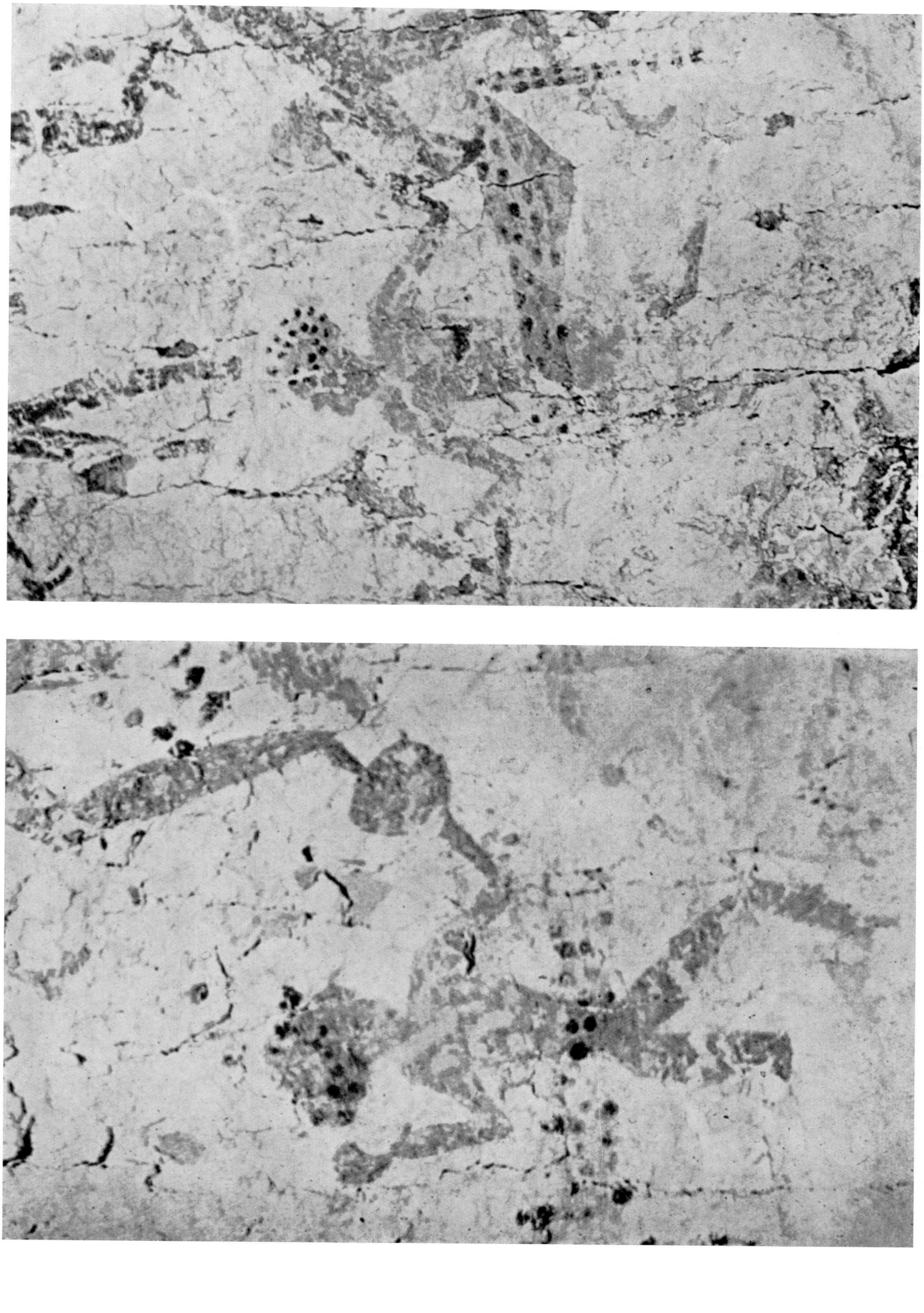

Pl. 297. Çatal Hüyük, Anatolia, Turkey, rock paintings of dancers, with costumes accentuated by black dots.

Pl. 298. *Above*: Kilwa, Jordan, rock engraving. *Below*: Adamgarh quarry, Hoshangabad, Madhya Pradesh, India, rock painting of archers.

Pl. 299. F. Overbeck, Odoardo and Gildippe, 1817–27. Fresco. Rome, Casino of the former Villa Giustiniani Massimo.

Pl. 300. P. Veit, L'Empireo, 1818–24. Fresco. Rome, Casino of the former Villa Giustiniani Massimo.

Pl. 301. D. G. Rossetti, Beata Beatrix, ca. 1863. Canvas, 34×26 in. London, Tate Gallery.

Pl. 302. J. E. Millais, Ophelia, 1852. Canvas, 30×44 in. London, Tate Gallery.

Pl. 303. *Left*: P. Tenerani, St. Benedict, 1845. Marble. Rome, S. Paolo fuori le Mura. *Right, above*: T. Minardi, Apparition of the Virgin, 1824. Canvas. Rome, S. Andrea al Quirinale. *Below*: F. Podesti, Proclamation of the Dogma of the Immaculate Conception, 1858. Fresco. Rome, Vatican, Sala dell'Immacolata.

Pl. 304. *Above*: J. E. Millais, Christ in the House of His Parents (The Carpenter's Shop), 1850. Canvas, $33\frac{1}{2} \times 54$ in. *Below, left*: W. H. Hunt, Claudio and Isabella, 1850. Panel, $30\frac{1}{2} \times 18$ in. Both, London, Tate Gallery. *Right*: W. H. Hunt, illustration for Tennyson's "Oriana" (Moxon ed., 1857). Woodcut.

Pl. 305. *Above*: P. Puvis de Chavannes, The Sacred Grove, 1884. Canvas, 3 ft., $^{1}/_{2}$ in. × 6 ft., 11 in. Chicago, Art Institute. *Below*: E. Burne-Jones, The Garden of Pan. Canvas, 4 ft., 11$^{1}/_{2}$ in. × 6 ft., 1$^{1}/_{2}$ in. Melbourne, National Gallery of Victoria.

Pl. 306. E. Burne-Jones, King Cophetua and the Beggar Maid, 1884. Canvas, 9 ft., 7½ in. × 4 ft., 5½ in. London, Tate Gallery.

Pl. 307. France. *Above, left*: Fréjus, Var, Baptistery, interior, late 4th–early 5th cent. *Right*: Jouarre, Seine-et-Marne, Abbey, crypt, 7th cent. *Below*: Vienne, Isère, St-Pierre (now Musée Lapidaire Romain), interior, 9th–10th cent.

Pl. 308. France. *Above, left*: Saint-Martin-du-Canigou, Pyrénées-Orientales, Abbey church, nave, 1001–26. *Right*: Tournus, Saône-et-Loire, St-Philibert, nave, 11th cent. *Below*: Dijon, Côte d'Or, St-Bénigne, crypt of rotunda, 1001–18 (rebuilt 1858).

Pl. 309. France. Vignory, Haute-Marne, Priory church, nave, early 11th cent.

Pl. 310. France. *Above*: Monthou-sur-Cher, Loir-et-Cher, church, detail of *chevet*, 11th cent. *Below, left*: Orléans, Loiret, St-Aignan, capital in crypt, 989–1029. *Right*: Maillezais, Vendée, Abbey church, capital on north wall, 11th cent.

Pl. 311. France. *Above*: Tympanum with interlace, 11th cent. Stone. Bessuéjouls, Aveyron, church. *Left, center*: Sarcophagus, 5th cent. Marble. Narbonne, Aude, *in situ*. *Below*: Lintel, with Christ in Majesty, angels, and apostles, 1020–21. Marble. Saint-Genis-des-Fontaines, Pyrénées-Orientales, church. *Right*: Crucifixion, 9th cent. Fresco (facsimile from Paris, Musée National des Monuments Français). Saint-Pierre-les-Églises, Vienne, church.

Pl. 312. Germany. *Above*: Regensburg, St. Emmeram, western crypt, 1052 (restored). *Below, left*: Essen, Münster, western choir, ca. 1000. *Right*: Gernrode, St. Cyriakus (founded 961), nave.

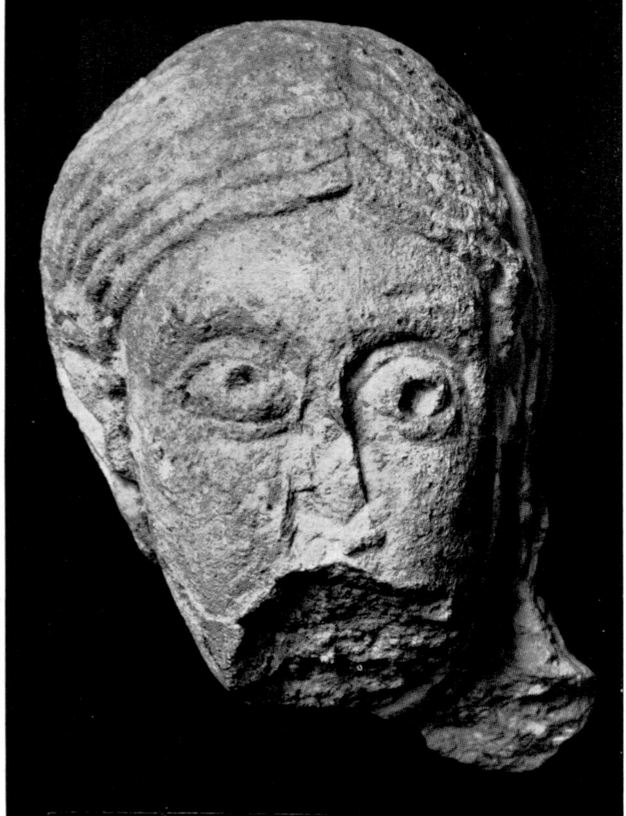

Pl. 313. *Above, left*: Reliquary found near Tiel, Netherlands, 8th cent. Gilded copper with stones and enamel, l., 2 1/2 in. Utrecht, Netherlands, Aartsbisschoppelijk Museum. *Right*: Enger purse reliquary, Germany, late 8th cent. Gold and silver repoussé on wood with enamel. Berlin, Staatliche Museen. *Below, left*: Covers of the so-called "Psalter of Dagulf," Lower Rhine, late 8th cent. Ivory, ht., 6 1/2 in. Paris, Louvre. *Right*: Head, from the *Westwerk* of St. Pantaleon, Cologne, 10th cent. Stone, ht., 16 1/2 in. Cologne, Diözesan-Museum.

Pl. 314. Germany. Illuminations. *Left*: Gospel book. The Four Evangelists, New Palace School, early 9th cent. Aachen, Cathedral Treasury. *Right*: Book of Pericopes of Henry II. The Marys at the Sepulcher, Reichenau, early 11th cent. Munich, Bayerische Staatsbibliothek (Clm. 4452, fol. 116v).

Pl. 315. England. *Left:* Beresford-Hope cross, 8th or early 9th cent. Enamel. London, Victoria and Albert Museum. *Right:* Arundel Psalter. Crucifixion, Winchester, ca. 1060. Illumination. London, British Museum (Arundel Ms. 60, fol. 52v).

Pl. 316. Spain. *Above and below, left*: San Pedro de Nave, Zamora, church, 7th cent. (reconstructed). *Below, right*: Baños de Cerrato, Palencia, S. Juan Bautista, 661.

Pl. 317. Spain. Pola de Lena, Oviedo, S. Cristina, 842–50.

Pl. 318. Spain. *Above*: Oviedo, S. Julián de los Prados ("Santullano"), nave, 812–42. *Below*: Naranco, Oviedo, S. Miguel de Lillo, 842–50.

Pl. 319. England. New Minster Charter. King Edgar offering the Charter to Christ, 966. Illumination, 8 1/8 × 6 3/8 in. London, British Museum (Cotton Vespasian A.VIII, fol. 2v).

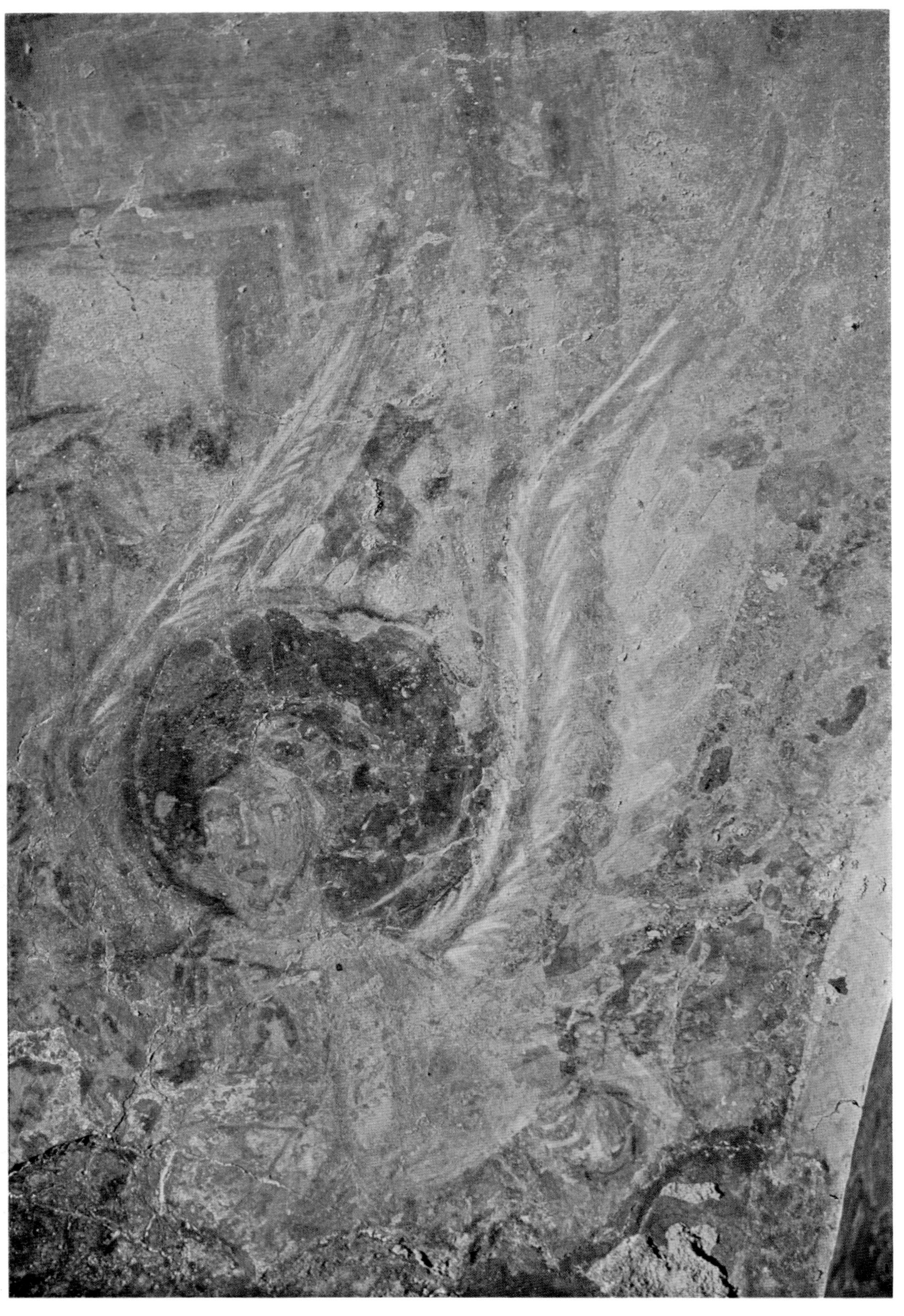

Pl. 320. Italy. Detail of an angel, 826–43. Fresco. Abbey of San Vincenzo al Volturno, Abruzzi, crypt of S. Lorenzo.

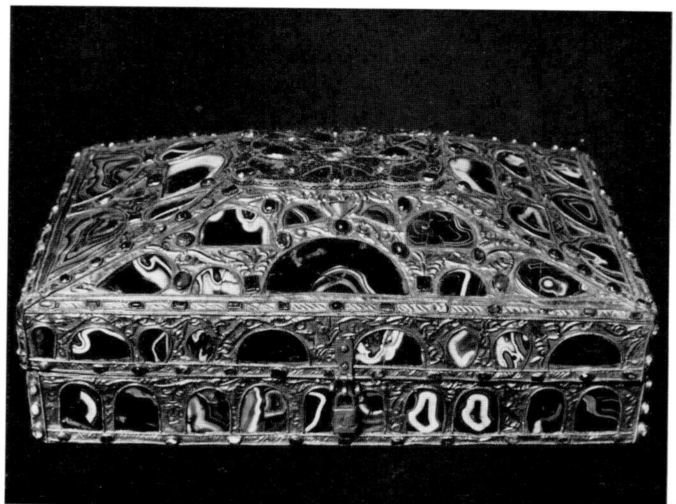

Pl. 321. Spain. *Above*: Capitals, 7th cent. Stone. San Pedro de Nave, Zamora, church. *Left*: Daniel in the Lions' Den. *Right*: Sacrifice of Abraham. *Center*: Impost with relief representing the sun, 7th cent. Stone. Quintanilla de las Viñas, Burgos, S. María. *Below, left*: Caja de las Reliquias, 9th cent. Gilded silver with repoussé on wood. Astorga, León, Cathedral. *Right*: Caja de las Ágatas, 910. Gold with repoussé, agate, and enamel. Oviedo, Cathedral.

Pl. 322. Spain. Bible. Scenes from the life of the Virgin, 920. Illumination. León, Cathedral Archives (Cod. 6).

Pl. 323. *Above, left*: Brescia, Italy, S. Salvatore, nave, second half of 8th cent. (photographed during restoration). *Right*: Zadar (formerly Zara), Yugoslavia, St. Donat, probably early 9th cent. (doc. 949). *Below, left*: Stilo, Calabria, Italy, La Cattolica, 10th cent. (?). *Right*: Galliano, Lombardy, Italy, Baptistery, early 11th cent.

Pl. 324. Italy. *Left, above*: Castelseprio, Lombardy, S. Maria, 7th–8th cent. *Below*: Galliano, Lombardy, S. Vincenzo, early 11th cent. *Right*: Cividale del Friuli, S. Maria in Valle ("Tempietto"), interior with stucco frieze of female figures, of disputed date.

Pl. 325. Italy. *Left*: Ciborium, early 9th cent. Marble. Ravenna, S. Apollinare in Classe. *Right*: Ciborium, of disputed date. Gilded stucco, on porphyry columns. Milan, S. Ambrogio.

Pl. 326. Italy. *Left*: Diptych, from Rambona, late 9th cent. Ivory, $12^{1}/_{4} \times 10^{3}/_{4}$ in. Rome, Vatican Museums. *Right, above*: Pluteus, 9th cent. Marble. Rome, S. Sabina. *Below*: Christ in Majesty with angels, detail of the Altar of Ratchis, 744–49. Marble. Cividale del Friuli, Cathedral. (Cf. VII, PL. 373.)

Pl. 327. Italy. *Left*: Crucifixion, 8th cent. Fresco. Rome, S. Maria Antiqua. *Right*: Pope Paschal I, St. Agatha, and St. Paul, 817–24. Mosaic. Rome, S. Cecilia in Trastevere, apsc.

Pl. 328. Sweden. *Above, left*: Crucifix, found in a 10th cent. grave at Birka, Uppland. Silver with filigree and granulation, ht., 1 3/8 in. *Right*: Crucifix encolpion, from Gåtebo, Bredsätra, Öland, ca. 1100 or earlier. Silver with design in niello, ht., 3 3/8 in. *Below, left*: Bowl, from Älvkarleby, Uppland, 12th cent. Silver repoussé, diam., 5 3/4 in. *Right*: Sarcophagus, from Eskilstuna, Södermanland, mid-11th cent. Limestone, ht., 7 ft., 3 3/8 in. (Reconstructed.) All, Stockholm, Statens Historiska Museum.

Pl. 329. Ex-voto, Italy, 16th cent. Panel. Madonna dell'Arco, near Naples, sanctuary.

Pl. 330 H. Rousseau, The Snake Charmer, 1907. Canvas, 5 ft., $6^{1}/_{2}$ in. × 6 ft., $2^{1}/_{2}$ in. Paris, Louvre.

Pl. 331. Examples of prehistoric, folk, and child art. *Left, above*: Two men carrying an animal, rock engraving in the Valcamonica, Lombardy, Italy. *Below*: J. J. Heuscher, An Inn, ca. 1865, Appenzell canton, Switzerland. Water color on paper with crayon. Basel, Schweizerisches Museum für Volkskunde. *Right*: Children's drawings, Italy.

Pl. 332. Assimilation of primitive forms. *Left, above*: A. Modigliani, Head of a Woman, ca. 1912. Stone, ht., 22 7/8 in. Paris, Musée d'Art Moderne. *Below*: C. Brancusi, The Kiss, 1908. Stone, ht., 28 in. Philadelphia, Museum of Art. *Right*: Examples of primitive, medieval, and prehistoric art. *Above*: Dan mask, Liberia. Wood. Paris, Musée de l'Homme. *Center*: Bracket, 12th cent. Stone. Montoire-sur-le-Loir, France, Chapel of St-Gilles. *Below*: So-called "stele-menhir," Lunigiana area, Italy. Stone, ht., ca. 39 3/8 in. La Spezia, Italy, Museo Archeologico Lunense.

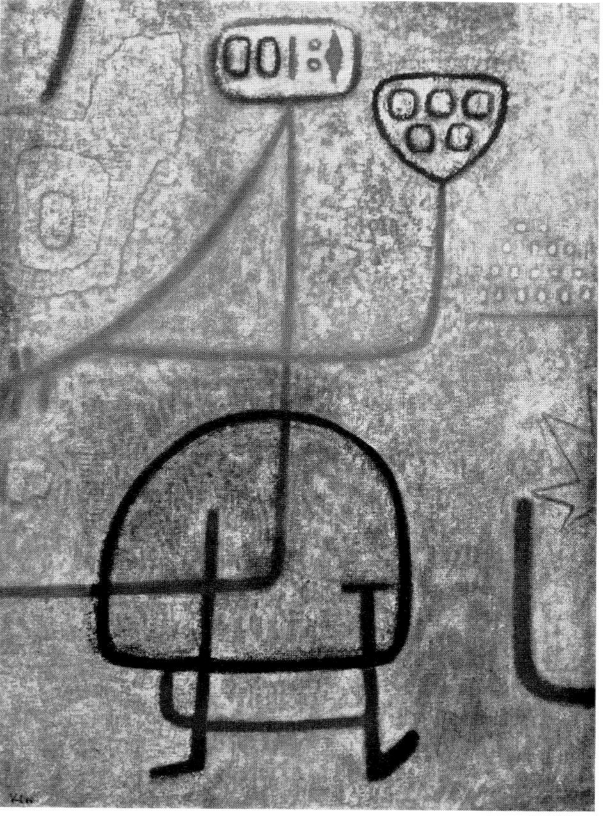

Pl. 333. Adoption of an ingenuous style. *Above*: J. Miró, Daphnis and Chloë, 1933. Etching, $10^{1}/_{4} \times 12^{1}/_{2}$ in. *Below*: P. Klee. *Left*: Fruit, 1938. Gouache, $14 \times 10^{5}/_{8}$ in. *Right*: La Belle Jardinière, 1939. Oil and tempera on canvas, $37^{3}/_{8} \times 27^{5}/_{8}$ in. Last two, Bern, Kunstmuseum, Paul-Klee-Stiftung.

Pl. 334. *Above, left*: W. Chandler, Reverend Ebenezer Devotion, 1770. Canvas, $55 \times 43\frac{3}{4}$ in. Brookline, Mass., Historical Society. *Right*: J. W. Stock, William Howard Smith, 1838. Canvas, $52\frac{1}{4} \times 37\frac{7}{8}$ in. Williamsburg, Va., Abby Aldrich Rockefeller Folk Art Coll. *Below*: E. R. Elmer, Mourning Picture: The Artist, His Wife, and Deceased Daughter, 1889. Canvas, 28×36 in. Northampton, Mass., Smith College Museum of Art.

Pl. 335. *Above*: H. Rousseau, War, 1894. Canvas, 3 ft., 9 in. × 6 ft., $4^3/_4$ in. Paris, Louvre. *Below*: A. Bauchant, Cleopatra's Barge, 1939. Canvas, $32 \times 39^3/_8$ in. New York, Museum of Modern Art.

Pl. 336. *Above, left*: T. A. Hoyer, Inside a Barn, 1937. Canvas, 30 1/8×24 1/8 in. New York, Museum of Modern Art (loan from U.S. WPA Art Program). *Right*: M. Hirshfield, Girl in a Mirror, 1940. Canvas, 40 1/8×22 1/4 in. New York, Museum of Modern Art. *Below*: J. Becker, The Snowsheds on the Central Pacific Railroad in the Sierra Nevada, 19th cent. Canvas, 18 1/2×25 1/2 in. Tulsa, Okla., Thomas Gilcrease Institute of American History and Art.

Pl. 337. *Above*: I. Večenaj, The Tillers, 1960. Canvas. Zagreb, Yugoslavia, Gallery of Primitive Art. *Below*: O. Metelli, Morning Performance. Canvas. Private coll.

Pl. 338. *Above*: M. Urteaga, Burial of an Illustrious Man, 1936. Canvas, $23\times32^{1}/_{2}$ in. New York, Museum of Modern Art. *Below*: H. Pippin, John Brown Going to His Hanging, 1942. Canvas, $24\times29^{7}/_{8}$ in. Philadelphia, Pennsylvania Academy of the Fine Arts.

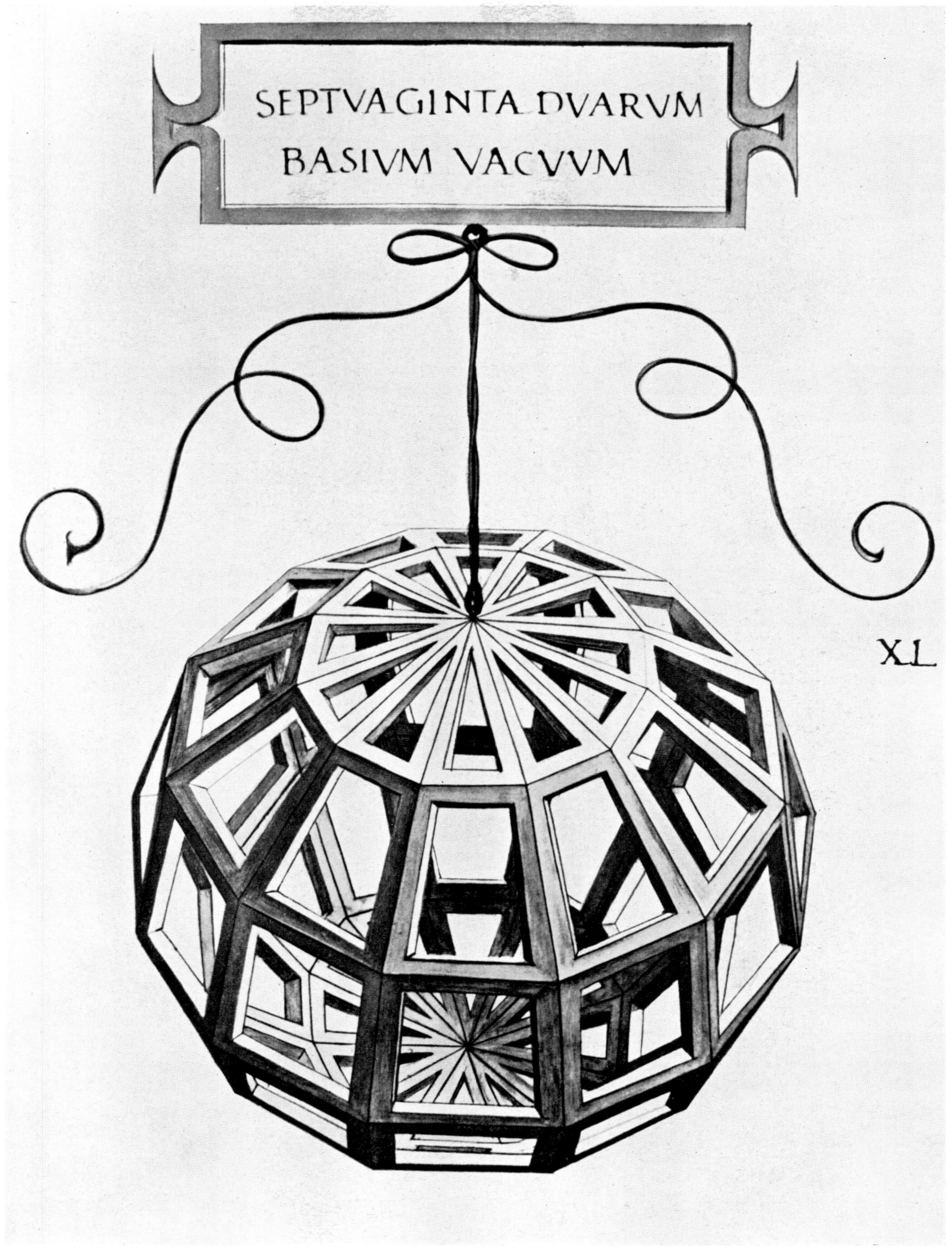

Pl. 339. A 72-faceted geometrical figure, from L. Pacioli, *De divina proportione*, Venice, 1509.

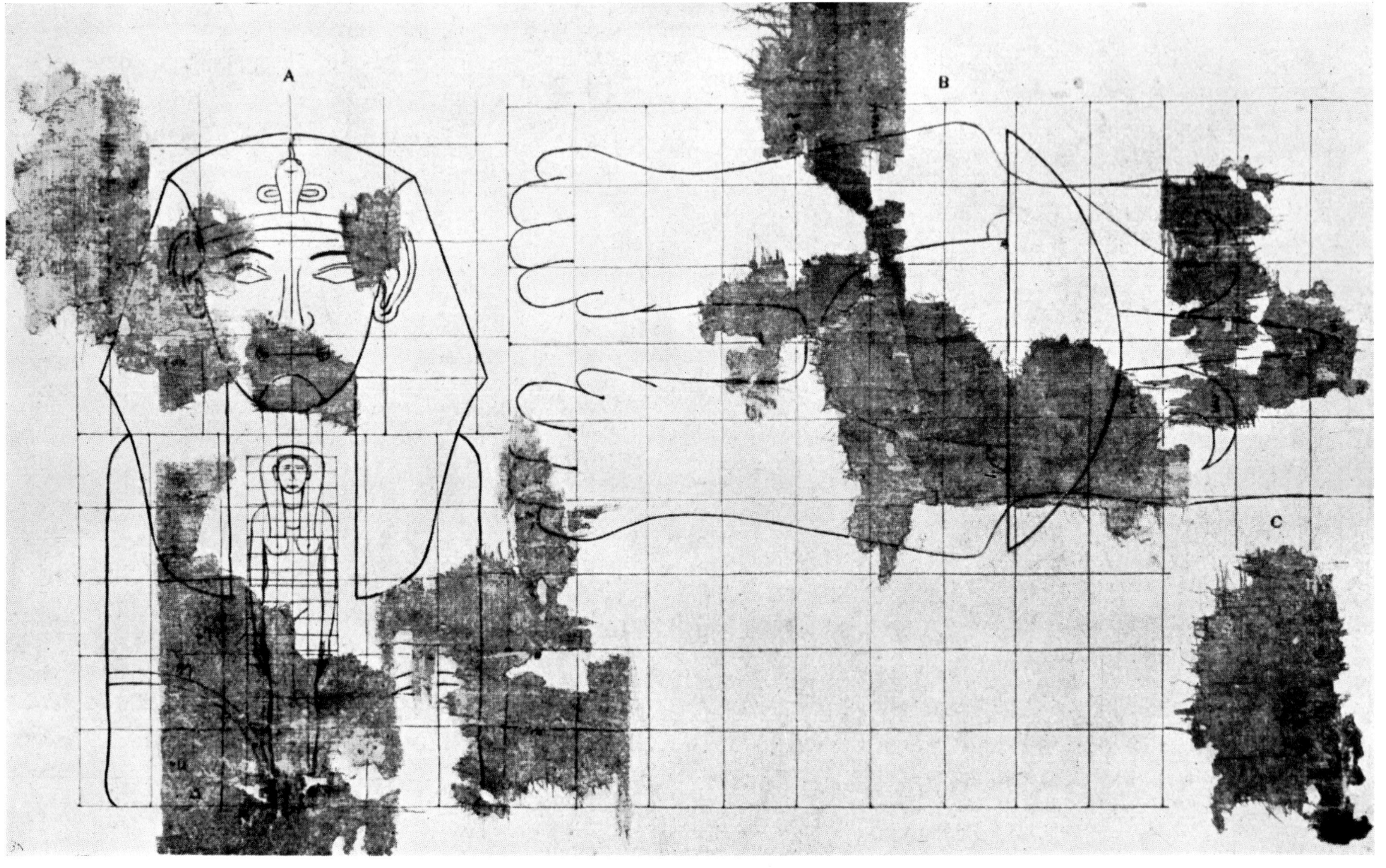

Pl. 340. Egyptian sculptor's working drawing for the execution of a sphinx, showing (A) frontal elevation, (B) ground plan, and (C) fragment of profile elevation, from Mellawi, 3d–2d cent. B.C. Papyrus, 13×21 5/8 in. Berlin, Staatliche Museen.

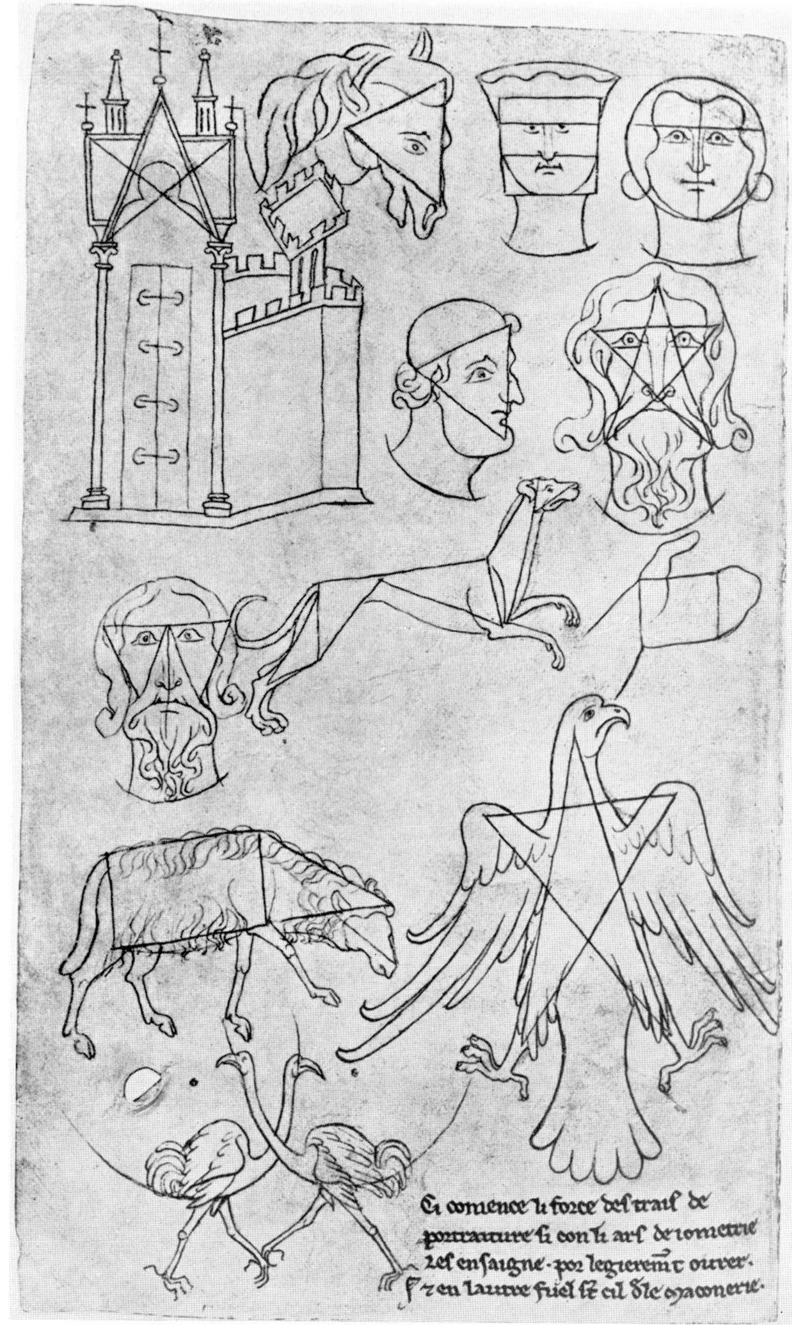

Pl. 341. Villard de Honnecourt, sketches illustrating the use of lines and geometrical figures to facilitate drawing, first half of 13th cent. Pen on parchment. Paris, Bibliothèque Nationale (Ms. fr. 19093).

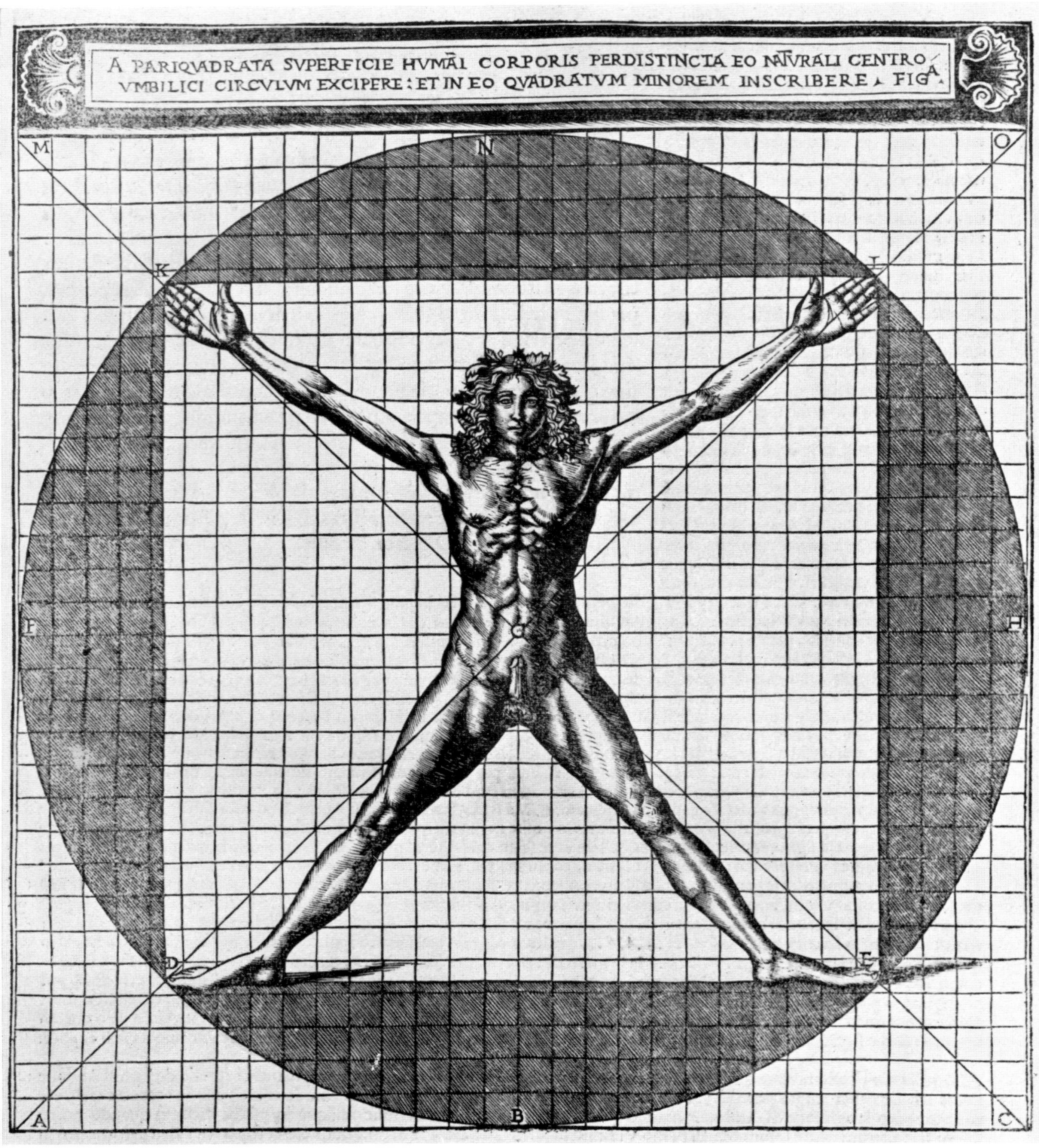

Pl. 342. The human figure inscribed in a square and a circle, from C. Cesariano's edition of Vitruvius, Como, 1521.

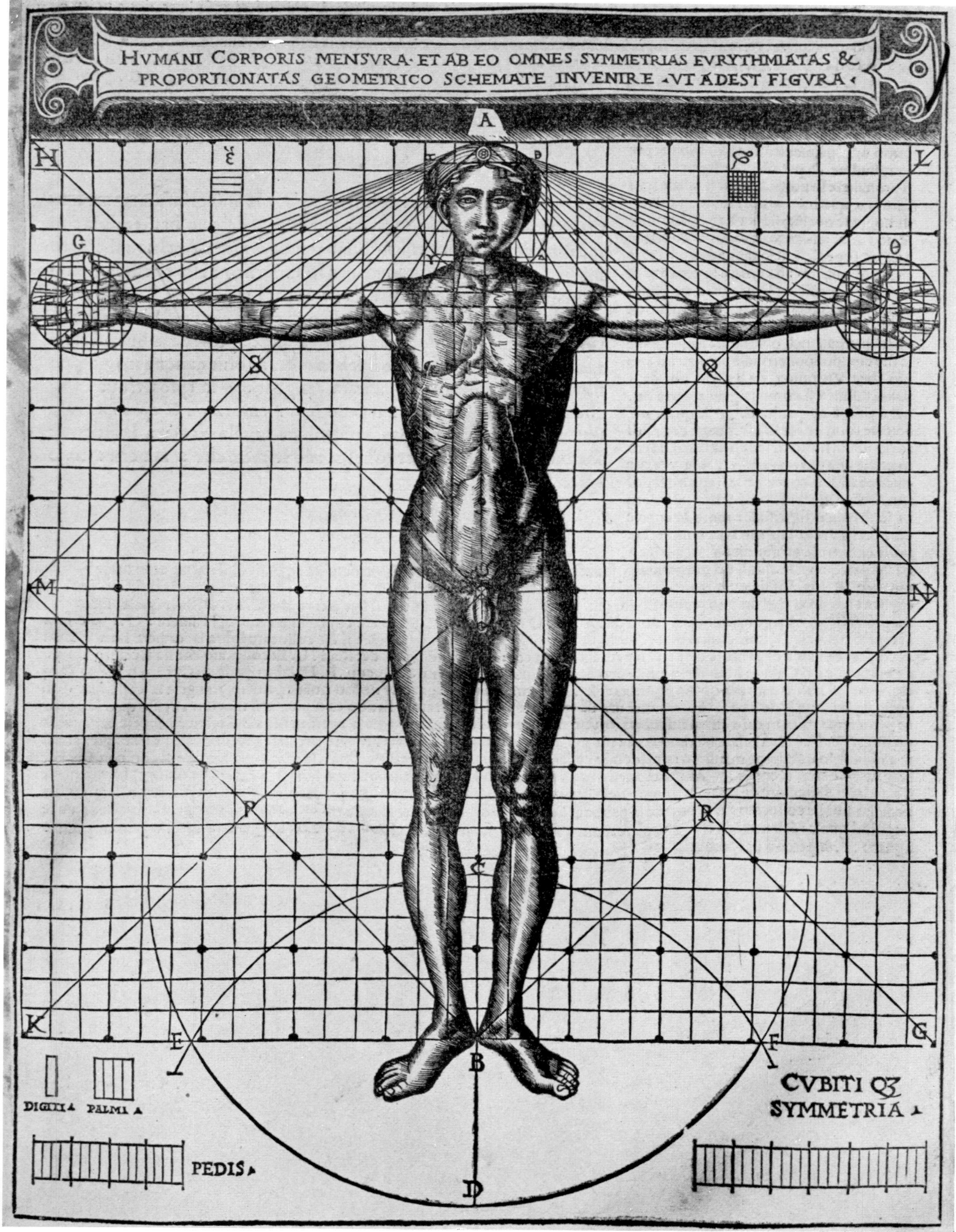

Pl. 343. Study of anatomical proportion, from C. Cesariano's edition of Vitruvius, Como, 1521.

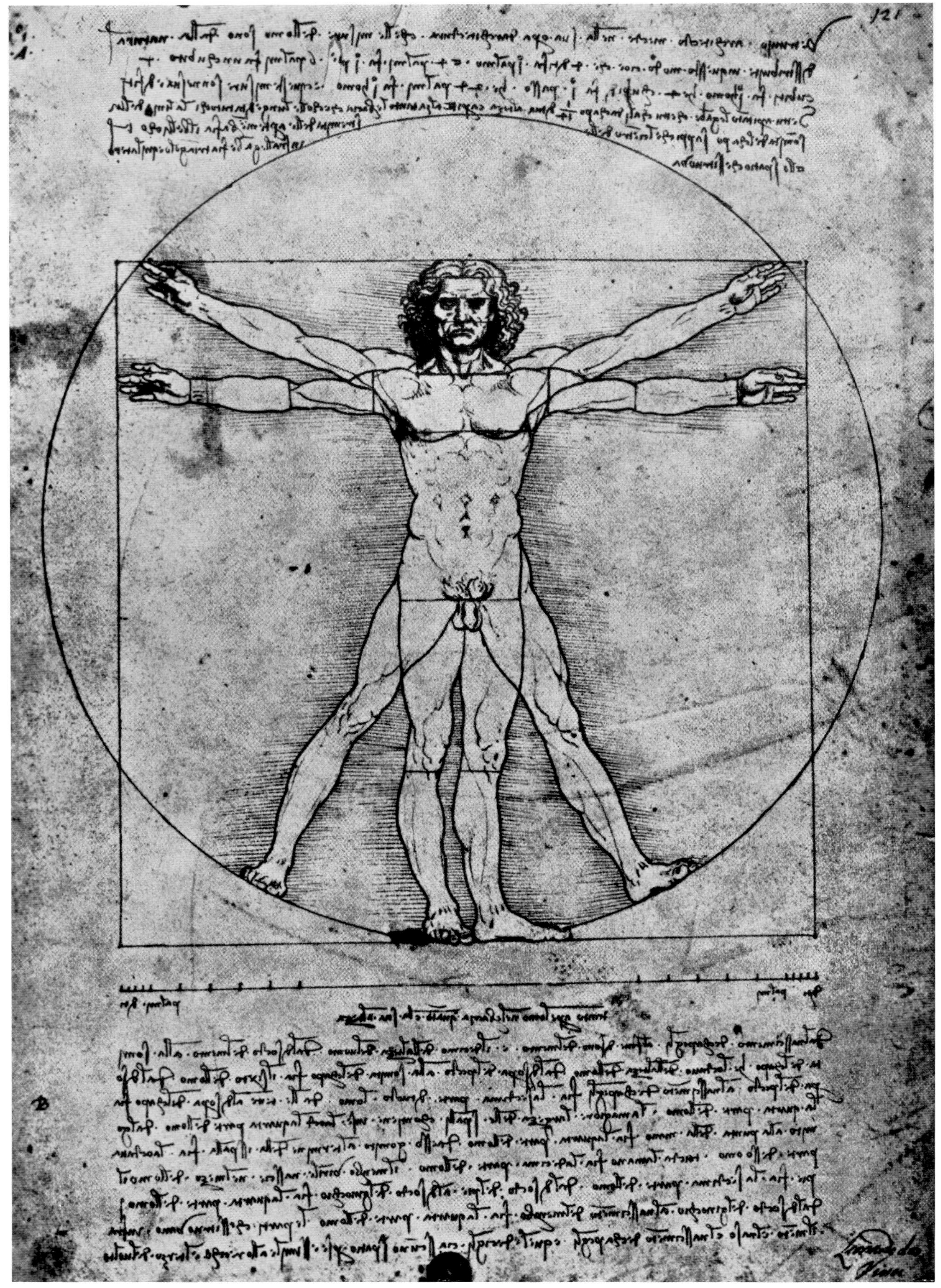

Pl. 344. Leonardo da Vinci, study of human proportions according to Vitruvius. Drawing, 13 1/2×9 3/4 in. Venice, Accademia.

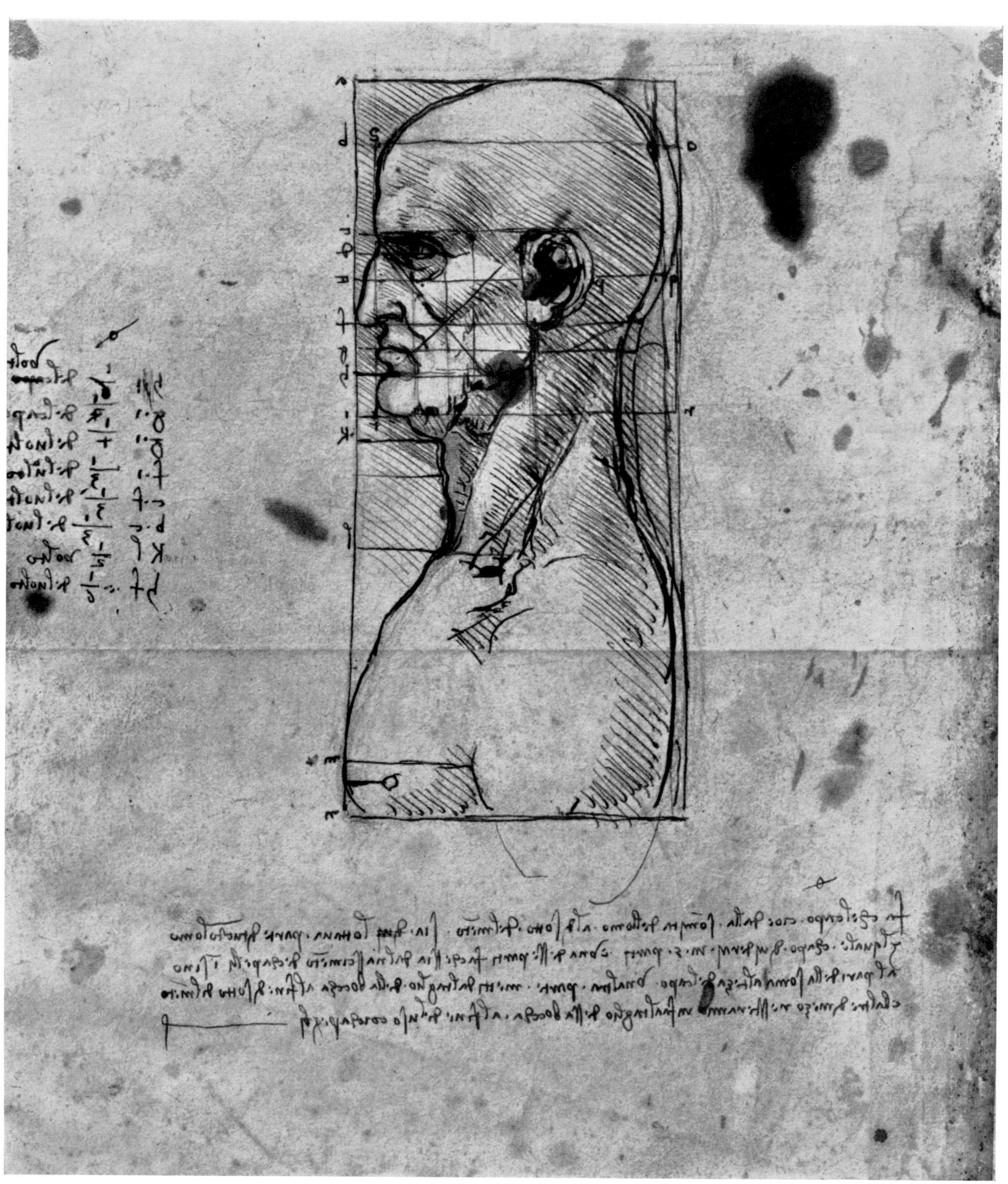

Pl. 345. Leonardo da Vinci, study of facial proportions. Drawing. Venice, Accademia.

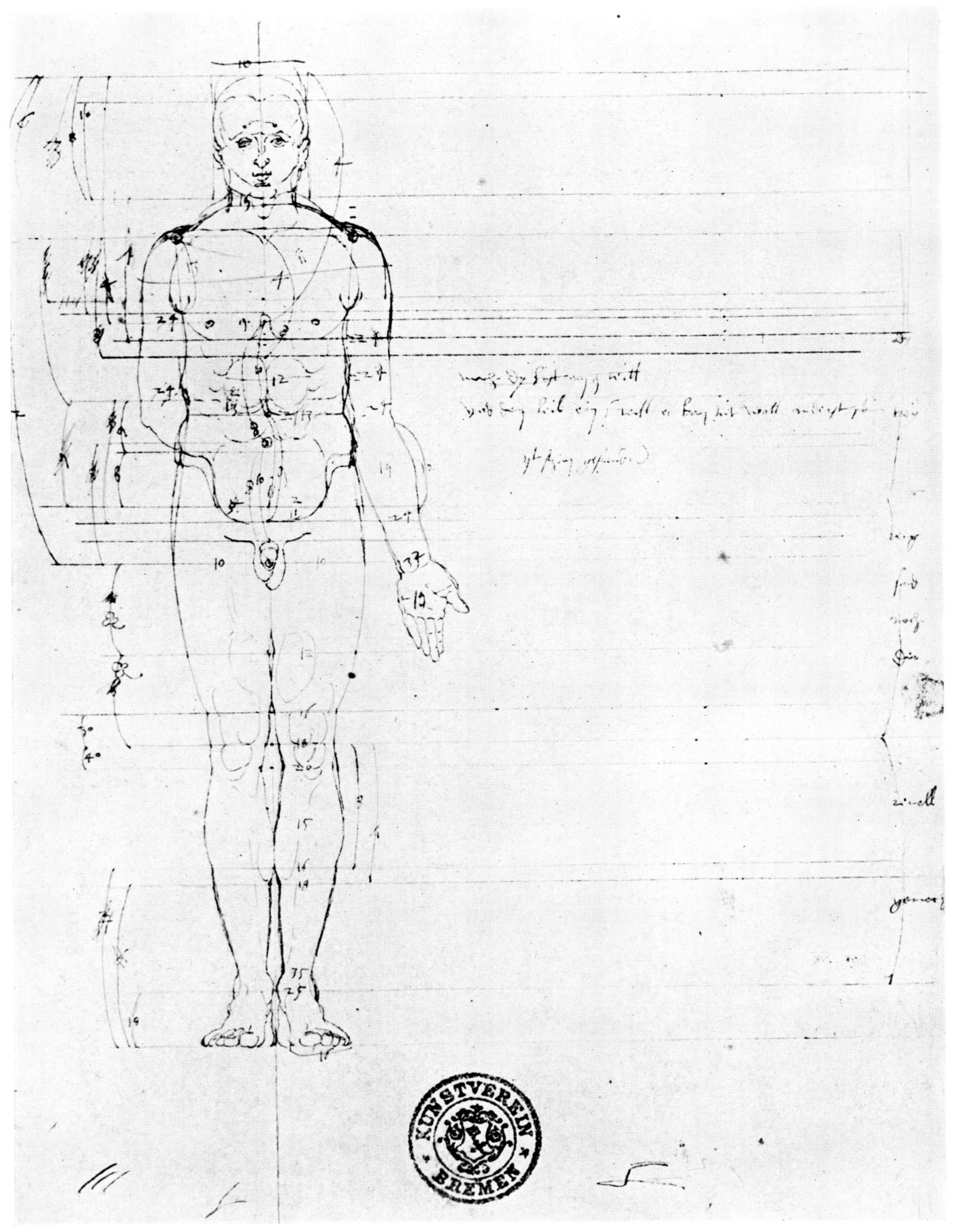

Pl. 346. A. Dürer, male figure, 1513, preparatory study for his treatise on human proportions. Pen drawing, 11×8 1/8 in. Bremen, Kunsthalle.

Pl. 347. A. Dürer, female figure with shield and lamp, ca. 1500. Pen drawing, $11\,^{3}/_{4} \times 8\,^{1}/_{8}$ in. Berlin, Kupferstichkabinett.

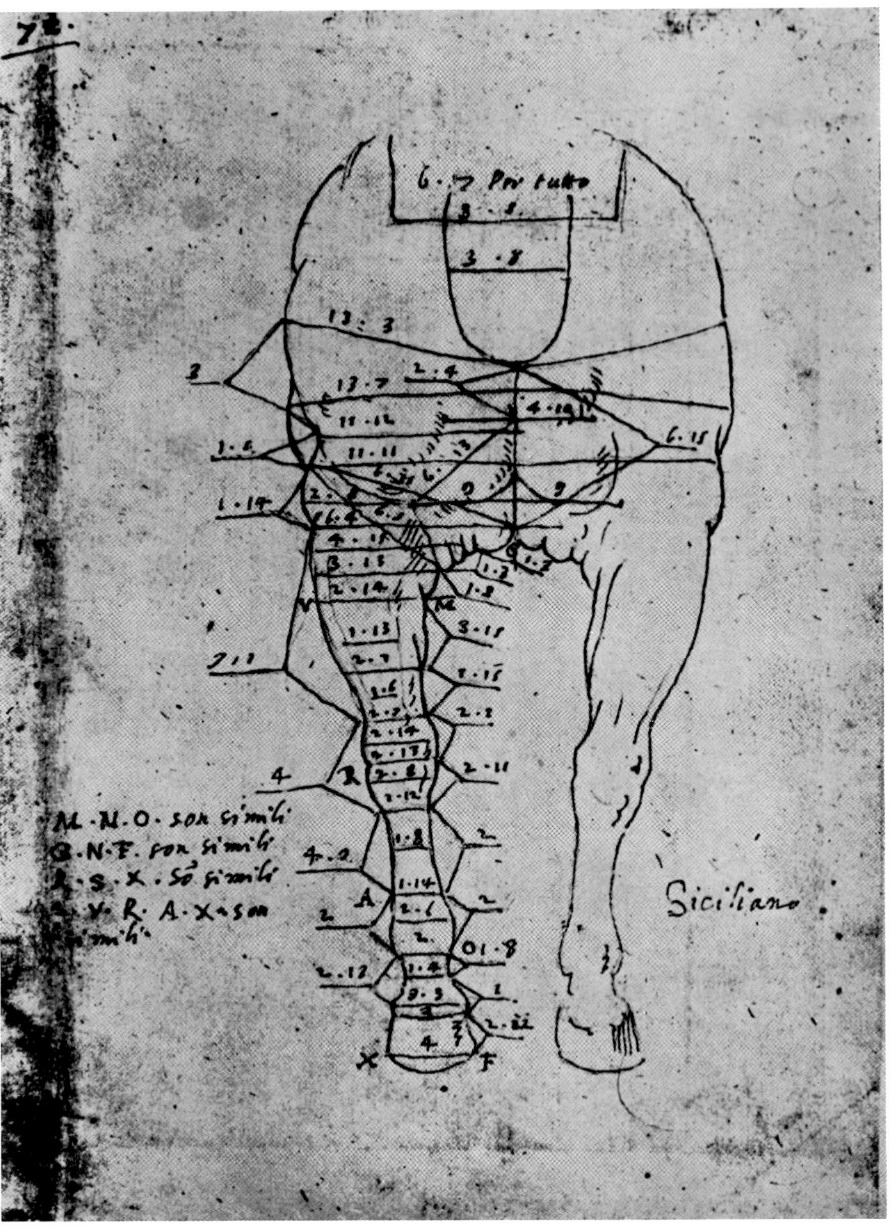

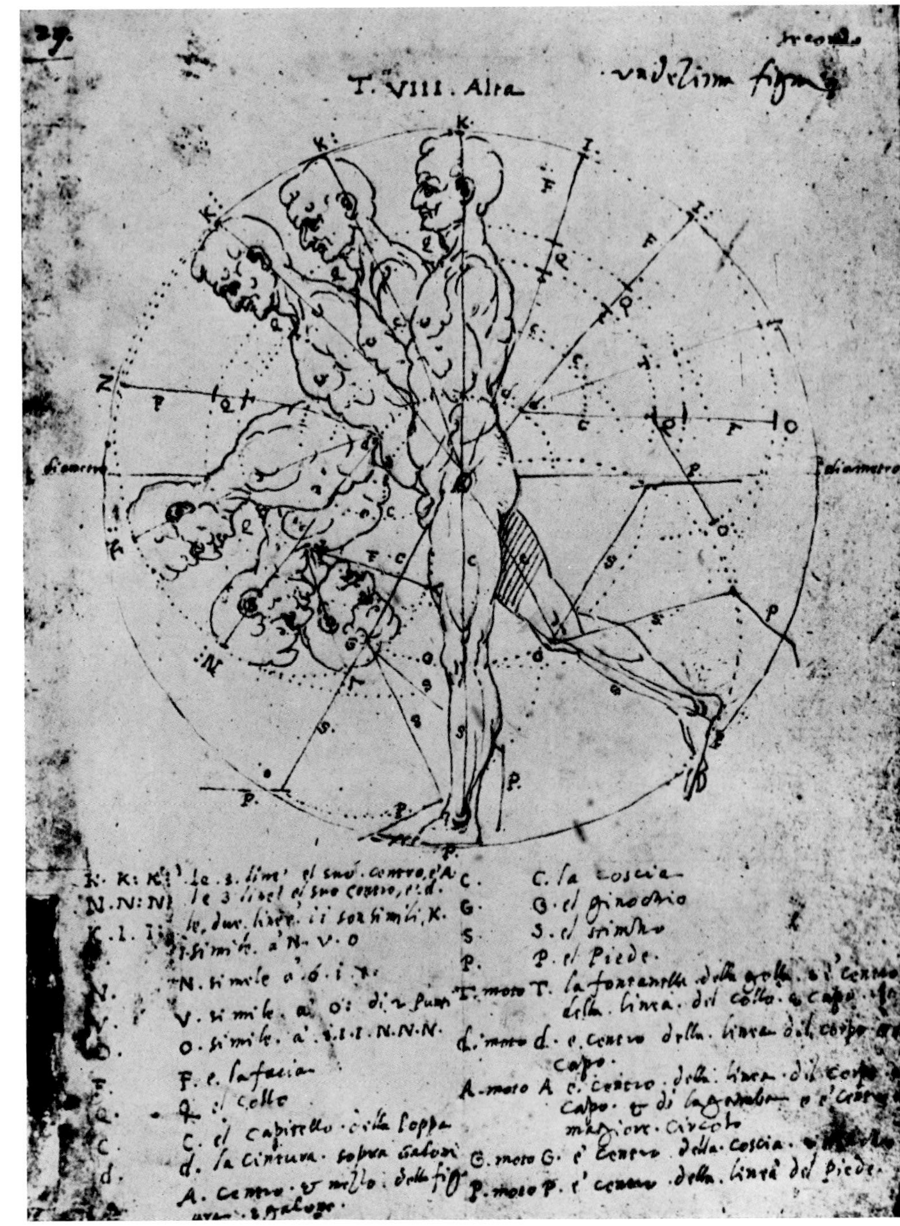

Pl. 348. Codex Huygens, by a Lombard artist, ca. 1570. Pen drawings; each, ca. $5^1/_4 \times 7^1/_4$ in. *Left*: Horse, shown without head, in frontal elevation (fol. 72). *Right*: Analysis of a figure bending forward and backward (fol. 29). New York, Pierpont Morgan Library.

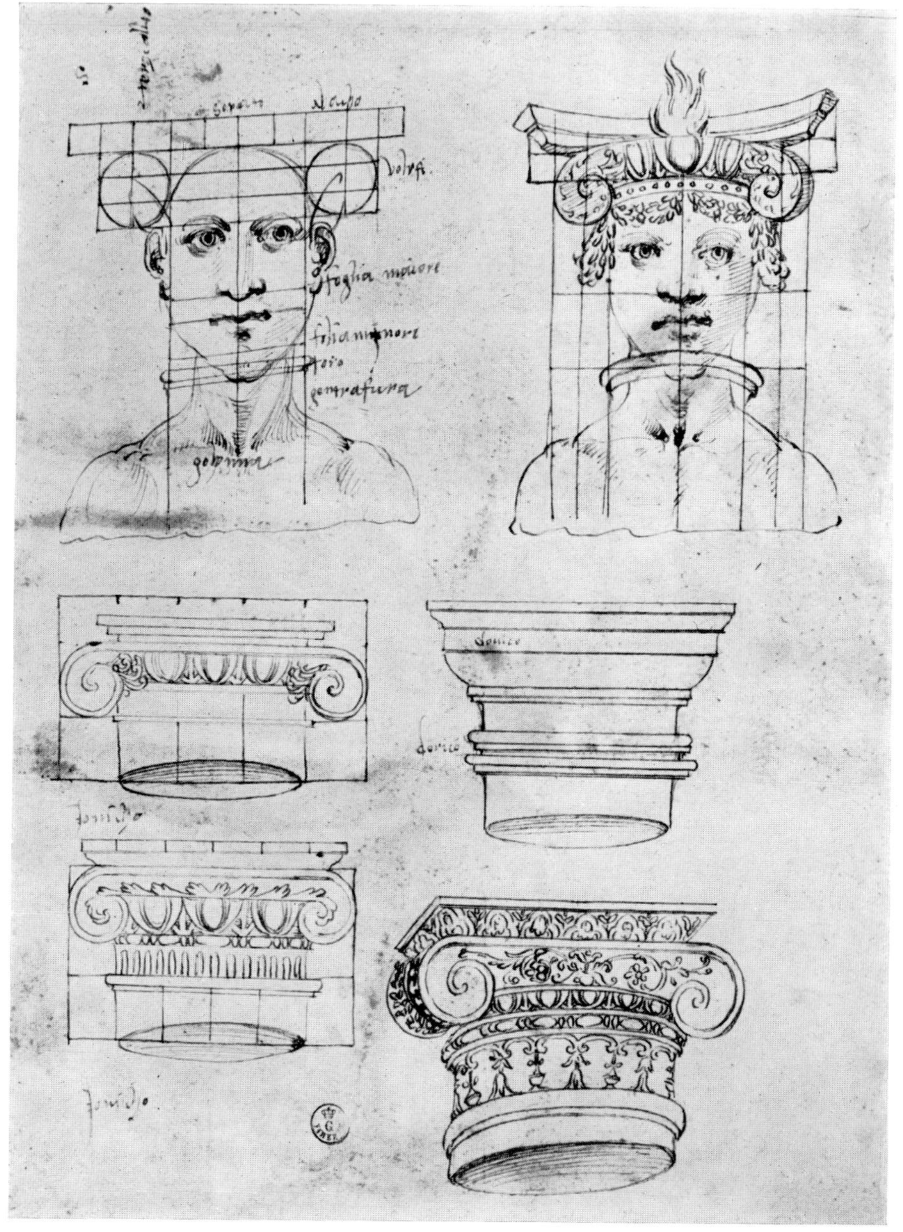

Pl. 349. Fra Giocondo (ca. 1433–1515), drawings relating human and architectural proportions. Florence, Uffizi, Gabinetto dei Disegni e Stampe.

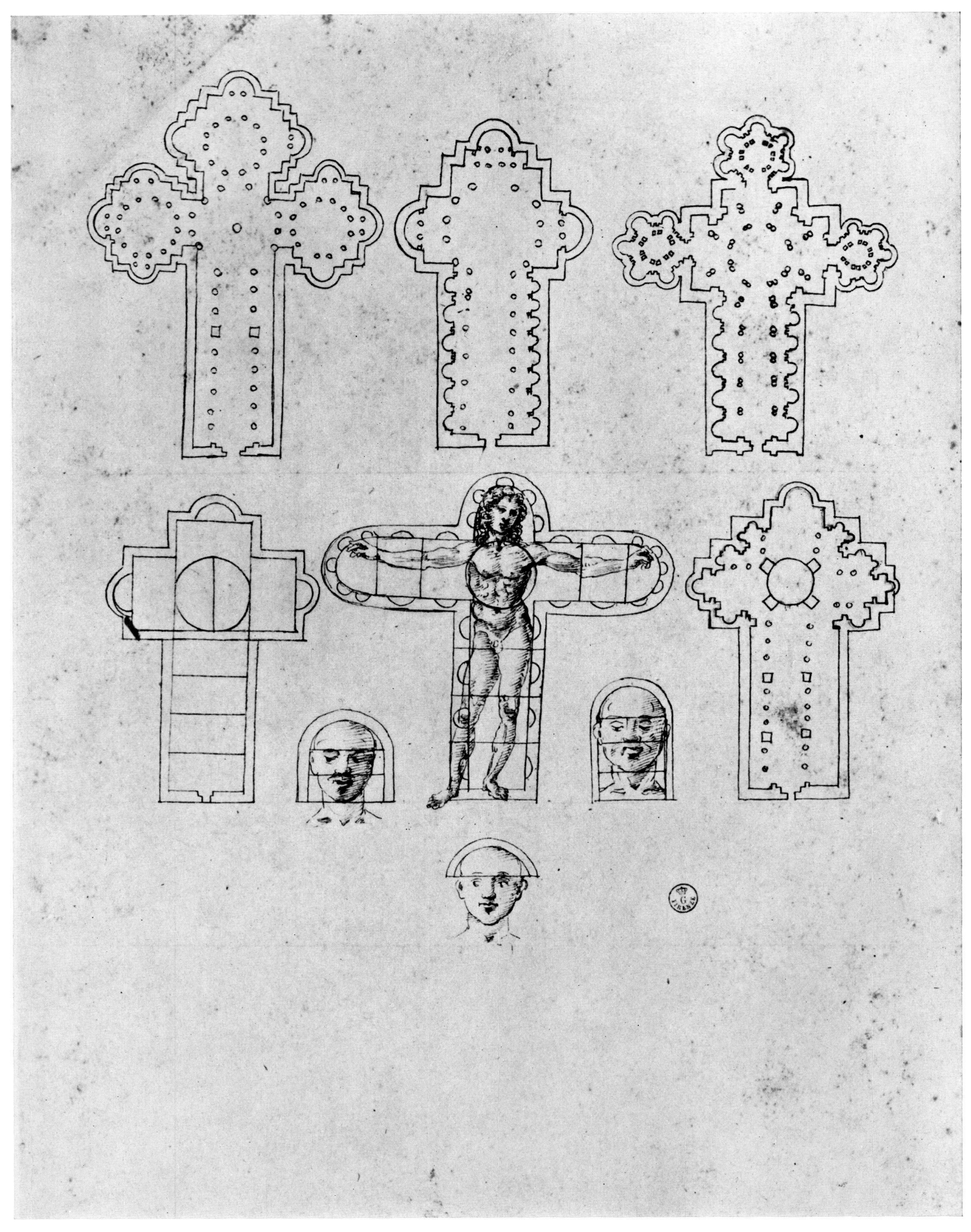

Pl. 350. P. Cataneo (d. 1569), six church plans with human figure and details. Drawing. Florence, Uffizi, Gabinetto dei Disegni e Stampe.

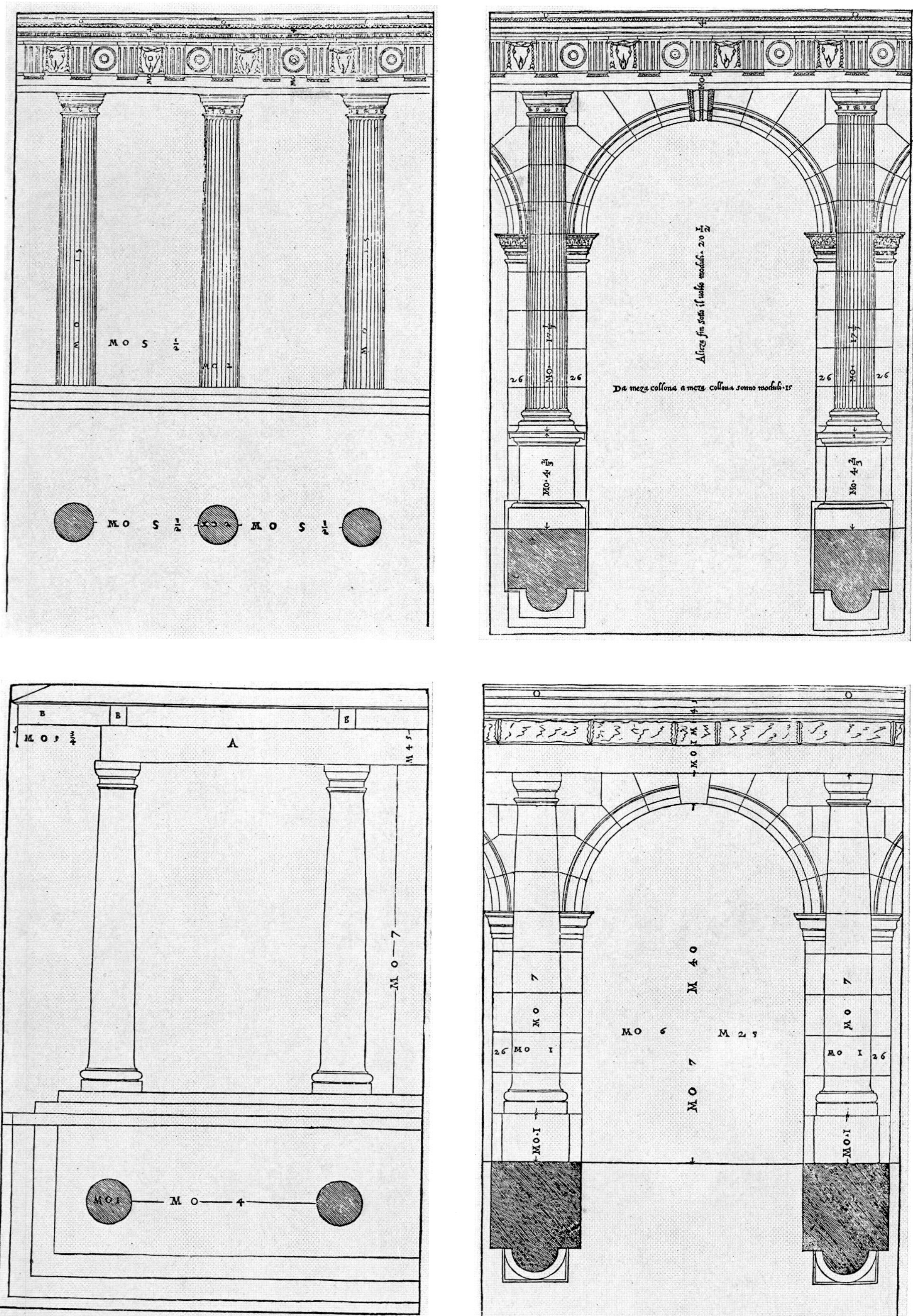

Pl. 351. A. Palladio, the Doric order, from *I quattro libri dell'architettura*, Venice, 1570.

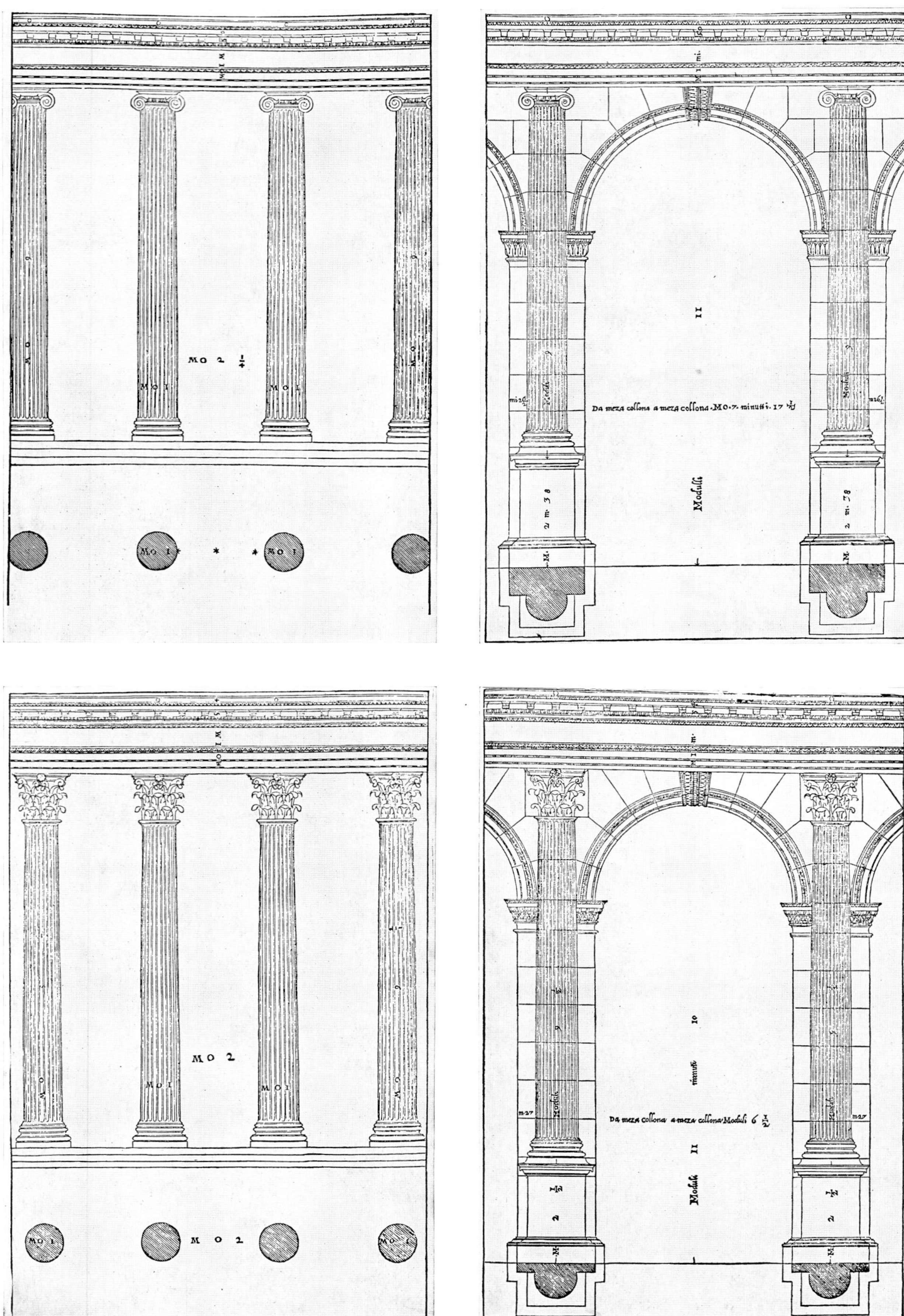

Pl. 352. A. Palladio, the Ionic order (*above*) and the Corinthian order (*below*), from *I quattro libri dell'architettura,* Venice, 1570.

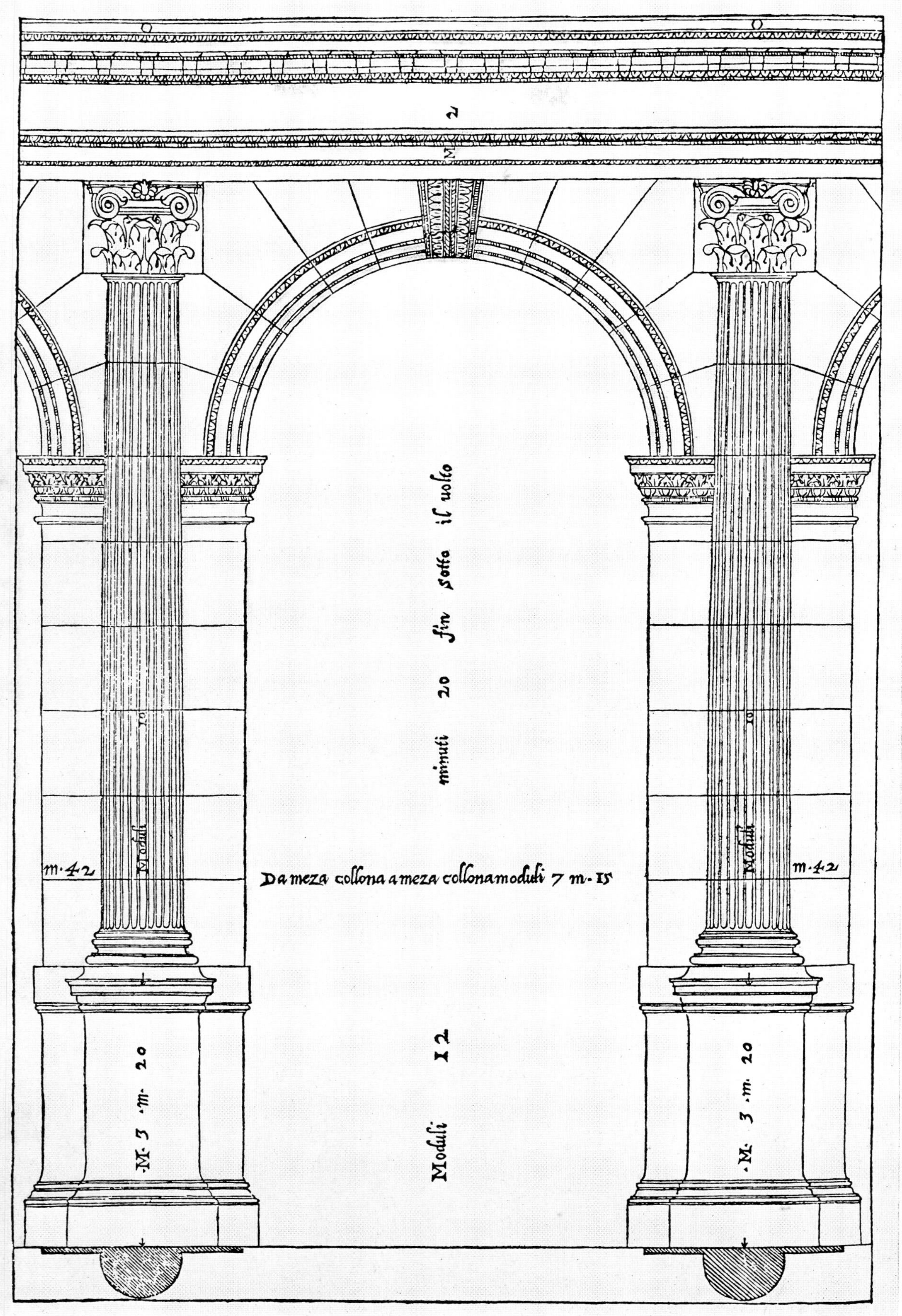

Pl. 353. A. Palladio, the composite order, from *I quattro libri dell'architettura,* Venice, 1570.

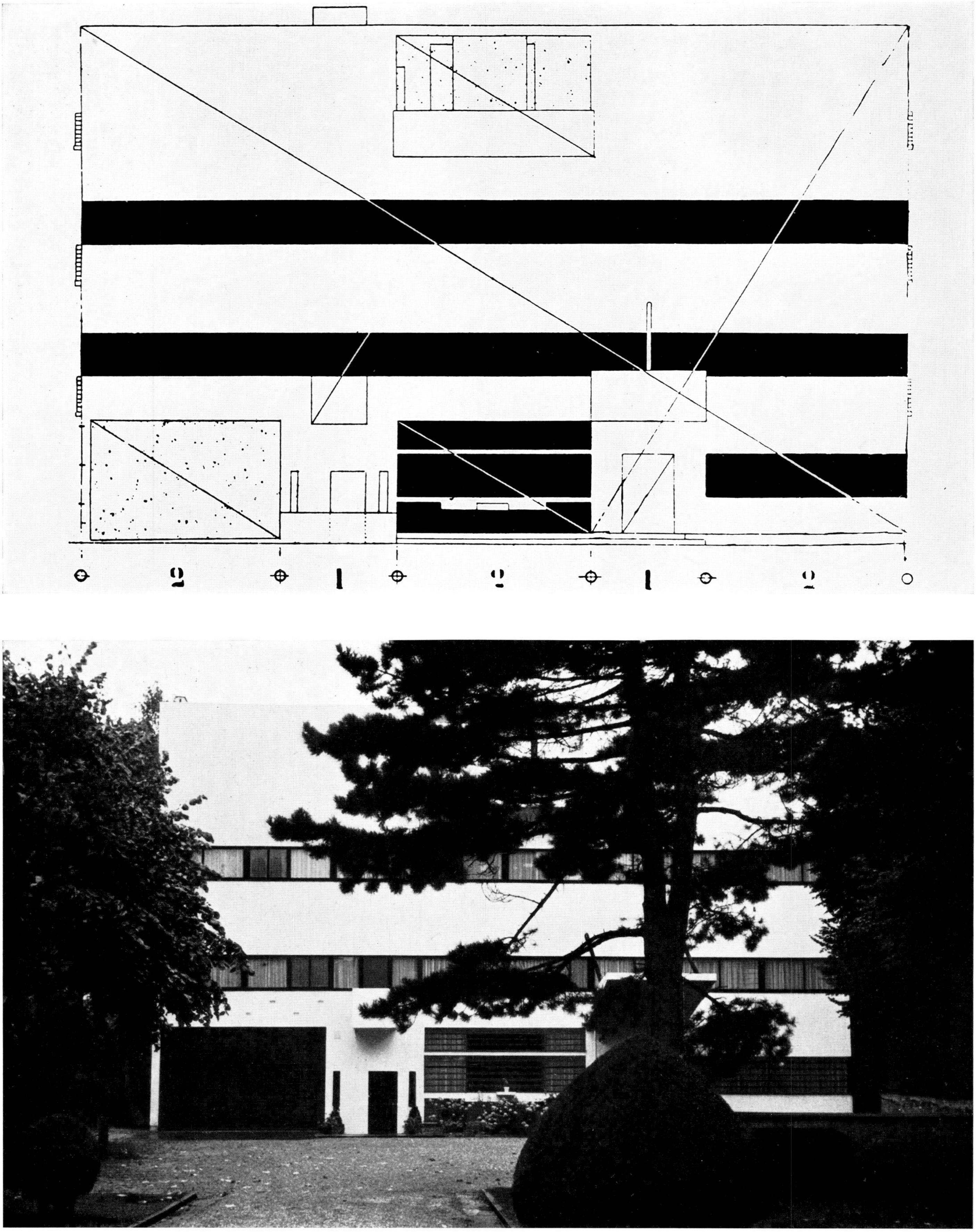

2 1 2 1 2

Pl. 354. Le Corbusier, Les Terrasses, a villa at Garches, Seine-et-Oise, France, 1927, north façade.

Pl. 355. Egyptian, First Intermediate Period. *Above, left*: Stele with a man, woman, and offerings. Painted limestone, $15^3/_4 \times 9^1/_8$ in. Leiden, Rijksmuseum van Oudheden. *Right*: Wall painting showing the carrying of grain, from the tomb of Ity, Gebelein. Turin, Museo Egizio. *Below*: Nubian, 1st cent. *Left*: King Natakamani (?), at Argo, Sudan. Granite. *Right*: Lion god, relief of west wall. Naga, Sudan, Lion Temple.

Pl. 356. *Above*: Greek. *Left*: Nuptial vase with Peleus and Thetis, from Eretria, mid-6th cent. B.C. Athens, National Museum. *Right*: Head of a young man, from Skiathos, ca. 480 B.C. Marble, ht., ca. 8 in. Halmyros, Greece, Museum. *Below, left*: Greco-Italic fragment with head of Acheloos (reconstructed as a disk), from Fratte, near Salerno, Italy, 5th cent. B.C. Terra cotta; diam. of disk, 25⁵/₈ in. Salerno, Museo Provinciale. *Right*: Greco-Bactrian medallion, first half of 2d cent. B.C. Silver, diam., 4³/₄ in. Leningrad, The Hermitage.

Pl. 357. Roman provincial. *Above, left*: Fragment of relief with a mourning barbarian woman, from Mogontiacum (Mainz, Germany), 1st cent. Marble, ht., 28 in. Mainz, Altertumsmuseum. *Right*: Stele with portraits, from Noricum, 3d cent. Marble, ht., 4 ft., 1 5/8 in. Celje, Yugoslavia, St. Maximilian. *Below, left*: Metope showing a Roman legionary in battle, from Trajan's monument at Adamklissi, Dobruja, Romania, first half of 2d cent. Limestone, ht., 4 ft., 9 7/8 in. Bucharest, National Museum. *Right*: Stele, from Phrygia, 3d cent. Marble, ht., 4 ft. Istanbul, Archaeological Museums.

Pl. 358. Roman provincial. *Above*: Stele with toilette scene, from Neumagen, Germany, 2d–3d cent. Marble, w., 4 ft., 7 in. Trier, Germany, Rheinisches Landesmuseum. *Below*: Triumphal procession from the Arch of Septimius Severus, Leptis Magna, ca. 203. Marble, ht., 5 ft., 7 in. Tripoli, Libya, Archaeological Museum.

Pl. 359. North Africa, Roman period. *Above*: Funerary relief with portraits, from Ghirza, Tripolitania, 4th–5th cent. Limestone, $24\frac{3}{8} \times 29\frac{1}{8}$ in. *Below*: Polychrome mosaic with Nile scene, from the Villa of the Nile at Leptis Magna, 2d cent. Both, Tripoli, Libya, Archaeological Museum.

Pl. 360. Byzantine style, Italy. *Left*: Messengers prostrating themselves before St. Silvester, ca. 1246. Fresco. Rome, Santi Quattro Coronati, Chapel of St. Silvester. *Right*: Madonna and Child enthroned, early 14th cent. Panel, 54×23 5/8 in. Tagliacozzo, Abruzzi, S. Maria d'Oriente.

Pl. 361. Peripheral Byzantine schools. *Left, above*: Armenian Gospel. The Entombment. Illumination. Washington, D.C., Freer Gallery of Art (32.18). *Below*: Embroidered Russian war banner showing the apparition of the Archangel Michael to Christ, Stroganov school, late 16th–early 17th cent. Moscow, Tretyakov Gallery. *Right*: Christ and St. Mary Magdalene, icon from Crete, ca. 1600. Ht., 23 3/4 in. London, National Gallery.

Pl. 362. Renaissance. *Left:* D. Cappelli (?), Madonna and Child with saints and donors, 1494. Fresco. Retrosi, near Amatrice, Italy, S. Maria delle Grazie, called the "Icona." *Right:* St. Sebastian, 15th cent. Fresco. Aosta, Italy, Collegiata di S. Orso.

Pl. 363. *Left*: Breton Calvary, detail of platform and frieze with scenes of the life of Christ, 1581–88. Granite. Guimiliau, Finistère, France, parish close. *Right*: Altar with carved wooden figures, 1609. Atri, Abruzzi, Italy, S. Domenico.

Pl. 364. *Above*: Palazzolo Acreide, Sicily, Church of the Annunziata, detail of façade, 18th cent. *Below*: S. Ignacio, province of Corrientes, Argentina, ruins of the Jesuit church, two views.

Pl. 365. Lecce, Apulia, Italy, SS. Nicolò e Cataldo, built 1180, remodeled 1716, baroque façade with original Romanesque portal.

Pl. 366. Nineteenth-century painting. *Above*: T. W. Dewing, The Recitation, 1891. Canvas, 30×55 in. Detroit, Institute of Arts. *Below, left*: R. Fontana, Gyspy, 1886. Canvas, $15^{3}/_{8}\times7^{7}/_{8}$ in. Formerly, Florence, E. Checcucci Coll. *Right*: Raja Ravi Varma, Hamsa Damayanti. Canvas. Trivandrum, India, Sri Chitra Art Gallery.

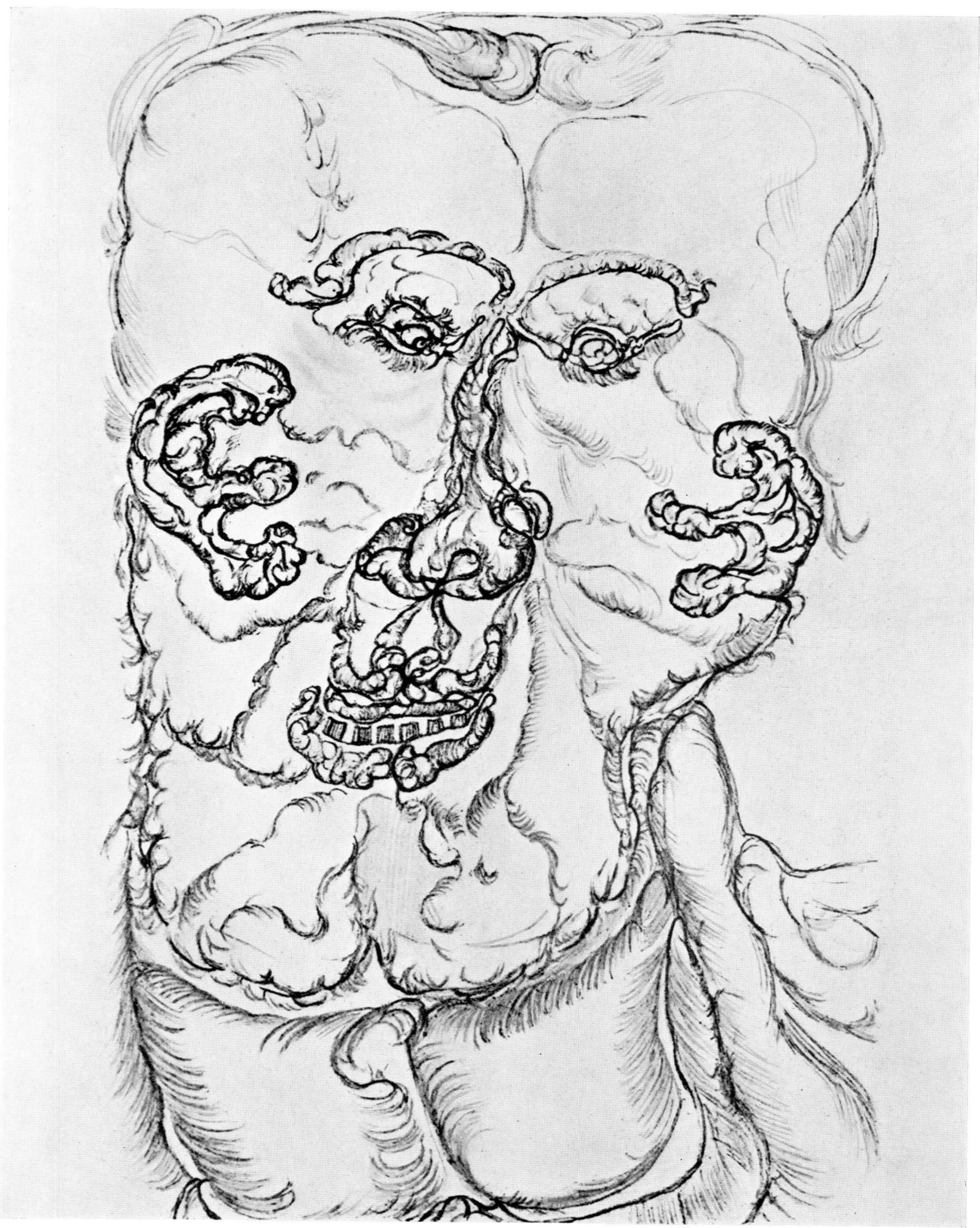

Pl. 367. Abnormal psychology. Pencil drawing by Case 26, from H. Prinzhorn, *Bildnerei der Geisteskranken,* Berlin, 1922. Ht., 13 in. (Original, Psychiatric Clinic of Heidelberg University, archives.)

Pl. 368. Child art. *Above*: Portrait by a child, age three and one-half, 1961. *Below*: Paintings on ceramics with dancers and gondolas, 1962–63. Both, Trieste, Scuola Materna Comunale di S. Sabba.

Pl. 369. Child art. Drawings. *Above*: Sports event, by E. Tallone, age nine. *Below*: Nativity, by P. Privileggi, age five, 1962.

Pl. 370. Observation rearranged by imagination. *Above*: Il Rosso, Skeletons, 1517. Sanguine. Florence, Uffizi, Gabinetto dei Disegni e Stampe. *Below*: F. de Nomé ("Monsù Desiderio"), imaginary scene, first half of 17th cent. Canvas. Private coll.

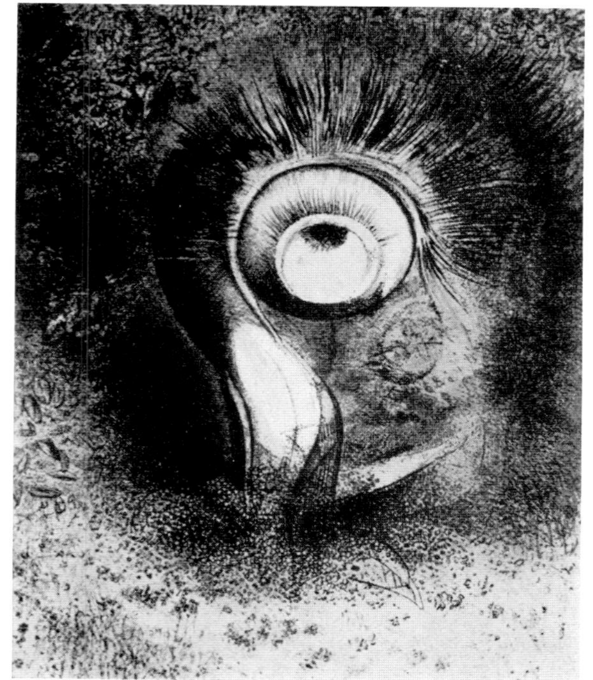

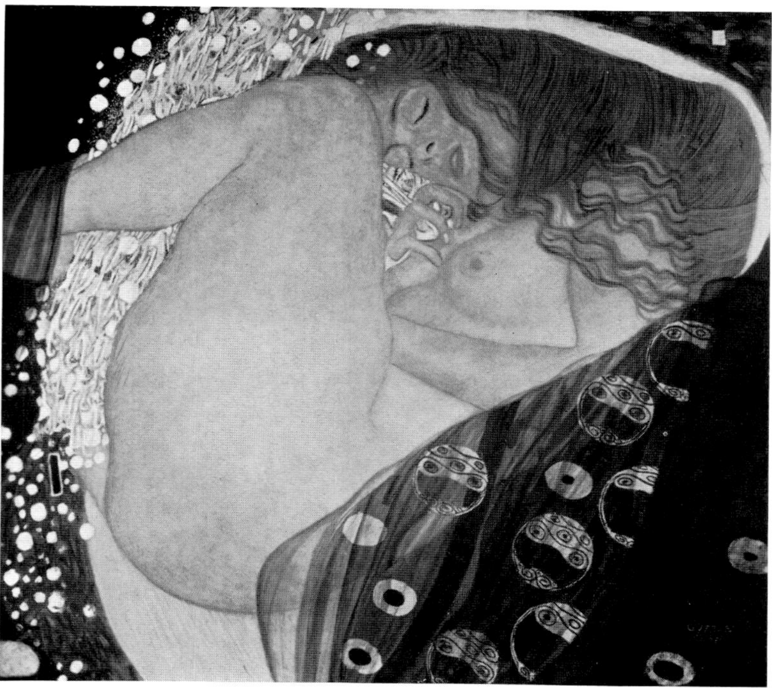

Pl. 371. Projection of emotion into forms. *Above, left*: O. Redon, lithograph from *Les Origines*, 1883. *Right*: G. Klimt, Danae, 1908. Canvas. Graz, Austria, F. Böck Coll. *Below*: V. van Gogh, Cypresses and Stars, 1889. Drawing, $18\frac{1}{2} \times 24\frac{1}{2}$ in. Bremen, Kunsthalle.

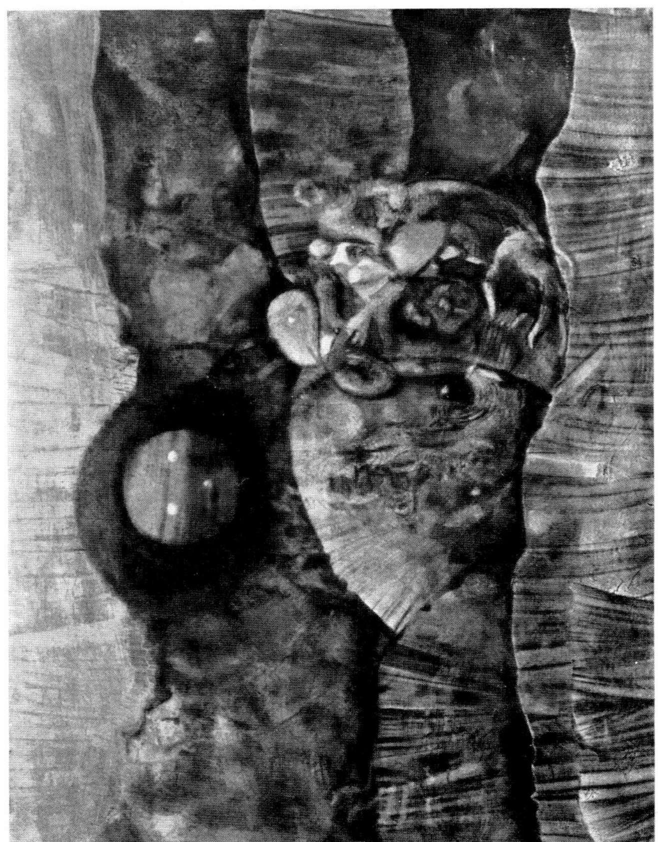

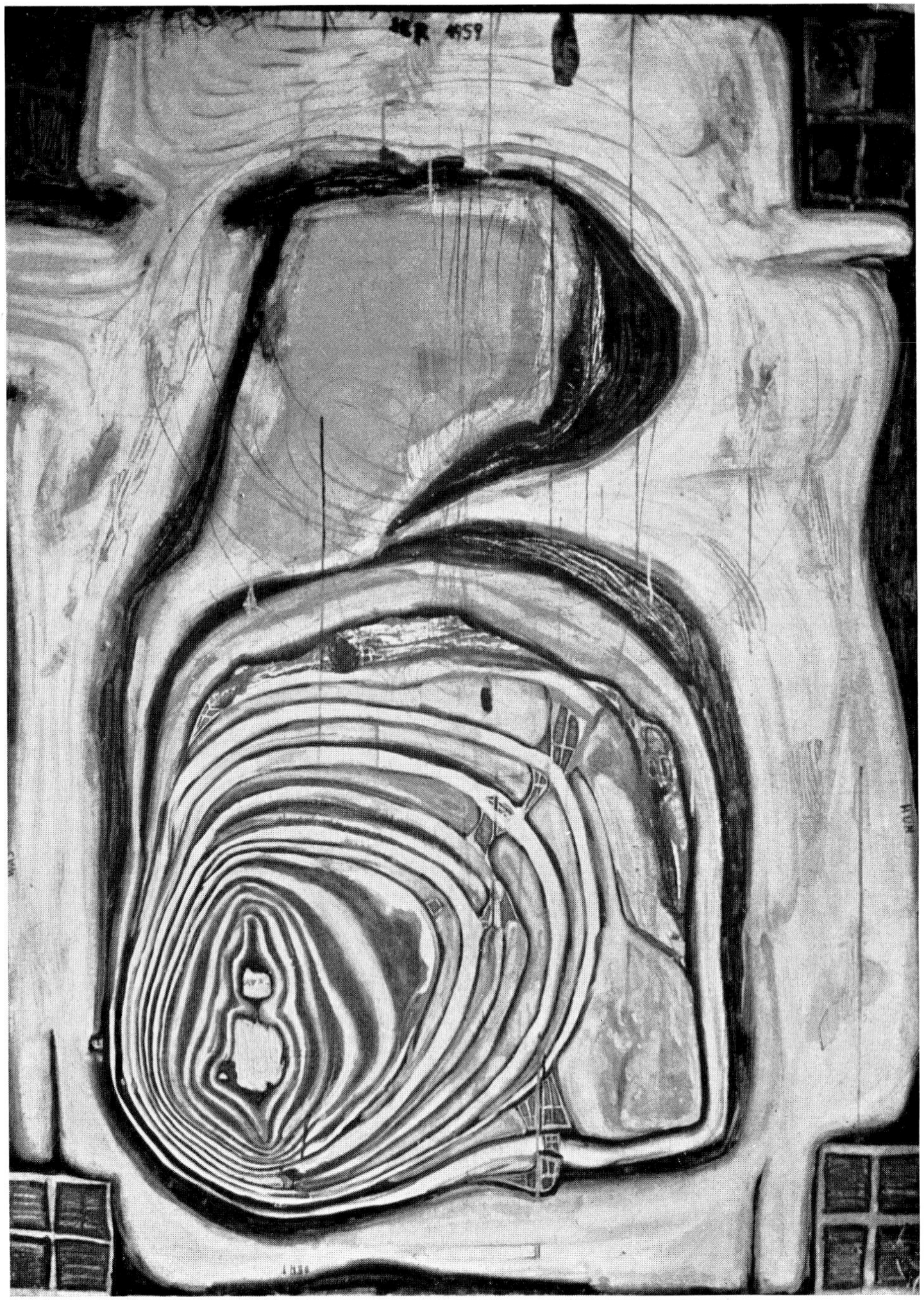

Pl. 372. Exploration of subconscious levels. *Left, above:* M. Ernst, Lune en Bouteilles. Canvas. London, private coll. *Below:* Y. Tanguy, Multiplication of the Arcs, 1954. Canvas, 40×60 in. New York, Museum of Modern Art. *Right:* Hundertwasser (F. Stowasser), The Wall, 1959. Composite media on paper, $51^5/_8 \times 39^3/_8$ in. São Paulo, Brazil, Museu de Arte Moderna.

Pl. 373. Child art. Water color, by L. Bartoleschi, age six, Rome.

Pl. 374. Surrealism. S. Dali, Metamorphoses of Narcissus, detail, 1936–37. Canvas; full size, $20^{1}/_{8} \times 30^{3}/_{4}$ in. London, Tate Gallery (on loan from E. James).

Pl. 375. *Above*: Announcements of elections and spectacles, painted on the wall of a house in Via dell'Abbondanza, Pompeii, 1st cent. (Destroyed, 1943.) *Center*: Trademark of a maritime trading corporation, floor mosaic in Piazzale delle Corporazioni, Ostia, 2d cent. *Below*: Mosaic with chariot race, publicizing the names of the competing horses, Barcelona, 2d cent. Ht., 11 ft., $8^1/_2$ in. Barcelona, Museo Arqueológico.

Pl. 376. *Above*: Food display, painted on the wall of the Thermopolium (XIII, PL. 262), Ostia, 2d cent. *Center*: Inn sign, prior to 15th cent., now affixed to Swan Inn, Clare, near Sudbury, Suffolk, England. Oak, l., ca. 10 ft. *Below, left*: Inn sign, the Shepherd and Dog, London Road, Brighton, England, 19th cent. Wrought iron. *Right*: Shop sign of an unidentified Parisian *boutique*, early 18th cent. Wrought iron, partially painted, $42^{1}/_{8} \times 37^{3}/_{8}$ in. Paris, Musée Carnavalet.

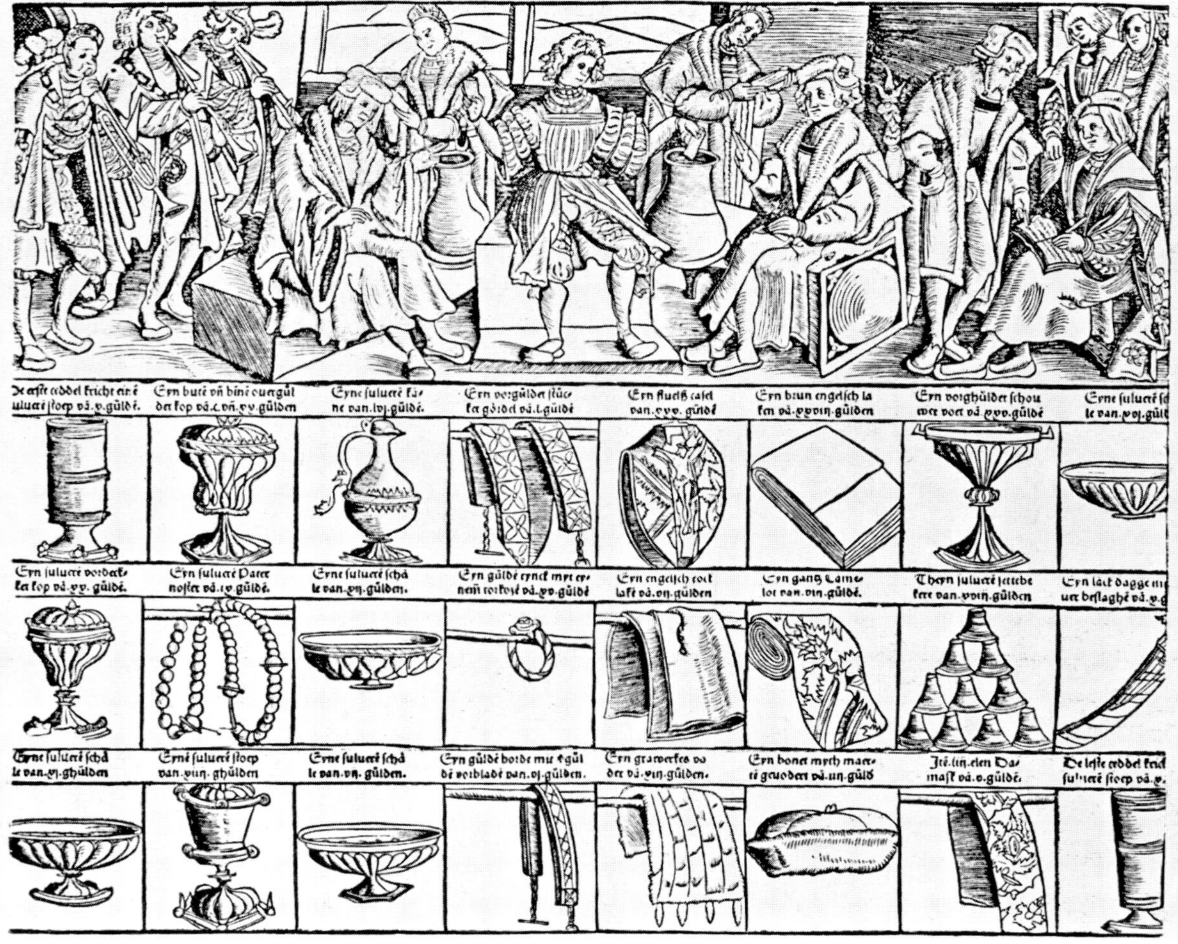

Pl. 377. *Above*: Placard for a lottery at Rostock, Germany, detail, attributed to E. Altdorfer, 1518. Woodcut. *Below, left*: Trade card of Francis Dodsworth, a gold and silver wiredrawer and thread manufacturer, London, ca. 1680. Engraving, 6 3/4 × 5 7/8 in. *Right*: Tobacco label for Thomas Marshall of Wapping, England, ca. 1675. Engraving, ca. 3 1/4 × 2 1/4 in. Last two, Cambridge, England, Pepysian Library.

MEssieurs & Dames,

Il est arriué en ceste Ville, vn tres-honneste Homme lequel à amené deux Animaux desquelz voyez la vraye Figure cy dessus, lesquelz Animaux ont esté presenté au Roy, dans le Ieu de Paulme du Louure à Paris, à la veuë de plusieurs grands Princes & Seigneurs, lesquels n'ont iamais veu leurs semblables, lesdicts animaux ont esté pris en Affrique, les noms desquels ne sont pas bien cogneus, quelque-vns les appellent Mago, les autres Tartarins, les Holandois qui les amenent des Indes, les appellent Bleuf-nez, lesdicts Animaux font plusieurs belles exercices, comme sauter, dancer, tenant espée & verre en main, &c. si bien que celuy qui les conduict desire & promet renuoyer tous ceux qui luy auront faict l'honneur de les veoir auec vn tres-parfaict contentement, lesdicts Animaux se montrent en la rue

à l'Enseigne

PARAPLUYES
ET PARASOLS
A PORTER DANS LA POCHE.

LES Parapluyes dont Mr Marius a trouvé le secret, ne pesent que 5. à 6. onces: ils ne tiennent pas plus de place qu'une petite Ecritoire, & n'embarassent point la poche; ainsi chacun peut sans s'incommoder en avoir un sur soy par précaution contre le mauvais temps. Ils sont cependant aussi grands, plus solides, resistent mieux aux grands vents, & se tendent aussi vite que ceux qui sont en usage.

C'est le témoignage que Messieurs de l'Academie Royale des Sciences en ont rendu.

Cette nouvelle Invention a paru avoir été bien reçuë du Public par le grand debit qui s'en est fait, ce qui a excité l'Auteur à la perfectionner, au point qu'il ne laisse plus rien à souhaiter du côté de la solidité.

A l'égard de ceux qui sont ornez, l'on conviendra qu'il ne s'est encore rien vû en Parassols de plus agréable pour le goût & la legereté, & que l'on peut contenter en ce genre les Curieux les plus difficiles, pour la richesse des montures & des ornemens. *Ils auront tous sa marque.*

Ils se font & se vendent à Paris chez Mr MARIUS, demeurant ruë des Fossez Saint Germain, aux trois Entonnoirs.

Par l'autorité d'un Privilege du Roy, portant deffense par toute l'étenduë du Royaume de les contrefaire, à peine de mille livres d'amende.

Zu Ihro Röm. Apostolis. K. K. Majest.
Von Fürsten zu Anhalt Zerbst. Neu errichteten Infanterie Regiment.

Pl. 378. *Left, above*: Handbill for a traveling monkey show, France, 1625. Woodcut. *Below*: Stationer's label used for the cover of a packet of playing cards, London, ca. 1680. Engraving, $7^{3}/_{4} \times 3^{3}/_{4}$ in. Cambridge, England, Pepysian Library. *Right, above*: Placard advertising pocket umbrellas, Paris, 1715. Woodcut. *Below*: Recruiting placard for an infantry regiment, Germany, mid-18th cent. Woodcut. Nürnberg, Germanisches National-Museum.

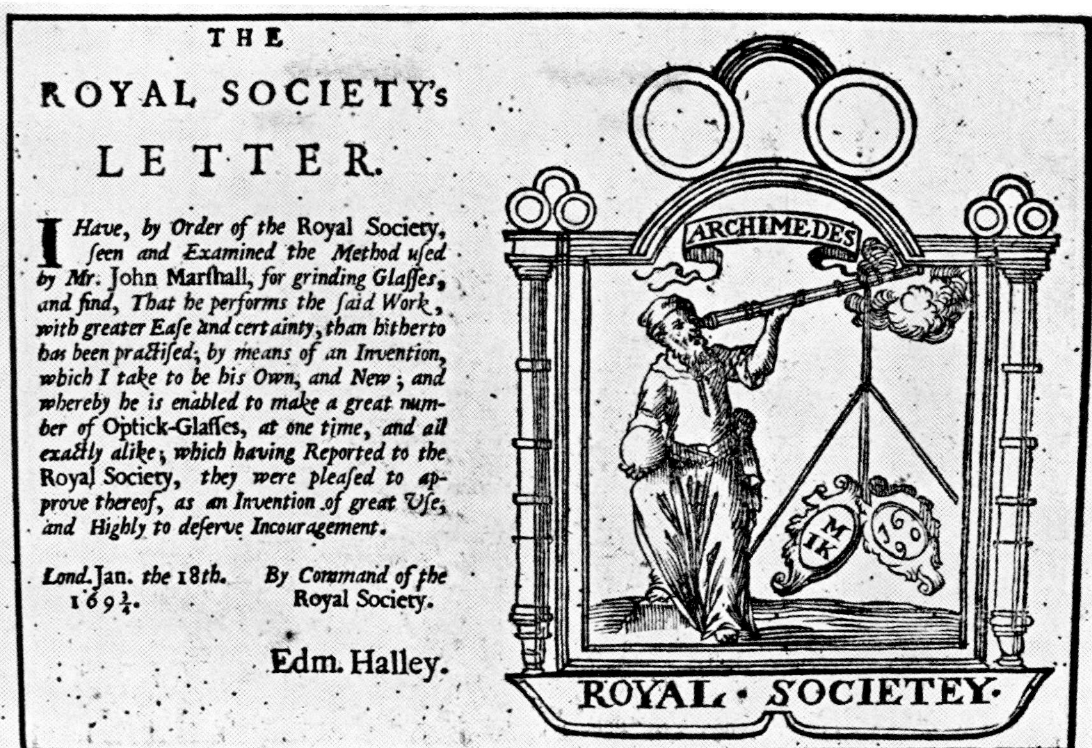

THE
ROYAL SOCIETY's
LETTER.

I Have, by Order of the Royal Society, seen and Examined the Method used by Mr. John Marshall, for grinding Glasses, and find, That he performs the said Work, with greater Ease and certainty, than hitherto has been practised; by means of an Invention, which I take to be his Own, and New; and whereby he is enabled to make a great number of Optick-Glasses, at one time, and all exactly alike; which having Reported to the Royal Society, they were pleased to approve thereof, as an Invention of great Use, and Highly to deserve Incouragement.

Lond. Jan. the 18th. 1693/4.

By Command of the Royal Society.

Edm. Halley.

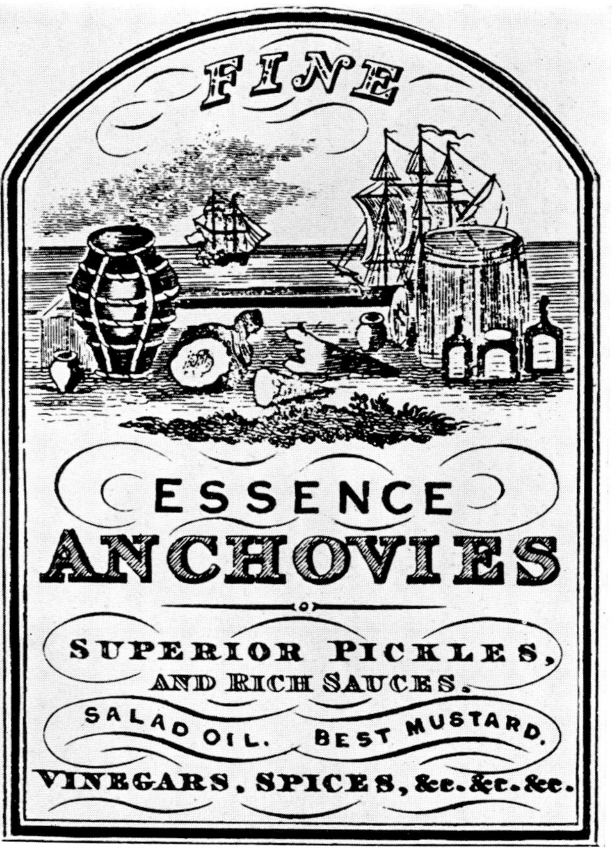

Pl. 379. *Above*: Endorsement by the Royal Society for a manufacturer of optical instruments, London, 1694. *Below, left*: Advertisement for W. Baker and Co., Boston, 19th cent. *Right*: Engraved label, London, ca. 1845. W., 2 1/4 in.

Pl. 380. J. Chéret, poster advertising ice skating at the Palais de Glace, Paris, 1893. Color lithograph.

Pl. 381. J. Toorop, poster for a salad-oil firm in Delft, Netherlands, late 19th cent. Color woodcut.

Pl. 382. Photographic travel poster, view of Arcos de la Frontera in Andalusia, issued by Ministerio de Información y Turismo, Spain, mid-20th cent.

Pl. 383. H. de Toulouse-Lautrec, poster for Jane Avril, Paris, 1893. Color lithograph.

Pl. 384. T. A. Steinlen, advertisement for Motocycles Comiot, Paris, early 20th cent. Color lithograph.

Pl. 385. A. Games, poster for *The Times*, London, 1961. Silk-screen print, reproduced in various sizes.

Pl. 386. *Left*: Z. Chotěnovský, poster for an opera by B. Smetana, Czechoslovakia, 1961, reproduced by offset. *Right*: C. Cagli, cinema poster, Italy, 1962.

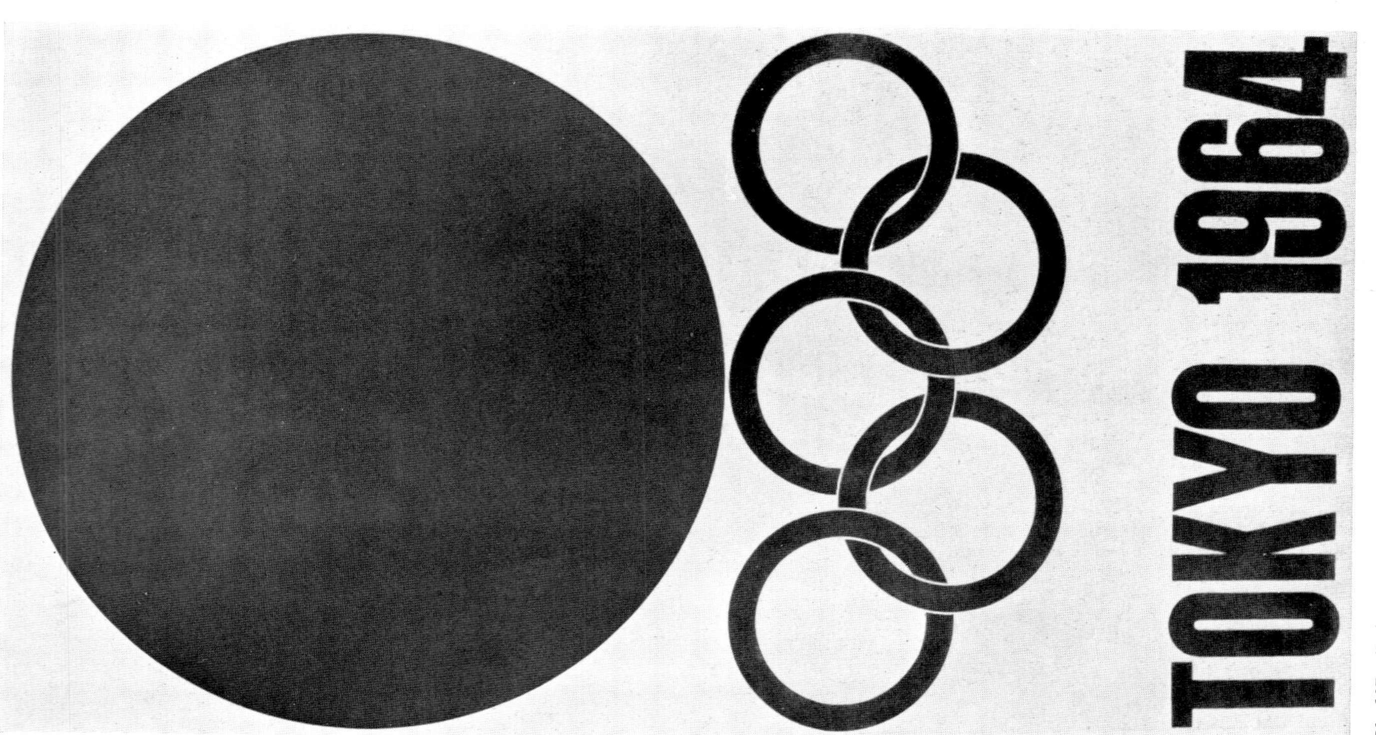

Pl. 387. *Left*: Y. Kamekura, poster for the Olympic games in Tokyo, 1964, combining the sun symbol of Japan and the interlocked circles of the Olympic games motif. *Right*: Y. Yoshimura, poster for the motion picture *Lover's Forest*, Japan, ca. 1961.

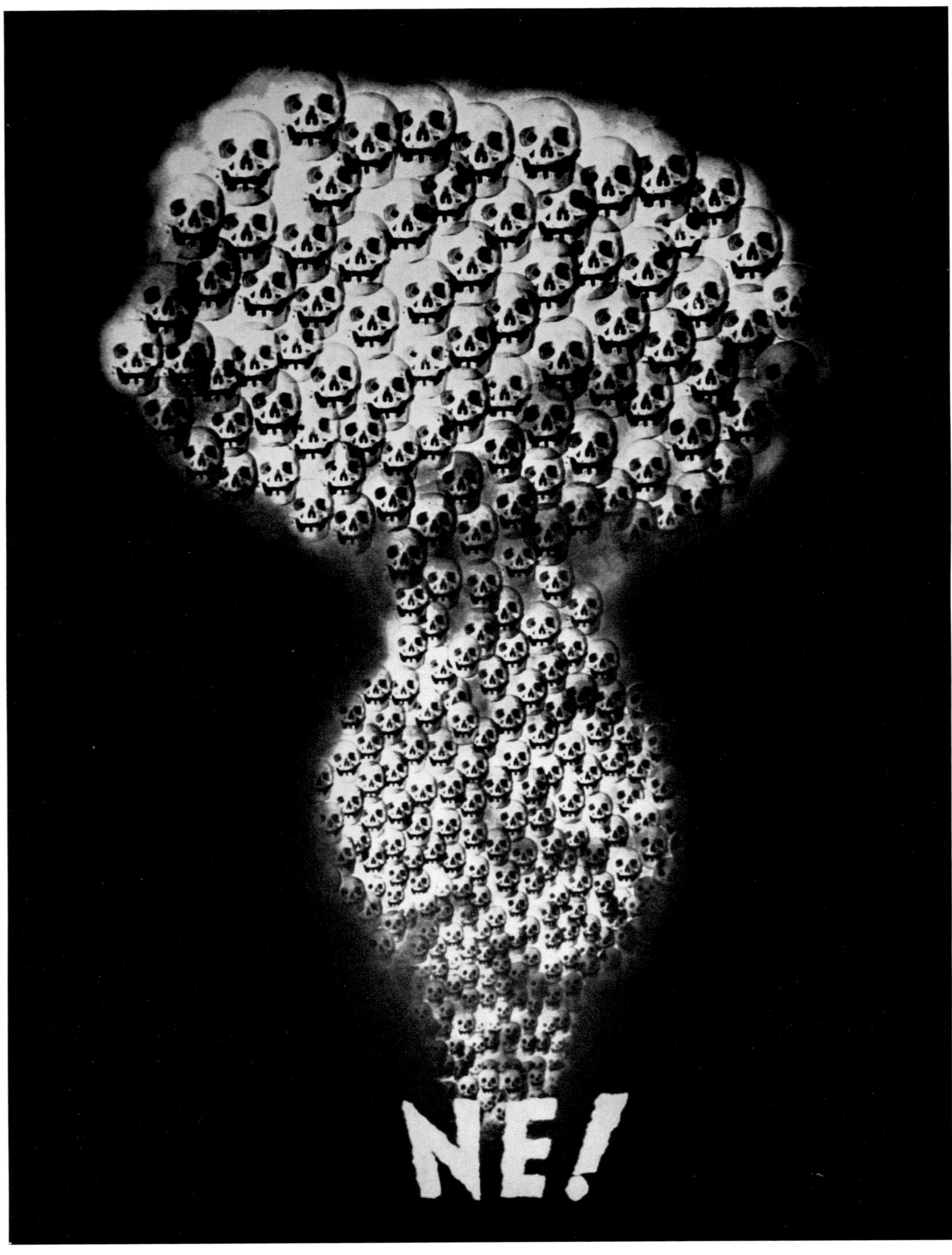

Pl. 388. V. Ševčík, poster for an exhibition on the theme of nuclear disarmament held in Prague, 1961.

Pl. 389. M. Jacno, design for publicity of the Théâtre National Populaire, France, 1954, variously used. Painting on paper; poster, 60×40 in.

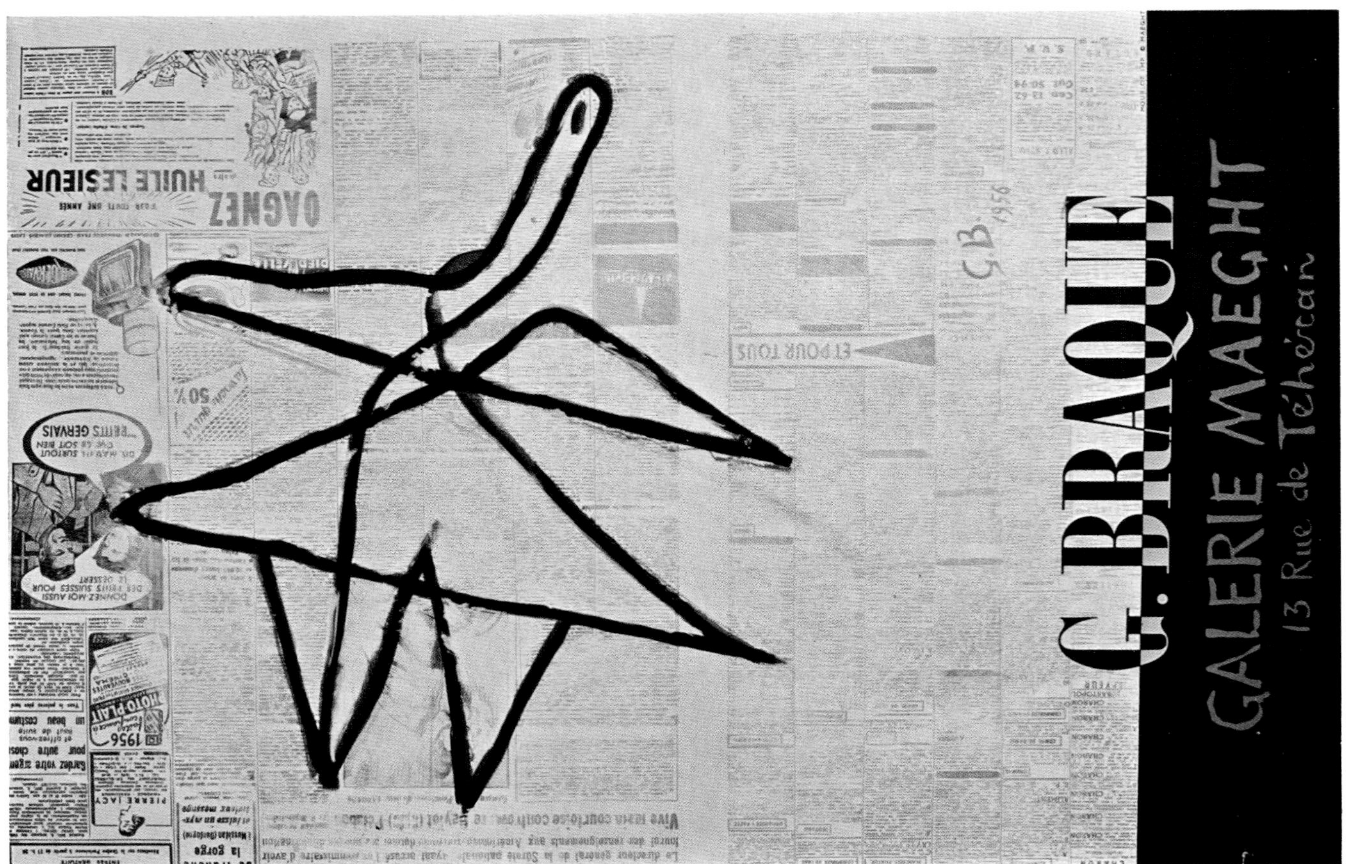

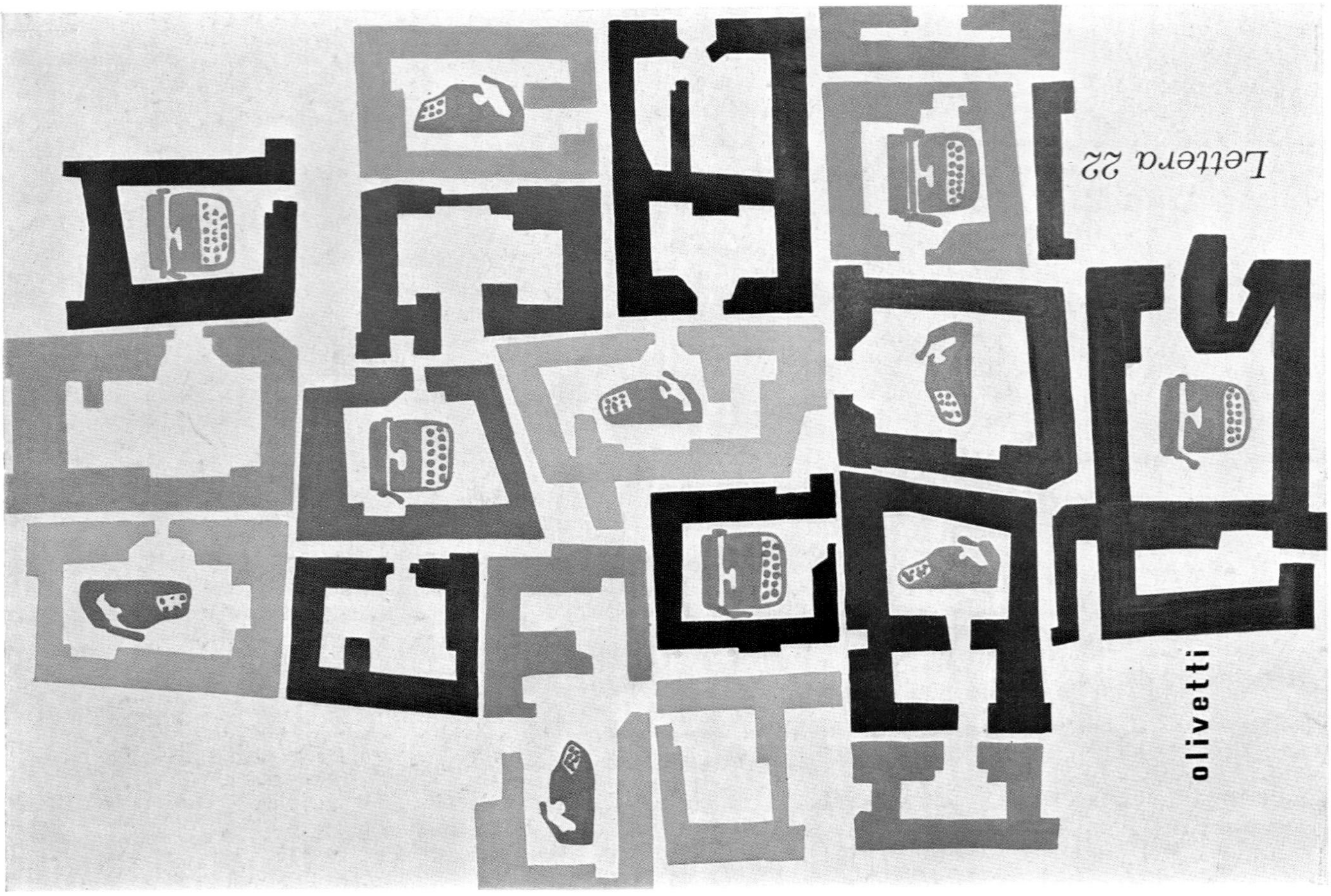

Pl. 390. *Left*: G. Braque, poster for an exhibition of his works, 1956. *Right*: G. Pintori, poster for the Olivetti typewriter "Lettera 22," 1954. W., 27 1/2 in.

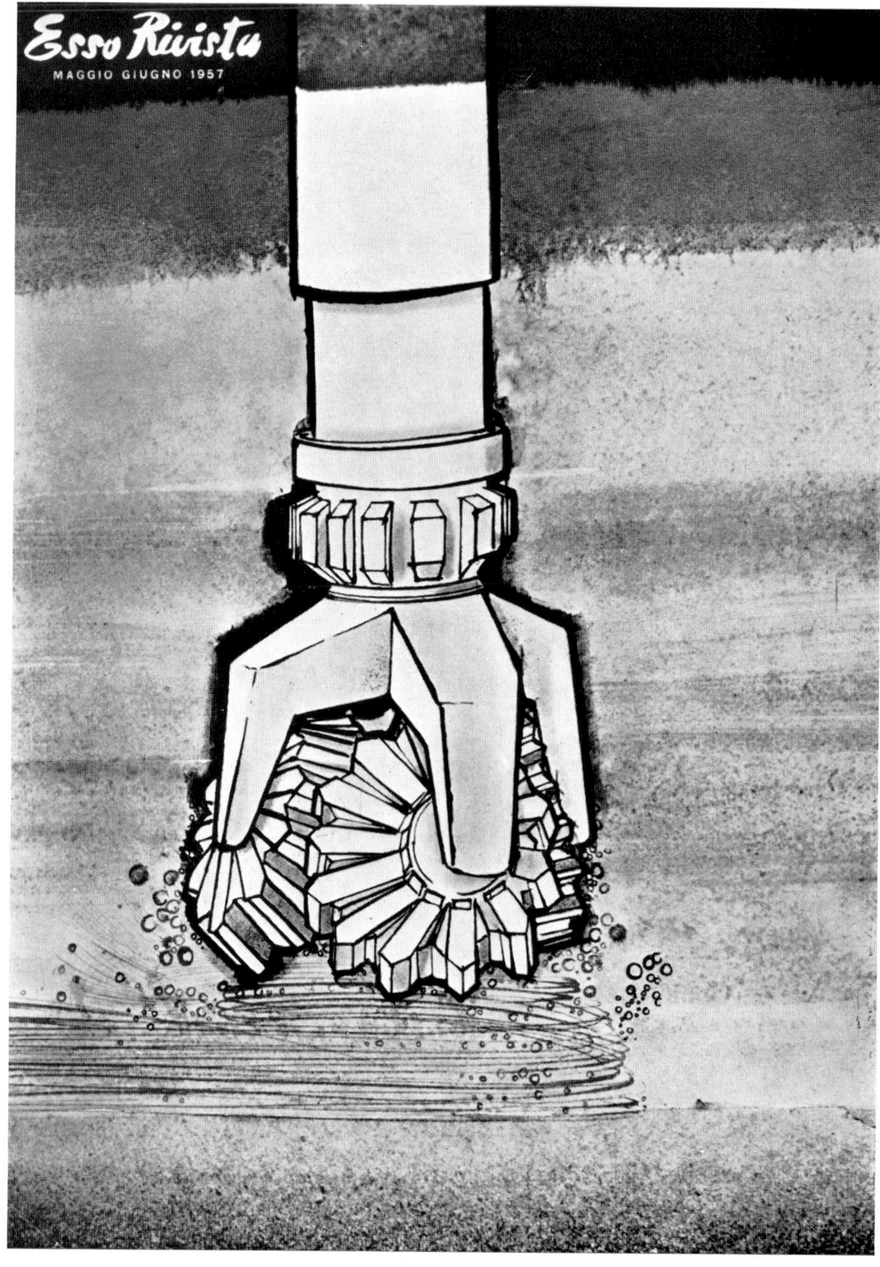

Pl. 391. Cover design. *Left*: A. François, for the magazine *Punch*, England, 1960. *Right*: B. Caruso, for the trade magazine *Esso Rivista*, Italy, 1957.

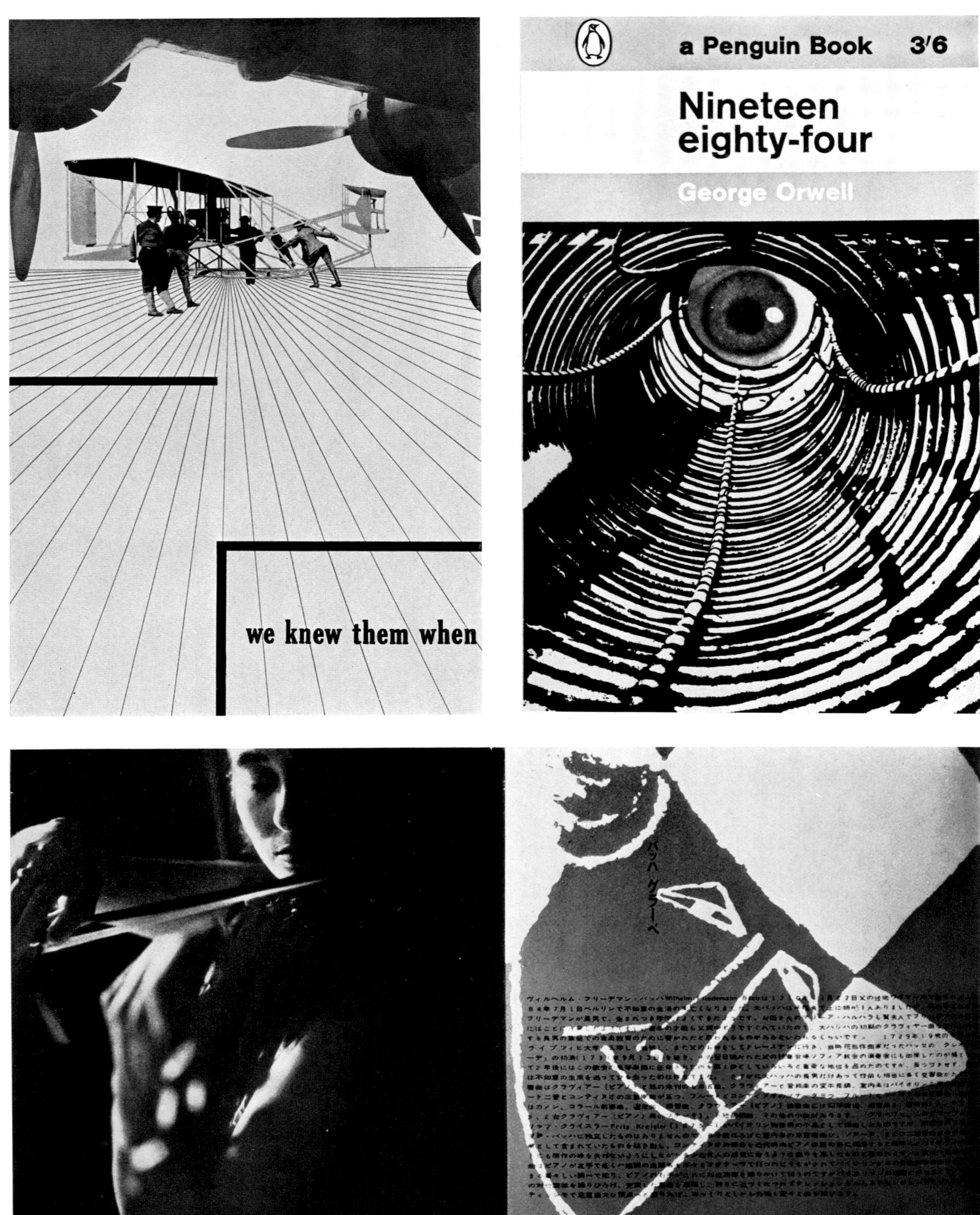

Pl. 392. Cover design. *Above, left*: L. Beall, broadside designed for *Collier's Magazine*, United States, 1941. Original art work, $15^1/_2 \times 10^1/_2$ in. *Right*: G. Facetti, paperback edition of G. Orwell, *Nineteen eighty-four*, Penguin Books, Harmondsworth, England, 1954. *Below*: A. Anzai, designer, with art work by T. Yokoo and H. Haashi, combined record album and booklet, Seirin Publishing Co., Tokyo, ca. 1962.

Pl. 393. *Above*: A. M. Cassandre, advertisement for Dubonnet, Paris, 1934. *Center*: L. Broggini, trademark for AGIP Supercortemaggiore gasoline, Italy, 1952. *Below*: Trademarks, ca. 1960–62. *Top row, left to right*: For Polar International Brokerage Corp., by Eckstein-Stone, Inc., New York; for Lanificio di Somma, by A. Calabresi, Milan; for Chicago Pharmacal Co., by J. Massey, Chicago. *Bottom row*: For Shinko Electric Co., by K. Yosheoka, Japan; for Dominion Linoleum, by E. Roch, Montreal.

Pl. 394. *Above*: Window display, for Hermes, Paris, Christmas, 1960, designed by A. Beaumel. *Below*: Packaging, mid-20th cent. *Left*: Terra-cotta bottle for gin, Amsterdam. *Center*: Wooden container for kobu-nuts, Japan. *Right*: Cork container for a bottle of sherry, Spain.

Pl. 395. Outdoor advertising. *Above*: Billboard for Morton Salt, a parody of the painting by A. M. Willard called "The Spirit of '76," designed by R. L. Dion, Chicago, appeared 1964. *Below*: Illuminated signs and advertisements on the Via Veneto, Rome, ca. 1962. (Cf. Piccadilly Circus, London, III, PL. 16.)

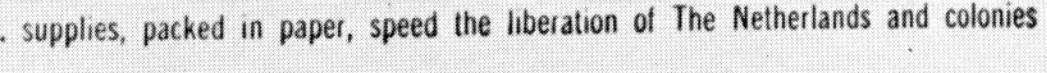

Pl. 396. Advertising and the fine arts. *Left*: Advertisement for Hosiery by Bryan, created by the surrealist painter S. Dali, reproduced in *Vogue*, May 15, 1944. (Photographed from archive material, damaged by pasting.) *Right*: Water color by W. de Kooning, 1943, a work in the commissioned art collection of Container Corporation of America, reproduced and exhibited as corporate advertising. Painting, $10^{1}/_{2} \times 8^{1}/_{2}$ in.

Pl. 397. Advertising and art trends. *Left*: Assemblage by "pop" artist A. Warhol, making use of commercial packaging and its design elements, arranged for an exhibition at the Stable Gallery, New York, April, 1964. *Right*: Cover design for the magazine *Holiday*, October, 1964, by G. Giusti, making use of assemblage.

Pl. 398. Architecture as institutional advertising. Trans World Airlines Flight Center at Kennedy International Airport, New York, designed by Eero Saarinen "to express the drama and wonder of air travel," opened May 28, 1962.

Pl. 399. Neapolitan *presepio* figure, one of the Three Kings, attributed to G. Picano, 18th cent. Wood; ht. (kneeling), 4 ft., 3 in. Naples, S. Maria in Portico.

Pl. 400. Sicilian theatrical *pupo,* character in the stories of the Paladins of France, 19th cent. Palermo, Museo Pitrè.

Pl. 401. *Above, left*: Boeotian bell-shaped figure with separately modeled legs, 8th–7th cent. B.C. Terra cotta, ht., 11 3/4 in. Boston, Museum of Fine Arts. *Center*: Roman jointed doll. Bone, ht., 5 7/8 in. Milan, Museo Teatrale alla Scala. *Right*: Doll made of seed pods, with red cloth and beads, Natal. Ht., 16 1/2 in. Milan, Coll. M. Strudthoff Brissoni. *Below, left*: Wax doll, Augsburg, second half of 18th cent. Ht., 21 1/4 in. Munich, Bayerisches Nationalmuseum. *Right*: Lead soldier, Stuttgart, 19th cent.

Pl. 402. *Above*: Doll kitchen, southern Germany, 18th cent. W., $25\frac{5}{8}$ in. *Below*: Shoemaker group from a *presepio*, by G. Bongiovanni and G. Vaccaro, Caltagirone, Sicily, early 19th cent. Terra cotta, w., $15\frac{3}{8}$ in. Both, Munich, Bayerisches Nationalmuseum.

Pl. 403. *Above*: Symbolic scene depicting a marionette battle of knights before King Solomon, engraving after Herrad von Landsberg, *Hortus Deliciarum* (fol. 215, pl. 55; orig., second half of 12th cent., destroyed). Paris, Bibliothèque Nationale. *Below*: Marionette theater, Venice, first half of 18th cent. London, Victoria and Albert Museum.

Pl. 404. *Above, left*: Rod puppet representing a Negro prince, Venice, 18th cent. Ht., 18 in. *Right*: Marionette representing Old Mother Shipton, England, ca. 1700. Ht., 34 in. *Below*: Guignol, Madelon, and Gnafron, hand puppets for the Guignol Lyonnais of Laurent Mourguet, Lyons, early 19th cent. Ht. of heads, 4 3/4 in. All, Lyons, Musée Historique.

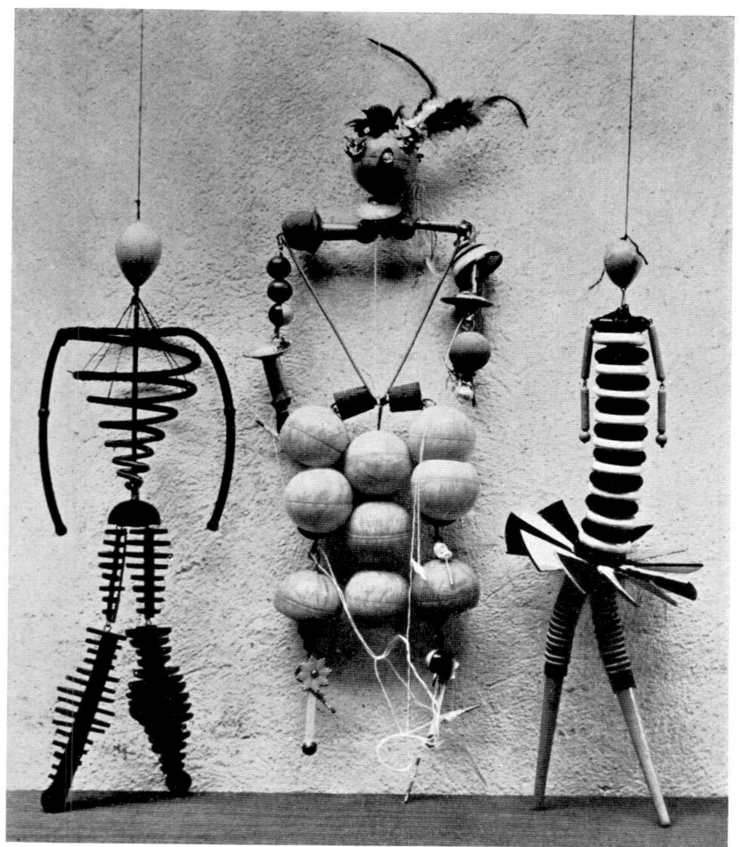

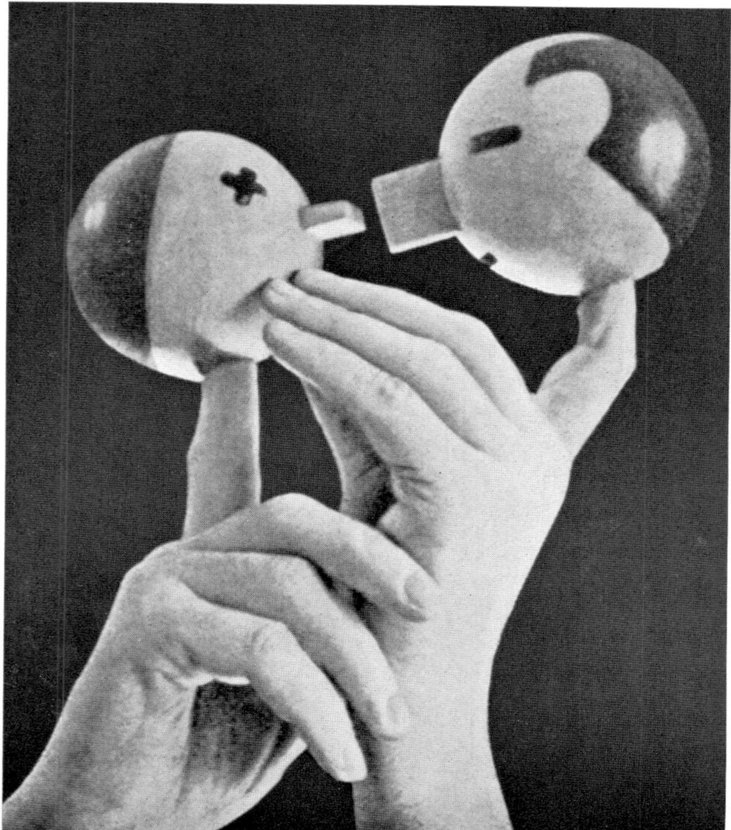

Pl. 405. *Above, left*: Marionettes of the Versuchsbuehne am Bauhaus, by L. Scheper, Dessau, ca. 1923. Munich, Theater-Museum. *Right*: Marionette representing the devil, by A. Altherr (1875–1945), founder of the Schweizerische Marionetten Theater, Zurich. *Below, left*: Demonstration of the manipulation of hand puppets with heads by S. V. Obraztzov (b. 1901, U.S.S.R.). *Right*: Two characters from the puppet film "Princ Bajaja," by J. Trnka, Czechoslovakia, 1950.

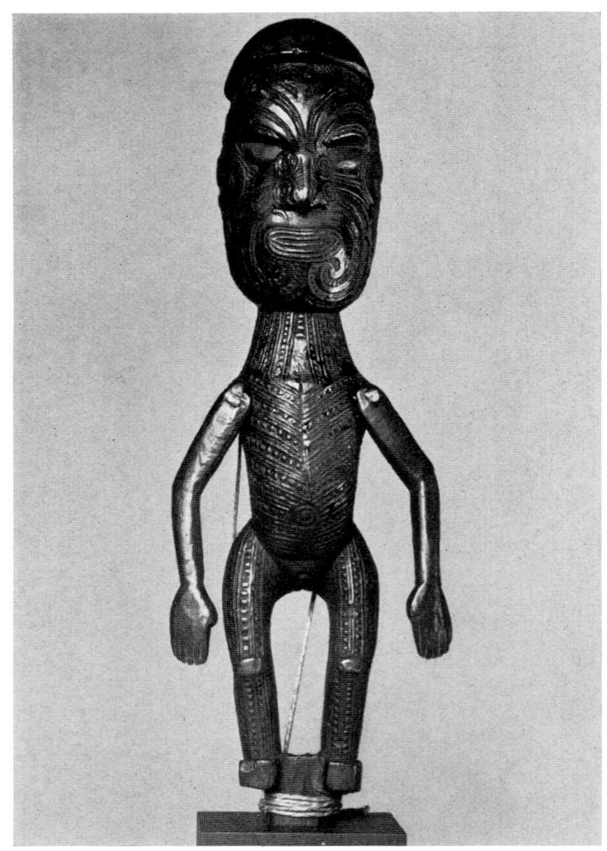

Pl. 406. *Above, left*: Japanese print demonstrating a shadow figure, 19th cent. W., 5 7/8 in. Milan, Coll. Brissoni. *Right*: General Tsao-tsao, character in the classic repertory of the shadow theater, China. Parchment. *Below, left*: Indonesian wayang, 20th cent. Leather, manipulated by rods; ht., 21 5/8 in. Milan, Coll. Brissoni. *Right*: Maori *karetao,* from New Zealand. Carved wood, manipulated by cords. London, British Museum.

Pl. 407. Neo-Sassanian reliefs on a mountainside at Rayy (Rhages), near Teheran, depicting Fatḥ 'Alī Shah killing a lion (*above*) and enthroned (*below*), first half of 19th cent.

Pl. 408. *Left*: Parts of a water pipe, early 19th cent. Polychrome enamels on silver, ht., $8^{1}/_{4}$ in. Teheran, Archaeological Museum. *Right*: Dagger carved with a prince and a legendary king, from Shiraz, ca. 1800. Steel blade, ivory handle; l. of handle, $4^{3}/_{8}$ in. Teheran, Coll. Gen. M. H. Firouz.

Pl. 409. Surgical instrument case, Iran, 19th cent. Painted papier-mâché box, $15\frac{5}{8} \times 10\frac{1}{4}$ in. Oxford, England, Ashmolean Museum.

Pl. 410. *Above*: Box inscribed with the name of Fatḥ 'Alī Shah, early 19th cent. Steel inlaid with gold, l., 11 1/2 in. *Below*: Enlargement showing decoration on a tube of a water pipe. Enameled silver, actual l., 1 1/8 in. Both, Teheran, Archaeological Museum.

Pl. 411. Head of Kṛṣṇa, cartoon of a *Rās Līlā* composition for a mural painting in Jaipur Palace Library, 18th cent. Color on paper, 26 × 18 in. New York, Metropolitan Museum.

Pl. 412. View of a city, 18th cent. Miniature. Benares, India, A. Boner Coll.

Pl. 413. *Above*: Rāma, Sītā, Lakṣmaṇa, and the golden deer, scene from the *Rāmāyaṇa*, Malwa school, mid-17th cent. Miniature. *Below*: Kṛṣṇa playing the flute, scene from the *Bhāgavata Purāṇa*. Basohli school, early 18th cent. Miniature, 6×8 in. Both, New Delhi, National Museum.

Pl. 414. *Left*: Illustrations to the *Rasikapriyā* by Keśava Dās, Mewar school. Miniatures. Bikaner, India, Coll. Maharaja of Bikaner. *Above*: Ca. 1670. *Below*: Ca. 1710–20. *Right, above*: Rādhā and Kṛṣṇa, Rajasthani school, late 18th cent. Miniature. *Below*: *Gopīs* approaching the temple of Kṛṣṇa, left half of wall hanging from a Vallabhacharya temple, Kishangarh school, ca. 1830. Tempera; full size, 6 ft., 3 in. × 4 ft., 5 in. Last two, Baroda, India, Museum and Picture Gallery.

Pl. 415. *Kumbha*, illustrating a musical mode, late Kangra style in Jammu, early 19th cent. Miniature. New Delhi, National Museum.

Pl. 416. *Above*: Datia, Govind Mandir, built by Raja Bīr Siṅgh Deo, ca. 1620. *Below*: Udaipur, Maharana's Palace, 16th–19th cent.

Pl. 417. *Above*: Amber, Jagat Śiromani temple, 1555–1646. *Below*: Orchha, Caturbhuja temple, built by Raja Bīr Siṅgh Deo (1605–28).

Pl. 418. *Above*: Nurpur, frieze of the Kṛṣṇa temple, built by Raja Bāsu Deo (1580–1613). *Below*: Devikund, funerary shrine (*chattrī*) of Maharaja Anūp Siṅgh, 1698 (repaired 1836).

Pl. 419. Jaisalmer, Dēwān Salīm Siṅgh's Palace, late 18th–early 19th cent.

Pl. 420. *Above*: Bikaner, Karan Mahal, ca. 1690 (renovated ca. 1755). *Below, left*: Kolayat, funerary shrine (*chattrī*) of Śrī Paraśurām Giri, detail of ceiling reliefs, 1692. *Right*: Tira Sujanpur, Narbadeśvara temple, detail of painted wall, ca. 1800.

Pl. 421. St. George and the Dragon. Panel, $12^{1}/_{4} \times 10^{5}/_{8}$ in. Paris, Louvre.

Pl. 422. The Tempi Madonna. Panel, 29 1/2×20 1/8 in. Munich, Alte Pinakothek.

Pl. 423. The Marriage of the Virgin. Panel, 5 ft., 6 7/8 in. × 3 ft., 10 1/2 in. Milan, Brera.

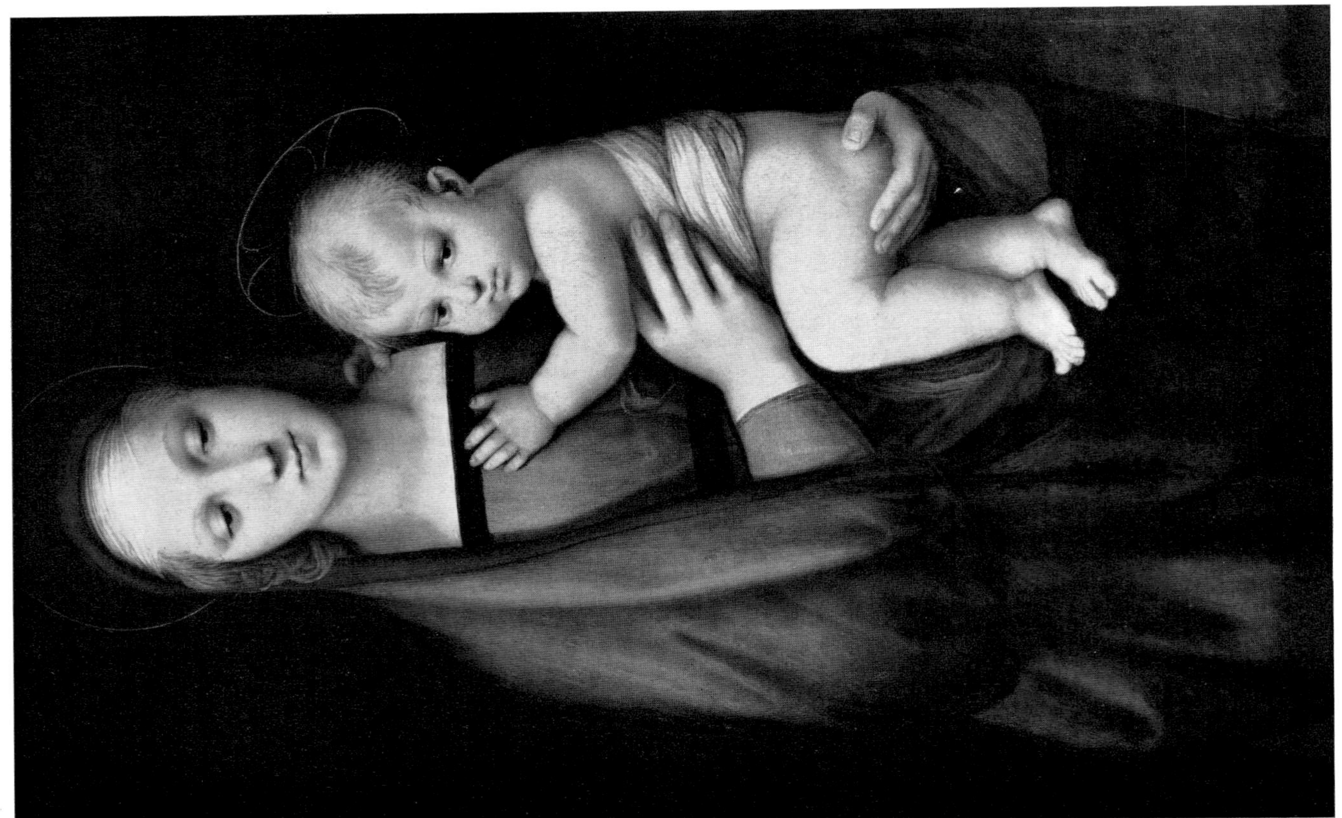

Pl. 424. *Left*: Madonna del Granduca. Panel, $33^{1}/_{8} \times 21^{5}/_{8}$ in. Florence, Pitti. *Right*: Portrait known as "La Muta." Panel, $25^{1}/_{4} \times 18^{7}/_{8}$ in. Urbino, Galleria Nazionale delle Marche.

Pl. 425. *Left*: Madonna del Prato. Panel, $44\frac{1}{2} \times 34\frac{3}{4}$ in. Vienna, Kunsthistorisches Museum. *Right*: The Canigiani Holy Family. Panel, $51\frac{5}{8} \times 42\frac{1}{8}$ in. Munich, Alte Pinakothek.

Pl. 426. Rome, Vatican, frescoes in the Stanze. *Above*: The School of Athens. *Below*: The Expulsion of Heliodorus.

Pl. 427. The Mass of Bolsena, detail. Fresco. Rome, Vatican, Stanza d'Eliodoro.

Pl. 428. Portrait of Leo X with Cardinals Giulio de' Medici and Luigi de' Rossi. Panel, 5 ft., 5/8 in. × 3 ft., 10 7/8 in. Florence, Uffizi.

Pl. 429. The Fire in the Borgo, detail of pl. 430.

Pl. 430. Rome, Vatican, frescoes in the Stanze. *Above*: The Repulse of Attila. *Below*: The Borgo.

Pl. 431. Rome, Vatican, the Logge of Raphael, right wing. (Cf. II, pl. 291; V, pl. 239.)

Pl. 432. *Above*: Sibyls and angels. Fresco. Rome, S. Maria della Pace. *Below*: The Death of Ananias, cartoon for a tapestry, 11 ft., 2 3/4 in. × 17 ft., 5 1/2 in. London, Victoria and Albert Museum.

Pl. 433. The Sistine Madonna. Canvas, 8 ft., 8 3/8 in. × 6 ft., 5 1/8 in. Dresden, Gemäldegalerie.

Pl. 434. *Above*: The Madonna of Foligno, detail (cf. IV, PL. 314). Canvas; full size, 10 ft., 6 in × 6 ft., 4 3/8 in. Rome, Vatican Museums. *Below*: Detail of PL. 435.

Pl. 435. The Transfiguration. Panel, 15 ft., 1 1/2 in. × 9 ft., 1 1/2 in. Rome, Vatican Museums.

Pl. 436. Rome, S. Maria del Popolo, view of the Chigi Chapel.

Pl. 437. J. F. Millet, The Washerwoman, ca. 1861. Panel, $17^{3}/_{8} \times 13$ in. Paris, Louvre.

Pl. 438. G. Courbet, Women Winnowing Corn, 1853–54. Canvas, 4 ft., $3^{5}/_{8}$ in. × 5 ft., $5^{3}/_{4}$ in. Nantes, France, Musée des Beaux-Arts.

Pl. 439. *Left*: J. F. Millet, Woman Sewing by Lamplight, 1872. Canvas, $39^{5}/_{8} \times 32^{1}/_{4}$ in. New York, Frick Coll. *Right*: H. Daumier, The Washerwoman, ca. 1861–63. Panel, $19^{1}/_{4} \times 13^{1}/_{8}$ in. Paris, Louvre.

Pl. 440. *Above*: G. Michel (1763–1843), The Storm. Canvas, 19×25 in. Strasbourg, Musée des Beaux-Arts. *Below*: T. Rousseau, Oak Trees at Apremont, 1852. Canvas, $25^{1}/_{4}\times 39^{3}/_{8}$ in. Paris, Louvre.

Pl. 441. *Above*: C.-F. Daubigny, The Seine at Bezons, ca. 1851. Panel, $14^{1}/_{2} \times 22^{1}/_{2}$ in. Paris, Louvre. *Below*: A. Chintreuil, Space, view near La Queue-les-Yvelines, 1869. Canvas, 3 ft., 4 in. × 6 ft., 7 in. Paris, Louvre.

Pl. 442. *Above*: C. Meunier, Au Pays Noir. Saint-Étienne, France, Musée d'Art et d'Industrie (*en dépôt*). *Below*: G. Segantini, Girl Knitting near Savognin, in the Engadine Valley, 1888. Canvas, $21^{1}/_{4} \times 34^{5}/_{8}$ in. Zurich, Kunsthaus.

Pl. 443. *Above*: C. de Groux (1825–70), Grace before the Meal. Ghent, Belgium, Musée des Beaux-Arts. *Below*: P. Szinyei-Merse, Picnic in May, 1873. Canvas, $48^{1}/_{2} \times 63^{3}/_{8}$ in. Budapest, National Museum.

Pl. 444. *Above*: G. Toma, Luisa Sanfelice in Prison, 1877. Canvas, 24×30 3/4 in. Rome, Galleria Nazionale d'Arte Moderna. *Below*: S. Lega, The Pergola, 1860. Canvas, 36 1/4×28 3/8 in. Milan, Brera.

Pl. 445. A. von Menzel, Room with Balcony, 1845. Oil on cardboard, $22\,^{7}/_{8} \times 18\,^{1}/_{2}$ in. Berlin, Staatliche Museen.

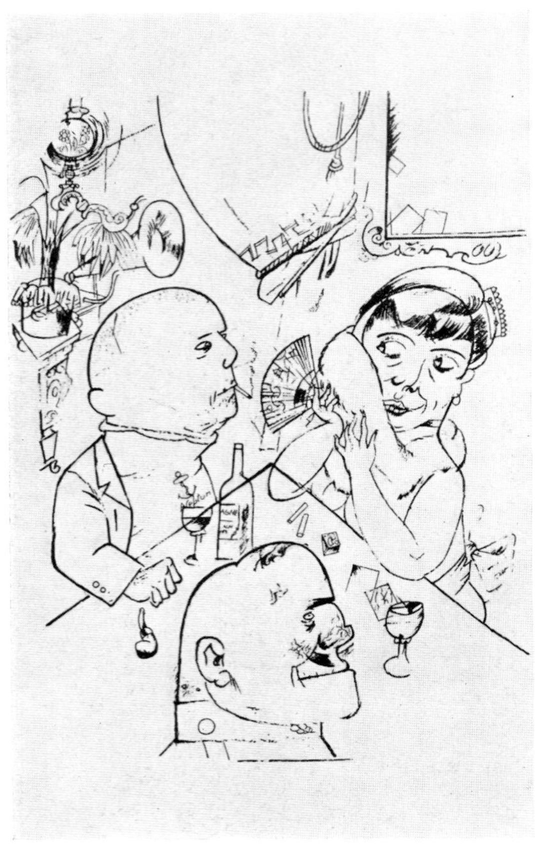

Pl. 446. *Left, above*: G. Grosz, Trio, 1919, from *Ecce Homo*, Berlin, Malik Verlag, 1923. Pen and ink. *Below*: B. Shahn, Blind Botanist, 1954. Tempera, 52×31 in. Wichita, Kans., Art Museum, R. P. Murdock Coll. *Right, above*: D. Rivera, Flower Seller, 1943. Oil on masonite, 59¾×47¼ in. Mexico City, M. Aleman Valdes Coll. *Below*: R. Guttuso, Nude. Drawing with water color. Coll. of the artist.

MARTHIN LVTHER, CATHARINA

Pl. 447. *Above, left*: Wenzel von Olmütz, *Roma caput mundi* ("the popish ass"), 1496. Engraving. *Right*: Cardinal-jester, ca. 1540. Woodcut. *Below*: The expulsion of the reformist preachers, showing Martin Luther — followed by his wife, Katharina von Bora — carrying the ousted preachers on his shoulders and pushing the Protestant reformers Melanchthon, Justus Jonas, and Carlstadt in a wheelbarrow. Engraving.

Pl. 448. *Above*: Christ and the pope, second half of the 16th cent. Woodcut. *Below*: L. Cranach the Younger (1515–86), The Reformation Triumphant, showing Martin Luther in the pulpit, with the Protestants receiving Communion and the pope and Catholic clergy in the mouth of Hell. Woodcut.

Pl. 449. P. Cati, The Council of Trent, 1588–89. Fresco. Rome, S. Maria in Trastevere.

Pl. 450. Rome. *Above*: Church of the Gesù (begun 1568), by Giacomo da Vignola and others, nave. *Below, left*: S. Maria Maggiore, tomb of Pius V, ca. 1585–87. *Right*: S. Maria sopra Minerva, tomb of Paul IV, 1566–72.

Pl. 451. P. Saenredam, Interior of St. Janskerk, Utrecht, 1645. Panel, $19^{1}/_{2} \times 16^{1}/_{2}$ in. Utrecht, Netherlands, Centraal Museum.

Pl. 452. *Above, left*: G. Celio, on designs by G. B. Fiammeri (G. Valeriano), Christ Blindfolded and Christ Exposed, ca. 1597. Both, canvas. Rome, Church of the Gesù. *Right*: N. Circignani (Pomarancio), scene from a cycle showing early Christian martyrs, 1582. Fresco. Rome, S. Stefano Rotondo. *Below, left*: S. Pulzone, portrait of Paul III. Canvas. Rome, Galleria Nazionale. *Right*: L. Carracci, Martyrdom of St. Margaret, 1616. Canvas, 10 ft., 2 in. × 6 ft., $6^3/_4$ in. Mantua, Italy, S. Maurizio.

Pl. 453. *Above*: Follower of Caravaggio (G. A. Galli?), The Virgin and St. Anne. Canvas, 39 3/4×51 1/2 in. Rome, Galleria Spada. *Below*: G. B. Crespi (Il Cerano), P. F. Mazzucchelli (Il Morazzone), and G. C. Procaccini, The Martyrdom of SS. Rufina and Secunda. Canvas, 6 ft., 3 5/8 in. × 6 ft., 3 5/8 in. Milan, Brera.

Pl. 454. G. L. Bernini, The Blood of Christ, 1671. Engraving by F. Spierre.

Pl. 455. Christ at Emmaus. Panel, 26 3/4 × 25 5/8 in. Paris, Louvre.

Pl. 456. *Left*: Self-portrait. Panel, $14^{3}/_{4} \times 11^{3}/_{8}$ in. *Right*: Self-portrait. Canvas, $23^{1}/_{4} \times 20^{1}/_{8}$ in. Both, The Hague, Mauritshuis.

Pl. 457. Danae. Canvas, 6 ft., 7/8 in. × 6 ft., 7 7/8 in. Leningrad, The Hermitage.

Pl. 458. Landscape: Stormy Weather. Panel, $20^1/_2 \times 28^3/_8$ in. Brunswick, Germany, Herzog-Anton-Ulrich-Museum.

Pl. 459. The Night Watch. Canvas, 11 ft., 9 3/8 in. × 14 ft., 4 1/2 in. Amsterdam, Rijksmuseum.

Pl. 460. *Above*: The Blinding of Samson. Canvas, 7 ft., 9 3/4 in. × 9 ft., 5 in. Frankfort on the Main, Städelsches Kunstinstitut. *Below, left*: The Parting of David and Jonathan. Panel, 28 3/4 × 24 1/4 in. Leningrad, The Hermitage. *Right*: Woman Bathing in a Stream (Hendrickje Stoffels?). Panel, 24 5/16 × 18 1/2 in. London, National Gallery.

Pl. 461. *Above*: Saul and David. Canvas, 4 ft., $3^{3}/_{8}$ in. × 5 ft., $4^{5}/_{8}$ in. The Hague, Mauritshuis. *Below, left*: Man in a Golden Helmet. Canvas, $26^{3}/_{8} \times 20^{1}/_{4}$ in. Berlin, Staatliche Museen. *Right*: Portrait of Titus. Canvas, $30^{3}/_{8} \times 24^{3}/_{4}$ in. Rotterdam, Museum Boymans-Van Beuningen.

Pl. 462. *Above*: Jacob Blessing the Sons of Joseph. Canvas, 5 ft., 8 1/2 in. × 6 ft., 10 1/4 in. Kassel, Germany, Gemäldegalerie. *Below*: The Denial of Peter. Canvas, 5 ft., 5/8 in. × 5 ft., 6 1/2 in. Amsterdam, Rijksmuseum.

Pl. 463. *Above*: The Conspiracy of the Batavians. Canvas, 6 ft., $5^1/_8$ in. × 10 ft., $1^5/_8$ in. Stockholm, Nationalmuseum. *Below*: The Bridal Couple (The Jewish Bride). Canvas, 3 ft., $11^3/_4$ in. × 5 ft., $5^1/_2$ in. Amsterdam, Rijksmuseum.

Pl. 464. Etchings. *Above*: The so-called "Hundred Guilder Print." *Below, left*: The Death of the Virgin. *Right*: The Raising of Lazarus.

Pl. 465. Christ and the Woman of Samaria. Panel, 18 1/4 × 15 3/8 in. Berlin, Staatliche Museen.

Pl. 466. Butchered Ox. Panel, 37×26$^{3}/_{8}$ in. Paris, Louvre.